1B	2B	3A	4A	5A	6A	7A	8A
							Hydrogen 1 **H** 1.0079	Helium 2 **He** 4.0026
			Boron 5 **B** 10.811	Carbon 6 **C** 12.011	Nitrogen 7 **N** 14.0067	Oxygen 8 **O** 15.9994	Fluorine 9 **F** 18.9984	Neon 10 **Ne** 20.1797
			Aluminum 13 **Al** 26.9815	Silicon 14 **Si** 28.0855	Phosphorus 15 **P** 30.9738	Sulfur 16 **S** 32.066	Chlorine 17 **Cl** 35.4527	Argon 18 **Ar** 39.948
Nickel 28 **Ni** 58.6934	Copper 29 **Cu** 63.546	Zinc 30 **Zn** 65.39	Gallium 31 **Ga** 69.723	Germanium 32 **Ge** 72.61	Arsenic 33 **As** 74.9216	Selenium 34 **Se** 78.96	Bromine 35 **Br** 79.904	Krypton 36 **Kr** 83.80
Palladium 46 **Pd** 106.42	Silver 47 **Ag** 107.8682	Cadmium 48 **Cd** 112.411	Indium 49 **In** 114.82	Tin 50 **Sn** 118.710	Antimony 51 **Sb** 121.757	Tellurium 52 **Te** 127.60	Iodine 53 **I** 126.9045	Xenon 54 **Xe** 131.29
Platinum 78 **Pt** 195.08	Gold 79 **Au** 196.9665	Mercury 80 **Hg** 200.59	Thallium 81 **Tl** 204.3833	Lead 82 **Pb** 207.2	Bismuth 83 **Bi** 208.9804	Polonium 84 **Po** (209)	Astatine 85 **At** (210)	Radon 86 **Rn** (222)

Europium 63 **Eu** 151.965	Gadolinium 64 **Gd** 157.25	Terbium 65 **Tb** 158.9253	Dysprosium 66 **Dy** 162.50	Holmium 67 **Ho** 164.9303	Erbium 68 **Er** 167.26	Thulium 69 **Tm** 168.9342	Ytterbium 70 **Yb** 173.04	Lutetium 71 **Lu** 174.967

Americium 95 **Am** (243)	Curium 96 **Cm** (247)	Berkelium 97 **Bk** (247)	Californium 98 **Cf** (251)	Einsteinium 99 **Es** (252)	Fermium 100 **Fm** (257)	Mendelevium 101 **Md** (258)	Nobelium 102 **No** (259)	Lawrencium 103 **Lr** (260)

DATE DUE

SEP 04 2012			

GENERAL CHEMISTRY

1000 ml
25 ml
A
100 mL
KIMAX
200 ml

GENERAL CHEMISTRY

CLIFFORD W. HAND
UNIVERSITY OF ALABAMA

WITH CONTRIBUTIONS FROM

ELLI S. HAND
UNIVERSITY OF ALABAMA

SAUNDERS GOLDEN SUNBURST SERIES
Saunders College Publishing
Harcourt Brace College Publishers
Fort Worth Philadelphia San Diego
New York Orlando San Antonio
Toronto Montreal London
Sydney Tokyo

Text Typeface: Berkeley Book
Compositor: Progressive Typographers
Acquisitions Editor: John J. Vondeling
Developmental Editor: Sandra Kiselica
Managing Editor: Carol Field
Project Editor: Margaret Mary Anderson
Copy Editor: Zanae Rodrigo
Manager of Art and Design: Carol Bleistine
Art Director: Anne Muldrow
Art Assistant: Caroline McGowan
Text Designer: Nanci Kappel, Kappel Design
Cover Designer: Lawrence R. Didona
Text Artwork: Rolin Graphics
Layout Artist: Nanci Kappel, Kappel Design
Director of EDP: Tim Frelick
Production Manager: Charlene Squibb
Marketing Manager: Marjorie Waldron

Cover Credit: © Erich Schrempp/Photo Researchers, Inc.

Printed in the United States of America

GENERAL CHEMISTRY

0-03-074172-6

Library of Congress Catalog Card Number: 92-051016

3456 69 987654321

To all of those from whom we have learned:
parents, children, teachers, and friends.

PREFACE

General Chemistry is intended for use in the first chemistry course taken by students planning further study in science or engineering. It is comprehensive in its coverage, and students who master the material will be well prepared for subsequent courses. Since the level of math and science preparedness among today's students entering college is variable, no background other than a moderate facility with algebra is assumed.

Approach and Philosophy

I began writing this book because none of the existing texts closely matched my approach to teaching. Specifically, I felt the need for improved treatments of both problem-solving and descriptive chemistry.

Problem-Solving

Acquisition of the ability to solve both word and quantitative problems in chemistry is a challenging task. Raised on the plug-and-chug approach, many students have not learned the more advanced skill of problem *analysis.* To those who lack the technique of breaking down a problem into simple steps, each problem seems new, and often insolvable unless a specific method for solving it has been learned previously.

My approach to this central teaching/learning difficulty has two components. First, a separate section in Chapter 1 is devoted to problem-solving. Techniques for making progress toward the solution of an unfamiliar problem are set out in an orderly fashion. These techniques, which include identification and specification of the characteristics of the answer, organizing the known information, working backward from the answer, and using the factor-label method, do not differ greatly from those used by any competent problem-solver. What sets this text apart from others, however, is the greater amount of space given to the discussion of problem-solving, the organization of techniques into a procedure, and the emphasis on the idea that problem-solving is a learnable skill rather than a mysterious innate ability. Although the presentation in Chapter 1 is quite different, the material was inspired by Polya's book, *How to Solve It.*

The second component of the approach is embodied in the detailed, and, equally importantly, the *consistent* way in which example problems are worked throughout the text. The procedure outlined in Chapter 1 is followed in every example, in every chapter. The inescapable message is, "Here is a procedure that works. Use it!" Every example is followed by an exercise, almost always the same problem with different wording, numbers, or compounds. The answer to each exercise is given immediately —no page turning is necessary.

The examples are generally longer and more fully explained than in other texts. However, there is some condensation as students progress; explanations, although still present, are more terse in later chapters. The layout and visual appearance of the examples are always the same. This allows students to take short-cuts as their skills improve. For instance, they can ignore the listing of known information, or skip the development of the solution plan and go directly to the actual working of the problem. For students whose skills develop more slowly, however, the information is always there.

Descriptive Chemistry

Balancing the treatment of chemical principles and descriptive chemistry has been a persistent problem in introductory textbooks. Many instructors have found that large doses of descriptive chemistry can be boring—small chunks are more easily learned and retained. The approach taken in this textbook is the parallel presentation of small units of descriptive chemistry together with chemical principles. The descriptive chemistry of one of the main groups forms the final section of most of the early chapters. Inclusion of this material within, rather than between, the chapters emphasizes its importance and underscores the inseparability of theory and observation. A valuable bonus of this approach is that the example problems throughout a chapter can be chosen from the descriptive chemistry that will follow, giving a double exposure that facilitates learning.

Two guidelines were used in choosing the topics for the descriptive chemistry sections. Most importantly, these sections had to make sense on the basis of principles already presented. For example, the chemistry of boron hydrides cannot be presented effectively prior to a reasonably comprehensive discussion of bonding. Since most students, presumably, will have acquired some familiarity with the topic before entering college, the choice for Chapter 1 is a basic overview of metals. More advanced aspects of metal chemistry, such as the redox nature of metallurgical operations, are covered in later chapters. The choice for Chapter 2 was the noble gases, because their chemistry is simpler and more limited than that of any other group.

The secondary guideline was the desirability of an orderly, progressive presentation. The order generally follows the periodic table, so Chapters 3, 4, and 5 cover the chemistry of the metals of Groups 1A, 2A, and 3A. Boron chemistry appears later, in part because it seemed to fit better with the other nonmetals, and in part because of the complexities of bonding in boron compounds. Halogen chemistry comes next, because it is simpler and more coherent than the chemistry of Groups 4A, 5A, and 6A. The chemistry of oxygen is given separate coverage, in Chapter 8, both because of its importance and its major differences from the chemistry of the other elements in Group 6A. The last descriptive chemistry to be covered in a separate section of a principles chapter is hydrocarbons, in Chapter 9. For students who take only one year of chemistry, this choice provides some exposure to organic chemistry, which is, by most measures, the largest discipline within chemistry.

Organization

General Chemistry is organized into three parts of approximately equal lengths.

The fundamental concepts, techniques, and vocabulary of chemistry, as well as some basic descriptive chemistry, are covered in the first third of the text. Problem-solving techniques are discussed in Chapter 1. The structure of atoms and molecules, chemical equations and stoichiometry, and heats of reaction are taken up in Chapters 2 through 4, followed by electronic structure, periodicity, and bonding in Chapters 6 through 9. Gases are discussed in Chapter 5. The descriptive chemistry in this part of the text includes metals, noble gases, Groups 1A, 2A, 3A, 7A, and oxygen.

The second part of the book is devoted to some important interactions that occur between ions and molecules: intermolecular forces, and processes involving the transfer of a proton or an electron. Entropy and free energy are introduced in Chapter 10, so that they may be used in the discussion of liquids and solids, solutions, equilibrium, and kinetics that follows (Chapters 11 through 14). Acids and bases are treated qualitatively in Chapter 15 and quantitatively in Chapter 16. Solubility and complex-ion equilibria are covered in Chapter 17. This part concludes with redox reactions, Chapter 18, and electrochemistry, Chapter 19.

The third part discusses the chemical features of our world. Organic and biochemistry, Groups 4A through 6A, the transition elements, and nuclear chemistry are covered here. Except for biochemistry (Chapter 21), which builds directly on organic chemistry (Chapter 20), these chapters are free-standing and may be presented in any order.

Features

- A detailed, *consistent* approach to problem-solving is used in the 260 in-text examples.
- A closely similar exercise, together with its answer, follows each example.
- A summary exercise, with answers, is included in each of the first 19 chapters. Summary exercises usually describe a single chemical substance, reaction, or process, and, by asking suitable questions, test student's learning of the entire chapter. In addition, these exercises reinforce some of the material in the preceding two or three chapters.
- End-of-chapter problems include general review questions and questions classified by type. The questions are paired, so that each odd-numbered problem is followed by an almost-identical repetition. More difficult problems are marked with an asterisk.
- Answers to all odd-numbered, end-of-chapter problems appear in the back of the text. Complete solutions to the odd-numbered problems, worked out in the same format used for the examples, are available in the *Student Solutions Manual.* Worked-out solutions to even-numbered, end-of-chapter problems are available in the *Instructor's Resource Guide.*
- Over 350 full-color photographs illustrate chemistry in the laboratory, in industry, and in everyday life.
- In most chapters, "Chemistry Old and New" essays discuss topics of historical or current interest. The topics range from army food to art forgeries. The essays are intended to be read for interest and pleasure, and they do not require a high level of chemical knowledge on the part of the reader. Each essay ends with discussion questions to stimulate students' critical thinking.
- Brief biographical notes, scattered throughout the text, bring to life many of the well-known names in chemistry.
- Marginal notes are used to emphasize important points, to mention interesting facts, or to help the reader locate related topics.
- The end-of-chapter summaries are longer and more detailed than in most texts. This feature drew very favorable comments from students during three years of class testing at the University of Alabama.
- A glossary (separate from the index) provides definitions of all important terms, keyed by page reference to the definition in the text.
- Ten Appendixes are included. These give expanded coverage to mathematical skills, including scientific notation, significant digits, and logarithms; tables

listing thermodynamic, solubility, electrochemical, and acid ionization data; and, for the curious, some derivations from kinetic theory, chemical kinetics, thermodynamics, and hybridization.

Support Package

- *Student Solutions Manual,* prepared by Leslie Kinsland of the University of Southwestern Louisiana. This supplemental text contains worked-out solutions to half of the end-of-chapter problems. All odd-numbered questions, both numerical and word problems, are included. Special care has been taken to present each solution in the same format as used in *General Chemistry,* to reinforce the problem-analysis component of the text.
- *Instructor's Resource Guide,* by Clifford Hand. Organized by chapter, this guide contains the following features: a brief chapter overview, a detailed chapter outline (with the author's annotations), suggestions for classroom demonstrations, worked-out solutions to the summary exercises, and worked-out solutions to the even-numbered end-of-chapter questions.
- *Test Bank for General Chemistry,* by Harry L. Blewitt of the University of Alabama, contains more than 2800 multiple-choice test questions, organized by textbook chapter and section. The test bank is available in both printed and computerized forms (IBM PC and Macintosh).
- *Laboratory Experiments for General Chemistry, 2nd ed.*, by Harold Hunt and Toby Block, Georgia Institute of Technology. This manual contains 42 experiments, designed with careful regard for safety in the general chemistry laboratory. An instructor's manual is available.
- *Qualitative Analysis and the Properties of Ions in Aqueous Solutions,* by Emil Slowinski, Macalester College, and William Masterton, University of Connecticut. A paperback supplement that encourages students to develop their own schemes for qualitative analysis. [A brief scheme for metal analysis is presented in Chapter 17 of *General Chemistry.*]
- *Transparencies.* Approximately 125 full-color transparencies of tables, diagrams and graphs from the text are available for classroom use with an overhead projector.
- *Student Study Guide,* by William P. Jensen of South Dakota State University, contains chapter outlines, example problems, lists of important terms, self-tests with answers, and sample problems with answers.
- *Chemical Demonstration Videotapes,* by Bassam K. Shakhashiri of the University of Wisconsin. These tapes contain fifty 3 to 5 minute demonstration experiments for classroom use.
- *Videodisk and Barcode Manual.* This disk contains all the Shakhashiri demonstrations and over 600 images drawn from various Saunders sources. A barcode book provides easy access to any image or demonstration.
- *Periodic Table Videodisc: Reactions of the Elements,* by Alton J. Banks of North Carolina State University. This videodisc is a visual compilation of information about elements and their reactions with air, water, acids, and bases.
- *Saunders Chemistry Updates.* A periodic newsletter, sent to users of Saunders chemistry texts, contains new and newsworthy information that professors can share with students.

ACKNOWLEDGMENTS

Many people spent many hours reviewing the several drafts of the manuscript. Their suggestions, incisive and often polite, have resulted in significant improvement in the quality of the text. We are grateful to them all.

Linda Atwood, California Polytechnic State University
Harry L. Blewitt, University of Alabama
John A. Collins, Broward Community College
John M. DeKorte, Northern Arizona University
Thomas A. Eckman, University of North Alabama
Roy G. Garvey, North Dakota State University
L. Peter Gold, Pennsylvania State University
Wyman K. Grindstaff, Southwest Missouri State University
Anthony W. Harmon, University of Tennessee at Martin
Larry W. Houk, Memphis State University
William P. Jensen, South Dakota State University
Ronald C. Johnson, Emory University
Stanley N. Johnson, Orange Coast College
Leslie N. Kinsland, University of Southwestern Louisiana
John C. Lai, Livingston University
Richard C. Legendre, University of South Alabama
Gilbert J. Mains, Oklahoma State University
Richard J. Margelis, Kalamazoo Valley Community College
William L. Masterton, University of Connecticut
Charles A. Trapp, University of Louisville
Russel F. Trimble, Southern Illinois University at Carbondale
Archie S. Wilson, University of Minnesota
William V. Wyatt, Arkansas State University

I am grateful also to Carl Salter of Birmingham-Southern University for the software used to generate some of the atomic orbital plots. Steven Albrecht of Ball State University read galleys and page proof for accuracy. Fred Juergens of the University of Wisconsin made some useful suggestions for illustrations. William Vining of Hartwick College offered both good advice and good facilities for producing some of the photographs. William R. Sandel of Harvard Apparatus provided some of the molecular models. Harry L. Blewitt was a frequent and reliable sounding-board for matters of pedagogy. Lisa Boffa and Ginger Pinholster contributed the essays on Chemistry Old and New.

Many thanks to the staff at Saunders College Publishing: to Sandi Kiselica, whose sure touch as Developmental Editor kept the project moving in the right direction; to Margaret Mary Anderson, the Project Editor, who competently pulled the pieces together and kept it all on schedule; to Photo Researcher Dena Digilio-Betz, who found many of the photographs; to Charles Winters, not only a top-notch photographer but also an imaginative student of chemistry; and to Publisher John Vondeling, who brought strong vision and strong demands to the project.

CONTENTS OVERVIEW

CONTENTS

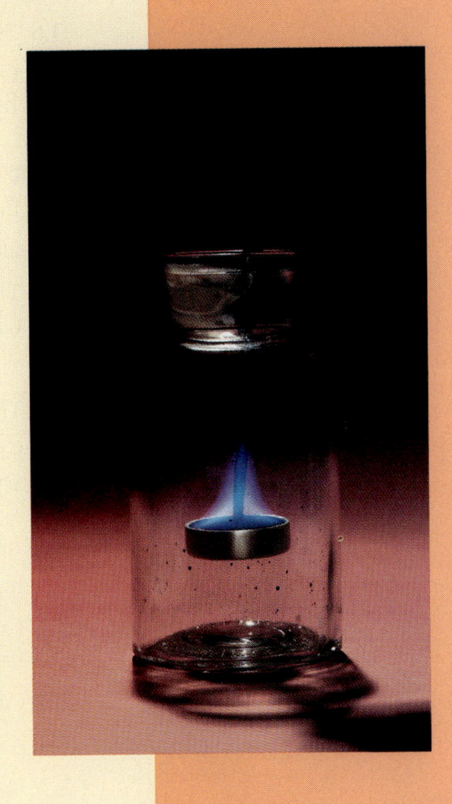

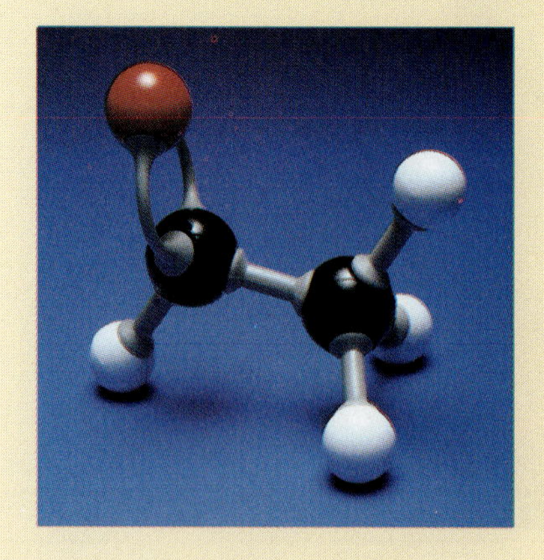

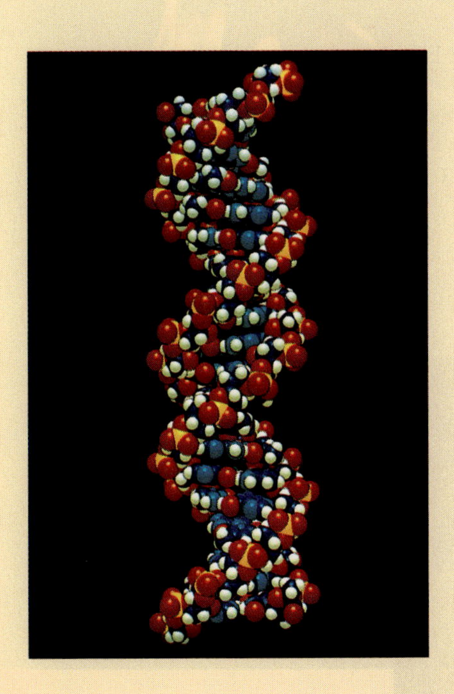

CHAPTER 1

SCOPE AND VOCABULARY OF CHEMISTRY

Bromine is one of the few substances that can be seen simultaneously in the gas, liquid, and solid states.

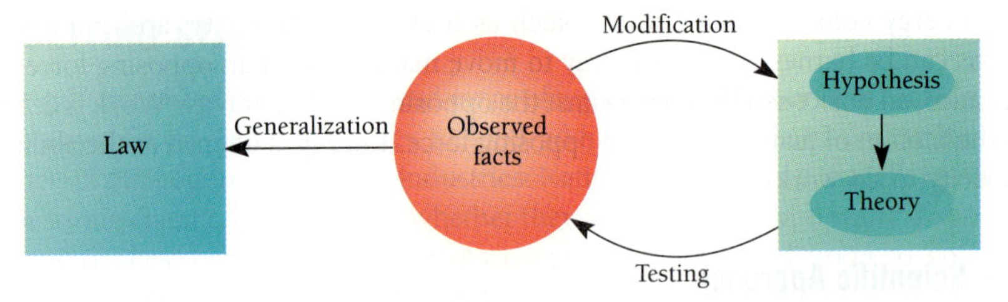

Figure 1.2 Observations of properties or behavior are the central, fundamental part of the scientific method. A group of related facts is sometimes summarized and generalized as a law. A group of facts usually stimulates the formation of a hypothesis, which is a tentative explanation of the facts. A hypothesis in turn stimulates testing. Often the new facts suggest modifications or refinement of the hypothesis, which in turn suggests additional experimental or observational tests. When a modified hypothesis has survived a great deal of testing without being disproven, it becomes known as a theory.

While a theory is the best available explanation of the facts, a theory is not "proved" by the facts, and cannot be regarded as being "true" in some absolute sense. There is always the possibility that new facts, which do not fit the theory, will be observed. A single reliable observation (a counterexample) can cause a theory to be discarded. You might, for example, develop a theory, based on countless observations, that metals are solid substances. Yet, *one* observation of liquid mercury is enough to disprove the theory. The history of science is replete with discarded theories, such as the geocentric model of the universe. For a scientist, a theory is neither true nor false. Rather, it is a good or useful theory if it explains all of the known facts and correctly predicts the outcomes of experiments not yet undertaken, and it is a bad, incomplete theory, or of limited use, if it does not. One difficulty, which we shall encounter from time to time in chemistry, is that scientific terms can have different meanings (sometimes strikingly different) when used in a nonscientific context. In ordinary speech, the word "theory" in a phrase like "It's only a theory," carries the connotation of shaky, irrelevant, and not to be taken seriously. In a scientific context, on the other hand, a theory has the highest degree of reliability: no better explanation of the facts is available.

The scientific method is an important tool in the practice of chemistry, but it is not always the means by which scientific knowledge is advanced. Often (some would say usually) new hypotheses and new insights into the workings of nature are reached through an intuitive process, and only later are the procedures of the scientific method applied. The role of intuition in science is certainly important, but hard to study since it is hidden from the conscious mind. The history of science abounds with anecdotes emphasizing the importance of intuitive thought.

A famous anecdote (almost certainly a myth) is that Newton formulated the theory of gravitation after watching an apple fall from a tree (or after being struck by a falling apple, depending on which version of the story you hear).

1.2 CLASSIFICATION OF MATTER

Physical State

Matter appears in many forms, differing in behavior as well as appearance. It is useful to describe or classify matter according to its physical state, its composition or purity, the ways in which it interacts with energy and other kinds of matter, and the nature of the tiny particles of which all matter is composed.

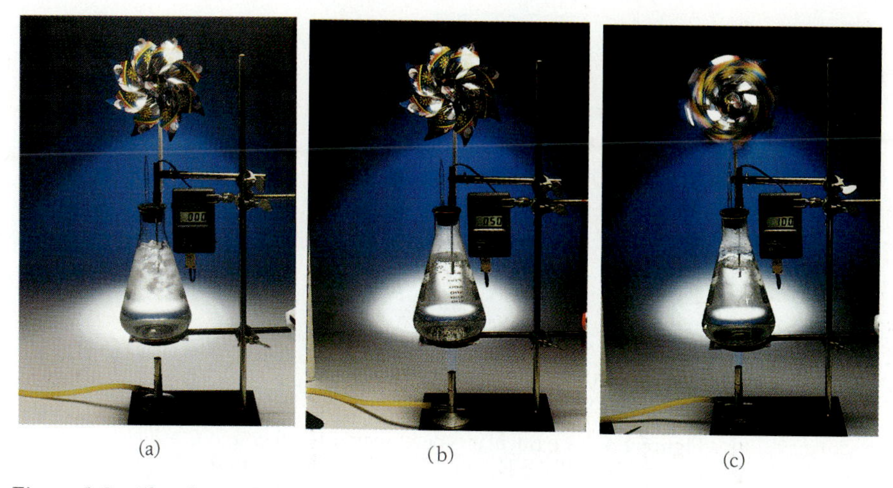

Figure 1.3 The phase transformations of H_2O are familiar to everyone. (a) At or below 0 °C, H_2O exists as solid ice. (b) Water is a liquid between 0 and 100 °C. (c) Under ordinary conditions, water boils and becomes steam at 100 °C.

The word "**substance**" is used to denote a particular type of matter. For example, air, water, and sand are substances. Substances exist in three different **physical states—solid, liquid,** and **gas.** A solid substance has both definite shape and volume; a liquid has a definite volume, but its shape is determined by its container. A gas, on the other hand, assumes not only the shape but also the volume of its container. Whereas the atoms composing a liquid or a solid substance are close enough to touch one another, the tiny particles of a gaseous substance are separated from one another, leaving considerable empty space in between. Sometimes the distinction between solid and liquid is not immediately clear. For example, sand assumes the shape of its container, but close examination reveals that the individual grains of sand have a fixed shape, unaffected by the shape of the container. Most solids, when heated, melt and become liquids. Liquids, in turn, evaporate or boil and become gases. Such changes in physical state are called **phase transformations.** Although different names are sometimes used (ice/water/steam), the different forms are the same substance (Figure 1.3). The word "phase" is sometimes used as a synonym for "physical state." Strictly speaking, however, phase refers to a component of a heterogeneous mixture (see below), as in the sentence, "The liquid phase of fog consists of water droplets."

Elsewhere in the universe, for instance in the sun's corona, in the core of neutron stars, or in black holes, other states of matter exist. However, where we live, substances are either solid, liquid, or gaseous.

The volume of a liquid or solid is fixed only if the temperature is unchanged. Virtually all liquids and solids expand slightly when heated.

Mixtures

A substance is either pure, or it is a **mixture** of two or more pure substances (Figure 1.4). Pure substances like water always have the same composition and properties, regardless of the origin or history of the particular sample of the substance. The **composition** of a mixture, that is, the relative amounts of the different substances, is not fixed. For example, a mixture of salt and sand may be 1 part salt and 10 parts sand or the other way around; both are called salt/sand mixtures. Mixtures can be made up of any combination of liquid, solid, or gaseous substances. Mixtures are either **homogeneous,** which means they have the same composition throughout the body of the sample, or **heterogeneous,** which means that within the sample there are regions of different composition. Using tweezers, for instance, you can locate and remove a grain of pure salt or of pure sand from a sand/salt mixture. Such a mixture is heterogeneous. No tweezers are delicate enough, however, to remove a piece of pure salt or pure water

A chunk of solid bromine has a definite size and shape. As a liquid, bromine has a definite volume but assumes the shape of the container. In the gaseous state, bromine assumes both the shape and volume of the container.

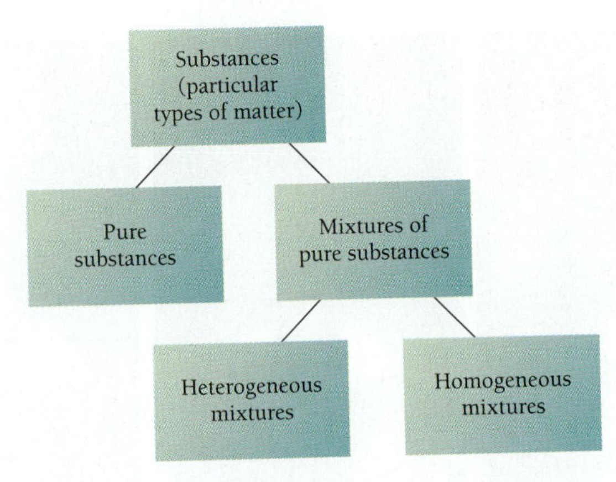

Figure 1.4 A particular type of matter is either a pure substance or a mixture. Heterogeneous mixtures contain regions of differing composition, while the composition of homogeneous mixtures is uniform.

from a sample of seawater. Homogeneous mixtures such as salt water are also called **solutions.** As seen in Figure 1.5, some solid substances are solutions. The air we breathe is a gaseous solution.

Figure 1.5 Brass is a homogeneous mixture, a solid solution of copper and zinc. Granite is a heterogeneous mixture of several different minerals.

Chemical and Physical Changes

Sometimes, when a substance undergoes a change, other, different substances are formed. For example, when charcoal burns in the backyard barbecue, it is converted to the new substances carbon dioxide and ashes. A **chemical change** is any change in which a new substance is produced. A **chemical reaction** is a specific chemical change that involves only a small number of different substances. Reactions are described by **chemical equations.** For example, the burning of carbon is represented by the following equation:

$$C + O_2 \longrightarrow CO_2$$

Chemical equations are discussed in some detail in Chapter 3.

Other processes, for example the grinding of sugar into a fine powder or the melting of ice, change the form or condition of a substance but do not produce new substances. These changes are called **physical changes.**

Chemical and Physical Properties

Some of the properties by which a substance is recognized and described are called **chemical properties,** because they can only be observed by inducing a *chemical change* in the substance. For example, the fact that a substance burns or neutralizes acids is determined only by igniting the substance or treating it with acid—both of these procedures generate new substances. Other properties, such as hardness, color, or melting temperature, can be observed without producing any new substances. Such properties are called **physical properties.** The components of a mixture can often be separated by "physical means," that is, by exploiting some difference in the physical properties of the pure substances (Figure 1.6).

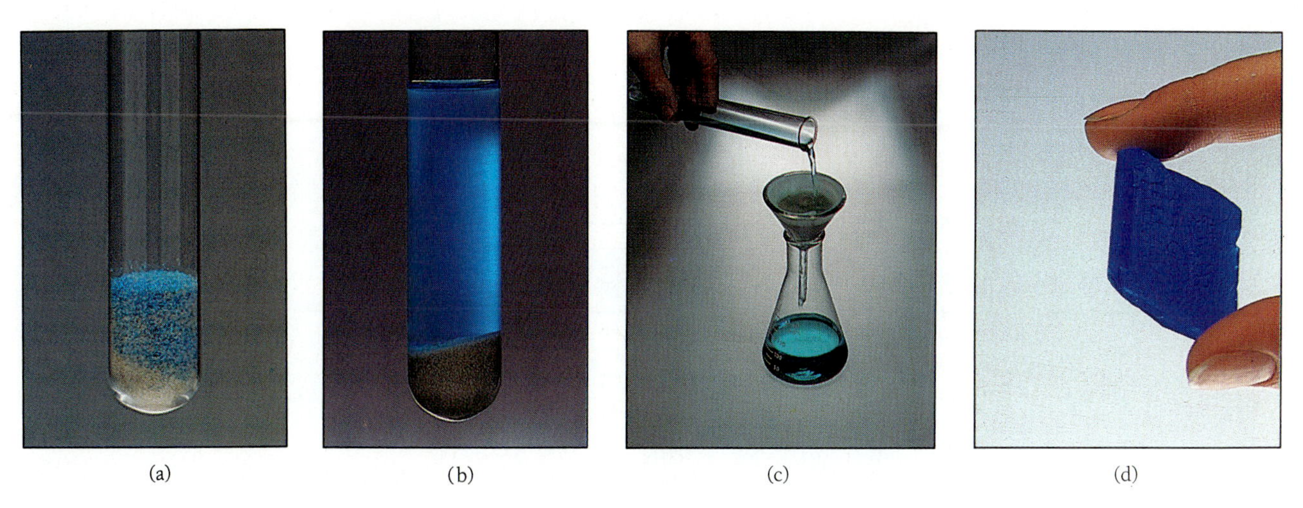

Figure 1.6 (a) A dry mixture of sand and copper sulfate can be separated by exploiting the difference in solubilities of these substances. (b) The copper sulfate dissolves in water, but the sand does not. (c) The mixture is filtered, leaving the solid sand in the filter paper. (d) When the water evaporates, crystals of pure copper sulfate are formed.

Elements and Compounds

The classification of matter can be extended to include elements and compounds (Figure 1.7). A pure substance is called an **element** if it is composed of only one type of atom. If a pure substance is composed of two or more different types of atoms, it is called a **compound.** There are 109 different types of atoms, hence there are 109 different elements. By contrast, millions of different compounds are known. The possibilities are virtually endless, because the 109 different elements can be combined in so many different ways. Each element has a name and an abbreviation of that name called the **elemental symbol.** An alphabetical list of the elements, together with their symbols and other information, appears on the inside front cover of this text. Most symbols come directly from the name. For example, the symbol for platinum is Pt. The symbols of nine of the elements are derived from their Latin names (Table 1.1). The symbol for tungsten (W) is derived from its German name, *Wolfram.* Elements

TABLE 1.1 Elements with Latin Symbols

English Name	Symbol	Latin Name
Antimony	Sb	*Stibium*
Copper	Cu	*Cuprum*
Gold	Au	*Aurum*
Iron	Fe	*Ferrum*
Lead	Pb	*Plumbum*
Mercury	Hg	*Hydrargyrum*
Potassium	K	*Kalium*
Silver	Ag	*Argentum*
Sodium	Na	*Natrium*

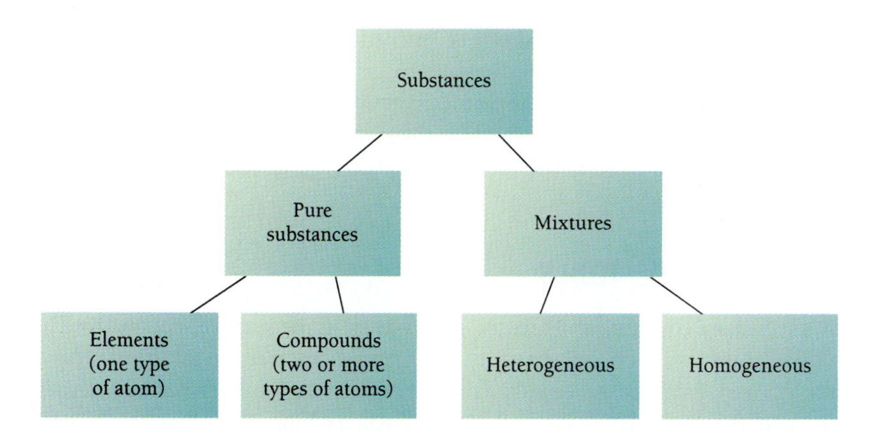

Figure 1.7 The classification of matter can now be extended by describing a pure substance as either an element or a compound.

1A	2A	3B	4B	5B	6B	7B	8B	8B	8B	1B	2B	3A	4A	5A	6A	7A	8A
1 H																1 H	2 He
3 Li	4 Be											5 B	6 C	7 N	8 O	9 F	10 Ne
11 Na	12 Mg											13 Al	14 Si	15 P	16 S	17 Cl	18 Ar
19 K	20 Ca	21 Sc	22 Ti	23 V	24 Cr	25 Mn	26 Fe	27 Co	28 Ni	29 Cu	30 Zn	31 Ga	32 Ge	33 As	34 Se	35 Br	36 Kr
37 Rb	38 Sr	39 Y	40 Zr	41 Nb	42 Mo	43 Tc	44 Ru	45 Rh	46 Pd	47 Ag	48 Cd	49 In	50 Sn	51 Sb	52 Te	53 I	54 Xe
55 Cs	56 Ba	57 *La	72 Hf	73 Ta	74 W	75 Re	76 Os	77 Ir	78 Pt	79 Au	80 Hg	81 Tl	82 Pb	83 Bi	84 Po	85 At	86 Rn
87 Fr	88 Ra	89 †Ac	104 Unq	105 Unp	106 Unh	107 Uns	108 Uno	109 Une									

*Lanthanide series	58 Ce	59 Pr	60 Nd	61 Pm	62 Sm	63 Eu	64 Gd	65 Tb	66 Dy	67 Ho	68 Er	69 Tm	70 Yb	71 Lu
†Actinide series	90 Th	91 Pa	92 U	93 Np	94 Pu	95 Am	96 Cm	97 Bk	98 Cf	99 Es	100 Fm	101 Md	102 No	103 Lr

Figure 1.8 The periodic table. Elements listed in the same vertical column have similar chemical properties.

Chemical nomenclature is overseen and approved by an international group known as IUPAC, the International Union of Pure and Applied Chemistry.

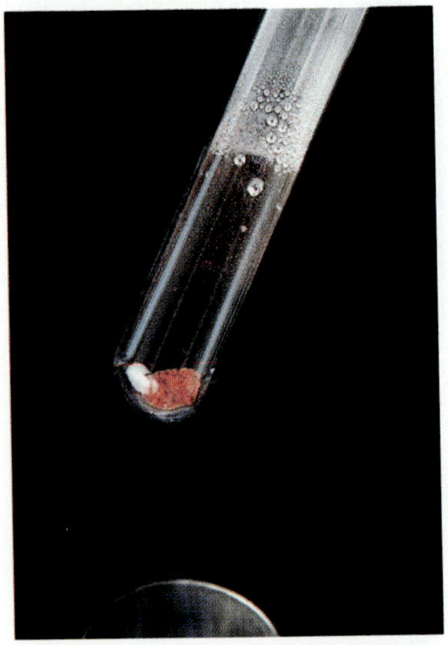

Figure 1.9 Heating of mercury(II) oxide results in the production of mercury, visible as silvery droplets near the mouth of the test tube, and oxygen, which passes unseen into the atmosphere.

106–109 have Latin names, assigned provisionally because of international disputes as to where and by whom the elements were first discovered.

The elements are commonly listed in a row-and-column format called the **periodic table,** shown in Figure 1.8 and reproduced in greater detail on the inside front cover of this text. The periodic table is arranged so that elements having similar chemical properties are in the same vertical column, called a **group.** The elements in the second column, for instance, are metals that react with water to form "bases," substances that neutralize the corrosive properties of acids. Each horizontal row in the table is called a **period.**

In the laboratory, compounds can be distinguished from elements on the basis of chemical properties. A compound can be broken down into simpler substances by means of chemical changes, but an element cannot. For example, heating induces the chemical transformation of the compound mercury(II) oxide into the elements mercury and oxygen (Figure 1.9). The fact that the elemental mercury produced has only 92.6% of the original mass of the mercury(II) oxide is evidence that mercury is the simpler substance.

1.3 SOME USEFUL TOOLS

Many specialized concepts and techniques are used in the study of chemistry. These include scientific notation and the units by which we describe the results of measurements, the use of significant digits to convey the accuracy and precision of measurements, and the factor-label method for unit conversions and other mathematical operations.

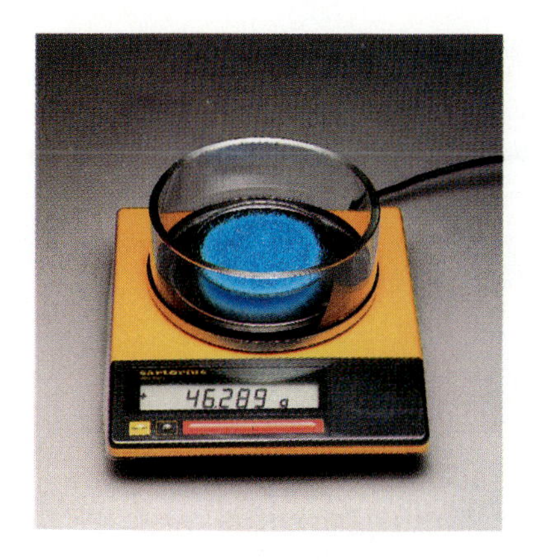

Figure 1.10 The physical quantity being measured here is mass. The result of the measurement is 46.289 grams.

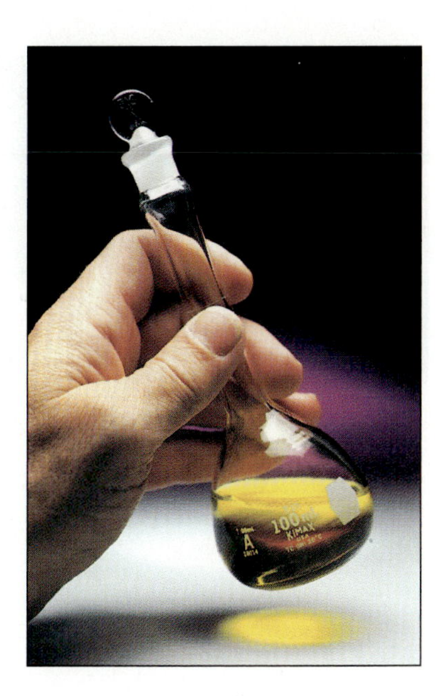

Figure 1.11 The quantities of liquids used in many laboratory experiments are measured in milliliters. This flask contains 50 mL of a solution of potassium chromate.

Numerical Quantities

A **physical quantity** in science supplies the answer to two questions: "How much?", and "Of what?" (Figure 1.10). Correspondingly, a quantity consists of two parts: the numerical part tells how much, and the **unit** tells of what. Units usually are abbreviated. For instance "s" stands for "seconds" or "mL" for milliliters (Figure 1.11). **Dimension** is often used as a synonym for unit, but there is a difference: *dimension* describes the type of quantity, while *unit* is a specific measure of a *quantity*. For example, seconds, days, and years are units of the dimension *time,* and centimeters, inches, and miles are units of the dimension *length.*

Units are an essential part of quantities, and must be included both in speaking and in writing. The world record for the 100 m sprint is 9.86 seconds; not "9.86," not "seconds," but "9.86 seconds." For many students inclusion of the units is a difficult habit to form, because in ordinary conversation we tend to omit them whenever they are implied by the context. We omit the time unit in a statement like, "My boyfriend is 22," because everyone who hears it automatically assumes, correctly, that the unit is "years." On the other hand, no one would say, "Last year my vacation lasted 3," since it is not clear from the context whether "days," "weeks," or "months" is intended. In science, it is not correct to omit the units, even when they can be inferred from the context.

Whenever two measured quantities having the same units are combined into a ratio, that ratio has no units. Suppose I weigh 116 lbs and my brother weighs 189 lbs. The statement, "My brother weighs 1.63 times as much as I do" contains a measured quantity that is *dimensionless,* that is, one that has *no units.* Other common dimensionless quantities are logarithms, exponentials, and trigonometric functions.

SI Units

A self-consistent set of units, called **SI (Système Internationale) Units,** is used to express almost all quantities in science. SI units are based on, and in many cases identical to, the units of the **metric system,** used by almost all countries of the world. In SI, seven quantities are regarded as fundamental and are referred to as **base quantities.** These base quantities, together with the symbols used to denote them, are listed in Table 1.2.

TABLE 1.2 Base Quantities in SI

Quantity	Symbol
Mass	m
Length	l
Time	t
Temperature	T
Amount of substance	n
Electric current	I
Luminous intensity	I_v^*

* I_v is seldom needed in chemistry.

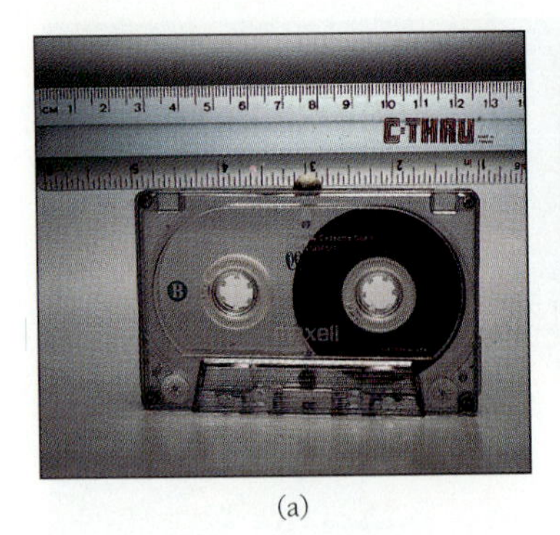

(a)

(b)

(c)

(d)

(a) The size of many familiar objects is measured in centimeters. This cassette tape is 10.0 cm long and 6.3 cm high. (b) The volume of this box is 1 cubic meter, about 30% greater than 1 cubic yard. (c) The length of each side of the cube is 10 cm. The pencil is 0.8 cm in diameter. (d) The same 10-cm cube has a volume of 1 liter (1000 cm^3). The mass of 1 L water is 1 kg. A half-filled 2-L soft drink bottle weighs about 1 kilogram.

Other physical quantities, such as speed, volume, or density, are products or quotients of the base quantities. Such physical quantities are known as "compound" or "derived" quantities. For example, the volume of a rectangular container is the product of three lengths (length, width, and height), and is denoted by $l \cdot l \cdot l$ or l^3. Speed is the quotient length divided by time. We will represent this quotient by lt^{-1}, and use similar notation for other quotients. The representation of speed by l/t is equally acceptable. However, problems can arise if there is more than one quantity in the denominator of a compound unit. A representation such as $m/l^3/t$ could mean either $(m/l^3)/t$ or $m/(l^3/t)$. Such ambiguity must be avoided.

The same ambiguity exists with numerical quotients: the value of 18/3/2 could be either 3 or 12.

Energy, an important and frequently encountered quantity in chemistry, is also a compound quantity. The dimensions of energy are ml^2t^{-2}, that is, mass times the square of length divided by the square of time.

The symbols m, l, and t are used for a generalized reference to a quantity, but the numerical value or *magnitude* of a specific quantity is expressed in particular units. For example, a particular object might have a length of 3.52 *meters* and a mass of 17.3 *kilograms*. The units of a compound quantity such as 0.97 cm s^{-1} are pronounced "centimeters per second." "Per" designates the unit in the denominator, as in the more familiar "miles per hour" (mi h^{-1}). The SI units for the five most important base quantities, together with the abbreviation of the unit, are listed in Table 1.3. Other derived quantities used in chemistry, together with their symbols, SI units, and relationships to base units, will be discussed as they arise.

TABLE 1.3 Units, Names, and Abbreviations

Quantity	Unit	Abbreviation
Mass	kilogram	kg
Length	meter*	m
Time	second	s
Temperature	kelvin	K
Amount of substance	mole	mol

* The internationally accepted spelling is "metre"; but, in the United States, "meter" is more common.

Scientific Notation and SI Prefixes

In the practice of science we encounter many physical quantities, ranging in size from the very large to the very small. For example, the height of an average man is about 1.8 m, while the diameter of an average atom is about 0.00000000035 m and the average distance from the earth to the sun is 150,000,000,000 m. These quantities are not only inconvenient to write, they are almost impossible to decipher or pronounce. The difficulties are resolved by the use of **scientific** (or **exponential**) **notation.** Using scientific notation, the atom diameter is written as 3.5×10^{-10} m, and the distance to the sun is written as 1.5×10^{11} m. The rules and procedures of scientific notation are summarized in Appendix A1.

TABLE 1.4 SI Prefixes

Name	Symbol	Multiplier	Name	Symbol	Multiplier
Femto-	f	10^{-15}	Deka-	da	10^{+1}
Pico-	p	10^{-12}	Hecto-	h	10^{+2}
Nano-	n	10^{-9}	**Kilo-**	k	10^{+3}
Micro-	μ	10^{-6}	**Mega-**	M	10^{+6}
Milli-	m	10^{-3}	Giga-	G	10^{+9}
Centi-	c	10^{-2}	Tera-	T	10^{+12}
Deci-	d	10^{-1}			

The magnitude of any SI unit can be modified by the use of a *prefix,* which becomes part of the name of the unit. Prefixes imply a multiplication by some power of ten. For example, *kilo-* means, "multiply the unit by one thousand (10^3)," while *centi-* means, "multiply the unit by 1/100 (10^{-2})." Thus, 1 kilometer (1 km) = 1000 meters, and 1 centimeter (1 cm) = 1/100 or 0.01 meter. SI prefixes, together with their abbreviations and corresponding multiplicative meaning, are given in Table 1.4. The more common prefixes are listed in boldface type.

Quantities are most often expressed using a prefix such that the number lies between 1 and 1000. If the quantity is 6.3×10^{-5} g, for example, it is more common to express it as 63 μg rather than 0.063 mg; however, all three forms are correct. There is a slight preference for multipliers whose exponents are divisible by 3; thus centi-, deci-, deka-, and hecto- are less common than the others in Table 1.4.

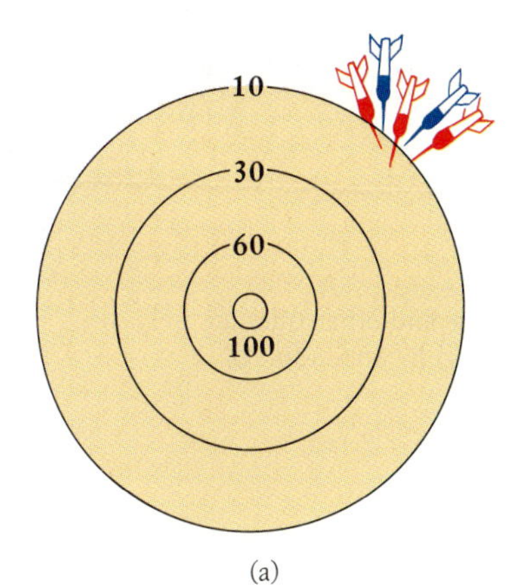

(a)

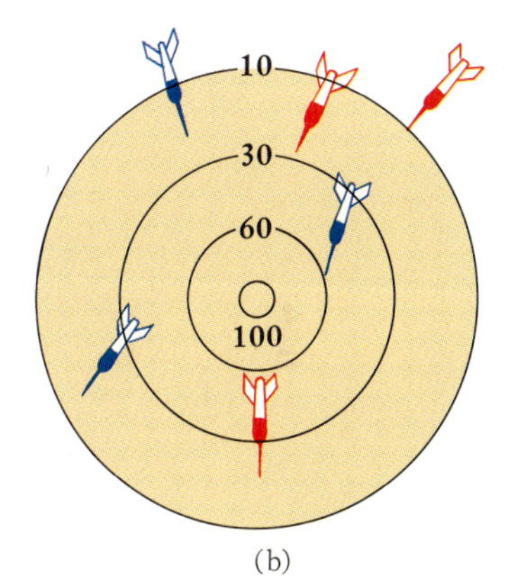

(b)

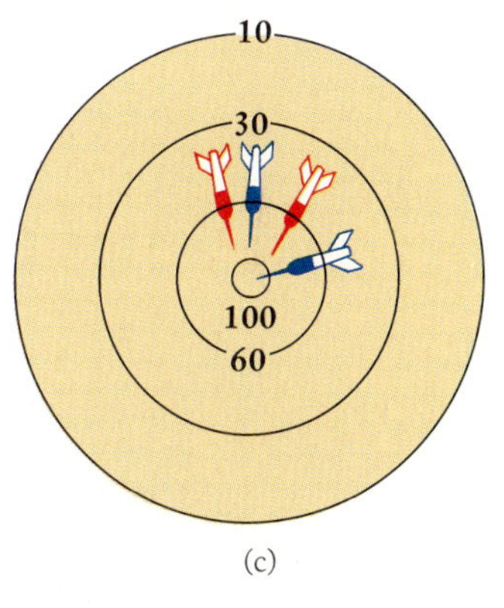

(c)

Figure 1.12 (a) The darts have been thrown with precision, since they are closely grouped. However, since the throws did not hit near the center of the target, the accuracy is poor: there are many possible reasons for this. (b) The second player is neither accurate nor precise, since the throws are not close to the center and are not closely grouped. (c) The third player is both accurate and precise.

Accuracy, Precision, and Significant Digits

A measured quantity can never be known *exactly,* since all measurements are subject to some degree of uncertainty. For example, following a visit to a friend you might check your odometer and see that you had driven 78.3 miles. You would be unable to tell if the actual distance was 78.316 or 78.292, however, because automobile odometers measure only to the nearest 1/10 mile. Also, there is the possibility that the odometer is in error, and that the actual distance is some totally different number such as 85 miles. In addition to magnitude and dimensions, any physical quantity has associated with it a precision and accuracy to which it is known.

Although often used as synonyms, the words "accuracy" and "precision" actually have different meanings. **Precision** indicates how well repeated measurements of the same quantity agree with one another; the word is meaningless when applied to a single measurement. **Accuracy** refers to how well a measurement agrees with the "true value." Often this true value is unknown, and the accuracy of a measurement can only be inferred. The word **"reliability"** is used in a general sense to mean accuracy, or precision, or both. Figure 1.12 illustrates the meaning of these three terms.

The way in which a quantity is written reveals the writer's knowledge (or opinion) of the reliability of the quantity. All the digits in a number are considered correct, except for the right-most digit, which has some uncertainty and is usually taken to be within ± 1 of the true value. Thus, "3.5 kg" means a mass that is greater than 3.4 kg but less than 3.6 kg. The notation "3.461 kg" means a mass that is greater than 3.460 kg and less than 3.462 kg. The number 3.5 has two **significant digits** (also called significant figures), while the number 3.461 has four significant digits. The greater the number of significant digits used to represent a particular quantity, the narrower the range within which the quantity lies; that is, the more reliably its magnitude is known. It is important to use a quantity's full reliability in calculations, and it is equally important to avoid overstating it. A multidigit number such as 3.4615385 often turns up in a calculator display, for instance as a result of dividing 4.5 by 1.3. Since these two factors are reliable only to two significant digits, however, it is misleading to write the quotient with eight digits. The rules for operations with significant digits include the following.

These and other rules are fully discussed in Appendix A2.

1. *Addition and subtraction.* The only decimal places that are significant in the answer are those having significant digits in *all* the addends or subtrahends.
2. *Multiplication and division.* The product or quotient is written with the same number of significant digits as the less reliable (that is, having the lesser number of significant digits) of the two factors.

The concept of significant digits does not apply to all quantities in chemistry. *Exact* quantities (integers, for example) are not the result of a measurement, and are not subject to error. Exact quantities do not affect the reliability of a calculation. For instance, if the measured mass of a new penny is 2.49 g, the mass of 3 pennies is $3 \cdot 2.49\ \text{g} = 7.47\ \text{g}$. The integer 3, although it has only a single digit, does not decrease the reliability with which the quantity 7.47 g is known. Exact quantities arise both from counting operations, for example 3 pennies, and from conversions between related units. The quantities in the statements, "1 centimeter equals 10 millimeters" and "1 inch equals 2.54 centimeters" are exact because they are the result of a *definition,* not a measurement.

Unit Conversions and the Factor-Label Method

Often it is necessary to express the value of a quantity in other than the given units. For instance, we might wish to express a distance of 86 kilometers in miles. Such operations are known as *unit conversions,* and they are performed by multiplying a quantity by a *unit conversion factor.* A **unit conversion factor** is a quotient whose numerator is a quantity expressed in one unit, and whose denominator is the same quantity expressed in another unit. The unit conversion factor connecting miles and kilometers is (1 mi/1.61 km), and the conversion of 86 km to miles is accomplished as follows.

$$?\ \text{mi} = 86\ \text{km} \cdot \frac{(1\ \text{mi})}{(1.61\ \text{km})}$$

The numerical part and the dimensional part of these quantities are treated separately.

$$?\ \text{mi} = \frac{86 \cdot 1}{1.61} \cdot \frac{\text{km} \cdot \text{mi}}{\text{km}}$$

Note that in the dimensional part the same factor (km) appears in both the numerator and the denominator, so that it may be cancelled.

$$?\ \text{mi} = \frac{54\ \cancel{\text{km}}\ \text{mi}}{\cancel{\text{km}}}$$

The result of this operation is the quantity expressed in the desired units.

Unit conversion factors such as (1 mi/1.61 km) are generated from more fundamental relationships called equivalence statements. An **equivalence statement** is a relationship between the magnitudes of the same quantity, expressed in different units. For example, the equivalence statement relating miles and kilometers is

$$1 \text{ mi} \Longleftrightarrow 1.61 \text{ km}$$

We use the symbol "⇔" rather than "=" as a reminder that this is a statement of *equivalence* rather than of *equality*. Clearly, 1 does not equal 1.61.

A unit conversion factor is formed by dividing one side of an equivalence statement into the other.

$$1 \text{ mi} \Longleftrightarrow 1.61 \text{ km}$$

$$\frac{1 \text{ mi}}{1.61 \text{ km}} \Longleftrightarrow \frac{1.61 \text{ km}}{1.61 \text{ km}}$$

Both the numerical and dimensional parts of the two quantities on the right-hand side of the equation cancel, leaving a dimensionless quantity whose value is 1.

$$\frac{1 \text{ mi}}{1.61 \text{ km}} \Longleftrightarrow 1$$

Since the resulting conversion factor is equivalent to 1, it may be inserted into any algebraic equation as a multiplier or divisor without affecting the meaning or value of the equation. Every equivalence statement generates two conversion factors, depending on which side is chosen as the divisor. Either conversion factor should be used, depending on the requirements of the problem.

$$1 \text{ mi} \Longleftrightarrow 1.61 \text{ km}$$

$$\frac{1 \text{ mi}}{1.61 \text{ km}} \Longleftrightarrow \frac{1.61 \text{ km}}{1.61 \text{ km}} \qquad \frac{1 \text{ mi}}{1 \text{ mi}} \Longleftrightarrow \frac{1.61 \text{ km}}{1 \text{ mi}}$$

$$\frac{1 \text{ mi}}{1.61 \text{ km}} \Longleftrightarrow 1 \qquad \text{and} \qquad 1 \Longleftrightarrow \frac{1.61 \text{ km}}{1 \text{ mi}}$$

EXAMPLE 1.1 Forming Conversion Factors From Equivalence Statements

a. Use the equivalence statement 2.54 cm ⇔ 1 inch to generate two conversion factors, then

b. Use the appropriate conversion factor to express the length 10.0 cm in inches.

The problem is solved as follows.

a.

$$2.54 \text{ cm} \Longleftrightarrow 1 \text{ inch}$$

$$\frac{2.54 \text{ cm}}{1 \text{ inch}} \Longleftrightarrow \frac{1 \text{ inch}}{1 \text{ inch}} \qquad \frac{2.54 \text{ cm}}{2.54 \text{ cm}} \Longleftrightarrow \frac{1 \text{ inch}}{2.54 \text{ cm}}$$

$$\frac{2.54 \text{ cm}}{1 \text{ inch}} \Longleftrightarrow 1 \qquad \text{and} \qquad 1 \Longleftrightarrow \frac{1 \text{ inch}}{2.54 \text{ cm}}$$

b. The quantity to be evaluated (the "target") has known units ("inches") but unknown magnitude ("?"), so we begin with the following incomplete equation.

$$? \text{ in} = (10.0 \text{ cm})$$

We next ask, "What multiplier is required so that the desired units are obtained?" Clearly, this quantity must have "cm" in the denominator, to cancel the original "cm," and "inch (in.)" in the numerator. Multiplication by the appropriate conversion factor then gives the quantity in the desired units.

$$? \text{ in.} = (10.0 \text{ cm}) \cdot \frac{(1 \text{ in.})}{(2.54 \text{ cm})} = 3.94 \text{ in.}$$

Many people prefer to emphasize the fact that a simple quantity such as 10.0 cm is actually a numerator, by writing it together with a dimensionless denominator of 1. Using this notation, the solution is

$$? \text{ in.} = \frac{(10.0 \text{ cm})}{(1)} \cdot \frac{(1 \text{ in.})}{(2.54 \text{ cm})} = \frac{(3.94 \text{ in.})}{(1)} = 3.94 \text{ in.}$$

Exercise

Use the equivalence statement 1 lb ⇔ 0.454 kg to generate two conversion factors. Then use the appropriate one to convert 156 lb to kg.

Answer: $\frac{0.454 \text{ kg}}{1 \text{ lb}}, \frac{1 \text{ lb}}{0.454 \text{ kg}}$; 70.8 kg.

In many problems a single unit conversion factor does not bring about the desired units, and two or more conversion factors must be used in a linked fashion. Since the value of all unit conversion factors is the dimensionless number 1, any number of them may be inserted as multipliers (or divisors) in an expression without affecting its value.

EXAMPLE 1.2 Conversion of One Mass Unit to Another

Following laboratory instructions calling for the use of 2.5 to 3.0 g powdered zinc metal in an experiment, a student weighed out 2.77 g. How many ounces is this? The equivalence statement for pounds and ounces (oz) is

$$16 \text{ oz} \Longleftrightarrow 1 \text{ lb}$$

The problem is solved as follows.

Using the approach of Example 1.1, we write down the target and observe that a conversion factor whose units are ounces per gram would achieve the correct cancellation of units.

$$? \text{ oz} = \frac{(2.77 \text{ g})}{(1)} \cdot \frac{(? \text{ oz})}{(1 \text{ g})}$$

Unfortunately, no such conversion factor is given in the problem. However, in order to cancel the *unwanted* unit (g) in the *numerator*, we must use a conversion factor having g in the *denominator*. This is generated from the equivalence statement 1 lb ⇔ 454 g (which must be either looked up or remembered), and its use in this problem will bring about a *partial* conversion of units. This is a step in the right direction.

$$? \text{ oz} = \frac{(2.77 \text{ \cancel{g}})}{(1)} \cdot \frac{(1 \text{ lb})}{(454 \text{ \cancel{g}})}$$

Now, what is needed is a conversion factor whose units are ounces per pound. This is generated from the equivalence statement given in the problem, so that

$$? \text{ oz} = \frac{(2.77 \cancel{\text{g}})}{(1)} \cdot \frac{(1 \cancel{\text{lb}})}{(454 \cancel{\text{g}})} \cdot \frac{(16 \text{ oz})}{(1 \cancel{\text{lb}})} = 0.0976 \text{ oz} = 9.76 \times 10^{-2} \text{ oz}$$

Multistep conversions are handled in this way, by seeking and using conversion factors that will cancel unwanted units and insert desired ones.

Exercise

Certain elements exist in several different forms, called *allotropes*. For example, one of the allotropes of the element carbon is graphite and another is diamond. Diamonds produced synthetically in a high-temperature, high-pressure process are usually small, flawed, and unsuitable for gems. Due to their hardness, however, they are valuable as abrasives.

The mass of diamonds and other gemstones is normally given in units of *carats*. Using the following conversion factors, and others as needed from the table on the inside back cover, calculate the mass in ounces and milligrams of a ¼ carat diamond. 1 carat ⇔ 3.09 grains, 7000 grains ⇔ 1 pound, and 16 ounces ⇔ 1 pound.

Answer: 1.77×10^{-3} ounces, 50.1 mg.

(a)

(b)

(c)

Graphite and diamond are two allotropic forms of carbon. (a) 0.2 grams of powdered graphite. (b) A 1-carat (about 0.2 g) gem-quality diamond. (c) Assorted industrial diamonds, whose total weight is 1 carat.

Unit conversions are a part of a broader approach to quantitative problems in science known as the **factor-label method.** In this method, units are treated algebraically in the same way as pure numbers—they can be multiplied, divided, cancelled, squared, and inverted. The only difference is that following such operations, units cannot always be reduced to a single value, while pure numbers can. Some examples of operations with units are given in Table 1.5.

Exponentiation (taking the square, cube, or square root, for example) of units deserves special mention because it is often incorrectly done.

TABLE 1.5 Operations with Pure Numbers and Units

Operation	Pure Numbers	Units
Multiplication	$3 \cdot 2 = 6$	(kilowatt) · (hour) = kilowatt-hour
Division	$\frac{3}{2} = 1.5$	$\frac{\text{mile}}{\text{hour}} = \frac{\text{mi}}{\text{h}} = \text{mi h}^{-1} = \text{mph}$
Cancellation	$\frac{\cancel{3} \cdot 4}{\cancel{3}} = 4$	$\frac{\text{m s}^{-1}}{\text{m}} = \text{s}^{-1} = \frac{1}{\text{s}}$
Exponentiation	$5^3 = 5 \cdot 5 \cdot 5 = 125$	$(\text{cm})^3 = \text{cm} \cdot \text{cm} \cdot \text{cm} = \text{cm}^3$
Reciprocation	$\frac{1}{5} = 5^{-1} = 0.2$	$\frac{1}{\text{kg}} = \text{kg}^{-1}$
	$\frac{1}{1/4} = \frac{1}{4^{-1}} = 4$	$\frac{1}{1/\text{mL}} = \frac{1}{\text{mL}^{-1}} = \text{mL}$
	$\frac{5}{5/3} = \frac{\cancel{5}}{\cancel{5} \cdot 3^{-1}} = 3$	$\frac{\text{mi}}{\text{mi/h}} = \frac{\cancel{\text{mi}}}{\cancel{\text{mi}} \cdot \text{h}^{-1}} = \text{h}$

EXAMPLE 1.3 Conversion of Cubic Units

The cubic centimeter (cm^3 or cc) is an SI unit of volume. In the laboratory, however, liters or milliliters are more commonly used to measure volume. The relevant equivalence statements are

$1\ cm^3 \Longleftrightarrow 1\ mL$

$1\ mL \Longleftrightarrow 0.001\ L$

$1\ qt \Longleftrightarrow 0.946\ L$

The SI base unit of volume is the cubic meter (m^3). Since 1 m^3 is a rather large volume (about 280 gallons), however, the liter (L) is more commonly used in laboratory operations. A liter is a little more than a quart. Convert the quantity 0.00264 m^3 first to (a) cubic centimeters, then to (b) liters using the equivalence statement 1 $cm^3 \Leftrightarrow$ 1 mL.

The problem is solved as follows.

a. ***Target:*** A volume in cubic centimeters, ? cm^3.

Starting Point: Volume in cubic meters, 0.00264 m^3.

Known Information: The equivalence statement 1 mL $\Leftrightarrow$ 1 cm^3 is given. Furthermore, the equivalence statement 1 cm $\Leftrightarrow$ 0.01 m is implicit in the metric prefix *centi-*, and 1 mL $\Leftrightarrow$ 0.001 L is implied by *milli-*.

Work

$$? \text{ cm}^3 = \frac{(0.00264 \text{ m}^3)}{(1)} \cdot \frac{(?)}{(?)}$$

The centimeter/meter conversion factor must be applied *three times,* because the quantity being converted is a volume (l^3) rather than a distance (l).

$$? \text{ cm}^3 = \frac{(0.00264 \text{ m}^3)}{(1)} \cdot \frac{(1 \text{ cm})}{(0.01 \text{ m})} \cdot \frac{(1 \text{ cm})}{(0.01 \text{ m})} \cdot \frac{(1 \text{ cm})}{(0.01 \text{ m})}$$

This operation is usually written in abbreviated form,

$$? \text{ cm}^3 = \frac{(0.00264 \text{ m}^3)}{(1)} \cdot \left(\frac{1 \text{ cm}}{0.01 \text{ m}}\right)^3 = 2.64 \times 10^3 \text{ cm}^3$$

Note that if the cm/m conversion factor is used only once (a common error), it can be seen from the units alone that the result is incorrect.

$$? \text{ cm}^3 = \frac{(\text{m}^3)}{1} \cdot \frac{(\text{cm})}{(\text{m})} = \frac{(\text{m}^2 \text{ cm})}{1}$$

b. The second part of the problem is also achieved in a two-step conversion.

$$? \text{ L} = \frac{(2.64 \times 10^3 \text{ cm}^3)}{(1)} \cdot \frac{(1 \text{ mL})}{(1 \text{ cm}^3)} \cdot \frac{(1 \text{ L})}{(1000 \text{ mL})} = 2.64 \text{ L}$$

Exercise

Trace amounts of impurities often impart an unpleasant taste or odor to drinking water. These offensive substances can be removed by passing the water through powdered charcoal (a form of carbon) contained in a small unit mounted on a water faucet. The method is effective primarily because of the large total surface area of a finely divided solid. Each gram of powdered charcoal, depending on how it was prepared, can have a total surface area of several million square inches. Convert 2.0×10^6 square inches to SI units.

Answer: 1.3×10^7 cm^2 or 1300 m^2.

(a)

(b)

(a) A tap-mounted filter is a convenient and effective means of water purification. (b) Inside the filter cartridge, impurities are captured and held on the surface of small particles of charcoal.

A second important feature of the factor-label method is that the units identify which of several possible solutions is correct, especially if the units are unfamiliar. This is illustrated in Example 1.4.

EXAMPLE 1.4 Conversion of Unfamiliar Units

In an unnamed foreign country, one kep of blarg costs 1.99 ziltags. Given that the exchange rate is 1 ziltag ⇔ \$0.37 and that there are 2.92 kilograms in one kep, calculate the cost in dollars of a kilogram of blarg.

Analysis and Work The target (\$) and the starting point [1 kg (of blarg)] are clear from the problem, but the unfamiliarity of the units makes what should be done with the other numbers far from obvious. Nonetheless the problem is stated in a way that resembles more familiar unit conversions, so we treat it as such and begin with the following.

$$? \$ = \frac{[1 \text{ kg (of blarg)}]}{(1)} \cdot \frac{(?)}{(?)}$$

Now try different combinations of the remaining numbers, *looking only at the units,* until the correct cancellation is achieved. These trials can be done systematically, as in Example 1.2, or by pure guesswork. The only combination that works is the following.

$$? \$ = \frac{[1 \text{ kg (of blarg)}]}{(1)} \cdot \frac{(1.99 \text{ ziltag})}{[1 \text{ kep (of blarg)}]} \cdot \frac{(\$0.37)}{(1 \text{ ziltag})} \cdot \frac{(1 \text{ kep})}{(2.92 \text{ kg})}$$
$$= \$0.25$$

Exercise

Express the price of blarg in ziltags per kilogram.

Answer: 0.68 ziltags per kg.

1.4 AN APPROACH TO PROBLEM-SOLVING

In the preceding section, we solved several numerical problems. The ability to solve problems is a very important skill in science in general and in chemistry in particular. Although there is more to chemistry than problem-solving, it is usually true that good problem-solvers do well in chemistry. But problem-solving is not an inborn ability possessed by some and not by others. The ability to solve problems is a skill that can be learned. As with other skills, practice is very important. Imagine the results of trying to learn other skills without practice—"Here's a tennis racquet and a rule book, the tournament starts in five minutes!" Yet anyone can learn to play tennis well enough to participate, with pleasure and satisfaction, in the sport. In the same way, anyone can learn problem-solving well enough to succeed in college chemistry. It's worth the effort, both because the alternative is to close yourself off from chemistry, a field of human activity that has far-reaching effects on our daily lives, and because problem-solving skills can be applied to many aspects of our lives.

The single most important skill in problem-solving is the careful analysis of a complicated problem into a sequence of simple steps. It is also the hardest to learn, because we all have a strong urge to solve problems by a single lunge to the correct answer. The lunge method is seductively simple; "Find the right formula and plug in the numbers given in the problem." Now, although this method sometimes works very well, in this imperfect world there are many problems that cannot be solved with a lunge. To try to do so leads to frustration and failure, which in turn create the false feeling that science is impossibly difficult and beyond the reach of ordinary people.

Our method will be to develop a stepwise procedure or recipe for analyzing and solving problems. The recipe cannot be followed mechanically, without any thought on the part of the problem-solver. However, it will show you how to approach and solve problems that on first glance seem impossible. After you have become accustomed to using the procedure, you will no doubt develop some shortcuts. However, at the beginning you should be patient until, through practice, you develop a level of skill that allows you to solve chemistry problems with confidence and satisfaction.

We illustrate the approach with the following problem: How much time should you allow for a trip of 190 miles, if your average speed is 49 mph and you expect to waste 45 minutes getting lost?

1. *Identify the target.* The first step is to *identify the target.* Describe, as completely as you can, what will constitute a satisfactory answer. Is the answer a time, a speed, a temperature, a mass? An important part of this step is to *write down the units* of the target, using a question mark for the numerical part of the quantity. Although it may seem strange to write down the answer before solving the problem, it is important to do so because it gives direction to your effort. The more clearly the target is described, the easier it will be to reach.

   The target of the example problem is a period of time. The units of the answer could be any time unit such as seconds, hours, or days, but since minutes and hours occur in the problem, choose one of these for the target.

   Target: A period of time.

   $$? \text{ h} =$$

2. *Examine the data.* The statement of a problem always contains information that is to be used in reaching the answer. The information is called the "knowns" or the "givens," or sometimes the "conditions."

a. *Identify the explicit knowns.* Carefully reread the problem with the intention of locating *all* the information given explicitly. In our example, the *explicit knowns* are 190 miles, 49 mph, and 45 additional minutes.

b. *Look for implicit knowns.* Some of the required information may be implied rather than stated explicitly in the problem. Necessary equivalence statements are often omitted because they are commonly known or easily looked up, and the problem should be examined to see if any such information is clearly missing. As you first read a problem, it may not be obvious that a particular piece of missing information is necessary. The possibility of implicit knowns should be kept in mind throughout the problem-solving process. Our example problem refers to both hours and minutes, and it is likely that a unit conversion will be required. The equivalence statement 1 hour ⇔ 60 minutes is thus an implicit known. So far we have the following.

Target: A length of time.

$$? \text{ h} =$$

Knowns: 190 mi, 49 mph, 45 min; 1 h ⇔ 60 min (implicit).

3. *Discover the relationship between the knowns and the target.* This is the most important step in problem-solving; without it, the problem cannot be solved. The relationship may be simple, such as a mathematical formula, or it may be a complex chain of reasoning that must be approached stepwise. Finding the relationship is also the most difficult step in problem-solving, and the only one that may call for creative thinking. There are several possibilities for finding the relationship; they are arranged in order of increasing complexity, and should be tried sequentially until success is achieved. The easiest is the lunge.

a. *Do you know a formula* that will lead you directly from the knowns to the target? In our example, there is no single formula. But consider a simpler example: "Calculate the volume of a sphere of radius $r = 5$ cm." If you remember that the formula for the volume of a sphere is $V = 4\pi r^3/3$, what remains is only the calculation of the value of the quotient $4(3.14)(5 \text{ cm}^3)/3$ (note that $\pi = 3.14$ is an implicit known in this problem). The major pitfall in this approach is the use of an incorrect or inappropriate formula, such as $V = \pi r^2$. Inspection of the units, however, immediately shows that the second formula *cannot* be correct: the right-hand side has the dimensions of area (l^2) and the target is a volume (l^3). Always check that a formula is dimensionally correct. A formula can be wrong for less obvious reasons: $V = r^3$ has the correct dimensions, but gives the volume of a cube rather than a sphere. If you cannot remember the formula, look it up; don't guess. If a formula won't solve the problem, other tactics must be used.

b. *Do the units of the knowns suggest a route to the target?* The factor-label method is an *invaluable* aid in problem-solving. In many problems, the given quantities have units that can be combined in only one way to yield a quantity having the dimensions of the target. Write down the knowns and try different ways of combining them. This trial-and-error procedure may well stimulate your memory and lead you to a useful formula. At the very least, it will allow you to discard some unproductive approaches. For instance, in our example, the knowns could be combined in the form

$$\frac{(190 \text{ miles})(49 \text{ miles hour}^{-1})}{(45 \text{ minutes})}$$

Since this combination has units of miles^2 hour^{-1} minutes^{-1}, and the target has units of hours, we may at once discard it.

c. *Have you solved a similar problem before?* If you have previously solved a problem such as, "How long does it take to walk a distance of 25 km, at an average pace of 20 minutes per kilometer and allowing 1 hour for resting?", it may be possible to adapt the previous approach to the present problem. Compare the two:

	Present Problem	Previous Problem
Target:	? (time) =	? (time) =
Distance:	190 mi	25 km
Interruption:	45 min	1 h
Speed:	49 *mi h*$^{-1}$	20 *min km*$^{-1}$

Both problems have the same target, a total time, and both give information as to distance, speed, and duration of interruptions. The two problems are similar, although not identical. The difference in magnitude and units of distance (km, mi) and time (hours, minutes) do not affect the *method* of reaching the answer. A more important difference, which requires a modification of procedure, lies in the way the speed information is given: one problem gives distance per unit time (49 miles per hour) and the other gives time per unit distance (20 minutes per kilometer). If the speed information of one of the problems is inverted, [1/(49 mi h^{-1}) = 0.020 h mi^{-1}], the problems are dimensionally identical and solved in exactly the same way. Problem-solvers must be able to follow the same general approach while allowing sufficient flexibility to accommodate the differences.

d. *The plan ("If only I knew . . .")*: If the lunge does not work, and the solution to a similar problem cannot be recalled, it becomes necessary to simplify the problem by analyzing it as a series of simple steps. The stepwise approach builds a logical chain connecting the target to the knowns. As always, a clear notion of direction, that is, a clear specification of the target, is required.

Sometimes the result of careful thinking about the target is the realization that the problem could be solved if only some missing bit of information were available. In our example, you might say to yourself, "*If only I knew* how long the actual driving took, I could add it to the 45 minutes spent in getting lost, and that would be the answer."

We have just accomplished two things that are essential to solving the problem: we have *specified the missing information,* and we have *planned the final step* of the solution. In so doing we have created the first link in a chain leading *backward* from the target to the knowns. It is much easier to work problems in this backwards fashion than to try to build a chain forward from the givens to the answer. Just as finding your way home from an unfamiliar place is easier than finding an unfamiliar place from your home, in problem-solving it is easier to work from the unknown to the knowns.

We have now identified, as part of the original problem, a *related, but simpler problem:* "Calculate the time it takes to drive 190 miles at an average speed of 49 mph." In doing so, we have also formulated a *plan* for working the problem. The steps are written in reverse order to emphasize that the plan was developed by working backwards from the target.

Second step. Add the lost time to the driving time.

First step. Calculate the driving time.

In more complicated problems it may be necessary to repeat this process several times. That is, it might turn out that the new, simpler problem still cannot be easily solved. Then we must think of the simpler problem as a new "original" problem and continue. This approach leads to another related but simpler problem, that is, another step in the plan. The end result of this repetitive process is the analysis of a complex problem, which cannot be solved directly, into a sequence of simpler steps. An easily followed trail will have been laid from the answer back to the starting point. An important aspect of the "if only I knew" approach is that it allows us to get started doing something constructive, after the lunge fails. We have now progressed to the following.

Analysis

Target: A length of time.

$$? \text{ h} =$$

Knowns: 190 mi, 49 mph, 45 min; 1 h = 60 min (implicit).

Relationship: No formula, but "if only I knew" suggests the following steps.

Plan

Step 2. Add the lost time to the driving time.

Step 1. Calculate the driving time.

4. *Carry out the steps of the solution.* After the formulation of the stepwise plan, only the mathematical operations remain. Perform each of the steps of the solution, in a *forward* direction this time. Do not round off the intermediate result, but do round the final answer to the correct number of significant digits.

Step 1.

Target: Driving time.

$$? \text{ h} =$$

Knowns: 190 mi, 49 mi h^{-1}.

Relationship: Is there a formula? Yes, there is: time = (distance)/(speed). This formula could have been reached by (a) remembering it or looking it up; (b) realizing that only one combination of knowns, namely (190 mi)/(49 mi h^{-1}), is dimensionally correct; or (c) remembering the related formula, distance = (speed) · (time), and rearranging it so as to isolate the unknown on the left.

Plan: Insert the knowns into the formula and evaluate.

Work:

$$\begin{aligned} ?\ \text{h} &= (\text{distance})/\text{speed}) \\ &= (190\ \text{mi})/(49\ \text{mi h}^{-1}) = 3.88\ \text{h}^{*} \end{aligned}$$

Note: Only two significant digits would be used if this quotient were the final answer to the problem. Following the rules given in Appendix A2, we carry extra digits in the intermediate stages of a calculation and round only the final answer to the correct number of significant digits.

Step 2. Add the two times

$$\begin{aligned} ?\ \text{h} &= \text{driving time} + \text{lost time} \\ &= 3.88\ \text{h} + 45\ \text{min} \end{aligned}$$

Since only quantities having the same units may be added or subtracted, a unit conversion must be performed.

$$\begin{aligned} &= 3.88\ \text{h} + \frac{(45\ \text{min})}{(1)} \cdot \frac{(1\ \text{h})}{(60\ \text{min})} \\ &= 3.88\ \text{h} + 0.75\ \text{h} \\ &= 4.63\ \text{h} \end{aligned}$$

The final answer, rounded to the correct number of significant digits, is

$$4.6\ \text{h}$$

5. *Check your work.* The final step in the procedure is to check your work. Although checking is essential, it is often omitted (even by experienced problem-solvers) in the mistaken belief that it's not worth the effort. It is. Your work should be checked after each step is performed, and again after the final answer is obtained.

There are several aspects to checking: the answer should (a) have the correct units; and (b) have a reasonable, sensible value; which is (c) in accord with an approximate, mental estimate. Use of the factor-label method is essential for keeping track of units, but units cannot be checked unless they were originally written down. If you try to save time by omitting the units, you

* This result could also have been reached by the factor-label method. The speed 49 mi h^{-1} is in fact a conversion factor derived from the equivalence statement 49 mi ⇔ 1 h. The other conversion factor, 1 h/49 mi, is used to solve the problem. As in Example 1.4, only one combination of these quantities gives the desired unit (h).

$$?\ \text{h} = \frac{(190\ \text{mi})}{(1)} \cdot \frac{(1\ \text{h})}{(49\ \text{mi})} = 3.88\ \text{h}$$

will have difficulty checking the results. In that case you may as well save more time by omitting the check, and resign yourself to a more-or-less steady diet of incorrectly worked problems.

a. Check the units. In checking the first step of the example problem, we find that, as they should, the units cancel to give the units of the target:

$$h = \frac{(\text{mi})}{(\text{mi h}^{-1})} = (\text{mi}) \cdot (\text{mi}^{-1}\ \text{h}) = \text{h}$$

In the second step as well, the units are correct, and there has been no addition or subtraction of quantities having different units.

$$h = h + \cancel{\text{min}} \cdot \frac{(\text{h})}{(\cancel{\text{min}})}$$

b. Is the answer reasonable? Numerical quantities do not exist by themselves; they are representations of some measurable aspect of the world. If we are told that after 3 weeks of illness Sydney's weight dropped to 4383 pounds, we know immediately that someone has made an error. The value is not sensible, because it is outside the reasonable range of human weight. Numbers in chemistry have reasonable ranges, too, although at first we are not as familiar with them as we are with everyday numbers like weights, distances, or prices. After studying chemistry for a while, you should know that 30 g cm^{-3} is an unreasonably high value for a density, or that 0.75 g mol^{-1} is an impossibly low value for a molar mass. You should try to develop a feel for the reasonable range as you learn new concepts in chemistry; it will put you in a much better position for checking the results of problem-solving. In checking the example problem, we find that the answers in both steps, 3.9 and 4.6 hours, are reasonable for a trip of 190 miles.

The possibility exists that Sydney might be a rhinoceros. If so, we may not be certain that an error has been made. We should make sure that we have accomplished the first step, *"identify the target."*

c. Estimate the answer. The procedure is to round off the numbers to 1 or 2 significant digits, and do the calculation mentally: 190 is about 200, and 49 is about 50; driving time is therefore about 200/50 or 4; 45 minutes is about 1 hour, which added to 4 is about 5; these estimates, ≈ 4 and ≈ 5, are close to what we calculated. Thus there is no obvious numerical error, and unless you wish to repeat the calculator operations as a double check there is nothing further to be done.

Note: Many problem-solvers prefer to estimate the answer before the calculation, rather than after.

As we proceed in this text we will have many occasions to solve problems, and there will be many examples of worked problems to study. We will make consistent use of the systematic approach. Here is a summary of the steps involved:

1. Identify the target.
  a. Describe the target as precisely as you can.
  b. Write down the units of the target.
2. Examine the data.
  a. What quantities are given in the problem?
  b. What equivalence statements or other facts are implicit in the problem?

3. Discover the relationship and plan the solution.
   a. Is there a formula? Is it dimensionally correct?
   b. Can the units of the knowns be combined to suggest an approach?
   c. Have you solved a similar problem before?
   d. "If only I knew. . . ."
      i. Missing information.
      ii. Simpler but related problem.
      iii. The plan is a backward chain.
4. Carry out the plan.
5. Check the work (stepwise and final answer).
   a. Are the units correct?
   b. Is the size of the number reasonable?
   c. Is the answer close to the estimate?

1.5 SOME QUANTITIES IN CHEMISTRY

One of the most important physical quantities in chemistry is the temperature. The temperature determines whether a substance is in the solid, liquid, or gaseous state. The chemical properties of a substance are also strongly influenced by temperature. Two other commonly encountered physical properties are density and specific gravity.

Temperature

Daniel Gabriel Fahrenheit (1686–1736) was a German scientist who moved to Amsterdam at the age of 31. There, although he was not associated with a university, he gave lectures in optics and other branches of physics.

Anders Celsius (1701–1744), a Swedish scientist whose main interest was in astronomy, developed his temperature scale in 1742. The Celsius scale was formerly called the centigrade scale.

William Thompson, First Baron Kelvin (1824–1907), was professor of natural philosophy at the University of Glasgow for more than 50 years. He worked primarily in the fields of heat and electricity, and was responsible for the widespread acceptance of the law of conservation of energy.

The concept of temperature is a familiar one, because we can easily sense the difference between hot and cold with our bodies. In day-to-day life, citizens of the United States use a numerical scale, devised by Fahrenheit, for measurement of temperature. Fahrenheit assigned the value 0 to the lowest temperature attainable by mixing salt and ice, and the value 100 to the temperature of the human body. On the **Fahrenheit scale** water freezes at 32 °F and boils at 212 °F. Since these two temperatures can be reproduced more easily, they have supplanted Fahrenheit's original defining points for the Fahrenheit scale. Outside the United States, the **Celsius scale** is used to measure temperature. On the Celsius scale the freezing point of water is assigned a temperature of 0 °C, and the boiling point is assigned the value 100 °C. Since 180 Fahrenheit degrees (F°) separate the freezing and boiling points of water, while only 100 Celsius degrees (C°) separate them, the Celsius degree is larger than the Fahrenheit degree. The equivalence statement is 1 C° $\Leftrightarrow$ 1.8 F° (exact). A third scale, the **Kelvin scale,** was developed later as a consequence of experimental studies of the behavior of gases and of theoretical insights into the nature of heat. The Kelvin degree is the same size as the Celsius degree, but the scale is shifted so that water freezes at 273.15 K and boils at 373.15 K. Note that the unit is K, not °K. Because 0 K is the lowest possible temperature, this scale is sometimes called the "absolute temperature" scale. Only the Celsius and Kelvin scales are used in science. Figure 1.13 shows the relationships between the three temperature scales.

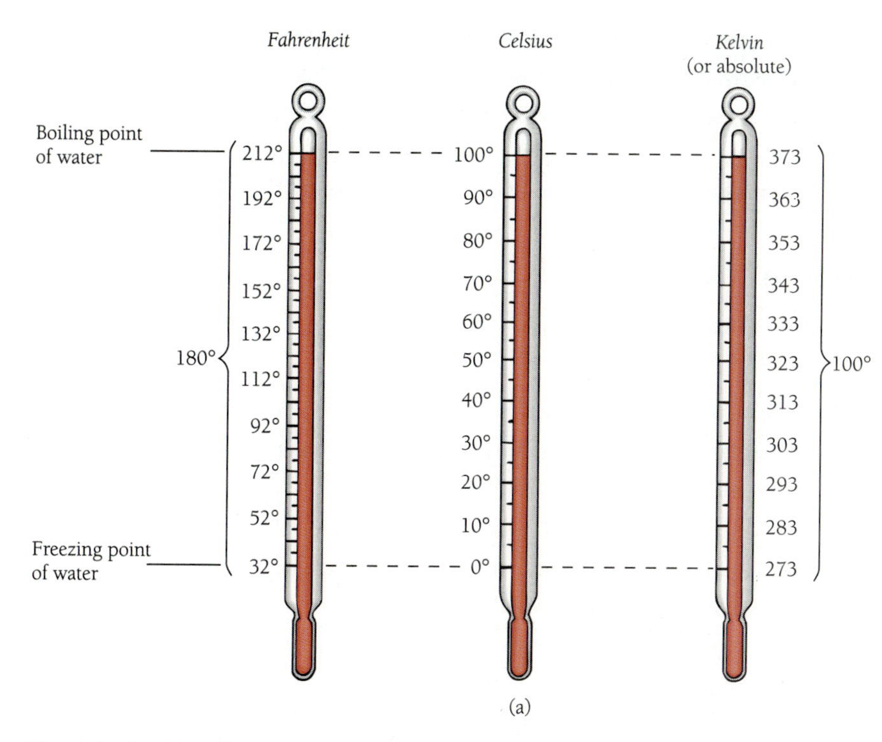

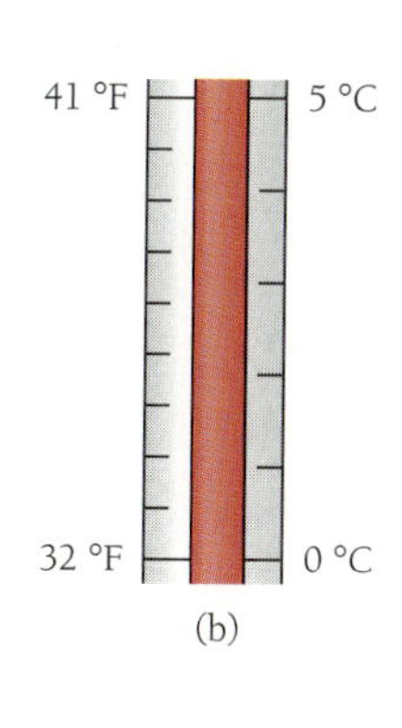

Figure 1.13 (a) Different temperature scales can be related by comparing the values each scale assigns to the fixed points at which water freezes and boils. (b) The C° and the K degree are the same size, but the F° is smaller by (100/180). It takes 9 F° to span the 5 C° range between 0 °C and 5 °C.

A temperature value in one scale can be expressed in another by means of the following relationships.

$$°F = 1.8(°C) + 32; \ °C = \frac{°F - 32}{1.8} \qquad (1\text{-}1)^*$$

$$K = °C + 273.15; \ °C = K - 273.15 \qquad (1\text{-}2)^*$$

* The relatively simple form of these temperature conversion equations has been achieved by omission of the units of the conversion factor 1.8, derived from the equivalence statement 1.8 F° ⇔ 1 C°. The dimensionally complete forms are:

$$T\ °F = (T\ °C) \cdot \frac{1.8\ F°}{1\ C°} + 32\ °F \quad \text{and}$$

$$T\ °C = (T\ °F - 32\ °F) \cdot \frac{1\ C°}{1.8\ F°} \qquad (1\text{-}1a)$$

$$T\ K = (T\ °C + 273.15\ °C) \cdot \frac{1\ K}{1\ C°} \quad \text{and}$$

$$T\ °C = (T\ K - 273.15\ K) \cdot \frac{1\ C°}{1\ K} \qquad (1\text{-}2a)$$

EXAMPLE 1.5 Temperature Conversions

Laboratories for chemical and biological research often have a "constant temperature room" to provide a stable environment for experiments and for storage of materials. A common temperature for such rooms is 68 °F. Express 68 °F in (a) °C and (b) in K.

Analysis (a)

Target: A temperature in °C.

$$? \, ^\circ\text{C} =$$

Knowns: A temperature in °F.

Relationship: Several formulas are available. The correct choice is Equation (1-1), since it contains both the target and the known.

Plan/Work Use the form of Equation (1-1) that has this problem's target on the left, substitute the known value, and solve.

$$? \, ^\circ\text{C} = \frac{^\circ\text{F} - 32}{1.8} \tag{1-1}$$

$$? \, ^\circ\text{C} = \frac{68 - 32}{1.8} = \frac{36}{1.8} = 20 \, ^\circ\text{C}$$

Check Positive Fahrenheit temperatures are always greater than the corresponding value in the Celsius scale. Furthermore, the Celsius value is about half the Fahrenheit value for temperatures greater than about 100 °F. You might find it useful to memorize the fact that 20 °C = 68 °F, for convenience in checking conversions near 70 °F.

Analysis (b) Using similar reasoning, select the appropriate formula and solve.

$$\text{K} = ^\circ\text{C} + 273.15$$

$$\text{K} = 20 \, ^\circ\text{C} + 273.15 = 293 \text{ K}$$

Note the use of significant figures in this problem.

Check A temperature in K is about 300 more than the same temperature in °C.

Exercise

a. A person with a body temperature of 104 °F is usually regarded as quite sick. Convert 104 °F to °C and K.

b. The value T = 298 K is often used to represent "room temperature," a convenient, approximate value of the temperature in many laboratories. Convert 298 K to °C and °F.

Answer: (a) 104 °F = 40 °C = 313 K; (b) 298 K = 25 °C = 77 °F.

(a)

Figure 1.14 (a) The balance shows that the block of wood is "heavier" than the class ring. (b) The behavior with respect to water shows that the class ring is "heavier" than the block of wood. The terms "heavy" and light" are ambiguous, and when used in science must be put in an unambiguous context.

(b)

Density

A children's playful riddle asks, "Which weighs more—a pound of feathers or a pound of lead?" Of course the answer is that they both weigh the same (have the same mass). The riddle is intended to illuminate and teach the difference between the concepts of mass (how much does it weigh?) and density (what volume does the mass occupy?). In ordinary speech, it is common to use the words "heavy" and "light" to refer to either density or mass. For example, "Alex is heavier (has more mass) than Peggy," or "Wood floats because it is lighter (lower density) than water" (Figure 1.14). Such ambiguous terms should be avoided in science. **Density** is formally defined as mass per unit volume, and most commonly expressed in units of g mL^{-1} for solids and liquids*, or g L^{-1} for gases. The SI unit "cubic centimeter (cm^3 or cc)" is often used as a synonym for milliliter: 1 cc = 1 cm^3 ⇔ 1 mL. Density is an intrinsic characteristic of a substance, independent of the amount (Table 1.6). For example, the density of water is 1 g mL^{-1}, whether the quantity is one drop or one bucket.

TABLE 1.6 Densities of Some Common Substances at 25 °C

Substance	Density
Dry air	0.0012 g mL^{-1}
Alcohol	0.79 g mL^{-1}
Seasoned oak	0.6–0.9 g mL^{-1}
Water	1.0 g mL^{-1}
Granite	2.7 g mL^{-1}
Iron	7.9 g mL^{-1}
Lead	11.3 g mL^{-1}

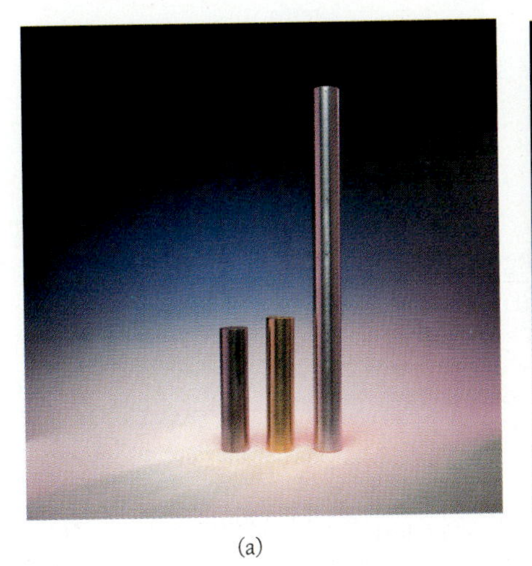

(a)

(b)

(a) These samples of steel, bronze, and aluminum all have the same mass (410 g). Because its density is the least, the aluminum bar has the largest volume. (b) These samples all have the same volume (50 cm^3), but their masses are different. The mass of the aluminum is 136 g, that of the bronze is 366 g, and that of steel is 393 g.

* The less commonly used units kg L^{-1} are numerically equal to g mL^{-1}; the density of gold, for instance, is 19.3 g mL^{-1} = 19.3 kg L^{-1}:

$$\frac{(19.3\text{ g mL}^{-1})}{(1)} \cdot \frac{(1\text{ kg})}{(1000\text{ g})} \cdot \frac{(1\text{ mL})}{(0.001\text{ L})} = 19.3\text{ kg L}^{-1}$$

A quantity, like density, that is independent of the amount of substance is called an **intensive quantity.** Temperature and pressure are two other commonly used intensive quantities. On the other hand, quantities whose value depends on the amount of substance are called **extensive quantities:** both mass and volume are extensive.

The relationship between mass, volume, and density is

$$\text{density} = \frac{\text{mass}}{\text{volume}}, \qquad \text{or } D = \frac{m}{V} \tag{1-3}$$

Equation (1-3) is used to solve several commonly encountered problems in chemistry.

EXAMPLE 1.6 Calculation of the Density of a Solid

The metallic element lead is one of the softest (most malleable) metals. It has a relatively high density. If 1 pound (0.454 kg) occupies a volume of 40.2 mL, what is the density of lead?

Analysis

Target: The target, a density, is clearly stated in the problem. We know from the definition that the units of density for a solid substance are g mL^{-1}.

$$? \text{ g mL}^{-1} =$$

Knowns: Mass and volume are explicit in the problem statement. Implicit knowns that may be useful are the equivalence statements 1000 g ⇔ 1 kg and 1000 mL ⇔ 1 L.

Relationship: The definition of density, Equation (1-3), may be used directly.

$$\text{density} = \frac{\text{mass}}{\text{volume}}, \qquad \text{or } D = \frac{m}{V}$$

Plan In this problem, there is only one step. Substitute the given values for the symbols in the formula.

Work

$$(\text{g mL}^{-1}) = \frac{0.454 \text{ kg}}{40.2 \text{ mL}} = 0.0113 \text{ kg mL}^{-1}$$

Check The units of the numerical answer are not the same as the units of the target—use a unit conversion factor.

$$(\text{g mL}^{-1}) = \frac{(0.0113 \text{ kg mL}^{-1})}{(1)} \cdot \frac{(1000 \text{ g})}{(1 \text{ kg})} = 11.3 \text{ g mL}^{-1}$$

Check 0.454 kg is about 500 g, and 500/40 is a little more than 10, so the math is correct. The densities of solid substances range from a little less than 1 to a little more than 22 g mL^{-1}, so the answer is reasonable.

Exercise

Although all metals can be melted if heated to a high enough temperature, the element mercury is the only metal that is a liquid at room temperature. If 0.00500 L has a mass of 68.0 g, what is the density of liquid mercury?

Answer: 13.6 g mL^{-1}.

EXAMPLE 1.7 Calculation of Volume from Mass and Density

The metal that we encounter most frequently is steel, which is a mixture of elements rather than a pure substance. Steel is mostly iron, with some carbon, chromium, nickel, manganese, and other elements added. The properties of steel depend on its composition, and many types of steel are manufactured. Type 201, a particular type of stainless steel, contains 16% to 18% chromium; 3% to 5% nickel; 5% to 7% manganese; 0% to 1% silicon; and traces of carbon, phosphorus, and sulfur.

What is the volume of a 10.0 g sample of a type of steel whose density is 7.62 g mL^{-1}?

Analysis

Target: The target is a volume. The units are not specified, so we may use whatever we choose. Since the density is given in g mL^{-1}, choice of mL for the target units will minimize the amount of unit conversion required.

$$? \text{ mL} =$$

Knowns: Mass and density are explicit.

Relationship: A formula exists, but some algebraic manipulation is required before it can be used. Cross multiply to isolate the target on the left.

$$D = \frac{m}{V}$$

$$\frac{1}{D} \cdot (V) \cdot (D) = \frac{m}{V} \cdot (V) \cdot \frac{1}{D}$$

$$V = \frac{m}{D}$$

Work

$$(? \text{ mL}) = \frac{m}{D} = \frac{10.0 \text{ g}}{7.62 \text{ g mL}^{-1}} = 1.31 \frac{(1)}{(\text{mL}^{-1})} = 1.31 \text{ mL}^*$$

Check The units are correct. Since the density is ≈ 10 g mL^{-1}, the volume occupied by 10 g should be about 1 mL.

* This result could also have been reached by direct application of the factor-label method using the conversion factor 1 mL/7.62 g derived from the equivalence statement 7.62 g ⇔ 1 mL:

$$10.0 \text{ g} \cdot \frac{1 \text{ mL}}{7.62 \text{ g}} = 1.31 \text{ mL}$$

Exercise

All substances, including gases, have characteristic densities. If the density of sea-level air is 1.41 g L^{-1} at room temperature, what is the volume of 0.0100 g air?

Answer: 0.00709 L or 7.09 mL.

Specific Gravity

The **specific gravity** of a substance is the ratio of its density to that of water at the same temperature.

$$\text{Sp. Gr.} = \frac{D_{\text{substance}}}{D_{\text{water}}} \qquad (1\text{-}4)$$

Almost all substances expand when heated. That is, the density *decreases* as the temperature is raised. Water is unusual, however, in that it *contracts* slightly on being warmed from 0 °C to 3.98 °C (where it has its maximum density of 0.99997 g mL^{-1}).

Since in the range 0 °C to 30 °C the density of water is 1.00 g mL^{-1}, while from 30 °C to 75 °C the density decreases gradually to 0.98 g mL^{-1}, for the most part specific gravity is numerically equal to density in g mL^{-1}. However, since it is a ratio of quantities having the same units, specific gravity is a dimensionless number. Although specific gravity is not often used in chemistry, it is common in commerce, industry, and medicine.

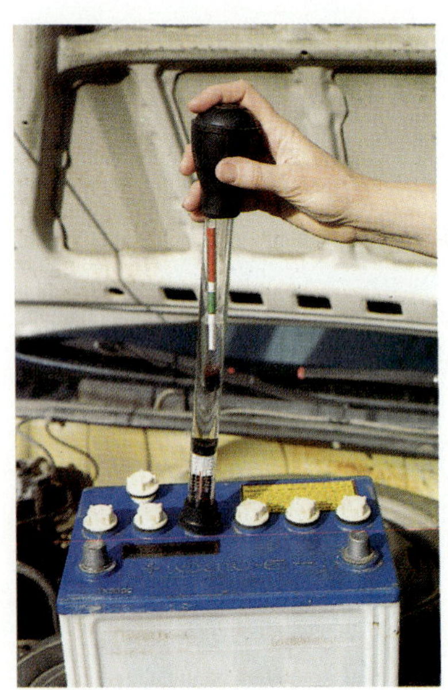

Figure 1.15 The extent of charge in an automobile battery is observed by measuring the specific gravity of the battery acid with a hydrometer.

EXAMPLE 1.8 Specific Gravity of a Liquid

The liquid ("battery acid") in an automobile storage battery is a solution of sulfuric acid in water. As the battery is discharged, sulfuric acid is consumed and the specific gravity of the solution decreases (see Figure 1.15). What is the density of the liquid in a partly discharged battery, if its specific gravity is 1.27?

Analysis

Target: A density.

$$? \text{ g mL}^{-1} =$$

Knowns: The specific gravity; the density of water is an implicit known.

Relationship: Equation (1-4), defining specific gravity, may be used directly to solve the problem.

Plan/Work

$$\text{Sp. Gr.} = \frac{D_{\text{substance}}}{D_{\text{water}}} \qquad (1\text{-}4)$$

$$\begin{aligned} D_{\text{substance}} &= (\text{Sp. Gr.}) \cdot (D_{\text{water}}) \\ &= (1.27) \cdot (1.00 \text{ g mL}^{-1}) \\ &= 1.27 \text{ g mL}^{-1} \end{aligned}$$

Check The units are correct for a density. Specific gravity and density are always numerically identical to 2 significant digits, and almost always to 3.

The chemistry of the charge/discharge cycle is discussed in Chapter 19.

Exercise

What is the specific gravity of the most dense metal, osmium, if its density is 22.5 g mL^{-1} at 25 °C?

Answer: 22.5.

1.6 INTRODUCTION TO METALS

We begin our study of the properties and behavior of specific types of substances with the metals, which, to most of us, are familiar and recognizable materials.

Properties

Of the 109 elements, approximately 85 (80%) are metals. They are distinguishable from nonmetals by their physical properties, which are given in Table 1.7. Metallic elements are found in the left and center of the periodic table, and nonmetals are found on the right. A diagonal line running approximately from boron to astatine separates the two. The distinction is not sharp, and some elements in the vicinity of the borderline—such as B, Si, Ge, As, Sb, and Te—do not fit neatly into either category. These elements have some properties of metals, and some of nonmetals. They are known as **metalloids,** or semimetals.

Metals are widely distributed over the surface of the earth. After oxygen and silicon, aluminum is the third most abundant element in the earth's crust. The earth's core is thought to consist primarily of iron, much of it in liquid form. Most metals are sufficiently reactive that they occur in nature as compounds rather than as pure elements. Compounds with oxygen and sulfur are common.

TABLE 1.7 Physical Properties of the Metallic Elements
Appearance: Opaque and highly reflective. Except for mercury, metals are solids at room temperature.
Color: Metals are silver in color, except for copper and gold.
Electrical: Variable, although all metals conduct electricity better than nonmetals. Silver is the best conductor.
Magnetic: Iron, cobalt and nickel are strongly attracted by a magnet; the others are not.
Thermal: Metals conduct heat better than most nonmetals.
Hardness: Variable; tungsten is very hard, while lead and some others are soft enough to be scratched with a fingernail.
Malleability: Most metals are malleable. That is, rather than breaking they are permanently deformed by bending or hammering.

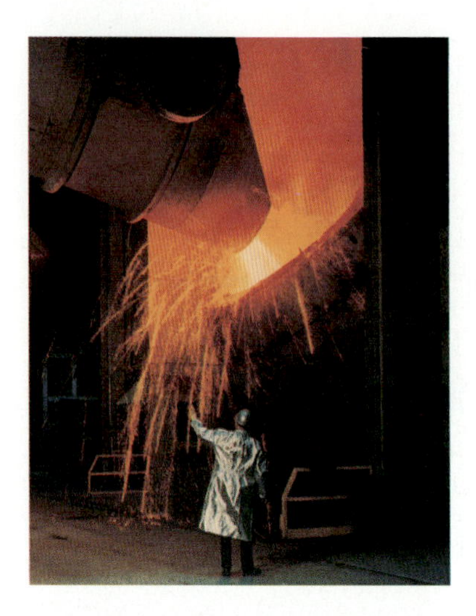

One stage in the process of steelmaking. Molten iron is being poured into the furnace, where impurities will be removed by reaction with oxygen and calcium carbonate.

Metallurgy

Rocks and soil containing usefully high amounts of a metallic element, or its compound(s), are called **ores.** The technology of extracting metals from their ores and refining them is called **metallurgy.** There are several stages to a metallurgical process. The first step in the production of steel, for instance, is to *concentrate* iron oxide-containing ore by removing most of the iron-free rock. Next the iron oxide is *reduced* to elemental iron by reaction with carbon monoxide, formed by heating coke with oxygen. Coke is a form of impure carbon made by heating coal in the absence of air. If pure metallic iron is the desired product, the remaining impurities are removed in a final *purification* step. However, the usual product of iron metallurgy is steel, so in the final step controlled amounts of carbon and other elements are added or allowed to remain.

Metallurgy of other metals is similar, but often involves other steps. Lead ore contains lead(II) sulfide (galena). It must be roasted (heated in air) to convert it to the oxide prior to reduction. Copper ore, containing copper(II) sulfide, is converted directly to elemental copper by roasting; no reaction with coke is necessary. Aluminum ore (bauxite) contains aluminum oxide. It is reduced by electrical power rather than by reaction with coke. Further details on metallurgy may be found in Chapters 18, 19, and 23.

Some less reactive metals are found in nature as elements rather than as compounds. The most familiar of these are silver and gold. Their availability in pure form, together with their scarcity, malleability, and natural beauty, has made them sought after throughout history for use in jewelery and coinage. Copper normally occurs as a compound, but on occasion deposits of elemental copper are found. Copper, silver, and gold are collectively known as the **coinage metals.**

Many useful metals are not pure elements, but are alloys. A metal **alloy** contains several metals, and may include small amounts of nonmetallic elements as well. Some alloys are compounds of two or more metals, some are homogeneous mixtures, and some are heterogeneous mixtures. Examples of common alloys are steel; bronze (copper, tin, and some phosphorus); brass (copper, zinc); solder (lead, tin); sterling silver (silver, copper); 14-carat gold (gold, silver, copper); and tinfoil (tin, lead, copper).

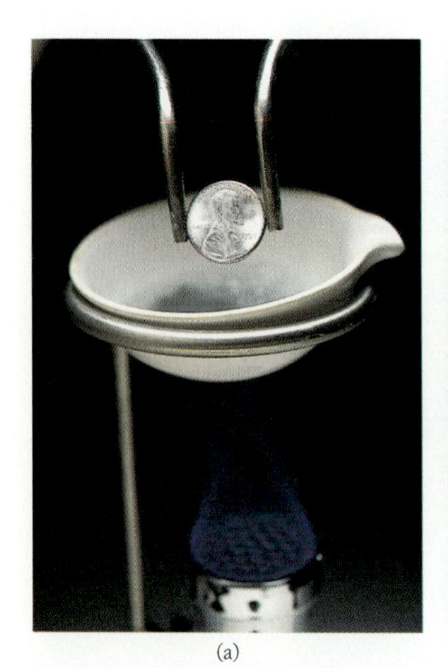

(a)

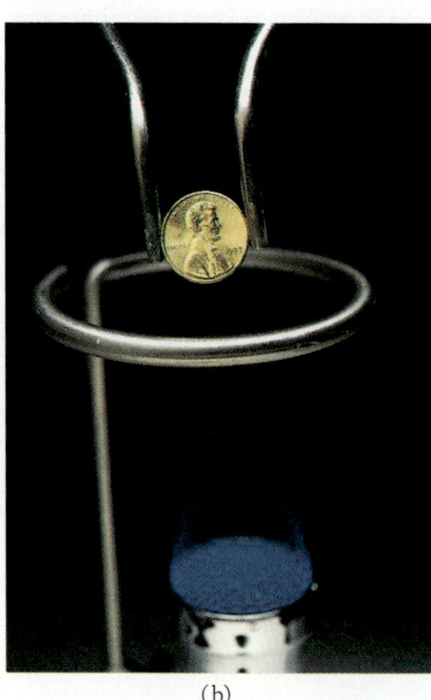

(b)

(c)

(a) This copper penny has been coated with zinc, by dipping it into a hot, concentrated solution of sodium zincate. (b) The zinc-coated penny is heated in the flame of a laboratory burner. At this high temperature, copper and zinc mix together to form the alloy, brass. (c) The characteristic color of copper is visible in the untreated penny (left), while the colors of zinc (middle) and brass (right) are quite different.

CHEMISTRY OLD AND NEW

MEASUREMENT AND SCIENCE

We use measurement systems so often in science and in our daily lives that we are hardly aware of them as an achievement in their own right. Like other branches of knowledge, the science of measurement and standardization—*metrology*—evolved from a few rudimentary ideas and techniques to the conveniently structured metric and SI systems in use today. Without accurate methods of determining mass, pressure, temperature, and a multitude of other quantities, experimental science would be impossible. In the past, the work of many scientists has been limited not by lack of information or technical skill, but by instruments that could not supply sufficiently precise determinations of physical quantities. The tick of a clock is satisfactory for timing most events in normal life, but a physicist working with femtosecond (10^{-15} s) laser pulses needs an incredibly accurate timing device. The official definitions of fundamental quantities such as the second, meter, and kilogram have been refined many times to allow for greater accuracy in research.

The first measuring device used by man was his own body. Units such as the cubit (the distance from the elbow to the tip of the middle finger) and the digit (the width of the thumb) were used to describe short to moderate lengths. Longer distances were measured by travel time or by the number of steps taken. For example the Roman term "mille passum," meaning a thousand paces, gave rise to the word *mile*. There is considerable variation in the length of peoples' strides and the width of their thumbs, however, and ultimately it became necessary to define a *standard* unit value. The British monarch Henry I solved the problem of conflicting definitions of the *yard* by decreeing that it was to be equal to the length of his own arm!

The English system of measurement—based on the foot, pound, quart, and bushel—evolved from these early standards and was tailored for commerce. Unfortunately its units of mass, length, and volume were not related in a clear way. The metric system, which was developed and adopted for use in France in the late 1700s, solved such problems by using decimal subunits and defining base units in terms of each other (1 mL = 1 cm^3; 1 gm = 1 mL water at conditions of maximum density). The inventors of the metric system set the length of the meter equal to one ten-millionth of the distance from the north pole to the equator along a certain longitude line.

A problem in metrology that persisted until the second half of this century stemmed from the fact that the definitions of units of length and mass referred to specific objects, such as King Henry's arm. Since measuring the earth each time a measuring stick was to be made was not feasible, even the meter was defined as the length of a particular metal rod. These length and mass standards were inaccessible, hard to reproduce precisely, and not very durable. The brass bar that served as the standard English yard from 1588 until 1824 was broken in half at some point, and mended with a crude dovetail joint described by one metrologist as "nearly as loose as a pair of tongs." Its replacement—another brass bar—was destroyed by fire in 1834 along with the rest of Britain's measurement standards. When scientists tried to reconstruct the lost standard pound weight, they found that most of the existing reproductions were so coarse and worn that they were practically useless. Today the meter is defined as the distance traveled by light in a vacuum during an interval of 1/299,792,458 of a second. Since the speed of light is invariant, this definition is natural, indestructible, and *very* precise.

Although a worldwide conversion to the metric and SI systems has taken place, in the United States there is widespread resistance to "metrification" as the changeover is called. Americans are so used to the English system that even simple day-to-day tasks, such as measuring cooking ingredients, would be difficult until people became used to the new units. By 1996, however, all of our highways will have signs giving distances in km and speed limits in km/h. Our journey towards full adoption of the metric system will have begun.

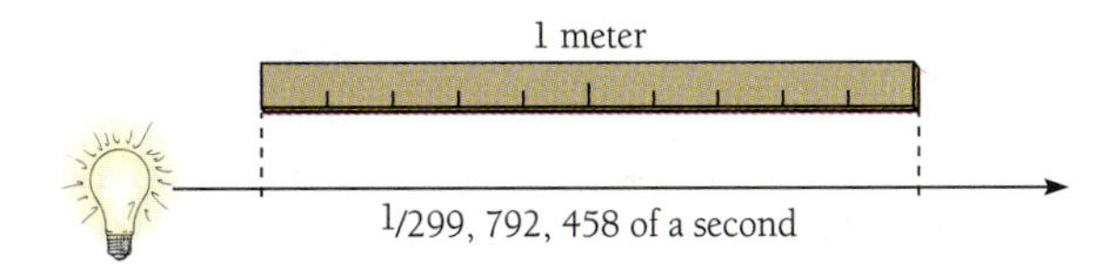

Discussion Questions

1. Which metrification strategy would cause the least inconvenience in the United States: An overnight changeover, or a gradual change, with a dual system existing for years while the English system is phased out?
2. What are some advantages and disadvantages of our present system of time units (1 day = 24 hours, 1 hour = 60 minutes, 1 minute = 60 seconds) compared to a new, decimal system (1 day = 10 hours, 1 hour = 100 minutes, 1 minute = 100 seconds)?

SUMMARY EXERCISE

In 1988, 21 hundred tons of gold were mined and purified worldwide. The market price of gold fluctuates, but for this exercise use a value of \$405 per troy ounce. Gold is a relatively heavy metal, with a density of 19.29 g mL^{-1} (measured at 17.3 °C; at this temperature the density of water is 0.9987 g mL^{-1}). Gold melts to a liquid at 1945 °F. Some useful equivalence statements are 1 ton ⇔ 2000 lb, 1 lb ⇔ 453.6 g, 1 troy ounce ⇔ 31.10 g. In solving the following problems, be sure to use the appropriate number of significant digits.

a. Express the annual production of gold in kg and troy ounces.

b. What is the dollar value of the annual production of gold?

c. Express the melting temperature in °C and K.

d. What is the specific gravity of gold at 17.3 °C?

e. Calculate the volume of gold that weighs one troy ounce.

f. Calculate the volume of \$175 worth of gold.

g. What is the mass of a gold nugget whose volume is 48.2 mL?

h. An empty beaker is found to weigh 35.5039 g. When a small flake of gold is put in it, the beaker weighs 35.5051 g. If this information, together with the density, is used to calculate the volume of the gold flake, what is the number of significant digits that should be used to express the answer? What is the most appropriate unit for the answer?

i. The U.S. national debt is about 4.0×10^{12} dollars. How many tons of gold would be required to pay off the debt? Assuming the 1988 annual production rate continues, how many years would it take to mine this amount?

Answers **(a)** 1.9×10^6 kg, 6.1×10^7 troy oz; **(b)** 2.5×10^{10} \$ (\$25 billion); **(c)** 1063 °C, 1336 K; **(d)** 19.32; **(e)** 1.612 mL; **(f)** 697 μL; **(g)** 930 g; **(h)** 2 significant digits, μL or nL; **(i)** 340,000 tons (3.4×10^5 tons), 161 years (1.6×10^2 years).

SUMMARY AND KEYWORDS

Chemistry is the study of matter and its changes. **Matter** has **mass** and occupies space, and is composed of **atoms.** The study of chemistry follows the **scientific method: observations,** which if repeatable are known as **facts;** facts lead to **hypotheses,** which are tentative explanations of facts. A **law** is a generalization of related facts. After a hypothesis has been sufficiently tested it becomes a **theory,** the best available explanation of facts.

A **substance** (a particular type of matter) can exist in different **physical states: gas, liquid,** or **solid.** Solids have definite shape and volume, liquids have definite volume, and gases have neither. When a substance changes from one state to another, it is said to undergo a **phase transformation.** The composition of an *impure* substance, called a **mixture,** is described by listing the identity and relative amounts of the pure substances it contains. **Homogeneous mixtures,** also called **solutions,** have the same composition throughout the body of the sample; **heterogeneous mixtures** do not. A pure substance is an **element** if it is composed of like atoms, or a **compound** if it is composed of atoms of two or more different types. There are 109 different elements, each with a unique name and **symbol.** Some elements exist in several different forms called **allotropes;** diamond and graphite are allotropic forms of carbon. Elements are commonly listed in the **periodic table,** in which elements in the same vertical column, the **group,** have similar chemical properties. Elements in the same horizontal row are in the same **period.**

Any change resulting in the formation of a new substance is a **chemical change** or **chemical reaction.** Otherwise, it is a **physical change,** which often is a phase transformation or change in particle size (of a solid). Characteristics describing a substance are **chemical properties** if their observation involves any chemical change. Otherwise, they are **physical properties.**

A **physical quantity** has a numerical part, which tells how much, and a **dimensional** part (the **units**), which tells of what. Some numbers are ratios of quantities, and are dimensionless. The self-consistent set of units used in chemistry is the **SI (Système Internationale).** The seven **base quantities** in the SI are mass (m), length (l), time (t), temperature (T), amount of substance (n), electric current (I), and luminous intensity (I_v). Other quantities ("derived") are products and/or quotients of these. Metric prefixes, of which the most common are **nano-** (n, 10^{-9}), **micro-** (μ, 10^{-6}), **milli-** (m, 10^{-3}), **kilo-** (k, 10^{3}), and **mega** (M), are used to modify the magnitude of a quantity, as in "1 ms (millisecond) = 10^{-3} s (second)."

Quantities are normally expressed in **scientific notation,** in which a number in the range 1 to 10 is multiplied by a power of 10. For example, 0.00538 in scientific notation is 5.38×10^{-3}. There are two aspects to the **reliability** of a quantity—it is **accurate** if it is close to the true value, and **precise** if repeated measurements cluster closely about the same value. The reliability of a quantity is represented by the number of **significant digits** used to write it. The results of a calculation can be no more accurate than the quantities entering into it. Therefore, in a quantity resulting from addition or subtraction, only those *decimal places* are significant that are significant in *all* the addends/subtrahends. Any result of multiplication or division has the same *number* of significant digits as the least reliable of the factors.

Two quantities may be related by an **equivalence statement,** for instance "1 inch ⇔ 2.54 cm," from which two **unit conversion factors** may be generated: (1 in./2.54 cm) ⇔ 1 ⇔ (2.54 cm/1 in.). Any quantity may be multiplied (or divided) by any conversion factor at any time; the operation changes the quantity's units but not its intrinsic value. The **factor-label method** requires that, in any mathematical operation, the units of a quantity be treated in the same way as the numerical part.

Solving an unfamiliar problem for which a one-step formula is not available involves careful specification of the **target,** *with its units,* and the identification of both explicit and hidden knowns. Missing information is identified ("If only I knew") and a stepwise chain leading *backwards* from the target to the knowns is constructed. Intermediate and final answers are checked for correct units and reasonable magnitude, and by estimating the answer.

Temperature is measured on either the **Celsius** scale, defined by assigning the value 0 °C to the freezing temperature of water and 100 °C to the boiling point, or on the **Kelvin** scale, which assigns values of 273.15 K and 373.15 K to these same two reference points. The size of the degree is the same on both scales. The **Fahrenheit** scale is still in use for nonscientific purposes in the United States. The conversion equations are: °F = 1.8(°C) + 32; °C = (°F − 32)/1.8; K = °C + 273.15; °C = K − 273.15.

Density is defined as the mass of a given amount of substance divided by its volume; the units are g mL^{-1} (= kg L^{-1}), or g L^{-1} (gases only). **Specific gravity** is the ratio of a material's density to that of water at the same temperature. The value of an **intensive property** (such as density, specific gravity, and temperature) is independent of the amount of substance, while that of an **extensive property** (such as mass or volume) is proportional to the amount.

Approximately 85 elements are metals, 18 are nonmetals, and the remainder (B, Si, Ge, As, Sb, and Te) are **metalloids.** Metals are opaque, reflective, and good conductors of heat and electricity. Most are **malleable.** Their hardness, magnetic properties, and chemical reactivity vary considerably. They are distributed widely in the earth's crust, where aluminum is the third most abundant element. Iron occurs in the earth's core as well as in the crust. Most metals occur naturally as compounds rather than as elements, although gold and silver and sometimes copper (the **coinage metals**) are found in elemental form.

In **metallurgy**, naturally occurring **ores** are *concentrated, reduced* to elemental form if necessary, and *purified.* **Alloys** are combinations of metallic elements, sometimes with nonmetals as well. Steel is an iron alloy containing carbon and several other elements. Steels containing relatively large amounts of nickel and/or chromium resist rusting and are known as stainless steel.

PROBLEMS

More difficult problems are marked with an asterisk. Answers to odd-numbered problems are given at the end of the text.

General

1. Give a concise definition of "matter." Does the term "substance" have a different meaning?
2. Distinguish between work and energy.
3. Describe the components of the scientific method. Name some areas of human concern, outside of science, for which the scientific method is a valuable approach. Name some areas of human concern in which the scientific method is not useful.
4. Distinguish between fact and theory.
5. Comment on the statement, "You needn't take it too seriously; after all, it's only a theory."
6. Describe each of the following as observation, fact, hypothesis, theory, and/or law. There may be some differences of opinion as to the correct descriptions, so be prepared to defend your choices.
   a. The phases of the moon repeat after approximately one month.
   b. The diameter of the moon is 3476 km.
   c. The moon travels in an orbit around the earth.
   d. The moon was formed when a passing star pulled a large amount of the earth's substance from what is now the Pacific Ocean.
   e. The moon is made of green cheese.
7. Describe each of the following as observation, fact, hypothesis, theory, and/or law. There may be some differences of opinion as to the correct descriptions, so be prepared to defend your choices.
   a. Objects are attracted to the earth with a force proportional to their mass.
   b. The force of gravity depends on the color of an object.
   c. Some objects, like hot air balloons, rise rather than fall when released.
   d. The weight of an astronaut on the moon is less than on the earth.
8. To what physical property or properties do the terms "light" and "heavy" refer?

Classification of Matter

9. Name, describe, and give examples of the three most common physical states of matter.
10. Give examples of substances that cannot easily be classified as solids, liquids, or gases.
11. Identify each of the following as a pure substance or a mixture.
    a. milk b. air
    c. compressed oxygen d. salt
    e. sugar
12. Identify each of the following as a pure substance or a mixture.
    a. beer b. water
    c. wood d. brass
    e. lead
13. What are the two types of impure substances?

*14. Describe two ways in which a mixture of salt and sand might be separated into the pure components.

15. Identify each of the following mixtures as homogeneous or heterogeneous.
    a. gasoline b. salad oil
    c. mayonnaise d. chicken noodle soup
    e. milk

16. Identify each of the following mixtures as homogeneous or heterogeneous.
    a. air
    b. seawater
    c. fog
    d. smoke
    e. the substance shown in Figure 1.5.
17. Define and give some examples of physical change.
18. Define and give some examples of chemical change.
19. Describe each of the following as either a chemical or physical change, or both.
    a. A wet towel dries in the sun.
    b. Lemon juice is added to tea, causing a color change.
    c. Hot air rises over a radiator.
    d. Coffee is brewed by passing hot water through ground coffee.
20. Describe each of the following as either a chemical change, a physical change, or both.
    a. Powdered sulfur is heated, first melting and then burning.
    b. Alcohol is evaporated by heating.
    c. Transparent rock candy (pure crystalline sugar) is finely ground into an opaque white powder.
    d. Chlorine gas is bubbled through brine (concentrated seawater), releasing liquid bromine.
    e. Electricity is passed through water, resulting in the evolution of hydrogen and oxygen gases.
21. Which of the following are chemical properties, and which are physical properties?
    a. Baking powder gives off bubbles of carbon dioxide when added to water.
    b. A particular type of steel consists of 95% iron, 4% carbon, and 1% miscellaneous other elements.
    c. The density of gold is 19.3 g mL^{-1}.
    d. Iron dissolves in hydrochloric acid with the evolution of hydrogen gas.
    e. Fine steel wool burns in air.
22. Which of the following are chemical properties, which are physical properties, and which are both?
    a. Metallic sodium is soft enough to be cut with a knife.
    b. When sodium metal is cut, the surface is at first shiny; after a few seconds' exposure to air, it turns a dull gray.
    c. The density of sodium is 0.97 g mL^{-1}.
    d. Cork floats on water.
    e. When sodium comes in contact with water, it melts, evolves a flammable gas, and eventually disappears altogether.
23. Which of the following pure substances are elements, and which are compounds?
    a. iron
    b. water
    c. carbon dioxide
    d. hydrogen
    e. mercury
24. Which of the following pure substances are elements, and which are compounds?
    a. aluminum
    b. sodium chloride
    c. sodium
    d. hydrochloric acid
    e. iron oxide
25. Identify each of the following as a pure substance or as a mixture. Further identify each substance as an element or compound if pure, or as heterogeneous or homogeneous if a mixture.
    a. dew
    b. maple syrup
    c. concrete
    d. paint
    e. table sugar
26. Identify each of the following as a pure substance or as a mixture. Further identify each substance as an element or compound if pure, or as heterogeneous or homogeneous if a mixture.
    a. tungsten
    b. granite
    c. diamond
    d. blood
    e. beer

Units and Scientific Notation

27. Identify each of the following units as SI or other. If SI, tell whether it is a base or a derived unit.
    a. cubic centimeter
    b. liter
    c. gallon
    d. meter
    e. gram
28. Identify each of the following units as SI or other. If SI, tell whether it is a base or derived unit.
    a. kilogram
    b. kilogram per liter
    c. second
    d. miles per hour
    e. Kelvin
29. Identify the numerical multiplier corresponding to each of the following metric prefixes, or vice versa. Do as many as you can from memory.
    a. nano-
    b. 10^3
    c. mega-
    d. centi-
    e. tera-
30. Identify the multiplier corresponding to each of the following metric prefixes, or vice versa.
    a. 10^{-3}
    b. pico-
    c. 10^1
    d. 10^{-6}
    e. deci-
31. Convert each of the following from arithmetic to scientific notation, or vice versa.
    a. 7325
    b. 0.83
    c. 18,236
    d. 7.5×10^6
    e. 0.000972
32. Convert each of the following from arithmetic to scientific notation, or vice versa.
    a. 0.000000000012
    b. 1234
    c. 12.6
    d. 6.34×10^{-6}
    e. 3.90×10^{-2}

33. Convert each of the following from arithmetic to scientific notation, or vice versa. (Do not change the units.)
    a. 692 m
    b. 0.003151596 g
    c. 2.54 cm $inch^{-1}$
    d. 3.00×10^8 m s^{-1}
    e. 0.000972 m^3

34. Convert each of the following from arithmetic to scientific notation, or vice versa. (Do not change the units.)
    a. 0.013 cm^3
    b. 0.946 L $quart^{-1}$
    c. 454 g $pound^{-1}$
    d. 65 mph
    e. 3.90×10^{-2} kg

35. Express the following quantities using a more appropriate SI prefix. Choose a prefix corresponding to an exponent that is divisible by three, and such that the value of the quantity is in the range 1 to 999. For example, 0.0050 L is better expressed as 5.0 mL.
    a. 0.083 kg
    b. 1.5×10^8 m
    c. 15,626 K
    d. 8372 mg
    e. 0.946 L

36. Express the following quantities using a more appropriate SI prefix.
    a. 3.00×10^8 g
    b. 0.76 m
    c. 0.35×10^{-8} L
    d. 0.96 km
    e. 6.2×10^{-6} s

37. Express the following quantities using a more appropriate SI prefix.
    a. 7,853 nm
    b. 0.000285 kPa
    c. 0.00183 °C
    d. 4.00×10^5 J
    e. 4.72×10^{-11} g

*38. Express the following quantities using a more appropriate SI prefix.
    a. 1.033×10^5 Pa
    b. 0.454 kg
    c. 792×10^3 mN
    d. 0.0793 μmol

*39. Express the following derived quantities using the most appropriate SI prefix. It is customary to express the denominator without a prefix. For example, the speed 3,458 m s^{-1} is normally expressed as 3.458 km s^{-1} rather than 3.458 m ms^{-1}, although the latter is not incorrect.
    a. 4.82×10^3 cm s^{-1}
    b. 0.00328 mol L^{-1}
    c. 50.3×10^{-6} L s^{-1}
    d. 3803 J m^{-3}

*40. Express the following derived quantities using a more appropriate SI prefix.
    a. 76.3×10^5 N m^{-2}
    b. 0.454 kg s^{-1}
    c. 7.95×10^3 J mol^{-1}
    d. 0.093 kg L^{-1}

*41. Express the following derived quantities using the indicated SI prefix(es). These conversions are more difficult than those in the preceding problems, and they are very important (see Example 1.3).
    a. 1.5×10^4 cm^2 (express in m^2)
    b. 9.03×10^2 kg m^{-3} (express in g cm^{-3})

*42. Express the following derived quantities using the indicated SI units.
    a. 135.5×10^5 cm^3 (express in m^3)
    b. 0.0000362 mol mL^{-1} (express in mmol L^{-1})

Significant Figures

43. How many significant figures are there in each of the following numbers?
    a. 105
    b. 0.82
    c. 0.0082
    d. 1500.

44. How many significant figures are there in each of the following numbers?
    a. 250.3
    b. 0.820
    c. 1500
    d. 150.0

45. How many significant figures are there in each of the following quantities?
    a. 108 nm
    b. 180 nL
    c. 018 Gg
    d. 180. pg

46. How many significant figures are there in each of the following quantities?
    a. 0.18 μL
    b. 1.08 pm
    c. 1.80 mL
    d. 18.0 μg

*47. How many significant figures are there in the quantities in each of the following statements?
    a. The length of a meter is 100 centimeters.
    b. The flask holds 125 mL.
    c. The flask holds 500 mL.
    d. The English pound is equivalent to 454 grams.
    e. Most dogs have 2 ears.

*48. How many significant figures are there in the numbers in each of the following statements?
    a. ". . . then add 100 g sodium chloride. . . ."
    b. This bottle holds exactly 500 mL.
    c. The girl weighs 100 pounds.
    d. They turned away 300 people.

49. Criticize the following statement: "According to the weather report, the high tomorrow will be 75 °F or 23.89 °C."

50. One of the arguments against adoption of the metric system in England was put forth by a prominent politician as follows: "I, for one, will not require an honest labourer, taking his ease at his favourite pub, to ask for his 0.45459 litres of ale." Criticize this statement.

51. Perform the following calculations, and use the proper number of significant digits to express the answer.
    a. $(185) \cdot (26)$
    b. $(0.032) \cdot (6.4 \times 10^2)$
    c. $(8.02 \times 10^{-3})/(2.218 \times 10^{-7})$

52. Perform the following calculations, and use the proper number of significant digits to express the answer.
    a. 79.2/39.6
    b. 0.9043×0.826
    c. $(6.02 \times 10^{23}) \cdot (1.61 \times 10^{-19})$

53. Perform the following calculations, and use the proper number of significant digits to express the answer.
    a. 75 + 103
    b. 752 + 103
    c. 752 + 10.3
    d. 752 − 10.3
    e. 75.2 − 10.3
    f. 0.752 + 10.3

54. Perform the following calculations, and use the proper number of significant digits to express the answer.
a. $0.752 - 10.3$ b. $0.7520 - 10.3$
c. $(75.2 \times 10^4) - 10.3$ d. $(75.2 \times 10^{-4}) + 10.3$
e. $750 - 103$

*55. Perform the following calculations, and use the proper number of significant digits to express the answer.
a. $[(6.91 \times 10^3) + 276]/185$
b. $(2.54 - 454) + (9.46 - 0.70)$
c. $72.53 - (0.0322 \cdot 213)$

*56. Perform the following calculations, and use the proper number of significant digits to express the answer.
a. $(3.142 + 92.62)/(72.62 - 4.625)$
b. $(6.02 \times 10^{23})/(7.92 + 35.3)$
c. $[(7.82 \times 10^{-2}) + (8 \times 10^{-4})]/(22.4 + 0.00194)$

57. The mass of substances used in the laboratory are often determined by weighing the substance in a container of some sort, then subtracting the mass of the empty container. If a chemist finds the mass of an empty beaker to be 37.603 g, and to be 37.682 g after a sample of titanium is placed in it, to how many significant digits does he know the mass of the titanium?

*58. Pure water freezes at 0.0 °C, but when water contains dissolved substances it freezes at a lower temperature. For example, if 5.85 g of table salt is dissolved in one liter of water, the freezing point is decreased by 0.372 °C. What is the freezing point of the solution?

59. Perform the indicated calculations and express the result to the correct number of significant figures. Note that the number of significant digits in the answers to (a) and (b) are different, and that (a) is therefore a much more reliable number. Often there are several different ways to go about making a measurement, and you should, if possible, avoid ways that require taking the difference between two large numbers.
a. $60725 + 60703$ b. $60725 - 60703$

*60. Perform the indicated calculations and express the result to the correct number of significant figures.
a. $(9.00352 \times 10^{-16}) + (90.0342 \times 10^{-17})$
b. $(9.00352 \times 10^{-16}) - (90.0342 \times 10^{-17})$

Temperature Scales

61. Convert the following Celsius temperatures to the Kelvin scale.
a. 25 °C b. −78 °C
c. 500 °C d. 1234 °C

62. Convert the following Celsius temperatures to the Kelvin scale.
a. 6215 °C b. −196 °C
c. −40 °C d. −10 °C

63. Convert the following Fahrenheit temperatures to the Celsius scale.
a. 68 °F b. 40 °F
c. 1234 °F d. 805 °F

64. Convert the following Fahrenheit temperatures to the Celsius scale.
a. 0 °F b. −40 °F
c. 350 °F d. 150 °F

65. Convert the following Celsius temperatures to the Fahrenheit scale.
a. 10 °C b. −30 °C
c. −196 °C d. 500 °C

66. Convert the following Celsius temperatures to the Fahrenheit scale.
a. −200 °C b. 220 °C
c. −12 °C d. 1001 °C

67. Liquefied gases have boiling points well below room temperature. On the Kelvin scale the boiling points of the following gases are: He, 4.2 K; H_2, 20.4 K; N_2, 77.4 K; O_2, 90.2 K. Convert these temperatures to the Celsius and the Fahrenheit scales.

68. Convert the temperatures at which the following metals melt to the Celsius and Fahrenheit scales: Al, 933.6 K; Ca, 1112 K; Ga, 302.93 K; Ag, 1235.1 K.

69. Bored with sightseeing on a trip to England, you wish to bake some cookies. You remember the recipe, and that the baking temperature should be 375 °F. Unfortunately, the oven in your flat is calibrated in °C. To what temperature should it be set? (The kitchen is not equipped with a calculator, so do this one in your head, or with pencil and paper.)

70. While visiting England you decide to take a walk, and on the telly the weatherman says that the temperature is 16 °C. Of course you will need an umbrella, but should you take a sweater as well? (Do this one in your head: you forgot to pack the calculator.)

Specific Gravity and Density

You will need the following information for some of the problems in this section: 1 $cm^3 \Leftrightarrow 1$ mL; the density of water at 30 °C is 0.9956 g mL^{-1}.

71. The density of cork varies somewhat from one sample to the next, but is usually about 0.21 g mL^{-1} at 30 °C. What is the specific gravity of cork?

72. The density of ordinary table salt is 2.16 g mL^{-1} at 30 °C. What is its specific gravity?

73. What is the density of copper at 30 °C, if its specific gravity is 8.92?

74. Calculate the density of cold rolled steel, given that its specific gravity is 7.88 at 30 °C.

*75. Does the specific gravity of water vary as the temperature is changed? Explain.

*76. Does the specific gravity of substances other than water change with temperature? Explain.

77. What is the density of cadmium metal, if 10.0 g occupies exactly 1.16 mL?

78. What is the density of silicon, if 50.6 g occupies exactly 21.72 mL?

*79. A convenient method of determining the density of an object is to submerge the object in water and measure the volume of displaced water. If a small bar of zinc weighing 24.3 g is dropped into a partially filled graduated cylinder and the water level rises from 40.8 mL to 44.2 mL, what is the density of zinc?

*80. Calculate the density of copper, if a piece weighing 0.625 kg causes the water level in a graduated cylinder to rise from 10.3 mL to 80.2 mL.

81. What is the volume of a kilogram of nickel, whose density is 8.90 g mL^{-1}?

82. Calculate the volume of a kilogram of magnesium, whose density is 1.74 g mL^{-1}.

*83. The density of dry air at sea level and at 22 °C is 1.12 g L^{-1}. What is the volume of one metric ton (1000 kg ⇔ 1 metric ton) of air? Express your answer in (a) liters, (b) cubic meters, and (c) cubic kilometers.

*84. Using the data from the preceding problem, calculate the mass of one cubic mile of air.

*85. Calculate the mass of one cubic meter (a "stere") of seawater, whose density is 1.025 g mL^{-1}. Owing to dissolved salt, the density of seawater is greater than that of fresh water. [Since the amount of dissolved salt (the "salinity") varies from place to place in the earth's oceans, so does the density.]

*86. What is the mass of a stere (preceding problem) of air at 22 °C? Use the data in problem 83.

*87. One cubic foot of lead weighs 708 pounds. Calculate its density in g mL^{-1}. You will need the equivalence statements 1 lb ⇔ 454 g, 1 inch ⇔ 2.54 cm, and 1 foot ⇔ 12 inches.

*88. What is the volume in cubic feet of 5.00 tons of lead? (1 ton ⇔ 2000 lb. See the preceding problem for other data and conversion factors.)

*89. The average density of dry balsa wood is 0.13 g mL^{-1}, that of Douglas fir is about 0.48 g mL^{-1}, and that of a typical hardwood such as white oak is 0.71 g mL^{-1}. The most dense of the common woods is ironwood, whose density when dry is 1.08 g mL^{-1}.
   a. Calculate the volume occupied by 5.00 lb of each of these woods.
   b. Why did Thor Heyerdahl choose balsa rather than ironwood (a far stronger material) for his raft Kon-Tiki?

*90. The mass of the Kohinoor Diamond is 108 carats. If the density of diamond is 3.51 g mL^{-1}, what is the volume of this gem? (1 carat ⇔ 3.09 grains, 7000 grains ⇔ 1 pound, 16 ounces ⇔ 1 pound, and 453.6 g ⇔ 1 pound.)

91. Oil floats on water because it is less dense and it does not dissolve. Oil also does not dissolve in alcohol, but does it float? Explain. The density of salad oil is typically about 0.93 g mL^{-1}, and a 5.00 mL sample of ethanol (grain alcohol) weighs 3.95 g.

*92. Pure gold has a density of 19.3 g mL^{-1} at room temperature. If a 0.779 mL sample of gold is found to have a mass of 12.3 g, what can be concluded about the composition of the sample?

*93. Aluminum foil is sold in supermarkets in long rolls measuring 66⅔ yards by 12 inches, in a thickness of 0.00065 inch. If the specific gravity of aluminum at 22 °C is 2.70, calculate the mass of a roll. The density of water at 22 °C is 0.998 g mL^{-1}.

*94. Plastic wrap sold in supermarkets is 12 inches wide and 66⅔ yards long. If the mass of such a roll is ¾ lb, and the specific gravity of the plastic is 1.43 at 22 °C, calculate the thickness of the material. The density of water at 22 °C is 0.998 g mL^{-1}. Express the answer in thousandths of an inch.

CHAPTER 2

ATOMS AND MOLECULES

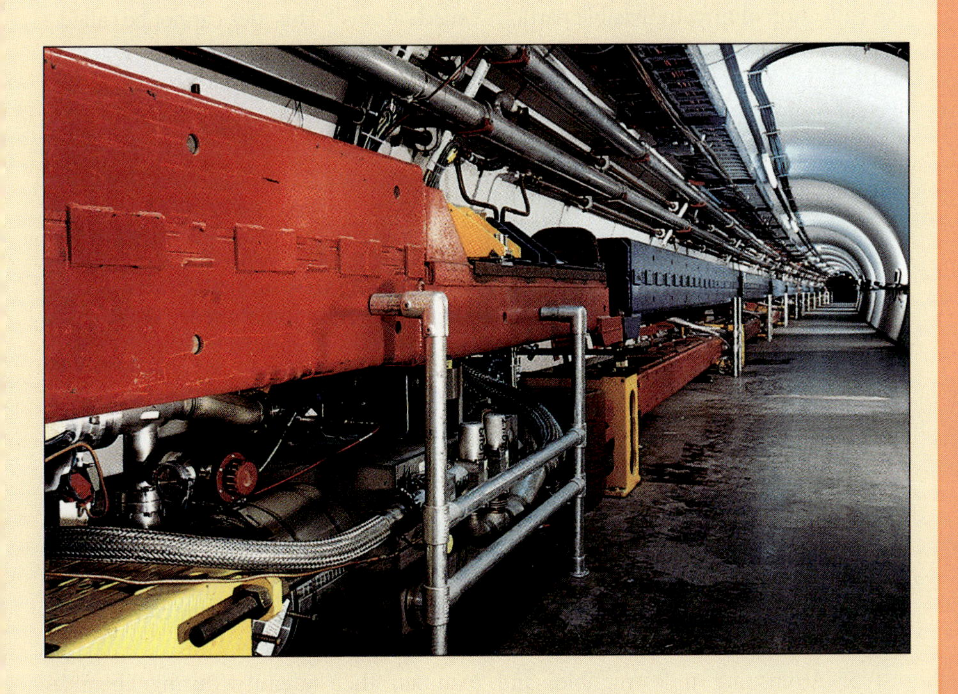

At the Fermi National Laboratory, liquid helium is used to cool the superconducting magnets that guide a beam of elementary particles into a circular path some four miles in diameter. Studies carried out with this gigantic research apparatus are concerned with the fundamental nature of matter.

2.1 ATOMIC THEORY

The Greeks felt that pure truth is reached through pure reason.

The basis for modern atomic theory was laid more than 2000 years ago by the Greek philosophers Leucippus and Democritus. They held that matter is not continuous, but is a collection of tiny, indivisible particles called "atoms." This idea cannot be called a scientific theory, because it was not based on observations or experiments. Over the years, however, enough experimental observations were made to provide a firm factual basis for the existence of atoms.

John Dalton's Ideas

The modern theory of the structure of matter begins with John Dalton, an English scientist of the early 19th century. His assessment of the existing evidence led him, in 1808, to publish the conclusions now known as Dalton's atomic theory. These conclusions included the following.

1. All matter consists of invisibly small particles, called **atoms.**
2. All atoms of the same element are identical to one another.
3. Atoms of different elements differ from one another, and, in particular, have different masses.
4. Atoms of different elements combine with one another, forming stable clusters of atoms called **molecules.** All molecules of a particular compound are identical. For example, all carbon dioxide molecules contain one carbon atom and two oxygen atoms—no more, no less.
5. Atoms are indestructible, and maintain their identity during chemical changes. Chemical reactions are rearrangements of atoms into new combinations, that is, new molecules.

With some modifications and additions, as discussed in later sections of this chapter, Dalton's theory stands today as essentially correct.

One of the diagrams used by John Dalton in his lecture on the atomic theory.

Components of Atoms

Although it is true that during chemical processes only the arrangement of the atoms changes, and not the atoms themselves, it is *not* true that atoms are indestructible.

> **John Dalton (1766–1844)**
>
> John Dalton was a Professor of Physics and Natural Philosophy at the University of Manchester. In addition to his theoretical work on the atomic structure of matter, he is known for his careful measurements of the behavior of gases (Dalton's law, page 201). Dalton was color blind, and, in 1794, wrote the first description of this deficiency. For some time thereafter, color blindness was known as "Daltonism."

Under conditions harsher than those of ordinary chemical reactions, atoms can be broken into smaller, subatomic units.

Elementary Particles

Atoms contain three kinds of subatomic units, collectively known as **elementary particles: electrons, protons,** and **neutrons.** Atoms of all 109 elements are composed of the same three elementary particles. Atoms of different elements differ only in the *number* of the elementary particles they contain.

Using large, expensive machines called particle accelerators (popularly known as "atom smashers"), it is possible to show that these elementary particles are made up of smaller, more fundamental entities (quarks). But the substructure of elementary particles does not directly influence chemical change, so this topic lies outside the realm of chemistry.

Mass and Charge; Atomic Units

The proton and the electron each carry an *electrical charge* of 1.6×10^{-19} coulombs. The SI unit, **coulomb (C),** is roughly equal to the amount of electrical charge that flows through a 100-watt light bulb every second. The charge of the proton and the charge of the electron are equal in magnitude but opposite in sign. The electron's charge is negative, while that of the proton is positive. The neutron's charge is zero. Although the neutron is slightly heavier than the proton, the two have approximately the same mass: 1.67×10^{-27} kg. The mass of the electron is 9.1×10^{-31} kg, only about 5×10^{-4} times the mass of the proton or neutron.

Numbers as small as these are hard to grasp, and the mass and charge of atoms and elementary particles are more often discussed in terms of *atomic units.* The atomic unit of electrical charge, the **elementary charge unit (ecu),** is defined so that the charge of the electron and that of the proton is 1 unit; that is, the equivalence statement is 1 ecu $\Leftrightarrow$ $1.6021892 \times 10^{-19}$ C. The **atomic mass unit (amu)** has a value such that the masses of the proton and the neutron are both approximately 1 amu (see page 59 for the exact definition of the amu). The equivalence statement is 1 amu $\Leftrightarrow$ 1.660540×10^{-27} kg. The essential characteristics of the elementary particles are shown in Table 2.1.

In tables of numerical data, the heading of a column gives the identity of the quantity and, following a slash, its units. Each number in a column is to be combined with the units in the heading. Thus, in the third column of Table 2.1, the proton mass is 1.6726 *in units of* 10^{-27} *kg;* that is, the proton mass is 1.6726×10^{-27} kg.

2.2 ATOMS AND ELEMENTS

We begin our discussion of atoms and elements with a description of the simplest of all atoms, hydrogen.

TABLE 2.1 Elementary Particles

Name	Symbol	Mass/10^{-27} kg	Charge/10^{-19} C	Mass/amu	Charge/ecu
Proton	p	1.6726	+1.6022	1.0073	+1
Neutron	n	1.6750	0	1.0087	0
Electron	e	0.00091095	−1.6022	0.00055	−1

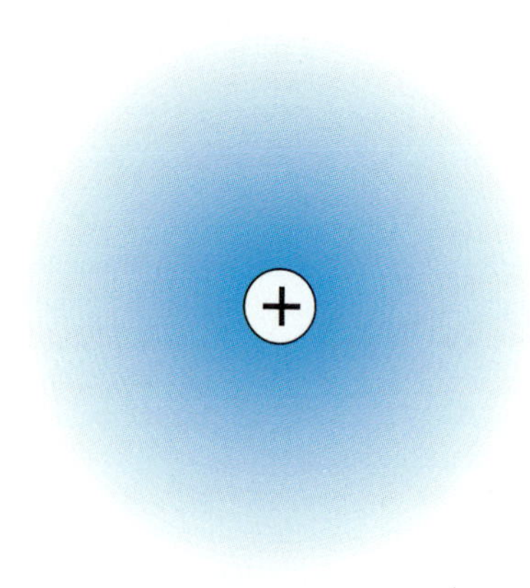

Figure 2.1 Almost all atoms of the simplest element, hydrogen, consist of an electron in motion around a proton. The motion is complex, and the atom is more aptly described as a proton located at the center of a spherical "electron cloud," a fog of negative charge.

The Hydrogen Atom

The hydrogen atom is composed of one proton and one electron*. Because these particles have equal and opposite charge, the hydrogen atom is electrically neutral. The electron moves rapidly around the much heavier proton, and it was at one time thought that the motion was similar to the motion of the earth around the sun. This hypothesis is called the "planetary model." It is now realized that the electron is not confined to a simple orbit around the proton, but that its motion brings it to all parts of a spherical region around the proton. The best way to visualize the atom is to imagine the result of a time exposure photograph, with the camera's shutter held open long enough so that the electron's location cannot be pinpointed. In such a photograph, the electron would appear to be smeared out over a spherical region about 1×10^{-10} m in diameter. The proton would appear in the center of a fog of negative electrical charge. The name **electron cloud** is used to describe the result of that motion (Figure 2.1).

Heavier Atoms

Atoms of elements other than hydrogen contain more than one proton and electron, and neutrons as well. However, their structure is similar to that of the hydrogen atom. The protons and neutrons are packed tightly together, forming the **nucleus,** a dense object containing all of the positive charge and more than 99.9% of the mass of the atom (Figure 2.2). The nucleus is the center about which the electrons move. As with the hydrogen atom, the motion of the electrons cannot be simply described. Electrons in atoms are best thought of as collectively forming an electron cloud. The number of electrons in an atom is the same as the number of protons in its nucleus, so that the total positive charge is equal to the total negative charge. In other words, all atoms are electrically neutral.

The **atomic number** (*Z*) of an atom is the number of protons in its nucleus. Each element has a unique atomic number: $Z = 1$ for hydrogen, and $Z = 109$ for the most recently discovered element. The periodic table lists the elements in order of increasing atomic number, which is written immediately above the symbol of the element. As mentioned in Chapter 1, the periodic table is further arranged so as to place elements of similar properties in the same vertical column.

Figure 2.2 Atoms consist of a dense core called the nucleus, which contains protons and neutrons. Electrons, equal in number to the protons, move about the nucleus in a complex way, forming a thin cloud of negative charge. Although the nucleus contains more than 99.9% of the mass, it occupies only about one hundred trillionth (10^{-14}) of the volume of the atom.

EXAMPLE 2.1 Atomic Number

a. What are the atomic numbers of the following elements: neon, neodymium, nobelium?

b. Which elements have the following atomic numbers? $Z = 18, 32, 92$.

* Although Dalton did not know this, there are actually three varieties of hydrogen atoms (see discussion of isotopes on page 45). Over 99.98% of all hydrogen atoms consist of one electron and one proton. About 0.02% consist of an electron, a proton, and one neutron; this variety is called *deuterium*. *Tritium* atoms, which are exceedingly rare, consist of an electron, a proton, and two neutrons. All three varieties have essentially the same chemical properties.

Analysis

a. ***Target:*** Atomic number of the element.

Known: Name of the element.

Relationship: The periodic table gives the atomic number corresponding to each elemental *symbol,* but the *name* of the element is not given. Also in the inside front cover, however, is an alphabetical list of elements together with their symbols and atomic numbers. All the required information is there.

Answer Neon(Ne), $Z = 10$; neodymium(Nd), $Z = 60$; nobelium(No), $Z = 102$.

b. ***Plan/Work*** Use the periodic table to find the symbol of the element with the given atomic number, and if necessary refer to the alphabetical list to find the same.

argon(Ar), $Z = 18$; germanium(Ge), $Z = 32$; uranium(U), $Z = 92$.

Exercise

Find the symbols and atomic numbers of all the members of Group 8A (the right-most column) of the periodic table: helium, neon, argon, krypton, xenon, and radon.

Answer: He, 2; Ne, 10; Ar, 18; Kr, 36; Xe, 54; Rn, 86.

The word **nucleon** means "nuclear particle" and refers to both protons and neutrons. Since both protons and neutrons have a mass of about 1 amu, and the electron mass is very much less, the mass of an atom in amu is about the same as the number of nucleons it contains. This number, an integer, is called the **mass number** of the atom.

The composition of a nucleus is compactly indicated by adding a left superscript and a left subscript to the elemental symbol. The notation $^{235}_{92}U$ means an atom of uranium, having atomic number 92 and mass number 235. Since 235 is the total number of neutrons plus protons, and 92 is the number of protons, the number of neutrons is $(235 - 92) = 143$. Only one element has a particular atomic number, so inclusion of both the elemental symbol and Z is redundant notation. An alternate form, for example Kr-84, gives only the symbol and *mass number.* The number of protons is implicit in the symbol of the element, and the number of neutrons must be found by difference. In Kr-84, for example, the number of protons (36) is found by looking up the atomic number in the periodic table. The number of neutrons (48) is found by subtracting the atomic number (36) from the mass number (84).

Isotopes

All atoms of a given element have the same number of *protons.* However, most elements exist in several variations, called **isotopes,** which are characterized by different numbers of *neutrons.* The number of neutrons has a negligible effect on chemical properties, so all isotopes of a given element show essentially the same

TABLE 2.2 Natural Abundance of the Isotopes of Hydrogen and Xenon

Isotope	Number of Atoms*	Natural Abundance (%)
$^{1}_{1}H$	9998	99.98
$^{2}_{1}H$	2	0.02
$^{3}_{1}H$	Very small	Very small
$^{124}_{54}Xe$	9	0.094
$^{126}_{54}Xe$	9	0.088
$^{128}_{54}Xe$	191	1.91
$^{129}_{54}Xe$	2624	26.24
$^{130}_{54}Xe$	405	4.05
$^{131}_{54}Xe$	2124	21.24
$^{132}_{54}Xe$	2693	26.93
$^{134}_{54}Xe$	1052	10.52
$^{136}_{54}Xe$	893	8.93

* In a representative sample of 10,000 atoms.

chemical behavior; they differ only in mass. The isotopes of an element are not all equally common. Representative samples of 10,000 atoms of hydrogen and xenon, for instance, have the composition shown in Table 2.2.

The relative amount of a particular isotope, expressed as a percentage, is called its isotopic abundance or **natural abundance.** When measured in naturally occurring samples, abundances of the isotopes of most elements are found to be essentially constant. Natural abundances do not depend on where the sample comes from or on its form. This constancy is a result of the near-identical chemical behavior of isotopes. As a particular element undergoes chemical change in natural geological or biological processes, there is little tendency for its isotopes to behave differently and become separated. The present isotopic abundances of the nonradioactive elements are unchanged from the original values established when the earth was formed.

Isotopic abundances of *radioactive* elements such as uranium have changed over the course of time. When the earth was formed, 17% of the uranium was U-235. Today the abundance of U-235 is only 0.7%. Reasons for this change are discussed in Chapter 24.

Ions

In some circumstances, an atom (or molecule) acquires one or more additional electrons and incorporates them into its electron cloud. Since the atom then contains more electrons than protons, it is no longer electrically neutral, but negatively charged. Such a species is called an **ion,** which means an atom (or molecule) having an electrical charge. An atom (or molecule) can also *lose* one or more electrons from its electron cloud, leaving it with *fewer* electrons than protons. Such a species, also called an ion, carries a positive charge. Positively charged ions are called **cations** (pronounced *cat' ions*), and negatively charged ions are called **anions** (*an' ions*). Both types of ions play important roles in chemical reactions.

Ions are designated by their elemental symbol and a superscript giving the charge, in units of elementary charge. For example, Cl^- is the symbol of the chloride ion, a chlorine atom that has acquired an additional electron and thus has a net negative charge of 1 ecu. Similarly, Ca^{2+} represents the calcium ion, a calcium atom that has lost two of its electrons and as a result has more protons than electrons; it has a net positive charge of 2 ecu.

EXAMPLE 2.2 Elementary Particles in Ions

How many protons, neutrons, and electrons are there in a $^{56}Fe^{3+}$ ion?

Analysis All the required information is contained within the symbol of the ion.

Work According to the periodic table, the atomic number of Fe (iron) is 26; that is, there are 26 protons.

The mass number, 56, is the number of protons plus the number of neutrons; therefore, there are $(56 - 26) = 30$ neutrons.

There are 26 electrons in the neutral *atom*. The +3 charge of the ion means that the atom has lost 3 electrons. The ion therefore has $(26 - 3) = 23$ electrons.

Exercise

How many protons, neutrons, and electrons are there in a $^{80}Se^{2-}$ ion?

Answer: 34 protons, 46 neutrons, and 36 electrons.

2.3 OBSERVATIONAL BASIS

We have seen that matter consists of atoms, invisibly small, neutral particles characteristic of each element, and that atoms are composed of smaller elementary particles of known mass and charge. In atoms, a very dense, positively charged nucleus containing protons and neutrons is surrounded by electrons, which form a cloud of negative charge. Since atoms contain equal numbers of electrons and protons, they are electrically neutral. Most elements have several isotopes, differing in mass number but not atomic number. We will now summarize some of the laboratory observations on which these conclusions are based.

Facts and Theories

Our presentation of the facts and theories of the structure of atoms begins with three laws of nature that apply to chemical compounds and their reactions. We then discuss the measurements of the charge and mass of the electron, and the experiment in which the nucleus was discovered. Finally we describe the observations that revealed the existence of isotopes.

Three Laws

Many scientists contributed, but it was in great measure due to the work of the French chemist Lavoisier that three generalities about compounds and chemical change were established. By the end of the 18th century, the following laws were generally accepted.

1. The total mass of the substances involved in a chemical change is unaffected by that change. When carbon burns, for example, the mass of the carbon dioxide produced is the same as the sum of the masses of carbon and oxygen consumed. This is called the **law of conservation of mass.**

Antoine Lavoisier (1743–1794)

Antoine Lavoisier had a strong sense of the importance of precise measurements, and developed accurate techniques for weighing chemical substances. He published the law of conservation of mass in his textbook, *Elementary Treatise on Chemistry,* and was among the first to distinguish clearly between elements and compounds. His career was cut off when he was guillotined during the French Revolution.

2. The composition of a particular chemical compound is always the same, whatever the source of that compound. For example, carbon dioxide contains 27% carbon and 73% oxygen, whether it is produced in combustion, respiration, or beer brewing. This is known as the **law of constant composition** or the **law of definite proportions.**
3. When two elements combine to form more than one compound, the compositions are related to one another by small integers. This is the **law of multiple proportions.** Carbon monoxide (57% O, 43% C) contains $57/43 = 1.333$ g oxygen for each gram carbon. Carbon dioxide contains 2.667 g O for each gram C. The ratio of these compositions, $2.667/1.333 = 2$, is a small integer.

These three laws form the base from which Dalton developed his atomic theory. The idea that molecules consist of definite numbers of atoms is clearly consistent with these laws.

The Ratio of Electron Mass to Electron Charge

Among the early (1848) experiments with electricity was Michael Faraday's application of high voltages to low-pressure gases contained within glass tubes. This results in the formation in the gas of an electrical *discharge,* of which the most obvious effect is the emission of light. The "neon sign" is based on this property of gases, although gases other than neon are also used in order to control the color of the emitted light (Figure 2.3).

Faraday's discovery stimulated much interest, and other scientists undertook similar investigations. It was discovered that "rays" other than light are given off, not from the discharge itself but from the electrodes, the metal structures used to conduct the electricity into the glass tube. Although these rays are invisible, they cause a visible

Michael Faraday (1791–1867)

Michael Faraday was a bookbinder's apprentice who educated himself in science through voracious reading. Ultimately he became director of the laboratory of the Royal Society of London and professor of the Royal Institution. His experimental findings in electricity laid the groundwork for a vast industry, because, among other things, he invented the first electrical generator. He was a successful popularizer of science, giving fascinating and well-attended Saturday lectures on a wide range of topics at the Royal Institution.

Figure 2.3 Application of electrical current to a tube containing neon, or other gases, creates an electrical discharge and causes the emission of light.

glow when they strike either the walls of the glass tube or a specially constructed target, coated with fluorescent material, within the tube. Two types of rays are observed: **cathode rays** are emitted from the negative electrode, whereas the region around the positive electrode is the source of **canal rays.** Both types of rays can be deflected by an electric field or by a magnetic field. Such deflection occurs only if the "rays" are not radiation but streams of fast-moving charged particles. Furthermore, the direction of deflection reveals the sign of the charge; canal rays are positively charged, and cathode rays are negatively charged.

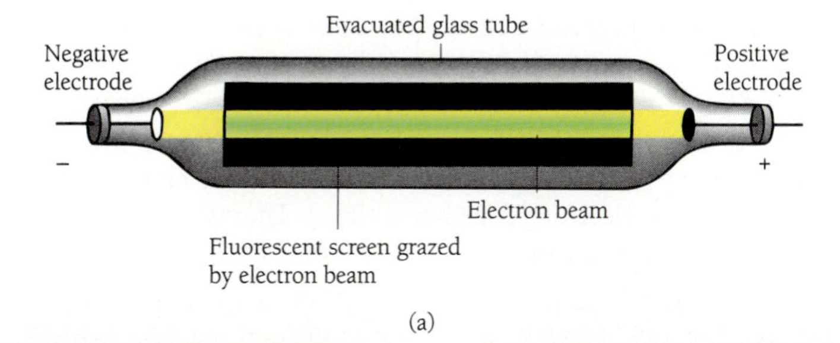

(a)

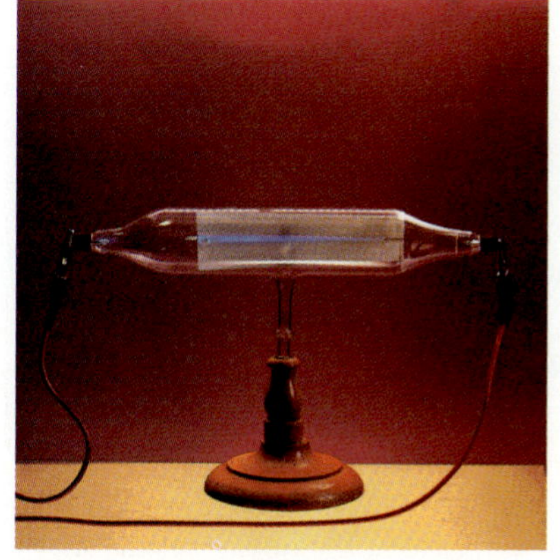

(b)

(c)

(a) A beam of electrons, created in an electrical discharge in a low-pressure gas, is made visible when it grazes along a fluorescent screen. (b) The beam travels in a straight line between the negative and positive electrodes. (c) The path traveled by the electrons becomes curved in the presence of a magnetic field.

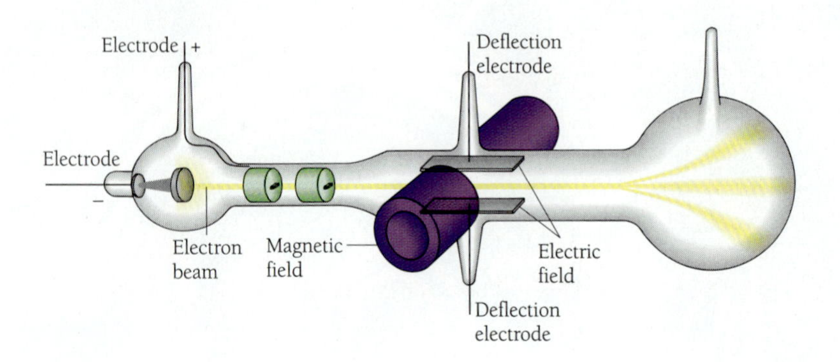

Figure 2.4 A schematic diagram of the apparatus used by J. J. Thomson to determine the mass-to-charge ratio of the electron. The m/e ratio is calculated from measurements of the effect of electric and magnetic fields on the path of electrons in the tube.

Using a *cathode ray tube* similar to the one shown in Figure 2.4 (the picture tube in today's TV sets is a modern version), J. J. Thomson investigated the mass and charge of cathode rays, or "electrons" as they had by then come to be called. Although he could not measure the quantities separately, he was able to determine accurately the value of m/e, the *ratio* of the mass to the charge of an electron. This ratio (5.7×10^{-12} kg/C) was always the same, independent of the chemical composition of the electrodes or the gas in the cathode ray tube. He concluded from this independence that the atoms of all elements contain electrons. Thomson further hypothesized that atoms are uniform spheres of positive charge with electrons embedded in them (Figure 2.5).

The Electron Charge and Mass

In 1912 Robert Millikan of the University of Chicago devised an ingenious method for measuring the charge of the electron. When an oil mist is exposed to the radiation from certain radioactive substances (see the following section), some of the individual oil droplets acquire an electrical charge. As a result the motion of the droplets, and in particular their rate of fall, can be controlled by an electric field. Millikan measured the charge carried by the droplets and discovered that it was always a *whole-number* multiple of 1.6×10^{-19} C. This is, charge $= n \cdot (1.6 \times 10^{-19}\ \text{C})$, where $n = 1, 2, 3,$ and so on, but never a value in between. He reasoned that since a droplet could only acquire a whole number of electrons, this smallest observable charge must be the charge e carried by a single electron. Together with J. J. Thomson's value of m/e, this established the mass of the electron.

$$m = (1.6 \times 10^{-19}\ \text{C}) \cdot (5.7 \times 10^{-12}\ \text{kg C}^{-1}) = 9.1 \times 10^{-31}\ \text{kg}$$

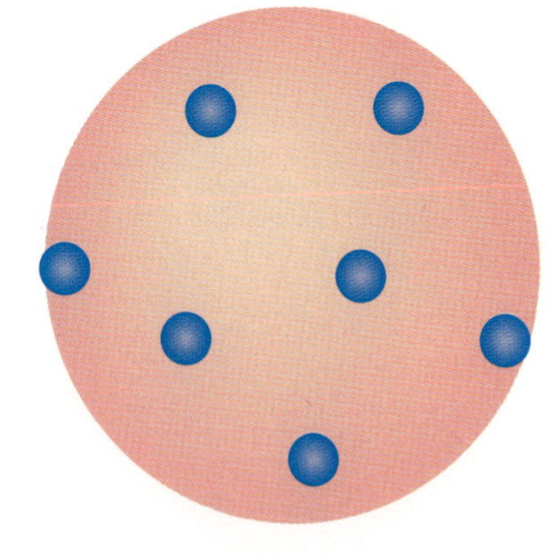

Figure 2.5 The Thomson model of the atom. An otherwise uniform sphere of positive charge has electrons embedded in it. The electrons can escape under the right circumstances, forming cathode rays.

Joseph John Thomson (1856–1940)

Joseph John Thomson was appointed Cavendish Professor of Experimental Physics at Cambridge in 1884 and held the position for almost 35 years. He contributed a great deal to our knowledge of electrical and magnetic fields and made extensive studies of electrical discharges in gases. He was awarded the 1906 Nobel Prize in physics.

Robert Millikan (1868–1953)

Robert Millikan, best known for his "oil-drop" experiment, also studied the photoelectric effect following suggestions advanced by Einstein. He was very interested in the relationship of science to other human activities, and wrote *Evolution of Science and Religion* (1928), and *Science and the New Civilization*. He was awarded the 1923 Nobel Prize in physics.

The Nuclear Atom

Around the turn of the century it was discovered that certain elements, notably uranium and radium, emitted "radiation" continuously, without being subjected to any electrical discharge. This phenomenon became known as **radioactivity,** and, in time, three different types of radioactive emissions were recognized. **Alpha (α) rays** are positively charged particles having the same mass as helium nuclei, **beta (β) rays** are fast-moving electrons, and **gamma (γ) rays,** rather than being particles, are radiation similar to visible light. The topic of radioactivity will be treated further in Chapter 24. For now, its importance lies in the fact that α-rays gave Rutherford and his coworkers, Geiger and Marsden, the tools they needed for a crucial experiment that established the existence of a nucleus within the atom. The experiment was to direct a stream of α-particles at a sheet of metal foil, and observe the results (Figure 2.6).

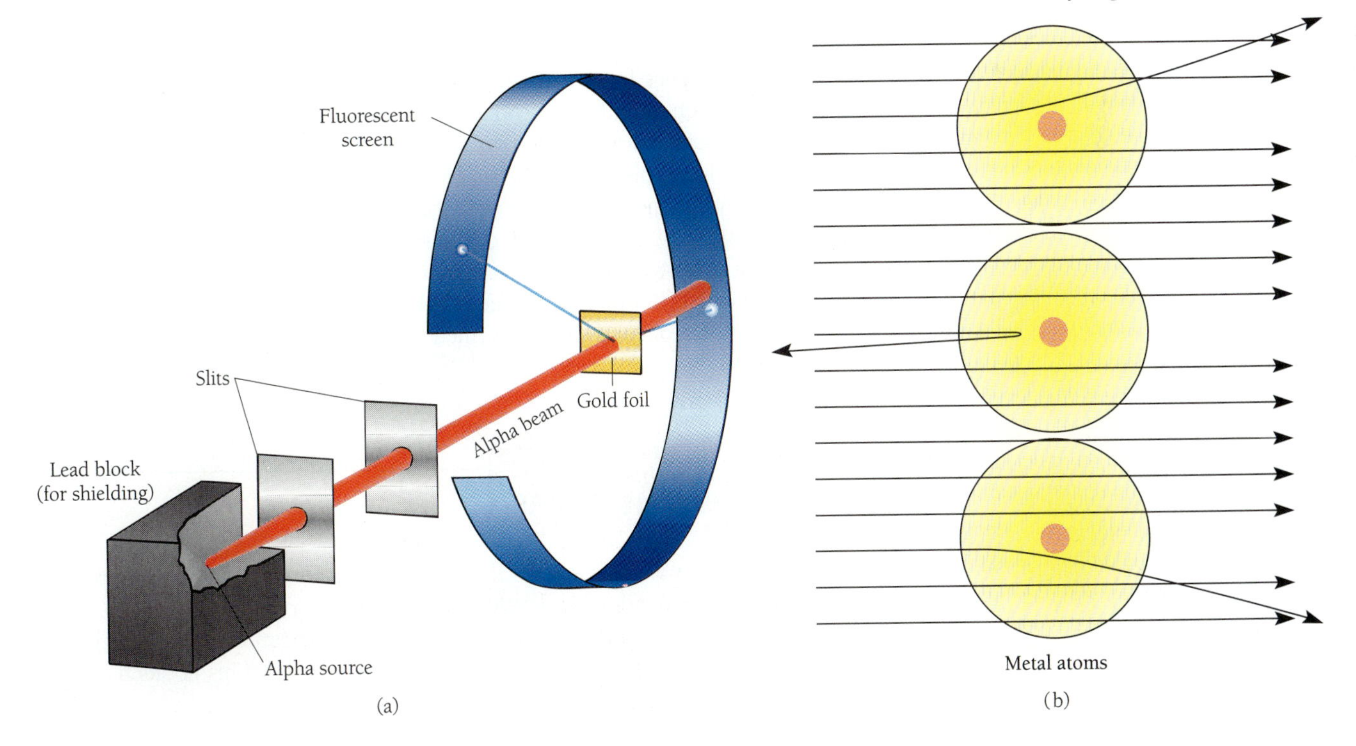

Figure 2.6 (a) In Rutherford's experiment, most α-particles pass through the metal foil undeflected while a few are strongly deflected. (b) Large deflections can occur only if the mass of the metal foil atoms is concentrated into a very small fraction of the total atomic volume, and that small volume also bears a relatively large electrical charge.

Ernest Rutherford (1871–1937)

Ernest Rutherford was a native New Zealander who spent most of his professional life in Canada and England. A student of J. J. Thomson, he succeeded him as Cavendish Professor at Cambridge in 1919. He was among the first to suggest that α-rays come from the disintegration of atomic nuclei. He won the Nobel Prize (1908), only two years after his teacher Thomson.

Instead of being blocked by the foil, the majority of the particles passed through undeflected, as if the foil were not there at all. This was unexpected, but even more so was the astonishing result that a few of the α-particles bounced back in the direction of their radioactive source. This can happen only if the mass of the foil atoms is concentrated into a very small fraction of the total atomic volume, and that small volume (the *nucleus*) also bears a relatively large electrical charge. Rutherford's experiment proved the existence of a small nucleus that is surrounded by a large region having virtually no mass. Thomson's earlier hypothesis of uniform spheres could not explain this behavior and was therefore discarded. As often happens in science, a major advance came from a totally unexpected experimental observation.

The Nature of Isotopes

The prefix *iso-* means "the same."

The existence of several different types of radioactive emissions from the same element led Frederick Soddy to suggest that an element can exist in different forms. Such forms, called isotopes, would have identical chemical properties but different physical properties.

Additional evidence for the existence of isotopes came from the *canal rays* of a discharge tube, which differ from cathode rays in that they are positively charged and have a much higher mass-to-charge ratio. The mass-to-charge ratio depends on the identity of the gas in the discharge tube, and canal rays are in fact beams of cations. The cations are produced when the discharge strips one or more electrons from the atoms of the gas. In 1912, J. J. Thomson noted that some very pure gases produced ions of several different mass-to-charge ratios. In neon, for example, 90% of the canal rays had a m/e ratio of 20 (in units of amu/ecu), while 10% had $m/e = 22$. He identified these as isotopes of neon, and thus confirmed that isotopes of the same element have different masses.

Frederick Soddy (1877–1956)

Frederick Soddy was Professor of Chemistry at Oxford when in 1921 he received the Nobel Prize for his development of the theory of isotopes. Although most of his scientific career was spent in investigations of radioactivity, he was also interested in politics and economics. One of his books has the unusual title, *Money Versus Man.*

A few years later (1918) the Canadian chemist A. J. Dempster constructed an improved discharge tube apparatus called a **mass spectrometer,** and used it to make precise measurements of the masses and relative abundances of isotopes. He discovered many isotopes and established that most elements exist in several isotopic forms. About one fifth of the elements have only one stable isotope: they are **monoisotopic.** Sodium, aluminum, and fluorine are the most abundant of the monoisotopic elements. About one fifth of the elements, including all those with an atomic number greater than 83, have *no* stable, nonradioactive isotopes. Several of these radioactive elements—radon, uranium, and plutonium—are frequently in the news. Precise values for isotopic abundances, such as those in Table 2.2, are the result of mass spectrometric measurements by Dempster and those who followed him.

Figure 2.11 illustrates a modern version of Dempster's apparatus.

2.4 MOLECULES, IONS, AND COMPOUNDS

The vast majority of pure substances are compounds rather than elements. That is, they contain more than one type of atom. The fundamental particles of compounds are molecules or ions, rather than atoms. The atoms in a molecule are joined by chemical bonds, and the composition of a molecule is indicated by its formula.

Molecules and Bonds

A molecule was earlier defined as a "stable cluster of atoms," and it was noted that molecules of a particular compound are identical to one another. The atoms in a molecule can be of the same element or of several different elements. There must be at least two, but there is no upper limit to the number of atoms in a molecule. A molecule is "stable," which means that, in the absence of chemical change, its atoms remain indefinitely in a fixed geometrical arrangement. Atoms in a molecule are held to their neighboring atoms by a relatively strong attractive force called a **chemical bond.** Molecules have no electrical charge, since they consist of electrically neutral atoms. Figure 2.7 shows some common molecules and their geometrical arrangement. Any compound that consists of molecules (for example water or carbon dioxide) is called a **molecular compound.** A sample of water consists of a large number of water molecules, all exactly alike (Figure 2.8). The molecule is the smallest particle of such a compound.

Figure 2.7 Molecular models are useful tools for visualizing the shape of molecules. The "ball-and-stick" models provide good representations of the locations of atoms and the way in which they are joined by chemical bonds. In the "space-filling" models, the electron clouds of the atoms are represented as solid objects. Space-filling models better represent the actual sizes of different atoms and the way in which they adjoin and overlap one another.

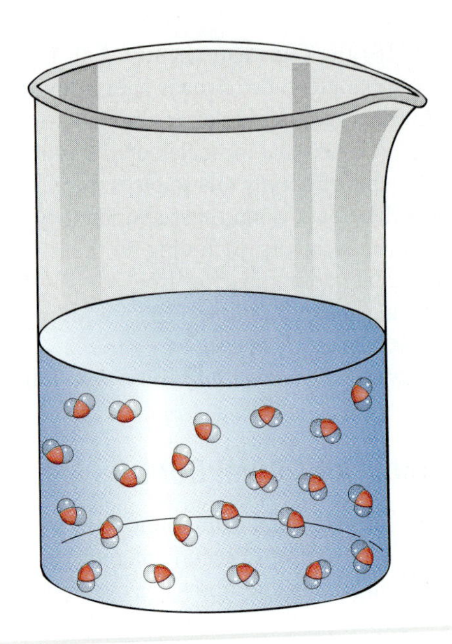

Figure 2.8 Liquid water consists of a large number of identical H_2O molecules.

Ionic Compounds

Ionic compounds containing OH^- are referred to as "hydroxides" rather than salts. Those containing O^{2-} are simply called "oxides."

Not all compounds consist of molecules; in some, like sodium chloride (table salt), the smallest particles are ions. These are called **ionic compounds** to distinguish them from the molecular compounds described in the preceding section. Unless they contain the OH^- ion or the O^{2-} ion, ionic compounds are also called **salts.** In sodium chloride, shown in Figure 2.9, the particles are sodium cations and chloride anions. The total positive charge is equal to the total negative charge, so that ionic compounds, like molecular compounds, are electrically neutral. The forces holding the oppositely charged ions together are called **ionic bonds.**

The terms "diatomic" and "polyatomic" are used to describe molecules as well as ions.

The anions and cations of sodium chloride are both derived from single atoms and are called **monatomic ions.** Larger ions, containing more than one atom, are called **polyatomic.** Chalk consists primarily of powdered calcium carbonate ($CaCO_3$), which contains monatomic calcium ions (Ca^{2+}, charge = +2 ecu), and polyatomic

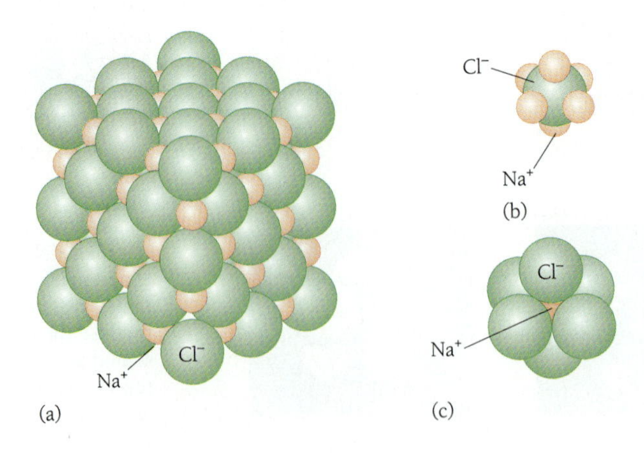

Figure 2.9 (a) Part of a crystal of sodium chloride. There are no molecules here, only ions. (b) Each chloride anion has 6 nearest-neighbor sodium cations, to which it is equally attracted. (c) Similarly, each sodium cation is equally attracted to its 6 nearest-neighbor chloride anions.

Figure 2.10 A carbonate ion consists of a centrally located carbon atom bonded to 3 oxygen atoms, plus 2 additional electrons. All 4 atoms lie in the same plane.

carbonate ions (CO_3^{2-}, charge = −2 ecu). Carbonate ions contain one carbon atom, three oxygen atoms, and two additional electrons (Figure 2.10). Like molecules, polyatomic ions are stable geometric arrangements of atoms held together by covalent bonds. They maintain their identity in many (but not all) chemical changes, and they differ from molecules only in that they carry a nonzero electrical charge. Polyatomic anions are more common than polyatomic cations. Some of the more common polyatomic ions are listed in Table 2.3.

The classification of compounds as ionic or molecular is a useful one, although there are borderline cases that have characteristics of both types. In general, however, a compound is ionic if it contains (1) an element from Group 1A or 2A of the periodic table, or ammonium ion, and (2) an element from Group 7A or a polyatomic ion. Most other compounds are molecular. We will discuss borderline cases as they arise, and from time to time we will mention other, less common, polyatomic ions.

TABLE 2.3 Names and Formulas of Common Polyatomic Ions

Anions			
OH^-	Hydroxide	ClO_3^-	Chlorate
CN^-	Cyanide	BrO_3^-	Bromate
ClO^-	Hypochlorite (also written OCl^-)	IO_3^-	Iodate
HS^-	Hydrogen sulfide	ClO_4^-	Perchlorate
NH_2^-	Amide	MnO_4^-	Permanganate
NO_2^-	Nitrite	HCO_3^-	Hydrogen carbonate (bicarbonate)
NO_3^-	Nitrate	HSO_4^-	Hydrogen sulfate
O_2^{2-}	Peroxide (*oxide* is monatomic, O^{2-})	$CH_3CO_2^-$	Acetate
CO_3^{2-}	Carbonate	HPO_4^{2-}	Hydrogen phosphate
SO_3^{2-}	Sulfite	$C_2O_4^{2-}$	Oxalate
SO_4^{2-}	Sulfate	$Cr_2O_7^{2-}$	Dichromate
PO_4^{3-}	Phosphate	$S_2O_3^{2-}$	Thiosulfate
Cations			
NH_4^+	Ammonium		
Hg_2^{2+}	Mercurous or mercury(I)*		

* Most metals form only monatomic cations. Mercury is an exception.

Formulas

The composition of a compound is described by a chemical formula, which indicates the relative amount of each element in the compound. Several different types of formula are in common use by chemists.

Molecular Formula

The composition of a molecule or molecular compound is described by listing the atoms it contains. A **molecular formula** is a list of elemental symbols, with numerical subscripts designating how many of that type atom are in the molecule. If an elemental symbol has no subscript, the molecule contains only one such atom. For example, NH_3 (ammonia) contains one N atom and three H atoms; H_2SO_4 (sulfuric acid) contains two H atoms, one S atom, and four O atoms; and N_2O_4 (dinitrogen tetroxide) contains two N atoms and four O atoms.

Empirical Formula

Ionic compounds, which do not consist of molecules, are described by an **empirical formula** whose subscripts give the *relative* numbers of ions in the compound. The subscripts must be whole numbers, and they must be the *smallest* set of integers that give the correct relative numbers of ions. The cation is always written before the anion. For example, $CaCl_2$ is the empirical formula of a compound having twice as many Cl^- ions as Ca^{2+} ions. The formula Ca_2Cl_4 is incorrect because the subscripts (2 : 4) are not the smallest that give the correct ratio (1 : 2). When the formula of a *compound* is written, the charges of its ions are omitted. When the formula of an ion by itself is written, the charge *must be included*. The charge of an ion is part of its formula. For example, sulfur trioxide (SO_3) and sulfite ion (SO_3^{2-}) are two distinct species with totally different chemical behaviors.

If the compound contains a polyatomic ion, parentheses are used when that ion's subscript is greater than one. In $CaSO_4$ the ratio of Ca^{2+} ions to SO_4^{2-} ions is 1 : 1, while in $Sn(SO_4)_2$ there are two SO_4^{2-} ions for every Sn^{2+} ion, and in $(NH_4)_2S$ there are two NH_4^+ ions for every S^{2-} ion. Subscripts to parentheses *multiply* the subscripts within parentheses, including the omitted subscript 1. For example, in barium nitrate, $Ba(NO_3)_2$, the relative numbers of the different atoms are Ba : N : O = 1 : 2 : 6. The **formula unit** of an ionic compound is the group of atoms listed in the empirical formula. That is, a formula unit of barium nitrate consists of one Ba atom, two N atoms, and six O atoms. In this context, it is usual to ignore the ionic status and speak instead of the "atoms" in a formula unit.

EXAMPLE 2.3 Composition of an Ionic Compound

Describe (a) the ionic composition and (b) the composition of the formula unit of the ionic compound $Al_2(SO_4)_3$ (aluminum sulfate).

Analysis "Ionic composition" means the identity and relative number of the ions making up the compound. In our example, these are seen to be aluminum and sulfate ions in the ratio of 2 : 3. The parentheses provide a clue that the (SO_4) unit is a polyatomic ion. Uncertainties as to the correct charge or name of ions can be resolved by consulting Table 2.3.

"Composition of the formula unit" means the identity and relative number of the atoms. We can obtain this information directly from the formula, keeping in mind the significance of subscripts and parentheses.

Work

a. Ionic composition: two aluminum ions for every three sulfate ions.

b. Composition of the formula unit: two Al atoms, three S atoms, and twelve O atoms.

Exercise

How many atoms are there of each type in the formula units of (a) $Ca(CN)_2$ (calcium cyanide); (b) $(NH_4)_2CO_3$ (ammonium carbonate), and (c) $Mg(NO_3)_2$ (magnesium nitrate)?

Answer: (a) $Ca(CN)_2$: one Ca atom, two C atoms, two N atoms; (b) $(NH_4)_2CO_3$: two N atoms, eight H atoms, one C atom, three O atoms; and (c) $Mg(NO_3)_2$: one Mg atom, two N atoms, six O atoms.

Empirical formulas are also used to describe molecular compounds, in order to show the smallest *relative* numbers of atoms. The molecular formula of acetic acid is $C_2H_4O_2$, while its empirical formula is CH_2O.

Dots in Formulas

Many solid compounds contain water, loosely held within the crystal structure. Such compounds are called **hydrates,** or **hydrated compounds,** and the water is referred to as **water of crystallization,** or **water of hydration.** Since the water is loosely held, it can often be driven off by heating, leaving an **anhydrous** compound (one lacking water). Water of hydration is indicated by a dot in the empirical formula. For example, hydrated magnesium sulfate contains 7 water molecules for each magnesium ion. This ratio is indicated by the formula $MgSO_4 \cdot 7H_2O$. Hydrated aluminum sulfate, $Al_2(SO_4)_3 \cdot 9H_2O$, contains 9 water molecules per formula unit.

Dots are also used in the formulas of minerals that are mixtures of compounds. For example, the mineral *perovskite* ($CaO \cdot TiO_2$) is a 1:1 mixture of calcium and titanium oxides. Dots are used only in formulas of solid substances that are homogeneous mixtures of two or more compounds.

Structural Formula

Often it is desirable to convey more information about a molecule than the number and type of its atoms. In a **structural formula,** the symbols of all atoms are written, and those atoms bonded to one another are connected by short lines. This information, that is, which atoms are bonded to which, is called the atom-to-atom **connectivity** of the molecule. Information about shape of the molecule is also frequently included in a structural formula. For example, both H—O—H and H╱O╲H are structural formulas for water because they both show the connectivity of the molecule. However, the latter conveys the additional information that the three atoms do not lie in a straight line. Table 2.4 gives the three types of formulas for some representative compounds.

TABLE 2.4 Comparison of Different Formula Types

Type	Molecular Compound	Molecular Compound	Molecular Compound	Ionic Compound	Polyatomic Ion
Name	Ethanol	Hydrogen peroxide	1-Butene	Ammonium sulfate	Ammonium ion
Molecular Formula	C_2H_6O or C_2H_5OH*	H_2O_2	C_4H_8	Not applicable	NH_4^+
Empirical Formula	C_2H_6O	HO	CH_2	$(NH_4)_2SO_4$ or $N_2H_8SO_4$	NH_4^+
Structural Formula	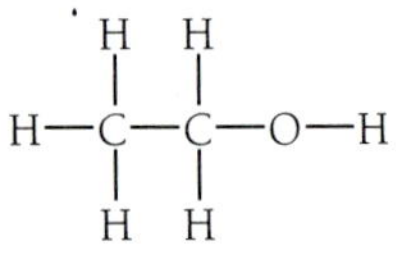 H H | | H—C—C—O—H | | H H	H—O—O—H	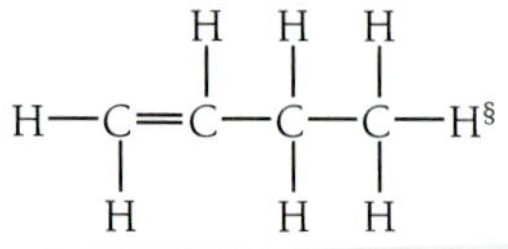 H H H | | | H—C=C—C—C—H§ | | | H H H	Not applicable‖	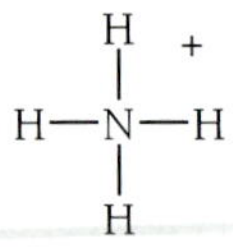 H + | H—N—H | H
	or H H | | H—C—C—H† | | H O—H 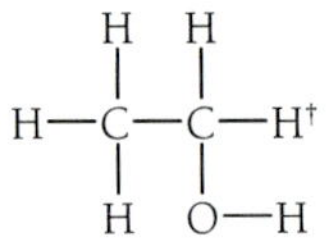	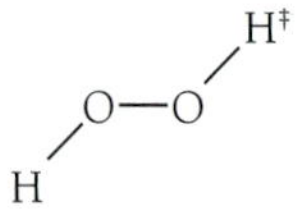 H‡ / O—O / H	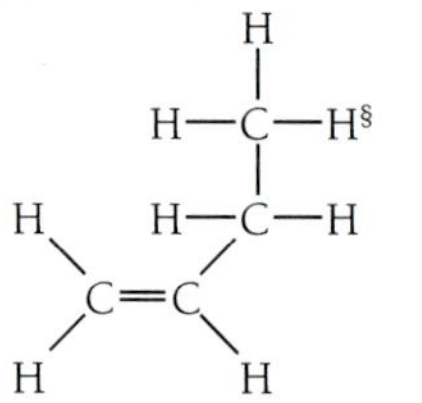 H | H—C—H§ | H H—C—H \ / C=C / \ H H		
Space-Filling Model				Not applicable‖	

* This version of the molecular formula is often used because it conveys the additional information that the molecular subgroup —OH is present.
† This structural formula is equivalent to the one immediately above it. They both convey the same information, namely connectivity. Both structures show that one of the carbon atoms is bonded to two hydrogen atoms, one —OH group, and one other carbon atom.
‡ Again, both structures shown are acceptable, but the second structure shows that the molecule is not a linear array of four atoms.
§ A double line is used in the structural formula to indicate a second type of bond (such "double bonds" are discussed in Chapter 8).
‖ A structural formula or space-filling model for an ionic compound would be a sketch or drawing such as that for sodium chloride, Figure 2.9. Such drawings are rarely used.

Modified Molecular Formula

The molecular formula of a compound is frequently written in a form that stresses part of the information carried by a complete structural formula. The intent of such a modified molecular formula is to indicate which *groups* of atoms are present in a molecule. For example, the molecular formula of acetic acid, $C_2H_4O_2$, is usually written as CH_3COOH to show the presence of the $—CH_3$ group and the —COOH group. Such submolecular groups of atoms determine the chemical properties of the compound.

2.5 THE MOLE, A COUNTING UNIT FOR CHEMISTRY

In almost all situations of chemical interest the *number* of atoms, molecules, or ions is an important quantity. For example, there are 2 ammonium ions in a formula unit of ammonium sulfate [$(NH_4)_2SO_4$]. Or, if *50* hydrogen molecules (H_2) react with *25* oxygen molecules (O_2), the result is *50* molecules of water (H_2O). Although atoms and molecules are **microscopic** objects, too small to see and too many to count, their number can be determined in another way. Suppose you were a thumbtack manufacturer, and received an order for 10 million thumbtacks. Knowing that a thumbtack weighs 0.15 g, rather than count them you could simply send the customer $(10{,}000{,}000) \cdot (0.15\text{ g}) = 1.5 \times 10^3$ kg (about 1½ tons) of tacks. Chances are, the number would not be *exactly* 10 million, but it would be close enough so that both you and the customer would be satisfied. The principle works just as well with atoms. Since their masses are known, they can be counted by weighing.

Chemists use the word *microscopic* to describe objects no larger than molecules. Ordinary quantities of matter, say anything large enough to weigh, are described as *macroscopic.*

We often make use of *counting units* when dealing with small items. For example, we buy eggs by the dozen (12), paper clips by the gross (144), paper by the ream (500), and so forth. The counting unit used for atoms, ions, and molecules is called the **mole,** abbreviated as **mol.** The mole is the SI unit for the *amount* of an element or compound. The number of particles in a mole is called Avogadro's number.

This abbreviation seems barely worthwhile, since it's only one letter shorter. However, "mol" is the approved form, and we shall use it.

Avogadro's Number and the Atomic Mass Unit

The value of Avogadro's number (N_A), as we will see in the following section, is about 6×10^{23}. The equivalence statement is *1 mol* $\Leftrightarrow$ *N_A particles,* and depending on *what* is being counted, "particles" can refer to atoms, molecules, ions, or subatomic particles. **Avogadro's number** is defined as the number of atoms in exactly 12 g of the pure isotope C-12. In other words, the mass of Avogadro's number of C-12 atoms is exactly 12 g. Closely related is the definition of the atomic mass unit: **1 amu ≡ ¹⁄₁₂ of the**

Actually, *any* object can be counted by the mole. But a mole of shoes, for instance, would have a mass and volume greater than that of the earth. Use of the mole is normally limited to elementary particles—atoms, molecules, and ions.

One mole of atoms of some common elements. Back row (left to right): bromine, aluminum, mercury, and copper. Front row (left to right): sulfur, zinc, and iron.

One-mole quantities of some compounds. The white compound is NaCl (58 g). The others are $CuSO_4 \cdot 5H_2O$ (deep blue, 250 g), $NiCl_2 \cdot 6H_2O$ (green, 238 g), $CoCl_2 \cdot 6H_2O$, (deep red, 166 g), and K_2CrO_7 (orange, 294 g).

From the definitions of the mole and the amu, the mass of 1 mol C-12 atoms is (N_A atoms) · (12 amu atom^{-1}) ⇔ 12 g. Dividing both sides of this equation by 12 leads to the relationship N_A amu ⇔ 1 g.

mass of *one* atom of C-12. That is, the mass of one C-12 atom is exactly 12 amu. Carbon-12 is the *standard* on which the atomic mass scale is based. These definitions of N_A and the amu have several important consequences.

1. The equivalence statement relating amu to grams is *N_A amu ⇔ 1 g*.
2. The mass, in grams, of one mole of any type of atom is numerically equal to the mass, in amu, of one of those atoms.
3. Avogadro's number is an *experimental quantity* whose value must be measured.

The Value of Avogadro's Number

There are numerous ways of measuring Avogadro's number. The most straightforward way is the measurement of the mass, in grams, of a single atom of C-12. Such measurements are made in a modern version of Dempster's mass spectrometer (Figure 2.11). The measurement yields the value 1.99×10^{-23} g for the mass of one C-12 atom, which establishes the equivalence statement 1.99×10^{-23} g ⇔ 12 amu. Di-

Count Amadeo Avogadro (1776–1856)

Amadeo Avogadro was Professor of Physics at the University of Turin in Italy. In 1811 he published the hypothesis that equal volumes of different gases contain equal numbers of molecules. He also suggested that certain elements (such as oxygen or hydrogen) exist as molecules rather than as single atoms, an idea that was not generally accepted until about 50 years later. Count Avogadro began his career as a lawyer, and turned to science later in life.

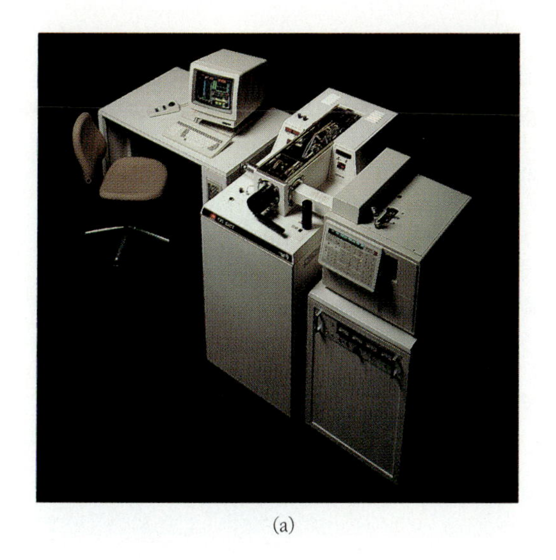

(a)

Figure 2.11 (a) A modern mass spectrometer. (b) Positive ions are formed when gas molecules are hit by a beam of electrons. The ions are accelerated by an electric field, and follow a path whose curvature depends on ion mass and the strength of a magnetic field. Changing the magnetic field brings ions of different mass into the detector, where their relative numbers are measured. (c) The output of the instrument is a *mass spectrum,* which, in this case, shows the relative amounts of ^{12}C and ^{13}C in a carbon-containing gaseous substance.

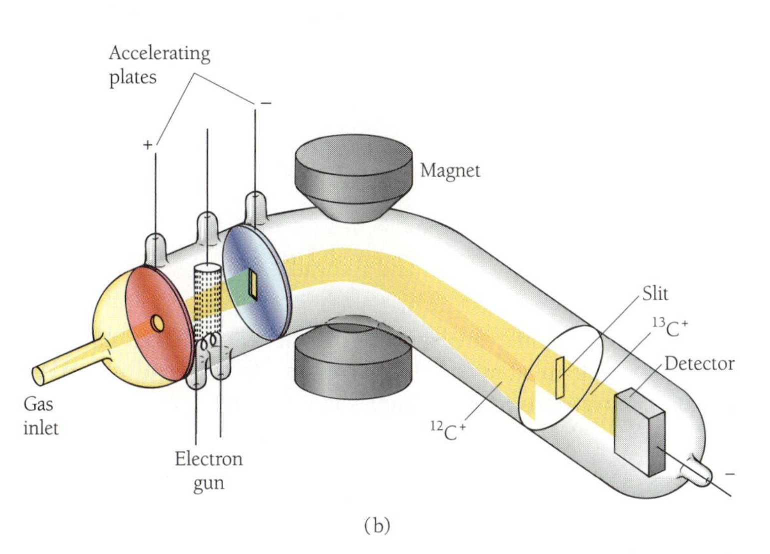

(b)

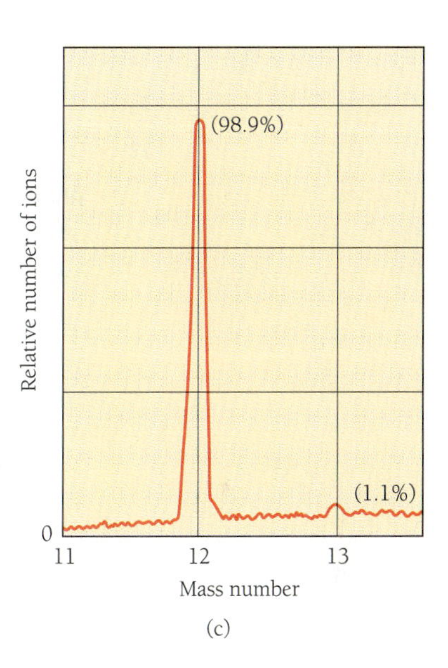

(c)

viding both sides of this equivalence statement by 1.99×10^{-23} gives the relationship $1 \text{ g} \Leftrightarrow 6.02 \times 10^{23}$ amu, which, when compared with the equivalence statement $1 \text{ g} \Leftrightarrow N_A$ amu, shows directly that $N_A = 6.02 \times 10^{23}$.

Some other means for measuring N_A involve measurement of the charge of the electron (Section 2.3) and measurement of the density and interatomic distances in a crystalline substance. The most accurate measurements lead to the value $N_A = 6.022137 \times 10^{23}$. The reciprocal of N_A appears in the equivalence statement $1 \text{ amu} \Leftrightarrow 1.660540 \times 10^{-24}$ g. Chemical calculations rarely require so many significant digits, and Avogadro's number should be memorized as 6.022×10^{23}.

Avogadro's number is almost unimaginably large (because atoms are almost unimaginably small), but some comparisons may convey a feeling of its immensity. Hydrogen atoms are so small that it would take 5×10^8 (five hundred million) of them placed side by side to make a length of one inch. Avogadro's number of H atoms, placed side by side, would stretch across the solar system and back. A computer capable of performing a million operations per second would require 20 billion years to add 6.02×10^{23} numbers. (According to the "Big Bang" theory, the age of the universe is less than that, about 12 to 15 billion years.)

The mole is a counting unit that permits the use of numbers of comprehensible size to describe ordinary quantities of chemical substances. Although it is true that 180 g water (about a cupful) contains about 6×10^{24} molecules, it is more meaningful to express this amount as 10 moles. Conversion between moles and number of particles is achieved by substituting the value of N_A into the equivalence statement 1 mol $\Leftrightarrow$ N_A particles. This leads to the two unit conversion factors

$$\frac{(6.022 \times 10^{23} \text{ particles})}{(1 \text{ mole})} \Longleftrightarrow 1 \Longleftrightarrow \frac{(1 \text{ mole})}{(6.022 \times 10^{23} \text{ particles})}$$

EXAMPLE 2.4 Atoms and Moles

a. How many krypton atoms are there in 1.5 moles?

b. How many moles contain 4.00×10^{24} radon atoms?

Analysis

Target: A number of atoms of a particular type.

$$? \text{ atoms} =$$

Knowns: The number of moles is given, and Avogadro's number is an implicit known.

Relationship: Moles and atoms are related by the equivalence statement 1 mol $\Leftrightarrow$ N_A atoms. Mole to atom conversions are solved the same way as the more familiar unit conversions in the examples in Chapter 1.

Work

a.

$$? \text{ Kr atoms} = \frac{(1.5 \text{ mol Kr})}{(1)} \cdot \frac{(6.02 \times 10^{23} \text{ atoms})}{(1 \text{ mol})}$$

$$= 9.0 \times 10^{23} \text{ Kr atoms}$$

Check $1\frac{1}{2} \times 6 = 9$, so there has been no math error. A chemically significant number of atoms or molecules is large, usually with an exponent in the range of 15 to 25. If the exponent is not in this range, either the problem is unusual or you have made a mistake.

b. As in part (a), this is a unit conversion. As always, we need to pay attention to the units.

Work

$$? \text{ mol Rn} = \frac{(4.00 \times 10^{24} \text{ Rn atoms})}{(1)} \cdot \frac{(1 \text{ mol})}{(6.02 \times 10^{23} \text{ atoms})}$$

$$= 6.64 \text{ mol Rn}$$

Check 4×10^{24} is somewhat more than 6×10^{23}, and so it corresponds to somewhat more than one mole. The answer is reasonable.

Exercise

Calculate the number of sugar molecules in 150 moles, and the number of moles in 8.0×10^{18} ammonium ions.

Answer: 9.03×10^{25} sugar molecules, 1.3×10^{-5} mol NH_4^+ ions.

EXAMPLE 2.5 Atoms in Compounds

How many F atoms are there in 2.3 mol XeF_4?

Analysis

Target: A number of F atoms.

$$? \text{ F atoms} =$$

Knowns: 2.3 mol XeF_4 is explicit, Avogadro's number is implicit.

Relationship: If we knew the number of XeF_4 molecules we could solve the problem because the molecular formula tells us that each molecule contains 4 F atoms.

Plan Leading *backwards* from the target, the two-step plan is:

2. Find the number of F atoms from the number of XeF_4 molecules.
1. Find the number of XeF_4 molecules in 2.3 mol XeF_4.

Work

1.

$$? \text{ XeF}_4 \text{ molecules} = \frac{(2.3 \text{ mol XeF}_4)}{(1)} \cdot \frac{(6.02 \times 10^{23} \text{ molecules})}{(1 \text{ mol})}$$
$$= 13.8 \times 10^{23} \text{ XeF}_4 \text{ molecules}$$

2.

$$? \text{ F atoms} = \frac{(13.8 \times 10^{23} \text{ XeF}_4 \text{ molecules})}{(1)} \cdot \frac{(4 \text{ F atoms})}{(1 \text{ XeF}_4 \text{ molecule})}$$
$$= 5.5 \times 10^{24} \text{ F atoms}$$

Check The units cancel correctly in both steps, and the number of atoms in the final answer is appropriate for a quantity of a few moles.

Exercise

How many atoms of each type are there in 0.337 moles of $Ca(NO_3)_2$?

Answer: 2.03×10^{23} Ca atoms; 4.06×10^{23} N atoms; 12.2×10^{23} O atoms.

2.6 MOLAR MASS: ATOMIC WEIGHT, MOLECULAR WEIGHT, AND FORMULA WEIGHT

The mass of an atom is one of the properties that distinguish elements from each other. Many chemical calculations are based on element masses. In practice we deal with the mass of one mole of atoms (the "molar mass") rather than the mass of one atom. Determination of molar masses requires information about isotope abundances, as well as accurate measurements of atomic masses.

Masses and Natural Abundances of Isotopes

It is difficult to make *absolute* measurements of atomic masses. Modern mass spectrometers are normally designed to yield accurate but *relative* information such as, "The mass of C-13 atoms is 1.083613 times the mass of C-12 atoms." Since the mass of the C-12 atom is defined to be exactly 12 amu, this establishes the mass of the C-13 atom as $12 \times 1.083613 = 13.00335$ amu. The mass of one mole of C-13 is 13.00335 g. The mass-spectrometric measurement also yields the relative abundance of different isotopes of an element. A sample of carbon, for instance, is seen to consist of 98.892% C-12 atoms and 1.108% C-13 atoms.

Molar Mass of Elements (Atomic Weight)

A mole of an element is a mixture of isotopes of slightly different mass. The mass of a mole is the *weighted average* of the molar masses of the isotopes. This average is found by summing the product of the mass and *fractional abundance* (percent abundance/100) of all the isotopes. The result is called the **molar mass** of the element. Out of long-standing habit among chemists, this quantity is often called by its former name, the **atomic weight.**

EXAMPLE 2.6 Isotopes, Percent Abundance, and Molar Mass

Use the mass and percent abundance data given in the preceding section to calculate the molar mass of carbon.

Work

$(12.00000 \text{ g})(0.98892) = 11.867 \text{ g}$	(mass of 98.892% of 1 mol C-12)
$+(13.00335 \text{ g})(0.01108) = 0.1441 \text{ g}$	(mass of 1.108% of 1 mol C-13)
12.011 g	(mass of 1 mol natural carbon)

Exercise

Calculate the molar mass of neon. The isotopic masses and natural abundances are: Ne-20, 19.9924 amu and 90.51%; Ne-21, 20.9940 amu and 0.27%; Ne-22, 21.99138 amu and 9.22%.

Answer: 20.18 g mol^{-1}, or (average) 20.18 amu $atom^{-1}$.

Although relative molar masses of two elements can be determined with a mass spectrometer, they can also be determined by chemical means. For example, accurate analysis shows that water consists of 88.810% (by mass) oxygen and 11.190% hydrogen. That is, for each gram of hydrogen there are 88.120/11.190 = 7.9366 grams of oxygen. Since the empirical formula of water is H_2O, there are two H atoms for each O atom, and the mass of an average O atom must be 2(7.9366) = 15.873 times the mass of an average H atom. This was the method used by Dalton and all those who followed him, until the advent of the mass spectrometer.

Determination of molar mass by analysis of compounds suffers from the disadvantage that the correct empirical formula of the compounds must be known. Dalton believed the formula of water to be HO, and therefore that the atomic mass of O was 8 rather than 16 times that of hydrogen. This was the source of considerable confusion over molar mass, and the situation was not satisfactorily resolved until the middle of the 19th century.

Modern values for molar masses, given in the alphabetical list of elements on the inside front cover, are based on both types of measurement. The values are listed without units, so that either g mol^{-1} or amu (for an "average" atom) may be used. The same values are also given beneath the symbols of each element in the periodic table.

Gram–Mole Conversions

Molar mass is used to determine the number of moles in a given mass of a substance. In essence, this is the "counting" of atoms referred to on page 59. The ability to convert between grams and moles, quickly and correctly, is an important skill in chemistry.

EXAMPLE 2.7 Gram to Mole Conversion

How many moles are there in 50.0 g argon?

Analysis The target and known are clear, and the problem has the form of a unit conversion. The required conversion factor, which must connect moles and grams, comes from the table of molar masses or the periodic table: for argon, 1 mol ⇔ 39.9 g.

Work

$$? \text{ mol Ar} = \frac{(50.0 \text{ g Ar})}{(1)} \cdot \frac{(1 \text{ mol Ar})}{(39.9 \text{ g Ar})}$$
$$= 1.25 \text{ mol Ar}$$

Check 50 g is a little more than the molar mass, so the answer should be a little more than one mole.

Note: Conversion between grams and moles *always* requires the use of the molar mass.

Exercise

How many moles are there in 75 g He?

Answer: 19 mol.

EXAMPLE 2.8 Conversion of Grams to Atoms

How many atoms are there in 50 g xenon?

Analysis If we knew the number of moles in 50 g, we could use Avogadro's number to find the number of atoms. The problem is a two-stage unit conversion, using the equivalence statements 1 mol Xe ⇔ 131.3 g and 1 mol ⇔ 6.02×10^{23} atoms.

Work

$$? \text{ atoms} = (50 \text{ g}) \cdot \frac{(1 \text{ mol})}{(131.3 \text{ g})} \cdot \frac{(6.02 \times 10^{23} \text{ atoms})}{(1 \text{ mol})}$$
$$= 2.3 \times 10^{23} \text{ atoms}$$

Check 50 is about ⅓ of 131, so the answer should be about ⅓ of Avogadro's number.

Exercise

Calculate the number of atoms in a pound of lead.

Answer: 1.32×10^{24} atoms.

EXAMPLE 2.9 Mass of a Given Number of Atoms

Calculate the mass of 1,000,000 sulfur atoms.

Analysis We can solve the problem if we know the mass of 1 atom, which can be found from the molar mass and Avogadro's number.

Plan

2. Multiply the mass of 1 S atom by 1,000,000.
1. Find the mass of 1 S atom from the molar mass and Avogadro's number.

Work Once the plan has been made, it is simpler to combine the two steps into a single calculation.

$$? \text{ g} = \frac{(32.06 \text{ g})}{(1 \text{ mol})} \cdot \frac{(1 \text{ mol})}{(6.022 \times 10^{23} \text{ atoms})} \cdot (1.000 \times 10^6 \text{ atoms})$$
$$= 5.324 \times 10^{-17} \text{ g}$$

Check The mass of 1 atom is about $30/(6 \times 10^{23}) = (30/6) \times 10^{-23} = 5 \times 10^{-23}$; 1 million ($10^6$) times this is 5×10^{-17}.
Note: Since the number of significant digits in 1,000,000 is not clear, this quantity should not limit the precision of the result.

Exercise

What is the mass of 4.54×10^{13} rubidium atoms?

Answer: 6.44 ng.

Molar Mass of Compounds (Molecular Weight)

The mass of 1 mole of a compound, its *molar mass*, has units of g mol^{-1}. This quantity is often referred to by its former name, **molecular weight.** The molar mass of a molecular compound is equal to the sum of the molar masses of all the atoms in the molecule. Unless the situation suggests otherwise, it is usual to compute and use molar masses accurate to the nearest 0.1 g mol^{-1} rather than the full accuracy available from the periodic table.

EXAMPLE 2.10 Calculation of Molar Mass of a Molecular Compound

The first xenon-containing compound to be isolated was $XePtF_6$. Calculate its molar mass.

Analysis/Plan The molar mass is found by summing the molar masses of the atoms.

Work Multiply the molar mass of each element by its subscript, and add.

$$\begin{array}{lr} \text{Xe: } 1 \text{ mol} \cdot 131.3 \text{ g mol}^{-1} = & 131.3 \text{ g} \\ \text{Pt: } 1 \text{ mol} \cdot 195.1 \text{ g mol}^{-1} = & 195.1 \text{ g} \\ \text{F: } 6 \text{ mol} \cdot 19.00 \text{ g mol}^{-1} = & 114.0 \text{ g} \\ \hline \text{XePtF}_6\text{:} \qquad 1 \text{ mol} = & 440.4 \text{ g} \\ \text{molar mass} = & 440.4 \text{ g mol}^{-1} \end{array}$$

Check The most common error is to forget that the molar mass of each element must be multiplied by the subscript of that element in the formula. Has this been done?

Exercise

Calculate the molar mass of PtO_2F_6.

Answer: 341.1 g mol^{-1}.

One mole of an ionic compound is equal to 6.02×10^{23} *formula units,* since such compounds do not consist of individual molecules. The older name **formula weight** is sometimes used for the molar mass of an ionic compound. This distinction is purely formal, and has no effect on the calculation of molar mass.

EXAMPLE 2.11 Molar Mass of an Ionic Compound

Calculate the molar mass of the ionic compound calcium acetate, whose empirical formula is $Ca(CH_3CO_2)_2$.

Analysis The relationship between molar mass and formula is the same for ionic compounds as it is for molecular compounds.

Plan Proceed as with a molecular compound, taking care that the number of each kind of atom in the formula unit is correctly counted.

Work

1. Rewrite the formula to show more clearly the number of each kind of atom in the formula unit.

$$Ca(CH_3CO_2)_2 \longrightarrow Ca(C_2H_3O_2)_2 \longrightarrow CaC_4H_6O_4$$

2. Multiply the molar mass of each element by its subscript, and add.

$$\begin{array}{lrr} & \text{Ca: } 1\text{ mol} \cdot 40.1\text{ g mol}^{-1} = & 40.1\text{ g} \\ & \text{C: } 4\text{ mol} \cdot 12.0\text{ g mol}^{-1} = & 48.0\text{ g} \\ & \text{H: } 6\text{ mol} \cdot 1.008\text{ g mol}^{-1} = & 6.05\text{ g} \\ & \text{O: } 4\text{ mol} \cdot 16.0\text{ g mol}^{-1} = & 64.0\text{ g} \\ \hline Ca(CH_3CO_2)_2: & 1\text{ mol} = & 158.2\text{ g} \\ & \text{molar mass} = & 158.2\text{ g mol}^{-1} \end{array}$$

Note: Since many compounds have a relatively large number of hydrogen atoms, errors in the tenths place of molar mass can occur if the rounded-off value of 1.0 g mol^{-1} is used for the molar mass of hydrogen. It is better to use the value 1.008 g mol^{-1} and round the result to the nearest tenth *after* the calculation.

Exercise

Xenon forms a small number of polyatomic ions, such as the perxenate ion XeO_6^{4-}. Calculate the molar mass of aluminum perxenate, $Al_4(XeO_6)_3$.

Answer: 789.9 g mol^{-1}.

2.7 COMPOSITION AND CHEMICAL FORMULAS

There is a close relationship between the formula and the composition of a compound. If the formula is known, the percent composition can be calculated. If the percent composition is known, the empirical formula can be determined.

Percent Composition

The **percent composition** of a compound is a list of the mass percents of each of the elements present in the compound. **Mass percent** or "percent by mass" is the number of grams (or pounds, or tons) of something in a sample whose total mass is 100 grams (or pounds, or tons). Since percent is a *ratio,* any mass unit may be used in place of grams. Also, since weight is proportional to mass, weight percent (percent by weight) has the same numerical value as mass percent. Common abbreviations for mass percent and weight percent are %(m/m) and %(w/w). Percent composition can be calculated from the empirical or molecular formula and the table of molar masses of each element. The percent of an element Y in a compound is found from Equation (2-1). Example 2.12 gives the procedure.

$$\%(\text{element Y}) = \frac{(\text{molar mass of Y}) \cdot (\#\text{ of Y atoms in formula})}{(\text{molar mass of compound})} \times 100 \qquad (2\text{-}1)$$

EXAMPLE 2.12 Percent Composition of a Compound

XeF_6 is a colorless crystalline substance that reacts violently with water, producing xenon trioxide (XeO_3) and hydrogen fluoride (HF). Calculate the percent composition of xenon trioxide. Report the result to the nearest 0.1%.

Analysis

Target: The %(m/m) of Xe and O in XeO_3.

Knowns: The formula is given, and the molar masses are implicit knowns.

Plan Calculate the molar mass of XeO_3, then apply Equation (2-1) to find the mass percent of Xe and O.

Work

$$\begin{array}{lr} \text{Xe: } 1\text{ mol} \cdot 131.3\text{ g mol}^{-1} = & 131.3\text{ g} \\ \text{O: } 3\text{ mol} \cdot 16.0\text{ g mol}^{-1} = & \underline{48.0\text{ g}} \\ \text{XeO}_3\text{: } \quad 1\text{ mol} = & 179.3\text{ g} \end{array}$$

$$\text{Xe: } \frac{(131.3\text{ g mol}^{-1}) \cdot (1)[\#\text{ Xe atoms in formula}]}{(179.3\text{ g mol}^{-1})} \times 100 = 73.2\%$$

$$\text{O: } \frac{(16.0\text{ g mol}^{-1}) \cdot (3)[\#\text{ O atoms in formula}]}{(179.3\text{ g mol}^{-1})} \times 100 = 26.8\%$$

Check The sum of the percentages of all elements must be 100:

$$73.2 + 26.8 = 100$$

Note: Although the calculation above is correct, because it is based on the definition of mass percent [Equation (2-1)], there is a convenient shortcut available. Calculation of the mass of oxygen in one mole of XeO_3 has already been done as part of the calculation of molar mass. The mass of O is 48.0 g, and there is no need to repeat the step $[(16.0\text{ g mol}^{-1}) \cdot 3 = 48.0\text{ g mol}^{-1}]$. The value may be used directly.

$$\%\text{ O} = \frac{48.0\text{ g}}{179.3\text{ g}} \cdot 100 = 26.8\%$$

Exercise

Find the percent oxygen in $Mg(NO_3)_2$.

Answer: Magnesium nitrate contains 64.7% oxygen (by mass).

Empirical Formula from Composition

An unknown or new compound can be analyzed by a variety of laboratory techniques to determine its percent composition experimentally. The empirical formula can be found by converting the mass percents of the elements into mole percents and then into the mole ratios that are the element subscripts in the formula.

Recall that the empirical formula gives the relative numbers of atoms of each element in a compound, expressed as the set of smallest whole numbers.

EXAMPLE 2.13 Calculation of an Empirical Formula

A compound of xenon and oxygen is known to contain 26.8% O. What is its empirical formula?

Analysis

Target: The empirical formula, that is, a list of the relative number of atoms of each element in the compound, expressed as the smallest whole numbers that indicate the correct atom ratio.

Knowns: The identity of the elements, and the mass percent of one of them. The percent of the other is an implicit known, as are the molar masses of the elements.

Relationship: Because the mole is a counting unit, the relative number of *atoms* is the same as the relative number of *moles*. Further, if we knew the relative number of *grams*, we could find the relative number of moles.

Consider a sample of 100 g of compound: because its mass percent is 26.8, there are 26.8 g O in that sample. *In a 100 g sample, the number of grams of each element is equal to its mass percent,* which is given in the problem.

Plan The stepwise plan, leading backwards from the target, is as follows.

3. Find the ratio of the molar amounts. This is the same as the ratio of atoms in the molecule, and, when converted to a whole-number ratio, gives the subscripts in the empirical formula.
2. Find the amounts in moles from the gram amounts and the molar masses.
1. Find the number of grams of each element in 100 g of the compound.

Work

1. In 100.0 g of the compound, the mass of oxygen is

$$26.8\% \text{ of } 100.0 \text{ g} = (100.0 \text{ g})(0.268) = 26.8 \text{ g O}$$

and the mass of xenon is

$$100.0 \text{ g} - 26.8 \text{ g} = 73.2 \text{ g Xe}$$

2.

$$\text{O: } (26.8 \text{ g O}) \cdot \frac{(1 \text{ mol O})}{(16.0 \text{ g O})} = 1.675 \text{ mol O}$$

$$\text{Xe: } (73.2 \text{ g Xe}) \cdot \frac{(1 \text{ mol Xe})}{(131.3 \text{ g Xe})} = 0.5575 \text{ mol Xe}$$

(Since these are intermediate results, we delay rounding off to the correct number of significant digits.)

3. The mole ratio 1.675 : 0.5575 is converted to a whole-number ratio by dividing each number by the smaller of the two.

$$\text{O: } 1.675/0.5575 = 3.004$$
$$\text{Xe: } 0.5575/0.5575 = 1$$

As a result of the limited number of significant digits, as well as the limited accuracy of chemical analysis, mole ratios are rarely found to be exact integers (except for the smallest, which is always exactly one). But since the *atom* ratio must be an integer, the result is rounded off:

- there are three moles O per mole Xe, or three O atoms per Xe atom;
- the empirical formula is XeO_3.

Exercise

Xenon forms another compound with oxygen, a colorless, explosive gas resulting from the action of concentrated sulfuric acid on barium perxenate. This xenon oxide contains 32.8% O. What is its empirical formula?

Answer: XeO_4.

The empirical formula of a compound of more than two elements is found from the percent composition by a similar procedure. The mass in grams of each element (which, in a 100 g sample, is numerically the same as its mass percent) is converted to the amount in moles. The smallest molar amount is divided into each of the others to find the mole ratios, which, after rounding, are the subscripts of the empirical formula. If one or more of the resulting subscripts is a fractional number, the *entire* set must be multiplied by a small integer to produce a set of whole-number subscripts. For example, if the fraction is ½, multiply by 2; if the fraction is ⅓ or ⅔, multiply by 3; if the fraction is ¼ or ¾, multiply by 4; and so on.

EXAMPLE 2.14 Another Empirical Formula

The most concentrated form of nitrogen available to home gardeners for fertilizer is ammonium nitrate, which contains 35.0% N, 5.0% H, and O. What is the empirical formula?

(a)

(b)

(a) A small-scale explosion of ammonium nitrate. (b) In 1947, a fire aboard a ship in a harbor 10 miles north of Galveston, Texas, caused the cargo of ammonium nitrate to explode. The force of the explosion was about the same as that of the Nagasaki atom bomb two years earlier. People up to two miles away were killed by flying debris, and homes within one mile were leveled. Nearby chemical plants and gas storage tanks caught fire and blazed, out of control, for days. When it was over, 550 people had been killed, and more than 4000 were injured.

Analysis

Target: The empirical formula, which is found from the mole ratios.

Knowns: The percent composition (except for oxygen). The molar masses are implicit knowns.

Relationship: The percent composition can be converted to mole amounts using the molar masses.

Plan

2. Divide the gram amount of each element (= percent) by its molar mass to convert to moles, and convert the mole amounts to integer ratios.
1. Find the missing percentage of oxygen.

Work

1. % O = 100.0 − (35.0 + 5.0) = 60.0
2. It is very helpful to use a structured worksheet to keep track of the necessary conversions.

Element	Grams (= %)	Molar Mass	Moles	Mole Ratio (moles ÷ smallest)	Atom Ratio (rounded mole ratio)	Integer Ratio
N	35.0	÷14.0	= 2.50	$\frac{2.50}{2.50} = 1$	1	× 2 = 2
H	5.0	÷1.01	= 4.95	$\frac{4.95}{2.50} = 1.98$	2	× 2 = 4
O	60.0	÷16.0	= 3.75	$\frac{3.75}{2.50} = 1.50$	$\frac{3}{2}$	× 2 = 3

Note that direct use of the rounded-off mole ratios in the next-to-last column would lead to an empirical formula with a fractional subscript: $NH_2O_{3/2}$. Since fractional subscripts are not used in empirical formulas, these ratios must be multiplied by 2 to convert them into whole numbers. This operation leaves the *relative* numbers of atoms in the formula unit unchanged. The empirical formula of the compound is $N_2H_4O_3$. (The formula of the ionic substance ammonium nitrate is normally written as NH_4NO_3.)

Exercise

When a small amount of water is cautiously added to xenon hexafluoride (XeF_6), a compound is formed that contains Xe, F, and O, and has the following percentage composition: Xe, 50.3%; F, 43.6%. What is its empirical formula?

Answer: $XeOF_6$.

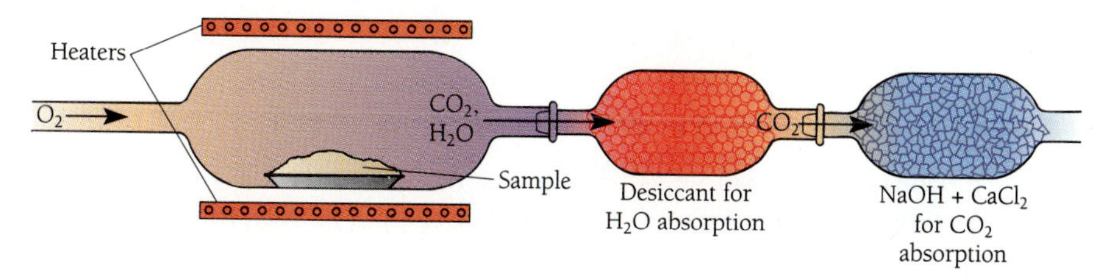

Figure 2.12 Combustion analysis. The CO_2 and H_2O resulting from burning of a sample containing C and H are absorbed, and the amount of each of them is determined from the mass gains of the respective absorbents.

Combustion Analysis

Combustible compounds are routinely analyzed by burning a small sample at high temperature in a stream of pure oxygen (Figure 2.12). Because each carbon atom in the original sample becomes part of a CO_2 molecule, and each H atom becomes part of a H_2O molecule, analysis of the combustion products for CO_2 and H_2O permits the calculation of the amounts of C and H in the original sample. Since oxygen is added to support the combustion, the oxygen content of the original sample cannot be determined by this procedure. However, after the amounts of all other elements are found, the amount of oxygen can be determined by difference.

EXAMPLE 2.15 Empirical Formula from Combustion Analysis

Suppose 9.572 mg of a compound known to contain only C, H, and O is subjected to combustion analysis and found to yield 17.08 mg CO_2 and 6.986 mg H_2O. Determine the empirical formula.

Analysis

Target: The empirical formula, that is, the whole-number mole ratios of the elements.

Knowns: Masses of CO_2 and H_2O produced, and sample mass. The molar masses are implicit knowns.

Relationship: C in the sample is related to the amount of CO_2 produced. H in the sample is related to the amount of H_2O. O in the sample must be determined by difference.

Plan

3. As in Example 2.14, the empirical formula is given by the rounded-off mole ratios (after converting to whole numbers if necessary).

2. Find the amount of O in the sample by difference, from the known amounts of C and H.

1. Find the amount of C in the sample from the amount of CO_2 collected, and the amount of H from the amount of H_2O collected. This step consists of the following parts. To find the mass of H in the original sample,

a. convert the given mass of H_2O to moles using the molar mass (18.02 g mol^{-1}),

b. convert moles H_2O to moles H in the original sample,

c. convert moles H to mass H using the molar mass (1.008 g mol^{-1}). Similarly, find the mass of C in the original sample via the sequence

$$\text{mass } CO_2 \longrightarrow \text{moles } CO_2 \longrightarrow \text{moles C in sample} \longrightarrow \text{mass C}$$

Work

1. Follow the steps of the plan.

a. Find the moles of H_2O produced.

$$? \text{ mmol } H_2O = (6.986 \text{ mg}) \cdot \frac{(1 \text{ mmol})}{(18.02 \text{ mg})} = 0.3877 \text{ mmol}$$

Note that conversion of the data from mg to g is unnecessary if the units *mg* and *mmol* are used rather than g and mol. The molar mass of H_2O is 18.02 g mol^{-1}. It is easily verified (try it) that 1 mmol ⇔ 18.02 mg.

b. Since there are *two* H atoms in a water molecule, each mole of water produced requires *two* moles of H in the sample. The number of millimoles of H in the sample is therefore $2 \times 0.3877 = 0.7754$.

c. The mass of H in the sample is

$$? \text{ mg H} = (0.7754 \text{ mmol H}) \cdot \frac{(1.008 \text{ mg H})}{(1 \text{ mmol H})} = 0.7816 \text{ mg H}$$

Next find the amount of carbon. The amount of CO_2 collected is

$$? \text{ mmol } CO_2 = (17.08 \text{ mg}) \cdot \frac{(1 \text{ mmol})}{(44.01 \text{ mg})} = 0.3881 \text{ mmol}$$

Since each C atom in the sample gives rise to *one* CO_2 molecule, the original sample also contained 0.3881 mmol C. The mass of C in the sample is

$$? \text{ mg C} = (0.3881 \text{ mmol C}) \cdot \frac{(12.01 \text{ mg C})}{(1 \text{ mmol C})} = 4.661 \text{ mg C}$$

2. Find the mass of oxygen by subtracting the mass of C and H from the sample mass.

9.572	mg sample
−4.661	mg C
−0.7816	mg H
4.129	mg O

Convert this to millimoles of O in the sample.

$$? \text{ mmol O} = (4.129 \text{ mg O}) \cdot \frac{(1 \text{ mmol O})}{(16.00 \text{ mg O})} = 0.2581 \text{ mmol O}$$

3. The millimolar amounts are C, 0.3881; H, 0.7752; O, 0.2581. Put these into the fourth column of the worksheet format of Example 2.14 and continue.

Element	Grams	Molar Mass	Milli-moles	Mole Ratio (millimoles ÷ smallest)	Atom Ratio (rounded mole ratio)	Integer Ratio
C	—	—	0.3881	$\frac{0.3881}{0.2581} = 1.504$	$\frac{3}{2}$	$\times 2 = 3$
H	—	—	0.7754	$\frac{0.7754}{0.2581} = 3.004$	3	$\times 2 = 6$
O	—	—	0.2581	$\frac{0.2581}{0.2581} = 1$	1	$\times 2 = 2$

The empirical formula is $C_3H_6O_2$.

Exercise

Suppose 4.237 mg of a compound containing only C, H, and O is analyzed, producing 6.237 mg CO_2 and 2.551 mg H_2O. Determine the empirical formula.

Answer: CH_2O.

Molecular Formula

The methods of the preceding section lead to the *empirical formula.* In many cases the *molecular formula,* which shows the actual number of atoms in a molecule of the compound, is different. Since the elements are in the same proportions in both formulas, the subscripts of the molecular formula are a multiple of the subscripts in the empirical formula. For example, the empirical formula of cyclopropane is CH_2, while its molecular formula is C_3H_6. If the molar mass of a compound is known, even approximately, the molecular formula can be determined from the empirical formula.

EXAMPLE 2.16 Molecular Formula from Empirical Formula

Combustion analysis of glucose yields the empirical formula CH_2O. In a separate experiment, the molar mass of glucose is determined to be in the range of 175 – 200 g mol^{-1}. What is the molecular formula?

Glucose is one of many compounds that are called *carbohydrates* because their empirical formulas suggest they are composed of C and H_2O. Glucose is formed from starch (and other compounds) in the digestive process. Energy used by muscles and other organs is supplied by *glycogen,* which is formed from glucose in the liver.

Analysis

Target: The molecular formula.

Knowns: The empirical formula and the approximate molar mass.

Relationship: The number of atoms of each element in the molecule is a whole number multiple of the subscript in the empirical formula. That is, the molecule consists of either 1, 2, 3, . . . CH_2O formula units and the molar mass of the compound is either 1, 2, 3, . . . times the molar mass of the CH_2O formula unit.

Plan Calculate the molar mass of a compound consisting of 1, 2, 3, . . . formula units, and choose the one that is within the given range.

Since the molar mass of the compound is only approximately known, there is no need to calculate the molar mass of the formula unit to tenth-gram accuracy.

Work The molar mass of the formula unit CH_2O is

$$\begin{array}{rr} \text{C:} & 1\ \text{mol} \cdot 12\ \text{g mol}^{-1} = 12\ \text{g} \\ \text{H:} & 2\ \text{mol} \cdot 1\ \text{g mol}^{-1} = 2\ \text{g} \\ \text{O:} & 1\ \text{mol} \cdot 16\ \text{g mol}^{-1} = 16\ \text{g} \\ \hline \text{CH}_2\text{O:} & 1\ \text{mol} = 30\ \text{g} \end{array}$$

Possible molecules having the empirical formula CH_2O, and their molar masses, are as follows.

Multiplier	Molecular Formula	Molar Mass/g mol^{-1}
1	CH_2O	30
2	$(CH_2O)_2$ or $C_2H_4O_2$	60
3	$(CH_2O)_3$ or $C_3H_6O_3$	90
4	$(CH_2O)_4$ or $C_4H_8O_4$	120
5	$(CH_2O)_5$ or $C_5H_{10}O_5$	150
6	$(CH_2O)_6$ or $C_6H_{12}O_6$	*180*
7	$(CH_2O)_7$ or $C_7H_{14}O_7$	210
⋮		

$C_6H_{12}O_6$ is the only possibility whose molar mass is within the given range. It is therefore the correct molecular formula for the glucose molecule.

Another way to solve the problem is to divide the molar mass of the formula unit into the approximately known molar mass of the compound (in this case, use a value near the midpoint of the approximate range) and find the multiplier by rounding the quotient to the nearest integer.

$$\frac{190}{30} = 6.33 \longrightarrow 6, \text{ so the molecular formula is } (CH_2O)_6 \text{ or } C_6H_{12}O_6$$

Exercise

What is the molecular formula of a compound whose empirical formula is C_2H_3O and whose molar mass is around 180 g mol^{-1}?

Answer: $C_8H_{12}O_4$.

2.8 CHEMICAL NOMENCLATURE

Millions of different chemical compounds have been isolated from natural sources or synthesized in the laboratory, and the number of known compounds is increasing every day. In earlier times, when it was possible for a chemist (or alchemist) to be familiar with the properties of most of the known substances, it was feasible to use names such as "wood alcohol," "Prussian blue," or "Glauber's salt." The name of a substance often reflected the source, or the place where it was first made, or the person who discovered it. Today, names are chosen to indicate the structural formula of the compound. Such names, called **systematic names,** follow internationally agreed-upon rules of systematic chemical nomenclature. Nonsystematic names (water, ammonia, nicotine, and aspirin, for example) are called **trivial names** and are still in use. Systematic nomenclature is designed so that a chemist, on seeing the name of a compound for the first time, can draw its structural formula.

IUPAC again. See the note in the margin of page 8.

Chemical compounds are divided into two classes—**organic** and **inorganic.** These categories have been in use since the early days of chemistry. They reflect the belief, now known to be false, that certain compounds (organic) occur only in living organisms. The modern definition of an organic compound is that it contains carbon and hydrogen, usually in combination with oxygen, nitrogen, or sulfur. All other compounds are classified as inorganic. An introduction to the nomenclature of organic compounds is given in Chapters 8 and 19. In the following sections, we present some of the rules of the systematic chemical nomenclature of inorganic compounds and ions.

There are a few exceptions to this classification. For example, compounds containing the hydrogen carbonate ion, HCO_3^-, are regarded as inorganic because their chemical and physical properties are similar to those of inorganic substances.

Binary Compounds

Compounds containing only two elements are called **binary compounds.** They take their names from the elements that compose them. Some binary compounds are ionic, for example calcium oxide (CaO) and cesium chloride (CsCl). Some binary compounds are molecular, like nitrogen trifluoride (NF_3) and carbon monoxide (CO). Recall (page 55) that unless a binary compound consists of an element in Group 1A or 2A and an element in Group 7A, it is probably a molecular compound. We will discuss molecular and ionic compounds separately.

Molecular Compounds

Four rules govern the systematic naming of binary molecular compounds.

1. The element named first lies to the left of the second-named element in the periodic table (*carbon* tetrachloride, CCl_4).
2. If both elements are in the same group (vertical column) of the periodic table, the element with the greater atomic number is named first (*sulfur* dioxide, SO_2).

   These two rules have some exceptions. Oxygen is named as if it appeared between Cl and F in the periodic table (dichlorine monoxide, Cl_2O, but oxygen difluoride, OF_2). Elements in Group 8A are named first (krypton difluoride, KrF_2). Some binary molecular compounds, particularly those containing hydrogen, have *trivial names.*

   Except for compounds with trivial names, the left to right ordering of elements is the same in both the name and the formula of a compound. Therefore, the name can be obtained by reading the formula from left to right.
3. The name of the second element is modified to end in *-ide.*
4. The number of each type of atom in the molecule is indicated by using one of the following prefixes:

one: *mono-*	two: *di-*	three: *tri-*	four: *tetra-*
five: *penta-*	six: *hexa-*	seven: *hepta-*	eight: *octa-*

   The final vowel of a two-syllable prefix is omitted before *-oxide.* The prefix "mono-" is not used with the first-named element.

TABLE 2.5 Binary Acids

Group	Molecular Formula	Name (when pure)	Name (when dissolved in water)
6A	H_2S	Hydrogen sulfide	Hydrosulfuric acid
	H_2Se	Hydrogen selenide	Hydroselenic acid
	H_2Te	Hydrogen telluride	Hydrotelluric acid
7A	HF	Hydrogen fluoride	Hydrofluoric acid
	HCl	Hydrogen chloride	Hydrochloric acid
	HBr	Hydrogen bromide	Hydrobromic acid
	HI	Hydrogen iodide	Hydroiodic acid
	HCN*	Hydrogen cyanide	Hydrocyanic acid

* HCN is included in this list because its behavior is similar to the true binary acids. Cyanide was originally thought to be an element.

Some additional examples of the application of these rules are

CBr_4	carbon tetrabromide	N_2O_4	dinitrogen tetroxide
CO	carbon monoxide	NF_3	nitrogen trifluoride
PCl_5	phosphorus pentachloride	CS_2	carbon disulfide

Exercise

Give the names of the following molecular compounds: SO_3, N_2O, Cl_2O_7.

Acids

Some molecular compounds (and all ionic compounds) separate into cations and anions when dissolved in water. This process is called **dissociation.** The name **acid** is given to any compound of hydrogen that dissociates into a hydrogen cation and an anion when dissolved in water. For example, HCl dissociates to H^+ and Cl^- in water. The hydrogen cation (H^+) is called the *hydrogen ion,* or the *proton.* The chemistry of acids is considered in Chapters 15–17. There are only a few binary acids, and all of them are molecular compounds. According to long-established practice, the binary acids bear different names when pure and when dissolved in water. Their names and formulas are given in Table 2.5.

The hydrogen atom consists of one proton and one electron. When a H^+ atom loses its electron to become a H^+ ion, all that is left is the proton. Many chemists refer to the H^+ ion as a "proton."

Ions and Ionic Compounds

Most elements form monatomic ions, whose charge depends on the position of the element in the periodic table. The charges of monatomic ions formed by elements in the "A" groups of the periodic table are given in Table 2.6.

Monatomic cations have the same name as the element, while an anion name is modified to end in *-ide*. For example, Na^+ is the *sodium* ion and Br^- is the *bromide* ion. Ions with a single positive or negative charge are called **monovalent** (mahn′ oh vale′ ent) ions. Doubly charged ions are **divalent** (dye′ vale ent), and triply charged ions are **trivalent** (try′ vale ent).

The names of ions are often verbalized by spelling them: Na^+ is "en-ay-plus" and Br^- is "bee-are-minus."

Compounds containing polyatomic ions are not true binary compounds because they contain more than two elements. However, they consist of only two different *ions,* and they are named the same way as compounds consisting of two monatomic ions. The names of polyatomic ions are related to their structure, but less obviously so than is the case for monatomic ions. In many cases, an older name has been incorporated into systematic nomenclature. For instance, SO_4^{2-} bears the name "sulfate ion," which does not explicitly indicate the presence of 4 oxygen atoms and a charge of −2. Table 2.3 (page 55) gives the names and formulas of some common polyatomic ions; these should be memorized.

The ending "*-ide,*" which usually refers to a monatomic anion, also appears in a few polyatomic anions. These names are historical relics, left over from the early days when it was thought that hydroxide (OH^-) and cyanide (CN^-) consisted of only one element.

Formulas of ionic compounds are written with the cation first, then the anion. The charges of the ions are *not* written in the formula of a compound. The name of an ionic compound is easily found by reading the formula from left to right because the name and formula are written in the same order. For example, *potassium iodide* (KI), *lithium carbonate* (Li_2CO_3), *magnesium sulfate* ($MgSO_4$), and *ammonium chloride* (NH_4Cl).

Since the cation-forming elements lie to the left of the anion-forming elements in the periodic table, the "cation first" rule for the formula of an ionic compound is, in effect, the same as the "left element first" rule for molecular compounds (see page 77).

Ionic compounds, except those whose anion is O^{2-} or OH^-, are called *salts.*

Hydrides

In contrast to other elements, hydrogen forms both monovalent cations (by loss of its single electron), and monovalent anions (by gain of an electron). The hydrogen anion, H^-, is called the *hydride ion.* In combination with the metals of Groups 1A and 2A, it forms ionic compounds called **hydrides,** such as rubidium hydride (RbH) and calcium hydride (CaH_2). The chemical properties of hydrides are very different from those of other hydrogen-containing compounds.

Empirical Formulas, Names, and Stoichiometry of Ionic Compounds

Chemical compounds are electrically neutral. In ionic compounds, therefore, the sum of the positive charges of the cations must be equal to the sum of the negative charges of the anions. This fact fixes the empirical formulas of ionic compounds. For example, the compound formed from a *divalent* cation such as Ca^{2+} and a *monovalent* anion such as Br^- must contain twice as many anions as cations, and its empirical formula must be $CaBr_2$. The atom or mole ratio of the elements in a compound is called its **stoichiometry.** The stoichiometry of calcium bromide is Ca : Br = 1 : 2. Since the stoichiometry

TABLE 2.6 Periodic Variation of the Charge of Monatomic Ions*

Group						
1A	2A	3A	4A	5A	6A	7A
Li^+	Be^{2+}			N^{3-}	O^{2-}	F^-
Na^+	Mg^{2+}	Al^{3+}		P^{3-}	S^{2-}	Cl^-
K^+	Ca^{2+}				Se^{2-}	Br^-
Rb^+	Sr^{2+}				Te^{2-}	I^-
Cs^+	Ba^{2+}					At^-
Fr^+	Ra^{2+}					

* The elements not listed either do not form monatomic ions or, if they do, the charge of the ion does not correlate simply with position in the periodic table.

of an ionic compound is established by the charges of its component ions, the relative numbers of the ions need not be included in the name. Some examples are sodium chloride (NaCl), magnesium chloride ($MgCl_2$), and aluminum chloride ($AlCl_3$); potassium nitride (K_3N), potassium oxide (K_2O), and potassium fluoride (KF); ammonium sulfate [$(NH_4)_2SO_4$], magnesium sulfate ($MgSO_4$), and aluminum sulfate [$Al_2(SO_4)_3$].

The empirical formula of an ionic compound can be determined from its name and a knowledge of the charges of its ions given the following rules. If the cation and anion have equal (but opposite) charges, neither ion has a subscript in the formula. Otherwise, the stoichiometric subscript of each ion is equal to the absolute value of the charge of the other ion.

EXAMPLE 2.17 Names and Formulas of Ionic Compounds

a. Give the empirical formulas of strontium iodide, aluminum oxalate, and magnesium carbonate.

b. Give the names of Na_2S and $(NH_4)_2Cr_2O_7$.

c. Give the name and formula of the ionic compound formed between chlorine and barium.

Analysis

a. ***Target:*** An empirical formula, which expresses the stoichiometry of the compound.

Relationship: The compound must be electrically neutral.

Required Information: The charges of the ions present in the compounds.

Work Consultation of Table 2.6 shows that the ionic charges are: Sr, +2 and I, −1. Two I^- ions are required to balance the charge of one Sr^{2+} ion, so the stoichiometry must be Sr : I = 1 : 2. The formula is SrI_2. For aluminum oxalate, the ionic charges are aluminum, +3 (Table 2.6) and oxalate, −2 (Table 2.3). The subscript of aluminum is therefore the absolute value of (−2) and that of oxalate is the absolute value of (+3). The formula is $Al_2(C_2O_4)_3$. For magnesium (+2) and carbonate (−2) ions, the rule produces the formula $Mg_2(CO_3)_2$; when the subscripts are divided by 2 (the common divisor), the correct formula is obtained: $MgCO_3$.

b. ***Target:*** The names of ionic compounds.

Knowns: The empirical formulas.

Relationship: The ions of a compound are named in the order they appear in the formula (cation first), the stoichiometry is implicit rather than explicit in the name, and the names of monatomic anions end in *-ide*.

Answers Sodium sulfide and ammonium dichromate.

c. ***Target:*** The empirical formula, from which the name is then obtained.

Work Barium is a Group 2A element and forms divalent cations (charge +2). Chlorine is a Group 7A element and forms monovalent anions (charge −1). The formula is $BaCl_2$ and the name is barium chloride.

Exercise

Give the name and formula of the ionic compounds formed between (a) oxygen and calcium, (b) aluminum and hydroxide ions, and (c) magnesium and nitrogen.

Answer: (a) Calcium oxide, CaO; (b) aluminum hydroxide, $Al(OH)_3$; (c) magnesium nitride, Mg_3N_2.

One Element, Two Ions

Many metallic elements (but not any of those in Table 2.6) form more than one type of monatomic cation. For example, iron forms both Fe^{2+} and Fe^{3+}, and copper forms Cu^+ and Cu^{2+}. A simple name like "iron sulfate" is ambiguous, because iron forms both $FeSO_4$ and $Fe_2(SO_4)_3$. In such cases, the stoichiometry of the compound must be indicated explicitly in the name. In the **Stock system** of nomenclature, the cation charge is indicated by parenthetical Roman numerals immediately following the English name of the element: $Fe(NO_3)_2$ is iron(II) nitrate (pronounced "iron-two-nitrate") and $Fe(NO_3)_3$ is iron(III) nitrate (pronounced "iron-three-nitrate"). Note that iron(III) nitrate does not mean $Fe_3(NO_3)$. Other examples of this nomenclature are manganese(II) carbonate, $MnCO_3$, and manganese(IV) oxide, MnO_2.

There is an older, outmoded system that is still used enough so that chemists need to be familiar with it. In this system the element name is modified to end in *-ous* to name the cation with the lower charge, and to end in *-ic* to name the cation with the greater charge. In most cases the Latin name for the element is used rather than the English name. Fe^{2+} is *ferrous* ion, and Fe^{3+} is *ferric ion.* With these suffixes, compound names are no longer ambiguous: ferrous nitrate is $Fe(NO_3)_2$, and ferric nitrate is $Fe(NO_3)_3$. Similarly, copper forms two bromides: cuprous bromide ($CuBr$) and cupric bromide ($CuBr_2$). Keep in mind that the endings *-ous* and *-ic* do not specify the charge of the ion, but only which of two ions has the lower charge.

Mercury is an exception to the Latin rule. $HgCl_2$ is *mercuric* chloride, not *hydrargyric* chloride, and Hg_2Cl_2 is *mercurous* chloride. The correct formula is Hg_2Cl_2 (not $HgCl$) because the cation is Hg_2^{2+} (see Table 2.3). An even older name for Hg_2Cl_2 is *calomel.*

Almost all of the elements that form several different ions are in the "B" groups of the periodic table. These groups are the subject matter of Chapter 23.

Oxoacids, Oxoanions, and Salts

There are many acids that contain oxygen in addition to hydrogen and one other element. Nitric acid, HNO_3, is a common example. Such compounds are called **oxoacids.** Like a binary acid, an oxoacid dissociates into H^+ and an anion when dissolved in water. These polyatomic anions are called **oxoanions** because they contain oxygen. Some of them have already been listed in Table 2.3.

The names of oxoacids and oxoanions are derived from the third element (neither H nor O) they contain, and the same name is used for both pure and dissolved acid. The name indicates (indirectly) the number of oxygen atoms. For elements forming only two oxoacids, the ending *-ic* is used for the acid with the larger number of oxygen atoms, while the ending *-ous* denotes the acid with the smaller number of oxygen atoms. These endings are changed when describing the ion or salt containing the ion: *-ic* → *-ate* and *-ous* → *-ite*. For elements forming more than two oxoacids, a combination of prefixes and endings is used to indicate the number of O atoms. The prefix *per-* means "more, thoroughly, complete," while the prefix *hypo-* means "lower, less." The nomenclature of these compounds is summarized in Table 2.7.

TABLE 2.7 Names of Some Oxoacids, Oxoanions, and Salts

Molecular Formula and Acid Name	Anion Formula and Name	Formula and Name of Typical Salt
H_2CO_3 Carbonic acid	CO_3^{2-} Carbon*ate* ion	Li_2CO_3 Lithium carbon*ate*
HNO_2 Nitr*ous* acid	NO_2^- Nitr*ite* ion	$Ba(NO_2)_2$ Barium nitr*ite*
HNO_3 Nitr*ic* acid	NO_3^- Nitr*ate* ion	KNO_3 Potassium nitr*ate*
H_2SO_3 Sulfur*ous* acid	SO_3^{2-} Sulf*ite* ion	K_2SO_3 Potassium sulf*ite*
H_2SO_4 Sulfur*ic* acid	SO_4^{2-} Sulf*ate* ion	K_2SO_4 Potassium sulf*ate*
HClO *Hypo*chlor*ous* acid	ClO^- *Hypo*chlor*ite* ion	NaClO Sodium *hypo*chlor*ite*
$HClO_2$ Chlor*ous* acid	ClO_2^- Chlor*ite* ion	$NaClO_2$ Sodium chlor*ite*
$HClO_3$ Chlor*ic* acid	ClO_3^- Chlor*ate* ion	$NaClO_3$ Sodium chlor*ate*
$HClO_4$ *Per*chlor*ic* acid	ClO_4^- *Per*chlor*ate* ion	$NaClO_4$ Sodium *per*chlor*ate*

2.9 THE NOBLE GASES

The year 1888 saw the first reports, from William Ramsay and also from Lord Rayleigh, of a previously unknown gaseous constituent of the atmosphere. The new substance was named "argon," after a Greek word meaning inactive, because it did not react with any other known material. Like all advances in science, this discovery had its roots in the prior work of others. About one hundred years previously, Henry Cavendish noticed that a small amount of gas (a little less than 1% by volume) remained after he had removed the nitrogen, oxygen, and water vapor from a sample of air. Neither he nor any other scientist followed up on this observation at the time.

Years later, Rayleigh carefully measured the density of nitrogen prepared by removing oxygen, water, and carbon dioxide from air. When he compared this value with the density of nitrogen prepared chemically from ammonia, he found a discrepancy. The density of "atmospheric" nitrogen was 1.2672 $g\ L^{-1}$, while that of "chemical" nitrogen was 1.2505 $g\ L^{-1}$. He knew that the difference was not due to error, and challenged chemists to explain it.

The existence of helium on the sun had been postulated several decades earlier, to explain certain features of the solar spectrum observed during the eclipse of 1868. Ramsay was the first to obtain a sample of the new element in an earthbound laboratory.

The Scottish chemist Ramsay accepted the challenge, and in 1894 he separated pure argon from Rayleigh's "atmospheric" nitrogen. He determined that the concentration of argon in air is 0.934%, so that the measurement of Cavendish, made more than 100 years earlier, was remarkably close. The year following his preparation of pure argon, Ramsay isolated the gaseous element helium from *cleveite* (an ore of

Sir William Ramsay (1852–1916)

Sir William Ramsay was Professor of Chemistry at University College, London. Although he is not remembered for any work other than that with the noble gases, he is the only scientist ever to discover an entire Group of new elements. In 1904 he received the Nobel Prize in chemistry, thus sharing the credit with Lord Rayleigh.

Henry Cavendish (1731–1810)

Henry Cavendish was a wealthy and somewhat eccentric British chemist and physicist. He was exceedingly taciturn, had few friends, and spent his time, money, and emotional effort in his laboratory. As a chemist, he showed that water is produced by the combustion of hydrogen. As a physicist, he measured the gravitational constant and the mass of the earth, and he discovered what is today known as Coulomb's law (the force between two electrical charges is inversely proportional to the square of the distance between them). He preferred not to publish the results of his studies, however. Years later Coulomb also characterized the electrical force and got the credit. Cavendish also prediscovered Ohm's law (electrical current is proportional to voltage). Since he had no better alternative, he used his own body to judge the strength of the current by the discomfort it caused. Some 100 years after his death, his family, still wealthy, endowed the Cavendish Chair of Experimental Physics at Cambridge. As can be seen from other biographical notes in this chapter, holders of this chair have made many important contributions to the development of modern chemistry.

uranium). Helium is not a constituent of any chemical compound in cleveite, but is physically trapped in the mineral and released when the ore is crushed.

Ramsay believed that helium and argon were members of a new group in the periodic table, now known as the **noble gases.** He further believed that there had to be additional, similar elements of greater atomic number. In three short months in 1898 he isolated neon, krypton, and xenon from a large amount of liquified air. In 1910 he discovered the final member of the group, radon.

Occurrence and Uses

The noble gases are sometimes called the *rare gases*. This is a misleading name, however, because argon, which constitutes 1%(v/v) of the atmosphere, is a rather common element. The group was earlier known as the *inert gases* because, for 65 years after their discovery, it was believed that they were incapable of forming compounds with other elements.

Some properties of the noble gases are given in Table 2.8. Note that both melting point and boiling point of the noble gases increase with atomic number. This is an example of a **periodic trend**—a regular variation in physical and chemical properties

John William Strutt, Third Baron Rayleigh (1842–1919)

John William Strutt was Professor of Physics at Cambridge. He was convinced of the value of truly accurate measurements in science. The discovery of argon, for which he was awarded the 1904 Nobel Prize in physics, grew out of his efforts to determine a more accurate value of the atomic weights of oxygen and nitrogen. In addition to this work, he made lasting contributions to both acoustics and optics.

Figure 2.13 One of the liquid helium-cooled magnets in the particle accelerator at the Fermi National Laboratory in Illinois. This research instrument is housed in an underground circular tunnel that is four miles in circumference.

TABLE 2.8 The Noble Gases

Name	Symbol	Atomic Number	Molar Mass/g mol^{-1}	Melting Point/K	Boiling Point/K	% (v/v) of Atmosphere
Helium	He	2	4.0026	—*	4.2	5.24×10^{-4}
Neon	Ne	10	20.180	24.5	27.3	1.82×10^{-3}
Argon	Ar	18	39.948	84.0	87.5	9.34×10^{-1}
Krypton	Kr	36	83.80	116.6	120.9	1.14×10^{-4}
Xenon	Xe	54	131.29	161.2	166.1	8.7×10^{-6}
Radon	Rn	86	222	202	211.4	Trace

* Helium is a unique substance because, at ordinary pressure, it does not freeze to a solid no matter how low the temperature. If the pressure is increased to about 25 times the normal atmospheric pressure, however, helium freezes at about 1 K.

as atomic number increases within a group of related elements. Periodic trends are very common and very useful in classifying and predicting the behavior of elements and compounds.

With the exception of radon, all the noble gases are permanent constituents of the atmosphere. Radon, a product of natural radioactive processes, is continuously formed in the earth's crust. Although a small amount percolates into the atmosphere, it disappears by radioactive decay within a few days.

Helium is commercially recovered as a minor constituent of natural gas. It is continuously formed in the decay of the radioactive elements in the earth's crust, and it tends to collect in deposits of natural gas. An important use of helium is as an exceedingly low-temperature refrigerant. Superconducting magnets, for example, are kept at 4.2 K by immersion in liquid helium (Figure 2.13). Another use is in "heliarc welding," in which the area being welded is flooded with helium (Figure 2.14). This protects the hot metal from contact with atmospheric oxygen, thus preventing the formation of joint-weakening oxides. Finally, helium is used as a buoyant gas in weather balloons and airships.

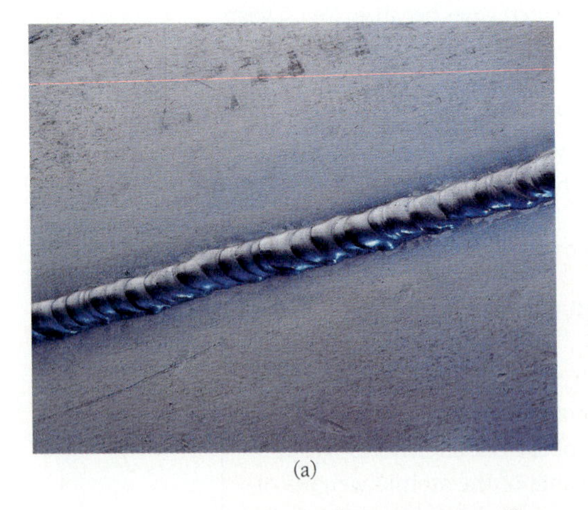

(a)

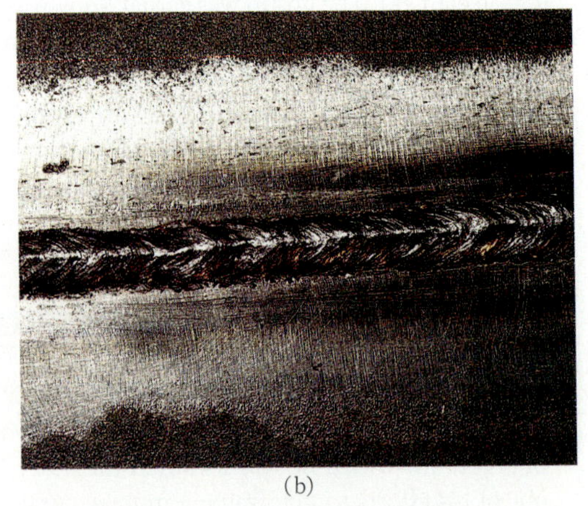

(b)

Figure 2.14 Comparison of two steel joints: one (a) welded with helium, the other (b) welded without helium.

Neon, argon, krypton, and xenon are all byproducts of the commercial separation of nitrogen and oxygen from air, but only Ar is produced in substantial quantity (some 0.25 million tons annually in the United States). Argon is used primarily to provide an inert atmosphere in some metallurgical processes and in light bulbs, where it prolongs filament life. Like helium, it is used in welding. Neon is used only in neon signs, and krypton and xenon have essentially no uses outside of chemical research. Although it is rarely used as such, xenon is an anesthetic gas.

Radon is a radioactive element, and it has been experimentally used in radiation treatment of cancer. In the past several years there has been increasing concern about the potential hazard from radon inside buildings and residences. Radon tends to accumulate in houses because it seeps up from the ground. This is particularly true in "energy efficient" dwellings that minimize leakage of air to the outside. The health hazards of radiation are discussed in Chapter 23.

Chemistry of Xenon and Krypton

The failure of early attempts to observe reactions of the noble gases led to the hypothesis that they are inert. This conclusion in turn had a formative influence on the developing ideas of the relationship between chemical properties and position in the periodic table (discussed in Chapter 8 in connection with the octet rule and noble-gas structure in ion formation). Linus Pauling was among the few who felt that the evidence did not confirm the hypothesis. Beginning in the 1930s he made occasional suggestions that experimental chemists should try harder to prepare noble gas compounds.

In 1962, Neil Bartlett, a Canadian, combined his accidental discovery of an unexpected reaction between O_2 and PtF_6 with his considerable chemical skills and showed that xenon also reacts with PtF_6. He and other chemists soon prepared numerous compounds in which xenon is bonded to either oxygen or fluorine. Far from being inert, in fact, xenon spontaneously forms XeF_2 when mixed with fluorine and exposed to sunlight. Some other compounds of xenon are XeF_4, XeF_6, $XeOF_4$, XeO_2F_2, XeO_3, and XeO_4 (Figure 2.15).

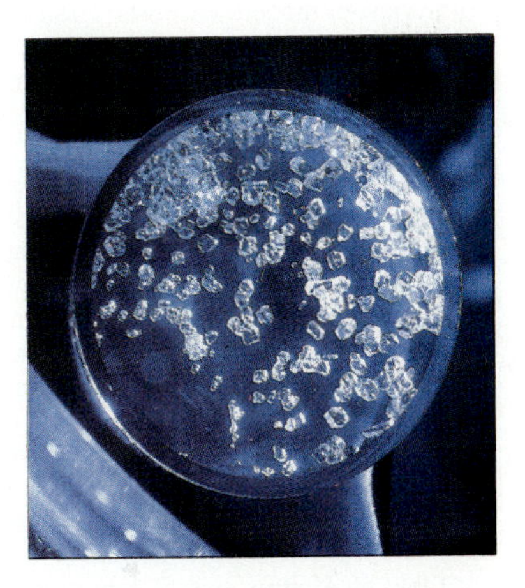

Figure 2.15 Crystals of XeF_4 formed by direct reaction of Xe and F_2 at 400 °C.

Helium, neon, and argon still deserve the name "inert" since no compounds have yet been prepared from them. Krypton forms KrF_2 and KrF_4. It is surmised, but not yet demonstrated in the laboratory, that the reactions of radon are similar to those of xenon.

Linus Pauling (1901–)

Linus Pauling spent a long and productive career at the California Institute of Technology. His early contributions to the theory of chemical bonding earned him the 1954 Nobel Prize in chemistry. Later he became interested in the structure of proteins. Along with many other scientists he became passionately concerned with world peace, and in 1962 he was awarded the Nobel peace prize. He is also known for his advocacy, against prevailing medical opinion, of vitamin C as an effective preventative of colds.

CHEMISTRY OLD AND NEW

ELEMENTAL SYMBOLS THROUGHOUT HISTORY

The modern representation of the chemical elements by one-, two-, and three-letter symbols, which originated in the early nineteenth century, conveys a great deal of information at a glance. For example, "Al_2O_3" tells us the type of substance (a compound), its composition (aluminum and oxygen), stoichiometry (two Al for every three O), and its molar mass ($2 \times 27.0\ g\,mol^{-1} + 3 \times 16.0\ g\,mol^{-1}$). This representation is much more concise than other ways of describing the substance and tells us a lot more. "Aluminum oxide," while expressing composition, does not give any information about elemental proportions. "Alumina" is vague on all points. Prior to the nineteenth century, however, several less satisfactory systems (at least from the view of modern chemists) existed. Ancients and alchemists developed shorthand notation for the known elements even before they were distinguishable from more complex substances, such as acids and salts.

Already in classical times, seven pure metals known to the Greeks and Romans had become associated with the seven known celestial bodies through their shared characteristics. For example, heavy, dense lead was associated with the slow-moving Saturn, and gold was allied to the yellow Sun. Since mythological gods were also related to planets, the metals acquired the deities' insignia as well. These signs (left table, below) were the first elemental symbols. In the middle ages the alchemists expanded the list by adding representations for chemical equipment, and for processes such as oxidation (♈) and sublimation (⏏). The codes served not only as useful laboratory abbreviations, but as a safeguard against outsiders' knowledge of alchemical practices.

In 1718, Geoffroy the Elder published a chart called the "Table of Affinity," which represented several classes of substances and the chemicals that reacted with them. This chart utilized the established alchemical signs, along with many new ones. In 1795, a chart of commonly used symbols from a scientific dictionary included + ⊕ (sugar), + ⏢ (borax), ₒ°ₒ (ether), ♆⚬—⚬ (arsenic), and ⚶ (sulfur). The plus signs denoted an acidic substance and the "pitchfork," a metallic one. Yet another system of representation popular at this time, authored by J. H. Hassenfratz and P. A. Adet, depicted acids and metals as letters enclosed by squares and circles (right table below). Other elements took the form of lines or arcs in particular orientations. John Dalton expanded on this idea in 1806 by using detailed circles for all of the elements. A new feature of his system was that the circles could be combined to show the stoichiometric and structural makeup of compounds—for example, he represented carbon dioxide by ○●○. While appropriate for simple substances, Dalton's method would be nightmarish when used to depict a large molecule such as Vitamin $B_{12}(C_{63}H_{88}CoN_{14}P)$!

In 1814, the Swedish chemist Jons Jakob Berzelius made public his idea of using simple representative letters to denote the elements in compounds. His proposal also included superscripts showing the number of each type of atom involved (for example, $CaCl^2$). He later modified his plan to make it even more concise. Oxygen and sulfur, among other elements, were given special symbols which could be placed above other components. A horizontal bar drawn across a letter indicated a pair of atoms: Al_2O_3, in his notation, would be $\overset{\cdots}{\text{A}\!\!\!-\!\text{l}}$. However, these modifications did not gain wide acceptance. Apart from the change of position of numerals from superscripts to subscripts, we use Berzelius' original system to this day. Our method of representing the elements was invented long before many of them were discovered, and a full 55 years before the periodic table, where they are most often seen, was published by Dmitri Mendeleev.

Discussion Question

Do you think the relatively late introduction of the modern system of elemental symbols delayed the development and growth of chemical science?

Alchemical Symbols for Metallic Elements Known before 1861

Metal	Planet/Associated Mythological Figure	Symbol
Copper	Venus	♀
Iron	Mars	♂
Mercury	Mercury	☿
Tin	Jupiter	♃
Lead	Saturn	♄
Gold	Sun/Sol	☉
Silver	Moon/Luna	☽

Some Elemental Symbols and Their Inventors

Modern	Alchemical	Hassenfratz and Adet	Dalton	Berzelius
O	Not known	—	○	·
C	Not known	(	●	C
S	⏝	◡	⊕	I
Hg	☿	Ⓗ	◌	Hg

SUMMARY EXERCISE

The masses (amu) and natural abundances of the isotopes of krypton are Kr-78, 77.945(0.35%); Kr-80, 79.9164(2.25%); Kr-82, 81.9135(11.6%); Kr-83, 82.9141(11.5%), Kr-84, 83.9115(57.0%); and Kr-86, 85.9106(17.3%). When a gaseous mixture of krypton and fluorine is exposed to an electric discharge or β-radiation from a radioactive source, a compound containing Kr and F(32.0%) is obtained. The molar mass of this compound is in the range 100 to 150 g mol^{-1}. The compound reacts with antimony pentafluoride to yield a material best described by the empirical formula $KrF_2 \cdot 2SbF_5$. This material is stable at room temperature, but decomposes at its melting point (122 °F).

a. Calculate the molar mass of krypton. Report the result using one more significant digit than is justified by the data. Which quantity(ies) limit(s) the number of significant digits in the result?

b. What is the empirical formula of the krypton-fluorine compound? What is its name? What is the molecular formula?

c. What is the number of electrons, protons, and neutrons in the fluoride ion? (Fluorine is a monoisotopic element.)

d. What is the percentage of fluorine in the substance $KrF_2 \cdot 2SbF_5$?

e. What is the melting point of $KrF_2 \cdot 2SbF_5$ in °C? In K?

Answers **(a)** 83.80 g mol^{-1}; the quantities 57.0% and 17.3% limit the reliability of the result to ± 0.1 amu. **(b)** KrF_2, krypton difluoride, KrF_2. **(c)** 10 electrons, 9 protons, 10 neutrons. **(d)** 41.1%. (e) 50 °C, 323 K.

SUMMARY AND KEYWORDS

The laws of **conservation of mass**, **constant composition**, and **multiple proportions** led Dalton to modern **atomic theory.** The theory holds that matter consists of **atoms,** that atoms of a given **element** are identical, that atoms of different elements have different masses, and that chemical reactions involve regrouping rather than destruction or creation of atoms. Atoms consist of a **nucleus,** a small dense core of **neutrons** and **protons,** surrounded by a larger, diffuse cloud of **electrons.** Protons and electrons have equal and opposite charge, so that atoms, having equal numbers of each, are electrically neutral. Neutrons are uncharged. Neutrons and protons, collectively called **nucleons,** have nearly the same mass, which is almost 2000 times the mass of electrons. The number of protons in the atoms of an element is the **atomic number** (Z), and the number of nucleons is the **mass number.** Most elements have several **isotopes,** variant forms having the same chemical properties and atomic number but different mass numbers. Isotopes are readily observed, and their masses accurately measured in a **mass spectrometer.** *Radioactive* isotopes are unstable, and undergo changes in which they emit **α-rays** (He nuclei), **β-rays** (electrons), or **γ-rays** (high energy radiation).

Molecular compounds consist of **molecules,** which are stable clusters of two or more atoms held together by **chemical bonds.** The **molecular formula** gives the number of atoms of each type in the molecule. The **structural formula** shows the **connectivity** of the molecule, that is, which atoms are bonded together. **Ions** are formed when an atom or molecule either gains one or more electrons to become a negatively charged **anion** or loses one or more electrons to become a positively charged **cation.** Ions can be **monatomic** or **polyatomic.** The charges of monatomic ions of the elements in the "A" groups vary regularly with position in the periodic table. Elements in the "B" groups typically form more than one type of monatomic ion. **Ionic compounds** consist of anions and cations, and, if the anion is not OH^- or O^{2-}, the compound is a **salt.** The **empirical formula** gives the type and relative number of atoms of each element in a compound. The **water of hydration** (loosely bound in the crystal structure) of a solid **hydrate** is separated by a dot from the formula of the **anhydrous** compound. The group of atoms described by the empirical formula of an ionic compound is called the **formula unit.**

In counting **microscopic** objects (atoms, molecules, or ions) chemists use a **counting unit** called the **mole (mol),** equal in value to **Avogadro's number,** $N_A = 6.022 \times 10^{23}$. N_A is the number of atoms in exactly 12 g of pure C-12. This fixes not only the number of particles in a mole but also the numerical value of the conversion factor between grams and **atomic mass units (amu).** The **molar mass (atomic weight)** of an element is the mass(g) of one mole of the naturally occurring isotopic mixture of an element. The atomic weight, in amu, is the mass of an atom averaged over the **natural abundances** of the isotopes. The mass of one mole of a compound is also called the molar mass (or molecular weight). The conversion factor between mass and moles is derived from equivalence statements such as, in the case of CO_2, 1 mol $CO_2 \Leftrightarrow 44.0$ g CO_2 (or 1 mmol $\Leftrightarrow$ 44.0 mg).

The empirical formula is used to calculate the **percent composition** of a compound. Laboratory measurement of composition, for instance through *combustion analysis,* is used to determine the empirical formula. Determination of the *molecular formula* requires the additional measurement of molar mass.

With a few exceptions, compounds are classified as **organic** if they contain carbon and hydrogen; otherwise they are **inorganic.** Compounds and ions are identified either by historically used **trivial** names such as "water," or by **systematic names,** which exactly describe the structural formula. Ionic compounds are named with the ion names, cation first. The names of monatomic anions, as well as those of ionic and molecular **binary compounds,** end in *-ide.* With some exceptions, the first-named element lies to the left or below the second in the periodic table. When a metal forms (only) two different cations, the ending *-ic* modifies the (Latin) name to identify the ion having the higher charge; the ending *-ous* denotes the lower. Alternatively (and always when there are more than two types of cation), parenthetical Roman numerals after the English name give the cationic charge **(Stock system).** In binary molecular compounds the number of atoms of each type is indicated by the prefixes *mono-, di-, tri-, tetra-, penta-, hexa-, hepta-,* or *octa-.* These prefixes are not used in naming ionic compounds, whose **stoichiometry** (relative number of each type of ion) is fixed by the requirement of electrical neutrality and the known charges of the ions. The **hydride ion** is H^-, while the *hydrogen ion* or *proton* is H^+.

Compounds that **dissociate** into anions and H^+ on dissolving in water are called **acids. Oxoanions** contain oxygen and one other element, and combine with H^+ to form **oxoacids.**

The **noble gases** (Group 8A) are generally unreactive. They are monatomic gases whose boiling points and melting points increase with atomic number (a **periodic trend**). Helium is obtained from natural gas, while the others (except for radon, which is radioactive), are isolated from air. Krypton forms two molecular compounds with fluorine, and xenon forms a variety of compounds with fluorine and oxygen. No compounds of other noble gases have been isolated.

PROBLEMS

General

1. Describe, in your own words, what the word "atom" meant to each of the following: Democritus, Dalton, Thomson, Rutherford.
2. Rank the following in order of increasing mass: neutron, electron, proton, nucleus.
3. What is a nucleon?
4. Distinguish between the mass, mass number, and atomic number of an atom.
5. What are isotopes?
6. What is meant by the term "polyatomic ion"? Give examples.
7. Distinguish between an empirical formula and a molecular formula.
8. Of the following, which are empirical formulas, which are molecular formulas, and which could be either: $C_6H_{12}O_6$, LiCl, C_2H_4, KI, NO, H_2O?
9. What information is given by a structural formula and not given by a molecular formula?
10. What is a mole, and how many items does it correspond to?
11. What is the definition of Avogadro's number?
12. Suppose the mole were redefined as, "The number of atoms in 12 *pounds* of pure C-12". Would the value of Avogadro's number be affected. If not, why not? If so, what would be the new value?
13. What experimental technique can be used to determine the precise masses of isotopes?
14. In what way(s) are the isotopes of a given element like one another? In what way(s) are they different?
15. What is meant by the term "natural abundance"? How is it measured?
16. Why is the isotopic composition of most elements essentially the same, no matter where the sample comes from?
17. What is meant by a monoisotopic element? Name a common one.
18. Is a trivial name unimportant? Explain.
19. In what order are the components listed in the name of an ionic compound?
20. In what order are the components listed in the name of a binary molecular compound?

Isotopes

21. Give the name of the following monoisotopic elements. How many neutrons are in the nuclei of each?
 a. $^{23}_{11}Na$ b. $^{27}_{13}Al$
 c. $^{59}_{27}Co$ d. $^{19}_{9}F$
 e. $^{55}_{25}Mn$ f. $^{75}_{33}As$
22. Each of the following is a monoisotopic element. Give the name of the element and the number of neutrons in the nuclei of each.
 a. $^{133}_{55}Cs$ b. $^{9}_{4}Be$
 c. $^{209}_{83}Bi$ d. $^{197}_{79}Au$
 e. $^{103}_{45}Rh$ f. $^{181}_{73}Ta$
23. Identify each of the isotopes in Problem 21 using the other abbreviated form, as, for example, C-12.
24. Identify each of the isotopes in Problem 22 using the other abbreviated form, as, for example, C-12.
25. Give the name and symbol of the element represented by X in each of the following isotopes. Each, incidentally, is the most plentiful of its element.
 a. $^{16}_{8}X$ b. $^{24}_{12}X$
 c. $^{39}_{19}X$ d. $^{40}_{18}X$
 e. $^{28}_{14}X$ f. $^{138}_{56}X$
26. Each of the following is the most abundant of the isotopes of element X. Give the name and symbol of X.
 a. $^{14}_{7}X$ b. $^{35}_{17}X$
 c. $^{63}_{29}X$ d. $^{20}_{10}X$
 e. $^{56}_{26}X$ f. $^{195}_{78}X$
27. Determine the number of protons, neutrons, and electrons in each of the following species.
 a. $^{24}_{12}Mg$ b. $^{45}_{21}Sc$
 c. $^{91}_{40}Zr$ d. $^{27}_{13}Al^{3+}$
 e. $^{65}_{30}Zn^{2+}$ f. $^{108}_{47}Ag^{+}$
28. Determine the number of protons, neutrons, and electrons in each of the following species.
 a. $^{52}_{24}Cr$ b. $^{93}_{41}Nb$
 c. $^{137}_{56}Ba$ d. $^{63}_{29}Cu^{+}$
 e. $^{56}_{26}Fe^{2+}$ f. $^{55}_{26}Fe^{3+}$

*29. Determine the number of protons, neutrons, and electrons in each of the following species.
 a. N-15 b. Os-188
 c. W-184 d. $^{37}_{17}Cl$
 e. S-34, doubly charged anion
 f. Rb-85, singly charged cation

*30. Determine the number of protons, neutrons, and electrons in each of the following species.
a. 2^+ cation of Be-9
b. Pb-206
c. $^{81}_{35}Br^-$
d. singly-charged cation of He-3
e. 3^- ion of N-15
f. 1^- ion of I-127

31. What is the symbol of the species composed of each of the following sets of subatomic particles?
a. 25p, 30n, 25e
b. 20p, 20n, 18e
c. 33p, 42n, 33e
d. 53p, 74n, 54e

32. What is the symbol of the species composed of each of the following sets of subatomic particles?
a. 94p, 150n, 94e
b. 79p, 118n, 76e
c. 34p, 45n, 36e
d. 54p, 77n, 54e

Molecules, Ions, and Compounds

33. How many oxygen atoms are there in one molecule of each of the following?
a. O_2
b. P_4O_{10}
c. $C_{12}H_{22}O_{11}$
d. $HClO_2$

34. How many oxygen atoms are there in each of the following nitrogen oxide molecules?
a. NO
b. NO_2
c. N_2O_4
d. N_2O_5

*35. How many hydrogen atoms are there in one molecule of each of the following?
a. $(CH_3)_2O$
b. CH_3CH_2Cl
c. CH_3CH_2OH
d. $CH_3(CH_2)_6CH_3$

*36. How many hydrogen atoms are there in one molecule of each of the following?
a. $CH_3CH_2CO_2H$
b. $CH_2ClCOCH_3$
c. $(C_2H_5)_2NH$
d. CH_2OHCH_2OH

*37. How many oxygen atoms are there in each of the following formula units?
a. K_2SO_4
b. $(NH_4)_2CO_3$
c. $FeSO_4$
d. $Ca(OH)_2$
e. $Al_2(CO_3)_3$
f. $Fe_2(SO_4)_3$

*38. How many oxygen atoms are there in each of the following formula units?
a. $BaCO_3$
b. $Al(NO_3)_3$
c. $Cr(OH)_2$
d. $Fe(ClO_2)_3$
e. $Pb(BrO_3)_2$
f. $Cr(OH)_3$

39. How many hydrogen atoms are there in each of the following formula units?
a. $(NH_4)_2S$
b. NH_4HCO_3
c. $BaCl_2 \cdot 2H_2O$
d. $CuSO_4 \cdot 5H_2O$

40. How many oxygen atoms are there in each of the following formula units?
a. $Na_2CO_3 \cdot 7H_2O$
b. $Ag_4Fe(CN)_6 \cdot H_2O$
c. $KCH_3CO_2 \cdot 3H_2O$
d. $S_2O_3^{2-}$

Molar Mass and Natural Abundance

41. Given that the masses and natural abundances of the isotopes of magnesium are as follows, calculate the molar mass.

$^{24}_{12}Mg$	23.985042 amu	78.99%
$^{25}_{12}Mg$	24.985837 amu	10.00%
$^{26}_{12}Mg$	25.982593 amu	11.01%

42. The three stable isotopes of silicon have masses and natural abundances as follows. Calculate the molar mass.

$^{28}_{14}Si$	27.976927 amu	92.23%
$^{29}_{14}Si$	28.976495 amu	4.67%
$^{30}_{14}Si$	29.973770 amu	3.10%

43. Approximately 50% of naturally occurring bromine is composed of the isotope Br-79, while the other half is Br-81. Without looking up the exact masses of the isotopes, estimate the molar mass of bromine.

*44. Approximately 75% of naturally occurring chlorine is composed of the isotope Cl-35, while the rest is Cl-37. Without looking up the exact masses of the isotopes, estimate the molar mass of chlorine.

*45. There are two stable isotopes of bromine, having masses of 78.918336 and 80.916289 amu, respectively. The molar mass of the naturally occurring mixture of these isotopes is 79.904 g mol^{-1}. Determine the natural abundance of the isotopes, and report the answer to three significant digits.

*46. There are two stable isotopes of chlorine, having masses of 34.968852 and 36.965903 amu, respectively. The molar mass of the naturally occurring mixture of these isotopes is 35.453 g mol^{-1}. Determine the natural abundance of the isotopes, and report the answer to three significant digits.

Moles, Molar Mass, and Grams

47. Express the following amounts in moles (or millimoles).
a. 3.01×10^{23} F atoms
b. 999 Ca atoms
c. 1.59×10^{22} NO_3^- ions
d. 7.50×10^{22} electrons

48. Express the following amounts in moles (or millimoles).
a. 6.02×10^{24} U-235 atoms
b. 9999 Mn^{2+} ions
c. 1.5×10^{19} protons
d. 0.035×10^{26} shoes

49. How many atoms are there in each of the following amounts?
a. 3.98 mmol He
b. 6.02 mol tungsten
c. 8.83×10^{-20} mol Sr-90
d. 1.234 mol In

50. How many atoms are there in each of the following amounts?
a. 10.0 mol Fe
b. 0.00037 mol Ag
c. 62.8 μmol Rn
d. 12.4×10^{-12} mol W

51. How many molecules or ions are there in each of the following amounts?
a. 5.00×10^{-6} mol water
b. 1.9×10^{-23} mol C_2H_4
c. 165 mmol HCO_3^- ions
d. 25 mol NH_3

52. How many molecules or ions are there in each of the following amounts?
a. 683 mmol S^{2-} b. 12 kmol Li
c. 32.6 μmol $HClO_4$ d. 1.6×10^{-4} mol XeF_2

53. Calculate the molar mass of the following compounds.
a. CaH_2 b. N_2O_5 c. BaF_2 d. CuCl

54. Calculate the molar mass of the following compounds.
a. K_2SO_4 b. $Co(NO_3)_3$ c. $Mg(OH)_2$ d. $KMnO_4$

*55. Calculate the molar mass of the following compounds.
a. carbon dioxide b. dinitrogen tetroxide
c. sulfur hexafluoride

*56. Calculate the molar mass of the following compounds.
a. strontium iodide b. potassium carbonate
c. sodium acetate d. calcium phosphate

57. Calculate the mass of each of the following.
a. 0.500 mol $NiSO_4$ b. 605 mmol $CuSO_4 \cdot 5H_2O$
c. 0.875 mol H_2CO_3 d. 6.02×10^{-6} mol XeF_4

58. Calculate the mass of each of the following.
a. 10.3 mol $(NH_4)_2SO_4$ b. 28 mmol HNO_3
c. 0.429 mol MnO_2 d. 5.25×10^{-8} mol $TiCl_4$

59. What amount (how many moles) of substance is in each of the following quantities?
a. 7.5 g $Sr(OH)_2$ b. 100. g SiO_2
c. 23 mg UF_6 d. 10.0 g N_2O_4

60. What amount of substance is in each of the following quantities?
a. 39.1 g SO_2 b. 0.115 g $MgBr_2$
c. 5.97 g LiCl d. 65.2 mg $CsNO_3$

*61. Calculate the number of moles in a half pound of each of the noble gases.

*62. The summary exercise in the previous chapter referred to 21 hundred tons (1 ton ⇔ 2000 pounds) of gold. How many moles is this?

63. A laboratory recipe calls for 0.15 mol $MgSO_4$ (molar mass = 120.4 g mol^{-1}). How many grams should be weighed out?

64. If you need a solution that contains 0.834 mol NaCl for each liter of water, how much NaCl should you weigh out in order to prepare 2 L of solution?

*65. If each mole of mercury(II) oxide (HgO) yields one-half mole of oxygen (O_2) on decomposition, how many moles of oxygen are produced by the decomposition of 100 g HgO?

*66. One mole of any group 1A metal will react with water to produce one-half mole of hydrogen (H_2). How many moles of H_2 are produced by the reaction of 100 g Li? Of 100 g Rb?

*67. How many hydrogen atoms are there in each of the following?
a. 1.5 mol $HClO_2$ b. 65 mmol H_2SO_4
c. 0.094 mol $(NH_4)_2CO_3$ d. 0.750 mol $BaCl_2 \cdot 2H_2O$

*68. How many oxygen atoms are there in each of the following?
a. 1.5 mol $HClO_2$ b. 65 mmol H_2SO_4
c. 0.094 mol $(NH_4)_2CO_3$ d. 0.750 mol $BaCl_2 \cdot 2H_2O$

*69. How many nitrogen atoms are there in 50 g (2 significant digits) of each of the following?
a. NH_3 b. $NaNO_2$
c. C_2N_2 d. $(NH_4)_2CO_3$

*70. How many carbon atoms are there in 50 g (2 significant digits) of each of the following?
a. K_2CO_3 b. $KHCO_3$
c. C_2H_6 d. CH_3CO_2H

Composition

71. Calculate the mass percent of the Group 1A element in each of the following: LiOH, NaOH, KOH.

72. Calculate the percent oxygen (by mass) in each of the four oxoacids of chlorine.

73. Which compound has the higher precent carbon: carbon monoxide or carbon dioxide?

74. Calculate the percent nitrogen in each of the following.
a. KNO_3 b. $Mg(NO_3)_2$ c. NH_4NO_3

Empirical Formula

75. A xenon oxofluoride contains 6.12% O, 43.6% F, and the remainder is Xe. What is the empirical formula?

76. One of the two compounds of krypton that has been prepared is a fluoride that contains 31.2% F (the remainder is Kr). What is its empirical formula?

77. A 5.682 g sample of an oxide of phosphorus contains 2.480 g P. What is its empirical formula?

*78. A 1.650 g sample of a mercury chloride contains 1.402 g Hg. What is the empirical formula?

*79. Analysis of a 6.3200 g sample yields the following amounts of nitrogen and oxygen: N, 1.923 g; O, 4.397 g. No other elements are present. What is the empirical formula of the compound? If the molar mass is known to be in the range 80 to 100 g mol^{-1}, what is the molecular formula?

*80. Analysis of a hydrated salt gives the following amounts: Cu, 3.99 mg; S, 2.01 mg; O, 9.06 mg; H, 0.635 mg. What is the empirical formula? Assuming that the salt contains the sulfate ion, SO_4^{2-}, write the empirical formula with a dot showing the water of hydration.

*81. Zinc chloride is an ionic compound that contains 48.0% Zn. What is the charge of the zinc cation?

*82. An iron chloride is analyzed and found to contain 65.6% Cl. What is the charge of the iron ion?

83. One of many compounds containing only carbon and hydrogen (hydrocarbons) is cyclohexane, whose empirical formula is CH_2. If its molar mass is in the range 80 to 90 g mol^{-1}, what is its molecular formula?

84. Ethane contains 79.9% C, and the remainder is H. What is the empirical formula of ethane? If its molar mass is known to be greater than 25 g mol^{-1} but less than 35 g mol^{-1}, what is its molecular formula?

*85. When 6.25 mg of the anesthetic ether, which contains C, H, and O, is subjected to combustion analysis, 14.84 mg CO_2 and 7.56 mg H_2O are obtained. What is the empirical formula? If the molar mass is known to be between 50 and 100 g mol^{-1}, what is the molecular formula?

*86. When 35.19 mg ethylene glycol (antifreeze), which contains C, H, and O, is subjected to combustion analysis, 49.90 mg CO_2 and 30.64 mg H_2O are obtained. What is the empirical formula of ethylene glycol? If its molar mass is known to be in the range 50 to 70 g mol^{-1}, what is its molecular formula?

Chemical Nomenclature

87. Write the formulas of each of the following ionic compounds.
    a. rubidium fluoride
    b. ammonium sulfate
    c. potassium sulfide
    d. barium sulfide
    e. magnesium hydrogen carbonate
    f. sodium nitrite

88. Write the formulas of each of the following ionic compounds.
    a. iron(II) chloride
    b. magnesium perchlorate
    c. cerium(IV) fluoride
    d. copper(II) bromide
    e. cesium sulfite
    f. cupric iodide

89. Write the molecular formulas of each of the following molecular compounds.
    a. sulfur trioxide
    b. chlorous acid
    c. carbon tetrachloride
    d. phosphorus pentachloride
    e. ammonia
    f. hydrogen iodide

90. Write the molecular formulas of each of the following molecular compounds.
    a. iodine heptafluoride
    b. dinitrogen pentoxide
    c. nitrogen trifluoride
    d. dinitrogen oxide
    e. sulfur hexafluoride
    f. xenon difluoride

*91. Give the formula of each of the following compounds, and label each as a molecular or empirical formula. If the empirical and molecular formulas are different, give both. If you can, identify each compound as ionic or molecular.
    a. manganese(IV) oxide
    b. sulfur dioxide
    c. ferric chloride
    d. dinitrogen tetroxide
    e. cobalt(III) dichromate
    f. hydrogen peroxide

*92. Give the formula of each of the following compounds, and label each as a molecular or empirical formula. If the empirical and molecular formulas are different, give both. If you can, identify each compound as ionic or molecular.
    a. aluminum carbonate
    b. mercurous chloride
    c. dinitrogen pentoxide
    d. tin(II) bromide
    e. bromic acid
    f. lead(IV) oxide

93. Name the following ionic compounds.
    a. NH_4HCO_3
    b. $Na_2C_2O_4$
    c. Mg_3N_2
    d. $K_2Cr_2O_7$
    e. CaH_2
    f. $Ca(HSO_4)_2$

94. Name the following ionic compounds.
    a. $(NH_4)_3PO_4$
    b. $Al_2(C_2O_4)_3$
    c. K_2O_2
    d. K_2S
    e. $NaHCO_3$
    f. LiCl

95. Name the following molecular compounds.
    a. $HClO_3$
    b. Cl_2O_7
    c. N_2O_5
    d. CO
    e. SO_3
    f. CI_4

96. Name the following molecular compounds.
    a. NO
    b. O_2F
    c. H_2Te
    d. NO_2
    e. XeO_3
    f. H_2O

97. Name the following compounds.
    a. $FeCl_2$
    b. HClO
    c. $GeCl_4$
    d. CaH_2
    e. Na_2O
    f. $MgCl_2$

98. Name the following compounds.
    a. $(NH_4)_3N$
    b. $HgCl_2$
    c. KH
    d. PCl_3
    e. HF
    f. ClF_3

Noble Gases

99. Give the names and elemental symbols of all the noble gases in order of increasing molar mass.

100. Give the names and formulas of all the known noble-gas fluorides.

101. Identify the noble gas that is
    a. the most common (most abundant).
    b. the least abundant.
    c. the most reactive.

CHAPTER 3

CHEMICAL EQUATIONS AND STOICHIOMETRY

The reaction of potassium with water releases enough heat to ignite the reaction product, hydrogen gas.

3.1 CHEMICAL EQUATIONS

A particular chemical change, involving the formation of new substances, is described by a chemical equation. In its simplest form, a **chemical equation** is a list of the formulas of all the species (elements and/or compounds) initially present, followed by an arrow and a list of all the species present after the change. The species initially present are called the **reactants,** the arrow is read as "reacts to give," "goes to," or "yields," and the final species are called the **products.** Optionally, the state of each species may be indicated in parentheses following the formula. The most frequently used state designations are (s) = solid, (ℓ) = liquid, (g) = gaseous, and (aq) = dissolved in water **(aqueous solution).** Equation (3-1a) describes the reaction in which aqueous ammonium sulfate is formed by bubbling gaseous ammonia through aqueous sulfuric acid.

Recall that a solution is a homogeneous mixture of two or more pure substances.

$$H_2SO_4(aq) + NH_3(g) \longrightarrow (NH_4)_2SO_4(aq) \quad (3\text{-}1)$$

Equations are used as compact descriptions of chemical change.

EXAMPLE 3.1 Writing an Equation for a Reaction

Metallic lithium reacts with water to produce lithium hydroxide, which remains dissolved in the water, and gaseous hydrogen, which escapes to the atmosphere. Write the equation for this reaction.

Analysis

Target: A chemical equation, which is an ordered list of chemical formulas; an arrow separates the products from the reactants.

Knowns: The names of the reactants and the products.

Missing Information: Chemical formulas rather than names are required.

Plan

3. Write the equation using Equation (3-1a) as a model.
2. Read the problem carefully to distinguish the products from the reactants.
1. Determine the formulas from the names, and include the state designation if known and if desired.

Work

1. Lithium is Li(s) (most metals are solids). Water is indicated by $H_2O(\ell)$. Dissolved lithium hydroxide is LiOH(aq), and gaseous hydrogen is $H_2(g)$.
2. Li and H_2O are reactants, while LiOH and H_2 are products.
3. The equation is

$$Li(s) + H_2O(\ell) \longrightarrow LiOH(aq) + H_2(g)$$

The stable form of hydrogen is the diatomic molecule H_2. Similarly, N_2, O_2, F_2, Cl_2, Br_2, and I_2 are the normal forms of these elements.

Exercise

Sodium reacts with water in the same way as lithium. The heat liberated in the reaction is so great that a second reaction often occurs. The hydrogen spontaneously ignites, combining with the oxygen of the air to form steam. Write the chemical equation for this second reaction.

Answer: $H_2(g) + O_2(g) \longrightarrow H_2O(g)$.

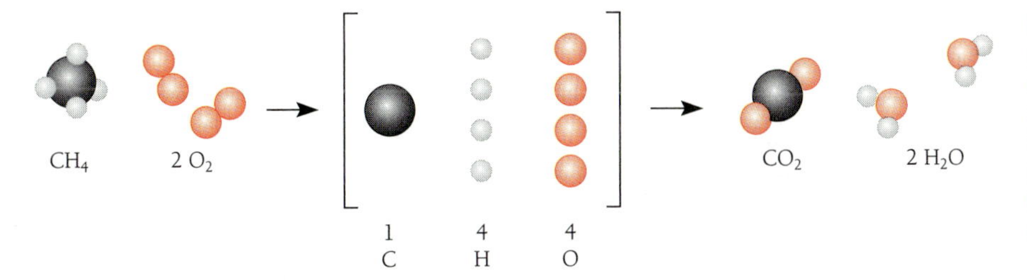

Figure 3.1 In a chemical reaction a group of atoms changes from an arrangement known as "reactants" to an arrangement known as "products." Since no atoms are created or destroyed during such rearrangements, their numbers are the same on both sides of the equation.

Balanced Equations

Equation (3-1a) identifies the participants in the chemical change, but it is not balanced. A **balanced equation** gives not only the identity, but also the relative number of molecules (or formula units) of all the participating species. Equation (3-1b) is the balanced equation for the sulfuric acid/ammonia reaction.

$$H_2SO_4(aq) + 2\ NH_3(g) \longrightarrow (NH_4)_2SO_4(aq) \qquad (3\text{-}1b)$$

This equation shows that twice as many molecules of NH_3 as H_2SO_4 are required, and that the number of $(NH_4)_2SO_4$ formula units produced is the same as the number of H_2SO_4 molecules consumed. These relationships are indicated by putting an integer, the **stoichiometric coefficient,** immediately in front of the formula of each species. If the stoichiometric coefficient is 1, as it is for both H_2SO_4 and $(NH_4)_2SO_4$ in Equation (3-1b), it is not written explicitly in the equation.

A balanced equation is a direct expression of the law of conservation of mass because individual atoms are unchanged by chemical reactions. Only the arrangement of the atoms changes. This fact provides the criterion for deciding whether or not an equation is balanced. *If the number of atoms of each element in the reactants is the same as that in the products, the equation is balanced* (Figure 3.1).

EXAMPLE 3.2 Is the Equation Balanced?

When ignited with a burner, ammonium dichromate decomposes rather spectacularly, with a shower of sparks. This reaction can be used to prepare Cr_2O_3, and it is sometimes used as a demonstration of a colorful chemical change. The reaction proceeds according to the equation

$$(NH_4)_2Cr_2O_7(s) \longrightarrow N_2(g) + 4\ H_2O(g) + Cr_2O_3(s)$$

Is this a balanced chemical equation?

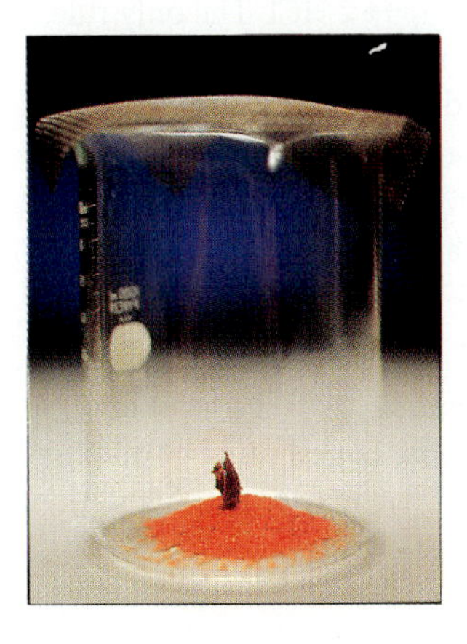

The decomposition of ammonium dichromate leaves gray-green chromium oxide as the only solid product.

Analysis/Plan Count and compare the number of atoms of each element on each side of the equation.

Work

Element	Left Side	Right Side
Cr	The formula unit contains 2 Cr atoms, and there is 1 formula unit. Total Cr: $2 \times 1 = 2$.	Each Cr_2O_3 contains 2 Cr atoms, and there is 1 formula unit. No other product species contains Cr. Total Cr: $2 \times 1 = 2$.
N	Each NH_4^+ ion has 1 N atom, and there are 2 NH_4^+ ions in the formula unit. There is 1 formula unit. Total N: $1 \times 2 \times 1 = 2$.	Each N_2 molecule has 2 N atoms, and there is 1 N_2 molecule in the equation. No other species contains N. Total N: $2 \times 1 = 2$.
H	Each NH_4^+ ion has 4 H atoms, and there are 2 NH_4^+ ions in the formula unit. There is 1 formula unit. Total H: $4 \times 2 \times 1 = 8$.	Only H_2O among the products contains H. Each molecule has 2 H's, and there are 4 H_2O molecules. Total H: $2 \times 4 = 8$.
O	There are 7 O atoms per formula unit, and 1 formula unit. Total O: $7 \times 1 = 7$.	There are 4 H_2O's, each containing 1 O atom. Also, there is 1 Cr_2O_3 with 3 O atoms. Total O: $(4 \times 1) + (1 \times 3) = 7$.

The number of atoms on each side of the equation is the same for all elements, so the equation is balanced.

Exercise

Show that (3-1b) is balanced, and that (3-1a) is not.

The Balancing Act

Choosing the correct stoichiometric coefficients, that is, balancing a previously unbalanced equation, can frequently be done with an educated guess. This is called *balancing by inspection.* It is easily seen that the unbalanced equation $H_2 + Cl_2 \rightarrow HCl$ becomes balanced by placing a 2 in front of HCl: $H_2 + Cl_2 \rightarrow 2\,HCl$. The only rule is that new species may not be put into an equation and no species may be removed. The equation $H_2 + O_2 \rightarrow H_2O$ cannot be balanced by adding O, a new species, to the products. (The equation $H_2 + O_2 \rightarrow H_2O + O$ falsely indicates that monatomic oxygen is a product.) The *subscripts* in a formula may not be changed, because to do so would be to add a new species. $H_2 + O_2 \rightarrow H_2O_2$ is not a balanced form of $H_2 + O_2 \rightarrow H_2O$—it is a different reaction. When inspection does not quickly produce a balanced equation, a systematic approach is used.

1. Balance the different elements one by one, beginning with the most complicated species (that is, the compound containing the largest number of different elements). Leave until later any pure elements that may be in the equation.

2. Within the most complicated species, begin if possible with an element that appears in only one species on each side of the equation. Insert stoichiometric coefficients as needed to show the same number of atoms of that element on each side of the equation.
3. If the balancing of one element causes the *un*balancing of a previously balanced element, restore its balance before proceeding.

EXAMPLE 3.3 Balancing an Equation

When potassium burns in oxygen or air, the product is not the oxide but the superoxide KO_2. Potassium superoxide is used in self-contained, closed-circuit breathing equipment which supplies oxygen and absorbs carbon dioxide. Moisture in the breath reacts with the superoxide, producing oxygen and potassium hydroxide. Carbon dioxide is removed from the breath by reaction with the potassium hydroxide. Balance the following equations.

$$\text{a. } H_2O(g \text{ or } \ell) + KO_2(s) \longrightarrow KOH(s) + O_2(g)$$

$$\text{b. } CO_2(g) + KOH(s) \longrightarrow KHCO_3(s)$$

Plan Follow the systematic guidelines.

Work

a. H_2O and KO_2 seem equally complicated. O appears in both compounds, but, because it appears also as the pure element O_2, O should be balanced *after* K and H. Since K is already in balance, 1 atom on each side of the equation, we begin with H.

Hydrogen: The coefficient 2 placed in front of KOH brings H into balance with 2 atoms on each side.

$$H_2O(g \text{ or } \ell) + KO_2(s) \longrightarrow \mathbf{2}\ KOH(s) + O_2(g)$$

This change throws potassium out of balance; restore the balance by placing a 2 in front of KO_2.

$$H_2O(g \text{ or } \ell) + \mathbf{2}\ KO_2(s) \longrightarrow 2\ KOH(s) + O_2(g)$$

At this point both K and H are in balance (check to make sure). The remaining element, O, is easily balanced because it appears in one place as a pure element.

Oxygen: There are $1 + (2 \times 2) = 5$ O atoms on the left, and $2 + 2 = 4$ O atoms on the right. Balance is achieved by increasing the number on the right from 4 to 5. This can be done only by using the *fractional coefficient* $1\frac{1}{2}$ for O_2.

$$H_2O(g \text{ or } \ell) + 2\ KO_2(s) \longrightarrow 2\ KOH(s) + 1\tfrac{1}{2}\ O_2(g)$$

Oxygen is now balanced, since the left side still shows $1 + (2 \times 2) = 5$ atoms, and the right side now also shows $2 + [(1\frac{1}{2}) \times 2] = 5$ atoms.

Fractional stoichiometric coefficients such as $1\frac{1}{2}$ are perfectly correct, but whole numbers are normally used. The mole ratios are unchanged when *all* stoichiometric coefficients are multiplied by the same factor, and 2 is the smallest factor that clears the fraction 1½. The final step in writing the balanced equation is to multiply all coefficients by 2, giving

$$2\ H_2O(g\ \text{or}\ \ell) + 4\ KO_2(s) \longrightarrow 4\ KOH(s) + 3\ O_2(g)$$

Check Always, after balancing an equation, check that it is actually balanced.

H: $2 \times 2 = 4$ on the left, $4 \times 1 = 4$ on the right

K: $4 \times 1 = 4$ on the left, $4 \times 1 = 4$ on the right

O: $(2 \times 1) + (4 \times 2) = 10$ on the left, $(4 \times 1) + (3 \times 2) = 10$ on the right

b. A check shows that $CO_2(g) + KOH(s) \rightarrow KHCO_3(s)$ is already balanced.

Exercise

Potassium oxide, K_2O, can be produced by the reaction of potassium with potassium nitrate: $K(s) + KNO_3(s) \rightarrow K_2O(s) + N_2(g)$. Balance this equation with whole number coefficients.

Answer: The *sum* of all stoichiometric coefficients in the balanced equation is 19.

3.2 STOICHIOMETRY

The term **stoichiometry** refers to the amounts of the various species participating in a chemical reaction. Typically, a problem in stoichiometry asks, "How much of *this* is required to react with *that*?" or "How much of *this* is produced when a given amount of *that* reacts?" The amounts can be in moles, mass, or volume units. Stoichiometry problems may *appear* to be complicated, but in fact they are nothing more than unit conversions. They are solved in the same way as the kilometers-to-miles and other conversions in Chapter 1.

Mole Relationships

The stoichiometric coefficients of a balanced chemical equation give the *relative* amounts of reacting species. Equation (3-lb) shows that all of the following amounts react completely, with none of either reactant left over.

Amount of H_2SO_4 +	Amount of NH_3	React to Give	Amount of $(NH_4)_2SO_4$
1 molecule	2 molecules		1 formula unit
24 molecules	48 molecules		24 formula units
6.022×10^{23} molecules	12.04×10^{23} molecules		6.022×10^{23} formula units
1 mole	2 moles		1 mole
0.35 mol	0.70 mol		0.35 mol

Any set of numbers in the ratio 1 : 2 : 1 may be used, provided they refer to amounts in moles (or molecules) rather than in units of mass.

The stoichiometric coefficients of a balanced chemical equation imply a set of equivalence statements (Figure 3.2). In the case of Equation (3-lb), these are: 1 mol $H_2SO_4 \Leftrightarrow$ 1 mol $(NH_4)_2SO_4$, 2 mol $NH_3 \Leftrightarrow$ 1 mol $(NH_4)_2SO_4$, and 1 mole

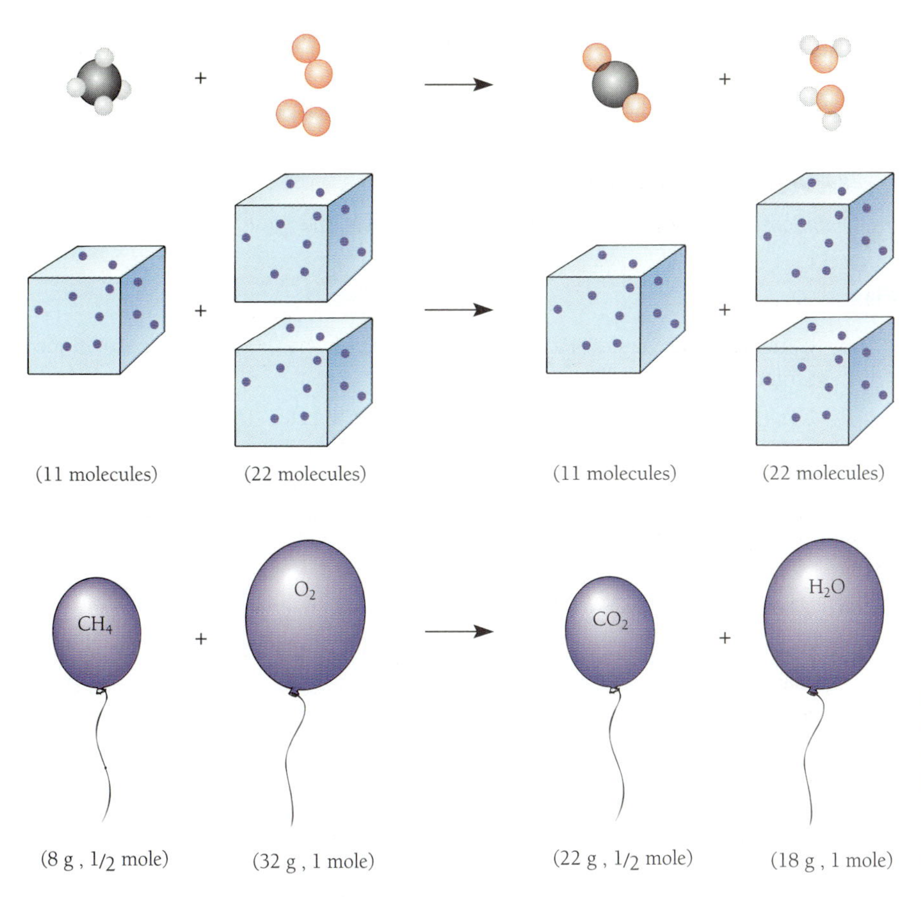

Figure 3.2 The stoichiometric coefficients of a balanced chemical equation give the relative amounts (ratios) of the participating species, both reactants and products. The ratios refer to amounts given either as molecules or moles.

$H_2SO_4 \Leftrightarrow 2$ mol NH_3. Like other equivalence statements, these give rise to conversion factors, known as **stoichiometric ratios** or **mole ratios,** such as

$$\frac{1 \text{ mol } H_2SO_4}{2 \text{ mol } NH_3} \quad \text{or} \quad \frac{1 \text{ mol } H_2SO_4}{1 \text{ mol } (NH_4)_2SO_4}$$

Of course these mole ratios apply only to this particular reaction. Other reactions have different ratios.

Mole ratios are used to determine the amounts of species consumed or produced in a reaction. First, the coefficients are read from the *balanced equation* and put in the form of an equivalence statement. Then, a unit conversion factor is generated from the equivalence statement. Finally, the conversion factor is applied to express the molar amount of one substance in terms of another.

EXAMPLE 3.4 The Mole Ratio in Stoichiometry

Sodium cyanide (which, like all cyanides, is extremely poisonous) is used in the extraction of gold and silver from their ores. Gold reacts with cyanide and atomspheric oxygen to form the complex salt sodium dicyanoaurate(I). This salt is soluble in water and can be washed out of the crushed ore. The equation for the reaction is

$$4\,Au(s) + 8\,NaCN(aq) + 2\,H_2O(\ell) + O_2(g) \longrightarrow 4\,NaOH(aq) + 4\,NaAu(CN)_2(aq)$$

How many moles of NaCN are required in order to extract 0.0285 moles of gold from its ore?

Analysis

Target: An amount (moles) of NaCN.

$$? \text{ mol NaCN} =$$

Knowns: The amount of Au.

Relationship: A balanced chemical equation gives the mole ratios of *all* species.

Plan

3. Generate a conversion factor, and use it to convert 0.0285 moles Au to the target, moles of NaCN.
2. Use the stoichiometric coefficients of the two species (target and known) to write an equivalence statement.
1. Make sure the equation is balanced. If it isn't, balance it.

Work

1. The equation is balanced as given.
2. The equivalence statement connecting Au and NaCN is 4 mol Au ⇔ 8 mol NaCN.
3. The conversion factor is (8 mol NaCN)/(4 mol Au).

$$? \text{ mol NaCN} = (0.0285 \text{ mol Au}) \cdot \frac{(8 \text{ mol NaCN})}{(4 \text{ mol Au})}$$
$$= 0.0570 \text{ mol NaCN}$$

Exercise

In the metallurgical process, metallic gold is later recovered from the solution by treatment with zinc.

$$NaAu(CN)_2(aq) + Zn(s) \longrightarrow Na_2Zn(CN)_4(aq) + Au(s)$$

How many moles of zinc must be used to recover 1.33 moles of gold?

Answer: 0.665 mol Zn.

Mass Relationships: The Standard Stoichiometry Problem

The amount of a substance is usually determined by weighing, so the known quantity in a problem like Example 3.4 is normally expressed in mass units rather than moles. Similarly, it is usually desirable to have the answer (the quantity of target substance) expressed in grams. The balanced equation and the mole ratios remain of central importance in such problems. It is only necessary to include gram-mole conversions in the solution (see Example 3.5 and Figure 3.3). These problems, with the mass of one substance given and the mass of another as the target, occur so frequently in chemistry that we will give them a special name: the **standard stoichiometry problem (SSP).** The SSP is always solved in the same way, whether the two substances involved are reactants, products, or one of each.

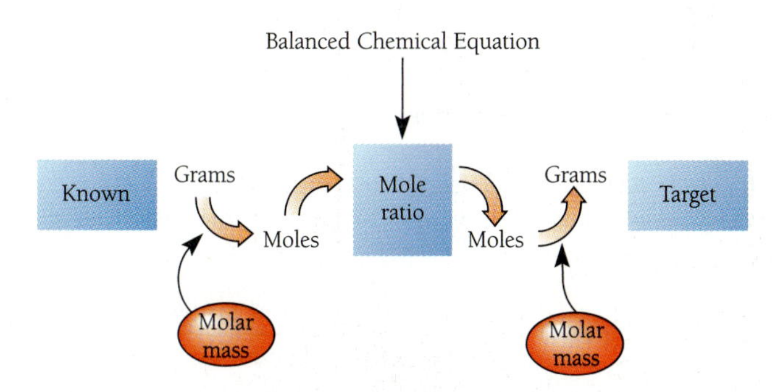

Figure 3.3 The basic problem in stoichiometry is the calculation of the mass of one substance from the mass of another. It is solved by a three-stage unit conversion requiring two molar masses and a stoichiometric or mole ratio of the two species.

EXAMPLE 3.5 Stoichiometry in Terms of Mass: The Standard Stoichiometry Problem

A reactant often used in the synthesis of new compounds is lithium aluminum hydride, $LiAlH_4$, which is produced by the action of lithium hydride on aluminum chloride. Finely powdered LiH and $AlCl_3$ are shaken up (suspended) in liquid ether, in which they are insoluble. The unimportant product, LiCl, is likewise insoluble in ether. The desired product, $LiAlH_4$, dissolves in ether, and is recovered from the filtered solution by evaporation of the ether. The reaction is

$$4\ \mathrm{LiH(s)} + \mathrm{AlCl_3(s)} \xrightarrow{\text{ether}} 3\ \mathrm{LiCl(s)} + \mathrm{LiAlH_4}$$

How much $LiAlH_4$ can be produced from 14.8 g LiH?

Analysis The problem is similar to one we have solved before (Example 3.4) and is solved in the same way. However, we must also use gram-mole conversion factors.

Unless otherwise specified or clear from the context, the quetion "How much?" refers to the mass in g, mg, or kg.

Target: ? g $LiAlH_4$ =

Knowns: The mass of LiH, and the chemical equation.

Missing Information: The molar masses of the two species; these are implicit in the empirical formulas.

Plan Convert the given mass of LiH to moles. Then make the mole to mole conversion as in Example 3.4 (make sure the equation is balanced). Finally, convert moles $LiAlH_4$ to grams.

Work

1. The equivalence statement is 1 mol LiH ⇔ 7.949 g, and the unit conversion is

$$?\ \text{mol LiH} = (14.8\ \text{g LiH}) \cdot \frac{(1\ \text{mol LiH})}{(7.949\ \text{g LiH})} = 1.862\ \text{mol}$$

2. From the stoichiometric coefficients of the balanced chemical equation, 1 mol $LiAlH_4$ ⇔ 4 mol LiH.

$$?\ \text{mol LiAlH}_4 = (1.862\ \text{mol LiH}) \cdot \frac{(1\ \text{mol LiAlH}_4)}{(4\ \text{mol LiH})} = 0.4655\ \text{mol}$$

3. The final conversion, from moles $LiAlH_4$ to grams, is

$$? \text{ g LiAlH}_4 = (0.4685 \text{ mol LiAlH}_4) \cdot \frac{(38.0 \text{ g LiAlH}_4)}{(1 \text{ mol LiAlH}_4)} = 17.69 \text{ g}$$

Rounding to the correct number of significant digits gives 17.7 g $LiAlH_4$.

In practice, the SSP is more conveniently treated as a single, three-stage, conversion operation.

$$? \text{ g LiAlH}_4 = (14.8 \text{ g LiH}) \cdot \frac{(1 \text{ mol LiH})}{(7.95 \text{ g LiH})} \cdot \frac{(1 \text{ mol LiAlH}_4)}{(4 \text{ mol LiH})} \cdot \frac{(38.0 \text{ g LiAlH}_4)}{(1 \text{ mol LiAlH}_4)}$$

$$= 17.7 \text{ g LiAlH}_4$$

Note: The mole-to-mole conversion factor *always* has the target compound in the numerator, whether it is a reactant or product.

Exercise

Lithium aluminum hydride must be used with care and in a moisture-free atmosphere because it reacts with water.

The considerable amount of heat liberated in this reaction is often enough to set the product, hydrogen, on fire.

$$LiAlH_4(s) + 4\ H_2O(\ell) \longrightarrow LiOH(aq) + Al(OH)_3(aq) + 4\ H_2(g)$$

If 0.144 g hydrogen is produced by the reaction of $LiAlH_4$ with water, how much $Al(OH)_3$ is also produced? (*Hint:* Remember that in an SSP no distinction is made between reactants and products; there is only the target compound and the known compound.)

Answer: 1.39 g.

3.3 PERCENTAGES IN STOICHIOMETRY

Percentages are used in two ways in stoichiometry. If the amount of product obtained in a reaction is less than expected, the actual amount can be described by the *percent yield.* If a particular reactant is not pure, its composition can be described by the *percent purity.*

Percent Yield of Product

The maximum possible amount of a product that can be obtained from given amounts of reactants is called the **theoretical yield.** The amount actually obtained is called the **actual yield,** and it is usually less than the theoretical yield (Figure 3.4). Some of the reasons for the discrepancy are, insufficient time allowed for the reaction to go to completion; further reaction or decomposition of the desired product, once formed; losses during isolation and purification of the product; and other reactions, called **side reactions,** of one or more of the reactants that lead to undesired **side products.** The ratio of the actual yield to the theoretical yield is commonly expressed as the **percent yield.**

$$\text{percent yield} = \frac{\text{actual yield}}{\text{theoretical yield}} \times 100 \qquad (3\text{-}2)$$

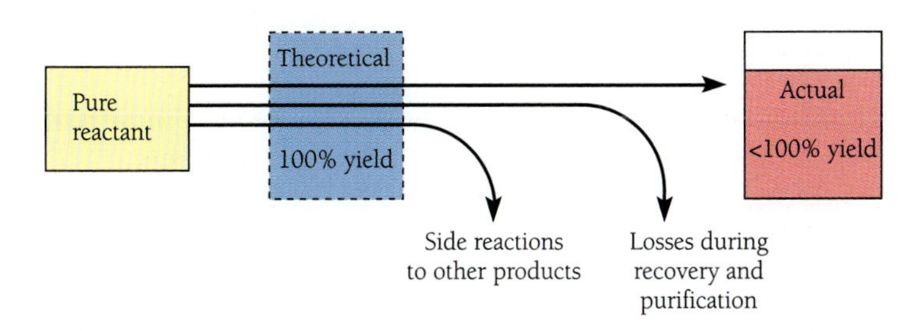

Figure 3.4 The actual yield in a real reaction is less than the theoretical yield, although, for many reactions, careful techniques can reduce the difference between the two to a negligible amount.

Calculation of the theoretical yield from a given amount of **starting mateial** (the reagent from which a desired product is prepared) is a standard stoichiometry problem. In Example 3.5, the theoretical yield of $LiAlH_4$ from 14.8 g of LiH, assuming no side reactions, further reaction of any of the $LiAlH_4$, nor losses during purification, is 17.7 g. If in an actual laboratory procedure only 13.1 g $LiAlH_4$ is isolated from the reaction, the percent yield calculated by Equation (3-2) is (13.1 g/17.7 g)100 = 74.0%. Note that the percent yield is a dimensionless number.

EXAMPLE 3.6 Calculation of Percent Yield

The hydroxides of Group 1A elements react with chlorine to produce hypochlorite salts, which are used as industrial and household bleaching agents. In particular, the household bleach Clorox is a dilute solution of sodium hypochlorite, made by the reaction of dilute sodium hydroxide with chlorine.

$$NaOH(aq) + Cl_2(g) \longrightarrow NaClO(aq) + NaCl(aq) + H_2O(\ell)$$

Calculate the percent yield if 91.3 g NaClO is obtained from the reaction of 105 g Cl_2.

Analysis

Target: A percent yield.

Knowns: The chemical equation, the actual yield, and the amount of starting material are known. Since the empirical formulas are given, the molar masses are implicit knowns.

Relationship: Equation (3-2) can be used directly.

Missing Information: The theoretical yield.

Plan

2. Use Equation (3-2) to find the percent yield.
1. Following Example 3.5, solve the SSP to find the theoretical yield.

Work

1. After balancing, the equation is

$$2\ NaOH(aq) + Cl_2(g) \longrightarrow NaClO(aq) + NaCl(aq) + H_2O(\ell)$$

The theoretical yield is

$$? \text{ g NaClO} = (105 \text{ g Cl}_2) \cdot \frac{(1 \text{ mol Cl}_2)}{(70.9 \text{ g Cl}_2)} \cdot \frac{(1 \text{ mol NaClO})}{(1 \text{ mol Cl}_2)} \cdot \frac{(74.4 \text{ g NaClO})}{(1 \text{ mol NaClO})}$$
$$= 110.18 \text{ g NaClO}$$

2.
$$\text{percent yield} = \frac{\text{actual yield}}{\text{theoretical yield}} \times 100$$
$$= \frac{91.3\text{ g}}{110.18\text{ g}} \times 100 = 82.9\%$$

Exercise

One of the chemicals used in the development of photographic film is sodium thiosulfate, $Na_2S_2O_3$. This substance (called "hypo" by photographers) is manufactured by the reaction of sodium sulfite with sulfur according to the equation $Na_2SO_3(aq) + S(s) \rightarrow Na_2S_2O_3(aq)$. Calculate the percent yield if 2.19 kg $Na_2S_2O_3$ is isolated from the reaction of 2.50 kg Na_2SO_3.

Answer: 69.8%.

Composition of Reagents and Mixtures

Sometimes a reagent is not a pure substance but is a mixture of the reactant and one or more other substances. Often this is by plan because some procedures are more efficient, safe, or economical when mixtures are used. A **reagent** is any mixture, solution, or pure substance used to bring about chemical change. A **reactant,** on the other hand, is a specific compound or element present before a reaction and converted during the reaction into a product. For example, hydrochloric acid, a reagent, is an aqueous solution of the reactant hydrogen chloride. Stoichiometric calculations must take into account the composition of the reagents. Suppose we had determined that the reaction in Example 3.6 is more economical if the chlorine is mixed with 10% nitrogen (by mass). Then the calculation of theoretical yield should use (105 g)·90% = 94.5 g for the mass of chlorine, rather than 105 g.

When the percent composition of a reagent is unknown, as is often the case, it can be determined by the methods of stoichiometry. With careful attention to technique, many reactions can be carried out in an essentially loss-free manner, giving the full theoretical yield of products. If such a reaction produces only 90% of the expected yield, we may conclude that the reagent is only 90% pure. Measurement of the amount of a particular element or compound in a sample, called **quantitative analysis,** is an important aspect of the branch of chemistry known as *analytical chemistry*. Percent composition of a reactant in a mixture is defined by Equation (3-3).

$$\text{percent composition} = \frac{\text{reactant mass}}{\text{mixture mass}} \times 100 \qquad (3\text{-}3)$$

Example 3.7 is a typical analytical procedure, in which a mixture is treated with a suitable reagent to yield a product that is carefully isolated and weighed (Figure 3.5). The mass of the *reactant species* in the mixture is calculated (SSP) from the mass of the product, and the percent of the reactant in the mixture is found with Equation (3-3).

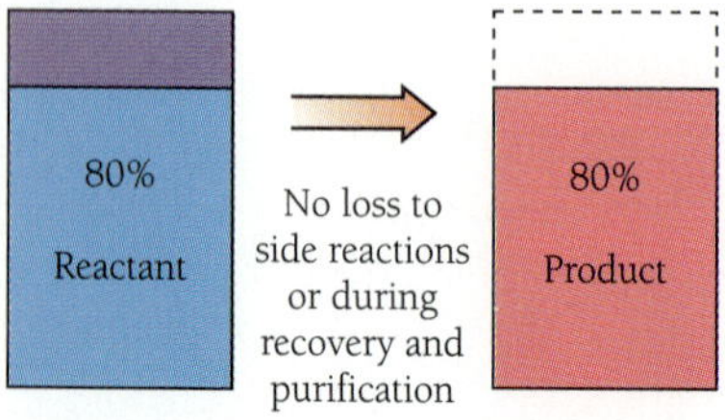

80% Reactant

No loss to side reactions or during recovery and purification

80% Product

Figure 3.5 Impure reagents can also cause the actual yield to be less than expected, even if there are no losses.

EXAMPLE 3.7 Percent Composition from Reaction Stoichiometry

If 1.00 kg of gold-bearing quartz rock is crushed and treated with sodium cyanide according to

$$4\,Au(s) + 8\,NaCN(aq) + 2\,H_2O(\ell) + O_2(g) \longrightarrow 4\,NaOH(aq) + 4\,NaAu(CN)_2(aq),$$

and 88.8 mg $NaAu(CN)_2$ is isolated, what is the percentage of gold in the ore?

Analysis

Target: Percent gold, a dimensionless number between 0 and 100.

Knowns: The product mass, the sample mass, and the balanced equation.

Relationship: Equation (3-3) can be used, but not without more information.

Missing Information: The amount of gold that yields 88.8 mg $NaAu(CN)_2$.

Plan

2. Apply Equation (3-3) to calculate percent composition.
1. Solve the SSP to find the mass of gold required to produce 88.8 mg $NaAu(CN)_2$.

Work

1. The stoichiometric coefficients of the balanced chemical equation lead to the mole ratio

$$\frac{(4 \text{ mol } NaAu(CN)_2)}{(4 \text{ mol Au})}$$

SSP

$$? \text{ g Au} = (88.8 \times 10^{-3} \text{ g } NaAu(CN)_2) \cdot \frac{(1 \text{ mol } NaAu(CN)_2)}{(272.0 \text{ g } NaAu(CN)_2)} \cdot \frac{(4 \text{ mol Au})}{(4 \text{ mol } NaAu(CN)_2)} \cdot \frac{(197.0 \text{ g Au})}{(1 \text{ mol Au})}$$
$$= 64.3 \times 10^{-3} \text{ g} = 64.3 \text{ mg}$$

2. The percent composition is

$$\text{percent composition} = \frac{\text{reactant mass}}{\text{mixture mass}} \times 100$$
$$= \frac{64.3 \times 10^{-3} \text{ g}}{1000 \text{ g}} \times 100 = 6.43 \times 10^{-3}\%$$

Check A rough calculation, (100 mg)(200/300) $\approx$ 70 mg, shows the answer to be reasonable. This may seem like a very small percentage, but bear in mind that the 88.8 mg of gold contained in our 1-kg sample is worth almost $1.

Exercise

A chemist in an industrial quality control laboratory was asked to check the purity of some potassium superoxide destined for use in breathing equipment (see Example 3.3). She added water to a 5.29 g sample of KO_2 and observed that 1.56 g O_2 was produced. Assuming that none of the impurities liberated oxygen, what was the purity of the sample?

Answer: The sample contained 87.4% KO_2.

3.4 LIMITING REACTANT

Reactions are often carried out with the reactants present in **stoichiometric amounts**—amounts exactly proportional to their mole ratio in the balanced chemical equation. In come circumstances, for instance if one of the reactants is very expensive or difficult to obtain, it is desirable to supply *more* than the stoichiometric amount of the other reactant. This is called adding a **reactant in excess.** The reactant present in *less* than the stoichiometric amount is called the **limiting reactant.** The limiting reactant controls the amount of product that can be obtained, because after the limiting reactant has been consumed, no additional product(s) can be formed. Thus, either the two reactants are present in *stoichiometric amounts,* or, one is *limiting* and the other is *in excess.* Note that "stoichiometric amounts" does *not* mean "the same number of moles of each."

As an example of these concepts, consider the reaction (3-1b), with which this chapter began.

$$H_2SO_4(aq) + 2\ NH_3(g) \longrightarrow (NH_4)_2SO_4(aq) \qquad (3\text{-}1b)$$

Suppose 0.9 mol H_2SO_4 and 1.3 mol NH_3 are allowed to react. The coefficients show that each mole of H_2SO_4 requires two moles NH_3 for complete reaction. For 0.9 mol H_2SO_4, the stoichiometric amount of NH_3 is 2×0.9 mol $=$ 1.8 mol. Since less than this amount is present, NH_3 is the limiting reactant, even though it is present in the greater molar amount. No more than 0.65 mol $(NH_4)_2SO_4$ can be formed from 1.3 mol NH_3. The concept of the limiting reactant is illustrated in another context in Figure 3.6.

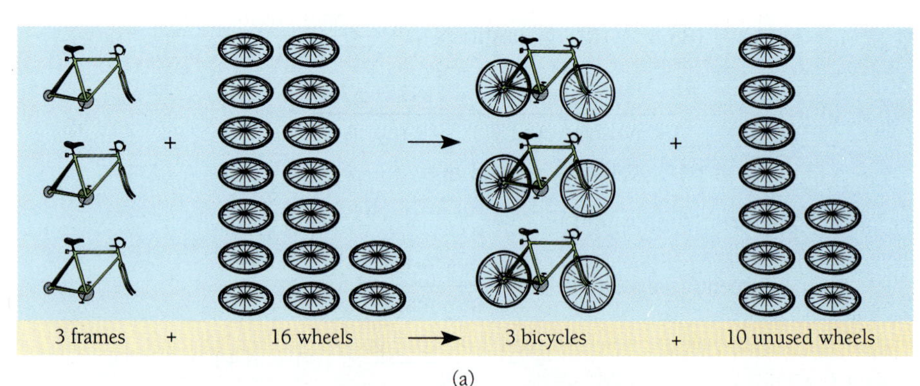

(a)

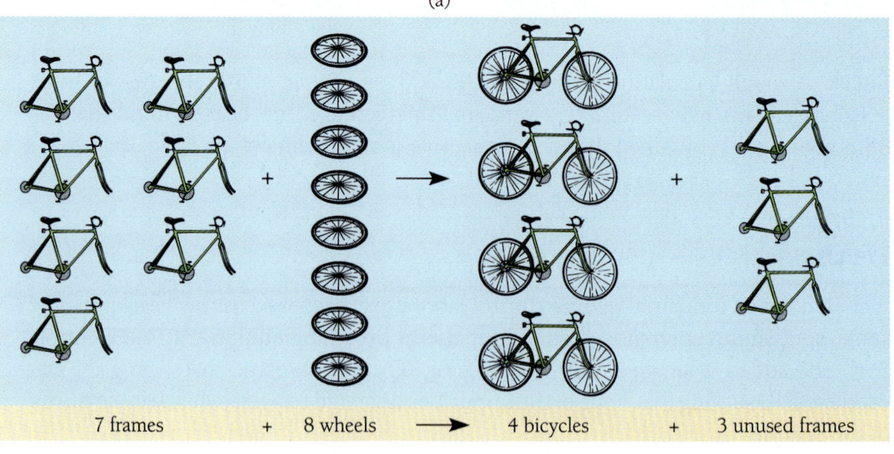

(b)

Figure 3.6 In the "reaction" producing bicycles from frames and wheels, the number produced can be controlled by either frames or wheels. In (a) the 3 frames are the limiting reactant, so that no more than 3 bikes are produced, no matter how many wheels are available. In (b) the 8 wheels limit the production to 4 bikes, in spite of the availability of more frames. Note that the number by itself does not necessarily determine which "reactant" is limiting. For example in (b) there are fewer frames but the number of wheels limits the number of bikes that can be produced.

Example 3.8 is a typical limiting reactant problem, in which the known quantities are in mass units rather than in moles.

EXAMPLE 3.8 Limiting Reactant

The only Group 1A element to react directly with nitrogen is lithium, which forms the nitride according to the equation

$$6\ Li(s) + N_2(g) \longrightarrow 2\ Li_3N(s)$$

Suppose 51 g Li is allowed to react with 39 g N_2. Determine (a) which reactant is limiting, (b) the amount of Li_3N formed, and (c) how much of the reactant in excess is left over after the reaction has gone as far as it can.

a. **Analysis** After the limiting reactant has been completely consumed, no more product can be formed. Suppose we calculate, in turn, the amount of product that would result from the complete consumption of each reactant. The smaller of these two amounts must be the one that is actually obtained, that is, must correspond to the limiting reactant.

Plan Solve the SSP for each of the reactants.

Work Check that the equation is balanced (it is), and calculate the amount of Li_3N obtainable from 51 g Li.

$$?\ g\ Li_3N = (51\ g\ Li) \cdot \frac{(1\ mol\ Li)}{(6.94\ g\ Li)} \cdot \frac{(2\ mol\ Li_3N)}{(6\ mol\ Li)} \cdot \frac{(34.8\ g\ Li_3N)}{(1\ mol\ Li_3N)} = 85\ g\ Li_3N$$

Calculate the amount of Li_3N obtainable from 39 g N_2.

$$?\ g\ Li_3N = (39\ g\ N_2) \cdot \frac{(1\ mol\ N_2)}{(28.0\ g\ N_2)} \cdot \frac{(2\ mol\ Li_3N)}{(1\ mol\ N_2)} \cdot \frac{(34.8\ g\ Li_3N)}{(1\ mol\ Li_3N)} = 97\ g\ Li_3N$$

The reaction proceeds until 85 g, the smaller of these two amounts of Li_3N, has been produced. At this point the reaction must stop, since no more Li is available. Li is the limiting reactant. Since enough N_2 was originally present to produce 97 g Li_3N, when 85 g Li_3N has been produced there is still some unreacted N_2. That is, N_2 is the reactant in excess.

b. **Analysis/Plan/Work** All the work has been done. The amount of Li_3N formed is 85 g.

c. **Plan** Solve the SSP for the mass of N_2 required to react with 51 g Li. Then subtract this amount from the amount originally present to get the mass of unreacted N_2.

Work

$$?\ g\ N_2 = (51\ g\ Li) \cdot \frac{(1\ mol\ Li)}{(6.94\ g\ Li)} \cdot \frac{(1\ mol\ N_2)}{(6\ mol\ Li)} \cdot \frac{(28.0\ g\ N_2)}{(1\ mol\ N_2)} = 34\ g\ N_2$$

Since 39 g was originally present, $(39 - 34) = 5$ g remains unreacted.

Exercise

Identify the limiting reactant, calculate the amount of $(NH_4)_2SO_4$ formed, and determine the amount of excess reactant remaining after 55 g each of NH_3 and H_2SO_4 are allowed to react according to Equation (3-1b).

Answer: H_2SO_4 is limiting, 74 g $(NH_4)_2SO_4$ is produced, and 36 g NH_3 remains unreacted.

Any problem that specifies the amount of more than one reactant is a limiting reactant problem. Problems in which the amounts of more than two reactants are specified are solved in the same way. The SSP is solved for each reactant whose amount is given. The least of these possible yields is the actual yield, and the corresponding reactant is the limiting reactant.

3.5 SOLUTIONS: COMPOSITION AND STOICHIOMETRY

Reactions taking place in solution are common (two are used in Example/Exercise 3.6), and the methods of stoichiometry are easily applied to them. In order to do so, however, it is necessary to know how much reactant is present in a given amount of solution, that is, to know the *composition* of the solution.

The term **"solvent"** is used to describe whichever component of a solution has the same state as the solution itself. In a sugar/water solution, water is the solvent because it is a liquid and the solution is a liquid. If more than one pure component has the same state as the solution, the solvent is the one present in the largest amount. For example, in a solution of 5 mL water and 50 mL alcohol, the solvent is alcohol, while in a solution of 50 mL water and 5 mL alcohol, the solvent is water. Any component of the solution other than the solvent is a **solute.** "Solvent" and "solute" are terms of convenience, to be used only when helpful. For example, it is not particularly useful to describe either component of a 50:50 mixture of alcohol and water as the solvent, so it is not generally done.

Solutions in water are called aqueous solutions.

A solution can be solid, liquid, or gaseous. In Chapter 1 we saw that some alloys are solutions of one metal in another. Air is a gas-phase solution containing nitrogen, oxygen, and small quantities of other gases. However, most of the solutions that a chemist deals with are liquids, and the most common solvent in the chemical laboratory is water.

Composition of Solutions

The relative amounts of solute and solvent are collectively known as the *composition* of the solution. Solutions containing a relatively small quantity of solute are called **dilute,** and those containing a relatively large amount of solute are called **concentrated.** Several different ways to describe composition are in general use, and chemistry students should be familiar with all of them.

Percent Composition

The use of *percent* to describe composition of solutions is common in chemistry, health careers, industry, and everyday life. The household bleach Clorox is labeled as

"5.25% sodium hypochlorite," for example. Some care must be taken in the interpretation of such numbers, however. Although a percent is a dimensionless ratio meaning "per hundred," the numerator and denominator of the ratio do have dimensions. For instance, 5.25% NaClO might mean 5.25 *grams* NaClO/100 *grams* solution, 5.25 *mL* NaClO/100 *mL* solution, or 5.25 *moles* NaClO per 100 *moles* solution. These are the most common dimensions, although other possibilities exist. All three of these solutions are 5.25% NaClO; but in fact each has a different composition. Therefore it is necessary, when using percent composition to describe a solution, to specify the units of measure. This is done as follows.

The **mass percent [%(m/m)]** of a solution, also called "percent by mass," is defined by Equation (3-4). Since weight is proportional to mass, any ratio of weights is numerically equal to the ratio of the corresponding masses. Percent is a ratio, and the "weight percent" or "percent by weight" of any solution is equal to the mass percent.

$$\%(\text{m/m}) = \frac{(\text{mass of solute})}{(\text{mass of solution})} \times 100 \qquad (3\text{-}4)$$

Mass percent is the most commonly used form of percent composition. If percentage units are not specified, it is *assumed* that %(m/m) is meant. For example, if 155 g of a solution contains 26.3 g solute, it is a (26.3/155)100 = 17.0% solution.

Volume [%(v/v)] percent, which is also called "percent by volume," is defined by Equation (3-5).

$$\%(\text{v/v}) = \frac{(\text{volume of solute})}{(\text{volume of solution})} \times 100 \qquad (3\text{-}5)$$

Volume percent is used when both solute and solvent are liquids, but not when the solute is a solid. For example, a 10%(v/v) solution of alcohol in water consists of 10 mL alcohol plus enough water to bring the total volume to 100 mL. Volume percent is often used (particularly in medical laboratories) because liquid volumes are easily measured. Since volume percent rarely appears in stoichiometry problems, however, we will not discuss it further in this chapter.

EXAMPLE 3.9 Obtaining a Specified Amount of Solute

In order to increase their effectiveness in decomposing grease, liquid drain cleaners for home use normally contain sodium hydroxide as a solute. If the composition of such a solution is 7.50%(m/m), how much solution would you have to weigh out in order to have a sample containing 10.0 g NaOH?

Analysis

Target: The mass of solution, in grams, sufficient to contain 10 g solute.

Knowns: Explicit knowns are 7.50%, the type of percentage (m/m), and the desired mass of solute.

Relationship: Equation (3-4) connects the known information with the target.

Plan Rearrange Equation (3-4) to isolate the target on the left. Then insert the known information and solve.

Work

$$\%(\text{m/m}) = \frac{(\text{mass of solute})}{(\text{mass of solution})} \times 100$$

$$(\text{mass of solution}) = \frac{(\text{mass of solute})}{\%(\text{m/m})} \times 100$$

$$= (10.0\ \text{g}/7.50) \times 100 = 133\ \text{g}$$

Alternatively, we could have solved this problem by the factor-label method. The equivalence statement appropriate for a 7.50%(m/m) solution is

$$7.50\ \text{g solute} \Longleftrightarrow 100\ \text{g solution}$$

in which the quantity 100 g is *exact.*

$$?\ \text{g solution} = (10.0\ \text{g solute}) \cdot \left(\frac{100\ \text{g solution}}{7.50\ \text{g solute}}\right)$$

$$= 133\ \text{g}$$

Check An approximate mental calculation might go as follows: "100 grams of a 7.50% solution contains 7 to 8 g solute, so if I want 10 g, which is a little more than 7 to 8, I need a little more than 100 g solution. The answer, 133, is a little more than 100, so it is reasonable."

Note: Since neither the identity of the solvent nor that of the solute was used in reaching the answer, both could have been omitted from the problem.

Exercise

How much of a 0.35%(m/m) aqueous solution of $LiNO_3$ should be weighed out in order to obtain an amount containing 5.00 g lithium nitrate?

Answer: 1.4 kg.

EXAMPLE 3.10 Preparing a Solution of Specified Composition

Lithium carbonate is a helpful medication for moderating certain types of mental illness, chiefly manic depression. Give laboratory instructions for the preparation of 750 grams of a 0.900% aqueous solution of Li_2CO_3.

Analysis

Target: Two quantities are required, the amounts of Li_2CO_3 and H_2O.

Knowns: Total mass of the solution and the percent composition. We assume that %(m/m) is meant, so that the units of the targets are grams.

Relationship: Equation (3-4) connects the masses of solute and solution, but does not contain the mass of solvent. However, once the masses of solute and solution are known, the mass of solvent can be found by difference.

Plan

2. Find the mass of solvent.
1. Rearrange Equation (3-4), insert the known quantities, and solve for the mass of solute.

Work

1.
$$\%(\text{m/m}) = \frac{(\text{mass of solute})}{(\text{mass of solution})} \times 100$$

$$(\text{mass of solute}) = \frac{\%(\text{m/m}) \cdot (\text{mass of solution})}{100}$$

$$= \frac{(0.900) \cdot (750 \text{ g})}{100} = 6.75 \text{ g}$$

2.
$$\text{mass of solution} = \text{mass of solvent} + \text{mass of solute}$$

$$750 \text{ g} = \text{mass of solvent} + 6.75 \text{ g}$$

$$\text{mass of solvent} = 750 \text{ g} - 6.75 \text{ g} = 743 \text{ g}$$

The laboratory instructions are, therefore, "Dissolve 6.75 g Li_2CO_3 in 743 g H_2O."

Check 1% of 750 is 7.5, so we need a little less than 7.5 g solute. All but ≈ 1% of the solution is solvent, so we need 750 − (1% of 750) ≈ 750 − 8 g water.

Exercise

How would you prepare two kilograms of a 1.12% aqueous solution of potassium permanganate?

Answer: Dissolve 22.4 g $KMnO_4$ in 1.98×10^3 g H_2O.

Molarity

The most common unit for describing the amount of solute in a solution is the molarity. The **molarity** of a solution is defined as the number of moles of solute per liter of solution.

$$\text{molarity} = \frac{(\text{moles of } \textit{solute})}{(\text{liters of } \textit{solution})} \tag{3-6}$$

If a solution contains 0.25 moles of solute in a volume of 1 liter, it is called a 0.25 **molar** solution. The symbol for molarity is **M**, and the units of M are **mol L^{-1}**. The molarity of a solution is a conversion factor linking moles of solute with solution volume. The equivalence statement, for a 0.25 M solution, is 0.25 mol solute ⇔1 L solution. The number of moles of solute in a stated volume of solution, a frequently required quantity, is found from this relationship. For example, the number of moles of solute in 1.8 L of a 0.25 M solution is

$$? \text{ moles solute} = (1.8 \text{ L}) \cdot \frac{(0.25 \text{ mol solute})}{(1 \text{ L solution})} = 0.45 \text{ mol}$$

The relationship among the number of moles of solute (n), the volume (V) in liters, and the molarity (M) in moles per liter, of a solution, is conveniently memorized in the form of Equation (3-7).

$$n = \text{M} \cdot V \tag{3-7}$$

Sometimes the word "concentration" is used as a synonym for molarity. However, concentration means only "amount of solute in a solution," and can refer to percent as well as molarity. In order to avoid ambiguity, it is best to use the word "molarity" when referring to moles per liter.

EXAMPLE 3.11 Using a Specified Amount of Solute

A laboratory procedure calls for the addition of 0.025 moles of KOH to a reaction mixture, and you have available a 0.100 M KOH solution. How much (what volume) should you add?

Analysis

Target: A volume, in liters:

$$? \text{ L} =$$

Knowns: The desired quantity of solute, 0.025 moles, and the molarity, 0.100 M, of the solution.

Relationship: Equation (3-7) connects the knowns with the target.

Plan Rearrange Equation (3-7) to isolate the target on the left, insert the known values, and solve.

Work The equation $n = \text{M}V$ is rearranged to

$$V = \frac{n}{\text{M}}$$

$$V = \frac{(0.025 \text{ mol})}{(0.100 \text{ M})}$$

To see that the units cancel correctly, M should be written out as mol L^{-1}.

$$V = \frac{(0.025 \text{ mol})}{(0.100 \text{ mol L}^{-1})} = 0.25 \text{ L}$$

Note: If you prefer, this problem may be solved as a unit conversion using the equivalence statement 0.1 mol ⇔ 1 L.

$$? \text{ L} = (0.025 \text{ mol}) \cdot \frac{(1 \text{ L})}{(0.100 \text{ mol})} = 0.25 \text{ L}$$

Exercise

What volume of a 0.335 M RbI solution contains exactly 1 mole rubidium iodide?

Answer: 2.99 L.

Figure 3.7 Volumetric flasks are used in the preparation of solutions of accurately known molarity. These flasks are available in many sizes.

To prepare solutions of a specified molarity you need to know the mass of the solute, which is calculated from its molar mass and the desired molarity. You also need to know the required volume of solution, which, except for very dilute solutions, is *not* the same as the volume of solvent. Proceed as follows: (a) weigh out the correct amount of solute, (b) dissolve it in an amount of solvent that is *less than* the desired volume of solution, and (c) carefully add solvent until the desired volume is reached. A variety of laboratory glassware, calibrated in volume units, is available for such procedures (Figure 3.7).

The volume of a solution is *not* equal to the sum of the volumes of the solvent and the solute. For example, the volume of 0.800 mol K_2CO_3 is 48 mL. When this amount is dissolved in 1000 mL water the resulting solution has a volume of 1018 mL. Its molarity is (0.800 mol/1.018 L) = 0.786 M.

EXAMPLE 3.12 Preparation of a Solution of Specified Molarity

Give instructions for the preparation of 500 mL of a 0.650 M solution of sodium carbonate.

Sodium carbonate is a household chemical called "washing soda." Years ago it was added to the tub along with soap when the clothes were washed. Nowadays it is added to detergents by the manufacturer.

Analysis

Target: ? g Na_2CO_3 =

Knowns: The identity of the solute, and the desired molarity and volume. The molar mass of sodium carbonate is implicit.

Relationship: If we knew the number of *moles* of solute in 500 mL of solution, we could calculate the mass in grams. Equations (3-6) and (3-7) both connect the number of moles of solute with the knowns, and either one can be used.

Plan The two-step chain leading backward from the target is as follows:

2. Calculate the number of grams from the number of moles and the molar mass.
1. Calculate the number of moles using Equation (3-7).

Work

1.

$$n = MV$$

$$? \text{ mol} = (0.650 \text{ mol L}^{-1})(500 \text{ mL}) = 325$$

Check The units of this answer are *mL mol L*$^{-1}$ rather than mol; apparently an error has been made. Equation (3-7) requires volume units of *liters*, not milliliters, and the mistake was to omit the volume unit conversion factor. This is an instance of an error that cannot be detected unless the units are written down and some attention is paid to them. The error is set right with the conversion factor connecting L and mL.

$$? \text{ mol} = (0.650 \text{ mol L}^{-1}) \cdot (500 \text{ mL}) \cdot \frac{(0.001 \text{ L})}{(1 \text{ mL})} = 0.325 \text{ mol}$$

2. The molar mass of Na_2CO_3 is 106.0 g mol^{-1}, so that the required mass of sodium carbonate is

$$? \text{ g} = (0.325 \text{ mol}) \cdot \frac{(106.0 \text{ g})}{(1 \text{ mol})} = 34.5 \text{ g}$$

The laboratory instructions are "Dissolve 34.5 g Na_2CO_3 in less than 500 mL solvent, say about 400 to 450 mL. Then add more solvent until the solution volume is 500 mL."

Note: The problem does not give the identity of the solvent, nor does the calculation depend on it. When the solvent is not specified it is usually assumed that water is intended, particularly if the solute is an ionic compound.

Exercise

Give a procedure for preparing 1.5 L of 0.0100 M potassium chloride.

Answer: Dissolve 1.1 g KCl in some water, then add more water until the solution volume is 1.5 L.

Among those with a family history of high blood pressure, there is a strong correlation between excessive sodium intake and the development of hypertension. Potassium chloride has a flavor similar to that of sodium chloride, and is widely used as a salt substitute in cooking.

Dilutions

An alternate method for the preparation of a solution of desired molarity is by **dilution,** that is, by the addition of solvent to a more concentrated **stock solution.** Calculation of the molarity resulting from dilution depends on the fact that, when pure solvent is added to a solution, the amount of *solute* remains the same even though the volume increases (Figure 3.8).

EXAMPLE 3.13 Dilution of a Stock Solution

How would you prepare 500 mL of 1.25 M $NaNO_2$ from a stock solution whose molarity is 6.00 M?

Sodium nitrite, added to meat products such as cold cuts and hot dogs, prevents the growth of the deadly *botulinus* microorganism and produces an attractive pink color. However, some evidence links the nitrite ion with the formation of carcinogenic compounds, and its use in foods is under study.

Analysis

Target: The volume of stock solution required (mL or L).

Knowns: The molarity of the stock solution, and the volume and molarity of the desired dilute solution.

Relationship: We have not seen a formula that may be used in a "lunge" at the problem. However, we do know that the number of moles of solute is the same before and after the addition of more solvent.

Missing Information: If we knew the amount of solute in the diluted solution, we also would know the amount of solute that must be provided by the stock solution.

Plan

1. Determine the amount of solute in the desired volume of dilute solution, using Equation (3-7).
2. Find the volume of stock solution that contains the desired amount of solute, using Equation (3-7).

Work

1. The amount of solute in the desired dilute solution is

$$n = MV$$

$$? \text{ mol} = (1.25 \text{ mol L}^{-1})(0.500 \text{ L}) = 0.625 \text{ mol}$$

2. Solve Equation (3-7) for the volume of stock solution that contains 0.625 moles solute.

$$V = \frac{n}{M}$$

$$? \text{ L} = \frac{(0.625 \text{ mol})}{(6.00 \text{ mol L}^{-1})} = 0.104 \text{ L} = 104 \text{ mL}$$

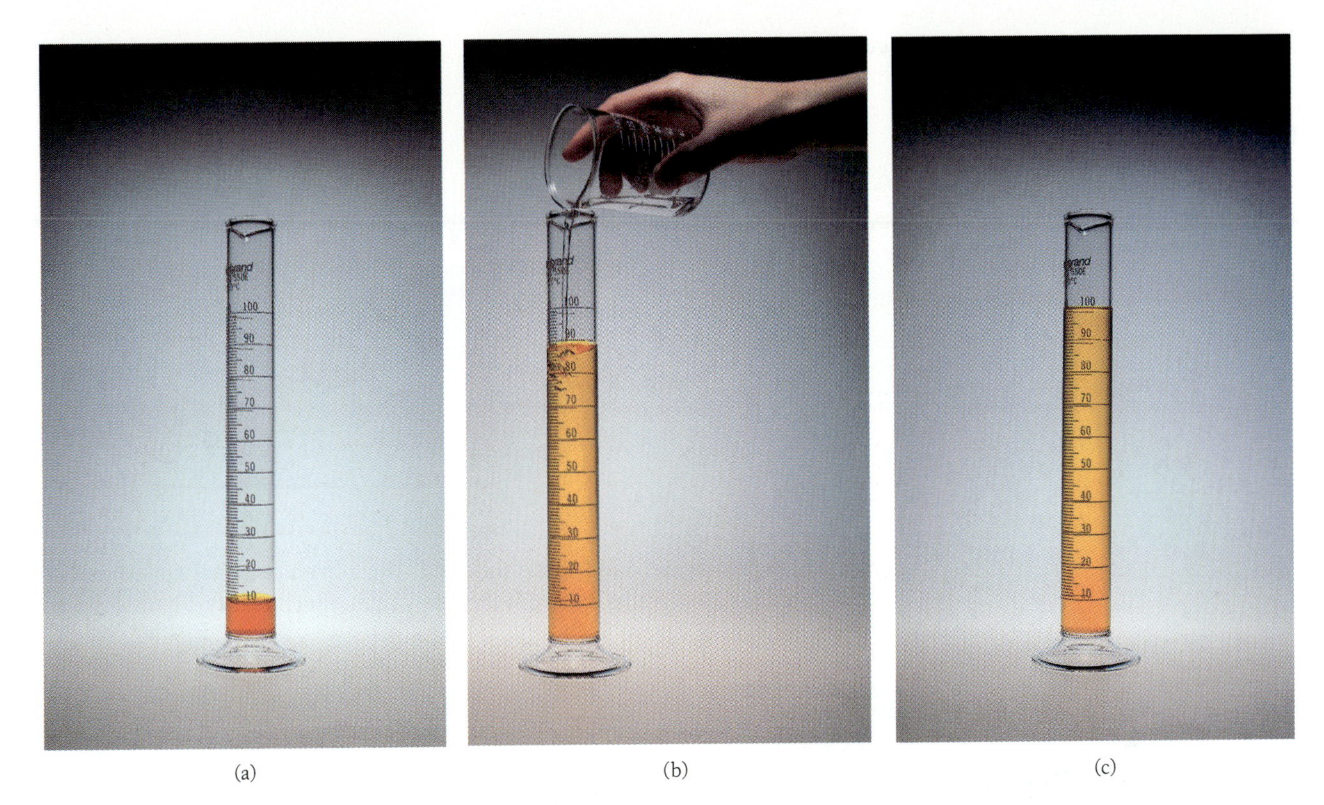

Figure 3.8 (a) 10 mL of 0.10 M $K_2Cr_2O_7$ in a graduated cylinder. (b) When 90 mL H_2O is added, the amount of solute (0.0010 mol or 0.29 g) present in the container stays the same. (c) Since the solution volume has increased, however, the molarity has decreased from 0.10 M to 0.010 M.

The dilute solution is prepared by adding water to 104 mL of stock solution until the volume reaches 500 mL.

A one-step procedure is derived as follows:

$$n_{\text{stock}} = n_{\text{dilute}}$$

$$\text{M}_{\text{stock}} \cdot V_{\text{stock}} = \text{M}_{\text{dilute}} \cdot V_{\text{dilute}}$$

$$V_{\text{stock}} = V_{\text{dilute}} \cdot \frac{\text{M}_{\text{dilute}}}{\text{M}_{\text{stock}}}$$

This form is convenient when many such calculations must be made.

Exercise

How would you prepare 1 liter of 0.100 M aqueous KOH from a 3.00 M stock solution?

Answer: To 33.3 mL of the stock solution, add sufficient water to bring the volume to 1 L.

Stoichiometry of Reactions in Solution

Stoichiometric calculations are in principle the same whether the reaction takes place between pure substances or between dissolved substances. When a reaction takes place in solution, the *mole ratio* is still the heart of the calculation, and the unit

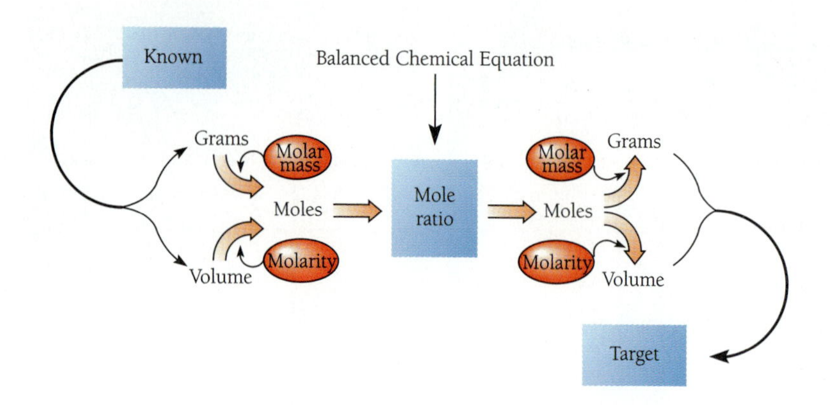

Figure 3.9 The standard stoichiometry problem has different entry and exit routes, depending on whether amounts are specified as grams of a pure substance or as volume and molarity of a solution.

conversion approach of the SSP is still followed. The difference is that mole amounts are calculated not from grams, but from solution volume and molarity. Figure 3.9 is an expanded version of Figure 3.3.

EXAMPLE 3.14 Stoichiometry of Solution Reactions

Sodium hydrogen carbonate (sodium bicarbonate) is also known as "baking soda." It is a common remedy for heartburn or acid indigestion because it reacts with and consumes HCl. HCl is the predominant acidic component of stomach fluid.

Sodium hydrogen carbonate reacts with hydrochloric acid according to the equation $NaHCO_3(aq) + HCl(aq) \rightarrow NaCl(aq) + H_2O(\ell) + CO_2(g)$. What volume of 0.752 M $NaHCO_3$ is required to react with 127 mL of 0.143 M HCl?

Analysis

Target: ? L $NaHCO_3$ =

Knowns: The mole ratio $NaHCO_3$:HCl is implicit in the chemical equation, which is given. The amount of HCl(moles) is implicit in the given volume and molarity.

Plan Use the three-stage conversion of the SSP.

HCl volume, molarity → moles HCl → moles $NaHCO_3$ → liters $NaHCO_3$

Work Make sure the chemical equation is balanced.

$$\begin{array}{ccccc} & \textbf{liters} & \rightarrow \textbf{moles HCl} & \rightarrow \textbf{moles NaHCO}_3 & \rightarrow \textbf{liters NaHCO}_3 \\ ?\text{ L NaHCO}_3 = & (0.127\text{ L}) \cdot & \dfrac{(0.143\text{ mol HCl})}{(1\text{ L})} \cdot & \dfrac{(1\text{ mol NaHCO}_3)}{(1\text{ mol HCl})} \cdot & \dfrac{(1\text{ L})}{(0.752\text{ mol NaHCO}_3)} \end{array}$$

$$= 0.0242\text{ L} = 24.2\text{ mL}$$

Exercise

What volume 0.0863 M $NaHCO_3$ is required to react with 35.3 mL of 0.109 M H_2SO_4? (The products of the reaction are H_2O, CO_2, and Na_2SO_4.)

Answer: 89.2 mL.

In other problems of stoichiometry, one species is a pure substance while the other is a solute. The SSP is used here also.

EXAMPLE 3.15 Stoichiometry when Both Solutions and Pure Substances Are Involved

The reaction of concentrated sulfuric acid with solid potassium perchlorate produces perchloric acid and potassium sulfate. What volume of 9.50 M $H_2SO_4(aq)$ is required to react with 100 g $KClO_4(s)$?

In the commercial production of perchloric acid, $HClO_4$ is removed from the reaction vessel in the gas phase. Perchloric acid is dangerously explosive (probably owing to trace quantities of the impurity Cl_2O_7), but it can be safely stored and handled in aqueous solution.

Analysis

Target: ? L H_2SO_4 =

Knowns: The mass of $KClO_4$ and the molarity of H_2SO_4, the identities of the products, and the molar masses of all species are implicit.

Missing Information: The balanced chemical equation is not given.

Plan

1. Write and balance the chemical equation.
2. Solve the SSP to find the volume of H_2SO_4.

Work

1. All participating species are given, so the complete but unbalanced equation can be written from the problem statement.

$$KClO_4(s) + H_2SO_4(aq) \longrightarrow HClO_4(g) + K_2SO_4(aq)$$

Cl, S, and O are already in balance; we may begin with either K or H. The coefficient 2 in front of $KClO_4$ brings K into balance.

$$\mathbf{2}\ KClO_4(s) + H_2SO_4(aq) \longrightarrow HClO_4(g) + K_2SO_4(aq)$$

The Cl balance is restored with another coefficient 2, in front of $HClO_4$.

$$2\ KClO_4(s) + H_2SO_4(aq) \longrightarrow \mathbf{2}\ HClO_4(g) + K_2SO_4(aq)$$

On checking the other elements (H, S, and O), it is found that all are in balance.

2. The molar mass of $KClO_4$ is 138.5 g mol^{-1}, and the SSP is

$$\begin{array}{cccc} & \rightarrow \textbf{moles } \mathbf{KClO_4} & \rightarrow \textbf{moles } \mathbf{H_2SO_4} & \rightarrow \textbf{liters } \mathbf{H_2SO_4} \\ ?\ \text{L } H_2SO_4 = (100\text{ g } KClO_4) \cdot & \dfrac{(1\text{ mol } KClO_4)}{(138.5\text{ g } KClO_4)} \cdot & \dfrac{(1\text{ mol } H_2SO_4)}{(2\text{ mol } KClO_4)} \cdot & \dfrac{(1\text{ L})}{(9.50\text{ mol } H_2SO_4)} \end{array}$$

$$= 0.0380\text{ L} = 38.0\text{ mL}$$

Check The units cancel correctly. $100/140$ is a little more than $2/3$ mol $KClO_4$, and since the mole ratio is $1/2$, about $1/3$ mol acid is required. 0.1 L (100 mL) of $\approx$ 10 M acid contains one mole, so the answer should be close to $100/3 \approx 35$ mL.

Exercise

It was found that 78.3 mL of a sulfuric acid solution was required to react completely with 100 g potassium perchlorate. What was the molarity of the acid?

Answer: 4.61 M.

Other stoichiometry problems are occasionally seen in which the amount in moles must be obtained from the mass in pounds or tons rather than grams. Also, the mass of a pure liquid might have to be determined from its volume and density. Similarly, the molar amount of a gas can be obtained from its pressure, temperature, and volume (see Chapter 4). All of these, however, are simply variations of the same problem, the SSP.

3.6 THE ALKALI METALS

At sufficiently high pressure hydrogen may exist as a metal. So far it has not been possible to attain high enough pressure in the laboratory to observe this allotrope. The core of the planet Jupiter may consist of metallic hydrogen.

The elements below hydrogen in the left-most column of the periodic table, Group 1A, are known as the **alkali elements or alkali metals.** Hydrogen is excluded from the classification not only because it is not a metal, but also because its chemical properties are quite different. Sodium and potassium are the most abundant of the alkali metals, and their compounds together make up about 5% of the mass of the earth's crust. Francium is radioactive, and even its longest-lived isotope disintegrates in only a few hours.

Properties of the Alkali Metals

A diatomic molecule or ion is one that consists of two atoms.

The alkalis are not as dense as other metals (Figure 3.10). Li, Na, and K are even less dense than water. The alkalis are also unusual among metals for their softness, and they are easily cut by a steel knife blade (Figure 3.11). They melt at relatively low temperatures, and also have relatively low boiling points. In the gas phase they are primarily monatomic, but about 1% of the vapor consists of the *diatomic* molecules Li_2, Na_2, and so on. They do not form molecules in the solid or liquid phase (the structure of metals is discussed in Section 9.4).

Table 3.1 lists some physical properties and also illustrates the periodic trends characteristic of the alkali metals. Melting point and boiling point decrease as atomic number increases, while the size of the atoms increases. The density increases and the hardness decreases, but in both of these properties sodium and potassium are out of order. Minor irregularities in periodic trends are typical of most groups in the periodic table.

Figure 3.10 The density of lithium is low enough so that it floats on oil, which in turn floats on water. Most other metals, like the copper pellets shown here, are dense enough to sink to the bottom of the water layer.

Figure 3.11 Sodium can easily be cut with a knife. The freshly cut surface has a silvery appearance, but turns gray almost immediately when exposed to air. The tarnish is due to the formation of sodium peroxide, sodium hydroxide, and sodium carbonate, by reaction with the oxygen, water, and carbon dioxide of the atmosphere.

TABLE 3.1 Physical Properties of the Alkali Metals

Element	Symbol	Atomic Number	Molar Mass	Atomic Radius/pm	Melting Point/°C	Boiling Point/°C	Density /g mL^{-1}	Hardness*
Lithium	Li	3	6.941	152	186	1342	0.534	0.6
Sodium	Na	11	22.99	186	97.8	883	0.971	0.4
Potassium	K	19	39.10	227	63.3	760	0.862	0.5
Rubidium	Rb	37	85.47	248	38.9	686	1.53	0.3
Cesium	Cs	55	132.9	265	28.4	669	1.87	0.2
Francium†	Fr	87	223					

* Hardness is given on the *Mohs Scale,* developed to compare the hardnesses of minerals. The scale is set by assigning a hardness of 1 to the very soft mineral talc ($3MgO \cdot 4SiO_2 \cdot H_2O$, also called soapstone; talcum powder is finely ground talc) and a hardness of 10 to the hardest known mineral, diamond. For comparison, the Mohs hardness of lead is 1.5, that of brass is 3 to 4, and that of steel ranges from 5 to 8.

† It is difficult and dangerous to assemble and work with large quantities of radioactive substances. Consequently the properties of francium are not well known. The number listed as molar mass is the mass number of the most stable isotope.

The chemistry of the alkali metals is dominated by two properties that are shared by all members of the group.

1. As elements, the metals are highly reactive. Alkali metals react with essentially all nonmetallic elements. Their reactivity increases with atomic number so that, for instance, lithium reacts slowly with water while the reaction between cesium and water is explosive.
2. Alkali metals form only ionic compounds in which the charge of the metal cation (M^+) is +1.

It is often convenient to use the symbol "M" to mean *any* metallic element. In the context of this section of the text, M refers to any or all of the alkali metals.

Reactions of the Alkali Metals

Some reactions of the alkali metals are as follows:

Hydrogen: The metals react with H_2 to form ionic compounds called metal **hydrides.**

$$2\ M(s) + H_2(g) \longrightarrow 2\ MH(s)$$

The hydrides themselves react vigorously with water, forming the hydroxide and liberating gaseous hydrogen.

$$MH(s) + H_2O(\ell) \longrightarrow MOH(aq) + H_2(g)$$

Halogens: The elements of Group 7A, known collectively as **halogens,** all react with alkali metals to form salts (Figure 3.12). The reaction of potassium with fluorine is typical.

$$2\ K(s) + F_2(g) \longrightarrow 2\ KF(s)$$

Oxygen: Several different types of oxygen compounds are known for the alkali metals. The oxides, M_2O, can be prepared from the nitrates according to the reaction in Exercise 3.3. Alkali oxides react with water to produce hydroxides.

$$M_2O(s) + H_2O(\ell) \longrightarrow 2\ MOH(aq)$$

Direct reaction with oxygen produces the oxide only in the case of lithium.

$$4\ Li(s) + O_2(g) \longrightarrow 2\ Li_2O(s)$$

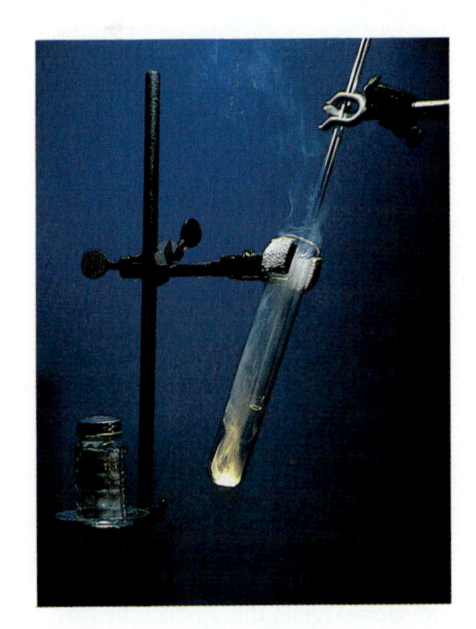

Figure 3.12 The reaction of sodium (an alkali metal) with chlorine (a halogen) is fast and vigorous. Sodium burns readily in a chlorine atmosphere, producing a white smoke of tiny sodium chloride particles.

When sodium reacts with oxygen, the principal product is the *peroxide*.

$$2\ Na(s) + O_2(g) \longrightarrow Na_2O_2(s)$$

Peroxides react with water to produce hydrogen peroxide, a useful bleaching agent.

$$Na_2O_2(s) + 2\ H_2O(\ell) \longrightarrow H_2O_2(aq) + 2\ NaOH(aq)$$

The principal product arising from the reaction of K, Rb, or Cs with oxygen is the *superoxide,* MO_2. Potassium superoxide liberates oxygen on contact with water (see Example 3.3).

Water: All alkali metals react with water. Li does so slowly, whereas Na does so vigorously (Figure 3.13). The heat generated by the reaction with sodium is usually enough to melt the metal and ignite the hydrogen. The reactions with potassium, rubidium, and cesium are even more violent, and always result in a hydrogen explosion. This increase in reactivity is an excellent example of the way in which chemical properties vary regularly within a group.

$$2\ M(s) + 2\ H_2O(\ell) \longrightarrow 2\ MOH(aq) + H_2(g)$$

Ammonia: Sodium reacts with ammonia to produce the ionic compound sodium amide (trivial name: sodamide).

$$2\ Na(s) + 2\ NH_3(\ell) \longrightarrow 2\ NaNH_2(s) + H_2(g)$$

The reaction is complex, and one of the intermediate species formed is a "free" electron, that is, an electron that is not incorporated into the electron cloud of any ion or molecule. Apparently, free electrons fit nicely into the cavities between the molecules of liquid ammonia. Dissolved electrons impart a deep blue color to the solutions, which can be prepared from any of the alkali metals.

Substances that in some way participate in, but are not themselves changed by, a chemical reaction are often written above or below the arrow.

$$M(s) \xrightarrow{NH_3(\ell)} M^+(\text{dissolved}) + e^-(\text{dissolved})$$

Salts of the Alkali Metals

Salts of the alkali metals are, in general, highly soluble in water. Notable exceptions are LiF and Li_2CO_3, which are only slightly soluble. Alkali salts all have relatively high melting points.

The behavior of lithium is not unique. The first member of a group often departs from characteristic group behavior to a greater extent than the heavier elements. Another example is boron, which is a metalloid. All the other members of Group 3A are metals.

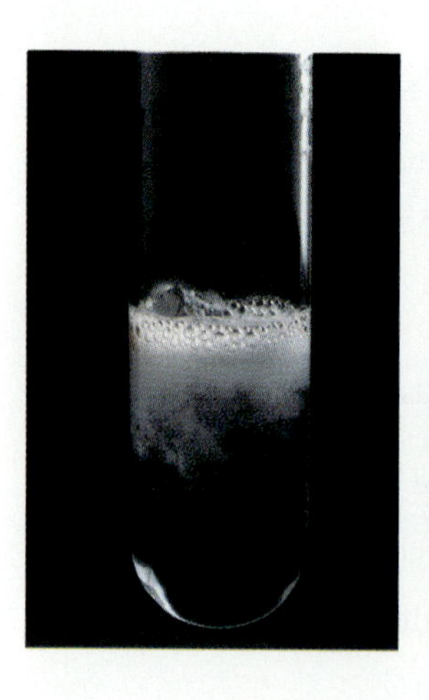
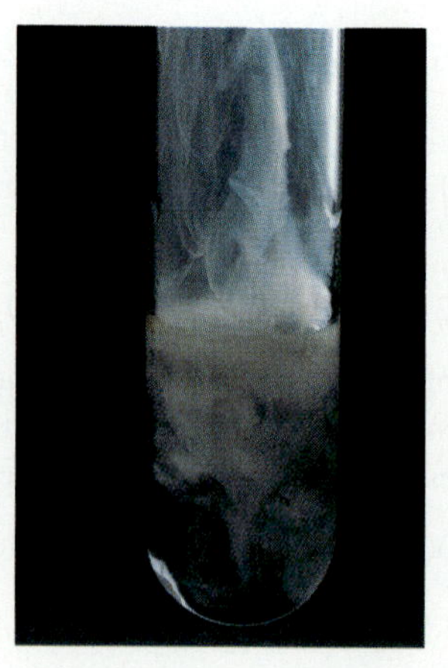

Figure 3.13 (a) Lithium metal reacts on contact with water. (b) The reaction of sodium with water is more vigorous, while (c) potassium reacts with enough violence to ignite the product, hydrogen. (See the first page of this chapter for a more colorful view of this reaction.) All three metals are less dense than water, and can be seen floating at the surface.

Figure 3.14 Alkali salts containing colored anions such as permanganate and dichromate take on the color of the anion. If the anion is colorless, as most are, the alkali salt is also colorless. These salts are potassium permanganate (dark purple), potassium dichromate (orange), and potassium nitrate (white).

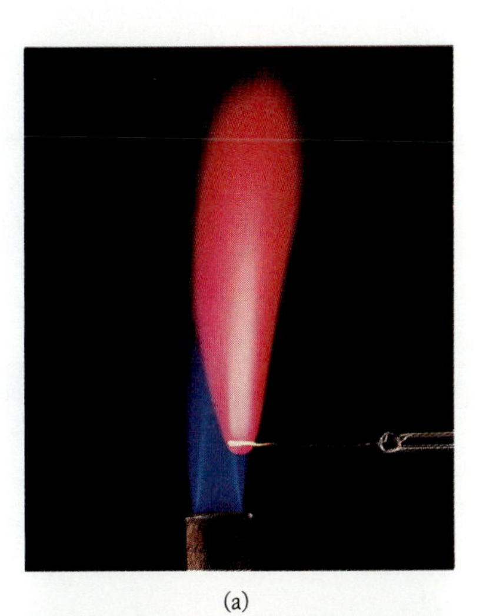

(a)

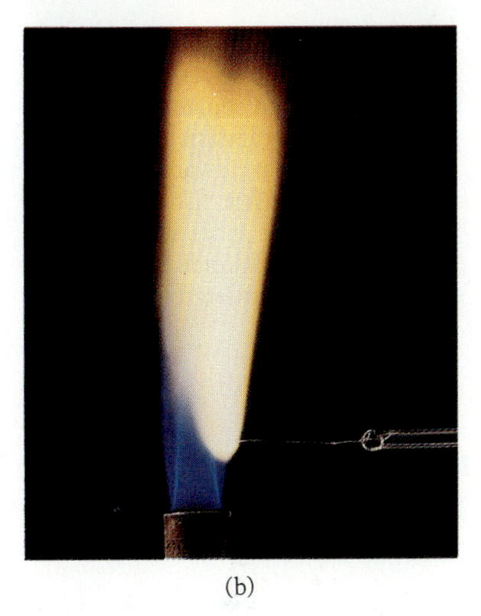

(b)

Figure 3.15 The flame test is used as a quick means of demonstrating the presence of an alkali metal in a compound. Each Group 1A metal emits a characteristic color when any one of its compounds is strongly heated. (a) Lithium. (b) Sodium.

The cations of Group 1A are colorless. Salts containing a colored anion such as permanganate(MnO_4^-) or dichromate($Cr_2O_7^{2-}$), are colored (Figure 3.14); otherwise, the salts of alkali metals are colorless.

Group 1A elements and their compounds emit colored light when strongly heated. Since the color of the emitted light is different for each element, this property may be used to identify which alkali element is present in a compound (Figure 3.15). Sodium vapor lights, used to illuminate some urban highways, give off a characteristic yellow glare produced by the passage of an electric current through high temperature, gaseous sodium.

Salts containing the *ammonium* ion, NH_4^+, are very similar to alkali salts in solubility and color. Because of this, NH_4^+ is sometimes called a "pseudoalkali."

Sources and Production of the Metals and Some Compounds

Metallic sodium is produced by passing an electric current through molten sodium chloride. This process, called *electrolysis,* is discussed in Chapter 18. Chlorine, the other product of the reaction, is also an important industrial compound. The raw material, sodium chloride, is mined in relatively pure form from deposits that are the residues of ancient seas.

$$2\ NaCl(\ell) \xrightarrow{\text{electrolysis}} 2\ Na(\ell) + Cl_2(g)$$

Similarly, lithium is produced by electrolysis of $LiCl(\ell)$. The raw material is obtained from *brine* (a concentrated salt solution), primarily from Searles Lake in California. Several lithium-containing minerals such as *spodumene,* $Li_2O \cdot Al_2O_3 \cdot 4SiO_2$ are also used in lithium production.

Potassium can also be produced by electrolysis, but the product is both purer and cheaper if made by passing molten sodium through molten potassium chloride. Mining of KCl deposits provides the raw material. Potassium is obtained as a vapor from this high temperature process.

The symbol "Δ" below a reaction arrow means that the reactants must be heated for the reaction to take place.

$$Na(\ell) + KCl(\ell) \xrightarrow[\Delta]{} NaCl(\ell) + K(g)$$

(a)

(b)

Figure 3.16 (a) The residues of ancient seas are good sources of alkali compounds. This ancient seabed in the Mojave desert near Edwards, California provides an extensive, flat, hard surface for landings of the space shuttle. (b) Most sodium chloride is mined from nearly pure underground deposits.

Cesium can be extracted from the mineral *pollucite,* $2Cs_2O \cdot 2Al_2O_3 \cdot 9SiO_2 \cdot 2H_2O$. Some cesium, however, and all rubidum, is obtained as a byproduct in the manufacture of Li. Lithium ores usually contain small amounts of Cs and Rb as impurities.

Alkali metal compounds are important industrially. Two of them, NaOH and Na_2CO_3, are among the top ten in production. About 20 billion pounds of each are produced annually in the USA. Sodium hydroxide is obtained from the electrolysis of aqueous sodium chloride.

$$2\ NaCl(aq) + 2\ H_2O(\ell) \xrightarrow{\text{electrolysis}} 2\ NaOH(aq) + Cl_2(g) + H_2(g)$$

Sodium carbonate (also called soda ash or washing soda, is obtained from large deposits of *trona,* $NaHCO_3 \cdot Na_2CO_3 \cdot 2H_2O$, in Wyoming. Heating of crushed trona drives off CO_2 and H_2O, leaving a residue of Na_2CO_3.

$$2\ NaHCO_3 \cdot Na_2CO_3 \cdot 2H_2O(s) \xrightarrow[\Delta]{} 3\ Na_2CO_3(s) + CO_2(g) + 5\ H_2O(g)$$

Some Uses of Alkali Metals and Compounds

Compounds of the alkali metals find numerous uses in our society. Only a few of these uses are given here.

Lithium: Li_2CO_3 is used in the treatment of mental illness. Li is a principal component of hydrogen bombs. Some special-purpose batteries use lithium (Chapter 18), and some lubricating greases contain lithium compounds.

Sodium: Metallic sodium is used as a heat transfer fluid in some industrial heat exchangers. NaOH, Na_2CO_3, and NaCl are extensively used as raw materials in the chemical industry, for example in the manufacture of glass, plastics, paper, and detergents.

Potassium: KNO_3 is used as fertilizer because both K^+ and NO_3^- are important nutrients for plants. KNO_3 is also used in gunpowder.

Rubidium and Cesium: Their extreme reactivity makes them useful as a "getter"—a small amount of metal placed inside a TV picture tube to scavenge the last traces of harmful oxygen from the not-quite-perfect vacuum. Cesium is also used as a light-sensitive component of some photoelectric cells (Section 6.2).

CHEMISTRY OLD AND NEW

THE CHEMICAL INDUSTRY

We sometimes think of chemical reactions as processes that are carried out inside of a scientific labortory. However, the production of chemicals is also performed by business people, resulting in over 80 billion dollars of sales and 600 billion pounds of materials per year in the United States alone. The chemical industry is one of the nation's largest and most diverse trades. It employs over one million people (accounting for 5.8% of all manufacturing jobs), and has a notable influence on the economy. While the United States had import–export trade deficits in other technological areas in 1991, there was a chemical trade surplus for that year of $18.8 billion.

Many aspects of modern life depend on some sector of the chemical business. Drugs, plastics, fuels, soaps and detergents, food additives, dyes, and synthetic fabrics are just some of our useful chemicals. Raw material for paints, fertilizers, adhesives, household cleaners, and floor coverings are also made in chemical plants. In addition to specialty items, huge amounts of organic and inorganic reagents—such as ethylene, benzene, sulfuric acid, ammonia, and sodium hydroxide—are produced each year and used in the manufacture of other chemicals. One of the most important processes carried out in industry is the separation of crude petroleum into gasoline, fuel oils, and lubricants. Production of commercially important polymers like polyvinyl chloride (PVC), polystyrene, and Nylon is another major operation.

In each 24-hour period this chemical manufacturing plant produces up to 7000 tons of ammonia, much of which is used as fertilizer.

Processes by which industrial chemicals are made are often very different from those used in laboratories. The typical preparation of an organic compound involves several steps: combining reactants, isolating the desired product from excess starting materials and byproducts, recycling the unused reactants, and preparing any waste for safe disposal. If a chemist in a lab wants to make a large amount of a compound, he or she does so by successive preparation of a number of smaller batches. But industrial plant utilize *continuous* processes. Reactants are constantly fed into a reaction vessel, where they are mixed. By the time the flow reaches the outlet of the vessel, the product has been formed. The mixture is then piped to a distillation tower or other apparatus for purification. Unused starting materials are separated and fed back into the reactor, while the product is collected and stored. Wastes are pumped to treatment centers to be prepared for disposal. Plants can operate 24 hours a day in this fashion without any idle or "down" time, an important factor when the high cost of equipment is considered. Since reactions carried out in this manner must be fast, *catalysts*—substances that speed up the rate of a chemical reaction without being consumed are often utilized. Industrial chemists and engineers are always working to improve the efficiency of their processes. Due to the large scale of manufacture, even a tiny percentage increase in product yield or decrease in waste production can save millions of dollars.

There is a great deal of controversy about safety hazards, threats to the environment, and the responsibility of the chemical industry in these matters. The Occupational Health and Safety Act of 1970 (OSHA) set stringent standards for working conditions at all plants. Many rules for the storage of chemicals, exposure limits to noise and harmful substances, and preparedness for emergencies have been established. These rules make chemical plants surprisingly safe places to work. While monetary concerns cause chemical businesses to move more slowly than the public would like in adopting new methods for treating and recycling waste, increasingly stringent laws limiting pollutants are passed each year. The ultimate goal of these regulations is the reduction of discharges of air and water contaminants to the point where we and our environment are not harmed in any way.

Discussion Questions

1. How would your life be affected if production of synthetic polymers in this country were halted?
2. Two of our prominent environmental problems are acid rain and water pollution. To what extent can the chemical industry be held responsible for these problems? If chemical industry does not bear primary responsibility, then who does?

SUMMARY EXERCISE

Sodium carbonate has been an important raw material in the chemical industry for many years. Prior to the discovery of high-grade ores, sodium carbonate was produced by the *Solvay process* from sodium chloride and limestone ($CaCO_3$). The other product of the process is calcium chloride.

a. Write a balanced overall equation for the Solvay process, which is carried out in the following five-step series.

1. Ammonia and carbon dioxide are bubbled through aqueous sodium chloride, resulting in the formation of solid sodium hydrogen carbonate and aqueous ammonium chloride. In this reaction, water is a reactant as well as the solvent.
2. Sodium hydrogen carbonate is decomposed by heating, yielding sodium carbonate, water, and carbon dioxide. The carbon dioxide is recycled, and used in step 1.
3. The remainder of the carbon dioxide needed in step 1 is obtained by heating limestone. The other product is calcium oxide.
4. The calcium oxide from step 3 is treated with water to produce solid calcium hydroxide.
5. The aqueous ammonium chloride produced in step 1 is treated with solid calcium hydroxide, yielding aqueous calcium chloride, water, and gaseous ammonia. The ammonia is recycled and used in step 1.

b. Write balanced chemical equations for each of the five steps of the Solvay process.

c. Name all compounds in the process that have an alkali or pseudoalkali cation.

d. How much $CaCO_3(s)$ is required to produce 416 metric tons (416×10^3 kg) of sodium carbonate? Express your answer in metric tons and pounds.

e. How much sodium carbonate can be produced by reacting 1.00 g sodium chloride with 1.00 g calcium carbonate? Which reactant is limiting?

f. If 822 g sodium carbonate is obtained from each kilogram of sodium chloride, what is the percent yield?

g. Suppose that 2.03 g CO_2 is obtained by heating 4.98 g limestone (step 3). What is the percentage of calcium carbonate in the limestone?

h. Suppose that the water is evaporated from a 5.75 g sample of the aqueous solution of ammonium chloride produced in step 1. If the residue (solid ammonium chloride) weighs 0.352 g, what is the concentration of ammonium chloride in the sample? Express the answer as percent by mass.

i. Calcium hydroxide reacts with hydrochloric acid to produce calcium chloride and water. What volume of 0.0985 M HCl is required to react with 0.500 L of 0.0100 M calcium hydroxide?

j. How would you prepare one liter of a 6.62 M stock solution of NaCl?

k. How would you prepare 650 mL of 0.100 M NaCl from this stock solution?

l. What mass of NaCl is dissolved in 1.5 L of 0.100 M NaCl?

Answers **(a)** 2 NaCl + $CaCO_3 \rightarrow Na_2CO_3 + CaCl_2$; **[b(1)]** NaCl(aq) + NH_3(g) + CO_2(g) + $H_2O(\ell) \rightarrow NaHCO_3$(s) + NH_4Cl(aq); **[b(2)]** 2 $NaHCO_3(s) \xrightarrow[\Delta]{} Na_2CO_3 + H_2O + CO_2$; **[b(3)]** $CaCO_3(s) \xrightarrow[\Delta]{}$ CaO + CO_2; **[b(4)]** CaO + $H_2O \rightarrow Ca(OH)_2$(s); **[b(5)]** 2 NH_4Cl(aq) + $Ca(OH)_2(s) \rightarrow CaCl_2$(aq) + 2 H_2O + 2 NH_3(g); **(c)** sodium chloride, sodium hydrogen carbonate, sodium carbonate, and ammonium chloride; **(d)** 393 tons, 8.66×10^5 lbs; **(e)** 0.907 g, NaCl; **(f)** 90.7%; **(g)** 92.7%; **(h)** 6.12%; **(i)** 102 mL; **(j)** dissolve 387 g NaCl in less than 1 L water, then add water until the solution volume is 1 L; **(k)** to 9.82 mL stock solution, add sufficient water to bring the volume to 650 mL; **(l)** 8.8 g.

SUMMARY AND KEYWORDS

A **chemical equation** lists formulas of *reactants* and **products** of a reaction, separated by an arrow, and optionally indicates the physical states and reaction conditions. In a **balanced equation,** the number of atoms of each element is the same on both sides. Equations are balanced by adjusting the **stoichiometric coefficients**—never by adding new species or changing the subscripts in a formula. Fractional coefficients may be used, but more commonly an equation is balanced with the smallest possible whole number coefficients. Stoichiometric coefficients give the **mole ratios** of participating species. Any two stoichiometric coefficients may be written as an equivalence statement relating the molar amounts of the two substances.

The **standard stoichiometry problem** requires a three-stage unit conversion: mass of known → moles of known → moles of target → mass of target. Known and target may be two reactants, two products, or one of each. When the target is a product, the result is called the **theoretical yield,** the amount resulting from complete conversion of the starting material. The **actual yield** is often less than the theoretical yield due to losses during product isolation, incomplete reaction, or **side reactions** leading to **side products.** The **percent yield** is 100 times the ratio of actual to theoretical yield. A **reactant** is a single chemical species, while a **reagent** is a mixture or solution containing the reactant. Determination of the composition or purity of reagents by stoichiometric techniques is an aspect to **quantitative analysis.**

When reactants are not present in **stoichiometric amounts,** one is the **limiting reactant** and the other(s) is(are) the **reactant(s) in excess.** The amount of product is determined by the amount of the limiting reactant and does not depend on amount(s) of the reactant(s) in excess.

The solution component which, when pure, is in the same physical state as the solution, is called the **solvent;** other components are **solutes.** The solvent in **aqueous solutions** is water. The **concentration** (amount of solute) of a solution can be described in several ways, including **mass (or weight) percent, %(m/m)** or **%(w/w); volume percent, %(v/v);** and **molarity (M),** moles solute per liter solution. Dilute solutions may be prepared as needed by the **dilution** of more concentrated solutions, called **stock solutions.** Calculations in dilution problems are based on the relationship $n = MV$. Unit conversions based on this relationship are also used in the stoichiometry of reactions taking place in solution.

The elements of Group 1A (other than H) are called the **alkali elements** or **alkali metals.** They are less dense, softer, and more reactive than other metals. The diatomic molecules M_2 exist in the gas phase but not in the liquid or solid. Alkali elements form exclusively ionic compounds, as the monovalent cations M^+. They react with most nonmetallic elements to form salts, almost all of which are soluble in water. Reaction with oxygen produces oxides, peroxides, and superoxides. The smallest alkali metal, lithium, differs from the others in some respects. For example, LiF and Li_2CO_3 are insoluble, and Li is the only alkali that reacts with nitrogen. Alkali metals can be prepared by electrolysis of their molten salts, but potassium is prepared commercially by reaction of KCl with sodium. The major sources of alkali metal ores are salt lakes, salt deposits, and mineral springs.

PROBLEMS

General

1. Explain in your own words what information is carried by an unbalanced chemical equation.
2. What additional information is gained when an unbalanced chemical equation is balanced?
3. How can a balanced equation be distinguished from an unbalanced one?
4. What is the relationship between a balanced equation and the atomic theory of Dalton?
5. What is the relationship between the law of conservation of mass and a balanced chemical equation?
6. Define stoichiometry.
7. What is a stoichiometric ratio?
8. How is the theoretical yield of a product determined?
9. Give some reasons why the actual yield might be less than the theoretical yield.
10. Distinguish between reactant and reagent.
11. What is a limiting reactant?
12. What is meant by the sentence, "The reactants were present in stoichiometric amounts?"
13. Distinguish between solvent and solute.
14. Define mass percent, weight percent, and volume percent.
15. Distinguish between the terms "concentration" and "molarity."
16. What is meant by the term "dilute solution"?

Chemical Equations

17. Which of the following equations are balanced, and which are not?
    a. $Zn + 2\ HCl \longrightarrow ZnCl_2 + H_2$
    b. $HNO_3 + 2\ HCl \longrightarrow NOCl + Cl_2 + 2\ H_2O$
    c. $H_2O + Cl_2 \longrightarrow 2\ HCl + O_2$
18. Which of the following equations are balanced, and which are not?
    a. $2\ ClO_2 + H_2O \longrightarrow HClO_3 + HClO_2$
    b. $NaOH + Al + H_2O \longrightarrow NaAlO_2 + H_2$
    c. $2\ NaHCO_3 \longrightarrow Na_2CO_3 + H_2O + 2\ CO_2$
19. Balance those equations in Problem 17 that are unbalanced.
20. Balance those equations in Problem 18 that are unbalanced.
21. Which of the following equations are balanced, and which are not?
    a. $K_4Fe(CN)_6 \longrightarrow 4\ KCN + Fe + 2\ C + N_2$
    b. $Na + H_2O \longrightarrow NaOH + H_2$
    c. $2\ KNO_3 + S + 2\ C \longrightarrow K_2S + N_2 + 2\ CO_2$
22. Which of the following equations are balanced, and which are not?
    a. $Fe_3O_4 + 3\ C \longrightarrow 3\ Fe + 3\ CO_2$
    b. $6\ NaOH + 4\ S \longrightarrow 2\ Na_2S + Na_2S_2O_3 + 3\ H_2O$
    c. $2\ CO_2 + 2\ K \longrightarrow K_2CO_3 + C$
23. Balance those equations in Problem 21 that are unbalanced.

*24. Balance those equations in Problem 22 that are unbalanced.

25. Balance the following equations.
    a. $Xe + F_2 \longrightarrow XeF_4$
    b. $Mg + O_2 \longrightarrow MgO$
    c. $N_2 + H_2 \longrightarrow NH_3$
    d. $Al + HCl \longrightarrow AlCl_3 + H_2$
    e. $Al + Cr_2O_3 \longrightarrow Al_2O_3 + Cr$
    f. $Zn + HNO_3 \longrightarrow Zn(NO_3)_2 + H_2$
26. Balance the following equations.
    a. $Fe + O_2 \longrightarrow Fe_2O_3$
    b. $Fe + O_2 \longrightarrow Fe_3O_4$
    c. $I_2O_5 + CO \longrightarrow I_2 + CO_2$
    d. $C_2H_4 + O_2 \longrightarrow CO + H_2O$
    e. $CH_4 + O_2 \longrightarrow CO_2 + H_2O$
    f. $CH_4 + O_2 \longrightarrow CO + H_2O$

27. Balance the following equations.
   a. $CS_2 + O_2 \longrightarrow CO_2 + SO_2$
   b. $BiCl_3 + H_2S \longrightarrow Bi_2S_3 + HCl$
   c. $RbOH + SO_2 \longrightarrow Rb_2SO_3 + H_2O$
   d. $HBF_4 + H_2O \longrightarrow H_3BO_3 + HF$
   e. $PCl_3 + Cl_2 \longrightarrow PCl_5$
   f. $Na_2SO_4 + C \longrightarrow Na_2S + CO_2$

28. Balance the following equations.
   a. $P_4O_{10} + C \longrightarrow P_4 + CO$
   b. $AgNO_3 + Na_2C_2O_4 \longrightarrow Ag_2C_2O_4 + NaNO_3$
   c. $KOH + H_3PO_4 \longrightarrow K_3PO_4 + H_2O$
   d. $UO_2 + HF \longrightarrow UF_4 + H_2O$
   e. $UF_4 + Mg \longrightarrow U + MgF_2$
   f. $N_2 + O_2 \longrightarrow N_2O$

*29. Balance the following equations.
   a. $P_4O_{10} + H_2O \longrightarrow H_3PO_4$
   b. $P_2H_4 \longrightarrow PH_3 + P_4$
   c. $PbO + NH_3 \longrightarrow Pb + N_2 + H_2O$
   d. $Mg_3N_2 + H_2O \longrightarrow Mg(OH)_2 + NH_3$

*30. Balance the following equations.
   a. $CaCN_2 + H_2O \longrightarrow CaCO_3 + NH_3$
   b. $Cu + HNO_3 \longrightarrow Cu(NO_3)_2 + NO + H_2O$
   c. $Cu(NO_3)_2 + NaHCO_3 \longrightarrow CuCO_3 + NaNO_3 + H_2O + CO_2$
   d. $SO_2Cl_2 + HI \longrightarrow H_2S + H_2O + HCl + I_2$

*31. Balance the following equations.
   a. $AsF_3 + PCl_5 \longrightarrow PF_5 + AsCl_3$
   b. $NH_3 + O_2 \longrightarrow NO + H_2O$
   c. $PbO_2 + Pb + H_2SO_4 \longrightarrow PbSO_4 + H_2O$

*32. Balance the following equations.
   a. $C_4H_{10} + O_2 \longrightarrow CO_2 + H_2O$
   b. $CsOH + H_3PO_4 \longrightarrow Cs_2HPO_4 + H_2O$
   c. $KClO_3 \longrightarrow KCl + O_2$

*33. Balance the following equation (challenging).

$$HMnO_4 + MnCl_2 + H_2O \longrightarrow MnO_2 + HCl$$

*34. Balance the following equation (very difficult).

$$S_2Cl_2 + NH_3 \longrightarrow N_4S_2 + NH_4Cl + S_8$$

35. Write balanced chemical equations for each of the following processes.
   a. Calcium phosphate reacts with sulfuric acid to produce calcium sulfate and phosphoric acid.
   b. Calcium phosphate reacts with water containing dissolved carbon dioxide to produce calcium hydrogen carbonate and calcium hydrogen phosphate.

36. Write balanced chemical equations for each of the following processes.
   a. When heated, nitrogen and oxygen combine to form nitrogen monoxide.
   b. Heating of a mixture of lead(II) sulfide and lead(II) sulfate produces metallic lead and sulfur dioxide.

37. Write balanced chemical equations for each of the following processes.
   a. Copper reacts with water, oxygen, and carbon dioxide to produce $Cu_2(OH)_2CO_3$.
   b. When heated, a mixture of potassium hydroxide and ammonium bromide produces potassium bromide, water, and ammonia.

38. Write balanced chemical equations for each of the following processes
   a. Gaseous chlorine reacts with liquid water to produce aqueous hypochlorous acid and aqueous hydrochloric acid.
   b. Potassium amide (KNH_2) reacts with water to produce aqueous potassium hydroxide and gaseous ammonia.

39. Write balanced chemical equations for each of the following processes.
   a. Potassium chlorate reacts with table sugar ($C_{12}H_{22}O_{11}$) to produce potassium chloride, carbon dioxide, and water.
   b. Lead(IV) oxide reacts with sulfur dioxide, producing lead(II) sulfate.

40. Write balanced chemical equations for each of the following processes.
   a. Heating of ammonium carbonate causes decomposition to ammonia, carbon dioxide, and water.
   b. Heating of ammonium dichromate produces chromium(III) oxide, nitrogen, and water.

Stoichiometry

41. How many moles of Cl_2 are formed when 0.0538 mol HCl reacts with nitric acid according to $HNO_3 + 3\ HCl \rightarrow NOCl + Cl_2 + 2\ H_2O$?

42. How many moles of iron are produced when 5.52 mol C is consumed in the reaction $Fe_3O_4 + 2\ C \rightarrow 3\ Fe + 2\ CO_2$?

43. How many moles of water are required to react with 0.0446 mol chlorine dioxide in the reaction $2\ ClO_2 + H_2O \rightarrow HClO_3 + HClO_2$?

44. How many moles of sodium sulfide are formed when 10 mol water is produced according to $6\ NaOH + 4\ S \rightarrow 2\ Na_2S + Na_2S_2O_3 + 3\ H_2O$?

45. What mass of H_2 is produced when 5.28 g Zn reacts according to $Zn + 2\ HCl \rightarrow ZnCl_2 + H_2$?

46. What mass of H_2O is produced when 10.0 g $NaHCO_3$ is decomposed according to $2\ NaHCO_3 \rightarrow Na_2CO_3 + H_2O + CO_2$?

47. What mass of ClO_2 is required to produce 8.36 kg $HClO_3$ according to $2\ ClO_2 + H_2O \rightarrow HClO_3 + HClO_2$?

48. What mass of CO_2 is produced when 75.3 mg C_8H_{18} burns according to $2\ C_8H_{18} + 25\ O_2 \rightarrow 16\ CO_2 + 18\ H_2O$?

49. What mass of anhydrous copper(II) sulfate remains when 150 g copper(II) sulfate pentahydrate ($CuSO_4 \cdot 5H_2O$) is heated sufficiently to drive off all its water of hydration?

50. What mass of anhydrous barium chloride remains when 35.0 g barium chloride dihydrate is heated sufficiently to drive off all its water of hydration?

51. If 0.328 g O_2 is produced in the reaction of Cl_2 with water, $2\ H_2O + 2\ Cl_2 \rightarrow 4\ HCl + O_2$, how much HCl is produced at the same time?

52. One of the combustion products of ammonia is nitrogen monoxide, produced in the reaction $4\ NH_3 + 5\ O_2 \rightarrow 4\ NO + 6\ H_2O$. What mass of NO is produced when sufficient ammonia burns to produce 175 g water?

*53. The density of N_2 at room temperature and atmospheric pressure is 1.15 g L^{-1}. How much lithium is required to react with 10.0 L N_2 according to $6\ Li + N_2 \rightarrow 2\ Li_3N$?

*54. At room temperature and atmospheric pressure, the densities of H_2, Cl_2, and HCl are 0.08244 g L^{-1}, 2.907 g L^{-1}, and 1.494 g L^{-1}, respectively.
   a. What volume of H_2 is required to react with 1.000 L Cl_2 according to $H_2(g) + Cl_2(g) \rightarrow 2\ HCl(g)$?
   b. What volume of HCl is formed?

Limiting Reactant

55. When 50 g each of Zn and S react according to $Zn + S \rightarrow ZnS$, how much ZnS is formed?

56. When 110.3 g Fe_3O_4 reacts with 32.8 g carbon according to the reaction $Fe_3O_4 + 2\ C \rightarrow 3\ Fe + 2\ CO_2$, how much Fe is produced?

*57. How much water is produced when 55 g C_3H_8 reacts with 155 g O_2? (The other product is carbon dioxide.)

*58. Boron trifluoride reacts with water to produce hydrogen fluoride and B_2O_3. How much HF is formed when 75.0 g BF_3 reacts with 25.0 g H_2O?

*59. Dinitrogen pentoxide reacts with water to produce nitric acid.
   a. How much HNO_3 is formed when 100.0 g N_2O_5 reacts with 20.0 g H_2O?
   b. Which is the limiting reactant?

*60. Hydrogen sulfide reacts with oxygen to form sulfur (S_8) and water.
   a. How much sulfur is formed when 100.0 g H_2S and 50.0 g O_2 react?
   b. Which reactant is limiting?

*61. Answer the following questions.
   a. How much Na_2O_2 is formed when 2.63 g Na reacts with 4.00 g O_2?
   b. How much of the reactant in excess remains unreacted?

*62. Answer the following questions.
   a. When 75.0 mg ammonia reacts with 165 mg oxygen, how much nitrogen monoxide is formed? (The other product is water.)
   b. How much of the reactant in excess remains unreacted?

*63. Answer the following questions.
   a. If 195.8 g KNO_3, 60.1 g S, and 41.9 g C react according to $2\ KNO_3 + S + 3\ C \rightarrow K_2S + N_2 + 3\ CO_2$, how much K_2S is formed?
   b. Which is the limiting reactant?

*64. Iron(II) chloride reacts with ammonia and water to produce iron(III) hydroxide and ammonium chloride.
   a. How much ammonium chloride is formed when 78 g $FeCl_3$, 25.0 g NH_3, and 25.0 g H_2O react?
   b. Which is the limiting reactant?

Percent Yield and Percent Purity

65. If 15 g sodium carbonate is obtained from the thermal decomposition of 50 g sodium hydrogen carbonate, $2\ NaHCO_3 \rightarrow Na_2CO_3 + H_2O + CO_2$, what is the percent yield?

66. What is the percent yield if 106 mg SO_2 is obtained from the combustion of 78.1 mg carbon disulfide according to the reaction $CS_2 + 3\ O_2 \rightarrow CO_2 + 2\ SO_2$?

*67. Disulfur dichloride and carbon tetrachloride are the products of the reaction of carbon disulfide and chlorine. What is the percent yield if 75 kg S_2Cl_2 is obtained from the reaction of 145 kg Cl_2?

*68. Suppose 5.72 g LiH is reacted with excess $AlCl_3$ and 4.03 g $LiAlH_4$ is isolated. What is the percent yield?

69. If, from a 50.0 g sample of an iron ore containing Fe_3O_4, 2.09 g of Fe is obtained by the reaction $Fe_3O_4 + 2\ C \rightarrow 3\ Fe + 2\ CO_2$, which is the percent of Fe_3O_4 in the ore?

*70. A 18.23 g sample of barium chloride dihydrate ($BaCl_2 \cdot 2H_2O$), contaminated with sand, was heated to drive off the water of hydration; 1.563 g H_2O was collected. Assuming no losses of H_2O during the procedure, what was the percent purity of the sample?

71. Aqueous silver nitrate forms solid silver chloride when mixed with aqueous potassium chloride (or any other dissolved chloride salt), according to $AgNO_3(aq) + KCl(aq) \rightarrow AgCl(s) + KNO_3(aq)$. Suppose 0.546 g of silver nitrate, known to contain some sodium nitrate as an impurity, is dissolved in water. It is then treated with excess KCl(aq), and 0.444 g AgCl is obtained. What is the percent purity of the silver nitrate?

*72. Aqueous barium chloride reacts with aqueous sodium sulfate (or any other dissolved sulfate salt) to form solid barium sulfate. If a 6.07 g sample of barium chloride, contaminated with sodium chloride, is dissolved in water, then treated with excess $Na_2SO_4(aq)$, and 2.89 g barium sulfate is obtained, what is the purity of the barium chloride?

*73. Suppose your company wishes to prepare 150 kg sodium peroxide by the direct reaction of sodium with oxygen. You know from past experience that the reaction proceeds in 92% yield. How much sodium should you use?

*74. Tin(IV) chloride is produced in 85% yield by the reaction of tin with chlorine. How much tin is required to produce a kilogram of tin(IV) chloride?

Composition of Solutions

75. How much solute is contained in each of the following solutions? Express each answer in both moles and grams.
   a. 75 mL of 0.036 M $RbBrO_3$
   b. 500 g of 2.00% (m/m) $K_2Cr_2O_7$
   c. 1.625 L of a 0.998 M solution of CsCl

76. How much solute is contained in each of the following solutions? Express each answer in both moles and grams.
   a. 0.667 L of 1.03 M LiCl
   b. 100 g of a 0.25% (m/m) solution of Na_2S
   c. 0.254 mL of 0.0412 M $NaHCO_3$

77. Solve the following problems.
   a. What volume of 0.563 M solution contains 3.25 mol solute?
   b. What volume of 0.0979 M KOH contains exactly 1 mol KOH?
   c. What mass of 1.93% (m/m) solution contains 10.0 g solute?
   d. What mass of 5.25% (m/m) NaClO contains exactly 1 mol NaClO?

78. Answer the following questions.
   a. What volume of 0.00269 M solution contains 0.00541 mol solute?
   b. What volume of 0.991 M CsI contains exactly 1 mol CsI?
   c. What mass of 0.75% (m/m) solution contains 5.9 g solute?
   d. What mass of 1.25% (m/m) RbCl contains exactly 1 mol RbCl?

79. Give directions for the preparation of each of the following solutions.
   a. 500 g of 2.5% aqueous $KClO_4$
   b. 1 L of 5% (v/v) alcohol in water
   c. 2 L of 0.0200 M KOH

80. Give directions for the preparation of each of the following solutions.
   a. 3.00 L of 0.0155 M K_2CO_3
   b. 500 mL of 0.100 M K_2SO_4
   c. 350 g of 2.00% alcohol in water

81. Give laboratory instructions for the preparation of each of the following solutions.
   a. 250 mL 0.100 M NaOH from 3.04 M stock solution
   b. 750 mL 0.25 M KCl from 6.0 M stock solution
   c. 300 mL 0.20 M H_2SO_4 from 4.00 M stock solution

82. Give laboratory instructions for the preparation of each of the following solutions.
   a. 25 mL 1.00 M LiCl from 3.00 M stock solution
   b. 50 mL 3.0 M K_2CO_3 from 6.0 M stock solution
   c. 3 L 0.020 M H_2SO_4 from 3.00 M stock solution

Reactions in Solution

83. What volume of 0.0496 M $HClO_4$ is required to react with 25.0 mL 0.505 M KOH according to $KOH + HClO_4 \rightarrow KClO_4 + H_2O$?

84. How much 0.746 M NaOH is required to react with 50.0 mL 1.007 M H_2SO_4? The reaction products are Na_2SO_4 and H_2O.

85. If 36.2 mL KOH(aq) are required to react with 25.0 mL 0.0513 M HNO_3 according to $KOH + HNO_3 \rightarrow KNO_3 + H_2O$, what is the molarity of the KOH?

86. If 1.035 mL H_2SO_4 are required to react with 1.000 L 3.55×10^{-3} M CsOH according to $H_2SO_4 + 2\ CsOH \rightarrow Cs_2SO_4 + 2\ H_2O$, what is the molarity of the H_2SO_4?

87. What volume of 0.0974 M HCl is required to react with 0.250 g Zn according to $Zn + 2\ HCl \rightarrow ZnCl_2 + H_2$?

88. What mass of Al can be consumed by the reaction of 355 mL 0.500 M NaOH according to $2\ NaOH + 2\ Al + 2\ H_2O \rightarrow 2\ NaAlO_2 + 3\ H_2$?

89. If 36.1 mL of KOH is required to react with 0.247 g oxalic acid ($H_2C_2O_4$) according to $2\ KOH + H_2C_2O_4 \rightarrow K_2C_2O_4 + 2\ H_2O$, what is the molarity of the KOH?

90. Suppose 0.113 g Na reacts with excess water so that the final volume of the resulting NaOH solution is 100 mL.
   a. Calculate the molarity of the solution.
   b. What mass of H_2 will be formed?

Alkali Metals

91. Describe the reactions of the alkali metals with oxygen, and write balanced equations for each.

92. Which of the alkali metals is (are) unreactive toward nitrogen?

93. Which of the Group 1A elements is most reactive toward water? Which is the least reactive? Write the general equation.

94. Describe, with equations, the use of potassium superoxide in breathing equipment.

95. What are the trends in melting point, boiling point, and density in Group 1A?

96. What raw materials are used in the manufacture of the alkali metals?

97. Give two methods by which sodium carbonate is commercially produced.

98. Describe a commercial use of sodium cyanide.

99. What polyatomic ion is similar in its chemical behavior to the alkali ions?

100. What is a "getter"?

101. How is sodium hydroxide produced, and what is an important byproduct of its manufacture?

102. List some uses of each of the Group 1A elements.

103. Name two sodium compounds used in the curing and preserving of meat.

104. How may a solution of electrons be prepared?

CHAPTER 4

ENERGY, HEAT, AND CHEMICAL CHANGE

The famous White Cliffs near Dover, England, consist of almost pure chalk—one of the common forms of calcium carbonate.

(a) (b)

Figure 4.1 (a) The kinetic energy of falling water is converted to electrical energy in a hydroelectric power plant. (b) The water wheel is an earlier means of harnessing the kinetic energy of moving water.

4.1 ENERGY AND HEAT

In Chapter 1, we placed the observable phenomena of the universe into the two categories of matter and energy. The nature of matter, specifically its organization into atoms and molecules, was discussed in Chapter 2. In this chapter, we begin our study of energy and its interaction with matter.

The Forms, Units, and Constancy of Energy

Radiant energy is discussed in Chapter 6, energy in the form of work is discussed in Chapter 10, some aspects of electrical energy are covered in Chapter 19, and nuclear energy is part of the material in Chapter 24.

Energy appears in many forms, which, for the most part, can be freely converted to one another. *Kinetic energy* (energy of motion) is often converted to other forms of energy. For example, when you rub your hands together to warm them, you are converting kinetic energy to *heat*. For many years, our society has used the kinetic energy of moving water (Figure 4.1). *Light* or *radiant energy* from the sun is converted to *heat* when it is absorbed by matter. *Chemical energy* is converted to *heat* in a fire and to *electrical energy* in a flashlight battery; *electrical energy* is converted to *work* in a motor, or to *light* or *heat* in common household appliances (Figure 4.2).

(a)

(b)

(c)

Figure 4.2 (a) A lightning stroke is the transfer of electrical energy between the earth and a cloud. (b) Chemical energy is converted into heat as wood burns. (c) A hair dryer converts electrical energy into heat and the kinetic energy of moving air.

Although it is of central importance in science and technology, energy was not chosen to be one of the *base* quantities in the SI system. Energy is a *compound* quantity, whose dimensions are $m \cdot \ell^2 \cdot t^{-2}$. The SI unit is the **joule (J)**, defined in terms of the base quantities by the equivalence statement 1 joule $\Leftrightarrow$ 1 kg $\cdot$ 1 m^2 $\cdot$ 1 s^{-2}. All forms of energy have the same dimensions and can be measured in joules.

Base quantities in SI are discussed in Chapter 1, page 9.

Although energy may be converted from one form to another, the *amount* of energy does not change during the conversion. This fact, confirmed by countless observations, is called the **law of conservation of energy;** energy is neither created nor destroyed in chemical or physical processes (Figure 4.3).

Figure 4.3 Radiant energy from the sun is converted into heat and electricity when it is absorbed by a photovoltaic or *solar panel*. No energy is lost, since the sum of the heat energy and the electrical energy is exactly equal to the initial radiant energy.

Heat

Any substance contains more energy at high temperature than at low temperature. This property is put to use in a generating plant, which converts the energy of high temperature steam to electrical energy. Energy associated with high temperature is called *thermal energy* or *heat*. Thermal energy is the sum of the kinetic energies of all of a substance's atoms, which are in constant motion. There is a direct relationship between atomic motion and temperature: the greater the vigor of atomic motion, the higher the temperature of the substance.

The microscopic motion of atoms in substances is discussed further in Section 5.4.

Since it is difficult to observe the motion of atoms, heat is formally defined in terms of other, measureable quantities: **heat** or **thermal energy** is the form of energy that moves or flows from one place or object to another because of a difference in temperature (Figure 4.4). Heat flows spontaneously from a warm object to a cooler object. If the objects are in contact with one another, heat flows by *conduction*. If the objects do not touch each other, heat flows by radiation or *convection*. Like other forms of energy, heat is measured in joules.

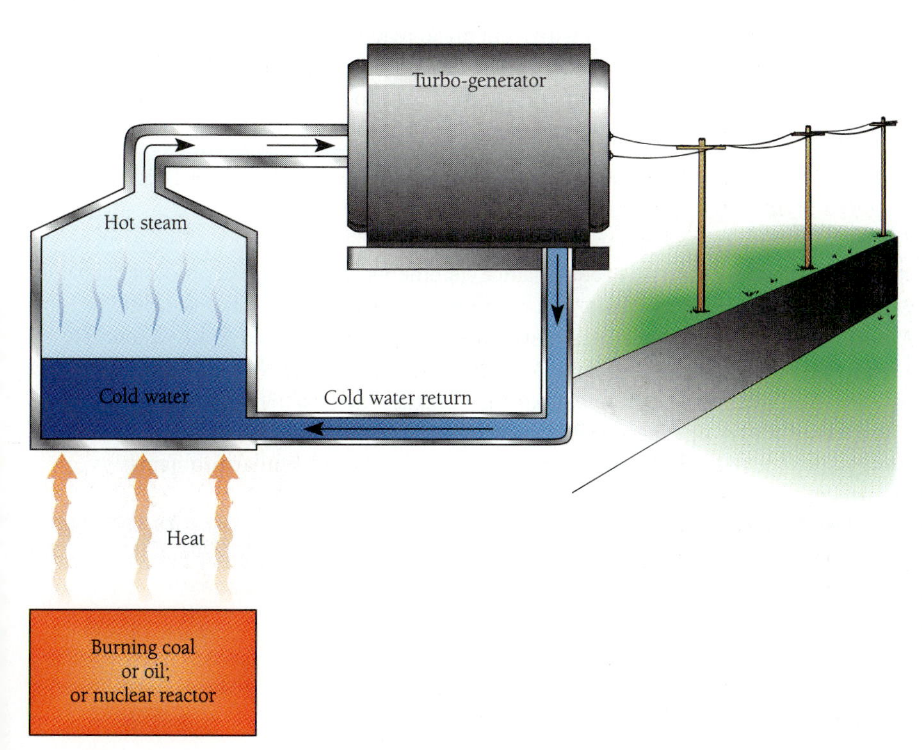

Figure 4.4 In an electrical generating plant, heat flows from a heat source, often a nuclear reaction or burning coal or oil, to a boiler where it is transferred to water. The thermal energy of the resulting high temperature steam is converted to electrical energy in a turbine/generator. As the steam loses its energy it becomes cold water, which is returned to the boiler.

4.2 HEAT IN CHEMICAL REACTIONS

A reaction taking place in fireflies releases energy in the form of light, and the chemical reaction in a flashlight battery releases electrical energy. However, most chemical reactions release or consume energy in the form of heat.

As we noted on the first page of this text, changes in matter are often accompanied by the release or consumption of energy. Almost all chemical reactions involve energy, and, most often, the energy is in the form of *heat.* The study of heat involved in chemical change is called **thermochemistry.** The quantity of heat involved in a reaction, like the quantity of product formed, is directly proportional to the quantity (mass) of reactants consumed.

Exothermic and Endothermic Reactions

Some reactions *consume* heat. For example, no decomposition of mercury(II) oxide to mercury and oxygen can occur without absorption of heat. Such reactions are described as **endothermic.**

Other reactions, described as **exothermic,** *generate* heat and release it to the environment. A combustion reaction, such as the burning of natural gas (methane, CH_4), is an example of an exothermic reaction. **Combustion** is a rapid, self-sustaining exothermic reaction of a substance with air or oxygen, often giving off some of its energy in the form of light.

In discussions of the heat generated or consumed in a chemical reaction, the amount of heat is commonly expressed in **kilojoules (kJ).** Some feeling for the size of this unit can be obtained from familiar situations. It takes 60 kJ to heat a cup of water (about 200 mL) from room temperature to 100 °C, and another 500 kJ to boil it away. Combustion of one charcoal briquet (about 40 g) releases 1300 kJ to the environment.

Heat of Reaction

The amount of heat associated with a chemical reaction is called the **heat of reaction,** and is given the symbol ΔH or ΔH_{rxn}. The heat of reaction for an endothermic reaction is a positive quantity, while, for exothermic reactions, ΔH is negative. The heat of reaction is conveniently written as part of the chemical equation.

$$2\ HgO(s) \longrightarrow 2\ Hg(\ell) + O_2(g); \qquad \Delta H_{rxn} = +181.7\ kJ \quad (4\text{-}1)$$

$$CH_4(g) + 2\ O_2(g) \longrightarrow CO_2(g) + 2\ H_2O(\ell); \qquad \Delta H_{rxn} = -890.4\ kJ \quad (4\text{-}2)$$

The amount of heat generated or consumed by a chemical reaction is proportional to the amount of reacting substances.

This notation implies a set of equivalence statements connecting the amount of heat, in kJ, with the amount, in moles, of a reactant or product species. For the HgO decomposition, the equivalence statements are 2 mol HgO $\Leftrightarrow$ 181.7 kJ, 2 mol Hg $\Leftrightarrow$ 181.7 kJ, and 1 mol $O_2 \Leftrightarrow$ 181.7 kJ. For Equation (4-2) they are 1 mol $CH_4 \Leftrightarrow$ −890.4 kJ, 2 mol $O_2 \Leftrightarrow$ −890.4 kJ, 1 mol $CO_2 \Leftrightarrow$ −890.4 kJ, and 2 mol $H_2O \Leftrightarrow$ −890.4 kJ. These statements are used to form the unit conversion factors necessary for the solution of Examples 4.1 and 4.2, and similar problems.

EXAMPLE 4.1 Heat of Combustion

How much heat is given off when 175 g CH_4 burns according to Equation (4-2)?

Analysis

Target: ? kJ =

Knowns: The balanced equation, the heat of reaction, and the mass of CH_4.

Relationship: The equivalence statements 1 mol $CH_4 \Leftrightarrow$ 16.0 g and 1 mol $CH_4 \Leftrightarrow -890.4$ kJ connect the target with the knowns.

Plan Use the factor-label method for a two-stage unit conversion.

Work

$$? \text{ kJ} = (175 \text{ g } CH_4) \cdot \frac{(1 \text{ mol } CH_4)}{(16.0 \text{ g } CH_4)} \cdot \frac{(-890.4 \text{ kJ})}{(1 \text{ mol } CH_4)}$$
$$= -9.74 \times 10^3 \text{ kJ}$$

Check 175 g is close to 10 moles, so the heat given off should be close to 10 times the heat of reaction.

Exercise

How much heat is given off when 55 g water is produced by the combustion of methane according to Equation (4-2)?

Answer: -1.4×10^3 kJ (or simply 1.4×10^3 kJ; see the following paragraph).

It is not necessary to include the algebraic sign when *speaking* about heats of reaction, because the same information is carried by the words that are used. A reaction that "gives off" (or "generates," "liberates," "produces," and so on) heat is clearly exothermic, and the sign of its ΔH_{rxn} must be negative. A reaction that "requires" (or "absorbs," "consumes," "uses," and so on) heat is clearly endothermic, and its ΔH_{rxn} is a positive quantity. It is common (and perfectly acceptable) to say, in giving the answer to Example 4.1, that "175 g CH_4 gives off 9.74×10^3 kJ of heat when it burns." It is *not* acceptable to say, "The heat of reaction is 9.74×10^3 kJ," because in this sentence it is not clear whether $+9.74 \times 10^3$ or -9.74×10^3 is meant. When a ΔH_{rxn} value is used in a calculation, it is good practice to include the sign even if it will be ignored when the answer is spoken.

EXAMPLE 4.2 Heat Absorbed in an Endothermic Reaction

How much HgO is decomposed by the absorption of 33.3 kJ of heat according to Equation (4-1)?

Analysis

Target: ? g HgO =

Knowns: The balanced equation, and the heat of reaction; the molar mass of HgO is implicit.

Relationship: The equivalence statements for ΔH_{rxn} and the molar mass of HgO connect the target with the knowns.

Plan/Work Use the factor-label method.

$$? \text{ g HgO} = (33.3 \text{ kJ}) \cdot \frac{(2 \text{ mol HgO})}{(+181.7 \text{ kJ})} \cdot \frac{(216.6 \text{ g})}{(1 \text{ mol})}$$
$$= +79.4 \text{ g}$$

Exercise

How much heat is required to prepare 10.0 g O_2 by decomposing HgO?

Answer: 56.8 kJ.

4.3 THERMOCHEMICAL EQUATIONS AND CALCULATIONS

A chemical equation, together with its associated ΔH_{rxn}, is called a **thermochemical equation.** Equations (4-1) and (4-2) are thermochemical equations, and Equations (4-3) through (4-7) are additional examples.

$$CH_4(g) + 2\ O_2(g) \longrightarrow CO_2(g) + 2\ H_2O(g); \qquad \Delta H_{rxn} = -802\ \text{kJ} \qquad (4\text{-}3)$$

$$H_2O(\ell) \longrightarrow H_2O(g); \qquad \Delta H_{rxn} = +44.0\ \text{kJ} \qquad (4\text{-}4)$$

$$2\ Na(s) + 2\ H_2O(\ell) \longrightarrow 2\ NaOH(s) + H_2(g); \qquad \Delta H_{rxn} = -367.6\ \text{kJ} \qquad (4\text{-}5)$$

$$H_2(g) + \tfrac{1}{2}\ O_2(g) \longrightarrow H_2O(\ell); \qquad \Delta H_{rxn} = -285.85\ \text{kJ} \qquad (4\text{-}6)$$

$$\tfrac{1}{2}\ H_2(g) + \tfrac{1}{2}\ Cl_2(g) \longrightarrow HCl(g); \qquad \Delta H_{rxn} = -92.3\ \text{kJ} \qquad (4\text{-}7)$$

These examples illustrate the following features of thermochemical equations.

1. The physical state of each substance (solid, liquid, or gaseous) is indicated in parentheses. This is important, as can be seen by comparing Equations (4-2) and (4-3).
2. In contrast to ordinary chemical equations, thermochemical equations often include fractional stoichiometric coefficients.
3. ΔH_{rxn} is expressed in kJ.
4. ΔH_{rxn} can be positive or negative, and the sign is included in the equation.
5. A physical process such as the vaporization of a liquid [Equation (4-4)] can also be described by a thermochemical equation.

As mentioned, a thermochemical equation contains several implicit equivalence statements linking a stated amount of heat with definite molar amounts of products and reactants. A given reaction, however, can be correctly described by more than one thermochemical equation. For example, Equations (4-7) and (4-8) both describe the exothermic formation of HCl(g).

$$H_2(g) + Cl_2(g) \longrightarrow 2\ HCl(g); \qquad \Delta H_{rxn} = -184.6\ \text{kJ} \qquad (4\text{-}8)$$

Note that the two different equivalence statements implied by the given values of ΔH_{rxn}, 1 mol HCl(g) $\Leftrightarrow$ -92.3 kJ from Equation (4-7) and 2 mol HCl(g) $\Leftrightarrow$ -184.6 kJ from Equation (4-8), lead to the same amount of heat *per mole of HCl(g) produced,* namely -92.3 kJ mol^{-1}. Equations (4-7) and (4-8) are both correct, and, depending on the context, either may be used.

Often it is necessary to know the heat produced/consumed when a reaction is run in reverse. If the production of HCl from H_2 and Cl_2 *produces* 92.3 kJ per mole HCl ($\Delta H_{rxn} = -92.3$ kJ), then the decomposition of one mole HCl into H_2 and Cl_2 must *consume* (receive from the surroundings) 92.3 kJ per mole HCl.

$$HCl(g) \longrightarrow \tfrac{1}{2} H_2(g) + \tfrac{1}{2} Cl_2(g); \qquad \Delta H_{rxn} = +92.3 \text{ kJ}$$

These considerations lead to two rules for the manipulation of thermochemical equations.

1. If the stoichiometric coefficients of a thermochemical equation are multiplied by a common factor, ΔH_{rxn} for the reaction must also be multiplied by the same factor.
2. When a thermochemical equation is reversed, the sign of ΔH_{rxn} must also be reversed.

Hess's Law

Consider the combustion of carbon, Equation (4-9).

$$C(s) + O_2(g) \longrightarrow CO_2(g); \qquad \Delta H_{rxn} = -393.5 \text{ kJ} \tag{4-9}$$

When the reaction is carried out in the presence of an excess of oxygen, carbon dioxide is the only product. If the oxygen supply is limited, however, the major reaction is the formation of CO [Equation (4-10)]. Since carbon monoxide itself is combustible [Equation (4-11)], the production of CO_2 may actually be carried out in two separate steps.

$$\text{Step a:} \quad C(s) + \tfrac{1}{2} O_2(g) \longrightarrow CO(g); \qquad \Delta H_a = -110.5 \text{ kJ} \tag{4-10}$$

$$\text{Step b:} \quad CO(g) + \tfrac{1}{2} O_2(g) \longrightarrow CO_2(g); \qquad \Delta H_b = -283.0 \text{ kJ} \tag{4-11}$$

It is easily seen that the heat of reaction in Equation (4-9), the *overall* reaction, is the sum of the heats of reaction of the steps in Equations (4-10) and (4-11). That is, $\Delta H_{rxn} = \Delta H_a + \Delta H_b$. This equality is an example of a general relationship known as **Hess's law:** the value of ΔH_{rxn} is the same whether a reaction is carried out in a single step or as a sequence of separate steps.

Sequential reaction steps are combined to obtain the chemical equation for an overall reaction, according to the following procedure. First write a "preliminary sum equation," with all the reactants from all the steps on the left and all the products from all the steps on the right. The preliminary sum of Equations (4-10) and (4-11) is

$$C(s) + \tfrac{1}{2} O_2(g) \longrightarrow CO(g) \tag{4-10}$$

$$CO(g) + \tfrac{1}{2} O_2(g) \longrightarrow CO_2(g) \tag{4-11}$$

$$\overline{C(s) + CO(g) + (\tfrac{1}{2} + \tfrac{1}{2})\, O_2(g) \longrightarrow CO(g) + CO_2(g)}$$

Next, combine the coefficients of the same species; do this separately on each side of the equation.

$$C(s) + CO(g) + \tfrac{2}{2}\,O_2(g) \longrightarrow CO(g) + CO_2(g)$$

The final step is the cancellation of stoichiometric coefficients of species appearing on both the left and the right of the arrow. Provided their stoichiometric coefficients are the same on both sides, these species are simply removed from the equation. The result of these operations

$$C(s) + \cancel{CO(g)} + \tfrac{2}{2}\,O_2(g) \longrightarrow \cancel{CO(g)} + CO_2(g)$$

is the overall equation

$$C(s) + O_2(g) \longrightarrow CO_2(g)$$

The chemical meaning of this cancellation is that a species produced in one step and then consumed (in equal amounts) in a later step does not survive as a product of the overall reaction, and therefore it does not appear in the chemical equation. In some reactions a compound appears on both sides of the preliminary sum equation, but with different coefficients. In that case it remains in the overall equation, with a coefficient equal to the difference between the left- and right-side coefficients in the preliminary sum. For example, if the steps are

$$Na_2O + H_2O \longrightarrow 2\ NaOH \qquad \text{and}$$

$$2\ NaOH + H_2SO_4 \longrightarrow Na_2SO_4 + 2\ H_2O$$

the preliminary sum is

$$Na_2O + H_2O + 2\ NaOH + H_2SO_4 \longrightarrow Na_2SO_4 + 2\ H_2O + 2\ NaOH$$

NaOH cancels completely, but only one of the H_2O's on the right cancels. The overall reaction is

$$Na_2O + H_2SO_4 \longrightarrow Na_2SO_4 + H_2O$$

EXAMPLE 4.3 Finding the Overall Reaction

Write the overall reaction for the stepwise reaction of potassium hypochlorite with chlorine to form potassium perchlorate.

$$KClO + Cl_2 + 2\ KOH \longrightarrow KClO_2 + 2\ KCl + H_2O$$

$$KClO_2 + Cl_2 + 2\ KOH \longrightarrow KClO_3 + 2\ KCl + H_2O$$

$$KClO_3 + Cl_2 + 2\ KOH \longrightarrow KClO_4 + 2\ KCl + H_2O$$

Analysis/Plan This problem is an application of the addition principles just discussed.

Work Add the equations to obtain the preliminary sum of all the reactants and all the products. Combine the coefficients of the same species.

$$\begin{array}{l} KClO + Cl_2 + 2\,KOH \longrightarrow KClO_2 + 2\,KCl + H_2O \\ KClO_2 + Cl_2 + 2\,KOH \longrightarrow KClO_3 + 2\,KCl + H_2O \\ KClO_3 + Cl_2 + 2\,KOH \longrightarrow KClO_4 + 2\,KCl + H_2O \\ \hline \end{array}$$

$$KClO + KClO_2 + KClO_3 + (1 + 1 + 1)\,Cl_2 + (2 + 2 + 2)\,KOH \longrightarrow$$
$$KClO_2 + KClO_3 + KClO_4 + (2 + 2 + 2)\,KCl + (1 + 1 + 1)\,H_2O$$

$$KClO + KClO_2 + KClO_3 + 3\,Cl_2 + 6\,KOH \longrightarrow$$
$$KClO_2 + KClO_3 + KClO_4 + 6\,KCl + 3\,H_2O$$

In this reaction, all species that appear on both sides have the same stoichiometric coefficients. Identify and cancel them,

$$KClO + \cancel{KClO_2} + \cancel{KClO_3} + 3\,Cl_2 + 6\,KOH \longrightarrow$$
$$\cancel{KClO_2} + \cancel{KClO_3} + KClO_4 + 6\,KCl + 3\,H_2O$$

leaving as the overall reaction

$$KClO + 3\,Cl_2 + 6\,KOH \longrightarrow KClO_4 + 6\,KCl + 3\,H_2O$$

Check Always check to make sure that the overall reaction is balanced.

Exercise

Write the overall reaction that proceeds by the following steps.

$$\text{(a)}\quad S + O_2 \longrightarrow SO_2;$$
$$\text{(b)}\quad SO_2 + \tfrac{1}{2}\,O_2 \longrightarrow SO_3$$

Answer: $S + \frac{3}{2}\,O_2 \rightarrow SO_3$.

Hess's law is often used to deduce ΔH_{rxn} of a new reaction, or one whose ΔH_{rxn} is difficult to measure. If ΔH_{rxn} for the reaction in Equation (4-10) were not known, for instance, it could be calculated from the known values for the reactions in Equations (4-9) and (4-11). The procedure is as follows. Write Equation (4-9), and underneath it, write Equation (4-11) in reverse. Be careful to reverse the sign of ΔH as well. Add the two equations to get the equation for the overall reaction, and add the two ΔH's to get the corresponding value of ΔH_{rxn}.

$$C(s) + O_2(g) \longrightarrow CO_2(g); \qquad \Delta H = -393.5\text{ kJ} \qquad (4\text{-}9)$$
$$CO_2(g) \longrightarrow CO(g) + \tfrac{1}{2}\,O_2(g); \qquad \Delta H = +283.0\text{ kJ} \qquad (4\text{-}11r)$$

$$\text{Overall: } C(s) + \tfrac{1}{2}\,O_2(g) \longrightarrow CO(g); \qquad \Delta H = -110.5\text{ kJ} \qquad (4\text{-}10)$$

As a matter of fact, the heat of this reaction is difficult to measure, because in any real combustion experiment it is not possible to produce pure CO; some CO_2 is always formed as well.

EXAMPLE 4.4 Determination of the Heat of an Unknown Reaction from Hess's Law

Determine the heat of reaction for $SiH_4(g) + 2\ O_2(g) \rightarrow 2\ H_2O(\ell) + SiO_2(s)$, given the following thermochemical equations.

a. $Si(s) + 2\ H_2(g) \longrightarrow SiH_4(g);\qquad \Delta H_a = 34\text{ kJ}$

b. $H_2(g) + \frac{1}{2}\ O_2(g) \longrightarrow H_2O(\ell);\qquad \Delta H_b = -286\text{ kJ}$

c. $O_2(g) + Si(s) \longrightarrow SiO_2(s);\qquad \Delta H_c = -911\text{ kJ}$

Analysis Since three thermochemical equations are given, and a fourth is asked for, this is almost certainly a Hess's law problem requiring the summation of several reactions to achieve a specified overall reaction.

Plan Manipulate each of the three given equations by reversing, and/or by multiplying by a small factor, so that when summed, the correct overall equation is obtained. This can and should be done systematically, as follows.

Work Write down the target reaction.

$$SiH_4(g) + 2\ O_2(g) \longrightarrow 2\ H_2O(\ell) + SiO_2(s)$$

We could have chosen to begin with $SiO_2(s)$, because it too appears only once in the given equations. Both choices lead to the same answer.

Select from the products one that appears only once in the given equations. In this case, $H_2O(\ell)$ appears only in Equation (b). It follows that (b) is one of the equations in the Hess's law sum, since that is the only way $H_2O(\ell)$ can appear as a product in the target reaction. However, 2 mol H_2O appears in the target, and only *1* mol is in (b): (b) must therefore be multiplied by 2, and the first reaction in the sum is

(1) $2\ H_2(g) + O_2(g) \longrightarrow 2\ H_2O(\ell);\qquad \Delta H_1 = 2\Delta H_b = 2(-286) = -572\text{ kJ}$

Now do the same with all the other products (in this case there is only one other) in the target reaction. $SiO_2(s)$ appears as a product in Equation (c), so (c) must be a contributor to the Hess's law sum. No multiplication is necessary since the stoichiometric coefficient is the same in both the target and (c). The second reaction in the sum is therefore

(2) $O_2(g) + Si(s) \longrightarrow SiO_2(s);\qquad \Delta H_2 = \Delta H_c = -911\text{ kJ}$

When each product in the target equation has been considered in this way, it is useful to do an intermediate summation.

(1) + (2) $2\ H_2(g) + 2\ O_2(g) + Si(s) \longrightarrow SiO_2(s) + 2\ H_2O(\ell)$

This is close to the target, but still lacks $SiH_4(g)$ as a reactant. Therefore, an equation containing this missing species must be found. Equation (a) contains $SiH_4(g)$, but as a product rather than a reactant: (a) must therefore be reversed before it can contribute correctly to the sum.

(3) $SiH_4(g) \longrightarrow Si(s) + 2\ H_2(g);\qquad \Delta H_3 = -\Delta H_a = -34\text{ kJ}$

Continue this procedure until all reactants in the target reaction have been accounted for. In this example there are no further reactants to consider, and the target reaction is the sum of (1), (2), and (3). Hess's law can now be applied, giving

$$
\begin{array}{llll}
(1) & 2\ H_2(g) + O_2(g) \longrightarrow 2\ H_2O(\ell); & \Delta H_1 = 2\ \Delta H_b = -572\ \text{kJ} \\
(2) & O_2(g) + Si(s) \longrightarrow SiO_2(s); & \Delta H_2 = \Delta H_c = -911\ \text{kJ} \\
(3) & SiH_4(g) \longrightarrow Si(s) + 2\ H_2(g); & \Delta H_3 = -\Delta H_a = -34\ \text{kJ} \\
\hline
& SiH_4(g) + 2\ O_2(g) \longrightarrow 2\ H_2O(\ell) + SiO_2(s); & \Delta H_{rxn} = \Delta H_1 + \Delta H_2 + \Delta H_3 = -1517\ \text{kJ}
\end{array}
$$

Check Make sure the overall reaction is balanced, and that the signs of all ΔH values have been properly used.

Exercise

The following heats of combustion are known.

$$C_2H_5OH(\ell) + 3\ O_2(g) \longrightarrow 2\ CO_2(g) + 3\ H_2O(\ell); \qquad \Delta H = -1366.6\ \text{kJ}$$

$$H_2(g) + \tfrac{1}{2}\ O_2(g) \longrightarrow H_2O(\ell); \qquad \Delta H = -285.8\ \text{kJ}$$

$$C(s) + O_2 \longrightarrow CO_2(g); \qquad \Delta H = -393.5\ \text{kJ}$$

Use this information to calculate the heat of the reaction

$$2\ C(s) + 3\ H_2(g) + \tfrac{1}{2}\ O_2(g) \longrightarrow C_2H_5OH(\ell)$$

Answer: -277.8 kJ.

Heat of Formation

There is usually more than one reaction that can be used to prepare a given compound. In particular, it is always possible to write a chemical equation that shows a compound being produced from its elements. A **formation equation** is defined as a balanced thermochemical equation in which one mole of a compound is produced from its constituent elements in their standard states. The **standard state** of a substance is its most stable form at 25 °C and one atmosphere (1 atm) pressure. Some standard states are: mercury, liquid; sodium chloride, solid; carbon dioxide, gas; and carbon, solid (graphite). Formation reactions, like other reactions, can be endothermic or exothermic. The heat of reaction ΔH_{rxn} of a formation equation is called the *heat of formation* of the compound. If the formation reaction is carried out at 1 atm pressure, the heat of reaction is called the **standard heat of formation** of the compound, and given the symbol ΔH_f°. The superscript "o" means, "standard state and 1 atm pressure." Examples of formation reactions include the following.

Pressure, and the units used to describe pressure, are discussed in Chapter 5.

Diamond is also a stable form of solid carbon, but since graphite is slightly more stable, it is designated as the standard state of carbon.

$$\tfrac{1}{2}\ H_2(g) + \tfrac{1}{2}\ Cl_2(g) \longrightarrow HCl(g); \qquad \Delta H^\circ_{rxn} = -92.3\ \text{kJ}, \qquad (4\text{-}12)$$

and

$$\Delta H_f^\circ\,[HCl(g)] = -92.3\ \text{kJ mol}^{-1}$$

$$H_2(g) + \tfrac{1}{2}\ O_2(g) \longrightarrow H_2O(\ell); \qquad \Delta H^\circ_{rxn} = -285.\ 85\ \text{kJ}, \qquad (4\text{-}13)$$

and

$$\Delta H_f^\circ\,[H_2O(\ell)] = -285.85\ \text{kJ mol}^{-1}$$

The units of ΔH_{rxn} and ΔH_f° are not identical. The definition of a formation equation specifies that *one mole* of compound appears as the product, so the amount of heat is specific to one mole and the units of ΔH_f° are kJ *mol*$^{-1}$.

The following points about formation reactions and standard heats of formation should be noted.

1. It is a consequence of the definition of "formation equation" that the heat of formation of any *element* in its standard state is zero.
2. It is not necessary that a formation *reaction* be experimentally convenient or, for that matter, even experimentally possible. The formation *equation* can always be written, and its ΔH determined by indirect means if necessary.
3. The heat of any reaction, including a formation reaction, may change if the temperature changes. Therefore, a ΔH_f° value refers to a particular temperature, which, if desired, may be indicated in the subscript: $\Delta H_{f,500}^\circ [CO_2(g)]$ is the heat of formation of carbon dioxide at 500 K. The most common reference temperature is 298 K (25 °C), and ΔH_f° values lacking a temperature indication in the subscript may be assumed to refer to 298 K.

The paragraph following Example 4.3 outlines an indirect means of determining the heat of formation of carbon monoxide.

All ΔH_f° values used in this text refer to 25 °C.

Each chemical compound has its own heat of formation, listed in tables such as Table 4.1 and Appendix G.

Calculation of Heat of Reaction

Provided that values of ΔH_f° for all reactants and products are known, heats of formation can be used to calculate the heat of any chemical reaction. It is not necessary to *measure* the heat of reaction. Consider the formation of ethylene glycol from acetylene and water in Equation (4-14).

$$C_2H_2(g) + 2\ H_2O(\ell) \longrightarrow C_2H_4(OH)_2(\ell) \qquad (4\text{-}14)$$

The heat of reaction is calculated as follows. The formation equation of each of the *product* compounds (there is only one in this example) is written as a thermochemical equation, and multiplied by its stoichiometric coefficient in Equation (4-14). Any *elements* formed in the reaction are ignored, since their heats of formation are zero. For each *reactant* compound, the formation equation is reversed and multiplied by its stoichiometric coefficient in Equation (4-14). Again, elements are ignored. Remember

TABLE 4.1 Standard Heats of Formation of Selected Compounds at 25 °C

Compound	ΔH_f°/kJ mol^{-1}	Compound	ΔH_f°/kJ mol^{-1}
$Al_2O_3(s)$	−1675.7	$CaO(s)$	−635.09
C(diamond)	1.895	$CaCO_3(s)$	−1206.92
$CH_4(g)$	−74.81	$HCl(g)$	−92.307
$C_2H_2(g)$	226.73	$H_2O(\ell)$	−285.830
$C_2H_4O(\ell$, acetaldehyde)	−192.3	$H_2O(g)$	−241.818
$C_2H_5OH(\ell$, ethanol)	−277.8	$H_2O_2(\ell)$	−187.78
$C_2H_4(OH)_2(\ell$, glycol)	−255	$HNO_3(\ell)$	−174.10
$C_4H_{10}O(\ell$, ethyl ether)	−279.3	$HgO(s)$	−90.83
$C_3H_4(g)$	190.5	$MnO_2(s)$	−520
$C_3H_8(g)$	−103.8	$NO(g)$	90.25
$C_6H_{12}O_6$(s, glucose)	−1273.3	$NO_2(g)$	33.18
$CO(g)$	−110.525	$NaOH(s)$	−425.609
$CO_2(g)$	−393.509	$SiH_4(g)$	34.3
		$SiO_2(s)$	−910.94

that the sign of the heat of reaction must be changed when the direction of an equation is reversed.

$$2\ C(s) + 3\ H_2(g) + O_2(g) \longrightarrow C_2H_4(OH)_2(\ell);$$
$$\Delta H_a = \Delta H_f^\circ(C_2H_4(OH)_2) = \quad -255\ kJ \quad (4\text{-}14a)$$
$$C_2H_2(g) \longrightarrow 2\ C(s) + H_2(g);$$
$$\Delta H_b = -\Delta H_f^\circ(C_2H_2) \quad = \quad -(+227)\ kJ \quad (4\text{-}14b)$$
$$2\ H_2O(\ell) \longrightarrow 2\ H_2(g) + O_2(g);$$
$$\Delta H_c = -2\Delta H_f^\circ(H_2O) \quad = -2(-286)\ kJ \quad (4\text{-}14c)$$

$$C_2H_2(g) + 2\ H_2O(\ell) \longrightarrow C_2H_4(OH)_2(\ell); \qquad \Delta H^\circ_{rxn} = \Delta H_a + \Delta H_b + \Delta H_c$$

The sum of these reactions is the overall reaction in Equation (4-14), whose heat we wish to determine. By Hess's law the heat of the overall reaction, Equation (4-14), is the sum of the heats of the steps.

$$\begin{aligned}\Delta H^\circ_{rxn} &= [\Delta H_f^\circ(C_2H_4(OH)_2)] + [-\Delta H_f^\circ(C_2H_2)] + [-2\Delta H_f^\circ(H_2O)] \quad (4\text{-}15)\\ &= -255 + (-227) + (-2)\cdot(-286)\\ &= +90\ kJ\end{aligned}$$

This algebraic procedure corresponds to the chemical process of dismantling all the reactant molecules into their elements, and then constructing the product molecules from the elements. Although chemical reactions do not take place in this artificial manner, Hess's law assures us that the heat of reaction calculated in this way is the same as the heat of the actual reaction.

After you have studied this calculation and understood the reasoning behind it, it will not be necessary to write down the individual steps [(4-14a) through (4-14c)]. The desired heat of reaction may be obtained directly from a generalization of Equation (4-15). To find ΔH°_{rxn} for the desired reaction, add up the heats of formation of all the products. Each ΔH_f° in this sum must be multiplied by the stoichiometric coefficient n of that product in the balanced chemical equation. Call this sum $\Sigma\ [n \cdot \Delta H_f^\circ$ (products)]. Then add up all the heats of formation of the reactants, each ΔH_f° multiplied by its stoichiometric coefficient. Call this sum $\Sigma\ [n \cdot \Delta H_f^\circ$(reactants)]. The heat of the reaction is given by the difference between these sums:

The Greek uppercase sigma, Σ, is the symbol used to mean "the sum of."

$$\Delta H^\circ_{rxn} = \Sigma\ [n \cdot \Delta H_f^\circ(\text{products})] - \Sigma\ [n \cdot \Delta H_f^\circ(\text{reactants})] \qquad (4\text{-}16)$$

EXAMPLE 4.5 Heat of Combustion

Calculate ΔH°_{rxn} for the combustion of methane.

$$CH_4(g) + O_2(g) \longrightarrow CO_2(g) + H_2O(\ell)$$

Analysis

Target: ΔH°_{rxn} for a given reaction.

Relationship: In this case the lunge works. Equation (4-16) is used for ΔH°_{rxn} calculations.

Plan

1. Balance the chemical equation if necessary.
2. Look up the heats of formation of all compounds (see Table 4.1).
3. Insert the ΔH_f° values into Equation (4-16) and solve.

Work

1. The balanced equation is

$$CH_4(g) + 2\,O_2(g) \longrightarrow CO_2(g) + 2\,H_2O(\ell)$$

2. The heats of formation are

products: $CO_2(g)$, -393.5 kJ mol^{-1}; $H_2O(\ell)$, -285.8 kJ mol^{-1}
reactants: $CH_4(g)$, -74.8 kJ mol^{-1}; ΔH_f° for O_2 is zero.

3.

$$\begin{aligned}\Delta H^\circ_{rxn} &= \Sigma\,[n \cdot \Delta H_f^\circ\,(\text{products})] - \Sigma\,[n \cdot \Delta H_f^\circ\,(\text{reactants})] \qquad (4\text{-}16)\\ &= [\Delta H_f^\circ(CO_2) + 2\Delta H_f^\circ(H_2O)] - [\Delta H_f^\circ(CH_4) + 2\Delta H_f^\circ(O_2)]\\ &= [-393.5 + 2(-285.8)] - [-74.8 + 2(0)]\\ &= -890.3\text{ kJ}\end{aligned}$$

Check ΔH°_{rxn} is negative, as it must be, since all combustion reactions are exothermic.

Exercise

Calculate the heat of combustion of acetylene.

$$C_2H_2(g) + \tfrac{5}{2}\,O_2(g) \longrightarrow 2\,CO_2(g) + H_2O(\ell)$$

Answer: -1299.5 kJ (per mole of acetylene).

Standard heats of formation are used in this method, and therefore the heat of reaction found from them is the *standard heat of reaction* ΔH°_{rxn}. The standard heat of reaction is the heat produced or consumed when the reaction takes place at normal atmospheric pressure. The symbol ΔH_{rxn}, without the superscript "o", refers to reactions taking place at any other pressure.

In addition to finding a reaction heat without resorting to experiment, we can also use Equation (4-16) to determine an unknown heat of formation. In this case the heat of reaction must be measured, and all other heats of formation must be known.

EXAMPLE 4.6 Calculation of Heat of Formation

Nitrogen dioxide reacts with water to produce nitric acid and nitrogen monoxide according to the equation

$$3\,NO_2(g) + H_2O(\ell) \longrightarrow 2\,HNO_3(\ell) + NO(g); \qquad \Delta H^\circ_{rxn} = -71.7\text{ kJ}$$

Values of ΔH_f° (kJ mol^{-1}) are NO(g), 90.3; $NO_2(g)$, 33.2; $H_2O(\ell)$, -285.8. Calculate the heat of formation of $HNO_3(\ell)$.

Analysis The relationship in Equation (4-16) applies. In contrast to the previous example, however, this time one of the heats of formation is unknown.

Plans After checking that the equation is balanced (it is), set up Equation (4-16) with all the given information and solve for the missing term.

Work

$$\Delta H^\circ_{rxn} = \Sigma\,[n \cdot \Delta H^\circ_f(\text{products})] - \Sigma\,[n \cdot \Delta H^\circ_f(\text{reactants})] \quad (4\text{-}16)$$

$$-71.7 = [2\Delta H^\circ_f(HNO_3) + \Delta H^\circ_f(NO)] - [3\Delta H^\circ_f(NO_2) + \Delta H^\circ_f(H_2O)]$$

Again, note that the stoichiometric coefficients from the balanced chemical equation must be used as multipliers.

$$= [2\Delta H^\circ_f(HNO_3) + 90.3] - [3(33.2) + (-285.8)]$$
$$= 2\Delta H^\circ_f(HNO_3) + 90.3 - 99.6 + 285.8$$

$$-2\Delta H^\circ_f(HNO_3) = 90.3 - 99.6 + 285.8 + 71.7 = +348.2 \text{ kJ}$$

$$\Delta H^\circ_f(HNO_3) = -\frac{348.2}{2} = -174.1 \text{ kJ mol}^{-1}$$

Exercise

The decomposition of $H_2O_2(\ell)$ into $H_2O(\ell)$ and $O_2(g)$ liberates 98.1 kJ heat to the environment for each mole of hydrogen peroxide decomposed. Using this fact and the value $\Delta H^\circ_f[H_2O(\ell)] = -285.8$, calculate $\Delta H^\circ_f[H_2O_2(\ell)]$.

Answer: -187.8 kJ mol^{-1}.

4.4 MEASUREMENT OF HEAT—CALORIMETRY

There are three types of process that involve heat transfer: (1) chemical reactions; (2) phase transformations such as melting or vaporization that take place without change of temperature; and (3) heating or cooling of a substance, without change of physical state. Although these processes can occur together, for example chemical reactions are often accompanied by a rise in temperature, calculations are made as if they occurred separately. Heat resulting from chemical reactions was treated in the previous section, and in the remainder of this section we consider the other two processes.

Heat in Physical Processes

A change in physical state of a substance can be written as a chemical equation.

$$H_2O(s) \longrightarrow H_2O(\ell); \qquad \Delta H^\circ = \Delta H^\circ_{fus} = 6.0 \text{ kJ mol}^{-1} \quad (4\text{-}17)$$

$$H_2O(\ell) \longrightarrow H_2O(g); \qquad \Delta H^\circ = \Delta H^\circ_{vap} = 44.0 \text{ kJ mol}^{-1} \quad (4\text{-}18)$$

The heat of reaction is called **heat of fusion, ΔH_{fus},** for the transition solid → liquid, and **heat of vaporization, ΔH_{vap},** for the transition liquid → vapor. These heats vary from one substance to another, but they are *always* positive; melting of a solid and evaporation (or boiling) of a liquid *always* require the input of heat. Heats of vaporization also vary with temperature. It takes 40.7 kJ to evaporate a mole of water at 100 °C, compared to the 44.0 kJ required at 25 °C.

When water evaporates from the skin, the necessary heat is withdrawn from the body. This cools the skin, and is an important part of the body's temperature control mechanism.

Heats of reaction for the reverse processes are of course negative quantities, and freezing and condensation are *always* exothermic processes.

$$H_2O(\ell) \longrightarrow H_2O(s); \qquad \Delta H^\circ = -\Delta H^\circ_{fus} = -6.0 \text{ kJ mol}^{-1} \quad (4\text{-}17r)$$

$$H_2O(g) \longrightarrow H_2O(\ell); \qquad \Delta H^\circ = -\Delta H^\circ_{vap} = -44.0 \text{ kJ mol}^{-1} \quad (4\text{-}18r)$$

For phase transitions, heat = $n \cdot \Delta H_{fus}$ or heat = $n \cdot \Delta H_{vap}$.

Calculation of the amount of heat transferred in a phase transition is similar to the calculation for a true chemical reaction: ΔH°_{fus} or ΔH°_{vap} is multiplied by the number of moles undergoing the transition. Often the amount of substance is given in grams, so that conversion to moles must be included in the calculation.

The symbol "q" is commonly used to represent an unknown quantity of heat.

The symbol "q" is used to represent the heat transferred in a chemical or physical process. In keeping with the fact that ΔH_{rxn} is positive for an endothermic reaction, which requires the absorption or gain of heat from the surroundings, we note that a positive value of q means a *gain* of heat, while if q is negative, heat is *lost*.

EXAMPLE 4.7 Heat of Fusion

How much heat is required to melt 15 g of ice?

Analysis

Target: An amount of heat

$$q = ?\ \text{J}$$

Knowns: The mass of ice to be melted.

Relationship: The heat can be calculated from the number of moles if the molar heat of fusion is known.

Missing Information The molar heat of fusion (6.0 kJ mol^{-1}) is an implicit known, because it can be looked up in a table or obtained from Equation (4-17). The molar mass of water is another implicit known.

Plan/Work Set the problem up as a unit conversion, proceeding from grams → moles → kilojoules.

$$q = (15\ \text{g}) \cdot \frac{(1\ \text{mol})}{(18.0\ \text{g})} \cdot \frac{(6.0\ \text{kJ})}{(1\ \text{mol})}$$

$$= 5.0\ \text{kJ}$$

Note: q is a positive quantity, indicating that ice must absorb heat in order to melt.

Exercise

Calculate the heat absorbed when a cup (250 g) of water evaporates at 25 °C.

Answer: 611 kJ.

Melting of a solid is *always* an endothermic process. Enough heat is absorbed from the hand to melt this bar of gallium metal.

Temperature Change, Heat Capacity, and Specific Heat

Whenever there is neither a phase change nor a chemical reaction, a transfer of heat causes a change of temperature. The amount of heat necessary to increase the temperature of an object by 1 °C is known as the **heat capacity (*C*)** of the object. Heat capacity is proportional to the mass of the object, it is always positive, and its units are J °C^{-1}. The heat transfer that accompanies temperature changes is given by

$$q = C(T_{\text{final}} - T_{\text{initial}})$$

or, using the abbreviation $\Delta T = T_{\text{final}} - T_{\text{initial}}$

$$q = C\Delta T \tag{4-19}$$

The symbol "Δ" is commonly used to refer to the *difference* between the value of a quantity before and after some change takes place.

Since the size of the kelvin and the Celsius degree is the same, any ΔT has the same numerical value when expressed in K or °C. Either scale may be used in heat capacity problems, and either J °C^{-1} or J K^{-1} may be written for the units of heat capacity.

EXAMPLE 4.8 Heat Capacity and Temperature Change

Calculate the heat required to increase the temperature of an iron bar from 25 °C to 600 °C (red heat, approximately). The heat capacity of the bar is 8.0 kJ °C^{-1}.

Analysis The process is a temperature change without phase change, so Equation (4-19) is appropriate. Any problem based on this equation will supply values for two of the quantities q, C, and ΔT, and ask for the calculation of the third. In this case, C and $\Delta T = (T_{\text{final}} - T_{\text{initial}})$ are explicit knowns.

Plan Insert the known values into Equation (4-19) and solve.

Work

$$\begin{aligned} q &= C\Delta T \\ &= (8.0 \text{ kJ °C}^{-1})[(600 - 25)(\text{°C})] \\ &= 4.6 \times 10^3 \text{ kJ} \end{aligned}$$

Exercise

What is the heat capacity of an iron bar if it releases 5.0×10^3 kJ heat to the environment as it cools from 800 °C to 350 °C?

Answer: $C = 11$ kJ °C^{-1} or 11 kJ K^{-1}.

Heat capacity information for different substances is normally given by the **specific heat (c_s)**, the heat capacity per gram of substance. Specific heat is defined as the heat required to increase the temperature of one gram of substance by one 1 °C (or 1 K). The units of specific heat are J °C^{-1} g^{-1} or J K^{-1} g^{-1}, and, like heat capacity, it is by definition a positive quantity. Each substance has a characteristic value of c_s, a few of which are given in Table 4.2.

The heat capacity of an object is given by the object's mass times the specific heat of the material.

$$C(\text{J K}^{-1}) = m(\text{g}) \cdot c_s(\text{J K}^{-1}\text{ g}^{-1})$$

For calculations involving c_s rather than C, Equation (4-20) is used in place of (4-19)

$$q = mc_s\Delta T \tag{4-20}$$

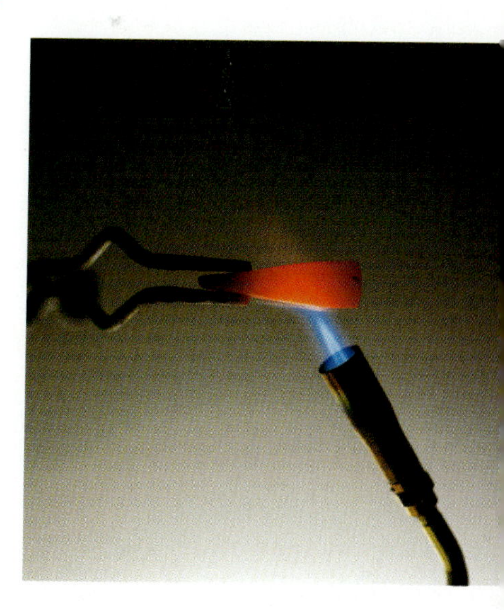

Metals can easily be brought to red heat (7600 °C) in the flame of a propane torch.

TABLE 4.2 Specific Heats and Molar Heat Capacities of Selected Substances

Substance	c_s/J K^{-1} g^{-1}	c_P/J K^{-1} mol^{-1}
$H_2O(\ell)$	4.184	75.31
$H_2O(s)$	2.09	37.7
$H_2O(g)$	2.03	36.6
$C_2H_5OH(\ell$, ethanol)	2.46	113.0
$SiO_2(s)$	0.74	44.3
$CaCO_3(s)$	0.81	81.3
Fe(s)	0.448	25.0
Cu(s)	0.385	24.5
Pb(s)	0.129	26.7
Air	1.01	
Glass	0.78	
Granite	0.80	
Polyethylene	2.3	

EXAMPLE 4.9 Specific Heat and Temperature Rise

How much heat is required to increase the temperature of a cup of water (200 g) from 25 °C to 97 °C?

Analysis This problem involves temperature change without phase change, so Equation (4-19) or (4-20) should be used. Heat capacity is not given, but specific heat can be looked up in tables such as Table 4.2.

Plan Insert the given data into Equation (4-20) and solve.

Work

$$\begin{aligned} q &= mc_s\Delta T \\ &= (200\text{ g}) \cdot (4.184\text{ J °C}^{-1}\text{ g}^{-1}) \cdot (97\text{ °C} - 25\text{ °C}) \\ &= 60 \times 10^3\text{ J} = 60\text{ kJ} \end{aligned}$$

Exercise

What is the specific heat of copper metal, if 290 J are required to heat a 15 g sample from 25 °C to 75 °C?

Answer: $c_s = 0.39$ J °C^{-1} g^{-1} or 0.39 J K^{-1} g^{-1}.

Often, heat capacity information is expressed as per mole quantities rather than per gram quantities. These are called **molar heat capacity** and given the symbol c_P;

the units are J K^{-1} mol^{-1} or J °C^{-1} mol^{-1}*. Molar heat capacity is equal to specific heat times the molar mass of the substance: $c_P = c_s \cdot \mathcal{M}$. Both specific heat and molar heat capacity are shown in Table 4.2.

Other Units—Calories and Foods

Years ago, before the equivalence of all forms of energy was realized, heat was measured in a different unit, called the **calorie (cal).** This unit was originally defined so that 1 calorie is the amount of heat required to increase the temperature of 1 gram H_2O by 1 °C. Today, however, the value of the calorie is defined in terms of the SI energy unit: the equivalence statement is 1 cal ⇔ 4.184 J. Many textbooks, including some recent ones, continue to use calories as a measure of heat. Older handbooks and other sources list thermochemical quantities, including heats of combustion, standard heats of formation, specific heats, and so on, in calories rather than joules.

The definition 1 cal ⇔ 4.184 J is exact. Its use in unit conversions does not limit the precision of the result to four significant digits, any more than the use of the exact equivalence 1 yard = 3 feet limits a height conversion to one significant digit.

EXAMPLE 4.10 The Units of Specific Heat

The specific heat of gold at 0 °C is 0.0302 cal g^{-1} °C^{-1}. Convert this value to SI units, and express it also in molar units.

Analysis

Target: A specific heat in SI units, J g^{-1} K^{-1} and J mol^{-1} K^{-1}.

Knowns: The value in cal g^{-1} °C^{-1}; the conversion factors are implicit, and can be looked up if not remembered.

Relationship: The heat capacity value 0.0302 cal g^{-1} °C^{-1} is another way of writing the equivalence statement

$$0.0302 \text{ cal} \Longleftrightarrow 1 \text{ g} \cdot 1 \text{ °C}$$

Also, the units °C and K are interchangeable in heat-capacity calculations, because they refer to a temperature *difference.*

Plan Use the factor-label approach to unit conversions.

Work

$$c_s = \frac{(0.0302 \text{ cal})}{(1 \text{ g °C})} \cdot \frac{(4.184 \text{ J})}{(1 \text{ cal})}$$
$$= 0.126 \text{ J g}^{-1} \text{ °C}^{-1} = 0.126 \text{ J g}^{-1} \text{ K}^{-1}$$

* The subscript "P" refers to normal heating or cooling processes in which the pressure does not change. If the process takes place without change in *volume,* the symbol c_v is used for the molar heat capacity. c_v is hardly ever used for liquids and solids, which rarely undergo constant-volume temperature changes. But it is used for gaseous substances, and, for gases, c_v is about 8 J mol^{-1} K^{-1} less than c_P.

The further conversion to molar units is achieved by

$$c_P = \frac{(0.126 \text{ J})}{(1 \text{ g °C})} \cdot \frac{(197.0 \text{ g})}{(1 \text{ mol})}$$
$$= 24.9 \text{ J mol}^{-1} \text{ °C}^{-1} = 24.9 \text{ J mol}^{-1} \text{ K}^{-1}$$

Check Since 1 cal $\Leftrightarrow$ 4.184 J, any quantity expressed in joules is about four times larger than when expressed in calories. The heat capacity is $\approx 4(0.03) = 0.12$, and, in molar units, it is about 200 times that: $200(0.12) = 24$.

Exercise

An older handbook gives the standard heat of formation of CaF_2(s) as −291.49 kcal mol^{-1}. Convert this quantity to SI units.

Answer: −1219.6 kJ mol^{-1}.

Another non-SI unit that continues to be widely used, especially in connection with foods and nutrition, is the nutritional Calorie (C), equal to 1000 cal. Either of the equivalence statements 1 C $\Leftrightarrow$ 1000 cal or 1 C $\Leftrightarrow$ 4184 J may be used in unit conversions.

The energy required for the functioning of our bodies is obtained from the reaction of the components of food with oxygen. The process is, in effect, a slow combustion, and food is the fuel. Much of the energy from this exothermic process is used as heat, to maintain body temperature. Some is used in muscular activity, and some is used in tissue growth and the synthesis of the many compounds needed to maintain life and health.

The energy released by the combustion of one gram of food, called its *fuel value*, can be measured in a bomb calorimeter (Section 4.4). Typical fuel values of carbohydrates are about 4 C g^{-1} (17 kJ g^{-1}); of fats, about 9 C g^{-1} (38 kJ g^{-1}); and of proteins, about 4 C g^{-1} (17 kJ g^{-1}). Table 4.3 lists a few foods, together with their fuel values and approximate compositions. For comparison, the listing includes two actual fuels.

TABLE 4.3 Fuel Values of Selected Substances

	Fuel Value		Composition (%)			
	C g^{-1}	kJ g^{-1}	Carbohydrate	Fat	Protein	Water
Banana	0.9	3.9	23	0.7	1.3	74
Bread, white	2.7	11	50	4.0	8.4	37
Cheese, Swiss	3.4	14	4	26	26	44
Chocolate bar	4.1	17	76	11	3	11
Eggs, boiled	1.6	6.7	2	12	12	75
Fish sticks	2.5	10	14	11	21	52
Hamburger	2.9	12	0	21	26	54
Milk (2%)	0.5	2.1	4.9	2.0	3.3	89
Peanut butter	5.9	25	19	50	31	1
Gasoline	11	48				
Natural gas	12	49				

Measurement Apparatus and Techniques

Measurement of heat involved in a chemical or physical process is carried out with an apparatus called a **calorimeter.**

Specific Heat and Temperature Change

Calorimeters can in some cases be assembled from surprisingly simple parts. An ordinary disposable styrofoam coffee cup provides a well-insulated reaction vessel for studying heat changes in liquids. Using such a "coffee-cup calorimeter," unknown specific heats and heats of reaction can be determined by measurements of mass, volume, and temperature change (Figure 4.5).

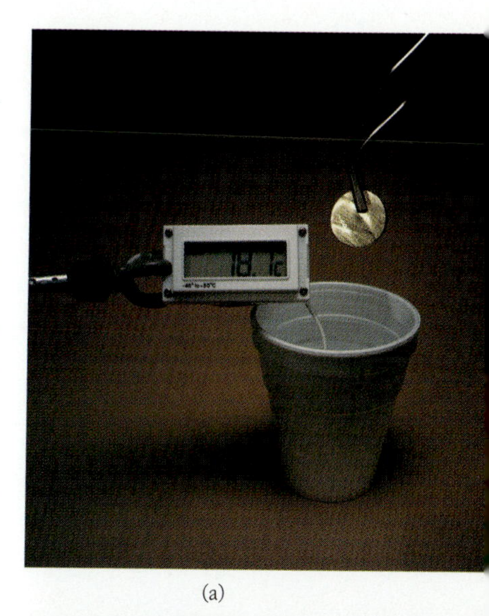

(a)

(b)

Figure 4.5 (a) A heated piece of metal is dropped into a styrofoam cup containing a known amount of water at a known temperature. (b) From the observed rise in temperature, the specific heat of the metal can be calculated.

EXAMPLE 4.11 Measurement of Specific Heat

A 57.2 g chunk of metal, whose specific heat is unknown, is heated to 90 °C and dropped into a coffee cup containing 50.0 g (= 50 mL) water at 25.0 °C. Heat is transferred from the hot metal to the cooler water, and, as a result, the metal temperature falls and the water temperature rises. The heat transfer continues until the water and the metal are both at the same temperature, 31.2 °C. Calculate the specific heat of the metal.

Analysis

Target: The specific heat of a metal

$$c_s = ?\ \text{J g}^{-1}\ {}^\circ\text{C}^{-1}$$

Knowns: Initial and final temperatures, and the masses of water and metal. The specific heat of water is an implicit known.

Relationship: The heat lost by the metal as it cools is equal in magnitude to the heat gained by the water as it warms to the final temperature. However, since heat lost is a negative quantity, while heat gained is positive, the algebraic statement of this relationship must be written with a negative sign.

$$q_{\text{metal}} = -q_{\text{water}}$$

This equation, together with Equation (4-20), is sufficient to solve the problem.

Plan

1. Use the initial and final temperatures, mass, and specific heat of water to calculate the heat gained by the water, q_{water}.
2. Use $q_{\text{metal}} = -q_{\text{water}}$ and the known mass and temperature change to determine the specific heat of the metal.

Work

1.

$$\begin{aligned} q_{\text{water}} &= mc_s\Delta T \\ &= (50.0\text{ g})(4.184\text{ J }{}^\circ\text{C}^{-1}\text{ g}^{-1})[(31.2 - 25.0)({}^\circ\text{C})] = 1297\text{ J} \end{aligned}$$

2.

$$q_{metal} = mc_s\Delta T$$

$$c_s = \frac{q_{metal}}{m\Delta T}, \quad \text{and} \quad q_{metal} = -q_{water} = -1297\text{ J}$$

$$c_s = \frac{-1297\text{ J}}{(57.2\text{ g})(31.2\text{ °C} - 90\text{ °C})} = 0.39\text{ J °C}^{-1}\text{ g}^{-1}$$

Note that in step 2, $\Delta T = T_{final} - T_{initial}$ is a negative quantity, as is the corresponding quantity of heat. ΔT and q are always negative in a cooling process.

Exercise

A small sample of metal at 77.3 °C is dropped into a coffee cup containing 51.8 g H_2O at 23.2 °C, and the water temperature rises to 28.1 °C. If the specific heat of the metal is known to be 0.228 J g^{-1} °C^{-1}, what is the mass of the metal?

Answer: 94.7 g.

EXAMPLE 4.12 Calculation of Final Temperature

A piece of lead weighing 100 g is heated to 135 °C and dropped into a styrofoam cup containing 75.0 g H_2O at 26.3 °C. What is the final temperature of the system?

Analysis This is a heat transfer problem involving temperature change with no phase transformation. (Even though the initial temperature of the metal exceeds the boiling point of water, the metal will quickly cool. The amount of water that evaporates is insignificant and may be ignored.)

Target: The final temperature, which is the same for both the lead and the water.

Knowns: Specific heats will have to be looked up, but all other required information is given.

Relationship: The heat gained by the water, q_{water}, is equal to the heat lost by the metal as it cools, $-q_{metal}$. The relationship $q = mc_s\Delta T$ holds for both the water and the metal.

Plan

1. Write the equality between the two heats as $q_{water} = -q_{lead}$.
2. Represent each of these heats by Equation (4-20), using T_f for the unknown final temperature.
3. Solve the resulting equation for T_f, the only unknown quantity.

Work

1.

$$q_{water} = -q_{lead}$$

2.

$$(mc_s\Delta T)_{water} = -(mc_s\Delta T)_{lead}$$

Recall that $\Delta T = (T_{\text{final}} - T_{\text{initial}})$, and insert all known values.

$$(75.0\text{ g})(4.184\text{ J }°\text{C}^{-1}\text{ g}^{-1})(T_f - 26.3)(°\text{C}) = -(100\text{ g})(0.128\text{ J }°\text{C}^{-1}\text{ g}^{-1})(T_f - 135)(°\text{C})$$

3.

$$313.8(T_f) - 8253 = -12.8(T_f) + 1728$$

$$(313.8 + 12.8)(T_f) = 1728 + 8253$$

$$326.6(T_f) = 9981$$

$$T_f = \frac{9981}{326.6} = 30.6°\text{C}$$

Check The final temperature is intermediate between the two initial temperatures, as it must be.

Exercise

What is the final temperature after 35 g H_2O at 20 °C is mixed with 65 g H_2O at 31 °C?

Answer: 27 °C.

The expression $q_{\text{water}} = -q_{\text{metal}}$, which was an important part of the solution of Examples 4.11 and 4.12, can be interpreted in a way that enlarges the simple relationship "heat gained = (−1)heat lost." Since the coffee cup is insulated, no heat enters it from the laboratory. That is, the *total* heat gain of the calorimeter, which is the sum of the heat gained by the water and the metal, is zero. This relationship follows from the law of conservation of energy. The algebraic form of this equality is $q_{\text{water}} + q_{\text{metal}} = 0$, which is of course equivalent to $q_{\text{water}} = -q_{\text{metal}}$. Either equation, and either interpretation, may be used in solving problems.

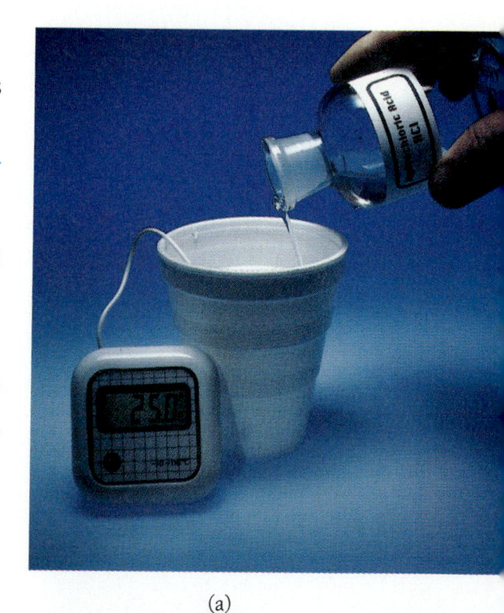

(a)

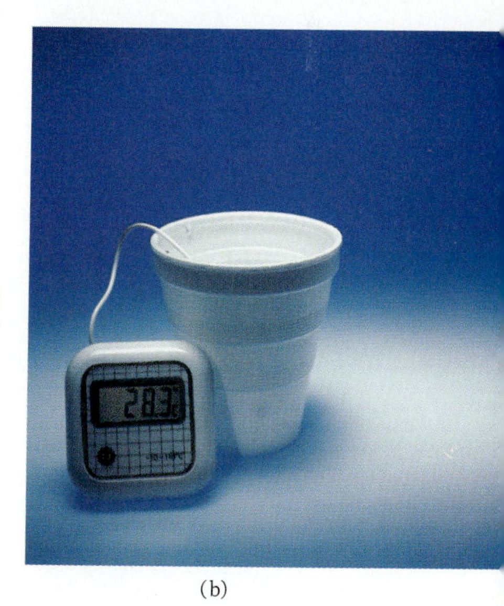

(b)

(a) When HCl(aq) is added to NaOH(aq), (b) the temperature rises.

Heat of Reaction

The coffee-cup calorimeter can also be used to determine the heat change in a chemical reaction. A common experiment in the introductory chemistry laboratory is the determination of the heat of the reaction between an acid and a base.

EXAMPLE 4.13 Measurement of Heat of Reaction

A 25.0 mL sample of 0.500 M hydrochloric acid at 25.0 °C is added to 25.0 mL of 0.500 M sodium hydroxide, also at 25.0 °C, in a coffee-cup calorimeter. After thorough mixing, the temperature of the solution is found to have increased to 28.3 °C. Calculate ΔH_{rxn}, expressing the result as kJ per mole HCl.

Analysis The exothermic reaction released heat that was absorbed by the solution, causing its temperature to rise.

Target: A heat of reaction, $\Delta H_{\text{rxn}} = ?$ kJ

Knowns: The volume and molarity of the reagents, and the temperature rise.

Relationship: The amount of heat can be calculated from $q = mc_s\Delta T$, provided that the specific heat and mass of the solution are known.

Missing Information: The mass and specific heat are not given. Mass can be calculated if the density is known, but that is not given either. The best we can do is to assume that, since the solution is not highly concentrated, the density and specific heat have approximately the same values as for pure water. Then, for the solution,

$$c_s \approx 4.2 \text{ J g}^{-1}\text{ °C}^{-1}$$

$$m \approx 50.0 \text{ mL} \cdot \frac{1.0 \text{ g}}{1.0 \text{ mL}} = 50 \text{ g}$$

Note: The errors introduced by these approximations are not severe, and probably of no greater importance than the imprecision (two significant figures) of the measured temperature rise. The calculated result should be within a few percent of the true value.

Plan

1. From the temperature rise and Equation (4-20), calculate the heat gained by the solution. This heat gain is equal to the heat released in the reaction (but with a negative sign, since it is heat *lost by the reactants*).
2. Find the number of moles of acid reacted from the volume and molarity of the solution.
3. Calculate the heat for one mole of acid, and check that the sign is correct.

Work

1.

$$\begin{aligned} q_{\text{solution}} &= mc_s\Delta T \\ &= (25 \text{ g} + 25 \text{ g})(4.2 \text{ J °C}^{-1}\text{ g}^{-1})(28.3 - 25.0)(\text{°C}) \\ &= 693 \text{ J}, \qquad \text{and} \end{aligned}$$

$$q_{\text{rxn}} = -q_{\text{solution}} = -693 \text{ J}$$

2.

$$n = V \cdot M = (0.0250 \text{ L}) \cdot (0.500 \text{ mol L}^{-1}) = 0.0125 \text{ mol}$$

3.

$$\Delta H_{\text{rxn}} = \frac{q_{\text{rxn}}}{n} = \frac{-693 \text{ J}}{0.0125 \text{ mol}} = -5.5 \times 10^4 \text{ J mol}^{-1}$$

$$= -55 \text{ kJ mol}^{-1}$$

Check Since the reaction causes a temperature rise, it *produces* heat: that is, it is exothermic and therefore ΔH_{rxn} is a negative quantity.

Exercise

Suppose 0.0225 mol of a substance reacts in a coffee-cup calorimeter containing 75 g H_2O, causing a temperature rise of 2.39 °C. What is ΔH_{rxn}?

Answer: $\Delta H_{\text{rxn}} = -33 \text{ kJ mol}^{-1}$

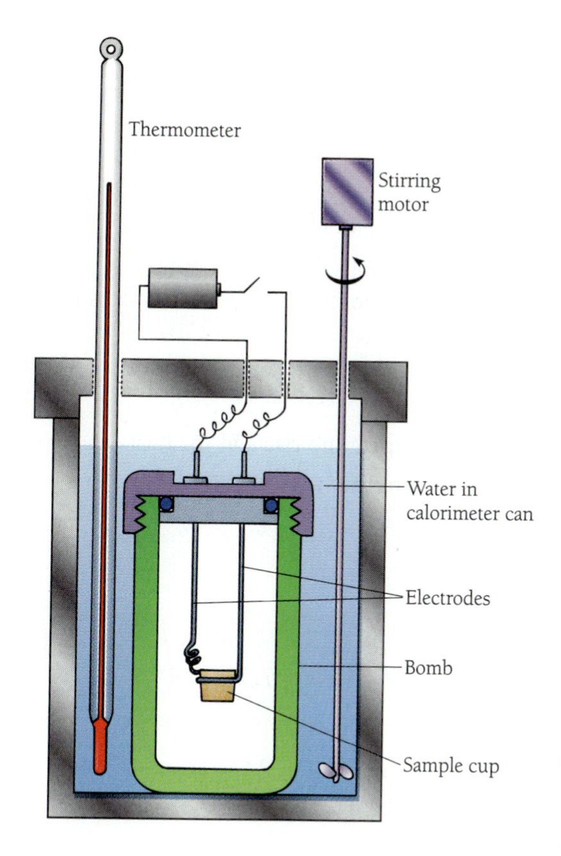

Figure 4.6 (a) The bomb calorimeter. A combustible sample is ignited by electrical heating and burns in an oxygen atmosphere inside a steel bomb. The heat given off by the reaction causes a rise in the temperature of the bomb and the water bath. The heat of reaction can be calculated from the temperature rise. (b) A disassembled calorimeter bomb.

Heat of Combustion

The heat generated when a substance burns is called the **heat of combustion (ΔH_c).** The units of ΔH_c are usually kJ per mol (of substance burned), although occasionally kJ g^{-1} is used. Many of the values appearing in tables of heats of formation are derived from heats of combustion. Measurement of the heat of combustion of a substance is carried out in an apparatus known as a **bomb calorimeter** (Figure 4.6). This apparatus consists of the bomb (a strong steel screw-top jar) and a water bath. In operation, a small amount (0.1 to 1.0 g) of sample, either liquid or solid, is placed in the bomb, which is then tightly closed. A high pressure of pure O_2 is admitted through a one-way valve in the lid, and the bomb is immersed in the water bath. The sample is ignited (by an electric current), and the heat evolved in the reaction causes a rise in the temperature of both the bomb and the water bath. The heat evolved is calculated from the measured temperature rise and the known heat capacity of the apparatus. This heat capacity, called the "calorimeter constant," depends not only on the bomb but also on the amount of water in the bath, and it must be remeasured each time the water is changed. Usually, the calorimeter constant is determined by burning a sample with a known heat of combustion just before the experiment with an unknown sample.

EXAMPLE 4.14 Calibration of a Bomb Calorimeter

A bomb calorimeter is calibrated by burning 0.332 g benzoic acid ($C_7H_6O_2$), which causes a temperature rise from 23.13 to 24.61 °C. Calculate the heat capacity of the calorimeter (that is, the total heat capacity of the bomb and the water). The heat of combustion of benzoic acid is −3227 kJ mol^{-1}.

The heat capacity of the calorimeter is the sum of the heat capacities of its parts, and specifically includes both the bomb and the water bath.

Analysis

Target: The heat capacity C of a calorimeter

$$C = ?\ \mathrm{J\ °C^{-1}}$$

Knowns: The mass and heat of combustion of the sample are given, as is the temperature rise. The molar mass of benzoic acid is an implicit known.

Relationship: If we knew the amount of heat evolved, we could evaluate C using Equation (4-19). The heat can be calculated from the amount of sample and its heat of combustion.

Plan The stepwise plan leading backward from the target is:

3. Calculate the heat capacity from the temperature rise and the heat.
2. Calculate the heat from the sample mass, molar mass, and heat of combustion.
1. Find the molar mass.

Work

1.

$$\text{molar mass} = 7(12) + 6(1) + 2(16) = 122\ \mathrm{g\ mol^{-1}}$$

2.

$$q = (0.332\ \mathrm{g}) \cdot \frac{(1\ \mathrm{mol})}{(122\ \mathrm{g})} \cdot \frac{(-3227\ \mathrm{kJ})}{(1\ \mathrm{mol})} = -8.78\ \mathrm{kJ}$$

3.

$$q = C\Delta T$$

$$C = \frac{q_{\text{calorimeter}}}{\Delta T} = \frac{+8.78\ \mathrm{kJ}}{(24.61 - 23.13)(°\mathrm{C})} = 5.93\ \mathrm{kJ\ °C^{-1}} = 5930\ \mathrm{J\ °C^{-1}}$$

Note that the heat calculated in step 2 is a negative quantity. It is negative because it is heat *lost* by the benzoic acid/oxygen system. In step 3, however, the same heat is entered as a positive quantity. The sign must be changed because it is heat *gained* by the calorimeter, whose heat capacity is the goal of the measurement. Remember that heat capacities are, by definition, positive quantities.

Exercise

By means of an electric heater, 1.00×10^4 J of thermal energy is added to a bomb calorimeter, resulting in a temperature rise of 1.63 °C. What is the heat capacity of the calorimeter?

Answer: 6.13 kJ °C^{-1}.

EXAMPLE 4.15 Measurement of Heat of Combustion

A 0.622 g sample of caffeine is burned in a bomb calorimeter whose measured heat capacity is 7910 J °C^{-1}. During the combustion the temperature of the water bath rises from 23.92 to 25.64 °C. Calculate the heat of combustion per gram of caffeine, and express the result in kJ mol^{-1} as well. (The molar mass is 194 g mol^{-1}.)

Analysis

Target: The heat of combustion per gram of sample, (? J g^{-1}), and per mole, (? J mol^{-1}).

Knowns: The heat capacity of the calorimeter, the temperature rise, the sample mass, and the molar mass are given.

Relationship: If we knew the amount of heat released in the combustion of the sample (0.622 g), we could use unit conversion factors to evaluate the combustion heat per gram and per mole. The heat released in the reaction must be equal to (the negative of) the heat gained by the calorimeter, which can be calculated from $q = C\Delta T$.

Plan The stepwise plan leading back from the target to the given information is therefore:

3. Calculate the heat per mole from the heat per gram and the molar mass.
2. Calculate the heat per gram from the total heat and the sample mass.
1. Calculate the total heat from the heat capacity of the calorimeter and the temperature rise. Change the sign to get the heat generated (lost) by the reaction.

Work

1.

$$\begin{aligned} q &= C\Delta T \\ &= (7910\ \text{J}\ ^\circ\text{C}^{-1})(25.64 - 23.92)(^\circ\text{C}) \\ &= 13{,}605\ \text{J} \\ q_{\text{rxn}} &= -13{,}605\ \text{J} \end{aligned}$$

2.

$$\begin{aligned} \text{heat of reaction} &= \frac{-13{,}605\ \text{J}}{0.622\ \text{g}} = -21{,}873\ \text{J g}^{-1} \\ &= -21.9\ \text{kJ g}^{-1} \end{aligned}$$

3. Also,

$$\begin{aligned} \text{heat of reaction} &= -21.9\ \text{kJ g}^{-1} \cdot \frac{194\ \text{g}}{1\ \text{mol}} \\ &= -4.24 \times 10^{3}\ \text{kJ mol}^{-1} \end{aligned}$$

Some minerals from Group 2A. Clockwise, from upper left, they are: dolomite, strontianite, beryl, barite, and emerald.

Note: As in the previous example, the algebraic sign of the heat must be changed. In step 1, 13,605 J is positive because it is heat gained by the calorimeter. In step 2, −13,605 J is negative because it is the heat lost by the reagents in an exothermic reaction. Any reaction, like combustion, that causes a *rise* in the temperature of the calorimeter is necessarily an *exothermic* reaction with a negative heat of reaction.

Exercise

What is the heat of combustion (per gram) of a substance if the burning of 0.494 g leads to a temperature rise of 2.15 °C in a bomb calorimeter whose heat capacity is 6.07 kJ $°C^{-1}$?

Answer: -26.4 kJ g^{-1}.

4.5 THE ALKALINE EARTH ELEMENTS—GROUP 2A

Beryl is the principal constituent of the gemstone emerald.

The metals of Group 2A, Be, Mg, Ca, Sr, Ba, and Ra, are known as the **alkaline earths.** They are quite reactive, and therefore do not occur in elemental form in nature. Beryllium is relatively rare; its main ore is *beryl* [$Be_3Al_2(SiO_3)_6$]. Magnesium and calcium compounds are abundantly distributed, both as minerals in the earth's crust and as salts dissolved in seawater. Magnesium occurs in the minerals *dolomite* ($MgCO_3 \cdot CaCO_3$), *carnallite* ($KCl \cdot MgCl_2 \cdot 6H_2O$), and quite a few others. Among the common minerals composed primarily of calcium carbonate are *chalk, limestone, calcite* and *marble. Gypsum* ($CaSO_4 \cdot 2H_2O$) and *fluorspar* (CaF_2) are also common. Barium occurs as the sulfate in the mineral *barite*, and strontium, much less abundant, occurs principally in the mineral *strontianite* ($SrCO_3$). The last member of the group, radium, is radioactive and exists in only minute quantities in nature. Some physical properties of the elements are given in Table 4.4.

Radium occurs as an impurity in uranium ore, from which it was first isolated by Marie and Pierre Curie. They earned the 1903 Nobel Prize in physics for this work.

Production and Uses of the Alkaline Earths

All of the Group 2A metals can be produced by reaction of the chloride salt with molten sodium.

$$MCl_2(s) + 2\,Na(\ell) \longrightarrow M(s) + 2\,NaCl(s)$$

The tires of the Hummer vehicles used in the Gulf War were made to be blow-out proof by placing light-weight, strong magnesium inserts within them. The Hummer pictured here is a civilian vehicle and its tires do not contain magnesium. The blow-out proof tires are available only for military use.

TABLE 4.4 Physical Properties of the Alkaline Earth Metals

Element	Symbol	Atomic Number	Molar Mass	Atomic Radius/pm	Melting Point/°C	Boiling Point/°C	Density /g mL^{-1}	Hardness (Moh's)
Beryllium	Be	4	9.012	112	1278	2970	1.85	
Magnesium	Mg	12	24.31	160	649	1090	1.74	2.0
Calcium	Ca	20	40.08	197	839	1484	1.55	1.5
Strontium	Sr	38	87.62	215	768	1384	2.63	1.8
Barium	Ba	56	137.33	222	725	1640	3.62	
Radium	Ra	88	226.03		700	1140	5.5	

In commercial operations, however, it is more common to electrolyze the molten salt.

$$MCl_2(\ell) \xrightarrow[\text{elect.}]{} M(\ell) + Cl_2(g)$$

Although calcium carbonate is a raw material for a large number of industrial processes, very little metallic calcium is produced. Of all the alkaline earth metals, only magnesium is produced in quantity.

The raw material for the production of magnesium is seawater. Addition of calcium hydroxide to seawater causes the separation of solid magnesium hydroxide, which is removed by filtration and converted to the water-soluble chloride by treatment with hydrochloric acid.

$$MgCl_2(aq) + Ca(OH)_2(s) \longrightarrow Mg(OH)_2(s) + CaCl_2(aq)$$

$$Mg(OH_2)(s) + 2\ HCl(aq) \longrightarrow MgCl_2(aq) + 2\ H_2O$$

Evaporation of the water leaves solid $MgCl_2$, which is then melted and electrolyzed to give elemental magnesium.

$$MgCl_2(\ell) \xrightarrow[\text{elect.}]{} Mg(\ell) + Cl_2(g)$$

Overall, it is an efficient process: the chlorine produced in the electrolysis is converted to HCl and recycled, and the calcium hydroxide is produced from seashells, whose primary constituent is $CaCO_3$.

The production of $Ca(OH)_2$ from $CaCO_3$ is described on page 162.

Beryllium finds increasing use as an alloying material. Addition of 2% to 3% Be to copper produces a *bronze* that is harder, stronger, and more elastic than copper, but retains the good electrical conductivity of copper. Some beryllium is used in x-ray apparatus, because it is the most transparent (to x-rays) of all metals.

Magnesium is a light-weight, relatively strong metal, and its strength and corrosion resistance can be improved by alloying with various other metals. Such alloys are extensively used in the aircraft industry. Another use is in the protection of steel (ship hulls and underground pipelines) from corrosion. Magnesium is also used in the nuclear power industry, both as a jacketing material for fuel elements (Chapter 24) and as a reagent in one of the steps of uranium processing.

The mechanism by which magnesium protects steel from corrosion is discussed in Chapter 19.

Calcium is used primarily in the form of lime (see below). Lime is important in steelmaking, pollution control, chemical manufacture, water treatment, papermaking, and (as concrete) the construction industry.

Strontium hydroxide is used in sugar refining to extract further quantities of sugar from the residue left from an earlier stage of the refining process. $Sr(NO_3)_2$ is used in fireworks. Barium sulfate is sometimes used as a whitener in paint and paper. Radium was once used as a component of luminous paint used for wristwatch dials, before it

(a)

(b)

(c)

Figure 4.7 When heated in a flame, compounds of the alkaline earth elements emit intensely colored light. (a) Calcium, (b) strontium, and (c) barium.

was realized that its radioactivity could cause cancer. But radiation also cures cancer, and $RaBr_2$ was once extensively used in cancer therapy. Other radioactive elements are now displacing radium from this use (Chapter 24).

Figure 4.8 The impressive colors of fireworks displays are produced by adding certain salts to the explosive charges in the rockets: barium salts give a green color, strontium salts give red, sodium salts give yellow, and copper salts are used for blue. Magnesium metal gives off a brilliant white light as it burns.

Some Compounds and Reactions

The Group 2A metals are quite reactive, although less so than the alkali metals. The chemistry of Group 2A is dominated by the ready formation of ionic compounds containing the doubly charged cations M^{2+}. Beryllium, however, forms molecular compounds only, and there is no evidence indicating the existence of Be^{2+} in any compound. Alkaline earth compounds, like those of the alkali metals, impart rich colors to a flame (Figure 4.7). This property is exploited in the manufacture of colorful fireworks (Figure 4.8).

Hydrides, Oxides, and Hydroxides

All Group 2A hydrides are known, and most can be produced by reaction of the metal with hydrogen.

$$M(s) + H_2(g) \xrightarrow{300-700\ ^\circ C} MH_2(s)$$

Beryllium does not react with hydrogen, but BeH_2 can be prepared by thermal decomposition (heating) of $Be[C(CH_3)_3]_2$.

Group 2A metals react with oxygen to produce the oxides.

$$2\ M(s) + O_2(g) \longrightarrow 2\ MO(s)$$

The reaction of beryllium is slow unless the metal is strongly heated. Magnesium ribbon burns readily in air, producing a brilliant white light. At room temperature, however, magnesium metal becomes coated with a thin layer of MgO that effectively seals the surface and prevents further reaction. The remaining group members react slowly at room temperature. All oxides can be prepared by heating the carbonates. The oxides undergo exothermic reactions with water, producing hydroxides.

$$MCO_3(s) \xrightarrow{\Delta} MO(s) + CO_2(g)$$

$$MO(s) + H_2O \longrightarrow M(OH)_2(aq)$$

The reactions of the Group 2A metals with water occur with increasing vigor as the atomic number increases. Be is unreactive, even at red heat; Mg reacts very slowly at room temperature (but readily with hot steam); Ca, Sr, and Ba react rapidly but not violently.

The increasing vigor of the reactions of the Group 2A oxides with water is another example of a group trend.

$$M(s) + 2\ H_2O(\ell) \longrightarrow M(OH)_2(aq) + H_2(g)$$

Some Other Compounds

All the Group 2A metals react with nitrogen at high temperature to form the nitrides. With the exception of Be_3N_2, these are ionic compounds that react with water to give ammonia.

$$3\ M(s) + N_2(g) \xrightarrow{\Delta} M_3N_2(s)$$

$$M_3N_2(s) + 6\ H_2O(\ell) \longrightarrow 3\ M(OH)_2(s \text{ or } aq) + 2\ NH_3(aq)$$

Similarly, direct reaction with carbon, sulfur, and the elements of Group 7A leads in most cases to binary ionic compounds such as CaC_2, MgS, and $BaCl_2$. The reaction of calcium with carbon produces calcium acetylide.

$$Ca(g) + 2\ C(s) \xrightarrow{2000\ °C} CaC_2(s)$$

A formerly important use of calcium acetylide was based on its rapid reaction with water to give the fuel gas acetylene (C_2H_2).

Calcium acetylide is also known by its older name, calcium carbide.

$$CaC_2(s) + 2\ H_2O(\ell) \longrightarrow Ca(OH)_2(s) + C_2H_2(g)$$

Years ago, helmets worn by miners and spelunkers had compartments holding water and CaC_2, separated by a valve for controlled addition of water to the reagent. A tube led to a headlamp in which the acetylene burned with a brilliant white flame (Figure 4.9). When heated with nitrogen, CaC_2 forms calcium cyanamide, $CaCN_2$. Calcium cyanamide is used both as fertilizer and as a raw material in the plastics industry.

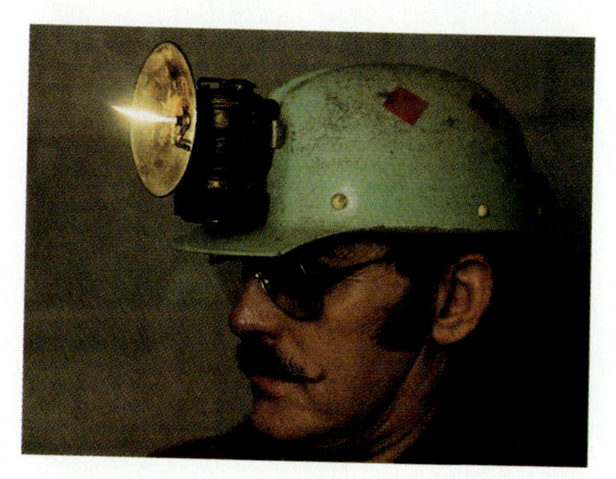

Figure 4.9 In a miner's lamp, the reaction of water dripping onto solid calcium acetylide produces acetylene, which is conducted to a jet where it burns with a brilliant flame.

A few drops of water sprinkled onto dry calcium oxide produces considerable heat and steam.

Lime

The two forms of lime are *quicklime* (calcium oxide, CaO) and *slaked lime* (calcium hydroxide, $Ca(OH)_2$). Together, the annual production of CaO and $Ca(OH)_2$ in the USA is about 20 million tons, which makes the lime industry the fourth most important in the country. Quicklime is produced by the thermal decomposition of limestone

$$CaCO_3(s) \xrightarrow{1200\ ^\circ C} CaO(s) + CO_2(g)$$

Addition of water produces slaked lime

$$CaO(s) + H_2O(\ell) \longrightarrow Ca(OH)_2(s)$$

About 8 million tons of lime are used annually by the steel industry to react with and remove the SiO_2 (sand) that contaminates iron ore. Slaked lime is an important material in the control of air pollution, since it reacts with the SO_2 formed in the combustion of sulfur-containing coal.

$$Ca(OH)_2(s) + SO_2(g) \longrightarrow CaSO_3(s) + H_2O(g)$$

The product of the reaction, calcium sulfite, is a solid that cannot escape into the atmosphere. Lime is also used in papermaking and as a raw material in the production of at least 150 important chemicals.

Lime making is an ancient process. The Romans mixed slaked lime with sand to produce the *mortar* used in their construction industry. *Portland cement* is a clay-limestone mixture that has been heated to convert the limestone to quicklime. When Portland cement is mixed with water, it quickly hardens or "sets" to a stone-like consistency. The added water forms $Ca(OH)_2$, and is also incorporated as water of hydration. The result is a strong matrix of interconnected mineral crystals. *Concrete* is a mixture of Portland cement, sand, and gravel.

Roman masons produced the mortar and cement required to build the structures which still stand today. Pictured is the Basilica of Constantine, a part of the Roman Forum in Rome, Italy.

Chemistry Old and New

FOOD TECHNOLOGY FOR OPERATION DESERT STORM

Food preservation technology is an important application of chemistry, from which people benefit each day. Almost every item found in the supermarket—from frozen TV dinners to canned vegetables to enriched flour—has been improved upon in some way with additives and preservatives. While processed foods are a major part of our diet, they are often a convenience rather than a necessity. On the other hand, the extreme circumstances of combat operations render food technology critical for the success of a military mission. Combat units, which are required to carry their own food, are often out of touch with base camp for many days at a time. Military food must be lightweight, have long storage lives, and need no refrigeration. Meals must also be nutritious (and appealing) enough to keep troops in good health and spirits. The enormous task of feeding thousands of soldiers is no less important than designing weapons or planning attack routes. It is no wonder that they say "an army travels on its stomach."

Until recently, the standard combat meal for army troops was the infamous C-ration—an unappealing tin of cold hash. In 1991, during Operation Desert Storm, however, troops were treated to "MRE"s (Meal, Ready-To-Eat), developed by food scientists at the U.S. Army Natick Research, Development, and Engineering Center in Natick, Massachusetts. MREs provide about 1300 calories and can serve as a soldier's only food for ten days. Sealed inside flexible polypropylene/aluminum foil pouches, MREs are designed to be used with Flameless Ration Heaters (FRHs). The FRH consists of a polymer sleeve containing a perforated box filled with a mixture of powdered magnesium (40%) and powdered iron (60%). To obtain a hot meal, soldiers need only to slip the MRE pouch into the FRH and add an ounce of water. The exothermic reaction of the water and metal powders to form oxides and hydroxides heats the MRE to about 160° F in less than 15 minutes.

Another Natick product developed for the Gulf War is preserved bread. Stable in storage for three years or more, this war-zone luxury is possible only through sophisticated chemical technology. Necessary additives include a sucrose (sugar) ester which acts as an emulsifying agent and stabilizer, a *humectant* to keep the bread from becoming brittle, and oxygen scavengers. Humectants are hygroscopic ("water-loving") chemicals that absorb water vapor from the atmosphere, allowing food to remain moist even in a dry environment. One such humectant is glycerin, which can absorb up to 50% of its mass in water vapor. Oxygen scavengers are important as preservatives because many chemical components of food react with O_2 to form peroxides, which can then degrade to undesirable substances. For example, the characteristic rancid odor and taste that develops in aging

meat and butter is due to a reaction with oxygen. By adding a chemical scavenger that removes O_2 from sealed food packages, such reactions can be delayed.

Current army research involves the development of products with even greater storage stabilities; one goal is a 1500 calorie meal with a shelf life of 10 years. *Freeze-drying* will play a large role in the success of the new ration. This technique involves freezing the water contained in foods to below 0° C and then placing the frozen food in a vacuum. Under these conditions, the H_2O molecules pass directly from ice crystals to the vapor phase (the process is called *sublimation*) where they are removed by the vacuum pump. The chemicals responsible for flavor are not harmed by freeze-drying, and no significant change of taste results. In addition, the elimination of water decreases the weight of food, which is important for troops carrying their own provisions. Since most bacteria require aqueous environments to grow, freeze-dried foods are also decay-resistant and may be stored for years without refrigeration.

Discussion Questions

1. What other uses might there be for the FRH or for the Mg/Fe/H_2O reaction as a source of heat?
2. Would powdered sodium be a suitable substitute for magnesium in the FRH?

SUMMARY EXERCISE

In a reaction known as the *thermite reaction*, powdered aluminum reacts with iron(III) oxide to produce aluminum oxide and molten iron. The reaction is highly exothermic, and releases 3.70 kcal for each gram of aluminum consumed.

a. Write a balanced equation for the thermite reaction.

b. What is ΔH_{rxn} for the reaction in (a)? Express the result in kJ.

c. The heat of fusion of metallic iron is 14.9 kJ mol^{-1}. What is ΔH for the reaction in (a) when the product is solid iron rather than liquid?

d. The standard heat of formation of aluminum oxide is −1676 kJ mol^{-1}. What is the standard heat of formation of iron(III) oxide?

e. How much iron is produced when 50.0 g iron(III) oxide reacts with 50.0 g aluminum? How much heat(kJ) is generated?

f. Suppose 75.0 mL Hg(ℓ) is warmed by addition of 1.50 kJ heat, and the temperature of the mercury is observed to rise from 25.0 °C to 96.1 °F. Determine the heat capacity of the mercury sample, the specific heat of mercury, and the molar heat capacity of mercury. The density of mercury at 25 °C is 13.6 g mL^{-1}.

Answers **(a)** $2\ Al(s) + Fe_2O_3(s) \rightarrow Al_2O_3(s) + 2\ Fe(\ell)$; **(b)** $\Delta H_{rxn} = -835$ kJ; **(c)** $\Delta H = -865$ kJ; **(d)** $\Delta H_f^\circ = -811$ kJ mol^{-1}; **(e)** 35.0 g Fe, 262 kJ produced; **(f)** $c = 141$ J °C^{-1}, $c_s = 0.139$ J g^{-1} °C^{-1}, $c_P = 27.8$ J mol^{-1} °C^{-1}.

SUMMARY AND KEYWORDS

Energy appears in many forms: heat, work, radiant energy, electrical energy, and so on. The SI unit of energy is the joule (J). The **law of conservation of energy** states that, although it may be changed to another form, energy is neither created nor destroyed in a chemical or physical process. Any object has **kinetic energy** by virtue of its motion, and the sum of the kinetic energies of all of a substance's atoms is called **thermal energy** or **heat.** Heat moves or flows spontaneously from a region of high temperature to a region of low temperature.

Exothermic chemical reactions produce heat, and **endothermic** reactions consume heat. In either case the amount of heat is proportional to the amount of material reacting. The **heat of reaction** (**ΔH_{rxn}** or simply **ΔH**) is negative for exothermic reactions and positive for endothermic process. A **thermochemical equation** gives the value of ΔH_{rxn} in addition to the stoichiometric coefficients. **Hess's law** states that the heat of reaction is the same regardless of whether the reaction takes place all at once or as a sequence of steps. A **formation** reaction is described by a thermochemical equation in which one mole of a compound is the sole product, and all reactants are pure elements in their **standard states** (the most stable state at 25 °C and 1 atm pressure). The heat of a formation reaction taking place at 1 atm pressure is called the **standard heat of formation (**ΔH_f°**)** of the product compound. Each chemical compound has its own ΔH_f° and the standard heat of formation of an element is zero. The heat of any chemical reaction can be calculated from the equation $\Delta H^\circ_{rxn} = \Sigma\ [n \cdot \Delta H_f^\circ\ (\text{products})] - \Sigma\ [n \cdot \Delta H_f^\circ\ (\text{reactants})]$.

The **heat of fusion ΔH_{fus}** is the heat input required to melt one mole of a solid substance, and the **heat of vaporization ΔH_{vap}** is the heat required to vaporize one mole of a liquid. Melting and evaporation are always endothermic.

When heat is added to an object or substance that undergoes no change of phase or chemical reaction, the temperature rises. The amount of heat added is $q = C\Delta T$, where C, the **heat capacity** of the object, is proportional to the mass, and ΔT is the temperature increase ($T_{final} - T_{initial}$). **Specific heat** c_s (J g^{-1} °C^{-1} or J g^{-1} K^{-1}), the heat capacity per gram, and c_P (J mol^{-1} K^{-1}), the molar heat capacity, are defined so that $q = mc_s\Delta T$ and $q = nc_P\Delta T$, where m = mass in grams and n = amount in moles.

Heat capacities and heats of reaction are measured in a **calorimeter.** A coffee-cup calorimeter is suitable for low-accuracy measurements of heat capacities and heats of reactions in solution, while **heats of combustion** are measured in a **bomb calorimeter.** Since a calorimeter is insulated and cannot gain heat from the environment, $q_{calorimeter} + q_{object}$ (or $q_{reaction}$) $= 0$, or $q_{calorimeter} = -q_{object}$ (or $-q_{reaction}$).

PROBLEMS

General

1. What is meant by heat of reaction?
2. If a reaction is characterized by $\Delta H_{rxn} = -500$ kJ mol^{-1}, does it absorb heat from the surroundings or release heat to them?
3. What is Hess's law?
4. Give a precise definition of formation equation and ΔH_f°.
5. What is meant by standard state?
6. Show that it is a consequence of the definition of heat of formation that $\Delta H_f^\circ = 0$ for any element in its standard state.
7. How may heats of reaction be determined from heats of formation?
8. Define exothermic and endothermic.
9. Is the freezing of water an exothermic or endothermic process?
10. Identify each of the following phase transitions as endo- or exothermic.
    a. solid → liquid
    b. gas → liquid
    c. liquid → solid
    d. liquid → gas
    e. solid → gas
    f. gas → solid
11. Define and give units for the following.
    a. Heat capacity
    b. Specific heat
    c. Molar heat capacity at constant pressure
12. For processes not involving chemical reaction or phase transformation, give equations relating heat and temperature change to (a) heat capacity, (b) specific heat, and (c) molar heat capacity.
13. Describe how a coffee-cup calorimeter would be used to determine the specific heat of a 10 g piece of metal.
14. Describe the operation of a bomb calorimeter.

Heat of Reaction

15. How much heat is required to decompose 33.9 g HBr by the following reaction?

$$2\,HBr(g) \longrightarrow H_2(g) + Br_2(\ell); \qquad \Delta H_{rxn} = +72.8\text{ kJ}$$

16. How much heat is required to decompose 694 g HCl by the following reaction?

$$2\,HCl(g) \longrightarrow H_2(g) + Cl_2(g); \qquad \Delta H_{rxn} = +184.6\text{ kJ}$$

17. How much heat is liberated when 0.0426 moles of sodium reacts with excess water according to the following equation?

$$2Na(s) + 2\,H_2O(\ell) \longrightarrow H_2(g) + 2\,NaOH(aq); \qquad \Delta H_{rxn} = -368\text{ kJ}$$

*18. How much heat is liberated by the reaction of 195 mL 0.962 M HCl with an excess of aqueous NaOH according to the following equation?

$$HCl(aq) + NaOH(aq) \longrightarrow NaCl(aq) + H_2O(\ell); \qquad \Delta H_{rxn} = -55.9\text{ kJ}$$

19. How much heat is required for the production of 1 metric ton (1000 kg) of iron by the following reaction?

$$FeO(s) + CO(g) \longrightarrow Fe(s) + CO_2(g); \qquad \Delta H_{rxn} = 9.0\text{ kJ}$$

20. How much heat is liberated per gram of sulfuric acid formed in the following reaction?

$$SO_3(g) + H_2O(\ell) \longrightarrow H_2SO_4(\ell); \qquad \Delta H_{rxn} = -133\text{ kJ}$$

*21. What is the heat of the reaction $2\,H_2(g) + O_2(g) \rightarrow 2\,H_2O(g)$ if 672 kJ is released when 50.0 g water is produced?

*22. What is the heat of the reaction $PbO(s) + C(s) \rightarrow Pb(s) + CO(g)$ if 23.8 kJ must be supplied in order to convert 49.7 g lead(II) oxide to lead?

74. A menthol ($C_{10}H_{20}O$) sample weighing 0.199 g was burned in a bomb calorimeter whose heat capacity was 4602 J K^{-1}, causing the temperature to rise from 23.04 to 24.79 °C. Calculate the heat of combustion of menthol in kJ mol^{-1}.

75. Naphthalene ($C_{10}H_8$) has a heat of combustion of 5.157×10^3 kJ mol^{-1}. What would be the temperature rise of a bomb calorimeter having a heat capacity of 4532 J K^{-1} if 0.25 g naphthalene were burned in it?

76. A 0.483 g sample of butter was burned in a bomb calorimeter whose heat capacity was 4572 J K^{-1}, and the temperature was observed to rise from 24.76 to 27.93 °C. Calculate the fuel value of butter in
    a. kJ g^{-1}
    b. Nutritional Calories per gram
    c. Calories per 5 g pat.

Alkaline Earths

77. Which, if any, of the Group 2A metals occur(s) naturally in elemental form?

78. Which, if any, of the Group 2A elements is produced on a large scale and used as the metal rather than as a compound?

79. What is the chief source of (a) calcium and (b) magnesium for commercial use?

80. Identify a commercial process that uses seashells as a raw material.

81. In commercial operations, how are the pure Group 2A metals prepared from their compounds? Write a general equation, using M to represent the element, that describes the process.

82. Which metal is most transparent to x-rays?

83. Name a Group 2A compound that is used in sugar refining.

84. What are some of the ways in which beryllium differs from the other Group 2A elements?

85. What are the chemical name and formula of "lime"?

86. What are some uses of lime?

87. Write general equations, using M to represent the metal, for the reaction of the alkaline earth elements with (a) oxygen and (b) nitrogen.

88. Write general equations, using M to represent the metal, for the reaction of the alkaline earth elements with (a) hydrogen and (b) water.

89. Write a general equation, using M to represent the metal, for the reaction of the Group 2A oxides with water. Are the reactions of the Group 2A oxides with water endo- or exothermic?

90. Although magnesium is quite reactive toward oxygen, the metal can be exposed to the atmosphere indefinitely without significant extent of reaction. Explain this apparent paradox.

91. What is the mass of $Ca(OH)_2$ formed when 750 g CaO reacts with excess water?

92. At high temperature, barium reacts with pure oxygen to form barium peroxide. Write a balanced equation for this reaction.

93. How much $Ca(OH)_2$ is required to react with one ton (2000 pounds) of SO_2?

94. Like those of the alkali metals, the hydrides of Group 2A elements react vigorously with water. One of the products of the reaction is gaseous hydrogen. Write a balanced equation for the reaction of strontium hydride with water.

CHAPTER 5

THE GASEOUS STATE

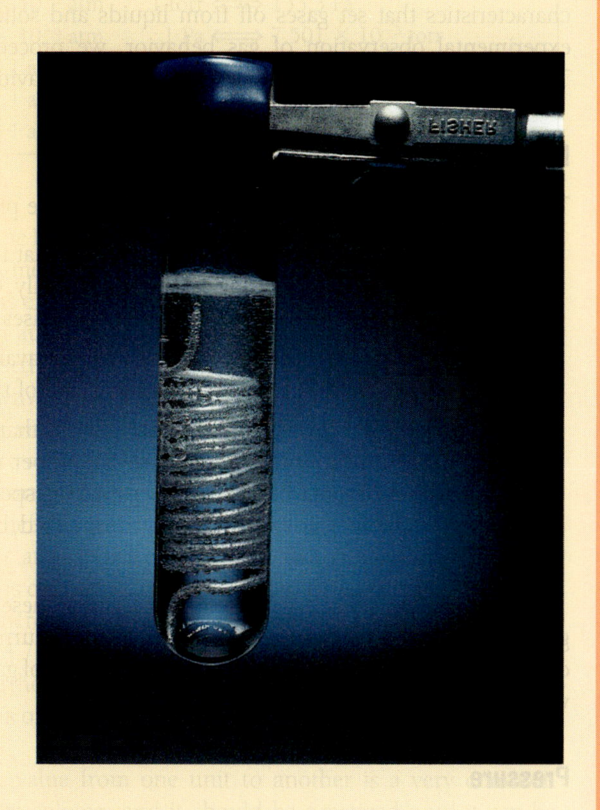

The reaction of aluminum with aqueous sodium hydroxide, in which sodium aluminate and bubbles of hydrogen gas are formed, is one of the many reactions that yield products in the gaseous state.

$$Al(s) + 6\ NaOH(aq) \longrightarrow 2\ Na_3AlO_3 + 3\ H_2(g)$$

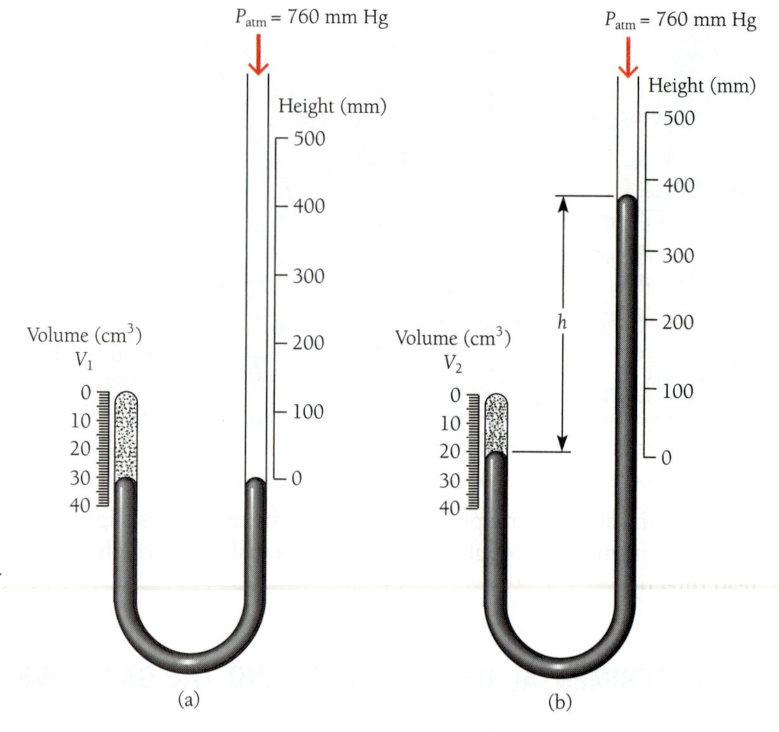

Boyle used a simple device called a "J-tube" in his studies. In (a) the mercury level is the same in the open and closed ends of the tube, so the gas pressure must be the same in each side (atmospheric pressure). When more mercury is added to the tube (b) the gas in the closed side is compressed to a smaller volume. The new volume is easily read from the scale, and the new pressure is 760 torr plus h, the height difference between the mercury levels.

TABLE 5.2 Pressure-Volume Relationship for 1 Gram Dry Air at 25 °C

Pressure/atm	Volume/mL
0.5	1686
1.0	843
2.0	422
5.0	169
10.0	84

helium for the air and make measurements at 25 °C, the value of the product changes to PV = 6113 mL atm. These relationships are shown in graphical form in Figure 5.5.

This state of affairs is commonly described by saying that the product of pressure and volume is a constant, provided that the temperature and amount of gas do not change.

$$P \cdot V = \text{constant}^* \tag{5-1a}$$

When Equation (5-1a) is rearranged to (5-1b), an additional feature becomes clear.

$$V = \text{constant} \cdot \left(\frac{1}{P}\right) \tag{5-1b}$$

When V is plotted against the reciprocal of pressure, as has been done in Figure 5.6, the result is a straight line. The slope of the line depends on the temperature and the amount of gas in the sample. Such plots are generally held to be more informative than the curved lines of Figure 5.5, because a straight-line plot implies a particularly simple relationship between the variables. (Recall the equation of a straight line, $y = mx + b$, and note that in a relationship of *proportionality* such as (5-1b) the y-intercept "b" has the value zero.)

* This does *not* mean that $P \cdot V$ is always the same. If the amount of gas or the temperature is changed, the product PV will have a different, but still constant, value. It is convenient to use the notation of Equation (5-1a), but we must be careful to understand and maintain the distinction between a constant number such as 843 mL atm, which may change in value if the conditions change, and a number such as the electrical charge on the electron, which is truly constant and has the same value under all circumstances.

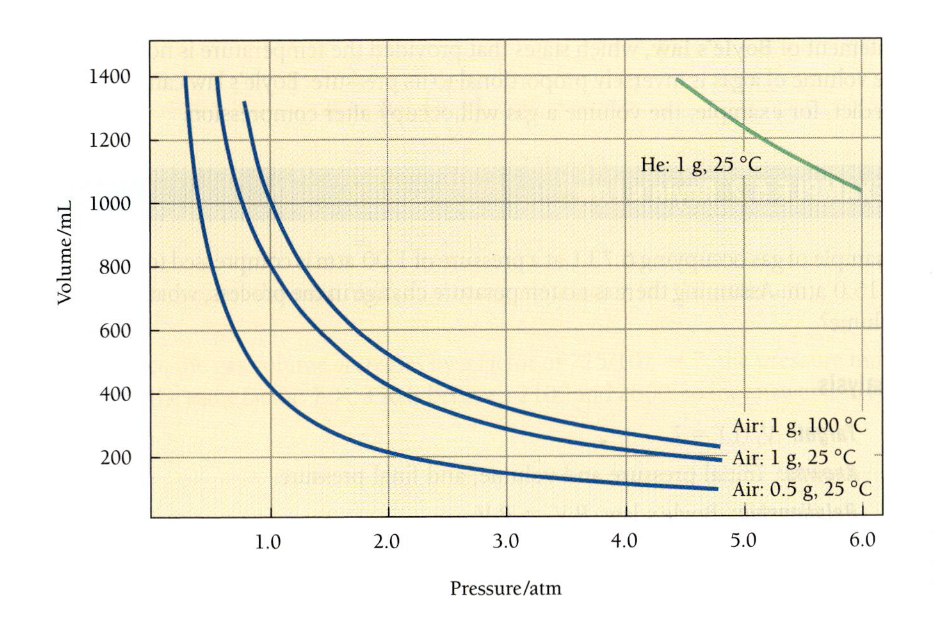

Figure 5.5 Pressure-volume relationships observed in helium and air. For air, the effect of changing the temperature and the sample size is shown.

If a sample of gas undergoes a change from an initial pressure P_i to a final pressure P_f, its volume changes from the initial value V_i to the final value V_f.

$$P_i \text{ and } V_i \xrightarrow[\text{temperature}]{\text{same}} P_f \text{ and } V_f$$

Provided that the temperature does not change during the process, the two sets of conditions are related through Equation (5-2).

$$P_iV_i = P_fV_f \tag{5-2}$$

Equation (5-2) applies to a particular sample of gas; it is not valid if the *amount* of gas (number of moles) changes.

Either form of Equation (5-1), or Equation (5-2), may be taken as the algebraic

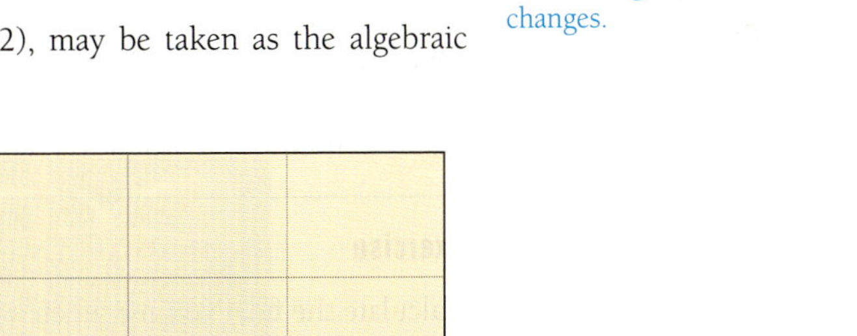

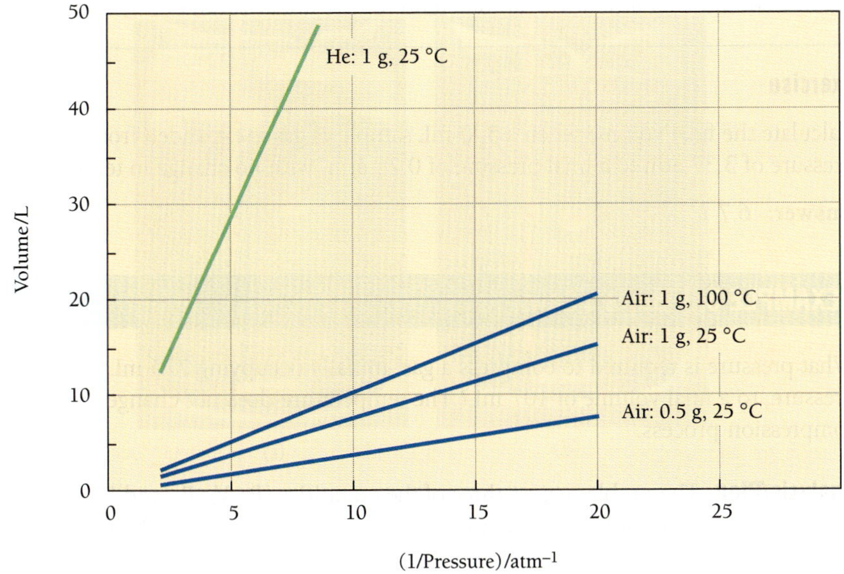

Figure 5.6 When volume is plotted vs. the reciprocal of pressure, a straight line is obtained. The slope is given by the value of "constant" in the equation $V = \text{constant} \cdot (1/P)$. The four lines represent the same data plotted in Figure 5.5.

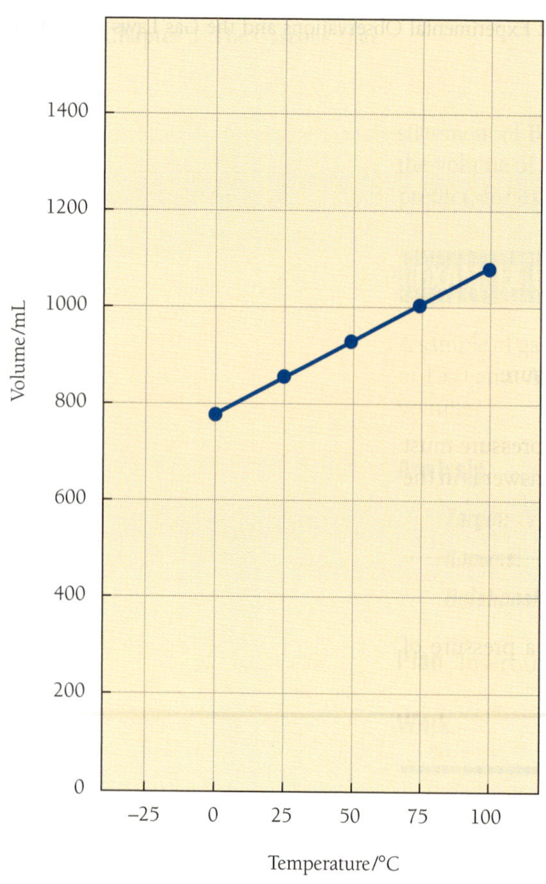

Figure 5.8 The volume of 1 gram air at 1 atm is plotted as a function of Celsius temperature (data taken from Table 5.3). The volume and temperature of a gas are linked by a straight-line or *linear* relationship.

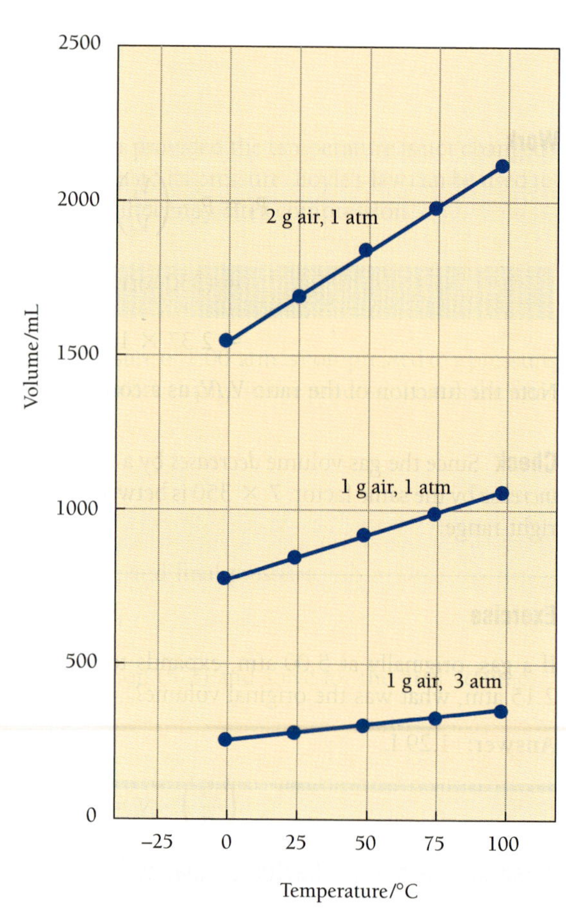

Figure 5.9 Volume-temperature relationship for different samples of air: top line, 2 grams air at 1 atm pressure; middle line, 1 g at 1 atm; bottom line, 1 g at 3 atm. The equation of the top line is $V = 5.66 \cdot T + 1546$ (where V is in mL and T is in °C), and the x-intercept is $-1546/5.66 = -273$ (°C).

The dependence of volume on Celsius temperature is linear but *not* proportional. A proportional relationship would have the form V = const · T, and hence would have V = 0 when T = 0.

The x-intercept of the line $y = mx + b$ is found from the formula

$$x\text{-intercept} = \frac{-b}{m}$$

When the data in Table 5.3 are presented graphically (Figure 5.8), it is seen that volume depends linearly on temperature. Such relationships are described by the generalized equation $y = mx + b$, which in this case is $V = mt + b$. Figure 5.9 shows volume vs. temperature plots for air under several different conditions. A striking fact emerges when the x-intercepts of the lines in Figure 5.9 are calculated. The x-intercept of all three lines is the same. In fact, no matter how much gas is used or what the pressure is, the x-intercept of a plot of volume vs. Celsius temperature is *always* −273 °C.

It was not possible for Charles to extend his measurements to more than a few degrees below 0 °C, but he could extrapolate his data into this experimentally inaccessible region and reach the tentative conclusion that the volume of any gas would be

Jacques Charles (1746–1823)

Jacques Charles was a Professor of Physics in Paris. His diverse interests led him into the field of electricity, where he confirmed many of Benjamin Franklin's experiments, and into aeronautics. He was the first to use hydrogen in balloons, and in 1783 made an ascent to an altitude of almost two miles.

zero at a temperature of −273 °C. "Absolute zero" is an appropriate name for this temperature, because it seems to be the lower limit to temperature. Below this temperature a gas would have a negative volume, and that is clearly impossible. The lines in Figure 5.9 are redrawn in Figure 5.10, with dotted lines indicating the extrapolations to low temperature. The temperature axis in Figure 5.10 is labeled both in degrees Celsius and in kelvins. The equation of the top line in Figure 5.10 is $V(\text{mL}) = 5.66 \cdot T(°\text{C}) + 1546$ when the temperature is expressed in degrees Celsius. When the temperature is expressed in kelvins, the equation becomes much simpler: $V(\text{mL}) = 5.66 \cdot T(\text{K})$.

In general, the proportional relationship given by Equation (5-3) holds for any gas when the temperature is expressed in kelvins.

$$V = \text{constant} \cdot T \qquad \text{or} \qquad (V/T) = \text{constant} \tag{5-3}$$

When a sample of gas undergoes any change at constant pressure,

$$V_i \text{ and } T_i \xrightarrow[\text{pressure}]{\text{same}} V_f \text{ and } T_f$$

the final values of volume and temperature are linked to the initial values by Equation (5-4). Equations (5-3) and (5-4) are algebraic forms of **Charles's law,** which states that, provided that the pressure is not changed, the volume of a sample of gas is proportional to its absolute (Kelvin scale) temperature.

$$\frac{V_i}{T_i} = \frac{V_f}{T_f} \tag{5-4}$$

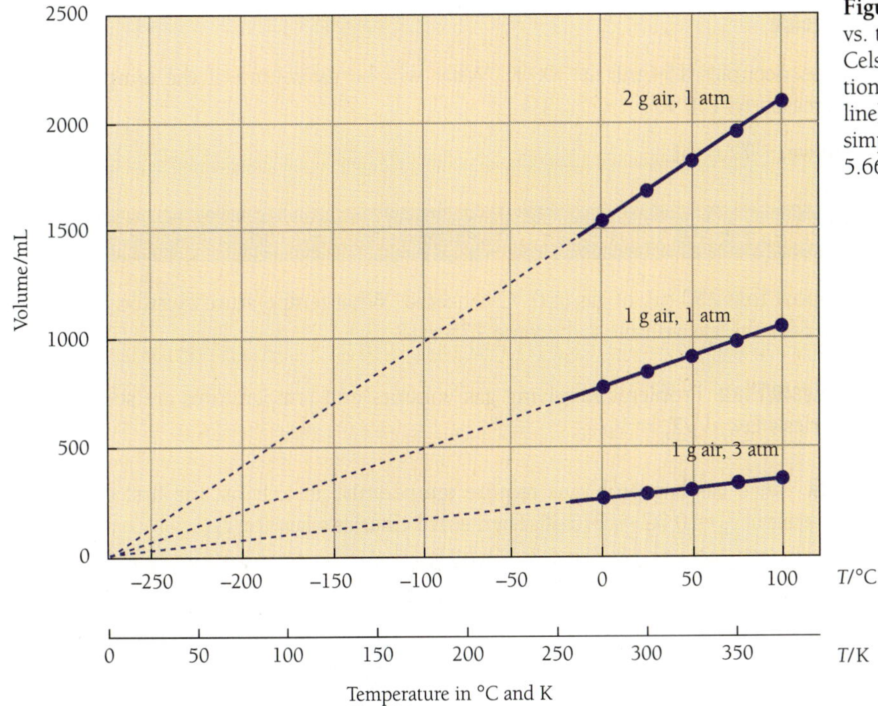

Figure 5.10 Data from Figure 5.9 plotted vs. temperature in kelvins as well as degrees Celsius. Use of °C leads to the linear relationship $V(\text{mL}) = 5.66 \cdot T(°\text{C}) + 1546$ (top line), while use of kelvins leads to the simpler, *proportional*, relationship $V(\text{mL}) = 5.66 \cdot T(\text{K})$.

A word of caution: in all gas-law problems, the temperature *must* be expressed in kelvins.

EXAMPLE 5.4 Charles's Law

A sample of gas occupying 0.273 L at a temperature of 300 K is heated to 400 K. Assuming there is no pressure change in the process, what is the final volume?

Analysis

Target: $V_f = ?$ L

Knowns: T_i, V_i, and T_f are known, as is the fact that the pressure does not change.

Relationship: Charles's law, $V_i/T_i = V_f/T_f$, connects the target with the knowns.

Plan Solve Equation (5-4) for V_f, insert the known values, and evaluate.

Work

$$V_f = \frac{T_f V_i}{T_i} = V_i \cdot \left(\frac{T_f}{T_i}\right)$$

$$= (0.273 \text{ L}) \cdot \frac{(400 \text{ K})}{(300 \text{ K})}$$

$$= 0.364 \text{ L}$$

Check Since gases expand when heated, the final volume should be larger. The temperature ratio is a conversion factor, which increases the volume by $(400/300) \approx 1.3$.

Exercise

A gas occupies 154 mL at 300 K. What will be its volume if the temperature is decreased to 150 K?

Answer: 77.0 mL.

EXAMPLE 5.5 Charles's Law

Suppose that 250 mL of gas at 0 °C is heated. What temperature must be reached in order for the gas volume to become 275 mL?

Analysis/Plan Problems involving gas volumes and temperatures are solved using Charles's law, $V_i/T_i = V_f/T_f$.

To convince yourself of the necessity of converting temperatures to the kelvin scale, try solving the problem in Example 5.5 using the temperature as given, in °C.

Work Since gas-law problems require temperature in kelvins, the first step is the conversion $T_i = 0\text{ °C} = 273$ K. Then solve Charles's law for T_f.

$$T_f = T_i \cdot \left(\frac{V_f}{V_i}\right)$$

$$= (273 \text{ K}) \cdot \frac{(275 \text{ mL})}{(250 \text{ mL})}$$

$$= 300 \text{ K} = 27 \text{ °C}$$

Exercise

The temperature of a 1000-mL gas sample is 25 °C. At what temperature would its volume be 800 mL?

Answer: −35 °C.

Combined Gas Law: *PV/T* = Constant

Boyle's and Charles's laws can be combined into one algebraic statement.

$$PV = (\text{constant})T \qquad \text{or} \qquad \frac{PV}{T} = \text{constant} \tag{5-5}$$

When the temperature is held constant, as it must be for Boyle's law to hold, Equation (5-5) is the same as Equation (5-1). When the pressure is held constant, as it must be for Charles's law to hold, Equation (5-5) is the same as Equation (5-3). Equation (5-5) is known as the *combined gas law*. The numerical value of the constant depends on the number of moles in the gas sample, but not on the pressure, temperature, or volume. Example problems based on the combined gas law are given in Section 5.3.

Avogadro's Principle

The relationship between the amount of gas (number of moles) and the volume it occupies is equally straightforward. If the amount of gas in a sample is doubled, and the temperature and pressure are left unchanged, the volume of the gas doubles. That is, the volume occupied by a gas is proportional to the amount of gas. If a sample contains n moles, the relationship is

$$V = (\text{constant}) \cdot n \tag{5-6}$$

This fact of gas behavior is related to **Avogadro's principle**, which states that equal volumes of different gaseous substances contain equal numbers of molecules (provided that the temperature and pressure are the same). It follows from Avogadro's principle that the constant in Equation (5-6) has the same value for all gases, *regardless of their chemical nature.*

Ideal Gas Law: *PV* = *nRT*

When Avogadro's principle is incorporated into the combined gas law, Equation (5-5), the result is a new equation.

$$PV = (\text{constant}) \cdot n \cdot T \tag{5-7}$$

In this equation, the constant is a true constant; that is, its value is the same for all gases, in any amount and under all conditions of temperature, pressure, and volume. It is called the **gas constant** (or occasionally the "universal" gas constant or the "ideal" gas constant), and it is given the symbol $\mathbf{R}$. In this form the combined gas law is known as the **ideal gas law.** It is expressed algebraically by Equation (5-8).

$$(\text{Pressure}) \cdot (\text{Volume}) = (\text{number of moles}) \cdot R \cdot (\text{Kelvin Temperature}) \qquad \text{or}$$

$$PV = nRT \tag{5-8}$$

Remember that, in order for Equation (5-8) to hold, the temperature must be expressed in kelvins. The value of the gas constant is readily measured, as shown in Example 5.6.

EXAMPLE 5.6 Evaluation of the Gas Constant

A gas sample of 0.3277 mol is placed into a 0.9703-L container held at a temperature of 273.16 K. The pressure is measured and found to be 7.570 atm. What are the numerical value and the units of the gas constant R?

Analysis

Target: The gas constant

$$R = ?$$

Knowns: P, V, n, and T are all given.

Relationship: The ideal gas law, Equation (5-8), connects the knowns with the target.

Plan Rearrange Equation (5-8) and substitute the known information.

Work

$$PV = nRT$$

$$R = \frac{PV}{nT}$$

$$= \frac{(7.570 \text{ atm})(0.9703 \text{ L})}{(0.3277 \text{ mol})(273.16 \text{ K})}$$

$$= 0.08206 \text{ L atm mol}^{-1} \text{ K}^{-1}$$

Check It is difficult to check this result, since we do not know beforehand what the units should be or what would be a reasonable numerical value. The best we can do is go back and make sure we have not incorrectly copied the digits or the units.

Exercise

A gas sample of 0.175 mol is confined to a 2500-mL container at 298 K, and the pressure is found to be 1300 torr. What is the value of R?

Answer: 6.23×10^4 mL torr mol^{-1} K^{-1}.

The Units of *R*

The gas constant is a true constant, independent of the value of any other quantity. It can be expressed in any of several different units, for example those obtained in Example 5.6; $R = 0.08206$ L atm mol^{-1} K^{-1}. In some texts and handbooks, R is given as 82.06 mL atm mol^{-1} K^{-1}. This is not a different value, however, it is the result of a unit conversion operation.

$$R = 0.08206 \text{ L atm mol}^{-1} \text{ K}^{-1} \left(\frac{1 \text{ mL}}{0.001 \text{ L}}\right)$$

$$R = 82.06 \text{ mL atm mol}^{-1} \text{ K}^{-1}$$

Similarly, if pressure is expressed in pascals and volume in cubic meters, the result is the gas constant R in SI units:

$$R = 8.314 \text{ m}^3 \text{ Pa mol}^{-1} \text{ K}^{-1}$$

Pressure is defined as force/area, so that the SI equivalence statement is $1 \text{ Pa} \leftrightarrow 1 \text{ N}/1 \text{ m}^2$. But since energy is force · distance, $1 \text{ J} \leftrightarrow 1 \text{ N} \cdot 1 \text{ m}$. Combining these two equivalence statements leads to a relationship between energy, pressure, and volume, $1 \text{ J} \leftrightarrow 1 \text{ Pa} \cdot 1 \text{ m}^3$, so that the gas constant may also be written

$$R = 8.314 \text{ J mol}^{-1} \text{ K}^{-1}$$

Since the most common units for gas pressures and volumes are atm and L (or mL), the most useful value of R to memorize (together with its units, of course) is $R = 0.08206 \text{ L atm mol}^{-1} \text{ K}^{-1}$.

5.3 CALCULATIONS INVOLVING GAS LAWS

There are many occasions in chemistry, for instance physical measurements of gas properties or stoichiometry, when it is necessary to make calculations using the gas laws. Two broad categories of gas-law problems may be distinguished, corresponding to the combined gas law (5-5) and to the ideal gas law (5-8). We will treat these two categories separately.

Process Problems

If the problem is concerned with a *process* taking place, in which the amount of gas is unchanged, then we may call it a **process problem.** In such problems a gas sample is converted from an initial state of given pressure, volume, and temperature (P_i, V_i, T_i) to a final state (P_f, V_f, T_f) in which either the pressure, volume, or temperature is unknown. The solution of process problems involves the use of the combined gas law in algebraic form, as expressed in Equation (5-9).

$$\frac{P_i V_i}{T_i} = \frac{P_f V_f}{T_f} \qquad (5\text{-}9)$$

This equation follows directly from Equation (5-3).

EXAMPLE 5.7 The Combined Gas Law: A Process Problem

The anesthetic gas N_2O (nitrous oxide or "laughing gas," often used by dentists) is stored at high pressure in steel tanks. Suppose a tank holds 2.00 liters, and that the pressure in the tank is 15.0 atm when the temperature is 30 °C. What volume would the gas occupy at a pressure of 1.00 atm and a temperature of 22 °C?

Analysis The target is the volume of a gas sample after a change takes place. It is a process problem, and therefore Equation (5-9) is the appropriate relationship.

Plan/Work As we study the problem, it is helpful to arrange the known quantities in a way that reminds us of the physical nature of the process.

initial state		final state
$P_i = 15.0$ atm		$P_f = 1.00$ atm
$V_i = 2.00$ L	⟶	$V_f = ?$ L
$T_i = 30$ °C		$T_f = 22$ °C

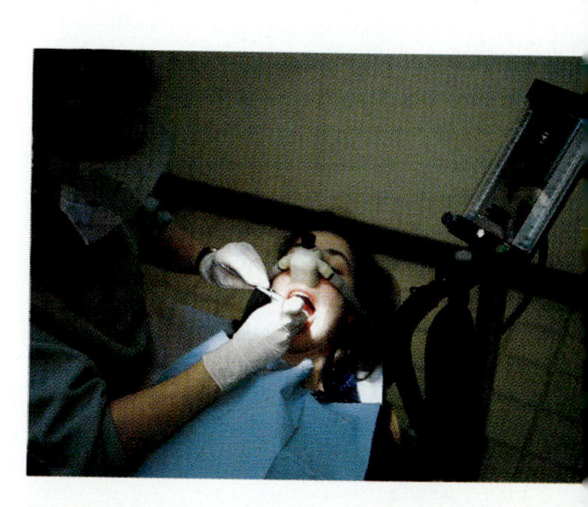

In working gas-law problems, temperature must be expressed in kelvins. After this conversion is made, the knowns and target are as follows.

$$\begin{array}{l} P_i = 15.0 \text{ atm} \\ V_i = 2.00 \text{ L} \\ T_i = 303 \text{ K} \end{array} \longrightarrow \begin{array}{l} P_f = 1.00 \text{ atm} \\ V_f = ? \text{ L} \\ T_f = 295 \text{ K} \end{array}$$

We solve Equation (5-9) for the target V_f, insert the known quantities, and evaluate.

$$V_f = V_i \cdot \left(\frac{P_i}{P_f}\right) \cdot \left(\frac{T_f}{T_i}\right)$$

$$= (2.00 \text{ L}) \frac{(15.0 \text{ atm})}{(1.00 \text{ atm})} \frac{(295 \text{ K})}{(303 \text{ K})}$$

$$= 29.2 \text{ L}$$

Check See the paragraph following the Exercise.

Exercise

If 5.00 L of gas at 25 °C and 0.500 atm is compressed to a volume of 1.00 L, and the temperature remains unchanged, what will be the new pressure?

Answer: 2.50 atm.

There is a special danger, in any process problem, of inverting one of the factors in the formula. Since the units cancel, the factor-label method does not guard against the use of (P_f/P_i) when (P_i/P_f) is intended. Checking a process problem is done qualitatively, by thinking about the effect of changes in P, V, and T. Example 5.7 is checked as follows.

- There is a decrease in pressure during the process, and since (according to Boyle's Law) the volume depends inversely on pressure, the volume must increase. Therefore, the pressure factor must be (15 atm/1 atm) rather than (1 atm/15 atm).
- The temperature decreases during the process, and since, according to Charles's law, the volume is proportional to temperature, the volume must decrease. Therefore, the temperature factor must be (295 K/303 K) rather than (303 K/295 K).

A common abbreviation encountered in gas-law problems is **standard temperature and pressure (STP).** Standard temperature and pressure are defined as 0 °C and 1 atm. The phrase "STP volume" means the volume a sample of gas would occupy if the pressure and temperature were 1 atm and 0 °C. The volume of one mole of gas at STP is 22.4 L. This number, the "molar volume at STP," can be calculated from the ideal gas law using the quantities $n = 1$ mol, $P = 1$ atm, $T = 273$ K, and $R = 0.08206 \text{ L atm mol}^{-1} \text{ K}^{-1}$.

$$V = \frac{nRT}{P} = \frac{(1.00 \text{ mol})(0.08206 \text{ L atm mol}^{-1} \text{ K}^{-1})(273 \text{ K})}{(1.00 \text{ atm})}$$

$$V = 22.4 \text{ L}$$

If you cannot remember the value of the gas constant, you can calculate it from the STP molar volume:

$$R = \frac{PV}{nT} = \frac{(1 \text{ atm})(22.4 \text{ L})}{(1 \text{ mol})(273 \text{ K})}$$

$$= 0.0821 \text{ L atm mol}^{-1} \text{ K}^{-1}$$

The molar volume at STP is the same for all gases, and it is a useful number to commit to memory.

EXAMPLE 5.8 A Process Problem Involving STP

The volume of a hot air balloon is about 150 m^3, and the pressure is the same as the pressure of the atmosphere surrounding it, that is, 1 atm. What is the temperature of the air in the balloon if its STP volume is 120 m^3?

Analysis

Target: A temperature

$$? \text{ K} =$$

Knowns: The volume at STP, and the volume under other, non-STP conditions.

Relationship: This is a process problem, although it may not be obvious that a process is involved. The clue is supplied by the phrase "STP volume" in the problem. We must *imagine* that the process

$$\text{air(STP)} \longrightarrow \text{air(balloon conditions)}$$

has taken place. Now we can set up the problem as we did in Example 5.7, using STP for the initial conditions.

Work

Initial (STP)		Final
$P_i = 1.00$ atm		$P_f = 1.00$ atm
$V_i = 120$ m^3	$\longrightarrow$	$V_f = 150$ m^3
$T_i = 273$ K		$T_f = ?$ K

(STP)

Equation (5-9) works for all process problems. After rearrangement, the form is

$$T_f = T_i \cdot \left(\frac{P_f}{P_i}\right) \cdot \left(\frac{V_f}{V_i}\right)$$

$$= (273 \text{ K}) \frac{(1.00 \text{ atm})}{(1.00 \text{ atm})} \frac{(150 \text{ m}^3)}{(120 \text{ m}^3)}$$

$$= 341 \text{ K} (= 68 \text{ °C})$$

Check An increase in volume is associated with an increase in temperature. The volume factor must be (150 m^3/120 m^3) rather than (120 m^3/150 m^3).

Note: Since the pressure remains the same, the simpler relationship $V_i/T_i = V_f/T_f$ (Charles's law) holds. The problem could have been solved as in Example 5.5.

Exercise

What is the STP volume of a gas that occupies 250 mL at 1.5 atm and 50 °C?

Answer: 3.2×10^2 mL or 0.32 L.

Example 5.8 contains two features worth remembering. A problem can be a process problem even though it does not explicitly refer to a process. What is required is that a particular sample, that is, an *unchanging amount* of gas, is described under two different sets of PVT conditions. Whether or not the problem contains the word "process" is irrelevant.

The volume units (cubic meters) in Example 5.8 are different from the usual liters. This presents no difficulty, since it is the dimensionless ratio (V_f/V_i) that appears in the solution. The same is true for pressure. Since it is the ratio (V_f/V_i) and/or (P_f/P_i) that enters the calculation, the units cancel and the only requirement is that the same units be used consistently throughout the problem. Temperature, however, must be expressed in kelvins; °C or °F must not be used.

EXAMPLE 5.9 *P-V-T* Changes Expressed as Ratios

Gaseous hydrogen chloride, used in some laboratory procedures, is stored under pressure in steel tanks. What will be the final pressure if HCl is compressed to one tenth of its initial volume, while its temperature increases by 10%? The initial pressure is 1.00 atm.

Analysis The target (P_f) is clear enough, but there seems to be less known information than in the preceding example. However, the word "compressed" implies a process, so the combined gas law is applicable.

Work We proceed as before by arranging the known information.

$$\begin{array}{lcl} P_i = 1.00 \text{ atm} & & P_f = ? \\ V_i = ? & \longrightarrow & V_f = ? \\ T_i = ?\text{ K} & & T_f = ?\text{ K} \end{array}$$

At first glance it appears that there is not enough information to solve this problem. However, there are two bits of information that we have not used. The temperature increases by 10%, so that $T_f = T_i + 0.10\ T_i = 1.10\ T_i$. Also, the final volume is one tenth of the initial volume, so that $V_f = 0.100\ V_i$. Now, when we collect the given information, the outlook is considerably brighter.

$$\begin{array}{lcl} P_i = 1.00 \text{ atm} & & P_f = ? \\ V_i = V_i & \longrightarrow & V_f = 0.100\ V_i \\ T_i = T_i & & T_f = 1.10\ T_i \end{array}$$

When the combined gas law equation is solved for the final pressure, volume and temperature appear as ratios. The unknown quantities V_i and T_i cancel, and the problem can be solved with the information at hand.

Work

$$P_f = P_i \cdot \left(\frac{T_f}{T_i}\right) \cdot \left(\frac{V_i}{V_f}\right)$$

$$= (1.00 \text{ atm}) \frac{(1.10\ \cancel{T_i})}{(\cancel{T_i})} \frac{(\cancel{V_i})}{(0.100\ \cancel{V_i})}$$

$$= 11.0 \text{ atm}$$

Check Since the volume decreases, the pressure must increase. Also, an increase in temperature brings about an increase in pressure.

Exercise

Suppose the pressure applied to 7.5 L of gas is doubled, while the temperature is unchanged. What is the new volume?

Answer: 3.75 L.

Single-State Problems

Often a problem does not specify or imply a change in the conditions of a gas sample. Rather, the problem asks for information about a sample in a specified state or condition. Such problems may be called **single-state problems,** and their solution always involves the ideal gas law, $PV = nRT$.

EXAMPLE 5.10 A Single-State Problem and the Ideal Gas Law

What is the volume of 1.5 moles of gas at 1000 K and 10.0 atm pressure?

Analysis The problem does not describe, even implicitly, a process. It is a single-state problem, to be solved by use of the ideal gas law.

Work

$$PV = nRT$$

$$V = \frac{nRT}{P}$$

$$= \frac{(1.5\ \text{mol})(0.08206\ \text{L atm mol}^{-1}\ \text{K}^{-1})(1000\ \text{K})}{10.0\ \text{atm}}$$

$$= 12\ \text{L}$$

Check The units are correct. R is about 0.1, so the numerical factor $R(1000)/(10)$ is about 10, and the answer should be close to $10(1.5) = 15$.
Note: The identity of the gas is not given, used, or needed.

Exercise

What is the volume of 1.0 mol of gas at 1000 torr and 500 K?

Answer: 31 L. (If your answer was 0.041, check your units and you will find that they did not cancel correctly.)

Example 5.10 is the model for several types of closely related problems. The ideal gas law $PV = nRT$ contains the gas constant R and four measurable quantities. The problem statement gives three of these quantities explicitly, and asks for the value of the fourth. The answer is found by inserting the given values into the ideal gas law and solving for the unknown.

Some single-state problems are more complicated, and require a stepwise procedure.

When treated with concentrated acid, limestone reacts to produce a foam of tiny bubbles of carbon dioxide.

EXAMPLE 5.11 Measuring the Amount of Gas Collected

Carbon dioxide is conveniently produced in the laboratory by the action of dilute acid on limestone, which is largely composed of calcium carbonate. (Limestone is actually a mixture of calcium and magnesium carbonates, but carbon dioxide can be produced from either material.)

$$CaCO_3(s) + 2\ HCl(aq) \longrightarrow CO_2(g) + H_2O + CaCl_2(aq)$$

In one such experiment, the carbon dioxide collected had a volume of 250 mL, measured at STP. How many grams of CO_2 were collected?

Analysis

Target: ? g CO_2 =

Knowns: T, P, V are explicit. Implicit knowns are the gas constant and the known properties of CO_2.

Relationship: Either $PV = nRT$ or $P_iV_i/T_i = P_fV_f/T_f$, depending on whether this is a single-state or process problem. There is in fact a change taking place, as a solid is transformed to a gas, *but it is not a process in which a gas changes from one condition to another.* Therefore it is a single-state problem, and the equation $PV = nRT$ is applicable.

Plan

2. Calculate the number of grams using the molar mass.
1. Calculate the number of moles of CO_2 using $PV = nRT$.

Work

1.

$$n = \frac{PV}{RT}$$

$$= \frac{(1.00\ \text{atm})(0.250\ \text{L})}{(0.08206\ \text{L atm mol}^{-1}\ \text{K}^{-1})(273\ \text{K})}$$

$$= 0.01116\ \text{mol}$$

2.

$$? \text{ g } CO_2 = (0.01116 \text{ mol } CO_2)\left(\frac{44.0 \text{ g } CO_2}{1 \text{ mol } CO_2}\right)$$
$$= 0.491 \text{ g } CO_2$$

Check Since 0.25 L is about 1/100 of 22.4 L, the volume occupied by 1 mol at STP, the amount is about 1/100 mol. The molar mass is 44 g, and 44(1/100) = 0.4.

Exercise

Freon-12 (CCl_2F_2) is widely used as the "working fluid" in refrigerators and air conditioners. What mass of this compound occupies 5.0 L at STP?

Answer: 27 g.

EXAMPLE 5.12 Calculation of Gas Density

Uranium hexafluoride is one of the heaviest known gases. Calculate its density at 85 °C and 1.00 atm pressure.

Analysis

Target: Since the units of density are g L^{-1}, the target is the mass of 1 L of gas.

Knowns: Temperature and pressure; implicit knowns are the gas constant and the molar mass of UF_6.

Relationship: The problem concerns a gas. No process is specified, so it is a single-state problem to be solved with the ideal gas law.

Plan Find the number of moles in one liter, and convert this value to grams using the molar mass.

Work

1.

$$n = \frac{PV}{RT} \quad \text{or}$$

$$\frac{n}{1 \text{ L}} = \frac{P}{RT}$$
$$= \frac{1.00 \text{ atm}}{(0.08206 \text{ L atm mol}^{-1} \text{ K}^{-1})(358 \text{ K})}$$

(Once more, note the use of K rather than °C.)

$$= 0.03404 \text{ mol L}^{-1}$$

2. The molar mass of UF_6 is first calculated and found to be 352 g mol^{-1}.

$$\frac{? \text{ g}}{1 \text{ L}} = (0.03404 \text{ mol L}^{-1})\left(\frac{352 \text{ g}}{1 \text{ mol}}\right) = 12.0 \text{ g L}^{-1}$$

The number of moles in one liter is, of course, the *molarity* of the gas. However, the term "molarity" is usually applied to solutions—it is rarely used for pure substances (although it is not incorrect to do so).

Since by definition the density of a substance is the mass of one liter, the answer to the problem is

$$D = 12.0 \text{ g L}^{-1}$$

Check Combined into one step, the mathematical operations were $\mathcal{M}P/RT$, where $\mathcal{M}$ represents molar mass. Using $R \approx 0.1$, this may be estimated as (350)(1)/(0.1)(350) = 10, which is close to the answer obtained.

Exercise

Calculate the STP density of air. For the purposes of this problem, air may be regarded as a gas whose molar mass is 29 g mol^{-1}.

Answer: 1.3 g L^{-1}.

EXAMPLE 5.13 Calculation of the Amount of a Gaseous Reaction Product

If the measurement is made at 1.00 atm and 25 °C, what will be the volume of gas produced when 1.00 g $KClO_3$ decomposes into KCl and O_2?

Analysis This is a single-state problem, because, even though a reaction is taking place, the gas itself undergoes no change. Calculation of the volume of oxygen using the ideal gas law requires values for P, T, and n. All are given except n, which is related to the amount of $KClO_3$ decomposed.

Plan

2. Calculate V from $V = nRT/P$.
1. Calculate n by the stoichiometric methods of Chapter 3.

Work

1. The balanced equation is $2\ KClO_3 \rightarrow 2\ KCl + 3\ O_2$, giving the stoichiometric ratio 3 mol $O_2 \Leftrightarrow$ 2 mol $KClO_3$. The molar mass of $KClO_3$ is 122.6 g mol^{-1}.

$$n = (1.00 \text{ g } KClO_3)\left(\frac{1 \text{ mol } KClO_3}{122.6 \text{ g } KClO_3}\right)\left(\frac{3 \text{ mol } O_2}{2 \text{ mol } KClO_3}\right)$$

$$= 0.01223 \text{ mol } O_2$$

2. The volume of O_2 is

$$V = \frac{nRT}{P} = \frac{(0.01223 \text{ mol})(0.08206 \text{ L atm mol}^{-1} \text{ K}^{-1})(298 \text{ K})}{1.00 \text{ atm}}$$

$$= 0.299 \text{ L}$$

Check The units are correct in each step. Also, 0.3 L is approximately what should be expected: 1 g $KClO_3$ is a little less than 1/100 of a mole, and (by the stoichiometric ratio 3/2) should produce a little more than 1/100 of a mole of O_2. At STP 1/100 of a mole occupies 22.4/100 = 0.2 L.

Exercise

What STP volume of CO_2 is produced when 1.00 g carbon is burned?

Answer: 1.87 L.

5.4 A THEORETICAL MODEL OF GAS BEHAVIOR

The observed behavior of gases summarized by the gas laws is well explained by a model known as the **kinetic molecular theory (KMT).** As noted in Chapter 1, a theory is a conceptual model that explains the observed facts. Usually the explanation is based on a set of starting statements, called postulates. The smaller the number of postulates, the more satisfactory the theory. A good theory correctly predicts the results of observations that have not yet been made. A good theory is simple and readily grasped.

Some theories, for example the theory of relativity, require a great deal of study on the part of the grasper before they can be readily grasped.

The Postulates of the Kinetic Molecular Theory

The word "kinetic" in the name kinetic molecular theory is derived from the Greek word for motion. The word "molecular" in the name reinforces Dalton's idea that matter is composed of small particles. The name implies that the motion of molecules is important to the theory. The postulates of the kinetic molecular theory of gases are:

1. Gases, like solids and liquids, consist of small particles called molecules (or atoms).
2. The particles that make up a sample of gas are separated, on the average, by distances that are rather large in comparison to the actual size of the particles themselves.

   For example, the molecules of a gas at STP are separated by an average distance of about 3 nm, while the radius of an oxygen molecule is about 0.15 nm. Since the average separation is about 20 times the radius, the volume of empty space is about $20^3 = 8000$ times the actual volume of the oxygen molecules.
3. The particles experience no attractive (or repulsive) forces among themselves. Consequently each particle behaves for the most part as if the others were not there.
4. The particles are in constant motion, which is random with respect to both direction and speed.
5. An individual particle continues in straight-line motion until it undergoes a collision with another particle or with a wall of the container. Such collisions are elastic, which means that no energy of motion is lost during a collision.
6. Although individual particles travel with different velocities, and thus have different kinetic energies, the *average* energy of the particles is not random; it is proportional to absolute temperature. Furthermore the relationship between average kinetic energy and temperature is the same for all gases, independent of pressure, volume, molar mass or any chemical property specific to the type of gas.

Elastic collisions are not observed with macroscopic objects. A tennis ball dropped from shoulder height would rebound to exactly the same height and continue bouncing indefinitely if its collisions with the ground were elastic.

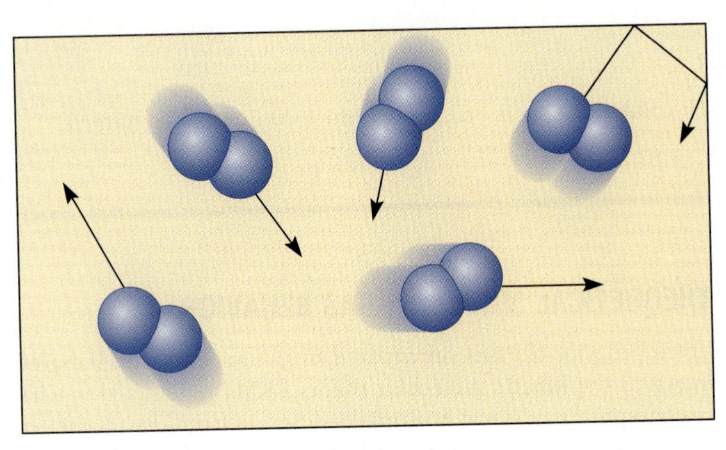

Figure 5.11 According to the kinetic molecular theory, a gas consists of widely spaced molecules in random motion, undergoing collisions with one another and with the walls of the container.

Most of the postulates of the KMT are illustrated in Figure 5.11, which depicts a diatomic gas such as oxygen or nitrogen. A gas consists mostly of empty space, punctuated only occasionally by a molecule.

The postulates of the KMT may be used to predict that the pressure of a gas sample is proportional to its absolute temperature and inversely proportional to its volume. That is, the KMT correctly predicts the set of observations known as the combined gas law. Except for the numerical value of the gas constant, it predicts the ideal gas law as well. The derivation of this prediction is given in Appendix B.

Distribution of Molecular Velocities

A basic formula from physics is $E = mv^2/2$, which relates the kinetic energy of a moving object to its mass and velocity.

One of the cornerstones of the KMT is that the average energy of molecular motion is proportional to the absolute temperature. Since energy of motion is related to velocity, the average velocity is also determined by the temperature. The motion of the molecules is random, however. They do not all travel with the same velocity. Some move faster than the average, some move slower. Furthermore, the velocity of an individual molecule changes as it undergoes collisions with other molecules. Even so, the average velocity is well defined at any given temperature. The average velocity $<v>$ of molecular motion is given by Equation (5-10), in which R is the gas constant in J mol^{-1} K^{-1}, T is the temperature in K, and $\mathcal{M}$ is the molar mass in *kilogram mol*$^{-1}$.

$$<v> = \left(\frac{8RT}{\pi \mathcal{M}}\right)^{1/2} \qquad (5\text{-}10)$$

EXAMPLE 5.14 Average Velocity of Gas Molecules

Calculate the average velocity of a nitrogen molecule on a very cold day, when the temperature is −5 °F.

Analysis The problem is solved by inserting known and given values into the appropriate formula, which is Equation (5-10).

Work First convert the given temperature to kelvins.

$$-5\ °F \longrightarrow -21\ °C \longrightarrow 252\ K$$

The molar mass of N_2 is 0.0280 kg mol^{-1}. Inserting these values into the equation, we find

$$<v> = \left(\frac{8RT}{\pi \mathcal{M}}\right)^{1/2}$$
$$= \left[\frac{(8)(8.314\ J\ mol^{-1}\ K^{-1})(252\ K)}{(3.142)(0.0280\ kg\ mol^{-1})}\right]^{1/2}$$
$$= 437\ m\ s^{-1}$$

Note: All quantities on the right-hand side of the equation are in SI units. This *ensures* that the answer will have the SI units of velocity, m s^{-1}, and will be correct provided no calculator mistake has been made. If any non-SI quantity is used, of course, the calculated value will be incorrect. Two very common errors are the use of molar mass in the more familiar *grams* per mole rather than the SI unit *kilograms* per mole, and the use of some value of R other than 8.314 J mol^{-1} K^{-1}.

The units in this calculation cancel correctly, since the dimensions of energy are $m\ell^2 t^{-2}$. The appropriate equivalence statement for SI units is

$$1\ J \Leftrightarrow 1\ kg \cdot 1\ m^2/1\ s^2.$$

Exercise

At what temperature do nitrogen molecules have an average velocity of 1000 m s^{-1}?

Answer: 1323 K.

The way in which the molecular velocities vary from one molecule to another is called the "distribution of molecular velocities," or the **Maxwell-Boltzmann distribution.** The Maxwell-Boltzmann distribution can be deduced from the postulates of the KMT.

James C. Maxwell (1831–1879) and Ludwig Boltzmann (1844–1906)

James Maxwell contributed to many different fields of physics. His early work on the KMT led to the distribution function which, in part, bears his name. He is perhaps better known for his studies of light and other forms of electromagnetic radiation. He made significant contributions to thermodynamics, and also studied color and color blindness.

Ludwig Boltzmann originally worked in the field of thermodynamics. He played an important role in the development of the new science of statistical mechanics. Later he made important discoveries in the area of electromagnetic radiation.

TABLE 5.4 Distribution of Molecular Velocities in Nitrogen at STP

Velocity Range/m s^{-1}	Percentage of Molecules Having Velocities in this Range
0–200	7
200–400	32
400–600	36
600–800	19
800–1000	5
>1000	1

Velocity distribution information can be presented in a table of values showing the percentage of molecules having velocities within specified *ranges* of velocities. For example, the distribution of velocities in gaseous nitrogen at STP is given in Table 5.4.

For most purposes it is more convenient to present this information in graphical form, as in Figure 5.12. One advantage of the graphical form is that the most probable velocity, which corresponds to the maximum of the curve, can be seen at a glance. In the case of nitrogen at room temperature, the most probable velocity is about 420 m s^{-1}. When you consider that 420 meters is about 4½ times the length of a football field, you realize that molecular motion is very rapid indeed. Graphs such as Figure 5.12 take on a slightly different appearance depending on the temperature and the molar mass: the effect of these parameters is shown in Figures 5.13 and 5.14.

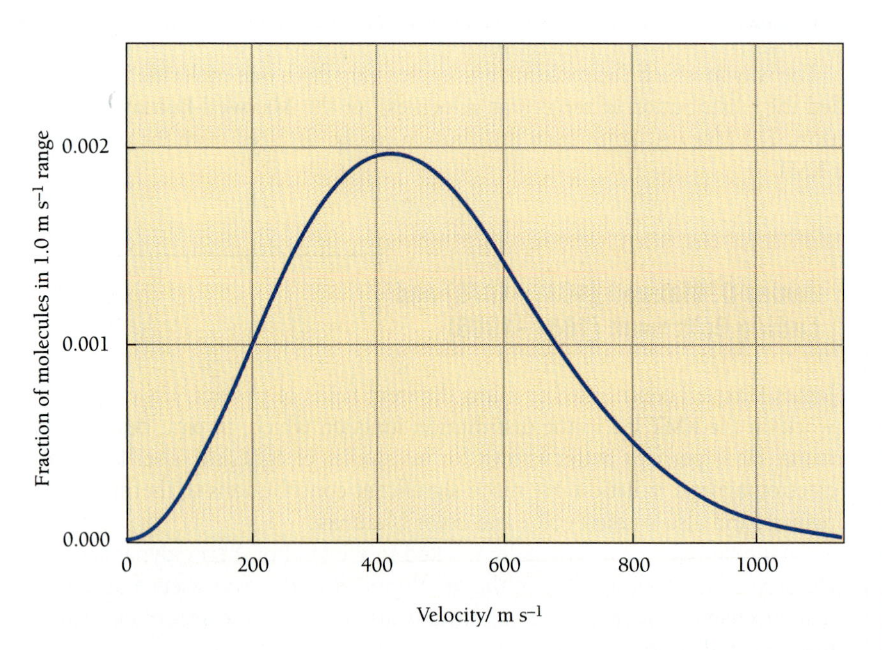

Figure 5.12 Distribution of molecular velocities in nitrogen at room temperature. The abscissa is a velocity *range* 1 m s^{-1} wide rather than a discrete velocity. For example, an abcissa value of 200 m s^{-1} means the range of velocities from 199.5 m s^{-1} to 200.5 m s^{-1}. Correspondingly, the meaning of the ordinate is the fraction of molecules having velocities within the 1 m s^{-1} range centered on 200 m s^{-1}.

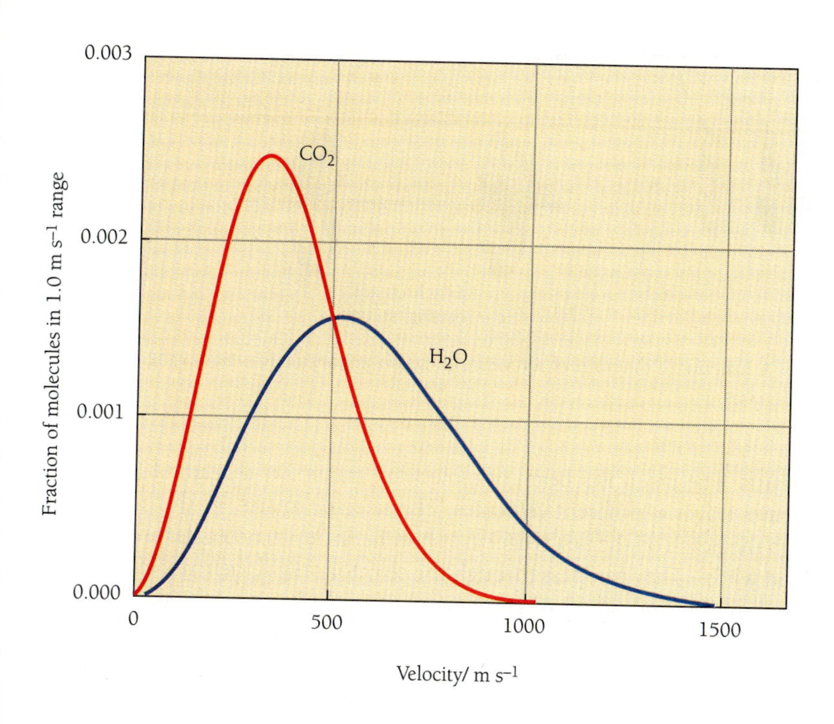

Figure 5.13 Distribution of velocities in gases of different mass. The molar mass of carbon dioxide is about two and a half times that of water vapor. Correspondingly, CO_2 molecules move more sluggishly. The temperature is 300 K in both curves.

Recently it has become possible to make direct measurements of the velocities of gas molecules. These measurements are carried out in what is called a "molecular beam" apparatus, shown in schematic form in Figure 5.15. In such an apparatus, gas leaks through a small hole in the source into a near-vacuum, and the collimating slits block the passage of all molecules that are not traveling in a straight line along the axis

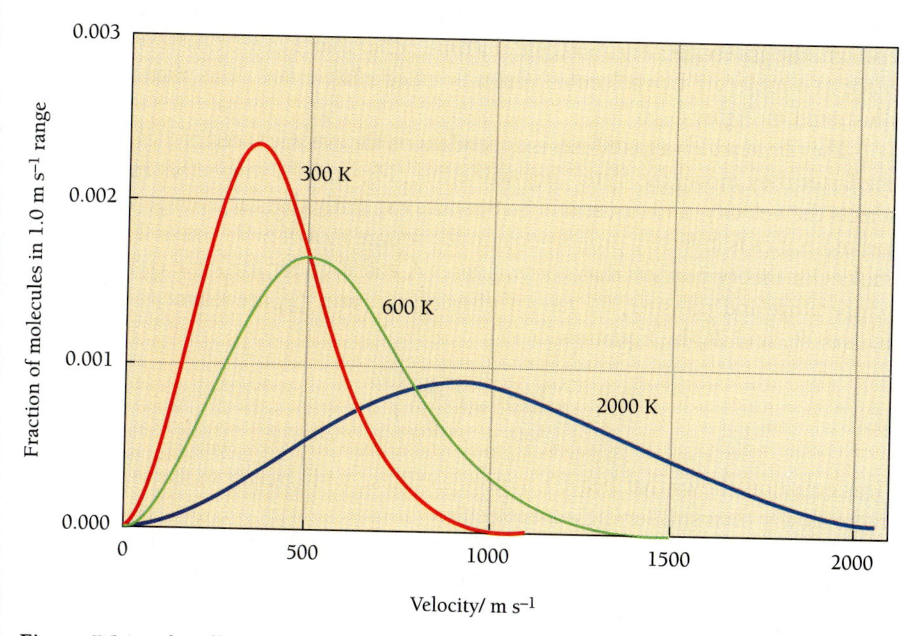

Figure 5.14 The effect of temperature on the molecular velocity distribution. The gas is argon, and the three temperatures are 300 K, 600 K, and 2000 K.

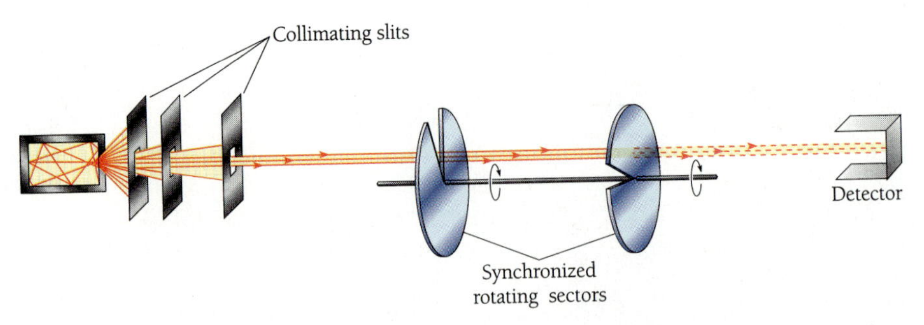

Figure 5.15 A functional diagram of a molecular beam apparatus for measuring the velocity of gas molecules.

of the apparatus. The result is a beam of molecules, much like a beam of light. The beam molecules travel at different velocities. The selector rotates, so that molecules enter through the notch on the first wheel in a short burst. Most molecules are blocked by the second wheel. They can pass through the selector only if their velocity is such that they reach the second wheel when its notch is in the beam position.

Molecules passing through the selector are counted by the detector. The rotational speed of the selector can be varied, so that it can be "tuned" to pass molecules of any chosen velocity. In this way, the number of molecules in each velocity range can be determined. The results of such experiments prove that the predictions of the KMT, as shown in Figures 5.12 through 5.14, are correct.

Effusion and Diffusion of Gases

Molecular beam studies are comparatively recent, having been carried out only in the past 30 years or so. Long before that, however, other features of gas behavior had provided ample verification of the KMT. One such feature is the phenomenon of **effusion,** which is the escape of gas through a very small hole. If the hole is small enough, the gas escapes one molecule at a time rather than in a jet. Gas flow that takes place in a molecule-by-molecule fashion is called *effusive flow.* The phenomenon is illustrated in Figure 5.16.

The rate at which a gas effuses is dependent on the average speed of the molecules; the faster they move, the faster they will effuse. Figure 5.13 shows that molar mass affects the velocity, and Equation (5-10) states explicitly that the average velocity of gas molecules is inversely proportional to the square root of molar mass. The kinetic molecular theory predicts that, for two gases A and B under the same conditions of temperature and pressure, the rates of effusion $r_{\text{eff,A}}$ and $r_{\text{eff,B}}$ are related to the molar masses $\mathcal{M}_A$ and $\mathcal{M}_B$ by Equation (5-11).

$$\frac{r_{\text{eff,A}}}{r_{\text{eff,B}}} = \left(\frac{\mathcal{M}_B}{\mathcal{M}_A}\right)^{1/2} \tag{5-11}$$

This equation shows that heavier molecules (those with greater molar mass) effuse more slowly than smaller, lighter molecules.

The relationship between effusion rate and molar mass is known as Graham's law. Of course, the rate of effusion depends on the size of the hole as well as the molar mass, so that comparisons between different gases have to be made in the same apparatus. Also, since molecular velocities depend on temperature (Figure 5.14), comparisons must be made at the same temperature. As shown in the following example, Graham's law provides a method for measuring the molar mass of a gas.

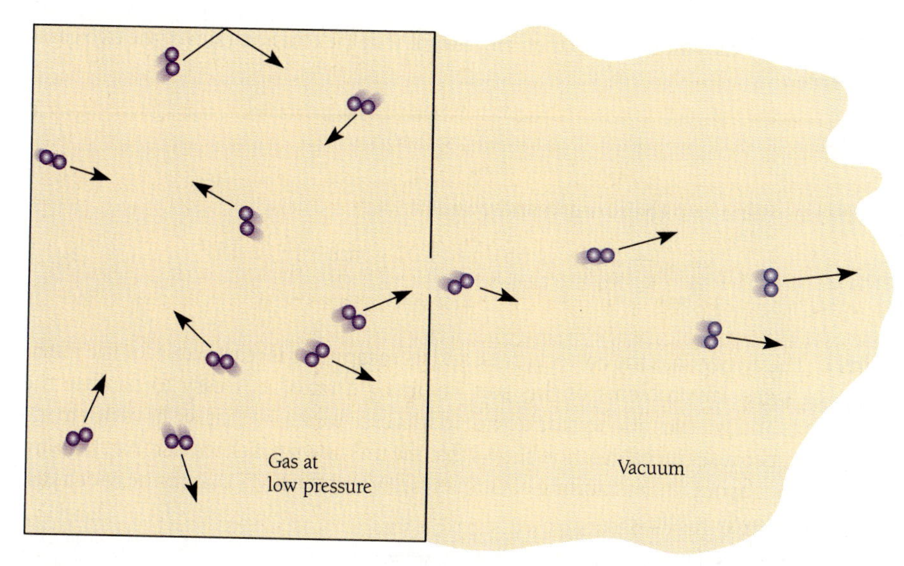

Figure 5.16 In the process of effusion, gas molecules escape one by one through a small hole. The hole must be smaller than the average distance between molecules, otherwise the gas rushes out in a jet.

EXAMPLE 5.15 Rate of Gas Effusion

An unknown gas effuses 2.65 times more slowly than helium, when the measurements are made at the same temperature and pressure in the same apparatus. What is the molar mass of the unknown gas?

Analysis

Target: A molar mass

$$? \text{ g mol}^{-1} =$$

Knowns: The effusion rate ratio = 2.65, and, implicitly, the molar mass of helium (4.00 g mol^{-1}).

Relationship: The word "effusion" in the problem points directly to Graham's law, Equation (5-11).

Plan Insert the knowns into Equation (5-11) and evaluate. Be sure that the ratio is correctly used.

Work Let helium be gas A and the unknown gas be gas B. Then Equation (5-11) becomes

$$\frac{r_{\text{eff,He}}}{r_{\text{eff,unk}}} = \left(\frac{\mathcal{M}_{\text{unk}}}{\mathcal{M}_{\text{He}}}\right)^{1/2}$$

Square both sides and rearrange

$$\left(\frac{r_{\text{eff,He}}}{r_{\text{eff,unk}}}\right)^2 = \frac{\mathcal{M}_{\text{unk}}}{\mathcal{M}_{\text{He}}}$$

$$\mathcal{M}_{\text{unk}} = \mathcal{M}_{\text{He}} \cdot \left(\frac{r_{\text{eff,He}}}{r_{\text{eff,unk}}}\right)^2$$

"Effuses 2.65 times more slowly than He" means that the effusion rates of the two gases are related by

$$\frac{r_{\text{eff,He}}}{r_{\text{eff,unk}}} = 2.65$$

Insert this value into the formula and evaluate.

$$\mathcal{M}_{\text{unk}} = \mathcal{M}_{\text{He}} \cdot (2.65)^2 = (4.00 \text{ g mol}^{-1}) \cdot (2.65)^2$$
$$\mathcal{M}_{\text{unk}} = 28.1 \text{ g mol}^{-1}$$

Possible identities of the gas in Example 5.15 are N_2, CO, and C_2H_4 (ethene). All these compounds are gases, and all have a molar mass of 28.0 g mol^{-1}.

Check The answer has the correct units, and its magnitude is reasonable for the molar mass of a gas. In problems of this sort involving a ratio, it is easy to confuse the subscripts and get the ratio upside down. In this case we know that we have not made that error, because our calculation shows that the unknown gas is heavier than helium. We know from our understanding of Graham's law and the KMT that the heavier a gas, the more slowly it effuses.

Exercise

What is the molar mass of a gas that effuses twice as fast as methane, CH_4?

Answer: 4 g mol^{-1}.

One potential pitfall in Graham's law problems is the use of effusion *times* rather than effusion *rates*. The time is the reciprocal of the rate. A rate such as 0.05 mol s^{-1} corresponds to a time of 1/0.05 or 20 s mol^{-1}. (Note the inversion of the units as well as the digits.) It follows that a ratio of rates is the reciprocal of a ratio of times. Example 5.15 might have given the information that the effusion time for He was 100 s and the effusion time for the unknown gas was 265 s. The ratio of *times* is 100/265 = 0.377; the ratio of *rates* is 265/100 = 2.65. Such problems must be read with great care.

Diffusion is another phenomenon that can be predicted and understood from the point of view of the KMT. **Diffusion** is the transport of one substance through another on a molecule-by-molecule basis. It occurs readily in gases, slowly in liquids, and almost imperceptibly slowly in solids. Diffusion is a common event in our environment. If someone is cooking something delicious in the kitchen, it will be some time before the occupants of the more distant parts of the house become aware of it. The molecules responsible for the aroma do not instantly travel to other locations. The motion of an individual gas molecule is unpredictable, because it is determined by random collisions which leave it with random speed and direction. A molecule will eventually visit all parts of the container, but it takes a long time to do so. This behavior is illustrated in Figure 5.17.

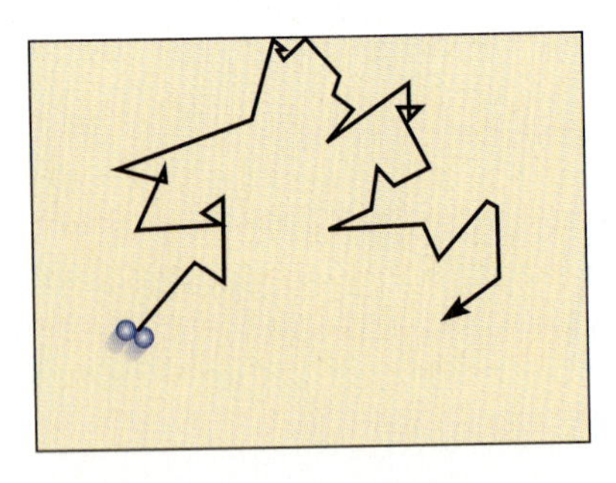

Figure 5.17 A molecule's journey through a container is punctuated by collisions, and so becomes a sequence of steps that are random in length, direction, and speed. Eventually the molecule visits all parts of the container.

If molecules of a different type are released in one corner of a vessel containing a gas, they diffuse through the gas at a rate that is quite slow compared to the average velocity of molecular motion between collisions. The process of diffusion continues until, much later, the mixture of the two gases becomes uniform throughout the container.

Diffusion rate is slower than, but nonetheless proportional to, the average velocity of molecular motion. The average velocity in turn depends inversely on the square root of molar mass (Equation 5-10), so that the overall diffusion rate is inversely proportional to the square root of molar mass. If we compare the diffusion rates of two different gases A and B under the same experimental conditions, we find that they follow the same behavior as do effusion rates, namely Graham's law.

$$\frac{r_{\text{diff,A}}}{r_{\text{diff,B}}} = \left(\frac{\mathcal{M}_B}{\mathcal{M}_A}\right)^{1/2} \tag{5-12}$$

This dependence on the square root of molar mass can be nicely demonstrated in a simple laboratory experiment, shown in Figure 5.18. A plug of cotton soaked in aqueous HCl is inserted into one end of a long piece of glass tubing, and a cotton plug soaked in aqueous NH_3 is placed in the other end. Gaseous HCl and NH_3 diffuse toward each other through the air in the tube, and when they meet they react.

$$HCl(g) + NH_3(g) \longrightarrow NH_4Cl(s)$$

The product is a solid, visible as a white dust that coats the inside of the tube at the meeting point. Ammonia has a molar mass of 17 g mol^{-1}, and it diffuses faster than the heavier HCl ($\mathcal{M}$ = 36.5 g mol^{-1}). The distance traveled by each is proportional to its diffusion rate, so the ratio of the distances (d_{HCl}/d_{NH_3}) equals the ratio of the rates. This ratio, $r_{\text{diff,HCl}}/r_{\text{diff,NH}_3}$, is given by Equation (5-12).

$$d_{HCl}/d_{NH_3} = \frac{r_{\text{diff,HCl}}}{r_{\text{diff,NH}_3}} = \left(\frac{\mathcal{M}_{NH_3}}{\mathcal{M}_{HCl}}\right)^{1/2}$$
$$= \left(\frac{17 \text{ g mol}^{-1}}{36.5 \text{ g mol}^{-1}}\right)^{1/2} = 0.68$$

If the experiment is done in a 100-cm tube, it is found that the white powder first appears at a point 59 cm from the NH_3 end and 41 cm from the HCl end. The ratio is 41/59 = 0.69, in agreement with the prediction of Graham's law.

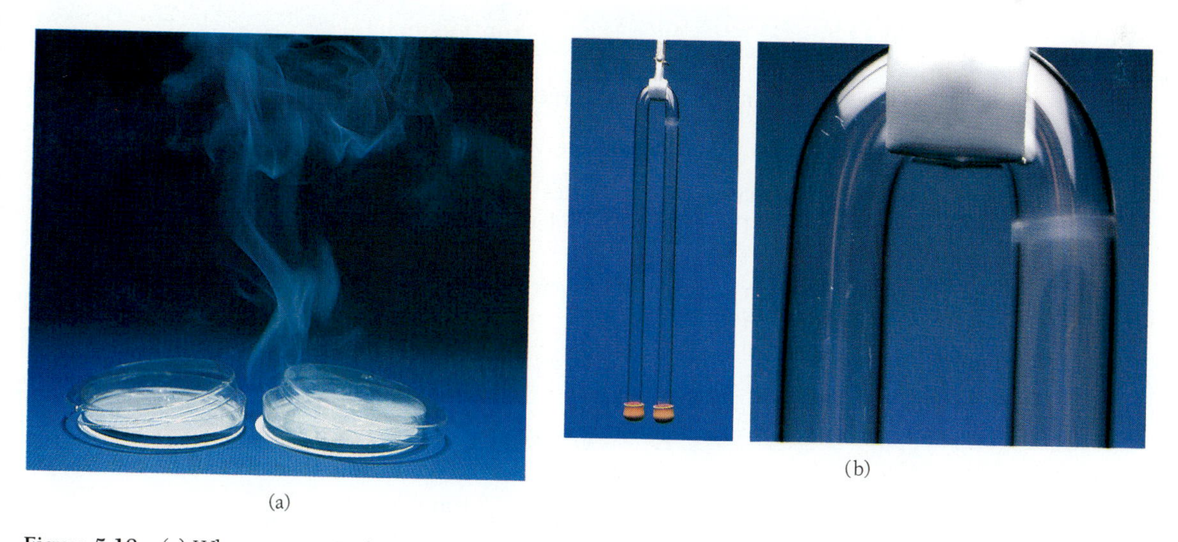

(a) (b)

Figure 5.18 (a) When gaseous HCl and NH_3 meet, a white smoke of solid NH_4Cl forms. (b) A rough measure of the relative molar masses of the two reactants can be obtained by injecting small amounts of NH_3 (left) and HCl (right) into opposite ends of a glass tube. The two gases do not meet in the middle, because HCl, having the greater molar mass, diffuses more slowly than NH_3.

5.5 MIXTURES OF GASES

The kinetic molecular theory asserts that there are no significant attractive forces between the molecules of a gas, or, put another way, that gas molecules move independently of one another. This being so, it should not matter to a molecule whether its neighbors in the gas are the same chemical species or a different chemical compound. Thus, the KMT predicts that mixtures of several different gases should behave in the same way as pure gases. Experimentally this is found to be the case.

Composition of Gaseous Mixtures: Mole Fraction

Although the composition of mixtures of gaseous substances can be described in terms of mass percent or molarity, for many purposes it is more useful to use a different measure called the *mole fraction*. Some of the advantages of mole fraction are that it does not require the identification of one component as the solvent, it is well suited to mixtures of more than two components, and, unlike molarity, it is independent of temperature.

The **mole fraction** of a component in a mixture is the number of moles of that substance divided by the total number of moles of all substances in the mixture. Consider a mixture of gases containing n_A moles of type A, n_B moles of type B, n_C moles of type C, and so on. The mole fraction X_A of component A is given by

$$X_A = \frac{n_A}{n_{total}} = \frac{n_A}{(n_A + n_B + n_C + \cdots)} \tag{5-13}$$

Similar definitions hold for the mole fractions $X_B, X_C, \ldots$ of the other species. Several things follow immediately from the definition of mole fraction: the mole fraction of any component is a number between 0 and 1; the mole fraction of a pure substance is 1; the sum of the mole fractions of all components of a mixture is 1.

EXAMPLE 5.16 Mole Fractions in Gas Mixtures

Calculate the mole fraction of oxygen in a mixture that contains 10.0 g each of H_2, N_2, and O_2.

Analysis Equation (5-13) relates mole fraction to the amounts, in moles, of all components. These amounts are found from the given masses and the molar masses.

Plan For each component, convert grams to moles. Then find the mole fraction of O_2 from Equation (5-13).

Work

$$H_2 \quad (10.0\ g)\left(\frac{1\ mol}{2.02\ g}\right) = 4.95\ mol$$

$$N_2 \quad (10.0\ g)\left(\frac{1\ mol}{28.0\ g}\right) = 0.357\ mol$$

$$O_2 \quad (10.0\ g)\left(\frac{1\ mol}{32.0\ g}\right) = 0.313\ mol$$

$$\text{total} = 5.62\ mol$$

Mole fraction oxygen:

$$X_{O_2} = \frac{n_{O_2}}{n_{total}}$$

$$= \frac{0.313}{5.62} = 0.0557$$

Check By definition, a mole fraction must be a positive number less than 1. *Note:* Mole fractions are dimensionless ratios.

Exercise

Calculate the mole fractions of the other two gases in the mixture and confirm that the sum of the mole fractions is 1.00.

Answer: $X_{N_2} = 0.0635$, $X_{H_2} = 0.881$; $X_{O_2} + X_{N_2} + X_{H_2} = 1.00$.

Partial Pressure and Dalton's Law

Since the gases of a mixture behave independently, the n_A moles of component A independently exert pressure on the walls of the container. This pressure, symbolized by P_A, is called the **partial pressure** of component A; it is numerically equal to the pressure that would exist in the container if n_A moles of A were the only gas present. The total pressure, which is the measured pressure, is the sum of the partial pressures of all the components.

Partial pressures must be determined indirectly, since instruments such as the barometer measure only the total pressure.

$$P_T = P_A + P_B + P_C + \cdots \qquad (5\text{-}14)$$

Equation (5-14) is known as **Dalton's law** of partial pressures. Partial pressure and mole fraction are closely related to one another. The KMT predicts that the ideal gas law holds for any type of gas, and therefore it should hold for a gas consisting of several different components.

$$P_T = \frac{n_{total}RT}{V} \qquad (5\text{-}15)$$

$$P_T = (n_A + n_B + n_C + \cdots)\frac{RT}{V}$$

$$P_T = \frac{n_A RT}{V} + \frac{n_B RT}{V} + \frac{n_C RT}{V} + \cdots \qquad (5\text{-}16)$$

The ideal gas law holds as well for the individual components of the mixture, so we may write

$$P_A = \frac{n_A RT}{V}, \qquad P_B = \frac{n_B RT}{V}, \qquad P_C = \frac{n_C RT}{V}, \qquad \text{etc.} \qquad (5\text{-}17)$$

Substitution of Equation (5-17) into Equation (5-16) gives Dalton's law.

$$P_T = P_A + P_B + P_C + \cdots$$

The relationship between mole fraction and partial pressure becomes clear if we take one of the identities in Equation (5-17) and divide it by Equation (5-15).

$$\frac{P_A}{P_T} = \frac{\left(\frac{n_A RT}{V}\right)}{\left(\frac{n_{total} RT}{V}\right)}$$

$$\frac{P_A}{P_T} = \frac{n_A}{n_{total}} = X_A \tag{5-18}$$

The relationship in Equation (5-18) is usually expressed in the rearranged form of Equation (5-19).

$$P_A = X_A P_T \tag{5-19}$$

That is, the partial pressure of any gas in a mixture is given by its mole fraction times the total pressure.

EXAMPLE 5.17 Mixture Composition in Parts per Million

Since the number of *molecules* is proportional to the number of moles, a pollutant level of 2 ppm may also be thought of as 2 pollutant *molecules* per million molecules of mixture.

In air quality analysis, the concentration of pollutants is often expressed in units of parts per million (ppm). A pollutant level of 2 ppm, for instance, means that out of every million moles in the mixture, there are 2 moles of pollutants. What is the partial pressure of ozone in the air, if the atmospheric pressure is 755 torr and the ozone level is 6.3 ppm?

Analysis

Target: A partial pressure, to be expressed in any acceptable pressure units but most sensibly in torr, the units given in the problem for the total pressure of the mixture.

Knowns: The atmospheric(total) pressure, and the pollutant level in ppm. The molar mass of the pollutant is an implicit known, which may or may not be needed.

Relationship: Equation (5-19), $P_{O_3} = X_{O_3} P_T$, could be used here if we knew the mole fraction of ozone.

New Target: The mole fraction of ozone.

Relationship: We have not yet seen a formula appropriate to this new problem. However, there must be a relationship between ppm and mole fraction, since they both describe composition in fractional terms.

$$X_{O_3} = \text{moles} \frac{O_3}{\text{total moles}}, \qquad \text{or}$$

$$X_{O_3} = \text{moles of } O_3 \text{ per mole of mixture.}$$

If the mole fraction is multiplied by 10^6, the product gives the number of moles O_3 per million moles of mixture.

$$10^6 \cdot X_{O_3} = \text{moles } O_3 \text{ per million total moles.}$$

That is, the desired relationship is $10^6 \cdot X_{O_3} = \text{ppm } O_3$. The problem is solved as follows.

Work The mole fraction of ozone is found from the relationship just derived.

$$10^6 \cdot X_{O_3} = \text{ppm } O_3$$

$$X_{O_3} = \text{ppm } \frac{O_3}{10^6}$$

$$= \frac{6.3 \text{ ppm}}{10^6} = 6.3 \times 10^{-6}$$

The partial pressure of ozone is

$$P_{O_3} = X_{O_3} P_T$$
$$= (6.3 \times 10^{-6})(755 \text{ torr})$$
$$= 4.8 \times 10^{-3} \text{ torr}$$

Exercise

On a day when the atmospheric pressure was 765 torr, the partial pressure of oxygen was found to be 160 torr. What is the mole fraction of oxygen in the atmosphere?

Answer: 0.209.

Vapor Pressure of Water

When water (or any other liquid that evaporates readily) is placed in a stoppered bottle, some of it evaporates and mixes with the air above the surface of the liquid. The process does not continue indefinitely, but stops when the partial pressure of the water vapor reaches a limiting value called the *equilibrium vapor pressure*. The value of the equilibrium vapor pressure increases as the temperature increases, and moreover it is different for different liquids. The topic of the vapor pressures of liquids is discussed at greater length in Section 11.4.

The equilibrium vapor pressure of water at 25 °C is 24 torr. Any gas, stored or handled in contact with liquid water in a closed container, is therefore a mixture of gases in which the partial pressure of water is $P_{H_2O} = 24$ torr. This fact is relevant to a

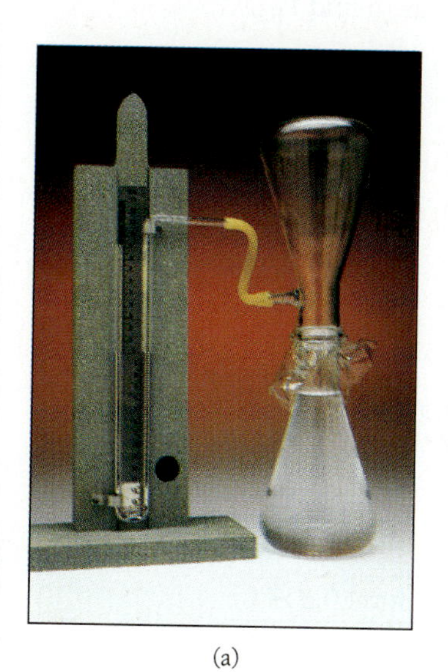

(a)

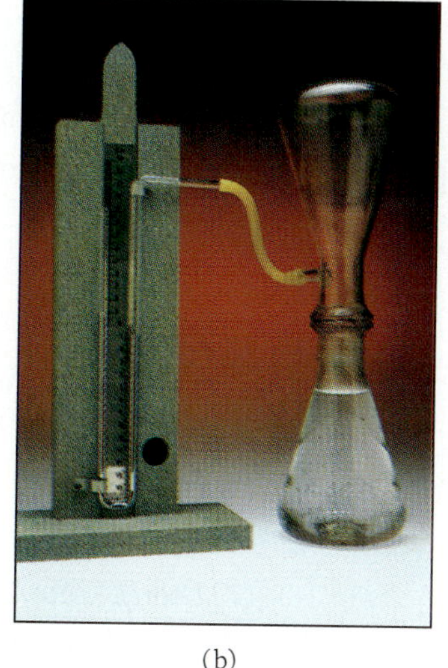

(b)

(a) The lower flask is filled with water, while the inverted flask contains nothing but dry air. The two are separated by a thin sheet of plastic wrap. The manometer shows that the pressure in the upper flask is equal to atmospheric pressure. (b) When the barrier is removed, some water evaporates into the upper flask, increasing the pressure by an amount equal to the equilibrium vapor pressure of water (approximately 20 mmHg at room temperature).

Figure 5.19 Carbon dioxide is generated by the action of dilute acid on a mineral containing the carbonate ion, for instance limestone, and collected in a pneumatic trough. A bottle is completely filled with water, then inverted with its mouth under the surface of water contained in a large trough or dish so that no air is admitted. The gas to be collected is sent through tubing to the inverted bottle, where it displaces the water. The result is the desired gas, free from air but containing a small amount of water vapor.

common laboratory technique for collecting small amounts of gaseous products of a reaction. The gas is collected in an apparatus known as a pneumatic trough, which is shown in Figure 5.19. Often this operation is described as "collecting a gas over water." Of course gases such as ammonia, which are soluble in water, cannot be collected in this way.

If the amount of gaseous reaction product is to be determined by pressure, volume, and temperature measurements, the presence of the water vapor must be taken into consideration. The procedure is illustrated in Example 5.18.

EXAMPLE 5.18 Collection of a Gas Over Water

Gaseous hydrogen can be produced in the laboratory by the action of dilute acid on a moderately active metal such as zinc.

$$\mathrm{Zn + 2\ HCl(aq) \longrightarrow H_2(g) + ZnCl_2(aq)}$$

Suppose the hydrogen is collected over water in a pneumatic trough. If the laboratory temperature is 25 °C and the atmospheric pressure is 755 torr, how many moles of H_2 have been produced when the volume of gas collected is 337 mL? The vapor pressure of water is 25 °C is 24 torr.

Analysis

Target: The amount of hydrogen

$$? \text{ mol } \mathrm{H_2} =$$

Knowns: Atmospheric pressure, $P = 755$ torr; vapor pressure of water, $P_{\mathrm{H_2O}} =$ 24 torr; and the volume and temperature of the mixture of hydrogen and water vapor. An implicit known is the fact that the pressure inside the collection bottle (P_T) is the same as the atmospheric pressure in the laboratory.

Relationship: V and T are given, so if we knew $P_{\mathrm{H_2}}$ we could find $n_{\mathrm{H_2}}$ from the ideal gas law. Dalton's law, $P_\mathrm{T} = P_{\mathrm{H_2}} + P_{\mathrm{H_2O}}$, allows us to find $P_{\mathrm{H_2}}$.

Plan

2. Calculate the number of moles of H_2 from the ideal gas law.
1. Find the partial pressure of H_2 from Dalton's law.

Work

1.

$$P_{\mathrm{H_2}} + P_{\mathrm{H_2O}} = P_\mathrm{T}$$
$$P_{\mathrm{H_2}} = P_\mathrm{T} - P_{\mathrm{H_2O}}$$
$$P_{\mathrm{H_2}} = 755 - 24 = 731 \text{ torr}$$

2.

$$n_{\mathrm{H_2}} = \frac{P_{\mathrm{H_2}}V}{RT}$$

$$n_{\mathrm{H_2}} = \frac{(731 \text{ torr})(1 \text{ atm}/760 \text{ torr})(0.337 \text{ L})}{(0.08206 \text{ L atm mol}^{-1} \text{ K}^{-1})(298 \text{ K})}$$
$$= 0.0133 \text{ mol}$$

Note that several unit conversions have been included in the solution: 337 mL → 0.337 L, 25 °C → 298 K, and P(torr) → P(atm). The units of pressure and volume *must* be atm and L if the numerical value 0.08206 is used for the gas constant.

Check The units have correctly cancelled. At STP, 0.01 mole of gas occupies 0.224 L. We are near STP and have a little more than 0.337 L, so we expect a little more than 0.01 moles. Our answer, 0.0133 mol, is about right in magnitude.

Exercise

750 mL of gas was collected over water when the laboratory temperature and pressure are 22 °C and 758 torr. The vapor pressure of water at that temperature is 20 torr. What amount of gas was collected?

Answer: 0.0301 mol.

5.6 STOICHIOMETRY OF GAS REACTIONS

Chemical reactions that involve gaseous species as reactants and products, such as the reactions shown in Equations (5-20) and (5-21), have particularly simple relationships among the volumes of the participating gases.

$$2\ C(s) + O_2(g) \longrightarrow 2\ CO(g) \qquad (5\text{-}20)$$

$$3\ H_2(g) + N_2(g) \longrightarrow 2\ NH_3(g) \qquad (5\text{-}21)$$

Gay-Lussac's Law

The law of combining volumes, better known as **Gay-Lussac's law,** states that the relative volumes of gaseous substances participating in chemical reactions, as products as well as reactants, are in the ratio of small whole numbers. It is understood that the volumes are to be measured under the same conditions of temperature and pressure. The law applies to all reactions involving at least two gaseous substances, whether or not there are any nongaseous participants.

Combining Volumes and the Ideal Gas Law

Gay-Lussac's law can be derived from the ideal gas law and the principles of stoichiometry. The ideal gas law tells us that the volume of a gas sample is proportional to the number of moles it contains. It follows that the ratio of the amounts, in moles, of two

Both Gay-Lussac's law and Avogadro's principle (equal volumes of gases contain equal numbers of molecules) follow directly from the ideal gas law, $PV = nRT$. Historically, the ideal gas law was formulated after the laws of Boyle, Charles, Gay-Lussac, and Avogadro were known.

Joseph Louis Gay-Lussac (1778–1850)

Joseph Louis Gay-Lussac was a Professor of Chemistry and Physics in Paris, and, in later life, he was a member of France's Chamber of Deputies (similar to the U.S. House of Representatives). In addition to studying gas-phase reactions, he made a balloon ascent to over 7 km to study the composition of the atmosphere. He also discovered, independently of Charles and some 15 years later, the gas behavior now known as Charles's law.

gases is equal to the ratio of the volumes (provided the measurements are made at the same temperature and pressure). Since the stoichiometric coefficients in most reactions are small whole numbers, the mole ratios and volume ratios of the reactants are whole numbers or simple fractions. Thus, in Reaction (5-20) one volume of O_2 reacts to produce two volumes of CO. Similarly in Reaction (5-21) three volumes of H_2 react with one volume of N_2 to produce two volumes of NH_3. Gay-Lussac's law was an important step in our understanding of gas behavior. It is consistent with the atomic theory of matter, and historically it led to the concept of the mole.

Problems in stoichiometry of gas reactions are often solved by combining the ideal gas law with the techniques presented in Chapter 3. The quantities of the gaseous reactants and/or products of such reactions are expressed as volumes (measured at the same temperature and pressure) rather than as grams.

EXAMPLE 5.19 Volume Relationships in Gas-Phase Reactions

How much ammonia can be produced by the reaction of 5.00 L of hydrogen with an excess of nitrogen?

Analysis In Chapter 3 we learned to solve stoichiometry problems by a standard approach that begins with the balanced chemical equation, $N_2 + 3\,H_2 \rightarrow 2\,NH_3$. The given amount of one species (in moles) is then converted to the amount of the other species using the mole ratio obtained from the chemical equation.

$$\frac{2 \text{ mol } NH_3}{3 \text{ mol } H_2}$$

It is implicit in the ideal gas law, however, that a volume ratio is equal to the corresponding mole ratio.

$$\frac{2 \text{ L } NH_3}{3 \text{ L } H_2} = \frac{2 \text{ mol } NH_3}{3 \text{ mol } H_2}$$

Work

$$? \text{ L } NH_3 = (5.00 \text{ L } H_2)\left(\frac{2 \text{ L } NH_3}{3 \text{ L } H_2}\right) = 3.33 \text{ L } NH_3$$

Check The stoichiometric coefficients show that the amount of NH_3 formed is somewhat less than the amount of H_2 consumed.

Warning: This approach requires that the pressure and temperature at which the volumes are measured be the same; this is a safe assumption unless the problem specifically states otherwise. If the problem involves different pressures and/or temperatures, a three-step solution is used.

$$\text{volume of A} \longrightarrow \text{moles of A} \longrightarrow \text{moles of B} \longrightarrow \text{volume of B}$$

Exercise

What volume of CO_2 is produced in the combustion of 3 L CO?

Answer: 3 L.

If a problem in stoichiometry involves both a gas and a liquid or solid, Gay-Lussac's law cannot be used. Instead we must rely on the methods presented in Chapter 3.

EXAMPLE 5.20 Stoichiometry in Gas Reactions

The first step in the commercial production of sulfuric acid is the combustion of sulfur, $S(s) + O_2(g) \rightarrow SO_2(g)$. How many liters of O_2, measured at STP, are required for the combustion of 35 tons of sulfur?

Analysis This is an ordinary problem in stoichiometry, but it requires more than the usual number of unit conversions. Watch the units carefully.

Work

$$? \text{ mol S} = (35 \text{ ton S})\left(\frac{2000 \text{ lb}}{1 \text{ ton}}\right)\left(\frac{454 \text{ g}}{1 \text{ lb}}\right)\left(\frac{1 \text{ mol S}}{32.0 \text{ g S}}\right)$$

$$= 9.93 \times 10^5 \text{ mol S}$$

$$? \text{ mol } O_2 = (9.93 \times 10^5 \text{ mol S})\left(\frac{1 \text{ mol } O_2}{1 \text{ mol S}}\right)$$

$$= 9.93 \times 10^5 \text{ mol } O_2$$

$$? \text{ L } O_2 = n_{O_2} \cdot \frac{RT}{P}$$

$$= \frac{(9.93 \times 10^5 \text{ mol } O_2)(0.08206 \text{ L atm mol}^{-1} \text{ K}^{-1})(273 \text{ K})}{1 \text{ atm}}$$

$$= 2.22 \times 10^7 \text{ L } O_2$$

Check $35 \cdot 2000 \cdot 454/32.0 \approx 2000 \cdot 500 = 10 \times 10^5$ mol S = mol O_2; $10 \times 10^5 \cdot 22.4$ L/mol at STP $\approx 200 \times 10^5 = 2 \times 10^7$.

Exercise

The density of liquid ethanol is 0.789 g mL^{-1}. How many liters of CO_2 (measured at STP) are produced in the combustion of 1.00 L ethanol according to the following equation?

$$C_2H_5OH(\ell) + 3\ O_2(g) \longrightarrow 2\ CO_2(g) + 3\ H_2O(\ell)$$

Answer: 768 L.

5.7 REAL GASES AND THEIR DEVIATIONS FROM IDEAL BEHAVIOR

With careful measuring technique and good measuring instruments, it is possible to show that most gases do not follow the ideal gas law with perfect accuracy. In practice, the pressure measured for a given n, V, and T is not quite equal to nRT/V. This behavior, known as "deviation from ideality," is more pronounced at high pressure and/or low temperature. Deviations from ideality can be expressed in terms of the quotient $Z = PV/nRT$. This quotient, called the **compressibility factor,** is a dimensionless number whose magnitude indicates the extent to which a real gas departs

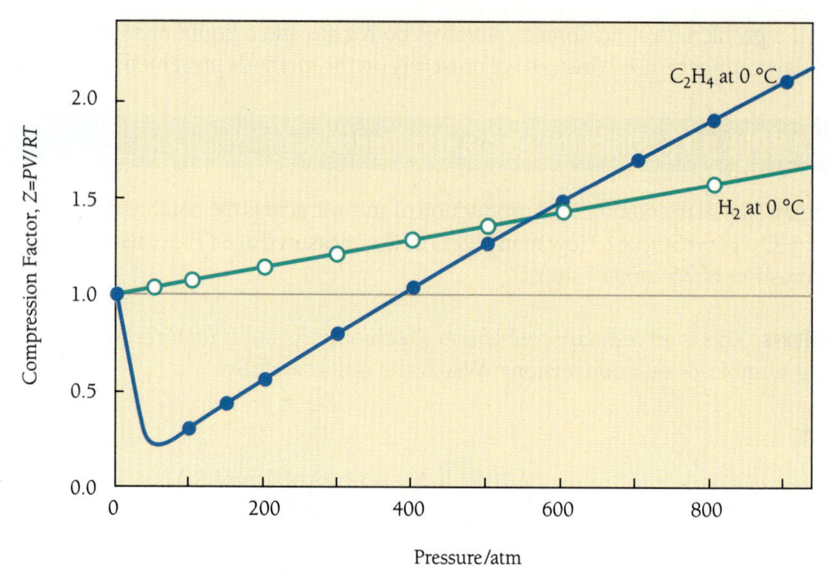

Figure 5.20 The ratio *PV/nRT* is plotted as a function of pressure for C_2H_2 and H_2 at 0 °C.

from ideal behavior. The compressibility of an ideal gas (for which $PV = nRT$) is $Z = 1$. To the extent that a real gas deviates from ideality, the value of the compressibility factor deviates from 1. Figure 5.20 shows how the compressibility factor of two different gases changes as the pressure and temperature change.

It is not hard to understand why such deviations occur. The ideal gas law is derived from the postulates of the KMT and can be no more reliable than those postulates. Two of the postulates in particular are only approximately correct. First, although the actual volume of the molecules of a gas is small with respect to the volume of the container, it is not zero. Moreover, as the pressure increases and the molecules come closer together, the actual molecular volume takes up an increasing fraction of the volume of the container. The volume occupied by one molecule is unavailable to others for their random motion. The available or *effective volume,* V_{eff}, is the volume that is unoccupied by other molecules, and it is less than the measured volume of the container. This is illustrated in Figure 5.21.

Second, the KMT asserts that there are no forces between molecules. This may be so at low pressure, when the molecules are far apart, but at close distances molecules do attract one another. At high pressures, attractive forces can bind several molecules together in loose clusters that are constantly forming and breaking up. Although the clusters are temporary, at any given time some small fraction of the molecules are tied up in this way. Clustering reduces the total number of independent particles in the gas, and it thus reduces the total force exerted by the gas on the walls of its container (Figure 5.22). The measured pressure is less than it would be if the gas were ideal. In other words, the *effective pressure* of a real gas is greater than the measured pressure.

Attractive forces are responsible for the fact that gases condense to liquids when the pressure is increased sufficiently. If there were no attractive forces, molecules would never cluster together in the liquid state.

These ideas suggested the **van der Waals equation,** which makes possible accurate calculations even when the gas behavior is far from ideal. The van der Waals equation is similar to the ideal gas law, but modified to incorporate the concepts of *effective pressure* and *effective volume.* These quantities differ slightly from the measured pressure, P_{m}, and the measured volume, V_{m}.

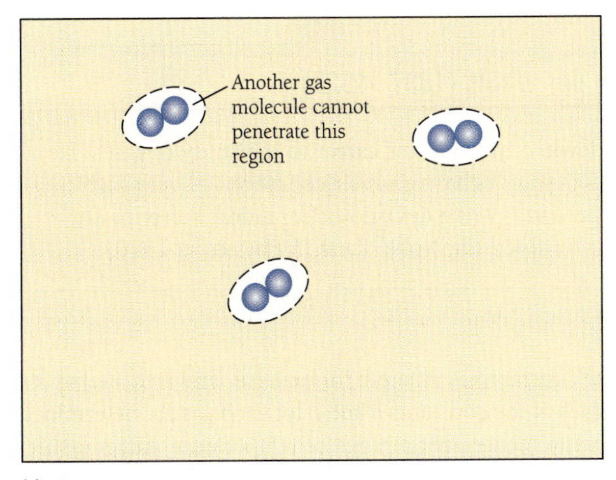

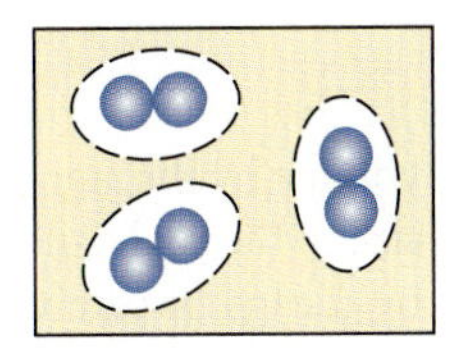

(a) Low pressure

(b) High pressure

Figure 5.21 (a) Because molecules cannot overlap, each of these diatomic molecules is surrounded by a region into which the center of another molecule cannot penetrate. At low pressure, however, the volume in which molecules are free to move about is nearly as great as the actual volume of the container. (b) At high pressure, the available (effective) volume is a much smaller fraction of the actual container volume.

$$PV = nRT \qquad \text{(ideal gas)}$$

$$P_{\text{eff}}V_{\text{eff}} = nRT \qquad \text{(real gas)}$$

$$\left(P_{\text{m}} + \frac{an^2}{V^2}\right)(V_{\text{m}} - nb) = nRT \qquad \text{(Van der Waals gas)} \qquad (5\text{-}22)$$

In the van der Waals equation, the effective pressure is greater than the measured pressure by an amount $a(n/V)^2$. As the quantity n/V (whose units are moles per liter) increases, molecules are on the average closer to one another, the attractive forces are stronger, and the discrepancy between the measured pressure and the effective pressure increases. The constant a is different for each gas, and is related to the strength of the attractive forces between molecules. The effective volume is less than the measured

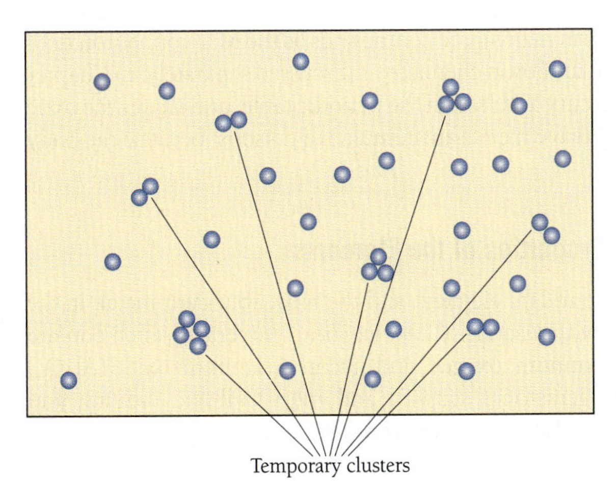

Figure 5.22 At high pressures, gas particles form temporary clusters that move as a unit. Cluster formation reduces the number of independent particles in a gas, thereby reducing the measurable pressure the gas can exert. Cluster formation occurs in polyatomic gases as well as the monatomic gas pictured here.

Johannes Diderik van der Waals (1837–1923)

Johannes van der Waals devoted much of his career to the study of gases. He was particularly interested in their behavior under conditions of very high pressure or very low temperature, when deviations from ideal behavior are most pronounced. He was awarded the Nobel Prize in Physics in 1910.

volume by the amount *nb*. The constant *b* is different for each gas, and is related to the size of the molecules. The values of the constants *a* and *b* for each gas are determined by means of careful *PVT* measurements over a range of temperature and pressure. Table 5.5 gives values of these constants, as measured for several common gases.

TABLE 5.5 Van der Waals Constants for Some Common Gases

Gas	Molar mass/g mol^{-1}	a/L^2 atm mol^{-2}	b/L mol^{-1}
Ar	40	1.35	0.0322
O_2	32	1.36	0.0318
H_2O	18	5.46	0.0305
CH_4	16	2.25	0.0428
C_2H_6	30	5.49	0.0638
C_3H_8	44	8.66	0.0845
C_4H_{10}	58	14.5	0.123

5.8 THE METALS OF GROUP 3A

The elements of Group 3A are boron, aluminum, gallium, indium, and thallium. With the exception of boron, they are metals. Boron is a metalloid, and its chemical properties are quite different from those of the other Group 3A elements. For instance, although the other elements form many ionic compounds as the M^{3+} ion, boron does not: its compounds are molecular. Boron chemistry is discussed separately, in Chapter 22. Thallium lies between mercury and lead in the periodic table, and shares the toxic properties of these elements. Heavy metals are cumulative poisons, because the body has no way of excreting them.

Recall that beryllium, the first element in Group 2A, also forms molecular compounds in contrast to the other alkaline earth elements. The first member of a group is usually different from the others.

Occurrence and Physical Properties of the Elements

Aluminum is the third most abundant element, and the most abundant metal, in the earth's crust. There are literally thousands of *aluminosilicate* minerals, which contain other metals in addition to aluminum, oxygen, and silicon. Aluminum oxide (Al_2O_3) occurs as *corundum* and the gemstones *sapphire* and *ruby*. Gallium, indium, and thallium are rare, and are found primarily as trace-level impurities in the ores of aluminum and zinc.

TABLE 5.6 Physical Properties of the Group 3A Elements

Element	Symbol	Atomic Number	Molar Mass	Atomic Radius/pm	Melting Point/°C	Boiling Point/°C	Density/g mL^{-1}	Hardness*
Boron	B	5	10.81	88	2300	2550	2.34	9.5
Aluminum	Al	13	26.982	143	660	2327	2.70	2.5
Gallium	Ga	31	69.72	122	29.8	2403	5.91	1.5
Indium	In	49	114.82	162	157	2000	7.31	1.2
Thallium	Tl	81	204.383	171	304	1457	11.85	

* Mohs's hardness. See footnote to Table 3.1, page 119, for an explanation.

Some physical properties of the Group 3A elements are listed in Table 5.6. Several group trends in properties are evident, but melting and boiling points do not vary regularly. The very low melting point (approximately room temperature) of gallium is unusual, particularly since its boiling point is so high. Gallium remains liquid over a wider temperature range than any other substance.

Production and Uses

About four million tons of aluminum are produced in the United States each year by electrolysis of the ore *bauxite,* which contains a mixed oxide-hydroxide of aluminum. Purified bauxite is dissolved in *cryolite* (Na_3AlF_6), and an electrical current is passed through the solution. The temperature is held at about 1000 °C in order to keep the bauxite/cryolite mixture in the liquid state. Liquid aluminum collects in the bottom of the vessel and is periodically drawn off, in a relatively pure state. The Hall-Heroult process, as it is called, is energy-intensive—it consumes 4% to 5% of the electrical power produced in the United States. A more detailed description of the process is given in Section 19.5.

Charles Hall was an undergraduate at Oberlin College in Ohio when he discovered how to produce aluminum by electrolysis of bauxite. Independently, and in the same year (1886), the process was discovered by Heroult in France.

Gallium, indium, and thallium can be prepared by electrolysis of aqueous solutions of their soluble salts.

The melting point of gallium is 29.9 °C, a few degrees less than body temperature.

Electrical wiring is a major use for aluminum.

LED stands for "light-emitting diode."

Aluminum is a versatile metal whose many uses stem from three of its properties: (1) high electrical conductivity, (2) low density, and (3) good resistance to corrosion. Conductivity of different metals is often compared by referring to the current-carrying capacity of wires having the same diameter. On this basis aluminum is not quite as good a conductor of electricity as copper, because an aluminum wire can carry only 84% of the current carried by a copper wire of the same diameter. However, because of its much lower density, an aluminum wire can carry twice the current of a copper wire of the same *mass*. As copper becomes scarcer and more expensive, aluminum will be increasingly used in both long-distance power transmission and local wiring.

The low density of aluminum makes it a useful metal in the transportation industry, particularly in air transport where light weight is so important. Although pure aluminum is a soft, low-strength material, it can be alloyed with other elements (chiefly Cu, Mn, Mg, and Si) to increase both hardness and strength. Most alloys retain the corrosion resistance of pure aluminum, thus increasing their usefulness.

Aluminum sulfate is used in making paper for books and magazines, as an additive that increases the hardness and decreases the water-absorbing tendency of untreated paper.

About 95% of the gallium produced in the United States is used in the electronics industry. Alloyed with phosphorus, antimony, or arsenic, gallium is fabricated into a wide range of semiconductor devices. One common example is the LED displays of stereo equipment and other items of consumer electronics. Indium is used in the electronics industry, not only as a semiconductor but also, because of its low melting point, as a solder. A trace of thallium improves the sensitivity of the sodium iodide crystals used in radiation detectors, but the element has few other uses because of its scarcity and its toxicity.

Some Compounds and Reactions

Although the elements of Group 3A (except for boron) are metals, they are near the borderline that separates metals from nonmetals in the periodic table. Consequently they share some properties with nonmetals, most notably that they form molecular as well as ionic compounds. In almost all of its ionic compounds, aluminum exists as the Al^{3+} ion. Gallium and indium form compounds as M^+ as well as M^{3+}, and the ionic compounds of thallium almost always contain Tl^+ rather than Tl^{3+}.

Many metals, including those of Group 3A, react with acids with the formation of hydrogen.

$$2\,Al(s) + 6\,HCl(aq) \longrightarrow 2\,AlCl_3(aq) + 3\,H_2(g)$$

Aluminum and gallium share the less common property of reacting with base as well as with acid.

$$2\,Al(s) + 6\,NaOH(aq) \longrightarrow 2\,Na_3AlO_3(aq) + 3\,H_2(g)$$

Oxides and Hydroxides

All the Group 3A metals react with oxygen at high temperature to produce oxides of formula M_2O_3. The reaction of aluminum is typical.

$$4\,Al(\ell) + 3\,O_2(g) \xrightarrow[\Delta]{} 2\,Al_2O_3\,(s)$$

Thallium forms Tl_2O as well as Tl_2O_3. There are several forms of naturally-occurring aluminum oxides. They are generally unreactive, hard substances. One of them, *corundum,* has been used in sandpaper and other abrasives, while others, *ruby* and *sapphire,* are valuable gemstones. The colors of these gems are due to impurities, and pure aluminum oxide is colorless.

Manufactured Al_2O_2 is increasingly used for abrasives, replacing natural corundum. Synthetic ruby is used in lasers.

When exposed to air at room temperature, aluminum quickly forms a thin surface film of Al_2O_3. This hard, impervious layer protects the metal from further contact with oxygen, and is responsible for the excellent corrosion resistance of aluminum.

The hydroxides $M(OH)_3$ form when aqueous solutions of M^{3+} salts are treated with base.

$$Ga(NO_3)_3(aq) + 3\ NaOH(aq) \longrightarrow Ga(OH)_3(s) + 3\ NaNO_3(aq)$$

These hydroxides, when heated, lose water to form first the mixed or *hydrous oxides,* and then, on stronger heating, the anhydrous oxides. Many of the naturally occurring Group 3A minerals, including bauxite, are hydrous oxides.

$$Al(OH)_3(s) \xrightarrow[\Delta]{} AlO(OH)(s) + H_2O(g)$$
$$2\ AlO(OH)(s) \xrightarrow[\Delta]{} Al_2O_3(s) + H_2O(g)$$

The hydroxides $Al(OH)_3$ and $Ga(OH)_3$ react with both acids and bases. (These reactions are discussed in detail in Chapter 23.) In contrast, $B(OH)_3$ does not react with acids, and neither $In(OH)_3$ nor $Tl(OH)_3$ reacts with bases.

Substances that react with both acids and bases are called "amphoteric." This behavior will be explored further in Chapter 15.

$$Al(OH)_3(s) + 3\ HCl(aq) \longrightarrow AlCl_3(aq) + 3\ H_2O$$
$$Al(OH)_3(s) + KOH(aq) \longrightarrow KAlO_2(aq) + 2\ H_2O$$

Compounds with Other Nonmetals

Aluminum reacts with chlorine to produce aluminum chloride, whose empirical formula is $AlCl_3$. Aluminum chloride melts at 192 °C to form Al_2Cl_6, a molecular compound whose structural formula is as follows.

```
Cl     Cl     Cl
  \   /  \   /
   Al     Al
  /   \  /   \
Cl     Cl     Cl
```

Gallium forms a similar molecular chloride, Ga_2Cl_6, whereas the chlorides of indium and thallium are ionic compounds. Aluminum chloride reacts vigorously with a small amount of water, forming the hydroxide and hydrochloric acid.

$$AlCl_3(s) + 3\ H_2O(\ell) \longrightarrow Al(OH)_3(s) + 3\ HCl(g)$$

The Group 3A elements react with Group 5A elements, most commonly P, As, and Sb, to form 1:1 compounds like GaP and InSb. These solid compounds are semiconductors, and many of them are used in the electronics industry. Their electrical properties can be varied by the carefully-controlled addition of other elements.

Some aluminum-containing minerals are (clockwise from upper right) ruby, sapphire, and corundum. These samples are resting on a surface of clay, which is an aluminosilicate mineral.

Ionic Compounds

The M^{3+} ions of the Group 3A metals form a variety of salts with most of the common anions. Nitrates, carbonates, sulfates, and so on, are all well known. As atomic number increases within this group, M^+ ions are increasingly common. In the case of thallium, Tl^+ compounds such as TlCl and Tl_2SO_4 are much more common than Tl^{3+} compounds.

CHEMISTRY OLD AND NEW

CHEMISTRY IN SPACE

Aboard the space shuttle *Discovery* (September 1988), NASA astronaut George Nelson prepares a crystal-growing experiment.

The Group 3A element, aluminum, is one of the most important commercial metals. Aluminum is not a particularly strong metal, but its strength and usefulness can be increased by alloying it with magnesium and small amounts of copper and silicon. However, due to its low specific gravity (about 2.7), aluminum does not readily mix with other, more dense elements. When combined and melted together, such a blend separates into layers, with the aluminum floating on top of the heavier metal. It was not until 1983 that scientists were able to prepare the first alloy of aluminum and zinc. Their method did not involve any special processing or complex mixtures of ingredients. It simply eliminated the basic problem: gravity. When the two metals were melted together in space shuttle *Columbia's* Spacelab research module, the large difference in their densities did not prevent mixing, as it does on earth.

A spaceship may seem an unlikely place to conduct research in metallurgy. But since the inception of the space shuttle program in 1982, many important experiments in chemistry, materials science, and biology have been carried out under conditions of zero gravity. The absence of gravity means that certain processes—such as alloying and crystallization—may work better in space than on earth. Experiments in these areas could lead to new space-manufactured materials with a variety of beneficial uses.

The exposure of animals and plants to zero gravity has yielded some startling results. Pituitary cells from rats flown aboard a shuttle were found to produce less growth hormone than earth-bound cells. Human lymphocytes (the disease-fighting white blood cells) placed in orbit showed stunted division, and bacterial spores exposed to solar radiation and microgravity exhibited a mutation rate ten times greater than normal. Chicken embryos fertilized on the ground but allowed to undergo first trimester development on the shuttle *Discovery* did not hatch from their eggs even when returned to an incubator on earth. Not all living things seem to be adversely affected by space flight, though: tomato seeds that spent over five years aboard NASA's unmanned Long Duration Exposure Facility (LDEF) lab sprouted faster, produced plants with more chlorophyll, and had fewer mutations than control seeds.

Some of the most exciting Spacelab experiments have examined the formation of crystals. Whether grown from solution or by a sublimation process in a sealed tube, crystals normally begin their growth at the wall of the container. The influence of the wall often results in smaller, deformed, imperfect crystals. Aboard the *Challenger,* however, sublimation of germanium selenide produced crystals which—due to the zero-gravity environment—grew from the *center* of the tube. These crystals were more than ten times larger than similarly-grown earth specimens. In addition, lead iodide crystals deposited from solution on *Discovery* proved to be purer and much more symmetrical than earth-produced ones. The use of ultrapure lead iodide could improve the sensitivity of x-ray films, allowing a proportional reduction in the amount of radiation received by patients undergoing medical and dental x-rays.

Space shuttle experiments show that gravity has a detrimental effect on some other processes as well. Tiny polystyrene spheres (used by the National Bureau of Standards as size references) made on a *Challenger* flight came out rounder and more uniform than any produced before. These little beads—only 9.89 microns in diameter—are important for calibrating machines that measure particle size in flour, talcum, and other ground powders. Sold by NBS in 25 gram quantities for $384, polystyrene reference spheres became the first commercial product manufactured in space.

NASA officials are now making plans for the construction of space station *Freedom,* which is to be fully staffed by early in the next century. This permanent, international facility will pave the way for space science to evolve from what is often seen as a novelty to a full-fledged division of research.

Discussion Questions

1. Can you think of any reasons why "space tomatoes" should have fewer mutations than "earth tomatoes," while "space bacteria" should have more?
2. Heat transfer by *convection* depends on the fact that warm gases and liquids have lower density than the same substances at lower temperature, and on the effect of gravity on these density differences. In what ways would the absence of convective heat transfer affect people and processes in space?

SUMMARY EXERCISE

1. The pressure of O_2 inside a 103.6 L storage tank is measured at 73 °F and found to be 12.2 atm. Express this pressure in (a) torr, (b) Pa, and (c) bars. Assuming that the gas behaves ideally, calculate (d) the mass of oxygen in the tank, (e) the density of oxygen under these conditions, (f) the volume the gas would occupy if it were further compressed to 55.5 atm with no change in temperature, and (g) the STP volume of the gas. (h) What is the average velocity of the oxygen molecules in the tank?
2. 250 mL of N_2 produced by the thermal decomposition of sodium azide (NaN_3) is collected over water when the laboratory temperature is 25 °C and the atmospheric pressure is 752 torr. The equilibrium vapor pressure of water at 25 °C is 24 torr. Calculate (i) the partial pressure of N_2, and (j) the mole fraction of water vapor in the collected gas, and (k) the mass of sodium azide that decomposed (the other product of decomposition is sodium metal).
3. (l) What is the molar mass of a gas whose density at 30.0 °C and 845 torr is 2.60 g L^{-1}? (*Hint:* See Example 5.12.) (m) What volume of fluorine (F_2) is required to react with 379 mL Xe to form XeF_4? (n) Using the van der Waals equation and data from Table 5.5, calculate the pressure exerted by 12.0 mol O_2 confined to a volume of 1.75 L at −85 °C. (o) What would the pressure be if oxygen behaved ideally under these conditions?

Answers **(a)** 9.27×10^3 torr; **(b)** 1.24×10^6 Pa; **(c)** 12.4 bars; **(d)** 1.67 kg; **(e)** 16.1 g L^{-1}; **(f)** 22.8 L; **(g)** 1.17×10^3 L; **(h)** 442 m s^{-1}; **(i)** 728 torr; **(j)** 0.0319; **(k)** 424 mg; **(l)** 58.1 g mol^{-1}; **(m)** 758 mL; **(n)** 71.3 atm; **(o)** 106 atm.

SUMMARY AND KEYWORDS

All gases are *compressible:* their volume is inversely dependent on the applied pressure. **Boyle's law,** which holds at constant temperature, states this relationship as $PV =$ constant. Pressure is measured with a **barometer** and is commonly expressed in units of **atmospheres (atm)** or **torr (= mmHg)**; 760 torr ⇔ 1 atm. The SI unit is the **Pascal (Pa)** (1.01×10^5 Pa ⇔ 1 atm). Like other substances, gases expand when heated: **Charles's law,** which holds at constant pressure, states that $V/T =$ (constant). The *combined gas law* says that the ratio PV/T is a constant for a given sample of gas. **Avogadro's principle** states that equal volumes of different gases contain equal numbers of molecules, and when this is incorporated into the combined gas law the result is the **ideal gas law,** $PV = nRT$. Except for unit conversions, the value of the gas constant R is always the same.

In a **process problem,** a sample of gas undergoes some change from a definite initial state to a definite final state; the combined gas law is used in the solution. A gas may be produced or consumed in a chemical reaction, but it is not a process problem unless the *gas* undergoes a change in state. In a **single-state problem** the gas is in only one defined condition, and the ideal gas law is used.

The conceptual model of gas behavior is the **kinetic molecular theory (KMT),** whose postulates are: gases consist of atoms or molecules that are small in comparison to the distance between them; they are in rapid, random motion, undergoing occasional collisions with one another and the container walls (the pressure exerted by a gas is the result of these collisions with the walls); the molecules exert no forces on one another, and the average energy of molecular motion is proportional to absolute temperature. The ideal gas law may be derived from these postulates. At any temperature, molecules in a gas travel with a distribution of velocities, called the **Maxwell-Boltzmann distribution,** which can be measured in a molecular beam. The phenomena of **effusion** (one-by-one passage of molecules through a tiny orifice) and **diffusion** (molecule-by-molecule mixing of several different gases) are explained by the KMT, which correctly predicts the effect of both temperature and molar mass on effusion and diffusion.

Components of gas mixtures behave independently, each exerting a **partial pressure** equal to the pressure it would exert if alone in the container. According to **Dalton's law,** the total pressure of a gas mixture is the sum of the partial pressures of the components. The **mole fraction** of a mixture component is the ratio of the number of moles of the component to the total number of moles of all components; this ratio in turn is equal to the ratio of the partial pressure of the component to the total pressure.

The law of combining volumes, or **Gay-Lussac's law,** states that in chemical reactions the relative volumes of the participating gases are in the ratio of small whole numbers.

Real gases deviate from ideal behavior at high pressures and/or low temperatures, so that calculations based on the ideal gas law are only approximately correct. Deviations are due to the oversimplified nature of the postulates of the KMT. Real molecules have small but nonzero volume, and exert small but nonzero forces on one another. The result is that the **compressibility factor,** $Z = PV/nRT$, is not equal to 1 for all gases under all conditions. The **van der Waals equation,** $(P + an^2/V^2)(V - nb) = nRT$, incorporates the concepts of *effective volume* and *effective pressure.* The van der Waals constants a and b are related to the strength of intermolecular attractive forces and to the size of the molecules.

PROBLEMS

General

1. What are the three states of matter and their distinguishing characteristics?
2. Whereas pure substances can almost always be classified as solid, liquid, or gaseous, mixtures are often difficult to describe. How would you classify whippped cream? Mud? Clouds?
3. Define pressure. Give a precise scientific definition, and also one that could be understood by someone without any scientific training.
4. What are some of the units used to measure pressure? How do they relate to one another? What is the value of the average pressure of the atmosphere in these various units? Why do we have more than one unit for pressure?
5. Describe the preparation and operational principles of a mercury barometer.
6. Suppose some liquid other than mercury were used to make a barometer. Would normal atmospheric pressure support a liquid column 760 mm high? Explain your answer.
7. One statement of Boyle's law is that for a given sample of gas at a given temperature, the product of pressure and volume is a constant. What is the meaning of the word "constant" in this context?
8. What is the relationship between gas volume and temperature when pressure is constant?
9. What is the ideal gas law? To what gases does it apply? Are there any limitations to the validity of the law?
10. "The gas constant $R = 0.08206$ L atm mol^{-1} K^{-1}." "The gas constant $R = 82.06$ mL atm mol^{-1} K^{-1}." Are both of these statements true? If not, which is correct? If so, explain the apparent contradiction of $0.08206 = 82.06$.
11. What is meant by the "absolute zero" of temperature? How can it be measured?
12. Explain the meaning of the following terms: theory, postulate, axiom, assumption, model, hypothesis.
13. List the postulates of the kinetic molecular theory. Are these postulates true? Are they reasonable? Are they useful?

14. What is meant by the term "velocity distribution"?

15. Is there a relationship between temperature and molecular velocity? What is it?

16. What is the relationship between temperature and average energy of molecular motion? Does the same relationship hold between temperature and average velocity of molecular motion?

17. What experimental evidence supports the validity of the KMT?

18. Distinguish between effusion and diffusion. What relationship exists between effusion rate and temperature? Between effusion rate and molar mass? Do the same relationships hold among diffusion, temperature, and molar mass? Explain your answer.

19. Describe the procedure for collecting gases over water.

20. What is meant by partial pressure? What is the relationship between partial pressure and total pressure? Can a pure substance have a partial pressure, or does the term apply only to mixtures of gases?

21. How do real gases differ from ideal gases? What is the relationship between these differences and the validity of the postulates of the KMT?

22. Explain the meaning of the acronym STP.

23. What aspects of gas behavior are associated with the following names: Boyle, Charles, Avogadro, Dalton, Torricelli, van der Waals, and Graham?

24. Define mole fraction. Does the mole fraction of a component of a gas mixture change when the temperature of a sample of the mixture changes? When the pressure changes?

Pressure

25. Express the pressure 5.00 atm in torr, Pa, kPa, and bars.

26. A pressure of 500 Pa equals how many torr?

27. The average pressure of the atmosphere at an altitude of 20 km is only 5.5% of the sea-level value. Express the pressure at 20 km in torr.

28. The pressure beneath the surface of the sea increases by about 1 atm for each 5 fathoms depth (1 fathom = 6 feet). What is the approximate pressure at a depth of 25 fathoms?

*29. The downward force exerted by a column of liquid is given by the formula $f = DghA$, where h and A are the height and cross-sectional area of the column, D is the density of the liquid, and g is the acceleration of gravity, 9.819 m s^{-2}. The *pressure* (force per unit area) is $P = f/A = Dgh$. If SI units (kg, m, s) are used, the calculated pressure is in pascals. Calculate the pressure exerted by a column of mercury 0.7600 m high, given that the density of mercury is 13.58 g mL^{-1}.

*30. The normal pressure of the atmosphere will support a column of mercury 760 mm high in a barometer. Given that the density of water is 13.6 times less than the density of mercury, how high a column of water will atmospheric pressure support in a water barometer?

31. Steel tanks for storage of gases are capable of withstanding pressures greater than 150 atm. Express this pressure in psi.

32. Automobile tires are normally inflated to a pressure of 28 psi above atmospheric pressure. Express this pressure difference in atmospheres, and also calculate the absolute pressure inside a tire.

Boyle's Law

33. A gas occupies 5.00 L at 25 °C and 1.00 atm. What is the volume if the gas is compressed to 5.00 atm with no change in temperature?

34. A gas occupies 256 mL at 75 °C and 6.21 atm. What is the volume if the pressure is reduced to 1.00 atm with no change in temperature?

35. A gas fills a 7.5 L container at 100 °C and 25 atm. What is the new volume if the gas undergoes a constant temperature expansion until its pressure is 1.00 atm?

*36. A weather balloon, when inflated to 1 atm pressure with helium, has a volume of 9.59 m^3. How large a tank is required to store, at a pressure of 120 atm, the helium needed to fill ten balloons? Assume the storage tank and balloons are at the same temperature.

37. By what factor does the volume of a gas increase if the pressure is halved?

38. By what factor does the volume of a gas decrease if the pressure is tripled?

39. A gas is initially at 5.3 L and 75 kPa. What pressure must be exerted in order to compress the gas to 4.0 L? Assume that the temperature is unchanged during the compression.

40. A gas is initially at 703 mL and 755 torr. What pressure must be exerted in order to compress the gas to 48.9 mL? Assume that the temperature is unchanged during the compression.

41. If 0.500 L of a gas at 2.77 atm pressure is expanded into a vessel whose volume is 5.00 L, what will the pressure be? Assume no change in temperature.

*42. If a 48.2 L storage tank is to hold enough gas to fill a 5.51 m^3 chamber to a pressure of 0.963 bar, what must the pressure be?

43. If all the gas in a container is transferred to a new container having ⅓ the volume of the original, by what factor does the pressure change?

44. If a gas at 2.69×10^3 torr is expanded into a vessel whose volume is 3.62 times that of the original container, what will the pressure be? Assume no change in temperature.

Charles's Law

45. If 500 mL of gas is heated from 300 to 400 K in a constant pressure process, what is the new volume?
46. What is the new volume if 5.00 L of gas is heated at constant pressure from 25 °C to 50 °C?
47. A 375 mL sample of gas is cooled from STP to 150 K in a constant pressure process. What is the new volume?
*48. On a hot summer day when the outside temperature is 93 °F, a customer enters an air-conditioned store where the temperature is 72 °F. If 8.33 L of air enters the store with the customer, what volume will that air occupy after it reaches the temperature of the store?
49. A sample of gas occupying 2.58 L at 300 K is allowed to expand into a 4.00 L vessel. If the pressure is unchanged during the process, what is the final temperature?
50. A 936 mL sample of gas (measured at STP) is compressed into a 500 mL vessel. To what temperature must the gas be cooled so that its final pressure is 1 atm?
51. A sample of oxygen occupying 876 mL at 20 atm and 50 °C undergoes a constant-pressure expansion to a final volume of 1.60 L. What is the final temperature?
52. If 2.00 L of gas at 450 torr and 303 K is put into a 1.5 L vessel at 450 torr, what is the final temperature?
53. A sample of gas occupies 0.893 L at 500 K and 1 atm. If the pressure remains unchanged, to what volume must the gas be reduced in order to decrease the temperature by 20%?
54. A 500 mL sample of gas at 500 torr and 500 K is cooled so that its temperature drops to 80% of the original value. If there is no change in pressure, what is the new volume?
55. At what temperature will a gas have exactly half the volume it occupies at room temperature (25 °C)? The pressure is the same at both temperatures.
*56. A precocious child played with a balloon on a hot summer day when the temperature at the park reached 93 °F. In the evening she noticed that the balloon's volume had decreased to 96.5% of its afternoon size. Assuming the atmospheric pressure to be the same at both times, what was the evening temperature?

Combined Gas Law—Process Problems

57. A 5.00 L sample of gas at 3.35 atm and 307 K is compressed and cooled to 6.73 atm and 290 K. What is the new volume of the gas?
58. A 250 mL sample of gas at 0.357 atm and 260 K is compressed and cooled to 0.873 atm and 190 K. What is the new volume of the gas?
59. A gas sample occupying 15.0 L at 2.00 atm and 298 K is compressed to 10.0 L. During the process the temperature rises to 333 K. What is the new pressure?
60. A gas sample occupying 75.0 mL at 650 torr and 373 K is compressed to 60.6 mL. During the process the temperature rises to 433 K. What is the new pressure?
61. If 0.958 L of gas at 500 torr and 35 °C is expanded into a 1.00 L vessel while the pressure drops to 400 torr, what is the new temperature?
62. If 1500 mL of gas at 0.500 atm and 15 °C is compressed into a 1.00 L vessel and the pressure rises to 0.850 atm, what is the new temperature?
63. 15 L of gas at STP is expanded to a pressure of 0.75 atm and a temperature of 100 °C. What is its volume under these conditions?
64. What is the STP volume of 350 mL of gas at 24 °C and 755 torr?
65. Suppose the absolute temperature of a gas, and its volume, are both doubled. By what factor does the pressure increase?
66. If the temperature (in K) of a gas increases by 10%, and at the same time the pressure increases by 10%, what is the net effect on the volume?
67. A gas at 1500 torr and 30 °C is compressed to 75% of its initial volume and heated to 100 °C. What is the new pressure?
68. If the volume of a gas sample, initially at 125 kPa and 298 K, is doubled while the pressure drops to 100 kPa, what is the new temperature?
69. If a gas is initially at 2.00 atm and its temperature increases from 300 K to 400 K, what must the new pressure be in order to keep the volume the same?
70. Suppose a gas at STP in a 1.00 L vessel is heated to 75 °C. What is the final pressure?

Ideal Gas Law—Single-State Problems

71. What is the pressure if 0.115 mol gas is confined to a 2.20 L volume at 150 °C?
72. What is the pressure if 325 mg CO_2 gas is confined to a 1.50 L vessel at 10 °C?
73. What volume is occupied by 2.00 mol gas at 200 K and 130 kPa?
74. What is the STP volume of 15.0 g SF_6?
75. A gas sample occupies 30.0 L at 0 °C and 2.75 atm. How many moles of gas are there?
76. How many moles of gas does it take to fill a 7.38 L container to a pressure of 3.86 bar at 35 °C?
77. What must the temperature be to accommodate 0.0750 mol of gas in a volume of 1.50 L at 700 torr pressure?
78. It is determined that 0.134 moles of gas in a 1.500 L container exerts a pressure of 0.95 atm. What is the temperature?

79. It takes 2.61 g NO to fill a 5.00 L vessel to a pressure of 913 torr. What is the temperature?

80. 37 mg O_2 is placed into a 100 mL bottle at 0 °C. What is the pressure?

81. What is the mass of 945 mL Xe at 650 torr and 230 °C?

82. What mass of argon is contained in a 150 L tank if the pressure is 1875 psi and the temperature is 298 K?

83. What is the STP density of gaseous SF_6?

84. What is the density of steam at 100 °C and 1.00 atm?

85. What must the pressure be in order for CO_2 to have a room temperature (25 °C) density of exactly 1.00 g L^{-1}?

86. At what pressure does neon have a density of 1.09 g L^{-1} when the temperature is 100 °C?

87. If 1.563 g of an unknown gas is placed in a 2.509 L vessel at 25.00 °C and the measured pressure is 677.8 torr, what is the molar mass?

88. If the mass of an 836 mL sample of gas at 25.00 °C and 0.869 atm is 521 mg, what is the molar mass of the substance?

89. What is the molar mass of a gas if a 789 mg sample has a volume of 751 mL at 280 torr and 30.0 °C?

90. If the mass of 1.00 L of gas is 5.64 g when the pressure is 1250 torr and the temperature is 298 K, what is the molar mass?

91. What is the molar mass of a gas whose STP density is 2.86 g L^{-1}?

92. What is the molar mass of a gas whose STP density is 6.13 g L^{-1}?

93. What is the density of acetylene (C_2H_2) at 5.00 atm and 25 °C?

94. What is the density of propane (C_3H_8) at 500 torr and 25 °C?

Kinetic Molecular Theory and Effusion

95. Calculate the average velocity of nitrogen molecules at STP. (*Hint:* Recall that, in SI, the unit of mass is the kg; the molar mass of N_2 is 0.028 kg mol^{-1}.)

96. Calculate the average velocity of molecules in gaseous sulfur dioxide at 100 °C.

97. Suppose conditions are such that the molecules of a certain gas have an average velocity of 500 m s^{-1}. What will the average velocity be if the pressure is doubled with no change in temperature?

98. Suppose conditions are such that the molecules of a certain gas have an average velocity of 500 m s^{-1}. What will the average velocity be if the volume is doubled with no change in temperature?

99. What is the molar mass of a gas whose average molecular velocity at STP is 514 m s^{-1}?

100. What is the molar mass of a gas whose average molecular velocity at 100 °C is 2.9×10^2 m s^{-1}?

101. The velocity an object (or a molecule) must have in order to overcome the earth's gravitational field and escape into space is called the "escape velocity"; it is 11.2 km s^{-1}. What must the temperature be for an average H_2 molecule to escape from the earth?

102. What temperature is required for an average H_2 molecule to escape from the planet Mercury, whose escape velocity is 3.5 km s^{-1}?

*103. What is the ratio of the average speeds of O_2 and N_2 molecules at 298 K? What is the ratio of their average energies at this temperature?

104. What is the molar mass of a gas whose molecules travel, on the average, four times as fast as molecules of O_2?

105. An unknown gas effuses from a container at a rate of 6.46×10^{-5} mol s^{-1}. Under the same conditions argon effuses at 1.38×10^{-4} mol s^{-1}. What is the molar mass of the unknown gas?

106. If it takes 578 s for 10 μmol gas to effuse from a certain container and only 280 s for He to effuse under the same conditions, what is the molar mass of the unknown gas?

Gas Mixtures

107. A gas is collected over water when the atmospheric pressure is 745 torr and the temperature is such that the vapor pressure of water is 25 torr. What is the partial pressure of the unknown gas?

108. What is the mole fraction of the unknown gas in the preceding problem?

109. The mole fraction of helium in a certain gas mixture is 0.974. What is the partial pressure of helium when the total pressure of the mixture is 2.28 atm?

110. The mole fractions of the three most abundant gases in dry air are N_2, 0.78; O_2, 0.21; and Ar, 0.010. Calculate the partial pressure of each component when the atmospheric pressure is 755 torr.

111. A mixture of 1.00 mol He and 2.00 mol N_2 is put into a 10.0 L container at 298 K. Calculate the total pressure, and the partial pressure and mole fraction of each gas. Calculate the mass percent of He in the mixture.

112. A mixture of 10 g H_2 and 10 g O_2 is placed in a 5-L container at −50 °C. Calculate the mole fractions, the partial pressures, and the total pressure.

*113. The STP density of a mixture of hydrogen and oxygen is 1.00 g L^{-1}. Calculate the mole fraction and partial pressure of each gas.

*114. The mass of a 780 mL sample (measured at 150 °C and 805 torr) of a mixture of neon and argon is 686 mg. Calculate the mole fraction of neon in the mixture.

Stoichiometry of Gas Reactions

115. What volume of CO_2, measured at 30 °C and 745 torr, is produced by the action of acid on 1.00 g $CaCO_3$? The reaction is

$$CaCO_3 + 2\ HCl \longrightarrow CaCl_2 + H_2O + CO_2$$

116. What volume of H_2, measured at 25 °C and 755 torr, is produced when 1.32 g Zn reacts with HCl(aq)?

117. If 15.0 L (STP) of SO_2 is formed when S reacts with O_2, how much S was consumed in the reaction?

118. How much Fe_2O_3 is formed when 2.95 L O_2, measured at 25 °C and 500 torr, reacts with excess Fe?

119. How many liters of H_2, measured at 600 K and 7.5 atm, are required to react with 100 L N_2 (also measured at 600 K and 7.5 atm) to form NH_3? What volume of NH_3 is produced?

*120. 1.5 L H_2 and 1.0 L O_2, both measured at STP, are allowed to react. What mass of H_2O is produced? What is the identity and STP volume of the unreacted gas?

*121. A certain type of coal contains 1.89% (w/w) of sulfur. When the coal burns, all the sulfur is converted to gaseous SO_2. What STP volume of SO_2 is produced when a 500-mg sample of coal is burned?

*122. If a 500-mg sample of coal yields 7.50 mL SO_2 when burned, what is the percentage of sulfur (w/w) in the coal? The volume measurement was made at 23.9 °C and 758 torr.

*123. A gaseous compound containing carbon and hydrogen is analyzed and found to contain 85.7% C. The mass of 100 mL of this gas, measured at 100 °C and 100 torr, is 0.04 g. What is the molecular formula?

*124. Certain metals react with carbon monoxide to form carbonyl compounds of formula $M(CO)_x$. A 500-mL sample of nickel carbonyl, $Ni(CO)_x$, is found to have a pressure of 357 torr when held at 30.0 °C. If the sample weighs 1.6 g, what is the empirical formula of nickel carbonyl?

Real Gases

125. Given that constants in the van der Waals equation have values of $a = 3.59\ L^2\ atm\ mol^{-2}$ and $b = 0.043\ L\ mol^{-1}$ for carbon dioxide, calculate the pressure exerted by 5.0 mol CO_2 when confined to a 1.00-L container at 300 K. Compare this to the pressure that would be exerted if the gas behaved ideally.

126. Repeat the calculation of Problem 125, but using a volume of 400 mL.

Group 3A Elements

127. One of the Group 3A elements is the third most abundant element on Earth. Which one?

128. Which of the Group 3A elements does not form ionic compounds?

129. Which of the Group 3A elements is very toxic?

130. Gallium has a unique physical property. What is it?

131. What properties of aluminum metal account for its widespread use in our society?

132. Which of the Group 3A elements are used in the electronics industry? What are the natural sources of these elements?

133. Each of the Group 3A metals is produced by electrolysis of a solution of one of its compounds. For each element, identify the solvent.

134. Like aluminum hydroxide, gallium hydroxide is amphoteric. Write balanced equations illustrating this behavior.

135. All the metals of Group 3A react with chlorine to form chlorides of empirical formula MCl_3. Which two elements form *molecular* compounds having the molecular formula M_2Cl_6?

136. What type of compounds form when the hydroxides of the Group 3A metals are mildly heated? Write an equation for this process.

*137. Suppose that the electrolysis of 1.00 kg of an impure sample of bauxite yields 416 g of aluminum metal. What is the percentage of AlO(OH) in the bauxite sample?

138. How much pure gallium(III) nitrate must be dissolved in water and electrolyzed in order to produce 1.00 kg of metallic gallium?

139. What volume (STP) of hydrogen is produced when 15.4 g Al dissolves in hot NaOH(aq)?

*140. What volume of HCl(g), measured at 750 torr and 22 °C, is produced when 1.00 mL $H_2O(\ell)$ reacts with an excess of $AlCl_3(s)$?

*141. Aluminum reacts with the oxides of other metals, for example iron, to produce aluminum oxide and the other metal. This reaction, called the *thermite* reaction, is sufficiently exothermic that it can be used for welding of steel rails and pipes.
   a. Write a balanced equation for the reaction of aluminum with iron(III) oxide.
   b. Calculate the standard enthalpy change of the reaction in (a).
   c. How much heat is released per gram of aluminum that reacts?
   d. Given that the heat of vaporization of water at 100 °C is 540 cal g^{-1}, how much water could be boiled by the heat released when one pound (454 g) of aluminum reacts with iron oxide?

PARTICLES, WAVES, AND THE STRUCTURE OF ATOMS

Wave motion is everywhere in nature. If you were at this scene in Florida Bay, where light waves from the setting sun are reflected from waves on the water's surface, your ears would have registered the sound waves created by the splashing oars.

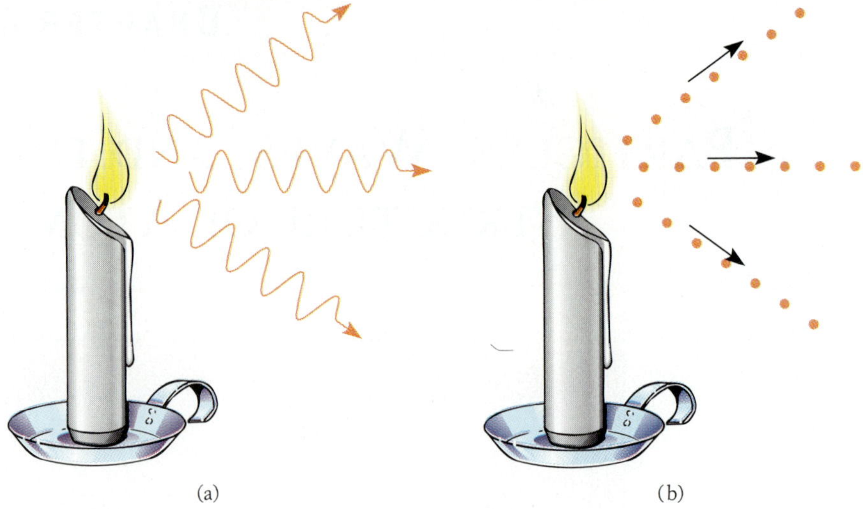

Figure 6.1 In the early days of modern science, there were two competing explanations of the nature of light: (a) light is radiant energy, streaming away from the source as waves; (b) a light ray is a stream of small particles, travelling rapidly in a straight line.

Our ideas about the fundamental nature of atoms have changed considerably since Dalton published his atomic theory almost 200 years ago. In this chapter we present the modern view of atomic structure, together with some of the observations and discoveries leading up to it.

We have already discussed heat of reaction and heat capacity (Chapter 4), which are two aspects of the general topic of the interaction of matter and energy. We now broaden the discussion to include radiant energy, also called electromagnetic radiation or light. Study of the interaction of electromagnetic radiation and matter is the experimental base on which our knowledge of the structure of atoms rests.

6.1 ELECTROMAGNETIC RADIATION

Many properties of light are familiar and easily observed: the casting of shadows, reflection at smooth surfaces, and **refraction,** the bending of a light beam at the interface between two different transparent media. By the late 17th century it had become known (from astronomical observations of planetary eclipses) that light travels at a high but not infinite speed. Two fundamentally different theories had been developed.

1. The *wave theory* held that light is a kind of fast-moving wave whose motion is analogous to that of waves visible on the surface of the sea.
2. The *corpuscular* or *particle theory* held that light consists of invisibly small, fast-moving "corpuscles" or particles, and that a beam of light is similar to a stream of bullets from a machine gun (Figure 6.1).

Both theories were consistent with reflection, refraction, and rapid straight-line motion (Figure 6.2).

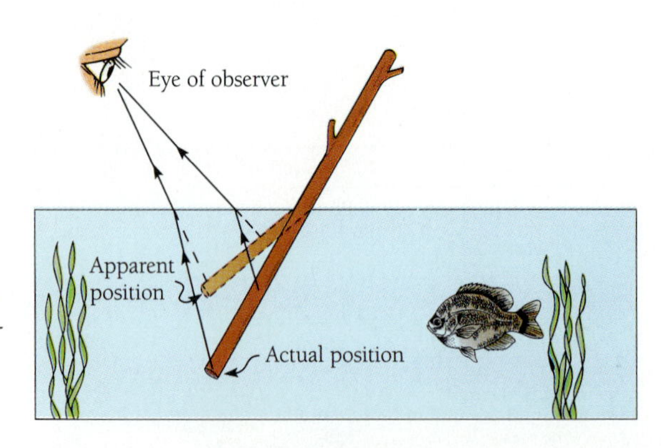

Figure 6.2 A partly submerged branch appears bent at the interface between air and water. The effect occurs because the brain expects light to travel in straight lines, and interprets the refracted (bent) rays as if they were straight.

Properties of Light

Some aspects of light, like reflections and shadows, are obvious and inescapable. Other properties, like the colors hidden in white light and the way in which two light beams interact with one another, are best studied in the laboratory.

Refraction and the Colors of Light

Isaac Newton was among the first modern scientists to make systematic observations of the refraction of light. His careful experiments with white light showed that it is a composite, which on passage through a prism is not only refracted, but is also *analyzed* (separated) into its component colors (Figure 6.3). (Previously it had been believed

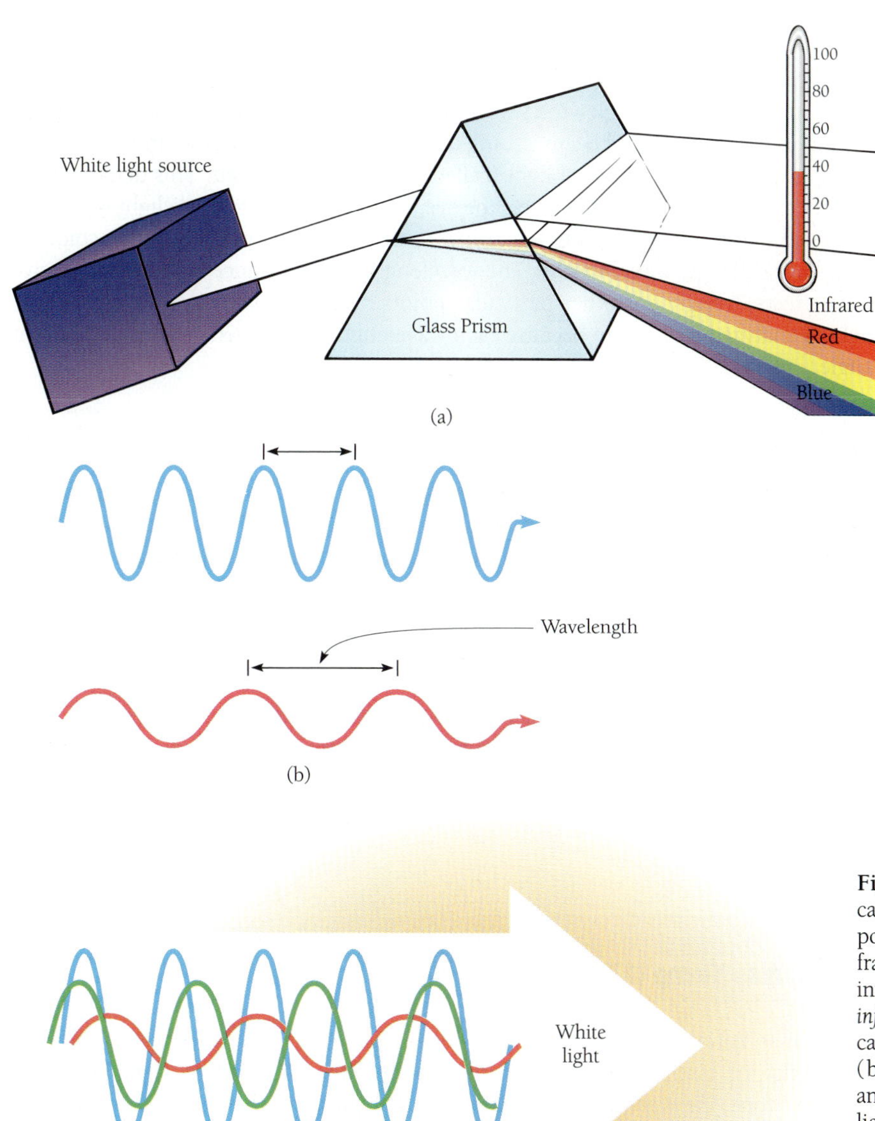

Figure 6.3 (a) Passage through a prism causes white light to separate into its component colors, because each color is refracted by a different amount. There is an invisible color adjacent to the red, called the *infrared,* which carries enough energy to cause a temperature rise in a thermometer. (b) According to the wave theory, the peaks and troughs are more closely spaced in blue light than in red; that is, blue light has a shorter wavelength. (c) Light that is a mixture of all colors is perceived as colorless or white by the human eye.

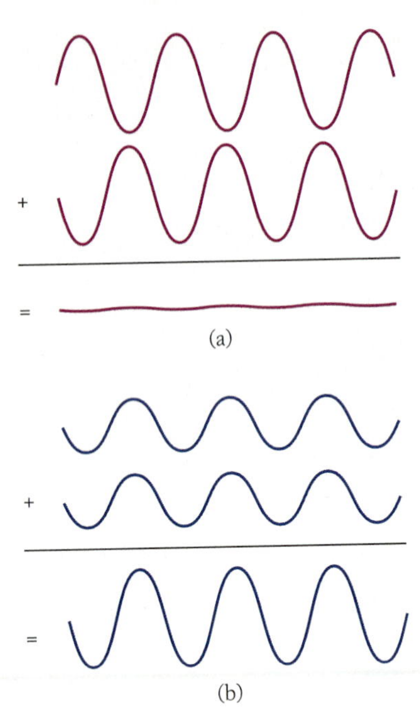

Figure 6.4 Wave interference. (a) Destructive interference occurs when the peaks of one wave coincide with the troughs of another, identical wave. The result is the cancellation and disappearance of both waves. (b) Constructive interference occurs when the peaks of two identical waves coincide. The result is a single wave of doubled intensity.

Sir Isaac Newton (1642–1727)

Sir Isaac Newton is widely regarded as one of the most outstanding scientists of the modern era. He made influential and sometimes revolutionary contributions to optics and theories of planetary motion and gravitation, and he was one of the inventors of calculus. He was greatly respected by his contemporaries, but was often involved in bitter controversy with other scientists. Because of his great reputation, his belief in the particle theory of light delayed acceptance of the wave theory for many years.

that the colors were not inherent in light, but were somehow created by the prism.) More than a century after Newton's work, it was discovered that a thermometer placed next to, but not in, the red beam issuing from a prism experiences a temperature rise. There is an invisible "color" of light (called *infrared*) beyond red. This demonstrates that there is more to light than meets the eye, and suggests that visible light is just one aspect of a larger phenomenon. According to the wave theory, the different colors of light have different **wavelengths.** The wavelength is the distance between successive wave crests, troughs, or other particular points of a repetitive wave pattern. White light is analyzed by a prism because each wavelength is refracted through a different angle.

Interference and Diffraction

By 1817 Thomas Young's experiments had shown that, under the right circumstances, two parallel rays of light could cancel each other out and result in no light at all. This phenomenon is called **destructive interference. Constructive interference** occurs when two waves add to, rather than subtract from, one another (Figure 6.4). The shifting colors seen in a thin film of oil on a puddle or on the plumage of some birds are due to interference, which, depending on the angle of view, cancels some colors (wavelengths) and reinforces others.

Diffraction, the spreading of waves into a shadow area, also occurs with light. The extent to which visible light spreads into shadows is small, and cannot be seen without special apparatus. For the most part shadows appear quite sharp. Diffraction also occurs when light is scattered by an array of close, regularly spaced grooves such

Thomas Young (1773–1829)

Thomas Young was a physician with broad interests. A knowledgeable Egyptologist, he made an important contribution to the deciphering of the Rosetta Stone. He was the first to point out that the wave theory could explain the phenomenon of polarized light. He was not believed at first, because contemporaries found it unthinkable that Newton could have been wrong.

Sir James Clerk Maxwell (1831–1879)

Sir James Clerk Maxwell, after holding a number of academic posts, became in 1871 the first Cavendish Professor at Cambridge. He made fundamental contributions to thermodynamics and to the kinetic theory of gases (Section 5.4b) as well as to the theory of radiation. In an 1861 lecture to the Royal Society, he showed one of the first color photographs ever made.

Figure 6.5 White light is spread out into its component colors when it undergoes *diffraction* by the closely spaced grooves of a compact disk. Diffraction involves both constructive and destructive interference of light waves.

as those on a compact disk. The angle at which the diffracted light leaves the grooves depends on the wavelength, and, as a result, faint colors can be seen (Figure 6.5).

The phenomena of interference and diffraction, common to all wave motion, cannot be explained by the particle theory. Because of this, the wave theory of light became firmly established in the latter half of the 19th century.

Electromagnetic Waves

In a series of papers in the early 1860s, James Clerk Maxwell laid out a comprehensive and mathematically rigorous theory that explained essentially all that was known about light. **Electromagnetic radiation,** according to Maxwell, is a fluctuating electric and magnetic field, travelling at a very high speed. Since the speed of electromagnetic waves, calculated from Maxwell's equations, was very close to the measured speed of light, Maxwell concluded that light is a form of electromagnetic radiation. It is the sensitivity of the human eye that seems to set light apart from other electromagnetic waves.

In ocean waves, the height of the surface regularly increases and decreases, so that a floating object such as a cork describes an oscillating motion as the wave crests pass. In electromagnetic waves, the strength of an electric field increases and decreases, so that a charged object such as an electron describes an oscillating motion as the wave crests pass. Perpendicular to the electric field, and oscillating with the same frequency, is a magnetic field (Figure 6.6). The waves travel in the direction perpendicular to

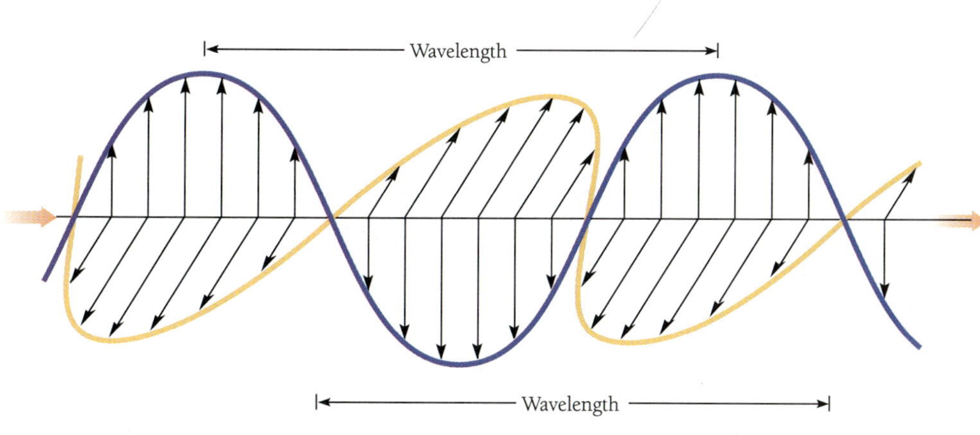

Figure 6.6 Light and other forms of electromagnetic radiation consist of perpendicular electric and magnetic fields, oscillating synchronously and travelling at over 10^8 m s^{-1} in straight-line motion.

Wavelength is almost always represented by the symbol λ (lowercase Greek "lambda"), and frequency is represented by ν (lowercase "nu").

both fields. The distance between adjacent peaks is the wavelength, and the number of peaks that pass a stationary point each second as the wave travels past is the **frequency.** The SI unit of frequency is the **hertz (Hz),** related to the base units by the equivalence statement 1 Hz $\Leftrightarrow$ 1 s^{-1}. Wavelength (λ) and frequency (ν) are related to the speed (c) of the wave by Equation (6-1).

$$c = \lambda\nu \tag{6-1}$$

The speed of electromagnetic radiation through a vacuum is the same for all wavelengths: 2.997925×10^8 m s^{-1}. This constant of nature is usually called the *speed of light.* Light travels somewhat more slowly through transparent substances such as air or glass. Moreover its speed through glass varies with wavelength, and this variation causes different colors to be refracted through different angles. The set of all wavelengths, from the very short to the very long, is called the **electromagnetic spectrum** (Figure 6.8).

Human eyes respond to electromagnetic radiation with a wavelength between about 400 nm and 700 nm, so this region of the spectrum is called **visible light.** Other portions of the electromagnetic spectrum have different names and are used for a variety of purposes (Figure 6.7). Although not an SI unit, the **Angstrom (Å) unit**

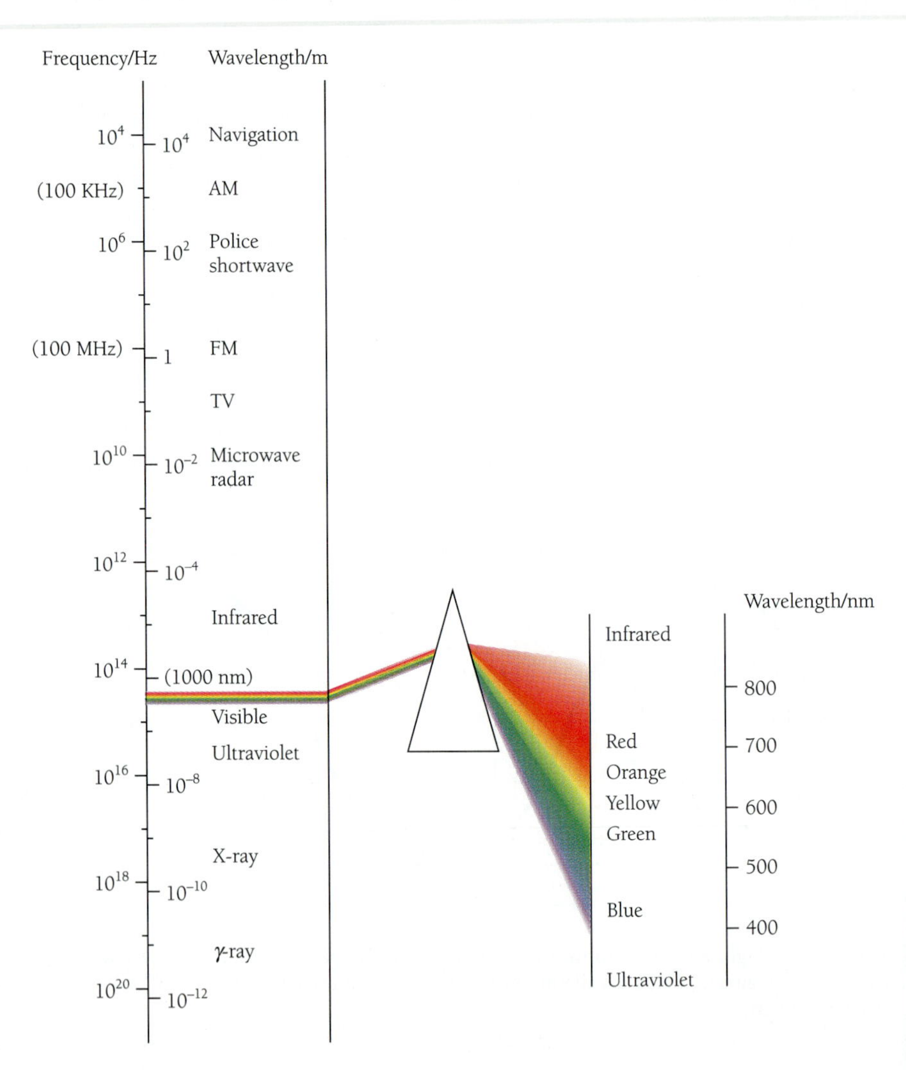

Figure 6.7 The electromagnetic spectrum encompasses a large range of useful radiations, differing from one another only in wavelength and frequency.

(10 Å ⇔ 1 nm) is still in limited use, particularly for the wavelengths of x-rays. Radiation of wavelength longer than a few centimeters, often collectively called "radio waves," is usually identified by frequency rather than wavelength. Whatever the spectral region, conversion between frequency and wavelength is often necessary.

EXAMPLE 6.1 Frequency and Wavelength

The yellow light emitted by heated sodium, in a flame or in a streetlight, has a wavelength of 589 nm. What is the frequency of this radiation?

Analysis

Target: A frequency, in Hz (s^{-1}).

Knowns: The wavelength; the speed of light, a constant, is an implicit known.

Relationship: Equation (6-1) supplies a direct algebraic link between target and knowns.

Plan Rearrange Equation (6-1), insert the known quantities (including appropriate unit conversion factors), and evaluate.

Work

$$c = \lambda\nu$$

$$\nu = c/\lambda$$

$$= \frac{3.00 \times 10^{8}\ \text{m s}^{-1}}{589\ \text{nm}} \cdot \frac{1\ \text{nm}}{10^{-9}\ \text{m}}$$

$$= 5.09 \times 10^{14}\ \text{s}^{-1} = 5.09 \times 10^{14}\ \text{Hz}$$

The *dimensions* of frequency are s^{-1} ("per second" or "reciprocal seconds"), but the *name* of the unit is hertz. Frequencies should be written with the unit symbol "Hz." A similar situation exists with solutions: the *name* of the concentration unit is molarity, M, but its *dimensions*, mol L^{-1}, are used in algebraic manipulations.

Check Although 10^{14} oscillations per second may seem like a high number, the answer is correct. Visible light is characterized by very short wavelengths and very rapid oscillations.

Exercise

Calculate the wavelength of the radiation broadcast by an AM station operating at 880 kHz and by an FM station at 92.3 MHz.

Answer: AM, 341 m; FM, 3.25 m.

Interaction of Light with Matter

Of all the ways that light and matter interact, two stand out as being particularly important to our understanding of atoms and molecules. These two are the absorption of light and the emission of light.

Absorption

When light passes through a substance, it is common for part of the light to be absorbed. Furthermore, the amount absorbed usually depends on wavelength. When white light passes through the blue glass of a stained-glass window, for instance, most

Figure 6.8 Heated objects emit blackbody radiation over a broad band of wavelengths. Both the intensity and the color of the light depend on the temperature of the object. As an object cools, the light it emits becomes dimmer and redder.

of the longer wavelength (yellow, green, red) components are absorbed. Only the blue passes through. A piece of red glass, on the other hand, selectively absorbs the blue, green, and yellow components, and allows only red light to pass through.

Emission

Heated objects emit light, and the intensity of the light at different wavelengths depends on the temperature of the object. Therefore the apparent color of the object also depends on temperature, and ranges from the dull red glow of cooling lava to the blue-white brilliance of a hot star (Figure 6.8). Such radiation, which also includes components of ultraviolet, infrared, and other invisible spectral regions, is called **blackbody radiation** (Figure 6.9). The name comes from the fact that black (nonreflective) objects are the most efficient emitters of light.

Spectroscopy

The detailed study of the wavelength dependence of the absorption or emission of electromagnetic radiation by matter is called **spectroscopy.** A large part of our knowledge of atoms and molecules comes from spectroscopic measurements.

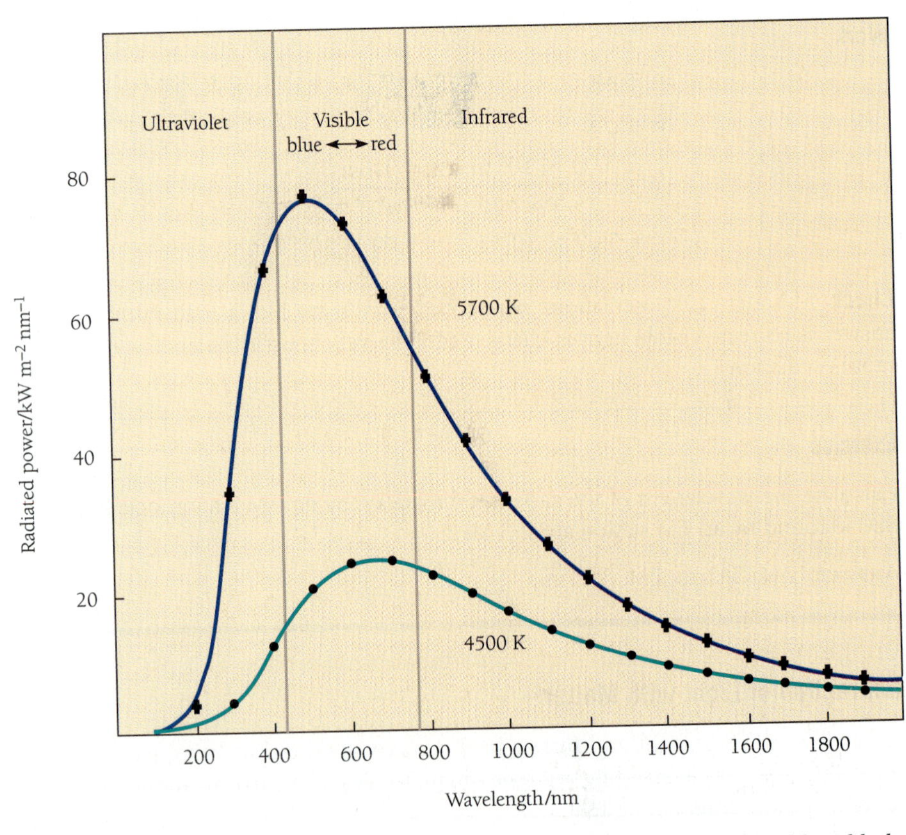

Figure 6.9 The amount of light (the radiated power or the "intensity") emitted by a black body depends on temperature, and on wavelength as well. Unless the temperature is quite high, most of the light is emitted in the infrared region of the spectrum. The mix of wavelengths emitted by an object at 5700 K, the temperature of the sun, is usually called "white" light. Cooler stars, whose temperature is, say, 4500 K, emit relatively less blue light and thus appear reddish. Very hot stars appear blue.

(a)

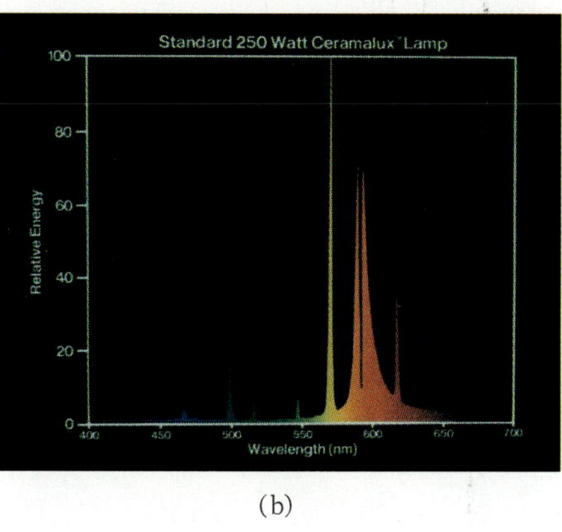

(b)

Figure 6.10 (a) Sodium-vapor street lights emit yellow light, which affects the apparent colors of objects. (b) Unlike the blackbody spectra in Figure 6.9, the sodium-lamp emission spectrum consists of several wavelength bands, with most of the light appearing in a broad band centered at about 600 nm. (Courtesy Philips Lighting Company.)

Continuous and Line Spectra

A solid body, when heated, emits radiation in a **continuous spectrum** (plural: spectra). This means that light of all wavelengths is emitted, and that intensity varies smoothly with wavelength as in Figure 6.9. A low pressure gas, on the other hand, when heated or energized by an electrical discharge, emits a spectrum that is **discrete** rather than continuous. Light is emitted at a few distinct wavelengths, called spectral lines, and no light is emitted in the wavelength intervals between the lines (Figure 6.10). The emission spectrum of a gas is called a **discrete spectrum** or **line spectrum.** These same differences appear in absorption as well. Solid materials tend to absorb light over a broad band of wavelengths, while gases absorb only at discrete spectral lines. Both absorption and emission spectra are observed by means of an instrument called a **spectrometer** (Figure 6.11). Apart from the source of the light to be investigated, the basic elements of an emission spectrometer are (a) a set of slits and/or lenses that produce a narrow beam of light travelling in only one direction; (b) a prism or other device that *disperses* the light, that is, sends its different wavelengths in different directions; and (c) a detector to measure the intensity of light emitted at different wavelengths. If the detector is a photographic film (as opposed to an electronic light sensor), the instrument is sometimes called a spectro*graph* rather than a spectro*meter*. In an absorption spectrometer, the source is a light bulb or other device that emits a

The work "discrete" means distinct, separate, or not connected.

In a spectro*scope*, once widely used for the study of emission spectra, the human eye is the detector.

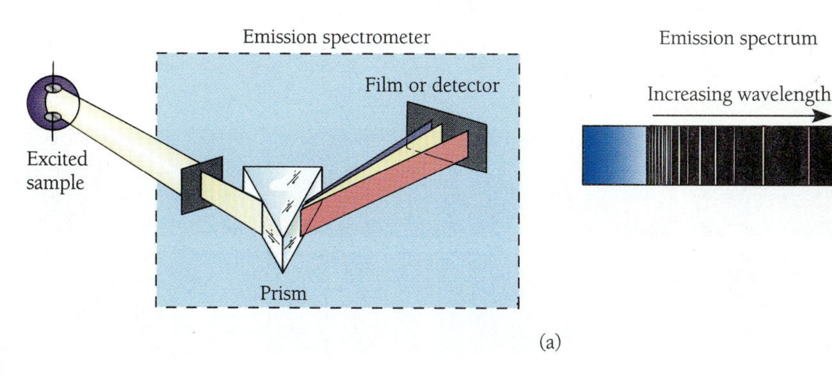

(a)

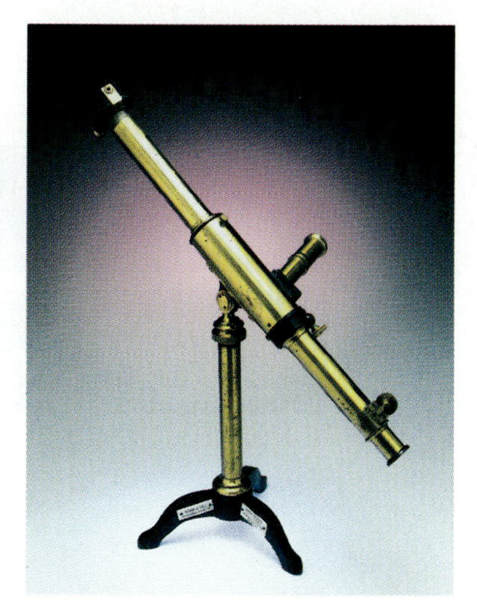

(b)

Figure 6.11 (a) The essential features of an emission spectrometer are the light source, the slits that allow passage of only a narrow beam, the prism, and the detector. (b) An early spectroscope.

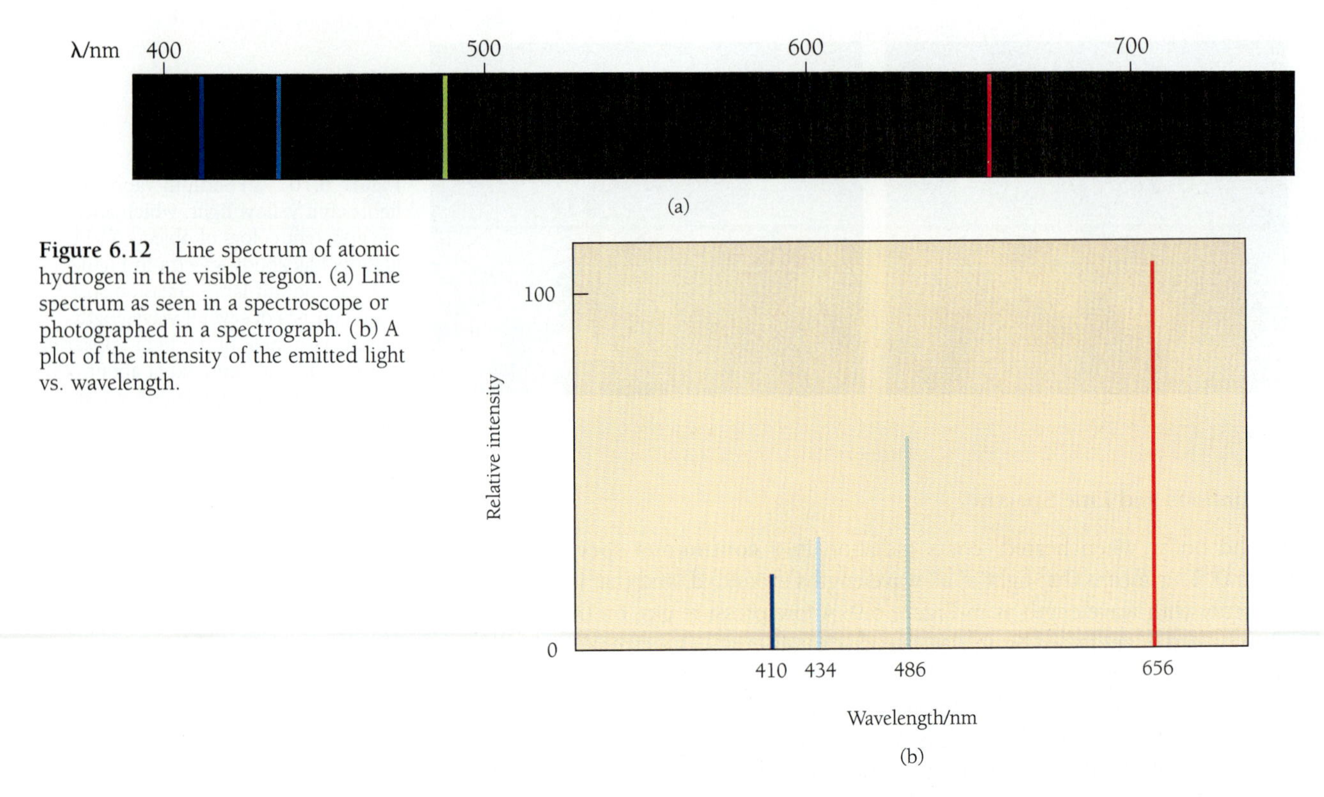

Figure 6.12 Line spectrum of atomic hydrogen in the visible region. (a) Line spectrum as seen in a spectroscope or photographed in a spectrograph. (b) A plot of the intensity of the emitted light vs. wavelength.

continuous spectrum, and the sample to be studied is placed in the light path (usually between the prism and the detector). Certain wavelengths are absorbed by the sample, so that light reaching the detector has some wavelengths missing or attenuated.

Emission Spectrum of Hydrogen

Emission spectra are often presented in the form of a graph of relative intensity of light vs. frequency or wavelength, as in Figures 6.9 and 6.10. Figure 6.12 shows part of the spectrum of light emitted from hydrogen energized by an electrical discharge. The high-energy conditions within a discharge cause much of the gas to be present as atoms rather than as diatomic molecules, and it is the emission spectrum of *atomic hydrogen* rather than H_2 that is shown in Figure 6.12. These lines form a related group called the "Balmer series," after their discoverer. Other lines, not in the Balmer series,

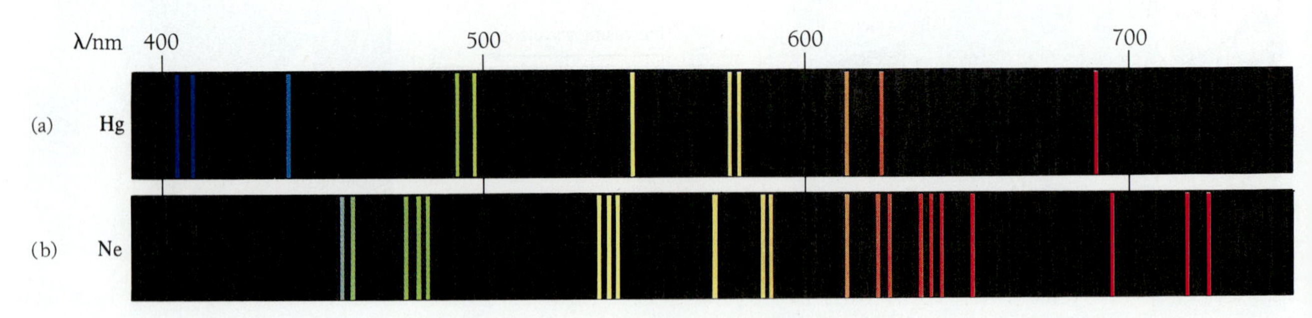

Figure 6.13 Each element has a characteristic line spectrum, different from the spectra of other elements. (a) The mercury spectrum, a major component of the light from fluorescent lighting. (b) The neon spectrum. The preponderance of lines in the red region of the spectrum gives neon signs their characteristic red color.

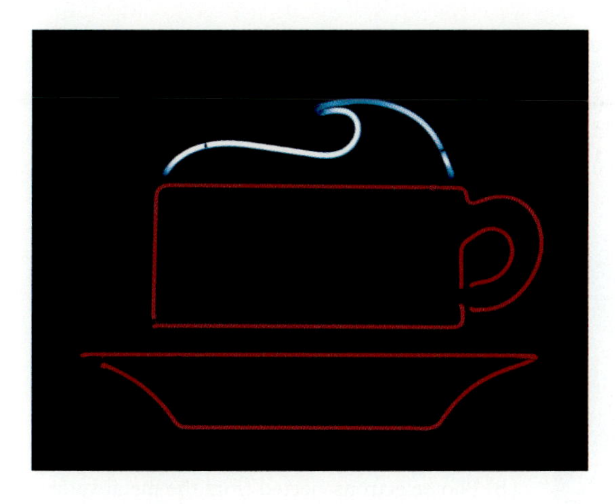

Figure 6.14 A neon sign owes its red color to the characteristic emission spectrum of neon shown in Figure 6.13. Other elements are used to produce different colors.

occur in other, invisible, regions of the spectrum. A well-known example is the "21 cm line," used in radio astronomy to observe hydrogen atoms in the far reaches of the universe.

The number, wavelength, and relative intensity of the lines in a spectrum depend on the identity of the gas in the discharge tube. Each element emits a characteristic line spectrum by which it can be identified (Figures 6.12 and 6.13). There is considerable regularity in both the intensity and the wavelength of atomic emission lines. Atomic spectra are closely related to the details of atomic structure, which will be developed in later sections of this chapter.

6.2 THE MODERN VIEW OF LIGHT

Near the close of the 19th century physicists were in possession of a large body of experimental facts and a set of mathematically rigorous theories that were very successful in explaining these facts. This theoretical structure, now known as **classical physics,** included (among others) Newton's laws of mechanical motion and Maxwell's wave theory of radiation. There were some minor difficulties. For example, classical physics could not explain *why* gases emitted line spectra rather than continuous spectra, but at least this behavior was not inconsistent with theory.

The Problem

It was not long, however, before several more serious cracks appeared in the structure of classical physics. Specifically it became clear that classical physics could not explain the details of the emission of light by solids. The newly discovered photoelectric effect presented another puzzle.

The Blackbody Spectrum

Calculations, based on classical physics, of the spectrum of the light emitted by a black body yielded unfortunate results. Rather than the experimentally observed smooth decrease in emitted light intensity as the wavelength decreases toward zero, as seen in Figure 6.9, classical physics predicts that the intensity should *rise* throughout the ultraviolet region and beyond. This is not only irreconcilable with experimental fact, but leads to the absurd prediction of infinite intensity at zero wavelength.

(a)

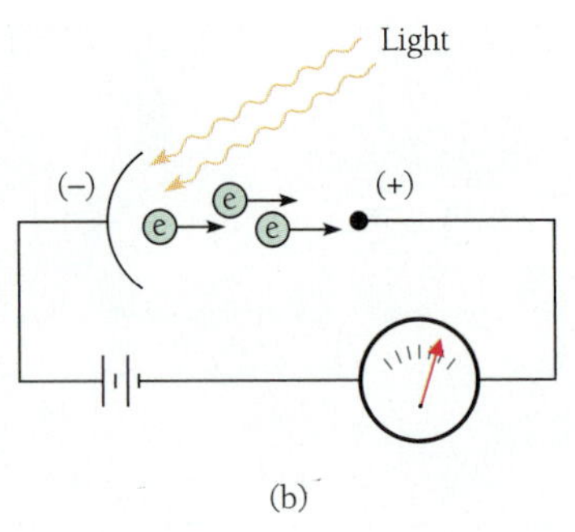

(b)

Figure 6.15 (a) A typical phototube, a device for sensing and measuring the intensity of light. (b) Photoelectrons ejected from the metal surface are attracted to and collected by the positive electrode. The magnitude of the resulting electric current is proportional to the light intensity.

The Photoelectric Effect

The phototube serves as the light detector in some spectrometers.

When light strikes a metal surface, it can happen that an electron is ejected from the metal. This is called the **photoelectric effect.** In a *phototube* (Figure 6.15) such **photoelectrons** are collected and become the current in the associated electric circuit. The following facts can be observed (Figure 6.16).

Because of the inverse relationship ($\lambda = c/\nu$) between frequency and wavelength, a *minimum* frequency corresponds to a *maximum* wavelength.

1. For each metal, there is a minimum frequency below which light does not cause the emission of photoelectrons. That is, there is a frequency *threshold.*
2. The kinetic energy of the ejected electrons increases in proportion to the increase in the frequency of the light, once the threshold is reached.
3. The kinetic energy of the ejected electrons does not increase when the intensity of light increases; the *number* of electrons does.

According to wave theory, however, the energy carried by light is proportional to its intensity but independent of frequency. On this basis we expect the following.

1. There should be no threshold. Sufficiently intense light of any frequency should cause the ejection of electrons.
2. Kinetic energy of electrons should be independent of frequency.
3. Kinetic energy of electrons should increase as light intensity increases.

These predictions are opposite to the facts, and it is clear that classical physics was seriously lacking.

The Solution: The Dual Nature of Light

The questions raised by the blackbody spectrum and the photoelectric effect were answered in a totally unexpected way. The resolution of these difficulties, in the early years of the 20th century, marks the beginning of the era of modern physics.

Quantization

The first part of the new physics was supplied by Max Planck in 1900. He showed that the correct blackbody spectrum is calculable if it is assumed that a heated object emits radiant energy in discrete amounts rather than continuously. He called this discrete

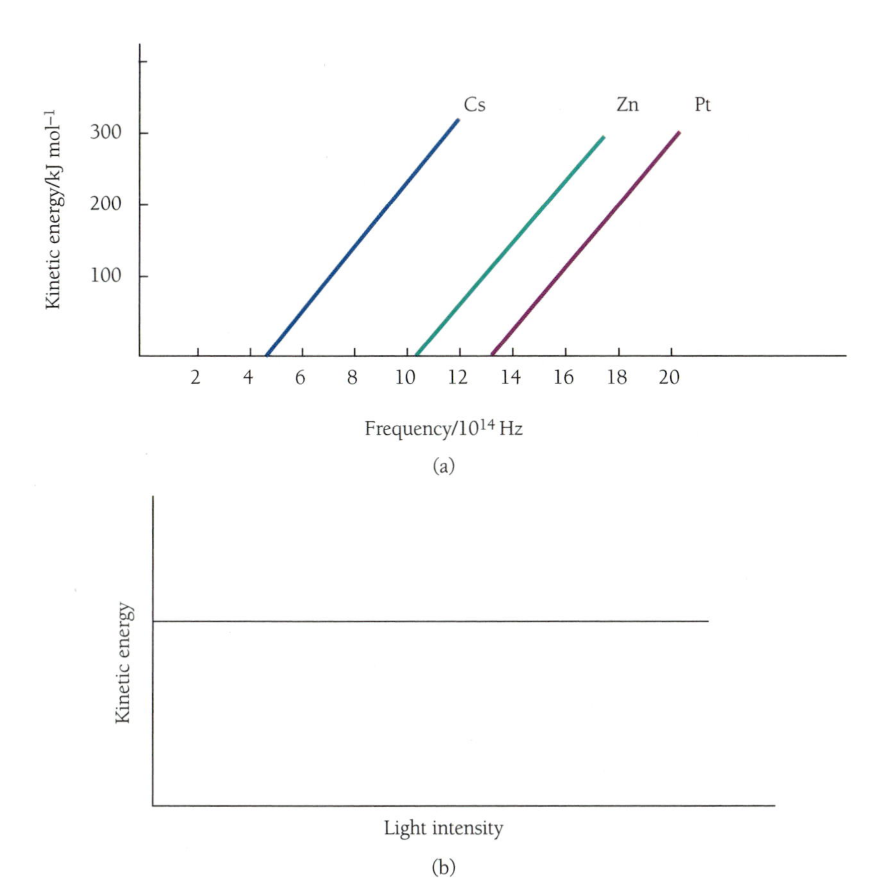

Figure 6.16 (a) The kinetic energy of photoelectrons increases linearly as the frequency of the light causing the effect increases, but below the threshold frequency no electrons are observed at all. The threshold frequency is different for each metal, but the slope is the same for all. This suggests that the slope is a property of the radiation rather than the metal. (b) The kinetic energy of photoelectrons is independent of the intensity of the light.

amount a **quantum** (Latin: how much; plural, **quanta**), defined as the smallest possible amount of a quantity. He stated that the energy of blackbody emission is **quantized,** which means discrete rather than continuous. He further showed that the energy of a quantum is proportional to the frequency of the emitted light.

$$E = h\nu \qquad (6\text{-}2)$$

The proportionality constant, $h = 6.626 \times 10^{-34}$ J s, is now known as Planck's constant.

The dimensions of Planck's constant are energy · time, so that when h multiplies a frequency (time^{-1}), the product is an energy in joules.

Several years later Albert Einstein enlarged on these ideas, and showed that all features of the photoelectric effect can be explained if it is assumed that quantization is inherent in the nature of light, rather than just a feature of blackbody emission. He proposed that light consists of individual *particles,* separate from one another and each carrying a quantum of energy $E = h\nu$. Today these particles of light are called **photons.**

Max Planck (1858–1947)

Max Planck, Professor of Physics in Berlin, is remembered primarily as a theoretical physicist. He was awarded the 1918 Nobel Prize for his explanation of the spectrum of blackbody radiation.

Albert Einstein (1879–1955)

Albert Einstein is one of the most important scientists of the 20th century. Working essentially alone, and in his spare time from his job as a patent office clerk, he published three important papers in a single year (1905). These were in widely separated fields of physics, and each would have been enough to establish his reputation as a leading physicist of the day. One explained that Brownian motion (the random jerky motion of tiny dust particles suspended in the air, visible in a sunbeam) is the result of the impact of fast-moving molecules; another dealt with the particulate nature of light and the photoelectric effect; the third was the theory of special relativity, which revolutionized current ideas about time and space. Almost immediately he was offered a professorship at Zurich, and a few years later moved to Berlin. In 1921, he was awarded the Nobel Prize (for the photoelectric effect rather than for relativity, which was then the subject of considerable controversy). He was an accomplished violinist as well as a scientist.

Forced by his religion as well as by his strong pacifist ideals to leave Germany in the 1930s, he moved to the United States and spent his remaining years at the Institute for Advanced Study in Princeton, N.J. Much of his effort, throughout his life, was spent in an attempt to unite electromagnetism and gravitation in a single theory. This goal, still elusive decades after his death, continues to motivate theoretical physicists.

The photoelectric effect is explained as follows. When a metal surface is struck by and absorbs a photon, the energy carried by the photon is transferred to one of the electrons in the metal. If the electron acquires enough energy to overcome the attractive force of the positively charged nuclei in its vicinity, it may escape from the surface as a photoelectron. The minimum energy necessary for the escape varies from one metal to another. Since the energy of a photon is proportional to the frequency of the light, there is a corresponding minimum or *threshold* frequency below which the photoelectric effect cannot occur. If the frequency is above the minimum, any additional photon energy is carried away in the form of kinetic energy of the photoelectron.

Since the wavelength of light is inversely proportional to frequency, the photoelectric threshold is a *minimum* required frequency but a *maximum* effective wavelength.

EXAMPLE 6.2 Threshold Energy and the Photoelectric Effect

The energy required for photoelectrons to escape from the surface of barium metal is 240 kJ mol^{-1}. Calculate the minimum frequency, and maximum wavelength, of light capable of producing the photoelectric effect in barium.

Analysis

Target: A frequency (Hz) and a wavelength (m).

Knowns: Energy requirements per mole of electrons. The speed of light, Planck's constant, and Avogadro's number are implicit knowns.

Relationship: Einstein's theory relates the photoelectric energy requirement to the energy of a photon. The equations $E = h\nu$ and $c = \lambda\nu$ allow conversion of photon energy to frequency and wavelength.

Plan From the minimum energy per mole of electrons, determine the energy required for one electron. Find the frequency, and then the wavelength, of a photon carrying this energy.

Work

1.

$$E_{\min} = (240 \times 10^3 \text{ J mol}^{-1}) \cdot \frac{1 \text{ mol}}{6.022 \times 10^{23} \text{ electrons}}$$
$$= 3.985 \times 10^{-19} \text{ J electron}^{-1}$$
$$= 3.99 \times 10^{-19} \text{ J electron}^{-1}$$

2. Each photon must have at least 3.99×10^{-19} J, and since $E = h\nu$, the minimum frequency is

$$\nu_{\min} = \frac{E}{h}$$
$$= \frac{3.985 \times 10^{-19} \text{ J}}{6.626 \times 10^{-34} \text{ J s}}$$
$$= 6.015 \times 10^{14} \text{ s}^{-1} = 6.01 \times 10^{14} \text{ Hz}$$

Use at least one extra digit in a calculation, and do not round to the appropriate number of significant digits until the final answer is reached.

This frequency corresponds to a (maximum) wavelength of

$$\lambda_{\max} = \frac{c}{\nu_{\min}}$$
$$= \frac{2.998 \times 10^8 \text{ m s}^{-1}}{6.015 \times 10^{14} \text{ s}^{-1}}$$
$$= 4.984 \times 10^{-7} \text{ m} = 498 \text{ nm}$$

Check 498 nm corresponds to blue-green light, and most photoelectric thresholds are in the visible region of the spectrum.

Exercise

To what threshold energy does light of $\lambda = 600$ nm correspond?

Answer: 199 kJ mol^{-1}.

Wave-Particle Duality

The experimental facts are that in certain circumstances (diffraction, interference) the behavior of light is wave-like, whereas in others (photoelectric effect, blackbody radiation), it is particle-like. Light is more complex than simple waves or simple particles, yet has some of the properties of each. This concept is known as **wave-particle duality.** An important aspect is that, like mass and charge, energy is quantized rather than continuous. Quantization is the rule rather than the exception in nature.

6.3 THE PLANETARY MODEL OF ATOMIC STRUCTURE

Following Rutherford's discovery of the nucleus, and also the measurement of the mass of the electron, it was perhaps inevitable that people began to think of an atom as a miniature solar system, with the electrons travelling in well-defined orbits around the small, heavy nucleus. This idea, the **planetary model,** was put into coherent mathematical form by Niels Bohr. The planetary model is no longer used (although it correctly accounts for the emission spectrum of hydrogen), because it gives a false picture of atomic structure. Electrons do not, in fact, travel in orbits around atomic nuclei. However, Bohr's theory is worth study because it was an essential step in the development of the modern view, and because it contained several valid insights into the natural laws that apply to very small systems.

Bohr's Key Assumptions

Like the atomic theory of Dalton, the theory of the **Bohr atom** begins with a set of assumptions or postulates.

1. Electrons travel in circular orbits around the nucleus. The orbital motion of the electrons gives to the atom a certain energy, calculable from the electron's speed, mass, and charge.
2. The energy of an atom is quantized rather than continuous. An atom can exist only with certain definite amounts of energy, that is, with its electrons in certain definite orbits. These "allowed states" are called **energy levels.**
3. When an atom drops from a higher energy level to a lower energy level, it emits a photon whose energy is equal to the energy difference between the two levels.
4. An atom in a particular energy level does not emit radiation. That is, the atom's energy is constant as long as it remains in a single level. (Because of this stability, energy levels are also called **stationary states.**)

The first of these postulates was part of and consistent with classical physics, but the others were new. Postulates 2 and 3, taken together, explained why atoms such as hydrogen have discrete rather than continuous spectra: if only certain energies are possible for an atom, then only certain energies can be emitted.

Niels Bohr (1885–1962)

The Danish physicist Niels Bohr studied under both Thomson and Rutherford. In 1920, he became the first director of the Institute for Theoretical Physics in Copenhagen. He began publishing his ideas on the planetary model in 1913, two years after the existence of the nucleus became known. In 1922, he earned the Nobel Prize. He fled occupied Denmark in 1943 and spent two years in Los Alamos, N.M., where he contributed to the development of the atomic bomb. He was convinced that control over nuclear weapons could only be achieved through the free exchange of people and ideas among countries.

According to classical theory, an electrically charged object that undergoes a change in speed or direction of motion emits electromagnetic radiation. An electron in a planetary orbit undergoes continuous change of direction, and ought to continually lose energy by radiation. If it did lose energy, however, it would inevitably spiral in and come in contact with the nucleus. An atom with orbiting electrons should therefore exist only temporarily. Bohr was forced to introduce postulate 4 in order to account for the continued existence and stability of atoms. This postulate was not only new, it was a direct contradiction of classical theory.

5. Bohr's final postulate was that an electron's angular momentum in an allowed orbit must be a whole number multiple of $h/2\pi$. That is, $mvr = nh/2\pi$, where m, v, and r are the electron's mass, velocity, and orbital radius, while $n = 1, 2, 3, \ldots$ (any integer).

   The fifth postulate allows the calculation of the energy of the atom in each of the stationary states. For the hydrogen atom, Bohr found that the energy of a state is given by Equation (6-3)

$$E_n = -hc\mathcal{R}\frac{1}{n^2}; \qquad n = 1, 2, 3, \ldots \qquad (6\text{-}3)$$

The symbol E_n in this equation represents the energy of the nth level in the series E_1, E_2, E_3, and so on. The subscript integer n is called the **quantum number** of the energy level: n is equal to 1 in the level of lowest energy, 2 in the next level up, and so on. The quantities h and c are Planck's constant and the speed of light, and $\mathcal{R}$ is the Rydberg constant (after Johannes Rydberg, an early observer of atomic spectra): $\mathcal{R} = 1.097 \times 10^7\ \text{m}^{-1}$.

Atomic energies could be described either by reference to the lowest level, as "so many joules above the lowest," or by reference to the highest level, as "so many joules below the highest." In the latter instance, which is the one used today, atomic energies such as those calculated from Equation (6-3), are always negative numbers.

This system of measuring from the highest level is also used to describe the depth of the ocean, or of a well or mine shaft.

The Hydrogen Spectrum

Frequencies and wavelengths of the lines in the spectrum of hydrogen can be calculated from Equation (6-3), together with the Planck/Einstein relationship $E = h\nu$ between the energy and frequency of a photon.

Emission

If E_U and E_L are the energies of the upper and lower states, then the frequency of the light emitted when an atom drops from the upper to the lower state is given by Equation (6-4).

$$\Delta E = E_U - E_L = h\nu = \frac{hc}{\lambda} \qquad (6\text{-}4)$$

Applying Equation (6-3) separately to the two energies E_U and E_L, we find

$$\Delta E = E_U - E_L = -hc\mathcal{R}\left(\frac{1}{n_U^2} - \frac{1}{n_L^2}\right) \qquad (6\text{-}5)$$

where the integers n_U and n_L are the quantum numbers of the upper and lower levels. (Note that n_U is larger than n_L, so that ΔE is a positive number.) The energy carried by the emitted photon is equal to the energy difference between the levels.

$$E_{\text{photon}} = \Delta E_{\text{atom}}$$

Since $E_{\text{photon}} = h\nu = hc/\lambda$, and ΔE_{atom} is given by Equation (6-5), the wavelength of the emitted light can be calculated from Equation (6-6).

$$\frac{hc}{\lambda} = -hc\mathcal{R} \cdot \left(\frac{1}{n_U^2} - \frac{1}{n_L^2}\right) \quad \text{or} \quad \frac{1}{\lambda} = -\mathcal{R} \cdot \left(\frac{1}{n_U^2} - \frac{1}{n_L^2}\right) \qquad (6\text{-}6)$$

We have just shown that Equation (6-6) follows from Bohr's postulates. But in fact the same equation, with the same numerical value of $\mathcal{R}$, was obtained from experimental data by Rydberg in 1885. At that time it was known only that if small integers were used, the equation correctly predicted the observed wavelengths of the lines in the hydrogen spectrum. The significance of the integers n_U and n_L remained a mystery until Bohr's explanation some 30 years later. The exact agreement between the calculated wavelengths and those observed experimentally in the emission spectrum of hydrogen was a stunning success of the Bohr theory.

EXAMPLE 6.3 Emission Wavelengths and Energy Levels in Hydrogen

The emission lines in the Balmer series of the hydrogen spectrum (Figure 6.12) correspond to transitions in which $n_L = 2$ and $n_U = 3, 4, 5, \ldots$ (a) Calculate the wavelength of the first Balmer line (that is, the line corresponding to $n_U = 3$). (b) What is the difference in energy between these two states of the hydrogen atom?

Analysis/Relationship/Plan Equation (6-6) relates quantum numbers to photon wavelength, and the relationship $E = hc/\lambda$ allows conversion of wavelength to energy. The values of $\mathcal{R}$, h, and c are implicit knowns.

Work

a. Insert the given values of n_U and n_L into Equation (6-6) and solve for λ.

$$\begin{aligned}\frac{1}{\lambda} &= -\mathcal{R} \cdot \left(\frac{1}{n_U^2} - \frac{1}{n_L^2}\right)\\ &= -(1.097 \times 10^7 \text{ m}^{-1}) \cdot \left(\frac{1}{3^2} - \frac{1}{2^2}\right)\\ &= 1.524 \times 10^6 \text{ m}^{-1}\end{aligned}$$

$$\lambda = \frac{1}{1.524 \times 10^6 \text{ m}^{-1}} = 6.563 \times 10^{-7} \text{ m} = 656.3 \text{ nm}$$

b. The energy carried by a 6.563×10^{-7} m photon is

$$\begin{aligned}E &= \frac{hc}{\lambda}\\ &= (6.626 \times 10^{-34} \text{ J s}) \cdot \frac{2.998 \times 10^8 \text{ m s}^{-1}}{6.563 \times 10^{-7} \text{ m}}\\ &= 3.027 \times 10^{-19} \text{ J}\end{aligned}$$

This is the value of the energy difference between the two states of the hydrogen atom, that is,

$$\Delta E = 3.027 \times 10^{-19}\ \text{J}$$

The energy difference between atomic states is frequently given on a per mole rather than a per atom basis, so

$$\Delta E = (3.027 \times 10^{-19}\ \text{J atom}^{-1}) \cdot \frac{6.022 \times 10^{23}\ \text{atom}}{1\ \text{mol}}$$

$$= 1.823 \times 10^{5}\ \text{J mol}^{-1} = 182.3\ \text{kJ mol}^{-1}$$

Check

a. ($\frac{1}{9} - \frac{1}{4}$) is about $(0.1 - 0.25) = -0.15$; multiplied by 1×10^{7} is 0.15×10^{7}, and the reciprocal is about 7×10^{-7}, so the wavelength calculation is correct.

b. For light in the uv and visible spectral regions, the energies (per mole of photons) are several hundred kJ mol^{-1}.

Exercise

Calculate the wavelength of the second line in the Balmer series ($n_U = 4$), and the energy difference between the upper and lower states.

Answer: 486.2 nm; 246.1 kJ mol^{-1}.

Absorption

The Bohr model of atomic structure also explains the *absorption* of light by atoms. Initially in a state of energy E_L, an atom jumps to a new state of higher energy E_U when it absorbs a photon. Only a photon whose energy $h\nu$ is exactly equal to the energy difference $\Delta E = E_U - E_L$ can be absorbed, so the absorption spectrum is discrete rather than continuous. Equations (6-4), (6-5), and (6-6) hold for the energies, wavelengths, and frequencies of lines in the absorption as well as the emission spectrum (Figure 6.17).

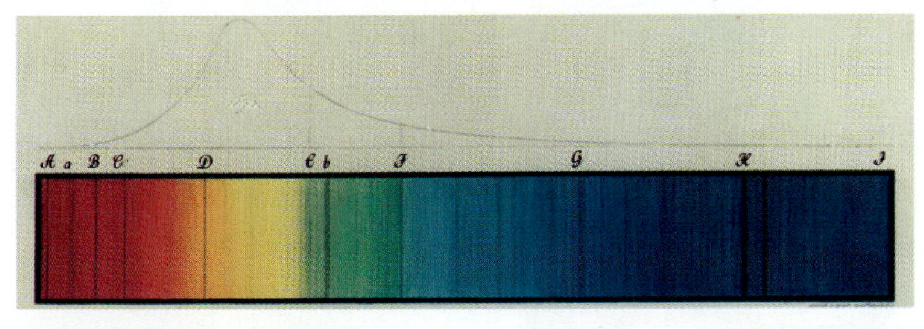

Figure 6.17 An early emission spectrum of the sun. The many dark lines, called the *Fraunhofer lines,* are due to absorption of certain wavelengths by atoms in the outer layers of the sun. First observed in the early 19th century, almost all of these lines could be correlated with the spectra of known elements, but some could not. The inability of physicists to identify these unexplained lines led to the hypothesis that a new element, helium, unknown on earth, was present in the sun. The hypothesis was later confirmed by Ramsay's discovery of helium in certain minerals.

Legacy of a Superseded Theory

Attempts were made, by Bohr and others, to extend the planetary model to atoms containing more than one electron. Although there was some progress, the extended theories were complicated and were eventually superseded by a quite different theory. Bohr's first and fifth postulates were discarded, and it is no longer believed that electrons travel in orbits around the nucleus. However, the other three postulates remain as key features of the new theory: atoms exist in discrete, *stationary states* and change their energies only by absorption or emission of a photon. Equation (6-6) is still used to calculate the wavelengths of lines in the hydrogen spectrum, and its counterpart, Equation (6-4), still gives the correct values for the energy levels of the hydrogen atom. Although superseded, and based in part on incorrect postulates (1 and 5), the Bohr theory was an important advance in that it established a new principle: the microscopic world of atoms and electrons has its own set of rules, and does not necessarily obey the same laws as macroscopic objects.

6.4 QUANTUM MECHANICS

The new theory came to be called *quantum mechanics,* because of the central importance of the concept of quantized energy levels. In addition to quantization, however, two other revolutionary ideas were involved in the new physics. These new ideas were that particles have wavelike properties, and that there is an inherent blurring or uncertainty associated with the motion of particles.

Matter Waves

Louis Victor de Broglie (1892–1977), a member of the French nobility, published his ideas in his PhD dissertation. Five years later he was awarded the Nobel Prize (1929), and subsequently became Professor of Physics at the Sorbonne (University of Paris).

In 1924 the French physicist Louis Victor de Broglie suggested that, since under certain circumstances light waves behaved as particles, it ought to be true that particles such as electrons should, on occasion, behave as **matter waves.** Wave-particle duality ought to apply impartially to both. Put in mathematical form, this notion leads to the **de Broglie relationship,** Equation (6-7), which gives the wavelength associated with a moving particle in terms of its momentum p [momentum = mass(m) · velocity(v)].

$$\lambda = \frac{h}{p} = \frac{h}{mv} \tag{6-7}$$

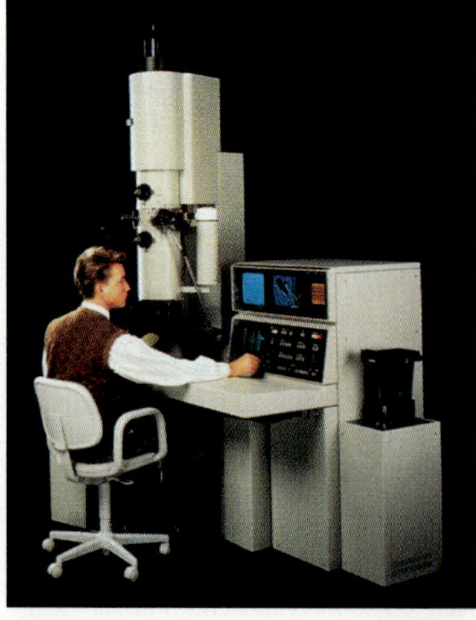

(a)

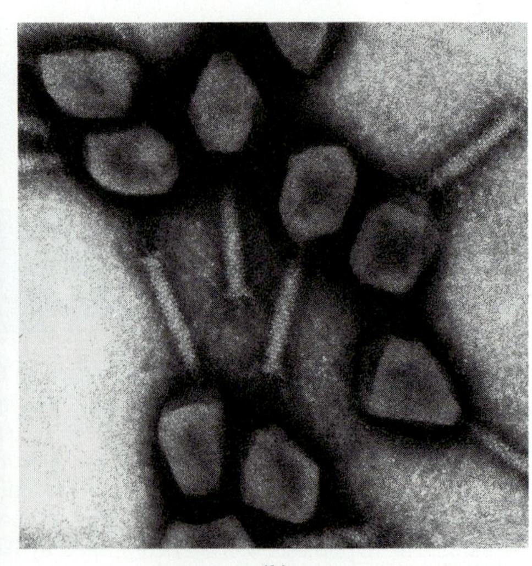

(b)

Figure 6.18 (a) The electron microscope depends for its action on the wave properties of a beam of electrons. It is capable of imaging much smaller structures than a light microscope, because the wavelength of a high velocity electron beam is much smaller than the wavelength of visible light. (b) This image of the T4 bacteriophage virus was obtained with an electron microscope. The head-to-tail length of the virus is only 2×10^{-7} m.

De Broglie's concept was soon demonstrated experimentally. In 1927, Davisson and Germer observed diffraction in an electron beam reflected from a crystal, and G. P. Thompson observed diffraction in an electron beam passed through a thin metal foil. Since diffraction is a property of waves, not particles, de Broglie's hypothesis of matter waves was confirmed. Furthermore, the wavelength of the electrons in the beam was exactly that predicted by Equation (6-7). The electron microscope (Figure 6.18) is a practical application of the wave properties of an electron beam.

EXAMPLE 6.4 The Wavelength of an Electron

According to the Bohr theory, the angular momentum of the electron in the lowest-energy stationary state of the hydrogen atom is $h/2\pi = 1.06 \times 10^{-34}$ kg m s^{-1}, which, supposing the electron were actually in an orbit, would correspond to a velocity of 2.19×10^{6} m s^{-1}. Calculate the de Broglie wavelength of an electron moving with this velocity.

Analysis

Target: A wavelength, in meters.

Knowns: The electron velocity; electron mass and Planck's constant are implicit knowns.

Relationship: Equation (6-7) may be used directly.

Work

$$\begin{aligned}\lambda &= \frac{h}{mv} \\ &= \frac{6.6 \times 10^{-34}\ \text{J s}}{(9.1 \times 10^{-31}\ \text{kg}) \cdot (2.2 \times 10^{6}\ \text{m s}^{-1})} \\ &= 3.3 \times 10^{-10}\ \text{J s}^2\ \text{kg}^{-1}\ \text{m}^{-1}\end{aligned}$$

Check The units of the answer do not appear to be correct, and yet no mistake has been made. But recall (Section 1.3) that the dimensions of energy, a compound unit, are $(m\ell^2 t^{-2})$. In SI, the equivalence statement is $1\ \text{J} \Leftrightarrow 1\ \text{kg m}^2\ \text{s}^{-2}$. Using this relationship as a conversion factor, we find

$$\begin{aligned}\lambda &= (3.3 \times 10^{-10}\ \text{J s}^2\ \text{kg}^{-1}\ \text{m}^{-1}) \cdot \frac{1\ \text{kg m}^2\ \text{s}^{-2}}{1\ \text{J}} \\ &= 3.3 \times 10^{-10}\ \text{m} = 330\ \text{pm}\end{aligned}$$

Exercise

Calculate the de Broglie wavelength of a 150 g baseball moving at 40 m s^{-1} (about 90 mph, the speed of a fast pitch).

Answer: 1.1×10^{-34} m, or 1.1×10^{-22} pm.

Keep the results of this Example and Exercise in mind, as they are relevant to the comments in the following section.

Werner Heisenberg (1901–1976)

Werner Heisenberg, best known for his work in quantum mechanics, also made contributions to nuclear physics and to theories of turbulent flow. After studying under Niels Bohr in Copenhagen, he spent the remainder of his career in Germany. He loved music, and felt that there were strong connections between music and physics. His lifelong interest in philosophy resulted in a 1958 book, *Physics and Philosophy: The Revolution in Modern Science.* He was awarded the 1932 Nobel Prize in physics (for work in quantum mechanics, not the principle that bears his name).

Uncertainty and the Impossibility of Orbits

In 1927, Werner Heisenberg proved that the position and the momentum of a moving object cannot both be known with perfect accuracy: if the momentum is accurately known, then the position cannot be, and vice versa. This concept is known as the **uncertainty principle.** According to Heisenberg, the uncertainty does not arise from the limitations of any particular measuring device, but is inherent in the nature of motion. Even with hypothetically "perfect" instruments there remains an irreducible uncertainty in our knowledge of moving objects. The mathematical form of the uncertainty principle is

$$\Delta x \cdot \Delta p > h \qquad \text{(6-8a)}$$

in which Δx (meters) is the uncertainty in position, and Δp (kg m s^{-1}) is the uncertainty in momentum. As shown in Equation (6-8a), the product of these uncertainties must be at least as great as Planck's constant. Since momentum is the product of mass and velocity, the uncertainty principle can also be written in the form

$$\Delta x \cdot m\Delta v > h \qquad \text{(6-8b)}$$

where m (kg) is the mass of the object and Δv (m s^{-1}) is the uncertainty in its velocity.

EXAMPLE 6.5 The Uncertainty Principle

Suppose it is known that the speed of an electron cannot be greater than 3×10^6 m s^{-1} nor less than 1×10^6 m s^{-1}. What is the minimum uncertainty with which its position can be known?

Analysis

Target: A positional uncertainty, in meters.

Knowns: The limits to the electron's speed; the mass of the electron and Planck's constant are implicit knowns.

Relationship: Problems of position, momentum, or velocity uncertainty are solved by direct application of Equation (6-8).

Plan

1. Find the uncertainty in velocity from the known information.
2. Find the uncertainty in position from Equation (6-8).

Work

1. If the speed is known to be in the range $1 \times 10^6 - 3 \times 10^6$ m s^{-1}, the uncertainty is

$$\Delta v = (3 \times 10^6 - 1 \times 10^6) = 2 \times 10^6 \text{ m s}^{-1}$$

2. From Equation (6-8b), the uncertainty in position must be

$$\Delta x \cdot m\Delta v > h$$

$$\Delta x > \frac{h}{m\Delta v}$$

$$> \frac{6.6 \times 10^{-34} \text{ J s}}{(9.1 \times 10^{-31} \text{ kg}) \cdot (2 \times 10^6 \text{ m s}^{-1})}$$

$$> 4 \times 10^{-10} \text{ J s}^2 \text{ kg}^{-1} \text{ m}^{-1}$$

As in Example 6.4, the units are rewritten with the equivalence statement $1 \text{ J} \Leftrightarrow 1 \text{ kg m}^2 \text{ s}^{-2}$

$$\Delta x > (4 \times 10^{-10} \text{ J s}^2 \text{ kg}^{-1} \text{ m}^{-1}) \left(\frac{1 \text{ kg m}^2 \text{ s}^{-2}}{1 \text{ J}}\right)$$

$$> 4 \times 10^{-10} \text{ m} = 400 \text{ pm}$$

That is, the position of the electron cannot be determined more precisely than ≈ 400 pm.

Exercise

Calculate the positional uncertainty of a 0.15 kg baseball travelling between 28 and 32 m s^{-1}.

Answer: 1×10^{-33} m.

The numerical results of the two preceding Examples and Exercises are instructive, for they reveal a clear distinction between microscopic and macroscopic objects. The baseball wavelength, $\approx 10^{-22}$ pm, is about $1/10^{20}$ of the diameter of an atomic nucleus. It is so small as to be unobservable. Since they are unobservable, "wave properties" of large objects have no physical reality. Macroscopic objects are particles, not waves, and they obey the laws of classical physics. The wavelength of an electron whose velocity is 2×10^6 m s^{-1}, on the other hand, is comparable to the diameter of an atom. Its wave properties are no more difficult to observe than atoms themselves.

The measured diameter of the hydrogen atom (Section 7.5) is about 75 pm, less than one-fifth the wavelength of the electron "matter wave" of Example 6.4. Since the wavelength of the electron is larger than the atom that contains it, is it appropriate to describe the electron as being in a "definite orbit" around the nucleus? Heisenberg's answer to this question is a clear "no": the uncertainty principle renders the planetary model of the atom useless. The description of an orbit requires knowledge of both position and momentum of the orbiting object. These are not knowable for electrons, because, as demonstrated in Example 6.5, the positional uncertainty of an "orbiting" electron can be larger than the atom itself. Electron motion cannot be described in terms of orbits, and another model, more appropriate to the experimental facts, must be used.

Electron Waves and the Schrödinger Equation

The new model was supplied by Erwin Schrödinger in 1926. He reasoned that, since electrons had wave properties, **electron waves** in an atom should be describable by mathematical techniques similar to those used in describing ocean waves, sound waves, and so on. He modified a generalized wave equation by including the de Broglie relationship, then applied it to the motion of the electron in a hydrogen atom. The modified equation, now known as the **Schrödinger equation,** is the base upon which the modern view of chemistry rests. It was an immediate success, because it allows the calculation of the exact frequencies of all the lines in the H-atom spectrum. Also, it offered immediate relief from the two problems raised by the planetary model.

1. Since an electron wave is not an electrical charge in a circular orbit, there is no reason why it should radiate energy. Bohr's unexplainable postulate 4 is no longer necessary.
2. Discrete (rather than continuous) states occur naturally with all types of wave motion, so there is no need to introduce quantization as an artificial, unexplained postulate.

Because quantization is one of its central attributes, and to distinguish it from classical mechanics, Schrödinger's approach to the problems of atomic and molecular structure is called **quantum mechanics.** It has been elaborated in the 60-odd years since its introduction, but it has not been changed in any fundamental way.

Stationary Waves

The natural occurrence of quantized wave motion is easily seen in waves on a rope or string. A snap of the wrist while holding the end of a rope produces a disturbance that moves quickly to the far end of the rope. Regularly repeated hand motion produces a train of travelling crests and troughs called a **travelling wave,** much like ocean waves. The wavelength of this travelling wave depends on the frequency of the motion that produces it and can have almost any value.

In common with all wave motion, the relationship between speed (S), frequency (ν), and wavelength (λ) of travelling waves on a rope is $S = \lambda\nu$.

If, however, the wave motion of a string is confined by fixing the two ends, as in a guitar or other stringed instrument, the situation is quite different. The waves no longer travel the length of the string, but they vibrate in place; they are called

Erwin Schrödinger (1887–1961)

Erwin Schrödinger published his famous equation in 1926, and seven years later shared the Nobel Prize with Dirac. He was Professor of Physics in Berlin until 1933, when he resigned and left Germany in protest against the national policy of racism. He wandered Europe for seven years, and finally settled at the Institute for Advanced Studies in Dublin. There he remained until he retired, in 1956, to his native Austria. He was an accomplished linguist, philosopher, and historian of science. In 1944, he wrote a book entitled, *What Is Life.* In 1954, he published a study of Greek philosophy of science, *Nature and the Greeks.*

stationary or **standing waves.** Furthermore, only certain wavelengths of a standing wave can persist in a fixed string. If the length of the string between the fixed points is 50 cm, for instance, then stable stationary waves of wavelength 100 cm, 50 cm, 33⅓ cm, 25 cm, and so on, are possible, but waves of intermediate length cannot occur. In addition to the clamped end points, certain points on the string, called **nodes,** do not move at all (Figure 6.19). The allowed wavelengths of the string motion are given by Equation (6-9)

$$\lambda = \frac{2D}{n + 1}; \qquad n = 0, 1, 2, 3, \ldots \tag{6-9}$$

in which D is the distance between the two fixed endpoints of the string and n is the number of nodes. Although the musicians who play stringed instruments do not call it that, n is in fact a quantum number because it characterizes the discrete or quantized states of the wave motion.

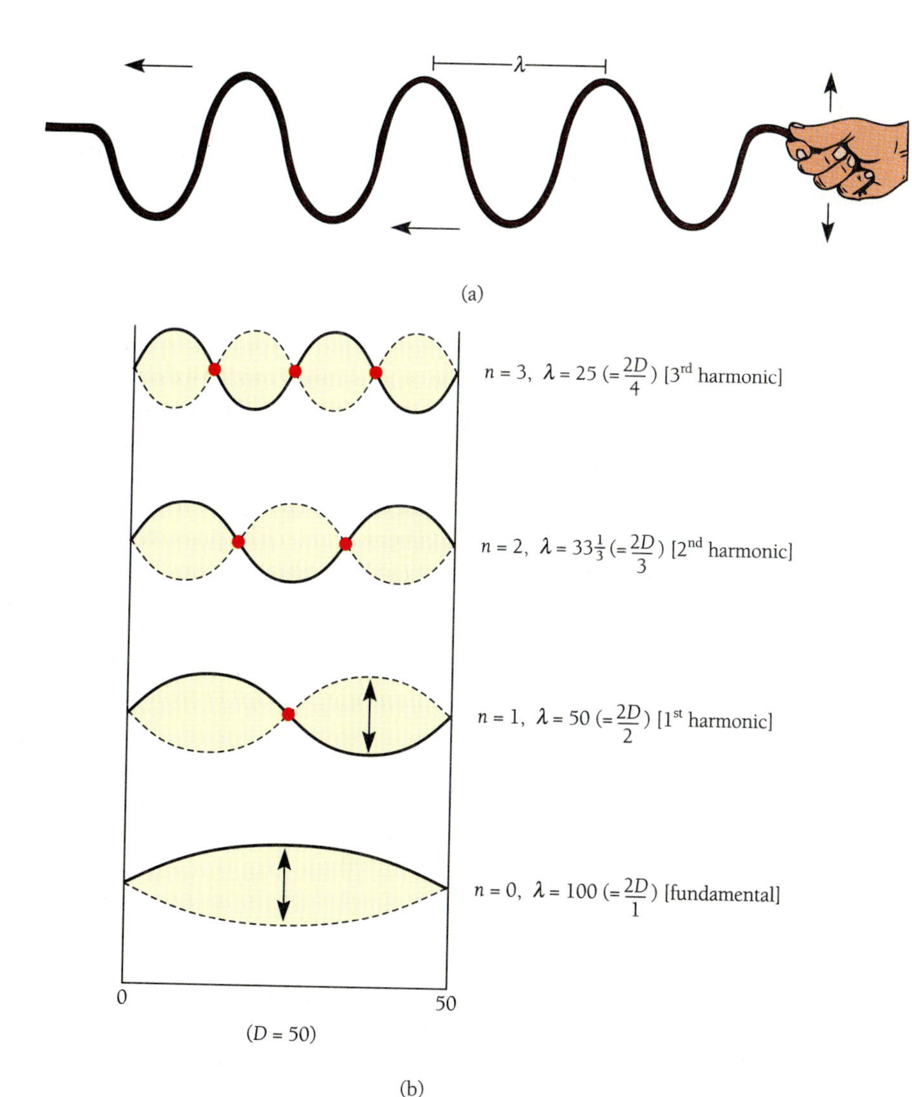

Figure 6.19 (a) A rope with free ends can support travelling wave motion of any wavelength, depending on the frequency of the oscillation that produces the wave. (b) A string that is fixed at both ends allows only certain wavelengths of stationary wave motion: others are "forbidden." In a string of 50 cm length, the allowed wavelengths are 100, 50, 33⅓, 25, 20 cm, and so on. Note that in all but the first of these vibrating states, there are nodes—points on the string that do not move at all.

Standing waves, nodes, allowed wavelengths, quantization, and quantum numbers all occur whenever wave motion of any sort is confined to a limited region of space. Thus an electron wave, confined by the attractive force between opposite charges to the region near the nucleus, is naturally quantized. Electron waves are difficult to visualize, but one thing is clear: we shall have to give up any notion of an electron "being at" a particular place in an atom. Just as a sound wave exists throughout the vicinity of a siren, an electron wave exists throughout the vicinity of the nucleus.

Electronic Wave Functions

An *algebraic equation* has for its solution one or more *numbers*. For example, the solutions to the equation $x^3 + 2x^2 - x = 2$ are the numbers $(-2, -1, +1)$. A *differential equation*, on the other hand, has for its solution one or more *functions*. (A function is a recipe or procedure for calculation of numbers.) The first two of the (many) solutions to the differential equation describing the standing waves of a 50 cm guitar string clamped at both ends (at $x = 0$ and $x = 50$ cm) are the functions $y = \sin[180(x/50)]$ and $y = \sin[360(x/50)]$. The solutions all have the general form $y = \sin[180(n + 1)(x/50)]$, where n is the number of nodes. Each of the different modes of string movement in Figure 6.19 is described by a different function.

The Schrödinger equation for the hydrogen atom is also a differential equation, and its solutions are called electronic **wave functions.** Because of their mathematical complexity, wave functions are rarely written out in full. Instead they are represented (abbreviated) by the Greek letter ψ(psi). But like the wave functions of a guitar string, ψ is a recipe for the calculation of the properties of the electron wave. Wave functions are sometimes called "stationary states," following the terminology introduced by Bohr.

We have previously described an electron in an atom as forming a cloud or fog of negative charge. Although ψ itself has no direct physical significance, ψ^2 gives the thickness or density of the cloud. That is, numerical evaluation of the square of the wave function at any point within the three-dimensional space of the atom gives the concentration of negative charge, called the **electron density,** at that point.

The meaning of ψ^2 can be described in another way, which emphasizes a different aspect and is useful in other contexts. According to Heisenberg, our knowledge of the location of the electron within the atom cannot be *exact*. The value of ψ^2 at any point in an atom, however, is the *probability* that a measurement will discover the electron at that point. This interpretation emphasizes another difference between classical physics and quantum mechanics: the laws of the former are exact statements, while the latter deals in probabilities.

These two interpretations of the meaning of ψ^2 are not at odds with one another. Rather, they are two ways to describe in words a reality that can be precisely described only in abstract mathematical terms. Most of the time, chemists think of ψ^2 as describing the shape, size, and density of the electron cloud.

6.5 THE HYDROGEN ATOM

The methods of quantum mechanics provide us with a detailed description of electrons in atoms. The hydrogen atom, which contains only one electron, is the simplest of all atoms. Therefore, hydrogen is a good starting point for a description of the modern view of the behavior of electrons in atoms.

Orbitals

Chemists have frequent need to refer to wave functions, and the word **orbital** was coined to mean "a wave function ψ, one of many related functions that are solutions to the Schrödinger equation of an electron in an atom." A good way to think of an orbital is as a region around the nucleus, having a particular size and shape, and containing the diffuse negative charge of an electron. An electron occupying an orbital has a particular energy, characteristic of that orbital.

Quantum Numbers

Quantization and quantum numbers arise naturally when wave motion is confined to a limited region of space. There is one quantum number in the functions that describe the waves of a one-dimensional guitar string. The functions that describe the vibrations of a two-dimensional drumhead contain two quantum numbers. The electron waves of a hydrogen atom exist in three-dimensional space, and their wave functions contain three quantum numbers. All three are dimensionless whole numbers, and their values are interrelated.

Names and Rules

The values that the three quantum numbers can have are governed by the following rules.

- **Principal quantum number, n:** $n = 1, 2, 3, \ldots$ Although n can be *any* positive integer, orbitals with n larger than 7 are not very important in chemistry.
- **Angular momentum quantum number, ℓ:** ℓ can be 0 or any positive integer less than the principal quantum number n. That is, the possible values are $\ell = 0, 1, 2, \ldots (n - 1)$.
- **Magnetic quantum number, m_ℓ:** m_ℓ can be 0, or a positive or negative integer whose absolute value is not greater than ℓ. That is, for an orbital with a given value of ℓ, the possible values of m_ℓ are $-\ell$, $-(\ell - 1), \ldots, 0, \ldots, (\ell - 1), \ell$.

These rules are mathematical in origin. A simple analogy might be the more familiar rule, "the square of an integer must be a positive integer."

Any set of particular values of these three quantum numbers (n, ℓ, m_ℓ), for example the set (2, 1, 1), corresponds to a unique state of the electron or atom, and describes an orbital having a particular size, shape, orientation, and energy.

Orbitals whose quantum numbers follow the rules given above are called **allowed** orbitals. Orbitals whose quantum numbers do not follow these rules are **forbidden,** and do not exist. That some orbitals exist and others do not is a law of nature, like the laws of conservation of mass and energy.

The law of conservation of energy can be stated in a similar way, as "processes in which energy is created or destroyed are forbidden, and do not take place."

EXAMPLE 6.6 The Rules for Quantum Numbers

Describe (by giving the values of the three quantum numbers) all possible orbitals having principal quantum number $n = 2$.

Analysis/Plan The value of the principal quantum number is given, and stepwise application of the rules yields the allowed values of the angular momentum and magnetic quantum numbers.

Work

1. Since ℓ must be less than n (but cannot be negative), possible values are $\ell = 0$ or 1. Thus, only the sets (2, 0, x) and (2, 1, x) are possible.
2. For the possible values of m_ℓ, consider first the case $\ell = 1$. Since the absolute value of m_ℓ cannot exceed ℓ, the only possibilities are $m_\ell = -1$, 0, or $+1$. Possible orbitals are (2, 1, -1), (2, 1, 0), and (2, 1, $+1$).
3. When $\ell = 0$, the only possibility is $m_\ell = 0$. Therefore, the only possible orbital is (2, 0, 0).

   There are a total of four orbitals having $n = 2$.

Exercise

Describe and give the total number of orbitals having (a) $n = 3$; (b) $n = 4$ and $\ell = 1$; and (c) $n = 3$ and $m_\ell = -1$.

Answer: (a) (3, 2, 2), (3, 2, 1), (3, 2, 0), (3, 2, -1), (3, 2, -2), (3, 1, 1), (3, 1, 0), (3, 1, -1), (3, 0, 0): 9 different orbitals have $n = 3$. (b) (4, 1, 1), (4, 1, 0), (4, 1, -1): three orbitals can have $n = 4$ and $\ell = 1$. (c) (3, 2, -1) and (3, 1, -1): two orbitals can have $n = 3$ and $m_\ell = -1$.

Orbital Descriptions and Classification: Shells and Subshells

The values of both n and ℓ are important to the chemical behavior of an atom. However, the value of m_ℓ is less important, and it is common to describe an orbital by n and ℓ only. The numerical value of n is used, while a kind of shorthand or code is used for ℓ: the letter *s* stands for $\ell = 0$, *p* for $\ell = 1$, *d* for $\ell = 2$, and *f* for $\ell = 3$. Thus a "1*s*" orbital has $n = 1$ and $\ell = 0$, a "3*d*" orbital has $n = 3$ and $\ell = 2$, and so on.

EXAMPLE 6.7 Orbitals and Subshells

How many 2*p* orbitals are there?

Analysis/Relationship The combination of the letter codes for ℓ and the quantum number rules leads to the answer.

Work "2*p*" means $n = 2$ and $\ell = 1$: for an orbital with $\ell = 1$, there are three possible values of m_ℓ.

$$m_\ell = -1, 0, \text{ and } +1$$

Therefore there are three 2*p* orbitals, (2, 1, -1), (2, 1, 0), and (2, 1, $+1$).

Exercise

For a particular value of n (equal to or greater than 3), how many *d* orbitals are there?

Answer: 5.

The set of all orbitals having the same value of the principal quantum number is called a **shell:** the first or $n = 1$ shell, the second or $n = 2$ shell, and so on. Within a shell, the set of orbitals having the same value of ℓ is called a **subshell.** In the $n = 3$ shell, for example, the s subshell consists of 1 orbital, the p subshell consists of 3 orbitals, and the d subshell consists of 5 orbitals. The second shell contains only the s (1 orbital) and p (3 orbitals) subshells, and the $n = 1$ shell has only the s subshell.

Significance of Quantum Numbers

Each of the three quantum numbers carries a definite physical meaning, as follows.

- n: The principal quantum number determines the energy of the orbital, its size, and the number of nodes.
- ℓ: The angular momentum quantum number gives the shape of the electron cloud and its angular momentum.
- m_ℓ: The magnetic quantum number gives the spatial orientation of the orbital.

Orbital Sizes, Shapes, Nodes, and Orientations

The density of the electron cloud is found by calculating ψ^2 at all points in the neighborhood of the nucleus. The size, shape, and orientation of the electron cloud depend on the values of the quantum numbers as follows.

s Orbitals

Orbitals having $\ell = 0$ are spherical in shape. The electron density ψ^2 is large at the nucleus, and drops as the distance from the nucleus increases. There are several different pictorial representations of an orbital. The *stipple plot* (Figure 6.20a) places the nucleus at the origin of a Cartesian (x, y, z) coordinate system, and represents the changing value of electron density in a cross-section of the orbital by the density of dots on the graph. Alternatively, a *contour line* enclosing, say, 90% or 95% of the total negative charge, can be drawn (Figure 6.20b).

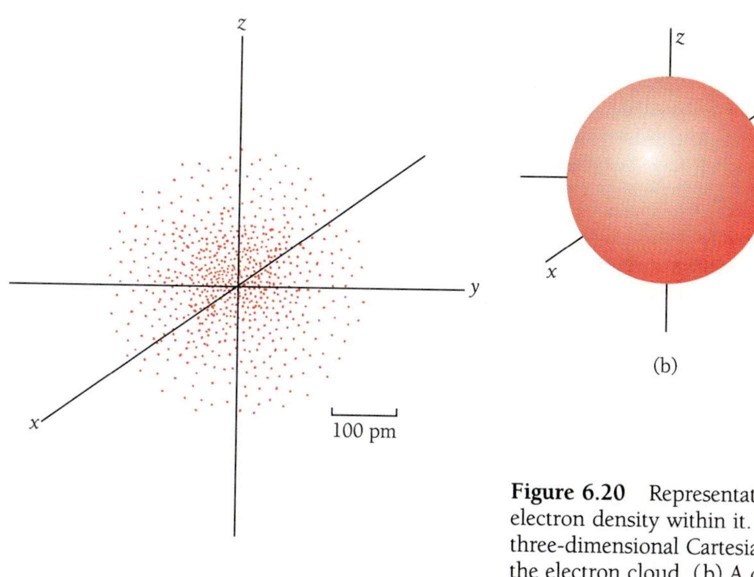

Figure 6.20 Representations of the shape of a 1s orbital and the variations of electron density within it. (a) Stipple plot with the nucleus at the origin of a three-dimensional Cartesian coordinate system, showing the varying density of the electron cloud. (b) A contour line, actually a contour *surface* in three dimensions, encloses most (90%) of the negative charge.

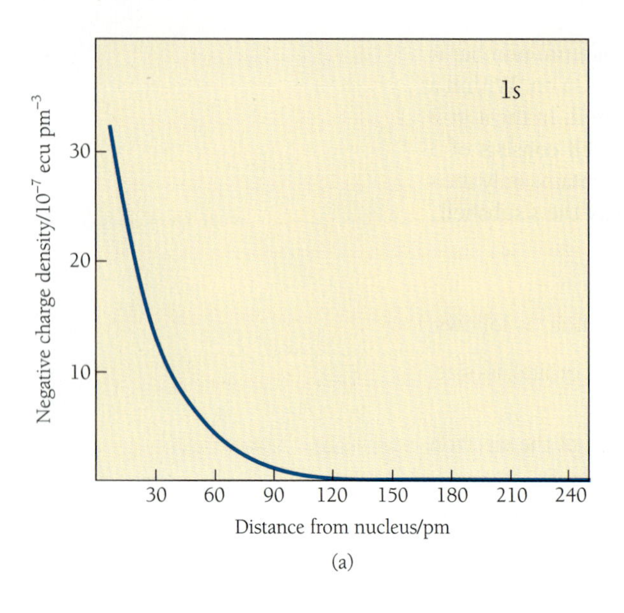

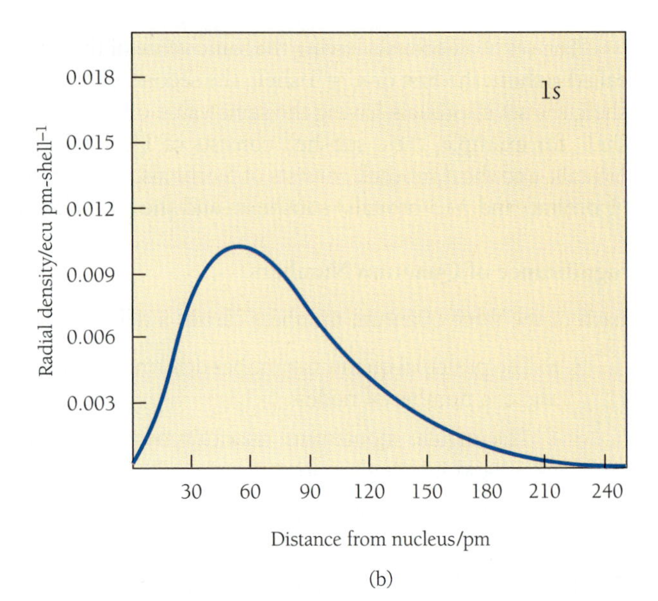

Figure 6.21 Two graphs of the 1*s* orbital. (a) Electron density (ψ^2) vs. distance from the nucleus. (b) Radial probability function, or the probability of finding the electron at a particular distance from the nucleus. The abbreviation "ecu" in the *y*-axis label stands for "elementary charge unit" (1.6×10^{-19} C).

A graph of numerical information can be made by plotting the value of electron density as a function of distance from the nucleus (Figure 6.21a). Although the *density* is high, there is not much total negative charge in the small region near the nucleus. There is a higher probability of finding the electron at a somewhat greater distance from the nucleus; this is shown in Figure 6.21b, a graph of what is called the "radial probability function." The most likely distance from the nucleus for the electron in a 1*s* orbital in the hydrogen atom (53 pm) is, interestingly enough, just the distance that Bohr calculated as the radius of the lowest-energy "orbit" in his model of the atom. The difference between the two graphs in Figure 6.21 is one of emphasis: (a) gives the negative charge (or probability of finding the electron) in a cubic picometer at a particular *point* in the atom, while (b) gives the negative charge (or probability) in a spherical shell 1 pm thick located at a particular *distance from the nucleus.*

The 1*s*, 2*s*, and other *s* on orbitals differ both in number of nodes and in size. Nodes in three-dimensional waves, rather than being points, are *surfaces* over which the electron density is zero. A *nodal surface* of an *s* orbital is the surface of a sphere. The 2*s* orbital has one node, the 3*s* has two, and so on (Figure 6.22). An *s* orbital whose principal quantum number is *n* has $(n - 1)$ spherical nodes. The *size* of the orbital also depends on the principal quantum number. Orbitals with larger values of *n* have significant electron density further from the nucleus.

p Orbitals

In the *p* orbitals ($\ell = 1$), a planar node separates two **lobes** of electron density, which are shaped like slightly flattened spheres. Since $\ell = 1$, there are three possible values of m_ℓ and hence three orbitals in a *p* subshell. These have different orientations in space, and are called the p_x, p_y, and p_z orbitals, depending along which coordinate axis they lie (Figure 6.23). The labels of the coordinate axes are chosen arbitrarily, and

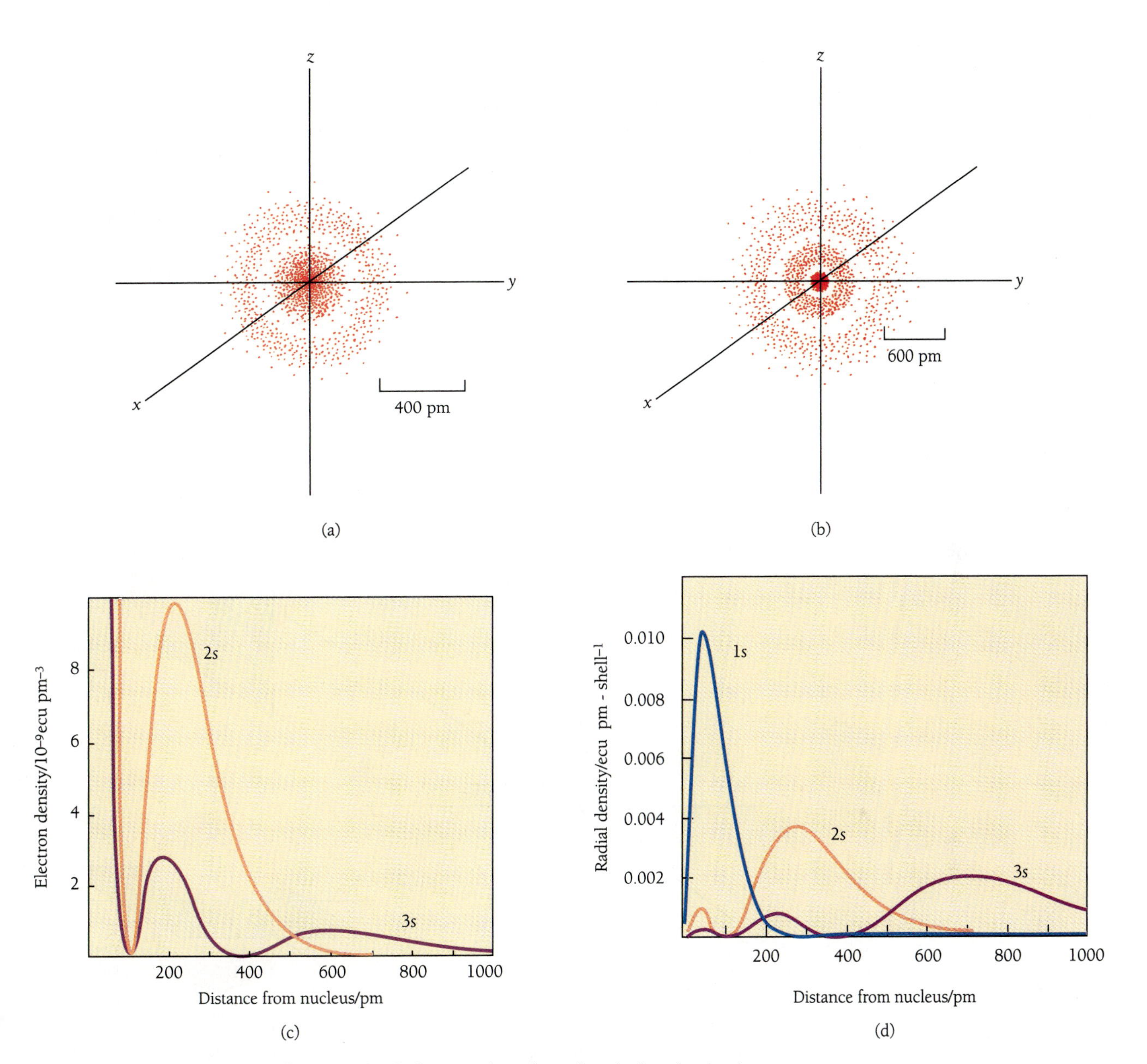

Figure 6.22 (a) Stipple plot of the 2*s* orbital, showing the spherical node that divides the orbital into two parts. (b) The 3*s* orbital has two nodes. (c) Variation of ψ^2 with distance from the nucleus in the 2*s* and 3*s* orbitals, showing the increasing size and number of nodes as the principal quantum number increases. There are $(n - 1)$ spherical nodes in *s* orbitals. (d) Radial probability function. As *n* increases, the electron is more likely to be found further from the nucleus.

there is no correspondence between the labels (x, y, z) and the values $(-1, 0, +1)$ of m_ℓ. The significant aspect of the designations p_x, p_y, and p_z is that the three *p* orbitals are perpendicular to one another.

A 3*p* orbital is larger than a 2*p*, and it has a spherical node in addition to the planar node (Figure 6.24). Each increase in the principal quantum number brings increased size and an additional spherical node to an orbital.

The different orientation of the p_x, p_y, and p_z orbitals takes on added significance when the atom is in a magnetic field. One of the orbitals is aligned parallel to the magnetic field, while the other two are perpendicular to it.

Figure 6.23 (a) Stipple plot of the $2p_z$ orbital. (b) Contour representation of the three $2p$ orbitals. They are identical in shape, but differ in orientation.

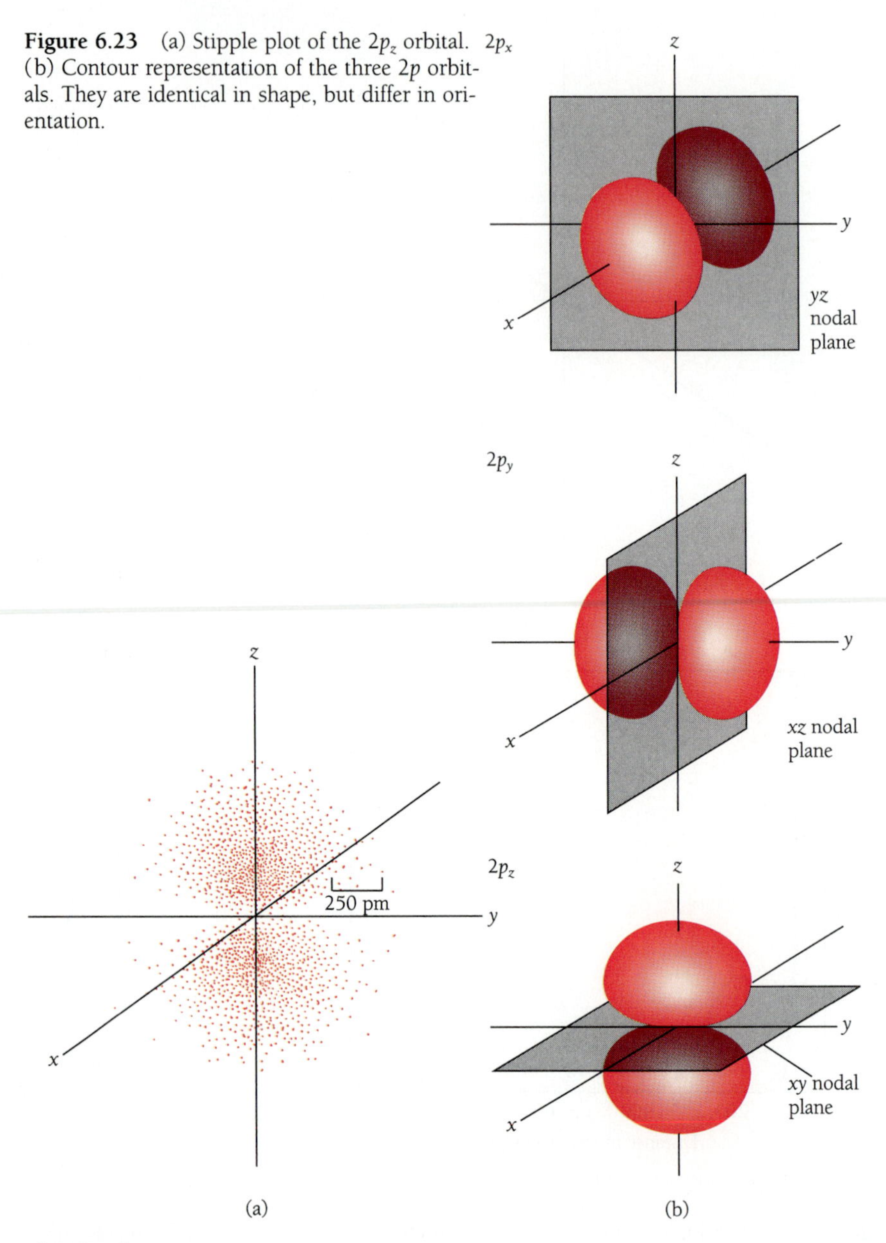

d Orbitals

When $\ell = 2$ there are five values of m_ℓ. That is, there are five orbitals in a *d* subshell. Contour plots of these orbitals are given in Figure 6.25. The d_{xy}, d_{yz}, d_{xz}, and $d_{x^2-y^2}$ orbitals each have four lobes and two planar nodes. The d_{z^2} orbital has three lobes and two conical nodes. Just as for *s* and *p* orbitals, an increase in principal quantum number of a *d* orbital brings additional spherical nodes and increased size (but no change in overall shape).

f Orbitals

As shown in Figure 6.26, the seven orbitals in the 4*f* subshell have more complicated shapes than the *s*, *p*, and *d* orbitals.

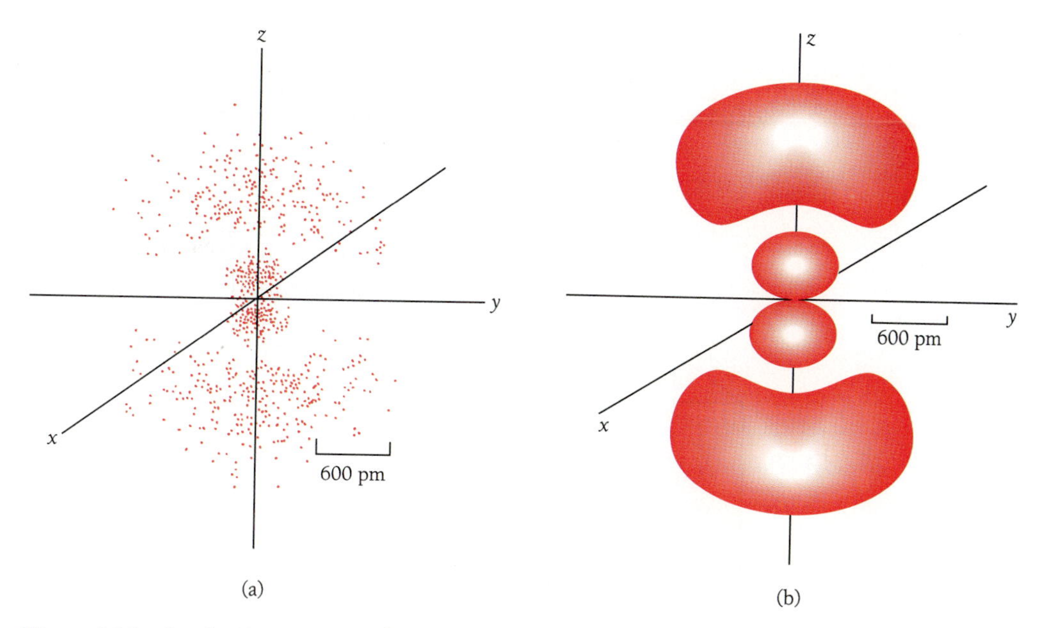

Figure 6.24 Stipple (a) and contour (b) plots of a hydrogen 3*p* orbital, showing planar and spherical nodes. The overall shape is similar to that of a 2*p* orbital, but the electron density is divided among more lobes.

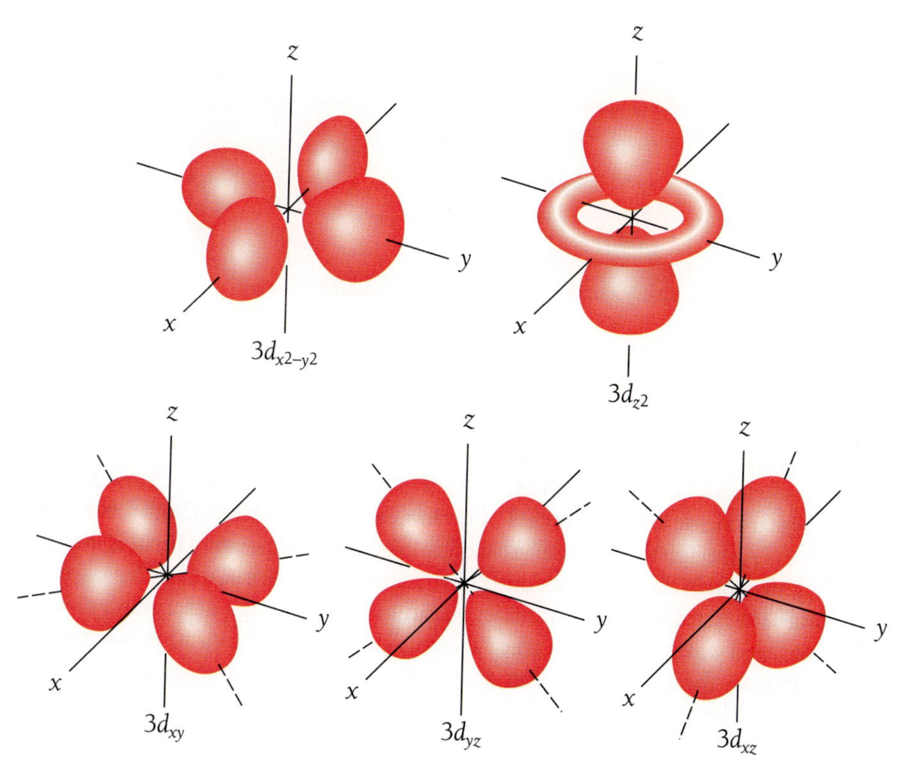

Figure 6.25 The orbitals of the 3*d* subshell. The d_{xy}, d_{yz}, and d_{xz} orbitals have four lobes, located between the coordinate axes. The center lines of these lobes (dotted lines) lie in the *xy* plane for the d_{xy} orbital, and so on. Each has two planar nodes; for the d_{xy} orbital the nodes are the *xz* and the *yz* planes. The $d_{x^2-y^2}$ orbital is like the d_{xy}, but rotated 45° so that the lobes lie along the axes rather than between them. The d_{z^2} orbital has three regions of electron density, and its two nodes are conical rather than planar.

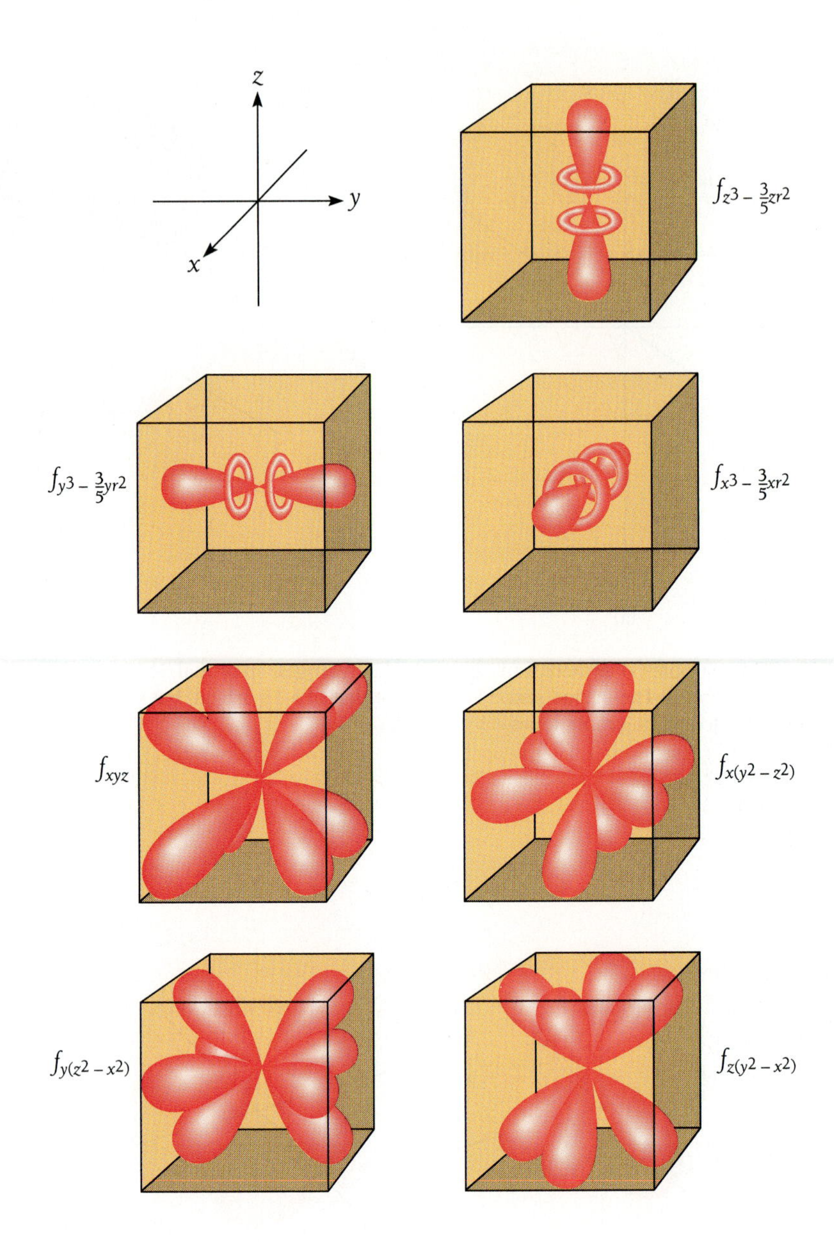

Figure 6.26 Contour plots of the seven orbitals in the *f* subshell.

Where Is the Electron?

Students often ask questions such as, "Which lobe is the electron in?", "How does an electron get from one lobe to another?", or "How can an electron be in both lobes at the same time?" Although these are sensible questions, and discussion of them is instructive, they all contain the hidden assumption that an electron is a *particle*. This assumption makes the questions unanswerable: quantum mechanics is a description of *wave* properties, and waves are not "located" at a particular point. The electron wave exists in the atom in the same sense that a sound wave exists in a room containing a loudspeaker. The sound wave exists simultaneously throughout the room, and the electron wave exists simultaneously throughout the atom.

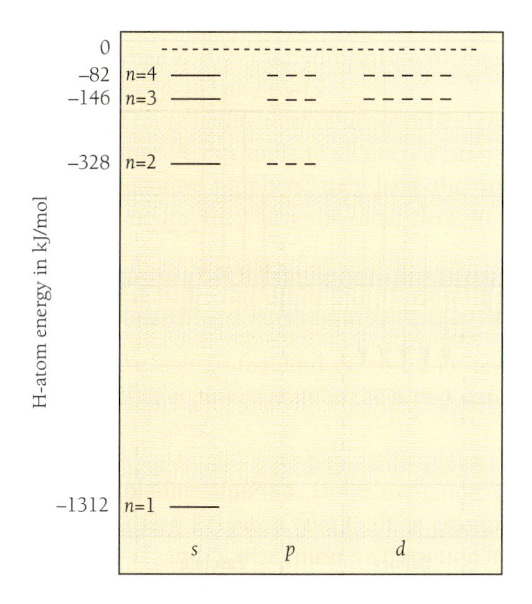

Figure 6.27 Energies of the hydrogen atom orbitals. All the orbitals in the same shell (that is, those having the same principal quantum number) have the same energy, given by $E_n = -1312/n^2$ kJ mol^{-1}. States having $n > 4$ are not shown because their energies are all crowded together between $E = -82$ kJ and 0 kJ mol^{-1}. The dashed line at $E = 0$ corresponds to the energy of a proton and electron that are separated by such a large distance that they do not interact at all.

Orbital and Electron Energies in the Hydrogen Atom

The energy associated with a hydrogen atom orbital, or equivalently, the energy of an electron in a particular orbital, is inversely proportional to the square of n. In the hydrogen atom, orbital energy does not depend on the values of ℓ or m_ℓ. Equation (6-10), in which E_n is the energy of any orbital with principal quantum number n, gives the numerical relationship.

We will see in the following chapter that when there is more than one electron in an atom, orbital energies depend on the value of both n and ℓ.

$$E_n = \frac{-2.18 \times 10^{-18}\text{ J}}{n^2} \qquad (6\text{-}10)$$

Multiplication of the constant in Equation (6-10) by Avogadro's number leads to Equation (6-11), which gives orbital energies per mole of H-atoms (Figure 6.27). The energy *difference* between two levels, the upper level with quantum number n_U and the lower with quantum number n_L is found from Equation (6-12).

$$E_n = \frac{-1312}{n^2}\text{ kJ mol}^{-1} \qquad (6\text{-}11)$$

There are many problems for which molar units are more appropriate than "per atom" units.

$$\Delta E = E_U - E_L = -1312\left(\frac{1}{n_U^2} - \frac{1}{n_L^2}\right)\text{ kJ mol}^{-1} \qquad (6\text{-}12)$$

Note that, except for the symbols used for the factor that multiplies ($1/n^2$), these equations are the same as those previously derived by Bohr for the planetary model. A hydrogen atom with its electron in the lowest possible level ($n = 1$) is said to be in the **ground state.** If the electron is in any other level, the atom is in an **excited state.**

Emission Spectrum and Ionization

The quantum mechanical interpretation of atomic spectra is essentially the same as that based on the Bohr model. An electrical discharge in gaseous hydrogen produces atoms in excited states, and a photon of light is emitted whenever an electron undergoes a transition from an orbital of higher energy to one of lower energy. Since a number of different excited states are produced in the discharge, a number of different

CHEMISTRY OLD AND NEW

SPECTROSCOPY AND THE DETECTION OF ART FORGERIES

The usefulness of spectroscopy as an analytical tool extends far beyond the reaches of laboratory and industrial chemistry. Because they are nondestructive and can reveal information about both the types of atoms and the kinds of chemical functionalities present in a material, spectroscopic methods are used in many diverse fields such as forensic science, art analysis, medicine, and the monitoring of environmental quality.

The detection of forged paintings is an activity that has benefited greatly from the use of spectroscopy. Traditional methods of uncovering forgeries relied on the presence of stylistic or technical inconsistencies, or anachronisms in material use. For example, a modern painter trying to imitate a seventeenth-century artist might use an improper brush stroke or inaccurately represent clothing from that period. The forger might use a type of canvas backing or pigment that was unavailable at the supposed date of the painting's completion. While most evidence of forgery is based solely on the judgment of an art historian, pigment "mistakes" can be scientifically detected.

There have been many changes in the tools and techniques used by artists over the past 500 years. The dyes used in making paints have evolved from natural plant- and mineral-based colorants to synthetic pigments with superior permanence and fade resistance. The inorganic compounds chrome yellow ($PbCrO_4$) and Rinmann's green ($ZnCo_2O_4$) are examples of modern pigments. Since each dye is derived from a different natural or synthetic source, it has a unique composition that can be positively identified by spectroscopy. A knowledge of the dates at which various pigments first came into use can help to determine if a work of art is authentic or if it is a recently produced forgery. One famous example of art fraud uncovered by using infrared spectroscopy is the case of the *Virgin and Child.* This painting, attributed to an unknown fifteenth century Italian artist, had long troubled experts: details of the Christ Child's face and halo and certain features of the paint surface did not match well with other works from that period. The infrared spectrum of a blue pigment used in the background showed a wide absorption band at a wavelength of 4.78 μm. This band is characteristic of the dye Prussian Blue (ferric ferrocyanide, $Fe_4[Fe(CN)_6]_3$). Since Prussian Blue was not available to artists until 1704, when it was first synthesized by the German chemist Diesbach, the *Virgin and Child* was deemed a forgery.

The *Virgin and Child* was exposed as an art forgery through spectroscopy.

A characteristic Prussian Blue infrared band was also found in the analysis of pigment taken from a Medieval Byzantine manuscript known as the *Archaic Mark,* a transcription of the Gospel of that disciple. This document is now thought to be a Greek forgery created during the 1920s.

Besides infrared spectroscopy, the identification of pigments is often accomplished with another spectroscopic technique called x-ray fluorescence. This tool reveals qualitative and quantitative information about the chemical elements (rather than the compounds) present in a sample. X-ray fluorescence is extremely sensitive; a complete dye analysis can be performed on a microscopic chip of paint, which can be removed inconspicuously from a work of art with a tiny needle.

Discussion Question

Would spectroscopy be a valuable tool for detecting forged sculptures?

▶ SUMMARY EXERCISE

A sample of lithium was heated to its boiling point (1342 °C) and an electrical discharge was established in the vapor. When the light emitted from the discharge was analyzed with a spectrometer, it was found that electron transitions between states of the Li^{2+} ion (a one-electron species) were responsible for many of the lines in the spectrum. One of these lines had a wavelength $\lambda = 208.4$ nm. **(a)** What is the frequency of light of this wavelength? **(b)** In what spectral region does the light lie? **(c)** How much energy is carried by each photon? **(d)** When it emits a 208.4 nm photon, the electron in Li^{2+} drops from an upper to a lower energy level. What is the energy difference, in kJ mol^{-1}, between these two energy levels? **(e)** What are the principal quantum numbers of the two levels? [*Hint:* This question can be answered by trial-and-error substitution of pairs of integers into Equation (6-14).]

In another experiment it was found that the threshold frequency for the photoelectric effect in gold was 1298 THz. **(f)** What is the minimum energy (kJ mol^{-1}) necessary for the removal of an electron from metallic gold? **(g)** When struck by high-energy x-rays, gold emits photoelectrons having considerable kinetic energy. If such an electron is travelling at 7.5×10^3 m s^{-1}, what is its de Broglie wavelength?

(h) Describe the shape of an orbital with quantum numbers $n = 3$ and $\ell = 1$. **(i)** Explain why the 2*d* subshell is forbidden.

Numerical prefixes for SI units are listed in Section 1.3.

Answers **(a)** 1.439×10^{15} Hz; **(b)** ultraviolet; **(c)** 9.532×10^{-19} J; **(d)** 574.0 kJ mol^{-1}; **(e)** $n_U = 4$, $n_L = 3$; **(f)** 8.601×10^{-19} J per electron, or 517.9 kJ mol^{-1}; **(g)** 97 nm; **(h)** two major lobes, directed along the *x*-, *y*-, or *z*-axis and separated by a node in the plane of the other two axes—a spherical node cuts through the lobes and separates them into two parts; **(i)** the notation "2*d*" means $n = 2$ and $\ell = 2$, which violates the rule that *n* must be greater than ℓ.

SUMMARY AND KEYWORDS

Wave properties of light are characterized by **wavelength** (λ; distance between successive crests), **frequency** (ν; number of crests passing a fixed point each second), and **velocity** ($c = \lambda\nu$). Light waves are periodic fluctuations of both electric and magnetic fields. The **electromagnetic spectrum** encompasses γ- and x-rays, ultraviolet, visible, and infrared light, microwaves, and various radio and TV waves. All of these are **electromagnetic radiation,** differing only in wavelength and frequency. The velocity of electromagnetic radiation through empty space, called the *speed of light* (c), is 3×10^8 m s^{-1}. The **photoelectric effect,** ejection of an electron from an illuminated metal surface, and the **blackbody radiation** emitted by a heated object can only be explained by the *particle theory,* which is that light consists of discrete particles called **photons** or **quanta,** each carrying energy $E = h\nu$. **Diffraction,** the spreading of waves into a shadow region, and **destructive interference,** the cancellation of light intensity, can only occur with waves. Other properties of light, color, reflection, **refraction** (change in direction on entering a new medium), and the casting of shadows, can be explained by either theory. **Wave-particle duality** asserts that light has properties of both.

A **spectrum** is a display of frequencies (or wavelengths) at which a sample interacts with light, together with the intensity of interaction. **Emission** spectra of heated objects are **continuous,** while emission spectra of heated gases are **discrete.** Discrete spectra consist of a number of *lines,* with no light emitted in the wavelength intervals between the lines. Spectra are observed by means of a **spectrometer,** an instrument that **disperses** light into its component wavelengths.

Quantum mechanics replaced the **planetary model** of atoms, which was inconsistent with Heisenberg's **uncertainty principle** ($\Delta x \Delta p > h$) and could not explain the wave properties of electrons (**de Broglie wavelength,** $\lambda = h/mv$). The **Schrödinger** equation describes the electron(s) in an atom on the basis of wave rather than particle properties. Light waves are **travelling waves,** while electron waves in an atom are **stationary waves, quantized** into a set of discrete energies.

A **wave function (ψ)** of the hydrogen atom (a solution of the Schrödinger equation) describes the size, shape, energy, orientation, and number of nodes of an **orbital,** the region near the nucleus that contains the diffuse mass and charge of the electron. The numerical value of ψ^2 is interpreted either as the charge density of the electron cloud or as the probability of finding the electron at a particular point. An orbital is identified by its **quantum numbers: *n* (principal), ℓ (angular momentum),** and **m_ℓ (magnetic).** Only orbitals having $n = 1, 2, \ldots, \ell = 0, 1, 2, \ldots (n - 1)$, and $m_\ell = 0, \pm 1, \pm 2, \ldots \pm \ell$ are **allowed;** others are **forbidden,** that is, they do not exist. The value of *n* determines the energy, size, and number of **nodes** (surfaces on which the charge density is zero) of the orbital; ℓ determines its shape, and m_ℓ determines its orientation. A **shell** is a set of orbitals having the same value of *n*. Within a shell, the orbitals in a **subshell** have the same value of ℓ. Subshells are coded as *s*, *p*, *d*, or *f*, when $\ell = 0, 1, 2$, or 3. An *s* orbital is spherical, the three mutually perpendicular *p* orbitals have two **lobes** separated by a planar node, and the five *d* orbitals have three or four lobes, separated by two planar or conical nodes.

An atom whose electron is in the level of lowest energy is in its **ground state;** otherwise it is in an **excited state.** The Lyman (ultraviolet), Balmer (visible), and Paschen (infrared) series in the spectrum of hydrogen are sets of lines emitted when electrons drop from higher levels to those with $n = 1, 2$, and 3, respectively. Excitation from the ground state to a level of very large *n* ($n = \infty$) corresponds to removal of the electron, that is, to *ionization;* the energy required is the **ionization energy.**

Atoms or ions having only one electron are **one-electron species.** The orbitals of all one-electron species are identical except in size and energy, which for principal quantum number *n* and atomic number *Z* is given by $E_n = -1312(Z^2/n^2)$ kJ mol^{-1}.

PROBLEMS

General

1. Distinguish between reflection and refraction of light.
2. Is refraction of radio waves possible?
3. Other than its effect on the eye, what are two differences between blue light and green light?
4. What is meant by dispersion of light?
5. Describe the photoelectric effect.
6. What is the relationship between wavelength and frequency of light? Of any type of wave?
7. What is meant by diffraction?
8. What is blackbody radiation?
9. Distinguish between two forms of interference.
10. Describe electromagnetic radiation.
11. Criticize the statement, "The speed of light is 3×10^8 m s^{-1}."
12. Compared to visible light, which of the following have greater wavelength: microwaves, x-rays, infrared, radar, FM radio?
13. What is a spectrum? Distinguish between emission and absorption spectra.
14. What is a discrete spectrum?
15. What type of light source emits a continuous spectrum? A discrete spectrum?
16. What is a spectrometer?
17. Name two characteristics of light that cannot be fully explained by the particle theory, and two that cannot be fully explained by the wave theory.
18. What is a photon?
19. Define the term "wave-particle duality."
20. What is meant by the term "quantized"? What are quanta?
21. List the feature(s) of the Bohr theory that survived to become a part of quantum theory, and the feature(s) that did not survive.
22. What is a stationary wave? What is a stationary state?
23. What is a matter wave?
24. Explain why the planetary model of the atom is incompatible with the uncertainty principle.

25. What is a wave function? What is its physical significance?

26. What is an orbital?

27. What are quantum numbers?

28. Describe and sketch the shapes of *s*, *p*, and *d* orbitals.

29. What is meant by shell and subshell?

30. What is an excited state?

31. Give the formulas for three one-electron species with atomic number greater than ten.

32. Define ionization energy. How is it calculated for a one-electron species?

Electromagnetic Radiation

33. The visible region of the spectrum extends over the approximate wavelength range 400 to 700 nm. What is this range in frequency units?

34. The wavelength of a prominent line in the emission spectrum of mercury is 254 nm. What is the frequency?

35. What is the frequency of x-rays having a wavelength of 1.224 Å (Angstroms)?

36. What is the frequency of γ-rays having a wavelength of 0.093 Å?

37. What is the wavelength of a television signal that is broadcast at 600 MHz?

38. What is the wavelength of the radiation in a microwave oven, if the frequency is 6.00×10^{10} Hz?

39. Sound travels through air at about 330 m s^{-1}. What is the wavelength of the note "A" (440 Hz)?

40. The speed of sound in seawater is 1531 m s^{-1}. What is the wavelength of the note "A" (440 Hz) if you are listening to underwater stereo?

*41. How long did it take for TV signals to return to earth from the Explorer landing craft on Mars, 100 million km away?

*42. The average distance of the earth from the sun is 93 million miles. How long does it take sunlight to reach the earth?

43. It takes radio waves about 1.5 seconds to travel to the moon (which introduces a three-second delay in spoken communication with astronauts on the moon). Use this information to estimate the distance to the moon.

44. It takes sunlight a little more than four hours (250 minutes), on the average, to reach the planet Neptune. What is the average radius of Neptune's orbit?

45. Calculate the energy carried by one photon of the following electromagnetic radiations.
    a. Blue light (450 nm) b. FM broadcast (92.7 MHz)
    c. Lyman α-line (121 nm)

46. Repeat the calculations in Problem 45, but express the answers in kJ mol^{-1} of photons.

47. Calculate the energy carried by one photon of the following electromagnetic radiations.
    a. Red light (692 nm) b. AM broadcast (880 kHz)
    c. Radar (1 cm)

48. Repeat the calculations in Problem 47, but express the answers in J mol^{-1} or kJ mol^{-1} of photons.

49. When compounds of barium are heated in a flame, green light ($\lambda = 554$ nm) is emitted. How much energy does one mole of these photons carry?

50. What is the energy of 2.53 moles of photons of wavelength 1.06 μm?

*51. Chemical changes stimulated by radiation are called *photochemical reactions*. How much energy per mole, approximately, is available from light in the following spectral regions? (In each case choose a wavelength that is somewhere in the middle of the region, for example 550 nm for visible light.)
    a. X-ray b. Visible

52. How much energy (per mole) is available from light in the following spectral regions?
    a. Ultraviolet b. Infrared

53. Suppose a particular photochemical reaction requires 350 kJ mol^{-1} to proceed. What is the maximum wavelength of light that will be effective? In what region of the spectrum is this?

54. A photochemical reaction requires 225 kJ mol^{-1} to proceed. What is the maximum wavelength of light that will be effective? In what region of the spectrum is this?

*55. If a microwave oven emits 10 W of 60 GHz radiation, how many photons are emitted per second? Express this number also in moles s^{-1} (1 W $\Leftrightarrow$ 1 J s^{-1}).

56. A 60-W light bulb consumes energy at the rate of 60 J s^{-1}. Much of the light is emitted in the infrared region, and less than 5% of the energy appears as visible light. Make the simplifying assumptions that 5% of the light is visible and that all visible light has a wavelength of 550 nm (yellow/green), and calculate the number of visible photons emitted per second.

Matter Waves and Uncertainty

57. The "light" in an electron microscope consists of a beam of electrons travelling with a velocity around 20% of the speed of light. What is the wavelength of this "matter wave"? How does this compare with the diameter of a hydrogen atom, which is about 75 pm?

58. Calculate the de Broglie wavelength of a 1500 kg satellite travelling in earth orbit at 18,500 mph. Comment on the significance of your answer.

59. The wave properties of a beam of neutrons are used in the technique of *neutron diffraction* to investigate the geometrical structure of molecules. What velocity must a neutron have in order to have a de Broglie wavelength of 12 pm?

60. What velocity must an α-particle (a helium nucleus) have in order to have a de Broglie wavelength of 0.529 Å?
61. Suppose a 150 g baseball is travelling with a speed between 89 and 91 km per hour. What would be the minimum possible uncertainty in knowing its location? Comment on the magnitude of the answer.
62. The momentum of the electrons in an electron microscope is usually around 1.2×10^{-22} kg m s^{-1}, with an uncertainty of about 0.01%. What is the uncertainty in the position of an object viewed by an electron microscope? How does this compare with the diameter (several hundred pm) of a typical atom?

Photoelectric Effect and Spectroscopy

63. The threshold energy for the photoelectric effect in platinum is 545 kJ mol^{-1}. What is the threshold wavelength (the maximum wavelength that causes the effect)? Will a platinum phototube respond to visible light?
64. The threshold energy for the photoelectric effect in cesium is about 190 kJ mol^{-1}. What is the threshold wavelength? Will a cesium phototube respond to infrared radiation?
65. Calculate the difference in energy (kJ mol^{-1}) between two atomic states, if light of wavelength 850 nm is emitted in the transition between the levels. In what spectral region is the light?
66. The "21 cm line" is used by radio astronomers to observe interstellar hydrogen. What is the energy difference (per mole) between the levels involved in this transition? (*Note:* These levels, very close in energy, arise from subtle aspects of atomic behavior that are not discussed in this text.)
67. What wavelength of light is emitted when a hydrogen atom drops from the $n = 6$ to the $n = 4$ state? In what region of the spectrum does this occur?
68. Calculate the wavelengths of the first two lines (those of lowest frequency) in the Lyman series of the hydrogen spectrum.
69. What wavelength is emitted when a He^+ ion undergoes a transition from the $n = 2$ level to the $n = 1$ level?
70. Repeat the calculation of the preceding problem for the Li^{2+} and Be^{3+} ions, and comment on the trend.
71. What wavelength of light is required to excite a hydrogen atom from the $n = 2$ to the $n = 6$ level? In what region of the spectrum is this light?
72. What wavelength of light is required to excite a He^+ ion from the $n = 2$ to the $n = 3$ level? In what region of the spectrum is this light?

*73. One of the lines in the Lyman series of the hydrogen spectrum has a wavelength of 93.8 nm. What is the upper level of this transition? (*Note:* Since λ is given to only three significant digits, the calculated value of n_U will not be an exact whole number.)

*74. One of the lines in the Balmer series has a wavelength of 486 nm. What is the quantum number of the upper level? What color is this light?

*75. Infrared radiation of 1876 nm is observed in emission from an electrical discharge in hydrogen. What are the quantum numbers of the upper and lower levels? *Hint:* The most practical way to solve this is by trial and error, using pairs of small integers.

*76. A line of wavelength 365 nm is observed in emission from an electrical discharge in hydrogen. What are the quantum numbers of the upper and lower levels?

Quantum Mechanics

77. Which of the following sets of quantum numbers (n, ℓ, m_ℓ) are not allowed for hydrogen and other one-electron species? In each case, state why not.
    a. (1, 1, 1) b. (3, 1, 0)
    c. (1, 0, 1) d. (0, −1, 0)
    e. (3, 2, 1) f. (3, 2, −2)
78. Which of the following sets of quantum numbers (n, ℓ, m_ℓ) are not allowed for hydrogen and other one-electron species? In each case, state why not.
    a. (0, 0, 0) b. (3, 1, −2)
    c. (3, 2, −2) d. (−1, 0, 0)
    e. (6, 0, 5) f. (6, 5, 0)
79. How many different orbitals can have the designation 4*d*?
80. How many orbitals can have $n = 4$?
81. How many different orbitals can have $m_\ell = -2$ and n less than 5?
82. How many different orbitals can have $\ell = 1$ and n less than 6?
83. Sketch each of the following orbitals.
    a. 2*s* b. 2*p*

*84. Sketch each of the following orbitals.
    a. 3*p* b. 4*s*

85. Identify the values of the principal and the angular momentum quantum number in each of the following: (a) 3*d*, (b) 3*f*, (c) 2*p*. Identify each as allowed or forbidden.
86. Identify the values of the principal and the angular momentum quantum number in each of the following: (a) 1*s*, (b) 5*s*, (c) 2*d*. Identify each as allowed or forbidden.
87. What is the maximum value of m_ℓ in an *f* orbital? In a *p* orbital?
88. Do the designations 1*s*, 3*d*, and so on, refer to shells, subshells, or orbitals?
89. What is the principal quantum number of an *s* orbital with 3 spherical nodes?
90. Describe the nodes in a 3*p* orbital.
91. A fellow student, having some conceptual troubles, asks you, "If the probability of an electron being on a nodal surface is truly zero, how can an electron get from one 2*p* orbital lobe to the other?" Your job: Answer this question in your own words.

CHAPTER 7

ELECTRON CONFIGURATION AND PERIODIC PROPERTIES

The reaction between liquid bromine and aluminum metal is so vigorous that the metal glows white-hot when the two elements are brought into contact. The fumes coming from the beaker consist of vaporized Br_2 (reddish-brown) and fine particles of the product $Al_2Br_6(s)$.

7.1 MANY-ELECTRON ATOMS

The energy levels and electron orbitals of the hydrogen atom (and one-electron ions) were described in the preceding chapter. These descriptions are also generally accurate for atoms having more than one electron—the so-called **many-electron species.**

Electron Orbitals in Many-Electron Atoms

The orbitals of many-electron atoms are often called **hydrogenlike orbitals,** because they are nearly the same as those of the hydrogen atom. There are some differences, however, which are important because they affect the chemical properties of the various elements.

Orbital Energy

The energy of an orbital in the hydrogen atom depends only on the principal quantum number n. All orbitals in the same shell, for instance in the $3s$, $3p$, and $3d$ subshells, are at the same energy (Figure 7.1). A set of orbitals having the same energy is said to be **degenerate.** In atoms containing more than one electron, the degeneracy of the subshells is *removed* or *split.* Each subshell in the shell has a different energy, increasing as the value of ℓ increases. Figure 7.2 shows that, in such an atom, the energy of the $3d$ subshell ($\ell = 2$) is greater than that of $3p$ ($\ell = 1$), which in turn is greater than that of $3s$ ($\ell = 0$). Orbital energy does *not* depend on the magnetic quantum number m_ℓ: the three orbitals in a p subshell lie at exactly the same energy. Similarly, the five orbitals of a d subshell form a degenerate set, as do the seven f orbitals. Orbital energies depend on the magnitude of the positive charge of the nucleus, and are therefore different for every element. However, the general *pattern* of energy levels in Figure 7.2 is common to almost all elements other than hydrogen.

The reason that subshell energy increases in the order $s < p < d$ . . . is discussed in Section 7.5, page 289.

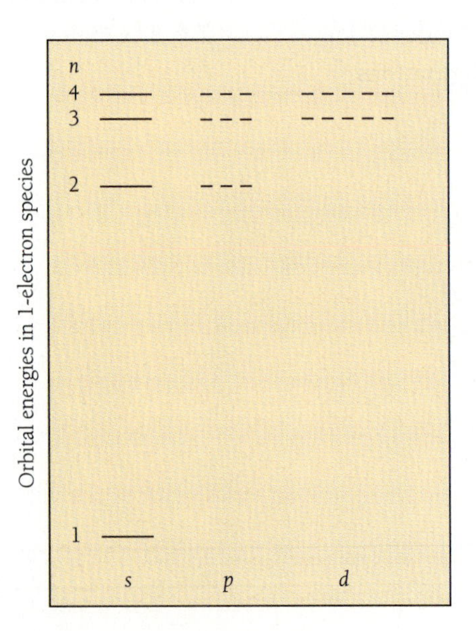

Figure 7.1 Orbitals in an energy level of a one-electron species are degenerate. All orbitals having the same value of the principal quantum number have the same energy.

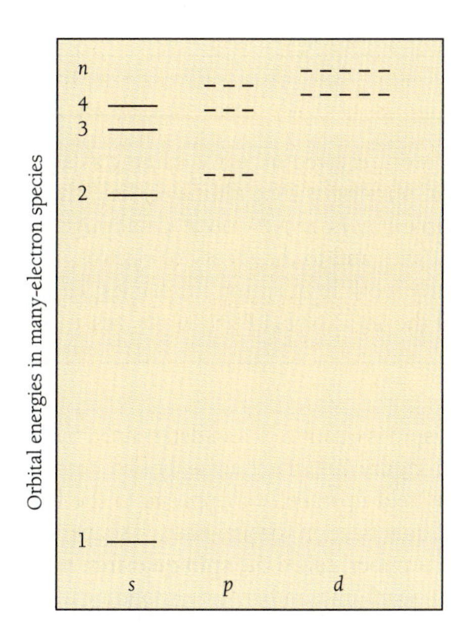

Figure 7.2 In atoms having more than one electron, orbital energy depends on the angular momentum quantum number (ℓ) as well as on the principal quantum number. For a given n, the energy increases as ℓ increases. No scale appears on the ordinate of this graph, because the energy of a given level depends on the number of protons in the nucleus and is different for each element. The general pattern of energy increasing with ℓ, however, is the same for almost all elements.

Orbital Size, Shape, and Orientation

Apart from their energy, many-electron orbitals do not differ significantly from hydrogen orbitals. They have the same characteristic shapes, as determined by the value of the angular momentum quantum number ℓ (Figures 6.20 for s, 6.23 for p, and 6.25 for d), and they have the same number and type of nodes, depending on the values of both n and ℓ. Within a given subshell the orbitals differ from one another in orientation in the same way as in the hydrogen atom (Figures 6.23 and 6.25). The overall size of the electron cloud increases as the principal quantum number increases, as it does in hydrogen.

Electron Spin

Thorough examination of the spectra of hydrogen and other atoms reveals some structural details that are too small to be seen in Figures 6.12 and 6.13 (page 230). This *fine structure* can be accounted for by introducing the postulate that an electron possesses an intrinsic angular momentum, a property called "spin." **Spin** means that the electron behaves as if it were spinning on its axis, much the same way that the earth spins daily on its axis. Not long after the discovery of electron spin, Paul Dirac showed that spin need not be introduced as an additional *postulate* of quantum theory, but that it appears explicitly in wave functions when the theory of relativity is combined with the Schrödinger equation.

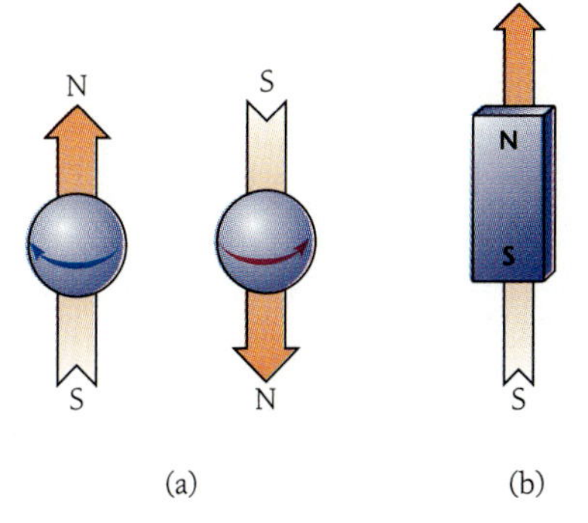

Figure 7.3 (a) An electron generates a magnetic field, whose direction is governed by the direction of the spin. (b) The magnetic field is weaker, but otherwise the same, as that created by a bar magnet.

A spinning, negatively charged electron generates a magnetic field, behaving like a tiny bar magnet. Figure 7.3 shows that the north (N) to south (S) direction of the magnetic field depends on the direction of the spin.

Spin is an intrinsic property of electrons, whether they are part of an atom or by themselves (as in a cathode ray tube). However, spin is observable only when the electron is in an *external* magnetic field (that is, a field other than the one created by

Paul A. M. Dirac (1902–1984)

Paul Dirac, the British son of a schoolmaster, received his early training as an engineer. He went on to study theoretical physics, and he ultimately held the same mathematics chair previously occupied by Isaac Newton at Cambridge. He made significant contributions to statistics and to the theory of electromagnetic radiation, in addition to his early work in quantum mechanics. He and Schrödinger were joint recipients of the 1933 Nobel Prize in physics.

The effect of electron spin can be observed in atomic spectra, because, among other reasons, the nucleus creates a weak magnetic field around itself. The topic of nuclear magnetic fields is not discussed in this text.

the electron itself). The direction of electron spin is quantized, and determined by the direction of the external field. As shown in Figure 7.4, there are only two possible orientations: spin is aligned either with the field (parallel) or opposite to the field (antiparallel). As with the other examples of quantization (orbital energy, shape, and orientation), there is an associated quantum number, called the **spin quantum number (m_s).** The two possible values of the spin quantum number, corresponding to the two allowed orientations, are $m_s = +1/2$ and $m_s = -1/2$. The two orientations are also called **spin up** and **spin down.** This "up/down" terminology does not imply any kind of directionality with respect to the earth, as it does in words like "uphill" and "downhill." It is simply a convenient way to refer to two opposite orientations. "Left" and "right" or "east" and "west" could also have been used.

The relative orientation of two electrons in the same vicinity is limited to two possibilities. If the two spins are *opposite,* the spins and their magnetic fields cancel

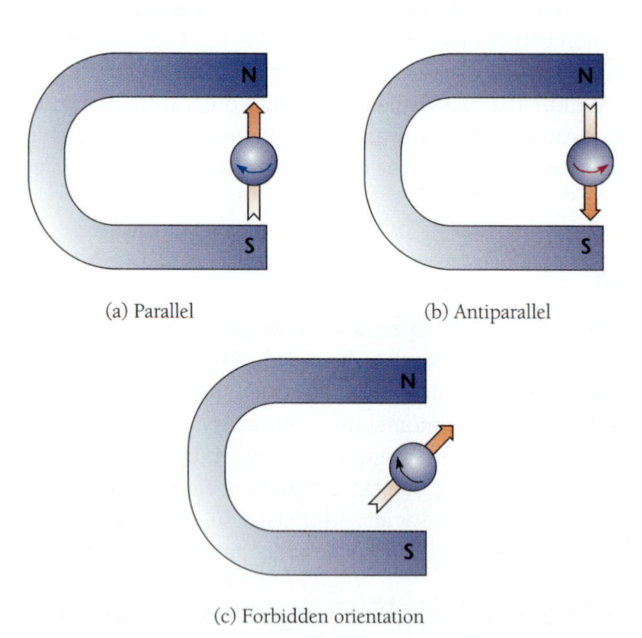

(a) Parallel

(b) Antiparallel

(c) Forbidden orientation

Figure 7.4 Electron spin is quantized, and can assume only two orientations with respect to an external magnetic field. The spin axis is in both cases aligned with the field, and the two possibilities are (a) parallel, and (b) antiparallel. All other orientations, such as (c), are forbidden.

each other. The net result is zero spin and zero magnetic field. In this condition, the electrons, and their spins, are said to be **paired.** If the two spins are aligned in the *same* direction, the net magnetic field is not zero: the electrons and the spins are said to be **unpaired.** Paired electrons have one spin up and one down. One has $m_s = +½$ while the other has $m_s = -½$, and the total spin is $(+½) + (-½) = 0$. Electron spin is represented by arrows, so that paired electrons are indicated by ⇅. Unpaired electrons have both spins up or both down, total spin $(+½) + (+½) = 1$ or $(-½) + (-½) = -1$, and can be represented by ↑↑ or ↓↓ (Figure 7.5).

It can be misleading to rely too heavily on analogies between microscopic particles and objects of macroscopic size. The picture of an electron "spinning on its axis" is such an analogy, but it is nevertheless useful because it is easily grasped. It is more accurate to say that an electron has both angular momentum and a magnetic moment, two properties that arise when a charged *macroscopic* object spins on its axis. These properties may well arise through other means in electrons.

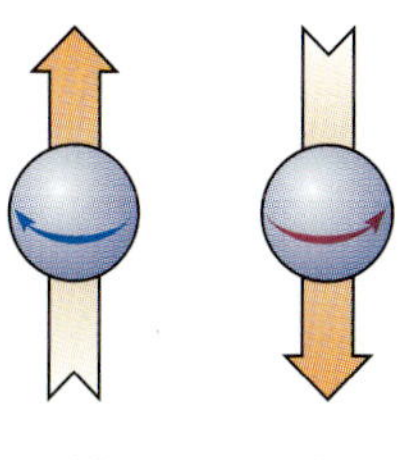

(a) Spins paired

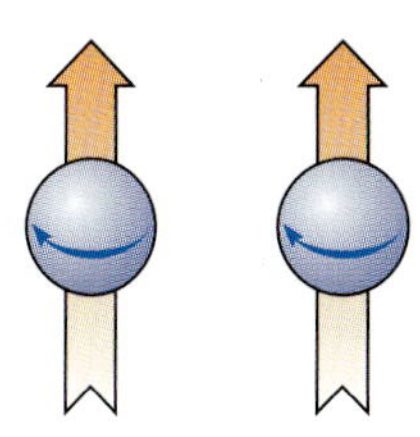

(b) Spins unpaired

Figure 7.5 Possible orientations for two electrons in the same vicinity are (a) spins paired (⇅), resulting in a net magnetic field of zero, and (b) spins unpaired (↑↑ or ↓↓), resulting in a nonzero net magnetic field.

Orbital Occupancy and the Exclusion Principle

The existence of electron spin means that four quantum numbers are needed to fully describe an electron in an atom. The first three (n, ℓ, and m_ℓ) describe a particular orbital. They are also used to describe an electron occupying that orbital. The fourth quantum number, m_s, describes the electron but not the orbital. The orbitals available to electrons in a many-electron atom are governed by a law of nature called the **Pauli exclusion principle,** which states that *no two electrons in an atom may have the same values of all four quantum numbers.* Each electron has a unique set of values $\{n, \ell, m_\ell, m_s\}$, not duplicated by any other electron, regardless of the total number of electrons in the atom.

Each orbital in an atom is described by a unique set of three quantum numbers $\{n, \ell, m_\ell\}$. For example, the 3*s* orbital has $n = 3$, $\ell = 0$, and $m_\ell = 0$, which we can abbreviate as $\{3, 0, 0\}$. If any one of these three quantum numbers has a different value, the set describes some orbital other than 3*s* (for instance, the set $\{3, 1, 0\}$ describes a 3*p* orbital). The two allowed spin states of an electron in the 3*s* orbital are described by $m_s = +½$ and $m_s = -½$, and the only possible sets of four quantum numbers for a 3*s* electron are $\{3, 0, 0, +½\}$ and $\{3, 0, 0, -½\}$. Since each electron in an atom must have a different set of quantum numbers, it follows directly from the Pauli exclusion principle that *no more than two electrons may occupy a given orbital.* Furthermore, *if an orbital is occupied by two electrons, their spins are paired.* The electrons in many-electron atoms are distributed so that each orbital is either (a) **filled,** or fully occupied by two electrons, *paired* (⇅); (b) **half-filled,** having one electron (↑ or ↓); or (c) **vacant** or **empty,** having no electrons.

Wolfgang Pauli (1900–1958)

Wolfgang Pauli was born in Austria and emigrated to the United States. He became a U.S. citizen during World War II. After the war he resumed his professorship in theoretical physics at the Federal Institute of Technology in Zurich, Switzerland. He was awarded the 1945 Nobel Prize in physics for his development of the exclusion principle.

7.2 ELECTRON CONFIGURATION OF THE ELEMENTS

The most important influence on the physical and chemical properties of elements is the way in which the electrons of the atoms are distributed in the various orbitals. This distribution is known as the **electron configuration** of the atom. Since the atoms of each element have a different number of electrons, the electron configuration of each element is different. In the following sections, we present a systematic method by which the electron configuration of most elements can be predicted. The method describes the electron configuration of a many-electron atom by listing the orbitals and their occupancy, in order of increasing energy. The procedure is called the **aufbau** method, or simply aufbau.

"Aufbau" is a German word meaning "a building up." It refers to the building up of a *description* of an atom, not to the building up of the atom itself. There is no implication that aufbau is an actual *process* by which atoms are formed.

The aufbau method leads to the electron configuration called the **ground state** of the atom, meaning the state or configuration of lowest energy. All other configurations have higher energy and are called **excited states.** Many different excited-state configurations are possible, but they last for only a short time following absorption of energy by an atom.

Aufbau and the First-, Second-, and Third-Period Elements

The ground-state electron configuration of many-electron atoms is governed by three factors related to orbital energy and electron spin. The first two factors, which are sufficient for understanding the configuration of the first five elements, are as follows.

1. Only the *lowest lying* (that is, having the lowest energy) available orbitals are occupied by electrons. Relative energies of orbitals are shown in Figure 7.2.
2. Orbital availability is determined by the Pauli principle, that is, an orbital already containing two electrons is full and not available for any more electrons. Note that each type of subshell has a different capacity:
    - **a.** an *s* subshell has 1 orbital and can hold up to 2 electrons;
    - **b.** a *p* subshell has 3 orbitals and can hold up to 6 electrons;
    - **c.** a *d* subshell has 5 orbitals and can hold up to 10 electrons;
    - **d.** an *f* subshell has 7 orbitals and can hold up to 14 electrons.

The aufbau method is the assignment of electrons to orbitals in a manner consistent with these factors. The ground-state electron configurations of elements of atomic number $Z = 1$ through 5 are as follows.

First Period

- $Z = 1$: Hydrogen has one electron, which occupies (and half fills) the orbital of lowest energy. According to Figure 7.1, the 1*s* orbital has the lowest energy. The number of electrons occupying an orbital is indicated by a superscript, and the electron configuration of hydrogen is written as H: $1s^1$.
- $Z = 2$: Helium has two electrons, both occupying the lowest orbital, 1*s*. The electron configuration is He: $1s^2$.

Configurations are often shown by **orbital diagrams,** which describe orbital occupancy pictorially. Orbitals are represented by boxes, and each electron is represented by an arrow. The quantum numbers *n* and ℓ of the orbitals are indicated below the boxes. These diagrams are used because they indicate electron spin as well as orbital occupancy. The orbital diagram for helium is He: [↑↓] (1s). Note that the arrows point in opposite directions, indicating the paired spins required by the Pauli exclusion principle.

Second Period

- $Z = 3$: Lithium's three electrons cannot all occupy the 1s orbital—the third electron must be in the next higher orbital, the 2s. The electron configuration and orbital diagram for lithium are as follows.

  Li: $1s^2 2s^1$ Li: [↑↓] [↑]
  1s 2s

- $Z = 4$: Beryllium has four electrons, which occupy the two lowest orbitals.

  Be: $1s^2 2s^2$ Be: [↑↓] [↑↓]
  1s 2s

- $Z = 5$: Four of boron's electrons fill the two lowest orbitals. The fifth must be accommodated in the next higher orbital, the 2p.

  B: $1s^2 2s^2 2p^1$ B: [↑↓] [↑↓] [↑][ ][ ]
  1s 2s 2p

Since the three orbitals in the p subshell have the same energy, and the two possible spins of the electron are not distinguishable from one another in the absence of a magnetic field, there is only one way for a p subshell to contain a single electron. That is, the six conceivable diagrams for the 2p subshell of boron, [↑][ ][ ] (2p), [ ][↑][ ] (2p), [ ][ ][↑] (2p), [↓][ ][ ] (2p), [ ][↓][ ] (2p), and [ ][ ][↓] (2p), are entirely equivalent. They refer to a single physical situation, namely a p subshell containing one electron.

Elements with $Z > 5$ have more than one electron in the 2p subshell. Carbon ($Z = 6$) has two 2p electrons. In order to decide which of the many possible orbital diagrams is correct for carbon, we need an additional criterion. The third factor governing electron configuration, which applies to all partly filled subshells, is known as Hund's rule.

3. **Hund's rule** states that the orbitals in a partly filled subshell are occupied in the way that *maximizes total electron spin,* that is, so as to maximize the sum of the spin quantum numbers m_s. Another way of stating Hund's rule is that, as far as possible, electrons in a subshell occupy separate orbitals and have parallel spins. It follows from Hund's rule that partly filled subshells are occupied in the way that maximizes the number of unpaired electrons.

Friedrich Hund (1896–) wrote several textbooks on theoretical physics, and made important contributions to our understanding of chemical bonding. His other interests have resulted in a book, *History of Physics Terminology* (1976) and more recent contributions (1990) to *Isis,* a journal for the history of science.

A partial explanation of Hund's rule is that electrons in different orbitals occupy different regions of the atom. They are therefore farther apart than if they were in the same orbital. Since the distance between them is greater, the repulsion energy of the negative charges is less, and configurations obeying Hund's rule therefore have lower energy than others.

According to Hund's rule, the configuration and orbital diagram for carbon are

$Z = 6$ C: $1s^2 2s^2 2p^2$ C: [↑↓] [↑↓] [↑][↑][ ]
1s 2s 2p

Note that the subshell diagrams [↑][↑][ ] ($2p$), [↑][ ][↑] ($2p$), [ ][↑][↑] ($2p$), [↓][↓][ ] ($2p$), [↓][ ][↓] ($2p$), and [ ][↓][↓] ($2p$) all refer to the *same* configuration, namely two electrons, spins parallel, in the p subshell. Ground-state configurations such as [↑][↓][ ] ($2p$) are prohibited by Hund's rule.

EXAMPLE 7.1 Electron Configuration and Orbital Diagram

For ground-state nitrogen atoms,

a. give the electron configuration, and

b. draw an orbital diagram.

Analysis

Target/Knowns: The two targets are clearly specified, and the implicit known information is the number of electrons, that is, the atomic number ($Z = 7$) of nitrogen.

Relationship: Electrons occupy the lowest available orbitals, whose energies increase in the order $1s < 2s < 2p$. The maximum occupancies of subshells are $s(2)$, $p(6)$, $d(10)$, and $f(14)$. Total spin is maximized in partly filled subshells.

a. Plan Beginning with $1s$, assign electrons to each subshell until it is full. Then move to the next higher subshell and continue until all seven electrons have been assigned.

Work Place two electrons in the $1s$ subshell, two in the $2s$ subshell (total of four electrons assigned so far), and the remaining three in the $2p$ subshell. The electron configuration of ground-state nitrogen is

$$\text{N: } 1s^2 2s^2 2p^3$$

b. Plan

1. For any *filled* subshells, put an up-down arrow pair in each of the subshell's boxes in the orbital diagram.
2. For a partly filled subshell, place "spin-up" arrows one-by-one into the orbitals. Pair the spins only if there are more electrons than orbitals.

Work

1. The filled-subshell portion of the configuration is $1s^2 2s^2$, for which the diagram is

   [↑↓] ($1s$) [↑↓] ($2s$)

2. The three remaining electrons singly occupy the three orbitals of the $2p$ subshell, and the complete orbital diagram for ground-state nitrogen is

   N: [↑↓] ($1s$) [↑↓] ($2s$) [↑][↑][↑] ($2p$)

Exercise

Give the electron configuration and orbital diagram of ground-state oxygen.

Answer: See below.

The $2p$ subshell can accommodate a total of six electrons, two in each of the orbitals. The elements $Z = 7$ through $Z = 10$ have the following electron configurations.

		Orbital diagram	1s	2s	2p
$Z = 7$	N: $1s^2 2s^2 2p^3$	N:	[↑↓]	[↑↓]	[↑][↑][↑]
$Z = 8$	O: $1s^2 2s^2 2p^4$	O:	[↑↓]	[↑↓]	[↑↓][↑][↑]
$Z = 9$	F: $1s^2 2s^2 2p^5$	F:	[↑↓]	[↑↓]	[↑↓][↑↓][↑]
$Z = 10$	Ne: $1s^2 2s^2 2p^6$	Ne:	[↑↓]	[↑↓]	[↑↓][↑↓][↑↓]

In neon, the final member of the second period, the orbitals in the $n = 1$ and $n = 2$ shells are filled. This configuration ($1s^2 2s^2 2p^6$) is called the **neon core.** Orbitals increase in size as the principal quantum number increases, so that electrons with $n = 1$ or 2 are on the average closer to the nucleus than those with larger values of n. Thus, the neon *core* is part of the inner electron cloud of all elements heavier than neon. It is common to represent this as [Ne] when writing electron configurations or orbital diagrams for the elements sodium through chlorine. For example, the configuration of sodium can be written as either $1s^2 2s^2 2p^6 3s^1$ or $[Ne]3s^1$. Either of the corresponding orbital diagrams [↑↓] (1s) [↑↓] (2s) [↑↓][↑↓][↑↓] (2p) [↑] (3s) or [Ne][↑] (3s) may be used.

The **noble-gas core** notation is sometimes used for second-period elements as well. The **helium core** is $1s^2$, and the corresponding notation for nitrogen, for instance, is $[He]2s^2 2p^3$.

One advantage of the core notation is that it clearly distinguishes between inner and outer electrons. Electrons in the core, called **core electrons,** are all paired, and all occupy completely filled subshells. Core electrons are also called **inner electrons.** Electrons in the highest occupied level, outside the core, are called **outer electrons.**

The notation [He] is actually *longer* than $1s^2$, so it is not an *abbreviation.* Instead, its purpose is to make a clear distinction between core electrons and outer electrons.

Third Period

The aufbau method gives the following electron configurations for the third-period elements.

		Orbital diagram	1s	2s	2p	3s	3p
$Z = 11$	Na: $1s^2 2s^2 2p^6 3s^1$	Na:	[↑↓]	[↑↓]	[↑↓][↑↓][↑↓]	[↑]	[ ][ ][ ]
$Z = 12$	Mg: $1s^2 2s^2 2p^6 3s^2$	Mg:	[↑↓]	[↑↓]	[↑↓][↑↓][↑↓]	[↑↓]	[ ][ ][ ]
$Z = 13$	Al: $1s^2 2s^2 2p^6 3s^2 3p^1$	Al:	[↑↓]	[↑↓]	[↑↓][↑↓][↑↓]	[↑↓]	[↑][ ][ ]

			1s	2s	2p	3s	3p
$Z = 14$	Si: $1s^22s^22p^63s^23p^2$	Si:	[↑↓]	[↑↓]	[↑↓][↑↓][↑↓]	[↑↓]	[↑][↑][ ]
$Z = 15$	P: $1s^22s^22p^63s^23p^3$	P:	[↑↓]	[↑↓]	[↑↓][↑↓][↑↓]	[↑↓]	[↑][↑][↑]
$Z = 16$	S: $1s^22s^22p^63s^23p^4$	S:	[↑↓]	[↑↓]	[↑↓][↑↓][↑↓]	[↑↓]	[↑↓][↑][↑]
$Z = 17$	Cl: $1s^22s^22p^63s^23p^5$	Cl:	[↑↓]	[↑↓]	[↑↓][↑↓][↑↓]	[↑↓]	[↑↓][↑↓][↑]
$Z = 18$	Ar: $1s^22s^22p^63s^23p^6$	Ar:	[↑↓]	[↑↓]	[↑↓][↑↓][↑↓]	[↑↓]	[↑↓][↑↓][↑↓]

EXAMPLE 7.2 Quantum Numbers in Ground-State Atoms

How many electrons have quantum number $\ell = 0$ in ground-state chlorine?

Analysis

Target: The total number of electrons with $\ell = 0$, that is, the total number of electrons in all the *s* subshells.

Knowns: The name of the element, and therefore its atomic number $Z = 17$.

Relationship: The aufbau principle provides the electron configuration so that the number of *s* electrons can be counted. (The answer can also be obtained by application of the quantum number rules. See Example 6.6, page 247.)

Plan

1. Write the electron configuration.
2. Count the electrons in subshells with $\ell = 0$ (*s* electrons).

Work Following the procedure of Example 7.1 gives $1s^22s^22p^63s^23p^5$ for the ground-state configuration of an atom with 17 electrons. There are two *s* electrons in each of the shells with principal quantum number 1, 2, and 3, so the total number of electrons with $\ell = 0$ is six.

Exercise

How many electrons with $\ell = 1$ are there in the ground state of chlorine?

Answer: 11.

Electron Configuration of the Heavier Elements

Prediction of ground-state electron configurations by aufbau requires knowledge of the relative energies of subshells, which determines the order in which the subshells are to be filled. So far we have obtained this information from Figure 7.2, but there are several other schemes by which energy-level ordering can be recalled. We could simply memorize the order (1*s*,2*s*,2*p*,3*s*,3*p*. . .), which is simple enough in the first three periods. However, the filling order is more complicated in the heavier elements.

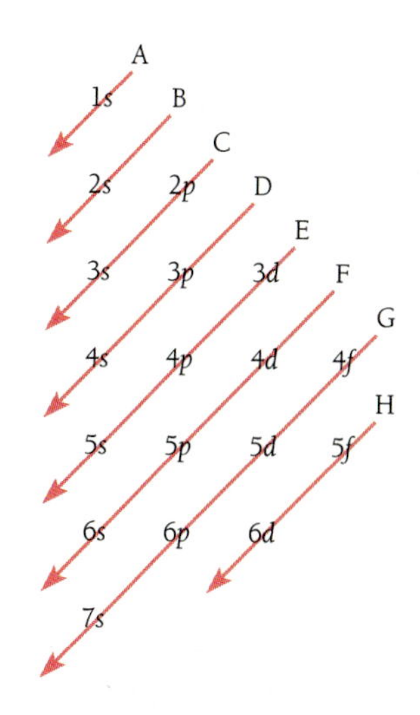

Figure 7.6 When the symbols for the electron subshells of many-electron atoms are arranged in this pattern, the arrows beginning at A, B, C, and so on, pass successively through the orbitals in aufbau filling order. A few atoms, discussed in the text, do not follow this pattern.

Energy Level Order

Close examination of Figure 7.2 shows that the energy of the 4*s* subshell is slightly less than that of the 3*d* subshell. This means that, in the aufbau scheme for fourth-period elements, the 4*s* orbital must be filled before any electrons can be assigned to the 3*d* orbitals. Figure 7.6 is a mnemonic device that gives the aufbau filling order for elements in *all* periods. The arrows labeled A, B, C, and so on, pass successively through the subshell symbols in the correct aufbau filling order. The result, obtained in this way from the diagram, is 1*s*,2*s*,2*p*,3*s*,3*p*,4*s*,3*d*,4*p*,5*s* and so on. Note that this scheme correctly shows that the 4*s* level fills before the 3*d* level. As we shall see, there are a few elements that do not follow this pattern.

An additional method is to use the periodic table. As shown in Figure 7.7, the periodic table can be subdivided into blocks corresponding to the type of subshell to which aufbau assigns the last electron of an element. In Groups 1A and 2A, the last electron is assigned to an *s* subshell. In Groups 3A through 8A, the last electron is assigned to a *p* subshell (helium is an exception). The assignment is to a *d* subshell in the "B" groups, and to an *f* subshell in the lanthanides and actinides.

Fourth Period

The electron configurations in the fourth period are given on page 274. The notation [Ar] represents the configuration of the core electrons, $1s^2 2s^2 2p^6 3s^2 3p^6$. The configurations of the two elements marked with an asterisk (Cr and Cu) do not follow the aufbau rules.

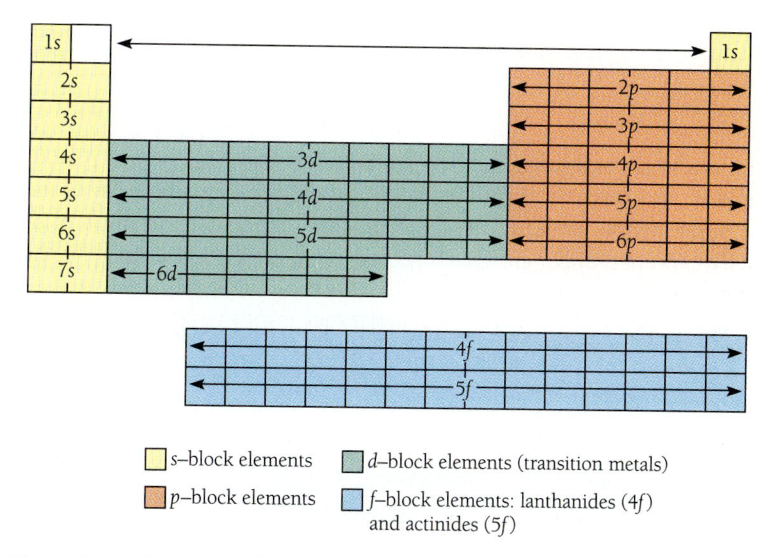

Figure 7.7 Electron configurations and the block structure of the periodic table. In the aufbau method, the final electron is assigned to the subshells listed in the blocks.

Atomic Number	Element Symbol	Ground-State Configuration	Orbital Diagram 4s	3d	4p
19	K	[Ar]$4s^1$	[Ar] [↑]	[ ][ ][ ][ ][ ]	[ ][ ][ ]
20	Ca	[Ar]$4s^2$	[Ar] [↑↓]	[ ][ ][ ][ ][ ]	[ ][ ][ ]
21	Sc	[Ar]$4s^2 3d^1$	[Ar] [↑↓]	[↑][ ][ ][ ][ ]	[ ][ ][ ]
22	Ti	[Ar]$4s^2 3d^2$	[Ar] [↑↓]	[↑][↑][ ][ ][ ]	[ ][ ][ ]
23	V	[Ar]$4s^2 3d^3$	[Ar] [↑↓]	[↑][↑][↑][ ][ ]	[ ][ ][ ]
24	Cr*	[Ar]$4s^1 3d^5$	[Ar] [↑]	[↑][↑][↑][↑][↑]	[ ][ ][ ]
25	Mn	[Ar]$4s^2 3d^5$	[Ar] [↑↓]	[↑][↑][↑][↑][↑]	[ ][ ][ ]
26	Fe	[Ar]$4s^2 3d^6$	[Ar] [↑↓]	[↑↓][↑][↑][↑][↑]	[ ][ ][ ]
27	Co	[Ar]$4s^2 3d^7$	[Ar] [↑↓]	[↑↓][↑↓][↑][↑][↑]	[ ][ ][ ]
28	Ni	[Ar]$4s^2 3d^8$	[Ar] [↑↓]	[↑↓][↑↓][↑↓][↑][↑]	[ ][ ][ ]
29	Cu*	[Ar]$4s^1 3d^{10}$	[Ar] [↑]	[↑↓][↑↓][↑↓][↑↓][↑↓]	[ ][ ][ ]
30	Zn	[Ar]$4s^2 3d^{10}$	[Ar] [↑↓]	[↑↓][↑↓][↑↓][↑↓][↑↓]	[ ][ ][ ]
31	Ga	[Ar]$4s^2 3d^{10} 4p^1$	[Ar] [↑↓]	[↑↓][↑↓][↑↓][↑↓][↑↓]	[↑][ ][ ]
32	Ge	[Ar]$4s^2 3d^{10} 4p^2$	[Ar] [↑↓]	[↑↓][↑↓][↑↓][↑↓][↑↓]	[↑][↑][ ]
33	As	[Ar]$4s^2 3d^{10} 4p^3$	[Ar] [↑↓]	[↑↓][↑↓][↑↓][↑↓][↑↓]	[↑][↑][↑]
34	Se	[Ar]$4s^2 3d^{10} 4p^4$	[Ar] [↑↓]	[↑↓][↑↓][↑↓][↑↓][↑↓]	[↑↓][↑][↑]
35	Br	[Ar]$4s^2 3d^{10} 4p^5$	[Ar] [↑↓]	[↑↓][↑↓][↑↓][↑↓][↑↓]	[↑↓][↑↓][↑]
36	Kr	[Ar]$4s^2 3d^{10} 4p^6$	[Ar] [↑↓]	[↑↓][↑↓][↑↓][↑↓][↑↓]	[↑↓][↑↓][↑↓]

The electron configuration of an atom in the fourth or higher period can be described in two ways. In the above list, subshells are given in order of increasing energy. Some people prefer to list them in shell order, with subshells of the same principal quantum number grouped together: $1s$, $2s2p$, $3s3p3d$, $4s4p$. . . In the case of iron, this choice leads to

Fe: [Ar]$3d^6 4s^2$ and [Ar] [↑↓][↑][↑][↑][↑] (3d) [↑↓] (4s) [ ][ ][ ] (4p)

EXAMPLE 7.3 Forbidden Configurations

Two of the following electron configurations are impossible. Identify them and explain why: $1s^2 1p^6 2s^2 2p^6 3s^1$; $1s^1 2s^2 2p^6$; $1s^2 2s^2 2p^7 3s^2 4s^1$.

Analysis Identification of an impossible configuration requires recognition that a rule has been broken. Examine each configuration carefully, keeping in mind the Pauli principle and the rule limiting the quantum number ℓ to a value less than n. Since m_ℓ and m_s are not indicated in the notation, the rules on their values need not be considered.

Work $1s^2 1p^6 2s^2 2p^6 3s^1$: According to the quantum number rules, ℓ must be less than n. Since $\ell = 1$ for a p subshell, there is no such thing as a $1p$ orbital. This configuration is impossible.

$1s^1 2s^2 2p^6$: Since there is only one electron in the $1s$ orbital, this configuration does not follow the aufbau principle of occupying the lowest energy orbitals. However, it violates neither the Pauli principle nor the $\ell < n$ rule. It is thus a possible configuration, representing one of the *excited states* of the fluorine atom.

$1s^2 2s^2 2p^7 3s^2 4s^1$: According to the Pauli principle, no more than six electrons can occupy a p subshell. This configuration is impossible.

Exercise

Explain why the configuration $1s^2 2s^2 2p^6 3s^3 3p^5$ is impossible.

Answer: An s subshell has only one orbital, which can hold at most two electrons. The configuration $3s^3$ does not exist.

Some Exceptions

The energy-level ordering of Figures 7.2 and 7.6 is generally valid, but there are some elements whose electron configurations do not fit this scheme in all detail. Subshell energies depend on the nuclear charge, that is, on the atomic number, and for atomic numbers 21 through 30 the $3d$ and $4s$ levels are very close in energy. The relative energies also depend on whether the levels are occupied by electrons. The result is that the fourth period contains two elements, Cr and Cu, whose electron configuration is not what is expected on the basis of Figure 7.6. Each of these elements has only one electron in the $4s$ orbital.

The deviation of the ground-state configurations of Cr and Cu from the normal pattern illustrates an additional principle, namely that the presence of a half-filled (Cr) or completely filled (Cu) d subshell gives a measure of additional stability to an atom.

The Transition Elements

The elements in the "B" groups of the periodic table are known as **transition elements** or **transition metals.** The classification is based primarily on chemical properties. Chemical properties are determined by electron configuration, and most of the transition elements are characterized by a *partly* filled d or f subshell in the ground state. This feature is not suitable as a definition for the transition elements, however, because it would exclude the Group 1B and 2B elements (whose outermost d subshells are *completely* filled). Once the block structure of the periodic table (Figure 7.7) is understood, the transition elements are defined most simply as the d-block and f-block elements.

Some chemists do not regard the elements of Group 2B as transition metals, and choose instead to treat them as a class by themselves.

The transition elements are strong, hard, high-melting and high-boiling metals that conduct heat and electricity well. Most are capable of forming more than one cation. Transition elements in the fourth period have one or more electrons in the $3d$ subshell, and are members of the *first transition series.* Those in the fifth period, with one or more $4d$ electrons, constitute the *second transition series.* The *third transition series* contains those sixth-period elements having one or more electrons in the $5d$ subshell. The d-block elements are the *main* transition elements, while the f-block elements constitute the **inner transition series.** The inner transition series are usually subdivided into the **lanthanides,** with partly filled $4f$ subshells, and the **actinides,** with partly filled $5f$ subshells.

The electron configurations in the fifth period are as follows. (The notation [Kr] represents the configuration of the core electrons, $1s^2 2s^2 2p^6 3s^2 3p^6 4s^2 3d^{10} 4p^6$.)

Atomic Number	Element Symbol	Ground-State Configuration	Orbital Diagram	5s	4d	5p
37	Rb	[Kr]$5s^1$	[Kr]	[↑]	[ ][ ][ ][ ][ ]	[ ][ ][ ]
38	Sr	[Kr]$5s^2$	[Kr]	[↑↓]	[ ][ ][ ][ ][ ]	[ ][ ][ ]
39	Y	[Kr]$5s^2 4d^1$	[Kr]	[↑↓]	[↑][ ][ ][ ][ ]	[ ][ ][ ]
40	Zr	[Kr]$5s^2 4d^2$	[Kr]	[↑↓]	[↑][↑][ ][ ][ ]	[ ][ ][ ]
41	Nb	[Kr]$5s^1 4d^4$	[Kr]	[↑]	[↑][↑][↑][↑][ ]	[ ][ ][ ]
42	Mo	[Kr]$5s^1 4d^5$	[Kr]	[↑]	[↑][↑][↑][↑][↑]	[ ][ ][ ]
43	Tc	[Kr]$5s^2 4d^5$	[Kr]	[↑↓]	[↑][↑][↑][↑][↑]	[ ][ ][ ]
44	Ru	[Kr]$5s^1 4d^7$	[Kr]	[↑]	[↑↓][↑↓][↑][↑][↑]	[ ][ ][ ]
45	Rh	[Kr]$5s^1 4d^8$	[Kr]	[↑]	[↑↓][↑↓][↑↓][↑][↑]	[ ][ ][ ]
46	Pd	[Kr]$5s^0 4d^{10}$	[Kr]	[ ]	[↑↓][↑↓][↑↓][↑↓][↑↓]	[ ][ ][ ]
47	Ag	[Kr]$5s^1 4d^{10}$	[Kr]	[↑]	[↑↓][↑↓][↑↓][↑↓][↑↓]	[ ][ ][ ]
48	Cd	[Kr]$5s^2 4d^{10}$	[Kr]	[↑↓]	[↑↓][↑↓][↑↓][↑↓][↑↓]	[ ][ ][ ]
49	In	[Kr]$5s^2 4d^{10} 5p^1$	[Kr]	[↑↓]	[↑↓][↑↓][↑↓][↑↓][↑↓]	[↑][ ][ ]
50	Sn	[Kr]$5s^2 4d^{10} 5p^2$	[Kr]	[↑↓]	[↑↓][↑↓][↑↓][↑↓][↑↓]	[↑][↑][ ]
51	Sb	[Kr]$5s^2 4d^{10} 5p^3$	[Kr]	[↑↓]	[↑↓][↑↓][↑↓][↑↓][↑↓]	[↑][↑][↑]
52	Te	[Kr]$5s^2 4d^{10} 5p^4$	[Kr]	[↑↓]	[↑↓][↑↓][↑↓][↑↓][↑↓]	[↑↓][↑][↑]
53	I	[Kr]$5s^2 4d^{10} 5p^5$	[Kr]	[↑↓]	[↑↓][↑↓][↑↓][↑↓][↑↓]	[↑↓][↑↓][↑]
54	Xe	[Kr]$5s^2 4d^{10} 5p^6$	[Kr]	[↑↓]	[↑↓][↑↓][↑↓][↑↓][↑↓]	[↑↓][↑↓][↑↓]

Note that there are more irregular configurations in the fifth period: of the ten elements Y through Cd of the second transition series, only Y, Zr, Tc, and Cd have the predicted configuration. Clearly, aufbau is not a reliable guide to outer-electron configuration among the transition elements.

EXAMPLE 7.4 Abbreviating with the Noble-Gas Core

Identify the noble-gas core of each of the following: Pt, Nd, Zr.

Analysis The relationship between target and knowns is implicit in the definition of the term "noble-gas core," namely the configuration of the noble gas immediately preceding the element in question.

Work Consultation of the periodic table reveals the answers as Pt, [Xe]; Nd, [Xe]; and Zr, [Kr].

The electron configurations of the remaining elements are given in Figure 7.13 (page 284). Although aufbau-based prediction of electron configurations in the transition series often fails, it is good practice to *try*—the attempt requires careful thought about the relevant factors.

EXAMPLE 7.5 Electron Configurations in the Transition Elements

Write the expected configurations of silver and hafnium, and suggest reasonable alternatives.

Analysis Aufbau supplies the *expected* configuration, while reasonable alternatives may be suggested by the stability of filled and half-filled subshells.

Ag: The rules of aufbau yield the configuration [Kr]$5s^2 4d^9$ for the element with 47 electrons. Knowing the stability of a filled *d* subshell, we might suggest the configuration [Kr]$5s^1 4d^{10}$ as an alternative. This suggestion (which is in fact correct) is supported by the fact that copper, in the same group as silver, has the corresponding outer-shell configuration—Cu:[Ar]$4s^1 3d^{10}$.

Hf: With atomic number 72, hafnium is in the sixth period and so has the [Xe] core of 54 electrons. All levels through 5*p* are filled. Figure 7.6 shows that the order of the subsequent levels is 6*s*, 4*f*, 5*d*, 6*p*. Since there are $(75 - 54) = 18$ electrons outside the [Xe] core, the aufbau prediction is Hf: [Xe]$6s^2 4f^{14} 5d^2$. Hafnium is beyond the lanthanide series, so it does not have a partly filled *f* subshell. $4f^{14}$ must be correct, and the only question is the distribution of the four remaining electrons in the 5*d* and 6*s* subshells. Five electrons are required to half-fill a *d* subshell, so no stability can be gained by hafnium in this way. The expected configuration is correct.

Exercise

Suggest possible configurations for Mo ($Z = 42$).

Answer: The correct configuration is given on page 276.

Electron Configuration of Ions

The aufbau method is not limited to atoms, but can be used to predict electron configurations in positive and negative ions as well. The ground-state configuration of a singly or doubly charged *anion* is obtained from that of the parent atom by adding one or two electrons to the next available orbital. The ground-state configurations of the cations of Groups 1A, 2A, and 3A elements are found by *removing* one, two, or three electrons, respectively, from the highest occupied orbitals of the parent atom.

Cations of the elements of the "B" groups do not follow this simple behavior. In the first place, they tend to form more than one cation—Cu^+ and Cu^{2+}, for example. Secondly, the electrons lost in the formation of cations are *not* those that were assigned last to the atom in the aufbau method. Instead, the +1 and +2 ions of the fourth period "B" elements Sc through Zn are formed by loss of the 4*s* electrons from the neutral atom. Electrons are lost from the 3*d* subshell only if the element forms ions of charge greater than the number of 4*s* electrons in the neutral atom. For example, the configurations of iron and its two common cations are Fe:$[Ar]4s^23d^6$, Fe^{2+}:$[Ar]3d^6$, and Fe^{3+}:$[Ar]3d^5$. The configurations of copper and its two common cations are Cu:$[Ar]4s^13d^{10}$, Cu^+:$[Ar]3d^{10}$, and Cu^{2+}:$[Ar]3d^9$.

EXAMPLE 7.6 Configuration of Monatomic Ions

Write the ground-state electron configuration of Mg^{2+}.

Analysis

Target: An electron configuration.

Knowns: The atomic number (12) of the parent atom, and the charge (+2) of the cation.

Relationship: Compared to the atom, the ion has lost two electrons. Magnesium is an "A" group element, so the electrons are lost from the highest (last-filled) level.

Work The configuration of the magnesium atom, expressed as an orbital diagram, is

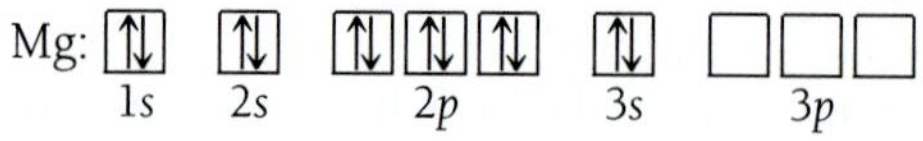

Mg:	[↑↓]	[↑↓]	[↑↓][↑↓][↑↓]	[↑↓]	[ ][ ][ ]
	1*s*	2*s*	2*p*	3*s*	3*p*

The highest occupied subshell is 3*s*, and removal of the two electrons from this subshell gives, as the ground-state configuration of the magnesium ion,

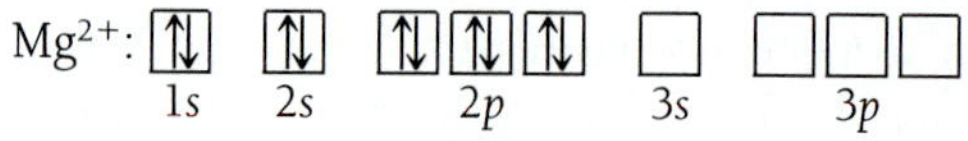

Mg^{2+}:	[↑↓]	[↑↓]	[↑↓][↑↓][↑↓]	[ ]	[ ][ ][ ]
	1*s*	2*s*	2*p*	3*s*	3*p*

Exercise

Write the ground-state configuration of the fluoride ion, F^-.

Answer: $1s^22s^22p^6$.

7.3 EXPERIMENTAL EVIDENCE FOR ELECTRON CONFIGURATIONS

Our knowledge of the energies of the orbitals in many-electron atoms comes from several different sources. Energies may be calculated directly from the Schrödinger equation, but accurate calculations, especially for larger elements, require a great deal of computer time. Emission spectra of the element yield information about energy *differences* among levels, but spectra of the heavier elements contain many lines and must be interpreted very carefully. *Photoelectron spectroscopy* offers a means for simple and direct measurements of orbital energies.

Photoelectron Spectroscopy

Photoelectron spectroscopy (PES) is related to the photoelectric effect described in Chapter 6, but there is an important difference. In the photoelectric effect the absorption of light causes the ejection of an electron from a metal *surface,* while in PES the absorption of a photon by a single *atom* causes the ejection of an electron. Either x-ray or short-wavelength ultraviolet light is used in PES, so there is more than enough energy to eject an electron from the atom. The excess energy is carried away by the electron as kinetic energy, which can be measured accurately. The difference between the energy $h\nu$ carried in by the photon and the kinetic energy carried away by the electron is the energy required to eject an electron *from a particular energy level.* The actual energy of the level is the negative of this ejection energy. These relationships are shown in Figure 7.8, and a schematic diagram of a photoelectron spectrometer is shown in Figure 7.9.

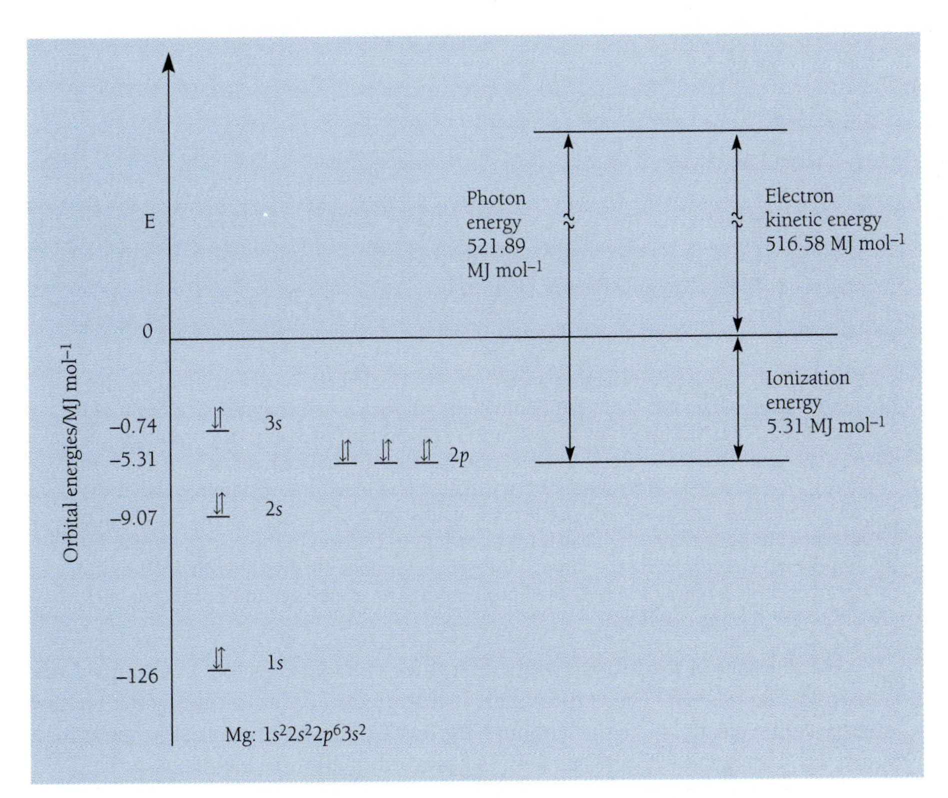

Figure 7.8 When an x-ray photon carrying energy $h\nu$ = 522 MJ mol^{-1} is absorbed by a ground-state magnesium atom, an electron can be ejected from any one of the occupied orbitals. The energy in excess of the 5.31 MJ mol^{-1} required to remove an electron from the 2*p* subshell, for instance, appears as kinetic energy of the ejected electron.

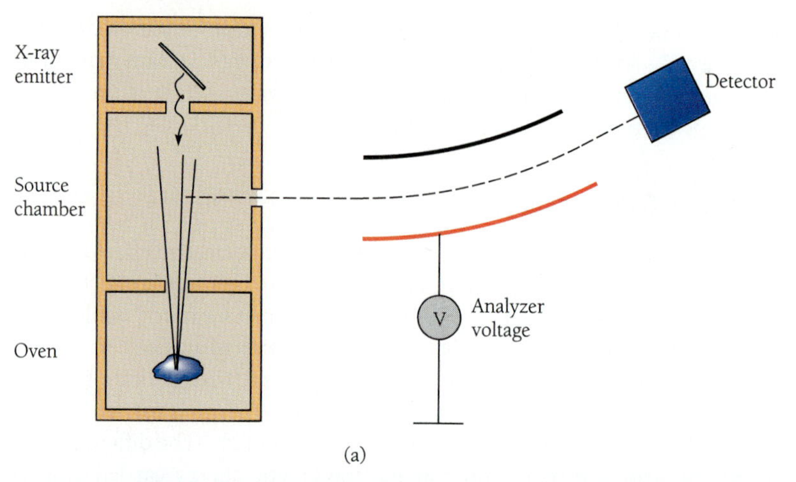

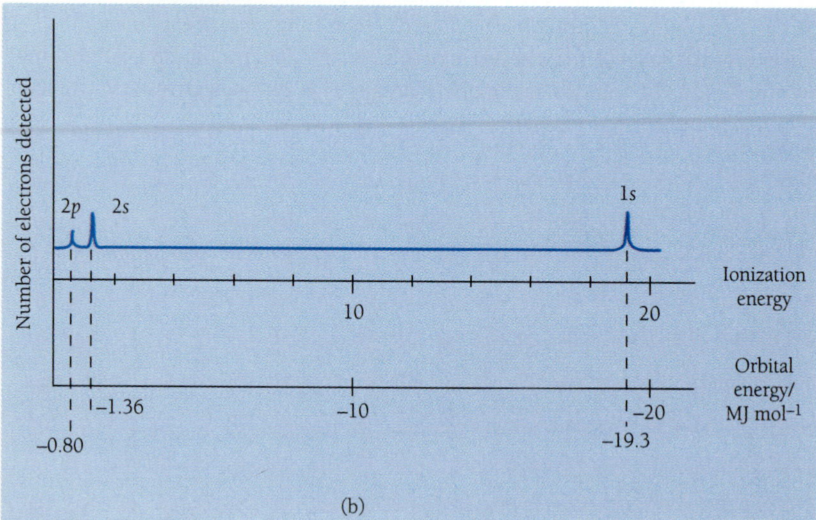

Figure 7.9 (a) A photoelectron spectrometer. An element is vaporized in a high temperature oven, and the vapor is directed into the source chamber as a beam of individual atoms. Irradiation by x-rays causes the ejection of an electron from one of the energy levels of the atom. The electron passes into the analyzer, where its kinetic energy is measured. Many atoms are irradiated, and electrons arising from all of the occupied levels are observed. (b) Typical results for a sample of boron. The number of electrons received at the detector is shown as a function of the energy required to remove an electron from the atom. Three peaks or lines clearly show the presence of three occupied energy levels in the boron atom. The ionization and orbital energies are given in units of MJ mol^{-1}.

The energies of the orbitals in the first three metallic elements in Group 1A are shown in Figure 7.10. The vertical scale is distorted in order to display the relative energies more clearly. Note that although the levels differ widely in energy, the aufbau *order,* $1s < 2s < 2p < 3s < 3p < 4s$, is observed in all three elements.

Magnetic Behavior of Atoms

One of the physical properties of materials is their behavior in a magnetic field. It is a familiar fact that steel paper clips can be picked up by a magnet, while copper pennies cannot (Figure 7.11). Three types of behavior can be distinguished, and all are related to electron spin.

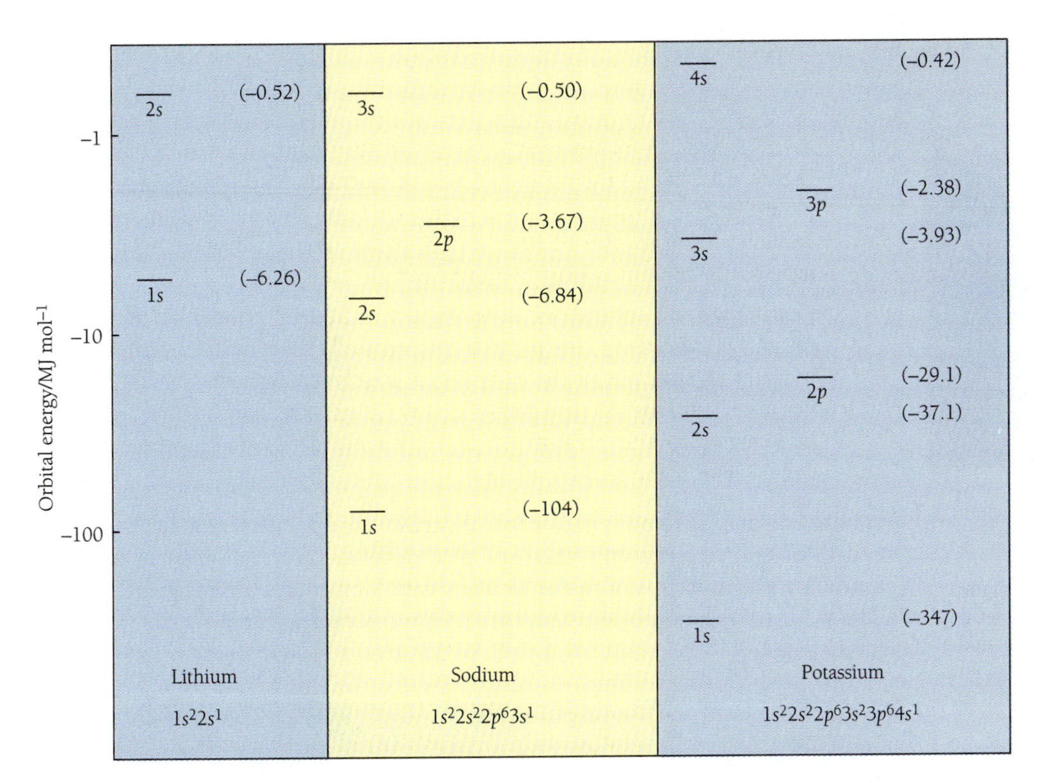

Figure 7.10 Energy levels in lithium, sodium, and potassium, as determined by photoelectron spectroscopy. The vertical scale is distorted, but retains the correct ordering of the levels. (The vertical scale is in fact proportional to the logarithm of energy. Logarithmic plots are frequently used when the range of the number to be plotted is large.) The numbers in parentheses give the measured values, in MJ mol^{-1}, of the orbital energies.

1. **Diamagnetic** substances are very weakly repelled by a magnet. All electrons in such substances are paired, so the substance has no magnetic field arising from unpaired spins.
2. **Paramagnetic** substances which contain at least one unpaired electron are weakly attracted by a magnet. The effect occurs regardless of the physical state or chemical combination—elements and compounds, solids, liquids, and

Figure 7.11 A mixture of copper and iron can be separated easily with a magnet. Iron is strongly attracted by a magnet, and copper is not.

Figure 7.12 Paramagnetic substances are attracted by a magnetic field, whether they are solids, liquids, or gases. Oxygen, shown here in liquid form at −188 °C, is a paramagnetic substance that can be held by a sufficiently strong magnet. The force of attraction is much weaker than that experienced by more familiar *ferromagnetic* substances such as steel.

gases, can all be paramagnetic, provided they contain unpaired electrons. Figure 7.12 shows the effect of a strong magnetic field on liquid oxygen, a paramagnetic substance. All substances whose atoms, ions, or molecules contain an *odd number* of electrons are paramagnetic, since such substances necessarily contain at least one unpaired electron.

3. **Ferromagnetic** substances are *strongly* attracted by a magnetic field. This property is also due to the presence of unpaired electrons in the material, but in ferromagnetic metals the attraction is greatly strengthened by the *cooperative behavior* of the electron spins of many atoms, spread throughout relatively large regions in the metal structure. Only a few metals are ferromagnetic: Fe, Co, Ni, and some of their alloys.

Paramagnetism is an important clue to the electron configuration of a substance. The strength of the attraction in a magnetic field depends on the number of unpaired electrons in a formula unit of a substance, and measurements of paramagnetic strength allow the determination of that number. Such measurements confirm the validity of Hund's rule.

EXAMPLE 7.7 Paramagnetism

Which of the fourth-period atoms are paramagnetic in the ground state?

Analysis

Target: Identity of atoms that are weakly attracted by a magnet.

Knowns: Electron configurations of all fourth-period atoms are implicit knowns.

Relationship: Paramagnetic species have at least one unpaired electron.

Work Examination of the orbital diagrams shows that, in the fourth period, all elements but Ca, Zn, and Kr have unpaired electrons.

Exercise

Name the paramagnetic elements of the third period.

Answer: All elements but Mg and Ar are paramagnetic.

7.4 THE PERIODIC TABLE OF THE ELEMENTS

The arrangement of elements in the periodic table is closely connected with electron configuration. For instance, all the elements of Group 1A have the outer-electron configuration ns^1, where n is the value of the principal quantum number. However, the table was developed long before the concept of electron configurations was known.

Early Form

One of the earliest versions of the periodic table was published by Dmitry Mendeleev in 1872. He arranged the elements in order of increasing molar mass, and further placed them into groups having similar chemical and physical properties.

Dmitry Mendeleev (1834–1907)

Dmitry Mendeleev was for many years Professor of Inorganic Chemistry at St. Petersburg University in Russia. He developed his ideas about the periodicity of chemical properties while writing what would become a widely used textbook of chemistry. Later, he became interested in the utilization of natural resources, chiefly salt, coal, and oil.

Others had previously grouped elements by molar mass, but Mendeleev recognized that such an ordering of the known elements led to groupings having *dis*similar properties. He resolved this dilemma by assuming the existence of as-yet-undiscovered elements. For instance, titanium forms a *dioxide* TiO_2, as do carbon and silicon. Boron and aluminum form oxides with M_2O_3 stoichiometry, so by *assuming* the existence of an element with molar mass ≈ 44, he was able to group Ti with C and Si rather than with B and Al. Furthermore, knowing in which group the unknown element must fit, he was able to predict some of its properties, such as melting point, density, and stoichiometry of its compounds. Scandium (molar mass 45) was discovered seven years later, and it had the predicted properties. Mendeleev left other gaps in his table, wherever necessary to maintain the similarity of chemical properties within a group. The missing elements were soon discovered: gallium (1875), germanium (1886), and so on.

The Modern Periodic Table

Mendeleev's gaps are now filled in, and the fundamental importance of electron configuration is now recognized. Also, we now think of the table as a list in order of atomic *number* rather than molar mass. A modern table is given in Figure 7.13. Molar masses, densities, and other properties are often added to such tables, depending on their intended use, but here we limit the data to the atomic number and the all-important outer-shell electron configuration. Modern American usage designates the groups in the periodic table with a numeral and a letter, while IUPAC recommends using only the numerals 1 through 18. There is some resistance to the IUPAC numbering, and it is likely that both systems will coexist in the future. IUPAC has also recommended the Latin-based names unnilquadium (one-zero-four), and so on, for elements heavier than Lawrencium. This was done largely to quell international disputes as to where these elements were first discovered, and by whom.

As described in Section 7.2 and Figure 7.7, certain blocks of elements correspond to the subshells that are filled as the atomic number increases. These blocks are named as follows.

- *Main-Group* or *Representative Elements: s* and *p* subshells fill. These are the "A" groups.
- *Transition Metals* or *Transition Elements: d* subshells fill. These are the "B" groups.
- *Inner Transition Elements* (also called the *lanthanides* and *actinides*): *f* subshells fill.

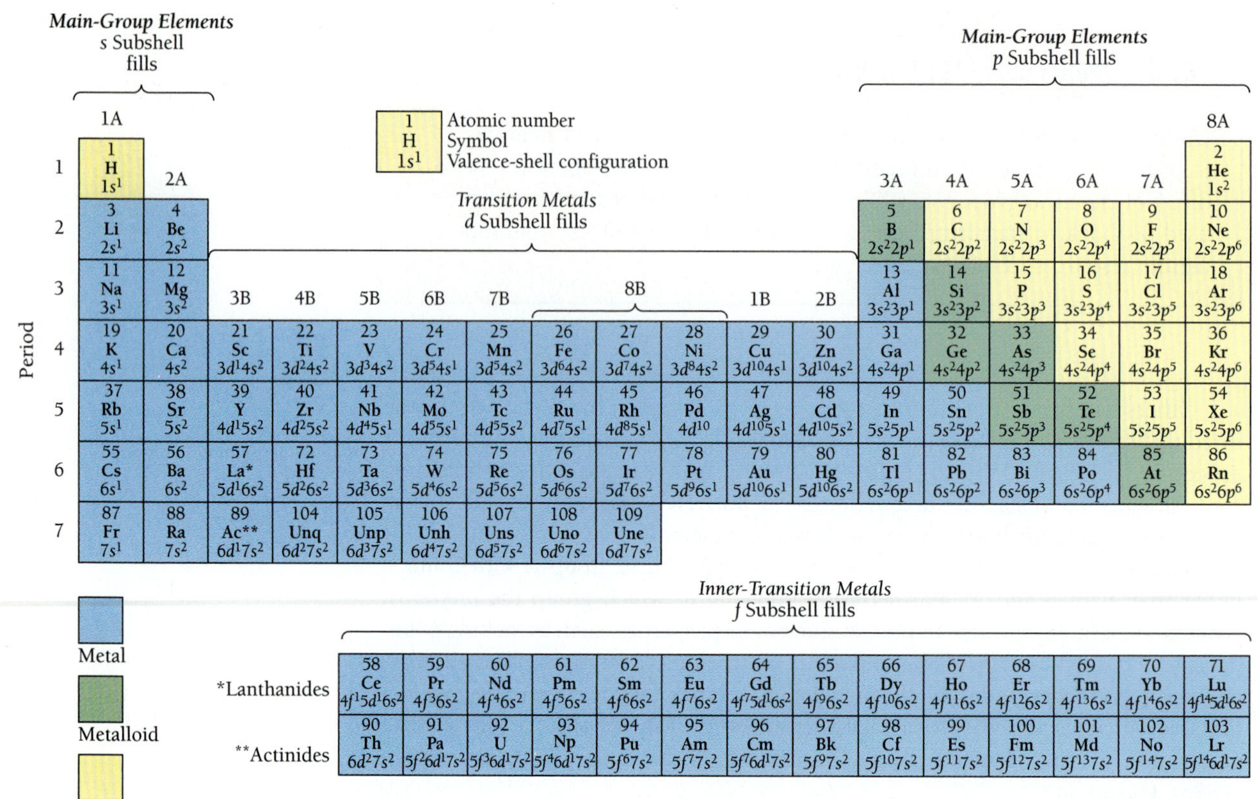

Period	1A	2A	3B	4B	5B	6B	7B	8B	8B	8B	1B	2B	3A	4A	5A	6A	7A	8A
1	1 H $1s^1$																	2 He $1s^2$
2	3 Li $2s^1$	4 Be $2s^2$											5 B $2s^22p^1$	6 C $2s^22p^2$	7 N $2s^22p^3$	8 O $2s^22p^4$	9 F $2s^22p^5$	10 Ne $2s^22p^6$
3	11 Na $3s^1$	12 Mg $3s^2$											13 Al $3s^23p^1$	14 Si $3s^23p^2$	15 P $3s^23p^3$	16 S $3s^23p^4$	17 Cl $3s^23p^5$	18 Ar $3s^23p^6$
4	19 K $4s^1$	20 Ca $4s^2$	21 Sc $3d^14s^2$	22 Ti $3d^24s^2$	23 V $3d^34s^2$	24 Cr $3d^54s^1$	25 Mn $3d^54s^2$	26 Fe $3d^64s^2$	27 Co $3d^74s^2$	28 Ni $3d^84s^2$	29 Cu $3d^{10}4s^1$	30 Zn $3d^{10}4s^2$	31 Ga $4s^24p^1$	32 Ge $4s^24p^2$	33 As $4s^24p^3$	34 Se $4s^24p^4$	35 Br $4s^24p^5$	36 Kr $4s^24p^6$
5	37 Rb $5s^1$	38 Sr $5s^2$	39 Y $4d^15s^2$	40 Zr $4d^25s^2$	41 Nb $4d^45s^1$	42 Mo $4d^55s^1$	43 Tc $4d^55s^2$	44 Ru $4d^75s^1$	45 Rh $4d^85s^1$	46 Pd $4d^{10}$	47 Ag $4d^{10}5s^1$	48 Cd $4d^{10}5s^2$	49 In $5s^25p^1$	50 Sn $5s^25p^2$	51 Sb $5s^25p^3$	52 Te $5s^25p^4$	53 I $5s^25p^5$	54 Xe $5s^25p^6$
6	55 Cs $6s^1$	56 Ba $6s^2$	57 La* $5d^16s^2$	72 Hf $5d^26s^2$	73 Ta $5d^36s^2$	74 W $5d^46s^2$	75 Re $5d^56s^2$	76 Os $5d^66s^2$	77 Ir $5d^76s^2$	78 Pt $5d^96s^1$	79 Au $5d^{10}6s^1$	80 Hg $5d^{10}6s^2$	81 Tl $6s^26p^1$	82 Pb $6s^26p^2$	83 Bi $6s^26p^3$	84 Po $6s^26p^4$	85 At $6s^26p^5$	86 Rn $6s^26p^6$
7	87 Fr $7s^1$	88 Ra $7s^2$	89 Ac** $6d^17s^2$	104 Unq $6d^27s^2$	105 Unp $6d^37s^2$	106 Unh $6d^47s^2$	107 Uns $6d^57s^2$	108 Uno $6d^67s^2$	109 Une $6d^77s^2$									

*Lanthanides	58 Ce $4f^15d^16s^2$	59 Pr $4f^36s^2$	60 Nd $4f^46s^2$	61 Pm $4f^56s^2$	62 Sm $4f^66s^2$	63 Eu $4f^76s^2$	64 Gd $4f^75d^16s^2$	65 Tb $4f^96s^2$	66 Dy $4f^{10}6s^2$	67 Ho $4f^{11}6s^2$	68 Er $4f^{12}6s^2$	69 Tm $4f^{13}6s^2$	70 Yb $4f^{14}6s^2$	71 Lu $4f^{14}5d^16s^2$
**Actinides	90 Th $6d^27s^2$	91 Pa $5f^26d^17s^2$	92 U $5f^36d^17s^2$	93 Np $5f^46d^17s^2$	94 Pu $5f^67s^2$	95 Am $5f^77s^2$	96 Cm $5f^76d^17s^2$	97 Bk $5f^97s^2$	98 Cf $5f^{10}7s^2$	99 Es $5f^{11}7s^2$	100 Fm $5f^{12}7s^2$	101 Md $5f^{13}7s^2$	102 No $5f^{14}7s^2$	103 Lr $5f^{14}6d^17s^2$

Figure 7.13 The modern form of the periodic table showing the outer-shell electron configurations of the elements. The color shadings identify the elements as metals, metalloids, and nonmetals. The group identifications 1A, 2A, 2B, ... 7A, 8A correspond to modern American usage. IUPAC has recommended the adoption of a scheme in which the groups are simply numbered 1–18.

7.5 PERIODICITY OF ELEMENT PROPERTIES

The **periodic law** is the formal statement of the relationship between atomic number and the properties of elements. It states that *when the elements are arranged in order of atomic number, there is a periodic recurrence of similar properties.* In modern usage the term **periodicity** has come to refer more generally to the trends in physical and chemical properties that occur in the groups and periods of the periodic table. We have already discussed some of these trends, for example the boiling temperatures of the noble gases (Section 2.9). In the following sections we discuss the trends in the size of atoms and ions, the energy involved in the loss or gain of an electron, and the extent to which the elements exhibit chemical properties characteristic of metals.

Figure 7.14 Large blocks of elements are named according to which subshells of the outermost shell are filled or partly filled. The chemistry of the main-group elements is dominated by *s* and *p* electrons, and that of the transition elements is dominated by *d* or *f* electrons.

Molar Volume

One of the first discussions of periodicity was offered by Lothar Meyer, who used the physical property of molar volume to illustrate the periodic law. The molar volume (mL mol^{-1}) of an element is found by dividing the density (g mL^{-1}) of the solid into the molar mass (g mol^{-1}). The regular recurrence of high volume elements is immediately apparent when molar volume is graphed as a function of atomic number (Figure 7.15).

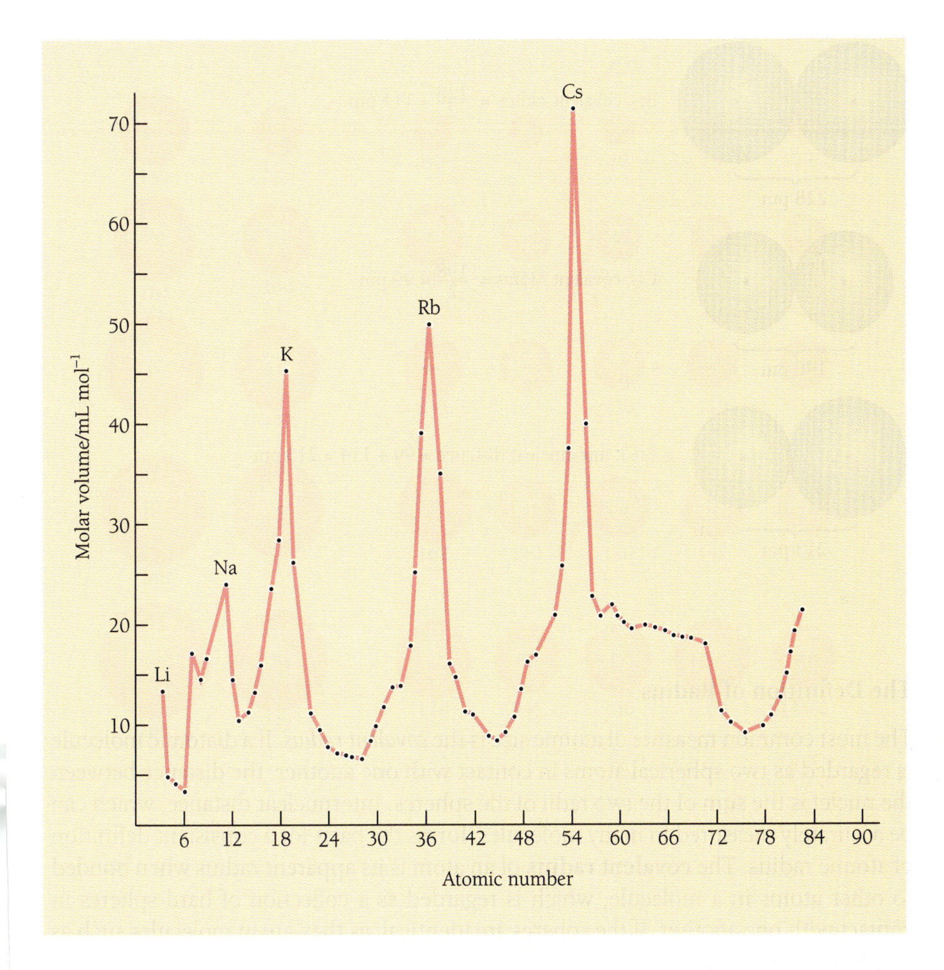

Figure 7.15 A modern version of Meyer's molar volume plot. In its original form (published in 1870), the abscissa was atomic weight (molar mass) rather than atomic number, and many as-yet-undiscovered elements such as the noble gases were missing. Nonetheless the results were a convincing demonstration of the periodic law.

Covalent Radius

Because it refers to individual atoms rather than to bulk properties, atomic *radius* is considered a more useful property than molar volume. However, an electron cloud does not have a definite boundary or edge, and so an atom does not have a definite radius. Nonetheless, some atoms are larger than others, and useful comparison can be made provided the term "atomic radius" is consistently and carefully defined.

Lothar Meyer (1830–1895)

Lothar Meyer received a medical education, and later was Professor of Physics at Tübingen University in Germany. Like Mendeleev, he developed a version of the periodic table while writing a textbook. Unlike Mendeleev, he was reluctant to make predictions about undiscovered elements. Therefore, he did not claim priority for his periodic table, even though it was published several years prior to Mendeleev's.

dot in the lower left corner identifies the largest atom (Cs) and reminds us that covalent radius increases both down a group and from right to left across a period.

Periodic trends in atomic size offer more detailed insight into the electronic structure of atoms.

Shielding

The increase in covalent radius within a given *group* is simply explained. As the atomic number increases, the principal quantum number increases. The size of the outer-shell orbitals increases as the principal quantum number increases. Therefore, since it is the size of the outer orbitals that determine the size of the atom, atomic radius increases as atomic number increases within a group.

In order to understand the decrease in radius as atomic number increases within a given *period,* we must take into account both the positive charge of the nucleus and the effect that the negatively charged electrons have on each other. Consider the configuration of the sodium atom, with its [Ne] core electrons and its single outer electron:

$$\underset{\text{core}}{1s^22s^22p^6} \;\Big|\; \underset{\text{outer}}{3s^1}$$

Since its principal quantum number is larger than that of the core electrons, much of the 3*s* electron density lies outside the core. The 3*s* electron is attracted by the positively charged nucleus, and simultaneously repelled by the negatively charged electrons of the core. The outer electron is thus screened or **shielded** by the core electrons from the full effect of the nuclear charge. As a result, the 3*s* electron experiences an **effective nuclear charge (Z_{eff})** that is considerably less than the actual nuclear charge. For sodium ($Z = 11$), Z_{eff} is ≈ 2.5. The effective nuclear charge is not a measureable property, and there are several different methods for assigning numerical values. The *trends* in Z_{eff}, which are more important than the actual values, are the same no matter which scheme is used to obtain them.

Some methods of assigning effective nuclear charge are based on measurements of ionization energy, which is discussed in Section 7.5.

Compared to the sodium atom, the magnesium atom has an additional proton in the nucleus and an additional outer electron in the 3*s* orbital.

$$\underset{\text{core}}{1s^22s^22p^6} \;\Big|\; \underset{\text{outer}}{3s^2}$$

Although there is some shielding of the outer electrons by each other, the effect of the increased number of protons in the nucleus is more important, and Z_{eff} is greater for magnesium than for sodium. For magnesium ($Z = 12$), $Z_{eff} \approx 3.3$.

The elements in the third period, beginning with sodium, all have the same core, and their outer electrons all have the same principal quantum number ($n = 3$). As the atomic number increases, the effective nuclear charge increases from ≈ 2.5 in Na to ≈ 6.8 in Ar. As Z_{eff} increases, the outer electrons are more strongly attracted and hence lie closer to the nucleus. This accounts for the regular decrease in atomic radius as atomic number increases. As seen in Figure 7.18, the effect is observed in all periods, although it is less regular in the transition elements.

EXAMPLE 7.8 Relative Size of Atoms

Which element in the following pairs has the larger covalent radius? (Ar,Rb); (K,Cu)

Analysis The tablet (Figure 7.19) indicates general trends in atomic size. Application of the trend to specific elements normally requires a complete periodic table.

Work

(Ar,Rb): Rb is larger because it is both lower and further to the left in the periodic table than Ar. Alternatively it can be reasoned that Rb is larger than Kr because the first member of a new period is always larger than the preceding noble gas, and Kr is larger than Ar because within any group, size increases with atomic number. Therefore Rb is larger than Ar.

(K,Cu): K is larger since it lies to the left of Cu in the same period. K has the smaller Z, hence the smaller Z_{eff} and the larger size.

Exercise

Consulting only the periodic table, arrange the elements Ge, Sn, and In in order of increasing size.

Answer: See Figure 7.17.

The concept of shielding also provides an explanation of why the energies of the subshells in many-electron atoms increase in the order $s < p < d < f$. The core electrons do not form a perfect shield, since part of the outer-shell electron cloud is found near the nucleus (see Figure 6.22, page 251). That is, there is some **penetration** of the core by the outer electrons. The extent to which an outer electron penetrates the core depends on the shape of the orbital. As shown in Figure 7.20 for the 3*s*, 3*p*, and 3*d* orbitals, the electron density near the nucleus is different for *s*, *p*, and *d* electrons. The result of these differences in core penetration is that each type experiences a different value of Z_{eff}. Much of the negative charge of a 3*s* electron lies close to the nucleus, but this is not the case for 3*p* or 3*d* electrons. Consequently a 3*s* electron is less shielded, and experiences a larger value of Z_{eff} than a 3*p* and 3*d* electron *in the same atom*. The larger effective nuclear charge means a stronger attractive force, hence a lower energy, for *s* electrons. Similarly, the *p* subshell has a lower energy than the *d* subshell, which in turn lies at lower energy than the *f* subshell.

The order of increasing subshell energies ($s < p < d < f$) is one of the cornerstones of our ideas about electron configuration. No explanation was given when this ordering was shown in Figure 7.2 and used in the aufbau scheme. The explanation is shielding.

Both *p* and *d* orbitals have a node at the nucleus.

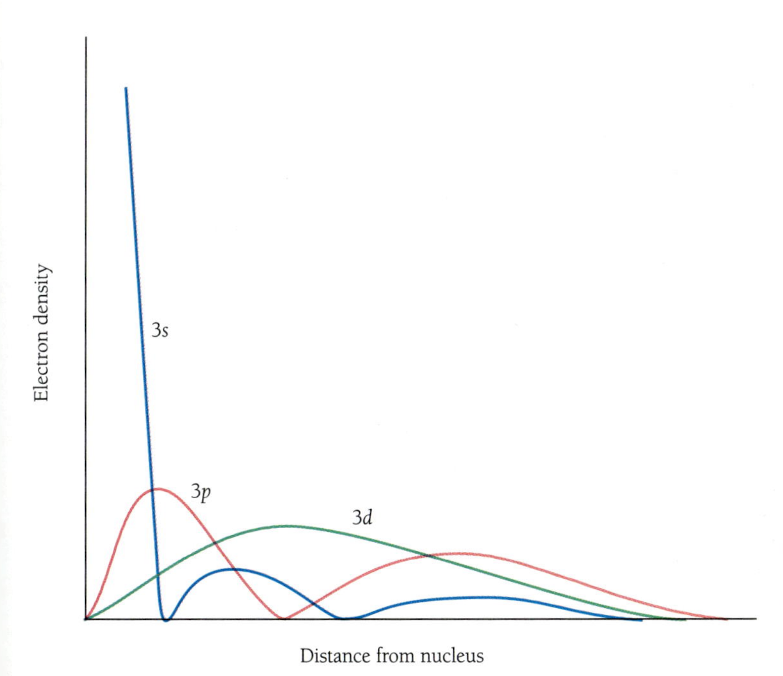

Figure 7.20 Negative charge density as a function of distance from the nucleus for orbitals in the 3*s*, 3*p*, and 3*d* subshells. These graphs are not drawn to the same scale, but are sketches intended to illustrate the differences more clearly.

Ionization Energy

Ionization energy is called "ionization potential" in some older texts.

The energy required to remove an electron from the highest occupied orbital of a gas-phase atom is called the **ionization energy (IE)** of that atom. When the process is written as a thermochemical equation for an element M, ionization energy is heat of reaction

The symbol ΔE is used for heat of reaction (rather than ΔH) when referring to reactions or processes taking place in isolated atoms or molecules.

$$M(g) \longrightarrow M^+(g) + e^-; \qquad \Delta E = IE$$

Ionization is always an endothermic process, so IE is always a positive number. The energy required for ionization can be supplied by a number of means, including heat, absorption of a photon, and impact by high-energy particles such as α-particles or electrons. Ionization energies can be determined experimentally with a photoelectron spectrometer or a mass spectrometer. With sufficient energy input, it is possible to remove one, two, three, four, or more electrons from an atom, and the additional energies required at each successive step are referred to as the second, third, fourth, and so on, ionization energies. Unless otherwise specified, the term "ionization energy" always refers to the *first* ionization energy. Data for the ionization energies of some elements are given in Table 7.1 and Figure 7.21.

Ionization energies are often given in units of *electron volts (eV)*, related to SI units by the equivalence statement 1 eV ⇔ 96.485 kJ mol^{-1}.

Several trends in ionization energies are evident from the data.

1. From Table 7.1, it is clear that the successive ionization energies of an element increase in the order first < second < third, and so on. The ionization energies of neon, for example, increase from 2 to well over 100 MJ mol^{-1} as more and more electrons are removed. To ionize a neutral atom, energy input is required to overcome the attractive force between the (negative) electron being removed and the (positive) ion left behind. When the second electron is removed, the remaining ion has a 2+ charge, so the attractive force opposing the removal is greater. Each successive step increases by 1+ the charge of the ion, so each successive step requires more energy. This is true for all atoms, and has little to do with specific electron configuration.

2. In each of the second-period elements Li through Ne, there is an unusually large jump between certain successive ionization energies. The jump, set off by bold type in Table 7.1, marks the difference between core and outer

TABLE 7.1 Successive Ionization Energies/MJ mol^{-1}

Element	1st	2nd	3rd	4th	5th	6th	7th	8th	9th	10th
H	1.312									
He	2.372	5.250								
Li	0.520	**7.298**	**11.82**							
Be	0.899	1.757	**14.85**	**21.01**						
B	0.801	2.427	3.660	**25.03**	**32.83**					
C	1.086	2.353	4.620	6.222	**37.83**	**47.28**				
N	1.402	2.857	4.578	7.475	9.445	**53.27**	**64.36**			
O	1.314	3.388	5.300	7.469	10.99	13.33	**71.33**	**84.08**		
F	1.681	3.374	6.020	8.407	11.02	15.16	17.87	**92.04**	**106.43**	
Ne	2.081	3.952	6.122	9.370	12.18	15.24	20.00	23.07	**115.38**	**131.48**

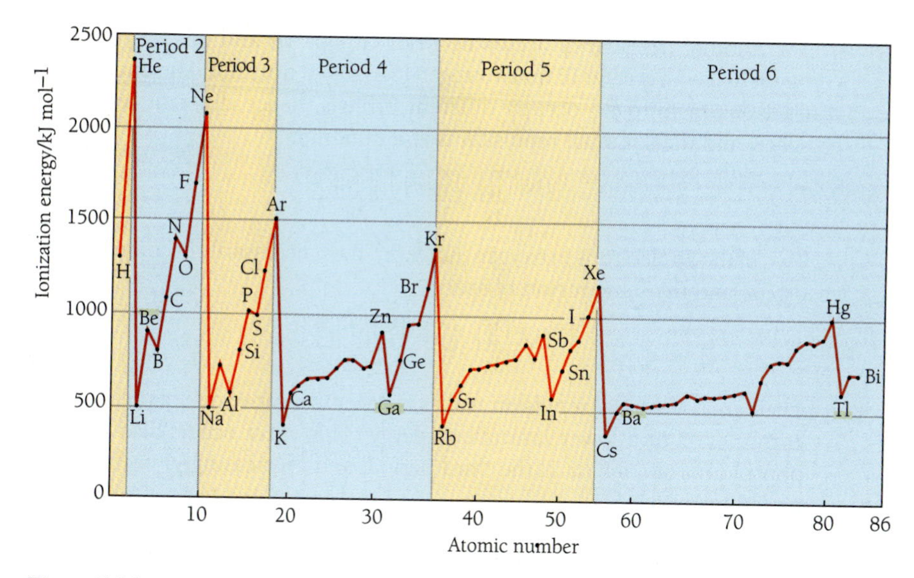

Figure 7.21 First ionization energy of the elements. Note the decrease as atomic number increases within a group, and the general but irregular increase within a period.

electrons. Nitrogen ([He]$2s^2 2p^3$) has five electrons in its outer shell. The first five ionization steps correspond to the removal of these electrons, and the ionization energies, although increasing, are all less than 10 MJ mol^{-1}. The last two steps correspond to the removal of the core electrons (1*s*), which requires much more energy. Core electrons, lying closer to the nucleus, are much more tightly held than outer electrons.

3. There is a general increase in the value of the *first* ionization energy as atomic number increases in the second period from Li to Ne. This is to be expected, because the effective nuclear charge generally increases with atomic number in a period. In addition, the atomic radius decreases. Other things being equal, the higher the value of Z_{eff} and the smaller the size of the atom, the greater the amount of energy needed to remove an electron. Figure 7.21 shows that this trend is generally followed in all rows of the periodic table.

Coulomb's law states that the energy required to remove an electron located at a distance r from a positive charge Z_{eff} is proportional to the quotient (Z_{eff}/r).

4. Within any group, for example the alkaline earths or the noble gases, ionization energy decreases as atomic number increases. Although Z_{eff} increases slightly, the electron that is being removed by ionization lies at a larger distance from the nucleus in the heavier atoms of the group. Therefore, the attractive force holding it to the nucleus is less, and it is more easily removed. These trends in first ionization energy are summarized in the tablet, Figure 7.22.

5. Certain irregularities are apparent in Figure 7.21. The Group 3A element boron ought to have a greater IE than its Group 2A predecessor (beryllium), whereas in fact it is less. This may be explained by the fact that a 2*p* electron is lost when boron ionizes, while a 2*s* electron is lost when beryllium ionizes. Since core penetration is less for *p* electrons than for *s* electrons in the same shell, Z_{eff} is less for *p* electrons. A lesser Z_{eff} means that boron's outer electron is more loosely held, and less energy is required to remove it. A similar irregularity occurs, for a similar reason, between Mg and Al in the third period. The effect is less noticeable in higher periods.

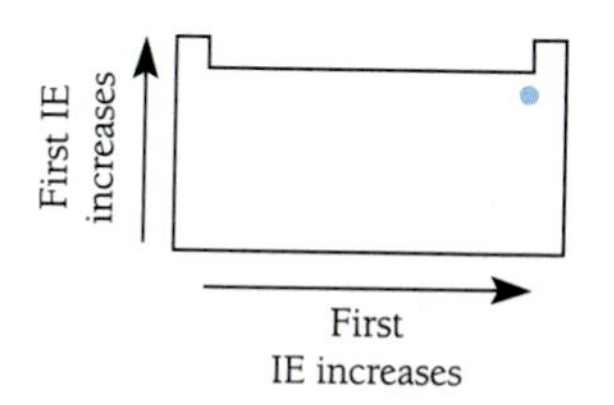

Figure 7.22 First ionization energy increases from left to right within a period, and decreases from top to bottom within a group. The element with the greatest ionization energy is helium.

Another irregularity occurs between Groups 5A and 6A. Oxygen (6A) ought to have a greater IE than nitrogen (5A), but it does not. The unpaired $2p$ electrons in nitrogen occupy different orbitals, hence different regions of space, and their mutual repulsion is at a minimum

(N: [↑↓] [↑↓] [↑][↑][↑]).
$1s$ $2s$ $2p$

Two of the $2p$ electrons in oxygen, however, must necessarily occupy the same orbital and the same region of space

(O: [↑↓] [↑↓] [↑↓][↑][↑]).
$1s$ $2s$ $2p$

This discussion should remind you of the explanation of Hund's rule, given on page 269.

Their greater mutual repulsion makes it easier to remove one of them, which corresponds to a lower ionization energy. This effect occurs also between phosphorus and sulfur in the third period, and (decreasingly) in the higher periods as well.

EXAMPLE 7.9 Relative Ionization Energies

Without consulting Figure 7.21, predict which element of the following pairs has the greater ionization energy: (Br, I); (Mg, Al); (Mg, P).

Analysis/Work The relationship between target and knowns is the general trend of ionization energy in the periodic table, but modified, as discussed above, by the effects of outer-electron configuration.

(Br, I): Both have the same outer-shell configuration (ns^2np^5), so that modification of the general trend is not appropriate. Bromine has the higher IE because it is smaller, so the outer electrons are more tightly bound.

(Mg, Al): By their relative location in the periodic table, we expect the order IE(Al) > IE(Mg). However, since p electrons are better shielded, it is easier to remove the p electron from Al ([Ne] [↓↑] [↑][ ][ ]) ($3s$, $3p$) than to remove an s electron from Mg ([Ne] [↓↑] [ ][ ][ ]) ($3s$, $3p$). Mg has the higher IE.

(Mg, P): Both are in the same period, so phosphorus, with the larger atomic number, should have the greater Z_{eff} and the greater IE. On the other hand, the greater shielding of p electrons tends to decrease Z_{eff} and IE. The two effects work in opposite directions. Generally, when Z differs by more than 1, increased atomic number is more important than the change from s to p. The ionization energy of P is greater than that of Mg.

Exercise

Without looking at Figure 7.21, place the following elements in order of increasing ionization energies: N, Si, P.

Answer: Now look at Figure 7.21.

Ionic Radius

Ionic radius can be measured by the technique of x-ray diffraction, which is described in Chapter 11.

The **ionic radius** of monatomic anions and cations is defined in terms of internuclear distances in *ionic* compounds rather than in covalent molecules. Ionic radius is best understood in relation to the radius of the atom from which the ion is formed. A

TABLE 7.2 Comparison of Ionic and Covalent Radii/pm

	Group				
	1A	2A	3A	6A	7A
Anion radius				Se^{2-}, 198	Br^-, 195
Covalent radius	K, 231	Ca, 197	Ga, 122	Se, 117	Br, 114
Cation radius	K^+, 133	Ca^{2+}, 99	Ga^{3+}, 62		

cation, having lost one or more electrons, is smaller than the parent atom. An anion, having gained one or more electrons, is larger than the parent atom. Some examples of this relationship in the main-group elements of the fourth period are given in Table 7.2.

Table 7.3 compares the radii of species having different atomic numbers but the same number of electrons. Such species are called **isoelectronic**, whether they are charged or neutral. Ionic radius always decreases as atomic number increases in an isoelectronic series, because Z_{eff} increases (since Z itself increases), so the electrons are more strongly attracted to the nucleus.

TABLE 7.3 Variation of Radius in an Isoelectronic Series of Ions

Species	N^{3-}	O^{2-}	F^-	Ne	Na^+	Mg^{2+}
Radius/pm	171	140	136	—	95	65

Neon forms no compounds, hence its "ionic radius" cannot be measured.

EXAMPLE 7.10 Relative Size of Ions

Which is larger, F^- or Cl^-?

Work Since the *covalent* radius of Cl is larger than that of F, the same relationship is expected in the singly charged anions: Cl^- is larger than F^-.

Exercise

Which of the following are pairs of isoelectronic species? (a) (O^{2-}, S^{2-}); (b) (S^{2-}, Ar); (c) (Zn^{2+}, Co^{3+}); (d) (Br^-, Rb^+); (e) (Cl^-, Sr^{2+}).

Answer: (b) and (d).

Metallic Character

In each of the Groups 3A through 6A, the smallest member is a nonmetal and the largest member is a metal. In each group, at least one of the elements of intermediate size is a metalloid. It is clear that **metallic character** increases with atomic number in these groups. Also, metallic character decreases as atomic number increases within a period. Metallic character is not precisely defined, but is associated both with chemical properties, such as the tendency to form ionic compounds, and with physical properties, such as the ability to conduct heat and electricity. The trend exists in other groups

Not only does iodine *look like* a metal, it actually becomes a metal when subjected to sufficiently high pressure.

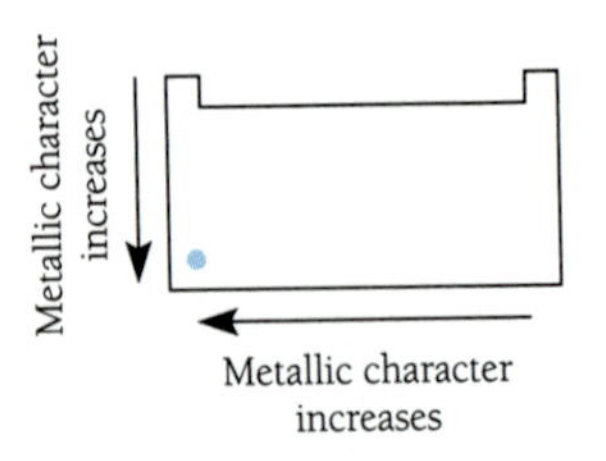

Figure 7.23 Metallic character increases with atomic number within a group, and decreases with atomic number within a period. The most highly metallic elements are found in the lower left of the periodic table.

as well: in Group 1B, silver is a better conductor than copper, and in Group 7A, crystals of iodine have an almost metallic sheen. No one has ever seen a crystal of astatine, but it can be presumed to have even more metallic character than iodine.

Trends in metallic character can be kept in mind with the aid of the tablet in Figure 7.23: as is the case for atomic radius, the dot for the maximum is in the lower left. Only the ionization energy is largest in the upper right.

7.6 THE HALOGENS

The elements of Group 7A—fluorine, chlorine, bromine, iodine, and astatine—are known collectively as the **halogens.** The stable form of these elements is the diatomic molecule: F_2, Cl_2, Br_2, and I_2 (Figure 7.24). The name halogen, which indicates a major source of these elements, derives from a Greek word meaning "sea salt." Chloride ion constitutes two thirds of the anionic component of blood plasma, and is involved in many bodily functions. Sodium chloride is an essential nutrient: each of us requires several grams per day. Iodine is a constituent of the hormone *thyroxin,* also essential. Bromide ion is a central nervous system depressant, and as such was once a popular component of over-the-counter pain pills. However, bromide ion is toxic, and in high doses causes delirium.

A great deal of evidence links fluoride ion with a reduction in the incidence of tooth decay, especially in children. Tooth enamel consists primarily of *hydroxyapatite,* $Ca_{10}(PO_4)_6(OH)_2$. If a reasonable level of fluoride ion is present in the diet during the growing phase of teeth, a significant amount of *fluoroapatite,* $Ca_{10}(PO_4)_6F_2$, is incorporated in the enamel in place of hydroxyapatite. Fluoroapatite is less soluble in mouth acids, hence fluoride-containing teeth are less susceptible to decay. Adult teeth

Figure 7.24 Pictured elements are chlorine (greenish-yellow), bromine (reddish-brown) and iodine (violet). At room temperature, chlorine is gaseous, bromine exists both as liquid and gas, and iodine both as solid and gas. A small amount of liquid bromine and solid iodine can be seen at the bottom of the center and right-hand flasks, respectively. Pure fluorine is not pictured here, because it is so reactive that it can be handled with safety only in specialized apparatus.

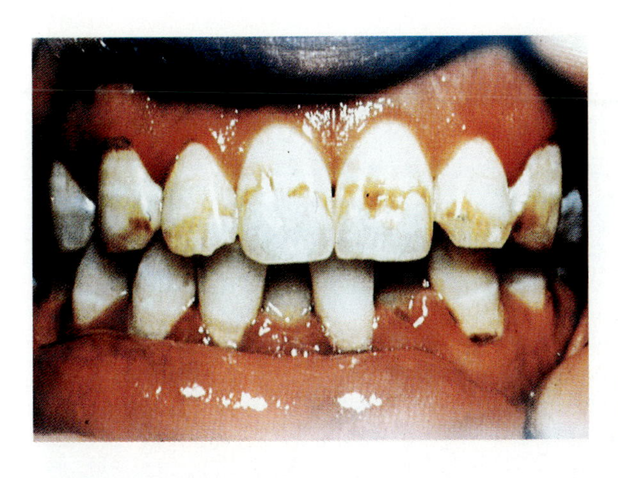

Figure 7.25 These teeth show varying degrees of discoloration caused by ingestion of fluorides. The amount of fluoride present in municipal water supplies is unlikely to cause noticeable damage.

undergo demineralization, as Ca^{2+} and PO_4^{3-} ions pass from the tooth to the plaque, and remineralization, as the reverse process occurs. Demineralization is accelerated by acids produced by the bacteria in the plaque. Too much fluoride is harmful: it weakens and discolors teeth (Figure 7.25), and it accumulates in the bones, causing pain and possible crippling.

Many communities augment the fluoride content of natural water by adding fluoride to the municipal water supply, and a few decrease it by removing fluoride. About half the population of the U.S. and Canada drinks artificially fluoridated water. Most other industrial nations do not fluoridate their drinking water. Artificial fluoridation has been practiced in this country since 1945. During all this time it has been a political controversy, arousing strong passions among both proponents and opponents. It has been a model of how *not* to use science in public policy. Both sides have deliberately, and essentially continuously, suppressed and distorted scientific evidence that does not support their beliefs.

Occurrence, Preparation, and Properties of the Elements

In discussions of the chemistry of the Group 7A elements, the symbol X is used to represent any halogen.

Halogens are widely distributed in nature. All occur as the **halide ion (X^-)** in seawater, although the concentrations of F^- and I^- are considerably less than those of Cl^- and Br^-. Fluorine occurs in the minerals *cryolite* (Na_3AlF_6), *fluorspar* (CaF_2), and *fluoroapatite* [$Ca_{10}(PO_4)_6F_2$]. Cl, Br, and I occur not only in seawater, but also in salt deposits that are the remnants of ancient seas. Certain forms of marine life, notably seaweed, have the ability of concentrating iodine within their tissues. Astatine does not occur naturally, because it has no stable isotopes. Small quantities can be artificially prepared in a nuclear reactor. Because it lasts only hours before disappearing by radioactive decay, however, less is known about the chemistry of astatine than about the other halogens.

Fluorine is prepared by electrolysis of hydrogen fluoride dissolved in potassium fluoride. The mixture contains 25% to 50% HF, and is molten at 100 °C.

$$2\ HF(solution) \xrightarrow[\text{electricity}]{\text{KF, 100 °C}} H_2(g) + F_2(g)$$

A useful laboratory preparation for small quantities of chlorine is the reaction of manganese dioxide with hydrochloric acid.

$$MnO_2(s) + 4\ HCl(aq) \longrightarrow MnCl_2(aq) + Cl_2(g) + 2\ H_2O$$

Figure 7.26 A commercial plant for the production of chlorine by the electrolysis of brine. The chemistry of this process is discussed in Section 19.5.

Chlorine is prepared commercially by the electrolysis of aqueous sodium chloride obtained from seawater or salt deposits (Figures 7.26 and 7.27). Chlorine is an important industrial chemical: about 10 million metric tons are produced annually in the U.S. alone.

Bromine is prepared by bubbling chlorine through seawater, which contains dissolved sodium bromide.

$$Cl_2(g) + 2\ NaBr(aq) \longrightarrow Br_2(g) + 2\ NaCl(aq)$$

Bromine vapor is swept out of the reaction vessel by a current of air. About 2500 L of seawater is required to produce one mole (160 g) of bromine.

Figure 7.27 A great deal of sodium chloride is dissolved in the waters of the Great Salt Lake in Utah. During dry seasons, when the water level drops, salt deposits can be found on the shore.

Iodine can be prepared by bubbling chlorine through an aqueous solution of an iodide salt, in a reaction similar to that used in the production of bromine.

$$Cl_2(g) + 2\ NaI(aq) \longrightarrow I_2(g) + 2\ NaCl(aq)$$

A more convenient laboratory preparation involves treatment of an iodide salt with manganese dioxide under acidic conditions.

$$2\ NaI(aq) + 2\ H_2SO_4(aq) + MnO_2(s) \longrightarrow I_2(s) + MnSO_4(aq) + Na_2SO_4(aq) + 2\ H_2O$$

Some natural sources contain iodine in the form of the iodate ion (IO_3^-). Treatment of aqueous sodium or potassium iodate with sodium hydrogen sulfite yields iodine.

The H_2SO_4/MnO_2 reaction can also be used to prepare Br_2 and Cl_2 from the corresponding alkali halides.

$$2\ NaIO_3(aq) + 5\ NaHSO_3(aq) \longrightarrow I_2(s) + 3\ NaHSO_4(aq) + 2\ Na_2SO_4(aq) + H_2O$$

Table 7.4 lists some of the physical properties of the halogens.

The most striking chemical property of the halogens is their great reactivity. They react, forming many different compounds, with both metals and nonmetals. The first member of the group, fluorine, reacts readily, often violently, with all elements except He, Ne, and Ar. The reactivity of the other halogens decreases in the order Cl > Br > I. Binary compounds of halogens with nonmetallic elements are molecular. Reactions in which a halogen atom acquires an electron to become a halide ion occur readily. This is particularly true if the atom from which the electron is acquired has a low ionization energy, as do most metals. A reaction in which an electron is transferred from one species to another is called an **oxidation.** The species that gives up the electron is said to be **oxidized,** and the reactant that acquires the electron is called an **oxidizing agent.** The halogens are powerful oxidizing agents, their oxidizing power increasing in the order I < Br < Cl < F.

Oxidation reactions are discussed in detail in Chapter 18.

Metal Halides

Halogens react with the metals of Groups 1A, 2A, and 3B (including the lanthanides), to form halides with formulas MX, MX_2, and MX_3, respectively. With the exception of the beryllium halides, all of these are ionic compounds. Some of the transition metals form ionic halides, for example AgCl, while others form molecular halides like $TiCl_4$ and UF_6. Some metal halides are difficult to classify as either ionic or molecular.

TABLE 7.4 Physical Properties of the Halogens

Element	Symbol	Atomic Number	Molar Mass	Covalent Radius/pm	Melting Point/°C	Boiling Point/°C	Ionization Energy /kJ mol^{-1}
Fluorine	F	9	19.00	64	−220	−188	1681
Chlorine	Cl	17	35.45	99	−103	−34.6	1250
Bromine	Br	35	79.90	114	−7.2	58.8	1139
Iodine	I	53	126.91	133	113.5	184.4	1008
Astatine	At	85	(210)	—	302	—	

The value listed in the table for the molar mass of astatine is the mass *number* of the element's most stable isotope.

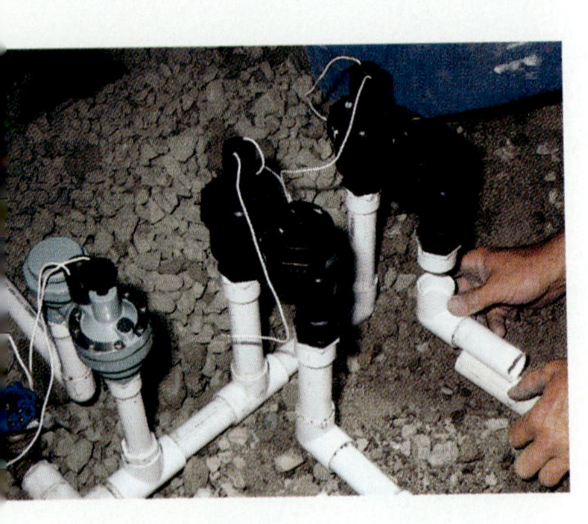

Figure 7.28 Polyvinyl chloride (PVC) can be easily molded to desired shapes. Molded PVC pipes are widely used in both residential and industrial applications.

Metal halides are common materials, and the most common, NaCl, is mined in many locations in essentially pure form. For thousands of years, sodium chloride has been of great importance to society. Besides being an essential nutrient, salt has been used as money, that is, as a medium of exchange. It has even caused wars! Today it is a cornerstone of modern chemical industry. Over half of the NaCl mined in the United States is converted to NaOH and Cl_2, both of which are raw materials for many manufacturing processes.

Hydrogen Halides

Halogens react with hydrogen to form the **hydrogen halides** HF, HCl, HBr, and HI. All of these are molecular compounds, gaseous at room temperature.

$$X_2 + H_2(g) \longrightarrow 2\ HX(g)$$

In keeping with the reactivity trend generally observed in Group 7A, (F > Cl > Br > I), the reaction of hydrogen with F_2 is explosive, and that with Cl_2 is less vigorous. Br_2 and I_2 react only at high temperature (and then incompletely). These reactions can be speeded up by exposing the hydrogen/halogen gas mixture to strong light. HF and HCl can be more conveniently prepared in the laboratory by treatment of a fluoride or chloride salt with concentrated sulfuric acid.

$$NaF(s) \text{ or } NaCl(s) + H_2SO_4(conc) \longrightarrow NaHSO_4(s) + HF(g) \text{ or } HCl(g)$$

Hydrogen bromide and HI can be prepared by treatment of the halogen with aqueous hydrogen sulfide.

$$H_2S(aq) + Br_2(\ell) \text{ or } I_2(s) \longrightarrow S(s) + 2\ HBr(g) \text{ or } 2\ HI(g)$$

Although there is very little pain at first, burns caused by HF are slow to heal. HF reacts with calcium in the skin to form CaF_2, which impedes the healing process.

The halogen halides are all very soluble in water, and the solutions are acids. Hydrogen fluoride is unusual because it reacts with silicon dioxide, the primary constituent of glass.

$$SiO_2(s) + 4\ HF(aq) \longrightarrow SiF_4(g) + 2\ H_2O(\ell)$$

Hydrogen fluoride is one of the very few common reagents that attacks glass.

Hydrochloric acid is an important industrial chemical, produced principally as a byproduct in the manufacture of chlorine-containing compounds such as PVC (polyvinyl chloride, a plastic material used for water and sewer pipes; Figure 7.28). HCl in turn is used in the manufacture of other chemicals and pharmaceuticals, as well as in food processing. About half the HCl manufactured in the United States is used for metal "pickling," a process for cleaning of metal surfaces.

An older name, still used in industry, for hydrochloric acid is "muriatic acid."

Halogen Oxides

Numerous binary compounds of oxygen and the halogens are known: all are molecular, and (with the exception of I_2O_5) all are unstable, decomposing spontaneously to the elements. The decomposition is slow with some, but with many, particularly the fluorine and chlorine oxides, it occurs unpredictably with explosive violence. All are powerful oxidizing agents. The known oxides are listed in Table 7.5, in order of increasing oxygen-to-halogen mole ratio.

The only halogen oxide of commercial importance is ClO_2, used to bleach wood pulp during one stage of paper manufacture. Because of its explosiveness, it is neither stored nor shipped. Instead, it is prepared as needed from sodium chlorate.

$$2\ NaClO_3(aq) + SO_2(g) + H_2SO_4(aq) \longrightarrow 2\ ClO_2(g) + 2\ NaHSO_4(aq)$$

TABLE 7.5 Oxygen Compounds of the Halogens*

(O/X ratio)	½	1	2	2	2.5	3	3.5
Fluorine	OF_2	O_2F_2					
Chlorine	Cl_2O		ClO_2	Cl_2O_4		Cl_2O_6	Cl_2O_7
Bromine	Br_2O		BrO_2				
Iodine				I_2O_4	I_2O_5		I_2O_7

* Some additional compounds have been reported, but are poorly characterized: Cl_2O_3, Br_2O_7, Br_3O_8, I_4O_9.

Oxoacids and Their Salts

The oxoacids of the halogens are listed in Table 7.6. Most are unstable species that cannot be isolated in pure form, but are prepared and used as aqueous solutions. *Salts* of the oxoacids are much more stable. The chlorine oxoacids are hypochlorous, $HClO$; chlorous, $HClO_2$; chloric, $HClO_3$; and perchloric, $HClO_4$. The anions of the bromine oxoacids are hypobromite, BrO^-; bromite, BrO_2^-; bromate, BrO_3^-; and perbromate, BrO_4^-. The fluorine- and iodine-containing species are correspondingly named. All the halogen oxoacids and their salts are good oxidizing agents.

The hypohalous acids in Table 7.6 are listed by their *structural formulas,* to show the arrangement of the atoms. According to the rules given in Chapter 2, the *empirical formulas* are written HClO, HBrO, and HIO. Both forms (HOX and HXO) are in common use. Except for HOF, which reacts with water, hypohalous acids are formed by the reaction of the halogen with water.

$$X_2 + H_2O \longrightarrow HOX(aq) + HX(aq)$$

Similar reactions occur in the presence of sodium hydroxide or other bases. Under these conditions the products are the halide and hypohalite salts.

$$X_2 + 2\ NaOH(aq) \longrightarrow NaOX(aq) + NaX(aq) + H_2O$$

TABLE 7.6 Halogen Oxoacids and their Anions

Acid	Hypohalous	Halous	Halic	Perhalic
Fluorine	HOF*			
Chlorine	HOCl†	$HClO_2$†	$HClO_3$†	$HClO_4$
Bromine	HOBr†	$HBrO_2$‡	$HBrO_3$†	$HBrO_4$†
Iodine§	HOI		HIO_3	HIO_4
Anion	**Hypophalite**	**Halite**	**Halate**	**Perhalate**
	OX^- or XO^-	XO_2^-	XO_3^-	XO_4^-

* Reacts with water, hence does not exist in solution.
† Stable in aqueous solution, but cannot be isolated in pure form.
‡ Existence doubtful.
§ H_5IO_6 and $H_4I_2O_9$ are also known.

Sodium hypochlorite is used as a bleach, both industrially and in the household (Clorox).

Solutions of hypohalites are unstable, and gradually decompose. The reaction of hypobromite to form the bromide and bromate salts is typical.

$$3\ NaBrO(aq) \longrightarrow 2\ NaBr(aq) + NaBrO_3(aq)$$

This reaction, in which a single species reacts with itself to form two different products, is an example of a **disproportionation** reaction.

Chlorite salts can be made from the reaction of chlorine dioxide with base, in a reaction that also produces chlorate salts.

$$2\ ClO_2(g) + 2\ KOH(aq) \longrightarrow KClO_2(aq) + KClO_3(aq) + H_2O$$

Halate salts are produced as indicated above, by the disproportionation of hypohalites. These salts are thermally unstable, and gentle heating of potassium chlorate in the presence of manganese dioxide is a convenient method for producing small quantities of oxygen in the laboratory.

$$2\ KClO_3(s) \xrightarrow[\Delta]{MnO_2} 2\ KCl(s) + 3\ O_2(g)$$

Potassium perchlorate, a powerful bleaching agent, is produced commercially by the electrolysis of aqueous potassium chlorate.

$$KClO_3(aq) + H_2O \xrightarrow[\text{elect.}]{} KClO_4(aq) + H_2(g)$$

Interhalogen Compounds

A number of molecular compounds, called **interhalogens,** are formed from two different halogens. All of these are fluorides except BrCl, ICl, ICl_3, and IBr. The following interhalogen fluorides have been prepared.

Chlorine:	ClF	ClF_3	ClF_5	
Bromine:	BrF	BrF_3	BrF_5	
Iodine:		IF_3	IF_5	IF_7

Like halogens themselves, the interhalogens are all highly reactive, powerful oxidizing agents. In addition, several of the halogen fluorides are useful as fluorinating agents, that is, species capable of introducing one or more fluorine atoms into another molecule. The diatomic interhalogens tend to have properties intermediate between those of the two parent halogens. For example, ClF is a corrosive oxidizing agent that is more reactive than Cl_2 but less so than F_2. ClF boils at −100 °C, between the boiling points of Cl_2 (−35 °C) and F_2 (−188 °C). Most interhalogens are produced by reaction of the elements under various conditions of temperature, pressure, and stoichiometric ratio of the reactants.

Chemistry Old and New

INFAMOUS SCIENTIFIC "MAYBES"

Many radical theories and experiments have found their way into the scientific literature, only to be proven untrue at a later time. Some of these "discoveries," which promised to revolutionize chemical thought, attracted great numbers of supporters, and generated significant public interest before being ultimately debunked. Dubbed "the science of things that aren't so" by Nobel laureate Irving Langmuir, these failed ideas illustrate the danger of drawing hasty conclusions from inadequate data.

In 1902, shortly after the discovery of x-rays, the French scientist Rene Blondlot announced that he had detected another kind of ray. He reported that a dimly illuminated piece of paper in a dark room would become brighter if a hot metal wire, encased in an iron tube, was allowed to "radiate" through an aluminum window at the paper. Blondlot's so-called "N-rays" revealed a number of bizarre properties upon further investigation. For example, a brick, wrapped in black paper and left in the street for a period of time, would somehow gather and store N-rays. When the brick was later held against a subject's head, the rays would pass through the skull and brighten the paper. Blondlot also claimed that people gave off their own N-rays. The British physicist R. W. Wood was skeptical. Visiting Blondlot's laboratory, he secretly removed a crucial aluminum prism from an experiment in progress. When the French professor's results did not change, Wood knew that the claims for the existence of N-rays were unsound. All support for the existence of the new radiation disappeared after Wood described his visit, and his own secret "experiment," in a 1904 letter to the journal *Nature*.

A more recent example of a deflated scientific theory is the great polywater controversy of the 1960s. First reported in the Soviet Union, polywater was proposed to be a viscous, polymeric type of H_2O produced when water vapor was condensed in very narrow glass tubes. Polywater resembled petroleum jelly, with properties vastly different from those of regular water. These properties indicated that it was the most stable form of H_2O, and some feared that all water on the planet could be converted into polywater. Even the prestigious National Bureau of Standards was convinced of the existence of this new form of water, and issued a report of its spectroscopic properties in 1969. Three years later, conclusive evidence showed that polywater was nothing but regular H_2O with impurities.

Perhaps the most memorable modern example of a scientific "maybe" is the cold fusion frenzy. On March 23, 1989, chemists B. Stanley Pons and Martin Fleischmann called a press conference to announce that they had induced *nuclear fusion* in a test tube at room temperature. Their process, which utilized a palladium electrode immersed in deuterated or "heavy" water to fuse deuterium atoms, promised to supply the planet with a "clean, virtually inexhaustible source of energy." Chemists and physicists in all parts of the world rushed to confirm these astonishing results. While some investigators were able to observe the production of heat or fusion products with similar setups, most were not. No two researchers obtained results in substantial agreement with one another. Numerous experimental errors and unknown variables complicated things even further. Pons and Fleischmann, in their eagerness to make public what might have been the most important scientific discovery of the twentieth century, bypassed important checks and balances normally present in research. Instead of submitting a peer-reviewed paper to a scholarly journal, they took their findings straight to the press. They kept certain details of their apparatus secret (to protect patent rights) instead of sharing them with colleagues. They promised incredible benefits for mankind from a process that had yet to be repeated by another scientist.

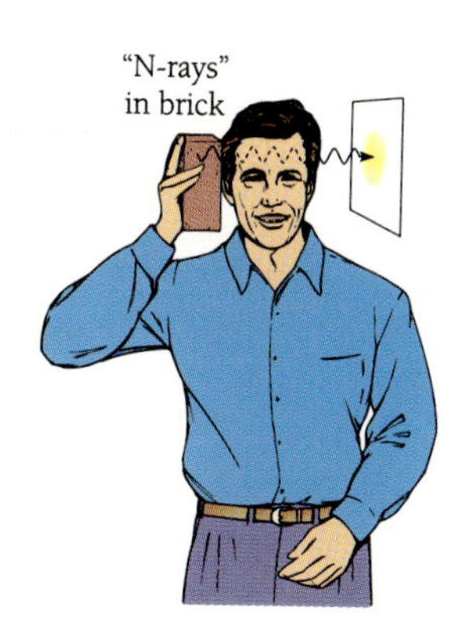

According to French scientist, Rene Blondet, if a brick was wrapped in black paper and left in a street for a period of time, it gathered and stored "N-rays." If the brick was later held to a person's head, the rays would pass through the skull and brighten a piece of paper.

Now, after four years, the reality of cold fusion is doubtful at best. Most scientists believe that, even though a few experimental observations have not been satisfactorily explained, there is little or no reason to suspect that nuclear fusion has ever taken place in a test tube. The most important lesson to be learned from the fusion fiasco, and other cases of controversial chemistry, is that the scientific method, although often slow and unwieldy, is essential when great discoveries are at stake.

Discussion Questions

1. Which, if any, of the "maybes" outlined above do you think came about through conscious, deliberate fraud?
2. Some people think that occasional outbreaks of the "science of things that aren't so" are beneficial because they cause scientists to be more careful. Do you agree?

70. Which class of elements have a partly filled *d* subshell? Name several elements in this class.
71. Elements having a partly filled 5*d* subshell are known as __________.
72. Elements having a partly filled 5*f* subshell are known as __________.
73. What outer-shell configuration(s) is/are characteristic of the main-group elements?
74. The elements in Groups 1A and 2A are sometimes called the "*s*-block" elements. Why?
75. What groups in the periodic table are called the "*p*-block" elements? The "*f*-block" elements?
76. If a third inner-transition series were to exist, what subshell would be partly filled?
77. In which group of the periodic table would a hypothetical element of atomic number $Z = 117$ belong? What is the name of this group?
78. In which group of the periodic table would a hypothetical element of atomic number $Z = 120$ belong? What is the name of this group?

Periodic Trends

79. Arrange the following elements in order of increasing covalent radius: N, Si, P.
80. Arrange the following elements in order of increasing covalent radius: Sc, Sr, Cs.
81. Identify the larger of each of the following pairs of species.
    a. (Mg, Na) b. (F, Cl)
82. Identify the larger of each of the following pairs of species.
    a. (Rh, Ag) b. (La, Ac)
83. Identify the larger species in each of the following pairs.
    a. (S^{2-}, Cl^-) b. (Se^{2-}, Se^+)
84. Identify the larger species in each of the following pairs.
    a. (S^{2-}, S) b. (Ca^{2+}, Ba^{2+})
85. In each of the following pairs, identify the species with the larger ionization energy.
    a. (S, O) b. (Rb, Sr)
86. In each of the following groups, identify the species with the largest ionization energy.
    a. (Te, Se, Br) b. (K, Ca, Rb)
87. Rank the elements Ne, Cl, Se in order of increasing ionization energy.
88. Rank the species Na, Na^+, Mg^{2+} in order of increasing ionization energy.
89. Which of the following processes requires the larger energy input? Explain your answer. $Co^{2+}(g) \rightarrow Co^{3+}(g) + e^-$; $Cu^+(g) \rightarrow Cu^{2+}(g) + e^-$.
90. Why is the second ionization energy of hydrogen not listed in Table 7.1?
91. Are there any elements for which the second ionization energy is less than the first? If so, name them. If not, explain.
92. The first ionization energy of Ca is greater than that of K, but the second ionization energy of Ca is less than that of K. Explain.
93. How much energy is required to ionize 5.00 mg neon?
*94. How much energy is required to remove all the outer electrons from one mole of boron atoms?
95. Compared to others in the same period, which elements have the highest ionization energy?
96. Compared to others in the same period, which elements have the largest covalent radius?
97. Estimate the internuclear distances in HBr and HI.
98. Estimate the internuclear distances in IBr and ICl.
99. Predict the value of the ionization energy of the (hypothetical) next higher halogen after astatine.
100. Predict the value of the ionization energy of the (hypothetical) next higher noble gas after radon.
101. Consider the elements B, C, Al, and Si. Which is largest? Which has the greatest ionization energy?
102. Rank the following species in order of increasing size and ionization energy: Na^+, Mg^{2+}.
103. Use Figure 7.15 to estimate the molar volume (which has never been measured) of francium. Estimate its covalent radius.
*104. Estimate the density of titanium.
105. Write the symbols for five species having 25 electrons.
106. The radii of the isoelectronic species K^+ and Ca^{2+} are 133 and 99 pm, respectively. Comment on the relative magnitudes of these radii.
107. While periodic trends offer a relatively easy way to predict that the ionization energy of Cl is greater than that of Se, it is not so easy to predict whether S or Br has the greater ionization energy. Why is the latter comparison more difficult? Which actually has the larger IE?
108. Which has the greater metallic character, Mg or Ca? How would you test the correctness of your prediction?

Halogens

109. Name the halogens and describe each as solid, liquid, or gas at room temperature.
110. Rank the halogens in order of increasing size, ionization energy, and metallic character.
111. What is the major source of the halogens?

112. What is an oxidizing agent? Which halogen is the best oxidizing agent?

113. Rank the hydrogen halides in order of increasing internuclear distance. Which, if any, are soluble in water? Which, if any, are gaseous at room temperature?

114. Are the binary oxides of the halogens molecular or ionic compounds? Which halogen forms the largest number of oxides?

115. How is bromine produced? Give any relevant chemical equations.

116. Give names and formulas of the oxoacids of chlorine. Give the formula and name of a salt of each of these acids.

117. Which, if any, of the hypohalous acids reacts with water?

118. Which halogen forms the most interhalogen compounds? Are interhalogens ionic or molecular?

119. Define disproportionation.

Chapter 8

The Chemical Bond

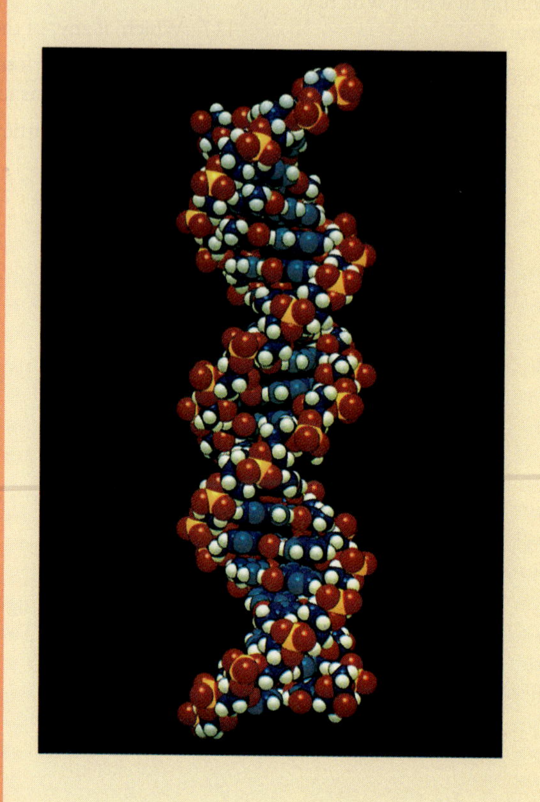

A computer-generated model of a short section of a DNA molecule. The shape of molecules, especially among compounds that participate in reactions within living organisms, has a great influence on chemical properties.

8.1 INTRODUCTION

In this chapter we take a step beyond atoms and elements, and turn our attention to the way in which atoms are organized into the larger units of chemistry: molecules and compounds. What is it that binds atoms together into the stable structures of molecules? Why are some compounds ionic and others molecular? What are the shapes of molecules, and what factors determine these shapes?

To begin with, it is the *electrons* that, in one way or another, create the force that binds atoms together into larger structures. This binding force is called the **chemical bond.** The electrons that participate in chemical bonding are called the **valence electrons.** The word **"valence"** is an adjective meaning having to do with chemical bonding. Only the outer electrons of an atom are valence electrons, because only the outer electrons are able to interact with electrons of other atoms. The inner electrons of the filled-shell core are in a sense hidden from the outside world, and are not utilized in bonding. The outer shell of electrons is often called the **valence shell.**

Used as a noun, "valence" means the combining power of an element, or the number of bonds an atom can form.

Our discussion of chemical bonding covers three types of bonds: the *ionic* bond, the *covalent* bond, and the *metallic* bond. Ionic bonding and covalent bonding are discussed in this chapter, while the metallic bond is dealt with in Chapter 9.

8.2 THE IONIC BOND

Ionic substances, such as rocks and minerals, are common materials. These substances consist of ions that are held together by forces arising from their electrical charge.

Coulomb Forces

A strong force, called the **Coulomb force,** exists between particles that bear an electrical charge. When one particle is positive and the other is negative, the force is attractive, but if the sign of the charge is the same for both particles, the force is repulsive. In order to separate two oppositely charged particles, energy must be supplied to overcome the attractive force. The amount of energy is independent of the mass of the particles, but does depend on the magnitude of the charge and on the distance between the particles. The relationship among energy (E), distance (d) between the centers, and the charges (q_1 and q_2) of the two particles is given by Equation (8-1).

When the particles bear *unlike* charges, the value of E is negative. This in turn means that energy must be supplied in order to separate the particles, that is, that the force between them is attractive.

$$E = \frac{Kq_1q_2}{d} \qquad (8\text{-}1)$$

K is a constant whose numerical value depends on the units in which charge and distance are expressed. The SI unit of electrical charge is the coulomb (C), and in SI units, $K = 8.98 \times 10^9$ J m C^{-2}. The *inverse* dependence on distance means that, as two particles of opposite charge approach one another, the attractive force between them becomes stronger and more energy is required to separate them. Equation (8-1) is known as **Coulomb's law,** after the French physicist whose careful measurements established its validity.

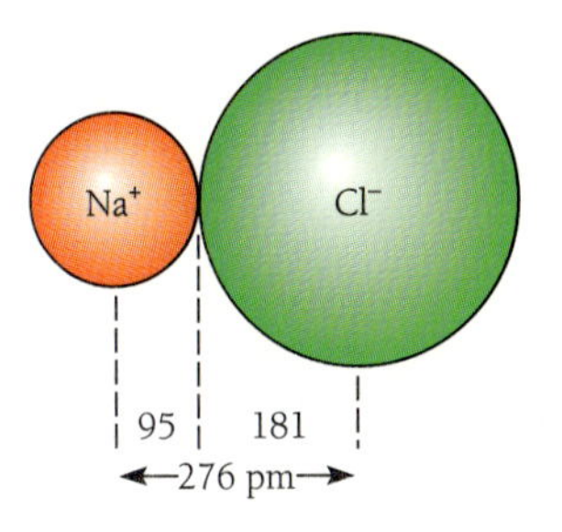

Figure 8.1 The attractive force between ions of opposite charge and the energy required to separate them depends on the distance (that is, the sum of the ionic radii). For Na^+ and Cl^-, the distance is 95 pm + 181 pm and the required energy is ≈ 500 kJ mol^{-1}.

Ions are charged particles of microscopic size, and the distance between the centers of adjacent ions is quite small (Figure 8.1). Therefore, the denominator in Equation (8-1) is small, and the energy of attraction between two oppositely charged ions is large. For a Na^+ ion and a Cl^- ion (whose radii are 95 pm and 181 pm, respectively), the energy of attraction is $\approx -8 \times 10^{-19}$ J, or about -500 kJ mol^{-1}. The *Coulomb force* that binds such ions together is called the **ionic bond.**

Moving *one* dot from the O atom with the −1 formal charge to the N atom still leaves one atom with only seven electrons, but the resulting structure is preferable because all formal charges are zero.

$$:\ddot{O}=\ddot{N}-\dot{\ddot{O}}:$$

This structure, too, has an equivalent form with the double bond on the right. Nitrogen dioxide is best described as a resonant species with two contributing structures.

$$:\ddot{O}=\ddot{N}-\dot{\ddot{O}}: \longleftrightarrow :\dot{\ddot{O}}-\ddot{N}=\ddot{O}:$$

Check All 17 valence electrons are used, there are no formal charges, and the octet rule is satisfied for two of the three atoms.

Exercise

Draw the Lewis structure of the CO^+ ion.

Answer: $\cdot C{\equiv}O{:}^+ \longleftrightarrow {:}C{\equiv}O{\cdot}^+$

Expanded Valence Shells

Phosphorus forms two chlorides, PCl_3 and PCl_5. The Lewis structure of phosphorus trichloride presents no problems.

```
          ..
         :Cl:
   ..    /
  :Cl—P:
   ..    \
         :Cl:
          ..
```

However, in PCl_5, the phosphorus atom is covalently bonded to *five* Cl atoms. Therefore, it has (at least) 10 valence electrons, which are said to occupy an **expanded valence shell.** Another example occurs in XeF_4, in which the central atom is surrounded by 12 electrons.

```
       ..
      :Cl:                 ..        ..
  ..   |   ..              :F        F:
 :Cl\  |  /Cl:              ..\     /..
  ..  \P/  ..                  :Xe:
     /   \                  ../     \..
  :Cl:   :Cl:              :F        F:
   ..     ..                ..        ..
```

Note that expanded valence shells can include lone pairs as well as bonding electrons.

Elements of the first and second periods do not violate the octet rule in this way. *Expanded valence shells are found only with elements in the third and higher periods.* Furthermore, expanded valence shells occur only in central atoms, not terminal atoms. Central atoms can have 10, 12, or 14 electrons.

The $M = 6N + 2 - V$ (page 317) rule is useful for identifying species with expanded valence shells. If M is negative, the central atom of the molecule (or ion) has an expanded valence shell. Furthermore, the absolute value of M is the number of additional electrons (above eight) in the valence shell. For PCl_5, $M = -2$, and the valence shell of the phosphorus atom has $8 + |-2| = 10$ electrons. For XeF_4, $M = -4$: there are $8 + |-4| = 12$ electrons in the valence shell of the central atom.

EXAMPLE 8.10 An Expanded Valence Shell

Draw the Lewis structure of IF_5.

Analysis The molecular formula suggests five terminal fluorine atoms bonded to a central iodine atom, that is, an expanded valence shell. Evaluation of $M = (6 \cdot 6) + 2 - 42 = -4$ confirms the suggestion.

Work Draw the suggested skeletal structure and complete the valence shells of the terminal fluorine atoms. The resulting structure, in which 40 of the 42 valence electrons have been used, is

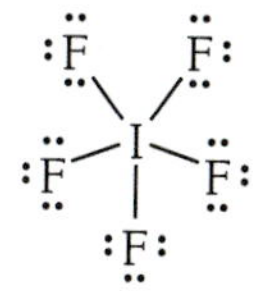

Since an expanded valence shell can accommodate up to 14 electrons, place the remaining 2 as a lone pair on the iodine atom.

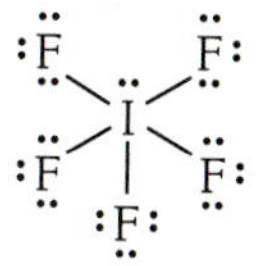

Check All of the valence electrons have been used, and all terminal atoms conform to the octet rule.

Exercise

Draw the Lewis structure of SF_6.

Answer:

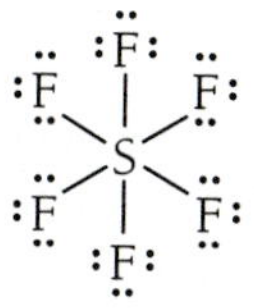

Expanded valence shells occur in polyatomic ions, such as SF_5^-, as well as in neutral molecules.

All the rules for drawing Lewis structures, and the exceptions to the rules, are summarized in Table 8.2.

8.4 ELECTRONEGATIVITY AND POLAR BONDS

The paired electrons of a covalent bond are usually not equally shared by the two bonded atoms. As we will see, the atom with the greater *electronegativity* gets the lion's share.

TABLE 8.2 Assignment of Lewis Structures to Molecules and Ions

1. Add the group numbers of all the "A" group atoms in the molecule. Add 1 for each unit of anionic charge, or subtract 1 for each unit of cationic charge. The result is the total number of valence electrons.
2. Arrange the atoms as a *central* atom surrounded by *terminal* atoms.
   a. If the molecule contains only one atom of a certain element (other than hydrogen), place it in the center.
   b. If more than one element is present as a single atom, the central atom is further down or to the left in the periodic table.
3. Insert a pair of electrons, indicating a covalent bond, between the central atom and each terminal atom.
4. Arrange the remaining valence electrons pairwise so as to satisfy the octet rule (duet rule for hydrogen) for as many atoms as possible; do the central atom last. Molecules with an odd number of electrons (for example NO and NO_2) cannot possibly satisfy the octet rule for all atoms.
5. If there are too few electrons to complete the central-atom octet, indicate multiple bonds by moving one or more lone pairs from one or more terminal atoms into bonding locations. *Electron-deficient* compounds have only four electrons in the valence shell of a central beryllium atom, or six in a central boron atom.
6. If there are too many electrons, add them as lone pairs to the central atom, indicating an expanded valence shell.
7. If there are several possible Lewis structures, equally good by the octet rule, choose the one that minimizes the formal charges.
8. If there are several equally good structures that differ only in placement of electrons, show them as resonance contributors.

The Polar Covalent Bond

One example of unequal electron sharing is provided by hydrogen chloride. In the diatomic HCl molecule, the Cl atom attracts electrons more strongly than the H atom does. Consequently the shared electrons of the bond lie closer to the chlorine end of the molecule, which thereby acquires a negative **partial charge.** The hydrogen end of the molecule, having lost part of its share of electrons, is left with a partial positive charge. The HCl molecule is a neutral species (as are all molecules), but it is electrically asymmetrical. It has a positive end and a negative end, and its two partial charges are equal in magnitude but opposite in sign. The term "partial charge" is used because the amount of charge on each end of the molecule is less than that which would result from the complete *transfer* of an electron that characterizes ionic bonding. Partial charges in a molecule are indicated by a lowercase Greek delta (δ), and the electrical asymmetry of the *bond* is often shown by an arrow with its crossed tail lying over the positive end of the bond.

$$\overset{\delta+}{\mathrm{H}}-\overset{\delta-}{\ddot{\underset{\cdot\cdot}{\mathrm{Cl}}}}: \qquad \overset{+\!\!\longrightarrow}{\mathrm{H}-\ddot{\underset{\cdot\cdot}{\mathrm{Cl}}}:}$$

Such bonds are called **polar** or **polar covalent** bonds. The electrical asymmetry of a polar bond is called its **bond moment.** Bond moments can be measured accurately in diatomic molecules, and in some polyatomic molecules as well.

						H 2.1										
Li 1.0	Be 1.5											B 2.0	C 2.5	N 3.0	O 3.5	F 4.0
Na 0.9	Mg 1.2											Al 1.5	Si 1.8	P 2.1	S 2.5	Cl 3.0
K 0.8	Ca 1.0	Sc 1.3	Ti 1.5	V 1.6	Cr 1.6	Mn 1.5	Fe 1.8	Co 1.9	Ni 1.9	Cu 1.9	Zn 1.6	Ga 1.6	Ge 1.8	As 2.0	Se 2.4	Br 2.8
Rb 0.8	Sr 1.0	Y 1.2	Zr 1.4	Nb 1.6	Mo 1.8	Tc 1.9	Ru 2.2	Rh 2.2	Pd 2.2	Ag 1.9	Cd 1.7	In 1.7	Sn 1.8	Sb 1.9	Te 2.1	I 2.5
Cs 0.7	Ba 0.9	La-Lu 1.0-1.2	Hf 1.3	Ta 1.5	W 1.7	Re 1.9	Os 2.2	Ir 2.2	Pt 2.2	Au 2.4	Hg 1.9	Tl 1.8	Pb 1.9	Bi 1.9	Po 2.0	At 2.2
Fr 0.7	Ra 0.9	Ac 1.1	Th 1.3	Pa 1.4	U 1.4	Np-No 1.4-1.3										

Figure 8.5 Electronegativities of the elements (Pauling scale).

Partial charge should not be confused with formal charge. *Partial* charge is a measurable property of molecules. *Formal* charge, while useful, is an abstract concept not subject to measurement.

Electronegativity

The existence of polar bonds means that the atoms of different elements differ in their power to attract electrons. We define **electronegativity** as the relative tendency of a covalently bonded atom to attract the bonding electron pair to itself. When two atoms are joined by a covalent bond in a molecule, the shared electron pair lies closer to the atom that has the greater electronegativity. That atom thereby acquires a partial negative charge, and the atom of lesser electronegativity is left with a partial positive charge. Electronegativity is related to electron affinity and ionization energy, but it refers to bonded rather than isolated atoms.

Electronegativity is a qualitative concept that cannot be measured. However, it can be estimated by several methods, the most widely used of which is the **Pauling electronegativity scale.** Electronegativity values on this scale run from 0.7 for francium to 4.0 for fluorine, the element with the greatest tendency to attract bonding electrons. Pauling electronegativities for the elements, given in Figure 8.5, are obtained from bond energies.

Bond energies are discussed in Section 8.5.

An earlier electronegativity scale, devised by Mullikan in 1934, is based on the difference between an element's ionization energy and its electron affinity. The slight numerical differences that exist among the various scales are not important, as electronegativities are not used for precise calculations. Generally speaking, metals have electronegativities less than 2.0, and the metalloids and nonmetals have electronegativities greater than 2.0. Electronegativity generally increases as atomic number increases in a period, and decreases as atomic number increases in a group (Figure 8.6).

The values in Figure 8.5 may be used in a qualitative way to predict bond moments. The larger the electronegativity *difference* between two elements, the more polar the bond between them. The negative end of the bond moment is at the atom with the greater electronegativity.

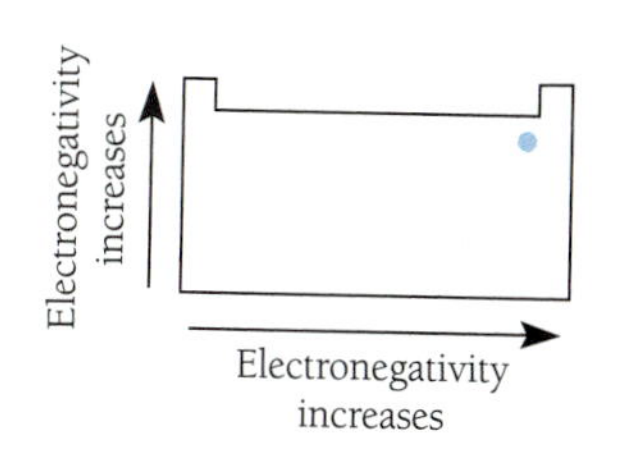

Figure 8.6 Periodic trends in electronegativity.

EXAMPLE 8.11 Bond Polarity

Rank the following bonds in order of increasing polarity, and identify the direction of the bond moment: H—Cl, Br—Br, I—Br, C—H, C—F.

Analysis The bond moment is related to the electronegativity *difference*. The atom with the greater electronegativity bears the partial negative charge.

Plan Look up the electronegativities in Figure 8.5, and calculate the difference for each bond. Arrange the bonds in order of increasing difference, and draw an arrow over them pointing to the element with the greater electronegativity.

Work

Bond	H—Cl	Br—Br	I—Br	C—H	C—F
EN	2.1 3.0	2.8 2.8	2.5 2.8	2.5 2.1	2.5 4.0
ΔEN	0.9	0	0.3	0.4	1.5

Results

Order of Polarity	Br—Br	<	I—Br	<	C—H	<	H—Cl	<	C—F
Polarity Direction	Br—Br		$\overset{+\!\!\longrightarrow}{\text{I—Br}}$		$\overset{\longleftarrow\!\!+}{\text{C—H}}$		$\overset{+\!\!\longrightarrow}{\text{H—Cl}}$		$\overset{+\!\!\longrightarrow}{\text{C—F}}$

Note that the Br—Br bond lacks an arrow: since the electronegativity difference is zero, there is no bond moment.

Exercise

Which of the following bonds is most polar? In which bond(s) does carbon lie at the negative end: Cl—C, H—B, Be—H, C—F, C—P, N—S?

Answer: The C—F bond is the most polar. Only in the C—P bond does carbon lie at the negative end.

Electronegativity is an additional guide to the choice of the central atom when Lewis structures are drawn. Because of the periodic trends in electronegativity, the "lower or to the left" rule (page 315) is generally equivalent to the rule "choose the least electronegative element as the central atom." Either rule may be used, but neither is absolutely reliable. Doubtful cases should still be checked by evaluating formal charges.

EXAMPLE 8.12 Electronegativity and Lewis Structures

Using electronegativity to determine the most likely geometrical arrangement of atoms, draw the Lewis structure of thionyl chloride, $SOCl_2$.

Plan Choose the central atom, then follow the usual procedure. Use formal charges to check the result.

Work

There are two unique atoms in the molecule, hence there are two possible beginning arrangements of the nuclei.

O—S(—Cl)(—Cl) and S—O(—Cl)(—Cl)

A B

Of these, A is preferable because the electronegativity of the central atom (S, EN = 2.5) is lower than that of the terminal atoms (O, EN = 3.5; and Cl, EN = 3.0). The molecule has 26 valence electrons, which can be used to draw the bonds and complete all octets as follows.

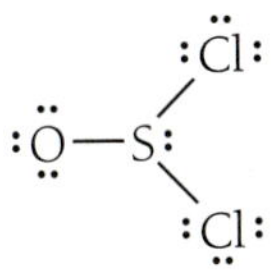

:Ö—S:(—:Cl:)(—:Cl:)

This structure has nonzero formal charges (oxygen, −1 and sulfur, +1), and is therefore suspect. Structure B is no improvement, not only because the most electronegative element is in the center, but because it too has +1 and −1 formal charges. However, as a third-period element, sulfur often has an expanded valence shell. With this in mind, we can write a better structure in which all atoms have a formal charge of zero.

:Ö—S:(—:Cl:)(—:Cl:) ⟶ Ö=S:(—:Cl:)(—:Cl:)

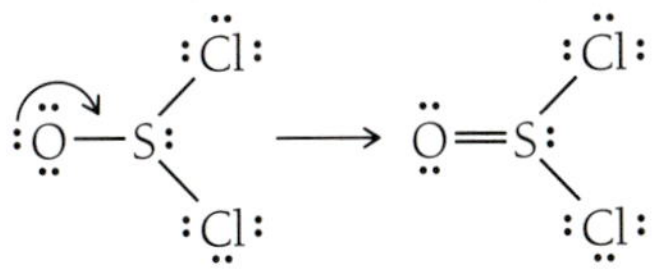

Exercise

Show that no structure can be drawn for arrangement B, above, in which all rules are satisfied and there are no formal charges.

Answer: Trial and error, or note from Table 8.1 that oxygen cannot have a formal charge of zero when bonded to three other atoms.

An additional use of electronegativity is in chemical nomenclature. In naming and writing the empirical formulas of binary compounds, the more electronegative element is mentioned last. The binary oxygen-fluorine compound is oxygen difluoride, OF_2, but the chlorine counterpart is dichlorine oxide, Cl_2O.

Bond Character

In the preceding sections, ionic and covalent bonding have been presented as if they were as clearly distinguishable as black and white. Either the solid consists of ions or it does not. Either an electron is transferred from one atom to another or it is not. But, as is often the case, the real world is not as neat and simple as the theories we use to explain it. Some compounds are in the borderline area between the two categories.

Compounds that separate into H^+ and an anion when dissolved in water are called *acids*, and their properties are discussed in Chapters 15 and 16.

Although clearly a polar covalent molecule, HCl nonetheless separates into H^+ and Cl^- ions when dissolved in water. Even NaCl, our prototype ionic compound, is not completely ionic. In the gaseous state, sodium chloride exists as ion pairs rather than as independent ions. If you were asked to write the Lewis structure of a covalently bonded NaCl molecule, you would very quickly come up with Na:C̈l:.

The bonding in the NaCl molecule/ion pair may be represented as a resonance hybrid, to which the ion pair A contributes much more importantly than the polar molecule B.

$$\underset{\textbf{A}}{Na^+\text{---}:\ddot{Cl}:^-} \longleftrightarrow \underset{\textbf{B}}{\overset{+\!\!\longrightarrow}{Na{-}\ddot{Cl}:}}$$

If the sodium atom's single valence electron is *completely* transferred to the chlorine atom, the result is called an ionic bond. If the electron is only *partially* transferred, the result is a polar covalent bond. In fact, there is no sharp dividing line between the two situations. That is, there is no sharp dividing line between ionic and covalent bonding. Instead there is a continuous range of bond types, often numerically expressed on a scale of **percent ionic character.** Percent ionic character in diatomic molecules can be calculated from measured values of bond moment and internuclear distance. For the molecules LiF, HF, and HCl the ionic characters are 90%, 40%, and 20%, respectively.

Internuclear distance, also called *bond length,* is discussed in Section 8.5.

The amount of ionic character in a bond depends on the difference between the electronegativities of the two bonded atoms: the greater the electronegativity difference, the greater the ionic character (Figure 8.7). The largest electronegativity differences are found in the alkali halides, but even in these compounds the bonds are not 100% ionic. In contrast, bonds having 100% *covalent character* do exist, in compounds such as H_2, O_2, Br_2, and so on. In these molecules the electronegativity difference, and the percent ionic character, is truly zero.

It is only in the solid state and the liquid state (discussed in Chapter 11) that a significant qualitative difference between covalent and ionic compounds exists. Liquid and solid ionic compounds consist of charged particles, while in the gaseous state they are better described as ion pairs or clusters whose overall charge is neutral. Covalent compounds, when pure, consist of neutral particles in the gas, liquid, and solid states (Figure 8.8).

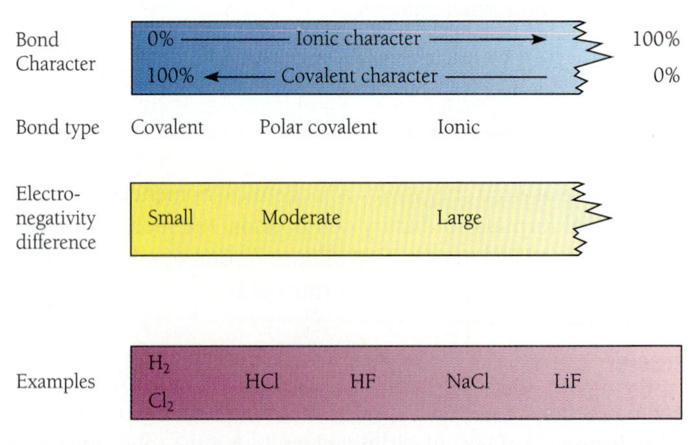

Figure 8.7 As the electronegativity difference between two atoms increases, the percent ionic character of the bond joining them rises from zero in molecules like H_2 and Cl_2 to about 90% (in LiF).

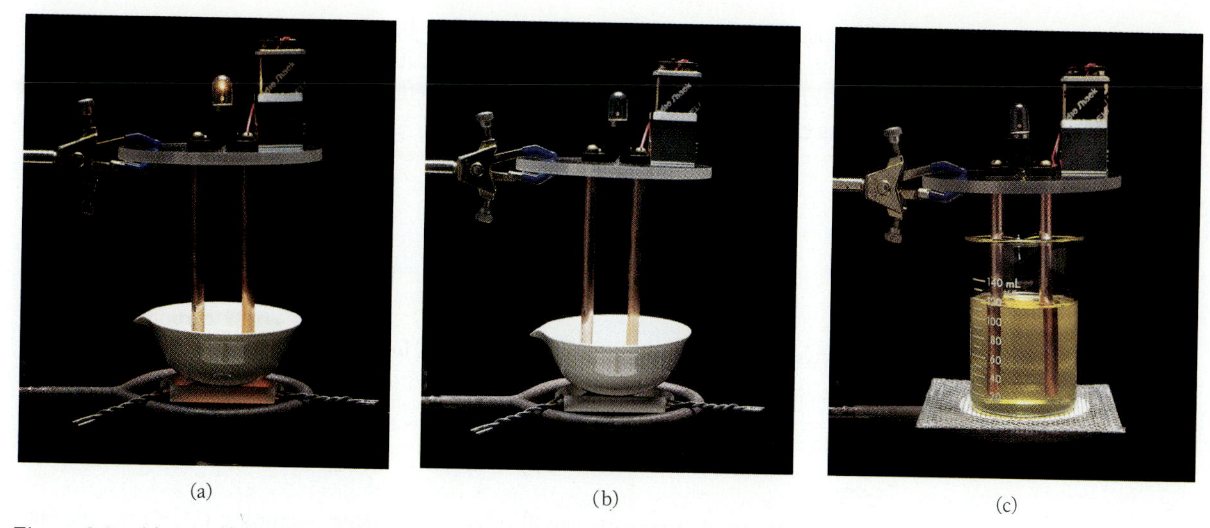

Figure 8.8 (a) At sufficiently high temperature, ionic compounds melt to form electrically conducting liquids. (b) At room temperature, ionic compounds are nonconducting solids. Although they consist of ions rather than neutral molecules, ionic solids do not conduct electricity because their ions are not free to move. (c) Since it is a mixture of molecular compounds and contains no ions, oil does not conduct electricity.

8.5 PROPERTIES OF COVALENT BONDS

In addition to their order and polarity, covalent bonds can be characterized by their length and strength. *Bond lengths* are useful in understanding the size and shape of molecules, and *bond strengths* have an important influence on chemical reactivity of molecules.

Bond Length

The attractive force of the covalent bond pulls atoms together until an opposing force, the mutual repulsion of the electron clouds, becomes strong enough to prevent closer approach. At this point of balance of the forces, the distance between the nuclei has a characteristic value called the **bond length.** Normally a few hundred picometers, the length of the covalent bond depends on bond order, the nature of the two atoms, and to a lesser extent on other atoms in the molecule. As we saw in Chapter 7, the bond length is equal to the sum of the *covalent radii* of the bonded atoms. Bond lengths can be measured by x-ray diffraction (Chapter 11), and also, very accurately, by microwave spectroscopy.

The effect of bond order on bond length can be seen in three molecules having C—C bonds of different order.

```
     H  H          H       H
     |  |           \     /
  H—C—C—H            C=C          H—C≡C—H
     |  |           /     \
     H  H          H       H
```

Molecule	ethane	ethene	ethyne
C—C Bond Order	1	2	3
Length (pm) of C—C Bond	154	134	120

An atom's covalent radius depends on whether the atom is singly, doubly, or triply bonded.

Compared to the single bond, the double bond is shorter because the stronger bonding force of two electron pairs pulls the atoms closer together. The triple bond, with three electron pairs, is shorter still. The effect of bond order is similar with other atoms capable of multiple bonding.

Molecule	methanol	formaldehyde	carbon monoxide
Structure	$H_3C{-}O{-}H$	$H_2C{=}O$	:C≡O:
C—O Bond Order	1	2	3
Length (pm) of C—O Bond	143	123	113

The effect of atom type can also be seen in these examples: C—O bonds, both single and multiple, are somewhat shorter than C—C bonds. Finally, there is a slight influence from other parts of the molecule: the length of the C—H bond is 109 pm in methane (CH_4), 110 pm in ethane, 107 pm in ethene, and 106 pm in ethyne. The similarity of these C—H bond lengths suggests that an *average* value for a certain bond type would be a generally useful concept. Such average values are collected in Table 8.3.

TABLE 8.3 Average Length(pm) of Covalent Bond

Single Bonds

Period	1	2				3				4	5
	H	C	N	O	F	Si	P	S	Cl	Br	I
H	58	110	98	94	92	145	138	132	127	142	161
C		154	147	143	141	194	187	181	176	191	210
N			140	136	134	187	180	174	169	184	203
O				132	130	183	176	170	165	180	199
F					128	181	174	168	163	178	197
Si						234	227	221	216	231	250
P							220	214	209	224	243
S								208	203	218	237
Cl									198	213	232
Br										228	247
I											266

Multiple Bonds

C=C	134	N=N	123	C=N	127	C=O	123	N=O	119
C≡C	120	N≡N	109	C≡N	116	C≡O	113	N≡O	108

EXAMPLE 8.13 Using the Table of Average Bond Lengths

Estimate the length of (a) the C—N bonds in cyanogen, C_2N_2, and (b) the C—O bond in the carbonate ion.

Analysis/Plan Since bond length depends on bond order, first draw the Lewis structure. Then use the values of bond lengths given in Table 8.3.

Work

a. The Lewis structure of cyanogen, :N≡C—C≡N:, shows that both C—N bonds are of order 3. Therefore the value in Table 8.3, 116 pm, is a good estimate.

b. The carbonate ion is a resonance hybrid of the structures shown on page 324. The average C—O bond order is $4/3$, for which there is no corresponding entry in Table 8.3. There is no strict arithmetic relationship between bond order and bond length, so we can only say that the length should be intermediate between 143 pm (for a single bond) and 123 pm (for a double bond). The measured value (in $CaCO_3$) is 129 pm.

Exercise

Estimate the length of the C—O bond in phosgene, $COCl_2$.

Answer: 123 pm.

Bond Energy

A diatomic molecule can be caused to break apart or *dissociate* into its two component atoms by several means, including heating and absorption of a photon.

$$O_2(g) \longrightarrow 2\ O(g); \qquad \Delta H = 498\ kJ \qquad (8\text{-}8)$$

$$CO(g) \longrightarrow C(g) + O(g); \qquad \Delta H = 1076\ kJ \qquad (8\text{-}9)$$

The process is always endothermic, since energy is required to overcome the attractive force of the covalent bond joining the two atoms. Conversely, energy is *released* when a bond is *formed.* The energy required to break a covalent bond is called the **dissociation energy** or **bond energy (kJ mol^{-1})**; the term **bond strength** is also used. The notation **D(X—Y)** is used to denote the energy of the X—Y bond. Equations (8-8) and (8-9) show that D(O═O) = 498 kJ mol^{-1} in oxygen and D(C═O) = 1076 kJ mol^{-1} in carbon monoxide. Like bond length, bond energy varies with bond order and the type of atoms bonded together. Bond energies are measured by spectroscopy or by the calorimetric methods for heats of reaction discussed in Chapter 4. Numerical values of bond energies always refer to dissociation of isolated molecules, that is, to gas-phase dissociation reactions.

EXAMPLE 8.14 Bond Dissociation Energy

It is found that the maximum wavelength of light capable of dissociating gas-phase HCl is 227 nm. What is the energy of the H—Cl bond?

Analysis

Target: An energy, in $kJ\ mol^{-1}$.

Relationship: The energy carried by a photon is $E = h\nu = hc/\lambda$. Since energy decreases as wavelength increases, the maximum effective wavelength corresponds to the minimum energy necessary to break the bond.

Plan Convert the given wavelength to energy, then multiply by Avogadro's number to express the bond energy in molar units.

Work

The energy carried by light is discussed in Chapter 6.

1. The energy of one photon is

$$E = \frac{hc}{\lambda} = \frac{(6.63 \times 10^{-34}\ J\ s) \cdot (3.00 \times 10^{8}\ m\ s^{-1})}{277 \times 10^{-9}\ m}$$

$$= 7.181 \times 10^{-19}\ J$$

Thus, the energy to dissociate one molecule is

$$= 7.18 \times 10^{-19}\ J\ molecule^{-1}$$

2. The dissociation energy or bond energy per mole is

$$D(H{-}Cl) = \left(\frac{7.181 \times 10^{-19}\ J}{1\ molecule}\right) \cdot \left(\frac{6.02 \times 10^{23}\ molecules}{1\ mol}\right)$$

$$= 4.32 \times 10^{5}\ J\ mol^{-1} = 432\ kJ\ mol^{-1}$$

Check The units cancel correctly, and the magnitude of the answer is close to the average value given in Table 8.4.

Exercise

Calculate the maximum wavelength of light capable of causing dissociation of N_2, for which $D(N{\equiv}N) = 945\ kJ\ mol^{-1}$.

Answer: 127 nm.

The relationship between bond energy and thermochemical heats of reaction is more complicated for polyatomic molecules. For instance, the energy required for *atomization* (complete dissociation) of one mole of water is

$$H_2O(g) \longrightarrow O(g) + 2\ H(g); \qquad \Delta H = 920\ kJ \qquad (8\text{-}10)$$

Since in this reaction *two* O—H bonds are broken, it might be assumed that each had a strength of $920/2 = 460\ kJ\ mol^{-1}$. Such is not the case. Breaking the *first* bond requires 489 kJ, so $D(HO{-}H) = 489\ kJ\ mol^{-1}$. Breaking the *second* bond is easier,

TABLE 8.4 Average Bond Energies in kJ mol^{-1}*

Single Bonds

Period		2				3			4	5
	H	C	N	O	F	Si	S	Cl	Br	I
H	436	413	391	467	565	323	347	427	363	295
C		347	305	358	485	301	259	339	276	240
N			159	201	272			200	243	
O				146	190	368		203		234
F					154		327	253	237	
Si						176	226	360	289	213
S							266	253	218	
Cl								239	218	208
Br									193	175
I										149

Multiple Bonds

C=C	615	N=N	418	C=N	615	C=O	750	O=O	498
C≡C	840	N≡N	945	C≡N	891	C≡O	1076		

* Missing entries correspond to values that are not well established.

and D(O—H) = 431 kJ mol^{-1}. The total **atomization energy,** the energy required to break *all* bonds in a molecule, of course conforms to Hess's law of reaction heat summation.

$$
\begin{array}{ll}
H_2O(g) \longrightarrow OH(g) + H(g); & \Delta H_1 = 489 \text{ kJ} \\
OH(g) \longrightarrow H(g) + O(g); & \Delta H_2 = 431 \text{ kJ} \\
\hline
H_2O(g) \longrightarrow O(g) + 2\,H(g); & \Delta H = \Delta H_1 + \Delta H_2 = 920 \text{ kJ}
\end{array}
$$

These two dissociation energies of the O—H bonds in water are different because the processes are different. More energy is required to remove a hydrogen atom from a relatively stable water molecule than from OH, a less stable fragment of a molecule. Unlike the successive ionization energies discussed in Section 7.5, however, these successive dissociation energies show no regular increase or decrease. The four successive dissociation energies of the C—H bonds in methane (CH_4) are 435, 453, 425, and 339 kJ mol^{-1}. For most molecules these individual values are unknown, and an *average* value (413 kJ mol^{-1} in the case of methane) is used to characterize the bond energies of a polyatomic molecule.

Average bond energies for a given type of bond vary somewhat from one compound to another. For instance the average C—H bond energy is 401 kJ mol^{-1} in C_2H_2, 409 kJ mol^{-1} in C_2H_4, and 413 kJ mol^{-1} in C_2H_6. For most bond types the spread is not more than 10 to 20 kJ mol^{-1} among different compounds. Values for some common bond types are given in Table 8.4.

Average bond energies are useful quantities in thermochemical calculations. They must be used with caution, however, as the bonds in particular molecules can differ from the average by 10 to 20 kJ mol^{-1}.

EXAMPLE 8.15 Calculating Bond Energies from Heats of Reaction

For the atomization reaction $NH_3(g) \rightarrow N(g) + 3\,H(g)$, $\Delta H = 1173$ kJ mol^{-1}. Calculate the average value of D(N—H) in ammonia.

Analysis/Relationship/Work The given reaction consists of the breaking of all three N—H bonds. The average bond energy is one third of the total.

$$D(N{-}H)[avg] = \frac{1173}{3} = 391 \text{ kJ mol}^{-1}$$

Exercise

The heat of atomization of ozone is 604 kJ mol^{-1}. Calculate the average bond energy and compare it with the values for D(O—O) and D(O═O) given in Table 8.4.

Answer: 302 kJ mol^{-1}. The value is intermediate between the values for O—O and O═O, as expected since ozone is a resonant species with a bond order of 3/2.

Estimation of Thermochemical Quantities

Bond dissociation is an endothermic process, so ΔH is a positive quantity. Bond formation is an exothermic process, with a negative value of ΔH.

The average bond energies of Table 8.4 are useful in estimating heats of reaction. Regardless of the actual mechanism, any chemical reaction can be *thought of* as proceeding stepwise. In the first step, all reactants are completely dissociated into their component atoms. The heat of reaction for this step is the total bond energy of all the bonds in the molecule. In the second step, the atoms are reassembled into the products. The total bond energy of all the newly formed bonds, *expressed as a negative quantity*, is the heat of reaction of the second step. The overall heat of the reaction is found from Hess's law.

$$\begin{array}{lll} \text{Step 1} \quad \text{reactants} \longrightarrow \text{atoms;} & \Delta H_1 \text{ (a positive quantity)} \\ \text{Step 2} \quad \text{atoms} \longrightarrow \text{products;} & \Delta H_2 \text{ (a negative quantity)} \\ \text{overall reaction:} \quad \text{reactants} \longrightarrow \text{products;} & \Delta H = \Delta H_1 + \Delta H_2 \end{array}$$

As an example, consider the reaction $H_2 + F_2 \rightarrow 2\,HF$.

$$\begin{array}{llll} \text{Step 1} & H_2(g) \longrightarrow 2\,H(g); & \Delta H_1 = D(H{-}H) & = 436 \text{ kJ} \\ & F_2(g) \longrightarrow 2\,F(g); & \Delta H_2 = D(F{-}F) & = 154 \text{ kJ} \\ \text{Step 2} & 2\,H + 2\,F(g) \longrightarrow 2\,HF(g); & \Delta H_3 = -2 \cdot D(H{-}F) & = -1130 \text{ kJ} \\ \hline \text{overall} & H_2(g) + F_2(g) \longrightarrow 2\,HF(g); & \Delta H = \Delta H_1 + \Delta H_2 + \Delta H_3 & \\ & & = D(H{-}H) + D(F{-}F) & \\ & & \quad - 2 \cdot D(H{-}F) & \\ & & = (436 + 154 - 1130) \text{ kJ} & \\ & & = -540 \text{ kJ} & \end{array}$$

The heat of formation of HF(g) is *one half* of the heat of the reaction in which *two* moles of HF are formed from $H_2(g)$ and $F_2(g)$. $\Delta H_f^\circ \approx (-540/2) = -270$ kJ mol^{-1}. The exact value, given in Appendix G, is $\Delta H_f^\circ[HF(g)] = -271.1$ kJ mol^{-1}.

The use of average bond energies to estimate the heat of a reaction involving somewhat larger molecules is illustrated in Example 8.16.

EXAMPLE 8.16 Estimating a Heat of Reaction

Estimate the heat of hydrogenation of ethene in the reaction $C_2H_4(g) + H_2(g) \rightarrow C_2H_6(g)$. Compare this to the actual value, obtained from tabulated heats of formation by the methods presented in Chapter 4.

Analysis/Relationship The heat of reaction is the energy cost of breaking all the reactant bonds, less that regained from making all the product bonds. We need to know the type (single, double, or triple) of each bond in the reactants and in the products.

Plan

1. Draw the Lewis structures of reactants and products to determine the type and order of each bond.
2. Using Table 8.4, find the sum of the bond energies of all the broken bonds, and subtract from it the sum of the energies of all the newly formed bonds.

Work

1. The reaction, written with Lewis structures, is

$$\begin{matrix} H & & H \\ & C{=}C & \\ H & & H \end{matrix} + H{-}H \longrightarrow H{-}\overset{\displaystyle H}{\underset{\displaystyle H}{C}}{-}\overset{\displaystyle H}{\underset{\displaystyle H}{C}}{-}H$$

2. Bond breaking requires energy as follows.

4 C—H bonds	$4 \cdot 413$ kJ mol^{-1} = 1652 kJ	
1 C=C bond	615	
1 H—H bond	436	
	Total 2703 kJ	$(= \Delta H_1)$

Bond formation liberates energy as follows.

6 C—H bonds	$6 \cdot 413$ kJ mol^{-1} = 2478 kJ	
1 C—C bond	347	
	Total 2825 kJ	$(= -\Delta H_2)$

By Hess's law, the heat of reaction is

$$\Delta H = \Delta H_1 + \Delta H_2$$
$$= 2703 + (-2825) = -122 \text{ kJ per mole of } C_2H_4$$

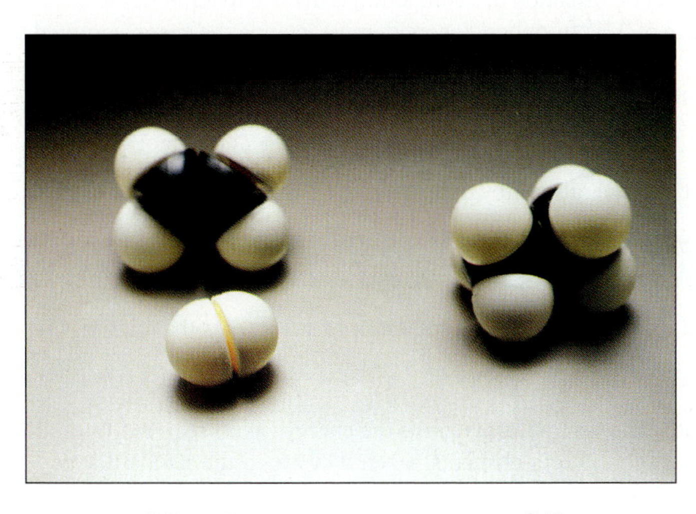

$C_2H_4 + H_2 \longrightarrow C_2H_6$

In these estimation problems, the final step can also be expressed as

$$\Delta H_{rxn} = \sum [\text{energy of broken bonds}] - \sum [\text{energy of formed bonds}]$$
$$= 2703 - 2825 = -122 \text{ kJ}$$

The hydrogenation reaction is estimated to be exothermic by 122 kJ. The actual heat of reaction, obtained from tabulated heats of formation, is

$$\Delta H_{rxn} = \sum [n \cdot \Delta H_f(\text{products})] - \sum [n \cdot \Delta H_f(\text{reactants})]$$
$$= \Delta H_f(C_2H_6) - [\Delta H_f(C_2H_4) + \Delta H_f(H_2)]$$
$$= -69.1 - [52.3 + 0]$$
$$= -121.4 \text{ kJ}$$

Note: The same result is obtained more simply by ignoring all bonds that are unchanged in the reaction, and considering only those that are broken or formed. In this hydrogenation reaction, for example,

	Broken Bonds	Formed Bonds
	C═C, H—H	C—C, 2 C—H
Total Energy	615 + 436 = 1051	347 + 2(413) = 1173

$$\Delta H_{rxn} = \sum [\text{energy of broken bonds}] - \sum [\text{energy of formed bonds}]$$
$$= 1051 - 1173 = -122 \text{ kJ}$$

Exercise

Estimate the heat of the gas-phase reaction

$$CH_4 + 2\ Br_2 \longrightarrow CH_2Br_2 + 2\ HBr$$

Answer: Exothermic by 66 kJ per mole of CH_4.

This method of estimating heats of reaction applies to *gas-phase* reactions, because it is based on bond energies in individual, isolated molecules. When appropriate heats of vaporization are known, however, the method can be more widely applied. For example, approximate bond energies can be used to estimate unknown heats of formation.

The heat of formation of a compound is defined in terms of the *standard states,* so an estimate of ΔH_f° from bond energies must include the energy required to produce gas-phase atoms from the standard states of the elements. For carbon-containing compounds, the heat of vaporization is needed.

$$C(s) \longrightarrow C(g); \qquad \Delta H^\circ = +717 \text{ kJ mol}^{-1}$$

Note that this quantity is, by definition, the heat of formation of gaseous carbon.

EXAMPLE 8.17 Estimating a Heat of Formation

Use bond energies to estimate the heat of formation of ethanol, $CH_3CH_2OH(g)$.

Analysis

Target: The heat of formation is the energy (heat) required to produce ethanol from its constituent elements, when they are in the standard state.

Relationship: Here, as in similar problems in Chapter 4, Hess's law is the central relationship. Since average bond energies are to be used, we need to know what bonds the molecule contains. This information is obtained from the Lewis structure.

Plan

1. Devise a sequence of reactions, the sum of which is the *formation* reaction of ethanol.

$$2\ C(s) + \tfrac{1}{2}\ O_2(g) + 3\ H_2(g) \longrightarrow [\text{atoms}] \longrightarrow CH_3CH_2OH(g)$$

2. Draw the Lewis structure. From it, determine what bonds are formed and how much energy is released in their formation. The total of the ΔH values for all steps is the heat of formation of ethanol.

Work

The first part of the sequence, the formation of atoms, is

(a) $2\ C(s) \longrightarrow 2\ C(g); \quad \Delta H_a = 2 \cdot 717 = 1434\ kJ$

(b) $\tfrac{1}{2}\ O_2(g) \longrightarrow O(g); \quad \Delta H_b = \dfrac{D(O{=}O)}{2} = 249\ kJ$

(c) $3\ H_2(g) \longrightarrow 6\ H(g); \quad \Delta H_c = 3 \cdot [D(H{-}H)] = 1308\ kJ$

The subtotal of (a), (b), and (c) is

(1) $C(s) + \tfrac{1}{2}\ O_2(g) + 3\ H_2(g) \longrightarrow \text{atoms}; \quad \Delta H_1 = \Delta H_a + \Delta H_b + \Delta H_c = 2991\ kJ$

The remaining step is the formation of the bonds in the Lewis structure of ethanol,

The heats of reaction are *negative* quantities in reactions of bond *formation*.

```
                  H   H
                  |   |
(2) atoms ---> H--C---C--O--H;
                  |   |
                  H   H
```

$\Delta H_2 = (-1) \cdot (\text{sum of bond energies})$

$$\begin{aligned}\Delta H_2 &= -1 \cdot \{5 \cdot [D(C{-}H)] + D(C{-}C) + D(C{-}O) + D(O{-}H)\} \\ &= -\{5 \cdot [413] + 347 + 358 + 467\} = -3237\ kJ\end{aligned}$$

The overall reaction is

$$2\ C(s) + \tfrac{1}{2}\ O_2(g) + 3\ H_2(g) \longrightarrow CH_3CH_2OH(g); \quad \begin{aligned}\Delta H &= \Delta H_1 + \Delta H_2 \\ &= 2991 + (-3237) \\ &= -246\ kJ\end{aligned}$$

The estimated heat of formation is $-246\ kJ\ mol^{-1}$, compared with the measured value of $-235\ kJ\ mol^{-1}$. The difference, although not large, is due to the use of *average* bond energies.

Exercise

There are two good Lewis structures that can be written for the molecular formula C_2H_6O. One is for ethanol, above, while the other is

```
     H     H
     |     |
  H—C—O—C—H
     |     |
     H     H
```

This structure also corresponds to an existing compound, methyl ether. (Two compounds having the same molecular formula but different structural formulas are known as *isomers*. More will be said about this in later chapters.)

Estimate the heat of formation of methyl ether.

Answer: -203 kJ mol^{-1}.

8.6 MOLECULAR GEOMETRY

A great deal of experimental evidence indicates that molecules have a characteristic, definite **molecular shape**—a well-defined spatial arrangement of the atoms with respect to one another. Furthermore, molecular shape strongly influences the chemical properties of a compound. Subtle differences in shape can profoundly affect chemical reactivity, particularly in reactions taking place in living organisms. Although modern physical measurements (such as spectroscopy) provide the best detailed information, the concept of molecular shape was deduced from chemical evidence more than a century ago.

Consider the molecular compound dichloromethane, CH_2Cl_2. The Lewis structure is a central C atom with two H atoms and two Cl atoms occupying terminal positions. If the shape is such that all five atoms lie in the same plane, there are two possible arrangements. These might be called "chlorines opposite" and "chlorines adjacent."

```
Cl   H            Cl   Cl
  \ /               \ /
   C       and       C
  / \               / \
H    Cl            H   H
```

Because of the difference in location of the chlorine atoms, these structures represent two different compounds with different properties. On the other hand, if the C atom occupies the center of a tetrahedron, and the other four atoms lie at the vertices, only one arrangement is possible. The two drawings shown at the top of p. 345 may at first sight *seem* to represent different arrangements of atoms, but in fact they are just two different views of the same structure. Unlike a square, a tetrahedron has no "opposite" vertices—each terminal atom in CH_2Cl_2 is adjacent to the other three. The experimental *fact* is that all samples of pure CH_2Cl_2, prepared by a variety of means, have identical chemical and physical properties. The inescapable conclusion is that there is only one substance, and that its *molecular shape* is tetrahedral rather than planar.

A tetrahedron is a four-sided solid. Each of the four identical sides is an equilateral triangle. It might also be described as a three-sided pyramid.

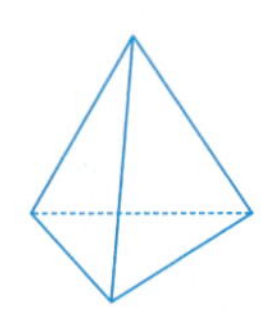

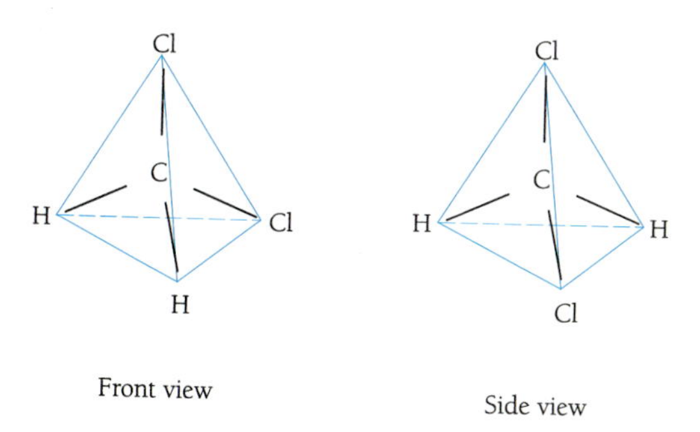

The blue lines show the outline of the tetrahedron, but have no physical significance. The black lines represent covalent bonds. The lone pairs on the chlorine atoms are not shown.

In other cases, certain molecular formulas were known to describe substances having two or more different sets of properties. For example, there are three different compounds having the molecular formula $C_2H_2Cl_2$.

A: Cl(H)C=C(Cl)H — B: Cl(H)C=C(H)Cl — C: $Cl_2C=CH_2$

In each of these molecules, all six atoms lie in the same plane. The different arrangements of atoms in the plane produce differences in the properties of these compounds.

Representation of Three-Dimensional Shapes

Visualization of molecular shapes is easy enough when all atoms lie in the same plane. As shown above for $C_2H_2Cl_2$, we need only to draw the shape with approximately correct bond lengths and angles between the bonds. A **bond angle** is the angle formed by imaginary lines joining the nucleus of an atom to the nuclei of two other atoms to which it is bonded. For example, the C=C—H bond angles in the $C_2H_2Cl_2$ molecules (A, B, and C, above) are all approximately 120°. Visualization is more difficult for three-dimensional molecules, however, and several methods are in common use. *Ball-and-stick* models show molecular shapes simply and accurately. Perspective drawings of these models do almost as well (Figure 8.9). More elaborate *space-filling* models, with balls scaled to the different covalent radii of different atoms, can be used to give a better representation of molecular volume (Figure 8.10). Both types of model

Figure 8.9 (a) Ball-and-stick model of dichloromethane. The carbon atom is black, the hydrogens are white, and the chlorines are green. (b) Perspective drawing of (a). These models indicate the relative positions of the nuclei well, but do not give a good representation of the relative sizes of the atoms.

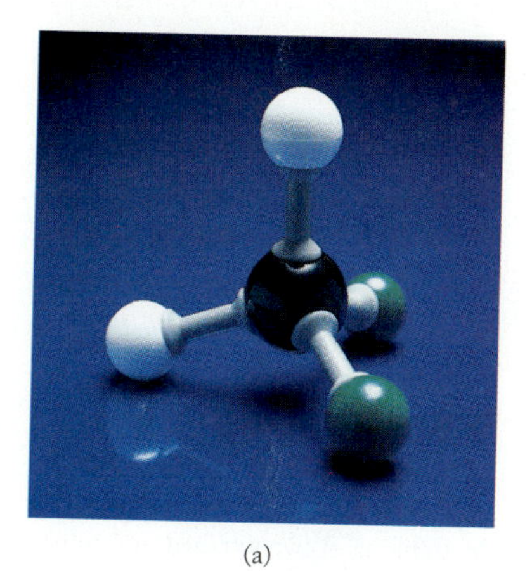

(a)

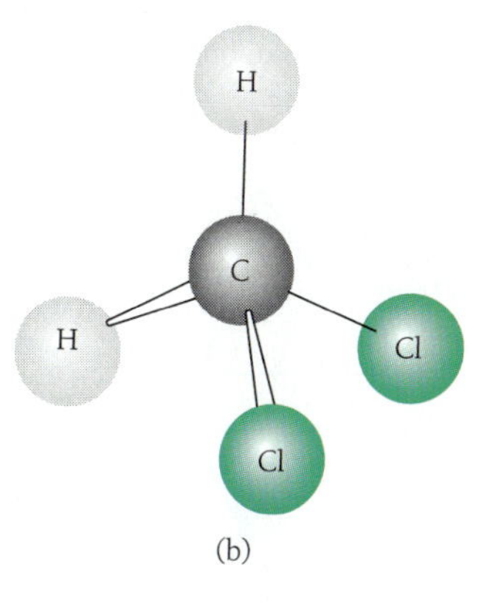

(b)

Figure 8.10 Space-filling model of ethanol, CH_3CH_2OH. Such models give a good idea of molecular volume, but the angular relationships among the atoms are not as obvious.

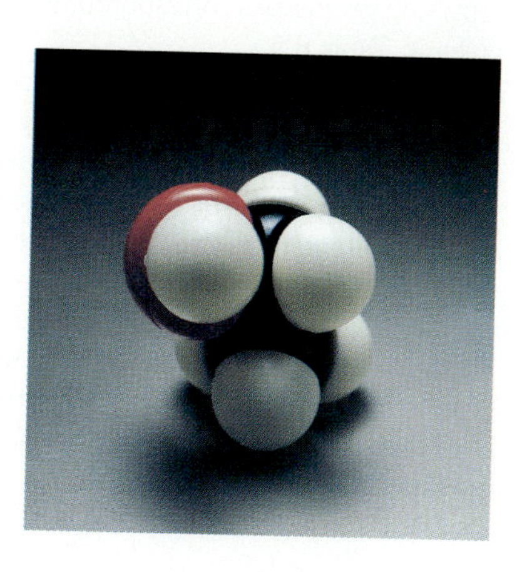

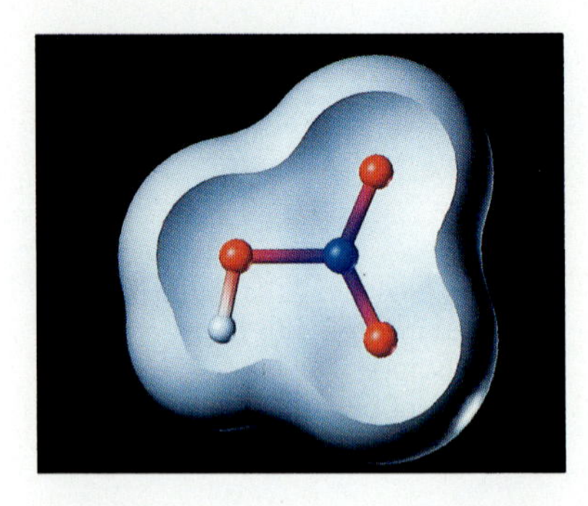

Figure 8.11 Molecular visualization is greatly enhanced by computer techniques, as shown by this ball-and-stick model of nitric acid inside a model of the molecular surface. (Model generated by J. Weber, University of Geneva, Switzerland.)

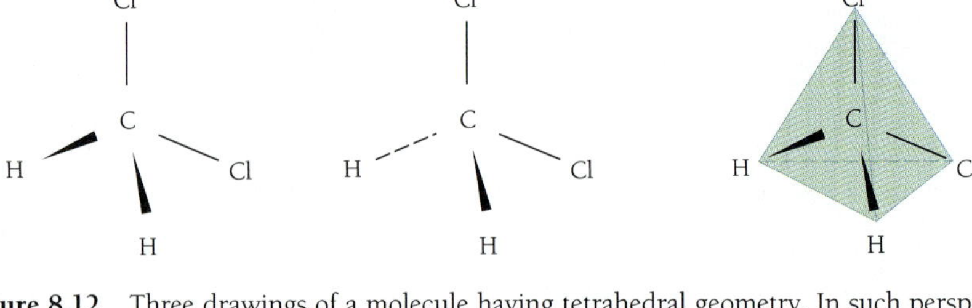

Figure 8.12 Three drawings of a molecule having tetrahedral geometry. In such perspective drawings, a wedge or dashed line indicates a bond to an atom below the plane of the paper, and an oppositely directed wedge represents a bond to an atom above the paper. The outline shape of the molecule is sometimes indicated by thin or colored lines.

can be generated by "molecular modelling" computer programs as well as by actual construction from a model kit (Figure 8.11).

A common pencil-and-paper representation uses wedges (to indicate perspective) for bonds pointing above the plane of the paper (toward the reader) or below the plane of the paper, and solid lines for bonds lying in the plane. Alternatively, a dashed line can be used to represent bonds pointing away from the reader (Figure 8.12). Thin or colored lines may be added to indicate the overall geometry.

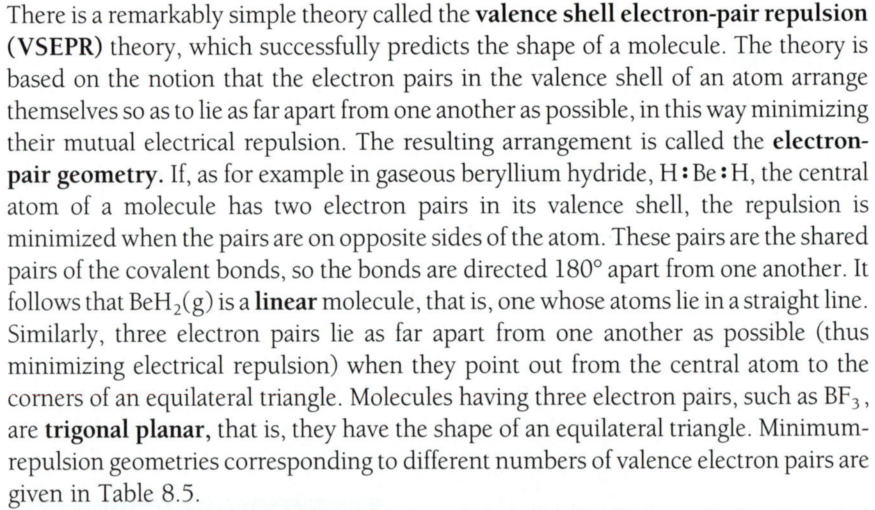

Electron-Pair Geometry and VSEPR

There is a remarkably simple theory called the **valence shell electron-pair repulsion (VSEPR)** theory, which successfully predicts the shape of a molecule. The theory is based on the notion that the electron pairs in the valence shell of an atom arrange themselves so as to lie as far apart from one another as possible, in this way minimizing their mutual electrical repulsion. The resulting arrangement is called the **electron-pair geometry.** If, as for example in gaseous beryllium hydride, H**:**Be**:**H, the central atom of a molecule has two electron pairs in its valence shell, the repulsion is minimized when the pairs are on opposite sides of the atom. These pairs are the shared pairs of the covalent bonds, so the bonds are directed 180° apart from one another. It follows that $BeH_2(g)$ is a **linear** molecule, that is, one whose atoms lie in a straight line. Similarly, three electron pairs lie as far apart from one another as possible (thus minimizing electrical repulsion) when they point out from the central atom to the corners of an equilateral triangle. Molecules having three electron pairs, such as BF_3, are **trigonal planar,** that is, they have the shape of an equilateral triangle. Minimum-repulsion geometries corresponding to different numbers of valence electron pairs are given in Table 8.5.

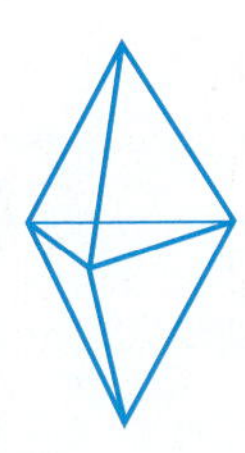

A trigonal bipyramid may be visualized by thinking of two tetrahedrons, joined together at their bases. The six faces of the trigonal bipyramid are equilateral triangles, identical to one another.

These geometries can be visualized with the aid of balloons, which, when tied together at the necks, tend to point in approximately the same directions as the electron pairs in the valence shell of the central atom in a molecule (Figure 8.13). Like valence electrons, the balloons assume the least crowded orientations. Indeed, VSEPR is often presented in terms of the least *crowding,* rather than the repulsion, of electron pairs.

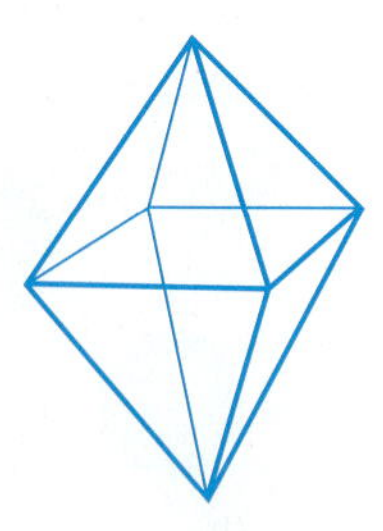

An octahedron is an eight-sided solid, which may be thought of as two four-sided pyramids placed base-to-base. The eight faces of an octahedron are equilateral triangles, identical to one another.

Some aspects of electron-pair geometries are not immediately obvious from the drawings in Table 8.5. The octahedral geometry of six electron pairs is completely symmetrical. All S—F bond lengths in the SF_6 molecule are the same, and, in chemical reactions, the behavior of each of the fluorine atoms is the same. Perspective drawings sometimes conceal this symmetry, but molecular models (Figure 8.14) show it clearly.

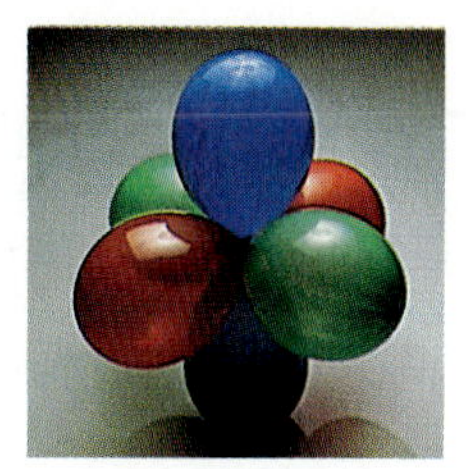

Figure 8.13 The electron-pair geometries described in Table 8.5 can be visualized by balloons.

The trigonal bipyramidal structure of five-pair molecules, like PCl_5, is not as symmetrical. It may be thought of as two tetrahedrons with their bases joined, with the P atom lying in the center of the equilateral triangle formed by the joined bases. Three of the Cl atoms lie in the same plane as the central P atom. They are trigonally arranged so that these three Cl—P—Cl angles are all 120°. These three bonds, and the

TABLE 8.5 Relationship Between Electron-Pair Geometry and Number of Valence Electron Pairs

Number of Electron Pairs in Valence Shell	Electron-Pair Geometry (Angle)	Example Molecule	Drawing
2	Linear (180°)	BeH_2	H — Be — H (180°)
3	Trigonal planar (120°)	BF_3	BF_3 (120°, 120°, 120°)
4	Tetrahedral (109.5°)	CH_4	CH_4 (109.5°)
5	Trigonal bipyramidal (90°, 120°, 180°)	PCl_5	PCl_5 (90°, 120°, 120°, 120°)
6	Octahedral (90°, 180°)	SF_6	SF_6 (90°, 90°, 90°)

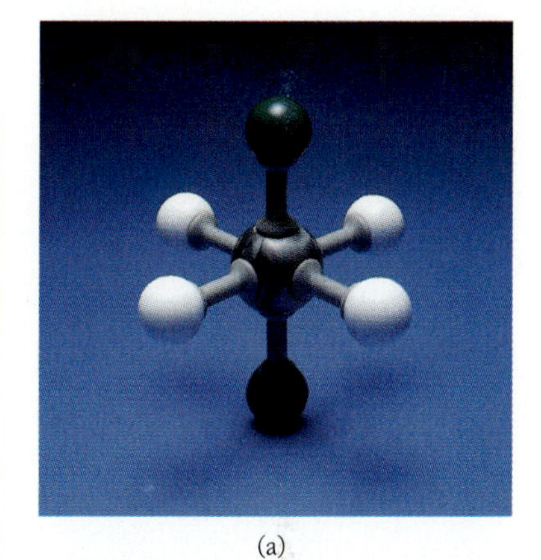

(a)

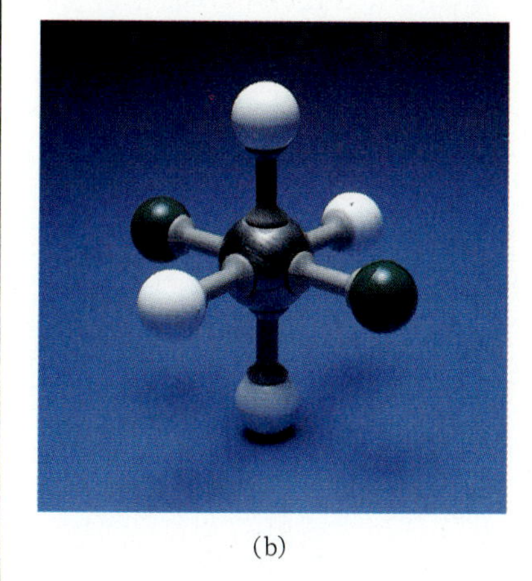

(b)

Figure 8.14 Two of the fluorine atoms in a model of SF_6 have been given a different color. The same model is photographed in different orientations to show that all six vertices of an octahedron are equivalent.

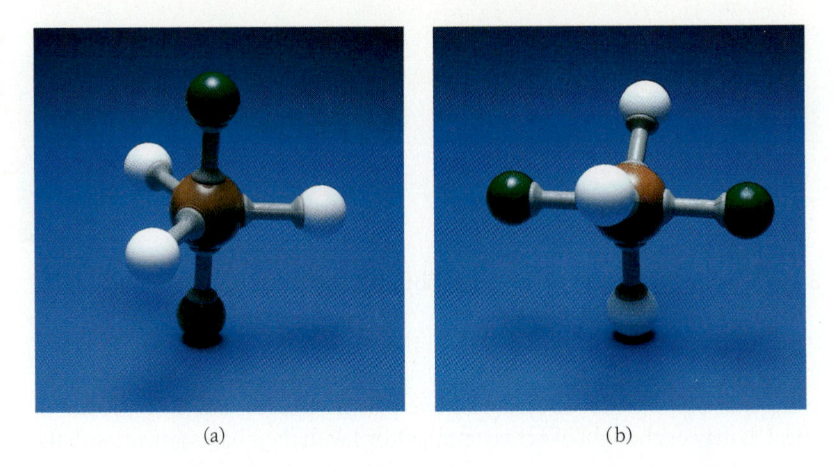

(a) (b)

Figure 8.15 (a) A model of PCl_5 with the axial and equatorial atoms in different colors. (b) The same model, turned on end. Axial and equatorial bonds in the trigonal bipyramidal structure are different from one another, both in their physical and chemical properties.

locations of the corresponding terminal atoms, are called **equatorial.** The other two Cl atoms lie in **axial** positions directly above and below the equatorial plane, forming a straight line with the P atom in the middle. The Cl(axial)—P—Cl(equatorial) angles are 90°. Trigonal bipyramidal molecules have two types of bonds: two equivalent axial bonds, and three equatorial bonds which are mutually equivalent but different from the axial bonds. In general, axial bonds are somewhat longer than equatorial bonds (219 and 204 pm, respectively, in PCl_5). The ball-and-stick model (Figure 8.15) illustrates this geometry.

Effect of Multiple Bonds

The two electron pairs of a double bond are both confined to the same region of space, between the two bonded atoms. These four electrons act together in repelling the other pairs in the valence shell. The carbon dioxide molecule has four electron *pairs* in the valence shell of carbon, but only two *electron regions* of mutually repelling negative charge. VSEPR theory therefore predicts (correctly) that O═C═O is a linear molecule, with a bond angle of 180°. In applying VSEPR theory, then, we must count *bonds* rather than electron pairs, because a bond, whether single, double or triple, is a single region of charge. Three examples of this way of counting are given below.

Compound	**hydrogen cyanide**	**formaldehyde**	**phosphorus oxychloride**
Lewis Structure	H—C≡N:	$H_2C=\ddot{O}$: (H, H bonded to C═O)	$Cl_3P=O$ (three :Cl: bonded to P═O, lone pairs shown)
Central-Atom Electron Pairs	4	4	5
Number of *Electron Regions* (number of bonds)	2	3	4
Electron-Pair Geometry	**linear**	**trigonal planar**	**tetrahedral**

The electron-pair geometry of resonant species is that predicted from *any one* of the resonance contributors, all of which have the same geometry. The nitrate ion is a resonance hybrid of three equivalent structures, each of which has a double bond and two single bonds. The geometry, with three electron regions, is trigonal planar.

Resonance contributors differ in placement of electrons, not nuclei.

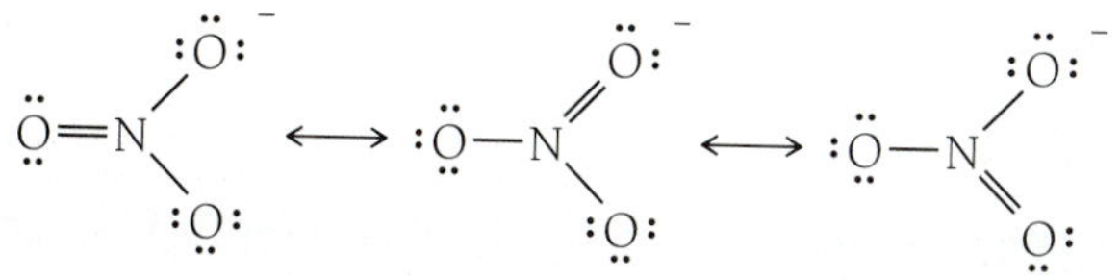

EXAMPLE 8.18 Electron-Pair Geometry

Predict the electron-pair geometry of the carbonate ion, CO_3^{2-}.

Analysis/Relationship According to VSEPR, the electron-pair geometry of a species is determined solely by the number of electron regions of the central atom.

Missing Information The Lewis structure is the *essential starting point* for VSEPR.

Plan

1. Determine the Lewis structure.
2. Count the electron regions of the central atom.
3. Read the geometry from Table 8.5.

Work

1. Placing the three O atoms symmetrically around the unique C atom and distributing the $[4 + (3 \cdot 6) + 2] = 24$ valence electrons, we find that the octet rule is satisfied for all atoms by the structure

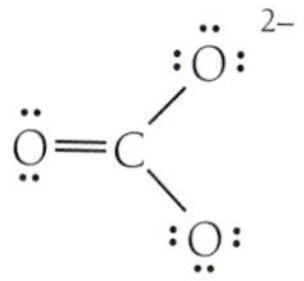

2. Carbonate ion is a resonance hybrid of three equivalent structures (Section 8.3), but since any one may be used to determine the geometry, we need not draw the others. The central atom has three electron regions: one double bond and two single bonds.
3. According to Table 8.5, structures with three electron regions are trigonal planar.

Exercise

Determine the electron-pair geometry of phosgene, $COCl_2$.

Answer: Trigonal planar.

Molecular Shape

Many molecules have nonbonding electrons (lone pairs) in the valence shell of the central atom. Both lone pairs and bonded pairs are regions of negative charge that repel one another, and they are treated alike in VSEPR theory. That is, the electron-pair

TABLE 8.6 Relationship between Electron-Pair Geometry and Molecular Shape for Central Atoms Having 2, 3, and 4 Electron Regions

Number of Electron Regions	Bonds	Lone Pairs	Electron Geometry	Molecular Shape	Example	Drawing
2	2	0	Linear	Linear	BeH_2	H : Be : H
	2	0	Linear	Linear	CO_2	O : : C : : O
3	3	0	Trigonal planar	Trigonal planar	BF_3	F, F, B, F
	3	0	Trigonal planar	Trigonal planar	CH_2O	H, O, C, H
	2	1	Trigonal planar	Bent	GeH_2	Ge, H, H
4	4	0	Tetrahedral	Tetrahedral	CH_4	H, C, H, H, H
	3	1	Tetrahedral	Pyramidal	NH_3	N, H, H, H
	2	2	Tetrahedral	Bent	H_2O	H, O, H

geometry is the same regardless of whether the electron regions are bonds or lone pairs. However, there is a difference in the way the *molecular geometry* is described. It is usual to describe the geometry of a molecule by the locations of its atoms, while ignoring the orientations of lone pairs. We will use the term *molecular shape* to mean the spatial arrangement of the atomic nuclei, and the term *electron-pair geometry* to mean the orientation of electron regions around the central atom. In general, molecular shape and electron-pair geometry are different in molecules having lone pairs.

Consider the water molecule: with two bonds and two lone pairs, the central oxygen atom has four electron regions, and the electron-pair geometry is therefore tetrahedral. The four electron pairs point outward to the four vertices of the tetrahedron, but only two of the vertices are occupied by hydrogen atoms. The three atomic nuclei all lie in the same plane, and the *molecular shape* is *bent* rather than tetrahedral.

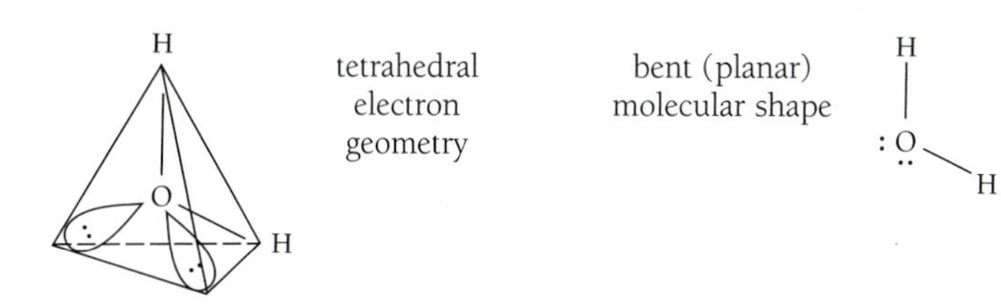

It does not matter which of the vertices are considered to be occupied by H atoms, as all vertices of a tetrahedron are equivalent to one another. The molecule is planar in any case, because any three points in space lie in one plane. Additional descriptions of molecular shape are gathered in Table 8.6.

EXAMPLE 8.19 Electron-Pair Geometry and Molecular Shape

What is the electron-pair geometry of the hydronium ion, H_3O^+? What is the molecular shape?

Analysis Both the electron-pair geometry and molecular shape are required. These are different if the central atom has lone pairs in its valence shell, and the same otherwise.

Plan Draw the Lewis structure, then look up the electron-pair geometry in Table 8.5 and the molecular shape in Table 8.6.

Work

1. Placement of the $[6 + (3 \cdot 1) - 1] = 8$ valence electrons so as to complete the valence shells of all three atoms leads to the structure

```
        +
      H
     /
H—O:
     \
      H
```

2. There are three bonds and one lone pair for a total of four electron regions, so the electron-pair geometry is tetrahedral. According to Table 8.6 the molecular shape is pyramidal.

Exercise

Describe the electron-pair geometry and molecular shape of the amide ion, NH_2^-.

Answer: Tetrahedral; bent.

Shapes of Expanded Valence Shell Molecules

An additional guide is needed to predict the shapes of molecules whose central atom has one or more lone pairs in an expanded valence shell. VSEPR theory includes the further postulate that nonbonding lone pairs repel each other more strongly and require somewhat more room than bonding pairs. The axial positions of the trigonal bipyramid are somewhat more crowded than the equatorial positions. An axial electron pair has *three* other pairs at right angles to it, while an equatorial pair has only *two* others at right angles (and two at 120°, which contribute much less to crowding). Thus, when the central atom has five pairs, any *lone pairs* occupy the less crowded equatorial positions.

All six vertices of an octahedron are geometrically equivalent, so there is only one possible shape for a molecule whose central atom has five bonded pairs and one lone pair. When there are four bonded pairs and two lone pairs, the least crowding occurs when the lone pairs are opposite to one another. The shapes of expanded-valence shell molecules are given in Table 8.7.

EXAMPLE 8.20 Expanded-Valence Shells and Molecular Shape

What is the electron-pair geometry of the central atom, and what is the molecular shape of the triiodide ion, I_3^-?

Analysis/Plan The approach is the same as in Example 8.19

molecular formula ⟶ Lewis structure ⟶ electron-pair geometry ⟶ molecular shape

Work

1. The triiodide ion has $(3 \cdot 7) + 1 = 22$ valence electrons. Either by trial and error, or by evaluating $M = (6 \cdot N) + 2 - V = -2$, we realize that the central atom has an expanded valence shell. The Lewis structure is

$$[:\ddot{\underset{..}{I}}-\ddot{\underset{..}{I}}{}^{\cdot\cdot}-\ddot{\underset{..}{I}}:]^-$$

With five electron regions, the electron-pair geometry is trigonal bipyramidal.

2. In the trigonal bipyramid, lone pairs occupy the equatorial positions. The two terminal atoms occupy the axial positions, and the molecular shape is linear.

Exercise

Determine the electron-pair geometry and the shape of the SF_5^- ion.

Answer: Octahedral; square pyramid.

TABLE 8.7 Relationship between Electron-Pair Geometry and Molecular Shape for Central Atoms Having 5 or 6 Electron Regions

Number of Electron Regions	Bonds	Lone Pairs	Electron Geometry	Molecular Shape	Example	Drawing
5	5	0	Trigonal bipyramidal	Trigonal bipyramidal	PCl_5	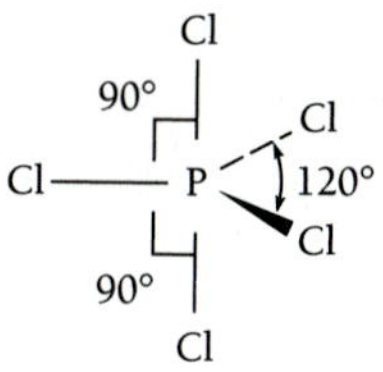 Cl, Cl, Cl, Cl, Cl around P; 90°, 90°, 120°
	4	1	Trigonal bipyramidal	See-saw	SF_4	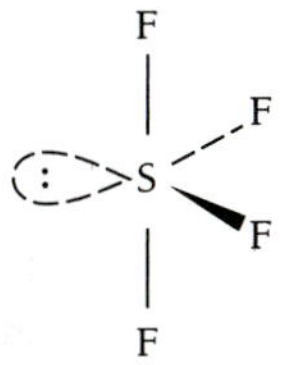 F, F, F, F around S
	3	2	Trigonal bipyramidal	T-shaped	$BrCl_3$	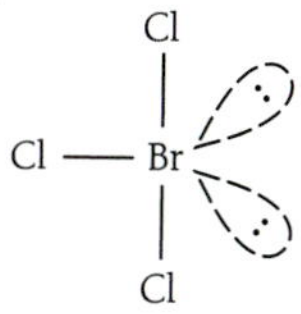 Cl, Cl, Cl around Br
	2	3	Trigonal bipyramidal	Linear	XeF_2	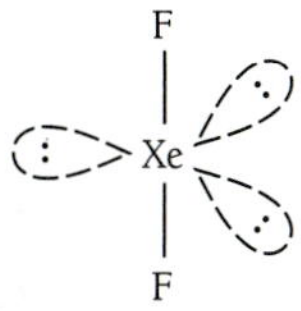 F, F around Xe
6	6	0	Octahedral	Octahedral	SF_6	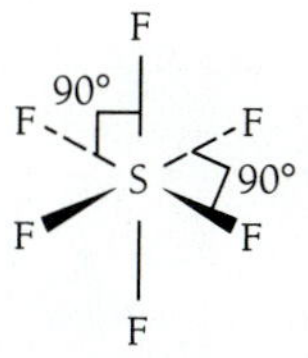 F, F, F, F, F, F around S; 90°, 90°
	5	1	Octahedral	Square pyramidal	IF_5	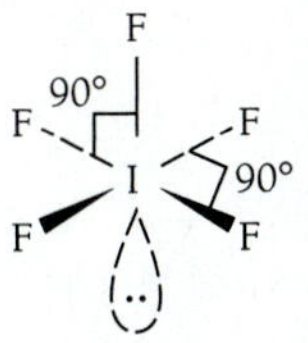 F, F, F, F, F around I; 90°, 90°
	4	2	Octahedral	Square planar	XeF_4	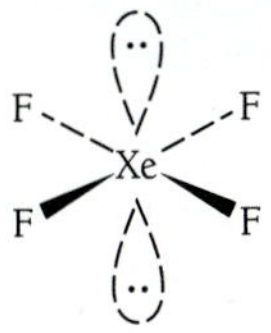 F, F, F, F around Xe

Bond Angles

The presence of lone pairs in a molecule causes the bond angles to deviate somewhat from those in the ideal electron-pair geometry. Because of its larger space requirement, a lone pair induces a slight *decrease* in the angles between the atoms bonded to the central atom. Thus, the two lone pairs in H_2O squeeze the bonding pairs closer together, so that the H—O—H bond angle is 105°, about 4° less than the perfect tetrahedral angle of 109.5°. Similarly, in SF_4 and $BrCl_3$ the axial bonds are bent slightly back from the lone pairs.

The effect of a multiple bond is similar to that of a lone pair. When one of the terminal atoms is joined to the central atom by a double bond, the larger space requirement of the two electron pairs in the double bond causes the singly-bonded terminal atoms to be crowded together. For example, in phosphorus oxyfluoride (PF_3O) the F—P—F angles are about 7° less than the perfect tetrahedral angles predicted by VSEPR.

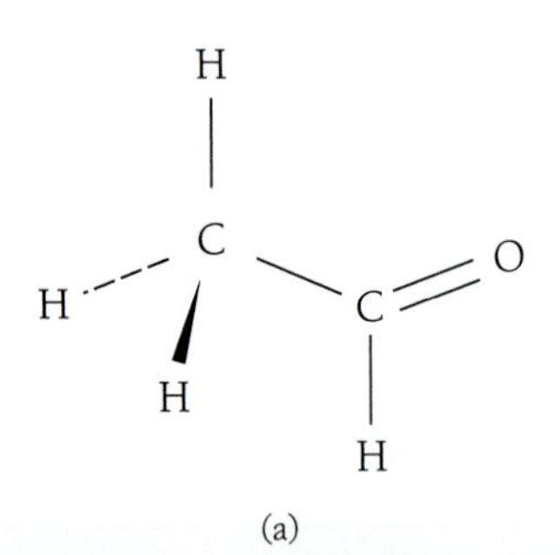

(a)

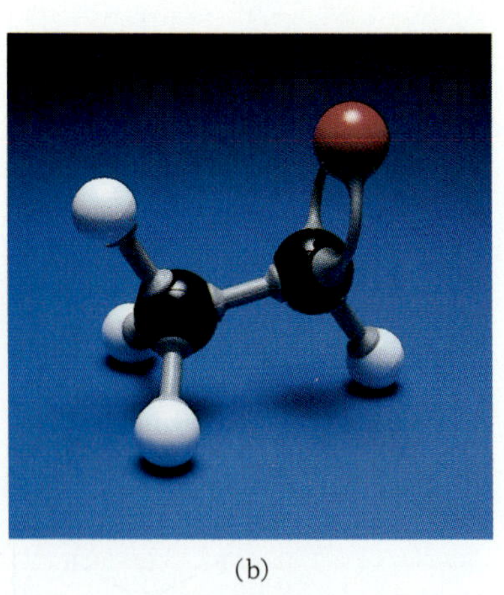

(b)

Figure 8.16 (a) Perspective drawing and (b) ball-and-stick model of acetaldehyde, a molecule with two central atoms. Note that the trigonal planar carbon atom and the three atoms to which it is bonded all lie in the same plane.

Multicenter Molecules

VSEPR can be used to predict the shape of molecules having more than one central atom, by applying the rules to each atom in turn. In acetaldehyde, whose Lewis structure is

the carbon atom bonded to four other atoms is tetrahedral, and the carbon atom bonded to three other atoms is trigonal planar. Figure 8.16 shows the molecular shape.

8.7 DIPOLE MOMENTS

The distribution of electrical charge in a molecule is often unsymmetrical, as a result of the differing electronegativities of the atoms. This electrical asymmetry is described by a measurable quantity called the *dipole moment.*

Polar Bonds and Polar Molecules

Most bonds are polar, that is, they have a bond moment greater than zero. Only if both ends of a bond are exactly the same, as for instance the C=C bond in ethene, is a bond completely nonpolar. A polyatomic molecule is, among other things, an array of

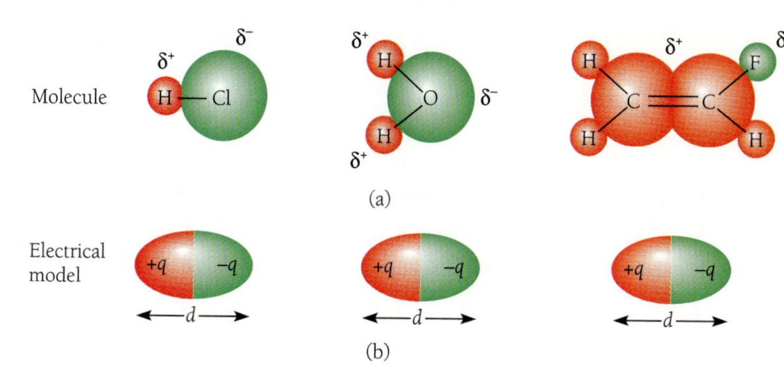

Figure 8.17 (a) Polar molecules such as HCl, H_2O, and CH_2CHF have a positive end and a negative end. (b) All polar molecules behave as if they consisted of charges $+q$ and $-q$ (equal and opposite) separated by a distance d. The value of the product qd is different in different molecules.

individual bond moments having a definite spatial relationship to one another. These bond moments add together to produce an overall *molecular moment* called the **dipole moment** of the molecule. Molecules that have a nonzero dipole moment are called **polar molecules.** Bond moments are *vector* quantities, meaning that they have both magnitude and direction. A molecule's bond moments can, in some cases, cancel each other out, leaving a **nonpolar** molecule with a dipole moment of zero. This can happen even though the molecule contains individual bond moments that are quite large. The dipole moment of a molecule is an easily measured quantity, and one that has considerable influence on the chemical and physical properties of the compound.

Measurement of Dipole Moments

Regardless of their shape, size, or the type of atoms they contain, polar molecules have a positive end and a negative end, but are neutral overall. The electrical behavior of all polar molecules is similar: it is as if they consisted of equal and opposite (partial) charges separated by a distance about the same as the molecular diameter (Figure 8.17).

Polar molecules tend to become aligned with an electric field, created for example by connecting two metal plates to a battery (Figure 8.18). The greater the field, the greater the alignment tendency. The energy required to align polar molecules is proportional to the magnitude of the dipole moment.

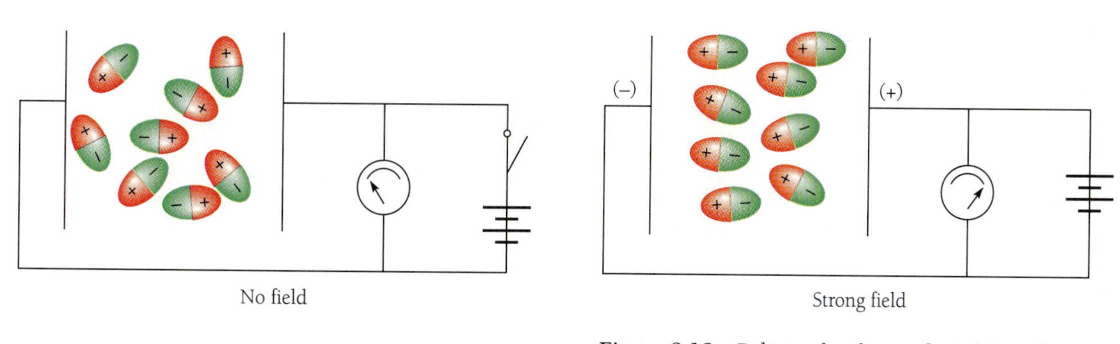

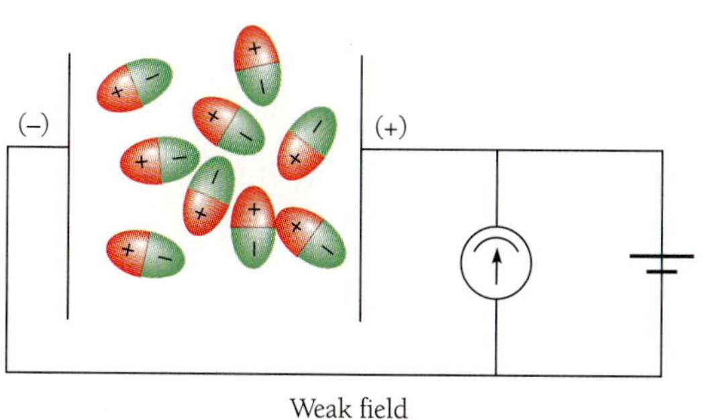

Figure 8.18 Polar molecules are forced into alignment by an electric field. An arrangement of two metal plates separated by a polar substance is called a capacitor or a condenser.

Electrical measurements cannot separately determine the magnitude of the partial charges and the distance separating them—only their *product* is measurable. (This product ($\mu = q \cdot d$) is the dipole moment (μ) of the molecule. The measured value of the dipole moment of water is 6.48×10^{-30} Coulomb-meters. Chemists usually discuss dipole moments in **Debye units (D),** defined as $1 \text{ D} \Leftrightarrow 3.34 \times 10^{-30}$ C m. For water, $\mu = 1.94$ D. Most polar molecules have dipole moments a few D in magnitude.

Molecular Polarity and Symmetry

Even though the C—O bond moments are relatively large (because the electronegativity difference between C and O is relatively large), CO_2 is nonpolar. It is a symmetrical linear molecule in which the equal and opposite bond moments cancel each other. Similarly, the trigonal planar BF_3 molecule is nonpolar even though the B—F bond moments are large. Indeed, provided that all the terminal locations are occupied by the same type of atom, *all* linear, trigonal planar, tetrahedral, trigonal bipyramidal, and octahedral molecules are nonpolar. This cancellation of bond moments is shown in Figure 8.19.

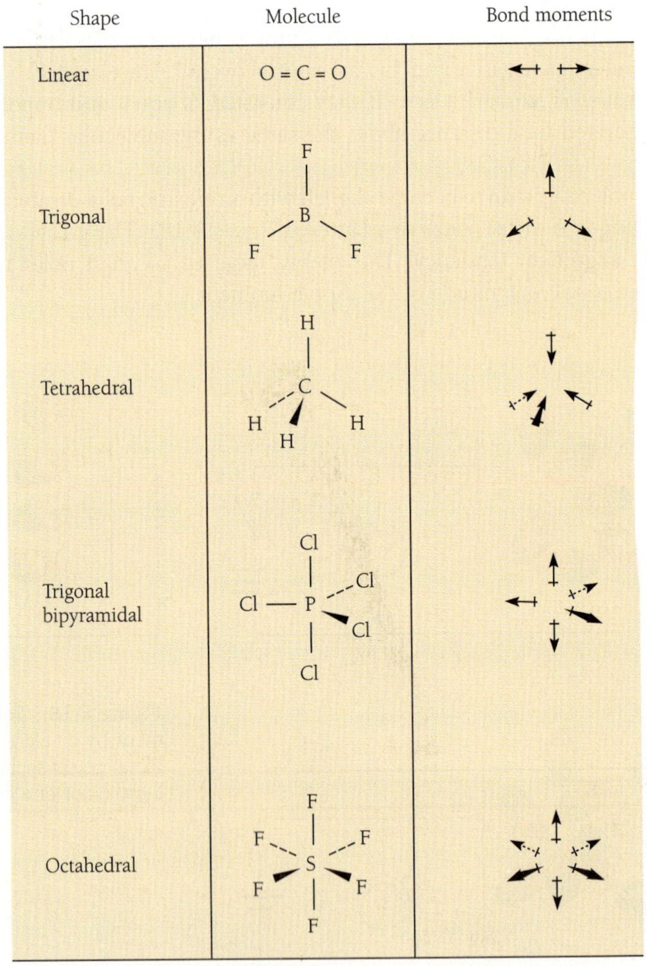

Shape	Molecule	Bond moments
Linear	O = C = O	
Trigonal	BF_3	
Tetrahedral	CH_4	
Trigonal bipyramidal	PCl_5	
Octahedral	SF_6	

Figure 8.19 Cancellation of bond moments is complete when all terminal positions of a given electron-pair geometry are occupied by the same type atom.

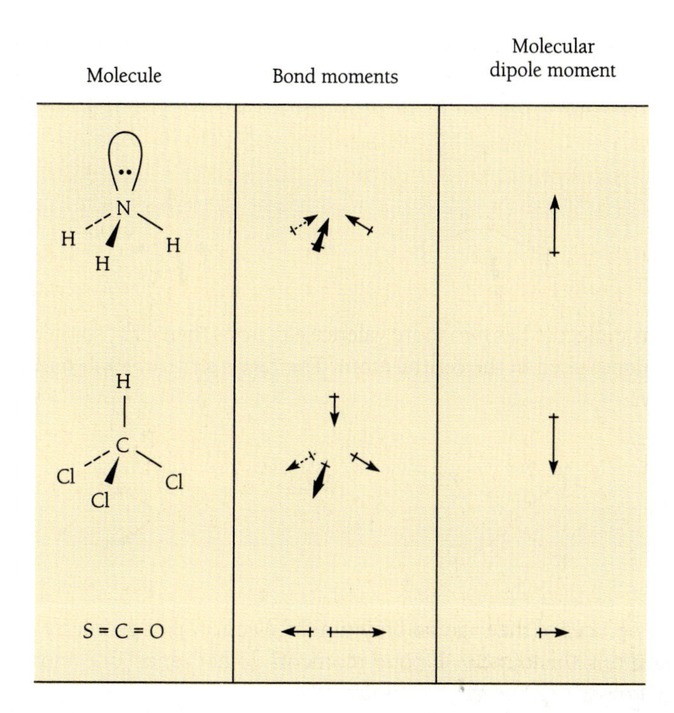

Figure 8.20 Vector addition of bond moments in polar molecules. The dipole moment in COS arises because the electronegativity of S is less than that of O. Hence, the C—S bond moment does not completely cancel the oppositely directed C—O bond moment.

However, when the terminal atoms are not all the same, cancellation of bond moments does not necessarily occur. Both NH_3 and $CHCl_3$ have tetrahedral electron-pair geometry, but both are polar. The symmetry can be destroyed by lone pairs as well as by different atoms (Figure 8.20).

EXAMPLE 8.21 Polarity of Molecules

Describe each of the following as polar or nonpolar, and explain the reasons for the choice: CF_4, SF_4.

Analysis Since virtually all covalent bonds have a nonzero bond moment, the question to be answered is whether or not the molecular shape is such that the bond moments cancel out to zero.

Plan Do the following for each molecule.

1. Determine the molecular shape from the Lewis structure and VSEPR.
2. Decide if the molecule has the correct symmetry for cancellation of its bond moments.

Work

CF_4. The Lewis structure is

```
    F
    |
F—C—F
    |
    F
```

Lone pairs on terminal atoms do not affect the cancellation of bond moments.

Molecules having four bonds and no lone pairs on the central atom are tetrahedral. Since all terminal atoms are the same in CF_4, the bond moments cancel. CF_4 is nonpolar.

SF_4. This molecule has two more valence electrons than CF_4, so there must be an expanded valence shell in the central atom. The Lewis structure and molecular shape are as follows.

Since not all vertices of the trigonal bipyramidal electron-pair geometry are occupied by the same atom, the four S—F bond moments do not cancel one another out. The molecule is polar.

Exercise

Is XeF_4 polar?

Answer: No. The electron-pair geometry is octahedral, and the two lone pairs occupy opposite positions. The molecular shape is square planar, and the four S—F bond moments cancel.

8.8 OXYGEN

There are good arguments for describing oxygen as the most important element. Certainly all life on earth depends on it, and there are few chemical reactions taking place in living organisms that do not involve oxygen or its compounds. Almost all the mass of the solar system is contained in the sun, and, as the sun consists primarily of hydrogen and helium, these are the most abundant elements in the solar system. Excluding hydrogen and helium, however, oxygen is about twice as abundant in the solar system as all other elements combined. Oxygen is the most abundant element on earth: it makes up about 25% (by mass) of the atmosphere, 89% of the hydrosphere (the earth's seas, lakes and rivers), and about 50% of the lithosphere (the earth's crust and mantle). Calcium carbonate, in the form of chalk, limestone, and marble, is 48% oxygen, while the oxygen content of sand and silicate minerals like flint is over 50%.

Properties and Preparation

As shown in Figure 7.12, page 282, oxygen is attracted by a magnetic field.

There are three naturally occurring isotopes of oxygen, with the following abundances: O-16, 99.76%; O-17, 0.04%; O-18, 0.20%. Elemental oxygen occurs as a diatomic, paramagnetic gas whose physical properties are given in Table 8.8. Liquid oxygen has a pale blue color, and the solid is a somewhat deeper blue. Dissolved oxygen supports an enormous amount of marine life.

TABLE 8.8 Molecular Oxygen, O_2

Molar Mass	Bond Length	Bond Energy	Melting Point	Boiling Point	Density of Solid
32.00 g mol^{-1}	120.8 pm	489.34 kJ mol^{-1}	−218.8 °C	−183 °C	1.43 g mL^{-1}

Oxygen Atoms

Covalent Radius	Ionization Energy	Electronegativity
66 pm*	1314 kJ mol^{-1}	3.5

* The covalent radius refers to *singly* bonded atoms. Hence it is somewhat larger than half the internuclear distance in O_2, a *doubly* bonded species.

Not surprisingly, oxygen is a good oxidizing agent, and reacts readily with most other elements. Only the noble gases and a few of the transition elements do not react directly with oxygen, and even some of these have stable oxides that can be prepared by indirect means. Much of the chemical behavior of oxygen can be explained in terms of its high electronegativity, second only to fluorine.

Elemental oxygen also exists in the allotropic form **ozone (O_3)**, a deep blue substance melting at −250 °C and boiling at −112.3 °C. It is an unstable substance that, after a time, decomposes to O_2. Ozone is prepared by passing oxygen through an electrical discharge: the effluent gas contains up to 10% O_3. Ozone has a strong odor, and in small concentrations it is responsible for the pleasant, fresh smell detectable after a thunderstorm (Figure 8.21). Ozone is a highly reactive, powerful oxidizing agent, and because of this it is used as a disinfectant in some municipal water systems. Because it screens out harmful ultraviolet radiation from the sun, the presence of a significant concentration of ozone in the stratosphere (the "ozone layer") is a major environmental factor for terrestrial life.

In the United States, it is more common to use chlorine in water systems.

Although we are accustomed to, and dependent upon, the presence of oxygen in the atmosphere, the primordial atmosphere of the earth probably contained no oxygen at all. During the first 3 billion years of the earth's history, the amount of oxygen in the atmosphere rose gradually. Solar radiation caused the decomposition of upper-atmospheric water vapor into H and O. Oxygen was heavy enough to be retained by the earth's gravity, while the lighter hydrogen leaked away into space. Minerals present in early rock strata show little evidence of reaction with oxygen, while more recently formed minerals have undergone extensive oxidative weathering (such as rusting of iron). Such geologic evidence suggests that, about 1.9 billion years ago, the oxygen concentration of the atmosphere had finally risen to about 2% of today's value. After the origin of life on the earth, photosynthesis by plants became an increasingly important contributor to the rising concentration of atmospheric oxygen.

Oxygen is obtained on an industrial scale from air. The temperature of air is reduced by successive cycles of compression (which heats the air), cooling, and expansion (which further cools the air), until finally the temperature decreases to the point at which air condenses to a liquid. Since its boiling point is somewhat higher than that of nitrogen, the oxygen condenses first. "Liquid air" produced this way consists mostly of oxygen. If required, pure oxygen is obtained by distillation of liquid air. The noble gases (Section 2.9) are by-products of this industrial process. Some 1.7 million tons of O_2 were produced in the United States in 1989, making it one of the more important substances in the chemical industry (Figure 8.22).

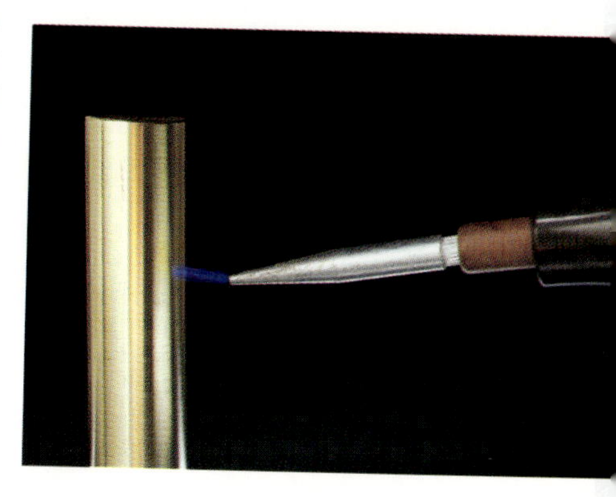

Figure 8.21 Even small sparks will produce enough ozone from oxygen in the air for its characteristic odor to be noticeable.

Covalent Compounds of Oxygen

Oxides of Group 7A elements are discussed in Section 7.6.

Many organic molecules contain covalently bonded oxygen, but the most common covalent oxide is water. The nonmetals of Groups 4A through 7A form covalent binary oxides, many by direct reaction of the elements (Figures 8.23, 8.24, and 8.25). For example,

$$C(s) + O_2(g) \longrightarrow CO_2(g) \quad (8\text{-}22)$$

$$S(s) + O_2(g) \longrightarrow SO_2(g) \quad (8\text{-}23)$$

$$4\ P(s) + 5\ O_2(g) \longrightarrow P_4O_{10}(s) \quad (8\text{-}24)$$

Many nonmetals form more than one oxide, for instance CO and CO_2, SO_2 and SO_3, and N_2O, NO_2, N_2O_4, N_2O_3, and N_2O_5. In their reactions with water, the oxides of the nonmetals behave quite differently from the metal oxides. The resulting solutions are *acids,* not bases, and the compounds are called **acidic oxides.**

$$CO_2(g) + H_2O(\ell) \longrightarrow H_2CO_3(aq) \quad (8\text{-}25)$$

$$3\ NO_2(g) + H_2O(\ell) \longrightarrow 2\ HNO_3(aq) + NO(g) \quad (8\text{-}26)$$

$$P_4O_{10}(s) + 6\ H_2O(\ell) \longrightarrow 4\ H_3PO_4(aq) \quad (8\text{-}27)$$

$$SO_3(g) + H_2O(\ell) \longrightarrow H_2SO_4(aq) \quad (8\text{-}28)$$

The nomenclature of oxoacids and oxoanions is discussed in Section 2.8.

The **oxoacids** thus formed react with bases such as the Group 1A and 2A hydroxides to form (ionic) salts containing **oxoanions** or **hydrogen oxoanions.**

$$NaOH(aq) + H_2CO_3(aq) \longrightarrow NaHCO_3(aq) + H_2O \quad (8\text{-}29)$$

$$2\ KOH(aq) + H_2SO_4(aq) \longrightarrow K_2SO_4(aq) + 2\ H_2O \quad (8\text{-}30)$$

The **peroxo anions** are a related group of polyatomic ions in which an —O— atom is replaced by the —O—O— group. For example,

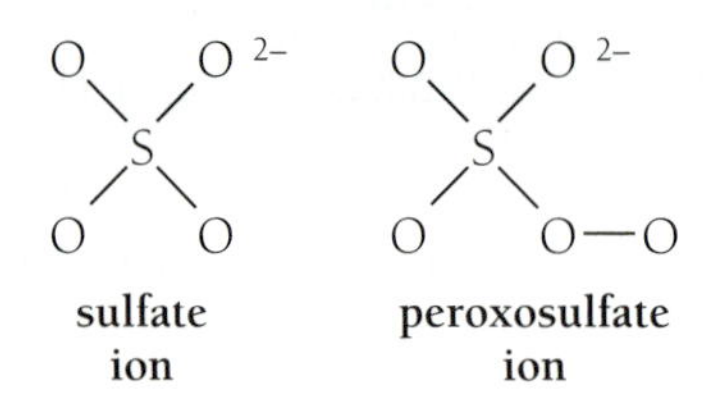

Salts containing peroxo ions tend to be unstable.

(a)

(b)

Figure 8.23 (a) At room temperature, carbon does not react with the oxygen in the air. (b) At high temperature, the reaction between carbon and atmospheric oxygen is vigorous.

(a) (b)

Figure 8.24 (a) Atmospheric oxygen does not react with sulfur at room temperature. (b) At high temperature, sulfur burns readily.

Another well-known covalent compound of oxygen is H_2O_2. Hydrogen peroxide is an oxidizing agent, widely used as a disinfectant and bleach. It is prepared commercially by the electrolysis of concentrated sulfuric acid or solutions containing the $S_2O_8^{2-}$ ion. Pure H_2O_2, a liquid at room temperature, is very soluble in water. When heated, pure H_2O_2 decomposes explosively before its normal boiling point is reached.

$$2\ H_2O_2(\ell) \xrightarrow[\Delta]{} 2\ H_2O + O_2 \qquad (8\text{-}31)$$

Because of its instability, hydrogen peroxide is normally handled in aqueous solutions in concentrations not exceeding 30%. Such solutions decompose slowly, but not explosively.

Oxygen-containing organic compounds are discussed in Section 20.2.

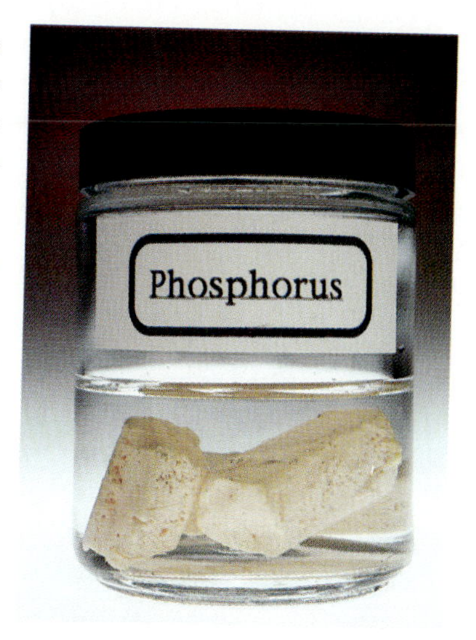

Figure 8.25 Phosphorus must be stored under water to protect it from oxygen.

The Oxygen Cycle

Since the beginning of life on earth, oxygen has circulated continuously through the atmosphere, the hydrosphere, the lithosphere, and the *biosphere* (the total mass of living organisms), in a pattern called the *oxygen cycle* (Figure 8.26). Both terrestrial and aquatic plants absorb atmospheric or dissolved carbon dioxide and, with energy for the endothermic reactions supplied by sunlight, incorporate it into large, complicated organic molecules called *carbohydrates,* chiefly starch and cellulose. The oxygen (and carbon) of carbohydrates is ultimately reconverted to carbon dioxide through animal and bacterial metabolic processes, and reenters the atmosphere. Marine organisms incorporate oxygen as calcium carbonate or silicates into their shells, which eventually become limestone or other forms of oxygen-containing sedimentary rock. Continental uplift brings this rock to the surface, from which weathering brings the carbonate ions back to the sea.

Carbohydrates are discussed in Section 21.3.

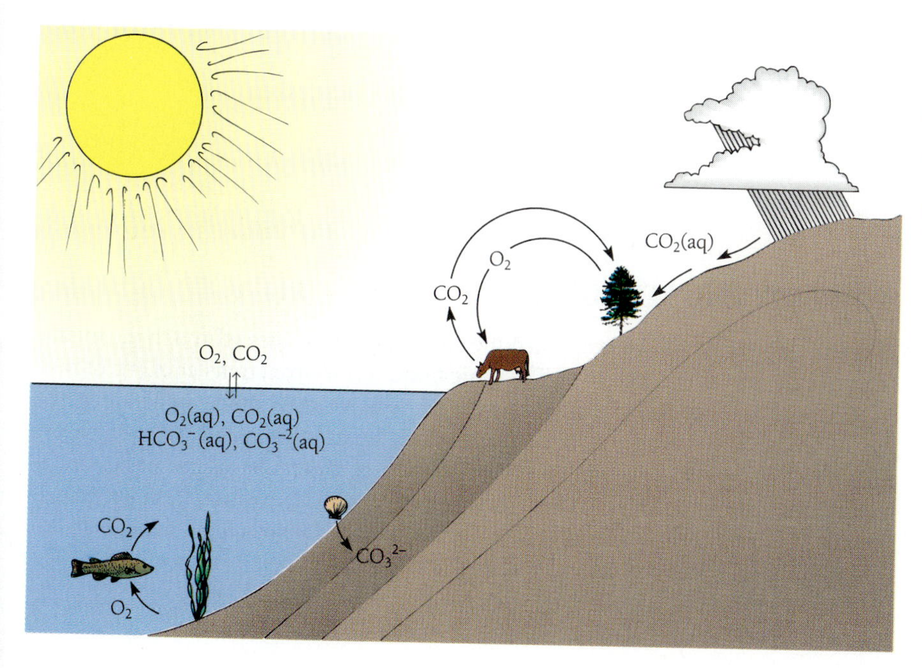

Figure 8.26 The oxygen cycle. Oxygen, as O_2 and CO_2, circulates through plant and animal life in the atmosphere and hydrosphere. Over the longer times of geologic processes, it also circulates (as CO_2 and the carbonate ions, HCO_3^- and CO_3^{2-}) through the marine biosphere and sedimentary rocks.

SUMMARY EXERCISE

Heating of aluminum fluorosulfate [$Al(SO_3F)_3$] to 500 °C causes its decomposition to aluminum sulfate and sulfuryl fluoride [SO_2F_2].

a. Write a balanced equation for the reaction.

b. Calculate the mass of aluminum fluorosulfate required to produce 10.0 g SO_2F_2.

c. How many *p* electrons are there in an aluminum *atom?*

d. What is the charge of the fluorosulfate ion?

e. What is the angular momentum quantum number of the highest-energy electron in the aluminum *ion?*

f. Draw the Lewis structure of SO_2F_2.

g. What are the orders of all bonds in SO_2F_2?

h. What is the total number of valence electrons in the fluorosulfate ion?

i. What shape does VSEPR theory predict for the SO_2F_2 molecule?

j. There is another oxofluoride of sulfur, SOF_2 (sulfonyl fluoride). Is this compound polar?

k. Draw resonance structures for the nitrite ion (NO_2^-) and determine the formal charges on each atom.

Answers **(a)** $2\ Al(SO_3F)_3 \xrightarrow[500°]{} 3\ SO_2F_2 + Al_2(SO_4)_3$; **(b)** 21.2 g; **(c)** 7; **(d)** −1;

(e) 1; **(f)** (Lewis structure: S bonded to two F by single bonds and to two O by double bonds; each F with three lone pairs, each O with two lone pairs) :F̈—S(=O:)(—F:)=Ö:; **(g)** S—F bonds are of order 1, S—O bonds are of order 2; **(h)** 32; **(i)** tetrahedral; **(j)** yes;

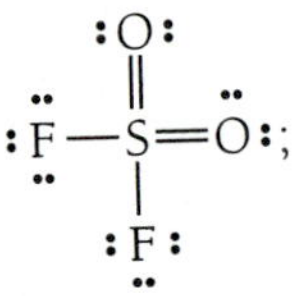

(k) (resonance structures: O=N—O⁻ ⟷ ⁻O—N=O, N with one lone pair; the singly bonded O carries formal charge −1)

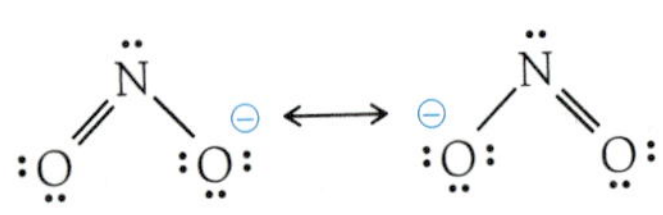

SUMMARY AND KEYWORDS

The interatomic binding force, called the **chemical bond,** arises through the interaction of **valence** (outer) electrons. The **ionic bond** results from transfer of one or more electrons to another atom, giving positive and negative ions, which are held together by *electrostatic* or *Coulomb forces.* In the solid state each ion is attracted by several nearest-neighbor ions, forming a **crystal lattice** of indefinite extent. There are no molecules in an ionic solid. In the gaseous state, ionic compounds form *ion pairs* or *clusters.* In forming ions, atoms gain or lose electrons so as to achieve a completed or filled valence shell. Group 1A and 2A elements form M^+ and M^{2+}, respectively, Group 7A elements form X^-, and so on. Ion charges determine the stoichiometry of ionic compounds.

The binding force in a **covalent bond** results from the sharing of a **pair** of valence electrons by two atoms. Covalent bonding usually follows the **octet rule:** atoms share one or more pairs of valence electrons so as to achieve a filled valence shell of eight electrons. Exceptions to the octet rule include hydrogen, **electron-deficient** atoms such as boron, and the heavier, **expanded valence shell** atoms (third and higher periods). **Dot** or **Lewis structures** are a convenient representation of covalent bonding in molecules and polyatomic ions. **Single** and **multiple bonds (double** and **triple)** are formed through sharing of one, two, or three electron pairs; the number of shared pairs is called the **bond order.** Unshared *(nonbonded)* valence pairs are called **lone pairs.**

The **formal charge** of each atom provides a rough guide to the location of charge in a molecule or polyatomic ion. Lewis structures with the least formal charge are the best representation of the actual bonding in molecules and ions. In **resonance,** bonding electrons are **delocalized** over several bonds, a situation that cannot be represented by a single Lewis structure. A **resonance hybrid** is an average of several Lewis structures, called **resonance contributors.** Resonant species have fractional bond orders, and are more stable than comparable nonresonant species.

The strength of an atom's attraction for bonding electrons is the **electronegativity** of the element. **Polar covalent** bonds (between atoms of different electronegativity) have a positive and a negative end, which creates a **bond moment.** Only the bonds in homonuclear diatomic molecules and certain other symmetrical species (for example the C—C bond in C_2H_6) are nonpolar. There is no sharp dividing line between highly polar and ionic bonds. All polar bonds have partially ionic and partially covalent character. Bond moments in a molecule combine to produce an overall molecular **dipole moment,** measured in **Debyes (D)** ($1\ D \Leftrightarrow 3.34 \times 10^{-30}$ C m). A typical **polar molecule** has a dipole moment of a few D. Molecules with high symmetry can be nonpolar, because the *vector* sum of nonzero bond moments can be zero.

The sum of the covalent radii of two bonded atoms is the **bond length,** which decreases as bond order increases. **Bond strength** or **bond energy** is the energy required to break the bond (that is, to *dissociate* a species into two fragments). Bond energy increases with bond order. Average values of bond energy may be used to estimate thermochemical quantities such as heats of formation and heats of reaction.

VSEPR theory predicts that the most common spatial arrangements of valence electrons, **linear, trigonal planar, tetrahedral, trigonal bipyramidal,** and **octahedral,** are assumed by central atoms in whose valence shell the sum of the number of bonds and lone pairs is 2, 3, 4, 5, and 6, respectively. For molecules whose central atom has one or more lone pairs, the **molecular shape,** which describes the spatial arrangement of the atoms only, is different from the **electron-pair geometry.** The terms *bent, pyramidal, see-saw,* and *T-shaped* are used to describe molecular shape. Lone pairs are bulkier than bonded pairs, hence occupy the less crowded **equatorial** rather than the **axial** positions in the trigonal bipyramidal geometry.

Oxygen is the most abundant element in the solar system (excluding the sun). It has three naturally occurring isotopes and two allotropic forms (O_2 and O_3, *ozone*). Ordinary oxygen (O_2) is paramagnetic. *Oxide* (O^{2-}) and *peroxide* (O_2^{2-}) are common anions, while *superoxide* (O_2^-) and *ozonide* (O_3^-) are less common. A number of *oxo-* and *peroxo-* anions, such as SO_4^{2-} and SO_5^{2-}, are known. Oxygen readily forms covalent bonds, and a large number of oxygen-containing organic compounds are known. The *oxygen cycle* describes the circulation of oxygen, in the form of O_2, carbohydrates, CO_2, HCO_3^- and CO_3^{2-}, through the atmosphere, lithosphere, hydrosphere, and biosphere.

PROBLEMS

General

1. Define valence.
2. Which elements are more likely to enter into ionic bonding as positive ions, those with high or low ionization energy?
3. Discuss the statement, "Ionic compounds do not consist of molecules."
4. Does the octet rule apply to ionic bonding as well as to covalent bonding? Explain.
5. Distinguish between formula unit and empirical formula.
6. In a polyatomic ion such as $Cr_2O_7^{2-}$, where do the extra electrons come from?
7. Define each of the following characteristics of the covalent bond: order, length, and energy.
8. What is the relationship between bond order and bond length?
9. What is the relationship between bond order and bond energy?
10. Under what circumstances can a bond have fractional order?
11. What is formal charge? Is the actual charge on an atom different from the formal charge?
12. How is formal charge used to evaluate Lewis structures?
13. What is resonance? How can you tell whether or not a species is resonant?
14. Criticize the statement, "The carbonate ion spends one third of its time in each of three equivalent contributing structures."
15. List some observable characteristics that could be used to distinguish resonant from nonresonant species.
16. What is meant by a hybrid structure?
17. What is meant by electron delocalization?
18. Criticize the statement, "The octet rule is not very useful, since there are so many exceptions."
19. What is bond polarity and how does it arise?
20. Which is the most electronegative element? The least electronegative?
21. Describe the general trend of electronegativity within the periodic table.
22. Why is bond moment called a vector quantity?
23. "Bond moments cannot be measured." True or false? Explain.
24. Distinguish between partial charge and formal charge.
25. Comment on the statement, "Most covalent bonds are polar."
26. Comment on the statement, "A polar covalent bond is a covalent bond with some ionic character."
27. What is electronegativity, and how is it measured?
28. Why is the electronegativity of neon not given in Figure 8.5?
29. Alkali and alkaline earth halides are all ionic compounds. Use this fact, together with the electronegativity values in Figure 8.5, to suggest a minimum value of ΔEN below which ionic bonding is unlikely.
30. If it is true that each type of atom has a characteristic covalent radius, how can the lengths of covalent bonds between the same two atoms differ in different molecules? Or in the same molecule, for that matter?
31. Explain, in general terms, the relationship between bond energy and heat of reaction.
32. Explain why it is neither incorrect nor inconsistent to describe water as both tetrahedral and bent.
33. For which molecular shape(s) is it appropriate to use the terms "axial" and "equatorial"? Illustrate your answer with sketches.
34. Distinguish between the terms "dipole moment" and "polar."
35. Which of the following are measurable quantities: bond moment, dipole moment, and distance between the centers of positive and negative charge in a polar molecule?
36. In what sense is the dipole moment of a molecule the sum of its bond moments? How can the sum be zero if none of the bond moments is zero?

Ionic Bonding

37. What monatomic ions would you expect to form from each of the following atoms?
    a. Li   b. Sr   c. S   d. Al
38. What monatomic ions would you expect to form from each of the following atoms? In each case, would you expect ion formation to be endothermic or exothermic?
    a. Rb   b. Se   c. Ba   d. N
39. Write the electron configurations of each of the following.
    a. Mg and Mg^{2+} b. Li and Li^{+}
    c. I and I^{-} d. Al and Al^{3+}
40. Name the noble gas whose electron configuration is the same as that of the monatomic ion of each of the following elements.
    a. Te   b. Ca   c. Br   d. At
41. Give the empirical formula of the ionic compound consisting of each of the following.
    a. Ba and O   b. Al and O
    c. Cs and S   d. sodium and $Cr_2O_7^{2-}$

42. Give the empirical formula of the ionic compound consisting of each of the following.
    a. Mg and N
    b. Li and I
    c. NH_4^+ and CO_3^{2-}
    d. potassium and peroxide ion

Lewis Structure

43. Write the Lewis structures of the following molecules.
    a. CH_2F_2 b. COClBr c. CO d. NH_2Cl

44. Write the Lewis structures of the following molecules.
    a. SCl_2 b. BCl_3 c. COS d. H_2S

45. Write the Lewis structures of the following species. If there is resonance, show all contributing structures.
    a. XeF_2 b. N_3^- c. ClO_3^- d. HN_3

46. Write the Lewis structures of the following species. If there is resonance, show all contributing structures.
    a. SO_3^{2-} b. HCO_3^- c. BeF_2 d. N_2H_2

47. Write the Lewis structures of the following species. If there is resonance, show all contributing structures.
    a. HO_2^-
    b. formate ion, HCO_2^-
    c. acetic acid, CH_3CO_2H (this molecule has three C—H bonds and one O—H bond)
    d. H_2SO_4 (H is bonded to O, and there are no O—O bonds)

48. Write the Lewis structures of the following species. If there is resonance, show all contributing structures.
    a. POF_3
    b. CH_3NH_2
    c. KrF_2
    d. propanol, $CH_3CH_2CH_2OH$

Formal Charge

In answering the questions in this section, it is not necessary to include formal charges of zero in lists or diagrams.

49. Determine the formal charges on all atoms in the following species.
    a. CH_2F_2 b. CO

50. Determine the formal charges on all atoms in the following species.
    a. $COCl_2$ b. BrF_3

51. Determine the formal charges on the atoms in the ammonium ion. Which atom carries the bulk of the positive charge?

52. Determine the formal charges on the atoms in the hydronium ion, H_3O^+. Which atom carries the bulk of the positive charge?

53. Determine the formal charges on all atoms in the following species. If there is resonance, show formal charges on all contributors.
    a. SO_2 b. HCO_3^- c. $BeCl_2$ d. N_2H_2

54. Determine the formal charges on all atoms in the following species. If there is resonance, show formal charges on all contributors.
    a. O_2^-
    b. NO_2^-
    c. acetic acid, CH_3CO_2H
    d. ethanol, CH_3CH_2OH

Bond Order

55. Determine the bond order for all bonds in the following species. Watch for resonance and its effect on bond order.
    a. CH_2F_2 b. COClBr c. CO d. NH_2Cl

56. Determine the bond order for all bonds in the following species. Watch for resonance and its effect on bond order.
    a. SCl_2 b. HCN c. COS d. H_2S

57. Determine the bond order for all bonds in the following species. Watch for resonance and its effect on bond order.
    a. IF_5
    b. NO_2^-
    c. acetic acid, CH_3CO_2H
    d. H_2SO_4

58. Determine the bond order for all bonds in the following species. Be on the lookout for resonance and its effect on bond order.
    a. SO_2 b. HCO_3^- c. BeF_2 d. N_2H_2

Bond Length and Energy

59. Why are some bond energies different from the values given in Table 8.4?

60. Suppose you learn through experiment that a particular bond, say a C—Br single bond, is shorter in one molecule than another. What would you predict about the relative bond energies in these two compounds?

61. In which compound, SO_2 or SO_3, would you expect to find the shorter S—O bond?

62. Comment on the strengths of the C—O bonds in the acetate ion ($CH_3CO_2^-$) relative to the C—O bonds in acetic acid (CH_3CO_2H).

63. Estimate the heat of reaction for the following gas-phase reactions.
    a. $H_2C{=}CH_2 + Br_2 \longrightarrow BrH_2C{-}CH_2Br$
    b. $H_2O_2 \longrightarrow H_2O + \frac{1}{2}\,O_2$

64. Estimate the heat of reaction for the following gas-phase reactions.
    a. $N_2 + 3\,H_2 \longrightarrow 2\,NH_3$
    b. $CH_4 + Cl_2 \longrightarrow CH_3Cl + HCl$
    c. $CO + H_2O \longrightarrow CO_2 + H_2$

65. Use the table of average bond energies to estimate the heats of formation of the following species.
    a. $ClF_3(g)$
    b. $N_2H_2(g)$

66. Use the table of average bond energies to estimate the heats of formation of the following species. [Note that the heat of formation of C(g) is 717 kJ mol^{-1}.]
   a. $H_2O_2(g)$ b. $CH_3CO_2H(g)$

Electronegativity and Bond Moments

67. Without looking at a list of electronegativities, identify the more electronegative element in the following pairs.
   a. N, O b. K, Rb c. Mg, K d. O, S
68. Looking at a periodic table rather than a list of electronegativities, identify the most electronegative element in each of the following sets.
   a. B, C, Al b. O, P, S c. Be, Ge, F d. O, F, Cl
69. Which is the more polar bond in each of the following? Try to answer this question without referring to Figure 8.5.
   a. C—F, H—F b. O—H, S—H
   c. S—Cl, S—Br d. C—H, O—H
70. Which is the more polar bond in each of the following? Try to answer this question without referring to Figure 8.5.
   a. C—N, N—H b. Si—O, Si—Cl
   c. N—O, C—O d. Br—F, O—F
71. In each of the following pairs, which bond has the greater ionic character?
   a. C—Br, Si—O b. N—O, Cl—F
   c. C—B, C—H d. Li—Br, C—Cl
72. In each of the following pairs, which bond has the greater ionic character?
   a. C—F, H—F b. O—Si, Cl—Si
   c. Cl—Br, Br—S d. Cl—O, F—O

Molecular Geometries

73. Give the electron-pair geometry and the molecular shape of each of the following.
   a. IF_5 b. CCl_2F_2 c. NO_2^- d. CH_2O
74. Give the electron-pair geometry and the molecular shape of each of the following.
   a. NH_2Cl b. H_2S c. SO_3^{2-} d. BH_4^-
75. Give the electron-pair geometry and the molecular shape of each of the following.
   a. SF_4 b. $POClBr_2$ c. OCN^- d. XeF_2
76. Give the electron-pair geometry and the molecular shape of each of the following.
   a. SiF_6^{2-} b. HNO_3 c. ClF_3 d. $XeOF_4$
77. What value does VSEPR predict for the indicated bond angle in each of the following molecules?
   a. H_2Te; <H—Te—H b. H_2CO; <H—C—H
78. For each of the following molecules, give an approximate value of the indicated bond angle.
   a. HCO_2H; <O—C—O
   b. CH_3OH; <C—O—H
79. What are the bond angles in the PCl_5 molecule?
   a. <Cl(axial)—P—Cl(equatorial)
   b. <Cl(equatorial)—P—Cl(equatorial)
   c. <Cl(axial)—P—Cl(axial)
80. For each of the following molecules, give an approximate value of the indicated bond angle.
   a. XeF_4; <F—Xe—F
   b. $NHCl_2$; <H—N—Cl
   c. CCl_2F_2; <Cl—C—F and <F—C—F
   d. ClF_3; <F—Cl—F

Dipole Moments

81. Identify each of the following as polar or nonpolar. If polar, which end of the molecule is negative?
   a. XeF_4 b. CS_2 c. H_2S d. H_2CO
82. Identify each of the following as polar or nonpolar. If polar, which end of the molecule is negative?
   a. PCl_5 b. PCl_4F
   c. $H_2C{=}CH_2$ d. $CHCl_3$
83. In which of the following is the dipole moment equal to zero?
   a. HF b. SF_4
84. In which of the following is the dipole moment equal to zero?
   a. $H_2C{=}CHF$ b. NH_3

*85. In which of the following is the dipole moment greater than zero?
   a. SO_4^{2-} b. SF_2 c. HCN d. BeF_2
86. In which of the following is the dipole moment greater than zero?
   a. $GeCl_2$ b. XeF_4 c. BCl_3 d. SF_6

Oxygen

87. Which of the following contains the lowest percentage of oxygen: the atmosphere, the hydrosphere, or the lithosphere?
88. By what mechanism could O_2 have entered the earth's atmosphere prior to the appearance of photosynthesizing plants?
89. Describe the process by which O_2 is produced commercially.
90. Give balanced chemical equations for two methods of producing small amounts of oxygen in the laboratory.
91. Give the names and formulas of three ions that consist only of oxygen.
92. Give the names and formulas of the compounds that result from the direct reaction of oxygen with each of the Group 1A metals.
93. Give the names and formulas of six anions that contain oxygen and one other element.

94. Some elements react with oxygen to form more than one binary compound.
    a. Give the names and formulas of two binary compounds of oxygen with the same metallic element.
    b. Give the names and formulas of two binary compounds of oxygen with the same nonmetallic element.
95. What type of binary oxide reacts with water to form acids?
96. What type of binary oxide reacts with water to form basic solutions?
97. How much oxygen is formed when 50.0 g of sodium peroxide reacts completely with water?
98. How much potassium chlorate must be decomposed to produce 1.00 L of oxygen, measured at STP?
99. Complete and balance the following equations.
    a. $P_4O_{10}(s) + 6\ H_2O(\ell) \longrightarrow$
    b. $3\ NO_2(g) + H_2O(\ell) \longrightarrow$

CHAPTER 9

FURTHER CONSIDERATIONS OF BONDING

When water is added to small pieces of calcium acetylide in the bottom of the graduated cylinder, a rapid reaction takes place. The reaction product, gaseous acetylene (C_2H_2) burns with a smoky flame.

9.1 INTRODUCTION

In the preceding chapter we discussed the *Lewis* model of chemical bonding: the octet rule, transferred electrons in ionic bonding, and shared electron pairs in covalent bonding. We saw the success of the model in accounting for the stoichiometry of compounds, and (together with the VSEPR theory) molecular geometry as well. But however useful as a conceptually simple model of bonding, the Lewis ideas do not provide any details of "electron-pair sharing." Furthermore, since it predates the advent of quantum mechanics, the Lewis model is not grounded in our modern views of the electronic structure of atoms and molecules.

Present-day approaches to understanding the covalent bond have followed two parallel tracks, each with its advantages and disadvantages. Neither approach should be regarded as "correct," but rather as more or less useful for understanding different aspects of bonding in molecules. In the **valence bond** approach, a covalent bond consists of *overlapping orbitals* on adjacent atoms, resulting in enhanced electron density in the region between the nuclei. The valence bond approach is an extension, in the mathematical language of quantum mechanics, of the Lewis model. Its great advantage is that bonding can be discussed in terms of the familiar dot structures, because molecules are regarded as being held together by localized bonds between adjacent atoms.

In the **molecular orbital** approach, molecules are held together whenever electrons occupy *bonding orbitals*. These orbitals extend over the entire molecule, rather than being associated with a single atomic nucleus (as they are in valence bond theory). Electron delocalization is thus a natural part of the molecular orbital approach, and there is no need to introduce resonance as a supplemental concept to overcome the limitations of Lewis structures. Both approaches offer useful explanations of bond order, length, strength, and angles. Since it deals with electron *pairs*, valence bond theory is less useful than molecular orbital theory in discussions of paramagnetism and molecules with an odd number of electrons. However, valence bond theory provides a simpler and more useful way of discussing molecular shapes.

9.2 VALENCE BOND THEORY

Although many chemists and physicists have contributed to its development, the name most often associated with the valence bond approach is that of Linus Pauling (historical box, page 85). Valence bond theory is in essence a combination of the Lewis concept of a bond with the modern model of quantized electron orbitals in atoms.

Overlap of Atomic Orbitals: σ and π Bonds

Whereas the covalent bond in H_2 is described in the Lewis model as the sharing of the two valence electrons, in the valence bond approach the bond is formed by the **overlap** of two half-filled 1*s* orbitals, one on each of the H atoms (Figure 9.1). Overlap

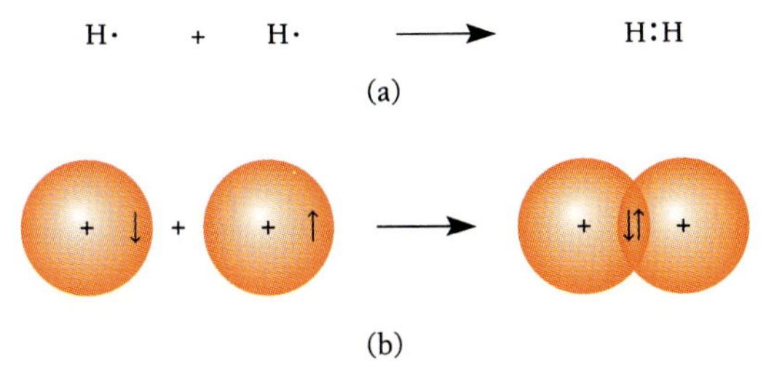

Figure 9.1 In the Lewis model (a) the covalent bond in H_2 consists of the shared pair of valence electrons. In the valence bond model (b), the bond is an internuclear region of enhanced electron density arising from the overlap of half-filled orbitals.

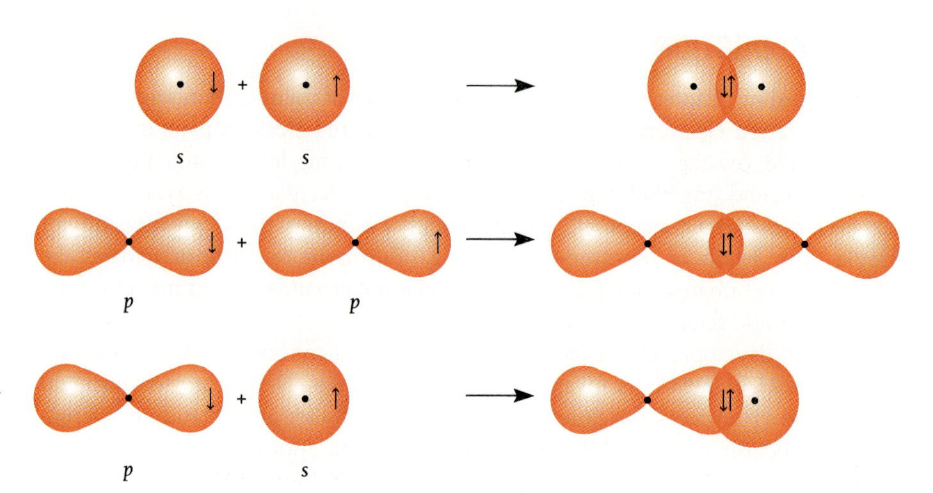

Figure 9.2 Covalent bonds, such as these σ bonds, can form through overlap of half-filled atomic orbitals having the same or different values of the angular momentum quantum number.

of atomic orbitals is similar to the constructive interference of waves, and it enhances the electron density between the nuclei. This region of enhanced electron density is the covalent bond. The greater the extent to which the orbitals overlap, the stronger the bond. Orbital overlap can be regarded as an addition of two charge distributions, but it is probably more correct to think of it as a form of constructive interference. The Pauli exclusion principle applies, so that the two electrons that occupy these orbitals must have paired spins. Filled (doubly occupied) orbitals cannot overlap to form bonds. If they did, more than two electrons would occupy the same region of space, in violation of the Pauli principle.

Overlap and bonding can occur with atomic *p* orbitals as well as with *s* orbitals. Furthermore, the overlapping orbitals can be of different types. Figure 9.2 illustrates some common forms of bonds. Regardless of their origin from *s* or *p* orbitals, these bonds are called **sigma (σ) bonds.** The σ bond is a single bond, or, in the Lewis model, one shared pair of electrons. The Greek "s" in this notation is intended to emphasize some aspects common to σ bonds and *s* orbitals. These common features, illustrated in Figure 9.3, include the following.

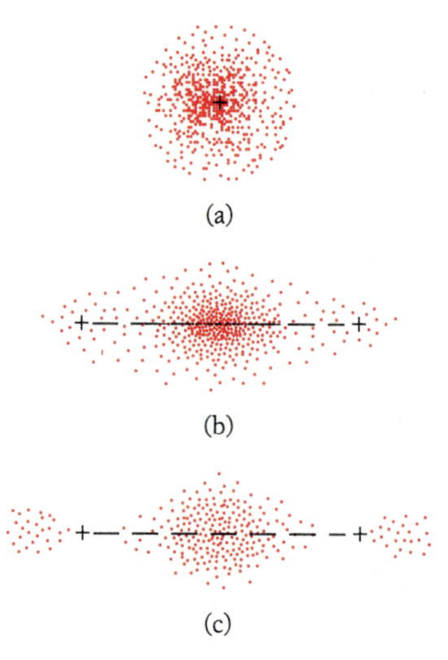

Figure 9.3 Stipple plots showing distribution of electron density and symmetry of (a) an *s* orbital of an atom, (b) a σ bond formed from two atomic *s* orbitals, and (c), a σ bond formed from two atomic *p* orbitals. Note the shape, the location of maximum electron density, and the absence of nodes in the internuclear region.

Figure 9.4 Footballs, ice cream cones, and pencils all have cylindrical symmetry. That is, their shape appears the same when they are rotated about the axis of symmetry.

1. *The greatest electron density is at the center.* The center of an *s* orbital is the nucleus, while the center of a σ bond is the **internuclear axis,** the line passing through the nuclei of the bonded atoms.
2. *There is only one region of electron density, and there are no nodes.* (Although σ bonds formed from *p* orbital overlap [Figure 9.3c] retain nodes near the nuclei, there are no nodes in the internuclear region, that is, in the *bond* itself.)

A σ bond has **cylindrical symmetry** about the internuclear axis. (Recall that an *s* orbital has **spherical symmetry** about the nucleus.) Some other objects with cylindrical symmetry are illustrated in Figure 9.4.

Another type of bond is formed by *side-to-side* overlap of *p* orbitals, as illustrated in Figure 9.5. These bonds, called **π bonds,** differ from σ bonds in shape, number of nodes, and symmetry. The electron density of a π bond, like that of an atomic *p* orbital,

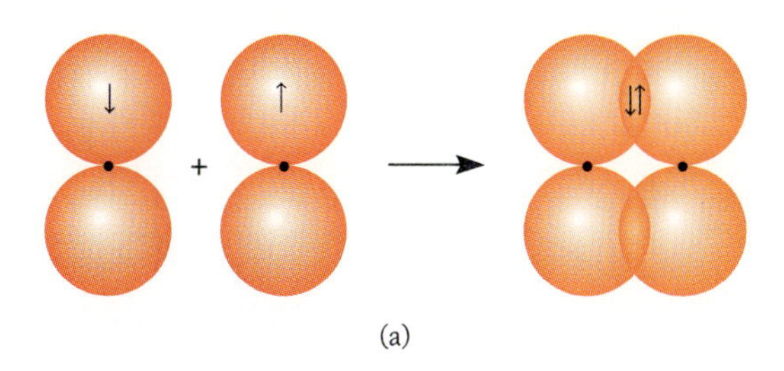

(a)

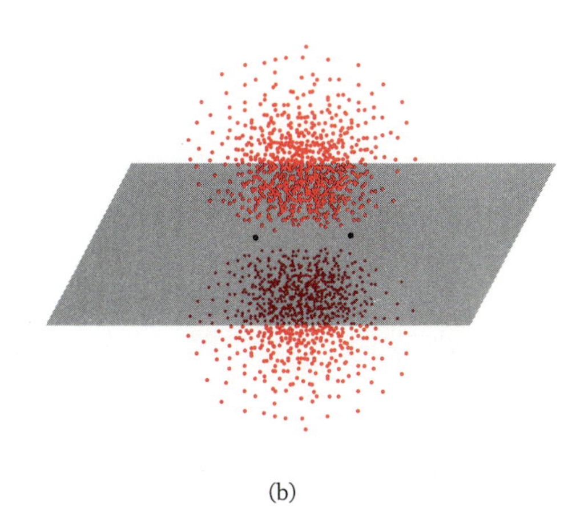

(b)

Figure 9.5 (a) Formation of a π bond by side-to-side overlap of atomic *p* orbitals. (b) Stipple plot of electron density in a π bond, showing that it consists of two lobes separated by a nodal plane passing through the two nuclei.

occurs in two lobes separated by a nodal plane. The electron density of a π bond is zero along the internuclear axis, which lies in the nodal plane of the bond. A π bond does not have cylindrical symmetry about the internuclear axis. As with σ orbitals, however, the greater the overlap of the atomic *p* orbitals, the greater the strength of the π bond.

A π bond always occurs in combination with a σ bond. A double bond consists of a σ bond and a π bond between the same two nuclei, while a triple bond is one σ bond and *two* π bonds. Figure 9.6 illustrates the bonding in molecular nitrogen.

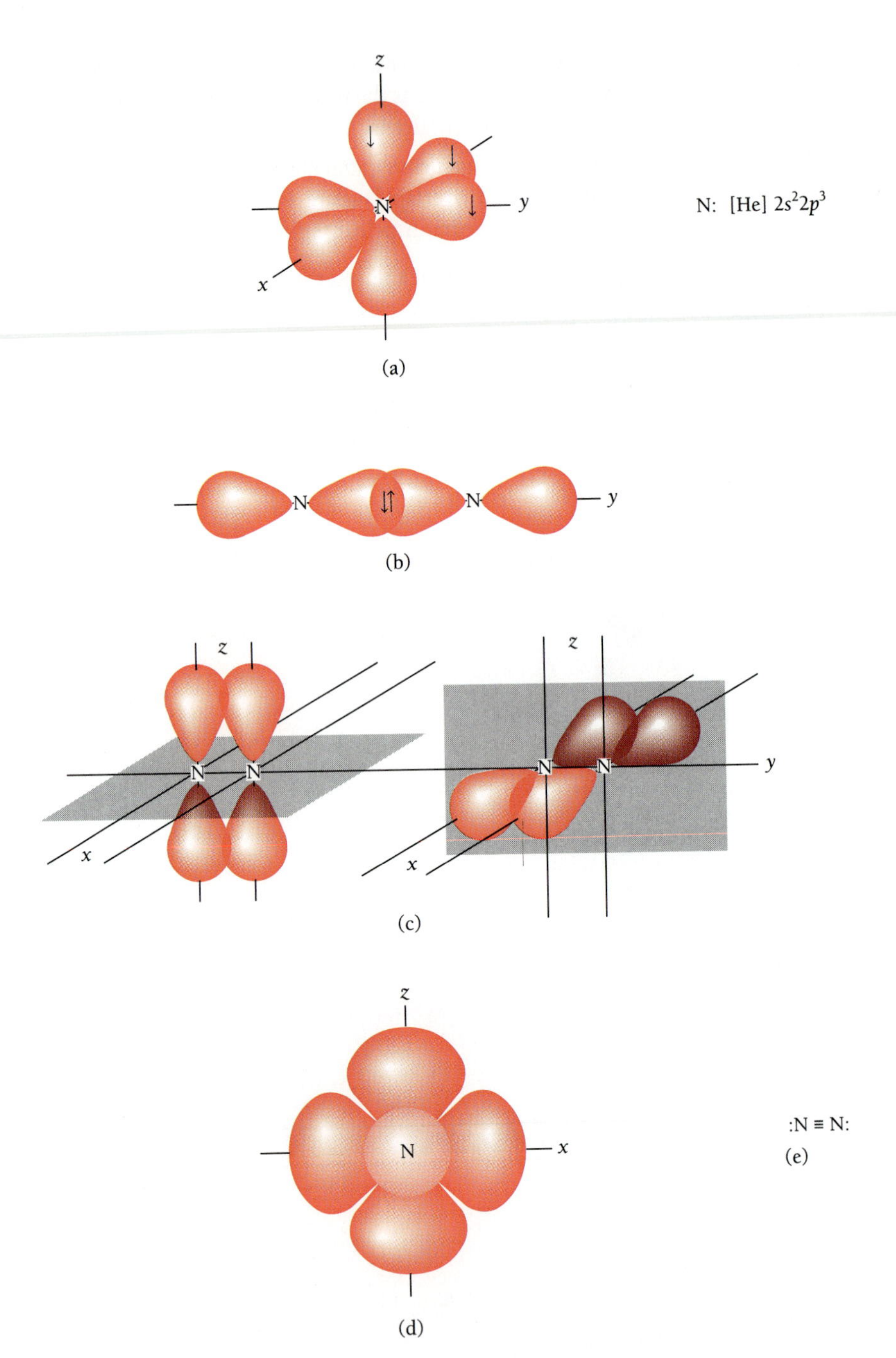

Figure 9.6 Multiple bonding in the N_2 molecule. (a) A nitrogen atom has three unpaired electrons, occupying the $2p_x$, and $2p_y$, and $2p_z$ orbitals. (b) The p_y orbitals overlap to form a σ bond. (c) The p_z orbitals form a π bond whose node is the *x-y* plane, while the p_x orbitals overlap to form another π bond whose node is the *y-z* plane. (d) An end-on view down the *y*-axis shows the different lobes of the σ and the two π bonds. (e) The dot structure of N_2, although less detailed, also indicates a triple bond.

The Valence State: Promotion

Although valence bond theory offers an excellent description of σ and π bonds as overlapping orbitals, there are many molecules whose stoichiometry cannot be explained by this description alone. The beryllium atom, for example, has the electron configuration [⇅] [⇅] [ ][ ][ ] (1s 2s 2p). Since it has no half-filled orbitals in the valence shell, beryllium ought to be incapable of forming bonds by overlap with half-filled orbitals on other atoms. In fact beryllium forms compounds, for example BeH_2 and BeF_2, that have two covalent bonds. Clearly, some modification of the theory is required.

Suppose beryllium enters into bonding not from its ground-state configuration [He] [⇅] [ ][ ][ ] (2s 2p), but from the *valence state* [He] [↑] [↑][ ][ ] (2s 2p). A **valence state** describes the electron configuration of an atom just before it forms bonds, rather than when it is isolated from other atoms. To reach this state, an electron must be excited from the 2*s* to the 2*p* subshell. This is called **promotion** in valence bond theory, and the energy required is the *promotion energy*. Now, having two half-filled orbitals, a beryllium atom can form two covalent bonds. Furthermore, the energy released in bond formation is greater than that required for promotion. These energy relationships are diagrammed in Figure 9.7.

It must not be supposed that promotion is an actual *process* that takes place prior to bonding. Rather, it is a conceptual model—an analysis of a complex phenomenon in terms of simpler parts. It is easier for us to think about bond formation as if it took place in the sequential steps of promotion and overlap, particularly if we wish to

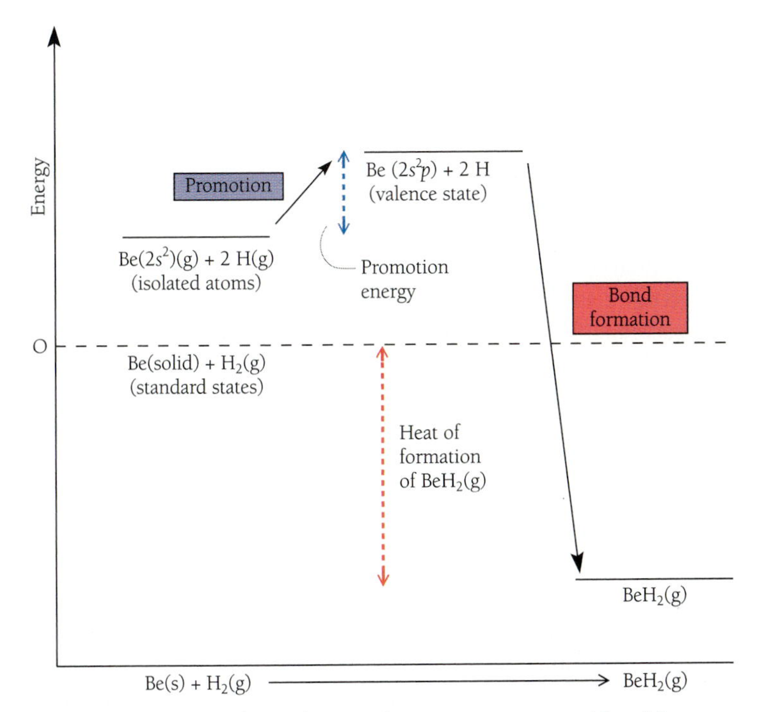

Figure 9.7 Energy relationships in electron promotion and bond formation. The $2s \rightarrow 2p$ promotion energy (blue) can be estimated from atomic spectroscopy or calculations based on the Schrödinger equation. For beryllium, it is less than 500 kJ mol^{-1}. Since formation of two Be—H bonds supplies more energy than this, BeH_2 is a stable molecule.

understand the energies involved in bonding. But there is no evidence to suggest that bonding is other than a smooth, continuous process, or that the hypothetical "promoted state" is ever achieved in reality.

Promotion to a valence state that is different from the ground state of an isolated atom also explains the stoichiometry of other second-period compounds. Boron forms trihalides such as BCl_3, indicating a valence state with three unpaired electrons. This configuration can be achieved by $s \rightarrow p$ promotion from the ground state $[\mathrm{He}]\underset{2s}{\boxed{\uparrow\downarrow}}\;\underset{2p}{\boxed{\uparrow}\boxed{}\boxed{}}$ to the promoted state $[\mathrm{He}]\underset{2s}{\boxed{\uparrow}}\;\underset{2p}{\boxed{\uparrow}\boxed{\uparrow}\boxed{}}$. Similarly, carbon forms compounds such as methane, CH_4, containing four covalent bonds. The $s \rightarrow p$ promotion from the ground state $[\mathrm{He}]\underset{2s}{\boxed{\uparrow\downarrow}}\;\underset{2p}{\boxed{\uparrow}\boxed{\uparrow}\boxed{}}$ to the valence state $[\mathrm{He}]\underset{2s}{\boxed{\uparrow}}\;\underset{2p}{\boxed{\uparrow}\boxed{\uparrow}\boxed{\uparrow}}$ satisfactorily accounts for the observed empirical formula.

Although the additional feature of electron promotion improves the theory, certain details of bond characteristics and molecular shape remain unexplained. From a valence state {*s*, *p*} beryllium would form a BeH_2 molecule having one $Be_{2s}—H_{1s}$ bond and one $Be_{2p}—H_{1s}$ bond. These two bonds are different, and should lead to different chemical behavior of the two hydrogen atoms. Such a difference ought to be observable by some chemical or physical means. The experimental facts are otherwise: the two Be—H bonds are identical in every way, as are the three bonds in BCl_3 and the four bonds in CH_4. No theory predicting bonds of two different types in these compounds can be correct. As to molecular shape, we would expect the three $C_{2p}—H_{1s}$ bonds in methane to have the same geometrical orientation as the three 2*p* orbitals, namely to be mutually perpendicular. No prediction can be made about the orientation of a $C_{2s}—H_{1s}$ bond, since the spherical 2*s* orbital has no directional characteristics. Again, the experimental facts are otherwise: CH_4 is known to be tetrahedral, with four identical C—H bonds making an angle of 109.5° with one another.

Hybrid Orbitals

There is in fact no reason to believe that the valence orbitals of a beryllium atom, when *bonded* to two other atoms in a molecule, are the same as those in an *isolated* atom. From the observed molecular geometry (linear, with equal bond lengths) and chemical properties of BeH_2, we know that the two bonding orbitals of beryllium must be identical and oriented at 180° to one another. These bonding orbitals are called **hybrid orbitals** because, like biological hybrids, they are distinct from and yet have some of the characteristics of the **parent orbitals** of an isolated beryllium atom.

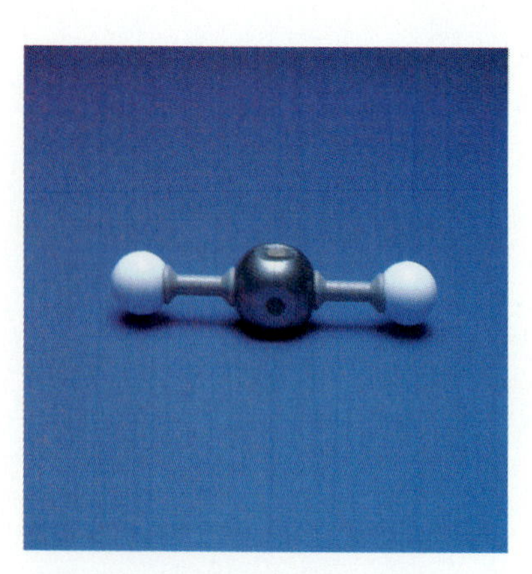

Beryllium dihydride, BeH_2.

sp Hybrids

The valence state of an isolated Be atom consists of one *s* and one *p* orbital (the parent orbitals), each singly occupied. The bonding state consists of two *sp hybrid orbitals,* each singly occupied. Although we will not present it here, the mathematical description of hybrid orbitals leads directly to their shape and relative orientation, shown for *sp* hybrids in Figures 9.8 and 9.9. The term **"hybridization"** refers both to the mathematical process of describing hybrid orbitals and to the condition of the atom whose orbitals are described. In the latter sense, we say, for instance, that "the hybridization of the Be atom in BeH_2 is *sp*."

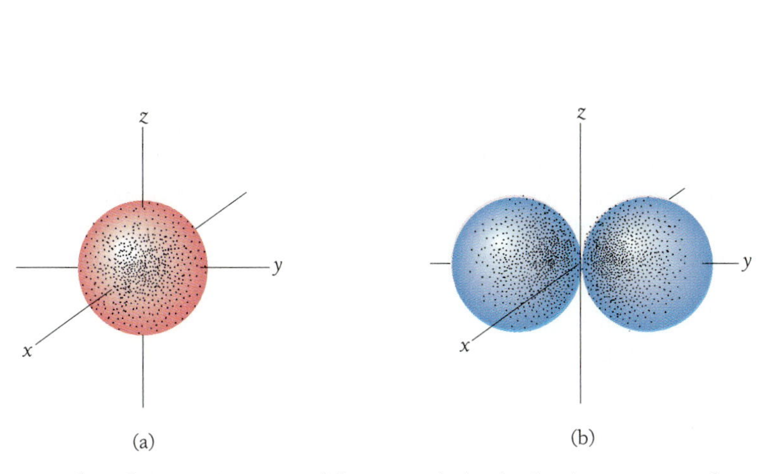

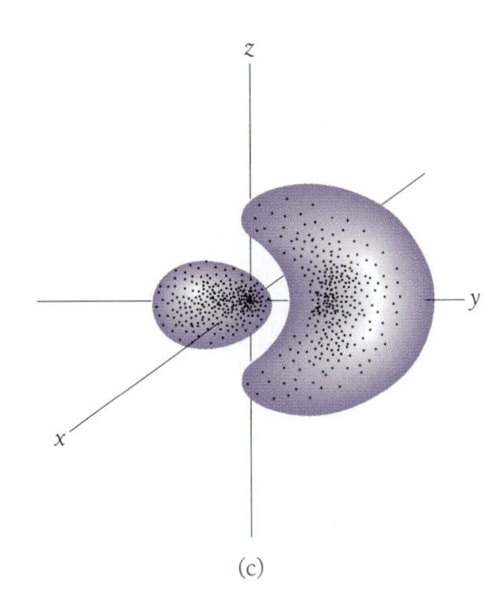

Figure 9.8 Hybrid orbitals retain some characteristics from both parent orbitals. (a) An *s* orbital of an isolated atom has its electron density in a single lobe. (b) A *p* orbital of an isolated atom has two lobes separated by a planar node. (c) An *sp* hybrid orbital has a curved nodal surface separating a major and a minor lobe. The hybrid retains the cylindrical symmetry of the parent *p* orbital.

The relative orientation of the two *sp* hybrid orbitals is more easily appreciated in a drawing that distorts the shape somewhat. For this reason hybrid shapes are frequently represented as in Figure 9.9, with the minor lobe very small or missing, and the major lobe drawn more slender than it actually is. Just as for hydrogenlike orbitals, the Pauli exclusion principle limits hybrid orbital occupancy to two electrons, spins paired. In valence bond theory, the valence state of beryllium is described by the orbital diagram [He] [↑][↑] (*sp*). From this state a Be atom can complete both of its

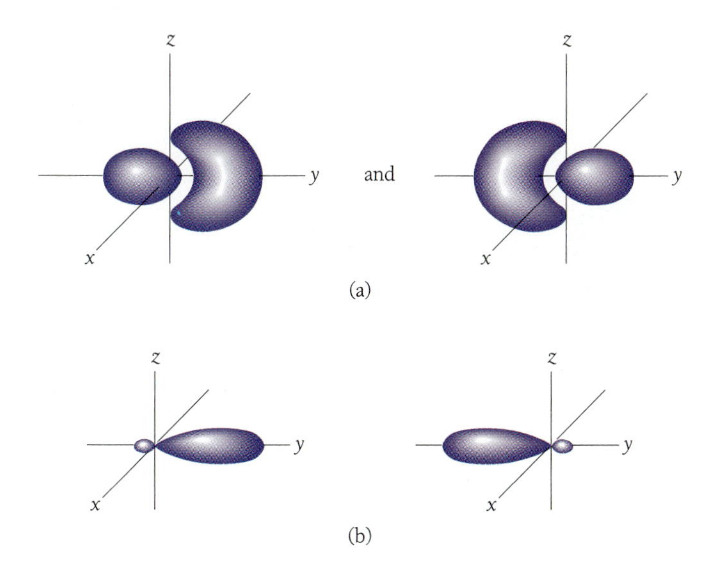

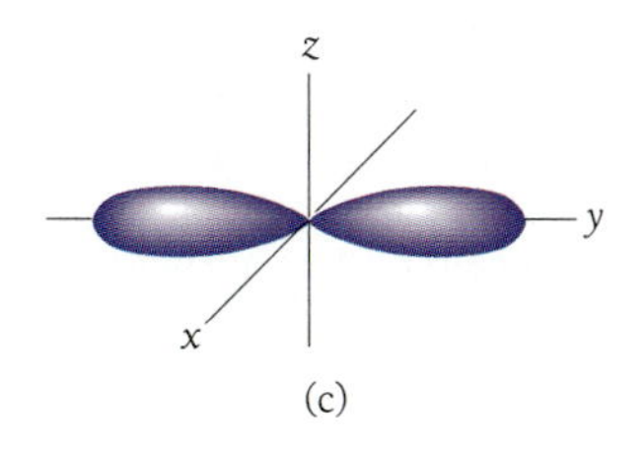

Figure 9.9 Contour drawings of hybrid orbitals are often distorted so as to emphasize their directional characteristics rather than their shapes. (a) Undistorted drawing of the two *sp* hybrids. (b) Distorted drawing, emphasizing the cylindrical symmetry and minimizing the representation of the minor lobes. (c) Relative orientation of the major lobes of the two *sp* hybrid orbitals, emphasizing the 180° bond angle in a molecule having *sp* bonding.

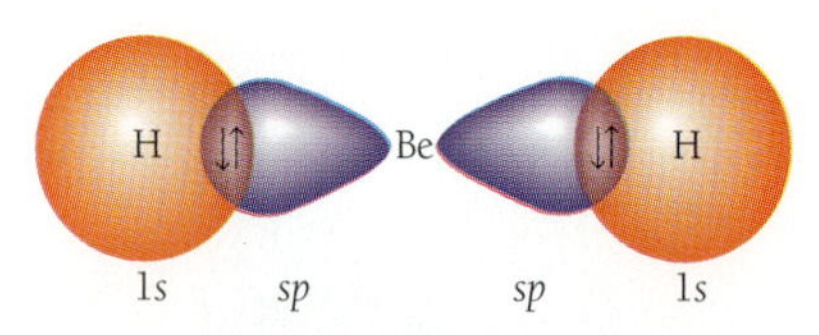

Figure 9.10 Valence bond theory describes each of the two covalent bonds in BeH_2 as resulting from the overlap of a half-filled 1*s* orbital on a hydrogen atom with a half-filled *sp* hybrid orbital on Be.

half-filled valence orbitals by overlap with half-filled orbitals on adjacent atoms, forming BeX_2 molecules such as BeH_2 (Figure 9.10) or BeF_2. The two *sp* bonding orbitals are oriented at a 180° angle to one another, and the H—Be—H bond angle is 180°.

While hybridization is a useful model for preserving and extending the Lewis ideas, it is important to remember that it is not a *process* that takes place within an electron cloud preliminary to bond formation. Rather, hydrogenlike and hybrid orbitals are alternative ways of *describing* the distribution of negative charge density in the electron cloud surrounding the nucleus. The hydrogenlike orbitals discussed in Chapter 7 are a good framework for discussing and understanding the energy levels and spectra of *atoms,* whereas the hybrid orbitals of valence bond theory are a good framework for understanding covalent *bonding.* This and other aspects of hybridization are discussed in more detail in Appendix D.

Boron trichloride, BCl_3.

sp^2 Hybrids

A similar approach is used to describe the bonding of atoms that form three covalent bonds. In boron, for example, the isolated-atom state is $[\text{He}]\underset{2s}{\boxed{\uparrow\downarrow}}\ \underset{2p}{\boxed{\uparrow}\boxed{}\boxed{}}$; promotion of an *s* electron gives the valence state $[\text{He}]\underset{2s}{\boxed{\uparrow}}\ \underset{2p}{\boxed{\uparrow}\boxed{\uparrow}\boxed{}}$. This valence-electron configuration is equally well described in terms of three equivalent *sp hybrid* orbitals: $[\text{He}]\ \underset{sp^2}{\boxed{\uparrow}\boxed{\uparrow}\boxed{\uparrow}}$. The shape of an sp^2 orbital is similar to that of an *sp* hybrid (Figure 9.11). Boron forms three σ bonds by overlap of its three half-filled sp^2 orbitals with half-filled orbitals on adjacent atoms. Since the three sp^2 hybrid orbitals

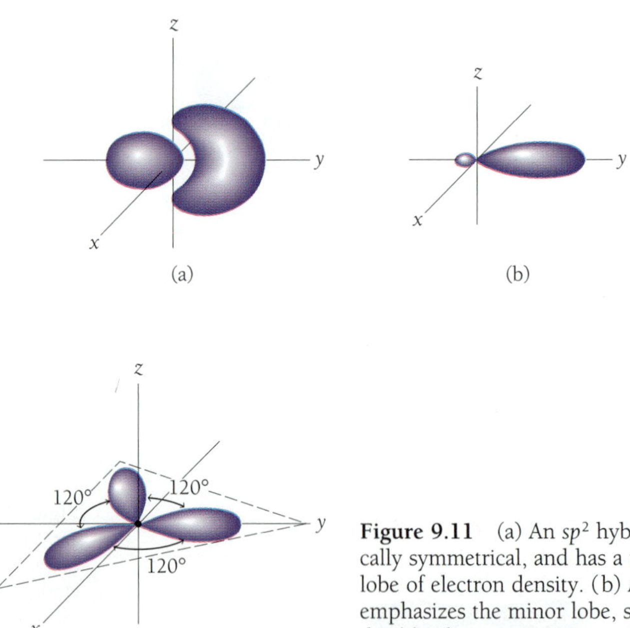

Figure 9.11 (a) An sp^2 hybrid orbital is cylindrically symmetrical, and has a major and a minor lobe of electron density. (b) A distorted sketch de-emphasizes the minor lobe, so as to better show the (c) relative orientations of the three sp^2 hybrids. The axes are coplanar, and point outward to the corners of an equilateral triangle.

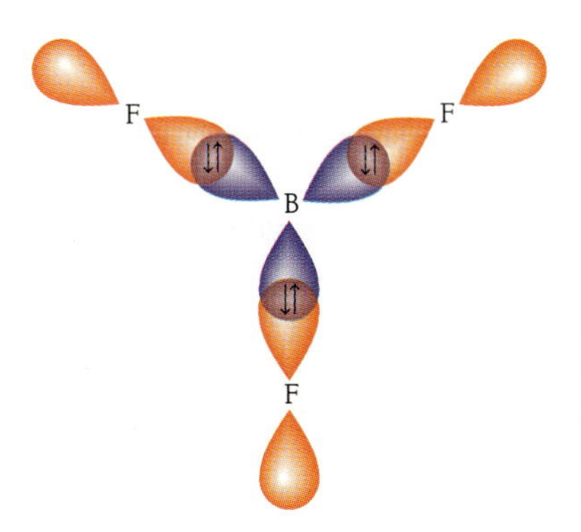

Figure 9.12 Each covalent bond in the BF_3 molecule is a σ bond formed by overlap of a half-filled 2*p* orbital on a fluorine atom with one of three half-filled sp^2 hybrid orbitals on the boron atom.

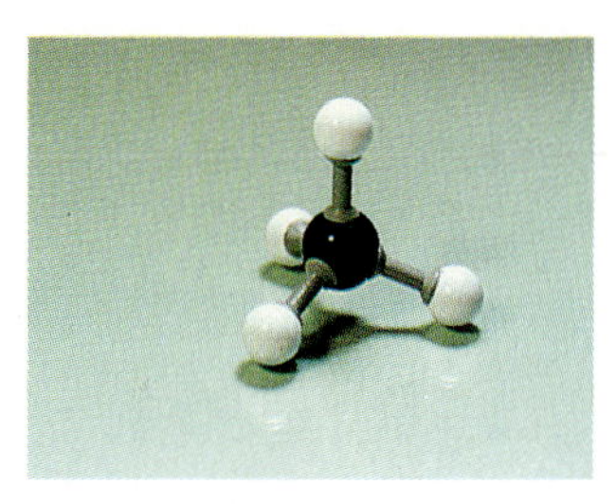

Methane, CH_4.

are oriented at angles of 120° to one another, so are the covalent bonds. The shape of BCl_3 and other molecules with sp^2 hybridization of the central atom is *trigonal planar* (Figure 9.12).

sp^3 Hybrids

The tetrahedral shape and bonding in molecules, such as CH_4, whose central atoms form four covalent bonds, is similarly described in terms of hybrid orbitals. The isolated-atom electron configuration of carbon is [He] [↑↓] (2s) [↑][↑][↑] (2p); promotion of an *s* electron gives the valence state [He] [↑] (2s) [↑][↑][↑] (2p). This valence-electron density is equally well described in terms of sp^3 *hybrid* orbitals, [He] [↑][↑][↑][↑] (sp^3). The shape of an sp^3 orbital is similar to that of *sp* and sp^2 hybrids (Figure 9.13). Carbon forms four σ bonds by overlap of its four half-filled sp^3 orbitals with half-filled orbitals on adjacent atoms. The four sp^3 hybrid orbitals are oriented at angles of 109.5° to one another, and so are the covalent bonds. The shape of CH_4 and other molecules with sp^3 hybridization of the central atom is *tetrahedral.*

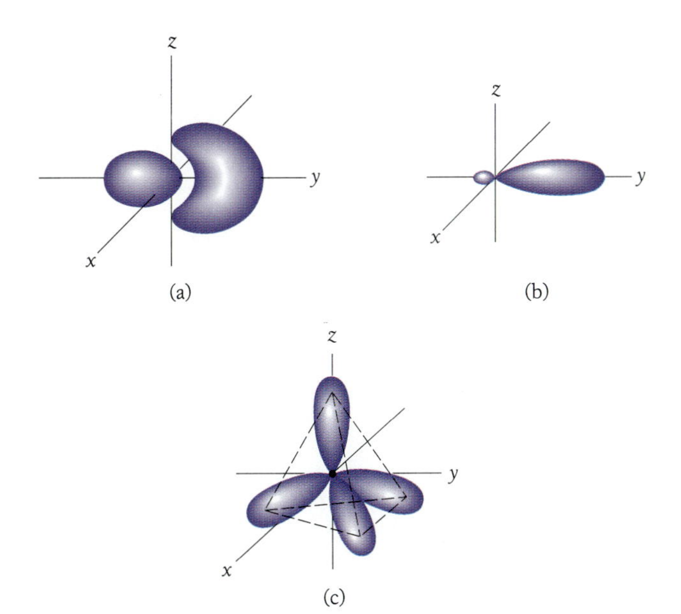

Figure 9.13 (a) An sp^3 hybrid orbital is cylindrically symmetrical, and has a major and a minor lobe of electron density. (b) A distorted sketch de-emphasizes the minor lobe, so as to better show the (c) relative orientation of the four sp^3 hybrids. The axes point outward to the corners of a tetrahedron, so that the bond angles associated with an "sp^3 carbon" are 109.5°.

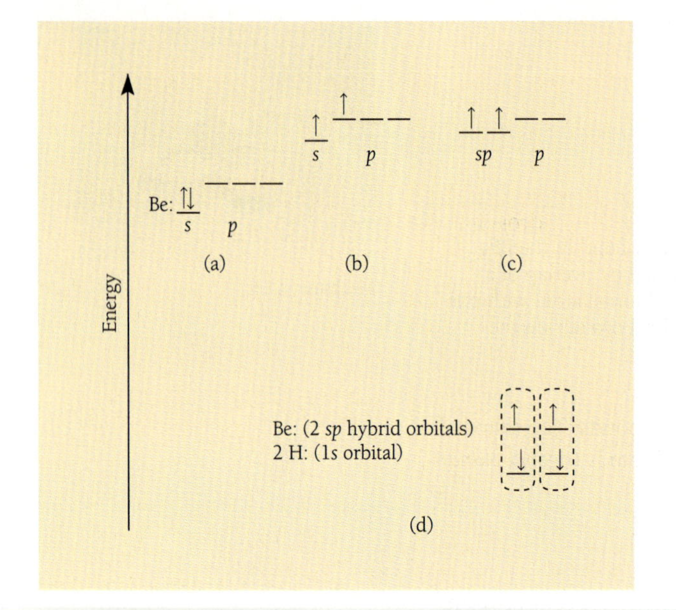

Figure 9.14 Energy relationships in the promotion/hybridization/overlap model of valence bond theory. (a) An isolated Be atom has the valence electron configuration {*s*, *s*}, which in the promotion step becomes (b) the valence state {*s*, *p*}. (c) The valence state is better described as two half-filled hybrid orbitals {*sp*, *sp*}, which by overlap with atomic orbitals on adjacent hydrogen atoms form (d) the two equivalent covalent bonds in the linear BeH_2 molecule.

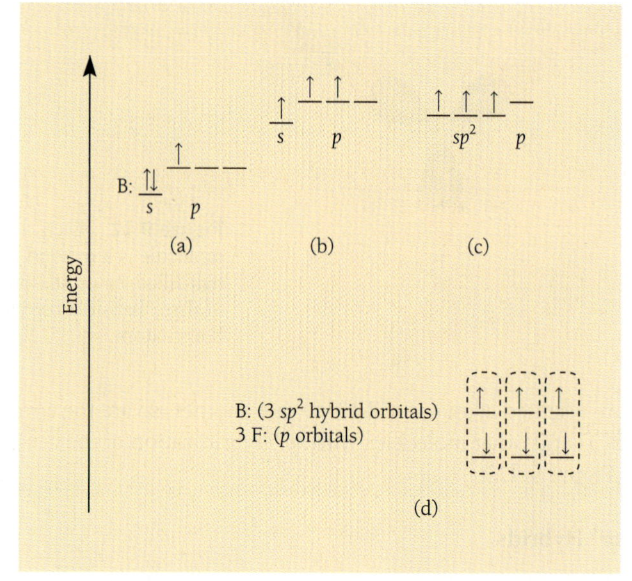

Figure 9.15 Energy relationships in the promotion/hybridization/overlap model of valence bond theory. (a) An isolated B atom has the valence electron configuration {*s*, *s*, *p*}, which in the promotion step becomes (b) the valence state {*s*, *p*, *p*}. (c) The valence state is better described as three half-filled hybrid orbitals {sp^2, sp^2, sp^2}, which by overlap with *p* atomic orbitals on adjacent fluorine atoms form (d) the three equivalent covalent bonds in the trigonal planar BF_3 molecule.

Note that in all three types of hybridization (*sp*, sp^2, and sp^3), the number of equivalent hybrid orbitals used to describe the valence shell is the same as the number of hydrogenlike parent orbitals.

The concepts of promotion and hybridization make valence bond theory useful for explaining the stoichiometry, bond characteristics, and molecular shape of covalently bonded molecules. The explanations are made *as if* covalent bond formation were the three-step process diagrammed in Figures 9.14, 9.15, and 9.16. We repeat

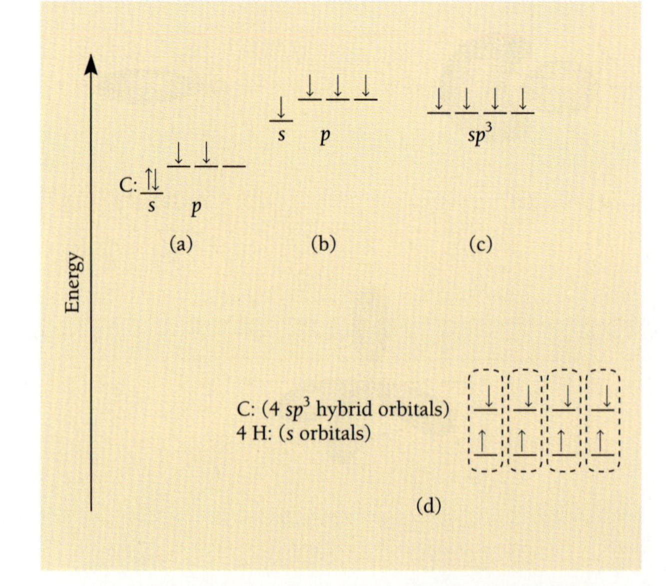

Figure 9.16 Energy relationships in the promotion/hybridization/overlap model of valence bond theory. (a) An isolated C atom is described by the valence electron configuration {*s*, *s*, *p*, *p*}, which in the promotion step becomes (b) the valence state {*s*, *p*, *p*, *p*}. (c) The valence state is better described as four half-filled hybrid orbitals {sp^3, sp^3, sp^3, sp^3}, which by overlap with atomic orbitals on adjacent hydrogen atoms form (d) the four equivalent covalent bonds in the tetrahedral CH_4 molecule.

that this is a conceptual model only. Bonding is not a stepwise process, nor is the actual electron density described by the orbitals of a hybrid set different from that described by the corresponding hydrogenlike set.

Participation of *d* Orbitals in Hybrids

The bonding and shape of a molecule whose central atom forms more than four bonds cannot be explained by reference to a hybrid set based only on *s* and *p* hydrogenlike orbitals. Hybrid orbitals in such molecules include a contribution from *d* orbitals as well. Specifically, hybridization of an *s* orbital, three *p* orbitals, and a *d* orbital produces a set of five *sp^3d hybrids.* In the phosphorus pentachloride molecule, for example, the P—Cl σ bonds are formed by overlap of orbitals on the chlorine atoms with phosphorus sp^3d orbitals. These have a shape similar to that of *s*, *p* hybrids, but are directed outward to the corners of a trigonal bipyramid (Figure 9.17). Similarly, involvement of two *d* orbitals produces a set of six *sp^3d^2 hybrids,* octahedrally arranged, as in sulfur hexafluoride (Figure 9.18). The bonding of atoms with expanded valence shells (Section 8.6) is explained in valence bond theory in terms of these *d*-containing

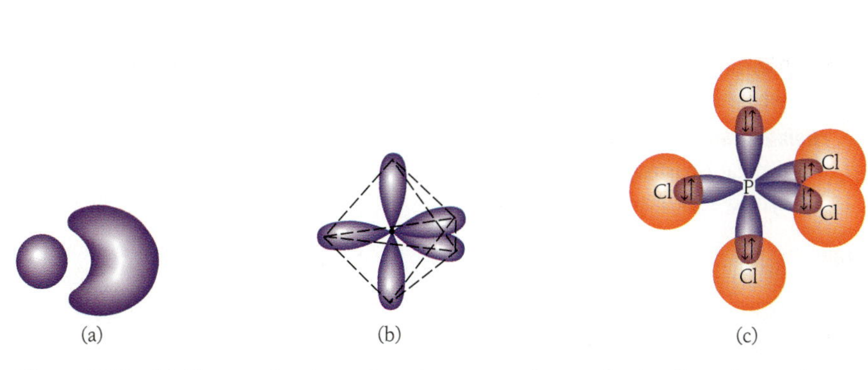

Figure 9.17 (a) Electron density in the valence state of atoms having five valence electrons is described by the set of five sp^3d hybrid orbitals. (b) The orientation of these orbitals is trigonal bipyramidal, which (c) corresponds to the shape of five-pair molecules such as PCl_5.

Phosphorus pentachloride, PCl_5.

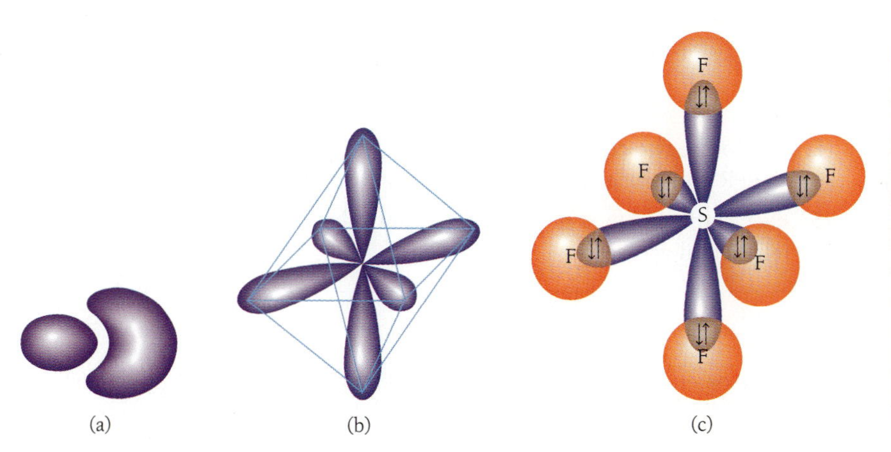

Figure 9.18 (a) Electron density in the valence state of atoms having six valence electrons is described by the set of six sp^3d^2 hybrid orbitals. (b) The orientation of these orbitals is octahedral, which (c) corresponds to the shape of six-pair molecules such as SF_6.

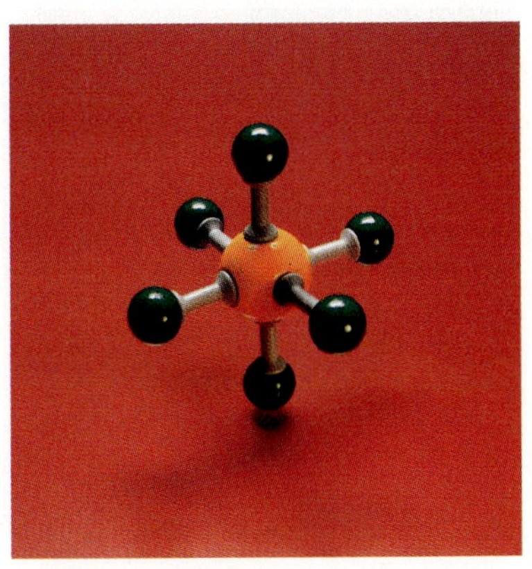

Sulfur hexafluoride, SF_6.

hybrid orbital sets. Such atoms are regarded as exceptions in the Lewis theory, because they violate the octet rule. In the valence bond approach they are not exceptions to a rule—they are simply a different type of hybridization.

Hybrid Orbital Geometry

There is a straightforward relationship between the shape of a molecule and the hybrid set used to describe the bonding. For instance, if the shape is known to be tetrahedral, the appropriate hybrid set is sp^3, and the central atom in tetrahedral molecules is said to be "sp^3 hybridized." Since molecular shape can be predicted by VSEPR theory, knowledge of the Lewis structure leads directly to both shape and hybridization.

EXAMPLE 9.1 Geometry Determines Hybridization

Predict the shape of ethane, C_2H_6. What hybrid set is most appropriate to describe the bonding in ethane? That is, what is the hybridization of the central atom(s) in this molecule?

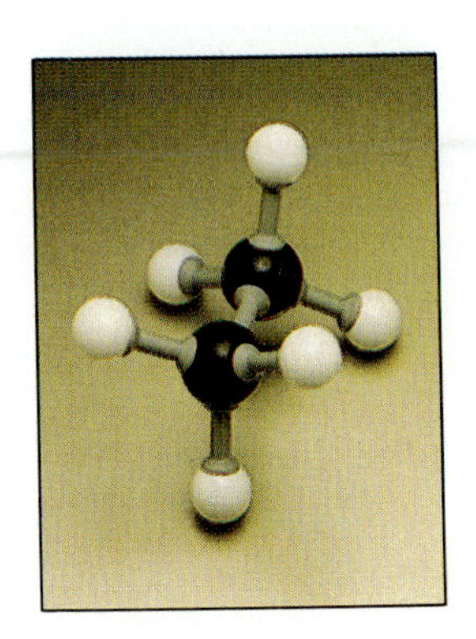

Ethane, C_2H_6.

Analysis

Target: Description of the geometrical configuration ("trigonal planar," "bent," and so on) around the central atom(s) is sufficient to describe the shape of a molecule.

Relationship: The geometry is related to the Lewis structure by VSEPR, and the hybridization is related to geometry through the known orientation of the orbitals in hybrid sets.

Plan

1. Draw the Lewis structure.
2. Use VSEPR to determine the shape around each central atom.
3. Identify the hybrid orbital set having the same orientation.

Work

1. The only Lewis structure satisfying the octet rule is

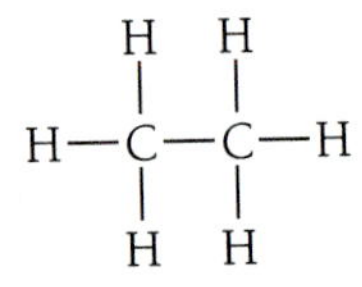

2. Each of the two central atoms has four electron pairs in its valence shell, so by VSEPR the geometry around each is tetrahedral.
3. Each carbon atom is sp^3 hybridized, since that is the only hybrid set having tetrahedral geometry.

Exercise

Predict the geometry of silane, SiH_4. What is the hybridization of the Si atom?

Answer: Tetrahedral, sp^3.

Silane, SiH_4.

The orientation of orbitals depends only on the *number* of hybrid orbitals in the set. The relationship between electron-pair geometry and hybridization also holds true even when one or more of the orbitals are occupied by lone pairs rather than used to form bonds with adjacent atoms.

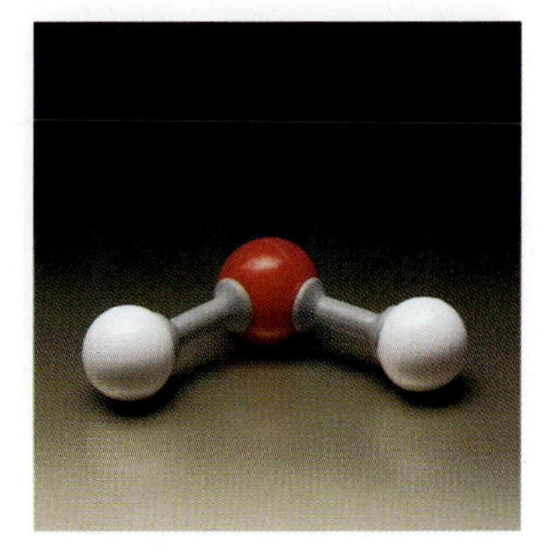

Water, H_2O.

EXAMPLE 9.2 Hybridization in the Water Molecule

Describe the hybridization and bonding of the oxygen atom in the water molecule.

Analysis Once the geometry is known, the hybrid type follows immediately. To "describe the bonding" means to identify the atomic orbitals that overlap to form the bonds, and to identify the bonds as σ or π.

Plan As in the preceding example, use VSEPR to obtain the electron-pair geometry from the Lewis structure. Then identify the hybrid type and describe the bonds.

Work The Lewis structure of water is

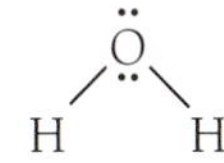

With four electron pairs around the central atom, the electron-pair geometry is tetrahedral. (Since there are only two bonding pairs, the *molecular shape* is described as "bent.")

Tetrahedral atoms have sp^3 hybridization. Each of the two O—H bonds is a σ bond formed by $O_{sp^3} - H_{1s}$ orbital overlap. Each of the two remaining sp^3 orbitals is occupied by a lone pair (Figure 9.19a).

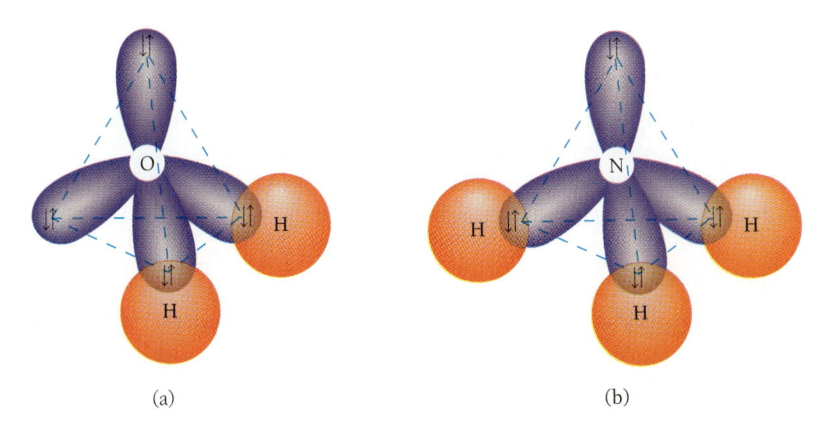

Figure 9.19 (a) With sp^3 hybridization of the central O atom, the electron-pair geometry in the water molecule is tetrahedral. Two of the four equivalent sp^3 orbitals do not participate in overlap and bonding, but are occupied by lone pairs. The molecule is bent, with an H—O—H bond angle of 104½°. (b) The hybridization in ammonia is sp^3. One of the four equivalent hybrid orbitals is occupied by a lone pair. The *shape* of the molecule (as opposed to the electron-pair geometry) is pyramidal, with H—N—H angles of 107°.

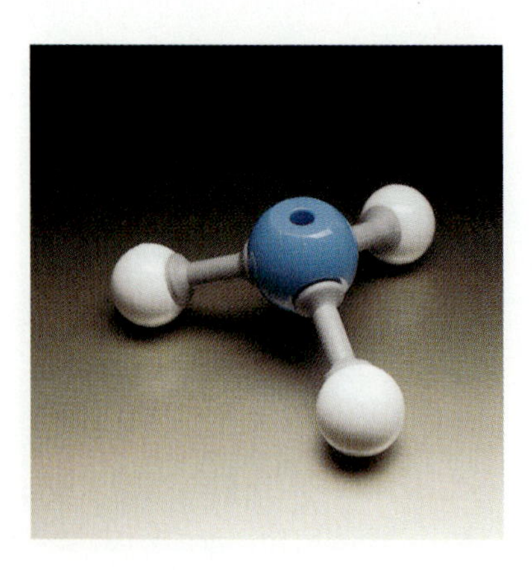

Ammonia, NH_3.

Differences between the actual bond angles in a molecule and those predicted by VSEPR were discussed in Section 8.6.

Exercise

Describe the hybridization and bonding in ammonia.

Answer: sp^3: three $N_{sp^3} - H_{1s}$ σ bonds and an sp^3 lone pair.

The bonding in the H_2O molecule could also be described without reference to hybrid orbitals. In this description, the valence state of oxygen is taken as the ground (nonpromoted) configuration [He] $\underset{2s}{[\uparrow\downarrow]}$ $\underset{2p}{[\uparrow\downarrow][\uparrow][\uparrow]}$. Two σ bonds are formed by overlap of the half-filled *p* orbitals with the 1*s* orbitals of two H atoms, which implies an H—O—H bond angle of 90° (the angle between the atomic *p* orbitals). Since this is quite different from the actual angle of 104.5°, the description in terms of hybrid orbitals is preferable. The actual shape of most molecules correlates better with hybrid orbitals than with the unhybridized atomic orbitals.

The relationships between electron-pair geometry and hybridization are summarized in Table 9.1.

TABLE 9.1 VSEPR and Hybridization

Number of Electron Regions	Pair Geometry from VSEPR	Hybrid Type
2	Linear	sp
3	Trigonal planar	sp^2
4	Tetrahedral	sp^3
5	Trigonal bipyramidal	sp^3d
6	Octahedral	sp^3d^2

Multiple Bonding

In the valence bond approach, a multiple bond is described as a σ bond plus one or two π bonds. A double bond consists of one σ and one π bond, and a triple bond consists of one σ and two π bonds. The π bonds are formed by overlap of unhybridized hydrogenlike atomic orbitals, not from hybrids.

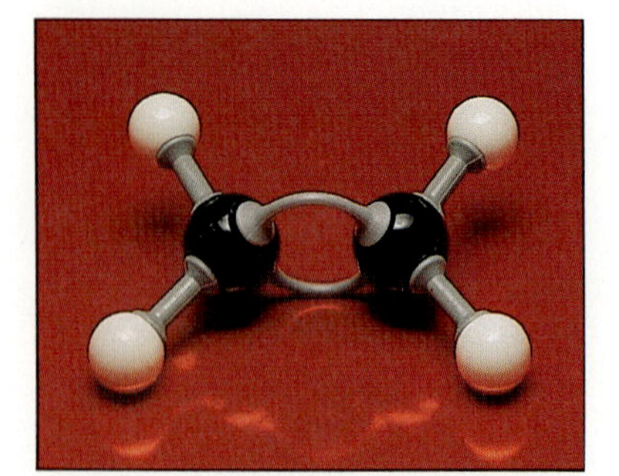

Ethene, C_2H_4.

Bonding in Ethene

It is firmly established by experiment, as well as clearly predicted by VSEPR, that the ethene molecule, $H_2C{=}CH_2$, is a planar species whose H—C—H and H—C—C bond angles are close to 120°. Although C has four valence electrons, the observed geometry precludes sp^3 hybridization and suggests sp^2 instead. The bonding is explained in valence bond terms as follows. The *s* orbital of the isolated carbon atom mixes with *two* of the *p* orbitals to form a set of three sp^2 hybrid orbitals. The remaining *p* orbital is unhybridized. The axis of the unhybridized *p* orbital is perpendicular to the plane containing the axes of the sp^2 hybrid set (Figure 9.20). The valence orbitals of carbon in ethene are $\{sp^2, sp^2, sp^2, p\}$: each of the four valence electrons occupies one of

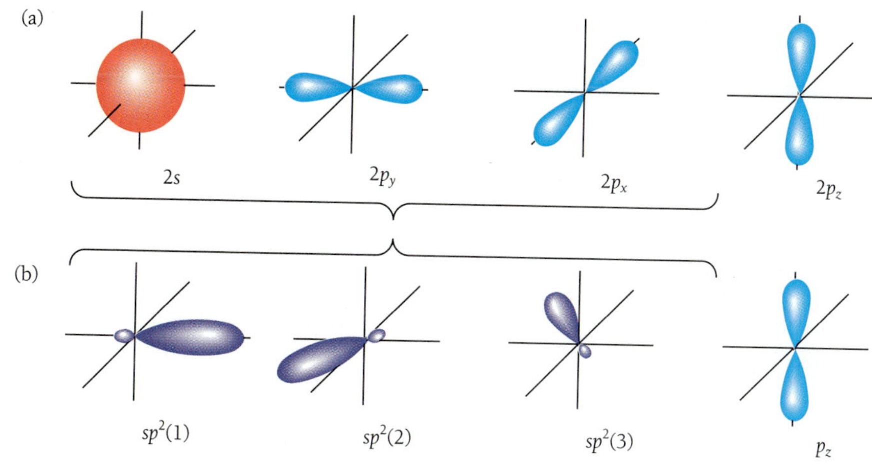

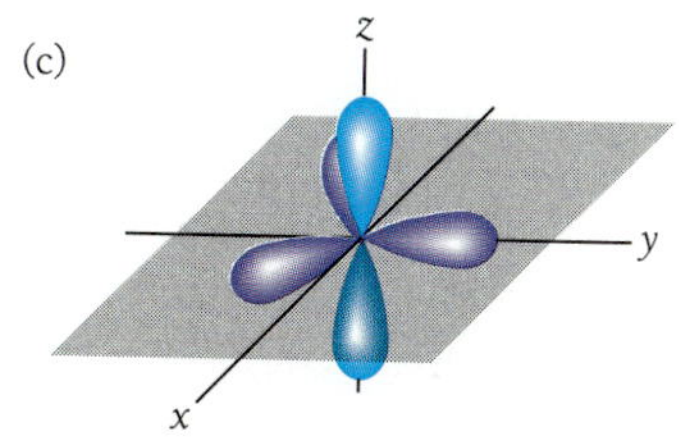

Figure 9.20 The electron density in the valence shell of carbon can be described in several ways. (a) One description uses the four orbitals $\{2s, 2p_x, 2p_y, 2p_z\}$. (b) Another description, which leads to the same overall electron density, uses three sp^2 hybrids and one unhybridized p orbital. (c) The axes of the sp^2 hybrid orbitals lie in the same plane, and the axis of the p orbital is perpendicular to this plane.

these, and the orbital diagram is [He] [↑][↑][↑] (sp^2 sp^2 sp^2) [↑] (p). Each of the sp^2 orbitals overlaps and forms a σ bond with an orbital on an adjacent atom, forming two $C_{sp^2} - H_{1s}$ and one $C_{sp^2} - C_{sp^2}$ bond. This results in a **σ framework** or "skeletal structure," as shown in Figure 9.21, having sp^2 or trigonal planar geometry about each of the central carbon atoms. The half-filled p orbitals on the carbon atoms (Figure 9.22) are adjacent to one another and form a π bond by side-to-side overlap. Overlap

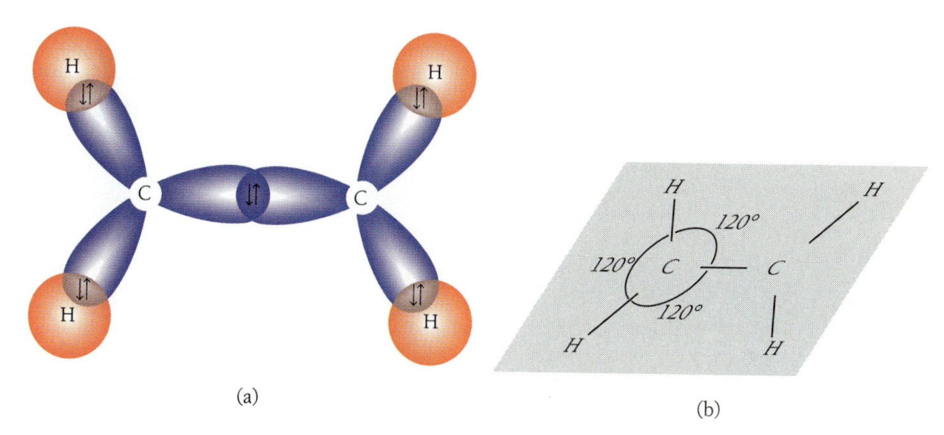

Figure 9.21 (a) Each of the half-filled sp^2 hybrid orbitals on the central carbon atoms of the ethene molecule overlaps with a half-filled orbital on an adjacent atom to form a σ bond. (b) The resulting σ framework is characterized by trigonal planar geometry around each carbon atom.

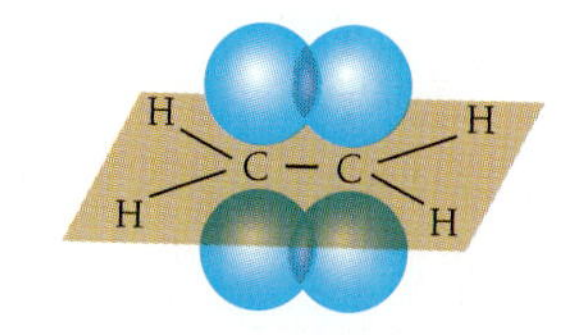

Figure 9.22 The half-filled p orbitals, unused in forming the σ framework of ethene, enter into side-by-side overlap to form a π bond.

is maximized, and the π bond is strongest, if the axes of the two p orbitals are parallel to one another. Since each p orbital is perpendicular to its own set of sp^2 hybrids, all six hybrid orbitals lie in the same plane. It follows that all six atoms in the ethene molecule lie in one plane.

Bonding in Ethyne

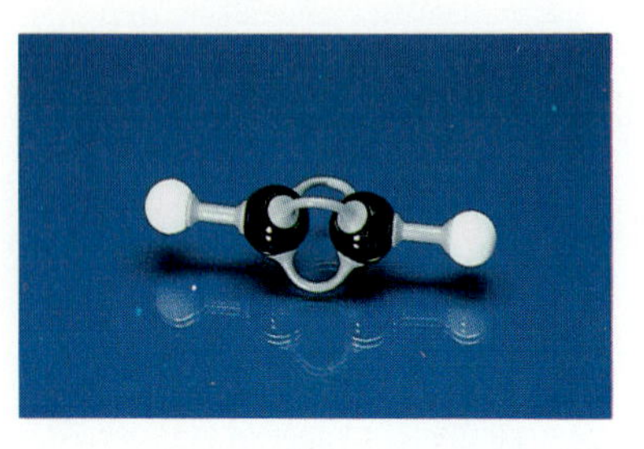

Ethyne, C_2H_2.

The bonding in ethyne, H—C≡C—H, is similarly explained in valence bond terms. Its linear structure indicates sp hybridization of the central carbon atoms, so that the valence shell is described as consisting of two sp hybrids and two unhybridized p orbitals (Figure 9.23). Each of these contains one electron, so the orbital diagram is [He] $\underset{sp}{\boxed{\uparrow}}\underset{sp}{\boxed{\uparrow}}\ \underset{p}{\boxed{\uparrow}}\underset{p}{\boxed{\uparrow}}$. The σ framework of ethyne is formed by two $C_{sp} - H_{1s}$ bonds and one $C_{sp} - C_{sp}$ bond (Figure 9.24a). *Both* p orbitals of one carbon atom are correctly positioned for side-by-side overlap with the p orbitals of the other carbon atom. Two π bonds are formed, with nodal planes at right angles to one another (Figure 9.24b). Note the close similarity between the bonding in C_2H_2 and that in N_2 (Figure 9.6).

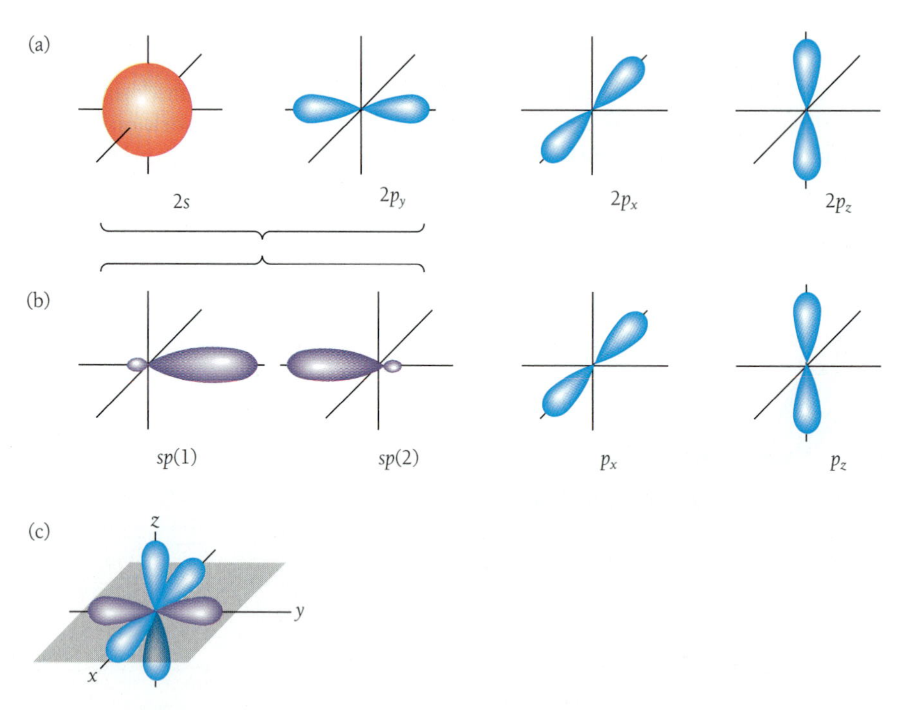

Figure 9.23 The electron configuration in the valence shell of second-period elements can be described either by (a) the four orbitals $\{2s, 2p_x, 2p_y, 2p_z\}$ or (b) two sp hybrids and two unhybridized p orbitals. (c) The axes of the sp hybrids are colinear and perpendicular to the axes of the remaining p orbitals.

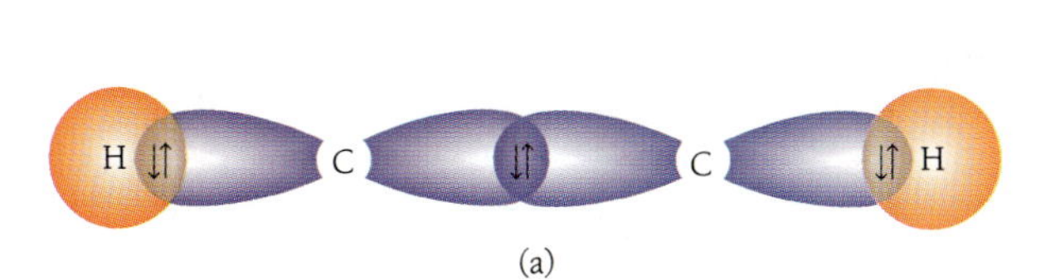

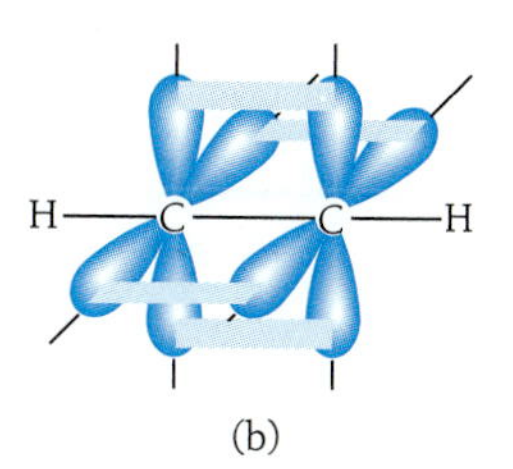

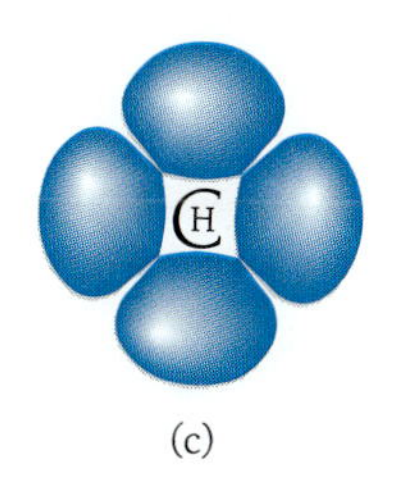

Figure 9.24 (a) The σ framework of ethyne is formed by overlap of the *sp* hybrid orbitals on the central carbon atoms with each other, and with the 1*s* orbitals of the terminal hydrogen atoms. (b) The two sets of half-filled *p* orbitals, unused in σ bonding, form two π bonds by side-by-side overlap. (c) An end-on view shows more clearly that each π bond has two lobes, symmetrically located on opposite sides of the axis containing the four atoms of the molecule.

Resonance

The concepts of resonance and electron delocalization, added to Lewis theory to explain bonding in species like ozone and the carbonate ion, are preserved in essentially unchanged form in the valence bond approach. Moreover, since valence bond theory is based on mathematical calculations, it provides a numerical assessment of the relative importance of the several structures contributing to a resonance hybrid. These quantitative details, however, are not often used by chemists.

9.3 MOLECULAR ORBITALS

The concepts of the preceding section are not the only way to describe bonding and structure. Whereas the valence bond approach uses hydrogenlike atomic orbitals to form combinations (hybrid orbitals) that are *localized on a particular atom,* an alternative model called the *molecular orbital* approach uses the same hydrogenlike atomic orbitals to form different combinations (molecular orbitals or **MO's**) that are not localized but are spread over the entire molecule. In this model, a bond results from electron occupancy of a molecular orbital rather than from overlap of atomic orbitals. In qualitative terms, the two models are quite different from one another. However, when numerical calculations are made (of such properties as bond angles, lengths, and strengths), the differences are less apparent. Each model has its strengths, and both are widely used. The MO method offers a more satisfactory explanation of the electron delocalization characteristic of resonant species and a better understanding of paramagnetism. The valence bond picture retains the simplicity of the Lewis shared-pair concept.

Diatomic Molecules: H_2, He_2, Li_2, and Be_2

MO theory is particularly well suited for discussion of the bonding in **homonuclear diatomic molecules,** which are those molecules consisting of two identical atoms. We begin by applying MO theory to hydrogen, the simplest molecule of them all.

Bonding in H_2

When two hydrogen atoms are close to one another, their electron clouds interact. According to molecular orbital theory, the 1*s* orbitals of the atoms combine, either by constructive or destructive interference, to form two *molecular orbitals (MO's)* (Figure 9.25). MO's like these, that are cylindrically symmetrical about the internuclear axis, are designated **σ MO's.** The orbital of higher energy is called the **antibonding MO** and

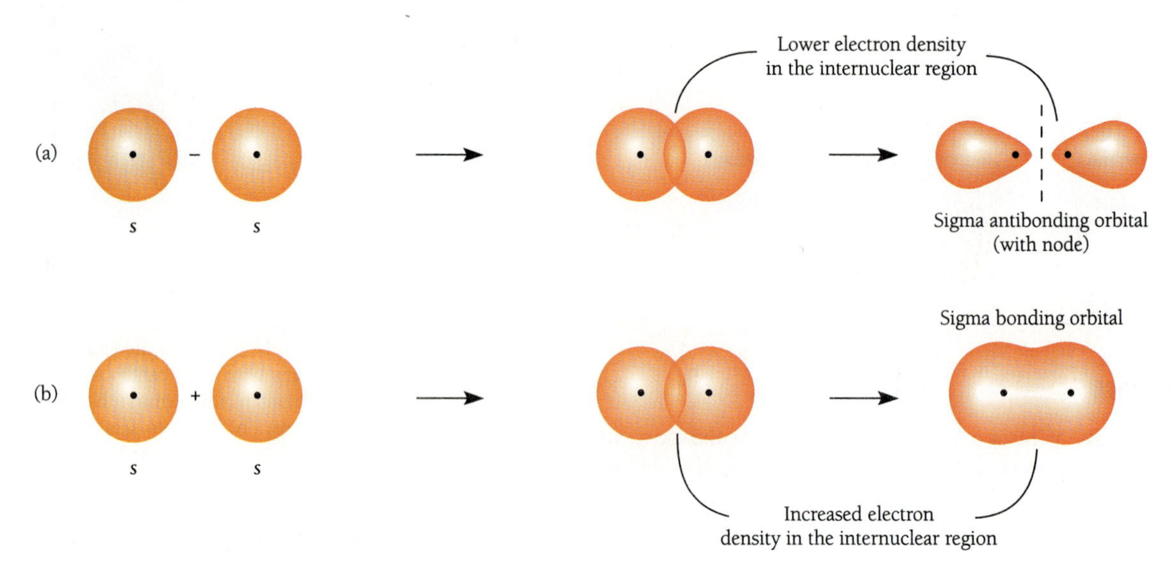

Figure 9.25 The 1s orbitals of two isolated hydrogen atoms mix together to form two types of molecular orbital. (a) The antibonding orbital, the result of destructive interference of the two 1s orbitals, is designated σ^*; it has two lobes separated by a nodal plane bisecting the internuclear axis. (b) The bonding orbital, the result of constructive interference, is designated σ; it has a single lobe.

given the symbol σ^*. The two lobes of a σ^* orbital are separated by a nodal plane that bisects the internuclear axis. The orbital of lower energy, denoted σ, is called the **bonding MO;** it consists of a single lobe (that is, it has no nodes). Antibonding molecular orbitals are always higher in energy (Figure 9.26), and more spread out or delocalized than the corresponding bonding MO's. The notation is often expanded to include the parent atomic orbitals (AO's) as a subscript, especially in more complicated molecules. The bonding MO in hydrogen may be designated as σ, σ_{1s}, or σ_{1s-1s}.

The two electrons of the hydrogen molecule occupy the lower energy orbital, spins paired. Since from this state the separated-atom condition can be restored only by input of energy, the two atoms remain together as a molecule (Figure 9.27). In MO theory, then, a *covalent bond is formed when a bonding MO is occupied.* Description of bonding in molecular orbital terms involves the following principles, which apply to more complicated molecules as well as to H_2.

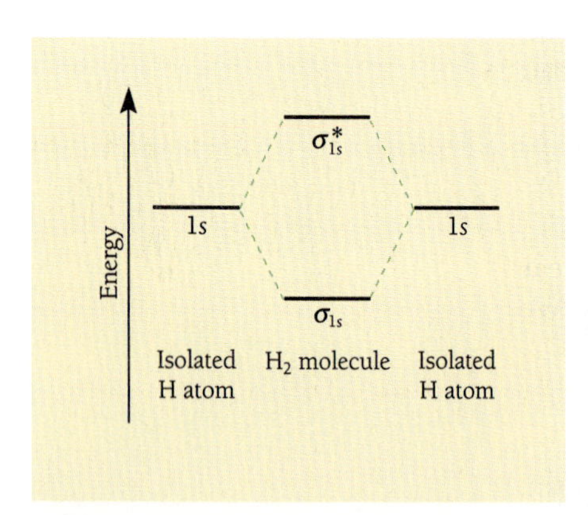

Figure 9.26 The energy of the bonding MO is less than the energies of the AO's from which it is formed, and the energy of the antibonding MO is greater.

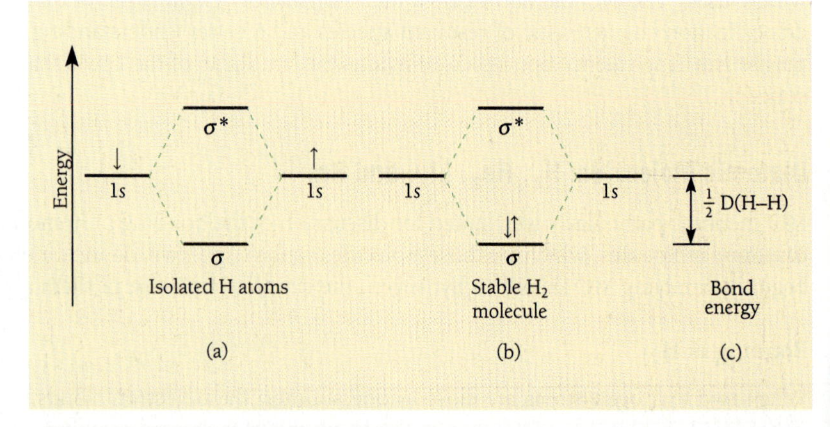

Figure 9.27 A stable H_2 molecule exists because the energy of a pair of electrons in the bonding MO (b) is less than when they occupy the two separate AO's (a). The energy difference (c) is *half* the bond energy D(H—H), because bond breaking involves restoring *both* electrons to their AO's. The dashed lines connect the MO's with the parent AO's from which they are formed.

1. Hydrogenlike atomic orbitals are combined to produce a set of molecular orbitals, which extend over the entire molecule. (Only valence-shell AO's are used—the basic ideas of bonding can be understood without considering core electrons.) The number of MO's is the same as the number of component AO's.
2. Half of the molecular orbitals are bonding MO's, with energy less than that of the AO's. The other half are antibonding, with greater energy.
3. The valence electrons from all atoms are assigned to the MO's according to the same aufbau procedure used in assigning electron configuration in atoms (Chapter 7) as follows:
   a. lowest energy MO's are filled first,
   b. the Pauli principle limits orbital occupancy to two electrons, and
   c. Hund's rule maximizes electron spin in a set of MO's of the same energy.

In some more complicated molecules, which we do not discuss in this text, there are *nonbonding* orbitals as well as bonding and antibonding orbitals.

Occupancy of a bonding MO by a pair of electrons constitutes a covalent bond, equivalent to a shared pair in the Lewis theory and to overlap of half-filled orbitals in the valence bond description.

Bond Order and the Non-Existence of He_2

Consider the interaction of two helium atoms. Although it contains two electrons rather than one, the valence orbital in atomic helium is the same as in hydrogen: 1*s*. Since it is the atomic *orbitals* (rather than the electrons in them) that combine to form MO's, the energy-level scheme of Figure 9.26 applies to He_2 as well as to H_2. Since He_2 has four valence electrons, however, the *occupancy* of the MO's is not the same as in H_2. With two electrons in a higher-energy orbital and two in a lower-energy orbital, the total energy of He_2 would be essentially the same as that of two separate He atoms. The "bond energy" is zero, or, in other words, there is no net force that could hold the two atoms together. The He_2 molecule does not exist (Figure 9.28). In general, MO theory holds that covalent bonding exists whenever the number of electrons in bonding MO's exceeds the number in antibonding MO's.

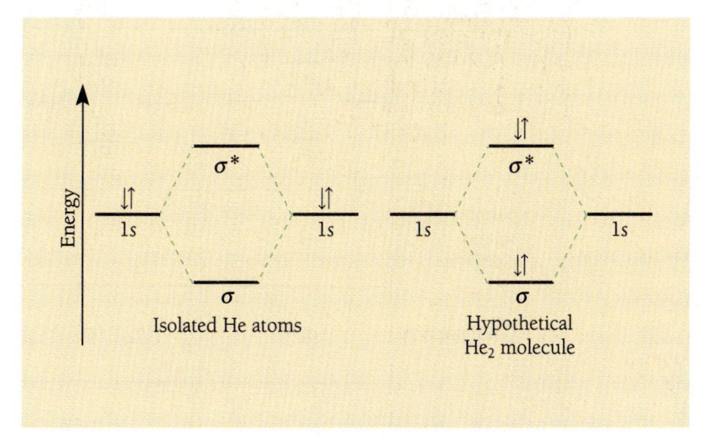

Figure 9.28 In a (hypothetical) He_2 molecule, the lowered energy of the two electrons in the σ bonding orbital is counteracted by the increased energy of two electrons in the σ* antibonding orbital. The result is a total bond energy of zero, that is, no bond at all. He_2 does not exist as a stable molecule.

Although in most cases the result is the same, the definition of **bond order** in MO theory is different from its definition in Lewis theory:

$$\text{bond order} = \frac{1}{2}\left[\begin{pmatrix}\text{total number}\\ \text{of electrons in}\\ \text{bonding MO's}\end{pmatrix} - \begin{pmatrix}\text{total number}\\ \text{of electrons in}\\ \text{antibonding MO's}\end{pmatrix}\right] \tag{9-1}$$

According to this definition the bond order in H_2 is one, and that in He_2 is zero. That is, the H_2 molecule has a single bond, and the He_2 molecule has no bond.

A molecule is often described by giving its electron *configuration,* which, just as for atoms, is a listing of the occupancy of its molecular orbitals. The hydrogen molecule is H_2: $(\sigma_{1s})^2$, and the helium molecule is He_2: $(\sigma_{1s})^2(\sigma^*_{1s})^2$.

Li_2 and Be_2

Core orbitals do not have an important effect on bonding. Since H and Li both have the same valence-shell configuration (s^1), they form MO's in essentially the same way. Be and He, with the s^2 configuration, are also alike. The MO description of the bonding in Li_2 and Be_2 is summarized in Figure 9.29. MO theory correctly predicts that Li_2 exists and Be_2 does not.

MO's with π Symmetry

Two different types of MO can result from the combination of atomic p orbitals, depending on their relative orientation. As illustrated in Figure 9.30, these MO's correspond to the σ and π bonds of valence bond theory, and they are similarly named.

The energy of bonding π orbitals, like that of σ orbitals, lies below that of the component AO's. Therefore occupation of π bonding orbitals by electrons creates a force holding a molecule together, that is, a covalent bond.

Molecule	Li_2	Be_2
Energy levels	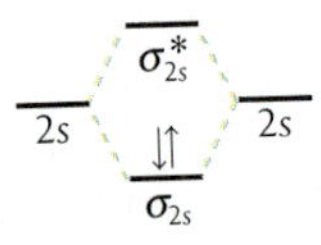 2s, σ^*_{2s}, ⥯ σ_{2s}, 2s	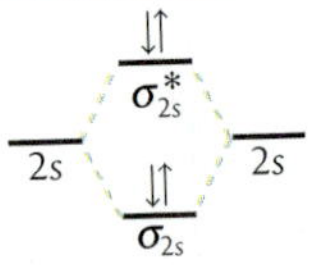 2s, ⥯ σ^*_{2s}, ⥯ σ_{2s}, 2s
Number of valence electrons	2	4
Configuration	$(\sigma_{2s})^2$	$(\sigma_{2s})^2\,(\sigma^*_{2s})^2$
Bond order	1	0
Exists as a stable molecule?	Yes	No

Figure 9.29 The two 2s AO's on adjacent Li or Be atoms combine to form two MO's having σ symmetry: the bonding MO σ_{2s} and the antibonding σ^*_{2s}. The bonding MO lies at lower energy than the parent AO's, and the antibonding orbital lies at higher energy. A lithium atom has a single valence electron, so the diatomic lithium molecule has two: these occupy the bonding MO, resulting in a stable Li_2 molecule with a bond order of 1. Each beryllium atom has two valence electrons, so the diatomic beryllium molecule would have four: these would occupy both the σ and σ^* MO's resulting in a Be_2 species with a bond order of zero. That is, MO theory predicts (correctly) that Be_2 does not exist as a stable molecule.

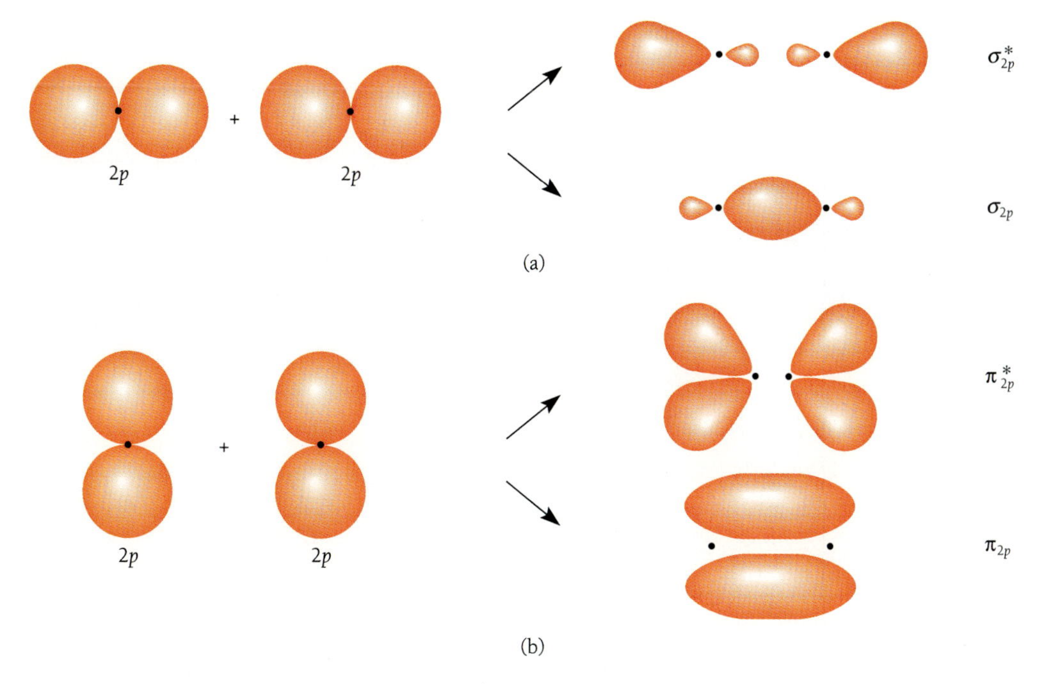

Figure 9.30 (a) Two $2p$ AO's can combine in an end-to-end fashion to form two MO's that have cylindrical symmetry about the internuclear axis. The higher-energy MO has a node bisecting the internuclear axis, and is the antibonding σ^*_{2p} orbital. The lower-energy orbital has a single lobe in the internuclear region, and is the bonding σ_{2p} orbital. (b) If the p orbitals are aligned in side-to-side fashion, the resulting combinations are not cylindrically symmetrical. There is a node along the internuclear axis, and the MO's have π symmetry. The lower-energy bonding orbital (π_{2p}) has two lobes, lying off-axis but between the nuclei, while the higher-energy π^*_{2p} orbital has four lobes, and an additional node bisecting the internuclear axis.

Other Second-Period Diatomic Species

The atomic orbitals of two second-period atoms are combined to produce molecular orbitals having the relative energies given in Figure 9.31. These orbitals have the shapes and symmetries previously illustrated in Figures 9.25 and 9.30. The atomic 2*s* orbitals combine to give σ/σ^* MO's, while the higher-energy 2*p* orbitals combine by end-to-end overlap to give σ/σ^* MO's and by side-by-side overlap to give π/π^* MO's.

Neutral Molecules

The electronic structure and bonding of all second-period homonuclear diatomic molecules can be predicted from the energy-level scheme shown in Figure 9.31. Application of the rules given in Section 9.3 leads to the orbital diagrams in Figure 9.32. The magnetic behavior of the molecules is an important check on the usefulness of the MO model. We saw in Section 7.2 that atoms with one or more unpaired electrons are *paramagnetic*. The same is true for molecules. MO theory correctly predicts that B_2 and O_2 are paramagnetic. Note that the valence bond model, which predicts the Lewis structure $\ddot{\underset{\cdot\cdot}{O}}=\ddot{\underset{\cdot\cdot}{O}}$, fails to account for the presence of unpaired electrons in O_2. Magnetic behavior, bond strength, and bond length of the second-period homonuclear diatomic molecules are included in Figure 9.32.

Paramagnetic species are weakly attracted by a magnetic field, while diamagnetic species are weakly repelled.

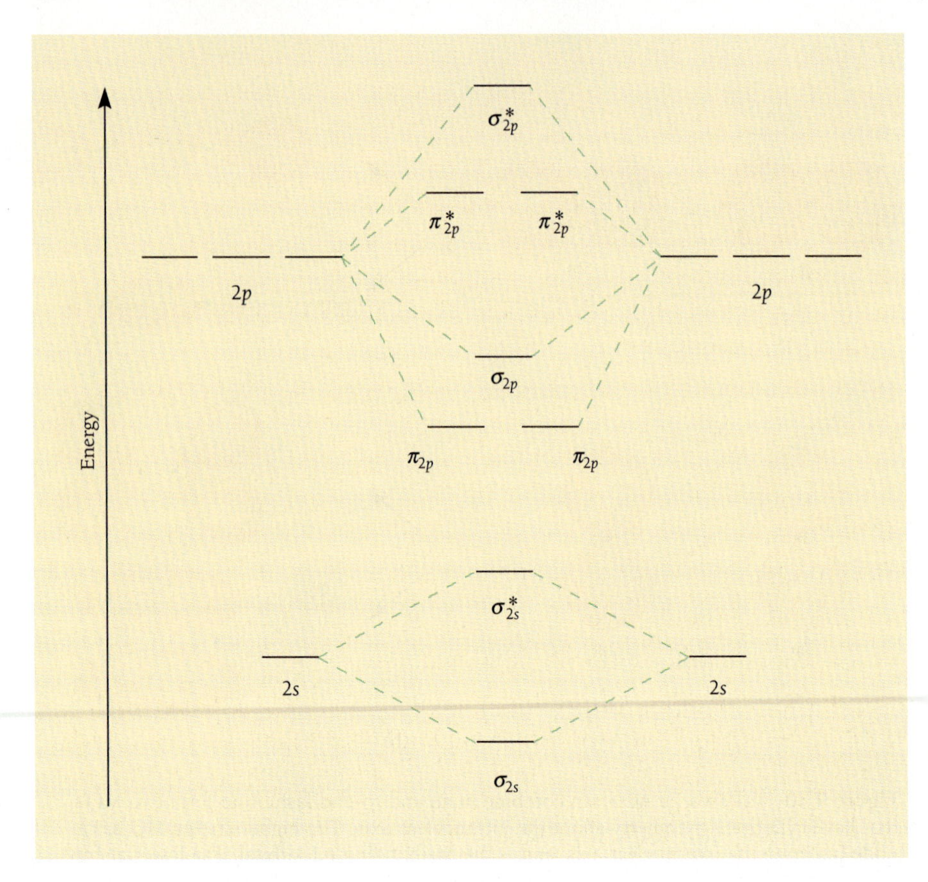

Figure 9.31 Relative energies of the molecular orbitals formed from the 2*s* and 2*p* atomic orbitals of two second-period atoms. Note that the MO's having π symmetry are degenerate. That is, they occur as sets of two orbitals having the same energy.

Ionic Species

Molecular orbital occupancy, and the related physical properties, can be determined for polyatomic ions as well as for neutral molecules. The MO aufbau scheme uses the total number of electrons, and does not depend on whether that total is equal to the number of protons in the nuclei.

EXAMPLE 9.3 Paramagnetism

Is the peroxide ion (O_2^{2-}) paramagnetic?

Analysis

Target: By the definition of "paramagnetism," the problem can be restated in the equivalent form, "Does O_2^{2-} have unpaired electron(s)?"

Relationship: The aufbau scheme leads to the electronic configuration, from which the number of unpaired electrons may be counted.

Plan

1. Count the number of valence electrons. If this number is odd, at least one electron is unpaired.
2. If the number is even, apply the aufbau scheme and see if any electrons are unpaired.

Work

1. The two Group 6A atoms of the O_2 *molecule* together have 6 + 6 = 12 valence electrons. The O_2^{2-} *ion* has an additional two electrons, making a total of 14. Since this is not an odd number, the ion may or may not have unpaired electrons.
2. The aufbau scheme of 14 electrons in the MO levels of Figure 9.31 leads to the configuration

$$O_2^{2-}: (\sigma_{2s})^2(\sigma_{2s}^*)^2(\pi_{2p})^4(\sigma_{2p})^2(\pi_{2p}^*)^4$$

All electrons are paired, so O_2^{2-} is diamagnetic.

Exercise

Does He_2^+ exist? If so, what is its bond order?

Answer: He_2^+ exists and has a bond order of ½.

	Li_2	(Be_2)	B_2	C_2	N_2	O_2	F_2	(Ne_2)
σ_{2p}^*	—	—	—	—	—	—	—	↑↓
π_{2p}^*	— —	— —	— —	— —	— —	↑ ↑	↑↓ ↑↓	↑↓ ↑↓
σ_{2p}	—	—	—	—	↑↓	↑↓	↑↓	↑↓
π_{2p}	— —	— —	↑ ↑	↑↓ ↑↓	↑↓ ↑↓	↑↓ ↑↓	↑↓ ↑↓	↑↓ ↑↓
σ_{2s}^*	—	↑↓	↑↓	↑↓	↑↓	↑↓	↑↓	↑↓
σ_{2s}	↑↓	↑↓	↑↓	↑↓	↑↓	↑↓	↑↓	↑↓
Total valence e's	2	4	6	8	10	12	14	16
Bond order	1	0	1	2	3	2	1	0
Magnetism	dia–	—	para–	dia–	dia–	para–	dia–	—
Bond length (pm)	267	—	159	131	110	121	143	—
Bond strength (kJ mol^{-1})	105	—	290	620	942	495	154	—

Figure 9.32 Molecular orbital occupancy in the second-period homonuclear diatomic molecules. By predicting the presence of unpaired electrons in B_2 and O_2, MO theory correctly accounts for the observed paramagnetism of these species. The nonexistence of Be_2 and Ne_2 are also well explained by MO theory. Note the pattern of the variation of bond length and bond energy with bond order.

The fact that an ionic species such as O_2^{2-} or He_2^+ has a bond order greater than zero does not *necessarily* mean that it forms stable ionic compounds. The peroxide ion is stable, since ionic peroxides such as BaO_2 are common compounds, but no compounds of He_2^+ have been prepared. Although He_2^+ is formed in electrical discharges in gaseous helium by the reaction

$$He + He^+ \longrightarrow He_2^+$$

it is quickly destroyed by electron attachment

$$He_2^+ + e^- \longrightarrow He + He$$

and no He_2^+ remains after the discharge is turned off.

Other Diatomic Molecules

The molecular orbital model can also be applied to the bonding and structure of the homonuclear diatomic molecules of the third and higher periods. The energy-level scheme of Figure 9.31 applies as well to the MO's of the Cl_2 and Br_2 molecules. However, the principal quantum number is not 2 (as in F_2), but 3 (in Cl_2) or 4 (in Br_2). Similarly, MO theory nicely explains the nonexistence of diatomic molecules of all the rare gases, not just helium and neon.

EXAMPLE 9.4 Bond Order in 4th-Period Diatomic Molecules

Use MO theory to predict the bond order and magnetism of Ca_2.

Analysis Both targets could be reached if we had an energy-level scheme showing which orbitals are occupied. The relationship to known information is supplied by the discussion of the preceding paragraph, which indicates that the MO's formed from atomic 3*s* and 3*p* orbitals have the same relative energies shown in Figure 9.31 for 2*s* and 2*p* AO's.

Plan

1. Count the electrons in the valence shells of two Ca atoms.
2. Follow the aufbau rules in placing these electrons in the energy levels of Figure 9.31 (strictly speaking, the figure should be relabeled with 4's rather than 2's for the principal quantum number).
3. Determine bond order and magnetism from the MO occupancy.

Work

1. Each Ca atom has two valence electrons, therefore Ca_2 has four.
2. Four electrons are enough to fill the bottom two levels: the electron configuration is $(\sigma_{4s})^2(\sigma_{4s}^*)^2$.
3. There are two bonding electrons and two antibonding electrons, so the bond order is 0.
4. Since there is no net bonding force, the molecule does not exist. Questions as to its magnetic properties are therefore irrelevant.

Exercise

Predict the bond order and magnetism of Rb_2.

Answer: BO = 1; diamagnetic.

The bonding and properties of **heteronuclear diatomic molecules,** that is, those consisting of two different atoms, can also be discussed in terms of MO theory. For instance, using Figure 9.31 and the fact that NO has 11 valence electrons, we predict an electron configuration of $(\sigma_{2s})^2(\sigma^*_{2s})^2(\pi_{2p})^4(\sigma_{2p})^2(\pi^*_{2p})^1$, a bond order of 2.5, and paramagnetic behavior for nitric oxide. Similarly, carbon monoxide is predicted to have a bond order of 3 and to be diamagnetic. Note that for carbon monoxide, the same predictions are made by the valence bond approach.

The chief difficulty with the application of MO theory to heteronuclear diatomic molecules is that the *atomic orbitals,* from which the MO's are constructed, lie at different energies in different atoms. This difference, if great enough, can affect the order of the energies of the molecular orbitals. In carbon monoxide, for instance, the relative energies of the MO's (Figure 9.33) are not the same as those for the homonu-

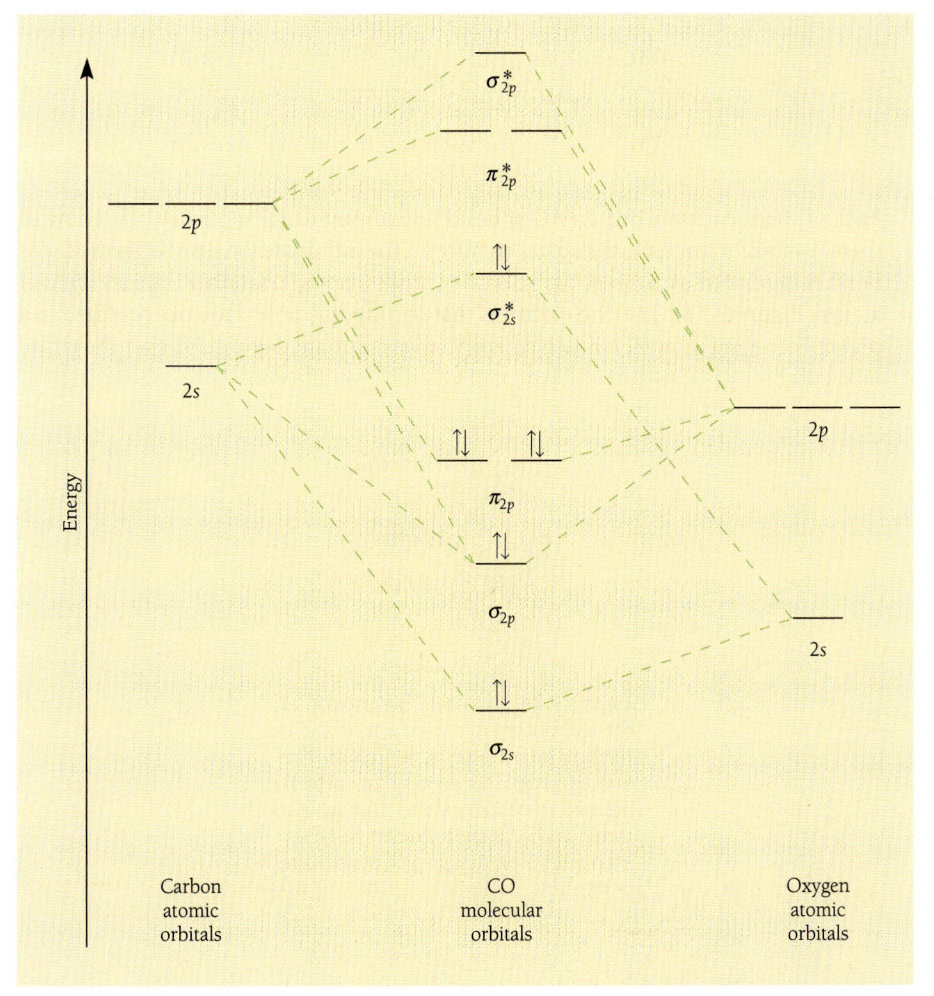

Figure 9.33 Occupancy and relative energies of the molecular orbitals in CO. Note that the energies of the *atomic* orbitals are not the same for the two atoms, in contrast to the situation in homonuclear molecules (Figure 9.31). The ten valence electrons of the CO molecule are shown occupying the five molecular orbitals of lowest energy, according to the aufbau principle.

clear diatomic species (Figure 9.31). Although the relative energies of molecular orbitals can be calculated, accurate calculations are difficult and require a large computer. For heteronuclear molecules, MO theory is not a simple, easily pictured model.

Polyatomic Molecules

Calculational difficulty increases when polyatomic molecules or ions, are considered, and even simple polyatomics, such as water or ozone require large investments of computer time. Nevertheless there are significant advantages to the use of MO theory.

Resonance

The concept of resonance, invented in Lewis and valence bond theory to explain the intermediacy of observed molecular properties between those of several Lewis structures, is not needed when the molecular orbital description of bonding is used. The valence bond description of the carbonate ion, for instance, is a resonance hybrid of three contributing structures.

$$\left[\ddot{O}{=}C(\ddot{O}{:})(\ddot{O}{:})\right]^{2-} \longleftrightarrow \left[{:}\ddot{O}{-}C({=}\ddot{O}{:})(\ddot{O}{:})\right]^{2-} \longleftrightarrow \left[{:}\ddot{O}{-}C(\ddot{O}{:})({=}\ddot{O}{:})\right]^{2-}$$

Each of these contains one C—O π bond in addition to the σ framework. None of them, by itself, represents the actual bonding in the carbonate ion. In MO theory the π bond is described as a molecular orbital of π symmetry, extending over all three O atoms (Figure 9.34). The two electrons that occupy this orbital are not confined to a region between the nuclei of two adjacent atoms, but are delocalized over the entire molecule.

Electron delocalization, which increases the stability of a molecule, is a natural part of MO theory. In valence bond theory, delocalization must be introduced as the additional postulate of *resonance*.

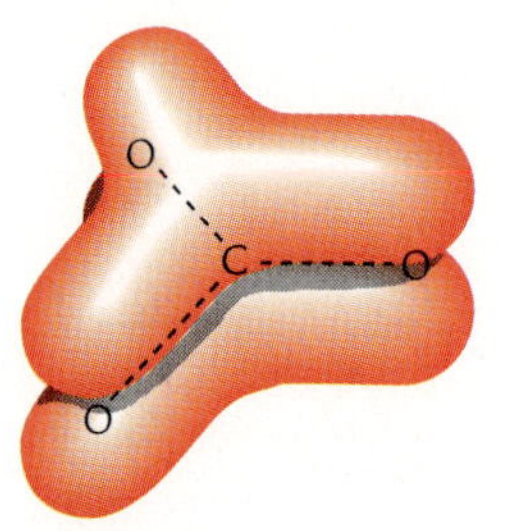

Figure 9.34 The MO description of the carbonate ion shows a π orbital extending over the entire structure. The two electrons that occupy this orbital are said to be delocalized, that is, associated with the entire molecule rather than with a particular pair of adjacent atoms.

Thinking with Both VB and MO

Most chemists are reluctant to give up the simplicity and clarity of the valence bond model, with its electron-pair bonds localized between two atoms and the clear imagery of Lewis dot structures and other forms of structural formulas. However, the naturalness of the MO description of multiple bonding and resonance is also attractive. Consequently the simpler features of both models are usually combined in descriptions of resonant molecules. For example, consider benzene, C_6H_6. The observed hexagonal shape and the fact that all six C—C bonds are the same length (140 pm) lead to the VB description as a resonance hybrid of the two equivalent structures

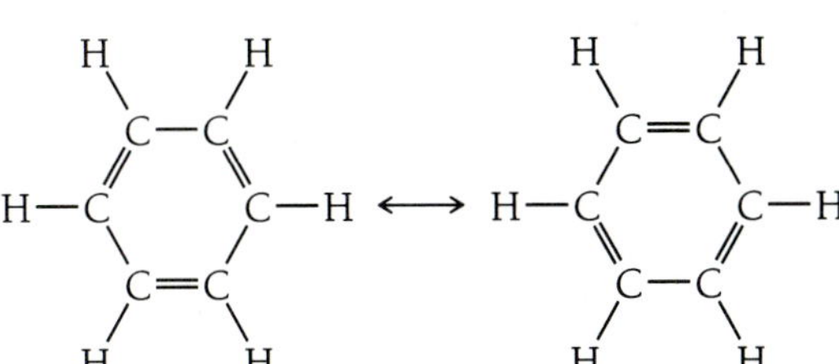

Figure 9.35 σ framework of the benzene molecule. The carbon atoms are sp^2 hybridized, and each forms σ bonds by overlap with two adjacent C_{sp^2} orbitals and one H_{1s} orbital.

The observed 120° bond angles and the fact that all 12 atoms of the molecule lie in the same plane indicate sp^2 hybridization of each carbon atom. The σ framework of C_{sp^2}—C_{sp^2} and C_{sp^2}—H_{1s} bonds is shown in Figure 9.35.

Three electrons from the valence shell of each carbon atom participate in the σ framework. The fourth electron occupies the unhybridized atomic p orbital, which is oriented with its axis perpendicular to the plane of the sp^2 hybrid orbitals (Figures 9.20 and 9.36). These p AO's combine to form bonding molecular orbitals, of π symmetry, that extend over the entire molecule (Figure 9.36). (Actually, six AO's combine to form six MO's, three low-energy bonding orbitals and three antibonding orbitals of higher energy. Only the lowest-energy orbital is shown in Figure 9.36.)

Other molecules are described in similar fashion. The σ framework is described by hybrid orbitals and localized bonds, while delocalization, if any, is described by MO's formed from unhybridized atomic p orbitals.

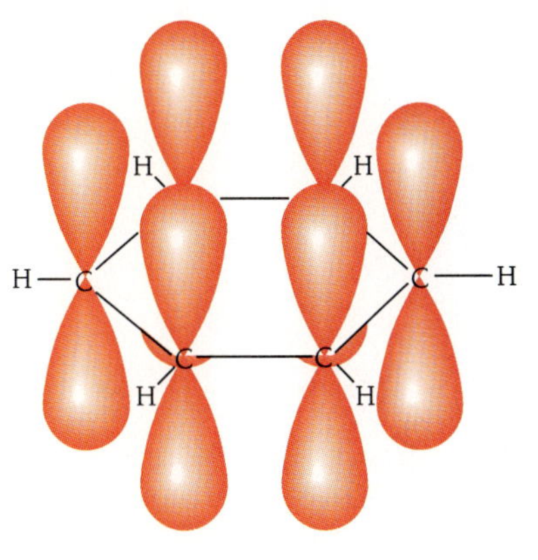

Component AO's

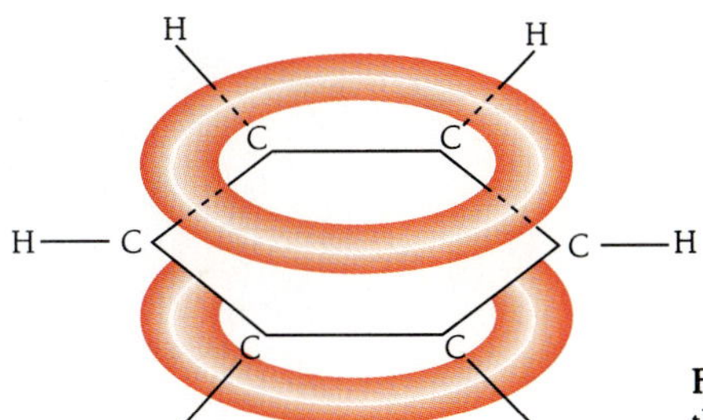

π molecular orbital

Figure 9.36 The atomic p orbitals of the six carbon atoms in benzene, unused in the σ framework, are combined to form a molecular orbital. The π electrons that occupy the orbital are delocalized over the entire molecule.

9.4 BONDING IN SOLID MATERIALS

Among the molecules included in the discussion of the previous section is dilithium, the diatomic molecule that is the predominant species when the element is in the gas phase. However, the boiling point of lithium is over 1300 °C, so it exists in the gas phase only under extreme conditions. What can our theories of bonding tell us about lithium when it is in its usual condition as a metallic solid? A useful theory of metallic bonding should be able to explain the characteristics of metals: high electrical and thermal conductivity, malleability, and appearance.

The "Sea of Electrons" Theory of Metallic Bonding

One early idea about the electronic structure of metals suggested that a crystal of a metal can be described as a collection of metal cations immersed in a swarm of valence electrons. The valence electrons are taken to be very loosely held by an individual metal ion, and thus free to wander throughout the entire piece of metal. A common metaphor for this model is "metal cations awash in a sea of electrons" (Figure 9.37). The model provides a good qualitative explanation of some characteristic properties of metals. If a battery is connected to a metal wire, for example, electrons will be attracted to and move toward the positive terminal. Electrical current, defined as the *motion of electrical charge,* flows easily in metals. Metals are good conductors of electricity because the *charge carriers,* that is, the valence electrons, are freely mobile. When the temperature of one end of a piece of metal is increased, the thermal energy entering the system appears as increased kinetic energy of the electrons. The faster-moving electrons spread freely and rapidly throughout the sample, depositing some of their increased energy and causing local heating wherever they go. Thermal energy moves readily from one part of a sample to another, so metals are good conductors of heat.

Heat flows easily from your hand to a metal, and consequently metals feel cold to the touch. A piece of wood at the same temperature does not feel cold. Because it is a poor conductor, wood draws much less heat from the skin.

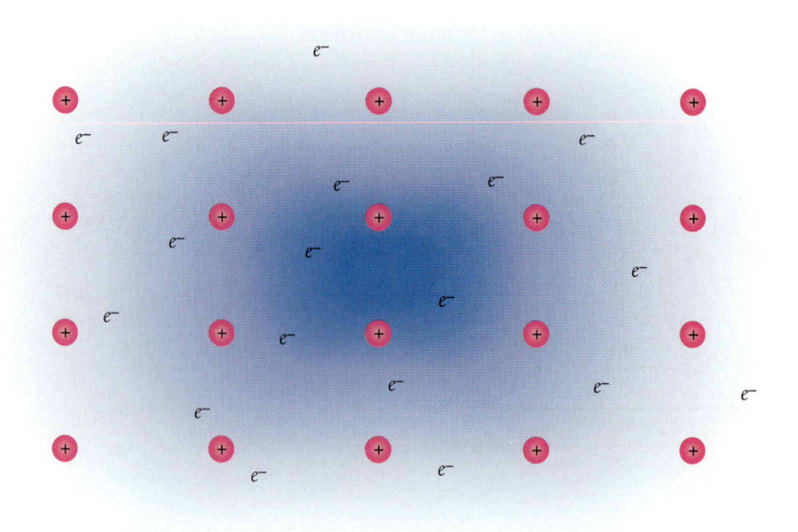

Figure 9.37 An early model describes a metal as a collection of positive ions immersed in a sea of loosely bound valence electrons. These electrons are not associated with a particular cation, and can move freely throughout the entire piece of metal.

Most metals are malleable, because, having only loosely bound valence electrons, individual bonds between adjacent atoms are weak and easily broken. This allows the structure to be rearranged, and a wire to be bent or otherwise deformed, by a relatively weak push or impact.

The *sea of electrons* model is easily visualized, and provides a simple explanation for several properties of metals. However, explanation of other properties requires a somewhat more sophisticated model.

Band Theory of Metals

The atomic orbitals of two adjacent lithium atoms combine to form two molecular orbitals, as shown in Figure 9.29. If three atoms are close enough to interact, three MO's are formed, and so on, with the number of MO's always equal to the number of AO's from which they are formed. In a crystal of metallic lithium, which has perhaps 10^{20} atoms, the number of MO's is so large that it is not appropriate to describe them as discrete energy levels. Instead, they are merged into a continuous **energy band** (Figure 9.38). The lower part of the band is identified with bonding orbitals that are *localized,* that is, confined to the immediate regions between adjacent nuclei. The upper part of the band consists of antibonding orbitals that are *delocalized,* that is, spread out through the entire piece of metal. There are two valence electrons in the Li_2 molecule, enough to half-fill its two MO's. Similarly, in Li metal, enough valence electrons are available to half-fill the s band. The orbitals in the lower part of the band are, for the most part, fully occupied by electrons. Since these occupied localized orbitals are the bonds that hold the metal atoms together, the lower part of the band is called the **valence band.**

The delocalized orbitals in the upper part of the band are largely unoccupied. However, because the levels are so closely spaced, it takes very little energy for an electron at the top of the valence band to move to an unoccupied orbital. Even at very low temperature, electrons have enough kinetic energy for this. Once in the upper, unoccupied, part of the band, an electron can move easily to any orbital in any part of the solid. Electrons in the upper part of the band are free to move throughout the

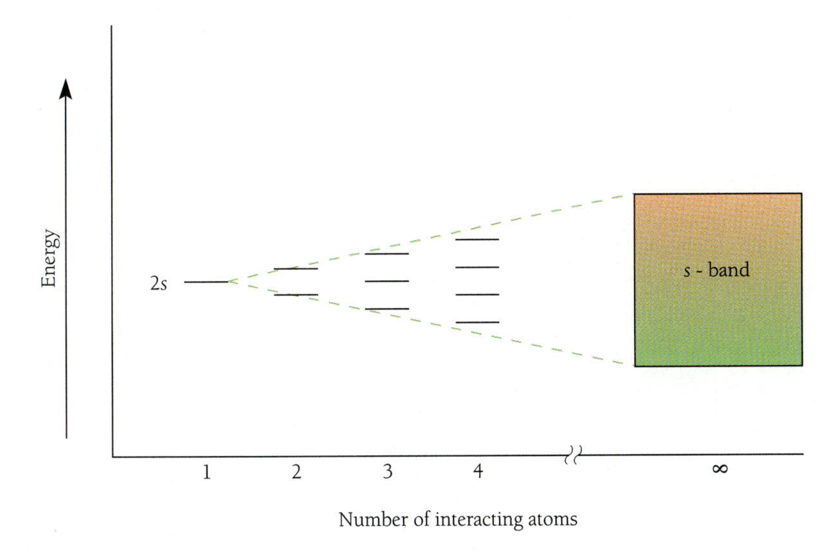

Figure 9.38 The number of MO's is equal to the number of component atomic orbitals. For the huge number of interacting atoms in a metal crystal, the MO's merge into a continuous energy band. Since it is the 2*s* AO's that form the band, it is referred to as the "s" or "2s" band.

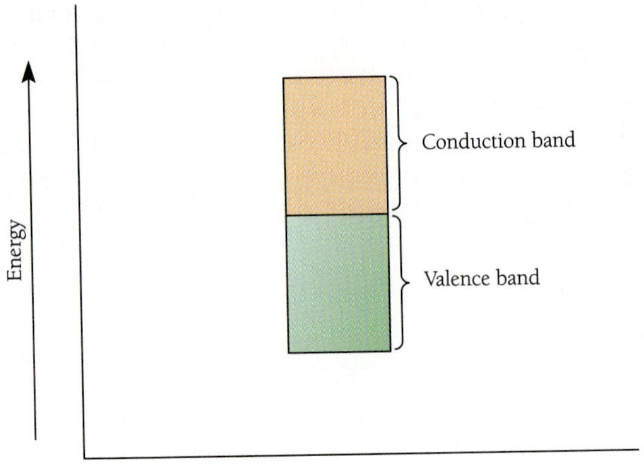

Figure 9.39 The merged MO's of Figure 9.38 are divided into two parts: the upper "conduction" band, and the lower "valence band." Electron occupancy of the valence band accounts for the bonding forces holding the crystal together, while electrons in the conduction band account for the electrical and thermal conductivity of the metal.

entire crystal, and this mobility accounts for both the electrical and thermal conductivity of metals (just as it does in the sea of electrons model). The upper part of the band is therefore called the **conduction band** (Figure 9.39).

This model accounts nicely for the fact that metals are opaque and lustrous, and for the fact that (with several exceptions) they exhibit no particular color. Since there is a continuous range of energies available, an electron can be excited to the conduction band by absorption of a photon of *any* energy (that is, light of any wavelength). All wavelengths are absorbed at the metallic surface, and no light passes through. Instead, the absorbed photons are immediately re-emitted, giving rise to the lustrous appearance of a metal surface.

Insulators and Semiconductors

It is only in metals that the conduction band overlaps or adjoins the valence band. In other materials the conduction band lies at higher energy, and is separated from the valence band by a **band gap.** If the gap is sufficiently large, no electrons have enough energy to move into the conduction band from the valence band. Furthermore, valence electrons cannot move through such materials because there are no vacant

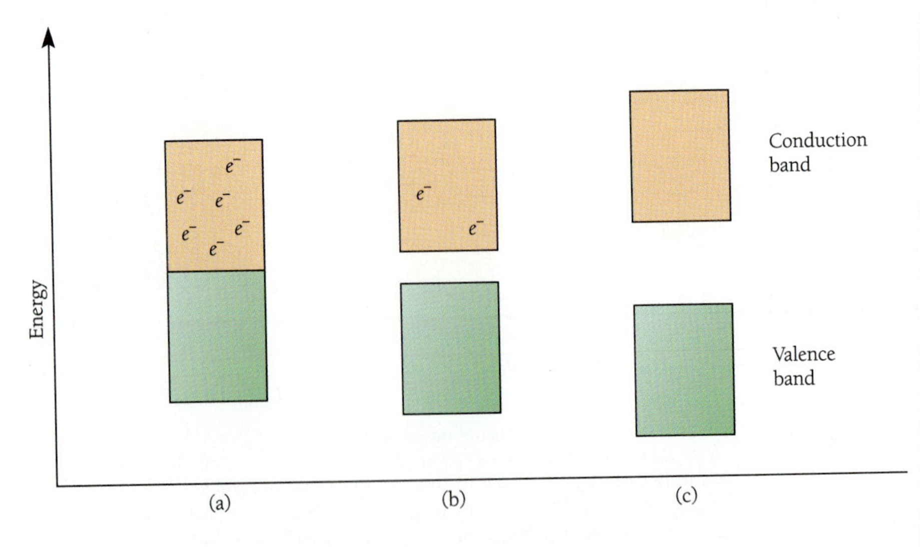

Figure 9.40 (a) In metals, the top of the valence band and the bottom of the conduction band lie at the same energy. Even at very low temperatures there are some electrons in the conduction band. (b) A small gap separates the two bands in semiconductors, and a moderate increase in temperature increases the occupancy of the conduction band. (c) The band gap in insulating materials is so large that no significant number of electrons can ever reach the conduction band. All remain in the valence band, localized between atoms as they are in individual molecules.

(a)

(b)

(c)

Figure 9.41 The electrical conductivity of semiconductor materials is affected by temperature, light intensity, electrical fields, and magnetic fields. These properties are exploited in devices such as (a) digital thermometers; (b) light-measuring devices in cameras; (c) computers and calculators, which utilize semiconductors for the conversion of light energy to electrical energy as well as in their circuitry and displays.

orbitals in the valence band. Since electrons cannot move freely, the substance is an **insulator**—a poor conductor of heat and electricity. If the band gap is small enough, however, and provided the temperature is not too low, a few electrons can reach the conduction band by absorbing thermal energy. Therefore, there are a few electrons in the conduction band and a few vacancies in the valence band. Materials having small band gaps are characterized by a moderate amount of electron mobility, and are called **semiconductors.** The relative energies of the bands in metals, semiconductors, and insulators are compared in Figure 9.40.

Pure silicon and pure germanium are semiconductors.

Semiconductors are insulators at low temperature, but become increasingly conductive as the temperature is increased and thermal energy allows more electrons to jump the band gap. Such a material can be made into a temperature-measuring device (Figure 9.41). In addition, absorption of light by a semiconductor can cause the excitation of electrons from the valence band to the conduction band. Therefore the conductivity depends on light intensity, and semiconductors can be used for photocells and light meters.

Semiconductors are useful materials, whose versatility comes in part from the fact that the magnitude of the band gap can be changed. In a manufacturing procedure called **doping,** the band gap of a semiconductor such as silicon or germanium is adjusted by the addition of carefully controlled amounts of other materials, often Group 3A or 5A elements. Semiconductor research and technology is an active and exciting border area between chemistry, physics, and engineering.

9.5 HYDROCARBONS

In our discussion of chemical bonding in this chapter, we have used CH_4, C_2H_2, C_2H_4, C_2H_6, and C_6H_6 as examples. These molecules, called **hydrocarbons,** are members of a large and important class of organic compounds containing only carbon and hydrogen. Hydrocarbons are further classified by their bonding, or equivalently by their chemical properties, which are determined by the bonding.

Alkanes

Older usage refers to alkanes as *paraffins.*

Hydrocarbons having only sp^3-hybridized carbon atoms are called **alkanes.** Each carbon atom in an alkane is σ-bonded to four other atoms. Four is the maximum number of bonds that a carbon atom can form, since it has only four valence electrons. Therefore, alkanes are also called **saturated** hydrocarbons. The element carbon is unique in the extent to which it can bond it itself, forming *chain compounds*—molecules with structures like the following.

```
         H   H   H   H   H   H   H
         |   |   |   |   |   |   |
  . . . —C — C — C — C — C — C — C . . .
         |   |   |   |   |   |   |
         H   H   H   H   H   H   H
```

Other elements, notably S and Si, behave similarly. But the extent to which they do so is limited, and *long* chains are not found in S or Si compounds.

There seems to be no limit to the length of chains that can be formed from carbon atoms. The empirical formulas of alkanes are $\mathbf{C_nH_{2n+2}}$.

In the IUPAC system the names of saturated hydrocarbons all end in *-ane,* and a prefix is used to indicate the number of carbon atoms the compound contains. There are many alkanes with more than ten carbon atoms, but only the first ten prefixes should be memorized. These, together with their numerical significance, are given in Table 9.2.

Many alkanes are widely used as fuels. Natural gas, bottled gas, kerosene, gasoline, and heating oil all are alkanes or mixtures of alkanes.

Saturated hydrocarbons are generally unreactive at room temperature. At higher temperature they react with oxygen, in combustion reactions whose products are carbon dioxide and water. Equation (9-2) describes the combustion of butane.

$$2\ C_4H_{10} + 13\ O_2 \longrightarrow 8\ CO_2 + 10\ H_2O \qquad (9\text{-}2)$$

If oxygen is the limiting reactant, some carbon monoxide is formed as well. Since carbon monoxide is odorless and highly toxic, precautions (such as good ventilation and good design of burners) must be taken when alkanes are burned.

Hydrocarbons react with chlorine (and other halogens) in the presence of light, forming several different **chlorinated hydrocarbons** in a stepwise sequence.

$$CH_4 + Cl_2 \xrightarrow{h\nu} CH_3Cl + HCl \qquad (9\text{-}3)$$

$$CH_3Cl + Cl_2 \xrightarrow{h\nu} CH_2Cl_2,\ CHCl_3,\ \text{etc.} \qquad (9\text{-}4)$$

TABLE 9.2 Alkane Nomenclature

Prefix	Number of Carbon Atoms	Molecular Formula	Alkane Name	Prefix	Number of Carbon Atoms	Molecular Formula	Alkane Name
Meth-	1	CH_4	Methane	Hex-	6	C_6H_{14}	Hexane
Eth-	2	C_2H_6	Ethane	Hept-	7	C_7H_{16}	Heptane
Prop-	3	C_3H_8	Propane	Oct-	8	C_8H_{18}	Octane
But-	4	C_4H_{10}	Butane	Non-	9	C_9H_{20}	Nonane
Pent-	5	C_5H_{12}	Pentane	Dec-	10	$C_{10}H_{22}$	Decane

Equations (9-3) and (9-4) are overall reactions for a stepwise sequence beginning with the formation of atomic chlorine by the *photodissociation* of molecular chlorine.

$$Cl_2 + \text{light} \longrightarrow 2\ Cl\cdot \quad (9\text{-}5)$$

$$CH_4 + Cl\cdot \longrightarrow CH_3\cdot + HCl \quad (9\text{-}6)$$

$$CH_3\cdot + Cl_2 \longrightarrow CH_3Cl + Cl\cdot \quad (9\text{-}7)$$

A dot is often added to an atomic or molecular formula to indicate the presence of an unpaired electron. Most odd-electron species are highly reactive, and few exist as stable compounds.

Only a few $Cl\cdot$ atoms are needed initially, since although they are consumed in step (9-6) they are regenerated in step (9-7). Chlorinated hydrocarbons are widely used in chemical technology: PVC (polyvinyl chloride) pipe is used for residential water and sewer systems; CH_2Cl_2 (dichloromethane), $CHCl_3$ (trichloromethane, also called chloroform), and $C_2H_4Cl_2$ (dichloroethane) are excellent solvents, used for paint remover, extraction of caffeine from coffee, and a wide variety of cleaning and degreasing tasks.

The sequence 9-5 through 9-7 is an example of a *chain reaction*. Chain reactions are discussed in Section 14.6.

Structural Isomers

Part of the richness and diversity of organic chemistry stems from the possibility that two or more distinct compounds can share the same molecular formula. In alkanes containing four or more carbon atoms, several different arrangements of the carbon chains are possible. For example, the following structural formulas can be written for C_6H_{14}.

```
    H  H  H  H  H  H
    |  |  |  |  |  |
 H—C—C—C—C—C—C—H
    |  |  |  |  |  |
    H  H  H  H  H  H
```

hexane

```
    H  H  H  H  H              H  H  H  H  H
    |  |  |  |  |              |  |  |  |  |
 H—C—C—C—C—C—H          H—C—C—C—C—C—H
    |  |  |  |  |              |  |  |  |  |
    H  |  H  H  H              H  H  |  H  H
       |                             |
    H—C—H                         H—C—H
       |                             |
       H                             H
```

2-methylpentane **3-methylpentane**

The names of these compounds are discussed in the following paragraph.

```
          H                       H
          |                       |
       H—C—H                   H—C—H
          |                       |
    H  H  |  H                H  |  H  H
    |  |  |  |                |  |  |  |
 H—C—C—C—C—H          H—C—C—C—C—H
    |  |  |  |                |  |  |  |
    H  |  H  H                H  |  H  H
       |                         |
    H—C—H                     H—C—H
       |                         |
       H                         H
```

2,3-dimethylbutane **2,2-dimethylbutane**

These compounds are all known, and all have different properties.

Different molecules sharing the same molecular formula are called **isomers.** When, as in this case, the difference is in the *connectivity,* they are called **structural isomers.** (Recall from Chapter 2 that a structural formula designates the *connectivity* of a molecule, that is, shows which atoms are bonded to one another.) Compounds in which no carbon atom is bonded to more than two other carbon atoms (for example hexane, above) are called **normal** or **straight-chain** hydrocarbons. If at least one carbon atom is bonded to three or four other carbon atoms, the compound is a **branched** (or branched-chain) hydrocarbon.

These isomeric variations are named according to IUPAC rules so that the structural formula is unambiguously indicated by the name. Isomeric alkanes are given a base name (Table 9.2) corresponding to the *longest* continuous chain of carbon atoms in the molecule. Prefixes to the base name are used to designate the identity and location of substituents. A **substituent** is an atom (other than hydrogen) or *group* of atoms that replaces a hydrogen atom in a molecule. A **group** is a portion or fragment of a molecule, usually named for its parent molecule. Examples of *alkyl groups* (fragments of alkanes) are *methyl* (CH_3—) and *ethyl* (C_2H_5—). The dash emphasizes that a group is not a complete molecule. *Halo* substituents are *fluoro* (F—), *chloro* (Cl—), and so on. When several different substituents are present, they are named in alphabetical order. Each carbon atom in the continuous chain is numbered, and the name of each substituent is preceded by the number of the carbon atom to which it is bonded.

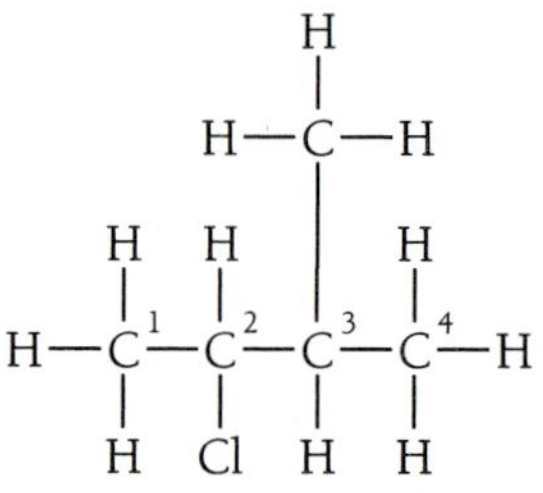

```
             H
             |
          H—C—H
             |
     H   H   |   H
     |1  |2  |3  |4
  H—C——C——C——C—H
     |   |   |   |
     H   Cl  H   H
```

2-chloro-3-methylbutane

If the same substituent occurs 2, 3, 4, and so on, times in a compound, the substituent name is prefixed with *di-*, *tri-*, *tetra-*, and so on. For example, the compound depicted in Figure 9.42 is called 2,4-dimethylpentane.

Often several drawings *seem* to represent different structural formulas, as shown below for C_5H_{12}.

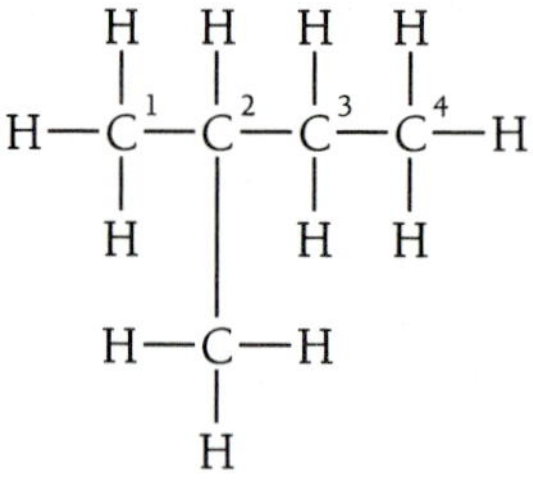

```
                                                          H
                                                          |4
                                                       H—C—H
                                                          |
     H   H   H   H          H   H   H   H          H   H  |
     |1  |2  |3  |4         |1  |2  |3  |4         |1  |2 |3
  H—C——C——C——C—H       H—C——C——C——C—H       H—C——C——C—H
     |   |   |   |          |   |   |   |          |   |  |
     H   |   H   H          H   H   |   H          H   |  H
         |                          |                  |
      H—C—H                      H—C—H              H—C—H
         |                          |                  |
         H                          H                  H
         A                          B                  C
```

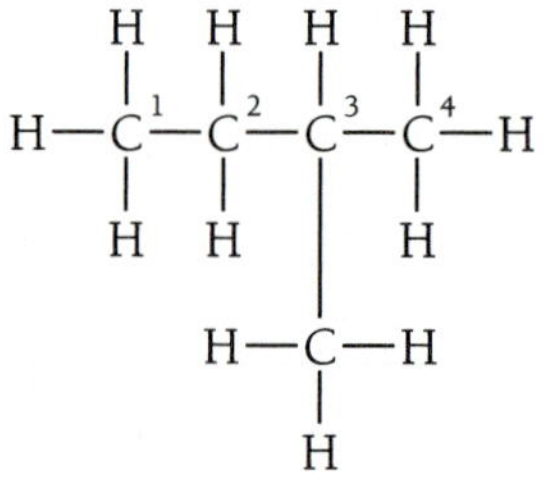

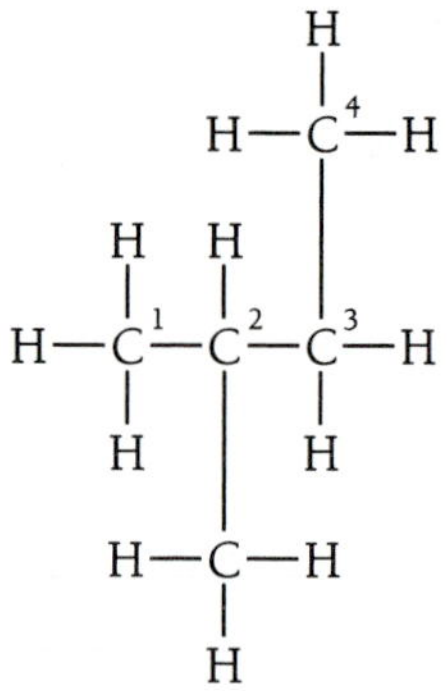

However, all three of these have the same connectivity, and represent the same molecule. Drawings A and B might be thought of as "front" and "back" views. The compound could be named 2-methylbutane or 3-methylbutane, depending on the

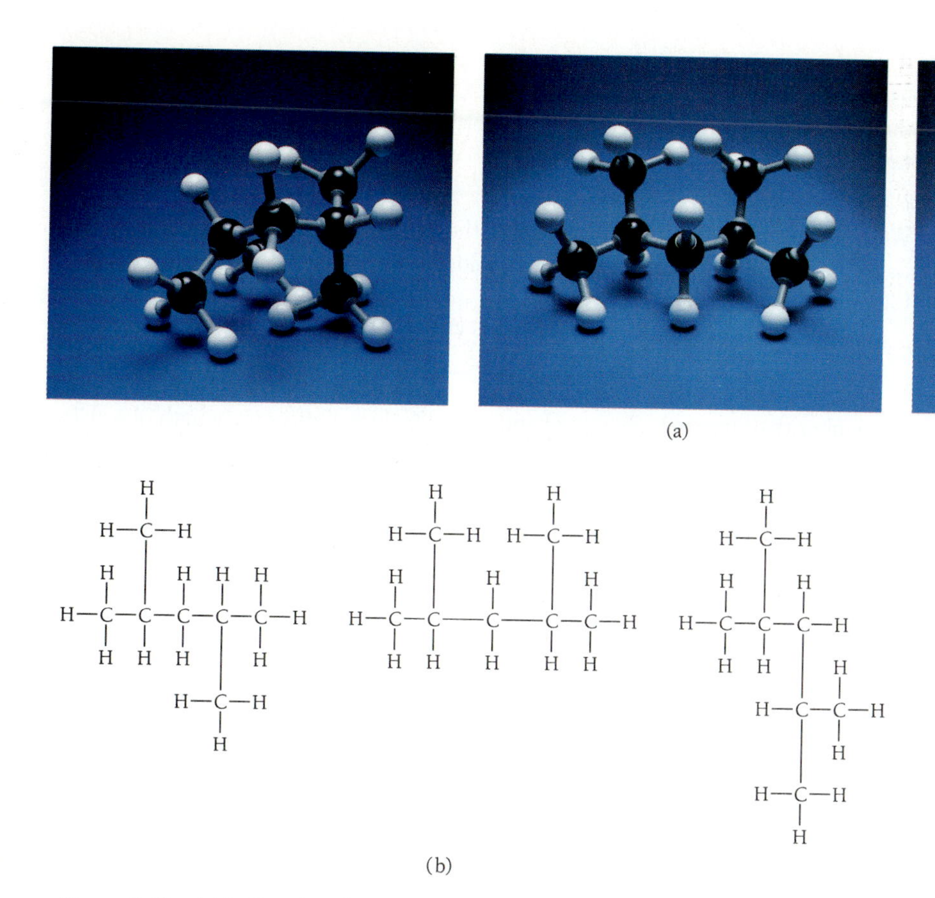

Figure 9.42 (a) Ball-and-stick models of 2,4-dimethylpentane, in three of the many different conformations that may arise through rotation about C—C single bonds. (b) Three ways of writing the structural formula of 2,4-dimethylpentane. All represent the same molecule.

vantage point. The name with the lowest possible number for the substituent is correct. Drawing C exemplifies another reason why different-looking structural formulas can be drawn for the same compound. Molecules have an inherent flexibility, since their parts can freely undergo internal rotation about single (but not multiple) bonds. A molecule is said to adopt different **conformations** as it rotates about its single bond(s). The drawings and ball-and-stick models of Figure 9.42 do not represent different molecules, but simply different conformations of 2,4-dimethylpentane.

EXAMPLE 9.5 Isomers of Heptane

Draw the structural formulas of all possible isomers of heptane.

Analysis The known information is implicit in the IUPAC rules: *-ane* means a saturated hydrocarbon of molecular formula C_nH_{2n+2}, and *hept-* means $n = 7$. The empirical formula is C_7H_{16}. Almost *any* way of putting together seven C atoms, with four bonds each, is the structural formula of a heptane isomer. What is needed is a systematic approach that examines all possibilities, and rejects those that are simply different views of the same isomer.

Plan Begin with the longest possible chain, and progress to the shortest. At each stage use the remaining C atoms as substituents, trying all possible locations. Check for duplications by imagining the structural formulas that would result from bond rotations or different vantage points.

Work The task is simpler if we draw only the carbon *skeleton* of the structures under consideration. If desired, H atoms can be added later, as needed, to give each C atom a total of four bonds.

Only one isomer having a continuous chain of seven C atoms is possible.

```
C—C—C—C—C—C—C
```

Remember that drawings showing bends or folds do not affect the length of a chain. For example,

```
C—C—C—C—C—C       and     C—C—C—C
|                         |
C                         C—C—C
```

both represent the same 7-carbon chain as

```
C—C—C—C—C—C—C
```

A 6-carbon chain may have a methyl substituent in two different positions.

```
C—C—C—C—C—C     C—C—C—C—C—C
  |                 |
  C                 C
```

Note:

```
C—C—C—C—C—C     is the same as     C—C—C—C—C—C
      |                                |
      C                                C
```

A 5-carbon chain has the following four possibilities for two methyl substituents.

```
C—C—C—C—C,     C—C—C—C—C,
  | |            |   |
  C C            C   C

  C                   C
  |                   |
C—C—C—C—C,   and  C—C—C—C—C
  |                   |
  C                   C
```

A 5-carbon chain has one possibility for an *ethyl* substituent.

```
C—C—C—C—C
    |
    C
    |
    C
```

A 4-carbon chain has only a single possibility for three methyl substituents.

```
  C C
  | |
C—C—C—C
  |
  C
```

There are no 3-carbon chains that have not already been drawn. For instance, in the following two structures the longest chains consist of four and five carbon atoms, not three.

```
     C            C—C
     |            |
  C—C—C        C—C—C
     |            |
     C—C          C
     |            |
     C            C
```

Nine structures with different connectivity can be drawn, so there are nine structural isomers of C_7H_{16}. The number of structural isomers increases quite rapidly as the number of carbon atoms increases.

Exercise

How many different dimethylhexanes (C_8H_{18}) exist?

Answer: Five.

EXAMPLE 9.6 Naming Isomeric Alkanes

Name the nine isomeric alkanes in Example 9.5.

Analysis The known information is the set of structural formulas having the carbon skeletons developed in the solution to Example 9.5. The relationship between target and knowns is established by the following IUPAC rules.

a. Select the longest continuous chain to obtain the base name.

b. Then identify each substituent by name and number along the chain. Use the lowest numbers if several possibilities exist.

Plan Working systematically down from the longest chain, name all structures.

Work

- 7-C chain: heptane
- 6-C chain: 2-methylhexane, 3-methylhexane
- 5-C chain: 2,3-dimethylpentane, 2,4-dimethylpentane, 2,2-dimethylpentane, 3,3-dimethylpentane, 3-ethylpentane
- 4-C chain: 2,2,3-trimethylbutane

Check The number of carbons in the chain plus those in the substituents must add up to seven (*hept*ane).

Exercise

Draw the carbon skeletons and name the isomeric C_4H_{10} compounds.

Answer: Butane, 2-methylpropane.

Alkenes

In older terminology, alkenes were known as *olefins.* The first two members of the series were called *ethylene* and *propylene,* names still frequently seen.

Hydrocarbons containing a double bond bear the general name **alkene.** These compounds are described as **unsaturated,** because the two carbon atoms with only three bonds to other atoms could form additional bonds. The molecular formula of alkenes is C_nH_{2n}. Individual compounds are named with the same prefixes given in Table 9.2 but with the ending *-ene.* The first few members of this group include the following.

```
Name          ethene           propene               1-butene

               H  H           H  H  H              H  H  H  H
               |  |           |  |  |              |  |  |  |
Structural  H—C=C—H        H—C=C—C—H          H—C=C—C—C—H
Formula                             |                    |  |
                                    H                    H  H
```

Name	**ethene**	**propene**	**1-butene**
Modified Molecular Formula	$CH_2{=}CH_2$	$CH_2{=}CHCH_3$	$CH_2{=}CHCH_2CH_3$

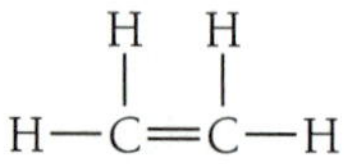

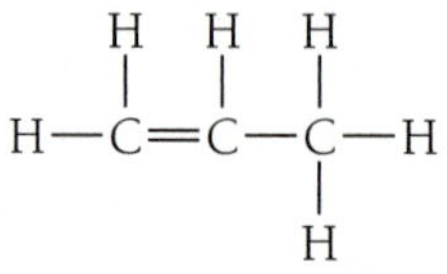

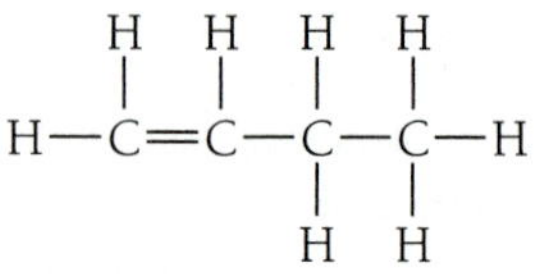

A doubly bonded carbon atom has sp^2 hybridization, and structural formulas are often drawn to indicate its approximately 120° bond angles.

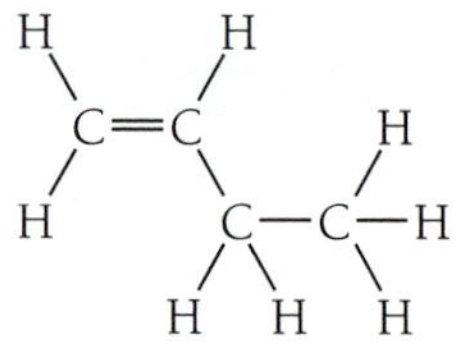

```
 H       H                    H       H
  \     /                      \     /
   C=C       H                  C=C
  /   \     /                  /   \
 H     C—C—H              H—C       C—H
      / \  \                / \     / \
     H   H  H              H   H   H   H

     1-butene                  2-butene
```

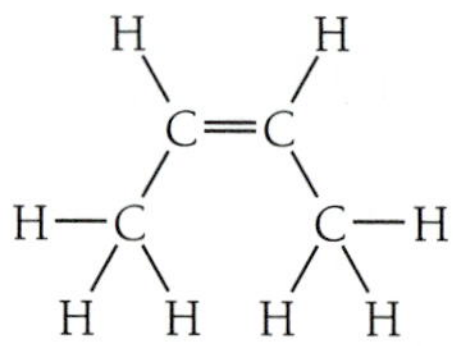

These two isomers of butene are sometimes called **positional isomers,** because they differ only in the position of the double bond. Position of the double bond is indicated in the name by prefixing with the lowest possible number of one of the doubly bonded carbon atoms.

Rotation about a double bond can sometimes result from the absorption of an ultraviolet photon, which carries enough energy to break the π part of the double bond.

Rotation about a *double* bond does not occur under normal conditions. Such rotation would eliminate the side-by-side overlap of the atomic p orbitals, in effect breaking the π part of the double bond. This would require much more energy than is available at ordinary temperatures. The fact that rotation does not occur, together with the 120° bond angles of doubly bonded carbon atoms, creates the possibility of a more subtle form of isomerism. **Geometrical isomers** occur in pairs, with substituents either on the same side or on opposite sides of a double bond. Thus there are two forms of 2-butene, and although they have the same connectivity, they are different compounds with different chemical and physical properties.

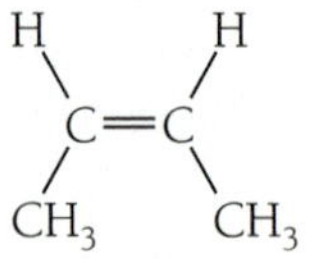

```
  H       H           H       CH3
   \     /             \     /
    C=C                 C=C
   /   \               /   \
 CH3    CH3          CH3     H

  cis-2-butene      trans-2-butene
```

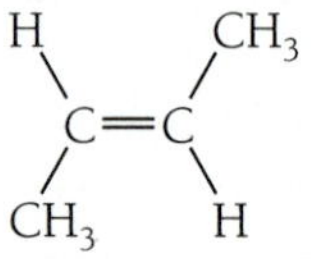

Geometrical isomers are identified by the Latin prefixes *cis-*, which means near or on the same side, and *trans-*, which means across or on the far side.

$$\begin{array}{c} \text{H} \quad\quad \text{CH}_3 \\ \diagdown \quad \diagup \\ \text{C}=\text{C} \\ \diagup \quad \diagdown \\ \text{H} \quad\quad \text{CH}_3 \end{array}$$

methylpropene

A fourth isomer of C_4H_8 is methylpropene: this is a *structural* rather than a geometrical isomer of the others. Its name does not specify the position of the double bond or the methyl substituent, because there is only one possible bonding arrangement. Also, since geometrical isomerism is not possible in methylpropene, there is no need for a *cis-* or *trans-* prefix.

The double bond in an alkene undergoes characteristic **addition reactions,** in which the two parts of a reactant molecule are said to "add across the double bond." For example, bromine readily adds to propene, forming an alkyl dibromide.

$$\begin{array}{c} \text{H} \quad\quad \text{H} \\ \diagdown \quad \diagup \\ \text{C}=\text{C} \\ \diagup \quad \diagdown \\ \text{CH}_3 \quad\quad \text{H} \end{array} + Br_2 \longrightarrow \text{CH}_3-\overset{\displaystyle \text{Br}}{\underset{\displaystyle \text{H}}{\overset{|}{\underset{|}{\text{C}}}}}-\overset{\displaystyle \text{Br}}{\underset{\displaystyle \text{H}}{\overset{|}{\underset{|}{\text{C}}}}}-\text{H} \quad (9\text{-}8)$$

propene **1,2-dibromopropane**

The reaction in Equation (9-8) occurs so readily that it is used as a quick laboratory test for the presence of an alkene (Figure 9.43). Hydrocarbons containing more than one double bond are known as *dienes, trienes,* and so on. For example, $CH_2{=}CHCH_2CH{=}CH_2$ is 1,4-pentadiene, and $CH_2{=}CHCH{=}CHCH{=}CHCH_3$ is 1,3,5-heptatriene.

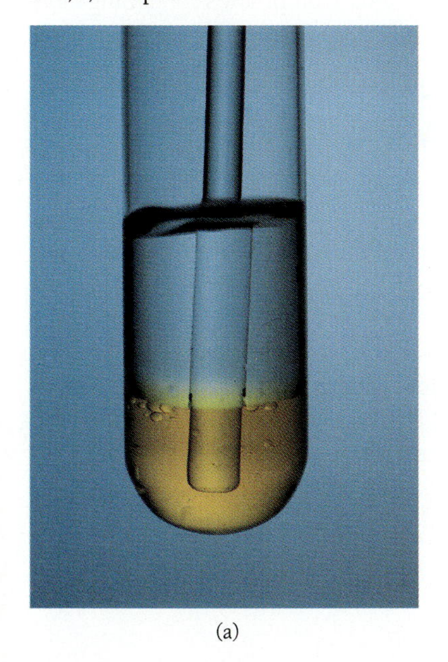

(a)

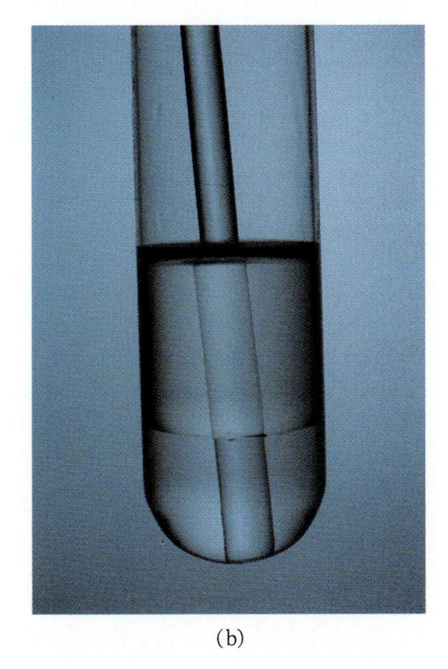

(b)

Figure 9.43 (a) Hexene (colorless) is added to a test tube containing a brown, aqueous solution of bromine ("bromine water"). The insoluble hexene floats on top of the aqueous layer, which is more dense. (b) When the mixture is stirred, Br_2 is quickly consumed by an addition reaction with the alkene. The decolorization of bromine water is a specific test for the presence of unsaturation in a hydrocarbon.

Alkynes

Hydrocarbons containing a triple bond are known as **alkynes.** These unsaturated compounds contain two adjacent carbon atoms that are *sp* hybridized. Although the first member of the class is more frequently called by its older name, *acetylene,* the IUPAC names use the prefixes of Table 9.2 and the ending *-yne*:

Name	**ethyne (acetylene)**	**propyne**	**1-butyne**
Structural Formula	H—C≡C—H	H—C≡C—C(H)(H)—H	H—C≡C—C(H)(H)—C(H)(H)—H
Modified Molecular Formula	HC≡CH	$HC{\equiv}CCH_3$	$HC{\equiv}CCH_2CH_3$

Alkynes also undergo addition reactions, but, since the initial product is an alkene, it can undergo further addition.

$$H{-}C{\equiv}C{-}H \xrightarrow{Br_2} H{-}C(Br){=}C(Br){-}H \xrightarrow{Br_2} H{-}C(Br)_2{-}C(Br)_2{-}H$$

Compounds containing more than one triple bond are called *diynes, triynes,* and so on. Some hydrocarbons have both double and triple bonds. Compounds containing more than one multiple bond are generally referred to as *polyunsaturated.*

EXAMPLE 9.7 Naming a Polyunsaturated Molecule

Name the compound $CH_3{-}CH{=}CH{-}CH{=}CH_2$.

Analysis/Knowns The compound is a 5-carbon unsaturated hydrocarbon with two double bonds at known positions. All necessary rules have been given.

Work The longest continuous chain is five carbon atoms, and the compound has two double bonds—the base name must be pentadiene. The double bonds are at positions 1 and 3, so the full name is 1,3-pentadiene. (The possible name 2,4-pentadiene is incorrect because it does not use the lowest possible numerical prefixes.)

Exercise

Write the structural or modified molecular formula of 3-methyl-1,4-pentadiyne.

Answer: $CH{\equiv}C{-}CH(CH_3)C{\equiv}CH$.

Alkanes, alkenes, and alkynes are collectively known as **aliphatic** hydrocarbons, to distinguish them from the *aromatic* hydrocarbons discussed in the following section.

Aromatic Hydrocarbons

A large class of compounds, called **aromatic** hydrocarbons, is characterized by extensive delocalization of π electrons over a *cyclic* molecular framework, or **ring system.** The simplest member of this class, benzene, was introduced in Section 9.3 as an example of resonance. Others, such as *naphthalene,* have two or more ring systems fused together.

Many aromatic hydrocarbons are carcinogenic (cancer-inducing).

benzene **naphthalene**

The Lewis structures of aromatic compounds are characterized by alternating single and double bonds in the ring(s). Because of resonance, however, all the bonds in a six-membered aromatic ring are identical and intermediate in character between single and double bonds. Since there are no double bonds, aromatic compounds do not undergo the addition reactions characteristic of alkenes and alkynes. Instead, aromatic compounds undergo **substitution** reactions, in which a hydrogen atom is replaced by another atom or group.

$$C_6H_6 + Br_2 \xrightarrow{FeCl_3} C_6H_5Br + HBr \qquad (9\text{-}9)$$

benzene **bromobenzene**

The delocalization of electrons in an aromatic ring system is not disturbed in substitution reactions, and the additional stability associated with resonance is preserved in the product compound.

Chemistry Old and New

Oil Recovery

Whether you call it "black gold" or "Texas tea," the thick liquid known as petroleum or crude oil is essential to modern industrial society. Petroleum is a complex mixture of compounds, but its major constituents are saturated, unbranched hydrocarbons. It is found in porous rock strata, and is believed to have formed by the slow decay of plant and animal matter in oxygen-poor marine sediments. Small amounts of nitrogen, oxygen, and sulfur compounds are also present in petroleum.

Processing of crude oil begins with *distillation,* which separates the crude oil into low-boiling fuels such as gasoline and diesel oil, high-boiling oils and greases used as lubricants, and even tar and asphalt for roads and construction. In a further step called *cracking,* high molecular weight compounds are converted to the more volatile, low molecular weight compounds found in high-quality gasoline.

Oil is a precious, but non-renewable resource, and we have already used up a substantial portion of the earth's known petroleum deposits or *reservoirs.* Each year, consumption of oil in the United States rises, while production falls. Oil imports, with their attendant political and military implications, are on the rise. Since most of the oil that can be extracted easily from shallow wells has already been consumed, oil producers are digging deeper and using more sophisticated recovery methods.

Three phases of recovery mark the life of an oil reservoir. When a well is first tapped, the "primary recovery" methods use natural underground conditions to bring the oil to the surface. A typical reservoir contains a layer of oil sandwiched between water and natural gas, all trapped beneath a layer of impermeable rock. Such deposits are naturally pressurized, and drilling a hole through the overlying rock is enough to bring the oil bubbling to the surface. The pressure drops as oil is removed from the reservoir, and ultimately becomes too low to power the natural flow of oil to the surface.

In the "secondary recovery" phase, pressurized gas or water is injected through another bore hole, and once again the oil flows. Eventually, all of the oil in the liquid layer can be extracted. But this can leave behind as much as two-thirds of the original oil, trapped as tiny droplets in the water or impregnating the porous rock.

Tapping the remaining oil requires the techniques of "tertiary recovery," which fall into two broad categories: (1) thermal methods, and (2) flooding. Thermal methods, including steam injection, use heat to reduce the viscosity of the trapped oil, or to vaporize it, allowing it to flow freely out of the porous rock matrix. In flooding, on the other hand, a substance that alters the physical or chemical properties of the oil or the rock is injected into the reservoir. For example, carbon dioxide injection both swells and thins the trapped oil, driving it out of the rock. Flooding with a surfactant decreases surface tension and prevents the blockage of rock pores by bubbles. Flooding with sodium hydroxide dissolves the acidic components of petroleum, and sweeps the hydrocarbons out of the rock.

Techniques were simpler in the early days of oil recovery. "Gushers" like this well near Long Beach, California, were a common sight when the natural pressure of an underground reservoir was sufficient to bring oil to the surface.

On the cutting edge of advanced technology are some methods still under development. For example, heating devices, to replace steam injection, are being tried. Another approach being investigated is the passing of a low-voltage electrical current through the rock. The use of specially bred microbes shows promise, both for generation of surfactant molecules to loosen the oil and for producing carbon dioxide to increase the pressure.

Discussion Questions

1. Which should be more effective for flushing oil out of older oil wells, pressurized water or pressurized carbon dioxide?
2. How do the formation and consumption of petroleum fit into the earth's "carbon cycle?"

SUMMARY EXERCISE

About 5 million tons of polyvinyl chloride (PVC) are produced annually in the United States. Its most familiar use is the "plastic pipe" of residential water and drainage systems. PVC is a polymer made from *vinyl chloride* (an older name for chloroethene), which in turn requires large amounts of chlorine for its manufacture. Another familiar polymer is natural rubber, formed from 2-methyl-1,3-butadiene (C_5H_8, older name *isoprene*).

Polymers, discussed in Chapter 20, are very large molecules made up of long chains of repeating, identical subunits.

a. Draw the structural formula of vinyl chloride. Are there any isomers of this compound? If so, draw their structures.

b. Use VSEPR to determine the electron-pair geometry around both carbon atoms of vinyl chloride. Is the molecular shape the same around each carbon? If not, what is it?

c. What is the hybridization of the carbon atoms in vinyl chloride?

d. What is the molecular orbital electron configuration of the Cl_2 molecule?

e. Is Cl_2 paramagnetic? What is its bond order?

f. Compare the bond energy of Cl_2 to that of Cl_2^+ and Cl_2^-. Is either of these ions paramagnetic?

g. Draw the structural formula of isoprene.

h. Draw structural formulas (carbon skeletons only) of all *structural* isomers of C_5H_8. In which of these structural isomers is *geometrical* isomerism possible?

Answers **(a)** $(H)_2C{=}C(Cl)(H)$ [H and H on the first carbon; Cl and H on the second]; there are no isomers. **(b)** Trigonal planar; since there are no lone pairs in the Lewis structure, the molecular shape is the same. **(c)** Both carbon atoms are sp^2 hybridized. **(d)** $(\sigma_{3s})^2(\sigma^*_{3s})^2(\pi_{3p})^4(\sigma_{3p})^2(\pi^*_{3p})^4$. **(e)** Cl_2 is not paramagnetic, and the bond order is one. **(f)** The bond order in Cl_2^+ is 3/2, so its bond energy is greater than that of Cl_2; the bond order of Cl_2^- is 1/2, so its bond energy is less; since both ions have an odd number of electrons, they are both paramagnetic. **(g)** $H_2C{=}C(CH_3){-}C(H){=}CH_2$. **(h)** In addition to isoprene itself, the following structural isomers exist: **(i)** C=C—C—C=C, **(ii)** C=C—C=C—C, and **(iii)** C—C=C=C—C; only **(ii)** exists as a pair of *cis/trans* isomers.

SUMMARY AND KEYWORDS

In *Lewis* theory, a covalent bond is a shared pair of electrons. In **valence bond** theory, a bond is formed from overlapping atomic orbitals. In **molecular orbital** theory, bonding forces arise from occupancy of bonding MO's. In a **σ bond** the electron density lies in a single lobe on, and has cylindrical symmetry about, the **internuclear axis (bond axis).** In **π bonds,** there are two lobes of electron density symmetrically located above and below the nodal plane containing the internuclear axis. Multiple bonds consist of a σ bond and one or two π bonds.

The **valence state,** which describes the electron configuration of an atom as it forms bonds, can be either the ground state or a *promoted state* of the isolated atom. The required **promotion energy** is supplied by the formation of bonds with other atoms. Since the *hydrogenlike orbitals* {*s*, *p*, *d* . . . } do not have the geometrical orientation observed in most molecules, the valence state is better described in terms of a set of **hybrid orbitals.** The valence shell of an isolated second-period atom has four orbitals {*s*, *p*, *p*, *p*}. Depending on the molecular geometry, one of the hybrid sets {*sp*, *sp*, *p*, *p*} (linear), {sp^2, sp^2, sp^2, *p*} (trigonal planar), or {sp^3, sp^3, sp^3, sp^3} (tetrahedral) is used to describe the bonding. Depending on which set is used, the bonded atom is said to be *sp*-, sp^2-, or sp^3 **hybridized.** Molecular geometry, which determines the hybridization, is observed experimentally or predicted from VSEPR. The *sp*, sp^2, and sp^3 hybrid orbitals have cylindrical symmetry, and consist of a major and a minor lobe. The major lobe forms a covalent bond by overlap with an orbital on an adjacent atom. In third- and fourth-row elements the *d* orbitals of the valence shell may also participate, resulting in sp^3d (trigonal bipyramidal) and sp^3d^2 (octahedral) hybridization.

A double bond consists of a σ bond formed from an sp^2 hybrid and a π bond formed from an unhybridized *p* orbital. A triple bond consists of a σ bond formed from an *sp* hybrid and two π bonds formed from *p* orbitals. The σ bonds of a molecule are collectively referred to as the **σ framework.**

In **molecular orbital** theory, which offers a more satisfactory description of bonding in diatomic molecules and resonant molecules, hydrogenlike orbitals (AO's) from all the atoms in a molecule are combined to form molecular orbitals (MO's). Both **bonding MO's** (low energy) and **antibonding MO's** (high energy) are formed. σ MO's have cylindrical symmetry, and π MO's have a planar node. Both σ^* and π^* orbitals have zero electron density in the internuclear region. The set of MO's in a molecule is filled according to the molecular *aufbau* rules. Like AO's, MO's can accommodate a maximum of two electrons. Orbital occupancy determines the bond order in the **homonuclear diatomic** molecules and ions: BO = (# *e*'s in bonding orbitals − # *e*'s in antibonding orbitals)/2. The hypothetical molecules He_2, Be_2, Ne_2, and so on, which have BO = 0, have a bond energy of zero and therefore do not exist as stable species.

The language of VB and MO theory is combined when describing resonant species. The σ framework of a molecule is described using hybrid orbitals, while resonance *delocalization* of electrons is described by MO's constructed from unhybridized atomic *p* orbitals.

In the *sea of electrons* model, a metal is described as a collection of cations immersed in a delocalized cloud of valence electrons. The high electrical and thermal conductivity of metals results from the ability of the loosely bound electrons to move freely throughout the entire crystal. In *band theory,* AO's from all the atoms in a crystal combine to form MO's, so numerous that they are better regarded as a continuous **energy band** than as discrete energy levels. The lower part of the band, called the **valence band,** corresponds to localized bonding orbitals. The upper part, the **conduction band,** corresponds to delocalized antibonding orbitals. Even the small amount of thermal energy available at low temperature is sufficient to excite some electrons to the low-lying conduction band of metals. These delocalized electrons account for the conductivity of metals. Light of any wavelength is absorbed by a metal surface, and immediately re-emitted. This accounts for the opacity, lustre, and lack of color of metals.

A large **band gap** separates the valence and conduction bands in **insulators,** so that electrons can never populate the conduction band; thus insulators cannot conduct electricity or heat. The band gap is of intermediate size in **semiconductors,** and the conduction band is increasingly populated as the temperature is increased. Semiconductor band gaps can be adjusted by **doping,** the addition of controlled amounts of certain impurities to the substance.

Compounds containing only C and H are **hydrocarbons.** If all C atoms are sp^3 hybridized the compound is an **alkane.** It is called **saturated** because each carbon atom is bonded to the maximum possible number (four) of other atoms. Saturated hydrocarbons are relatively unreactive, although they do burn. An **alkene** contains a double bond, and, because the sp^2 carbon atoms are bonded to only three other atoms, alkenes are called **unsaturated. Alkynes,** also unsaturated, contain a triple bond. Unsaturated hydrocarbons undergo **addition reactions,** becoming saturated in the process.

Aromatic hydrocarbons contain a six-membered ring of sp^2 carbon atoms. The Lewis structure shows alternating single and double bonds, but actually all bonds are identical and of intermediate order. The p electrons of carbon are **delocalized** into π orbitals that extend over the entire ring. Aromatic hydrocarbons undergo **substitution** reactions rather than the addition reactions characteristic of **aliphatic** (nonaromatic) unsaturated hydrocarbons.

Larger hydrocarbons occur in sets of **structural isomers,** which share the same molecular formula but differ in connectivity: one isomer has a *straight chain* of carbon atoms, while the others contain **branched chains.** Unsaturated hydrocarbons also have **positional isomers,** which differ in the location of their multiple bond(s). Alkenes can also exist as pairs of *cis-trans* or **geometrical isomers,** which have the same connectivity but differ in spatial arrangement near the double bond. Geometrical isomerism occurs because rotation about a double bond is not possible. Because rotation about single bonds occurs freely, long-chain hydrocarbons are flexible molecules, able to adopt different **conformations.**

PROBLEMS

General

1. Using as few words as possible, define what is meant by the term "covalent bond" in (a) Lewis theory, (b) valence bond theory, and (c) molecular orbital theory.
2. Describe a σ bond. Describe a π bond.
3. Name some objects, other than those in Figure 9.4, that have cylindrical symmetry.
4. What is meant by the term "valence state"? Give some examples.
5. Criticize the statement, "Before it can form any bonds, a carbon atom must absorb sufficient energy to reach the valence state."
6. Define promotion and promotion energy.
7. What are hydrogenlike orbitals? What is a parent orbital?
8. Criticize the statement, "Under the influence of approaching neighbor atoms, an atom's parent AO's are reshaped into a set of hybrid orbitals."
9. What is the relationship between the number of hybrid orbitals in a set and the number of corresponding parent orbitals?
10. Describe the similarities in the shapes of the hybrid orbitals sp, sp^2, and sp^3.
11. Comment on the statement, "Resonance is an artificial concept in Lewis theory and an unnecessary one in molecular orbital theory."
12. What is the difference between resonance and delocalization?
13. What type of hybridization is associated with double bonds? With triple bonds?

*14. Describe the shape, including the location of important nodal surfaces, of σ and σ^* orbitals.

15. Describe the shape, including the location of important nodal surfaces, of π and π^* orbitals.
16. What properties of diatomic molecules are better explained by molecular orbital theory than by valence bond theory?

17. Criticize the statement, "σ orbitals have a single lobe, while π orbitals have two."
18. What is the relationship among bond order, bond energy, and dissociation energy?
19. Why is sp^3d hybridization not used to describe the valence states of second-period elements?
20. How are thermal and electrical conductivity related?
21. What aspects, if any, of the sea of electrons model are contradicted by the band theory of metals?
22. What is an energy band in a metal? Distinguish between valence band and conduction band.
23. Compare the way in which each of the theories of metals accounts for high electrical conductivity.
24. What is a band gap? How is it related to conductivity?
25. How do the orbitals in the conduction band differ from those in the valence band?
26. Distinguish between alkanes, alkenes, and alkynes. For each class of compounds give an example that is *not* given in the text.
27. What is the meaning of the term "saturated," when used to describe hydrocarbons?
28. By what aspects of chemical behavior may alkenes be distinguished from aromatic compounds?
29. Distinguish between aromatic and aliphatic hydrocarbons in terms of their bonding.
30. What is the fundamental reason for the difference in reactivity between alkanes and alkenes?
31. Define, with examples, the term "isomer."
32. Use examples to illustrate several types of isomerism.

Hybrid Orbitals

33. Describe the hybridization of the atom designated by boldface type in each of the following species.
    a. $\mathbf{N}H_3$ b. $\mathbf{N}H_4^+$

*34. Describe the hybridization of the atom designated by boldface type in each of the following species.
    a. $\mathbf{N}O_2$ b. $\mathbf{N}O_3^-$

35. Describe the hybridization of the atom designated by boldface type in each of the following species.
    a. $\mathbf{S}F_4$ b. $\mathbf{S}F_6$

*36. Describe the hybridization of the atom designated by boldface type in each of the following species.
    a. $\mathbf{Si}H_4$ b. $\mathbf{Al}Cl_4^-$

*37. Describe the hybridization of the atom designated by boldface type in each of the following species.
    a. $H_2\mathbf{N}NH_2$ b. N_3^- (**central** atom)

*38. Describe the hybridization of the atom designated by boldface type in each of the following species.
    a. $\mathbf{B}F_3$ b. $H_3N{-}\mathbf{B}F_3$

39. Describe the hybridization of the atom designated by boldface type in each of the following species.
    a. $\mathbf{P}F_5$ b. $\mathbf{H}CCl_3$

*40. Describe the hybridization of the atom designated by boldface type in each of the following species.
    a. $\mathbf{I}Cl_3$ b. $\mathbf{C}O_2$

41. Give the hybridization of each of the carbon atoms in the molecule whose structural formula is

```
      H   H   H   H
      |   |   |   |
   H—C—C=C—C—H
      |           |
      H           H
```

42. Give the hybridization of each of the carbon atoms in the molecule whose structural formula is

```
        H   H      O
        |   |     //
   H—C=C—C—C        H
      |      |     \     |
      H      H      O—C—H
                         |
                         H
```

*43. What is the hybridization of each of the carbon atoms in the molecule whose modified molecular formula is CH_2CH_2?

*44. Describe the hybridization of each of the carbon atoms in the molecule whose modified molecular formula is $CH_2CHCH_2CHCHCl$.

*45. Describe the hybridization of each of the carbon atoms in the following molecules.
    a. CH_3CHO b. $CCl_2CHCH_2CHCH_2$

*46. Give the hybridization of each of the carbon atoms in the following molecules.
    a. CH_3CH_2OH b. $CH_3CH_2OCH_2CH_3$

Valence-Bond Theory: Bonding and Resonance

47. For each of the following molecules, give the hybridization of the atom in bold type. Then describe each of the bonds formed by that atom as σ, π, or some combination of σ and π.
    a. $HCCH$ b. F_2CCF_2

48. For each of the following molecules, describe the hybridization of the atom in bold type. Also, identify each of the bonds formed by that atom as σ, π, or some combination of σ and π.
    a. $CH_3{-}CHO$ b. $CH_3{-}CH_2OH$

49. What is the hybridization of the sulfur atom in each of the following compounds? Identify each of the bonds as σ, π, or some combination of σ and π.
    a. SO_2 b. SO_3

*50. What is the hybridization of the xenon atom in each of the following compounds? Identify each of the bonds as σ, π, or some combination of σ and π.
    a. XeF_4 b. XeF_2 c. $XeOF_4$

*51. Give the hybridization of the carbon atom in each of the following molecules. Then describe each of the bonds formed by that atom as σ, π, or some combination of σ and π.

a. CS_2 b. $COCl_2$

52. For each of the following compounds, give the hybridization of the atom in bold type, and describe each of the bonds formed by that atom as σ, π, or some combination of σ and π.

```
     H   O                H   O
     |  //                |  //
a. H—C—C           b. H—C—C
     |   \                |   \
     H    N—H             H    N—H
          |                    |
          H                    H
```

*53. Give the hybridization of the carbon atoms, resonance structures (if any), and bond types (σ, π, or combination of σ and π), of the allene molecule, whose σ framework is

```
  H       H
  |       |
H—C—C—C—H
```

*54. Give the hybridization of the nitrogen atoms, resonance structures (if any), and bond types (σ, π, or combination of σ and π), of the N_2O_3 molecule, whose σ framework is

```
          O
         /
O—N—N
         \
          O
```

Molecular Orbitals

55. Describe the bond order and magnetism of each of the following diatomic ions.

a. H_2^+ b. H_2^- c. He_2^+

56. Describe the bond order and magnetism of each of the following diatomic ions.

a. Li_2^+ b. Li_2^- c. Be_2^-

57. Use MO theory to predict the relative bond energy in the following pairs of species.

a. F_2, F_2^+ b. C_2, C_2^-

58. Use MO theory to predict the relative bond energy in the following pairs of species.

a. N_2, N_2^+ b. Be_2^-, Be_2^+

59. Use MO theory to predict the relative bond energy in the following pairs of species.

a. CO, CO^+ b. NO, NO^+

60. Use MO theory to predict the relative bond energy in the following pairs of species.

a. NO, CO b. CO^+, NO^+

61. A number of diatomic oxygen species are known. Describe the magnetic behavior of each of the following, and rank them in order of increasing bond energy: O_2, O_2^+, O_2^-, O_2^{2-}.

62. Describe the magnetic behavior of each of the following. Also, rank them in order of increasing bond energy: N_2, N_2^-, N_2^+.

63. Which of the following species would you expect to increase in stability on the addition of one electron: C_2, N_2, O_2, F_2, CO, NO?

64. Which of the following species would you expect to increase in stability on the removal of one electron: C_2, N_2, O_2, F_2, CO, NO?

65. When carbon vaporizes, at extremely high temperatures, among the species present in the vapor is the diatomic molecule C_2. Is C_2 paramagnetic? Which would you expect to have the greatest bond energy: C_2, C_2^-, or C_2^{2-}?

66. Although the molecule S_8 is the predominant form of sulfur in the liquid phase, the species S_2 is also known (in the gas phase). Is S_2 paramagnetic? Which would you expect to have the greatest bond energy: S_2, S_2^-, or S_2^{2-}?

Hydrocarbons

67. Name each of the following hydrocarbons.

```
              H
              |
            H—C—H
              |
     H   H    |   H
     |   |    |   |
a. H—C—C—C—C—H
     |   |    |   |
     H   H    H   H

              H
              |
            H—C—H
              |
     H   H    |   H
     |   |    |   |
b. H—C—C—C—C—H
     |   |    |   |
     H   |    H   H
         |
       H—C—H
         |
       H—C—H
         |
         H

         H
         |
       H—C—H
         |
     H   |    H   H
     |   |    |   |
c. H—C—C—C—C—H
     |   |    |   |
     H   |    H   H
         |
       H—C—H
         |
       H—C—H
         |
         H
```

68. Name each of the following hydrocarbons.

```
      H  H  H  H  H
      |  |  |  |  |
a. H—C—C—C—C=C—H
      |  |  |
      H  H  H
```

```
      H  H  H  H  H
      |  |  |  |  |
b. H—C=C—C—C—C—H
            |  |  |
            H  H  H
```

```
      H  H  H  H  H
      |  |  |  |  |
c. H—C—C=C—C—C—H
      |        |  |
      H        H  H
```

69. What is the total number of σ bonds in each of the molecules in Problem 68? The number of π bonds?

70. What is the total number of σ bonds in the following hydrocarbon? The number of π bonds?

```
              H  H
              |  |
     H—C≡C—C=C—H
```

71. Draw the structural formulas of each of the following compounds.
    a. 2-methylhexane
    b. 2,3-dimethylpentane
    c. 2-chloro-3-methylnonane
    d. 2-bromo-4-chloro-4,5,7-trimethyldecane

72. Draw the structural formulas of each of the following compounds.
    a. 3-methyl-1-pentene
    b. propene (Why is this not called 1-propene?)
    c. 1,3-butadiene
    d. 4-methyl-2-pentyne

73. Draw complete structural formulas corresponding to each of the following modified molecular formulas.
    a. $CH_3CH_2CHCHCH_2Cl$
    b. $CH_3CH{=}CHCH_2CHO$
    c. $CH_2{=}CHCH_2CH_2CH_2OH$
    d. $CH_3CH_2(CH_2)_3CH_3$

74. Draw complete structural formulas corresponding to the following modified molecular formulas.
    a. $CH_3CH(CH_3)CH_2CH_3$
    b. $CH_3C(CH_3)_2CH_2CH_3$
    c. $C(CH_3)_4$
    d. CH_3COOH

CHAPTER 10

THERMODYNAMICS

Because its ions are no longer confined to tiny crystals, the entropy of $KMnO_4$ increases when it dissolves and mixes with water. Entropy always increases during spontaneous processes.

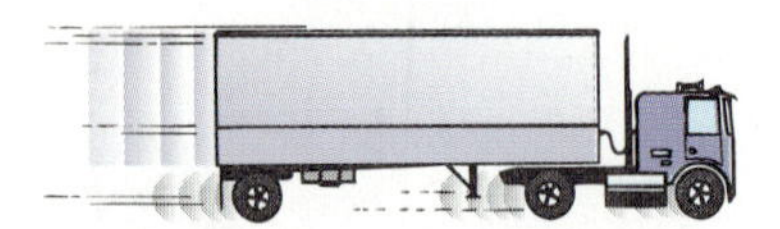

Figure 10.1 A speeding car has a great deal of kinetic energy. Because it is heavier, a truck travelling at the same speed has more kinetic energy.

10.1 ENERGY, HEAT, AND WORK

In Chapter 4, we discussed *thermochemistry,* which concerns the role of heat in chemical reactions. Thermochemistry is part of a broader subject called **thermodynamics,** which is the study of heat, work, and energy in chemical and physical processes. In this chapter, we consider the transfer of energy from one place to another and the transformation of energy from one form to another, and we begin to address the question of *why* chemical reactions take place.

Kinetic and Potential Energy

Energy is classified into two types: potential and kinetic. A moving object has energy by virtue of the fact that it is in motion: energy of motion is called **kinetic energy** (Figure 10.1). Equation (10-1) shows that the kinetic energy of an object depends on its mass and its velocity.

$$KE = \tfrac{1}{2}mv^2 \qquad (10\text{-}1)$$

A heavier and/or faster-moving object has more energy than a lighter, slower one. The kinetic energy of atoms and molecules is very important in thermochemistry and thermodynamics.

Potential energy is the energy an object has by virtue of its *position* rather than its motion (Figures 10.2 through 10.4). Snow on a mountaintop has more potential energy than snow in the valley, simply because it is higher, and water behind a dam has more potential energy than water in the river below the dam. Compressed air, because of its position (occupancy of a small volume), has more potential energy than the same mass of air after it has expanded. Conversion of potential energy to kinetic energy is a common process, as illustrated in Figures 10.2 through 10.4.

In a swinging pendulum (Figure 10.5), there is a continual back-and-forth exchange of kinetic and potential energy. As it changes direction at the end of its swing, the pendulum is momentarily motionless: the kinetic energy is zero. At the same time, because the pendulum is in its highest position, the potential energy is at a maximum. At the lowest point (the midpoint of the swing), the potential energy is at a

Figure 10.2 Snow on a mountain has a great deal of potential energy, because of its altitude. An avalanche converts this potential energy into kinetic energy, sometimes with catastrophic results.

Figure 10.3 The gravitational potential energy of water is converted first to kinetic energy and then to electrical energy as water moves from a high position behind the dam to a lower position in the river below.

minimum, but the kinetic energy is maximized because the pendulum is moving at its maximum velocity. Potential energy has been converted to kinetic energy. At all other positions the energy of the pendulum is both kinetic and potential, in varying amounts.

Potential energy can also be thought of as "stored energy." In addition to the gravitational energy of the snow in Figure 10.2 or the water in Figure 10.3, other forms of stored energy are common. For example, *electrical* energy is stored in batteries and used by motors and lightbulbs; *chemical* energy is stored in fuels and released during combustion; *nuclear* energy is stored in atomic nuclei, and can be released in bombs or nuclear reactors.

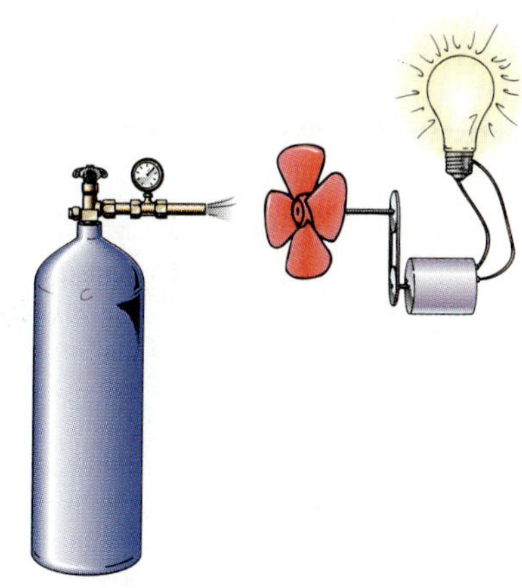

Figure 10.4 The potential energy of compressed air can be converted into other forms of energy: first into kinetic energy as the air and pinwheel are set into motion, then into electrical energy as the pinwheel turns a generator.

Work and Force

Most of us use the word "work" in phrases such as, "You have to work very hard to learn a foreign language properly," or "It was a lot of work hauling all those books to the third floor!" Like many common words and concepts, however, "work" must be defined carefully and restrictively when used in a scientific context. In science, **work** is motion against an opposing force.

$$\text{work} = (\text{force}) \cdot (\text{distance}) \qquad (10\text{-}2)$$

Hauling books to the third floor is work, because it is (upward) motion against the (downward) force of gravity. Mental activity, however difficult it may be, is not work in the scientific sense. The SI unit of work is the same as that of energy: the joule (J).

Force is a common concept used to describe the strength of a push, or the strength of gravitational or electrical attraction. The *definition* of force is in terms of its effect on the velocity of an object. Newton's second law of motion states that the acceleration (*change* in velocity) of an object is equal to the *force* applied to the object divided by its mass. This law is usually seen in the form of Equation (10-3).

$$\text{force} = (\text{mass}) \cdot (\text{acceleration}) \qquad (10\text{-}3)$$

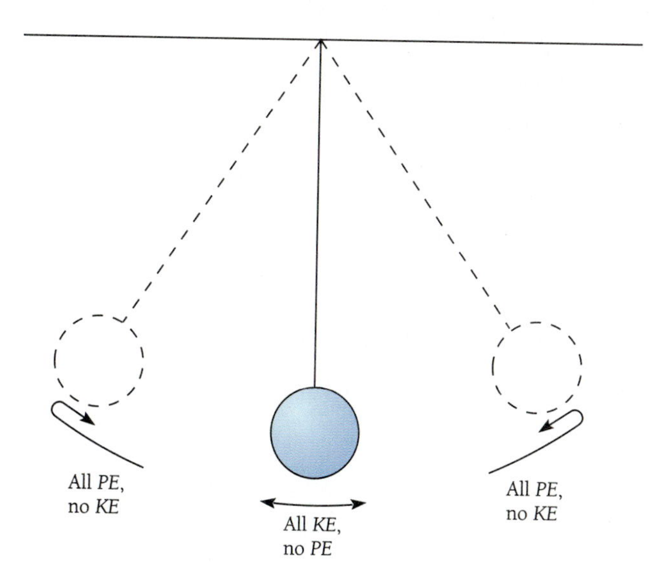

Figure 10.5 The energy of a swinging pendulum is all kinetic at the midpoint of the swing, and all potential at the extreme points.

The work required to lift a 1 kg mass a distance of 1 m against the force of gravity is about 10 J.

The dimensions of acceleration are velocity ($\ell \cdot t^{-1}$) divided by time, or $\ell \cdot t^{-2}$. Equation (10-3) can be used to define the SI unit of force, the **newton (N)**: $1\ \text{N} \Leftrightarrow \text{kg} \cdot 1\ \text{m} \cdot 1\ \text{s}^{-2}$. The definitions presented in Equations (10-2) and (10-3) of work and force can be combined to give a more useful equivalence statement, $1\ \text{J} \Leftrightarrow 1\ \text{N} \cdot 1\ \text{m}$.

EXAMPLE 10.1 Gravitational Work

Calculate the work done when a 150-lb person climbs an 11-foot flight of stairs. The gravitational force acting on an object is proportional to its mass, and is 9.807 N kg^{-1}.

Analysis

Target: Work done against the force of gravity:

$$?\ \text{J} =$$

Knowns: The mass of the object, the distance moved, and the force of gravity are explicit; unit conversion factors are implicit knowns.

Relationship: Equation (10-2), the definition of work, relates the target with one of the knowns (distance).

Missing Information: However, the *total* force must be calculated before Equation (10-2) can be used; the total force in turn can be calculated from the knowns and the statement that gravitational force is proportional to mass.

Plan

2. Calculate the work from Equation (10-2).
1. Calculate the gravitational force acting on the person, using unit conversion factors as appropriate.

Work

1. Both the word "proportional" and the dimensions of the given quantities indicate that the force on the object is found as follows:

$$\begin{aligned} \text{force} &= [\text{mass (kg)}] \cdot [\text{force per kg (N kg}^{-1})] \\ &= (150\ \text{lb}) \cdot \frac{0.454\ \text{kg}}{1\ \text{lb}} \cdot \frac{9.807\ \text{N}}{1\ \text{kg}} \\ &= 668\ \text{N} \end{aligned}$$

2.

$$\text{work} = (\text{force}) \cdot (\text{distance}) \tag{10-2}$$

$$= (668\ \text{N}) \cdot (11\ \text{ft}) \cdot \frac{12\ \text{in.}}{1\ \text{ft}} \cdot \frac{2.54\ \text{cm}}{1\ \text{in}} \cdot \frac{1\ \text{m}}{100\ \text{cm}}$$

$$= 2.24 \times 10^3\ \text{N m} \equiv 2.24 \times 10^3\ \text{J} = 2.2\ \text{kJ}$$

Check A careful check for correct cancellation of units is very important, especially when working with unfamiliar quantities.

Exercise

The force of gravity on the surface of the moon is about one sixth of that on earth, or 1.6 N kg^{-1}. Calculate the work required for a 200-lb astronaut to climb a 30-ft ladder on the moon.

Answer: 1.3 kJ.

Although motion against a gravitational force is a common form of work, there are others. A bicyclist, for instance, must do work to overcome the forces of friction and wind resistance. The stretching of a spring or rubber band involves motion against the restoring force of elasticity.

PV Work

One of the most common forms of work in chemistry is ***PV* work,** which is the work involved when a substance (usually a gas) expands or is compressed. When, for instance, oxygen or acetylene is compressed for storage in a small, high-pressure cylinder, work is done on the gas. The *motion* is inward, as the gas is gathered into a much smaller volume, and the *opposing force* is the outward pressure of the confined gas. In other processes, for instance fuel combustion within an automobile engine, the outward motion of an *expanding* gas is harnessed to perform the useful work of propelling a car.

Energy and work are not the same, but they are closely related. A compressed gas can be made to do work (Figure 10.6), and therefore has more energy than if it had not been compressed. The work expended in compressing the gas goes to increase its energy. Although work is often called a *form of energy,* it is more accurate to say that work is a *means* by which energy can be transferred from one object or place to another.

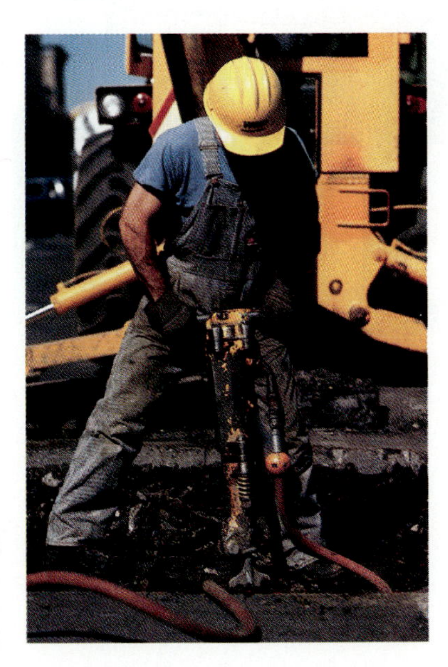

Figure 10.6 A compressor does work when it squeezes air into a smaller volume; the air hammer in turn uses the potential energy of the compressed air to do work when it breaks rock into fragments against the cohesive forces that hold the rock together.

Heat

As we saw in Chapter 4, *heat* or *thermal energy* is the sum of the kinetic energy of all the atoms in a substance. According to the kinetic molecular theory (Chapter 5), there is a direct relationship between the energy of atomic motion and the temperature of a substance: the greater the energy, the greater the temperature. In thermodynamics, **heat** is defined as the form of energy that moves from one place to another because of a difference in temperature. Heat flows spontaneously from a warm object to a cooler object. If the objects are in contact with one another, heat flows by *conduction*. If the objects do not touch each other, heat flows by *radiation*. Heat flow, like work, is a means of energy transfer (Figure 10.7).

10.2 BASIC CONCEPTS OF THERMODYNAMICS

As with most topics, the study of thermodynamics begins with a review of vocabulary. In this section we define some important terms, and outline some concepts that are central to the subject.

The System and Its Environment

Discussions of energy in chemistry are normally concerned with processes that take place within specified boundaries, that is, within a limited and well-defined portion of the universe. The contents of any such region are called a **system.** For instance, if we are interested in the temperature change when water and concentrated acid are mixed, we define the system as the water plus the acid. In this way, we confine our attention to the contents of the beaker and do not worry about whatever else may be going on in

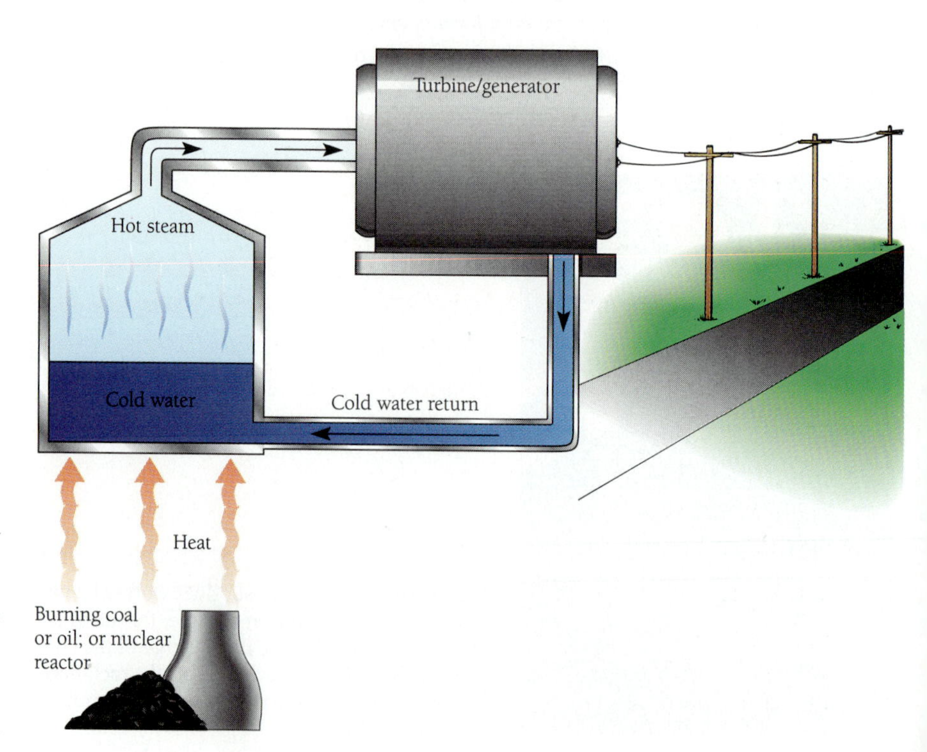

Figure 10.7 In an electrical generating plant, heat flows from a heat source, often a nuclear reaction or the burning of coal or oil, to a boiler, where it is transferred to water. The thermal energy of the resulting high temperature steam is converted, first to mechanical energy and then to electrical energy, in a turbine/generator. As the steam loses its energy it becomes cold water, which returns to the boiler.

(a)

(b)

(c)

Figure 10.8 (a) A beaker containing liquid bromine is an *open* system. (b) Eventually the bromine evaporates completely, leaving the system and passing into the surroundings. (c) A stoppered flask is a *closed* system. No substance can enter or leave it, and its mass cannot change.

the laboratory. A particular system may be defined in any way we wish, provided that its boundaries and contents are specified. The rest of the universe, all that lies outside the boundaries of the system, is known as the **surroundings** or the environment.

An **open system** is one that can freely exchange matter with the surroundings. For example, water in an uncovered beaker transfers water vapor to the room as it evaporates. A **closed system** is one that cannot gain or lose matter: no water vapor passes into or out of a stoppered bottle (Figure 10.8). A closed system can interact with the surroundings *only* through the interchange of heat and/or work (Figure 10.9).

The State of a System

A complete description of a system includes the values of its physical properties, such as temperature, pressure, density, and electrical conductivity. Consider two systems containing the same kind and amount of matter, for instance two 10 g samples of water: if the measured value of each property is the same in both systems, the systems are said to be in the same **state.** If the value of one or more properties is different in the two systems, they are in different states. Since the word "state" is also used to refer to the *physical state* of a substance (that is, solid, liquid, or gas), confusion with the broader definition is possible. When necessary for clarity, the phrase "physical state" or "phase" is used to refer to the solid, liquid, or gaseous condition.

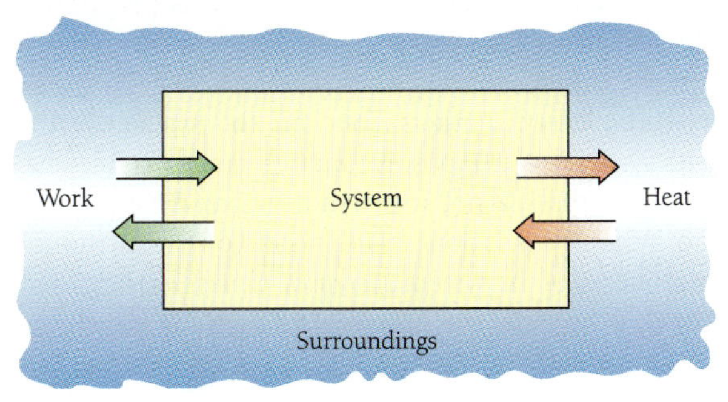

Figure 10.9 A closed system interacts with its surroundings when heat and/or work flows across the boundaries. In general, heat or work flow is accompanied by some change within the system.

Figure 10.10 An unattended solar-powered remote water level measuring station. Radiant energy from the sun is converted into heat and electricity when it falls on the photovoltaic panel. No energy is lost, since the sum of the heat energy and the electrical energy is exactly equal to the amount of radiant energy absorbed.

Internal Energy

The amount of energy contained in a system usually changes when the state changes. When the addition of heat to a substance causes a rise in temperature (which means that it is in a different state), the energy of the substance increases. **Internal energy** is the name given to the energy contained in a system. Internal energy is an *extensive* quantity, that is, its numerical value is proportional to the amount of substance in the system. Internal energy is equal to the sum of the kinetic and potential energies of all the atoms in the system. The potential energy of atoms arises from their relative position, that is, the way in which they are grouped. In a system consisting of one mole of carbon atoms and two moles of oxygen atoms, the potential energy is larger when the atoms are arranged as one mole C(s) + one mole O_2(g) than when they are arranged as one mole CO_2(g). In the former state, the system can undergo combustion and give off energy, while in the latter it cannot.

The internal energy of a system has a definite value, but it cannot be measured. However, it is the *changes* in internal energy that are important, and these changes are readily measured. (The notion that a quantity has a definite but nonmeasurable value may seem strange at first, but exactly the same situation holds for another, more familiar quantity. The elevation of the city of Atlanta is listed as 1050 feet. This quantity is *not* the altitude, which has a definite value but cannot be measured—it is the *change* in altitude between sea level and the city.)

The First Law of Thermodynamics

We have previously referred to the law of *conservation of energy,* which states that energy is neither created nor destroyed in a chemical or physical process (Figure 10.10). A specialized statement of this law, appropriate to chemical and physical processes involving energy in the form of work and heat, is used in thermodynamics. In this form, the law of conservation of energy becomes the **first law of thermodynamics:** "the change in a system's internal energy is equal to the sum of the heat and work transferred to it during a change of state." In algebraic terms, the first law is

The symbol "Δ" always means a *change:* the final value minus the initial value of some quantity.

$$E_{\text{final}} - E_{\text{initial}} = q + w$$
$$\Delta E = q + w \qquad (10\text{-}4)$$

The heat added to the system, that is, the heat transferred from the surroundings to the system, is denoted by q. The work done on the system, that is, the work transferred from the surroundings to the system, is denoted by w. All three quantities, E, q, and w, are measured in joules (Figure 10.11).

Both q and w can be positive or negative quantities, and the distinction is important. When a system absorbs heat from the surroundings, q is positive, and the internal energy of the system *increases.* Provided that w is not large and negative, ΔE is positive when heat is absorbed. When the system acts as a source of heat, and transfers (loses) heat to the surroundings, q is negative, and the internal energy of the system *decreases.* Similar relationships hold for work. When work is done on the system *by* the surroundings (work added to the system), w is a positive quantity, and the internal energy of the system *increases.* Provided that q is not large and negative, ΔE is positive when work is done on the system. When work is done *by* the system on the surroundings, w is negative and the internal energy of the system *decreases.* Unless q is large and positive, ΔE is negative when work is done by the system. Although the transfer of heat or work involves both the system and the surroundings, the algebraic sign of q or w is determined by whether the *system* loses or gains heat or work.

If $q > 0$, the *system absorbs heat* from the surroundings.
If $q < 0$, the surroundings absorb heat *from the system.*
If $w > 0$, the surroundings do work *on the system.*
If $w < 0$, the *system does work* on the surroundings.

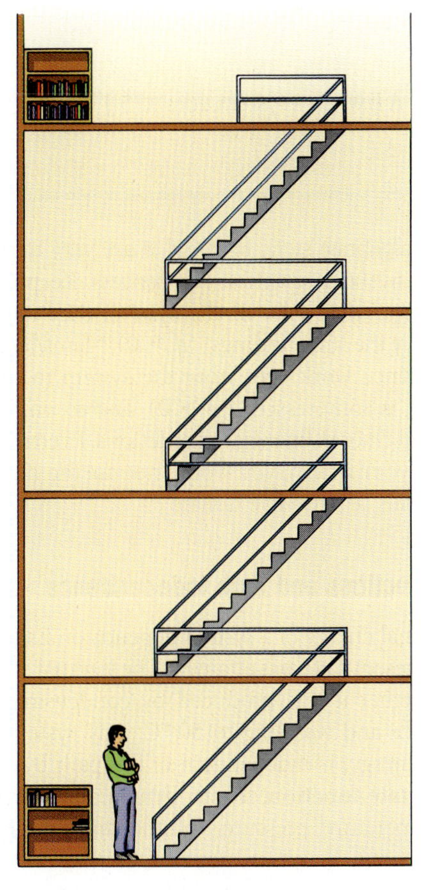

(a) Heating 250mL H_2O from 25 °C to 97 °C

(b) Carrying 680 kg of books up 4 flights (11m)

Figure 10.11 The energy required for typical operations involving heat often seems larger than for those involving work. (a) It takes 75 kJ of energy, in the form of heat, to raise the temperature of a cup of water (about 250 mL) from room temperature to near the boiling point (25 °C to 97 °C). (b) The same amount of energy (75 kJ), in the form of work, is enough to carry three-quarters of a ton of books (680 kg) up to a fourth-floor apartment (11.2 m).

EXAMPLE 10.2 Heat, Work, and Internal Energy

What is the change in internal energy when 200 J of heat is added to a system and the system does 150 J of work on the surroundings?

Analysis

Target: The change in internal energy, ΔE, in joules.

Knowns: The magnitudes of q and w are given explicitly, but their signs are implicit knowns that must be inferred from the context. The sign of q is positive, because heat is "added to a system." The sign of w is negative, since the system "does work *on* the surroundings."

Relationship: The first law, $\Delta E = q + w$.

Plan Insert the given values into the first law and solve.

Work

$$\begin{aligned}\Delta E &= q + w \\ &= 200\text{ J} + (-150\text{ J}) = 50\text{ J}\end{aligned}$$

Exercise

Calculate ΔE when a system does 175 J work and loses 50 J heat to the surroundings.

Answer: $\Delta E = -225$ J.

The algebraic signs of q and w are very important, yet often they are not given explicitly in the description of a process. Keeping in mind that the signs are defined from the point of view of the system, they may be inferred from phrases like "is warmed by the surroundings" ($q > 0$), "gains/loses heat" ($q > 0/q < 0$), "does work" ($w < 0$, since work goes from the system to the surroundings), "produces work" ($w < 0$), "is compressed" ($w > 0$), and so on. We are accustomed to using similar contextual clues in financial transactions. Even though a debt is negative, that is, a debt *decreases* your wealth, you would not use the sign explicitly by saying, "I still owe the bank 'minus seventy-five dollars' ".

State Functions and Path Independence

The internal energy of a system depends on its state, but a system has no memory of how its present state was attained. The internal energy of a mole of water at 50 °C is the same whether it was produced by condensing steam and cooling to 50 °C or by melting ice and warming to 50 °C. Any system property whose value, like that of internal energy, is independent of the path by which the system attains its state, is called a **state function,** thermodynamic property, or property of state. Internal energy, temperature, pressure, volume, and physical properties such as density are all state functions. However, not all quantities in thermodynamics are state functions: heat and work are not. You can warm your hands by rubbing them together (doing work on them) or by holding them up in front of a fire (adding heat). The initial and final states are the same (cold hands $\rightarrow$ warm hands) and therefore the internal energy is the same, but in the first process w is positive and q is zero, while by the second pathway q is positive and w is zero. Note that although q and w individually are not state functions, their *sum* $(q + w) = \Delta E$ is path-independent because E is a state function.

10.3 ENTHALPY—A NEW STATE FUNCTION

Although internal energy is a quantity of fundamental importance in thermodynamics, it is not always the most convenient one for discussions of chemical reactions. In this section we introduce a new quantity, called *enthalpy,* which is more closely related to heat and work in the ordinary chemical processes that take place in the laboratory or in the environment.

PV Work

The work associated with chemical processes is almost always *PV* work. For instance, when a gasoline/air mixture burns within the cylinder of an automobile engine, the system (here the system is defined as the reacting gases) *expands,* driving the piston

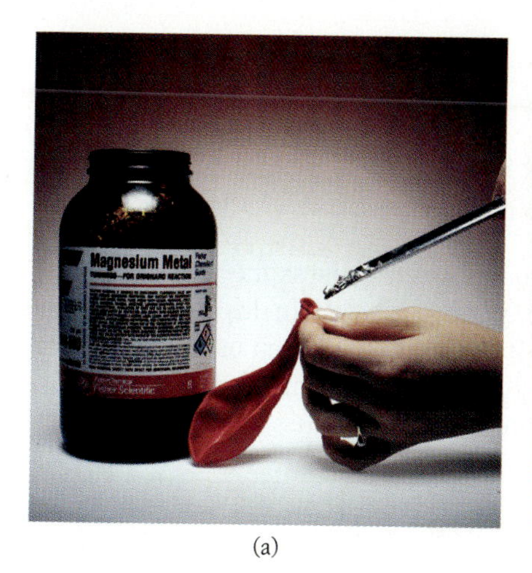

(a)

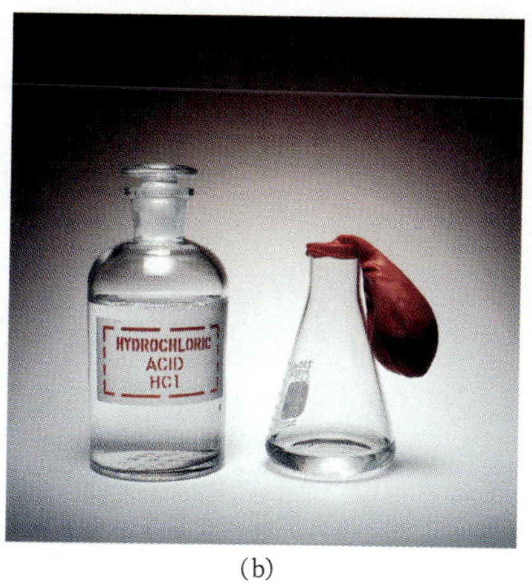

(b)

(c)

Figure 10.12 (a) A balloon containing magnesium turnings is (b) placed over the mouth of a flask containing hydrochloric acid. (c) When the balloon is inverted, the reaction $Mg(s) + 2\ HCl(aq) \rightarrow H_2(g) + MgCl_2(aq)$ takes place and the system volume increases. The system does work on the surroundings.

and ultimately doing the useful work of propelling the car. The system does work on the surroundings, and it is the expansion, that is, the volume increase, of the system that is responsible. The first law is obeyed whether or not work is put to a *useful* purpose. If some magnesium metal is added to a dilute acid solution in an open beaker, hydrogen is produced according to $Mg(s) + 2\ HCl(aq) \rightarrow MgCl_2(aq) + H_2(g)$. Since a gas is evolved, the volume of the system increases during the reaction. Work is done on the surroundings by the system, although the work is dissipated in the useless shoving around of the laboratory air (Figure 10.12).

Work is not always done *by* the system. When the volume of a system *decreases* during a chemical change, the system is compressed and the surroundings have done work on it. The quantity of work done on the system (added to it) is given by Equation (10-5).

$$w = -P\Delta V \tag{10-5}$$

[Equation (10-5) is derived in Appendix C.] Take careful note of the minus sign, which is required for consistency with the distinction between work done *on* the system and work done *by* the system. When a system (for instance a sample of gas) is compressed, its volume decreases. The change in volume, $\Delta V = V_{\text{final}} - V_{\text{initial}}$, is negative, but $w = -P\Delta V$ is positive. When the system expands, ΔV is positive and $-P\Delta V$ is negative. This is correct, because a system does work on the surroundings when it expands.

EXAMPLE 10.3 *PV* Work and the First Law

a. Suppose 85.0 J heat is added to a system, causing it to expand by 6.97×10^{-4} m^3 against a pressure of 9.33×10^4 Pa (about 700 torr). Calculate the change in the internal energy of the system.

b. If the addition of 85.0 J heat to a system causes it to expand by 0.150 L against a pressure of 0.750 atm, what is ΔE?

Analysis/Relationship/Plan Both (a) and (b) require the first law, $\Delta E = q + w$. The work must be calculated from $w = -P\Delta V$, and unit conversion may be necessary.

Work

a.
$$\begin{aligned}\Delta E &= q + w\\ &= q - P\Delta V\\ &= 85.0\text{ J} - (9.33 \times 10^4\text{ Pa}) \cdot (6.97 \times 10^{-4}\text{ m}^3)\\ &= 85.0\text{ J} - 65.0\text{ Pa m}^3\end{aligned}$$

Since quantities cannot be added or subtracted unless they have the same units, something must be done before we can proceed. In this case we can rely on SI: if the relationship $w = -P\Delta V$ is correct (it is), then use of SI units for pressure (Pa) and volume (m^3) *must* result in the SI unit of work (J). That is, use of SI units throughout guarantees the correctness of the equivalence statement $1\text{ J} \Leftrightarrow 1\text{ Pa m}^3$.

$$\Delta E = 85.0\text{ J} - 65.0\text{ J} = 20.0\text{ J}$$

b.
$$\begin{aligned}\Delta E &= q + w\\ &= q - P\Delta V\\ &= 85.0\text{ J} - (0.750\text{ atm}) \cdot (0.150\text{ L})\\ &= 85.0\text{ J} - (0.1125\text{ L atm})\end{aligned}$$

Since liters and atmospheres are not SI units, a unit conversion is necessary before the problem is completed. The appropriate equivalence statements are $1\text{ atm} \Leftrightarrow 1.013 \times 10^5\text{ Pa}$, $1\text{ m}^3 \Leftrightarrow 10^3\text{ L}$, and $1\text{ J} \Leftrightarrow 1\text{ Pa m}^3$.

$$\begin{aligned}\Delta E &= 85.0\text{ J} - (0.1125\text{ L atm}) \cdot \frac{1\text{ m}^3}{10^3\text{ L}} \cdot \frac{1.013 \times 10^5\text{ Pa}}{1\text{ atm}} \cdot \frac{1\text{ J}}{1\text{ Pa m}^3}\\ &= 85.0\text{ J} - 11.4\text{ J} = 73.6\text{ J}\end{aligned}$$

Exercise

Suppose a system is exposed to a pressure of 5.00 atm, which causes the system's volume to decrease by 0.225 L. If there is no change in internal energy ($\Delta E = 0$), how much heat is transferred? Does the system absorb heat from the surroundings, or give heat off to the surroundings?

Answer: $q = -114$ J; since q is negative, heat is transferred from the system to the surroundings.

Constant-Volume Processes

One laboratory situation in which reactions take place at constant volume is the measurement of heats of reaction in a bomb calorimeter (Section 4.4).

During any change of state that takes place without change in volume, for instance if a reaction takes place in a closed, rigid container, ΔV is 0 and no PV work is exchanged with the surroundings. Therefore, $w = 0$. According to the first law, then, if a process takes place with no change in volume, the entire change in internal energy is due to the heat gained or lost by the system.

$$\begin{aligned}\Delta E &= q + w = q + 0\\ \Delta E &= q_v \qquad (10\text{-}6)\end{aligned}$$

The subscript "v" indicates that q_v is the heat transferred in a constant-volume process. The quantity ΔE may be described either as "change in internal energy" or "heat at constant volume."

Constant-Pressure Processes and the Definition of Enthalpy

Many chemical processes occur without change in pressure. For example, any reaction carried out in an open container is subject to the unchanging atmospheric pressure of the laboratory. ΔV is usually not zero in such processes, and therefore w is not zero either. According to the first law, the heat transferred at constant pressure is $q_P = \Delta E - w$. This expression is more cumbersome than that for constant-volume heat ($q_V = \Delta E$), and it is convenient to define a new thermodynamic property so that heat involved in a constant-pressure process can be as simply described as constant-volume heat.

We define the **enthalpy (*H*)** so that the change in enthalpy, ΔH, is equal to the heat added to a system during a process taking place at constant pressure

An older name for enthalpy is "heat content."

$$\Delta H \equiv q_P \tag{10-7}$$

Substitution of ΔH for q_P and $-P\Delta V$ for w in Equation (10-4) gives an expression for the enthalpy change in a constant-pressure process.

$$\Delta E = \Delta H - P\Delta V \qquad \text{or} \qquad \Delta H = \Delta E + P\Delta V \tag{10-8}$$

The thermodynamic properties enthalpy and internal energy should be viewed as a pair: $\Delta H = q_P$ and $\Delta E = q_V$. Both H and E are related to the energy content of a system, and both have the same units (joules). Either internal energy or enthalpy should be used, depending on whether the process takes place at constant volume or at constant pressure. Like internal energy, enthalpy is a state function, that is, its value in a particular state is independent of the path by which the state was reached. Like internal energy, the enthalpy of a system has a definite but nonmeasurable value. That is, only *changes* in enthalpy are measurable.

Since changes in E, P, and V are all path-independent, the right-hand side of Equation (10-8) is path-independent. Therefore, the left-hand side is also path-independent, and enthalpy is a state function.

The fact that enthalpy is a state function leads immediately to a principle of great usefulness in thermochemistry. The enthalpy change in any process depends only on the initial and final states, and is the same for *all* paths leading from the initial to the final state. In a *chemical* process, the initial state is the reactants and the final state is the products. Therefore, ΔH_{rxn} *is the same whether a reaction is carried out all at once or as a sequence of individual steps.* This principle was introduced and used in Chapter 4, under the name *Hess's law.*

10.4 HEAT OF REACTION

From a thermodynamic point of view, a chemical reaction is a process in which the system changes from its initial state (the reactants) to its final state (the products). As with most thermodynamic processes, there usually is a change in internal energy (ΔE_{rxn}) and enthalpy (ΔH_{rxn}) of the system.

Reaction Energy and Reaction Enthalpy

The quantities ΔE_{rxn} and ΔH_{rxn} that are associated with each chemical reaction are called **energy of reaction** and **enthalpy of reaction.** Since chemical reactions rarely take place under conditions of constant volume, ΔE_{rxn} is rarely used. For any reaction

(a)

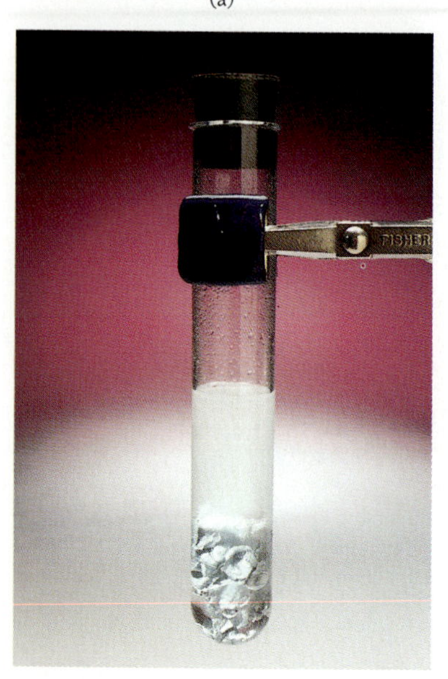

(b)

(a) This exothermic gas-forming reaction, $Zn(s) + HCl(aq) \longrightarrow ZnCl_2(aq) + H_2(g)$, releases an amount of heat ΔH_{rxn} to the environment when it takes place in an open container at the unchanging atmospheric pressure of the laboratory. (b) When the same reaction takes place in a corked test tube, the pressure rises but the volume remains the same (provided the cork does not pop). Under constant-volume conditions the heat transferred to the environment is ΔE_{rxn}, which in this reaction is slightly greater than ΔH_{rxn}.

taking place under constant pressure (that is, any reaction not confined to a closed, rigid vessel), ΔH_{rxn} is equal to the heat generated or consumed in the reaction. In Chapter 4 we followed the common practice of referring to ΔH_{rxn} as *heat* of reaction and ΔH_f° as *heat* of formation, but we now see that the more precise terms for these quantities are *enthalpies* of reaction and formation. Unless there is danger of confusion with the corresponding energies, however, we will continue to use the more familiar terms, heat of reaction and heat of formation.

The reason that an *exothermic* reaction, which is a source of heat, should have a *negative* value of ΔH_{rxn}, is now apparent. The process is considered from the point of view of the *system,* which is defined as the collection of reacting substances. When a reaction gives off heat to the surroundings, the enthalpy of the *system* decreases and ΔH is a negative quantity. Therefore, ΔH_{rxn} *is negative for exothermic reactions* and ΔH_{rxn} *is positive for endothermic reactions.* These relationships are summarized in Figures 10.13 and 10.14.

The Difference between ΔE_{rxn} and ΔH_{rxn}

The numerical difference between ΔE and ΔH is negligibly small in any reaction not producing or consuming gaseous substances. Even in gas reactions the difference may be safely ignored, since it rarely exceeds a few tenths of a percent. For example, when an alkali metal reacts with water to produce gaseous hydrogen and a dilute hydroxide solution, ΔH is about 200 kJ per mole of metal consumed and ΔE is only 0.6 kJ less. If the value of ΔE_{rxn} is needed in some calculation requiring great accuracy, it can be obtained from Equation (10-9).

$$\Delta E_{rxn} = \Delta H_{rxn} - \Delta n \cdot RT \tag{10-9}$$

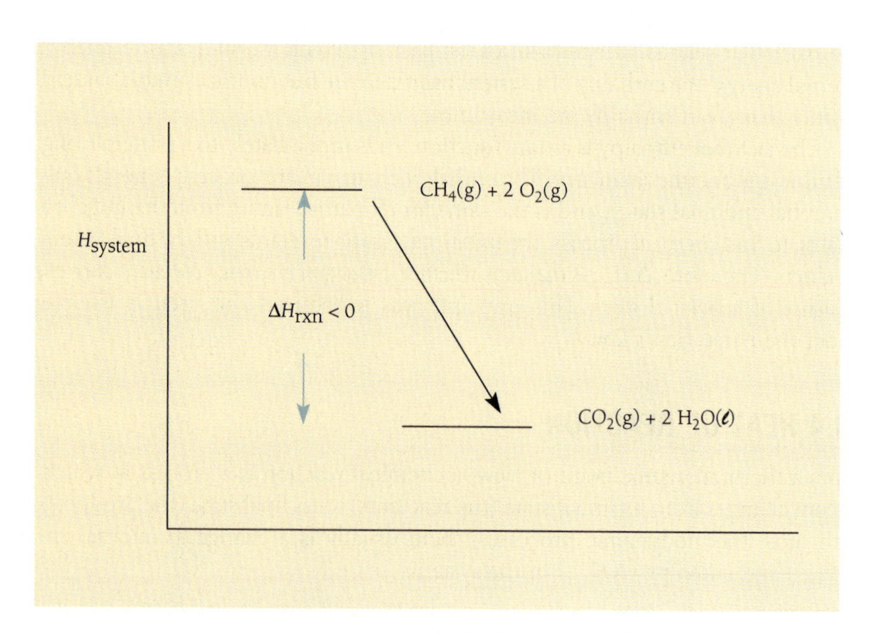

Figure 10.13 In this exothermic reaction, the initial state of the system is the reactants CH_4 and 2 O_2, which, taken together, have a higher enthalpy than the products CO_2 and 2 H_2O, the final state of the system. During the process of reaction the system produces heat and *loses* it to the surroundings, so that H_{system} decreases. ΔH_{rxn} is a negative quantity in exothermic reactions.

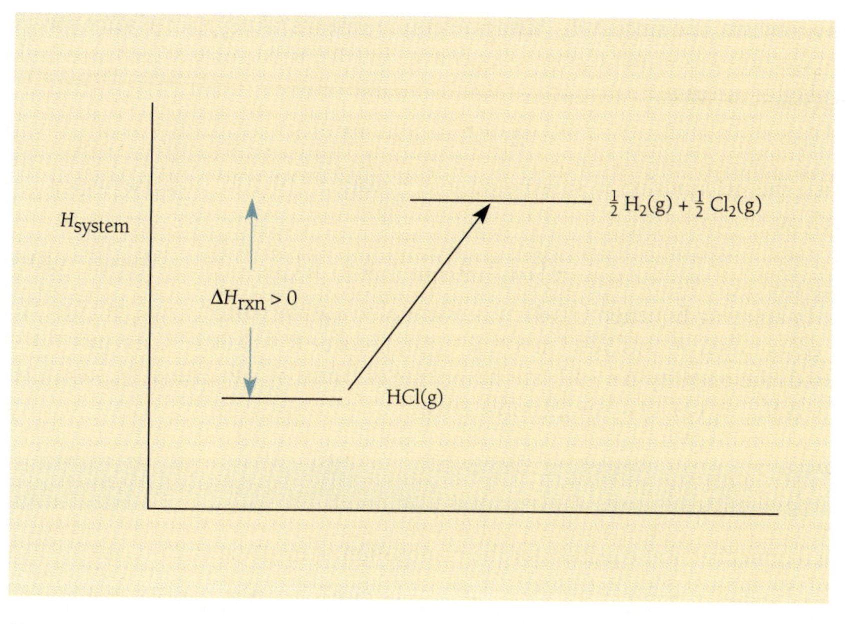

Figure 10.14 In this endothermic reaction, the initial state of the system is the reactant HCl, which has a lower enthalpy than the final state of the system, the products ½ H_2 and ½ Cl_2. During the process of reaction the system *absorbs* the heat it needs from the surroundings, so that H_{system} increases. ΔH_{rxn} is a positive quantity in endothermic reactions.

The quantity Δn, called the "*(gas) mole number change,*" equals the sum of the stoichiometric coefficients of all the *gaseous* products minus the sum of the stoichiometric coefficients of all the *gaseous* reactants. For example, in the reaction

$$2\ C_2H_6(g) + 7\ O_2(g) \longrightarrow 4\ CO_2(g) + 6\ H_2O(\ell)$$

there are 4 moles of gaseous products and (2 + 7) moles of gaseous reactants.

$$\Delta n = (4) - (2 + 7) = -5 \text{ mol}$$

Equation (10-9) is obtained as follows. The relationship $\Delta H = \Delta E + P\Delta V$ (10-8) was derived for changes taking place at constant pressure. The more general form, applicable to processes in which both pressure and volume may change, is

$$\Delta H = \Delta E + \Delta(PV) \qquad (10\text{-}10a)$$

The notation $\Delta(PV)$ means the change in the quantity PV, that is, $\Delta(PV) = (P_{final} \cdot V_{final} - P_{initial} \cdot V_{initial})$. If the process is a chemical reaction, $\Delta(PV) = (P_{final} \cdot V_{products} - P_{initial} \cdot V_{reactants})$, $\Delta H = \Delta H_{rxn}$, and $\Delta E = \Delta E_{rxn}$, so that

$$\Delta H_{rxn} = \Delta E_{rxn} + \Delta(PV) \qquad (10\text{-}10b)$$

Compared to the volume of gaseous species, the volume of solid and liquid substances in the reaction is negligibly small and may be ignored. For gaseous species, the quantity $\Delta(PV)$ can be estimated from the ideal gas law.

$$PV = nRT$$

so that

$$\Delta(PV) = \Delta(nRT)$$

R is a constant, and since ΔH_{rxn} always refers to reactions taking place at some specific temperature, T does not change. Any change in (PV) is due only to a change in the number of moles n.

$$\Delta(PV) = \Delta n \cdot RT$$

When this relationship is substituted into Equation (10-10b) and rearranged, the result is

$$\Delta E_{rxn} = \Delta H_{rxn} - \Delta n \cdot RT \tag{10-9}$$

In applying Equation (10-9), it is essential to use the correct value and units for the gas constant: $R = 8.314\ \text{J mol}^{-1}\ \text{K}^{-1}$. (If the value $R = 0.0821\ \text{L atm mol}^{-1}\ \text{K}^{-1}$ is used, the quantity $\Delta n \cdot RT$ will have units of L atm, which cannot be subtracted from ΔH_{rxn} in J or kJ—see Example 10.3.)

EXAMPLE 10.4 Energy of Reaction

The thermochemical equation for the combustion of methane at 25 °C is

$$CH_4(g) + 2\ O_2(g) \longrightarrow CO_2(g) + 2\ H_2O(\ell); \qquad \Delta H_{rxn} = -890.4\ \text{kJ}$$

Calculate the energy of reaction.

Relationship/Plan The target, ΔE_{rxn}, is obtained from Equation (10-9). The first step is the determination of Δn from the balanced equation.

Work Check that the equation is balanced (it is). Use the stoichiometric coefficients of the gaseous product (CO_2) and the gaseous reactants (CH_4 and O_2) to determine Δn.

$$\Delta n = (1) - (1 + 2) = -2\ \text{mol}$$

The energy of reaction is

$$\begin{aligned}
\Delta E_{rxn} &= \Delta H_{rxn} - \Delta n \cdot RT \\
&= -890.4\ \text{kJ} - (-2\ \text{mol}) \cdot (8.314\ \text{J mol}^{-1}\ \text{K}^{-1}) \frac{1\ \text{kJ}}{1000\ \text{J}} \cdot (298\ \text{K}) \\
&= -890.4\ \text{kJ} - (-4.96\ \text{kJ}) \\
&= -885.4\ \text{kJ}
\end{aligned}$$

Note: Units are critical. You must (a) use $R = 8.314\ \text{J mol}^{-1}\ \text{K}^{-1}$; (b) convert the product $\Delta n \cdot RT$ from J to kJ in order to subtract from ΔH_{rxn}; and (c) use the temperature in kelvins, not °C.

Check ΔE_{rxn} does not normally differ from ΔH_{rxn} by more than a few kJ, but it may be larger or smaller (or exactly the same if $\Delta n = 0$). Check that the correct sign of Δn was used.

Exercise

When the product of the reaction is gaseous water, $\Delta H_{rxn} = -802.3$ for the combustion of methane at 25 °C. What is ΔE_{rxn}?

Answer: -802.3 kJ.

One situation in which the difference between ΔE_{rxn} and ΔH_{rxn} must be taken into account is the measurement of heats of combustion in a bomb calorimeter (Section 4.3). Because it is done at constant volume, a bomb calorimeter measurement yields ΔE_{rxn}, which must be converted to ΔH_{rxn} before it can be related to the heats (enthalpies) of formation ordinarily used by chemists.

EXAMPLE 10.5 Reaction Energy and Reaction Enthalpy

The energy of combustion of phenol, $C_6H_5OH(s)$, measured in a bomb calorimeter at 25 °C, is −3051.0 kJ per mol of phenol burned.

Solutions of phenol have been used as an antiseptic in dentistry since the 19th century.

a. Calculate the enthalpy of combustion.

b. Use the result to determine the heat of formation of phenol.

Analysis

Relationship:

a. If the value of Δn is known, the enthalpy of combustion can be found from Equation (10-9) and the given energy.

b. The heat of formation can be found by the method discussed in Chapter 4 and Equation (4-16):

$$\Delta H_{rxn} = \Sigma[n \cdot \Delta H_f(\text{products})] - \Sigma[n \cdot \Delta H_f(\text{reactants})]$$

Plan (part a)

1. Write a balanced equation for the combustion reaction in order to determine Δn.

2. Use Equation (10-9) to determine ΔH_{rxn} from Δn and the given ΔE_{rxn}.

Plan (part b)

3. Evaluate ΔH_f for phenol.

Work

1. The products of a combustion reaction are CO_2 and H_2O, and in a calorimeter at 25 °C the water is produced in liquid form. The balanced equation is

$$C_6H_5OH(s) + 7\ O_2(g) \longrightarrow 6\ CO_2(g) + 3\ H_2O(\ell)$$

and the gas mole number change is

$$\Delta n = (6) - (7) = -1\ \text{mol}$$

2. After rearranging Equation (10-9), find the enthalpy of the combustion reaction by

$$\begin{aligned}\Delta H_{rxn} &= \Delta E_{rxn} + \Delta n \cdot RT \\ &= -3051.0\ \text{kJ} + (-1\ \text{mol}) \cdot (8.314\ \text{J mol}^{-1}\ \text{K}^{-1}) \\ &\quad \cdot \frac{1\ \text{kJ}}{1000\ \text{J}} \cdot (298\ \text{K}) \\ &= -3051.0\ \text{kJ} + (-2.48\ \text{kJ}) \\ &= -3053.5\ \text{kJ}\end{aligned}$$

3. $$\Delta H_{rxn} = \Sigma[n \cdot \Delta H_f(\text{products})] - \Sigma[n \cdot \Delta H_f(\text{reactants})]$$

$$-3053.5 \text{ kJ} = [6 \cdot \Delta H_f(CO_2) + 3 \cdot \Delta H_f(H_2O)] - [\Delta H_f(\text{phenol}) - 7 \cdot \Delta H_f(O_2)]$$

The heat of formation of O_2, a pure element, is zero. Values for $CO_2(g)$ and $H_2O(\ell)$ are found in Appendix G. Inserting these values, we have

$$-3053.5 \text{ kJ} = [6 \cdot (-393.51) + 3 \cdot (-285.83)] - \Delta H_f(\text{phenol})$$

$$-3053.5 \text{ kJ} = -3218.6 \text{ kJ} - \Delta H_f(\text{phenol})$$

so that

$$\Delta H_f(\text{phenol}) = -165.1 \text{ kJ mol}^{-1}$$

Check ΔH_f values, including signs, have been correctly copied from Appendix G. The correct value and units of R and T have been used.

Exercise

Combustion of aluminum wire in a bomb calorimeter at 25 °C results in the formation of $Al_2O_3(s)$, and liberates 836.0 kJ per mole of aluminum burned. Write the equation, calculate ΔH for the combustion reaction, and calculate $\Delta H_f(Al_2O_3)$.

Answer: $2\ Al(s) + 3/2\ O_2(g) \rightarrow Al_2O_3(s)$; $\Delta H_{rxn} = -1675.7$ kJ; $\Delta H_f(Al_2O_3) = -1675.7 \text{ kJ mol}^{-1}$.

10.5 SPONTANEITY, DIRECTIONALITY, AND ENTROPY

Some processes, both chemical and physical, occur spontaneously. A **spontaneous process** is one that has an intrinsic tendency to take place, without any external stimulation or influence. For example, a drop of food coloring added to water will, in time, spread out and form a homogeneous solution (Figure 10.15). Stirring will hasten the process, but it is not essential. Sodium reacts spontaneously with water, forming hydrogen and aqueous sodium hydroxide. These processes occur naturally in one direction, but not in reverse. We never observe the spontaneous separation of a

(a)

(b)

Figure 10.15 (a) A drop of food coloring added to water begins mixing immediately. Mixing continues spontaneously, even without stirring. (b) Hours or days later, mixing is complete.

colored solution into pure water and a droplet of color, nor can metallic sodium ever be produced by bubbling hydrogen through aqueous sodium hydroxide. In general we can designate a forward/reverse *directionality* to a chemical or physical process, and say that under specific conditions of temperature, pressure, and concentration, the process occurs spontaneously in one direction and not in the other. We now proceed to a discussion of the reasons for reaction spontaneity and directionality. What factors can be identified as the *driving forces* behind spontaneous processes? Why do spontaneous processes take place, and why do they take place in one direction and not the other?

Driving Forces: Enthalpy and Disorder

Physical processes often take place spontaneously in the direction that is accompanied by a decrease in the energy of the system. For example, apples fall from the tree to the ground, thus decreasing their potential energy. Similarly, spontaneous *chemical* processes often are accompanied by a decrease in the energy or enthalpy of the system. The reaction of sodium and water is an exothermic chemical process leading to a decrease in the enthalpy of the ($Na + H_2O$) system, and an increase in the enthalpy of the surroundings. Many spontaneous chemical reactions are exothermic, and enthalpy is one of the driving forces determining reaction spontaneity. *Other things being equal, processes occur spontaneously in the direction that decreases the enthalpy of the system,* that is, processes for which $\Delta H < 0$.

However, enthalpy does not tell the whole story. In many spontaneous processes, the enthalpy is unchanged or even increased. Water left in an open dish spontaneously evaporates, even though its enthalpy increases by an amount equal to the heat of vaporization. Two pure gases placed in the same container spontaneously form a homogeneous mixture, even though the enthalpy change is zero for the process. It is clear that there is another driving force, operating independently of enthalpy change, that affects the spontaneity and directionality of chemical and physical processes. In both of these examples, the process occurs in the direction that increases the amount of *disorder* or *randomness* in the system. A sample of water, prior to evaporation, is relatively well ordered, because all of its molecules are confined to the small volume of the liquid. After evaporation the water vapor is disordered, because the molecules are then scattered throughout the atmosphere. Similarly, a mixture of substances is more disordered, that is, more randomly arranged, than the unmixed components.

Entropy

Disorder is a property of a chemical or physical system, just as are temperature, volume, density, enthalpy, and so on. The name given to the thermodynamic property of disorder (or randomness, or chaos) is **entropy (S)**. Entropy is an extensive state function, and its SI units are $J\ K^{-1}$. A large value of entropy means that the system has a great deal of disorder, and a small value means that the system is well ordered. The entropy of a system is affected by the temperature, composition, volume, and physical state of the system.

Effect of Temperature According to the kinetic molecular theory, as the temperature of a substance increases, its atoms move with increasing velocity and energy. This is true for solids and liquids as well as for gases. As their energy increases, the atoms move with increasing freedom and their positions within the substance become more random. In all substances, *entropy increases as the temperature increases.*

Effect of Composition Consider two pure gaseous substances contained in a vessel at the same temperature and pressure, but separated from one another by a removable partition (Figures 10.16 and 10.17). When the partition is removed, the gases diffuse into one another and form a homogeneous mixture that is more disordered than the separate components. The entropy of the system has increased as a result of the mixing. Similarly, when a solid or liquid dissolves (Figure 10.18), its molecules are no longer confined to one small region. Instead, they are randomly distributed throughout the solution, and the entropy of the system has increased. Almost invariably, *entropy increases when two or more pure substances form a homogeneous mixture.*

"Solution" is another name for a homogeneous mixture.

Effect of Volume When a gas expands it occupies a larger volume. Its molecules are arranged in a less ordered fashion, simply because they can travel farther in their random motion. *The entropy of a gas increases as its volume increases.* Since the volume of a gas is inversely proportional to the pressure, it is also true that *the entropy of a gas decreases as the pressure increases.*

Effect of Physical State The particles (atoms, molecules, or ions) of a solid substance are locked in place, free to vibrate but not to travel from one location to another. When the solid melts to a liquid, its particles are no longer confined to one place, but can

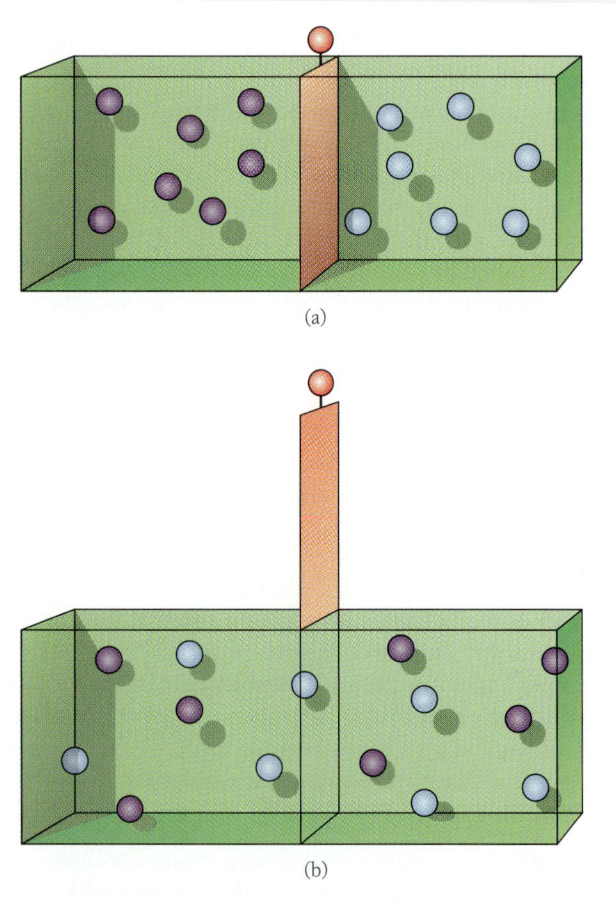

Figure 10.16 When the partition separating the two pure gases (a) is removed, the gases mix and the arrangement of their molecules becomes more random (b).

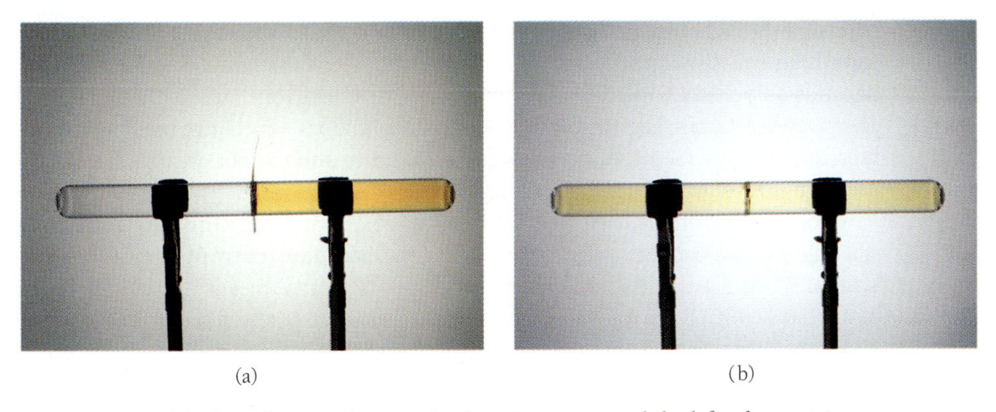
(a) (b)

Figure 10.17 (a) The right test tube contains bromine vapor, and the left tube contains air. (b) When the partition separating the two gases is removed, mixing is spontaneous and complete within a few minutes.

move throughout the entire liquid. This is a state of greater disorder, hence greater entropy. Similarly, when a liquid vaporizes, its atoms or molecules are free to move throughout the whole volume of the container. Again, this is a state of greater disorder. For a given substance, *entropy in the gaseous state is greater than in the liquid state, which in turn is greater than in the solid state.*

In the absence of significant enthalpy changes, processes occur spontaneously in the direction of increased entropy, that is, in the direction for which ΔS is a positive quantity.

Standard Entropy

Because molecular motion decreases, entropy decreases as the temperature decreases. There is no molecular motion, and no entropy, at absolute zero. This aspect of entropy is stated more precisely as the **third law of thermodynamics:** at absolute zero, the entropy of a perfectly ordered crystal of a pure substance is zero. Unlike enthalpy and internal energy, therefore, entropy has a natural reference level from which it can be measured. The entropies of a large number of pure elements and compounds have

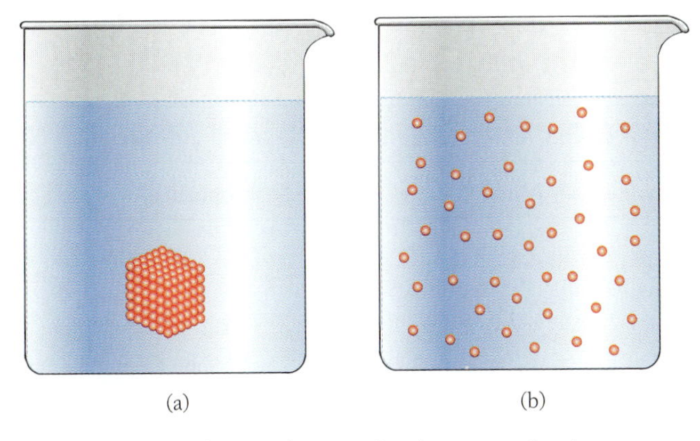
(a) (b)

Figure 10.18 When a substance dissolves, its molecules are no longer confined to a small region. The solution (b) has greater disorder than the separate pure solid and pure liquid (a).

been determined, and are listed in thermodynamic tables such as Appendix G. Entropy is an *extensive* quantity (that is, S is proportional to the amount of substance) that depends on pressure, and the values given in tables refer to one mole of the element or compound at 1 atm. The entropy of a liquid or solid does not change much when the pressure changes, because the *volume* of a liquid or solid is not greatly affected by pressure. The effect is much larger for gases.

The standard state of an element or compound is the most stable state at 25 °C and 1 atm.

The entropy of one mole of an element or compound, measured in its standard state at one atmosphere pressure, is called the **standard entropy,** and given the symbol $S°$. As usual in thermodynamics, the superscript "o" means "at 1 atm." Values of standard entropy of a few substances are listed in Table 10.1. Since entropy depends on temperature, these values are correct only for 25 °C.

Entropy Change in Chemical Reactions

The effect of physical state on entropy, noted above, can in some cases be used to predict whether entropy increases or decreases during a chemical reaction. Since the entropy of gaseous substances is greater than that of solids or liquids, reactions that produce gases from solid or liquid reactants generally result in an increase in entropy. For example, in the decomposition of calcium carbonate, the products have greater entropy than the reactant, and ΔS for the reaction is a positive quantity.

$$CaCO_3(s) \longrightarrow CaO(s) + CO_2(g)$$

When metallic magnesium burns, on the other hand, a gas is consumed in the reaction and the entropy decreases: ΔS for the reaction is negative.

A suspension of finely divided MgO in water, called "milk of magnesia," is a common remedy for upset stomach.

$$2\ Mg(s) + O_2(g) \longrightarrow 2\ MgO(s)$$

The entropy change in a chemical reaction is called the **entropy of reaction** (**ΔS_{rxn}**), and it is found from the relationship $\Delta S_{rxn} = S_{products} - S_{reactants}$. For the calcium carbonate decomposition and the combustion of magnesium, the entropies of reaction are $+161\ J\ K^{-1}$ and $-217\ J\ K^{-1}$, respectively, when the reactions take place at 1 atm pressure and 25 °C.

The **standard entropy of reaction,** $\Delta S°_{rxn}$ (or simply $\Delta S°$), is calculated from tabulated standard entropies in a way that parallels the calculations of $\Delta H°_{rxn}$ (Chapter 4).

$$\Delta S°_{rxn} = \Sigma[n \cdot S°(\text{products})] - \Sigma[n \cdot S°(\text{reactants})] \qquad (10\text{-}11)$$

TABLE 10.1 Selected Values of Standard Entropy at 25 °C

Substance	$S°$/J mol^{-1} K^{-1}	Substance	$S°$/J mol^{-1} K^{-1}
$H_2O(\ell)$	69.91	$H_2O(g)$	188.825
$H_2O_2(\ell)$	109.6	$H_2(g)$	130.684
$CaO(s)$	39.75	$O_2(g)$	205.138
$CaCO_3(s)$	92.9	$CO_2(g)$	213.74
$Mg(s)$	32.68	$C_2H_2(g)$	200.94
$MgO(s)$	26.94	$C_2H_4(g)$	219.56
$Mg(OH)_2(s)$	63.18	$C_2H_6(g)$	229.60

The stoichiometric coefficient (n) of each reactant and product is obtained from the balanced chemical equation, and values of $S°$ are obtained from the table. However, there is one important difference between the calculations of $\Delta S°_{rxn}$ and $\Delta H°_{rxn}$. In calculations of $\Delta H°_{rxn}$, elements are ignored because $\Delta H_f°$ for any element is, by definition, zero. In contrast, elements cannot be ignored in calculation of $\Delta S°_{rxn}$, because $S°$ is not equal to zero for *any* substance.

$S°$ is a nonzero, positive quantity for all substances, elements, and compounds alike.

EXAMPLE 10.6 Entropy of Reaction

Calculate the standard entropy of reaction for the addition of H_2 to acetylene (a gas, C_2H_2) to form ethane (a gas, C_2H_6).

Analysis The target, $\Delta S°_{rxn}$, is calculated from Equation (10-11). The necessary values of $S°$ are found in Table 10.1, and the stoichiometric coefficients are obtained from the balanced chemical equation.

Work The balanced equation is

$$C_2H_2(g) + 2\ H_2(g) \longrightarrow C_2H_6(g)$$

In doing calculations of this sort, it is convenient to write the standard entropies, multiplied by the appropriate stoichiometric coefficient, underneath the species in the equation.

$$\begin{array}{llll} & C_2H_2(g) & +\ 2\ H_2(g) \longrightarrow & C_2H_6(g) \\ n \cdot S°\ \text{J mol}^{-1}\ \text{K}^{-1}: & 200.94 & 2(130.684) & 229.60 \end{array}$$

$$\begin{aligned} \Delta S°_{rxn} &= \Sigma[n \cdot S°(\text{products})] - \Sigma[n \cdot S°(\text{reactants})] \\ &= [1 \cdot S°(C_2H_6)] - [1 \cdot S°(C_2H_2) + 2 \cdot S°(H_2)] \\ &= [1 \cdot 229.60 - (1 \cdot 200.94 + 2 \cdot 130.684)](\text{mol} \cdot \text{J mol}^{-1}\ \text{K}^{-1}) \\ &= -232.71\ \text{J K}^{-1} \end{aligned}$$

Check Correct stoichiometric coefficients have been used, and correct values of entropy have been copied from the table ($S°$ values are always positive). The negative value of ΔS is expected for a process in which the number of moles of gaseous species decreases.

Exercise

Verify the value quoted on page 440 for the standard entropy of combustion of magnesium according to $2\ Mg(s) + O_2(g) \rightarrow 2\ MgO(s)$.

Answer: $\Delta S°_{rxn} = -216.62\ \text{J K}^{-1}$.

As with $\Delta H°_{rxn}$, if a reaction is written in reverse, the sign of $\Delta S°_{rxn}$ changes.

$$C_2H_2(g) + 2\ H_2(g) \longrightarrow C_2H_6(g); \qquad \Delta S° = -232.71\ \text{J K}^{-1}$$
$$C_2H_6(g) \longrightarrow C_2H_2(g) + 2\ H_2(g); \qquad \Delta S° = +232.71\ \text{J K}^{-1}$$

Keep in mind that, although $S°$ for an element or compound is always a positive quantity, ΔS°_{rxn} can be positive or negative. The sign of ΔS°_{rxn} can often be predicted from the following generalizations.

1. ΔS°_{rxn} is positive if the equation shows a greater number of moles of gaseous products than gaseous reactants. It is negative if the converse is true.
2. For reactions that do not involve gases, ΔS°_{rxn} is usually positive if the number of moles of products is greater than the number of moles of reactants, and negative if the converse is true.
3. If the number of moles of gas does not change during the reaction, and if the total number of moles of all species also remains the same, ΔS°_{rxn} is generally small, and may be positive or negative.

When these rules are applied to the hydrogenation reaction of Example 10.6, it is seen that ΔS_{rxn} should indeed be negative.

10.6 FREE ENERGY

Enthalpy and entropy are both important to the understanding of chemical and physical processes. In addition, however, it is very useful to combine these two state functions into a third function, called *free energy*. Free energy is closely related to spontaneity, directionality, and equilibrium.

Reaction Directionality and the Definition of Free Energy

Both ΔH and ΔS affect the direction of spontaneous chemical reactions and physical changes. If ΔH is zero or very small, a reaction occurs spontaneously in the direction of greater entropy. That is, ΔS is *positive* in a low-enthalpy spontaneous reaction. If, on the other hand, ΔS is zero or very small, a reaction occurs spontaneously in the direction of lower enthalpy. That is, ΔH is *negative* in a low-entropy spontaneous reaction.

The two driving forces of enthalpy and entropy can be combined into a single thermodynamic property, called **free energy, *G*.** Changes in free energy are related to changes in enthalpy and entropy by

$$\Delta G = \Delta H - T\Delta S \qquad (10\text{-}12)$$

Free energy is related to a law of nature that governs the directionality of spontaneous processes. The **second law of thermodynamics** states that *the total entropy of a system and its surroundings always increases in a spontaneous change.* That is,

$$\Delta S_{total} = \Delta S_{system} + \Delta S_{surroundings} > 0$$

Total entropy increases when heat flows from a warm object to a cool object, for example, or when a gas expands into a vacuum, or when two pure gases mix by diffusion.

Since it is not always convenient to consider changes in the surroundings, the concept of free energy was developed to allow application of the second law to changes in the system only. In this form, the second law states that *the free energy of a system always decreases during a spontaneous process.* That is,

$$\Delta G_{system} < 0$$

The relationship between these two forms of the second law is outlined in Appendix C.

Several things should be noted about the definition $\Delta G = \Delta H - T\Delta S$. First, ΔH, T, and ΔS can all be measured, so that ΔG, which is determined from them, is also a measurable quantity. Second, ΔS has units of J K^{-1} and cannot be subtracted from ΔH(J or kJ) unless its units are changed to J through multiplication by the temperature. Third, the negative sign of the $T\Delta S$ term is required to make ΔH and ΔS act in the same sense. We saw above that when ΔS is small or zero, processes occur spontaneously in the direction for which ΔH is a negative quantity. Since $\Delta S \approx 0$, by Equation (10-12) $\Delta G \approx \Delta H$ and ΔG is negative. On the other hand, when ΔH is small or zero the process is spontaneous in the direction of *increased* entropy, and ΔS is positive. For these processes Equation (10-12) shows that $\Delta G \approx -T\Delta S$, so that ΔG is negative in this case as well. In other words, ΔG has been defined so that free energy decreases in a spontaneous process. If ΔG is zero or positive, the process *cannot* occur spontaneously.

Although a negative value of ΔG means that a process will occur spontaneously, it says nothing about the *rate* at which the process takes place. A spontaneous process may occur at an imperceptibly slow rate, so that for all practical purposes it does not take place at all. For example, ΔG for the process C(diamond) $\rightarrow$ C(graphite) is -2.9 kJ mol^{-1}, but in all the centuries since the discovery of diamonds, no spontaneous change to graphite has ever been observed (Figure 10.19). The rate of chemical change, and the factors that affect it, are the subject of Chapter 14.

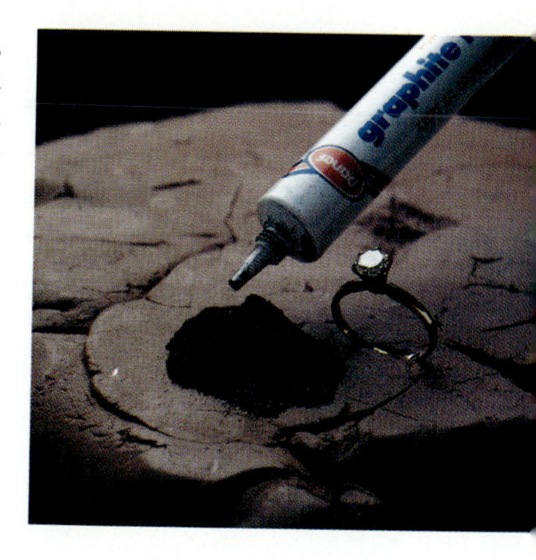

Figure 10.19 Even though ΔG for the process is negative, diamonds do not spontaneously change into graphite at a perceptible rate.

Diamonds are forever.

Temperature Dependence of ΔG

The free energy change for a particular process depends on temperature, in a way that follows directly from the defining equation $\Delta G = \Delta H - T\Delta S$. If ΔS is positive, then ΔG decreases as the temperature increases. Conversely, ΔG increases as temperature increases for those processes having a negative value of ΔS. Even the *direction* of spontaneous change can depend on temperature. Suppose that for a particular process ΔH and ΔS are both negative: it follows that ΔG is negative at low temperature, and becomes positive at sufficiently high temperature. These relationships are summarized in Table 10.2.

The importance of these relationships, in chemistry, is that a reaction may be spontaneous in one direction at low temperature, and spontaneous in the other direction at high temperature.

Standard Free Energies of Formation and Reaction

The **standard free energy of formation, ΔG_f°,** of a compound is defined as the free energy change when 1 mole of the compound is produced in a formation reaction. This parallels the definition of the standard *heat (enthalpy)* of formation given in

TABLE 10.2 Effect of Temperature on ΔG

ΔH	ΔS	ΔG
+	−	Positive at all temperatures, and increases as temperature increases.
−	+	Negative at all temperatures, and decreases as temperature increases.
−	−	Negative at low temperature, and becomes positive at high temperature.
+	+	Positive at low temperature, and becomes negative at high temperature.

Section 4.2. Recall that in a *formation reaction* 1 mole of a compound is produced from its component elements, all of which are in their standard states, at a pressure of 1 atmosphere. Like the heat of formation, ΔG_f° of a compound may be positive or negative, but ΔG_f° *is zero for all pure elements in their standard states.* Values of ΔG_f° for selected compounds are given in Table 10.3, and additional values are listed in Appendix G. The **standard free energy of reaction, ΔG_{rxn}°,** is found by using Equation (10-13).

$$\Delta G_{rxn}^\circ = \Sigma[n \cdot \Delta G_f^\circ(\text{products})] - \Sigma[n \cdot \Delta G_f^\circ(\text{reactants})] \qquad (10\text{-}13)$$

The direction of spontaneity of a reaction, *when all species are at the standard pressure (1 atm),* can be assessed from the sign of ΔG_{rxn}°. If ΔG_{rxn}° is a negative quantity, the reaction is spontaneous as written. If $\Delta G_{rxn}^\circ = 0$, the reaction does not proceed in either direction. If ΔG_{rxn}° is positive, the reaction is spontaneous in the reverse direction. Methods for assessing the direction of spontaneous reaction under nonstandard conditions are developed in Chapter 13.

EXAMPLE 10.7 Free Energy of Reaction

Calculate the change in standard free energy for the incomplete combustion of ethene (C_2H_4). The products of the reaction are CO(g) and H_2O(g). In which direction does the reaction proceed spontaneously under standard pressure?

Analysis/Plan The calculation of ΔG_{rxn}° is essentially the same as that for ΔS_{rxn}° and ΔH_{rxn}°. Balance the equation, look up values of ΔG_f° for all species, and use Equation (10-13). The direction of spontaneity is predicted from the sign of ΔG_{rxn}°.

Work

	$C_2H_4(g)$ +	$2\ O_2(g)$	$\longrightarrow$	$2\ CO(g)$ +	$2\ H_2O(g)$
$n \cdot \Delta G_f^\circ/\text{kJ mol}^{-1}$:	+68.15	2(0)		$2 \cdot (-137.168)$	$2 \cdot (-228.572)$

$$\begin{aligned}\Delta G_{rxn}^\circ &= \Sigma[n \cdot \Delta G_f^\circ(\text{products})] - \Sigma[n \cdot \Delta G_f^\circ(\text{reactants})] \\ &= [2 \cdot (-137.168) + 2 \cdot (-228.572)] - [1 \cdot (+68.15)]\ (\text{mol} \cdot \text{kJ mol}^{-1}) \\ &= -799.63\ \text{kJ}\end{aligned}$$

Since ΔG_{rxn}° is negative, the reaction is spontaneous as written when the partial pressure of each gas is 1 atm.

Note: Elements in their standard states may be ignored in these calculations, because their free energies of formation are zero. Also, it is good practice to include the sign of positive values: write +68.5 rather than 68.5.

The incomplete combustion of ethene yields carbon monoxide and water.

TABLE 10.3 Selected Values of Free Energy of Formation at 25 °C

Compound	ΔG_f°/kJ mol^{-1}	Compound	ΔG_f°/kJ mol^{-1}
$H_2O(\ell)$	−237.129	$H_2O(g)$	−228.572
$H_2O_2(\ell)$	−120.35	$CO(g)$	−137.168
$CaO(s)$	−604.03	$CO_2(g)$	−394.359
$CaCO_3(s)$	−1128.79	$C_2H_2(g)$	+209.20
$Mg(OH)_2(s)$	−833.51	$C_2H_4(g)$	+68.15
$PbO(s)$	−187.89	$C_2H_6(g)$	−38.82
$P_4O_{10}(s)$	−2697.7	$NH_3(g)$	−16.45

Exercise

Calculate the standard free energy for the reaction of lead(II) oxide with ammonia to form lead, nitrogen, and liquid water.

Answer: −114.82 kJ.

EXAMPLE 10.8 Calculation of Free Energy of Formation

The standard free energy of reaction for the thermal decomposition of calcium carbonate, which yields calcium oxide and carbon dioxide, is +130.40 kJ. Using other values of ΔG_f° from Table 10.3, confirm that the value given for $\Delta G_f^\circ(CO_2)$ is correct.

Analysis/Plan Write the balanced chemical equation, then use Equation (10-13) to determine $\Delta G_f^\circ(CO_2)$.

Work

$$\begin{array}{llll} & CaCO_3(s) \longrightarrow & CaO(s) + CO_2(g); & \Delta G^\circ_{rxn} = +130.40 \text{ kJ} \\ n \cdot \Delta G_f^\circ/\text{kJ mol}^{-1}: & -1128.79 & -604.03 \quad ? & \end{array}$$

$$\begin{aligned} \Delta G^\circ_{rxn} &= \Sigma[n \cdot \Delta G_f^\circ(\text{products})] - \Sigma[n \cdot \Delta G_f^\circ(\text{reactants})] \\ +130.40 &= [-604.03 + \Delta G_f^\circ(CO_2)] - [-1128.79] \\ -\Delta G_f^\circ(CO_2) &= -604.03 - (-1128.79) - (+130.40) \\ \Delta G_f^\circ(CO_2) &= -394.36 \text{ kJ mol}^{-1} \end{aligned}$$

Here again, liberal use of parentheses and explicit signs is good practice.

This result is consistent with the value in Table 10.3.

Exercise

When $P_4O_{10}(s)$ reacts with a small amount of water, solid H_3PO_4 is formed. [If too much water is used, the product is $H_3PO_4(aq)$ rather than $H_3PO_4(s)$.] Given that the standard free energy change for this reaction is −355.9 kJ, calculate ΔG_f° of $H_3PO_4(s)$.

Answer: −1119.1 kJ mol^{-1}.

The relationship $\Delta G = \Delta H - T\Delta S$ holds for *any* process that takes place at constant temperature. If the temperature and any two of ΔG, ΔH, and ΔS are known, the third is easily calculated. In particular, it is true that $\Delta G^\circ_{rxn} = \Delta H^\circ_{rxn} - T\Delta S^\circ_{rxn}$.

Be careful! $\Delta G^\circ_{rxn} = \Delta H^\circ_{rxn} - T\Delta S^\circ_{rxn}$, but $\Delta G^\circ_f \neq \Delta H^\circ_f - TS^\circ$

Note that ΔG°_f *does not* equal $(\Delta H^\circ_f - TS^\circ)$. This is easily seen for elements, for which ΔG°_f and ΔH°_f are both zero, and S° is a positive, nonzero, quantity. The reason for this is that ΔG°_f and ΔH°_f are the free energy and enthalpy change in the *formation reaction,* while S° is an absolute quantity unrelated to any reaction.

EXAMPLE 10.9 Use of ΔG°_{rxn} and ΔS°_{rxn} to Find the Heat of Reaction

Use the information in Tables 10.1 and 10.3 to determine the value of ΔH°_{rxn} for the reaction $Mg(s) + 2\ H_2O(g) \rightarrow Mg(OH)_2(s) + H_2(g)$ at 25 °C.

Analysis/Plan The required relationship is $\Delta G^\circ_{rxn} = \Delta H^\circ_{rxn} - T\Delta S^\circ_{rxn}$, and values of ΔS°_{rxn} and ΔG°_{rxn} are obtained by the methods of Examples 10.6 and 10.7.

Work

1. Standard entropy.

	Mg(s) +	$2\ H_2O(g) \longrightarrow$	$Mg(OH)_2(s)$ +	$H_2(g)$
$n \cdot S^\circ$/J mol^{-1} K^{-1}:	32.68	2 · (188.825)	63.18	130.684

$$\begin{aligned}\Delta S^\circ_{rxn} &= \Sigma[n \cdot S^\circ(\text{products})] - \Sigma[n \cdot S^\circ(\text{reactants})]\\ &= [63.18 + 130.684] - [32.68 + 2(188.825)]\ \text{J K}^{-1}\\ &= -216.47\ \text{J K}^{-1}\end{aligned}$$

Check ΔS is large and negative, as it should be for a reaction in which the number of gas moles decreases.

2. Standard free energy.

	Mg(s) +	$2\ H_2O(g) \longrightarrow$	$Mg(OH)_2(s)$ +	$H_2(g)$
$n \cdot \Delta G^\circ_f$/kJ mol^{-1}:	0	2 · (−228.572)	−833.51	0

$$\begin{aligned}\Delta G^\circ_{rxn} &= \Sigma[n \cdot \Delta G^\circ_f(\text{products})] - \Sigma[n \cdot \Delta G^\circ_f(\text{reactants})]\\ &= [-833.51] - [2 \cdot (-228.572)]\ \text{kJ}\\ &= -376.37\ \text{kJ}\end{aligned}$$

Note: The negative value indicates that, under standard conditions, this reaction is spontaneous as written.

3. Standard enthalpy.

$$\begin{aligned}\Delta G^\circ_{rxn} &= \Delta H^\circ_{rxn} - I\Delta S^\circ_{rxn},\ \text{or}\\ \Delta H^\circ_{rxn} &= \Delta G^\circ_{rxn} + T\Delta S^\circ_{rxn}\\ &= -376.37\ \text{kJ} + (298\ \text{K}) \cdot (-216.47\ \text{J K}^{-1}) \cdot \frac{1\ \text{kJ}}{1000\ \text{J}}\\ &= -440.88\ \text{kJ}\end{aligned}$$

Note: It is absolutely necessary (and frequently forgotten) that the units of ΔH, ΔG, and $T\Delta S$ must be the same. This usually requires conversion of $T\Delta S$ from J to kJ.

Exercise

Calculate the standard enthalpy change for the thermal decomposition of liquid hydrogen peroxide (H_2O_2) to produce gaseous water and oxygen.

Answer: −54.0 kJ per mole of H_2O_2 decomposed.

Pure liquid hydrogen peroxide is a highly corrosive and dangerously explosive substance. The "peroxide" available in drugstores is a 3% (by mass) aqueous solution, and it is not dangerous.

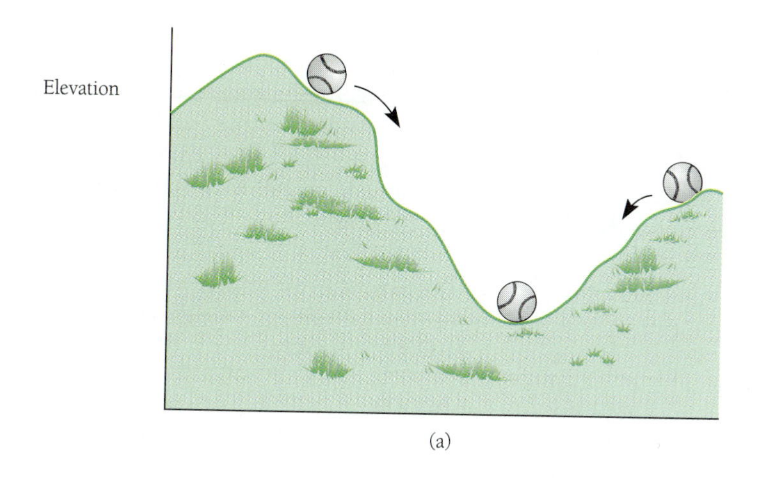

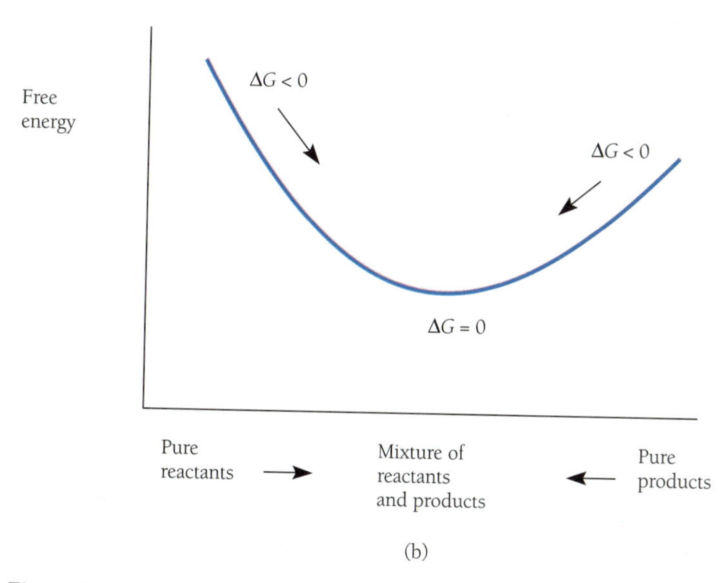

Figure 10.20 (a) A ball stops rolling when it reaches a minimum elevation. (b) A chemical reaction stops when it reaches a minimum in free energy.

Equilibrium

ΔG is always negative in a spontaneous process. As a spontaneous chemical reaction proceeds, the free energy of a system decreases, and, for many reactions, it reaches a minimum *before* all the reactants are consumed. At this point $\Delta G = 0$, and the reaction stops because further reaction (in either direction) would cause an increase in free energy (Figure 10.20).

Processes in which $\Delta G = 0$ are very important in chemistry. If $\Delta G = 0$, then $\Delta H = T\Delta S$. The enthalpy driving force is exactly balanced by the entropy driving force, and the net result is that the process or reaction has no tendency to proceed in either direction. Under these conditions a reaction is said to be *at equilibrium*. The topic of equilibrium is developed in detail in later chapters of this book.

SUMMARY EXERCISE

Gaseous *thionyl chloride* ($SOCl_2$) reacts vigorously with liquid water to produce gaseous sulfur dioxide and HCl(g). The standard entropy and free energy of this reaction are $\Delta S^\circ_{rxn} = 242\ J\ K^{-1}$ and $\Delta G^\circ_{rxn} = -55.4$ kJ at 25 °C. The standard enthalpy of formation of *liquid* thionyl chloride at 298 K is −246 kJ mol^{-1}. Other relevant thermodynamic quantities at 25 °C are

	$H_2O(\ell)$	$SO_2(g)$	HCl(g)
ΔH°_f/kJ mol^{-1}	−285.830	−296.83	−92.307
S°/J K^{-1} mol^{-1}	69.91	248.22	186.908
ΔG°_f/kJ mol^{-1}	−237.129	−300.194	−95.299

a. Write the balanced equation for the reaction of thionyl chloride with water.

b. Are there any lone pairs of electrons on the S atom in $SOCl_2$? If so, how many?

c. Describe the shape of the $SOCl_2$ molecule.

d. Characterize the order and symmetry (σ or π) of all bonds in $SOCl_2$.

e. Determine the value of ΔH°_{rxn}.

f. Using only the data given in this exercise, calculate ΔH°_f, ΔG°_f, and S° for $SOCl_2(g)$ at 298 K.

g. What is the standard heat of vaporization of $SOCl_2(\ell)$?

Answers **(a)** $SOCl_2(g) + H_2O(\ell) \rightarrow SO_2(g) + 2\ HCl(g)$; **(b)** yes, one; **(c)** pyramidal; **(d)** the two S-Cl bonds are σ bonds of order 1, and the S-O bond is a double bond with a σ and a π component; **(e)** +16.7 kJ; **(f)** $\Delta H^\circ_f = -212.3$ kJ mol^{-1}, $\Delta G^\circ_f = -198.3$ kJ mol^{-1}, and $S^\circ = 310$ J K^{-1} mol^{-1}; **(g)** 34 kJ mol^{-1}.

SUMMARY AND KEYWORDS

Motion against an opposing force is called **work.** An important form of work in chemistry is *compressive* or **PV work**, that is, volume change against the opposing force of pressure. The *PV* work done *on* a system is $w = -P\Delta V$, and the work done *by* a system on the surroundings is $w = +P\Delta V$. Work is the form of energy transferred from one place to another as a result of motion, while **heat** is the form of energy transferred as a result of a difference in temperature. The study of heat and work in chemical and physical processes is called **thermodynamics.** The study of heat produced in or consumed by chemical reactions is *thermochemistry*. An object has **kinetic energy** by virtue of its motion, and **potential energy** by virtue of its position. The law of *conservation of energy* states that energy is neither created nor destroyed in a chemical or physical process.

Any limited portion of the universe may be defined as a thermodynamic **system,** and the rest of the universe is the **surroundings.** A system is in a well-defined condition known as a **state,** which changes as the system exchanges heat and/or work with the surroundings. A change in a system's state is usually accompanied by a change in its **internal energy** E (the sum of the kinetic and potential energies of all the atoms in the system). The **first law of thermodynamics** states that a system's internal energy change is the sum of the heat and work exchanged with the surroundings: $\Delta E = q + w$. A positive value of q means that the system absorbs heat from the surroundings, and w is positive when the surroundings do work on the system.

Internal energy is a **state function,** which means that its numerical value does not depend on the path by which the system reaches that particular state. Physical properties of a system are state functions, but q and w are not: heat and work do depend on the path. When a system changes to a new state having the same volume as the former state, the increase in internal energy is equal to the heat absorbed: $\Delta E = q_V$. The state function **enthalpy (H)** is defined so that $\Delta H = q_P$, the heat exchanged with the surroundings when a change of state occurs without change in pressure. The **enthalpy of reaction (ΔH_{rxn})** is the heat absorbed or generated by a system undergoing chemical reaction at constant pressure. ΔE and ΔH are *extensive* quantities, that is, their values are proportional to the amount of material reacting.

Chemical and physical processes usually occur spontaneously in one direction, but not in the reverse direction. This directionality is controlled both by the drive toward lower enthalpy and the drive toward greater entropy. **Entropy, S** (J K^{-1}), is a state function that measures the amount of chaos, disorder, or randomness in a system. Entropy increases as the temperature or volume of a system increases, on mixing pure substances together, and on melting or vaporization. The **second law of thermodynamics** states that the total entropy of a system and its surroundings increases during any spontaneous process.

The **third law of thermodynamics** states that, at absolute zero, the entropy of a perfectly ordered crystal of a pure substance is zero. The **standard entropy,** $S°$, is the entropy of one mole of substance at 1 atm pressure. Standard entropy is a positive quantity for all elements and compounds. The **standard entropy of reaction** is found from $\Delta S°_{rxn} = \Sigma[n \cdot S° \text{ (products)}] - \Sigma[n \cdot S° \text{ (reactants)}]$.

The state function **free energy, G,** is defined so as to combine enthalpy and entropy: $\Delta G = \Delta H - T\Delta S$. The second law requires that the free energy of a system decreases during any spontaneous change. A chemical process for which $\Delta G_{rxn} < 0$ proceeds spontaneously (although possibly at an undetectably slow rate). If $\Delta G_{rxn} > 0$, the reaction cannot take place spontaneously. If $\Delta G_{rxn} = 0$, the reaction is in *equilibrium.* The **standard free energy of formation,** $\Delta G_f°$, of a compound is the free energy change of its *formation reaction,* in which one mole of a compound is the sole product, and pure elements in their **standard states** (the most stable state at 25 °C and 1 atm) are the only reactants. $\Delta G_f° = 0$ for elements, and may be positive or negative for compounds. The **standard free energy of reaction** is found from $\Delta G°_{rxn} = [\Sigma n \cdot \Delta G_f° \text{(products)}] - [\Sigma n \cdot \Delta G_f° \text{(reactants)}]$.

PROBLEMS

General

1. What is the difference between thermodynamics and thermochemistry?
2. Define the term "work" as it is used in chemistry and physics.
3. What is the most common type of work in chemical processes?
4. Define the terms "kinetic" and "potential" energy. Can they be interconverted? If so, give an example. If not, why not?
5. Define heat. What units are commonly used to measure heat?
6. Define the following terms, as they are used in thermodynamics: "system," "surroundings."

7. The phrase, "states of matter," is used to refer to solid, liquid, and gas. Distinguish between this usage and the usage of the term "state" in thermodynamics.
8. What is a state function? Give examples.
9. What is Hess's law, and how is it related to the fact that enthalpy is a state function?
10. Define internal energy. In what unit is it measured?
11. State the first law of thermodynamics, in words rather than as an equation.
12. Give a qualitative description of enthalpy. Why are both ΔE and ΔH used to describe changes in a system?
13. Suppose the work in some process is given as -200 J. What meaning should be attached to the negative sign?
14. What is the relationship between pressure, volume, and the work done on a system?
15. Give an algebraic statement of the first law of thermodynamics.
16. What is the common symbol for the phrase "heat required by an endothermic reaction taking place at constant volume"? At constant pressure?
17. If a reaction is characterized by $\Delta H_{rxn} = -500$ kJ mol^{-1}, does it absorb heat from the surroundings or release heat to them?
18. What two driving forces underlie all spontaneous chemical or physical processes? What thermodynamic property (state function) is associated with each of them?
19. What property of systems is described by entropy?
20. What is free energy, and how is it defined?
21. What is meant by the term "standard state"?
22. Give a precise definition of "formation" equation and ΔG_f°.
23. Show that it is a consequence of the definition of "free energy of formation" that $\Delta G_f^\circ = 0$ for any element in its standard state.
24. How may standard free energy of reaction be determined from standard free energies of formation?
25. What is the relationship between free energy change, spontaneity, and equilibrium?
26. What are the units of energy, enthalpy, entropy, and free energy?

Heat, Work, and the First Law

27. How much work is involved in moving an object a distance of 5.00 meters against an opposing force of 3.30 N?
28. If a rubber band is stretched so that its length increases by 2.00 cm, and the force required is 4.00 N, how much work is done on the band?
29. How far can an object be moved against a force of 0.753 N if 100 J is available?
30. Suppose an object is moved a distance of 6.00 feet, and the work done is 100 J. What is the magnitude of the force that opposes the motion?

*31. Calculate the work required to move a 150-lb person up one flight of stairs (9.0 feet). The force of gravity is 9.8 N kg^{-1}.

*32. Suppose an astronaut, who together with his space suit weighs 240 lbs on the earth, climbs a 50 m-high hill on the moon. If the astronaut does 9.1 kJ work, what is the force of gravity on the moon? Express your answer in N kg^{-1}.

*33. One liter (1 kg) of water acquires 0.57 kJ of kinetic energy when it drops over a 190-foot waterfall (the height of Niagara Falls). If all this energy is converted to heat, by how much does the temperature of the water rise? The specific heat of water is 4.184 J g^{-1} K^{-1}. (It is said that Joule, while honeymooning in the Alps, attempted an experimental measurement of this effect.)

34. If the energy required to heat one gram of water from 25 °C to 50 °C were used instead (as work) to lift the water against the force of gravity, to what height could the water be lifted? The specific heat of water is 4.184 J g^{-1} K^{-1}.
35. What is the change in internal energy of a system if 100 J heat is added and the system does 50 J work on the surroundings?
36. A heated bar of metal transfers 75 J heat to the surroundings, and does no work. What is the value of ΔE for the bar?
37. A gas is compressed by the expenditure of 50 J work, and at the same time 50 J heat is added. What is ΔE for the gas?
38. A liquid substance is vaporized by the addition of 40.4 kJ heat. Because the vapor has a greater volume, the substance does 1.14 kJ work on the surroundings when it vaporizes. What is the change in internal energy of the substance?
39. A gas expands against a pressure of 50 kPa, in the process increasing its volume by 0.015 m^3. What is ΔE of the gas?
40. A gas is compressed by a pressure of 5.00 atm, and in the process its volume decreases by 6.35 L. What is ΔE of the gas?

Standard Entropy, Entropy of Reaction, Free Energy of Formation, and Free Energy of Reaction

41. Use data from Appendix G to determine ΔS°_{rxn} when acetaldehyde ($S^\circ = 38.3$ J K^{-1}) burns according to the reaction

$$CH_3CHO(\ell) + 3\ O_2(g) \longrightarrow 2\ CO_2(g) + 2\ H_2O(\ell)$$

42. Use data from Appendix G to determine ΔS°_{rxn} for the reaction

$$C(s) + H_2O(g) \longrightarrow CO(g) + H_2(g)$$

*43. Use data from Appendix G to calculate the change in entropy when 10.0 g of HgO(s) decomposes, under standard conditions, according to $2\ HgO(s) \rightarrow 2\ Hg(\ell) + O_2(g)$.

*44. Given that

$$N_2(g) + 2\ O_2(g) \longrightarrow 2\ NO_2(g); \quad \Delta S^\circ = -83.45\ J\ K^{-1}$$

$$2\ NO(g) + O_2(g) \longrightarrow 2\ NO_2(g); \quad \Delta S^\circ = -146.54\ J\ K^{-1}$$

calculate the entropy of the reaction $N_2(g) + O_2(g) \rightarrow 2\ NO(g)$. Do not use data from Appendix G.

45. The standard entropy of reaction for $PCl_3(g) + Cl_2(g) \rightarrow PCl_5(g)$ is -170.27 J K^{-1}. What is the standard entropy of $PCl_5(g)$?

46. The standard entropy of reaction of hydrogen with allene, $C_3H_4(g)$, to give propane, $C_3H_8(g)$, is -49.72 J K^{-1}. What is the standard entropy of allene?

47. Use data from Appendix G to calculate ΔG°_{rxn} for the thermite reaction, $2\ Al(s) + Fe_2O_3(s) \rightarrow 2\ Fe(s) + Al_2O_3(s)$.

48. Use data from Appendix G to calculate ΔG°_{rxn} for the reaction $CaO(s) + H_2O(\ell) \rightarrow Ca(OH)_2(aq)$.

49. Use data from Appendix G to calculate ΔG°_f for $SnO_2(s)$, given that, for $SnO_2(s) + 2\ CO(g) \rightarrow 2\ CO_2(g) + Sn(s)$, $\Delta G^\circ_{rxn} = 5.2$ kJ.

50. Use data from Appendix G to calculate ΔG°_f for NOCl(g), given the fact that NOCl dissociates to NO(g) and $Cl_2(g)$ when heated, with a standard free energy change of 20.47 kJ per mole of NOCl dissociated.

51. For the reaction $PbO(s) + C(s) \rightarrow Pb(s) + CO(g)$, $\Delta H^\circ_{rxn} = +106.8$ kJ and $\Delta S^\circ_{rxn} = +188.0$ J K^{-1}. What is the value of ΔG°_{rxn}? Do not use values of ΔG°_f from Appendix G.

52. For the reaction $N_2H_4(\ell) + O_2(g) \rightarrow N_2(g) + 2\ H_2O(\ell)$, $\Delta H^\circ_{rxn} = -622.29$ kJ and $\Delta S^\circ_{rxn} = +5.08$ J K^{-1}. What is the value of ΔG°_{rxn}? Do not use values of ΔG°_f from Appendix G.

53. The standard entropy of the reaction $PbS(s) + \frac{3}{2}\ O_2(g) \rightarrow PbO(s) + SO_2(g)$ is -81.99 J K^{-1}, and ΔG°_{rxn} is -389.4 kJ. Calculate the value of the standard enthalpy of reaction, without using values of ΔH°_f from Appendix G.

54. For the reaction of sodium and liquid water to form hydrogen and aqueous sodium hydroxide, ΔS°_{rxn} and ΔG°_{rxn} are -7.68 J K^{-1} and -182.02 kJ, respectively, per mole of sodium. How much heat is liberated when 0.100 g Na reacts with excess water? Do not use any values from Appendix G.

55. If the enthalpy change for a reaction is 24.9 kJ mol^{-1} and the entropy change is 75 J mol^{-1} K^{-1} at 298 K, what is the free energy change at 298 K? At 100 K? (Assume that neither ΔH nor ΔS changes with temperature. For most reactions, this assumption is only approximately valid.)

56. If the entropy change of a reaction is -150 J mol^{-1} K^{-1} and the enthalpy change is -50.0 kJ mol^{-1} at 298 K, what is the free energy change at 298 K? At 500 K? (Assume that neither ΔH nor ΔS changes with temperature. For most reactions, this assumption is only approximately valid.)

Spontaneity, Entropy, and Free Energy

57. For each of the following processes, determine the sign of ΔS and explain your reasoning.
    a. Antifreeze is added to the water in an automobile radiator.
    b. A film of ice forms on a puddle of water.

58. For each of the following processes, determine the sign of ΔS and explain your reasoning.
    a. $N_2O_4 \rightarrow 2\ NO_2$
    b. A cloud forms in the sky.

59. Suppose a process occurs at 25 °C with $\Delta H = 100$ kJ and $\Delta S = 100$ J K^{-1}. Is the reaction spontaneous at this temperature? At what temperature will the reaction be in equilibrium?

60. Suppose a process occurs at 25 °C with $\Delta G = 10.0$ kJ and $\Delta S = 200$ J K^{-1}. Is the reaction spontaneous at this temperature? At what temperature will the reaction be in equilibrium?

*61. Suppose a reaction is at equilibrium at 298 K.
    a. For which of the possible combinations of algebraic signs of ΔS and ΔH will the reaction be spontaneous in the forward direction when the temperature is increased?
    b. For which of the combinations of signs will the reaction become spontaneous if the temperature is decreased?

CHAPTER 11

LIQUIDS AND SOLIDS

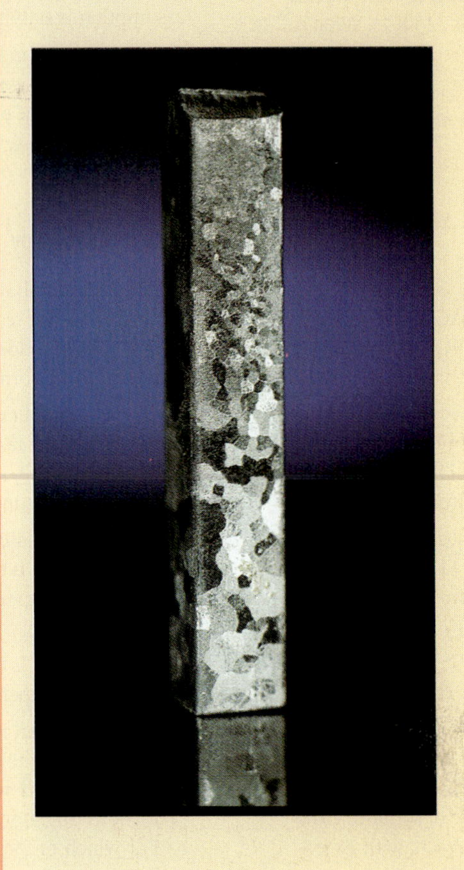

Each of the irregularly shaped regions in this bar of very pure aluminum consists of a single crystal. The different orientation of the crystals gives rise to differences in the way they reflect and scatter light.

11.1 STATES OF MATTER

We begin our study of the liquid and solid states by asking some fundamental questions about states of matter. Why is matter sometimes gaseous and sometimes liquid or solid? What are the properties of liquids and solids, and how can they be explained? Why and under what circumstances does matter undergo changes from one physical state to another? Our answer to these questions begins with the forces that molecules exert on each other.

Intermolecular Attractions

Gases consist of widely separated molecules whose attraction for one another is too weak to have a significant effect on gas behavior. But the strength of these intermolecular attractive forces depends on the distance between the molecules—if the pressure on a gas is increased, pushing the molecules closer together, the forces increase in strength. Considerable departures from the ideal gas law are observed in real gases at high pressure.

Similar effects occur if the temperature of a gas changes. At high temperature the average kinetic energy of the molecules is great enough so that their motion is unhindered by the weak intermolecular forces. At low temperature the average kinetic energy is less, and molecular motion is affected by intermolecular forces. Departures from ideal-gas behavior increase as the temperature decreases.

Condensation

When a gas is cooled, the average energy of molecular motion gradually decreases. At some point an abrupt change takes place in the nature of the collisions that occur between gas molecules. Instead of bouncing away, colliding molecules begin to stick together, because the average kinetic energy is insufficient to overcome the attractive intermolecular forces. Small clusters of molecules form, and rapidly grow in size as colliding gas molecules continue to stick. Clusters coalesce with one another, and ultimately most of the molecules of the sample are grouped together in a single mass. This process is called **condensation.** The substance is no longer a gas, but has become what is called a **condensed phase:** the term refers to both liquids and solids.

The arrangement of molecules in liquids and solids is more orderly than in gases. The *entropy* of condensed phases is less than that of the gas phase.

Essentially the same train of events takes place if, instead of being cooled, a gas undergoes an increase in pressure. Because the strength of the intermolecular forces increases as the molecules approach one another, there comes a point where the forces are sufficient to bind the molecules together even though the average energy of motion may still be relatively high. Clusters form, grow, and coalesce, and the result is the formation of the condensed phase.

Condensation of a gas occurs on either compression or cooling. The condensed state is characterized by tightly packed molecules, in contrast to the widely separated molecules of a gas. The forces responsible for the existence of solids and liquids are called **cohesive forces.**

Several types of cohesive forces are discussed in Section 11.3.

Condensed Phases: The Kinetic Molecular Theory View

In both the liquid and solid state, the particles (which may be atoms, molecules, or ions) of a substance are in close contact with one another. Moreover, the molecules are in motion. They have an average kinetic energy that is proportional to the temperature, just as it is in gases. In liquids, the particles are free to move from one location to

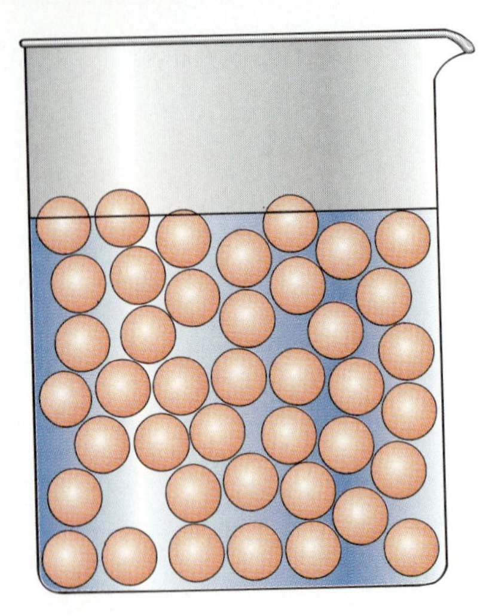

Figure 11.1 The molecules of a liquid are in contact with one another, but randomly arranged. This imperfect packing results in occasional holes. Random sliding motion carries molecules throughout the container.

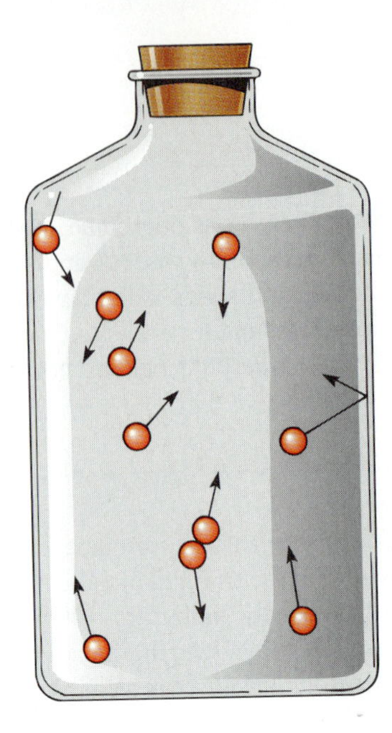

Figure 11.2 Gas molecules are widely separated from one another. They move rapidly in random, straight-line motion that is interrupted by occasional collisions.

another, and their motion is random as they slip and slide past one another. This is quite different from gases, in which molecules move in straight lines and only occasionally contact other molecules. The way in which liquid molecules are packed together is random, and there are some locations, called "holes," that are momentarily unoccupied by any molecules at all. Liquid structure is illustrated in Figure 11.1. For comparison, the structure of the gaseous state is illustrated in Figure 11.2.

Liquids have some holes in them.

The entropy of solids is less than that of liquids.

By contrast, the particles of a solid substance are locked in place, and there is almost no movement of molecules from one location to another. Yet the particles of a solid also have an average kinetic energy proportional to the temperature. The motion in solids is almost entirely *vibrational*—the particles vibrate back and forth, but very seldom leave their original location. In some solids the particles are neatly packed in regular arrays, while in others the packing is random. Figure 11.3 illustrates the molecular structure of a solid.

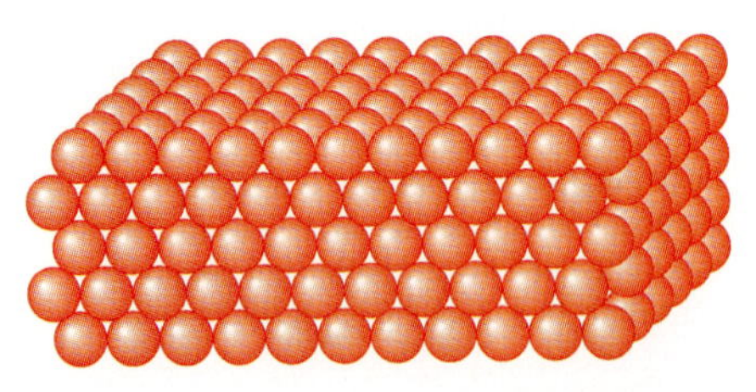

Figure 11.3 The molecules of a solid are in close contact with one another. Their only motion is vibrational, and they are confined to a specific location. Only rarely do they move to another location within the solid.

11.2 PROPERTIES OF LIQUIDS

The general physical properties of liquids—fluidity, viscosity, and surface tension—are a direct consequence of the properties of the molecules themselves. Molecular size and shape, the strength of the intermolecular forces, and the way in which the molecules are packed together, all help determine the behavior of liquids.

Fluidity

The most obvious characteristic of liquids is that they are **fluids,** which means that they can be stirred, splashed, passed through a sieve, and poured from one container to another. The word "fluid" is derived from the Latin word meaning "to flow." The molecules of a liquid are in contact with one another at all times, and there is not much free space between them. The total volume occupied by a given amount of liquid is approximately equal to the sum of the volumes of the individual molecules, and is not affected by the shape of its container. The molecules of a liquid occupy all the nooks and crannies of the container.

Gases are fluids, because they also can flow.

Liquids are not very compressible, but their volume does decrease slightly when they are subjected to a large increase in pressure. Also, in common with almost all substances, they expand somewhat when their temperature is raised.

Viscosity

Although all liquids are fluid, they vary in their degree of fluidity. This variation is noticeable in the ease with which they may be poured. The property of a liquid that determines its resistance to flow is its **viscosity.** Highly viscous liquids resist flow, while liquids of low viscosity flow easily. Common examples can be found at the breakfast table. Syrup and honey have high viscosity, and they flow over the pancakes very slowly. In contrast, low-viscosity liquids such as milk, juice, or coffee, when spilled, spread out over the table all too rapidly.

The viscosity of a liquid is affected by the intermolecular forces that hold the molecules together, and by the size and shape of its molecules as well. Small, compact molecules like H_2O and CH_3OH slide easily over one another, and these liquids have relatively low viscosities. Long, flexible molecules (for example heavy oils and greases) become entangled with one another and strongly resist flow.

Measurement of viscosity provides information about the shape of molecules.

For all but a few liquids, viscosity decreases as temperature increases. Since the energy of molecular motion increases as temperature increases, it becomes easier for molecules to move about. A practical consequence of this temperature dependence is that lubricating oils and greases for machines and vehicles must be designed for specific operating temperatures and environmental conditions.

Surface Tension

When a liquid such as water is in an open dish, the molecules at the surface are in an environment quite different from the environment of the molecules in the bulk of the liquid below the surface. As shown in Figure 11.4, surface molecules are strongly attracted to the molecules just beneath them, and only slightly attracted to the few gas molecules that happen to be near the surface. The intermolecular forces are unbalanced, and the surface molecules are drawn more tightly toward the bulk of the liquid. The result is a surface layer, only a few molecules thick, that is denser and more cohesive than the bulk of the liquid. A liquid behaves as if it were covered with a tough, elastic skin. This property of liquids is called **surface tension.** Liquids (like water) with strong intermolecular forces have a high surface tension.

It takes energy to increase the surface area of a liquid, just as it takes energy to increase the surface area of a rubber balloon. Surface tension tends to pull the surface into the shape that minimizes the surface area. There are no irregularities in the surface of a liquid at rest, because any mounds, bulges or ripples would increase the surface area. Soap bubbles and small water droplets are spherical (rather than cubical, for instance) because a sphere has the smallest possible surface area for a given volume

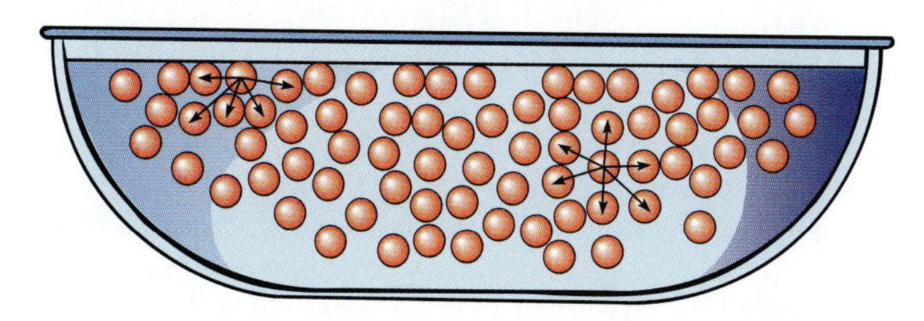

Figure 11.4 Molecules in the bulk of the liquid are subject to forces that, on average, are the same in all directions. But because of the unbalanced forces that exist at the surface of a liquid, the molecules in the surface layer are drawn strongly inward. This results in a surface layer which is more dense and less easily penetrated, as if the surface were covered by a tough skin.

Figure 11.5 (a) The surface tension of a soap solution pulls the surface of soap bubbles into the shape having the least surface area for a given volume—the sphere. (b) Tiny dewdrops on a blade of grass or spiderweb also assume a spherical shape.

(Figure 11.5). The high surface tension of water is utilized by a class of insects that live and move on the surface of ponds and puddles (Figure 11.6). Water striders weigh so little that their feet never penetrate the surface "skin" of the water. For these insects, water is as sturdy as a concrete pavement is for us. Another demonstration of the existence of a surface film is that, with careful technique, a needle or a razor blade can be made to float on water (Figure 11.7).

In quantitative discussions of surfaces, the *coefficient of surface tension* is defined as the energy required to increase the surface area of a liquid by 1 m^2.

Another effect that is governed by intermolecular forces is liquid *wetting* of a surface. Water does not wet the surface of a freshly waxed car (Figure 11.8). Water molecules are only weakly attracted to wax molecules, so surface tension, unhindered, pulls the water into spherical droplets. Other surfaces, notably clean glass and the old paint of an unwaxed car, exert a stronger force on water molecules. The intermolecular forces between two different materials are called **adhesive** forces. The *cohesive* forces within the water droplets are less than the *adhesive* forces between water and the unwaxed surface. As a result the water droplets spread out and wet the surface.

Wetting of a surface occurs when adhesive forces are stronger than cohesive forces.

Figure 11.6 Because of surface tension, water striders never get their feet wet. The surface layer of a pond is so tough that, for a light-weight insect, it is as solid as concrete.

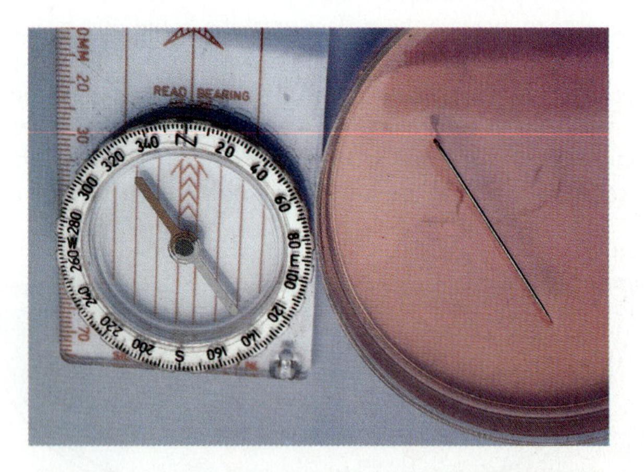

Figure 11.7 If carefully placed, a needle can be floated on the surface of water. If the needle is first magnetized by rubbing it with a magnet, it will align itself in a north–south direction, forming a compass.

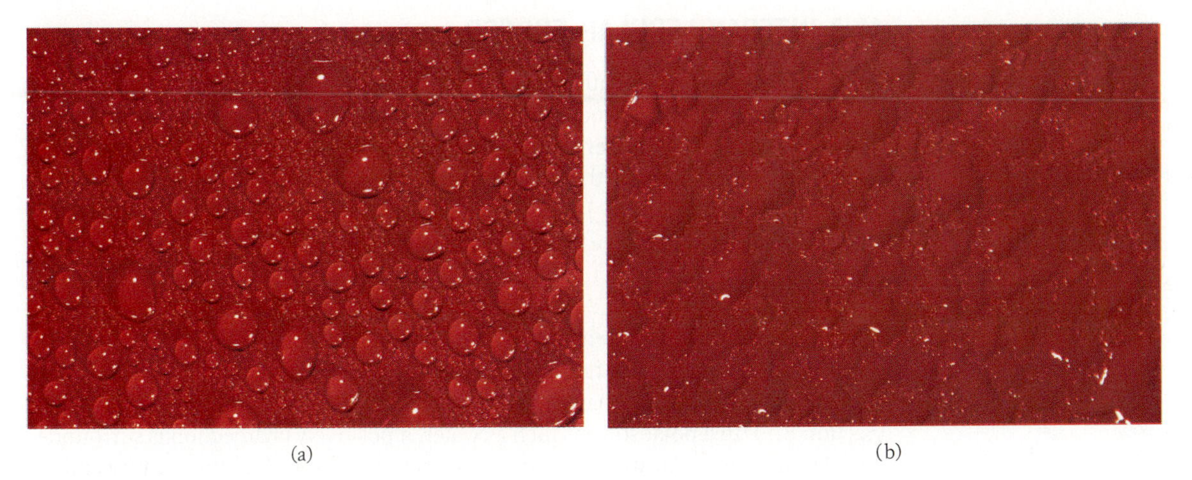
(a) (b)

Figure 11.8 (a) Water droplets do not adhere well to a waxed surface. Surface tension pulls the droplets into a near-spherical shape. Larger drops are flattened by gravitational force. (b) The stronger adhesive forces of an unwaxed surface causes water droplets to spread out, thus maximizing the attraction and "wetting" the surface.

Capillary action is another effect that depends on intermolecular forces. Water rises in a narrow glass tube (a capillary) because the adhesive forces to glass are stronger than the cohesive forces of water (Figure 11.9). This can be seen even in large-diameter glass tubes, such as a laboratory buret. The surface forms a *meniscus* (Latin: *lens*) as the edges of the water surface, attracted to the glass by adhesive forces, climb up the wall of the tube. Not all liquids behave this way. In mercury, the adhesive forces are weaker than the cohesive forces, so the edges of a glass-enclosed mercury surface cannot rise as high as the center. A mercury meniscus curves downward at the edge. Another example of capillary action is the absorption of water by a cloth or paper towel, which may be thought of as containing many small bundles of tiny tubes (Figure 11.10).

Figure 11.9 The strong adhesive forces between water and glass pull the surface up against the force of gravity (left). In mercury, the cohesive forces predominate, so the center of the column rises higher than the edges (right).

Figure 11.10 Strong adhesion between water and cloth or paper fibers pulls water into a towel.

11.3 INTERMOLECULAR FORCES

The name van der Waals first appeared in our discussion of the behavior of real gases (page 210).

Molecules of the same, or different, type are attracted to one another by several kinds of forces. These forces, which are responsible for both *adhesion* and *cohesion*, are known collectively as **van der Waals forces.** Although all are electrical forces between opposite charges, each of the several types of van der Waals force arises from a different cause.

Charge-Dipole Forces

The polar bonds between atoms of different electronegativity have *bond moments* whose vector sum is the *dipole moment* of the molecule.

Molecules composed of atoms of different elements often have an unsymmetrical distribution of electrical charge, or *dipole moment.* Molecules of such *polar* compounds are electrically neutral, but have a positively charged end and a negatively charged end (Section 8.7). In a polar solvent such as water, a positively charged ion is surrounded by polar molecules. An attractive force exists between the ion and the negative end of a water molecule. The positive end of the water molecule is repelled by the ion, but the force is weaker because the distance is greater. The net result, called the **charge-dipole force,** is attractive. These interactions are diagrammed in Figure 11.11. Similarly, there is an attractive force between a negative ion and the positive end of a water molecule. The charge-dipole force, although weaker than the force between oppositely charged ions, is the strongest of the van der Waals forces. The properties of aqueous solutions of ionic compounds are determined in large measure by charge-dipole forces.

Dipole-Dipole Forces

As shown in Figure 11.12, the oppositely charged portions of two polar (but uncharged) molecules give rise to an attractive force when the molecules are aligned in a head-to-tail fashion. The attractive force, called a **dipole-dipole force,** is weaker than

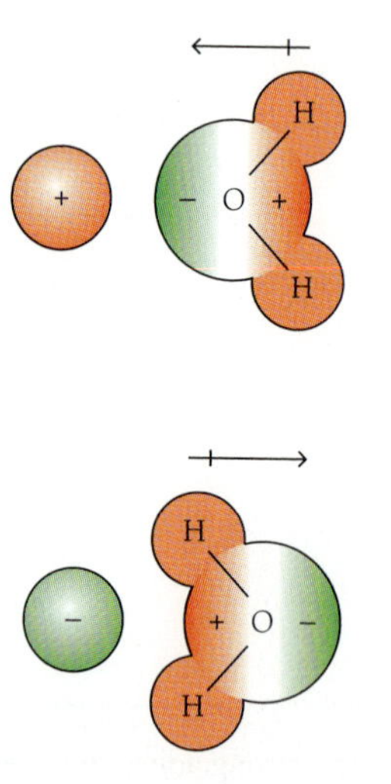

Figure 11.11 A net attractive force exists between an ion and a polar molecule. The ion is strongly attracted to one end of the molecule, and weakly repelled by the more distant end.

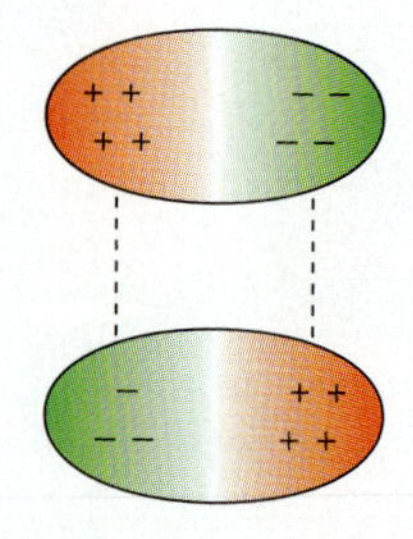

Figure 11.12 Oppositely charged ends of polar molecules attract one another, leading to a net attractive force between polar molecules. This attractive force exists even though the molecules carry no net electrical charge.

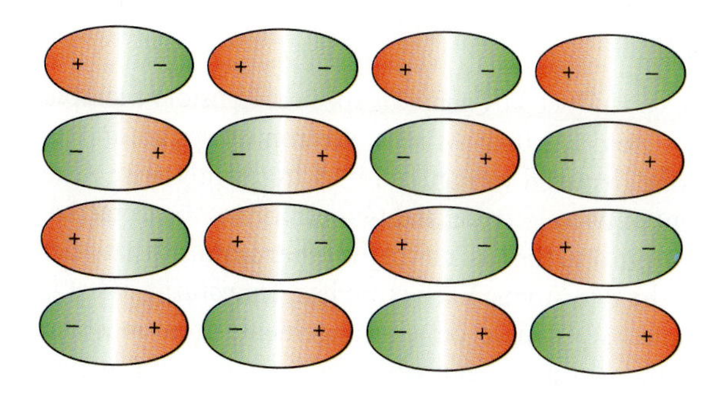

Figure 11.13 Molecules of a polar solid are arranged in a head-to-tail fashion that maximizes dipole-dipole interactions and creates a cohesive force.

a charge-dipole force. In the condensed state, particularly in a solid, polar molecules are arranged in a way that maximizes dipole-dipole attractions (Figure 11.13).

Electrical forces exist between uncharged polar molecules as well as between ions.

The arrangement of molecules in a polar liquid is irregular, and constantly changing, but the head-to-tail alignment of adjacent molecules occurs much more frequently than the head-to-head alignment. Thus on the average there is a net attraction, due to dipole-dipole forces, which causes polar substances to cohere strongly, both in the solid and the liquid state.

The words *cohesion, cohesive, cohere,* and *coherent* all come from the Latin verb cohærere, meaning "to stick together."

Induced Dipole Forces

A single atom is spherical in shape and has a symmetrical distribution of electrical charge. It is *nonpolar*. However, if it is brought near a charge, for instance a negative ion or the negative end of a dipolar molecule, the atom's charge distribution will *become* unsymmetrical. The atom's electron cloud will be repelled by and move away from the negative charge, while the atom's positive nucleus will be attracted and move closer (Figure 11.14). The originally nonpolar atom *acquires* a dipole moment, and it becomes polar. Such dipoles, caused by influences from outside the molecule, are called **induced dipoles.** The term is meant to distinguish these dipoles from the **permanent dipoles** of polar molecules like HCl. However, an induced dipole is still a dipole, and it is attracted by a neighboring ion by a *charge-induced dipole* force or to a neighboring polar molecule by a *dipole-induced dipole* force.

Induced dipoles arise in solutions of nonpolar substances in polar solvents, for example oxygen or argon dissolved in water. The forces are weaker than those arising from permanent dipoles, and, precisely because the forces are weak, nonpolar compounds are not very soluble in polar solvents.

The factors affecting solubility are discussed more thoroughly in Chapter 12.

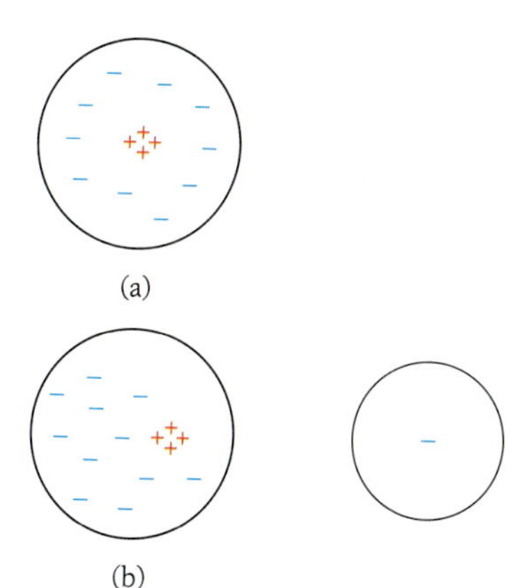

Figure 11.14 The presence of a nearby charge can induce a dipole in an otherwise nonpolar species. (a) An isolated atom, with symmetrical distribution of charge. (b) The same atom when near a charge.

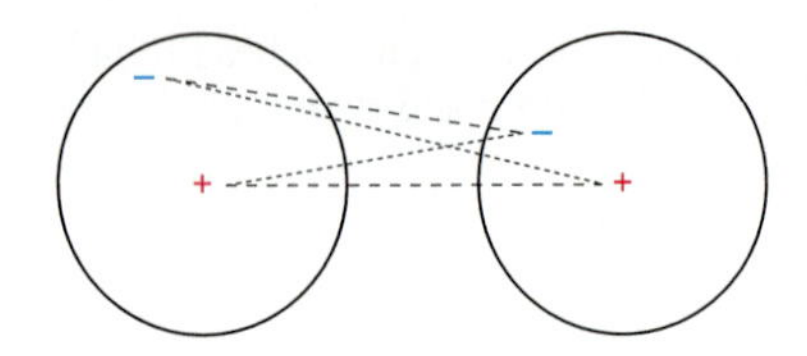

Figure 11.15 This instantaneous, temporary arrangement of electrons in adjacent hydrogen atoms results in a net attraction, because the attractive forces (dotted lines) are stronger than the repulsive interactions (dashed lines). Averaged over all possible arrangements, the result is the *dispersion* force, a weak attraction between the two atoms.

The dispersion force is named for **Fritz London** (1900–1954), the German physicist who first explained it.

Dispersion or London Forces

Surprisingly, dipole moments are also induced by *nonpolar* species. Consider a simple case, two hydrogen atoms that are close together. As each electron moves around its nucleus, arrangements like the one depicted in Figure 11.15 frequently arise. The strongest force in this arrangement of two electrons and two protons is the attractive force between the electron in one atom and the nucleus of the other atom, so that a net force of attraction exists between the two atoms. Each atom has, *temporarily,* what is called an *instantaneous* dipole moment. Of course, other arrangements occur in which there is a repulsive force between the two atoms. But when the force is averaged over all possible locations of the electrons in both atoms, the result is a small net attractive force. This attraction is called the **dispersion** or **London force.** The dispersion force accounts for the cohesion of all nonpolar liquids. It is also responsible for the condensation, at low temperature, of nonpolar gases such as Cl_2, O_2, N_2, and the noble gas elements. Since the dispersion force arises from electrons, it is generally stronger in compounds or elements with many electrons. Other things being equal, the dispersion force increases with molar mass.

Hydrogen Bonding

A special type of dipole-dipole interaction exists in certain hydrogen-containing compounds. When a hydrogen atom is bonded to a small atom of large electronegativity, the bonding electrons are pulled so far away from the H atom as to leave it almost bare. Under these circumstances the hydrogen atom behaves almost like an isolated proton, and as such is attracted to the nonbonded electrons (lone pairs) of an atom on a neighboring molecule. This attraction results in a close association known as a **hydrogen bond.** Despite the name, hydrogen bonds are not true covalent bonds, because they do not involve the sharing of an electron pair between two atoms. Dashed or dotted lines are used to indicate hydrogen bonding (Figure 11.16). Normally stronger than both London and dipole-dipole forces, H-bonds have perhaps one-twentieth to one-tenth the strength of an ordinary covalent bond.

There is some evidence that Cl and S also are capable of hydrogen bonding. If so, the interaction is very much weaker than that with N, O, and F.

Hydrogen bonding occurs only in compounds in which a hydrogen atom is bonded to N, O, or F. A hydrogen bond can be thought of as a unit in which a hydrogen atom is shared between any two of N, O, or F atoms. The sharing is not equal, and the H atom is closer to the atom to which it is covalently bonded than to the atom to which it is hydrogen bonded.

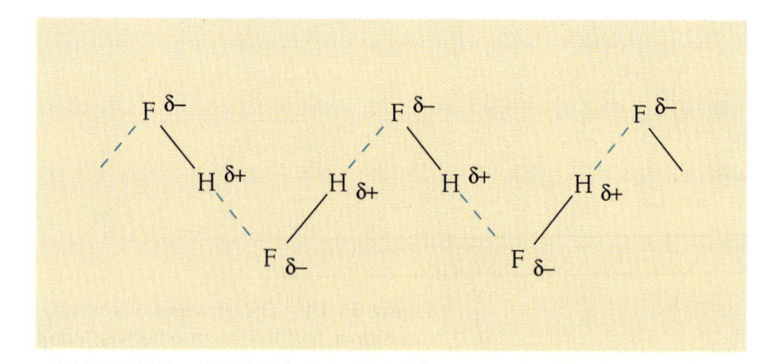

Figure 11.16 Solid hydrogen fluoride consists of zig-zag chains in which individual molecules are held together by hydrogen bonds.

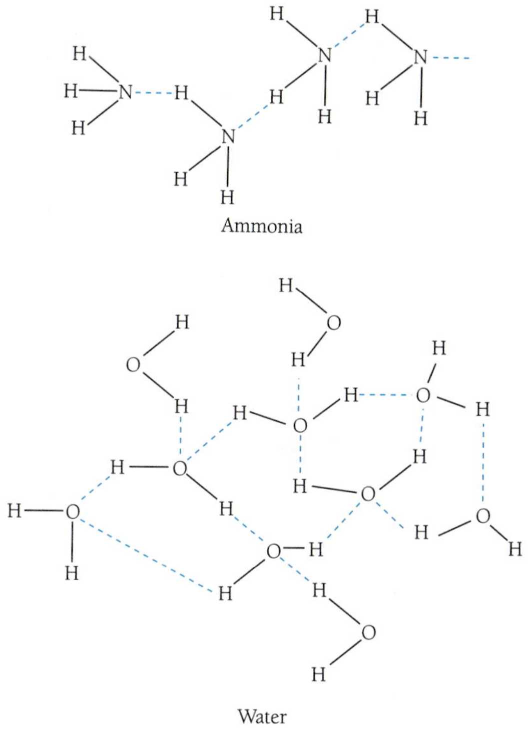

Ammonia

Water

Figure 11.17 The cohesive forces in liquid NH_3 and H_2O are due primarily to the existence of hydrogen bonding, in which a hydrogen atom, covalently bonded to N or O, is in addition strongly attracted to another N or O atom on a neighboring molecule. The effect is particularly strong in H_2O, because each molecule can be H-bonded to as many as four others.

Hydrogen bonding is responsible for the relatively strong cohesion of liquid H_2O and NH_3, as shown in Figure 11.17. The cohesion is particularly strong in H_2O, because, having two lone pairs and two H atoms, most water molecules are hydrogen bonded to four others. The cohesion of NH_3 is weaker, because the average molecule interacts with only two others.

Hydrogen bonding is an important feature of the structure of DNA, which is responsible for the preservation and transmission of genetic information from one generation of plants and animals to the next (Chapter 21).

A Summary of van der Waals Forces

Nonbonding intermolecular forces arise from electrical interactions between charges, dipoles, and induced dipoles. The magnitude of the dipole moment induced in an atom or molecule increases with the number of electrons, and it depends also on the type and proximity of the species responsible for the induced dipole. The strongest forces are exerted by ions, and the weakest forces are exerted by induced dipoles. Van der Waals forces, and typical systems in which they dominate the overall cohesive forces of liquids, are summarized in Table 11.1.

As a general rule, the strength of intermolecular forces decreases in the order of their listing in Table 11.1. This order does not hold in all cases, because the actual strength of the forces depends on the value of the permanent dipole moment and on molar mass. The dipole-dipole forces are stronger than the London forces in small, highly polar molecules like H_2O and HCN.

TABLE 11.1 Intermolecular Forces of Attraction

Type	Description	Typical Systems
Ion-dipole	Force between charged particles and molecules with a permanent dipole moment	Ionic compounds dissolved in polar solvents, such as NaCl in water
Hydrogen bonding	Especially strong dipole-dipole interaction between molecules containing at least one H atom bonded to N, O, or F	H_2O, NH_3, CH_3OH, HF, and mixtures of these substances
Dispersion or London force	Force between temporary dipoles induced in each other by otherwise nonpolar compounds	All substances; it is the only cohesive force in nonpolar substances like C_2H_6, O_2, and Ne
Dipole-dipole	Force between molecules having a permanent dipole moment	HBr, CCl_2F_2
Induced dipole	Force between nonpolar species and charged or polar species	Nonpolar compounds dissolved in polar solvents; I_2 in H_2O; also ionic compounds dissolved in nonpolar solvents: NaCl in CCl_4

CCl_2F_2 is one of the CFC's (chlorinated fluorocarbons) responsible for the threat to the earth's ozone layer.

EXAMPLE 11.1 Identification of Forces

What forces account for the cohesion in CHF_3?

Analysis Table 11.1 gives the forces found in different species. It is necessary to know whether the compound is ionic, and, if not, whether the molecules have a dipole moment and/or are capable of hydrogen bonding. This information can be obtained from the Lewis structure and the VSEPR theory.

Work CHF_3 cannot be ionic, because it contains no Group 1A or 2A elements (or ammonium ion). The structural formula of CHF_3 is

```
     F
     |
 H—C—F
     |
     F
```

The molecule is tetrahedral, and has a nonzero dipole moment. Hydrogen bonding is not possible, since the H-atom is not bonded to N, O, or F. From Table 11.1, therefore, the only forces are dipole-dipole and dispersion.

Criteria for deciding whether or not a molecule has a dipole moment are given in Section 8.7.

Exercise

What cohesive forces exist in MgF_2?

Answer: Charge-induced dipole and London forces are present, but are very much weaker than the main cohesive force, the ionic bond.

11.4 PHASE TRANSITIONS

When a substance changes from one physical state to another, as in melting or boiling, the process is called a **phase transition.** The theory of phase transitions is closely related to the kinetic molecular theory.

Melting and Freezing

When a liquid is cooled, the average kinetic energy of its molecules decreases and at a certain temperature becomes insufficient to overcome the attractive forces exerted by neighboring molecules. The sliding motion of the molecules ceases, and the substance becomes a solid, incapable of flow (Figure 11.18). The phenomenon is known as **freezing,** and the temperature at which it occurs is called the **freezing point.**

When a solid substance melts, its molecules become free to move past one another.

When a solid substance is warmed, it eventually reaches a temperature at which the vibrational motion of its molecules is sufficiently energetic to overcome the forces holding the particles in their fixed locations. Further addition of heat does *not* result in a temperature increase, but causes an increasing number of particles to break free and begin to move past one another. The solid becomes a liquid, in a process called **melting** or **fusion.** The temperature at which this occurs is called the **melting point (mp).** The melting point and the freezing point are the same for a given substance. (It is usually called the melting point when it is above room temperature, and the freezing point when it is below room temperature.) After the phase transition is complete, further addition of heat once again causes a temperature increase. The process is diagrammed in Figure 11.19, which also indicates the boiling process discussed below.

Figure 11.18 It is not strictly true that solids cannot flow. Subjected to the extreme forces generated by continental drift, solid rock can be deformed, and over a long period of time, flow into a new shape, These rock strata originally were in flat, parallel layers.

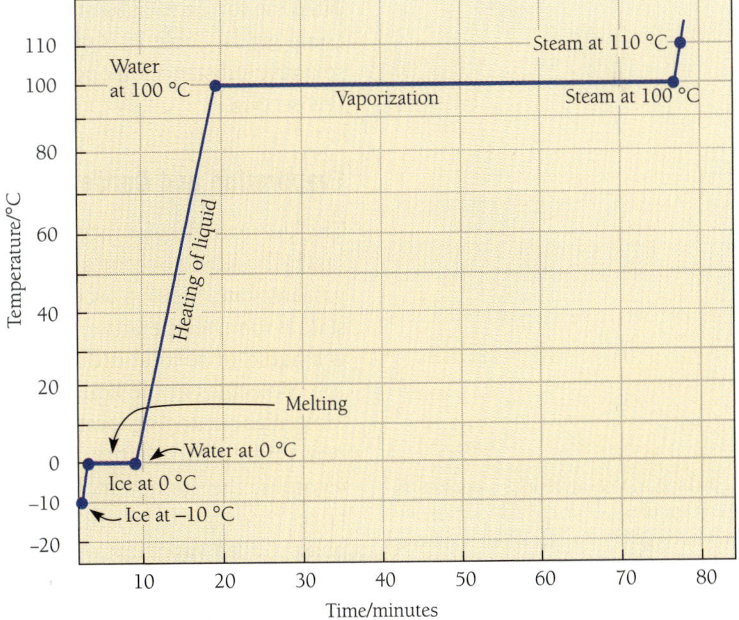

Figure 11.19 Temperature change caused by the addition of heat at a constant rate of 40 J min^{-1} to 1 gram of ice at −10 °C. The temperature remains constant during melting and vaporization.

TABLE 11.2 Melting and Boiling Points

Substance	Type of Force*	mp/°C	bp/°C
H_2O	HDL	0	100
CH_3CH_2OH (ethanol)	HLD	−117	79
$(CH_3)_2C{=}O$ (acetone)	LD	−95	56
C_6H_6 (benzene)	L	6	80
NH_3	HDL	−78	−33
HF	HDL	−83	19
HCl	DL	−114	−85
CCl_4	L	−23	77
Ar	L	−189	−186
H_2	L	−259	−253
$TiCl_4$†	L	−25	136
O_2	L	−218	−183

* The types of intermolecular force are indicated as follows: H, hydrogen bonding; D, dipole-dipole; L, London (dispersion). When there is more than one, the first force listed is the predominant cohesive force.

† Even though it is a metal halide, titanium tetrachloride is a nonpolar, covalent compound.

Strong cohesive forces result in high melting points.

The melting point of a solid is a physical property that depends on the cohesive forces that hold the particles in place. If the forces are strong, higher average kinetic energy, and thus higher temperature, must be attained before the substance melts. Conversely, weak cohesive forces result in low melting points. Melting point is an important indicator of the strength of intermolecular attractive forces. Because it is different for each substance, melting point is used to recognize and characterize compounds. Table 11.2 gives the melting point and type of cohesive force for representative substances, along with the normal boiling points (bp) to be discussed later in this section.

Evaporation and Condensation

The kinetic molecular theory also provides a framework for understanding the process in which a liquid becomes a vapor. The terms **vapor** and **gas** both refer to matter in the gaseous state. "Gas" is used to describe a substance, like oxygen, that is in the gaseous state at room temperature, while "vapor" is used to describe a substance, like H_2O, that is a liquid or solid at ordinary temperature and pressure.

Molecules in the liquid are continually jostled and bumped by their neighbors. Molecules occupying locations at the surface of a liquid are on occasion bumped from beneath by a particle having well above average energy. If the collision is sufficiently vigorous, the surface molecule acquires enough energy to overcome the cohesive forces binding it to its neighbors. It is literally bumped out of the liquid into the vapor phase. Called **vaporization** or **evaporation,** the process occurs more frequently as the temperature is increased, because the average energy of molecular collisions is proportional to temperature.

A surface molecule is like a person at the edge of a dense crowd—interested in what's going on, yet often separated from the crowd by its sometimes vigorous movement.

The reverse process, *condensation,* occurs when gas molecules strike the surface of the liquid and stick there rather than bouncing back into the vapor phase. Condensation rate is proportional to the rate at which vapor-phase molecules collide with the surface, which in turn increases proportionally as the pressure increases.

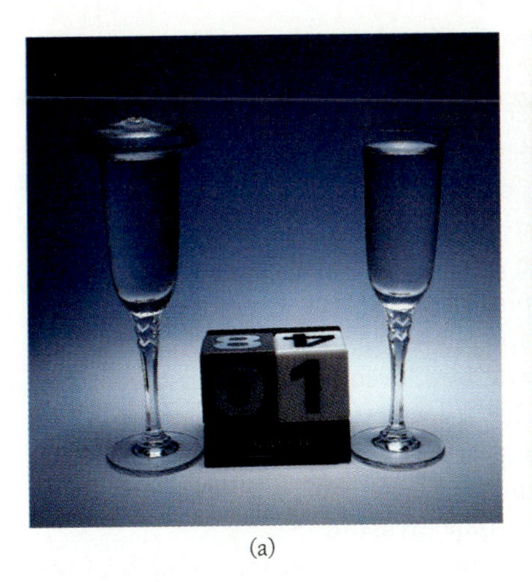
(a)

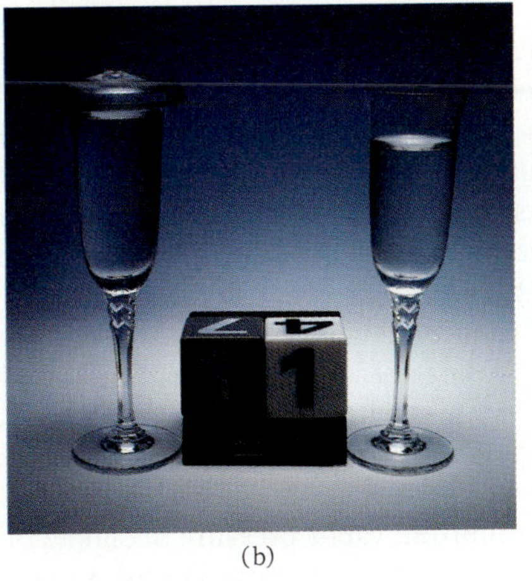
(b)

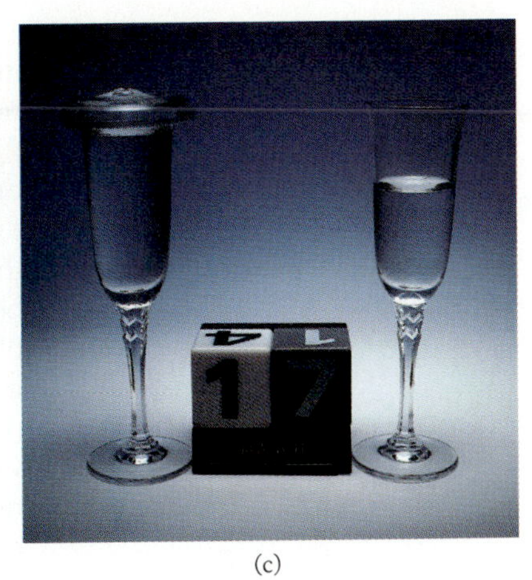
(c)

Figure 11.20 The water level in a covered container remains the same day after day. There is no net transfer of water from the liquid to the vapor phase, and, of course, no loss of water from the container. The open container loses water by evaporation, at a rate that depends on temperature, humidity, and air currents.

Vapor Pressure of Liquids

If water is left in an open container, eventually it all evaporates. Liquids, like water, that evaporate readily are called **volatile.** Not all liquids are volatile—some evaporate so slowly that they can be left in an open dish for years without noticeable loss.

Motor oil is an example of a nonvolatile liquid.

Suppose that a volatile liquid is put into a *closed* container. The level of the liquid in a partly filled, stoppered bottle does not change as time passes, there is no apparent evaporation (Figure 11.20). According to the KMT, however, the processes of vaporization and condensation take place continually. If the amount of liquid remains the same, it can only mean that the processes of vaporization and condensation take place at exactly the same rate. The liquid and the vapor are in **equilibrium.** This condition is called **dynamic equilibrium,** to emphasize the fact that there is a great deal of activity even though there is no observable change (Figure 11.21).

The surface seems calm, but there is continual traffic as the molecules pass back and forth between the liquid and vapor phases.

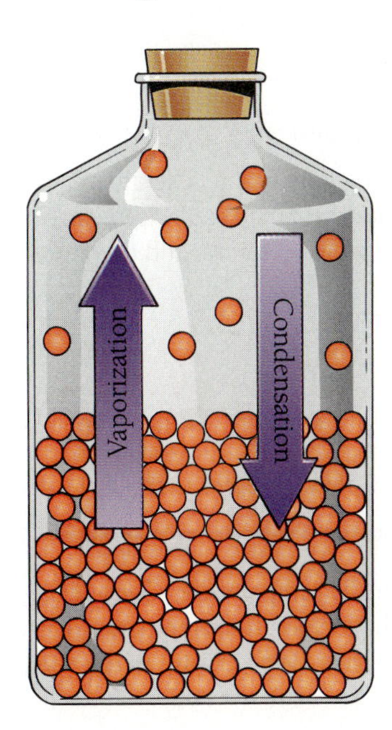

Figure 11.21 According to the KMT, the level of the liquid in a closed container does not change, because the rate at which molecules enter the vapor phase (rate of vaporization) is the same as the rate at which molecules leave the vapor phase (rate of condensation).

The rate of vaporization of the molecules of a liquid increases as the temperature increases, but is not affected by the pressure. On the other hand, the rate of condensation of the vapor molecules is not greatly affected by the temperature, but increases as the pressure of the vapor increases. Consider what happens when a liquid is introduced into an otherwise empty, but closed, container. It begins to evaporate, and at first there is no condensation because there are no molecules in the vapor phase. But as the evaporation proceeds, the pressure in the vapor phase increases and condensation begins. As the process continues, the vapor-phase pressure increases as long as the rate of condensation is less than the rate of evaporation. But as long as the pressure is increasing, the rate of condensation increases as well. Eventually the pressure reaches the point at which the rate of condensation has reached the same value as the rate of vaporization, and *there can be no further change in pressure*—the system is at equilibrium. The pressure at which a liquid is in equilibrium with its own vapor is called its **equilibrium vapor pressure** or simply the **vapor pressure.**

The value of the equilibrium vapor pressure depends on the strength of the cohesive forces in the liquid. Liquids with strong cohesive forces have low vapor pressures, that is, they are less volatile. Study of vapor pressure is a means of learning about intermolecular forces, and vapor pressure was an important clue in the discovery of hydrogen bonding (see page 470).

EXAMPLE 11.2 Vapor Pressures and Forces

Of the two compounds, formic acid (HCOOH) and acetic acid (CH_3COOH), which would you expect to have the higher vapor pressure (at a given temperature)?

Analysis Since compounds with weak cohesive forces have higher vapor pressures, this problem (like Example 11.1) involves identification of the forces in each compound and estimating their relative strength.

Work The structural formulas of the two compounds

```
        O                   H     O
       //                   |    //
  H—C           and     H—C—C
       \                    |    \
        O—H                 H     O—H

   formic acid              acetic acid
```

show that both are polar, and both are capable of hydrogen bonding. The difference lies in London forces, which generally increase as molar mass increases. Acetic acid, with the greater molar mass, should have the stronger cohesive forces, and therefore the smaller vapor pressure at any given temperature.

Exercise

Which has the lower vapor pressure, NO or O_2?

Answer: NO. (It is polar, and therefore has stronger cohesive forces than O_2.)

Equilibrium vapor pressure depends strongly on the temperature. Consider a closed container containing a liquid in equilibrium with its vapor. If the temperature is increased, the rate of evaporation increases while at first there is no change in the rate of condensation. This imbalance of rates results in an increase in the number of molecules in the vapor phase, that is, an increase in the pressure. The increased pressure is accompanied by an increase in the rate of condensation. The system soon reaches a new state of dynamic equilibrium, characterized by a higher pressure as well as increased temperature. Some values of the vapor pressure of water are given in Table 11.3. The equilibrium vapor pressure of *all* liquids increases as the temperature is increased. This behavior is illustrated in Figure 11.22.

TABLE 11.3 Equilibrium Vapor Pressure of Water Near Room Temperature

T/°C	P_{eq}/torr
5	6.5
10	9.2
15	12.8
20	17.5
25	23.8
30	31.8
35	42.2
40	55.3

Relative Humidity

There are two important differences between vaporization of a liquid in open and closed containers. First, equilibrium is rarely attained in an open container, as the vapor molecules simply diffuse into the open atmosphere, in effect leaving the system. Vaporization occurs, but there is little or no condensation. Second, the gas phase above the surface is a *mixture* of air and vapor, whether or not equilibrium is attained. If there is equilibrium, it is not the total pressure but the *partial pressure* of the vapor in the gas mixture that is equal to the equilibrium vapor pressure of the liquid. For example, at a temperature of 20 °C (68 °F) the equilibrium vapor pressure of water is 17.5 torr. If the barometric pressure that day is 755 torr, then atmospheric air *in equilibrium with liquid water* consists of water vapor at a partial pressure of 17.5 torr and other gases at (755 − 17.5) = 737.5 torr. When the partial pressure of water vapor in the atmosphere is equal to the equilibrium vapor pressure, the air is said to be "saturated," or to have a relative humidity of 100%.

The relative humidity of the atmosphere is usually less than 100%, because in the natural environment, atmospheric water vapor is never in equilibrium with lakes, rivers, and oceans. **Relative humidity (RH)** is defined in terms of the actual or measured partial pressure (P) of water in the atmosphere and the equilibrium vapor pressure (P_{eq}) as

$$RH = \left(\frac{P}{P_{eq}}\right) \times 100 \tag{11-1}$$

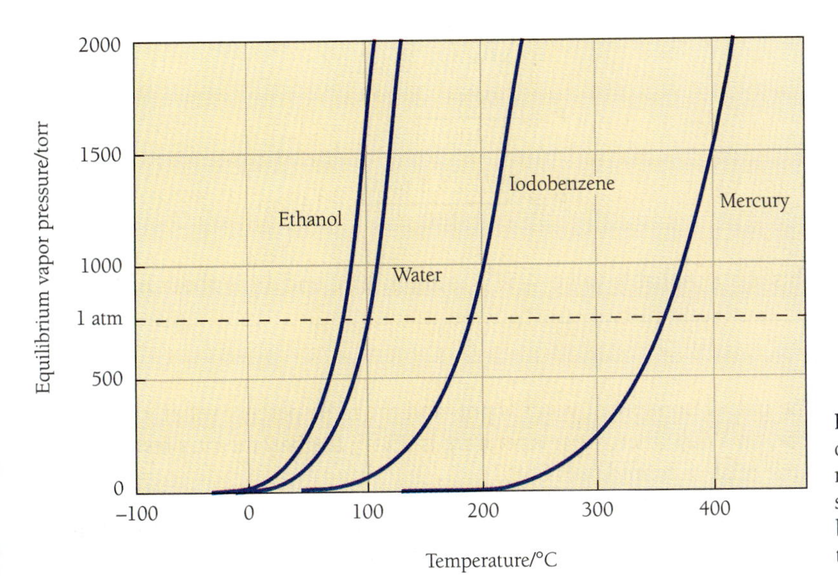

Figure 11.22 The equilibrium vapor pressures of ethanol, water, iodobenzene (C_6H_5I), and mercury are plotted vs. temperature. The general shape of these plots is the same for all liquids, but the pressures corresponding to the same temperature vary widely.

EXAMPLE 11.3 Relative Humidity and Partial Pressure of Water Vapor

On a day when the relative humidity is 57% and the temperature is 77 °F, what is the partial pressure of water vapor in the atmosphere?

Analysis

Target: A partial pressure

$$? \text{ torr} =$$

Knowns: *RH* and temperature.

Relationship: Equation (11-1) relates the target to the knowns, but the given information is insufficient.

Missing Information: P_{eq}, the equilibrium vapor pressure of water at the given temperature.

Plan Leading backward from the target to the knowns, the stepwise solution plan is as follows.

3. Insert the values of *RH* and P_{eq} into Equation (11-1) and solve for *P*.
2. Look up the value of P_{eq} in Table 11.3, but first complete Step 1.
1. Convert 77 °F to temperature in °C.

Work

1.

$$T\ °\text{C} = (T\ °\text{F} - 32)\left(\frac{5}{9}\right)$$

$$T\ °\text{C} = (77 - 32)\left(\frac{5}{9}\right) = 25$$

2. From Table 11.3, P_{eq} at 25 °C is 23.8 torr.
3. Equation (11-1) is rearranged, the known values are inserted, and the pressure is evaluated:

$$RH = \left(\frac{P}{P_{eq}}\right) \times 100$$

$$P = \frac{(RH)(P_{eq})}{100}$$

$$P = \frac{(57)(23.8 \text{ torr})}{100} = 14 \text{ torr}$$

Check This is a reasonable value—57% is about half, and 14 is about half of 24.

Exercise

What is the partial pressure of water vapor, and the mole percent water vapor in the atmosphere, on a day when the temperature is 95 °F, the barometric pressure is 747 torr, and the relative humidity is 85%?

Answer: $P(H_2O) = 36$ torr, and atmospheric air contains 4.8 mol% water vapor.

EXAMPLE 11.4 Calculation of Relative Humidity

When the outside temperature is 59 °F and the partial pressure of water vapor is 10 torr, what is the relative humidity?

Analysis As in the preceding example, Equation (11-1) supplies the relationship between the target (RH) and the knowns [T(°F) and $P(H_2O)$]. Again, Table 11.3 must be consulted to find the appropriate value of P_{eq}.

Plan

1. Convert the temperature to °C.
2. Look up the corresponding value of P_{eq}.
3. Calculate RH from Equation (11-1).

Work

1.

$$T\,°C = (T\,°F - 32)\left(\frac{5}{9}\right)$$

$$T\,°C = (59 - 32)\left(\frac{5}{9}\right) = 15$$

2. From Table 11.3, at 15 °C P_{eq} = 12.8 torr.

3.

$$RH = \left(\frac{P}{P_{eq}}\right) \times 100$$

$$= \left(\frac{10 \text{ torr}}{12.8 \text{ torr}}\right) \times 100 = 78\%$$

Exercise

What is the relative humidity on a day when the partial pressure of water vapor is 6.5 torr and the temperature is 41 °F?

Answer: 100%. (It's a cold, rainy day.)

The *dew point* of moist air is closely related to the relative humidity. When air with RH less than 100% is cooled, the partial pressure of water vapor remains the same. But because P_{eq} decreases as the temperature decreases, RH ($= 100P/P_{eq}$) *increases* as the temperature drops. Ultimately a temperature is reached at which $P_{eq} = P$, and any further cooling results in the condensation of water vapor as dew. The temperature at which condensation begins is called the **dew point.** As found in Example 11.3, for instance, the partial pressure of water vapor is 14 torr at 25 °C and RH = 57%. Table 11.3 shows that P_{eq} = 14 torr when the temperature is somewhere in the range 15 °C to 20 °C. The exact dew point is 16.4 °C (61.5 °F) for this particular sample of air.

Dew point and relative humidity both describe the water content of moist air.

Boiling

When a volatile liquid in an open container is heated, evaporation proceeds more and more rapidly as the temperature increases. At some specific temperature the evaporation suddenly becomes violent and much more rapid. Bubbles of vapor form below the surface of the liquid, grow in size, and rise to the surface of the liquid where they burst, releasing large quantities of the substance into the vapor phase. The phenomenon is known as **boiling,** and the temperature at which it takes place is called the **boiling point.** Boiling of a liquid in an open container occurs when the equilibrium vapor pressure of the liquid is equal to the atmospheric pressure. Boiling points of different liquids vary widely: nitrogen boils at −196 °C, lead boils at 1620 °C.

Boiling is a constant-temperature process.

A liquid cannot be heated to a temperature greater than its boiling point. Heat added to a liquid at its boiling point is used to overcome the cohesive forces, and causes more liquid to vaporize, but the temperature of the remaining liquid does not increase (Figure 11.19).

The KMT provides an explanation for this marked change in the evaporation process when a liquid reaches its boiling point. Consider a bubble, or cavity, beneath the surface of the liquid. Vaporization and condensation take place across the surface of the cavity, just as at the main surface above. The gas inside the cavity contains no air—it is pure vapor, and its pressure is the equilibrium vapor pressure of the liquid at that temperature. If this vapor pressure is less than the confining pressure, that is, the pressure of the atmosphere bearing down on the liquid, the cavity will collapse and be squeezed out of existence. Consequently, cavities do not form when the vapor pressure is low. If the vapor pressure is equal to or greater than the confining pressure, however, cavities grow rather than collapse. Vaporization takes place throughout the body of the liquid, not just at the surface—the liquid **boils.**

Boiling takes place throughout the bulk of a liquid, while evaporation occurs only at the surface.

Because the equilibrium vapor pressure of a liquid increases as the temperature increases, the temperature at which boiling takes place depends on the confining pressure experienced by the liquid. For a liquid in an open container, the confining pressure is the atmospheric pressure. If the confining pressure is increased, a liquid's temperature must be increased before its equilibrium vapor pressure becomes equal to the confining pressure. If the pressure is decreased, the equilibrium vapor pressure becomes equal to the confining pressure at a lower temperature. At a pressure of 1520 torr (2 atm), for instance, water does not boil until its temperature reaches 120 °C. At a pressure of 400 torr, on the other hand, water boils at 83 °C. In Denver, Colorado, which lies 1600 m above sea level, the average atmospheric pressure is 630 torr rather than 760 torr. Water boils at 95 °C in Denver, and because the temperature is lower, it takes somewhat longer to cook foods by boiling.

The temperature at which the vapor pressure of a liquid is equal to 760 torr is called the **normal boiling point.** Inclusion of the word "normal" distinguishes it from the *boiling point,* the temperature at which the vapor pressure is equal to the confining pressure (whatever that may be).

Intermolecular Forces and Boiling Points; Hydrogen Bonding

Melting points and boiling points are both high when strong cohesive forces are present.

The strength of the cohesive forces in a liquid affects both the vapor pressure and the normal boiling point. Liquids with strong forces have a low vapor pressure and a high boiling point. Look again at Table 11.2 and note the relationship between boiling point and the type of cohesive force of a compound. Because the strength of London dispersion forces increases as the number of electrons in a species increases, it is generally true that boiling point increases with molar mass. When the boiling points of

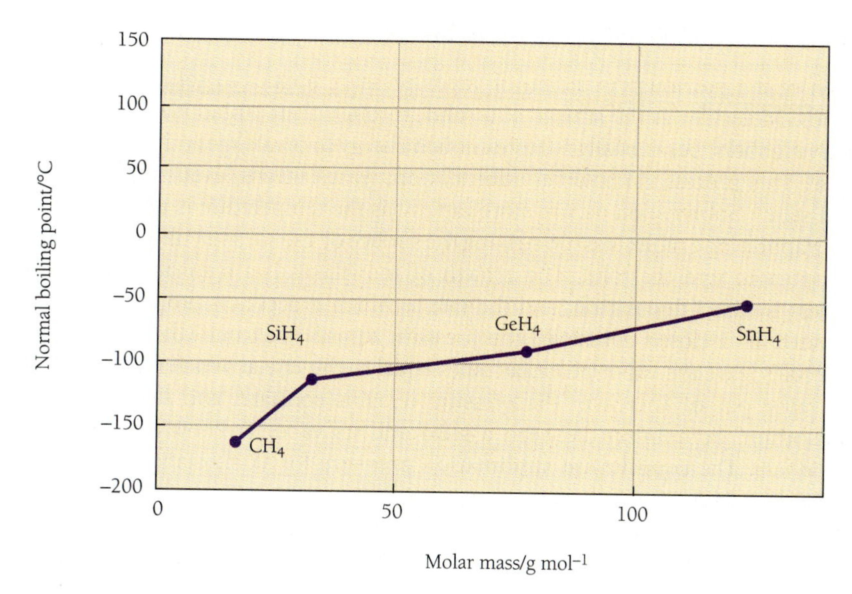

Figure 11.23 Boiling points of the Group 4A hydrides increase as molar mass increases.

the Group 4A hydrides CH_4, SiH_4, GeH_4, and SnH_4 are plotted against molar mass, as in Figure 11.23, this trend is apparent. The same trend exists in other sets of similar substances, such as the halogens or the noble gases.

Molecules capable of forming hydrogen bonds are important exceptions to this trend. When boiling point is plotted against molar mass of the Group 5A, 6A, and 7A hydrides NH_3, PH_3, AsH_3, and SbH_3; H_2O, H_2S, H_2Se, and H_2Te; HF, HCl, HBr, and HI (Figure 11.24), it is apparent that NH_3, H_2O, and HF do not fit the pattern. Their boiling points are higher than expected, and it can be inferred that the cohesive forces in these liquids are anomalously strong. The difference is hydrogen bonding, found only in compounds containing N, O, or F. Trends in boiling points supply very strong evidence for the existence of hydrogen bonding.

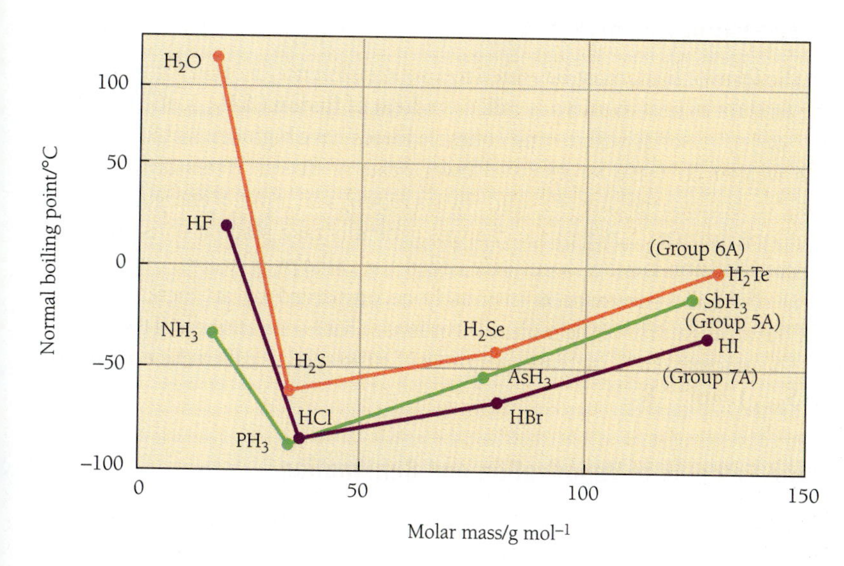

Figure 11.24 Boiling point versus molar mass among the Group 5A, 6A, and 7A hydrides. The smallest molecule in each group has an unexpectedly high boiling point, as a result of strong hydrogen bonding.

Sublimation and Vapor Pressure of Solids

"Sublimation" has a different meaning in psychology courses.

"Deposition" means something different to lawyers.

Ionic solids have exceedingly low sublimation pressures because their cohesive forces are very strong.

Much of the previous discussion of vapor pressure applies to solids as well as to liquids. Molecules at the surface of a solid, through their vibrational contacts with their neighbors, occasionally acquire enough energy to break free and enter the vapor phase. This process is known as **sublimation,** which means "evaporation from the solid state." Sublimation occurs more rapidly as the temperature is increased.

Vapor-phase molecules, on striking the surface of a solid, sometimes stick and are incorporated into the solid. The growth of snowflakes is a familiar example. The process is called **deposition,** and the rate at which it occurs is proportional to the pressure. In a closed container, the pressure adjusts itself until the rate at which molecules leave the vapor phase becomes equal to the rate at which molecules enter the vapor phase. Thereafter the pressure remains constant, and the system is at equilibrium. Equilibrium between a solid and its vapor is known as *sublimation equilibrium.* The **equilibrium sublimation pressure** for all substances increases as temperature increases.

When a solid in an open container is heated, the equilibrium sublimation pressure increases until it becomes equal to the confining pressure of the atmosphere. Further heat input increases the *rate* of sublimation without increasing the temperature of the solid. This temperature is known as the **sublimation temperature,** or the **normal sublimation temperature** if the atmospheric pressure is 760 torr.

Solid carbon dioxide, or "dry ice," is a familiar example of a substance that sublimes. At 1 atm pressure, the temperature of dry ice cannot rise above the normal sublimation temperature, −78 °C, so that it is an excellent refrigerant for frozen foods. The variation of equilibrium sublimation pressure of dry ice with temperature is given in Table 11.4.

TABLE 11.4 Equilibrium Sublimation Pressure of CO_2

T/°C	P_{eq}/torr
−134	1
−120	10
−109	40
−100	100
−86	400
−78	760
−69	1520

11.5 THERMODYNAMICS OF PHASE TRANSITIONS

The thermodynamic concepts of enthalpy, entropy, and free energy were developed in the preceding chapter. These concepts are immediately applicable to processes in which a substance changes from one physical state to another.

Enthalpy and Entropy of Fusion

In some texts and handbooks, ΔH_{fus} values are given in cal g^{-1} or kcal mol^{-1}.

The distinctions between heat, enthalpy, and energy are discussed in Section 10.3.

Both enthalpy and entropy of fusion are always positive quantities.

Energy, in the form of heat, must be added to a solid substance in order for melting to occur. The amount of heat required is called the **heat of fusion (ΔH_{fus})**, and the units are J mol^{-1} or J g^{-1}. Solids with strong cohesive forces have high heats of fusion, and solids with weaker cohesive forces have lower heats of fusion. Heats of fusion are always positive, because melting of a solid always requires an input of energy, that is, it is an *endothermic* process. Since phase changes generally take place at constant pressure, the heat of fusion is more accurately called the *enthalpy of fusion.*

Melting of a solid substance is invariably accompanied by an increase in the disorder of the arrangement of the molecules or ions. That is, the entropy of the system increases. The amount of the increase, per mole of substance, is called the **entropy of fusion (ΔS_{fus}, J mol^{-1} K^{-1}).**

Entropy and Enthalpy of Vaporization and Sublimation

Like the fusion process, vaporization of a liquid is invariably accompanied by an increase in disorder. The entropy of the system increases, and **entropy of vaporization (ΔS_{vap}, J mol^{-1} K^{-1})** is always a positive quantity.

TABLE 11.5 Phase Transition Data

Substance	Type of Force*	mp/°C	ΔH_{fus}/kJ mol^{-1}	bp/°C	ΔH_{vap}/kJ mol^{-1}
H_2O	HDL	0	6.00	100	40
CH_3CH_2OH (ethanol)	HLD	−117	4.80	79	39
$(CH_3)_2C{=}O$ (acetone)	LD	−95	5.68	56	30
C_6H_6 (benzene)	L	6	9.90	80	30
NH_3	HDL	−78	7.69	−33	23
HF	HDL	−83	4.58	19	28
HCl	DL	−114	2.00	−85	16
CCl_4	L	−23	3.28	77	30
Ar	L	−189	6.70	−186	6
H_2	L	−259	0.12	−253	1
$TiCl_4$	L	−25	9.37	136	34
O_2	L	−218	0.44	−183	7

* The types of intermolecular force are indicated as follows: H, hydrogen bonding; D, dipole-dipole; L, London (dispersion). The first force listed in the predominant cohesive force.

Evaporation of a liquid requires energy, in the form of heat. The **heat of vaporization (ΔH_{vap})**, has units of J g^{-1} or J mol^{-1}. Like fusion, vaporization usually takes place at constant pressure, and so it is more correctly called *enthalpy of vaporization.* For all liquids, ΔH_{vap} is a positive quantity. In the inverse process, when a vapor condenses to a liquid, heat is given off. This is why burns caused by steam are more severe than burns caused by boiling water. It is not so much that steam can be at a higher temperature, but that on contact with skin the steam condenses to liquid water, in the process liberating some 2.2 kJ of heat for each gram of steam. This extra heat causes the greater damage to the skin.

In common with melting points, normal boiling points, and vapor pressures, ΔH_{fus} and ΔH_{vap} increase as the strength of the cohesive forces increases. Some representative values are given in Table 11.5.

The situation is essentially the same in the process of sublimation. Energy is required for a molecule to break free of solid-state cohesive forces and enter the vapor phase, and sublimation is always endothermic. The required energy is known as the **heat of sublimation (ΔH_{sub}**, in J mol^{-1} or J g^{-1}). Similarly, ΔS_{sub} (J mol^{-1} K^{-1} or J g^{-1} K^{-1}) is always positive.

Free Energy of Phase Transitions

We noted earlier (Section 10.6) that the free energy change ΔG associated with any change is zero when the system is at equilibrium. *Phase transitions are equilibrium processes,* and therefore $\Delta G = 0$. Free energy is defined by the equation $\Delta G = \Delta H - T\Delta S$, so that Equation (11-2) is true for *any* phase transition.

$$\Delta S = \Delta H/T \tag{11-2}$$

This relationship provides a convenient means of calculating ΔS_{fus}, ΔS_{vap}, or ΔS_{sub} from the value of the corresponding ΔH. As always in thermodynamics, temperature must be expressed in kelvins. Some values of ΔS_{vap} calculated from Equation (11-2) are given in Table 11.6.

TABLE 11.6 Entropies of Vaporization at the Normal Boiling Point

Substance	ΔH_{vap}/kJ mol^{-1}	nbp/K	ΔS_{vap}/J mol^{-1} K^{-1}
H_2O	40.0	373	107
C_6H_6	30	353	85
HCl	16	188	85
CCl_4	30	350	86
$TiCl_4$	34	409	83
$(CH_3)_2CO$	30	329	91

Exceptions are often more interesting than the rules they break.

It is immediately apparent that there is little variation in ΔS_{vap}, whether the compounds are polar or not. Except for water, all values are in the range 87 ± 4 J mol^{-1} K^{-1}. Apparently, a comparable amount of disorder is generated when *any* liquid vaporizes. The generalization that $\Delta S_{vap} \approx 87$ J^{-1} mol^{-1} K^{-1} is called *Trouton's rule,* and it holds for most compounds whose normal boiling points are in the range 200 to 400 K. ΔS_{vap} of water is much greater than that of other liquids because the molecules of liquid water are linked by extensive H-bonding into networks of many molecules. Although the networks are temporary and shift as the molecules move about, they are always present, creating a considerable degree of order. The entropy of liquid water is therefore *less* than that of liquids incapable of hydrogen bonding, and consequently the entropy *increase* on vaporization is greater.

Anomalously high entropies of vaporization are found in other hydrogen-bonded liquids as well. For both HF and NH_3, $\Delta S_{vap} = 96$ J mol^{-1} K^{-1}.

The following two examples illustrate some of the calculations involving phase transitions.

EXAMPLE 11.5 Heat of Fusion and Vaporization

How much heat is required (a) to melt one gram of ice, and (b) to vaporize one gram of water?

Analysis This problem is a review of some of the examples in Chapter 4.

Target: An amount of heat,

$$? \text{J} =$$

Knowns: The substance and the processes; the molar mass is an implicit known.

Relationship: The amount of heat required is proportional to the amount of substance

$$q = (n \text{ moles}) \cdot (\Delta H \text{ J mol}^{-1})$$

Missing Information: The values of ΔH; these may be obtained from Table 11.5.

Work

a.

$$\begin{aligned} q &= n \cdot \Delta H_{fus} \\ &= (1.0 \text{ g}) \cdot \left(\frac{1 \text{ mol}}{18 \text{ g}}\right) \cdot (6.0 \text{ kJ mol}^{-1}) \\ &= 0.33 \text{ kJ} = 330 \text{ J} \end{aligned}$$

b.

$$q = n \cdot \Delta H_{vap}$$
$$= (1.0 \text{ g}) \cdot \left(\frac{1 \text{ mol}}{18 \text{ g}}\right) \cdot (40 \text{ kJ mol}^{-1})$$
$$= 2.2 \text{ kJ}$$

Check Estimation of the answers gives (a) 6/18 = 1/3 = 0.33; and (b) 40/18 ≈ 2. The units cancel correctly.

Note: Look again at Figure 11.19, which shows temperature changes caused by addition of heat to 1 g water. The lengths of the lines corresponding to vaporization and melting are in the ratio 2.2 : 0.33, or ≈ 6.7 : 1. In general, the heat required for vaporization is much greater than for melting.

Exercise

Calculate the heat required to vaporize 125 g benzene (C_6H_6).

Answer: 48 kJ.

EXAMPLE 11.6 Entropy of Vaporization

At its normal boiling point of 58 °C, the heat of vaporization of 2,3-dimethylbutane (C_6H_{14}) is 29.8 kJ mol^{-1}. Calculate the entropy change when 50.0 g of this compound is vaporized at 58 °C.

Analysis This example is similar to the preceding problem, with one difference—the required value of ΔS_{vap} for 2,3-dimethylbutane is not available in any table in this book.

Relationship: Vaporization is an equilibrium process, which means that $\Delta G_{vap} = 0$ and $\Delta S_{vap} = \Delta H_{vap}/T_{nbp}$.

Work

$$\Delta S_{vap} = \frac{\Delta H_{vap}}{T_{nbp}}$$
$$= \frac{(29.8 \text{ kJ mol}^{-1})}{(58 + 273)\text{K}}$$
$$= 0.0900 \text{ kJ mol}^{-1} \text{ K}^{-1}$$
$$= 90.0 \text{ J mol}^{-1} \text{ K}^{-1}$$

The entropy change accompanying the vaporization of 50.0 g is

$$\Delta S = (50.0 \text{ g}) \cdot \left(\frac{1 \text{ mol}}{86.2 \text{ g}}\right) \cdot (90.0 \text{ J mol}^{-1} \text{ K}^{-1})$$
$$= 52.2 \text{ J K}^{-1}$$

Check Trouton's rule ($\Delta S_{vap} \approx 87$ J mol^{-1} K^{-1}) indicates that 90 J mol^{-1} K^{-1} is a reasonable value. Since 50 g is about half a mole, 52 J K^{-1} is about right.

Exercise

Calculate ΔS for the vaporization of 1.00 gram of metallic sodium. The heat of vaporization is 89.6 kJ mol^{-1} at the normal boiling point, 883 °C.

Answer: 3.37 J K^{-1}.

11.6 PHASE DIAGRAMS

We have seen that phase transitions occur, and equilibria exist, between the solid and liquid, the solid and vapor, and the liquid and vapor. Figure 11.25 shows some examples of these transitions. In the following section we will examine some additional aspects of phase transitions and equilibria.

More on Vapor Pressure Plots

Graphs of vapor pressure vs. temperature, such as Figure 11.26 for water, carry a great deal of information about the substance. Each point on the diagram, whether on or off the line, is a specific pair of (P, T) values, and designates a distinct thermodynamic state. All points on the line represent states in which liquid and vapor are in equilibrium, and the two phases can coexist *only at these points.*

Of course, point (A) can be reached from *any* point on the line by an appropriate combination of changes in pressure and temperature.

Beginning from an arbitrary point on the line, a change in pressure or temperature (or both) will bring the system to a different point (state), which in general does *not* lie on the line. States like point (A) in Figure 11.26 can be reached by an increase in pressure (from point 1) or a decrease in temperature (from point 2). Either of these changes causes the vapor to condense, so that all points, like point (A), lying above and to the left of the line, represent a one-phase system: pure liquid, with no vapor. Similarly, point (B) can be reached by a pressure decrease (from point 2) or a temperature increase (from point 1). Either of these changes results in vaporization. Points like (B), that lie below and to the right of the line, also correspond to one-phase states: pure vapor, with no liquid present. Figure 11.26, called a **phase diagram,** shows which state exists at a given temperature and pressure. A phase diagram is divided into single-phase regions by the *equilibrium line.*

(a)

(b)

Figure 11.25 (a) Photomicrograph of ascorbic acid (vitamin C) crystals growing from the pure liquid. The colors are produced by polarizing filters in the microscope. Viewed without these filters, the crystals are colorless and harder to see. (b) Frost crystals on a winter window. Like the liquid-solid transition in (a), direct deposition of water vapor tends to produce tree-like structures known as dendrites.

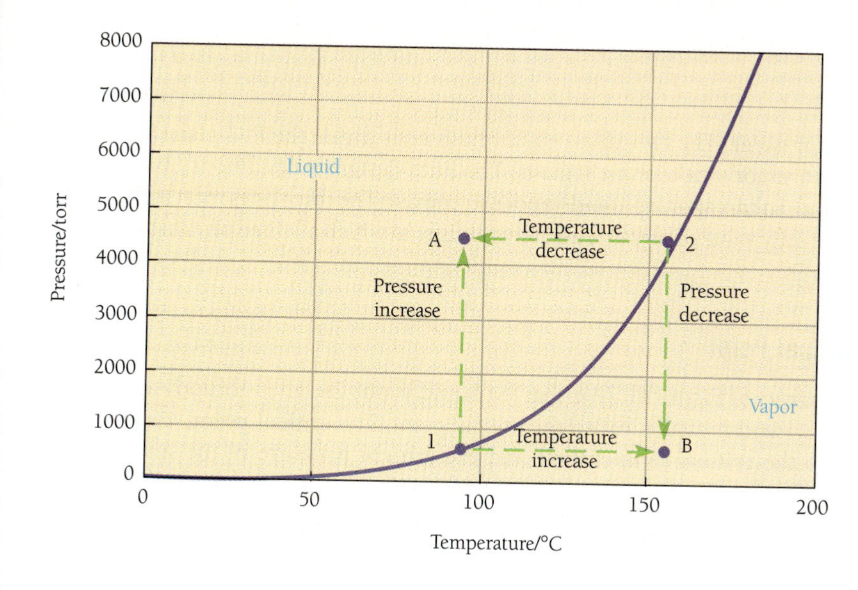

Figure 11.26 A plot of vapor pressure vs. temperature, or phase diagram, for water. Points (A), (B), (1), and (2) are explained in the text.

Lines, Regions, and Points

A similar plot can be drawn to display the liquid and solid phases together. A melting/freezing equilibrium line designates pairs of P, T values at which solid and liquid can coexist. The line separates the one-phase solid region from the one-phase liquid region. A third diagram can be drawn for the sublimation equilibrium. Normally the three plots are combined into one, as in Figure 11.27, the complete phase

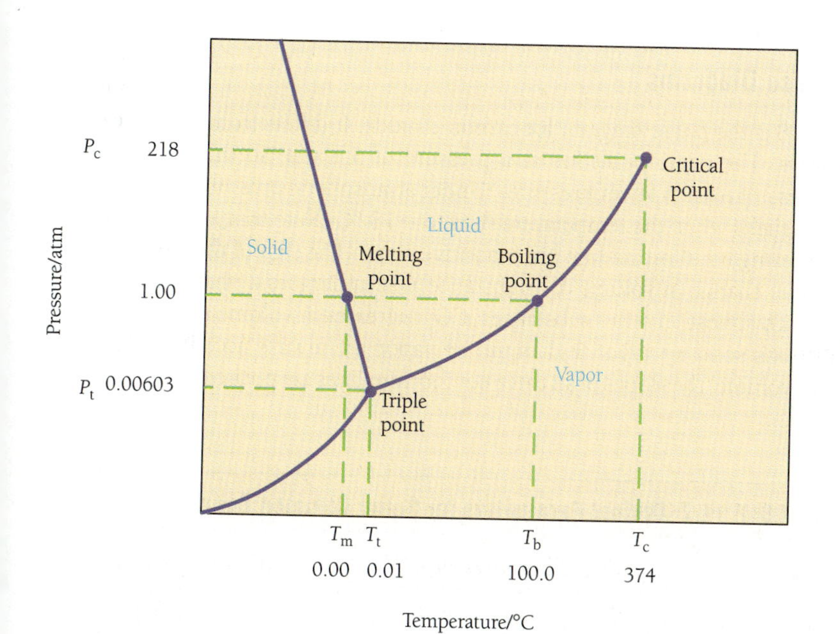

Figure 11.27 Phase diagram of water. Equilibrium lines separate one-phase regions, and meet at the triple point. The triple point, $T =$ 0.01 °C and $P =$ 0.00603 atm (4.58 torr), is the only state in which all three physical states can coexist. The normal melting point (T_m) and boiling point (T_b) are the points where the appropriate equilibrium line intersects the $P = 1$ atm (dashed) line. The liquid/vapor line ends at the critical point. The solid/vapor and solid/liquid lines do not end in specific points, but extend into experimentally inaccessible ranges of pressure or temperature.

diagram for water. The two-phase equilibrium lines divide the P, T plane into solid, liquid, and vapor single-phase regions. The lines designate conditions under which solid/liquid, solid/vapor, or liquid/vapor can coexist. The three lines meet at the **triple point,** the only value of pressure and temperature at which all three physical states can coexist.

The Critical Point

Another aspect of Figure 11.27 is that the liquid/vapor line ends abruptly at a certain temperature and pressure, called the critical point. The **critical point,** which corresponds to the **critical temperature** and the **critical pressure** of the substance, is usually defined as the temperature above which a vapor cannot be liquified, no matter how much pressure is applied. In the region above or to the right of the critical point, the substance is a **supercritical fluid.** It is neither a liquid nor a gas, but has some of the properties of both. Like a liquid, a supercritical fluid is essentially incompressible. Its molecules are tightly packed, in contact with one another, and it takes an enormous increase in pressure to decrease the volume of a sample. Like a gas, it fills its container —no amount of applied pressure is enough to confine it to the bottom part of a flask. The critical parameters for some substances are given in Table 11.7.

The solid/vapor and solid/liquid equilibrium lines extend into experimentally inaccessible regions of pressure and temperature, rather than ending abruptly at a critical point.

There are no liquids with critical pressures less than 2 atm. Since we spend our lives at 1 atm, supercritical fluids and their properties are not part of our common experience. Nevertheless they are used in the laboratory as high-temperature solvents, and in some industrial operations as well.

Supercritical carbon dioxide is used as a nontoxic solvent in one process for decaffeinating coffee.

Using Phase Diagrams

Figure 11.28 is the phase diagram for carbon dioxide. It differs from that of water in several ways. The triple point lies at a pressure of 5.1 atm, so that liquid carbon dioxide does not exist at lower pressures. Under atmospheric pressure, CO_2 sublimes rather than melts when the temperature rises to −78 °C. Note that, unlike water, the melting/freezing equilibrium line has a positive slope. The sign of the slope is determined by the change in volume of a substance as it undergoes a phase transition. In common with almost all other substances, CO_2 increases in volume as it melts. Since the transitions solid → vapor and liquid → vapor are always accompanied by an increase in volume, the slopes of all three equilibrium lines are positive. The volume of

TABLE 11.7 Critical Parameters for Some Common Substances

Substance	Critical Temperature/°C	Critical Pressure/atm
Ammonia	132.5	112.5
Carbon dioxide	31	72.9
Ethanol	243	63
Ethene	9.9	50.5
Helium	−267.9	2.3
Methane	−82.1	45.8
Mercury	1492	1490
Water	374.1	218.3

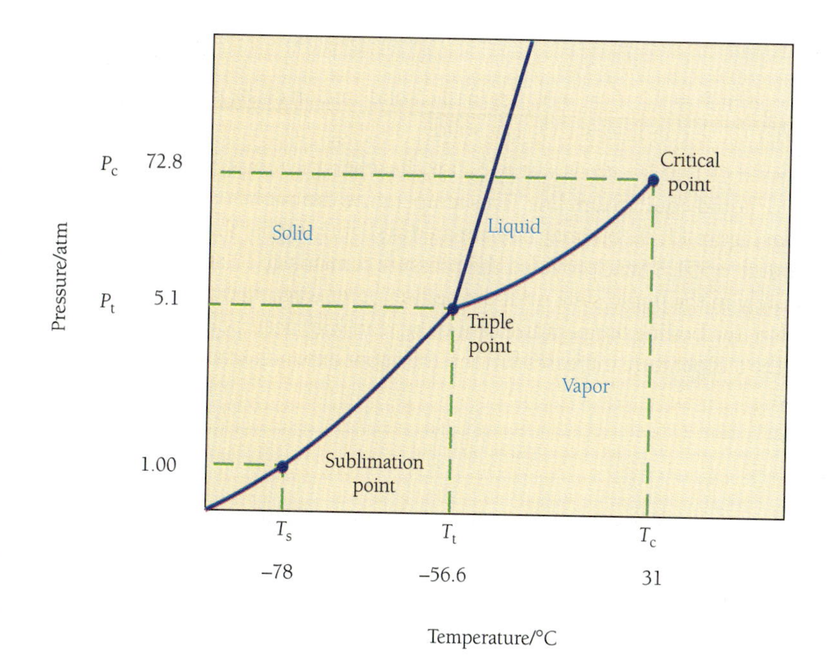

P_c = critical pressure
P_t = triple point pressure
T_s = normal sublimation temperature
T_t = triple point temperature
T_c = critical temperature

Figure 11.28 Phase diagram for carbon dioxide. Except for the slope of the melting equilibrium line (see text), the general shape of the diagram is similar to that of water. The triple-point pressure is 5.1 atm, so that liquid CO_2 cannot exist unless the pressure is more than this.

H_2O, on the other hand, decreases as it melts. This is directly observable from the fact that ice, having a larger volume per gram and hence a smaller density, floats on water rather than sinks to the bottom, as a more dense substance does. In consequence, the slope of the solid/liquid equilibrium line in Figure 11.27 is negative. Only one other substance, gallium metal, is known to decrease in volume on melting.

Phase diagrams are useful not only in determining what physical state exists at a given temperature and pressure, but also in predicting what changes occur as conditions change.

This unusual property of water is critical to aquatic life. When a lake freezes in winter, ice floats to the top, forming a solid cover that serves as an insulating blanket. The lower regions of the lake remain liquid, allowing fish to go about their normal business.

EXAMPLE 11.7 Interpretation of Phase Diagrams

a. Determine the state of CO_2 at −60 °C and 10 atm.

b. Describe what changes occur as a sample of CO_2 is warmed from −60 °C to +60 °C, while the pressure remains constant at 10 atm.

Analysis/Plan The known information and the necessary relationships are all contained in Figure 11.28. Identify the physical state at the starting point ($P = 10$, $T = -60$), then move in the indicated direction—horizontally for a temperature increase with no change in pressure. Describe what happens as dividing lines are reached and crossed.

Work

a. The point $P = 10$, $T = -60$ lies in the "solid" region of the phase diagram, so under these conditions CO_2 is a solid.

b. As solid CO_2 at 10 atm is warmed, it melts when the temperature reaches the solid/liquid equilibrium line. This temperature cannot be read with accuracy from Figure 11.28, but may be estimated at slightly higher than −56.6 °C (say around −55). Then, provided the pressure remains at 10 atm, CO_2 remains entirely in the liquid state until the boiling point is reached at perhaps 25 °C. Again, the boiling temperature cannot be read with accuracy from this graph. As the temperature increases further, the vapor expands according to Charles' law.

Exercise

Describe what occurs when H_2O at −25 °C and 1 torr is warmed to +25 °C, then compressed to 2 atm.

Answer: Ice sublimes to vapor (at some temperature less than 0.01 °C). Then, as the pressure increases, the vapor condenses to liquid water. The condensation takes place at 23.8 torr (from Table 11.3).

11.7 THE SOLID STATE

In this section we discuss the geometrical arrangement of the atoms, molecules, or ions in a solid substance, and the experimental means for obtaining this information. We also consider the relationship between the physical properties of a solid and the type of cohesive force that binds its particles together.

Microscopic and Macroscopic Structure

Most solid substances are **crystalline**—their microscopic particles lie in a definite, regular, and often simple geometrical pattern. Although the particles making up a real crystal may be polyatomic, for instance NO_2^- ions, it is often helpful to regard a crystal as a three-dimensional array of featureless dots, or **lattice points.** Such an idealized array is called a **crystal lattice.** The lattice points of the crystal lattice corresponding to a $NaNO_2$ crystal could represent either Na^+ ions or NO_2^- ions, but not both. Each lattice point must have precisely the same environment and orientation.

Visualization of geometrical concepts is easier in two dimensions than three, so by way of introduction Figure 11.29 shows two different *two-dimensional* lattices. Each lattice may be described as being built up from a number of identical **unit cells,** in the same sense that a floor is built of a number of identical tiles. Although a unit cell might be defined in different ways (Figure 11.29), it is most useful to define it as the smallest possible parallelogram that has the same symmetry as the lattice. The size and shape of a two-dimensional unit cell are defined by the four lattice points at the cell corners. A lattice of any desired size can be formed by lining up unit cells next to one another, in the same way that a floor of any size can be covered by lining up adjacent tiles.

A parallelogram is a four-sided figure having two pairs of parallel sides.

In three dimensions, not surprisingly, there are more possibilities. However, there are only 14 different forms that a *three-dimensional* crystal lattice may take

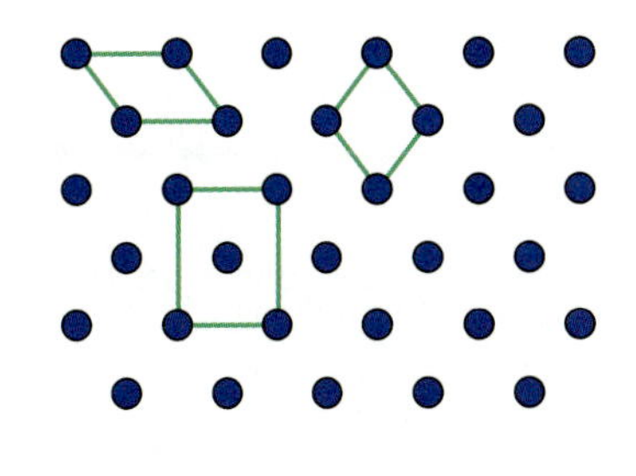

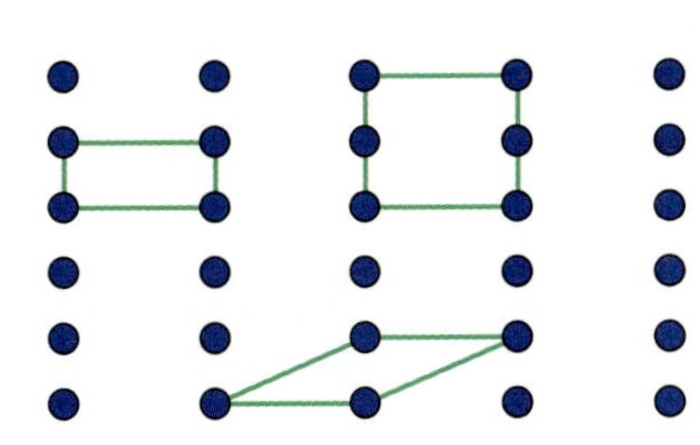

Figure 11.29 Two forms of two-dimensional lattice. The lines indicate several possible ways to define a two-dimensional "unit cell."

Figure 11.30 These naturally occurring crystalline minerals are (clockwise from upper left) beryl, calcite, fluorite, garnet, and barite (center).

(Section 11.8). This is a mathematical limitation, having nothing to do with chemical or physical properties of substances. It is analogous to saying that there are only two types of four-sided figures with 90° angles: squares and rectangles. Each lattice is built up of identical, brick-like, unit cells. The unit cell of a three-dimensional crystal lattice is a *parallelepiped,* a three-dimensional figure having eight corners and three pairs of parallel sides or faces.

A parallelepiped

One of the most striking consequences of the regular microscopic structure is that macroscopic chunks of crystalline solids, called **crystals**, take on precise geometric shapes, as shown in Figure 11.30. Each distinct type of crystal lattice can give rise to crystals of numerous different shapes.

Some substances do not form crystals. When they solidify, their particles do not line up in the precise repetitive pattern of a crystal lattice. Instead they are more randomly arranged, and the structure, on the microscopic level, looks more like a frozen liquid. The particles are packed tightly together and are in contact, but they do not move around as they do in the liquid state. Such substances, lacking crystal structure, are called **amorphous solids.**

X-Ray Crystallography

Although inferences about a crystal lattice can be made from the shape of a crystal, all of our *direct* knowledge comes from the measurement technique called x-ray crystallography. In common with other forms of electromagnetic radiation, x-rays undergo diffraction (Section 6.1). Diffraction modifies the intensity of an x-ray beam reflected from the surface of a crystal in a way that depends directly and simply on the distance between particles in the crystal lattice. The initial discovery of x-ray diffraction by crystals was made in 1912 by the German physicist von Laue. The development of the theory and technique of **x-ray crystallography** (the elucidation of crystal structure) was done at Cambridge University by William and Lawrence Bragg.

Max Theodor Felix von Laue (1879–1960) was awarded the 1914 Nobel Prize for this discovery.

William Henry Bragg (1862–1942) and **Lawrence William Bragg** (1890–1971) began work in 1912, a few months after von Laue's discovery. They shared the 1915 Nobel Prize in physics, the only father-and-son team ever to do so.

The fundamental measure in x-ray crystallography is the *diffraction angle*—the change in direction of a beam of x-rays that occurs when the beam strikes a crystal. Consider two parallel rays of an x-ray beam directed toward the surface of a crystal.

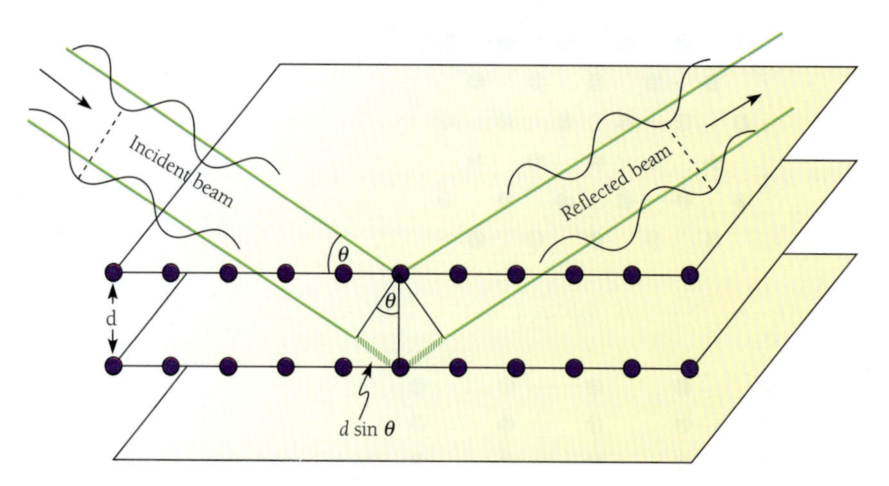

Figure 11.31 Diffraction of x-rays from the surface of a crystal: a two-dimensional sketch of a three-dimensional problem. The angle of reflection θ at which constructive interference takes place is related to the wavelength of the x-rays and to the distance d between planes in the crystal lattice.

Constructive and destructive interference are discussed on page 224.

In the simplified, two-dimensional sketch of Figure 11.31, the atoms of the crystal are represented by a set of regularly spaced points. A distance d separates the planes or layers of atoms, and Figure 11.31 is a side-on view of the top two planes of a crystal surface. The x-rays are initially in phase, so that they interfere constructively. After reflection, the two rays are still in phase only if the distances they have travelled differ by a whole number of wavelengths $n\lambda$. This condition holds only at certain values of the angle θ. The additional distance travelled by the lower ray, determined by trigonometry, is $2d \sin \theta$. The requirement for maximum intensity of the outgoing beam, known as the *Bragg equation,* is

$$\text{path length difference} = 2d \sin \theta = n\lambda \qquad (11\text{-}3)$$

EXAMPLE 11.8 X-Rays and Crystal Dimensions

X-radiation of wavelength 1.544 Å (angstroms) is found to be reflected at an angle of 13.95° from a crystal of magnesium. What is the interplanar distance in the crystal lattice?

Analysis

Target: A crystal lattice spacing,

$$? \text{ Å} =$$

(In the absence of other instructions, choose the units of the target to match units given in the problem.)

Knowns: X-ray wavelength and reflection angle.

Relationship: The Bragg equation, (11-3). All the information required to solve the problem is given, except the following.

Missing Information: The value of the integer n in the Bragg equation.

Plan There is no way that the missing value of n can be found from the available information. Therefore we must *assume* a value, or calculate the values of d corresponding to all possible values of n.

Work Solving the Bragg equation for d, we find

$$d = \frac{n\lambda}{2 \sin \theta}$$

$$= \frac{n \cdot 1.554 \text{ Å}}{2 \cdot \sin(13.95°)} = n \cdot 3.202 \text{ Å}$$

Thus, $d = 3.202, 6.404, 9.606, 12.81$ Å, . . . when $n = 1, 2, 3, 4$, . . . The distance between *adjacent* planes in the crystal structure is 3.202 Å; 6.404 Å is the distance between *alternate* planes; 9.606 Å is the distance between every third plane, and so on.

Exercise

X-radiation of wavelength 1.544 Å is found to be reflected at an angle of 15.56° from a crystal of silver bromide. What is the interplanar distance in the crystal lattice?

Answer: 2.878 Å.

Crystal lattices are very symmetrical, so that many different sets of parallel planes, each with a different inter-plane spacing, can be drawn through their points (Figure 11.32). Since each set of planes reflects x-rays, a crystal reflects x-rays at many different angles. X-ray crystallography involves measurement of the angles and intensities of all beams diffracted from the surface of a crystal. Analysis of the data yields a

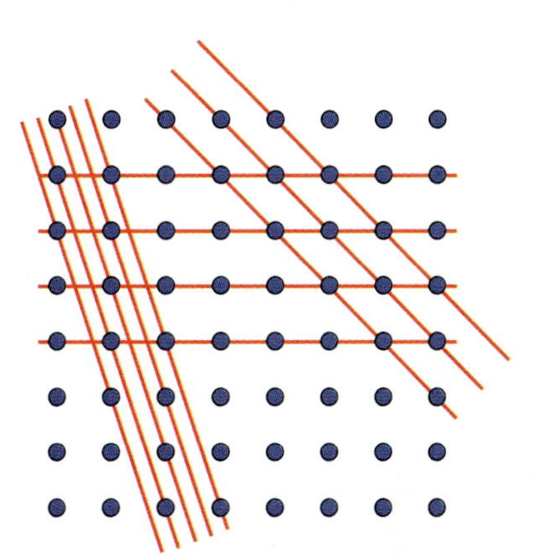

Figure 11.32 Three of the many different sets of parallel lines that can be drawn through the "lattice points" of a (hypothetical) two-dimensional crystal. Each has a different spacing between the lines. In a three-dimensional crystal, many sets of parallel planes can be drawn through the lattice points. Each set has a different spacing, and so reflects x-rays by a different angle according to the Bragg equation.

complete three-dimensional map of the unit cell, showing the locations of all atoms. Figure 11.33 is a diagram of x-ray diffraction equipment, and it shows the diffraction pattern produced by a crystal.

Types of Solids

Solids exhibit a wide range of properties—a thin sheet of aluminum foil is flexible enough to be used as a wrapper, for instance, but a thin sheet of glass is not. Solids are sometimes classified according to their properties, but it is more common to use a classification scheme that takes into account the nature of the particles and their cohesive forces. We distinguish four types of solids: ionic, molecular, covalent or macromolecular, and metallic.

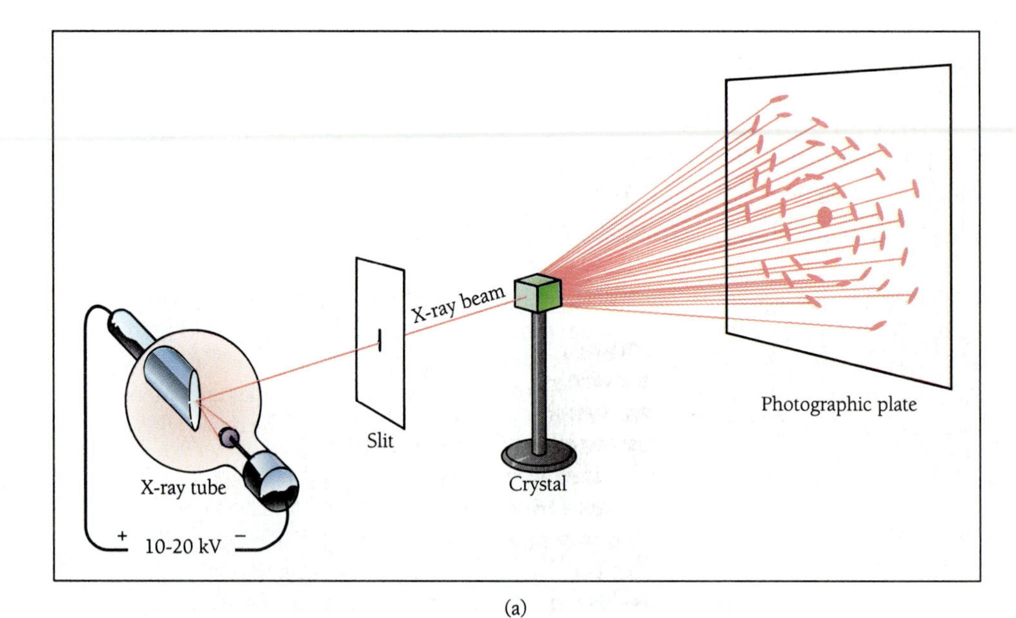

(a)

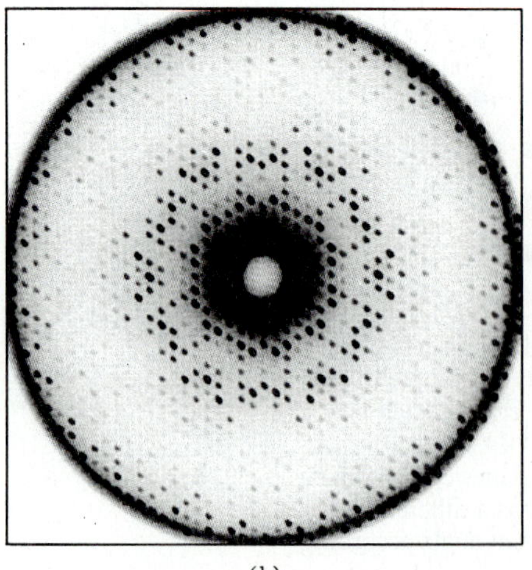

(b)

Figure 11.33 (a) A crystal mounted in a beam of x-rays gives rise to a set of diffracted beams (called "reflections" by crystallographers). These can be recorded on film, as shown, but most modern apparatus uses electronic detectors. (b) Each spot on the exposed film corresponds to diffraction by a different set of planes in the crystal lattice. The central spot is due to the undiffracted beam, which has passed undisturbed through the crystal.

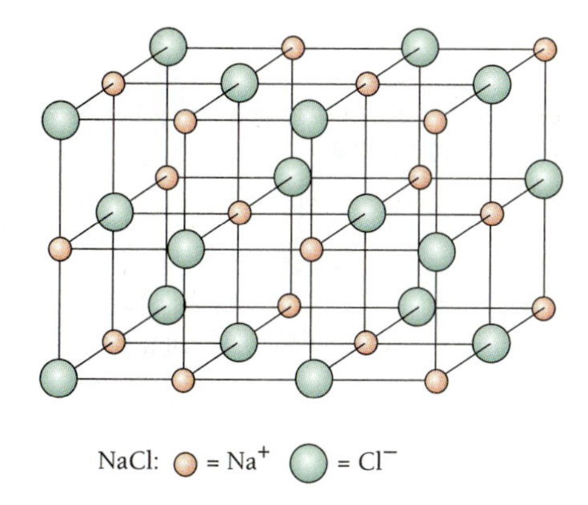

Figure 11.34 The ions in sodium chloride are arranged so that each sodium ion is surrounded by and equidistant from six chloride ions, and each chloride ion is similarly surrounded by six sodium ions.

Ionic Solids

The particles of an **ionic solid** are positive and negative ions, and the force that binds them together into a cohesive unit is the coulomb attraction between unlike charges. Sodium chloride is an ionic solid whose constituent particles are the ions Na^+ and Cl^-. The structure of solid sodium chloride is shown in Figure 11.34. Each sodium ion is surrounded by and equidistant from six chloride ions, and each chloride ion is surrounded by and equidistant from six sodium ions. Imaginary lines drawn through the centers of a Na^+ ion and its six nearest-neighbor Cl^- ions would intersect at right angles. Discrete "molecules" of NaCl do not exist in this structure, because there are no clear boundaries between one group of ions and another. Solids held together by a network of bonds, yet not having discrete molecules, are called **network solids.** All ionic compounds form crystalline solids. In general, ionic solids are hard and brittle, have high melting points, and are poor conductors of both heat and electricity. At sufficiently high temperatures they melt to form electrically conducting liquids having high boiling points and high heats of vaporization.

Molecular Solids

The particles that make up the crystal lattice of a **molecular solid** are molecules. They are held together by one or more of three types of intermolecular cohesive forces that exist between electrically neutral species: dispersion forces, dipole-dipole forces, and hydrogen bonding. The noble gases are included in this category even though they exist as atoms rather than as molecules, because their cohesive force is the dispersion force. Figures 11.35 and 11.36 show the geometrical arrangements of the molecules in crystals of benzene and water. Benzene is a symmetrical molecule with no permanent dipole moment, so its cohesive forces are dispersion only. Ice, on the other hand, is held together by dispersion forces, dipole-dipole forces, and hydrogen bonds.

Because the forces of dispersion, dipole-dipole, and hydrogen bonding are relatively weak, molecular solids tend to have low melting points. Many of them are soft, even waxy, solids. They melt to form liquids having low boiling points and low heats of vaporization. There are no molecular substances with boiling points higher than a few hundred degrees, and there are many that are liquid or gaseous even below room temperature. Since the strength of dispersion forces increases with size, the melting

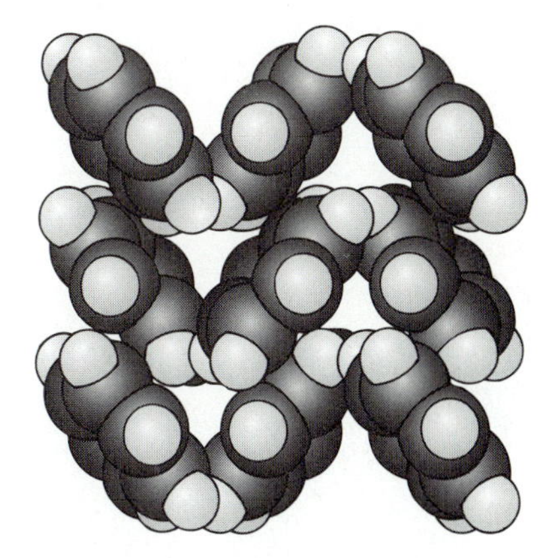

Figure 11.35 Arrangement of benzene molecules in the crystal lattice. The shaded areas represent carbon atoms and the white areas represent hydrogen atoms.

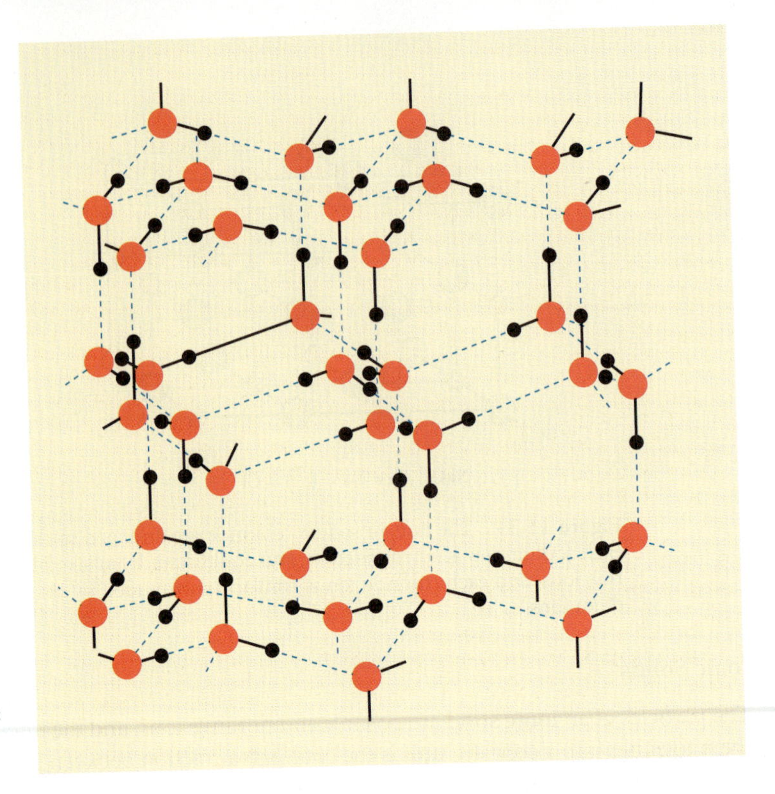

Figure 11.36 Arrangement of water molecules in the crystal lattice. Hydrogen bonding is indicated by dashed blue lines.

and boiling points of chemically similar molecular substances tend to increase with molar mass. The noble gases illustrate this trend, as do the normal (unbranched) alkanes (Table 11.8).

Covalent Solids

Both ionic solids and macromolecular solids are *network solids.*

In its usual sense, the word "molecule" refers to a group of atoms joined together in a fixed geometrical array by covalent bonds. In certain solids, the atoms are held together in a vast three-dimensional array by covalent bonds. Such solids are called **covalent** or **macromolecular** solids. In a covalent solid there are no boundaries that divide the crystal into separate molecules. In a sense each piece of the solid is one gigantic molecule.

Diamond, one of the forms of elemental carbon, is an example of a covalent solid. As shown in Figure 11.37, each carbon atom is joined to four neighbors through

TABLE 11.8 Variation in Melting and Boiling Points with Size

Element	Melting Point/°C	Boiling Point/°C	Compound	Melting Point/°C	Boiling Point/°C
He	—*	−269	CH_4	−182	−16
Ne	−249	−246	C_2H_6	−183	−8
Ar	−189	−186	C_3H_8	−190	−4
Kr	−157	−153	C_4H_{10}	−138	−1
Xe	−112	−107	C_5H_{12}	−130	36
Rn	−71	−62	C_6H_{14}	−95	69

* The cohesive forces in helium are so weak that it cannot be solidified at atmospheric pressure.

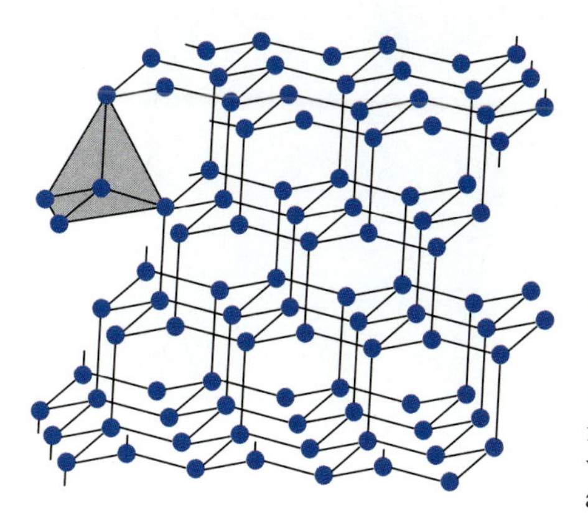

Figure 11.37 Diamond is a crystalline solid in which the cohesive forces are covalent bonds. Each carbon atom is tetrahedrally bonded to four others. There are no molecules in this solid, since there is no way to tell where one "molecule" leaves off and the next begins.

ordinary covalent bonds. The diamond lattice repeats indefinitely in three dimensions, so that diamond is a crystalline solid. Other examples are silicon carbide (SiC) and natural quartz (pure SiO_2).

Covalent bonds are stronger than dipole-dipole interactions or hydrogen bonds. Therefore covalent solids have high melting points, and are usually extremely hard and brittle substances. When they melt, they form liquids with exceedingly high boiling points.

Covalent substances need not be crystalline. Sometimes the geometrical arrangement of the atoms and the bonding are irregular, and not repeated throughout the crystal. Ordinary glass, which consists primarily of SiO_2, has such a structure (Figure 11.38). Any solid lacking a regular crystal lattice is called *amorphous.* A *covalent* amorphous solid is called a **glass.** The properties of glasses are quite different from those of crystals (Figure 11.39). For example, glasses do not exhibit a melting point in

Window glass is transparent, but not because it is the type of solid called a "glass." The mineral *obsidian* is a glass, but it is black and not at all transparent. On the other hand, many crystals, like diamond, are transparent.

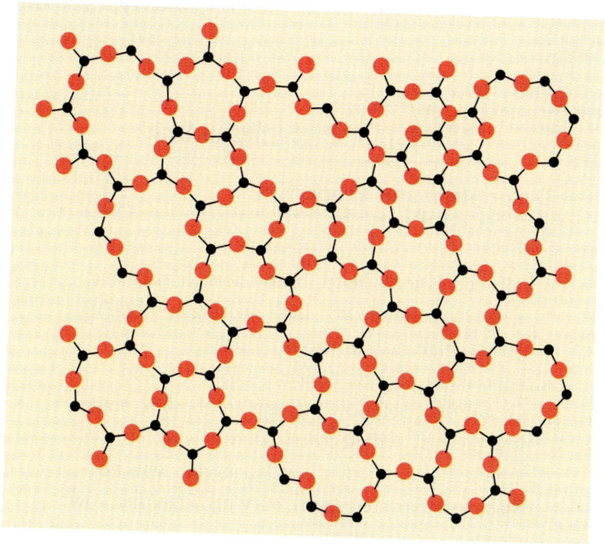

Figure 11.38 Glass is a covalent network solid, primarily SiO_2, that lacks the regularity of a crystal lattice. Silicon atoms are represented in black, oxygen atoms in red. This is a two-dimensional representation of the bonding in SiO_2, but in the actual material, the network of covalent bonds extends in three dimensions.

(a)

(b)

Figure 11.39 Glasses (a) differ from crystals (b) in form and appearance as well as in physical properties.

the same sense that crystals do. Glasses can be liquefied by heating, but the transition from solid to liquid takes place gradually, over a wide range of temperature. When heated, a glass first becomes soft and bendable—it loses the brittleness that characterizes it at lower temperatures. As the temperature is increased it becomes softer and softer, then gooey, then a viscous liquid, and finally an ordinary liquid. Crystals, on the other hand, make a sharp transition, at a specific and reproducible temperature, from the solid state to the liquid state.

Metallic Solids

Metals are distinguished by their physical and chemical properties: they are opaque, reflective, malleable, and have high thermal and electrical conductivities (Figure 11.40). They have low ionization potentials and electronegativities, often form ionic halides, and do not normally form negative ions. **Metallic solids** are usually crystalline, and their cohesive force is often called the "metallic bond." Metallic bonding is best understood from the point of view of band theory, discussed in Section 9.4.

The description and physical properties of the various types of solids are summarized in Table 11.9.

Figure 11.40 All metals have a lustrous, reflecting surface, but there are some color differences among them. The sphere is steel, the cylinder is aluminum, and a gold chain is nestled in a coil of copper wire.

TABLE 11.9 Properties of Solids

Type	Description and Properties	Example
Crystalline		
Ionic	Composed of ions. Hard, brittle, low thermal and electrical conductivity, high melting and boiling points, high heat of vaporization.	NaCl
Molecular	Composed of discrete molecules (or atoms, in Group 8A). Soft, low melting and boiling points, low thermal and electrical conductivity.	H_2O, wax
Covalent	Composed of covalently bonded atoms, with no discrete molecules. Hard, very high melting, low electrical conductivity.	Diamond, SiC
Metallic	Composed of atoms, with no discrete molecules. Shiny, malleable, high thermal and electrical conductivity.	Fe, Cr, Cu, Ag
Glass	Irregular arrangement of atoms. Softening range rather than sharp melting temperature. Brittle below the softening temperature.	Window glass

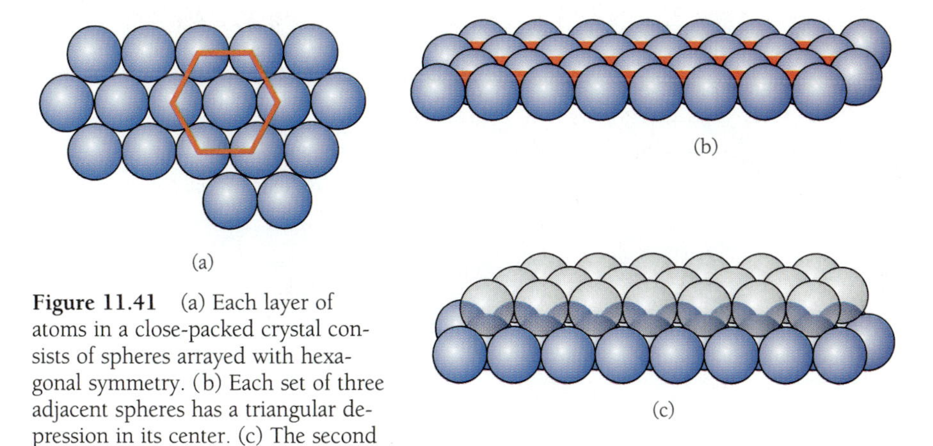

Figure 11.41 (a) Each layer of atoms in a close-packed crystal consists of spheres arrayed with hexagonal symmetry. (b) Each set of three adjacent spheres has a triangular depression in its center. (c) The second layer of spheres fits into the depressions of the first.

11.8 CRYSTAL STRUCTURES

Solids are classified, as we have seen, by the nature of their cohesive forces. An additional classification scheme, for *crystalline* solids, is based on the geometrical details of the crystal lattice.

Metals and the Close-Packed Structures

If the particles of a solid are spherical and identical in size, as is the case in metals, they tend to be arranged so that each sphere is in contact with as many other spheres as possible. This arrangement, called **close packing,** maximizes the cohesive forces and minimizes the amount of free space between the atoms. There are two different ways that close packing of identical spheres can be achieved. Both consist of identical layers that are packed in an array that can be described as hexagonal (Figure 11.41). Each set of three adjacent spheres has a triangular depression in its center, and the spheres of the next layer fit in these depressions.

Although two such layers can fit together in only one way, there are two possibilities for assembling a stack of three or more layers. In one, called *hexagonal close-packed* (hcp), each sphere of the third layer is directly above a sphere in the first layer. Each sphere of the fourth layer is directly above one in the second, and the layer pattern repeats as ABABABAB . . . (Figure 11.42a). Magnesium, Ti, Co, and Zn are among the metals having the hcp structure. In the other, called *cubic close-packed* (ccp), it is not until the fourth layer that any sphere lies directly over one in a lower layer, and the pattern of layers repeats itself as ABCABCABC. . . . Calcium, Cu, Pb, Ag, and Au are among the metals that crystallize in the ccp lattice (Figure 11.42b). In both the hcp and the ccp structures, each atom is in contact with 12 nearest-neighbor atoms.

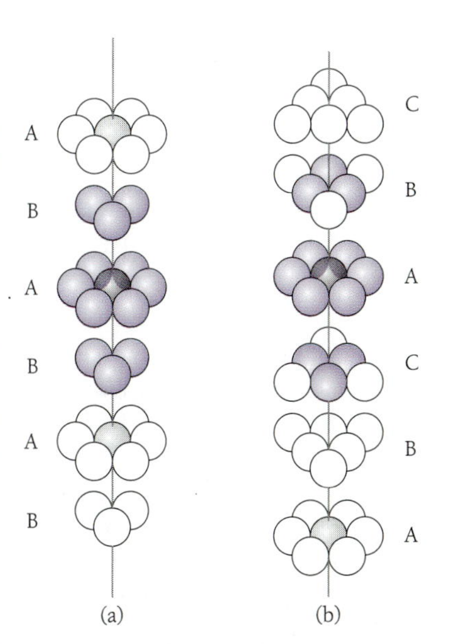

Figure 11.42 Hexagonal and cubic close-packed structures differ in the way in which identical layers are stacked. (a) An exploded view of the hcp structure, showing two different relative positions of the layers. These alternate in the hcp structure, giving rise to a pattern that repeats as ABABAB. . . . (b) There are three different relative positions of the layers in the ccp structure, giving rise to an ABCABCABC . . . pattern. In both the hcp and the ccp structures, each atom has 12 nearest-neighbor atoms that it touches. The 12 nearest neighbors of one representative atom (dark shading) are shown lightly shaded in both structures.

The Cubic Lattices

In some crystals, the layers of atoms consist of a *square* array. This structure is more open, that is, has more free space, than the ccp and hcp lattices. Several different crystal lattices can result, depending on how the layers are stacked together. In the

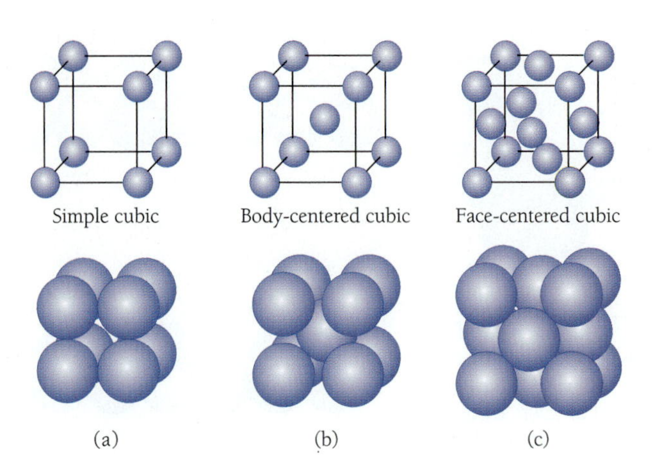

Figure 11.43 Units cells in the (a) primitive cubic, (b) body-centered cubic, and (c) face-centered cubic lattices. In each structure, the unit cell is defined by lattice points lying at the eight corners of a cube. The body-centered lattice includes an additional point in the center of the unit cell, while the face-centered cubic has six additional points at the center of each of the six faces of the cube.

simple or **primitive** cubic lattice, each atom is directly over an atom in the layer beneath. The unit cell is a cube containing no lattice points other than the eight corner points (Figure 11.43a). Each atom has six nearest neighbors, three in its own cell and three in adjacent cells. In the **body-centered** cubic lattice, each atom sits in a depression formed between four adjacent atoms in the layer beneath. The unit cell contains a central lattice point in addition to the eight points at the corners of the cube (Figure 11.43b). Each atom has eight nearest neighbors. The alkali metals, and also Cr, Mo, and W, are among the metals that have body-centered cubic structures. A third type of cubic lattice, called **face-centered** cubic, has eight lattice points lying at and defining

In geometry, a cube is a cube; in crystal structure, there are three kinds of cubes.

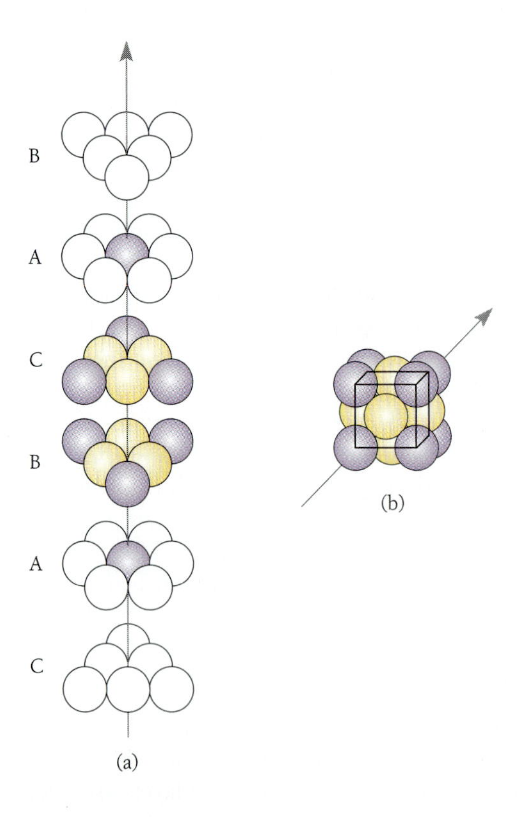

Figure 11.44 (a) Six layers of the cubic close-packed structure of Figure 11.42b are redrawn, in an exploded view that exaggerates the vertical spacing between the layers. The arrow shows which atoms or depressions lie directly over one another. (b) The alignment arrow is tilted, the correct interplane spacing is restored, only the shaded atoms from (a) are redrawn, and lines have been added to better show that this is the same as the face-centered cubic unit cell of Figure 11.43c.

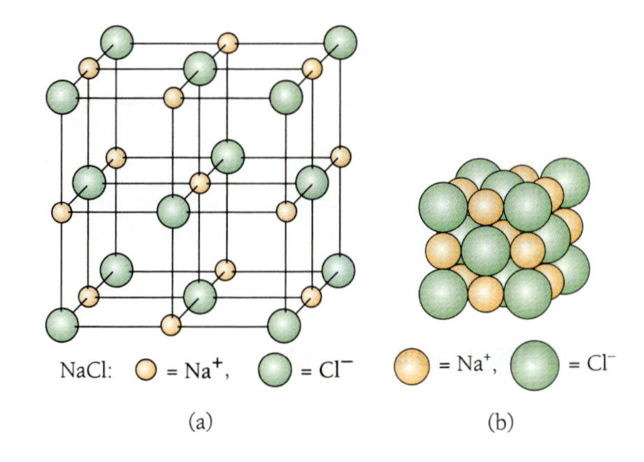

Figure 11.45 The NaCl structure is face-centered cubic. (a) The locations of the Na^+ ions are shown by orange circles, and the Cl^- ions by green circles. Either the anions or the cations may be regarded as defining the eight corners of the unit cell; here the anions are chosen. (b) In the actual structure the ions touch one another. Note that the Cl^- ions are larger than the Na^+ ions.

the corners of a cubic unit cell. It has an additional lattice point centered on each of the six faces of the cube (Figure 11.43c). This lattice was described above (Figure 11.42b) but tilted at a different angle—it is identical to the cubic-close packed structure. Figure 11.44 shows the relationship between the two descriptions.

Many ionic compounds crystallize in the cubic system. In sodium chloride, the 14 lattice points of the face-centered cubic structure are occupied by Cl^- ions, and the Na^+ ions lie halfway between them (Figure 11.45). Another common structure among ionic compounds is the primitive cubic, found in CsCl, CsBr, and CsI (Figure 11.46). Superficially, the structure of these compounds *looks* like the body-centered cubic lattice, but it is not, because the unit cell must be defined by *identical* lattice points. Although cations may be used, anions are customarily chosen to define the unit cell.

One of the several factors that influence the crystal structure of an ionic compound is that the anions and cations are usually of different sizes. The relative ionic size determines the **coordination number,** which is the maximum number of oppositely charged species that can crowd around an ion. If the cation is approximately the same size as the anion (or only slightly smaller), eight anions can fit around it and the coordination number is 8. These compounds (like CsCl, Figure 11.46) crystallize in the simple cubic lattice. If the cation is smaller, fewer anions can fit around it. The coordination number is 6, and these compounds, like NaCl, crystallize in the face-centered cubic lattice. Finally, if the cation is much smaller, say less than half the size of the anion, as in ZnS, the coordination number is 4.

In most ionic compounds of 1 : 1 stoichiometry, the cation is smaller than the anion.

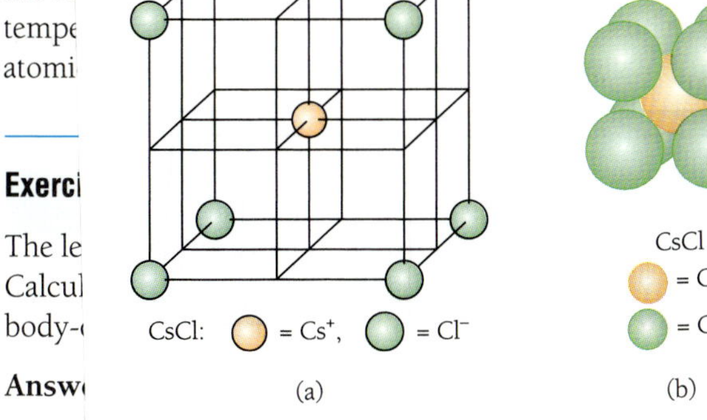

Figure 11.46 The cesium chloride structure. (a) Eight anions define the unit cell of the simple cubic lattice, and a single cation lies in the center. In this structure, each ion has eight nearest-neighbor ions of opposite charge. (b) In cesium chloride, the anions and cations are more nearly equal in size than in sodium chloride.

CHAPTER 12

SOLUTIONS

Once begun by the addition of a tiny "seed," crystal growth from a supersaturated solution of sodium acetate proceeds rapidly.

Figure 12.1 A tidal pool contains a heterogeneous mixture of sand, seawater, and life forms such as seaweed and the anemones and mussels in this picture. The seawater itself is a homogeneous mixture (solution) of sodium chloride (and other salts) and water. The surrounding rock is a heterogeneous mixture of several types of minerals.

12.1 DISSOLUTION, SOLVATION, AND SOLUBILITY

In Chapter 1 we classified mixtures of two or more pure substances as either *heterogeneous* or *homogeneous.* In a heterogeneous mixture (Figure 12.1), there are regions of varying composition. In a homogeneous mixture, all regions, no matter how small, have the same composition. Homogeneous mixtures are also called **solutions** (Figure 12.2). Solutions have some properties in common with their pure components, and some properties that occur only in the mixture. Solutions, as well as their pure components, can be solid, liquid, or gaseous substances. The component that has the same physical state (when pure) as the solution is called the **solvent.** The other components are called **solutes.** The composition of a solution can be described in various ways, the most common of which are *molarity, mass %, mole fraction,* and *mole %.* In this chapter, we discuss some of the properties of solutions, and we develop some ideas about **dissolution,** the *process* in which a solution is formed from the components. We will also be concerned with **solubility,** the *extent* to which the process takes place.

If more than one pure component is in the same physical state, as in homogeneous mixtures of two liquids, the one present in the greater amount is usually called the solvent.

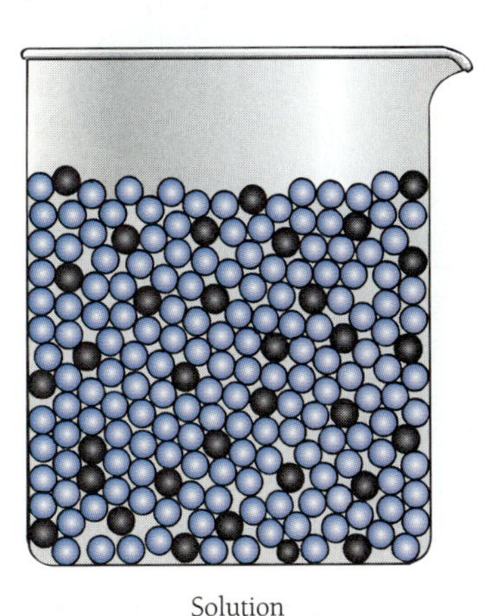

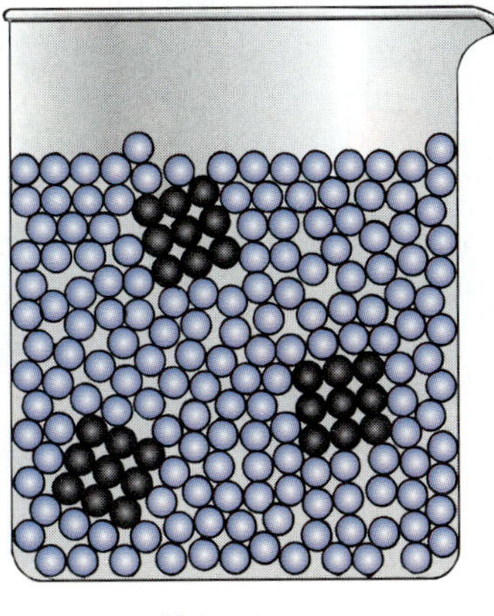

Figure 12.2 In a solution, the molecules (or ions) of the solute are evenly dispersed throughout the sample. In a heterogeneous mixture, there are distinct regions of differing composition.

Supersaturated solutions, a third category, are discussed below.

of solute is added, no amount of stirring will make it all dissolve. When a solution has accepted all the solute it can, it is called a **saturated** solution. A solution that can accept additional solute is **unsaturated** or **undersaturated.** A saturated solution is in dynamic equilibrium with undissolved solute.

The amount of solute that can dissolve in a given amount of solvent is called the **solubility** of the solute. Substances of high solubility are called *soluble,* substances of low to moderate solubility are called *sparingly soluble,* and substances of very low solubility are called *insoluble.* Numerical values of solubility can be expressed in several different units, the most common of which are grams of solute per liter of solution, and grams of solute per 100 mL of solution. Solubility can also be expressed as moles of solute per liter of solution. In general, solubility is different for each solute/solvent pair. Solubility depends on the temperature, and most solid substances become more soluble as the temperature increases. Solubility can be affected by the presence of other solutes in the solution. For example, the amount of NaCl that will dissolve in water containing some dissolved HCl is less than the amount that will dissolve in the same volume of pure water.

The effect of temperature on solubility is discussed below.

Table 12.1 gives the solubilities of several substances in water at 25 °C. These examples show that solubilities vary over a wide range.

Heat of Solution

The dissolution of a pure substance can be thought of as a chemical reaction, and described by a chemical equation. When writing such an equation, it is important to specify the *state* of the substance as solid (s), liquid (ℓ), gas (g), or as solvated (sol) or (aq) for aqueous solutions. The solvent is normally written under the arrow rather than as a reactant. If the substance is an ionic compound, the solvated ions are indicated separately.

$$KCl(s) \xrightarrow[H_2O]{} K^+(aq) + Cl^-(aq) \quad (12\text{-}1)$$

TABLE 12.1 Solubility in Water at 25 °C

Substance	Solubility/g L^{-1}
NaCl(s)	3.6×10^2
AgCl(s)	1.6×10^{-3}
$CaCO_3$(s) (limestone)	9×10^{-4}
CO_2(g)*	1.5
$C_2H_5OH(\ell)$ (ethanol)†	Infinite
$C_{12}H_{22}O_{11}$(s) (sugar)	2×10^3
C_2H_5Cl(g) (chloroethane)	6
$C_6H_6(\ell)$ (benzene)	0.9

* The solubility of carbon dioxide, like that of all gases, depends on the pressure at which the measurement is made. The value given in the table was measured at a gas pressure of 1 atm.

† Some pairs of liquids do not exhibit any limits to solubility; they are said to be "completely miscible" (mixable).

If the substance is an element or molecular compound, dissolution is simply shown as a change from the pure state to the dissolved state.

$$O_2(g) \xrightarrow[H_2O]{} O_2(aq) \qquad (12\text{-}2)$$

As is the case for other chemical reactions, solution reactions generally have a nonzero heat of reaction. Some are quite strongly exothermic. The heat released when a drop of water is added to sulfuric acid, for example, can be enough to cause the drop to boil and spatter. (Since the acid has a higher boiling point than water, the danger is less when drops of acid are mixed with water.) For this reason it is good laboratory practice, when preparing acid solutions, to add acid to water rather than to add water to acid.

A good rule in the laboratory is AAA: always add acid.

The heat transferred to or from the solution as a substance dissolves is called the **heat of solution (ΔH_{soln})**, or the **enthalpy of solution.** Like enthalpies of other reactions, it is a negative quantity for exothermic dissolution and positive for solution processes that require heat input. Heat of solution is sometimes indicated by writing "heat" into the equation as if it were an actual chemical substance. The endothermic reaction (12-1) can be written as

$$KCl(s) + heat \xrightarrow[H_2O]{} K^+(aq) + Cl^-(aq) \qquad (12\text{-}3)$$

Heat is shown as a product in an exothermic reaction.

$$O_2(g) \xrightarrow[H_2O]{} O_2(aq) + heat \qquad (12\text{-}4)$$

These two reactions are typical: generally, heat is absorbed from the environment when an ionic solid dissolves in water, and released to the environment when gases dissolve in water. The relationships between the enthalpies of the solute when pure and when dissolved are shown in Figure 12.7.

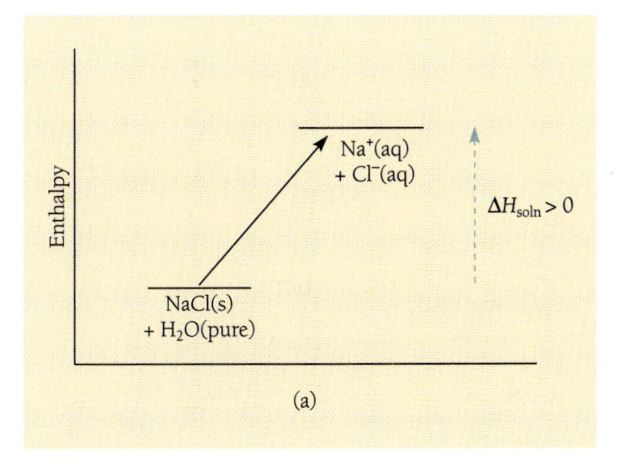

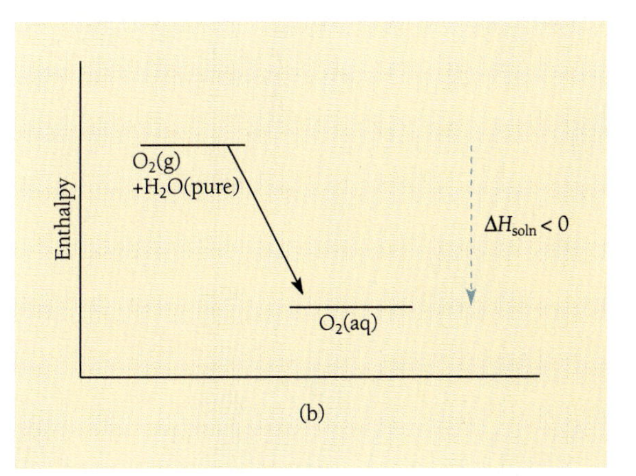

Figure 12.7 (a) When the dissolution of a solute is endothermic, heat must be absorbed from the surroundings as the process takes place. (b) When the dissolution of a solute is exothermic, the process is accompanied by the release of heat to the environment. The enthalpy of the solute is less when dissolved than when pure, and ΔH, the enthalpy of solution, is a negative quantity.

Effect of Temperature on Solubility

For almost all substances, solubility depends on temperature. Some substances become more soluble as the temperature is increased, while others become less soluble. The fact that a state of *equilibrium* exists between a saturated solution and the pure, undissolved solute allows the effect of temperature on the solubility of a substance to be predicted. According to **Le Châtelier's principle,** if a system in equilibrium is disturbed by an outside influence (for example a change in temperature), the system will respond in such a way as to counteract and reduce the effect of the disturbance. When the temperature of a mixture of solid KCl and its saturated solution is increased, by adding heat, the equilibrium system responds by using up part of the added heat.

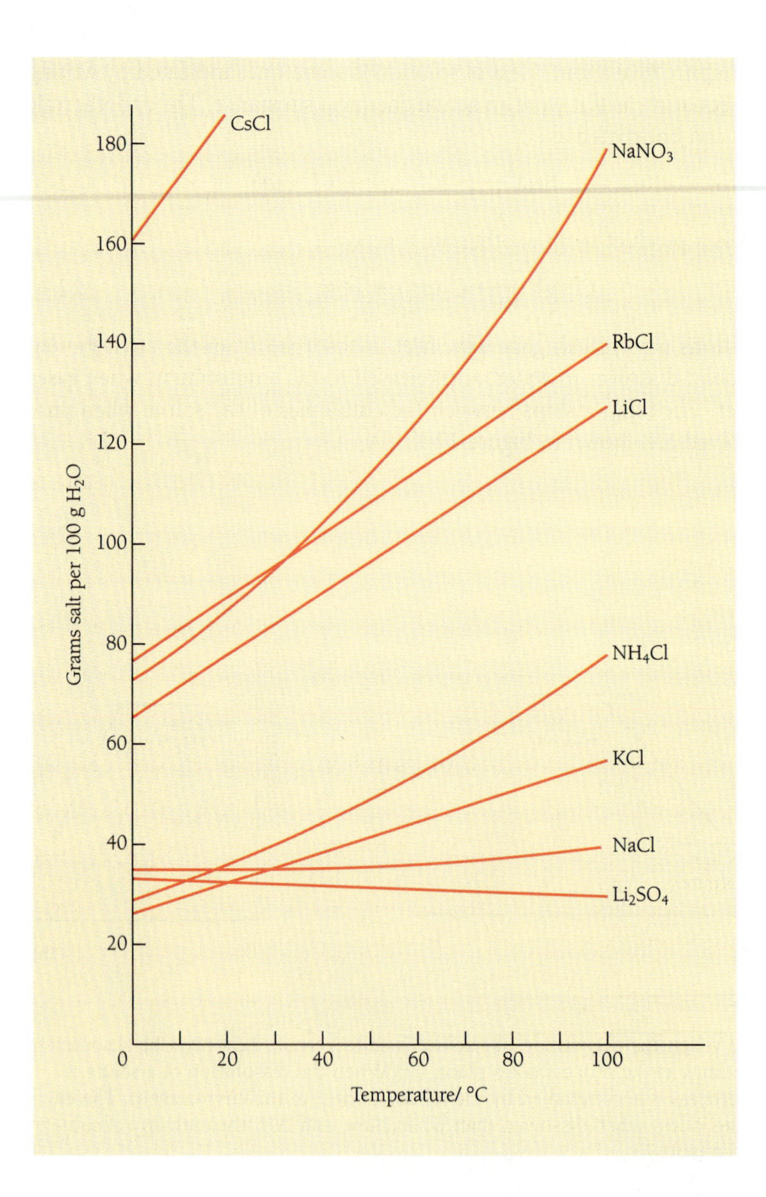

Figure 12.8 The solubility of most ionic solids increases with increasing temperature. This behavior is predicted from Le Châtelier's principle, since the solution reactions of most ionic solids are endothermic.

Henri Louis Le Châtelier (1850–1936)

This French chemist worked with ceramics, glasses, and cements, in a field now known as "materials science." In the course of his research he developed the platinum-rhodium thermocouple, a device still used for the measurement of very high temperatures. The principle of equilibrium that bears his name was published in 1886.

According to Equation (12-3) heat is consumed when KCl(s) dissolves, so the response to the added heat is the dissolution of more KCl(s). This response is summarized in the phrase, "the equilibrium shifts to the right"—that is, there is an increase in the amount of KCl(aq), a substance to the right of the arrow in the chemical equation. The general rule is, *for all substances whose solution reactions are endothermic, solubility increases as temperature increases.*

The opposite is true for substances like oxygen, whose solution reactions are exothermic. If the equilibrium system described by Equation (12-4) is cooled by removing heat, more O_2 dissolves. The system responds so as to generate heat in order to replace part of the loss. *For all substances whose solution reactions are exothermic, solubility increases as temperature decreases.* The behavior of soft drinks is a familiar example. A cold soft drink contains a high concentration of dissolved carbon dioxide. As the drink warms up, the solubility of carbon dioxide decreases, bubbles of gas escape, and the drink begins to taste "flat" as the concentration of dissolved carbon dioxide decreases. Figures 12.8 and 12.9 show the solubility of some common substances at different temperatures.

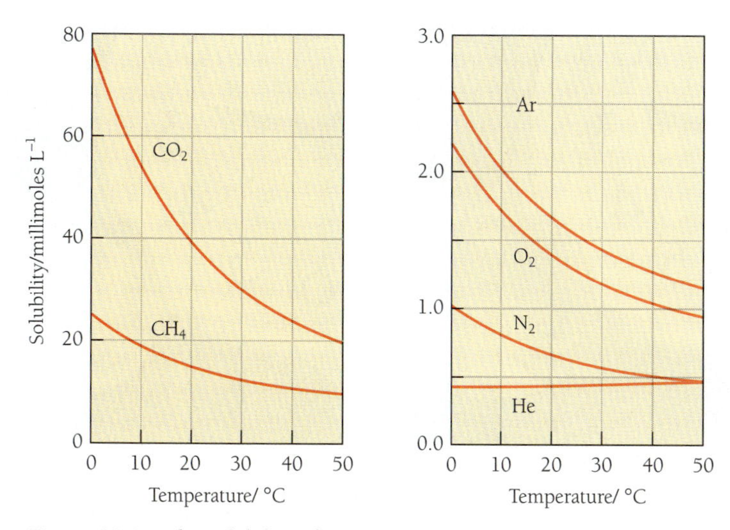

Figure 12.9 The solubility of most gases decreases as the temperature increases. The solution reactions are exothermic.

Supersaturated Solutions

When the solubility of a substance increases with temperature, it is possible to prepare a **supersaturated solution,** in which the concentration of solute is *greater* than in a saturated solution. A supersaturated solution cannot be prepared by simply adding more solute and stirring, for once the solute concentration reaches its saturation value, no further solute can dissolve. However, if the temperature is increased, more solute can dissolve, up to the new, higher, saturation concentration. If the solution is then cooled, the solubility decreases and crystallization of excess solute should occur. But crystallization takes time—it does not begin immediately, and once the process starts, it is not completed instantaneously. Thus, for a time, the concentration of the solute in the solution is *greater* than the solubility.

Supersaturation is a temporary condition. The system is not in equilibrium, and sooner or later crystallization of excess solute will occur, and equilibrium will be re-established. However, it is difficult for crystallization to take place unless some solid material is already present. In systems of ordinary purity there are usually some microscopic solid particles present (dust, for instance), which serve as nuclei for crystal growth. If such particles are excluded, and the hot solution is cooled very slowly in a stoppered container that prevents solvent evaporation, and provided the degree of supersaturation is not too great, supersaturated solutions can be prepared. The onset of crystallization can be hastened by the addition of a tiny crystal of pure solute, called a **seed,** to serve as a nucleus. The process is illustrated in Figure 12.10.

For most solids, dissolution is endothermic, while the reverse process, crystallization, is exothermic. Consequently, when a supersaturated solution crystallizes, heat is released to the environment and the mixture becomes noticeably warmer. This effect is put to use in a cordless heating pad available at drug stores. The Heat Solution® (Figure 12.11) is a pouch containing a supersaturated solution of sodium acetate, in which crystallization can be initiated by clicking a small metal button within the pouch. During the crystallization, which takes about 10 seconds, the pouch warms up to perhaps 50 °C. It cools gradually to room temperature over about a half

(a)

(b)

(c)

Figure 12.10 A supersaturated solution of potassium acetate is shown at various times after the introduction of a tiny "seed" crystal of pure potassium acetate. Crystal growth on the seed is rapid, and continues until the amount of solute remaining in solution has decreased to its equilibrium, or saturated, value.

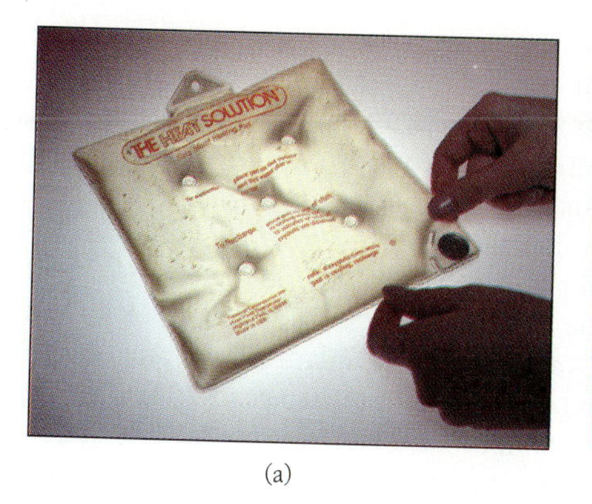

(a)

(b)

Figure 12.11 (a) The Heat Solution®, a supersaturated solution of sodium acetate, is stable until mechanically disturbed. (b) When the clicker button is pressed, crystallization begins and spreads rapidly throughout the pouch. The temperature rises to over 50 °C during crystallization.

hour, providing soothing heat to sore muscles and joints. Heating the pouch in boiling water redissolves the crystals, and the heating pad can be used repeatedly.

The cold pack often used at sporting events for first-aid treatment of bruises and sprains is another practical use of heat of solution. These "chemical" cold packs contain solid NH_4NO_3 and a sealed pouch of water. The pouch breaks when squeezed, and the ammonium nitrate dissolves in an endothermic process. The necessary heat can come only from the water, which cools to about 5 °C as a result.

Engine exhaust contains large amounts of $H_2O(g)$, a product of fuel combustion. Because the equilibrium vapor pressure of water is much less at low temperature, aircraft exhaust gases become supersaturated with water vapor as they cool. Deposition from the vapor, and the growth of a visible trail of ice crystals, begins either at a speck of soot from the exhaust or at a pressure fluctuation caused by turbulence.

In "cloud seeding," tiny crystals of AgI(s) are dispersed into a large region of supersaturated air, in the hope of creating rain clouds. Seeding almost always produces clouds, but the clouds do not always produce useful rain.

Skeptics say that seeding hardly ever produces rain.

12.2 SOLUBILITY TRENDS

In Chapter 10 we discussed the two driving forces behind spontaneous chemical and physical processes, namely the drive toward lower energy and the drive toward greater disorder or chaos. These driving forces are quantitatively described by changes in *enthalpy* (ΔH) and *entropy* (ΔS), and combined into *free energy* by the relationship $\Delta G = \Delta H - T\Delta S$. In spontaneous processes, such as the formation of a solution from the pure components, ΔG is a negative quantity. ΔG is large (and negative) when very soluble substances dissolve, and small when slightly soluble substances dissolve. Entropy almost always increases during dissolution, while ΔH can be either positive or negative. In the following sections, we discuss the factors influencing both ΔS and ΔH in solution formation.

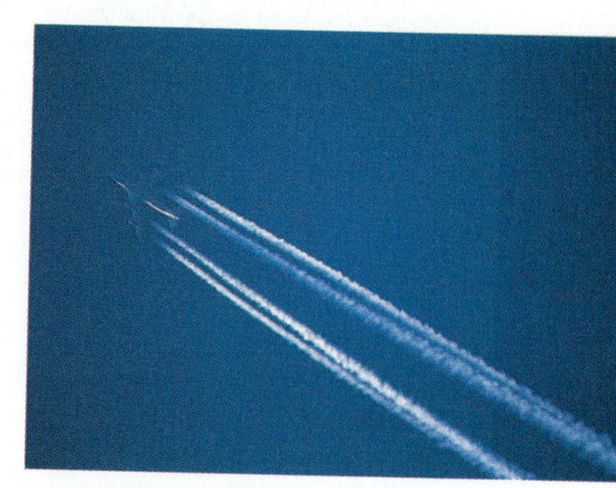

Figure 12.12 The "vapor trails" that sometimes form behind aircraft are a striking example of the deposition of a solid from air that is supersaturated with respect to water vapor.

The Role of Entropy

In the pure state, the particles of a solid are confined to a relatively small region, and in crystalline solids they are lined up in neat, orderly rows. The orderly arrangement is lost when the substance dissolves, and the particles of the solute are free to wander randomly throughout the much larger volume of the solution. The larger volume in the dissolved state contributes more to the entropy increase than the loss of crystalline order, and dissolution of crystals, amorphous solids, and liquids are all accompanied by a large increase in entropy. One indication of the relative importance of volume and crystalline order can be obtained by comparing the entropy changes accompanying melting and vaporization of a pure substance. Melting is a loss of crystalline regularity with little or no volume increase, while vaporization of a liquid is a volume increase with little or no change in regularity. For benzene, $\Delta S_{fus} = 36\ J\ mol^{-1}\ K^{-1}$ and $\Delta S_{vap} = 113\ J\ mol^{-1}\ K^{-1}$. For zinc, the values are $\Delta S_{fus} = 10\ J\ mol^{-1}\ K^{-1}$ and $\Delta S_{vap} = 110\ J\ mol^{-1}\ K^{-1}$. [These ΔS_{vap} values are not standard ($P = 1$ atm) entropies. Instead, they are for vaporization resulting in a vapor *volume* of 20 L mol^{-1}.]

The drive toward greater disorder is an important factor in almost all dissolution processes. In considering the different solubilities of various substances, therefore, we should be asking, "Why do some substances *not* dissolve?" The answer lies in the other driving force, enthalpy.

The Role of Enthalpy

The enthalpy difference between the solute, when pure, and the solute, when dissolved, is the heat or enthalpy of solution. The strength of the attractive forces between a solute particle and the surrounding solvent molecules is an important part of enthalpy of solution. Other things being equal, the stronger the forces the greater the (negative) value of the enthalpy of solution and the more exothermic the dissolution process. As indicated in Section 12.1, these forces are relatively strong for ionic and polar solutes dissolved in polar solvents, and they are even stronger if there is hydrogen bonding. As a consequence, ionic and polar compounds generally dissolve well in polar solvents. The process is driven by an enthalpy decrease as well as by the entropy increase. These observations are summarized in the aphorism, "like dissolves like," in which "like" refers primarily to polarity.

However, there is more to enthalpy of solution than the solute-solvent interactive forces. The nature of the pure materials is involved as well. Recall from Chapter 4 that the enthalpy change in a process is the same whether the process occurs all at once or stepwise. We can gain a better understanding of enthalpy of solution if we divide the process into three hypothetical steps.

The energy required to separate the particles of a solid is called the *lattice energy*.

1. The molecules or ions of the pure solute are pulled apart from one another, to a distance great enough for the cohesive forces to become negligibly small. In effect, this is the vaporization of the solute, with the hypothetical "vapor" occupying the same volume it will ultimately have in the solution. Since this step requires energy input to overcome the cohesive forces in the pure solute, the enthalpy change, ΔH_1, is positive.
2. The pure solvent is expanded slightly, creating throughout its volume many little "holes," each large enough to accommodate one solute particle. This step is also endothermic ($\Delta H_2 > 0$), because the solvent-solvent intermolecular forces must be overcome in order to create the holes.
3. The solute "vapor" is allowed to occupy the solvent "holes" to form the solution. Energy is released as the solute-solvent forces become established, and the enthalpy change for this step (ΔH_3) is negative.

The total enthalpy of these steps is the enthalpy of solution. The individual enthalpies are not measurable, because they are hypothetical. In real solution processes, the steps are not sequential, they are simultaneous and inseparable. These enthalpy relationships are diagrammed in Figures 12.13 and 12.14.

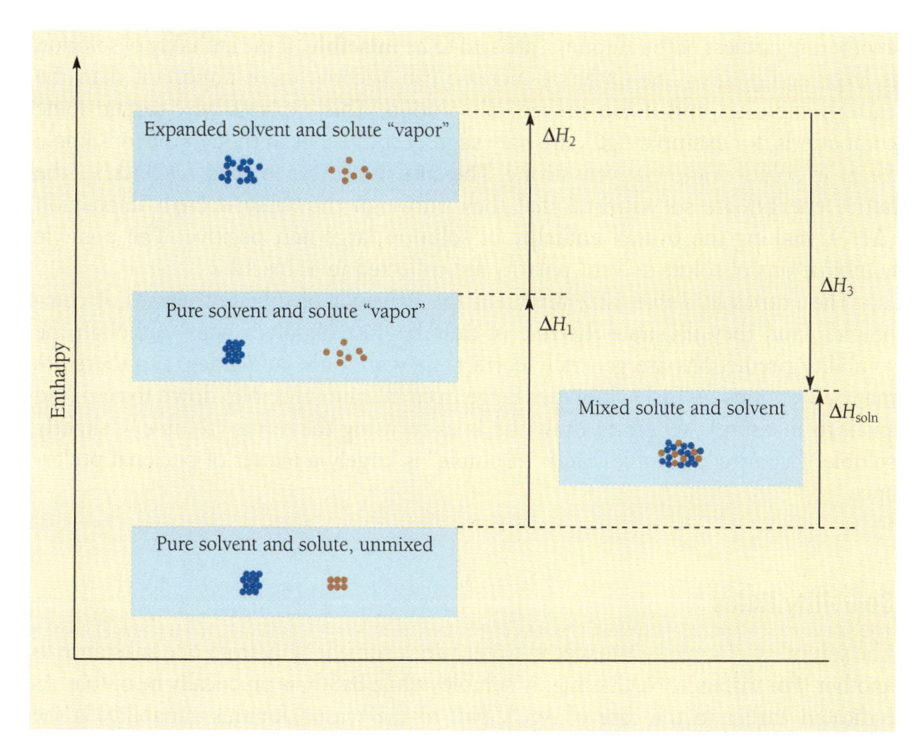

Figure 12.13 Enthalpy relationships in solvation. ΔH_1 is determined by the energy required to overcome the cohesive forces of the pure solute, ΔH_2 by the energy required to overcome enough of the solvent's cohesive forces to accommodate the solute particles, and ΔH_3 by the energy regained when the solute-solvent intermolecular forces are established. The enthalpy of solution is the sum $\Delta H_{soln} = \Delta H_1 + \Delta H_2 + \Delta H_3$. In this example all three components are relatively large, which is typical of ionic or polar solutes and polar solvents. The overall process for ionic solutes and aqueous solvent is most often endothermic, that is, ΔH_{soln} is a positive quantity.

It does not matter whether the enthalpy of solution is positive or negative, provided that it is small. In either case, dissolution is spontaneous, driven by the entropy increase. In other words, the substance tends to be soluble. If the enthalpy of solution is large and *negative,* entropy and enthalpy work together to make the substance very or completely soluble. Substances that dissolve completely in one

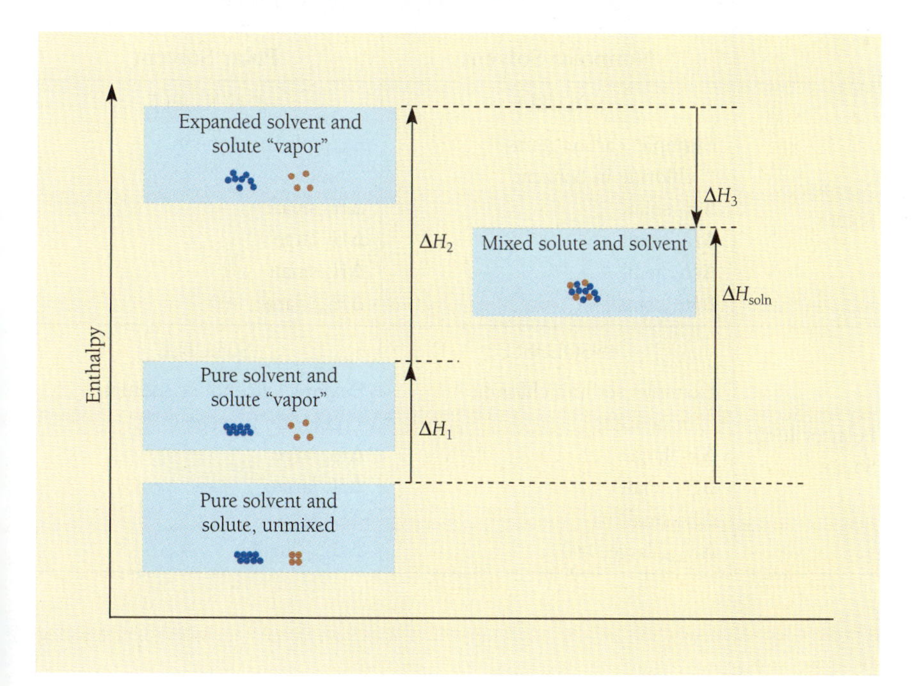

Figure 12.14 Enthalpy relationships in the solvation of a nonpolar solute by a polar solvent. ΔH_1 is small for nonpolar solutes, because the cohesive forces are limited to the relatively weak London dispersion forces. Solute-solvent force is likewise small. The overall ΔH_{soln} is determined primarily by the solvent expansion term, which is large and positive. Since a large energy input is required, nonpolar solutes are not very soluble in polar solvents.

another, regardless of the amounts, are said to be **miscible.** If the enthalpy of solution is large and *positive,* the enthalpy increase can become more important than the entropy increase, and it can prevent dissolution. This explains why certain ionic compounds, for example AgCl, are only slightly soluble, even though the ion-dipole forces in solution are relatively strong. The cohesive forces in solid AgCl (ΔH_1, the lattice energy) are so powerful that they outweigh the solute-solvent interaction (ΔH_3), making the overall enthalpy of solution large and positive. The possible combinations of solute-solvent polarity are collected in Table 12.2.

The entries in Table 12.2 represent the behavior of different classes of compounds, and they illustrate the rule of thumb, "like dissolves like." Although the solubility predictions are generally correct, they are only guidelines, and there are many exceptions. Actual solubilities range from infinite (miscible) down to zero (too small to measure). Where to draw the lines in using the terms "soluble," "slightly soluble," "sparingly soluble," and "insoluble" is largely a matter of personal preference.

Solubility Rules

The solubility of ionic substances in water varies considerably from one substance to another. For instance, NaCl is highly soluble, while $BaSO_4$ is practically insoluble. As indicated earlier in the case of AgCl, part of the reason for this variability is the difference in strength of the cohesive forces in the crystalline solids. Most of the observed solubility behavior of ionic compounds can be summarized in the set of solubility rules given in Table 12.3. These rules should be memorized, because the solubility of substances is an important part of a chemist's background knowledge.

TABLE 12.2 Solvation Enthalpy, Polarity, and Solubility

	Nonpolar Solvent	Polar Solvent
Nonplar Solute	SOLUBLE Example: carbon tetrachloride in benzene ΔH_1 small ΔH_2 small ΔH_3 small ΔH_{soln} small	INSOLUBLE Example: benzene in water ΔH_1 small ΔH_2 large ΔH_3 small ΔH_{soln} large, >0
Polar or Ionic Solute	INSOLUBLE Example: sodium chloride in benzene ΔH_1 large ΔH_2 small ΔH_3 small ΔH_{soln} large, >0	SOLUBLE Example: alcohol or sodium chloride in water ΔH_1 large ΔH_2 large ΔH_3 large ΔH_{soln} small

TABLE 12.3 Solubility* of Ionic Compounds in Water

Salts of Group 1A cations and NH_4^+ are soluble†
Salts of the nitrate(NO_3^-), acetate($CH_3CO_2^-$), chlorate(ClO_3^-), and perchlorate(ClO_4^-) anions are soluble
Chlorides, bromides, and iodides‡ are soluble, except:
  slightly soluble—$PbCl_2$; $PbBr_2$; $HgBr_2$
  insoluble—AgCl, AgBr, AgI; Hg_2Cl_2, Hg_2Br_2, Hg_2I_2; PbI_2; HgI_2
Sulfates(SO_4^{2-}) are soluble, except:
  slightly soluble—$CaSO_4$; Ag_2SO_4
  insoluble—$SrSO_4$; $BaSO_4$; $PbSO_4$; Hg_2SO_4
Hydroxides are insoluble, except:
  $Ba(OH)_2$ and Group 1A hydroxides are soluble
  $Ca(OH)_2$ and $Sr(OH)_2$ are slightly soluble
Fluorides, sulfides, carbonates(CO_3^{2-}), and phosphates(PO_4^{3-}) are insoluble, except:
  Group 1A and NH_4^+ salts are soluble

* "Soluble" means that the solubility is greater than 0.1 M. "Insoluble" means that the solubility is less than 0.01 M. "Slightly soluble" means a solubility in the range 0.01 M to 0.1 M.
† Lithium chemistry differs somewhat from the other Group 1A metals. LiF, Li_2CO_3, and Li_3PO_4 are insoluble.
‡ Salts of the cyanide(CN^-) and thiocyanate(SCN^-) anions have solubilities similar to those of iodides.

Precipitation Reactions

The rules given in Table 12.3 are useful in predicting whether or not a solid product will result from a reaction taking place in solution. A **precipitate** is the solid product of *a precipitation reaction,* in which two dissolved substances react to form an insoluble product.

EXAMPLE 12.1 Formation of a Precipitate

What precipitate, if any, will form when an aqueous solution of NaI is mixed with an aqueous solution of $Pb(NO_3)_2$?

Analysis Solutions of ionic compounds consist of separate anions and cations. When two such solutions are mixed, the solution contains up to four different ionic species. If any two ions are the components of an insoluble compound, a precipitate will form.

Plan Determine the ionic constituents of the combined solution, and list the possible compounds that could be formed. Then check Table 12.3 to see which (if any) of the possible products is insoluble.

Work When NaI(aq) and $Pb(NO_3)_2$ are mixed, the solution contains all four ions: Na^+, I^-, Pb^{2+}, and NO_3^-. These ions can be combined to form two different compounds: $NaNO_3$ and PbI_2. [The two original compounds, NaI and $Pb(NO_3)_2$, can be ignored. These *must* be soluble, since according to the statement of the problem, they were originally dissolved.]

Table 12.3 shows that $NaNO_3$ is soluble, while PbI_2 is insoluble. Therefore, a precipitate of $PbI_2(s)$ will form when the solutions are mixed.

Exercise

What precipitate, if any, will form when an aqueous solution of NaF is mixed with an aqueous solution of $CaCl_2$?

Answer: Calcium fluoride.

12.3 VOLATILE SOLUTES AND VAPOR PRESSURE

The presence of dissolved material affects the properties of a solution. For instance, the vapor pressure of a volatile liquid changes when a solute is dissolved in it. If the solute is more volatile than the solvent, the vapor pressure of the solution is greater than that of the solvent alone. In this section, we discuss the properties of solutions in which both solvent and solute are volatile substances.

Ideal Solutions and Raoult's Law

Equation (12-5) incorporates Dalton's law of partial pressures, which was discussed in Chapter 5.

The vapor pressure of a solution consisting of two volatile liquids is the sum of the *partial pressures* of the components. For example, the vapor pressure of a solution of benzene (C_6H_6) and toluene(C_7H_8) is

$$P_{\text{soln}} = P_{\text{benzene}} + P_{\text{toluene}} \quad (12\text{-}5)$$

François Marie Raoult (1830–1901) was professor of chemistry at the University of Grenoble in France. Most of his scientific effort was spent in the investigation of the behavior of solutions.

The partial pressures of the components depend on the composition of the solution. **Raoult's law** states that the vapor pressure of a volatile solution component is equal to its mole fraction in the solution times its vapor pressure when pure. Raoult's law for a benzene/toluene solution is given by Equations (12-6a and b),

$$P_{\text{benzene}} = (X_{\text{benzene}})(P^\circ_{\text{benzene}}) \quad (12\text{-}6a)$$

$$P_{\text{toluene}} = (X_{\text{toluene}})(P^\circ_{\text{toluene}}) \quad (12\text{-}6b)$$

Mole fraction was defined in Section 5.5. If a mixture contains n_A moles of substance A and n_B moles of substance B, the mole fractions of A and B are $X_A = n_A/(n_A + n_B)$ and $X_B = n_B/(n_A + n_B)$, respectively.

in which X_{benzene} and X_{toluene} are the mole fractions of the two components of the solution, and P°_{benzene} and P°_{toluene} are the vapor pressures of the pure substances. Equation (12-6c) is the general, algebraic statement of Raoult's law.

$$P = XP^\circ \quad (12\text{-}6c)$$

Combining Equations (12-5) and (12-6a and b) yields Equation (12-7), which may be used to calculate the vapor pressure of a benzene/toluene solution of any composition.

$$P_{\text{soln}} = (X_{\text{benzene}})(P^\circ_{\text{benzene}}) + (X_{\text{toluene}})(P^\circ_{\text{toluene}}) \quad (12\text{-}7)$$

Although there are solutions whose vapor pressures are accurately given by Raoult's law, there are many more for which the law is only approximately correct. Those that *do* follow the law are known as **ideal solutions.** Mixtures of liquids whose chemical and physical properties are very similar to one another are more likely to be ideal. Figure 12.15 illustrates Raoult's law, for a pair of miscible, volatile liquids.

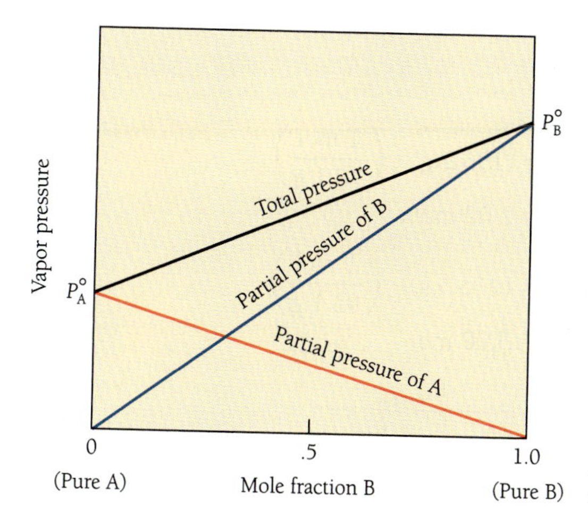

Figure 12.15 Raoult's law for ideal solutions predicts the pressure and composition of the vapor in equilibrium with a solution of two volatile liquids, A and B. The partial pressure of each component is proportional to its mole fraction in the liquid, and the total pressure is the sum of the partial pressures. P°_A and P°_B are the equilibrium vapor pressures of the pure liquids A and B, respectively.

Ideal solutions are important for the same reason that ideal gases are important: they provide an easily understood model, which is followed by real systems to a greater or lesser extent. We will have more to say about nonideal solutions shortly.

EXAMPLE 12.2 Vapor Pressure of Liquid-Liquid Solutions

At 20 °C, the vapor pressure of benzene(C_6H_6) is 75 torr and that of toluene(C_7H_8) is 22 torr. Consider a solution containing 156.2 g benzene and 276.3 g toluene. Calculate the partial pressure of each component in the equilibrium vapor, and the total pressure of the vapor.

Analysis

Targets:

$$? \text{ torr} = P_{\text{benzene}}$$

$$? \text{ torr} = P_{\text{toluene}}$$

$$? \text{ torr} = P_{\text{total}}$$

Knowns: Temperature, P°_{benzene}, P°_{toluene}, and the amounts of each compound. Molar masses are implicit knowns.

Relationship: Raoult's law provides a direct link between the knowns and the partial pressures of benzene and toluene in the vapor. Dalton's law allows the calculation of the total pressure when the partial pressures are known.

Missing Information: The mole fractions of the components are required for the application of Raoult's law, and the amounts (in moles) of each compound are required for calculation of mole fractions.

Plan

3. Calculate the total pressure from the partial pressures.
2. Calculate the partial pressures from the mole fractions and the pressures of the pure liquids.
1. Calculate the mole fractions of benzene and toluene.

Work

1.

$$n_{\text{benzene}} = (156.2\ \text{g}) \cdot \left(\frac{1\ \text{mol}}{78.1\ \text{g}}\right)$$
$$= 2.00\ \text{mol}$$
$$n_{\text{toluene}} = (276.3\ \text{g}) \cdot \left(\frac{1\ \text{mol}}{92.1\ \text{g}}\right)$$
$$= 3.00\ \text{mol}$$
$$X_{\text{benzene}} = \frac{n_{\text{benzene}}}{(n_{\text{benzene}} + n_{\text{toluene}})}$$
$$= \frac{2.00}{(2.00 + 3.00)}$$
$$= 0.400$$

Although the mole fraction of toluene could be calculated similarly, there is a simpler way. By definition, the sum of the mole fractions of all the components of a mixture is 1. In this case,

$$X_{\text{benzene}} + X_{\text{toluene}} = 1$$
$$X_{\text{toluene}} = 1 - X_{\text{benzene}}$$
$$= 1 - 0.400 = 0.600$$

2.

$$P_{\text{benzene}} = (X_{\text{benzene}})(P^{\circ}_{\text{benzene}}) \tag{12-6a}$$
$$= (0.400) \cdot (75\ \text{torr}) = 30\ \text{torr}$$
$$P_{\text{toluene}} = (X_{\text{toluene}})(P^{\circ}_{\text{toluene}})$$
$$= (0.60)(22\ \text{torr}) = 132\ \text{torr} \quad (!!)$$

3.

$$P_{\text{total}} = P_{\text{benzene}} + P_{\text{toluene}}$$
$$= 30\ \text{torr} + 132\ \text{torr} = 162\ \text{torr}$$

Check The total pressure of an ideal solution must be intermediate between the two pure-liquid pressures. Since this is *not* the case here, we know that an error has been made. Repeating the calculations, we locate the error in step 2. The correct calculation is

$$P_{\text{toluene}} = (0.60)(22\ \text{torr}) = 13\ \text{torr} \quad \text{and} \quad P_{\text{total}} = P_{\text{benzene}} + P_{\text{toluene}}$$
$$= 30\ \text{torr} + 13\ \text{torr} = 43\ \text{torr}$$

Note: The temperature is not used in the calculation. Its value is given only because the vapor pressures of the pure substances vary with temperature.

Exercise

At the normal boiling point of benzene (80 °C), the equilibrium vapor pressure of toluene is 350 torr. What is the vapor pressure at 80 °C of a benzene/toluene solution in which $X_{\text{toluene}} = 0.330$?

Answer: 625 torr.

EXAMPLE 12.3 Solution Composition and Vapor Pressure

At 50 °C, the vapor pressure of methanol is 400 torr and that of ethanol is 215 torr. What is the composition of a solution of these two liquids, if the vapor pressure of the solution at 50 °C is 300 torr?

Analysis

Target: The composition of the solution, expressed most conveniently in terms of the mole fractions of the components.

Knowns: $P°$ of the pure liquids, P of the solution, and the temperature.

Relationship: Dalton's law relates the total pressure to the component vapor pressures, Raoult's law provides the link between component mole fraction and vapor pressure, and the sum of the mole fractions must be 1.

Plan It may not be immediately obvious what steps are necessary to calculate the composition from the solution pressure. However, we have just solved a very similar problem, in which we calculated the pressure from composition. Let us assign the value y to one of the mole fractions, and follow the method of Example 12.2 to calculate the solution pressure in terms of y. This should lead to a relationship from which the numerical value of y can be obtained.

Work

1. Let $X_{\text{methanol}} = y$, so that $X_{\text{ethanol}} = 1 - y$
2. Raoult's law gives

$$P_{\text{methanol}} = (y) \cdot (400 \text{ torr}) \quad \text{and} \quad P_{\text{ethanol}} = (1 - y) \cdot (215 \text{ torr})$$

3. Dalton's law gives

$$P = P_{\text{methanol}} + P_{\text{ethanol}}$$

$$300 = (y)400 + (1 - y)215$$

This equation is the desired result, because it may be solved for y.

$$300 = 400y + 215 - 215y$$

$$300 - 215 = 400y - 215y$$

$$85 = 185y$$

$$y = \frac{85}{185} = 0.46$$

The mole fraction of methanol is 0.46, that of ethanol is $1 - 0.46 = 0.54$.

Exercise

Use the data of Example 12.2 to determine the composition of a benzene/toluene solution whose vapor pressure (at 20 °C) is 50 torr.

Answer: $X_{\text{benzene}} = 0.53$, $X_{\text{toluene}} = 0.47$.

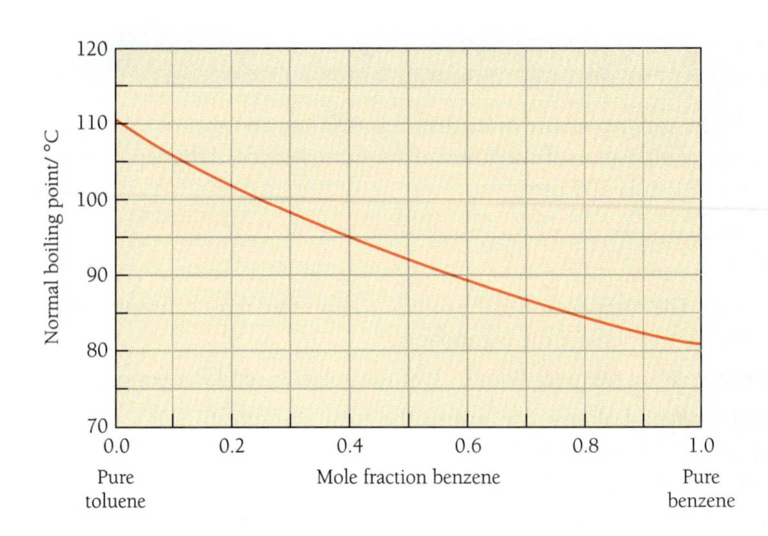

Figure 12.16 The normal boiling point of an ideal solution depends on composition. It varies smoothly (but not linearly) between the boiling points of the pure components. The normal boiling point of pure toluene is 110.6 °C, while that of pure benzene is 80.1 °C.

The composition in the vapor can be found from

$$X_{\text{benzene}} = \frac{P_{\text{benzene}}}{P_{\text{total}}} = \frac{30 \text{ torr}}{43 \text{ torr}}$$
$$= 0.70 = 70\% \text{ (mole)}$$

The composition of the vapor is not the same as the composition of the liquid. The vapor is always "richer in the more volatile component." This means that if component B is more volatile (has the higher vapor pressure when pure), as is the case in Figure 12.15, the mole fraction of B in the vapor phase will always be larger than the mole fraction of B in the liquid phase. For example, the solution in Example 12.2 is 40 mole% benzene and 60 mole% toluene. The more volatile of the two is benzene, and vapor in equilibrium with this solution is 70 mole% benzene and 30 mole% toluene.

The normal boiling point of a liquid is the temperature at which its equilibrium vapor pressure equals 1 atm. Since the vapor pressure of a solution depends on its concentration (Figure 12.15), it follows that the normal boiling point of a solution also depends on composition. The boiling point of an ideal solution varies between the boiling points of the two components, as shown in Figure 12.16.

Distillation

The new composition (89 mole%) can be calculated by repeating the procedure of Examples 12.2 and 12.4.

The fact that the vapor above a solution has a composition different from that of the liquid makes it possible to separate the components of a solution. This can be done in a process called **distillation.** Consider the solution discussed in Example 12.2. The liquid was 40 mole% benzene, while the vapor was 70 mole% benzene. If some of the vapor is removed and condensed to a liquid, it remains 70% benzene. When some of this condensate evaporates, the composition of *its* vapor would be 89% benzene. If the process of condensation and re-evaporation is repeated again and again, the percentage of benzene in the vapor steadily increases: 70%, 89%, 96%, 99%, and so on. In this way, a small amount of pure benzene can be obtained from the mixture. (Since only part of the condensate is evaporated, the amount of material decreases at each step.) The number of repetitive steps necessary for this purification depends on the initial composition and on the *difference* in the volatilities of the two substances. If the initial mole fraction of one of the components is relatively large, and the other component is much less volatile, a single distillation may produce a substance of high purity. If the two substances have similar volatilities, many steps are necessary. The use of distillation to purify liquids is widespread both in industrial processes (Figure 12.17) and in the research laboratory (Figure 12.18).

Figure 12.17 Separation of liquids by distillation is achieved on a large scale with an industrial distillation column. Each repetitive condensation and re-evaporation of the solution takes place in one of many identical chambers, called "plates," of the column. In operation, the more volatile of the components of the solution moves toward the top of the column, while the residual liquid at the bottom becomes progressively richer in the less volatile component. Many plates are required to separate liquids that have similar vapor pressures.

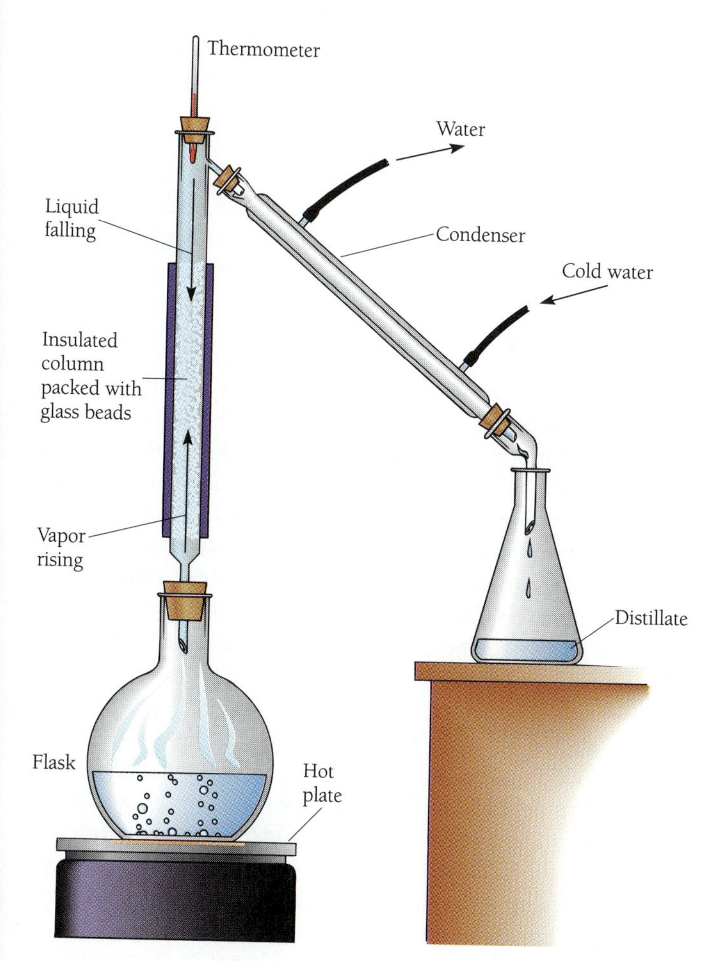

Figure 12.18 Distillation is carried out on a much smaller scale in routine laboratory operations. The "plates" of the laboratory column are not physically separated as they are in an industrial column. Instead, the column is loosely packed with small glass beads or helices, which provide sites for the repetitive condensation and re-evaporation necessary for good separation of the solution components.

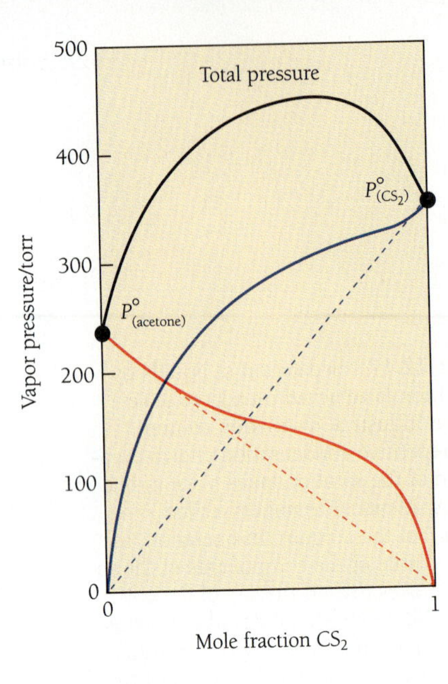

Figure 12.19 Solutions of acetone and carbon disulfide show large deviations from ideal behavior. The vapor pressure of such mixtures can be quite different from that predicted by Raoult's law. The solid lines show the measured partial pressures of the components and the total pressure, while the dashed lines represent the partial pressures expected if the solution were to behave ideally.

Real Solutions

Deviations from ideal behavior are common, and the vapor pressure of some solutions is greater than that of either component by itself. This behavior is illustrated in Figure 12.19, for solutions of acetone (CH_3COCH_3) and carbon disulfide. As with pure liquids, the equilibrium vapor pressure of a solution is controlled by the strength of the intermolecular attractive forces. If these forces are strong, molecules tend to remain in the liquid phase, and the vapor pressure is low. If the forces are weak, the vapor pressure is high. The cohesive forces in pure acetone are due primarily to dipole-dipole attraction. When nonpolar CS_2 is mixed with acetone, part of the dipole-dipole attraction is disrupted and replaced by the weaker dipole-*induced dipole* attraction. Both acetone and carbon disulfide molecules experience weaker cohesive forces in the solution than in the pure state, so their tendency to vaporize is greater.

Gaseous Solutes and Henry's Law

Ideal solutions form only between pairs of compounds that have very similar chemical and physical properties. This is not the case when one of the components is a liquid and the other is a gas, and such solutions deviate extensively from ideality. Nontheless, there are certain regularities in their behavior. For instance, the solubility of a gaseous solute usually decreases as the temperature increases (Section 12.1), and always increases as the gas pressure increases. The English chemist William Henry made a thorough investigation of the dependence of solubility on pressure. **Henry's law** (Equation 12-8) states that the molar solubility (S) of a gas in a liquid solvent is directly proportional to the partial pressure (P) of the gas above the liquid. The proportionality constant k_h is called the Henry's law constant.

William Henry (1774–1836) was a practicing physician as well as a chemist. This combination of careers is rare today.

$$S = k_h \cdot P \qquad (12\text{-}8)$$

The units in Equation (12-8) are S, mol L^{-1}; P, atm; and k_h, mol L^{-1} atm^{-1}. (Other units of solubility, such as grams per liter, mole fraction, or moles per kilogram, are sometimes used in Henry's law. Sometimes the law is written in inverse form, as $P = k'S$, where $k' = 1/k_h$. Be on the lookout for these variations if you use other texts or data sources.) The numerical value of the Henry's law constant is different for each gaseous solute and each liquid solvent, and it also changes if the temperature is changed. Values of k_h for several common gases dissolved in water at 25 °C are given in Table 12.4.

The modern trend toward standardized usage has not yet conquered personal preference in expressions of Henry's law.

EXAMPLE 12.4 Solubility of a Gas

Use the Henry's law constant given in Table 12.4 to calculate the aqueous solubility of CO_2 (g L^{-1}) when the carbon dioxide pressure is 5.0 atm and the temperature is 25 °C.

Analysis

Target: A solubility,

$$? \text{ g L}^{-1} =$$

Knowns: The Henry's law constant and the gas temperature and pressure. The molar masses of the solution components are implicit knowns.

Relationship: Equation (12-8) connects the solubility with the knowns. The units in (12-8) do not match the target units, so some conversion is required.

Plan Use Equation (12-8) to find the solubility in mol L^{-1}, then convert the units from mol L^{-1} to g L^{-1}.

Work

1.
$$\begin{aligned} S &= k_h \cdot P \qquad (12\text{-}8) \\ &= (0.034 \text{ mol L}^{-1} \text{ atm}^{-1}) \cdot (5.0 \text{ atm}) \\ &= 0.17 \text{ mol L}^{-1} \end{aligned}$$

2. The molar mass of CO_2 is 44.0 g mol^{-1}, so

$$\begin{aligned} \text{solubility} &= (0.17 \text{ mol L}^{-1}) \cdot \left(\frac{44.0 \text{ g}}{1 \text{ mol}}\right) \\ &= 7.5 \text{ g L}^{-1} \end{aligned}$$

Check The units are correct. We do not yet have enough experience with gaseous solubilities to know if the answer is numerically reasonable, but a mental estimate ($5 \cdot 0.03 = 0.15 \approx 0.2$; $0.2 \cdot 40 = 8$) shows that our result has the correct magnitude.

Exercise

Calculate the mass of carbon dioxide that will dissolve in 350 mL water at 25 °C when the CO_2 pressure is 17 atm. **Answer:** 8.9 g.

TABLE 12.4 Henry's Law Constants for Gases Dissolved in Water at 25 °C

Gas	k_h/mol L^{-1} atm^{-1}
H_2	0.00079
N_2	0.00065
O_2	0.0013
CH_4	0.0013
CO_2	0.034

12.4 NONVOLATILE SOLUTES AND COLLIGATIVE PROPERTIES

The components of an ideal solution behave more or less as they do when in the pure state. Although not many solutions behave ideally as far as both components are concerned, the behavior of the *solvent* is often close to ideal. The reason is that, in a dilute solution, the solvent is *almost* pure. The properties of dilute solutions therefore depend primarily on the properties of the solvent. Indeed, there are several interrelated solution properties, called **colligative properties,** that are independent of the nature of the solute. These properties depend on the concentration but not the identity of the solute, provided that (a) the solute is nonvolatile, and (b) the solution is dilute. In the context of colligative properties, a dilute solution is one whose solute concentration is less than about 0.05 M.

The word "colligative" comes from the Latin *colligatus,* which means "bound together."

The four colligative properties, discussed in the following sections, are vapor pressure decrease, boiling point increase, freezing point decrease, and osmotic pressure.

Vapor Pressure Decrease

The vapor in equilibrium with a solution of a nonvolatile substance contains *solvent* molecules only.

The effect of the addition of a small amount of a nonvolatile solute on the vapor pressure of a liquid can be predicted from the kinetic molecular theory. The vapor pressure of a liquid is controlled primarily by the rate at which molecules, located in the surface layer of the liquid, pass into the vapor phase. This rate in turn depends on the rate at which surface molecules experience high-energy collisions from below. If the liquid is not pure, but contains some solute molecules, some of the surface locations will be occupied by solute molecules. Energetic collisions with *solute* molecules in surface locations do not affect the vapor pressure, because the solute is nonvolatile. Solute molecules do not enter the gas phase no matter how hard they are bumped. Suppose the solution contains 1 mole percent solute. This means that 1% of the surface locations are occupied by solute molecules, and 99% of the surface locations are occupied by solvent molecules. Therefore, only 99% of the energetic collisions are effective in bumping molecules into the gas phase, and the vapor pressure is 99% of the pressure of the pure solvent. The solute physically blocks some surface sites, but the chemical properties of the solute are not involved. This effect is the colligative property known as the **vapor pressure decrease** (Figure 12.20).

The extent of vapor pressure decrease can be predicted from Raoult's law. For the *solvent* in an ideal solution, the general form of Raoult's law (Equation 12-6c) becomes

$$P_{\text{solvent}} = X_{\text{solvent}} P^{\circ}_{\text{solvent}} \qquad (12\text{-}6d)$$

If the solute is nonvolatile, P_{solution} is equal to P_{solvent}, and the vapor pressure of the *solution* is given by Equation (12-9).

$$P_{\text{solution}} = X_{\text{solvent}} P^{\circ}_{\text{solvent}} \qquad (12\text{-}9)$$

It is more convenient to think about the properties of solutions in terms of the amount of solute present rather than the amount of solvent. When only two components are present, $X_{\text{solvent}} = (1 - X_{\text{solute}})$. Substitution of this relationship into Equation (12-9) and rearrangement gives an expression for the vapor pressure decrease in terms of the mole fraction of solute.

$$\Delta P = (P^{\circ}_{\text{solvent}} - P_{\text{solution}}) = X_{\text{solute}} P^{\circ}_{\text{solvent}} \qquad (12\text{-}10)$$

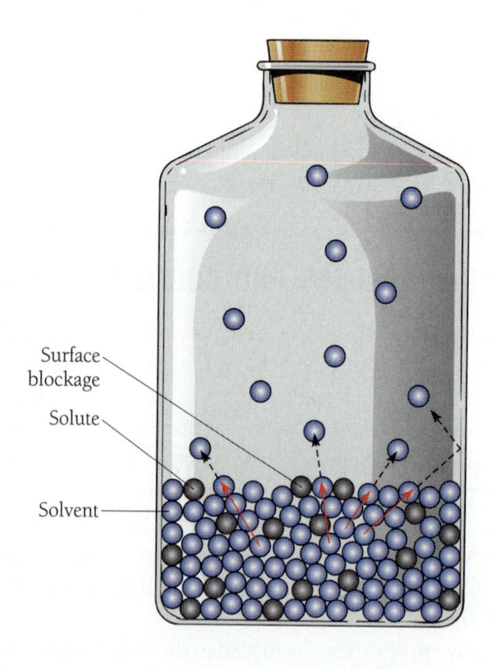

Figure 12.20 Only those solvent molecules lying at the surface can evaporate into the gas phase. If some surface sites are occupied by nonvolatile solute species, collisions experienced at those locations do not result in evaporation. The number of *solvent* molecules entering the gas phase, and therefore the vapor pressure, is decreased in proportion to the extent of surface blockage.

EXAMPLE 12.5 Vapor Pressure Decrease

What is the composition of an aqueous solution whose vapor pressure at 50 °C is 90.0 torr? The vapor pressure of pure water at this temperature is 92.5 torr, and the solute is nonvolatile.

Analysis

Target: The composition of a solution is asked for. We are not told *which* measure of composition (molarity, mole fraction, mass %, and so on) should be used, so any of these is satisfactory.

Knowns: The identity of the solvent, its vapor pressure $P°$, and the solution vapor pressure P are explicit in the problem statement.

Relationship: Is there a formula that connects the target and the knowns? Yes. In fact, there are two. Raoult's law (Equation 12-6d) relates the vapor pressure of the solvent to the mole fraction of the *solvent,* and Equation (12-10) relates the vapor pressure decrease to the mole fraction of the *solute.* Since the problem asks only for the composition, the mole fraction of either component is a satisfactory answer, and either equation can be used. We choose Equation (12-10) and solve the problem.

Work

$$\Delta P = (P^{\circ}_{\text{solvent}} - P_{\text{solution}}) = X_{\text{solute}} P^{\circ}_{\text{solvent}} \qquad (12\text{-}10)$$

$$X_{\text{solute}} = \frac{\Delta P}{P^{\circ}} = \frac{(P^{\circ}_{\text{solvent}} - P_{\text{solution}})}{P^{\circ}_{\text{solvent}}} = \frac{(92.5 \text{ torr} - 90.0 \text{ torr})}{92.5 \text{ torr}} = 0.027$$

Check The answer is a dimensionless number between 0 and 1, as a mole fraction must be, and it is a small number, as is appropriate for the mole fraction of the solute in a dilute solution.

Exercise

Pure toluene (molar mass 92.1 g mol^{-1}) has a vapor pressure of 350 torr at 80 °C. What is the vapor pressure of a solution of 1.00 g of the nonvolatile compound phenanthrene (molar mass 178 g mol^{-1}) in 100 g toluene?

Answer: 348 torr.

Changes in Boiling Point and Freezing Point

The fact that the vapor pressure of a solution is less than the vapor pressure of the pure solvent means that the boiling point is different also. The normal boiling point of a liquid is the temperature at which its vapor pressure becomes equal to one atmosphere. As shown in Figure 12.21, if the vapor pressure is decreased by the presence of a nonvolatile solute, the temperature must be increased in order to bring the vapor pressure back up to 1 atm. That is, the solution boils at a somewhat higher temperature. This increase in boiling point is called the **boiling point elevation (ΔT_b)**. It is a colligative property, because it depends on the concentration and not the identity of the solute.

A property closely related to ΔT_b is the change in the *freezing point* of a liquid when a nonvolatile solute is dissolved in it. The solution always freezes at a temperature less than the freezing point of the pure solvent. The decrease, called the **freezing point depression (ΔT_f)**, is also a colligative property. Figure 12.22 shows the relationship between freezing point and vapor pressure in solutions of nonvolatile solutes.

Molality and Calculations of ΔT_f and ΔT_b

The values of both ΔT_b and ΔT_f are directly proportional to solution composition. Although it is possible to describe composition in terms of the molarity of the solute, there are two disadvantages to the use of molarity in discussions of ΔT_b and ΔT_f.

1. Molarity does not describe the quantity of *solvent*, since it is defined in terms of the volume of *solution*. The volume of a solution, particularly of a concentrated solution, is not the same as the volume of solvent used to prepare it.
2. The molarity of a solution changes if the temperature is changed. This happens because solutions, like other substances, expand when heated. For example, suppose an aqueous solution is exactly 1.000 M when the temperature is 20 °C. If the temperature rises to 30 °C, which can easily happen in a non-air-conditioned laboratory in summertime, the solution expands by 0.3% and the volume becomes 1.003 L. The molarity changes to 1.000/1.003 = 0.997 M at the higher temperature. This variability is small enough to be unimportant in many laboratory operations, but must be taken into account in precise work.

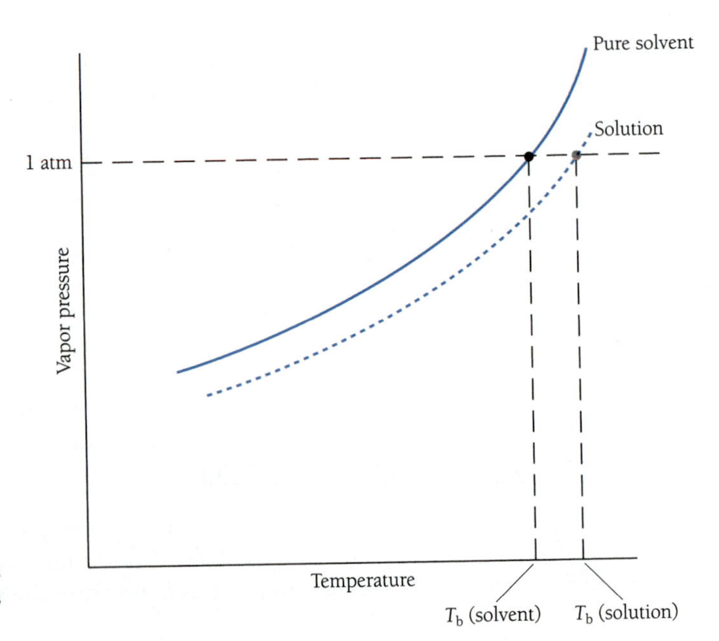

Figure 12.21 Regardless of the temperature, the vapor pressure of a solution of a nonvolatile solute (dotted line) is less than that of the pure solvent (solid line). The normal boiling point, T_b, is the temperature at which the lines cross the P = 1 atm mark (dashed line). The solution boils at the higher temperature.

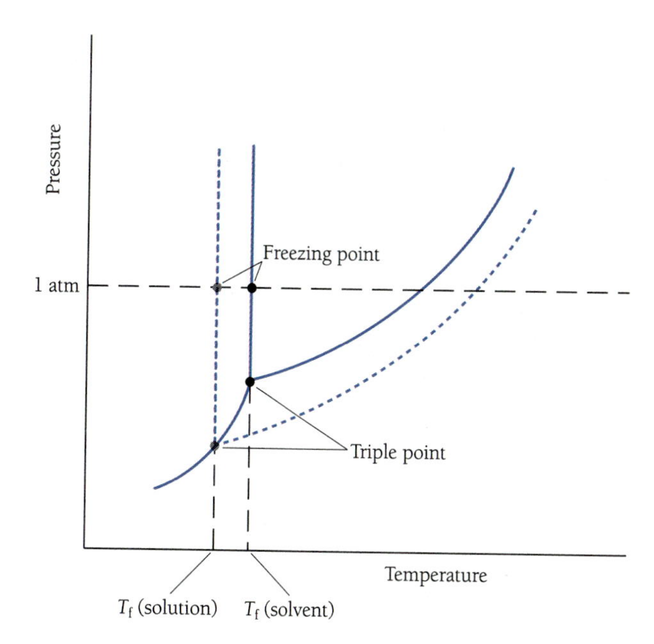

Figure 12.22 A portion of the phase diagram of a volatile substance when pure (solid lines) and when containing a non-volatile solute (dashed lines). As in Figure 12.21, there is a decrease in vapor pressure at all temperatures. Such solutes are normally insoluble in the *solid* solvent, so there is no corresponding decrease in the vapor pressure of the solid. Both the triple point and the freezing point of the solution are less than those of the pure solvent.

The **molality** of a solution is defined as the number of moles of solute per kilogram of *solvent.*

$$\text{molality} = \frac{\text{(moles solute)}}{\text{(kg solvent)}} \tag{12-11}$$

The symbol for molality is "*m*," and its units are mol kg^{-1}. A 1 *m* solution is said to be a "one *molal* solution." Since neither the amount of solute nor the mass of solvent depends on temperature, the molality of a solution is not affected by a temperature change. To prepare a solution of a given molality, both solute and solvent are weighed. (Note that if the solvent density is known, and it usually is, volume measurement can take the place of weighing of the solvent.)

A lowercase "*m*" is always used for molality, while an uppercase "M" is used to abbreviate molarity.

EXAMPLE 12.6 Calculation of Molality

Because fluoride ion is effective in reducing the incidence of cavities in teeth, small quantities of sodium fluoride are added to municipal water supplies in some localities. What is the molality of a solution of 18.4 g NaF in 350 mL H_2O at 25 °C?

The chemistry of tooth protection by fluoride is discussed in Section 7.6.

Analysis

Target: Solution molality,

$$? \text{ mol (solute) kg}^{-1} \text{ (solvent)} =$$

Knowns: Mass of solute and volume of solvent; the molar mass of the solute is implicit in its identity; the density of pure water at 25 °C, 1.00 kg L^{-1}, is another implicit known.

Relationship: Equation (12-11) is used in calculations of molality.

Missing Information: The mass of solvent is needed; it can be obtained from the density and the given volume.

Plan

3. Use Equation (12-11) to find the molality.
2. Calculate the number of moles of NaF.
1. Find the solvent mass from its volume and density.

Work

1.

$$\text{Solvent mass} = (350 \text{ mL } H_2O) \cdot (1.00 \text{ g mL}^{-1}) \cdot \left(\frac{1 \text{ kg}}{1000 \text{ g}}\right)$$
$$= 0.350 \text{ kg } H_2O$$

2. The molar mass of the solute is 23.0 (Na) + 19.0 (F) = 42.0 g mol^{-1}, so

$$n = (18.4 \text{ g NaF}) \cdot \left(\frac{1 \text{ mol}}{42.0 \text{ g}}\right)$$
$$= 0.438 \text{ mol NaF}$$

3.

$$\text{molality} = \frac{(\text{moles solute})}{(\text{kg solvent})} \qquad (12\text{-}11)$$
$$= 1.25 \text{ mol kg}^{-1} = 1.25\ m$$

Exercise

Another health-related use for a sodium salt is the preparation of "normal saline," a sterile, 0.85% (w/w) aqueous solution of sodium chloride. The molality of this solution is 0.155 mol kg^{-1}. Normal saline has approximately the same ion concentration as blood plasma, and it is a useful solvent for intravenously administered medication. How much NaCl must be added to 500 g water to prepare a 0.155 m solution?

Answer: 4.53 g NaCl.

Be careful to distinguish between "molar" and "molal" in speech as well as in writing.

The relationship between molarity and molality involves the molar mass ($\mathcal{M}$) of the solvent as well as the density (D) of the solution.

$$\text{molarity} = \frac{(\text{molality}) \cdot D}{1 + (\text{molality}) \cdot \left(\frac{\mathcal{M}}{1000}\right)}$$

One disadvantage of the use of molality is that there is no simple relationship between the easily measured volume of a solution and the quantity of solute it contains. The amount of solute in, say, 500 mL of 0.6 *molar* solution is (0.6 mol L^{-1}) · (0.500 L) = 0.300 mol. The amount of solute in 500 mL of 0.6 *molal* solution, on the other hand, cannot be determined without weighing the solution or knowing its density. The disadvantage disappears when dealing with dilute aqueous solutions, however. In dilute (less than about 0.1 M or 0.1 m) aqueous solutions, the value of molality is almost the same as that of molarity. This is so because the density of a dilute aqueous solution is about the same as that of pure water: 1.0 kg L^{-1}. That is, "kg solvent" and "L solution," the denominators of the equations defining molality (m = moles/kg H_2O) and molarity (M = moles/L solution), have essentially the same numerical value. For example, a 2%(w/w) aqueous solution of NH_4NO_3 has a molarity of 0.252 mol L^{-1} and a molality of 0.255 mol kg^{-1}. The discrepancy between molarity and molality becomes less as an aqueous solution is made more dilute.

TABLE 12.5 Some Boiling Point and Freezing Point Constants

Solvent	Molar mass/g mol^{-1}	T_f^*/°C	K_f/°C kg mol^{-1}	T_b^*/°C	K_b/°C kg mol^{-1}
Water	18.0	0.00	1.86	100.00	0.512
Ethanol	46.1	−114.6	1.99	78.4	1.22
Acetone	58.1	−95.8	2.68	56.1	1.73
Benzene	78.1	5.5	5.12	80.1	2.53
Naphthalene	128.2	80.2	6.8	218	6.4

* Values of T_f and T_b are for the pure solvent.

We now return to our discussion of the colligative properties, boiling point elevation and freezing point depression. Both ΔT_b and ΔT_f are proportional to the solute molality (m). The relationships are

$$[T_b(\text{solution}) - T_b(\text{pure})] \equiv \Delta T_b = K_b \cdot m \qquad (12\text{-}12)$$

and

$$[T_f(\text{solution}) - T_f(\text{pure})] \equiv \Delta T_f = -K_f \cdot m \qquad (12\text{-}13)$$

Note that ΔT_f is a negative quantity, because the freezing point of a solution is *less* than that of the pure solvent. K_b and K_f are called the **boiling point constant** and the **freezing point constant,** respectively, and both are positive quantities. The values of K_f and K_b are different from each other, and they differ from one solvent to another. Values of K_f and K_b for several common solvents are listed in Table 12.5.

EXAMPLE 12.7 Boiling Point Elevation

What is the boiling point of a 10.0 %(w/w) aqueous solution of sucrose ($C_{12}H_{22}O_{11}$)?

Analysis

Target: A boiling point,

$$?\ T_b\ (°\text{C or K}) =$$

Knowns: Identity of the solute and solvent, and the composition, are explicit. The boiling point (100 °C) and the boiling point constant (0.512 °C kg mol^{-1}) of the solvent, and the molar mass of the solute (342.1 g mol^{-1}) are implicit knowns.

Clearly there are other implicit knowns, for instance the molar mass and density of water. Since it is often difficult to decide in advance which hidden knowns will be useful in a particular problem, a good strategy is to begin planning the solution without making an exhaustive list. Then, as the plan or the work proceeds, be on the lookout for missing information that might be supplied by implicit knowns.

Relationship: Is there a formula? No, there is not. Equation (12-12) relates solution composition and boiling point, but it is in terms of molality, not weight percent. If we knew the molality, we could use it in Equation (12-12) to solve the problem.

This "if only we knew" statement in fact defines a new, subsidiary problem that must be solved first. The approach is the same.

PROBLEM

What is the molality of a 10% aqueous solution of sucrose?

Analysis

Target: A molality,

$$? \ m \text{ mol kg}^{-1} =$$

Knowns: Same as in the main problem

Relationship: Both molality and mass % involve the amount or mass of the components.

$$m = \frac{\text{moles solute}}{\text{kilograms solvent}} \quad \text{and}$$

$$\text{mass \%} = \left(\frac{\text{mass solute}}{\text{mass solution}}\right) \cdot 100$$

Plan Use the mass % to calculate the mass of sucrose and water in some specific amount (100 g is a good choice) of solution. Convert the mass of sucrose to moles, and divide by the mass of water in kilograms.

Work

1. The mass of sucrose in 100.0 grams of a 10% solution is

$$(\text{grams sucrose}) = \left(\frac{\text{mass \%}}{100}\right) \cdot (\text{grams solution})$$

$$= \left(\frac{10.0}{100}\right) \cdot (100.0 \text{ g}) = 10.0 \text{ g}$$

The mass of water in 100 g of solution containing 10.0 g sucrose is

$$100.0 \text{ g} - 10.0 \text{ g} = 90.0 \text{ g}$$
$$= 0.0900 \text{ kg } H_2O$$

2. The molar mass of sucrose ($C_{12}H_{22}O_{11}$) is 342.1 g mol^{-1}; the molality of the solution is

$$m = \frac{(\text{moles sucrose})}{(\text{kg water})}$$

$$= \frac{(10.0 \text{ g}) \cdot (1 \text{ mol}/342.1 \text{ g})}{(0.0900 \text{ kg})}$$

$$= 0.325 \text{ mol kg}^{-1}$$

We can now formulate a plan for the solution to the *main* problem.

Plan

1. Use Equation (12-12) and the molality to find the boiling point change.
2. Calculate T_b from ΔT_b and the boiling point of pure solvent.

Work

1. The boiling point elevation is

$$\begin{aligned} \Delta T_b &= m \cdot K_b \qquad (12\text{-}12) \\ &= (0.325 \text{ mol kg}^{-1}) \cdot (0.512 \text{ °C kg mol}^{-1}) \\ &= 0.166 \text{ °C} \end{aligned}$$

2. The boiling point of the solution is

$$\begin{aligned} T_b &= 100.0 \text{ °C} + \Delta T_b \\ &= (100.0 + 0.166) \text{ °C} \\ &= 100.2 \text{ °C} \end{aligned}$$

Exercise

Calculate the freezing point of a 1.54% solution of dichlorobenzene ($C_6H_4Cl_2$) in naphthalene ($C_{10}H_8$).

Answer: 79.5 °C.

Example 12.7 shows that, even for a fairly concentrated solution, the change in boiling point is small. For a solvent like benzene, whose value of K_b is five times larger, the change would be five times larger. However, it would still be only about 1 degree. Accurate measurement of temperature changes this small require specialized equipment. Nevertheless, a good-quality laboratory thermometer, accurate to 0.1 °C, can be used for the determination of *approximate* values of molar mass.

EXAMPLE 12.8 Freezing Point Depression

Pure benzene freezes at 5.5 °C. A solution prepared by dissolving 900 mg of an unknown substance in 27.3 g benzene is found to freeze at 4.2 °C. What is the approximate molar mass of the unknown substance?

Analysis

Target: A molar mass,

$$? \text{ g mol}^{-1} =$$

Knowns: Solute and solvent masses, and the freezing points of the pure solvent and the solution. The freezing point constant and molar mass of benzene are implicit knowns.

Relationship: Equation (12-13) seems relevant because it deals with freezing point depression, but insertion of the known values would lead to a value of molality, not molar mass. However, knowing the molality and the mass of solute, we can calculate its molar mass.

Plan

3. Calculate the molar mass from the number of moles and mass of the dissolved material.

2. Calculate the number of dissolved moles from the molality and the amount of solvent.
1. Calculate the molality from the known information and Equation (12-13).

Work

1.

$$\Delta T_f = -m \cdot K_f \qquad (12\text{-}13)$$

$$m = \frac{-\Delta T_f}{K_f}$$

$$= -\frac{(4.2\ °C - 5.5\ °C)}{5.12\ °C\ kg\ mol^{-1}}$$

$$= +0.254\ mol\ kg^{-1} = 0.254\ m$$

2. Rearrange the equation defining molality, and solve.

$$m = \frac{(\text{mol solute})}{(\text{kg solvent})}$$

$$\begin{aligned}(\text{mol solute}) &= (\text{molality}) \cdot (\text{kg solvent}) \\ &= [0.254\ \text{mol (kg solvent)}^{-1}] \cdot (0.0273\ \text{kg solvent}) \\ &= 0.00693\ \text{mol}\end{aligned}$$

3.

$$\text{molar mass} = \frac{\text{grams}}{\text{moles}} = \frac{900\ \text{mg}}{0.00693\ \text{mol}}$$

$$= \frac{0.900\ \text{g}}{0.00693\ \text{mol}} = 1.3 \times 10^2\ \text{g mol}^{-1}$$

Check The units are consistent throughout. Furthermore, 100 g mol^{-1} or so is not an unreasonable value for a molar mass.

1.03 g of an unknown substance was added to 97.3 g naphthalene ($C_{10}H_8$), and the mixture was found to have a freezing point of 78.6 °C. What is the molar mass of the unknown?

Answer: 45 g mol^{-1}.

Molar masses determined by this method are of limited accuracy. The equation $\Delta T_f = m \cdot K_f$ is exact in dilute solutions, but a ΔT of less than a few tenths of a degree is difficult to measure accurately. Larger ΔT's can be obtained in more concentrated solutions, but the equation $\Delta T_f = m \cdot K_f$ is then only approximately correct. Freezing point depression is a convenient method for determining molar mass to an accuracy of no more than two significant digits. Furthermore, the method is limited to compounds of low molar mass. A molality of ≈ 0.1 is required to produce a freezing point change large enough to measure accurately, and most substances with high molar mass have solubilities less than 0.1 *m*. The fourth colligative property, osmotic pressure, offers a method of measuring high molar masses.

Osmosis and Osmotic Pressure

Consider the experimental setup shown in Figure 12.23. In the right-hand flask there is an aqueous solution of a nonvolatile solute. The left flask contains pure water, and the two flasks are connected by an airtight tube. Initially the amounts of liquid in the two flasks are the same, but gradually there is a transfer of water from the pure solvent to the solution.

The reason for this change lies in the effect of the solute on the vapor pressure. Suppose the equilibrium vapor pressure of the pure solvent ($P°$) is 25 torr while the vapor pressure of the solution (P) is 24 torr. The *pure solvent* tries to maintain a vapor pressure of 25 torr by evaporation of liquid, while the *solution* tries to maintain a vapor pressure of 24 torr, by condensation of any solvent vapor in excess of 24 torr. The result is the gradual but inevitable transfer of solvent molecules from the pure solvent to the solution. The effect of this is that the solution becomes more dilute. Dilution is a spontaneous process because it is accompanied by an increase in entropy. Solute molecules are more disordered when they are spread out in the larger volume of a more dilute solution.

A more general description of the result of the experiment in Figure 12.23 is, *whenever a solution is separated from pure solvent by a region through which solvent but not solute can pass, there is a spontaneous flow of solvent into the solution, which thereby becomes more dilute.* The vapor phase is not the only, nor even the most common, such region. A thin sheet of cellophane, the skin of a grape, or a cell wall in plant or animal tissue, is such a region. A thin layer of material separating regions of different composition is referred to as a **membrane,** and, if the membrane allows the free passage of only certain types of molecules, it is called a **semipermeable membrane.** One way of thinking about the structure of a semipermeable membrane is as a thin sheet perforated with microscopic holes that are large enough for small molecules like water to pass through, but too small to allow passage of larger molecules that may be dissolved in the water. While this may be an adequate description of an artificial membrane such as cellophane, naturally occurring membranes such as cell walls are considerably more complex. Transport of substances through membranes is incompletely understood, and is the subject of current research.

(a)

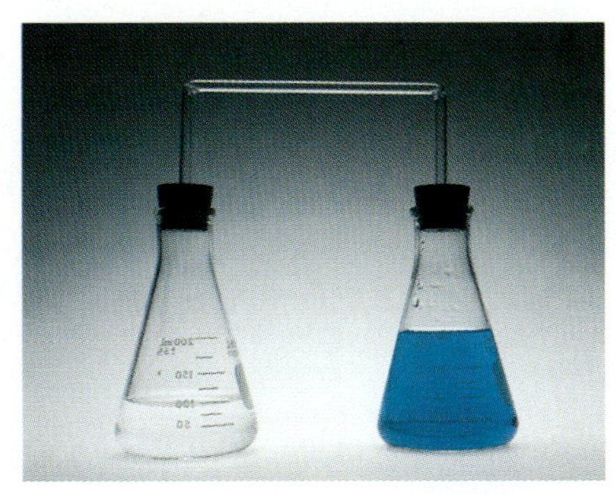

(b)

Figure 12.23 (a) An aqueous solution of copper sulfate, a nonvolatile solute, is connected to a flask containing pure solvent by a tube through which vapor, but not liquid, can freely pass. (b) After several weeks, the volume of the solution has increased and the volume of the pure solvent has decreased. A gradual transfer of solvent through the vapor phase has taken place.

Solvent flow takes place spontaneously (but rather slowly) across semipermeable membranes, always in the direction that causes the solution to become more dilute. This type of flow is called **osmosis.** The experimental setup in Figure 12.24 can be used to demonstrate osmotic flow.

The process depicted in Figure 12.24 eventually comes to a halt. As the liquid level in the right-hand flask rises, the pressure on the right-hand side of the membrane becomes increasingly larger than the pressure on the left-hand side. This pressure difference induces a flow of water through the membrane from right to left. When the pressure difference reaches the point where the right-to-left pressure-induced flow is equal to the left-to-right osmotic flow, equilibrium has been reached and no further change of the liquid level in either flask takes place. The pressure difference at equilibrium is called the **osmotic pressure,** and it is one of the colligative properties of solutions. The magnitude of the osmotic pressure is independent of the nature of the membrane, the solvent, and the solute (provided that it cannot pass through the membrane). Osmotic pressure (π) may be calculated by Equation (12-14), in which M is the molarity, R is the gas constant, and T is the temperature in kelvins.

$$\pi = \mathrm{M}RT \tag{12-14}$$

The units of M (mol L^{-1}), R (L atm mol^{-1} K^{-1}), and π (atm) must be consistent: use $R = 0.08206$ L atm mol^{-1} K^{-1}, and convert π to atm if it is given in some other pressure unit.

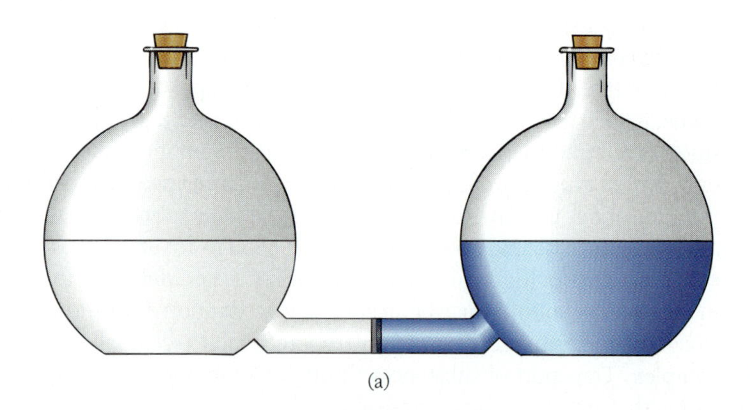

(a)

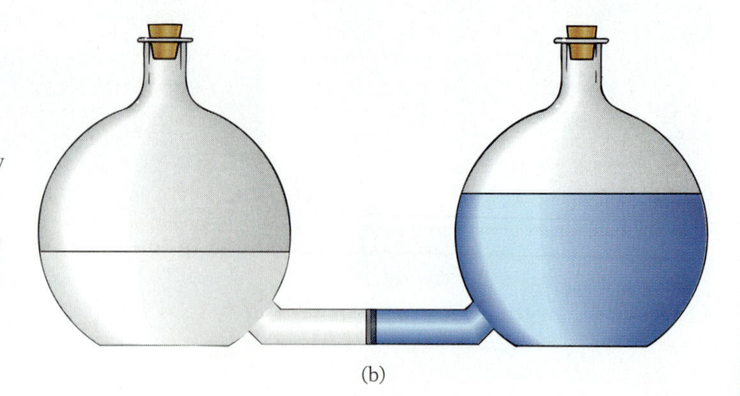

(b)

Figure 12.24 When a solution is separated from pure solvent by a semipermeable membrane through which only solvent can pass, there is a gradual transfer of solvent to the solution. The experiment begins in (a) with equal volumes of liquid in the two containers; (b) shows the situation some time later, after osmotic flow has taken place. The result is an increase in volume and a decrease in concentration of the solution. As far as the solution and solvent are concerned, there is very little difference between the experiments depicted in Figures 12.23 and 12.24.

Osmotic pressure can be determined by measurement of the height of the liquid levels after equilibrium has been attained in an apparatus similar to the one in Figure 12.24. Osmotic pressure is directly proportional to the height difference Δh. For aqueous solutions, the equivalence statement, from which the conversion factor may be generated, is

$$1 \text{ atm} \Longleftrightarrow 1.034 \times 10^4 \text{ mm } H_2O \qquad (12\text{-}15)$$

This equivalence statement is derived as follows. A pressure difference of 1 atm will support a column of mercury 760 mm high. The density of mercury is 13.6 times the density of water, so a column of water supported by 1 atm would be $(13.6) \cdot (760) = 1.034 \times 10^4$ mm high.

One of the applications of osmosis is the measurement of the very high molar masses of proteins.

The chemistry of proteins is discussed in Chapter 21.

EXAMPLE 12.9 Determination of a Large Molar Mass

Cytochrome C, an enzyme involved in the release of energy in muscle activity, is a protein of relatively high molar mass. At 25 °C a solution containing 0.0100 g in 100 mL H_2O has an osmotic pressure that is large enough to support a height difference of 1.90 mm between the solution and pure water. What is the molar mass of the material?

Analysis

Target: A molar mass,

$$? \text{ g mol}^{-1} =$$

Knowns: Temperature, Δh (water), and the concentration in g/100 mL.

Relationship: Equation (12-14) relates osmotic pressure to concentration, but in most cases cannot be used directly because of incompatible units. The height difference must be converted to a pressure [Equation (12-15)], and the correct units for R must be used. The result of using Equation (12-14) will be a concentration in moles per liter, and molarity can be combined with the composition in g L^{-1} to obtain the desired molar mass.

Plan

3. Calculate the molar mass from the concentration in g L^{-1} and the molarity.
2. Calculate the molarity from the given information and Equation (12-14), taking care that the units are correct.
1. Convert the height differential to osmotic pressure with Equation (12-15).

Work

1.

$$\pi = [\Delta h \text{ (mm } H_2O)] \cdot \left(\frac{1 \text{ atm}}{1.034 \times 10^4 \text{ mm } H_2O}\right) \qquad (12\text{-}15)$$

$$= (1.90 \text{ mm } H_2O) \cdot \left(\frac{1 \text{ atm}}{1.034 \times 10^4 \text{ mm } H_2O}\right)$$

$$= 1.838 \times 10^{-4} \text{ atm}$$

2.

$$\pi = MRT \qquad (12\text{-}14)$$

$$M = \frac{\pi}{RT}$$

$$= \frac{1.838 \times 10^{-4}\ \text{atm}}{(0.08206\ \text{L atm mol}^{-1}\ \text{K}^{-1})(298\ \text{K})}$$

$$= 7.514 \times 10^{-6}\ \text{mol L}^{-1}$$

3. One liter of solution contains 7.514×10^{-6} mol solute. According to the problem statement, 0.100 L (100 mL) contains 0.0100 grams solute. This establishes the equivalence statement for the amount contained in 1 L:

$$0.100\ \text{g} \Longleftrightarrow 7.514 \times 10^{-6}\ \text{mol}$$

Dividing both sides by 7.514×10^{-6}, we find

$$\left(\frac{0.100\ \text{g}}{7.514 \times 10^{-6}}\right) \Longleftrightarrow \left(\frac{7.514 \times 10^{-6}\ \text{mol}}{7.514 \times 10^{-6}}\right)$$

$$1.33 \times 10^{4}\ \text{g} \Longleftrightarrow 1\ \text{mol}$$

That is, the molar mass is 1.33×10^{4} g mol^{-1}.

Exercise

Calculate the molar mass of human hemoglobin, given that at 25 °C a solution containing 0.0314 g in 100 mL H_2O has an osmotic pressure that will support a column of water 1.23 mm high.

Answer: 64,500 g mol^{-1}.

Osmosis plays an important role in the chemistry of living systems. For instance, suppose some cells are removed from the organism and placed in pure water. A cell is, in a sense, an aqueous solution surrounded by a semipermeable membrane. To be sure, it is a very complex solution containing a large number of solutes, but it is nonetheless a solution. Osmosis occurs, and water passes through the membrane into the cell. The cell expands, and the membrane stretches to accommodate the increasing volume. Ultimately the limit of elasticity of the membrane is reached, and it bursts, destroying the cell. This process is easily observed with red blood cells. Osmotic rupture of red blood cells is called *hemolysis*.

Osmotic flow takes place when two solutions are separated by a semipermeable membrane. It is not necessary that one of the liquids be pure solvent, only that the two solutions have different concentrations. Osmotic flow takes place in the direction that tends to equalize the concentrations of the solutions. Thus, in hemolysis, water passes into the cell, making it more dilute so that its concentration becomes more like that of the liquid on the other side, pure water. If a cell is placed in a concentrated salt solution rather than in pure water, osmotic flow takes place in the opposite direction. Water passes out of the cell and into the concentrated solution. In this process, called *crenation*, the cell shrivels and shrinks in size. If the concentration of the solution surrounding the cell is just right, there is no net flow in either direction. Two solutions having the same osmotic pressure are called *isotonic*. A solution having a higher osmotic pressure (than a cell or another solution to which it is compared) is called *hypertonic.* If it has a lower concentration it is called *hypotonic.* A "normal saline" solution (0.9% w/w NaCl) is isotonic to blood. Normal saline is used, among other things, in the treatment of burns, shock, and dehydration, and as a solvent for medicines that must be administered intravenously.

The selectivity of membranes to pass some substances and block others varies greatly from one type of membrane to another. Some membranes, for example, allow the passage not only of water but also of small dissolved ions and molecules, while blocking larger molecules. The process that takes place when different solutions are connected by such a membrane is called *dialysis* rather than osmosis. Dialysis is used in research to purify, that is, remove the ions from, large molecules isolated from living systems. It is also the principle on which the artificial kidney is based (Figure 12.25). The function of the kidneys is the removal of metabolic wastes from the bloodstream. An artificial kidney is a machine containing a pump, filters, and a tubular semipermeable membrane. In use, a portion of the patient's blood flow is diverted from the body to the artificial kidney, where it makes contact through the membrane with the "dialysate," a solution containing all the essential dissolved components of the blood *except* the harmful wastes. The waste products, primarily urea (NH_2CONH_2, a small molecule), gradually pass from the blood to the dialysate. The essential solutes are not lost from the blood, either because their concentrations are the same on both sides of the membrane, or because they are too large to pass through the membrane.

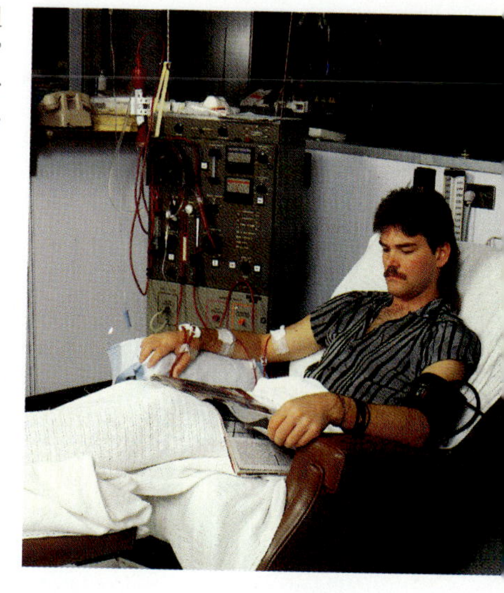

Figure 12.25 A patient undergoing dialysis with an artificial kidney.

Another practical use of osmosis is *desalination,* or the removal of salt from seawater (Figure 12.26). Seawater is forced at high pressure through semipermeable membrane tubes immersed in pure water. Because the mechanical pressure is greater than the osmotic pressure, the net flow of water through the membrane is from the seawater to the pure water side. The seawater becomes a more concentrated brine (salt solution), while pure water collects on the other side of the membranes.

The flow of water *from* the seawater, which leaves it more concentrated, is opposite to normal osmotic flow. Therefore, the desalination process is sometimes called "reverse osmosis."

The major problems associated with desalination by this method are that the process is relatively slow, and that the membrane tubes are easily clogged and subject to rupture. The required pressure may be estimated from Equation (12-14): the composition of seawater corresponds roughly to 1.1 moles of dissolved ions (Na^+, Mg^{2+}, Cl^-, and so on) per liter, which produces an osmotic pressure of $\pi \approx (1.1 \text{ mol L}^{-1}) \cdot (0.082 \text{ L atm mol}^{-1} \text{ K}^{-1}) \cdot (280 \text{ K}) \approx 25$ atm. This is the minimum pressure that must be applied to halt normal osmotic flow—flow *reversal* for desalination requires higher pressure.

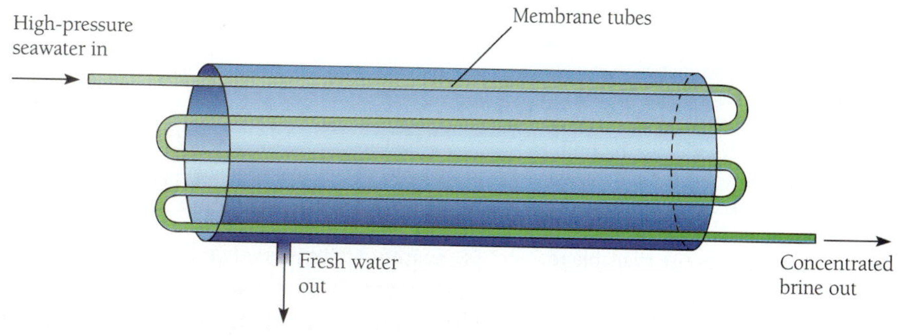

Figure 12.26 Inside the desalination unit, high-pressure seawater passes through long lengths of tubing made from a semipermeable membrane. Pure water is squeezed through the pores of the membrane, but the salts dissolved in the seawater remain behind.

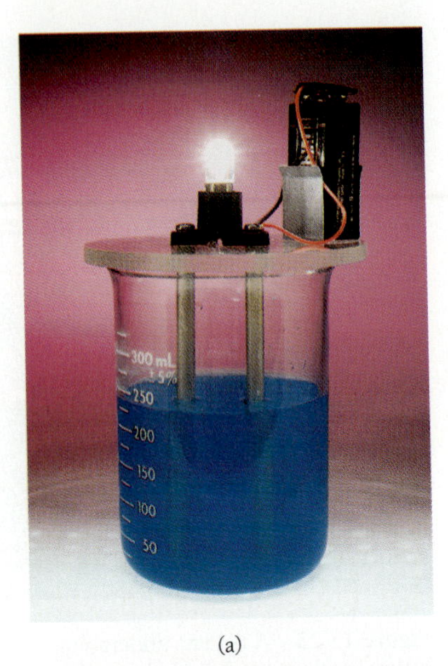

(a)

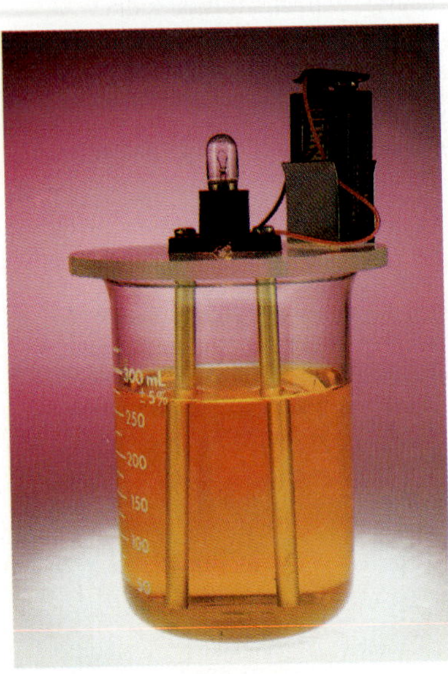

(b)

Figure 12.27 Electrolyte solutions, whose dissolved particles are ions rather than neutral molecules, are capable of carrying an electrical current. (a) In a solution of copper sulfate, the dissolved ions Cu^{2+} and SO_4^{2-} complete the circuit and allow the bulb to light. (b) In a solution of iodine, the dissolved particles are neutral I_2 molecules. The light does not glow because no charged particles are available to carry the current.

12.5 SOLUTIONS OF ELECTROLYTES

When an ionic substance dissolves in a polar solvent such as water, the particles that enter the solution are positive and negative ions rather than neutral molecules. Solutions whose solute particles bear an electrical charge are called *electrolyte solutions,* and the compounds that form them are called **electrolytes.** Solutions of electrolytes are quite different in behavior from solutions of nonelectrolytes. The most noticeable property of electrolyte solutions, and the reason for the name, is that they are conductors of electricity (Figure 12.27).

Weak and Strong Electrolytes

The electrical conductivity of a solution depends on the concentration of dissolved ions, and it is a property that is easily measured. Such measurements show that there are two broad classes of electrolytes. One group of substances, when dissolved in water, produces solutions that are good conductors of electricity. These substances are called **strong electrolytes.** The other type produces solutions that, compared to strong electrolytes at the same concentration, are poor conductors of electricity. These substances are called **weak electrolytes.** The extensive experimental and theoretical work of the Swedish chemist Svante Arrhenius provides an explanation for this difference in behavior. When a *strong electrolyte* such as NaCl dissolves (Equation 12-16),

$$NaCl(s) \xrightarrow[H_2O]{} Na^+(aq) + Cl^-(aq) \tag{12-16}$$

the dissociation is complete. The only particles that enter the solution are ions—there is no such thing as a dissolved "NaCl molecule." *All ionic compounds are strong electrolytes.* Some ionic compounds are only slightly soluble, and some dissolve scarcely at all. However, to the extent that they do dissolve, they are completely dissociated and are therefore classified as strong electrolytes. In *weak electrolytes,* on the other hand, the process of dissociation takes place *incompletely.* When the weak electrolyte acetic acid dissolves in water (Equation 12-17),

$$CH_3CO_2H(\ell) \xrightarrow[H_2O]{} H^+(aq) + CH_3CO_2^-(aq) + CH_3CO_2H(aq) \tag{12-17}$$

neutral undissociated molecules as well as ions enter the solution. Many acids are weak electrolytes. A common way of describing the two types of electrolyte is to say that strong electrolytes are completely or "100% dissociated," while weak electrolytes are incompletely or "less than 100% dissociated."

Colligative Properties

As we have seen, the colligative properties of solutions are independent of the chemical nature of the solute. It is the *number* of solute particles, not their type, that determines the magnitude of the effect. Therefore, we would expect a 0.1 m solution

> **Svante Arrhenius (1859–1927)**
>
> Svante Arrhenius is well known for his work on acids and bases (Chapter 15), which grew out of a much broader study of the nature of electrolyte solutions and the process of electrolytic dissociation. He received the 1903 Nobel Prize for this work. He also studied the nature and mechanism of chemical toxicity.

of a 1 : 1 strong electrolyte such as NaCl to have a freezing point depression or osmotic pressure twice as great as a 0.1 *m* solution of a nonelectrolyte such as ethanol. For each mole of electrolyte that dissolves, *two* moles of particles, anions and cations, enter the solution. As far as colligative properties are concerned, a 0.1 *m* solution of NaCl behaves the same as a 0.2 *m* solution of a nonelectrolyte. The effect is more striking for more complicated salts such as $MgCl_2$, which yields three moles of dissolved particles per mole of dissolved salt, or $Al_2(SO_4)_3$, which yields five moles of dissolved particles per mole of dissolved salt.

The properties of electrolyte solutions were first studied systematically by the Dutch chemist Jacobus van't Hoff. His work led to a modified form of the basic equations describing the colligative properties [Equations (12-10) and (12-12) through (12-14)]. The modified equations, which include a numerical factor (i) related to the type of solute, are

$$\Delta P = i[X_{\text{solute}} \cdot P^{\circ}_{\text{solvent}}] \qquad (12\text{-}10a)$$

$$\Delta T_b = i[K_b \cdot m] \qquad (12\text{-}12a)$$

$$\Delta T_f = -i[K_f \cdot m] \qquad (12\text{-}13a)$$

$$\pi = i[MRT] \qquad (12\text{-}14a)$$

The factor i, called the **van't Hoff factor,** is the number of moles of *particles* produced when 1 mole of a substance dissolves. For strong electrolytes with a 1 : 1 stoichiometry (for example, LiBr), $i = 2$. For 1 : 2 and 2 : 1 salts (for example, $CaCl_2$ and K_2SO_4), $i = 3$. For nonelectrolytes, $i = 1$.

EXAMPLE 12.10 The van't Hoff Factor

Which of the following aqueous solutions has the lowest freezing point: 0.20 *m* NaCl, 0.15 *m* K_2SO_4, or 0.30 *m* sucrose (a nonelectrolyte)?

Analysis

Target: The relative freezing points of three solutions.

Knowns: The identities and molalities of the solutes. K_f for water is an implicit known.

Relationship: The freezing point depression Equation (12-13a) is a direct link between the knowns and the target.

Plan: One approach (a sure thing) would be to calculate the actual freezing points of all three solutions, and compare them to one another. However, since we need only to choose the lowest of the three, it is probably unnecessary to obtain numerical results. We will *examine* the equation and try to reason out the answer.

Work The equation is

$$\Delta T_f = -i \cdot K_f \cdot m$$

i (the van't Hoff factor) and m (the molality) are different for the three solutions, but, since K_f is the same, it must be that the solution with the lowest freezing point is the one with the largest value of $i \cdot m$.

for NaCl: $i = 2,\ m = 0.20,\ i \cdot m = 0.40$
for K_2SO_4: $i = 3,\ m = 0.15,\ i \cdot m = 0.45$
for sucrose: $i = 1,\ m = 0.30,\ i \cdot m = 0.30$

The solution with the largest $i \cdot m$, and therefore the largest ΔT_f and the lowest freezing point, is that containing potassium sulfate.

Exercise

Calculate the freezing points of each of the three solutions in the example.

Answer: NaCl, −0.74 °C; K_2SO_4, −0.84 °C; sucrose, −0.56 °C.

If the solute is a weak electrolyte, the van't Hoff factor is not easily determined. For a 1 : 1 *weak* electrolyte such as acetic acid, i depends on the extent to which the electrolyte is dissociated, and can range from 1 (very slight dissociation) to 2 (near 100% dissociation). The value of i differs for each weak electrolyte, and in addition depends on concentration and temperature. The effect of concentration on the extent of dissociation of weak electrolytes will be examined in detail in Chapter 16.

A further complication arises when the van't Hoff factor is measured in concentrated solutions. In such cases it is found that, even for strong electrolytes, the value of i is not exactly equal to the integer indicated by the stoichiometry of the salt. This deviation from the predicted behavior is due to a number of different causes, usually grouped together and called simply "solution nonidealities."

Activity

For many purposes it is unnecessary to identify the factors responsible for nonideal behavior of solutions. In order to predict the vapor pressure of a solution, for example, you do not need to know *why* the solution deviates from Raoult's law. All you need to know is *by how much* it deviates. The concept of solution *activity* was developed in order to deal with nonideal solutions without reference to the cause of the nonideality. **Activity** is an "effective concentration." If the actual concentration of a solution is 0.15 M, for instance, and the solution behaves (with respect to some property such as osmotic pressure) as if its concentration were 0.14 M, its effective concentration or activity is 0.14 M. The **activity coefficient** (γ) is defined as the ratio of the activity to the actual concentration: γ is a dimensionless number. In our example, the activity coefficient is $\gamma = (0.14 \text{ M})/(0.15 \text{ M}) = 0.93$. In ideal solutions, the activity is the same as the concentration, and the activity coefficient is therefore 1.00. As a solution departs more and more from ideal behavior, the activity coefficient differs more and more from 1.00.

In *concentrated* solutions of electrolytes, the activity coefficient can be greater than 1. In *dilute* solutions, γ is less than 1. In *very dilute* solutions, γ is equal to 1.

The activity coefficient of a solute must be measured for each different solute and solvent, and for each different concentration. The results of such measurements are collected in tables of chemical data, but there are many solutions for which the measurements have not been made. For many solutions, particularly dilute solutions, the behavior is nearly ideal, and the activity is nearly equal to the concentration. In the remainder of this text, we will assume that this is true for all solutions, and leave the study of activity for a more advanced course in chemistry.

12.6 COLLOIDAL DISPERSIONS

A mixture of sand and water is clearly a mixture and not a solution. We can stir or shake it vigorously enough so that the sand is uniformly dispersed throughout the liquid, but as soon as we stop stirring the sand separates from the liquid and settles to

Jacobus Hendricus van't Hoff (1852–1911)

Jacobus van't Hoff made detailed investigations of the electrical conductivity and osmotic pressure of solutions. The very first Nobel Prize in chemistry (1901) was awarded to van't Hoff for this work. Later in his career he became interested in the geometrical structure of molecules.

the bottom. Suppose we were to prepare a series of sand/water mixtures, each time grinding the sand into smaller particles. What would be the result? The most obvious effect would be that, the smaller the particle size, the longer would be the time required for the sand to settle to the bottom after we stopped stirring. Eventually the particles would be so small that they would not settle out, but would remain homogeneously dispersed throughout the liquid. Particles that do not settle out are said to be *suspended* in the liquid.

Nature of Colloids

Mixtures in which *very* small particles are homogeneously dispersed and suspended in a supporting medium are called **colloids,** or *colloidal dispersions.* The small particles of a colloid are called the **dispersed phase,** and the medium that supports them is called the **dispersion medium** or the *continuous phase.* The particles of the dispersed phase are too small to be seen, but nonetheless, they may contain billions of atoms. They usually have diameters in the range 10 to 1000 nm. Although colloidal dispersions are often completely transparent, they are not true solutions. Many of their properties are intermediate between those of mixtures and true solutions. They are stable against spontaneous settling into two phases, and they display colligative properties similar to those of true solutions.

For comparison, the diameter of an I_2 molecule is ≈ 0.55 nm. A cube of solid iodine 100 nm on a side contains about 100 million molecules.

Colloids are named and classified according to the solid, liquid, or gaseous nature of the dispersed phase and the dispersion medium. The several types of colloids are listed in Table 12.6.

TABLE 12.6 Types of Colloidal Dispersions

Dispersed Phase	Dispersion Medium	Type	Examples
Solid	Gas	Aerosol	Smoke
Solid	Liquid	Sol	Mud, blood
Solid	Solid	Solid sol	Ruby, stained glass
Liquid	Gas	Aerosol	Fog, mist
Liquid	Liquid	Emulsion	Milk, mayonnaise
Liquid	Solid	Gel	Jelly
Gas	Gas	*	
Gas	Liquid	Foam	Whipped cream, soapsuds
Gas	Solid	Foam	Plastic foam, pumice

* Does not exist—mixtures of gases are true solutions.

CHEMISTRY OLD AND NEW

THE IMPORTANCE OF EXPERIMENTAL ACCURACY

Few things are as frustrating as searching for an error made while solving a complex, multistep problem. For a research scientist, correcting mistakes is not always easy. Even the most innovative experiments depend on the results of previous investigators, and a researcher trying to fix a problem may find himself examining the work of many others in addition to his own. A famous example of this interdependency involves a mistake in the groundbreaking studies of Robert Millikan, who determined the charge of the electron. His error lay unnoticed for many years until a group of scientists, working in a seemingly unrelated field, sought to explain a small difference between their results and those obtained in earlier investigations. In the process of tracking down the source of the discrepancy, the experiments of five Nobel laureates were scrutinized, several important fundamental constants were redefined, and international arguments broke out over the origin of the problem.

In the early years of x-ray investigations, wavelengths were measured by observing the angle of diffraction from a crystal of known interatomic distances. An improved method, introduced by Compton in 1925, used a "grating"—a large number of parallel grooves machined on the surface of a metal plate—rather than a crystal. Since the spacings (d) between ruled grating lines were known more precisely than those between crystal planes, x-ray wavelengths (λ) calculated from them using the Bragg equation ($n\lambda = 2d \sin \theta$) were more precise. Experimenters rushed to check their λ values with this new method. Before long, a disturbing trend emerged: although the difference was small, grating-measured wavelengths seemed to be consistently larger than crystal-measured wavelengths. Erik Backlin, a student of Nobel laureate Karl Siegbahn, estimated the discrepancy at about 0.2% (or one part in 500). After double-checking his results, he examined the crystal-based calculations of his teacher and found them to be flawless. Even Compton was puzzled. In a 1929 paper he conceded, "We have been diligently searching for . . . a hidden error for the past six months without success, and now it seems necessary to appeal for help. . . ." Theories poured in from all over, but none offered a satisfactory explanation.

The procedure for determining d spacings in crystals had been derived by William and Lawrence Bragg, who received the 1915 Nobel Prize for this work. Their method, which used the mass and density of crystalline substances to find the volume and dimensions of unit cells, was sound. There was nothing left to do but dig deeper into the past. Unit cell masses were found by summing the molar masses of a crystal's constituent atoms and then dividing this total by Avogadro's number. Could the entire molar mass scale be off? But careful scrutiny of molar masses, most of which had been measured by another Nobelist (T. W. Richards), revealed no errors. The only possible remaining source of error was Avogadro's number, which had been determined from electrolysis measurements and thus was related to the charge of the electron (e).

K. Shiba was the first to suggest that the accepted value of the electronic charge was wrong. He proposed that Millikan had used an inaccurate value for the viscosity of air in his 1910 oil-drop experiment, which led to evaluation of e (and, yes, to a Nobel Prize). Millikan, who had rechecked his viscosity value in 1917 and deemed it "altogether unique in its reliability and

Properties of Colloids

We have seen (in Section 11.2) that the forces existing at the surface of a liquid are quite strong, and that they lend to the surface certain properties not characteristic of the bulk of the liquid. Similar surface effects occur at solid surfaces as well, although it is not usual to speak of the "surface tension" of solids. The forces at the surface of a solid attract and hold small atoms, molecules, or ions to the surface. This phenomenon is known as **adsorption.** If a colloid particle adsorbs ions, for instance OH^- ions, that are dissolved in the dispersing medium, the particle acquires an electrical charge. The electrical charge then repels other colloid particles, which bear a similar charge. The mutual repulsion helps to keep the dispersed phase from settling out.

Colloids that are kept in suspension by electrical repulsion can be precipitated by removing or reducing the repulsion. There are several ways in which this can be achieved. Adsorbed ions can be removed by *dialysis* of a colloid. The particles, no longer kept apart by electrical repulsion, cluster together, grow in size, and eventually settle to the bottom. Alternatively, a concentrated salt solution can be added. The increased concentration of ions in the dispersion medium tends to neutralize the charge on the colloid particles, allowing them to cluster and grow. Precipitation of a colloid is called **coagulation.**

Dialysis was described in Section 12.4.

Curdling of milk is a familiar example of coagulation.

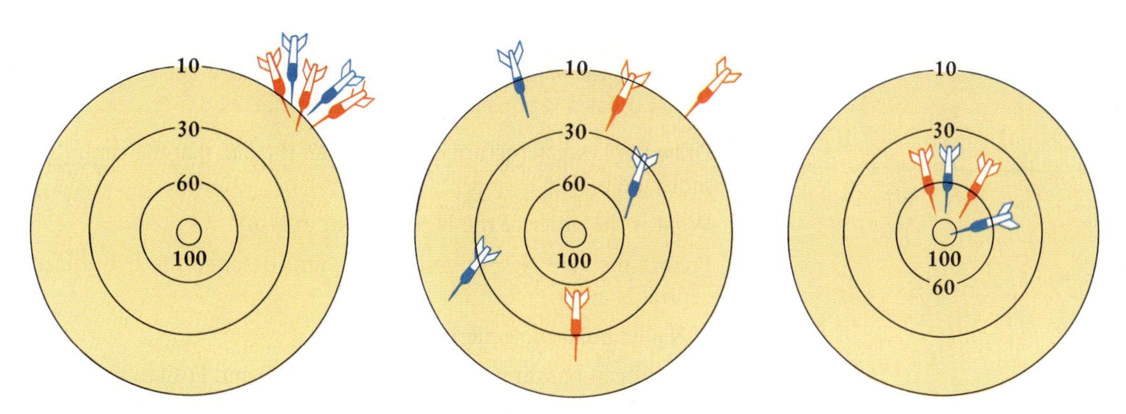

Avogadro's number had been *precisely* measured, but the average of the measurements was not accurate.

precision," was reluctant to accept this explanation. However, in 1936 G. Kellstrom, using an improved technique, remeasured the viscosity of air and found that Millikan's result was in error by about 0.2%. The x-ray wavelength discrepancies could now be explained! Unfortunately, all relevant work completed in the 26-year interval between the 1910 oil-drop experiment and Kellstrom's correction was now off by 0.2%. The charge of the electron was recomputed, as was Avogadro's number and all of the crystal-determined x-ray wavelengths. The crystal- and grating-measured wavelengths were finally in agreement, and the mystery of the one-part-in-500 error was solved. The two-fold moral of the story is, that application of the scientific method can resolve seemingly hopeless difficulties, and that even our heroes can make mistakes.

Discussion Questions

1. The numerical value of the 0.2% error in Avogadro's number is about 1×10^{21}. How large is an error of 0.2% in the following measured quantities? Your own height and weight; the length of a football field; the cost of a $140,000 house.
2. Suggest a situation, other than the one discussed in this essay, in which a measurement error of 0.2% would be of importance.

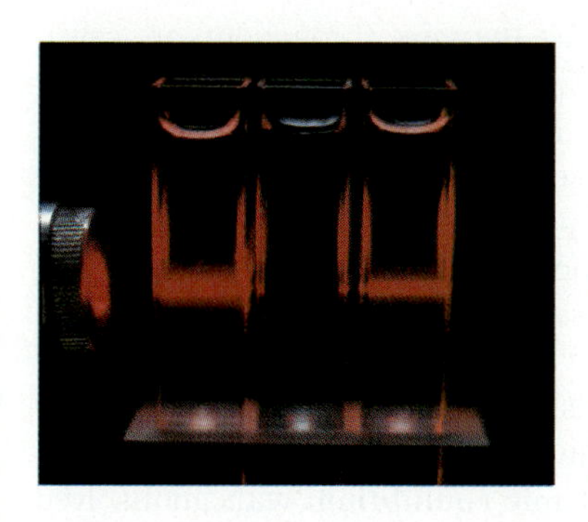

Figure 12.28 The beam from a small laser is shown passing successively through a colloid, a solution of sodium chloride, and a second sample of the colloid. The beam is scattered only by the colloid, which otherwise appears the same as a true solution.

One readily observable property of colloids is their ability to scatter light. A colloid that to the eye seems perfectly transparent may nonetheless scatter light much more effectively than a solution or a pure liquid. Light scattering by colloids is called the **Tyndall effect** (Figure 12.28). Another example of scattering of light by a colloid may be seen when a beam of sunlight passes through a darkened room. The dust particles dancing in the beam are the dispersed phase of an aerosol. Normally they are invisible because they are so small, but they are large enough to scatter light by the Tyndall effect.

John Tyndall (1820–1893) was an Irish-born British physicist who studied light, sound, and radiant heat. He was an authority on glaciers, and a popular lecturer on scientific topics for nonscientific audiences.

SUMMARY EXERCISE

Thermal decomposition of ammonium nitrate produces dinitrogen oxide (N_2O), also called "laughing gas." This relatively unreactive gas has a solubility of 2.228 g L^{-1} in water at 25 °C when the partial pressure of N_2O is 2.00 atm. At its normal boiling point (−88.5 °C) the vapor pressure of N_2O is of course 760 torr. For comparison, propene (C_3H_6) has a vapor pressure of 54.5 torr at −88.5 °C.

a. Draw the Lewis structure of N_2O, given the fact that the central atom in the molecule is N.

b. What is the order of the N—O bond in N_2O?

c. Predict the N—N—O bond angle, and describe the hybridization of the central N atom.

d. Is N_2O a polar molecule?

e. Is hydrogen bonding possible between N_2O and H_2O in aqueous solutions?

f. Calculate the Henry's law constant for aqueous solutions of N_2O.

g. Which compound, NH_4NO_3 or N_2O, would you expect to have the greater solubility in water? Why?

h. What is the freezing point of a solution of 1.5 g NH_4NO_3 in 125 g H_2O?

i. What is the boiling point of a solution of 9.8 g $Ba(NO_3)_2$ in 304 g H_2O?

j. If $Ba(NO_3)_2(aq)$ is mixed with $K_2SO_4(aq)$, what precipitate, if any, will form?

k. What is the vapor pressure of a solution of 10.0 g N_2O in 100.0 g propene at −88.5 °C? (Assume the solution obeys Raoult's law.)

Answers **(a)** $:\ddot{N}=N=\ddot{O}:$; **(b)** 2; **(c)** 180°, *sp*; **(d)** yes; **(e)** yes; **(f)** 0.0253 L mol^{-1} atm^{-1}; **(g)** NH_4NO_3, because it is ionic and hence more strongly solvated than N_2O; **(h)** −0.56 °C; **(i)** 100.19 °C; **(j)** $BaSO_4(s)$; **(k)** 116 torr.

SUMMARY AND KEYWORDS

Homogeneous mixtures, whose compositions are the same throughout, are called **solutions,** and may be solid, liquid, or gaseous. The **solvent** is that component which, when pure, has the same state (solid, liquid, or gas) as the solution. The other component(s) is/are the **solute(s).** Water is the solvent in an *aqueous* solution. Solute particles experience attractive forces from nearby solvent molecules, which are themselves rearranged by the forces. This interaction is called **solvation.** Ionic solutes, which undergo **dissociative ionization** on dissolution, are more strongly hydrated than molecular solutes.

A solution that can dissolve no more solute is **saturated,** and if excess solid solute is present the opposing processes of **dissolution** and **crystallization** occur at equal rates. A saturated solution is in a state of **dynamic equilibrium** with undissolved solute. Solutions can be **undersaturated,** and in some cases unstable **supersaturated solutions** can be prepared. The amount of solute that can dissolve in a given amount of solvent is called the **solubility** of the solute. Many solids become more soluble in water as the temperature increases, while most gaseous solutes become less soluble.

The major driving force that makes dissolution a spontaneous process is the entropy change, which is always positive. In general, there is also an enthalpy change associated with the process. The **heat of solution, ΔH_{soln}**, may be positive or negative, although dissolution is endothermic for most solid solutes. **Le Châtelier's principle,** which applies to all systems in equilibrium, may be used to predict solubility trends. If heat is required to make a substance dissolve ($\Delta H_{soln} > 0$), then an increase in temperature will increase the solubility. If the dissolution process is exothermic, then increasing the temperature will decrease the solubility.

Solute charge and *solvent polarity* are the two most important factors determining solubility. The principle *like dissolves like* means that ionic substances are more soluble in polar than nonpolar solvents, while nonpolar molecular substances are more soluble in nonpolar solvents. The solubilities of ionic compounds follow certain trends (Table 12.3).

The vapor pressure of **ideal solutions** is described by **Raoult's law.** Most *real* solutions obey Raoult's law only approximately. The process of **distillation,** used to separate the volatile components of a solution from one another, is based on the fact that the vapor in equilibrium with a solution is richer in the more volatile component. Gaseous solutes obey Henry's law, that is, have solubilities proportional to pressure.

Colligative properties of solutions depend on the concentration but not on the chemical nature of the (nonvolatile) solute. They are **vapor pressure decrease, boiling point elevation, freezing point depression,** and **osmotic pressure.** Each solvent has a characteristic **boiling point constant** and **freezing point constant.**

The process of **osmosis** is the flow of solvent through a **semipermeable membrane,** which blocks the passage of solute particles. Solvent flow in osmosis occurs in the direction that tends to equalize the concentrations of the solutions on the two sides of the membrane. The direction of osmotic flow can be reversed by the application of a pressure greater than the osmotic pressure. This principle is used in the *desalination* of seawater by *reverse osmosis. Dialysis* is similar to osmosis, but a dialysis membrane is permeable to smaller solute ions and molecules as well as to solvent. Only the larger solute species are blocked by a dialysis membrane.

Electrolytes are substances that *dissociate* into ions as they dissolve, forming solutions capable of carrying electrical current. **Strong electrolytes** dissociate completely, and **weak electrolytes** dissociate partially. Solutions of weak electrolytes contain undissociated solute molecules as well as their component ions. Ionic compounds are strong electrolytes, while many acids are weak electrolytes. The **van't Hoff factor** i is the number of moles of particles produced when 1 mole of solute dissolves.

Solution behavior departs increasingly from ideal behavior as the concentration increases. **Activity** is an "effective concentration." The **activity coefficient,** equal to 1.00 in ideal solutions, is the dimensionless ratio of activity to concentration.

A **colloid** or *colloidal dispersion* occupies the borderline between mixtures and true solutions. The **dispersed phase** consists of particles (from 10 to 1000 nm in diameter) which are too small to be seen but are much larger than the solute molecules of a true solution. Colloidal particles are homogeneously distributed and suspended in the **dispersion medium** or *continuous phase,* which may be solid, liquid, or gaseous. Colloids often scatter light (the **Tyndall effect**). Colloid particles can acquire an electrical charge by **adsorption** of ions. If the charge is decreased, by dialysis for instance, the particles can cluster more readily and the colloid **coagulates.**

PROBLEMS

General

1. Differentiate among the terms "mixture," "solution," and "colloid."
2. What terms describe solutions whose solute concentrations are less than, equal to, and greater than the solubility?
3. What factors influence the enthalpy of solution?
4. What effect does pressure have on the solubility (in a liquid solvent) of (a) a gaseous solute, (b) a liquid solute, and (c) a solid solute?
5. Is it possible for a solution to be saturated and yet not be in equilibrium with pure solute?
6. If a substance is more soluble in cold water than in hot water, what can you conclude about the enthalpy of solution?
7. Define the terms "deposition," "crystallization," and "precipitation."
8. Is it possible for the *rate* of dissolution to increase as the temperature increases, while the *solubility* of the substance decreases?
9. If the definition of a saturated solution is "a solution that can dissolve no more solute," how is it possible for a supersaturated solution to exist?
10. Why is entropy important in solubility?
11. What is an ideal solution?
12. Suppose the vapor in equilibrium with a solution had the same composition as the solution. Would it be possible to separate the two components by distillation? Explain your answer. (Incidentally, such solutions are not at all rare: they are called *azeotropes.*)
13. To what systems does Henry's law apply?
14. What is the relationship between Henry's law and Le Châtelier's principle?
15. Define the term "colligative property."
16. Name four colligative properties of solutions.
17. If a small amount of ethanol (bp 78 °C) is dissolved in water, would you expect the boiling point of the solution to be greater or less than 100 °C?
18. What is meant by reverse osmosis?
19. Osmosis is a spontaneous process, and therefore involves either an increase in entropy, a decrease in enthalpy, or both. Which do you think is the more important factor? Explain.
20. With respect to colligative properties, what is the important difference between ionic and molecular solutes?
21. What is an activity coefficient?
22. Distinguish between crystallization and coagulation.
23. How might you define a colloid in terms of particle size?
24. In what way can adsorption affect the properties of a colloid?

Solubility Rules and Precipitations

25. Account for the fact that ethanol (C_2H_5OH) is completely miscible with water.
26. Carbon tetrachloride is soluble in benzene but not in water. Why?
27. Classify each of the following ionic substances as soluble, slightly soluble, or insoluble in water.
    a. barium hydroxide
    b. silver carbonate
    c. potassium carbonate
    d. silver acetate
28. Classify each of the following ionic substances as soluble, slightly soluble, or insoluble in water.
    a. barium chloride
    b. silver sulfate
    c. silver chloride
    d. potassium chlorate
29. Classify each of the following ionic substances as soluble, slightly soluble, or insoluble in water.
    a. magnesium sulfate
    b. magnesium sulfide
    c. calcium fluoride
    d. ammonium chloride
30. Classify each of the following ionic substances as soluble, slightly soluble, or insoluble in water.
    a. aluminum hydroxide
    b. magnesium iodide
    c. cesium fluoride
    d. barium acetate
31. In each of the following, both compounds are water-soluble. Predict whether a precipitate will form when solutions of the two are mixed, and if so, identify the compound that precipitates.
    a. NH_4Cl, $Ba(NO_3)_2$
    b. $Ba(NO_3)_2$, Na_2SO_4
    c. NH_4NO_3, $AgNO_3$
32. In each of the following, both compounds are water-soluble. Predict whether a precipitate will form when solutions of the two are mixed, and if so, identify the compound that precipitates.
    a. $Pb(NO_3)_2$, NaI
    b. $Ba(NO_3)_2$, KCl
    c. $(NH_4)_2S$, $AgNO_3$
33. In each of the following, both compounds are water-soluble. Predict whether a precipitate will form when solutions of the two are mixed, and if so, identify the compound that precipitates.
    a. KCN, $Pb(NO_3)_2$
    b. $Mg(NO_3)_2$, Na_2S
    c. NaOH, KBr
34. In each of the following, both compounds are water-soluble. Predict whether a precipitate will form when solutions of the two are mixed, and if so, identify the compound that precipitates.
    a. NH_4Br, $Hg_2(NO_3)_2$
    b. KOH, Na_2S
    c. CsF, $MgCl_2$

Composition of Solutions

35. If 2.5 mol O_2 is mixed with 6.3 mol N_2, what is the composition of the mixture? Express the answer as a mole fraction and mole percent of each component.
36. If 0.00325 mol Na_2SO_4 is mixed with 0.00759 mol KCl, what is the composition of the mixture? Express the answer as mole fraction and mole percent of each component.
37. What is the mole fraction of carbon tetrachloride in a mixture of 5.00 g CCl_4 and 250 g C_6H_6?
38. What is the mole fraction of acetone (C_3H_6O) in a mixture of 50.0 g acetone and 50.0 g H_2O?
*39. A certain mixture of NaCl and NaI is 1.00% (w/w) NaI. What is the mole% NaI?
40. Ethanol is usually handled as a 95% (w/w) aqueous solution rather than as a pure liquid. What is the mole fraction of ethanol (C_2H_6O) in this solution?
41. Calculate the molality of a solution made by dissolving 1.00 g NaCl in 100.0 g H_2O.
42. Calculate the molality of a solution of 65.9 mg naphthalene ($C_{10}H_8$) in 18.3 g benzene (C_6H_6).
*43. An aqueous solution contains 2.07% (by mass) Na_2S. What is the molality of the solute?
*44. What is the molality of a solution that is 50.0% (by mass) each of benzene (C_6H_6) and toluene (C_7H_8)? In solutions of this sort, the distinction between solute and solvent is arbitrary. Solve this problem in two ways, first regarding benzene as the solvent, then toluene.
45. How much sucrose ($C_{12}H_{22}O_{11}$) must be dissolved in 75.0 g H_2O to produce a 0.100 *m* solution?
46. How much water must be added to 2.55 g NaCl in order to produce a 0.100 *m* solution?

Raoult's Law

47. The vapor pressure of octane (C_8H_{18}) is 390 torr at 100 °C, while octadecane ($C_{18}H_{38}$) has a negligibly small vapor pressure at that temperature. What is the vapor pressure of a solution of 10.0 g octadecane in 500 g octane at 100 °C?
48. What is the composition of a solution of octadecane in octane, if the solution has a vapor pressure of 350 torr at 100 °C? Use the data in the preceding problem, and express your answer in mole percent.
*49. The normal boiling point of acetone (C_3H_6O) is 56.5 °C, at which temperature the equilibrium vapor pressure of cyclohexanone ($C_6H_{10}O$) is 24.0 torr. Calculate the vapor pressure (at 56.5 °C) of a cyclohexanone/acetone mixture having 75 mole% cyclohexanone.
*50. Diethyl ether (usually called simply "ether") is a volatile liquid whose vapor pressure at 0 °C is 178 torr. D*imethyl* ether is even more volatile, having a boiling point well below room temperature. At 0 °C, its vapor pressure is 2.50 atm. What is the vapor pressure of a mixture of 3.00 mol diethyl ether and 1.00 mol dimethyl ether at 0 °C?
51. At 30 °C, the vapor pressure of 1-bromobutane (C_4H_9Br) is 51.3 torr while that of 1-chlorobutane (C_4H_9Cl) is 129.9 torr. What is the vapor pressure of a mixture of these two halobutanes that is 50% (by mass) 1-bromobutane?
52. Using the data of the preceding problem, calculate the vapor pressure of a mixture of 1-bromobutane and 1-chlorobutane that contains 50 mole% of each compound.
53. The vapor pressure of ethyl acetate ($C_4H_8O_2$) is 276.1 torr at 50.0 °C, at which temperature the vapor pressure of propyl acetate ($C_5H_{10}O_2$) is 110.3 torr. What is the partial pressure of each component above a mixture of 2.00 mol propyl acetate and 1.00 mol ethyl acetate? What is the composition of the vapor?
*54. Using the data of the preceding problem, calculate the vapor pressure of each component of a mixture of ethyl acetate and propyl acetate that contains 50.0% (by mass) of each component. What is the composition of the vapor?

Henry's Law

55. For carbon dioxide, the Henry's law constant is 0.039 mol L^{-1} atm^{-1} at room temperature. What is the maximum molarity of CO_2 in an aqueous solution in equilibrium with CO_2 at 5.00 atm?
56. The value of the Henry's law constant for argon in water at 0 °C is 2.5×10^{-3} mol L^{-1} atm^{-1}. Calculate the solubility, in mol L^{-1} and g L^{-1}, if the partial pressure of argon is 3.62 atm.
57. Assume that a soft drink is carbonated during manufacture by allowing the drink to come to equilibrium with 4.00 atm CO_2 at 0 °C, at which temperature the Henry's law constant is 0.077 mol L^{-1} atm^{-1}. What is the CO_2 concentration in a soft drink?
*58. A soft drink quickly goes stale as its CO_2 content, initially saturated or supersaturated, declines to a new equilibrium value. What is the CO_2 molarity in a room-temperature (25 °C) soft drink that has reached equilibrium with air, which is approximately 0.033 mole% CO_2?
59. If 0.256 g argon will dissolve in 1 liter of hot water when the argon pressure is 5.00 atm, what is the Henry's law constant at this temperature?
60. An argon pressure of 4.00 atm is sufficient to dissolve 1.00 g Ar in 3.56 L of warm water. Evaluate the Henry's law constant at this temperature.

Colligative Properties

Table 12.5 should be consulted for data needed in some of the following problems.

61. Calculate the freezing point of a 0.46 *m* aqueous solution of ethylene glycol, a nonelectrolyte.
62. Calculate the freezing point of a 0.013 *m* solution of iodine in benzene.

63. What is the molality of an aqueous solution that freezes at −0.53 °C?

64. Calculate the molality of a solution of phenanthrene in naphthalene that freezes 0.86 °C below the freezing point of pure naphthalene.

65. What are the freezing points of the following aqueous solutions?
    a. 0.25 *m* sucrose
    b. 0.50 *m* $SrBr_2$

66. What are the freezing points of the following aqueous solutions?
    a. 0.25 *m* NaCl
    b. 0.50 *m* K_2CO_3

*67. Calculate the freezing point of a 1.00% (by mass) solution of bromobenzene (C_6H_5Br) in each of the following.
    a. benzene
    b. naphthalene

68. Calculate the freezing point of a 1.00 mole% solution of 1,4-dichlorobenzene ($C_6H_4Cl_2$) in each of the following.
    a. benzene
    b. naphthalene

*69. Suppose 1.0 g of a 50% (by mass) mixture of $MgCl_2$ and $BaCl_2$ is dissolved in 50.0 g H_2O. Calculate the freezing point of the solution.

*70. Suppose 0.100 gram of a 50.0% (by mass) mixture of NaBr and CsBr is dissolved in 10.0 g H_2O. Calculate the freezing point of the solution.

*71. Dissolution of 0.500 g of a mixture of NaCl and CsBr in 100.0 g H_2O is found to decrease the freezing point by 0.125 °C. Calculate the percent (by mass) of NaCl in the mixture.

*72. Suppose 1.00 gram of a mixture of CCl_4 and CBr_4 is dissolved in 10.0 g benzene, and the freezing point is found to decrease by 0.24 °C. Determine the composition of the solution, and express it as %(w/w) CCl_4.

*73. 0.331 g of an unknown crystalline substance is ground together with 10.0 g camphor, and the resulting powder is found to melt to a clear liquid at 168.0 °C. Given that K_f for camphor is 40 °C kg mol^{-1}, and that pure camphor melts at 176.0 °C, what is the molar mass of the unknown? (Assume it to be a molecular compound.)

*74. 500 mg of an unknown substance is dissolved in 50.0 g benzene, and the solution is found to freeze at 4.3 °C. Calculate the molar mass of the substance. In solving this problem, you must make an assumption about the value of the van't Hoff factor. What is your assumption, and what did you base it on?

75. Calculate the boiling point of a solution of 12.3 g urea (NH_2CONH_2) in 105 g water. Urea is a nonvolatile molecular substance.

*76. What is the boiling point of an 8.7% (by mass) solution of benzoic acid ($C_6H_5CO_2H$) in benzene?

*77. When 125 g of dextrose (grape sugar) is dissolved in 850 g water, the normal boiling point of the resulting solution is found to be 100.42 °C. What is the molar mass of dextrose?

*78. When 546 mg benzimidazole is dissolved in 3.00 g ethanol, the boiling point is found to increase by 1.88 °C. What is the molar mass of benzimidazole?

79. The osmotic pressure of an aqueous solution of a nonvolatile nonelectrolyte was found to be 8.38 atm at 25 °C. What is the concentration of the solution? What would the osmotic pressure of this solution be at 0 °C?

*80. What solute concentration would be required to produce an osmotic pressure capable of supporting a column of water equal in height to a good basketball center, say 7 feet 2 inches? Assume room temperature.

*81. Human hemoglobin has a molar mass of 6.84×10^4 g mol^{-1}. How much hemoglobin must be dissolved to produce 1.00 L of an aqueous solution with a room-temperature osmotic pressure of 1.00 torr? What is the height of a water column this pressure can support?

*82. The molar mass of bovine insulin is 5,700 g mol^{-1}. How high a column of water could be supported by the osmotic pressure of a 0.10% (by mass) aqueous solution of insulin? Assume room temperature.

*83. The sheath protein of the tobacco mosaic virus has a huge molar mass. Dissolution of 10 mg of this protein in 1.00 mL water gives a solution whose osmotic pressure at room temperature is sufficient to support a column of water only 0.062 mm high, which is barely noticeable and hard to distinguish from capillary rise. What is the molar mass of this protein? (Other methods are available for the accurate determination of such high molar masses.)

*84. Calculate the molar mass of human serum albumin, given that 13.8 mg dissolved in 5.00 mL H_2O has an osmotic pressure of 0.748 torr at room temperature.

Other Problems

*85. 35.00 g NH_4Cl is heated with 65.0 g H_2O and stirred until all the solid dissolves, at which point the solution is cooled to 30 °C. Eventually some NH_4Cl crystallizes. The crystals are separated, dried, and found to weigh 5.62 g. Express the solubility of NH_4Cl at 30 °C as grams solute per 100 g solution (mass %) and as molality.

86. The concentration of CO_2 in a soft drink is 0.31 M. What is the total mass of CO_2 in a 12-ounce drink? (1 oz ⇔ 29.6 mL)

*87. The main component of automotive antifreeze is ethylene glycol ($C_2H_6O_2$). Although the mathematical relationships of the colligative properties are accurate only in dilute solution, Equation (12-13) may be used for rough estimates in concentrated solutions. Estimate the molality of ethylene glycol required to decrease the freezing point of water to 0 °F. Also express the composition as weight percent.

**88. Estimate the molality of a $NaCl/H_2O$ mixture that freezes at -10 °C (this is an estimate only, because the solution is not dilute). An ice/*water* mixture is in equilibrium at 0 °C, but when NaCl(s) is mixed with ice, the temperature must fall in order to maintain equilibrium in the ice/*NaCl(aq)* mixture. This principle is put to use in the preparation of homemade ice cream. Estimate the amount of NaCl(s) that must be added to 10 pounds of ice in order to decrease the temperature to -10 °C.

89. If 0.28 g of an ionic compound with empirical formula Hg_2SO_4 is dissolved in 100.0 g water, the resulting solution has a freezing point depression of 0.021 °C. How many ions are formed when one formula unit of this substance dissolves?

90. A solution of 0.883 g I_2 in 50.0 g CCl_4 is found to boil 0.35 °C above the normal boiling point. What is the boiling point elevation constant of carbon tetrachloride?

*91. Bromobenzene normally boils at 156.43 °C, but when 1.008 g benzoic acid ($C_6H_5CO_2H$) is dissolved in 65.32 g bromobenzene, the solution boils at 157.22 °C. When 2.3 g of an unknown molecular substance is dissolved in 50 g bromobenzene, the solution boils at 157.81 °C. What is the molar mass of the unknown?

92. At what temperature would a 1.00 M solution of a nonelectrolyte have an osmotic pressure of 1.00 atm? In what solvent could this experiment be performed?

93. Glutamine synthetase, an enzyme involved in energy transport in animal metabolism, has a molar mass of 592,000 g mol^{-1}. How many grams of this material must be dissolved to produce one liter of an aqueous solution whose (room temperature) osmotic pressure is 1.5 atm? Is this a reasonable value?

CHAPTER 13

EQUILIBRIUM

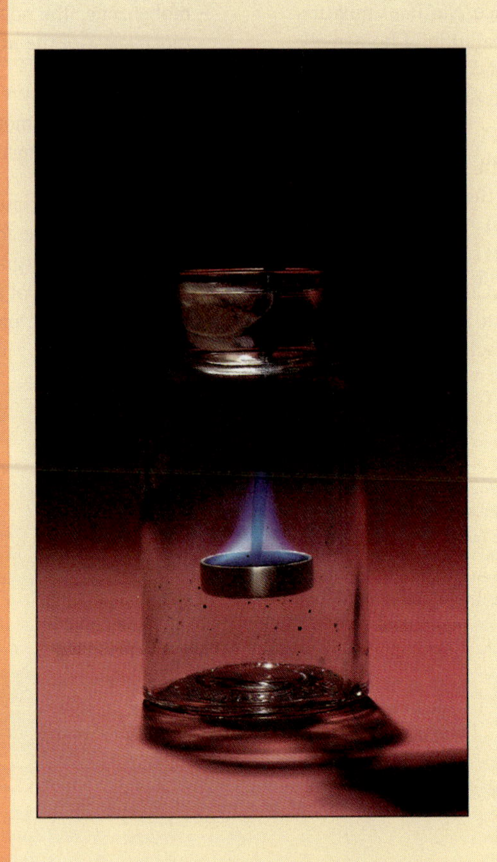

Sulfur burning in an atmosphere of pure oxygen. The major product is sulfur dioxide, some of which reacts further with oxygen to produce sulfur trioxide. The second reaction is reversible, and, under the right conditions, can reach equilibrium.

$$2\ SO_2 + O_2 \rightleftharpoons 2\ SO_3.$$

(a)

(b)

(c)

Figure 13.1 (a) When the stopper is removed from the bottle containing liquid and vapor bromine, (b) bromine vapor begins to escape. (c) After a time, the partial pressure of bromine in the enlarged volume regains its equilibrium value. Attainment of equilibrium always takes time, and it is always a smooth process, as shown in Figure 13.2.

13.1 EQUILIBRIUM IN PHYSICAL PROCESSES

We have already encountered several examples of equilibrium in physical processes: vapor pressure (Section 11.4) and other forms of phase equilibrium, solubility (Section 12.1), and osmotic pressure (Section 12.4). These systems may be used to illustrate certain features common to all equilibria, namely the *reversibility* of equilibrium systems, the *dynamic* nature of equilibrium processes, the *smooth approach* to equilibrium, and the fact that equilibrium values do not depend on the *direction* of approach.

Characteristics of Equilibrium

A **reversible** process is one that takes place readily in either direction in response to a small change in conditions. Addition of a small amount of heat to an ice/water mixture at 0 °C causes some ice to melt, while removal of a small amount of heat causes some water to freeze. Melting/freezing is a reversible physical process, as are all phase equilibria. Not all physical processes are reversible, however. For example, once crystallization of a supersaturated solution begins, no small change in the surroundings can reverse the process.

The vapor pressure exerted by a liquid in a closed container is maintained at a constant value through *dynamic equilibrium*. Evaporation and condensation both take place, but at equal and opposite rates. There is no net change in the amount of either liquid or vapor, and no change in the vapor pressure.

Dynamic equilibrium is not established instantaneously. When liquid evaporates into a closed container, the vapor pressure rises steadily in a *smooth approach* to its equilibrium value (Figures 13.1 and 13.2).

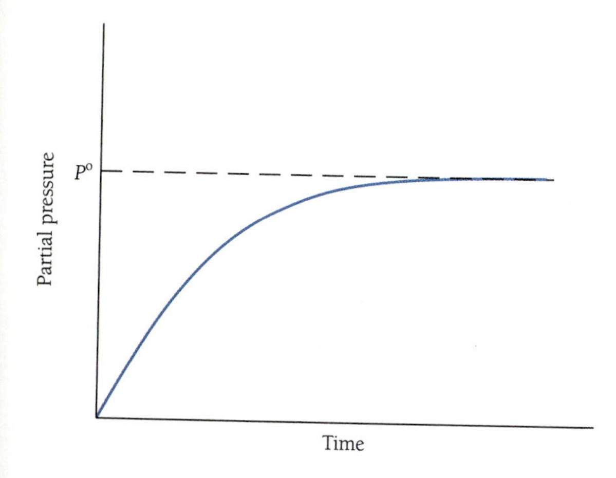

Figure 13.2 Partial pressure of bromine vapor, plotted as a function of time after the opening of the bottle in Figure 13.1a. The pressure rises smoothly and ever more slowly as the equilibrium value ($P°$) is approached. For bromine at room temperature, $P°$ is about 200 torr.

The lack of dependence on the *direction of approach* is sometimes expressed by the statement, "A system at equilibrium has no memory of how the equilibrium was attained."

Equilibrium vapor pressure does not depend on the *direction of approach* to equilibrium. Suppose a liquid/vapor system is in equilibrium at a pressure of 100 torr. If the volume of the system is decreased (Figure 13.3), the pressure is momentarily forced above 100 torr. However, the pressure soon returns to 100 torr as some of the vapor condenses to the liquid. Conversely, if the volume above the liquid is increased, the pressure decreases. Additional liquid evaporates, and the pressure returns to 100 torr.

Other Physical Equilibria

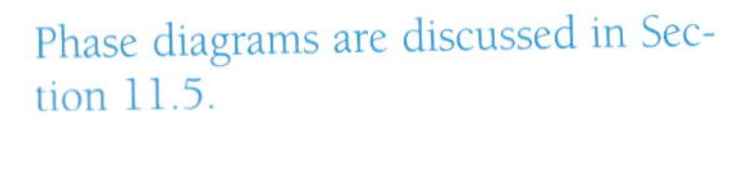

Phase equilibrium is not limited to vaporization, sublimation, and melting. Different solid forms of a substance, such as the allotropic forms of sulfur, phosphorus, tin, or carbon, can be in equilibrium. The specific values of temperature and pressure, at which phase equilibrium is possible, are indicated by *phase diagrams*. Figure 13.4 is a phase diagram for carbon.

Another common form of equilibrium, called solute **distribution** equilibrium, occurs when a substance is soluble in each of two immiscible liquids. Iodine, for example, is soluble in both carbon tetrachloride and water. These solvents are mutu-

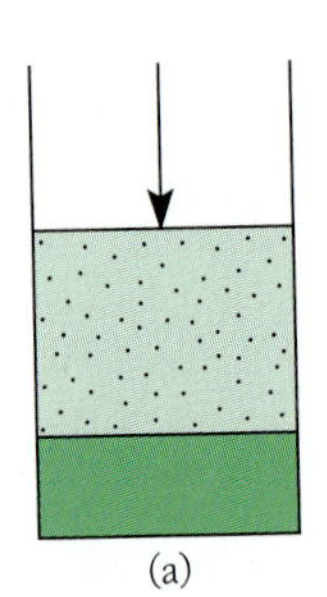

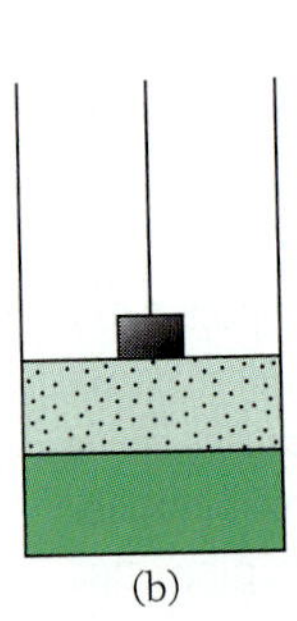

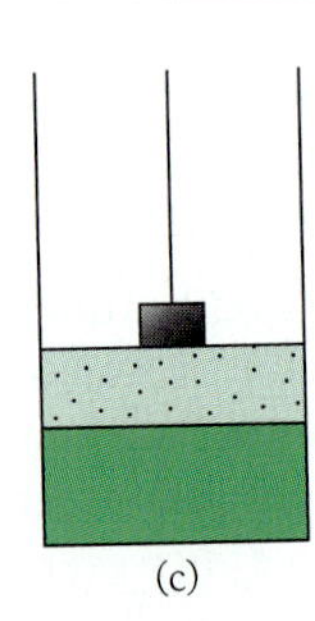

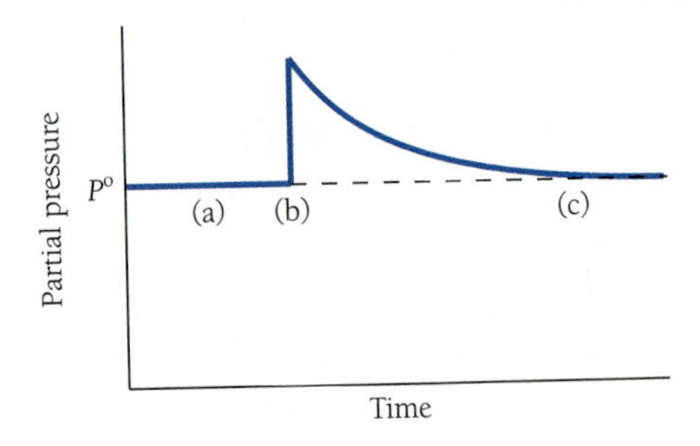

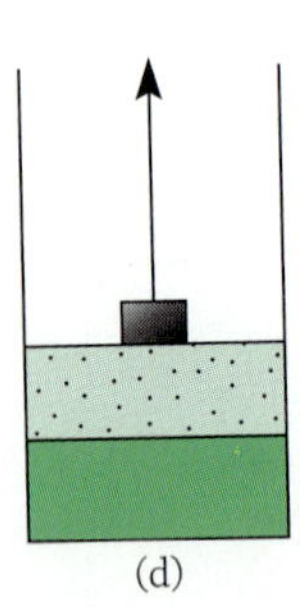

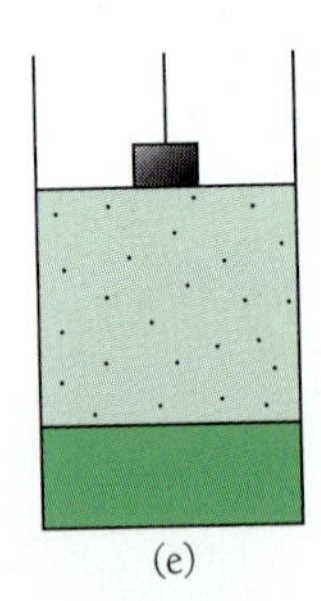

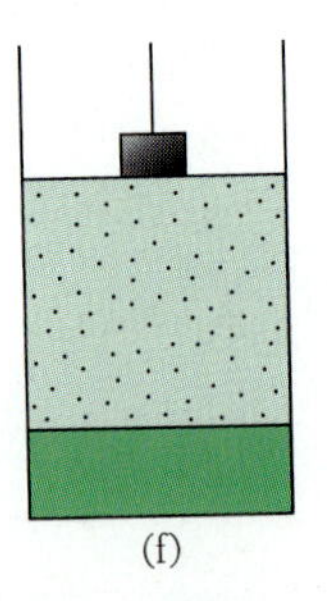

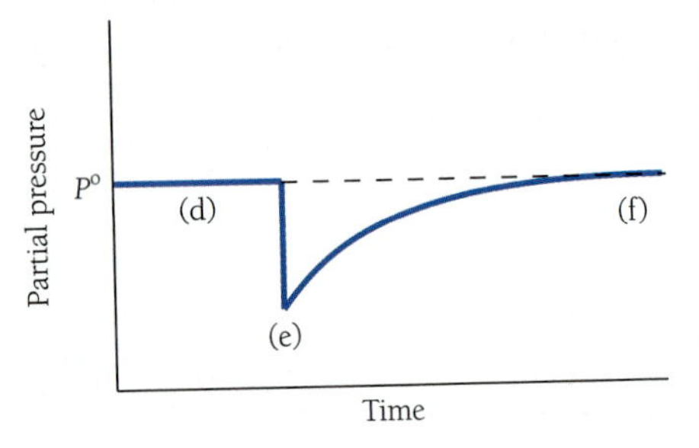

Figure 13.3 (a) When the volume of the cylinder is decreased by pushing the piston down, (b) the vapor pressure increases above $P°$, its equilibrium value. (c) Condensation occurs, and the pressure soon returns to its equilibrium value. (d) The opposite process takes place if the piston is partially withdrawn. (e) The pressure is momentarily less than its equilibrium value, (f) but it soon returns to it.

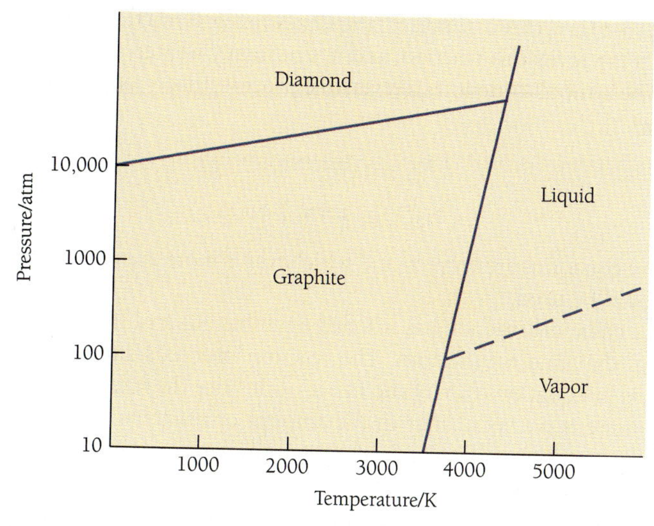

Figure 13.4 Phase diagram of carbon. Graphite is the stable solid form at low temperature and pressure. At high temperature, and pressures greater than 10,000 atm, graphite and diamond are in equilibrium.

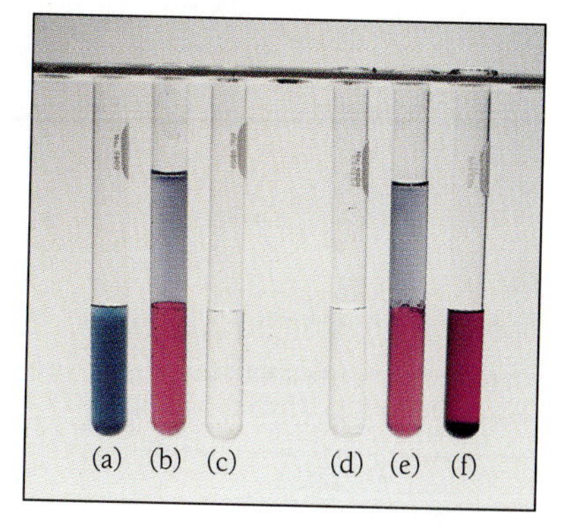

Figure 13.5 The molecular solid I_2 is distributed between the immiscible solvents water and carbon tetrachloride. A drop of aqueous 2% (w/w) starch solution has been added to produce a blue color in the aqueous iodine solutions. The carbon tetrachloride solutions are purple. In the left-hand group, (a) contains a few mg I_2 dissolved in 10 mL water, (c) contains 10 mL pure CCl_4, and (b) shows the result after thorough mixing of (a) and (c). In the right-hand group, (d) contains 10 mL pure water, (f) contains a few mg I_2 dissolved in 10 mL CCl_4, and (e) shows the result of mixing (d) and (f). Test tubes (b) and (e) each contain 10 mL H_2O, 10 mL CCl_4, and I_2. They are identical, because the equilibrium composition of a system does not depend on the direction of approach.

ally insoluble, and form two layers when mixed. Water lies on top of the CCl_4 because its density (1.0 g mL^{-1}) is less than that of CCl_4 (1.6 g mL^{-1}). When I_2 is added to a mixture of H_2O and CCl_4, part of it dissolves in the water and part dissolves in the CCl_4. The ratio of solute concentrations in the two layers is equal to the ratio of solubilities of I_2 in each solvent. This ratio of concentrations is constant, and it holds for saturated solutions and undersaturated solutions alike. Figure 13.5 shows that in distribution equilibria, as in all equilibria, the same concentrations are reached regardless of the direction of approach.

13.2 EQUILIBRIUM IN CHEMICAL REACTIONS: THE DISSOCIATION OF N_2O_4

Like physical processes, chemical reactions can be reversible. A reversible chemical reaction is one that can take place in either direction, depending on the conditions. For example, the gas nitrogen dioxide undergoes an **association** reaction, in which two molecules combine to form a molecule of dinitrogen tetroxide

$$2\ NO_2(g) \longrightarrow N_2O_4(g) \qquad (13\text{-}1)$$

The NO_2 molecule is a product of automobile exhaust, and it greatly impacts our environment. Its interaction with sunlight releases oxygen atoms, a primary cause of the smog that plagues some of our larger cities (Figure 13.6).

$$NO_2 + \text{sunlight} \longrightarrow NO + O \longrightarrow \text{smog}$$

In the other direction, dinitrogen tetroxide undergoes a **dissociation** reaction, in which it splits into two identical NO_2 molecules.

$$N_2O_4(g) \longrightarrow 2\ NO_2(g) \qquad (13\text{-}2)$$

Figure 13.6 The formation of nitrogen dioxide in automobile engines is the first step in a series of reactions that results in urban smog.

Equations (13-1) and (13-2) represent the same chemical reaction, written in opposite directions. It is a reversible reaction, and it is normally written as in Equations (13-3) *or* (13-4). The symbol $\rightleftharpoons$ denotes a reversible process, that is, one capable of reaching equilibrium.

$$2\ NO_2(g) \rightleftharpoons N_2O_4(g) \tag{13-3}$$

$$N_2O_4(g) \rightleftharpoons 2\ NO_2(g) \tag{13-4}$$

If you're thinking about *association,* use Equation (13-3). If you're thinking about *dissociation,* use Equation (13-4).

Both of these equations are correct, and whichever is more appropriate to a particular context should be used.

Rates of chemical reactions are discussed in Chapter 14.

When a chemical reaction has reached equilibrium, it is, like a physical process, in a state of dynamic equilibrium. The reaction proceeds simultaneously in both directions, as in Equations (13-1) and (13-2), but since the *rates* are the same in both directions, there is no net change in the amount of reactants or products. There is furious activity on the molecular level, but from a macroscopic point of view the reactions are unnoticeable.

Consider a container that is filled initially with pure N_2O_4, say at a pressure of 0.5 atm. The N_2O_4 begins to dissociate by reaction (13-2), and at first there is no reverse reaction (13-1) because there is no NO_2 to associate. As the partial pressure of NO_2 builds up, however, an increasing amount of association takes place. Ultimately the two rates become equal, and equilibrium is achieved. The total pressure in the container, and the partial pressures of NO_2 and N_2O_4, change during this process, as shown in Figure 13.7. Just as in physical changes, the approach to chemical equilibrium is smooth and continuous. It is rapid at first, and slower as the condition of equilibrium is more closely approached. If the experiment starts with the same mass of pure NO_2 rather than N_2O_4 in the container, the result is the same. Figure 13.8 shows

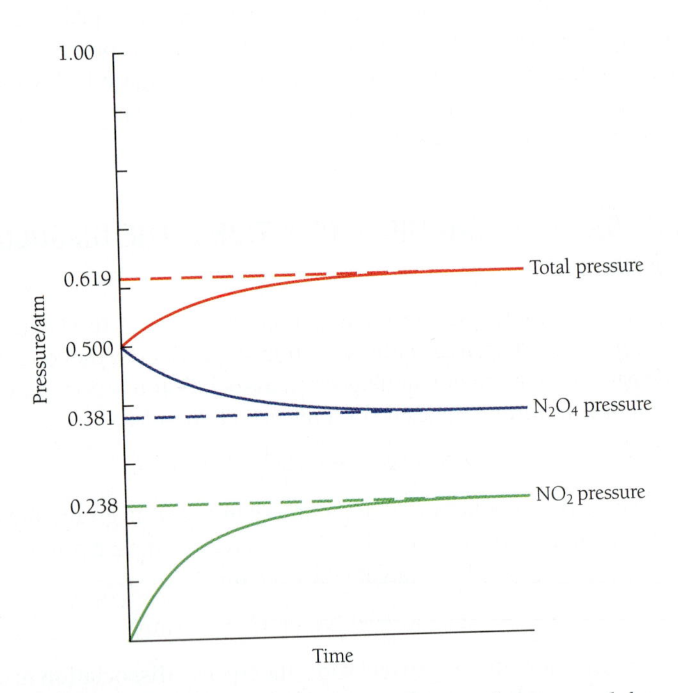

Figure 13.7 As N_2O_4 dissociates, its pressure decreases and the pressure of the product, NO_2, increases. The total pressure increases as well, because for each molecule of N_2O_4 that dissociates, *two* NO_2 molecules are formed. Both partial pressures, and the total pressure, reach their equilibrium values at the same time.

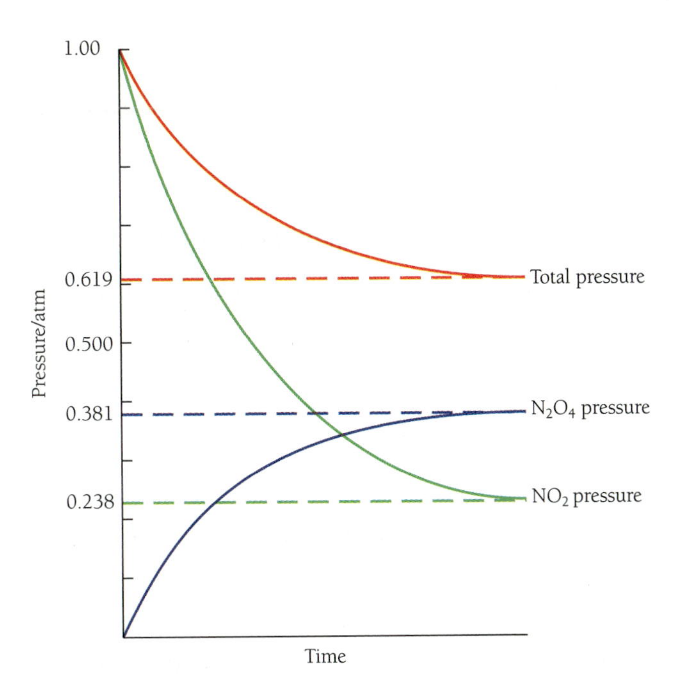

Figure 13.8 NO_2 is initially present at a pressure of 1 atm. As it associates, its pressure decreases and the pressure of the product N_2O_4 rises. The total pressure decreases, because for each pair of NO_2 molecules that reacts, only one N_2O_4 molecule is produced. The equilibrium pressures are the same as in Figure 13.7.

what happens when 1.0 atm NO_2 is allowed to react. Comparison with Figure 13.7 shows that in each case the equilibrium pressures, both total and partial, are identical. Equilibrium in chemical reactions does not depend on the direction of approach.

Some Numerical Results

In both of the experiments just discussed, the pressure of the equilibrium mixture is 0.619 atm, while the partial pressures of N_2O_4 and NO_2 are 0.381 atm and 0.238 atm, respectively. When similar experiments are performed, at a variety of different starting pressures of N_2O_4 and NO_2, the results in Table 13.1 are obtained. From the data in columns three and four it can be seen that, regardless of the initial conditions, the equilibrium mixture contains less NO_2 than N_2O_4. Therefore, as shown in column

Significant figure notation implies an uncertainty of ± 1 in the right-most digit. Therefore, all quantities in the right-hand column of Table 13.1 have the same value, to three significant digits.

TABLE 13.1 Partial Pressures in NO_2/N_2O_4 Equilibria*

Initial Pressures		Equilibrium Pressures		Equilibrium Ratios	
$P_{N_2O_4}$	P_{NO_2}	$P_{N_2O_4}$	P_{NO_2}	$P_{NO_2}/P_{N_2O_4}$	$P_{NO_2}^2/P_{N_2O_4}$
0.50	0.00	0.381	0.238	0.625	0.149
0.00	1.00	0.381	0.238	0.625	0.149
0.00	5.00	2.21	0.572	0.259	0.148
0.00	20.0	9.41	1.18	0.125	0.148
5.00	0.00	4.59	0.824	0.180	0.148
15.0	0.00	14.3	1.45	0.101	0.147
3.33	2.79	4.33	0.800	0.185	0.148

* Pressures, in atmospheres, in a rigid container (fixed volume) at 25 °C.

five, the ratio of their pressures $P_{NO_2}/P_{N_2O_4}$ is always less than one. But there is no obvious trend to the value of this ratio as the initial conditions are varied. Much more striking is the fact that the quotient $P_{NO_2}^2/P_{N_2O_4}$ is equal to 0.148 atm for *all* starting conditions.

Reaction Quotient, Equilibrium Constant, and Equilibrium Expression

The particular combination of partial pressures $P_{NO_2}^2/P_{N_2O_4}$, symbolized by Q, is called the **reaction quotient** of the $N_2O_4 \rightleftharpoons 2\ NO_2$ reaction.

$$Q = \frac{P_{NO_2}^2}{P_{N_2O_4}} \tag{13-5}$$

The numerator of Q is the pressure of the product, raised to the power of its coefficient (2) in the balanced chemical Equation (13-4). The denominator is the pressure of the reactant, also raised to the power of its stoichiometric coefficient (1). The reaction quotient can have any positive value, or it can be zero, depending on what the pressures are. In the first row of Table 13.1, the *initial* value of Q is $(0.00\ \text{atm})^2/0.50\ \text{atm} = 0\ \text{atm}$. The value of Q rises to 0.148 atm when the reaction reaches equilibrium. In the second, third, and fourth rows, the initial value of Q is undefined because its denominator is zero. The final value of Q is 0.148 atm. In the last row, Q falls from $(2.79\ \text{atm})^2/3.33\ \text{atm} = 2.34\ \text{atm}$ to 0.148 atm. It is apparent that, whatever value the reaction quotient may have initially, its value is 0.148 atm when the reaction reaches equilibrium at 25 °C.

When Q is evaluated by inserting the *equilibrium* values of the partial pressures, the result is called the **equilibrium constant, K_{eq}**, of the reaction.

$$Q_{eq} = K_{eq} = \frac{P_{NO_2,eq}^2}{P_{N_2O_4,eq}} \tag{13-6}$$

The right-hand side of Equation (13-6) is called the **equilibrium expression.** Since K_{eq} always refers to a system at equilibrium, the "eq" subscripts to the pressures are unnecessary. The equilibrium expression is written simply as (13-7).

$$K_{eq} = \frac{P_{NO_2}^2}{P_{N_2O_4}} \tag{13-7}$$

The "eq" subscript to K serves as a reminder that the pressures on the right-hand side are *equilibrium* pressures.

The ideal gas law was discussed in Section 5.2.

The amount of a gaseous species is sometimes expressed by its molarity rather than by its partial pressure. The ideal gas law gives the relationship between the molarity of a substance A and its partial pressure P_A, as $[A] = P_A/RT$. If the equilibrium expression is written with concentration rather than pressure units, the equilibrium constant will have a different numerical value (just as a person's weight might be 150 in pound units and 68 in kilogram units). For the $N_2O_4 \rightleftharpoons 2\ NO_2$ equilibrium at 298 K, the equilibrium expression and the equilibrium constant in the two different systems of units are

$$K_{eq} = \frac{P_{NO_2}^2}{P_{N_2O_4}} = 0.148\ \text{atm} \tag{13-8a}$$

and

$$K_{eq} = \frac{[NO_2]^2}{[N_2O_4]} = 0.00605\ \text{mol L}^{-1} \tag{13-8b}$$

General Form of the Equilibrium Expression

An equilibrium expression like that in Equation (13-7) can be written for any reaction that reaches equilibrium. The numerical value of the equilibrium expression, that is, the equilibrium constant, is always the same for a particular reaction at a given temperature. When, as is often the case, a reaction involves dissolved substances rather than gases, concentrations rather than partial pressures are used in the reaction quotient and K_{eq}.

Consider the reaction in which the substances A, B, and C are in equilibrium with the substances D, E, and F. Assume that some of these are dissolved substances and others are gases. When the stoichiometric coefficients a, b, c, d, e, and f are included, the equation might appear as

$$\text{a A(soln)} + \text{b B(soln)} + \text{c C(g)} \rightleftharpoons \text{d D(g)} + \text{e E(g)} + \text{f F(soln)} \quad (13\text{-}9)$$

The equilibrium expression for this generalized reaction is

$$K_{eq} = \frac{P_D{}^{d} \cdot P_E{}^{e} \cdot [F]^{f}}{[A]^{a} \cdot [B]^{b} \cdot P_C{}^{c}} \quad (13\text{-}10)$$

The square brackets, as usual, indicate concentration in moles per liter. Concentrations or partial pressures of products are in the numerator, and those of reactants are in the denominator. The exponents are the stoichiometric coefficients in the balanced chemical equation.

EXAMPLE 13.1 The Equilibrium Expression

Write the equilibrium expression for the reaction used in the commercial production of bromine from the bromide ions present in seawater

$$Br^-(aq) + Cl_2(g) \rightleftharpoons Cl^-(aq) + Br_2(aq)$$

Analysis

Target: An equilibrium expression

$$K_{eq} = \{\text{a quotient of concentrations and/or pressures}\}$$

Knowns: Reactants and products are identified.

Relationship: Equation (13-10) is appropriate, but its use requires a *balanced* chemical equation.

Missing Information: Stoichiometric coefficients in the equation.

Plan

2. Use Equation (13-10) to obtain the expression.
1. Balance the equation.

Work

1. The balanced equation is

$$2\ Br^-(aq) + Cl_2(g) \rightleftharpoons 2\ Cl^-(aq) + Br_2(aq)$$

2. For this specific reaction, the general form [Equation (13-10)] becomes

$$K_{eq} = \frac{[Cl^-(aq)]^2[Br_2(aq)]}{[Br^-(aq)]^2 P_{Cl_2}}$$

Bromine can be prepared by bubbling chlorine gas through an aqueous solution containing bromide ion.

Exercise

Write the equilibrium expression for the reaction of CO with O_2 to produce CO_2.

Answer: $K_{eq} = \dfrac{P_{CO_2}{}^2}{P_{CO}{}^2 P_{O_2}}$.

Forward and Reverse Reaction

The chemical equation for an equilibrium reaction such as $2\ NO_2 \rightleftharpoons N_2O_4$ can be written in several ways, two of which are

$$2\ NO_2(g) \rightleftharpoons N_2O_4\ (g) \qquad (13\text{-}3)$$

$$N_2O_4(g) \rightleftharpoons 2\ NO_2(g) \qquad (13\text{-}4)$$

One of these is called the *forward* reaction, and the other is called the *reverse* reaction. The choice is arbitrary and often signifies nothing more than which reaction was written first. One equilibrium expression [Equation (13-7)], corresponds to Reaction (13-4), and a different expression, Equation (13-11), corresponds to Reaction (13-3).

$$K_{eq,forward} = \frac{P_{NO_2}{}^2}{P_{N_2O_4}} = 0.148\ \text{atm} \qquad (13\text{-}7)$$

$$K_{eq,reverse} = \frac{P_{N_2O_4}}{P_{NO_2}{}^2} = \frac{1}{0.148\ \text{atm}} = 6.76\ \text{atm}^{-1} \qquad (13\text{-}11)$$

Since the roles of reactant and product are interchanged between Reactions (13-3) and (13-4), the numerator and denominator are interchanged between Equations (13-7) and (13-11). $K_{eq,reverse}$ is the *reciprocal* of $K_{eq,forward}$. Note the inversion of the units as well as the numerical part of the constant. Equation (13-12) is true for all chemical reactions.

$$K_{eq,reverse} = \frac{1}{K_{eq,forward}} \qquad (13\text{-}12)$$

Recall from Chapter 4 that chemical equations with fractional coefficients are often used in thermochemistry.

Another way of writing the chemical equation uses fractional coefficients.

$$\tfrac{1}{2}\ N_2O_4(g) \rightleftharpoons NO_2(g) \qquad (13\text{-}13)$$

The equilibrium expression and constant appropriate to this form are

$$K'_{eq} = \frac{P_{NO_2}}{P_{N_2O_4}{}^{1/2}} = [0.148\ \text{atm}]^{1/2} = 0.385\ \text{atm}^{1/2} \qquad (13\text{-}14)$$

Note the square root relationship that comes about when the coefficients of Equation (13-4) are multiplied by one-half: $K'_{eq} = (K_{eq,forward})^{1/2}$. The general rule, applicable to all equilibrium reactions, is as follows: if all stoichiometric coefficients are multiplied by some factor x, the equilibrium constant must be raised to the power x.

EXAMPLE 13.2 K_{eq} for Related Chemical Equations

At 500 °C, $K_{eq} = P_{SO_2}{}^2 P_{O_2}/P_{SO_3}{}^2 = 2.9 \times 10^{-4}$ atm for the reaction $2\ SO_3(g) \rightleftharpoons 2\ SO_2(g) + O_2(g)$. For each of the following chemical equations, write the equilibrium expression and evaluate the equilibrium constant.

a. $SO_3(g) \rightleftharpoons SO_2(g) + \frac{1}{2}\ O_2(g)$

b. $2\ SO_2(g) + O_2(g) \rightleftharpoons 2\ SO_3(g)$

c. $SO_2(g) + \frac{1}{2}\ O_2(g) \rightleftharpoons SO_3(g)$

Analysis

Targets: Equilibrium expressions, and values of K_{eq}.

Knowns: The equilibrium expression and K_{eq} for a related chemical equation.

Relationship/Plan: Two relationships have been discussed.

1. If an equation is reversed, invert the equilibrium expression and constant.
2. If all stoichiometric coefficients are multiplied by a common factor, raise the equilibrium expression and constant to that power.

Work

The starting point is the equation

$$2\ SO_3(g) \rightleftharpoons 2\ SO_2(g) + O_2(g); \qquad K_{eq} = 2.9 \times 10^{-4}\ \text{atm}$$

a.

$$SO_3(g) \rightleftharpoons SO_2(g) + \tfrac{1}{2}\ O_2(g)$$

All stoichiometric coefficients have been multiplied by ½, so that the equilibrium expression and the corresponding equilibrium constant become

$$K_{eq} = \frac{P_{SO_2} P_{O_2}{}^{1/2}}{P_{SO_3}}$$

$$K_{eq} = (2.9 \times 10^{-4}\ \text{atm})^{1/2} = 1.7 \times 10^{-2}\ \text{atm}^{-1/2}$$

b.

$$2\ SO_2(g) + O_2(g) \rightleftharpoons 2\ SO_3(g)$$

The equation has been reversed, so that the equilibrium expression and the corresponding equilibrium constant are

$$K_{eq} = \frac{P_{SO_3}{}^2}{P_{SO_2}{}^2 P_{O_2}}$$

$$K_{eq} = \frac{1}{2.9 \times 10^{-4}\ \text{atm}} = 3.4 \times 10^{3}\ \text{atm}^{-1}$$

c.

$$SO_2(g) + \tfrac{1}{2}\ O_2(g) \rightleftharpoons SO_3(g)$$

The equation has been reversed *and* its coefficients have been multiplied by ½, so we must invert *and* take the square root of the equilibrium constant. The result is

$$K_{eq} = \frac{P_{SO_3}}{P_{SO_2}P_{O_2}^{1/2}},$$

and

$$K_{eq} = \left[\frac{1}{2.9 \times 10^{-4}\ \text{atm}}\right]^{1/2} = 59\ \text{atm}^{-1/2}$$

Exercise

Be sure you understand why no units are given for K_{eq} in this exercise.

The equilibrium constant for the reaction $H_2(g) + I_2(g) \rightleftharpoons 2\,HI(g)$ is 6.2×10^2 at 298 K. What is K_{eq} for the reaction $HI(g) \rightleftharpoons \frac{1}{2}\,H_2(g) + \frac{1}{2}\,I_2(g)$ at this temperature?

Answer: 0.040.

13.3 INTERPRETATION OF THE EQUILIBRIUM CONSTANT

Equilibrium is not observable in all reaction systems. When carbon burns, for example, it is completely consumed. No carbon remains after the reaction ceases. Such reactions are said to "go to completion," and their equations are written with a single-headed arrow: $C + O_2 \rightarrow CO_2$. Other reactions do not take place to a measurable extent. For instance, water does not spontaneously decompose into hydrogen and oxygen. Of such reactions it is simply said that they "do not go." Many chemical reactions fall between these extremes of behavior. They "go," but reach equilibrium and stop before going to completion. If the equilibrium mixture contains mostly products (those species on the right of the arrow in the chemical equation) and only small amounts of reactants (the species on the left of the arrow), it is said that the equilibrium *"lies to the right."* If, on the other hand, the equilibrium mixture contains mostly reactants and only small amounts of products, the equilibrium is said to *"lie to the left."*

The Magnitude of K_{eq}

Since $K_{eq} = \dfrac{\text{numerator}}{\text{denominator}}$,

if $K_{eq} \ll 1$, then the numerator (products) is much less than the denominator (reactants); if $K_{eq} \gg 1$, then the numerator (products) is much greater than the denominator (reactants).

The numerical value of the equilibrium constant of a reaction indicates whether the equilibrium lies to the left or to the right. If the equilibrium constant is large ($\gg 1$), then the amount (either concentration or partial pressure) of the products in the numerator of the equilibrium expression must be greater than the amount of the reactants in the denominator, and the equilibrium therefore lies to the right. If K_{eq} is small ($\ll 1$), the amount of the products must be less than that of the reactants, and the equilibrium lies to the left. The magnitude of K_{eq} indicates how far the reaction proceeds.

If K_{eq} is very large, say greater than 1000, the reaction goes essentially to completion. If K_{eq} is very small, say less than 0.001, the reaction proceeds only very slightly. These numbers are approximate guidelines only, and should not be taken as rigorous statements of fact. Nonetheless they are useful in answering the question, "Does this reaction go or not?"

Relationship between Q, K_{eq}, and Reaction Directionality

Comparison of the value of K_{eq} with the value of Q for some particular set of concentrations or pressures indicates the direction in which the reaction will proceed. In the $N_2O_4 \rightleftharpoons 2\ NO_2$ dissociation, for example, the first row of Table 13.1 describes an experiment in which the initial value Q is zero. As the dissociation reaction proceeds to the right, Q rises smoothly to its equilibrium value of $Q = K_{eq} = 0.148$ atm. At all times when the reaction is proceeding to the right, Q is less than K_{eq}. In the experiment described in the bottom row of Table 13.1, Q decreases from its initial value of 2.34 atm to its equilibrium value of 0.148 atm. As long as Q is greater than K_{eq}, the reaction proceeds to the left. This behavior is characteristic of all chemical reactions, not just the N_2O_4 dissociation. As indicated in Table 13.2, if conditions are such that Q is less than K_{eq}, the reaction tends to proceed to the right. If Q is greater than K_{eq}, the reaction tends to proceed to the left. If Q is equal to K_{eq}, the reaction is at equilibrium.

TABLE 13.2 Relationship between Reaction Quotient, K_{eq}, and Direction of Reaction

$Q < K_{eq}$	Reaction proceeds to the right
$Q > K_{eq}$	Reaction proceeds to the left
$Q = K_{eq}$	Reaction is at equilibrium; does not proceed in either direction

EXAMPLE 13.3 Direction of Reaction

If NO_2 and N_2O_4 are introduced into a reaction vessel such that $P_{NO_2} = 1.00$ atm and $P_{N_2O_4} = 2.00$ atm, will the reaction $N_2O_4 \rightleftharpoons 2\ NO_2$ proceed to the left, to the right, or not at all? The equilibrium constant is $K_{eq} = 0.148$ atm.

Analysis Prediction of reaction directionality requires comparison of Q with K_{eq}. The problem statement contains all information necessary for the calculation of Q.

Work

$$Q = \frac{P_{NO_2}{}^2}{P_{N_2O_4}} = \frac{(1.00\text{ atm})^2}{2.00\text{ atm}} = 0.500\text{ atm}$$

Since $K_{eq} = 0.148$ atm, $Q > K_{eq}$ and the reaction proceeds to the left.

Exercise

Predict the reaction directionality when the *total* pressure is 0.500 atm and $P_{NO_2} =$ 0.300 atm.

Answer: Under these conditions $Q = 0.450$ atm, and the reaction proceeds to the left.

Equilibrium Constant and Free Energy of Reaction

The relationship between reaction directionality and the magnitude of K_{eq} calls to mind the discussion of free energy and the direction of spontaneous reaction. In Chapter 10 it was noted that if ΔG_{rxn} is negative, the reaction proceeds spontaneously to the right. If ΔG_{rxn} is positive, the reaction proceeds spontaneously to the left. If ΔG_{rxn} is zero, the reaction is at equilibrium. These relationships are summarized in Table 13.3.

TABLE 13.3 ΔG_{rxn} and K_{eq} in Equilibrium Processes

ΔG_{rxn} and Directionality		K_{eq} and Position of Equilibrium	
ΔG_{rxn}	Direction of Spontaneous Reaction	K_{eq}	Position of Equilibrium
Negative	Right	$\gg 1$	Right
Zero	Neither	1	Intermediate
Positive	Left	$\ll 1$	Left

"ln K_{eq}" is the symbol for the natural logarithm of K_{eq}. Use the LN or ln x key on your calculator.

When the partial pressure of each reactant and product in a gas-phase reaction is 1 atm, it follows from the definition that $\Delta G_{rxn} = \Delta G^\circ_{rxn}$.

Use of tabulated values of ΔG°_f to calculate K_{eq} is considerably faster and often more accurate than making measurements on equilibrium systems in the laboratory.

Table 13.3 indicates that there is a connection between the free energy change and the equilibrium constant of a reaction. The relationship is

$$\Delta G^\circ_{rxn} = -RT \cdot \ln K_{eq} \qquad (13\text{-}15)$$

in which T is the temperature (K) and R is the gas constant (8.314 J mol^{-1} K^{-1}). Note that it is the *standard* free energy change (ΔG°_{rxn}) that is related to K_{eq} by Equation (13-15). The value of ΔG°_{rxn} is independent of pressure, while the value of ΔG_{rxn} depends on the partial pressures of the reactants and products. If for a certain reaction K_{eq} is >1, then ln K_{eq} is a positive number and ΔG°_{rxn} is negative. For example, if $K_{eq} = 10^6$ at 25 °C, $\Delta G^\circ_{rxn} = -34.2$ kJ mol^{-1} by Equation (13-15). Both the large K_{eq} and the negative ΔG°_{rxn} show that the equilibrium lies to the right. If on the other hand the value of K_{eq} is <1, say 10^{-5}, its logarithm is negative and ΔG°_{rxn} is positive: $\Delta G^\circ_{rxn} = +28.5$ kJ mol^{-1}. The small K_{eq} and the positive ΔG°_{rxn} both show that the equilibrium lies to the left.

For many reactions, the value of ΔG°_{rxn} can be obtained by the methods of Chapter 10, using known values of ΔG°_f (Appendix G). K_{eq} can then be calculated from Equation (13-16), a rearranged form of Equation (13-15).

$$\ln K_{eq} = -\Delta G^\circ_{rxn}/RT \qquad (13\text{-}15a)$$

$$K_{eq} = e^{(-\Delta G^\circ_{rxn}/RT)} \qquad (13\text{-}16)$$

EXAMPLE 13.4 Free Energy and K_{eq}

Given that $\Delta G^\circ_f(NO_2) = +57.6$ kJ mol^{-1} and $\Delta G^\circ_f(N_2O_4) = +128.4$ kJ mol^{-1} at 400 K, calculate K_{eq} for the reaction $N_2O_4(g) \rightleftharpoons 2\ NO_2(g)$ at 400 K.

Analysis

Target: An equilibrium constant,

$$K_{eq} = ?$$

Knowns: The balanced equation, together with ΔG°_f data for all reaction participants.

Relationship: Equation (13-16) is appropriate, but it requires ΔG°_{rxn} data rather than ΔG°_f.

Missing: ΔG°_{rxn}

Plan

2. Use Equation (13-16) to obtain K_{eq}.
1. Use the methods of Chapter 10 to obtain ΔG°_{rxn} from the given ΔG°_f data.

Work

1.

$$\begin{aligned}\Delta G^\circ_{rxn} &= \Sigma[n \cdot \Delta G^\circ_f(\text{products})] - \Sigma[n \cdot \Delta G^\circ_f(\text{reactants})]\\ &= [2 \cdot \Delta G^\circ_f(NO_2)] - [1 \cdot \Delta G^\circ_f(N_2O_4)]\\ &= (2\text{ mol}) \cdot (57.6\text{ kJ mol}^{-1}) - (1\text{ mol}) \cdot (128.4\text{ kJ mol}^{-1})\\ &= -13.2\text{ kJ} = -13{,}200\text{ J}\end{aligned}$$

Note on Units: Recall from Chapters 4 and 10 that the units of ΔH°_{rxn} and ΔG°_{rxn}, normally abbreviated as J or kJ, arise from equivalence statements relating joules to specific quantities of substances. We have just determined that $\Delta G^\circ_{rxn} = -13.2$ kJ for the reaction consuming *one mole* of N_2O_4 and producing *two* moles of NO_2. That is, $\Delta G^\circ_{rxn} = -13.2$ kJ *per mole of* N_2O_4, and $\Delta G^\circ_{rxn} = -13.2$ kJ *per two moles of* NO_2. The dimensions of ΔG°_{rxn} are (energy mol^{-1}), although the "mol^{-1}" part is normally omitted. In calculations involving logarithms or exponents, however, including the present use of the relationship $K_{eq} = e^{(-\Delta G^\circ_{rxn}/RT)}$, the "mol^{-1}" part of the units cannot be ignored. Therefore, before continuing, we must write the result of step 1 as

$$\Delta G^\circ_{rxn} = -13{,}200\text{ J mol}^{-1}$$

2.

$$\begin{aligned}K_{eq} &= e^{(-\Delta G^\circ_{rxn}/RT)} = e^{-\{(-13{,}200\text{ J mol}^{-1})/[(8.314\text{ J mol}^{-1}\text{ K}^{-1})(400\text{ K})]\}}\\ &= e^{-(-3.969)}\\ &= 53\end{aligned}$$

Use the "e^x" or "inv(erse) ln x" key on your calculator.

Note on Units: Although we have been using equilibrium constants having a variety of units, the result of any ln x or e^x operation is dimensionless. This inconsistency is resolved in advanced chemistry courses by dividing each partial pressure in an equilibrium expression by 1 atm, and each concentration by 1 mol L^{-1}. This tactic converts all equilibrium constants to dimensionless quantities, regardless of reaction stoichiometry.

This text uses a different approach: we simply insert the correct units after evaluating the exponential term. Since the equilibrium expression for this reaction is $K_{eq} = P_{NO_2}{}^2/P_{N_2O_4}$, the correct units are (atm)^2/(atm) = atm. We write the answer to this example not as 53, but as

$$K_{eq} = 53\text{ atm}$$

Exercise

Calculate K_{eq} at 298 K for the reaction $2\ H_2O(g) \rightleftharpoons 2\ H_2(g) + O_2(g)$.

Answer: 7×10^{-81} atm. An equilibrium constant this small means that the reaction does not take place to a measurable extent.

In addition to the comments on units and the calculation itself, Example 13.4 illustrates two features worth noting. First, both ΔG°_f and ΔG°_{rxn} vary as the temperature changes. The values quoted in Example 13.4 for ΔG°_f of NO_2 and N_2O_4 are valid at

The superscript "°" in ΔG°_f and ΔG°_{rxn} means "standard pressure." These quantities do not vary as the pressure changes, because no matter what the reactant and product pressures are, the *standard* pressure is always 1 atm. ΔG varies with pressure, but ΔG° does not.

The temperature dependence of equilibrium constants is discussed in Section 13.5.

400 K, but they are *not* the same as the values given in Appendix G for 298 K. Advanced courses in chemistry discuss the temperature dependence of ΔG and other thermodynamic functions. The second feature is that the value of an equilibrium constant generally varies with temperature. Earlier in this chapter we used the value $K_{eq} = 0.148$ atm for the reaction $N_2O_4(g) \rightleftharpoons 2\ NO_2(g)$ at 25 °C (298 K). Example 13.4 shows that, at 400 K, the value is 53 atm.

The relationship between ΔG°_{rxn} and K_{eq} is also useful in determining unknown values of ΔG°_f, as shown in Example 13.5.

EXAMPLE 13.5 Free Energy and K_{eq}

K_{eq} for the reaction $F_2(g) + Cl_2(g) \rightleftharpoons 2\ ClF(g)$ is 4.08×10^{19} at 298 K. Calculate ΔG°_{rxn}, and also $\Delta G^\circ_f(ClF)$.

Analysis

Target: Both ΔG°_{rxn} and $\Delta G^\circ_f(ClF)$ are required.

Knowns: The balanced reaction, K_{eq}, and the temperature.

Relationship: Equation (13-15) is the needed relationship between the knowns and the target ΔG°_{rxn}; the other target $\Delta G^\circ_f(ClF)$ can then be determined by the methods of Chapter 10.

Plan

1. First find ΔG°_{rxn}, then
2. Use ΔG°_{rxn} to find $\Delta G^\circ_f(ClF)$.

Work

1.

$$\begin{aligned}\Delta G^\circ_{rxn} &= -RT \cdot \ln K_{eq} = -(8.314\ \text{J mol}^{-1}\ \text{K}^{-1})(298\ \text{K})(\ln 4.08 \times 10^{19}) \\ &= -111.9\ \text{kJ mol}^{-1}\end{aligned}$$

2. The desired free energy of formation is found from the relationship

Recall that the standard free energy of formation of any element in its standard state is zero.

$$\begin{aligned}\Delta G^\circ_{rxn} &= \Sigma[n \cdot \Delta G^\circ_f(\text{products})] - \Sigma[n \cdot \Delta G^\circ_f(\text{reactants})] \\ &= [2 \cdot \Delta G^\circ_f(\text{ClF})] - \{1 \cdot \Delta G^\circ_f[\text{F}_2(g)] + 1 \cdot \Delta G^\circ_f[\text{Cl}_2(g)]\} \\ -111.9\ \text{kJ mol}^{-1} &= [2 \cdot \Delta G^\circ_f(\text{ClF})] - \{1 \cdot 0 + 1 \cdot 0\} \\ 2 \cdot \Delta G^\circ_f(\text{ClF}) &= +111.9\ \text{kJ mol}^{-1} \\ \Delta G^\circ_f(\text{ClF}) &= \frac{+111.9\ \text{kJ mol}^{-1}}{2} \\ &= +56.0\ \text{kJ mol}^{-1}\end{aligned}$$

Exercise

K_{eq} for the reaction $N_2(g) + 3\ H_2(g) \rightleftharpoons 2\ NH_3(g)$ is $5.85 \times 10^5\ \text{atm}^{-2}$ at 298 K. Calculate $\Delta G^\circ_f(NH_3)$, and compare your answer with the value in Appendix G.

Answer: The value listed in Appendix G for $\Delta G^\circ_f(NH_3)$ is $-16.5\ \text{kJ mol}^{-1}$.

13.4 SOLUTION REACTIONS AND HETEROGENEOUS EQUILIBRIA

So far our discussion has focused on reactions taking place in the gaseous state. These, while instructive, represent only a small fraction of all chemical reactions. Others take place in (liquid) solutions, or at the interface between two different phases. Reversible reactions that reach equilibrium are common in these systems.

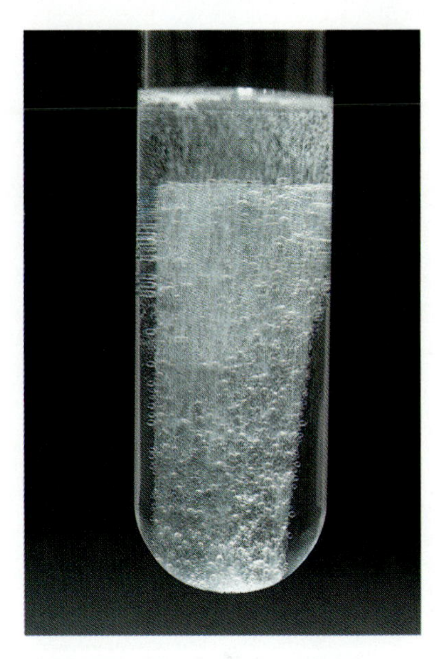

Figure 13.9 Active metals like zinc react with dilute aqueous acids such as HCl to produce gaseous hydrogen and a metal salt. This heterogeneous reaction occurs only at the metal surface, that is, at the interface between the solid and the liquid phase.

Reactions in Solution

Many common chemical reactions occur between solutes in a solution, for example the precipitation of insoluble salts and the complex biochemical interactions within living organisms. Another example is the reaction of acetic acid with ethanol to produce ethyl acetate and water.

$$CH_3CO_2H + C_2H_5OH \rightleftharpoons CH_3CO_2C_2H_5 + H_2O \qquad (13\text{-}17)$$

The reaction has an equilibrium constant of about 4 at room temperature. This means that the equilibrium lies to the right, but not very far to the right. The products, $CH_3CO_2C_2H_5$ and H_2O, predominate in the equilibrium mixture, but there are also appreciable concentrations of the reactants, CH_3CO_2H and C_2H_5OH. The equilibrium expression and equilibrium constant are given in Equation (13-18).

$$K_{eq} = \frac{[CH_3CO_2C_2H_5][H_2O]}{[CH_3CO_2H][C_2H_5OH]}; \qquad K_{eq} = 4 \qquad (13\text{-}18)$$

Note that K_{eq} is dimensionless: although the units of the individual species concentrations are mol L^{-1}, they cancel in the equilibrium expression.

Heterogeneous Reactions

Reactions in which some of the species are in different phases are called **heterogeneous reactions.** Since they are in different phases, the reactants cannot mix thoroughly with one another, so that heterogeneous reactions take place at the interface between two phases. The combustion of carbon is an example of a heterogeneous reaction. Carbon and oxygen can come into contact with one another only at the surface of a solid particle of carbon. Reaction takes place at the surface rather than throughout the body of the solid (Figure 13.9).

The thermal decomposition of calcium carbonate, (Equation 13-19),

$$CaCO_3(s) \underset{\Delta}{\rightleftharpoons} CaO(s) + CO_2(g) \qquad (13\text{-}19)$$

is an example of a heterogeneous reaction that reaches equilibrium when carried out in a closed container. (If the container is open, carbon dioxide escapes and the decomposition goes to completion.) If we follow the general rule (Equation 13-10), we would write the equilibrium expression for reaction (13-19) as (13-20).

The production of lime (CaO) from limestone ($CaCO_3$), discussed in Section 4.4, is carried out in open containers called kilns.

$$K_{eq} = \frac{P_{CO_2}[CaO(s)]}{[CaCO_3(s)]} \qquad (13\text{-}20)$$

However, some simplification of this expression is possible. CaO and $CaCO_3$ are pure solids, so their *concentrations* are fixed regardless of the size or number of particles of each that may be present. Since the terms [CaO(s)] and [$CaCO_3$(s)] do not change in value as the reaction proceeds, they are not included in the reaction quotient or the equilibrium expression. They are simply omitted, the correct equilibrium expression for reaction (13-19) is

$$K_{eq} = P_{CO_2} \qquad (13\text{-}21)$$

The *concentration* of a solid can be calculated from its density. For $CaCO_3$(s), whose density is 2.7 g mL^{-1}, the molarity of the solid is

$$(2.71\text{ g mL}^{-1})\cdot\frac{1\text{ mol}}{100.1\text{ g}}\cdot\frac{1000\text{ mL}}{1\text{ L}} = 27.1\text{ mol L}^{-1}$$

All equilibrium expressions involving pure solid substance(s) are treated in this way. *The concentrations of pure solid substances are not included in the equilibrium expression or the reaction quotient.*

EXAMPLE 13.6 Equilibrium Expression for Heterogeneous Reactions

Write the equilibrium expression for the decomposition of sodium hydrogen carbonate ($NaHCO_3$), which yields water vapor, carbon dioxide, and sodium carbonate (Na_2CO_3).

Analysis

Target: An equilibrium expression,

$$K_{eq} = ?$$

Knowns: Reactant and products are identified, and the gaseous state of water is specified.

Implicit: The physical state of $NaHCO_3(s)$, $Na_2CO_3(s)$, and $CO_2(g)$ are implicit knowns.

Formula: Equation (13-10) is appropriate, but its use requires a *balanced* chemical equation.

Missing: Stoichiometric coefficients in the reaction.

Plan

2. Use Equation (13-10) to obtain the expression for K_{eq}.
1. Balance the chemical equation.

Work

1. The balanced equation is

$$2\ NaHCO_3(s) \rightleftharpoons Na_2CO_3(s) + CO_2(g) + H_2O(g)$$

2. For this reaction the general form [Equation (13-10)] becomes

$$K_{eq} = P_{CO_2} \cdot P_{H_2O}$$

Solid species are omitted, and partial pressures are used for gaseous species.

Exercise

Write the equilibrium expression for the reaction $C(s) + 2\ H_2O(g) \rightleftharpoons CO_2(g) + 2\ H_2(g)$.

Answer: $K_{eq} = P_{CO_2} \cdot P_{H_2}{}^2/P_{H_2O}{}^2$.

Simplification of the equilibrium expression is appropriate for liquids as well as solids. If a liquid is pure, or if it is a solvent containing only a small amount of solute, its concentration does not change as the reaction proceeds. *The only terms included in an equilibrium expression or reaction quotient are those corresponding to gases or dissolved substances.* Terms corresponding to pure liquids or solids, or to the solvent of a dilute solution, are not included.

These crystals of sodium chloride are in equilibrium with the surrounding liquid, a saturated solution of sodium chloride in water.

Solubility Equilibria

A saturated solution in contact with some pure, undissolved solute is in a state of dynamic equilibrium. This is one of the few examples of dynamic equilibrium in which some changes in the system can be readily observed. Although the solute concentration does not change, and the mass of undissolved solute is likewise fixed, the *shape* of the crystals undergoes some change as time passes. The smallest crystals tend to disappear, while the larger ones grow. Irregular chunks tend to assume simpler, more regular shapes. Dissolution and crystallization continue, at equal and opposite rates, but they take place at different regions of the solid/liquid interface. The process is described by chemical equations, such as (13-22) for the ionic solute silver sulfate.

$$Ag_2SO_4(s) \rightleftharpoons 2\,Ag^+(aq) + SO_4^{2-}(aq) \qquad (13\text{-}22)$$

The equilibrium expression is

$$K_{eq} = K_{sp} = [Ag^+]^2[SO_4^{2-}] \qquad (13\text{-}23)$$

Solubility equilibria are so common that their equilibrium constants are given a special name and symbol: $\mathbf{K_{sp}}$ is called the **solubility product.** The solubility product is the equilibrium constant for the equilibrium between an ionic substance and its dissolved ions. Numerical calculations involving solubility products are discussed in Chapter 17.

13.5 SHIFTS IN EQUILIBRIA: LE CHÂTELIER'S PRINCIPLE

Le Châtelier's principle, introduced in Section 12.1 in the discussion of the effect of temperature on solubility, is applicable to all cases of chemical and physical equilibrium. Indeed, good arguments can be put forth to apply it to equilibrium situations in politics, economics, psychology, ecology, and so on. Le Châtelier's principle is that *if an equilibrium system is disturbed by an outside influence, for example a change in temperature, volume, pressure, or concentration of a reactant or product, the system will respond in such a way as to reduce the effect of the disturbance.* The principle is especially useful in predicting the effect of various changes on the position of chemical equilibrium.

Effect of Changes in Concentration

The equilibrium system

$$CH_3CO_2H + C_2H_5OH \rightleftharpoons CH_3CO_2C_2H_5 + H_2O \qquad (13\text{-}17)$$

is a homogeneous mixture in which none of the four components is singled out as the solvent. The reaction quotient is

$$Q = \frac{[CH_3CO_2C_2H_5][H_2O]}{[CH_3CO_2H][C_2H_5OH]}$$

and at equilibrium, $Q = K_{eq}$. If an additional amount of one of the *products* of this reaction, say water, is added to an equilibrium mixture, its concentration increases.

Therefore, the value of the reaction quotient increases, and it becomes greater than K_{eq}. As we have seen, whenever $Q > K_{eq}$, a reaction proceeds to the left. As this occurs, some of the additional water is consumed, decreasing its concentration. The equilibrium is said to "shift to the left." In terms of Le Châtelier's principle, the disturbance is the increase in concentration of one of the species, and the response of the equilibrium system is the decrease, by reaction, of the concentration of that species.

On the other hand, if an additional amount of one of the *reactants*, say CH_3CO_2H, is added to the reaction mixture, the reaction quotient decreases. In this case $Q < K_{eq}$, and the reaction proceeds to the right, consuming some of the added reactant. The equilibrium is said to "shift to the right." In terms of Le Châtelier's principle, the system has again acted to reduce the effect of a disturbance. These effects are summarized as follows.

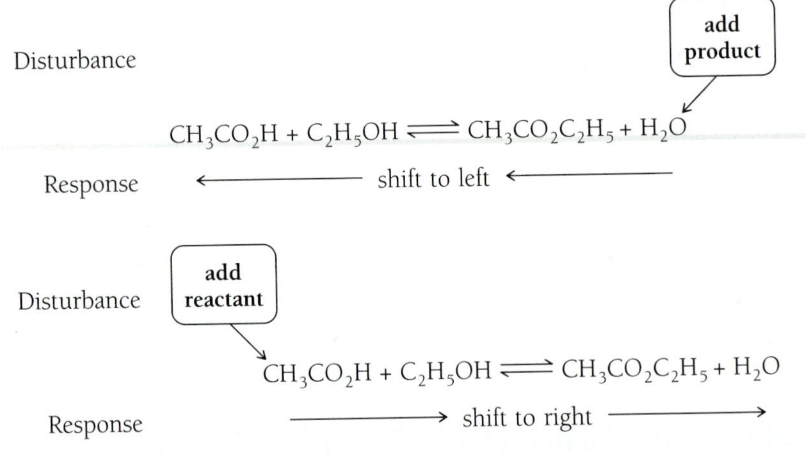

The principle also applies when a concentration is *decreased*. For instance, the weak electrolyte hydrofluoric acid dissociates in aqueous solution according to

The chemistry of acids such as HF is discussed in Chapters 15 and 16.

$$HF(aq) \rightleftharpoons H^+(aq) + F^-(aq) \tag{13-24}$$

Addition of NaOH(aq) decreases the concentration of $H^+(aq)$ by reaction with $OH^-(aq)$.

$$H^+(aq) + OH^-(aq) \longrightarrow H_2O$$

As a result, the equilibrium in Equation (13-24) shifts to the right: more HF(aq) dissociates, increasing the $H^+(aq)$ concentration and reducing the effect of the disturbance.

Reactions involving gases are similarly affected by additional quantities of reactants or products. Suppose the reaction

$$H_2(g) + I_2(g) \rightleftharpoons 2\ HI(g) \tag{13-25}$$

is at equilibrium in a rigid container, and an additional quantity of the reactant $H_2(g)$ is introduced. The partial pressure P_{H_2} increases, Q decreases, and the equilibrium shifts to the right.

As shown in Figure 13.10, the effect of concentration change is also apparent in solute distribution equilibria such as

$$I_2(\text{in } H_2O) \rightleftharpoons I_2(\text{in } CCl_4)$$

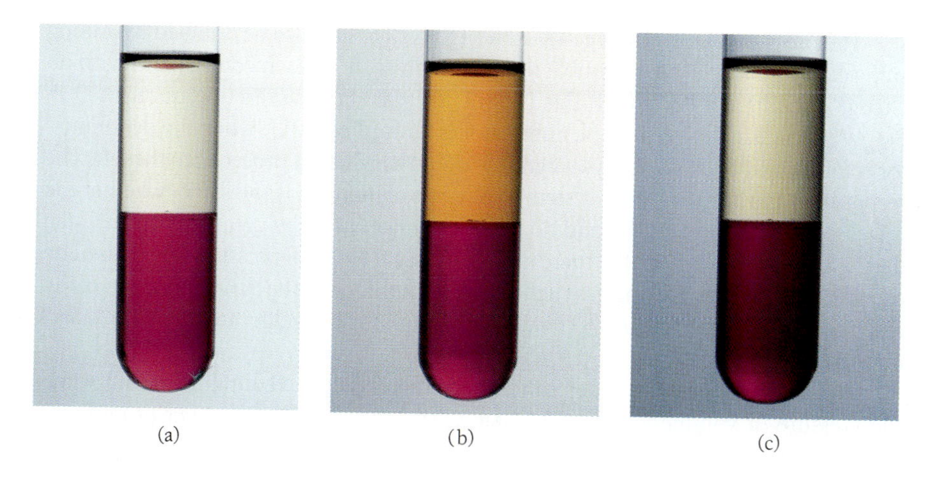

Figure 13.10 Dissolved iodine is distributed between the two immiscible solvents, water and carbon tetrachloride. In (a) the system is in equilibrium. In (b) a small amount of I_2 has been added to the water layer, which deepens its color. No other change has taken place. Some I_2 then passes down into the CCl_4 layer in order to reduce the extent of the disturbance, and in (c) the system is once again in equilibrium. The color of the water layer, and hence the concentration of iodine dissolved in it, is intermediate between (a) and (b). The color of the carbon tetrachloride layer has deepened, showing that some I_2 has passed to it in response to the disturbance.

Effect of Change in Volume or Pressure

Suppose that a mixture of NO_2 and N_2O_4 is at equilibrium in a cylinder whose volume can be changed by a movable piston. If the piston is moved so as to decrease the volume to half of its original value, the immediate effect is a doubling of the pressure. In response, the equilibrium

$$N_2O_4(g) \rightleftharpoons 2\ NO_2(g)$$

shifts to the left, which converts some NO_2 to N_2O_4. Since this conversion decreases the total number of moles of gaseous species, the total pressure in the cylinder decreases. In terms of Le Châtelier's principle, the disturbance is a decrease in volume, and the response is a decrease in the *number of moles of gas* (Figure 13.11).

If the piston had been moved so as to *increase* the volume, the pressure would have decreased, and the reaction would have proceeded to the right so as to increase the total number of moles of gas (Figure 13.12).

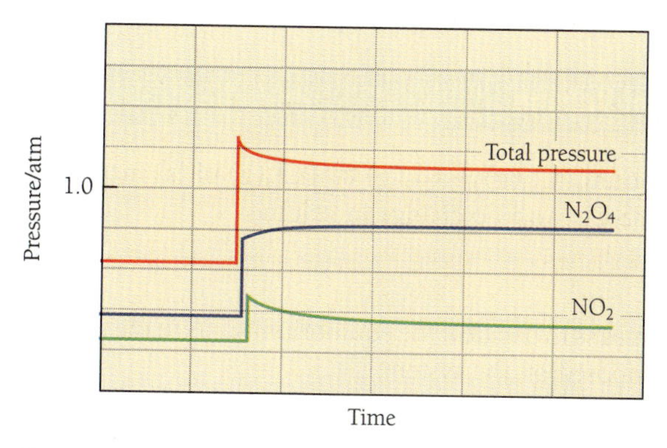

Figure 13.11 When an equilibrium system of NO_2 and N_2O_4 at a total pressure of 0.619 atm is compressed to half its original volume, the pressure initially doubles to 1.24 atm. The system responds so as to reduce the effect of the disturbance, by converting NO_2 to N_2O_4, and the pressure falls to the new equilibrium value of 1.17 atm. During the process of re-equilibration, the partial pressure of NO_2 falls and that of N_2O_4 rises.

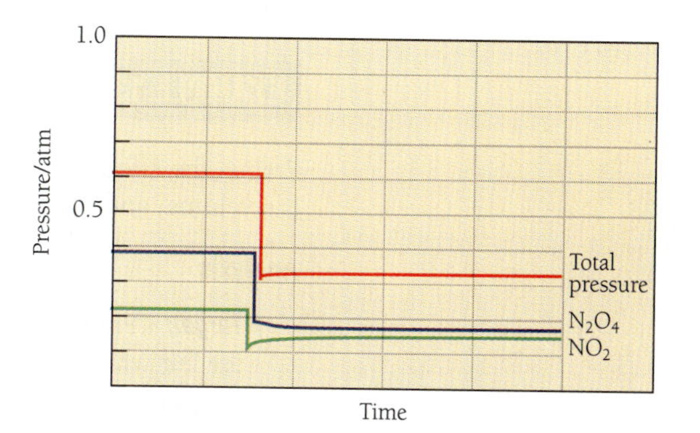

Figure 13.12 When an equilibrium system of NO_2 and N_2O_4 at a total pressure of 0.619 atm is expanded to double its original volume, the pressure initially drops to 0.310 atm. The system responds so as to reduce the effect of the disturbance, by converting N_2O_4 to NO_2, and the pressure rises to the new equilibrium value of 0.330 atm. During the process of re-equilibration, the partial pressure of NO_2 increases, and that of N_2O_4 decreases.

The pressure of the $N_2O_4(g) \rightleftharpoons 2\ NO_2(g)$ system can also be doubled by adding sufficient NO_2 and N_2O_4 to double the partial pressure of each component, with no change in container volume. If this is done, the value of the reaction quotient $Q = P_{NO_2}{}^2/P_{N_2O_4}$ is also doubled—the numerator is multiplied by 2^2, while the denominator is multiplied by 2. As always when $Q > K_{eq}$, the reaction proceeds to the left. The response of the $N_2O_4 \rightleftharpoons NO_2$ system to a pressure increase is the same, whether the increase is brought about by addition of more material or by compression (volume decrease). Note, however, that there is no response if the total pressure is increased by the addition of a nonreacting gas such as argon. In this case, the *partial pressures* of the reacting species do not change, Q remains at its equilibrium value, and there is *no effect* on the equilibrium.

The controlling factor in determining the direction of an equilibrium shift when the pressure or volume is changed is $\boldsymbol{\Delta n_{gas}}$, the change in the number of moles of gaseous species in the balanced chemical equation. Equation (13-4) shows 2 moles of gaseous species on the right and only 1 mole on the left: $\Delta n_{gas} = (2 - 1) = +1$.

$$N_2O_4(g) \rightleftharpoons 2\ NO_2(g); \qquad \Delta n_{gas} = +1 \tag{13-4}$$

Equations (13-25), (13-26), and (13-27) describe some other chemical equilibria that involve gaseous substances.

$$H_2(g) + I_2(g) \rightleftharpoons 2\ HI(g); \qquad \Delta n_{gas} = 0 \tag{13-25}$$

$$PCl_3(g) + Cl_2(g) \rightleftharpoons PCl_5(g); \qquad \Delta n_{gas} = -1 \tag{13-26}$$

$$2\ NaHCO_3(s) \rightleftharpoons Na_2CO_3(s) + CO_2(g) + H_2O(g); \qquad \Delta n_{gas} = +2 \tag{13-27}$$

Equation (13-27) illustrates that, in determining the value of Δn_{gas} from the stoichiometric coefficients of the balanced chemical equation, only the gaseous species are considered. For reactions like (13-4) and (13-27), in which Δn_{gas} is a positive number, increasing the volume (or decreasing the pressure) shifts the equilibrium to the right, because that creates more moles to fill up the increased volume (or to increase the pressure). If Δn_{gas} is negative, as it is in Equation (13-26), increasing the volume (decreasing the pressure) shifts the equilibrium to the left. If there is no change in the number of moles of gas ($\Delta n_{gas} = 0$), pressure/volume changes have no effect on the position of equilibrium. It follows from this that equilibrium reactions involving only solids or liquids are essentially unaffected by changes in pressure.

EXAMPLE 13.7 The Effect of Volume

Predict the effect on the equilibrium $PCl_3(g) + Cl_2(g) \rightleftharpoons PCl_5(g)$ of (a) increasing the volume, and (b) adding some Cl_2 with no change in volume.

Analysis

Target: "Predict the effect" means to decide if the equilibrium shifts to the right or the left when a change is imposed on the system.

Knowns: Each of the disturbances to the system, and the balanced chemical equation.

Implicit: The value of Δn_{gas} can be determined from the balanced equation.

Relationship: Le Châtelier's principle.

Plan Balance the equation, and determine the sign of Δn_{gas}. Apply Le Châtelier's principle to each of the disturbances separately.

Work

a. The number of gas moles on the right is 1, and that on the left is 2: $\Delta n_{gas} = (1 - 2) = -1$, a negative number. When the volume is increased, the system responds by increasing the number of moles: the equilibrium shifts to the left.

b. When chlorine is added, the system responds by consuming part of the added chlorine: the equilibrium shifts to the right.

Exercise

Predict the effect of (a) increasing the volume, and (b) adding some CO_2 to an equilibrium mixture of $NaHCO_3(s)$, $Na_2CO_3(s)$, $CO_2(g)$, and $H_2O(g)$.

Answer: The equilibrium shifts to the (a) right, (b) left.

Effect of Change in Temperature

The effect of temperature on chemical equilibrium can be understood by regarding heat as a substance that participates in a reaction, that is, as a "reactant" or "product." The chemical equation for an endothermic reaction can then be written in the form of Equation (13-28).

Until the mid-19th century it was believed that heat was a substance, called "caloric" or "phlogiston." This hypothesis was disproven by careful measurements of mass changes in chemical reactions, which showed that phlogiston had a zero or negative mass!

$$\text{heat} + \text{chemical reactants} \rightleftharpoons \text{chemical products} \qquad (13\text{-}28)$$

One such reaction is the dissociation of N_2O_4.

$$\text{heat} + N_2O_4(g) \rightleftharpoons 2\ NO_2(g) \qquad (\Delta H^\circ_{rxn} = +57.2\ \text{kJ})$$

Both tubes contain an equilibrium mixture of N_2O_4 (colorless) and NO_2 (reddish brown). The left-hand tube, immersed in a water bath at 50 °C, clearly has a higher partial pressure of NO_2 than the right-hand tube, immersed in an ice bath at 0 °C. This is in accord with Le Châtelier's principle, which predicts that, for the *endothermic* reaction $N_2O_4 \rightleftharpoons 2\ NO_2$, the partial pressure of NO_2 should increase as the temperature is increased.

TABLE 13.4 Effect of Changes on Equilibrium

Change	Direction of Equilibrium Shift	Change in K_{eq}
Increase in Concentration of		
Reactant	Right	None
Product	Left	None
Increase in Volume		
$\Delta n_{gas} < 0$	Left	None
$\Delta n_{gas} = 0$	None	None
$\Delta n_{gas} > 0$	Right	None
Increase in Pressure		
$\Delta n_{gas} < 0$	Right	None
$\Delta n_{gas} = 0$	None	None
$\Delta n_{gas} > 0$	Left	None
Increase in Temperature		
$\Delta H^\circ_{rxn} < 0$	Left	Decrease
$\Delta H^\circ_{rxn} > 0$	Right	Increase

If heat is added to this equilibrium system, by raising the temperature, Le Châtelier's principle predicts a shift to the right. Part of the added heat is consumed as N_2O_4 is converted to NO_2. The concentration of NO_2 increases and that of N_2O_4 decreases, so that the value of Q increases. But the reaction remains in equilibrium—it is Q_{eq} that has increased. Since $K_{eq} = Q_{eq}$, the value of K_{eq} must be larger at the higher temperature. K_{eq} for the N_2O_4 dissociation increases from 0.11 atm at 25 °C to 0.78 atm at 50 °C.

The opposite effect occurs in an exothermic reaction such as

$$PCl_3(g) + Cl_2(g) \rightleftharpoons PCl_5(g) + \text{heat} \qquad (\Delta H^\circ_{rxn} = -87.9 \text{ kJ})$$

In this reaction, heat is a "product." When the temperature is increased, the system consumes part of the heat by shifting the equilibrium to the left. The concentrations of PCl_3 and Cl_2 both increase, while that of PCl_5 decreases. The reaction quotient, and the equilibrium constant, are both smaller at the higher temperature. For this reaction, K_{eq} is 3.3×10^6 atm^{-1} at 25 °C and 0.46×10^6 atm^{-1} at 100 °C.

When $\Delta H^\circ > 0$ (endothermic), K_{eq} increases as T increases. When $\Delta H^\circ < 0$ (exothermic), K_{eq} decreases as T increases.

In summary, when the temperature is increased, the equilibrium constant of an endothermic reaction increases and that of an exothermic reaction decreases.

Remember that when a reversible chemical reaction is disturbed by a change in temperature, both the position of equilibrium and the value of K_{eq} change. However, when the equilibrium is disturbed by a change in concentration of a reactant or product, or by a change in pressure or volume, the equilibrium shifts but the value of K_{eq} remains the same. The conclusions are summarized in Table 13.4.

EXAMPLE 13.8 Effect of Pressure and Temperature

Predict the effect of (a) increasing the pressure and (b) increasing the temperature on the equilibrium in the Haber process,

$$N_2(g) + 3\,H_2(g) \rightleftharpoons 2\,NH_3(g)$$

Analysis/Relationship Prediction of the effect of a pressure change requires knowledge of Δn_{gas}, obtained from the balanced equation. Predicting the effect of temperature requires knowing whether the reaction is endo- or exothermic.

Plan

a. Make sure the equation is balanced, and determine Δn_{gas}. Then predict the effect of pressure from the sign of Δn_{gas}.

b. Calculate ΔH°_{rxn} from data in Appendix G, and use the sign of ΔH°_{rxn} to predict the effect of temperature.

Work

a. The number of moles of gas on the right is 2, and that on the left is 4: $\Delta n_{gas} = (2 - 4) = -2$. If the pressure is increased, the system responds by reducing the number of moles: the equilibrium shifts to the right.

This result is important in the commercial synthesis of ammonia. For a given amount of nitrogen and hydrogen, more ammonia is produced when the reaction is run at high pressure.

b. From the thermochemical data we find

$$\begin{aligned}\Delta H^\circ_{rxn} &= \Sigma[n \cdot \Delta H^\circ_f(\text{products})] - \Sigma[n \cdot \Delta H^\circ_f(\text{reactants})]\\ &= (2\text{ mol}) \cdot \Delta H^\circ_f(NH_3)\\ &\quad - [(3\text{ mol}) \cdot \Delta H^\circ_f(H_2) + (1\text{ mol}) \cdot \Delta H^\circ_f(N_2)]\\ &= (2\text{ mol}) \cdot (-46.11\text{ kJ mol}^{-1}) - [0 + 0]\\ &= -92.22\text{ kJ}\end{aligned}$$

The reaction is exothermic, so K_{eq} *decreases* when the temperature increases, and the equilibrium shifts to the left.

Exercise

Predict the effect of increasing the (a) temperature, and (b) pressure on the equilibrium $PCl_3(g) + Cl_2(g) \rightleftharpoons PCl_5(g)$.

Answer: (a) The reaction is exothermic, so K_{eq} decreases and the shift is to the left; (b) shifts to right.

The methods of thermodynamics can be used to derive a relationship describing the effect of temperature on the equilibrium constant of a reaction. Let K_1 be the value of K_{eq}, at temperature T_1, and K_2 be its value at some other temperature T_2. If the standard enthalpy of reaction is ΔH°_{rxn}, then

As usual in thermodynamics, temperature is expressed in kelvins.

$$\ln (K_2/K_1) = -\left(\frac{\Delta H^\circ_{rxn}}{R}\right) \cdot \left(\frac{1}{T_2} - \frac{1}{T_1}\right) \qquad (13\text{-}29a)$$

R is the gas constant, $8.314\text{ J mol}^{-1}\text{ K}^{-1}$.

or, in its equivalent form,

$$\ln\left(\frac{K_2}{K_1}\right) = \left(\frac{\Delta H^\circ_{rxn}}{RT_1T_2}\right) \cdot (T_2 - T_1) \qquad (13\text{-}29b)$$

EXAMPLE 13.9 Variation of K_{eq} with Temperature

At sufficiently high temperature, most diatomic molecules undergo partial dissociation into atoms. For example, at 1000 K, $K_{eq} = 9.1 \times 10^{-12}$ atm for the reaction $Cl_2(g) \rightleftharpoons 2\,Cl(g)$. Given that ΔH°_{rxn} for this reaction is 240 kJ, what is the value of K_{eq} at 1500 K?

A large part of the damage to the earth's ozone layer is caused by chlorine atoms.

Analysis/Plan Calculations involving the effect of temperature on equilibrium constants are solved using Equation (13-29a or 13-29b). The target is an equilibrium constant, and all required information is given in the problem (except the value of R, an implicit known).

Work Let the subscripts 1 and 2 designate the two temperatures. The known information is

$$T_1 = 1000\ \text{K}; \qquad K_1 = 9.1 \times 10^{-12}$$

$$T_2 = 1500\ \text{K}; \qquad K_2 = ?\ \text{(target)}$$

$$\Delta H^\circ_{rxn} = 240\ \text{kJ mol}^{-1}$$

Insert these quantities, and the gas constant, into Equation (13-29b). Note that, just as in Example 13.4, the "mol^{-1}" part of the units of ΔH°_{rxn} is used.

$$\ln\left(\frac{K_2}{K_1}\right) = \left(\frac{\Delta H^\circ_{rxn}}{RT_1T_2}\right) \cdot (T_2 - T_1)$$

$$= \left[\frac{240{,}000\ \text{J mol}^{-1}}{(8.314\ \text{J mol}^{-1}\ \text{K}^{-1}) \cdot (1000\ \text{K}) \cdot (1500\ \text{K})}\right] \cdot (1500\ \text{K} - 1000\ \text{K})$$

$$\ln\left(\frac{K_2}{K_1}\right) = 9.62$$

Use the "e^x" or "inv ln" key on the calculator.

$$\frac{K_2}{K_1} = e^{9.62} = 1.51 \times 10^4$$

$$K_2 = 1.51 \times 10^4 \cdot K_1$$
$$= 1.51 \times 10^4 \cdot 9.1 \times 10^{-12}\ \text{atm}$$
$$= 1.4 \times 10^{-7}\ \text{atm}$$

Exercise

Calculate the value of K_{eq} at $T = 2000$ K for the dissociation of chlorine.

Answer: 1.7×10^{-5} atm.

EXAMPLE 3.10 Standard Enthalpy of Reaction and K_{eq}

The equilibrium constant for the dissociation of fluorine, $F_2(g) \rightleftharpoons 2\ F(g)$, is 3.4×10^{-7} atm at 727 °C and 5.8×10^{-4} atm at 1727 °C. What is ΔH°_{rxn} for this reaction?

Analysis/Relationship/Plan The relationship between enthalpy of reaction, equilibrium constant, and temperature is given by Equation (13-29).

Work Let

$$T_1 = 727\ °\text{C} = 1000\ \text{K}; \qquad K_1 = 3.4 \times 10^{-7}\ \text{atm}$$

$$T_2 = 1727\ °\text{C} = 2000\ \text{K}; \qquad K_2 = 5.8 \times 10^{-4}\ \text{atm}$$

$$\ln\left(\frac{K_2}{K_1}\right) = \left(\frac{\Delta H^\circ_{rxn}}{RT_1T_2}\right) \cdot (T_2 - T_1)$$

$$\Delta H^\circ_{rxn} = \left[\ln\left(\frac{K_2}{K_1}\right)\right] \cdot \frac{(RT_1T_2)}{(T_2 - T_1)}$$

$$= \left[\ln\left(\frac{5.8 \times 10^{-4}\ \text{atm}}{3.4 \times 10^{-7}\ \text{atm}}\right)\right] \cdot (8.314\ \text{J mol}^{-1}\ \text{K}^{-1}) \cdot (1000\ \text{K}) \cdot \frac{(2000\ \text{K})}{(2000\ \text{K} - 1000\ \text{K})}$$

$$= [\ln(1.71 \times 10^3)] \cdot \frac{(1.66 \times 10^7\ \text{J mol}^{-1}\ \text{K})}{(1000\ \text{K})}$$

$$= 1.2 \times 10^5\ \text{J mol}^{-1}$$

$$= 120\ \text{kJ mol}^{-1} \qquad (\text{or } 120\ \text{kJ})$$

Use of the "per mole" part of the units is optional when reporting an enthalpy *of reaction.*

Exercise

The equilibrium constant for the dissociation of bromine, $Br_2(g) \rightleftharpoons 2\ Br(g)$, is 2.71×10^{-8} atm at 843 °C and 1.80×10^{-7} atm at 979 °C. What is ΔH°_{rxn} for this reaction?

Answer: 162 kJ.

13.6 CALCULATIONS OF EQUILIBRIUM QUANTITIES

The equilibrium expression is a numerical relationship among the concentrations and/or partial pressures of the products and reactants of a reversible chemical reaction. In this section, we discuss several different classes of common and important calculations based on this relationship.

The Equilibrium Constant

The experimental determination of an equilibrium constant requires measurement of the equilibrium concentrations or partial pressures of the reactants and products. These quantities are then substituted into the equilibrium expression, and K_{eq} is evaluated.

EXAMPLE 13.11 Calculation of K_{eq}

Measured at 500 °C, a mixture of gases is found to consist of 0.500 M $[H_2]$, 1.00 M $[N_2]$, and 0.085 M $[NH_3]$. Assuming that the reaction $N_2(g) + 3\ H_2(g) \rightleftharpoons 2\ NH_3(g)$ is in equilibrium, evaluate the equilibrium constant at 500 °C.

Analysis

Target: An equilibrium constant,

$$K_{eq} = ?$$

Knowns: Concentrations of $[H_2]$, $[N_2]$, and $[NH_3]$ at equilibrium, the chemical equation, and the temperature.

Relationship: K_{eq} is equal to the reaction quotient at equilibrium.

Missing: The equilibrium expression.

Plan Write the equilibrium expression, insert the numerical values of the concentrations, and evaluate.

Work The equilibrium expression, determined from the balanced equation, is

$$K_{eq} = \frac{[NH_3]^2}{[H_2]^3[N_2]}$$

Insertion of the given data yields

$$K_{eq} = \frac{(0.085\ M)^2}{(0.500\ M)^3(1.00\ M)} = 0.058\ M^{-2}$$

This answer is not incorrect, but gas-phase equilibrium constants are normally expressed in pressure units rather than in concentration units. A better solution would be to convert all molarities to atmospheres, by using the relationship $P = (n/V)RT = MRT$, before calculating the equilibrium constant.

Be sure you understand why, in this application, the value $R = 0.08206$ L atm mol^{-1} K^{-1} must be used.

$$P_{H_2} = (0.500\ \text{mol L}^{-1})(0.08206\ \text{L atm mol}^{-1}\ \text{K}^{-1})(773\ \text{K}) = 31.7\ \text{atm}$$

$$P_{N_2} = (1.00\ \text{mol L}^{-1})(0.08206\ \text{L atm mol}^{-1}\ \text{K}^{-1})(773\ \text{K}) = 63.4\ \text{atm}$$

$$P_{NH_3} = (0.085\ \text{mol L}^{-1})(0.08206\ \text{L atm mol}^{-1}\ \text{K}^{-1})(773\ \text{K}) = 5.39\ \text{atm}$$

$$K_{eq} = \frac{P_{NH_3}{}^2}{P_{H_2}{}^3 P_{N_2}}$$

$$K_{eq} = \frac{(5.39\ \text{atm})^2}{(31.7\ \text{atm})^3(63.4\ \text{atm})}$$

$$K_{eq} = 1.4 \times 10^{-5}\ \text{atm}^{-2}$$

Exercise

The equilibrium concentrations of CH_3CO_2H, C_2H_5OH, $CH_3CO_2C_2H_5$, and H_2O are 0.50 M, 0.25 M, 3.6 M, and 0.14 M, respectively. What is K_{eq} for the reaction $CH_3CO_2H + C_2H_5OH \rightleftharpoons CH_3CO_2C_2H_5 + H_2O$?

Answer: 4.0.

Often the equilibrium constant is known, and the problem asks for the concentration or pressure of one or another species at equilibrium.

EXAMPLE 13.12 Partial Pressure at Equilibrium

The equilibrium constant for the synthesis of ammonia is 1.4×10^{-5} atm^{-2} at 500 °C. If the partial pressures of NH_3 and N_2 are 5.39 atm and 63.4 atm, respectively, what is the equilibrium partial pressure of H_2?

Analysis/Plan All necessary numerical data are given in the problem, and the required equilibrium expression is available from the preceding example.

Work

$$K_{eq} = \frac{P_{NH_3}{}^2}{P_{H_2}{}^3 P_{N_2}}$$

$$P_{H_2}{}^3 = \frac{P_{NH_3}{}^2}{K_{eq} P_{N_2}}$$

$$= \frac{(5.39 \text{ atm})^2}{(1.4 \times 10^{-5} \text{ atm}^{-2})(63.4 \text{ atm})}$$

$$= 3.3 \times 10^4 \text{ atm}^3$$

$$P_{H_2} = (3.3 \times 10^4 \text{ atm}^3)^{1/3} = 32 \text{ atm}$$

Use the y^x key on your calculator.

Exercise

The equilibrium concentrations of CH_3CO_2H, C_2H_5OH, and H_2O are 0.50 M, 0.25 M, and 0.14 M, respectively, and the equilibrium constant for the reaction $CH_3CO_2H + C_2H_5OH \rightleftharpoons CH_3CO_2C_2H_5 + H_2O$ is 4.0 near room temperature. What is the equilibrium concentration of $CH_3CO_2C_2H_5$?

Answer: 3.6 M.

Extent of Reaction

The initial concentrations or pressures of species prior to reaction are usually known, and the problem is to determine the amount of reactant(s) consumed, or the amount of product(s) formed, when the system has reached equilibrium. These amounts, which are of course interrelated by the stoichiometry of the reaction, are described by a quantity called the extent of reaction (x). **Extent of reaction** is defined as the *change* in partial pressure (or concentration) of a reaction product, divided by the stoichiometric coefficient of that product in the balanced equation, as a particular reaction moves from an initial condition to a later (usually equilibrium) condition.

For example, suppose that both N_2 and O_2 are initially present at a partial pressure of 1.00 atm, and that after reaction (13-30) has reached equilibrium, it is found that exactly 0.010 atm O_2 has reacted.

$$N_2(g) + O_2(g) \rightleftharpoons 2\,NO(g); \qquad K_{eq} = 4.1 \times 10^{-4} \text{ at } 2000\text{ °C} \quad (13\text{-}30)$$

The amount of O_2 reacted, 0.010 atm, defines the *extent of reaction* in this example: $x = 0.010$ atm. Since the stoichiometric coefficients of O_2 and N_2 are the same, exactly 0.010 atm of N_2 has reacted as well. Furthermore, according to Equation (13-30) *two* moles of NO are produced for each mole of O_2 that reacts, so that at equilibrium the partial pressure of NO must be 0.020 atm. A summary of this information, in a structured format beneath the chemical equation, is a valuable problem-solving tool:

	$N_2(g)$	+	$O_2(g)$	$\rightleftharpoons$	$2\ NO(g)$
Initial pressure	1.00 atm		1.00 atm		0 atm
Change in pressure	$-x$ (= −0.010 atm)		$-x$ (= −0.010 atm)		$+2x$ (= +0.020 atm)
Equilibrium pressure	$1.00 - x$ (= 0.99 atm)		$1.00 - x$ (= 0.99 atm)		$0 + 2x$ (= 0.020 atm)

EXAMPLE 13.13 Extent of Reaction

N_2 and H_2 are put into a reactor at 500 °C, each at a partial pressure of 20.0 atm. After the reaction has proceeded for a time, it is found that the partial pressure of the product, NH_3, is 1.6 atm. What is the partial pressure of H_2?

Analysis

Target: P_{H_2}(atm), after some unspecified extent of reaction.

Knowns: The initial pressures $P_{N_2,\text{init}}$ and $P_{H_2,\text{init}}$ are given; implicit information includes the balanced chemical equation and the fact that $P_{NH_3,\text{init}} = 0$.

Relationship: P_{H_2} can be found from the extent of reaction, which in turn can be found from the initial and final pressures of NH_3.

Plan

2. Find P_{H_2} from the extent of reaction.
1. Use the structured format to summarize the known and unknown quantities and to evaluate the extent of reaction.

Work

1.

	N_2	+	$3\ H_2$	$\rightleftharpoons$	$2\ NH_3$
Initial	20.0		20.0		0

The changes from the initial to the final pressures are given by the product of the extent of reaction and the stoichiometric coefficients.

Change	$-x$	$-3x$	$+2x$
Final	$(20.0 - x)$	$(20.0 - 3x)$	$(0 + 2x)$

We know that the final pressure of NH_3 is 1.6 atm. That is, in terms of the extent of reaction x,

$$(0 + 2x) = 1.6 \text{ atm}$$

$$x = \frac{1.6 \text{ atm}}{2} = 0.80 \text{ atm}$$

2. The final pressure of H_2 is

$$P_{H_2} = (20.0 - 3x)$$
$$= 20.0 \text{ atm} - 3(0.80 \text{ atm})$$
$$= 17.6 \text{ atm}$$

Exercise

The initial concentrations of CH_3CO_2H, C_2H_5OH, $CH_3CO_2C_2H_5$, and H_2O are 0.64 M, 0.39 M, 3.44 M and 0.00 M, respectively. When the reaction $CH_3CO_2H + C_2H_5OH \rightleftharpoons CH_3CO_2C_2H_5 + H_2O$ reaches equilibrium, the concentration of water is found to be 0.14 M. What is the extent of reaction and what are the equilibrium concentrations of CH_3CO_2H, C_2H_5OH, and $CH_3CO_2C_2H_5$?

Answer: $x = 0.14$ M; 0.50 M, 0.25 M, 3.58 M.

Sometimes the extent of reaction is the target of a problem, but more often it is calculated in a substep that must be carried out in order to reach a further target.

EXAMPLE 13.14 Extent of Reaction and Equilibrium Constant

A mixture initially containing 0.400 atm SO_2 and 0.100 atm O_2 is found to contain 0.086 atm SO_3 after the reaction $2\,SO_2(g) + O_2(g) \rightleftharpoons 2\,SO_3(g)$ reaches equilibrium. What is the extent of reaction, and what is the value of the equilibrium constant?

Analysis

Target: The extent of reaction, x, and the equilibrium constant, K_{eq}, are the two targets of this problem.

Knowns: Initial pressures of SO_3(zero), SO_2, and O_2; the equilibrium pressure of SO_3; the balanced chemical equation.

Relationship: The extent of reaction can be found from the known initial and equilibrium pressures of SO_3: the formula K_{eq} = (equilibrium expression) applies.

Missing: The *equilibrium* pressures of SO_2 and O_2 are needed; these can be found from the *initial* pressures and the extent of reaction.

Plan

3. Insert the equilibrium pressures of all species into the equilibrium expression and solve for K_{eq}.
2. Find the equilibrium pressures of SO_2 and O_2 from their initial pressures and the extent of reaction.
1. Find the extent of reaction from the initial and equilibrium pressures of SO_3.

Work

1. Set up the known information under the species in the balanced equation using x for the unknown extent of reaction.

	$2\ SO_2$	+	O_2	$\rightleftharpoons$	$2\ SO_3$
Initial	0.400		0.100		0
Change	$-2x$		$-x$		$+2x$
Equilibrium	$0.400 - 2x$		$0.100 - x$		$0 + 2x$

Since the equilibrium pressure of SO_3, equal to twice the extent of reaction, is known to be 0.086 atm, x can be calculated.

$$(0 + 2x) = 0.086 \text{ atm}$$

$$x = \text{extent of reaction}$$

$$= \frac{0.086 \text{ atm}}{2} = 0.043 \text{ atm}$$

2. At equilibrium,

$$P_{SO_2} = 0.400 - 2x$$
$$= 0.400 - 0.086 = 0.314 \text{ atm}$$

$$P_{O_2} = 0.100 - x$$
$$= 0.100 - 0.043 = 0.057 \text{ atm}$$

3. Substitute these pressures into the equilibrium expression and evaluate the equilibrium constant.

$$K_{eq} = \frac{P_{SO_3}{}^2}{P_{SO_2}{}^2 P_{O_2}}$$

$$= \frac{(0.086 \text{ atm})^2}{(0.314 \text{ atm})^2 \cdot (0.057 \text{ atm})} = 1.3 \text{ atm}^{-1}$$

Exercise

A mixture initially containing 1.000 atm N_2 and 1.000 atm O_2 is found to contain 0.990 atm O_2 after the reaction $N_2(g) + O_2(g) \rightleftharpoons 2\ NO(g)$ reaches equilibrium. What is the value of the equilibrium constant?

Answer: 4.1×10^{-4}.

Equilibrium Concentrations or Pressures

A common type of problem arises when all initial concentrations or pressures and the value of the equilibrium constant are known, but the extent of reaction is not. Such problems are best solved by means of a standard, structured approach, patterned on the information summary of Examples 13.13 and 13.14. Successful problem-solvers almost always use this format, or a similar one that is equally well structured.

EXAMPLE 13.15 Equilibrium Pressures from K_{eq} and Initial Conditions

A vessel is filled with $COCl_2$ (phosgene) at 450 °C and an initial pressure of 0.75 atm. Calculate (a) the extent of reaction, and (b) the partial pressures of CO, Cl_2 and $COCl_2$, after equilibrium has been reached. The equilibrium constant for the reaction $COCl_2(g) \rightleftharpoons CO(g) + Cl_2(g)$ is 0.222 atm at 450 °C.

Analysis

Target: (a) The extent of reaction, and (b) partial pressures at equilibrium.

Knowns: Initial pressures of $COCl_2$, CO(zero), and Cl_2(zero), and the equilibrium constant.

Relationship: Problems having known K_{eq} and initial pressures are *always* solved with the structured approach given here.

Work

Step 1. Write the *balanced* chemical equation.

(1)		$COCl_2$	$\rightleftharpoons$	CO	+	Cl_2

Step 2. On the next line write the initial pressure of *every* species in the equation. In most problems, the units are obvious and may be omitted.

(1)		$COCl_2$	$\rightleftharpoons$	CO	+	Cl_2
(2)	**Initial**	0.75		0		0

Step 3. Next write the *changes* in pressure, in terms of x, the (unknown) extent of reaction, that occur as the reaction reaches equilibrium. The multiplier of x is always equal to the stoichiometric coefficient. The sign of x is negative for species consumed and positive for species formed.

(1)		$COCl_2$	$\rightleftharpoons$	CO	+	Cl_2
(2)	**Initial**	0.75		0		0
(3)	**Change**	$-x$		$+x$		$+x$

Step 4. In the next line write the equilibrium pressures. These are found by adding the "change" line to the "initial" line.

(1)		$COCl_2$	$\rightleftharpoons$	CO	+	Cl_2
(2)	**Initial**	0.75		0		0
(3)	**Change**	$-x$		$+x$		$+x$
(4)	**Equilibrium**	$(0.75 - x)$		$(0.0 + x)$		$(0.0 + x)$

Step 5. Using the balanced chemical equation, write the equilibrium expression and insert the pressure terms from the "equilibrium" line.

$$K_{eq} = \frac{P_{CO}P_{Cl_2}}{P_{COCl_2}}$$

$$K_{eq} = \frac{(0.0 + x)(0.0 + x)}{0.75 - x}$$

A quadratic equation has the form $Ax^2 + Bx + C = 0$, in which x is the unknown and A, B, and C may be positive or negative numbers. There are *two* solutions to a quadratic equation:

$$x = \frac{-B + \sqrt{B^2 - 4AC}}{2A} \quad \text{and}$$

$$x = \frac{-B - \sqrt{B^2 - 4AC}}{2A}$$

Step 6. Solve the algebraic expression and obtain the numerical value of x. In this example the result is a quadratic equation, but reactions of different stoichiometry lead to different forms.

$$K_{eq} = \frac{x^2}{0.75 - x}$$

$$0.75\,K_{eq} - K_{eq}x = x^2$$

$$x^2 + K_{eq}x - 0.75\,K_{eq} = 0$$

$$x = \frac{-K_{eq} \pm \sqrt{K_{eq}{}^2 - 4 \cdot (-0.75K_{eq})}}{2}$$

Inserting the value $K_{eq} = 0.222$ atm, we obtain

$$x = \frac{-0.222 \pm \sqrt{(0.222)^2 - 4 \cdot (-0.75) \cdot (0.222)}}{2}$$

$$x = \frac{-0.222 \pm 0.846}{2}$$

The two solutions are

$$x = +0.312 \text{ atm} \quad \text{and}$$

$$x = -0.534 \text{ atm}$$

Whenever a problem in chemistry is solved by means of a quadratic equation, one of the two answers is always physically impossible.

As we defined it in the problem setup, x is equal to the partial pressures of CO and Cl_2 at equilibrium. A negative pressure is impossible, so x cannot be a negative number. The correct solution is $x = 0.31$.

Step 7. After the extent of reaction x is found, the equilibrium pressures of all species are evaluated by means of the expressions in the "equilibrium" line of the problem setup.

		$COCl_2$	$\rightleftharpoons$	CO	+	Cl_2
(1)						
(4)	**Equilibrium**	$(0.75 - x)$		$(0.0 + x)$		$(0.0 + x)$

$$P_{COCl_2} = (0.75 - 0.312) = 0.44 \text{ atm}$$

$$P_{CO} = P_{Cl_2} = (0.00 + 0.31) = 0.31 \text{ atm}$$

Check Evaluate the reaction quotient for the calculated pressures: Q must equal K_{eq} at equilibrium.

$$Q = \frac{P_{CO}P_{Cl_2}}{P_{COCl_2}}$$

$$Q = \frac{(0.31 \text{ atm})(0.31 \text{ atm})}{0.44 \text{ atm}} = 0.22 \text{ atm}$$

The problem has been solved correctly.

Exercise

Calculate the extent of reaction and the equilibrium pressures of NO_2 and N_2O_4 in a vessel initially containing pure N_2O_4 at 0.25 atm. $K_{eq} = 0.11$ atm for the reaction $N_2O_4 \rightleftharpoons 2\ NO_2$.

Answer: $x = 0.070$ atm; $P_{N_2O_4} = 0.18$ atm, $P_{NO_2} = 0.14$ atm.

Example 13.15 involved gases, whose amounts are normally specified as partial pressures. Exactly the same structured approach is used to solve problems in which the amounts are specified as concentrations. These problems may involve gas mixtures or liquid solutions.

EXAMPLE 13.16 Equilibrium Concentrations from K_{eq}

A mixture containing 2.00 moles each of carbon dioxide and hydrogen is admitted to a 5.00 L reaction vessel maintained at 980 °C. For the reaction $CO_2(g) + H_2(g) \rightleftharpoons CO(g) + H_2O(g)$ at this temperature, $K_{eq} = 1.67$. What are the equilibrium molarities of all species?

Analysis/Plan K_{eq} and the initial conditions are given, and the targets are the concentrations at equilibrium. All such problems are solved by the structured approach used in Example 13.15: first calculate the extent of reaction, then evaluate the equilibrium concentrations.

Work The initial molarities of CO_2 and H_2 are 2.00 mol/5.00 L = 0.400 M.

	CO_2	+	H_2	$\rightleftharpoons$	CO	+	H_2O
Initial	0.400		0.400		0		0
Change	$-x$		$-x$		$+x$		$+x$
Equilibrium	$(0.400 - x)$		$(0.400 - x)$		x		x

The equilibrium expression is

$$K_{eq} = \frac{[CO][H_2O]}{[CO_2][H_2]}$$

$$K_{eq} = \frac{(x)(x)}{(0.400 - x)(0.400 - x)} = \frac{x^2}{(0.400 - x)^2}$$

Now, although this equation can be rearranged to the quadratic form and solved as in Example 13.15, the right-hand side is a perfect square, and a simpler solution is available. Take the square root of both sides:

This short-cut can be used whenever the equilibrium expression is a perfect square.

$$\sqrt{K_{eq}} = \frac{x}{0.400 - x}$$

$$0.400 \cdot \sqrt{K_{eq}} - x \cdot \sqrt{K_{eq}} = x$$

$$0.400 \cdot \sqrt{K_{eq}} = x + x \cdot \sqrt{K_{eq}} = x \cdot (1 + \sqrt{K_{eq}})$$

$$x = \frac{0.400 \cdot \sqrt{K_{eq}}}{1 + \sqrt{K_{eq}}}$$

Substitute the value $\sqrt{K_{eq}} = \sqrt{1.67} = 1.29$ to get

$$x = \frac{0.400(+1.29)}{1 + 1.29} = 0.225 \text{ M}$$

for the extent of reaction.

The equilibrium concentrations are found from the expressions in the "equilibrium" line of the problem setup:

$$[H_2] = [CO_2] = (0.400 - x) = 0.175 \text{ M}$$

$$[H_2O] = [CO] = x = 0.225 \text{ M}$$

After solving equilibrium problems, you should always check that $Q_{eq} = K_{eq}$.

Check $Q_{eq} = [(0.225)(0.225)]/[(0.175)(0.175)] = 1.65$, which, within significant-figure error, is the same as the given value of K_{eq}.

Exercise

The initial concentrations of CH_3CO_2H and C_2H_5OH are 1.00 M each. If the equilibrium constant for the reaction $CH_3CO_2H + C_2H_5OH \rightleftharpoons CH_3CO_2C_2H_5 + H_2O$ is 4.0, what are the equilibrium concentrations of all species?

Answer: $x = 0.67$ M; $[CH_3CO_2H] = [C_2H_5OH] = 0.33$ M, $[CH_3CO_2C_2H_5] = [H_2O] = 0.67$ M.

The structured approach to equilibrium problems is also used when the starting concentration of products is other than zero.

EXAMPLE 13.17 Both Reactants and Products Initially Present

At equilibrium, a certain mixture of gases contains H_2, I_2, and HI at the following partial pressures: H_2, 0.100 atm; I_2, 0.200 atm; and HI, 1.00 atm. Suppose the temperature changes, so that the equilibrium constant of the reaction $H_2(g)$ +

$I_2(g) \rightleftharpoons 2\ HI(g)$ changes also. If the value of the new equilibrium constant is 30, calculate the equilibrium pressures of all species.

Analysis/Plan Equilibrium pressures are to be calculated from the given initial pressures and equilibrium constant, so the structured approach is appropriate. The fact that a condition of equilibrium pre-existed at another temperature is irrelevant.

Work

	H_2	+	I_2	$\rightleftharpoons$	2 HI
Initial	0.100		0.200		1.00
Change	$-x$		$-x$		$+2x$
Equilibrium	$(0.100 - x)$		$(0.200 - x)$		$(1.00 + 2x)$

The change in HI pressure is $2x$, because the stoichiometric coefficient of HI is 2.

$$K_{eq} = \frac{P_{HI}^2}{(P_{H_2}) \cdot (P_{I_2})}$$

$$= \frac{(1.00 + 2x)^2}{(0.100 - x)(0.200 - x)}$$

Rearrangement gives the quadratic form

$$(K_{eq} - 4.00) \cdot x^2 - (0.300 \cdot K_{eq} + 4.00) \cdot x - (1.00 + 0.02 \cdot K_{eq}) = 0$$

Substitution of $K_{eq} = 30$ and use of the quadratic formula gives the two roots

$$x = -0.0291 \text{ atm} \qquad \text{and} \qquad x = +0.529 \text{ atm}$$

The extent of reaction x is a *change* in a pressure or concentration, and as such it can be either positive or negative. In the "equilibrium" line of the problem setup, the pressure of HI is shown as $(1.00 + 2x)$. If x is positive this represents an increasing pressure as the new equilibrium is approached, while a negative value of x means that the pressure of HI decreases. There is nothing in the problem statement that tells us which of these changes will occur. In order to solve the problem, we must try *both* possibilities and see what values of the equilibrium pressures are obtained from each value of x.

	Equilibrium pressures	**if $x = -0.029$**	**if $x = +0.53$**
P_{H_2}	$(0.100 - x)$	0.13 atm	*−0.43 atm*
P_{I_2}	$(0.200 - x)$	0.23 atm	*−0.33 atm*
P_{HI}	$(1.00 + 2x)$	0.94 atm	*+2.06 atm*

A negative value for a pressure is a logical impossibility. The italicized values are wrong, and the other choice is the correct one.

Check

$$Q = \frac{(0.94)^2}{(0.13)(0.23)} = 30 = K_{eq}.$$

Exercise

Hint: the shortcut works in this problem.

Calculate the equilibrium pressures if 1.5 atm each of H_2, I_2, and HI are put into a container at a temperature such that the equilibrium constant is 50.

Answer: $x = 1.0$ and 2.4; $P_{H_2} = P_{I_2} = 0.50$ atm; $P_{HI} = 3.5$ atm.

SUMMARY EXERCISE

Sulfuryl chloride, SO_2Cl_2, is a volatile liquid used in laboratory syntheses of some organic compounds. At its normal boiling point of 69.1 °C, its heat of vaporization is 223 J g^{-1}.

a. What is the entropy of vaporization at the normal boiling point? Is this an unusual value?

b. What cohesive forces exist in liquid SO_2Cl_2?

Sulfuryl chloride decomposes to sulfur dioxide and chlorine at elevated temperatures. In an experiment carried out at 200 °C, sulfuryl chloride was admitted to a flask at a pressure of 1.650 atm. After equilibrium was attained, the partial pressure of SO_2 in the flask was found to be 1.397 atm.

c. Write the balanced chemical equation for the decomposition.

d. Calculate the partial pressure of Cl_2, and the total pressure in the flask, at equilibrium.

e. Calculate the value of the equilibrium constant at 200 °C.

f. $\Delta H^\circ_{rxn} = +67.2$ kJ for this reaction. Calculate the value of the equilibrium constant at 298 K.

g. Calculate the standard free energy of reaction at 298 K.

h. Using any needed quantities from Appendix G, calculate the standard free energy of formation of sulfuryl chloride.

i. At 180 °C, the equilibrium constant for the decomposition is 3.42 atm. If a flask is filled with SO_2Cl_2 at 180 °C and 405 torr, what is the partial pressure of Cl_2 after equilibrium is reached?

Answers **(a)** $\Delta S^\circ_{vap} = 87.9$ J mol^{-1} K^{-1}, not an unusual value; **(b)** dipole-dipole and London forces; **(c)** $SO_2Cl_2(g) \rightleftharpoons SO_2(g) + Cl_2(g)$; **(d)** $P_{Cl_2} = 1.397$ atm, $P_{total} = 3.047$ atm; **(e)** 7.714 atm; **(f)** 3.38×10^{-4} atm; **(g)** +19.8 kJ; **(h)** −320.0 kJ mol^{-1}; **(i)** 0.47 atm.

SUMMARY AND KEYWORDS

Reversible processes can take place in either direction. A reversible process can attain a state of **dynamic equilibrium**, in which the process occurs simultaneously in both directions. Both rates are equal, so that there is no overall change in a system at equilibrium. **Evaporation/condensation** and **freezing/melting** are examples of reversible physical processes. A **phase diagram** indicates the conditions of temperature and pressure at which the solid, liquid, and gaseous forms of a substance are in equilibrium with one another. Other examples of physical equilibria are (a) **sublimation,** in which a solid is in equilibrium with its vapor; (b) **solubility,** in which a pure substance is in equilibrium with its saturated solution in some solvent; (c) **distribution,** in which a solute is partitioned between two immiscible solvents; and (d) **osmotic pressure,** in which equilibrium is maintained across a semipermeable membrane by the opposition of pressure-driven and concentration-driven solvent flow.

Many chemical reactions are reversible, for example the **dissociation/association** of dinitrogen tetroxide/nitrogen dioxide. Approach to equilibrium is smooth and steady, and the final equilibrium is independent of the direction of approach. For the generalized equation $aA + bB \rightleftharpoons cC + dD$, the **reaction quotient** Q is $[C]^c[D]^d/[A]^a[B]^b$. For gaseous species, partial pressures (atm) rather than molarities are used in a reaction quotient, and terms corresponding to pure solids and liquids, and to the solvent in dilute solutions, are not included. A reaction quotient formed from equilibrium pressures or concentrations is called the **equilibrium expression;** it is numerically equal to K_{eq}, the **equilibrium constant.** $Q_{eq} = K_{eq} = [C_{eq}]^c[D_{eq}]^d/[A_{eq}]^a[B_{eq}]^b$. The equilibrium constant is different for each chemical reaction, and K_{eq} for a given reaction changes if the temperature changes. When an equilibrium constant is large, the equilibrium *lies to the right.* When the equilibrium constant is small, the equilibrium *lies to the left.*

Heterogeneous reactions involve two phases, for instance gas/liquid or gas/solid, and they take place only at the interface between phases.

The effect of a change in temperature, pressure, volume, or reagent concentration on a chemical equilibrium can be predicted using **Le Châtelier's principle:** "A system in equilibrium responds to a change in conditions so as to minimize the effect of the change."

The change in gas moles, Δn_{gas}, of an equilibrium reaction is the sum of the stoichiometric coefficients of the gaseous products minus the sum of those of gaseous reactants. If Δn_{gas} is negative, the equilibrium shifts to the right when the reaction mixture is compressed (an increase of pressure and a decrease in volume). If Δn_{gas} is positive, the equilibrium shifts to the left when the system is compressed.

When the concentration of one of the species in a chemical equation is increased, the equilibrium shifts to the other side.

If a reaction is *exothermic,* its equilibrium constant decreases when the temperature increases, and the equilibrium shifts to the left. The equilibrium constant of an *endothermic* reaction increases as the temperature is increased.

Changes in pressure, volume, and/or composition affect the position of equilibrium but not the value of the equilibrium constant, while a change in temperature affects both. The units of an equilibrium constant are determined not only by the units (atm or mol L^{-1}) of the chemical species, but also by the form of the equilibrium expression. If the sum of the exponents in the numerator is the same as the sum of the exponents in the denominator, the equilibrium constant is dimensionless.

The change in concentration (or partial pressure) of any product species, as the reaction proceeds from an initial condition to a final condition, is given by the **extent of reaction,** x, multiplied by the stoichiometric coefficient of that species in the balanced chemical equation (and multiplied by -1 for reactant species). Equilibrium calculations should be solved by the structured approach outlined in Example 13.15.

PROBLEMS

GENERAL

1. Name three features that describe all equilibrium systems.
2. List some reversible physical processes, and some irreversible ones.
3. What is meant by the term "dynamic equilibrium"?
4. What is the relationship between equilibrium and the rates of opposing processes?
5. Explain the statement, "A system at equilibrium has no memory of how the condition of equilibrium was attained."
6. Is it possible to distinguish between water prepared by melting ice and warming to 25 °C and water prepared by condensing steam and cooling to 25 °C? If so, how? If not, why not?
7. What is meant by the phrase "solute distribution equilibrium"?
8. In what ways are chemical and physical equilibria alike? In what ways are they different?
9. What can be said about the magnitude of the equilibrium constant in a reaction whose equilibrium lies far to the right? To the left?
10. If the equilibrium constant for a reaction is large, what can be said about ΔG°_{rxn}?

Equilibrium Expressions

11. Write the equilibrium expression for each of the following reactions
    a. $2\ HgO(s) \rightleftharpoons 2\ Hg(\ell) + O_2(g)$
    b. $C(s) + H_2O(g) \rightleftharpoons CO(g) + H_2(g)$
    c. $HCO_2H(aq) + H_2O(\ell) \rightleftharpoons HCO_2^-(aq) + H_3O^+(aq)$
12. Write the equilibrium expression for each of the following reactions.
    a. $2\ CO(g) + O_2(g) \rightleftharpoons 2\ CO_2(g)$
    b. $2\ SO_2(g) + O_2(g) \rightleftharpoons 2\ SO_3(g)$
    c. $2\ O_3(g) \rightleftharpoons 3\ O_2(g)$
13. Write the equilibrium expression for each of the following reactions.
    a. $Cl_2(g) + 2\ Br^-(aq) \rightleftharpoons Br_2(aq) + 2\ Cl^-(aq)$
    b. $CuO(s) + H_2(g) \rightleftharpoons Cu(s) + H_2O(g)$
    c. $H_2(g) + S(s) \rightleftharpoons H_2S(g)$
14. Write the equilibrium expression for each of the following reactions.
    a. $2\ H_2(g) + O_2(g) \rightleftharpoons 2\ H_2O(g)$
    b. $COCl_2(g) \rightleftharpoons CO(g) + Cl_2(g)$
    c. $H_2(g) + I_2(g) \rightleftharpoons 2\ HI(g)$
15. Identify each reaction in Problems 11 and 13 as heterogeneous or homogeneous.
16. Identify each reaction in Problems 12 and 14 as heterogeneous or homogeneous.
17. The equilibrium constant for the reaction $COCl_2 \rightleftharpoons CO + Cl_2$ is 6.12×10^{-9} atm. What is the value of the equilibrium constant for the reaction $CO + Cl_2 \rightleftharpoons COCl_2$?
18. The equilibrium constant for the reaction $CO_2(g) + H_2(g) \rightleftharpoons CO(g) + H_2O(g)$ is 0.11 at 700 K. What is the value of the equilibrium constant for the reaction $CO(g) + H_2O(g) \rightleftharpoons CO_2(g) + H_2(g)$?
19. The equilibrium constant for the reaction $2\ SO_2 + O_2 \rightleftharpoons 2\ SO_3$ is 3.4 atm^{-1} at 1000 K. What is the value of the equilibrium constant for each of the following reactions?
    a. $2\ SO_3 \rightleftharpoons 2\ SO_2 + O_2$
    b. $SO_2 + \frac{1}{2}\ O_2 \rightleftharpoons SO_3$
20. The equilibrium constant for the reaction $2\ Cl_2(g) + 2\ H_2O(g) \rightleftharpoons 4\ HCl(g) + O_2(g)$ is 0.0750 atm at 750 K. What is the value of the equilibrium constant for each of the following reactions?
    a. $4\ HCl(g) + O_2(g) \rightleftharpoons 2\ Cl_2(g) + 2\ H_2O(g)$
    b. $Cl_2(g) + H_2O(g) \rightleftharpoons 2\ HCl(g) + \frac{1}{2}\ O_2(g)$

Reaction Quotient and Reaction Directionality

21. The dissociation reaction $COCl_2 \rightleftharpoons CO + Cl_2$ has an equilibrium constant of 6.12×10^{-9} atm at 100 °C. For each of the following mixtures, indicate whether there is equilibrium, or, if not, in which direction reaction will occur.

	P_{COCl_2}/atm	P_{CO}/atm	P_{Cl_2}/atm
a.	5×10^{-4}	2×10^{-4}	3×10^{-2}
b.	5×10^{-6}	3×10^{-5}	0.32
c.	2×10^{-3}	5×10^{-8}	6×10^{-5}

22. For the reaction $2\ SO_2 + O_2 \rightleftharpoons 2\ SO_3$ at 1000 K, the value of the equilibrium constant is 3.4 atm^{-1}. For each of the following mixtures, indicate whether there is equilibrium, or, if not, in which direction the reaction will proceed.

	P_{SO_2}/atm	P_{O_2}/atm	P_{SO_3}/atm
a.	3	2	1
b.	1	2	3
c.	2.4	3.4	3.4

23. The reaction $CO_2 + H_2 \rightleftharpoons H_2O + CO$ has an equilibrium constant of 0.043 at 600 K. For each of the following mixtures, indicate whether there is equilibrium, or, if not, in which direction the reaction will proceed.

	P_{CO_2}/atm	P_{H_2}/atm	P_{H_2O}/atm	P_{CO}/atm
a.	0.5	0.5	0.25	0.25
b.	0.25	0.25	0.005	0.005
c.	0.15	0.15	0.15	0.3

24. The equilibrium constant for the reaction $H_2(g) + I_2(g) \rightleftharpoons 2\ HI(g)$ is 1.3 at 298 K. For each of the following mixtures, indicate whether there is equilibrium, or, if not, in which direction the reaction will proceed.

	P_{H_2}/atm	P_{I_2}/atm	P_{HI}/atm
a.	1	1	1
b.	0.625	3.18	1.61
c.	1.05×10^{-3}	9.6×10^{-2}	0.0314

Free Energy and Equilibrium Constant

25. The standard free energy change for the hydrogenation of ethyne, $C_2H_2(g) + H_2(g) \rightleftharpoons C_2H_4(g)$, is −141 kJ at 298 K. Evaluate K_{eq} and comment on its magnitude.

26. For the reaction $H_2(g) + I_2(g) \rightleftharpoons 2\ HI(g)$, the standard free energy change is $-33.3\ kJ\ mol^{-1}$ at 1000 K. What is the value of the equilibrium constant at this temperature?

*27. The equilibrium constant for the reaction $H_2(g) + I_2(g) \rightleftharpoons 2\ HI(g)$ is 620 at 298 K. Calculate ΔG°_{rxn} at 25 °C.

*28. For the reaction $CH_4(g) + Cl_2(g) \rightleftharpoons CH_3Cl(g) + HCl(g)$, the equilibrium constant is 1.4×10^{12} at 500 K. Calculate ΔG°_{rxn} at this temperature.

29. At room temperature, the equilibrium constant for the reaction $CH_4(g) + Cl_2(g) \rightleftharpoons CH_3Cl(g) + HCl(g)$ is 7.42×10^{17}. Calculate ΔG°_{rxn} and $\Delta G^\circ_f(CH_3Cl)$, given that $\Delta G^\circ_f(CH_4) = -50.72\ kJ\ mol^{-1}$ and $\Delta G^\circ_f(HCl) = -95.30\ kJ\ mol^{-1}$. Check your answer by looking up $\Delta G^\circ_f(CH_3Cl)$ in Appendix G.

30. At room temperature, the equilibrium constant for the reaction $2\ SO_2 + O_2 \rightleftharpoons 2\ SO_3$ is 6.98×10^{24}. Calculate ΔG°_{rxn} and $\Delta G^\circ_f(SO_3)$, given the additional information that $\Delta G^\circ_f(SO_2) = -300.194\ kJ\ mol^{-1}$. Check your answer by looking up $\Delta G^\circ_f(SO_3)$ in Appendix G.

31. The standard free energies of formation at 25 °C are 68.2 and $-32.8\ kJ\ mol^{-1}$ for gaseous C_2H_4 and C_2H_6, respectively. Calculate K_{eq} at 298 K for the hydrogenation reaction $C_2H_4(g) + H_2(g) \rightleftharpoons C_2H_6(g)$.

32. The standard free energies of formation at 25 °C for $S_2(g)$ and $S_8(g)$ are 79.3 and 49.6 $kJ\ mol^{-1}$, respectively. Calculate K_{eq} for the reaction $4\ S_2(g) \rightleftharpoons S_8(g)$ at room temperature.

Shifts in Equilibrium and Le Châtelier's Principle

33. The reaction $2\ SO_2 + O_2 \rightleftharpoons 2\ SO_3$ is exothermic. What is the effect on the equilibrium of (a) increasing the temperature, (b) adding SO_3 to the system, (c) adding O_2 to the system, and (d) adding Ar to the system (at constant volume)?

34. The reaction $CO_2 + H_2 \rightleftharpoons CO + H_2O$ is endothermic. What is the effect on the equilibrium of (a) increasing the temperature, (b) adding CO to the system, (c) adding H_2 to the system, and (d) removing CO_2 from the system (this can be accomplished by adding solid KOH, which consumes CO_2 via $KOH + CO_2 \rightarrow KHCO_3(s)$ and does not otherwise interact with the system)?

35. For each of the following reactions, tell in which direction the equilibrium shifts on (i) decreasing the volume, and (ii) increasing the temperature.
    a. $3\ O_2(g) \rightleftharpoons 2\ O_3(g)$; endothermic
    b. $2\ NO_2(g) \rightleftharpoons 2\ NO(g) + O_2(g)$; endothermic

36. For each of the following reactions, describe the effect of (i) decreasing the volume, and (ii) increasing the temperature.
    a. $2\ HI(g) \rightleftharpoons H_2(g) + I_2(g)$; exothermic
    b. $N_2O_4(g) \rightleftharpoons 2\ NO_2(g)$; exothermic

37. What is the effect on the equilibrium $CaCO_3(s) \rightleftharpoons CaO(s) + CO_2(g)$ of adding (a) CO_2, and (b) CaO [careful]?

38. What is the effect on the equilibrium $C(s) + CO_2(g) \rightleftharpoons 2\ CO(g)$ of adding (a) CO_2, and (b) C(s)?

*39. The equilibrium constant for the reaction $N_2O_4 \rightleftharpoons 2\ NO_2$ is 0.148 atm at 298 K, and ΔH°_{rxn} for the reaction is 57.2 kJ. Calculate the value of the equilibrium constant at 473 K.

*40. The equilibrium constant for the reaction $CO + H_2O \rightleftharpoons CO_2 + H_2$ is 1.04×10^5 at 25 °C, and the standard heat of the reaction is −41.66 kJ. Calculate the value of the equilibrium constant at 100 °C.

*41. The equilibrium constant for the reaction $N_2O_4 \rightleftharpoons 2\ NO_2$ is 0.148 atm at 298 K, and ΔH°_{rxn} for the reaction is 57.2 kJ. At what temperature will the equilibrium constant be 1.52 atm?

*42. The equilibrium constant for the reaction $CO + H_2O \rightleftharpoons CO_2 + H_2$ is 1.04×10^5 at 25 °C, and the standard heat of the reaction is −41.66 kJ. At what temperature will the equilibrium constant equal 1.04×10^4?

*43. The equilibrium constant for the reaction $PCl_5 \rightleftharpoons PCl_3 + Cl_2$ is 3.0×10^{-7} atm at 298 K and 7.9×10^{-4} at 373 K. What is the value of ΔH°_{rxn} for this reaction?

*44. The equilibrium constant for the reaction $SO_2 + \frac{1}{2}\ O_2 \rightleftharpoons SO_3$ is $2.6 \times 10^{12}\ atm^{-1/2}$ at 25 °C, and at 550 °C, $K_{eq} = 23\ atm^{-1/2}$. Evaluate ΔH°_{rxn} for this reaction.

*45. At its normal boiling point of 100 °C, the heat of vaporization of water is $40.66\ kJ\ mol^{-1}$. What is the equilibrium vapor pressure of water at 50 °C? (*Hint:* Treat the vaporization process as a chemical reaction. Recall that, by definition, the vapor pressure of a liquid at its normal boiling point is 1 atm.)

*46. Use the data in the preceding problem to calculate the temperature at which the vapor pressure of water is 2.00 atm.

Extent of Reaction

47. A reaction vessel is filled with H_2 and CO_2, each at a partial pressure of 0.75 atm. Later it is found that the pressure of H_2O is 0.25 atm.
    a. What is the extent of the reaction $CO_2 + H_2 \rightleftharpoons CO + H_2O$.
    b. What are the partial pressures of the other species present?

48. CO, CO_2, and O_2 are allowed to react according to $2\ CO(g) + O_2(g) \rightleftharpoons 2\ CO_2(g)$. If the initial partial pressure of each gas is 0.500 atm, and later the pressure of CO has dropped to 0.300 atm, calculate the extent of reaction and the partial pressures of CO_2 and O_2.

49. SO_2 and O_2 are put into a reaction vessel and allowed to react to produce SO_3. Initially, the pressure of SO_2 is 0.750 atm and that of O_2 is 0.350 atm. Later the pressure of SO_2 is determined to be 0.500 atm. Calculate the partial pressures of the other species, and the total pressure.

50. N_2, H_2, and NH_3, all at 1.0 atm, are placed in a reaction vessel and allowed to react according to $N_2(g) + 3\,H_2(g) \rightleftharpoons 2\,NH_3(g)$. If at a later time the pressure of NH_3 has dropped to 0.8 atm, (a) what is the extent of reaction? (b) What are the partial pressures of the other species? (c) What is the total pressure?

Equilibrium Constant

51. The following equilibrium partial pressures were measured at 750 °C: $P_{H_2} = 0.387$ atm, $P_{CO_2} = 0.152$ atm, $P_{CO} = 0.180$ atm, and $P_{H_2O} = 0.252$ atm. What is the value of the equilibrium constant for the reaction $H_2 + CO_2 \rightleftharpoons CO + H_2O$?

52. The equilibrium vapor pressure of water is 150 torr at 60 °C. What is the value of the equilibrium constant for the reaction $H_2O(\ell) \rightleftharpoons H_2O(g)$?

*53. Initially 0.750 atm, the partial pressure of NO decreased to 0.634 atm after the reaction $2\ NO(g) \rightleftharpoons O_2(g) + N_2(g)$ reached equilibrium. What is the value of the equilibrium constant?

*54. A reaction vessel is filled with a mixture of 0.80 atm NO and 0.25 atm Cl_2. At equilibrium, the pressure of NO has fallen to 0.65 atm. What is the value of the equilibrium constant for the reaction $2\ NO(g) + Cl_2(g) \rightleftharpoons 2\ NOCl(g)$?

55. If the equilibrium pressure of CO_2 over a mixture of solid CaO and $CaCO_3$ is 0.25 atm, what is the value of the equilibrium constant for the reaction $CaCO_3(s) \rightleftharpoons CaO(s) + CO_2$?

*56. Ammonium carbamate decomposes according to $NH_4NH_2CO_2(s) \rightleftharpoons 2\ NH_3(g) + CO_2(g)$. If the partial pressure of CO_2 is 0.300 atm after a sample of pure ammonium carbamate reaches equilibrium in a closed container, what is the value of the equilibrium constant?

Equilibrium Concentrations

57. K_{eq} for the reaction $COCl_2(g) \rightleftharpoons CO(g) + Cl_2(g)$ is 4.25 atm at 400 °C. If a vessel initially contains $COCl_2$ at a pressure of 1.60 atm, what are the partial pressures of all species after equilibrium has been attained?

58. Suppose the initial pressure of $COCl_2$ in the preceding problem had been 5.00 atm. What would the equilibrium pressures be?

*59. For the reaction $H_2(g) + I_2(g) \rightleftharpoons 2\ HI(g)$ at 450 °C, the equilibrium constant is 65. Calculate the equilibrium pressures of all species in a vessel initially containing 0.500 atm HI.

*60. If in the preceding problem the initial pressure of HI had been 0.100 atm, what would the equilibrium pressure of HI be?

61. At 600 °C the equilibrium constant for the reaction $CO(g) + H_2O(g) \rightleftharpoons CO_2(g) + H_2(g)$ is 1.85. If 0.100 atm each of CO and H_2O is placed in a vessel and allowed to come to equilibrium, what are the partial pressures of all species?

62. Suppose an equimolar mixture of CO(g) and $H_2O(g)$ is placed in a flask at 600 °C and a *total* pressure of 0.500 atm. Using the value of K_{eq} from the preceding problem, calculate the equilibrium partial pressures of all species.

*63. Suppose 0.100 atm each of CO, H_2, CO_2, and H_2O is placed in a reaction vessel and allowed to come to equilibrium. Use the value of K_{eq} from Problem 61 and calculate the equilibrium pressures of all species.

*64. Repeat the preceding problem, using an initial pressure of 2.0 atm for all species.

65. K_{eq} for the reaction $N_2O_4 \rightleftharpoons 2\ NO_2$ is 0.148 atm. What is the equilibrium pressure of NO_2 in a vessel originally containing 0.500 atm N_2O_4?

66. Repeat the preceding problem, using an initial pressure of 1.50 atm for N_2O_4.

67. For the reaction $H_2(g) + Br_2(g) \rightleftharpoons 2\ HBr(g)$, the equilibrium constant is about 3500 at 1800 K. If the equilibrium partial pressure of H_2 is 0.00500 atm and that of Br_2 is 0.0500, what is the partial pressure of HBr and what is the total pressure?

68. The dissociation of bromine has an equilibrium constant of 8.9×10^{-5} atm at 1300 °C. If the equilibrium partial pressure of Br_2 is 0.500 atm, (a) what is the pressure of Br? (b) What percentage of the bromine exists as diatomic molecules, and what percentage exists as atoms?

*69. At 700 °C, K_{eq} is 1.50 atm for the reaction $C(s) + CO_2(g) \rightleftharpoons 2\ CO(g)$. If neither CO_2 nor C is present initially, and the initial pressure of CO is 1.75 atm, what is the equilibrium pressure of CO_2?

*70. Suppose the initial pressures of CO and CO_2 were both 1.00 atm in the preceding problem. What are the equilibrium pressures?

*71. The reaction $N_2 + O_2 \rightleftharpoons 2\ NO$ has an equilibrium constant of 0.025 at 2400 K. If a vessel initially contains 0.0100 atm each of N_2 and O_2, what are the equilibrium pressures of all species?

*72. Repeat the preceding problem using an initial pressure of 0.500 atm for both N_2 and O_2.

*73. The reaction $N_2 + O_2 \rightleftharpoons 2\ NO$ has an equilibrium constant of 0.025 at 2400 K. If the initial concentrations of N_2 and O_2 are both 0.0100 M (and no NO is present), what are the equilibrium concentrations of all species? Note this is similar to Problem 71, but the amounts are given as concentrations rather than as partial pressures.

*74. Repeat the preceding problem using an initial concentration of 0.500 M for both N_2 and O_2.

*75. At 25 °C, K_{eq} for the reaction $N_2O_4 \rightleftharpoons 2\ NO_2$ is 0.148 atm. What is the equilibrium concentration of NO_2 in a vessel originally containing 0.500 M N_2O_4? Note that the units of the equilibrium constant are different from the units in which the amounts of the chemical species are given. One way of dealing with this is to convert all concentrations to pressures using the ideal gas law, $P = MRT$.

*76. K_{eq} for the reaction $COCl_2(g) \rightleftharpoons CO(g) + Cl_2(g)$ is 4.92 atm at 600 °C. If a vessel initially contains $COCl_2$ at a concentration of 1.60 M, what are the concentrations of all species after equilibrium has been attained? Instead of converting molarity to partial pressures by the expression $P = MRT$, try converting the *equilibrium constant* from pressure to concentration units. Since K_{eq} has units of atm, it can be converted to concentration units by the same equation (in rearranged form), $M = P/RT$. That is, $K(\text{mol L}^{-1}) = K(\text{atm})/(RT)$.

Other Problems

77. Suppose $\Delta G^\circ_{rxn} = 0$ for some reaction. What is the value of K_{eq}?

78. (a) Write the equilibrium expression for the reaction $H_2O(\ell) \rightleftharpoons H_2O(g)$; $\Delta H = 44$ kJ. (b) What is the effect of an increase in temperature on the equilibrium constant? (c) Since all vaporizations are endothermic, what can be generally concluded about the relationship between equilibrium vapor pressure and temperature?

*79. Suppose SO_2, SO_3, and O_2 are in equilibrium according to $2\ SO_2(g) + O_2(g) \rightleftharpoons 2\ SO_3(g)$ and a quantity of inert gas such as helium is added to the system. What is the effect on the equilibrium if (a) He is added without changing the total pressure? In this case the volume of the vessel must increase as the inert gas is added. (b) He is added without changing the volume of the vessel? In this case the total pressure must rise as He is added.

80. For the reaction $H_2(g) + I_2(g) \rightleftharpoons 2\ HI(g)$ at 1500 K, the equilibrium constant is 30.0. Calculate the equilibrium concentrations of all species in a vessel initially containing 0.500 M each of H_2 and I_2.

*81. At 700 °C, K_{eq} is 1.50 atm for the reaction $C(s) + CO_2(g) \rightleftharpoons 2\ CO(g)$. Suppose the total gas pressure at equilibrium is 1.00 atm. What are the partial pressures of CO and CO_2?

82. Choose the initial pressures from any two rows of Table 13.1 and verify the calculation of the equilibrium pressures.

CHAPTER 14

CHEMICAL KINETICS

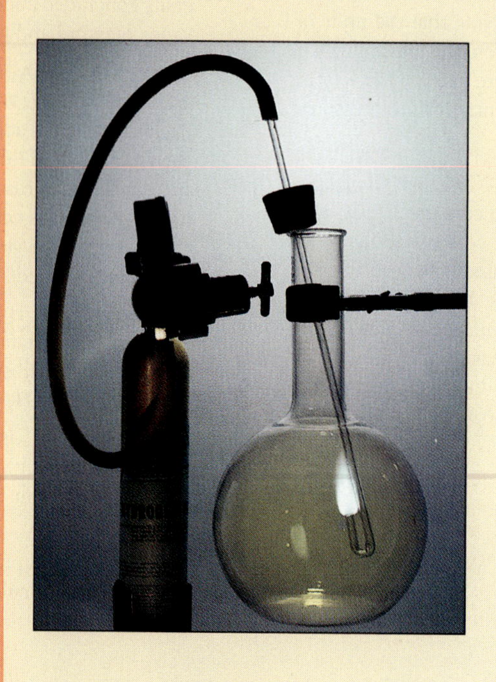

Hydrogen burns rapidly in an atmosphere of chlorine. The reaction product is gaseous hydrogen chloride.

14.1 INTRODUCTION

It is a matter of common experience that chemical reactions take place at different rates: a natural gas (methane) or hydrogen explosion can occur in a fraction of a second, while the complete conversion of a steel hoe to rust (iron oxide) is a process that requires years. The rate at which a reaction takes place has little to do with the concepts of spontaneity and equilibrium discussed in previous chapters. Mixtures of methane and oxygen are stable for years under ordinary conditions of temperature and pressure, because oxidation takes place so slowly as to be scarcely detectable. If the mixture is ignited with a spark or flame, however, the oxidation reaction takes place, literally, in a flash. Yet in each case the overall reaction is the same, it is spontaneous, and the same value of the equilibrium constant applies regardless of the rate (Figure 14.1).

The study of the rates of chemical reactions, and of the factors that influence the rates, is called **chemical kinetics.** There are a number of reasons for studying kinetics. On the practical side, process control, planning, and scheduling, both in the laboratory and in the chemical manufacturing industry, require knowledge of the time required for reactions to take place. On the theoretical side, chemical kinetics provides considerable insight into the mechanism of a reaction. "Mechanism" refers to the details of the interactions between the atoms, ions, or molecules that participate in the reaction. The mechanism of a reaction is not always apparent from the chemical equation. The explosive oxidation of methane occurs by a complex mechanism involving species, such as oxygen atoms (O) and hydroxyl radicals (OH), which do not even appear in the chemical equation. One of the goals of chemical kinetics is to obtain a complete and accurate description of a complex reaction mechanism.

Another goal is to understand the factors that influence chemical reaction rates. Among these factors are temperature, concentration of reactants, and, for solid reactants, particle size. The most important factor is the chemical nature of the reactants.

(a)

(b)

Figure 14.1 (a) When hydrogen-filled balloons are ignited by a candle flame, the explosive reaction between hydrogen and oxygen is complete within milliseconds. (b) After many years of exposure to the environment, the reaction between iron and oxygen is far from complete.

Why do some substances react rapidly and others slowly? There is no single answer to this important question, but we will discuss some of the reasons in Sections 14.7 (mechanisms), 14.8 (chain reactions), and 14.9 (catalysis).

14.2 COLLISION THEORY

We will begin our discussion of kinetics with a theoretical approach, by examining a model that explains most of the experimental observations. The model, known as the **collision theory** of reaction rates, is based on three postulates.

1. *Collisions between reacting molecules are necessary before reaction can occur.*
2. *Only those collisions having sufficient energy are effective in bringing about reaction.*
3. *Colliding molecules must be properly oriented with respect to one another in order for reaction to take place.*

The first postulate implies that the rate at which a chemical reaction occurs is controlled by, and can be no greater than, the rate at which collisions take place between the reaction partners. Since collision rates can be calculated from the kinetic molecular theory, the KMT is an important part of any quantitative application of the collision theory of reaction rates. According to the KMT, the rate of collisions between two types of molecule A and B is given by

$$\text{collision rate} = (\text{const})[\text{A}][\text{B}], \qquad (14\text{-}1)$$

where [A] and [B] are the concentrations in moles per liter. The numerical value of the constant depends on the temperature and on the mass and diameter of the A and B molecules. For methane and oxygen at room temperature, for instance, the equation is

$$\text{collision rate} = 7.4 \times 10^{34}\ [CH_4][O_2]\ \text{collisions L}^{-1}\ \text{s}^{-1} \qquad (14\text{-}2)$$

When experimentally determined rates of chemical reactions are compared with collision rates, as calculated by equations such as (14-2), it is almost always found that the reaction rates are much less than the collision rates. This can only mean that, in general, reactions do not take place every time a collision occurs between potentially reactive species. Some collisions are ineffective, so it is clear that other factors are involved (Figure 14.2).

One clue to this behavior is provided by the temperature dependence of reaction rates. As we will see in more detail in Section 14.6, the rate of most reactions increases sharply as the temperature is raised. This cannot be entirely explained by an increase in the rate of collisions, because collision rate increases only slightly as the temperature increases. However, the average *energy* of molecular motion increases considerably as the temperature increases. Therefore, we postulate that reaction does not take place unless the colliding species have at least some minimum amount of energy. This minimum energy required for a reactive collision to occur is called the **activation energy (E_a)**, and it is often thought of as an energy barrier that the colliding reactant molecules must surmount in order to form products.

According to the kinetic molecular theory, the average energy of molecular collisions is directly proportional to the absolute temperature.

When numerical calculations are made and compared with experiment, however, it is found that reaction does not always take place even if the collision is sufficiently energetic to overcome the activation energy barrier. There is another requirement to be met, one of orientation rather than energy. Consider the reaction $CO_2 + NO \rightarrow CO + NO_2$. The reaction takes place when an oxygen atom is transferred from CO_2 to

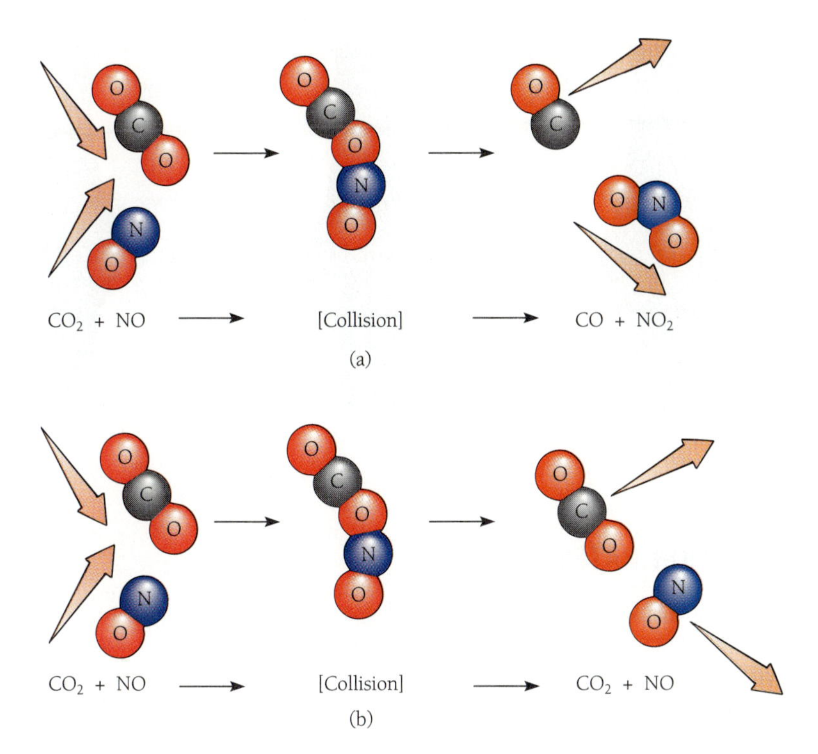

Figure 14.2 (a) Two molecules collide, and then separate into a new arrangement of atoms. This is a reactive collision. In order for a reactive collision to occur, certain conditions of energy and orientation must be fulfilled. (b) The same two molecules can collide, under slightly different conditions, and separate into the original configuration of atoms. This is an unreactive collision.

NO during the brief duration of the collision, as shown in Figure 14.2. Transfer cannot occur unless the colliding species are oriented so that the NO molecule can receive the oxygen atom. The structure of NO_2 is O—N—O rather than O—O—N, and reaction cannot take place if, at the moment of the collision, the O end of the NO molecule is adjacent to one of the oxygen atoms of CO_2. Figure 14.3 illustrates a non-reactive collision of this type.

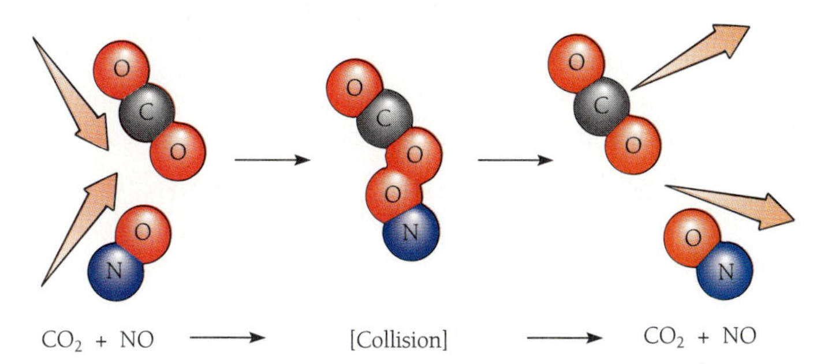

Figure 14.3 This collision between CO_2 and NO is unreactive because the colliding molecules are not oriented properly for reaction to occur. Such collisions are unsuccessful even though they may have more than enough energy to surmount the barrier.

Figure 14.4 The progress of a reaction is immediately apparent when one of the reactants is colored. As the reaction $Br_2(aq) + HCO_2H(aq) \rightarrow 2\,Br^-(aq) + 2\,H^+(aq) + CO_2$ proceeds, the yellow color of the solution slowly fades. The color is due to $Br_2(aq)$, which is consumed by reaction with formic acid.

14.3 REACTION RATES

The rate at which a chemical reaction takes place is measured by observing the rate of change in the concentrations of the compounds involved (Figure 14.4). **Reaction rate** is defined as the *rate of increase in concentration of a product* or as the *rate of decrease in concentration of a reactant,* whichever is more convenient. For the generalized reaction $A \rightarrow B$, the rate is given by Equation (14-3).

$$\text{rate} = \frac{\Delta[B]}{\Delta t} \quad \text{or} \quad r = -\frac{\Delta[A]}{\Delta t} \tag{14-3}$$

The symbol "Δ" means, as usual, "change in:" Δt denotes the time interval during which the change in concentration $\Delta[A]$ or $\Delta[B]$ was measured*. The rate is defined to be a positive quantity. If the rate is measured in terms of a *decrease* in reactant concentration, as in the second part of Equation (14-3), a negative sign is used. The units of the reaction rate are concentration/time, most often expressed as moles per liter per second (mol $L^{-1}\,s^{-1}$). It is permissible to use other units. For instance, moles per liter per hour (mol $L^{-1}\,h^{-1}$) might be more appropriate for a slower reaction.

There is only one rate of a reaction, even though products may appear at a rate different from the rate of disappearance of reactants. For example, in the Haber synthesis of ammonia the stoichiometry of the reaction

$$N_2 + 3\,H_2 \longrightarrow 2\,NH_3 \tag{14-4}$$

is such that for every mole of N_2 consumed, *three* moles of H_2 are required and *two* moles of NH_3 are produced. That is, the rate of N_2 consumption is only one-third the rate of H_2 consumption, and one-half the rate of NH_3 production.

$$r = -\frac{\Delta[N_2]}{\Delta t} = -\frac{1}{3}\left(\frac{\Delta[H_2]}{\Delta t}\right) = +\frac{1}{2}\left(\frac{\Delta[NH_3]}{\Delta t}\right) \tag{14-5}$$

* Students who have had a semester of calculus may be interested to know that the fundamental definition uses the *derivative* of concentration rather than a finite difference: rate = $d[B]/dt = -d[A]/dt$. This distinction is important only in a more advanced treatment of kinetics, for instance in a course in physical chemistry. Appendix E summarizes some of the applications of calculus to the equations of chemical kinetics.

Equation (14-5) is easily generalized for a reaction whose chemical equation (14-6) has stoichiometric coefficients $(a, b, \ldots) \rightarrow (c, d, \ldots)$.

$$aA + bB + \cdots \longrightarrow cC + dD + \cdots \tag{14-6}$$

For this reaction, the rate is

$$\begin{aligned} r &= -\frac{1}{a}\left(\frac{\Delta[\mathrm{A}]}{\Delta t}\right) = -\frac{1}{b}\left(\frac{\Delta[\mathrm{B}]}{\Delta t}\right) = \cdots \\ &= +\frac{1}{c}\left(\frac{\Delta[\mathrm{C}]}{\Delta t}\right) = +\frac{1}{d}\left(\frac{\Delta[\mathrm{D}]}{\Delta t}\right) = \cdots \end{aligned} \tag{14-7}$$

Defined in this way, the rate has the same numerical value regardless of which species is used to measure it.

(a)

(b)

Figure 14.5 Reactant concentration has a strong influence on reaction rate. (a) When held above the flame of a laboratory burner, steel wool reacts with atmospheric oxygen. The reaction is not especially fast, because the concentration of oxygen is low. (b) When the oxygen concentration is increased by blowing in pure O_2 from the side, the reaction is much faster.

Influence of Concentration

There are several factors that influence the rate of a particular chemical reaction. The most obvious of these is the concentrations of the reactants, because these concentrations determine the rate at which collisions occur between reacting species. For example, when nitric oxide reacts with ozone at room temperature [Equation (14-8)],

$$NO(g) + O_3(g) \longrightarrow NO_2 + O_2 \tag{14-8}$$

the rate is $r = 1 \times 10^7[NO][O_3]$ mol L^{-1} s^{-1}. That is, the rate is directly proportional to the concentrations of both of the reactants (Figure 14.5). The rate is obtained by measurement of any of the concentration changes:

$$-\frac{\Delta[NO]}{\Delta t} = 1 \times 10^7[NO][O_3] \text{ mol L}^{-1}\text{ s}^{-1}$$

$$-\frac{\Delta[O_3]}{\Delta t} = 1 \times 10^7[NO][O_3] \text{ mol L}^{-1}\text{ s}^{-1}$$

$$+\frac{\Delta[NO_2]}{\Delta t} = 1 \times 10^7[NO][O_3] \text{ mol L}^{-1}\text{ s}^{-1}$$

$$+\frac{\Delta[O_2]}{\Delta t} = 1 \times 10^7[NO][O_3] \text{ mol L}^{-1}\text{ s}^{-1}$$

Two other example reactions, with their measured rates, are given as Equations (14-9) and (14-10).

$$\begin{aligned} &OH^-(aq) + HCO_3^-(aq) \longrightarrow CO_3^{2-}(aq) + H_2O; \\ &r = 6 \times 10^9[OH^-][HCO_3^-] \text{ mol L}^{-1}\text{ s}^{-1} \end{aligned} \tag{14-9}$$

$$\begin{aligned} &NOCl(g) \longrightarrow NO(g) + \tfrac{1}{2} Cl_2(g); \\ &r = 2 \times 10^{-8}[NOCl]^2 \text{ mol L}^{-1}\text{ s}^{-1} \end{aligned} \tag{14-10}$$

The Rate Law

The rate may depend on the concentration by a direct proportionality, as in Equations (14-8) and (14-9), or it may depend on the concentration raised to some exponent other than one, as it does in Equation (14-10). For the reaction

TABLE 14.1 Units of Rate Constants

Reaction Order	Units of k
0	mol L^{-1} s^{-1}
1	s^{-1}
2	L mol^{-1} s^{-1}
3	L^2 mol^{-2} s^{-1}

aA + bB + · · · → products, the general expression for the rate is given by Equation (14-11),

$$r = k[\text{reactant}_1]^x[\text{reactant}_2]^y[\text{reactant}_3]^z \ldots \quad (14\text{-}11)$$

in which k is a constant and x, y, and z represent the exponents to which the molarities of the reactants are raised. Expressions such as Equation (14-11), which summarize the dependence of the rate of a reaction on the concentration of the reactants, are known as **rate laws** (or **rate expressions**). Do not confuse the exponents (x, y, . . .) in the *rate law* with the stoichiometric coefficients (a, b, . . .) in the *chemical equation*—in general, they are different quantities. In some reactions [for example as in Equation (14-9)], the rate-law exponents do have the same values as the stoichiometric coefficients, while in other reactions [for example as in Equation (14-10)] they have different values. In all reactions, the values of the concentration exponents must be determined by experiment.

A rate law consists of a set of concentration terms, each raised to the appropriate exponent, and a number k, called the **rate constant.** Each chemical reaction has its own value of the rate constant. It follows from the definition of Equation (14-11) that k is equal to the reaction rate (r) when the concentrations of all reactants are 1 M. The **order** of a reaction is defined as the sum of the exponents (x, y, z, . . .) in the rate law. Reactions (14-8) through (14-10) are all *second order,* because in each case the sum of the concentration exponents in the rate law is 2. The word "order" can also be applied to an individual reactant. Reaction (14-8) is "first order in NO," "first order in O_3," and "second order overall," while reaction (14-10) is "second order in NOCl."

Remember that, if an exponent is 1, it is implied and usually not written.

The rate constant k provides a means for comparing the *inherent* rates of different reactions. Since the rate of a reaction changes when the concentrations are changed, a reaction that is inherently fast can be made to proceed quite slowly by reducing the concentration of reactants. Therefore, comparisons of the *rates* of different reactions can be misleading unless the rates refer to the same conditions of reactant concentrations. On the other hand, the *rate constant* is independent of reactant concentrations. When two reactions of the same order are compared, the one with the larger rate constant is inherently the faster of the two.

Rate constants are always positive quantities, and the units of a rate constant depend on the order of the reaction (Table 14.1).

The units of a rate constant vary, but the units of a reaction *rate* are always the same: mol L^{-1} s^{-1}.

Determination of the Rate Law: Method of Initial Rates

Both the order of a reaction and the numerical value of its rate constant must be determined experimentally, by measurements of the reaction rate. One technique for determining reaction order and rate constant, known as the method of **initial rates,** is based on measurements made soon after the start of the reaction. Reaction order is determined separately for each reactant species, by measurement of the initial rate in at least two experiments in which the initial concentration of one species is varied while that of all other species is held constant. For example, suppose that the initial rate of a reaction is 1×10^{-4} mol L^{-1} s^{-1} under one set of conditions, and rises to 2×10^{-4} mol L^{-1} s^{-1} when the concentration of species [A] is doubled. If all other concentrations are the same in the two experiments, the change in rate must be due entirely to the change in [A]. Since the rate increases twofold when [A] increases twofold, the relationship is the direct proportionality, rate $\propto$ [A]. The reaction is first order with respect to [A]. If a twofold increase in [A] results in *no* increase in rate, the reaction is zero order with respect to A. If a twofold increase in [A] causes a fourfold (2^2) increase in rate, the reaction is second order with respect to [A]. If a twofold increase in [A] results in an eightfold (2^3) increase in rate, the reaction is third order with respect to [A]. For two experiments in which $[A]_1 \neq [A]_2$ and all other concentrations are the same, the general relationship is

$$(\text{Rate}_1/\text{Rate}_2) = \{[A]_1/[A]_2\}^x, \tag{14-12}$$

where x is the reaction order with respect to [A].

When working numerical problems involving the rates of chemical reactions, it is essential to distinguish between the *rate* and the *rate constant.*

EXAMPLE 14.1 Method of Initial Rates

The following data were collected at 175 °C during three trial runs of the reaction $NO + ½\ Cl_2 \rightarrow NOCl$.

Experiment	[NO]/M	$[Cl_2]$/M	Initial Rate of Increase in [NOCl]/mol L^{-1} s^{-1}
1	0.0035	0.0100	9.0×10^{-6}
2	0.0105	0.0100	8.0×10^{-5}
3	0.0105	0.0200	1.6×10^{-4}

a. Determine the order with respect to each reactant, and the overall reaction order.

b. Write the rate law and give the value of the rate constant.

Analysis

Target: The problem asks for the order of the reaction with respect to each reactant. Once these are determined, the overall order and the value of the rate constant must be obtained.

Knowns: The balanced chemical equation; initial rates at particular reactant concentrations.

Relationship: Equation (14-12) relates order and initial rate, while Equation (14-11) relates rate, rate constant, and concentration.

Plan

1. Seek pairs of experiments in which the initial concentration of only one species is different. From the observed effect of the difference on the rate, determine the order with respect to the reactant whose concentration changes. Do this successively for all reactants, and combine the results into a rate law.
2. Use Equation (14-11) to evaluate the rate constant.

Work

1. Compare experiments 1 and 2: [NO] increases by a factor of 0.0105/0.035 = 3, while $[Cl_2]$ is unchanged. The rate, however, increases by a factor of $8.0 \times 10^{-5}/9.0 \times 10^{-6} = 8.9 \approx 9$. The rate increase ($9 = 3^2$) is thus

proportional to the *square* of the concentration increase of NO, so the reaction is second order with respect to NO:

$$\text{rate} = (\text{a constant}) \cdot [\text{NO}]^2$$

Compare runs 2 and 3: [NO] is unchanged, while $[Cl_2]$ increases by a factor of 0.0200/0.0100 = 2. The rate is also doubled: $1.6 \times 10^{-4}/8.0 \times 10^{-5} = 2 = 2^1$. The rate increase is the *first* power of the concentration increase. The reaction is therefore first order with respect to chlorine

$$\text{rate} = (\text{a constant}) \cdot [Cl_2]$$

Combining these results, we find the overall rate law to be

$$r = k[\text{NO}]^2[Cl_2]^1 \qquad \text{or} \qquad r = k[\text{NO}]^2[Cl_2]$$

or, *second order* with respect to NO, *first order* with respect to Cl_2, and *third order* overall.

2. The rate constant is evaluated by inserting the initial concentrations and rate from *any* of the experiments into the rate law. Using the results from run 3, we find

$$r = k[\text{NO}]^2[Cl_2]$$

$$k = \frac{r}{[\text{NO}]^2[Cl_2]} = \frac{1.6 \times 10^{-4}\ \text{mol L}^{-1}\ \text{s}^{-1}}{(0.0105\ \text{M})^2(0.0200\ \text{M})}$$

$$= 73\ \text{M}^{-2}\ \text{s}^{-1} = 73\ \text{L}^2\ \text{mol}^{-2}\ \text{s}^{-1}$$

Exercise

The following data were collected during trial runs of the reaction $C_2H_5Br + OH^- \rightarrow C_2H_5OH + Br^-$

Run	$[C_2H_5Br]$/M	$[OH^-]$/M	Initial Rate of OH^- Disappearance/mol L^{-1} s^{-1}
1	0.150	0.200	4.8×10^{-5}
2	0.300	0.200	9.6×10^{-5}
3	0.300	0.400	19.2×10^{-5}

Determine the order of the reaction with respect to each reactant, and the overall order. Write the rate expression and give the numerical value and units of the rate constant.

Answer: First order with respect to each reactant, and second order overall; $r = k[C_2H_5Br][OH^-]$; $k = 1.6 \times 10^{-3}$ L mol^{-1} s^{-1}.

14.4 FIRST-ORDER REACTIONS

First-order reactions are common, both in laboratory studies and in the chemical industry. In this section we present a detailed discussion of the kinetics of first-order reactions, using as an example the conversion of methyl isocyanide to its isomer methyl cyanide (acetonitrile). This isomerization reaction takes place in the gas phase at elevated temperatures.

$$\begin{array}{c}\text{H}\\|\\\text{H—C—N}\equiv\text{C}\\|\\\text{H}\end{array} \longrightarrow \begin{array}{c}\text{H}\\|\\\text{H—C—C}\equiv\text{N}\\|\\\text{H}\end{array} \qquad (14\text{-}13)$$

methyl isocyanide **methyl cyanide**

TABLE 14.2 Isomerization of Methyl Isocyanide at 350 °C

Time/seconds	$[CH_3NC]$/ mol L^{-1}
0	0.01000
120	0.00432
240	0.00186
360	0.000805
480	0.000348
600	0.000150
720	0.0000648

Change in Concentration with Time

The isomerization of methyl isocyanide follows a first-order rate law in which the rate constant has the value $k = 7.00 \times 10^{-3}\ s^{-1}$ at 350 °C.

The temperature is usually mentioned when the value of a rate constant is given, because for most reactions, the rate constant increases when the temperature increases.

$$CH_3NC \longrightarrow CH_3CN; \qquad (14\text{-}14)$$

$$r = -\frac{\Delta[CH_3NC]}{\Delta t} = 7.00 \times 10^{-3}[CH_3NC]\ \text{mol L}^{-1}\ \text{s}^{-1}$$

When CH_3NC is introduced into a reaction vessel at 350 °C, it begins to react, and its concentration decreases continuously. When the initial concentration (C_0) is 0.0100 M, measurement of the concentration of CH_3NC at two-minute intervals gives the values shown in Table 14.2.

The relationship between concentration and time is more quickly grasped when the data in Table 14.2 are presented in graphical form: Figure 14.6 shows a plot of $[CH_3NC]$ vs. time. The rate of reaction, by definition a positive number, is equal to (−1) times the *slope* (a negative quantity) of the curve. Two such slopes are drawn, showing that late in the reaction (after 320 seconds have elapsed), the rate is considerably less than the earlier rate (after 110 seconds). This effect is predicted by the rate

The slope of a curved line is defined as the slope ($\Delta y/\Delta x$) of the tangent to the curve.

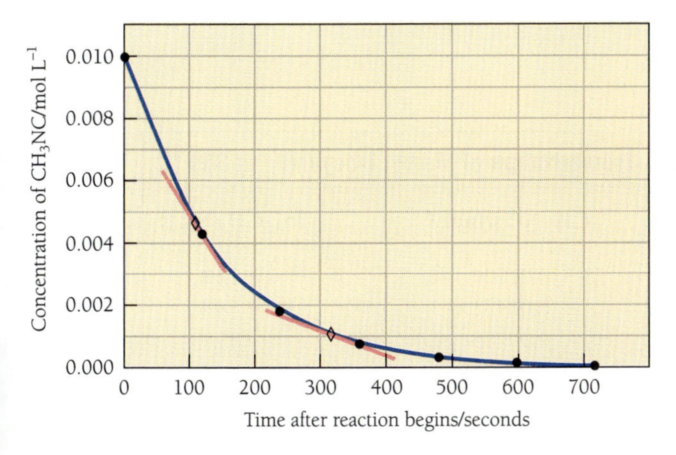

Figure 14.6 The concentration of unreacted methyl isocyanide decreases as the isomerization proceeds. Its rate of decrease is the rate of the reaction, $-\Delta[CH_3NC]/\Delta t$, which in turn is equal to (−1) times the slope of the tangent to the curve. Two such tangents are shown, one after 110 seconds of reaction and one after 320 seconds. The slopes, and hence the reaction rates, are different at these two times. At $t = 110$ s, the slope is -3.24×10^{-5} mol $L^{-1}\ s^{-1}$ (so that the rate is $+3.24 \times 10^{-5}$ mol $L^{-1}\ s^{-1}$). After 320 seconds, the rate has decreased to 7.45×10^{-6} mol $L^{-1}\ s^{-1}$.

TABLE 14.3 Calculation of the Rate Constant from Figure 14.6

Elapsed Time/s	Slope/ $mol\ L^{-1}\ s^{-1}$	Rate/ $mol\ L^{-1}\ s^{-1}$	$[CH_3NC]$/ $mol\ L^{-1}$	Rate Constant/s^{-1} ($k = r/[CH_3NC]$)
110	-3.24×10^{-5}	3.24×10^{-5}	4.63×10^{-3}	0.0070
320	-7.45×10^{-6}	7.45×10^{-6}	1.06×10^{-3}	0.0070

The usual reason for measuring a reaction rate is to determine the value of a rate constant.

law, $r = k[CH_3NC]$: as the concentration of CH_3NC decreases, so does the rate. The rate constant k is determined by dividing the rate r [$= (-1) \cdot$ (slope)] by the concentration, since $k = r/[CH_3NC]$. This calculation is outlined in Table 14.3.

Change in Logarithm of Concentration with Time

Although an unknown rate constant can be evaluated from a plot such as Figure 14.6, it is difficult to draw an accurate tangent to a curved line. A more accurate method is to plot the *natural logarithm (ln)* of the reactant concentration vs. time. The first two columns of Table 14.4 repeat the information of Table 14.2, and the third column gives the logarithm of the concentration.

Once again, columns of numbers are more easily interpreted if they are presented in graphical form. Figure 14.7 shows that the relationship between the logarithm of reactant concentration and time is quite different from that between the concentration itself and time. The most important feature of the graph is that it is a straight line, with an easily measured and constant slope. This behavior is common to all first-order reactions.

The algebraic equation for the line in Figure 14.7 may be found as follows. Begin with the general form for a straight line

$$y = m \cdot x + b,$$

where m is the slope and b is the y-intercept. Since the graph shows ln C on the y-axis vs. time on the x-axis, the general equation becomes

$$\ln C = m \cdot t + b \qquad (14\text{-}15)$$

When the time $t = 0$, the concentration is at its initial value, C_0, so that

$$\ln C_0 = m \cdot (0) + b = b \qquad (14\text{-}16)$$

TABLE 14.4 Isomerization of Methyl Isocyanide at 350 °C

Time/s	$[CH_3NC]/mol\ L^{-1}$	$\ln\{[CH_3NC]\}$
0	0.01000	−4.61
120	0.00432	−5.44
240	0.00186	−6.29
360	0.000805	−7.12
480	0.000348	−7.96
600	0.000150	−8.81
720	0.0000648	−9.64

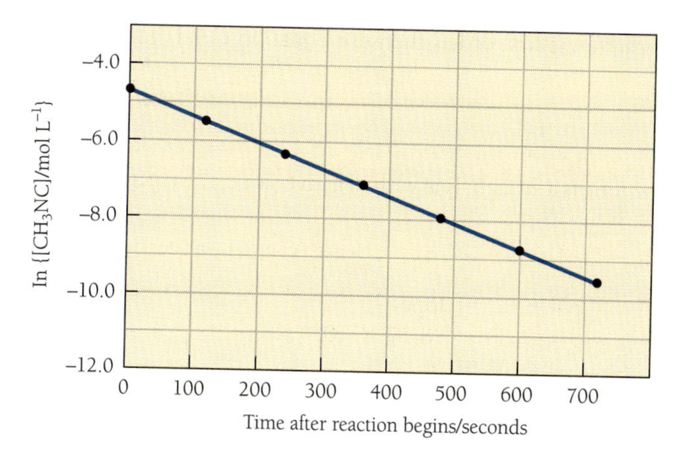

Figure 14.7 In the isomerization of methyl isocyanide, as in all first-order reactions, the logarithm of the concentration of reactant decreases linearly as the reaction proceeds. The slope of the line, which may be determined either from the graph or the data from which the graph is drawn, is directly proportional to the rate constant of the reaction.

It is shown in Appendix E that the slope m is related to the rate constant k by Equation (14-17)

$$m = \text{slope} = (-1) \cdot k \tag{14-17}$$

The complete equation is therefore

$$\ln C = -k \cdot t + \ln C_0 \tag{14-18}$$

Since $(\ln C - \ln C_0) = \ln(C/C_0)$, Equation (14-18) can be rearranged to yield

$$\ln\left(\frac{C}{C_0}\right) = -k \cdot t \quad \text{or} \tag{14-19a}$$

$$\ln\left(\frac{C_0}{C}\right) = +k \cdot t \tag{14-19b}$$

The calculus-based derivation of Equation (14-19) from the rate law of a first-order reaction is given in Appendix E.

In both forms of Equation (14-19), C_0 is the concentration at time 0 ("initial concentration"), and C is the concentration at any time, t, thereafter. Equation (14-19) is known as the *integrated rate law* for a first-order reaction. If (and only if) the reaction is *first order,* a plot of the logarithm of reactant concentration vs. time is a straight line *whose slope is equal to* (-1) *times the rate constant* [Equation (14-17)]. This relationship is used to calculate first-order rate constants from experimental data.

Be sure to use the *natural* logarithm (ln) of reactant concentration in graphs and calculations of first-order reactions.

EXAMPLE 14.2 Rate Constant of a First-Order Reaction

The concentration of a reactant is measured at regular time intervals and the results are plotted in the form of ln (concentration) vs. time. The graph is found to be a straight line with slope $= -8.57 \times 10^{-4}\ \text{s}^{-1}$. Determine the order of the reaction and the value of the rate constant.

Analysis

Target: Reaction order; reaction rate constant (the units of the rate constant are not known until the reaction order is known).

Knowns: A plot of ln C vs. time is a straight line; the numerical value of the slope is known.

Relationship: Equation (14-17) relates the slope and the rate constant, *provided the reaction is first order.*

Plan Confirm that the reaction is first order, then use Equation (14-17) to find the rate constant.

Work

1. If a plot of ln C vs. time is a straight line with negative slope, the reaction is first order.
2.

$$\begin{aligned} k &= (-1) \cdot (\text{slope}) \qquad (14\text{-}17) \\ &= (-1) \cdot (-8.57 \times 10^{-4}\ \text{s}^{-1}) \\ &= (+)8.57 \times 10^{-4}\ \text{s}^{-1} \end{aligned}$$

Note: k is a positive quantity, as it must be.

Exercise

If a plot of ln C vs. time has a slope of $-5.00 \times 10^{-3}\ \text{s}^{-1}$, what is the rate constant of the reaction?

Answer: $5.00 \times 10^{-3}\ \text{s}^{-1}$.

EXAMPLE 14.3 Evaluation of a First-Order Rate Constant

Using the data plotted in Figure 14.7, calculate the rate constant for the isomerization of methyl isocyanide.

Analysis

Target: A first order rate constant,

$$k = ?\ \text{s}^{-1}$$

Knowns: Values of ln $[CH_3NC]$ vs. time are explicit; the reaction order (first) is a hidden known, implicit in the fact that Figure 14.7 is a straight line.

Relationship: Equation (14-17).

Missing: The slope of the line.

Plan Evaluate the slope from the given data, then use Equation (14-17) (that is, change the sign of the slope) to obtain the rate constant.

Work

1. Choose any two times and read the corresponding values of ln C from the graph. For instance at t = 240 s the logarithm of the concentration is −6.29, while at t = 600 s ln C is −8.81.

$$\begin{aligned} \text{slope} &= \frac{\text{change in ln } C}{\text{change in time}} \\ &= \frac{-8.81 - (-6.29)}{600\ \text{s} - 240\ \text{s}} \\ &= \frac{-2.52}{360\ \text{s}} \\ &= -7.00 \times 10^{-3}\ \text{s}^{-1} \end{aligned}$$

2.

$$k = (-1) \cdot (\text{slope}) = 7.00 \times 10^{-3}\ \text{s}^{-1}$$

Exercise

Choose a different pair of times and recalculate k.

Answer: The answer is the same, $7.00 \times 10^{-3}\ \text{s}^{-1}$.

The following two examples illustrate the use of the integrated rate law [Equation (14-19)] in calculating an unknown time or concentration.

EXAMPLE 14.4 Elapsed Time in a First-Order Reaction

The isomerization of methyl isocyanide follows a first-order rate law with rate constant $k = 0.00700\ \text{s}^{-1}$ at 350 °C. How long will it take for an initial concentration $C_0 =$ 0.25 M to be reduced to 0.010 M?

Analysis

Target: An elapsed time,

$$t = ?\ \text{s}$$

Knowns: Reaction order and rate constant; initial and final concentrations.

Relationship: The integrated rate law [Equation (14-19)] relates concentration, rate constant, and elapsed time.

Plan Rearrange Equation (14-19) to isolate the unknown on the left, insert the known information, and solve.

Work

$$\ln\left(\frac{C}{C_0}\right) = -k \cdot t \qquad (14\text{-}19)$$

$$\begin{aligned} t &= -\left(\frac{1}{k}\right) \cdot \ln\left(\frac{C}{C_0}\right) \\ &= -\left(\frac{1}{0.00700\ \text{s}^{-1}}\right) \cdot \ln\left(\frac{0.010\ \text{M}}{0.25\ \text{M}}\right) \\ &= -(143\ \text{s}) \cdot \ln(0.040) = -(143\ \text{s}) \cdot (-3.22) \\ &= 460\ \text{s} \end{aligned}$$

Note: Elapsed time is always a positive quantity.

Exercise

For the same reaction, calculate the time required for the concentration to decrease from 0.010 M to 0.0010 M.

Answer: 3.3×10^{2} s.

EXAMPLE 14.5 Reactant Concentration After an Elapsed Time

If the initial concentration of methyl isocyanide is 0.500 M, what will it be after 5.00 minutes have elapsed?

Analysis This is similar to the preceding problem, and Equation (14-19) is applicable to all first-order processes. All required information is available, either in this or in the preceding problem.

Plan Insert the known information into the integrated rate law, and solve for the unknown concentration.

Work

$$\ln\left(\frac{C}{C_0}\right) = -k \cdot t$$
$$= -(0.00700\ \text{s}^{-1})(5.00\ \text{min})$$
$$= -0.0350\ \text{min s}^{-1}$$

At this point, if we have been careful about writing down the units, we should realize that there has been an error. Logarithms, by definition, are dimensionless quantities. Yet the logarithm we have obtained has units of min s^{-1}. The mistake was to use two different time units in the same equation. The error is corrected by using the appropriate conversion factor, 60 s ⇔ 1 min.

$$\ln\left(\frac{C}{C_0}\right) = (-0.0350\ \text{min s}^{-1})\left(\frac{60\ \text{s}}{1\ \text{min}}\right)$$
$$= -2.10$$

Experienced problem-solvers are aware of this pitfall, and make sure at the start that all quantities in a formula are expressed in compatible units.

The solution proceeds:

Use the "inv ln" or "e^x" key on the calculator.

$$\text{antilog}\left[\left(\ln\frac{C}{C_0}\right)\right] = \text{antilog}\,(-2.10) = e^{-2.10}$$
$$\frac{C}{C_0} = 0.122$$
$$C = C_0 \cdot (0.122)$$
$$= (0.500\ \text{M}) \cdot (0.122) = 0.0612\ \text{M}$$

Check Be sure that the result is *less than* the starting concentration.

Exercise

If the initial concentration of methyl isocyanide is 0.100 M, what is it after 2.00 minutes?

Answer: 0.0432 M.

One consequence of the form of Equation (14-19) is that the value of the concentration need not be measured directly—only the ratio (C/C_0) must be measured. Calculations may be done in terms of concentration ratios rather than actual values of concentration. This is an important feature, because a concentration ratio often can be measured with greater ease and accuracy than the concentration itself.

EXAMPLE 14.6 Percentage of Remaining Reactant

What percentage of the original concentration of methyl isocyanide will remain after it has been allowed to react for 8.5 minutes at 350 °C?

Analysis This is another variation on Equation (14-19). In this example, the target is a percentage (a ratio). All the required information is available in the problem statement or in Example 14.4.

Plan

1. Calculate the ratio C/C_0 from the known information and Equation (14-19).
2. Calculate the percentage from C/C_0 and the definition of percent:

$$\% = \left(\frac{C}{C_0}\right) \cdot 100$$

Work

1. Insert the given values into Equation (14-19), making sure that the units cancel correctly.

$$\ln\left(\frac{C}{C_0}\right) = -k \cdot t$$

$$= -(0.00700\ \text{s}^{-1}) \cdot (8.5\ \text{min})\left(\frac{60\ \text{s}}{1\ \text{min}}\right)$$

$$= -3.57$$

$$\frac{C}{C_0} = \text{inv ln}\ (-3.57) = e^{-3.57}$$

$$= 0.028$$

2. The *fraction* remaining is 0.028; the *percentage* remaining is 100 times this, or 2.8%.

Exercise

Calculate the percent remaining after 4.00 minutes.

Answer: 19%.

Half-Life

A very useful way of characterizing the rates of first-order reactions is by the **half-life**, or **half-time, $t_{1/2}$**. The half-life is the time required for the consumption of half of the reactant present initially. In the example of Table 14.4, $t_{1/2}$ is the time required for the

The decay of radioactive isotopes, discussed in Chapter 24, is a first-order reaction. These nuclear reactions are invariably characterized by their half-lives rather than by their rate constants.

initial concentration to decrease from 0.010 M to 0.0050 M. The half-life can be estimated, either from the table or from Figure 14.4 or 14.5, as being about 100 seconds. Equation (14-19) can be used to derive a simple relationship between $t_{1/2}$ and the value of the rate constant of a first-order reaction. If the initial concentration of reactant is C_0, then the concentration C when half of it has been consumed is $C_0/2$. An expression for the half-life is obtained when these values are inserted into Equation (14-19).

$$\ln\left(\frac{C}{C_0}\right) = -kt \qquad (14\text{-}19)$$

$$\ln\left[\frac{(C_0/2)}{C_0}\right] = -kt_{1/2}$$

$$\ln\left(\frac{1}{2}\right) = -kt_{1/2}$$

$$-0.693 = -kt_{1/2}$$

The relationship may be expressed in either of two forms [Equation (14-20a, b)],

$$t_{1/2} = \frac{0.693}{k} \quad \text{or} \qquad (14\text{-}20a)$$

$$k = \frac{0.693}{t_{1/2}} \qquad (14\text{-}20b)$$

The equation $k = 0.693/t_{1/2}$ applies *only* to first-order reactions.

and it holds true for all first-order reactions.

Equation (14-20) allows an easy conversion between a half-life and a first-order rate constant.

EXAMPLE 14.7 Half-Life and Rate Constant

The thermal decomposition of dinitrogen pentoxide

$$2\ N_2O_5 \longrightarrow 2\ N_2O_4 + O_2$$

is a first-order reaction whose half-life, measured at 27 °C and 300 torr pressure, is 0.134 minutes. What is its rate constant?

Analysis

Target: A first-order rate constant,

$$k = ?\ s^{-1}$$

Knowns: The chemical equation, half-life, temperature, and pressure.

Relationship: Equation (14-20) may be used directly.

Work

$$k = \frac{0.693}{t_{1/2}} \qquad (14\text{-}20)$$

$$= \frac{0.693}{(0.134\ \text{min})}$$

$$= 5.17\ \text{min}^{-1}$$

The units of the rate constant are determined by the units of time used to express the half-life. The rate constant can be used in these units, or if desired it can be converted to other units.

$$k = (5.17\ \text{min}^{-1}) \cdot \left(\frac{1\ \text{min}}{60\ \text{s}}\right)$$
$$= 0.0862\ \text{s}^{-1}$$

Exercise

If the rate constant of a first-order reaction is $7.5 \times 10^{-3}\ \text{s}^{-1}$, what is its half-life?

Answer: 92 s.

Note that some of the given information was not used in the solution of Example 14.7. It is customary to specify the temperature, and often the pressure as well, at which a rate constant or equilibrium constant is measured, even if these data are not necessary for the particular problem. The chemical equation was also given, but it served only to identify the reaction.

EXAMPLE 14.8 Amount of Reactant After Successive Half-Lives

What fraction of the amount originally present remains after a first-order reaction has proceeded for a time equal to (a) one half-life, and (b) four half-lives?

Analysis Part (a) is just a restatement of the definition: after a time equal to one half-life, half of the reactant has been consumed and half remains. The answer to part (a) is 0.5. Part (b) requires that the definition be applied four times.

Work After the first half-life, half remains; after the second, half of that or $(1/2)(1/2) = 1/4$ remains; after the third half-life, the fraction remaining is $(1/2)(1/2)(1/2) = 1/8$. The general form is that after n half-lives, the fraction remaining is $(1/2)^n$. In this problem, the time was given as $n = 4$ half-lives, so that the fraction remaining $= (1/2)^4 = 1/(2^4) = 1/16 = 0.0625$.

Exercise

How much remains after 10 half-lives?

Answer: 0.1%.

EXAMPLE 14.9 Fraction of Unreacted Substance After a Given Time

The half-life of the first-order decomposition of sulfuryl chloride, $SO_2Cl_2 \rightarrow SO_2 + Cl_2$, is 3.3×10^4 s at 600 K. What fraction of a sample of sulfuryl chloride remains after an elapsed time of 30.0 minutes?

Analysis

Target: A fraction of remaining reactant.

Knowns: The reaction order and half-life, and the temperature and pressure.

Relationship: There is no formula connecting the fractional consumption of reactant with the half-life, but "if only I knew" the rate constant, the problem and the solution would be essentially the same as Example 14.6.

New Target: The rate constant,

$$k = ?\ s^{-1}$$

Knowns: The half-life.

Relationship: Equation (14-20) may be used directly.

Plan Use Equation (14-20) to calculate the rate constant, then calculate the fraction remaining, that is, the ratio (C/C_0), using the method of Example 14.6.

Work

1.

$$k = \frac{0.693}{t_{1/2}} = \frac{0.693}{3.3 \times 10^4\ s} = 2.1 \times 10^{-5}\ s^{-1}$$

2.

$$\ln\left(\frac{C}{C_0}\right) = -kt = -(2.1 \times 10^{-5}\ s^{-1})(30.0\ min)\left(\frac{60\ s}{1\ min}\right) = -0.0378$$

$$\frac{C}{C_0} = e^{-0.0378}$$

$$\text{fraction remaining} = \frac{C}{C_0} = 0.96 \qquad \text{(or 96\%)}$$

Check The elapsed time, 1800 seconds, is considerably less than the half-life. Therefore, considerably less than half is consumed. A 4% consumption seems reasonable.

Exercise

If the half-life of a reaction is 45 seconds, how much of the material originally present remains after 5.0 minutes?

Answer: 1.0%.

14.5 REACTIONS OF OTHER ORDERS

Many chemical reactions take place according to rate laws other than first order. Reactions of order 0, 2, or 3 are common, and even fractional orders, such as 3/2, are known. For each different order, there is a different equation, giving the relationship among concentration, rate constant, and elapsed time. In each case, the data may be plotted in a way that produces a straight line whose slope is determined by the value of the rate constant.

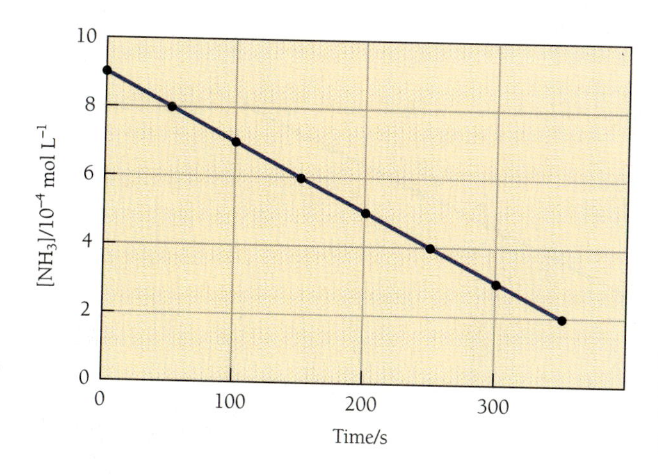

Figure 14.8 The reaction $2\ NH_3 \rightarrow N_2 + 3\ H_2$ follows zero-order kinetics. The concentration of ammonia decreases linearly with time, independent of the amount of ammonia present, when the gas is exposed to a hot metal surface.

Zero-Order Reactions

The high-temperature, low-pressure decomposition of ammonia,

$$2\ NH_3(g) \longrightarrow N_2(g) + 3\ H_2(g); \quad r = k[NH_3]^0 \quad \text{or} \quad r = k \quad (14\text{-}21)$$

Anything raised to the zero power equals 1.

which takes place on a hot metal surface, is an example of a zero-order reaction. The rate law is simply

$$r = k$$

The units of a zero-order rate constant are mol $L^{-1}\ s^{-1}$. Although the value of the rate constant depends on the type of metal, its surface area and cleanliness, and the temperature, the rate is independent of the *concentration* of ammonia. The concentration of ammonia remaining at any time (t) after the beginning of the reaction is given by the *integrated rate law*

The rate law $r = k$ applies to zero-order reactions only. Note that the rate of a zero-order reaction is independent of the concentrations of the reactants.

$$[NH_3] = [NH_3]_0 - kt \quad (14\text{-}22)$$

A plot of $[NH_3]$ vs. time gives a straight line whose slope is equal to the rate constant times (-1) (Figure 14.8).

Second-Order Reactions

The reaction of hydrogen carbonate ion with hydroxide ion

$$OH^- + HCO_3^- \longrightarrow CO_3^{2-} + H_2O; \quad r = k[OH^-][HCO_3^-]\ \text{mol}\ L^{-1}\ s^{-1} \quad (14\text{-}9)$$

is a second-order reaction (first order with respect to both hydrogen carbonate ion and hydroxide ion) whose rate constant is 6×10^9 L $mol^{-1}\ s^{-1}$ at 25 °C. Note that the units of a second-order rate constant are L $mol^{-1}\ s^{-1}$. When the initial concentrations of both reactants are the same*, reactant concentration decreases according to the second-order integrated rate law

The units of a *rate constant* are different for reactions of different order. The units of a reaction *rate (mol $L^{-1}\ s^{-1}$)* are the same for all orders.

$$\frac{1}{C} = \frac{1}{C_0} + kt \quad (14\text{-}23)$$

* When the initial concentrations of the reactants are different, the more complicated Equation (14-23a) is used. We will not encounter such situations in this text.

$$\frac{1}{[HCO_3^-]_0 - [OH^-]_0} \cdot \ln \frac{[HCO_3^-]/[HCO_3^-]_0}{[OH^-]/[OH^-]_0} = kt \quad (14\text{-}23a)$$

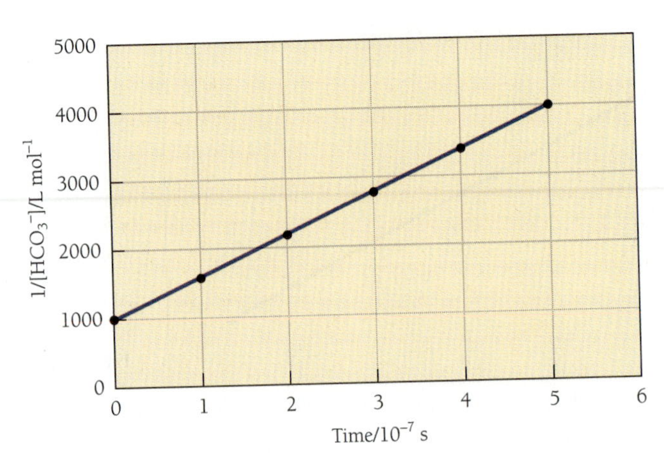

Figure 14.9 In second-order reactions, a plot of the reciprocal of concentration vs. time gives a straight line. The decomposition of hydrogen carbonate ion by the action of base, shown here, is a very fast reaction that goes essentially to completion in less than a thousandth of a second. Experimental measurements of such rapid reaction rates are difficult, and require sophisticated and expensive equipment.

In this equation, C is the concentration of either reactant at time t, and C_0 is its initial concentration: $C_0 = [HCO_3^-]_0 = [OH^-]_0$. A plot of the *reciprocal* of the concentration vs. time (Figure 14.9) gives a straight line whose slope is equal to the rate constant.

Half-life

The half-life may be defined for a reaction of any order, but the relationship between the half-life and the rate constant is different for each order. For zero-order reactions, $t_{1/2} = C_0/2k$; for first-order reactions, $t_{1/2} = 0.693/k$; for second-order reactions, $t_{1/2} = 1/C_0k$. The half-life of a reaction depends on concentration in all reactions except those of first order. Since the concentration of reactant decreases steadily during a reaction of any order, it follows that, except for first-order reactions, the half-life changes as a reaction proceeds. Figure 14.10 illustrates this behavior.

These relationships are obtained from the integrated rate laws, following the method used in Section 14.4 for first-order reactions.

The fact that it is independent of concentration makes the half-life of a first-order reaction a useful quantity. Half-lives of zero, second, and other orders are rarely used.

14.6 EFFECT OF TEMPERATURE ON REACTION RATE

The rate at which a chemical reaction takes place is influenced by temperature, to an extent that varies from one reaction to another. An increase in temperature brings about an increase in the rate of almost all reactions (Figure 14.11). There are two reasons, both related to the kinetic molecular theory, for this behavior. As the temperature increases, (a) the rate at which reactant molecules collide with one another increases, and (b) the average kinetic energy of the colliding molecules increases. The effect of temperature can be quite large. For many reactions, the rate can double if the temperature is increased by as little as 10 °C. For all reactions, the effect of temperature on rate can be calculated from the *activation energy* of the reaction.

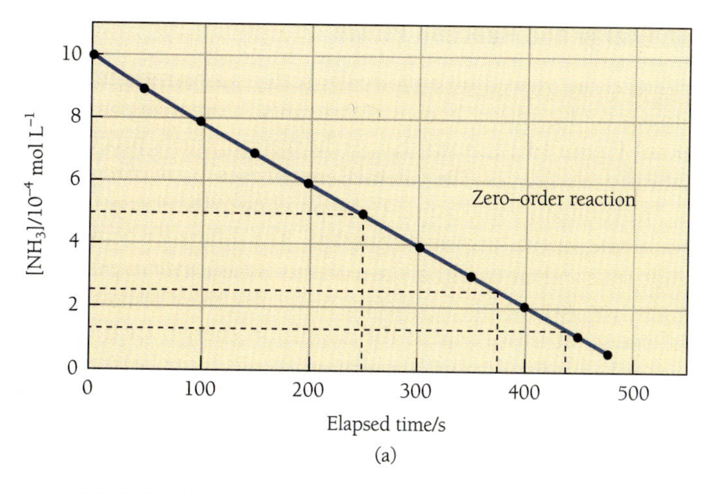

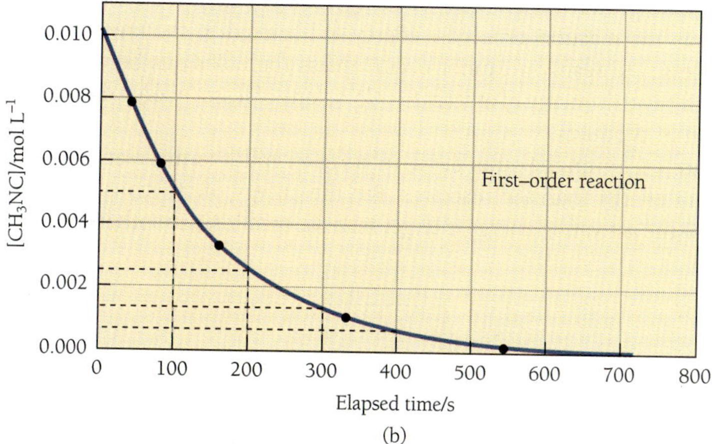

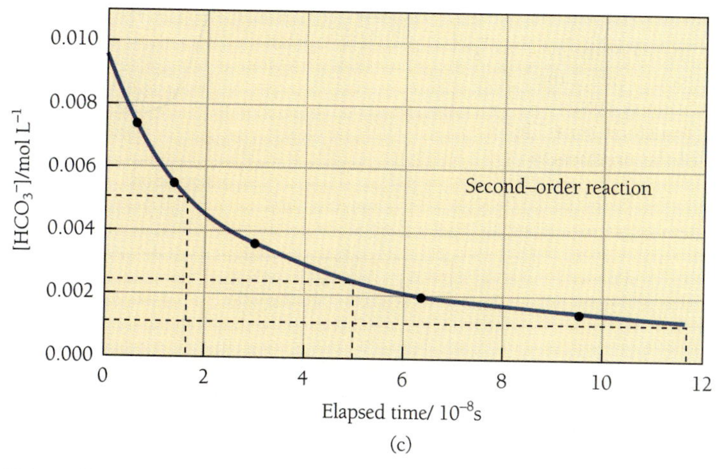

Figure 14.10 Concentration of reactant vs. time for reactions of (a) zero, (b) first, and (c) second order. The horizontal dashed lines show reactant concentrations equal to ½, ¼, and ⅛ of the starting concentration (and 1⁄16, in the first order plot). The vertical dashed lines indicate the times required for these successive 50% decreases, that is, the half-life of the reaction. In all but first-order reactions, half-life changes as the reaction proceeds.

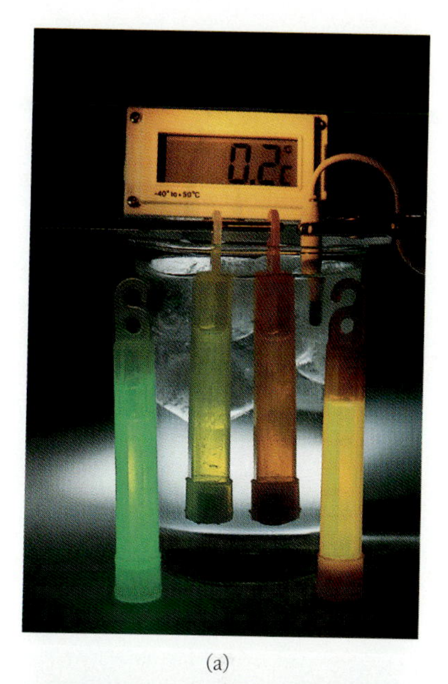

(a)

(b)

Figure 14.11 The light-producing reactions in Cyalume® light-sticks proceed more rapidly, and give off more light, as the temperature increases. In (a) the two room-temperature sticks outside the beaker are brighter than those inside, at 0.2 °C. In (b) the sticks at 59 °C are brighter than the room-temperature sticks.

Activation Energy and Reaction Profile

Potential energy is "energy of position." During a reactive collision, the potential energy changes as the positions of the atoms change with respect to one another.

A chemical reaction, viewed on a molecular scale, is a rearrangement of a group of atoms from one configuration, known as "reactants," to another configuration, known as "products." Figure 14.12 shows that, as the rearrangement proceeds, the potential energy of the system changes. The potential energy rises to a maximum value, and then decreases. The configuration of atoms corresponding to the energy maximum is called the *transition state,* or the "activated complex." The difference between the energy of the transition state and the energy of the reactants is the *activation energy*. Recall that, in order for reaction to take place, reactant molecules must bring to the collision an amount of energy at least equal to the activiation energy. Each different chemical reaction has a characteristic value of activation energy. It can be large or small, or even equal to zero for a few special reactions. But for many reactions, E_a is in the range 50 to 250 kJ mol^{-1}.

The energy changes shown in Figure 14.12 are appropriate for an exothermic reaction. The products have a lower energy content than the reactants, and consequently, as the reaction proceeds, energy is liberated to the environment as heat. Nonetheless, the reaction cannot occur unless the collisions between the reactant molecules have an energy of at least E_a. Collisions of lesser energy do not result in the formation of products.

Figure 14.13, on the other hand, refers to an endothermic reaction: the products have more energy than the reactants. But the energy of the transition state is even greater, so the activation energy is larger than the reaction's endothermicity. Collisions

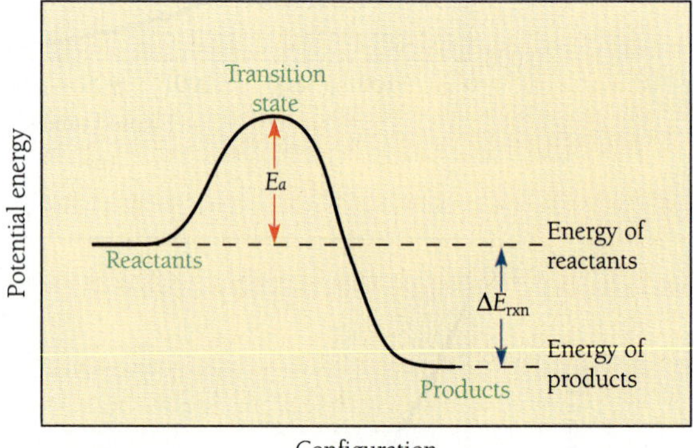

Figure 14.12 As a reaction proceeds, the potential energy of the system changes. In general, the energy increases by an amount called the "activation energy," and then decreases. The configuration of atoms corresponding to the maximum energy is called the "transition state." In an exothermic reaction, such as this one, the energy of the products is less than that of the reactants.

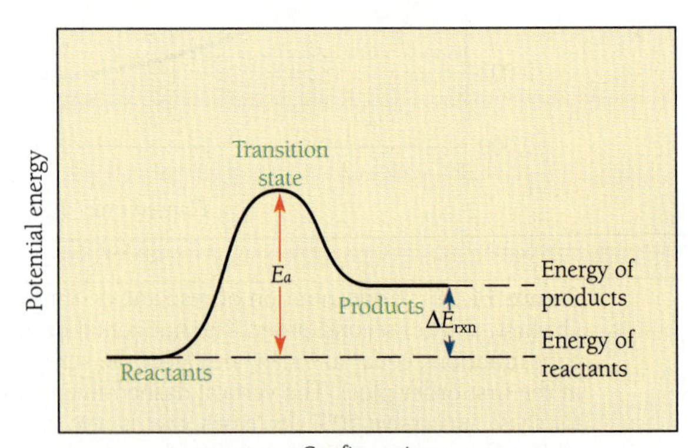

Figure 14.13 In an endothermic reaction, potential energy rises as the transition state is approached, and then decreases as the products are formed. The activation energy of an endothermic reaction is greater than the energy of reaction.

that supply energy greater than the endothermicity, *but less than E_a*, are nothing more than collisions: no reaction takes place.

Diagrams such as Figures 14.12 and 14.13 are called **reaction profiles** or **energy profiles.** Although the word "configuration" is used in these figures, the abscissa (x-axis) of an energy profile is normally called the *reaction coordinate.* **Reaction coordinate** denotes the change in geometrical relationship among the atoms as they proceed from the reactant configuration to product configuration. For example, in the dissociation reaction $Br_2 \rightarrow Br + Br$, the reactant configuration, molecular bromine, is a bonded pair of bromine atoms with an internuclear spacing of 230 pm, the normal bond length of the bromine molecule. The product configuration is again two bromine atoms, but these are separated by a much larger distance and no longer bonded to one another. For bromine dissociation, then, the "reaction coordinate" has a simple interpretation: it is the distance separating the two atoms.

In the energy profile of a more complicated reaction, for instance Equation (14-24), the interpretation of the reaction coordinate is less obvious.

$$H_2C{=}CH_2 + H_2 \longrightarrow H{-}\underset{H}{\overset{H}{C}}{-}\underset{H}{\overset{H}{C}}{-}H \qquad (14\text{-}24)$$

Several different bond lengths and angles change as a reactive collision proceeds, and the reaction coordinate cannot be identified as a single length or angle. In complicated reactions, the exact geometrical course of events may not even be known. It is best to think of the reaction coordinate as an unspecified, but definitely geometrical, description of the change from reactant molecule(s) to product molecule(s). It is important to realize that the reaction coordinate describes a *single collisional event,* and does not refer to the number or percentage of molecules that have reacted.

The number of molecules or moles that have reacted is described by the extent of reaction, defined in Chapter 13. Don't confuse reaction coordinate with extent of reaction.

The ordinate (y-axis) of the reaction profile is labeled "potential energy" in Figures 14.9 through 14.13. Since there are several different kinds of energy in chemistry, we should be clear as to which one is meant here. The reaction profile refers to a single molecular event, and the energy plotted is the *potential energy* associated with bond strengths and angles. The *kinetic energy* of the colliding species is converted to potential energy during the collision. The *difference* in potential energy between reactants and products is the energy of the reaction *per molecule.* Since it is often more useful to think in terms of molar quantities, the y-axis of a reaction profile is sometimes labeled "(ΔE)", the reaction energy *per mole.*

Most chemists do not distinguish between ΔE_{rxn} (energy of reaction) and ΔH_{rxn} (enthalpy of reaction) in qualitative thinking, nor is it necessary to do so unless precise measurements or calculations are called for. Indeed, many textbooks contain diagrams similar to Figures 14.9 and 14.10, in which the reaction energy (the difference in energy between reactants and products) is labeled ΔH rather than ΔE. In Chapter 10 we saw that, except for certain gas-phase reactions, ΔE_{rxn} and ΔH_{rxn} have the same numerical value (within the accuracy of routine measurements).

The difference between ΔE_{rxn} and ΔH_{rxn} was discussed in Section 10.4.

The particular geometrical configuration corresponding to the maximum of the potential energy curve in the reaction profile is called the **transition state.** The transition state is not an actual, stable molecule. It exists only as a fleeting configuration through which the collision partners must pass in order to become products. The transition state is not necessarily halfway between reactants and products, any more than our continental divide must be exactly halfway between the Atlantic and Pacific

Oceans. The significance of the transition state lies in the fact that it is the point of no return. Once a pair of colliding molecules have passed the transition state, they go on to become products.

The Arrhenius Equation

A note about the Swedish chemist, Svante Arrhenius, appears on p. 540.

The effect of temperature on reaction rate is given by the **Arrhenius equation** (14-25),

$$k = Ae^{-E_a/RT} \tag{14-25}$$

which relates the rate constant k to the absolute temperature T, the gas constant R, and the activation energy E_a. The quantity A is called the **frequency factor** or the **pre-exponential factor.** Its value, like that of the activation energy, is different for each reaction. Once the reaction is specified, the values of A and E_a are fixed: they do not depend on temperature. If the activation energy of a reaction is large (only highly energetic collisions are successful), the quantity $-E_a/RT$ is a large negative number, and $e^{-E_a/RT}$ is a small number. If so, the rate constant k is small and the reaction is slow. If the activation energy is small, the quantity $-E_a/RT$ is close to zero, and $e^{-E_a/RT}$ is only slightly less than one. Then, provided the frequency factor is not too small, the rate constant k is large and the reaction is fast. Another feature revealed by the Arrhenius equation is that reactions characterized by a large E_a experience a greater *change* in rate for a given temperature increase than reactions having a low value of E_a.

If E_a is large, e^{-E_a} and k are small. If E_a is small, e^{-E_a} and k are large.

The Arrhenius equation can be used to relate the rate constant at two different temperatures to the activation energy, even when the frequency factor is unknown. Equation (14-25) is written for two different temperatures, T_1 and T_2, at which the corresponding values of the rate constant, k_1 and k_2, are known.

$$k_1 = Ae^{-E_a/RT_1} \tag{14-26}$$

$$k_2 = Ae^{-E_a/RT_2} \tag{14-27}$$

Logarithms of each side of Equations (14-26) and (14-27) are found as follows.

$$\ln k_1 = \ln A - \frac{E_a}{RT_1} \tag{14-28}$$

$$\ln k_2 = \ln A - \frac{E_a}{RT_2} \tag{14-29}$$

When Equation (14-29) is subtracted from Equation (14-28), the result is

$$\ln k_1 - \ln k_2 = -\frac{E_a}{RT_1} + \frac{E_a}{RT_2}$$

$$= -\left(\frac{E_a}{R}\right)\cdot\left(\frac{1}{T_1} - \frac{1}{T_2}\right) \tag{14-30}$$

$$\ln\left(\frac{k_1}{k_2}\right) = -\left(\frac{E_a}{R}\right)\cdot\left(\frac{1}{T_1} - \frac{1}{T_2}\right) \tag{14-31a}$$

If you prefer, use the equivalent form

$$\ln\left(\frac{k_2}{k_1}\right) = \frac{E_a}{R}\cdot\frac{T_2 - T_1}{T_1T_2} \tag{14-31b}$$

to compare reaction rates at two different temperatures. The temperature must be expressed in kelvins, and, because the units of E_a are J mol^{-1} (or kJ mol^{-1}), the value $R = 8.314$ J mol^{-1} K^{-1} *must* be used for the gas constant.

The "mol^{-1}" part of the units means that E_a is the energy required to produce one mole of transition state from the reactants.

Numerical Calculations

Many problems in kinetics require familiarity with the relationship between reaction rate constant and temperature. Some examples of such problems follow.

EXAMPLE 14.10 Calculation of Activation Energy

An important member of the complex set of reactions by which the earth's ozone layer is established and maintained is the disappearance of ozone by reaction with atomic oxygen.

$$O + O_3 \longrightarrow O_2 + O_2$$

This reaction, like many in which an atom rather than a molecule is a reactant, is a relatively rapid one. At 0 °C the rate constant is 1.76×10^6 L mol^{-1} s^{-1}, while, at 72 °C, the rate constant is 1.11×10^7 L mol^{-1} s^{-1}. Calculate the activation energy of the reaction.

Analysis

Target: An activation energy,

$$E_a = ?\ \text{J mol}^{-1}$$

Knowns: Rate constants at two different temperatures.

Relationship: Target and knowns are related by Equation (14-31). Most problems involving reaction rates at more than one temperature are solved by means of this equation.

Plan Solve Equation (14-31) for E_a, substitute the known information, and evaluate. Whenever a complicated numerical expression is to be evaluated, it is helpful to organize the work by writing down all the quantities, together with their values if known. When the information given in this problem is systematically arranged, it is seen that four out of the five variable quantities in Equation (14-31) are known, and the fifth is the target.

$$T_1 = 0\ °\text{C} = 273\ \text{K}; \qquad k_1 = 1.76 \times 10^6\ \text{L mol}^{-1}\ \text{s}^{-1}$$

$$T_2 = 72\ °\text{C} = 345\ \text{K}; \qquad k_2 = 1.11 \times 10^7\ \text{L mol}^{-1}\ \text{s}^{-1}$$

$$E_a(\text{J mol}^{-1}) = ?$$

Work Solve Equation (14-31b) for the target quantity.

$$\ln\left(\frac{k_2}{k_1}\right) = \left(\frac{E_a}{R}\right) \cdot \frac{(T_2 - T_1)}{(T_1 T_2)} \qquad (14\text{-}31\text{b})$$

$$\left(\frac{E_a}{R}\right) \cdot \frac{(T_2 - T_1)}{(T_1 T_2)} = \ln\left(\frac{k_2}{k_1}\right)$$

$$\left(\frac{E_a}{R}\right) = (T_1 T_2) \cdot \frac{[\ln(k_2/k_1)]}{(T_2 - T_1)}$$

$$E_a = R \cdot (T_1 T_2) \cdot \frac{[\ln(k_2/k_1)]}{(T_2 - T_1)}$$

Substitute the known values, being careful to express temperature in K and to use $R = 8.314 \text{ J mol}^{-1} \text{ K}^{-1}$.

$$E_a = (8.314 \text{ J mol}^{-1} \text{ K}^{-1}) \cdot [(273 \text{ K}) \cdot (345 \text{ K})] \times \frac{[\ln(1.11 \times 10^7 \text{ L mol}^{-1} \text{ s}^{-1}/1.76 \times 10^6 \text{ L mol}^{-1} \text{ s}^{-1})]}{(345 \text{ K} - 273 \text{ K})}$$

$$E_a = 20{,}000 \text{ J mol}^{-1} = 20 \text{ kJ mol}^{-1}$$

Check Activation energies are positive quantities, rarely more than 300 kJ mol^{-1}. If the answer turns out negative, k_1 and k_2 or T_1 and T_2 have been interchanged.

Exercise

Calculate the activation energy of a reaction whose rate constant is $1.50 \times 10^3 \text{ s}^{-1}$ at 20 °C and $7.50 \times 10^3 \text{ s}^{-1}$ at 50 °C.

Answer: 42 kJ mol^{-1}.

If the activation energy and the value of the rate constant at one temperature are known, the value of the rate constant at any other temperature can be found from Equation (14-31).

EXAMPLE 14.11 Rate Constant at a Different Temperature

One of the reactions taking place in the Earth's upper atmosphere is the oxygen-atom transfer reaction, $O_3 + NO \rightarrow O_2 + NO_2$, whose second-order rate constant is $1.25 \times 10^7 \text{ L mol}^{-1} \text{ s}^{-1}$ at 50 °C. Given that the activation energy is 10.3 kJ mol^{-1}, calculate the value of the rate constant at −50 °C.

Analysis

Target: The rate constant at −50 °C,

$$k = ? \text{ L mol}^{-1} \text{ s}^{-1}$$

Knowns: Activation energy, and the rate constant at 50 °C. The chemical equation and the order of the reaction are given but are not used in the solution.

Relationship: Equation (14-31) applies to all such problems, regardless of reaction order.

Plan Tabulate all known information, convert to appropriate units if necessary, and solve for the unknown rate constant. It is somewhat easier to keep track of the units if this is done in two steps. First solve for the rate constant ratio k_1/k_2, and then solve for the target k_2.

Work

1.

$$T_1 = 50\ °C = 323\ K; \qquad k_1 = 1.25 \times 10^7\ L\ mol^{-1}\ s^{-1}$$

$$T_2 = -50\ °C = 223\ K; \qquad k_2 = ?$$

$$E_a = 10.3\ kJ\ mol^{-1} = 10{,}300\ J\ mol^{-1}$$

2.

$$\ln\left(\frac{k_1}{k_2}\right) = -\left(\frac{E_a}{R}\right)\cdot\left(\frac{1}{T_1} - \frac{1}{T_2}\right) \qquad (14\text{-}31a)$$

We used Equation (14-31a) instead of Equation (14-31b) just for variety.

$$= -\left(\frac{10{,}300\ J\ mol^{-1}}{8.314\ J\ mol^{-1}\ K^{-1}}\right)\cdot\left(\frac{1}{323\ K} - \frac{1}{223\ K}\right)$$

$$= +1.72$$

Note that all units have disappeared from the right-hand side. Logarithms are *dimensionless* quantities.

Exponentiate (find the antilogarithm of) both sides of the equation.

$$\text{antilog}\left[\ln\left(\frac{k_1}{k_2}\right)\right] = \text{antilog}\ (+1.72)$$

$$\frac{k_1}{k_2} = e^{1.72} = 5.585$$

3. Substitute the known value of k_1 and solve for k_2.

$$k_2 = \frac{k_1}{5.585}$$

$$= \frac{1.25 \times 10^7\ L\ mol^{-1}\ s^{-1}}{5.585}$$

$$= 2.24 \times 10^6\ L\ mol^{-1}\ s^{-1}$$

Check A rate constant is a positive quantity whose value decreases as the temperature decreases.

Exercise

What is the rate constant of the ozone-nitric oxide reaction at 0 °C?

Answer: $6.2 \times 10^6\ L\ mol^{-1}\ s^{-1}$.

Relationship between Activation Energy and Reaction Energy

Reversible reactions are characterized by *two* values of activation energy—one for the forward (to the right) reaction ($E_{a,f}$), and one for the reverse reaction ($E_{a,r}$). These two quantities have different numerical values, which, however, are related to one another

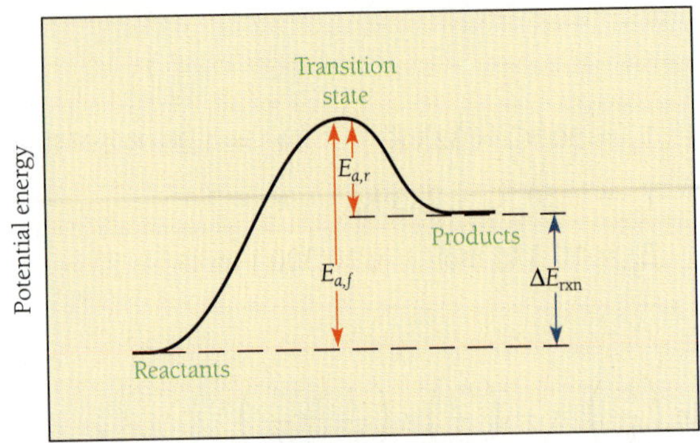

Figure 14.14 A reversible reaction has two activation energies: $E_{a,f}$ for the forward direction and $E_{a,r}$ for the reverse direction. In endothermic reactions, such as this one, $E_{a,f}$ is greater than $E_{a,r}$. For exothermic reactions $E_{a,f}$ is less than $E_{a,r}$. The transition state is the same in both directions.

through the energy of reaction (ΔE_{rxn}). The relationship is shown in Figure 14.14, from which it can be seen that

$$\Delta E_{rxn} + E_{a,r} = E_{a,f} \qquad \text{or} \qquad \Delta E_{rxn} = E_{a,f} - E_{a,r} \tag{14-32}$$

Equation (14-32) holds true for all reactions, whether ΔE_{rxn} is positive, negative, or zero.

EXAMPLE 14.12 Activation Energies and Energy of an Endothermic Reaction

The O-atom transfer reaction $CO_2 + NO \rightarrow CO + NO_2$, discussed on page 596 and pictured in Figure 14.2a, has a forward activation energy $E_{a,f} = 342$ kJ mol^{-1} and a reverse activation energy $E_{a,r} = 116$ kJ mol^{-1}. What is its energy of reaction?

Analysis

Target: An energy of reaction,

$$\Delta E_{rxn} = ?\ \text{kJ mol}^{-1}$$

Knowns: $E_{a,f}$ and $E_{a,r}$; the chemical equation is given, but not needed in the solution.

Relationship: Both Figure 14.14 and Equation (14-32) describe the relationship between target and knowns.

Work

$$\Delta E_{rxn} = E_{a,f} - E_{a,r} \tag{14-32}$$

$$\Delta E_{rxn} = 342\ \text{kJ mol}^{-1} - 116\ \text{kJ mol}^{-1}$$

$$\Delta E_{rxn} = 226\ \text{kJ mol}^{-1}$$

Note that the reaction is endothermic and requires a net input of heat from the environment.

Exercise

If a reaction has $E_{a,f} = 275$ kJ mol^{-1} and $E_{a,r} = 50$ kJ mol^{-1}, what is its ΔE_{rxn}? Is the reaction endothermic or exothermic?

Answer: $+225$ kJ mol^{-1}; endothermic.

Many people find it helpful, in working problems of this sort, to make a quick sketch of the reaction profile. For the reaction in Example 14.12, the sketch would look like Figure 14.15c. The sketch gives a quick pictorial confirmation of the sign relationships among the three energies. Since ΔE_{rxn}, $E_{a,f}$, and $E_{a,r}$ are all positive quantities in an endothermic reaction, it is easy to see that $E_{a,f}$ is the sum of the other two.

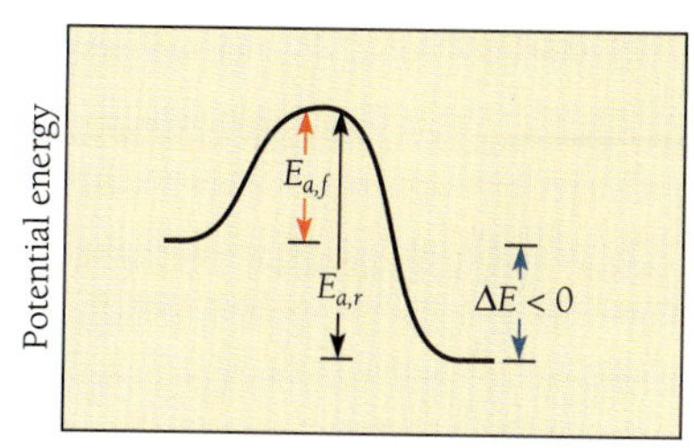

Exothermic: $E_{a,f} < E_{a,r}$
(a)

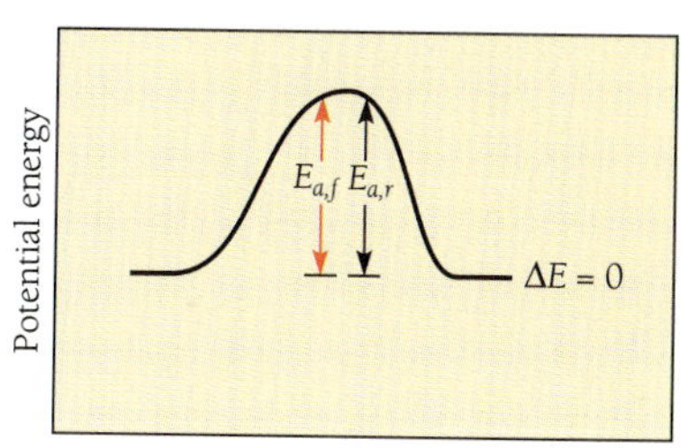

Thermoneutral: $E_{a,f} = E_{a,r}$
(b)

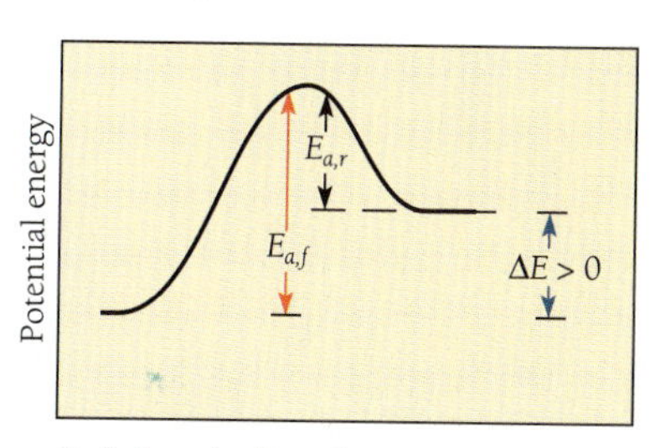

Endothermic: $E_{a,f} > E_{a,r}$
(c)

Reaction coordinate

Figure 14.15 Quick sketches like these are useful for remembering the relationship among $E_{a,f}$, $E_{a,r}$, and ΔE_{rxn} in reactions that are (a) exothermic, (b) thermoneutral ($\Delta E_{rxn} = 0$), and (c) endothermic.

EXAMPLE 14.13 Activation Energies and Energy of an Exothermic Reaction

Another O-atom transfer reaction that occurs among the oxides of carbon and nitrogen is $CO + N_2O \rightarrow CO_2 + N_2$. This reaction is exothermic, with $\Delta E_{rxn} = -365$ kJ mol^{-1}. Its activation energy is 96 kJ mol^{-1}. What is the activation energy of the reverse reaction?

Analysis/Relationship/Plan Like Example 14.12, this is a substitution into Equation (14-32).

Work Make a sketch showing the knowns and the target.

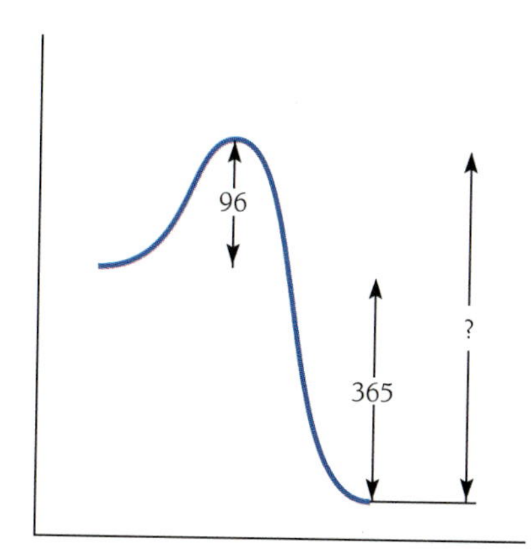

Rearrange Equation (14-32) and substitute the known values to obtain the value of $E_{a,r}$.

$$\begin{aligned} E_{a,r} &= E_{a,f} - \Delta E_{rxn} \\ &= 96 \text{ kJ mol}^{-1} - (-365 \text{ kJ mol}^{-1}) \\ &= 461 \text{ kJ mol}^{-1} \end{aligned}$$

Check E_a is always a positive quantity, and, for exothermic reactions, $E_{a,r}$ is greater than $E_{a,f}$.

Exercise

An endothermic reaction has $\Delta E_{rxn} = 125 \text{ kJ mol}^{-1}$ and a reverse activation energy of 70 kJ mol^{-1}. What is the activation energy in the forward direction?

Answer: 195 kJ mol^{-1}.

14.7 THE MECHANISM OF CHEMICAL REACTIONS

A balanced chemical equation such as Equation (14-33),

$$2\,O_3(g) \longrightarrow 3\,O_2(g) \tag{14-33}$$

which describes the thermal decomposition of ozone, contains a great deal of interesting and useful information. It provides a list of reactants, a list of products, and the relative molar amounts of each. Also, since the relative molar amounts are known, the relative gram amounts of the various species involved can be calculated. However, a stoichiometric equation tells us nothing about the *way* in which the transformation from reactants to products is achieved. That is, the equation contains no information as to the pathway or *mechanism* of the reaction.

Mechanism is a term that means a detailed, step-by-step description of the molecular events by which reactant molecules become product molecules. The mechanism of a chemical reaction describes which molecules collide and in what sequence they do so. It describes which chemical bonds are formed, which are broken, and in what sequence.

The decomposition of ozone does not result from a single collision of two O_3 molecules. Instead, it occurs in two separate elementary steps. An **elementary step** is a chemical reaction that takes place as a single event, often a collision between two molecules. The first step in ozone decomposition is the disintegration of one O_3 molecule to form an oxygen molecule and an oxygen atom.

$$O_3 \longrightarrow O_2 + O \tag{14-34}$$

In the second step, another ozone molecule collides with the oxygen atom and transfers an atom of oxygen to it, forming a new O═O bond. The collision partners separate in a new configuration, namely as two O_2 molecules.

$$O_3 + O \longrightarrow O_2 + O_2 \tag{14-35}$$

The net result of this activity is the consumption of two ozone molecules and the production of three oxygen molecules. This result, called the *overall reaction,* is found by adding the elementary steps.

$$\begin{array}{rcll} O_3 & \longrightarrow & O_2 + O & (14\text{-}34) \\ O + O_3 & \longrightarrow & O_2 + O_2 & (14\text{-}35) \\ \hline \cancel{O} + 2\,O_3 & \longrightarrow & 3\,O_2 + \cancel{O} & \text{or} \\ 2\,O_3 & \longrightarrow & 3\,O_2 & (14\text{-}33) \end{array}$$

The reaction *stoichiometry* is described by Equation (14-33), while the *mechanism* is described by Equations (14-34) and (14-35).

Any reaction that proceeds via a sequence of elementary steps is called an **overall** or complex reaction. The mechanism of a reaction is described by listing its elementary steps, along with some information as to the rate of each. Elementary steps such as the one represented by Equation (14-34), which involve a single molecule rather than a collision, are called **unimolecular.** Most complex reactions proceed via **bimolecular** steps, in which two molecules or atoms collide. Very rarely, an elementary step may involve the simultaneous collision of three species. Such steps are called **trimolecular** (or *termolecular* in older books).

No elementary steps involving more than three atoms or molecules are known.

The collision theory of reaction rates allows us to predict the order of an elementary step. The rate of a unimolecular step depends on the concentration of only one species, and the step is first order. A bimolecular step is second order because its rate is determined by the collision rate between two species. A trimolecular step involves a triple collision, the rate of which depends on the concentration of three species. Trimolecular steps are third order. There is an important difference between the rate law of a complex reaction and the rate law of an elementary step. Whereas *the rate law (order) of a complex reaction can only be determined by observation and experiment, the rate law (order) of any elementary step is determined by its stoichiometric coefficients.* Thus, since it is known that Equation (14-35) represents an elementary step, it is also known that it is second order and that its rate law is given by Equation (14-36).

$$O_3 + O \longrightarrow O_2 + O_2 \qquad (14\text{-}35)$$

$$\text{rate} = k[O_3][O] \qquad (14\text{-}36)$$

It is not obvious whether a particular chemical equation describes a complex reaction or an elementary step. This can only be established by experiment. Reaction (14-33) *could* take place as a single step, although in fact it does not. It is the responsibility of the person who is discussing a chemical equation to identify it as a complex reaction or an elementary step.

The understanding of reaction mechanisms is an important part of chemical kinetics. The approach is to postulate a hypothetical mechanism, and to predict from it the order of the overall reaction. Then, laboratory experiments are done to see if the proposed mechanism is consistent with the observed facts. If, as is often the case, several different proposed mechanisms are consistent with the facts, other experiments must be performed in order to distinguish between them.

Prediction of the overall rate law from a proposed sequence of elementary steps is aided by the concept of the **rate-determining step.** If one elementary step in a sequence is much slower than the others, the rate of the overall reaction will be determined by, and in fact be the same as, the rate of the slow step. A reaction cannot proceed faster than its slowest step. A few examples will help clarify the application of this principle.

EXAMPLE 14.14 Rate-Determining Step

The reaction $NO_2(g) + CO(g) \rightarrow NO(g) + CO_2(g)$, when carried out at low temperature, proceeds by the mechanism

(1) $$NO_2 + NO_2 \longrightarrow NO + NO_3 \qquad \text{(slow)}$$

(2) $$NO_3 + CO \longrightarrow NO_2 + CO_2 \qquad \text{(fast)}$$

What rate law would be observed in an experiment?

Analysis

Target: A rate law, in which the identity of the compounds and the concentration exponents are to be determined,

$$k = [?]^?[?]^? \ldots$$

Knowns: The steps of the mechanism, identified as "fast" or "slow."

Relationship: The rate law of any elementary step is implied by the stoichiometry and may be written immediately. The critical fact is that, in this mechanism, the first step, labeled "slow," is rate determining—the rate of the overall reaction is the same as the rate of the slow step.

Plan Identify the rate-determining step and write its rate law.

Work The rate-determining step is

(1) $$NO_2 + NO_2 \longrightarrow NO + NO_3 \quad \text{(slow)}$$

and, from the stoichiometry, its rate law is

$$\text{rate} = k_1[NO_2][NO_2] \quad \text{or} \quad \text{rate} = k_1[NO_2]^2$$

The subscript "1" means that the rate constant is for the first step in the reaction mechanism.

Since the second step is fast, there is no delay in the production of CO_2. The overall rate of consumption of NO_2 is equal to the rate of appearance of CO_2. The rate law may be expressed as

$$r = -\frac{\Delta[NO_2]}{\Delta t} = k_1[NO_2]^2 \quad \text{or} \quad r = +\frac{\Delta[CO_2]}{\Delta t} = k_1[NO_2]^2$$

Exercise

What overall rate law is predicted from the following proposed mechanism for the reaction of hydrogen peroxide with iodide ion in acid solution?

$$I^- + H_2O_2 + H_3O^+ \longrightarrow 2\,H_2O + HOI \quad \text{(slow)}$$

$$I^- + HOI \longrightarrow I_2 + OH^- \quad \text{(fast)}$$

$$OH^- + H_3O^+ \longrightarrow 2\,H_2O \quad \text{(fast)}$$

$$I_2 + I^- \longrightarrow I_3^- \quad \text{(fast)}$$

Answer: $r = k[H_2O_2][H_3O^+][I^-]$.

EXAMPLE 14.15 Rate-Determining Step

The decomposition of ozone follows first-order kinetics. Show that the mechanism

(1) $$O_3 \longrightarrow O_2 + O \quad \text{(slow)}$$

(2) $$O + O_3 \longrightarrow 2\,O_2 \quad \text{(fast)}$$

is consistent with a first-order rate law.

Analysis

Target: A true/false comment, "is the mechanism first order or not?" Since the order of a reaction is explicit in the rate law, the rate law can also be regarded as the target.

Knowns: The mechanism.

Relationship: The mechanism contains a slow step, so the concept of the rate-determining step is appropriate.

Work The rate of the slow step, and therefore of the overall reaction, is

$$\text{rate} = k_1[O_3]$$

Since the order of a reaction is the sum of the concentration exponents in the rate law, which in this case is 1, the reaction is first order. That is, the mechanism is consistent with the observation of first-order kinetics.

Exercise

The reaction $Cl_2(aq) + H_2S(aq) \rightarrow S(s) + 2\ HCl(aq)$ is known to proceed by a second-order rate law. Show that the following mechanism *cannot* be the correct one.

$$Cl_2 \longrightarrow 2\ Cl \qquad \text{(slow)}$$

$$Cl + H_2S \longrightarrow HCl + SH \qquad \text{(fast)}$$

$$SH + Cl \longrightarrow HCl + S(s) \qquad \text{(fast)}$$

Answer: The proposed mechanism leads to a first-order rate law, $r = k[Cl_2]$. The correct mechanism *must* lead to a second-order law, since that is what is observed.

Often there are surprises in the study of reaction mechanisms. The well-studied reaction

$$H_2(g) + I_2(g) \longrightarrow 2\ HI(g) \tag{14-37}$$

was thought for many years to take place in a single bimolecular elementary step, because the observed rate law is second order.

$$\text{rate} = k[H_2][I_2] \tag{14-38}$$

More than 40 years after the reaction was first characterized, it was realized that another, more complicated mechanism, given by Equations (14-39) and (14-40), is also consistent with the observed rate law.

$$I_2(g) \rightleftharpoons 2\ I(g) \qquad \text{(fast, in equilibrium: } K_{eq}\text{)} \tag{14-39}$$

$$H_2(g) + I(g) + I(g) \longrightarrow 2\ HI(g) \qquad \text{(slow, rate constant } k_2\text{)} \tag{14-40}$$

There is some evidence that the trimolecular step [Equation (14-40)] may actually proceed by two bimolecular steps.

$$H_2(g) + I(g) \longrightarrow H_2I(g) \tag{14-40a}$$

$$H_2I(g) + I(g) \longrightarrow 2\ HI(g) \tag{14-40b}$$

The question is not yet completely resolved, although it is accepted that the mechanism is considerably more complicated than the single bimolecular step, Equation (14-37). Occam's razor fails to cut this problem.

William of Occam ("the Invincible Doctor") taught philosophy at Oxford until 1323. He was responsible for the test, now known as *Occam's razor*, by which the simplest of several competing hypotheses is regarded as most likely to be true.

We now show that the mechanism consisting of two elementary steps, Equations (14-39) and (14-40), leads to a second-order rate law for the overall reaction. Step two, Equation (14-40), is very slow because it requires a comparatively infrequent three-body collision. Step two is rate determining. The rate of the overall reaction is therefore the same as the rate of step two, which is

$$\text{rate} = k_2[H_2][I][I] = k_2[H_2][I]^2 \tag{14-41}$$

Since the second step is comparatively slow, the first step [Equation (14-39)] has plenty of time to reach equilibrium. A state of equilibrium is maintained throughout the entire course of the reaction, so it is always true that

$$K_{eq} = \frac{[I]^2}{[I_2]} \qquad \text{or} \qquad [I]^2 = K_{eq}[I_2] \tag{14-42}$$

When Equation (14-42) is substituted into Equation (14-41), a second-order rate law is obtained.

$$\text{rate} = k_2 K_{eq}[H_2][I_2]$$

The product of the rate constant k_2 and the equilibrium constant K_{eq} is also a constant, which we may call, simply, k. That is, we make the substitution $k = k_2K_{eq}$. Thus, the rate law predicted from the mechanism given by Equations (14-39) and (14-40) is

$$\text{rate} = k[H_2][I_2]$$

It is identical to the experimentally observed rate law (14-38).

Two different mechanisms, one consisting of the single elementary step [Equation (14-37)] and the other consisting of the two-step sequence [Equations (14-39) and (14-40)], predict exactly the same rate law. In the absence of any additional information, there is no way to tell which mechanism is operating. This is a central problem in the study of chemical kinetics. Clever experimentation and careful interpretation of results are required to establish the correctness of a mechanism.

The observed rate law [Equation (14-38)] does not include the concentration of iodine *atoms,* although they play a prominent part in the mechanism. This is often the case with overall rate laws, and it is a simplifying feature rather than a complicating feature. Atomic iodine is a highly reactive species, which reacts almost as soon as it is formed. It is consumed either in reaction (14-40) or in the right-to-left reaction of the equilibrium [Equation (14-39)]. The iodine atom is so reactive that its concentration remains very small and therefore very hard to measure. A rate law including a nonmeasurable concentration of atomic iodine is useless for calculation of reaction rates.

In the hydrogen-iodine reaction we have seen an example of a new feature of mechanisms, the **pre-equilibrium.** An elementary step will be in equilibrium if its reverse reaction is faster than succeeding steps. The equilibrium expression may be used in deducing the rate law. Species such as atomic iodine, which are generated in a complex reaction but are neither reactants nor final products, are called **intermediates.** Since they are always more highly reactive than the other species involved in the reactions, they are sometimes called "reactive intermediates," or "unstable intermediates."

14.8 CHAIN REACTIONS

On the basis of the principles of periodicity and the chemical similarity of adjacent elements in the periodic table, one might expect the mechanism of the reaction of hydrogen with bromine to be similar to the mechanism of the hydrogen-iodine reaction. The fact that it is not is one of the many unexpected and interesting results in the study of chemical kinetics. The mechanism of the hydrogen-bromine reaction is

(initiation) $$Br_2 \longrightarrow 2\ Br \qquad (14\text{-}43)$$

(propagation) $$Br + H_2 \longrightarrow HBr + H \qquad (14\text{-}44)$$

(propagation) $$H + Br_2 \longrightarrow HBr + Br \qquad (14\text{-}45)$$

(termination) $$Br + Br \longrightarrow Br_2 \qquad (14\text{-}46)$$

Addition of the four steps of this mechanism gives the correct overall reaction, $H_2 + Br_2 \rightarrow 2\ HBr$.

There are two intermediates in this mechanism, H and Br. Both are highly reactive, and consequently reactions (14-44) to (14-46) are comparatively fast. Addition of these four steps gives the observed overall reaction,

$$H_2(g) + Br_2(g) \longrightarrow 2\ HBr(g)$$

Note the action of the two steps labeled "propagation." Together these steps are responsible for the formation of the final product, HBr. Both steps involve a reactive intermediate, and the intermediate used in one propagation step is regenerated in the other. Such mechanisms are called **chain mechanisms** or **chain reactions,** and the intermediates are given the special name **chain carriers.** In a chain reaction, chain carriers are formed in one or more **initiation steps.** In this case, bromine atoms are formed in Reaction (14-43). There follows a set of **propagation steps** in which the highly reactive chain carriers react with the stable reactants to produce the stable product(s) and another chain carrier. The propagation steps continue until all the reactants are consumed, and/or the chain carriers are destroyed in one or more steps called **termination steps.** The rate law for the overall reaction of bromine with hydrogen is

$$\text{rate} = k[H_2][Br_2]^{1/2} \qquad (14\text{-}47)$$

The reaction is first order in hydrogen, *one-half* order in bromine, and *three-halves* order overall. Although the study of chemical kinetics would be simpler if reaction orders were always integers, there is nothing in the definition of order that requires this to be so. Chain reactions are often of fractional order, and this is one of the experimentalist's first clues that a reaction involves a chain mechanism.

It is now commonly accepted that chlorine atoms are the primary cause of the decrease in the amount of ozone in the stratosphere. The chlorine atoms that threaten our ozone layer originate from manufactured compounds called "CFC"s (chlorinated fluorocarbons), used primarily in refrigeration and air conditioning. Ozone decrease proceeds by a chain reaction whose propagation steps are

Because they do not lead to the ultimate products, initiation and termination steps are sometimes omitted when the mechanism of chain reaction is written.

$$O_3 + Cl \longrightarrow O_2 + ClO \qquad (14\text{-}48)$$

$$ClO + O_3 \longrightarrow 2\ O_2 + Cl \qquad (14\text{-}49)$$

The overall reaction of this chain mechanism is simply

$$2\ O_3 \longrightarrow 3\ O_2$$

in which the chain carriers are Cl and ClO.

Much of the chemistry of the earth's atmosphere involves chain reactions, some of which are very complicated. Study of atmospheric reactions is an active field in

Figure 14.16 Hydrogen burns in an atmosphere of pure chlorine. The reaction $H_2 + Cl_2 \rightarrow 2\ HCl$ takes place by a chain mechanism involving the chain carriers H and Cl. Most flames involve chain reactions.

chemistry, with experiments being carried out in airplanes and satellites as well as in earth-bound laboratories. In addition, a great deal of data is gathered by balloon- and rocket-borne instruments.

Other examples of chain reactions are flames and explosions (Figure 14.16) and the industrial synthesis of most of our modern plastics and fabrics.

14.9 CATALYSIS

It is a relatively common experimental observation that the addition of certain substances can increase the rate of a chemical reaction. For example, pure potassium chlorate decomposes to potassium chloride and oxygen when heated. The process is slow and uneven at moderate temperature, and takes place with explosive rapidity at high temperature. If some manganese dioxide is thoroughly mixed with the potassium chlorate, however, the reaction takes place rapidly and smoothly at a temperature well below the danger point.

Production of O_2 by the catalytic decomposition of $KClO_3$ is a common experiment in introductory laboratory courses.

$$2\ KClO_3 \xrightarrow[MnO_2]{} 2\ KCl + 3\ O_2 \tag{14-50}$$

Manganese dioxide is said to "catalyze the thermal decomposition of potassium chlorate." A **catalyst** is any substance that increases the rate of a chemical reaction without itself being consumed. The action of a catalyst is called **catalysis.**

The Mechanism of Catalysis

The means by which catalysts achieve their effect can be understood in terms of the collision theory of reaction rates. We saw in Section 14.2 that a reaction rate is controlled by the activation energy and also by a geometrical factor, the requirement for proper orientation of the reacting species. When a catalyst is present, the reaction proceeds by a different pathway. The catalyzed pathway is faster because (1) the activation energy is less, and/or (2) the required geometrical alignment of the reactants is facilitated. The energy profile of a typical catalyzed reaction is compared with the

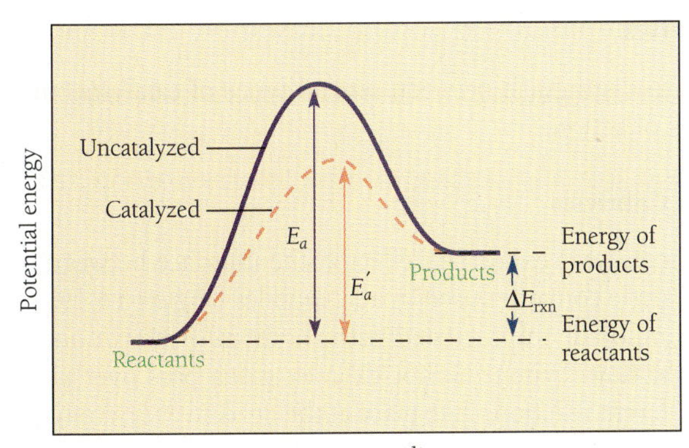

Figure 14.17 In many reactions, a catalyst increases the rate of a reaction by decreasing its activation energy. The lower the activation energy, the greater the fraction of collisions that can surmount the barrier and the faster the reaction. A catalyst does not affect the energy of the reactants or of the products, only that of the intermediate configurations. See also Figure 14.20.

profile for the same reaction when no catalyst is present in Figure 14.17. As shown by the Arrhenius equation (14-25), a decrease in activation energy causes an increase in reaction rate.

Figure 14.17 illustrates a very important feature of catalysis: the energies of the reactants and products are not changed when a catalyst is added to a reaction. Only the energy of the intermediate configurations along the reaction coordinate, including the energy of the transition state, are affected by a catalyst. Consequently a catalyst has no effect on the heat of reaction, nor on the value of the equilibrium constant of a reversible reaction. *A catalyst changes the rate at which a reaction reaches equilibrium, but does not change the position of equilibrium.*

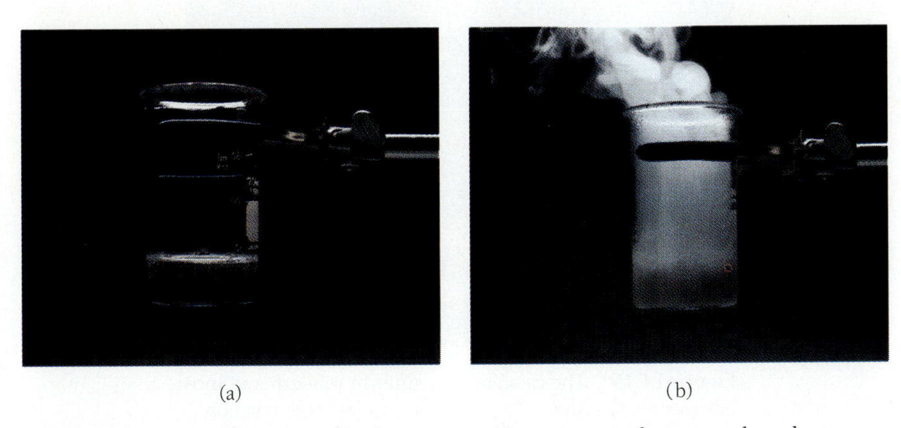

(a) (b)

Figure 14.18 Decomposition of hydrogen peroxide to water and oxygen takes place slowly in the absence of a catalyst. (a) Warm aqueous H_2O_2 (30% by mass) slowly releases bubbles of O_2. (b) After addition of a small amount of the catalyst $MnO_2(s)$, the evolution of oxygen is much more rapid. Some steam is generated in this exothermic process.

Types of Catalysis

It is convenient to distinguish between several types of catalysts, and several corresponding modes of action.

Heterogeneous Catalysis

Heterogeneous catalysts exert their effect at the interface between two phases. For example, the catalytic converter used in automobiles (Figure 14.19) consists of small beads of aluminum oxide impregnated with rhodium or with a noble metal such as platinum or palladium. When the hot exhaust gases pass over and through these beads, several different reactions take place at the bead surface. Among these reactions is (14-51).

$$2\ CO(g) + O_2(g) \longrightarrow 2\ CO_2(g) \quad (14\text{-}51)$$

Another example of heterogeneous catalysis is found in the Haber synthesis of ammonia, in which a mixture of hydrogen and nitrogen is compressed, heated, and exposed to a catalyst consisting primarily of finely divided iron.

$$N_2(g) + 3\ H_2(g) \xrightarrow[Fe]{} 2\ NH_3(g) \quad (14\text{-}52)$$

Homogeneous Catalysis

If a catalyst is present in the same phase (solid, liquid, or gas) as the reactants, it is called a **homogeneous catalyst.** This type of catalysis is often observed in aqueous solutions. Many reactions are acid-catalyzed, for example the hydrolysis of methyl formate.

$$HCO_2CH_3(aq) + H_2O \xrightarrow[HCl]{} HCO_2H(aq) + CH_3OH(aq) \quad (14\text{-}53)$$

The reaction between $Ce^{4+}(aq)$ and $Tl^{+}(aq)$ is catalyzed by $Mn^{2+}(aq)$ in a homogeneous process.

$$2\ Ce^{4+}(aq) + Tl^{+}(aq) \xrightarrow[Mn^{2+}]{} 2\ Ce^{3+}(aq) + Tl^{3+}(aq) \quad (14\text{-}54)$$

Figure 14.19 The catalytic converter in modern automobiles serves to reduce the amount of undesirable reaction products in the exhaust gases. The metal-impregnated beads catalyze the conversion of CO to CO_2 and of NO to N_2 and O_2. In addition, any unburned gasoline is converted to CO_2 and H_2O.

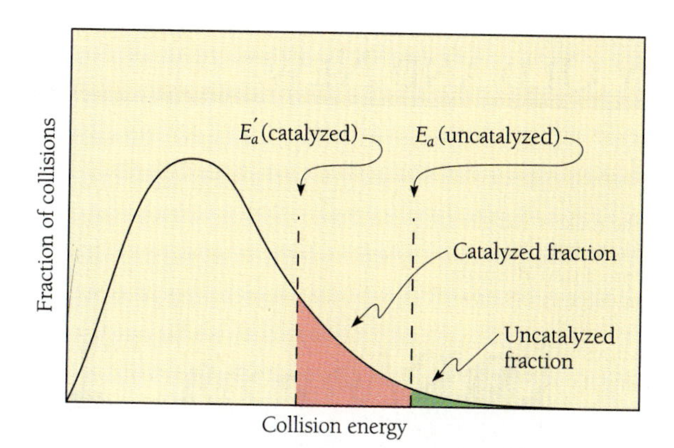

Figure 14.20 The curve shows the distribution of collision energies according to the kinetic molecular theory. An uncatalyzed reaction has some activation energy E_a, and a corresponding fraction of collisions having energy greater than E_a (green area). The addition of a catalyst decreases the activation energy to some lower value E_a', with the result that a greater fraction of collisions can lead to successful reaction (green area plus red area).

Homogeneous catalysis also takes place in the gaseous state. As we have seen, chlorine atoms catalyze the decomposition of ozone in the atmosphere.

$$O_3(g) + Cl(g) \longrightarrow O_2(g) + ClO(g) \quad (14\text{-}48)$$

$$ClO(g) + O_3(g) \longrightarrow 2\ O_2(g) + Cl(g) \quad (14\text{-}49)$$

Atomic chlorine is first consumed, and then regenerated in this two-step mechanism. Since there is no net consumption of the catalyst, the reaction is often written simply as

$$2\ O_3(g) \xrightarrow[Cl]{} 3\ O_2(g) \quad (14\text{-}55)$$

In both hererogeneous and homogeneous catalysis, the catalyzed reaction proceeds more rapidly than the uncatalyzed reaction because the catalyst provides an alternative reaction pathway having a lower activation energy. In terms of the collision theory of reaction rates, a lower activation energy means that a larger fraction of collisions has sufficient energy for successful reaction, so that the reaction proceeds more rapidly (Figure 14.20).

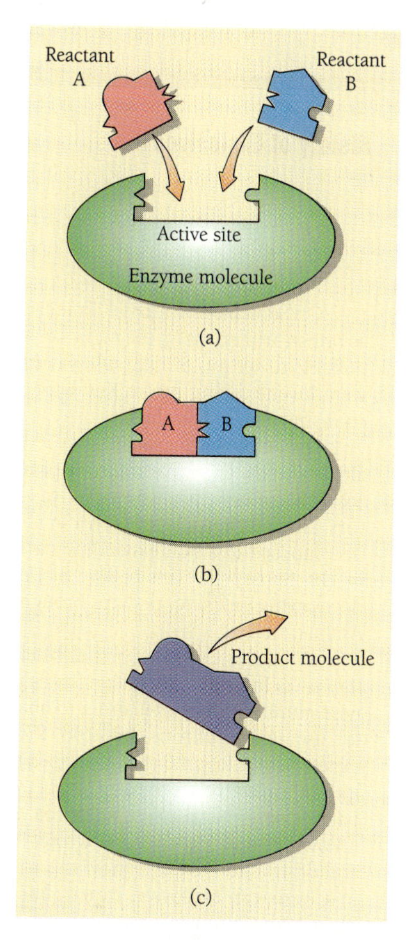

Figure 14.21 An enzyme is a large protein molecule having on its surface a particularly shaped region known as the "active site." Reactant molecules fit precisely into the active site, just as a key fits into a lock. The reaction is facilitated both by improving the geometrical factor and by decreasing the activation energy.

Biological Catalysis

Enzymes are large protein molecules that catalyze almost all of the reactions taking place in living organisms. These catalysts are very *efficient,* that is, they increase the reaction rate by a large factor. They are also highly *specific,* since they catalyze only one particular reaction (or at most a small group of similar reactions). The mechanism of enzyme catalysis is complex, but the "lock and key" analogy is a good model to remember. As shown in Figure 14.21, an enzyme molecule has a specific region, called the **active site,** at which the catalytic action takes place. The active site has a particular shape, and reactant molecules can fit into it in only one way—in the orientation most suitable for reaction. The fit is both precise and specific, in the same sense that a certain key fits precisely into one lock and no others. Most other reactant molecules do not fit, and hence most other reactions are not catalyzed. Enzymes lower the activation energy of reactions, in addition to optimizing the geometrical factor.

Not all catalysts are specific; literally thousands of reactions are catalyzed by acids.

CHEMISTRY OLD AND NEW

History of Explosives

"And the rockets' red glare,
The bombs bursting in air,
Gave proof thro' the night
that our flag was still there."

On September 14, 1814, Francis Scott Key wrote these lines as British artillery lit up the night sky at Fort McHenry, making it possible to see a U.S. flag, still waving. But chemical explosives were invented long before Key wrote the poem that was to become our national anthem.

No one knows exactly who invented the first explosives, although scholars have traced a primitive form of gunpowder (more accurately, black powder), called "huo yao," to 11th Century China. When ignited with a small flame, the powder rapidly decomposed to a hot, expanding gas—perfect for colorful fireworks. Some historians believe black powder might have been developed elsewhere, perhaps brought to China by European or Arabian traders.

Whatever its precise origin, black powder soon found military uses. In 1233, Mongolian forces reportedly captured a Chinese munitions factory in order to obtain supplies of gunpowder for use against Europe. In 1248, the English philosopher and experimenter Roger Bacon described an early black powder recipe that called for seven parts "saltpetre" (potassium nitrate), five parts "young hazel twigs" (charcoal), and five parts sulfur. By following this recipe, Bacon wrote, "thou wilt call up thunder and destruction, if thou know the art." Bacon probably was the first European explosives expert, but Berthold Schwartz, possibly a 14th Century German scholar, may have been the first to use black powder in munitions.

Cannons loaded with black powder were first used in Europe around 1325, and handguns arrived on the scene about 100 years later. Black powder also found peaceful uses in the mines of Hungary, Germany, and Sweden throughout the 1600s. By the 1700s, worldwide production of gunpowder was dominated by France. The renowned chemist Antoine Lavoisier was appointed chairman of the French gunpowder commission in 1775.

On the other side of the Atlantic, black powder was so important to early American settlers that landowners were ordered to build a compost shed for saltpeter production (1642). Mixed with limestone and water or urine, the compost was boiled, then cooled to produce saltpeter crystals. Full-scale U.S. development of high-quality gunpowder usually is credited to Eleuthère Irénée du Pont, who had apprenticed under Lavoisier in France. In 1802 the young du Pont built a powder mill on the Brandywine river near his home in Wilmington, Delaware. In 1804, its first year of commercial production, du Pont's mill manufactured about 11,000 lbs of black powder. The Du Pont powder company is often credited as the birthplace of the chemical industry in the United States.

New and more powerful explosives such as mercury fulminate were developed in the 1800s. Nitroglycerin was discovered in 1846 by the Italian chemist, Ascanio Sobrero. About the same time, German scientists Christian Friedrich Schonbein and Rudolf Christian Bottger reported the discovery of guncotton (nitrocellulose). In 1864, Swedish scientist Alfred Bernhard Nobel developed the first blasting cap—composed of mercury fulminate and ignited by a black powder fuse—as a safe method for detonating nitroglycerin. Then, in 1867, Nobel invented dynamite, a potent explosive that is safer to handle than nitroglycerin. This invention was soon commercialized, and the profits continue to fund the annual Nobel Prizes.

The military potential of explosives has been realized for a long time. This illustration from a 15th century manuscript shows cannons in use during the Hundred Years' War between England and France (1337–1453).

While U.S. production of munitions increased dramatically in the 19th century to support various military conflicts, including the Civil War, explosives also were needed for digging the Erie Canal, building railroads, and blasting in the mining of coal and metal ores. The use of explosives for both peaceful purposes and for weaponry continues to this day.

Discussion Questions

1. Gunpowder is a mixture of three substances: charcoal, potassium nitrate, and sulfur. How do you think it might have been discovered a thousand years ago?
2. Which use of explosives, in weapons or in more peaceful applications, has had the greater impact on human society?

▸ SUMMARY EXERCISE

Dinitrogen pentoxide is a volatile white solid that is stable below 0 °C. In the gaseous state at higher temperatures, it decomposes according to the overall reaction $N_2O_5 \rightarrow 2\ NO_2 + \frac{1}{2}\ O_2$. The following mechanism has been proposed,

$$N_2O_5 + N_2O_5 \longrightarrow N_2O_4 + N_2O_4 + O_2 \qquad \text{(slow)}$$

$$N_2O_4 \longrightarrow NO_2 + NO_2 \qquad \text{(fast)}$$

and the following data were collected in a series of experiments at 25 °C.

Run	$[N_2O_5]$/M	Initial Rate of $[N_2O_5]$ Decrease/mol L^{-1} s^{-1}
1	0.010	3.44×10^{-7}
2	0.030	1.03×10^{-6}
3	0.060	2.06×10^{-6}

a. Determine the order of the reaction and write the rate expression.

b. Give the numerical value and units of the rate constant.

c. Calculate the half-life of the reaction at 25 °C.

d. What fraction of the initial concentration of N_2O_5 remains after overnight storage (16 hours) at 25 °C?

e. The activation energy for the reaction is 103 kJ mol^{-1}. What is the half-life of the reaction at 125 °C?

f. Does the proposed mechanism lead to the correct stoichiometry of the overall reaction?

g. Does the proposed mechanism lead to the correct rate law?

h. The reaction $N_2O_5 \rightarrow 2\ NO_2 + \frac{1}{2}\ O_2$ reaches equilibrium rather than proceeding to completion. At 298 K, an equilibrium mixture was found to consist of the following partial pressures of the three substances: N_2O_5, 0.0178 atm; NO_2, 0.324 atm; O_2, 0.00801 atm. Calculate (i) K_{eq}; (ii) $\Delta G°$; and, using any needed values from Appendix G, (iii) $\Delta G^\circ_{f,298}[N_2O_5(g)]$.

Answers **(a)** First order, $r = k[N_2O_5]$; **(b)** $3.44 \times 10^{-5}\ s^{-1}$; **(c)** 2.01×10^4 s; **(d)** 0.138 or 13.8%; **(e)** 0.586 s; **(f)** yes, the overall reaction of the proposed mechanism is $2\ N_2O_5 \rightarrow 4\ NO_2 + O_2$, or $N_2O_5 \rightarrow 2\ NO_2 + \frac{1}{2}\ O_2$; **(g)** no, it leads to a second-order rate law; **[h(i)]** $K_{eq} = 0.528\ atm^{3/2}$; **[h(ii)]** +1.58 kJ (per mole N_2O_5); **[h(iii)]** 101.04 kJ mol^{-1}.

SUMMARY AND KEYWORDS

The study of the rates and mechanisms of chemical reactions is called **chemical kinetics.** The **collision theory** states that reactions occur only when the reacting species collide with the proper orientation and with sufficient energy to overcome the

activation energy (E_a). Activation energy is different for each chemical reaction, but it is not affected by temperature. The effect of concentration on reaction rate is summarized by a **rate law,** an expression of the form $r = k[A]^x[B]^y$ . . . The overall **order** of the reaction is the sum of the exponents. The order with respect to reactant A is x, and y is the order with respect to the reactant B, and so on. The **rate constant** k, always a positive number, is different for each reaction, and generally increases as the temperature is raised. The extent of the increase is controlled by the **Arrhenius equation,** $k = Ae^{-E_a/RT}$. A **reaction rate** is expressed in units of mol L^{-1} s^{-1}, while the units of the rate constant are different for each order of reaction. Reaction order can be determined by the method of **initial rates:** the rate at the start of a reaction is measured as a function of the concentration of one reactant, while keeping all other concentrations constant.

The decreasing reactant concentration in a *first-order* reaction is found from the *integrated rate law* $\ln(C/C_0) = -kt$. The **half-life** or **half-time** of a reaction is the time required for one-half of the amount originally present to react. For first-order reactions, the half-life is related to the rate constant by $t_{1/2} = 0.693/k$, but for all other reaction orders the half-life depends on the concentration. Second-order reactions having a single reactant, and those having two reactants with equal concentrations, follow the integrated rate law $1/C = 1/C_0 + kt$.

A **reaction profile** or **energy profile** is a plot of the *potential energy* of the reaction system vs. the **reaction coordinate.** The reaction coordinate represents changes in the geometrical configuration of the participating species during the course of the reaction. The geometrical configuration of atoms at the maximum energy of the reaction profile is called the **transition state.** A reaction profile provides information about the activation energy and ΔE_{rxn}, the *energy of reaction.* These are related by $\Delta E_{rxn} = E_{a,f} - E_{a,r}$, where $E_{a,f}$ is the activation energy of the forward reaction, and $E_{a,r}$ is the activation energy of the reverse reaction.

A reaction **mechanism** is the set of **elementary steps** that together make up the **overall** or **complex** reaction. Each elementary step describes a single molecular event. In contrast to a complex reaction, the rate law of an elementary step is determined by its stoichiometry. Elementary steps may be **unimolecular, bimolecular,** or **trimolecular.** The rate of an overall reaction is the same as the rate of its slowest or **rate-determining step.** The rate-determining step may be preceded by a reversible, equilibrium step. Mechanisms often indicate the participation of **intermediates** (sometimes called **reactive intermediates**) that do not appear in the equation for the overall reaction. Intermediates are produced, then later consumed, during the course of the overall reaction.

A chain reaction involves several reactive intermediates, called **chain carriers.** The mechanism of a chain reaction consists of **initiation, propagation,** and **termination steps.** The products of a chain reaction are formed in the propagation steps. Chain reactions are often of nonintegral order.

The rate of a reaction can be increased by the addition of a **catalyst,** which is not itself consumed in the reaction. The phenomenon is called **catalysis.** Compared to the uncatalyzed reaction, the *catalyzed* reaction has a lower activation energy and is therefore faster in both the forward and reverse directions. A catalyst does not affect the value of the equilibrium constant. Catalysis that occurs when the catalyst, the reactants, and the products are all in the same phase is called **homogeneous catalysis.** A common example is acid catalysis of a reaction in aqueous solution. If catalysis takes place when the catalyst is in a different phase from the reactants and products, it is called **heterogeneous catalysis.** Platinum and palladium, as well as many other transition metals, are good catalysts. Chemical reactions in living organisms are usually catalyzed by large protein molecules called **enzymes.** Enzymes facilitate reactions by arranging the correct orientation of reactants as well as by lowering the activation energy. Enzyme action takes place at a specific region on the enzyme molecule known as the **active site.**

PROBLEMS

General

1. List the three postulates of the collision theory of reaction rates.
2. What is activation energy, and what is its relationship to the thermochemical energy of reaction?
3. What specific factors influence the rate of a chemical reaction?
4. What is meant by the term "rate law"?
5. What is the "order" of a chemical reaction? Is it possible for a reaction to be of $1\frac{1}{2}$ order? If not, why not?
6. Distinguish between the rate of a reaction and the rate constant of the reaction.
7. What are the units of a reaction rate? What are the units of a rate constant? Do the units of either rate or rate constant depend on reaction order? If so, give examples.
8. Define the term "half-life." For which order reaction is the concept most useful? Why is it less useful for other orders?
9. What is the numerical relationship between the rate constant of a first-order reaction and its half-life?
10. Describe what is meant by the term "reaction profile," and draw sketches appropriate to both exothermic and endothermic reactions.
11. Discuss the interpretation of the term "reaction coordinate," being careful to distinguish it from the term "extent of reaction" used in equilibrium problems.
12. What is a catalyst? What is the difference between a homogeneous catalyst and a heterogeneous catalyst?
13. Using a simple model, describe the catalysis of a biochemical reaction by an enzyme.
14. Distinguish between complex reactions and elementary steps.
15. What is the specific meaning of the term "reaction mechanism"?
16. Describe the significance of the rate-determining step in reaction mechanisms.
17. Describe the significance of intermediates in reaction mechanisms.
18. What is a chain reaction? What is a chain carrier?

Rate Laws, Order, and Initial Rates

19. Identify the overall order and the order with respect to each reactant in the following reactions.
    a. $C_2H_5Br + OH^- \longrightarrow C_2H_5OH + Br^-$; $r = k[OH^-][C_2H_5Br]$
    b. $2\ O_3 \longrightarrow 3\ O_2$; $r = k[O_3]$
    c. $ClCO_2C_2H_5 \longrightarrow CO_2 + ClC_2H_5$; $r = k[ClCO_2C_2H_5]$
    d. $HI \longrightarrow \frac{1}{2}\ H_2 + \frac{1}{2}\ I_2$; $r = k[HI]^2$
20. Identify the overall order and the order with respect to each reactant in the following reactions.
    a. $NOCl \longrightarrow NO + \frac{1}{2}\ Cl_2$; $r = k[NOCl]^2$
    b. $2\ N_2O_5 \longrightarrow 2\ N_2O_4 + O_2$: $r = k[N_2O_5]$
    c. $C_6H_5CHO + HCN \longrightarrow C_6H_5CH(OH)CN$; $r = k[C_6H_5CHO][HCN]$
    d. $2\ NO_2 \longrightarrow 2\ NO + O_2$: $r = k[NO_2]^2$
21. Write the rate of appearance of each product and the rate of disappearance of each reactant in Problem 19. For example, in (a) the rates are:

$$C_2H_5Br: \quad \Delta[C_2H_5Br]/\Delta t = -k[C_2H_5Br][OH^-]$$
$$OH^-: \quad \Delta[OH^-]/\Delta t = -k[C_2H_5Br][OH^-]$$
$$C_2H_5OH: \quad \Delta[C_2H_5OH]/\Delta t = +k[C_2H_5Br][OH^-]$$
$$Br^-: \quad \Delta[Br^-]/\Delta t = +k[C_2H_5Br][OH^-]$$

22. For each reactant and product in Problem 20, write the rate of its appearance or disappearance.
23. Assume the rate law for a reaction $A \rightarrow B$ is $r = k[A]^y$. Suppose the concentration of A were doubled. What would be the effect on the rate if $y = 0$? If $y = 1$, 2, or 3?
24. Assume the rate law for a reaction $A \rightarrow B$ is $r = k[A]^y$. Suppose the concentration of A were tripled. What would be the effect on the rate if $y = 0$? If $y = 1$, 2, or 3?

25. For the reaction $2\ NO + 2\ H_2 \rightarrow N_2 + 2\ H_2O$, the rate law is $r = k[H_2][NO]^2$. What would be the effect on the rate if
    a. $[H_2]$ were doubled and [NO] were unchanged?
    b. $[H_2]$ and [NO] were both doubled?
26. For the reaction $2\ NO + 2\ H_2 \rightarrow N_2 + 2\ H_2O$, the rate law is $r = k[H_2][NO]^2$. What would be the effect on the rate if
    a. [NO] were doubled and $[H_2]$ were unchanged?
    b. $[H_2]$ were doubled and [NO] were tripled?
27. For the second-order reaction $2\ NO_2 \rightarrow 2\ NO + O_2$, $k = 1.8 \times 10^{-8}\ L\ mol^{-1}\ s^{-1}$ at 100 °C. What is the rate when $[NO_2] = 0.15$ M?
28. The decomposition of bromoethane, $CH_3CH_2Br \rightarrow C_2H_4 + HBr$, is first order and has a rate constant of $1.1 \times 10^{-7}\ s^{-1}$ at 300 °C. What is the rate when the concentration of the decomposing substance is 0.00500 M?
29. At sufficiently high temperatures, nitric oxide reacts with hydrogen to form water according to $H_2 + NO \rightarrow H_2O + \frac{1}{2} N_2$.
    a. If doubling the concentration of NO causes a fourfold increase in the rate, what is the order of the reaction with respect to NO?
    b. If doubling the concentration of H_2 causes a twofold increase in the rate, what is the order with respect to H_2?
    c. What is the overall order of the reaction?
    d. Write the rate law.
30. Iodine monochloride reacts with hydrogen according to $ICl + \frac{1}{2} H_2 \rightarrow HCl + \frac{1}{2} I_2$.
    a. If doubling the concentration of ICl causes a fourfold increase in the rate, what is the order of the reaction with respect to ICl?
    b. If doubling the concentration of H_2 causes a twofold increase in the rate, what is the order with respect to H_2?
    c. What is the overall order of the reaction?
    d. Write the rate law.
    e. What are the units of the rate constant?
31. The following data were collected during trial runs of a reaction having stoichiometry $A + B + C \rightarrow D$.

Run	[A]/M	[B]/M	[C]/M	Initial Rate/mol $L^{-1}\ s^{-1}$
1	0.05	0.05	0.05	0.80×10^{-3}
2	0.10	0.05	0.05	1.6×10^{-3}
3	0.05	0.10	0.05	3.2×10^{-3}
4	0.05	0.05	0.10	0.80×10^{-3}

    a. Determine the order of the reaction with respect to each reactant, and the overall order.
    b. Write the rate expression and give the numerical value and units of the rate constant.
32. The following data were collected at 400 °C during trial runs of the reaction $NO_2 \rightarrow NO + \frac{1}{2} O_2$

Run	$[NO_2]$/M	Initial Rate of $[NO_2]$ Decrease/mol $L^{-1}\ s^{-1}$
1	0.010	3.1×10^{-4}
2	0.020	1.2×10^{-3}
3	0.030	2.8×10^{-3}

    a. Determine the order of the reaction and write the rate expression.
    b. Give the numerical value and units of the rate constant.
33. The following data were collected during trial runs of a reaction having stoichiometry $A + B \rightarrow C$.

Run	[A]/M	[B]/M	Initial Rate/mol $L^{-1}\ s^{-1}$
1	0.05	0.05	3×10^{-5}
2	0.10	0.05	6×10^{-5}
3	0.10	0.10	1.2×10^{-4}

    a. Determine the order of the reaction with respect to each reactant, and the overall order.
    b. Write the rate expression and give the numerical value and units of the rate constant.

*34. In the presence of nitrogen at 1000 K, bromine atoms recombine according to $Br + Br \rightarrow Br_2$. The following data were collected during a series of trial runs.

Run	[Br]/M	$[N_2]$/M	Initial Rate of Br_2 Appearance/ mol $L^{-1}\ s^{-1}$
1	6.4×10^{-4}	8.0×10^{-2}	1.0×10^3
2	3.2×10^{-4}	8.0×10^{-2}	2.5×10^2
3	6.4×10^{-4}	1.6×10^{-1}	2.0×10^3

    a. Determine the order of the reaction and write the rate expression.
    b. Give the numerical value and units of the rate constant.

First-Order Reactions

35. The half-life for the first-order decomposition of cyclobutane, $C_4H_8 \rightarrow 2\ C_2H_4$, is 89 s at 498 °C. What is the rate constant at this temperature?

36. The half-life for the first-order isomerization reaction cyclopropane $\rightarrow$ propene is 2.6 s at 600 °C. What is the rate constant at this temperature?

37. The rate constant for the first-order decomposition of chloroethane, $CH_3CH_2Cl \rightarrow C_2H_2 + HCl$, is $0.13\ s^{-1}$ at 100 °C. What is its half-life?

38. In the discussion of geometrical isomerism (page 408), it was noted that rotation about double bonds does not occur under normal conditions. High temperatures are not "normal," however, and at 400 °C the isomerization of *cis*-2-butene to *trans*-2-butene takes place with a rate constant of 2.2×10^{-7} s^{-1}. What is the half-life of this first-order reaction?

39. If the initial concentration of a sample of cyclobutane is 0.500 M, how much remains after an elapsed time of 5.0 minutes? The rate constant for the first-order decomposition is $0.0078\ s^{-1}$.

40. The rate constant for the first-order reaction $N_2O_5 \rightarrow 2\ NO_2 + \frac{1}{2}\ O_2$ is $1.0 \times 10^{-5}\ s^{-1}$ at 45 °C, and the initial concentration of N_2O_5 is 0.00500 M.
    a. How long will it take for the concentration to decrease to 0.00100 M?
    b. How much longer will it take for a further decrease to 0.000900 M?

41. If the initial concentration of a N_2O_5 sample is 0.337 M, how much remains after one day of decomposition at 45 °C? (Use the data from Problem 40.)

42. Sulfuryl chloride decomposes to sulfur dioxide and chlorine according to $SO_2Cl_2 \rightarrow SO_2 + Cl_2$, in a first-order reaction whose rate constant is $2.10 \times 10^{-5}\ s^{-1}$ at 600 K. If the initial concentration of a sample of SO_2Cl_2 is 0.0100 M, what would be its concentration after storage at 600 K for one week?

43. The half-life of the first-order decomposition of sulfuryl chloride is 3.30×10^4 s at 600 K. If the initial concentration is 0.0250 M, what is the concentration after one hour at this temperature?

44. The half-life for the decomposition of cyclobutane is 89.0 s at 771 K.
    a. How much time is required for an initial concentration of 0.0200 M to decrease to 0.00500 M?
    b. How much additional time is required for the concentration to drop further to 0.00200 M?

45. If the concentration of a substance decreases from 0.0125 M to 0.00993 M in 176 s, and the process is known to be first order, what is the value of the rate constant?

46. It is found that 47.0 minutes is required for the concentration of a substance to decrease from 0.75 M to 0.25 M. What is the rate constant for this first-order decomposition?

47. Emission of radioactivity and decay of unstable nuclei follow first-order kinetics. These nuclear processes are not affected by temperature or the state of chemical combination. For example, radium chloride, whether pure or in solution, decays at the same rate as radium metal. For this reason it is customary to write the first order rate law not in terms of concentration, but in terms of the number of moles of radioactive nuclei. The rate law for the decay of Ra-226 nuclei is $-\Delta n/\Delta t = k \cdot n$, where n is the number of moles and $k = 1.36 \times 10^{-11}\ s^{-1}$.
    a. What is the half-life?
    b. How much of a sample, originally containing 1.00 mmol, remains after 1000 years?

48. The rate constant for the decay of $^{235}_{92}U$ is 3.1×10^{-17} s.
    a. What is the half-life?
    b. What fraction of the original number of $^{235}_{92}U$ nuclei remains after two billion (2.0×10^9) years?

49. How long a time must elapse before the radioactivity of a sample of $^{239}_{94}Pu$ decreases to 5.0% of its initial value? This is another way of wording the question, "How much time must elapse for the number of moles remaining to be equal to 5.0% of the original amount?" The half-life of this isotope of plutonium is 24,100 years.

50. A prominent component of the radioactive fallout produced by the weapons testing during the 1950s is $^{90}_{38}Sr$, whose half-life is 25 years. What fraction of this material remains in the environment today, after about 40 years?

Reaction Profiles

51. Sketch the reaction profile for a reaction that is exothermic by 125 kJ mol^{-1} and has an activation energy of 75 kJ mol^{-1}.

52. Sketch the reaction profile for a reaction that is endothermic by 75 kJ mol^{-1} and has an activation energy of 125 kJ mol^{-1}.

53. If an exothermic reaction has $\Delta E_{rxn} = -250$ kJ mol^{-1} and $E_a = 100$ kJ mol^{-1}, what is E_a for the reverse reaction?

54. What is ΔE_{rxn} for a reaction with a forward activation energy of 185 kJ mol^{-1} and a reverse activation energy of 35 kJ mol^{-1}?

55. Sketch the reaction profile of a reaction in which $E_{a,f} = E_{a,r}$. Is the reaction exothermic or endothermic?

*56. Some reactions, most notably atom recombination reactions such as $2\ Br \rightarrow Br_2$, occur with an activation energy that is essentially zero. If the bond energy of Br_2 is 193 kJ mol^{-1}, sketch the reaction profile of the dissociation reaction $Br_2 \rightarrow 2\ Br$.

57. What is the reverse activation energy of a reaction having $\Delta E_{rxn} = -108$ kJ mol^{-1} and $E_{a,f} = 15$ kJ mol^{-1}?

58. What is the forward activation energy of a reaction having $\Delta E_{rxn} = +108$ kJ mol^{-1} and $E_{a,r} = 15$ kJ mol^{-1}?

Arrhenius Equation and Temperature Dependence

*59. For the reaction $2\ N_2O_5 \rightarrow 2\ N_2O_4 + O_2$, the rate constant is $2.4 \times 10^{-4}\ s^{-1}$ at 293 K and $9.2 \times 10^{-4}\ s^{-1}$ at 303 K.
a. Calculate the activation energy.
b. Calculate the frequency factor.

*60. The rate constant for the second-order atom transfer reaction $NO + O_3 \rightarrow NO_2 + O_2$ is $5.16 \times 10^4\ L\ mol^{-1}\ s^{-1}$ at 400 K and 2.28×10^6 at 500 K.
a. What is the activation energy?
b. What is the value of the pre-exponential factor (A)?

61. The activation energy of the first-order decomposition of cyclobutane is 260 kJ mol^{-1}, and the rate constant is $6.08 \times 10^{-8}\ s^{-1}$ at 600 K. What is the value of the rate constant at 650 K?

*62. The activation energy of the second-order reaction $CH_3I + HI \rightarrow CH_4 + I_2$ is 140 kJ mol^{-1}, and at 200 °C the rate constant is 0.0132 L $mol^{-1}\ s^{-1}$. Calculate the value of the rate constant at 400 °C.

*63. At what temperature does the decomposition of cyclobutane proceed with a rate constant of $6.08 \times 10^{-6}\ s^{-1}$? Use the data from Problem 61.

*64. At what temperature will the rate constant of the reaction $CH_3I + HI \rightarrow CH_4 + I_2$ be 0.0500 L $mol^{-1}\ s^{-1}$? Use the data given in Problem 62.

*65. For many reactions, the rate constant is approximately doubled if the temperature is increased by 10 °C. This is a good rule of thumb, but it applies only near room temperature and only to those reactions whose activation energy is moderate rather than high or low. What is the activation energy of a reaction whose rate constant at 35 °C is exactly twice that at 25 °C?

*66. Suppose that, under otherwise identical conditions, a reaction is three times faster at 30 °C than at 20 °C. What is its activation energy?

Reaction Mechanisms

67. Write the rate law for each of the following elementary reactions:
a. $H + O_2 \longrightarrow OH + O$
b. $O + O_3 \longrightarrow 2\ O_2$
c. $2\ Cl + Cl_2 \longrightarrow 2\ Cl_2$

68. Write the rate law for each of the following elementary reactions:
a. $H + Br_2 \longrightarrow HBr + Br$
b. $NO + O_3 \longrightarrow NO_2 + O_2$
c. $CH_3NC \longrightarrow CH_3CN$

69. What is the overall reaction corresponding to the following mechanism?

$$N_2O_5 \longrightarrow NO_2 + NO_3$$
$$NO_3 + N_2O_5 \longrightarrow 3\ NO_2 + O_2$$

70. What is the overall reaction corresponding to the following mechanism?

$$Cl_2 \longrightarrow 2\ Cl$$
$$Cl + CO \longrightarrow COCl$$
$$COCl + Cl_2 \longrightarrow COCl_2 + Cl$$
$$2\ Cl \longrightarrow Cl_2$$

71. The reaction $NO_2(g) + CO(g) \rightarrow CO_2(g) + NO(g)$ proceeds at low temperatures via the mechanism
a.

$$NO_2 + NO_2 \longrightarrow NO + NO_3 \qquad \text{(slow)}$$
$$NO_3 + CO \longrightarrow NO_2 + CO_2 \qquad \text{(fast)}$$

b. At high temperatures, on the other hand, the mechanism is the direct transfer of an oxygen atom from NO_2 to CO:

$$NO_2 + CO \longrightarrow NO + CO_2$$

Write the rate laws corresponding to these two mechanisms.

72. Chlorine and nitric oxide react according to the overall equation $Cl_2 + 2\ NO \rightarrow 2\ NOCl$. The experimental rate law is $r = k[Cl_2][NO]$. Which of the following possible mechanisms fits the rate law? (Both may fit.)
a. $2\ NO + Cl_2 \longrightarrow 2\ NOCl$
b.

$$NO + Cl_2 \longrightarrow NOCl_2 \qquad \text{(slow)}$$
$$NOCl_2 + NO \longrightarrow 2\ NOCl \qquad \text{(fast)}$$

*73. The rate law for the reaction $2\ NO + O_2 \rightarrow 2\ NO_2$ is $r = k[NO]^2[O_2]$. Write a plausible mechanism for the reaction.

*74. Nitrogen dioxide decomposes according to the overall equation $2\ NO_2 \rightarrow 2\ NO + O_2$. Write a two-step mechanism that satisfies the experimentally determined rate law $r = k[NO_2]^2$.

Other Problems

75. A chemist investigating the first order decomposition of N_2O_5, for which the rate constant is $1.0 \times 10^{-5}\ s^{-1}$ at 45 °C, forgot to record the initial concentration in one experiment. However, the N_2O_5 concentration was 0.025 M after an elapsed time of 5.00×10^4 s. What was the original concentration?

76. Acetaldehyde decomposes according to $CH_3CHO \rightarrow CO + CH_4$. The reaction is second order, and the value of the rate constant is 0.55 L $mol^{-1}\ s^{-1}$ at 500 °C. What concentration of acetaldehyde remains after 60 seconds, if the initial concentration is 0.010 M?

77. The second-order decomposition, $NO_2 \rightarrow NO + \frac{1}{2} O_2$, has a rate constant equal to 3.1 L mol^{-1} s^{-1} at 400 °C.
   a. How long will it take for the NO_2 concentration to be reduced to ½ of its initial value of 0.030 M?
   b. Same question as (a), but use an initial concentration of 0.10 M.

*78. Show that the order of a reaction with respect to reactant A is given by the relationship

$$\text{order} = \frac{\ln(\text{Rate}_1/\text{Rate}_2)}{\ln\{[A]_1/[A]_2\}},$$

where the subscripts refer to two measurements of the initial rate when the concentrations of all reactants other than A were unchanged.

79. For a certain reaction, the reverse activation energy is greater than the forward activation energy. Is the reaction exothermic or endothermic?

80. Is a reaction endothermic or exothermic if the forward and reverse activation energies are the same?

81. What is the forward activation energy of a reaction having $\Delta E_{rxn} = -108$ kJ mol^{-1} and $E_{a,r} = 15$ kJ mol^{-1}?

82. What is the rate constant at 150 °C of a reaction whose frequency factor and activation energy are 9.3×10^{11} L mol^{-1} s^{-1} and 113 kJ mol^{-1}, respectively?

*83. Calculate the frequency factor for the decomposition of cyclobutane. Use the data from Problem 61.

84. Calculate the frequency factor of the reaction $CH_3I + HI \rightarrow CH_4 + I_2$. Use the data from Problem 62.

85. The exchange reaction between methyl bromide and chlorine has a rate constant of 5.9×10^{-3} L mol^{-1} s^{-1} at 25 °C. If the activation energy is 66.1 kJ mol^{-1}, at what temperature will the rate constant be 5.0×10^{-2} L mol^{-1} s^{-1}?

*86. The activation energy of the second-order gas-phase reaction between hydrogen iodide and ethyl iodide, $2\,HI + 2\,C_2H_5I \rightarrow 2\,C_2H_6 + I_2$, is 125 kJ mol^{-1}, and the rate constant is 4.91×10^{-8} L mol^{-1} s^{-1} at 25 °C. What is the value of the rate constant at 350 °C?

*87. By what factor does the rate of a reaction increase between 0 °C and 100 °C if the activation energy is 100 kJ mol^{-1}?

88. If the activation energy of a reaction is 100 kJ mol^{-1}, at what temperature will its rate constant be 75 times larger than the room temperature (25 °C) value?

89. What is the activation energy of a reaction whose rate constant increases fivefold when the temperature is increased from 50 °C to 100 °C? Can the frequency factor be determined from these data?

90. For a certain reaction, the rate constant decreases to one tenth of its original value when the temperature drops from 100 °C to 65 °C. What is its activation energy?

91. Sketch the reaction profile for a reaction having $E_a = 50$ kJ mol^{-1}, and $\Delta E_{rxn} = -75$ kJ mol^{-1}. Suppose that in the presence of a catalyst the activation energy is reduced to 25 kJ mol^{-1}. Sketch the reaction profile of the catalyzed reaction on the same sketch you drew for the uncatalyzed reaction.

*92. The following chain mechanism has been suggested for the photochlorination of methane.

(1) $Cl_2 \xrightarrow{\text{light}} Cl + Cl$

(2) $CH_4 + Cl \longrightarrow CH_3 + HCl$

(3) $CH_3 + Cl_2 \longrightarrow CH_3Cl + Cl$

(4) $CH_3 + Cl \longrightarrow CH_3Cl$

   a. Write the overall reaction.
   b. Identify the chain carrier(s).
   c. Identify each step as initiation, propagation, or termination.

*93. The mechanism of the acid-catalyzed hydrolysis of acetonitrile is

$$CH_3CN(aq) + H^+ \rightleftharpoons CH_3CNH^+ \quad \text{(fast)}$$

$$CH_3CNH^+ \longrightarrow \text{product} \quad \text{(slow)}$$

What is the rate law corresponding to this mechanism? (*Hint:* The mechanism is similar to that given on page 627 for the $H_2 + I_2$ reaction.)

*94. The rate law for the acid-catalyzed hydrolysis of acetonitrile (preceding problem) is $r = k[CH_3CN][H^+]$. What is the relationship between this rate constant, the equilibrium constant of the reaction $CH_3CN(aq) + H^+ \rightleftharpoons CH_3CNH^+$, and the rate constant (k_2) for the elementary step $CH_3CNH^+ \rightarrow$ product?

*95. The acid-catalyzed dehydration of ethanol,

$$CH_3CH_2OH \xrightarrow{H^+} CH_2{=}CH_2 + H_2O$$

follows the rate law $r = k[CH_3CH_2OH][H^+]$.
   a. Write a plausible mechanism for this reaction.
   b. Write the rate constant in terms of an equilibrium constant and an elementary step rate constant.

CHAPTER 15

ACIDS AND BASES

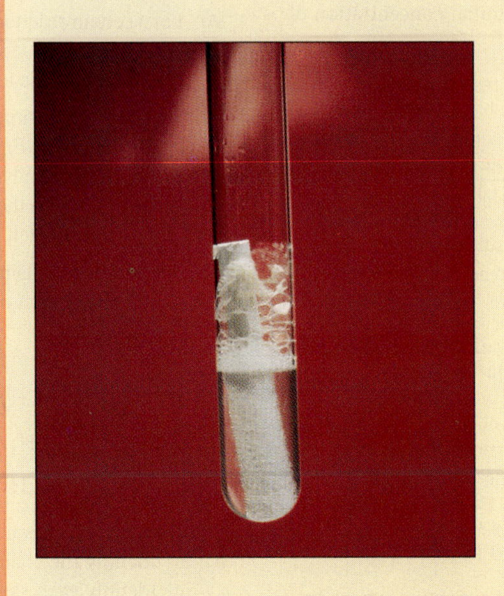

Hydrochloric acid reacts with blackboard chalk, a powdered form of calcium carbonate, to produce gaseous carbon dioxide and the salt calcium chloride. Calcium carbonate is the chief constituent of the minerals chalk, limestone, and marble; the ability to react with certain minerals is one of the well-known properties of acids.

15.1 INTRODUCTION

Interactions between compounds we call *acids* and their reaction partners, the *bases,* are central to the field of chemistry. The concepts presented in this chapter have broad applicability in all branches of chemistry, even in areas that, at first glance, seem to have little to do with acids or bases.

Acids were recognized as a distinct class of chemical substance a long time ago, much earlier than the emergence of chemistry as a science (Figure 15.1). The Egyptians were familiar with the properties of acetic acid (vinegar) and some of its uses other than as a natural foodstuff. Hundreds of years ago, in the days of the alchemists and in the early days of modern chemistry, substances were classified as acids if they exhibited certain properties or behavior. Some of these properties are sour taste, reaction with limestone to produce bubbles of gas, induction of a color change (from blue to red) in litmus, a vegetable dye, and ability to dissolve certain metals (Figure 15.2). If a substance could do all or most of these things, it was regarded as an acid. It was also recognized very early that there is another class of substances, having the power to reduce or entirely remove all of the properties of an acid. Such substances are called bases, and they have their own set of defining properties. If a substance has a bitter taste, a slippery or soapy feel, and turns litmus dye from red to blue, it is classified as a base. When a base reacts completely with an acid, the chemical properties of the base as well as those of the acid are lost (Figure 15.3).

The properties and behavior of acids and bases have been explained by three theories, developed by Arrhenius, Brønsted and Lowry, and Lewis. Each theory builds on a simple definition of acids and bases, and each addresses a different aspect of their properties and behavior.

Figure 15.1 The acidic properties of wine were well known in ancient times. These wine jars were recovered intact from the wreckage of a Roman cargo vessel that foundered in the Mediterranean centuries ago.

15.2 ARRHENIUS THEORY OF ACIDS AND BASES

In the late 19th century, the Swedish chemist Svante Arrhenius was beginning his study of the properties of electrolytes, a field of chemistry that would occupy much of his professional career. His research showed that acids and bases are electrolytes, and Arrhenius believed that their behavior could be understood in terms of the *dissociation* of electrolytes.

A brief note about Arrhenius appears on page 540 in the section on electrolytes. The Arrhenius equation (Chapter 14) describes the effect of temperature on reaction rates.

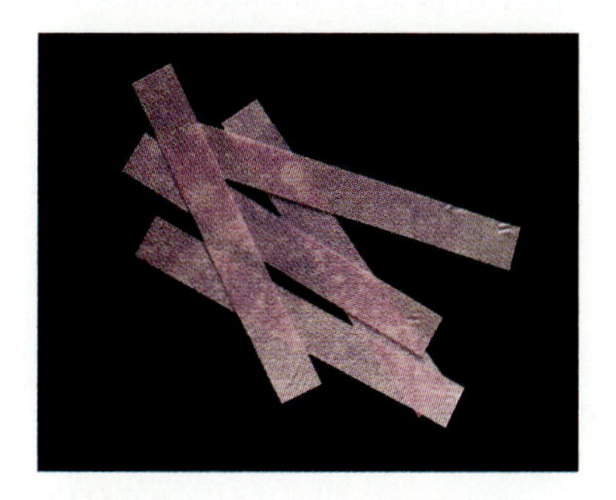

Figure 15.2 An easily noticeable property of acids is their effect on certain dyes. These paper strips, dyed blue with litmus, show the color changes resulting from being splashed with vinegar. Not all dyes respond in this way, and most dyes used in modern fabrics are resistant to such acid-induced color changes.

(a)

(b)

Figure 15.3 Acids and bases are familiar items in the home. (a) For example, vinegar (acetic acid) and vitamin C (ascorbic acid) are acids, as are most fruits and drinks. (b) Most cleaning products, such as ammonia and oven cleaner (sodium hydroxide), are bases.

The Role of H^+ and OH^-

Arrhenius proposed that the properties of acids are due to the presence of dissolved hydrogen ions, $H^+(aq)$. All solutions containing hydrogen ions, from any source, exhibit acid properties. An **acid** is any compound that, when dissolved in water, releases a hydrogen ion. For example, an acid such as hydrochloric acid *dissociates* into ions when it dissolves.

$$HCl(g) \longrightarrow H^+(aq) + Cl^-(aq) \qquad (15\text{-}1)$$

A **base**, on the other hand, is any substance that, when dissolved in water, releases hydroxide ions (OH^-). For example, when the base sodium hydroxide dissolves in water, it dissociates into sodium and hydroxide ions.

$$NaOH(s) \longrightarrow Na^+(aq) + OH^-(aq) \qquad (15\text{-}2)$$

In the Arrhenius view, the chemical properties of bases are due to the presence of OH^- in solution. Therefore, these properties are common to all solutions containing OH^-.

In the process of **neutralization,** an acid and a base react and mutually annihilate one another's properties. According to Arrhenius, the products of a neutralization are a salt and water.

$$HCl(g) + KOH(s) \longrightarrow KCl(aq) + H_2O(\ell) \qquad (15\text{-}3)$$

Most neutralizations take place in aqueous solutions, and the process is better represented by Equation (15-4).

$$K^+(aq) + OH^-(aq) + H^+(aq) + Cl^-(aq) \longrightarrow K^+(aq) + Cl^-(aq) + H_2O(\ell) \qquad (15\text{-}4)$$

K^+ and Cl^- appear on both sides of this equation, reflecting the chemical fact that they do not participate in, and are unchanged by, the reaction. Ions that are present but do not participate in a reaction are called **spectator ions.** A clearer picture of the chemical process is obtained by cancelling out the spectator ions, leaving a **net ionic equation** (15-5) that shows only the participants in the reaction.

$$H^+(aq) + OH^-(aq) \longrightarrow H_2O(\ell) \qquad (15\text{-}5)$$

The observable properties of acids and bases disappear in neutralization, because the H^+ and OH^- ions disappear when they react to form water.

The Arrhenius acid-base theory draws together many diverse properties and interactions of dissolved substances. It has stood the test of time and serves as the solid foundation on which more modern theories are built.

Acidity and Basicity

The term **acidity** refers to the extent to which a solution displays acid characteristics—strongly, moderately, or only faintly. Since acid characteristics are due to the presence of $H^+(aq)$, acidity also refers to the concentration of hydrogen ions. If $[H^+]$ is large, so is the acidity. If $[H^+]$ is small, the acidity is low, and the solution is said to be "slightly acid." The term **basicity** is similarly defined in terms of the concentration of hydroxide ion: a solution of high basicity has a large value of $[OH^-]$, while a slightly basic solution has a low $[OH^-]$. The range of numerical values of $[H^+]$ and $[OH^-]$ that occurs in commonly used laboratory reagents is enormous: in concentrated hydrochloric acid, for example, $[H^+]$ is about 12 mol L^{-1}. In household ammonia, on the other hand, $[H^+]$ is about 10^{-11} mol L^{-1}.

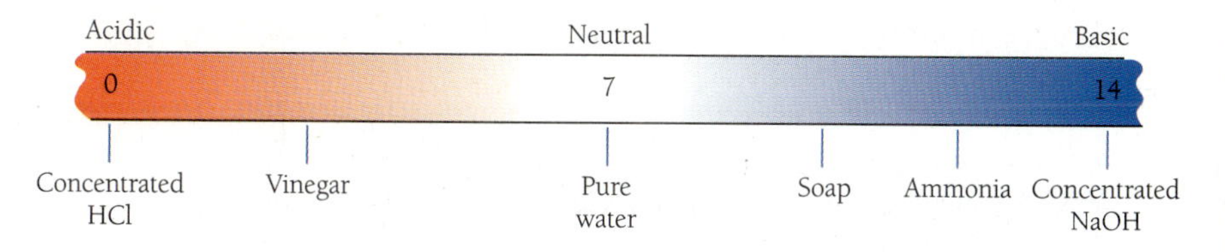

Figure 15.4 The pH scale is a convenient measure of acidity. The more acidic a solution, the *less* its pH. Aqueous solutions generally have pH values between 0 and 14. However, solutions having $[H^+] > 1$ M or $[OH^-] > 1$ M lie outside this range.

The pH Scale

As we have noted before, numbers that are very large or very small are distracting and inconvenient to use. Scientists often define new units so that such extra-large or extra-small numbers become more tractable. Just such a definition was set forth for acidity, when in 1909 the Danish chemist Søren Sørensen introduced the pH scale. According to this concept, acidity is expressed as the *logarithm(log)* of the hydrogen ion concentration. The **pH** of a solution is a dimensionless number defined in Equation (15-6), where $[H^+]$ is the concentration of hydrogen ion in mol L^{-1}.

$$pH = -\log[H^+] \tag{15-6}$$

The logarithm used in pH is the *common* or *base-10* logarithm(log), rather than the *natural* logarithm(ln) used in thermodynamics and kinetics. Logarithms are reviewed in Appendix A3.

The concentration of hydrogen ion is, most frequently, less than 1 mol L^{-1}. Its logarithm is accordingly a negative number, and the purpose of the negative sign in Equation (15-6) is to convert the pH (of most solutions) to a positive quantity. The negative sign creates an *inverse* relationship between acidity and pH: the greater the acidity, the lower the pH. The pH of acidic solutions is less than 7, while the pH of basic solutions is greater than 7 (Figure 15.4). Conversion of hydrogen ion concentration to pH is a very common operation in the practice and study of chemistry.

Strictly speaking, one can take the logarithm only of a dimensionless number. The dimensions of concentration, mol L^{-1}, are simply ignored in pH calculations. The pH itself, like all logarithms, is dimensionless.

EXAMPLE 15.1 Calculation of pH

Calculate the pH of the following solutions.

a. Any solution in which the hydrogen ion concentration is 10^{-2} mol L^{-1}.

b. Concentrated hydrochloric acid, in which the hydrogen ion concentration is 12 mol L^{-1}.

c. A solution of 0.0005 M HNO_3, in which the hydrogen ion concentration is 0.0005 mol L^{-1}.

Analysis

Target: The same in all three cases, pH = ?

Knowns: $[H^+]$.

Relationship: Equation (15-6).

Work

a. Since the concentration is exactly a power of 10, the calculation of pH is easy: the logarithm of 10^y, where y is *any* number, is simply y itself. Thus,

$$\log(10^{-2}) = -2 \qquad \text{and} \qquad pH = -\log[H^+] = -(-2) = 2$$

b. The concentration is not exactly a power of 10, so a calculator with a "log" key is needed.

$$\log(12) = 1.07918125$$

Since the pH is defined as the *negative* of the logarithm, we multiply by (-1) to get

$$\text{pH} = -1.07918125$$

Most calculators display eight or ten digits, which implies a precision far greater than that of even the most precise concentration measurements. In this problem, we should express the answer to two decimal places, as

pH is a dimensionless number.

$$\text{pH} = -1.08$$

The rule for pH (and other common logarithms) is that the number of digits to the right of the decimal point in the pH is the same as the number of significant figures in the concentration.

c. The problem is solved in the same way as (b). The only difference is that, since the concentration is less than 1 M, the pH is a positive number.

$$\text{pH} = -\log[\text{H}^+] = -\log(0.0005) = -(-3.3010300) = 3.3010300$$

After rounding to the proper number of significant figures, the answer is written as

$$\text{pH} = 3.3$$

Check The hydrogen ion concentration is between 10^{-3} and 10^{-4}, so the pH must be between 3 and 4.

Exercise

Calculate the pH of solutions in which the hydrogen ion concentration is (a) 1.0×10^{-5} M, (b) 4.7×10^{-3} M, and (c) 10 M.

Answer: (a) 5.00; (b) 2.33; (c) -1.0 or -1.00, depending on the ambiguous number of significant digits in "10 M."

Determination of the hydrogen-ion concentration in aqueous solutions is a very common procedure in the laboratory. Acidity is an important quantity, not only in chemistry, but also in geology, biology, ecology, and related sciences. Fast, accurate meters are available for measurement of pH (Figure 15.5). (The principles on which pH measurements are based are discussed in Chapter 19.)

The conversion of a pH value to a value of the hydrogen ion concentration is an equally common operation. Equation (15-6) gives the relationship between pH and the *logarithm* of hydrogen ion concentration. To find the relationship between pH and *concentration*, we must take antilogs of both sides of Equation (15-6)

$$\text{pH} = -\log[\text{H}^+] \qquad (15\text{-}6)$$

$$-\text{pH} = \log[\text{H}^+]$$

$$\text{antilog}(-\text{pH}) = \text{antilog}(\log[\text{H}^+])$$

$$10^{-\text{pH}} = 10^{\log[\text{H}^+]} = [\text{H}^+] \qquad (15\text{-}7)$$

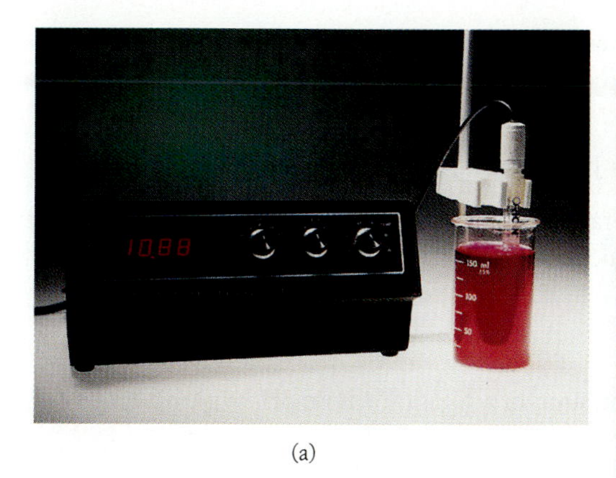

(a)

(b)

Figure 15.5 (a) This bench-top pH meter shows that the pH of a solution of sodium hydroxide is 10.88. The red color of the added phenolphthalein indicator provides visual confirmation that the pH is greater than 10. (b) A portable pH meter, used in environmental studies of the pH of natural waters.

EXAMPLE 15.2 Hydrogen Ion Concentration

Calculate the hydrogen ion concentration of solutions in which (a) pH = 5 and (b) pH = 9.70.

Analysis

Targets: A hydrogen ion concentration, $[H^+] = ?$ M

Knowns: The pH.

Relationship: Equation (15-7).

Plan Insert the given information into Equation (15-7) and solve.

Work

a.

$$[H^+] = 10^{-pH} \qquad (15\text{-}7)$$

$$[H^+] = 10^{-5} = 0.00001$$

b.

$$[H^+] = 10^{-pH}$$

$$[H^+] = 10^{-9.70}$$

Using the "10^x" function on the calculator we find

$$10^{-9.70} = 1.99526 \times 10^{-10}$$

On some calculators, the "10^x" key is labeled "inv log" or "antilog."

The change of sign in this operation must not be omitted: $10^{+9.70}$ is a very different number from $10^{-9.70}$. When rounded to the correct number of significant figures, the answer is

$$[H^+] = 2.0 \times 10^{-10}\ M$$

Note: The result of a 10^x or e^x operation is dimensionless, and appropriate units must be supplied. In the case of pH calculations, the units are always mol L^{-1}.

Exercise

What is the hydrogen ion concentration of a solution whose pH is (a) −1.2, (b) 3.0, and (c) 5.2?

Answer: (a) 2×10^1 M, (b) 0.001 M, and (c) 6×10^{-6} M.

The Sørensen or pH notation for $[H^+]$ has proven so convenient that its use has been extended to other widely ranging numbers. The basicity of a solution, that is, the OH^- molarity, is commonly expressed as pOH. The definition of pOH parallels that of pH.

$$\text{pOH} = -\log[OH^-] \quad (15\text{-}8)$$

15.3 THE HYDRONIUM ION

In the discussion of *solvation* in Chapter 12, it was pointed out that strong interactions exist between a dissolved ion and the molecules of a polar solvent. When the solvent is water, the interaction is called **hydration** rather than solvation. All ions, both cations and anions, are hydrated in aqueous solutions (Figure 15.6).

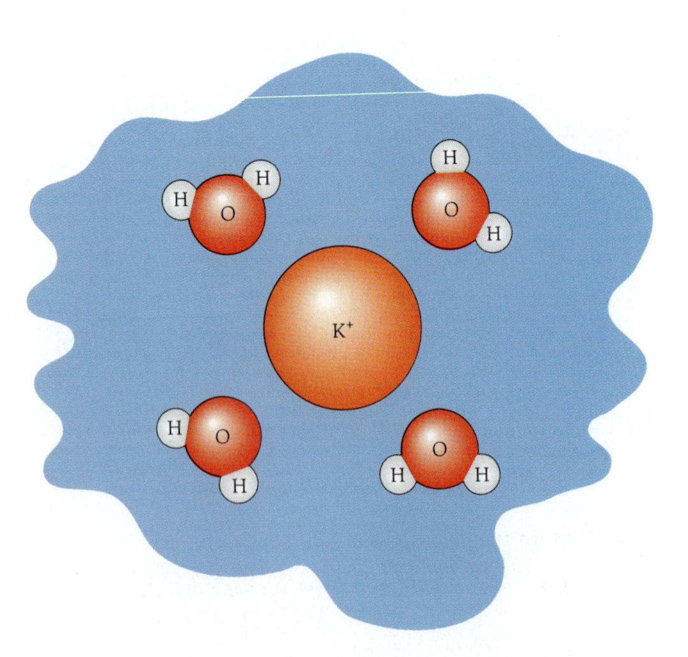

Figure 15.6 When present in polar liquids, cations such as K^+ (and anions as well) exert an orienting effect on nearby solvent molecules. The effect is caused by ion-dipole forces, and is called solvation. If the solvent is water, the effect is known as hydration.

The aqueous proton, that is, the hydrogen ion dissolved in water, is a special case of hydration. The interaction is considerably stronger than that for any other ion and results from a different cause. Whereas all ions are attracted to water by ion-dipole forces, the aqueous proton is joined to a water molecule by a strong covalent bond. The resulting species, H_3O^+, is called the **hydronium ion** (Figure 15.7). The three hydrogen atoms of H_3O^+ are indistinguishable from one another and identical in all respects. The formation of H_3O^+ is so favorable that *all* hydrogen ions interact with water in this way—there is no such thing as a "bare" or unhydrated proton in aqueous solution.

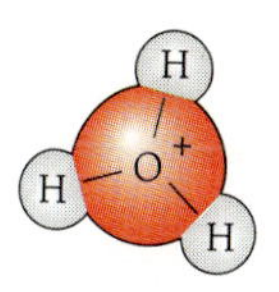

Figure 15.7 An aqueous hydrogen ion does not exist by itself, but is covalently bonded to a water molecule. The combination is known as a hydronium ion, H_3O^+.

Like other positive ions, H_3O^+ itself is hydrated. The hydration is particularly strong, because of hydrogen bonding with nearby water molecules. The exact structure of the hydrated proton is not known with certainty. One possible arrangement is shown in Figure 15.8.

In contrast, the hydroxide ion does *not* form a covalent bond with water. In solution, OH^- is simply a negative ion, which, like other oxygen-containing anions, interacts via hydrogen bonding and ion-dipole attraction to orient the solvent molecules in the immediate neighborhood (Figure 15.9).

One immediate consequence of the nonexistence of *bare* H^+ in aqueous solution is that the equation defining pH should probably be rewritten as

$$\mathrm{pH} = -\log[H_3O^+] \qquad \text{or} \qquad [H_3O^+] = 10^{-\mathrm{pH}} \tag{15-9}$$

For several reasons, however, including convenience and habit, many textbooks and many scientists continue to use the original notation, $\mathrm{pH} = -\log[H^+]$. Also in continued use are phrases like "aqueous proton," "dissolved hydrogen ion," and "H^+ concentration." These usages are acceptable, since it is understood that they refer to H_3O^+ rather than H^+.

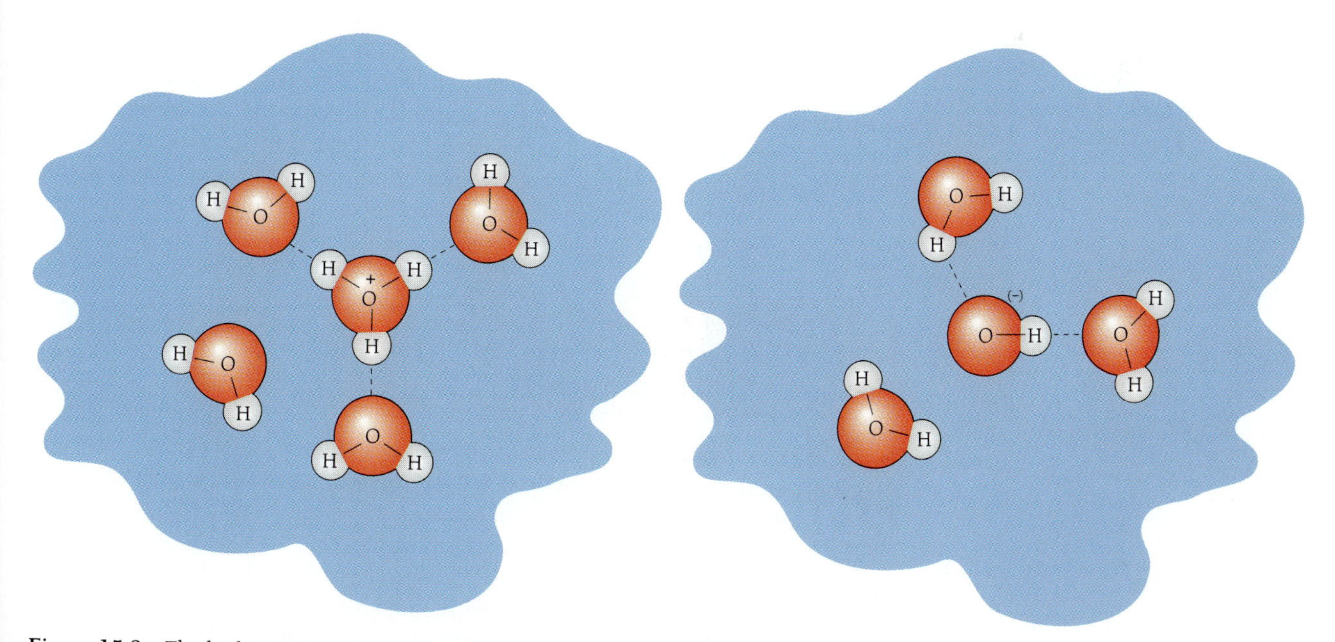

Figure 15.8 The hydronium ion is itself hydrated, linked to solvent molecules by hydrogen bonds as well as by ion-dipole forces.

Figure 15.9 The aqueous hydroxide ion is hydrated, but it is not covalently bonded to water. Hydrogen bonding is involved, and ion-dipole forces are also important.

15.4 THE BRØNSTED-LOWRY CONCEPT OF PROTON TRANSFER

In view of the fact that the true nature of the aqueous proton is H_3O^+ rather than H^+, it is clear that the process taking place when an acid dissolves in water is more complicated than the simple dissociation proposed by Arrhenius. It is much better described as a **proton transfer** reaction, in which an acid molecule donates an H^+ to a solvent molecule.

$$H_2\ddot{O}: + H:\ddot{Cl}: \longrightarrow H_2\ddot{O}^+:H + :\ddot{Cl}:^- \qquad (15\text{-}10)$$

In the process, the H—Cl covalent bond is broken and a new O—H covalent bond is formed. This new bond uses a lone pair of electrons in the valence shell of the oxygen atom in the water molecule.

In 1923, the Danish chemist Johannes Brønsted and the British chemist Martin Lowry expanded the concept of proton transfer into a new theory of acids and bases. Brønsted and Lowry worked independently of one another, and the theory is now known as the **Brønsted-Lowry theory** of acids and bases. The new theory incorporates much of the Arrhenius theory, but is based on the idea of proton transfer rather than acid dissociation. A word about nomenclature is in order here. Since the hydrogen ion has no electrons and its nucleus consists of a single proton, "H^+" and "proton" refer to the same species. Chemists tend to use the words "proton" and "hydrogen ion" interchangeably.

Proton Donors and Proton Acceptors

Arrhenius's term "dissociation" is inappropriate and should not be used in discussions of the Brønsted-Lowry theory.

Recall that the Arrhenius theory defines an acid as a source of H^+: when an acid dissolves in water, it dissociates into ions (Reaction 15-1). In the Brønsted-Lowry view, the process of dissolution of an acid is a chemical reaction, a proton transfer reaction such as (15-10). Since such reactions produce two ionic species from neutral reactants, they are also known as *ionization reactions.* A Brønsted-Lowry acid is defined as a **proton donor,** that is, a species that is capable of directly transferring a proton to another species (Figure 15.10). The Brønsted-Lowry definition of an acid is similar to the Arrhenius definition, since a "proton donor" is not greatly different from a "source of H^+." The Brønsted-Lowry definition of a base, however, is quite different from the Arrhenius definition.

Johannes Nicolaus Brønsted (1879–1947) and Thomas Martin Lowry (1874–1936)

Johannes Nicolaus Brønsted was a Professor at the University of Copenhagen. A great deal of his scientific effort was directed toward understanding the catalytic properties of acids and bases. In 1947 he was elected to the Danish parliament, but he died before taking office.

Thomas Martin Lowry became Professor of Physical Chemistry at Cambridge University in 1920. He studied the rotation of polarized light by various substances, and was also interested in the chemistry of sulfur chlorides. He was the author of a well-regarded book entitled *Historical Introduction to Chemistry.*

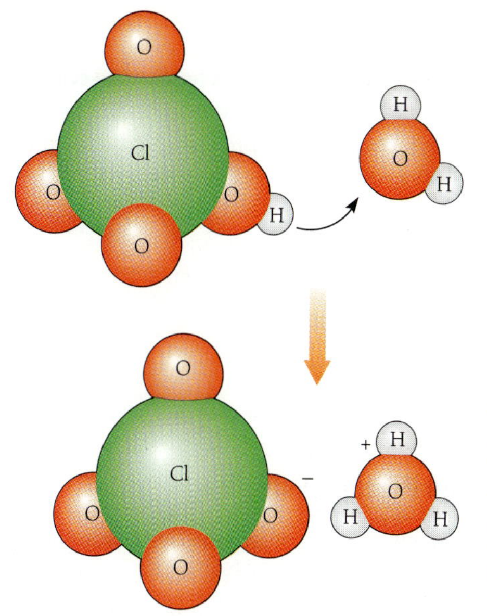

Figure 15.10 When an acid, such as perchloric acid, dissolves in water, it ionizes by transferring a proton to a nearby water molecule. Since the proton carries a positive charge, the residual perchlorate ion has a negative charge, and the newly formed hydronium ion is positively charged.

Brønsted and Lowry noted that when an acid reacts with a hydroxide base in a neutralization reaction, a proton is *transferred* from the acid to the base. Neutralization should therefore be described by Equation (15-11), in which a proton is transferred from a hydronium ion to a hydroxide ion.

$$H_3O^+ + OH^- \xrightarrow{} H_2O + H_2O \qquad (15\text{-}11)$$

Generalizing from this example of an acid-base interaction, Brønsted and Lowry defined a base as any species that is capable of accepting a proton directly from another species. More simply, a base is a **proton acceptor.** In order to form a bond with H^+, which has no electrons, a base must have an atom with a lone pair of electrons in its valence shell. The Brønsted-Lowry definition differs considerably from the more limited notion that a base is a source of hydroxide ions.

So far we have seen three important new features in the Brønsted-Lowry theory of acids and bases: (1) a base is a proton acceptor, (2) a reaction between an acid and a base is a proton transfer reaction, and (3) the *solvent* plays an intimate role—in Reaction (15-10), the solvent (H_2O) is a base because it accepts a proton from the acid HCl.

Brønsted-Lowry theory retains the idea that the common properties of acid solutions are due to the presence of hydrogen ions, although of course this refers to H_3O^+ rather than to H^+. Similarly, the properties of a basic solution are still considered to be due to the presence of OH^- ions. One immediate benefit of the Brønsted-Lowry concept is that it explains why ammonia is a base even though the ammonia molecule contains no hydroxide ion. When ammonia dissolves in water, it accepts a proton from a water molecule, which therefore behaves as an acid. A proton transfer reaction takes place between the acid H_2O and the base NH_3 (Figure 15-11).

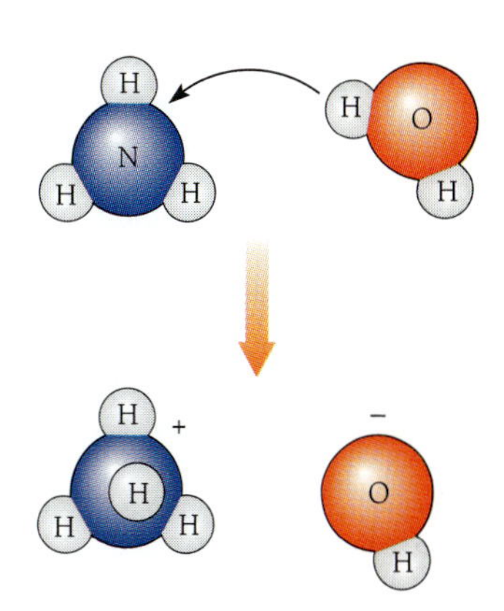

Figure 15.11 When ammonia dissolves, it acquires a proton from a nearby water molecule. As a result, the ammonium ion is positively charged and the residual hydroxide ion is negatively charged.

$$NH_3 + H_2O \longrightarrow NH_4^+ + OH^- \qquad (15\text{-}12)$$

Since hydroxide ions are produced in this Brønsted-Lowry acid-base reaction, the common basic properties, which are due to the presence of OH^-, are observable in the solution.

Equilibrium Nature of Acid-Base Interactions

Most acid-base reactions are reversible processes that proceed to a position of equilibrium rather than to completion. When equilibrium is reached, the solution contains significant concentrations of all species—reactants as well as products. When ammonia dissolves in water, the ionization reaction is described by Equation (15-13).

$$NH_3 + H_2O \rightleftharpoons NH_4^+ + OH^- \qquad (15\text{-}13)$$

The solution contains both NH_3 and NH_4^+ (as well as OH^-). When nitrous acid is dissolved in water, the ionization reaction is

$$HNO_2 + H_2O \rightleftharpoons H_3O^+ + NO_2^- \qquad (15\text{-}14)$$

and the solution contains HNO_2 as well as H_3O^+ and NO_2^-.

Strong and Weak Acids and Bases, and Percent Ionization

Equilibria that lie far to the right are represented by unsymmetrical equilibrium arrows ($\rightleftharpoons$) or by single reaction arrows ($\rightarrow$). For the strong acid HCl,

$$HCl + H_2O \rightleftharpoons H_3O^+ + Cl^- \quad \text{or}$$

$$HCl + H_2O \longrightarrow H_3O^+ + Cl^-$$

Weak-acid equilibria are represented by normal equilibrium arrows

$$HF + H_2O \rightleftharpoons H_3O^+ + F^-$$

With the exception of HNO_3, which is 98% ionized, all the acids listed in Table 15.1 are more than 99.9% ionized in a 1 M solution.

Other aspects of percent ionization are discussed in Chapter 16.

Acids are classified according to the extent to which the ionization equilibrium lies to the left or the right. If the equilibrium of an acid ionization reaction lies far to the right, so that essentially all of the acid molecules have donated their proton to a water molecule, the acid is a **strong acid.** The acid is said to be "completely ionized." If the equilibrium lies to the left, so that a significant number of the acid molecules are not ionized, the acid is a **weak acid.** Weak acids are "incompletely ionized." Most acids are weak—the only common strong acids are those listed in Table 15.1.

The distinction between strong and weak acids can be quantitatively expressed in terms of the **percent ionization**, defined as the concentration of the ionized form divided by the initial concentration, times 100. An acid is strong if it is close to 100% ionized; otherwise, the acid is weak. For example, in a 0.10 M solution of nitrous acid [which ionizes according to Equation (15-14)], the concentration of the ionized form, NO_2^-, is 0.0063 M. The percent ionization is

$$\% = 100 \cdot \frac{0.0063\ \text{M}}{0.10\ \text{M}}$$

$$= 6.3\%$$

Since the stoichiometry of Equation (15-14) requires that equal amounts of NO_2^- and H_3O^+ are produced, $[H_3O^+]$ may be substituted for $[NO_2^-]$ in the calculation. For any acid,

$$\%\ \text{ionization} = 100 \cdot \frac{[\text{ionized form}]}{[\text{initial acid}]} = 100 \cdot \frac{[H_3O^+]}{[\text{initial acid}]} \qquad (15\text{-}15)$$

EXAMPLE 15.3 Percent Ionization

The measured pH of a 0.055 M solution of an acid is 1.96. What is the percent ionization?

Analysis

Target: % ionization = ?

Knowns: Initial concentration and pH.

Relationship: Equation (15-15) applies, but requires $[H_3O^+]$ rather than pH.

Plan Calculate $[H_3O^+]$, then find the % ionization.

Work

$$[H_3O^+] = 10^{-pH} = 10^{-1.96} = 0.011\ M \quad (15\text{-}9)$$

$$\%\ \text{ionization} = 100 \cdot \frac{[H_3O^+]}{[\text{initial acid}]} \quad (15\text{-}15)$$

$$= 100 \cdot \frac{0.011\ M}{0.055\ M}$$

$$= 20\%$$

Exercise

When a 0.075 M solution of hydrofluoric acid is analyzed, it is found that $[HF] = 0.068$ M and $[F^-] = 0.0068$ M. What is the percent ionization of this weak acid?

Answer: 9.1%.

TABLE 15.1 Common Strong Acids

Formula	Name
HCl	Hydrochloric acid*
HNO_3	Nitric acid
H_2SO_4	Sulfuric acid
$HClO_4$	Perchloric acid
HBr	Hydrobromic acid*
HI	Hydriodic acid*

* Name used for aqueous solutions. When pure, the hydrohalic acids are called hydrogen chloride, hydrogen bromide, and hydrogen iodide.

The distinction between *strong* and *weak* applies to bases as well as acids. Because it is an ionic compound, sodium hydroxide is completely dissociated in aqueous solution. Since it is 100% ionized, NaOH is a **strong base.** All Group 1A and 2A hydroxides are strong bases. NaOH and KOH are most often used in the laboratory, because they are very soluble and relatively inexpensive. $Mg(OH)_2$ and $Ca(OH)_2$ are also inexpensive, but their low solubility limits their use to applications that do not require a dissolved base. For example, a well-stirred mixture of $Mg(OH)_2(s)$ and water forms a white suspension called *milk of magnesia,* popular as a remedy for excess acid in the stomach. Although a *soluble* strong base such as NaOH would cause great tissue damage, $Mg(OH)_2$ is safe—it dissolves in the stomach, as it reacts with stomach acid, but not in the mouth. Suspensions of $Ca(OH)_2$ are used in pollution control, to remove acidic components from the exhaust gases produced in the combustion of coal and oil.

Sodium hydroxide is also called *lye* or *caustic soda.*

Some other strong bases, not containing the hydroxide ion, are listed in Table 15.4.

The soluble hydroxide bases are $Sr(OH)_2$, $Ba(OH)_2$, and the Group 1A hydroxides.

In contrast to strong bases, **weak bases** are *incompletely ionized.* A weak base accepts a proton from water in a reversible reaction, such as Reaction (15-13),

$$NH_3 + H_2O \rightleftharpoons NH_4^+ + OH^- \quad (15\text{-}13)$$

in which the equilibrium lies to the left. The most common weak bases are ammonia and its derivatives. A "derivative" of ammonia is a compound, such as methylamine, in which one or more of ammonia's H-atoms has been replaced by another atom or group of atoms.

H—N(H)(H) — **ammonia**

H—N(H)(CH_3) — **methylamine**

The percent ionization of a weak base is defined and calculated in the same way as the percent ionization of weak acids.

$$\%\ \text{ionization} = 100 \cdot \frac{[\text{ionized form}]}{[\text{initial base}]} = 100 \cdot \frac{[OH^-]}{[\text{initial base}]} \quad (15\text{-}16)$$

Conjugate Pairs

Both the forward and reverse reactions of an equilibrium ionization such as Reaction (15-14) are proton transfer reactions. The ions H_3O^+ and NO_2^- satisfy the Brønsted-Lowry definitions of acid and base just as well as the compounds HNO_2 and H_2O.

$$\textbf{Forward} \quad \underset{\textbf{acid}}{HNO_2} + \underset{\textbf{base}}{H_2O} \longrightarrow H_3O^+ + NO_2^- \tag{15-14a}$$

$$\textbf{Reverse} \quad \underset{\textbf{acid}}{H_3O^+} + \underset{\textbf{base}}{NO_2^-} \longrightarrow H_2O + HNO_2 \tag{15-14b}$$

This situation is more compactly expressed by writing the equilibrium reaction and identifying each species as either an acid or a base.

$$\underset{\textbf{acid}_1}{HNO_2} + \underset{\textbf{base}_2}{H_2O} \rightleftharpoons \underset{\textbf{acid}_2}{H_3O^+} + \underset{\textbf{base}_1}{NO_2^-} \tag{15-14}$$

The species $\text{acid}_1/\text{base}_1$ and $\text{acid}_2/\text{base}_2$ are known as **conjugate pairs,** and a Brønsted-Lowry equilibrium reaction always involves *two* conjugate pairs. The conjugate base of an acid is the species that remains after the acid has donated its proton.

Examples of conjugate pairs, taken from Equations (15-10, 15-13, and 15-14) are HCl/Cl^-, H_3O^+/H_2O, H_2O/OH^-, NH_4^+/NH_3, and HNO_2/NO_2^-. To identify the acid conjugate to any base, add 1 H to the formula and increase the charge by 1. To identify the base conjugate to any acid, remove 1 H from the formula and decrease the charge by 1.

Conjugate pairs are always written *acid/base,* with the acid first.

EXAMPLE 15.4 Conjugate Pairs

a. What acid is conjugate to the base HSO_4^-?

b. What is the conjugate base of acetylsalicylic acid (aspirin), whose molecular formula is $C_9H_8O_4$?

Analysis

Target: A member of a conjugate pair.

Knowns: The chemical formula of one member of the pair.

Relationship: The chemical formulas of the members of a conjugate pair differ by one hydrogen ion.

Plan Add or subtract H^+ from the known formula, as appropriate.

Work

a. Adding 1 H to the formula HSO_4^- gives $H_2SO_4^-$; adding 1 to the charge gives $(-1) + 1 = 0$. The acid conjugate to HSO_4^- is H_2SO_4.

b. Removing 1 H from $C_9H_8O_4$ gives $C_9H_7O_4$; subtracting 1 from the charge gives $0 - 1 = -1$. The conjugate base of $C_9H_8O_4$ is $C_9H_7O_4^-$.

Exercise

What is the conjugate acid of the base $(CH_3)_2NH$?

Answer: $(CH_3)_2NH_2^+$.

Relative Strengths of Acids and Bases

In general, there are differences in strength among a group of weak acids—some are *relatively* weaker than others. For example, hydrocyanic acid (HCN) is considerably weaker than nitrous acid. This means that HCN has a lesser tendency than HNO_2 to release its proton, and the HCN equilibrium [Equation (15-17)] lies further to the left than the HNO_2 equilibrium [Equation (15-14)].

Weak Acid $$HNO_2 + H_2O \rightleftharpoons H_3O^+ + NO_2^- \quad (15\text{-}14)$$

Weaker Acid $$HCN + H_2O \rightleftharpoons H_3O^+ + CN^- \quad (15\text{-}17)$$

Provided the comparison is made between solutions of the same concentration, the pH will be greater (because the percent ionization is less) in HCN(aq) than in HNO_2(aq). Table 15.2 lists some acids, together with their conjugate bases, in order of decreasing strength. All the acids listed above H_3O^+ are strong acids, and all those listed below H_3O^+ are weak acids.

Reciprocal Relationship of Conjugate Pair Strengths

There is a simple relationship between the strength of an acid and the strength of its conjugate base: the stronger the acid, the weaker its conjugate base. This **reciprocal relationship** follows from the definition of acid strength in terms of an acid's tendency

TABLE 15.2 Relative Strength of Acids

(Increasing Acid Strength ↑)

Conjugate Acid		Conjugate Base	
Name	Formula	Formula	Name
Hydriodic acid	HI	I^-	Iodide ion
Hydrobromic acid	HBr	Br^-	Bromide ion
Perchloric acid	$HClO_4$	ClO_4^-	Perchlorate ion
Hydrochloric acid	HCl	Cl^-	Chloride ion
Sulfuric acid	H_2SO_4	HSO_4^-	Hydrogen sulfate ion
Nitric acid	HNO_3	NO_3^-	Nitrate ion
Hydronium ion	$\mathbf{H_3O^+}$	$\mathbf{H_2O}$	**Water**
Hydrogen sulfate ion	HSO_4^-	SO_4^{2-}	Sulfate ion
Phosphoric acid	H_3PO_4	$H_2PO_4^-$	Dihydrogen phosphate ion
Hydrofluoric acid	HF	F^-	Fluoride ion
Nitrous acid	HNO_2	NO_2^-	Nitrite ion
Formic acid	HCO_2H	HCO_2^-	Formate ion
Benzoic acid	$C_6H_5CO_2H$	$C_6H_5CO_2^-$	Benzoate ion
Acetic acid	CH_3CO_2H	$CH_3CO_2^-$	Acetate ion
Carbonic acid	H_2CO_3	HCO_3^-	Hydrogen carbonate ion
Hydrogen sulfide	H_2S	HS^-	Hydrogen sulfide ion
Dihydrogen phosphate ion	$H_2PO_4^-$	HPO_4^{2-}	Hydrogen phosphate ion
Ammonium ion	NH_4^+	NH_3	Ammonia
Hydrocyanic acid	HCN	CN^-	Cyanide ion
Phenol	C_6H_5OH	$C_6H_5O^-$	Phenolate ion
Water	$\mathbf{H_2O}$	$\mathbf{OH^-}$	**Hydroxide ion**
Ammonia	NH_3	NH_2^-	Amide ion
Hydrogen	H_2	H^-	Hydride ion

to donate a proton. If the acid loses a proton with relative ease (relatively strong acid), the conjugate base that remains behind has a relatively weak tendency to reacquire the lost proton. That is, the conjugate base is relatively weak. On the other hand, a very weak acid cannot easily transfer a proton to another species. This means that the proton is very tightly held. If the species does lose its proton, however, the resulting conjugate base has a strong tendency to reacquire the proton and hold it tightly. That is, the conjugate is a relatively strong base. As an example of these reciprocal relationships, recall that HNO_2 is a *stronger* acid than HCN. Therefore, NO_2^- is a *weaker* base than CN^-.

15.5 FURTHER ASPECTS OF THE BRØNSTED-LOWRY CONCEPT

We have seen that Brønsted-Lowry theory rests on the notion of the proton transfer reaction as the fundamental interaction between an acid and a base. Proton transfer is reversible, and differences in the position of equilibrium imply differences in acid strength. The reciprocal relationship provides a straightforward explanation of proton-transfer strengths in conjugate pairs. In the following sections, we discuss some additional aspects of acid-base behavior.

Autoionization of Water

Comparison of Reactions (15-10) and (15-13) reveals an interesting aspect of the behavior of water.

The single arrow means that the reaction goes to completion.

$$HCl + H_2O \longrightarrow H_3O^+ + Cl^- \qquad (15\text{-}10)$$

$$NH_3 + H_2O \rightleftharpoons NH_4^+ + OH^- \qquad (15\text{-}13)$$

In one reaction water is a base, that is, it accepts a proton from an acid. In the other it is an acid, donating a proton to a base. A substance that can both donate and accept protons is called an **amphiprotic*** substance (Figure 15.12).

The fact that water is amphiprotic has an important consequence: water can react with itself, in a process known as an **autoionization reaction.**

$$\underset{\textbf{acid}_1}{H_2O} + \underset{\textbf{base}_2}{H_2O} \rightleftharpoons \underset{\textbf{acid}_2}{H_3O^+} + \underset{\textbf{base}_1}{OH^-} \qquad (15\text{-}18)$$

In this reaction, one water molecule is a proton donor and another water molecule is a proton acceptor. There is a proton transfer between water molecules. Water is a very weak acid as well as a very weak base, so that the equilibrium does not proceed very far to the right. The autoionization equilibrium expression is

As usual, the concentration of the pure liquid water is omitted from the equilibrium expression.

$$K_w = [H_3O^+][OH^-] \qquad (15\text{-}19)$$

The equilibrium constant K_w is called the **autoionization constant** of water. Its value depends somewhat on temperature, and at 25 °C it is 1.0×10^{-14}. Table 15.3 gives the value of K_w at various temperatures.

It follows from the stoichiometry of Equation (15-18) that H_3O^+ and OH^- are produced in equal quantities in the autoionization of water. In pure water, therefore,

* The term "amphoteric" is sometimes used in place of "amphiprotic," although there is a slight difference in their meanings. "Amphoteric" means "capable of reacting with both acids and bases" (page 663). "Amphiprotic" means "capable of acting both as a proton donor and as a proton acceptor."

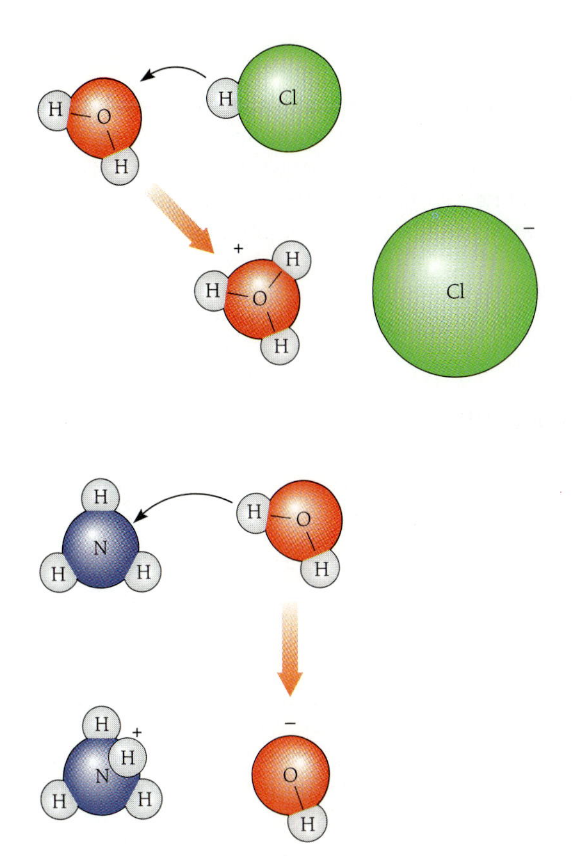

Figure 15.12 Depending on the chemical nature of its reaction partner, an amphiprotic substance such as water can either accept a proton or donate one. That is, it can enter into reactions either as an acid or a base.

TABLE 15.3 Temperature Dependence of K_w

Temperature/°C	K_w
0	0.114×10^{-14}
10	0.292×10^{-14}
20	0.681×10^{-14}
25	1.00×10^{-14}
30	1.47×10^{-14}
40	2.92×10^{-14}
50	5.47×10^{-14}

the concentrations of H_3O^+ and OH^- are the same. In pure water at 25 °C, $[H_3O^+] = [OH^-] = 1.0 \times 10^{-7}$ M, and the pH is 7.00. Since the concentration of $[OH^-]$ is the same, the pOH of pure water at 25 °C is also 7.00.

A useful relationship between pH and pOH can be developed by taking the logarithm of both sides of Equation (15-19), and then multiplying by -1.

$$K_w = [H_3O^+][OH^-] \qquad (15\text{-}19)$$

$$\log(K_w) = \log[H_3O^+] + \log[OH^-]$$

$$-\log(K_w) = -\log[H_3O^+] + (-\log[OH^-])$$

All terms in this equation correspond to the "p" notation, so that

$$pK_w = pH + pOH \qquad (15\text{-}20)$$

The "p" notation for any quantity X is $pX = -\log X$.

Since at 25 °C $K_w = 1.0 \times 10^{-14}$, $pK_w = -\log(1.0 \times 10^{-14}) = -(-14.00)$. At room temperature, therefore,

$$pH + pOH = 14.00 \qquad (15\text{-}21)$$

Equations (15-20) and (15-21) hold true even if the water is not pure, that is, even if there are molecular and/or ionic solutes dissolved in it. All aqueous solutions at 25 °C contain both H_3O^+ and OH^- in concentrations related by Equation (15-21).

In 1 M aqueous hydrochloric acid, $[H_3O^+] = 1$ M and $[OH^-] = 1 \times 10^{-14}$ M.

Ionic Acids and Bases

The Brønsted-Lowry definition of an acid as a species capable of donating a proton does not refer to the electrical charge of the species. Cations and anions, as well as uncharged molecules, can be acids (or bases). The species NH_4^+, HBr, and HPO_4^{2-} all are capable of donating a proton to a base, and all are acids. All undergo proton-transfer reactions with water, according to Equations (15-22) through (15-24).

$$NH_4^+ + H_2O \rightleftharpoons H_3O^+ + NH_3 \tag{15-22}$$

$$HBr + H_2O \rightleftharpoons H_3O^+ + Br^- \tag{15-23}$$

$$HPO_4^{2-} + H_2O \rightleftharpoons H_3O^+ + PO_4^{3-} \tag{15-24}$$

Similarly, the species CO_3^{2-}, F^-, NH_3, and $H_2NCH_2CH_2\overset{+}{N}H_3$ (ethylenediamine monopositive ion) all are capable of accepting a proton from an acid, and therefore all are bases. All undergo proton transfer reactions with water [Equations (15-25) through (15-28)].

$$CO_3^{2-} + H_2O \rightleftharpoons HCO_3^- + OH^- \tag{15-25}$$

$$F^- + H_2O \rightleftharpoons HF + OH^- \tag{15-26}$$

$$NH_3 + H_2O \rightleftharpoons NH_4^+ + OH^- \tag{15-27}$$

$$H_2NCH_2CH_2\overset{+}{N}H_3 + H_2O \rightleftharpoons H_3\overset{+}{N}CH_2CH_2\overset{+}{N}H_3 + OH^- \tag{15-28}$$

As far as proton-transfer reactions are concerned, there is no difference in behavior between ionic (charged) and molecular (uncharged) acids and bases. Note that several of the acids in Table 15.2 are ions.

Salts

Since chemical compounds are electrically neutral, an ionic acid or base is always accompanied by an ion of opposite charge, called a **counterion.** Thus the acid NH_4^+ occurs in such compounds as NH_4Cl, NH_4Br, and NH_4NO_3, in which the counterions are Cl^-, Br^-, and NO_3^-. The base F^- occurs in such compounds as NaF, CaF_2, and KF, in which the counterions are Na^+, Ca^{2+}, and K^+. All of these compounds are called "salts." A salt can be produced in a reaction between an uncharged acid and an uncharged base, as in Equations (15-29) and (15-30). When the base is OH^-, as is often the case in aqueous solution, water is also a product of the reaction.

$$\underset{\textbf{acid}}{HF(aq)} + \underset{\textbf{base}}{KOH(aq)} \longrightarrow \underset{\textbf{water}}{H_2O(\ell)} + \underset{\textbf{salt}}{KF(aq)} \tag{15-29}$$

Uncharged acids like HCl and HNO_3 are not salts, because they are not *ionic* compounds.

$$\underset{\textbf{acid}}{HCl(g)} + \underset{\textbf{base}}{NH_3(g)} \longrightarrow \underset{\textbf{salt}}{NH_4Cl(s)} \tag{15-30}$$

The formal definition of *salt* is, "an ionic compound whose anion is not OH^- or O^{2-}."

When a salt such as KF dissolves in water, the process may be thought of as taking place in two steps. First the salt dissociates into its ionic constituents K^+ and F^-.

$$KF(s) \longrightarrow K^+(aq) + F^-(aq) \tag{15-31}$$

After dissociation, these ions behave independently of one another. Salt dissociation is complete in the sense that no undissociated salt enters the solution: only the ionic constituents are present as solutes. Then, since the F^- anion is a base, it reacts with water in a Brønsted-Lowry proton-transfer reaction [Equation (15-26)]. The resulting solution contains OH^- ions, and therefore exhibits properties common to all basic solutions.

$$\underset{\textbf{base}_1}{F^-} + \underset{\textbf{acid}_2}{H_2O} \rightleftharpoons \underset{\textbf{acid}_1}{HF} + \underset{\textbf{base}_2}{OH^-} \quad (15\text{-}26)$$

Salts like KF, which form basic solutions in water, are called **basic salts.** The base F^- is conjugate to the weak acid HF, and the salt KF is formed by neutralization of HF with KOH. Because it can be formed by neutralization of a weak acid, a basic salt is also called a "salt of a weak acid."

The counterion, K^+, has no effect on the pH of a KF solution, because it does not undergo a proton transfer reaction with water. It cannot donate a proton, because it has none. It cannot accept a proton, because it has no valence shell lone pair with which to form a bond to a transferred proton.

A similar sequence takes place with salts whose cations are acids, for example NH_4Cl. As the salt dissolves, it dissociates completely into its component ions.

$$NH_4Cl(s) \longrightarrow NH_4^+(aq) + Cl^-(aq) \quad (15\text{-}32)$$

Since the NH_4^+ ion is an acid, it undergoes a proton-transfer reaction with water.

$$\underset{\textbf{acid}_1}{NH_4^+} + \underset{\textbf{base}_2}{H_2O} \rightleftharpoons \underset{\textbf{acid}_2}{H_3O^+} + \underset{\textbf{base}_1}{NH_3} \quad (15\text{-}22)$$

The resulting solution contains H_3O^+ ions, and therefore it exhibits the properties common to all acidic solutions. Salts like NH_4Cl, which form acidic solutions in water, are called **acid salts.** The acid NH_4^+ is conjugate to the weak base NH_3, and the

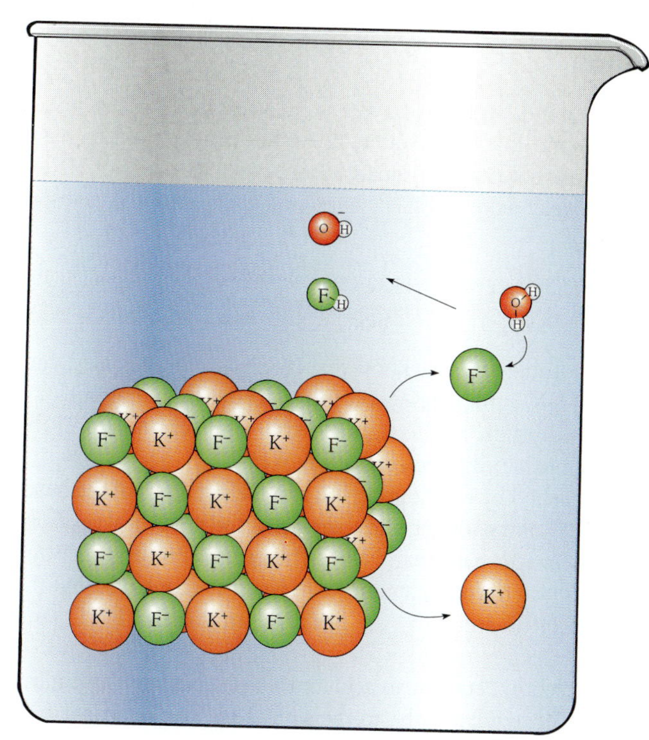

Figure 15.13 When a solid ionic substance such as KF dissolves, its ions are independently released into the solution. The fluoride ion is a weak base (the conjugate base of HF), and some F^- ions react by accepting a proton from a solvent molecule. The result is a solution containing some HF and OH^-, as well as K^+ and unreacted F^- ions.

salt NH_4Cl can be formed by neutralization of NH_3 with HCl. Because it can be formed by neutralization of a weak base, an acid salt is also called a "salt of a weak base."

The counterion Cl^- has no effect on the acidity of a solution of NH_4Cl, because it does not undergo a proton transfer reaction with water. It cannot be an acid, because it has no proton to donate. But Cl^- *is* a base, because it is the conjugate base of the acid HCl. Why then does it not react with water? The reason lies in the relative strength of the acid. HCl is a very strong acid, and according to the reciprocal relationship (Section 15.4) the conjugate base Cl^- is a very weak base. It is so weak a base that it does not react to a significant extent with water.

As noted above, neither K^+ nor Cl^- reacts with water. When KCl dissolves, it dissociates completely according to Equation (15-33). There is no further reaction, and no formation of either H_3O^+ or OH^-.

$$KCl(s) \longrightarrow K^+(aq) + Cl^-(aq) \qquad (15\text{-}33)$$

The solution remains neutral, and the salt KCl is an example of a **neutral salt.** A neutralization reaction of a strong acid with a hydroxide base produces water and a neutral salt.

The strong acids are HI, HBr, HCl, $HClO_4$, H_2SO_4, and HNO_3. The strong hydroxide bases are the Group 1A and 2A hydroxides.

$$HCl + KOH \longrightarrow KCl + H_2O \qquad (15\text{-}34)$$

Reactions of acid or basic salts with water are in principle no different from the reactions of molecular (uncharged) acids or bases with water. However, they cannot properly be called "ionization" reactions, because the acid or base portion of the salt is *already* an ionic species. Instead, reactions of salts with water are called **hydrolysis** reactions. The term "hydrolysis," in the context of Brønsted-Lowry theory, means "proton transfer reaction between an ion and water." The term is also used more generally to mean "reaction with water." Hydrolysis is a particular case of *solvolysis,* which means "reaction with solvent."

The Position of Equilibrium

Because of the central role of water in Brønsted-Lowry theory, the discussion so far has stressed proton-transfer equilibria involving water. But other proton-transfer reactions occur. Suppose an aqueous solution is prepared from equal amounts of HNO_2 and NaCN. Since HNO_2 is a weak acid and CN^- is a weak base, a proton-transfer equilibrium will be established.

$$\underset{\textbf{acid}_1}{HNO_2(aq)} + \underset{\textbf{base}_2}{CN^-(aq)} \rightleftharpoons \underset{\textbf{acid}_2}{HCN(aq)} + \underset{\textbf{base}_1}{NO_2^-(aq)} \qquad (15\text{-}35)$$

From their relative positions in Table 15.2, we know that HNO_2 is a stronger acid than HCN. That is, the tendency of HNO_2 to release a proton is greater than that of HCN. This difference drives the equilibrium to the right. In the solution, therefore, the concentration of HCN is greater than the concentration of HNO_2.

The same conclusion is reached by considering the relative strengths of the bases in Equation (15-35). Since HNO_2 is a stronger acid than HCN, the reciprocal relationship requires that CN^- be a stronger base than NO_2^-. The greater proton-accepting tendency of CN^- pulls the reaction to the right, and once again we conclude that the equilibrium concentrations of HCN and NO_2^- are greater than those of HNO_2 and CN^-.

In general, then, in any reaction between an acid and a base, the equilibrium lies to the side of the *weaker* acid (or, equivalently, to the side of the weaker base).

EXAMPLE 15.5 Position of Equilibrium

Equal volumes of 0.01 M acetic acid and 0.01 M potassium fluoride are mixed.

a. Write the equation for the equilibrium reaction between these two solutes.

b. When equilibrium is established, will the concentration of acetic acid be greater or less than the concentration of hydrofluoric acid?

Analysis/Relationships The reaction is a proton transfer from acetic acid to the base F^-. The equilibrium lies to the side of the weaker acid.

Work

a. The proton-transfer reaction between acetic acid and fluoride ion is

$$CH_3CO_2H + F^- \rightleftharpoons CH_3CO_2^- + HF$$

Potassium ion is neither an acid nor a base, and does not participate in proton-transfer reactions.

b. Their relative position in Table 15.2 indicates that CH_3CO_2H is a weaker acid than HF. Therefore, the equilibrium lies to the side of CH_3CO_2H. The equilibrium concentration of CH_3CO_2H is greater than that of HF.

Exercise

Equal molar amounts of benzoic acid and sodium acetate are dissolved in one liter of water. When equilibrium is established, what anion is present in the greatest concentration?

Answer: Benzoate ion, $C_6H_5CO_2^-$.

The Leveling Effect of Water

One aspect of the properties of acids is that no acid stronger than H_3O^+ can exist in aqueous solution. Suppose that HA represents an acid that is stronger than H_3O^+, that is, any acid *above* H_3O^+ in Table 15.2. In solution, HA donates its proton to water in Reaction (15-36),

$$HA + H_2O \rightleftharpoons A^- + H_3O^+ \qquad (15\text{-}36)$$

The boldface arrow emphasizes the fact that this equilibrium lies *far* to the right.

in which the equilibrium lies so far to the right that we say HA is *completely ionized*. The solution contains H_3O^+ and A^- rather than HA. This phenomenon is known as the **leveling effect of water.** All strong acids are completely ionized, and their solutions all contain H_3O^+ rather than the nonionized acid. Provided their concentrations are the same, all such solutions have the same pH. Therefore, even though strong acids differ among themselves as to the intrinsic strength of their tendency to donate protons, these differences cannot be observed in aqueous solution. The relative strengths of strong acids to donate a proton *can* be determined in the gas phase, however, or in highly acidic solvents such as concentrated sulfuric acid.

The leveling effect also ensures that no base stronger than OH^- can exist in aqueous solution. If a base B, stronger than OH^-, is present in water, it will react according to Equation (15-37).

$$B + H_2O \rightleftharpoons BH^+ + OH^- \qquad (15\text{-}37)$$

The equilibrium lies far to the right, and the solution contains OH^- rather than the base B. Just as for acids, the relative strengths of strong bases must be measured in the gas phase or in some solvent other than water. Some bases stronger than OH^- are listed in Table 15.4.

Polyprotic Acids and Polyhydric Bases

Lemon juice contains 5% to 8% citric acid.

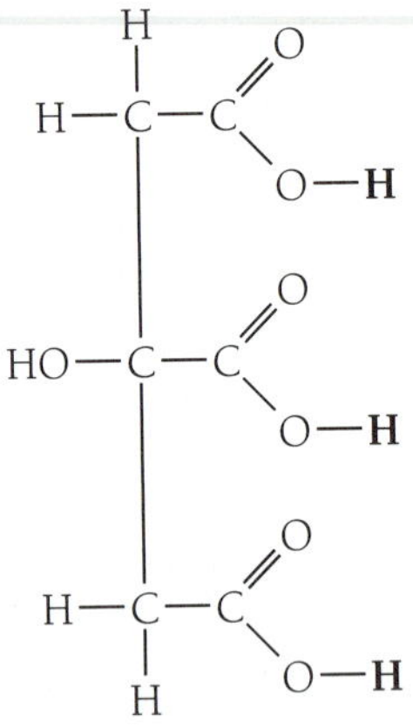

An acid that is capable of donating more than one proton in acid-base reactions is called a **polyprotic acid.** If it can donate two protons, it is **diprotic,** and if it can donate three protons, it is **triprotic.** Tetra- and higher-protic acids are known but are not common. Sulfuric acid (H_2SO_4), hydrosulfuric acid (H_2S), and carbonic acid (H_2CO_3) are common diprotic acids, while phosphoric acid (H_3PO_4) and citric acid ($C_6H_8O_7$) are triprotic. The behavior of citric acid illustrates an important feature of acid-base chemistry. Although the molecule contains eight hydrogen atoms, five of them are nonacidic. The other three are acidic, or *ionizable,* that is, they are available for proton transfer reactions. Other examples of molecules containing both ionizable and nonionizable protons are acetic acid and formic acid, both of which are monoprotic.

citric acid (triprotic)

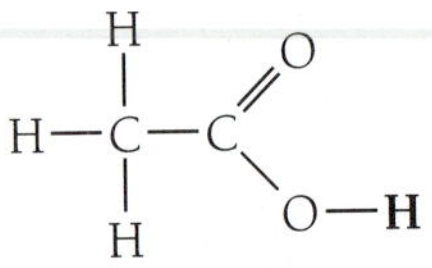

acetic acid (monoprotic)

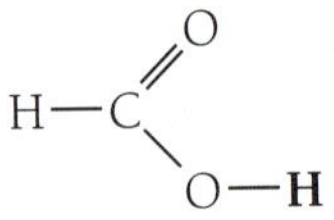

formic acid (monoprotic)

H-atoms bonded to carbon are generally not ionizable in aqueous solution (H—CN is an exception). H-atoms bonded to oxygen can be ionizable or not, depending on other nearby atoms. H-atoms in the —COOH group, as in citric, acetic, and formic acids, are ionizable.

TABLE 15.4 Bases Stronger than OH^-

Name	Formula	Conjugate Acid
Oxide ion	O^{2-}	OH^-
Sulfide ion	S^{2-}	HS^-
Amide ion	NH_2^-	NH_3
Hydride ion	H^-	H_2
Methoxide ion	CH_3O^-	CH_3OH

When a polyprotic acid ionizes, it does so sequentially in a series of monoprotic steps. For phosphoric acid the steps are

$$H_3PO_4 + H_2O \rightleftharpoons H_2PO_4^- + H_3O^+ \tag{15-38}$$

$$H_2PO_4^- + H_2O \rightleftharpoons HPO_4^{2-} + H_3O^+ \tag{15-39}$$

$$HPO_4^{2-} + H_2O \rightleftharpoons PO_4^{3-} + H_3O^+ \tag{15-40}$$

Polyprotic acids may be weak or strong. All three of the ionization equilibria in the phosphoric acid system lie to the left, that is, the acids H_3PO_4, $H_2PO_4^-$, and HPO_4^{2-} are all weak acids. In the carbonic acid system, both ionization equilibria lie to the left, so that the acids H_2CO_3 and HCO_3^- are weak.

$$H_2CO_3 + H_2O \rightleftharpoons HCO_3^- + H_3O^+$$

$$HCO_3^- + H_2O \rightleftharpoons CO_3^{2-} + H_3O^+$$

In sulfuric acid, however, the first equilibrium lies far to the right, so that H_2SO_4 is a strong acid. The second equilibrium lies to the left, and HSO_4^- is a weak acid.

$$H_2SO_4 + H_2O \longrightarrow HSO_4^- + H_3O^+$$

$$HSO_4^- + H_2O \rightleftharpoons SO_4^{2-} + H_3O^+$$

Regardless of the position of equilibrium of the first step, it is always true that the equilibria of the subsequent steps lie further and further to the left. That is, the acids become progressively weaker as the stepwise ionization proceeds. There is a good reason for this: due to the electrical attraction between unlike charges, it is harder to remove a positively charged proton from an anion like $H_2PO_4^-$ than from a neutral species like H_3PO_4. It is harder still to remove a proton from a doubly charged anion like HPO_4^{2-}.

Just as there are species that can donate more than one proton, there are species that can accept more than one proton. Such species are called **polyhydric bases.** Much of the discussion of polyprotic acids can be adapted to polyhydric bases with little difficulty. Ethylenediamine is a **dihydric base** that ionizes according to a stepwise process:

$$H_2NCH_2CH_2NH_2 + H_2O \rightleftharpoons H_2NCH_2CH_2\overset{+}{N}H_3 + OH^- \tag{15-41}$$

$$H_2NCH_2CH_2\overset{+}{N}H_3 + H_2O \rightleftharpoons H_3\overset{+}{N}CH_2CH_2\overset{+}{N}H_3 + OH^- \tag{15-28}$$

Both steps are weak-base ionizations, but the second step lies further to the left than the first. It is easier to transfer a positively charged proton to an electrically neutral species [Equation (15-41)] than to a positive ion [Equation (15-28)].

Since the reactions involved in polyprotic acid or polyhydric base ionization are reversible steps, it follows that some of the species involved are amphiprotic. Ethylenediamine monopositive ion, $H_2NCH_2CH_2NH_3^+$, donates a proton to hydroxide ion in the *right-to-left* reaction of the equilibrium (15-41), and accepts a proton from water in the *left-to-right* reaction of the equilibrium (15-28). Similarly, HPO_4^{2-} is a proton donor in (15-40) and a proton acceptor in (15-39). *The intermediate species in multistep acid or base ionizations are always amphiprotic.* In some cases, the first (uncharged) species is also **amphoteric,** that is, capable of reacting with either acids or bases. Aluminum hydroxide is an example of an amphoteric compound.

Amphoterism of metal hydroxides is discussed further in Chapter 23.

$$Al(OH)_3(s) + NaOH(aq) \longrightarrow Al(OH)_4^-(aq) + Na^+(aq) \tag{15-42}$$

$$Al(OH)_3(s) + HCl(aq) \longrightarrow Al(OH)_2(H_2O)^+(aq) + Cl^-(aq) \tag{15-43}$$

The group 2A hydroxides are also classified as dihydric bases, but for a different reason. Each mole dissociates (completely) into two moles of OH^- ions when it dissolves, and can accept two moles of protons from an acid.

$$Ca(OH)_2(s) \longrightarrow Ca^{2+}(aq) + 2\ OH^-(aq)$$

There are no *sequential* ionization steps, and no equilibrium proton transfers with water molecules.

15.6 MOLECULAR STRUCTURE AND ACID STRENGTH

What is it that makes one acid stronger than another? Like the rest of its chemical properties, the tendency of an acid to release a proton is influenced by its structure. The relationship between structure and acid strength is complex, but certain correlations can be made with confidence. The main feature that influences acidity is the strength and polarity of the bond joining the acidic proton to the remainder of the acid species. Bond strength and polarity are in turn affected by certain properties of the atom or group of atoms to which the acidic proton is bonded. These properties include electronegativity, atomic radius, and position within the periodic table.

The Binary Hydrides: Electronegativity and Atomic Radius

Electronegativity is discussed in Chapter 8.

The *binary acids,* also called binary hydrides, contain one element other than hydrogen. The second period binary hydrides, CH_4, NH_3, H_2O, and HF, form a series in which the acidity decreases from right to left. HF is a weak acid, H_2O is a very weak acid, NH_3 is weaker still, and CH_4 is the weakest acid of the four. The electronegativity (EN) of the element bonded to hydrogen also decreases from right to left, from F (EN = 4.0) to C (EN = 2.5). The larger the electronegativity, the greater the acid strength. This behavior is reasonable in terms of the electron distribution in these molecules. In a polar covalent bond, such as the bond in HF, the bonding electrons are pulled closer to the atom having the greater electronegativity. When a molecule acts as an acid by transferring a proton to an acceptor species, the electron pair of the covalent bond is left behind on the conjugate base.

$$\overset{\delta+}{H}-\overset{\delta-}{F} \longrightarrow H^+ \quad :F^-$$

When the molecule is highly polar, as is HF, the process of losing a proton is already well advanced. The electron pair is already associated primarily with the fluorine atom. It is relatively easy for the bond in HF to be extended still further and complete the loss of a proton. Therefore, other things being equal, acid strength should increase as the electronegativity of the atom to which hydrogen is bonded increases.

This trend is also evident in the acidities of the binary hydrides in other rows of the periodic table. In the third period the electronegativities increase in the order Si < P < S < Cl, and the acid strengths increase in the order $SiH_4 < PH_3 < H_2S <$ HCl. In the fourth period, the electronegativities increase in the order Ge < As < Se < Br. The acid strengths also increase in this order, $GeH_4 < AsH_3 < H_2Se <$ HBr.

However, the relationship between structure and acidity is more complex than the preceding remarks imply. For instance, HCl is a much stronger acid than HF, even though the electronegativity of F (EN = 4.0) is considerably greater than that of Cl (EN = 3.0). Clearly, electronegativity is not the only factor involved. Consider again the process of separation of a proton from the anionic portion of an acid molecule. In the separation of the proton from the partially negatively charged F or Cl atom, the force of electrical attraction between the atoms must be overcome. The attractive force is strongest at short distances, so the strength of the force decreases as the separation proceeds. It follows that the amount of work needed for the process is less if the initial

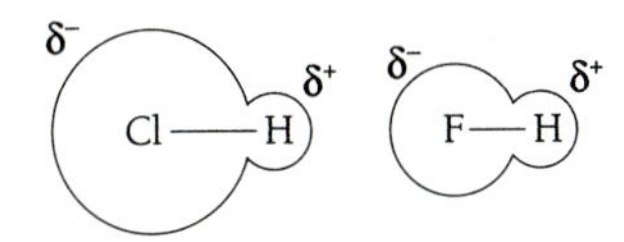

separation is already large. Other things being equal, it is easier to pull a proton off a large negatively charged species than a small one. In any *group* of the periodic table, therefore, the acidities of the binary hydrides should increase as the covalent radius of the element increases. Within any group, the covalent radius increases as the atomic number increases, so that the acid strength of the binary hydrides should increase with atomic number. This behavior is just what is observed. Since the length of a covalent bond is the sum of the covalent radii of the two atoms, it is also true that acid strength increases as bond length increases, other things being equal. In Group 7A the acidities increase in the order HF < HCl < HBr < HI, and in Group 6A the acidities increase in the order $H_2O < H_2S < H_2Se < H_2Te$.

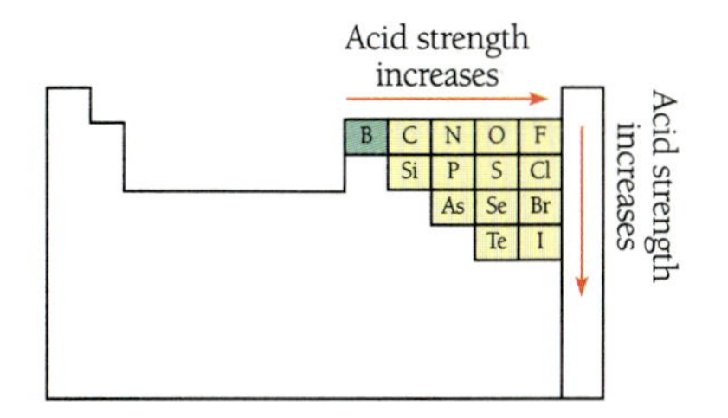

Figure 15.14 Within a period, the strength of the binary acids increases as the *electronegativity* of the atom to which hydrogen is bonded increases. Within a group, however, binary acid strength increases as the *size* of the atom to which hydrogen is bonded increases.

These two factors, electronegativity and atomic size, work in opposite directions in determining acid strength. As the atomic number increases within a group, the electronegativity decreases; this tends to decrease the acid strength. The size, however, increases with atomic number, and this tends to increase the acid strength. Since the acid strength in fact increases, we can conclude that the size factor is dominant within a group. This is reasonable inasmuch as chemical behavior is similar within a group, and size differences are quite large. Within a period, however, the electronegativity increases with atomic number, which tends to increase the acid strength. The size decreases slightly with increasing atomic number in a period, and this tends to decrease the acid strength. Since in fact the acid strength increases with atomic number, the electronegativity effect must be dominant. This is to be expected, since chemical behavior differs sharply within a period, while the size differences are small. Periodic trends in the strength of the binary acids are summarized in Figure 15.14.

Oxoacids: Electronegativity and Inductive Effect

Acids in which the ionizable proton is bonded to an oxygen atom are called **oxoacids.** The *ternary* (three-element) oxoacids, which contain one element in addition to oxygen and hydrogen, form a group that illustrates some further relationships between structure and acid strength. The *hypohalous acids* have the structure X—O—H, where X = Cl, Br, or I. These acids increase in strength in the order HOI < HOBr < HOCl, which means that the O—H bond is weaker and more easily broken in HOBr than in HOI, and weaker still in HOCl. This trend is a consequence of the **inductive effect,** which is a spreading out or transmission of the electron-withdrawing power of an electronegative atom to more distant portions of a molecule. Compare the weakest of these acids, HOI, with the strongest, HOCl. Since chlorine is more electronegative than iodine, the electrons in the O—X bond will be pulled more strongly toward the halogen in H—O—Cl than in H—O—I. This leaves the oxygen atom in H—O—Cl with somewhat lesser valence electron density. Some of the deficit is made up by the electrons of the *O—H* bond moving closer to the O atom. This weakens the O—H bond, to a greater extent in H—O—Cl than in H—O—I, causing HOCl to be a stronger acid than HOI. Since the electronegativity of Br is intermediate between that of Cl and I, the acid strength of HOBr is intermediate between that of HOCl and HOI.

Hypofluorous acid, HOF, is not stable in aqueous solution.

Chlorine forms four oxoacids, which differ only in the number of oxygen atoms that are bonded to the central chlorine atom. In this series of acids, the acid strength increases as the number of terminal oxygen atoms increases.

H—O—Cl
hypochlorous acid

H—O—Cl—O
chlorous acid

O
|
H—O—Cl—O
chloric acid

O
|
H—O—Cl—O
|
O
perchloric acid

Consider the hypochlorous acid molecule, H—OCl. The inductive withdrawal of electron density from the O—H bond will be increased if an additional electronegative atom is bonded to the chlorine atom. Oxygen is very electronegative (EN = 3.5), and so the O—H bond has less electron density (is weaker) in H—OClO than in H—OCl. Since its bond is weaker, the proton is more easily lost from H—OClO than from H—OCl. That is, H—OClO is the stronger acid. Each additional terminal oxygen atom bonded to the central atom further decreases the strength of the O—H bond. H—$OClO_2$ is a stronger acid than H—OClO, and H—$OClO_3$ is one of the strongest acids known.

The same pattern is followed by other oxoacids: nitric acid (H—ONO_2) is stronger than nitrous acid (H—ONO), sulfuric acid (H—OSO_2OH) is stronger than sulfurous acid (H—OSOOH), and so on. The behavior is quite general, and can be summarized as follows. The greater the number of terminal oxygen atoms, the greater the strength of the oxoacid, *regardless of the identity of the central atom.* For example, this rule correctly predicts that HNO_3 (two terminal oxygen atoms) is a stronger acid than $HClO_2$ (one terminal oxygen atom; Figure 15.15).

Two further examples of the inductive effect on acid strength are instructive. Chloroacetic acid is stronger than acetic acid,

Both acetic acid and chloroacetic acid are monoprotic.

H
|
Cl—C—C(=O)—O—H
|
H
chloroacetic acid

H
|
H—C—C(=O)—O—H
|
H
acetic acid

because the inductive withdrawal of electrons from the O—H bond is greater by chlorine than by the less electronegative hydrogen it replaces. 2-Chloropropionic acid is stronger than 3-chloropropionic acid,

H Cl
| |
H—C—C—C(=O)—O—H
| |
H H
2-chloropropionic acid

Cl H
| |
H—C—C—C(=O)—O—H
| |
H H
3-chloropropionic acid

because the inductive effect of the chlorine atom is stronger when the chlorine is nearer to the site of ionization.

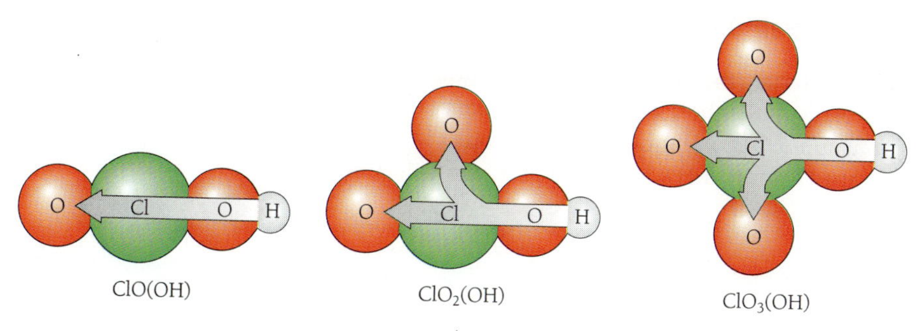

Figure 15.15 As the number of terminal oxygen atoms increases, there is an increasingly strong inductive withdrawal of electrons from the O—H bond, causing it to become progressively weaker. This in turn causes the bond to break more readily. That is, the acid is stronger because it donates its proton more readily.

In some circumstances, an atom or group of atoms can *donate* electron density to distant regions of the molecule. Chemists speak of "inductive donation" as well as "inductive withdrawal" of electrons. The inductive effect is an important and useful concept in chemistry, with implications beyond the area of acid-base strength.

15.7 LEWIS ACIDS AND BASES

When Brønsted and Lowry redefined a base as a proton acceptor, they offered new insight into acid-base interactions: namely, that acid-base reactions may be viewed as *proton transfers* from an acid to a base. About 10 years later (1932), the American chemist G. N. Lewis offered yet another definition, which extended acid-base concepts to a much broader range of chemical phenomena.

A **Lewis acid** is defined as a species that can *accept* a pair of electrons from another species, forming a coordinate covalent bond in the process. A Lewis acid contains an atom with an incomplete valence shell, for example the boron atom in BF_3, or an expandable valence shell, for example the phosphorus atom in PCl_3. A **Lewis base** is an electron pair *donor*. One of the atoms of a Lewis base must have a lone pair of valence electrons available for bonding with another species. Examples of Lewis bases are NH_3, H_2O, and OH^-. All Brønsted-Lowry bases are Lewis bases, and vice versa—any species that bonds to a proton can do so only by virtue of having an available electron pair. On the other hand, Lewis acids and Brønsted-Lowry acids are quite different. For example, the Lewis acid BF_3 cannot be a Brønsted-Lowry acid, since it has no proton to donate. The Brønsted-Lowry acid NH_4^+ is not a Lewis acid. Since the nitrogen atom in NH_4^+ already has four pairs of electrons in its valence shell, it cannot accept another. Table 15.5 summarizes the definitions of acids and bases in the three theories.

In forming an ordinary covalent bond, each atom contributes one electron. In a *coordinate bond,* both electrons are supplied by the same atom.

Recall that a Brønsted-Lowry base also has a lone pair, and that it also forms a coordinate bond to the transferred proton.

TABLE 15.5 Three Views of Acids and Bases

Theory	Acid	Base
Arrhenius	Source of H^+	Source of OH^-
Brønsted-Lowry	H^+ donor	H^+ acceptor, must have lone pair of electrons
Lewis	e-pair acceptor, has incomplete or expandable valence shell	e-pair donor, must have lone pair of electrons

Electron Pair Donation

In Lewis terms, the mechanism of the neutralization reaction between hydrogen and hydroxide ions is the donation and acceptance of an electron pair, with the consequent formation of a covalent bond. In Reaction (15-44) the Lewis base OH^- is the donor, and the Lewis acid H^+ is the acceptor, of an electron pair.

$$\underset{\text{acid}}{H^+} + \underset{\text{base}}{:\!O\!-\!H^-} \longrightarrow \underset{\text{adduct}}{H\!:\!O\!-\!H} \qquad (15\text{-}44)$$

When a Lewis acid and a Lewis base react to form a single, covalently bonded species, the product is often called an **adduct.**

Reaction (15-45) is an example of another type of Lewis acid-base interaction, called a *displacement* reaction.

$$\underset{\text{base}}{NH_3\!:} + \underset{\text{adduct}}{H\!:\!Cl} \longrightarrow \underset{\text{adduct}}{[H_3N\!:\!H]^+} + \underset{\text{base}}{:\!Cl^-} \qquad (15\text{-}45)$$

One Lewis base (NH_3) displaces another (Cl^-) from an adduct. The Lewis acid (H^+) appears as part of the adducts in this reaction rather than as a separate species.

The use of arrows to describe the movement of electrons is common in discussions of Lewis acid-base interactions. In Equation (15-44), the arrow indicates that a lone pair of electrons, initially in the valence shell of the oxygen atom, moves to form the bond between hydrogen and oxygen. In Equation (15-45), one arrow indicates the movement of an electron pair from the valence shell of the nitrogen atom to form a bond between nitrogen and hydrogen. The other arrow indicates movement in the opposite sense, from the covalent bond between the hydrogen atom and the chlorine atom to the valance shell of the chlorine atom to give the chloride ion.

Adduct Formation by Group 3A Compounds

Generally speaking, species having vacant orbitals (unfilled valence shells) can accept an electron pair. Many compounds of Group 3A elements are electron deficient. They can act as Lewis acids, because they have only six electrons in the valence shell of the Group 3A atom. The boron atom in boron trifluoride, for example, has a valence-shell vacancy in which a fourth electron pair could be accommodated: therefore it is a Lewis acid.

$$F\!:\!\underset{\displaystyle F}{\overset{}{\underset{\cdot\cdot}{B}}}\!:\!F$$

When BF_3 reacts with a Lewis base such as ammonia, an adduct is formed.

$$F_3B + :\!NH_3 \longrightarrow F_3B\!:\!NH_3 \qquad (15\text{-}46)$$

Aluminum chloride behaves in a similar fashion, forming the adduct $AlCl_4^-$ by reaction with the base Cl^-.

$$AlCl_3 + :Cl^- \longrightarrow [AlCl_4]^- \qquad (15\text{-}47)$$

Coordination Complexes of Metal Cations

The Lewis acid behavior of transition metal cations underlies a great deal of modern inorganic chemistry. These ions have vacant orbitals in the valence shell, so, as shown in Equations (15-48) and (15-49), they can undergo adduct-forming reactions with Lewis bases.

The electron configurations of the transition metals are discussed in Section 7.2.

$$Ag^+ + 2\,:C{\equiv}N^- \longrightarrow [Ag(:C{\equiv}N)_2]^- \qquad (15\text{-}48)$$

$$Cu^{2+} + 4\,:NH_3 \longrightarrow [Cu(:NH_3)_4]^{2+} \qquad (15\text{-}49)$$

Lewis adducts of cations are called **complexes** or **coordination complexes** (because they have coordinate covalent bonds), and they will be further discussed in Chapter 23 (Figure 15.16).

The term "adduct" is usually reserved for neutral compounds that, like $F_3B{:}NH_3$, are formed from a *neutral* Lewis acid and a *neutral* Lewis base.

Amphoterism of Certain Metal Hydroxides

Some metal hydroxides, for example those of aluminum, zinc, tin, and cobalt, react with either acids or bases. As hydroxides, they are Arrhenius bases that react readily with acids.

$$Zn(OH)_2(s) + 2\,H^+(aq) \longrightarrow Zn^{2+}(aq) + 2\,H_2O \qquad (15\text{-}50)$$

They are also Lewis acids, that can react with a Lewis base such as OH^-.

$$Zn(OH)_2(s) + 2\,OH^-(aq) \longrightarrow Zn(OH)_4^{2-}(aq) \qquad (15\text{-}51)$$

(a)

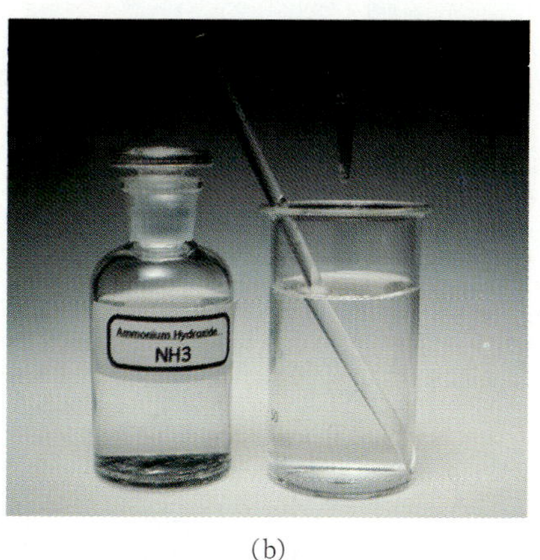

(b)

Figure 15.16 (a) A precipitate of AgCl forms immediately when $AgNO_3(aq)$ is added to NaCl(aq). (b) When $NH_3(aq)$ is added, the AgCl(s) dissolves. Silver ion, like many metal ions, is a good Lewis acid that forms complexes with ammonia and other Lewis bases. The silver-ammine complex ion is so stable that it permits AgCl, insoluble in pure water, to dissolve in aqueous solutions of ammonia.

The product of Reaction (15-51), the *zincate ion*, forms soluble salts with Group 1A cations. Zinc hydroxide reacts with, and dissolves in, both hydrochloric acid and potassium hydroxide. The molecular equations corresponding to (15-50) and (15-51) are

$$Zn(OH)_2(s) + 2\ HCl(aq) \longrightarrow ZnCl_2(aq) + 2\ H_2O \qquad (15\text{-}52)$$

$$Zn(OH)_2(s) + 2\ KOH(aq) \longrightarrow K_2Zn(OH)_4(aq) \qquad (15\text{-}53)$$

15.8 TITRATION: ANALYSIS FOR ACID (OR BASE) CONTENT

When an unknown sample of base is analyzed, the titrant is a strong acid of known concentration.

In laboratory work, it is often necessary to determine the amount of acid or base present in a given sample of material. We may wish to know the purity of the material, the molar mass of an unknown acid or base, or simply the number of moles present so that we may use the right amount in a laboratory procedure. Measurements of this sort are carried out by means of a procedure called **titration.** In the titration of an acid, a strong base of known concentration, called the **titrant,** is gradually added to the acid sample. The volume of base required to exactly neutralize the acid in the sample is carefully measured (Figure 15.17). The point of exact neutralization, called the

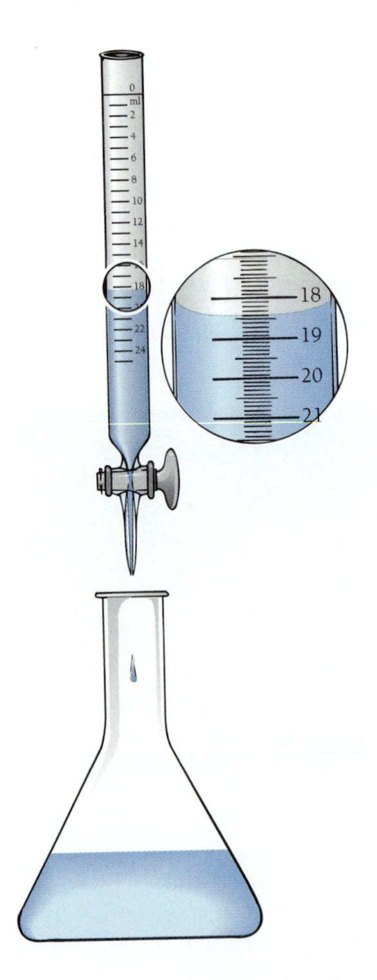

Figure 15.17 Titrant, a strong base of known concentration, is added from a buret to a sample containing an unknown amount of acid. The titrant level can be accurately read at any time (in this figure the titrant level is 18.40 mL), so that the volume of added titrant is known.

Figure 15.18 Indicators are compounds whose colors are strikingly different in acid and basic solutions. Three common indicators are phenolphthalein (colorless/pink), methyl violet (blue/violet), and methyl orange (red/yellow).

equivalence point, can be determined by measuring the pH. Initially the pH has the low (1 to 3) value characteristic of acid solutions. The pH increases gradually during the early stages of a titration, and rises abruptly as the equivalence point is reached. The pH need not be measured precisely in a titration. The color change of an *indicator* may be used to signal that the equivalence point has been reached. An **indicator** is a compound that has one color at low pH, and a different color at high pH (Figure 15.18).

The change of pH during a titration is discussed in detail in Chapter 16.

Similarly, the unknown base content of a sample can be determined by titrating with a strong acid of known molarity. The pH decreases during such a titration, and the equivalence point is detected either with a pH meter or an indicator.

Stoichiometry and Calculations

When a *monohydric* base such as NaOH is used to titrate a *monoprotic* acid, the number of moles of base required to reach the equivalence point is equal to the number of moles of acid originally present in the sample.

$$\text{moles acid} = \text{moles base}$$

$$n_a = n_b \tag{15-54}$$

The number of moles of base contained in V_b liters of titrant of molarity M_b (mol L^{-1}) is

$$n_b = M_b \cdot V_b$$

The relationship $n = M \cdot V$ was introduced in Chapter 3, page 111.

Many titrations are performed in order to determine the molarity of an acid solution. The number of moles of acid in a sample of known volume V_a and unknown molarity M_a is $n_a = V_a \cdot M_a$. At the equivalence point, therefore,

$$M_aV_a = M_bV_b \tag{15-55}$$

Equation (15-55) does not apply to titrations of a *diprotic* acid. A more generally applicable equation is developed on page 672.

Since V_a, V_b, and M_b are known, the unknown molarity M_a of an acid solution can be calculated. Equation (15-55) applies equally well to titrations of a monohydric base with a monoprotic acid.

Equations (15-54) and (15-55) apply only to titrations described by chemical equations in which the stoichiometric coefficients of the acid and the base are equal to one another. Following Example 15.6, we develop a more general method applicable to *all* titrations, including those in which these coefficients are not the same (for example, $H_2SO_4 + 2\ KOH \rightarrow K_2SO_4 + 2\ H_2O$).

EXAMPLE 15.6 Determination of an Unknown Acid Concentration

It is found that 29.28 mL of 0.100 M NaOH is required to reach the equivalence point in the titration of 25.0 mL of solution of hydrochloric acid. Calculate the molarity of the acid solution.

Analysis

Target: The molarity of an acid solution,

$$M_a = ?\ \text{mol L}^{-1}$$

Knowns: The volume of the acid sample, and the volume and molarity of the base titrant.

Implicit: The acid is monoprotic, as is the titrant.

Relationship: Equation (15-55) applies to titrations of dissolved monohydric bases and monoprotic acids.

Plan Insert the given values into Equation (15-55) and solve.

Work

$$M_a V_a = M_b V_b$$

$$M_a = \frac{M_b V_b}{V_a}$$

$$= \frac{(0.100\ \text{mol base L}^{-1})(0.02928\ \text{L})}{0.0250\ \text{L}}$$

$$= 0.117\ \text{mol L}^{-1}$$

Exercise

16.25 mL of 0.100 M HCl is required to reach the equivalence point in a titration of 50.00 mL of a monohydric base. What is the molarity of the base?

Answer: 0.0325 M.

Note that volumes were converted from mL to L in the work of Example 15.6. This was not an essential step. Whenever the units of V_a and V_b are the *same*, they cancel one another. Converting from mL to L, however, reinforces the habit of using a consistent set of units in solving numerical problems.

To convert from mL to L, move the decimal point three places to the left: 350 mL ⟶ 0.350 L. The equivalence statement is 1 mL ⇔ 10^{-3} L.

In practice, the titrant is always an aqueous solution of a strong monohydric base such as KOH or a strong monoprotic acid such as HCl. The *sample,* however, may be polyhydric or polyprotic. For instance, a barium hydroxide solution of unknown molarity might be titrated with aqueous HCl. The balanced equation for the titration reaction is

$$Ba(OH)_2(aq) + 2\ HCl(aq) \longrightarrow BaCl_2(aq) + 2\ H_2O \tag{15-56}$$

in which the stoichiometric mole ratio is

$$1 \text{ mol base} \Longleftrightarrow 2 \text{ mol acid} \quad (15\text{-}57)$$

In this titration, the simple relationship of Equation (15-55) must be modified to read

$$\begin{matrix}\text{moles base} \\ \text{in sample}\end{matrix} = \begin{pmatrix}\text{moles acid} \\ \text{in titrant}\end{pmatrix} \cdot \left(\frac{1 \text{ mol base}}{2 \text{ mol acid}}\right) \quad (15\text{-}58)$$

$$M_b V_b = M_a V_a \left(\frac{1 \text{ mol base}}{2 \text{ mol acid}}\right) \quad (15\text{-}59)$$

It is all too easy to put this factor 2 in the numerator rather than the denominator. This error can be avoided if the *full* units of the stoichiometric relationship are always written out, as they are in Equation (15-58).

Equation (15-58) is not new—it is a specific application of the general procedures of stoichiometric calculations developed in Chapter 3. As shown in the following example, Equation (15-58) is easily modified for use with titrations of a polyprotic acid.

EXAMPLE 15.7 Concentration of a Diprotic Acid

The liquid, called "battery acid," in an automobile storage battery is concentrated sulfuric acid. The concentration is about 5.2 M in a fully charged battery, and decreases to about 4.8 M as the battery runs down. When a 1.00-mL sample of battery acid was titrated, it was found that 34.53 mL of 0.300 M KOH was required to completely neutralize the sulfuric acid. Calculate the value of $[H_2SO_4]$ in the sample. Does the battery need to be recharged?

Analysis

Target: The molarity of an acid solution,

$$M_a = ?\ \text{mol L}^{-1}$$

Knowns: The volume of the sample, and the volume and molarity of the titrant; the sample is a diprotic acid, and the titrant is a monohydric base.

Relationship: $n_a = n_b \times$ (stoichiometric mole ratio) applies to *all* titrations, and $n = M \cdot V$ applies to all solutions.

Plan Determine the stoichiometric mole ratio from the balanced chemical equation, and use the relationships $n_a = n_b \cdot$ (mole ratio) and $n = M \cdot V$ to find the acid molarity.

Work The balanced chemical equation is $2\ KOH + H_2SO_4 \rightarrow K_2SO_4 + 2\ H_2O$, so the stoichiometric mole ratio is

$$1 \text{ mol acid} \Longleftrightarrow 2 \text{ mol base}$$

Modified to apply to the titration of an acid, Equation (15-58) becomes

$$\begin{matrix}\text{moles acid} \\ \text{in sample}\end{matrix} = \begin{pmatrix}\text{moles base} \\ \text{in titrant}\end{pmatrix} \cdot \left(\frac{1 \text{ mol acid}}{2 \text{ mol base}}\right)$$

or, in terms of solution molarity and volume,

$$M_a V_a = M_b V_b \cdot \left(\frac{1 \text{ mol acid}}{2 \text{ mol base}}\right)$$

$$M_a = \frac{M_b V_b}{V_a} \cdot \left(\frac{1 \text{ mol acid}}{2 \text{ mol base}}\right)$$

$$= \frac{(0.300 \text{ mol base L}^{-1}) \cdot (34.53 \text{ mL})}{(1.00 \text{ mL})} \cdot \left(\frac{1 \text{ mol acid}}{2 \text{ mol base}}\right)$$

$$= 5.18 \text{ mol L}^{-1}$$

The battery is fully charged, or nearly so.

Exercise

When a 25.0 mL sample of $Ba(OH)_2(aq)$ is titrated with 0.100 M HCl, it is found that 17.92 mL titrant is required to reach the equivalence point. What is the molarity of the barium hydroxide solution?

Answer: 0.0358 M.

In some titrations, the sample is a pure solid or liquid rather than a solution. The unknown is the amount, rather than the molarity, of the sample. This procedure is used in the determination of the unknown molar mass of an acid, as illustrated by Example 15.8.

EXAMPLE 15.8 Determination of Molar Mass

A pure sample of a monoprotic acid weighing 0.617 g is titrated and found to require 18.35 mL of 0.178 M KOH to reach the equivalence point. Calculate the molar mass of the acid.

Analysis

Target: A molar mass,

$$\mathcal{M} = ?\ \text{g mol}^{-1}$$

Knowns: Sample mass, and titrant molarity and volume. The stoichiometric ratio of 1 is implicit, since the problem specifies KOH and a monoprotic acid.

Relationship: The number of moles of a monoprotic acid is given by $n_a = n_b$, and the molar mass of a compound may be found from $\mathcal{M}$ = (mass in grams)/(number of moles).

Plan Find the number of moles of acid, then use n_a and the given mass to find the molar mass.

Work

1.
$$n_a = n_b = M_b V_b$$
$$= (0.178 \text{ mol L}^{-1}) \cdot (0.01835 \text{ L})$$
$$= 0.003266 \text{ mol}$$

2.
$$\mathcal{M} = \frac{\text{mass}}{\text{moles}}$$
$$= \frac{0.617 \text{ g}}{0.003266 \text{ mol}}$$
$$= 189 \text{ g mol}^{-1}$$

Exercise

A sample of base weighing 0.703 g is titrated with 0.500 M HCl, and 48.2 mL acid is required for complete neutralization. If other evidence shows that the base is dihydric, what is its molar mass?

Answer: 58.3 g mol^{-1}. If you reached a different answer, check your units, particularly the units of the stoichiometric ratio.

Normality

In earlier textbooks of chemistry, another method is used for titration calculations involving polyprotic acids or polyhydric bases. In this method, the terms *normality, equivalent,* and *equivalent weight* are defined and used in place of molarity, mole, and molar mass. The system is no longer widely used, so we do not discuss it here. However, since many older books, articles, and laboratory instructions refer to normality, a summary is given in Appendix F.

SUMMARY EXERCISE

One of the chemical products of muscle contraction is lactic acid ($CH_3CHOHCO_2H$), a monoprotic acid whose structure is

```
     H   OH
     |   |    //O
 H—C—C—C
     |   |    \
     H   H     O—H
```

Prolonged exercise can temporarily overload the body's capacity for elimination of this substance, and the resulting increase in lactic acid concentration in the muscles causes pain and stiffness. The pH of a 0.010 M solution of lactic acid is 2.96. For comparison, the pH of a 0.010 M solution of acetic acid is 3.38.

a. What are the concentrations of H_3O^+ and OH^- in 0.01 M lactic acid?

b. What is the percent ionization of lactic acid in a 0.010 M solution?

c. The lactic acid molecule has six H atoms, yet is monoprotic. Which of the H atoms is ionizable?

Chemisty Old and New

Acid Rain

Sometimes more acidic than lemon juice, acid rain is a threat to trees, rivers, lakes, fish, and wildlife. From the Rocky Mountains and the Ohio River Valley to Canada and Scandinavia, the problem has become all too common. Acid rain interferes with photosynthesis in trees and other green plants; it converts lakes into acid baths that cannot support life; it leaches toxic metals (primarily aluminum) from the soil and carries them to lakes, killing fish even when the acidity by itself is not dangerously high; it attacks marble and limestone and can literally deface statues and buildings that have suffered less harm in their first 2000 years of existence than in their last 50.

Natural rainfall is not pure water—it contains enough dissolved carbon dioxide to produce a slightly acidic solution with a pH of about 5.6. When the pH of rainfall drops to lower values, however, the environment is adversely affected. "Acid rain" is a generic term describing rainfall or snow whose pH is less than 5. Today, the average pH of rainfall in northeastern United States is 4.4, and values less than 2 have been recorded. For comparison, note that the pH of vinegar is 2.5.

Acid rain results from the presence of sulfur oxides, and to a lesser extent nitrogen oxides, in the atmosphere. Small amounts of these oxides arise naturally, from volcanic activity, lightning, and forest fires, but the major sources are associated with modern industrial societies. Although metal-ore smelting and chemical manufacturing contribute, the chief culprit is the combustion of coal, which can contain up to 5% sulfur. During the combustion process, the sulfur is converted to SO_2, which, unless it is removed from the flue gases, is released into the atmosphere.

Once in the atmosphere, SO_2 is converted to SO_3, which reacts with water to produce H_2SO_4, which falls as acid rain: $SO_3(g) + H_2O(\ell) \rightarrow H_2SO_4(aq)$. The mechanism of the conversion of SO_2 to SO_3 is complex and not fully understood. The direct reaction $SO_2 + \frac{1}{2} O_2 \rightarrow SO_3$ is too slow to account for the formation of acid rain. Dust particles, from both human and natural sources, may well provide a surface upon which the reaction can take place more rapidly. Another possibility is hydroxyl radical ($OH\cdot$), a molecular fragment produced by the action of sunlight on mixtures of NO_2, H_2O, and O_2. Sulfur dioxide reacts readily with $OH\cdot$: $SO_2(g) + 2\,OH\cdot(g) \rightarrow SO_3(g) + H_2O(g)$.

Tracing the source of acid rain is no easy task, since pollutants can travel far from their point of origin. The smokestacks of many power-generating plants are 1000 feet high, where prevailing winds can disperse pollutants over a wide area. Today, specially equipped aircraft—flying laboratories—are used to track chemically tagged smokestack plumes.

One of the most visible and unfortunate consequences of acid rain is the damage it does to forests. These trees near Most, Czech Republic, are dead or dying from acid rain.

Making our rainfall safe again will not be cheap. A great deal of sulfur can be removed from coal before it is burned, by several different techniques. Alternatively, SO_2 can be removed from the flue gases by reaction with CaO and O_2 to form $CaSO_4(s)$. In addition, it is possible to counteract the effect of acid rain on lakes by neutralizing them with massive annual doses of air-dropped $CaCO_3(s)$.

The problem continues to generate conflicts. Some say tough new restrictions are needed immediately to reduce SO_2 emissions. Others call for more research, arguing that effective environmental policies must be based on a clear understanding of the problem.

Discussion Questions

1. What problems, economic and environmental, are created by the formation of millions of tons of $CaSO_4(s)$ when SO_2 is removed from flue gases?
2. The waters of a lifeless lake are usually crystal clear, and often beautiful to see. Is this an important factor in the discussion of acid rain and how to deal with it?

d. Draw the structural formula, and give the name, of the conjugate base.

e. Which acid is weaker, acetic acid or lactic acid?

f. Suppose 0.01 mol each of lactic acid and sodium acetate are dissolved in a liter of water. When equilibrium is reached, which concentration will be higher, acetic acid or lactic acid?

g. If a small amount of KOH is dissolved in a 0.01 M solution of lactic acid, will the concentration of nonionized lactic acid increase, decrease, or stay the same?

h. Rather than search for some hydrochloric acid (a better choice for titrations), a student decides to use 0.100 M lactic acid for a titration of strontium hydroxide. The data show that 39.1 mL of lactic acid is required for complete neutralization when a 25.00 mL sample of $Sr(OH)_2(aq)$ is titrated. What is the concentration of the strontium hydroxide?

i. Suppose that one of the H atoms bonded to the terminal C atom of lactic acid were replaced by a Cl atom, giving an acid whose structure is

```
        H   OH
        |   |      O
  Cl—C—C—C//
        |   |      \
        H   H       O—H
```

Would this acid be stronger or weaker than lactic acid?

j. Look at the data in Table 15.3 and decide whether the following reaction is endothermic or exothermic.

$$H_2O + H_2O \rightleftharpoons H_3O^+ + OH^-$$

Answers **(a)** $[H_3O^+] = 0.0011$ M, $[OH^-] = 9.1 \times 10^{-12}$ M; **(b)** 11%; **(c)** only the $—CO_2H$ hydrogen is ionizable;

```
                         H   OH
                         |   |      O
(d) lactate ion, H—C—C—C//
                         |   |      \
                         H   H       O⁻
```

; **(e)** acetic acid; **(f)** acetic acid; **(g)** decrease; **(h)** 7.82×10^{-2} M; **(i)** stronger; **(j)** endothermic.

SUMMARY AND KEYWORDS

In the **Arrhenius** view, an acid is a *proton source,* while a base is a *hydroxide ion source.* Acids and bases *dissociate* when dissolved in water: **strong acids** are completely dissociated, while **weak acids** are only partially dissociated. Acidic properties of solutions are due to $H^+(aq)$, while basic properties are due to $OH^-(aq)$. Acids and bases react with one another in **neutralization** reactions, which yield two products: water and an ionic compound called a **salt. Spectator ions** do not participate in reactions, and are omitted from a **net ionic equation. Polyprotic acids** supply more than one proton; **diprotic** and **triprotic** acids are common. **Acidity** refers to the concentration of $H^+(aq)$, and **basicity** refers to the concentration of $OH^-(aq)$. Acidities are commonly expressed as **pH,** which is equal to $-\log[H^+(aq)]$: the smaller the pH the greater the acidity.

The **Brønsted-Lowry** view focuses on **proton transfer** from an acid **(proton donor)** to a base **(proton acceptor).** A Brønsted-Lowry base must have a lone pair, but need not contain hydroxide ions. An uncharged (molecular) acid or base, when dissolved, undergoes *ionization* by proton transfer to or from water. The *relative strength* of an acid refers to its tendency to transfer a proton to a base. In reversible proton-transfer reactions, the equilibrium lies to the side of the weaker acid and weaker base. Brønsted-Lowry acids and bases exist as **conjugate pairs.** A *conjugate acid* has one more proton than its *conjugate base.* When an ionic (rather than molecular) acid or base undergoes a proton transfer reaction with water, the process is called **hydrolysis.** Salts are classified as neutral, acidic, or basic, depending on the pH of their aqueous solutions. Salts of weak acids are basic, while salts of weak bases are acidic. Water is **amphiprotic,** that is, it can accept or donate a proton. Water and some other substances undergo **autoionization.** Water autoionizes to hydronium ion and hydroxide ion, and the pH of pure water at 25 °C is 7.00. The **autoionization constant** of water (K_w) is 1.0×10^{-14} at 25 °C. Regardless of the number and identity of the dissolved substances, in an aqueous solution at 25 °C it is always true that pH + pOH = 14.00.

In a **titration,** the amount of acid (or base) in a sample is determined by the amount of **titrant** required to reach the **equivalence point,** which is observed by an abrupt change in pH or by a change in color of an **indicator.**

The relative strength of an acid is affected by the electronegativity and the size of the atom to which the acidic hydrogen is bonded. Large electronegativity and large size both tend to increase the strength of binary acids. The strength of binary acids increases to the right within a period, and increases downwards within a group. The **inductive effect** is the transmission of electron-withdrawing or electron-donating power to a distant part of a molecule. The strength of the inductive effect falls off with distance. Inductive withdrawal of electron density from its O—H bond increases the strength of an acid.

A **Lewis base** is an electron pair donor, and a **Lewis acid** is an electron pair acceptor. All Brønsted-Lowry bases are Lewis bases, and vice versa. In a Lewis neutralization reaction, a coordinate covalent bond is formed between an acid and a base. Neutral compounds of Group 3A elements are good Lewis acids, which can add to a Lewis base to form an **adduct.** Metal cations, particularly of transition elements, add to Lewis bases to form **complexes.** In another type of reaction, one Lewis base *displaces* another from an acid-base adduct. Certain metal hydroxides are **amphoteric** because they are Lewis acids as well as Arrhenius bases.

PROBLEMS

General

1. What is meant by the terms, "Arrhenius acid" and "Arrhenius base"? Give examples of each.
2. Define the terms "neutralization" and "salt."
3. What is the hydronium ion? How does it differ from, say, a hydrated sodium ion?
4. What are polyprotic acids and polyhydric bases? Give examples of each.
5. What are Brønsted-Lowry acids and bases? Give examples of each.
6. Is it possible for a substance to be both an Arrhenius acid and a Brønsted-Lowry acid? If so, give an example.
7. Is it possible for a substance to be both an Arrhenius base and a Brønsted-Lowry base? If so, give an example.
8. Write the equation of an acid ionization in aqueous solution.
9. Write the equation of a base ionization in aqueous solution.

10. Distinguish between strong and weak acids. Give examples of each.

11. What is a conjugate pair? Give examples, including both charged and uncharged acids.

12. Define hydrolysis, and distinguish between hydrolysis and ionization.

13. What is meant by the leveling effect of water?

14. Explain the difference between the terms "amphoteric" and "amphiprotic."

15. What is meant by autoionization? Write the autoionization reaction for water.

16. What is a Lewis acid? Is it possible for one substance to be both a Lewis acid and a Brønsted-Lowry acid? If so, give an example.

Concentration and pH

17. Calculate the molarity of each of the following solutions.
 a. 15.18 g NaCl in 100 mL solution
 b. 50.0 g H_2SO_4 in 500 mL solution
 c. 0.100 g phenol (C_6H_5OH) in 1.00 L solution

18. Calculate the molarity of each of the following solutions.
 a. 0.038 g barium hydroxide in 50.0 mL solution
 b. 0.050 g iodine in 100 mL CCl_4
 c. 0.00055 g HCl in 1.00 mL solution

19. Calculate the mass of solute required to prepare each of the following solutions.
 a. 350 mL of 0.0100 M CsI
 b. 50 mL of 3.0 M KOH
 c. 500 mL of 0.0100 M glucose ($C_6H_{12}O_6$)

20. Calculate the mass of solute required to prepare each of the following solutions.
 a. 15 L of 0.050 M H_3PO_4
 b. 250 mL of 0.100 M H_2SO_4
 c. 100 mL of 0.100 M NaOH

21. How much water must be added to 25.0 mL of a 0.100 M solution in order to reduce its concentration to 0.0300 M?

22. It is desired to prepare 100 mL of a 0.0375 M solution by dilution of a stock solution whose concentration is 0.750 M. How much stock solution is required? How much water?

23. Calculate the pH of the following solutions of strong acids.
 a. 5.00×10^{-1} M HCl b. 0.030 M HNO_3
 c. 0.75 g L^{-1} $HClO_4$

24. Calculate the pH of the following solutions of strong acids.
 a. 7.5×10^{-2} M HBr
 b. 0.0062 M HI
 c. 2.84 g HNO_3 in 250 mL solution

25. To what molarity of HCl solution do the following pH values correspond: 3.05, 5.7, 0.0, −1.00, and 2.43?

26. To what molarity of KOH solution do the following pOH values correspond: 2.67, 4.2, 0.0, −1.00, and 1.243?

27. Calculate the pOH of solutions having the following values of pH: 4.19, 7.00, 8.25, and 12.2.

28. Calculate the pH of solutions having the following values of pOH: 3.25, 6.1, 7.00, 9.7, and 13.5.

29. Calculate the pH of the following solutions of strong bases:
 a. 3 M NaOH b. 1.3×10^{-4} M $Sr(OH)_2$
 c. 2.5 g L^{-1} LiOH

30. Calculate the pH of the following solutions of strong bases:
 a. 0.000178 M $Ba(OH)_2$ b. 5.8×10^{-2} M KOH
 c. 0.12 g $Sr(OH)_2$ in 500 mL of solution

Weak Acids and Bases, Percent Ionization

31. Calculate the percent ionization in each of the following solutions of weak, monoprotic acids.
 a. 0.015 M, $[H_3O^+] = 1.0 \times 10^{-4}$
 b. 9.95×10^{-2} M, $[H_3O^+] = 1.5 \times 10^{-3}$
 c. 4.5×10^{-3} M, pH = 5.52

32. Calculate the percent ionization in each of the following solutions of weak, monohydric bases.
 a. 7.5×10^{-3} M, pOH = 5.2
 b. 1.00 M, $[OH^-] = 1.5 \times 10^{-4}$
 c. 3.75×10^{-3} M, $[OH^-] = 3.5 \times 10^{-3}$

33. If a 0.503 M solution of a weak acid is 0.030% ionized, what is the hydrogen ion concentration? What is the pH?

34. If a 0.0375 M solution of a weak acid is 1.6% ionized, what is the hydrogen ion concentration? What is the pH?

*35. Calculate the hydroxide ion concentration and the pH of the following solutions of weak bases, given the percent ionization.
 a. 3.6×10^{-2} M (1.5%)
 b. 0.78 M (0.0032%)

*36. Calculate the hydroxide ion concentration and the pH of the following solutions of weak bases, given the percent ionization.
 a. 8.0×10^{-3} M (47%) b. 0.15 M (0.020%)

37. Classify each of the following acids as strong or weak: HF, HCl, HBr, HCN, H_2SO_4, HSO_4^-, HNO_3.

38. Classify each of the following acids as strong or weak: C_6H_5OH, HClO, $HClO_4$, H_2CO_3, H_2O, H_2S, NH_4^+.

39. Classify each of the following bases as strong or weak: CsOH, OH^-, H_2O, HCO_3^-, NH_2^-.

40. Classify each of the following bases as strong or weak: NH_3, NO_2^-, CN^-, HSO_4^-, Cl^-.

Brønsted-Lowry Acids and Bases

41. Identify all species in the following reactions as either an acid or base, in the Brønsted-Lowry sense.
a. $CN^- + H_2O \rightleftharpoons HCN + OH^-$
b. $HSO_4^- + H_2CO_3 \rightleftharpoons HCO_3^- + H_2SO_4$
c. $H_2C_2O_4 + NO_2^- \rightleftharpoons HNO_2 + HC_2O_4^-$

42. Identify all species in the following reactions as either an acid or base, in the Brønsted-Lowry sense.
a. $NH_4^+ + HCO_3^- \rightleftharpoons NH_3 + H_2CO_3$
b. $NH_2^- + H_2O \longrightarrow NH_3 + OH^-$
c. $O^{2-} + H_2O \longrightarrow OH^- + OH^-$

43. Arrange the species in the reactions of Problem 41 as Brønsted-Lowry conjugate pairs.

44. Arrange the species in the reactions of Problem 42 as Brønsted-Lowry conjugate pairs.

45. What is the conjugate base of each of the following acids?
a. H_2SO_4 b. H_2F^+
c. CH_3OH d. H_3O^+

46. What is the conjugate base of each of the following acids?
a. OH^- b. NH_4^+
c. C_2H_6 d. C_2H_2

47. What is the conjugate acid of each of the following bases?
a. H_2SO_4 b. CH_3OH c. I^-

48. What is the conjugate acid of each of the following bases?
a. CH_3^- b. $CH_3CH_2NH_2$ c. HSO_4^-

49. Write the aqueous ionization reaction for each of the following acids.
a. nitrous acid b. hydrobromic acid
c. acetic acid

50. Write the aqueous ionization reaction for each of the following acids.
a. phenol b. hydrosulfuric acid
c. chloric acid

51. Write the aqueous ionization reaction for each of the following bases.
a. nicotine ($C_{10}H_{14}N_2$) b. hydrazine (N_2H_4)
c. trimethylamine [$(CH_3)_3N$]

52. Write the aqueous ionization reaction for each of the following bases.
a. $(CH_3)_2NH$ b. pyridine (C_5H_5N)
c. hydroxylamine (NH_2OH)

*53. Classify each of the following solutes according to the pH of their aqueous solutions, as less than 7, approximately 7, or greater than 7.
a. KNO_3 b. K_2CO_3 c. NH_4Cl
d. Na_2S e. KNH_2

*54. Classify each of the following solutes according to the pH of their aqueous solutions, as less than 7, approximately 7, or greater than 7.
a. $NaNO_2$ b. KCl c. LiF
d. NaCN e. SrO

*55. Write the equation of the hydrolysis reaction(s) that occur(s) when each of the solutes in Problem 53 is dissolved.

*56. Write the equation of the hydrolysis reaction(s) that occur(s) when each of the solutes in Problem 54 is dissolved.

57. For each of the following amphiprotic species, write two different proton-transfer reactions with water.
a. HSO_4^- b. HCO_3^-

58. For each of the following amphiprotic species, write equations for the proton-transfer reaction(s) with water.
a. HS^- b. H_2O

59. Which is the stronger acid in each of the following pairs?
a. HCl, HNO_2 b. CH_3COOH, HF

*60. Which is the stronger acid in each of the following pairs?
a. H_3O^+, HCN b. HSO_4^-, HS^-

*61. Which is the stronger base in each of the following pairs?
a. Br^-, OH^- b. NO_2^-, F^-

*62. Which is the stronger base in each of the following pairs?
a. HSO_4^-, HS^- b. CN^-, OH^-

*63. Write the equation for the proton-transfer equilibrium that occurs between each of the following pairs of reactants. Indicate whether the equilibrium lies to the right or the left.
a. HF, NO_2^- b. HSO_4^-, HCO_3^-

*64. Write the equation for the proton-transfer equilibrium that occurs between each of the following pairs of reactants. Indicate whether the equilibrium lies to the right or the left.
a. $H_2PO_4^-$, HCN b. H_2CO_3, CO_3^{2-}

Polyprotic Acids and Polyhydric Bases

65. Write the chemical equations for the stepwise ionization of oxalic acid ($H_2C_2O_4$), a diprotic acid.

66. Write the chemical equations for the stepwise ionization of citric acid ($H_3C_6H_5O_7$), a triprotic acid.

67. Which is the stronger acid, $H_2PO_4^-$ or HPO_4^{2-}? Why?

68. When H_3AsO_4 is dissolved in water, several different arsenic-containing species are formed. What are they?

69. Sulfide ion is a dihydric base. Write the equations for its stepwise acceptance of two protons from water.

70. Both calcium hydroxide and ethylene diamine (H_2NCH_2-CH_2NH_2) are dihydric bases, but for different reasons. Explain.

Structure-Strength Correlation

71. Arrange the following binary hydrides in order of increasing acidity, and explain your reasoning: PH_3, NH_3, and AsH_3.
72. Arrange the following binary hydrides in order of increasing acidity, and explain your reasoning: PH_3, SiH_4, HCl, and H_2S.
*73. In each of the following pairs, identify the stronger acid and explain your reasoning.
    a. FCH_2COOH, $F_2CHCOOH$ b. H_2Se, H_2Te
    c. H_3PO_4, $H_2PO_4^-$
*74. In each of the following pairs, identify the stronger acid and explain your reasoning.
    a. H_2SO_3, H_2SO_4
    b. FCH_2CH_2COOH, $CH_3CHFCOOH$
    c. HCO_3^-, H_2CO_3
*75. Arrange the following acids in order of increasing strength, and explain your reasoning: $HClO_2$, HOCl, and HOBr.
*76. If HOF were stable in aqueous solution (it is not), would you expect it to be a stronger acid than HOCl? Explain.
*77. Why is NH_2^- a stronger base than OH^-? (*Hint:* Explain in terms of the relative strengths of the conjugate acids.)
*78. Which is the stronger base, PH_3 or NH_3? Why?

Lewis Acids and Bases

*79. In each of the following reactions, identify each species as a Lewis acid, base, or adduct.
    a. $AlCl_3 + Br^- \longrightarrow AlCl_3Br^-$
    b. $Be(OH)_2(s) + 2\ OH^-(aq) \longrightarrow Be(OH)_4^{2-}(aq)$
*80. In each of the following reactions, identify each species as a Lewis acid, base, or adduct. Rewrite the chemical equation using structural rather than molecular formulas, and show the movement of the electron pairs in each reaction.
    a. $SnCl_4 + 2\ Cl^- \longrightarrow SnCl_6^{2-}$
    b. $NH_3 + BF_3 \longrightarrow H_3N{:}BF_3$

Neutralization and Titration

81. What volume of 0.0998 M NaOH is required to reach the equivalence point in a titration of 33.5 mL 0.152 M HClO?
82. What volume of 0.109 M HNO_3 is required to reach the equivalence point in a titration of 25.00 mL 0.0750 M KOH?
83. If 23.07 mL of 0.1000 M NaOH is required in the titration of 16.51 mL aqueous hydrofluoric acid, what is the acid molarity?
84. If 9.81 mL of 0.1059 M HCl is required in the titration of 50.00 mL aqueous ammonia, what is the molarity of the ammonia?
85. What volume of 0.2473 M HCl is required to completely neutralize 25.00 mL of 0.0930 M $Ba(OH)_2$?
86. What volume of 0.1504 M KOH is required to completely neutralize 25.00 mL of 0.0750 M H_2SO_4?
87. What volume of 0.500 M KOH is required to neutralize 5.72 g benzoic acid ($C_6H_5CO_2H$)?
88. What volume of 0.500 M acetic acid is required to neutralize 5.72 g KOH?
89. What volume of 0.0200 M HCl is required to neutralize 3.48 g $Ca(OH)_2$?
90. What volume of 0.500 M H_2SO_4 is required to neutralize 0.762 g LiOH?
91. What volume of 0.500 M H_2SO_4 is required to neutralize 0.219 g $Mg(OH)_2$?
92. What volume of 0.01500 M $Ba(OH)_2$ is required to neutralize 0.219 g oxalic acid ($H_2C_2O_4$, diprotic)?
*93. If 19.7 mL of 0.1099 M KOH is required to neutralize a 25.0 mL sample of an acid, and the acid is known to be diprotic, what is its molarity?
*94. If 37.3 mL of 0.0996 M KOH is required to neutralize a 25.0 mL sample of an acid, and the acid is known to be triprotic, what is its molarity?
*95. It is found that a 0.482 g sample of a pure acid requires 18.3 mL of 0.103 M NaOH to reach the equivalence point in a titration. Assuming that the acid is monoprotic, what is its molar mass?
*96. A 486 mg sample of a pure acid requires 40.2 mL of 0.0991 M KOH to reach the equivalence point in a titration. Assuming that the acid is monoprotic, what is its molar mass?
*97. It is found that a 3.52 mg sample of a pure acid requires 6.63 mL of 0.0102 M KOH for complete neutralization. Assuming that the acid is diprotic, what is its molar mass?
*98. It is found that a 0.274 g sample of a pure acid requires 37.3 mL of 0.1006 M NaOH for complete neutralization. Assuming that the acid is diprotic, what is its molar mass?

Other Problems

99. Write proton-transfer autoionization equations for the following amphiprotic solvents.
    a. NH_3 b. NH_2OH
    c. H_2SO_4
*100. Calculate the pH of the following solutions of strong acids.
    a. 1.0% HCl (weight percent; assume density = 1.00 g mL^{-1})
    b. 0.050% HBr (weight percent; assume density = 1.00 g mL^{-1})
101. Two solutions of identical molarity are prepared. One is a weak acid, and the other is a strong acid. Which has the greater pH? Which reacts more vigorously with metallic zinc?

102. Classify each of the following reactions as ionization or hydrolysis.
 a. $CH_3COOH + H_2O \longrightarrow CH_3COO^- + H_3O^+$
 b. $(CH_3)_3N + H_2O \longrightarrow (CH_3)_3NH^+ + OH^-$
 c. $NH_2^- + H_2O \longrightarrow NH_3 + OH^-$
 d. $NH_3 + H_2O \longrightarrow NH_4^+ + OH^-$
 e. $CO_3^{2-} + H_2O \longrightarrow HCO_3^- + OH^-$

*103. All of the following equilibria lie to the right. Use this information to arrange the acids HSO_4^-, HBrO, HCN, HNO_2, and H_2S in order of increasing acid strength.
 a. $HSO_4^- + HS^- \rightleftharpoons H_2S + SO_4^{2-}$
 b. $HBrO + CN^- \rightleftharpoons BrO^- + HCN$
 c. $HSO_4^- + NO_2^- \rightleftharpoons SO_4^{2-} + HNO_2$
 d. $H_2S + BrO^- \rightleftharpoons HS^- + HBrO$
 e. $HNO_2 + HS^- \rightleftharpoons NO_2^- + H_2S$

104. Explain why Brønsted-Lowry acids are not Lewis acids. Are Brønsted-Lowry bases the same as Lewis bases? Explain.

105. If 39.27 mL of 0.01006 M KOH is needed to reach the first equivalence point in a titration of 25.00 mL H_2SO_4, what is the molarity of the acid?

The next four problems require an understanding of *normality* and other concepts found in Appendix F.

106. For each of the following solutions, change molarity to normality or normality to molarity, as appropriate, when the compound reacts with a strong acid or base.
 a. 0.5 M H_2SO_4
 b. 0.1 M HBr
 c. 1.5 M $Ba(OH)_2$
 d. 0.08 M NH_3
 e. 3.5×10^{-3} M HNO_3
 f. 0.75 N KOH
 g. 0.005 N H_2SO_4
 h. 3.5 N HCl
 i. 0.10 N H_3PO_4

107. How much 0.142 N acid will be required for the titration of 25.0 mL of 0.35 N base?

108. How much 0.0750 N acid is required to titrate 0.100 g $Ba(OH)_2$?

109. It is found that a 0.191 g sample of a pure acid requires 25.4 mL of 0.103 M NaOH to reach the equivalence point in a titration. What is the equivalent weight of the acid? If the acid is diprotic, what is its molar mass?

CHAPTER 16

IONIC EQUILIBRIUM: ACIDS AND BASES

Aluminum chloride, an acidic compound that contains no hydrogen, reacts vigorously with water: $AlCl_3(s) + 3\ H_2O \rightarrow Al(OH)_3(s) + 3\ HCl(aq)$. The HCl turns the indicator (methyl red) to its red form.

16.1 IONIZATION OF WEAK ACIDS AND BASES

The primary focus of the preceding chapter was the qualitative description of the chemistry of acids and bases. Our goal in this chapter is to develop the quantitative aspects of Brønsted-Lowry theory—that is, the quantitative treatment of proton-transfer reactions in aqueous solutions.

Ionization Constants

The ionization of an aqueous weak acid such as nitrous acid is described in Brønsted-Lowry terms as a reversible proton-transfer reaction between the acid and the solvent.

$$HNO_2 + H_2O \rightleftharpoons NO_2^- + H_3O^+ \tag{16-1}$$

Weak-acid ionization reactions reach a position of equilibrium in which appreciable concentrations of both the ionized form NO_2^- and the nonionized form HNO_2 are present. The system is described in the usual way by a reaction quotient, which at equilibrium is equal to the equilibrium constant of the reaction.

$$Q_{eq} = K_{eq} = \frac{[H_3O^+][NO_2^-]}{[HNO_2]} \tag{16-2}$$

The concentration of *water* does not appear in this expression. As discussed in Section 13.4, the concentration of the solvent in a dilute solution is not included in the equilibrium expression. This is true even if, as in weak acid or base ionization, the solvent is a participating reactant or product. The equilibrium constant for a weak acid is usually given the symbol K_a, called either the **ionization constant** of the weak acid, or the **acid ionization constant.**

Equilibrium expressions do not include species whose concentration does not change during the reaction.

We can represent the ionization equilibrium of *any* weak acid by Equation (16-3)

$$\underset{\textbf{acid}}{HA} + H_2O \rightleftharpoons H_3O^+ + \underset{\textbf{conjugate base}}{A^-} \tag{16-3}$$

where A^- stands for the *anionic portion* of the acid HA (that is, its conjugate base). The general expression for the ionization constant is

$$K_a = \frac{[H_3O^+][A^-]}{[HA]} \tag{16-4}$$

The ionization of weak bases is treated in the same way. An aqueous base such as ammonia ionizes according to

$$NH_3 + H_2O \rightleftharpoons NH_4^+ + OH^- \tag{16-5}$$

The corresponding equilibrium expression is

$$K_b = \frac{[NH_4^+][OH^-]}{[NH_3]} \tag{16-6}$$

in which K_b is the **base ionization constant.** For the generalized base ionization equilibrium

$$B + H_2O \rightleftharpoons BH^+ + OH^- \tag{16-7}$$

the ionization constant is

$$K_b = \frac{[BH^+][OH^-]}{[B]} \tag{16-8}$$

Once again, the concentration of *water* does not appear in Equation (16-8), even though water is a reactant.

The units of both K_a and K_b are mol L^{-1}. These units are normally not written out since they are always the same. In this text, we do not include units when quoting values of ionization constants, nor when using them in numerical calculations.

The Sørenson "p" notation, initially used to define pH, is also used to express the value of the ionization constant of a weak acid or base. The "pK" of an acid or base is defined by Equation (16-9).

In "p" notation, the symbol pX means $(-1) \cdot \log X$.

$$pK_a = -\log K_a; \qquad pK_b = -\log K_b \tag{16-9}$$

Conversions of pK values to values of ionization constants, and vice versa, are mathematically the same as conversions between pH and $[H_3O^+]$.

EXAMPLE 16.1 Conversion of K_a to pK_a

The ionization constant of benzoic acid ($C_6H_5CO_2H$) at room temperature is 6.3×10^{-5}. What is the pK_a of benzoic acid?

Analysis

Target: pK_a = ?

Knowns: The ionization constant. (The identity of the acid and the temperature are given, but not needed.)

Relationship: Equation (16-9) relates the pK_a to K_a.

Plan Substitute the known into the formula.

Work

$$\begin{aligned} pK_a &= -\log K_a \\ &= -\log (6.3 \times 10^{-5}) \\ &= -(-4.20) \\ &= 4.20 \end{aligned}$$

Check Most weak acids and bases have pK's that are in the range +1 to +10. A calculated value outside this range is not necessarily wrong, but the math should be double-checked.

Exercise

What is the pK_a of an acid whose ionization constant is 7.5×10^{-8}?

Answer: 7.12.

The inverse procedure of finding an acid or base ionization constant from a given pK is frequently needed, since data on acids and bases are often given as pK values rather than as K_a or K_b.

EXAMPLE 16.2 Conversion of pK_b to K_b

The pK_b of aniline ($C_6H_5NH_2$) is 9.42 at 25 °C. What is the base ionization constant?

Analysis

Target: K_b = ?

Knowns: The identity and pK_b of the base, and the temperature.

Relationship: Equation (16-9) connects the pK_b and K_b. Recall that the relationship $y = 10^{\log y}$ holds for any positive number y.

Plan Insert the given values into Equation (16-9), and solve by taking the antilogarithm. Do not omit the minus sign.

Use the "10^x," "inv log," or "2nd 10^x" key on the calculator.

Work

$$pK_b = -\log K_b \quad (16\text{-}9)$$

$$\log K_b = -pK_b$$

$$K_b = 10^{\log K_b} = 10^{-(pK_b)}$$

$$= 10^{(-9.42)} = 3.8 \times 10^{-10}$$

Exercise

What is the ionization constant of an acid whose pK_a is 6.2?

Answer: 6×10^{-7}.

Example 16.2 and its exercise differ in the precision of the quantities, as indicated by the number of significant digits used. As pointed out in the preceding chapter (Example 15.1), the number of significant digits in a quantity is the same as the number of digits *to the right of the decimal point* in its logarithm.

Relative Strengths of Weak Acids and Bases

The equilibrium expression $K_a = [H_3O^+][A^-]/[HA]$ relates the concentrations of solute species after a weak aqueous acid has reached ionization equilibrium. In common with all equilibrium constants, the magnitude of K_a is correlated with the position of equilibrium. If K_a is small, the equilibrium lies to the left: the concentration of the nonionized form HA (HNO_2, for example) is relatively large, and the concentration of the ionized form A^- (NO_2^-, for example) is relatively small. If, on the other hand, K_a is large, the substance exists predominantly in the ionized form A^-. We have previously defined a strong acid as one in which ionization is essentially complete. We now see that if K_a is large, the acid is strong. If K_a is small, the acid is weak. A similar correlation holds between the magnitude of K_b and the strength of bases.

In discussions of acid-base chemistry, the word "strength" refers to the tendency to lose (or gain) a proton, or it refers to the extent of ionization in aqueous solution. Outside this context, however, the word is sometimes used to refer to concentration, as in "strong coffee." The chemical properties of acidic solutions are due to H_3O^+ ions, and the greater the concentration of H_3O^+ the more pronounced the properties. For instance, active metals such as zinc react with and dissolve in acids:

$$Zn + 2\,H_3O^+(aq) \longrightarrow Zn^{2+}(aq) + 2\,H_2O + H_2(g)$$

(a)

(b)

(c)

Figure 16.1 Gaseous hydrogen is evolved rapidly and vigorously when zinc reacts with a strong acid. (a) Zinc reacts with 0.10 M HCl. (b) The reaction is slower when zinc reacts with 0.10 M acetic acid, a weak acid. The bubbles of hydrogen are smaller and fewer. The difference is due to the concentration of hydronium ion, which is 0.10 M in (a) and only 0.0013 M in (b). (c) A solution of 0.0013 M HCl reacts at the same rate as a 0.10 M solution of acetic acid, because the concentration of H_3O^+ is the same in both solutions.

The reaction is slow if $[H_3O^+]$ is low, and fast if $[H_3O^+]$ is large. Although it is natural to think of the high-$[H_3O^+]$ solution causing the fast reaction as being "stronger," it is incorrect. "Strength" refers to proton-transfer tendency, *not* to concentration (Figure 16.1).

There are large variations in strength among weak acids and bases. Acetic acid has $K_a = 1.8 \times 10^{-5}$, while phenol (C_6H_5OH) has $K_a = 1.0 \times 10^{-10}$. Acetic acid is a weak acid, but phenol is much weaker. Ammonia has $K_b = 1.9 \times 10^{-5}$, and pyridine has $K_b = 1.4 \times 10^{-9}$. Ammonia is a weak base, and pyridine is weaker. When acids are listed in order of decreasing ionization constants, the list also indicates the decreasing relative strength of the acids. Table 16.1 gives some representative acids, their ionization reactions, ionization constants, and pK_a's. The strongest acids are at the top of the table. Note particularly the inverse relationship between pK_a and acid strength: the weaker the acid, the larger the pK_a. A more complete table is given in Appendix J.

Such a list was given in the preceding chapter, although numerical values of K_a were not included (page 655).

Since there is a continuous range of K_a values among acids, the distinction between "strong" and "weak" is somewhat arbitrary. Most people would classify sulfuric and nitric acids as strong, trichloroacetic acid as borderline, and all the others in Table 16.1 as weak. These terms are matters of convenience, and not intended to serve as the basis for fine distinctions.

TABLE 16.1 Relative Strengths of Acids at 25 °C

(Increasing Acid Strength ↑)

Acid	Ionization Reaction	K_a	pK_a
Sulfuric acid	$H_2SO_4 + H_2O \rightleftharpoons H_3O^+ + HSO_4^-$	~1000	−3
Nitric acid	$HNO_3 + H_2O \rightleftharpoons H_3O^+ + NO_3^-$	44	−1.64
Trichloroacetic acid	$CCl_3CO_2H + H_2O \rightleftharpoons H_3O^+ + CCl_3CO_2^-$	0.22	0.66
Hydrofluoric acid	$HF + H_2O \rightleftharpoons H_3O^+ + F^-$	6.6×10^{-4}	3.18
Nitrous acid	$HNO_2 + H_2O \rightleftharpoons H_3O^+ + NO_2^-$	4.0×10^{-4}	3.40
Acetic acid	$CH_3CO_2H + H_2O \rightleftharpoons H_3O^+ + CH_3CO_2^-$	1.8×10^{-5}	4.74
Hydrosulfuric acid	$H_2S + H_2O \rightleftharpoons H_3O^+ + HS^-$	1×10^{-7}	7
Hypochlorous acid	$HOCl + H_2O \rightleftharpoons H_3O^+ + OCl^-$	3.0×10^{-8}	7.52
Hydrocyanic acid	$HCN + H_2O \rightleftharpoons H_3O^+ + CN^-$	4.0×10^{-10}	9.40
Phenol	$C_6H_5OH + H_2O \rightleftharpoons H_3O^+ + C_6H_5O^-$	1.0×10^{-10}	10.0
Ethanol	$C_2H_5OH + H_2O \rightleftharpoons H_3O^+ + C_2H_5O^-$	~10^{-16}	~16

Percent Ionization

When a neutral weak acid HA undergoes an ionization reaction, it loses a proton and becomes a negatively charged ion. The process can be thought of either as an ionization or as a *deprotonation*. The several pairs of terms commonly used to describe the species involved in the process are defined in Equations (16-3a) through (16-3c)

$$\text{HA} + H_2O \rightleftharpoons A^- + H_3O^+ \qquad (16\text{-}3)$$

$$\text{conjugate acid} + \text{water} \rightleftharpoons \text{conjugate base} + \text{hydronium ion} \qquad (16\text{-}3a)$$

$$\text{neutral form} + \text{water} \rightleftharpoons \text{ionized form} + \text{hydronium ion} \qquad (16\text{-}3b)$$

$$\text{protonated form} + \text{water} \rightleftharpoons \text{deprotonated form} + \text{hydronium ion} \qquad (16\text{-}3c)$$

As discussed in the preceding chapter, the extent to which the process takes place can be described by the *percent ionization.*

$$\%\ \text{ionization} = \frac{100 \cdot [A^-]}{[\text{initial acid}]} \qquad (16\text{-}10a)$$

Since the reaction produces equal amounts of A^- and H_3O^+, their concentrations are equal, and percent ionization may also be calculated by

$$\%\ \text{ionization} = \frac{100 \cdot [H_3O^+]}{[\text{initial acid}]} \qquad (16\text{-}10b)$$

Regardless of the extent to which ionization takes place, the total concentration of the protonated and deprotonated forms, ([HA] + [A^-]), must be equal to the initial concentration of the acid, that is, to the molarity of the solution. A third form of the definition, therefore, is

$$\%\ \text{ionization} = \frac{100 \cdot [A^-]}{[\text{HA}] + [A^-]} \qquad (16\text{-}10c)$$

The molarity of the solution (initial acid concentration) is usually known, so that determination of percent ionization requires a measurement or calculation of the equilibrium value of [A^-], [H_3O^+], or pH.

$$\text{fractional ionization} = \frac{\%\ \text{ionization}}{100}$$

Fractional ionization, defined as in Equations (16-10a, b, and c) but without the factor 100, is also used to describe the extent to which ionization takes place.

Calculation of Concentrations in Acid and Base Ionization Equilibria

Values of [H_3O^+] and other quantities are calculated by means of the structured approach set out in Chapter 13. The approach is applicable to all equilibrium problems, acid or otherwise.

EXAMPLE 16.3 Percent Ionization of a Weak Acid

The major acidic component of vinegar is acetic acid (CH_3CO_2H), whose ionization constant at 25 °C is 1.8×10^{-5}. The concentration of acetic acid in household vinegar is about 0.75 M. Calculate the percent ionization of the acetic acid.

Analysis

Target: % ionization = ?

Knowns: The identity of the acid, the temperature, the concentration, and the ionization constant.

Relationship: Equation (16-10a) can be used, but requires information not given: "if only I knew" the equilibrium concentration of acetate or hydronium ion, the percent ionization could be calculated.

New Target: The concentration of the ionized form.

Knowns: As before. Only the acid concentration and K_a are needed, the other information is extra.

Relationship: The structured approach of Chapter 13 supplies the needed pathway between the target and the knowns.

Plan Follow the structured approach to calculate the equilibrium concentration of acetate ion, then use Equation (16-10a) to find the percent ionization.

Work

1.

Balanced Equation	$CH_3CO_2H + H_2O \rightleftharpoons$	H_3O^+	$+ CH_3CO_2^-$
Initial	0.75	0	0
Change	$-x$	$+x$	$+x$
Equilibrium	$0.75 - x$	x	x

Since the concentration of H_2O does not appear in the equilibrium expression, it is ignored in the calculation.

$$K_a = \frac{[H_3O^+][CH_3CO_2^-]}{[CH_3CO_2H]} = \frac{x \cdot x}{0.75 - x}$$

$$K_a\,(0.75 - x) = x^2$$

$$x^2 + K_a x - 0.75\,K_a = 0$$

The quadratic equation is solved in the usual way, and the impossible root is discarded.

$$x = \frac{-K_a \pm \sqrt{K_a^2 + 4(0.75K_a)}}{2}$$

$$= 0.0037 \text{ mol L}^{-1}$$

Since x was defined so that at equilibrium $x = [CH_3CO_2^-] = [H_3O^+]$, we have found the missing information.

$$[CH_3CO_2^-] = 0.0037 \text{ M}$$

2. The initial concentration of the acid is given as 0.75 M, so

$$\%\text{ ionization} = \frac{100 \cdot [A^-]}{[\text{initial acid}]} \qquad (16\text{-}10a)$$

$$= \frac{100 \cdot [CH_3CO_2^-]}{[\text{initial } CH_3CO_2H]}$$

$$= \frac{100 \cdot (0.0037 \text{ M})}{(0.75 \text{ M})} = 0.49$$

The extent of ionization of the major acidic component of vinegar is about 0.5%.

Check The answer can be checked by evaluation of the equilibrium expression

$$K_a = \frac{(0.0037)^2}{0.75 - 0.0037} = 1.8 \times 10^{-5}$$

Exercise

What is the percent ionization in 0.100 M acetic acid?

Answer: 0.13%.

In working out Example 16.3, we used the quadratic equation. While this is straightforward, it is also tedious. There is an easier way that may almost always be used in acid-base ionization problems. In Example 16.3, the equation to be solved was

$$K_a = \frac{x^2}{0.75 - x} \tag{16-11}$$

If the percent ionization is small enough, the extent of reaction x is so small that the quantity $(0.75 - x)$ in the denominator is only slightly less than 0.75. Therefore, at the cost of a very slight inaccuracy in the result, we may substitute 0.75 for the quantity $(0.75 - x)$. That is, we may simply ignore the x in the denominator of Equation (16-11) and solve the simpler form, Equation (16-12).

$$K_a = \frac{x^2}{0.75} \tag{16-12}$$

$$x^2 = 0.75 \cdot K_a$$

$$x = \sqrt{x^2} = \sqrt{0.75 \cdot K_a}$$

In this example, the value of K_a is 1.8×10^{-5}, so that

$$x = \sqrt{0.75 \cdot 1.8 \times 10^{-5}}$$

$$= 0.0037 \text{ M}$$

The answer is the same as that reached in Example 16.3. Note that the given values (0.75 M and 1.8×10^{-5}) limit the precision of the result to two significant digits. This approximate method may be used in most calculations.

Laboratory measurements of concentrations are rarely accurate to more than three significant digits. Acid and base ionization constants, on the other hand, are usually known only to two significant digits. Therefore, the approximate method is usually valid as long as it introduces no error in the second digit of the result.

Let us say that an error of ± 5% is tolerable in most equilibrium calculations. The approximate formula used above for the extent of reaction is

$$x \approx \sqrt{KC} \tag{16-13}$$

in which C is the initial concentration of the acid or base, and K is either K_a or K_b. Values of $[H_3O^+]$ and other equilibrium concentrations calculated by Equation (16-13) are in error by no more than 5% whenever x is less than 10% of C. To solve this type of problem, first calculate x by using Equation (16-13) and compare the result to C. If you find that $x > 0.1 \cdot C$ (or, equivalently, if $x/C > 0.1$), recalculate x using the quadratic formula method of Example 16.3. Note that in Example 16.3, x is less than $0.1 \cdot C$, so the approximate result is within 5% of the result obtained by the quadratic formula.

Problem 97 at the end of the chapter asks you to prove that the error is less than 5% when x is less than $0.1 \cdot C$.

In Example 16.3, $x = 0.0037$, and $0.1 \cdot C = 0.075$. Therefore, x is much less than $0.1 \cdot C$.

In many acid-base equilibrium problems, it is the pH of the solution that is of interest. The following calculation is typical.

EXAMPLE 16.4 pH of a Weak-Acid Solution

Ant bites are painful in part because they leave small quantities of formic acid (HCO_2H, 46.0 g mol^{-1}) beneath the skin. If 50.0 mL of an aqueous solution contains 173 mg formic acid, what is the pH? The ionization constant of formic acid is 1.8×10^{-4}.

The name "formic acid" comes from *formica,* the Latin word for "ant."

Analysis

Target: $[H_3O^+]$, expressed as pH.

Knowns: The acid ionization constant, and information from which the initial concentration of the acid can be calculated.

Relationship: The structured approach to equilibrium problems, developed in Chapter 13, provides a pathway between the knowns and the target. In most cases the approximate formula, Equation (16-13), is appropriate.

Plan Calculate the initial acid concentration from the given data, use the structured approach to calculate the value of $[H_3O^+]$ at equilibrium, and calculate the pH.

Work

1.

$$\text{initial concentration} = (173\ \text{mg}) \cdot \frac{(1\ \text{mmol}/46.0\ \text{mg})}{(50.0\ \text{mL})}$$

$$= 0.0752\ \text{mmol mL}^{-1} = 0.0752\ \text{M}$$

2.

Balanced Equation	$HCO_2H + H_2O \rightleftharpoons$	H_3O^+ +	HCO_2^-
Initial	0.0752	0	0
Change	$-x$	$+x$	$+x$
At Equilibrium	$(0.0752 - x)$	x	x

$$K_a = \frac{[H_3O^+][HCO_2^-]}{[HCO_2H]} = \frac{x \cdot x}{0.0752 - x}$$

$$K_a \approx \frac{x^2}{0.0752}$$

$$x \approx \sqrt{K_a \cdot 0.0752}$$

$$\approx \sqrt{(1.8 \times 10^{-4}) \cdot 0.0752} = \sqrt{1.35 \times 10^{-5}}$$

$$\approx 0.00368$$

Check $x/C = 0.00368/0.0752 = 0.05$, so the approximation is valid.

$$[H_3O^+] = x = 0.00368\ \text{M}$$

3.

$$\text{pH} = -\log [H_3O^+]$$

$$= 2.43$$

Note: The final answer is rounded to two significant figures, in conformance with the two-digit value of K_a.

Exercise

Suppose an acid has a molar mass of 96 and a pK_a of 6.35. What is the pH of an aqueous solution containing 5.0 g of the acid per liter of solution?

Answer: Acid concentration = 0.052 M; $K_a = 4.5 \times 10^{-7}$, pH = 3.82.

The calculations for the equilibrium ionizations of weak bases are essentially the same. One must be careful in pH conversions, however, because in a base calculation the extent of reaction $x = [OH^-]$ rather than $[H_3O^+]$. The method leads to the pOH, not the pH, of the solution.

EXAMPLE 16.5 pH of a Solution of a Weak Base

Household ammonia is an aqueous solution of NH_3 at a concentration of about 2.0 M. Given that K_b for ammonia is 1.9×10^{-5} at 25 °C, calculate the pH.

Analysis

Target pH = ?; $[OH^-]$ or pOH is needed.

Knowns: The identity of the base, its concentration and its K_b.

Relationships: $[OH^-]$ is related to K_b by Equation (16-8). In aqueous solutions at 25 °C, pH + pOH = 14.00.

Plan Use the structured approach to find the equilibrium concentration of OH^-, calculate pOH from $[OH^-]$, and calculate pH from pOH.

Work

1.

Balanced Equation	NH_3	+ $H_2O \rightleftharpoons$	NH_4^+	+ OH^-
Initial	2.0		0	0
Change	$-x$		$+x$	$+x$
At Equilibrium	$2.0 - x$		x	x

$$K_b = \frac{[NH_4^+][OH^-]}{[NH_3]} = \frac{x \cdot x}{2.0 - x}$$

$$K_b = \frac{x^2}{2.0 - x} \approx \frac{x^2}{2.0}$$

$$x = \sqrt{K_b \cdot 2.0} = \sqrt{(1.9 \times 10^{-5}) \cdot 2.0}$$

$$\approx 0.00616$$

Check $x/C = 0.00616/2.0 = 0.003, < 0.1$.

In the case of base ionization, the quantity x is the equilibrium concentration of hydroxide ion rather than hydronium ion.

$$[OH^-] = x = 0.00616 \text{ M}$$

2.
$$pOH = -\log [OH^-] = -(-2.21)$$
$$= 2.21$$

3.
$$pH + pOH = 14.00$$
$$pH = 14.00 - pOH = 14.00 - 2.21$$
$$= 11.79$$

Exercise

If household ammonia is diluted by a factor of 10, so that its concentration is 0.20 M, what is its pH?

Answer: pOH = 2.71, pH = 11.29.

Determination of K_a from pH

The value of the ionization constant for a particular acid can be determined from measurements of equilibrium concentrations. Since the concentrations of hydronium ion and of the protonated and deprotonated forms are related through the stoichiometry, only one of these must be measured in order to determine the value of K_a. In practice, it is simplest to measure the pH.

EXAMPLE 16.6 Determination of K_a from Measurement of pH

A 0.0100 M solution of *o*-chlorophenol (C_6H_4ClOH) has a pH of 5.24 at 25 °C. What is the value of K_a for this acid?

Analysis K_a may be found from the equilibrium concentrations through Equation (16-4). Equilibrium concentrations can be found from pH, since they are related to $[H_3O^+]$ through stoichiometry.

Plan Use the standard approach for equilibrium problems. Note that in this problem, it is the ionization constant rather than the extent of reaction that is the unknown.

Work

Balanced Equation	$C_6H_4ClOH + H_2O \rightleftharpoons$	$C_6H_4ClO^-$	$+ H_3O^+$
Initial	0.0100	0	0
Change	$-x$	$+x$	$+x$
At Equilibrium	$(0.0100 - x)$	x	x

$$K_a = \frac{[H_3O^+][C_6H_4ClO^-]}{[C_6H_4ClOH]} = \frac{x \cdot x}{0.0100 - x}$$

The value of x is known, since $x = [H_3O^+]$ and $[H_3O^+] = 10^{-pH}$

$$x = 10^{-5.24} = 5.75 \times 10^{-6}$$

Substitute the value of x into the equilibrium expression and evaluate K_a.

$$K_a = \frac{(5.75 \times 10^{-6})^2}{0.0100 - 5.75 \times 10^{-6}}$$

$$= 3.3 \times 10^{-9}$$

Exercise

The pH of a 0.0500 M solution of an unknown acid is 3.75. What is the pK_a?

Answer: $K_a = 6.3 \times 10^{-7}$, $pK_a = 6.20$.

Very Dilute or Very Weak Acids

What is the pH of a 10^{-9} M solution of HCl? The quick (and incorrect) answer is, "The strong acid HCl is 100% ionized, so $[H_3O^+]$ is 10^{-9} and pH = 9." On careful consideration, we realize that $[H_3O^+]$ in pure water is 10^{-7}, which is 100 times greater than 10^{-9}. Since it is not possible to *decrease* $[H_3O^+]$ by the addition of any quantity of an *acid* to pure water, the quick answer must be wrong.

In very dilute solutions of acids, the amount of H_3O^+ supplied by the acid is negligible in comparison with the amount already present due to the autoionization of water. The same situation arises when the amount of H_3O^+ supplied by the acid is negligibly small, not because of low concentration, but because the acid is exceedingly weak. For instance, K_a for ethanol is about 10^{-16}, and according to Equation (16-13) the concentration of H_3O^+ in a 1 M solution should be $\approx \sqrt{K_a C} = 10^{-8}$. But this cannot be correct, because the concentration of H_3O^+ in pure water cannot be decreased by the addition of an acid, regardless of the strength of that acid. In practice, then, *the presence of exceedingly weak acids (or bases) has no effect on the pH of an aqueous solution.*

16.2 Hydrolysis of Salts

In Chapter 15 we described the proton-transfer reaction of a neutral acid (or base) as an *ionization,* while using a different word, *hydrolysis,* to describe the reaction of an ionic acid (or base).

Ionization $HNO_2 + H_2O \rightleftharpoons NO_2^- + H_3O^+$

Ionization $NH_3 + H_2O \rightleftharpoons NH_4^+ + OH^-$

Hydrolysis $NH_4^+ + H_2O \rightleftharpoons NH_3 + H_3O^+$

Hydrolysis $NO_2^- + H_2O \rightleftharpoons HNO_2 + OH^-$

An ionic acid or base is always associated with a counterion: that is, it is part of a compound. For example, ammonium ion is associated with Cl^- in NH_4Cl, and nitrite ion is associated with potassium ion in KNO_2. Because these compounds are salts, the term **"salt hydrolysis"** is often used to describe the proton-transfer reaction of an ionic acid or base with water.

Reciprocal Relationships

In any acid-base conjugate pair, there is a **reciprocal relationship** between the relative strengths: the stronger the acid, the weaker the conjugate base, and vice versa. HCN, for example, is a *very* weak acid—its conjugate base CN^-, while still a weak base, is relatively stronger than many weak bases. The acid ionization equilibrium, Equation (16-14), lies far to the left, while the equilibrium of the conjugate base hydrolysis, Equation (16-15), lies more to the center.

$$HCN + H_2O \rightleftharpoons H_3O^+ + CN^- \qquad (16\text{-}14)$$

$$CN^- + H_2O \rightleftharpoons OH^- + HCN \qquad (16\text{-}15)$$

The reciprocal relationship between the relative strengths of an acid and its conjugate base can be expressed in algebraic terms. Suppose we write the equilibrium expressions for Reactions (16-14) and (16-15)

$$K_a = \frac{[H_3O^+][CN^-]}{[HCN]} \quad \text{and} \quad K_b = \frac{[OH^-][HCN]}{[CN^-]}$$

and multiply them together, cancelling terms as appropriate.

$$K_a \cdot K_b = \frac{[H_3O^+]\cancel{[CN^-]}}{\cancel{[HCN]}} \cdot \frac{[OH^-]\cancel{[HCN]}}{\cancel{[CN^-]}} = [H_3O^+][OH^-]$$

The right-hand side of this expression is just the *autoionization constant of water,*

The ion product and autoionization constant of water were described in Section 15.5.

$$K_w = [H_3O^+][OH^-]$$

so that

$$K_a \cdot K_b = K_w \qquad (16\text{-}16)$$

$K_w = 1.0 \times 10^{-14}$ at 25 °C.

Equation (16-16) applies to *any* acid/base conjugate pair. When Equation (16-16) is written as $K_b = K_w/K_a$, it more clearly demonstrates the reciprocal relationship: the larger the value of K_a, the smaller the value of K_b, and vice versa.

All of the foregoing applies equally to weak bases and their conjugate acids, for instance ammonia and ammonium chloride.

$$NH_3 + H_2O \rightleftharpoons NH_4^+ + OH^-; \qquad K_b = [NH_4^+][OH^-]/[NH_3]$$

$$NH_4^+ + H_2O \rightleftharpoons NH_3 + H_3O^+; \qquad K_a = [NH_3][H_3O^+]/[NH_4^+]$$

$$K_b \cdot K_a = [OH^-][H_3O^+] = K_w$$

EXAMPLE 16.7 pH of a Solution of a Salt of a Weak Acid

Calculate the pH of a 0.100 M solution of NaCN. For the weak acid HCN, $K_a = 4.0 \times 10^{-10}$.

Analysis

Target: pH of a solution of the *salt* NaCN.

Knowns: Concentration and identity of the salt, and ionization constant of the *conjugate acid,* HCN.

Relationship: Because an ionic base (a salt) is involved rather than an uncharged base like NH_3, this problem may *look* different. Once the value of K_b is obtained from K_a and K_w, however, this problem is the same as Example 16.5.

Plan Calculate K_b from the given K_a, use the standard approach to find the equilibrium concentrations of all species, and obtain the pH from $[H_3O^+]$ or from the pOH.

Work

Since the values of K_w, K_a, and K_b all vary with temperature, some assumption must be made when a problem fails to specify the temperature. The usual assumption is 25 °C, at which $K_w = 1.0 \times 10^{-14}$.

1.

$$K_b = \frac{K_w}{K_a} = \frac{1.0 \times 10^{-14}}{4.0 \times 10^{-10}} = 2.50 \times 10^{-5}$$

2.

Balanced Equation	$CN^- + H_2O \rightleftharpoons$	$HCN +$	OH^-
Initial	0.100	0	0
Change	$-x$	$+x$	$+x$
At Equilibrium	$(0.100 - x)$	x	x

$$K_b = \frac{[HCN][OH^-]}{[CN^-]} = \frac{x \cdot x}{0.100 - x}$$

$$\approx \frac{x^2}{0.100}$$

$$x \approx \sqrt{K_b \cdot 0.100} = \sqrt{(2.50 \times 10^{-5}) \cdot 0.100}$$

$$\approx 0.00158$$

Check $x/C = 0.0016/0.100 = 0.016$, $< 10\%$. The approximation is valid.

$$[OH^-] = x = 0.00158 \text{ M}$$

3.

$$\begin{aligned} pOH &= -\log[OH^-] \\ &= -\log 0.00158 = -(-2.80) \\ &= 2.80 \end{aligned}$$

$$\begin{aligned} pH &= pOH - 14.00 \\ &= 11.20 \end{aligned}$$

This 0.100 M solution of a salt of a weak acid is far from neutral. It has the same $[OH^-]$ as 0.0016 M KOH.

Exercise

When a 0.100 M solution of acetic acid ($K_a = 1.8 \times 10^{-5}$) is neutralized with an equal volume of 0.100 M KOH, the result is a 0.0500 M solution of potassium acetate. What is the pH of 0.0500 M potassium acetate?

Answer: pOH = 5.28, pH = 8.72.

EXAMPLE 16.8 Relative Base Strengths of Salts

Which is the weaker base, NaF or $NaCH_3CO_2$?

Analysis Comparison of base strengths requires knowledge of the K_b's. In this case, since the bases are the *ions* F^- and $CH_3CO_2^-$, the K_b's must be obtained from the K_a's for the conjugate acids (Table 16.1).

Work On looking up the ionization constants of the conjugate acids, we discover that HF is a stronger acid than CH_3CO_2H. We know this both from the K_a values and from the fact that HF is listed higher in the table. We need not calculate anything: since HF is the stronger acid, F^- must be the weaker base.

Exercise

Arrange the following salts in order of decreasing base strength: sodium nitrite, potassium nitrate, potassium cyanide, and sodium trichloroacetate.

Answer: $KCN > NaNO_2 > NaCCl_3CO_2 > KNO_3$.

Hydrolysis of Metal Cations

Metal cations can hydrolyze because they are *Lewis* acids. For example, when a zinc salt dissolves, the Zn^{2+} ion forms a coordinate bond by accepting an electron pair from a water molecule, a Lewis base.

$$Zn^{2+} + :O(H)_2 \longrightarrow [Zn:O(H)_2]^{2+}$$

The sharing of the electron pair decreases the electron density at the oxygen atom, causing the electrons of the O—H bonds to move closer to the O atom. This weakens the O—H bond, allowing the coordinated water molecule to donate a proton more readily than a free water molecule does. That is, coordination with the metal cation increases the Brønsted-Lowry acid strength of water. Zinc ions actually bond to four water molecules, forming the complex ion $Zn(H_2O)_4^{2+}$. This species participates in a proton-transfer equilibrium with water, acting as a Brønsted-Lowry acid.

$$[Zn(OH_2)_4]^{2+} + H_2O \rightleftharpoons [Zn(OH_2)_3(OH)]^{+} + H_3O^+ \qquad (16\text{-}17)$$

Or, more simply,

$$Zn(H_2O)_4^{2+} + H_2O \rightleftharpoons Zn(H_2O)_3OH^+ + H_3O^+ \qquad (16\text{-}18)$$

Hydrated metal cations exhibit a wide range of acid strengths, as seen in Table 16.2.

TABLE 16.2 Acid Constants for Hydrated Cations

Cation	K_a at 25 °C
Fe^{3+}	3.6×10^{-3}
Al^{3+}	7.9×10^{-6}
Cu^{2+}	1.6×10^{-7}
Pb^{2+}	1.5×10^{-8}
Co^{2+}	1.3×10^{-9}
Fe^{2+}	3.2×10^{-10}
Ni^{2+}	2.5×10^{-11}

Acid Strength Increases ↑

Neutral Salts and Amphoteric Salts

Although salt hydrolysis is common, it is not universal. Chloride, nitrate, and other anions that are conjugate to strong acids are such weak bases that they do not hydrolyze at all. Since they do not accept protons from water, their presence has no effect on the pH of an aqueous solution. Although all metal ions coordinate with water, as discussed above, the resulting hydrated cations of Groups 1A and 2A (with the exception of Be^{2+}) are such weak acids that they cannot donate a proton to water. Therefore, Group 1A and Group 2A salts of strong acids, for example NaBr or $Ba(ClO_4)_2$, do not affect the pH when dissolved in water. They are **neutral salts.**

A salt of a strong acid and a strong base is a *neutral salt.*

If the anion of a salt is conjugate to a weak acid, and the cation is conjugate to a weak base, both ions have acid/base properties. These salts are *amphoteric,* which means that they react with both acids and bases. The pH of a solution of an amphoteric salt is determined by the opposing tendencies toward hydrolysis of the two ions. One such salt is ammonium fluoride. Since K_b for F^- is about 10^{-10} and K_a for NH_4^+ is about 10^{-9}, NH_4^+ is hydrolyzed to a slightly greater extent. A solution of NH_4F, although close to neutral, is slightly acidic.

16.3 BUFFER SOLUTIONS

Any solution that resists change in pH on the addition of either H_3O^+ or OH^- is known as a **buffer solution.** The ability of buffers to resist changes in pH is considerable. A 1 L solution containing 0.10 mol acetic acid and 0.18 mol sodium acetate is a buffer whose pH is 5.00 at 25 °C. If 5.0 mL of 1.0 M NaOH is added to 1 L of this buffer, the pH rises to only 5.03. By contrast, if the same amount of base is added to 1 L pure water, the pH rises from 7.0 to 11.7! Buffers are not only useful in the laboratory, they are essential to the function of living organisms. Human blood is a buffer solution whose pH remains very close to 7.4.

Buffer action depends in large part on a property of equilibrium systems called the *common ion effect.*

Common Ion Effect

A state of equilibrium exists when a weak acid such as HF is dissolved in water.

$$HF + H_2O \rightleftharpoons H_3O^+ + F^- \qquad (16\text{-}19)$$

This equilibrium is disturbed when a salt of the same acid, such as NaF, is added to the solution. Le Châtelier's principle predicts that the addition of NaF will shift the equilibrium to the left. The concentration of H_3O^+ will decrease and that of HF will increase. This shift in an equilibrium, caused by a substance containing the same ion as one of those involved in the equilibrium, is called the **common ion effect.** The common ion effect operates in any mixture of a weak acid and one of its salts, and also in mixtures of weak bases and their salts.

EXAMPLE 16.9 pH of a Common-Ion Mixture

a. Calculate the pH of a 1.00 L solution that contains 0.100 mol HF and 0.500 mol NaF. (The usual way of describing this solution is to say that it is 0.100 M *in* HF and 0.500 M *in* NaF.)

b. For comparison, calculate the pH of 0.100 M HF with no added salt.

Analysis (b) is the same as Example 16.4. For (a), we have the following.

Target: $[H_3O^+]$, expressed as pH.

Knowns: Initial concentrations of HF and F^-. K_a is an implicit known, to be obtained from Table 16.1.

Relationship: The equilibrium expression $K_a = [H_3O^+][F^-]/[HF]$ holds in all solutions of HF or F^-, regardless of what other solutes might be present. Specifically, it holds in this case, when there are two different sources of HF(aq) and F^-(aq).

Plan Use the structured approach—it is good strategy in almost all equilibrium problems. We will dispose of (b) first, since we have already solved such a problem in Example 16.4.

Work

b. $K_a(HF) = 6.6 \times 10^{-4}$.

Balanced Equation	HF	+	$H_2O \rightleftharpoons H_3O^+$	+ F^-
Initial	0.100		0	0
Change	$-x$		$+x$	$+x$
At Equilibrium	$(0.100 - x)$		x	x

$$K_a = \frac{[H_3O^+][F^-]}{[HF]} = \frac{x \cdot x}{0.100 - x}$$

$$K_a \approx \frac{x^2}{0.100}$$

$$\begin{aligned} x &\approx \sqrt{K_a \cdot 0.100} \\ &\approx \sqrt{(6.6 \times 10^{-4}) \cdot 0.100} \\ &\approx 0.00812 \end{aligned}$$

Check $x/C = 0.008/0.1 = 0.08$, and the approximation is valid.

$$[H_3O^+] = x = 0.00812 \text{ M}$$

$$\begin{aligned} \text{pH} &= -\log[H_3O^+] = -\log(0.00812) \\ &= -(-2.090) \\ &= 2.09 \end{aligned}$$

The pH of the pure acid is 2.09.

Work

a. When 0.100 mol HF is dissolved in 1 L of 0.500 M NaF, initially [HF] = 0.100 M and $[F^-]$ = 0.500 M. Following the structured approach, we write:

Balanced Equation	+ HF	$H_2O \rightleftharpoons H_3O^+$	+ F^-
Initial	0.100	0	0.500
Change	$-x$	$+x$	$+x$
At Equilibrium	$(0.100 - x)$	x	$(0.500 + x)$

$$K_a = \frac{[H_3O^+][F^-]}{[HF]} = \frac{x \cdot (0.500 + x)}{0.100 - x}$$

We saw in part (b) that, even in the absence of NaF, the extent of reaction was small enough so that the approximation $(0.100 - x) \approx 0.100$ was valid. When the acid anion is present from an additional source, as in this problem, Le Châtelier's principle ensures that the extent of reaction is even less. The approximations $[F^-]$ $(0.500 + x) \approx 0.500$ and [HF] $(0.100 - x) \approx 0.100$ are certainly valid.

$$K_a \approx \frac{x \cdot 0.500}{0.100}$$

$$x \approx \frac{K_a \cdot 0.100}{0.500}$$

$$\approx \frac{(6.6 \times 10^{-4}) \cdot 0.100}{0.500}$$

$$\approx 1.32 \times 10^{-4}$$

$$[H_3O^+] = x = 1.32 \times 10^{-4}$$

$$pH = -\log[H_3O^+] = 3.88$$

Check The pH of the pure acid (2.09) must rise when F^- (a base) is added.

Exercise

Calculate the pH of a solution that is 0.05 M in HNO_2 and 0.5 M in KNO_2.

Answer: pH = 4.4.

The preceding example shows that the common ion effect decreases the extent of ionization of a weak acid. In terms of extent of reaction, 0.100 M HF is 8% ionized in pure water, and only 0.13% ionized in 0.500 M NaF. The effect of the common ion is considerable.

Buffer Action

In Example 16.9 we calculated the pH of a HF/F^- buffer.

We introduced the concept of buffer solutions by describing what they *do*, namely, resist changes in pH. Now we describe buffer solutions by what they *are*, and how they work. Any solution containing significant and comparable concentrations of a weak acid and its conjugate base, such as HF/F^- or NH_4^+/NH_3, is a buffer.

To understand how buffers resist change in pH, consider a solution of HF and NaF. If a small amount of KOH (or other strong base) is added, hydroxide ion reacts with HF.

A single-headed arrow is used to indicate 100% consumption of reactants.

$$HF + OH^- \longrightarrow H_2O + F^- \qquad (16\text{-}20)$$

Hydroxide ion is a *strong* base, so the reaction goes essentially to completion—all the added OH^- is consumed. Since the added OH^- is consumed, it cannot affect the pH of the solution.

If a small amount of HCl (or other strong acid) is added to the solution, H_3O^+ is consumed by reaction with F^-

$$F^- + H_3O^+ \longrightarrow HF + H_2O \qquad (16\text{-}21)$$

H_3O^+ is a *strong* acid, so the reaction goes essentially to completion. Since added H_3O^+ is consumed, there can be no effect on pH. A buffer, then, is an *amphoteric* solution: it reacts with both acid and base.

The pH of Buffer Solutions

The equation used to calculate the pH of the buffer in Example 16.9,

$$K_a = [H_3O^+] \cdot \frac{[F^-]}{[HF]}$$

is a particular instance of the more general form, Equation (16-22).

$$K_a = [H_3O^+] \cdot \frac{[\text{conjugate base}]}{[\text{conjugate acid}]} \qquad (16\text{-}22)$$

Since buffers are commonly prepared by mixing a weak acid with its salt, Equation (16-22) can also be written in the following way.

$$K_a = [H_3O^+] \cdot \frac{[\text{salt}]}{[\text{acid}]} \qquad (16\text{-}23)$$

The arguments given in Example 16.9 for neglecting the extent of reaction are *always* valid in buffer calculations. Therefore, the concentration terms in Equations (16-22) and (16-23) refer to the *initial* concentrations of acid and conjugate base (salt) used to prepare the solution. The hydronium ion concentration of a buffer is found by rearranging Equation (16-22).

$$[H_3O^+] = K_a \cdot \frac{[\text{conjugate acid}]}{[\text{conjugate base}]} \qquad (16\text{-}24a)$$

Taking negative logarithms of both sides of Equation (16-24a) and using the definitions of pH and pK_a, we obtain

$$-\log[H_3O^+] = -\log K_a - \log \frac{[\text{acid}]}{[\text{base}]}$$

$$\text{pH} = pK_a - \log \frac{[\text{acid}]}{[\text{base}]}$$

$$\text{pH} = pK_a + \log \frac{[\text{base}]}{[\text{acid}]} \qquad (16\text{-}24b)$$

Equation (16-24b) is known as the Henderson-Hasselbalch equation. Equations (16-24a) and (16-24b) show that, in buffers having *equal concentrations* of the two components, $[H_3O^+] = K_a$ and $\text{pH} = pK_a$. This relationship is a useful guide in the preparation of buffers. These equations also show that, when the ratio [base]/[acid] is 10/1, the pH is 1 unit greater than the pK_a. Similarly, when the ratio [base]/[acid] is 1/10, the pH is 1 unit less than the pK_a.

EXAMPLE 16.10 pH of an Acid/Salt Buffer

Calculate the hydronium ion concentration in a buffer that is 0.100 M in acetic acid and 0.180 M in potassium acetate.

Analysis

Target: $[H_3O^+]$ in a buffer.

Knowns: The identity and concentrations of an acid and its conjugate base.

Implicit: The value $K_a = 1.8 \times 10^{-5}$ is found in Table 16.1. Use of this value requires the assumption that the temperature, unspecified in the problem, is 25 °C.

Relationship: Either form of Equation (16-24) may be used.

Work Since the known value is K_a rather than pK_a, and $[H_3O^+]$ rather than pH is called for, Equation (16-24a) is more direct.

$$[H_3O^+] = K_a \cdot \frac{[\text{conjugate acid}]}{[\text{conjugate base}]}$$

$$= (1.8 \times 10^{-5}) \cdot \frac{0.100 \text{ M}}{0.180 \text{ M}}$$

$$= 1.0 \times 10^{-5} \text{ M}$$

Exercise

Calculate $[H_3O^+]$ in a buffer solution that is 0.100 M in sodium acetate and 0.500 M in acetic acid.

Answer: 9.0×10^{-5}.

Buffers prepared from a weak base and one of its salts are less common, but Equation (16-24) is used to calculate pH in these systems, too.

EXAMPLE 16.11 pH of a Base/Salt Buffer

Calculate the hydronium ion concentration in a buffer that is 0.100 M in ammonia and 0.200 M in ammonium chloride. $K_b = 1.9 \times 10^{-5}$ for ammonia.

Analysis

Target: $[H_3O^+]$ in a buffer.

Knowns: The identity and concentrations of a base and its conjugate acid, and K_b for the base.

Relationship: Either form of Equation (16-24) may be used to find $[H_3O^+]$, but the value of K_a rather than K_b is needed. For any conjugate pair, $K_a \cdot K_b = K_w$.

Plan Find K_a from K_b, then use Equation (16-24a) to find $[H_3O^+]$.

Work

1.

$$K_a \cdot K_b = K_w$$

$$K_a = \frac{K_w}{K_b} = \frac{1.0 \times 10^{-14}}{1.9 \times 10^{-5}} = 5.26 \times 10^{-10}$$

2. NH_4^+ is the conjugate acid and NH_3 is the conjugate base.

$$[H_3O^+] = K_a \cdot \frac{[\text{conjugate acid}]}{[\text{conjugate base}]} = (5.26 \times 10^{-10}) \cdot \frac{0.200\ \text{M}}{0.100\ \text{M}} = 1.1 \times 10^{-9}\ \text{M}$$

Exercise

Calculate $[H_3O^+]$ in a buffer solution that is 0.250 M in ammonia and 0.500 M in ammonium chloride.

Answer: 1.1×10^{-9} M.

Buffer Capacity

The amount of acid (or base) a buffer can absorb without significant change in pH is called the **buffer capacity.** Buffer capacity is increased if the concentration of the components is increased. A buffer in which the conjugate acid and base are both 1.0 M has 10 times the capacity of a buffer in which the concentrations are both 0.10 M. Buffer capacity is greatest when the conjugate acid and base have equal concentrations, in which case $pH = pK_a$. For this reason, the components of a buffer for a particular task are normally chosen so that the pK_a of the acid is near the desired pH of the buffer.

Preparation of Buffer Solutions

There are two general methods for the preparation of buffer solutions. The first is to mix together appropriate quantities of a weak acid and one of its salts. As seen in Example 16.9, mixing 0.10 mol HF and 0.50 mol NaF in 1 L water produces a buffer of pH 3.88. In the second method, which is often more convenient in the laboratory, a solution of the weak acid alone is partially neutralized by the addition of strong base. Since the product of the neutralization reaction is a salt, the resulting solution contains both acid and salt and is therefore a buffer.

EXAMPLE 16.12 Buffer Preparation from a Weak Acid and its Salt

A chemist needs one liter of a buffer of pH 4.00. Following the general guideline of pH $\approx$ pK_a for buffers of good capacity, she selects formic acid/sodium formate, because the pK_a of formic acid is 3.74 at 25 °C. Calculate the amounts (moles) of acid and salt that must be added to 1 L water to produce the desired buffer.

Analysis The pH of a buffer is determined by the pK_a and the concentrations of the components, as related by Equation (16-24).

Plan Insert the given data into the equation and solve for the molar ratio [base]/[acid].

Work Since pK_a rather than K_a is given, and the target is pH, Equation (16-24b) is simpler to use.

$$\text{pH} = \text{p}K_a + \log\frac{[\text{base}]}{[\text{acid}]}$$

$$\log\frac{[\text{base}]}{[\text{acid}]} = \text{pH} - \text{p}K_a$$

$$\log\frac{[\text{base}]}{[\text{acid}]} = 4.00 - 3.74 = 0.26$$

$$10^{\log([\text{base}]/[\text{acid}])} = 10^{0.26}$$

$$\frac{[\text{salt}]}{[\text{acid}]} = 1.8$$

The problem is now solved, not for the individual concentrations, but for their ratio. The concentrations [HCOOH] = 1.00 and [HCOONa] = 1.8 M have the indicated ratio and will produce the desired buffer, but so will the concentrations [HCOOH] = 0.50 and [HCOONa] = 0.90 M. *Any* pair of concentrations with this ratio is an acceptable answer to the problem.

Of course if the problem specifies one of the concentrations, the other can be found. For instance if the concentration of sodium formate must be 0.25 M, then

$$\frac{[\text{base}]}{[\text{acid}]} = 1.8$$

$$[\text{acid}] = \frac{[\text{base}]}{1.8} = \frac{0.25\ \text{M}}{1.8} = 0.14\ \text{M}$$

Exercise

How many moles of sodium formate must be added to 1.00 liter of a 0.100 M solution of formic acid in order to produce a buffer of pH 3.50?

Answer: 0.057 mol.

EXAMPLE 16.13 Total Buffer

Suppose the total buffer, that is, the sum [HCOOH] + [HCOONa] in Example 16.12, had to be 0.500 M. What are the required concentrations of the acid and its salt?

Analysis We know the *ratio* of concentrations from Example 16.12, and the *sum* from the information given in this problem. Therefore we have two equations and two unknowns, and we can solve the problem algebraically.

Work

$$[\text{acid}] = \frac{[\text{base}]}{1.8} \qquad (1)$$

$$[\text{acid}] + [\text{base}] = 0.500 \text{ M} \qquad (2)$$

Substitute (1) into (2) and solve.

$$\frac{[\text{base}]}{1.8} + [\text{base}] = 0.500 \text{ M}$$

$$[\text{base}] + (1.8) \cdot [\text{base}] = (0.500 \text{ M}) \cdot (1.8)$$

$$(2.8) \cdot [\text{base}] = (0.500 \text{ M}) \cdot (1.8)$$

$$[\text{base}] = (0.500 \text{ M}) \cdot \frac{(1.8)}{2.8}$$

$$= 0.32 \text{ M}$$

The concentration of acid is found from either (1) or (2)

$$[\text{acid}] = \frac{[\text{base}]}{1.8} \qquad (1)$$

$$= \frac{0.32 \text{ M}}{1.8} = 0.18 \text{ M}$$

Exercise

What are the concentrations of sodium acetate and acetic acid in a buffer solution of pH 5.00? The total buffer must be 1.00 M, and the pK_a of acetic acid is 4.74.

Answer: [acid] = 0.35 M, [salt] = 0.65 M.

EXAMPLE 16.14 Buffer Preparation by Partial Neutralization

How much sodium hydroxide must be added in order to convert 1.00 L of 0.400 M acetic acid to a buffer of pH = 5.00? The pK_a of acetic acid is 4.74.

Analysis Partial neutralization of a weak acid with a strong base produces an amount of salt (*conjugate* base) equal to the amount of *strong* base added. The result is a mixture of an acid and its conjugate base—a buffer. Since the salt is produced from the acid, the total amount of acid + conjugate base must be equal to the original amount of acid

(0.400 mol), regardless of the extent of the partial neutralization. The Henderson-Hasselbach equation (16-24b) allows the calculation of the required concentration ratio [conjugate base]/[acid] from the pH and the pK_a.

Plan Calculate the required concentration ratio [salt]/[acid], and, from it and the total buffer concentration, find the molar amount of acid that must be converted to salt. This amount of acid is equal to the required amount of NaOH.

Work

$$\text{pH} = \text{p}K_a + \log \frac{[\text{conjugate base}]}{[\text{acid}]} \qquad (16\text{-}24b)$$

$$\log \frac{[\text{conjugate base}]}{[\text{acid}]} = \text{pH} - \text{p}K_a$$

$$\log \frac{[\text{conjugate base}]}{[\text{acid}]} = 5.00 - 4.74 = 0.26$$

$$\frac{[\text{conjugate base}]}{[\text{acid}]} = 10^{0.26} = 1.82$$

The *mole ratio* (mol conjugate base)/(mol acid) must also be 1.82, independent of the volume. Algebraically, the solution volume V cancels.

$$\frac{[\text{conjugate base}]}{[\text{acid}]} = \frac{(\text{mol conjugate base})/V}{(\text{mol acid})/V} = 1.82$$

We now have two equations and two unknowns.

$$\text{mol acid} = \frac{(\text{mol conjugate base})}{1.82} \qquad (1)$$

$$\text{mol acid} + \text{mol conjugate base} = 0.400 \text{ mol} \qquad (2)$$

Substitute (1) into (2)

$$\frac{(\text{mol conjugate base})}{1.82} + (\text{mol conjugate base}) = 0.400 \text{ mol}$$

$$(\text{mol conjugate base}) + 1.82 \cdot (\text{mol conjugate base}) = (0.400 \text{ mol}) \cdot 1.82$$

$$2.82 \cdot (\text{mol conjugate base}) = (0.400 \text{ mol}) \cdot 1.82$$

$$(\text{mol conjugate base}) = \frac{(0.400 \text{ mol}) \cdot 1.82}{2.82}$$

$$= 0.258 \text{ mol}$$

$$= 0.26 \text{ mol}$$

Note the use of significant figures.

Therefore, 0.26 mol NaOH must be added.

Note: Since the solution volume cancels out of the equation, the result is the same whether the 0.26 mol NaOH is added as 10.4 g pure solid, as 260 mL of 1.00-M solution, or as 2600 mL of 0.100 M solution. These three ways of partial neutralization yield different volumes of buffer, but all have the same pH.

Exercise

How much KOH must be added to 375 mL of 0.78 M propanoic acid ($pK_a = 4.88$) to produce a buffer of pH 4.00?

Answer: 1.9 g dry KOH or 34 mL of a 1.0 M solution.

Figure 16.2 In the titration of an unknown acid sample, a strong base such as KOH(aq) is added to the sample from a buret. The color of an added indicator changes when the acid has been neutralized.

16.4 Titrations

The laboratory procedure *titration,* for the determination of the precise amount of acid (or base) in a sample, was discussed in general terms in Section 15.8. Typically, a strong base of known concentration is gradually added to a sample containing an unknown amount of acid. The amount of acid in the sample is determined from the measured amount of base required to neutralize it (Figure 16.2).

Strong-Acid Titration Curves

During a titration of an acid sample, the pH continually rises as the basic titrant is added. When the point of exact neutralization, the **equivalence point**, is reached in the titration of a strong acid, the pH is 7. The pH then rises beyond 7 as more base is added. The process is conveniently summarized, in Figure 16.3, by a plot of pH vs. volume of titrant added. Such a plot is called a **titration curve.**

As shown in Figure 16.3, titration curves have a characteristic shape. There is a gradual rise in pH during the initial phase of the titration, when the sample is *undertitrated.* The curve rises more steeply as the equivalence point is approached and passed, and finally levels out again. The abrupt rise is called the **break** in the curve. The sample is *overtitrated* in the region beyond the equivalence point. The general shape of titration curves is modified by the concentration and strength of the acid being titrated. Figure 16.3 illustrates the effect of concentration on curve shape when the sample is a strong acid.

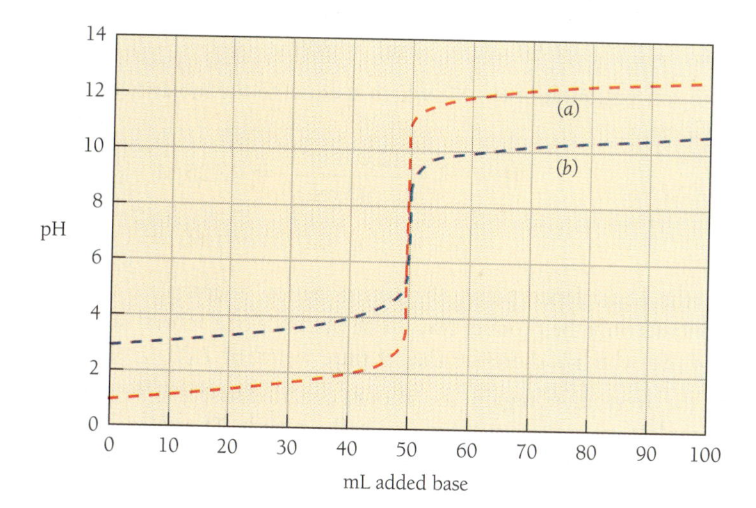

Figure 16.3 (a) Sample pH as a function of added base when 50 mL of 0.10 M HCl is titrated with 0.10 M NaOH. The pH changes only slightly until the equivalence point is closely approached. Then it rises abruptly before leveling out again. (b) Titration of 50 mL of 0.0010 M HCl with 0.0010 M NaOH. The break is sharp, but because the sample is dilute the pH varies over a narrower range.

Experimental titration curves are prepared by plotting the measured pH of the sample after the addition of various amounts of titrant. Alternatively, a theoretical curve can be constructed by calculating the pH at a few points in the under- and overtitrated regions. Table 16.3 lists some of the points from which the red line in Figure 16.3 was drawn. Four of the pH values in the table are calculated in Example 16.15.

TABLE 16.3 50 mL 0.10 M HCl Titrated with 0.10 M NaOH*

Base Added/mL	pH
0	*1.00*
10	1.18
20	1.37
30	*1.60*
40	1.95
50	*7.00*
60	11.96
70	*12.22*
80	12.36
90	12.46
100	12.52

* Italicized values are calculated in Example 16.15.

EXAMPLE 16.15 Strong-Acid Titration Curves

A 50 mL sample of 0.10 M HCl is titrated with 0.10 M NaOH. Calculate the pH (a) before any titrant has been added, (b) after the addition of 30 mL titrant, (c) at the equivalence point, and (d) after the addition of 70 mL titrant.

Analysis Since both acid and base are strong, equilibrium calculations are not required to determine $[H_3O^+]$. We need only the amounts, in moles, of acid and added base, and the volume of the solution.

Work

a. Initially the sample is 0.10 M strong acid, so

$$[H_3O^+] = 0.10 \text{ M, and pH} = -\log(0.10) = 1.00$$

The remaining calculations are easier to follow if the amounts are given in millimoles rather than moles. Recall that the units of molarity can be either mol L^{-1} or mmol mL^{-1}. That is, the solute concentration in a 1-M solution is 1 mmol mL^{-1}.

b. In the undertitrated region, all of the added base is consumed as it neutralizes an equal amount of acid. The pH is determined by the amount of (un-neutralized) acid remaining, and the new volume.

$$\begin{array}{lrl} & \text{initial acid} = (50 \text{ mL}) \cdot (0.10 \text{ mmol mL}^{-1}) = & 5.0 \text{ mmol HCl} \\ - & \text{added base} = (30 \text{ mL}) \cdot (0.10 \text{ mmol mL}^{-1}) = & 3.0 \text{ mmol NaOH} \\ \hline & \text{remaining acid} = & 2.0 \text{ mmol HCl} \end{array}$$

The new volume is

$$V = \text{initial acid volume} + \text{volume of added base}$$
$$= 50 \text{ mL} + 30 \text{ mL} = 80 \text{ mL}$$

The concentration of H_3O^+, which is equal to the molarity of the remaining HCl, is

$$[H_3O^+] = \frac{(2.0 \text{ mmol})}{(80 \text{ mL})} = 0.025 \text{ M}$$

$$\text{pH} = -\log(0.025) = 1.60$$

c. At the equivalence point, the neutralization is complete and the solution contains only the product NaCl. The presence of a neutral salt does not affect the pH, which is therefore that of pure water, or 7.00.

d. In the overtitrated region, the amount of added base is more than required for complete consumption of the acid in the sample. The pH is determined solely by the amount of excess base added, and the volume.

$$\begin{array}{lrl} & \text{added base} = (70 \text{ mL}) \cdot (0.10 \text{ mmol mL}^{-1}) = & 7.0 \text{ mmol NaOH} \\ - & \text{initial acid} = (50 \text{ mL}) \cdot (0.10 \text{ mmol mL}^{-1}) = & 5.0 \text{ mmol HCl} \\ \hline & \text{excess base} = & 2.0 \text{ mmol NaOH} \end{array}$$

The new volume is

$$V = \text{initial acid volume} + \text{volume of added base}$$
$$= 50 \text{ mL} + 70 \text{ mL} = 120 \text{ mL}$$

The base concentration is

$$[OH^-] = \frac{2.0 \text{ mmol}}{120 \text{ mL}} = 0.017 \text{ M}$$
$$pOH = -\log(0.017) = 1.78$$
$$pH = 14.00 - pOH = 12.22$$

Exercise

Verify two other values in Table 16.3 by calculating the pH after the addition of (a) 40 mL and (b) 60 mL of 0.10 M NaOH to 50 mL of 0.10 M HNO_3.

Answer: (a) pH = 1.95; (b) pH = 11.96.

Weak-Acid Titration Curves

A plot of pH changes during the titration of a weak acid with a strong base has an overall appearance similar to that of a strong acid titration curve, although there are some differences in detail. Compared to the curve for a strong acid of the same concentration, the weak acid curve has a higher pH and rises more steeply throughout the undertitrated region. In the overtitrated region, weak- and strong-acid curves are the same. The break is smaller for weak acids, because it begins at a higher pH. At the

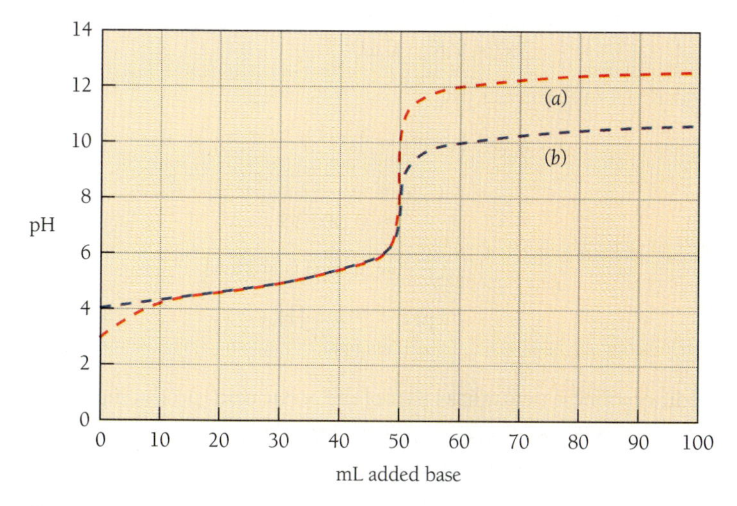

Figure 16.4 Sample pH as a function of added base during the titration of a weak acid with a strong base. (a) 50 mL of 0.10 M CH_3CO_2H is titrated with 0.10 M NaOH. The break is smaller than for strong acids, and the curve lies at higher pH in the undertitrated region. (b) Titration of 50 mL of 0.0010 M CH_3CO_2H with 0.0010 M NaOH. In both (a) and (b), the solution in the undertitrated region is a mixture of acetic acid and sodium acetate. Because of buffer action, the initial concentration of acid and the concentration of titrant have little effect on the shape of the curve in the undertitrated region.

TABLE 16.4 50 mL 0.10 M Acetic Acid Titrated with 0.10 M NaOH

Base Added/mL	pH
0	2.87
10	4.14
20	4.56
30	4.92
40	5.34
50	8.72
60	11.96
70	12.22
80	12.36
90	12.46
100	12.52

equivalence point in the titration of a weak acid, *the pH is greater than 7*. Calculated pH values during a weak-acid titration are given in Table 16.4, and the corresponding titration curve is shown as the red line in Figure 16.4.

EXAMPLE 16.16 Titration of a Weak Acid

A 50 mL sample of 0.10 M CH_3CO_2H ($pK_a = 4.74$) is titrated with 0.10 M KOH. Calculate the pH (a) before any titrant has been added, (b) after the addition of 30 mL titrant, (c) at the equivalence point, and (d) after the addition of 70 mL titrant.

Analysis These are really four separate problems. We must find the pH of (a) a weak acid (Example 16.4); (b) a buffer prepared by partial neutralization of a weak acid (Example 16.14); (c) a salt of a weak acid (Example 16.7); and (d) an overtitrated acid, which may also be regarded as a partially neutralized strong base.

Work

a. Initially the sample is a solution of a weak acid, so the pH is found by the standard approach.

Balanced Equation	$CH_3CO_2H + H_2O \rightleftharpoons$	$CH_3CO_2^-$ +	H_3O^+
Initial	0.10	0	0
Change	$-x$	$+x$	$+x$
At Equilibrium	$0.10 - x$	x	x

$$K_a = \frac{[H_3O^+][CH_3CO_2^-]}{[CH_3CO_2H]} = \frac{x \cdot x}{0.10 - x}$$

$$K_a \approx \frac{x^2}{0.10}$$

$$x \approx \sqrt{K_a \cdot 0.10}$$

Find K_a from the given pK_a:

$$K_a = 10^{-pK_a} = 10^{-4.74} = 1.82 \times 10^{-5}$$

$$x \approx \sqrt{(1.82 \times 10^{-5}) \cdot 0.10}$$
$$\approx 0.00135$$

x/C is less than 10%, so the approximation is valid.

$$[H_3O^+] = x = 0.00135 \text{ M}$$

$$pH = -\log [H_3O^+]$$
$$= 2.87$$

b. Addition of base neutralizes part of the acetic acid, producing acetate ion, the conjugate base.

It is very important to distinguish between acetate ion, the *conjugate* base, and NaOH, the *strong* base used as the titrant.

$$CH_3CO_2H + OH^- \longrightarrow CH_3CO_2^- + H_2O$$

Because OH^- is a strong base, this reaction proceeds essentially to completion—the amount of unreacted OH^- is negligibly small.

$$\text{initial acid} = (50 \text{ mL}) \cdot (0.10 \text{ mmol mL}^{-1}) = 5.0 \text{ mmol } CH_3CO_2H$$
$$- \quad \text{added } OH^- = (30 \text{ mL}) \cdot (0.10 \text{ mmol mL}^{-1}) = 3.0 \text{ mmol } OH^-$$
$$\text{remaining acid} = \quad 2.0 \text{ mmol } CH_3CO_2H$$

The amount of acetate ion produced is equal to the amount of added OH^-, so the buffer contains 2.0 mmol acid and 3.0 mmol conjugate base.

$$pH = pK_a + \log \frac{[\text{conjugate base}]}{[\text{acid}]} \quad (16\text{-}24b)$$

$$= 4.74 + \log \frac{(3.0 \text{ mmol}/V \text{ mL})}{(2.0 \text{ mmol}/V \text{ mL})}$$

$$= 4.92$$

Note: The volume can be ignored, because it cancels in Equation (16-24b).

c. The equivalence point is reached when 5.0 mmol (50 mL) KOH has been added. The acid has completely reacted, and the solution consists of 5.0 mmol potassium acetate in 100 mL H_2O. Since potassium acetate is a salt of a weak acid, it hydrolyzes, and the pH is greater than 7.

Balanced Equation	$CH_3CO_2^- + H_2O \rightleftharpoons$	CH_3CO_2H +	OH^-
Initial	0.050	0	0
Change	$-x$	$+x$	$+x$
At Equilibrium	$0.050 - x$	x	x

$$K_b = \frac{[OH^-][CH_3CO_2H]}{[CH_3CO_2^-]} = \frac{x \cdot x}{0.05 - x}$$

$$K_b \approx \frac{x^2}{0.05}$$

$$x \approx \sqrt{K_b \cdot 0.05}$$

Substitute the relationship $K_b = K_w/K_a$, and solve.

$$x \approx \sqrt{0.05 \cdot (K_w/K_a)}$$

$$x \approx \sqrt{0.05 \cdot 1.0 \times 10^{-14}/1.82 \times 10^{-5}}$$

$$\approx 5.24 \times 10^{-6}$$

$$[OH^-] = x = 5.24 \times 10^{-6} \text{ M}$$

$$pOH = 5.28$$

$$pH = 14.00 - pOH = 8.72$$

When a weak *base* like acetate ion hydrolyzes, the extent of reaction x is equal to the concentration of OH^-, not hydronium ion.

d. In the overtitrated region, the pH is controlled by the amount of strong base added in excess of that required for complete neutralization of the sample. Although the solution contains acetate ion, the hydrolysis equilibrium

$$CH_3CO_2^- + H_2O \rightleftharpoons CH_3CO_2H + OH^-$$

is forced to the left in the presence of excess OH^-, and makes a negligible contribution to the total $[OH^-]$.

$$\begin{array}{rl} \text{added } OH^- = (70 \text{ mL}) \cdot (0.10 \text{ mmol mL}^{-1}) = & 7.0 \text{ mmol } OH^- \\ - \quad \text{initial acid} = (50 \text{ mL}) \cdot (0.10 \text{ mmol mL}^{-1}) = & 5.0 \text{ mmol } CH_3CO_2H \\ \hline \text{excess hydroxide} = & 2.0 \text{ mmol } OH^- \end{array}$$

The new volume is

$$V = \text{initial acid volume} + \text{volume of added base}$$
$$= 50 \text{ mL} + 70 \text{ mL} = 120 \text{ mL}$$

The hydroxide concentration is

$$[OH^-] = \frac{2.0 \text{ mmol}}{120 \text{ mL}} = 0.017 \text{ M}$$

$$pOH = -\log(0.017) = 1.78$$

$$pH = 14.00 - pOH = 12.22$$

Note: In the overtitrated region, the calculation is the same for weak and strong acids. Compare this calculation with (d) of Example 16.15.

Exercise

Consider the titration of 50 mL 0.25 M HF ($pK_a = 3.18$) with 0.25 M KOH, and calculate the pH after the addition of (a) 25 mL, (b) 50 mL, and (c) 75 mL of titrant.

Answer: (a) pH = 3.18, (b) pH = 8.14, and (c) pH = 12.70.

Beyond the equivalence point, acid strength is irrelevant.

The calculations in (d) of Examples 16.15 and 16.16 are identical. It follows that, in the overtitrated region, titration curves for strong and weak acids are identical (provided the comparison is made at the same concentration).

The pH at the midpoint in a weak acid titration, that is, when half of the acid has been neutralized, is especially easy to calculate. At this point, the concentration of the salt is equal to the concentration of unreacted acid. The ratio [conjugate base]/[acid] = 1, so $K_a = [H_3O^+] \cdot 1$. *At the midpoint of a weak acid titration,* $pH = pK_a$. A final point to remember about weak acid titrations is that the pH at the equivalence point is *greater than* 7.

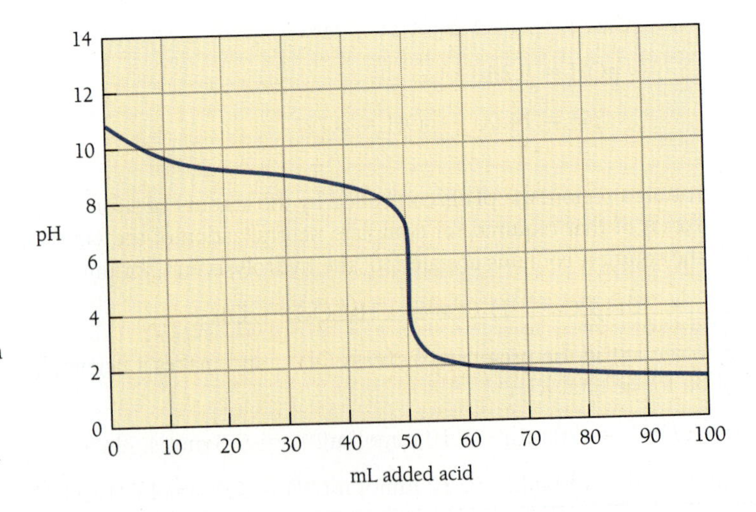

Figure 16.5 Sample pH as a function of added acid during the titration of 50 mL of 0.10 M NH_3, a weak base, with 0.10 M HCl. The curve is essentially a reflection of the red line in Figure 16.4.

Titration of Weak Bases

When a weak base is titrated with a strong acid, the titration curve is an upside-down reflection of a weak acid curve. The pH decreases during the titration, is somewhat less than 7 at the equivalence point, and is independent of base strength in the overtitrated region (Figure 16.5). Calculations for weak-base titration curves are generally the same as for weak-acid titrations.

EXAMPLE 16.17 Titration of a Weak Base

Consider the titration of 50 mL of 0.10 M NH_3 ($K_b = 1.9 \times 10^{-5}$) with 0.10 M HCl, and calculate the pH (a) before any titrant has been added, (b) after the addition of 30 mL titrant, (c) at the equivalence point, and (d) after the addition of 70 mL titrant.

Analysis There are no conceptual differences between weak-base and weak-acid titrations. Provided we are careful about the use of $[OH^-]$, pOH, $[H_3O^+]$, pH, K_b, and K_a, we can follow the method of Example 16.16.

Work

a. Initially the sample is a solution of a weak base, so the pH is found by the standard approach.

Balanced Equation	$NH_3 + H_2O \rightleftharpoons$	NH_4^+	$+ OH^-$
Initial	0.10	0	0
Change	$-x$	$+x$	$+x$
At Equilibrium	$0.10 - x$	x	x

$$K_b = \frac{[NH_4^+][OH^-]}{[NH_3]} = \frac{x \cdot x}{0.10 - x}$$

$$\approx \frac{x^2}{0.10}$$

$$x \approx \sqrt{K_b \cdot 0.10} = \sqrt{(1.9 \times 10^{-5}) \cdot 0.10} = 0.00138$$

$$[OH^-] = x = 0.00138\ \text{M}; \qquad \text{pOH} = 2.86; \qquad \text{pH} = 11.14$$

b. Addition of HCl neutralizes part of the ammonia,

$$NH_3 + H_3O^+ \longrightarrow NH_4^+ + H_2O$$

producing ammonium ion, the conjugate acid.

$$\begin{aligned} \text{initial base} &= (50\ \text{mL}) \cdot (0.10\ \text{mmol mL}^{-1}) = 5.0\ \text{mmol } NH_3 \\ -\ \text{added } H_3O^+ &= (30\ \text{mL}) \cdot (0.10\ \text{mmol mL}^{-1}) = 3.0\ \text{mmol } H_3O^+ \\ \hline \text{remaining base} &= \qquad\qquad 2.0\ \text{mmol } NH_3 \end{aligned}$$

The amount of ammonium ion produced is equal to the amount of added H_3O^+, so the buffer contains 2.0 mmol conjugate base and 3.0 mmol conjugate acid. The value of pK_a is required in order to find the pH from Equation (16-24b).

$$K_a = \frac{K_w}{K_b} = \frac{1.0 \times 10^{-14}}{1.9 \times 10^{-5}} = 5.26 \times 10^{-10}$$

$$pK_a = -\log K_a = 9.28$$

$$pH = pK_a + \log \frac{[\text{conjugate base}]}{[\text{conjugate acid}]} \quad (16\text{-}24b)$$

$$= 9.28 + \log \frac{2.0 \text{ mmol}/V \text{ mL}}{3.0 \text{ mmol}/V \text{ mL}}$$

$$= 9.10$$

c. The equivalence point is reached when 50 mL HCl has been added. The ammonia has completely reacted, and the solution consists of 5.0 mmol ammonium chloride in 100 mL H_2O. Ammonium ion is a weak acid, and the pH is found as usual.

When 5.0 mmol is dissolved in 100 mL, the concentration is 0.050 M.

Balanced Equation	NH_4^+ + $H_2O \rightleftharpoons$	NH_3 +	H_3O^+
Initial	0.050	0	0
Change	$-x$	$+x$	$+x$
At Equilibrium	$0.050 - x$	x	x

$$K_a = \frac{[H_3O^+][NH_3]}{[NH_4^+]} = \frac{x \cdot x}{0.050 - x}$$

$$K_a \approx \frac{x^2}{0.05}$$

$$x \approx \sqrt{K_a \cdot 0.05}$$

$$x \approx \sqrt{(5.26 \times 10^{-10}) \cdot 0.050}$$
$$\approx 5.13 \times 10^{-6}$$

$$[H_3O^+] = x = 5.13 \times 10^{-6} \text{ M}$$

$$pH = 5.29$$

d. In the overtitrated region, the pH is entirely controlled by the amount of HCl added in excess of that required for complete neutralization of the sample. The hydrolysis equilibrium

$$NH_4^+ + H_2O \rightleftharpoons NH_3 + H_3O^+$$

is forced to the left in the presence of excess H_3O^+, and makes a negligible contribution to the pH.

$$\text{added HCl} = (70 \text{ mL}) \cdot (0.10 \text{ mmol mL}^{-1}) = 7.0 \text{ mmol HCl}$$
$$-\ \text{initial } NH_3 = (50 \text{ mL}) \cdot (0.10 \text{ mmol mL}^{-1}) = 5.0 \text{ mmol } NH_3$$
$$\text{excess HCl} = 2.0 \text{ mmol } H_3O^+$$

The new volume is

$$V = \text{initial base volume} + \text{volume of added acid}$$
$$= 50 \text{ mL} + 70 \text{ mL} = 120 \text{ mL}$$

The acid concentration is

$$[H_3O^+] = \frac{2.0 \text{ mmol}}{120 \text{ mL}} = 0.017 \text{ M}$$

$$\text{pH} = -\log(0.017) = 1.78$$

Exercise

Calculate the pH after the addition of (a) 20 mL 0.10 M HCl, and (b) 100 mL 0.10 M HCl, to 50 mL of 0.10 M NH_3.

Answer: (a) pH = 9.46; (b) pH = 1.48.

Indicators

An acid/base **indicator** is a substance whose color depends on the pH of the solution in which it is dissolved. For example, phenolphthalein is an indicator that is colorless in acid solution and pink in basic solution. A small amount of phenolphthalein added to an acid titration sample causes the sample to turn pink when, at the break, the pH rises sharply into the basic region. The color change occurs at the **endpoint** of the titration. The pH at the *endpoint* is a chemical property of the indicator, while the pH at the *equivalence* point depends on the pK_a of the acid being titrated. The endpoint and the equivalence point are not necessarily the same. The difference need not lead to inaccurate titrations, however. As a titration nears the equivalence point, the pH rises so steeply that the addition of the final drop of titrant usually carries the titration through both the endpoint and the equivalence point. The inaccuracy introduced by assuming that the titration is complete when the endpoint has been reached is therefore no greater than one drop ($\approx$0.05 mL) of titrant, which is insignificant in most titrations.

Indicators are themselves weak acids or weak bases (both types work equally well). The general formula for an acid indicator is HIn, and its ionization reaction is

$$\text{HIn} + H_2O \rightleftharpoons H_3O^+ + \text{In}^-; \qquad K_{\text{In}} = \frac{[H_3O^+][\text{In}^-]}{[\text{HIn}]} \tag{16-25}$$

Any weak acid or base whose protonated and deprotonated forms have different colors could function as an indicator. A useful indicator must also be *strongly* colored, so that only a small amount need be added to the sample. Because its concentration is always very low, the indicator has a negligible effect on the pH of a solution being titrated.

The apparent color of an indicator depends on the relative amounts of the two forms present in the solution. In particular, color depends on the concentration ratio [HIn]/[In^-], which in turn is controlled by the acidity of the solution. When the equilibrium expression in Equation (16-25) is rearranged, it shows that the ratio [HIn]/[In^-] is directly proportional to [H_3O^+].

$$\frac{[\text{HIn}]}{[\text{In}^-]} = \frac{[H_3O^+]}{K_{\text{In}}} \tag{16-26}$$

In the case of phenolphthalein, HIn is colorless and In^- is red. At the beginning of the titration, [H_3O^+] is large (low pH), [HIn] is large, and there is very little In^- present. As the titration proceeds, the pH rises, [H_3O^+] drops, and the relative amount of In^-

increases. Eventually, $[In^-]$ becomes large enough to impart a color to the solution—the endpoint has been reached. The pH at which this occurs depends on the value of K_{In}. Therefore, an indicator is suitable for a particular titration only if the color change occurs at a pH close to the pH at the equivalence point (Figure 16.6).

Actually the color change occurs over an **indicator range** of about 2 pH units. Phenolphthalein is colorless below pH 8 and fully colored above pH 9.8. When a titration of 50 mL of 0.1 M HCl with 0.1 M NaOH approaches the equivalence point, the break is sharp enough so that two *drops* (about 0.05 mL each) of titrant is sufficient to change the pH from 4.3 to 9.7. Use of the indicator methyl red, with an indicator range of 4.2 to 6.2, would lead to essentially the same result, since the methyl red endpoint and the phenolphthalein endpoint are no more than one drop apart in this titration of a strong acid.

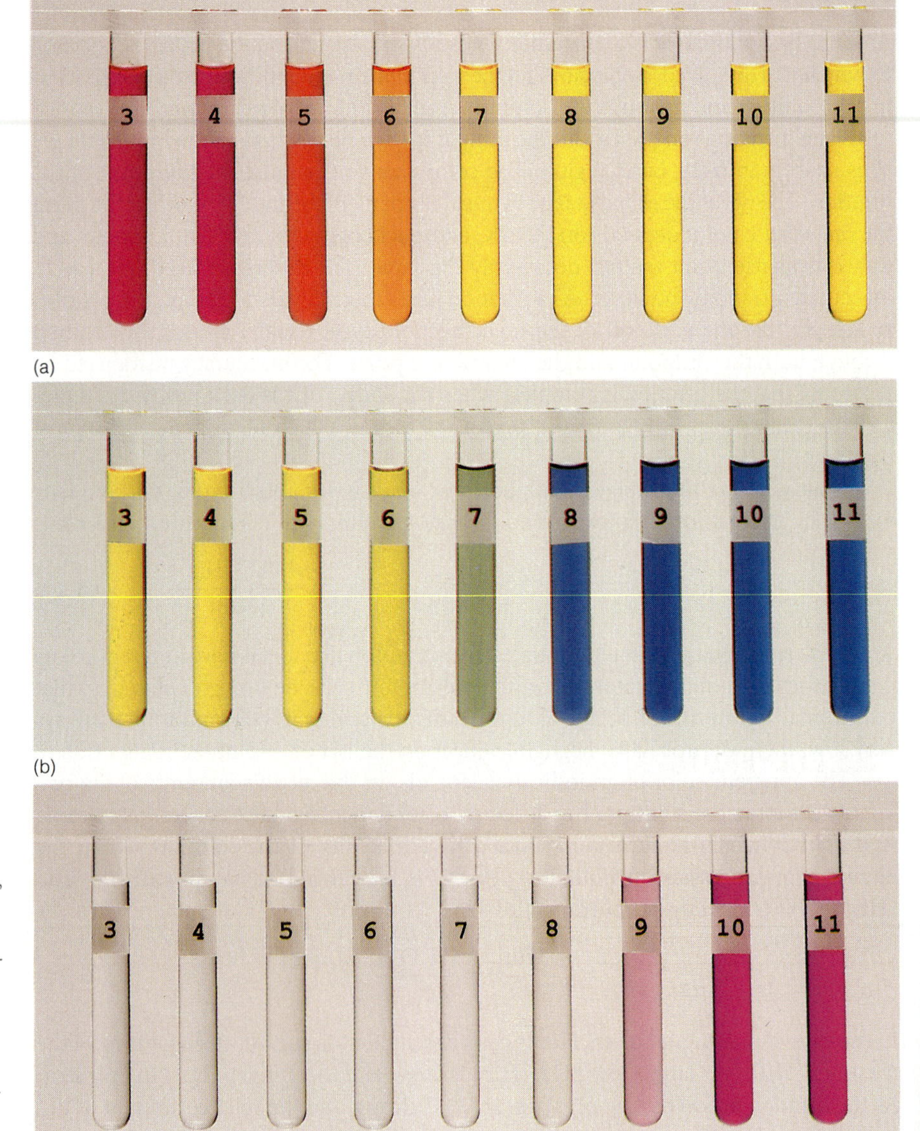

Figure 16.6 Each indicator responds differently to changing pH (given by the numbers on the test tubes). (a) *Methyl red* is red in acidic solutions, changes gradually through orange to yellow over the pH range 4.5 to 6.5, and is yellow in neutral and basic solutions. (b) *Bromthymol blue* is yellow in acid, changes to blue over the pH range 6 to 8, and is blue in basic solutions of pH 8 and greater. (c) *Phenolphthalein* is colorless in acidic and neutral solutions, and it does not achieve its full pink color until the pH rises to 10 or above.

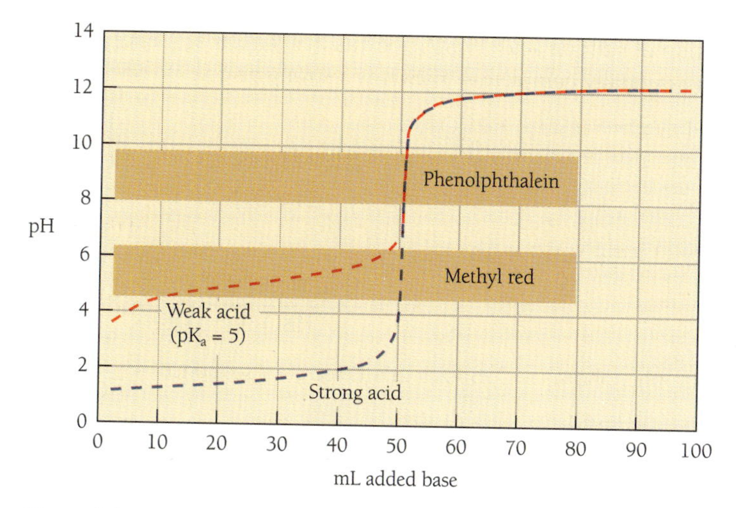

Figure 16.7 Titration of (a) 50 mL 0.10 M HCl and (b) 50 mL 0.10 M CH_3CO_2H with 0.1 M NaOH. The shaded bands show the pH ranges over which the indicators phenolphthalein and methyl red change colors. With phenolphthalein, both strong and weak acids show sharp endpoints at the equivalence points. With methyl red, the strong acid endpoint is sharp but the weak acid endpoint is gradual. Methyl red is useless as an indicator in the titration of weak acids.

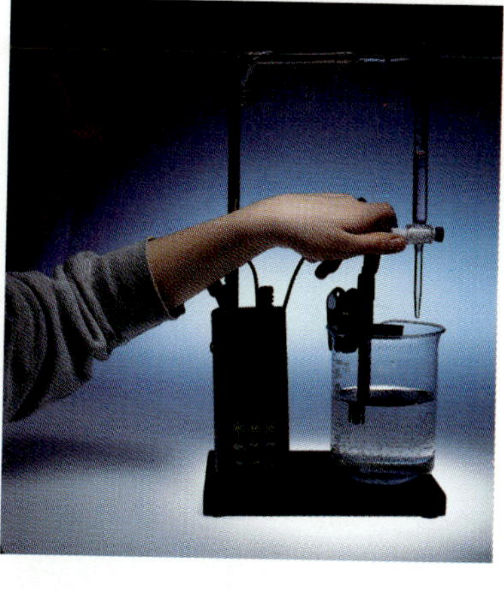

Figure 16.8 The changing pH during a titration can be measured with a pH meter. The equivalence point of any titration is the point at which the pH change is greatest for a given amount of added titrant.

The choice of indicator for the titration of a weak acid is more critical. Figure 16.7 shows that the color change is accomplished with the addition of a single drop of titrant for both methyl red and phenolphthalein in the case of a strong acid titration. In a weak acid titration, however, the pH is in the range 4.2 to 6.2 throughout most of the undertitrated region. Methyl red is not a suitable indicator for a weak acid titration, because the color change would occur gradually over many milliliters of titrant. Note that phenolphthalein, because its color change occurs at a pH higher than the pH at the equivalence point, is a good indicator for titrations of all but the weakest acids.

Although indicators are adequate for endpoint detection in the titrations of many different acids or bases, they are not the only means. A **pH meter** is a device for the direct measurement of the pH of a solution, and can be used in titrations to detect the equivalence point. Figure 16.8 shows a modern pH meter in use in a titration.

16.5 POLYPROTIC ACID EQUILIBRIA

Ionization of a polyprotic acid occurs in a series of steps: the protons are lost one at a time, rather than all at once. Each of these sequential steps is an equilibrium reaction, described by an ionization constant. Subscripts are added to the K's to designate the steps as first, second, or third. For example, the ionization of oxalic acid (diprotic) is described by Equations (16-27) and (16-28).

$$H_2C_2O_4 + H_2O \rightleftharpoons HC_2O_4^- + H_3O^+;$$

$$K_1 = \frac{[H_3O^+][HC_2O_4^-]}{[H_2C_2O_4]} = 5.9 \times 10^{-2} \quad (16\text{-}27)$$

$$HC_2O_4^- + H_2O \rightleftharpoons C_2O_4^{2-} + H_3O^+;$$

$$K_2 = \frac{[H_3O^+][C_2O_4^{2-}]}{[HC_2O_4^-]} = 6.4 \times 10^{-5} \quad (16\text{-}28)$$

The ionization of sulfuric acid, H_2SO_4, occurs in a similar sequence, except that the first step is essentially complete—H_2SO_4 is a strong acid.

$$H_2SO_4 + H_2O \rightleftharpoons HSO_4^- + H_3O^+; \quad K_1 = \frac{[H_3O^+][HSO_4^-]}{[H_2SO_4]} \sim 10^3 \quad (16\text{-}29)$$

$$HSO_4^- + H_2O \rightleftharpoons SO_4^{2-} + H_3O^+; \quad K_2 = \frac{[H_3O^+][SO_4^{2-}]}{[HSO_4^-]} = 1.2 \times 10^{-2} \quad (16\text{-}30)$$

The triprotic phosphoric acid ionizes in a three-step sequence.

$$H_3PO_4 + H_2O \rightleftharpoons H_2PO_4^- + H_3O^+; \quad K_1 = \frac{[H_3O^+][H_2PO_4^-]}{[H_3PO_4]} = 7.6 \times 10^{-3} \quad (16\text{-}31)$$

$$H_2PO_4^- + H_2O \rightleftharpoons HPO_4^{2-} + H_3O^+; \quad K_2 = \frac{[H_3O^+][HPO_4^{2-}]}{[H_2PO_4^-]} = 6.2 \times 10^{-8} \quad (16\text{-}32)$$

$$HPO_4^{2-} + H_2O \rightleftharpoons PO_4^{3-} + H_3O^+; \quad K_3 = \frac{[H_3O^+][PO_4^{3-}]}{[HPO_4^{2-}]} = 2.2 \times 10^{-13} \quad (16\text{-}33)$$

Ionization Constants

Structure-strength relationships were discussed in Section 15.6.

For oxalic acid, $HO_2C{-}CO_2H$, $K_1/K_2 \sim 900$. For adipic acid, $HO_2C(CH_2)_4CO_2H$, structurally similar but with a greater distance separating its ionizable hydrogens, $K_1/K_2 \sim 10$.

The values of the several K's for an acid always follow the order $K_1 > K_2 > K_3$. Because of the attractive force between unlike charges, removal of a positively charged proton from a negatively charged species like $H_2PO_4^-$ is relatively more difficult than from an uncharged species like H_3PO_4. Since removal of the second proton occurs less easily, the equilibrium in Reaction (16-32) does not lie as far to the right as that in Reaction (16-31). That is, $H_2PO_4^-$ is not as strong an acid as H_3PO_4. Polyprotic acids having only a small number of atoms, for example $H_2C_2O_4$, H_2SO_4, H_2S, H_2CO_3, H_3PO_4, and H_3AsO_4, have large *differences* between successive K's.

In some contexts, such as the stoichiometry of neutralization reactions, we are only mildly interested in the stepwise nature of polyprotic acid ionization. The total number of protons ultimately supplied by the acid is more important. We then write the overall ionization equation, and the corresponding overall ionization constant. Equations (16-27) and (16-28) for oxalic acid become

$$H_2C_2O_4 + 2\,H_2O \rightleftharpoons C_2O_4^{2-} + 2\,H_3O^+; \quad K = \frac{[H_3O^+]^2\,[C_2O_4^{2-}]}{[H_2C_2O_4]} = 3.8 \times 10^{-6} \quad (16\text{-}34)$$

Equation (16-34) cannot be used to calculate the pH of a solution of oxalic acid, because the relationship between the concentrations of H_3O^+ and $C_2O_4^{2-}$ is unknown. It is *not* true that one $C_2O_4^{2-}$ ion appears for each $H_2C_2O_4$ molecule that ionizes. A method for calculating the pH of polyprotic acid solutions is given in Example 16.19.

When the two equilibrium expressions are written out in full and multiplied together,

$$K_1 \cdot K_2 = \frac{[H_3O^+][\cancel{HC_2O_4^-}]}{[H_2C_2O_4]} \cdot \frac{[H_3O^+][C_2O_4^{2-}]}{[\cancel{HC_2O_4^-}]} = K$$

it is seen that the overall ionization constant K is related to the constants K_1 and K_2 for the individual steps by

$$K = K_1 \cdot K_2 \tag{16-35}$$

Similar relationships hold for all polyprotic acids, so that $K = K_1 \cdot K_2$ for H_2S, $K = K_1 \cdot K_2 \cdot K_3$ for H_3PO_4, and so on.

pH Dependence of Polyprotic Species

The relative amounts of the protonated and deprotonated forms of a weak acid are determined by the pH. The ionization equilibria for carbonic acid are

$$H_2CO_3 + H_2O \rightleftharpoons HCO_3^- + H_3O^+;$$

$$K_1 = \frac{[H_3O^+][HCO_3^-]}{[H_2CO_3]} = 4.0 \times 10^{-7} \tag{16-36}$$

$$HCO_3^- + H_2O \rightleftharpoons CO_3^{2-} + H_3O^+;$$

$$K_2 = \frac{[H_3O^+][CO_3^{2-}]}{[HCO_3^-]} = 4.0 \times 10^{-11} \tag{16-37}$$

The pH dependence is given by Equations (16-38) and (16-39), the rearranged forms of the equilibrium expressions.

$$\frac{[HCO_3^-]}{[H_2CO_3]} = \frac{K_1}{[H_3O^+]} \quad \text{or} \quad \log \frac{[HCO_3^-]}{[H_2CO_3]} = \text{pH} - \text{p}K_1 \tag{16-38}$$

$$\frac{[CO_3^{2-}]}{[HCO_3^-]} = \frac{K_2}{[H_3O^+]} \quad \text{or} \quad \log \frac{[CO_3^{2-}]}{[HCO_3^-]} = \text{pH} - \text{p}K_2 \tag{16-39}$$

Consider a solution of H_2CO_3, for which $\text{p}K_1 = 6.40$ and $\text{p}K_2 = 10.40$. If 1.00 mol H_2CO_3 is dissolved in a liter of water, the pH of the solution is 3.20, and the concentrations of the three forms are $[H_2CO_3] = 0.999$ M, $[HCO_3^-] = 6.3 \times 10^{-4}$ M, and $[CO_3^{2-}] = 4.0 \times 10^{-11}$ M. If a strong base is gradually added to the solution, the pH rises. At each new value of pH, the concentration *ratios* are easily found from Equations (16-38) and (16-39). The results of such calculations are plotted in Figure 16.9.

The pH is calculated by the method of Example 16.19.

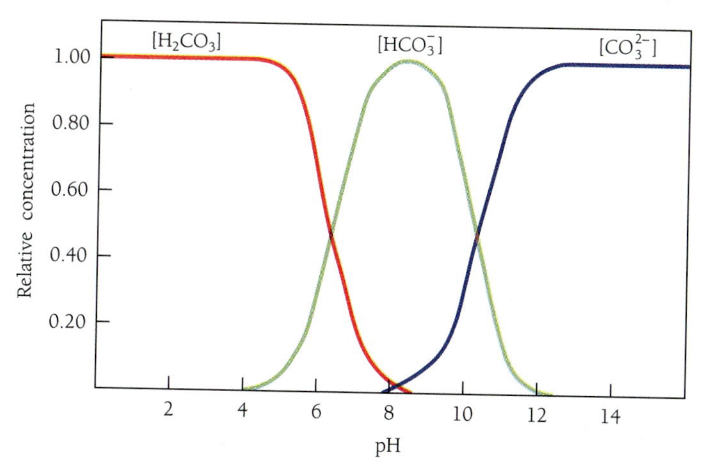

Figure 16.9 Relative concentrations of the three species in the carbonic acid system, as a function of pH. The dominant form at high pH is the fully deprotonated form CO_3^{2-}. There is *no* pH at which all three forms are present in significant concentrations.

Several conclusions may be drawn from Figure 16.9. First, there is *no* pH at which more than two forms are present in significant concentrations.* This fact is exceedingly useful, for it means that calculations of equilibrium concentrations in solutions of polyprotic acids can be carried out in the same way as for monoprotic acids. Second, at the two pH values where two curves cross one another, the concentrations of the two forms are equal. The ratio of their concentrations is 1, and the log of the ratio is 0. By Equations (16-38) and (16-39), then, pH = p*K*. That is, points of equal concentrations occur at $pH = pK_1 = 6.40$ for H_2CO_3 and HCO_3^-, and at $pH = pK_2 = 10.40$ for HCO_3^- and CO_3^{2-}. The crossing points are at pH = p*K* for all polyprotic acids.

EXAMPLE 16.18 pH at Equal Concentrations

The p*K*'s for phosphoric acid are 2.12, 7.21, and 12.66. At what pH will the concentrations of $H_2PO_4^-$ and HPO_4^{2-} be equal to one another?

Analysis A conjugate acid/base pair has equal concentrations when pH = p*K*. The only uncertainty in this problem is which of phosphoric acid's three p*K*'s must be used.

Work In a triprotic acid system, there are three equations analogous to Equations (16-38) and (16-39). For phosphoric acid they are

$$\log \frac{[H_2PO_4^-]}{[H_3PO_4]} = pH - pK_1$$

$$\log \frac{[HPO_4^{2-}]}{[H_2PO_4^-]} = pH - pK_2$$

$$\log \frac{[PO_4^{3-}]}{[HPO_4^{2-}]} = pH - pK_3$$

When the concentrations of $H_2PO_4^-$ and HPO_4^{2-} are equal, as required in this problem, the left-hand side of the second of these equations equals zero. Therefore,

$$pH = pK_2 = 7.21$$

After this sort of problem has been thoroughly understood, it is probably unnecessary to write out the equations—just choose the appropriate equilibrium and p*K*.

Exercise

At what pH will $[HAsO_4^{2-}] = [AsO_4^{3-}]$? For H_3AsO_4, $pK_1 = 2.25$, $pK_2 = 6.77$, and $pK_3 = 11.60$.

Answer: 11.60.

* This depends on the successive p*K*'s being separated by at least 3 to 4 units, and is generally true for acids containing only a few atoms (H_2CO_3, H_2S, $H_2C_2O_4$, H_3AsO_4, H_3BO_3, and so on). It is not true for many of the larger acids, whose successive p*K*'s are not as well separated. For example, at pH 5 a solution of citric acid ($H_3C_6H_5O_7$), with p*K*'s of 3.8, 4.8, and 5.4, contains about 20% $C_6H_5O_7^{3-}$, 50% $HC_6H_5O_7^{2-}$, and 30% $H_2C_6H_5O_7^-$.

Calculation of pH in Polyprotic Acids

According to Figure 16.9, no more than two of the three forms of carbonic acid can have significant concentrations in the same solution. If the pH is less than 8, $[CO_3^{2-}]$ is very small, while at pH greater than 8, $[H_2CO_3]$ is very small. When equilibrium calculations are made, only one of the two equilibrium expressions need be used. Equation (16-38) is used at low pH, and Equation (16-39) is used at high pH. Thus, although from a chemical point of view polyprotic acids are more complex than monoprotic acids, calculations are based on the same underlying algebra.

Calculations *are* more complicated for acids whose p*K*'s are not well separated. See the footnote on page 720.

EXAMPLE 16.19 pH of a Polyprotic Acid Solution

Calculate the pH of 0.0100 M H_2CO_3, for which $pK_1 = 6.40$ and $pK_2 = 10.40$.

Analysis Only one of the equilibrium expressions is required when $(pK_2 - pK_1) > \approx 3$. Initially the system contains only H_2CO_3, so the expression for K_2 cannot be useful because it does not involve $[H_2CO_3]$. Use the standard approach, and calculate the equilibrium concentrations from the K_1 expression.

Work

Balanced Equation	H_2CO_3 + H_2O ⇌	H_3O^+	+ HCO_3^-
Initial	0.0100	0	0
Change	$-x$	$+x$	$+x$
At Equilibrium	$0.0100 - x$	x	x

$$K_1 = \frac{[H_3O^+][HCO_3^-]}{[H_2CO_3]} = \frac{x \cdot x}{0.0100 - x}$$

$$K_1 \approx \frac{x^2}{0.0100}$$

$$x \approx \sqrt{0.0100 \cdot K_1} = \sqrt{0.0100 \cdot 10^{-pK_1}}$$
$$\approx 6.31 \times 10^{-5}$$

$x/C = 0.006$, so the approximation is valid.

$$[H_3O^+] = x = 6.31 \times 10^{-5}\ M$$

$$pH = -\log(6.31 \times 10^{-5})$$
$$= 4.20$$

The amount of H_3O^+ produced in the further ionization of HCO_3^- by

$$HCO_3^- + H_2O \rightleftharpoons H_3O^+ + CO_3^{2-}$$

is negligibly small, and does not affect the pH. Not only is K_2 very small, but also the H_3O^+ produced in the first step shifts the second equilibrium to the left. The second step can be safely ignored, and the pH, as calculated on the basis of the first step, is correct.

Exercise

Calculate the pH of a 0.050 M solution of dimethylmalic acid, for which $K_1 = 6.8 \times 10^{-4}$ and $K_2 = 8.7 \times 10^{-7}$.

Answer: 2.23.

Intermediate Salts

The intermediate ions (for example HCO_3^-) in a diprotic acid ionization are amphiprotic. Therefore, solutions of salts of these ions, called **intermediate salts** (for instance K_2HPO_4 or $NaHCO_3$), can react with either strong acids or strong bases. The pH of a solution of $NaHCO_3$ changes very little when a small amount of either a strong acid or a strong base is added, because the additional H_3O^+ or OH^- is consumed according to Equations (16-40) or (16-41).

$$HCO_3^- + H_3O^+ \longrightarrow H_2CO_3 + H_2O \tag{16-40}$$

$$HCO_3^- + OH^- \longrightarrow CO_3^{2-} + H_2O \tag{16-41}$$

That is, any solution of an intermediate salt of a diprotic acid is a *buffer*. The pH of such solutions is approximately equal to the average of pK_1 and pK_2, and is almost independent of the concentration of the salt. For $NaHCO_3(aq)$, pH $\approx$ 8.4. Intermediate salts of polyprotic acids behave in a similar fashion. $NaH_2PO_4(aq)$ is a buffer of pH $\approx (pK_1 + pK_2)/2$, and $Na_2HPO_4(aq)$ is a buffer of pH $\approx (pK_2 + pK_3)/2$.

Titration Curves for Polyprotic Acids

There are two equivalence points in the titration of a diprotic acid such as H_2CO_3. The first occurs when the acid H_2CO_3 is converted to HCO_3^-, and the second occurs when the acid HCO_3^- is converted to CO_3^{2-}. The two neutralizations do not occur simultaneously, because the stronger acid H_2CO_3 competes more successfully for the added OH^-. Neutralization of HCO_3^- does not begin until all of the H_2CO_3 has been consumed. Since K_2 is considerably less than K_1, the pH at the second equivalence point is noticeably higher than at the first. Figure 16.10 shows a titration curve for H_2CO_3.

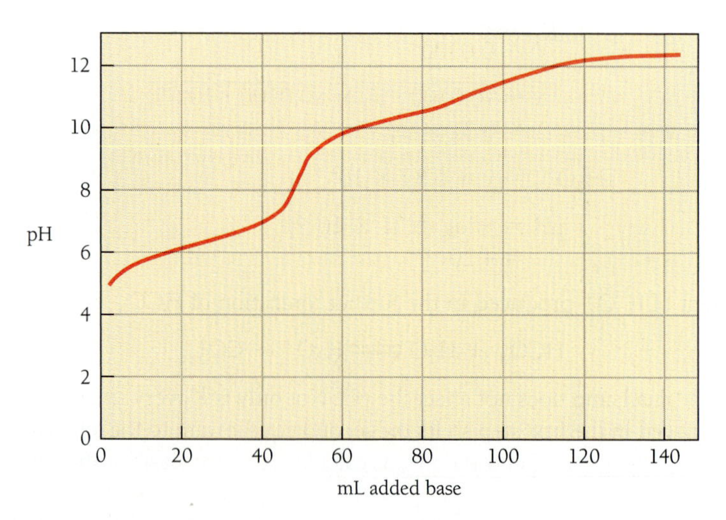

Figure 16.10 The titration of 50 mL of 0.10 M H_2CO_3 with 0.10 M KOH. Two breaks occur in the titration curve of diprotic acids, corresponding to the equivalence points of the two successive neutralization reactions. Since the acid HCO_3^- is very weak ($K_2 = 4 \times 10^{-11}$), the second break is very small and hard to see.

Titration of acids as weak or weaker than HCO_3^- poses special problems. The increase in pH during the break is less than the indicator range, and consequently the color change occurs over a relatively wide range of added titrant. A definitive endpoint cannot be observed. Even with a pH meter, the equivalence point is difficult to determine accurately.

16.6 CALCULATION SUMMARY

In this chapter, we have discussed the calculation of pH in a large variety of different aqueous systems. But apart from the definition of pH and the fact that pH + pOH = 14.00, only two equations have been used for these calculations: the Henderson-Hasselbalch equation, and the approximation $x \approx \sqrt{KC}$. Table 16.5 summarizes the types of acid-base systems and the way in which pH is calculated.

TABLE 16.5 pH Calculations in Aqueous Acid-Base Systems

Number of Solutes	Solute(s)	Required Equation	Remarks
One solute	Strong acid	$pH = -\log C$*	
	Strong base	$pOH = -\log C$	$pH = 14.00 - pOH$
	Weak acid	$pH = -\log \sqrt{K_a \cdot C}$	x/C must be <0.1†
	Weak base	$pOH = -\log \sqrt{K_b \cdot C}$	$pH = 14.00 - pOH$
	Salt of weak acid	$pOH = -\log \sqrt{K_b \cdot C}$	$K_b = K_w/K_a$
	Salt of weak base	$pH = -\log \sqrt{K_a \cdot C}$	$K_a = K_w/K_b$
Two solutes	Weak acid + salt	$pH = pK_a + \log \frac{[\text{base}]}{[\text{acid}]}$	["base"] = [salt]‡
	Weak base + salt	$pOH = pK_b + \log \frac{[\text{acid}]}{[\text{base}]}$	["acid"] = [salt]§
	or	$pH = pK_a + \log \frac{[\text{base}]}{[\text{acid}]}$	$K_a = K_w/K_b$

Titration of a Weak Acid with a Strong Base

Region	Solution Contains	Required Equation
Undertitrated	Weak acid + salt	$pH = pK_a + \log \frac{[\text{base}]^{\parallel}}{[\text{acid}]}$
Equivalence point	Salt of weak acid	$pOH = -\log \sqrt{(K_w/K_a) \cdot C}$
Overtitrated	Strong base#	$pOH = -\log$ [excess strong base]

* C is the concentration of a *single* solute: acid, base, or salt.
† x is the extent of reaction, equal to $[H_3O^+]$ in a weak-acid calculation and to $[OH^-]$ in a weak-base calculation.
‡ [base] is the concentration of the *conjugate base,* that is, the concentration of the salt of the weak acid.
§ [acid] is the concentration of the *conjugate acid,* that is, the concentration of the salt of the weak base.
|| [base] is the concentration of salt produced by the partial neutralization, and is equal to (moles of added OH^-)/volume. [acid] is the concentration of un-neutralized acid, and is equal to (original moles of acid − moles of added OH^-)/volume.
The solution also contains the salt of the weak acid, which in the presence of a *strong* base has a negligible effect on pH.

CHEMISTRY OLD AND NEW

HUMAN BODY BUFFERS

When you soothe an upset stomach by taking an antacid tablet, you are ingesting a base. Normal metabolic processes generate substantial amounts of acidic substances. Yet, most of the time, the pH of your blood remains very close to neutral, in the range 7.35 to 7.45. Buffer systems in the blood, and in other body fluids, quickly react with acidic and/or alkaline substances to maintain a neutral environment.

Buffer action is provided by numerous substances in the blood, including phosphates, citrates, proteins, and hemoglobin. However, the bicarbonate ion (HCO_3^-) serves as the body's primary blood buffer system. Bicarbonate ion is capable of consuming either excess acid (H_3O^+) or excess base (OH^-).

$$H_3O^+(aq) + HCO_3^-(aq) \longrightarrow H_2CO_3(aq) + H_2O(\ell)$$

$$OH^-(aq) + HCO_3^-(aq) \longrightarrow CO_3^-(aq) + H_3O(\ell)$$

When gaseous carbon dioxide dissolves, it reacts with water to produce carbonic acid, H_2CO_3.

$$CO_2(g) + H_2O(\ell) \rightleftharpoons H_2CO_3(aq)$$

In the body, this reaction is facilitated by the enzyme *carbonic anhydrase*. Carbonic acid is a weak acid that ionizes as follows.

$$H_2CO_3(aq) + H_2O(\ell) \rightleftharpoons H_3O^+(aq) + HCO_3^-(aq); pK_a = 6.1$$

The pH of blood depends on the relative concentrations of H_2CO_3 and HCO_3^-, and may be calculated by Equation (16-24b), the Henderson-Hasselbalch equation.

$$pH = pK_a + \log([HCO_3^-]/[H_2CO_3])$$

When the concentration of $HCO_3^-(aq)$ is about 20 times that of H_2CO_3, the pH is 7.4. The human body buffer maintains this pH by adjusting the relative concentrations of HCO_3^- and H_2CO_3.

Most of the body's energy needs are supplied by the exothermic oxidation of glucose to form carbonic acid (aqueous carbon dioxide).

$$C_6H_{12}O_6 + 6\ O_2 \longrightarrow 6\ H_2CO_3$$

Build-up of $[H_2CO_3]$, which would decrease the $[HCO_3^-]/[H_2CO_3]$ ratio and therefore decrease the pH, is prevented by a process that takes place in the lungs. There, the carbonic acid reverts to gaseous carbon dioxide, and passes from the body with each exhalation of the breath.

$$H_2CO_3 \longrightarrow CO_2 + H_2O$$

Many other processes within the body affect acidity. Strenuous exercise produces lactic acid, which increases acidity in muscle tissue and causes temporary soreness. Overdoses of acidic medications like aspirin, or of basic substances such as antacids, can alter the pH of blood.

Although a healthy human body naturally adjusts its pH level, two life-threatening conditions—acidosis and alkalosis—can occur whenever the pH drops below 6.7 or rises above 7.8. These conditions can result from disease, metabolic disorders, any serious disruption of breathing patterns, or as a side effect of certain medications.

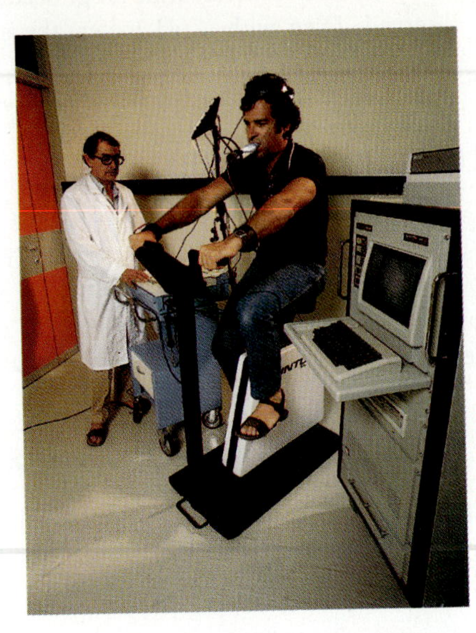

Heavy exercise increases the production of CO_2 in the muscles, and the acidity of the blood increases until the excess CO_2 can be exhaled.

For example, a person who hyperventilates during a mountain climb may expel too much CO_2, thus altering the ratio of HCO_3^- to CO_2 and causing *respiratory alkalosis*. Conversely, people with lung diseases or respiratory disorders, and patients under anesthesia, may develop respiratory acidosis if breathing is too shallow to expel excess CO_2. This condition can also be brought on by cardiac arrest.

A metabolic malfunction in diabetic patients sometimes results in the buildup of β-hydroxybutyric acid and acetoacetic acid in extracellular fluids, causing *metabolic acidosis*. Kidney malfunction or failure is usually followed by acidosis. Patients with metabolic or respiratory acidosis have in many cases been treated with sodium bicarbonate, administered intravenously. This quickly restores the proper 20:1 ratio of $[HCO_3^-]$ to $[H_2CO_3]$.

Discussion Questions

1. The inorganic phosphate ions $H_2PO_4^-$ and HPO_4^{2-} are closely associated with the reactions of adenosine triphosphate (ATP), an important intermediate in the distribution and use of energy in the body. What is the relationship between pH and the levels of $H_2PO_4^-$ and HPO_4^{2-} in the blood?
2. Patients with emphysema have difficulty breathing. What effect does this have on the pH of the blood?

▶ SUMMARY EXERCISE

Most mammals can synthesize ascorbic acid ($C_6H_8O_6$), but humans cannot. Since this compound, also called vitamin C, is a prominent component of most fruits and vegetables, we easily ingest the daily 0.05 g required for the prevention of scurvy. Many people believe that large doses of vitamin C prevent colds, although this view is not supported by a firm body of clinical evidence.

When a 500-mg tablet of vitamin C is dissolved in 50.0 mL water at 25 °C, the result is a 0.0568 M solution whose pH is 2.67.

a. Calculate the molar mass of ascorbic acid.

b. What is the percent ionization in this solution?

c. Calculate the acid ionization constant and the pK_a.

d. What are the values of K_b and pK_b for the ascorbate ion?

e. Calculate the pH of a 0.22 M solution of sodium ascorbate.

f. The pH of healthy human blood is in the range 7.35 to 7.45. What is the ratio of the concentration of ascorbic acid to the concentration of ascorbate ion when the pH is 7.40?

g. Suppose 50 mL of a 0.250 M solution of ascorbic acid is titrated with 0.250 M KOH. Calculate the pH after the addition of the following volumes of base: (i) 0 mL, (ii) 25.0 mL, (iii) 50.0 mL, and (iv) 100 mL.

h. The pH of blood is controlled by several buffer systems, the most important of which is H_2CO_3/HCO_3^-. The concentration of HCO_3^- in blood is about 0.0027 M when the pH is 7.40. Given that K_1 for carbonic acid is 4.0×10^{-7}, calculate the concentration of H_2CO_3 in blood.

Answers **(a)** 176 g mol^{-1}; **(b)** 3.76%; **(c)** $K_a = 8.4 \times 10^{-5}$, $pK_a = 4.08$; **(d)** $K_b = 1.2 \times 10^{-10}$, $pK_b = 9.92$; **(e)** 8.71; **(f)** 4.8×10^{-4}, which means that only about 0.05% of the total is present as the un-ionized acid; **[g(i)]** 2.34; **[g(ii)]** 4.08; **[g(iii)]** 8.59; **[g(iv)]** 12.92; **(h)** 2.7×10^{-4} M.

SUMMARY AND KEYWORDS

The ionization equilibrium of a weak acid is described by an equilibrium expression and K_a, the **(acid) ionization constant.** $HA + H_2O \rightleftharpoons A^- + H_3O^+$; $K_a = [H_3O^+][A^-]/[HA]$. Weak-base ionization equilibria are described by an equilibrium expression and K_b, the **(base) ionization constant.** $B + H_2O \rightleftharpoons BH^+ + OH^-$; $K_b = [BH^+][OH^-]/[B]$. Water appears in the chemical equations but not in the equilibrium expressions. K_a and K_b are often given in "p" notation as $pK_a = -\log(K_a)$ and $pK_b = -\log(K_b)$. The greater the relative strength of an acid, the greater its K_a and the smaller its pK_a.

The *structured approach* to calculation of equilibrium concentrations, pH, and *percent ionization* in weak acid or base systems involves (a) the balanced chemical equation; (b) the initial concentration(s); (c) the change in concentrations, or *extent of reaction x*; and (d) the equilibrium constant. The resulting quadratic equation has the approximate solution $x \approx \sqrt{KC}$, which may be used when $x/C \approx 0.1$ or less. If the acid is very weak and/or very dilute, the value of K_a or K_b does not control the pH because the major contributor to $[H_3O^+]$ and $[OH^-]$ is the autoionization of water.

The **reciprocal relationship** between the strengths of acid/base **conjugate pairs** is $K_a \cdot K_b = K_w$. A salt of a weak acid is a weak (ionic) base, and a salt of a weak base is a weak (ionic) acid. Salts of strong acids and bases are **neutral salts,** and they have no effect on pH. The anion of a weak acid and the cation of a weak base may combine to form an *amphoteric salt.*

In the **common ion effect,** predictable from Le Châtelier's principle, an acid (or base) ionization reaction is inhibited by the addition of a salt containing the anion of the acid (or cation of the base). **Buffer solutions** are prepared by mixing a weak acid (or base) and one of its salts, or by partial neutralization of a weak acid (or base). A buffer can react with and consume either acid or base, and therefore resists change in pH on addition of either H_3O^+ or OH^-. **Buffer capacity** is greatest when the pH is close to the pK_a of the buffer acid. The Henderson-Hasselbalch equation, $pH = pK_a + \log([base]/[acid])$, is used in buffer calculations.

A **titration curve** is a graph of sample pH vs. volume of added titrant. When an acid sample is titrated, the curve is characterized by a gradual rise in the *undertitrated* region, a sharp **break** when the sample is exactly neutralized (the **equivalence point**), and then a leveling out in the *overtitrated* region. The pH at the equivalence point is 7 for strong acids and greater than 7 for weak acids. Strong acids produce larger breaks than weak acids, and concentrated acids produce sharper breaks than dilute acids. At the midpoint of a titration of a weak acid, $pH = pK_a$.

An **indicator** is a weak acid whose conjugate base has a strikingly different color. A titration sample containing an indicator changes color when the titration nears the equivalence point. When the color changes, the titration is at its **endpoint** (close to but not necessarily the same as the equivalence point). Each indicator has a different **indicator range** (the 1.5 to 2.0 unit range of pH over which the color change takes place). The range of *phenolphthalein* is 8 to 9.8, which makes it suitable for the titration of all but very weak acids.

Each successive ionization step of a polyprotic acid is characterized by an equilibrium constant, smaller than that of the preceding step: $K_1 > K_2 > K_3$. The overall reaction has an equilibrium constant $K = K_1 \cdot K_2 \cdot K_3$. When the successive *K*'s differ greatly in magnitude (the usual case for acids containing only a few atoms), only two of the forms of a polyprotic acid can be present in significant quantities at a given pH. Except for sulfuric acid, the pH of a polyprotic acid solution is calculated by ignoring all but the first ionization. Titration curves show two breaks for diprotic acids, and three for triprotic acids. These breaks occur at successively higher pH. Breaks occurring at high pH are smaller and difficult to observe.

PROBLEMS

General

1. Distinguish between reaction quotient and equilibrium expression.
2. What is the justification for omitting the concentration of water from the reaction quotient of an acid or base ionization equilibrium, in light of the fact that Brønsted-Lowry theory clearly identifies water as a reaction participant?
3. Is the relationship between acid strength and pK_a value direct or inverse? Explain.
4. Define percent ionization.
5. Distinguish between equilibrium constant and ionization constant.
6. Distinguish between extent of reaction and fractional ionization.

*7. Derive the general expression $x = [-K + \sqrt{(K^2 + 4KC)}]/2$ for the extent of reaction in an equilibrium ionization of a weak acid or base. *C* is the initial molarity and *K* is K_a or K_b. What is the relationship between *x* and pH in this expression? Modify the expression for *x* to a form giving the fractional ionization in terms of *K* and *C*.

8. Under what circumstances is it appropriate to use the approximate equation $x \approx \sqrt{KC}$ rather than the quadratic form?
9. Why is the pH of an aqueous solution of a *very* weak acid, say one with $pK_a > 14$, independent of concentration?
10. What is meant by the term "reciprocal relationship," as applied to acids and bases? Illustrate your discussion with an algebraic expression.
11. Define hydrolysis, and give some example chemical equations of the hydrolysis of salts.
12. Can the same substance be both a base and a salt? Explain.
13. Distinguish between amphiprotic and amphoteric.
14. What is a buffer solution?
15. What is meant by the term "buffer capacity"?
16. What is the Henderson-Hasselbalch equation? Given an example of its use.
17. What is titration?
18. Consider the titration of a strong acid with NaOH, and describe the change in composition of the sample as it passes from the undertitrated region, through the equivalence point, and into the overtitrated region.
19. Repeat the preceding problem, but for a weak acid rather than a strong one.
20. What is meant by the "break" in a titration curve?
21. What is an indicator? What are the characteristics of a good indicator for an acid/base titration?
22. Distinguish between the endpoint and the equivalence point of a titration.
23. What is meant by the term "indicator range"?
24. What is the pH at the equivalence point of the titration of a strong acid? Of a weak acid?
25. In a description of the ionization of a diprotic acid, to what do the symbols K, K_1, and K_2 refer? Illustrate your answer with chemical and algebraic equations.
26. Is it possible that $K_1 < K_2$ for a diprotic acid? If so, give an example. If not, explain why not.

Ionization Constant and p*K*

27. Convert the following acid ionization constants to pK_a's, using the appropriate number of significant figures.
    a. fluoroacetic acid, $K_a = 2.2 \times 10^{-3}$
    b. benzenesulfonic acid, $K_a = 0.20$
    c. butanoic acid, $K_a = 1.51 \times 10^{-5}$
    d. *p*-chlorobenzoic acid, $K_a = 1 \times 10^{-4}$
28. Convert the following base ionization constants to pK_b's, using the appropriate number of significant figures.
    a. aniline, $K_b = 3.8 \times 10^{-10}$
    b. aziridine, $K_b = 1.1 \times 10^{-6}$
    c. benzamidine, $K_b = 5.4 \times 10^{-3}$
    d. pyridine, $K_b = 1.4 \times 10^{-9}$
29. Convert the following pK_a values to acid ionization constants, using the appropriate number of significant figures.
    a. bromoacetic acid, $pK_a = 2.86$
    b. thiophenol, $pK_a = 6.52$
    c. picric acid, $pK_a = 0.25$
    d. nitrous acid, $pK_a = 3.4$
30. Convert the following pK_b values to base ionization constants, using the appropriate number of significant figures.
    a. trimethylamine, $pK_b = 4.2$
    b. *N*-methylaniline, $pK_b = 9.16$
    c. hydrazine (K_1), $pK_b = 6.1$
    d. ethylamine, $pK_b = 3.37$

Equilibrium Concentrations, and Percent and Fractional Ionization

31. Calculate the pH of a 0.375 M solution of hypochlorous acid.
32. The pK_a of formic acid is 3.74. What is the pH of a 0.025 M solution?
33. Calculate $[OH^-]$ and pH in a 0.050 M solution of dimethylamine ($K_b = 5.9 \times 10^{-4}$).
34. Calculate the pH of a 1.00×10^{-3} M solution of pyridine.
35. Calculate the pH and percent ionization in a 0.0750 M solution of HCN.
36. What is the percent ionization in a 0.27 M solution of uric acid ($K_a = 1.3 \times 10^{-4}$)?
37. What is the percent ionization in a 0.100 M solution of formic acid? What is the pH?
*38. Calculate the percent ionization in a 0.0283 M solution of methylamine. What is the pH?
*39. The pH of a solution of formic acid is 3.00. What was the initial concentration of the acid? What are the equilibrium concentrations of the protonated and deprotonated forms?
40. What concentration of methylamine will produce a solution of $[OH^-] = 0.015$ M?

Ionization Constant and pH

41. The percent ionization in thiophenol is 0.53% when the concentration is 0.0100 M. What is the acid ionization constant of thiophenol?
42. The fractional ionization in 0.40 M iodoacetic acid is 0.0412. What is the acid ionization constant of iodoacetic acid?
43. Fluoroacetic acid is 4.6% ionized in a 1.00 M solution. What is its pK_a?
44. Pyridine is 0.053% ionized in a 0.00500 M solution. What is the pK_b of pyridine?
45. The pH of a 0.025 M solution of butanoic acid is 3.21. What is the acid ionization constant?
46. The pH of a 0.35 M solution of uric acid is 2.17. What is the pK_a of uric acid?

47. If the equilibrium concentration of formate ion is 0.0115 M in a 0.750 M solution of formic acid, what is the K_a of formic acid? Remember that the value 0.750 M refers to the total concentration of protonated and deprotonated forms.

48. The concentration of trimethylammonium ion is 7.9×10^{-4} M in a 0.0100 M solution of trimethylamine. What is the pK_b of trimethylamine?

49. The pH of a 5.00×10^{-3} M solution of *N*-methylaniline is 8.27. What is the value of the base ionization constant of *N*-methylaniline?

50. A 0.068 M solution of benzamidine has a pOH of 1.78. What is the value of pK_b for benzamidine?

Salt Hydrolysis

51. Write the chemical equation for the hydrolysis of KNO_2 and calculate the value of K_b.

52. The base ionization constant of ethylamine ($C_2H_5NH_2$) is 4.3×10^{-4}. What is K_a for ethylammonium iodide? Write the equilibrium expression that is equal to K_a.

53. Given that pK_a for HOCl is 7.52, calculate K_b for NaOCl. Write the chemical equation for the hydrolysis reaction.

54. Calculate K_a for pyridinium bromide (C_5H_5NHBr). What is the equilibrium expression?

*55. Classify each of the following aqueous solutions as acidic, neutral, or basic.
   a. hydrazinium bromide
   b. potassium sulfate
   c. potassium chloride
   d. trimethylammonium chloride

*56. Classify each of the following aqueous solutions as acidic, neutral, or basic.
   a. sodium nitrate
   b. pyridinium chloride
   c. potassium nitrite
   d. lithium hypochlorite

57. What is the pH of 0.0500 M pyridinium chloride?

58. Calculate the pH of 0.100 M potassium benzoate.

59. What is the pH of 0.0375 M ethylammonium chloride? What is the concentration of ethylamine in this solution?

60. Given that the pK_a of HF is 3.18, calculate the pH of a solution of 0.50 M NaF. What is the concentration of HF in this solution?

61. Given that the pH of a 0.044 M solution of hydroxylammonium chloride is 3.66, calculate K_a for the hydroxylammonium ion and K_b for hydroxylamine.

62. The pH of a 0.75 M solution of potassium propanoate is 9.38. Calculate the concentrations of propanoate ion and propanoic acid, K_b of the propanoate ion, and the ionization constant of propanoic acid.

Common Ion Effect

63. Calculate the equilibrium concentration of CN^- in a solution in which the equilibrium concentrations of HCN and H_3O^+ are 0.125 M and 0.830 M respectively.

64. Given that K_a of HNO_2 is 0.00040, calculate the concentration of NO_2^- in a solution in which the equilibrium concentrations of HNO_2 and H_3O^+ are 0.0924 M and 0.00100 M, respectively.

*65. What is the concentration of trimethylamine if 0.0348 mol trimethylammonium bromide is dissolved in 1.00 L of 0.300 M HBr? K_b for trimethylamine is 6×10^{-5}.

66. What is the concentration of NH_4^+ if 10.0 g KOH is added to 1.00 L of a 1.00 M solution of NH_3?

Buffer Solutions

67. The pK_a of HOCl is 7.52. Calculate the pH of a solution in which [HOCl] = 0.493 M and [OCl^-] = 0.0872 M.

68. What is the pH of a buffer consisting of 0.300 mol $NaNO_2$ and 0.100 mol HNO_2 in 1.500 L water?

69. What is the pH of a solution, 500 mL of which contains 10.0 g each of acetic acid and sodium acetate?

70. Calculate the pH of a buffer consisting of a 0.500 M solution of NH_3 to each liter of which has been added 7.82 g NH_4Cl.

71. What concentrations of *o*-ethylbenzoic acid (pK_a = 3.79) and potassium *o*-ethylbenzoate are required to prepare a buffer of pH = 4.00?

*72. What relative concentrations of aniline (pK_b = 9.42) and anilinium chloride are required to prepare a buffer of pH = 5.00?

*73. What mass ratio of trifluoroacetic acid (molar mass 114 g mol^{-1}) to sodium trifluoroacetate (molar mass 136 g mol^{-1}) must be added to water in order to prepare a buffer of pH = 1.00?

*74. What mass ratio of ethanolamine (61.6 g mol^{-1}) to ethanolammonium bromide (151.0 g mol^{-1}) must be added to water in order to prepare a buffer of pH = 9.50?

75. What acid might be a good choice for a buffer of pH = 5.00? What ratio of molar concentrations of acid/salt would produce this pH?

76. What acid might be a good choice for a buffer of pH = 7.50? What ratio of molar concentrations of acid/salt would produce this pH?

*77. How many moles KOH must be added to 1.00 L of 0.372 M acetic acid in order to prepare a buffer of pH = 5.00?

*78. How many grams NH_4Cl must be added to 400 mL of a 0.93 M solution of NH_3 in order to prepare a buffer of pH = 9.00?

Titration Curves

79. Calculate the pH at the equivalence point of a titration of acetic acid with 0.100 M KOH, assuming that the concentration of the acid is as follows.
a. 1.00 M b. 0.100 M
c. 0.0100 M

80. Calculate the pH at the equivalence point of the following titrations.
a. 0.0783 M HCl with 0.100 M KOH
b. 0.113 M HNO_2 with 0.100 M KOH
c. 0.218 M HOCl with 0.100 M KOH
d. 0.104 M NH_3 with 0.100 M HCl.

81. Calculate the pH during the titration of 50.0 mL 0.100 M HCl with 0.100 M KOH, after the addition of the following volumes of base.
a. 10.0 mL b. 25.0 mL
c. 49.0 mL d. 50.0 mL
e. 80.0 mL
Sketch a rough titration curve through these points.

*82. Repeat the preceding problem for the titration of 50.0 mL of 0.100 M HOCl with 0.100 M KOH.

83. Calculate the pH during the titration of 50.0 mL 0.100 M KOH with 0.100 M HCl, after the addition of the following volumes of base.
a. 10.0 mL b. 25.0 mL
c. 49.0 mL d. 50.0 mL
e. 80.0 mL
Sketch a rough titration curve through these points.

*84. Repeat the preceding problem for the titration of 50.0 mL of 0.100 M NH_3 with 0.100 M HCl.

Polyprotic Acids

85. Write the chemical equations and equilibrium expressions for the successive ionization steps of the following species.
a. H_2SeO_4 (diprotic) b. H_3AsO_4 (triprotic)

*86. Write the chemical equations and equilibrium expressions for the ionization of the dihydric base $H_2NCH_2CH_2NH_2$ (ethylenediamine).

The following pK_a values will be useful in solving the remaining problems in this chapter—H_2CO_3: $pK_1 = 6.40$, $pK_2 = 10.40$; H_3PO_4; $pK_1 = 2.12$, $pK_2 = 7.21$, $pK_3 = 12.66$.

87. (a) Write the overall chemical equation and equilibrium expression for the complete ionization of H_2CO_3. (b) What is the value of the overall ionization constant?

88. (a) Write the overall chemical equation and equilibrium expression for the complete ionization of H_3PO_4. (b) What is the value of the overall ionization constant?

89. Refer to Figure 16.9 and estimate the concentrations of H_2CO_3, HCO_3^-, and CO_3^{2-} when the pH is 7.00 and the total concentration of carbonate (that is, $[H_2CO_3] + [HCO_3^-] + [CO_3^{2-}]$) is 0.0500 M.

90. Refer to Figure 16.10 and estimate the pH at which the concentration of HCO_3^- is twice that of CO_3^{2-}.

91. Gaseous CO_2 is dissolved in a buffer solution of pH = 4.00, resulting in an equilibrium concentration $[H_2CO_3] = 0.0010$ M. What are the concentrations of HCO_3^- and CO_3^-?

*92. 1.50 g Na_2CO_3 is dissolved in water, and the pH is adjusted to 6.00 by the addition of HCl. The final volume of the solution is 500 mL. What are the concentrations of H_2CO_3, HCO_3^-, and CO_3^{2-}?

*93. What is the pH of a solution, 1.00 liter of which contains 10.0 g each of Na_2CO_3 and $NaHCO_3$?

*94. It is desired to have a buffer of pH = 7.00, using phosphoric acid and/or its salts. What species should be used, and what should be their concentration *ratio?*

*95. Calculate the pH if 70 mL 0.100 M KOH is added to 50 mL 0.100 M H_2CO_3. Note that this amount of base is more than enough to convert all H_2CO_3 to HCO_3^-, but not enough to convert all the HCO_3^- to CO_3^{2-}.

*96. Calculate the pH if 70 mL 0.100 HCl is added to 50 mL 0.100 M Na_2CO_3.

Other Problems

*97. Show that the error in using the expression $[H_3O^+] = x \approx \sqrt{K_a \cdot C}$ is no more than 5% whenever $x < 0.1 \cdot C$. *Hint:* Let $x_{approx} = \sqrt{K_a \cdot C}$, show that $x_{true} = \sqrt{K_a \cdot (C - x)}$ and then that $(x_{approx}/x_{true}) < 1.05$ when $x < 0.1 \cdot C$.

98. A 0.083 M solution of a certain acid is known to be 1.07% ionized. What is the pH?

99. The concentration of hydronium ion is 8.0×10^{-5} in a solution of HClO. What is the value of the concentration ratio $[ClO^-]/[HClO]$?

100. If the concentration of OH^- ion in a 0.15 M solution of nicotine is 3.97×10^{-4} M, what is K_b for nicotine?

*101. The pH of a 0.275 M solution of sodium *p*-methoxyphenolate is 11.85. What is the pK_a of *p*-methoxyphenol?

102. Using the information in Table 16.1, arrange the following ions in order of increasing strength as bases: CN^-, NO_3^-, HS^-, NO_2^-.

103. Calculate the pH of a solution that is 0.0753 M in HOCl and 0.156 M in NaOCl.

*104. A solution contains 7.50 g KNO_2 per liter. How much HNO_2 must be added to 1.00 L in order to prepare a buffer of pH = 4.00? (Assume no change in volume.)

105. Which of the following buffers has the greater resistance to change in pH (buffer capacity)? Explain your answer.
   a. [acid] = [salt] = 0.100 M
   b. [acid] = [salt] = 0.300 M

106. In the titration of an unknown acid, the pH is observed to be 3.64 at the point when exactly half of the acid has been neutralized. What is the ionization constant of the acid?

*107. Calculate the pH of the solution resulting from mixing equal volumes of 0.10 M HCl with each of the following.
   a. water
   b. 0.20 M NaOH
   c. 0.10 M sodium acetate
   d. 0.20 M sodium acetate
   e. 0.10 M NH_3
   f. 0.20 M NH_3

108. (a) What is the concentration of a solution of HCl if the titration of a 25.00 mL sample requires 37.53 mL of 0.09961 M KOH? (b) Would this answer change if the acid were hypochlorous acid rather than hydrochloric acid? If so, recalculate. If not, explain why not.

*109. 0.292 g of an unknown acid is dissolved in water and titrated with 0.100 M base. It is noted that the pH is 4.20 after the addition of 11.95 mL base and that the equivalence point is reached after the addition of 23.90 mL base. What are the molar mass and pK_a of the acid? (Assume that the acid is monoprotic.)

*110. (a) Write two different chemical equations, and their corresponding equilibrium expressions, for the proton transfer reactions of the bicarbonate ion (HCO_3^-) with water. (b) What is the value of the ionization constant (or hydrolysis constant) for each?

111. Two indicators are available for a titration: phenolphthalein, with a range of 8.0 to 9.8, and methyl red, with a range of 4.2 to 6.2. Which would be the better choice for each of the following titrations? Explain your choice.
   a. titration of acetic acid with KOH
   b. titration of hydrochloric acid with KOH
   c. titration of ammonia with HCl
   d. titration of sodium cyanide with HCl

CHAPTER 17

SOLUBILITY AND COMPLEX ION EQUILIBRIA

A precipitate of lead(II) chromate forms immediately when a solution of a soluble chromate salt is added to a solution of a soluble lead salt.

TABLE 17.1 Solubility Products of Some Slightly Soluble Salts at 25 °C

Ag_2SO_4	1.5×10^{-5}
$BaSO_4$	1.1×10^{-10}
$PbSO_4$	1.6×10^{-8}
$Mg(OH)_2$	7.3×10^{-12}
$Al(OH)_3$	4.5×10^{-33}
MgF_2	6.6×10^{-9}
CaF_2	3.9×10^{-11}
SrF_2	2.9×10^{-9}
$CaCO_3$	4.6×10^{-9}
Hg_2CO_3	8.9×10^{-17}
$PbCO_3$	7.4×10^{-14}
$AgCl$	1.8×10^{-10}
$PbCl_2$	1.7×10^{-5}

In the preceding chapter, the mathematical treatment of chemical equilibrium was applied to proton-transfer reactions in aqueous solutions. We saw that one of the underlying themes of acid-base chemistry is the unique position of water as a solvent, and as reactant, in the laboratory, the environment, and in living organisms. In this chapter we continue the theme, and discuss additional types of equilibria involving aqueous species.

17.1 SOLUBILITY AND SOLUBILITY PRODUCT

In the earlier discussion of solutions (Chapter 12), it was pointed out that the solubility of a substance depends on an equilibrium between the dissolved and undissolved (pure) forms of the substance. The important influence of enthalpy and entropy of solvation on solution equilibria was discussed. We now broaden the discussion to include the quantitative aspects of solubility, and consider some calculations based on solubility equilibrium constants.

Saturated Solutions

A solution in equilibrium with pure, undissolved solute is **saturated.** The equilibrium between undissolved and dissolved solute is described by a chemical equation, such as (17-1) for the ionic solute silver sulfate.

$$Ag_2SO_4(s) \rightleftharpoons 2\ Ag^+(aq) + SO_4^{2-}(aq) \tag{17-1}$$

The *reaction quotient* was first described in Section 13.2.

The reaction quotient for this reaction is given by Equation (17-2).

$$Q = [Ag^+]^2[SO_4^{2-}] \tag{17-2}$$

Recall that pure solids are always omitted from reaction quotients. A reaction quotient such as this one, which contains only ionic species in the numerator and whose denominator is 1, is called an **ion product.**

The equilibrium constant of *any* reaction is defined as Q_{eq}, the reaction quotient when the reaction is in equilibrium. When a system composed of pure (undissolved) solute and dissolved solute is in equilibrium, the ion product Q_{eq} is called the **solubility product,** K_{sp}. Equation (17-3) is called the *solubility product expression* for the salt Ag_2SO_4.

The notation K_{sp} (rather than K or K_{eq}) is always used for the equilibrium constant of a dissolution reaction.

$$K_{sp} = Q_{eq} = [Ag^+]_{eq}^2[SO_4^{2-}]_{eq} \tag{17-3}$$

Since the notation K_{sp} applies only to systems in equilibrium, the "eq" subscripts in the right-hand side of Equation (17-3) are unnecessary and normally omitted.

EXAMPLE 17.1 The Solubility Product Expression

Write the solubility product expression for the slightly soluble salt aluminum sulfate, $Al_2(SO_4)_3$.

Analysis A solubility product expression is an equilibrium constant expression. It is written using the stoichiometric coefficients of the balanced chemical equation as concentration exponents, but omitting solvent and pure solid species.

Work The balanced chemical equation is

$$Al_2(SO_4)_3(s) \rightleftharpoons 2\ Al^{3+}(aq) + 3\ SO_4^{2-}(aq)$$

The solubility product expression is

$$K_{sp} = [Al^{3+}]^2[SO_4^{2-}]^3$$

Exercise

Write the solubility product expression for CaF_2.

Answer: $K_{sp} = [Ca^{2+}][F^-]^2$.

The units of a solubility product depend on the stoichiometric coefficients in the dissolution equation, for instance Equation (17-1), and these coefficients in turn are the same as the ion subscripts in the formula of the compound. Since concentrations in solubility product expressions are always expressed in mol L^{-1}, the units of K_{sp} are usually not written down. Values of the solubility products for a few ionic compounds are given in Table 17.1, and a more extensive listing appears in Appendix H.

Values of K_{sp} larger than about 10^{-3} are rarely encountered, because solubility products are used only with salts that are not very soluble.

Although it is given a special name and symbol, the solubility product is in no way different from any other equilibrium constant. When K_{sp} is small, as it is for $BaSO_4$, it means that the dissolution reaction does not proceed very far to the right. The concentrations of both Ba^{2+} and SO_4^{2-} are very low in a saturated solution. This low concentration allows the use of $BaSO_4$ in certain medical procedures. Because it is opaque to x-rays, barium sulfate is often used in x-ray diagnoses of gastrointestinal difficulties. The patient drinks a "barium cocktail," a suspension of finely divided $BaSO_4(s)$ in water, after which the mysteries of the digestive tract are revealed by x-ray investigation (Figure 17.1). In spite of the fact that all *soluble* barium salts are toxic, the procedure is safe because $BaSO_4$ is essentially insoluble.

The **solubility** of a substance is the maximum amount that will dissolve. For example, 4.8 grams of silver sulfate is the most that can dissolve in one liter of water at 25 °C. The solubility of silver sulfate is therefore 4.8 g L^{-1}. The units of solubility should be carefully noted, because units other than g L^{-1} are sometimes used. Until fairly recently it was common practice to express solubility in "g/100 mL" or the equivalent "g/100 cc". If 4.8 g dissolves in 1 L, the solubility can also be expressed as 0.48 g/100 mL. The temperature must be specified when solubility values are quoted, since solubility usually changes with temperature.

Solubility is sometimes expressed in units of mol L^{-1}, and referred to as **molar solubility.** The conversion factor between *gram* solubility and *molar* solubility is the molar mass. For silver sulfate,

$$\text{molar solubility} = (4.8 \text{ g L}^{-1}) \cdot \left(\frac{1 \text{ mol}}{311.8 \text{ g}}\right) = 0.015 \text{ mol L}^{-1}$$

When an ionic substance such as silver sulfate dissolves, it is *completely* dissociated into its component ions—the only species to enter solution are Ag^+ and SO_4^{2-}. Note that the component ions of a salt can have different concentrations, as determined by the stoichiometry of the salt. If 0.015 mol of Ag_2SO_4 dissolves in one liter of water, the resulting ion concentrations are $[Ag^+] = 0.030$ M and $[SO_4^{2-}] = 0.015$ M.

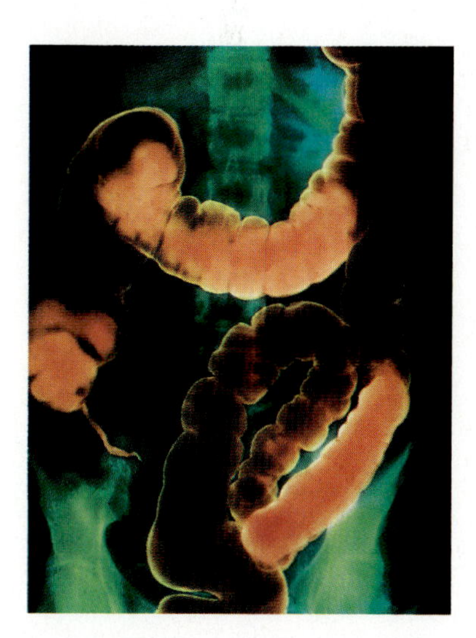

Figure 17.1 Barium and its compounds are opaque to x-rays. The intestinal tract is highlighted in an x-ray photograph after the patient swallows a suspension of insoluble $BaSO_4$.

Conversion between Solubility and K_{sp} for 1:1 Salts

Solubility and solubility product are related quantities. Since each describes the amount of solute in a saturated solution, it is possible to obtain the value of one from that of the other.

Figure 17.2 The mineral *gypsum* is $CaSO_4 \cdot 2H_2O$. One of the world's most spectacular deposits of gypsum occurs at White Sands National Monument in New Mexico. The "sands" are not sand, but pure gypsum.

EXAMPLE 17.2 Calculation of Solubility Product

The solubility of $CaSO_4$ is 0.67 g L^{-1} at 25 °C. What is its solubility product?

Analysis/Plan The target (K_{sp}) is obtained from the solubility product expression and the molarities of Ca^{2+} and SO_4^{2-} in a saturated solution. The molarities are calculated from the molar mass and the gram solubility.

Work

1. The molar mass of $CaSO_4$ is 136.1 g mol^{-1}, so that

$$\text{molar solubility} = \text{mol } CaSO_4\ L^{-1} = (0.67 \text{ g L}^{-1}) \cdot \left(\frac{1 \text{ mol}}{136.1 \text{ g}}\right)$$
$$= 4.9 \times 10^{-3} \text{ mol L}^{-1}$$

2. For each mole of $CaSO_4$ that dissolves, one mole each of Ca^{2+} and SO_4^{2-} appears in solution. The ion molarities in a saturated solution are therefore $[Ca^{2+}] = [SO_4^{2-}] = 4.9 \times 10^{-3}$ M.
3. The solubility product expression is $K_{sp} = [Ca^{2+}][SO_4^{2-}]$.
4. Substitution of the ion molarities gives for the solubility product $K_{sp} = [Ca^{2+}][SO_4^{2-}] = (4.9 \times 10^{-3})(4.9 \times 10^{-3}) = 2.4 \times 10^{-5}$.

Check The answer is reasonable, since solubility products are almost always less than 10^{-4}.

Exercise

Calculate the solubility product of AgCl, given that its solubility is 0.0019 g L^{-1} at 25 °C.

Answer: 1.8×10^{-10}.

Calculation of a solubility when the solubility product is known is frequently required, since many textbooks and handbooks list solubility products and not solubilities. Although problems of this sort can be solved by working Example 17.2 in reverse, it is more consistent to use the structured approach suitable for all equilibrium problems.

EXAMPLE 17.3 Calculation of Solubility from K_{sp}

One form of $CaSO_4$ is *alabaster*. It is a translucent stone, usually pure white but sometimes with reddish-brown streaks. Because it is very soft and easily carved, alabaster is used for statuary and other decorative purposes.

The solubility product of $CaSO_4$ is 2.4×10^{-5} at room temperature. What is its solubility in g L^{-1}?

Analysis Initially, when no $CaSO_4$(s) has dissolved, $[Ca^{2+}] = [SO_4^{2-}] = 0$. The molar solubility is related to the (unknown) extent of reaction. Follow the standard approach.

Work

Balanced Equation	$CaSO_4(s) \rightleftharpoons$	Ca^{2+} +	SO_4^{2-}
Initial		0	0
Change		$+x$	$+x$
At Equilibrium		x	x

$CaSO_4(s)$ does not appear in the standard equilibrium setup or calculation because, as a pure solid, it does not appear in the equilibrium expression.

$$K_{sp} = [Ca^{2+}][SO_4^{2-}]$$

$$2.4 \times 10^{-5} = x^2$$

$$x = \sqrt{2.4 \times 10^{-5}} = 4.9 \times 10^{-3}$$

Strictly speaking, we should write $\sqrt{2.4 \times 10^{-5}} = \pm 4.9 \times 10^{-3}$. As always with quadratic equations, one of the two roots is physically impossible. Since x is a concentration and cannot be negative, there should be no hesitation in discarding the negative root.

The molar solubility is 4.9×10^{-3} mol L^{-1}, and

$$\text{mass solubility} = \left(\frac{4.9 \times 10^{-3}\ \text{mol}}{\text{L}}\right) \cdot \left(\frac{136.1\ \text{g}}{1\ \text{mol}}\right) = 0.67\ \text{g L}^{-1}$$

Check A mental estimate of K_{sp} from the calculated molar solubility, $(5 \times 10^{-3})^2 = 25 \times 10^{-6}$, is very close to 2.4×10^{-5}, the value given in the problem. Also, salts whose molar solubility is greater than about 0.05 mol L^{-1} are usually regarded as "soluble," and rarely appear in solubility product problems. The calculated molar solubility of $CaSO_4$, 0.0049 mol L^{-1}, is low enough to be reasonable for an insoluble material.

Exercise

The solubility product of MgC_2O_4 (magnesium oxalate) at room temperature is 8.5×10^{-5}. Calculate its solubility in g L^{-1}.

Answer: 1.0 g L^{-1}.

Solubility/K_{sp} Conversion for Salts of More Complicated Stoichiometry

The square-root relationship of Examples 17.2 and 17.3 between the solubility product and the molar solubility holds true for all 1 : 1, 2 : 2, and 3 : 3 ionic compounds, that is, those whose anions and cations bear equal (but opposite) charges. Since the stoichiometric ratio of anions to cations is 1, solutions of these compounds have equal concentrations of anions and cations. Examples are AgCl, $BaSO_4$, and $AlPO_4$. However, for a salt not in this category, converting a solubility product to a solubility requires close attention to the stoichiometry of the salt.

EXAMPLE 17.4 Solubility from Solubility Product

The solubility product of HgI_2 is 1.1×10^{-28} at room temperature. What is its solubility in g L^{-1}?

Analysis This is an equilibrium problem with known initial conditions ($[Hg^{2+}] = 0$ and $[I^-] = 0$) and a given equilibrium constant. The molar solubility is related to the unknown extent of reaction. Follow the structured approach.

Work

Balanced Equation	$HgI_2(s) \rightleftharpoons$	Hg^{2+}	$+ 2\,I^-$
Initial		0	0
Change		$+x$	$+2x$
At Equilibrium		x	$2x$

$$K_{sp} = [Hg^{2+}][I^-]^2 = (x)(2x)^2$$

$$K_{sp} = 1.1 \times 10^{-28} = (x)(4x^2) = 4x^3$$

$$x^3 = \frac{1.1 \times 10^{-28}}{4} = 2.75 \times 10^{-29}$$

$$x = \sqrt[3]{2.75 \times 10^{-29}} = 3.02 \times 10^{-10}$$

The molar solubility is 3.02×10^{-10} mol L^{-1}, and therefore

$$\text{mass solubility} = (3.02 \times 10^{-10}\ \text{mol L}^{-1}) \cdot \left(\frac{454.4\ \text{g}}{1\ \text{mol}}\right)$$

$$= 1.4 \times 10^{-7}\ \text{g L}^{-1}$$

Exercise

The solubility product of Ag_2CO_3 is 6.4×10^{-12} at room temperature. Calculate its solubility in g L^{-1}.

Answer: 0.032 g L^{-1}. (Molar solubility = 1.2×10^{-4} mol L^{-1}.)

The stoichiometric subscript **2** in the formula HgI_2 appears *twice* in the solution to Example 17.4: once as the multiplier of the extent of reaction x, and once as the exponent of the concentration in the solubility product expression.

$$K_{sp} = [Hg^{2+}][I^-]^2 = (x)(\mathbf{2}x)^{\mathbf{2}}$$

This "**2**" means that $[I^-] = \mathbf{2}[Hg^{2+}]$. This is true when $HgI_2(s)$ is the sole source of $Hg^{2+}(aq)$ and $I^-(aq)$, as in solubility ↔ K_{sp} conversions. But it is not true when either of these ions arises from two *different* sources.

This "**2**" is part of the definition of the solubility product,

$$K_{sp} = [Hg^{2+}][I^-]^2.$$

The exponent is *always* used in numerical problems.

The stoichiometric subscripts are always used as *exponents* in solubility product calculations, because they are part of the expression for K_{sp}. However, they are used as *multipliers* of the unknown extent of reaction x only when conversions between solubility and K_{sp} are needed.

In Examples 17.8 through 17.16, we'll see some calculations in which stoichiometric subscripts are *not* used as multipliers.

EXAMPLE 17.5 Solubility from Solubility Product

The solubility product of $Sr_3(AsO_4)_2$ is 1.3×10^{-18} at room temperature. What is its solubility in g L^{-1}?

Analysis This is an equilibrium problem with known initial conditions ($[Sr^{2+}] = [AsO_4^{3-}] = 0$) and given equilibrium constant. The molar solubility is related to the unknown extent of reaction. Follow the structured approach.

Work

Balanced Equation	$Sr_3(AsO_4)_2(s) \rightleftharpoons$	$3\ Sr^{2+}$	$+\ 2\ AsO_4^{3-}$
Initial		0	0
Change		$+3x$	$+2x$
At Equilibrium		$3x$	$2x$

$$K_{sp} = [Sr^{2+}]^3[AsO_4^{3-}]^2 = (3x)^3(2x)^2$$

Note the dual use of the stoichiometric subscripts **3** and **2** from the formula $Sr_3(AsO_4)_2$. It is all too easy to forget and use these subscripts only once, so be sure you understand *why* this is so.

$$K_{sp} = 1.3 \times 10^{-18} = (27x^3)(4x^2) = 108x^5$$

$$x^5 = \frac{1.3 \times 10^{-18}}{108} = 1.20 \times 10^{-20}$$

$$x = (1.20 \times 10^{-20})^{1/5} = 1.04 \times 10^{-4}$$

The molar solubility is 1.04×10^{-4} mol L^{-1}, and therefore

$$\text{mass solubility} = (1.04 \times 10^{-4} \text{ mol L}^{-1}) \cdot \left(\frac{540.6 \text{ g}}{1 \text{ mol}}\right)$$

$$= 0.056 \text{ g L}^{-1}$$

Use the "x^y", "y^x", "$y^{1/x}$", or "$x^{1/y}$" key on your calculator. If you're unfamiliar with this operation, try it with something easy: $32^{1/5} = 32^{0.2} = 2$.

Exercise

The solubility product of PbI_2 is 7.9×10^{-9} at room temperature. Calculate its solubility in g L^{-1}.

Answer: 0.58 g L^{-1}. (Molar solubility = 0.0013 mol L^{-1}.)

EXAMPLE 17.6 K_{sp} from Solubility

The solubility of $Mg_3(PO_4)_2$ is 9.2×10^{-4} g L^{-1}. What is its solubility product?

Analysis/Plan The target, the numerical value of K_{sp}, is obtained from the solubility product expression and the molarities of Mg^{2+} and PO_4^{3-} in a saturated solution. The molarities can in turn be determined from the solubility (in grams per liter), the molar mass, and the stoichiometry.

Work

1. The molar mass of $Mg_3(PO_4)_2$ is 262.9 g mol^{-1}, so that

$$\text{molar solubility} = (9.2 \times 10^{-4}\ \text{g L}^{-1}) \cdot \left(\frac{1\ \text{mol}}{262.9\ \text{g}}\right)$$
$$= 3.5 \times 10^{-6}\ \text{mol L}^{-1}$$

2. For each mole of $Mg_3(PO_4)_2$ that dissolves, three moles of Mg^{2+} and two moles of PO_4^{3-} appear in solution. The ion molarities in a saturated solution are therefore

$$[Mg^{2+}] = 3(3.5 \times 10^{-6}\ \text{mol L}^{-1}) = 10.5 \times 10^{-6}\ \text{mol L}^{-1}$$
$$[PO_4^{3-}] = 2(3.5 \times 10^{-6}\ \text{mol L}^{-1}) = 7.0 \times 10^{-6}\ \text{mol L}^{-1}$$

3. The solubility product expression is

$$K_{sp} = [Mg^{2+}]^3[PO_4^{3-}]^2$$

and substitution of the ion molarities gives the solubility product.

$$K_{sp} = [Mg^{2+}]^3[PO_4^{3-}]^2 = (10.5 \times 10^{-6})^3(7.0 \times 10^{-6})^2$$
$$= 5.7 \times 10^{-26}$$

Check Make sure the correct exponents have been used, and the ion concentrations are in the correct ratio—$[Mg^{2+}]/[PO_4^{3-}] = 3/2$.

Exercise

Calculate the solubility product of $Ca(OH)_2$, given that its solubility is 0.87 g L^{-1} at 25 °C.

Answer: 6.5×10^{-6}.

17.2 PRECIPITATION AND CRYSTALLIZATION

In a *saturated* solution, the ion product Q is equal to the solubility product K_{sp}. If the ion product is less than the solubility product, the solute concentration is less than its equilibrium value, so that additional solute may be dissolved—the solution is **undersaturated.** As we will see, it is also possible to prepare solutions in which the ion product is greater than the solubility product. Such solutions are **supersaturated.**

$$Q < K_{sp}; \quad \text{undersaturated} \tag{17-4a}$$

$$Q = K_{sp}; \quad \text{saturated} \tag{17-4b}$$

$$Q > K_{sp}; \quad \text{supersaturated} \tag{17-4c}$$

Supersaturated solutions are unstable, and sooner or later the excess solute separates from the solution as a pure solid phase. As solute leaves the solution, the value of Q decreases, and deposition of the solid solute continues until $Q = K_{sp}$. The process (in the case of silver sulfate, for example) is the reverse of the dissolution reaction (17-1).

$$2\ Ag^+(aq) + SO_4^{2-}(aq) \longrightarrow Ag_2SO_4(s) \tag{17-5}$$

In the following sections, we discuss ways in which the supersaturation condition $Q > K_{sp}$ can arise: chemical reaction, evaporation of solvent, temperature change (which changes the value of K_{sp}), addition of another salt containing one of the same ions (common ion effect), and change in pH.

Figure 17.3 Addition of 0.2 M Na_2CO_3 to 0.2 M $Pb(NO_3)_2$ causes the ion product $[Pb^{2+}][CO_3^{2-}]$ to exceed the solubility product, $K_{sp}(PbCO_3) = 7.4 \times 10^{-14}$. Precipitation of solid $PbCO_3$ is immediate.

Precipitation Reactions

A chemical reaction that takes place when two solutions are mixed, and which results in the deposition of a solid product, is called a **precipitation reaction.** The solid product is called a *precipitate*. Suppose 100 mL each of 0.2 M $Pb(NO_3)_2$ and 0.2 M Na_2CO_3 are mixed together. The result is a solution of four ions, whose concentrations are $[Pb^{2+}] = 0.1$ M, $[Na^+] = 0.2$ M, $[NO_3^-] = 0.2$ M, and $[CO_3^{2-}] = 0.1$ M. Although both $Pb(NO_3)_2$ and Na_2CO_3 are soluble salts, *lead carbonate* is not. Its solubility product is only 7.4×10^{-14}, and in the mixed solution, the ion product $Q_{PbCO_3} = [Pb^{2+}][CO_3^{2-}] = 0.01$ greatly exceeds K_{sp}. The solution is supersaturated with respect to $PbCO_3$, and a precipitate of $PbCO_3(s)$ appears. Although nothing other than solubility is involved, it is correct to describe the process as a chemical reaction: a new substance, $PbCO_3(s)$, not found among the original reagents, is formed (Figure 17.3).

$$Pb(NO_3)_2(aq) + Na_2CO_3(aq) \longrightarrow PbCO_3(s) + 2\ NaNO_3(aq) \quad (17\text{-}6)$$

When Equation (17-6) is written in terms of the ionic species, it is seen that Na^+ and NO_3^- appear on both sides of the equation.

$$Pb^{2+}(aq) + 2\ NO_3^-(aq) + 2\ Na^+(aq) + CO_3^{2-}(aq) \longrightarrow PbCO_3(s) + 2\ Na^+(aq) + 2\ NO_3^-(aq)$$

That is, Na^+ and NO_3^- are *spectator ions,* and do not participate in the reaction. Cancelling out the spectator ions leaves a *net ionic equation,* which more clearly shows the nature of the precipitation reaction.

Spectator ions and net ionic equations were introduced in connection with acid-base reactions in Section 15.2.

$$Pb^{2+}(aq) + CO_3^{2-}(aq) \longrightarrow PbCO_3(s) \quad (17\text{-}6a)$$

The known solubilities of different types of ionic compounds provide guidelines for predicting whether a precipitation reaction will occur when two solutions are mixed. Conclusions based on these guidelines can be confirmed by calculation of the ion product and comparison with the K_{sp} value found in Table 17.1 or Appendix H.

The rules for the solubility of salts are summarized in Table 12.3 on page 517.

EXAMPLE 17.7 Precipitation Reactions

What precipitate, if any, forms when (a) KOH(aq) is mixed with $MgCl_2(aq)$, and (b) $Na_3PO_4(aq)$ is mixed with $CaBr_2(aq)$?

Analysis

Target: The identity of any insoluble salt that might be formed from the specified reagents.

Plan

1. List the ions present in the combined solutions, and the formulas of all possible salts that can be formed by combining these ions.
2. Determine whether any of the possible salts has an especially low solubility product (Table 17.1), or falls in one or another category of insoluble ionic compounds (Table 12.3).

Work

The problem refers to KOH(aq) and $MgCl_2$(aq). The state designation (aq) indicates that these compounds are already dissolved. Therefore, they need not be considered as candidates for precipitation.

a. The combined solutions contain the ions K^+, Mg^{2+}, Cl^-, and OH^-. In addition to the original compounds (potassium hydroxide and magnesium chloride), which are known to be soluble, these ions can join in the new combinations $Mg(OH)_2$ and KCl.

We predict $Mg(OH)_2$ will precipitate, since its solubility product is small ($\approx 10^{-11}$). Unless the original solutions were very dilute, $Q_{Mg(OH)_2}$ will exceed K_{sp} in the combined solutions.

b. The combined solutions contain the ions Na^+, Ca^{2+}, Br^-, and PO_4^{3-}, which can join in the new combinations NaBr and $Ca_3(PO_4)_2$. Neither of these appears in Table 17.1, but Table 12.3 indicates that phosphates (except alkali phosphates) are insoluble. We predict that a precipitate of calcium phosphate will form.

Exercise

What precipitate, if any, forms when the following pairs of solutions are mixed: (a) silver nitrate and sodium chromate? (b) NaBr and $FeCl_3$?

Answer: (a) Ag_2CrO_4; (b) no precipitate.

Of course if the mixture of solutions is very dilute, it may be that the solubility product of the insoluble salt is not exceeded. In doubtful cases, it is necessary to calculate the ion product and compare it to K_{sp}.

EXAMPLE 17.8 Use of K_{sp} to Predict Precipitation

What precipitate, if any, forms when 100 mL of 0.0100 M KF is added to 200 mL 0.0300 M $Mg(NO_3)_2$?

Analysis Possible product salts are KNO_3 and MgF_2. KNO_3 is soluble, but MgF_2 is only slightly soluble. Precipitation will occur if the ion product of MgF_2 exceeds its solubility product in the combined solutions.

Plan Calculate the concentrations of Mg^{2+} and F^- in the combined solution. Then evaluate the ion product of MgF_2 and compare it with the value of K_{sp} in Table 17.1.

Work The total amount of F^- ion in the original solution is

$$n = M \cdot V = (0.0100 \text{ mol L}^{-1}) \cdot (0.100 \text{ L}) = 1.00 \times 10^{-3} \text{ mol}$$

The volume of the combined solutions is (100 + 200) mL = 0.300 L, so the fluoride ion concentration in the mixture is

$$[F^-] = \frac{1.00 \times 10^{-3} \text{ mol}}{0.300 \text{ L}}$$
$$= 0.00333 \text{ M}$$

Similarly, the concentration of magnesium ion in the combined solutions is

$$[Mg^{2+}] = \frac{(0.200\ L)(0.0300\ mol\ L^{-1})}{0.300\ L}$$
$$= 0.0200\ M$$

The ion product of MgF_2 is

$$Q = [Mg^{2+}][F^-]^2 = (0.0200)(0.00333)^2$$
$$= 2.2 \times 10^{-7}$$

Since this is larger than the solubility product (6.6×10^{-9}), a precipitate of $MgF_2(s)$ will appear when these two solutions are mixed.

Exercise

Will precipitation take place when equal volumes of 0.020 M NaF(aq) and 0.020 M $BaCl_2$ are mixed together? (*Hint:* The actual volumes are not specified, and they are unimportant as long as they are equal. You may assume any volume you wish, for instance 1.00 L of each.)

Answer: $Q = 1.0 \times 10^{-6}$; since $Q < K_{sp}$ for BaF_2, no precipitate forms.

EXAMPLE 17.9 First Appearance of Precipitate

At what concentration of Cl^- does the first precipitate of $PbCl_2$ appear when concentrated aqueous NaCl is added dropwise to 0.750 M $Pb(NO_3)_2$?

Analysis Initially, $Q = [Pb^{2+}][Cl^-]^2 = 0$, since the solution contains no Cl^-. As Cl^- is added, $[Cl^-]$ and Q increase until $Q = K_{sp}$. Any further increase in $[Cl^-]$ results in the formation of a precipitate. The value of K_{sp} for $PbCl_2$ is 1.7×10^{-5} at room temperature.

Work

$$Q = [Pb^{2+}][Cl^-]^2 = K_{sp}$$
$$(0.750)[Cl^-]^2 = 1.7 \times 10^{-5}$$
$$[Cl^-]^2 = \frac{1.7 \times 10^{-5}}{0.750} = 2.27 \times 10^{-5}$$
$$[Cl^-] = \sqrt{2.27 \times 10^{-5}} = 0.0048\ M$$

The first precipitate appears when the chloride ion concentration is slightly greater than 0.0048 M.

Note: There was a *hidden assumption* in the method, namely that the concentration of Pb^{2+} was still 0.750 M after addition of the NaCl(aq). That is, we assumed no increase in volume. If the NaCl(aq) was sufficiently concentrated, say at least 1 M, the assumption is valid. Only 5 mL of 1 M NaCl(aq) is required to bring $[Cl^-]$ up to ≈ 0.005 M in one liter of solution. The volume increase 1000 mL $\rightarrow$ 1005 mL is not zero, but it is negligible when the solubility product is known to only two significant digits.

Exercise

At what concentration of Ag^+ does the first precipitate of AgCl appear when concentrated aqueous $AgNO_3$ is added to 0.00100 M NaCl?

Answer: $[Ag^+] = 1.8 \times 10^{-7}$ M. Very little $AgNO_3(aq)$ is required to cause precipitation of AgCl.

In more complicated systems, several different precipitates may be possible. Ion product calculations are used to decide which compound precipitates first.

EXAMPLE 17.10 Which Compound Precipitates First?

A solution is 0.50 M in silver ions (Ag^+) and 0.50 M in mercury(I) ions (Hg_2^{2+}). If aqueous sodium chloride is gradually added to the solution, which compound precipitates first? Assume that the volume does not increase significantly as the NaCl(aq) is added.

Analysis This is two separate problems packaged as one. Use the method of Example 17.9 to calculate the chloride ion concentration required for first appearance of AgCl(s), then repeat the procedure for the $[Cl^-]$ required for first appearance of Hg_2Cl_2(s). Whichever requires the lower value of $[Cl^-]$ is the first to separate from the solution.

Work

AgCl

$$K_{sp} = 1.8 \times 10^{-10} = [Ag^+][Cl^-]$$

$$[Cl^-] = \frac{1.8 \times 10^{-10}}{[Ag^+]} = \frac{1.8 \times 10^{-10}}{0.50} = 3.6 \times 10^{-10} \text{ M}$$

Hg_2Cl_2

$$K_{sp} = 1.2 \times 10^{-18} = [Hg_2^{2+}][Cl^-]^2$$

$$[Cl^-]^2 = \frac{1.2 \times 10^{-18}}{[Hg_2^{2+}]} = \frac{1.2 \times 10^{-18}}{0.50} = 2.4 \times 10^{-18}$$

$$[Cl^-] = \sqrt{2.4 \times 10^{-18}} = 1.5 \times 10^{-9} \text{ M}$$

The first precipitate to appear is AgCl, since its K_{sp} is exceeded at a lower concentration of chloride ions.

Exercise

A solution is 0.080 M in silver ions (Ag^+) and 0.020 M in mercury(I) ions. If KBr is gradually added to the solution, what is the composition of the precipitate that first appears? K_{sp} for AgBr is 5.0×10^{-13}, and K_{sp} for Hg_2Br_2 is 5.6×10^{-23}.

Answer: AgBr begins to precipitate when $[Br^-]$ exceeds 6.3×10^{-12} M. Hg_2Br_2 appears later, when $[Br^-]$ exceeds 5.3×10^{-11} M.

EXAMPLE 17.11 Fractional Precipitation

NaF(aq) is gradually added to a solution containing 0.025 M $Ca(NO_3)_2$ and 0.040 M $Mg(NO_3)_2$. The volume increase in this procedure is negligibly small.

a. What is the first compound to precipitate? What is the value of $[F^-]$ when this precipitate appears?

b. Calculate the value of $[F^-]$ when the second precipitate appears.

c. Determine the values of $[Ca^{2+}]$ and $[Mg^{2+}]$ when the second precipitate just begins to appear.

Analysis All targets are explicitly stated, and a check of Table 17.1 shows that both CaF_2 and MgF_2 are insoluble and expected to precipitate. The relationships between the numerical targets and knowns are the solubility products of MgF_2 and CaF_2. When solid CaF_2 and MgF_2 are both present, *both* solubility equilibria are established and both cation concentrations can be calculated from $[F^-]$ and the appropriate K_{sp}.

Plan

1. Calculate the value of $[F^-]$ required for each ion product to become equal to the solubility product. Precipitation of the compound begins when these concentrations are exceeded.
2. Using the K_{sp}'s and the $[F^-]$ at which the *second* precipitate forms, calculate the equilibrium values of both cation concentrations.

Work

1. Since the volume increase is negligible, both cation concentrations remain at their initial value until precipitation begins at $Q = K_{sp}$. The fluoride concentration required for precipitation of CaF_2 is

$$K_{sp} = [Ca^{2+}][F^-]^2$$

$$[F^-] = \sqrt{K_{sp}/[Ca^{2+}]} = \sqrt{(3.9 \times 10^{-11})/0.025}$$

$$= \sqrt{1.56 \times 10^{-9}} = 3.9 \times 10^{-5}\ \text{M}$$

The concentration required for precipitation of MgF_2 is

$$[F^-] = \sqrt{K_{sp}/[Mg^{2+}]} = \sqrt{(6.6 \times 10^{-9})/0.040}$$

$$= \sqrt{1.65 \times 10^{-7}} = 4.1 \times 10^{-4}\ \text{M}$$

As NaF(aq) is added gradually, the first solid, CaF_2(s), appears when $[F^-]$ reaches 3.9×10^{-5} M. The second solid, MgF_2, begins to precipitate later, when $[F^-]$ has increased to 4.1×10^{-4} M.

2. When the second precipitate (MgF_2) just begins to appear, $[Mg^{2+}]$ is still at its initial value, so

$$[Mg^{2+}] = 0.040$$

However, much of the Ca^{2+} originally present in the solution is now in the form of $CaF_2(s)$. The concentration of Ca^{2+} remaining in solution is calculated from the equilibrium expression

$$K_{sp} = [Ca^{2+}][F^-]^2$$

The value of $[F^-]$ to be used in this equation is its concentration when MgF_2 begins to precipitate, namely 4.1×10^{-4} M.

$$[Ca^{2+}] = \frac{K_{sp}}{[F^-]^2} = \frac{3.9 \times 10^{-11}}{(4.1 \times 10^{-4})^2}$$

$$= 2.3 \times 10^{-4} \text{ M}$$

Exercise

Suppose that $Na_2SO_4(aq)$ is added to a solution containing 0.035 M $AgNO_3$ and 0.015 M $Pb(NO_3)_2$. What is the *first* precipitate to appear, and what are the values of $[Ag^+]$ and $[Pb^{2+}]$ when the *second* precipitate just begins to appear?

Answer: $PbSO_4$; $[Ag^+] = 0.035$ M, $[Pb^{2+}] = 1.3 \times 10^{-6}$ M.

Example 17.11 shows that, when $[F^-] = 4.1 \times 10^{-4}$ M, virtually all the Mg^{2+} is in solution while virtually all (≈ 99%) of the Ca^{2+} is present in solid form as $CaF_2(s)$. If

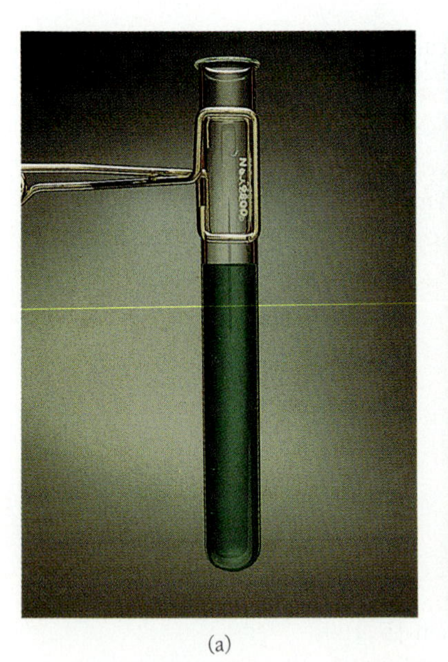

(a)

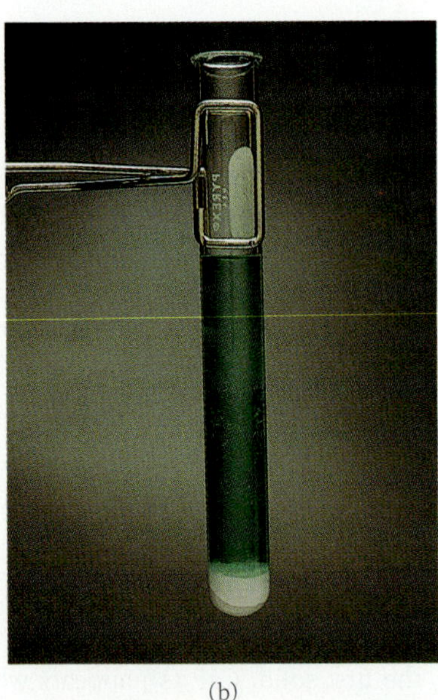

(b)

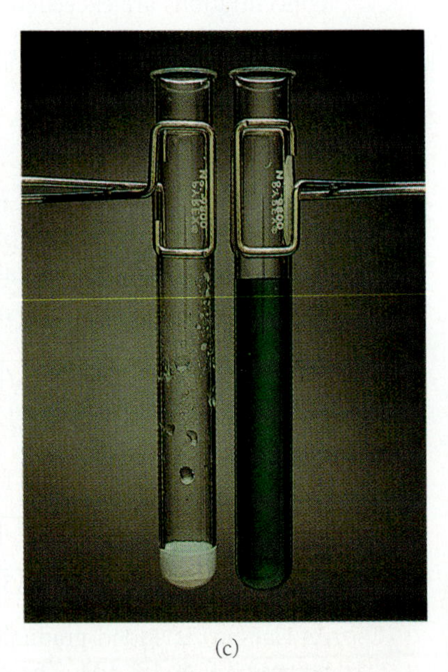

(c)

Figure 17.4 The technique of fractional precipitation is used to separate cations. The test tube in (a) contains Ag^+, Pb^{2+}, Cd^{2+} and Ni^{2+} ions. The color is due to the Ni^{2+} ions, which are green in aqueous solution. (b) After the addition of aqueous HCl, the insoluble compounds AgCl and $PbCl_2$ have precipitated and settled to the bottom. Since both $CdCl_2$ and $NiCl_2$ are soluble, the precipitate contains neither of these compounds. (c) The green solution, containing Cd^{2+} and Ni^{2+}, is simply poured off to complete the separation of Ag^+ and Pb^{2+} from Cd^{2+} and Ni^{2+}.

desired, $CaF_2(s)$ can be separated from Mg^{2+} by filtration. This separation and/or purification technique, based on differing solubilities of the components of a solution, is called **fractional precipitation** (Figure 17.4).

Crystallization

Consider an undersaturated solution, in which $Q < K_{sp}$. Apart from the addition of a reagent to cause a precipitation reaction, there are several ways to bring about a condition in which $Q > K_{sp}$. If some of the solvent evaporates, the solution volume decreases. Therefore the solute concentrations, and Q, increase. If the value of Q becomes greater than K_{sp}, deposition of pure, solid solute will eventually occur. Separation of a solid phase of pure solute from a supersaturated solution is called **crystallization** when it is caused by a physical change.

Crystallization caused by solvent evaporation happens in nature as well as in the laboratory. Limestone, which is primarily calcium carbonate, is soluble in cold water to the extent of about 1.5 mg L^{-1}. Although this is a low solubility, it is not zero, and over long periods of time large caverns are formed in limestone strata as the calcium carbonate dissolves in groundwater. Within these caverns, stalactites and stalagmites are formed by crystallization of calcium carbonate when some water evaporates from the saturated solution that drips from the ceiling of the cave (Figures 17.5 and 17.6).

The value of K_{sp}, like that of most equilibrium constants, depends on temperature. Solubility of most ionic substances increases with temperature, so that cooling is required to bring about a decrease in K_{sp}. If a temperature decrease of a particular solution is enough so that $K_{sp} < Q$, the solution becomes unstable and a solid phase

Figure 17.5 Cave structures such as these form by deposition of calcium carbonate in limestone caverns. The groundwater dripping through these caves is a saturated solution of calcium carbonate. As the water evaporates, $CaCO_3$ precipitates.

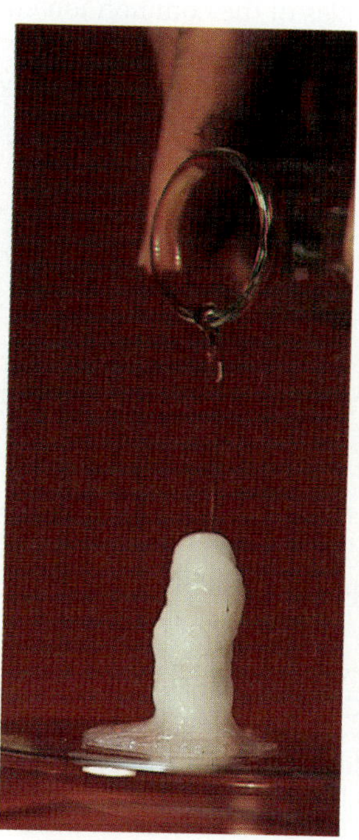

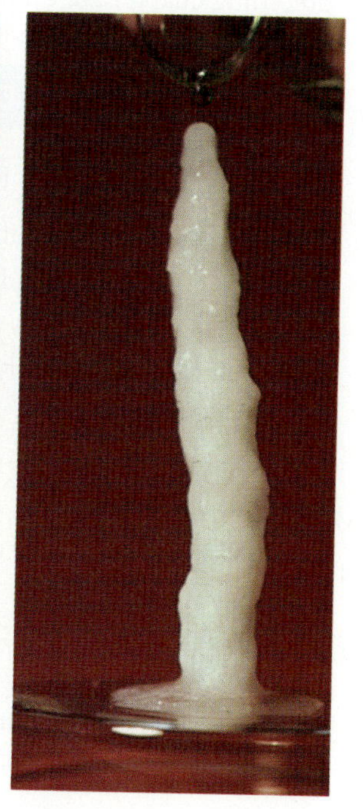

Figure 17.6 The natural process of stalagmite formation can be imitated in the laboratory by slowly pouring a supersaturated sodium acetate solution over a small pile of solid sodium acetate. Excess sodium acetate immediately precipitates when the solution comes in contact with the pure solid, and the pile grows rapidly.

separates from the liquid. That is, crystallization occurs. The process does not necessarily occur immediately, and under favorable conditions a supersaturated solution can persist for some time before crystallization occurs. A supersaturated solution is in an unstable, nonequilibrium state, however, and ultimately the excess solute will crystallize (Figure 17.7).

Some supersaturated solutions last for years.

The Common Ion Effect

The concentration of one of the species in an ion product can be increased by the addition of another salt containing that ion. Since that species is shared by the two salts, it is called a **common ion.** Suppose that a small amount of concentrated NaI solution is added to a saturated PbI_2 solution, in which $Q = K_{sp}$, so that the common ion (I^-) concentration increases. The increase in $[I^-]$ increases the ion product $Q = [Pb^{2+}][I^-]^2$ to a value greater than K_{sp}, thus creating a supersaturated solution. Le Châtelier's principle predicts that, in order to restore equilibrium, the additional iodide ion will cause the solution reaction to shift to the left.

$$PbI_2(s) \rightleftharpoons Pb^{2+}(aq) + 2\,I^-(aq)$$

That is, some solid PbI_2 is formed, and as a result, the concentration of $Pb^{2+}(aq)$ remaining in the solution decreases (Figure 17.8).

The presence of the common ion causes the solubility of PbI_2 to be less in NaI(aq) than in pure water. The **common ion effect** is the shifting of a chemical equilibrium when one of the ions involved in the equilibrium is present from an additional source. In this example, a *solubility* equilibrium is shifted. The buffer solutions discussed in the preceding chapter are examples of the common ion effect on *acid-base* equilibria.

Figure 17.7 When a saturated solution of sodium or potassium acetate is cooled, the solubility decreases and the solution becomes supersaturated. In this condition, it is unstable, and eventually solid solute crystallizes from the solution. The process can be hastened by the addition of a small "seed" crystal, which forms a nucleus for crystallization.

EXAMPLE 17.12 The Common Ion Effect on Solubility

The solubility product of PbI_2 is 7.9×10^{-9} at 25 °C.

a. Calculate the solubility (g L^{-1}) of PbI_2 in 0.100 M KI.

b. Compare the result with the solubility of PbI_2 in pure water.

Analysis Like all solubility problems, these are equilibrium problems. Use the structured approach.

Work

a.

	$PbI_2(s) \rightleftharpoons$	$Pb^{2+}(aq)$	+	$2\,I^-(aq)$
Balanced Equation				
Initial		0		0.100
Change		$+x$		$+2x$
At Equilibrium		x		$(0.100 + 2x)$

$$K_{sp} = [Pb^{2+}][I^-]^2$$

$$7.9 \times 10^{-9} = (x)(0.100 + 2x)^2$$

Even in the absence of KI, x is a very small number because K_{sp} is so small. Furthermore, the common ion effect shifts the equilibrium to the left and makes x even smaller. The amount of iodide ion contributed by PbI_2, $2x$, is insignificant compared to the 0.100 M pre-existing in the solution. That is, $0.100 + 2x \approx 0.100$. Thus, to an excellent approximation,

$$7.9 \times 10^{-9} = (x)(0.100)^2$$

Note: Even though the formula of the salt is PbI_2, we do *not* multiply the iodide concentration by 2 in this example. The *changes* in concentration are x and $2x$ (for Pb^{2+} and I^-), but the concentration of I^- *at equilibrium* is determined by the molarity of KI, which is unrelated to $[Pb^{2+}]_{eq}$. A factor-of-two relationship between $[Pb^{2+}]_{eq}$ and $[I^-]_{eq}$ holds *only* when $PbI_2(s)$ is the sole source of $Pb^{2+}(aq)$ and $I^-(aq)$.

$$x = \frac{7.9 \times 10^{-9}}{(0.100)^2} = 7.9 \times 10^{-7} \text{ mol L}^{-1}$$

Since x is the concentration of Pb^{2+} in the saturated solution, x is also equal to the amount of PbI_2 that has dissolved to produce the saturated solution. If the answer is to be expressed in g L^{-1}, the molar solubility must be multiplied by the molar mass.

$$\begin{aligned}\text{solubility} &= (7.9 \times 10^{-7} \text{ mol L}^{-1}) \cdot (461 \text{ g mol}^{-1}) \\ &= 3.6 \times 10^{-4} \text{ g L}^{-1}\end{aligned}$$

b. Solubility of PbI_2 in pure water.

Balanced Equation	$PbI_2(s) \rightleftharpoons$	$Pb^{2+}(aq)$ +	$2\ I^-(aq)$
Initial		0	0
Change		$+x$	$+2x$
At Equilibrium		x	$2x$

$$K_{sp} = [Pb^{2+}][I^-]^2$$

$$7.9 \times 10^{-9} = (x)(2x)^2 = 4x^3$$

$$x = \left(\frac{7.9 \times 10^{-9}}{4}\right)^{1/3} = 1.25 \times 10^{-3} \text{ M}$$

$$\text{solubility} = (1.25 \times 10^{-3} \text{ M}) \cdot (461 \text{ g mol}^{-1}) = 0.58 \text{ g L}^{-1}$$

Note: In this part of the problem, PbI_2 is the sole source of both lead and iodide ions, so the factor-of-two relationship holds at equilibrium: $[I^-]_{eq} = 2 \cdot [Pb^{2+}]_{eq} = 2x$.

Check The solubility of PbI_2 in KI(aq) is less than in pure water, as predicted by Le Châtelier's principle.

Exercise

The solubility product of AgCl is 1.8×10^{-10}. Calculate the solubility of AgCl in 0.100 M NaCl.

Answer: 1.8×10^{-9} mol L^{-1}, or 2.6×10^{-7} g L^{-1}.

Figure 17.8 Addition of a drop of concentrated sodium iodide to a saturated solution of lead iodide increases the value of the ion product and causes the immediate separation of solid PbI_2 from the solution.

17.3 INFLUENCE OF pH ON SOLUBILITY

The principles of the common ion effect on solubility are the same for all ionic compounds. However, hydroxides and salts of weak acids merit additional discussion, since their solubilities can easily be controlled by adjusting the pH of the solution.

Hydroxides

Many common hydroxide compounds are only moderately soluble. For example, at room temperature magnesium hydroxide has a solubility product of 7.3×10^{-12} and a molar solubility in pure water of 1.2×10^{-4} mol L^{-1}. In a saturated solution of $Mg(OH)_2$, $[OH^-] = 2.4 \times 10^{-4}$, pOH = 3.62, and pH = 10.38. If the pH is adjusted by the addition of either a strong acid or a strong base, the solubility of $Mg(OH)_2$ changes because the concentration of hydroxide ion changes. If OH^- is added,

$$Mg(OH)_2(s) \rightleftharpoons Mg^{2+}(aq) + 2\ OH^-(aq)$$

the equilibrium shifts to the left, decreasing the solubility of $Mg(OH)_2$. If H_3O^+ is added, it reacts with and removes OH^- from the solution, shifting the equilibrium to the right and increasing the solubility of $Mg(OH)_2$. Magnesium hydroxide, insoluble in pure water, dissolves readily in acid (Figure 17.9).

EXAMPLE 17.13 Maximum Mg^{2+} Concentration at High pH

Calculate the maximum concentration of Mg^{2+} that can exist in a solution of pH = 12.00.

Analysis/Relationship The maximum possible Mg^{2+} concentration occurs when $Q = K_{sp}$, because then the solution is saturated with respect to $Mg(OH)_2$. $[Mg^{2+}]_{max}$ can be calculated from the solubility product and $[OH^-]$.

Plan Calculate $[OH^-]$ from pH, insert the values of K_{sp} and $[OH^-]$ into the solubility product expression, and evaluate $[Mg^{2+}]$.

Work

$$pOH = 14.00 - pH = 2.00$$

$$[OH^-] = 1.00 \times 10^{-2}\ M.$$

$$K_{sp} = [Mg^{2+}][OH^-]^2$$

$$[Mg^{2+}] = \frac{K_{sp}}{[OH^-]^2}$$

$$= \frac{7.3 \times 10^{-12}}{(1.00 \times 10^{-2})^2} = 7.3 \times 10^{-8}\ M$$

Exercise

Calculate the maximum possible concentration of Zn^{2+} in a solution of pH = 10.00. K_{sp} for $Zn(OH)_2$ is 3.0×10^{-16}.

Answer: 3.0×10^{-8} M.

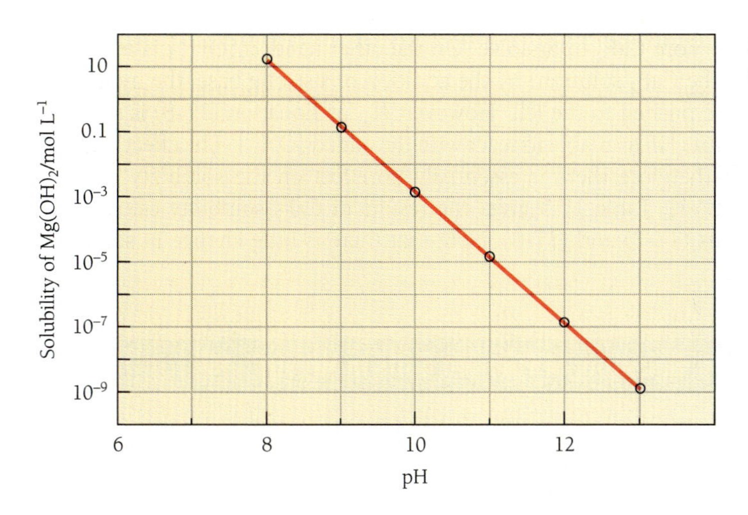

Figure 17.9 The effect of pH on the solubility of $Mg(OH)_2$.

Using the method of Example 17.13, we can calculate the maximum possible concentration, at any pH, of any ion that forms an insoluble hydroxide. The results of such calculations for Mg^{2+} and Zn^{2+} are collected in Table 17.2. The maximum possible concentrations of both Mg^{2+} and Zn^{2+} decrease as the pH increases ($[OH^-]$ increases).

Salts of Weak Acids

The solubilities of salts of weak acids, for instance CaF_2, depend on pH because the anion concentration is linked to pH through the weak acid equilibrium expression. When such a salt is dissolved, there are *two* equilibria involved.

$$CaF_2(s) \rightleftharpoons Ca^{2+} + 2\,F^-; \qquad K_{sp} = [Ca^{2+}][F^-]^2 = 3.9 \times 10^{-11} \tag{17-7}$$

$$F^- + H_2O \rightleftharpoons OH^- + HF; \qquad K_b = \frac{[OH^-][HF]}{[F^-]} = \frac{K_w}{K_a} = 1.5 \times 10^{-11} \tag{17-8}$$

If a saturated solution of CaF_2 is acidified by the addition of a strong acid such as HNO_3, the OH^- produced in the hydrolysis [Equation (17-8)] reacts with the added H_3O^+. This pulls the equilibrium to the right, which decreases the concentration

TABLE 17.2 pH-Dependence of Maximum Ion Molarities at 25 °C

pH	$[OH^-]$	$[Mg^{2+}]_{max}$	$[Zn^{2+}]_{max}$
6.00	1×10^{-8}		3.0
7.00	1×10^{-7}		3.0×10^{-2}
8.00	1×10^{-6}	7.3	3.0×10^{-4}
9.00	1×10^{-5}	0.073	3.0×10^{-6}
10.00	1×10^{-4}	7.3×10^{-4}	3.0×10^{-8}
11.00	1×10^{-3}	7.3×10^{-6}	3.0×10^{-10}
12.00	1×10^{-2}	7.3×10^{-8}	3.0×10^{-12}

TABLE 17.3 pH-Dependence of CaF_2 Solubility

pH	Solubility/g L^{-1}
0	2.2
2	0.11
4	0.018
6	0.017
8	0.015

of F^- and permits more CaF_2 to dissolve. On the other hand, if $[OH^-]$ is increased (from its value in the CaF_2 solution) by the addition of a strong base, the hydrolysis equilibrium will be pushed to the left. However, K_b for Equation (17-8) is so small that there is very little HF available to react with the added $[OH^-]$. The effect on $[F^-]$ is negligible, and therefore there is essentially no effect on the solubility equilibrium, Equation (17-7). Table 17.3 gives the results of CaF_2 solubility calculations based on Equations (17-7) and (17-8). Note that there is little change in solubility above pH 4.

EXAMPLE 17.14 Solubility of Fluoride Salts in HF(aq)

What is the molar solubility of CaF_2 in 0.100 M HF? The acid ionization constant of HF is 6.6×10^{-4}.

Analysis/Plan The molar solubility is equal to the maximum possible concentration of Ca^{2+}, and it can be calculated from K_{sp} once the value of $[F^-]$ is known. The fluoride ion concentration can be calculated from the HF molarity and K_a.

Work The fluoride ion concentration in 0.100 M HF is found by the methods of the preceding chapter.

$$[F^-] \approx \sqrt{K_a \cdot C} = \sqrt{(6.6 \times 10^{-4}) \cdot 0.100}$$
$$= 8.12 \times 10^{-3}\text{ M}$$

The maximum concentration of Ca^{2+} in this solution is found from the solubility product of CaF_2.

$$K_{sp} = [Ca^{2+}][F^-]^2$$
$$[Ca^{2+}]_{max} = \frac{K_{sp}}{[F^-]^2}$$
$$= \frac{3.9 \times 10^{-11}}{(8.12 \times 10^{-3})^2}$$
$$= 5.9 \times 10^{-7}\text{ M}$$

The solubility of CaF_2 in 0.100 M HF is 5.9×10^{-7} moles per liter.

Exercise

What is the molar solubility of SrF_2 in 0.085 M HF?

Answer: 5.2×10^{-5} moles per liter.

There was a hidden assumption in the method used in Example 17.14, namely that the concentration of $[F^-]$ established by the ionization of HF is unaffected by the subsequent dissolution of CaF_2. In fact, when 5.7×10^{-7} mol CaF_2 dissolves in 0.1 M HF, the concentration of $[F^-]$ increases by $2 \cdot (5.7 \times 10^{-7})$ M. Within the limits of two-digit precision (imposed by the values of K_{sp} and K_a), this increase is negligible. That is, $(8.1 \times 10^{-3} + 11.4 \times 10^{-7}) = 8.1 \times 10^{-3}$, and the accuracy of the calculation is not affected by this increase. But if K_a is very small, and/or K_{sp} is relatively large, this method can lead to erroneous results. In this textbook, however, we will not encounter any problems for which the method of Example 17.14 is unsatisfactory.

Other common salts whose solubilities are pH-dependent because of simultaneous weak-acid or weak-base equilibria are the acetates, carbonates, and phosphates, as well as all salts in which the cation is NH_4^+. Sulfides, too, are salts of a weak acid, but their solubilities are treated somewhat differently.

Sulfides

Hydrogen sulfide, a gaseous substance, is a weak diprotic acid with moderate solubility in water. At 25 °C a saturated solution contains 0.10 M H_2S. The first ionization step, with its corresponding equilibrium constant, is

$$H_2S(aq) + H_2O \rightleftharpoons H_3O^+ + HS^-; \qquad K_1 = \frac{[H_3O^+][HS^-]}{[H_2S]} = 1.0 \times 10^{-7} \qquad (17\text{-}9)$$

The second ionization

$$HS^- + H_2O \rightleftharpoons H_3O^+ + S^{2-}; \qquad K_2 = \frac{[H_3O^+][S^{2-}]}{[HS^-]} \qquad (17\text{-}10)$$

is very much weaker, and it is difficult to measure the value of K_2. For many years it was thought that $pK_2 \approx 13$, but recently it has been shown that $pK_2 = 19 \pm 2$. Now, if HS^- is such a weak acid, its conjugate base S^{2-} must be a very strong base—much stronger, in fact, than OH^-. Indeed, the reciprocal relationship requires that $K_b = K_w/K_2 = 10^{-14}/10^{-19} = 10^5$ for the base equilibrium

$$S^{2-} + H_2O \rightleftharpoons HS^- + OH^-; \qquad K_b \approx 10^5$$

Such a large value means that the equilibrium lies almost entirely to the right, which is another way of saying that the species S^{2-} does not exist in aqueous solution.

The sulfide ion does exist in the solid state, however, and the dissolution of an ionic sulfide such as ZnS(s) can be represented as the two-step process

$$ZnS(s) \rightleftharpoons Zn^{2+} + S^{2-}$$

$$S^{2-} + H_2O \longrightarrow HS^- + OH^-$$

or, more simply, as the overall equilibrium

$$ZnS(s) + H_2O \rightleftharpoons Zn^{2+} + HS^- + OH^- \qquad (17\text{-}11)$$

which has the corresponding solubility product expression

$$K'_{sp} = [Zn^{2+}][HS^-][OH^-] = 2 \times 10^{-24} \text{ at } 25\ °C \qquad (17\text{-}12)$$

The solubility product has been labeled with a prime (′) as a reminder that the anions appearing in the expression, HS^- and OH^-, are not those expected on the basis of the empirical formula of the salt.

The concentration of the HS^- ion depends on $[H_3O^+]$ through Reaction (17-9), and $[OH^-]$ depends on $[H_3O^+]$ through the autoionization of water. These relationships can be used to calculate the solubility of ZnS in acidic solution at any pH. The

TABLE 17.4 Solubility of ZnS

pH	In Acid	In Acid with 0.1 M H_2S	
	Solubility of ZnS/g L^{-1}	$[Zn^{2+}]_{max}$/ mol L^{-1}	Solubility of ZnS/g L^{-1}
0	4	0.02	2.0
1	0.4	2×10^{-4}	0.02
2	0.04	2×10^{-6}	2×10^{-4}
3	0.004	2×10^{-8}	2×10^{-6}
4	4×10^{-4}	2×10^{-10}	2×10^{-8}
5	4×10^{-5}	2×10^{-12}	2×10^{-10}
6	4×10^{-6}	2×10^{-14}	2×10^{-12}

results of several such calculations are given in Table 17.4. The values in the second column are the solubilities of ZnS in a strong acid, say HNO_3(aq), in which there is no common ion effect.

The maximum concentration of Zn^{2+} ions in aqueous solution can be controlled by saturating the solution with H_2S. A precipitate of ZnS appears when H_2S is added to a zinc-containing solution, regardless of the original source of Zn^{2+}. The solubility equilibrium, Equation (17-11), is established, and the equilibrium Zn^{2+} concentration, which is the maximum possible, is

$$[Zn^{2+}]_{max} = [Zn^{2+}]_{eq} = K'_{sp} \cdot \frac{1}{[HS^-]} \cdot \frac{1}{[OH^-]}$$

Substituting Equation (17-9) and the water autoionization equilibrium $K_w = [H_3O^+][OH^-]$, and inserting the values of K_1, K_w, and $[H_2S]_{sat} = 0.1$ M, we find

$$[Zn^{2+}]_{max} = K'_{sp} \cdot \frac{[H_3O^+]}{K_1[H_2S]} \cdot \frac{[H_3O^+]}{K_w}$$
$$= K'_{sp} \cdot 10^{22} \cdot [H_3O^+]^2$$

Insertion of K'_{sp} (2×10^{-24} for zinc sulfide) gives

$$[Zn^{2+}]_{max} = 0.02 \cdot [H_3O^+]^2 \qquad (17\text{-}13)$$

Some values of $[Zn^{2+}]_{max}$ and the corresponding solubility of ZnS in g L^{-1}, calculated from Equation (17-13), are given in the third and fourth columns of Table 17.4.

Group 1A sulfides are soluble, as are $(NH_4)_2S$ and SrS. BaS and CaS are slightly soluble.

Separation of metal cations by pH control of the solubilities of metal sulfides is discussed in Section 17.5.

Most metal sulfides are insoluble in pure water. Like ZnS, their solubility in saturated H_2S can be controlled by adjusting the pH with strong acid. This control can be used to separate aqueous metal ions from one another. If an acidic solution containing various metal ions is saturated with H_2S, metal sulfides having relatively small K'_{sp}'s will precipitate. Those with larger K'_{sp}'s precipitate only if the pH is increased by addition of base.

EXAMPLE 17.15 Maximum Metal Ion Concentration in Saturated H_2S

What is the maximum possible concentration of Fe^{2+}(aq) that can exist in a 0.100 M solution of H_2S whose pH has been adjusted (by the addition of HNO_3) to 3.00?

Analysis/Relationships The maximum Fe^{2+} concentration is found from the solubility product expression

$$K'_{sp} = [Fe^{2+}][HS^-][OH^-] = 6 \times 10^{-18}$$

$[OH^-]$ is found from the given pH, and $[HS^-]$ is found by evaluating the acid ionization equilibrium expression

$$K_a = \frac{[H_3O^+][HS^-]}{[H_2S]} = 1.0 \times 10^{-7}$$

using the given values of pH and $[H_2S]$.

Work The ionization equilibrium expression leads immediately to

$$[HS^-] = \frac{K_a[H_2S]}{[H_3O^+]}$$

$[H_3O^+] = 0.0010$ at pH $= 3.00$, and $[H_2S]$ is given as 0.100 M. Therefore,

$$[HS^-] = \frac{(1.0 \times 10^{-7})(0.100)}{(0.0010)}$$
$$= 1.0 \times 10^{-5}$$

At pH $= 3.00$, pOH $= 14.00 -$ pH $= 11.00$, and $[OH^-] = 1.0 \times 10^{-11}$. When the solubility product expression is evaluated with these values of $[OH^-]$ and $[HS^-]$, the result is

$$[Fe^{2+}][HS^-][OH^-] = K'_{sp}$$
$$[Fe^{2+}] = \frac{6 \times 10^{-18}}{[HS^-][OH^-]}$$
$$= \frac{6 \times 10^{-18}}{(1.0 \times 10^{-5})(1.0 \times 10^{-11})}$$
$$= 0.06 \text{ M}$$

Exercise

What is the maximum concentration of CdS that can exist in a 0.10 M solution of H_2S whose pH has been adjusted to 5.00?

Answer: 8×10^{-15} M.

17.4 COMPLEX ION EQUILIBRIA

We noted in our discussion of Lewis acids and bases that metal cations form Lewis adducts with electron-pair donors such as $:NH_3$. These adduct species are known as **coordination complexes**, because the bonds to the metal ion are donor-acceptor or "coordinate covalent" bonds. The chemistry of coordination complexes is considered in some detail in Chapter 23.

TABLE 17.5 Formulas of Some Complex Ions

$Ag(NH_3)_2^+$	$Fe(CN)_6^{3-}$
$Cu(NH_3)_4^{2+}$	$Ni(CN)_4^{2-}$
$Zn(NH_3)_4^{2+}$	$Fe(C_2O_4)_2^{2-}$
$Co(NH_3)_6^{2+}$	$Al(C_2O_4)_2^-$
$AgCl_2^-$	$Mn(en)_3^{2+}$*

* "en" is a common abbreviation for ethylenediamine, $H_2NCH_2CH_2NH_2$.

Rust stains can often be removed by treatment with a solution of oxalic acid or an oxalate salt. Rust consists largely of insoluble Fe_2O_3 and the iron in the rust reacts to form the complex ion $Fe(C_2O_4)_2^{2-}$. This converts the iron to a soluble form, and removes the stain.

Formation Constants of Complex Ions

Coordination complexes that carry an electrical charge are called **complex ions.** For example, silver ion reacts reversibly with aqueous ammonia to form the singly charged complex ion $Ag(NH_3)_2^+$.

$$Ag^+(aq) + 2\ NH_3(aq) \rightleftharpoons Ag(NH_3)_2^+(aq) \qquad (17\text{-}14)$$

A few other complex ions are listed in Table 17.5.

The equilibrium constants that characterize these reversible, complex-forming reactions are called **formation constants,** K_f. The equilibrium expression for the reaction of silver ion with ammonia is

$$K_f = \frac{[Ag(NH_3)_2^+]}{[Ag^+][NH_3]^2}; \qquad K_f = 1.6 \times 10^7 \text{ at } 25\ °C \qquad (17\text{-}15)$$

Note that K_f is large. This is typical of complex ion formation constants, and indicates that the equilibria of complex-forming reactions lie to the right. That is, most complex ions are very stable.

The Common Ion Effect in Complex Ion Equilibria

The existence of highly stable complex ions offers the possibility of controlling the concentration of metal ions and the solubility of slightly soluble salts. In a solution containing both ammonia and silver ions, for instance, the concentration of "free silver," $Ag^+(aq)$, can be made very much smaller than the concentration of "bound silver," that is, silver tied up in the complex ion $Ag(NH_3)_2^+$. The ratio of these concentrations depends on the amount of ammonia present, and is found by rearranging Equation (17-15).

$$\frac{[Ag^+]}{[Ag(NH_3)_2^+]} = \frac{1}{1.6 \times 10^7} \cdot \frac{1}{[NH_3]^2} = \frac{6.3 \times 10^{-8}}{[NH_3]^2}$$

When the NH_3 concentration is 0.5 M, for example, this ratio has the value 2.5×10^{-7}: $[Ag^+] \ll [Ag(NH_3)_2^+]$, and almost all of the silver is tied up in the complex ion.

Suppose some ammonia is added to a mixture of a saturated solution of AgCl and solid AgCl. Reaction (17-14) causes the concentration of $Ag^+(aq)$ to drop to a low value, which decreases the ion product $Q = [Ag^+][Cl^-]$ of silver chloride. The solution is no longer saturated, and more AgCl can dissolve. Silver chloride is considerably more soluble in ammonia than in pure water.

EXAMPLE 17.16 Increase in Solubility by Complex Ion Formation

a. Calculate the molar solubility of AgCl in 0.500 M NH_3.

b. Compare the result to the molar solubility in pure water. K_f for $Ag(NH_3)_2$ is 1.6×10^7.

Analysis

Target: Since the equilibrium concentration of Cl^- is equal to the number of moles of AgCl that dissolve in one liter—the molar solubility—the target is $[Cl^-]$. Note that we cannot use $[Ag^+]$ as the target, because most of the silver that dissolves is converted to the complex ion.

Relationship: Since both silver and chloride enter the solution by dissolution of AgCl, $[Cl^-]$ must be equal to the *total* silver concentration: $[Cl^-] = [Ag^+] + [Ag(NH_3)_2^+]$. Furthermore, both the relationship $K_{sp} = [Ag^+][Cl^-]$ and the expression for K_f, Equation (17-15), must hold at equilibrium.

Plan There are three relationships linking three unknowns, so the ordinary methods of algebra will lead to a solution.

Work

a. The three unknowns are $[Cl^-]$, $[Ag^+]$, and $[Ag(NH_3)_2^+]$, and the three equations are

(1) $$[Cl^-] = [Ag^+] + [Ag(NH_3)_2^+] \qquad \text{(mass balance)}$$

(2) $$K_{sp} = [Ag^+][Cl^-] \qquad \text{(solubility)}$$

(3) $$K_f = \frac{[Ag(NH_3)_2^+]}{[Ag^+][NH_3]^2} \qquad \text{(complexation)}$$

Rearrange (3) to $[Ag(NH_3)_2^+] = [Ag^+] \cdot K_f \cdot [NH_3]^2$ and substitute into (1) to obtain

$$\begin{aligned}[Cl^-] &= [Ag^+] + [Ag^+] \cdot K_f \cdot [NH_3]^2 \\ &= [Ag^+] \cdot (1 + K_f \cdot [NH_3]^2) \qquad \text{or}\end{aligned}$$

$$[Ag^+] = \frac{[Cl^-]}{1 + K_f \cdot [NH_3]^2}$$

Substitute this into (2) to obtain the expression

$$K_{sp} = \frac{[Cl^-]^2}{1 + K_f \cdot [NH_3]^2}$$

which, in final form, allows calculation of the value of the target $[Cl^-]$.

$$[Cl^-] = \sqrt{K_{sp} \cdot (1 + K_f \cdot [NH_3]^2)}$$

Insert the known values $K_{sp} = 1.8 \times 10^{-10}$, $K_f = 1.6 \times 10^7$, and $[NH_3] =$ 0.500 M and evaluate to get

$$\text{molar solubility of AgCl} = [Cl^-] = 0.027 \text{ mol L}^{-1}$$

b. For any 1 : 1 salt in pure water the molar solubility is equal to the square root of the solubility product. For AgCl, then,

This relationship was established in Example 17.3.

$$\begin{aligned}\text{molar solubility} &= \sqrt{K_{sp}} = \sqrt{1.8 \times 10^{-10}} \\ &= 1.3 \times 10^{-5}\end{aligned}$$

The presence of 0.50 M NH_3 increases the solubility of AgCl more than a thousandfold.

Exercise

The solubility product for silver iodide is 8.3×10^{-17}. Calculate the molar solubility of AgI in 1.0 M NH_3.

Answer: 3.6×10^{-5} mol L^{-1}. (Again, more than a thousand times its solubility in pure water.)

In solving Example 17.16, we assumed that $[NH_3]_{eq} = 0.500$ M. That is, we assumed that not enough AgCl would dissolve to reduce the ammonia concentration significantly. In fact, when 0.027 mol AgCl dissolves and forms the complex ion $Ag(NH_3)_2^+$, the concentration of NH_3 drops from 0.50 M to $0.50 - 2(0.027) = 0.45$ M. Whether or not this decrease is regarded as "significant" is a matter of choice, especially since K_f and K_{sp} are known only to two significant figures.

If desired, a more accurate result can be obtained as follows. Treat the equilibrium concentration of NH_3 as a fourth unknown, and introduce an additional mass balance requirement as a fourth equation: $[NH_3]_{eq} + 2 \cdot [Ag(NH_3)_2^+]_{eq} = [NH_3]_{initial} = 0.50$ M. Solving these four equations for the four unknowns leads to the result $[Cl^-] = 0.024$ M. The improvement in accuracy is probably not worth the trouble.

It follows from the results of Example 17.16 that ammonia can be used to dissolve precipitates of insoluble silver salts (Figure 17.10). Chapter 23 contains some further discussions of complex ion equilibria.

17.5 QUALITATIVE ANALYSIS OF METAL CATIONS

Suppose you were given a solution and told that it might contain either Ni^{2+} or Mn^{2+}, or both. What experiments could you perform in order to decide among these possibilities? This is a typical problem encountered in **qualitative analysis**—the identification of a substance, or of the components of a mixture.

Metal cations can often be separated, and identified, by fractional precipitation of their salts. There are several different schemes for the qualitative analysis of metal ions, depending on which ions are likely to be present. Most schemes separate ions into small groups according to the solubilities of several types of salt. Qualitative analysis of metal ions is a practical application of the concepts of ionic equilibria: solubilities and solubility products, the common ion effect, the effect of pH on solubility, and buffer action.

Table 17.4 gives the same information for $Zn^{2+}(aq)$.

We begin our discussion with a detailed consideration of the identification of Ni^{2+} and/or Mn^{2+} in a solution. A more general scheme for identification of about 20 different cations in aqueous solution is developed later.

(a)

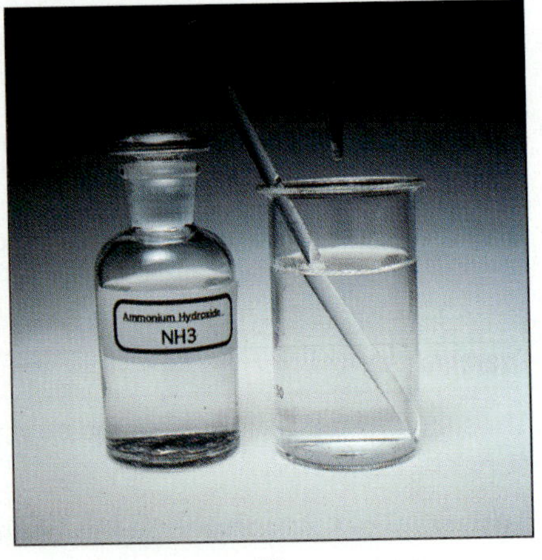

(b)

Figure 17.10 AgCl is quite insoluble in water, but it is more soluble in concentrated ammonia due to the formation of the complex ion $[Ag(NH_3)_2^+]$. The precipitate formed in (a) by the addition of aqueous $AgNO_3$ to a solution of NaCl can be redissolved by (b) the addition of concentrated ammonia.

TABLE 17.6 Maximum Ion Molarities in Saturated H_2S at 25 °C

pH	$[Ni^{2+}]_{max}/M$	$[Mn^{2+}]_{max}/M$
0	s†	s
1	s	s
3	3×10^{-3}	s
5	3×10^{-7}	s
7	3×10^{-11}	2×10^{-2}
9	3×10^{-15}	2×10^{-6}

* Calculated from equations like (17-13), using K'_{sp} (NiS) = 3×10^{-9} and K'_{sp} (MnS) = 2×10^{-10}.
† "s" means that the maximum metal ion concentration is greater than 1 M; that is, the sulfide is soluble.

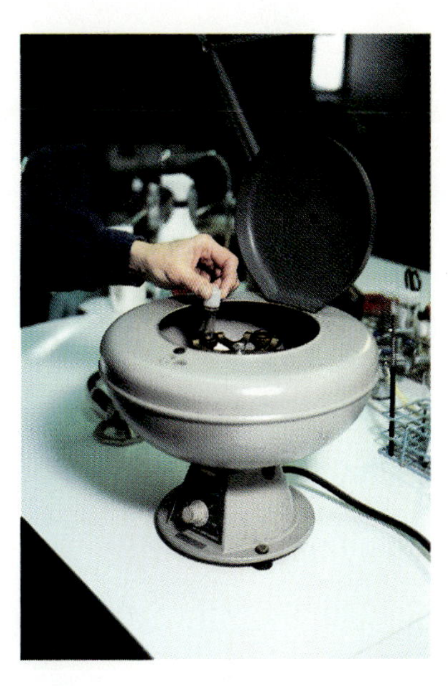

Figure 17.11 A bench-top centrifuge is used in qualitative analysis to hasten the settling of precipitates by substituting the much stronger centrifugal force for the force of gravity.

Separation of Ni^{2+} and Mn^{2+} by Fractional Precipitation

As discussed in Section 17.3, the solubility of metal sulfides is greatly affected by pH. If H_2S is added to a solution containing Ni^{2+} and/or Mn^{2+}, and the pH is low enough, no precipitate will form. If the pH is high, on the other hand, the metal sulfide(s) will precipitate when H_2S is added. The maximum possible concentration of a particular metal ion in a solution containing H_2S depends on pH, as shown in Table 17.6 for Ni^{2+} and Mn^{2+} ions.

The maximum Ni^{2+} and Mn^{2+} molarities listed in Table 17.6 suggest the following procedure for determining which ions are present.

1. Acidify the solution to pH $\approx$ 1 by adding concentrated HCl, then saturate with H_2S. No precipitate will appear, since both NiS and MnS are soluble at this pH.
2. Add NaOH(aq) gradually until the pH rises to about 5. If Ni^{2+} is present, and its concentration is at least 10^{-4} M, NiS will begin to precipitate at pH $\approx$ 2 to 4.
3. Remove the solid NiS (if formed), and continue adding OH^-. If Mn^{2+} is present, MnS will precipitate. If no precipitate has appeared when the pH has risen above 8 to 9, Mn^{2+} is not present at any concentration greater than 10^{-4} M.

A Scheme for Qualitative Analysis

About 20 cations can be identified by the following scheme, which is based on the varying solubilities of their chlorides, sulfides, and carbonates or phosphates. A few mL of a solution containing at least 10^{-4} M of any or all of the cations is treated as follows.

- *Step 1.* A few drops of concentrated HCl are added. If a precipitate forms, the solution must have contained Ag^+; Hg_2^{2+}; and/or Pb^{2+}, whose chlorides are insoluble. These ions constitute Analytical Group 1 (Figure 17.12). The chlorides of all other ions are soluble.

(a) (b)

Figure 17.12 A precipitate of the ions of Analytical Group 1, (a) after the addition of HCl to the unknown solution, and (b) after centrifugation.

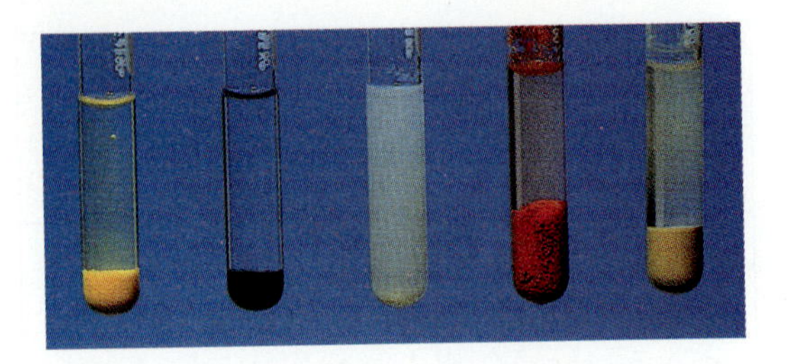

Figure 17.13 Qualitative analysis is a colorful operation, because many metal sulfides are brightly colored compounds. These are precipitates of (from left to right) CdS, HgS, As_2S_3, Sb_2S_3, and SnS_2.

- *Step 2.* Any precipitate formed in Step 1 is separated, and 0.3 M HCl saturated with H_2S is added to the residual solution. If a precipitate forms, it may contain any or all of the highly insoluble sulfides of ions in Analytical Group 2: As^{3+}, Bi^{3+}, Cd^{2+}, Cu^{2+}, Hg^{2+}, Sb^{3+}, Sn^{4+}, and Pb^{2+}. ($PbCl_2$ is slightly soluble, so some Pb^{2+} survives Step 1.)
- *Step 3.* After separation of any precipitate from Step 2, $NH_3(aq)$ is added to the solution. Ammonia reacts with HCl to form a NH_4^+/NH_3 buffer of pH $\approx$ 9. Formation of a precipitate indicates the presence of one or more of the ions of Analytical Group 3: Co^{2+}, Fe^{2+}, Mn^{2+}, Ni^{2+}, and Zn^{2+}. The sulfides of these metals are soluble at the lower pH used to precipitate the Analytical Group 2 sulfides. The precipitate also contains the insoluble hydroxides of Al^{3+} and Cr^{3+}.
- *Step 4.* The residual solution from Step 3 is treated with $(NH_4)_2CO_3(aq)$. Formation of a precipitate indicates the presence of one or more of the ions of Analytical Group 4: Ba^{2+}, Ca^{2+}, Mg^{2+}, and Sr^{2+}. [$(NH_4)_3PO_4$ may be used as

TABLE 17.7 Cation Groups in Qualitative Analysis

Analytical Group	Cations	Precipitating Reagent	pH	Insoluble Compounds
1	Ag, Pb^{2+}, Hg_2^{2+}	Conc. HCl	≈ 1	AgCl, $PbCl_2$, Hg_2Cl_2
2	As^{3+}, Bi^{3+}, Cd^{2+}, Cu^{2+}, Hg^{2+}, Sb^{3+}, Sn^{4+}, (Pb^{2+})	H_2S (sat.) in 0.3 M HCl	≈ 1	As_2S_3, Bi_2S_3, CdS, CuS, HgS, Sb_2S_3, SnS_2, (PbS)
3	Co^{2+}, Fe^{2+}, Mn^{2+}, Ni^{2+}, Zn^{2+}	Conc. NH_3	≈ 9	CoS, FeS, MnS, NiS, ZnS
4	Ba^{2+}, Ca^{2+}, Mg^{2+}, Sr^{2+}	$(NH_4)_2CO_3(aq)$	≈ 9	$BaCO_3$, $CaCO_3$, $MgCO_3$, $SrCO_3$
5	Na^+, K^+, (NH_4^+)	None		None

the precipitating reagent in this step, since the solubilities of metal carbonates and phosphates are similar.] The residual solution contains the ions of Analytical Group 5: NH_4^+, Na^+, and K^+. [NH_4^+ is present in any case, since it was added as $(NH_4)_2CO_3(aq)$. But it may also have been present in the original sample.]

Step 5. Each of the groups of precipitates is subjected to further treatment in order to separate and identify the individual ions within the group. For example, in Analytical Group 1, Pb^{2+} can be identified because $PbCl_2$ dissolves in boiling water. Ag^+ can be identified because AgCl dissolves in concentrated NH_3.

This scheme for qualitative analysis is summarized in Table 17.7, and a *flowchart* is given in Figure 17.14.

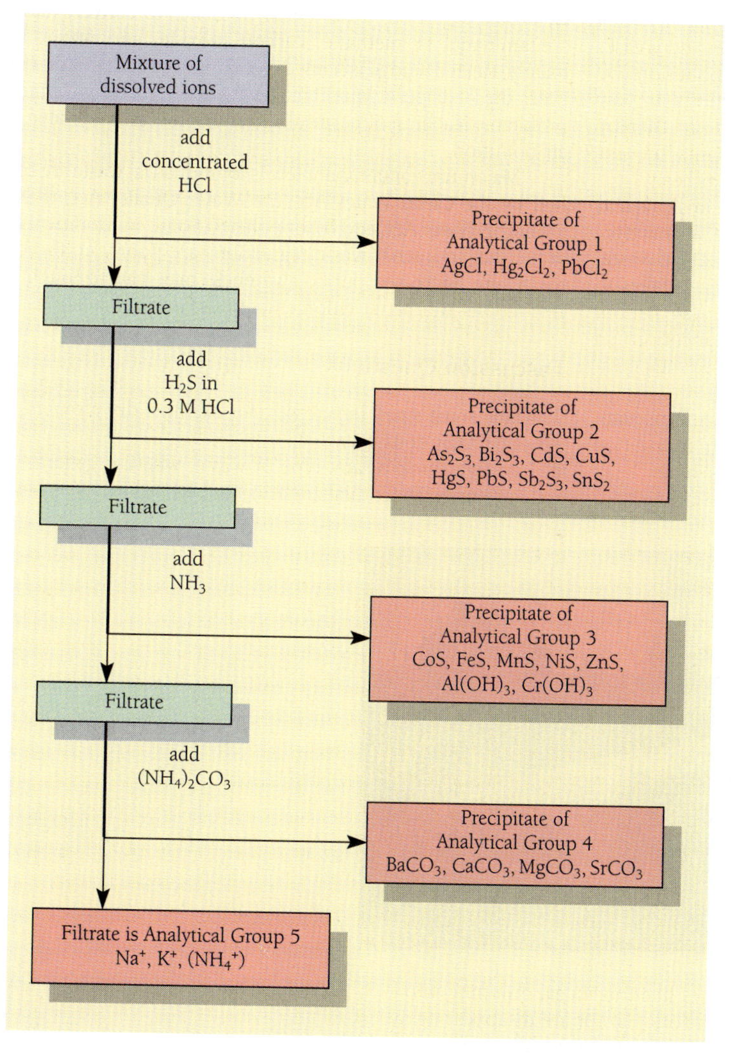

Figure 17.14 A scheme for the qualitative analysis of metal cations.

CHEMISTRY OLD AND NEW

INDUSTRIAL WASTE

From highly alkaline textile wastes to the radioactive by-products of nuclear fission, industrial wastes can threaten public health as well as the environment. Each year, U.S. industries generate far more waste than individuals. The U.S. Public Health Service has reported, in fact, that treated industrial wastewater carries over twice as much degradable organic matter into U.S. waterways than all city sewer systems combined.

Wastes are generated in the production of heat, electrical power, manufacturing chemicals, clothing, food and drugs, synthetic and natural materials, and many other activities. It is doubtful that *any* industrial operation is waste-free. Water pollutants range from suspended solids to inorganic salts, acids or alkalis, organic matter, toxic chemicals, microorganisms, radioactive materials, metals, and floatables including oil and grease.

Industrial processes can also cause "thermal pollution" by heating streams and rivers. Higher stream temperatures support increased bacterial activity, which depletes oxygen and thereby threatens wildlife. Non-toxic but colored pollutants prevent sunlight from entering a stream, reducing photosynthesis.

Effective treatment of wastewater begins with analysis and characterization of the contaminants. For example, it is important to know the biochemical oxygen demand (BOD) of waste. The BOD is based on how much dissolved oxygen it takes to decay organic material in wastewater over a five-day period. Contaminants that increase the BOD are of particular concern in marine estuaries because they cause suffocation of aquatic creatures.

The acidity or alkalinity of wastewater is also important, since a pH of less than 4.5 or more than 9.5 kills fish. In addition to the effect on fish, acid waters can corrode ships' hulls.

Treatment of industrial waste begins with settling-and-screening procedures, which use gravity and added chemical coagulants to separate large "settleable" solids and floatables. Any small contaminant (1 to 10 μm in diameter) still suspended in water may be removed via "flotation" techniques that use air bubbles to drive particles to the surface, where they are skimmed off. Secondary treatment involves "activating" the waste by bubbling oxygen into it. Microorganisms then oxidize some of the contaminants, converting them to a semi-solid sludge. Advanced treatment may involve physical, biological, or chemical processes. For example, disinfection by ultraviolet light or chlorine can kill harmful bacteria.

Sludge may also be subjected to aerobic composting or anaerobic digestion—the use of microbes to convert harmful contaminants to methane, carbon dioxide, and water. Ultimately, treated sludge is hauled to a landfill or incinerator. However, these disposal methods have their own environmental problems: leaching of harmful materials into groundwater, and air pollution.

Maximum contaminant levels for industrial waste are specified by the Clean Water Act, the Water Quality Act, the Safe Drinking Water Act, and other policies. Some environmentalists argue that industrial waste regulations are poorly enforced. In the future, increased public awareness may lead to tough new restrictions and better enforcement of existing regulations.

Discussion Questions

1. Name some small businesses, for example dry cleaners, that generate hazardous waste.
2. Can you think of any type of waste, whether generated by industry or by individuals, that is *not* harmful in some way?

Scenes like this illegal dumping into New York Harbor are disappearing as chemists find new and better ways of handling industrial waste.

SUMMARY EXERCISE

The amount of solute contained in 1.73 L of a saturated solution of PbI_2 is 1.0 g. The formation constant of the complex hydroxide ion $Pb(OH)_3^-$ is 3×10^{14}. All questions and data in this exercise refer to 25 °C.

a. Calculate the solubility (g L^{-1}) and the molar solubility of PbI_2.

b. What is the value of the solubility product of PbI_2?

c. How many grams of PbI_2 will dissolve in 1.00 L of a 0.030 M potassium iodide solution?

d. Will a precipitate of PbI_2 form if equal volumes of 1.0×10^{-4} M $Pb(NO_3)_2$ and 2.0×10^{-2} M NaI are mixed?

e. Would a small amount of $PbI_2(s)$ be more likely to dissolve in $HNO_3(aq)$ than in KOH(aq)?

f. What is the value of the ratio $[Pb^{2+}]/[Pb(OH)_3^-]$ in a solution of pH = 8.5?

g. How many grams of NH_4Cl must be added to 1.00 L of 0.0100 M $NH_3(aq)$ to produce a buffer solution of pH = 8.50? (K_b for NH_3 is 1.9×10^{-5}.)

Answers **(a)** 0.58 g L^{-1}, 1.3×10^{-3} mol L^{-1}; **(b)** 7.9×10^{-9}; **(c)** 4.0×10^{-3} g; **(d)** no; **(e)** KOH; **(f)** 100; **(g)** 3.2 g.

SUMMARY AND KEYWORDS

The **ion product** Q of a dissolved salt is the product of the concentrations of the constituent ions, each raised to the power equal to that ion's subscript in the formula of the salt. If it refers to a *saturated* solution, the ion product is called the *solubility product expression* and its numerical value is the **solubility product,** K_{sp}. The **solubility** of a substance is the number of grams that will dissolve in one liter of solvent, and the **molar solubility** is the same quantity expressed in mol L^{-1}. Molar solubility = $\sqrt{K_{sp}}$ for 1 : 1 salts like AgCl or $CuSO_4$, but the relationship is different for salts of other stoichiometry.

The solubility of a salt can be altered by the **common ion effect.** AgBr(s) is much less soluble in NaBr(aq) than in pure water because of the presence of the Br^- ion, common to both AgBr and NaBr.

In a saturated solution, $Q = K_{sp}$. If $Q < K_{sp}$, the solution is **undersaturated,** and if $Q > K_{sp}$, the *solution* is **supersaturated.** If $Q > K_{sp}$, the solution is unstable and **precipitation** takes place until $Q = K_{sp}$. The condition $Q > K_{sp}$ can be brought about in several ways: by decreasing the temperature, which (for most salts) decreases K_{sp}; by the addition of a common ion, which increases Q; or by the evaporation of solvent, which increases Q.

Since $[OH^-]$ is low in acid solution, metal hydroxides are more soluble at low pH. Salts of weak acids (for example HF, H_2CO_3, H_2S, or H_3PO_4) are also more soluble at low pH, because the anion is removed from the solution by reaction with H_3O^+. Most sulfides are insoluble, dissolving to a very slight extent according to $MS(s) + H_2O \rightleftharpoons M^{2+}(aq) + HS^-(aq) + OH^-(aq)$. The S^{2-} anion does not exist in aqueous solution.

Complex ions are **coordination complexes** bearing a nonzero electrical charge. Coordination complexes consist of a metal ion bonded (by coordinate covalent bonds) to one or more electron-donating molecules or anions (Lewis bases). Complex ions are in equilibrium with their constituent "free" cations and Lewis bases. The equilibrium constant is called the **formation constant** (K_f). The relative concentrations of free and bound cations in a solution depend on the concentration of the Lewis base. For example, the concentration ratio $[Ag^+]/[Ag(NH_3)_2^+]$ can be controlled by changing the concentration of ammonia.

Separation and **qualitative analysis** of mixtures of cations can be achieved by exploiting the different solubilities of the chlorides, sulfides, and carbonates, and the effect of pH on these solubilities. An analytical *scheme,* such as the one in Table 17.7, is a systematic procedure for separation of ions into analytical groups according to these solubilities.

PROBLEMS

General

1. Distinguish between the terms "ion product" and "solubility product expression."
2. What is the difference between a solubility product and a solubility product expression?
3. Define the terms "solubility" and "molar solubility."
4. What is meant by the common ion effect on solubility?
5. Define the terms "undersaturated," "saturated," and "supersaturated."
6. Describe a procedure for the preparation of supersaturated solutions.
7. How can the solubility of sparingly soluble hydroxides be manipulated?
8. How can the solubility of sparingly soluble sulfides be manipulated? Explain in terms of the basicity of the sulfide ion.
9. What is a complex ion? Give examples.
10. Is the sulfate ion a complex ion? Explain.
11. In the term "coordination complex," what does the word "coordination" refer to?
12. When a few drops of 1×10^{-5} M $AgNO_3$ are added to 0.01 M NaCl, a white precipitate appears immediately; when added to 5 M NaCl, no precipitate appears. Suggest an explanation for this behavior. (*Hint:* Consult Table 17.5.)
13. What is meant by the term "qualitative analysis"?

*14. Suppose a student subjects a sample to the entire qualitative analysis scheme, by successively adding HCl, H_2S, NH_3, and $(NH_4)_2CO_3$. She observes no precipitate at any stage, and concludes that the original sample contains no cations other than Analytical Group 5, Na^+, K^+, and/or NH_4^+. Suggest a means for distinguishing among these possibilities. (*Hint:* The answer is to be found in Chapters 6 and 15, not in Chapter 17.)

Solubility and Solubility Product

Appendix H should be consulted for needed values of solubility products.

15. Write the solubility product expression for the following salts.
    a. BaF_2 b. $Bi_2(SO_4)_3$
    c. CuBr d. $BaCO_3$
16. Write the solubility product expression for the following salts.
    a. $Co_3(AsO_4)_2$ b. Hg_2I_2
    c. HgI_2 d. $(NH_4)_2CO_3$
17. Calculate the molar solubility of CuCl from its K_{sp}.
18. Calculate the molar solubility of $BaCrO_4$ from its K_{sp}.

*19. Calculate the solubility, in mol L^{-1} and g L^{-1}, of $AuCl_3$.

20. Calculate the solubility, in mol L^{-1} and g L^{-1}, of SrF_2.
21. How many grams of PbC_2O_4 will dissolve in 1.00 L of water?
22. How many grams of $Ca(OH)_2$ will dissolve in 1.00 L of water? What is the pH of a saturated solution?

*23. Calculate the solubility, in g L^{-1}, of $AuBr_3$.

24. Calculate the solubility, in g L^{-1}, of CaF_2.
25. It is found that 1.0×10^{-7} moles of AgBr will dissolve in 141 mL of water. What is the value of K_{sp}?
26. A saturated solution of $AgIO_3$ contains 0.028 g L^{-1}. Calculate K_{sp}.
27. If 1.55×10^{-4} mol Ag_2SO_4 will dissolve in 10 mL H_2O, what is the value of K_{sp}?
28. A saturated solution of Ag_2CO_3 contains 0.032 g L^{-1}. Calculate K_{sp}.

Precipitation

29. What compound (if any) precipitates when equal volumes of 0.0020 M $CaCl_2$ and 0.0040 M Na_2SO_4 are mixed?
30. What compound (if any) precipitates when equal volumes of 0.0020 M $SrCl_2$ and 0.0040 M Na_2CO_3 are mixed?
31. Will a precipitate form when equal volumes of 0.010 M $BaCl_2$ and 1.5×10^{-4} M NaF are mixed? If so, what compound is it?

32. Will a precipitate form when equal volumes of 2.0×10^{-5} M $AgNO_3$ and 2.0×10^{-5} M $CaCl_2$ are mixed?

*33. How many moles of $Na_2CrO_4(s)$ must be dissolved in 500 mL of 0.010 M $CaCl_2$ in order to cause the appearance of a precipitate? (Assume no change in solution volume on the addition of the solid.)

34. How many grams of NaF must be added to 1.00 L of 1.0×10^{-4} M $BaCl_2$ in order to cause the appearance of a precipitate? (Assume no change in solution volume on the addition of the solid.)

35. If dilute aqueous $AgNO_3$ is added dropwise to a dilute equimolar solution of NaCl and NaBr, what is the composition of the first precipitate to appear?

36. If dilute aqueous KOH is added dropwise to a dilute equimolar solution of $BaCl_2$ and $Ca(NO_3)_2$, what is the composition of the first precipitate to appear?

37. If solid Na_2SO_4 is gradually added to and dissolved in a solution that is 0.010 M in $Ca(NO_3)_2$ and 0.010 M in $Pb(NO_3)_2$, in what order will solid $CaSO_4$ and $PbSO_4$ appear?

38. If solid Na_2CO_3 is added to a solution that is 1.0×10^{-3} M in $CuCl_2$ and 1.0×10^{-3} M in $CaCl_2$, in what order will solid $CuCO_3$ and $CaCO_3$ appear?

*39. In Problem 37, what is the remaining concentration of the ion that formed the first precipitate when the second precipitate just begins to appear?

*40. In Problem 38, what is the remaining concentration of the ion that formed the first precipitate when the second precipitate just begins to appear?

Common Ion Effect

41. What is the maximum concentration of $Ag^+(aq)$ that can exist in 0.0050 M NaBr?

42. What is the maximum concentration of $Ag^+(aq)$ that can exist in 0.015 M Na_2CrO_4?

43. What is the maximum concentration of $Hg^{2+}(aq)$ that can exist in 0.75 M KBr?

44. What is the maximum concentration of $Ba^{2+}(aq)$ that can exist in 0.022 M KOH?

*45. How many moles of $Mn(NO_3)_2$ will dissolve in 1.00 L of 0.10 M KOH?

*46. How many moles of $Cr(NO_3)_3$ will dissolve in 1.00 L of a solution whose pH is 5.00?

*47. What is the solubility (in g L^{-1}) of Ag_2CrO_4 in 0.0050 M Na_2CrO_4?

*48. What is the solubility (in g L^{-1}) of Ag_2CrO_4 in 0.063 M $AgNO_3$?

*49. What is the solubility of $SrCO_3$ (in g L^{-1}) in (a) pure water and (b) 0.033 M $Sr(NO_3)_2$?

*50. What is the solubility of $BaCO_3$ (in g L^{-1}) in (a) pure water and (b) 0.00089 M $BaCl_2$?

pH Control of Solubility

51. What is the maximum concentration of Mg^{2+} that can exist in a solution whose pOH is (a) 4.53, (b) 1.70?

52. What is the maximum concentration of Mn^{2+} that can exist in a solution whose pH is (a) 7.81, (b) 11.15?

53. To what value must the pH be adjusted (by addition of HCl) in order to dissolve 5.00 g $Mg(OH)_2$ in a liter of water?

54. To what value must the pH be adjusted in order to dissolve 100 g $Mg(OH)_2$ in 250 mL H_2O?

*55. Will a precipitate of $Mg(OH)_2$ appear if equal volumes of 1.0×10^{-4} M $MgCl_2$ and 0.010 M NH_3 are mixed? K_b for NH_3 is 1.9×10^{-5}.

*56. Will a precipitate of $AgC_2H_3O_2$ appear if equal volumes of 0.00100 M $AgNO_3$ and 0.010 M acetic acid are mixed? K_a for acetic acid is 1.8×10^{-5} and K_{sp} for $AgC_2H_3O_2$ is 4×10^{-3}.

*57. A solution contains 0.010 M each of $MgCl_2$ and $CaCl_2$. To what value must the fluoride ion concentration be adjusted (by addition of solid KF) in order to remove as much Ca^{2+} as possible [in the form of $CaF_2(s)$] while leaving the Mg^{2+} entirely in solution?

*58. Mg^{2+} and Ca^{2+} can be separated by fractional precipitation of the hydroxides. If a solution contains 0.010 M each of $MgCl_2$ and $CaCl_2$, to what value must the pH be adjusted in order to remove as much Mg^{2+} as possible [in the form of $Mg(OH)_2$] while leaving the Ca^{2+} entirely in solution?

*59. What is the maximum concentration of Zn^{2+} that can exist in saturated H_2S at pH 3.50?

*60. What is the maximum concentration of Pb^{2+} that can exist in saturated H_2S at pH 2.00?

*61. To what value must the pH of a saturated solution of H_2S be adjusted in order for 0.0050 mg FeS to dissolve in 1.0 L?

*62. A sample containing 0.100 g each of NiS and MnS is mixed with 1.00 L of 1 M NaOH, and the pH is reduced by the gradual addition of concentrated HCl. Which of the two solids will dissolve first, and at what pH will the dissolution of the first solid be complete?

Complex Ion Equilibria

63. Write the chemical equation describing the formation of the complex, and the formation constant expression, for each of the following complex ions.
a. $Ag(CN)_2^-$ b. $Cd(NH_3)_4^{2+}$

64. Write the chemical equation describing the formation of the complex, and the formation constant expression, for each of the following complex ions.
a. $Zn(OH)_4^{2-}$ b. $CoCl_6^{3-}$

65. Concentrated ammonia is added to a solution of silver nitrate, so that the final value of $[NH_3]$ is 0.35 M. Given that K_f for $Ag(NH_3)_2^+$ is 1.6×10^7, what is the value of the ratio $[Ag^+]/[Ag(NH_3)_2^+]$?

66. Given that K_f for $Zn(NH_3)_4^{2+}$ is 1.1×10^{12}, what is the value of the ratio $[Zn^{2+}]/[Zn(NH_3)_4^{2+}]$ in a solution of zinc nitrate and ammonia in which $[NH_3] = 0.0050$?

*67. Gaseous ammonia is added to a solution of 0.063 M $AgNO_3$ until the concentration of $NH_3(aq)$ rises to 0.18 M. Given that K_f for $Ag(NH_3)_2^+$ is 1.6×10^7, what are the concentrations of $Ag(NH_3)_2^+$ and Ag^+?

*68. Ammonia is added to a solution of 0.047 M $Zn(NO_3)_2$ until the concentration of NH_3 rises to 0.200 M. Given that K_f for $Zn(NH_3)_4^{2+}$ is 1.1×10^{12}, what are the concentrations of Zn^{2+} and $Zn(NH_3)_4^{2+}$?

Qualitative Analysis

69. What chemical property distinguishes Analytical Group 1 from Analytical Group 2?

70. What chemical property distinguishes Analytical Group 4 from Analytical Group 3?

*71. A precipitate from Analytical Group 1 is known to contain no $PbCl_2$, and therefore it is AgCl, Hg_2Cl_2, or a mixture of the two. Suggest a means for determining its composition.

72. What test might be performed in order to determine whether a solution contains Zn^{2+} or Cd^{2+}?

73. Lead ion is in Analytical Group 2 as well as Analytical Group 1, because $PbCl_2$ is slightly soluble. Suppose that, in the first step of the analysis procedure, hydrochloric acid is added to a sample until [HCl] rises to 0.15 M. What is the maximum concentration of Pb^{2+} that can be present *without* forming a precipitate?

74. Suppose you wish to know only whether or not a solution contains an ion from Analytical Group 3. Saturation of the solution with H_2S and adjusting the pH to approximately 9 yields a precipitate. Does this confirm the presence of a Group 3 ion? Explain.

Chapter 18

Oxidation-Reduction Reactions

Exploding fireworks supply a spectacular example of oxidation-reduction reactions.

18.1 REDOX REACTIONS

The word *redox* was coined as a contraction of "reduction-oxidation," and **redox reaction** is used to describe reactions in which oxidation and reduction occur. Such reactions are very common and very important. All of the following processes involve redox reactions: rusting of iron and steel, corrosion of other metals, extraction of metals from their ores, drying and weathering of paint, use of coal and oil for the production of power or heat, bleaching of clothes, disinfecting of wounds, spoiling of foods, photosynthesis in plants, respiration in animals, formation of acid rain, and many others.

Reaction with Oxygen

Following the discovery of oxygen in 1774 by Joseph Priestley and Karl Scheele, the word *oxidation* was used to describe the chemical changes induced in substances by exposure to and combination with oxygen. One form of oxidation is *combustion*, a self-sustaining exothermic reaction with oxygen that gives off useful light and heat. The combustion of gasoline, wood, and paper are familiar examples. Some metals also undergo combustion. A thin ribbon of magnesium metal burns in air, producing a hot, white flame and a smoke of finely divided MgO(s) particles. In the early days of photography, the rapid combustion of magnesium powder in air was used, when necessary, for additional lighting. The "flashbulb," a tangled strand of magnesium or aluminum wire enclosed in a bulb containing pure oxygen, was a later development (Figure 18.1). The oxidation of magnesium is described by Equation (18-1).

$$2\ Mg(s) + O_2 \xrightarrow[\Delta]{} 2\ MgO(s); \qquad \Delta H^\circ_{rxn} = -601.7\ kJ \tag{18-1}$$

Electron Loss or Gain

Magnesium also undergoes an exothermic reaction with chlorine, as illustrated in Figure 18.2.

$$Mg(s) + Cl_2 \longrightarrow MgCl_2(s); \qquad \Delta H^\circ_{rxn} = -641.3\ kJ \tag{18-2}$$

Joseph Priestley (1733–1804) and Karl Wilhelm Scheele (1742–1786)

Joseph Priestley was an English theologian, who wrote books on philosophy, government, and especially the history of the Christian Church. He was also interested in science. Although he prepared pure oxygen in his home laboratory, he failed to realize the significance of his discovery. His political views (in particular his support of the French revolution) were unpopular in England, and, after his house was wrecked by an angry mob, he emigrated to Pennsylvania.

Karl Wilhelm Scheele was a Swedish pharmacist. His discovery of oxygen occurred earlier than and independent of Priestley's, but he did not receive full credit because his published account came later. He also discovered nitrogen, independently and with little recognition.

The camera on the right is equipped with a flash bulb, which produces light by the rapid combustion of magnesium or aluminum wire. In more modern cameras (left) light is emitted by an electrical discharge.

Reactions (18-1) and (18-2) both give off light and heat, and both reaction products, MgO and $MgCl_2$, are *ionic* compounds. Product formation in Equations (18-1) and (18-2) can be represented by separate ion-forming steps, followed by a step showing the combination of ions.

Recall that ionic compounds are formed when a metal loses one or more of its valence electrons and becomes a positive ion, while at the same time a nonmetal gains one or more additional electrons in its valence shell and becomes a negative ion.

$$Mg \longrightarrow Mg^{2+} + 2\,e^- \qquad (18\text{-}1a)$$

$$\tfrac{1}{2}\,O_2 + 2\,e^- \longrightarrow O^{2-} \qquad (18\text{-}1b)$$

$$Mg^{2+} + O^{2-} \longrightarrow MgO(s) \quad \text{and} \qquad (18\text{-}1c)$$

$$Mg \longrightarrow Mg^{2+} + 2\,e^- \qquad (18\text{-}2a)$$

$$Cl_2 + 2\,e^- \longrightarrow 2\,Cl^- \qquad (18\text{-}2b)$$

$$Mg^{2+} + 2\,Cl^- \longrightarrow MgCl_2(s) \qquad (18\text{-}2c)$$

These reactions are very similar, since in each case Mg loses two electrons.

Chemists have found it useful to broaden the original meaning of the word "oxidation" to include reactions like that in Equation (18-2), even though O_2 is not involved. Today, any process in which an atom, molecule, or ion loses one or more

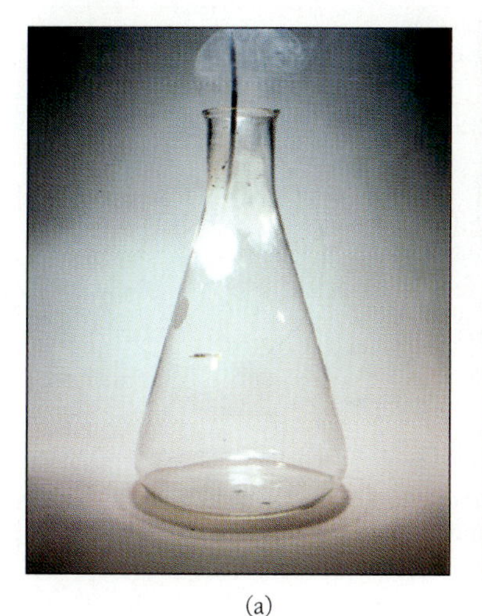

(a)

(b)

Figure 18.2 (a) Magnesium wire burns vigorously in an atmosphere of oxygen, and (b) equally well in an atmosphere of chlorine.

electrons is called an **oxidation.** In Reactions (18-1a) and (18-2a), Mg loses two electrons to become Mg^{2+}: Mg is *oxidized.* A substance like O_2 or Cl_2, which causes oxidation of another substance, is known as an **oxidizing agent** or **oxidant.**

"Reduction" was originally used to describe the process in which a metal ore was converted to the pure metal. This was called a "reduction" because the ore loses mass in the process.

The process in which a substance *gains* one or more electrons is known as **reduction.** In Reaction (18-1b), oxygen gains two electrons per atom to become O^{2-}: oxygen is *reduced.* Likewise, chlorine is reduced in Reaction (18-2b). Any substance, such as magnesium, that is capable of causing reduction of another substance, is called a **reducing agent** or **reductant.** Note carefully that, because electrons are negatively charged, when an atom *gains* electrons in a reduction, its electrical charge *decreases* (becomes more negative).

In any chemical reaction involving oxidation, the oxidizing agent becomes reduced, while the reducing agent becomes oxidized. Equation (18-3) summarizes the vocabulary of oxidation and reduction (Figure 18.3).

$$\underset{\substack{\textbf{is oxidized} \\ (\text{to } Mg^{2+})}}{\overset{\textbf{reductant}}{Mg(s)}} + \underset{\substack{\textbf{is reduced} \\ (\text{to } Cl^{-})}}{\overset{\textbf{oxidant}}{Cl_2}} \longrightarrow \underset{\substack{[Cl^{-}\text{--}Mg^{2+}\text{--}Cl^{-}] \\ (\text{an ionic solid})}}{MgCl_2(s)} \quad (18\text{-}3)$$

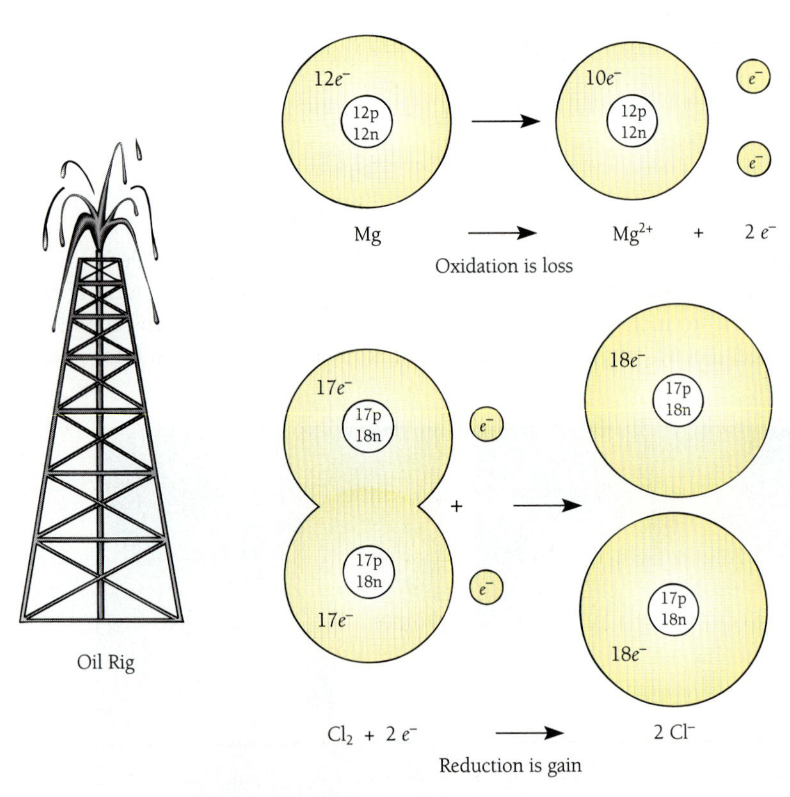

Figure 18.3 The processes of oxidation and reduction can be kept in mind by the use of the mnemonic "OIL RIG," which stands for "Oxidation Is Loss, Reduction Is Gain."

The electrons gained by Cl_2 in Reaction (18-2b) come directly from the Mg. There can be no gain of electrons by one reagent unless there is a simultaneous loss of electrons by another. *Oxidation is always accompanied by reduction, and vice versa.*

Reactions in which oxidation and reduction take place are called redox reactions. In many redox reactions, including almost all of those discussed in this chapter, the mechanism involves direct transfer of one or more electrons from the substance being oxidized to the substance being reduced. Reactions of this type are also called **electron transfer reactions.** Other redox reactions, including many important processes in living organisms, take place by mechanisms other than direct transfer of electrons.

Although the oxidative and reductive parts of a redox reaction cannot take place independently of one another, it is often convenient to *write* them separately. In this separate form, they are called **half-reactions.** The balanced half-reactions for the oxidation of magnesium by chlorine are

Oxidation $$Mg \longrightarrow Mg^{2+} + 2\,e^- \quad (18\text{-}4)$$

Reduction $$Cl_2 + 2\,e^- \longrightarrow 2\,Cl^- \quad (18\text{-}5)$$

Half-reactions are not balanced unless the electrons gained or lost are shown explicitly. Electrons appear on the right, as a product, in oxidation half-reactions, and on the left in reduction half-reactions.

18.2 OXIDATION NUMBERS

When studying electron transfer reactions, it is important to keep track of the number of electrons lost or gained. This is done by means of a system in which each atom in a molecule or ion is assigned an *oxidation number* (sometimes called an "oxidation state"). The **oxidation number** of an atom in a compound is a positive number equal to the number of valence electrons it loses in its *formation reaction* from its constituent elements, or a negative number equal to the number of electrons gained in the formation reaction. The oxidation number of an atom in an element is zero. A *change* in oxidation number of an atom during a chemical reaction indicates that one or more valence electrons have been lost or gained. Changes in oxidation number always accompany electron loss or gain. *Any reaction in which an atom undergoes a change in oxidation number is a redox reaction.* Conversely, *in any redox reaction the oxidation numbers of (at least) two atoms must change.* The modern definition of oxidation is a process in which the oxidation number of an atom increases. Reduction is defined as a process in which the oxidation number of an atom decreases.

Formation reactions were first introduced and used in Chapter 4, in discussions of thermochemistry.

Oxidation number is a concept useful in discussions of redox reactions, but it is not a physical or chemical property. Its numerical value is not measurable, because it is an *assigned* quantity.

Monatomic Ions

The assignment of oxidation numbers to elements in binary ionic compounds is straightforward. In KF, for example, the potassium atom loses one electron as it becomes the K^+ ion: therefore the oxidation number of K^+ is $+1$. The F^- ion has gained one valence electron (compared to neutral F), and its oxidation number is -1. In MgS, two electrons are lost when Mg^{2+} is formed from elemental magnesium, and

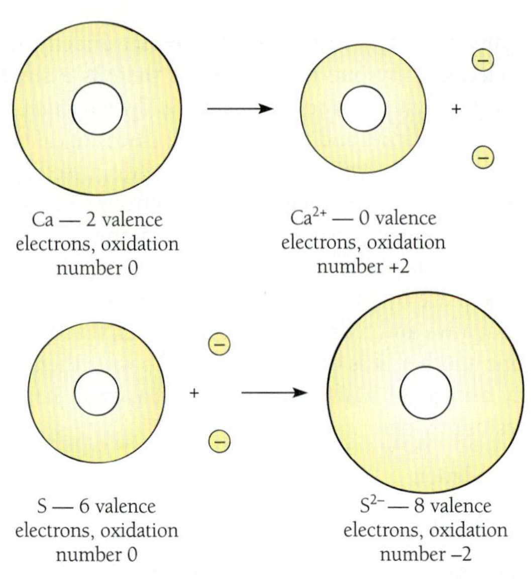

Figure 18.4 In the ionization process, a neutral metal atom loses electrons and becomes a cation with a positive charge equal to the number of electrons lost. This number is called the oxidation number, and its positive sign indicates that the element has been oxidized. A neutral atom of a nonmetal gains electrons during ionization, and becomes an anion having a negative charge equal to the number of electrons gained. Its oxidation number is negative, which signifies that the element has been reduced. The oxidation number of a monatomic ion is equal to its charge.

Ionic *charge* is written with the sign following the numeral: S^{2-}. *Oxidation number* is written with the sign preceding the numeral: -2.

the oxidation number of Mg^{2+} is $+2$. The S^{2-} ion, having gained two electrons in its formation from elemental sulfur, has an oxidation number of -2. For monatomic ions, then, the oxidation number is equal to the charge, *including the sign* (Figure 18.4).

Covalently Bonded Species

Do not confuse oxidation number with *formal charge* (Section 8.3), which is an aid in drawing Lewis structures. They are defined differently, and, except in elements and monatomic ions, formal charge and oxidation number usually have different values.

In contrast to the situation in ionic compounds, the atoms in covalently bonded species do not gain or lose electrons in the formation reaction. Instead, they acquire a greater or lesser proportion of the shared pair of electrons that constitutes a covalent bond. Nevertheless, oxidation numbers are assigned to atoms in covalent molecules *as if* electron transfer were complete rather than partial. That is, polar covalent bonds are regarded as being ionic, with the more electronegative atom "gaining" an electron (or more than one, in multiple bonds) from the less electronegative atom. In the covalently bonded CCl_4 molecule, for example, each Cl atom, having "gained" an electron, is assigned an oxidation number of -1. The C atom, having "lost" four electrons, is assigned an oxidation number of $+4$.

All polyatomic species, including elements, compounds, and polyatomic ions, contain covalent bonds. Because the procedure for assigning oxidation numbers is based on the fiction that polar covalent bonds are 100% ionic, we must be wary of attaching too much chemical significance to the values of oxidation numbers in covalent species.

Because it is equal to the ion charge, the oxidation number of a monatomic ion must be a whole number. This restriction does not apply in polyatomic ions or molecules, and *fractional* oxidation numbers are seen now and then. For example, in the triiodide ion (I_3^-) the oxidation number of iodine is $-1/3$.

In contrast to the oxidation number, the *charge* of a molecule or ion is a real physical quantity. The charge, which must be a whole number, is measurable in the laboratory.

Summary of Rules

Two principles are given in the preceding sections—that oxidation number refers to the valence electron *change* relative to the pure element, and that the electrons of a covalent bond are counted as belonging to the more electronegative element. These principles, together with the Lewis structure, are sufficient to assign oxidation numbers to each atom in any molecular or ionic species. In practice, the assignment of oxidation numbers is carried out by the successive application of the rules given in Table 18.1. These rules, which are derived from the two principles given above, should be memorized.

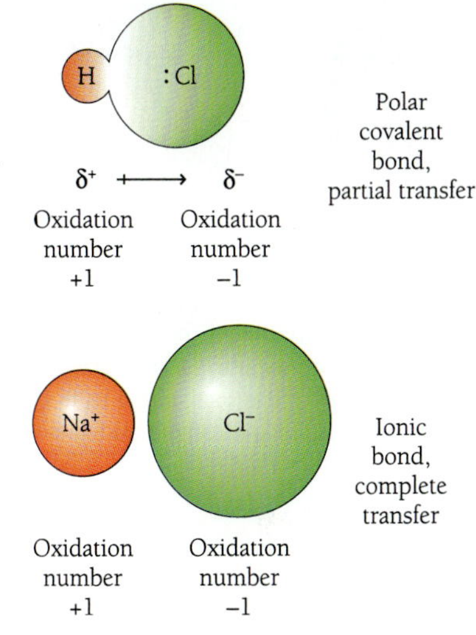

Figure 18.5 In the assignment of oxidation numbers, the partial electron transfer in a polar covalent bond is treated as if it were the complete transfer of an ionic bond.

TABLE 18.1 Assignment of Oxidation Numbers

Rule 1. The sum of the oxidation numbers of all the atoms in a molecule must equal zero, while in a polyatomic ion, this sum must equal the ionic charge. It follows that the oxidation number of an element is zero. This applies even if several different forms exist, so that the oxidation number of oxygen is zero in O_2 and also in O_3. It also follows that the oxidation number of a monatomic ion equals its charge.

Rule 2. In compounds, the *oxidation number of a Group 1A metal is +1, and that of a Group 2A metal is +2.*

Rule 3. Except in F_2 (when it is zero), *the oxidation number of fluorine is always −1.* The other Group 7A elements also have oxidation numbers of −1, unless bonded to oxygen or another halogen. In such molecules or ions, the oxidation number of a halogen can be positive.

Rule 4. When bonded to a nonmetal, *the oxidation number of hydrogen is +1.* No metal is more electronegative than hydrogen, so the oxidation number of hydrogen is −1 in metal hydrides like NaH or $LiAlH_4$.

Rule 5. The oxidation number of oxygen is −2, except in:

- molecules or ions containing an —O—O— bond. In peroxides such as BaO_2, Na_2O_2, or H_2O_2, Rule 2 or 4 takes precedence, and the oxidation number of oxygen is −1. In superoxides like KO_2, Rule 2 takes precedence and the oxidation number of oxygen is $-1/2$.
- fluorine-containing compounds. Rule 3 takes precedence, and oxygen has a positive oxidation number when bonded to F. The oxidation number of oxygen is +2 in OF_2 and +1 in O_2F_2.

Rule 6. Oxidation numbers of any atoms not covered by Rules 1 through 5 are determined by difference, using Rule 1. In applying Rule 1, it is essential to count each *atom*, rather than each element. For example, in H_2O the oxidation numbers are +1 for each of the H atoms, and −2 for the O atom. Their sum is $(+1) + (+1) + (-2) = 0$. In the sulfate ion, SO_4^{2-}, the oxidation numbers are +6 for S and −2 for each of the O atoms. Their sum is $(+6) + 4(-2) = -2$, the charge of the ion.

Many common plants, including nettles and other stinging weeds, contain oxalic acid and/or its conjugate base hydrogen oxalate ion. If the common houseplant "dumbcane" (Dieffenbachia picta) is chewed, the oxalic acid in it causes swelling of the tongue and loss of speech.

Careful, stepwise application of these rules will lead to the correct assignment of oxidation numbers in almost all cases. In the rare instances that the rules fail, oxidation numbers are assigned by referring directly to the electronegativities of covalently bonded atoms.

EXAMPLE 18.1 Oxidation Numbers in a Polyatomic Ion

What are the oxidation numbers of each of the atoms in the hydrogen oxalate ion, $HC_2O_4^-$?

Analysis All oxidation-number problems should be approached through a *stepwise* application of the rules in Table 18.1. You must know the charge of a polyatomic ion in order to assign oxidation numbers to its atoms.

Work Rule 4 assigns +1 to each H atom, and Rule 5 assigns −2 to each O atom. No rule specifically covers carbon, so its oxidation number must be obtained by difference through application of Rule 1. Set up a systematic list for the species, representing the unknown oxidation number by *x*.

Atom Type	Number of Atoms		Oxidation Number			
H	1	×	+1	=	+1	
O	4	×	−2	=	−8	
C	2	×	x	=	$2x$	
	Charge on species (= total)			=	−1	(Rule 6)

The resulting algebraic equation is solved for *x*.

$$(+1) + (-8) + (2x) = (-1)$$

$$2x = 6$$

$$x = 3$$

The oxidation number of carbon in the $HC_2O_4^-$ ion is +3.

Exercise

What are the oxidation numbers of the atoms in potassium permanganate, $KMnO_4$?

Answer: K, +1; Mn, +7; O, −2.

Elements with Variable Oxidation Numbers

Apart from fluorine and the metals of Groups 1A and 2A, most elements can exist in more than one oxidation state. These elements are often called **variable-valence elements.** In the compounds H_2S, $K_2S_2O_3$, SO_2, and H_2SO_4, for example, sulfur has oxidation numbers of −2, +2, +4, and +6, respectively. The oxidation number of mercury is +1 in Hg_2Cl_2 and +2 in $HgCl_2$. Figure 18.6 lists the common oxidation numbers of the elements.

1A	2A	3B	4B	5B	6B	7B	8B	8B	8B	1B	2B	3A	4A	5A	6A	7A	8A
H +1 −1																	**He**
Li +1	**Be** +2											**B** +3	**C** +4 +2 0 −1 −2 −3 −4	**N** +5 +4 +3 +2 +1 −3	**O** +2 $-\frac{1}{2}$ −1 −2	**F** −1	**Ne**
Na +1	**Mg** +2											**Al** +3	**Si** +4 −4	**P** +5 +3 −3	**S** +6 +4 +2 −2	**Cl** +7 +5 +3 +1 −1	**Ar**
K +1	**Ca** +2	**Sc** +3	**Ti** +4 +3 +2	**V** +5 +4 +3 +2	**Cr** +6 +5 +4 +3 +2	**Mn** +7 +6 +4 +3 +2	**Fe** +3 +2	**Co** +3 +2	**Ni** +2	**Cu** +2 +1	**Zn** +2	**Ga** +3	**Ge** +4 −4	**As** +5 +3 −3	**Se** +6 +4 −2	**Br** +5 +3 +1 −1	**Kr** +4 +2
Rb +1	**Sr** +2	**Y** +3	**Zr** +4	**Nb** +5 +4	**Mo** +6 +4 +3	**Tc** +7 +6 +4	**Ru** +8 +6 +4 +3	**Rh** +4 +3 +2	**Pd** +4 +2	**Ag** +1	**Cd** +2	**In** +3	**Sn** +4 +2	**Sb** +5 +3 −3	**Te** +6 +4 −2	**I** +7 +5 +1 −1	**Xe** +6 +4 +2
Cs +1	**Ba** +2	**La** +3	**Hf** +4	**Ta** +5	**W** +6 +4	**Re** +7 +6 +4	**Os** +8 +4	**Ir** +4 +3	**Pt** +4 +2	**Au** +3 +1	**Hg** +2 +1	**Tl** +3 +1	**Pb** +4 +2	**Bi** +5 +3	**Po** +2	**At** −1	**Rn**

Figure 18.6 The common oxidation numbers of the elements.

Since an atom acquires a positive oxidation number by loss of valence electrons, its maximum possible oxidation number is the number of valence electrons in the neutral atom. This number is a *periodic* property, varying regularly with atomic number. The largest (positive) oxidation number possible for an atom of any element is its group number. Although this aspect of oxidation numbers is contained in the information of Figure 18.6, it is more strikingly illustrated in Figure 18.7. The

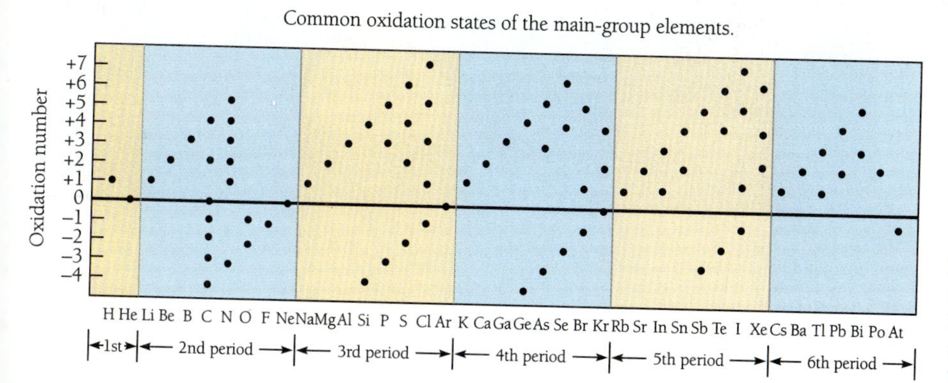

Figure 18.7 When the commonly observed oxidation numbers of the main-group elements are plotted against atomic number, the periodic trends are readily seen.

minimum (largest negative) oxidation number an element can have is its group number minus 8. Note that metallic elements do not commonly exhibit negative oxidation numbers.

Variable oxidation number is particularly evident among the transition metals. Iron, for example, forms many ionic compounds as Fe^{2+} and many as Fe^{3+}. Sometimes a particular color is associated with each oxidation number of a transition element (Figure 18.8), but in general, the anion also influences the color of a transition metal compound.

(a)

(b)

Figure 18.8 (a) Oxidation number has a striking effect on the color of manganese compounds. Salts of the Mn^{2+} ion (oxidation number +2) tend to be pale pink. MnO_2 (oxidation number +4) is an insoluble black solid. Salts of MnO_4^{2-} (oxidation number +6) are usually green, and salts of the MnO_4^- ion (oxidation number +7) have a strong purple color. (b) Oxidation state alone does not determine the color of a compound. The oxidation number of chromium is +3 in both $Cr(NO_3)_3$ (purple) and $CrCl_3$ (green), and it is +6 in both K_2CrO_4 (yellow) and $K_2Cr_2O_7$ (orange).

The best strategy for assigning oxidation numbers in an ionic compound of a variable-valence metal is to treat the anion and cation separately. In order to do this, you must be able to recognize the formulas of the common polyatomic anions, *and know their charges.*

A listing of the common polyatomic ions is given in Table 2.3, page 55.

EXAMPLE 18.2 Ionic Compounds of Variable-Valence Elements

What are the oxidation numbers of (a) iron and (b) sulfur in the ionic compound $FeSO_3$?

Analysis Since the compound contains more than one element (Fe and S) not covered by the rules, their oxidation numbers cannot be found from the requirement that the total must be zero in a neutral compound.

Plan Identify the anion in the formula, and, using the known charge of the anion, determine the cation charge. Treat the anion and cation separately, and apply the rules stepwise to determine the oxidation number(s) of the element(s) in them.

Work The cation appears first in the formula of an ionic compound, so the ions in $FeSO_3$ are

Cation Fe^{x+}

Anion SO_3^{x-}

The atom grouping SO_3^{x-} is recognizable as the doubly charged *sulfite* ion, SO_3^{2-}.

a. The compound stoichiometry (1 : 1) requires that the cation and anion charges be equal and opposite, so the cation is Fe^{2+}. Since this is a monatomic ion, the oxidation number is equal to the charge, +2.
b. The oxidation number of oxygen in the anion is −2, and that of sulfur is found by difference.

Atom Type	Number of Atoms		Oxidation Number		
O	3	·	−2	=	−6
S	1	·	x	=	x
	Charge on species (= total)			=	−2

$$x + (-6) = -2$$

$$x = -2 + 6 = +4$$

The oxidation number of sulfur in SO_3^{2-}, and therefore in $FeSO_3$ and all other compounds of SO_3^{2-}, is +4.

Exercise

What are the oxidation numbers of N and Cr in $(NH_4)_2Cr_2O_7$?

Answer: N, −3; Cr, +6.

18.3 BALANCING REDOX EQUATIONS

There are several general methods available for balancing redox equations. We will discuss two approaches, the *oxidation-number method* and the *ion-electron method.*

The Oxidation-Number Method

The most direct method of balancing redox equations is called the **oxidation-number method.** For those who have learned to assign oxidation numbers quickly and reliably, this method is faster and easier than the ion-electron method. The first step is the assignment of oxidation numbers. In subsequent steps, the stoichiometric coefficients in the equation are adjusted so that the increase in oxidation number of the oxidized species is equal to the decrease in oxidation number of the reduced species. We illustrate the method by balancing the equation describing the high temperature oxidation of ammonia

$$NH_3 + O_2 \longrightarrow N_2 + H_2O$$

Oxidation is accompanied by an *increase* in oxidation number.

1. Assign oxidation numbers to all elements in all reactants and products:

$$\overset{-3\ +1}{NH_3} + \overset{0}{O_2} \longrightarrow \overset{0}{N_2} + \overset{+1\ -2}{H_2O}$$

By looking for *changes* in oxidation numbers, determine which element is oxidized and which is reduced: nitrogen is oxidized from −3 in NH_3 to 0 in N_2, and oxygen is reduced from 0 in O_2 to −2 in H_2O. Adjust stoichiometric coefficients if necessary to ensure that these atoms, the *redox atoms,* are individually balanced.

$$2\ NH_3 + O_2 \longrightarrow N_2 + 2\ H_2O$$

2. Determine the values of the total increase and decrease in oxidation number. The increase is equal to the number of electrons lost in the oxidation. Similarly, the decrease in oxidation number is equal to the number of electrons gained in the reduction.

$2(-3 \longrightarrow 0) = +6$ — 6 electrons lost in oxidation

$$2\ \overset{-3}{N}H_3 + \overset{0}{O_2} \longrightarrow \overset{0}{N_2} + 2\ H_2\overset{-2}{O}$$

$2(0 \longrightarrow -2) = -4$ — 4 electrons gained in reduction

3. Using the least common multiple, adjust the stoichiometric coefficients of the species containing the redox atoms so as to make the total increase in oxidation number in the oxidation equal to the total decrease in oxidation number

in the reduction. Since each unit change in oxidation number corresponds to the transfer of one electron, this step ensures that the total number of electrons lost in oxidation equals the total number gained in reduction.

The least common multiple of 4 and 6 is 12, so we need to multiply the coefficient of NH_3 by 2 and that of O_2 by 3 in order to equalize the number of electrons lost and gained. We must also multiply the coefficients of the products by these factors to maintain the atom balance. The result is

$$\overset{-3}{4\,NH_3} + \overset{0}{3\,O_2} \longrightarrow \overset{0}{2\,N_2} + 6\,H_2\overset{-2}{O}$$

$4(-3 \longrightarrow 0) = +12$ — 12 electrons lost in oxidation

$6(0 \longrightarrow -2) = -12$ — 12 electrons gained in reduction

A check shows that the numbers of atoms of each element are the same on both sides of the equation. That is, the equation is now balanced.

4. Although it is not so in this case, sometimes there are additional product or reactant species that are not balanced by the first three steps. These must be balanced by inspection. If this further balancing is necessary, it must be done *without changing the coefficients of the oxidized and reduced species.* This step is illustrated in the following example.

EXAMPLE 18.3 Balancing by the Oxidation-Number Method

One laboratory procedure suitable for the production of small amounts of chlorine is the permanganate oxidation of chloride ion. Permanganate ion is usually reduced to Mn^{2+} when, as in this reaction, it is used as an oxidant in acid solution. Write a balanced equation for the reaction of potassium permanganate and sodium chloride in dilute sulfuric acid. The products of the reaction are gaseous chlorine, water, and the dissolved sulfate salts of sodium, potassium, and manganese(II).

Analysis/Plan Write the (unbalanced) equation from the given description, and balance it by the oxidation-number method outlined previously.

Work The unbalanced equation is

$$NaCl + KMnO_4 + H_2SO_4 \longrightarrow Cl_2 + MnSO_4 + Na_2SO_4 + K_2SO_4 + H_2O$$

1. The oxidation numbers are

$$\overset{+1}{Na}\overset{-1}{Cl} + \overset{+1}{K}\overset{+7}{Mn}\overset{-2}{O_4} + \overset{+1}{H_2}\overset{+6}{S}\overset{-2}{O_4} \longrightarrow \overset{0}{Cl_2} + \overset{+2}{Mn}\overset{+6}{S}\overset{-2}{O_4} + \overset{+1}{Na_2}\overset{+6}{S}\overset{-2}{O_4} + \overset{+1}{K_2}\overset{+6}{S}\overset{-2}{O_4} + \overset{+1}{H_2}\overset{-2}{O}$$

so the *redox atoms* are Mn and Cl. Mn is balanced, and balance of Cl is achieved by giving NaCl a coefficient of 2.

$$2\,Na\overset{-1}{Cl} + K\overset{+7}{Mn}O_4 + H_2SO_4 \longrightarrow \overset{0}{Cl_2} + \overset{+2}{Mn}SO_4 + Na_2SO_4 + K_2SO_4 + H_2O$$

2. The total changes in oxidation number are

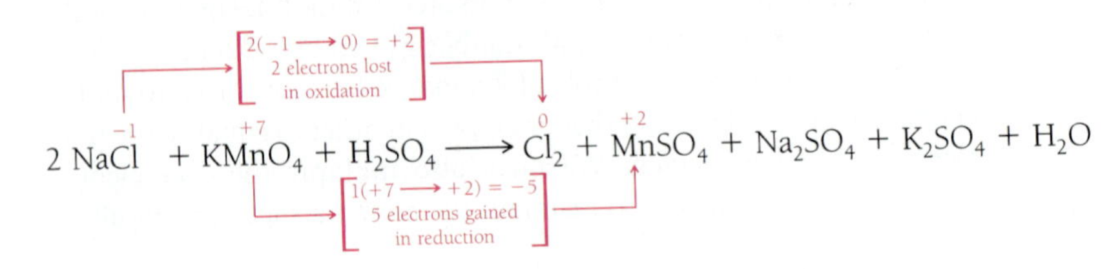

3. When the oxidation number changes are equalized by adjusting the stoichiometric coefficients, the equation becomes

10(−1 → 0) = +10
10 electrons lost
in oxidation

$$10\ \overset{-1}{\text{NaCl}} + 2\ \text{K}\overset{+7}{\text{Mn}}\text{O}_4 + \text{H}_2\text{SO}_4 \longrightarrow 5\ \overset{0}{\text{Cl}}_2 + 2\ \overset{+2}{\text{Mn}}\text{SO}_4 + \text{Na}_2\text{SO}_4 + \text{K}_2\text{SO}_4 + \text{H}_2\text{O}$$

2(+7 → +2) = −10
10 electrons gained
in reduction

4. The non-redox elements are balanced by inspection, but the coefficients of $NaCl$, $KMnO_4$, Cl_2, and $MnSO_4$ *must not be changed.*

   10 Na atoms on the left requires a coefficient of 5 for Na_2SO_4:

$$10\ \overset{-1}{\text{NaCl}} + 2\ \text{K}\overset{+7}{\text{Mn}}\text{O}_4 + \text{H}_2\text{SO}_4 \longrightarrow 5\ \overset{0}{\text{Cl}}_2 + 2\ \overset{+2}{\text{Mn}}\text{SO}_4 + 5\ \text{Na}_2\text{SO}_4 + \text{K}_2\text{SO}_4 + \text{H}_2\text{O}$$

   K and H are balanced, so no adjustment is required. Balance of S requires a coefficient of 8 for H_2SO_4:

$$10\ \overset{-1}{\text{NaCl}} + 2\ \text{K}\overset{+7}{\text{Mn}}\text{O}_4 + 8\ \text{H}_2\text{SO}_4 \longrightarrow 5\ \overset{0}{\text{Cl}}_2 + 2\ \overset{+2}{\text{Mn}}\text{SO}_4 + 5\ \text{Na}_2\text{SO}_4 + \text{K}_2\text{SO}_4 + \text{H}_2\text{O}$$

   The change in the coefficient of H_2SO_4 has disrupted the balance of hydrogen. Balance is restored with a coefficient of 8 for H_2O:

$$10\ \overset{-1}{\text{NaCl}} + 2\ \text{K}\overset{+7}{\text{Mn}}\text{O}_4 + 8\ \text{H}_2\text{SO}_4 \longrightarrow 5\ \overset{0}{\text{Cl}}_2 + 2\ \overset{+2}{\text{Mn}}\text{SO}_4 + 5\ \text{Na}_2\text{SO}_4 + \text{K}_2\text{SO}_4 + 8\ \text{H}_2\text{O}$$

   Oxygen is in balance, so no further changes are needed.

Check Each side has 10 Na's, 10 Cl's, 2 K's, 2 Mn's, 16 H's, 8 S's, and 40 O's.

Exercise

In basic solution, permanganate ion is often reduced to $MnO_2(s)$. Use the oxidation number method to balance the equation $KMnO_4 + Na_2SO_3 + H_2O \rightarrow MnO_2 + Na_2SO_4 + KOH$.

Answer: $2\ KMnO_4 + 3\ Na_2SO_3 + H_2O \rightarrow 2\ MnO_2 + 3\ Na_2SO_4 + 2\ KOH$.

The oxidation-number method is quite suitable for balancing equations in which the complete list of reactants and products is known. This is not always the case, as we will see in the following section.

Molecular Equations, Spectator Ions, and Net Ionic Equations

The reaction of aqueous iron(II) chloride with cerium(IV) chloride follows Equation (18-6).

$$FeCl_2(aq) + CeCl_4(aq) \longrightarrow FeCl_3(aq) + CeCl_3(aq) \qquad (18\text{-}6)$$

This equation is in a sense misleading, because it fails to indicate that ionic compounds, when dissolved in water, exist as separate anions and cations. Equations like (18-6) are called **molecular equations,** since all products and reactants are shown as neutral compounds rather than dissociated ions.

Molecular equations obscure the fact that *strong electrolytes,* including strong acids, strong bases, and ionic compounds, are fully dissociated in aqueous solution. An **ionic equation,** which shows strong electrolytes as their component aqueous ions, can be written for each molecular equation. The ionic equation corresponding to Equation (18-6) is

$$Fe^{2+}(aq) + 2\ Cl^-(aq) + Ce^{4+}(aq) + 4\ Cl^-(aq) \longrightarrow Fe^{3+}(aq) + 3\ Cl^-(aq) + Ce^{3+}(aq) + 3\ Cl^-(aq) \qquad (18\text{-}7a)$$

Since dissolved anions are not associated with any particular cation, this ionic equation can be written more compactly as

$$Fe^{2+}(aq) + Ce^{4+}(aq) + 6\ Cl^-(aq) \longrightarrow Fe^{3+}(aq) + Ce^{3+}(aq) + 6\ Cl^-(aq) \qquad (18\text{-}7b)$$

The chloride ions, while present in the solution, are unchanged in this reaction. All such ions, that undergo no change in a reaction, are called **spectator ions.** No essential information is lost when spectator ions are omitted from a chemical equation. An equation from which the spectator ions have been omitted is called a **net ionic equation.** The state designation (aq) is often omitted, because *all* ions appearing in a net ionic equation are in solution. The net ionic equation corresponding to Equation (18-7) is

Spectator ions and net ionic equations in acid-base reactions were discussed in Section 15.2.

$$Fe^{2+} + Ce^{4+} \longrightarrow Fe^{3+} + Ce^{3+} \qquad (18\text{-}8)$$

EXAMPLE 18.4 Writing Ionic Equations

Write the following (balanced) molecular equation first as (a) an ionic equation, then (b) as a *net* ionic equation.

$$2\ KMnO_4(aq) + 5\ K_2C_2O_4(aq) + 16\ HCl(aq) \longrightarrow 2\ MnCl_2(aq) + 8\ H_2O(\ell) + 10\ CO_2(g) + 12\ KCl(aq)$$

Analysis/Relationship The ionic equation is reached by knowing which species are strong electrolytes, that is, which ones dissociate on dissolution. Removal of spectator ions yields the *net* ionic equation.

Plan

a. Rewrite the equation, showing all strong electrolytes as separate ions.
b. Identify and remove all spectator ions from the equation.

Work

a. The strong electrolytes are either ionic compounds, strong acids, or strong bases.

- $KMnO_4$ is ionic, and should be written as $K^+ + MnO_4^-$
- $K_2C_2O_4$ is ionic, and should be written as $2\ K^+ + C_2O_4^{2-}$
- HCl is a strong acid, and should be written as $H^+ + Cl^-$
- $MnCl_2$ is ionic, and should be written as $Mn^{2+} + 2\ Cl^-$
- H_2O is molecular, and does not dissociate to a significant extent
- CO_2 is molecular, and does not dissociate
- KCl is ionic, and should be written as $K^+ + Cl^-$

The equation is rewritten. At this stage the designation (aq) may be dropped, since it applies to all ions.

$$2[K^+ + MnO_4^-] + 5[2\ K^+ + C_2O_4^{2-}] + 16[H^+ + Cl^-] \longrightarrow 2[Mn^{2+} + 2\ Cl^-] + 8\ H_2O(\ell) + 10\ CO_2(g) + 12[K^+ + Cl^-]$$

The indicated multiplications are carried out:

$$2\ K^+ + 2\ MnO_4^- + 10\ K^+ + 5\ C_2O_4^{2-} + 16\ H^+ + 16\ Cl^- \longrightarrow 2\ Mn^{2+} + 4\ Cl^- + 8\ H_2O(\ell) + 10\ CO_2(g) + 12\ K^+ + 12\ Cl^-$$

Next, coefficients of the same species are combined:

Left Side	$2\ K^+ + 10\ K^+ = 12\ K^+$
Right Side	$4\ Cl^- + 12\ Cl^- = 16\ Cl^-$

The result is the *ionic equation:*

$$12\ K^+ + 2\ MnO_4^- + 5\ C_2O_4^{2-} + 16\ H^+ + 16\ Cl^- \longrightarrow 2\ Mn^{2+} + 8\ H_2O(\ell) + 10\ CO_2(g) + 12\ K^+ + 16\ Cl^-$$

b. Spectator ions are cancelled out to obtain the *net* ionic equation.

$$2\ MnO_4^- + 5\ C_2O_4^{2-} + 16\ H^+ \longrightarrow 2\ Mn^{2+} + 8\ H_2O(\ell) + 10\ CO_2(g)$$

Check The total of the ionic charges must be the same on both sides.

$$2(-1) + 5(-2) + 16(+1) = 2(+2)$$

Exercise

Write the net ionic equation corresponding to the following molecular equation.

$$4\ Cr_2O_3(s) + 8\ H_2O(\ell) + 3\ HClO_4(aq) \longrightarrow 3\ HCl(aq) + 8\ H_2CrO_4(aq)$$

Answer: $4\ Cr_2O_3(s) + 8\ H_2O(\ell) + 3\ ClO_4^- \rightarrow 3\ Cl^- + 16\ H^+ + 8\ CrO_4^{2-}$.

$Cr_2O_3(s)$ is not shown as separate ions because it is not a dissolved substance.

Since net ionic equations are simpler and easier to interpret than molecular equations, the ability to write a molecular equation in net ionic form is useful. However, it is not necessary to be able to write a given net ionic equation in molecular form. Problems in stoichiometry, which require a balanced equation as a starting point, can be solved as easily with a net ionic equation as with a molecular equation. One such problem appears in Example 18.9, page 790.

The Ion-Electron Balancing Method

Molecular chemical equations are in balance when the *mass,* or the number of atoms of each element, is the same on both sides of the equation. Net ionic equations have the additional requirement that the total *electrical charge* must be the same on both sides of the equation. A balanced equation is the chemist's expression of the laws of conservation of mass and conservation of charge. The **ion-electron method** is a very useful procedure for balancing net ionic equations. In this method an unbalanced equation is written as separate half-reactions, which are separately balanced and then recombined to give the balanced overall equation.

Molecular redox equations are more often balanced by the oxidation-number method discussed previously.

The ion-electron procedure involves the following steps:

1. Separate the given reaction into half-reactions.
2. Balance each half-reaction separately:
   a. Balance all atoms other than O and H.
   b. Add H_2O to the O-deficient side to balance O.
   c. Add H^+ to the H-deficient side as needed to balance H; this completes the mass balance.
   d. Add electrons as needed to balance the charge.
3. Adjust the coefficients in one or both half-reactions so that each shows the same number of electrons, and add the two to get the balanced overall net ionic equation.

The big advantage of the ion-electron method is that, except for reactions in which the oxidation number of H or O changes, redox equations can be balanced *without reference to oxidation numbers.* We illustrate the procedure by balancing Equation (18-9).

Later we will show how to modify the procedure in order to use it for reactions in which the oxidation number of H or O changes.

$$H^+ + ClO_3^- + NO_2^- \longrightarrow Cl_2(g) + NO_3^- + H_2O \quad (18\text{-}9)$$

1. Separate the equation into two half-reactions. Choose one of the reactant species that contains an element other than H and O. Write it, together with the product species that contains the same element, as one of the half-reactions.

$$ClO_3^- \longrightarrow Cl_2$$

Ignoring H^+, OH^-, and H_2O (if any of them appear in the equation), write the remaining species as the other half-reaction.

$$NO_2^- \longrightarrow NO_3^-$$

Oxidation numbers can also be used to identify the half-reactions. The oxidation half-reaction includes the atom whose oxidation number *increases,* while the reduction half-reaction involves an oxidation number *decrease.*

2a. Balance all atoms *other than O and H* in each half-reaction by changing stoichiometric coefficients as needed. The chlorate/chlorine equation is balanced by adding the coefficient 2. The nitrite/nitrate half-reaction is already balanced, and needs no adjustment.

Never change the *subscripts* when balancing an equation.

$$2\ ClO_3^- \longrightarrow Cl_2 \qquad NO_2^- \longrightarrow NO_3^-$$

Water is a participant in many aqueous redox reactions, especially those involving oxoanions.

The redox atoms are now balanced in both of these half-reactions, but oxygen atoms are not. An imbalance in oxygen at this point indicates that water is a participant in the reaction. This chemical fact is reflected in the next step of the ion-electron method.

2b. Balance the oxygen in each half-reaction, if necessary, by adding water molecule(s) to whichever side of the equation is deficient in O atoms.

ClO_3^-/Cl_2 Half-Reaction $\quad 2\ ClO_3^- \longrightarrow Cl_2 + 6\ \mathit{H_2O}$

NO_2^-/NO_3^- Half-Reaction $\quad \mathit{H_2O} + NO_2^- \longrightarrow NO_3^-$

This procedure usually results in an imbalance of hydrogen, which is repaired in the next step.

2c. Balance the hydrogen in each half-reaction, if necessary, by adding H^+ *ion(s)* to whichever side of the equation is deficient in H atoms.

ClO_3^-/Cl_2 Half-Reaction $\quad \mathit{12\ H^+} + 2\ ClO_3^- \longrightarrow Cl_2 + 6\ H_2O$

NO_2^-/NO_3^- Half-Reaction $\quad H_2O + NO_2^- \longrightarrow NO_3^- + \mathit{2\ H^+}$

At this stage all elements should appear in equal numbers on both sides of the equations. That is, *mass balance* has been achieved for both half-reactions. The next step achieves the necessary *charge* balance, by including the electrons that are lost or gained by oxidation or reduction.

If, at this point, the charge is the same on both sides of the equation, the reaction is not a redox reaction.

This step involves the charge carried by the ions, not the oxidation number of atoms in the ion.

2d. Balance the charge, in each half-reaction, by adding electron(s) to whichever side has the larger total positive charge or the smaller total negative charge. The goal is to have the same total electrical charge on both sides of the half-reaction. Do not try to bring the total charge to zero, and do not try to give each half-reaction the same charge. Most especially, *do not confuse oxidation numbers of elements in a polyatomic ion with the electrical charge on the ion.*

NO_2^-/NO_3^- Half-Reaction $\quad H_2O + NO_2^- \longrightarrow NO_3^- + 2\ H^+$

Electrical Charges $\quad (0) + (-1) \longrightarrow (-1) + 2(+1)$

$$(-1) \longrightarrow (+1)$$

The charge is balanced by the addition of (-2), that is, 2 electrons, to the right side of the half-reaction.

$$H_2O + NO_2^- \longrightarrow NO_3^- + 2\ H^+ + 2\ e^-$$

Electrical Charges $\quad (0) + (-1) \longrightarrow (-1) + 2(+1) + 2(-1)$

Balanced Total Charge $\quad (-1) \longrightarrow (-1)$

ClO_3^-/Cl_2 Half-Reaction $\quad 12\ H^+ + 2\ ClO_3^- \longrightarrow Cl_2 + 6\ H_2O$

Electrical Charges $\quad (+12) + 2(-1) \longrightarrow 0 + 0$

$$(+10) \longrightarrow 0$$

The charge is balanced by the addition of (-10), that is, 10 electrons, to the left side of the half-reaction.

$$\mathit{10\ e^-} + 12\ H^+ + 2\ ClO_3^- \longrightarrow Cl_2 + 6\ H_2O$$

Balanced Total Charge $\quad 0 \longrightarrow 0$

The balancing of the half-reactions is now complete, and it should be checked. One of the half-reactions is an oxidation, involving electron loss. The lost electron(s) must appear as a product, on the right side of the equation. The other *must* be a reduction half-reaction, showing electrons on the left. The *number* of electrons in the two reactions need not be the same at this stage, but, unless one half-reaction shows electrons on the left, and the other shows them on the right, a mistake has been made. Each reaction should also be checked to make sure that the number of atoms of each element, and the total electrical charge, are the same on both sides.

OIL RIG

There can be no oxidation without an accompanying reduction.

3. The final step in the procedure is to add the two equations and cancel out redundant species. This cannot be done, however, until the number of electrons has been *equalized,* that is, made the same in both half-reactions. One or both of the equations must be multiplied by a small whole number so that the number of electrons is the same in each. In our example, the electron coefficients are 2 and 10. The least common multiple of these numbers is 10, and the coefficients of the NO_2^-/NO_3^- half-reaction must be multiplied by 5. The other half-reaction need not be changed in this example.

Oxidation Half-Reaction $$5\,H_2O + 5\,NO_2^- \longrightarrow 5\,NO_3^- + 10\,H^+ + 10\,e^-$$

Reduction Half-Reaction $$10\,e^- + 12\,H^+ + 2\,ClO_3^- \longrightarrow Cl_2 + 6\,H_2O$$

Adding these reactions and cancelling the redundant species gives the complete, balanced, net ionic equation.

$$\overset{0}{\cancel{5}}H_2O + 5\,NO_2^- + \overset{0}{\cancel{10}}\,e^- + \overset{2}{\cancel{12}}\,H^+ + 2\,ClO_3^- \longrightarrow$$
$$5\,NO_3^- + \overset{0}{\cancel{10}}\,H^+ + \overset{0}{\cancel{10}}\,e^- + Cl_2 + \overset{1}{\cancel{6}}\,H_2O \qquad \text{or}$$

$$5\,NO_2^- + 2\,H^+ + 2\,ClO_3^- \longrightarrow 5\,NO_3^- + Cl_2(g) + H_2O \qquad (18\text{-}10)$$

EXAMPLE 18.5 The Ion-Electron Method in Acidic Solution

Dichromate ion oxidizes nitrous acid according to the equation $Cr_2O_7^{2-}(aq) + HNO_2(aq) \rightarrow Cr^{3+}(aq) + NO_3^-(aq)$. Use the ion-electron method to balance this equation.

Plan Follow the steps outlined previously.

Work

1. The half-reactions are

$$Cr_2O_7^{2-} \longrightarrow Cr^{3+} \qquad \text{and} \qquad HNO_2 \longrightarrow NO_3^-$$

The state designation (aq) is unnecessary, since all reactants and products in this reaction are aqueous species.

2a. Balance all atoms other than H and O, to get

$$Cr_2O_7^{2-} \longrightarrow 2\,Cr^{3+} \qquad \text{and} \qquad HNO_2 \longrightarrow NO_3^-$$

2b. Add water to balance the oxygen.

$$Cr_2O_7^{2-} \longrightarrow 2\,Cr^{3+} + 7\,H_2O \qquad \text{and} \qquad H_2O + HNO_2 \longrightarrow NO_3^-$$

2c. Add H^+ to balance the hydrogen.

$$14\,H^+ + Cr_2O_7^{2-} \longrightarrow 2\,Cr^{3+} + 7\,H_2O \quad \text{and}$$

$$H_2O + HNO_2 \longrightarrow NO_3^- + 3\,H^+$$

2d. Add electrons to balance the charge.

Reduction $$6\,e^- + 14\,H^+ + Cr_2O_7^{2-} \longrightarrow 2\,Cr^{3+} + 7\,H_2O$$

Oxidation $$H_2O + HNO_2 \longrightarrow NO_3^- + 3\,H^+ + 2\,e^-$$

3. Multiplication of the oxidation half-reaction by three equalizes the electron count, and the overall redox reaction is

$$6\,e^- + 14\,H^+ + Cr_2O_7^{2-} + 3\,H_2O + 3\,HNO_2 \longrightarrow 2\,Cr^{3+} + 7\,H_2O + 3\,NO_3^- + 9\,H^+ + 6\,e^-$$

which, after cancelling redundant species, is

$$5\,H^+ + Cr_2O_7^{2-} + 3\,HNO_2 \longrightarrow 2\,Cr^{3+} + 4\,H_2O + 3\,NO_3^-$$

Check Each side has a total charge of +3, 8 H's, 2 Cr's, 13 O's, and 3 N's.

Exercise

Copper reacts with nitric acid but not with hydrochloric acid. Nitric acid is not only a strong acid, but it is also a powerful oxidant. Write a balanced net ionic equation for the reaction, $Cu(s) + NO_3^- + H^+ \rightarrow Cu^{2+}(aq) + NO(g)$.

The reaction of metallic copper with nitric acid produces brown fumes of NO_2. The solution takes on the green color of the other reaction product, copper nitrate.

Answer: $3\,Cu(s) + 2\,NO_3^- + 8\,H^+ \rightarrow 3\,Cu^{2+}(aq) + 2\,NO(g) + 4\,H_2O$.

Redox Reactions in Acid or Basic Solutions

The appearance of H^+ as a participating reactant in Equation (18-10) is an additional indication of the role of water and its ionization products in aqueous redox reactions. Moreover, it shows that the reaction takes place in acidic solution. Under basic conditions, however, the dominant ionic component of the solvent is OH^-. It is therefore inappropriate to show H^+ as a participant in a redox reaction in basic solution.

The participation of OH^- rather than H^+ in an aqueous redox reaction can be indicated by modifying the equation. Either or both of the half-reactions, or the overall equation, can be changed. For example, consider again the nitrite reduction of chlorate ion. Begin with the balanced equation,

$$5\ NO_2^- + 2\ H^+ + 2\ ClO_3^- \longrightarrow 5\ NO_3^- + Cl_2 + H_2O \qquad (18\text{-}10)$$

Add, to the side having the H^+ ions, an equal number of OH^- ions. Add the same number of OH^- ions to the other side, so as to maintain the equation in balance.

$$5\ NO_2^- + 2\ H^+ + 2\ \mathit{OH}^- + 2\ ClO_3^- \longrightarrow 5\ NO_3^- + Cl_2(g) + H_2O + 2\ \mathit{OH}^-$$

Write "$n\ H_2O$" in place of "$n\ H^+ + n\ OH^-$" and cancel the redundant H_2O molecules.

$$5\ NO_2^- + \overset{1}{\cancel{2}}\ \mathit{H_2O} + 2\ ClO_3^- \longrightarrow 5\ NO_3^- + Cl_2 + \overset{0}{\cancel{H_2O}} + 2\ OH^- \qquad \text{or}$$

$$5\ NO_2^- + H_2O + 2\ ClO_3^- \longrightarrow 5\ NO_3^- + Cl_2 + 2\ OH^- \qquad (18\text{-}11)$$

Equations (18-10) and (18-11) are both correctly balanced. Both are *overall* reactions, and neither describes the mechanism. Equation (18-10) is used when the solution is acidic, and Equation (18-11) is used when the solution is basic. Many redox reactions take place only in acidic solution, or only in basic solution, but not in both.

In summary, to balance a redox equation by the ion-electron method, follow the given stepwise procedure. The equation obtained by steps 1 through 3 is a net ionic equation that assumes that the reaction takes place in acid solution. If the reaction is known to take place in basic solution, carry out the two additional steps.

EXAMPLE 18.6 The Ion-Electron Method in Basic Solution

Write a balanced net ionic equation for the reaction

$$MnO_4^- + N_2O_4(g) \longrightarrow MnO_2(s) + NO_3^-$$

which takes place in basic solution.

Plan Following the stepwise procedure, balance the reaction first in acid solution, then convert to basic solution.

Work

1. Separate into half-reactions.

$$MnO_4^- \longrightarrow MnO_2 \qquad N_2O_4 \longrightarrow NO_3^-$$

It is not necessary to include state designations when balancing an equation, but, if present originally, they should be restored in the final answer.

2a. Balance all elements other than H and O.

$$MnO_4^- \longrightarrow MnO_2 \qquad N_2O_4 \longrightarrow 2\ NO_3^-$$

2b. Balance O with H_2O.

$$MnO_4^- \longrightarrow MnO_2 + 2\ \mathit{H_2O} \qquad 2\ \mathit{H_2O} + N_2O_4 \longrightarrow 2\ NO_3^-$$

2c. Balance H with H^+.

$$4\ \mathit{H}^+ + MnO_4^- \longrightarrow MnO_2 + 2\ H_2O$$

$$2\ H_2O + N_2O_4 \longrightarrow 2\ NO_3^- + 4\ \mathit{H}^+$$

2d. Balance charge with e^-.

$$3\,e^- + 4\,H^+ + MnO_4^- \longrightarrow MnO_2 + 2\,H_2O$$

$$2\,H_2O + N_2O_4 \longrightarrow 2\,NO_3^- + 4\,H^+ + 2\,e^-$$

Check One reaction shows electron loss, the other shows electron gain.

3. Equalize the electron count.

$$2 \cdot (3\,e^- + 4\,H^+ + MnO_4^- \longrightarrow MnO_2 + 2\,H_2O) \quad \text{or}$$

$$6\,e^- + 8\,H^+ + 2\,MnO_4^- \longrightarrow 2\,MnO_2 + 4\,H_2O$$

$$3 \cdot (2\,H_2O + N_2O_4 \longrightarrow 2\,NO_3^- + 4\,H^+ + 2\,e^-) \quad \text{or}$$

$$6\,H_2O + 3\,N_2O_4 \longrightarrow 6\,NO_3^- + 12\,H^+ + 6\,e^-$$

Add the half-reactions,

$$6\,e^- + 8\,H^+ + 2\,MnO_4^- + 6\,H_2O + 3\,N_2O_4 \longrightarrow 2\,MnO_2 + 4\,H_2O + 6\,NO_3^- + 12\,H^+ + 6\,e^-$$

and eliminate redundant species to get the overall equation, balanced as if it took place in acid solution.

$$2\,MnO_4^- + 2\,H_2O + 3\,N_2O_4 \longrightarrow 2\,MnO_2 + 6\,NO_3^- + 4\,H^+$$

Adjust the equation to show OH^- rather than H^+. Add OH^- to both sides

$$4\,OH^- + 2\,MnO_4^- + 2\,H_2O + 3\,N_2O_4 \longrightarrow 2\,MnO_2 + 6\,NO_3^- + 4\,H^+ + 4\,OH^-$$

combine H^+ and OH^- as H_2O

$$4\,OH^- + 2\,MnO_4^- + 2\,H_2O + 3\,N_2O_4 \longrightarrow 2\,MnO_2 + 6\,NO_3^- + 4\,H_2O$$

and cancel the redundant water. The overall equation, now balanced for basic solution and with state designations restored, is

$$4\,OH^- + 2\,MnO_4^- + 3\,N_2O_4(g) \longrightarrow 2\,MnO_2(s) + 6\,NO_3^- + 2\,H_2O$$

Exercise

Balance the following half-reaction, in basic solution.

$$NO_3^- \longrightarrow NH_3$$

Answer: $NO_3^- + 6\,H_2O + 8\,e^- \rightarrow NH_3 + 9\,OH^-$.

As mentioned earlier, the ion-electron procedure we have outlined fails when either O or H undergoes a change in oxidation number. With a slight modification of the procedure, however, the ion-electron method can be used to balance the equation for *any* redox reaction. Two changes are required.

Hydrogen changes its oxidation number in reactions involving H^- and/or H_2, for example $H^- + H_2O \rightarrow H_2 + OH^-$. Oxygen changes its oxidation number in reactions involving O_2^-, O_2^{2-}, and/or O_2, for example $4\,O_2^- + 2\,H_2O \rightarrow 3\,O_2 + 4\,OH^-$.

a. In step 1, use oxidation numbers to separate the reaction participants into oxidation and reduction half-reactions.

b. In step 2, O or H must be balanced *if its oxidation number changes.*

EXAMPLE 18.7 The Ion-Electron Method for Redox Reactions of H or O

Hydride ion (H^-) is a powerful reductant capable of reducing peroxide ion (O_2^{2-}) to OH^- in basic solution. In the process, hydride ion is oxidized to gaseous hydrogen. Write a balanced equation for the reaction.

Analysis Since it is clear from the problem that both O and H undergo oxidation or reduction, the modified procedure must be used.

Plan Write an oxidation half-reaction, in which the only change in oxidation numbers is an increase, and a reduction half-reaction, in which the only change is a decrease in oxidation number. Then follow the stepwise procedure of the ion-electron method.

Work

1. Assign oxidation numbers and note their changes.

$$\overset{-1}{O_2^{2-}} + \overset{-1}{H^-} \longrightarrow \overset{0}{H_2} + \overset{-2+1}{OH^-}$$

The half-reactions are

$$\textbf{Oxidation} \quad \overset{-1}{H^-} \longrightarrow \overset{0}{H_2}$$

$$\textbf{Reduction} \quad \overset{-1}{O_2^{2-}} \longrightarrow \overset{-2}{OH^-}$$

2a. After balancing H in the oxidation reaction and O in the reduction reaction, the equations are

$$\textbf{Oxidation} \quad 2\,H^- \longrightarrow H_2$$

$$\textbf{Reduction} \quad O_2^{2-} \longrightarrow 2\,OH^-$$

2b and 2c. No further balancing of O is needed in either equation. Balance H in the reduction equation in the usual way, by adding H^+ where needed.

$$\textbf{Oxidation} \quad 2\,H^- \longrightarrow H_2$$

$$\textbf{Reduction} \quad 2\,H^+ + O_2^{2-} \longrightarrow 2\,OH^-$$

2d. Charge is balanced with e^-.

$$\textbf{Oxidation} \quad 2\,H^- \longrightarrow H_2 + 2\,e^-$$

$$\textbf{Reduction} \quad 2\,e^- + 2\,H^+ + O_2^{2-} \longrightarrow 2\,OH^-$$

3. The electron count is already equal. The half-reactions are added, and redundant species are eliminated to get the overall equation, balanced as if the reaction took place in acid solution.

$$\textbf{Overall} \quad 2\,H^- + 2\,H^+ + O_2^{2-} \longrightarrow 2\,OH^- + H_2$$

To convert to a basic solution equation, add 2 OH^- to both sides.

$$2\,OH^- + 2\,H^- + 2\,H^+ + O_2^{2-} \longrightarrow 2\,OH^- + H_2 + 2\,OH^-$$

Then combine H^+ and OH^- and write as H_2O. Cancellation of redundant H_2O is not necessary in this case. In final form, the overall equation is

$$2\,H^- + O_2^{2-} + 2\,H_2O \longrightarrow 4\,OH^- + H_2$$

Check Each side has a total charge of −4 and the following atoms: 6 H's, 4 O's.

Lithium hydride, a gray solid, reacts vigorously with water to produce lithium hydroxide and gaseous hydrogen. When a small amount of LiH is dropped into water, the reaction takes place immediately. Phenolphthalein indicator has been added to the water to demonstrate the formation of OH^-.

Exercise

Hydride ion reacts with water to form hydroxide ion and gaseous hydrogen. Write a balanced net ionic equation for this process.

Answer: $H^- + H_2O \rightarrow H_2 + OH^-$

This equation can also be balanced by inspection, which remains a valid procedure for balancing simple equations of any sort.

Auto Oxidation-Reduction

It sometimes happens that a substance functions both as an oxidant and as a reductant, in the same reaction. That is, the substance is oxidizing and reducing itself, in a process called **auto oxidation-reduction.** Such reactions are also called **disproportionation reactions.** The hydrolysis of chlorine and the decomposition of hydrogen peroxide are examples.

$$Cl_2(aq) + H_2O \longrightarrow Cl^- + HClO \qquad (18\text{-}12)$$

$$H_2O_2 \longrightarrow H_2O + O_2 \qquad (18\text{-}13)$$

Such reactions can be balanced by the oxidation-number method, or, once the half-reactions are correctly identified, by the ion-electron method.

EXAMPLE 18.8 Auto Oxidation-Reduction Reactions

Write a balanced net ionic equation for the reaction of chlorine with water, producing hypochlorous acid and chloride ion. Use oxidation numbers to identify the half-reactions, and balance them by the ion-electron method.

Work

1. The oxidation numbers in Equation (18-12) are

$$\overset{0}{Cl_2} + \overset{+1\ -2}{H_2O} \longrightarrow \overset{-1}{Cl^-} + \overset{+1+1-2}{HClO}$$

The oxidation number of chlorine increases from 0 to +1 in HClO, and decreases from 0 to −1 in Cl^-. The half-reactions for these processes are

Oxidation $Cl_2 \longrightarrow HClO$

Reduction $Cl_2 \longrightarrow Cl^-$

2a. Balance for elements other than H and O.

Oxidation $Cl_2 \longrightarrow 2\ HClO$

Reduction $Cl_2 \longrightarrow 2\ Cl^-$

2b. Balance the oxygen by adding H_2O where needed.

Oxidation $2\ H_2O + Cl_2 \longrightarrow 2\ HClO$

Reduction $Cl_2 \longrightarrow 2\ Cl^-$

2c. Balance the hydrogen by adding H^+ where needed.

Oxidation $2\ H_2O + Cl_2 \longrightarrow 2\ HClO + 2\ H^+$

Reduction $Cl_2 \longrightarrow 2\ Cl^-$

2d. Balance the charge by adding electrons where needed.

Oxidation $2\ H_2O + Cl_2 \longrightarrow 2\ HClO + 2\ H^+ + 2\ e^-$

Reduction $2\ e^- + Cl_2 \longrightarrow 2\ Cl^-$

3. Since the same number of electrons appears in each, the half-reactions are added without further change.

$$2\ H_2O + 2\ Cl_2 \longrightarrow 2\ HClO + 2\ H^+ + 2\ Cl^- \quad \text{or}$$

$$H_2O + Cl_2(aq) \longrightarrow HClO(aq) + H^+ + Cl^- \qquad (18\text{-}12a)$$

Exercise

Manganate ion (MnO_4^{2-}) disproportionates readily in neutral or acidic solutions, forming $MnO_2(s)$ and MnO_4^-. Using the ion-electron method, write a balanced equation for this process.

MnO_4^{2-} is the formula of the manganate ion. The formula of the *per*manganate ion is MnO_4^-.

Answer: $3\ MnO_4^{2-} + 4\ H^+ \rightarrow 2\ MnO_4^- + MnO_2(s) + 2\ H_2O$.

18.4 REDOX TITRATIONS

The amount of a reducible substance present in a sample can be determined by titration with a strong reducing agent as the titrant. Similarly, if a sample contains an oxidizable substance, it can be titrated with a strong oxidizing agent. Potassium permanganate is often used in redox titrations. Not only is the MnO_4^- ion a strong oxidant, but also, because it is deeply colored, it serves as its own indicator (Figure 18.9). In the titration of a sample of sulfite ion, for instance, the titration reaction is

$$5\ SO_3^{2-} + 2\ MnO_4^- + 6\ H^+ \longrightarrow 2\ Mn^{2+} + 5\ SO_4^{2-} + 3\ H_2O \qquad (18\text{-}14)$$

As MnO_4^- is added to the SO_3^{2-} sample, the concentrations of Mn^{2+} and SO_4^{2-} build up while that of SO_3^{2-} decreases. No MnO_4^- is present as long as the sample is undertitrated. Both SO_3^{2-} and SO_4^{2-} are colorless, and Mn^{2+} is a very pale pink, unnoticeable in dilute solutions. The permanganate ion is so strongly colored, however, that the solution turns noticeably pink as soon as *one drop* of titrant is added beyond the equivalence point.

Figure 18.9 When $KMnO_4(aq)$ is the titrant in a redox titration, it serves as its own indicator. As soon as one drop of excess $KMnO_4$ is added to this sample of oxalic acid ($H_2C_2O_4$), the characteristic pink color of MnO_4^- signals that the endpoint has been passed.

Calculations in redox titrations are no different from other calculations in stoichiometry. At the equivalence point, the relationship is

$$\text{moles reductant} = (\text{moles oxidant}) \cdot (\text{stoichiometric ratio}) \qquad (18\text{-}15)$$

As usual, the number of moles in a sample is found from its molarity and volume, $n = M \cdot V$.

EXAMPLE 18.9 A Redox Titration

The "breathalyzer" used by police to confirm suspected drunken driving is based on the redox titration of alcohol (ethanol, C_2H_6O) by dichromate ion. The suspect breathes into an apparatus containing $K_2Cr_2O_7$ in acid solution. If there is enough ethanol in the breath to reach the endpoint and reduce all of the dichromate, the color of the solution changes from the characteristic orange of $Cr_2O_7^{2-}$ to the green of Cr^{3+}. In the laboratory, the titration is done more conventionally, using a buret to measure the amount of $K_2Cr_2O_7$. If 43.2 mL of 0.0100 M $K_2Cr_2O_7$ is required to reach the endpoint when a 25.0-mL sample of an ethanol-containing solution is titrated, what is the concentration of ethanol? The product of the oxidation of ethanol is acetaldehyde, C_2H_4O.

Analysis/Plan The starting point of any calculation in stoichiometry is a balanced chemical equation. Once the mole ratio has been obtained from the coefficients of the equation, Equation (18-15) is used to obtain the numerical result. The relationship $n = \text{M} \cdot V$ is used as needed.

Work

1. From the problem statement, the (unbalanced) half-reactions are seen to be

$$Cr_2O_7^{2-} \longrightarrow Cr^{3+} \qquad \text{and} \qquad C_2H_6O \longrightarrow C_2H_4O$$

Either the ion-electron or the oxidation-number method may be used to obtain the overall equation

$$8\ H^+ + 3\ C_2H_6O + Cr_2O_7^{2-} \longrightarrow 2\ Cr^{3+} + 3\ C_2H_4O + 7\ H_2O$$

which indicates the mole ratio

$$3 \text{ moles } C_2H_6O \Longleftrightarrow 1 \text{ mole } Cr_2O_7^{2-} \Longleftrightarrow 1 \text{ mole } K_2Cr_2O_7$$

2. $$\text{moles reductant} = (\text{moles oxidant}) \cdot (\text{stoichiometric ratio}) \qquad (18\text{-}15)$$

$$\text{moles } C_2H_6O = (\text{moles } Cr_2O_7^{2-}) \cdot \left(\frac{3 \text{ mol } C_2H_6O}{1 \text{ mol } Cr_2O_7^{2-}}\right)$$

$$= (0.0432 \text{ L}) \cdot (0.0100 \text{ mol L}^{-1}) \cdot \left(\frac{3 \text{ mol } C_2H_6O}{1 \text{ mol } Cr_2O_7^{2-}}\right)$$

$$= 0.001296 \text{ mol } C_2H_6O$$

3. Since the sample volume was 25.0 mL = 0.0250 L, the concentration of ethanol is

$$n = \text{M} \cdot V$$

$$\text{M} = \frac{n}{V}$$

$$= \frac{0.001296 \text{ mol } C_2H_6O}{0.0250 \text{ L}}$$

$$= 0.0518 \text{ M}$$

A 0.0518 M solution contains approximately 0.24% (by mass) of ethanol, which is well over the legal limit for drunken driving.

Exercise

A 50.0-mL sample of a solution containing an unknown concentration of Na_2SO_3 is titrated with 0.0500 M $KMnO_4$ in basic solution. If 63.2 mL titrant is required to reach the endpoint, what is the concentration of SO_3^{2-} ion in the unknown?

Answer: 0.158 M.

18.5 APPLICATIONS

Oxidation-reduction processes are very common in industry, in the laboratory, and in natural processes. Sulfuric acid is produced by oxidation of sulfur, ammonia is obtained by reduction of nitrogen, and chlorine serves as an oxidant in the manufacture of many useful chemicals. Oxidation produces heat and electrical power. Our drinking water is purified by the addition of oxidants (chlorine or ozone), our wastewater is treated by oxidation, and even in household chemistry we use oxidants (bleach) when washing clothes. Food is utilized by the body (or spoiled, if not consumed) in redox reactions. Green plants produce starch by reduction of CO_2, and the fall colors of trees are developed in redox reactions. A lightning stroke produces NO_2 from atmospheric gases in a redox reaction, and may also ignite a forest fire, one of nature's most spectacular redox processes.

Redox Reactions in Metallurgy

Most metallic elements are found in nature as compounds rather than as pure metals. For instance, tin occurs primarily as SnO_2, and is converted to the metallic state before it is used. This is done in a redox reaction, using carbon (coke) as the reducing agent.

$$SnO_2 + C(s) \xrightarrow[\Delta]{} Sn(\ell) + CO_2(g) \tag{18-16}$$

The reaction is carried out above the melting point of tin (232 °C), and the liquid metal is poured off the reaction mixture in a relatively pure state.

Some elements, like lead and zinc, occur primarily as sulfide ores. These are first converted to oxides by *roasting* (heating in air).

$$2\,MS(s) + 3\,O_2(g) \xrightarrow[\Delta]{} 2\,MO(s) + 2\,SO_2(g) \tag{18-17}$$

$$M = Pb \text{ or } Zn$$

Equation (18-17) is a redox reaction, although it is oxygen and sulfur whose oxidation numbers change, not the metal. Zinc also occurs as the carbonate, which decomposes to the oxide on heating.

$$ZnCO_3(s) \xrightarrow[\Delta]{} ZnO(s) + CO_2(g) \tag{18-18}$$

The oxides of both zinc and lead can be reduced to the metal by heating with carbon. For zinc, the process is carried out above the boiling point (907 °C),

$$2\,ZnO(s) + C(s) \xrightarrow[\Delta]{} 2\,Zn(g) + CO_2(g) \tag{18-19}$$

and the product is in the vapor phase. Very pure metal results from the subsequent deposition of the vapor.

The most common ore of copper is *chalcopyrite*, $CuFeS_2$. Part of the sulfur is oxidized when this ore is roasted, while the copper and iron components form separate compounds.

$$6\ CuFeS_2(s) + 13\ O_2(g) \longrightarrow 3\ Cu_2S(s) + 2\ Fe_3O_4(s) + 9\ SO_2(g) \quad (18\text{-}20)$$

Heated with limestone and silica, the iron component of the roasted ore forms a liquid of variable composition called *slag.*

$$CaCO_3(s) + Fe_3O_4(s) + 2\ SiO_2(s) \xrightarrow[\Delta]{} CO_2(g) + CaO\cdot FeO\cdot Fe_2O_3\cdot(SiO_2)(\ell) \quad (18\text{-}21)$$

Cu_2S is insoluble in slag, and so may be separated from it. When copper(I) sulfide is heated in air, it is reduced directly to the metal. No reaction with carbon or other reducing agent is required.

$$Cu_2S(\ell) + O_2(g) \xrightarrow[\Delta]{} 2\ Cu(\ell) + SO_2(g) \quad (18\text{-}22)$$

The chief ore of mercury is HgS. Like Cu_2S, this compound is reduced directly to the metal on roasting.

$$HgS(s) + O_2(g) \xrightarrow[\Delta]{} Hg(g) + SO_2(g) \quad (18\text{-}23)$$

Some metal ores require reducing agents more powerful than carbon.

$$KCl(\ell) + Na(g) \xrightarrow[\Delta]{} NaCl(\ell) + K(g) \quad (18\text{-}24)$$

Other metal ores are reduced by electrolysis (Section 19.5).

$$2\ NaCl(\ell) \xrightarrow[\text{elect.}]{} 2\ Na(\ell) + Cl_2(g) \quad (18\text{-}25)$$

All these processes, which produce a pure metal from one of its ionic compounds, are redox reactions. In the compounds, the metallic element has a positive oxidation number: +4 in SnO_2, +2 in ZnS and $ZnCO_3$, +1 in KCl, and so on. The metals gain one or more electrons in the reduction process, becoming uncharged atoms whose oxidation number is zero.

Redox Reactions in Living Organisms

Electron-transfer reactions are an important feature of metabolic processes. For instance, energy is produced in the body primarily through the exothermic oxidation of glucose. Reaction (18-26) is an overall reaction, whose mechanism includes dozens of steps.

Chapter 21 contains additional information about energy production and use in the body.

$$C_6H_{12}O_6 + 6\ O_2 \longrightarrow 6\ CO_2 + 6\ H_2O; \qquad \Delta H^\circ = -2800\ \text{kJ} \quad (18\text{-}26)$$

Green plants convert carbon dioxide and water to glucose and oxygen in the process of *photosynthesis,* for which the overall reaction is the reverse of Equation (18-26). In photosynthesis, the oxidation number of carbon is reduced from +4 in CO_2 to zero in $C_6H_{12}O_6$, while the oxidation number of O is increased from −2 in H_2O to zero in O_2. The overall process is endothermic: the required energy comes from sunlight absorbed by the plants. Photosynthesis is a very complicated process, involving many elementary steps and many intermediates. There is a great deal of current research aimed at understanding the fine details of redox reactions in living systems.

CHEMISTRY OLD AND NEW

PHOTOCHROMIC MATERIALS

Like a chameleon, photochromic glass can temporarily change color in response to environmental changes. Most often used in "photoray" eyeglasses that darken in sunlight and turn clear indoors, photochromic glass exploits a reversible photochemical reaction of silver chloride.

Photochromic glass is a "smart" material, whose "brain cells" are tiny crystals of silver chloride or other silver halides. These microcrystals are trapped within the solid structure of ordinary glass, a covalent network solid composed primarily of silica (SiO_2).

At night or indoors, photoray eyeglasses are clear, allowing up to 90% of visible light to pass. But when ultraviolet light from sunlight is absorbed by AgCl, an electron is ejected from a chloride ion.

$$Cl^- + \text{light} \longrightarrow Cl + e^-$$

The liberated electron enters the valence shell of a nearby silver ion, forming a neutral atom of silver.

$$e^- + Ag^+ \longrightarrow Ag$$

In pure samples of AgCl this process is reversible, and the chlorine atom immediately regains its lost electron from the silver atom to form the original silver and chloride ions. Photochromic glass, however, also contains copper(I) ions. These interfere with the process by providing an alternate reaction path for the chlorine atoms.

$$Cu^+ + Cl \longrightarrow Cl^- + Cu^{2+}$$

With no chlorine atoms left to accept electrons, silver atoms begin to accumulate. Eventually there are enough silver atoms to coat the surface of each microcrystal with a thin, opaque film of metallic silver. These coated crystals reflect and absorb up to 80% of the visible light, darkening the lenses of photoray glasses.

But walk back inside (away from ultraviolight light), and the dark color of the glasses gradually fades as Cu^{2+} ions accept electrons from silver atoms.

$$Cu^{2+} + Ag \longrightarrow Cu^+ + Ag^+$$

As a result, the microcrystals once more become transparent silver chloride.

Photochromic glass is prepared by adding silver chloride to the mixture of ingredients. Silver chloride is soluble in molten glass at high temperatures, but as the glass cools to about 600–700 °C, a precipitate of microcrystalline silver chloride forms. Roughly 5 nm in diameter, the crystals are spaced about 50 nm apart.

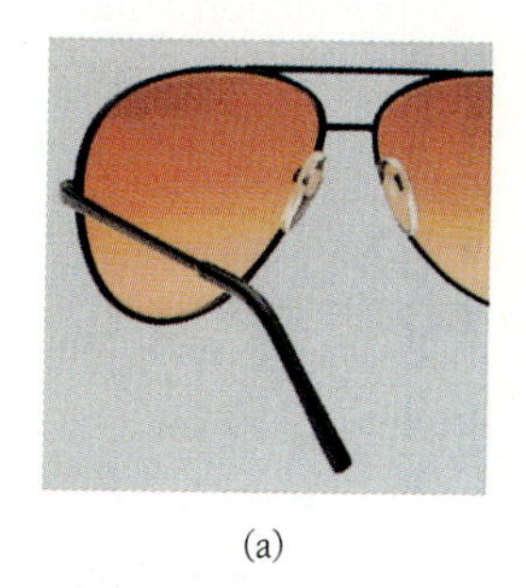

(a)

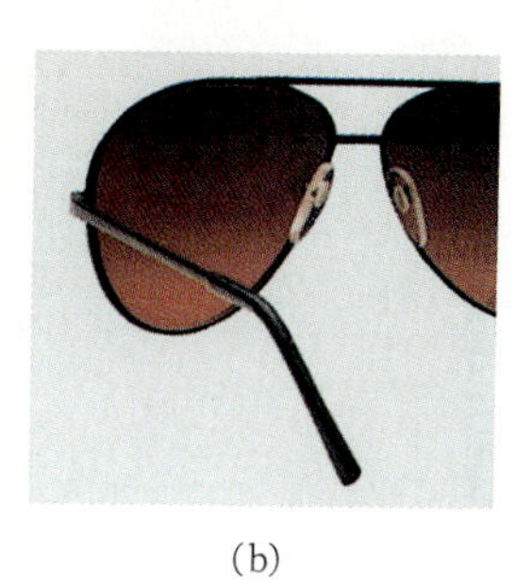

(b)

Photoray glasses are relatively transparent indoors (a), but will darken appreciably when exposed to direct sunlight (b).

Both in its composition and behavior, black-and-white photographic film is very similar to photochromic glass. In film, silver bromide microcrystals ("grains") are suspended in a gel rather than in a glass. Exposure to light results in the formation of silver atoms, just as in photochromic glass. But only a few atoms are formed (because photographic exposure times are so short), and the image must be developed, that is, enhanced by chemical treatment.

Exposed film is developed with a reducing agent that converts transparent AgBr to black Ag. The rate at which a particular grain is reduced depends critically on how many silver atoms are initially present—that is, on how much light the grain has absorbed. Development times and conditions are carefully controlled to ensure that only the exposed grains are reduced. Film is made insensitive to further exposure to light in a second processing step, called "fixing," which removes Ag^+ ions while leaving the black grains of silver in place.

Discussion Questions

1. How would installation of photochromic glass in the windows of a house or office building affect the cost of heating and cooling the building?
2. In which location would the effect be greater, Minnesota or Florida?

SUMMARY EXERCISE

Ammonia and oxygen are the raw materials in the Ostwald process for the commercial production of nitric acid. The (unbalanced) equations for the three steps of the process are:

(1) $$NH_3 + O_2 \longrightarrow NO + H_2O$$

(2) $$NO + O_2 \longrightarrow NO_2$$

(3) $$NO_2 + H_2O \longrightarrow HNO_3 + NO$$

a. Which, if any, of these steps are redox reactions?

b. What are the oxidation numbers of nitrogen in NH_3, NO, NO_2, and HNO_3?

c. Write balanced molecular equations for Reactions (1) and (2).

d. Assuming that step (3) takes place in acidic solution, write balanced net ionic equations for (i) the oxidation half-reaction, (ii) the reduction half-reaction, and (iii) the overall reaction.

e. Write the balanced molecular equation for step (3).

f. Combine your answers to (c) and (e) into a single balanced equation for the three-step process. *Hint:* one mole of water is produced along with each mole of HNO_3.

g. How much oxygen is required for the complete conversion of 1.00 metric ton (1 ton $\Leftrightarrow$ 10^6 g) of ammonia to nitric acid?

h. HNO_3 is an Arrhenius acid. Is it a strong acid? Is it also a Brønsted-Lowry acid? A Lewis acid?

Answers **(a)** All three; **(b)** −3, +2, +4, and +5; **(c)** $4\ NH_3 + 5\ O_2 \rightarrow 4\ NO + 6\ H_2O$ and $2\ NO + O_2 \rightarrow 2\ NO_2$; **(d)** oxidation: $H_2O + NO_2 \rightarrow NO_3^- + 2\ H^+ + e^-$; reduction: $2\ H^+ + NO_2 + 2\ e^- \rightarrow NO + H_2O$; overall: $3\ NO_2 + H_2O \rightarrow 2\ H^+ + 2\ NO_3^- + NO$; **(e)** $3\ NO_2 + H_2O \rightarrow 2\ HNO_3 + NO$; **(f)** $NH_3 + 2\ O_2 \rightarrow HNO_3 + H_2O$; **(g)** 3.76 tons; **(h)** HNO_3 is a strong Arrhenius and Brønsted-Lowry acid, but it is not a Lewis acid.

SUMMARY AND KEYWORDS

The mnemonic *OIL RIG* is a reminder that **oxidation** involves loss of one or more electrons, while **reduction** involves gain of one or more electrons. Any species capable of causing oxidation in another is called an **oxidant** or **oxidizing agent.** At the same time it causes oxidation, an oxidant is itself reduced. A species capable of causing reduction, while itself undergoing oxidation, is called a **reductant** or a **reducing agent.**

A reduction must be accompanied by an oxidation, and vice versa. Reactions involving oxidation/reduction are called **redox reactions.** These include **electron transfer reactions,** in which one or more electrons are transferred directly from one molecule or ion to another, as well as more complicated processes. In an **auto oxidation-reduction** or **disproportionation** reaction, the same substance is both oxidized and reduced. It is convenient to write the two parts of a redox reaction separately as an **oxidation half-reaction** and a **reduction half-reaction.** The electrons lost or gained are shown explicitly as products of an oxidation half-reaction and as reactants in a reduction half-reaction.

Each element in a compound is assigned an **oxidation number** that describes the number of electrons it has gained in (or lost from) its valence shell in the formation reaction of the compound from its constituent elements. The electron pair shared between covalently bonded atoms is counted as belonging to the more electronegative atom. These principles lead to the set of rules (Table 18.1) for the assignment of oxidation numbers.

Except for fluorine and the Group 1A and 2A metals, most elements exhibit **variable valence,** that is, they exist in more than one oxidation state. The maximum possible oxidation number of a main-group element is its group number, and the minimum is the group number minus 8.

Redox equations can be written as **molecular equations** or **ionic equations,** or as **net ionic equations** in which **spectator ions** are not shown. Spectator ions are those that are unchanged in the reaction. Water and its constituent ions, H^+ and OH^-, often participate as reactants or products in aqueous redox reactions. Half-reactions can be balanced using the **ion-electron method,** which first balances elements other than H and O, then O and H, and finally electrical charge. In the **oxidation number** balancing method, stoichiometric coefficients of the oxidant and reductant are adjusted to make the total change (increase + decrease) in oxidation number equal to zero.

PROBLEMS

General

1. Differentiate among the following terms: combustion, oxidation, and reaction with oxygen.
2. Define reduction and reducing agent.
3. Distinguish between a Lewis base and a reducing agent.
4. Invent another mnemonic (besides OIL RIG) to help remember whether electrons are lost or gained in oxidation/reduction.
5. What is a half-reaction? Can it occur by itself?
6. Why are some redox reactions called electron-*transfer* reactions?
7. Give a formal definition of oxidation number.
8. What is the maximum possible oxidation number of a given element? What is the most negative oxidation number?
9. What is a spectator ion?
10. Distinguish among the terms molecular equation, ionic equation, and net ionic equation. Write all three equations for a particular redox reaction. For a particular acid-base reaction, write all three equations.
11. What is meant by disproportionation? Give an example.
12. Is it possible for the same ion or compound to be both a Brønsted-Lowry acid and an oxidizing agent? If so, give an example. If not, explain why it is impossible.

Oxidation Numbers

13. For each of the following compounds, determine the oxidation number of the element in boldface type.
    a. $\mathbf{Sr}Cl_2$ b. $\mathbf{Fe}_2O_3$ c. $Na\mathbf{N}H_2$
14. For each of the following compounds, determine the oxidation number of the element in boldface type.
    a. $K\mathbf{Br}O_4$ b. $\mathbf{Hg}$ c. $\mathbf{Mn}O_4^-$
15. Determine the oxidation number of all elements in the following species.
    a. PCl_5 b. NF_3
    c. NO_2 d. N_2O_4
    e. BaO_2 f. Na_2SO_3
16. Determine the oxidation number of all elements in the following species.
    a. $S_2O_3^{2-}$ b. IF_5
    c. Hg_2Cl_2 d. H_3PO_4
    e. O_3 f. $PtCl_6^{2-}$
17. Determine the oxidation number of all elements in the following species.
    a. SF_6 b. XeF_4
    c. P_4O_{10} d. $HAsO_2$
    e. NH_4^+ f. CH_3OH
18. Determine the oxidation number of all elements in the following species.
    a. $Zn(OH)_4^{2-}$ b. $CuSO_4 \cdot 5H_2O$
    c. BaH_2 d. OF_2
    e. ClO^- f. SO_3

Reaction Types

19. Which of the following are redox reactions?
    a. $PCl_3 + Cl_2 \longrightarrow PCl_5$
    b. $H_2SO_3 + 2\ KOH \longrightarrow K_2SO_3 + 2\ H_2O$
    c. $SO_3 + H_2O \longrightarrow H_2SO_4$
    d. $3\ NO_2 + H_2O \longrightarrow 2\ HNO_3 + NO$
    e. $Mg + 2\ HCl \longrightarrow H_2 + MgCl_2$

20. Which of the following are redox reactions?
 a. $AgNO_3(aq) + HCl(aq) \longrightarrow AgCl(s) + HNO_3(aq)$
 b. $3\ NaClO \longrightarrow NaClO_3 + 2\ NaCl$
 c. $2\ HI(aq) + H_2O_2(aq) \longrightarrow I_2(s) + 2\ H_2O$
 d. $2\ Br^- + Cl_2 \longrightarrow 2\ Cl^- + Br_2$
 e. $C_2H_4 + H_2 \longrightarrow C_2H_6$

21. Which of the following are redox reactions?
 a. $Al_2S_3 + 6\ H_2O \longrightarrow 2\ Al(OH)_3 + 3\ H_2S$
 b. $Xe + 2\ F_2 \longrightarrow XeF_4$
 c. $Br_2 + H_2O \longrightarrow HBr + HBrO$
 d. $N_2O_4 \longrightarrow 2\ NO_2$
 e. $CaC_2 + 2\ H_2O \longrightarrow Ca(OH)_2 + C_2H_2$

22. Which of the following are redox reactions?
 a. $H_2CO_3 \longrightarrow CO_2 + H_2O$
 b. $3\ ZnS(s) + 2\ NO_3^-(aq) + 8\ H^+ \longrightarrow 3\ Zn^{2+}(aq) + 2\ NO(g) + 3\ S(s) + 4\ H_2O$
 c. $Cr_2O_7^{2-} + H_2O \longrightarrow 2\ CrO_4^{2-} + 2\ H^+$
 d. $IO_3^- + 2\ Cr(OH)_4^- \longrightarrow I^- + 2\ CrO_4^{2-}$
 e. $PbO(s) + NO_3^- \longrightarrow NO_2^- + PbO_2(s)$

Oxidant/Reductant Recognition

23. For each of the redox reactions in Problem 19, identify the species that is oxidized and the species that is reduced.

24. For each of the redox reactions in Problem 20, identify the species that is oxidized and the species that is reduced.

25. For each of the redox reactions in Problem 21, identify the oxidant and the reductant.

26. For each of the redox reactions in Problem 22, identify the oxidant and the reductant.

Ionic and Molecular Equations

27. Each of the following is a balanced molecular equation. Write each as an ionic equation, and then, by eliminating the spectator ions, write each as a net ionic equation.
 a. $3\ NaClO \longrightarrow NaClO_3 + 2\ NaCl$
 b. $3\ Cu(s) + 8\ HNO_3(aq) \longrightarrow 3\ Cu(NO_3)_2(aq) + 2\ NO(g) + 4\ H_2O(\ell)$

28. Each of the following is a balanced molecular equation. Write each as an ionic equation, and then, by eliminating the spectator ions, write each as a net ionic equation.
 a. $Br_2 + H_2O \longrightarrow HBr + HBrO$
 b. $3\ Na_2MnO_4(aq) + 2\ H_2O(\ell) \longrightarrow MnO_2(s) + 2\ NaMnO_4(aq) + 4\ NaOH(aq)$

29. Each of the following is a balanced molecular equation. Write each as an ionic equation, and then, by eliminating the spectator ions, write each as a net ionic equation.
 a. $8\ HCl(aq) + H_3AsO_4(aq) + 4\ Mg(s) \longrightarrow AsH_3(g) + 4\ MgCl_2(aq) + 4\ H_2O$
 b. $KNO_2(aq) + 5\ H_2O(\ell) + 2\ Al(s) \longrightarrow 2\ Al(OH)_3(aq) + NH_3(aq) + KOH(aq)$

30. Each of the following is a balanced molecular equation. Write each as a net ionic equation.
 a. $Mg + 2\ HCl \longrightarrow H_2 + MgCl_2$
 b. $3\ P_4(s) + 12\ KNO_3(aq) + 8\ HNO_3(aq) + 8\ H_2O(\ell) \longrightarrow 12\ KH_2PO_4(aq) + 20\ NO(g)$

Half-Reactions

31. For each of the following unbalanced redox reactions, write the unbalanced oxidation half-reaction and the unbalanced reduction half-reaction.
 a. $UO_2^{2+}(aq) + Sn^{2+}(aq) \longrightarrow U^{4+}(aq) + Sn^{4+}(aq)$
 b. $Hg(\ell) + H_2SO_4(aq) + Cr_2O_7^{2-} \longrightarrow Hg_2SO_4(s) + Cr^{3+}(aq)$
 c. $H_2O_2(aq) + Mn^{2+}(aq) \longrightarrow MnO_4^-(aq) + H_2O$
 d. $C_2O_4^{2-}(aq) + IO_3^-(aq) \longrightarrow CO_2 + I^-(aq)$

32. For each of the following unbalanced redox reactions, write the unbalanced oxidation half-reaction and the unbalanced reduction half-reaction.
 a. $Cu(s) + NO_3^-(aq) \longrightarrow Cu^{2+}(aq) + NO_2(g)$
 b. $H_2S(g) + MnO_4^-(aq) \longrightarrow S(s) + MnO_2(s)$
 c. $ClO^-(aq) + P_4(s) \longrightarrow PO_4^{3-} + Cl^-(aq)$
 d. $PbSO_4(aq) + H_2O \longrightarrow Pb(s) + PbO_2(s) + H_2SO_4(aq)$
 e. $S_2O_3^{2-}(aq) + I_2(aq) \longrightarrow I^-(aq) + S_4O_6^{2-}(aq)$

Half-Reaction Balancing

33. Balance each of the half-reactions obtained as answers to Problem 31. Assume that the reactions take place in acidic solution.

34. Balance each of the half-reactions obtained as answers to Problem 32. Assume that (a) and (d) take place in acidic solution. Balance (b) and (c) as if they took place in basic solution.

Balancing Redox Reactions

35. Complete the ion-electron balancing procedure, begun in Problem 33, by writing the overall net ionic equation for each of the reactions in Problem 31.

36. Complete the ion-electron balancing procedure, begun in Problem 34, by writing the overall net ionic equation for each of the reactions in Problem 32.

*37. Balance the following equations by the oxidation number method, assuming that the reaction takes place in acidic solution. Give the number of electrons transferred in the balanced equation.
 a. $I_2(s) + H_2S \longrightarrow SO_4^{2-} + I^-$
 b. $Cl_2 + H_2O \longrightarrow ClO^- + Cl^-$

*38. Balance the following equations by the oxidation number method, assuming that the reaction takes place in basic solution. Give the number of electrons transferred in the balanced equation.
a. $Bi(OH)_3(s) + Sn^{2+} \longrightarrow Sn^{4+} + Bi(s)$
b. $MnO_4^- + H_2O_2(aq) \longrightarrow MnO_2(s) + O_2(g)$
c. $NO_2(g) + OH^- \longrightarrow NO_3^- + NO(g)$
d. $Cr_2O_3(s) + ClO^- \longrightarrow CrO_4^{2-} + Cl^-$

39. Each of the following is an unbalanced net ionic equation. Balance each, using any method you choose.
a. $I^- + H_2O_2(aq) + H^+(aq) \longrightarrow I_2(s) + H_2O$
b. $IO_3^- + Cr(OH)_4^- \longrightarrow I^- + CrO_4^{2-} + H^+ + H_2O$
c. $Cu(s) + NO_3^- + H^+ \longrightarrow NO_2(g) + Cu^{2+} + H_2O$

*40. Balance the following equations, which range from tricky to very difficult. Use any method you choose. *Hint:* There may be more than two elements whose oxidation number changes.
a. $CrI_3(s) + Cl_2(aq) \longrightarrow CrO_4^{2-} + IO_4^- + Cl^-$
b. $Fe(CN)_6^{4-} + Ce^{4+} + H_2O \longrightarrow$
$Ce(OH)_3(s) + Fe(OH)_3(s) + CO_3^{2-} + NO_3^- + H^+$
c. $Fe_3O_4(s) + H_2(g) \longrightarrow Fe(s)$
d. $Cu_2S(s) + NO_3^- + H^+ \longrightarrow$
$Cu^{2+} + NO(g) + S_8(s) + H_2O$

*41. The following equation is easily balanced by the ion-electron method, but, because of uncertainties in oxidation numbers, less easily by the oxidation-number method.

$$CN^- + MnO_4^- \longrightarrow CNO^- + MnO_2(s)$$

The rules given in Table 18.1 do not help in assignment of oxidation numbers in CN^- or CNO^-. The general principle of assigning the bonding electron pairs to the more electronegative atom ($C < N < O$), however, together with reasonable Lewis structures, leads to an oxidation number of -3 for the N atom in CN^- and -1 for the N atom in CNO^-. Use these assignments to balance the equation by the oxidation-number method, in basic solution.

*42. In the preceding problem, you were asked to balance equations involving the CN^- and CNO^- ions, and told to assign the oxidation number of N in these ions as -3 in CN^- and -1 in CNO^-. A redox equation can be correctly balanced by the oxidation number method, however, even if you are not certain of the oxidation number assignment. As an experiment, try balancing the equation several times, assuming first that the oxidation number of N is -3 in both CN^- and CNO^-, and then that it is $+3$ in both CN^- and CNO^-. There is no uncertainty about oxygen, whose oxidation number is -2 in both ions. Explain why all three of these assignments lead to the same balanced equation.

CHAPTER 19

ELECTROCHEMISTRY

When copper wire is placed in a solution containing Ag^+ ions, a spontaneous redox reaction results in the growth of delicate whiskers of silver metal on the wire. $Cu(s) + 2\,Ag^+(aq) \rightarrow 2\,Ag(s) + Cu^{2+}$.

19.1 CONDUCTION OF ELECTRICAL CURRENT

Electricity and electrical current are familiar phenomena for most of us. We deal with them every day as we start our cars, light our homes, use battery-operated radios, or see lightning strokes. Electrical current often has a chemical basis, rooted in the electron-transfer reactions described in Chapter 18. **Electrochemistry** is the name given to those aspects of redox reactions that involve electrical current.

Electrical Current, Charge, and Potential

Electrical current is one of the *base quantities* (Section 1.3) whose formal definitions are the starting points of the SI system. Current is measured in units of **amperes** (A). One ampere is formally defined as the amount of current that produces a force of 2×10^{-7} newtons per meter of length between two parallel conductors placed one meter apart in a vacuum. The magnitude of this unit, however, can be better appreciated from familiar situations. The current in a 60-watt light bulb is about ½ A, and the current that must be supplied by an automobile battery in order to start the engine can be as high as several hundred A.

André Marie Ampère (1775–1836) was a French mathematician, philosopher, and physicist, who studied the relationship between electricity and magnetism.

Electrical *charge* and current are closely related quantities. Electrical current consists of the motion of charged particles, in much the same way that the current in a river consists of the motion of water particles (molecules). **Charge,** like mass, is a fundamental property of matter. Charge is best understood in terms of its characteristics. Charge exists in two varieties, called positive and negative, which attract one another. Charges of the same sign repel one another. Among the elementary particles, the proton has positive charge, the electron has negative charge, and the neutron has no charge. The charge of the electron or the proton is called the **elementary charge.** It is the smallest charge that has been observed, and its value is 1.60×10^{-19} *coulombs* (C).

Indirect evidence indicates that electrons, protons, and neutrons are composed of smaller entities known as *quarks,* which carry a charge that is ⅓ or ⅔ of the elementary charge. So far, quarks have not been directly observed.

Familiar uses of electricity involve charges very much larger than the elementary charge. When a 60-watt light bulb is operating, for instance, about 30 C flow through it each minute. A heavy-duty automobile battery is capable of storing about half a million coulombs. The **coulomb** is formally defined in SI units as the charge passing a fixed point in one second, in any conductor carrying a current of one ampere. That is, a current of one ampere, flowing in a circuit for one second, delivers a charge of one coulomb. Several equivalence statements, useful in electrical problems, can be generated from this definition.

Charles Augustin de Coulomb (1736–1806) retired from the French military at the age of 53 in order to pursue experimental physics. He invented several electrical and magnetic instruments, and measured the force exerted by electrically charged objects on one another.

$$1\ \text{C} \Longleftrightarrow 1\ \text{A} \cdot 1\ \text{s}; \qquad 1\ \text{A} \Longleftrightarrow \frac{1\ \text{C}}{1\ \text{s}}; \qquad \frac{1\ \text{C}}{1\ \text{A}} \Longleftrightarrow 1\ \text{s} \qquad (19\text{-}1)$$

Another quantity of importance in electrochemistry is *potential,* which is related to energy and capacity to do work. **Potential** may be thought of as the driving force behind the flow of electrical current. Electrical potential is analogous to gravitational potential. In preindustrial societies the gravitational potential energy in rivers and streams is often used to do work of grinding, boring, lifting, and so on. Conversion of this potential energy to useful work is done by means of a water wheel (Figure 19.1).

The amount of energy a given mass of water transfers to a water wheel depends on the height through which the water falls. As the mass (a large number of water molecules) moves from a region of high gravitational potential to a region of low gravitational potential, it acquires kinetic energy proportional to the potential difference, that is, to the height. In Figure 19.1 more work can be obtained in (b) from the same mass of water than in (a), because in (b) the water falls through a greater height.

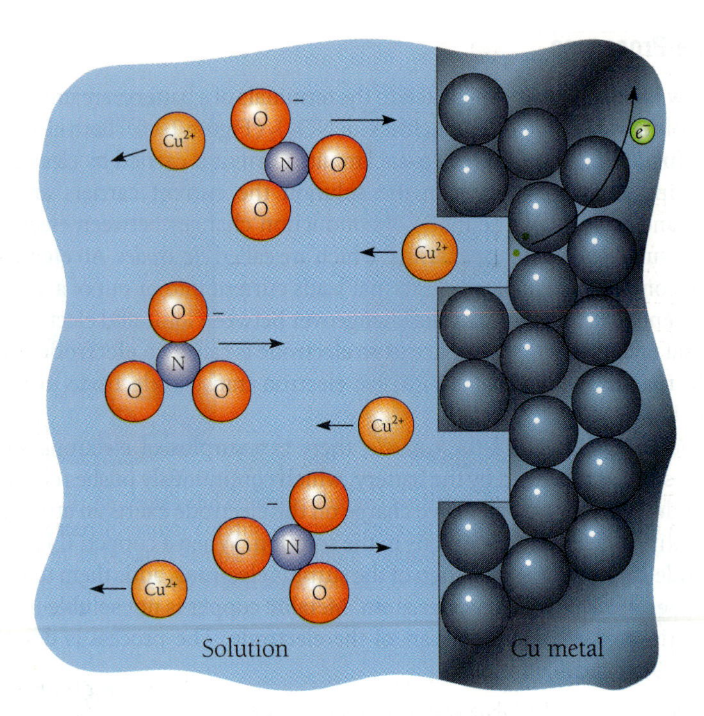

Figure 19.8 A surface copper atom gives up two electrons to the anode and moves out into the solution as a Cu^{2+} ion.

At the same time, the opposite process is occurring at the other electrode. Electrons are drawn out of the electrode by the battery, and the resulting deficit is made up when one of the surface copper atoms gives up two of its valence electrons and becomes a copper(II) ion. The water-soluble ion leaves the surface of the electrode and joins the other dissolved ions. The process is described by Equation (19-4) and in Figure 19.8.

$$Cu(s) \longrightarrow Cu^{2+}(aq) + 2\,e^- \qquad (19\text{-}4)$$

The phrase, "AN OX and a RED CAT" will remind you that oxidation takes place at the anode, and reduction takes place at the cathode.

Since this electrode process is a loss of electrons, it is an oxidation. An electrode at which oxidation takes place is called an **anode.** Electrode processes are reversible. The anode in Figure 19.6 can be converted to a cathode simply by interchanging the connections to the battery.

Electrochemical Cells

The apparatus depicted in Figure 19.6 is an **electrochemical cell.** Although they can be constructed in different ways using many different materials, all electrochemical cells have several features in common. A cell has two electrodes: an anode and a cathode. Each electrode is immersed in an **electrolyte,** a material (usually liquid, paste, or gel) in which ionic conduction takes place. Outside the cell, the metal wires of the *external circuit* connect the electrodes to batteries, motors, light bulbs, meters, and so on. There are two types of electrochemical cell.

Luigi Galvani (1737–1798)

Luigi Galvani was a noted surgeon and a professor of anatomy at the University of Bologna in Italy. He observed contractions in frog muscle touched simultaneously by the two ends of a bow or arc made of two metals. His belief that the muscle tissue was the source of electricity put him in direct conflict with Professor Alessandro Volta, who believed that the two-metal bow was the source. The controversy stimulated research in both physics and medicine.

1. In an **electrolytic cell,** an external source of electrical current, such as a battery, causes a chemical change to take place.
2. In a **voltaic cell,** also called a **galvanic cell,** a spontaneous chemical reaction generates the electrical current that flows in the cell and the external circuit.

A redox reaction, called the **cell reaction,** occurs in both electrolytic and voltaic cells. The copper/copper sulfate/copper cell in Figure 19.6 is an electrolytic cell. The external power source causes the oxidation of copper [Equation (19-4)] at the anode, and the reduction of copper ion [Equation (19-3)] at the cathode. The batteries in Figure 19.2, on the other hand, are voltaic cells. In them, a redox reaction (Section 19.4) produces electrical current that is put to use in the battery's external circuit.

19.2 VOLTAIC CELLS AND HALF-CELLS

The chemical process taking place within an electrochemical cell is an electron-transfer reaction. As such, it involves both an oxidation half-reaction and a reduction half-reaction. But although these half reactions must occur simultaneously, they do not occur in the same *place.* The anode and the cathode are in separate locations within an electrochemical cell.

Cell Configuration

Figure 19.9 shows a typical setup of a voltaic cell. The two vessels containing the solutions are called the zinc and copper **half-cells.** Half-cells are named for the electrode processes that take place in them. In the left half-cell, the electrode reaction is the oxidation half-reaction $Zn(s) \longrightarrow Zn^{2+}(aq) + 2\ e^-$: the cell is called the *oxidation* half-cell. It may also be called the anode half-cell or the *anode compartment.* In the other half-cell, the electrode process is the reduction half-reaction $Cu^{2+}(aq) + 2\ e^- \longrightarrow Cu(s)$: the cell is the *reduction* half-cell, the cathode half-cell, or the *cathode compartment.*

The *overall* or *cell reaction* in Figure 19.9 is $Cu^{2+}(aq) + Zn(s) \longrightarrow Cu(s) + Zn^{2+}(aq)$. As in all voltaic cells, the cell reaction proceeds spontaneously, provided the external circuit is complete. That is, there must be a metallic pathway so that electrons,

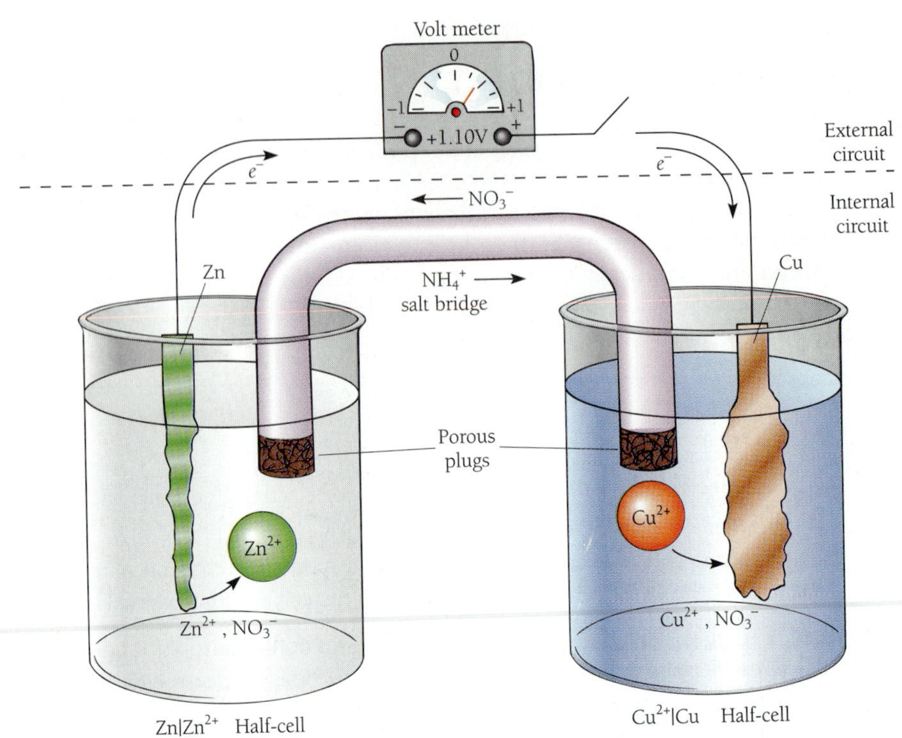

Figure 19.9 A voltaic cell consists of an oxidation half-cell and a reduction half-cell, separated by a porous barrier (salt bridge). A voltmeter is used in the external circuit to measure the difference in electrical potential between the two compartments.

generated at the anode, can pass through the external circuit to the cathode, where they are needed for the reduction half-reaction. If the switch in the external circuit of Figure 19.9 is opened, the circuit is broken and the overall cell reaction ceases.

In order for the cell to operate, the *internal circuit* must also be complete. The two half-cells must have an electrical connection between them. This cannot be a metallic wire, as it is in the external circuit, because immersion of a metallic wire into the solutions would constitute additional electrodes at which redox reactions would take place. The connection between the two half-cells must be an *ionic conductor*. The device shown in Figure 19.9 is one type of ionic-conducting connector, called a **salt bridge.** It consists of a glass tube filled with a concentrated salt solution—2 M NH_4NO_3 is often used. Usually the solvent in a salt bridge is a semirigid gel such as agar. This prevents convective flow and mixing, but allows individual ions to move. Current in the salt bridge is carried by NH_4^+ and NO_3^- ions, moving in opposite directions.

Use of a metallic conductor would bring the total number of electrodes to four. The result would be two interconnected electrochemical cells rather than a single cell.

The salt bridge performs two functions: (1) it provides an ionically conducting pathway between the half-cells and thus completes the electrical circuit, while at the same time (2) it prevents mixing of the different electrolytes in the half-cells. These two criteria can be met by other means. The tube connecting the half-cells can be blocked in the middle by a sheet of blotting paper or a piece of coarse, unglazed ceramic (like a flower pot). Either of these serves to prevent mixing while allowing passage of dissolved ions. **Porous barrier** is a general term denoting a salt bridge or any other device that performs these functions.

Electrode Types

A variety of different types of electrode, serving many purposes, are in common use. Several of these are described in this section.

Metal/Solution Electrode

Each electrode in Figure 19.9 consists of a metal in contact with a solution containing as solute a cation (oxidized form) of the metal. This is a common type called the **metal/solution electrode.** It is easily prepared, and it can be made from any metal that does not react with water.

Gas Electrode

Oxidation or reduction of a gaseous species can also take place in an electrochemical cell. Figure 19.10 shows the operation of a hydrogen electrode, a specific example of a **gas electrode.** Since gaseous hydrogen does not conduct electricity, some other means is necessary to permit the flow of electrons into or out of the cell. This function is carried out by a platinum wire or foil. Platinum is used because it is unreactive and does not participate in the redox reaction. In cases where cost is a factor, less expensive materials such as graphite are used instead of platinum. In gas electrodes, the electrode process occurs only at the place where the three phases meet. That is, at the junction between the platinum wire, the gaseous hydrogen, and the solution containing the dissolved hydrogen ions. Like other electrode types, the gas electrode is reversible and can be the site of either an oxidation half-reaction or a reduction half-reaction. The electrode reactions are represented by

Oxidation $$H_2(g) \longrightarrow 2\ H^+(aq) + 2\ e^- \quad (19\text{-}5)$$

Reduction $$2\ H^+(aq) + 2\ e^- \longrightarrow H_2(g) \quad (19\text{-}6)$$

The mechanism of an electrode process is generally complicated, and both of these equations represent the sum of several elementary steps. The reduction [Equation (19-6)] involves two one-electron processes, producing two hydrogen atoms adsorbed on the platinum surface. Two adsorbed atoms then combine to form molecular hydrogen.

$$H^+(aq) + e^- \longrightarrow H\ (\text{adsorbed}) \quad (19\text{-}6a)$$

$$H^+(aq) + e^- \longrightarrow H\ (\text{adsorbed}) \quad (19\text{-}6b)$$

$$2\ H\ (\text{adsorbed}) \longrightarrow H_2(g) \quad (19\text{-}6c)$$

The sum or net effect of these steps is

$$2\ H^+(aq) + 2\ e^- \longrightarrow H_2(g) \quad (19\text{-}6)$$

Metal/Salt Electrode

The oxidized form of the electrode element need not be a dissolved cation, as is the case with the metal/solution electrode. For instance, consider what happens when a silver electrode is operated in a solution containing chloride ions. Silver ions, oxidatively produced at a silver anode, cannot enter the solution as $Ag^+(aq)$ without exceeding the solubility product of the insoluble salt AgCl. Instead, they combine with a chloride ion without ever leaving the metal surface. The result is the formation of a layer of insoluble silver chloride on the surface of the silver anode. The electrode reaction is

$$Ag(s) + Cl^-(aq) \longrightarrow AgCl(s) + e^- \quad (19\text{-}7)$$

(a)

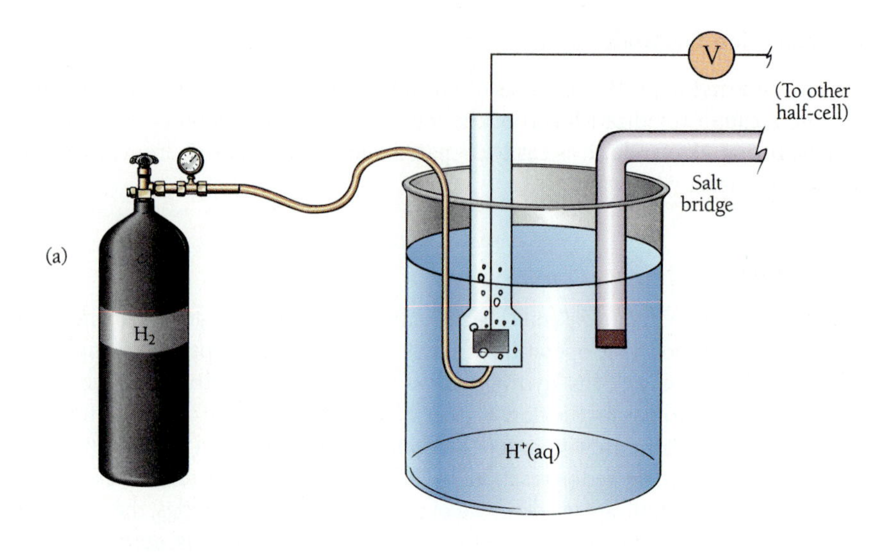

(b)

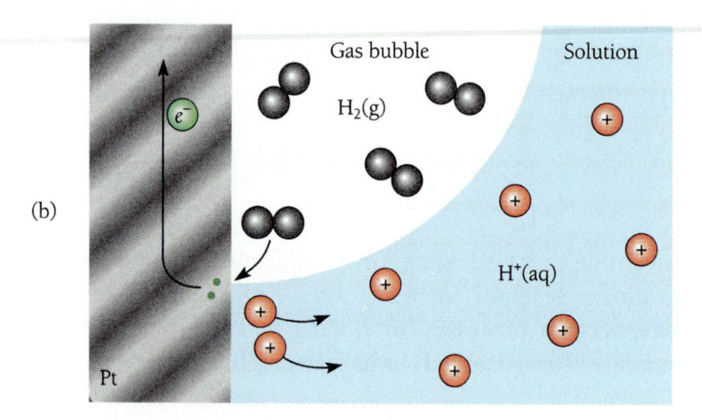

(c)

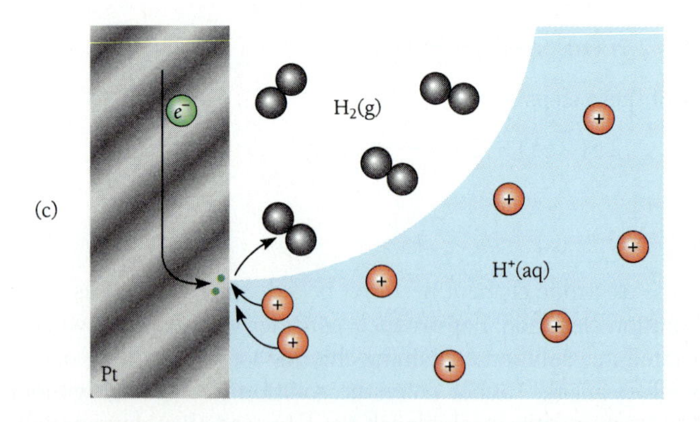

Figure 19.10 (a) The hydrogen electrode. Gaseous hydrogen and dissolved $H^+(aq)$ contact one another at the surface of a wire or foil made of platinum, an inert material that does not participate in the overall redox reaction. Gas electrodes are reversible and can be a source or a sink for electrons. (b) When the gas electrode acts as an anode, the electrode reaction is $H_2(g) \rightarrow 2\ H^+(aq) + 2\ e^-$. The reaction occurs only at the junction of the three phases—gas, solution, and metal. (c) When a gas electrode acts as a cathode, the electrode reaction is $2\ H^+(aq) + 2\ e^- \rightarrow H_2(g)$.

If the electrode reaction involves an insoluble salt rather than a dissolved species, the electrode at which it takes place is called a **metal/salt electrode.** The electrode of Equation (19-7) is called a Ag/AgCl electrode, and is shown schematically in Figure 19.11.

Redox Electrode

An electrode in which both the oxidized and the reduced forms are *solutes* is called a **redox electrode.** Such electrodes utilize variable-valence elements, that is, those that exhibit several different oxidation states in their compounds. For example, iron readily forms compounds in both the +2 and +3 oxidation states, and a length of platinum wire or foil immersed in a solution containing both $Fe^{2+}(aq)$ and $Fe^{3+}(aq)$ constitutes a redox electrode (Figure 19.12). Depending on the nature of the other half-cell, either the oxidation $Fe^{2+}(aq) \rightarrow Fe^{3+}(aq) + e^-$ or the reduction $Fe^{3+}(aq) + e^- \rightarrow Fe^{2+}(aq)$ can take place at the platinum surface. At such a redox anode, a $Fe^{2+}(aq)$ ion comes in contact with the platinum wire, deposits an electron, and drifts away in the oxidized form $Fe^{3+}(aq)$. The reverse process takes place when the electrode functions as a cathode. Redox electrodes are usually described by naming both forms, for instance a "Fe^{2+}/Fe^{3+} electrode," or a "Mn^{2+}/Mn^{3+} electrode" (Figure 19.13).

Platinum does not participate in the electrode reaction.

Redox electrodes need not involve monatomic metal cations. The only requirement is that the oxidized and reduced forms be soluble species. The reduction half-reaction

$$ClO_4^-(aq) + H_2O + 2\,e^- \longrightarrow ClO_3^-(aq) + 2\,OH^-$$

for example, can be used in a redox electrode.

Redox electrodes are widely used in biomedical research. A great deal of information concerning brain function and the transmission of nerve impulses in the body has been obtained using miniaturized redox electrodes inserted into nerve cells (Figure 19.14).

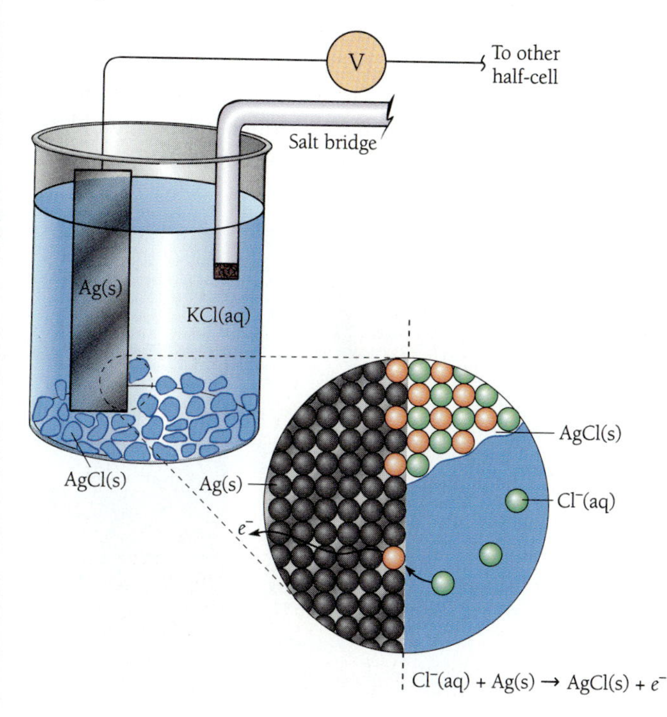

Figure 19.11 The Ag/AgCl electrode is designed to allow intimate contact of the three phases—Ag(s), AgCl(s), and $Cl^-(aq)$. The electrode reaction takes place at the junction of these three phases. The silver chloride layer is sufficiently porous to allow the infiltration of chloride ions. The reaction is $Ag(s) + Cl^-(aq) \rightarrow AgCl(s) + e^-$ when the electrode is an anode, and $AgCl(s) + e^- \rightarrow Ag(s) + Cl^-(aq)$ when the electrode is functioning as a cathode.

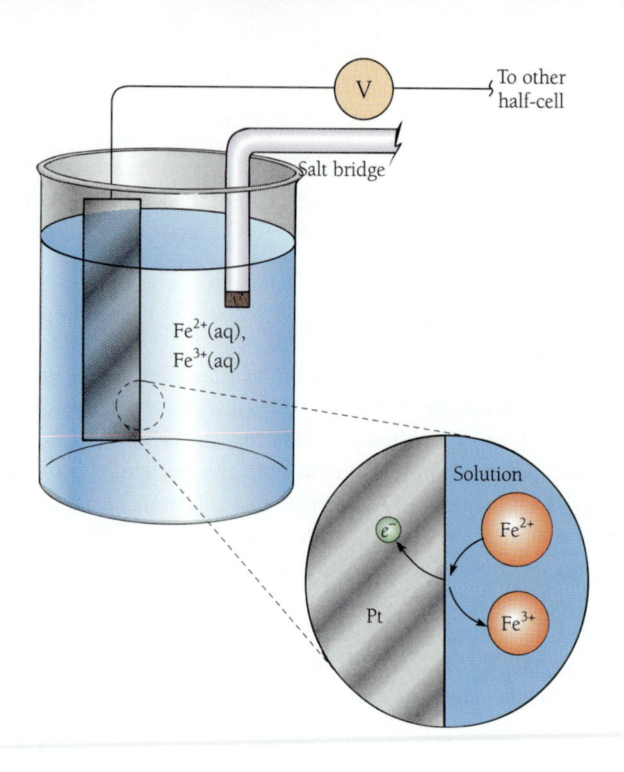

Figure 19.12 The Fe^{2+}/Fe^{3+} redox electrode consists of a platinum wire immersed in a solution containing both Fe^{2+}(aq) and Fe^{3+}(aq). The dissolved ions acquire or lose a valence electron at the surface of the inert wire.

Notation

Since the anion does not participate in the reaction at the Zn/Zn^{2+} electrode, it makes no difference whether $ZnCl_2$, $Zn(NO_3)_2$, or any other soluble zinc salt, is used to prepare the half-cell.

Diagrams of electrochemical cells, such as the one appearing in Figure 19.9, are very useful in discussions of cell processes and reactions. However, they take time to draw, and contain much more information than is necessary for most purposes. It is convenient to employ a shorthand notation, in which the *cell diagram* of Figure 19.9 is represented by $Zn|Zn^{2+}(1\ M)||Cu^{2+}(1\ M)|Cu$. In this notation, concentration of aqueous species is given in parentheses (concentration information may be omitted if it is not relevant to the discussion), and the vertical line | represents a phase boundary. $Zn|Zn^{2+}(1\ M)$ means metallic zinc (a solid) in contact with 1 M aqueous zinc ion (a liquid). Counterions are not shown unless they participate in the electrode reaction. The double vertical line || separating the aqueous species indicates a porous barrier, such as a salt bridge, connecting the two half-cells. The external circuit is not shown, although its wires are assumed to connect with the metals written on the left and the right of the shorthand notation. The anode, at which oxidation takes place, is written

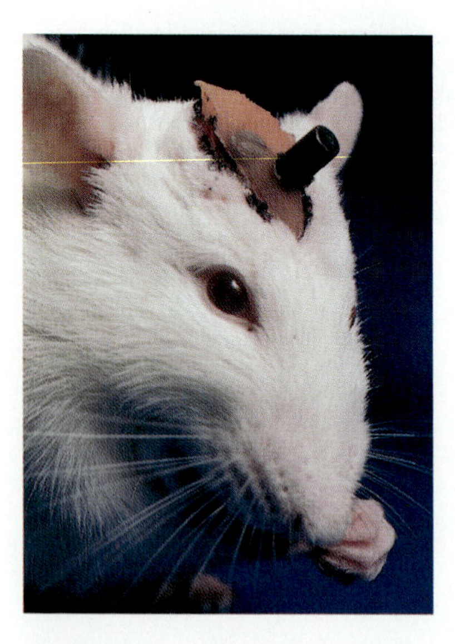

Figure 19.14 Redox electrodes can be made small enough to study redox reactions taking place within individual cells.

Figure 19.13 The pale pink Mn^{2+} ion is being oxidized to the brown Mn^{3+} form at the anode of this electrolytic cell. The anode compartment is connected by a salt bridge to the cathode compartment (not shown). The voltmeter shows that the applied potential is 5.00 V.

on the left. Thus, the notation $Zn|Zn^{2+}(1\ M)||Cu^{2+}(1\ M)|Cu$ implies that at the zinc electrode the half-reaction is $Zn(s) \rightarrow Zn^{2+}(1\ M) + 2\ e^-$. The half-reaction at the other electrode is the reduction $Cu^{2+}(1\ M) + 2\ e^- \rightarrow Cu(s)$, and the overall cell reaction is $Zn(s) + Cu^{2+}(1\ M) \rightarrow Cu(s) + Zn^{2+}(1\ M)$. For cells consisting only of metal/solution electrodes, the half-reactions (except for the electrons) can be read easily from the notation by mentally replacing the single vertical lines with arrows and the double vertical line with "and." $Zn|Zn^{2+}||Cu^{2+}|Cu$ becomes $Zn \rightarrow Zn^{2+}$ and $Cu^{2+} \rightarrow Cu$.

Not all cells need a porous barrier. If the electrolyte (the liquid, ionically conducting part of the cell) is the same for both half-cells, there is no reason to prevent the two from mixing, and no need for a porous barrier. The cells in Figures 19.6 and 19.10 operate without porous barriers: their shorthand notations are $Cu(s)|Cu^{2+}(aq)|Cu(s)$ and $Pt|H_2(1\ atm)|HCl(1\ M)|Cl_2(1\ atm)|Pt$. Note that pressure replaces concentration in the notation for a gas electrode. In the notation for redox electrodes, a comma separates the oxidized and reduced forms: $Pt|Fe^{2+}(aq), Fe^{3+}(aq)||Cu^{2+}(aq)|Cu$. The inert platinum metal is shown as Pt| or |Pt, because it *is* a necessary part of the cell and it *does* include a phase boundary.

Pt| and |Pt are ignored when writing the half-reactions from the shorthand notation.

EXAMPLE 19.1 Shorthand Notation for Electrochemical Cell Diagrams

a. Write the shorthand notation for a voltaic cell in which the overall cell reaction is $2\ Ag(s) + Br_2(\ell) \rightarrow 2\ Br^-(aq) + 2\ Ag^+(aq)$.

b. Write the half-reaction taking place at the cathode of the cell $Pt|Cl_2(g)|Cl^-(aq)||Br^-(aq)|AgBr(s)|Ag$.

Analysis/Plan Follow the rules for notation, given above, precisely and completely.

Work

a. Begin by identifying the oxidation and reduction half-reactions so that they may be written in the proper order in the notation. The easiest way to do this is to assign oxidation numbers

If you are unable to assign oxidation numbers, balance the half-reactions using the ion-electron method (Section 18.3). The oxidation half-reaction shows electrons as a *product*.

$$\overset{0}{Ag}(s) + \overset{0}{Br}_2(\ell) \longrightarrow \overset{+1}{Ag}{}^+(aq) + \overset{-1}{Br}{}^-(aq)$$

from which it is seen that the (unbalanced) half-reactions are

Oxidation	$Ag(s) \longrightarrow Ag^+(aq)$
Reduction	$Br_2(\ell) \longrightarrow Br^-(aq)$

Group the reactants and products with the oxidation reaction on the left, the reduction reaction on the right, and the dissolved species in the middle.

$$Ag(s) \qquad Ag^+(aq) \qquad Br^-(aq) \qquad Br_2(\ell)$$

Indicate a porous barrier between the two half-cells (unless the two electrolytes are identical), and indicate the phase boundaries.

In this case the need for the porous barrier is obvious: unless the two solutions are kept separate, $Ag^+(aq)$ and $Br^-(aq)$ will react to form an insoluble precipitate of AgBr.

$$Ag(s)|Ag^+(aq)||Br^-(aq)|Br_2(\ell)$$

Since the wires of the external circuit cannot be connected to $Br_2(\ell)$, an inert electrode must be indicated.

$$Ag(s)|Ag^+(aq)||Br^-(aq)|Br_2(\ell)|Pt$$

Check All phase boundaries are indicated by | ; the aqueous species are adjacent in the middle, separated by || ; and the left- and right-most species are metals.

RED CAT

b. The process taking place at the cathode is the reduction half-reaction, which is written on the right in shorthand notation. The participants in the right-hand reaction are read from the diagram as $Br^-(aq)$, $AgBr(s)$, and $Ag(s)$. When these are arranged into a balanced reduction half-reaction, the result is

$$AgBr(s) + e^- \longrightarrow Ag(s) + Br^-(aq)$$

Check Electrons appear as a reactant in a reduction half reaction; the equation is balanced.

Exercise

a. Write the shorthand notation for a voltaic cell in which the overall cell reaction is $Cu^{2+}(aq) + H_2S(aq) \rightarrow 2\ H^+(aq) + Cu(s) + S(s)$.

b. Write the half-reaction taking place at the anode of the cell $Pt|Hg(\ell)|Hg_2Cl_2(s)|Cl^-(aq)||Fe^{3+}(aq), Fe^{2+}(aq)|Pt$.

Answer: (a) $Pt|H_2S(aq)|S(s)||Cu^{2+}(aq)|Cu(s)$; (b) $2\ Hg(\ell) + 2\ Cl^-(aq) \longrightarrow Hg_2Cl_2(s) + 2\ e^-$.

19.3 POTENTIAL OF CELLS AND HALF-CELLS

When a voltmeter is connected between the two electrodes of a *voltaic* cell, it is found that a potential difference or voltage exists between them. This voltage is called the **cell potential** ($\mathscr{E}$), and typically has a value of a few volts or less. Like all electrical potential differences, a cell potential has a *polarity,* that is, a positive or negative sign. Voltage measurement of the cell $Zn|Zn^{2+}(aq)||Cu^{2+}(aq)|Cu$ shows that the zinc electrode is negative with respect to the copper electrode. The polarity of a voltaic cell indicates two things: (1) electrons move spontaneously through the external circuit from the negative electrode to the positive electrode, and (2) since the source of the electrons moving out of the negative electrode is the oxidation half-reaction, it follows that oxidation occurs at the negative electrode of a voltaic cell.

Standard Cell Potential

The potential of the cell $Zn|Zn^{2+}(aq)||Cu^{2+}(aq)|Cu$ is 1.1 V when the concentrations of $Zn^{2+}(aq)$ and $Cu^{2+}(aq)$ are both 1 M. Mention of the concentration is necessary because cell potentials are found to vary as the electrolyte concentrations change. Comparisons of cell potentials of different cells are therefore meaningful only if they refer to the same conditions. **Standard conditions** for electrochemical cells are, depending on the substance, *pure* solid, *1.00 M* solution, or *1.00 atm* gas. A potential measured under these conditions is called a **standard cell potential** ($\mathscr{E}°$).

Half-Cell Potential

Each different voltaic cell has its own characteristic standard cell potential. Furthermore, overall cell potential arises from the *independent* contributions of the two

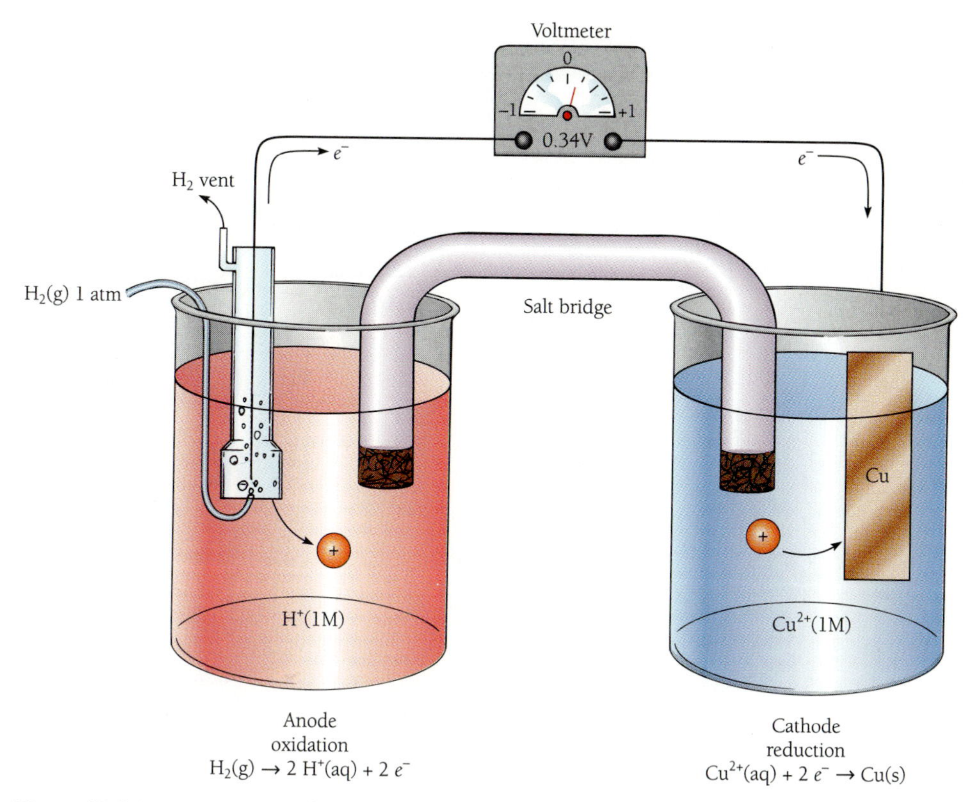

Figure 19.15 Measurement of the standard reduction potential of the Cu^{2+}/Cu half-cell. The direction of the electron flow in the external circuit shows that Cu^{2+} is being reduced at the cathode.

half-cells. Each different half-cell has a characteristic *half-cell potential,* independent of the half-cell to which it is coupled.

While the overall cell potential of a voltaic cell is readily measured, the potential of a *half-cell* cannot be measured at all. The reason for this is that "cell potential" is actually a potential *difference* between the two electrodes. One cannot measure the potential "difference" of a single electrode, any more than one can measure the difference in height of a single person. However, the question, "What is the difference in electrical potential between a $Cu(s)|Cu^{2+}(1\ M)$ half-cell and some *reference* half-cell?" can easily be answered by measurement. It has been agreed that the reference half-cell shall be the **standard hydrogen electrode (SHE).** The standard hydrogen electrode is a gas electrode (Figure 19.10) operated under standard conditions: hydrogen pressure 1.00 atm, $[H^+] = 1.00$ M, and 25 °C. The half-cell potential of the SHE is *assigned* a value of $\mathscr{E}° = 0.00$ V,* and all other half-cell potentials are expressed relative to this value. The symbol $\mathscr{E}$ used for electrochemical cell potential comes from an older synonym for voltage, electromotive force. As usual, the superscript "o" means that the quantity refers to standard conditions.

* This use of a reference standard, with an assigned value of zero, to provide a base for comparison of quantities whose absolute values are immeasurable, has been encountered before, in thermochemistry. The "heat content" of a substance cannot be measured, while *changes* in heat content are easily measurable. We choose to discuss enthalpy by assigning a value of zero to the heat content of pure elements, and *measuring* the heat of formation of a compound relative to elements as reference standards.

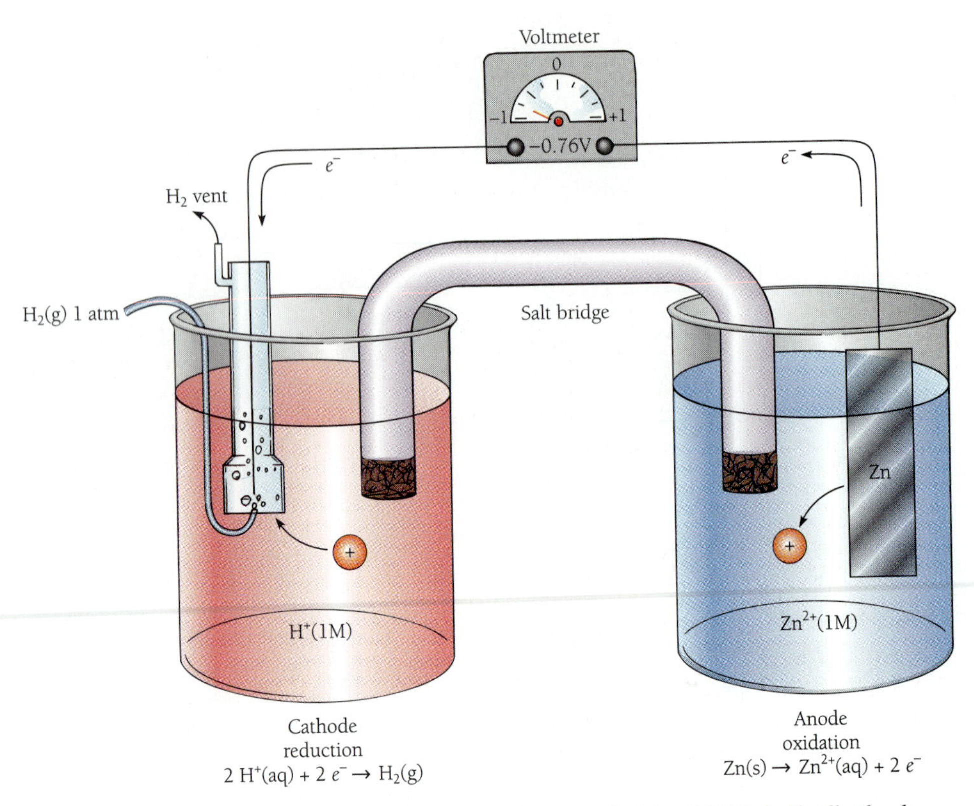

Figure 19.16 Measurement of the standard potential of the Zn^{2+}/Zn half-cell. The direction of the electron flow in the external circuit shows that Zn is being oxidized at the anode.

Standard Reduction Potential

The value of a half-cell potential is determined by means of a voltmeter connected to a voltaic cell that consists of the unknown half-cell as one electrode and the SHE as the other. Since a voltmeter indicates direction of electron flow as well as potential difference, the measurement determines not only the value of the standard cell potential but also identifies which electrode is the source of electrons. Figures 19.15 and 19.16 illustrate the measurement of the potential of the Cu^{2+}/Cu and the Zn^{2+}/Zn half-cells. The voltmeter reading in Figure 19.15 indicates (a) that the cell potential is 0.34 V, and (b) that electrons are flowing through the external circuit *into* the copper half-cell. Therefore, the copper half-cell is the cathode, and its electrode reaction is the reduction $Cu^{2+}(aq) + 2\ e^- \rightarrow Cu(s)$. The cell potential, measured under standard conditions against a SHE, is the **standard reduction potential (SRP)** of the Cu/Cu^{2+} half-cell. $\mathscr{E}^\circ_{red}\ (Cu^{2+}/Cu) = 0.34$ V.

Figure 19.16 shows a similar measurement of the Zn^{2+}/Zn half-cell. The meter shows a voltage of 0.76 V, and shows also that electrons are flowing through the external circuit *out of* the zinc half-cell. Since the zinc half-cell is the source of electrons, it must be the anode. The electrode half-reaction is the oxidation $Zn(s) \rightarrow Zn^{2+}(aq) + 2\ e^-$. The measured potential, +0.76 V, is therefore the standard *oxidation* potential of the Zn^{2+}/Zn half-cell.

Oxidation and reduction potentials for a given half-cell have the same absolute value but opposite signs. If the potential for the *oxidation* $Zn(s) \rightarrow Zn^{2+}(1\ M) + 2\ e^-$ is +0.76 V, the potential for the *reduction* $Zn^{2+}(1\ M) + 2\ e^- \rightarrow Zn(s)$ is −0.76 V.

$$\mathscr{E}^\circ_{ox} = -\mathscr{E}^\circ_{red} \qquad \text{(any half-reaction)} \qquad (19\text{-}8)$$

A *positive* value of the SRP for a half-cell means that, when connected to a hydrogen electrode under standard conditions, the half-cell undergoes a *reduction* half-reaction. If a half-cell's SRP is negative, its half-reaction when connected to a SHE is an oxidation.

Standard reduction potentials have been measured for a large number of different half-cells, and the results are listed in tables such as Table 19.2. These tables usually give the reduction half-reaction, so that oxidized and reduced forms are clearly identified, and the number of electrons involved in the reduction is unambiguously shown. Sometimes the term "standard *electrode* potentials" is used instead of "standard reduction potential."

An expanded version of Table 19.2, with many more entries, appears in Appendix I.

Relative Oxidizing Strength and Ease of Reduction

The positive sign of the standard reduction potential of Cu^{2+}/Cu means that the reduction of Cu^{2+} in the overall cell reaction $Cu^{2+}(1\ M) + H_2(1\ atm) \rightarrow 2\ H^+(1\ M) + Cu(s)$ takes place *spontaneously* when a Cu^{2+}/Cu half-cell is connected to a SHE. The *negative* sign of the standard reduction potential of Zn^{2+}/Zn, on the other hand, means that a positive sign is associated with the *oxidation* rather than the reduction. It is the oxidation of Zn(s) in the overall cell reaction $2\ H^+(1\ M) + Zn(s) \rightarrow Zn^{2+}(1\ M) + H_2(1\ atm)$ that occurs spontaneously. It follows that Cu^{2+} is a *stronger oxidizing agent* than H^+. (If H^+ were the stronger oxidizing agent, the overall cell reaction $2\ H^+ + Cu \rightarrow Cu^{2+} + H_2$ would occur.) Similarly, Zn^{2+} is a *weaker* oxidizing agent than H^+. A table of SRP's is in fact an ordered list of oxidants, with the strongest oxidizing agent (F_2) at the top and the weakest (Li^+) at the bottom. This aspect of the table is remembered more easily if you memorize the fact that F_2 is one of the strongest known oxidizing agents.

We can equally well conclude, from the fact that $Cu^{2+} + H_2 \rightarrow 2\ H^+ + Cu(s)$ occurs spontaneously, that H_2 is a *stronger reducing agent* than Cu. With this in mind, the table of SRP's can also be seen as an ordered list of reductants, with the strongest at the bottom. You should memorize the fact that metallic lithium is one of the strongest known reducing agents.

There is a *reciprocal relationship* in electron-transfer reactions that is analogous to the reciprocal relationship between conjugate acids and bases in proton-transfer reactions. A reducing agent is strong if it loses its electron(s) readily, that is, if it is easily oxidized. Since the oxidized form has no great tendency to regain those easily-lost electrons, it is a relatively weak oxidizing agent. The reciprocal relationship in electron-transfer strength can be summarized in diagrammatic form.

Conjugate pairs and the strong acid/weak base relationship are discussed in Section 15.4.

$$\text{strong oxidant} \underset{-e^-}{\overset{+e^-}{\rightleftarrows}} \text{weak reductant}$$

$$\text{strong reductant} \underset{+e^-}{\overset{-e^-}{\rightleftarrows}} \text{weak oxidant}$$

TABLE 19.2 Standard Reduction Potentials in Aqueous Solution at 25 °C

Reduction		Half-Reaction		$\mathcal{E}°$(V)
F_2(g)	+	2 e^-	$\longrightarrow$ 2 F^-(aq)	+2.87
Ce^{4+}(aq)	+	e^-	$\longrightarrow$ Ce^{3+}(aq)	+1.61
MnO_4^-(aq) + 8 H^+(aq)	+	5 e^-	$\longrightarrow$ Mn^{2+}(aq) + 4 H_2O	+1.51
Cl_2(g)	+	2 e^-	$\longrightarrow$ 2 Cl^-(aq)	+1.358
$Cr_2O_7^{2-}$(aq) + 14 H^+(aq)	+	6 e^-	$\longrightarrow$ 2 Cr^{3+}(aq) + 7 H_2O	+1.33
O_2(g) + 4 H^+(aq)	+	4 e^-	$\longrightarrow$ 2 H_2O	+1.229
IO_3^-(aq) + 6 H^+(aq)	+	5 e^-	$\longrightarrow$ ½ I_2(aq) + 3 H_2O	+1.195
ClO_4^-(aq) + 2 H^+(aq)	+	2 e^-	$\longrightarrow$ ClO_3^-(aq) + H_2O	+1.19
Br_2(ℓ)	+	2 e^-	$\longrightarrow$ 2 Br^-(aq)	+1.066
NO_3^-(aq) + 4 H^+(aq)	+	3 e^-	$\longrightarrow$ NO(g) + 2 H_2O	+0.96
NO_3^-(aq) + 3 H^+(aq)	+	2 e^-	$\longrightarrow$ HNO_2(aq) + H_2O	+0.94
2 Hg^{2+}(aq)	+	2 e^-	$\longrightarrow$ Hg_2^{2+}(aq)	+0.920
Ag^+(aq)	+	e^-	$\longrightarrow$ Ag(s)	+0.7994
F^{3+}(aq)	+	e^-	$\longrightarrow$ Fe^{2+}(aq)	+0.771
I_2(s)	+	2 e^-	$\longrightarrow$ 2 I^-(aq)	+0.535
Cu^+(aq)	+	e^-	$\longrightarrow$ Cu(s)	+0.521
O_2(g) + 2 H_2O(ℓ)	+	4 e^-	$\longrightarrow$ 4 OH^-(aq)	+0.40
Cu^{2+}(aq)	+	2 e^-	$\longrightarrow$ Cu(s)	+0.337
Hg_2Cl_2(s)	+	2 e^-	$\longrightarrow$ 2 Hg(ℓ) + 2 Cl^-	+0.27
AgCl(s)	+	e^-	$\longrightarrow$ Ag(s) + Cl^-	+0.222
Cu^{2+}(aq)	+	e^-	$\longrightarrow$ Cu^+(aq)	+0.153
S(s) + 2 H^+(aq)	+	2 e^-	$\longrightarrow$ H_2S(aq)	+0.14
AgBr(s)	+	e^-	$\longrightarrow$ Ag(s) + Br^-	+0.0713
2 H^+	+	**2 e^-**	$\longrightarrow$ **H_2(g)** [definition]	**0.000**
Fe^{3+}(aq)	+	3 e^-	$\longrightarrow$ Fe(s)	−0.037
Pb^{2+}(aq)	+	2 e^-	$\longrightarrow$ Pb(s)	−0.126
Sn^{2+}(aq)	+	2 e^-	$\longrightarrow$ Sn(s)	−0.14
Ni^{2+}(aq)	+	2 e^-	$\longrightarrow$ Ni(s)	−0.25
Cd^{2+}(aq)	+	2 e^-	$\longrightarrow$ Cd(s)	−0.403
Cr^{3+}(aq)	+	e^-	$\longrightarrow$ Cr^{2+}(aq)	−0.41
Fe^{2+}(aq)	+	2 e^-	$\longrightarrow$ Fe(s)	−0.44
Zn^{2+}(aq)	+	2 e^-	$\longrightarrow$ Zn(s)	−0.763
2 H_2O	+	2 e^-	$\longrightarrow$ H_2(g) + 2 OH^-(aq)	−0.8277
Al^{3+}(aq)	+	3 e^-	$\longrightarrow$ Al(s)	−1.66
Mg^{2+}(aq)	+	2 e^-	$\longrightarrow$ Mg(s)	−2.37
Na^+(aq)	+	e^-	$\longrightarrow$ Na(s)	−2.714
Li^+(aq)	+	e^-	$\longrightarrow$ Li(s)	−3.045

The *form* of the table of SRP's, and the information it contains, can be remembered as presented in Table 19.3.

Cell Potential and Cell Reaction

The table of SRP's supplies the information necessary to determine, by calculation and without measurement, the cell potential and overall cell reaction of many different voltaic cells. The procedure for this is laid out in Example 19.2.

TABLE 19.3 Standard Reduction Potentials in Aqueous Solution at 25°C

Oxidizing Agents			Reducing Agents	Approximate $\mathscr{E}°/V$
(strongest)			(weakest)	
$F_2(g)$	$+ 2\,e^-$	$\longrightarrow$	$2\,F^-(aq)$	+3
oxidizing power ↑ ease of reduction ↑			reducing power ↓ ease of oxidation ↓	
$Li^+(aq)$	$+ e^-$	$\longrightarrow$	Li(s)	−3
(weakest)			(strongest)	

EXAMPLE 19.2 Calculation of the Potential of a Voltaic Cell

Write the overall cell reaction and calculate the standard cell potential of a voltaic cell composed of a $Ni^{2+}(1\ M)|Ni(s)$ half-cell and a $Cu^{2+}(1\ M)|Cu(s)$ half-cell.

Analysis We know that cell potential arises from the independent contributions of the two half-cell potentials, and that one of these is a *reduction* potential while the other is an *oxidation* potential. From the relative positions of Ni^{2+} and Cu^{2+} in Table 19.2 we see that Cu^{2+} is a stronger oxidant than Ni^{2+}, so that Ni(s) must be oxidized by Cu^{2+}. In the process, of course, Cu^{2+} is reduced. Thus, the anode is the $Ni^{2+}(1\ M)|Ni(s)$ half-cell and the cathode is the $Cu^{2+}(1\ M)|Cu(s)$ half-cell.

Plan Use Table 19.2 to write down the $Cu^{2+} \rightarrow Cu$ reduction half-reaction, and the $Ni \rightarrow Ni^{2+}$ oxidation half-reaction, together with the standard potentials. Convert the SRP of the nickel half-reaction to an oxidation potential by changing its sign.

Add the two half-reactions to get the overall cell reaction, and add the two half-cell potentials to get the cell potential.

Work

$$\begin{array}{lrl}
\textbf{Reduction} & Cu^{2+}(1\ M) + 2\,e^- \longrightarrow Cu(s); & \mathscr{E}^\circ_{red} = +0.34\ V \\
\textbf{Oxidation} & Ni(s) \longrightarrow Ni^{2+}(1\ M) + 2\,e^-; & \mathscr{E}^\circ_{ox} = -\mathscr{E}^\circ_{red} = -[-0.25] = +0.25\ V \\
\hline
\textbf{Overall} & Ni(s) + Cu^{2+}(1\ M) \longrightarrow Ni^{2+}(1\ M) + Cu(s); & \mathscr{E}^\circ_{cell} = \mathscr{E}^\circ_{red} + \mathscr{E}^\circ_{ox} \\
 & & \mathscr{E}^\circ_{cell} = (+0.34\ V) + (+0.25\ V) \\
 & & \mathscr{E}^\circ_{cell} = +0.59\ V
\end{array}$$

Check Did you reverse the sign of the tabulated $\mathscr{E}°$ for the oxidation half-reaction? This must *always* be done when half-cell potentials are added to obtain an overall cell potential.

Exercise

What are the overall cell reaction and cell potential of a voltaic cell composed of the two half-cells, $Sn^{2+}(1\ M)|Sn(s)$ and $Cu^{2+}(1\ M)|Cu$?

Answer: $Sn(s) + Cu^{2+}(1\ M) \longrightarrow Sn^{2+}(1\ M) + Cu(s)$; $\mathscr{E}^{\circ}_{cell} = +0.48\ V$

Two very important relationships were used in the solution to Example 19.2. Equation (19-8) relates the oxidation and reduction potentials of any one half-cell, and Equation (19-9) relates the cell potential to the half-cell potentials.

$$\mathscr{E}^{\circ}_{ox} = -\mathscr{E}^{\circ}_{red} \qquad \text{(any half-reaction)} \qquad (19\text{-}8)$$

$$\mathscr{E}^{\circ}_{cell} = \mathscr{E}^{\circ}_{red} + \mathscr{E}^{\circ}_{ox} \qquad (19\text{-}9)$$

In many cases, the stoichiometric coefficients of one or both half-reactions must be multiplied by a small integer in order to equalize the number of electrons. This is the same procedure used in the ion-electron method for redox equation balancing (Section 18.3), and the half-reactions may be added only if the electrons cancel. The equation for the overall cell reaction must reflect the fact that electrons are neither produced nor consumed in cell reactions, they are *transferred.* However, the value of $\mathscr{E}^{\circ}_{red}$ (or $\mathscr{E}^{\circ}_{ox}$) must *not* be changed when equalizing the number of electrons. Voltage is an intensive quantity (like density), *independent of the amount of material.* Thus, for instance,

Recall the water wheel analogy of Figure 19.1. The height of the waterfall (the gravitational potential) does not depend on the amount of water passing through.

$$Ag^{+} + e^{-} \longrightarrow Ag(s); \qquad \mathscr{E}^{\circ}_{red} = +0.80\ V \qquad \text{and}$$

$$2\ Ag^{+} + 2\ e^{-} \longrightarrow 2\ Ag(s); \qquad \mathscr{E}^{\circ}_{red} = +0.80\ V$$

are both correct.

In calculating half-cell potentials it is convenient but not necessary to know in advance which half-cell is the cathode, as was done in Example 19.2. If it is not known, the *assumption* is made that one or the other is the cathode, and the calculation is carried out on the basis of that guess. If the resulting cell potential is positive, the guess was correct. If it is negative, the guess was incorrect. In the latter case, the correct results are obtained by simply inverting the sign of the potential and the direction of the arrow in the chemical equations.

EXAMPLE 19.3 Cell Potential with Uncertain Electrode Identification

Calculate the potential of the cell composed of the two half-cells $Ni(s)|Ni^{2+}(1\ M)$ and $Pt|Fe^{2+}(1\ M), Fe^{3+}(1\ M)$. What is the overall cell reaction?

Analysis The procedure given in the preceding Example may be followed, provided we can identify the anode and the cathode.

Plan Arbitrarily choose one or the other as the cathode, then proceed as in Example 19.2.

Work Assume that the nickel electrode is the cathode.

Reduction $Ni^{2+}(1\ M) + 2\ e^- \longrightarrow Ni(s);$ $\qquad \mathscr{E}^\circ_{red} = -0.25\ V$

Oxidation $Fe^{2+}(1\ M) \longrightarrow Fe^{3+}(1\ M) + e^-;$ $\qquad \mathscr{E}^\circ_{ox} = -\mathscr{E}^\circ{}_{red} = -(+0.771\ V)$

The iron half-reaction must be multiplied by 2, so that the number of electrons is the same in both half-reactions. The half-cell potential is *not* changed by this multiplication.

Oxidation $2\ Fe^{2+}(1\ M) \longrightarrow 2\ Fe^{3+}(1\ M) + 2\ e^-;$ $\qquad \mathscr{E}^\circ_{ox} = -(+0.771\ V)$

The two reactions and potentials are added.

Reduction	$Ni^{2+}(1\ M) + 2\ e^- \longrightarrow Ni(s);$	$\mathscr{E}^\circ_{red} = -0.25\ V$
Oxidation	$2\ Fe^{2+}(1\ M) \longrightarrow 2\ Fe^{3+}(1\ M) + 2\ e^-;$	$\mathscr{E}^\circ_{ox} = -0.771\ V$
Cell	$Ni^{2+}(1\ M) + 2\ Fe^{2+}(1\ M) \longrightarrow 2\ Fe^{3+}(1\ M) + Ni(s);$	$\mathscr{E}^\circ_{cell} = -1.02\ V$

The fact that the cell potential is negative means that the starting guess was incorrect. The *iron* electrode is actually the cathode, and it is Fe^{3+}, not Ni^{2+}, that is reduced in the cell reaction. The correct reaction and cell potential are the reverse of those obtained above.

The potential of a voltaic cell is *always* a positive quantity.

$$2\ Fe^{3+}(1\ M) + Ni(s) \longrightarrow Ni^{2+}(1\ M) + 2\ Fe^{2+}(1\ M); \qquad \mathscr{E}^\circ_{cell} = +1.02\ V$$

Exercise

Find the overall cell reaction and the potential of the voltaic cell composed of the half-cells $Ag^+(1\ M)|\ Ag(s)$ and $Cu^{2+}(1\ M)|\ Cu(s)$.

Answer: $2\ Ag^+(1\ M) + Cu(s) \longrightarrow 2\ Ag(s) + Cu^{2+}(1\ M);$ $\qquad \mathscr{E}^\circ_{cell} = +0.46\ V$

Spontaneity of Cell Reactions

The results of the preceding sections may be generalized as follows. The shorthand diagram of a voltaic cell is written (left half-cell: oxidation)||(right half-cell: reduction). The oxidation (anode) reaction is a source of electrons, which flow from left to right in the external circuit. The "oxidation-is-left" rule allows the half-reactions and overall reaction to be written from the cell diagram. Furthermore, cell potential is calculated on the basis of this rule. A positive result proves that the left–right assignments in the cell diagram were correct. If the result is negative, the cell diagram was incorrectly written: the overall cell reaction is *not* spontaneous as written, but is spontaneous in the *reverse* direction.

The relationship between the sign of cell potential and the direction of spontaneous reaction has implications beyond electrochemical cells, since redox reactions also take place outside the apparatus of a voltaic cell (Figure 19.17). A piece of metallic copper dropped into a solution of silver nitrate reacts spontaneously according to $Cu(s) + 2\ Ag^+ \rightarrow Cu^{2+}(aq) + 2\ Ag(s)$. However, a piece of silver dropped into a solution of copper nitrate does not react at all. The reaction $Cu^{2+}(aq) + 2\ Ag(s) \rightarrow Cu(s) + 2\ Ag^+$ is *not* spontaneous. The direction of spontaneity of a redox reaction is the same whether it takes place in an electrochemical cell or not. *If the calculated cell potential is positive, the reaction is spontaneous as written.*

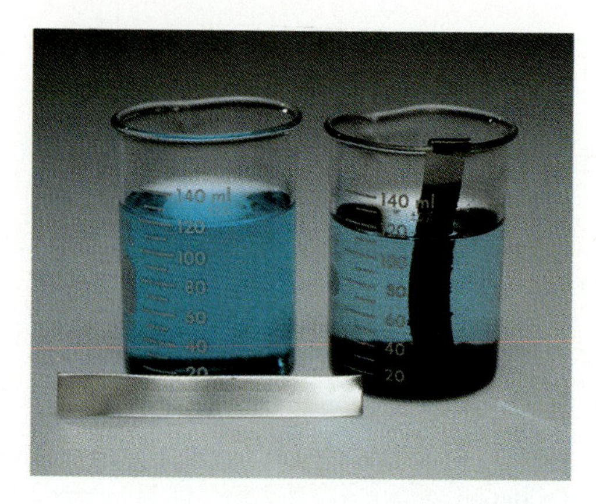

Figure 19.17 Redox reactions having a positive cell potential take place spontaneously when the reactants are mixed, regardless of whether or not they are in an electrochemical cell. A clean strip of zinc (seen in front of the beaker on the left) is placed in a dilute solution of $CuSO_4$. The reaction $Cu^{2+} + Zn \longrightarrow Zn^{2+} + Cu$, for which $\mathscr{E}° = 1.1$ V, begins immediately. Later, the zinc strip has become covered with a dark coating of finely divided copper, while the paler color of the solution indicates that the concentration of Cu^{2+} has decreased.

EXAMPLE 19.4 Reaction Spontaneity Outside of Voltaic Cells

Describe the results of the following two experiments.

a. A piece of zinc is dropped into a 1 M $AgNO_3$ solution.

b. A piece of silver is dropped into a 1 M $Zn(NO_3)_2$ solution.

Analysis Since the reactants in (a) are Zn(s) and Ag^+(1 M), the redox reaction $Zn(s) + 2\ Ag^+(1\ M) \longrightarrow Zn^{2+}(aq) + 2\ Ag(s)$ is a possible outcome. The reverse reaction is a possible outcome of experiment (b). One of these is spontaneous as written, the other is not. If the reaction in (a) is spontaneous, it can be made the basis of a voltaic cell having a *positive* cell potential. If it is not spontaneous, calculation of the corresponding cell potential yields a negative number.

Plan Calculate the potential of a cell having the target reaction (a) as its overall cell reaction, note the sign, and decide on spontaneity.

Work

a. Assume that Ag^+ is reduced in the cell reaction.

Reduction	$2\ Ag^+ + 2\ e^- \longrightarrow 2\ Ag(s);$	$\mathscr{E}^\circ_{red} = +0.80$ V
Oxidation	$Zn(s) \longrightarrow Zn^{2+}(aq) + 2\ e^-;$	$\mathscr{E}^\circ_{ox} = -[\mathscr{E}^\circ_{red}] = -[-0.763] = +0.76$ V
Cell	$Zn(s) + 2\ Ag^+ \longrightarrow 2\ Ag(s) + Zn^{2+}(aq);$	$\mathscr{E}^\circ_{cell} = +1.56$ V

Since the potential is positive, the reaction is spontaneous as written, and it occurs when metallic zinc comes in contact with aqueous silver ion.

b. Assume that Zn^{2+} is reduced in the reaction.

Reduction	$Zn^{2+}(aq) + 2\ e^- \longrightarrow Zn(s);$	$\mathscr{E}^\circ_{red} = -0.76$ V
Oxidation	$2\ Ag(s) \longrightarrow 2\ Ag^+ + 2\ e^-;$	$\mathscr{E}^\circ_{ox} = -[\mathscr{E}^\circ_{red}] = -[+0.80] = -0.80$ V
Cell	$2\ Ag(s) + Zn^{2+}(aq) \longrightarrow Zn(s) + 2\ Ag^+;$	$\mathscr{E}^\circ_{cell} = -1.56$ V

Since this potential is negative, the reaction is not spontaneous as written, and no reaction occurs when metallic silver comes in contact with aqueous zinc ion.

Note: It was not necessary to do both (a) and (b), because the result of one may be inferred from the result of the other.

Exercise

What reaction, if any, takes place when metallic copper contacts 1 M aqueous nickel(II)?

When a nickel bar is placed in contact with a solution containing Cu^{2+} ions (left), metallic copper plates out on the nickel. The color changes from blue to green as Cu^{2+} ions are consumed and replaced by newly formed Ni^{2+} ions in the solution (right). No reaction occurs when a bar of copper is placed in a solution containing Ni^{2+} ions (not shown).

Answer: No reaction.

Concentration Dependence

When a voltaic cell has electrolyte concentration(s) different from the standard value of 1 M, the voltage it generates is different from the standard cell potential. A familiar example is the flashlight battery, a voltaic cell whose voltage decreases continuously as the cell operates. The concentration of reactants decreases as the cell reaction proceeds. When the reactants are used up, the voltage has dropped to zero—the battery is dead.

The procedure for calculating nonstandard potentials requires the evaluation of the *reactant quotient* (Q). Recall that the numerator of the reaction quotient consists of the concentration (or pressure, if the species is a gas) of each product of the reaction, each raised to a power equal to its stoichiometric coefficient in the balanced equation. The denominator includes the concentrations (pressures) of the reactants, raised to the stoichiometric coefficients. If the substance is a pure solid, it is omitted from the reaction quotient. For example, the reaction quotient of the reaction

Reaction quotients were discussed in Chapter 13, beginning on p. 558.

$$2\ Fe^{3+}(aq) + Zn(s) \longrightarrow 2\ Fe^{2+}(aq) + Zn^{2+}(aq)$$

is

$$Q = \frac{[Fe^{2+}]^2[Zn^{2+}]}{[Fe^{3+}]^2}$$

When all concentrations are known, the cell potential under nonstandard conditions can be calculated from the **Nernst equation** (19-10).

$$\mathscr{E} = \mathscr{E}^\circ - \frac{RT}{n\mathscr{F}} \ln Q \qquad (19\text{-}10)$$

In this equation, R is the gas constant, 8.314 J mol^{-1} K^{-1}; T is the absolute temperature (K); n is the number of electrons transferred in the redox reaction; and $\mathscr{F}$ is the total charge on one mole of electrons. The charge carried by one electron is the elementary

Most calculations in electrochemistry are carried to no more than three significant figures. Use 96,500 or 9.65×10^4 C mol^{-1}.

charge, 1.602×10^{-19} C, so that the charge carried by one mole of electrons is easily calculated.

$$\mathcal{F} = (1.6021773 \times 10^{-19} \text{ C per electron}) \times (6.0221367 \times 10^{23} \text{ electrons per mole})$$

$$\mathcal{F} = 96{,}485.309 \text{ C mol}^{-1}$$

Some facts about Michael Faraday, the English scientist after whom this constant is named, are given on page 48.

This physical constant, electron charge per mole, is called the **faraday.** The relationship among the quantities potential, charge, and energy allows the faraday to be expressed either as 96,485 C mol^{-1} or, equivalently, as 96,485 J V^{-1} mol^{-1}.

A factor of 2.303 is customarily included in Equation (19-10) to convert from natural to common logarithms. Most calculations refer to a temperature of 25 °C, at which the factor $(2.303 \cdot R \cdot T/\mathcal{F})$ has the value 0.0592. Thus, in its most-used form, at 25 °C, the Nernst equation reduces to

$$\mathcal{E} = \mathcal{E}° - \frac{0.0592}{n}\log(Q) \qquad (19\text{-}11)$$

$\ln x = 2.303 \cdot \log_{10}x$

The number of electrons transferred, n, does not appear explicitly in a balanced redox equation. The value of n must be determined from the half-reactions or the change in oxidation numbers. Note that when *all* concentrations are 1 M (or 1 atm for gases), the value of Q is 1, $\log(Q) = 0$, and $\mathcal{E} = \mathcal{E}°$; that is, $\mathcal{E} = \mathcal{E}°$ under standard conditions.

EXAMPLE 19.5 Cell Potential Under Non-Standard Conditions

Calculate the cell potential (at 25 °C) of the following voltaic cell:

$$\text{Pt}|Cl_2(2.50 \text{ atm})|Cl^-(1.00 \text{ M})||Fe^{3+}(0.050 \text{ M}), Fe^{2+}(0.0010 \text{ M})|\text{Pt}$$

Analysis

Target: A nonstandard cell potential.

Known: Temperature, identity of reagents and their concentrations in the half-cells.

Relationship: The Nernst equation applies, but it cannot be used unless $\mathcal{E}°$ is known.

New Target: $\mathcal{E}°$ for the given cell.

Plan Find $\mathcal{E}°$ by the method of Example 19.3, then use it and the Nernst equation to calculate $\mathcal{E}_{cell}$.

Work Chlorine is on the left in the shorthand cell diagram, so we assume it to be the anode.

Oxidation	$2\,Cl^-(1.00\text{ M}) \longrightarrow Cl_2(2.50\text{ atm}) + 2\,e^-$;	$\mathcal{E}°_{ox} = -\mathcal{E}°_{red} = -(+1.358)$ V
Reduction	$2\,Fe^{3+}(0.050\text{ M}) + 2\,e^- \longrightarrow 2\,Fe^{2+}(0.0010\text{ M})$;	$\mathcal{E}°_{red} = +0.771$ V
Cell	$2\,Fe^{3+}(0.050\text{ M}) + 2\,Cl^-(1.00\text{ M}) \longrightarrow Cl_2(2.50\text{ atm}) + 2\,Fe^{2+}(0.0010\text{ M})$;	$\mathcal{E}°_{cell} = -0.587$ V

Since the temperature is 25 °C, the common version of the Nernst equation (19-11) can be used.

$$\mathcal{E} = \mathcal{E}° - \frac{0.0592}{n}\log(Q) \qquad (19\text{-}11)$$

The half-reactions show that $n = 2$, and Q is obtained from the balanced cell reaction.

$$\begin{aligned}\mathscr{E} &= \mathscr{E}^\circ - \frac{0.0592}{2}\log\frac{[Fe^{2+}]^2 \cdot P_{Cl_2}}{[Fe^{3+}]^2[Cl^-]^2} \\ &= -0.587 - \frac{0.0592}{2}\log\frac{(0.0010)^2(2.50)}{(0.050)^2(1.00)^2} \\ &= -0.587 - \frac{0.0592}{2}\log(0.0010) \\ &= -0.587 - \frac{0.0592}{2}(-3.00) \\ &= -0.587 + 0.089 = -0.498\text{ V}\end{aligned}$$

Use units of mol L^{-1} for dissolved species and atm for gaseous species when substituting quantities into the Nernst equation.

Note: Since the calculated value of $\mathscr{E}$ is *negative,* the cell reaction *does not* occur as written. Instead, it is spontaneous in the reverse direction.

The potential of a spontaneously operating voltaic cell is a *positive quantity*.

$$Cl_2(2.50\text{ atm}) + 2\ Fe^{2+}(0.0010\text{ M}) \longrightarrow 2\ Fe^{3+}(0.050\text{ M}) + 2\ Cl^-(1.00\text{ M})$$

The correct cell diagram is

$$Pt|Fe^{2+}(0.0010\text{ M}), Fe^{3+}(0.050\text{ M})||Cl^-(1.00\text{ M})|Cl_2(2.50\text{ atm})|Pt$$

and the voltage is +0.498, not −0.498.

Exercise

For the cell $Cu(s)|Cu^{2+}(0.500\text{ M})||Ag^+(0.00225\text{ M})|Ag(s)$, calculate the potential at 25 °C.

Answer: $\mathscr{E}_{cell} = +0.32$ V.

EXAMPLE 19.6 Finding an Unknown Concentration

For the cell of Example 19.3, whose reaction and standard potential are $2\ Fe^{3+}(aq) + Ni(s) \rightarrow 2\ Fe^{2+}(aq) + Ni^{2+}(aq)$; $\mathscr{E}^\circ = +1.020$ V, calculate the concentration of $Ni^{2+}(aq)$ if $[Fe^{3+}] = [Fe^{2+}]$ and the measured cell potential is 1.070 V at 25 °C.

Analysis Any cell potential problem having at least one concentration different from 1.00 M or having an unknown concentration requires the use of the Nernst equation. In this case, the value of $\mathscr{E}^\circ$ is known from previous work, but often $\mathscr{E}^\circ$ will also have to be calculated.

Plan Set up the Nernst equation, insert all known values, and evaluate the target quantity, $[Ni^{2+}]$.

Work

$$\mathscr{E} = \mathscr{E}^\circ - \frac{0.0592}{n}\log(Q)$$

The oxidation number of Ni changes from 0 to +2 in the reaction, $n = 2$. The form of Q is available from the balanced cell reaction.

$$\mathscr{E} = \mathscr{E}^\circ - \frac{0.0592}{2}\log\frac{[Fe^{2+}]^2[Ni^{2+}]}{[Fe^{3+}]^2}$$

The concentrations of Fe^{3+} and Fe^{2+} are the same, so these factors cancel, leaving only $[Ni^{2+}]$. Solve the equation for $\log[Ni^{2+}]$, insert the known values, and evaluate.

$$(\mathscr{E} - \mathscr{E}°) = -\frac{0.0592}{2}\log[Ni^{2+}]$$

$$\log[Ni^{2+}] = \frac{2}{0.0592}\cdot(\mathscr{E}° - \mathscr{E})$$

$$= \frac{2}{0.0592}\cdot(1.020 - 1.070)$$

$$= -1.69$$

Take the antilog to find $[Ni^{2+}]$ from $\log[Ni^{2+}]$

$$[Ni^{2+}] = 10^{-1.69} = 0.020\ M$$

Note: The difficulty with units in logarithmic and exponential quantities arises again in electrochemistry. We deal with it as we did in the chapters on equilibrium and kinetics. That is, we ignore units when taking logarithms, and, as we have done here, *insert* the correct units after taking an antilogarithm. The units of the concentrations in the Nernst equation are always moles per liter.

Exercise

For the cell $Sn(s)|Sn^{2+}(aq)||Cu^{2+}(aq)|Cu(s)$, calculate the standard cell potential and the Cu^{2+} molarity required to produce a potential $\mathscr{E} = +0.37$ V when the temperature is 25 °C and $[Sn^{2+}] = 1.00$ M.

Answer: $\mathscr{E}° = +0.48$ V, $[Cu^{2+}] = 1.9 \times 10^{-4}$ M.

Measurement of pH

The effect of concentration on the potential of a cell can be used to measure the pH of a solution. Consider a voltaic cell made up of a hydrogen anode of unknown H^+ molarity, and a SHE cathode: $Pt|H_2(1\ atm)|H^+\ (x\ M)||H^+(1.00\ M)|H_2(1\ atm)|Pt$. The half-reactions and standard potentials are

Anode	$H_2(1.00\ atm) \longrightarrow 2\ H^+(x\ M) + 2\ e^-$;	$\mathscr{E}°_{ox} = 0.00$ V
Cathode	$2\ H^+(1.00\ M) + 2\ e^- \longrightarrow H_2(1.00\ atm)$;	$\mathscr{E}°_{red} = 0.00$ V
Cell	$2\ H^+(1.00\ M) \longrightarrow 2\ H^+(x\ M)$;	$\mathscr{E}°_{cell} = 0.00$ V

Since $\mathscr{E}° = 0.00$ V, the Nernst equation for this cell reduces to

$$\mathscr{E} = -\frac{0.0592}{2}\cdot\log\frac{[H^+]^2}{1.00^2}$$

where $[H^+]$ is the unknown concentration in the anode half-cell,

$$\mathscr{E} = -\frac{0.0592}{2}\cdot\log[H^+]^2$$

$$\mathscr{E} = -0.0592\cdot\log[H^+]$$

The factor of 2 in the denominator cancels out, because of the relationship $\log(x^2) = 2\log(x)$ in the numerator.

Since by definition $pH = -\log[H^+]$,

$$\mathscr{E} = +0.0592\cdot pH \qquad \text{or} \qquad pH = \frac{\mathscr{E}}{0.0592} = 16.9\cdot\mathscr{E}$$

Measurement of the potential of this voltaic cell provides a direct determination of the pH of an aqueous solution. This type of voltaic cell is called a **concentration cell**, because its potential arises from a concentration difference between the electrolytes of otherwise identical electrodes.

Because it requires a gas storage tank and a regulating device for maintaining a constant gas pressure, and because hydrogen is a dangerously flammable gas, the hydrogen electrode is not used for routine pH measurements. Modern instruments for measuring pH are based on a different principle: they sense the potential difference that develops when a H^+ concentration difference exists across a very thin glass membrane. Figure 19.18 shows a small pH meter suitable for both laboratory and field use.

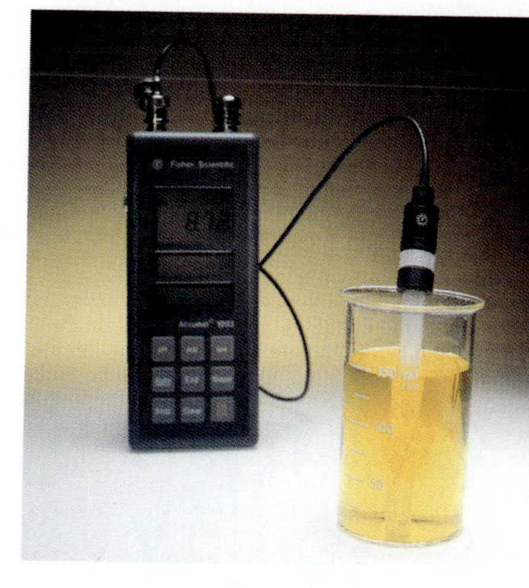

Figure 19.18 A portable pH meter with digital readout. The sensing probe contains both electrodes of a voltaic cell in a single structure. The beaker contains an aqueous solution of potassium chromate, which is basic because of the OH^- formed in the equilibrium $2\ CrO_4^{2-} + H_2O \rightleftharpoons Cr_2O_7^{2-} + 2\ OH^-$.

19.4 COMMONLY USED VOLTAIC CELLS

The reaction in a voltaic cell takes place spontaneously, causing an electrical current to flow in the external circuit. The current may then serve a variety of useful purposes, and voltaic cells used in this way are called *batteries*. Many different types of batteries have been developed, each with somewhat different characteristics and suited to a different use.

The Lead Storage Battery

The automobile battery, an indispensable part of the modern transportation system, uses lead grids for electrodes. The anode grid is filled with spongy, porous lead, and the cathode grid is filled with lead dioxide. The body of the cell is filled with $H_2SO_4(aq)$. The cell diagram is $Pb|PbSO_4(s)|H_2SO_4(aq)|PbO_2(s)|PbSO_4(s)|Pb(s)$. Note that there is no salt bridge or other porous barrier in this cell. The anode and cathode are in the same compartment and use the same electrolyte. When the battery discharges, supplying electrical power to the external circuit, the electrode reactions are

Anode $$Pb(s) + HSO_4^- \longrightarrow PbSO_4(s) + H^+ + 2\ e^-$$

Cathode $$PbO_2(s) + 3\ H^+ + HSO_4^- + 2\ e^- \longrightarrow PbSO_4(s) + 2\ H_2O(\ell)$$

The overall cell reaction is

$$Pb(s) + PbO_2(s) + 2\ H_2SO_4(aq) \longrightarrow 2\ PbSO_4(s) + 2\ H_2O(\ell)$$

During discharge, the sulfuric acid electrolyte becomes more dilute, and both electrodes become coated with lead sulfate. These electrode reactions are reversible, so that when the car's alternator supplies electrical current to the battery, $PbSO_4(s)$ is consumed and H_2SO_4 is produced. The fact that it is rechargeable is an important characteristic of the lead storage battery (Figure 19.19).

Dry Cell

The common flashlight battery or dry cell, patented by Georges Leclanché in 1866, is a convenient source of small amounts of electrical power (Figure 19.20). In the dry cell, a central graphite cathode is surrounded by a moist paste of MnO_2 and NH_4Cl. The anode, which serves also as the container for the electrolyte paste, is made of zinc. The cell potential is about 1.5 V, and the reactions are

Anode $$Zn(s) \longrightarrow Zn^{2+}(aq) + 2\ e^-$$

Cathode $$2\ NH_4^+ + 2\ MnO_2(s) + 2\ e^- \longrightarrow Mn_2O_3(s) + H_2O(\ell) + 2\ NH_3(aq)$$

Overall Cell Reaction $$2\ NH_4^+ + 2\ MnO_2(s) + Zn(s) \longrightarrow Zn^{2+}(aq) + Mn_2O_3(s) + H_2O(\ell) + 2\ NH_3(aq)$$

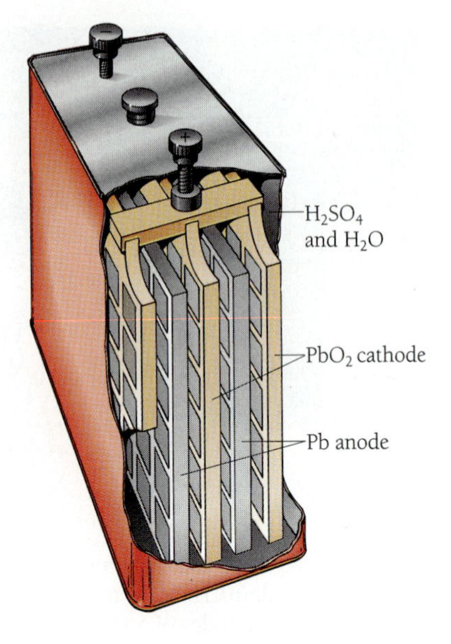

Figure 19.19 The lead storage battery. The cell potential, which decreases as the battery is discharged, is about 2 V. The electrical systems of most automobiles require 12 V, and an automobile battery contains six such cells.

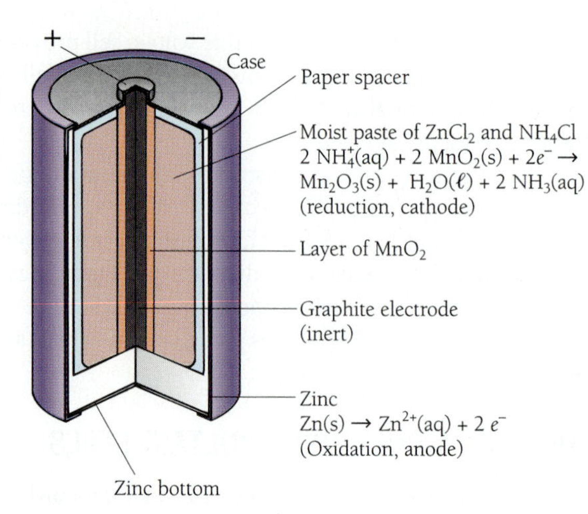

Figure 19.20 The dry cell, sometimes called the Leclanché cell after its inventor.

Unlike the lead storage battery, the dry cell cannot be recharged. As the cell discharges, the reaction product NH_3 diffuses away from the cathode. Application of electrical power to the cell cannot cause the cell reaction to proceed to the left, because the concentration of NH_3 *at the electrode* is too low. For a cell to be rechargeable, all species formed during discharge must remain in contact with the electrode at which they are produced. Otherwise, they will be unavailable as reactants during the reverse reaction that accompanies recharging.

The "alkaline battery" is similar to the dry cell, but uses NaOH in place of NH_4Cl in the electrolyte paste. Its voltage, also about 1.5 V, does not decrease during use as much as that of the dry cell. Furthermore, its voltage is less affected by temperature changes than that of the dry cell. Like the Leclanché cell, the alkaline battery cannot be recharged.

Other Rechargeable Batteries

A battery is called "dry" if its electrolyte is a moist paste rather than a true liquid solution.

Environmental considerations favor a reusable battery rather than a throwaway. The nickel-cadmium or *nicad* cell is a rechargeable, dry battery used in calculators, radios, electric razors, and portable power tools (Figure 19.21). Its electrode reactions are

Figure 19.21 The nicad cell is a convenient, rechargeable source of small amounts of electrical power, used in many small appliances.

Anode	$Cd(s) + 2\ OH^- \longrightarrow Cd(OH)_2(s) + 2\ e^-$
Cathode	$NiO(OH)(s) + H_2O(\ell) + e^- \longrightarrow Ni(OH)_2(s) + OH^-(aq)$
Overall	$2\ NiO(OH)(s) + Cd(s) + 2\ H_2O(\ell) \longrightarrow 2\ Ni(OH)_2(s) + Cd(OH)_2(s)$

Since all reactants and products of the overall reaction are pure solids (or solvent), there are no concentration terms in the reaction quotient: $Q = 1$. Therefore, $\mathscr{E} = \mathscr{E}°$, independent of the extent of reaction, and the cell potential is essentially unchanged as the cell discharges. The solid products, $Ni(OH)_2$ and $Cd(OH)_2$, cling to the electrodes and are available as reactants during recharging.

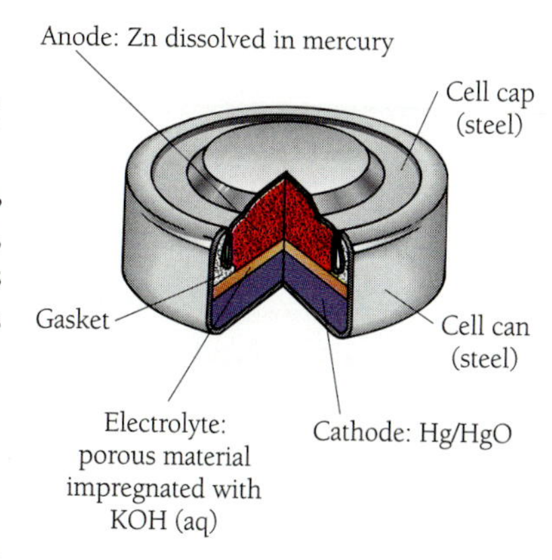

A mercury battery. The diameter of this cell is less than a centimeter.

Mercury and Silver Batteries

The mercury battery is characterized by a very stable voltage output (1.35 V) and a useful lifetime considerably longer than that of a Leclanché cell. Mercury batteries are widely used in wristwatches, hearing aids, and pacemakers. The reactions are

Anode	$Zn(s) + 2\ OH^- \longrightarrow ZnO(s) + H_2O(\ell) + 2\ e^-$
Cathode	$HgO(s) + H_2O(\ell) + 2\ e^- \longrightarrow Hg(\ell) + 2\ OH^-$
Cell	$HgO(s) + Zn(s) \longrightarrow ZnO(s) + Hg(\ell)$

A closely related battery has the same anode reaction, but uses silver rather than mercury in the cathode reaction.

$$AgO(s) + H_2O(\ell) + 2\ e^- \longrightarrow Ag(s) + 2\ OH^-$$

Both the mercury cell and the silver cell can be recharged, but are normally used as throwaways. The reaction quotient Q equals 1 in both the silver and the mercury batteries. Therefore, as these batteries discharge, their potential does not change significantly.

Fuel Cells

A **fuel cell** is a voltaic cell whose reactants are stored outside the cell and continuously replenished as they are consumed in the cell reaction. The chemical energy stored in fuels such as methane or hydrogen can be converted directly to electrical power in a fuel cell. Direct conversion is advantageous because it is more efficient than burning the fuel and using the heat to generate electrical power. High costs have prevented wide use of fuel cells, but they are ideal for certain applications, for example spacecraft power. Figure 19.22 is a diagram of the type of H_2-O_2 fuel cell used to produce electrical power in NASA manned space vehicles. The electrolyte is aqueous potassium hydroxide, and the only product of the cell reaction is H_2O. The cell is designed so that water is lost by evaporation as fast as it is produced. The fuel cell reactions are:

Anode	$2\ H_2(g) + 4\ OH^- \longrightarrow 4\ H_2O + 4\ e^-$
Cathode	$O_2(g) + 2\ H_2O(\ell) + 4\ e^- \longrightarrow 4\ OH^-(aq)$
Cell	$2\ H_2(g) + O_2(g) \longrightarrow 2\ H_2O(\ell)$

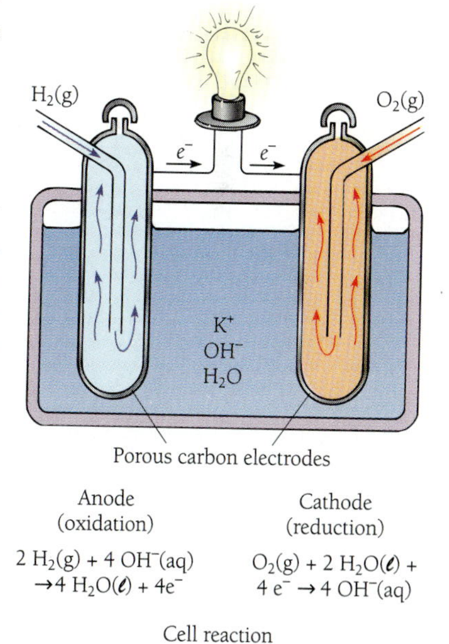

Figure 19.22 The hydrogen fuel cell is a nonpolluting source of power whose only product is water. Both anode and cathode are constructed of porous carbon impregnated with a small amount of Pt, Ag, or CoO catalyst.

19.5 ELECTROLYTIC CELLS

All of the electrochemical cells discussed so far have been voltaic cells, in which a spontaneous chemical reaction takes place with a consequent flow of electrical current

The rechargeable batteries discussed previously are voltaic cells while discharging, and electrolytic cells while being recharged.

in an external circuit. However, in many systems the process is reversible. An *electrolytic cell* is an electrochemical cell in which electrical energy, from an external source, causes a redox reaction to take place. The process is called **electrolysis** (Figure 19.23). For instance, if electrical energy is applied to water, gaseous hydrogen and oxygen are produced.

$$2\ H_2O(\ell) \xrightarrow[\text{energy}]{\text{electrical}} 2\ H_2(g) + O_2(g) \tag{19-12}$$

The electrode reactions are

$$\textbf{Anode} \quad 2\ H_2O(\ell) \longrightarrow O_2(g) + 4\ H^+(aq) + 4\ e^- \tag{19-13}$$

$$\textbf{Cathode} \quad 4\ H_2O(\ell) + 4\ e^- \longrightarrow 2\ H_2(g) + 4\ OH^-(aq) \tag{19-14}$$

$$\textbf{Cell} \quad 6\ H_2O(\ell) \longrightarrow 2\ H_2(g) + O_2(g) + 4\ H^+(aq) + 4\ OH^-(aq)$$

$$\text{or } 2\ H_2O(\ell) \longrightarrow 2\ H_2(g) + O_2(g) \tag{19-15}$$

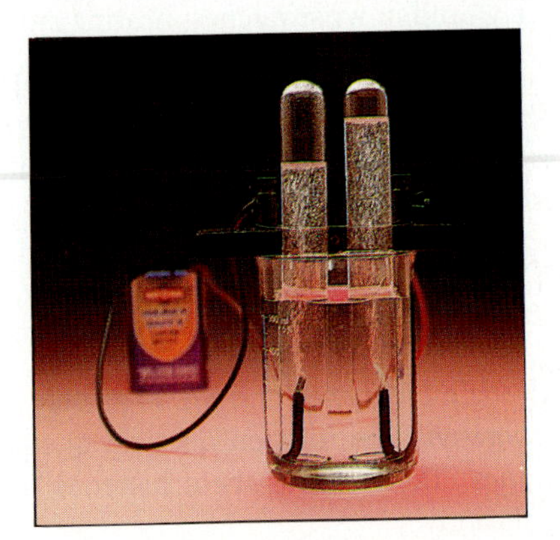

Figure 19.23 Hydrogen and oxygen are products of the electrolysis of water. The amount of hydrogen produced is twice that of oxygen.

RED CAT

The definitions of anode and cathode hold in electrolytic as well as voltaic cells—oxidation takes place at the anode, and reduction takes place at the cathode. The *polarity* of the electrodes, however, is different in the two types of cell. The anode of a voltaic cell has a surplus of electrons, produced by the oxidation half-reaction. The anode therefore is negatively charged relative to the cathode. In an electrolytic cell, however, the battery (or other external power source) forces a surplus of electrons to the *cathode,* where they are needed for the reduction half-reaction. The cathode of an electrolytic cell, therefore, is negative relative to the anode. These considerations are summarized in Table 19.4.

TABLE 19.4 Electrode Polarity in Electrochemical Cells

Cell Type	Electrode	Process	Polarity	Reason
Voltaic	Anode	Oxidation	Negative	Electron production creates surplus
	Cathode	Reduction	Positive	Electron use in reduction creates deficit
Electrolytic	Anode	Oxidation	Positive	Electrons drawn off by battery
	Cathode	Reduction	Negative	Battery forces needed electrons to cathode

Electrode Reactions

Since the anode and cathode of an electrochemical cell must be connected by a substance in which *ionic* conduction can take place, electrical power applied to *pure* water does not cause any significant amount of electrolysis. Pure water is an insulator, not a conductor. If a small amount of Na_2SO_4 (or NaOH, H_2SO_4, or some other strong electrolyte) is dissolved in it, water becomes a good conductor and is readily electrolyzed.

Autoionization in pure water produces H_3O^+ and OH^- at a concentration of about 10^{-7} M. This is too small to conduct much current, and electrolysis of pure water is practically impossible.

A new question arises if a sodium compound is used to improve the conductivity of water. The presence of $Na^+(aq)$ means that there are *two* possible cathode reactions.

$$2\ H_2O(\ell) + 2\ e^- \longrightarrow H_2(g) + 2\ OH^-(aq) \quad \text{and} \tag{19-14}$$

$$Na^+(aq) + e^- \longrightarrow Na(s) \tag{19-16}$$

The experimental fact is that water is reduced and sodium ion is not. Reaction (19-16) does not take place when aqueous solutions of NaCl are electrolyzed. The relationships in Tables 19.2 and 19.3 allow us to understand why this is so.

It is shown explicitly in Table 19.3 that species near the top of the list are more easily reduced than those further down. H_2O is above Na^+, and therefore it is more easily reduced. In terms of the standard reduction potential, we can say that the more positive the SRP, the more easily the species is reduced. Table 19.2 shows that the SRP of water (-0.83 V) is more positive than the SRP of Na^+ (-2.71 V), and therefore water is more easily reduced.

The rule for predicting the cathode reaction in an electrolytic cell is simple. Other things being equal, the species that *is* reduced is the one that is most easily reduced. Other conceivable cathode reactions do not take place. Ions whose reduction potential is more positive than ≈ -0.8 V are reduced at the cathode, and if no such ions are present, only water is reduced. This rule is only an approximate guide, for two reasons. First, it is $\mathscr{E}_{red}$ and not $\mathscr{E}^\circ_{red}$ that should be compared. The concentration of the metal cation affects its value of $\mathscr{E}_{red}$, and the pH affects the value of $\mathscr{E}_{red}$ for water. Second, and more vexing, the *rate* of electrode processes is an important factor. A particular reaction may be favored by comparison of SRP's, and yet occur so slowly that the *unfavorable* reaction predominates.

The reaction with the largest SRP is favored.

Similar considerations hold at the anode. Table 19.2 shows that $F^-(aq)$ is extremely difficult to oxidize. In aqueous solutions of fluoride salts, therefore, the anode reaction is the oxidation of water.

$$2\ H_2O(\ell) \longrightarrow O_2(g) + 4\ H^+(aq) + 4\ e^- \tag{19-13}$$

Chloride ion is more easily oxidized, however, and in aqueous solutions containing Cl^- the anode reaction is*

$$2\ Cl^-(aq) \longrightarrow Cl_2(g) + 2\ e^- \tag{19-17}$$

Quantitative Aspects of Electrolysis

The overall cell reaction in electrolysis is nonspontaneous, and it proceeds only as long as an external source of electrical current is connected to the cell. The amount of products generated is proportional to the number of electrons that pass through the

* Comparison of $\mathscr{E}^\circ$ values in Table 19.2 leads to the prediction that O_2, not Cl_2, should be produced at the anode during the electrolysis of aqueous solutions containing Cl^-. But the *rate* of the process controls the result in this case. Electrode reactions involving H_2 or O_2 are usually much slower than others, and so, even though it is favored by $\mathscr{E}^\circ$ arguments, very little O_2 is produced. The effect of reaction rate on electrode processes is known as "overvoltage."

In the half-reaction for oxygen formation, Equation (19-13), the stoichiometric coefficient of the transferred electrons is in plain sight.

cell, or equivalently, to the amount of electrical charge that passes through the cell. In the anode reaction of water electrolysis, for instance, the production of one mole of $O_2(g)$ requires that four moles of electrons or $4 \cdot 96{,}500 \approx 4 \times 10^5$ coulombs pass through the cell. Calculations of this sort are another aspect of reaction *stoichiometry*. It is useful to think of the equation for the cell reaction as having a hidden term giving the number of electrons transferred. In effect, this hidden term gives the stoichiometric coefficient of the transferred electrons.

$$2\ H_2O(\ell) \longrightarrow 2\ H_2(g) + O_2(g) \qquad (4\ e^- \text{ or } n = 4) \tag{19-15}$$

The faraday ($\mathcal{F} = 96{,}485$ C mol^{-1}) is the unit conversion factor that relates the amount of material produced in electrolysis to the electrical charge passing through the cell. Apart from the fact that the stoichiometric coefficient of electrons in equations like (19-15) is hidden, the only difference between the example problems that follow and those given in Chapter 3 (Stoichiometry) is that amounts of electrons are expressed in coulombs rather than grams or solution volume. Many problems in the stoichiometry of electrolysis reactions, like Example 19.7, involve the usual calculations of amounts of products or reactants. Others, like Example 19.8, are of a new type. In these, the target to be calculated is the hidden stoichiometric coefficient of electrons, or, equivalently, the charge of an ion.

EXAMPLE 19.7 Amount of Product from Current and Electrolysis Time

How many grams of H_2 are produced when water is electrolyzed for a period of 1.00 hour with a current of 0.358 A?

Analysis/Plan Although the target is clearly defined, it is not connected to the given information by any single formula. Let's try the "if only I knew" approach and work backward from the target.

- Grams H_2 can be calculated if we know *moles* H_2 and molar mass.
- Moles H_2 can be calculated if we know *moles of electrons* and the reaction stoichiometry. Knowledge of "reaction stoichiometry" in electrolysis problems includes knowledge of n, the number of electrons transferred.
- Moles of electrons can be calculated if we know the total *charge* flowing through the cell and the value of the faraday.
- Total charge can be calculated from the known relationship among *charge*, *current*, and *time* [Equation (19-1)]. Current and time are given in the problem statement.

Work

1. Calculate the total charge from the current and the time, using appropriate unit conversions.

$$1\ \text{C} \Longleftrightarrow 1\ \text{A} \cdot 1\ \text{s} \tag{19-1}$$

$$\begin{aligned} ?\ \text{C} &= (0.358\ \text{A})(1.00\ \text{h})(60\ \text{min h}^{-1})(60\ \text{s min}^{-1}) \\ &= 1.29 \times 10^3\ \text{A s} = 1.29 \times 10^3\ \text{C} \end{aligned}$$

2. Using the faraday as the conversion factor, find the number of moles of electrons.

$$\begin{aligned} ?\ \text{mol electrons} &= (1.29 \times 10^3\ \text{C}) \cdot \frac{1\ \text{mol electrons}}{96{,}500\ \text{C}} \\ &= 1.34 \times 10^{-2}\ \text{mol electrons} \end{aligned}$$

3. Using the stoichiometric relationship implied by the balanced chemical equation, find the mass of hydrogen produced.

$$2\ H_2O(\ell) \longrightarrow 2\ H_2(g) + O_2(g) \qquad (4\ e^- \text{ or } n = 4)$$

$$? \text{ g } H_2 = (1.34 \times 10^{-2} \text{ mol } e^-) \cdot \frac{2 \text{ mol } H_2}{4 \text{ mol } e^-} \cdot \frac{2.02 \text{ g } H_2}{1 \text{ mol } H_2}$$

$$= 0.0135 \text{ g } H_2$$

Note particularly the use of n, the number of electrons transferred, as a stoichiometric factor in step 3.

Exercise

Calculate the mass of metallic copper deposited on the cathode when a solution of Cu^{2+} is electrolyzed for 15 minutes with a current of 2.73 A.

Answer: 0.81 g Cu.

EXAMPLE 19.8 Calculation of Ionic Charge

A solution of an ionic compound of copper is electrolyzed by a current of 0.100 A, causing the deposition of metallic copper on the cathode. After the current has continued for 15.0 minutes, it is found that the mass of the cathode has increased by 29.6 mg. What is the oxidation number of copper in this ionic compound?

Analysis The cathode reaction in this electrolysis is

$$Cu^{n+} + n\ e^- \longrightarrow Cu$$

where n is the number of moles of electrons passing through the cell per mole of deposited copper. The value of n is also the charge on the copper ion, and, therefore, copper's oxidation number in the compound.

Copper compounds are known in both the +1 and +2 oxidation state.

Plan Use the current, time, and the faraday to determine the number of moles of electrons, and determine the number of moles of copper from the cathode mass increase and the molar mass. The ratio of these amounts is n, the number of electrons in the reduction half-reaction and the oxidation number of the copper ion.

Work The number of moles of electrons passing through the cell is

$$? \text{ mol } e^- = (0.100 \text{ A}) \cdot (15.0 \text{ min}) \cdot \frac{60 \text{ s}}{1 \text{ min}} \cdot \frac{1 \text{ mol } e^-}{9.65 \times 10^4 \text{ C}}$$

$$= 9.33 \times 10^{-4} \text{ mol } e^-$$

$1 \text{ C} \Longleftrightarrow 1 \text{ A s}$

The amount of copper deposited is

$$? \text{ mol Cu} = (0.0296 \text{ g}) \cdot \frac{1 \text{ mol Cu}}{63.5 \text{ g}}$$

$$= 4.66 \times 10^{-4} \text{ mol Cu}$$

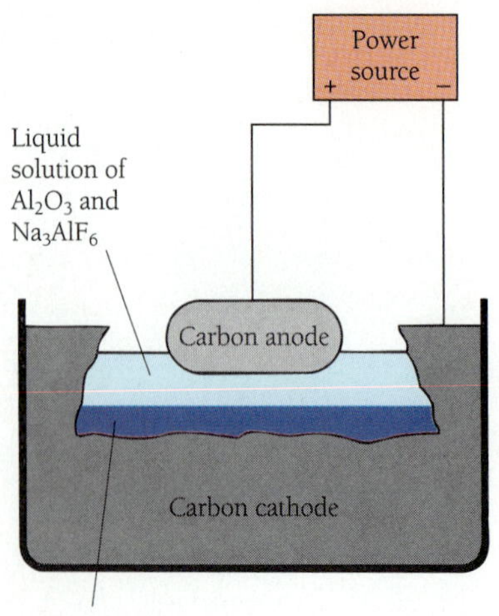

The ratio of these amounts is

$$n = \frac{9.33 \times 10^{-4} \text{ mol } e^-}{4.66 \times 10^{-4} \text{ mol Cu}} = \frac{2.00 \text{ mol } e^-}{1 \text{ mol Cu}}$$

The charge on the copper ions, and the oxidation number of copper, is +2.

Exercise

If the mass of the cathode increases by 21.6 mg during the passage of 0.200 A through a solution of a chromium salt for 10.0 min, what is the charge on the chromium ion?

Answer: +3.

Uses of Electrolysis

Electrolysis is a common industrial technique that is used in the production of metals and other substances, the application of protective and decorative coatings, and other processes. Some examples of these industrial processes are discussed in this section.

Production of Aluminum

The most abundant metal in the earth's crust is aluminum. It occurs in concentrated form as *bauxite* ($Al_2O_3 \cdot 2H_2O$), and it is more widely dispersed in a variety of alumino-silicate minerals such as the feldspars. Its oxidation state in nature is +3, and its reduction potential is sufficiently negative (−1.66 V) so that it cannot be reduced by carbon or other readily available chemical reducing agents. Large quantities of aluminum are obtained from a process, invented independently by Hall and Heroult in 1886, that uses electrical power to reduce aluminum ore to its metallic state.

The American **Charles M. Hall** (1864–1914) was only 22 years old when he showed that aluminum could be produced by electrolysis. The Frenchman **Paul Heroult** was only 23 when, in the same year, he made the same discovery.

The raw material of the *Hall-Heroult process* is bauxite. After treatment to remove metal-ion impurities (chiefly Fe^{3+}) and water, the purified Al_2O_3 is dissolved in the molten salt Na_3AlF_6 *(cryolite)*. Electrical current is passed through the solution in a cell with graphite electrodes. The cell must be operated at 1000 °C in order to keep the Al_2O_3/Na_3AlF_6 mixture molten. The electrode reactions are

Anode	$2\,O^{2-} \longrightarrow O_2(g) + 4\,e^-$
Cathode	$Al^{3+} + 3\,e^- \longrightarrow Al(\ell)$
Overall	$2\,Al_2O_3(\text{solution}) \longrightarrow 3\,O_2(g) + 4\,Al(\ell)$

Additional information about aluminum metal can be found in Section 5.8.

Aluminum is widely used for containers, electrical wiring, and metal parts for which light weight, low cost, and good corrosion resistance are important. A substantial fraction (4% to 5%) of the total electrical power produced in this country is consumed in the manufacture of aluminum.

Production of NaOH and Cl_2

Electrolysis of sodium chloride is the source of a number of industrially important materials. When pure NaCl is electrolyzed at a temperature above its melting point (800 °C), the products are gaseous Cl_2 and metallic Na.

Anode	$2\ Cl^- \longrightarrow Cl_2 + 2\ e^-$
Cathode	$Na^+ + e^- \longrightarrow Na(\ell)$
Overall	$2\ NaCl(\ell) \longrightarrow 2\ Na(\ell) + Cl_2(g)$

In the *chlor-alkali* process, the electrolysis of aqueous rather than molten NaCl leads to different products. Provided the salt concentration is kept high, Cl^- rather than H_2O is oxidized at the anode. The reactions are

Aqueous NaCl is called *brine.*

Anode	$2\ Cl^-(conc) \longrightarrow Cl_2 + 2\ e^-$
Cathode	$2\ H_2O + 2\ e^- \longrightarrow H_2(g) + 2\ OH^-$
Overall	$2\ H_2O + 2\ Cl^-(conc) \longrightarrow Cl_2 + H_2(g) + 2\ OH^-$

The electrolysis is carried out in a "diaphragm cell" (Figure 19.24). The molecular equation for the process is

$$2\ NaCl(aq) + 2\ H_2O(\ell) \longrightarrow 2\ NaOH(aq) + Cl_2(g) + H_2(g)$$

All three of the products are valuable industrial chemicals, although the sodium hydroxide is somewhat contaminated with sodium chloride. New diaphragm materials are being developed to lessen this problem.

A different type of cell, using a liquid mercury cathode, is also used in the chlor-alkali process. Although salt-free NaOH is produced in cells of this design, the inevitable loss of small amounts of mercury poses environmental hazards.

Electrolytic Refining

Purification of a substance prior to sale or use is often called *refining* (for instance, "refined sugar"). Metallic copper, obtained from its ore as described in Section 18.5, contains other metals as impurities. Much of the copper produced in this country is

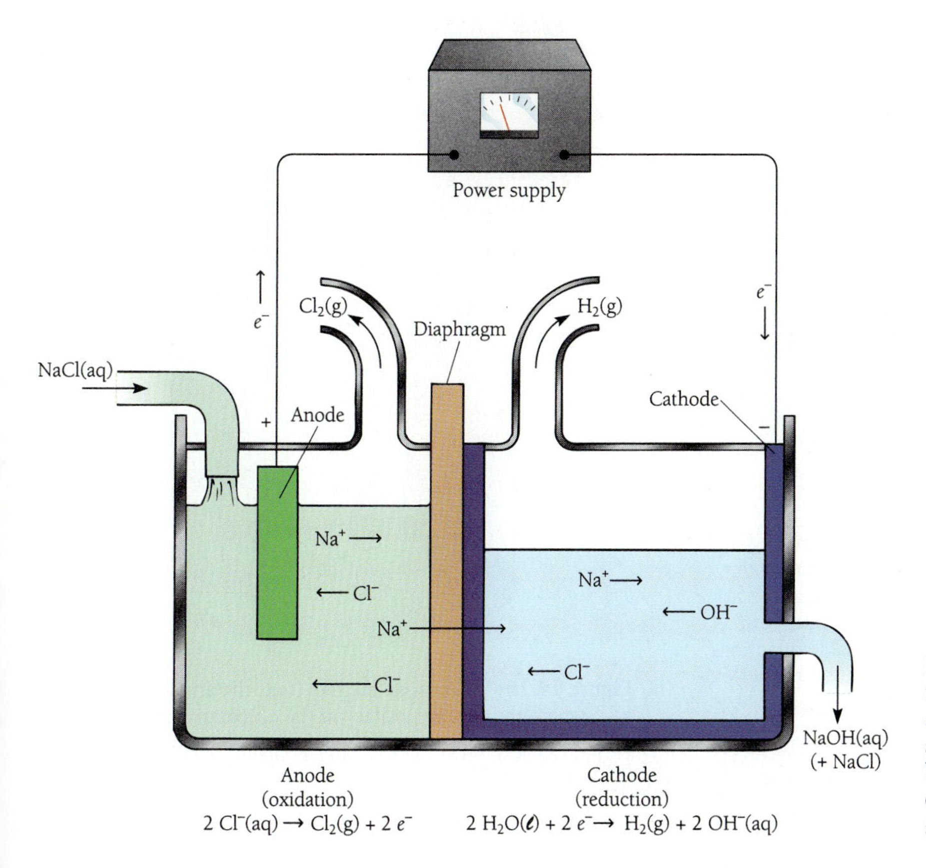

Figure 19.24 A diaphragm cell used in the electrolysis of NaCl(aq). The anode and cathode compartments are separated by a diaphragm, a porous barrier that prevents mixing of the gaseous products while allowing the passage of water and dissolved ions.

refined by an electrolytic method in a cell similar to that of Figure 19.25. The anode is a thick slab of impure copper, and the cathode is a thin sheet of purified copper. The electrolyte is an aqueous solution of $CuSO_4$ and H_2SO_4. At the anode, copper and active-metal impurities such as Fe and Zn are oxidized and enter the solution as aqueous ions. The cell voltage is kept low enough so that the noble-metal impurities Au, Pt, and Ag are not oxidized. Instead, as the anode gradually disappears, they fall to the bottom as the "anode sludge," from which they are later recovered, purified, and sold. Cations produced at the anode migrate to the cathode, where only Cu^{2+} is reduced—the cell voltage is not high enough to reduce Fe^{2+} or Zn^{2+}. The result is the buildup, on the cathode, of copper metal that is more than 99.5% pure.

Electroplating

Metal objects can be coated or "plated" with a thin layer of another metal in an electrolytic process called **electroplating.** The object to be coated is made the cathode of an electrolytic cell whose anode is the desired coating metal, for example silver. The cathode reaction is $Ag^+(aq) + e^- \rightarrow Ag(s)$, and, as current passes through the cell, a thin coating of silver builds up. The anode reaction is $Ag \rightarrow Ag^+ + e^-$, and the overall process is the transfer of silver from the anode to the cathode (Figure 19.26). Jewelery

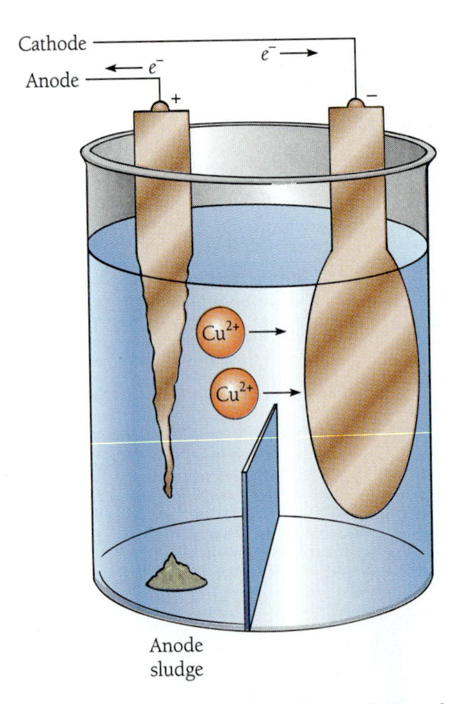

Figure 19.25 In the electrolytic refining of copper, the metal is transported through the electrolyte in the form of $Cu^{2+}(aq)$, and it is deposited on the cathode in very pure form. The anode sludge contains valuable quantities of silver, gold, and platinum which are not oxidized in the process.

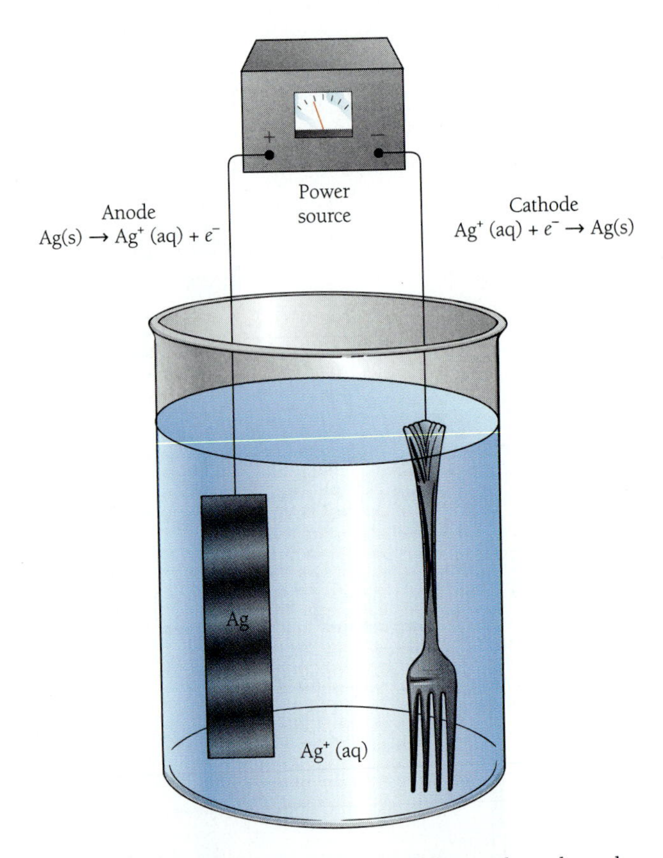

Figure 19.26 Silver ions migrate from the anode to the cathode, where they are deposited on the object to be silver plated. The electrolyte can be any soluble silver salt such as $AgNO_3$.

and fine electronic parts are gold plated in a similar process. Even plastic objects can be plated in this way, provided that the plastic is first coated with an electrically conductive material.

Electroplated coatings not only improve the appearance, but can also protect the objects against corrosion. Chromium-plated objects remain bright and do not tarnish when exposed to the atmosphere. Electroplating is also used to produce "tin" cans (tin deposited on steel) and galvanized iron (zinc deposited on steel).

19.6 CORROSION

The slow oxidation of metals by reagents present in the environment is called **corrosion.** The rusting of iron or steel, that is, the slow conversion of iron to iron oxide, is a familiar example. The process takes place only under moist conditions, and it is accelerated by the presence of ionic substances dissolved in the moisture. Rusting is an electrochemical process in which iron is oxidized at the anode of a natural electrochemical cell, and oxygen from the air is reduced at the cathode.

Most corrosion is due to H_2O and O_2 in the environment.

Rusting of automobile parts occurs much more rapidly in those regions of the country where salt is put on the roads in winter.

Rusting begins with the electrode half-reactions

Anode	$Fe(s) \longrightarrow Fe^{2+}(aq) + 2\,e^-$
Cathode	$2\,H_2O(\ell) + O_2(aq) + 4\,e^- \longrightarrow 4\,OH^-$
Overall	$2\,Fe(s) + 2\,H_2O(\ell) + O_2(aq) \longrightarrow 4\,OH^- + 2\,Fe^{2+}(aq)$

which occur beneath the surface of a water droplet. Oxidation occurs at a surface site where, by virtue of a defect, impurity, crystal boundary, or strain, iron is marginally easier to oxidize than at neighboring sites. A shallow depression called an *anode pit* develops as iron enters the solution. The electrons released in the oxidation travel through the body of the object (which therefore acts as the external circuit of the

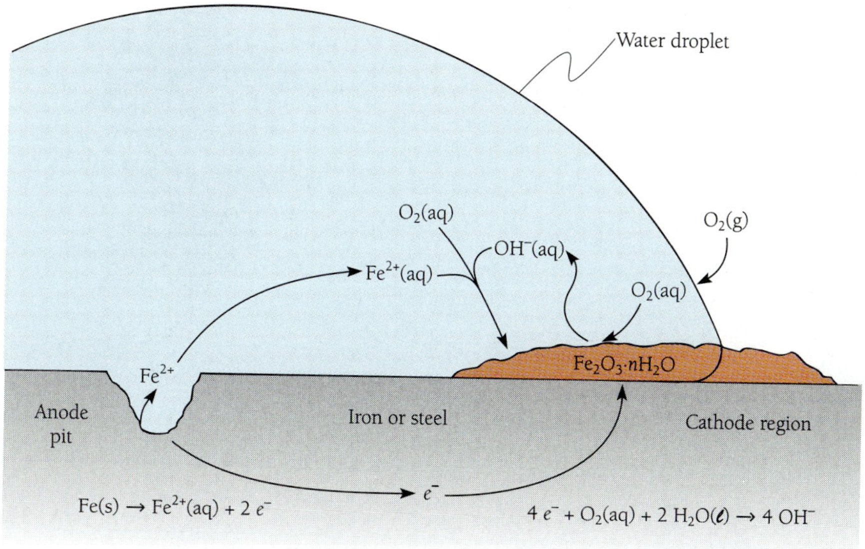

Figure 19.27 Rusting of iron. Fe(s) is oxidized to Fe^{2+} at the "anode pit," while O_2 from the air is reduced to OH^- near the edge of the water droplet. $Fe^{2+}(aq)$ is further oxidized by $O_2(aq)$ and deposited as Fe_2O_3. The result is the buildup of iron oxide and the eventual consumption of all available metallic iron.

The cathode region is often located near the edge of the droplet, where the concentration of $O_2(aq)$ is the highest.

voltaic cell) to the cathode region. A diagram of this naturally occurring electrochemical cell is given in Figure 19.27. The process is complete when $Fe^{2+}(aq)$ ions are further oxidized by dissolved oxygen, forming the precipitate called rust. This combined oxidation and precipitation is complicated, and is represented by the following overall reaction.

$$4\ Fe^{2+}(aq) + O_2(aq) + 8\ OH^- \longrightarrow 2\ Fe_2O_3 \cdot H_2O(s) + 2\ H_2O$$

The number of water molecules involved is variable, and the formula for rust is better represented by $Fe_2O_3 \cdot nH_2O(s)$.

Steel objects that must be exposed to corrosive environments can be prevented from rusting by connecting them to a more active metal such as magnesium or titanium. Buried pipelines and ship hulls are often protected by this technique, known as *cathodic protection* (Figure 19.28). The connection establishes a voltaic cell in which magnesium and steel are the electrodes, moist earth or seawater is the electrolyte, and the connecting wire or ship hull is the external circuit. Because it is so easily oxidized ($\mathcal{E}^\circ_{red}$ of Mg^{2+} is -2.37 V), magnesium is necessarily the anode of the cell. The pipeline or ship hull is therefore forced to be the cathode, and, since only reduction takes place at a cathode, no pitting and loss of iron can occur. Instead, dissolved oxygen is reduced to hydroxide ions at the cathode. The magnesium (or titanium) anode is gradually consumed by oxidation to Mg(II) [or Ti(IV)], and it must be periodically replaced.

Other types of oxidative corrosion occur with different metals. Silver becomes tarnished on exposure to sulfur compounds in the atmosphere or in certain foods (for example, eggs), forming black $Ag_2S(s)$. Copper roofs and gutters become coated with gray-green $CuSO_4 \cdot 2Cu(OH)_2$ following reaction with atmospheric O_2, H_2O, and SO_2. Sometimes the corrosion products form a coating that retards or prevents further oxidation. When aluminum or chromium is oxidized by exposure to air, a thin,

The steel hulls of ships can be cathodically protected by replaceable blocks of titanium. Four such blocks can be seen in this photo, as thin strips in line with the propeller.

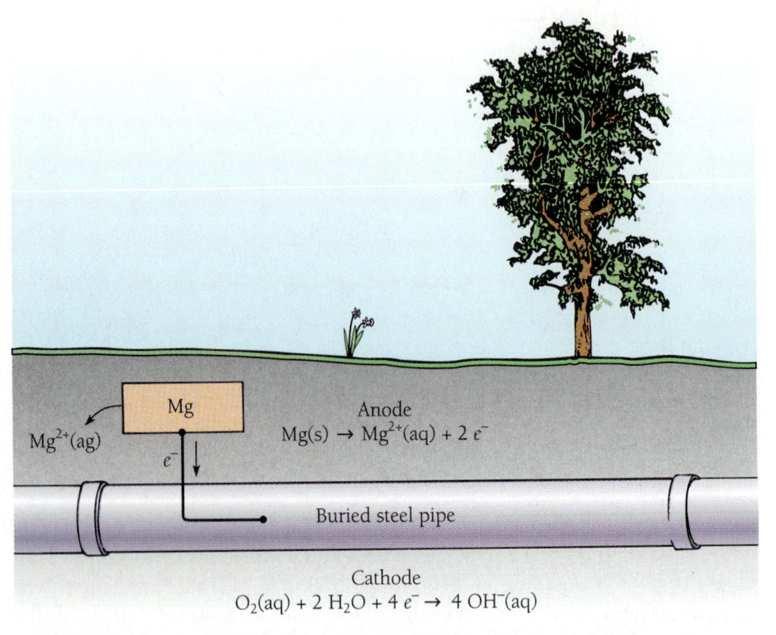

Figure 19.28 Cathodic protection of steel against rusting. The more active magnesium is a sacrificial anode, the moist soil is the electrolyte, and the iron pipe is the cathode at which reduction of O_2 rather than loss of iron occurs. The magnesium is placed near the surface so that it can be replaced as necessary.

TABLE 19.5

Spontaneity	Cell Potential	Free Energy
Spontaneous, as written	$\mathscr{E} > 0$	$\Delta G_{rxn} < 0$
At equilibrium	$\mathscr{E} = 0$	$\Delta G_{rxn} = 0$
Nonspontaneous (but spontaneous in reverse)	$\mathscr{E} < 0$	$\Delta G_{rxn} > 0$

Note that there are no "o" superscripts to these relationships: they apply to all pressures and concentrations, not just 1 atm and 1 M.

transparent, impermeable layer of Al_2O_3 or Cr_2O_3 builds up and seals the metal from further contact with air. Platinum and gold are very difficult to oxidize, and so do not corrode at all.

19.7 THERMODYNAMICS OF ELECTROCHEMICAL CELLS

It was noted in Section 19.3 that if the calculated potential of a voltaic cell has a positive value, the overall cell reaction is spontaneous as written. If the calculated potential is negative, the reaction is nonspontaneous as written but proceeds spontaneously in the reverse direction. Furthermore, we know that when a battery runs down, its cell reaction stops and its voltage drops to zero. In the chapter on thermodynamics (Section 10.6), we learned of another quantity related to reaction spontaneity, namely the free energy of reaction, ΔG_{rxn}. These relationships are summarized in Table 19.5. Clearly, there is an inverse relationship between $\mathscr{E}$ and ΔG. Put into algebraic form, that relationship is

$$\Delta G_{rxn} = -n\mathscr{F}\mathscr{E} \qquad (19\text{-}18)$$

Equation (19-18) holds under all conditions, including standard conditions, so it is also true that

$$\Delta G^\circ_{rxn} = -n\mathscr{F}\mathscr{E}^\circ \qquad (19\text{-}19)$$

In Equations (19-18) and (19-19), as in other equations dealing with electrochemical cells, the factor n is the number of electrons transferred from the reductant to the oxidant in the overall cell reaction.

The stoichiometric coefficient n of electrons in an *overall* reaction is a hidden variable.

Electrical measurements can be made with great accuracy, so that measurements of cell potential are often the best way to determine the value of the thermochemical quantity ΔG_{rxn}. Once the free energy is known, it can be combined with a measurement of ΔH_{rxn} and the relationship $\Delta G = \Delta H - T\Delta S$ to find ΔS_{rxn} as well.

EXAMPLE 19.9 Free Energy of a Cell Reaction

A voltaic cell consisting of a Ag(s) | Ag^+(aq) half-cell and a Ni(s) | Ni^{2+}(aq) half-cell has a measured potential of 1.05 V. The nickel electrode is negative with respect to the silver electrode. Write the overall cell reaction and calculate its free energy change.

Analysis

Target: An overall cell reaction and an associated free energy change.

Knowns: The identity of the half-cells and the cell polarity.

Relationship/Plan: ΔG can be obtained from Equation (19-18) once n and $\mathscr{E}$ are known; $\mathscr{E}$ is given, and n can be deduced from the stoichiometry of the overall cell reaction. The cell reaction can be written from the half-cell reactions once it is known which is oxidation and which is reduction.

Work Since the negative electrode of a voltaic cell is the anode, the half-reactions are

Oxidation	$Ni(s) \longrightarrow Ni^{2+}(aq) + 2\,e^-$	
Reduction	$Ag^+(aq) + e^- \longrightarrow Ag(s)$	
Overall	$2\,Ag^+ + Ni(s) \longrightarrow Ni^{2+}(aq) + 2\,Ag(s)$	$(n = 2)$

From the cell reaction, $n = 2$. Since the reaction in a voltaic cell is always spontaneous, the sign of $\mathscr{E}$ is positive. The free energy of the reaction is

$$\begin{aligned} \Delta G_{rxn} &= -n\mathscr{F}\mathscr{E} \qquad (19\text{-}18) \\ &= -(2\text{ mol } e^-) \cdot \frac{96{,}485\text{ C}}{1\text{ mol } e^-} \cdot (+1.05\text{ V}) \\ &= -2.03 \times 10^5\text{ V C} \\ &= -203\text{ kJ} \end{aligned}$$

1 joule $\Longleftrightarrow$ 1 volt-coulomb

The problem does not specify the concentrations, and they are not needed. However, if $[Ni^{2+}] = [Ag^+] = 1$ M, the result would be written as ΔG°_{rxn} rather than ΔG_{rxn}.

Check The sign of ΔG is negative, as it must be for a spontaneous reaction. According to Table 19.2, the SRP of Ag^+ is more positive than that of Ni^{2+}. Therefore, Ag^+ is more easily reduced than Ni^{2+}, and the reaction directionality deduced above is correct.

Exercise

The cell consisting of a $Pt|F_2(1.00\text{ atm})|F^-(0.00100\text{ M})$ half-cell and a $Pt|Cl_2(1.00\text{ atm})|Cl^-(10\text{ M})$ half-cell has a measured potential of 1.75 V. The chlorine electrode is negative with respect to the fluorine electrode. Write the overall cell reaction and calculate its free energy change.

Answer: $F_2(1.00\text{ atm}) + 2\,Cl^-(10\text{ M}) \rightarrow 2\,F^-(0.00100\text{ M}) + Cl_2(1.00\text{ atm})$; $\Delta G_{rxn} = -338$ kJ.

The relationship between $\mathscr{E}$ and ΔG_{rxn} can be combined with the Nernst equation to show why dry cell batteries, and most other voltaic cells, "run down" as they are used. The overall cell reaction of an ordinary flashlight battery is

The dry cell, or Leclanché cell, is described on page 825.

$$2\,NH_4^+ + 2\,MnO_2(s) + Zn(s) \longrightarrow Zn^{2+}(aq) + Mn_2O_3(s) + H_2O(\ell) + 2\,NH_3(aq)$$

The Nernst equation for this cell is

$$\mathscr{E} = \mathscr{E}^\circ - \frac{0.0592}{n}\log Q$$

$$\mathscr{E} = \mathscr{E}^\circ - 0.0296\log\frac{[Zn^{2+}][NH_3]^2}{[NH_4^+]^2}$$

The value of the reaction quotient $Q = [Zn^{2+}][NH_3]^2/[NH_4^+]^2$, and its logarithm, continually increases as Zn^{2+} and NH_3 are produced and NH_4^+ is consumed. The value of $\mathscr{E}$ therefore decreases steadily ("runs down") as the reaction proceeds. When $\mathscr{E}$ reaches zero, ΔG is also zero, so that the overall cell reaction is in equilibrium. The reaction has stopped, and the battery is dead.

A final comment about thermodynamics concerns the equilibrium constant of redox reactions. We saw in Section 13.3 that, for any reaction, the equilbrium constant is related to the standard free energy of reaction by

$$\ln K_{eq} = \frac{-\Delta G^\circ_{rxn}}{RT}$$

When this equation is combined with the relationship between cell potential and free energy, $\Delta G^\circ_{rxn} = -n\mathscr{F}\mathscr{E}^\circ$, the result is Equation (19-20), a general relationship between standard cell potential and equilibrium constant for redox reactions.

$$\ln K_{eq} = \frac{n\mathscr{F}\mathscr{E}^\circ}{RT} \tag{19-20}$$

EXAMPLE 19.10 Equilibrium Constant of a Cell Reaction

The cell reaction and standard potential for the cell $Ni(s)|Ni^{2+}(aq)||Pb^{2+}(aq)|Pb(s)$ are

$$Pb^{2+}(aq) + Ni(s) \rightleftharpoons Pb(s) + Ni^{2+}(aq); \qquad \mathscr{E}^\circ = +0.12\text{ V}; \qquad (n = 2)$$

What is the value of the equilibrium constant at room temperature?

Analysis/Plan The target, K_{eq}, is related to the given quantities by Equation (19-20). Insert the known values, assume $T = 298$ K, and evaluate K_{eq}.

Work

$$\ln K_{eq} = \frac{n\mathscr{F}\mathscr{E}^\circ}{RT}$$

$$= (2\text{ mol } e^-) \cdot \frac{96{,}485\text{ C}}{1\text{ mol } e^-} \cdot \frac{(0.12\text{ V})}{(8.314\text{ J mol}^{-1}\text{ K}^{-1}) \cdot (298\text{ K})}$$

$$= 9.3$$

$$K_{eq} = e^{9.3} = 1.1 \times 10^4$$

Exercise

The room temperature equilibrium constant for the reaction $Cl_2(g) + 2\,Br^-(aq) \rightleftharpoons Br_2(\ell) + 2\,Cl^-(aq)$ is 7.5×10^9 at 25 °C. What is the value of $\mathscr{E}^\circ$ for the cell $Pt|Cl_2(g)|Cl^-(aq)||Br^-(aq)|Br_2(\ell)|Pt$?

Answer: 0.29 V.

SUMMARY EXERCISE

Titanium, the fourth most abundant metal in the earth's crust, occurs primarily as $FeTiO_3$ in the ore *ilmenite*. Metallic titanium is stronger than steel, yet has only about half the density. Moreover, it is not corroded by moist air or seawater. Its strength, light weight, and corrosion resistance make titanium and its alloys valuable structural materials, particularly in aircraft and spacecraft. However, it is difficult to obtain metallic titanium from its ore. At the high temperatures needed for metallurgy, Ti metal is quickly oxidized by a wide variety of substances, including O_2, N_2, and even C. The oxidation number of titanium in its compounds can be +4 (most common), +3, or +2 (readily oxidized by water).

Coke is a naturally occurring material similar to coal, but having a somewhat lower carbon content.

Elemental titanium is prepared by a series of reactions called the *Kroll process*. Ilmenite is heated with solid carbon (coke) in an atmosphere of chlorine, resulting in the formation of the volatile molecular compound $TiCl_4$, along with CO and $FeCl_2(s)$. After separation from Cl_2 and CO by condensation, pure $TiCl_4$ is reacted with magnesium in an argon atmosphere at about 800 °C. The product, solid metallic titanium, is easily separated from unreacted magnesium and the other product, $MgCl_2$, both of which are liquid at this temperature.

a. The oxidation number of Fe in ilmenite is +2. What is the oxidation number of Ti in this ore?

b. Write a balanced molecular equation for the production of $TiCl_4$ from ilmenite. What element is oxidized? What element is reduced?

c. Write a balanced equation for the production of Ti from $TiCl_4$. What element is oxidized? What element is reduced?

d. Salts of Ti^{4+} are strongly hydrated in aqueous solution, and $Ti^{4+}(aq)$ is more realistically represented by the formulas $Ti(OH)_2^{2+}$ or $Ti(OH)_3^{+}$. Nonetheless $Ti^{4+}(aq)$ is electrochemically active, and the SRP of $Ti^{4+}(aq) + e^- \rightarrow Ti^{3+}(aq)$ is about 0.10 V. Calculate the cell potential of a voltaic cell consisting of a $Pt|Ti^{4+}(aq), Ti^{3+}(aq)$ redox electrode and a $Cu^{2+}|Cu(s)$ electrode.

e. What is the overall reaction of the cell described in (d)? Which electrode is the anode?

f. Give the cell diagram for the cell in (d). Use the shorthand notation.

g. If the temperature is 25 °C, $[Cu^{2+}] = 0.010$ M, and $[Ti^{4+}(aq)] = [Ti^{3+}(aq)]$, what is the potential of the cell in (d)?

h. Calculate ΔG°_{rxn} and K_{eq} at 298 K for the reaction in (e).

i. Suppose the reaction in (d) is reversed by attaching a battery to the external circuit, making the $Pt|Ti^{4+}(aq), Ti^{3+}(aq)$ electrode the cathode. For what length of time must a current of 1.013 A be passed through the cell in order to decrease the mass of the anode by 1.000 g? How many moles of ions will be reduced at the cathode during this time?

Answers **(a)** +4; **(b)** $FeTiO_3(s) + 3\ C(s) + 3\ Cl_2(g) \rightarrow FeCl_2(s) + 3\ CO(g) + TiCl_4(g)$; C is oxidized and Cl_2 is reduced; **(c)** $TiCl_4(g) + 2\ Mg(\ell) \rightarrow 2\ MgCl_2(\ell) + Ti(s)$; Mg is oxidized and Ti^{4+} is reduced; **(d)** +0.24 V; **(e)** $Cu^{2+}(aq) + 2\ Ti^{3+}(aq) \rightarrow 2\ Ti^{4+}(aq) + Cu(s)$; the $Pt|Ti^{4+}(aq), Ti^{3+}(aq)$ electrode is the anode; **(f)** $Pt|Ti^{3+}(aq), Ti^{4+}(aq)||Cu^{2+}(aq)|Cu$; **(g)** +0.18 V; **(h)** −46 kJ; 1.3×10^8; **(i)** 2998 s (50 min); 0.03147 mol.

SUMMARY AND KEYWORDS

Electric **charge** is a property of electrons, protons, and ions, and electric **current** is the motion of negatively and/or positively charged particles. Current equals charge per unit time: in SI units, 1 ampere = 1 coulomb/1 second. Electrical **potential,** expressed in the SI unit **volt** (V), is analogous to gravitational potential and measures the capacity of electric charge to do work; 1 joule = (1 volt) · (1 coulomb). Potential *difference,* also called **voltage,** is the driving force behind current. In **metallic** current, the moving valence electrons of the metal carry the current. In solutions and fused salts, **ionic** current is carried by anions and cations. No current can flow in an **insulator** because no charged particles are free to move in these materials.

Electrochemistry is the study of the interchange of electrical and chemical energy in redox reactions whose oxidation and reduction half-reactions take place in physically separate locations. An **electrode** is a material in which current is carried by electrons, in contact with an **electrolyte,** a material in which ionic conduction takes place. A reversible redox *half-reaction* called an **electrode reaction** takes place at the electrode surface. An electrode is called an **anode** if its electrode reaction is an oxidation, or a **cathode** if its reaction is a reduction. Some different types of electrodes are the *metal/solution, gas, metal/salt,* and *redox* electrodes.

An **electrochemical cell** consists of two interconnected electrodes. In a **voltaic cell,** a spontaneous **cell reaction** causes the flow of electrons through the *external circuit.* Such a device transforms chemical energy into electrical energy. Common voltaic cells are the *lead storage battery,* the *dry cell,* the *alkaline cell,* the *rechargeable* or *nicad* cell, the *mercury* cell, the *silver cell,* and the **fuel cell.** The *anode* compartment of a voltaic cell is often separated from the *cathode* compartment by a **porous barrier** such as a **salt bridge,** which prevents mixing of two different electrolytes while allowing the passage of ions. Cells are commonly represented by *cell diagrams,* either in full or in shorthand notation.

The difference in electrical potential, or **voltage,** between the two electrodes of a voltaic cell is called the **cell potential** or the **standard cell potential** if the measurement is made under standard conditions (1 M for solutes and 1 atm for gases). The *polarity* of a cell describes which electrode is at the higher potential: in voltaic cells, the negative electrode is the anode. Each half-cell *independently* contributes its **half-cell potential** to the overall cell potential. The **standard reduction potential (SRP)** of a reduction half-reaction is the cell potential when that reaction occurs (under standard conditions) at the cathode of a cell whose anode is the **standard hydrogen electrode (SHE).** The SRP of a reactant indicates its relative strength as an oxidant, or its relative ease of reduction by another reagent. Common SRP's range from $\approx +3$ V for F_2, the strongest oxidant, to ≈ -3 V for $Li^+(aq)$, the weakest. The *standard oxidation potential* of a half-reaction is equal to but opposite in sign to its SRP.

The cell potential of a voltaic cell is the sum of the oxidation and reduction potentials of the half-reactions: $\mathscr{E}_{cell} = \mathscr{E}_{red} + \mathscr{E}_{ox}$. If $\mathscr{E}_{cell}$ is positive, the cell reaction is spontaneous as written. If $\mathscr{E}_{cell}$ is negative, the reaction is spontaneous in the reverse direction. If $\mathscr{E}_{cell} = 0$, the cell is at equilibrium and no reaction occurs. Cell potential and free energy change of the cell reaction are related by $\Delta G = -n\mathscr{F}\mathscr{E}$, where n is the number of electrons transferred (the hidden stoichiometric coefficient) in the reaction. The **faraday** ($\mathscr{F}$) is the charge (96,485 C) of one mole of electrons. The dependence of cell potential on the *reaction quotient* (Q) of the cell reaction is given by the **Nernst equation,** $\mathscr{E} = \mathscr{E}° - (0.0592/n) \cdot \log(Q)$. The operation of a pH meter is based on this relationship.

In an **electrolytic cell,** externally supplied electrical energy is consumed, causing **electrolysis,** a nonspontaneous redox reaction, to occur. The reaction at the anode is an oxidation, but, in contrast to voltaic cells, the anode of an electrolytic cell is positive. For each mole of reactant oxidized at the anode (or reduced at the cathode), $n\mathcal{F}$ coulombs of charge passes through the cell; n is the stoichiometric coefficient of electrons in the oxidation (or reduction) half-reaction.

Aluminum is commercially produced by electrolysis of its ore *bauxite* (Al_2O_3). Large amounts of NaOH, Cl_2, and H_2 are produced by electrolysis of NaCl(aq). Copper is purified to a high degree by *electrolytic refining,* and various objects are commonly coated with silver, chromium, or other metals by **electroplating. Corrosion** is the oxidation of metals by oxidants in the environment, in a naturally occurring voltaic cell.

PROBLEMS

General

1. Define electrochemistry. How does its meaning differ from that of redox chemistry?
2. What is the relationship between electrical charge and current?
3. What is the smallest charge found in nature and where is it found?
4. Distinguish between charge and potential.
5. What are the units of electrical potential and work, and how is potential related to work?
6. Compare and contrast metallic and ionic conduction.
7. What differences, if any, exist between the mechanisms of conduction in dissolved sodium chloride and molten sodium chloride?
8. What is meant by the term "electrode"?
9. What is an electrode process?
10. Define anode and cathode.
11. What is an electrochemical cell? What are the two types of cell?
12. What is the function of a salt bridge?
13. How can the number of electrons transferred in a redox reaction be determined?
14. Name and describe some different types of electrode.
15. Describe in detail the meaning of the following notation.

$$Zn(s)|Zn^{2+}(0.3\ M)||Cl^-(0.1\ M)|AgCl(s)|Ag(s)$$

16. What is the significance of the equation $\mathcal{E}_{cell} = \mathcal{E}_{red} + \mathcal{E}_{ox}$?
17. What is the relationship between $\mathcal{E}_{red}$ and $\mathcal{E}_{ox}$ for a given voltaic cell?
18. What is meant by the acryonyms SHE and SRP?
19. How is SRP defined and measured?
20. If two half-reactions have SRP's of +0.5 and −0.5 V, which one involves the stronger oxidant?
21. What is the strongest commonly occurring oxidant? The weakest?
22. What is the strongest commonly occurring reductant? The weakest?
23. What are the relationships among cell potential, free energy of reaction, and spontaneity of cell reaction?
24. Write the Nernst equation, in two different forms, and explain its meaning.
25. Describe two commonly used voltaic cells.
26. Describe some industrial processes that involve electrochemistry.
27. What is the faraday? How is it related to Avogadro's number?
28. Describe the rusting of iron from an electrochemical point of view.

Electrical Units

29. Calculate the total charge passing through a circuit if a current of 0.25 A flows for 20 seconds. How much work is done by this amount of charge if the potential difference is 10 V?
30. How much charge is moved in a flashlight battery that operates for 15 minutes at a current of 20 mA? How much work is done if the battery voltage is 1.5 V?
31. What length of time must elapse for a current of 75 mA to deliver 15 coulombs of charge?
32. How long must a 5.00 A current run in order to move an amount of charge equal to the charge on 1 mole of electrons?

33. The watt (W) is the unit that measures electrical power or the rate of electrical work. The familiar measure of electrical work, used by the electrical power industry, is the *kilowatt-hour*. The relationship, in SI units, is 1 watt · 1 second ⇔ 1 joule. How much work is done when a 60-watt light bulb operates for one hour?

34. A portable electric heater is rated at 1.3 kilowatts. How much heat is produced during three hours of operation?

*35. If the voltage in the preceding problem is 117 V, what is the current? How much charge (expressed in faradays) passes through the circuit in three hours?

*36. An automobile starter draws 235 A from a 12-V battery for 30 s before the engine starts. How much work is done? Compare this with the thermal energy needed to heat a cup of water (250 g) from 25 °C to 90 °C.

Voltaic Cells

37. Using shorthand notation, describe the cell in Figure 19.9.

38. Using shorthand notation, describe the cells in Figures (a) 19.15, (b) 19.16, and (c) 19.22.

39. Make pictorial sketches, similar to Figure 19.9, of each of the following cells.
 a. $Sn(s)|Sn^{2+}(aq)||Ag^{+}(aq)|Ag(s)$
 b. $Zn(s)|Zn^{2+}(aq)||Fe^{3+}(aq),Fe^{2+}(aq)|Pt$

40. Make pictorial sketches, similar to Figure 19.9, of each of the following cells.
 a. $Pt|H_2(1\ atm)|H^{+}(aq)||Cl^{-}(aq)|AgCl(s)|Ag(s)$
 b. $Pt|H_2(1\ atm)|H^{+}(pH = 5)||H^{+}(pH = 3)|H_2(1\ atm)|Pt$

41. Write the anode and cathode half-reactions for the cell in Problem 37.

42. Write the anode and cathode half-reactions for each of the cells in Problem 38.

43. Write the overall cell reaction for each of the cells in Problem 39. What is the value of n in each case?

*44. Write the cell reaction for each of the cells in Problem 40. What is the value of n in each case?

Reaction Spontaneity and Relative Oxidant/Reductant Strength

45. Which of the following reactions occur(s) spontaneously?
 a. $2\ MnO_4^- + 5\ Hg_2^{2+}(aq) + 16\ H^+ \rightarrow 2\ Mn^{2+}(aq) + 10\ Hg^{2+}(aq) + 8\ H_2O$
 b. $NiCl_2(aq) + Pb(s) \rightarrow PbCl_2(aq) + Ni(s)$
 c. $KClO_4(aq) + 2\ HCl(aq) \rightarrow KClO_3(aq) + Cl_2(g) + H_2O$

46. Which of the following reactions occurs spontaneously?
 a. $SnCl_2(aq) + Zn(s) \rightarrow ZnCl_2(aq) + Sn(s)$
 b. $H_2(g) + Br_2(\ell) \rightarrow 2\ HBr(aq)$
 c. $Cu^{2+}(aq) + Cu(s) \rightarrow 2\ Cu^{+}(aq)$

47. Write the equation for the reaction that takes place when the following reagents are mixed. If no reaction takes place, write "NR." Assume standard conditions.
 a. $NaF(aq)$, $NaCl(aq)$ b. $Cl_2(g)$, $KBr(aq)$
 c. $I_2(s)$, $KBr(aq)$ d. $AgBr(s)$, $Ni(s)$, $H_2O(\ell)$
 e. $Cu(s)$, H^+

*48. Write the equation for the reaction that takes place when the following reagents are mixed. If no reaction takes place, write "NR." Assume standard conditions.
 a. $FeCl_3(aq)$, $FeCl_2(aq)$ b. $HNO_3(aq)$, $HCl(aq)$
 c. $HCl(aq)$, $Ni(s)$ d. $I_2(s)$, $Br_2(\ell)$
 e. $KMnO_4(aq)$, $KBr(aq)$

*49. Write the equation for any reaction that occurs in the following situations. If no reaction takes place, write "NR."
 a. A strip of tin is immersed in $Cu(NO_3)_2(aq)$.
 b. A strip of zinc is immersed in $Zn(NO_3)_2(aq)$.
 c. Gaseous oxygen is bubbled into water.

50. Write the equation for any reaction that occurs in the following situations. If no reaction takes place, write "NR."
 a. Powdered iron is added to an aqueous solution of $AgNO_3$.
 b. Solid iodine is added to aqueous hydrochloric acid.
 c. Solid iodine is added to aqueous nitric acid.

51. Use the position of the reduction half-reaction in Table 19.2 to determine which is the better oxidizing agent (under standard conditions) in each of the following pairs.
 a. Br_2, Cl_2 b. $Pb^{2+}(aq)$, $Cd^{2+}(aq)$
 c. $KMnO_4(aq)$, $Na_2Cr_2O_7(aq)$ d. $CuCl(aq)$, $CuCl_2(aq)$
 e. $Zn^{2+}(aq)$, $Al^{3+}(aq)$ f. $AgCl(s)$, $AgBr(s)$

52. In each of the following pairs, identify the species that is the stronger oxidizing agent under standard conditions.
 a. $Cu^{+}(aq)$, $Ag^{+}(aq)$ b. $Fe^{2+}(aq)$, $Fe^{3+}(aq)$
 c. H_2, O_2 d. I_2, Br_2
 e. $F^{-}(aq)$, $Ce^{4+}(aq)$ f. $Fe^{3+}(aq)$, $Al^{3+}(aq)$

53. Use the position of the reduction half-reaction in the table of SRP's to decide which is the better reducing agent (under standard conditions) in each of the following pairs.
 a. $Pb(s)$, $Ni(s)$ b. $Fe^{2+}(aq)$, $Hg_2^{2+}(aq)$
 c. $Ag(s)$, $Sn(s)$ d. $HNO_2(aq)$, $H_2S(aq)$
 e. $CrCl_3(aq)$, $MnSO_4(aq)$ f. $Fe(s)$, $H_2(g)$

54. In each of the following pairs, identify the species that is the stronger reducing agent under standard conditions.
 a. Sn, Pb b. $Br^{-}(aq)$, $Fe^{2+}(aq)$
 c. $NO_3^{-}(aq)$, $I^{-}(aq)$ d. HCl, H_2S
 e. HNO_2, HNO_3 f. Al, Mg

55. In each of the following pairs, identify the species that is the more easily reduced under standard conditions.
 a. Cu^{+}, Cu^{2+} b. Hg^{2+}, Fe^{3+}
 c. Al^{3+}, Fe^{3+} d. ClO_4^-, $Cr_2O_7^{2-}$

56. In each of the following pairs, identify the species that is the more easily oxidized under standard conditions.
   a. Hg, Ag [in the presence of $Cl^-(aq)$]
   b. Na, Li
   c. Cu^+, Ag^+
   d. Cr^{3+}, NO_3^-

Cell Potentials

57. Calculate the cell potential and identify the anode of the cell in Figure 19.9.

*58. Calculate the standard cell potential of the voltaic cell shown in Figure 19.22.

*59. Calculate the standard potential of the cells consisting of the following electrode pairs. When the cell operates spontaneously under standard conditions, what is the cell reaction? Describe the spontaneously operating cell in shorthand notation, with the anode on the left.
   a. $Pt|Hg^{2+}(aq)|Hg_2^{2+}$; $Pt|I_2(s)|I^-(aq)$
   b. $Cu^+(aq)|Cu(s)$; $Pt|S(s)|H_2S(aq)$
   c. $AgBr(s)|Ag(s)$; $Zn^{2+}(aq)|Zn(s)$
   d. $NO_3^-(aq),HNO_2(aq)|Pt$; $ClO_4^-(aq),ClO_3^-(aq)|Pt$

*60. Calculate the standard potential of the cells consisting of the following electrode pairs. When the cell operates spontaneously under standard conditions, what is the cell reaction? Describe the spontaneously operating cell in shorthand notation, with the anode on the left.
   a. $Pt|MnO_4^-(aq)|Mn^{2+}(aq)$; $NO_3^-(aq)|NO(g)|Pt$
   b. $Sn(s)|Sn^{2+}(aq)$; $Ag^+(aq)|Ag(s)$
   c. $Zn(s)|Zn^{2+}(aq)$; $Fe^{3+}(aq)|Fe^{2+}(aq)|Pt$
   d. $Pt|H_2(1\ atm)|H^+(aq)$; $Cl^-(aq)|AgCl(s)|Ag(s)$

*61. Calculate the standard potential of the following cells. When the cell operates spontaneously under standard conditions, what is the cell reaction? Does the cell diagram correctly identify the electrode on the left as the anode? If not, write the diagram correctly.
   a. $Zn(s)|Zn^{2+}(aq)||Sn^{2+}(aq)|Sn(s)$
   b. $Ag(s)|Ag^+(aq)||Cd^{2+}(aq)|Cd(s)$
   c. $Ag(s)|AgCl(s)|HCl(aq)|H_2(g)|Pt$
   d. $Pt|H_2(g)|H^+(aq)||Cl^-(aq)|Cl_2(g)|Pt$

*62. Calculate the standard potential of the following cells. When the cell operates spontaneously under standard conditions, what is the cell reaction? Does the cell diagram correctly identify the electrode on the left as the anode? If not, write the diagram correctly.
   a. $Pt|Cu^+(aq),Cu^{2+}(aq)||Pb^{2+}(aq)|Pb(s)$
   b. $Al(s)|Al^{3+}(aq)||Ce^{4+}(aq),Ce^{3+}(aq)|Pt$
   c. $Pt|NO(g)|HNO_3(aq)||Ni^{2+}(aq)|Ni(s)$
   d. $Pt|ClO_4^-(aq),ClO_3^-(aq)||OH^-(aq)|O_2(g)|Pt$

Concentration Dependence

63. Write the chemical equation for the spontaneous reaction that occurs when each of the following electrode pairs is coupled in a voltaic cell. Indicate the value of *n*, the hidden stoichiometric coefficient of the transferred electrons. Write the algebraic expression for the reaction quotient *Q*.
   a. $Pt|MnO_4^-(aq), Mn^{2+}(aq)$; $NO_3^-(aq)|NO(g)|Pt$
   b. $Sn(s)|Sn^{2+}(aq)$; $Ag^+(aq)|Ag(s)$
   c. $Zn(s)|Zn^{2+}(aq)$; $Fe^{3+}(aq), Fe^{2+}(aq)|Pt$
   d. $Pt|H_2(1\ atm)|H^+(aq)$; $Cl^-(aq)|AgCl(s)|Ag(s)$

*64. Write the chemical equation for the spontaneous reaction that occurs when each of the following electrode pairs is coupled in a voltaic cell. Indicate the value of *n*, the hidden stoichiometric coefficient of the transferred electrons. Write the algebraic expression for the reaction quotient *Q*.
   a. $Pt|Hg^{2+}(aq), Hg_2^{2+}(aq)$; $I_2(s)|I^-(aq)|Pt$
   b. $Cu^+(aq)|Cu(s)$; $S(s)|H_2S(aq)|Pt$
   c. $AgBr(s)|Ag(s)$; $Zn^{2+}(aq)|Zn(s)$
   d. $Pt|NO_3^-(aq), HNO_2(aq)$; $ClO_4^-(aq), ClO_3^-(aq)|Pt$

65. Write the chemical equation for the cell reaction in each of the following cells. Indicate the value of *n*, the hidden stoichiometric coefficient of the transferred electrons. Write the algebraic expression for the reaction quotient *Q*.
   a. $Zn(s)|Zn^{2+}(aq)||Sn^{2+}(aq)|Sn(s)$
   b. $Ag(s)|Ag^+(aq)||Cd^{2+}(aq)|Cd(s)$
   c. $Ag(s)|AgCl(s)|HCl(aq)|H_2(g)|Pt$
   d. $Pt|H_2(g)|H^+(aq)||Cl^-(aq)|Cl_2(g)|Pt$

66. Write the chemical equation for the cell reaction in each of the following cells. Indicate the value of *n*, the hidden stoichiometric coefficient of the transferred electrons. Write the algebraic expression for the reaction quotient *Q*.
   a. $Pt|Cu^+(aq),Cu^{2+}(aq)||Pb^{2+}(aq)|Pb(s)$
   b. $Al(s)|Al^{3+}(aq)||Ce^{4+}(aq),Ce^{3+}(aq)|Pt$
   c. $Pt|NO(g)|HNO_3(aq)||Ni^{2+}(aq)|Ni(s)$
   d. $Pt|ClO_4^-(aq),ClO_3^-(aq)||OH^-(aq)|O_2(g)|Pt$

*67. Calculate the cell potential of each of the following electrochemical cells at 25 °C.
   a. $Sn(s)|Sn^{2+}(4.5 \times 10^{-3}\ M)||Ag^+(0.100\ M)|Ag(s)$
   b. $Zn(s)|Zn^{2+}(0.500\ M)||$ $Fe^{3+}(7.2 \times 10^{-6}\ M), Fe^{2+}(0.15\ M)|Pt$
   c. $Pt|H_2(1\ atm)|HCl(0.00623\ M)|Cl_2(1\ atm)|Pt$

*68. Calculate the cell potential of each of the following electrochemical cells at 25 °C.
   a. $Pt|H_2(10.0\ atm)|H^+(1.00 \times 10^{-3}\ M)$ $||Ag^+(0.00496\ M)|Ag(s)$
   b. $Pt|H_2(1.00\ atm)|H^+(pH = 5.97)$ $||H^+(pH = 3.47)|H_2(1.00\ atm)|Pt$
   c. $Pt|H_2(0.0361\ atm)|H^+(0.0100\ M)$ $||H^+(0.0100\ M)|H_2(5.98 \times 10^{-4}\ atm)|Pt$
   d. $Pt|Hg(\ell)|Hg_2Cl_2(s)|Cl^-(2.50\ M)$ $||Br^-(1.28\ M)|Br_2(\ell)|Pt$

Because the next four problems involve exponentiation (antilogarithms), the numerical answers are more than usually sensitive to errors introduced by rounding off intermediate results. If your answer is within 20% to 30% of the value given in Appendix K, you should conclude that you have worked the problem correctly.

*69. For each of the following cells, calculate the ratio of concentrations required to produce the indicated cell potential.
   a. $Ni(s)|Ni^{2+}(aq)||Sn^{2+}(aq)|Sn(s)$; $[Ni^{2+}]/[Sn^{2+}] = ?$ when $\mathscr{E} = +0.27$ V
   b. $Sn(s)|Sn^{2+}(aq)||Ni^{2+}(aq)|Ni(s)$; $[Sn^{2+}]/[Ni^{2+}] = ?$ when $\mathscr{E} = +0.27$ V

*70. For the following cells, calculate the value of concentration or ratio of pressures required to produce the indicated cell potential.
   a. $Cu(s)|Cu^{2+}(0.015\ M)||Ag^{+}(aq)|Ag(s)$; $[Ag^{+}] = ?$ when $\mathscr{E} = +0.18$ V
   b. $Pt|Cl_2(g)|Cl^{-}(aq)||F^{-}(aq)|F_2(g)|Pt$; assume $[F^{-}] = [Cl^{-}]$ and calculate P_{F_2}/P_{Cl_2} when $\mathscr{E} = 1.65$ V

*71. For the following cells, calculate the concentration ratio or pressure required to produce the indicated cell potential. Remember that when the cell potential is zero, the cell reaction is at equilibrium.
   a. $Pt|Cr^{2+}(aq), Cr^{3+}(aq)||H^{+}(pH = 3)|H_2(1.00\ atm)|Pt$; calculate $[Cr^{2+}]/[Cr^{3+}]$ when $\mathscr{E} = 0$ V
   b. $Ag(s)|AgBr(s)|HBr(2.50\ M)|H_2(g)|Pt$; calculate P_{H_2} when $\mathscr{E} = +0.077$ V

*72. For the following cells, calculate the value or ratio of pressure required to produce the indicated cell potential. Remember that when the cell potential is zero, the cell reaction is at equilibrium.
   a. $Pt|H_2(g)|H^{+}(pH = 4)||H^{+}(pH = 3)|H_2(g)|Pt$; calculate $P_{H_2}(anode)/P_{H_2}(cathode)$ when $\mathscr{E} = 0$ V
   b. $Pt|H_2(2.00\ atm)|KOH(aq)|O_2(g)|Pt$; calculate P_{O_2} when $\mathscr{E} = 1.23$ V

Quantitative Aspects of Electrochemistry

73. For each of the following cations, how many faradays are required for reduction and deposition of one mole of metal at the cathode of an electrolytic cell?
   a. Fe^{3+} b. Ag^{+}
   c. Hg_2^{2+}

74. How much charge, expressed in faradays, is required for the anodic oxidation of one-half mole of metal to produce each of the following cations?
   a. Hg^{2+} b. Au^{3+}
   c. Na^{+}

75. How much charge, expressed in coulombs, is required to reduce each of the following ions to 1.00 g metal?
   a. Fe^{3+} b. Ag^{+}
   c. Hg_2^{2+}

76. How much charge, expressed in coulombs, is required to oxidize 1.00 g metal to each of the following ions?
   a. Hg^{2+} b. Au^{3+}
   c. Na^{+}

77. How many moles of $H_2(g)$ are produced when water is electrolyzed with a current of 0.553 A for a period of 15 min? What volume is occupied by this amount of gas at STP?

78. A $Cd^{2+}(aq)$ solution is electrolyzed with a current of 0.0846 A for a period of 275 s. What mass of Cd(s) is plated out on the cathode?

79. The mass of silver deposited on a spoon during electroplating was 0.634 mg. How much electrical charge passed through the cell?

*80. After 2 hours, 25 minutes, 15.3 seconds of operation, the mass of the cathode of a $Ag(s)/Ag^{+}(aq)$ electrolytic cell was found to have increased by 1.0035 g. How much charge passed through the cell? What was the average value of the current?

*81. How long does an electrolysis cell have to run at a current of 150 A in order to produce 1 pound (454 g) of aluminum?

*82. It is desired to fill a balloon with H_2 at a pressure of 1.05 atm and a temperature of 25 °C. The volume of the balloon, when filled, is 750 mL. How long must a current of 2.00 A be passed through the cell in order to produce this amount of H_2 by electrolysis of water?

*83. An electrolytic cell contains 50.0 mL of a 0.152 M solution of $FeCl_3$. A current of 0.775 A is passed through the cell, causing deposition of Fe(s) at the cathode. What is the concentration of $Fe^{3+}(aq)$ in the cell after this current has run for 20.0 min?

*84. Suppose 250 mL of a 0.333 M solution of $CuCl_2$ is electrolyzed. How long will a current of 0.75 A have to run in order to reduce the concentration of Cu^{2+} to 0.167 M? What mass of Cu(s) will be deposited on the cathode during this time?

Batteries and Corrosion

85. Write the anode and cathode half-reactions, and the cell reaction, of the following voltaic cells.
   a. dry cell
   b. lead storage battery (discharge cycle)
   c. silver cell

86. Write the anode and cathode half-reactions, and the cell reaction, of the following voltaic cells.
   a. nicad rechargeable battery
   b. mercury cell
   c. hydrogen fuel cell

87. Tarnished silver is coated with a patina of $Ag_2S(s)$. This coating can be removed by boiling the silverware in an aluminum pan, with some baking soda or salt added to make the solution conductive. Explain this from the point of view of electrochemistry.

88. Does the rusting of iron depend on pH? If so, describe the effect of pH.

Additional Problems

89. Calculate the standard free energy change of each of the reactions in Problem 45. What is the value of the equilibrium constant at 25 °C?

90. Calculate the standard free energy change of each of the reactions in Problem 46. What is the value of the equilibrium constant at 25 °C?

91. The standard free energies of formation of H^+ and Cl^- are 0.00 and -131 kJ mol^{-1}, respectively. Use these values to calculate the SRP of $Cl_2(g)$, and check your answer against the value in Table 19.2.

92. (a) What current is required to move 115 C in 1 minute? (b) If the circuit operates at 9 V, how much work is done? (c) What is the power (in watts)?

93. Which of Fe^{3+}, Cl^-, and Cu^+ is the most powerful oxidant?

94. Which of Na, I^-, and Ag is the most powerful reducing agent?

95. Write an algebraic expression of the form $\mathscr{E} = A + B \cdot \text{pH}$ for the potential of the cell $Zn(s)|Zn^{2+}(1.00\ M)||H^+(x\ M)|H_2(1\ atm)|Pt$. What are the numerical values of the constants A and B at 25 °C?

96. Certain ions are unstable in aqueous solution because they disproportionate, that is, undergo auto oxidation-reduction. For instance, if the reaction $2\,Cu^+(aq) \rightarrow Cu(s) + Cu^{2+}(aq)$ were spontaneous, $Cu^+(aq)$ would be unstable. Is it? Is Fe^{2+} stable in aqueous solution?

97. A student in a freshman laboratory performs an electrolysis experiment in which 1.952 g Ag metal is plated out from a $AgNO_3$ solution by running a 0.500 ampere current through the cell for exactly 1 hour. He uses the data to determine the value of the faraday. What is his result, and what is the percent error from the true value of 96,485 C mol^{-1}? Suggest a likely explanation for the discrepancy.

98. Iron and steel objects are commonly protected against rusting by coating with a layer of zinc (galvanized iron) or tin ("tin" cans). The mechanism by which these two coating materials offer protection is quite different, particularly near regions where the coating is incomplete, broken, or otherwise damaged so that the underlying iron is exposed. Discuss.

ORGANIC COMPOUNDS

Hexamethylenediamine is dissolved in water in the lower layer, and adipoyl chloride is dissolved in hexane to form the immiscible upper layer. The two solutes can react only at the interface between the layers. The reaction produces the polymer nylon, and with care, a long continuous nylon rope can be withdrawn from the beaker.

Most of the almost 15 million known chemical compounds are *organic.* Our basic needs are supplied by organic compounds—the food we eat, the clothes we wear, the wood we build our houses with, the medicines that cure our diseases, the fuel that keeps us warm and takes us from one place to another, all consist primarily of organic compounds. Of the 50 chemicals produced in the largest quantities in this country, about 30 are organic. The name "organic" stems from the *vital force* theory, which held that certain substances could be produced only in living organisms. The theory was abandoned after 1828, when Friedrich Wöhler showed, in an easily reproducible experiment, that the organic compound urea could be produced simply by heating the inorganic substance ammonium cyanate.

Urea is produced in animal metabolic processes and is excreted as a means of ridding the body of excess nitrogen.

$$\underset{\textbf{(inorganic)}}{\underset{\textbf{cyanate}}{\underset{\textbf{ammonium}}{NH_4OCN}}} \xrightarrow[\Delta]{H_2O} \underset{\textbf{(organic)}}{\underset{\textbf{urea}}{O{=}C(NH_2)_2}} \tag{20-1}$$

In modern usage an **organic** compound is one that contains carbon and hydrogen, usually in combination with oxygen, nitrogen, and/or sulfur. Salts (and acids) of the ions HCO_3^-, CO_3^{2-}, CN^-, OCN^-, and a few others, are treated as exceptions and classified as inorganic.

20.1 HYDROCARBONS, HETEROCYCLES, AND FUNCTIONAL GROUPS

The binary compounds of carbon and hydrogen, the *hydrocarbons,* were discussed in Chapter 9. The *alkanes* contain only single bonds, the *alkenes* have one or more C═C double bonds, and the *alkynes* have one or more C≡C triple bonds. In *aromatic* hydrocarbons, six carbon atoms are joined by bonds of 3/2 order into a hexagonal array. The prefixes *meth-*(1), *eth-*(2), *prop-*(3), *but-*(4), *pent-*(5), *hex-*(6), *hept-*(7), *oct-*(8), and

Figure 20.1 Many different hydrocarbons are separated from crude oil in large industrial distillation towers such as these.

Friedrich Wöhler (1800–1882)

Friedrich Wöhler, like many other scientists of his time, was trained as a physician. He soon returned to his childhood hobby, chemistry, and after studying and working in Stockholm he became professor of chemistry at the University of Göttingen in Germany. He contributed greatly to the theory of isomerism, and was the author of *Outlines of Organic Chemistry,* a widely used text.

so on of a hydrocarbon name are used to indicate the number of atoms in the longest continuous chain of carbon atoms in the molecule. The ending of a name indicates the absence of multiple bonds (*-ane*), or the presence of double (*-ene*) or triple (*-yne*) bonding. Numerical prefixes indicate the position of multiple bonds or of substituent atoms or groups. Because multiply bonded carbon atoms are bonded to fewer than the maximum number (four) of other atoms, alkenes and alkynes can undergo *addition* reactions and are said to be *unsaturated* (with respect to bonding capacity). Alkanes are normally unreactive at room temperature, and are *saturated.*

Except for the smallest molecules, a given molecular formula corresponds to two or more *structural isomers.* For example, C_4H_{10} is the molecular formula of both 2-methylpropane and butane.

$$H_3C-\underset{\displaystyle |}{\overset{\displaystyle CH_3}{CH}}-CH_3 \qquad H_3C-CH_2-CH_2-CH_3$$

2-methylpropane **butane**

Moreover, *alkenes* often exist as *geometrical isomers,* such as the two forms of 2-butene.

$$\begin{array}{ccc} H & & H \\ & C=C & \\ H_3C & & CH_3 \end{array} \qquad \begin{array}{ccc} H & & CH_3 \\ & C=C & \\ H_3C & & H \end{array}$$

***cis*-2-butene** ***trans*-2-butene**

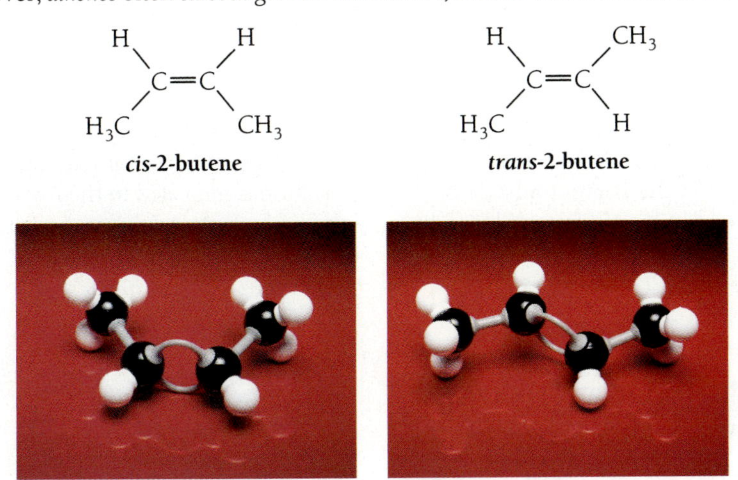

Geometrical isomers are named according to whether like atoms or groups are on the same side (*cis* isomer) or on opposite sides (*trans* isomer) of the double bond.

After this brief synopsis of Chapter 9, we now resume the discussion with a more detailed consideration of isomerism, and with the description of additional classes of hydrocarbons and related compounds.

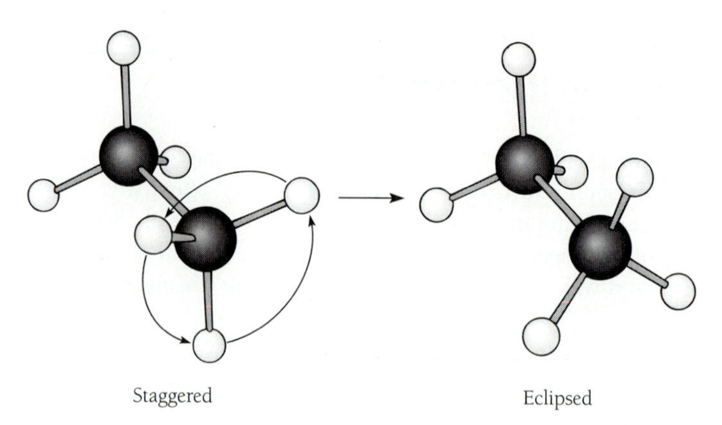

Figure 20.2 The *eclipsed* and *staggered* conformations of ethane alternate rapidly as the methyl groups freely rotate.

Conformation

Molecules are not rigid structures. They undergo a variety of internal motions, including *rotation* of one part of the molecule with respect to another. In ethane, for example, the two $—CH_3$ groups rotate freely against one another in a propeller-like motion (Figure 20.2). The different configurations that can be achieved through such rotations are called **conformations** or **conformational isomers** (Figure 20.3). As expected from the kinetic molecular theory, the atoms of a substance are in continuous motion. In general, conformational isomers of any compound are constantly being converted into one another. At normal temperatures, a given molecule goes through all possible conformations many times each second.

Internal rotations occur more slowly at low temperature. In special cases, at very low temperature, this motion can be "frozen" and conformational isomers can be distinguished from one another.

Rotation about double bonds is essentially impossible. Rotation about single bonds occurs freely and continually.

Bond rotation leading to different conformations of a molecule occurs only with single bonds. Rotation about *double bonds* does not occur. *Cis-* and *trans-* isomers of a molecule containing a double bond are *not* readily converted to one another. They are different compounds with different chemical and physical properties.

Cyclic Hydrocarbons

Organic compounds whose carbon atoms are bonded together to form a ring or loop are called **cyclic compounds.** This structural feature is indicated in the name of the compound, which is formed by addition of the prefix *cyclo-* to the name. Saturated

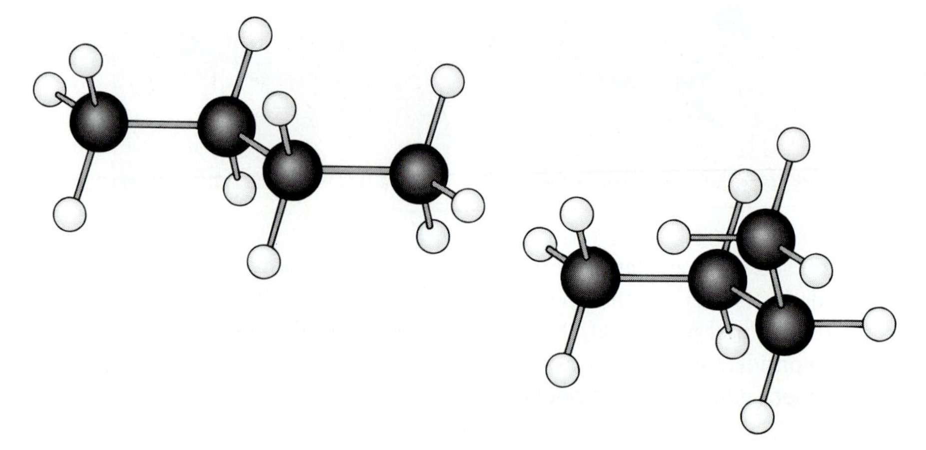

Figure 20.3 Two of the many possible conformations of butane resulting from free rotation about the central C—C bond.

cyclic hydrocarbons are called **cycloalkanes.** The smallest cycloalkane is cyclopropane (C_3H_6), in which the carbon atoms lie at the vertices of an equilateral triangle.

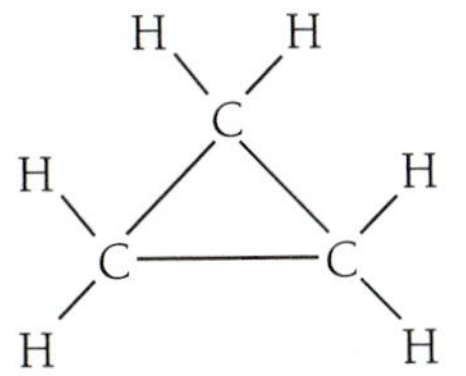

Note that cyclopropane is a structural isomer of prop*ene* (and not prop*ane,* as one might expect from the name). Whenever possible (without sacrificing clarity) the structural formulas of organic compounds are abbreviated by omitting C and H atoms. Cyclopropane is represented by a simple triangle,

The structural formula of propene is

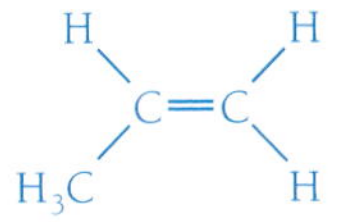

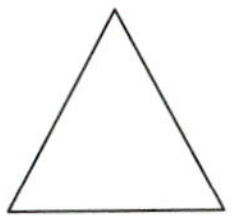

cyclopropane

in which each vertex is understood to represent the CH_2 group. Other common cyclic hydrocarbons are cyclobutane, cyclopentane, and cyclohexane.

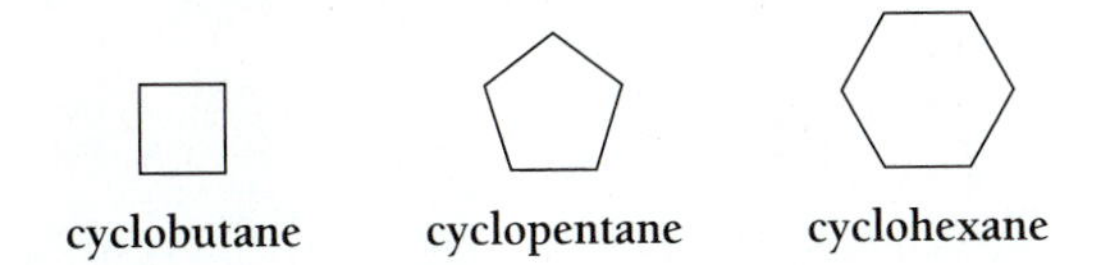

cyclobutane **cyclopentane** **cyclohexane**

The generic formula of the cycloalkanes is C_nH_{2n}, making them isomeric with the **acyclic** (not containing a ring) alk*enes.*

Cyclic hydrocarbons having a double bond in the ring are known as **cycloalkenes.** In these abbreviated structural formulas, a two-line vertex (⌃), which shows two bonds, represents an sp^3 carbon atom with two H atoms.

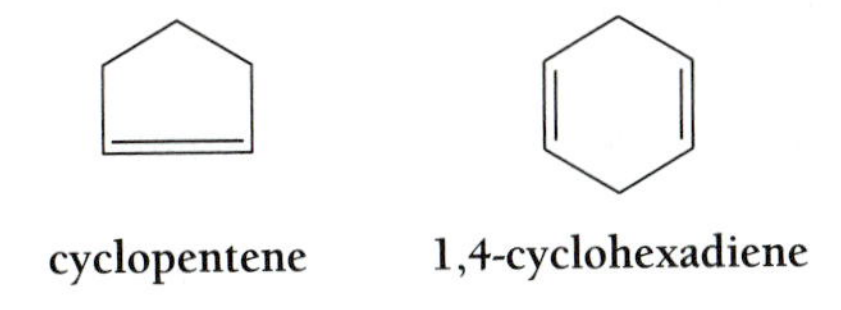

cyclopentene **1,4-cyclohexadiene**

A three-line vertex (⌃), which shows three bonds, represents an sp^2 carbon atom with only one H atom. The complete structural formula of cyclopentene, for example, is

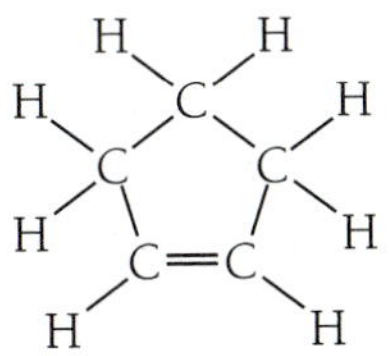

cyclopentene

Cyclic compounds may have rings of any size, but rings of more than seven atoms are less common.

You learned in geometry that *any* three points lie in one plane.

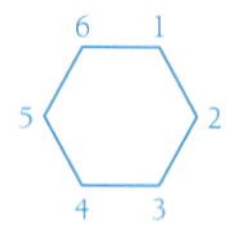

The cyclopropane ring is planar, that is, all three carbon atoms lie in one plane. The other cycloalkane rings are puckered, not planar. In cyclobutane, for example, one of the C atoms lies approximately 20° out of the plane formed by the other three. Furthermore, like other molecules, cyclic compounds have more than one possible conformation. In cyclohexane, for example, there is one stable conformation and several others that are less stable. These are easier to explain and visualize if the ring carbons are numbered, as in the sketch at the left. In the **chair** (stable) and the **boat** (unstable) conformations, the four central carbons 1, 3, 4, and 6 lie in one plane, while carbons 2 and 5 lie outside this plane. In the chair, 2 and 5 are on opposite sides of the plane, while in the boat, they are on the same side.

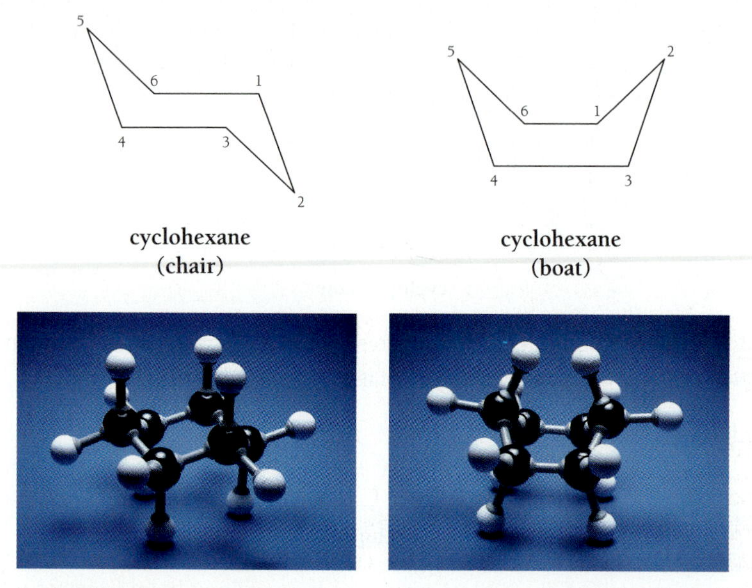

cyclohexane (chair) **cyclohexane (boat)**

Other conformations are intermediate between these two. Individual molecules switch rapidly among these conformations, in a flipping motion consisting of partial rotations about the C—C single bonds. The average cyclohexane molecule spends almost all of its time as a chair, since that conformation is the most stable.

Heterocyclic Compounds

A large class of compounds, called **heterocyclic compounds,** exists in which one or more of the ring carbon atoms of a cyclic hydrocarbon has been replaced by another atom. Here are some examples:

ethylene oxide **furan** **tetrahydrofuran**

Some heterocyclic compounds have extensive delocalization of π-electron density, and hence behave like aromatic compounds. Pyridine and pyrimidine, for example, have many reactions similar to those of benzene.

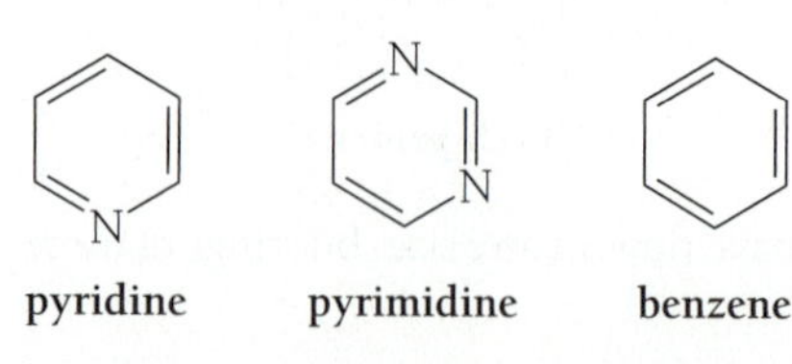

pyridine **pyrimidine** **benzene**

Geometrical Isomerism in Cyclic Compounds

Like a double bond, a ring provides a frame of reference that divides a molecule into two sides. This feature makes geometrical or *cis/trans* isomerism possible in substituted cycloalkanes. For example, there are two isomers of 1,2-dichlorocyclopropane.

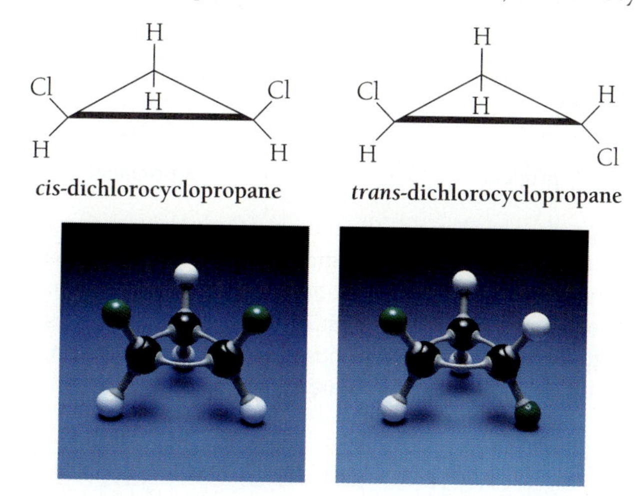

cis-dichlorocyclopropane *trans*-dichlorocyclopropane

Functional Groups

Alkanes are relatively less reactive than most organic compounds. They react with oxygen, but only at high temperature. They may be halogenated, but only when energized by ultraviolet light. The presence of a double bond, however, confers additional reactivity. For example, *alkenes* react readily with halogens to form alkyl dihalides (Section 9.5), which are said to be *derivatives* of hydrocarbons.

$$\mathrm{CH_3CH{=}CH_2} + \mathrm{Br_2} \longrightarrow \mathrm{CH_3{-}CHBr{-}CH_2Br} \qquad (9\text{-}8)$$

propene **1,2-dibromopropane**

A **derivative** is a compound prepared from another compound by some typical reaction. A derivative bears a close structural resemblance to its reference compound.

The double bond of an alkene is a **functional group**—a subset of atoms within the molecule that serves as the site of, and exerts a dominant influence on, chemical behavior of the compound. In the sections that follow, compounds characterized by a variety of functional groups are classified and discussed according to whether they contain oxygen, nitrogen, or other atoms.

20.2 OXYGEN-CONTAINING COMPOUNDS

Oxygen is one of the most abundant, and reactive, elements on earth. It is hardly surprising, therefore, that oxygen finds its way into so many different organic compounds. We begin by considering compounds in which oxygen is singly bonded to two other atoms.

There are many inorganic compounds of oxygen, also. Metal oxides and ionic compounds of CO_3^{2-}, NO_3^-, and PO_4^{3-}, to name a few, are very common.

Alcohols, Phenols, and Ethers

An **alcohol** is a compound in which an oxygen atom is singly bonded to a hydrogen atom and an sp^3-hybridized carbon atom. The functional group —OH is called the **hydroxyl** group. One general method of preparing an alcohol is the acid-catalyzed addition of water across the double bond of an alkene.

$$\underset{\textbf{propene}}{CH_2{=}CH{-}CH_3} + H{-}OH \xrightarrow{H^+} \underset{\textbf{2-propanol}}{CH_3{-}\underset{\substack{|\\ OH}}{CH}{-}CH_3} \qquad (20\text{-}2)$$

Alcohols are named by changing the ending of the hydrocarbon root name from *-e* to *-ol*, and, if necessary, indicating the position of the hydroxyl group with a numerical prefix. The 2-propanol produced in Reaction (20-2) could also be named 2-hydroxypropane. Some of the smaller alcohols are still known by earlier, nonsystematic names: 2-propanol, used as rubbing alcohol in hospitals for fever reduction, is often called isopropanol. Some other small alcohols are shown below.

Structure	CH_3OH	CH_3CH_2OH	$CH_3{-}\overset{\substack{OH\\ |}}{\underset{\substack{|\\ CH_3}}{C}}{-}CH_3$
IUPAC Name	**methanol**	**ethanol**	**2-methyl-2-propanol**
Older Name	**methyl alcohol**	**ethyl alcohol**	**tertiary butyl alcohol**
Even Older Name	**wood alcohol**	**grain spirits**	

The International Union of Pure and Applied Chemistry is the overseer of chemical terminology.

All small alcohols are liquid at room temperature, and they are often used as solvents in the laboratory and in industry. Methanol is used as fuel in some high-performance engines, and some brands of gasoline contain up to 10% ethanol.

Ethanol-containing gasoline is called "gasohol."

Alcohols react with most oxidizing agents. These reactions, and the products that are formed in them, are discussed later in this section (pages 857 and 858). Alcohols also react with hydrogen halides, to produce alkyl halides. Equation (20-3) is a typical example.

$$\underset{\textbf{3-pentanol}}{CH_3CH_2\overset{\substack{OH\\ |}}{C}HCH_2CH_3} \xrightarrow{HBr} \underset{\substack{\textbf{3-bromopentane}\\ \textbf{(major product)}}}{CH_3CH_2\overset{\substack{Br\\ |}}{C}HCH_2CH_3} + \underset{\substack{\textbf{2-bromopentane}\\ \textbf{(minor product)}}}{CH_3\overset{\substack{Br\\ |}}{C}HCH_2CH_2CH_3} \qquad (20\text{-}3)$$

Organic chemists frequently use unbalanced equations, and often show one or more of the reactants above or below the arrow.

This reaction does not proceed by a simple replacement of one functional group by another: if this were the case, there would be only one product. The fact is that in addition to 3-bromopentane, the major product, a small amount of 2-bromopentane is also produced in Reaction (20-3). Few organic reactions proceed by a single elementary step, and formation of more than one product is the rule rather than the exception.

The reactivity of alcohols is influenced by the other groups attached to the sp^3 carbon to which the hydroxyl group is bonded. It is often useful, when referring to alcohols, to differentiate among these possibilities. A **primary (1°)** alcohol has two H atoms bonded to the carbon to which the hydroxyl group is attached, a **secondary (2°)** alcohol has one such H atom, and a **tertiary (3°)** alcohol has none.*

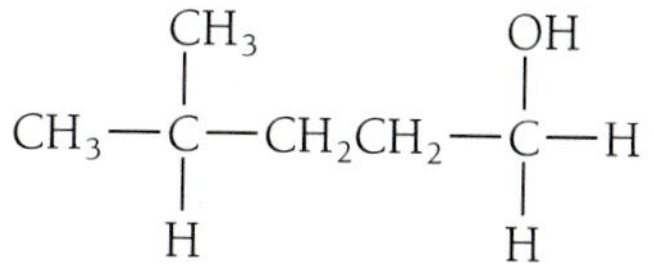

$$CH_3-\underset{H}{\overset{CH_3}{C}}-CH_2CH_2-\underset{H}{\overset{OH}{C}}-H \qquad CH_3-\underset{H}{\overset{CH_3}{C}}-CH_2-\underset{H}{\overset{OH}{C}}-CH_3 \qquad CH_3-\underset{CH_3}{\overset{OH}{C}}-CH_2-CH_2-CH_3$$

4-methyl-1-pentanol (a primary alcohol) 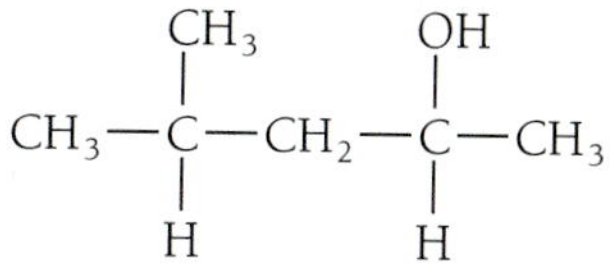　**4-methyl-2-pentanol (a secondary alcohol)** 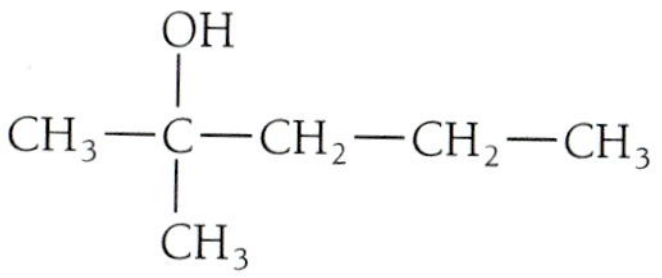　**2-methyl-2-pentanol (a tertiary alcohol)**

Generally, the rate at which alcohols react is different for 1°, 2°, and 3° alcohols.

Alcohols may be regarded as derivatives of water, in which one of the H-atoms has been replaced by an alkyl group "R." This point of view is helpful in understanding some of the physical and chemical properties of alcohols, which resemble those of water.

The abbreviation R— is used to represent an organic group, such as methyl ($-CH_3$), ethyl ($-CH_2CH_3$), isopropyl [$-CH(CH_3)_2$], and so on, without specifying a particular one.

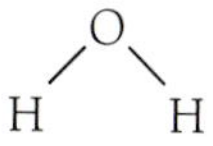

$$\underset{\textbf{water}}{H-O-H} \qquad \underset{\textbf{an alcohol}}{R-O-H}$$

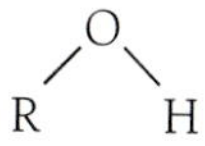

Like water, alcohols are very weak acids. Their K_a's are in the range 10^{-15} to 10^{-16}, so that they are essentially un-ionized in aqueous solutions. The extremely low acid strength means that the conjugate base, the *alkoxide ion* $R-O^-$, is a very strong base. In aqueous solution, the hydrolysis reaction (20-4) goes to completion.

$$\underset{\textbf{alkoxide ion}}{R-O^-(aq)} + H_2O \longrightarrow \underset{\textbf{alcohol}}{R-OH} + OH^-(aq) \qquad (20\text{-}4)$$

Like water, alcohols react with alkali metals to produce solutions that are strongly basic [Equation (20-5)].

$$Na(s) + \underset{\textbf{ethanol}}{CH_3CH_2OH(\ell)} \longrightarrow \tfrac{1}{2}H_2(g) + Na^+ + \underset{\textbf{ethoxide ion (a strong base)}}{CH_3CH_2O^-} \qquad (20\text{-}5)$$

Like water, alcohols are polar, and hydrogen-bonded in the liquid state. These strong intermolecular forces cause alcohols to have much higher boiling points than the corresponding hydrocarbon: ethane boils at −89 °C, while ethanol boils at +79 °C.

Hydrogen bonding (Section 11.3) is an especially strong form of dipole-dipole attraction that occurs when a hydrogen atom is bonded to N, O, or F.

* Another way of remembering the distinction is to regard all alcohols as related to methanol, CH_3OH. An alcohol is 1°, 2°, or 3° depending on whether 1, 2, or all 3 of methanol's C—H bonds have been replaced by C—C bonds.

This difference becomes less dramatic as the number of carbon atoms increases, because the —OH becomes a relatively smaller fraction of the molecule.

An aromatic compound with a hydroxyl substituent bonded to one of the carbon atoms of the ring is called a **phenol,** after the simplest such compound, C_6H_5OH. Many phenols are known by their trivial names rather than their systematic names.

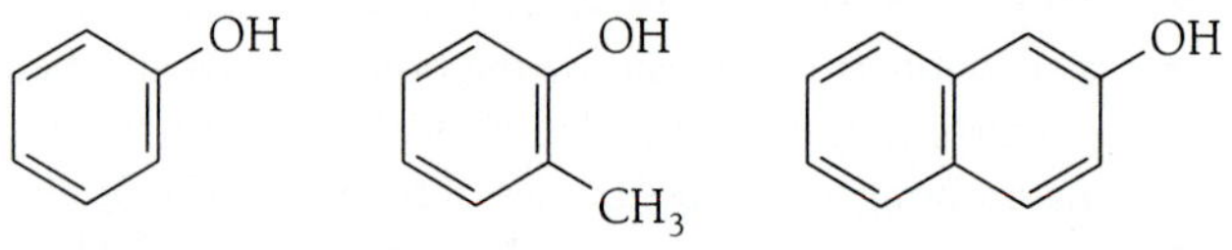

phenol *ortho*-**cresol** **β-naphthol**

phenyl group

"phen" + "ol" = "phenol"

The name *phenol* is related to **phenyl,** the name of the $—C_6H_5$ group derived from benzene. Compared to the *aliphatic* (nonaromatic) alcohols, phenols have greater acid strength. The ionization constant of phenol is about 10^{-10}, roughly comparable to that of HCN. Both these acids are about 0.01% ionized in 0.01 M solutions.

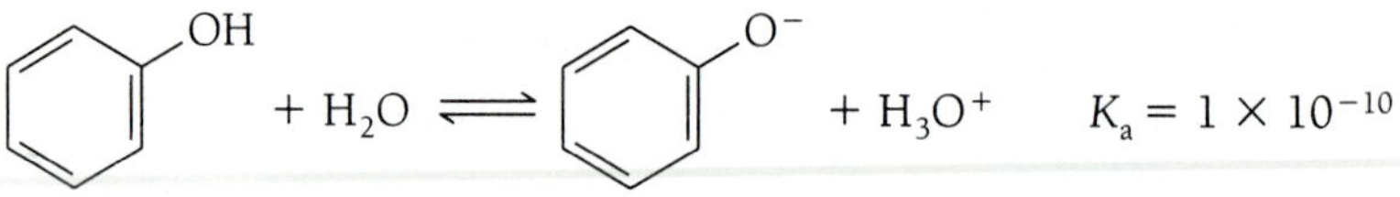

$$C_6H_5OH + H_2O \rightleftharpoons C_6H_5O^- + H_3O^+ \qquad K_a = 1 \times 10^{-10}$$

Phenol itself, formerly called *carbolic acid,* is a minor component of coal tar. It is produced on a large scale from benzene, by stepwise reaction with chlorine, sodium hydroxide, and hydrochloric acid.

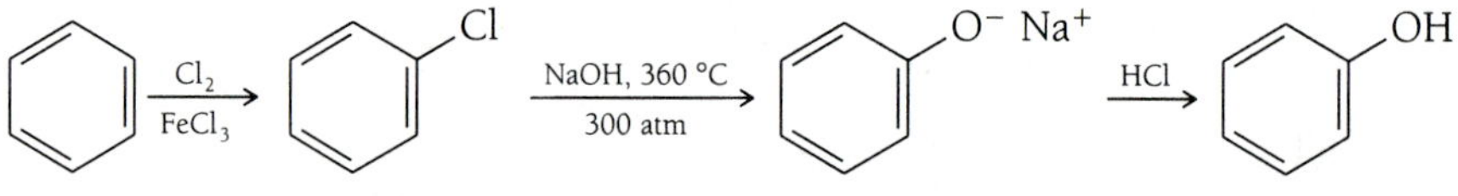

$$\text{benzene} \xrightarrow[FeCl_3]{Cl_2} \text{chlorobenzene} \xrightarrow[300\ \text{atm}]{NaOH,\ 360\ °C} \text{sodium phenolate } (C_6H_5O^-\ Na^+) \xrightarrow{HCl} \text{phenol } (C_6H_5OH)$$

The allergen in poison ivy and poison oak is a phenol.

Phenol is an important raw material in the chemical manufacturing industry. It has been used as an antiseptic and disinfectant, particularly in dentistry.

An **ether** is a compound that contains two alkyl groups joined by an oxygen atom. Alternatively, an ether may be described as a water molecule in which both hydrogen atoms have been replaced by alkyl groups. Ethers are named by specifying the alkyl groups, in alphabetical order.

$CH_3CH_2—O—CH_3$ **ethyl methyl ether**

$(CH_3CH_2)_2O$ **diethyl ether**

$CH_3—O—CH(CH_3)_2$ **isopropyl methyl ether**

Although they are polar substances, ethers lack the capability of hydrogen bonding. As a result, ethers have considerably higher vapor pressures and lower boiling points than the corresponding alcohols. Dimethyl ether boils at −46 °C, more than 100° lower than its isomer, ethanol. Diethyl ether is a well-known (but no longer widely used) anesthetic. It is prepared by treating ethanol with sulfuric acid.

$$2\ C_2H_5OH \xrightarrow[\Delta]{H_2SO_4} (C_2H_5)_2O + H_2O \tag{20-6}$$

Ethers are not very reactive, and, in part because of this, they are often used as solvents in the laboratory.

Aldehydes and Ketones

A **carbonyl** group consists of an sp^2 carbon atom doubly bonded to an oxygen atom. This group is present in **aldehydes,** in which a carbonyl group is bonded to an organic group and a hydrogen atom, and in **ketones,** in which a carbonyl group is bonded to two organic groups. Aldehydes are named by changing the *-e* ending of the hydrocarbon to *-al.*

$$R{-}CHO \qquad H{-}CHO \qquad CH_3{-}CHO \qquad CH_2{=}CH{-}CH_2{-}CHO$$

an aldehyde | **methanal (formaldehyde)** | **ethanal (acetaldehyde)** | **3-butenal**

Particularly for the smaller aldehydes, older names are widely used (Figure 20.4). Ketones are named by changing the hydrocarbon ending from *-e* to *-one,* or in older references, by identifying the groups bonded to the carbonyl group.

$$R{-}C({=}O){-}R' \qquad H_3C{-}C({=}O){-}CH_3 \qquad H_3C{-}C({=}O){-}CH(CH_3){-}CH_3$$

a ketone | **propanone (acetone)** | **3-methyl-2-butanone (isopropyl methyl ketone)**

The symbol R′ (R prime) is used to indicate that there are two different organic groups, R and R′, in the molecule.

The smallest ketone is called by its trivial name, *acetone.*

There are many naturally occurring aldehydes and ketones. Glucose is an aldehyde that is metabolized in animal bodies to produce energy. (Glucose also contains —OH groups, so it is an alcohol as well as an aldehyde.) The sex hormones progesterone and testosterone are ketones.

The flavor of vanilla is due to an aldehyde.

Ketones can be synthesized by the oxidation of secondary alcohols.

$$\text{cyclohexanol} \xrightarrow[H_2SO_4]{Cr_2O_7^{2-}(aq)} \text{cyclohexanone} \qquad (20\text{-}7)$$

cyclohexanol | **cyclohexanone**

Figure 20.4 *Formaldehyde* reacts with phenol to form a hard substance called *Bakelite.* Bakelite, one of the first synthetic plastics, is still used today, for combs, auto parts, and electrical fixtures.

Oxidation of a primary alcohol leads to an aldehyde. However, since aldehydes themselves are readily oxidized, it is often difficult to halt the reaction at the aldehyde stage.

$$CH_3{-}CH_2{-}CH_2{-}OH \xrightarrow[H_2SO_4]{Cr_2O_7^{2-}(aq)} \left[CH_3{-}CH_2{-}CHO\right] \longrightarrow CH_3{-}CH_2{-}C({=}O){-}O{-}H \qquad (20\text{-}8)$$

propanol | **propanal** | **propanoic acid**

The usual product of the oxidation of a 1° alcohol is a carboxylic acid (see below).

Both aldehydes and ketones can be reduced. The product is either the corresponding hydrocarbon or the alcohol, depending on the strength of the reducing agent.

$$CH_3{-}CH_2{-}\overset{\overset{\displaystyle O}{\|}}{C}{-}CH_3 \xrightarrow[H_2O]{NaBH_4} CH_3{-}CH_2{-}\overset{\overset{\displaystyle OH}{|}}{CH}{-}CH_3 \qquad (20\text{-}9)$$

An *amalgam* is an alloy of a metal with mercury.

The product of reduction by $LiAlH_4$, $NaBH_4$, or H_2 (in the presence of a catalyst) is the alcohol, while the stronger reductant zinc amalgam in acid solution yields the alkane.

$$CH_3CH_2CH_2CH_2C{\overset{\displaystyle O}{\underset{\displaystyle H}{\lessgtr}}} \xrightarrow[HCl]{Zn(Hg)} CH_3CH_2CH_2CH_2CH_3 \qquad (20\text{-}10)$$

pentanal **pentane**

Addition reactions across the double bond of the carbonyl group are characteristic of aldehydes and ketones. The addition of HCN is most conveniently carried out in two steps:

$$H_3C{-}C{\overset{\displaystyle CH_3}{\underset{\displaystyle O}{\lessgtr}}} \xrightarrow[2.\ H_2SO_4(aq)]{1.\ NaCN(aq)} CH_3{-}\overset{\overset{\displaystyle CH_3}{|}}{\underset{\underset{\displaystyle C{\equiv}N}{|}}{C}}{-}OH \qquad (20\text{-}11)$$

acetone **acetone cyanohydrin**

The strong polarity of the C=O bond is important to the mechanism of addition reactions. In the first step [Equation (20-11a)], a CN^- ion is attracted to and forms a bond to the positive end of the C=O bond, yielding a larger anionic species.

$$H_3C{-}\overset{\overset{\displaystyle CH_3}{|}}{\underset{\delta^+}{C}}{=}\underset{\delta^-}{O} + C{\equiv}N^- \longrightarrow H_3C{-}\overset{\overset{\displaystyle CH_3}{|}}{\underset{\underset{\displaystyle C{\equiv}N}{|}}{C}}{-}O^- \qquad (20\text{-}11a)$$

In the second step the anion, a base like CN^- itself, is neutralized by the strong acid.

$$H_3C{-}\overset{\overset{\displaystyle CH_3}{|}}{\underset{\underset{\displaystyle C{\equiv}N}{|}}{C}}{-}O^- + H_3O^+ \longrightarrow H_3C{-}\overset{\overset{\displaystyle CH_3}{|}}{\underset{\underset{\displaystyle C{\equiv}N}{|}}{C}}{-}OH + H_2O \qquad (20\text{-}11b)$$

Carboxylic Acids

Vinegar is a 5% aqueous solution of acetic acid.

The product of the oxidation of a 1° alcohol is a **carboxylic acid,** which is a compound containing the group $-CO_2H$. For example, treatment with the oxidizing agent potassium permanganate converts ethanol to *acetic acid,* $CH_3{-}CO_2H$.

$$CH_3CH_2OH \xrightarrow{KMnO_4(aq)} CH_3C{\overset{\displaystyle O}{\underset{\displaystyle O{-}H}{\lessgtr}}} \qquad (20\text{-}12)$$

ethanol **acetic acid**

The smallest carboxylic acid is *formic acid,* H—CO_2H. Larger carboxylic acids are named by replacing the *-e* of the parent hydrocarbon name with *-oic acid.* Carboxylic acids tend to have strong odors. The smell of rancid butter arises in part from butanoic acid, and caproic acid ($C_5H_{11}CO_2H$, IUPAC name hexanoic acid) got its trivial name because its odor is reminiscent of the odor of unwashed goats (from the sign and constellation Capricorn, the goat).

An ant bite stings because a small amount of formic acid is left in the tiny puncture-wound.

Carboxylic acids can be converted to 1° alcohols by reducing agents such as $LiAlH_4$.

$$R{-}CO_2H \xrightarrow{LiAlH_4} R{-}CH_2{-}OH \qquad (20\text{-}13)$$

Acid Strength

Carboxylic acids are stronger Brønsted acids than alcohols and phenols.

$$CH_3CH_2CH_2C(=O)OH + H_2O \rightleftharpoons CH_3CH_2CH_2C(=O)O^- + H_3O^+ \qquad (20\text{-}14)$$

butanoic acid **butanoate ion**

There are two reasons for this. First, the C=O group withdraws electron density from the adjacent O—H bond in a carboxylic acid, making it weaker and more easily broken than the O—H bond in an alcohol. Second, the conjugate base that is the product of ionization, the **carboxylate** anion R—CO_2^-, is a resonant species. Two equivalent Lewis structures can be drawn,

The relationships between molecular structure and the relative strengths of acids are discussed in Section 15.6.

$$CH_3CH_2CH_2C(=O)O^- \longleftrightarrow CH_3CH_2CH_2C(O^-)=O$$

resonance forms of the butanoate anion

and in the actual structure the two oxygen atoms are identical with respect to bonding and negative charge. As always, resonance or electron delocalization confers stability. Because it is *more* stable, the carboxylate anion accepts a proton *less* readily than an alkoxide ion R—O^-. That is, it is a *weaker* base. If R—CO_2^- is a weaker base than R—O^-, it follows that R—CO_2H is a stronger acid than R—OH. Carboxylic acids in which R is an unsubstituted alkyl group are of moderate strength, having pK_a values in the range of 4 to 5. Formic acid (H—CO_2H) is somewhat stronger (pK_a = 3.75). Substitution in the alkyl group, particularly with halogen atoms, has a marked effect on acid strength. Trichloroacetic acid (CCl_3CO_2H), for example, has a pK_a of 0.65, so it is almost as strong as nitric acid.

Remember the reciprocal relationship: the stronger the acid, the weaker the conjugate base.

The trends in relative acid strengths among the organic acids follow the principles discussed in Section 15.6. The most obvious factor contributing to greater acid strength is the inductive withdrawal of electron density from the bond joining the ionizable hydrogen atom to the rest of the acid molecule. In discussions of organic chemistry it is common to refer to *electron-withdrawing* and *electron-donating* substituent groups.

The term "electronegative," which also describes electron-withdrawing power, is applied only to single atoms.

If the acid contains an electron-withdrawing group, the O—H bond is weaker and more easily broken. The acid is stronger, because it donates its proton more readily. Conversely, the presence of an electron-donating group tends to weaken an

Figure 20.5 Several common household products contain ethyl acetate, because it is an excellent solvent with a pleasant odor.

acid. Some common electron-withdrawing groups are $-NO_2$, $-\overset{R}{\overset{|}{C}}=O$, $-F$, and $-Cl$. These groups differ in the extent to which they withdraw electrons. Fluorine is somewhat more effective than chlorine, for example, so that fluoroacetic acid is somewhat stronger than chloroacetic acid. Formic acid is a stronger acid than phenol, which is in turn stronger than methanol.

formic acid ($pK_a = 3.75$) **phenol** ($pK_a = 10.0$) **methanol** ($pK_a = 15.5$)

The carbonyl group is a good electron withdrawer, the phenyl group ($-C_6H_5$) is not as good, and the methyl group donates rather than withdraws electron density.

Esters

Carboxylic acids react with alcohols to form **esters,** in a reaction catalyzed by strong acids. Such a reaction, for example the formation of methyl benzoate,

$$C_6H_5COOH + CH_3OH \overset{H^+}{\rightleftharpoons} C_6H_5COOCH_3 + H_2O \quad (20\text{-}15)$$

benzoic acid **methanol** **methyl benzoate**

is called an *esterification.* Esters are named with the group name of the alcohol (*-yl* ending) and the acid name (*-ate* ending). The product of the reaction between ethanol and propanoic acid is *ethyl propanoate.* Some familiar ester-containing products are shown in Figure 20.5.

Esterifications are members of a general class of reactions, called **condensations,** in which two smaller molecules become coupled into a larger unit. It is characteristic of condensation reactions that a small molecule, usually water, is eliminated from the point where the coupling takes place. In esterifications, the eliminated water molecule is composed of OH from the acid and H from the alcohol. Esterifications are reversible reactions that reach equilibrium, but can be driven to completion by removing the product water as soon as it is formed.

Although their chemical equations *look* alike, an esterification is not an acid-base neutralization. Alcohols are not bases, and the mechanism of esterification is not a direct proton transfer.

$$R-C(=O)-O-H + H-O-R' \overset{H^+}{\rightleftharpoons} R-C(=O)-O-R' + H_2O \quad (20\text{-}16)$$

Because the reaction is reversible, esters can be *cleaved* to their acid and alcohol components in the presence of an excess of water and strong acid. Many esters have pleasant, fruity odors. For example, the odor of bananas is due to *amyl acetate* (*$CH_3CO_2CH_2CH_2CH_2CH_2CH_3$*). Since hydrogen bonding is not possible in esters, their boiling points are usually lower than those of the corresponding acids.

The boiling point of acetic acid is 118 °C, while that of ethyl acetate is 77 °C.

A summary of oxygen-containing compounds and functional groups is given in Table 20.1.

TABLE 20.1 Some Classes of Organic Compounds Containing Oxygen

Class Name	Functional Group	Examples		
Alcohols	R—OH ("hydroxyl" or "hydroxy" group)	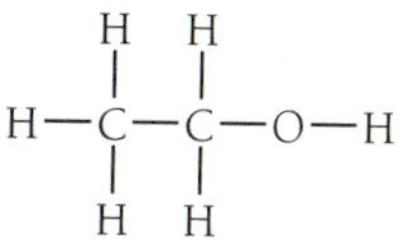 H–C(H)(H)–C(H)(H)–O–H **ethanol (1°)**	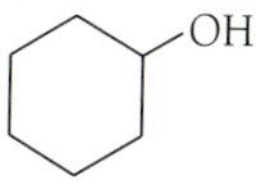 (cyclohexane ring)–OH **cyclohexanol (2°)**	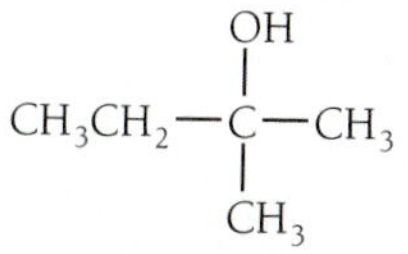 CH_3CH_2–C(OH)(CH_3)–CH_3 **2-methyl-2-butanol (3°)**
Phenols	Ar—OH	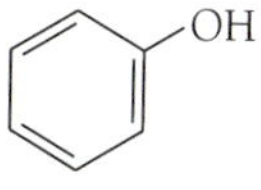 (benzene ring)–OH **phenol**	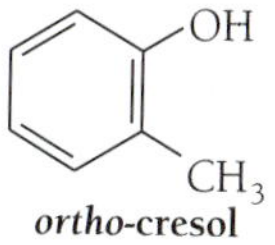 (benzene ring)(–OH)(–CH_3) ***ortho*-cresol**	
Ethers	R—O—R′	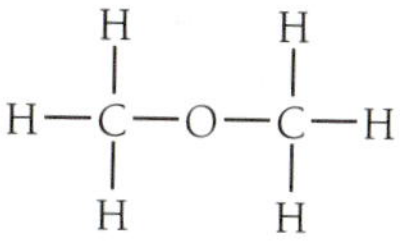 H–C(H)(H)–O–C(H)(H)–H **dimethyl ether**	C_2H_5—O—CH_3 **ethyl methyl ether**	
Aldehydes	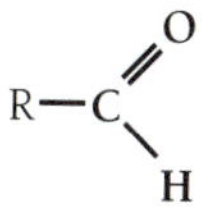 R–C(=O)–H (C=O is called the "carbonyl" group)	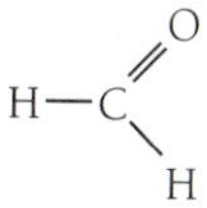 H–C(=O)–H **formaldehyde**	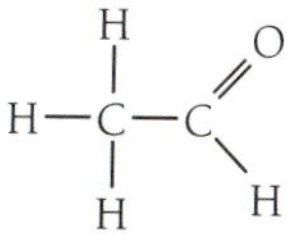 H–C(H)(H)–C(=O)–H **acetaldehyde (ethanal)**	
Ketones	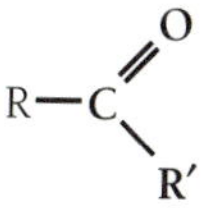 R–C(=O)–R′	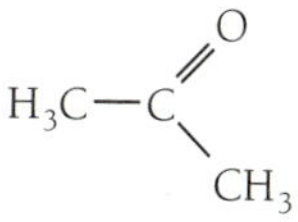 H_3C–C(=O)–CH_3 **acetone**	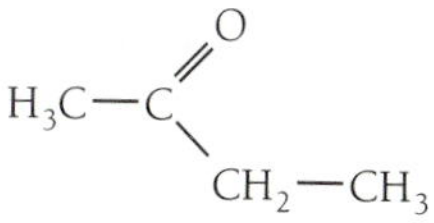 H_3C–C(=O)–CH_2—CH_3 **butanone**	
Carboxylic Acids	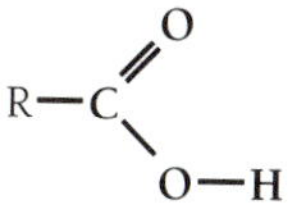 R–C(=O)–O—H	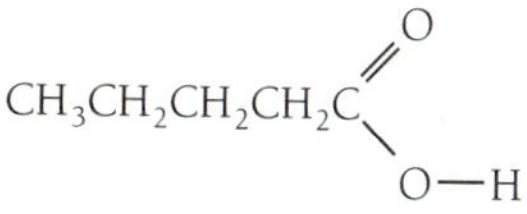 $CH_3CH_2CH_2CH_2$C(=O)–O—H **pentanoic acid**		
Esters	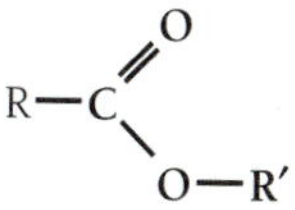 R–C(=O)–O—R′	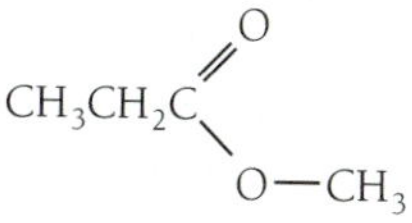 CH_3CH_2C(=O)–O—CH_3 **methyl propanoate**	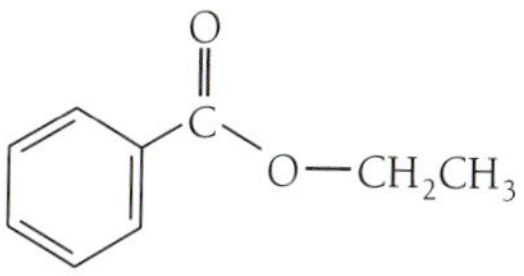 (benzene ring)–C(=O)–O—CH_2CH_3 **ethyl benzoate**	

"Ar" stands for "aromatic group," and it serves a function similar to "R" in general formulas.

Redox Reactions in Organic Chemistry

Several of the reactions discussed in this section have been identified as oxidations or reductions, and an additional redox reaction is illustrated in Figure 20.6. Calculation of oxidation numbers in large molecules is tedious, and although it can be used to

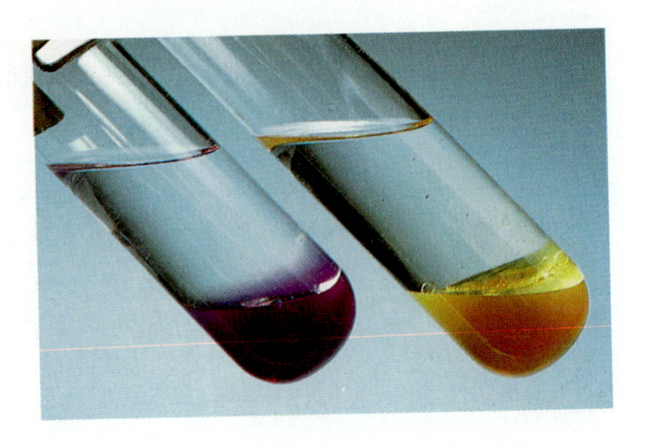

Figure 20.6 Alkenes react with oxidizing agents such as aqueous potassium permanganate. Two hydroxyl groups are added across the double bond: $\mathrm{R(H)C{=}C(H)R} \rightarrow \mathrm{H{-}C(OH)(R){-}C(OH)(R){-}H}$. This photo illustrates a simple laboratory test for the presence of an alkene. On the left, no reaction results from shaking $KMnO_4(aq)$ with hex*ane* (upper layer). On the right, hex*ene* is quickly oxidized by the $KMnO_4$, which is reduced to $MnO_2(s)$ in the process.

identify the participants in a redox reaction, there is often an easier way. A reaction is an oxidation if a compound gains oxygen (or loses hydrogen), and it is a reduction if the compound gains hydrogen (or loses oxygen). Gain of one oxygen atom involves the loss of two electrons, and it is sometimes called a *two-electron oxidation*. In contrast, loss of a hydrogen atom is a *one-electron process*. Example 20.1 demonstrates the relationship of these facts to the earlier definition (Chapter 18) of oxidation as an increase in oxidation number and reduction as a decrease.

EXAMPLE 20.1 Oxidation in Organic Reactions

Show that oxidation occurs in both of the following: (a) the conversion of methanol to formaldehyde, which involves loss of hydrogen, and (b) the conversion of propanal to propanoic acid, which is a gain of oxygen.

Analysis The goal is to demonstrate that oxidation occurs in these two reactions.

Plan Show that in each process there is at least one atom that undergoes an increase in oxidation number.

Work

a. The structural-formula representation of the conversion of methanol to formaldehyde is

$$\mathrm{H{-}C(H)_2{-}O{-}H} \longrightarrow \mathrm{H{-}C({=}O){-}H}$$

However, for the purpose of assigning oxidation numbers, empirical formulas are more convenient. Also, in this form it is easier to count H and O atoms and see if any have been lost or gained in the reaction.

$$CH_4O \longrightarrow CH_2O$$

Using the rules for assignment of oxidation numbers (Section 18.2),

$$\overset{-2}{C}H_4O \longrightarrow \overset{0}{C}H_2O$$

we find that the oxidation number of carbon has increased from −2 to 0. Therefore, oxidation has occurred. Two H-atoms have been lost, one for each increase of +1 in the oxidation number.

b. The oxidation numbers in propanal (CH_3CH_2CHO) and propanoic acid ($CH_3CH_2CO_2H$) are

$$\overset{-4/3\ +1\ -2}{C_3H_6O} \longrightarrow \overset{-2/3\ +1\ -2}{C_3H_6O_2}$$

The oxidation number of carbon has increased by +⅔, so there has been an oxidation. Since there are 3 C atoms, the total increase is 3(+⅔) = +2, associated with the gain of one O atom by the molecule.

Exercise

Show that the hydrogenation of propene is a redox reaction.

$$CH_3CH{=}CH_2 + H_2 \longrightarrow CH_3CH_2CH_3$$

Answer: The oxidation number of carbon decreases from −2 in propene to −8/3 in propane.

20.3 NITROGEN-CONTAINING COMPOUNDS

Several important classes of organic compounds contain nitrogen.

Amines

Amines are derivatives of ammonia, in which one or more of the H atoms of NH_3 have been replaced by organic groups. Amines are referred to as primary, secondary, or tertiary depending on whether one, two, or all three of the H atoms have been replaced. Simple amines are named by prefixing the word *-amine* with the names of the groups (in alphabetical order) that are bonded to the nitrogen atom.

$CH_3CH_2NH_2$	$(CH_3)_2NH$	$CH_3CH_2NHCH_3$	$(CH_3CH_2)_3N$
ethylamine	**dimethylamine**	**ethylmethylamine**	**triethylamine**
(1°)	(2°)	(2°)	(3°)

Phenyl amines are usually named as substituted **anilines,** after the simplest member of the group.

NH₂ H N CH₃

aniline **N-methylaniline**

The "N" in the name N-methylaniline indicates that the methyl group is bonded to the nitrogen atom, rather than to a carbon atom.

Anilines are used as raw materials in the manufacture of dyestuffs. About 50,000 tons of aniline are produced annually in the United States.

All amines are bases. Substitution of one or more of ammonia's hydrogen atoms with an organic group does not involve the lone pair of electrons on the N-atom, and therefore it does not impair the ability of the N-atom to accept a proton from a Brønsted acid.

$$H_3N\colon + H_2O \longrightarrow NH_4^+ + OH^- \qquad pK_b = 4.8 \qquad (20\text{-}17)$$

$$(CH_3)_3N\colon + H_2O \longrightarrow (CH_3)_3NH^+ + OH^- \qquad pK_b = 4.2 \qquad (20\text{-}18)$$

Salts produced by neutralization of amines with hydrochloric acid are usually named by changing the ending from *-amine* to *-ammonium,* and adding the word *chloride.* They are also called *amine hydrochlorides.* Like other salts, these are ionic compounds. Many of them are soluble in water, although the parent amines, particularly those of high molar mass, are often insoluble.

$$\text{C}_5\text{H}_9\text{NH}_2 + \text{HCl(aq)} \rightleftharpoons \text{C}_5\text{H}_9\text{NH}_3^+ \; \text{Cl}^- \qquad (20\text{-}19)$$

cyclopentylamine — **cyclopentylammonium chloride** or **cyclopentylamine hydrochloride**

Amines can be produced in several ways, for example from the reaction between ammonia and an alkyl halide.

$$CH_3I + 2\,NH_3 \xrightarrow[H_2O]{} CH_3NH_2 + NH_4^+ + I^- \qquad (20\text{-}20)$$

Simple amines, often with strong, unpleasant odors, are produced by bacterial action in decaying animal matter, such as dead fish.

Many plants produce complex amines called *alkaloids* (alkali-like), that have powerful effects on the human body. Quinine, for treatment of malaria, and morphine, for alleviation of pain, are alkaloids widely used in medicine. Thiamine, an essential vitamin, is an alkaloid. Some alkaloids have harmful effects: the toxic, highly addictive drugs nicotine and cocaine are alkaloids.

Amides

Imagine that the —OH group of a carboxylic acid is replaced by an amino group ($—NH_2$). The resulting structure is an **amide,** a molecule containing the amide group, $—CO—NH_2$. One way of making amides is by the condensation of a carboxylic acid with an amine.

$$CH_3—C(=O)—OH + H_2N—C_6H_5 \xrightarrow[(-H_2O)]{} CH_3—C(=O)—NH—C_6H_5 \qquad (20\text{-}21)$$

acetic acid **aniline** **acetanilide**

The simpler amides are named by replacing the *-ic* or *-oic acid* ending of the acid with *-amide:* propanoic acid becomes propanamide, and so on. The H-atoms of the simple amide group can be substituted by organic groups, forming *N*-substituted amides. *N,N*-Dimethylformamide, usually called "DMF," is a useful laboratory solvent.

$$H—C(=O)—N(CH_3)—CH_3$$

***N,N*-dimethylformamide**

Proteins are discussed in Section 21.1.

The substructure $—\overset{O}{\overset{\|}{C}}—\overset{H}{\overset{|}{N}}—$, the *amide linkage,* is a characteristic part of protein molecules.

20.4 OTHER FUNCTIONAL GROUPS

In this section, we discuss a few of the many types of organic compounds containing sulfur or phosphorus.

Sulfonic Acids

Benzene and other aromatic hydrocarbons react with concentrated sulfuric acid or sulfur trioxide to produce **sulfonic acids,** which are derivatives of sulfuric acid.

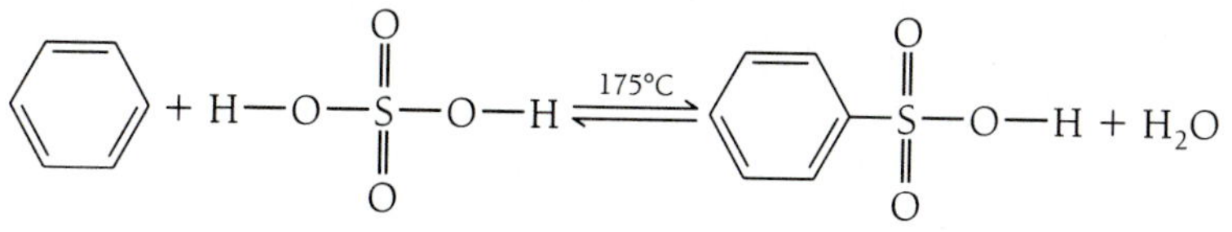

$$C_6H_6 + H{-}O{-}SO_2{-}O{-}H \overset{175°C}{\rightleftharpoons} C_6H_5{-}SO_2{-}O{-}H + H_2O \quad (20\text{-}22)$$

benzene **sulfuric acid** **benzenesulfonic acid**

Removal of water from the reaction mixture, by distillation, shifts the equilibrium of Reaction (20-22) to the right and improves the yield of the product *benzenesulfonic acid.* Reaction of sulfonic acids with NaOH gives salts such as sodium *benzenesulfonate.*

Benzenesulfonic acid is used in the manufacture of detergents, which are discussed in Section 21.4.

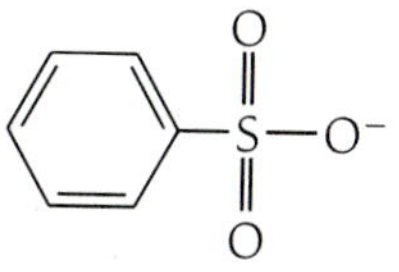

$$C_6H_5{-}SO_2{-}O^-$$

benzenesulfonate anion

Derivatives of sulfonic acids can be made in a variety of ways. In general, these derivatives have structures analogous to derivatives of carboxylic acids, for example sulfonic acid esters and *sulfonamides.*

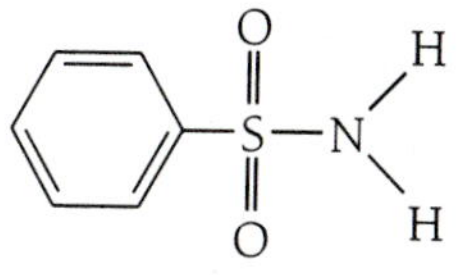

$$C_6H_5{-}SO_2{-}NH_2 \qquad H_2N{-}C_6H_4{-}SO_2{-}NH_2$$

benzenesulfonamide 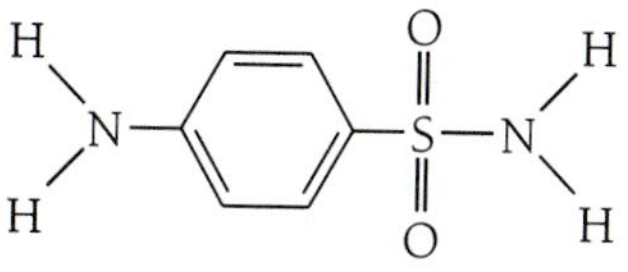 **sulfanilamide**

Sulfanilamide is the simplest member of a class of antibacterial compounds called *sulfa drugs,* discovered and used extensively during World War II.

Sulfonic acids are much stronger than the corresponding carboxylic acid. The pK_a of benzenesulfonic acid, Ph—SO_3H, is −0.6, while that of benzoic acid, Ph—CO_2H, is 4.2. This is to be expected since the parent compound sulfuric acid (HO—SO_3H, $pK_a \approx -3$) is so much stronger than the parent of the carboxylic acids (HO—CO_2H, $pK_a = 6.4$).

Ph— is a common abbreviation for the *phenyl group,* C_6H_5—

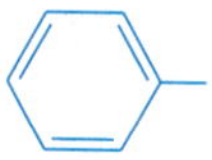

Phosphates

Phosphoric acid reacts with alcohols and other molecules containing the —OH group to form **phosphate esters.**

$$H-O-\overset{\overset{O}{\|}}{\underset{\underset{H}{|}}{\underset{O}{\underset{|}{P}}}}-O-H \qquad H-O-\overset{\overset{O}{\|}}{\underset{\underset{H}{|}}{\underset{O}{\underset{|}{P}}}}-O-R$$

phosphoric acid **a phosphate monoester**

These are central to the chemical reactions taking place in living systems—as components of DNA and RNA, they participate in the expression and transmission of genetic information from one generation to the next. Phosphate esters are involved in the storage, release, and utilization of energy in most organisms. Some of these reactions are discussed in Chapter 21.

Examples of organic groups and compounds containing N, S, and P are given in Table 20.2.

TABLE 20.2 Some Classes of Organic Compounds Containing Nitrogen, Sulfur, and Phosphorus

Class Name	Functional Group	Examples		
Amines	$R-\ddot{N}H_2$ (1°)	CH_3NH_2 **methylamine (1°)**	$(CH_3)_2NH$ **dimethylamine (2°)**	
	$R-\ddot{N}H-R$ (2°)			
	$R-\ddot{N}(R)-R$ (3°)			$(C_2H_5)_3N$ **triethylamine (3°)**
Amides	$R-C(=O)-NH_2$	$CH_3-CH_2-C(=O)-NH_2$ **propanamide**	$CH_3-C(=O)-NH-CH_3$ ***N*-methylacetamide**	
	$R-C(=O)-NHR$			$H-C(=O)-N(CH_3)-CH_3$ ***N,N*-dimethylformamide**
	$R-C(=O)-NR_2$			
Sulfonic acids	$R-S(=O)_2-O-H$	$C_6H_5-S(=O)_2-O-H$ **benzenesulfonic acid**	$H_3C-S(=O)_2-O-H$ **methanesulfonic acid**	
Phosphate esters	$O=P(O-)_2-O-$	$O=P(O-H)(O-CH_3)-O-CH_3$ **dimethyl phosphate**		$O=P(O-H)(O-H)-O-P(=O)(O-CH_3)-OH$ **methyl diphosphate**

20.5 POLYMERS

A very large molecule, or **macromolecule,** composed of repeating subunits, is called a **polymer.** The compounds from which polymers are prepared are called **monomers.** Polymers composed of two or more different monomers are called *copolymers* to distinguish them from *homopolymers,* prepared from a single monomer.

Polyfunctional Compounds

Many organic compounds are **polyfunctional,** that is, they have more than one functional group. A common example is 1,2-dihydroxyethane (ethylene glycol), which has two hydroxyl groups. The functional groups in a polyfunctional molecule need not be the same. An *amino acid,* for instance, contains an $—NH_2$ group and a $—CO_2H$ group. An important feature of polyfunctional molecules is that they are able to react with more than one other molecule. A bifunctional acid such as *malonic acid,* $HO_2C—CH_2—CO_2H$, can undergo esterification (or other reactions characteristic of carboxylic acids) at both ends.

Amino acids are essential nutrients. Their structures and properties are discussed in the following chapter.

$$CH_3O—H + HO—C(=O)—CH_2—C(=O)—OH + H—OCH_3 \longrightarrow H_3C—O—C(=O)—CH_2—C(=O)—O—CH_3 + 2\,H_2O \quad (20\text{-}23)$$

methanol **malonic acid** **methanol** (each esterification releases H_2O)

dimethyl malonate

Condensation Polymers

When a bifunctional acid undergoes esterification with an alcohol that is also bifunctional, the result is a polymeric compound with high but indefinite molar mass, known as a *polyester.* The structural formulas of polymers are drawn so as to show only a single subunit, repeated *n* times.

$$\cdots + HO—C(=O)—C_6H_4—C(=O)—OH + HOCH_2CH_2OH + HO—C(=O)—C_6H_4—C(=O)—OH + HOCH_2CH_2OH + \cdots \longrightarrow$$

$$\cdots—C(=O)—C_6H_4—C(=O)—OCH_2CH_2O—C(=O)—C_6H_4—C(=O)—OCH_2CH_2O—\cdots + n\,H_2O \quad (20\text{-}24)$$

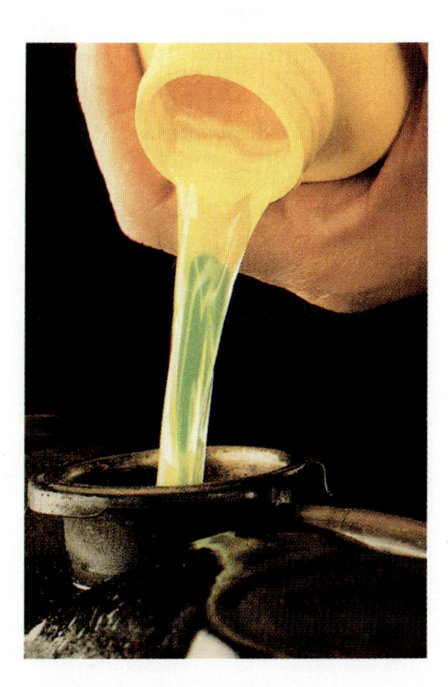

Ethylene glycol is used as antifreeze in automobile radiators.

The polyester produced in Reaction (20-24) and abbreviated in the sketch below is called Dacron when it is manufactured as a fiber or fabric, and Mylar when it is manufactured as a thin film.

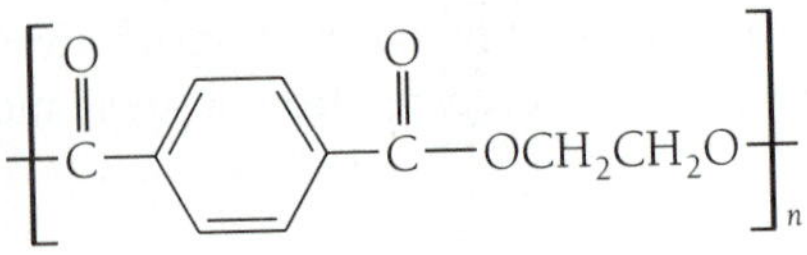

$$\left[-\overset{O}{\overset{\|}{C}}-C_6H_4-\overset{O}{\overset{\|}{C}}-OCH_2CH_2O- \right]_n$$

Many polymers contain thousands of monomeric units. For the most part these units are bonded together in a linear fashion, like the links of a chain. Indeed, the number of monomer units in a linear macromolecule is called the **chain length.** Polymeric materials, which, like Dacron, are produced in a condensation reaction, are called **condensation polymers.** Condensation polymerization proceeds by the stepwise linking of monomers, so it is sometimes called "step-reaction polymerization." Some common condensation polymers are listed in Table 20.3.

Addition Polymers

In **addition polymerization** the characteristic chain-building step is an addition across the double bond of an alkene. The simplest **addition polymer** is *polyethylene,* whose monomer, as the name indicates, is ethylene (ethene).

$$n\ CH_2{=}CH_2 \longrightarrow {-\!\!\!+}CH_2CH_2{+\!\!\!-}_n \qquad (20\text{-}25)$$

TABLE 20.3 Condensation Polymers

Monomers	Polymer	Uses	
$HO_2C{-}(CH_2)_4{-}CO_2H$ **adipic acid** and $H_2N{-}(CH_2)_6{-}NH_2$ **1,6-diaminohexane**	$\left[\overset{O}{\overset{\|}{C}}(CH_2)_4\overset{O}{\overset{\|}{C}}NH(CH_2)_6NH\right]_n$ **Nylon 66**	Fibers (textiles), small parts	 Nylon is used for strong, long-wearing fabrics.
 $HO_2C{-}C_6H_4{-}CO_2H$ **terephthalic acid** and $HOCH_2CH_2OH$ **ethylene glycol**	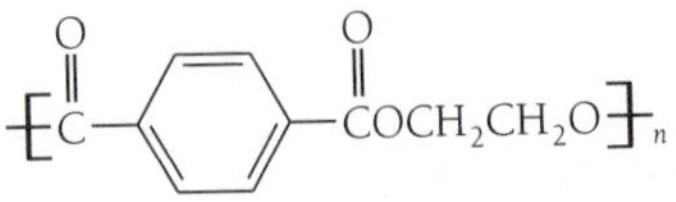 $\left[\overset{O}{\overset{\|}{C}}{-}C_6H_4{-}\overset{O}{\overset{\|}{C}}OCH_2CH_2O\right]_n$ **Dacron, Mylar**	Thin films, fibers	 High-quality sails are woven from dacron fiber.

TABLE 20.3 cont'd

Monomers	Polymer	Uses
$HO-C_6H_4-C(CH_3)_2-C_6H_4-OH$ **bisphenol A** and $Cl-C(=O)-Cl$ **phosgene**	$-[C_6H_4-C(CH_3)_2-C_6H_4-O-C(=O)-O]_n-$ **Lexan (a "polycarbonate")**	Shatter-resistant windows, glasses
	Polycarbonate polymers have replaced glass in many applications.	
$HO-(CH_2)_4-OH$ **1,4-butanediol** and $O=C=N(CH_2)_6N=C=O$ **1,6-hexanediisocyanate**	$-[(CH_2)_4OC(=O)NH(CH_2)_6NHC(=O)O]_n-$ **Polyurethane**	Elastic fibers, foams, varnishes
	Polyurethane varnish is hard, quick-drying, and scratch-resistant.	
$HO_2C-C_6H_4-CO_2H$ **terephthalic acid** and $H_2N-C_6H_4-NH_2$ **1,4-diaminobenzene**	$-[C(=O)-C_6H_4-C(=O)NH-C_6H_4-NH]_n-$ **Kevlar**	Tire cords, high strength composites
	Bulletproof vests have a thick layer of kevlar.	

A common mechanism of addition polymerization begins with a free radical *initiator*. A **free radical** is a fragment of a molecule, usually the result of a broken bond, that contains an unpaired electron. Organic peroxides, R—O—O—R, are common precursors of free radicals. The RO—OR bond is weak, and mild heating causes many peroxides to dissociate into two free radicals.

$$\mathrm{R{-}O{:}O{-}R} \xrightarrow[\Delta]{} \mathrm{R{-}O\cdot} + \mathrm{\cdot O{-}R} \qquad (20\text{-}26)$$

When an alkene reacts with a free radical [Equation (20-27)], the π bond breaks. One of the π electrons from the broken bond joins with the unpaired electron of the free radical to form a new covalent (σ) bond, and the other remains unpaired.

$$\mathrm{R{-}O\cdot} + \begin{array}{c}\mathrm{H\ \ H}\\ |\ \ \ |\\ \mathrm{C{=}C}\\ |\ \ \ |\\ \mathrm{H\ \ H}\end{array} \longrightarrow \mathrm{R{-}O{-}}\begin{array}{c}\mathrm{H\ \ H}\\ |\ \ \ |\\ \mathrm{C{-}C\cdot}\\ |\ \ \ |\\ \mathrm{H\ \ H}\end{array} \qquad (20\text{-}27)$$

The product, the radical-alkene adduct, is itself a free radical, and can react with an additional alkene monomer, producing yet another free radical.

$$\mathrm{R{-}O{-}}\begin{array}{c}\mathrm{H\ \ H}\\ |\ \ \ |\\ \mathrm{C{-}C\cdot}\\ |\ \ \ |\\ \mathrm{H\ \ H}\end{array} + \begin{array}{c}\mathrm{H\ \ H}\\ |\ \ \ |\\ \mathrm{C{=}C}\\ |\ \ \ |\\ \mathrm{H\ \ H}\end{array} \longrightarrow \mathrm{R{-}O{-}}\begin{array}{c}\mathrm{H\ \ H\ \ H\ \ H}\\ |\ \ \ |\ \ \ |\ \ \ |\\ \mathrm{C{-}C{-}C{-}C\cdot}\\ |\ \ \ |\ \ \ |\ \ \ |\\ \mathrm{H\ \ H\ \ H\ \ H}\end{array} \qquad (20\text{-}28)$$

In this way a polymer chain of indefinite length, incorporating thousands of monomers, can result from a single free radical initiator.

A variety of monomers, all of which may be thought of as substituted ethene, are used in the synthesis of addition polymers. Although occasionally two different alkenes are used to prepare a *copolymer*, most commercially important addition polymers are *homopolymers*. Table 20.4 gives some examples.

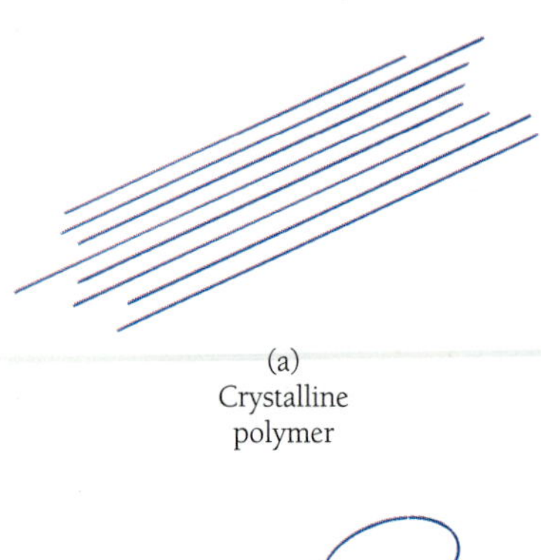

(a)
Crystalline
polymer

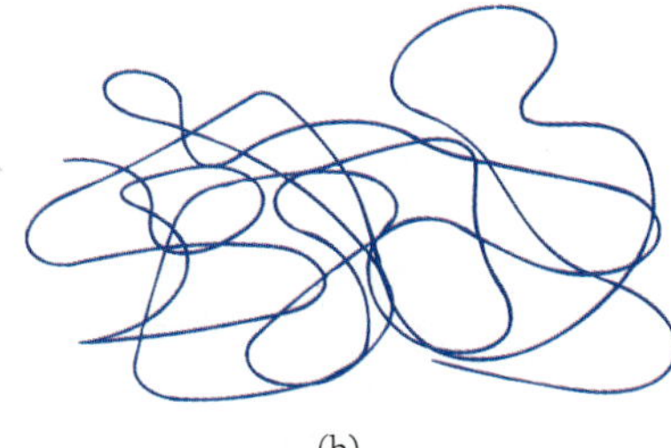

(b)
Amorphous
polymer

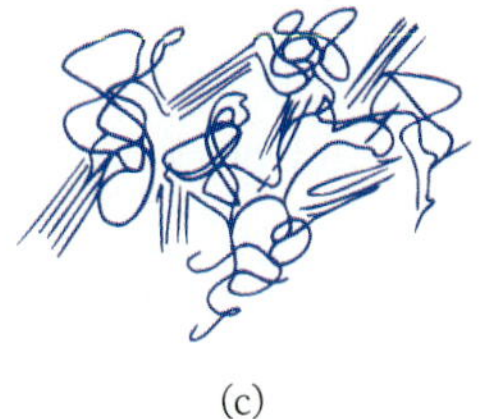

(c)
Partly
crystalline
polymer

Figure 20.7 Structural organization of linear polymers. (a) In crystalline polymers, the molecules are parallel, held together by van der Waals forces. (b) In amorphous polymers, the molecules are randomly arranged and tangled about one another. (c) Partly-crystalline polymers have small regions of crystallinity, separated by amorphous regions.

Structure of Polymers

Most polymers are solid materials whose properties are affected by the geometrical relationship between the chain-like strands. Polymers whose molecules lie parallel to one another (Figure 20.7a) are *crystalline*, and have the sharp melting point characteristic of crystals. If the strands form a tangled mass, as shown in Figure 20.7b, the polymer is called *random* or *amorphous* (lacking structure). Amorphous polymers undergo gradual softening rather than true melting as the temperature is increased, and they tend to be elastic rather than brittle. Both natural and synthetic rubber are amorphous polymers. In a third type of structure (Figure 20.7c), regions of oriented chains are separated by amorphous regions. Such polymers, characterized by a variable *degree of crystallinity*, have properties intermediate between amorphous and crystalline polymers.

Most polymers are relatively unreactive, since the reactive functional groups of the monomers were utilized in the polymerization reaction. They are stable materials useful for textiles, coatings (paints), adhesives, construction materials, and so on. The list is virtually endless. By proper choice of monomers and careful control of such variables as chain length and degree of crystallinity, polymer chemists can produce materials of widely different physical properties. Polymers are classified according to their physical behavior, in particular their response to mechanical or thermal stress. **Fibers,** which have a high degree of crystallinity, have high tensile strength in one

TABLE 20.4 Addition Polymers

Monomer	Polymer	Uses
$CH_2{=}CH_2$ **ethylene**	$+\!CH_2{-}CH_2\!+_n$ **polyethylene**	Films, containers, pipes, electrical insulation
	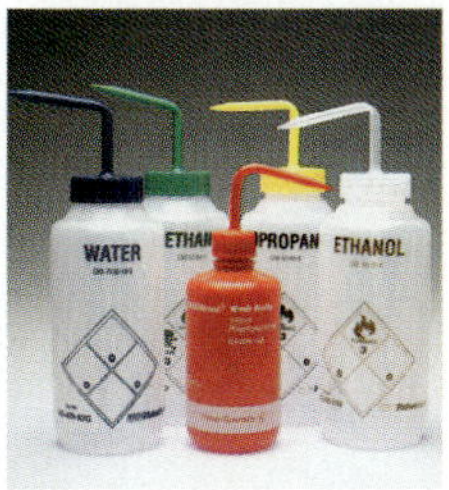	Polyethylene provides a nonbreakable, nonreactive container for liquids.
$CH_2{=}CHCH_3$ **propylene**	$+\!CH_2{-}\underset{}{\overset{CH_3}{\overset{\vert}{C}}}H\!+_n$ **polypropylene**	Rope, artificial turf
		Polypropylene rope is often used in water sports.
$CH_2{=}CHCl$ **vinyl chloride**	$+\!CH_2{-}\overset{Cl}{\overset{\vert}{C}}H\!+_n$ **polyvinyl chloride**	Plumbing pipe, floor tiles
	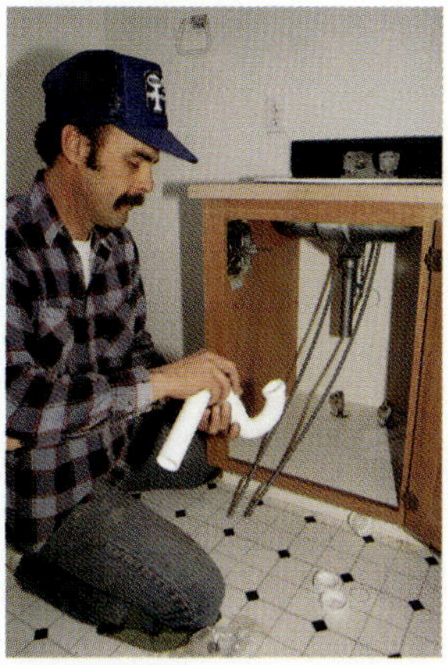	Polyvinyl chloride pipe is common in residential and commercial applications.

(continued on next page)

TABLE 20.4 cont'd

Monomer	Polymer	Uses
$CF_2{=}CF_2$ **tetrafluoroethene**	$-\!\!\!+CF_2-CF_2+\!\!\!-_n$ **Teflon**	Nonstick coatings
		This "nonstick" pan is coated with teflon.
$CH_2{=}CH-C_6H_5$ (benzene ring on CH) **styrene**	$+CH_2-CH(C_6H_5)+_n$ **polystyrene**	Packaging, coffee cups
		The low thermal conductivity of polystyrene foam is put to use in coffee cups.
$CH_2{=}CH-C{\equiv}N$ **acrylonitrile**	$+CH_2-CH(C{\equiv}N)+_n$ **Orlon, Acrilan**	Fibers
		Orlon has many desirable qualities for clothing.

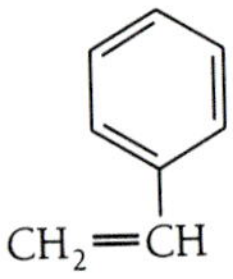

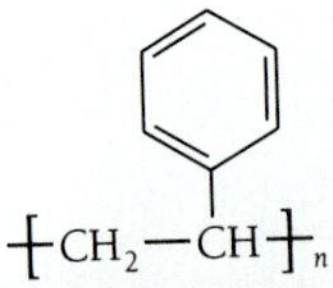

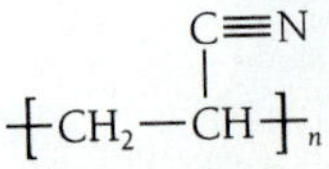

TABLE 20.4 cont'd

Monomer	Polymer	Uses
$CH_2{=}C(CH_3){-}CO_2CH_3$ **methyl methacrylate**	$\{CH_2{-}C(CH_3)(CO_2CH_3)\}_n$ **Lucite, Plexiglass**	Windows, lenses
		Plexiglass is strong, transparent, and easily formed into a variety of shapes.
$CH_2{=}CCl_2$ **1,1-dichloroethene**	$\{CH_2{-}CCl_2\}_n$ **Saran**	Packaging
		Practically everybody uses saran.
$CH_2{=}CH{-}CH{=}CH_2$ **1,3-butadiene**	$\{CH_2{-}CH{=}CH{-}CH_2\}_n$ **polybutadiene** 	Synthetic rubber*†

* Natural rubber is polyisoprene, an addition polymer of the monomer *isoprene,* $CH_2{=}C(CH_3){-}CH{=}CH_2$.

† Synthetic rubbers used for tires are *copolymers* of butadiene and styrene.

(a)

(b)

Figure 20.8 (a) Polymeric fibers are manufactured in large quantities. (b) Tennis balls made of elastomeric polymers are familiar substances.

dimension. **Elastomers** are readily deformed by mechanical stress, but return to their original shape when the stress is relieved (Figure 20.8). **Plastics** have a variety of properties intermediate between those of fibers and elastomers. In general, plastics can be softened or melted by heating and molded into a variety of useful shapes. If softening and reshaping can be done repeatedly, the polymer is **thermoplastic.** If the softening and shaping can be done only once, the polymer is **thermosetting.** A second heating of a thermosetting polymer results in chemical decomposition rather than softening.

A significant part of the chemical industry in the United States is devoted to polymers. More than half of the country's chemists work with polymers, and the mass of polymeric substances produced annually (60 billion pounds in 1989) is greater than the total mass of people in the country.

SUMMARY AND KEYWORDS

In a **cyclic compound,** some of the atoms are bonded so as to form a ring. **Heterocyclic compounds** contain at least one atom other than carbon in the ring. **Cycloalkanes** are isomeric with alkenes. The various shapes a molecule can assume through bending, twisting, or rotating about *single* bonds are called **conformations.** Such rotation does not occur about *double* bonds. Except for cyclopropane, cycloalkanes have nonplanar conformations. In cyclohexane, the **chair** conformation is more stable than the **boat.**

Molecules (other than alkanes) contain one or more **functional groups:** a subset of atoms that influences and serves as a site for chemical reactivity. Oxygen-containing functional groups are **hydroxyl** ($—OH$), which occurs in **1°, 2°, and 3° alcohols** and also in **phenols**; the $—O—$ group, which appears in **ethers**; the **carbonyl group** ($>C{=}O$), which characterizes **aldehydes** ($R—CO—H$) and **ketones** ($R—CO—R'$); and the group $—C(=O)—O—$, characteristic of **carboxylic acids** ($R—CO_2—H$) and **esters** ($R—CO_2—R'$).

The *amino* group ($—NH_2$) is characteristic of **amines, derivatives** of ammonia in which one or more of the H-atoms has been replaced by an organic group. **Amides** contain the ($—C(=O)—N<$) group. **Sulfonic acids** contain the $—SO_2—OH$ group. **Phosphates** (esters of phosphoric acid) are common molecules in living systems.

Oxidation (increase in oxidation number or loss of electrons) is associated with change in the relative amount of H or O in a compound. Loss of an H-atom from a molecule is a one-electron oxidation, gain of an O-atom is a two-electron oxidation. Alcohols can be oxidized to aldehydes or ketones, and aldehydes can be oxidized to carboxylic acids.

Polymers are chain-like **macromolecules** composed of many **monomer** units. **Chain lengths** of a thousand or more monomer units are common. **Addition** polymers are formed in **free radical** additions to the double bond in ethene or substituted ethenes. **Condensation** polymers are formed in condensation reactions of **polyfunctional** monomers. In a **condensation** reaction, two molecules are joined into a larger adduct. The reaction involves the splitting out and loss of a small molecule, usually water. The chain-like strands are arranged parallel to one another in *crystalline* polymers, while in *amorphous* polymers, the strands are randomly arranged. Some polymers have both crystalline and amorphous regions. **Fibers** have high tensile strength, **elastomers** are readily but reversibly deformed by mechanical stress, and **plastics** have intermediate properties. **Thermosetting** plastics can be shaped, once, by heating; a second heating causes decomposition. **Thermoplastic** materials can be repeatedly softened and molded by heating.

PROBLEMS

General

1. What is the distinction between organic and inorganic compounds?
2. Comment on the statement, "In organic compounds, the hybridization of the carbon atoms is sp^3."
3. What is a functional group? Give examples.
4. What types of isomers are common in organic compounds?
5. What evidence supports the statement, "Resonance confers stability on a molecule?"
6. Define the terms "monomer" and "polymer."
7. Identify the two major types of polymer. What feature differentiates them?
8. What is meant by the term "conformation"? Name two conformations of cyclohexane, and sketch them.
9. Is it possible for a nonplanar conformation of cyclopropane to exist? Explain.
10. In an organic molecule undergoing reduction, does the number of hydrogen atoms (relative to the number of carbon or oxygen atoms) increase, decrease, or remain the same?
11. Identify each of the following chemical changes as an oxidation, reduction, or neither.
    a. methyl ethyl ketone $\longrightarrow$ 2-butanol
    b. $CH_2{=}CHCH_3 \longrightarrow CH{\equiv}CCH_3$
    c. methylbenzene $\longrightarrow$ benzoic acid
    d. bromomethane $\longrightarrow$ fluoromethane

*12. Show that in general, for a compound $C_nH_xO_y$ undergoing a chemical change to a compound $C_nH_{x'}O_{y'}$, oxidation has taken place if $x' < x$ or if $y' > y$.

Hydrocarbons and Halogenated Hydrocarbons

13. What hybridization of carbon characterizes the bonding in (a) alkanes, (b) alkenes, (c) alkynes, (d) cycloalkanes, and (e) aromatic hydrocarbons?
14. Give a general empirical formula (of the type C_nH_m) for alkanes, alkenes, alkynes, and cycloalkanes.
15. Draw the structural formula of each of the following.
    a. 2,2-dimethylpropane b. 1,2,4-trichlorobenzene
    c. 2-chloro-2-pentene d. cyclopentene
16. Draw the structural formula of each of the following.
    a. 1,3-dibromopentane b. 1,3-dibromobenzene
    c. 1,3-pentadiene d. propyne
17. Name each of the following.
    a. $H_2C(CH_2{-}CH_3){-}CH_2CH_2CH_3$ (structure: H_2C bearing a $CH_2{-}CH_3$ group, bonded to $CH_2CH_2CH_3$)
    b. (benzene ring bearing Cl)
    c. $H_3C{-}CH(CH_3){-}CH_2CH_2CH(Br)_2$ (structure: $H_3C{-}CH{-}CH_2CH_2CH$ with CH_3 on the second carbon and two Br on the terminal CH)
    d. (benzene ring bearing Cl and, at the adjacent position, CO_2H)
18. Name each of the following.
    a. $CH_3{-}CH{=}CH{-}CH_2{-}CH{=}CH{-}CH_3$
    b. $F_2FC{-}CF_2F$ (structure: F–C–C–F with two F on each carbon)
    c. (five-membered ring: $HC{=}CH$ bonded to $HC{=}CH$... ring $HC{-}CH$, $HC{=}HC$, $CH{=}CH$, joined through CH_2)
    d.

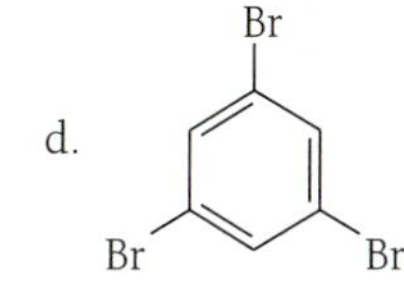

*19. Which of the following names carries enough information so that a unique structural formula can be drawn?
    a. hexane b. ethane
    c. pentenol d. dichlorobutane
    e. cyclopentene f. dichloromethane

*20. Which of the following names carries enough information so that a unique structural formula can be drawn?
    a. dibromobenzene b. *cis*-dibromopropene
    c. 2-fluoropropane d. propyne
    e. 1,2-dibromocyclohexane f. 1-iodo-3-bromobenzene

21. Explain why "3-methylpropane" is *not* a correct IUPAC name.

*22. Explain why "1-chlorocyclobutane" is *not* a correct IUPAC name.

*23. Explain why "*cis*-1,2-dichloroethane" is *not* a correct IUPAC name.

*24. Draw the complete structural formula corresponding to each of the following modified molecular formulas.
a. $CH_3CH_2CH_2CH{=}CH_2$ b. $CH_3CHClCH_2CH_3$
c. $CH_2{=}CHC(CH_3)_3$ d. $CH_3CH_2CH(C_2H_5)CH_3$

25. Draw the structural formula corresponding to each of the following.
a. *cis*-$CHCl{=}CHCl$ b. $CH_3CH_2C(CH_3)_3$
c. $(CH_3)_3CCH_2CH_2CH{=}CHCH_3$ d. $C(CH_3)_4$

26. Name each of the compounds in Problem 24.

27. Name each of the compounds in Problem 25.

28. Draw structural formulas and give names for five structural isomers of C_6H_{14}.

*29. Draw structural formulas and give names for as many structural and geometrical isomers of C_5H_{10} as you can. (*Hint:* Do not overlook cyclic structures. There are 12 possible isomers.)

*30. Draw structural formulas and give names for as many structural isomers of C_5H_8 as you can. (There are at least 19.)

31. Which of the following compounds can exist as *cis* and *trans* isomers?
a. 1,2-dichloroethane b. 1,2-dichloroethyne
c. 1,2-dichloroethene d. 1,1-dichloroethene

32. Which of the following compounds can exist as *cis* and *trans* isomers?
a. 1,2-dichlorocyclohexane b. 1,3-dichlorocyclohexane
c. 1,2-dichlorobenzene d. 1,3-dichlorobenzene

33. Draw structures and name isomers of $C_3H_4Cl_2$, other than the two already given in Section 20.1.

Functional Groups

34. Give the name of the class(es) of compounds containing each of the following functional groups.
a. —OH b. $—NH_2$
c. R—O—R′

35. Give the name of the class(es) of compounds containing each of the following functional groups.
a. $—CONH_2$ b. —NHR
c. >C=O

36. Name the classes of compounds containing the following groups.
a. 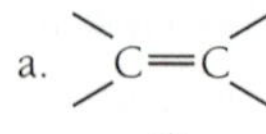 b. $—CO_2H$
c. R—C(=O)—R′

*37. Name the classes of compounds containing the following groups.
a. $R—CH_2OH$ b. $RCH(OH)CH_3$
c. RCHO d. $—SO_3H$

*38. Draw modified molecular formulas and structural formulas for the following compounds.
a. propanal
b. propanoic acid
c. isopropyl methyl ether
d. methyl propanoate (the methyl ester of propanoic acid)

*39. Draw modified molecular formulas and structural formulas for the following.
a. diethylamine b. acetic acid
c. acetamide d. acetone

*40. Draw modified molecular formulas and structural formulas for
a. 3-chlorobutanoic acid b. 1,2,3-trihydroxypropane
c. 6-aminohexanoic acid d. 1-chloro-1-phenylethane

41. Draw modified molecular formulas and structural formulas of three isomeric alcohols C_4H_9OH, and label them as 1°, 2°, or 3°.

*42. Draw structural formulas for propanal and butanal, and specify the hybridization of each of the carbon atoms in these molecules. Generalize your results to describe the hybridization of the carbon atoms in any aldehyde $CH_3(CH_2)_nCHO$.

*43. Draw structural formulas for acetone and 2-butanone, and specify the hybridization of each of the carbon atoms in these molecules. Generalize your results to describe the hybridization of the carbon atoms in any ketone $CH_3(CH_2)_nCOCH_3$.

44. Draw structural formulas for acetic acid and propanoic acid, and specify the hybridization of each of the carbon atoms in these molecules. Generalize your results to describe the hybridization of the carbon atoms in any carboxylic acid, $CH_3(CH_2)_nCO_2H$.

45. How many N—H bonds are there in a secondary amine?

46. Which, if any, of the functional groups discussed in this chapter are stabilized by resonance?

*47. What is the major organic product expected from each of the following reactions? (Do not try to balance the equations.)
a. $CH_2{=}CH_2 + H_2O \longrightarrow ?$
b. $CH_3(CH_2)_3CH_2OH + KMnO_4(aq) \longrightarrow ?$
c. $CH_3CH_2OH + Na \longrightarrow ?$
d. $CH_3CH_2OH + HCO_2H \longrightarrow ?$

*48. What is the major organic product to be expected from each of the following reactions?
a. $CH_3CH_2CHO + KMnO_4(aq) \longrightarrow ?$
b. $CH_3CH_2CHO + NaBH_4(aq) \longrightarrow ?$
c. $CH_3COCH_3 + NaBH_4(aq) \longrightarrow ?$
d. $(CH_3)_2NH + HCl(aq) \longrightarrow ?$

*49. What is the major organic product expected from each of the following reactions?

a. 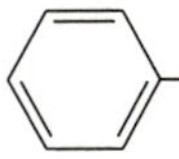 $C_6H_5-CO_2H\ (aq) + NaOH\ (aq) \longrightarrow ?$

b. cyclohexanone + $\xrightarrow[H_2O]{NaBH_4}$?

c. 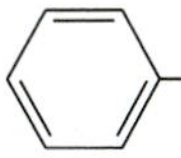 $C_6H_5-OH\ (aq) + NH_3\ (aq) \longrightarrow ?$

*50. When acetaldehyde (ethanal) reacts with $KMnO_4(aq)$ in basic solution, the inorganic product is $MnO_2(s)$. Identify the organic product, and use the half-reaction method to balance the net ionic half-reactions.

*51. Calculate the theoretical yield of the organic product if 15 g each of acetaldehyde and $KMnO_4$ react in aqueous solution. (You will need the balanced equation from Problem 50.)

Polymers

52. Give an example of a condensation reaction. What is the essential feature of monomers used in condensation polymerizations?

*53. The examples of condensation polymers given in the text are all copolymers, that is, they contain two different monomers. Is it possible for a single monomer to polymerize so as to form a condensation *homo*polymer? If so, suggest an example. If not, explain why not.

54. Is it possible to produce a copolymer by addition polymerization? If so, give an example. If not, explain why not.

*55. Nylon is decomposed by acids, while polyethylene is not. Suggest an explanation for this difference in behavior.

*56. How many moles of styrene are required to produce 15 g polystyrene?

57. Name two types of polymer you have used today. Give the type (addition or condensation) and the monomer(s).

58. What is the percentage (by mass) of oxygen in Plexiglas (polymethylmethacrylate)?

59. What is the percentage (by mass) of chlorine in Saran Wrap?

60. Give the structural formula of the monomer used in preparation of the polymer whose structure is as follows.

a. 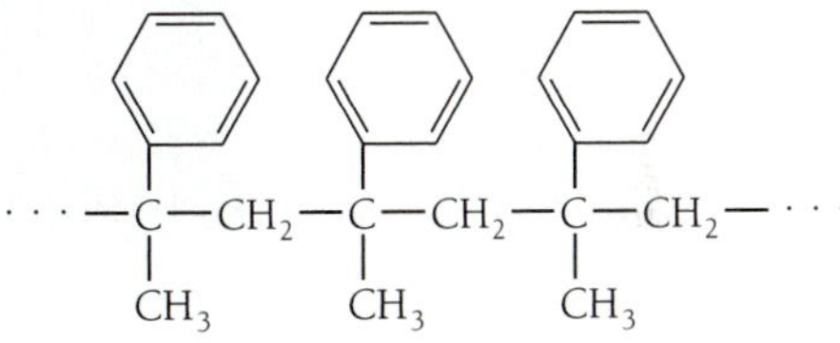 $\cdots-C(C_6H_5)(CH_3)-CH_2-C(C_6H_5)(CH_3)-CH_2-C(C_6H_5)(CH_3)-CH_2-\cdots$

b. $\cdots-C(CH_3)_2-CH_2-C(CH_3)_2-CH_2-C(CH_3)_2-CH_2-\cdots$

CHAPTER 21

MOLECULES OF LIFE

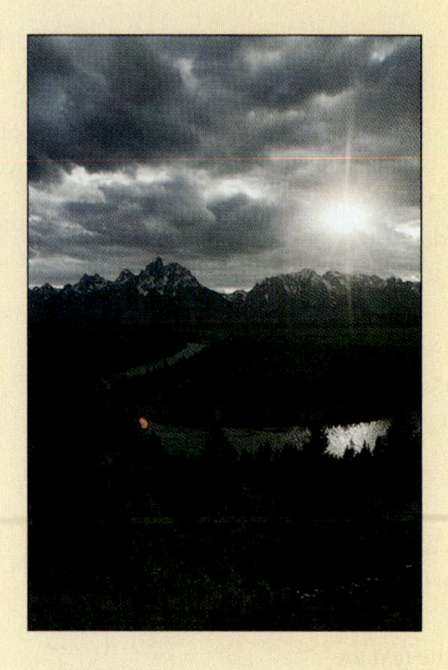

The molecules found in living organisms are of stunning complexity. Yet they have their beginnings in photosynthesis, the solar-powered coupling of two of the simplest compounds in nature, carbon dioxide and water.

The historical classification of "organic" compounds as those produced by living organisms has fallen by the wayside, ever since the laboratory synthesis of organic compounds became routine. Nonetheless, there is a large group of compounds that are present in, and important to the function of, living organisms. The study of these substances, their properties, and their reactions constitutes the field of **biochemistry.** Many biological molecules are quite large, and many are polymers. Many are characterized by a new type of isomerism, called *stereoisomerism* or *chirality*. In this chapter, we examine the structure, properties, and functions of four major classes—proteins, carbohydrates, nucleic acids, and lipids—of these "molecules of life."

21.1 PROTEINS

Proteins are polymeric molecules, ranging in molar mass from perhaps 5000 to 5,000,000. Synthesized as needed in plant and animal cells, proteins are utilized in various ways: as enzymes that catalyze the myriad reactions necessary to maintain the cell, for oxygen transport (hemoglobin), as connective tissue (cartilage and tendon), as the contractive element (myosin) in muscle, as hormones (insulin), and even as snake venom. The monomers of these versatile polymers are the *amino acids.*

Amino Acids

Compounds containing both a carboxylic acid function ($—CO_2H$) and an amino group ($—NH_2$) are called **amino acids.** Those in which the amino group and the carboxyl group are bonded to the same carbon atom are known as **α-amino acids.** Their general structure is

$$H_2N—CH(R)—C(=O)—O—H$$

Amino acids have a basic group ($—NH_2$) as well as an acidic group ($—CO_2H$). These groups interact by proton transfer, so that the structure of amino acids is better represented by

$$H_3N^+—CH(R)—C(=O)—O^-$$

Such structures, which have two ionized groups and yet are uncharged overall, are called **zwitterions** or *inner salts.* Amino acids have this inner salt structure both in aqueous solution and in the pure solid. In spite of this, many chemists prefer to represent amino acids by the uncharged structural formula $H_2N—CHR—CO_2H$.

Over 100 amino acids have been identified in nature. The ones that are monomers of proteins are quite common, and the others are rare. Living organisms synthesize proteins only from the 20 amino acids shown in Figure 21.1. All of these are α-amino acids, and they differ from one another only in the structure of the R-group. All are known by trivial names and/or three-letter abbreviations rather than by IUPAC systematic names. Eight of them (called the "essential" amino acids) must be present in

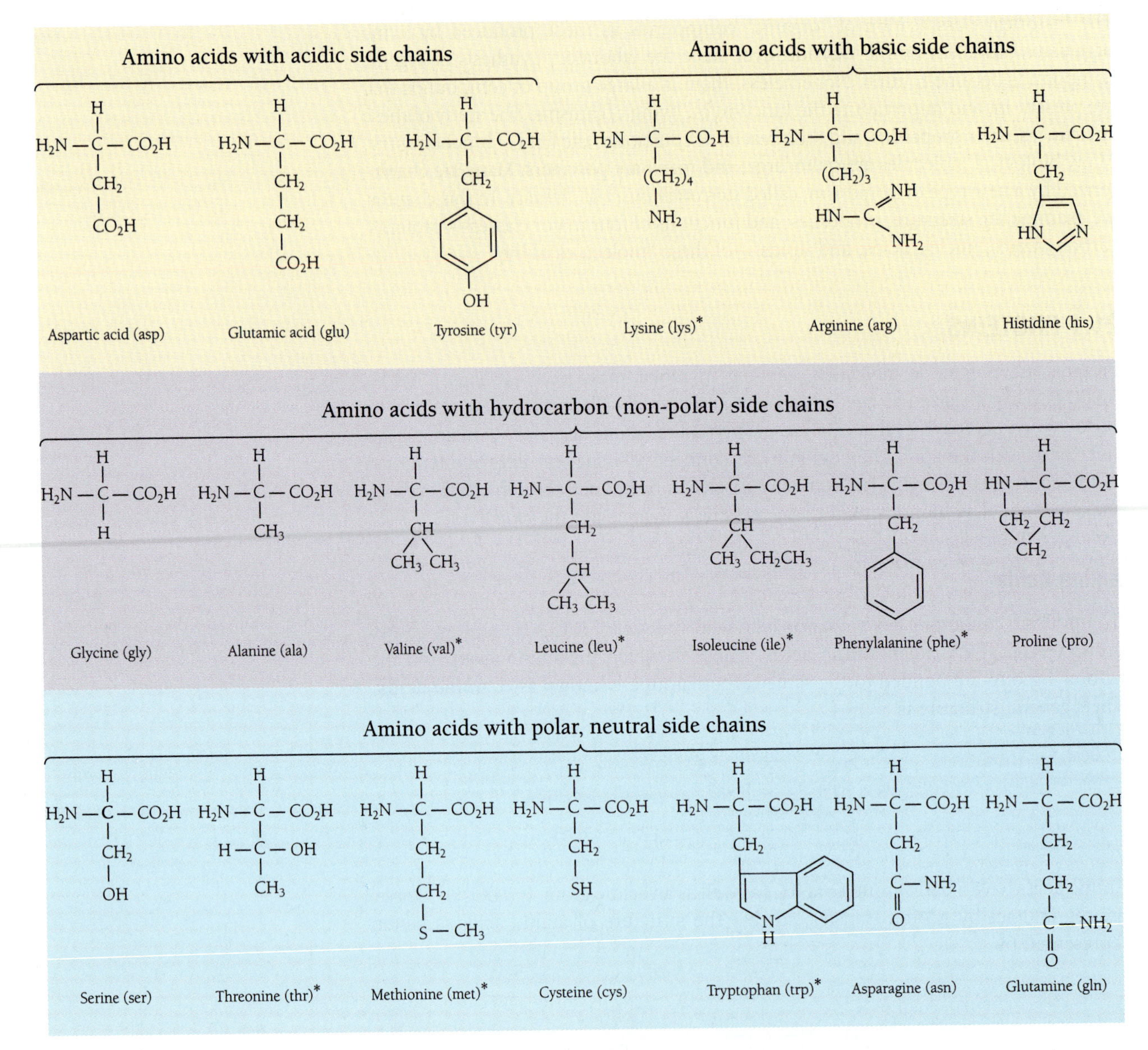

Figure 21.1 The 20 naturally occurring amino acids, arranged according to the character of the R-groups. All are present in the human body, but eight of them (marked with an asterisk) must be obtained from the diet because the body cannot synthesize them.

the diet because they cannot be synthesized in sufficient quantities by the human body.

The nature of the R-group (acid, base, polar, nonpolar, size) has a great influence on the properties of amino acids and the proteins that incorporate them. For example, the carboxylic acid groups in the side-chains of aspartic acid and glutamic acid are ionized at physiological pH (≈ 7.4), so that these amino acids bear a negative charge. Similarly, the (basic) amino groups in the side-chains of lysine and arginine are protonated at physiological pH, so these amino acids bear a positive charge.

Stereoisomerism

Amino acids exhibit a subtle but important form of isomerism. Whenever a carbon atom is bonded to four *different* atoms or groups, as is the case in all α-amino acids except glycine, the molecule exists in two isomeric forms. These differ from one another in the same sense that the right and left hands are different. The two isomers are called the L and D forms (Latin *laevo,* "left" and *dextro,* "right"). One form is superimposable on (identical to) the *mirror image* of the other. Molecules capable of this type of isomerism are said to be **chiral** (*ki′ ral,* rhymes with "spiral"), and the two forms are referred to as **enantiomers** (en an′ tee o mers) (Figure 21.2). This mirror-image relationship is a particular form of **stereoisomerism,** a more general term that implies the same atom-to-atom connectivity but a different spatial arrangement. Geometrical isomerism (Section 8.5) and conformational isomerism (Section 20.1) are other forms of stereoisomerism.

The physical properties (density, melting point, and so on) of a pair of enantiomers are identical, except for one: the interaction with *polarized light.* The photons or rays of ordinary light have all possible orientations of their electromagnetic fields, while in polarized light, all photons have the same orientation. The plane of the fluctuating electric field (the plane containing the vertical arrows in Figure 6.6, page 225) is called the *plane of polarization.* When polarized light passes through a sample of one of a pair of enantiomers, the plane of polarization is rotated clockwise by one enantiomer and counterclockwise by the other. A substance that rotates polarized light is called **optically active.**

Except in one situation, the chemical behavior of two enantiomers is also identical. Each has, for instance, the same value of K_a when functioning as an acid. The exception, which is of crucial importance to the chemistry of life, is that in reactions with *another* chiral molecule, one enantiomer reacts much more slowly than the other. Indeed, in many cases, one of the two does not react at all.

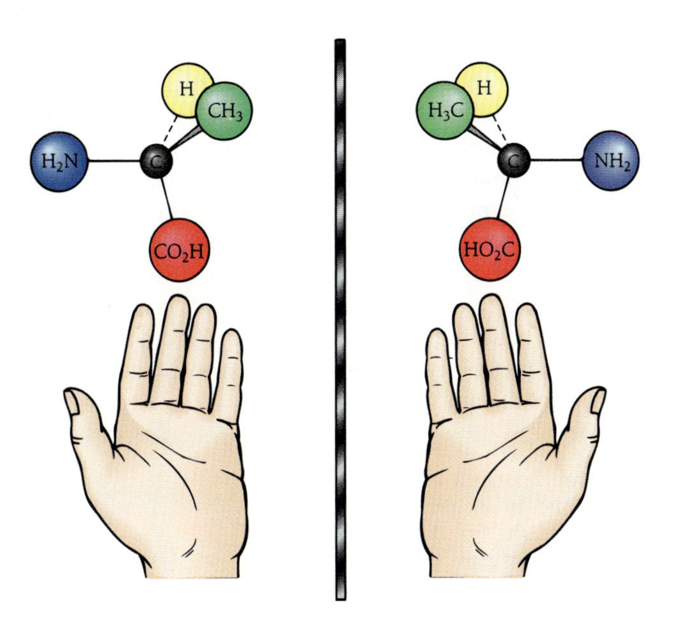

Figure 21.2 The two enantiomers of alanine are mirror images of one another. That is, they are related in the same way as the left and right hands.

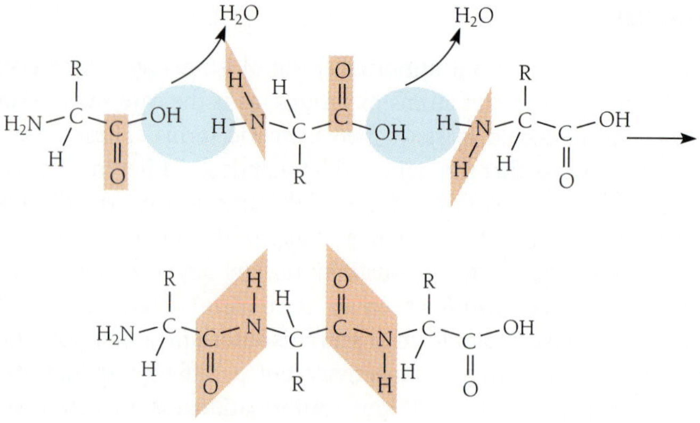

Figure 21.3 Formation of a tripeptide by condensation of three amino acids. Note the loss of two molecules of water.

The term *"racemic mixture"* applies to 50 : 50 mixtures of *any* pair of enantiomers, not just to amino acids.

When amino acids are synthesized in the laboratory from non-chiral reagents, both enantiomers are formed in equal amounts. The result is an equimolar mixture of the two, called a **racemic mixture.** Such a mixture is not optically active, and there is no physical or chemical clue that it is anything other than a single compound. When amino acids are produced in living organisms, however, the reactions are facilitated (catalyzed) by enzymes, which are themselves chiral molecules. The result of the participation of a chiral catalyst is that only one of the two enantiomers is produced. Proteins consist *entirely* of L amino acids. Why nature prefers this form, and not its mirror image, is one of the unanswered questions of evolution.

The Peptide Bond

The formation of an *amide* in a condensation reaction between the amino group (—NH_2) and the carboxylic acid group (—CO_2H) was described in Section 20.3. Since amino acids contain both of these groups, they are capable of forming condensation polymers (Figure 21.3). When incorporated in such a polymer, the amino acid monomers are called *residues.* The —CONH— group that links one residue to the next is called a **peptide bond.** Amino acid polymers having fewer than a dozen or so residues are called **polypeptides,** and larger, naturally occurring, polypeptides are called **proteins.** As shown in Figure 21.4, two resonance structures contribute to the —CONH— group. There is considerable double-bond character in the bond between the C and N atoms in amides and polypeptides. As a result, the molecule cannot rotate about the C—N bond and the adjacent atoms are forced to lie in the same plane. This rigidity limits the possible conformations of a polypeptide, and has a great effect on the overall shape of the molecule.

Figure 21.4 Electron delocalization in the peptide bond. In one of the resonance structures, the C—N bond is a double bond. This prevents rotation about the C—N bond, and forces the six atoms associated with the bond (marked in boldface type) to lie in the same plane.

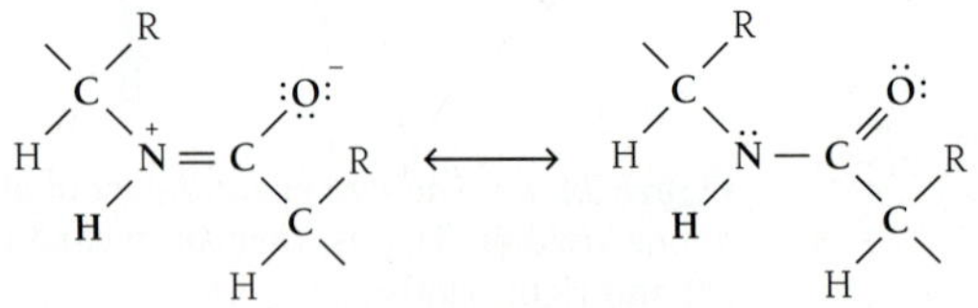

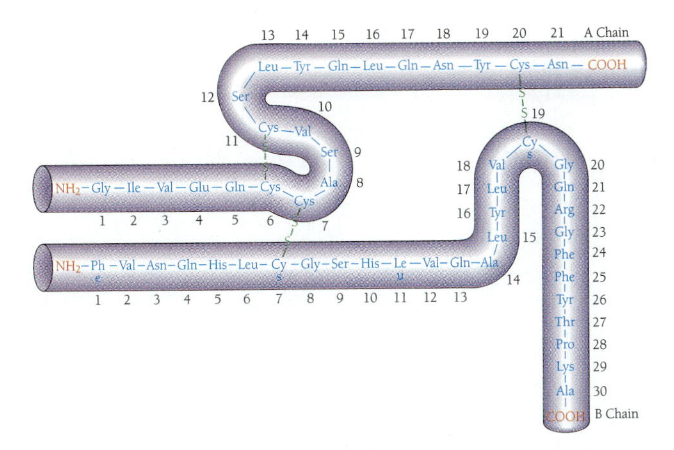

Figure 21.5 Primary structure of bovine insulin. The two chains, consisting respectively of 21 and 30 residues, are cross-linked in two places by S—S covalent bonds. Covalent bonding between protein chains is observed only at cysteine residues.

Protein Structure

The *sequence,* by which is meant the identity and order of the amino acid residues in a polypeptide, is called the **primary (1°) structure** of a protein. The 1° structure of the hormone *insulin,* important in regulating the metabolism of sugar in the body, is illustrated in Figure 21.5. The determination of this structure, a lengthy and difficult task, was accomplished in 1953 by Frederick Sanger. The method was the decomposition of the protein into smaller pieces, consisting of only a few residues, separating these, and determining the sequence of each. Each of the many protein molecules in the sample broke in different places, so the fragments contained overlapping sequences. The original sequence could be deduced from this information, rather like the assembly of a jigsaw puzzle. More recently, parts of this complex procedure have been automated and computerized, so that *protein sequencing* is no longer such a formidable task.

A long chain can coil or fold on itself in a variety of ways, and this aspect of protein conformation is called the **secondary (2°) structure.** A major force holding a protein in its 2° structure is the hydrogen bond, which occurs in most of the peptide bonds of the chain. Several types of secondary structure have been identified. In the **pleated sheet** (Figure 21.6), polypeptide chains lie parallel to one another, and are held in place by hydrogen bonding between the amide (NH) hydrogen in one chain and the

Frederick Sanger (1913–)

Frederick Sanger earned his doctorate at Cambridge University, where he remained to continue research in biochemistry. He spent 10 years on the structure of insulin, and in 1958 received the Nobel Prize for this work. In 1980 he shared a Nobel Prize for the determination of nucleic acid structure, becoming only the fourth person to earn two Nobel awards.

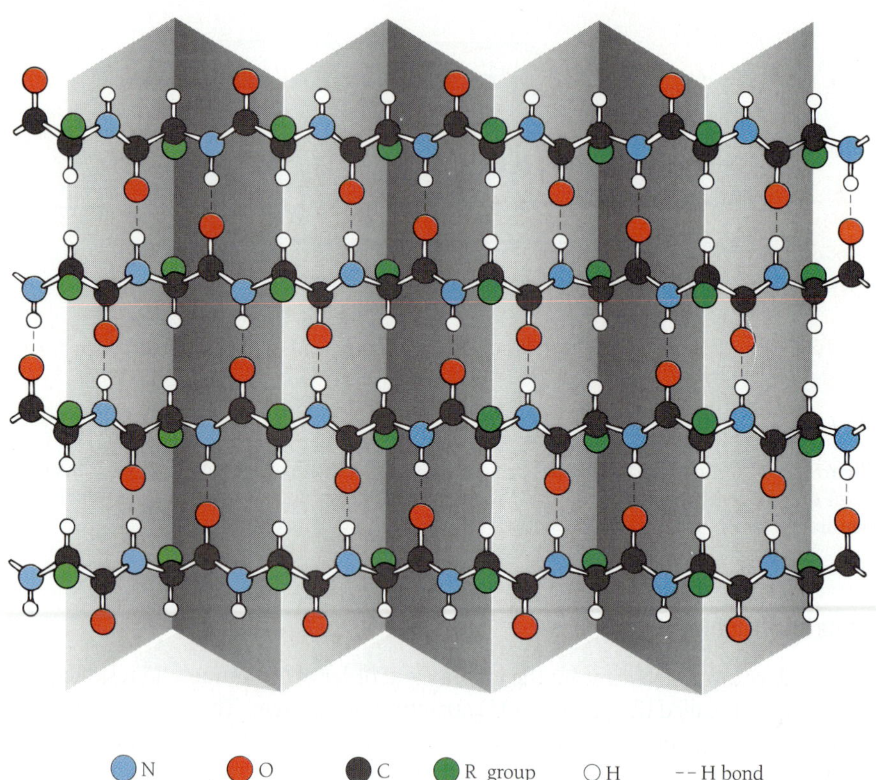

Figure 21.6 The pleated sheet secondary structure of proteins. The name comes from the fact that, viewed from the side (top or bottom of this drawing), the two-dimensional structure is buckled or folded, accordion style, along the backbone of each polypeptide chain. The six atoms of the peptide bonds lie in the sheet, while the R-groups extend above and below the sheet.

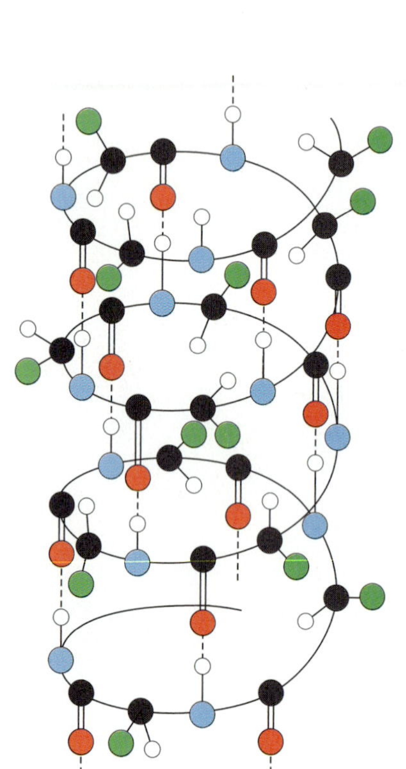

Figure 21.7 The α-helix structure of a polypeptide chain. The bond angles and lengths have been distorted for clarity in the diagram. In the actual structure, all bond angles and lengths are close to their normal values, and the coplanarity of the atoms in the peptide bond is maintained. The many H-bonds make the α-helix a very stable structure.

carbonyl (C═O) oxygen in the next. Most pleated-sheet proteins consist of distinct chains, but in a few cases, a single chain can fold on itself so that *part* of its 2° structure is a pleated sheet. Although H-bonds are individually weak, many of them acting together make the pleated sheet a strong, cohesive, stable structure. The pleated sheet is found in silk, and in other fibrous proteins used in structural parts of organisms.

Another type of 2° structure occurs when a polypeptide chain coils in such a way that H-bonding can occur between peptide bonds on the same chain. This structure, called the **α-helix,** was proposed by Pauling and Corey in 1951. In the α-helix, H-bonding occurs between the N—H part of a peptide bond and the C═O group of another peptide bond four residues down the chain (Figure 21.7). Since all R-groups lie on the outside of the resulting coil, different proteins present a different face to the surroundings. The many H-bonds confer considerable stability on the α-helix. However, certain amino acid residues tend to destabilize the structure. For example:

- too many charged (acidic or basic) residues in close proximity create an electrostatic repulsive force that prevents formation of a helix;
- too many large R-groups simply do not fit in the helical structure;
- alone among the amino acids, proline's N-atom is part of a ring and *cannot* reach the proper conformation; an α-helix therefore makes a sharp bend at each proline residue.

Thus, although 2° structure is quite different from 1° structure, it is in fact dictated by the primary structure.

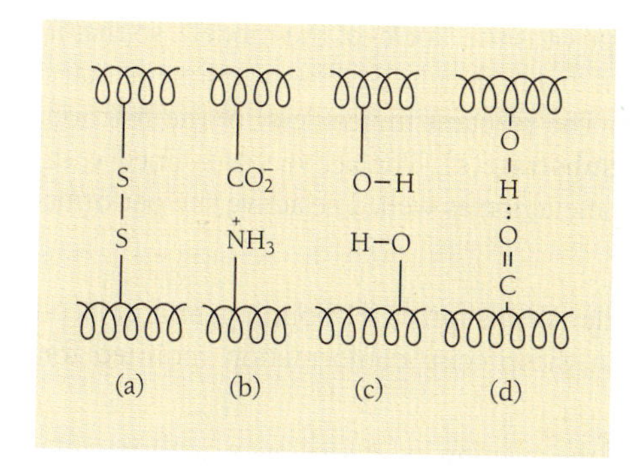

Figure 21.8 Two protein chains (or different segments of the same chain) are held together by a variety of forces between the R-groups of the amino acid residues. These forces include (a) covalent bonding between cysteine residues, (b) electrostatic attraction between acidic and basic residues, (c) dipole forces between polar residues, and (d) hydrogen bonds.

The α-helix itself is flexible rather than rod-like, and the particular shape it assumes is called the **tertiary (3°) structure** of the protein. In many proteins the 3° structure is determined and maintained by *cross-links* formed by —S—S— bonds between cysteine residues (such as in insulin, Figure 21.5), although other forces are also important. Figure 21.8 summarizes the main interactions that help to stabilize a protein's tertiary structure. Figure 21.9 shows the tertiary structure of *myoglobin,* a protein used for oxygen storage in muscle tissue.

Some proteins consist of several distinct chains, each with its 1°, 2°, and 3° structure, bound together in a unit. This level of organization is called the **quaternary structure (4°).** *Hemoglobin,* which carries oxygen from the lungs, consists of four protein molecules (each similar in structure to myoglobin).

Enzymes

In Chapter 14 we introduced the term **enzyme** to describe a biological catalyst, and discussed the function of enzymes in terms of the *lock-and-key* model (Figure 14.21, page 633). Enzymes are proteins whose tertiary (and sometimes quaternary) structure contains folds and convolutions that form a groove or pocket called the **active site.**

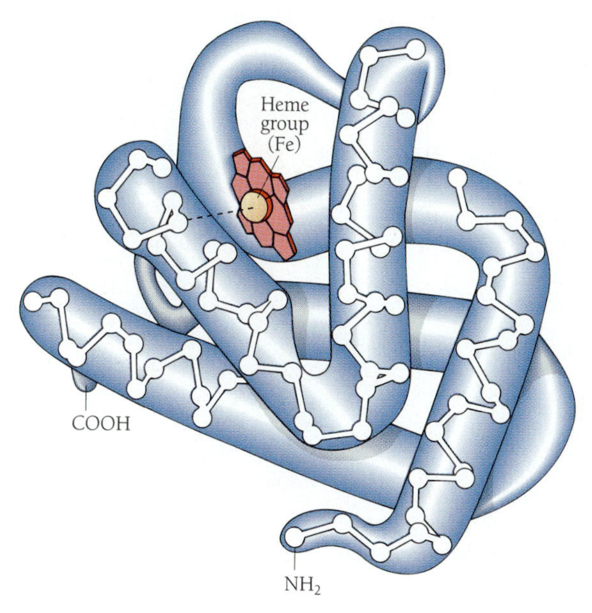

Figure 21.9 The tertiary structure of myoglobin. This protein contains 153 amino acid residues. The 2° structure consists of several short lengths of α-helix, arranged in a tertiary structure that enfolds a *heme* group. The heme group ($C_{34}H_{32}O_4N_4Fe$), a planar structure incorporating an iron atom (yellow), is capable of binding an oxygen molecule and releasing it when needed by the body.

The active site has a particular, precise shape (it is the "lock" of the model), so that it can accept and hold another molecule only if that molecule itself has a complementary shape, that is, if it fits into the active site. The reactant molecule(s) of the reaction catalyzed by the enzyme is/are called the **substrate(s).** The active site is lined with certain amino acid side-chains, and often a metal ion as well. The active site performs two main functions.

1. It *binds* the substrate through van der Waals forces, H-bonding, and/or electrostatic interactions, holding it in the precise conformation required for reaction.
2. It decreases the activation energy of the reaction by, for instance, withdrawing electron density from a substrate bond that is to be broken.

Enzyme-substrate interaction is illustrated in Figure 21.10 for the enzyme *carboxypeptidase A,* whose function is to split off one amino acid from the end of a polypeptide chain. Carboxypeptidase A contains 307 amino acid residues.

Other enzymes catalyze condensation reactions, by holding two substrate molecules together in the precise orientation required for formation of a bond between them. Such enzymes are necessary, for example, in protein synthesis. The usefulness of enzymes in biochemical reactions stems from two features. First, each enzyme is highly specific, catalyzing only one reaction (or a few reactions having similar substrates). Second, enzyme reactions can be faster than their noncatalyzed counterparts by a factor of a million or more: each enzyme molecule can service thousands of substrate molecules per second.

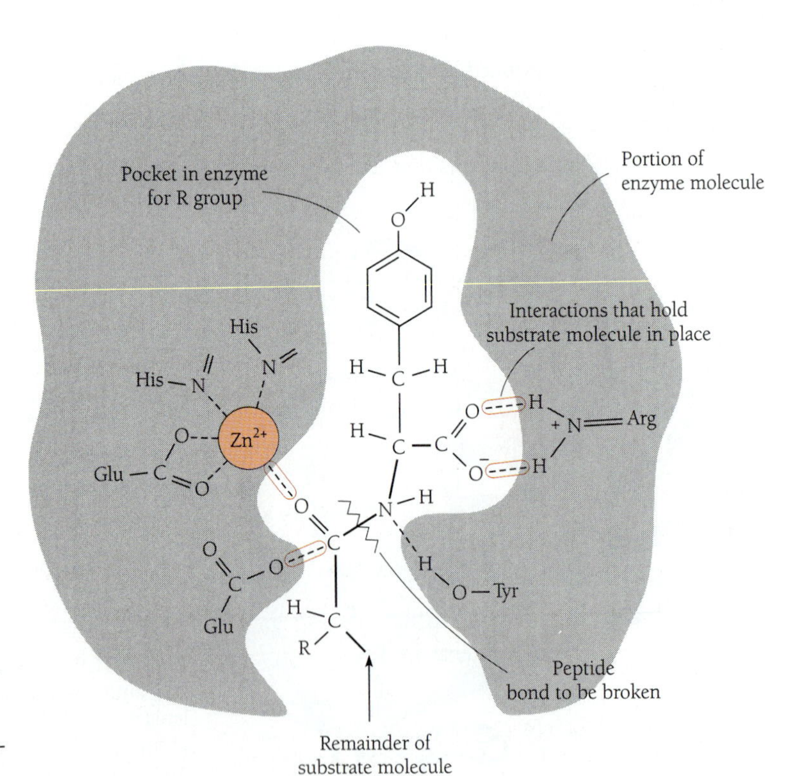

Figure 21.10 The active site of carboxypeptidase A is lined with specific amino acid side-chains that bind the end of a polypeptide substrate molecule. This enzyme contains a Zn^{2+} ion that helps to weaken the peptide bond that must be broken in order to split off (hydrolyze) the terminal amino acid. Other groups lining the active site aid in adding the molecule of water. Following the hydrolysis, the shortened polypeptide and the newly freed amino acid (which no longer fit the active site precisely) diffuse away. The enzyme is then ready to accept another polypeptide.

21.2 CARBOHYDRATES

Another group of biologically important polymers has the approximate empirical formula $[C(H_2O)]_n$. These polymers, together with their monomers, are called **carbohydrates,** because the formula corresponds to "hydrated carbon." Polymeric carbohydrates are produced by photosynthesis in green plants, and are used in structural material (wood) by plants and as primary sources of energy by both plants and animals. The smaller carbohydrates are called **sugars,** because most of them have a sweet taste. Sugars are polyhydroxy aldehydes or polyhydroxy ketones. As such, they can undergo reactions characteristic both of alcohols and of aldehydes or ketones.

Monosaccharides

Many of the naturally occurring sugars have 5 or 6 carbon atoms. These compounds are called **monosaccharides,** and their individual names all end in *-ose. Glucose,* of central importance in the energy metabolism of animals, is an aldohexose, or 6-carbon aldehyde. Fructose (fruit sugar) is a *ketohexose.* Ribose, a constituent of nucleic acids (Section 21.3), is an *aldopentose.*

```
   H—C=O            CH2—OH
     |                |
   H—C—OH            C=O          H—C=O         H—C=O
     |                |             |             |
  HO—C—H         HO—C—H           H—C—OH        H—C—H
     |                |             |             |
   H—C—OH          H—C—OH         H—C—OH        H—C—OH
     |                |             |             |
   H—C—OH          H—C—OH         H—C—OH        H—C—OH
     |                |             |             |
     CH2—OH          CH2—OH         CH2—OH        CH2—OH
   glucose          fructose       ribose       deoxyribose
```

In most monosaccharides, there is a hydroxyl group bonded to every carbon atom other than the carbonyl carbon. Note that several of these carbon atoms are *chiral,* that is, they have four different groups bonded to them. Therefore there are many possible **optical isomers,** which are compounds differing only in the bonding configuration of one or more chiral atoms. In an aldohexose such as glucose, each of the four chiral carbons can exist in one of two configurations, so that there are a total of $2 \cdot 2 \cdot 2 \cdot 2 = 16$ optical isomers. There are eight different mirror-image pairs (enantiomers) among these 16 optical isomers.

Only 3 of the 16 isomers of glucose occur widely in nature.

Since the C—C bonds in the backbone of these monosaccharides are single bonds that can freely rotate, many conformations are possible. In some of these, the carbonyl group comes quite close to the —OH group on the next-to-last carbon. A molecule in this conformation can undergo **cyclization,** by an *intra*molecular addition of the hydroxyl group to the C=O double bond (Figure 21.11). The cyclic form predominates in aqueous solution. An additional chiral center, not present in the linear form, is created at C-1 (the C=O carbon atom) when cyclization occurs. Therefore there are two stereoisomeric forms of cyclic monosaccharides, called α and β. In α-glucose, the newly formed hydroxyl group is *axial,* which means that it is approximately perpendicular to the "plane" of the six-membered ring. In β-glucose,

Since the six-membered ring is puckered rather than planar, the "plane" of the ring is only roughly defined.

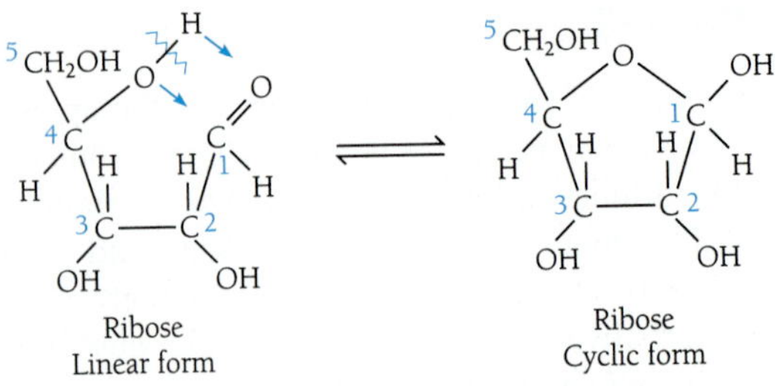

Figure 21.11 Cyclization of ribose comes about as a result of the addition of one of the hydroxyl groups across the C=O double bond. The numbers are used to identify specific carbon atoms or to groups bonded to specific carbon atoms.

this hydroxyl group is *equatorial,* meaning that it extends out from the ring (Figure 21.12). This slight structural difference has a profound effect on the chemistry of both plants and animals.

Disaccharides

The —OH group on the carbon atom adjacent to the ring oxygen atom of a monosaccharide readily undergoes condensation reactions with alcohols or amines, forming a product called a **glycoside.** A monosaccharide can react in this way with any of the —OH groups on another monosaccharide, forming a particular type of glycoside known as a **disaccharide.** In the reaction, a molecule of water is eliminated from two hydroxyl groups and the molecules become joined by a —O— bridge known as a

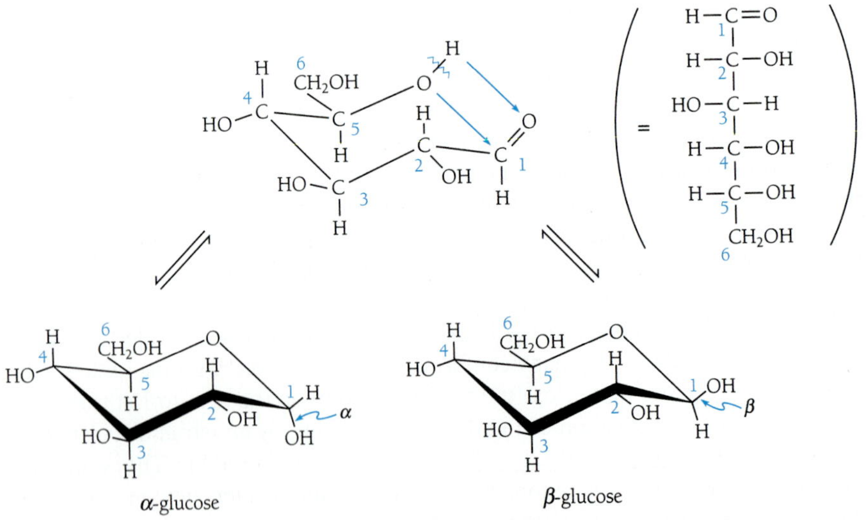

Figure 21.12 Cyclization of glucose. Since there is free rotation about the C—C bonds (in the straight-chain form), this cyclization can occur in two ways. There are two distinct cyclic forms of glucose. The α- and β-forms are stereoisomers differing only in the configuration at the number one carbon atom. The heavy lines in the structural formulas are used in perspective views to indicate the portion of the molecule closest to the viewer.

glycoside linkage. Ordinary table sugar, **sucrose,** is a disaccharide composed of α-glucose and β-fructose (Figure 21.13). Since a monosaccharide has several hydroxyl groups, there are numerous possibilities for formation of a glycoside linkage. These are distinguished, when necessary, by incorporating the carbon numbers into the name. Because C-1 of glucose is linked to C-2 of fructose, sucrose is a *1,2-glycoside.* Other common disaccharides are lactose ("milk sugar," β-galactose + glucose), maltose (α-glucose + glucose), and cellobiose (β-glucose + glucose). These are 1,4-glycosides, which are more common than 1,2-glycosides (Figure 21.14).

Mono- and disaccharides taste sweet, but not all equally so (Table 21.1). Candy manufacturers often use fructose, because, gram for gram, it is almost twice as sweet as sucrose. A sweet taste does not mean that a substance is a carbohydrate. The artificial sweetener *saccharin* ($C_7H_5NO_3S$) is about 500 times as sweet as sucrose, while the newer product, NutraSweet (a dipeptide) is 200 times as sweet as sucrose. Some naturally occurring proteins also taste sweet.

A sweet taste is no sign that a compound is safe to eat. Some compounds of beryllium are both poisonous and sweet-tasting, as is lead acetate. One of the early names for lead acetate was "sugar of lead."

Oligosaccharides

Some naturally occurring compounds, known as **oligosaccharides,** consist of more than two monosaccharide units. The prefix **oligo-** means "a few"—more than two, and perhaps as many as ten or twelve.

Polysaccharides

As its name implies, a **polysaccharide** is a polymer composed of many monosaccharide subunits. The most important naturally occurring polysaccharides are starch, glycogen, and cellulose. These are composed entirely of glucose monomers, and the differences between them are due to the way in which the monomers are joined. Plants and animals alike use the oxidation of glucose as a source of energy (for driving endothermic chemical reactions as well as for muscle activity). In animals, glucose,

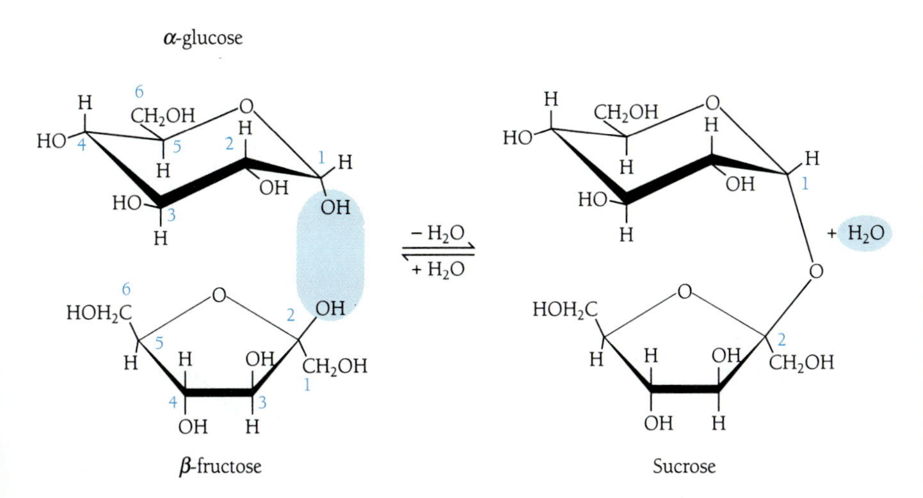

Figure 21.13 Formation of sucrose, a disaccharide, by condensation of an α-glucose molecule with a β-fructose molecule. Since the condensation takes place between the C-1 hydroxyl of glucose and the C-2 hydroxyl of fructose, sucrose is more informatively called a 1,2-glycoside.

TABLE 21.1 Relative Sweetness of Sugars

Sugar	Sweetness
Sucrose*	100
Glucose	74
Fructose	173
Galactose	32
Maltose	32
Lactose	1

* Sucrose is arbitrarily assigned a sweetness value of 100.

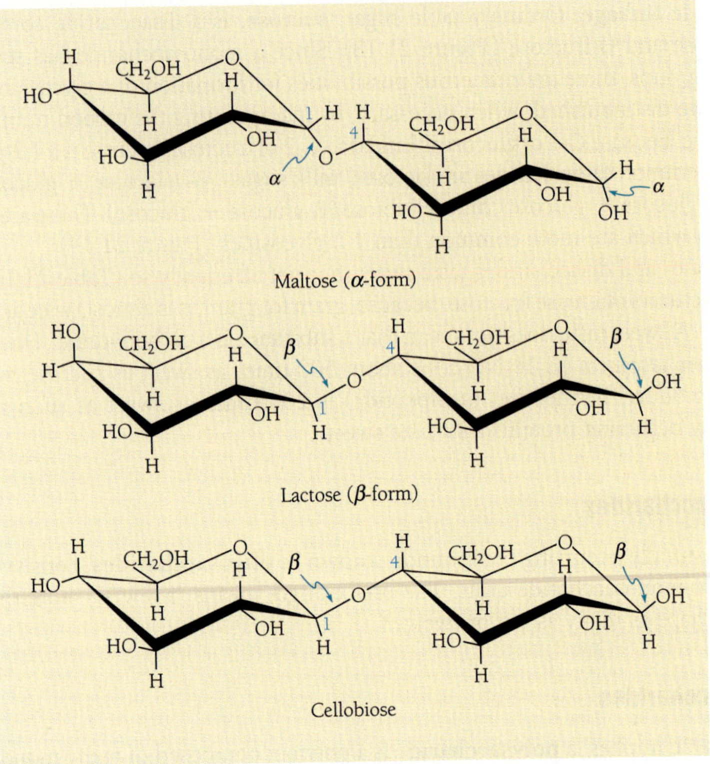

Maltose (α-form)

Lactose (β-form)

Cellobiose

Figure 21.14 Maltose, lactose, and cellobiose, three naturally occurring disaccharides. Both maltose and cellobiose consist of two glucose molecules. The only difference is that maltose is an α-glycoside while cellobiose is a β-glycoside.

and hence energy, is stored in glycogen polymers. In plants, the same function is performed by starch. Cellulose is used as a structural material by plants. It is a major component of wood, and it is by far the most abundant naturally occurring polymer.

Starch is a mixture of two polymers, *amylose* and *amylopectin*. About 20% to 30% of starch consists of amylose, in which all glycoside linkages are (α-1,4). Depending on the species of plant that produces the starch, amylose molecules can contain 25 to 1000 glucose units (Figure 21.15). Amylopectin also consists of glucose monomers

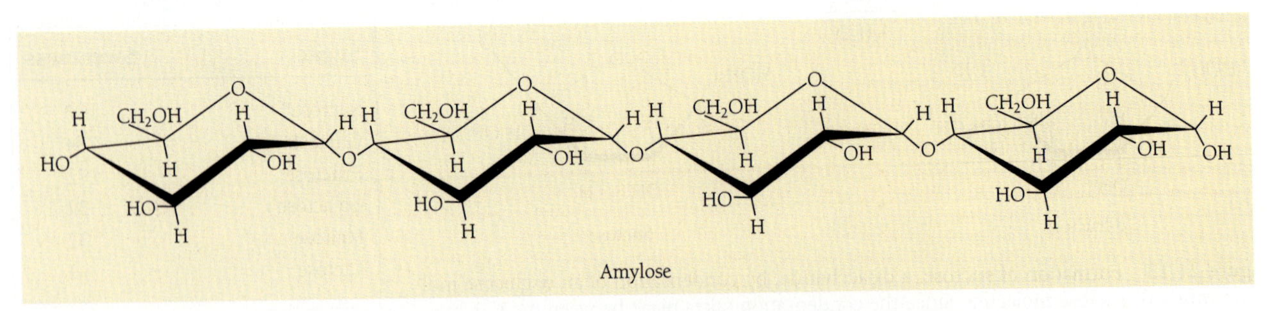

Figure 21.15 One of the components of starch is amylose, a linear polymer in which the glucose monomers are joined by (α-1,4) glycoside linkages. The molar mass of naturally occurring amyloses ranges from about 4×10^3 to 150×10^3.

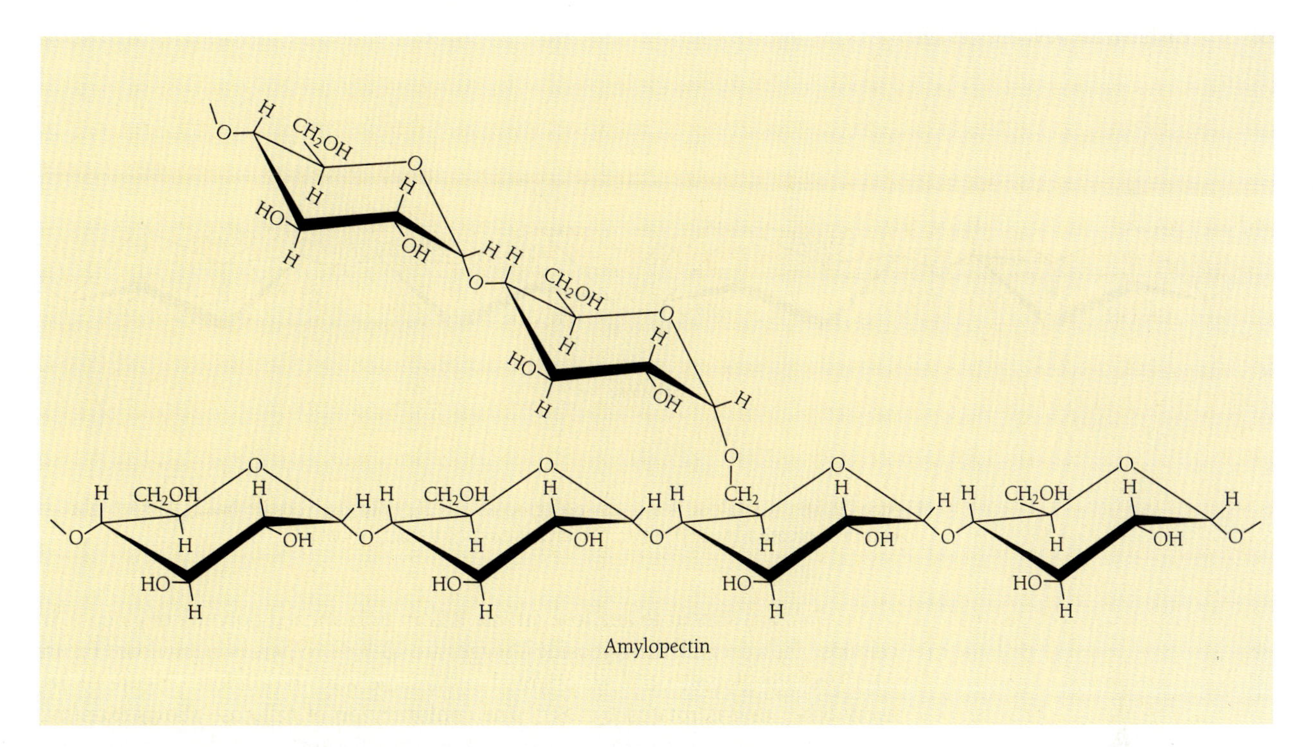

Figure 21.16 A branch point in amylopectin. About 4% of the glucose units have (α-1,6) as well as (α-1,4) glycoside linkages. Each branch point begins a new chain, and the result is a branched polymer incorporating about ten times as many glucose units as in amylose.

with (α-1,4) linkages, but with some *chain branching:* approximately every 20th to 25th glucose is linked at C-6 as well as at C-4. Amylopectins typically contain 300 to 5000 glucose units (Figure 21.16).

Glycogen is very similar to amylopectin, except that it has about twice as many (α-1,6) branch points, and is larger, with perhaps 30,000 glucose monomers (Figure 21.17).

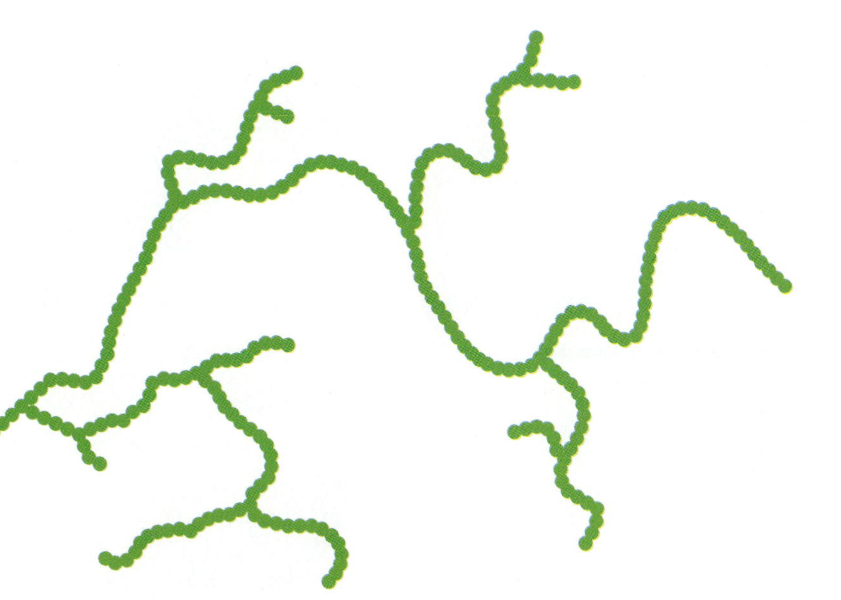

Figure 21.17 A typical amylopectin or glycogen molecule. Each dot represents an α-glucose unit.

Cellulose

Figure 21.18 Cellulose is a linear polyglucose having only β-glycoside linkages. Each glucose unit is rotated by 180° from the orientation of its neighbors.

Cellulose is similar to amylose, in that it is an unbranched polymer of glucose. In cellulose, however, all the glycoside linkages are (β-1,4). Whereas in amylose (Figure 21.15) all the glucose units are oriented in the same way, in cellulose (Figure 21.18) every other unit is rotated by 180°. The conformation of cellulose allows extensive hydrogen bonding between hydroxyl groups on adjacent molecules, resulting in a rigid two-dimensional structure (similar to the pleated sheet 2° structure of some proteins). This accounts for the great structural strength of woody plants. Extensive *inter*molecular hydrogen bonding is not possible in starch, however. Instead, the α-linkages are favorable for a helical structure in amylose, similar to the helical structure of proteins (Figure 21.19).

Possibly the most noticeable (to us) difference between the α- and β-linkages is that humans can use starch as food, while cellulose is not digestible. The first step in

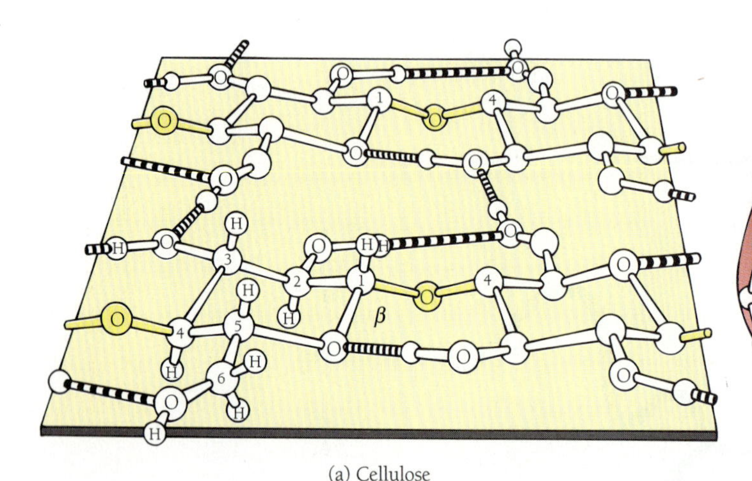

(a) Cellulose

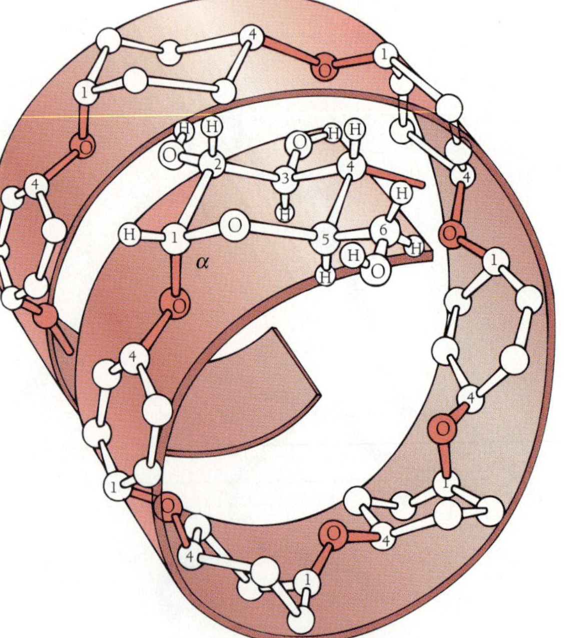

(b) Starch (amylose)

Figure 21.19 (a) Cellulose forms tough sheets of parallel chains, held together by *inter*molecular hydrogen bonding. (b) Starch (amylose) molecules assume a helical structure, with *intra*molecular hydrogen bonding.

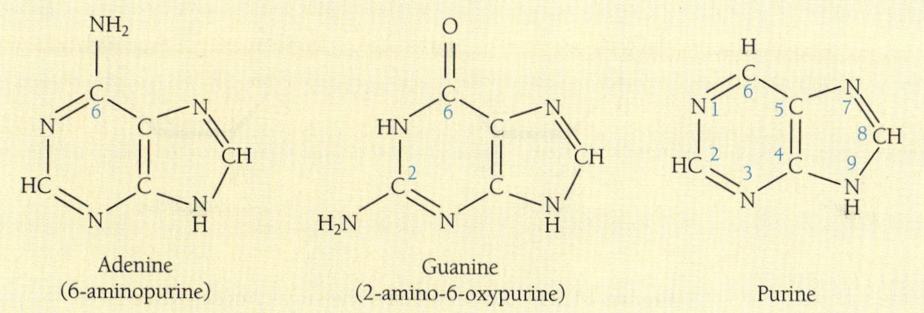

Figure 21.20 Purine and the purine bases.

the digestion of starch is *hydrolysis,* in which each glycoside linkage is broken apart by the addition of a water molecule. The result is the complete conversion of the polymer to its separate monomer molecules, glucose.

$$\text{-(glucose)-}_n + (n - 1)\ H_2O \longrightarrow n \text{ glucose}$$

The α-linkages of starch are easily hydrolyzed. The β-linkages of cellulose, protected from hydrolysis by the secondary structure of cellulose, are untouched by the digestive system. Certain bacteria, however, produce enzymes capable of hydrolyzing cellulose and utilizing the glucose it contains. Such bacteria inhabit the gut of termites and grazing mammals, allowing these and related species to prosper on a diet of cellulose. It is possible to hydrolyze cellulose by cooking with acid, but up to now it has not been feasible to use this process as a food source.

21.3 NUCLEIC ACIDS

The *nucleic acids,* DNA and RNA, are probably best known for the part they play in the transmission of genetic information from one generation to the next, although they are of central importance in other biological functions as well. Both occur in plants, animals, and viruses. Both are polymers: RNA is usually smaller than DNA, which in some species can have a molar mass of over 10^{12} g mol^{-1}.

Viruses contain either DNA or RNA, but not both.

The molar mass of human DNA is about 3×10^{12} g mol^{-1}, while that of some bacterial DNA is a thousand times less.

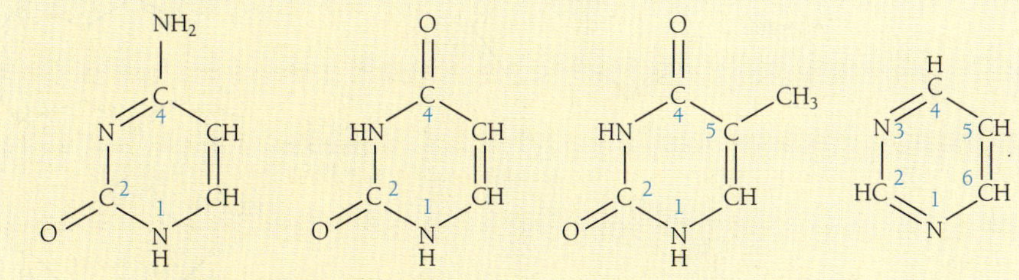

Figure 21.21 Pyrimidine and the pyrimidine bases.

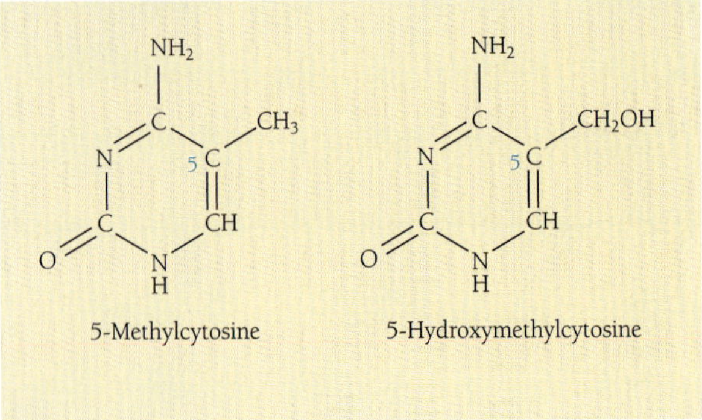

5-Methylcytosine

5-Hydroxymethylcytosine

Figure 21.22 Modified cytosine bases.

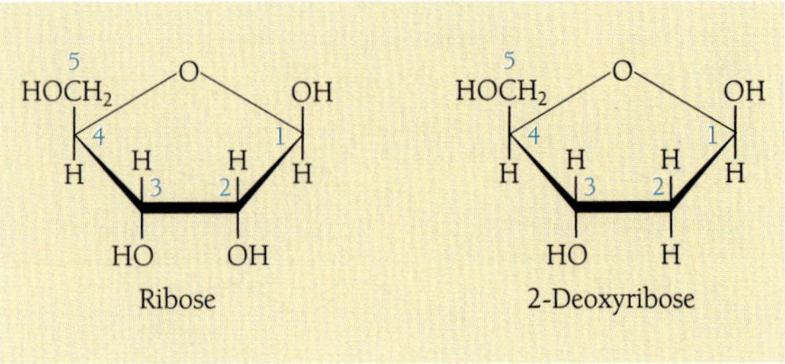

Ribose

2-Deoxyribose

Figure 21.23 RNA contains the sugar ribose, while DNA contains the closely related 2-deoxyribose.

Nucleosides, Nucleotides, and Primary Structure

A **nucleic acid** is a polymer made up of a large number of similar subunits. These subunits, called **nucleotides,** consist of three parts: phosphoric acid, a sugar, and a nitrogen-containing base. Two of these bases, *adenine* and *guanine,* are called the **purine bases.** They are derivatives of the simpler compound *purine,* and they occur in both DNA and RNA. Figure 21.20 gives their structure, along with that of purine. Three of the bases, *cytosine, uracil,* and *thymine,* are derivatives of the simpler compound *pyrimidine.* These are known as the **pyrimidine bases,** and their structures are shown in Figure 21.21. Thymine occurs only in DNA, uracil occurs only in RNA, and cytosine occurs in both. All five of the bases are plate-like, essentially planar, molecules.

Until recently it was thought that only these five bases occur in nucleic acids, but now it is known that other, closely related, bases occur as well. For example, the *modified bases methylcytosine* and *hydroxymethylcytosine* (Figure 21.22) are common in both plant and animal DNA, and they may play an important role in reproduction.

In the nucleic acids, each base is bonded to a sugar molecule. In **RNA (ribonucleic acid)** the sugar is ribose, while in **DNA (deoxyribonucleic acid)** the sugar is ribose in which the C-2 —OH group has been replaced by —H. Figure 21.23 shows the two sugars, and Figure 21.24 shows that they are bonded to the nitrogen base by a

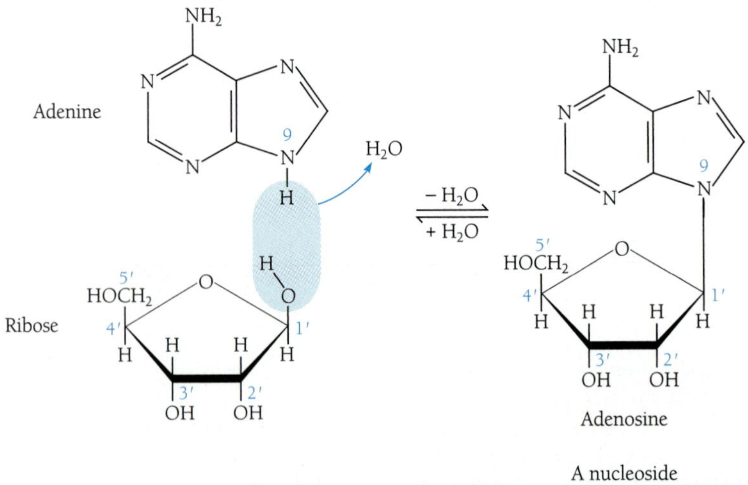

Adenine

Ribose

$-H_2O$

$+H_2O$

Adenosine

A nucleoside

Figure 21.24 A nucleoside is a sugar-base combination bonded at carbon C-1 of ribose or deoxyribose, and N-9 of purine bases or N-1 of pyrimidine bases. The two parts are joined in a condensation reaction.

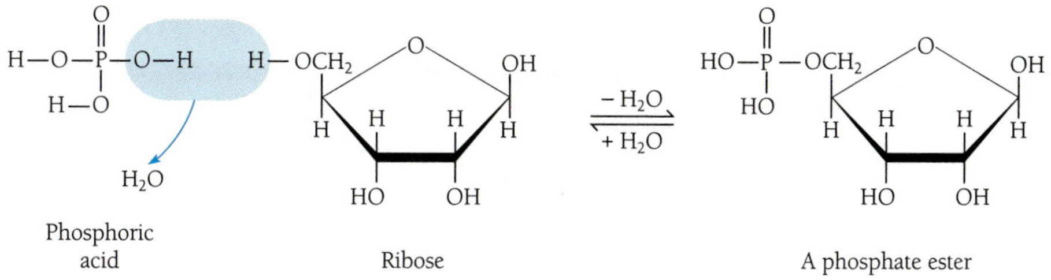

Phosphoric acid

Ribose

A phosphate ester

Figure 21.25 Condensation of phosphoric acid with an alcohol (one of the hydroxy groups of ribose or deoxyribose) leads to a phosphate ester.

condensation reaction involving the hydroxyl group at C-1. A sugar-base combination is called a **nucleoside**, and the bond joining the two is a *nucleoside bond.*

Nucleotide comes after nucleoside, both in size and in the alphabet.

The third component of a nucleotide is a phosphate group, bonded to the sugar. This bond, an *ester* linkage, comes about through condensation of phosphoric acid and the hydroxy group on C-5 of the sugar (Figure 21.25). The three-part structure of a typical nucleotide is shown in Figure 21.26.

Although best known as components of nucleic acids, nucleotides and nucleosides have other functions as well. Some of these are of central importance in the distribution and expenditure of energy in the organism, discussed in Section 21.5. The compound AZT (Figure 21.27), whose structure differs from that of deoxythymidine only in that it has an azide group ($—N_3$) rather than a hydroxyl at C-3 of the sugar, is used in the treatment of AIDS. AZT interferes with DNA synthesis, and the AIDS virus is more sensitive to it than is the human body. Although not a cure, it is a worthwhile treatment that delays the onset of symptoms of the disease.

Nucleotides are joined to one another by further sugar-phosphate condensation, at C-3 of ribose or deoxyribose. The resulting *polynucleotides* are the nucleic acids DNA and RNA. A short segment of an RNA molecule is shown in Figure 21.28. The sequence of nucleotides along the sugar-phosphate backbone is called the **primary (1°) structure** of nucleic acids. Since both DNA and RNA contain only one type of sugar, primary structure actually refers only to the sequence of bases.

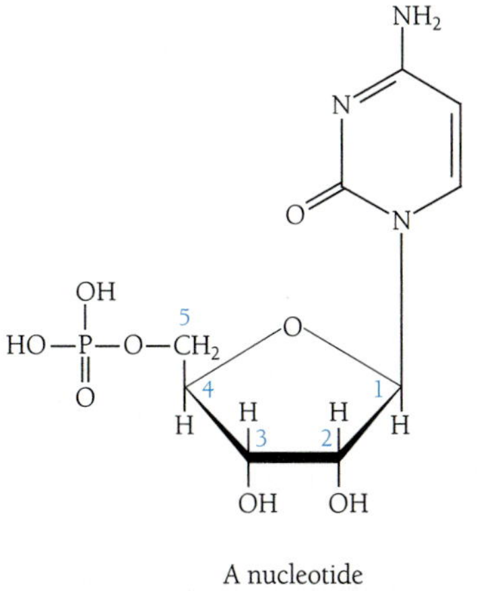

A nucleotide

Figure 21.26 Nucleotides consist of a phosphate group, ribose or deoxyribose, and a purine or pyrimidine base. The phosphate ester linkage occurs at C-5 of the sugar.

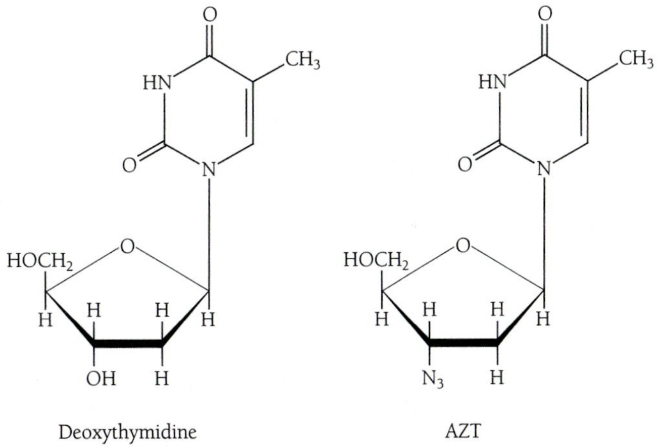

Deoxythymidine

AZT

Figure 21.27 Deoxythymidine and the synthetic analogue AZT. The difference between these two compounds is slight, but sufficient to disrupt the replication of the AIDS virus.

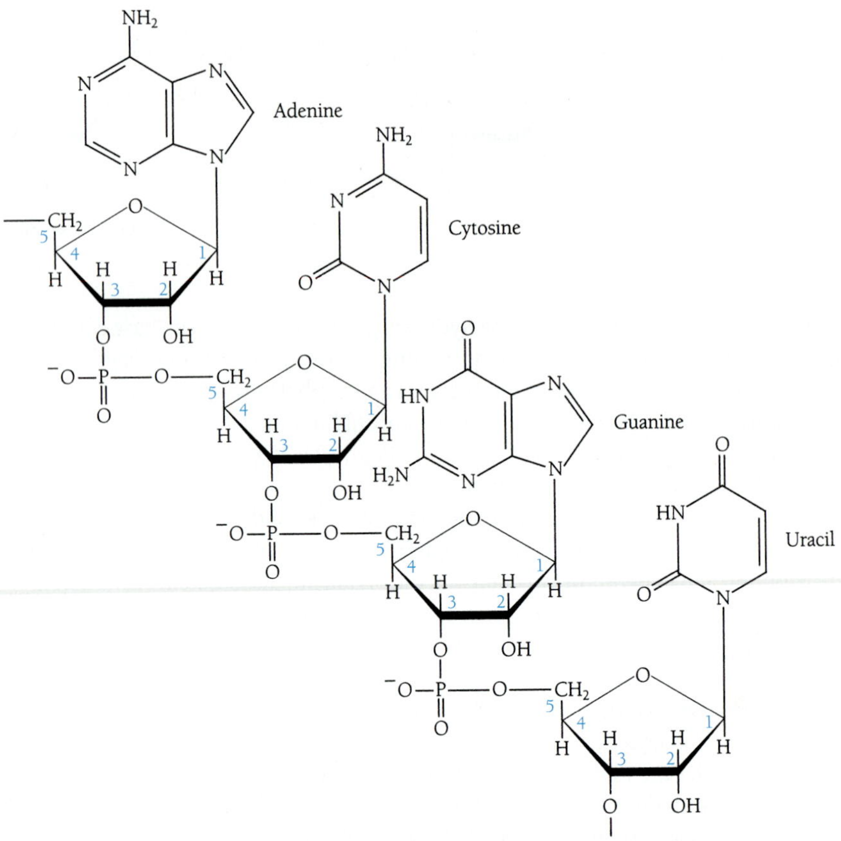

Figure 21.28 A typical segment of a molecule of ribonucleic acid. The phosphate ester linkages are at C-3 and C-5 of the sugar, while the base is attached to C-1.

Secondary Structure of Nucleic Acids

A nucleic acid molecule is a long, flexible chain or strand, which might be coiled or folded into a huge variety of overall shapes. However, only a few of these possibilities are found in living organisms. The way in which the strands are coiled is called the **secondary (2°) structure** of the nucleic acid.

Hydrogen Bonding and Base Pairing

The shapes of the purine and pyrimidine bases are particularly well suited for hydrogen bonding between them. Guanine and cytosine form three H-bonds, while adenine and thymine form two. In DNA, the bases always occur in H-bonded pairs, and the pairing is always the same.

- G≡C Guanine and cytosine are always paired together, with three hydrogen bonds.
- A=T Adenine and thymine are always paired together, with two hydrogen bonds.

No other combinations have the correct geometry for the formation of multiple hydrogen bonds (Figure 21.29).

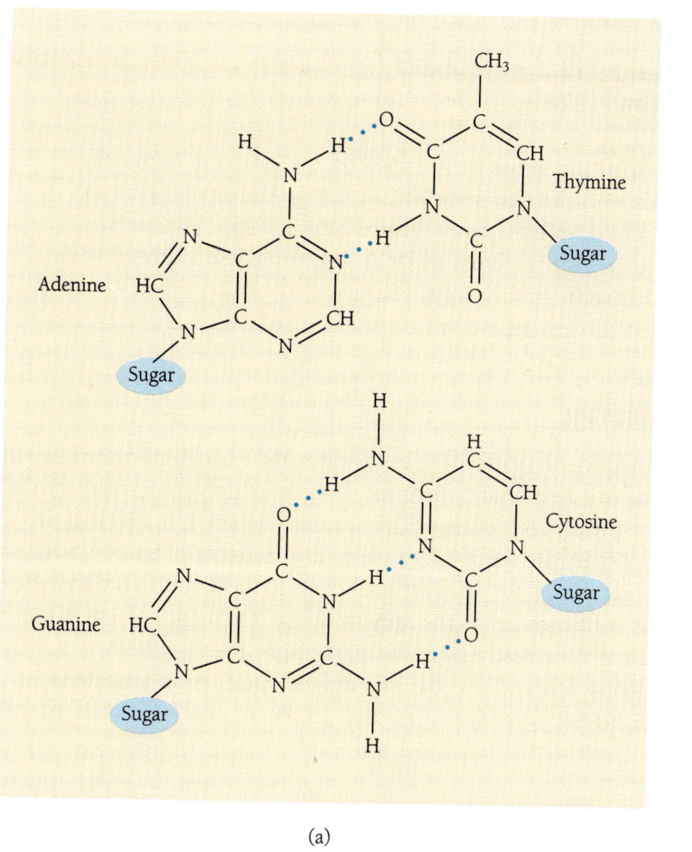

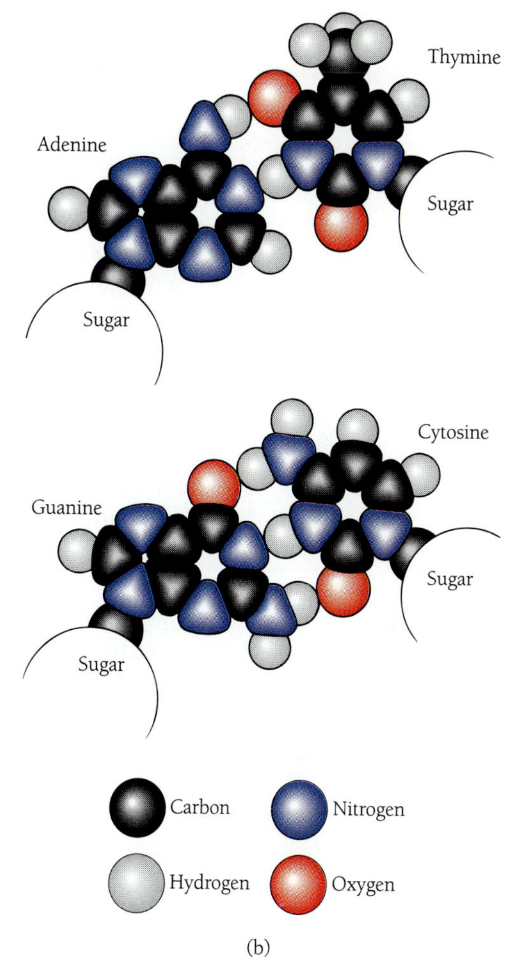

Figure 21.29 (a) Hydrogen bonding between pairs of bases. The interaction is specific, occurring only between the pairs of bases shown. (b) Space-filling models, showing the same interaction, emphasize how well the base pairs fit together. Note that the **—A⁝⁝⁝T—** (2 H-bonds) and the **—G⁝⁝⁝C—** (3 H-bonds) pairs have essentially the same size and shape.

The Double Helix

The secondary structure of DNA was deduced in 1953 by James Watson and Francis Crick, working from x-ray diffraction data provided by Maurice Wilkins and Rosalind Franklin. Many people regard the discovery of the DNA structure as one of the 20th century's most important advances in science. In this structure, "the *double helix*," two linear polynucleotides are joined together by hydrogen bonding between their bases. Although H-bonds are individually weak, the total bonding force holding the two strands together is quite strong. Since the bases are always paired in the same way, (A, T) and (G, C) are called **complementary pairs.** The extent of hydrogen bonding, and hence the strength and stability of the DNA molecule, is maximized if the two strands are wrapped around each other in a double helix. In this configuration *each base* can be hydrogen bonded to its complement on the other strand. The double helix can be visualized as a twisted ladder: the rungs or steps represent the hydrogen-bonded base pairs, while the curved uprights represent the sugar-phosphate backbones of the two polymer chains (Figures 21.30 and 21.31).

Crick's 1966 book, *Of Molecules and Men,* discusses the significance and implications of the discovery. Watson's popular book, *The Double Helix,* is a personal account of the history.

Francis H. C. Crick (1916–) and James D. Watson (1926–)

Francis Crick was working in the early 1950s on the structure of biological molecules at the Cavendish Laboratory of Cambridge University. James D. Watson, a young American biochemist, joined him in 1951. Within two years they had deduced the structure of DNA, and, nine years later, they were honored with a share in the Nobel Prize.

The Nucleic Acid Message

Nestled within the protective ramparts of the sugar-phosphate backbones, the bases of DNA are relatively safe from external influences. They react only under very special circumstances, and the primary structure of DNA remains intact. It is well that this is so, because the primary structure is not only a particular sequence of bases, it is also a *message*. The purine and pyrimidine bases are analogous to the letters of the alphabet, whose sequence spells out a message in the same way that a sequence of letters spells out the message you are now reading in this paragraph. Because DNA is a large polymer containing many nucleotides, the amount of information contained in a single molecule is enormous: comparable, in fact, to the amount contained in this

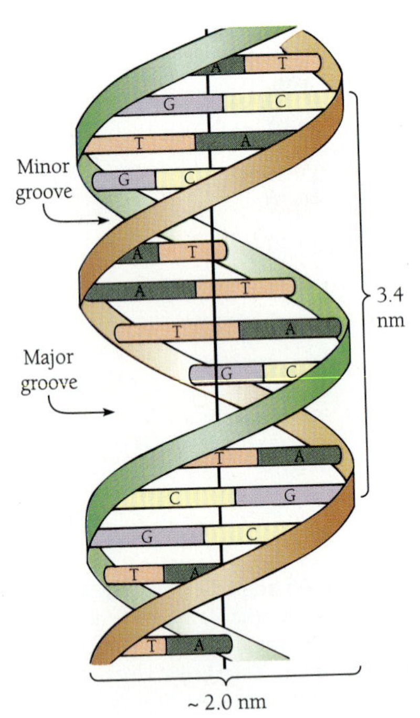

Figure 21.30 The secondary structure of DNA is a double helix, in which two linear polynucleotides are held together by extensive hydrogen bonding between complementary base pairs. The structure resembles a twisted ladder.

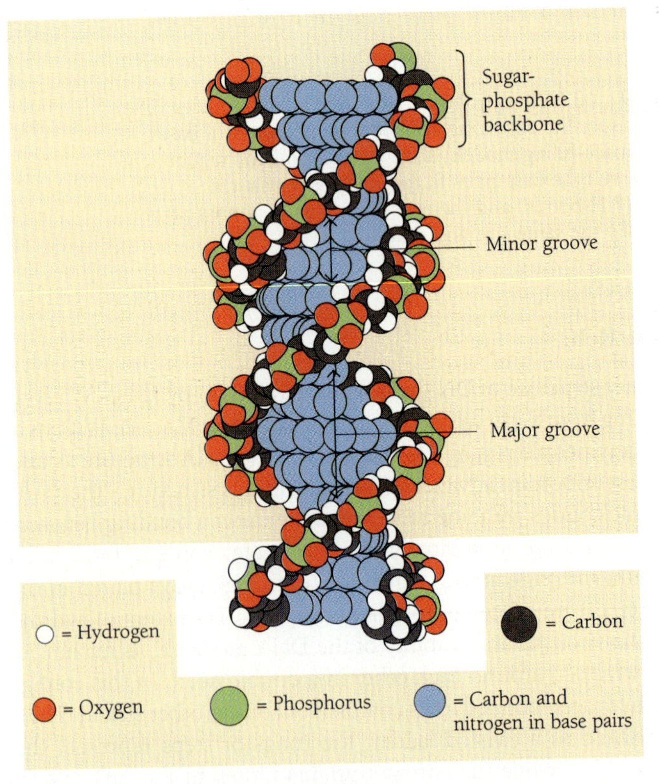

Figure 21.31 A space-filling model of a segment of DNA.

Maurice Wilkins (1916–) and Rosalind Franklin (1920–1958)

Maurice Wilkins, a transplanted New Zealander, and Rosalind Franklin worked together at the Biophysical Laboratory of Kings College in London. They were responsible for some of the advances in x-ray crystallography that made possible the structure determination of very large molecules. Wilkins shared the 1962 Nobel Prize with Watson and Crick, and, had she lived, Franklin might have shared it as well.

book! The content of the DNA message is nothing less than the laboratory instructions for all the complex chemistry of a living cell.

Problem 44 at the end of the chapter emphasizes the comparison between a DNA molecule and a book.

Protein Biosynthesis and the Genetic Code

The way in which a living cell "reads" the DNA message, and carries out the instructions it contains, is a fascinating and complicated process. A great deal is known about it, and a great deal remains to be learned.

Although DNA itself is not directly involved, the synthesis of proteins within a living cell follows instructions contained in the DNA message. Part of the DNA message is copied onto smaller molecules of ribonucleic acid, called *messenger RNA*. The information carried by *m*-RNA describes the length and the order of amino acids (the 1° structure) of one of the many proteins needed by a cell. As in DNA, these instructions are spelled out by the sequence of bases along the RNA backbone.

The message is written into the primary structure of RNA in three-letter "words" called **codons.** Each group of three bases (such as . . . CAG . . ., . . . AAC . . ., and so on) along the backbone forms one codon. Since the RNA alphabet contains only the four letters C, G, A, and U, only $4 \cdot 4 \cdot 4 = 64$ three-letter words are possible. The meanings of these words are now known, and are collectively referred to as the **genetic code.**

In a living cell, proteins are assembled by the stepwise addition of amino acids to a growing polymer chain. The process continues until a complete protein, having a specific and unique primary structure, is formed. Each codon directs the addition of a specific amino acid to the protein chain. The codon GAA, for instance, means "attach a glutamine monomer to the chain." Since there are 64 codons and only 20 amino acids, some codons are synonyms: both GAA and GAG, for example, code for glutamine. In addition, some codons are used for punctuation, as a "period," whose meaning is "this protein is finished—stop adding amino acids." In this way, the information contained in the primary structure of DNA results in the biosynthesis of proteins, many of which are enzymes that catalyze the chemical reactions of the cell. The biosynthesis of proteins is a multistep reaction, involving the intermediate participation of several different types of RNA and numerous enzymes.

Replication

Of critical importance to a living organism is the preservation and transmission of the DNA message during growth and reproduction. At the cellular level, this means that new DNA molecules, identical to the old, must be produced during cell division. Like

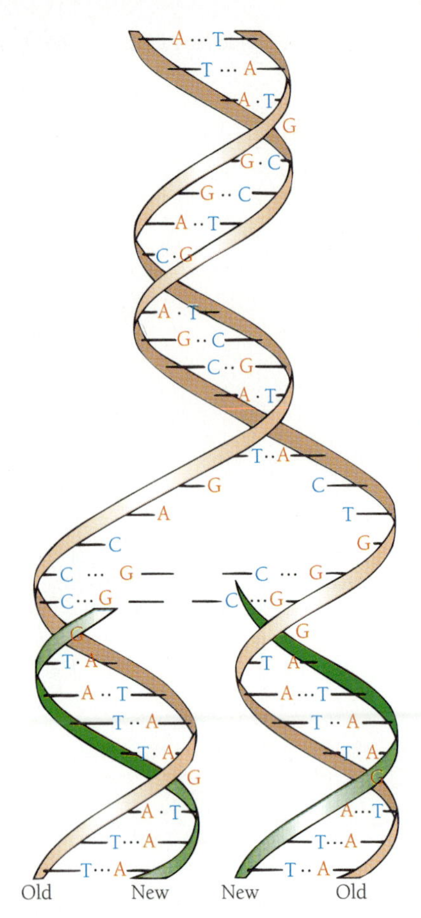

Figure 21.32 During DNA replication, the double helix unwinds, and each of the old strands becomes a template and site for the production of a new, complementary strand. When the process is complete, each of the two identical "daughter" molecules contains one of the old ("parent") strands.

protein synthesis, this is a multistep process involving RNA and many enzymes. The overall process is called DNA *replication.*

Each strand of the DNA double helix carries the same information, because of the complementary nature of base pairing: each adenine is paired with a thymine, and each guanine is paired with a cytosine. The relationship between the two strands is rather like that between a photograph and its negative—one can easily be produced from the other, because they are both the same picture.

If a DNA molecule is like a photo stuck to its negative, the replication process requires (1) separating the two, (2) making a photo from the negative and a negative from the photo, and (3) reassembly of the four parts as two identical photo-negative pairs. In replication, then, the DNA molecule unwinds from its helical conformation, separating the two strands. Each strand, as it unwinds, is a template for the assembly of a new, complementary strand. As the new strand is assembled, it and the original strand are wound into a new double helix. Many enzymes and RNA molecules are involved in bringing the proper nucleotides to the right place at the right time, so that the end result is two identical DNA molecules, each carrying the same genetic message. The process is diagrammed in Figure 21.32.

21.4 LIPIDS

Biological molecules that are insoluble in water but soluble in nonpolar solvents are called **lipids:** many lipids are esters, but the classification is based on solubility rather than molecular structure. Lipids perform a wide variety of functions within an organism.

Glycerides: Fats and Oils

Glycerol (1,2,3-propanetriol, older name "glycerin") is a polyfunctional alcohol that readily condenses with three moles of acid, forming esters called **triglycerides.** The acid components may be the same, or as in a *mixed* triglyceride, different. Natural *fats* and *oils* are mixtures of mixed triglycerides. Treatment of these compounds with hot NaOH(aq) or KOH(aq) causes hydrolysis and releases the sodium or potassium salts of the acids. Hydrolysis of an ester, when carried out in basic solution, is called *saponification.*

The word "saponification" is derived from the French word for soap, *sapon.*

$$\begin{array}{c} H_2C{-}O{-}\overset{O}{\overset{\|}{C}}{-}R_1 \\ | \\ HC{-}O{-}\overset{O}{\overset{\|}{C}}{-}R_2 \\ | \\ H_2C{-}O{-}\overset{O}{\overset{\|}{C}}{-}R_3 \end{array} \xrightarrow[H_2O]{KOH} \begin{array}{c} H_2C{-}O{-}H \\ | \\ HC{-}O{-}H \\ | \\ H_2C{-}O{-}H \end{array} + \begin{array}{c} K^+ \; {}^-O{-}C({=}O){-}R_1 \\ K^+ \; {}^-O{-}C({=}O){-}R_2 \\ K^+ \; {}^-O{-}C({=}O){-}R_3 \end{array}$$

a mixed triglyceride **glycerol** **three fatty acids (as potassium salts)**

The acids, called **fatty acids,** are straight-chain hydrocarbons containing 14 to 20 carbon atoms and having a carboxylic acid group (—COOH) at one end. Fatty acids may be *saturated* (containing no multiple bonds), *unsaturated* (containing one double bond), or *polyunsaturated* (having more than one double bond). The following are three examples of some 70 naturally occurring fatty acids.

$H_3C(CH_2)_{16}COOH$ — Stearic acid; mp 67–69 °C

$H_3C(CH_2)_5CH{=}CH(CH_2)_7COOH$ — Palmitoleic acid; mp 0.5 °C

$H_3C(CH_2)_4CH{=}CHCH_2CH{=}CH(CH_2)_7COOH$ — Linoleic acid; mp −5 °C

The extent of saturation affects the molecular shape, hence the extent of intermolecular attractive forces and the melting point of both the fatty acids (Figure 21.33) and the corresponding triglycerides. Triglyceride mixtures that are solid at room temperature are called **fats,** and are mostly of animal origin. The acid portions of fats are predominantly *saturated* fatty acids. **Oils** are triglyceride mixtures that are liquid at room temperature. Oils, which are mostly of plant origin, consist predominantly of *unsaturated* fatty acids.

Fats and oils are an important part of the diet, because they are good energy sources. Gram for gram they supply about twice as many calories as carbohydrates. Shortening and margarine are made by *saturating* vegetable oils, which increases their melting point enough to make them solid or semisolid at room temperature. The process, also called *hydrogenation,* involves the catalytic addition of hydrogen across the double bond(s) of unsaturated triglycerides. There is considerable evidence that excessive consumption of saturated fats increases the risk of heart disease and strokes, and that the risk is less with unsaturated oils. Recently, as a result of consumer concern, there has been a trend toward the use of shortening and spreads prepared from *partially* hydrogenated oils. These retain some degree of unsaturation, and therefore, presumably, pose less of a health risk.

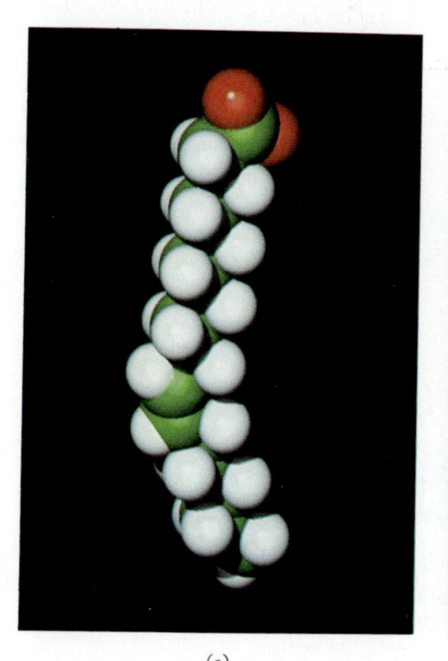

(a)

(b)

Figure 21.33 Fatty acids have a polar "head" (top) and a long, nonpolar, hydrocarbon "tail." Compared to the kinked tail of the unsaturated acids (a), the relatively straight tail of the saturated acids (b) allows closer packing in the solid and liquid states. This in turn leads to stronger intermolecular van der Waals forces and higher melting points.

Soaps

Remember the phrase, "like dissolves like."

The alkali salts of fatty acids, produced by saponification of fats, are called **soaps.** Much of the dirt ("soil") that becomes attached to our clothes, dishes, and bodies contains lipids, greases, and other nonpolar materials that are insoluble in water. Because they are similar to hydrocarbons, the nonpolar tails of fatty acid anions are called **hydrophobic** (water-hating). The tails readily penetrate and become embedded in particles of nonpolar soil. But the carboxylate heads, which are **hydrophilic** (water-loving) because they bear the negative charge of the ions, cannot enter a nonpolar medium. In this way, a soil particle acquires a coating of polar, water-soluble carboxylate anions, and is easily removed by rinsing in water (Figure 21.34).

So-called "hard water" contains Ca^{2+} and Mg^{2+} ions leached from limestone rocks. When hard water is used for washing, several annoying aspects of soap chemistry become evident. Unlike the sodium and potassium salts, the calcium and magnesium salts of fatty acids are insoluble in water. When soap is added to hard water, the Ca and Mg salts precipitate as a grayish scum. As a result, more soap must be added. Not until all the Ca^{2+} and Mg^{2+} ions are removed from the water does the soap become available for dissolving dirt. This problem does not arise with synthetic soaps, called **detergents,** because their calcium and magnesium salts are more soluble. The first detergents to be manufactured were the *alkyl benzene sulfonates.* These compounds, like soap, have long, nonpolar tails and an anionic head.

$$CH_3CH_2CH_2CH_2CH_2CH_2CH_2CH_2CH_2CH_2CH_2-C_6H_4-SO_3^-\ Na^+$$

As is often the case, solution of one problem created another. Although alkyl benzene sulfonate detergents work better in hard water, they are much more slowly degraded by bacteria when they are returned to the environment in waste water. This problem became known when wells in some rural areas began delivering foamy water, due to the build-up of detergent in the ground water. Substitute detergents were found, and today we use compounds that are readily biodegradable and that do not form insoluble calcium or magnesium salts.

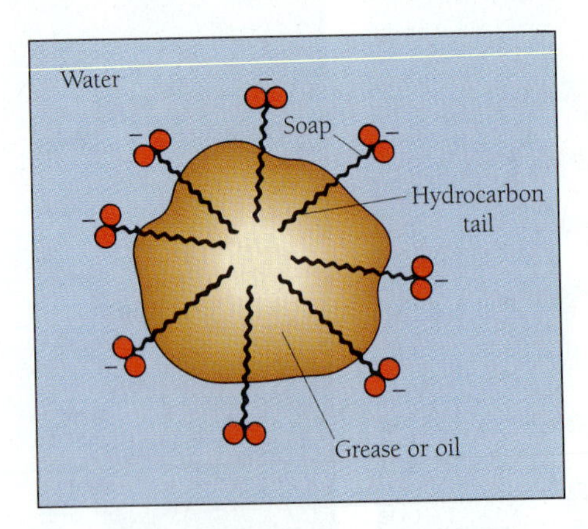

Figure 21.34 A nonpolar dirt particle becomes water soluble when the nonpolar hydrocarbon tails of soap anions penetrate it, leaving the negatively charged $-CO_2^-$ on the surface.

Waxes

Esters of a fatty acid with a long-chain primary alcohol (usually 24 to 28 carbon atoms) are called **waxes.** These are soft solids used for lubrication and protection by many plants and animals (Figure 21.35), for instance as waterproof coatings on fruits and leaves.

Phospholipids

Replacement of one of the fatty acid residues of a triglyceride with a phosphate ester leads to a compound called a *phosphoglyceride,* one of several classes of **phospholipids** that incorporate a phosphate group.

$$CH_3-CH_2-(CH_2)_{14}-CH_2-CH_2-CH_2-CH_2-\overset{\overset{\large O}{\|}}{C}-O-CH_2$$

$$CH_3-CH_2-(CH_2)_5-CH{=}CH-(CH_2)_7-CH_2-\overset{\overset{\large O}{\|}}{C}-O-CH$$

$$H_2C-O-\underset{\underset{\large O^-}{|}}{\overset{\overset{\large O}{\|}}{P}}-O-CH_2-CH_2-\underset{\underset{\large CH_3}{|}}{\overset{\overset{\large CH_3}{+|}}{N}}-CH_3$$

a typical phospholipid

These molecules have a polar head and two nonpolar tails, which causes them to cluster into large layered structures (Figure 21.36) used by many organisms as membranes surrounding cells and subcellular structures. In such a bilayer membrane, the hydrophilic portions of the molecules, on the outside, can freely interact with an aqueous environment. The nonpolar, hydrophobic tails are attracted to and interact with each other. Most cell membranes also contain a variety of protein molecules, imbedded in or attached loosely to the phospholipid bilayer structure. Membranes serve several functions. For example, they protect the enclosed structure from the

Figure 21.35 Bees use wax for construction.

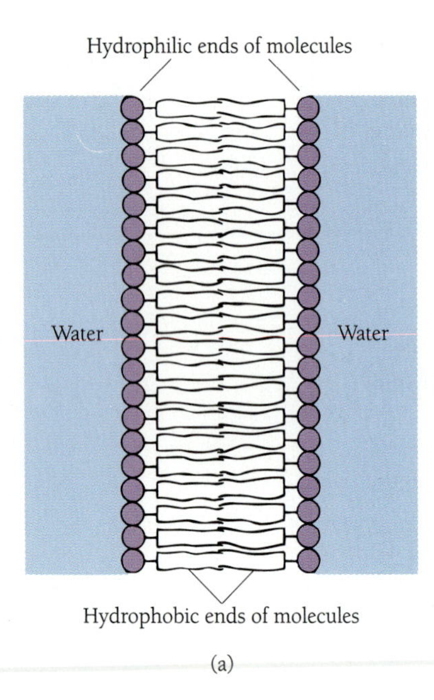

(a)

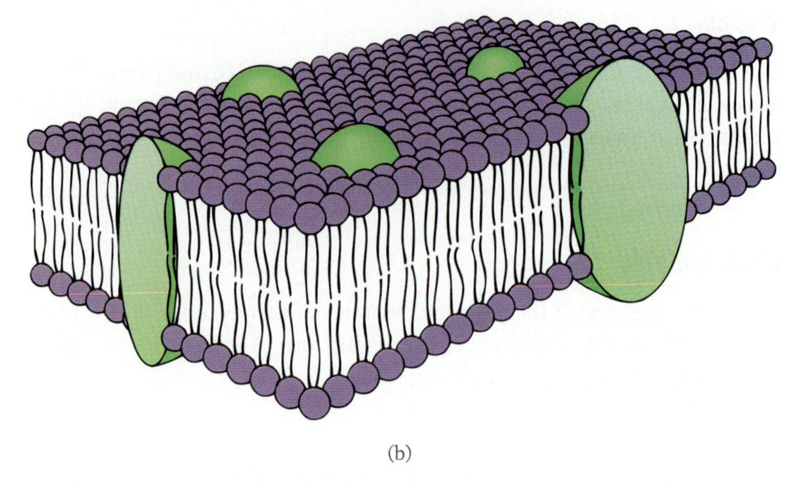

(b)

Figure 21.36 (a) Cross-sectional view of a phospholipid bilayer membrane. The polar phosphate heads are exposed to the aqueous environment, while the hydrocarbon tails cluster together in the central region. (b) Protein molecules, indicated by the green regions, are a part of the membrane structure. Proteins participate in the selective transfer of substances across membranes.

environment, which is normally quite different from the cell interior. They also exclude unwanted substances while passing (often through channels associated with the imbedded proteins) needed nutrients, and they aid in the excretion of waste material.

Steroids

All **steroids** share a common skeletal structure (Figure 21.37a) of three six-membered rings and one five-membered ring fused together. The best-known and most abundant steroid is *cholesterol* (Figure 21.37b), used by the body in a variety of ways.

CH_3 CH_3 CH_3 CH_3 CH_3 HO

(a) Steroid skeleton

(b) Cholesterol

Figure 21.37 (a) The four-ring structure common to all steroids. (b) Cholesterol, the most abundant steroid. The *-ol* suffix indicates the presence of the hydroxy group.

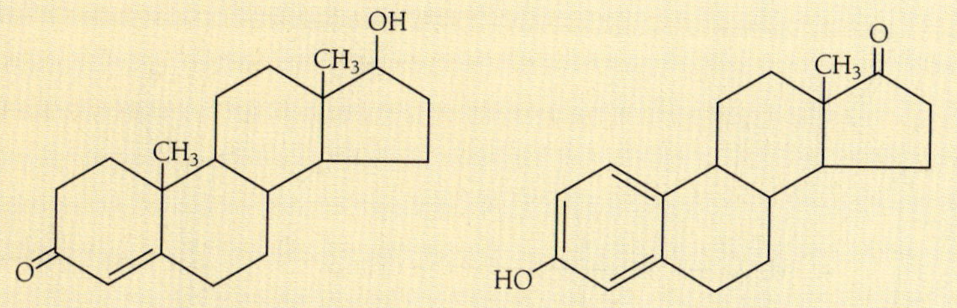

OH CH$_3$ CH$_3$ O (a) Testosterone

O CH$_3$ HO (b) Estrone

Figure 21.38 Two hormones that regulate the development of secondary sex characteristics in males (testosterone) and females (estrone).

Cholesterol is synthesized (in the liver) as needed, but in addition some is obtained from the meat and dairy products that we eat. It is believed that excessive dietary intake of cholesterol contributes to the build-up of fatty deposits in the arteries, which in turn leads to heart disease and strokes. Cholesterol is the starting material for the synthesis of numerous hormones, including the sex hormones, which regulate many body functions (Figure 21.38).

Synthetic hormones derived from the male sex hormone testosterone are used by some athletes to increase muscle mass. Because they interfere with the natural hormone levels in the body, there are some unfortunate side effects of this use of steroids. Among these are decreased sperm production and breast enlargement in males, baldness, growth of facial hair and menstrual irregularities in females, and disruption of normal puberty in children. Often a personality change (for the worse) accompanies steroid use.

21.5 BIOENERGETICS

A great deal of energy is necessary to operate a living organism. Energy is needed not only for muscle activity and keeping the body warm, but also for driving the complex biochemical machinery of the cell. This activity is collectively known as *cellular metabolism.* Plants, and most animals, obtain energy from the oxidation of carbohydrates. Some animals (including humans) are also able to derive energy from the oxidation of proteins and fats. In this section we will sketch some aspects of *bioenergetics,* including the part played by the nucleotide adenosine triphosphate (ATP).

ATP: The Energy Carrier

Cellular **metabolism** includes two types of activity: an **anabolic** process results in the formation of complex molecules from smaller species, while a **catabolic** process results in the breakdown of large, complex molecules into smaller ones. Generally speaking, catabolic processes are energy suppliers and anabolic processes are energy consumers. Because of the close connection between free energy and equilibrium, biochemists prefer to discuss bioenergetics in terms of ΔG, rather than ΔE or ΔH. A reaction that consumes free energy, that is, in which $\Delta G > 0$, is **endergonic.** A reaction that liberates free energy, that is, in which $\Delta G < 0$, is an **exergonic** process. A

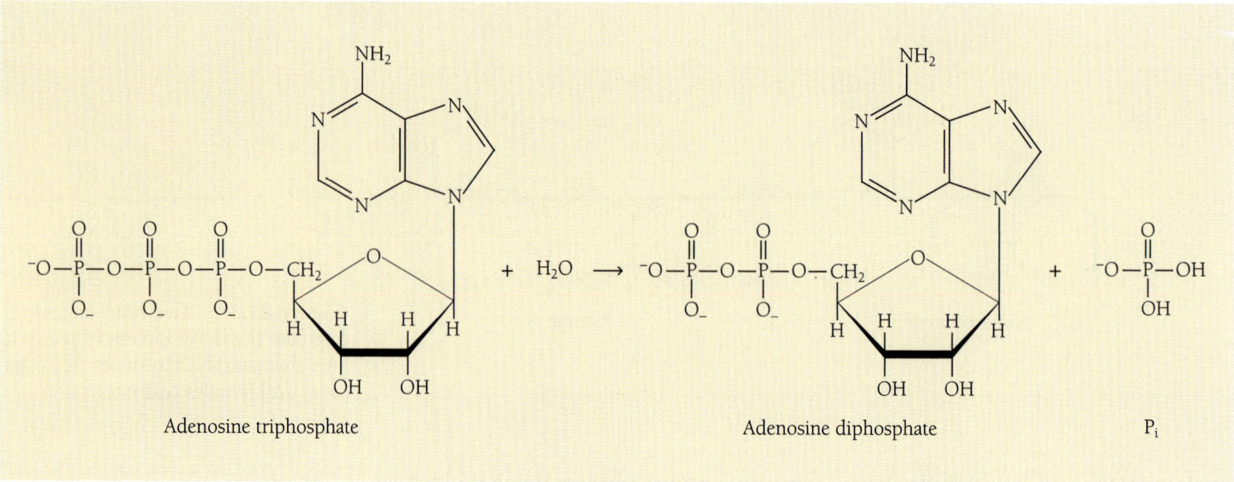

$$^{-}O-\overset{O}{\overset{\|}{P}}(O^-)-O-\overset{O}{\overset{\|}{P}}(O^-)-O-\overset{O}{\overset{\|}{P}}(O^-)-O-CH_2\text{-(adenosine)} + H_2O \longrightarrow {}^{-}O-\overset{O}{\overset{\|}{P}}(O^-)-O-\overset{O}{\overset{\|}{P}}(O^-)-O-CH_2\text{-(adenosine)} + {}^{-}O-\overset{O}{\overset{\|}{P}}(OH)-OH$$

Adenosine triphosphate — Adenosine diphosphate — P_i

Figure 21.39 The principal carrier of energy in living organisms is adenosine triphosphate. The hydrolysis of one of the energy-rich P—O bonds in ATP supplies the energy needed for other, endergonic reactions in the cell.

major source of energy is the oxidation of glucose, available from the breakdown of carbohydrates in food.

Note the use of ΔG rather than $\Delta G°$. Biochemical reactions normally take place under conditions other than the standard, 1 M, concentration.

$$C_6H_{12}O_6 + 6\ O_2 \longrightarrow 6\ CO_2 + 6\ H_2O; \qquad \Delta G = -2870\ \text{kJ} \qquad (21\text{-}1)$$

This oxidation takes place in the digestive system, from which energy, as well as nutrients, is transported to other locations in the body. The most important energy carrier is *adenosine triphosphate (ATP)* (Figure 21.39). When one mole of this compound loses the terminal phosphate group (P_i), the product is called *adenosine diphosphate (ADP)*. The loss of a phosphate group takes place in an exergonic hydrolysis reaction, and about 30 kJ of free energy becomes available. The reaction, shown in structural formulas in Figure 21.39, is usually written in shorthand form.

P_i stands for "inorganic phosphate." It can be $H_2PO_4^-$ or HPO_4^{2-}.

$$\text{ATP} + H_2O \longrightarrow \text{ADP} + P_i; \qquad \Delta G = -33\ \text{kJ} \qquad (21\text{-}2)$$

Coupled Reactions

Although the two processes take place in the same cell, the energy released in ATP hydrolysis is not immediately available for driving another, endergonic reaction: the reactions must be *coupled* in some closer way. One form of coupling is through a common intermediate. A typical endergonic reaction is the formation of the ester of a carboxylic acid.

Values vary from one ester to another, but 20 to 25 kJ is typical.

$$\text{R—COOH} + \text{HO—R}' \longrightarrow \text{R—COO—R}' + H_2O; \qquad \Delta G = 23\ \text{kJ} \quad (21\text{-}3)$$

Through coupling, the reaction can proceed by a mechanism in which each step is exergonic, that is, spontaneous.

$$(1) \qquad \text{R—COOH} + \text{ATP} \longrightarrow \text{R—CO—ADP} + P_i; \qquad \Delta G < 0\ \text{kJ} \qquad (21\text{-}4)$$

$$(2) \qquad \text{R—CO—ADP} + \text{HO—R}' \longrightarrow \text{R—COO—R}' + \text{ADP}; \qquad \Delta G < 0\ \text{kJ} \qquad (21\text{-}5)$$

$$\textbf{Overall} \quad \text{R—COOH} + \text{HO—R}' + \text{ATP} \longrightarrow \text{R—COO—R}' + \text{ADP} + P_i; \qquad \Delta G = -10\ \text{kJ} \quad (21\text{-}6)$$

The overall reaction is the same as if the two processes in Equations (21-2) and (21-3) had occurred independently. The participation of the common intermediate R—CO—ADP *couples* the reactions and allows all steps to proceed with $\Delta G < 0$. The mechanism given above is an idealization: in living systems, such reactions may involve numerous steps, all of which are catalyzed by different enzymes. Figures 21.40 and 21.41 show a convenient way to symbolize coupled reactions.

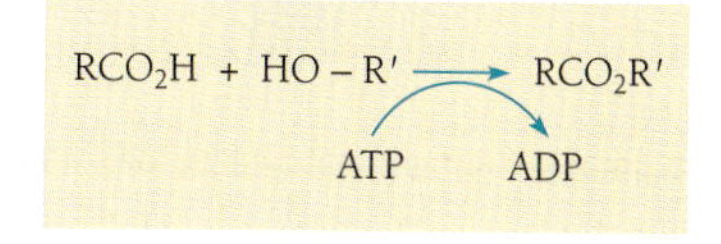

Figure 21.40 Coupled reactions are usually shown in symbolic form, without numerical values of ΔG.

The ATP consumed in cellular metabolism is produced during the oxidation of glucose. This is accomplished in a complex sequence of coupled reactions whose overall effect is the production of 36 molecules of ATP for every molecule of glucose consumed.

$$C_6H_{12}O_6 + 6\ O_2 \longrightarrow 6\ CO_2 + 6\ H_2O; \qquad \Delta G = -2870\ \text{kJ}$$

$$36\ \text{ADP} + 36\ P_i \longrightarrow 36\ \text{ATP} + 36\ H_2O; \qquad \Delta G = +1200\ \text{kJ}$$

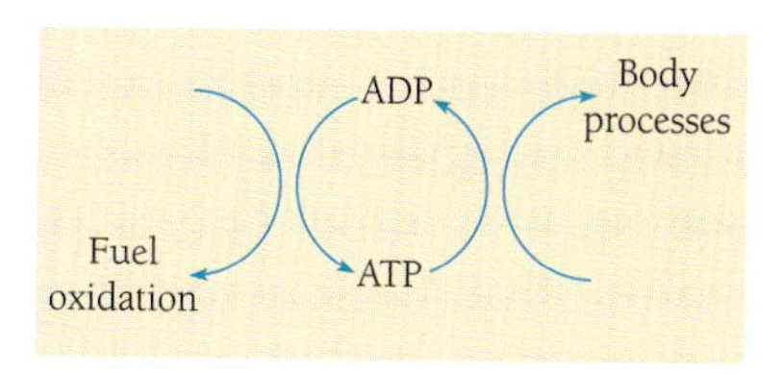

Figure 21.41 Energy released in the oxidation of fuels is stored and transported in the ADP/ATP cycle.

It is interesting to note that $1200/2870 \approx 40\%$ of the available energy is stored in ATP for later use. The remainder appears as heat necessary to maintain body temperature at the optimum value. The efficiency of energy conversion compares favorably with that in other machines such as automobile engines, which typically convert no more than 5% of the energy of gasoline into motive power. The first 10 steps of the glucose oxidation mechanism, which together produce 2 molecules of ATP, take place in the absence of O_2. This sequence is known as *anaerobic glycolysis,* and it operates during intense muscle activity when the supply of oxygen is insufficient for the complete oxidation to CO_2 and H_2O.

Photosynthesis

Regardless of how it is used, or by which organism, the story of biochemical energy begins with **photosynthesis**—the fixation of energy from the sun. Most of this activity is carried out in marine algae, although higher green plants make a significant contribution. The overall reaction of photosynthesis is

$$6\ CO_2 + 6\ H_2O \xrightarrow{\textbf{sunlight}} C_6H_{12}O_6(\text{glucose}) + 6\ O_2$$

One way to describe this reaction is as a reduction of carbon dioxide and an oxidation of water (to O_2). When plants and animals metabolize glucose in order to utilize the stored solar energy, the reverse reaction, the oxidation of glucose (to CO_2) and the reduction of oxygen (to H_2O), takes place.

The mechanism of photosynthesis is complicated, and much remains to be learned about the details of electron transport in the various redox steps. The process takes place in several stages, and, not surprisingly, the ADP/ATP reaction is intimately involved. In the primary stage, sunlight is absorbed by a light-sensitive pigment. The energy is used in a sequence of reactions whose overall effect is the conversion of ADP to ATP and the oxidation of water in the half-reaction

$$2\ H_2O \longrightarrow O_2 + 4\ H^+ + 4\ e^-$$

In the secondary stage, which can occur in the dark, carbon dioxide is reduced and assembled into a glucose molecule.

The study of photosynthesis is today an active field of research, not only for its intrinsic importance but also because it may lead to the discovery of artificial reactions that trap solar energy in similar ways. Researchers have made some progress along these lines, using inorganic catalysts or photosensitive electrochemical cells to oxidize water.

SUMMARY AND KEYWORDS

Biochemistry is the study of compounds and reactions that occur in living organisms. **α-Amino acids** have an $—NH_2$ group and a $—CO_2H$ group bonded to the same carbon atom. Both these groups are ionized, forming a **zwitterion,** but the molecule is overall uncharged. Amino acids are **chiral** molecules, which exhibit differences in bond directionality called **optical isomerism,** a form of **stereoisomerism.** An **optically active** substance rotates the plane of polarized light passing through it. **Enantiomers** are mirror-image stereoisomers of one another, like a pair of gloves. Enantiomers rotate the plane of polarized light in opposite directions. An equimolar mixture of enantiomers, a **racemic mixture,** is optically inactive. **Polypeptides** are polymers of amino acids, either synthetic or naturally occurring. **Proteins** are very large natural polypeptides. The **peptide bond** or amide linkage (—CONH—) joins amino acid *residues* in a polypeptide.

The **primary (1°) structure** of a protein is its sequence of amino acids. The **secondary (2°) structure,** determined primarily by hydrogen bonding between peptide bonds, is the configuration of the polymer chain(s). Secondary structure may involve a single molecule, as in the **α-helix,** or numerous molecules, as in the **pleated sheet.** The **tertiary (3°) structure,** the way in which an α-helix is twisted, bent, or coiled, is determined by specific interactions between amino acid residues. These interactions can include covalent bonds (—S—S— links) as well as electrostatic and dipole forces. Biochemical reactions are catalyzed by **enzymes,** proteins whose 3° structure produces a channel or cleft called the **active site.** The **substrate** (reactant molecule) binds to the active site and undergoes reaction, after which the product(s) is/are released.

Monosaccharides such as **glucose** are polyhydroxy aldehydes or ketones having five to six carbon atoms; their *cyclic* form predominates. Two monosaccharides can undergo condensation to form a **disaccharide** such as **sucrose** (table sugar: glucose + fructose). The link between monosaccharides is either an **α**- or a **β-glycoside linkage,** depending on the configuration at C-1 of the linkage. Mono-, di-, and **oligosaccharides** are called **sugars. Polysaccharides** are used in nature for energy storage and structural strength. *Cellulose* is poly (β-glucose) in which long, relatively *unbranched* chains lie parallel, held together in a secondary structure by hydrogen bonding. There are several naturally occurring forms of poly (α-glucose). *Amylose,* having up to about 1000 glucose units, is relatively small and unbranched; *amylopectin* is larger and branched; *starch* is a mixture of these two polymers. *Glycogen,* used by animals for energy storage, is highly branched and has perhaps 30,000 monomers. The α-glycoside linkage is easily hydrolyzed (addition of water with consequent bond rupture) to give monosaccharides, while the β-linkage is not. Because of this difference, starch is a digestible food and cellulose is not. Mono-, di-, oligo-, and polysaccharides are collectively called **carbohydrates.**

A **nucleotide** consists of a phosphate group, a sugar, and a nitrogen-containing base; a **nucleoside** contains only a sugar and a base. *Adenine* and *guanine* are **purine bases,** while *cytosine, uracil,* and *thymine* are **pyrimidine bases.** The nucleic acids **RNA (ribonucleic acid)** and **DNA (deoxyribonucleic acid)** are polynucleotides having one of these bases bonded to each sugar along a sugar-phosphate backbone. The sequence of bases is the **primary structure.** The **secondary structure** of DNA, the *double helix,* consists of two DNA molecules held together by multiple H-bonding between the *complementary base pairs* G≡C and A=T. The information inherent in the base sequence of DNA, when transferred to *m*-RNA, spells out a message in a four-letter alphabet {A, C, G, U}. The message gives instructions for the synthesis of proteins, and each three-letter word **(codon)** directs the addition of a particular amino acid to a growing polypeptide chain. The correspondence between codons and the 20 amino acids used in protein synthesis is called the **genetic code.** The integrity of the message is preserved during cell division by *replication,* in which a DNA molecule gives rise to two identical "daughter" molecules.

Fat-soluble biomolecules are called **lipids.** *Glycerol* forms triple esters **(triglycerides)** with 14- to 22-carbon unbranched carboxylic acids **(fatty acids). Fats** (triglycerides of animal origin) are higher melting and contain more *saturated* fatty acids. **Oils** (triglycerides of plant origin) are lower melting and contain more *unsaturated* fatty acids. **Waxes** are esters formed from fatty acids and long-chain alcohols. *Saponification* (treatment with aqueous base) of triglycerides gives **soap,** a mixture of glycerol and the alkali salts of fatty acids. A greasy dirt particle becomes soluble when the nonpolar **(hydrophobic)** ends of soap anions penetrate it, leaving the **hydrophilic** (water-soluble) carboxylate ends as a surface coating. A type of **phospholipid** found in cell membranes is a triglyceride containing a phosphate ester and two fatty acids. All **steroids** have a common structure containing four fused rings, but differ in the substituents on the rings. Many hormones are steroids.

Cellular **metabolism** includes **anabolism** (synthesis) and **catabolism** (breakdown). Free energy supplied in the **exergonic** oxidation of glucose (derived from carbohydrates in the diet) is stored in the energy-rich *adenosine triphosphate (ATP)* molecule. Energy is available as needed from the conversion of ATP to ADP: this reaction is *coupled* through a common intermediate to a variety of **endergonic** reactions. The ultimate source of biochemical energy is **photosynthesis,** in which sunlight is absorbed by green plants and used to drive the endergonic conversion of carbon dioxide and water to glucose.

PROBLEMS

Optical Activity

1. What is polarized light, and what is meant by the term "optical activity"?

2. Distinguish between the following types of isomerism: structural, geometrical, and optical.

3. Define the following terms: "chiral molecule or compound," "chiral carbon atom," "enantiomers," "racemic mixture."

4. What structural feature distinguishes optically active from optically inactive compounds?

5. Which of the following can exist as a pair of enantiomers?
   a. CHBrClF b. HClC=CClH
   c. CCl_2F_2 d. $CH_3—CH(NH_2)—CO_2H$

6. Which of the following can exist as a pair of enantiomers?
   a. $CHClFCH_3$ b. $H_2C(NH_2)—CO_2H$
   c. $H_2C=CH—CClBrCH_3$ d. $H—C(=O)—C(H)(OH)—CH_3$

7. How many chiral carbon atoms are there in 2-bromobutane? How many optical isomers of this molecule are there?

8. How many chiral carbon atoms are there in the linear form of glucose? How many optical isomers of this molecule are there?

9. How many chiral carbon atoms are there in the cyclic form of glucose? How many optical isomers of this molecule are there?

10. Draw the structure of cholesterol, and indicate each chiral carbon with an asterisk. How many optical isomers of this molecule are there?

11. Sketch the structures of the enantiomers of alanine.

12. All but 1 of the 20 amino acids listed in Figure 21.1 exist as pairs of enantiomers. Identify the one that does not, and explain why not.

Proteins

13. What type(s) of interaction can occur between the side-chains of glutamic acid and serine?
14. What type(s) of interaction can occur between the side-chains of glutamic acid and arginine?
15. Draw the structure of a typical α-amino acid. What do you think the structure of a β-amino acid might be?
16. What type of reaction results in the formation of a dipeptide from two amino acids? Illustrate with a balanced chemical equation.
17. Draw the structural formulas of the two dipeptides that can be formed from alanine and serine.
18. Draw the complete structural formula of a tripeptide containing three different amino acids.
19. How many different tripeptides can be formed from aspartic acid, serine, and tryptophan? Write their primary structures in the abbreviated form (asp—ser—trp, for instance).

*20. Consider glutamic acid, an amino acid having an acidic side-chain. Draw a structural formula indicating the form you expect this substance to have in basic solution, say when the pH is 10 or greater. What is its structure in acidic solution?

21. Consider lysine, an amino acid having a basic side-chain. Draw a structural formula indicating the form you expect this substance to have in strongly acid solution, say when the pH is 3 or less. What is its structure in basic solution?
22. Aqueous solutions of amino acids are buffers, that is, they resist change in pH on the addition of either strong acid or strong base. Explain.
23. What is an essential amino acid? How many are there?
24. How many amino acids are used by living organisms in the synthesis of proteins?
25. Name some functions performed by proteins in living organisms.
26. Distinguish between the 1°, 2°, and 3° structure of proteins.
27. What forces are responsible for creating or maintaining the 1° structure of proteins? The 2° structure? The 3° structure?

Carbohydrates

28. What is a carbohydrate, where does the name come from, and what functional group(s) is/are found in all carbohydrates?
29. Distinguish between *aldose* and *ketose*.
30. Define the terms *"hexose"* and *"pentose."*

*31. Why are carbohydrates generally soluble in water, but not in nonpolar organic solvents?

32. What is a glycoside? What is a disaccharide?
33. What is the difference between the α and β forms of a carbohydrate?
34. Which of the following do(es) *not* have α and β forms: linear monosaccharides, cyclic monosaccharides, disaccharides?

*35. Look at Figure 21.15, and explain why amylose (a component of starch) is sometimes described as a polymer of maltose rather than glucose.

*36. Look at Figure 21.28, and explain why cellulose is sometimes described as a polymer of cellobiose rather than glucose.

*37. How is the structure of starch different from that of cellulose?

*38. How is the structure of glycogen different from that of starch?

Nucleic Acids

39. List the components of the two types of nucleic acid.
40. Distinguish between a nucleotide and a nucleoside.
41. Distinguish between the 1° and 2° structure of DNA.
42. What is the relationship between the primary structures of proteins and nucleic acids?
43. Suppose one strand of a DNA double helix has a segment with the base sequence —GGACA—. What must be the sequence on the opposite, complementary strand?

*44. Estimate the total number of letters of the alphabet in this book. If each letter corresponds to a nucleotide of average molar mass 330 g mol^{-1}, what would be the molar mass of a DNA molecule containing this number of nucleotides?

Lipids

45. How are lipids defined?
46. Distinguish between fats and oils.
47. What is meant by the terms *"saturated"* and *"polyunsaturated"* when applied to fats and oils?
48. Distinguish between a fat and a fatty acid.
49. What is a triglyceride? What classes of naturally occurring substances are triglycerides?
50. To what class of organic compounds do waxes belong?
51. Why are steroids classified as lipids even though they are not triglycerides?
52. Distinguish between soap and detergent.
53. What is meant by the term "hydrophobic"? Give some examples of hydrophobic substances.
54. What is meant by the term "hydrophilic"? Give some examples of hydrophilic substances.
55. What distinguishes a phosphoglyceride from a fat or oil?
56. Draw the structural formula of glycerol.

*57. What products would be expected from the saponification of one mole of a phosphoglyceride?

Bioenergetics

58. Distinguish between *metabolism, catabolism,* and *anabolism.*

59. What is the difference between *exothermic* and *exergonic?* Is it possible for a reaction to be both exothermic and endergonic?

60. The energy stored in ATP cannot be used for driving endergonic reactions simply because energy release and energy consumption take place in the same cell or vicinity. Explain why not.

61. How many moles of ATP must be converted to ADP in order to drive a reaction whose free energy change is +962 kJ per mole of reaction?

*62. Suppose that, in Problem 61, 42 moles of ATP are converted to ADP in the actual process. What would be the efficiency of the mechanism? That is, what percent of the energy released by ATP is used to drive the metabolism, and how much is dissipated as heat?

*63. A second phosphate group can be lost from ADP, forming adenosine monophosphate (AMP): the reaction ADP + $H_2O \longrightarrow$ AMP + P_i is exergonic by about 30 kJ. How many moles of ATP must be converted to AMP in order to drive a reaction that requires 672 kJ, if the process is 62.2% efficient?

CHAPTER 22

THE MAIN-GROUP ELEMENTS

Elemental silicon has a shiny, almost metallic appearance.

22.1 INTRODUCTION

The **main-group** or **representative elements** are those designated "A" in the periodic table. They have *s*, or *s* and *p*, electrons in their valence shells, and for this reason are sometimes subdivided into the *s-block* (Groups 1A and 2A) and the *p-block* (Groups 3A through 8A) elements. The chemistry of some of these elements has already been discussed: Group 1A in Chapter 3, Group 2A in Chapter 4, the Group 3A metals in Chapter 5, Group 7A in Chapter 7, and Group 8A in Chapter 2. The chemistry of oxygen is sufficiently important to have received special attention in Chapter 8. An introductory study of main-group chemistry should include these sections as well as the material in this chapter.

Various aspects of the organic chemistry of carbon are covered in Chapters 9, 20, and 21.

All the elements in Groups 1A (except H) and 2A are metals, and most of the remaining main-group elements are nonmetals. Inasmuch as metallic character increases with atomic number within a group (Chapter 7), some of the elements in Groups 3A through 6A are metalloids or metals. Only the nonmetals and metalloids among the main-group elements are discussed in this chapter. The discussion is organized according to group number, in order to make use of the similarity in chemical behavior among the members of a given group.

22.2 BORON

Boron has two stable isotopes: B-10 (19%) and B-11 (81%). Although it is rare (0.0003% of the earth's crust), boron is relatively easy to obtain from the mineral deposits that remained after certain ancient lakes dried up. Some of these are almost pure *borax*, $Na_2B_4O_7 \cdot 10H_2O$. Large-scale mining of borax took place during the 1880s in Death Valley, California, but more recently other sites in the American Southwest have been used (Figure 22.1). Purification after mining is a simple matter of dissolving the borax in hot water, and separating the solution from the insoluble

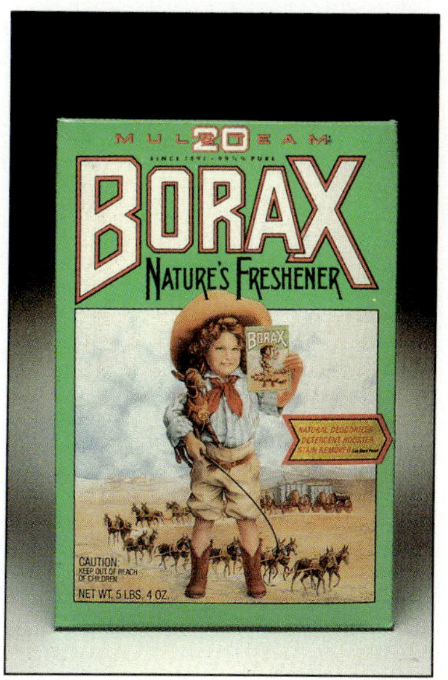

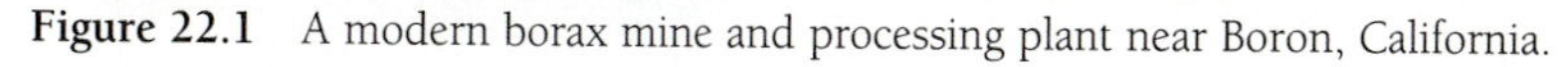
Figure 22.1 A modern borax mine and processing plant near Boron, California.

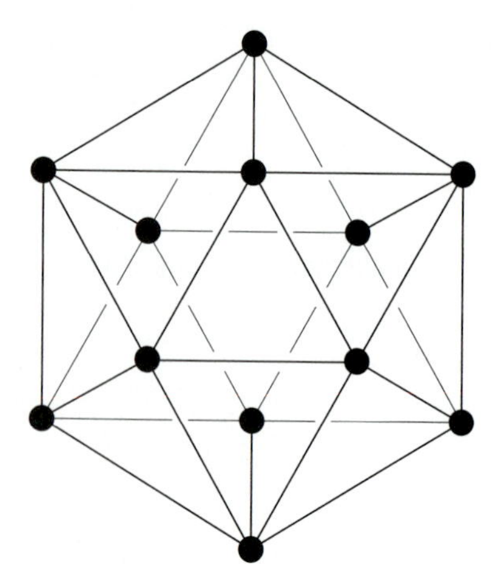

Figure 22.2 Several allotropic forms of boron contain 12 atoms at the vertices of an icosahedron, a 20-sided figure. The lines in this drawing represent geometrical relationships, not conventional 2-electron bonds: the bonding in boron and its compounds is complex and varied. Similar structural units occur in compounds of boron with hydrogen and carbon.

impurities. Evaporation of the water leaves relatively pure borax. Treatment of borax with sulfuric acid yields boric acid (H_3BO_3), which is converted to boric oxide (B_2O_3) by heating. An amorphous solid material containing 95% to 98% B can be obtained by reduction of boric oxide with magnesium.

$$B_2O_3(s) + 3\ Mg(s) \xrightarrow[\Delta]{} 2\ B(s) + 3\ MgO(s) \qquad (22\text{-}1)$$

Removal of impurities (chiefly oxides and borides) from this solid is very difficult. Small quantities of pure boron are prepared by thermal decomposition of BCl_3, BBr_3, or BI_3 in a hydrogen atmosphere.

Pure boron is a hard (almost as hard as diamond), crystalline, high-melting substance having very low electrical and thermal conductivity. It is a structurally complex material that has several allotropic forms. The fundamental structural element is an *icosahedron,* a geometrical figure having 20 triangular faces (Figure 22.2).

Boron and its compounds are widely used materials. Certain alloys and polymers used in jet engines are strengthened by the addition of fibers of elemental boron. Borax, a slightly alkaline compound, is added to detergents for pH control, and sodium perborate ($NaBO_3$) is a mild but effective bleach (Clorox II). Dilute solutions of boric acid are used as mild antiseptics and antifungal agents; boron nitride and boron carbide are used as abrasives; boric oxide is a raw material in the manufacture of glass. Sodium borohydride ($NaBH_4$), a good reducing agent, and BF_3, one of the strongest Lewis acids known, are important reagents in the synthesis of many organic and inorganic compounds.

Electron-Deficient Bonding

Electron-deficient bonding in boron trifluoride is discussed on page 326.

The ionization energy of boron is high enough so that it does not form ionic compounds as B^{3+}, as might be expected of a Group 3A element. Instead, boron reacts with nonmetals to form binary molecular compounds. Many of these are *electron deficient.* In the trihalides, for example, the boron atom has only six electrons in its valence shell.

$$\text{Cl:B}\begin{matrix}\cdot\cdot\text{Cl}\\ \cdot\cdot\text{Cl}\end{matrix}$$

Boron trichloride is a trigonal planar molecule, so that the hybridization of the boron atom is sp^2. As a consequence of their electron deficiency, boron trihalides are good electron acceptors (Lewis acids). They form adducts with compounds having lone pairs, such as ammonia derivatives and ethers. These adducts have tetrahedral geometry around the boron atom.

$$BF_3(g) + :NF_3(g) \longrightarrow F_3B:NF_3(s) \qquad (22\text{-}2)$$

$$BF_3 + R{-}\ddot{\underset{\cdot\cdot}{O}}{-}R \longrightarrow F_3B{:}O{:}\begin{matrix}R\\ R\end{matrix} \qquad (22\text{-}3)$$

Boron trihalides are volatile substances that are very reactive toward water, and in moist air they are converted to boric acid and the hydrohalic acid.

$$BBr_3(\ell) + 3\ H_2O(\ell) \longrightarrow H_3BO_3(aq) + 3\ HBr(aq) \qquad (22\text{-}4)$$

Oxygen Compounds of Boron

Treatment of borax with concentrated sulfuric acid gives boric acid, a moderately soluble solid.

$$Na_2B_4O_7 \cdot 10H_2O(s) + H_2SO_4(aq) \longrightarrow Na_2SO_4(aq) + 5\,H_2O + 4\,H_3BO_3(s) \quad (22\text{-}5)$$

The empirical formula of boric acid is usually written H_3BO_3 in order to emphasize that its solutions are acidic. However, its structure is better represented by $B(OH)_3$, because it is a trihydroxy compound with *no* ionizable hydrogens. That is, it is not a Brønsted acid. Boric acid is a monoprotic *Lewis acid* that ionizes by forming a coordinate bond with a hydroxide ion obtained from a water molecule.

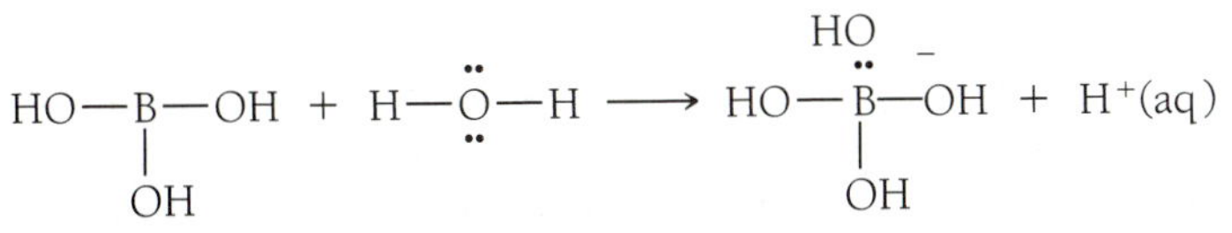

$$HO{-}B(OH){-}OH + H{-}\ddot{\underset{\cdot\cdot}{O}}{-}H \longrightarrow [HO{-}B(OH)_2{-}OH]^- + H^+(aq) \quad (22\text{-}6)$$

Dehydration of boric acid by heating leaves boric oxide (B_2O_3), a hard, glassy solid.

$$2\,H_3BO_3(s) \xrightarrow[\Delta]{} B_2O_3(s) + 3\,H_2O(g) \quad (22\text{-}7)$$

The structure of oxygen-containing compounds of boron is complex and varied, and includes both sp^2 and sp^3 hybridization. The anion in $Na_2B_4O_7 \cdot 10H_2O$, for instance, is $B_4O_5(OH)_4^{2-}$, in which two of the boron atoms (sp^3) are shared by six-membered rings. The hybridization of the other two boron atoms is sp^2.

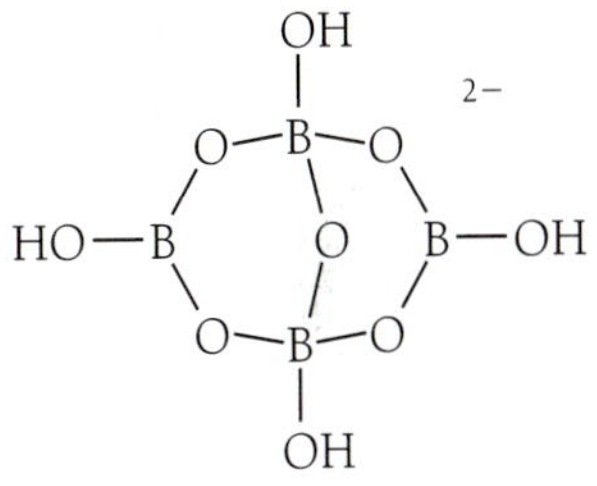

$B_4O_5(OH)_4^{2-}$ ion

Boranes

Binary compounds of boron with hydrogen are called **boranes.** The simplest of these is *diborane,* B_2H_6, and about 20 others are known. The bonding in diborane (and in the higher boranes as well) is unusual, because there are not enough valence electrons to form conventional 2-electron covalent bonds between all the atoms. Instead, an electron pair is shared among *three* atoms, resulting in a **three-center** or **bridge bond.** Curved lines are used in the structural formulas to represent three-center bonds. In many of the higher boranes, for instance *decaborane* ($B_{10}H_{14}$), the boron atoms occupy some or all of the vertices of a partial or complete icosahedron (such as in Figure 22.2). They are held together by several different types of three-center bonds, and they also include some boron atoms bonded to *five* other atoms. Clearly, Lewis dot structures are not adequate to describe these compounds. Boranes in which one or more of the boron atoms have been replaced by carbon are called *carboranes.*

The simplest borane is actually BH_3, but it is highly reactive and cannot be isolated as a pure compound.

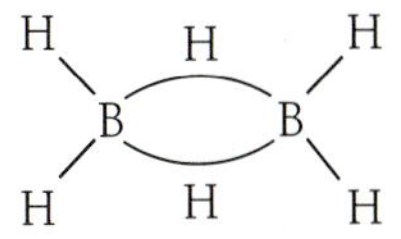

Diborane

Other Boron Compounds

Binary compounds of boron with metals are called **borides.** These exhibit a diversity of structure, bonding, and properties, but most are hard, high-melting, and chemically unreactive. Borides with metal-boron ratios from M_4B to MB_{12} are known.

Boron nitride (BN) is an extremely hard material whose bonding, structure, and physical properties are similar to those of elemental carbon. It is prepared by the reaction of boron with ammonia at very high temperature (white heat), and exists in a layered, graphite-like structure as well as in diamond-like crystals. It is used as an industrial abrasive. Boron carbide (B_4C), prepared by the high-temperature reduction of B_2O_3 with carbon, is also hard enough to be a useful abrasive.

22.3 CARBON, SILICON, AND OTHER ELEMENTS IN GROUP 4A

Group 4A is probably the most widely used group of elements in the periodic table. All the members of this group are well known, relatively common, and put to use in a variety of ways in our society.

Carbon

Coal consists primarily of carbon ($\approx 80\%$), but contains H, S, O, and N as well.

Carbon is both common and rare: it is widely distributed, but accounts for only 0.08% of the earth's crust. Except for rare deposits of graphite or diamonds, it does not occur naturally in elemental form. Marble, limestone, dolomite, and many other minerals contain the carbonate ion, often as $CaCO_3$ and/or $MgCO_3$. Chalk is almost pure $CaCO_3$. Carbon constitutes a considerable portion of the mass of living matter, and the atmosphere is about 0.03% (by mass) CO_2. There are two stable isotopes, C-12 (98.89%) and C-13 (1.11%). Radioactive C-14 is slowly and continuously formed in the atmosphere. However, it also slowly disintegrates, and its concentration remains at a very low level.

The formation and decay of radioactive carbon in the atmosphere is discussed in Section 24.8.

There are at least two allotropic forms of carbon. *Diamond,* one of the hardest substances known, is a network solid with a cubic crystal structure held together by covalent bonds between sp^3 carbon atoms. Its durability, transparency, and rarity make it a valuable gemstone, and its hardness is utilized in industrial cutting tools and abrasives. Synthetic diamonds are produced from graphite in a high-temperature, high-pressure process. Such crystals are often imperfect and not used as gems (Figure 22.3).

The well-known "sparkle" of diamonds is the result of artful cutting and shaping, which enhances the naturally high reflectivity of crystalline carbon.

(a)

(b)

Figure 22.3 (a) Industrial diamonds are produced in high-pressure furnaces like this. (b) Synthetic diamonds are plentiful, and used primarily as abrasives.

Graphite and diamond are the two allotropic forms of carbon.

Graphite is a soft network solid that has a layered crystal structure (Figure 22.4). Each layer consists of sp^2-hybridized carbon atoms joined by covalent bonds in a hexagonal network. Adjacent layers are held together weakly, by van der Waals forces. This layered structure results in some unusual physical properties. The weak forces between layers are easily overcome, so that layers can easily slide past one another (Figure 22.5). In consequence, graphite is a good "dry" lubricant, usable at very high temperatures. Electrons move easily within a layer, but cannot easily jump to an

Much like the individual sheets in a stack of paper, graphite layers can easily slide past one another.

(a)

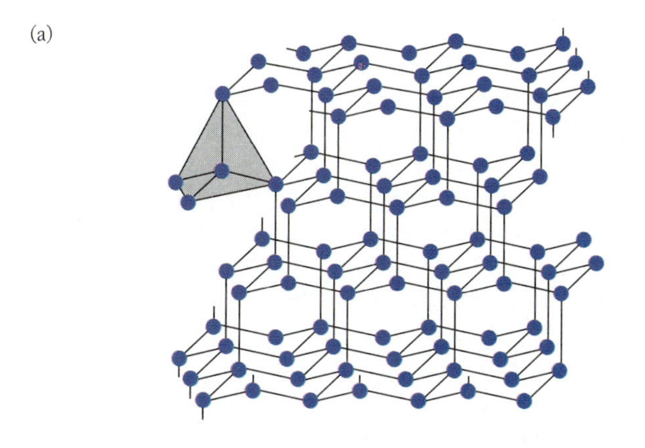

(b)

335 pm

142 pm

Figure 22.4 Crystal structures of elemental carbon: (a) diamond and (b) graphite. Both forms are network solids.

Figure 22.5 "Lead" pencils actually consist of powdered graphite, held together by a small amount of added glue. Only the slightest hand pressure is required to transfer a trace of the soft graphite to paper.

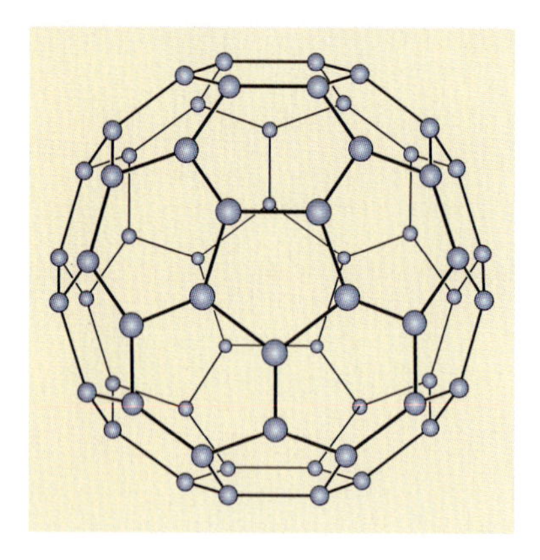

Figure 22.6 One of many possible "fullerenes," large clusters of carbon atoms bonded in a spherical, soccer-ball structure.

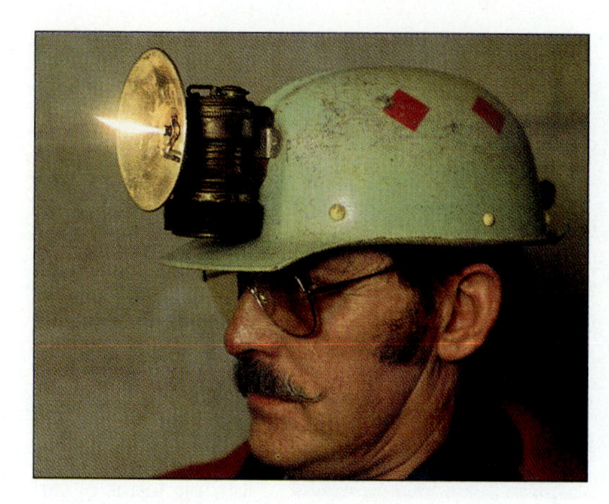

Figure 22.7 The combustion of acetylene in headlamps provides light for cave explorers. The acetylene is produced by reaction of CaC_2 with H_2O.

adjacent layer. Therefore, a graphite crystal is a good conductor of electricity and heat *parallel* to the layers, but it is a poor conductor perpendicular to them. Graphite electrodes are used in electrochemical cells, for instance in the production of aluminum (page 832).

High-temperature degradation of some organic polymers (polyesters, for example) produces strong graphite fibers. When imbedded in other polymers, these fibers form a *composite* material used in tennis racquets, spacecraft, and other applications requiring strength and light weight. Charcoal and soot contain very small crystals (*microcrystals*) of graphite. Soot formed by electrical heating of a graphite rod in a helium atmosphere contains large *cluster molecules* as well as microcrystalline graphite. These species have 40 to 80 C atoms arranged in a partial or complete spherical structure of linked pentagons and hexagons, much like the surface of a soccer ball (Figure 22.6).

These carbon cluster molecules are called "fullerenes." Buckminster Fuller is the architect who designed the geodesic dome which these molecules so strongly resemble.

Carbides

At high temperatures, direct combination of carbon with most metals produces **carbides.** Some of these, chiefly those formed with Groups 1A and 2A metals, are ionic compounds containing either the C^{4-} or C_2^{2-} ion. Compounds of the C^{4-} ion react with water to produce methane [Equation (22-8)],

$$Be_2C(s) + 4\ H_2O(\ell) \longrightarrow 2\ Be(OH)_2(s) + CH_4(g) \qquad (22\text{-}8)$$

while the product of C_2^{2-} hydrolysis is acetylene [Equation (22-9); Figure 22.7].

$$CaC_2(s) + 2\ H_2O(\ell) \longrightarrow Ca(OH)_2(s) + H{-}C{\equiv}C{-}H(g) \qquad (22\text{-}9)$$

A crystal of silicon carbide, also called carborundum.

With elements of similar electronegativity, carbon forms covalent carbides: SiC crystallizes in the diamond structure, with every other C atom replaced by Si. It shares many properties with diamond, and is a useful abrasive. Boron carbide has been mentioned (Section 22.2).

Binary compounds of carbon with other nonmetals are not classified as carbides, but are given other names. For example, the Group 7A elements are called halides, not carbides. *Carbon tetrachloride,* produced industrially by the reaction of methane with an excess of chlorine, was widely used as a solvent before its toxicity was fully appreciated.

Dry cleaners now use other grease-dissolving liquids, such as CH_3CCl_3, that do not share the toxic properties of carbon tetrachloride.

$$CH_4(g) + 4\,Cl_2(g) \longrightarrow CCl_4(\ell) + 4\,HCl(g) \qquad (22\text{-}10)$$

Carbon tetrachloride is an important raw material for the production of *CFC's* (chlorinated fluorocarbons), which are used in refrigerators and air conditioners.

CFC's ultimately reach the stratosphere, where absorption of ultraviolet light causes them to release Cl atoms. Atomic chlorine reacts with ozone by the mechanism discussed in Section 14.8.

$$CCl_4(\ell) + HF(g) \longrightarrow CCl_3F(g) + HCl(g) \qquad (22\text{-}11)$$

The widespread use of CFC's as refrigerant gases poses a threat to the earth's ozone layer.

Oxides

Carbon forms several oxides: CO, CO_2, C_3O_2 *(carbon suboxide),* and a few other rare and unstable compounds. *Carbon monoxide* is a colorless, odorless gas that is only slightly soluble in water. It reacts with aqueous hydroxide ion to produce the formate ion, the anion of formic acid (HCO_2H).

$$CO(g) + OH^-(aq) \longrightarrow HCO_2^-(aq) \qquad (22\text{-}12)$$

Carbon monoxide is a Lewis base that can donate electrons to some metals to form *metal carbonyls.*

$$Ni(s) + 4\,CO(g) \longrightarrow Ni(CO)_4(\ell) \qquad (22\text{-}13)$$

Carbon monoxide is infamous for its toxicity.

The mechanism of carbon monoxide poisoning is discussed in Section 23.10.

Carbon dioxide, the major product of the combustion of carbon and its compounds, is the *anhydride* of carbonic acid. That is, it reacts with water to form H_2CO_3.

$$CO_2(g) + H_2O(\ell) \longrightarrow H_2CO_3(aq) \qquad (22\text{-}14)$$

The structure of carbonic acid includes a carbonyl group and two hydroxyl groups.

```
      OH
     /
O=C
     \
      OH
```

Carbonic acid reacts with base to produce *hydrogen carbonates* (bicarbonates) and *carbonates.*

$$OH^-(aq) + H_2CO_3(aq) \longrightarrow HCO_3^-(aq) + H_2O \qquad (22\text{-}15)$$

$$2\,OH^-(aq) + H_2CO_3(aq) \longrightarrow CO_3^{2-}(aq) + 2\,H_2O \qquad (22\text{-}16)$$

Many marine organisms, from microscopic size on up, extract dissolved ions from seawater to produce calcium carbonate for their shells and/or skeletons. Eventually these hard parts settle to the sea bottom, where over eons of time they are converted to limestone and other types of carbonate minerals. The ocean itself is a vast H_2CO_3/HCO_3^- buffer system in which marine organisms, limestone sediments, and atmospheric CO_2 all play a part.

Increased atmospheric concentration of CO_2, caused by human activities, is decreasing the amount of infrared radiation emitted by the earth into space. The result, a slight but significant increase in the average temperature of the atmosphere, will have a noticeable effect on climate in the coming decades. Ultimately, the extra CO_2 in the

Global warming, caused by human introduction of CO_2 and other gases into the atmosphere, is called the greenhouse effect.

atmosphere will be absorbed by the ocean and converted to limestone. However, this happens very slowly, on a geological timescale, and it is not a factor in the short-term solution to the problem of global warming.

Cyanides and Other Anions

Hydrogen cyanide, a weak acid, is a toxic gas with an odor similar to that of crushed almonds. It is thought to have been a prominent constituent of the primordial atmosphere of earth, and has been identified as one of the gases that exists in interstellar space. It may have been important in the origin of life on earth, and even today there are marine organisms ("blue-green algae") that require HCN or its salts in their metabolism. One reaction is particularly suggestive of HCN's possible role in the earth's history: HCN polymerizes to form *adenine,* one of the nitrogen bases of DNA and RNA.

$$5\ HCN(g) \xrightarrow[NH_3,\ H_2O]{} \text{adenine} \qquad (22\text{-}17)$$

NH2, N, N, N, N, H

adenine

Salts of HCN, containing the **cyanide** ion (CN^-), resemble halogen salts in their properties. Because of this similarity, cyanide salts are sometimes called *pseudohalides.*

Cyanide ion is a good Lewis base, and forms Lewis adducts with metal ions. One of these, $Au(CN)_2^-$, is used to extract gold from crushed gold-bearing rock. When the rock is treated with $CN^-(aq)$ in the presence of air, gold leaches out of the rock in the form of the soluble ion $Au(CN)_2^-$. Atmospheric oxygen is required to oxidize the gold from its native oxidation state (0) to that in the complex ion, +1. After the $Au(CN)_2^-(aq)$ is separated from the crushed rock, it is reduced to metallic gold by reaction with powdered zinc.

The chemistry of complex ions such as $Au(CN)_2^-$ is considered in Chapter 23.

Other common inorganic carbon-containing ions include *cyanate,* (OCN^-), *thiocyanate,* (SCN^-), and *cyanamide,* (CN_2^{2-}).

Silicon

Silicon is the second most abundant (26%) element on earth. It crystallizes in the diamond structure, but there is no layered form analogous to graphite. Silicon is usually classified as a metalloid, because it is a semiconductor, but it shares many other properties with metals. Silicon is produced by the high-temperature reaction of carbon with ordinary sand, which is almost pure *silica* (SiO_2).

$$SiO_2(s) + 2\ C(s) \xrightarrow[3000\ °C]{} Si(\ell) + 2\ CO(g) \qquad (22\text{-}18)$$

Very pure silicon, used in the electronics industry for the manufacture of microchips, is obtained in a process that begins with the formation of silicon tetrachloride from silicon of ordinary purity.

$$Si(s) + 2\ Cl_2(g) \longrightarrow SiCl_4(\ell) \qquad (22\text{-}19)$$

After distillation, $SiCl_4$ is reduced by pure magnesium, and the water-soluble magnesium chloride is washed away.

$$SiCl_4(\ell) + 2\ Mg(s) \longrightarrow Si(s) + 2\ MgCl_2(s) \qquad (22\text{-}20)$$

Silicon is further purified by *zone refining*. In this technique, a moveable heating element creates a small zone of liquid in a bar of solid silicon. Impurities are more soluble in the liquid than the solid, so, as the liquid zone moves, the impurities are swept to the end of the bar. Repeated passes of the heater leave a bar of silicon containing less than one impurity atom per *billion* silicon atoms (Figure 22.8).

Covalent Compounds of Silicon

Many compounds of silicon are analogous to organic compounds of carbon. For example, silicon forms alkane-like **silanes** of formula Si_nH_{2n+2}. But there are differences as well. Unlike CH_4, *silane* (SiH_4) ignites spontaneously in air, and reacts briskly with water.

$$SiH_4(g) + 2\ O_2(g) \longrightarrow SiO_2(s) + 2\ H_2O(g) \qquad (22\text{-}21)$$

$$SiH_4(g) + 2\ H_2O(\ell) \longrightarrow SiO_2(s) + 4\ H_2(g) \qquad (22\text{-}22)$$

Only a handful of compounds containing Si═Si double bonds are known, and there are no Si-H compounds analogous to alkenes or alkynes. Although silanes up to Si_6H_{14} have been prepared, only SiH_4 and Si_2H_6 are stable.

The formation of covalent bonds between more than two or three atoms of the same element is called **catenation** (from Latin *catena*, chain). Carbon has this ability to a unique degree: the length of the $-(C-C)_n-$ chain in polyethylene (Section 20.5) is virtually unlimited, while in silanes the chain length is limited to 6 Si atoms. This difference in behavior may be understood in terms of bond energies. The C—C bond energy is 347 kJ mol^{-1} (Table 8.3, page 336), comparable to that of C—Cl(339), C—O(358), and C—H(413). In contrast, the Si—Si bond energy is only 176 kJ mol^{-1}, far less than that of Si—Cl(360), Si—O(368), or Si—H(323). Molecules containing Si—Si bonds, if formed, tend to rearrange into other configurations having stronger bonds.

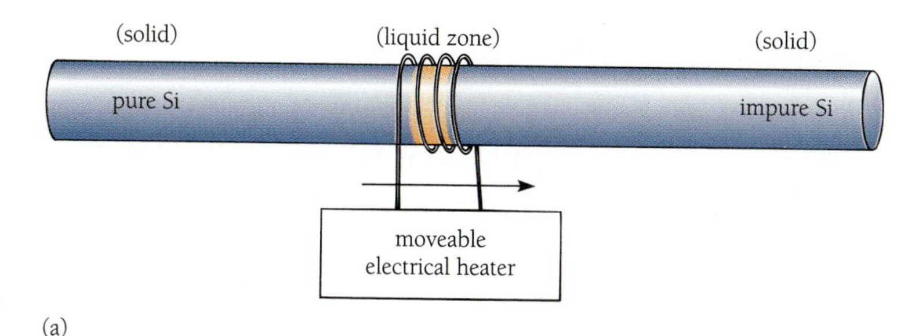

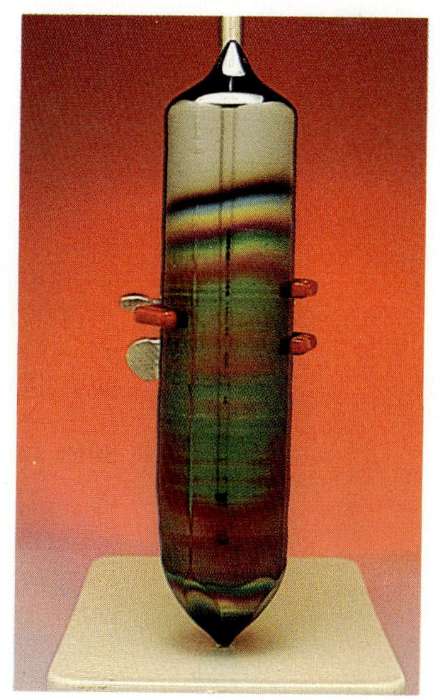

Figure 22.8 (a) In the technique of zone refining, impurities are trapped in the hot liquid zone and swept to the end of the bar, which is later cut off and discarded. (b) An ingot of ultrapure silicon, after zone refining.

Direct combination of silicon with halogens leads to the *tetrahalides,* which may be gaseous, liquid, or solid, depending on molar mass. $SiCl_4$, $SiBr_4$, and SiI_4 react with oxygen-containing molecules such as water to replace the halogen atom with oxygen.

$$SiBr_4(\ell) + 2\ H_2O(\ell) \longrightarrow SiO_2(s) + 4\ HBr(g) \qquad (22\text{-}23)$$

This reactivity is a consequence of the greater energy of the Si—O bond (compared to the energy of the Si—Cl, Si—Br, and Si—I bonds), which makes molecules with Si—O bonds more stable than similar molecules with Si—Cl, Si—Br, or Si—I bonds. The Si—F bond is stronger than that in the other halides, and because of this the reaction of SiF_4 with water [analogous to Equation (22-23)] reaches equilibrium rather than going to completion. SiF_4 reacts with HF to form *fluorosilicic acid,* in which six fluorine atoms are bonded to the silicon atom.

$$SiF_4(g) + 2\ HF(aq) \longrightarrow H_2SiF_6(aq) \qquad (22\text{-}24)$$

There are many compounds of tetravalent (sp^3) silicon with formulas SiR_nX_{4-n} (R is an organic group such as $—CH_3$, $—CH_2CH_3$, $—OCH_3$, and so on, and X is Cl or F). Silicon tetrachloride is a frequently used starting material in the synthesis of such compounds.

$$\underset{}{SiCl_4(\ell)} + 4\ \underset{\textbf{ethanol}}{C_2H_5OH(\ell)} \longrightarrow \underset{\textbf{tetraethoxysilane}}{Si(OC_2H_5)_4(\ell)} + 4\ HCl \qquad (22\text{-}25)$$

These compounds tend to be water-sensitive, reacting to produce SiO_2 as the final product.

Silicone Polymers

Although long-chain silicon analogues of hydrocarbons are not stable, polymers in which Si and O atoms alternate along a chain of indefinite length can be made by reaction of a suitably substituted chlorosilane with water.

$$(CH_3)_2SiCl_2(\ell) + H_2O(\ell) \longrightarrow \!\!+\!Si(CH_3)_2—O\!+_{\!n} + 2\ HCl(g) \qquad (22\text{-}26)$$

These polymers, which have the structure

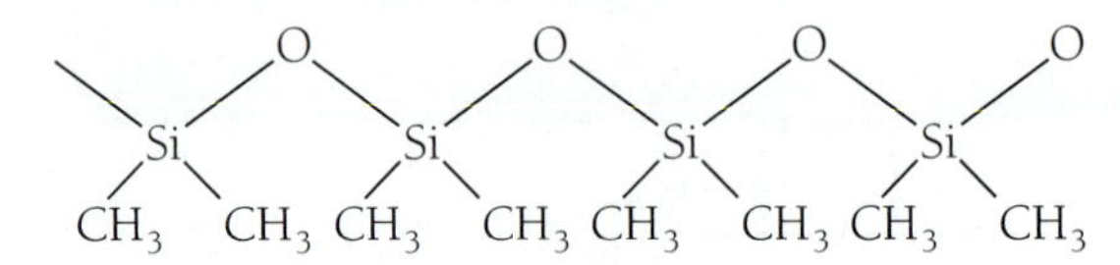

are called **silicones.** Use of different groups in place of the methyl group results in polymers with a wide variety of properties. Most are chemically inert, and stable at higher temperatures than hydrocarbon polymers. Some are used for high-temperature greases and oils, and some products ("Silly Putty" for example) have unusual elastic properties.

Silica and Silicate Minerals

Silicon and oxygen are the two most abundant elements on earth. In combination with other elements, mostly metals, Si and O form over 90% of the earth's crust as silica and the silicate minerals. The earth's mantle, beneath the crust, is believed to be composed chiefly of *olivine,* a mixed iron-magnesium silicate. Microscopic marine organisms called *diatoms* extract silicon compounds from seawater to form skeletons of silica. In some areas, the sediment at the sea bottom and the underlying rock are composed largely of compacted diatom skeletons.

(a)

(b)

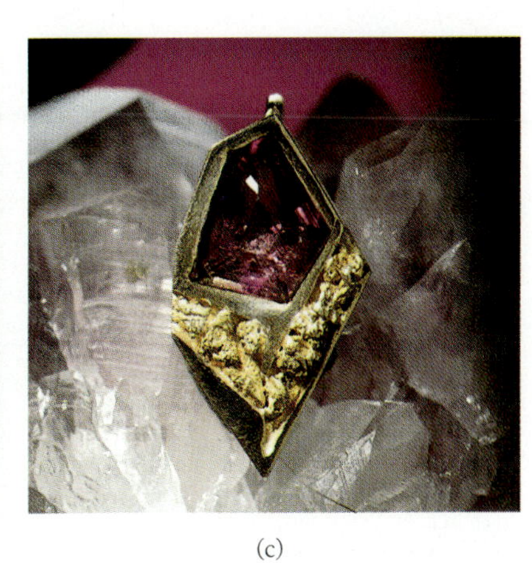

(c)

Figure 22.9 (a) A natural crystal of quartz; (b) amethyst crystals in a geode found in a Brazilian riverbed. This specimen, on display at the Alabama State Museum of Natural History, is 75 cm tall and weighs more than 150 kg; (c) an amethyst brooch on a bed of natural amethyst.

Silica occurs naturally as *quartz* (Figure 22.9), a hard, transparent crystalline substance; as sand, which has a variable composition but contains a large amount of weathered silica; and in about 20 other crystalline forms. Quartz is found as hexagonal crystals, often quite large. Some of these are beautifully colored by the presence of impurities, and are used as gemstones. "Rose quartz," "rock crystal," and "amethyst" are the common names of three such crystals. The structure of quartz consists of long, helical —Si—O—Si—O— chains, reminiscent of the α-helix found in proteins. The presence of helical chains introduces *stereoisomerism* (Section 21.2), and as a result natural quartz is optically active. An individual crystal rotates polarized light either to the right or to the left. In their visual appearance as well, crystals are mirror images of one another and can be separated by careful inspection and sorting.

The optical activity of quartz does not stem from the presence of left- or right-handed *molecules*. It comes instead from the fact that, like a rope or a garden hose, a —Si—O—Si—O— chain can be coiled in two mirror-image ways.

When crystalline quartz is melted and recooled, the resulting solid is a *glass* that does not retain the regularity of a crystal (Figure 22.10). Although an O atom still lies between each pair of Si atoms, the arrangement of Si atoms is random. *Flint* and *obsidian* are natural forms of silica having this random, glassy structure. As pointed out in Section 11.7, glasses lack a definite melting point. Instead of abruptly turning into a liquid, they become progressively softer and less viscous over a broad temperature range. Window glass and other types (for instance the Pyrex used in ovenware) are produced by addition of Na_2CO_3, B_2O_3, and other substances to silica. These addi-

The temperature at which a glass loses its rigidity and can be bent without breaking is called the "softening point." The softening "point" is actually a temperature *range* of a few degrees over which the change occurs.

The common mineral *flint* has a glassy structure, which causes it to break along curved surfaces rather than along the well-defined planes of a crystalline substance. This property, along with its hardness, makes flint a good raw material for cutting tools. These paleolithic spear points were found near Wenatchee, WA.

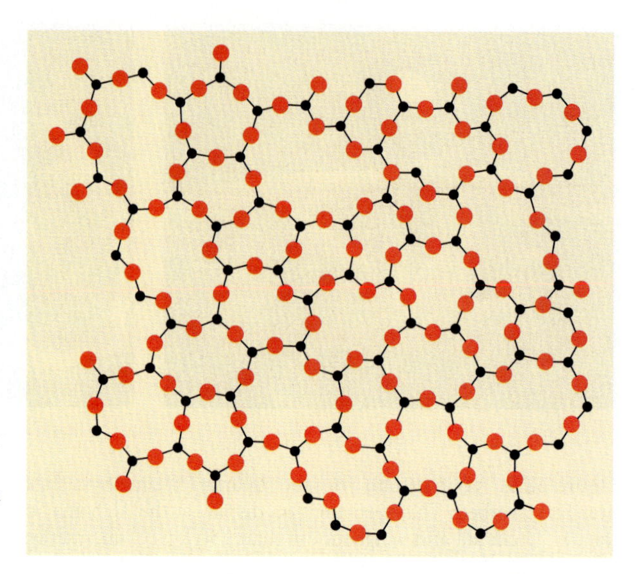

Figure 22.10 When molten quartz freezes, the result is a glass that lacks the regularity of a crystal lattice. Silicon atoms are shown in black, oxygen atoms in red. This is a two-dimensional representation of the bonding in SiO_2, but in the actual material, the network of covalent bonds extends in three dimensions.

The term "heat resistance," applied to glass, means that it does not expand much when the temperature is increased. Thus, heat-resistant glass does not crack when its temperature is suddenly changed.

tives lower the softening point and impart other desirable characteristics, such as heat resistance and color, to the product. Many different recipies are used, and glasses with many different properties are available.

Silicate minerals contain metal ions in addition to Si and O. Familiar names such as asbestos, mica, talc, and clay all describe silicate minerals. The fundamental structural units is SiO_4, in which an sp^3 Si atom occupies the center of a tetrahedron whose vertices are O atoms. The simplest silicates contain the *orthosilicate* ion, SiO_4^{4-}. Sodium orthosilicate and a few other minerals such as Ca_2SiO_4, a major component of cement, are simple ionic compounds of the orthosilicate ion. However, most silicates have more complex structures in which SiO_4 tetrahedra are joined by sharing an oxygen atom at one or more vertices. Each O-atom that is *not* shared by two tetrahedra carries a formal charge of −1. A huge variety of structures can be formed in this way (Figure 22.11).

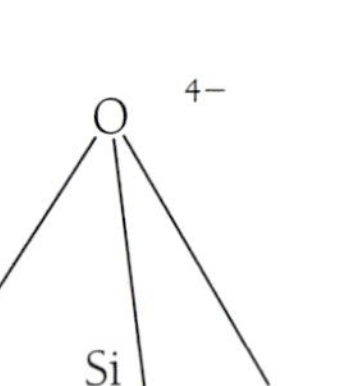

Orthosilicate ion

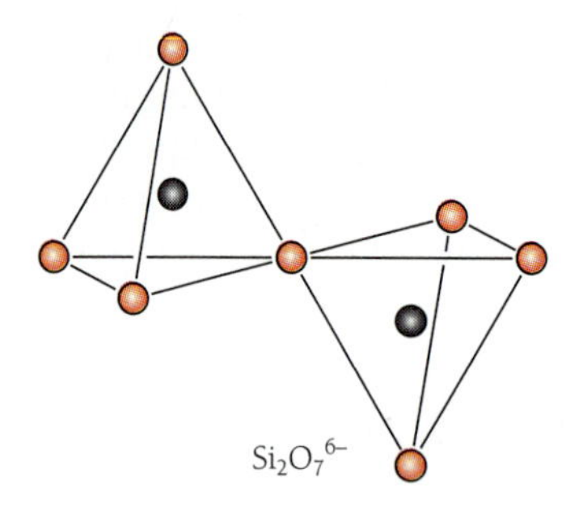

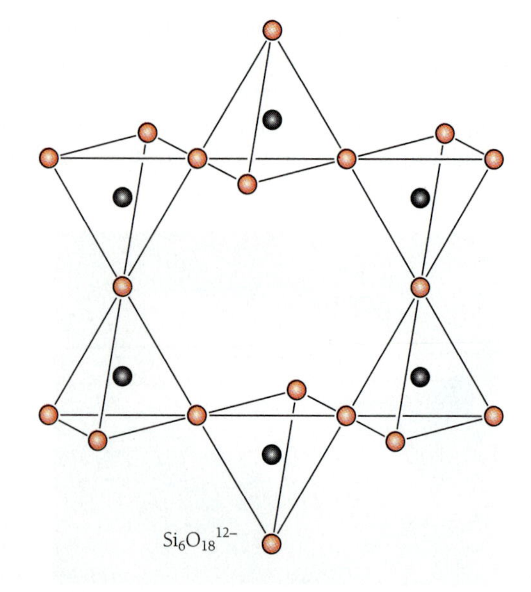

Figure 22.11 Two of the many different structures possible when SiO_4 units are linked together at their vertices.

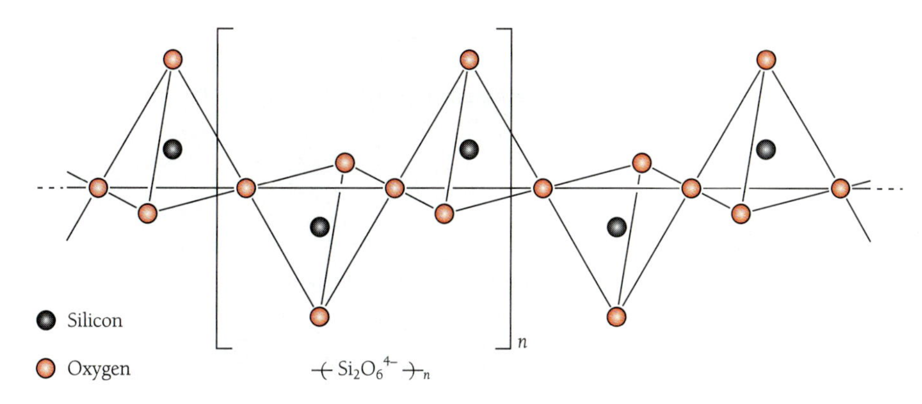

Figure 22.12 The *pyroxene* minerals contain long strands of linked tetrahedra. In this polymeric anion, the repeating unit is $Si_2O_6^{4-}$. The negative charge is balanced by a variety of metal ions.

The *pyroxene* minerals contain long chains of linked tetrahedra, ionically bonded to a variety of metal cations (Figure 22.12). Double chains of tetrahedra, of indefinite length (Figure 22.13), give a fibrous texture to the minerals containing them, the *amphiboles.* Among the amphiboles are the *asbestos* minerals, whose high tensile strength and heat resistance caused them to be widely used in cement, floor tiles, brake linings, fireproofing, and insulation. But inhalation of asbestos fibers over a prolonged period can cause *asbestosis,* a nonmalignant scarring of the lungs. Because workers in the industry are subject to this serious disease, asbestos is no longer widely used. However, there is no direct evidence that living or working in a building having some asbestos-containing constructional materials causes asbestosis.

Asbestos in its natural state has a fibrous structure.

The *micas* contain two-dimensional layers of SiO_4 tetrahedra, with metal ions lying between and separating the layers. Mica can easily be separated into very thin,

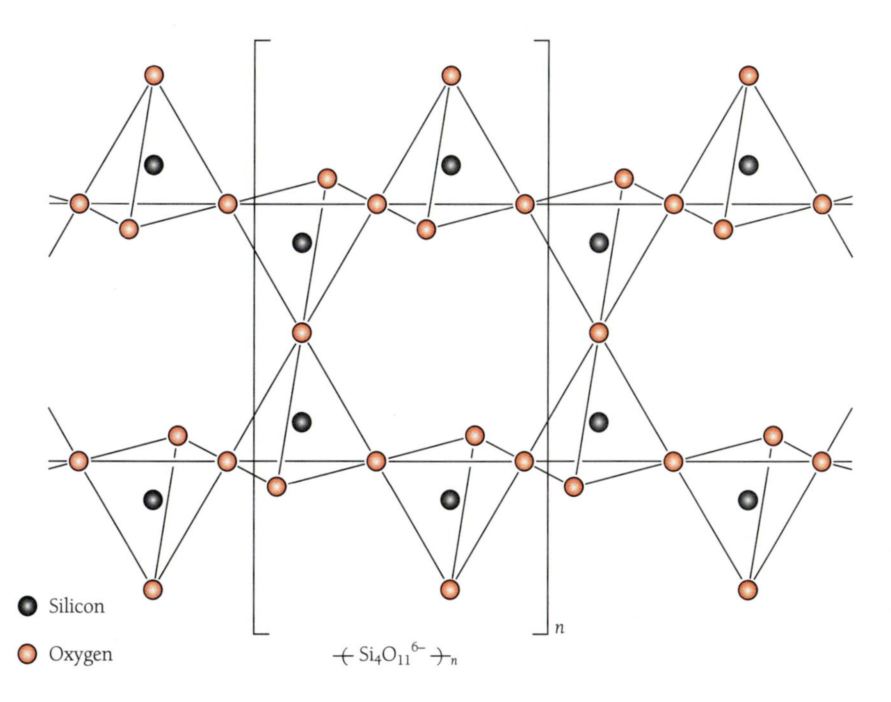

Figure 22.13 Indefinitely long, double chains of silicate tetrahedra occur in the *amphiboles.* The *asbestos* minerals are in this class. The negative charges of the silicate chains are balanced by a variety of +1, +2, and +3 metal ions.

Figure 22.15 Wet clay can be worked into almost any desirable shape. High temperature baking turns clay into a hard, brittle substance.

Figure 22.14 Naturally occurring mica can easily be separated into thin, transparent sheets.

transparent sheets that are used for furnace windows (Figure 22.14). In other substances, the silicate layers can slide easily over one another. This accounts for the slippery feel and lubricating properties of *talc* (soapstone), from which *talcum powder* is made.

Aluminosilicates form another large subclass of the silicate minerals. In these, some of the Si atoms in the structures discussed above are replaced by aluminum. Since the oxidation number of aluminum (+3) is less than that of silicon (+4), an additional source of positive charge is required in order to achieve electrical neutrality in aluminosilicates. Typically, the additional charge is supplied by Group 1A or 2A ions. *Feldspar,* $KAlSi_3O_8$, whose structure contains two-dimensional layers of linked aluminosilicate tetrahedra, is a very common component of the earth's crust. As aluminosilicate rocks are ground into powder by weathering, some of the exposed O-atoms react with water and are converted to —OH groups. These hydrated aluminosilicates include the *clay* minerals, which are an important component of soil.

Clay is the raw material for one of the oldest industries in human history, ceramics. A **ceramic** material is produced by strong heating of moist clay (Figure 22.15). This treatment drives water from the spaces between the layers of linked tetrahedra, leaving a rigid mass of microcrystals glued together by SiO_2 glass. The reddish color of many earthenware objects is due to the presence of impurities, chiefly iron. Ceramics are used for tiles, tableware, porcelain, and, increasingly, as substitutes for metal parts in high-temperature applications. The outer surfaces of space shuttles are coated with ceramic tiles.

When tetrahedra are joined in a three-dimensional array, relatively large cavities can remain in the crystal structure. Metal ions, usually Group 1A or 2A, occupy some of these cavities in the aluminosilicate minerals called **zeolites.** These are porous materials, and when water containing a high concentration of Ca^{2+} and Mg^{2+} ions (hard water) is passed through a sodium-containing zeolite, an *ion-exchange* process takes place. The Na^+ ions in the crystal structure are replaced by Ca^{2+} and Mg^{2+} ions. That is, hard water is converted to soft water that does not form a precipitate with soap. The exchange process is reversible, so that a used zeolite can be "recharged" by running salt water through it for a time. The equilibrium reaction is given by Equation (22-27), in which ZE^{2-} represents a negatively charged site within a zeolite cavity.

The chemistry of soaps, detergents, and hard water is discussed in Section 21.4.

$$Ca^{2+}(aq) + Na^+\cdot\cdot\cdot ZE^{2-}\cdot\cdot\cdot Na^+ \rightleftharpoons 2\,Na^+(aq) + Ca^{2+}\cdot\cdot\cdot ZE^{2-} \quad (22\text{-}27)$$

The equilibrium is driven to the right by hard water, in which the concentration of Ca^{2+} is high, and to the left by salt water, in which the concentration of Na^+ is high. Nowadays synthetic polymers, the *ion-exchange resins,* have displaced natural and synthetic zeolites in many water-softening applications, especially in smaller units that fit directly on water taps.

The Group 4A elements, clockwise from upper right. Silicon, lead, germanium, tin, and carbon. All of these samples have been processed except carbon, which is shown in one of its natural forms: graphite crystals in a quartz matrix.

Germanium, Tin, and Lead

The remaining members of Group 4A are germanium, tin, and lead, whose metallic character steadily increases in that order. All form covalent hydrides analogous to alkanes: the *germanes* GeH_4 through Ge_9H_{20} have been prepared, *stannane* (SnH_4) and *distannane* (Sn_2H_6) are known, while lead forms only *plumbane* (PbH_4). All of the covalent tetrahalides MX_4 (M = Ge, Sn, or Pb; X = F, Cl, Br, or I) have been prepared except $PbBr_4$ and PbI_4. Lead displays its metallic character by forming the ionic *dihalides* PbF_2 and $PbCl_2$. $PbBr_2$ and PbI_2 are also known, but, because of the low electronegativity of Br and I, these compounds have considerable covalent character. Both tin and lead, and to a lesser extent germanium, behave like other metals in forming ionic salts with many common anions: $Sn(NO_3)_2$, $PbSO_4$, and so on. Both tin and lead form salts as +2 cations.

22.4 NITROGEN, PHOSPHORUS, AND THE OTHER GROUP 5A ELEMENTS

Every one of us is in intimate contact with the first two members of this group. Each breath we draw is ≈80% nitrogen, and, as we saw in Chapter 21, phosphorus plays a central role in our metabolism and genetic inheritance. Arsenic is well known to murder mystery fans, but most of us know little about antimony and bismuth.

Nitrogen

Most of the matter in the universe is concentrated in the stars, and in them nitrogen is the fourth most abundant element (after H, He, and C). On earth, however, it is a different story. The earth's crust contains less than 0.03% N, concentrated for the most part in deposits of $NaNO_3$ and KNO_3. Furthermore, because they are highly soluble in water, these deposits survive only in very dry regions. The bulk of earth's nitrogen is in the atmosphere, which is 78% N_2. There are two stable isotopes: ^{14}N (99.63%) and ^{15}N (0.37%). N_2 itself is almost inert: the bond energy of $N{\equiv}N$ (945 kJ mol^{-1}) is so great that virtually all reactions of N_2 are endothermic enough to be possible only at high temperature. Indeed, the only element that reacts with N_2 at ordinary temperature is Li, which forms Li_3N (lithium nitride). Nonetheless, as one of the primary elements in proteins and nucleic acids, nitrogen is of utmost importance to us.

The entry point of nitrogen into the food chain is the production of protein by plants, but, because of its unreactivity, atmospheric N_2 cannot be used directly by plants. Nitrogen *compounds* are required. Nitrogen in any oxidation state other than zero is called "fixed nitrogen," and any process that converts N_2 into compounds is called "nitrogen fixation." There are only two natural, nonindustrial sources of fixed nitrogen: (1) lightning strokes, which momentarily increase atmospheric temperature high enough for the formation of nitrogen oxides from the elements, and (2) *nitrogen-fixing bacteria,* which live in symbiosis with certain plants (legumes) and convert atmospheric N_2 to compounds as part of their complicated metabolism. Bacterial decay of plant and animal matter (including animal wastes) is an important source of nutrients for plants. However, bacterial decay is not nitrogen *fixation:* it is merely the conversion of one form of fixed nitrogen to another.

Fritz Haber (1868–1934)

Fritz Haber, trained as an organic chemist, made important contributions in electrochemistry and in the kinetics of heterogeneous reactions. He developed his ammonia synthesis over a period of four years just prior to World War I. Important contributions to the choice of catalyst, reactor materials, and design were made by the engineer Carl Bosch. The process was immediately put to use by Germany for the production of munitions, and of badly needed fertilizer. Although there was opposition stemming from the military importance of the discovery, Haber received the 1918 Nobel Prize.

Each year, rainfall brings ≈5 kg of lightning-fixed nitrogen per acre to the soil (average in the United States), and careful cultivation of legume crops can fix perhaps 10 to 20 times as much. (Note, however, that the latter method makes the land unavailable for production of other foods during that growing season.) This amount of naturally fixed nitrogen is insufficient to feed today's global population, let alone the number of people we can expect in the future.

Fortunately, industrial means for large-scale fixation of nitrogen have been available for almost a century. The Haber process is the direct combination, at moderate temperature and pressure, of N_2 and H_2 to form ammonia.

$$N_2(g) + 3\,H_2 \xrightarrow[\text{Fe}]{\text{500 atm, 450 °C}} 2\,NH_3(g) \qquad (22\text{-}28)$$

The key to the process is the use of an iron catalyst. The annual production of ammonia in the United States is almost 20 million tons, about three-quarters of which is used in agriculture. Gaseous ammonia can be injected directly into the soil (Figure 22.16), or it can be used as the raw material for production of dry fertilizers such as NH_4NO_3.

Figure 22.16 Direct injection of gaseous ammonia is an efficient means of fertilizing soil.

Hydrides

Ammonia is a pungent gas whose normal boiling point is −33 °C. The vapor pressure of the liquid at 25 °C is only 10 atm, so that ammonia can be stored, shipped, and handled in liquid form, without refrigeration, in closed containers of moderate strength. Its lone pair of electrons allows ammonia to act as a Brønsted base, accepting a proton from an acid to form the ammonium ion, NH_4^+. The solubility behavior of ammonium salts is similar to that of the alkali metal salts.

Almost all salts of Group 1A cations, and of NH_4^+, are soluble in water.

Nitrogen forms another hydride, *hydrazine* ($H_2N—NH_2$), produced by reaction of sodium hypochlorite with excess ammonia.

$$3\ NH_3(g) + NaOCl(aq) \longrightarrow NaOH(aq) + NH_4Cl(aq) + N_2H_4(aq) \quad (22\text{-}29)$$

Pure hydrazine is unstable, and decomposes to N_2 and H_2 with the liberation of 49 kJ mol^{-1}. Hydrazine and its organic derivatives are good reducing agents. The main thrust engines of the lunar landing module that carried men to and from the surface of the moon were fueled by a mixture of methyl hydrazine [$CH_3NH—NH_2$] and dimethyl hydrazine [$(CH_3)_2N—NH_2$], reacting with the oxidizing agent N_2O_4.

$$(CH_3)_2N—NH_2(\ell) + 2\ N_2O_4(\ell) \longrightarrow 3\ N_2(g) + 4\ H_2O(g) + 2\ CO_2(g) \quad (22\text{-}30)$$

This reaction is highly exothermic, liberating about 1760 kJ of heat for each mole of methyl or dimethyl hydrazine. Equally importantly, the two reactants ignite on contact with one another, eliminating the need for an ignition system in the lunar lander.

Oxides

There are nitrogen oxides in all positive oxidation states up to +5: N_2O (+1), NO (+2), N_2O_3 (+3), NO_2 and N_2O_4 (+4), and N_2O_5 (+5). All are gases at room temperature, and at high temperature, all are more or less readily convertible to the others. They are unusual among oxides because they are *endothermic compounds*—that is, their heats of formation are positive quantities. The decomposition of an endothermic compound is an exothermic process.

Dinitrogen oxide (formerly called *nitrous oxide*) has the structure N=N=O. It decomposes to its elements at moderate temperature, although it is stable at room temperature. It supports combustion almost as easily as oxygen, because its loosely bound O-atom is available for combination with combustible substances. It is used as an anesthetic, primarily in dentistry, where it is sometimes called "laughing gas." Dinitrogen oxide is formed on gentle heating of ammonium nitrate.

$$NH_4NO_3(s) \longrightarrow N_2O(g) + 2\ H_2O(g) \quad (22\text{-}31)$$

Nitrogen oxide (NO, formerly called *nitric oxide*) is a colorless gas produced by the action of nitric acid on metals.

$$3\ Cu(s) + 8\ H^+(aq) + 2\ NO_3^-(aq) \longrightarrow 2\ NO(g) + 4\ H_2O(\ell) + 3\ Cu^{2+}(aq) \quad (22\text{-}32)$$

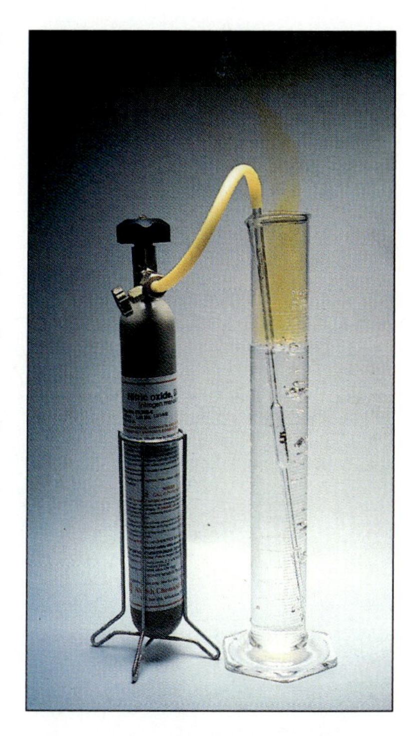

Figure 22.17 The colorless gas NO reacts quickly with atmospheric oxygen, forming a brown gas containing an equilibrium mixture of NO_2 and N_2O_4. The water in the graduated cylinder does not participate in the reaction—it is there only to demonstrate that the NO coming from the storage tank is colorless, and that the reaction does not occur until NO contacts the air.

Nitrogen oxide, like N_2O, decomposes to its elements at moderate temperature. At elevated temperature, however, the dissociation to N_2 and O_2 is a reversible reaction. Small amounts of NO are formed when air is heated above 700 °C. Because of this, NO is always a byproduct of combustion in air. On contact with oxygen, NO reacts quickly to form nitrogen dioxide (Figure 22.17).

$$2\ NO(g) + O_2(g) \longrightarrow 2\ NO_2(g) \quad (22\text{-}33)$$

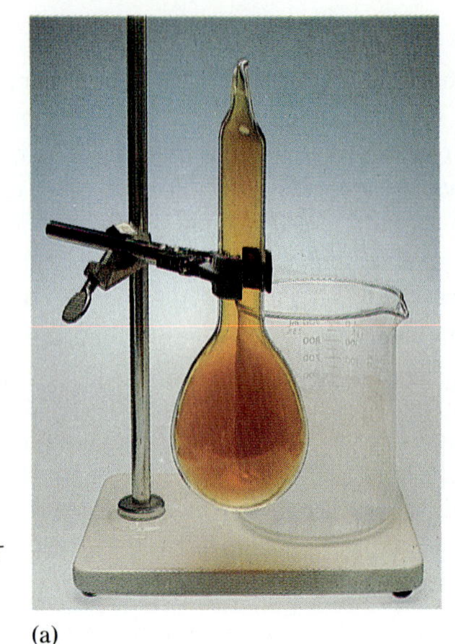

(a)

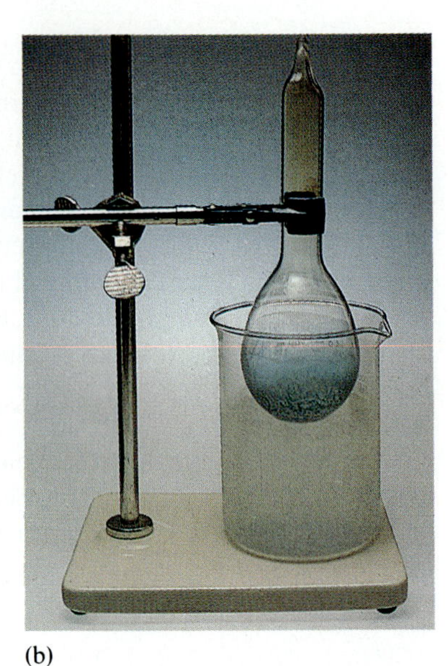

(b)

(a) The bulb contains an equilibrium mixture of NO_2 (brown) and N_2O_4 (colorless), plus a small amount of NO (colorless). (b) When cooled in liquid nitrogen (−196 °C), the NO and NO_2 in the mixture undergo an association reaction to form the blue solid N_2O_3.

A *dimer* is a polymer consisting of two monomer units.

Nitrogen dioxide (NO_2), is a pungent brown gas that readily associates to form the dimer *dinitrogen tetroxide* (N_2O_4). The equilibrium between nitrogen dioxide and dinitrogen tetroxide, $2\ NO_2 \rightleftharpoons N_2O_4$, was discussed in Chapter 13: these two compounds always exist as a mixture in the gaseous state. An NO_2/N_2O_4 mixture is a powerful oxidant, whose use in rocket propulsion was mentioned above.

Dinitrogen trioxide (N_2O_3) and *dinitrogen pentoxide* (N_2O_5) are much less stable than the other oxides, and they have no commercial use.

Acids

Several of the oxides of nitrogen, like oxides of other nonmetals, react with water to give acidic solutions.

$$N_2O_3(g) + H_2O(\ell) \longrightarrow 2\ HNO_2(aq) \tag{22-34}$$

$$N_2O_5(g) + H_2O(\ell) \longrightarrow 2\ HNO_3(aq) \tag{22-35}$$

$$2\ NO_2(g) + H_2O(\ell) \longrightarrow HNO_2(aq) + HNO_3(aq) \tag{22-36}$$

A disproportionation is an auto oxidation-reduction reaction.

$HNO_2(aq)$ is somewhat unstable and slowly disproportionates to NO and HNO_3.

$$3\ HNO_2(aq) \longrightarrow HNO_3(aq) + 2\ NO(g) + H_2O(\ell) \tag{22-37}$$

Because of this instability, Equation (22-36) is often combined with Equation (22-37) and written to indicate the overall reaction.

$$3\ NO_2(g) + H_2O(\ell) \longrightarrow 2\ HNO_3(aq) + NO(g) \tag{22-38}$$

Nitrous acid (HNO_2) is a weak acid whose aqueous solutions are stable, for a time, but which cannot be isolated in pure form. However its salts, the *nitrites*, are common. Sodium nitrite ($NaNO_2$) has been used for many years as a food preservative. Sodium nitrite imparts a desirable reddish color to meat and retards bacterial decay, but recently it has come under suspicion as a possible cause of cancer.

Nitric acid (HNO_3) is both a strong acid and an oxidizing agent. This combination of properties, together with its low cost, accounts for its wide use in the chemical industry. More than 7 million tons of nitric acid were produced in the United States in 1989 by the *Ostwald process,* which begins with the catalytic oxidation of ammonia.

$$4\ NH_3(g) + 5\ O_2(g) \xrightarrow[Pd]{} 4\ NO(g) + 6\ H_2O(g) \quad (22\text{-}39)$$

Gold, palladium, or rhodium can be used as the catalyst.

$$2\ NO(g) + O_2(g) \longrightarrow 2\ NO_2(g) \quad (22\text{-}33)$$

$$3\ NO_2(g) + H_2O(\ell) \longrightarrow 2\ HNO_3(aq) + NO(g) \quad (22\text{-}38)$$

The nitrogen oxide produced in Reaction (22-38) is recycled in Reaction (22-33), so that the overall reaction is

$$NH_3(g) + 2\ O_2(g) \longrightarrow HNO_3(aq) + H_2O(\ell) \quad (22\text{-}40)$$

Aqueous solutions of nitric acid decompose slowly on exposure to light and/or heat. Nitric acid itself is colorless, and the brown color acquired by $HNO_3(aq)$ in storage is due to the build-up of one of the decomposition products, $NO_2(aq)$. Because it is an oxidizing agent as well as an acid, HNO_3 can dissolve metals lying above hydrogen in the table of Standard Reduction Potentials (Chapter 19), for example copper [Equation (22-32); Figure 22.18].

The metals Pt, Au, Rh, and Ir are not attacked by concentrated nitric acid, but will dissolve in a mixture of HCl and HNO_3 *(aqua regia).*

Aqua regia is Latin for "royal water."

$$3\ Pt(s) + 22\ H^+(aq) + 4\ NO_3^-(aq) + 18\ Cl^-(aq) \longrightarrow 3\ H_2PtCl_6 + 4\ NO(g) + 8\ H_2O(\ell) \quad (22\text{-}41)$$

Halogen Compounds

The binary nitrogen halides NF_3, NCl_3, and NBr_3, as well as several mixed halides like NF_2Cl, are produced by indirect means rather than by direct reaction of the elements. NF_3, for example, is a product of electrolysis of NH_4F dissolved in $HF(\ell)$. The nitrogen halides are strong Lewis bases, and with the exception of NF_3, all are highly reactive, sometimes explosively so. Reaction of iodine with concentrated ammonia produces *nitrogen triiodide* ($NI_3 \cdot NH_3$), an unstable solid that detonates at the slightest mechanical shock.

Nitrides

Ionic nitrides containing the *nitride ion* (N^{3-}) and having empirical formulas of Li_3N, Ca_3N_2, and so on, are formed by direct combination of nitrogen with the active metals of Groups 1A and 2A, and some other metals. With the exception of Li, these metals require high temperatures for direct reaction with N_2. Metalloids and some nonmetals form covalent nitrides, giving compounds that are similar to the carbides of metalloids. Most covalent nitrides are hard, inert, covalent network solids.

Figure 22.18 Because it is an oxidizing agent as well as an acid, HNO_3 can react with copper (which does not react with non-oxidizing acids). The products of this reaction are $Cu^{2+}(aq)$ (blue) and NO_2 (brown gas).

Phosphorus

Phosphorus constitutes about 1% of the earth's crust, where it occurs primarily as *phosphate rock,* $Ca_3(PO_4)_2$. Phosphate rock, mined in Florida, can be heated with coke and sand to produce elemental phosphorus vapor.

$$2\ Ca_3(PO_4)_2(s) + 10\ C(s) + 6\ SiO_2(s) \longrightarrow 6\ CaSiO_3(s) + 10\ CO(g) + P_4(g) \quad (22\text{-}42)$$

Phosphate rock is treated with sulfuric acid to produce a mixture of $CaSO_4 \cdot 2H_2O$ and $Ca(H_2PO_4)_2 \cdot H_2O$ that is used as fertilizer. Phosphorus, as a major constituent of

Figure 22.19 *White phosphorus* (in test tube) is stored under water, because it ignites spontaneously in air, forming P_4O_{10}. The other allotropic form, *red phosphorus,* does not react with oxygen at room temperature.

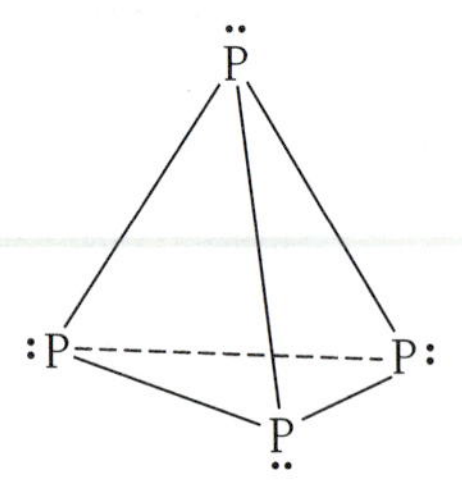

White phosphorus

nucleic acids (Section 21.4) and ATP (Section 21.6) is of course an essential nutrient for both plants and animals. There are several allotropic forms of phosphorus. *White phosphorus* is a waxy, low-melting solid that forms when gaseous P_4 condenses. It ignites spontaneously in air, and is normally stored under water (with which it does not react). The 4 atoms of the P_4 molecule, in both the gaseous and solid states, lie at the corners of a tetrahedron. As a nonpolar molecular substance, it is soluble in organic solvents such as benzene. White phosphorus slowly changes to a more stable form, *red phosphorus,* that does not ignite in air (Figure 22.19). Its structure is more complex, probably involving covalently bonded polymeric molecules having considerably more than 4 P atoms.

Phosphides, Hydrides, and Halides

Binary compounds of phosphorus with metals and metalloids are called *phosphides.* Most are covalently bonded, but a few, including Na_3P and Ca_3P_2, contain phosphorus in the -3 oxidation state, as the *phosphide ion,* P^{3-}. These react rapidly with water, yielding basic solutions and PH_3.

$$2\ P^{3-}(aq) + 6\ H_2O(\ell) \longrightarrow 2\ PH_3(g) + 6\ OH^-(aq) \qquad (22\text{-}43)$$

Most covalent phosphides are network solids, having no discrete molecules. These have a large range of properties and structures. A few of them (for example Fe_2P) are such good conductors of electricity that they are classified as metals.

Phosphine, PH_3, can be prepared by the hydrolysis of ionic phosphides [Equation (22-43)]. It is a toxic, flammable gas that has the same pyramidal structure as ammonia. Unlike ammonia, however, PH_3, is not soluble in water and is not a good Brønsted base. Provided no water is present, phosphine reacts with strong acids to form salts of the *phosphonium* ion, PH_4^+.

$$PH_3(g) + HI(g) \longrightarrow PH_4I(s) \qquad (22\text{-}44)$$

Phosphonium salts are inherently unstable. $PH_4Cl(s)$, for example, decomposes to PH_3 and HCl at any temperature above 0 °C.

The phosphonium ion is a very strong acid, and its salts react completely with water to give phosphine.

$$PH_4I(s) + H_2O(\ell) \longrightarrow H_3O^+(aq) + I^-(aq) + PH_3(g) \qquad (22\text{-}45)$$

Substituted PR_4^+ ions like $P(CH_3)_4^+$ are common in organic chemistry. The highly unstable *diphosphine* (P_2H_4) is similar to hydrazine (N_2H_4). No other phosphorus hydrides are known.

Phosphorus trichloride and *phosphorus pentachloride* were mentioned in the discussion of bonding (Section 8.3). PCl_5 is an example of the "expanded valence shell" class of exceptions to the octet rule. Other tri- and pentahalides, including mixed halides (molecules containing more than one type of halogen), are known. Phosphorus halides are prepared by direct reaction of white phosphorus with pure or mixed elemental halogens. The trihalide is the main product if phosphorus is in excess, and the pentahalide is produced with excess halogen.

$$P_4(s) + 6\ Br_2(\ell) \longrightarrow 4\ PBr_3(\ell) \qquad (22\text{-}46)$$

$$PBr_3(\ell) + Br_2(\ell) \longrightarrow PBr_5(\ell) \qquad (22\text{-}47)$$

All the halides are molecular compounds with pyramidal (PX_3) or trigonal bipyramidal (PX_5) shape, and most are gases or low-boiling liquids. PCl_3 reacts violently with water to yield phosphorous acid.

$$PCl_3(g) + 3\ H_2O(\ell) \longrightarrow 3\ HCl(aq) + H_3PO_3(aq) \qquad (22\text{-}48)$$

The noun "phosphorus" refers to the element, while the adjective "phosphorous" refers to combined phosphorus in the lower of two oxidation states.

The phosphorus atom in PCl_3 is readily oxidized from the +3 state to +5 on exposure to oxygen. The product, *phosphoryl chloride,* is a volatile liquid whose molecules have a tetrahedral structure.

$$2\ PCl_3(g) + O_2(g) \longrightarrow 2\ POCl_3(\ell) \qquad (22\text{-}49)$$

Analogous compounds are formed with the other phosphorus trihalides.

Phosphoryl chloride

Oxides, Acids, and Salts

Phosphorus forms a rich and complex group of compounds with oxygen. The product of the spontaneous combustion of white phosphorus is *phosphorus pentoxide* P_4O_{10}.

$$P_4(s) + 5\ O_2(g) \longrightarrow P_4O_{10}(s) \qquad (22\text{-}50)$$

(This compound has been known for many years, and its name corresponds to the empirical formula, P_2O_5, rather than to the molecular formula.) The reaction of P_4O_{10} with water is vigorous and highly exothermic, and if carelessly carried out can result in spattering and boiling.

$$P_4O_{10}(s) + 6\ H_2O(\ell) \longrightarrow 4\ H_3PO_4(aq) \qquad (22\text{-}51)$$

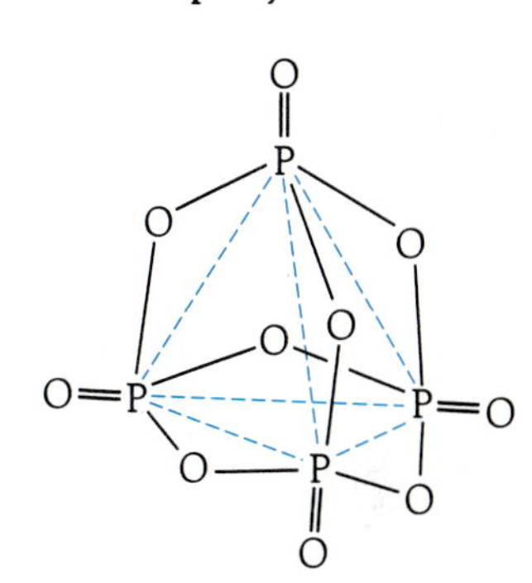

Phosphorus pentoxide

Because of this reactivity toward water, phosphorus pentoxide is an effective drying agent: it can even remove the elements of water from covalent molecules, for instance converting H_2SO_4 to SO_3. Reactions (22-50) and (22-51) together are used to prepare highly pure phosphoric acid, but larger quantities of H_3PO_4 are commercially produced by treatment of phosphate rock with an excess of sulfuric acid. The structure of P_4O_{10} is related to the P_4 tetrahedron, but with an oxygen atom lying close to each of the four edges of the P_4 tetrahedron. In addition, each P atom is doubly bonded to an O atom. Another oxide, P_4O_6, has the same overall structure but lacks the doubly bonded oxygen atoms.

Phosphorus trioxide

There are many different phosphorus oxoacids and salts, only a few of which are mentioned here. Oxoacids containing phosphorus in the +5 state are called *phosphoric acids,* and their salts are the *phosphates.* The simplest oxoacid of P(V) is H_3PO_4, a tetrahedral molecule. Although it is often simply called "phosphoric acid," the correct name for H_3PO_4 is *orthophosphoric acid.* All three H-atoms are acidic: that is, H_3PO_4 is a triprotic acid. In accord with the general trend for polyprotic acids, the third hydrogen is only slightly ionized, so that HPO_4^{2-} is a very weak acid ($K_3 = 2 \times 10^{-13}$). Salts of H_3PO_4, usually referred to as "phosphates" rather than *orthophosphates* (the correct name), contain the tetrahedral PO_4^{3-} ion. This ion is a resonance hybrid of four equivalent structures that differ only in the location of the double bond.

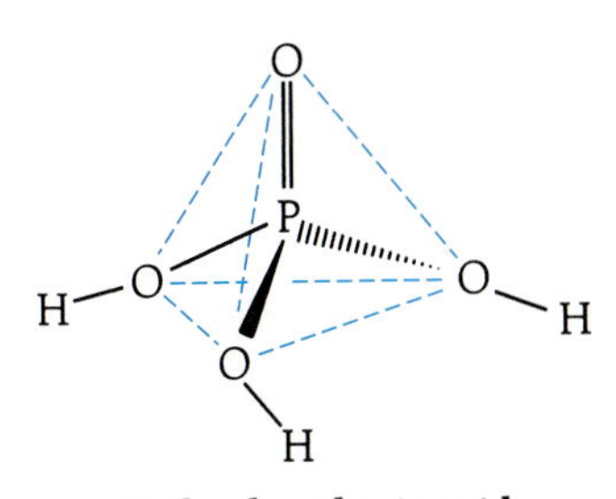

Orthophosphoric acid

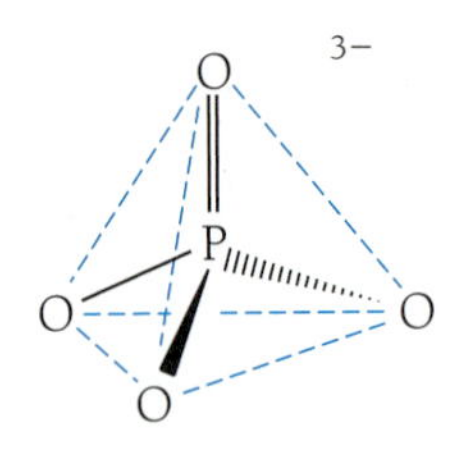

Phosphate ion

Other phosphoric acids and related salts exist due to a feature that the PO_4 group shares with the SiO_4 group: the ability to link one or more tetrahedra by sharing an oxygen atom. *Pyrophosphoric acid* ($H_4P_2O_7$) is prepared by heating orthophosphoric acid. Heating causes a condensation reaction in which two molecules are joined and a molecule of water is lost.

$$2\ H_3PO_4 \xrightarrow[\Delta]{} H_4P_2O_7 + H_2O \qquad (22\text{-}52)$$

Salts of pyrophosphoric acid are called *pyrophosphates*. Longer chains of phosphorus-oxygen tetrahedra exist in other compounds, for example in *adenosine triphosphate* (ATP, Section 21.5). *Metaphosphoric acid* ($H_3P_3O_9$) and its salts, the *metaphosphates*, contain three tetrahedra linked in a triangular array. Compounds containing a phosphorus atom bonded to four oxygen atoms are often referred to simply as "phosphates," regardless of the number of tetrahedral units they have.

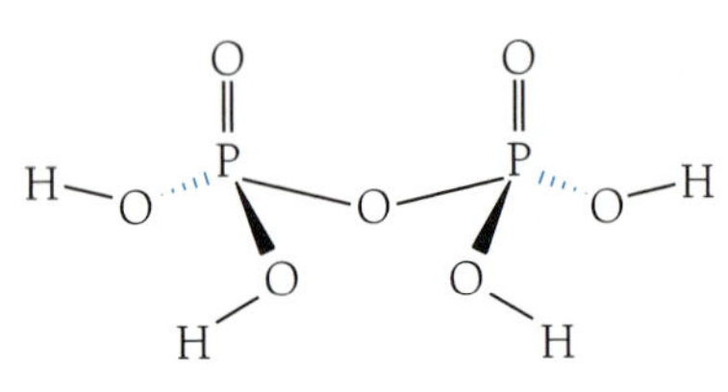

Pyrophosphoric acid, $H_4P_2O_7$

The use of phosphate rock as fertilizer, mentioned above, is just one of the many purposes served by phosphoric acids and their salts in our society. The automobile and ship industries make use of the fact that treatment of metal parts with phosphoric acid confers some degree of resistance to corrosion. A related use is "naval jelly," a consumer product for cleaning corroded metals. Phosphoric acid imparts a pleasant taste to soft drinks, and the acidic properties of the intermediate ion $H_2PO_4^-$ are put to use in baking powder. Na_3PO_4 is a relatively strong base (because HPO_4^{2-} is such a weak acid), and is used to adjust the pH of detergents. In this use, it also aids in water softening, tying up Mg^{2+} and Ca^{2+} in phosphate-containing polyatomic ions. After it is used, most of our water is purified and returned to the environment. Since normal wastewater treatment does not remove phosphates, lakes and rivers can receive heavy doses. Phosphorus is an essential nutrient, and overuse of phosphate-containing detergents is equivalent to adding fertilizer to natural waters. This causes a heavy growth of algae, the subsequent bacterial decay of which depletes dissolved oxygen and kills other forms of aquatic life. Growing awareness of this problem in the late 1960s led, relatively quickly, to legislation that eliminated or severly curtailed the use of phosphates in detergents.

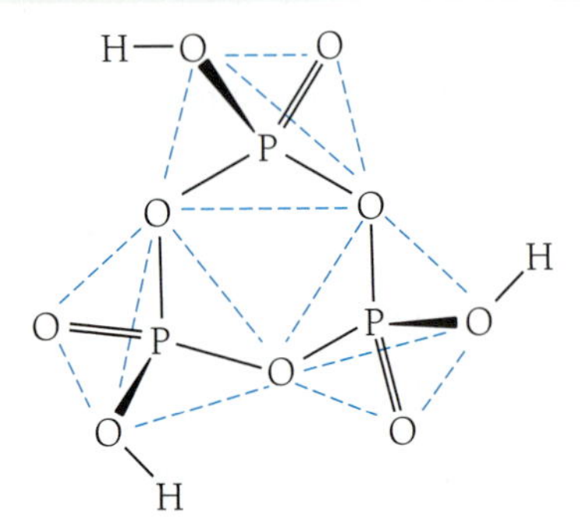

Metaphosphoric acid, $H_3P_3O_9$

Baking powder contains sodium dihydrogen phosphate and also sodium hydrogen carbonate. In the presence of water, the components of baking powder react to give small bubbles of carbon dioxide: $HCO_3^- + H_2PO_4^- \rightarrow CO_2(g) + H_2O + HPO_4^{2-}$. The bubbles impart desirable texture to cakes and other baked goods.

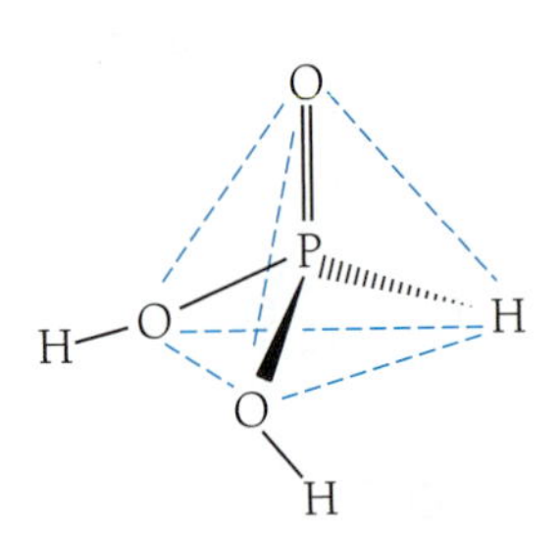

Phosphorous acid

One or two of the O—H groups in H_3PO_4 can be replaced by a single H-atom, leading to additional phosphorus oxoacids in lower oxidation states: H_3PO_3 (phosphorous acid, oxidation number +3) and H_3PO_2 (hypophosphorous acid, +1). These are unusual compounds, because in most oxoacids hydrogen is bonded only to oxygen. The P—H hydrogen is not ionizable, and these compounds are respectively diprotic and monoprotic (weak) acids. (Formic acid, H—COO—H, behaves similarly. It is monoprotic, because the C—H hydrogen is not ionizable.) Phosphorus in the +3 oxidation state is readily oxidized, so that, for instance, the *hydrogen phosphite* ion can reduce some metal ions to the free metal.

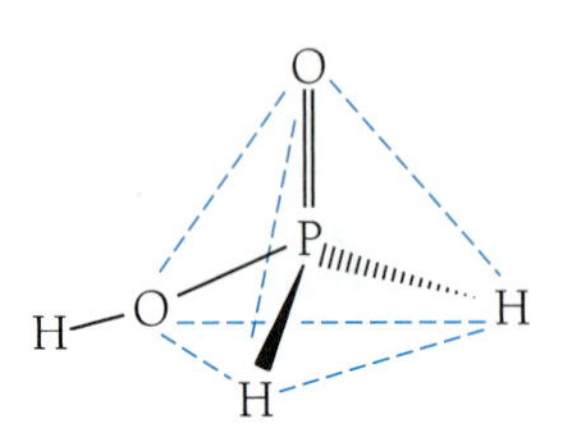

Hypophosphorous acid

$$HPO_3^{2-}(aq) + 2\ Ag^+(aq) + H_2O(\ell) \longrightarrow 2\ Ag(s) + PO_4^{3-}(aq) + 3\ H^+(aq) \quad (22\text{-}53)$$

Arsenic, Antimony, and Bismuth

The remaining elements in Group 5A are similar to phosphorus in many ways, although trends in their chemistry are affected by their increasing metallic character as atomic number increases. Thus bismuth, as Bi^{3+}, readily forms salts with common anions, while salts of As^{3+} and Sb^{3+} are less common. The molecular forms As_4 and Sb_4 exist, but the corresponding Bi_4 does not. All three elements have shiny, black allotropic forms that are metallic in character. Sb and As form *arsine* and *stibine* (AsH_3 and SbH_3), analogs of ammonia. All the Group 5A elements form trihalides: some of these are definitely molecular, and some are more ionic in character. All the Group 5A trihalides are strong Lewis bases. The oxides of As and Sb, and their corresponding oxoacids, are similar in structure and reactivity to those of phosphorus.

22.5 SULFUR, SELENIUM, AND TELLURIUM

Oxygen is such an important element that we discussed it separately, in Chapter 8. In this section, we consider all but one of the remaining elements in Group 6A. Polonium is not discussed, because it is so rare.

Sulfur

Sulfur constitutes about 0.06% of the earth's crust. It exists as the free element concentrated in geological formations associated with certain NaCl deposits. It exists also in combined form, as sulfate-containing minerals and sulfide ores such as FeS, CuS, and ZnS. Some SO_2 is emitted to the atmosphere during volcanic eruptions. Up to 5% of the mass of coal, natural gas, and petroleum is sulfur, present as H_2S or combined in a variety of organic compounds. This presents a serious environmental problem, for this sulfur is converted to SO_2 and released to the atmosphere when such fuels are burned. Once in the atmosphere, SO_2 is converted to sulfuric acid and returns to earth as acid rain. A great deal of effort goes into reducing such sulfur emissions, both by desulfurization of fuel and by removing SO_2 from the exhaust gases of electrical generating plants and other heavy users of fossil fuels. Sulfur dioxide and hydrogen sulfide are combined in the *Claus process* to produce elemental sulfur, thus converting pollutants to an important raw material.

Acid rain is discussed in Chapter 15, in Chemistry Old and New.

$$SO_2(g) + 2\ H_2S(g) \longrightarrow 3\ S(\ell) + 2\ H_2O(g) \quad (22\text{-}54)$$

About 40% of the sulfur used in the United States is a byproduct of the effort to reduce sulfur emissions from fossil fuel combustion and from the production of copper and zinc from their sulfide ores (Section 18.5).

Some sulfur is obtained by the *Frasch process* from deposits of pure sulfur, largely in Louisiana. Compressed air and superheated, high-pressure water are pumped into the deposits, and a frothy mixture of water, molten sulfur, and air is returned to the surface. Very pure (>99%) sulfur is obtained after cooling, so that no further processing is required (Figure 22.20).

The most stable form of the element is the cyclic molecule S_8. As a molecular substance it is soluble in many organic solvents, including *carbon disulfide* (analogous to CO_2). Other cyclic forms, containing up to 12 S-atoms, are less stable: all except S_6 quickly revert to S_8. This complexity, which arises from sulfur's well-developed ability

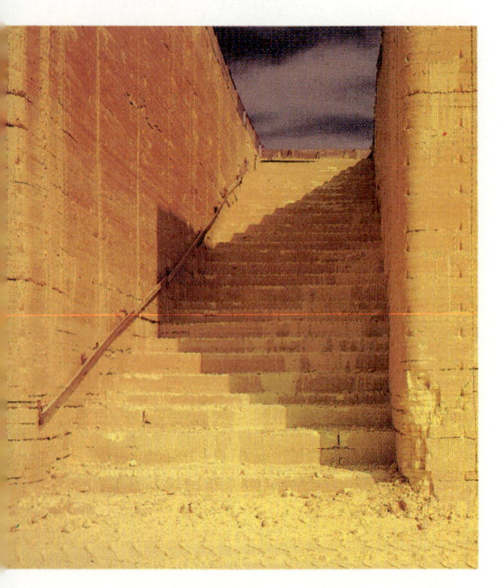

Figure 22.20 The sulfur obtained from deposits in Louisiana is almost pure. These steps are cut into blocks of sulfur stacked for shipment.

to catenate, results in the existence of numerous allotropic forms. S_8 rings can pack together in two different ways to form two crystalline allotropes: *orthorhombic* (mp 112 °C) and *monoclinic* (mp 119 °C; Figure 22.21). A third allotrope is formed when S_8 is heated. Above about 150 °C, S_8 rings begin to break apart and recombine as polymeric —S—S—S— chains of varying length. As a result of tangling of the chains, the viscosity *increases* as temperature increases. With almost all other liquids, the viscosity *decreases* as the temperature is raised. Further heating causes the chains to break into smaller fragments, and the viscosity decreases. If this liquid is cooled suddenly the result is a rubbery solid that slowly (over several days) reverts to the crystalline form as the more stable S_8 rings re-form. Hot sulfur vapor has a deep blue color, due to the presence of S_2 molecules in a complex mixture of atoms and molecules of various lengths.

Sulfides, Hydrides, and Halides

Sulfur combines with most metals at elevated temperatures to form *sulfides* such as HgS. Some of these are best described as ionic, but all have considerable covalent character. Many of them, particularly sulfides of the transition metals, have a variable composition. In these *nonstoichiometric compounds* some of the metal atoms are missing from the crystal lattice, so that the metal/sulfur ratio is better represented by a nonintegral empirical formula such as $Fe_{0.85}S$. These ratios are variable, and another sample of iron sulfide might have the composition $Fe_{0.95}S$. Probably *no* sample is truly stoichiometric, with an Fe/S ratio of 1.000. Although no mention of them has been made of them in earlier chapters, nonstoichiometric compounds are not a curiosity found only in metal sulfides. They occur, but less commonly, with other elements as well.

The sulfide ion *does* exist in solid compounds.

Alkali-metal sulfides are best thought of as stoichiometric ionic compounds, with two M^+ ions for each S^{2-} ion. The sulfide ion is such a strong base that it cannot exist in an aqueous environment, so the dissolution of an ionic sulfide involves the active participation of water (Section 17.3).

$$Na_2S(s) + H_2O(\ell) \longrightarrow 2\,Na^+(aq) + HS^-(aq) + OH^-(aq) \qquad (22\text{-}55)$$

Some *disulfide* compounds, containing the S_2^{2-} ion, are known. If aqueous sulfides or disulfides are boiled with sulfur, linear *polysulfide* ions S_n^{2-} having up to half a dozen or so sulfur atoms are formed. This comes about through the tendency of sulfur to catenate.

$$S_n^{2-}(aq) + S(s) \longrightarrow S_{n+1}^{2-}(aq) \qquad (22\text{-}56)$$

Hydrogen sulfide (H_2S) is an exceedingly toxic gas produced by the action of acids on many sulfides.

$$Na_2S(s \text{ or } aq) + 2\,HCl(aq) \longrightarrow H_2S(g) + 2\,NaCl(aq) \qquad (22\text{-}57)$$

Figure 22.21 Naturally occurring crystals of monoclinic sulfur (left) and orthorhombic sulfur (right). The two crystalline forms of S_8 are easily distinguishable by their shape.

Although formally a diprotic acid, its second dissociation constant is so small ($\approx 10^{-19}$) that essentially no S^{2-} ions are formed in its aqueous solutions. Small quantities of H_2S can be produced, and are useful in qualitative analysis of metal cations (Section 17.5), by hydrolysis of *thioacetamide.*

$$\underset{\textbf{Thioacetamide}}{CH_3{-}\overset{\overset{\displaystyle S}{\|}}{C}{-}NH_2(aq)} + 2\ H_2O + H^+(aq) \xrightarrow[\Delta]{} \underset{\textbf{Acetic acid}}{CH_3{-}\overset{\overset{\displaystyle O}{\|}}{C}{-}OH} + H_2S(aq) + NH_4^+(aq) \quad (22\text{-}58)$$

Hydrogen sulfide is also produced by the bacterial decay of sulfur-containing proteins, and as such is responsible for the odor of rotten eggs. Other sulfur hydrides, of general formula H_2S_n, incorporate catenated chains of S atoms. These hydrides are less stable than H_2S, and have no uses.

Many sulfur compounds have unpleasant odors.

A variety of molecular sulfur halides with formulas S_2X_2, SX_2, SX_4, and SX_6 are known. For example, *sulfur hexafluoride* (SF_6) is a gaseous compound prepared by direct reaction of sulfur with fluorine. It is inert toward most reagents, and is used as an insulator in high-voltage electrical equipment. In contrast, the gas *sulfur tetrafluoride* (SF_4), prepared from SF_2 and NaF, reacts readily with numerous compounds. On exposure to water, it hydrolyzes instantly:

$$SF_4(g) + 2\ H_2O(\ell) \longrightarrow SO_2(g) + 4\ HF(g) \quad (22\text{-}59)$$

In the absence of water, SF_4 is useful for the synthesis of organic fluoro-compounds, such as R_2CF_2 from $R_2C{=}O$ or RCF_3 from RCOOH, and in the preparation of metal fluorides from their oxides.

Oxides, Acids, and Salts

Oxides The only stable oxides of sulfur are SO_2 and SO_3. *Sulfur dioxide* is produced by the combustion of elemental sulfur or almost any sulfur-containing compound. *Sulfur trioxide* results from the reaction of SO_2 with oxygen. Reaction (22-60) is slow when pure reagents are used, but is much faster when catalyzed by surfaces.

$$2\ SO_2(g) + O_2(g) \longrightarrow 2\ SO_3(g) \quad (22\text{-}60)$$

The rapid conversion of SO_2 to SO_3 on the surfaces of airborne dust particles is an important step in the formation of acid rain. Both oxides react with water, SO_2 to produce H_2SO_3 and SO_3 to produce H_2SO_4. The resonance-stabilized structure of SO_2 has been mentioned (Exercise 8.7, page 324).

Oxoacids Sulfur dioxide dissolves readily in water to produce an acidic solution. The solute is usually described as *sulfurous acid,* H_2SO_3, although there is little evidence for the existence of a discrete H_2SO_3 molecule: it is probably more correct to describe it as $SO_2 \cdot nH_2O$ ($n = 6$ or 7). Whatever its structure, the solute is a diprotic acid of moderate strength.

$$H_2SO_3(aq)\ [\text{or } SO_2 \cdot nH_2O] \rightleftharpoons H^+(aq) + HSO_3^-(aq) \quad (22\text{-}61)$$

$$HSO_3^-(aq) \rightleftharpoons H^+(aq) + SO_3^{2-}(aq) \quad (22\text{-}62)$$

Sulfurous acid reacts with bases to form salts containing the *hydrogen sulfite* (HSO_3^-) ion (formerly called bisulfite) or the *sulfite* ion (SO_3^{2-}). The sulfur atom in both these ions is in the +4 oxidation state, and can easily be oxidized to +6. Therefore solutions of sulfite salts, particularly alkaline solutions, are good reducing agents.

$$SO_3^{2-}(aq) + 2\ OH^-(aq) + Cl_2(g) \longrightarrow SO_4^{2-}(aq) + 2\ Cl^-(aq) + H_2O \quad (22\text{-}63)$$

Sulfuric acid (H_2SO_4) is the number one industrial chemical, manufactured in greater quantity than any other substance. Older, less economical syntheses have given way to the *contact process,* used today in virtually all manufacture of H_2SO_4. The process begins with the oxidation of elemental sulfur to SO_2,

$$S(s) + O_2(g) \longrightarrow SO_2(g) \quad (22\text{-}64)$$

followed by the further (catalytic) oxidation to SO_3.

Many different catalysts are effective in this reaction. The most widely used is vanadium(V) oxide, V_2O_5.

$$2\ SO_2(g) + O_2(g) \xrightarrow[\text{cat}]{} 2\ SO_3(g) \quad (22\text{-}60)$$

Although sulfur trioxide reacts with water to produce sulfuric acid, the direct reaction produces a fine mist that is difficult to handle in a large-scale operation. Therefore, an indirect route is used: SO_3 is dissolved in part of the H_2SO_4 produced in a later step, where it reacts to form the intermediate *oleum* or "fuming sulfuric acid," whose major component is *pyrosulfuric acid,* $H_2S_2O_7$.

$$SO_3(g) + H_2SO_4(aq) \longrightarrow H_2S_2O_7(aq) \quad (22\text{-}65)$$

Subsequent reaction of oleum with water yields concentrated (18 M) sulfuric acid.

$$H_2S_2O_7(aq) + H_2O(\ell) \longrightarrow 2\ H_2SO_4(aq) \quad (22\text{-}66)$$

About half of the approximately 40 million tons of H_2SO_4 produced each year is used to convert phosphate rock to fertilizer. The remainder is used in a huge variety of industrial processes, including petroleum refining and the manufacture of drugs, paper, detergents, and metals. One familiar use is as the electrolyte in automobile storage batteries (Section 19.4).

Although sulfuric acid contains sulfur in its maximum oxidation state (+6), it is only a moderately strong oxidizing agent. Concentrated solutions have a strong affinity for water, so much so that drops of water added to concentrated acid will boil and spatter as a result of the high heat of solution. Like P_4O_{10}, H_2SO_4 is a sufficiently powerful dehydrating agent to remove the elements of water from many organic materials—often only a charred mass of elemental carbon remains (Figure 22.22).

Sulfuric acid

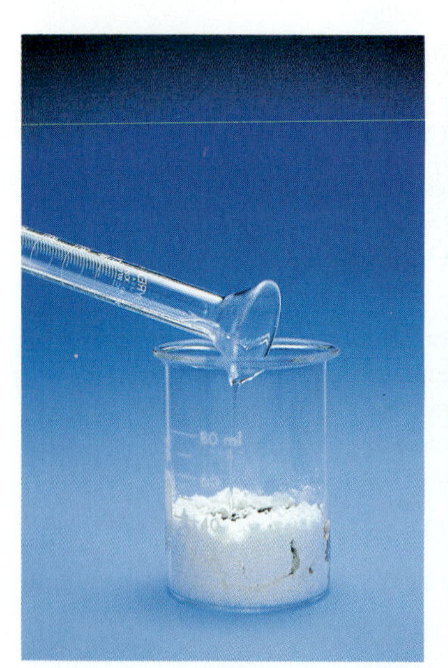

Figure 22.22 Ordinary sugar is dehydrated by concentrated sulfuric acid, leaving elemental carbon.

$$C_{12}H_{22}O_{11}(s) + 11\ H_2SO_4(conc) \longrightarrow 12\ C(s) + 11\ H_2SO_4 \cdot H_2O(aq) \quad (22\text{-}67)$$

Salts of sulfuric acid contain the *sulfate* ion (SO_4^{2-}), in which an sp^3-hybridized sulfur atom is tetrahedrally bonded to four oxygen atoms. In sulfuric acid, two of the oxygen atoms are bonded also to hydrogen. Metal sulfates are common, both as laboratory and industrial chemicals and as minerals.

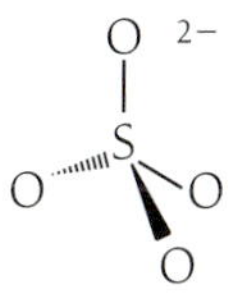

Sulfate ion

As with many of the XO_4 tetrahedral forms discussed in this chapter, numerous variations of the basic structure exist. The simplest of these is the *thiosulfate* ion ($S_2O_3^{2-}$), in which one of the oxygen atoms is replaced by a sulfur atom. Thiosulfates are prepared by boiling sulfite salts with sulfur, or alternatively by the oxidation of sulfide salts in aqueous solution.

$$SO_3^{2-}(aq) + S(s) \longrightarrow S_2O_3^{2-}(aq) \quad (22\text{-}68)$$

$$2\ HS^-(aq) + 2\ O_2(g) \longrightarrow S_2O_3^{2-}(aq) + H_2O(\ell) \quad (22\text{-}69)$$

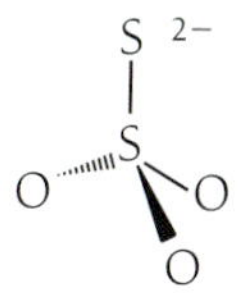

Thiosulfate ion

Sodium thiosulfate is used in the final step ("fixing") of black-and-white film processing, to remove AgBr(s) from the film. Silver bromide is the light-sensitive component of film, and its removal allows the finished photograph to be exposed to light without further darkening. The film to be fixed is washed in a solution of sodium thiosulfate, which removes silver in the form of a soluble *complex ion.*

Complex ions such as $Ag[S_2O_3]_2^{3-}$ are discussed in Chapter 23.

$$AgBr(s) + 2\ S_2O_3^{2-}(aq) \longrightarrow Br^-(aq) + Ag[S_2O_3]_2^{3-}(aq) \quad (22\text{-}70)$$

Thiosulfate ion is a good reducing agent, used in titrations of iodine.

$$2\ S_2O_3^{2-}(aq) + I_2(aq) \longrightarrow 2\ I^-(aq) + S_4O_6^{2-}(aq) \quad (22\text{-}71)$$

The $S_4O_6^{2-}$ ion produced in this reaction is the *tetrathionate ion,* one of many *polythionate* ions that can be formed as a result of sulfur's tendency to catenate. These have the structural formulas $O_3S{-}(S)_n{-}SO_3^{2-}$ (including $n = 0$: $O_3S{-}SO_3^{2-}$ is the *dithionate ion*). Polythionate ions are sulfur chains of various lengths, terminated at both ends by tetrahedral SO_3 groups.

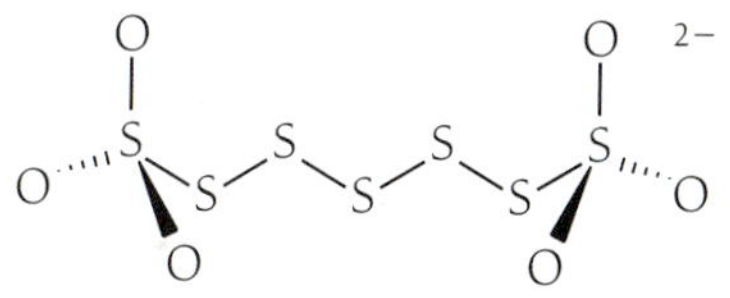

Polythionate ion

Sulfur-oxygen tetrahedra can also be joined by sharing a corner atom. Pyrosulfuric acid and its salts, the pyrosulfates, have this structure.

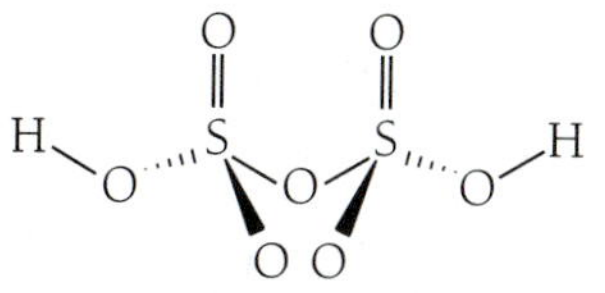

Pyrosulfuric acid

Selenium and Tellurium

Selenium and tellurium are much less common than sulfur, and they are not widely used. Their chemistry has not been studied as extensively as that of sulfur. Their general behavior parallels that of sulfur, with some differences. Although a cyclic, molecular form, Se_8, exists, it is unstable. The stable forms of both Se and Te are metallic-appearing crystals. Both are metalloids, and selenium's semiconductor properties made it useful in rectifiers (to convert ac current to dc) in the early days of the electronics industry. Polonium, the heaviest element of Group 6A, is definitely a metal. Both Se and Te form X^{2-} ions, and their salts, the *selenides* and *tellurides,* are often present in small amounts as impurities in metallic sulfide ores. The weak acids H_2Se and H_2Te are stronger and less stable than H_2S, and both are toxic, evil-smelling gases. A variety of molecular halides of Se and Te have been prepared.

Selenium and tellurium both form dioxides and trioxides analogous to SO_2 and SO_3, but they are solid rather than gaseous at room temperature. These oxides react with bases to form salts containing the ions XO_3^{2-}, HXO_3^-, XO_4^{2-}, and HXO_4^-. Selenium dioxide is a useful oxidizing agent in the laboratory. Selenic acid (H_2SeO_4) is structurally similar to sulfuric acid, but telluric acid has octahedral geometry corresponding to the molecular structure $Te(OH)_6$.

SUMMARY AND KEYWORDS

The **main-group (representative)** elements are divided into the *s-block* (Groups 1A and 2A) and the *p-block* (Groups 3A through 8A). Metallic character increases with atomic number within each group. The smallest member of each group is likely to depart from typical group properties more than the others.

Group 3A. Boron, the only nonmetal in Group 3A, occurs primarily as **borax,** $Na_2B_4O_7 \cdot 10H_2O$, which can be treated with H_2SO_4 to form *boric acid* [$B(OH)_3$]. Boric acid loses water on heating, leaving B_2O_3 (boric oxide), a raw material in the manufacture of glass. Boron forms *electron-deficient* molecular halides that are good Lewis acids. The **boranes** (boron hydrides) exhibit **three-center bonds** and have a variety of structures based on the *icosahedron.* Binary compounds with metals are **borides.** Boron nitride and carbide are high-melting abrasives.

Group 4A. Carbon occurs as *carbonate* minerals such as limestone. **Diamond,** a three-dimensional network solid, and **graphite,** with a two-dimensional layered structure, are two allotropic forms of carbon. Group 1A and 2A **carbides** contain either the C^{4-} or C_2^{2-} ion, which hydrolyze to CH_4 and C_2H_2, respectively. The most common carbon oxides are CO (toxic) and CO_2, both produced in the combustion of carbon-containing compounds. CO_2 reacts with water to give the weak diprotic *carbonic acid,* H_2CO_3. *Hydrogen cyanide* is another toxic gas, and its salts, the cyanides (CN^-), share this toxicity. NaCN is used in the metallurgy of gold. Carbon *halides* such as CCl_4 are usually classified as organic compounds.

Silicon, the second most common element, occurs as **silica** (SiO_2) and **silicate minerals.** Crude silicon is obtained by high-temperature reduction of SiO_2 with C, and high-purity Si is obtained by reduction of $SiCl_4$. **Silanes** (Si_nH_{n+2}) are molecular compounds analogous to alkanes, although not nearly as numerous. Silicon forms *tetrahalides* that react with water to produce SiO_2. One or more of the halogen atoms may be replaced by an organic group, giving rise to a large family of organo-silicon compounds. **Silicone** polymers, in which each Si atom of the alternating Si—O backbone bears two organic groups, are generally stable materials with good high-temperature stability and diverse physical properties. *Quartz* is one of several crystalline forms of SiO_2. The structures of the **silicate minerals** incorporate various derivatives of the *orthosilicate ion,* SiO_4^{4-}. Long one-dimensional chains of linked SiO_4 tetrahedra occur in the *pyroxene* minerals. Two-stranded chains are found in the *amphiboles,* a class that includes the *asbestos* minerals. In *mica* and *talc,* the tetrahedra are organized into two-dimensional sheets. Some of the Si atoms are replaced by Al in the **aluminosilicates,** which include *feldspar* and *clay.* Heating of moist clay produces a **ceramic,** a material consisting of aluminosilicate microcrystals imbedded in a glassy matrix of SiO_2. Three-dimensional structures of linked tetrahedra that have regularly spaced cavities are found in the **zeolites,** useful in water-softening.

The elements germanium, tin, and lead form some covalent compounds, such as the MH_4 hydrides *germane, stannane* and *plumbane*, and the *tetrahalides*. Their oxides are similar to SiO_2. Unlike the lighter elements of Group 4A, Sn and Pb form ionic salts in the +2 oxidation state.

Group 5A. Elemental nitrogen (N_2) constitutes 78% of the atmosphere, and is virtually unreactive due to its strong triple bond. *Fixed* nitrogen (compounds) is required for the growth of plants. Nitrogen is fixed (and used primarily for fertilizer) in the *Haber synthesis* of *ammonia* (NH_3) from N_2 and H_2. Ammonia and *hydrazine* (N_2H_4) are the two hydrides of nitrogen. Both are Brønsted bases, and hydrazine is a powerful reducing agent. All of the numerous nitrogen oxides are gaseous, *endothermic* compounds. Two nitrogen oxoacids are formed by the reaction of nitrogen oxides with water, *nitrous* (HNO_2) and *nitric* (HNO_3) acids. HNO_2 is an unstable weak acid whose salts, the *nitrites*, are stable. HNO_3 is a strong acid, and an oxidizing agent as well. It is produced in quantity from NH_3 and O_2 in the *Ostwald process*. Nitrogen forms highly reactive trihalides, which are strong Lewis bases.

Phosphorus is mined as *phosphate rock*, $Ca_3(PO_4)_2$, which, after treatment with sulfuric acid, is used for fertilizer. The allotrope called *red phosphorus* is somewhat more stable than *white phosphorus*, which consists of discrete P_4 molecules. The binary compounds of phosphorus with metals, the *phosphides*, react with water to produce *phosphine*, PH_3. Phosphine is more reactive than the analogous NH_3. Two types of halide molecules exist, PX_3 and PX_5. PCl_3 reacts with water to produce H_3PO_3 and with O_2 to produce $O{=}PCl_3$. Phosphorus burns in air to form P_4O_{10}, an avid dehydrating agent that combines with water to form *orthophosphoric acid*, H_3PO_4. In P_4 and P_4O_{10}, the P-atoms lie at the vertices of a tetrahedron, while in H_3PO_4 the P-atom occupies the center of a tetrahedron whose vertices are O and OH groups. In other forms of phosphoric acid and phosphate salts, several PO_4 tetrahedra are joined by sharing an O-atom at a common vertex. In living matter, phosphorus is contained in nucleic acids and ATP. The acids H_3PO_3 and H_3PO_2 are less common than H_3PO_4.

Arsenic, antimony, and bismuth exist in metallic-appearing allotropic forms: As and Sb are metalloids, while Bi is definitely a metallic element. The molecular forms As_4 and Sb_4 are very unstable, and Bi_4 does not exist. As and Sb form hydrides, halides, oxides, and oxoacids similar to the phosphorus compounds. Bi forms some ionic halides, and its oxides are unlike those of phosphorus.

Group 6A. Sulfur exists in nature both as the pure element and as metal sulfides and sulfate minerals. Combustion of coal and oil, which contain small amounts of sulfur, results in the formation of SO_2, a major cause of acid rain. The most stable form of the element is the cyclic S_8 molecule, which crystallizes in two distinct forms. Sulfur's ability to *catenate* (form covalent bonds to itself) permits the formation of a third allotrope, a rubbery solid containing long polymeric chains $\text{+}(S)_n\text{+}$. Some metallic *sulfides* are ionic (for example Na_2S), while some are *nonstoichiometric* compounds whose atomic composition is characterized by nonintegral mole ratios. Catenation also leads to linear *polysulfide* ions S_n^{2-} ($n = 2$ to 6), formed by boiling aqueous sulfides with sulfur. Catenated hydrides (HS_nH) exist, but the only common sulfur hydride is the toxic gas H_2S. In its molecular halides, sulfur exhibits a variety of oxidation states: $S_2X_2(+1)$, $SX_2(+2)$, $SX_4(+4)$, and $SX_6(+6)$. The oxides SO_2 and SO_3 react with water to form *sulfurous* and *sulfuric* acids, respectively. Sulfuric acid is made commercially by the *contact process*, in which sulfur is burned to SO_2, catalytically oxidized further to SO_3, and reacted with water. The *sulfate* ion (SO_4^{2-}) is tetrahedral, and in the related *thiosulfate* ion ($S_2O_3^{2-}$) one of the vertices is occupied by an S atom rather than an O atom. The *polythionate* anions consist of two thiosulfate tetrahedra linked by catenated sulfur atoms. Two SO_4 tetrahedra share a corner O-atom in the *pyrosulfate ion,* $S_2O_7^{2-}$.

Much of the chemistry of selenium and tellurium is similar to that of sulfur. Two significant differences are that Se and Te do not catenate to the extent that S does, and the stable allotropes are metals or metalloids.

PROBLEMS

Boron

1. How many neutrons are there in each of the stable isotopes of boron?
2. What is the chief ore of boron, and where is it found?
3. What reducing agent is used in the preparation of elemental boron from its ore?
4. What is the empirical formula of boron nitride, and what reagents are used to prepare it?
5. Boron compounds display two unusual types of bonding. Describe them, and give examples.
6. The fact that boron halides are volatile indicates that their bonding is covalent rather than ionic. What physical property of boron atoms also suggests that boron should form molecular rather than ionic compounds? Explain.
7. What is the empirical formula of the binary compound of boron and chlorine? What is the molecular shape of this and other boron halides?
8. Explain why the equation $H_3BO_3 + H_2O \rightarrow H_3O^+ + H_2BO_3^-$ is a poor representation of the ionization of boric acid.
9. The acid ionization constant of boric acid is 7.3×10^{-10}. Calculate the pH of a 0.050 M solution.
10. What are the elements contained in each of the following types of compounds: (a) borides, (b) boranes, and (c) carboranes.
11. Give the name and formula of a boron compound that (a) reacts with water to give HBr, (b) is a good reducing agent, (c) is a useful abrasive, and (d) results from the strong heating of boric acid.
12. How much diborane (B_2H_6) can be produced from the reaction of 25.0 g BCl_3 with excess $LiAlH_4$?

Carbon and Other Group 4A Elements

13. Carbon has two stable isotopes. Identify them, and give their approximate natural abundances.
14. Name the two main allotropic forms of carbon and describe their bonding and structures.
15. Which allotrope of carbon is useful as a lubricant? Explain.
16. What is a fullerene?
17. Give the name and formula of a compound that reacts with water to produce methane. Write a balanced equation for the reaction.
18. Give the name and formula of a compound that reacts with water to produce ethyne. Write a balanced equation for the reaction.
19. Give the name and formulas of three oxides of carbon. Are these molecular or ionic compounds? Which of them reacts with metals to form adducts?
20. Name and give the formulas of three carbon-containing anions (other than CO_3^{2-} and HCO_3^-). Which of them forms salts similar in properties to halides? Which of them is used in mining?
21. For each of the following species, draw the Lewis structure and indicate the hybridization of the carbon atom.
    a. HCO_3^- b. HCN
    c. HCO_2^-
22. For each of the following species, draw the Lewis structure and indicate the hybridization of the carbon atom.
    a. CCl_4 b. C_2H_2
    c. C_2H_3Cl
23. Rank the following in order of increasing acid strength: H_2CO_3, HCO_2H, HCO_3^-.
24. The pH of natural rainwater is less than 7, even in the absence of the SO_2 produced by combustion of sulfur-containing fuels. Explain.

*25. Would you expect the pH of water dripping from a stalactite in a limestone cavern to be less than 7, approximately 7, or greater than 7? Explain.

26. What class of carbon compounds is used as refrigerant gases?
27. What carbon compound is the monomer of one of the bases in nucleic acids?
28. Potassium hydroxide reacts with atmospheric carbon dioxide, producing $KHCO_3(s)$. After a stream of dry air was passed through a tube containing KOH(s), the mass of the tube was found to increase by 17.7 mg. How many moles of carbon dioxide were removed from the air?
29. How many liters (STP) of acetylene (ethyne) result from the reaction of 100 g CaC_2 with excess water?

30. What raw material, and what reducing agent, are used in the production of elemental silicon?

*31. Suggest a reason why no silanes with more than six catenated Si atoms are known.

32. Write an equation for the reaction of a silicon tetrahalide with water.

33. What is a silicone?

34. What are the principal naturally occurring compounds of silicon?

35. What structural feature is common to the silicate minerals? How do aluminosilicates differ from ordinary silicates?

36. Explain the difference between clay and ceramic.

37. What accounts for the optical activity of quartz?

38. What is the formula and structure of the orthosilicate ion, and what orthosilicate compound is used in cement?

39. How much SiO_2 is produced when 10.0 liters (STP) of silane reacts with an excess of water?

40. What is the hybridization of the silicon atoms in Si_2H_6?

41. What are the names of the compounds GeH_4, SnH_4, and PbH_4?

42. Glass, while not a pure substance, is nonetheless composed primarily of SiO_2. Quartz is pure SiO_2. What is the major structural difference between these two forms of silica?

43. Both tin and lead form many ionic compounds such as $PbSO_4$ and $Sn(NO_3)_2$, but there are no known compounds of Si^{+2}. Account for this difference in terms of periodic trends.

44. What evidence exists to indicate that tin has both metallic and nonmetallic character?

45. Write an equation for the reaction you would expect to occur between plumbane and water.

46. Metallic lead is produced from PbO by reduction with carbon. The other product of the reaction is carbon monoxide. How much PbO is required to produce 1 metric ton (1000 kg) of lead? How much carbon?

Nitrogen and Other Group 5A Elements

47. One of the two stable isotopes of nitrogen has a natural abundance of 0.37%. How many neutrons are contained in the nucleus of this isotope?

48. Suggest a reason why N_2 is so unreactive.

49. With what industrial process is the name Fritz Haber associated? Write the equation(s) for this process.

50. What is "fixed nitrogen," why is it important, and what are its two main natural sources?

51. Give the formulas and names of, and the hybridization of the nitrogen atom(s) in, the two hydrides of nitrogen.

52. Give an example of a binary compound of nitrogen (other than a hydride) that is (a) molecular and (b) ionic.

53. Draw the dot structures of the following nitrogen-containing ions. Describe the geometry of the triatomic ions.
a. N_3^- (azide ion)
b. CN^-
c. NCO^-
d. CNO^-

54. Give the names, formulas, and oxidation numbers of nitrogen in as many oxides of nitrogen as you can. Which of these are the least stable?

55. What is an endothermic compound, and which of the nitrogen oxides fit this description? Is either CO or CO_2 endothermic?

56. Which nitrogen oxide reacts spontaneously with oxygen? Write the equation for this reaction.

57. What are the two oxoacids of nitrogen? Are they strong or weak? Are they stable in aqueous solution? Which oxides are the anhydrides of these acids?

58. Why is it that some metals that do not react with HCl(aq) do react with HNO_3(aq)? Write an equation for the reaction of copper metal with HNO_3(aq).

59. Like many nitrogen-containing compounds, hydrazine is a base: in acid solution it exists primarily as the $N_2H_5^+$ ion. Hydrazine is also a reducing agent. Write a balanced equation for the redox reaction of hydrazine with IO_3^- in acid solution. The products of the reaction include N_2(g) and I_2(s).

60. What is the raw material used in the production of phosphorus? Write an equation for the reaction in which elemental phosphorus is produced from this compound.

61. In the allotropic form called "white phosphorus," the element occurs as P_4 molecules. Describe the geometry and hybridization in P_4.

62. Write an equation for the reaction of (a) sodium phosphide with water (net ionic equation) and (b) the preparation of phosphorus trichloride.

63. Write equations for the reaction of (a) the spontaneous combustion of white phosphorus and (b) the reaction of the combustion product with water.

64. Give the names and draw the geometric structures of the following oxoacids of phosphorus.
a. H_3PO_4
b. H_3PO_3
c. H_3PO_2
d. $H_4P_2O_7$
e. $H_3P_3O_9$

*65. Both H_3PO_2 and H_3PO_3, and their anions, are good reducing agents. In particular, $\mathscr{E}°$ for the half-reaction $H_3PO_4 + 2\ H^+ + 2\ e^- \rightarrow H_3PO_3(aq) + H_2O$ is −0.276 V at 25 °C. Write the equation for the reaction that takes place when, under standard conditions of concentration, H_3PO_3(aq) is mixed with (a) Pb^{2+}(aq) and (b) Zn^{2+}(aq).

*66. The pK_a's for H_3PO_4 are 2.12, 7.21, and 12.66. Calculate the pH of a 0.0150 M solution of orthophosphoric acid.

67. Which element(s) in Group 5A exist(s) as tetratomic molecules?

68. Both N and P form monatomic anions, while Bi forms a monatomic cation. Comment on this difference.

Sulfur, Selenium, and Tellurium

69. What are the two major sources of sulfur used in the United States?

70. Describe the allotropic forms of sulfur. Which of them is the most stable?

71. What anions are formed by sulfur in its binary ionic compounds?

72. Which is the stronger acid, H_2S or H_2Se? Why? (*Hint:* See Section 15.6.)

73. Complete and balance the following equations.
    a. $K_2S + H_2O \longrightarrow$
    b. $K_2S + HCl(aq) \longrightarrow$
    c. $SF_4 + H_2O \longrightarrow$

74. Draw the dot structure, describe the molecular geometry, and give the hybridization of the central atom in each of the following.
    a. SF_4
    b. SF_6
    c. S_3^{2-}

75. There are two common oxides of sulfur, and both are anhydrides of sulfur oxoacids. Name the oxides, and the corresponding oxoacids, and write the equation for the reactions of the oxides with water.

76. Draw the Lewis structure of sulfur dioxide. What is the order of the S—O bonds?

77. How much SO_2 is produced from the combustion of 1.00 ton (2000 lbs) of coal containing 5.2% (by mass) sulfur?

78. What raw materials are used in the production of sulfuric acid? What is the name of the process? Write equations for the process.

79. Name the following compounds.
    a. K_2SO_4
    b. $KHSO_4$
    c. K_2SO_3
    d. $KHSO_3$

*80. The second ionization of sulfuric acid is

$$HSO_4^- + H_2O \longrightarrow H_3O^+ + SO_4^{2-}; \qquad pK_2 = 1.92$$

Calculate the pH of a 0.60 M solution of $KHSO_4$.

81. Describe the geometry of, and give the hybridization of the central atom in, each of the following.
    a. sulfate ion
    b. thiosulfate ion

82. Describe the structure of a polythionate ion.

*83. In alkaline solution, sulfite salts react with permanganate salts to produce a precipitate of manganese dioxide. Write a balanced net ionic reaction for this process.

*84. The titration of an alkaline solution containing 0.9454 g K_2SO_3 required 42.8 mL of $KMnO_4(aq)$ to reach the equivalence point. What is the molarity of the $KMnO_4$ solution?

85. Write an equation for the reaction of gaseous SO_3 with concentrated $H_2SO_4(aq)$. Name the product.

86. Which of the Group 6A elements form(s) a monatomic, doubly charged anion?

87. What differences in structure exist among sulfuric, selenic, and telluric acids?

Other Problems

88. What valence electron configurations are found in the main-group elements?

89. Are the metalloid elements found in the main groups or the transition elements, or both? Which are the metalloid elements?

90. What are the two most common elements in the earth's crust?

91. Which would you expect to have the higher boiling point, H_2Se or H_2Te?

92. What is the empirical formula of a compound that contains only boron and hydrogen, and is 11.55% H by mass? If the molar mass is known to be between 100 and 150 g mol^{-1}, what is the molecular formula? What is the name of this compound?

*93. If 20.0 g of white phosphorus burns completely in air, and the combustion product is dissolved in 750 mL water, what is the molarity of the solution?

CHAPTER 23

THE TRANSITION ELEMENTS

Copper is one of the few metals that occurs naturally in elemental form.

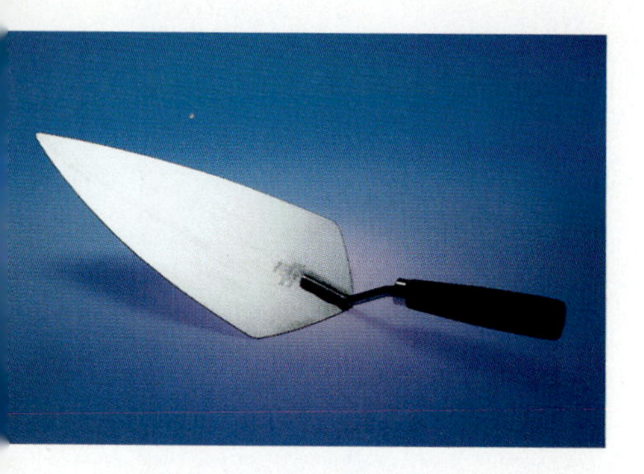

The most familiar transition-metal containing substance is steel.

23.1 INTRODUCTION

Members of the "B" groups of the periodic table are called **transition elements** or **transition metals.** This name was given because some of their chemical behavior is intermediate between that of the electropositive metals of Groups 1A and 2A, and the electronegative nonmetals on the right-hand side of the periodic table. Although there is of course some variation, many of the transition metals have similar physical properties. In addition, they show much similarity in their chemical behavior. A transition element typically forms compounds in several different oxidation states. Compounds of transition elements tend to be highly colored, and many are paramagnetic. All transition elements are metals, with metallic lustre and good electrical conductivity, and all form coordination complexes (discussed later in this chapter) with electron-pair donors. Transition elements vary widely in their abundance in the earth's crust.

Transition elements are put to a wide variety of uses. Steel is essentially an iron-carbon alloy, whose properties are controlled by the addition of one or more other transition elements. Vanadium, chromium, molybdenum, manganese, cobalt, nickel, and tungsten are commonly used in steelmaking. Titanium alloys are used where a strong, light-weight metal is needed. Paints are colored with transition-metal compounds. Metallic chromium and zinc are used for protective coatings. Platinum and palladium are among those transition elements that are used as industrial catalysts. The process of photographic film development is based on the redox properties of silver. In nature, many compounds of biochemical importance contain transition metals: certain vitamins, many enzymes, and hemoglobin, to name a few.

Although the "B group elements" include the lanthanides (Z = 58 to 71) and actinides (Z = 90 to 103), these are usually considered separately from the other transition elements. They are discussed in Section 23.5.

23.2 PHYSICAL PROPERTIES

Both physical and chemical properties of the elements are closely related to the shape, orientation, energy, and occupancy of the valence-electron orbitals. In this sense, electron configuration is the most fundamental of element properties. We begin our survey of transition metals, therefore, with a discussion of their electron configurations.

Electron Configuration

The electron configuration of the *d* subshell strongly influences the chemical behavior of transition elements. All transition elements, except Zn, Cd, and Hg, have a partially filled *d* subshell in one or more of the common oxidation states (including zero). The ground-state electron configurations of the fourth-period transition elements are given in Table 23.1, along with the values of selected physical properties. The obvious trend in electron configuration is the filling of the *d* subshell as atomic number increases in the period, but there are two exceptions: chromium and copper. Although these exceptions have sometimes been explained by referring to the "special stability" of the half-filled shell (Cr) or the completely filled shell (Cu), the reasons are probably more complex. The 3*d* and 4*s* subshells are very close in energy in *all* the fourth-period elements, so that the electron configurations occasionally depart from the *aufbau* guidelines discussed in Chapter 7.

TABLE 23.1 Selected Properties of the Fourth-Period Transition Elements

Property	Scandium	Titanium	Vanadium	Chromium	Manganese
Atomic Number	21	22	23	24	25
Valence Electron Configuration*	$4s^23d^1$	$4s^23d^2$	$4s^23d^3$	$4s^13d^5$	$4s^23d^5$
Atomic Radius (pm)†	161	145	132	127	124
Density (g cm^{-3})	2.99	4.54	6.11	7.20	7.43
Ionization Energy (kJ mol^{-1})					
First	631	658	650	653	718
Second	1235	1310	1414	1592	1509
Third	2389	2653	2828	2987	3249
Melting Point (°C)	1541	1660	1890	1857	1244
Boiling Point (°C)	2831	3287	3380	2672	1962
Std. Reduction Potential (V)‡	−2.08	−1.63	−1.2	−0.91	−1.18
Common Oxidation States§	+3	+2,+3,+4	+2,+3,+4,+5	+2,+3,+6	+2,+3,+4,+7
Electronegativity	1.3	1.5	1.6	1.6	1.5

Property	Iron	Cobalt	Nickel	Copper	Zinc
Atomic Number	26	27	28	29	30
Valence Electron Configuration*	$4s^23d^6$	$4s^23d^7$	$4s^23d^8$	$4s^13d^{10}$	$4s^23d^{10}$
Atomic Radius (pm)†	124	125	125	128	133
Density (g cm^{-3})	7.88	8.9	8.90	8.96	7.13
Ionization Energy (kJ mol^{-1})					
First	759	758	737	746	906
Second	1561	1646	1753	1958	1733
Third	2957	3232	3394	3554	3793
Melting Point (°C)	1535	1495	1453	1083	420
Boiling Point (°C)	2250	2870	2732	2567	907
Std. Reduction Potential (V)‡	−0.44	−0.28	−0.23	+0.34	−0.76
Common Oxidation States§	+2,+3	+2,+3	+2	+1,+2	+2
Electronegativity	1.8	1.9	1.9	1.9	1.6

* All have the argon core, so that the complete configuration of scandium, for example, is [Ar]$4s^23d^1$.
† Half the distance between adjacent nuclei in the pure crystalline metal.
‡ Except for scandium, the value quoted is for the process $M^{2+}(aq) + 2e^- \rightarrow M(s)$. Scandium has no +2 ion, so the value in this table refers to the process $Sc^{3+}(aq) + 3e^- \rightarrow Sc(s)$.
§ Compounds in other oxidation states are known, but are rare.

Atomic Radius and Density

The value of atomic radius given in Table 23.1 is the *metallic radius,* defined as half the distance between adjacent nuclei in the pure, crystalline metal. The listed values follow the usual periodic trend and generally decrease with atomic number. Shielding of one electron by others in the same subshell is not perfect, and the *effective nuclear*

Section 7.5 (page 286) includes a discussion of another way in which atomic radius can be defined and measured.

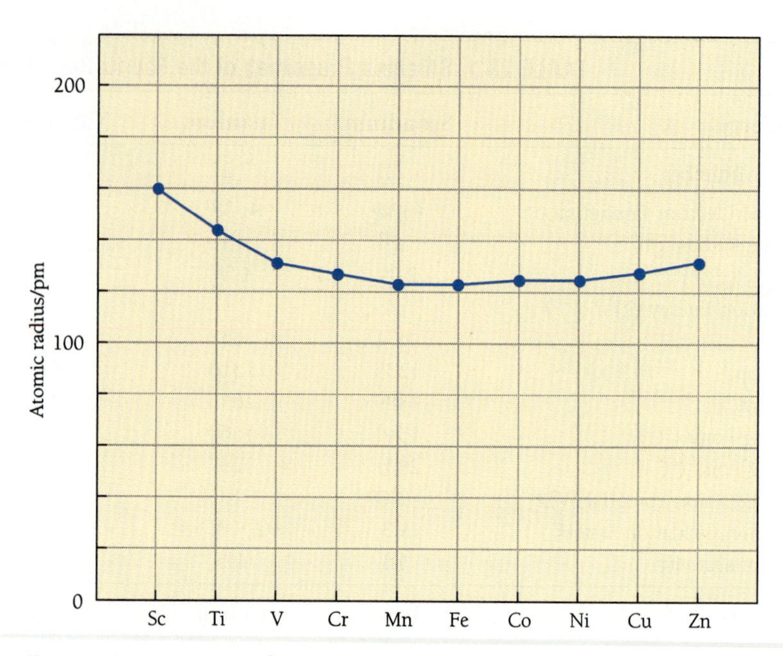

Figure 23.1 Variation of atomic radius (metallic radius) with atomic number in the fourth-period transition elements.

charge increases slightly as a proton is added to the nucleus and an electron is added to the 3*d* subshell. The electron cloud shrinks, therefore, because of the stronger attraction of the increased effective nuclear charge. This trend is more easily seen when the values are plotted against atomic number (Figure 23.1). The slightly increased radius shown by copper and zinc may also be explained in terms of electron shielding. The

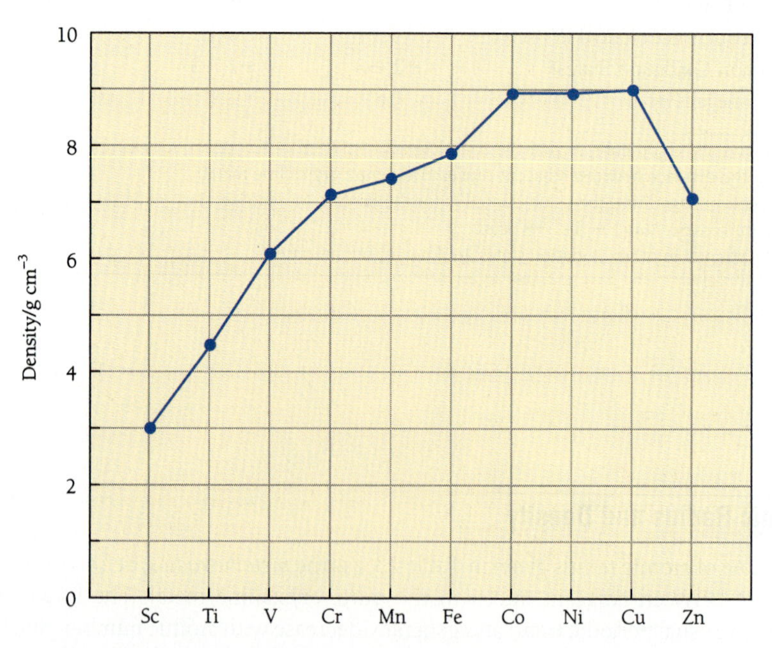

Figure 23.2 Variation of density of metals with atomic number in the fourth-period transition elements.

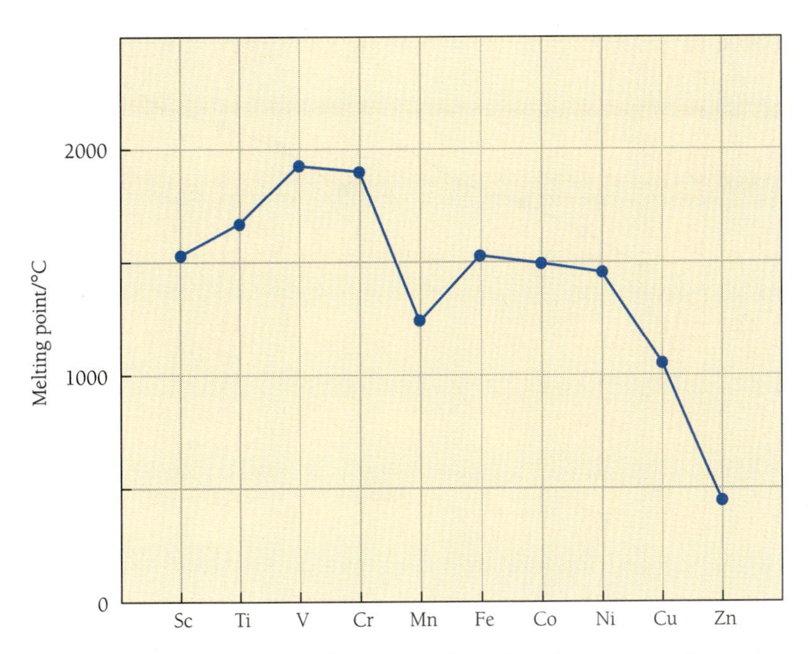

Figure 23.3 Variation of melting point of metals with atomic number in the fourth-period transition elements.

complete 3*d* subshell shields the 4*s* electrons from the nucleus more effectively than an incomplete *d* subshell, hence the effective nuclear charge in Cu and Zn is somewhat less than expected. Therefore, the 4*s* electrons are less strongly attracted, and lie at a greater distance from the nucleus.

See Section 7.5 for a discussion of the relationship between shielding, effective nuclear charge, and atomic radius.

The density of the metals increases as atomic number increases, as shown in Figure 23.2. The increase is due primarily to the fact that density is *inversely* proportional to atomic volume, which in turn varies as the cube of the atomic radius. The increasing mass of the metals also contributes to the trend.

Melting Point

Figure 23.3 shows the variation in melting points, and, except for manganese, there is a fairly regular trend. The melting point of a substance is a measure of the strength of its cohesive forces, so that the melting point of a metal increases as the strength of the metallic bond increases. The bond strength correlates roughly with the number of *un*paired *d* electrons, which rises from 1 to 5 and then falls to 0 as the atomic number increases from 21(Sc) to 30(Zn).

Ionization Energy, Electronegativity, and Oxidation States

When a fourth-period transition element ionizes, it loses electrons first from the 4*s* subshell, then from the 3*d* subshell. Ionization energy generally increases with atomic number (Figure 23.4) due to the increased nuclear charge. This is true not only for the first ionization energy (to M^+) but also for the subsequent ionizations in which cations of higher charge are produced.

Electronegativity also follows the general periodic trend, increasing as atomic number increases within a period. Manganese and zinc are exceptions to the trend, but the differences are in any case not large.

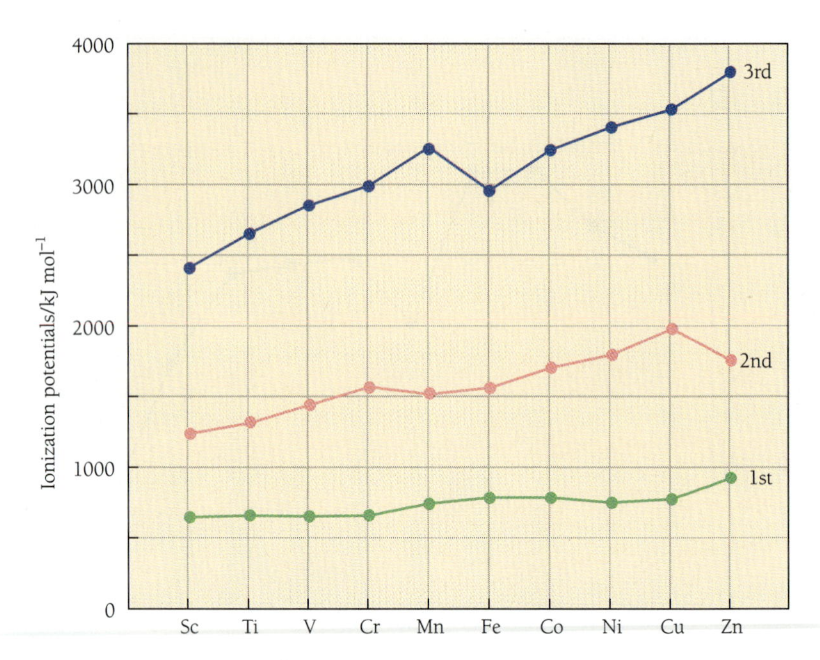

Figure 23.4 Variation of ionization energy with atomic number in the fourth-period transition-metal atoms.

A transition metal always uses its 4*s* electrons in compound formation. In its lowest oxidation state, a metal can form ionic compounds like $MnCl_2$ by *transfer* of the 4*s* electrons to another species. In higher oxidation states, the 3*d* electrons are also involved: bonding is by electron *sharing* to form covalent species like MnO_4^-. The number of electrons lost or shared is the oxidation state of the metal in the compound. Since one or more of the 3*d* electrons may be used, several oxidation states in transition metal compounds are possible. In the elements Sc through Mn, the maximum oxidation state corresponds to the use of *all* the 4*s* and 3*d* electrons. In the heavier elements Fe through Zn, the number of *unpaired* 3*d* electrons decreases. Loss or sharing of the remaining paired electrons to form bonds is less favorable, and the maximum oxidation number shown by these elements is less than the total number of valence electrons.

Standard Reduction Potential

With the exception of manganese and zinc, standard reduction potential of the doubly charged cation increases regularly as atomic number increases (Figure 23.5). Recall from Chapter 19 that any species (such as Fe^{2+}) whose reduction potential is negative cannot be reduced by H_2, and that its reduced form (Fe) *can* be oxidized by H^+. All fourth-period transition metals except copper can be oxidized by $H^+(aq)$. Therefore, these metals will dissolve in (that is, react with) dilute HCl, forming soluble chloride salts and liberating gaseous H_2. Dissolution of copper requires an oxidizing agent stronger than H^+, such as the nitrate ion [for the reaction $NO_3^-(aq) + 4\,H^+(aq) + 3\,e^- \rightarrow NO(g) + 2\,H_2O$, the standard reduction potential is +0.96 V]. Copper dissolves easily in nitric acid.

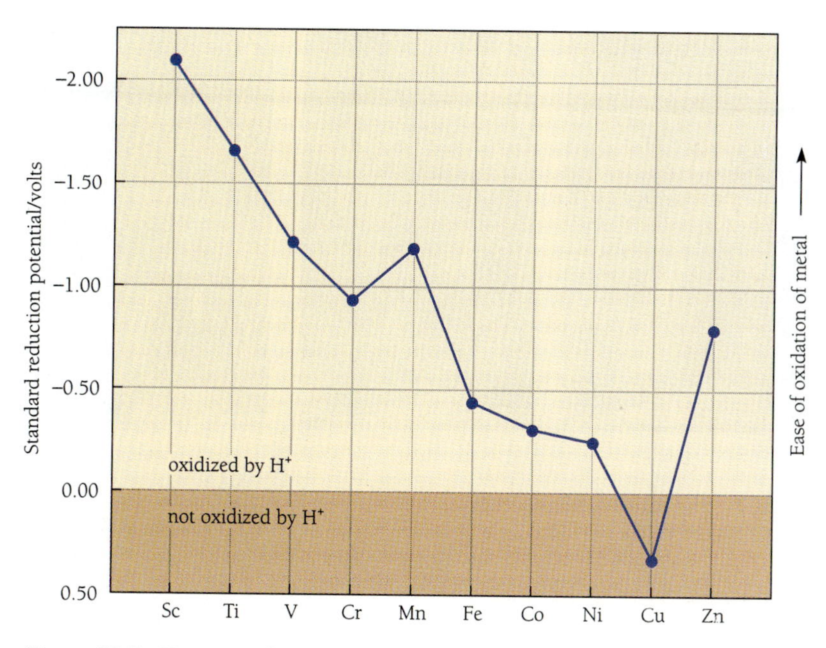

Figure 23.5 Variation of standard reduction potential with atomic number in the doubly charged cations of the fourth-period transition elements. Since scandium does not form a doubly charged ion, the SRP plotted here refers to the Sc^{3+} ion.

Color and Magnetism

Most transition metal compounds are colored, some very strongly (Figure 23.6). Their color is affected by their oxidation state. For instance, $FeCl_2$ is greenish-yellow and $FeCl_3$ is yellow-brown. Color is determined by the wavelength of the visible light that is absorbed, and this in turn depends on the spacing between electronic energy levels in the substance (Chapter 7). It is the spacing of the 3*d* levels that determines the color of fourth-period transition metal compounds.

More information about color is presented in Section 23.9.

The magnetic properties of transition metals and their compounds are determined by the configuration of the valence electrons. Many of these compounds are

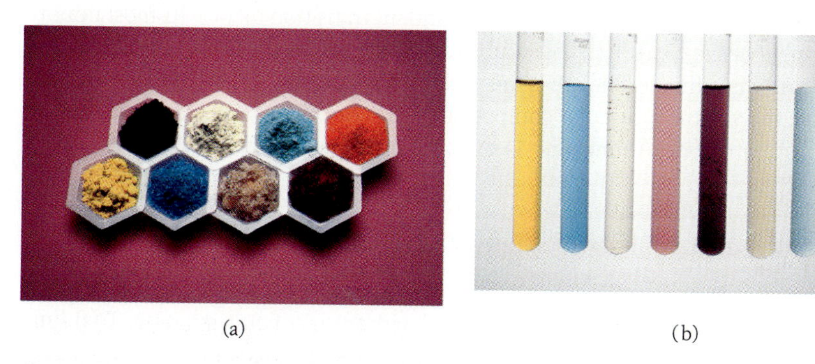

(a) (b)

Figure 23.6 (a) Many transition metal compounds are highly colored. Bottom row, left to right: $FeCl_3$, $CuSO_4$, $MnCl_2$, $CoCl_2$; top row, left to right: $Cr(NO_3)_3$, $FeSO_4$, $NiSO_4$, $K_2Cr_2O_7$. (b) The same compounds in aqueous solution. Left to right: $FeCl_3$, $CuSO_4$, $MnCl_2$, $CoCl_2$, $Cr(NO_3)_3$, $FeSO_4$, $NiSO_4$, $K_2Cr_2O_7$.

The yellow color of this 15th century Ming Dynasty vase is the result of the addition of $FeCl_2$ to the glaze.

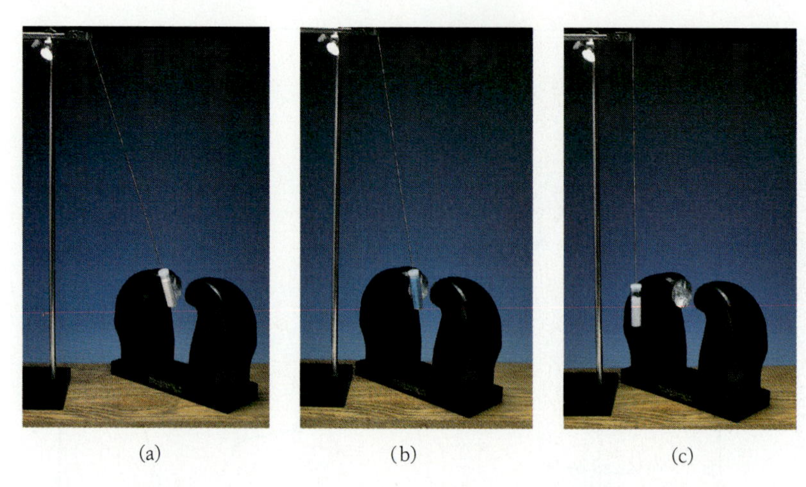

(a) (b) (c)

Figure 23.7 The paramagnetism of some transition-metal compounds is apparent in a strong magnetic field. (a) $MnSO_4$ is attracted to the magnet. (b) $CuSO_4$ is also attracted, but less so than $MnSO_4$ because its paramagnetism is less. (c) KCl is not attracted at all, because it is not a paramagnetic substance.

paramagnetic, meaning that they have at least one unpaired electron. Modern theories of bonding (Section 23.9) are able to make good predictions of the strength of paramagnetism (Figure 23.7).

23.3 SOME CHEMISTRY OF THE FOURTH-PERIOD TRANSITION ELEMENTS

Scandium is not a particularly rare element. At 5 ppm (0.0005%) of the earth's crust, it is as common as arsenic (5 ppm) and almost twice as common as boron (3 ppm). There are no concentrated deposits of scandium minerals, however, and the element is not mined or used in quantity. It is unusual among the transition elements in that it exhibits only one oxidation state, +3. The Sc^{3+} ion has no 3*d* electrons, and as a result its compounds are colorless and diamagnetic.

Titanium, at 5800 ppm (0.58%) of the earth's crust, is the ninth most abundant element (and, after iron, the second most abundant transition metal). Its most important ores are *ilmenite,* $FeTiO_3$, a constituent of granite, and *rutile,* TiO_2. Preparation of the pure metal is difficult. The ore is first treated with carbon and chlorine at red heat to produce the volatile molecular compound $TiCl_4$ (bp 136.4 °C). After some purification, $TiCl_4$ is reduced to metallic Ti by molten magnesium at 800 °C.

$$FeTiO_3(s) + 3\ C(s) + 3\ Cl_2(g) \longrightarrow FeCl_2(s) + 3\ CO(g) + TiCl_4(g) \quad (23\text{-}1)$$

$$TiCl_4(g) + 2\ Mg(s) \longrightarrow 2\ MgCl_2(\ell) + Ti(s) \quad (23\text{-}2)$$

Titanium metal is both strong and lightweight, and it is used in aircraft and other applications where weight must be minimized without sacrificing strength. Titanium does not corrode in air, because, like Al (and many other metals), it is quickly covered by a tough, impervious oxide coating. At high temperature, titanium forms compounds with most nonmetals. TiB, TiC, and TiN are hard, stable, high-melting substances. The most stable oxidation state is +4, and the most common compound is

Titanium dioxide is used as the white pigment in paint.

TiO_2. Titanium dioxide has a brilliant white color, and it is used as the white pigment in paint, paper, plastics, and other products. More than a million tons are produced annually. Both Ti(II) and Ti(III) compounds are unstable in aqueous solution and are readily oxidized to the +4 state.

The use of Roman numerals in the names of inorganic compounds is discussed in Section 2.8.

Vanadium is fairly common (150 ppm) in the earth's crust, occurring in low concentrations in many locations. It is also found in concentrated deposits of certain minerals, including *carnotite,* $K(UO_2)VO_4 \cdot 3/2H_2O$. Carnotite is mined and processed chiefly for its uranium content, and vanadium is extracted as a byproduct. Vanadium is also present in Venezuelan petroleum. This causes some problems, because some of the catalysts used in petroleum refining are deactivated ("poisoned") by vanadium. Like titanium, vanadium is sufficiently reactive to make preparation of the pure metal difficult. If necessary, metallic vanadium can be obtained by electrolysis of molten salts. For many applications, however, the pure metal is not required. Reduction of a mixture of V_2O_5 and Fe_2O_3 with aluminum yields *ferrovanadium,* which is added to iron to make vanadium steel, a tough, springy material with good corrosion resistance. About three-quarters of the vanadium produced in this country is used for this purpose. The other major use is as the catalyst (V_2O_5) for the conversion of SO_2 to SO_3 in the manufacture of sulfuric acid (Section 22.5). The most common oxidation state observed in vanadium compounds is +5. Examples are V_2O_5, VCl_5, and various compounds containing the *vanadate* ion, VO_4^{3-}. The VO^{2+} ion (+4 state) is also known.

Chromium is common (200 ppm) in the earth's crust, and its use as a protective and decorative coating makes it very well known. The chief ore is *chromite,* $FeCr_2O_4$, which is reduced with carbon to produce *ferrochromium,* a Cr-Fe alloy used directly in steelmaking.

$$FeCr_2O_4(s) + 4\ C(s) \xrightarrow[\Delta]{} Fe(\ell) + 2\ Cr(\ell) + 4\ CO(g) \qquad (23\text{-}3)$$

Pure Cr can be obtained by treatment of chromite with alkali and O_2 to produce the water-soluble Na_2CrO_4, which is separated and then reduced to Cr_2O_3 with carbon. The final step is the further reduction of Cr_2O_3 to Cr with aluminum. The common oxidation states are +2, +3, and +6. Cr(II) compounds are good reducing agents, while Cr(VI) compounds are powerful oxidizing agents. The oxides Cr_2O_3 (green) and CrO_3 (red-orange) are used as pigments in paint.

Chromium is used as a decorative and protective coating for metal parts.

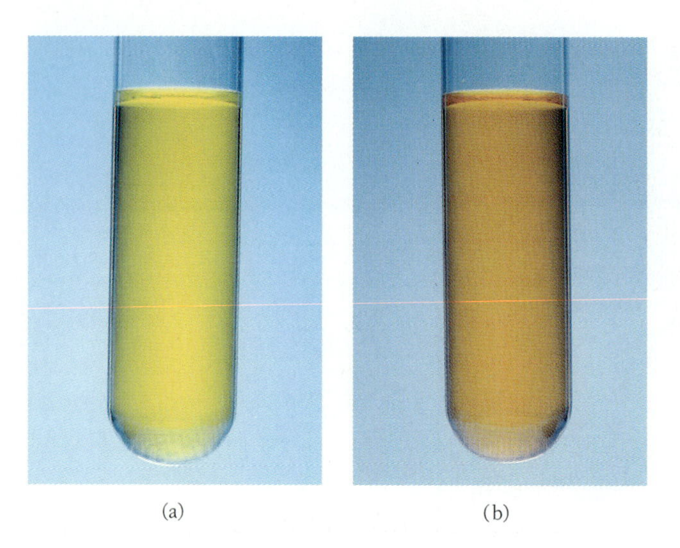

(a) (b)

Figure 23.8 There is a color change associated with the equilibrium between the chromate and dichromate ions. (a) The yellow color of CrO_4^{2-} predominates in basic solution. (b) Addition of a small amount of HCl(aq) shifts the equilibrium to the right, and the solution takes on the orange color of $Cr_2O_7^{2-}$.

Chromium(VI) oxide, CrO_3, dissolves in water to form one of several different oxygen-containing species. The *chromate* ion, CrO_4^{2-}, predominates in basic solution, while in acid solution the equilibrium favors the *dichromate* ion, $Cr_2O_7^{2-}$ (Figure 23.8).

$$2\ CrO_4^{2-}(aq) + 2\ H^+(aq) \rightleftharpoons Cr_2O_7^{2-}(aq) + H_2O \qquad (23\text{-}4)$$

The chromate ion is tetrahedral, and the dichromate ion consists of two tetrahedra sharing an oxygen atom at one vertex. The electrolyte and source of chromium in electroplating (Section 19.5) is a mixture of chromic acid (H_2CrO_4) and sulfuric acid.

Manganese metal.

Manganese is found primarily as *pyrolusite,* which contains MnO, MnO_2, and Fe_2O_3. It is a relatively abundant metal (1000 ppm or 0.1%), used primarily in steel. Pyrolusite is reduced with carbon to produce *ferromanganese,* a Mn-Fe-C alloy used directly in steelmaking. Manganese steels are very tough: their abrasion- and shock-resistance makes them useful for armor plate, bulldozer blades, and other products. The most common oxidation states are +2 and +7, but many compounds of the +3, +4, +5, and +6 states exist. Because of this diversity, manganese has an extensive and complicated oxidation-reduction chemistry. For example, the *manganate* ion (MnO_4^{2-}, oxidation number +6) disproportionates spontaneously in acid or neutral solution, yielding Mn(VII) and Mn(IV).

$$3\ MnO_4^{2-}(aq) + 4\ H^+(aq) \longrightarrow 2\ MnO_4^-(aq) + MnO_2(s) + 2\ H_2O \qquad (23\text{-}5)$$

The *permanganate* ion (MnO_4^-, oxidation number +7) is a powerful oxidant, but the nature of the reduced form depends on the pH of the solution. In acid, MnO_4^- is reduced to Mn^{2+}, while in base, the product is $MnO_2(s)$.

$$MnO_4^-(aq) + 8\ H^+(aq) + 5\ e^- \longrightarrow Mn^{2+}(aq) + 4\ H_2O; \qquad \mathscr{E}° = +1.51\ V \qquad (23\text{-}6)$$

$$MnO_4^-(aq) + 2\ H_2O + 3\ e^- \longrightarrow MnO_2(s) + 4\ OH^-(aq); \qquad \mathscr{E}° = +1.23\ V \qquad (23\text{-}7)$$

Acidic permanganate solutions are used by analytical chemists in redox titrations. Because of the intense purple color of MnO_4^-, such titrations usually do not require an added indicator.

Iron is the fourth most abundant element in the earth's crust (4.7%). The major ores are *hematite*, Fe_2O_3, and *magnetite,* Fe_3O_4. The metallurgy of iron has been discussed (Section 1.6). Pure iron, rarely seen, is easily oxidized in moist air to rust, a mixture of iron oxides. The common oxidation states of iron are +2 and +3. Essentially the only use of iron is in steel, but there is a bewildering variety of different steels and different uses. Iron is important in biochemistry, because as a component of *hemoglobin,* it is an essential feature of the oxygen-transport system of most higher animals.

The role of hemoglobin in oxygen transport is discussed in Section 23.10.

Cobalt (23 ppm) occurs most frequently as the sulfide, often in association with Ni or As. It is obtained in greatest quantity as a byproduct of the extraction of Ni, Cu, and Pb from their ores. The metal itself is relatively unreactive and does not combine with H_2 or N_2, even at elevated temperature. Exposure to high-temperature moist air gradually converts the metal to the oxide CoO. This oxide is used to produce a deep blue color ("cobalt blue") in glasses, ceramic glazes, and paints. Cobalt is used in some specialty alloys, for example *alnico,* an Al-Ni-Co-containing steel from which some corrosion-resistant "stainless steels" and strong permanent magnets are made. The chemistry of cobalt is dominated by the +2 and +3 oxidation states.

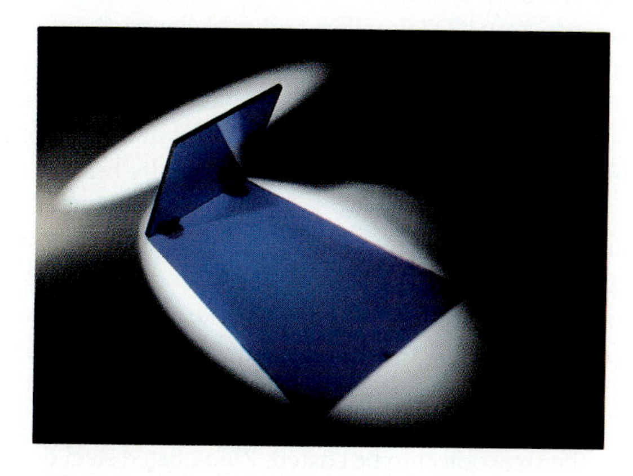

Glass can be given a deep blue color by adding a cobalt(II) salt to the molten glass.

Nickel occurs as 20 ppm of the earth's crust. It is often found as the sulfide, together with the sulfides of copper and iron. Roasting of the ore produces the oxide, which is then reduced with carbon. The result is a mixture of nickel, copper, and iron, from which the nickel is separated by a special method. On exposure to CO at 60 °C, nickel forms *nickel carbonyl,* $Ni(CO)_4$, a gaseous compound that is easily separated from the solid Cu and Fe. At somewhat higher temperature, $Ni(CO)_4$ dissociates, leaving very pure metallic nickel.

$$\mathrm{Ni(s)} + 4\,\mathrm{CO(g)} \underset{200\text{ °C}}{\overset{60\text{ °C}}{\rightleftharpoons}} \mathrm{Ni(CO)_4(g)} \qquad (23\text{-}8)$$

The coin we call the "nickel" is actually a Cu-Ni alloy containing only 30% nickel.

Nickel is used in a number of alloys. One of these, called Monel metal, is nickel that contains about 25% Cu and 3% Fe. This alloy is very resistant to corrosion, and is one of the few materials that can withstand direct exposure to fluorine. U.S. coins, including both the 5-cent piece and the 3-layer dimes and quarters, contain an alloy that is 30% Ni and 70% Cu. Although other, unstable, oxidation states have been observed, the +2 state of nickel is most important. Nickel(II) compounds, both pure and in aqueous solution, are often a deep emerald green in color.

Copper is moderately abundant (70 ppm) and is widely distributed as sulfides, arsenides, chlorides, and carbonates, often in combination with iron and other transition elements. The extraction of copper from its chief ore, *chalcopyrite* ($CuFeS_2$) is discussed in Section 18.5, and the electrolytic refining of impure copper is discussed

This sample contains native (elemental) copper as well as two different copper minerals. *Malachite* (green) is $Cu_2(OH)_2CO_3$; *azurite* (blue) is $Cu_3(OH)_2(CO_3)_2$.

in Section 19.5. The metal is soft and ductile, and it has very high thermal and electrical conductivity. About half of the copper produced is used for electrical purposes. Together with Zn, Sn, and Pb, copper forms a variety of alloys called *brass* and *bronze*. Both sterling silver and the 14-carat gold used for jewelry contain copper, which lends hardness and durability to these alloys. Although less expensive substitutes are available today, copper has been extensively used for roofing and water pipes. The chalky green color of weathered copper is due to the formation of $CuCO_3$ and "basic sulfates," $CuSO_4 \cdot (2 \text{ or } 3)Cu(OH)_2$. Copper forms compounds in the +1 and +2 oxidation states. Aqueous solutions of Cu(I) salts are unstable due to spontaneous disproportionation.

$$2\ Cu^+(aq) \longrightarrow Cu(s) + Cu^{2+}(aq) \qquad (23\text{-}9)$$

As a result the only Cu(I) compounds that persist in a moist environment are those that are insoluble, such as Cu_2O, Cu_2S, CuCl, and CuCN.

Several features account for the fact that copper was the material used in the first metallurgy in human history. (1) Like silver and gold, copper is sufficiently unreactive so that it is sometimes found as the metal—the only processing required before use is beating or bending to shape. (2) Rich veins of sulfide ores are often associated with deposits of elemental copper—there was no difficulty in finding or recognizing the ore. (3) The sulfide ore is reduced to the metal simply by heating in air—no other reagents are required ($CuS + O_2 \rightarrow Cu + SO_2$).

Zinc is about twice as abundant (130 ppm) as copper, and sometimes occurs together with copper, as sulfide ores. Reduction of the mixed sulfides (by heating with carbon) gives *brass:* this name is applied to alloys containing substantial amounts of both zinc and copper. Most of the zinc produced in this country is used in the manufacture of "galvanized iron," zinc-coated steel. The coating is applied by dipping steel into molten zinc, or electrolytically in a cell in which the electrolyte is $ZnSO_4(aq)$ and the cathode is the steel object to be coated. Zinc-coated steel does not rust, partly because of the formation of a thin, protective surface layer of zinc oxide. Additional protection arises from an electrochemical mechanism: zinc is more easily oxidized than iron, so rusting cannot take place until all the zinc is consumed (Section 19.6). Zinc is an important element in biochemistry and an essential nutrient in the human diet. Many of the enzymes that govern metabolic processes, in virtually all organisms, contain zinc.

23.4 HIGHER TRANSITION SERIES

There are additional series of transition elements beyond the fourth period. In the fifth period the elements Y($Z = 39$) through Cd($Z = 48$) are characterized by a partially or completely filled 4*d* subshell. In the sixth period, it is the 5*d* subshell that is filling in the elements La($Z = 57$) through Hg($Z = 80$). The three series of elements are referred to by several names: the **first, second**, and **third transition series**; the **fourth-, fifth-**, and **sixth-period transition elements**; or the **3*d*, 4*d***, and **5*d* transition elements.**

The elements in the second and third transition series are similar in many ways to the fourth-period transition elements. They are characterized by multiple oxidation states and highly colored compounds, many of which are paramagnetic. The trend of atomic size shown in Figure 23.1 for the fourth-period transition elements occurs also in the fifth and sixth periods. As shown in Figure 23.9, there is a decrease of atomic size as atomic number increases (with a slight upturn at the end of the series, in Groups 1B and 2B). The reason for the decrease has been discussed: the effective nuclear

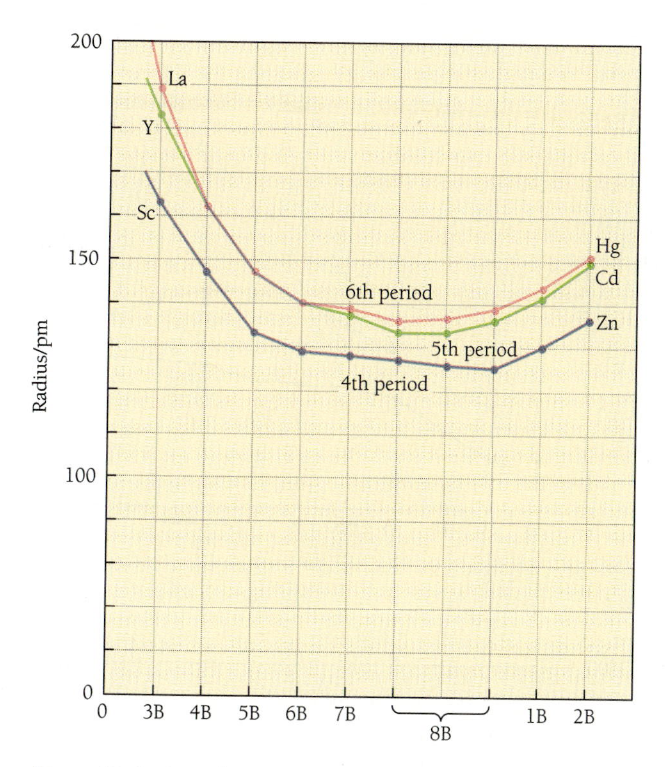

Figure 23.9 Trends in atomic radii of the transition elements in the fourth, fifth, and sixth periods.

charge is increasing, so that the electrons are pulled in more tightly. The fifth-period elements are about 10% to 15% larger than the corresponding fourth-period elements because of the additional shell of electrons. The sixth-period elements, however, are about the same size as their fifth-period counterparts. The size increase expected on going from the fifth- to the sixth-period transition series does not occur because of an effect known as the **lanthanide contraction.** There are 14 "extra" elements, the lanthanides, sandwiched in between elements #57 and #72 at the beginning of the series. The usual shrinkage with increasing atomic number occurs, and the cumulative effect is enough to offset the size increase due to the additional shell. Sixth-period hafnium, the first element after the lanthanides, is in fact slightly smaller than fifth-period zirconium. Table 23.2 shows that the normal size increase occurs with Cs, but, after the lanthanide contraction, not with Hf or Pt.

TABLE 23.2 Effect of the Lanthanide Contraction

Atomic Radius in pm			
Period	Group 1A	Group 4B	Group 8B
4	227 (K)	147 (Ti)	125 (Ni)
5	248 (Rb)	160 (Zr)	138 (Pd)
6	265 (Cs)	158 (Hf)	138 (Pt)

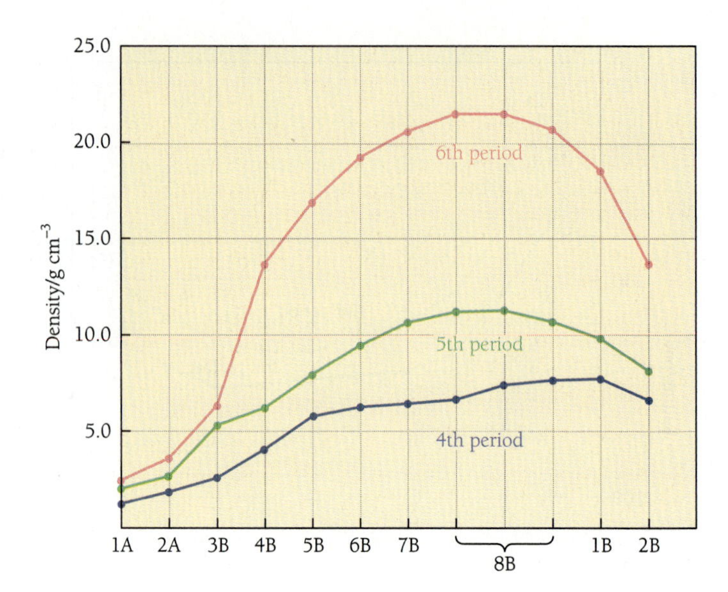

Figure 23.10 Density trends in the transition elements (and *s*-block elements) of the fourth, fifth, and sixth periods. The sixth-period densities are significantly larger because the mass is greater, while the atomic radii are about the same as in the fifth period.

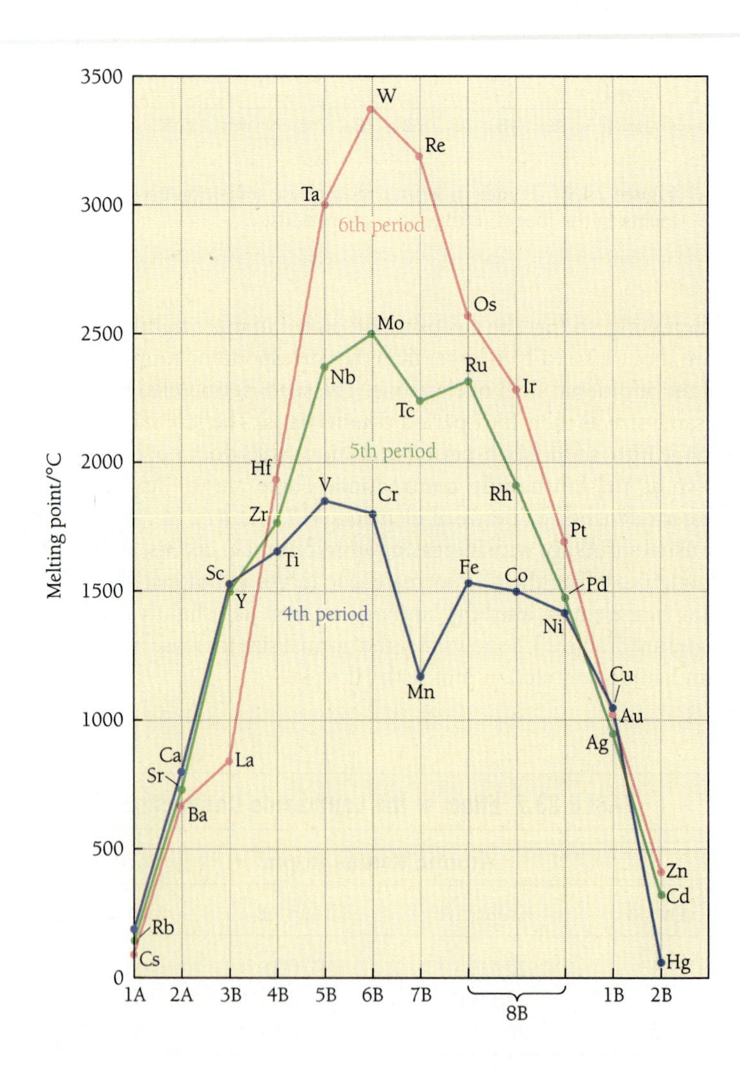

Figure 23.11 Trends in melting point among the transition metals (and the *s*-block elements) of the fourth, fifth, and sixth periods.

Trends in density (Figure 23.10) and melting point (Figure 23.11) are similar in all three periods. The fifth- and sixth-period transition elements are much less abundant in the earth's crust, but nonetheless several of them have important industrial uses. Tungsten is used in lamp filaments, high-temperature steel alloys, and cutting tools (as the extremely hard tungsten carbide). Platinum is an effective and widely used catalyst in a variety of industrial applications, silver is essential to photography, mercury is used in batteries and fluorescent lights as well as in thermometers and barometers.

Cutting and grinding tools made of tungsten carbide are used in metal-working.

23.5 INNER TRANSITION SERIES

There are two other transition series, whose behavior is affected by the presence of a partly filled *f* (rather than *d*) subshell.

The Lanthanides

The 4*f* subshell is slightly lower in energy than the 5*d* subshell, and it must be filled with 14 electrons before the 5*d* shell begins to fill in the sixth-period transition series. The 14 elements that follow lanthanum are characterized by an increasing number of 4*f* electrons. Together with lanthanum, these elements are known as the **lanthanides** or the **rare earths.** They are referred to collectively as an *inner transition series.* The name is given because they all have **5***p* and **6***s* electrons in the outer shells, while the 4*f* subshell, which is filling, is a smaller, "inner" shell. The lanthanides are remarkably similar to one another in physical and chemical behavior. As a result of this similarity they often occur together in nature, and they are difficult to separate from one another. The most common ores contain a mixture of oxides, M_2O_3. Compounds of the lanthanides are ionic, not covalent, and the only common oxidation state is +3. All of the M^{+3} ions except La^{+3} and Lu^{+3} are paramagnetic, and most of the lanthanide salts are colored. In spite of the name "rare earths," they are not particularly rare. In fact they are more abundant than tungsten, mercury, arsenic, and a number of other, more familiar elements. Although not used in large amounts in our society, the rare earths do have some specialty applications. The "flint" in cigarette lighters is a cerium-steel alloy, and a mixture of rare-earth oxides is added to the mantles of gas lanterns in order to increase their brightness. Neodymium is a component of some high-intensity lasers, and several other lanthanides are used as light emitters in color television screens.

The Actinides

The elements Ac($Z = 89$) through Lr($Z = 103$) in the seventh period form a second inner transition series, called the **actinides.** The 5*f* subshell is partially filled in most of these elements. All are radioactive, and only Th and U occur in useful quantity in the earth's crust. The others are produced in trace quantities in nuclear reactions (Section 24.4), both in nature and in the laboratory. Because of their radioactivity, actinides cannot be handled without heavy protective shielding. Many ingenious remote-handling techniques have been developed for the study of their chemistry (Figure 23.12).

All of the actinides are metals, and their compounds are predominantly ionic rather than covalent: the +3 and +4 oxidation states are most common. As with other transition elements, some covalent compounds are known in higher oxidation states. One example of a covalent actinide compound is $UF_6(g)$.

Figure 23.12 The chemistry of the actinides must be studied from a distance, using robot manipulators.

23.6 COORDINATION COMPLEXES

In a coordinate covalent bond, both of the shared electrons come from the same atom. Coordinate bonding is also called *donor-acceptor* bonding.

We have made heavy use of ideas about chemical bonding to explain the chemical behavior of compounds. Our understanding of the chemistry of the main-group metals begins with the ionic bond, whereas in organic chemistry, the covalent bond is by far the most important type. The chemistry of the transition elements, on the other hand, is dominated by *coordinate covalent* bonds. Transition metal compounds that are held together by coordinate bonds are called coordination compounds or coordination complexes.

Metal Ions, Lewis Adducts, and Ligands

Recall from Section 15.7 that *Lewis acids* are species capable of accepting an electron pair from a *Lewis base,* thus forming an *adduct* in which the two parts are joined by a coordinate covalent bond. Metal cations, because of their positive charge, are almost always able to accept an electron pair and function as Lewis acids. Transition metal ions, having vacancies in their *d* subshells, are well suited for this type of interaction. Electron-pair donor species that bond to metals in this way are called **ligands.** A ligand must have at least one unshared (lone) pair of electrons. A ligand may be a neutral molecule, such as NH_3 or H_2O, or a negative ion, such as Cl^- or CN^-. The adducts, which contain a metal ion and several ligands, are called **coordination compounds** if they are neutral species, and **coordination complexes** if they are ionic. The adduct itself may be a cation, an anion, or a neutral compound, depending on the charges of the metal ion and the ligands. For example, $[Cu(NH_3)_4]^{2+}$ is a cation, because the copper ion has a charge of +2 and the ligands are neutral species. On the other hand, $[Fe(CN)_6]^{3-}$ is an anion, because the ferric ion has a charge of +3 and each of the six ligands has a charge of −1. $Cr(OH)_3(H_2O)_3$ is a neutral compound, because the positive charge of its metal ion is balanced by the negative charge of three of its six ligands. It is customary to enclose the formula of a coordination complex (an ion) in square brackets—do not confuse this with the use of square brackets to indicate the concentration of a solute.

Coordination Number and the Shape of Coordination Complexes

The number of metal-ligand coordinate bonds in a coordination compound or complex is called the **coordination number** of the metal ion. Coordination numbers of 4 and 6 are most common, while 2, 5, 7, 8, and 9 occur less frequently. If the coordination number is 2, the complex is linear in shape: that is, the ligand-metal-ligand bond angle is 180°. One example is $[Ag(CN)_2]^-$, the complex formed in the extraction of silver from its ore.

The role of cyanide ion in mining is discussed in Section 22.3.

180°

$$[N{\equiv}C{-}Ag{-}C{\equiv}N]^-$$

If the coordination number is 6, the complex is octahedral in shape. That is, the six ligands lie at the vertices of a regular octahedron centered on the metal ion. The $[Co(NH_3)_6]^{3+}$ ion, shown in Figure 23.13, is an example of this geometry. Be sure you understand the meaning of the sketch in Figure 23.13b, because this representation will be used for octahedral complexes throughout the remainder of the text.

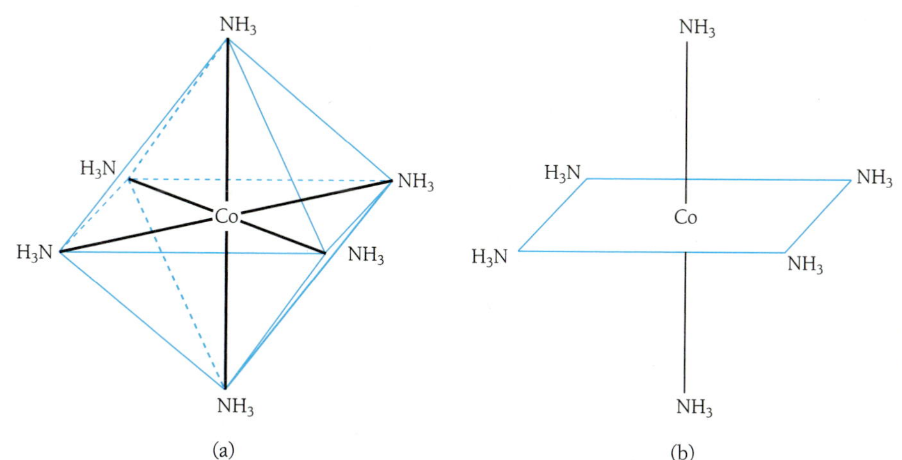

Figure 23.13 (a) The six ligands of the complex cation $[Co(NH_3)_6]^{3+}$ occupy the vertices of a regular octahedron. The shape is indicated by blue lines outlining the edges of the octahedron, while the ligand-metal coordinate bonds are shown by heavy black lines. (b) Octahedral complexes are more commonly represented by an abbreviated drawing that shows two bonds (the black lines) and the four edges of the octahedron that lie in the plane perpendicular to the indicated bonds.

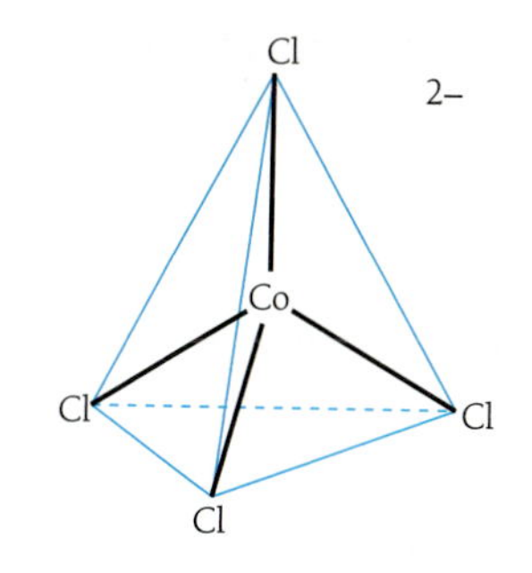

Figure 23.14 Tetrahedral geometry of the complex ion $[CoCl_4]^{2-}$.

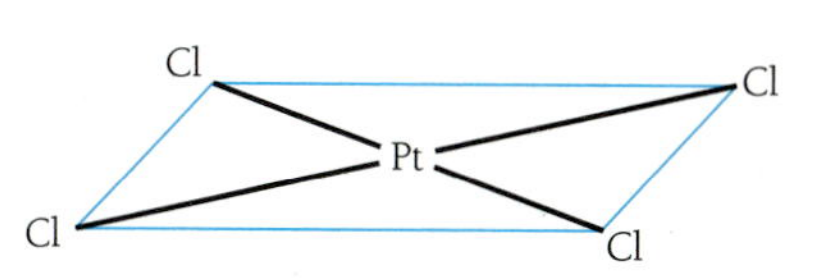

Figure 23.15 Square planar geometry of the complex ion $[PtCl_4]^{2-}$.

Two different shapes are possible when the coordination number is 4. In some complexes, such as $[CoCl_4]^{2-}$ and $[NiCl_4]^{2-}$, the ligands are tetrahedrally arranged around the metal (Figure 23.14). In others, such as $[PtCl_4]^{2-}$, the ligands lie at the corners of a square. This shape is called **square planar.** All five nuclei of the complex ion $[PtCl_4]^{2-}$ lie in a plane, as indicated in Figure 23.15. The set of locations or **binding sites** that are or can be occupied by ligands is called the **coordination sphere.**

The Valence Shell Electron Pair Repulsion theory was introduced in Chapter 8 to predict the shapes of covalently bonded molecules. The shapes of *coordination complexes* cannot be predicted reliably with this simple theory.

Types of Ligands

Some ligands have more than one atom with a lone pair of electrons, and can form more than one coordinate bond with a metal ion. Ligands that can form two bonds are called **bidentate,** meaning that they can grip the metal ion with "two teeth." Ligands that form only one bond, such as NH_3, H_2O, OH^-, and Cl^-, are called **monodentate.** *Ethylenediamine,* $H_2N{-}CH_2{-}CH_2{-}NH_2$, is an example of a bidentate ligand. The size and shape of this molecule is such that both of its nitrogen atoms can form a coordinate bond with the same metal ion. That is, one ligand molecule occupies *two* locations in the coordination sphere. Figure 23.16 illustrates this structure, in the complex cation $[Co(H_2N{-}CH_2{-}CH_2{-}NH_2)_3]^{3+}$. Ethylenediamine is a common ligand, and is frequently abbreviated as "en": $[Co(en)_3]^{3+}$.

Other **polydentate** ligands can form more than two bonds. *Ethylenediaminetetraacetate (EDTA)* can form six bonds, using the lone pairs indicated in the structure below: it is a *hexa*dentate ligand.

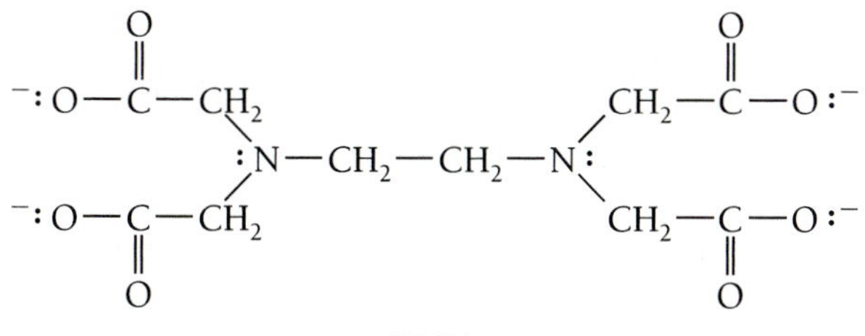

EDTA

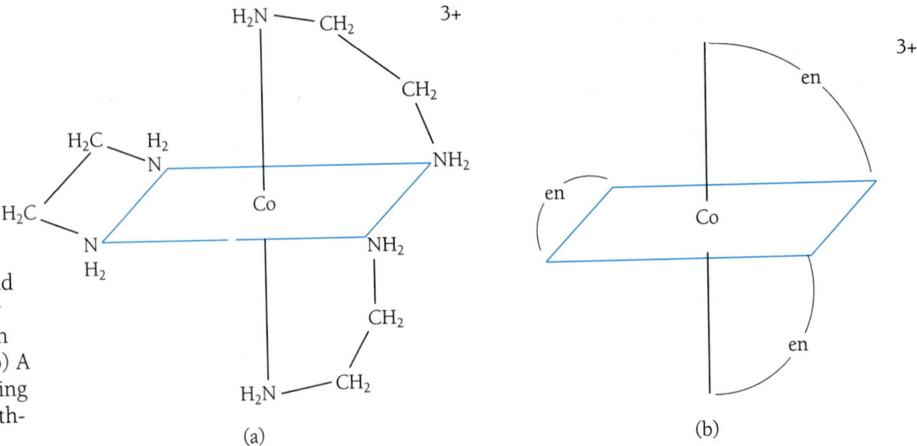

Figure 23.16 (a) A bidentate ligand such as ethylenediamine can occupy two binding sites in the coordination sphere of an octahedral complex. (b) A simpler diagram of the same ion, using the common abbreviation "en" for ethylenediamine.

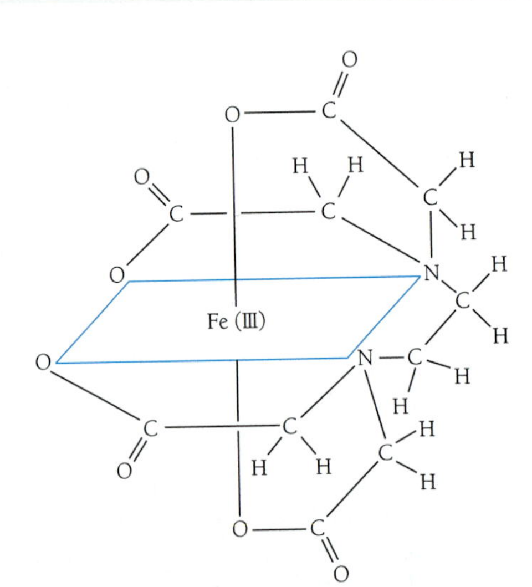

Figure 23.17 A single EDTA ion can occupy all six sites in a metal ion's coordination sphere.

Polydentate ligands are also called **chelating agents**, and the complexes they form are **chelates** (*kee' lates*, from the Greek *chele*, "claw"). In an EDTA chelate, (Figure 23.17), the single ligand completely enfolds the metal ion. Some common ligands are shown in Table 23.3.

23.7 FORMULA, OXIDATION STATE, AND NOMENCLATURE

Once the oxidation state of the metal ion is known, and a list of ligands is given, the charge on the complex can be deduced and an empirical formula can be written. The metal is written first, followed by the ligand(s). Ligands are listed in alphabetical order of their names: NH_3 ("ammine") precedes H_2O ("aquo").

TABLE 23.3 Some Common Ligands*

Type	Structure and Name
Monodentate	$:\ddot{Cl}:^-$, chloro⁺; $:C{\equiv}N:^-$, cyano; $:O(H)_2$, aquo;
	$:NH_3$ (N—H, H, H), ammine; $:\ddot{O}{-}H^-$, hydroxo; $:C{\equiv}O:$, carbonyl;
	$:\dot{N}{=}\ddot{O}$, nitrosyl; $:N(=O)(-O:^-)$, nitro; $:P(H)_3$ (P—H, H, H), phosphine
Bidentate	$^-:\ddot{O}{-}C(=O){-}C(=O){-}\ddot{O}:^-$, oxalato;
	$H_2\ddot{N}{-}CH_2{-}CH_2{-}\ddot{N}H_2$, ethylenediamine
Tridentate	$H_2\ddot{N}{-}CH_2{-}CH_2{-}\ddot{N}H{-}CH_2{-}CH_2{-}\ddot{N}H_2$, diethylenetriamine
Hexadentate	$(^-:\ddot{O}{-}C(=O){-}CH_2)_2\ddot{N}{-}CH_2{-}CH_2{-}\ddot{N}(CH_2{-}C(=O){-}\ddot{O}:^-)_2$ EDTA

* The atom that forms the coordinate bond to the metal is shown in bold type.
⁺ The other halogens, F^-, Br^-, and I^-, can also act as ligands.

EXAMPLE 23.1 Formula of a Coordination Complex

Write the formula of the nickel(II) complex that has three bromide ions and one water molecule as ligands.

Analysis

Target: An empirical formula. As always, the formula of a species includes the charge, if different from zero.

Relationship: The charge of the complex is the sum of the charges of the metal ion and the ligands. The order in which the components are to be written is given in the text.

Plan Determine the charge, then write the components in order. Be sure to include the square brackets, if the complex is an ion.

Work The sum of the component charges is

$$+2 \text{ (nickel ion)} + 3(-1) \text{ (bromide ion)} + 0 \text{ (water)} = -1$$

The order is: metal, aquo, bromo. The formula is $[Ni(H_2O)Br_3]^{-1}$.

Exercise

Write the formula of a complex of Cr^{3+} having as ligands three ammonia molecules and three hydroxide ions.

Answer: $Cr(NH_3)_3(OH)_3$.

Transition metals characteristically exhibit multiple oxidation states, any one of which may be found in different complex compounds or ions. As with simpler ionic species like $MgCl_2$, the oxidation number of the metal in a coordination complex is implicit in the empirical formula and need not be specified.

EXAMPLE 23.2 Oxidation Number of a Complexed Metal

Determine the oxidation number of cobalt in the complex cation $[Co(NH_3)_4(OH)_2]^+$.

Analysis

Target: The oxidation number of the cobalt ion.

Relationship: The charge on the complex must equal the sum of the charges of the metal ion and the ligands. The charge on the metal ion is equal to the oxidation number.

Plan Add up the charges on each of the six ligands, then subtract this total from the charge of the complex (+1) to obtain the charge on the metal ion.

Work

1. There are two OH^- ligands, each bearing a charge of −1, and four neutral NH_3 ligands: the total ligand charge is −2.
2. total charge of complex − ligand charge = metal ion charge

$$(+1) - (-2) = +3$$

The oxidation number of cobalt in $[Co(NH_3)_4(OH)_2]^+$ is +3.

Exercise

What is the oxidation number of the metal in (a) $[Fe(CN)_6]^{3-}$ and (b) $[Cr(en)_3]^{2+}$?

Answers: **(a)** +3, **(b)** +2.

TABLE 23.4 Rules for Naming Coordination Compounds

1. As with all ionic compunds, the cation is named before the anion.
2. Within each ion, ligands are named before the metal. If more than one type of ligand is present, they are named in alphabetical order.
3. The number of each kind of simple ligand is given by the prefixes *mono-* (optional), *di-*, *tri-*, *tetra-*, *penta-*, and *hexa-*. If the name of a more complicated ligand already includes one of these, the prefix *bis-*, *tris-*, *tetrakis-*, *pentakis*, or *hexakis-* is used instead (for example, *bis-dimethylamine*). The numerical prefix is ignored in the alphabetical listing: tetra*bromo* precedes di*chloro*.
4. The ending of an anionic ligand is replaced by *o*: iod*ide* becomes iod*o*, sulfate becomes sulfat*o*, nitrite becomes nitrit*o*, and so on.
5. The name of a neutral ligand is unchanged, although there are some common exceptions to this rule: H_2O is *aquo*, NH_3 is *ammine*, CO is *carbonyl*, NO is *nitrosyl*.
6. For those metals having variable oxidation state, the oxidation number is indicated by Roman numerals in parentheses following the metal name. The name of the metal is unchanged in neutral and positive complexes, but it adds the suffix *-ate* if the complex is negatively charged. Often the Latin name is used for the metal in complex anions: *ferrate* and *plumbate*, but *nickelate* and *zincate*.

Am' mine, with two m's, refers to the ligand NH_3. *A mine'*, with one m, refers to the class of organic compounds NR_3 (Section 20.3).

Some new rules are necessary for naming coordination complexes, but the same basic principle of chemical nomenclature is followed: the name must be unique, that is, it must correspond to only one possible structure. The rules for coordination complexes are given in Table 23.4.

EXAMPLE 23.3 Naming Coordination Complexes

Study each of the following formulas, and observe that the name conforms to the rules in Table 23.4.

$[Co(NH_3)_5Br]Cl_2$	pentaamminebromocobalt(III) chloride
$K_2[Fe(H_2O)_2(CN)_4]$	potassium diaquotetracyanoferrate(II)
$[Cr(en)_2(OH)_2]$	bis(ethylenediamine)dihydroxochromium(II)
$[Ru(NH_3)_3(CO)_3][Zn(OH)_4]$	triamminetricarbonylruthenium(II) tetrahydroxozincate(II)

Exercise

(a) Name the coordination compound $K_2[CoCl_4]$, and (b) give the formula of diammineethylenediaminediiodochromium(III) chloride.

Answers: (a) Potassium tetrachlorocobaltate(II); (b) $[Cr(NH_3)_2(en)I_2]Cl$.

23.8 ISOMERISM IN COORDINATION COMPOUNDS

The structure of coordination compounds permits several different types of isomerism: two of these (geometrical and optical) we have seen before in organic compounds, and two (coordination and linkage) are new.

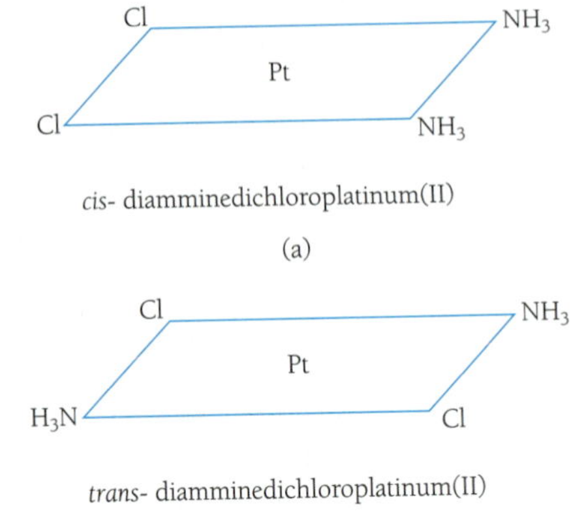

Figure 23.18 Square planar complexes with the general formula MX_2Y_2 or MX_2YZ exist as two geometrical isomers. (a) In *cis*-diamminedichloroplatinum(II), the identical ligands occupy adjacent corners of the square. (b) In the *trans* isomer, the identical ligands occupy opposite corners.

Stereoisomerism

Geometrical isomerism in alkenes is discussed in Section 9.5, and that of substituted cycloalkanes in Section 20.1.

Square planar complexes having two identical ligands (and two others) can exist as *geometrical* or *cis-trans isomers,* depending on whether the identical ligands occupy adjacent or opposite corners of the square. Such isomerism occurs in complexes of general formula MX_2YZ and MX_2Y_2, where X, Y, and Z represent different ligands. Figure 23.18 shows the two isomers of diamminedichloroplatinum(II), an MX_2Y_2 complex. Geometrical isomerism is a form of **stereoisomerism** in which compounds have the same atom-to-atom connectivity but differ in the spatial arrangement of atoms. Geometrical isomers are different compounds with different, often strikingly different, properties. For example *cis*-$Pt(NH_3)_2Cl_2$ inhibits cell division, and is thus a valuable, if somewhat toxic, antitumor agent. Its isomer, *trans*-$Pt(NH_3)_2Cl_2$, has no effect on cell division.

Six-coordinate (octahedral) complexes having two identical ligands can also exist as geometrical isomers. The two identical ligands are either adjacent to *(cis)* or across from *(trans)* one another, as shown in Figure 23.19. Geometrical isomerism can also

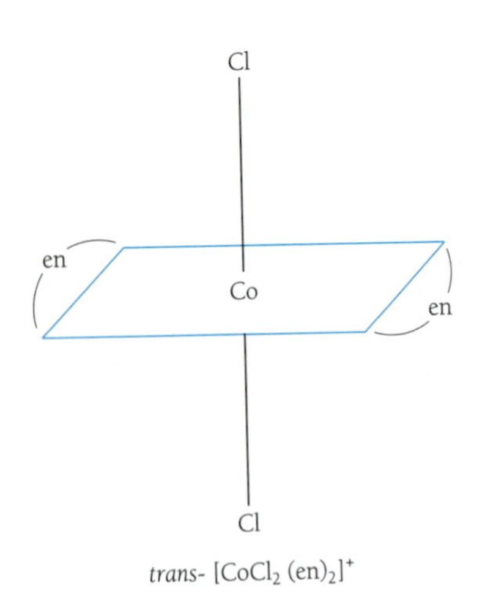

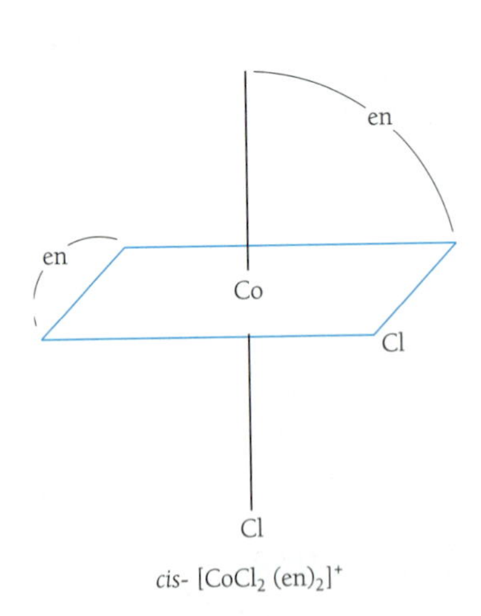

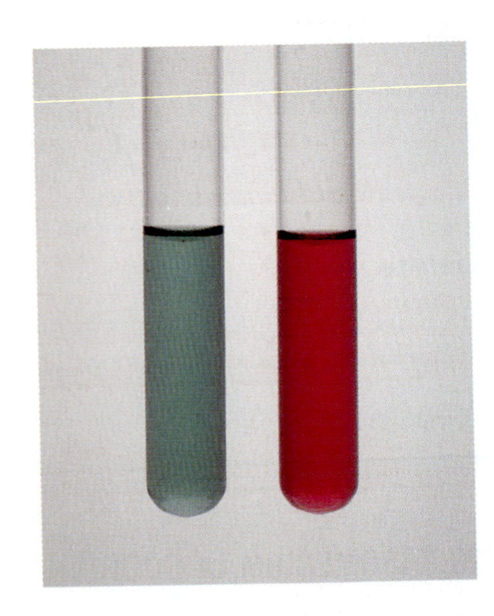

Figure 23.19 Geometrical isomerism in octahedral complexes. The complex ion $[CoCl_2(en)_2]^+$ in two isomeric forms. Many of the properties, including color, are different in these forms. The *cis* isomer is purple, while the *trans* isomer is green.

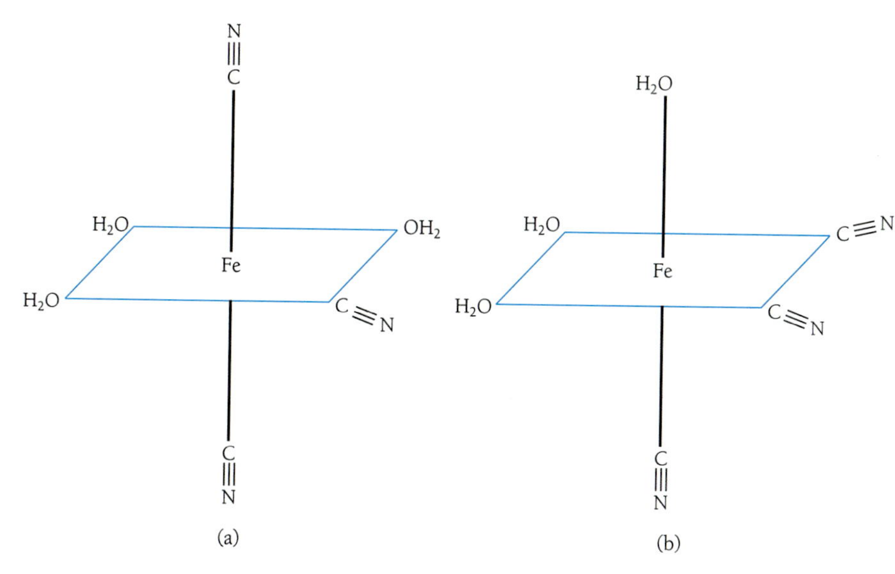

Figure 23.20 Geometrical isomerism in MX_3Y_3 complexes. (a) The *mer* isomer ("meridian" plane). (b) The *fac* isomer ("face" plane).

occur if an octahedral complex has three identical ligands, as for example in the neutral compound $Fe(H_2O)_3(CN)_3$. The three identical ligands can lie in the same plane as the metal ion, or they can lie in a plane that is one of the faces of the octahedron (Figure 23.20).

Another form of stereoisomerism, *optical isomerism,* was introduced in Chapter 21 in the discussion of amino acids. Any compound with a nonsuperimposable mirror image is optically active, that is, it causes rotation of the plane of polarized light passing through it. Such a compound and its mirror image are a pair of *enantiomers:* they have identical physical properties but rotate the plane of polarized light in opposite directions. Optical isomerism can also occur in octahedral coordination complexes. For example, the $[CrCl_2(en)_2]^+$ ion has a nonsuperimposable mirror image, as shown in Figure 23.21. No amount of rotation or viewing from another angle will allow superimposition of the mirror images—unquestionably there are two isomers. The possibility for optical isomerism exists also in tetrahedral complexes having four different ligands, but to date no examples of such complexes have been prepared and unambiguously identified.

Structural Isomerism

Several new types of structural isomers, not seen in molecular compounds, can occur in coordination compounds. (Recall that structural isomers differ in their atom-to-atom connectivity.) Certain ligands, for example NO_2^-, have more than one atom that can donate a lone pair and form a coordinate bond to a metal ion. The ligand NO_2^- can form a bond to a metal ion either with its N atom or with one of its O atoms. In the former case, it is called *nitro-,* and indicated in formulas as NO_2. In the latter case, it is called *nitrito,* and indicated by ONO in empirical formulas. These different modes of

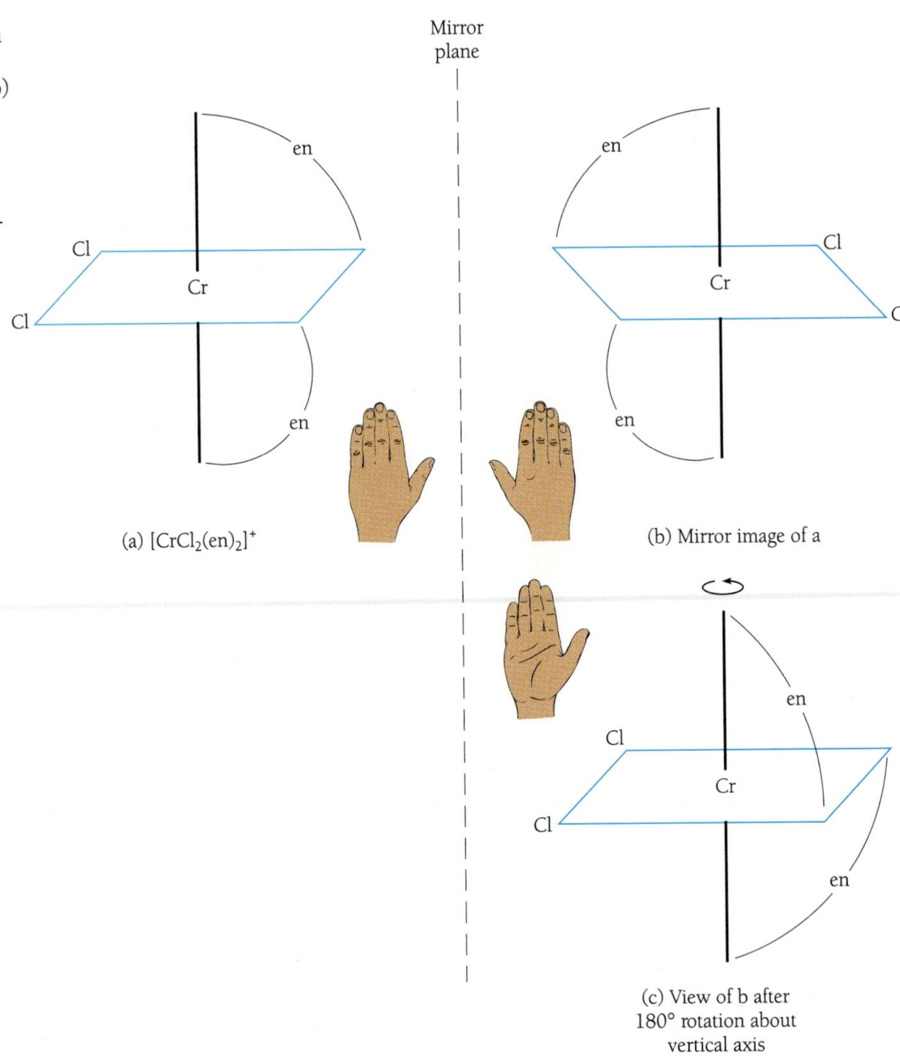

Figure 23.21 The $[CrCl_2(en)_2]^+$ ion cannot be superimposed on its mirror image. (a) The structure of the ion. (b) The mirror image of the structure in (a). (c) The ion in (b) after 180° rotation about a vertical axis. The ion in view (c) is clearly different from and cannot be superimposed on, the original in (a).

bonding result in **linkage isomerism** (Figure 23.22). Another example of this type of ligand is the SCN^- ion (thiocyanate), which can donate an electron pair from its S atom or its N atom, giving rise to linkage isomers in thiocyanate complexes.

Finally, an anionic species can serve either as a ligand or as the counterion in a coordination compound. In $[Pt(NH_3)_3Cl]Br$, Cl^- is a ligand and Br^- is the counterion. In $[Pt(NH_3)_3Br]Cl$, these roles are reversed. This is an example of **coordination isomerism,** a form of structural isomerism in which the composition of the complex ion is different, but the empirical formula of the compound is the same. The relationships among the various types of isomerism discussed in this text are summarized in Table 23.5.

23.9 BONDING IN COORDINATION COMPLEXES

Three major theories are available as frameworks for discussion of the electronic structure and bonding in coordination complexes. Two of these, the *valence bond* and the *molecular orbital* theories, we have seen before. Both theories are generally applica-

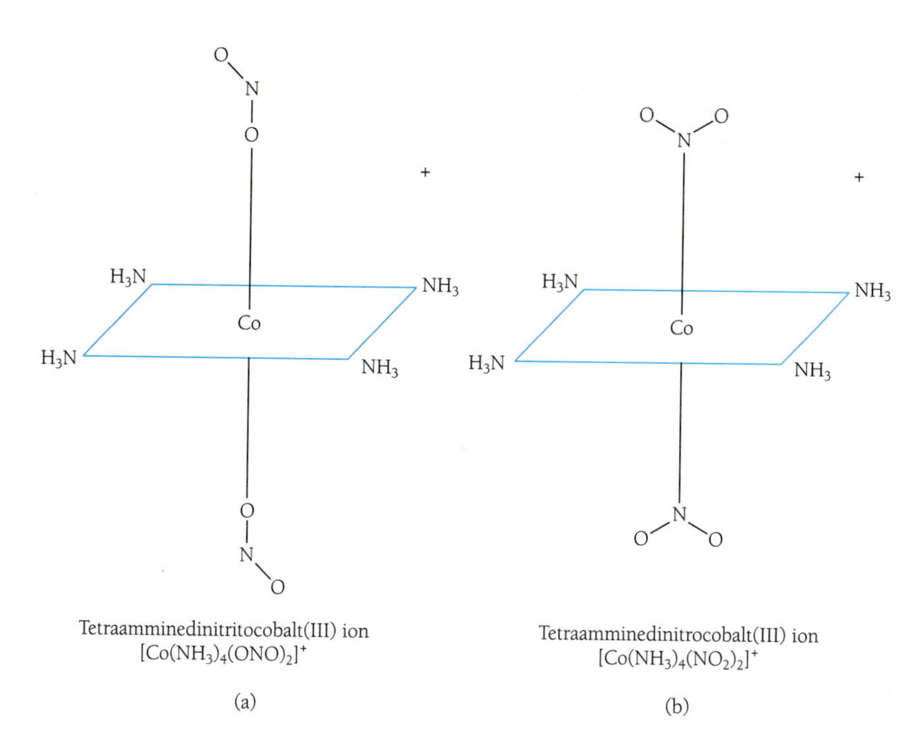

Figure 23.22 A pair of linkage isomers. They differ in the ligand atom that forms the coordinate bond.

ble to all kinds of molecules. The third, *crystal field* theory, is new. Crystal field theory is used primarily for compounds of transition metals. Each of the theories has its strengths and weaknesses, and each emphasizes different aspects of bonding and structure.

Valence Bond Theory

The *valence bond theory* treats the metal-ligand bond as resulting from the overlap of a filled ligand orbital with an empty metal orbital. The overlap region is occupied by the lone pair of electrons donated by the ligand. The bond is treated as entirely covalent, lacking any ionic character. The model begins with the observed geometry of the complex, because the geometry determines which of the metal orbitals are used in

TABLE 23.5 Classification of Isomers

Class	Type	Distinguishing Characteristic
Structural		Different atom-to-atom connectivity
	Positional	In organic compounds, different placement of multiple bonds or substituents
	Linkage	Same ligand, different bonding site
	Coordination	Interchange of ligand and counterion
Stereo		Same connectivity, different spatial arrangement
	Geometrical	Substituents or ligands on same or opposite side of double bond or metal ion
	Optical	Nonsuperimposable mirror images
Conformational		Same connectivity and spatial arrangement, different single-bond rotational angles

bonding. The hybrid set *sp* is used if the complex is linear, sp^3 if tetrahedral, dsp^2 if square planar, and d^2sp^3 is used for octahedral complexes. The distribution of the metal's valence electrons in the remaining (unhybridized) *d* orbitals determines whether the complex is diamagnetic or paramagnetic. As an example, we will consider the complex cation $[Cr(NH_3)_6]^{3+}$. The orbital diagrams for Cr($Z = 24$), Cr^{3+}, and d^2sp^3-hybridized Cr^{3+} are

Orbital diagrams are introduced and explained in Section 7.2.

			3*d*	4*s*	4*p*
Neutral atom	Cr:	[Ar]	[↑][↑][↑][↑][↑]	[↑]	[ ][ ][ ]
Ground-state ion	Cr^{3+}:	[Ar]	[↑][↑][↑][ ][ ]	[ ]	[ ][ ][ ]

			3*d*	d^2sp^3
Hybrid	Cr^{3+}:	[Ar]	[↑][↑][↑]	[ ][ ][ ][ ][ ][ ]
Bonded to six ligands	Cr^{3+}:	[Ar]	[↑][↑][↑]	[••][••][••][••][••][••]

After accepting six electron pairs from six ligands, the chromium ion still has three unpaired electrons in its 3*d* subshell. We therefore predict that the complex ion, and any of its compounds, such as $[Cr(NH_3)_6]Cl_3$, are paramagnetic. When the measurement is made, the prediction is found to be correct. Furthermore, the observed strength of the paramagnetism is that expected for a compound with three unpaired electrons. Cobalt ($Z = 27$) has three more electrons than chromium. The electron configuration of a d^2sp^3-hybridized Co^{3+} ion in an octahedral complex should differ from that of Cr^{3+}, above, only by having an additional three electrons in the 3*d* subshell:

			3*d*	4*s*	4*p*
Neutral atom	Co:	[Ar]	[↑↓][↑↓][↑↓][↑][↑]	[↑]	[ ][ ][ ]
Ground-state ion	Co^{3+}:	[Ar]	[↑↓][↑↓][↑↓][ ][ ]	[ ]	[ ][ ][ ]

			3*d*	d^2sp^3
Hybrid	Co^{3+}:	[Ar]	[↑↓][↑↓][↑↓]	[ ][ ][ ][ ][ ][ ]
Bonded to six ligands	Co^{3+}:	[Ar]	[↑↓][↑↓][↑↓]	[••][••][••][••][••][••]

This configuration leads to the prediction that cobalt complexes of stoichiometry $Co(L)_6$, where L is any ligand, should be diamagnetic. The observed fact is that some are, but some are not. $[Co(NH_3)_6]Cl_3$ is diamagnetic, with no unpaired electrons, but $K_3[CoF_6]$ is paramagnetic, with *four* unpaired electrons. The simple theory presented here cannot explain this fact.

Valence bond theory can be modified by allowing participation of **4***d* orbitals in hybridization, and when this is done the theory correctly predicts the paramagnetism of $K_3[CoF_6]$. This complication, however, together with the failure to predict the *colors* of coordination compounds and the fact that any ionic character in the bonding is ignored, has made the valence bond theory less popular than the others.

Crystal Field Theory

The **crystal field theory** of coordination complexes emphasizes the ionic character of the metal-ligand bond. In this sense it is complementary to the valence bond theory, which emphasizes the covalent character of the bonding. Crystal field theory is successful in predicting the color and paramagnetism of coordination compounds. Like the valence bond theory, crystal field theory begins with the observed geometry of the complex.

Electric Field of Ligands in Octahedral Complexes

A ligand is either an anion or a polar molecule with its negative end oriented toward the metal ion. In crystal field theory, both types are treated in the same way, as negative *point charges* located at the vertices of the octahedron (or square, or tetrahedron, depending on the type of complex). The complex is held together by the attractive force between these charges and the positive charge of the metal ion. In this model (Figure 23.23) the bonding is entirely ionic, and covalent character is ignored.

A point charge is infinitesimally small, having no extent in space.

The ligand point charges also affect the *d* orbitals of the metal ion: orbital energy is increased by the nearby negative charges. However, the effect is not the same for all five of the *d* orbitals. The d_{z^2} and $d_{x^2-y^2}$ orbitals, whose major lobes lie along the coordinate axes and thus point directly at the ligands, are affected more strongly than the d_{xy}, d_{xz}, and d_{yz} orbitals, whose lobes are directed in between the ligands (Figure

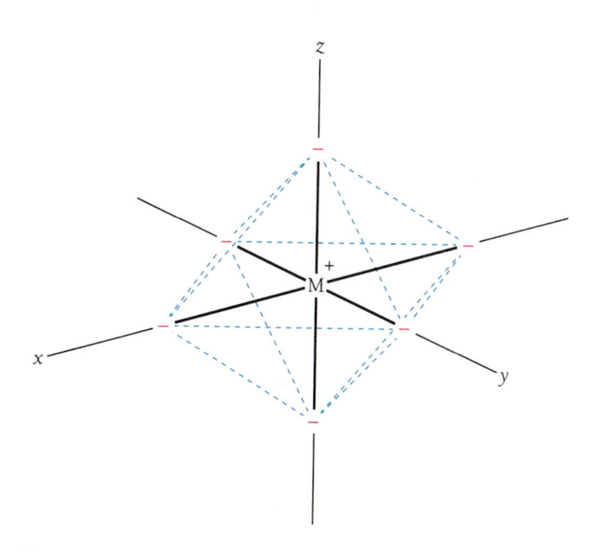

Figure 23.23 The crystal field model of an octahedral complex: ligands are treated as point negative charges.

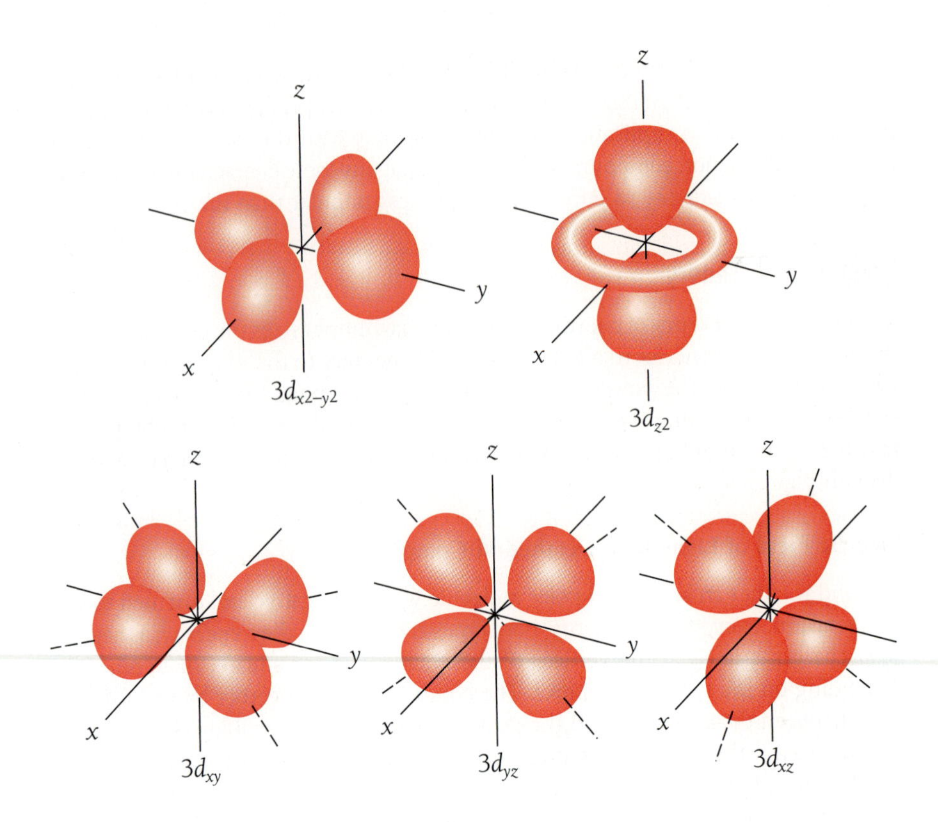

Figure 23.24 Shapes and orientations of the five *d* orbitals.

23.24). The set of *d* orbitals is split into two groups, differing in energy (Figure 23.25). The energy difference is called the **crystal field splitting,** and given the symbol **Δ**. The magnitude of the splitting depends on the nature of the ligand. If the ligand is a strong Lewis base, such as NH_3 or CN^-, **Δ** is large. Weak Lewis bases like halide ions give rise to smaller values of **Δ**. In general, the splitting caused by different ligands increases as follows.

weak $\quad Cl^- < F^- < OH^- < H_2O < NH_3 < en < NO_2^- < CN^- \quad$ strong

increasing **Δ** →

This ordering is called the **spectrochemical series.**

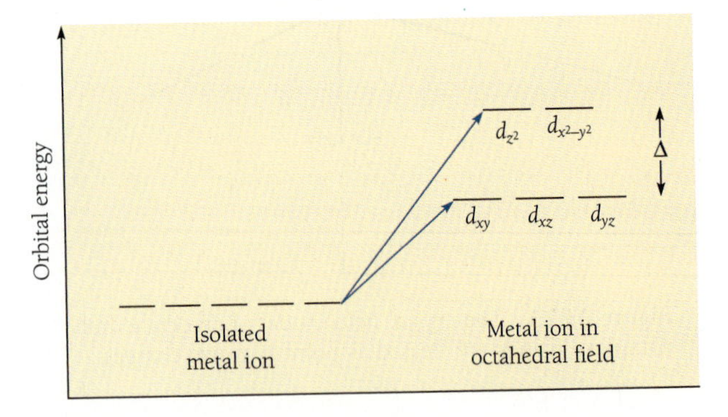

Figure 23.25 Splitting of the energies of the *d* orbitals in the crystal field of six octahedrally arranged ligands.

Electron Configuration and Magnetic Properties

We are now in a position to understand why some complexes of Co^{3+}, for instance $[Co(CN)_6]^{3-}$, are diamagnetic, while others, like $[CoF_6]^{3-}$, are paramagnetic. Cobalt(III) is a d^6 ion; that is, it has six electrons in its $3d$ subshell. In the absence of ligands all of the d orbitals have the same energy, and, in accordance with Hund's rule, electrons in them avoid pairing to the greatest possible extent. The orbital diagram is [↑↓][↑][↑][↑][↑] (3d). In the **weak-field case,** the ligands are low in the spectrochemical series. The splitting Δ is small, and the electron distribution is the same as in the isolated ion (in which there is no splitting). In the **strong-field case,** when the ligands are high in the series, Δ is large enough so that the energy benefit of having all electrons in the lower level overcomes the energy cost of pairing (Figure 23.26). There are four unpaired electrons in the weak-field case, so $K_3[CoF_6]$ is paramagnetic. There are no unpaired electrons in the strong-field case, so $K_3[Co(CN)_6]$ is diamagnetic. The weak-field case is sometimes called the **high-spin case,** and similarly, the strong-field case can be called the **low-spin case.**

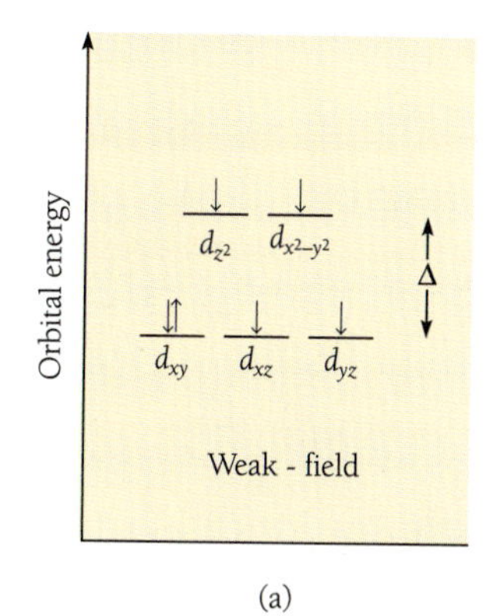

(a)

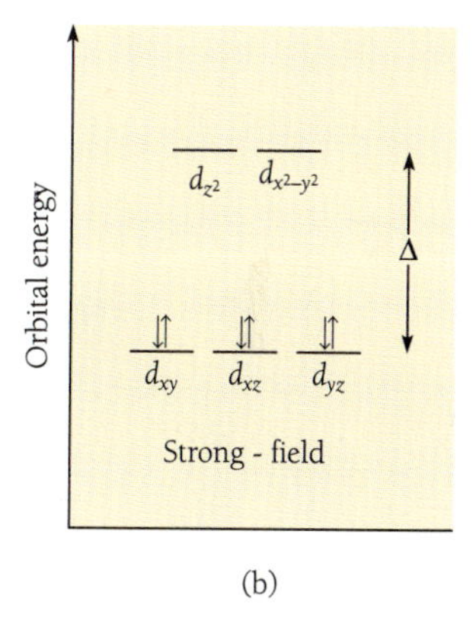

(b)

Figure 23.26 Electron configuration of a d^6 species such as Co^{3+}. In the weak-field case (a), with a ligand such as F^-, the splitting is small and pairing is minimized. In the strong-field case (b), with a ligand such as CN^-, the splitting is large and occupancy of the upper level is minimized.

EXAMPLE 23.4 Magnetic Properties of Coordination Compounds

Use the crystal field theory to predict the magnetic properties of $K_3[Fe(CN)_6]$: is it paramagnetic? If so, how many unpaired electrons are there?

Analysis

Target: The number of unpaired electrons in the $3d$ subshell of the iron ion.

Relationship: The target can be obtained from the orbital diagram, which in turn can be determined from the number of d electrons and the energy-level spacing. The number of electrons is known from the atomic number and/or the electron configuration of the atom, together with the charge of the metal ion of the complex. The energy-level spacing is assumed to be large or small, depending on the position of the ligand in the spectrochemical series.

This assumption may not be correct in all cases. The spectrochemical series is a useful guide, but it is not infallible.

Plan Determine the charge on the complex ion, and, from it, the oxidation state and d-electron configuration of the metal ion. Use the spectrochemical series to decide between the weak- and strong-field case, complete the orbital diagram, and determine the magnetic properties.

Work

1. Adding up the charges of 3 K^+ ions and 6 CN^- ions gives -3; since the compound $K_3[Fe(CN)_6]$ has a charge of 0, the charge of the iron ion must be $+3$.

2. The configuration of Fe($Z = 26$) is [Ar] $4s^2 3d^6$. Removing two electrons from the $4s$ shell and one from the $3d$ level leaves an ion with the configuration Fe^{3+}: [Ar] $3d^5$.

3. CN^- is high in the spectrochemical series, so this is almost certainly a strong-field case—Δ is large, and the five electrons fill the lower of the two levels first.

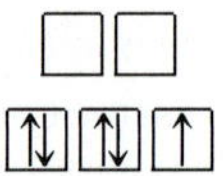

$K_3[Fe(CN)_6]$ is paramagnetic, with one unpaired electron.

Exercise

How many unpaired electrons are there in $K_3[Fe(Cl)_6]$?

Answer: Five.

Colors of Octahedral Complexes

Ordinary white light (sunlight, daylight, or from light bulbs) consists of a balanced mixture of all colors. All wavelengths in the visible region of the electromagnetic spectrum—from about 400 to 700 nm—are present. When white light passes through a substance, like water, that does not absorb light in this wavelength range, the substance appears colorless. But if the substance absorbs, say, all the blue light, what emerges is a mixture of red, green, and yellow light. Our eyes perceive this mixture as yellow-orange, and we say that the substance has a yellow or orange color. Similarly, substances that absorb red light appear green, and those absorbing yellow-green light appear purple. There is a direct relationship between the color of a substance and the wavelength of light that it absorbs.

The nature of light, and the process of light absorption, are discussed in Chapter 6.

When a molecule or ion absorbs a photon, one of the electrons of that species is boosted from a lower to a higher energy level. The equation relating the energy *difference* between the two levels and the wavelength of the absorbed photon is $\Delta E = hc/\lambda$. Now, in coordination complexes the energy levels involved in the absorption of visible radiation are the *d* orbitals, split into two groups by the electric field of the ligands. The energy difference is Δ, the crystal field splitting. When a $[CoF_6]^{3-}$ ion absorbs a photon, an electron moves from one of the (d_{xy}, d_{yz}, d_{xz}) group of orbitals to one of the orbitals in the $(d_{z^2}, d_{x^2-y^2})$ group, as shown in Figure 23.27. Fluoride ion is

The equation $\Delta E = hc/\lambda$, which relates the wavelength (λ) of a photon to energy levels, is discussed in Section 6.3. h is Planck's constant, and c is the velocity of light.

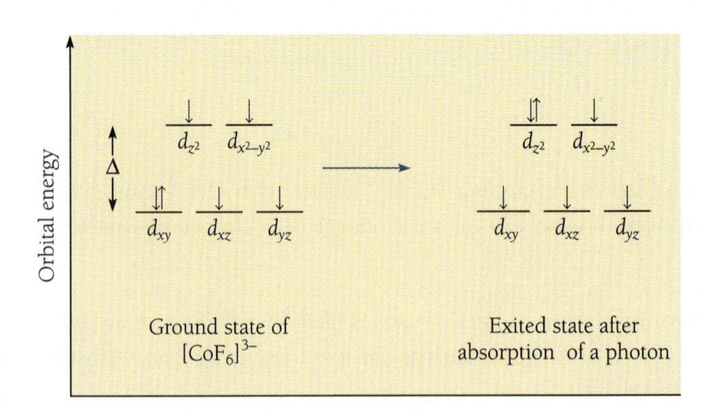

Figure 23.27 Absorption of a photon by a $[CoF_6]^{-3}$ ion causes excitation of an electron from the lower to the upper of the 3*d* levels. Since this can occur only with light of wavelength $\lambda = hc/\Delta = 700$ nm, which is red, compounds containing this ion appear green.

TABLE 23.6 Light Absorption and Color of Cobalt(III) Complexes

Complex	Wavelength Absorbed/nm	Color of Light Absorbed	Color of Complex
$[CoF_6]^{3-}$	≈ 700	red	blue-green
$[Co(H_2O)_6]^{3+}$	≈ 600	orange	blue
$[Co(NH_3)_5Cl]^{2+}$	≈ 535	yellow	purple
$[Co(NH_3)_5OH]^{2+}$	≈ 500	blue-green	red
$[Co(NH_3)_6]^{3+}$	≈ 475	blue	yellow-orange
$[Co(CN)_5Br]^{3-}$	≈ 415	violet	yellow

low in the spectrochemical series, and Δ for $[CoF_6]^{3-}$ is only about 170 kJ mol^{-1}. This energy is carried by light of wavelength $\lambda \approx 700$ nm. It is red light, and its absorption results in a blue-green color for compounds containing this ion. Table 23.6 lists the absorption wavelengths and resulting colors of some other cobalt complexes.

Values of Δ in the fifth- and sixth-period transition metal complexes, which involve the 4*d* and 5*d* orbitals, respectively, are similar. In particular, the ordering of the spectrochemical series is the same for the entire group of transition-metal complexes.

Other Geometries

The crystal field theory is easily extended to complexes having shapes other than octahedral—the approach is the same. Ligand point charges are placed at appropriate locations, and their effect on the energies of the *d* orbitals is calculated. The shape of the complex (linear, square planar, tetrahedral, and so on) determines the location of the point charges relative to the *d* orbitals. The extent and pattern of splitting depends on the geometry. When the point charges are tetrahedrally located, *none* of the *d* orbital lobes points directly at a ligand, hence none of the *d* orbitals moves to exceptionally high energy. Moreover, the d_{z^2} and $d_{x^2-y^2}$ orbitals are affected less than the others (rather than more as they are in the octahedral geometry). The crystal field splitting for tetrahedral complexes is shown in Figure 23.28.

There is another difference between tetrahedral and octahedral geometries, in addition to the different splitting pattern of the *d* orbitals. For any given ligand, the value of Δ in a tetrahedral complex is only about half what it is in an octahedral complex. In fact, there are *no* ligands high enough in the spectrochemical series to produce the strong-field (low-spin) case in tetrahedral complexes—all known complexes are high-spin. Consequently, the magnetic properties of tetrahedral complexes can be predicted without using the spectrochemical series. After counting the number of *d* electrons in the metal ion, simply place them into the five *d* orbitals according to Hund's rule of maximum spin, as if there were no splitting at all.

Each of the different geometries has its own characteristic *d* orbital splitting pattern, and deductions about magnetic properties and color can be made from them. The crystal field splitting for linear and square planar complexes is given in Figure 23.29.

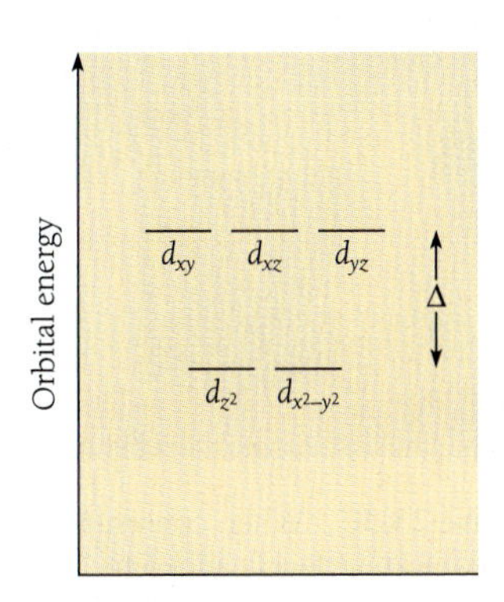

Figure 23.28 Splitting of *d* orbitals in a tetrahedral field.

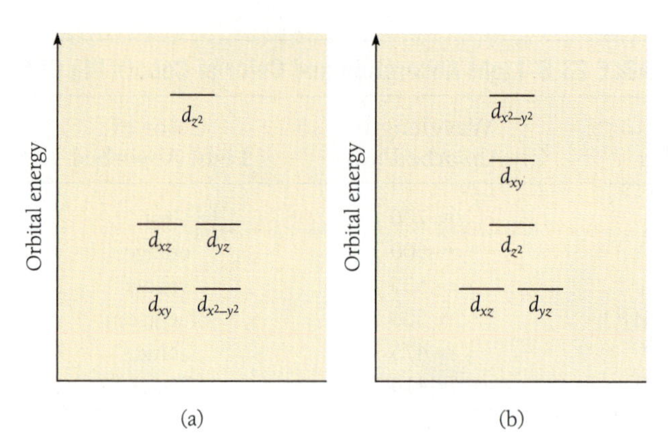

Figure 23.29 Crystal field splitting in (a) linear complexes and (b) square planar complexes.

Molecular Orbital Theory

The third bonding theory that may be applied to coordination complexes is the molecular orbital theory. As is the case when MO theory is applied to simpler molecules, the bonding and geometry of coordination complexes may be discussed in terms of occupancy of *bonding, antibonding,* and *nonbonding* orbitals, having σ or π characteristics. Atomic orbitals from both metal ion and ligands are used to construct the molecular orbitals, which are not localized to a particular "bond" but rather extend over the entire complex. MO theory does not address the question of ionic/covalent character of particular bonds. Although it gives a more realistic account of bonding, MO theory is more difficult to apply at the conceptual, nonmathematical level than is crystal field theory. There is no advantage to the use of MO theory in an introductory text.

Figure 23.30 When concentrated aqueous ammonia is added to $Ni(NO_3)_2(aq)$, an immediate ligand exchange reaction takes place in which $[Ni(H_2O)_6]^{2+}$ (green) is transformed into $[Ni(NH_3)_6]^{2+}$ (blue).

23.10 CHEMICAL PROPERTIES OF COMPLEXES

Water is an excellent ligand. It follows that a metal ion in aqueous solution, $M^{x+}(aq)$, is in fact the complex ion $[M(H_2O)_n]^{x+}$ (*x* is the oxidation state and *n* is the coordination number). There is no such thing as a "bare" metal ion in aqueous solution, and $Ni^{2+}(aq)$, for example, is actually $[Ni(H_2O)_6]^{2+}$. You might wonder why textbook authors and other chemists continue to use the notation $Ni^{2+}(aq)$ or even Ni^{2+}, knowing that it does not represent the structure of the hydrated ion. One reason is that the coordination number in the hydrated ion is not always known with certainty. Another reason is simplicity. Good chemical notation carries only the needed information, and unnecessary items are omitted, even if true. It would add nothing of importance to write a reduction half-reaction as, for example, $[Ni(H_2O)_6]^{2+} + 2\,e^- \rightarrow Ni(s) + 6\,H_2O$, so it is rarely done. On the other hand, when the information is needed or relevant, as it often is in this chapter, the presence of H_2O as a ligand is included in the notation.

Formation Constants

When aqueous ammonia is added to an aqueous solution of a Ni^{2+} salt, the process that takes place (Figure 23.30) is a **ligand exchange** reaction in which NH_3 replaces

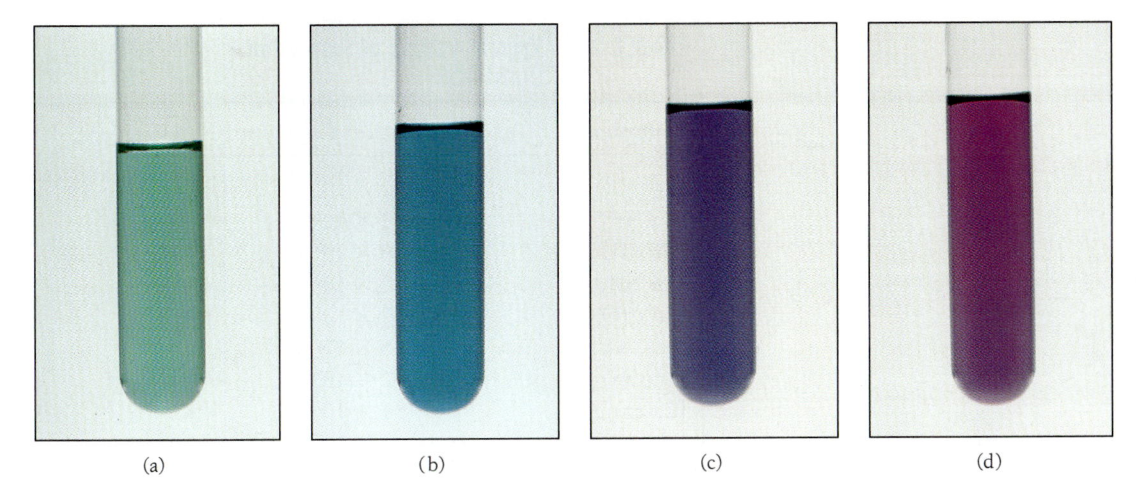

Figure 23.31 Ligand exchange is a stepwise process. (a) The green color of a solution of a nickel(II) salt is due to the hexaaquo complex $[Ni(H_2O)_6]^{2+}$. As the bidentate ligand ethylenediamine (en) is added to the solution, the color of the solution changes due to the sequential formation of (b) $[Ni(H_2O)_4en]^{2+}$, (c) $[Ni(H_2O_2)(en)_2]^{2+}$, and (d) $[Ni(en)_3]^{2+}$.

H_2O in the coordination sphere of the Ni^{2+} ion. The result is the formation of the complex $[Ni(NH_3)_6]^{2+}$.

$$[Ni(H_2O)_6]^{2+} + 6\ NH_3(aq) \rightleftharpoons [Ni(NH_3)_6]^{2+} + 6\ H_2O \qquad (23\text{-}10)$$

Such reactions are equilibrium processes, and furthermore they occur stepwise. That is, the ligands are exchanged one at a time, in a series of six stepwise equilibrium reactions (Figure 23.31). The first two steps are

$$[Ni(H_2O)_6]^{2+} + NH_3(aq) \rightleftharpoons [Ni(NH_3)(H_2O)_5]^{2+} + H_2O \qquad (23\text{-}10.1)$$

$$[Ni(NH_3)(H_2O)_5]^{2+} + NH_3(aq) \rightleftharpoons [Ni(NH_3)_2(H_2O)_4]^{2+} + H_2O \qquad (23\text{-}10.2)$$

The final step is

$$[Ni(NH_3)_5(H_2O)]^{2+} + NH_3(aq) \rightleftharpoons [Ni(NH_3)_6]^{2+} + H_2O \qquad (23\text{-}10.6)$$

For the overall reaction (23-10) the equilibrium expression and constant are

For many complex ions, only the overall equilibrium constant is known—the equilibrium constants for the individual steps have not been measured.

$$K_f = \frac{[[Ni(NH_3)_6]^{2+}]}{[[Ni(H_2O)_6]^{2+}][NH_3(aq)]^6} = 9 \times 10^8$$

As usual, the concentration of the solvent (H_2O) is omitted from the equilibrium expression, even though it is formally a reaction product. (*Note:* In equilibrium expressions involving complexes, it is not always clear whether square brackets refer to molarities or to the complex ion. We use the convention that boldface brackets **[ ]** refer to molarities, while brackets in normal type enclose the complex ion.) As noted above, the simpler notation $Ni^{2+}(aq)$ is often used to represent the fully hydrated ion $[Ni(H_2O)_6]^{2+}$, so that the equilibrium expression may be written

$$K_f = \frac{[[Ni(NH_3)_6]^{2+}]}{[Ni^{2+}(aq)][NH_3(aq)]^6}$$

TABLE 23.7 Typical Values of Formation Constants

Complex	Formation Constant
$[Ag(NH_3)_2]^+$	2×10^7
$[Cu(NH_3)_4]^{2+}$	2×10^{13}
$[Ni(NH_3)_6]^{2+}$	9×10^8
$[Zn(NH_3)_4]^{2+}$	3×10^9
$[Cd(CN)_4]^{2-}$	2×10^{17}
$[Fe(CN)_6]^{3-}$	8×10^{43}
$[Zn(OH)_4]^{2-}$	2×10^{20}
$[Cu(en)_2]^{2+}$	1×10^{20}
$[Ni(en)_3]^{2+}$	1×10^{18}
$[Ni(EDTA)]^{2-}$	4×10^{18}

Additional discussion of complex ion formation can be found in Section 17.4.

The equilibrium constant for the overall process [Equation (23-10)] is called the **formation constant** of the complex ion $[Ni(NH_3)_6]^{2+}$. Some representative values of formation constants are given in Table 23.7.

The formation constants in Table 23.7 are all very large, indicating that complex ions are very stable species. In addition, note that only one significant digit is used to express these quantities. In general, formation constants are not known as precisely as the other types of equilibrium constants discussed in this text.

Calculations based on formation constants are no different from those based on any other type of equilibrium constant.

EXAMPLE 23.5 Complex Ion Equilibrium

What is the ratio of concentration of $[Cu(NH_3)_4]^{2+}$ to that of $Cu^{2+}(aq)$ in a solution in which the concentration of free (unbound) NH_3 is 0.0100 M?

Analysis

Target: The concentration *ratio* $[[Cu(NH_3)_4]^{2+}]/[Cu^{2+}(aq)]$ (rather than an actual concentration *value*) is called for.

Knowns: The chemical equation, and therefore the equilibrium expression as well, is implicit in the fact that the complex ion contains *four* ligands. The value of the formation constant is in Table 23.7.

Relationship: The equilibrium expression.

Plan Write the equilibrium expression, solve for the target ratio, insert the known value of NH_3 concentration, and evaluate.

Work The reaction is

$$Cu^{2+}(aq) + 4\ NH_3(aq) \rightleftharpoons [Cu(NH_3)_4]^{2+}$$

and the equilibrium expression is

$$K_f = \frac{[[Cu(NH_3)_4]^{2+}]}{[Cu^{2+}(aq)][NH_3(aq)]^4}$$

Solving for the target ratio and inserting numerical values, we find

$$\frac{[[Cu(NH_3)_4]^{2+}]}{[Cu^{2+}(aq)]} = K_f \cdot [NH_3(aq)]^4$$
$$= (2 \times 10^{13}) \cdot (0.0100)^4$$
$$= 2 \times 10^5$$

The ratio is 2×10^5: the tetraammine copper(II) complex ion predominates, and very little $Cu^{2+}(aq)$ (the *hydrated ion*) remains in solution.

A related problem is solved in Example 17.16, page 754.

Exercise

Calculate the ratio of the concentration of bis(ethylenediamine) copper(II), $[Cu(en)_2]^{2+}$, to that of $Cu^{2+}(aq)$, in a solution in which the concentration of ethylenediamine is 0.010 M.

Answer: 1×10^{16}.

The concentration ratio [M(ligand)]/[M(aquo)] reflects the stability of the complex—the greater the ratio, the more stable the complex. From a comparison of the two ratios calculated in Example/Exercise 23.5, it is clear that the ethylenediamine-copper(II) complex is more stable than the ammonia-copper(II) complex. As it happens, this is a specific example of a general rule: the stability of *chelates* is normally greater than the stability of complexes having only mono dentate ligands, and the greater the number of bonding atoms possessed by a multidentate ligand, the greater the stability of its chelates. EDTA, a *hexa*dentate ligand, forms very stable complexes.

Several uses of EDTA depend on the great stability of its complexes. Trace quantities of metal ions can catalyze unwanted reactions in foods, leading to undesirable flavors. If the salt calcium disodium EDTA is added to the food, these metal ions are tied up in an EDTA chelate and rendered ineffective as catalysts. Therefore, the flavor is preserved. Look at the labels of some prepared foods, for instance salad dressing, to see how common this practice is. Another application is in the treatment of metal poisoning. If $Na_2[Ca(EDTA)]$ is ingested by the victim of lead poisoning, for instance, an exchange reaction results in the formation of the chelate $[Pb(EDTA)]^{2-}$. This complex is so stable that the lead cannot participate in other reactions, and, consequently, it is harmless. Ultimately the $[Pb(EDTA)]^{2-}$ is eliminated from the body, leaving behind the sodium and calcium (which are, in any case, essential nutrients).

One further generalization can be made about the stability of complexes. A complex containing a ligand that forms strong coordinate bonds to the metal ion is necessarily a stable complex with a relatively large value of K_f. It is to be expected that ligands forming strong bonds should have a large effect on the metal's *d* orbitals, that is, have large values of the crystal field splitting Δ. It is experimentally observed that low-spin complexes, formed from ligands high in the spectrochemical series, generally have large values of the formation constant.

It often happens that a particular metal ion is involved simultaneously in both a complex-ion formation reaction and a solubility equilibrium. For instance, consider the silver chloride solubility equilibrium. In the presence of ammonia, the species $Ag^+(aq)$ in Equation (23-11)

$$AgCl(s) \rightleftharpoons Ag^+(aq) + Cl^-(aq); \qquad K_{sp} = 1.8 \times 10^{-10} \qquad (23\text{-}11)$$

also participates in the ligand exchange reaction (23-12).

$$Ag^+(aq) + 2\ NH_3(aq) \rightleftharpoons [Ag(NH_3)_2]^+ \qquad (23\text{-}12)$$

Example 17.16, page 754, is a numerical calculation of the increase in solubility of silver chloride in aqueous ammonia, as compared to its solubility in pure water.

Le Châtelier's principle tells us that, if ammonia is added to a saturated solution of AgCl, formation of the complex ion by Equation (23-12) will decrease the concentration of $Ag^+(aq)$ and cause a shift of the solubility equilibrium in Equation (23-11) to the right. That is, more AgCl will dissolve. The conclusion is that AgCl is more soluble in ammonia than in pure water.

Kinetic Aspects

The ligand exchange reaction of Figure 23.30 takes place very rapidly, and is complete during the time required for mixing of the reagents. When ligand exchange takes place in less than a minute or so, as in this case, the complex is said to be **labile.** On the other hand, when concentrated $Cl^-(aq)$ is added to a solution of a cobalt salt, formation of $[CrCl_6]^{3-}$ from $[Cr(H_2O)_6]^{3+}$ takes hours or days. Such complexes, slow to exchange their ligands, are called **inert.** This common terminology is unfortunate, because it implies that inert complexes do not react. They do, but slowly. "Nonlabile" would be a better term. With the exception of those formed by Cr(III) and Co(III), virtually all the octahedral (coordination number = 6) complexes of the fourth-period transition elements are labile. Although there are many labile complexes, inert complexes predominate in the fifth- and sixth-period transition series.

Like most reactions, ligand exchange proceeds more rapidly as the temperature is increased. Conversion of $[Cr(H_2O)_6]^{3+}$ to $[CrCl_6]^{3-}$ takes only a few minutes if the solution is boiled.

Acid-Base Behavior

Ligands in the coordination sphere of a complex can do more than exchange with other ligands. They can also take part in reactions, often without becoming detached from the complex. Consider the hexaaquochromium(III) ion, formed when a Cr^{3+} salt dissolves in water. As shown in Figure 23.32, one of this ion's ligands can readily transfer a proton to a solvent molecule, generating an H_3O^+ ion and becoming a hydroxo ligand in the process. The $[Cr(H_2O)_6]^{3+}$ ion is therefore an ordinary Brønsted acid. The chemical equation for the proton transfer, and the associated K_a, are

$$[Cr(H_2O)_6]^{3+} + H_2O \rightleftharpoons H_3O^+ + [Cr(H_2O)_5(OH)]^{2+} \qquad (23\text{-}13)$$

$$K_a = 1 \times 10^{-4}$$

As a Brønsted acid, hexaaquochromium(III) ion is weaker than phosphoric acid ($K_1 = 7.6 \times 10^{-3}$) and stronger than acetic acid ($K_a = 1.8 \times 10^{-5}$).

Zinc hydroxide, $Zn(OH)_2$, is only slightly soluble in water: its solubility product is 3.0×10^{-16}. It is to be expected that $Zn(OH)_2$ is soluble in acid, because $H^+(aq)$ reacts with OH^-, removing it from solution and pulling the solubility equilibrium $Zn(OH)_2(s) \rightleftharpoons Zn^{2+} + 2\ OH^-$ to the right. However, zinc hydroxide is amphoteric—it also reacts with OH^-, and therefore is soluble in basic as well as acidic solutions. Amphoterism in zinc hydroxide can be explained by the proton-transfer behavior of ligands.

When zinc hydroxide, $Zn(OH)_2$, dissolves in water, part of it enters the solution as the four-coordinate complex $[Zn(H_2O)_2(OH)_2]$. This species is only slightly soluble (about 4×10^{-6} moles per liter), because, as a neutral molecule, it does not interact strongly with polar solvents. In basic solutions, however, one or both of the *aquo* ligands of $[Zn(H_2O)_2(OH)_2]$ can donate a proton to $OH^-(aq)$. Both the resulting tri- and tetra-hydroxo complexes are negatively charged, and as ionic species they are readily soluble in water. The mechanism by which zinc hydroxide reacts with and dissolves in base is given by Equations (23-14) and (23-15).

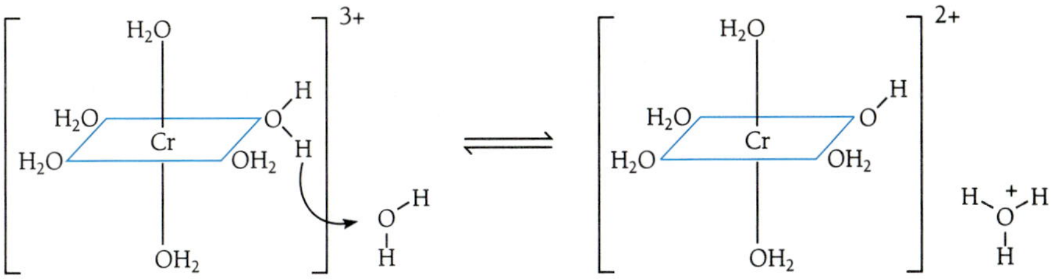

Figure 23.32 An *aquo* ligand of a hydrated metal ion can act as a proton donor. Aqueous solutions of such ions are acidic.

$$[Zn(H_2O)_2(OH)_2] + OH^- \rightleftharpoons [Zn(H_2O)(OH)_3]^- + H_2O \quad (23\text{-}14)$$

$$[Zn(H_2O)(OH)_3]^- + OH^- \rightleftharpoons [Zn(OH)_4]^{2-} + H_2O \quad (23\text{-}15)$$

The tetrahydroxozincate(II) ion, $[Zn(OH)_4]^{2-}$, is sometimes simply called "zincate ion."

In the presence of acid, on the other hand, one or both of the *hydroxo* ligands of $[Zn(H_2O)_2(OH)_2]$ can accept a proton. Both the tri- and tetra-aquo complexes are positively charged, and therefore soluble. The mechanism by which zinc hydroxide reacts with and dissolves in acid is given by Equations (23-16) and (23-17).

$$[Zn(H_2O)_2(OH)_2] + H_3O^+ \rightleftharpoons [Zn(H_2O)_3(OH)]^+ + H_2O \quad (23\text{-}16)$$

$$[Zn(H_2O)_3(OH)]^+ + H_3O^+ \rightleftharpoons [Zn(H_2O)_4]^{2+} + H_2O \quad (23\text{-}17)$$

Amphoteric behavior of hydroxides is not limited to transition elements. We have previously mentioned (Section 5.8) that $Al(OH)_3$ and $Ga(OH)_3$ are amphoteric.

Biological Complexes

Coordination compounds play a central role in many of the chemical processes that take place within living organisms. Although numerous ligands are found in nature, one type is of particular importance. The compound shown in Figure 23.33 is *porphine,* and if one or more of the hydrogen atoms on the outside of the ring is replaced

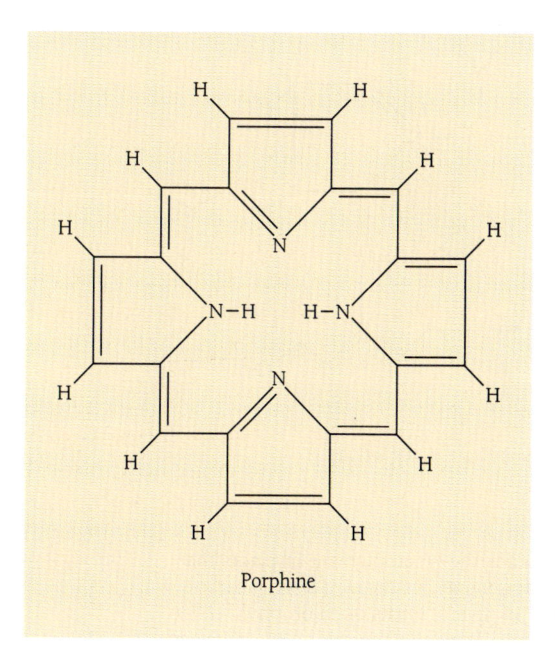

Figure 23.33 Porphine. Substitution of any or all of the H atoms on the outside of the ring gives rise to a class of tetradentate ligands called *porphyrins,* which are both common and important in living organisms.

The green pigment in plants is not for beauty, but for capturing the solar energy used in production of carbohydrates.

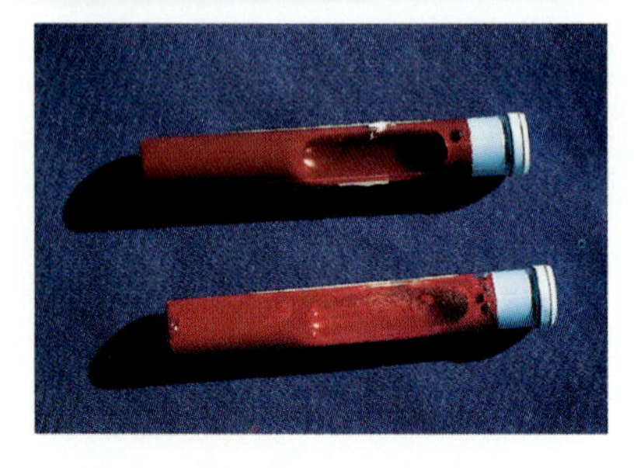

Blood taken from a vein (top) has a slightly purplish color due to the deoxygenated heme it contains. Arterial blood (bottom) contains a greater amount of oxygenated heme, which has a brighter red color.

by another group, the molecule is called a **porphyrin.** When the two *central* H-atoms are removed, any porphyrin becomes an excellent tetradentate ligand, capable of clamping a metal ion in its center by forming coordinate bonds with its four nitrogen atoms. Porphyrins are *planar* ligands—all of the atoms in Figure 23.33 lie in one plane. In a porphyrin complex, the chelated metal ion lies in the same plane. The complex is a neutral molecule when the metal ion is doubly charged, as with Fe^{2+} and Mg^{2+}.

Chlorophyll (Figure 23.34) is the green pigment in plants that absorbs sunlight and begins the complicated chain of events called *photosynthesis*. The overall process of photosynthesis is the conversion of water and carbon dioxide to carbohydrates and oxygen. Chlorophyll is somewhat unusual from the point of view of the chemistry of magnesium—normally, Group 2A elements form complexes only with oxygen ligands such as EDTA.

The function of absorbing O_2 in the lungs, transporting it throughout the body, and storing it until needed is carried out by the Fe(II) complex *heme* (Figure 23.35). Fe(II) has a coordination number of six, so that after binding to the four N atoms of the tetradentate porphyrin, there is room for two more ligands. One is a large protein whose tertiary structure (Section 21.1) allows it to enfold and protect the heme group. The heme-protein complex *myoglobin* (Figure 21.9) is used for oxygen storage in muscle tissue. The sixth ligand is either O_2 or H_2O, depending on whether or not oxygen is being stored. The molecule that transports oxygen is *hemoglobin,* which consists of four myoglobin molecules held together by hydrogen bonds. Oxygenated hemoglobin is red, and responsible for the color of arterial blood. Deoxygenated heme has a bluish or purplish tinge, which is imparted to the venous blood returning to the lungs.

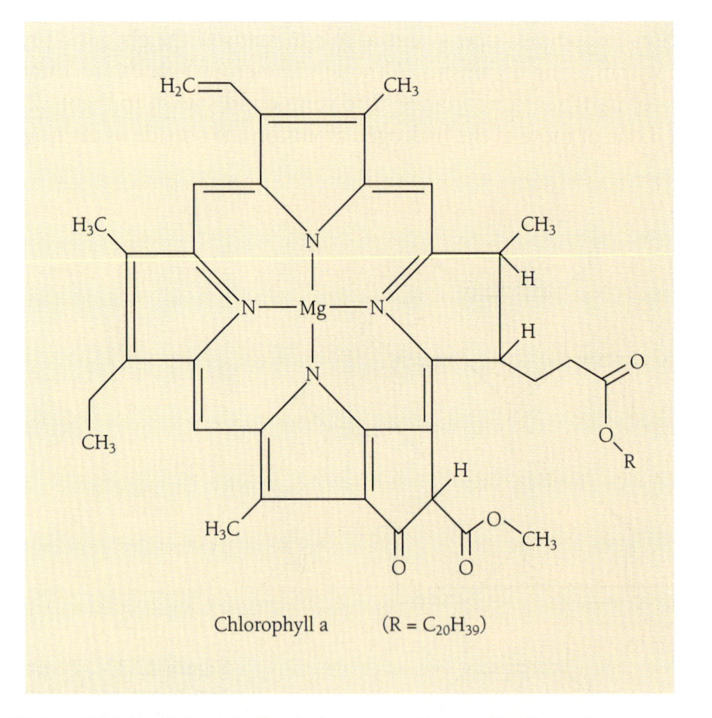

Figure 23.34 Chlorophyll a is the green pigment in plants that absorbs sunlight and provides energy for the synthesis of carbohydrates. Chlorophyll a is one of several forms of chlorophyll, differing slightly in the composition of the substituents on the porphyrin ring.

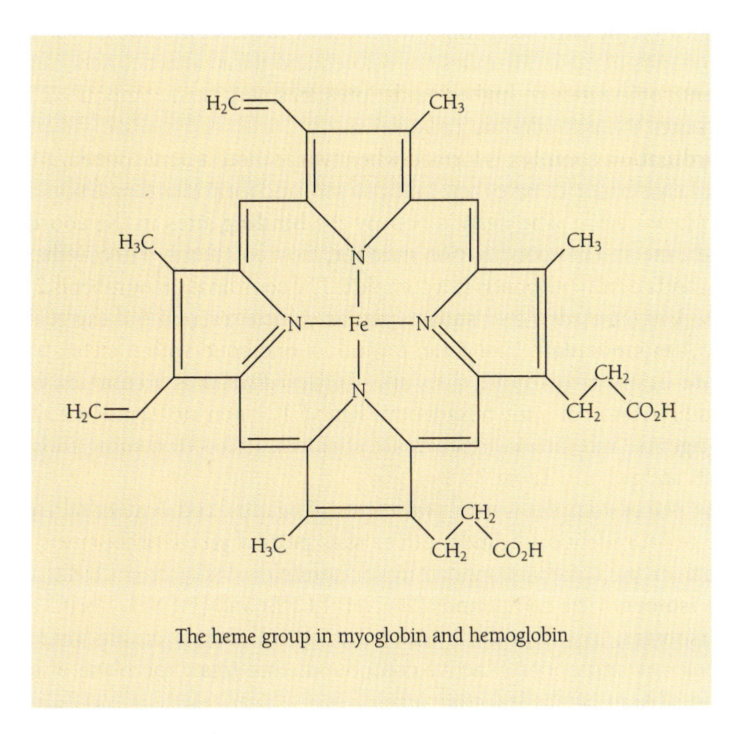

Figure 23.35 The Fe(II) complex *heme* is used in oxygen storage and transport in our bodies. Four of Fe's six coordinate bonds are to the tetradentate porphyrin ligand, and one is to a large protein molecule. The sixth site in the coordination sphere is available for a molecule of O_2.

Carbon monoxide is similar to oxygen in size and shape, and can also bind to the sixth site in heme. The CO-hemoglobin complex is about 250 times more stable than the O_2-hemoglobin complex, and, once formed, remains in the bloodstream for a long time. Carbon monoxide is poisonous because it prevents hemoglobin from carrying out its role in oxygen transport. Some animals have oxygen transport systems based on other metals. Horseshoe crabs, for example, use a copper-porphyrin-protein complex. It is presumed that the unusually large amount of vanadium in Venezuelan crude oil is the residue of long-extinct animals that used vanadium for this purpose.

SUMMARY AND KEYWORDS

The *transition metals,* in the B groups, are characterized (except for Zn, Cd, and Hg) by partly filled *d* subshells in at least one of their oxidation states. Their chemistry is dominated by the *d* orbitals: 3*d* in the **first series (fourth period)**, 4*d* in the **second series (fifth period)**, and 5*d* in the **third series (sixth period).** They have multiple oxidation states, and form both ionic (lower oxidation states) and covalent (higher oxidation states) compounds that are highly colored and frequently paramagnetic. Physical properties of transition elements exhibit regular trends as atomic number increases within a period: atomic radius decreases, density increases, melting point rises and then drops, ionization potential and electronegativity rise, and standard reduction potential (of the +2 ion) drops. Atomic radii in the second series are 10% to 15% larger than in the first. The **lanthanide contraction** explains why radii in the third series are about the same as those in the second series.

Elements of the *inner transition series* have partly filled *f* subshells: 4*f* in the sixth-period **lanthanides** ($Z = 57$ to 71), and 5*f* in the seventh-period **actinides**

(Z = 89 to 103). The most common oxidation state of the lanthanides is +3; most +3 ionic compounds of the lanthanides are colored and paramagnetic. Actinides form a few covalent compounds in higher oxidation states and ionic compounds in lower oxidation states. All actinides are radioactive.

A **coordination complex** is formed when two or more anionic or neutral **ligands** (which are Lewis bases) donate their lone pairs to, and form coordinate bonds with, a metal ion (Lewis acid). The ligands occupy the **binding sites** in the **coordination sphere** of the metal. The **coordination number,** most commonly 4 or 6, is the number of metal-ligand coordinate bonds in a complex. The coordination number is 2 in *linear* complexes, 4 in *tetrahedral* and **square planar** complexes, and 6 if the geometry is *octahedral.* A **monodentate** ligand can form only one bond with a metal, whereas a **polydentate** ligand forms more than one. *Ethylenediamine* is a common bidentate ligand, and *EDTA* is a common hexadentate ligand. Polydentate ligands are also called **chelating agents,** and their complexes are **chelates.** Rules for naming coordination compounds are given in Table 23.4.

Square planar complexes with two identical ligands, and octahedral complexes with three or four identical ligands, can exist as pairs of *geometrical isomers.* Ligands with more than one possible donor atom, for example —NO_2^- or —ONO^-, give rise to **linkage isomers.** The compounds $[Pt(NH_3)_3Cl]Br$ and $[Pt(NH_3)_3Br]Cl$ are **coordination isomers,** in which the roles of ligand and counterion are interchanged. *Optical isomerism* and optically active compounds that rotate the plane of polarized light are possible in octahedral complexes.

Coordination complexes can be described by *valence bond theory,* which ignores ionic character in metal-ligand bonds and correlates complex geometry with hybridization of d orbitals in the metal ion. Provided that $4d$ as well as $3d$ orbitals are used for hybridization of fourth-period metal ions, the theory correctly predicts magnetic properties. Valence bond theory fails to predict colors. *Molecular orbital theory* gives a more accurate picture of the bonding, but is more difficult to use.

Crystal field theory ignores the covalent character of the metal-ligand bonds, and declares the complex to be held together by electrostatic attraction between the ligands and the positive metal ion. Ligands are viewed as structureless negative "point charges" existing at locations determined by the observed geometry of the complex. The metal ion's d orbitals are split into several groups (two groups in octahedral complexes). That is, the five d orbitals acquire different energies by their interaction with the ligands. The **crystal field splitting** (Δ) is large for ligands high in the **spectrochemical series.** In such cases, called **strong-field** or **low-spin,** orbitals in the lower of the two groups are filled before electrons occupy orbitals in the upper group. In the **weak-field** or **high-spin** case, electrons occupy the five d orbitals so as to maximize the number of unpaired spins. The color of coordination compounds depends on the magnitude of Δ. The d-orbital splitting pattern depends on the geometry of the complex. In linear complexes, the d orbitals split into three groups; in square planar geometry, into four groups.

Ligands can displace one another from the coordination sphere in stepwise, reversible **ligand exchange** reactions. Such reactions are fast in **labile** complexes and slow in **inert** complexes. The **formation constant** is the equilibrium constant for the overall reaction in which a complex, initially having only *aquo* ligands, exchanges all of them for ligands of another type. Chelates have larger formation constants, that is, are more stable and less likely to undergo ligand exchange, than other complexes. Some ligands can participate in proton transfer reactions (act as a Brønsted-Lowry acid or base) while bonded to the metal. Many metal salts dissolve to give acidic solutions because one or more of the ligands in $[M(H_2O)_n]^{x+}$ can donate a proton, converting a solvent molecule to H_3O^+. Some metal hydroxides are amphoteric because of their proton-donating *aquo* ligand(s) and their proton-accepting *hydroxo* ligand(s).

Many compounds of biological importance are coordination complexes of **porphyrins,** which are large, planar tetradentate ligands incorporating four N-atoms. *Chlorophyll* is a magnesium(II)-porphyrin complex that absorbs solar energy in photosynthesis; *myoglobin* and *hemoglobin* are Fe(II)-porphyrin-protein complexes responsible for oxygen storage and transport in most animals.

PROBLEMS

Transition Elements: General

1. What features of electron configuration distinguish the transition elements from metals in the A groups of the periodic table?
2. What is meant by the terms "first," "second," and "third" transition series?
3. What is the lanthanide contraction?
4. Name the two inner transition series of elements.
5. Which elements in the second and third transition series have electron configurations that depart from the *aufbau* prediction?
6. Which of the actinides is/are radioactive?
7. Which of the actinides exist(s) naturally in other than trace quantities in the earth's crust?
8. In what order are electrons lost when transition metals form ions of successively higher charge?
9. Some chemists classify Zn, Cd, and Hg with the main-group elements rather than with the transition elements. What is the justification for this?
10. What type of bonding characterizes transition elements in their lower oxidation states? In their higher oxidation states?
11. Write electron configurations for the following elements.
    a. titanium b. nickel c. ruthenium
    d. molybdenum e. yttrium f. gold
12. Write electron configurations for the following species.
    a. V^{5+} b. Pt^{4+} c. Cu^+
    d. Cu^{2+} e. Ti^{4+} f. Hg^{2+}
13. Identify the oxidation state of the transition element in each of the following species.
    a. TiO_3^{2-} b. $AgNO_3$
    c. $TiCl_4$ d. $K_3Fe(CN)_6$
14. Identify the oxidation state of the transition element in each of the following species.
    a. MnO_2 b. K_2CrO_4
    c. Cr_2O_3 d. $K_2Cr_2O_7$

The First Transition Series

15. Which +2 ion among the fourth-period transition elements has the highest standard reduction potential?
16. Which element(s) in the first transition series has/have an electron configuration that departs from the *aufbau* prediction? Try to answer this without looking up the facts.
17. What is the highest oxidation number commonly found among the elements of the first transition series? Which element(s) exhibit(s) this oxidation state? Give examples.
18. Scandium, with only one 3*d* electron, has a greater atomic radius than all other elements in the first transition series, all of which have more than one 3*d* electron. Explain this.
19. Which of the metals Sc through Zn is most abundant in the earth's crust? Which has the second greatest abundance?
20. Which is the least abundant of the fourth-period transition elements?
21. With some exceptions, the melting point of the fourth-period transition elements rises from Sc to Cr, then falls. Explain this trend.
22. What is the relationship between electron configuration and maximum observed oxidation number in the first transition series?
23. Choose any two elements from the first transition series and describe how the metal is obtained from the ore.
24. Write one interesting chemical fact about each of the elements Sc through Zn.

Coordination Compounds: General

25. Define the terms "Lewis acid" and "Lewis base."
26. What is a coordinate bond?
27. Distinguish between coordination *compound* and coordination *complex.*
28. What is meant by the term "ligand"? What characteristic distinguishes ligands from species that are not ligands?
29. Define coordination number.
30. In what sense is a coordination compound an adduct?

31. What is a chelate? How is a chelate different from an ordinary coordination complex?

32. Name three chelating agents, and draw their structures.

33. Can the ammonium ion, NH_4^+, serve as a ligand in a coordination complex?

34. Would you expect BF_3 to be as good a ligand as NH_3? Explain your answer.

35. Classify each of the following ligands as monodentate, bidentate, etc.
    a. NO   b. pyridine, C_5H_5N
    c. oxalate, $C_2O_4^{2-}$ d. iodo

*36. Classify each of the following ligands as monodentate, bidentate, etc. (*Hint:* Use VSEPR to determine the shape of the azide and triiodide ions.)
    a. azide, N_3^- b. $CH_3—C≡N$ c. I_3^-
    d. acetylacetone: $CH_3—C(=O)—CH_2—C(=O)—CH_3$

37. The structure of *cyclam,* a saturated compound related to porphine, is shown below.

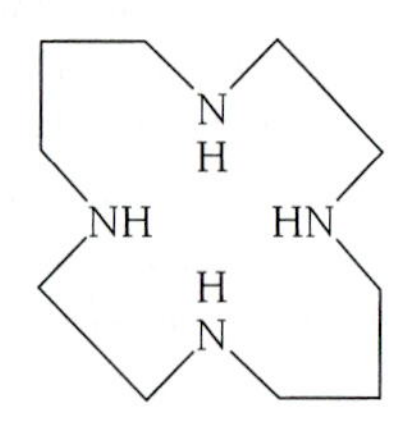

cyclam

How many coordinate bonds can cyclam form with a metal ion?

38. 1,10-Phenanthroline is a chelating agent, while its isomer 4,7-phenanthroline is not. Explain this difference.

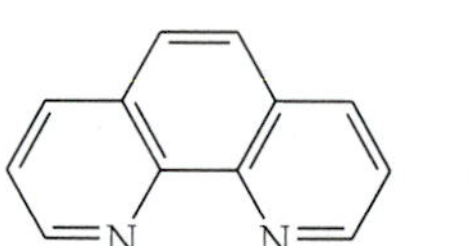

1,10-phenanthroline   **4,7-phenanthroline**

Oxidation State and Nomenclature

39. Determine the oxidation state of the transition metal in each of the following species.
    a. $[V(en)_3]^{3+}$ b. $[Ir(NH_3)_4Cl_2]^+$
    c. $[PtCl_4]^{2-}$ d. $[Cr(NH_3)_5SO_4]^+$

40. Name each of the species in Problem 39. (Ignore possible isomerism.)

41. Name each of the following complex ions. (Ox is the abbreviation for $C_2O_4^{2-}$, the oxalate ion.)
    a. $[Fe(H_2O)_2(CN)_4]^-$ b. $[Co(NH_3)_2(en)Cl_2]^+$
    c. $[Ni(NH_3)_2(ox)_2]^{2-}$ d. $[Zn(OH)_4]^{2-}$

42. Determine the oxidation state of the transition metal in each of the ions in Problem 41.

43. Determine the oxidation state of the transition metal in each of the following compounds.
    a. $K[Au(CN)_2]$ b. $Rb_4[Ni(ox)_3]$
    c. $[Cr(en)_2(NO)_2]Cl_2$ d. $[Ni(en)_3]Br_2$

44. Name each of the compounds in Problem 43.

45. Name each of the following compounds. (Phen is the abbreviation for 1,10-phenanthroline, whose structure is given above, in Problem 23.38.)
    a. $[Rh(CO)_5Cl]Cl_2$ b. $[Ru(phen)_3]Cl_2$
    c. $Na_2[Fe(EDTA)]$ d. $Pt(NH_3)_2Br_2$

46. Determine the oxidation state of the transition metal in each of the compounds in Problem 45.

47. Write the empirical formula of each of the following complex ions. Do not omit the ion charge from the formula.
    a. tetraaquodihydroxoaluminum(III)
    b. hexachloroferrate(II)
    c. triamminebromozinc(II)
    d. tris(ethylenediamine)nickel(II)

48. Write the empirical formula of each of the following complex ions. Do not omit the ion charge from the formula.
    a. amminetriiodoplatinate(II)
    b. tetracarbonyldinitrosomolybdenum(II)
    c. bis(diethylenetriamine)cobalt(III)
    d. dichloroaurate(I)

49. Write the empirical formula of each of the following compounds.
    a. iron(III) hexacyanoferrate(II)
    b. lithium diamminetetrabromochromate(II)
    c. chlorotrinitroplatinum(II) chloride
    d. calcium triaquotrichloronickelate(II)

50. Write the empirical formula of each of the following compounds.
    a. sodium dioxolatozincate(II)
    b. bis(ethylenediamine)dinitrosorhodium(III) nitrate
    c. bis(ethylenediamine)dichlorocobalt(III) sulfate
    d. potassium aquotrinitritocuprate(II)

Geometry and Isomerism

51. Distinguish between structural isomerism and stereoisomerism.

52. What type of isomerism can exist in (a) tetrahedral complex ions? (b) In square planar complex ions?

53. Let W, X, Y, and Z represent monodentate ligands in a square planar complex. Which of the following exist(s) as a pair of geometrical isomers: MWXYZ, MX_2YZ, MX_2Y_2, or MXY_3?

54. Let W, X, Y, and Z represent monodentate ligands in an octahedral complex. In which of the following is geometrical isomerism possible: $MX_2Y_2Z_2$, MW_3XYZ, or MWX_3Y_2? (Ignore the possibility of optical isomerism.)

*55. In which of the following is optical isomerism possible: $M(en)Cl_2$, $M(en)_3$, M(en)FClBrI, or $M(en)_2FCl$?

56. How many isomers are there for the complex ion dibromobis(ethylenediamine)cobalt(III)? Draw their structures.

57. Draw the structures of four isomers of the compound $Ni(NH_3)_2BrCl(en)$.

58. Draw structures for all isomers of each of the following species:
a. $[Pt(NH_3)_2(H_2O)_2]^{2+}$
b. $[Pt(NH_3)_2(H_2O)(SCN)]^{+}$
c. $[Pt(NH_3)_2(H_2O)(SCN)]Br$.

Bonding, Magnetism, and Color

59. Name and briefly describe each of the three theories that are used to describe the bonding in coordination complexes.

*60. In general, what changes in the physical properties of a complex might be expected when its ligands are replaced by ligands lower in the spectrochemical series (smaller value of Δ)?

*61. Compounds of Sc^{3+} are colorless, and compounds of Ti^{3+} are highly colored. Account for this difference.

62. Arrange the following ligands in order of increasing crystal field splitting: cyano, chloro, aquo.

63. For halogen ligands, crystal field splitting increases with electronegativity. Arrange the halide ions in order of increasing crystal field splitting.

*64. Solutions of Cu(II) salts are colored, whereas solutions of Zn(II) salts are not. Explain.

*65. In mystery stories, victims of cyanide poisoning are often described as having a blue color to the skin. Suggest a reason for this based on the similarity in size and shape between O_2 and CN^-.

66. It is observed that compounds containing the complex $[CoBr_6]^{3-}$ are paramagnetic, while those containing $[Co(py)_6]^{3+}$ are diamagnetic. What can you conclude about the relative positions of the ligands Br^- and py on the spectrochemical series? (py is an abbreviation of pyridine, C_5H_5N.)

*67. For each of the following complexes, determine the oxidation number of the metal, whether it is a high- or low-spin case, and predict the number of unpaired electrons.
a. $[VF_6]^-$ b. $[V(CN)_6]^-$
c. $[VF_6]^{2-}$ d. $[V(CN)_6]^{2-}$

*68. For each of the following complexes, determine the oxidation number of the metal, whether it is a high- or low-spin case, and predict the number of unpaired electrons.
a. $[CrF_6]^{3-}$ b. $[Cr(CN)_6]^{3-}$
c. $[NiF_6]^{4-}$ d. $[Ni(CN)_6]^{4-}$

69. Predict the number of unpaired electrons in each of the following complexes.
a. tetrafluoromanganese(IV)
b. hexafluorotungstate (V)
c. potassium hexachlororhenate(IV)

70. Predict the number of unpaired electrons in each of the following complexes.
a. hexafluororuthenium(IV)
b. hexacyanorhodate(III)
c. hexabromoniobate(V)

Chemical Properties

71. Is it possible for an "inert" complex to react? Explain.

72. When $H_2S(g)$ is bubbled through a solution of $Cu(NO_3)_2$, there is immediate formation of a black precipitate. If concentrated KCN(aq) is first added to the copper nitrate solution, no precipitate forms on addition of H_2S. Explain. (Don't try this experiment—both H_2S and KCN are dangerous poisons.)

73. Calculate the pH of a 0.00100 M solution of chromium trichloride, assuming that all chloride is present as counterion, not ligand. K_a for $[Cr(H_2O)_6]^{3+}$ is 1×10^{-4}. Compare this value to the pH of an acetic acid solution of the same molarity.

74. Calculate the ratio of the concentration of $[Cd(CN)_4]^{2-}$ to that of $[Cd(H_2O)_4]^{2+}$ in a solution in which $[CN^-] =$ 0.05 M.

75. Calculate the ratio of the concentration of $[Fe(CN)_6]^{3-}$ to that of $[Fe(H_2O)_6]^{3+}$ in a solution that is 0.050 M in CN^-.

76. Calculate the ratio of the concentration of $[Zn(NH_3)_4]^{2+}$ to that of $[Zn(H_2O)_4]^{2+}$ in a solution in which $[NH_3] =$ 0.100 M.

*77. A chemist wishes to remove from solution as much $Ni^{2+}(aq)$ as possible, to avoid possible interfering effects with a planned laboratory procedure. She has available 0.075 M solutions of ethylenediamine and sodium calcium EDTA. Which should she use? Explain your choice.

78. A concentrated aqueous solution of $CuCl_2$ is green in color, but when the solution is diluted by the addition of water, it turns blue. Explain.

79. When KCN(aq) is added to a solution of $ZnSO_4$, a white precipitate forms. The precipitate dissolves when more KCN is added. Explain.

CHAPTER 24

THE CHEMISTRY OF NUCLEAR PROCESSES

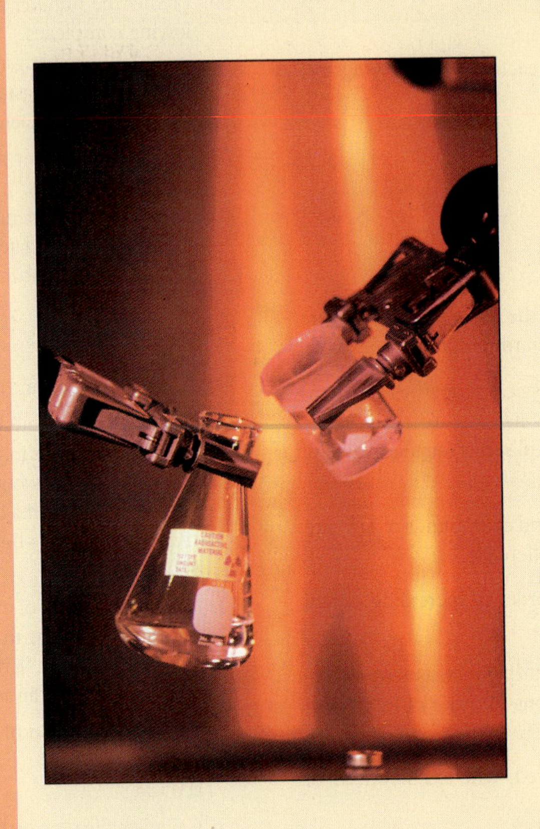

Chemists use precision robot manipulators to study the properties of radioactive substances from a safe distance.

Except for a brief description of radioactivity in Section 2.3, and mention of the fact that the actinides are radioactive, we have not yet discussed anything that can occur within the nucleus of an atom. In fact, the nucleus is much more than an inert repository of the atom's mass and positive charge, and important changes can take place there as well as in the atom's electron shells. Most of these processes result in the formation of different nuclei, having a different composition of protons and neutrons. Nuclear changes characteristically involve much greater energy than ordinary chemical processes. The study of nuclear events is a branch of science called *nuclear chemistry*.

24.1 SPONTANEOUS DISINTEGRATION OF NUCLEI

The study of the structure and behavior of the atomic nucleus began in 1896, when Antoine Becquerel discovered that uranium-containing substances emitted a previously unobserved form of radiation. The new phenomenon was named **radioactivity,** and the substances were described as **radioactive.** Soon after, Pierre and Marie Curie discovered two new radioactive elements, radium and polonium. Today we recognize that elements may have both radioactive and nonradioactive isotopes: the former are called **radioisotopes.** Recall that "isotope" usually refers to one of a small number of variant forms of a particular element, which have the same number of protons but a different number of neutrons. A more general term is **nuclide,** which is used to refer to any one of several thousand known nuclei, regardless of atomic number.

Marie Sklodowska Curie

Emission of α-, β-, and γ-"Rays"

Soon after Becquerel's discovery, Rutherford (historical box, page 52) suggested that radioactivity involves the spontaneous, highly exothermic *decay* or *disintegration* of atomic nuclei, and that the energy liberated in the process is carried away by the

Pierre Curie (1859–1906) and Marie Sklodowska Curie (1867–1934)

Pierre Curie was Professor of physics at the University of Paris (Sorbonne). His early work was in crystallography and the effects of temperature on magnetism. Together with his brother Jacques, he discovered *piezoelectricity,* the effect of pressure on the electrical properties of crystals. In 1895 he married Marie Sklodowska, a Polish student of chemistry.

Marie Sklodowska Curie continued her own research in chemistry following her marriage to Pierre. Her 1898 report of a probable new element in the uranium ore *pitchblende* was so exciting that her husband dropped his own research to work with her. They worked together until his death, and she later was appointed Professor of physics. Their joint work on radioactivity earned a share of the 1903 Nobel Prize. In 1911, Marie Curie was awarded a second Nobel Prize (the first person ever to be so honored) for the discovery of radium and polonium.

The nuclear-powered merchant vessel, *N. S. Savannah,* is no longer in service. It lies at anchor in Charleston, South Carolina, at the Patriots Point Naval and Maritime Museum.

Antoine Henri Becquerel (1852–1908)

Antoine Henri Becquerel was Professor of physics at the École Polytechnique in Paris. As the discoverer of radioactivity, he shared the 1903 Nobel Prize in physics with Marie and Pierre Curie. He also studied light absorption by crystals, phosphorescence, and the influence of magnetic fields on the rotation of polarized light.

radiation. He identified three types of rays as α, β, and γ. **Alpha (α) "rays"** are actually fast-moving composite particles consisting of two protons and two neutrons: an α-particle is identical to the nucleus of a helium atom. **Beta (β) "rays"** are fast-moving electrons, and only **gamma (γ) rays** are true electromagnetic radiation. Gamma rays have very short wavelengths, because their energy is so high. The relationship between energy and wavelength is discussed in Section 6.2. The equation $E = h\nu = hc/\lambda$ is valid for all forms of electromagnetic radiation, including γ-radiation.

Energy of Nuclear Decay

Most people are aware that nuclear-powered ships can cruise for months on relatively small quantities of fuel, and we are all familiar with the awesome power of nuclear weapons. The implication is clear: nuclear processes release vast amounts of energy. Nuclear energy can be expressed in the SI unit *joules,* but the **electronvolt (eV)** is also used by nuclear chemists. The equivalence statement for conversions to and from SI is $1\ \text{eV} \Leftrightarrow 1.602 \times 10^{-19}\ \text{J}$. Since nuclear energies are large, the mega-electronvolt is commonly used: $1\ \text{MeV} \Leftrightarrow 1.602 \times 10^{-13}\ \text{J}$.

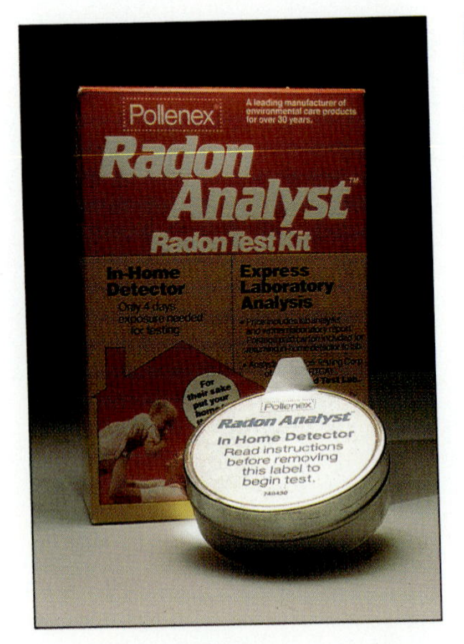

Relatively inexpensive kits are available for testing the radon content of the air in residences and other buildings.

EXAMPLE 24.1 Energy Released in Radioactive Decay

Radon undergoes α-decay, with an energy release of 5.49 MeV per atom. How much energy is released per mole of radon?

Analysis

Target: A molar energy, in J mol^{-1}.

Relationship: The energy conversion factor is known, and Avogadro's number converts atomic to molar units.

Plan Use the factor-label method of unit conversion.

Work

$$? \text{ J mol}^{-1} = \left(5.49\ \frac{\text{MeV}}{\text{nucleus}}\right)\left(\frac{10^6\ \text{eV}}{1\ \text{MeV}}\right)\left(\frac{1.602 \times 10^{-19}\ \text{J}}{1\ \text{eV}}\right) \times \left(\frac{6.022 \times 10^{23}\ \text{nuclei}}{1\ \text{mol}}\right)$$

$$= 5.30 \times 10^{11}\ \text{J mol}^{-1}$$

Check If this were a *chemical reaction,* you should conclude that an error had been made, because the number is so large. But the math is correct—nuclear reaction energies are perhaps a million times larger than the energies released in chemical reactions. (Combustion of methane releases about 9×10^5 J mol^{-1}.)

Exercise

Calculate the molar energy released in the 1.72 MeV β-decay of phosphorus-32.

Answer: 1.66×10^{11} J mol^{-1}.

Writing Nuclear Equations

A nuclide is identified by its elemental symbol, preceded by a subscript giving the positive charge of the nucleus (the atomic number) and a superscript equal to the mass number (the number of *nucleons,* or total number of protons and neutrons). Using this notation, nuclear processes are represented by equations similar to ordinary chemical equations. The equation for the formation of radon gas by the α-decay of radium nuclei in the earth's crust is

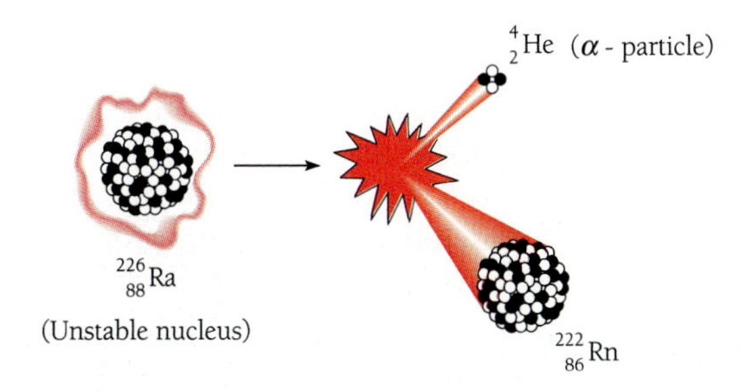

Sooner or later the inherent instability of the $^{226}_{88}$Ra nucleus results in spontaneous decomposition into an α-particle (a nucleus of $^{4}_{2}$He) and a $^{222}_{86}$Rn nucleus. The energy released in this disintegration appears as kinetic energy of the fragments.

in which the symbol $^{4}_{2}$He is used for the α-particle. The product of the decay, in this case ^{222}Rn, is called the **daughter** nuclide. Carbon-14, a β-emitter, is used to determine the age of objects in the technique of radiocarbon dating (Section 24.8). The equation for this process is

$$^{14}_{6}\text{C} \longrightarrow {}^{14}_{7}\text{N} + {}^{\ 0}_{-1}e$$

Either $^{\ 0}_{-1}e$ or $^{\ 0}_{-1}\beta$ may be used as the symbol of the β-particle. Note that the atomic number subscript, normally the same as the positive charge of the nucleus, is −1 for the electron. The superscript, which gives the total number of neutrons plus protons, is zero for the β-particle.

It often happens that α- or β-decay produces a **metastable** daughter, that is, a nucleus in an *excited state.* After a while an excited nucleus drops to the ground state, losing its energy in the form of a γ-particle (photon). This process is essentially the same as the emission of light by an excited atom or molecule (Section 6.3), except that excited *nucleons* rather than electrons are involved. An example of this sequential process is seen in the β-decay of ^{131}I, which produces ^{131}Xe in an excited state.

^{131}I is used in the diagnosis and treatment of thyroid disorders.

$$^{131}_{53}\text{I} \longrightarrow {}^{131}_{54}\text{Xe} + {}^{\ 0}_{-1}e$$

When the daughter drops to the ground state by γ-emission, there is no change in the number of neutrons or protons.

$$^{131}_{54}\text{Xe} \longrightarrow {}^{131}_{54}\text{Xe} + {}^{0}_{0}\gamma$$

Except for loss of energy, the original nucleus is unchanged. Symbols used for other common particles in nuclear chemistry are $^{1}_{0}n$ for the neutron and $^{1}_{1}p$ or $^{1}_{1}\text{H}$ for the proton.

When equations for nuclear processes are correctly written, they are balanced for mass and charge. The mass number (superscript) of the nuclide on the left is equal to the sum of the mass numbers of the particles on the right, and the positive charge (subscript) on the left is equal to the sum of the charges on the right.

EXAMPLE 24.2 Chemical Equations of Radioactive Decay

Write a balanced equation for the β-decay of ^{32}P.

Analysis

Target: A nuclear equation, balanced with respect to charge and mass.

Relationship: The sum of the superscripts on the right must equal that on the left, and the same is true for the subscripts.

Work The complete symbol for ^{32}P is $^{32}_{15}\text{P}$, so the target equation is

$$^{32}_{15}\text{P} \longrightarrow {}^{y}_{z}\text{X} + {}^{0}_{-1}e$$

where the element symbol X, the mass number y, and the atomic number z must be determined. The superscript (mass) equation is $y + 0 = 32$, or $y = 32$. The subscript (charge) equation is $z + (-1) = 15$, or $z = 16$. Insertion of these values gives the balanced equation.

$$^{32}_{15}\text{P} \longrightarrow {}^{32}_{16}\text{X} + {}^{0}_{-1}e$$

The atomic number of the daughter nuclide is 16, so the correct element symbol is S. The complete equation is

$$^{32}_{15}\text{P} \longrightarrow {}^{32}_{16}\text{S} + {}^{0}_{-1}e$$

Exercise

Write a balanced equation for the α-decay of ^{230}Th.

Answer: $^{230}_{90}\text{Th} \rightarrow {}^{226}_{88}\text{Ra} + {}^{4}_{2}\text{He}$.

Other Modes of Decay

In the years since the discovery of α-, β-, and γ-radiation, a number of other, less common, modes of radioactive decay of nuclei have been identified. One of these is the emission of a **positron,** a particle having the same mass as an electron but bearing a positive charge. Others include emission of a neutron, orbital electron capture, and spontaneous fission (the splitting of a nucleus into two nuclei of intermediate size and, usually, several neutrons as well). Some of these will be described in later sections of the chapter.

24.2 STABILITY OF NUCLEI

An important and interesting question is, "Why are some nuclei stable, while others undergo spontaneous disintegration?" The general answer lies in the forces that bind the nucleons together in the nucleus. To begin with, an electrostatic force of *repulsion* exists between positively charged protons. This force is inversely proportional to the square of the distance separating them. Since protons are packed into a nucleus whose diameter is about 100,000 times smaller than that of the atom, this force is very strong. All nuclei other than hydrogen contain more than one proton, so we might equally well ask, "Why are there *any* stable nuclei heavier than hydrogen?" There *must* be an attractive force, stronger than electrostatic repulsion, that prevents nuclei from flying apart.

Many experiments by physicists have confirmed the existence of an attractive force, called the nuclear **"strong force,"** between nucleons. This force does not depend on electrical charge, and it affects protons and neutrons equally. The strong force counters the electrostatic force, and it allows the nucleus to exist as a stable entity.

The strong force operates only within the nucleus, that is, at distances less than a few femtometers (1 fm $\Leftrightarrow 10^{-15}$ m). Outside the nucleus, at any distance greater than a few fm, the strong force is negligible.

Mass and Numbers

There are limits to the effectiveness of the nuclear strong force. In particular, it cannot overcome the mutual electrical repulsion of positive charges if there are too many protons in a nucleus. The first rule of stability, then, is that there are *no* stable nuclei with more than 83 protons. All elements beyond bismuth in the periodic table are unstable. This is usually thought of in terms of the mass of the nucleus, so that an equivalent statement is, "Elements heavier than bismuth are radioactive."

Another factor in nuclear stability is whether the numbers of neutrons and protons are even or odd. Table 24.1 shows at a glance that, of the 279 naturally occurring nonradioactive nuclides, odd numbers of nucleons are less common.

In addition to the odd-even aspect, certain numbers of nucleons confer extra stability. Nuclei having 2, 8, 20, 28, 50, or 82 protons or neutrons tend to be more stable than other nuclides. These so-called **magic numbers** suggest a structured nucleus in which nucleons are arranged in shells, similar to the electron structure of the outer atom.

The Neutron-to-Proton Ratio

The stability of nuclei is also related to the ratio of the number of neutrons to the number of protons, the *n/p ratio.* Among the lighter elements, stable nuclides have approximately the same numbers of neutrons and protons: $n/p \approx 1$. As the atomic number increases, however, the relative number of neutrons increases, and n/p rises to

TABLE 24.1 Naturally Occurring Stable Nuclides

Number of Nuclides	Number of Protons	Number of Neutrons
168	Even	Even
57	Even	Odd
50	Odd	Even
4	Odd	Odd

about 1.5 in the largest stable nuclei. Apparently additional neutrons, with their strong-force cohesiveness, are required to overcome the electrostatic repulsion of a larger number of protons in the heavier elements. An interesting feature emerges in Figure 24.1, where the number of neutrons is plotted against the number of protons in all *stable* nuclides. Stable nuclei lie in a **stability belt** above the line corresponding to $n/p = 1$.

When an unstable nucleus decays, it generally does so in a way that puts the daughter nuclide into or closer to the stability belt. The characteristic mode of decay of nuclides *above* the stability belt is β-emission. In β-decay, one of the neutrons in a nucleus is converted to a proton-electron pair

$$^{1}_{0}n \longrightarrow {}^{1}_{1}p + {}^{0}_{-1}e \qquad \text{(within the nucleus)}$$

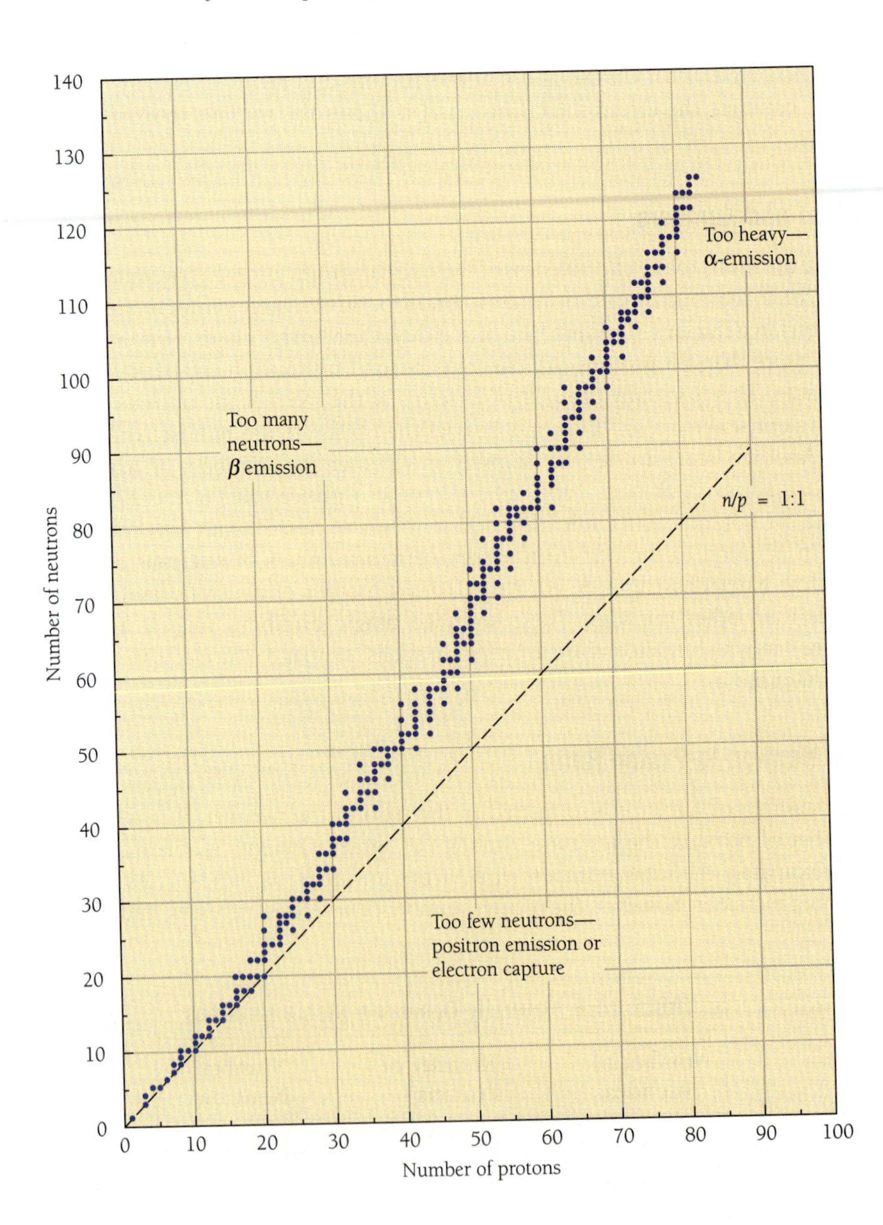

Figure 24.1 A plot of the number of neutrons versus the number of protons in all the stable nuclides. In all stable nuclides except hydrogen, the number of neutrons is at least as great as the number of protons, so that $n/p \geq 1$. The heavier the nuclide, the greater the ratio, and stable nuclei lie in an upward-curving "stability belt." There are no stable nuclei with more than 83 protons.

and the electron is ejected at high speed, while the proton remains in the nucleus. The daughter nucleus, having lost a neutron and gained a proton, has a lower *n/p* ratio. An example is ^{106}Ru.

$$^{106}_{44}\text{Ru} \longrightarrow {}^{106}_{45}\text{Rh} + {}^{\;0}_{-1}e$$

n/p	**1.41**	**1.36**

For those nuclei below the stability belt, both positron emission and, less commonly, electron capture produce daughter nuclei with an increased *n/p* ratio. In the former, a proton becomes a neutron plus a positron, which is ejected.

With many nuclides having low *n/p* ratios, both positron emission and electron capture are observed. It is not possible, from the value of *n/p* alone, to predict which of these processes is more likely.

$$^{1}_{1}\text{p} \longrightarrow {}^{1}_{0}\text{n} + {}^{\;0}_{+1}e \qquad \text{(within the nucleus)}$$

The daughter has more neutrons and fewer protons. For example, ^{25}Al decays by positron emission. (Either $^{\;0}_{+1}e$ or $^{\;0}_{+1}\beta$ may be used as the symbol of the positron).

$$^{25}_{13}\text{Al} \longrightarrow {}^{25}_{12}\text{Mg} + {}^{\;0}_{+1}e \qquad (\text{or } {}^{\;0}_{+1}\beta)$$

n/p	**0.92**	**1.08**

In *electron capture,* also called "K-capture," the nucleus absorbs an electron from a low-lying atomic orbital. The captured electron combines with a proton to produce a neutron, which remains in the nucleus.

The electron shell with $n = 1$ was originally called the "K" shell, hence the name "K-capture."

$${}^{\;0}_{-1}e + {}^{1}_{1}\text{p} \longrightarrow {}^{1}_{0}\text{n}$$

The *n/p* ratio increases, and energy is emitted in the form of an x-ray photon. The decay of ^{41}Ca is an example.

The wavelength of the emitted photon is determined by the energy levels of the daughter atom, not the parent.

$${}^{\;0}_{-1}e + {}^{41}_{20}\text{Ca} \longrightarrow {}^{41}_{19}\text{K} + h\nu$$

n/p	**1.05**	**1.16**

α-Decay, which is common among the heavier radioisotopes, is rare for nuclides of atomic number less than 83. The daughter nuclide has a slightly larger *n/p* ratio:

$$^{221}_{87}\text{Fr} \longrightarrow {}^{217}_{85}\text{At} + {}^{4}_{2}\text{He}$$

n/p	**1.540**	**1.553**

Emission of a γ-ray leaves the *n/p* ratio unchanged.

EXAMPLE 24.3 Modes of Nuclear Disintegration

Predict whether α-decay, β-decay, or positron emission is most likely when the radioisotope ^{50}Mn decays. Write a balanced equation for the process.

Analysis

Target: An equation describing a specific decay mode.

Relationship: The decay mode is suggested by the atomic number and by the *n/p* ratio, and the equation follows from the decay mode.

Plan Decide whether α-decay is likely for an isotope of manganese, and, if not, consult Figure 24.1 to decide whether the number of neutrons is greater than (β-decay) or less than (positron emission) expected for a stable nucleus.

Work The atomic number of ^{50}Mn is 25, so it is not heavy enough to decay by emission of an α-particle. There are 25 protons and (50 − 25) 25 neutrons in its nucleus, and Figure 24.1 shows that a stable nuclide with atomic number 25 has perhaps 26 to 30 neutrons. Since 25 is below this range, ^{50}Mn probably decays by positron emission. The equation is

$$\begin{array}{lccc} & {}^{50}_{25}\text{Mn} \longrightarrow & {}^{50}_{24}\text{Cr} + {}^{0}_{+1}e \\ n/p & 1.00 & 1.08 \end{array}$$

Exercise

Write the equation for the most likely decay mode of the radioisotope ^{41}Cl.

Answer: ${}^{41}_{17}\text{Cl} \rightarrow {}^{41}_{18}\text{Ar} + {}^{0}_{-1}e$

Nuclear Binding Energy

As discussed in Section 24.5, elements heavier than hydrogen are formed in the interior of stars and in supernova explosions.

Even among the nonradioactive nuclides—those that do not spontaneously disintegrate—there are differences in relative stability. A characteristic value of nuclear binding energy is associated with each nuclide. Nuclear binding energy is energy of cohesion, or the amount of energy that must be supplied to break a nucleus apart into its constituent nucleons. It is analogous to the atomization energy of a molecule, but much larger. As illustrated in Figure 24.2, the value for the helium nucleus is $\approx 3 \times 10^9$ kJ mol^{-1}, 10 million times greater than a typical molecular dissociation energy. It follows from the definition that the binding energy is the amount of energy released to the environment when a nucleus is formed from protons and neutrons. A large value of binding energy means that the total of the cohesive forces is large, and that the nucleus is very stable.

Careful measurements with mass spectrometers (Section 2.5) show that the mass of an atom is always less than the sum of the masses of its constituent protons, neutrons, and electrons, by an amount that is called the **mass defect.** During formation of nuclei, part of the nucleon mass is converted to energy and released to the environment. The relationship between the mass defect and the binding energy is given by the **Einstein equation,**

$$E = mc^2 \tag{24-1}$$

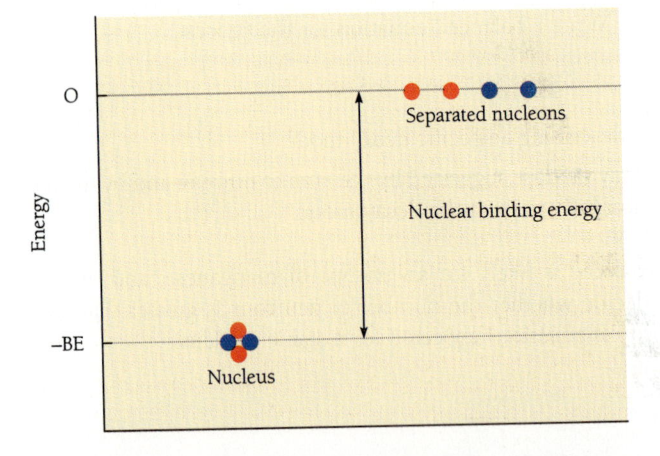

Figure 24.2 The binding energy of the helium nucleus, 2.73×10^9 kJ mol^{-1}, must be supplied to break one mole of helium nuclei into the constituent protons and neutrons.

in which c is the speed of light and m is the amount of mass converted to energy. This relationship provides a means of calculating the nuclear binding energy from the known mass of an atom.

Very precise values of the masses of all the known nuclides are given in the "Table of the Isotopes" in the *CRC Handbook*. Note that this table gives the mass of one *atom*, which includes the electrons, not the mass of the *nucleus*.

EXAMPLE 24.4 Nuclear Binding Energy

Calculate the mass defect and the binding energy of the $^{4}_{2}He$ isotope, whose atomic mass is 4.00260 amu.

Analysis

Target: Both mass defect and its energy equivalent, the binding energy, are called for.

Relationship: The mass defect is defined as the difference between the mass of the atom and the sum of the masses of its protons, neutrons, and electrons. The binding energy is the amount of energy released when this amount of mass disappears. Binding energy is related to mass defect by the Einstein equation.

Plan Subtract the given mass of the $^{4}_{2}He$ atom from the total mass of the individual protons, neutrons, and electrons. Then use the Einstein mass-energy relationship to convert the mass defect to energy units. Be careful to distinguish between amu and kg, and also between quantities that refer to one atom and those that refer to one mole. If all quantities are expressed in SI units (kg, m, s), the calculated energy will have units of joules.

The SI relationship between energy and the base quantities mass, length, and time is $1\ J \Leftrightarrow 1\ kg \cdot 1\ m^2 \cdot 1\ s^{-2}$.

Work The notation tells us that the $^{4}_{2}He$ atom contains 2 protons, 2 neutrons, and 2 electrons, whose masses are listed on the inside back cover of the text. The total mass of these components is

$$\begin{array}{lrl} & \text{protons: } 2(1.007276) & = 2.014552 \text{ amu} \\ + & \text{neutrons: } 2(1.008665) & = 2.017330 \\ + & \text{electrons: } 2(0.0005486) & = 0.001097 \\ \hline & \text{total:} & = 4.032979 \text{ amu} \end{array}$$

Subtracting the given mass of the atom from this total, we obtain

$$\begin{array}{lll} - \text{ atomic mass:} & 4.00260 & \\ \hline \text{mass defect:} & = 0.03038 \text{ amu} & \equiv 0.03038 \text{ g mol}^{-1} \end{array}$$

Convert the mass defect to kg (the SI *base* unit of mass), insert the appropriate values into the Einstein equation, and solve.

$$\begin{aligned} E &= mc^2 \\ &= \left(\frac{0.03038 \text{ g}}{1 \text{ mol}}\right)\left(\frac{1 \text{ kg}}{1000 \text{ g}}\right)\left(\frac{2.99792 \times 10^8 \text{ m}}{1 \text{ s}}\right)^2 \\ &= 2.730 \times 10^{12} \text{ kg m}^2 \text{ s}^{-2} \text{ mol}^{-1} \ (= \text{J mol}^{-1}) \\ &= 2.730 \times 10^{9} \text{ kJ mol}^{-1} \end{aligned}$$

Check This is a large number, as it should be for a nuclear binding energy. After use of the equivalence statement $1\ J \Leftrightarrow 1\ kg\ m^2\ s^{-2}$, the units are correct for an energy value.

Exercise

Calculate the mass defect and binding energy of the nuclide $^{35}_{17}Cl$, whose atomic mass is 34.9689 amu.

Answer: 0.3201 g mol^{-1}; 2.877×10^{13} J mol^{-1}.

Mass defect and binding energy increase as the number of nucleons increases, and as a result it might appear that large nuclei are more stable than small nuclei. This, however, is a hasty and incorrect conclusion. A better measure of the relative stability of nuclei is the **binding energy per nucleon,** which is obtained by dividing the binding energy by the total number of nucleons (the mass number). This gives the average energy of cohesion of the nucleons in a nucleus.

If electronvolt units are desired, the equivalence statement **931.4943 MeV ⇔ 1 amu** may be used for mass/energy conversions rather than the Einstein equation.

The equivalence statements 1 amu ⇔ 1.660540×10^{-27} kg and 1 eV ⇔ $1.6021773 \times 10^{-19}$ J lead directly to the energy equivalent of 1 amu:

$$\begin{aligned} E &= mc^2 \\ &= (1\text{ amu}) \cdot \left(\frac{1.660540 \times 10^{-27}\text{ kg}}{1\text{ amu}}\right) \cdot \left(\frac{2.99792458 \times 10^{8}\text{ m}}{1\text{ s}}\right)^2 \\ &= 1.492419 \times 10^{-10}\text{ J} \\ &= (1.492419 \times 10^{-10}\text{ J}) \cdot \left(\frac{1\text{ eV}}{1.6021773 \times 10^{-19}\text{ J}}\right) \\ &= 9.314943 \times 10^{8}\text{ eV} = 931.4943\text{ MeV} \end{aligned}$$

EXAMPLE 24.5 Binding Energy per Nucleon

The most stable nuclide is ^{56}Fe, whose mass is 55.93494 amu. Calculate the binding energy per nucleon in ^{56}Fe.

Analysis

Target: The binding energy per nucleon in MeV.

Relationship/Plan: Find the mass defect from the given isotope mass and the known proton, neutron, and electron masses; convert the mass defect to total binding energy; and divide by the number of nucleons.

Work The atomic number of iron is 26, so an atom of ^{56}Fe has 26 electrons, 26 protons, and 30 neutrons. The total mass of these components is

$$\begin{aligned} &\text{protons: } 26(1.007276) &&= 26.189176\text{ amu} \\ + \;&\text{neutrons: } 30(1.008665) &&= 30.259950 \\ + \;&\text{electrons: } 26(0.0005486) &&= 0.014264 \\ \hline &\text{total:} &&= 56.463390\text{ amu} \end{aligned}$$

Subtracting the given mass of the atom from this total, we obtain

− atomic mass:	55.93494	
mass defect:	= 0.52845 amu	= 0.52845 g mol^{-1}

When this is converted to energy units (MeV) and divided by the number of nucleons, the result is

$$\text{binding energy per nucleon} = \left(\frac{0.52845 \text{ amu}}{56 \text{ nucleons}}\right)\left(\frac{931.4943 \text{ MeV}}{1 \text{ amu}}\right)$$
$$= 8.7901 \text{ MeV}$$

Check All but the smallest nuclei have binding energies in the range 7 to 9 MeV.

Exercise

Calculate the binding energy per nucleon of ^{20}Ne, whose atomic mass is 19.99244 amu.

Answer: 8.0321 MeV.

Binding energy per nucleon is plotted *versus* atomic number, for all stable nuclides, in Figure 24.3. Nuclei of mass number 20 to 160 are relatively stable, while both heavier and lighter nuclei are less so (although most are stable enough so that they do not undergo radioactive decay). Most nonradioactive nuclides have values in the range 7 to 9 MeV.

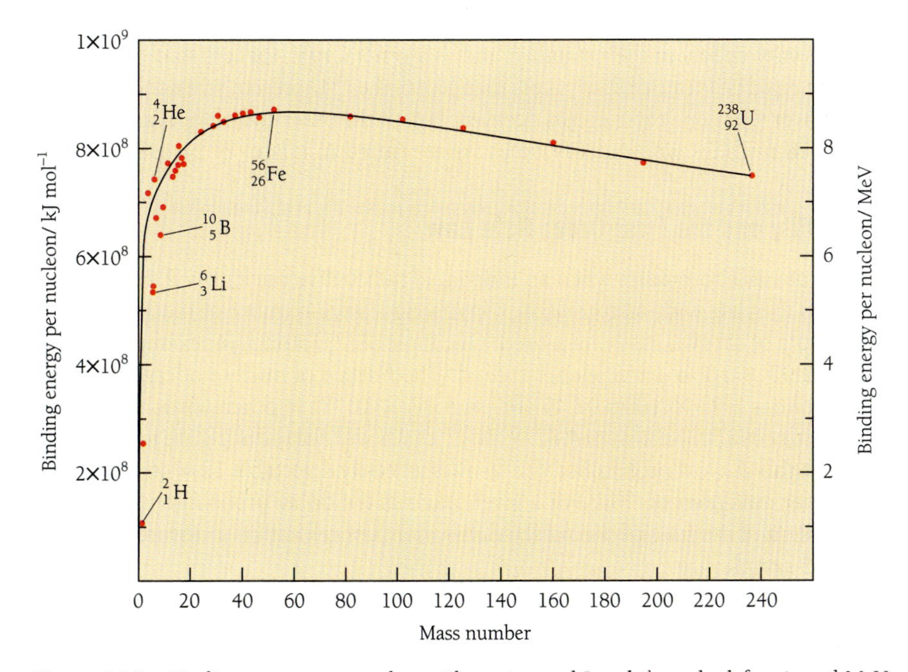

Figure 24.3 Binding energy per nucleon. The units are kJ mol^{-1} on the left axis and MeV per nucleon on the right. The greater the binding energy per nucleon, the more stable the nucleus. Both the light elements and the heavy elements are less stable than those of intermediate atomic number. The most stable nuclide is ^{56}Fe.

Fission and Fusion

Although it is a rare event, on occasion a large nuclide such as $^{235}_{92}U$ disintegrates by spontaneous **fission:** the breaking apart into two nuclides of intermediate mass number and one or more small particles as well.

$$^{235}_{92}U \longrightarrow ^{90}_{38}Sr + ^{143}_{54}Xe + 2\,^{1}_{0}n$$

The daughter nuclides of a fission event are more stable than the parent. That is, they have greater values of binding energy per nucleon (Figure 24.3). This greater stability means that nuclear fission is always an exothermic process.

Two very light nuclei can become more stable by joining together in a process called **fusion.** Two examples are

$$^{1}_{1}H + ^{2}_{1}H \longrightarrow ^{3}_{2}He$$

$$^{3}_{2}He + ^{1}_{1}H \longrightarrow ^{4}_{2}He + ^{0}_{+1}e$$

Fusion is also an exothermic process, since the product nucleus has a greater binding energy per nucleon than the reactant nuclei.

Nuclear fusion does not occur under ordinary conditions, because electrostatic repulsion between the nuclei keeps them apart. However, if the temperature is very high ($\geq 10^7$ K), some of the nuclei have enough kinetic energy to overcome the repulsion. If they can approach closely enough for the nuclear strong force to become important, fusion will occur. The core temperature of the sun (and other stars) is high enough so that fusion not only occurs, but is the source of the sun's energy. The liberated energy is radiated into space as the nuclear fuel "burns."

24.3 RATE OF NUCLEAR DECAY

Nuclides vary widely in their stability, and as a general rule, the less stable a nuclide, the more rapid is its rate of decay. Less than half of the ^{238}U present when the earth was formed over 4 billion years ago has decayed, but on the other end of the scale there are short-lived nuclides whose decay times are measured in micro- or picoseconds.

Probability and the First-Order Rate Law

It is not possible to predict when an individual radioactive nucleus will disintegrate, but (as is the case with many statistics involving a large number of individuals) the *average* rate of decay is easily measured. One mole of ^{238}U atoms, for example, emits a total of 2.93×10^6 α-particles per second. The *fraction* of nuclei undergoing decay each second is $2.93 \times 10^6 / 6.02 \times 10^{23} = 4.87 \times 10^{-18}$. In other words, an individual nucleus has a probability of 4.87×10^{-18} that it will decay during the next second. Although this is a low probability, the result is an easily detectable 12 α-emissions per second per milligram of ^{238}U. For a given nucleus the decay probability is constant, independent of the state of chemical combination, concentration, pressure, temperature, and sample size.

Let N be the number of ^{238}U atoms in the sample, and let ΔN be the number of nuclei undergoing decay (that is, the number of α-emissions) during the time period Δt. The *rate* of decay, $r = -\Delta N/\Delta t$, is given by the probability (k) that any one atom will decay during Δt, multiplied by the number of atoms in the sample.

$$r = -\frac{\Delta N}{\Delta t} = k \cdot N \qquad (24\text{-}2)$$

Equation (24-2) looks very much like the first-order rate laws [Equation (24-3)] discussed in Chapter 14 on chemical kinetics.

$$r = -\frac{\Delta C}{\Delta t} = k \cdot C \qquad (24\text{-}3)$$

Equation (14-14), page 603, has this form.

The only difference is that Equation (24-2) is based on the total number of reactant atoms, while Equation (24-3) is based on reactant concentration C (mol L^{-1}). If both sides of Equation (24-3) are multiplied by the sample volume and Avogadro's number, it becomes identical to Equation (24-2). That is, Equations (24-2) and (24-3) are two forms of the same relationship. The probability of decay (k) is in fact a first-order rate constant. It is a direct consequence of the probabilistic nature of the process that all nuclear decay follows first-order kinetics. Although Equations (24-2) and (24-3) are equivalent, nuclear chemists almost always use Equation (24-2).

EXAMPLE 24.6 Rate of Particle Emission

The radioactive isotope of hydrogen is *tritium,* which decays by β-emission with a rate constant of $1.76 \times 10^{-9}\ s^{-1}$: ${}^{3}_{1}H \rightarrow {}^{3}_{2}He + {}^{\ 0}_{-1}e$. How many β-particles are emitted by a 1.00 mg sample of tritium in one second?

Analysis

Target: The number of nuclear disintegrations in a 1 s time period.

Relationship: Equation (24-2) gives the rate of decay, but is based on the number rather than the mass of nuclei in the sample.

Plan Determine the number of nuclei present and apply Equation (24-2).

Work The exact atomic weight of ${}^{3}_{1}H$ is not given, but it is almost the same as the mass number, 3. This approximation is good to at least 2 significant figures, for any nuclide.

$$\text{Number of nuclei} = (1.00\ \text{mg})\left(\frac{0.001\ \text{g}}{1\ \text{mg}}\right)\left(\frac{1\ \text{mole}}{3.0\ \text{g}}\right)\left(\frac{6.02 \times 10^{23}\ \text{nuclei}}{1\ \text{mole}}\right)$$
$$= 2.0 \times 10^{20}\ \text{nuclei}$$

Insert the known values into Equation (24-2)

$$\text{rate of β-emission} = \text{decay rate}$$

$$r = -\frac{\Delta N}{\Delta t} = (1.76 \times 10^{-9}\ s^{-1}) \cdot N$$
$$= (1.76 \times 10^{-9}\ s^{-1}) \cdot (2.0 \times 10^{20})$$
$$= 3.5 \times 10^{11}\ \text{β-particles per second}$$

Exercise

There are three radioactive isotopes of oxygen. One of them, ${}^{14}O$, decays by positron emission with a rate constant of $0.00906\ s^{-1}$. How many positrons are emitted in 15 minutes by a 1.00-μg sample?

Answer: 3.5×10^{17}.

Half-Life

Nuclear chemists almost always discuss decay rate in terms of half-life rather than rate constant. The relationship between half-life and rate constant for first-order processes (Section 14.4) is

$$t_{1/2} = \frac{0.693}{k} \tag{24-4}$$

It is customary to express nuclear decay half-lives in seconds, minutes, hours, days, or years, using the abbreviations s, min, h, d, and y, depending on the magnitude of $t_{1/2}$.

EXAMPLE 24.7 Half-Life of Nuclear Decay

Calculate the half-life of ^{238}U.

Analysis

Target: A half-life, in convenient units.

Relationship: The equation $t_{1/2} = 0.693/k$.

Work The rate constant, given in the text above, is $4.87 \times 10^{-18}\ s^{-1}$. The half-life is therefore

$$t_{1/2} = \frac{0.693}{k}$$

$$= \frac{0.693}{4.87 \times 10^{-18}\ s^{-1}} = 1.42 \times 10^{17}\ s$$

A more customary answer requires conversion to a longer time unit

$$t_{1/2} = (1.42 \times 10^{17}\ s)\left(\frac{1\ \text{hour}}{3600\ s}\right)\left(\frac{1\ \text{day}}{24\ \text{hours}}\right)\left(\frac{1\ \text{year}}{365\ \text{days}}\right)$$

$$t_{1/2} = 4.50 \times 10^{9}\ y$$

Exercise

The rate constant for the β-decay of ^{72}Ga is $1.36 \times 10^{-5}\ s^{-1}$. Calculate its half-life.

Answer: 14.2 h.

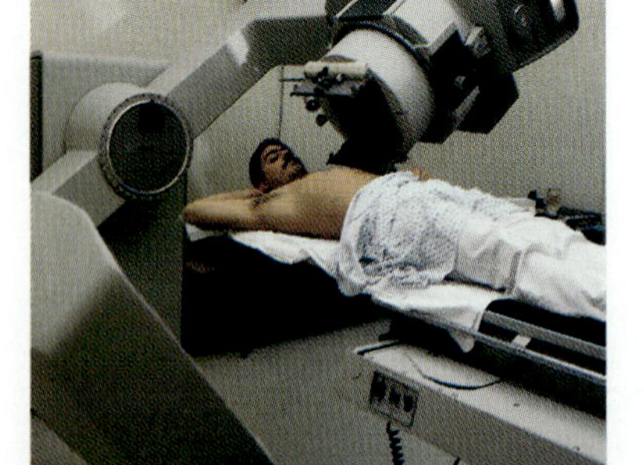

A cobalt-60 source for treating cancer with γ-radiation.

As illustrated in the following example, numerical problems involve the use of the *integrated* form of the rate law, Equation (24-5).

$$\ln (N/N_0) = -k \cdot t \tag{24-5}$$

In this equation N_0 is the original number of radioactive nuclei in the sample, N is the number remaining after an elapsed time t, and k is the rate constant.

EXAMPLE 24.8 Half-Life and the Rate of Decay

^{60}Co, a γ-emitter used in cancer therapy, has a half-life of 5.25 y. What fraction of the original number of Co nuclei remains after 2.00 years?

Analysis

Target: A dimensionless fraction; using the symbols defined in the text, this fraction is N/N_0.

Relationship: The integrated first-order rate law, Equation (24-5), connects the target with the knowns, and Equation (24-4) is used to obtain the rate constant from the half-life.

Work Find the rate constant from the half-life:

$$k = \frac{0.693}{t_{1/2}}$$
$$= \frac{0.693}{5.25\ \text{y}} = 0.132\ \text{y}^{-1}$$

The units of the rate constant (years) are appropriate to the data given in the problem, and should not be changed.

Insert the given information into Equation (24-5) and solve.

$$\ln(N/N_0) = -k \cdot t$$
$$= -(0.132\ \text{y}^{-1}) \cdot (2.00\ \text{y}) = -0.264$$

Exponentiate (take the antilogarithm of) both sides.

$$\text{fraction remaining} = N/N_0 = e^{\ln(N/N_0)} = e^{-0.264} = 0.768$$

A little more than ¾ of the original nuclei remain.

Check Since $t_{1/2} \approx 5$ y, after 5 years the fraction remaining is ≈ 0.5. After a lesser time, the fraction must be greater than 0.5.

Exercise

What fraction of the ^{238}U present when the earth was formed 4.3×10^9 years ago remains today? The half-life is 4.50×10^9 y.

Answer: 0.52.

24.4 INDUCED NUCLEAR REACTIONS

Radioactive decay is the spontaneous disintegration of a single nucleus. In contrast, fusion is the combination of two nuclei, and it takes place only when the nuclei meet in a high-energy collision. Sufficiently high kinetic energy is acquired naturally by nuclei in the high-temperature environment of stellar interiors. On earth, fusion and other reactions between two nuclear particles are studied in large machines called *particle accelerators.*

Alpha Bombardment and Neutron Capture

In 1919, Rutherford demonstrated that α-particles from radioactive decay have sufficient energy to induce nuclear reactions. When he bombarded a sample of nitrogen with α-particles, he observed the formation of an isotope of oxygen.

$$^{14}_{7}\text{N} + ^{4}_{2}\text{He} \longrightarrow ^{17}_{8}\text{O} + ^{1}_{1}\text{H}$$

Since that time, many other nuclear transformations have been induced by bombardment of stable nuclides with α-particles. Some of these reactions produce unstable nuclides that subsequently undergo spontaneous disintegration. For example, α-bombardment of ^{27}Al produces ^{30}P, which decays by positron emission with a half-life of 2.5 m. ^{30}P is an *artificial* radioisotope.

$$^{27}_{13}Al + ^{4}_{2}He \longrightarrow ^{30}_{15}P + ^{1}_{0}n$$

Nuclear reactions are also induced by neutrons.

$$^{6}_{3}Li + ^{1}_{0}n \longrightarrow ^{4}_{2}He + ^{3}_{1}H$$

Since the uncharged neutron is not repelled by a positively charged nucleus, high kinetic energy is not required for the reaction. Neutron *capture* is a more appropriate term than neutron *bombardment*.

Particle Accelerators

Particles having an electric charge, such as protons, electrons, or ions, can be accelerated to high kinetic energies by the action of electric fields. A diagram of a **cyclotron** is shown in Figure 24.4. Beams of high-energy particles can be produced and directed toward a suitable target material, where they interact with and induce nuclear transformations in target nuclei. The first cyclotron was constructed in the 1930s in California. It was a small machine, with a diameter of less than six inches. Modern versions of particle accelerators, in which particles usually travel in circular rather than spiral paths, have diameters of *miles*. In them, particles can be accelerated to more than 90% of the speed of light. The largest accelerator (Figure 24.5), operated by the European consortium (CERN), is 8.5 km (5.3 miles) in diameter. An even larger accelerator, the "superconducting supercollider," is under construction in Texas. It will have a circumference of 85 km (53 miles).

Figure 24.5 The LEP (large electron-positron) accelerator straddles the border between France and Switzerland. The Geneva (Switzerland) airport is in the foreground, while much of the rest of the photo lies in France. Completed in 1989, the LEP is the largest particle accelerator in use today. In it, electrons and positrons travel in opposite directions in an almost-circular path 26.7 km in circumference. After acceleration, the two types of particle meet head-on, bringing a total of more than 100 GeV (100×10^9 eV, about 10 million million kJ mol^{-1}) to the collision.

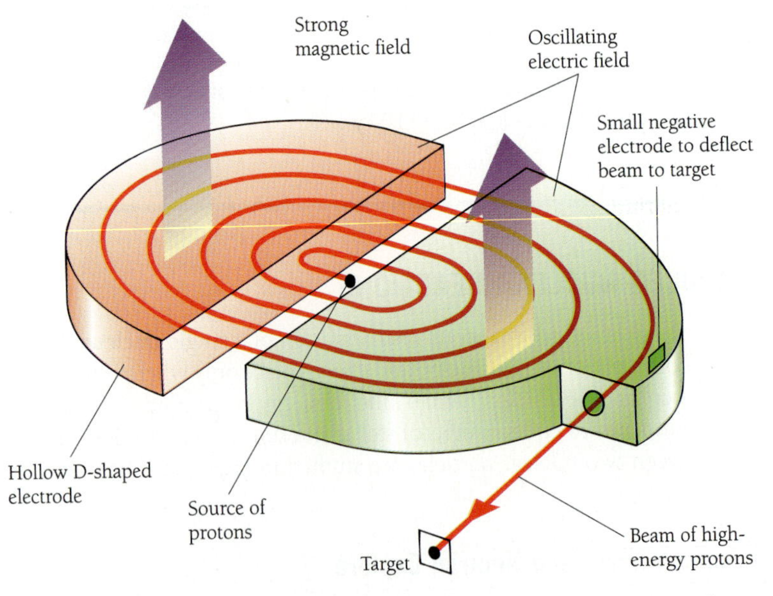

Figure 24.4 A cyclotron. Charged particles introduced at the center spiral outward toward the exit port, gaining energy on each circuit of the apparatus. An electric field is applied to the gap between the "dees" (named for their shape) so that the particle is accelerated when it crosses the gap. The polarity of the electric field is switched on each half-circuit, so that the particle always experiences an accelerating field, never a retarding field. The path followed by the particles is curved because powerful magnets (not shown) produce a magnetic field perpendicular to the plane of the dees.

The blue line in this photo is caused by light emitted from a beam of protons produced in a cyclotron at the Argonne National Laboratory.

Transuranium Elements

The heaviest naturally occurring element is uranium. The artificial elements having $Z = 93$ to 109 are called the **transuranium elements,** and they have all been discovered since 1940. They are produced either by neutron capture or by bombardment of suitable target nuclei with charged particles from an accelerator. *Neptunium* is prepared by exposing ^{238}U to neutrons. The unstable ^{239}U formed by neutron capture quickly decays by β-emission, forming ^{239}Np.

Since they are uncharged, neutrons cannot be accelerated in a cyclotron. A fission reactor (Section 24.6) is a convenient source of neutrons for synthesis of artificial radioisotopes.

$$^{1}_{0}n + {}^{238}_{92}U \longrightarrow {}^{239}_{92}U \longrightarrow {}^{239}_{93}Np + {}^{0}_{-1}e$$

The ^{239}Np isotope decays within a few days by β-emission, forming an isotope of *plutonium.*

$$^{239}_{93}Np \longrightarrow {}^{239}_{94}Pu + {}^{0}_{-1}e$$

A different isotope of plutonium can be produced by bombarding ^{238}U with *deuterons* (nuclei of deuterium, the second hydrogen isotope). Again, a short-lived neptunium intermediate is formed, which decays to the daughter ^{238}Pu.

$$^{2}_{1}H + {}^{238}_{92}U \longrightarrow {}^{238}_{93}Np + 2\,{}^{1}_{0}n$$

$$^{238}_{93}Np \longrightarrow {}^{238}_{94}Pu + {}^{0}_{-1}e$$

Heavier transuranium elements can be made by bombarding the lighter ones with protons, deuterons, α-particles, and so on. In this way, researchers have gradually worked their way out to element 109, synthesized in 1984. The main problem with this approach is that the stability, and the half-life, of these elements decreases dramatically as the atomic number increases. It is increasingly difficult to prepare elements of high atomic number in sufficient quantity to serve as targets for further bombardment, and there is less time to identify and characterize a suspected new element. Special techniques have been developed for investigating the chemical properties of a relatively small number of atoms.

A slightly different approach to the synthesis of high-Z transuranium elements is heavy-ion bombardment—the use of nuclei heavier than helium as projectiles. Rutherfordium ($Z = 104$) was first prepared in this way, by bombarding californium ($Z = 98$) with ^{12}C nuclei.

$$^{12}_{6}C + {}^{249}_{98}Cf \longrightarrow {}^{257}_{104}Rf + 4\,{}^{1}_{0}n$$

Elements 93 to 102 were first synthesized in the United States, primarily at the University of California. In the late 1950s, a Russian group began to achieve some new syntheses, and in the 1970s a German group successfully entered the field.

Energetics of Nuclear Reactions

The energy released or required by any nuclear reaction, spontaneous or induced, is calculated from the difference in mass between reactants and products. In a spontaneous reaction, the mass decreases as matter is converted to energy. Either gain or loss of mass can occur in induced reactions. Example 24.9 illustrates this calculation for two different spontaneous processes, α-decay and β-decay.

EXAMPLE 24.9 Energy Release in Nuclear Processes

Calculate the energy released in (a) the β-decay of $^{55}_{24}Cr$, and (b) the α-decay of $^{238}_{92}U$.

Analysis

Target: An energy, expressed in any convenient units, such as J per event or J per mole of events.

Relationship: The Einstein equation $E = mc^2$ relates the energy released to the mass difference between reactants and products.

Plan Write the nuclear equation for each process, look up the atomic masses of the parent and daughter nuclides, calculate the mass decrease, and convert to energy units.

Work

a. The equation is $^{55}_{24}Cr \rightarrow {}^{55}_{25}Mn + {}^{\ 0}_{-1}e$. The *CRC Handbook* gives the following masses for the nuclides,

$$^{55}_{24}Cr = 54.940842 \text{ amu}; \qquad ^{55}_{25}Mn = 54.938047 \text{ amu}$$

The mass change (products − reactants) is

$$54.938047 - 54.940842 = -0.002795 \text{ amu}$$

Note that this is negative, signifying a mass decrease: the lost mass appears as kinetic energy of the reaction products.

$$\begin{aligned} E &= mc^2 \\ &= \left(\frac{-0.002795 \text{ g}}{1 \text{ mol}}\right)\left(\frac{1 \text{ kg}}{1000 \text{ g}}\right)\left(\frac{2.9979 \times 10^8 \text{ m}}{1 \text{ s}}\right)^2 \\ &= -2.512 \times 10^{11} \text{ kg m}^2 \text{ s}^{-2} \text{ mol}^{-1} \\ &= -2.512 \times 10^{11} \text{ J mol}^{-1} \end{aligned}$$

Again, the negative sign denotes energy lost by the decaying nucleus. It is usual to refer to energy release as an unsigned quantity, as "β-decay of a ^{55}Cr atom releases 2.512×10^{11} J mol^{-1} to the environment." If desired, the result is converted to eV using the equivalence statement 1 eV ⇔ 1.602×10^{-19} J and Avogadro's number.

$$\begin{aligned} E &= \left(\frac{2.512 \times 10^{11} \text{ J}}{1 \text{ mol}}\right)\left(\frac{1 \text{ mol}}{6.022 \times 10^{23} \text{ decays}}\right)\left(\frac{1 \text{ eV}}{1.602 \times 10^{-19} \text{ J}}\right) \\ &= 2.604 \times 10^6 \text{ eV decay}^{-1} \end{aligned}$$

The "per decay" part of the units is usually omitted, and the quantity is written simply as

$$E = 2.604 \text{ MeV}$$

Check Energies of radioactive emissions and other nuclear reactions are usually in the MeV range, so this is a reasonable answer.

b. The balanced equation for the α-decay of uranium-238 is

$$^{238}_{92}\text{U} \longrightarrow {}^{234}_{90}\text{Th} + {}^{4}_{2}\text{He}$$

The nuclide masses listed in the *CRC Handbook* are

$$^{234}_{90}\text{Th: } 234.04359 \text{ amu}$$

$$^{4}_{2}\text{He: } 4.00260$$

$$^{238}_{92}\text{U: } 238.05078$$

The mass difference between products and reactants is

$$(234.04359 + 4.00260) - 238.05078 = -0.00459 \text{ amu}$$

The energy released to the environment is

$$E = (0.00459 \text{ amu})\left(\frac{931.4943 \text{ MeV}}{1 \text{ amu}}\right)$$

$$= 4.28 \text{ MeV}$$

Exercise

Calculate the energy released in the β-decay of ^{90}Sr. The *CRC Handbook* gives the following nuclide masses: ^{90}Sr, 89.907738 amu; ^{90}Y, 89.907152 amu.

Answer: 0.546 MeV.

24.5 NUCLEAR CHEMISTRY IN THE COSMOS

This section is concerned with processes occurring far outside the confines of the laboratory—within the earth itself, the sun, and the wider reaches of the universe. We will address the questions, "What is the source of the energy emitted by the sun?", "How did the 100-odd elements come into being?", and "What is the explanation for the natural radioactivity of the earth's crust?"

Stellar Energy and the Origin of the Elements

The origin of the universe can never be duplicated by experiment, and no scientific theory can explain it. It is possible, however, to learn some things about the early history of the universe, and to develop theories that are consistent with what we know from laboratory studies of nuclear events and from astronomical observations.

The Big Bang

Most scientists believe that the universe began as a dense cloud of matter/energy at an incredibly high temperature, expanding in a furious explosion called the *big bang*. As the cloud expanded, it cooled sufficiently to allow the separate existence of matter and radiation. The matter was present primarily as protons and electrons, in roughly equal

The recent (1992) satellite-based observations of fluctuations in the intensity of cosmic background radiation has provided additional support for the big bang theory.

numbers, and as neutrons. While protons are stable particles, neutrons (unless they are part of atomic nuclei) are unstable. Most of the neutrons in the early universe disintegrated into protons and electrons, which later combined, forming hydrogen atoms.

Laboratory studies of isolated neutrons show that this process takes place with a half-time of a little more than ten minutes.

$$^{1}_{0}n \longrightarrow {}^{1}_{1}H + {}^{0}_{-1}e$$

Some neutrons escaped this fate, by fusion with protons to form nuclei of deuterium and helium, but most did not. Most of the matter in the universe, then and now, is $^{1}_{1}H$.

The Formation and Energy Source of Stars

Gravitational attraction caused some of the newly formed hydrogen atoms to condense into clumps, at whose centers the density and temperature were high enough for hydrogen fusion to take place. The fusion of hydrogen nuclei produces helium, some deuterium, and a great deal of energy. The energy was radiated into space, and these clumps were the first stars. As their hydrogen "fuel" became depleted, gravitational force caused the stars to become smaller and hotter. The core temperature became high enough for the stars to maintain their energy by helium fusion. Heavier nuclei appeared—for example, ^{12}C formed in a two-step fusion of three helium nuclei.

$$^{4}_{2}He + {}^{4}_{2}He \longrightarrow {}^{8}_{4}Be; \qquad {}^{8}_{4}Be + {}^{4}_{2}He \longrightarrow {}^{12}_{6}C$$

Nuclear fusion reactions are the source of the radiant energy emitted by stars. A star's color indicates its temperature: hot stars are blue, while cooler stars are yellow.

Nuclei larger than ^{12}C were not formed in quantity, however, because the temperature and density cannot rise high enough in stars of small or moderate size. This is the end of stellar evolution for small stars. They gradually cool as the last of their fuel is expended, leaving a carbon core and a rich mix of lighter elements in their outer layers.

Stellar Explosions and the Synthesis of Heavy Elements

The fate of large stars is more dramatic. Hotter and denser than smaller stars, they maintain their energy by fusion of ever-larger nuclei, eventually forming a great deal of iron. The process stops with iron, because further increase in the mass of nuclei is endothermic. Their fuel exhausted, the stars collapse under gravitational pressure and explode in a *supernova* with unimaginable violence. During the explosion the temperature rises to the point where endothermic fusion of nuclei of intermediate mass can occur, forming all the heavier elements. These become part of the expanding debris from the explosion, and they result in the formation of interstellar dust clouds.

Iron-56 is the most stable nuclide; that is, it has the largest binding energy per nucleon.

Second-Generation Stars and Planets

Once again, gravitational force causes the formation of clumps, this time containing dust as well as hydrogen. The larger of these bodies are sufficiently hot and dense for fusion to take place, and they reignite as second-generation stars like our Sun. Smaller

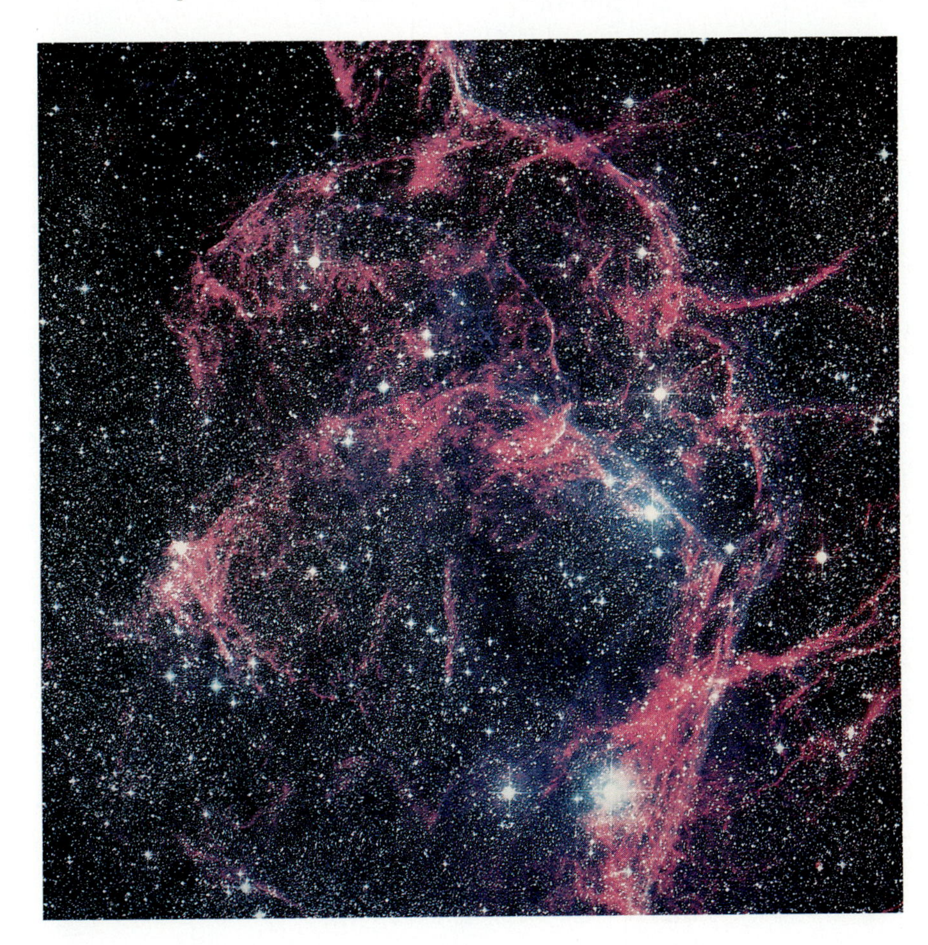

These red filaments are remnants of the Vela supernova, which took place about 12,000 years ago. Heavy elements are formed from smaller nuclei in the incredible fury of a supernova.

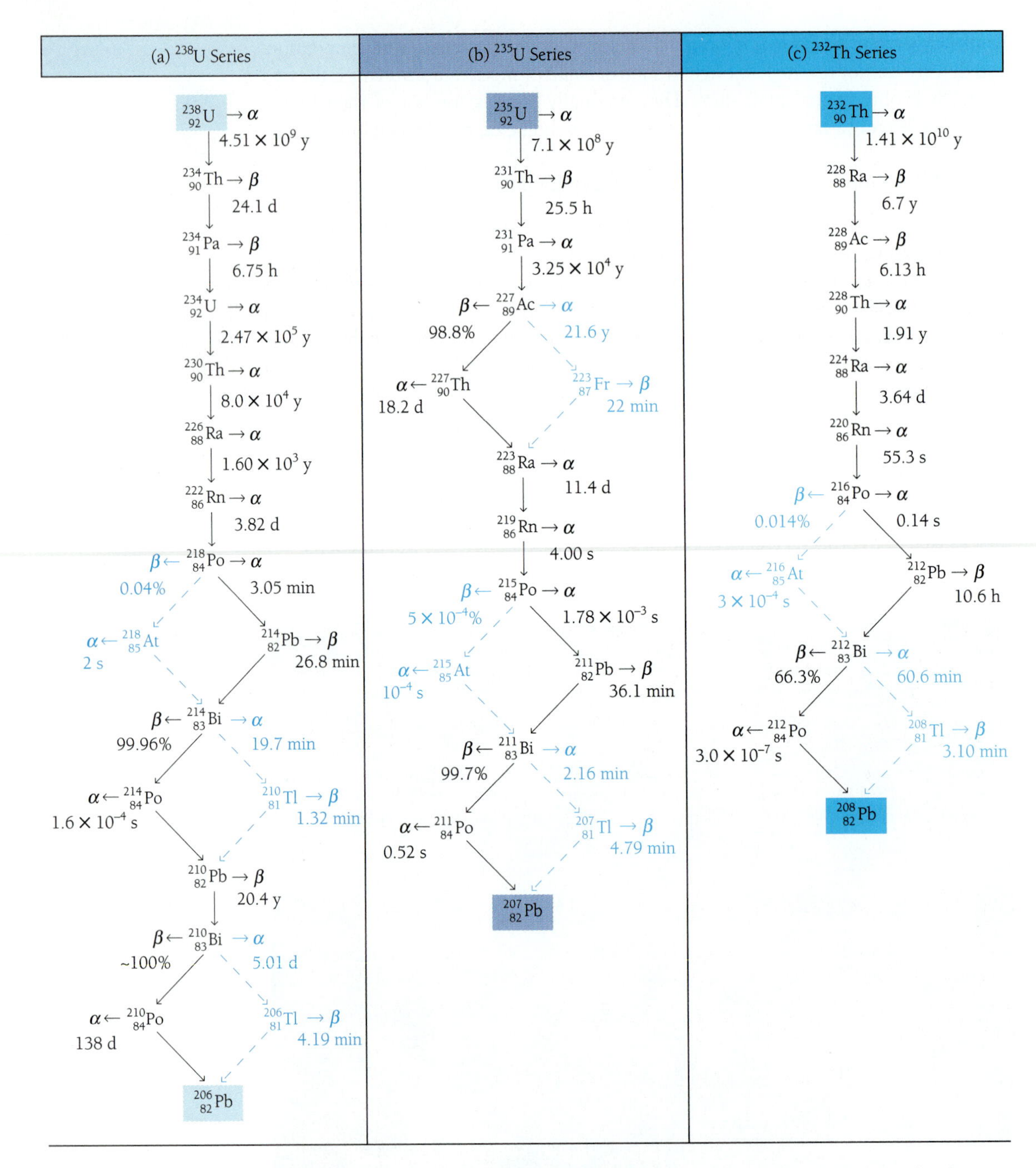

Figure 24.6 The three naturally occurring radioactive decay series. Alternate, but less common, pathways are shown by dashed lines.

accumulations of cosmic dust, containing all the stable elements (and some of the longer-lived radioactive ones, too) become planets like Earth. The big bang theory holds that *supernova* explosions, and the formation of second-generation stars and planets, are continuing processes in the universe today.

Radioactive Decay Series

About half of the uranium and almost all of the thorium present when the earth was formed remain today, and these elements are the source of the natural radioactivity of the earth's crust.

When $^{238}_{92}U$ decays, the daughter nuclide is $^{234}_{90}Th$.

$$^{238}_{92}U \longrightarrow {}^{234}_{90}Th + {}^{4}_{2}He$$

$^{234}_{90}Th$ is itself radioactive, decaying with a half-life of 24.1 days by β-emission to produce $^{234}_{91}Pa$, another unstable nuclide. The process continues through a total of 14 successive decays, until finally a stable nuclide, $^{206}_{82}Pb$, is reached. This sequential process is called a radioactive **decay series.** Three such series occur naturally in the earth's crust: two begin with uranium isotopes, and the third begins with thorium (Figure 24.6).

Most of the steps in all three series move the *n/p* ratio closer to the stability belt. Since α-emission produces a daughter nuclide with an increased *n/p* ratio, an occasional β-decay is required to keep *n/p* from rising further and further above the stability belt.

All of the intermediate nuclides in these series have much shorter half-lives than the parent species $^{235}_{92}U$, $^{238}_{92}U$, and $^{232}_{90}Th$. Intermediate nuclides disappear comparatively quickly and are not present to a significant extent in the earth's crust.

Amidst huge clouds of steam, a lava flow enters the Pacific Ocean at the coast of Hawaii. Some of the heat driving volcanic action comes from radioactivity in the interior of the earth.

24.6 NUCLEAR POWER

Ever since the realization, over 50 years ago, of the immense energies involved in nuclear transformations, means have been sought to put some of this power to use in industrial society. The energy of nuclear fission has been successfully applied to ship propulsion, weapons, and generation of electricity. To date, the power of nuclear fusion has been used only in weapons.

Critical Mass and the Bomb

The *spontaneous* fission of a nucleus (Section 24.2) is a rare event, but fission occurs readily when certain nuclei capture a neutron. When exposed to low-energy ("slow") neutrons, ^{235}U undergoes **induced fission.** Many different fission reactions take place following neutron capture. Three examples are:

$$^{1}_{0}n + {}^{235}_{92}U \longrightarrow {}^{90}_{38}Sr + {}^{143}_{54}Xe + 3\ {}^{1}_{0}n$$

$$^{1}_{0}n + {}^{235}_{92}U \longrightarrow {}^{137}_{52}Te + {}^{97}_{40}Zr + 2\ {}^{1}_{0}n$$

and

$$^{1}_{0}n + {}^{235}_{92}U \longrightarrow {}^{141}_{56}Ba + {}^{92}_{36}Kr + 3\ {}^{1}_{0}n$$

Over 200 different daughter nuclides have been identified in the fission of ^{235}U. Nuclear fission is an exothermic process, with an average energy release of 8.2×10^7 kJ per gram of ^{235}U.

Compared to the 56 kJ released when one gram of natural gas burns, 8.2×10^7 kJ per gram of ^{235}U "fuel" is a huge amount of energy.

Fission of a heavy nucleus produces several neutrons among the fragments. In ^{235}U, the average is 2.4 neutrons per fission event. Some of these neutrons cause fission in other nuclei, and a **branched chain reaction** results (Figure 24.7).

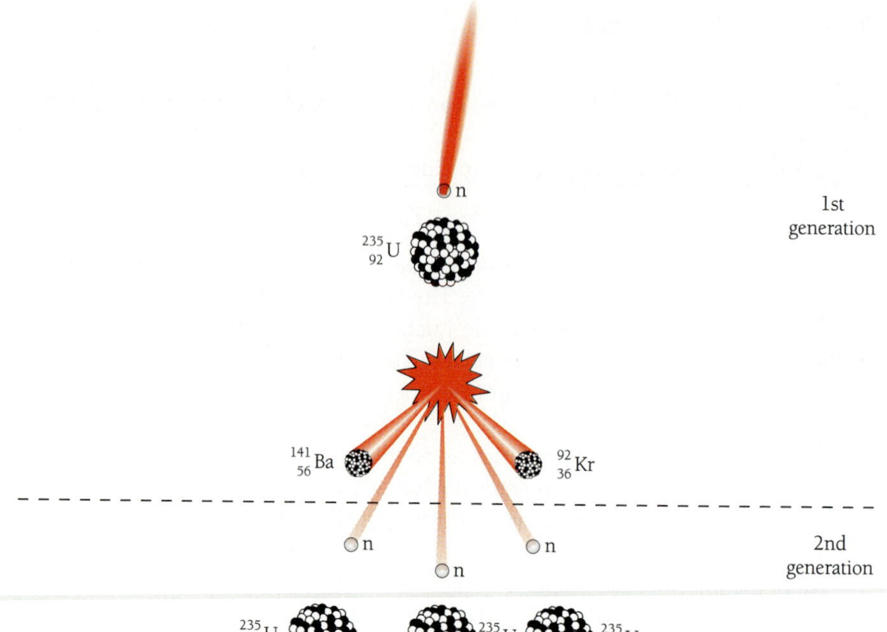

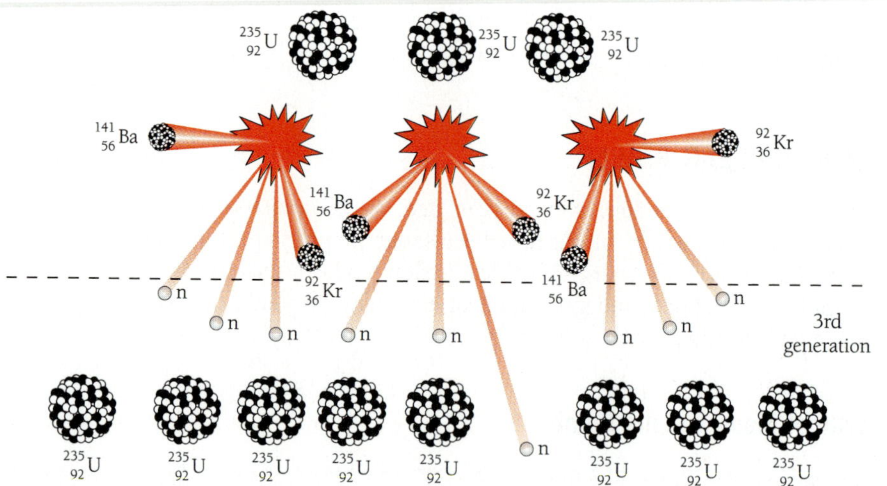

Figure 24.7 A branched chain reaction in a fissionable material such as ^{235}U. If each fission event produced two neutrons, and both were captured by uranium nuclei, the number of fission events would double in each generation. In the 10th generation, more than 1000 nuclei would undergo fission. In reality, the number is greater, because many fission events produce more than two neutrons: the average is 2.4. The result is a rapid release of a vast amount of energy, that is, a nuclear explosion.

The number of neutrons that are captured and induce further fission is less than the number produced. Some are absorbed by nonfissionable nuclei of other atoms, and some escape from the sample entirely. The number that escape depends on the size and shape, the density, and the mass of the sample. If, on the average, less than one neutron per fission induces a later fission event, the chain reaction dies out quickly, and only a minor amount of energy is released. The sample is said to have *subcritical* mass. The **critical mass** for a sample of given shape and density is the mass in which the average number of captured neutrons is exactly 1. In such samples, the chain reaction is sustained. The number of nuclei per second undergoing fission is constant, and energy is released at a constant and controlled rate. In larger samples, having *supercritical* mass, the average number of captured neutrons is greater than one. The number of fission events per second increases exponentially, and all the energy in the sample is released very quickly. In one type of atomic bomb, two subcritical masses of ^{235}U are brought together to form a supercritical mass, which quickly explodes (Figure 24.8).

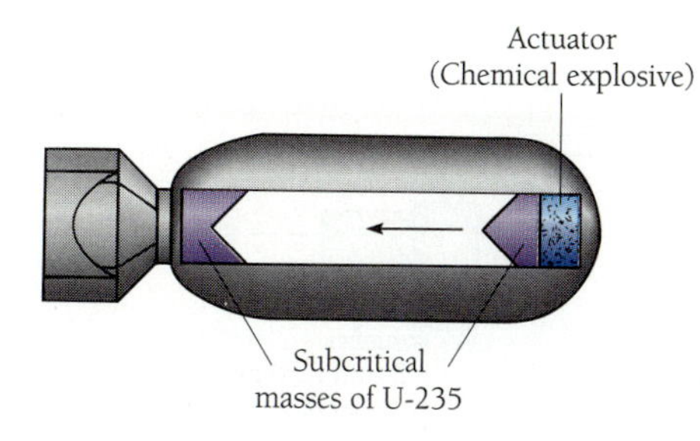

Figure 24.8 One type of atomic bomb contains two subcritical masses of ^{235}U or ^{239}Pu. A chemical explosive drives the two pieces together, forming a supercritical mass. A branched chain reaction follows, resulting in the almost instantaneous release of a huge amount of nuclear energy.

Fission Reactors

Today, about 15% of the electrical power used in the United States is generated from the heat produced by fission of ^{235}U in **nuclear reactors**—devices for maintaining a large-scale chain reaction at *exactly* the critical level, so that energy is released at a constant and manageable rate.

Small reactors are used to propel submarines and large surface vessels. Others are used primarily as neutron sources in the production of artificial radioisotopes. Most nuclear reactors, however, are designed for maximum production of electrical power. The first commercial generating plant began operating in 1957 in Pennsylvania.

An early problem was the low density of the fissionable isotope, ^{235}U, whose natural abundance is only 0.7% (the rest is ^{238}U). Even in the pure metal, ^{235}U atoms are too far apart to sustain fission in natural uranium, no matter how large the sample. However, natural uranium can be *enriched* in ^{235}U by *gaseous diffusion*. The first step is the production of uranium hexafluoride, a gaseous molecular compound. Since its slightly smaller molar mass makes it diffuse more rapidly, $^{235}UF_6$ can be separated from the heavier $^{238}UF_6$. The end product is not pure $^{235}UF_6$, but a mixture of isotopes in which the ^{235}U content has been increased to 2% to 4%.

The effect of molar mass on the rate of diffusion in gases is discussed in Section 5.4.

Laser separation is a promising alternative to gaseous diffusion for isotope enrichment. In this process, uranium-containing molecules absorb light and are raised to an excited electronic energy level. The chemical reactivity of excited states is usually different from that of the ground state, and treatment with a suitable reagent can provide a means of separating excited molecules from ground-state molecules. *Isotope separation is possible because the wavelength of light emitted by lasers can be "tuned" so as to be absorbed by molecules containing* ^{235}U, *but not those containing* ^{238}U.

Enrichment to 2% to 4% is satisfactory for nuclear reactors, but greater concentrations of ^{235}U are necessary for nuclear weapons.

The neutrons released by fission have a range of kinetic energies. Only slow neutrons can be captured by ^{235}U, while the faster ones either escape or are absorbed by other nuclides. Nuclear reactors require the use of a **moderator,** a substance that slows down fast neutrons. Neutrons are slowed most effectively by collisions with very light nuclei, and among the substances used as moderators by reactors of various designs are water, helium, graphite, and liquid sodium. Commercial reactors in the United States use water, an effective moderator but one that also absorbs some of the neutrons in the reaction

$$^{1}_{0}n + ^{1}_{1}H \longrightarrow ^{2}_{1}H$$

Enough neutrons are lost by this means so that uranium enriched to 3% ^{235}U is required. Canadian reactors, on the other hand, use "heavy water" (^2H_2O or D_2O). D_2O does not absorb neutrons, and heavy-water reactors are fueled with natural rather than enriched uranium. However, large amounts of D_2O must be separated from natural water. Separation is achieved by multistage distillation, using excess heat from the reactor that would otherwise be wasted.

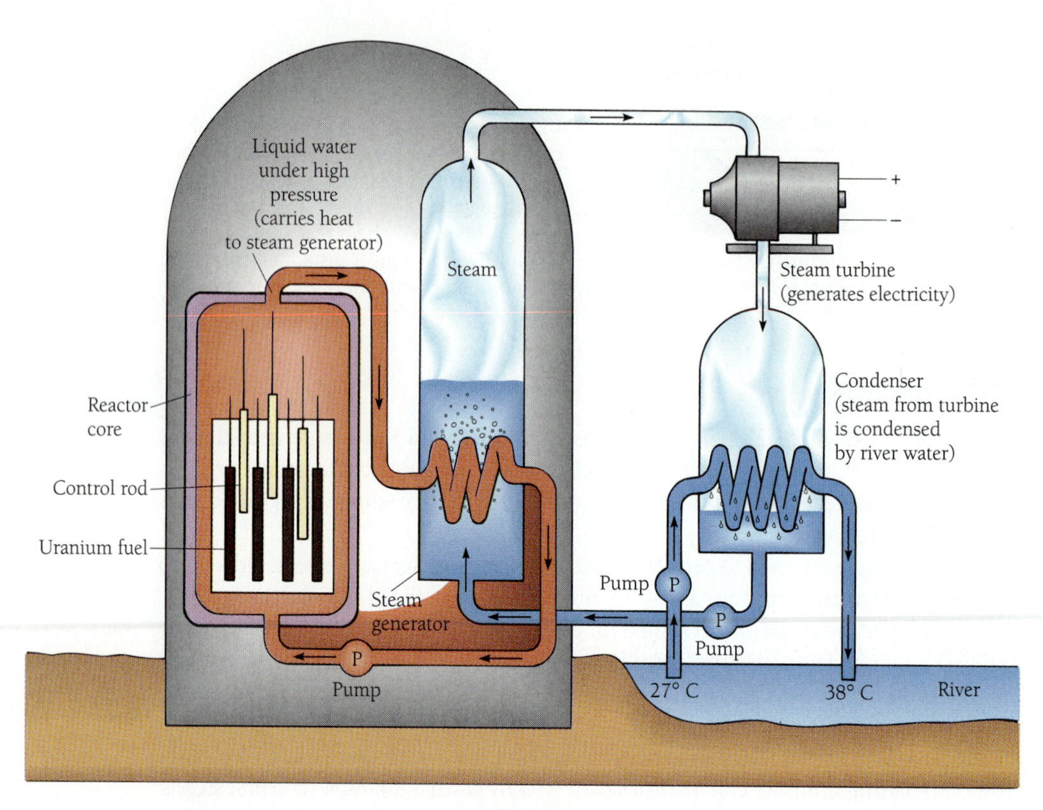

Figure 24.9 Central features of a light-water reactor (LWR). The core contains UO_2/U_3O_8 fuel rods, and boron- or cadmium-containing control rods whose position controls the level of the chain reaction. Ordinary water serves as both moderator and heat-transfer medium, and is continuously flushed through the core and the steam generator. A steam-driven turbine/generator produces electricity. A cooling source, often river water, is required to condense the steam after passage through the turbine.

The **core** of a reactor (Figure 24.9) contains an assembly of several hundred **fuel rods** interspersed with movable **control rods.** Fuel rods are zirconium tubes packed with uranium oxide pellets, and control rods contain effective neutron absorbers such as boron or cadmium. When fully inserted, the control rods absorb most of the neutrons and keep the core well below the critical level. When the control rods are partly withdrawn, fewer neutrons are absorbed and the chain reaction proceeds at a steady rate. As the fuel is consumed, and neutron-absorbing fission products build up, the position of the control rods must be adjusted to maintain criticality. The heat generated by fission is removed from the core by circulating water or sodium. Some distance away, the heat is used to produce steam, which drives an electrical generator.

In much of their design, nuclear power plants are similar to those fueled by coal or oil. The major differences are that in a conventional plant, the heat is produced in a burner, and different techniques are required for removal and disposal of the waste products.

Many experts believe the world's remaining supply of ^{235}U is so small that it can be a significant source of power for at most another 50 years. Consequently, some industrial countries have a different type of reactor, the **breeder reactor,** in varying stages of design, construction, or use. The central feature of the breeder reactor is that the core design maximizes neutron capture by the nonfissionable ^{238}U. The resultant ^{239}U then undergoes two β-decays to yield ^{239}Pu. Plutonium-239 is fissionable, and after it is collected and formed into rods, it can be used as reactor fuel. Because a breeder reactor produces additional fuel as well as electricity, present supplies of uranium can be used for centuries rather than decades.

The core of the Super Phenix, one of the breeder reactors used to generate electrical power in France.

Approaches to breeder reactors differ among industrial nations. The United States has essentially discontinued their development, out of popular concern about their safety and environmental impact. France has several in operation, and since the mid-1980s has filled a significant portion of the country's electrical needs with breeders. Japan and Russia have vigorous development programs in place.

We are all aware that there are pros and cons to the use of nuclear reactors, and that the continuing public debate is often strident. It is especially hard for citizens untrained in science, because there are truly knowledgeable people on each side of the many issues that make up the controversy. Today, most people realize that a nuclear reactor is not a bomb, and *cannot* be made to explode even through deliberate effort (by saboteurs, for instance). However, there are other matters of real concern.

In 1986, after the meltdown, the Chernobyl number 4 reactor was entombed in a concrete shell. Heavily protected workers entered the tomb for the first time in 1991.

One concern has to do with unavoidable release of radioactive material into the environment. This is minimal in mining, refining, enriching, and day-to-day operation of reactors, but cannot be shrugged off in the case of radioactive wastes. Disposal of radioactive wastes (and for that matter, any form of hazardous waste) is a problem with political as well as technical aspects. In this country, we have not yet decided how it should be done, and the wastes are piling up in temporary storage at the reactors where they are generated.

A second area of concern is the occurrence of an accident at a reactor site, resulting in the spread of a dangerous amount of radioactivity into the air, soil, and groundwater. Just such an accident occurred in 1986 in Chernobyl in the Soviet Union, as a result of equipment failure combined with a deliberate decision to disable certain critical, automatic safety features. Failure of a cooling system caused a heat buildup sufficient to ignite the thousand tons of graphite moderator. The fire burned for many days, and radioactive material from the breached reactor core was spread over thousands of square miles of Northern Europe. It was a chemical explosion, not a nuclear explosion, but the results were devastating. More than 30 people died immediately, and many more will suffer illness and death from the delayed effects of exposure to large amounts of radiation. A Chernobyl-style accident could not happen in a U.S. reactor. Here, reactor cores are secured inside large *containment vessels* and cannot spill their contents. Also, there are no large amounts of flammable materials in U.S. nuclear power plants, which use water, not graphite, as the moderator. But other accidents can, and do, happen. In 1979, at the Three Mile Island plant in Pennsylvania, a combination of carelessness and the failure of emergency backup cooling pumps resulted in overheating and partial melting of a reactor core. The situation was brought under control without release of harmful amounts of radioactivity to the environment, but the incident left lingering doubts as to the safety of nuclear power.

An interesting chapter in the earth's geological history occurred in West Africa about two billion years ago. The natural processes of erosion and runoff created a rich vein of uranium ore, and when a sufficient amount of uranium had accumulated, the vein "went critical." Groundwater served as the moderator in what was, in effect, a natural nuclear reactor. Such a reactor would operate sporadically because the heat

Since ^{235}U decays more rapidly than ^{238}U, its relative abundance decreases as time passes. When the earth was formed, the natural abundance of ^{235}U was 17%. By two billion years ago it had dropped to $\approx 3\%$, still enough to sustain fission in a light-water reactor.

generated by fission would vaporize the groundwater. Lacking a moderator, the chain reaction would stop, only to start again when the temperature dropped enough to allow groundwater to seep back in. During dry periods when the water table was low, reactor operation ceased entirely. This natural reactor operated on and off for perhaps 150,000 years. Our knowledge of this event comes from the isotope abundances in the remaining uranium and from the fission products, many of which are still there.

Fusion Reactors

The controlled fusion of hydrogen nuclei is a potential source of power, attractive for two main reasons: (1) there is a virtually inexhaustible supply of hydrogen nuclei in the oceans; (2) the major products of hydrogen fusion are nontoxic and nonradioactive, and so pose no threat to the environment or to public health. Some radioactive substances will be generated as parts of the reactors themselves absorb neutrons, but the problems of waste disposal will be substantially less than those posed by the operation of fission reactors. However attractive the possibility of nuclear fusion is, it presents enormous engineering problems. We will probably be well into the 21st century before fusion power is a reality. The difficulty is that fusion requires temperatures in excess of 10^7 K, and there are no construction materials that can withstand such conditions. Two approaches to nuclear fusion are being investigated.

Magnetic Containment

At such high temperature, electrons are stripped away from all atoms, and substances exist as a *plasma*—a gas of nuclei and electrons. The motion of these charged particles can be controlled magnetically, and a correctly shaped magnetic field can confine a high temperature plasma to a small region (a "magnetic bottle"). The hot plasma never contacts the container walls, which therefore can be kept cool enough to survive. Laboratory apparatus has been constructed to test this approach, and hydrogen fusion *has* been achieved. However, the fusion power output so far obtained is less than the power input required to maintain the magnetic field. Progress has been steady but painfully slow, and some people doubt that the method will ever produce useful amounts of power.

Inertial Confinement (by Lasers)

In this scheme, a small fuel pellet is irradiated on all sides by synchronized pulses from a dozen powerful lasers. The pellet is instantly compressed and heated to sufficiently high temperature and density for fusion to occur. Prototype reactors have not yet produced useful amounts of energy.

Although there are other possibilities, the most promising reaction for fusion power in either confinement scheme is the "DT," or deuterium-tritium reaction.

$$^{2}_{1}\text{H} + ^{3}_{1}\text{H} \longrightarrow ^{4}_{2}\text{He} + ^{1}_{0}\text{n}$$

Deuterium is readily available, and tritium is produced in the reactor by either

$$^{2}_{1}\text{H} + ^{2}_{1}\text{H} \longrightarrow ^{3}_{1}\text{H} + ^{1}_{1}\text{H} \quad \text{or} \quad ^{6}_{3}\text{Li} + ^{1}_{0}\text{n} \longrightarrow ^{3}_{1}\text{H} + ^{4}_{2}\text{He}$$

Modern nuclear weapons (the "hydrogen bomb") use uranium or plutonium fission to create temperatures high enough to initiate fusion in hydrogen.

24.7 EFFECTS OF RADIATION ON MATTER

The common feature of all radioactive emissions is their high energy, which is transferred to any substance that absorbs radiation. Energy transfer occurs gradually in a stepwise fashion as the incoming particle or photon interacts with the many atoms and molecules along its path. The most noticeable effects are the rupture of chemical bonds, creating pairs of free radicals, and the stripping of valence electrons, creating ions. High-energy radiation leaves a trail or "track" of ions in a substance, and therefore it is often referred to as **ionizing radiation.** The term is not limited to radioactivity, but includes x-rays produced by other means, and *cosmic rays* as well. Cosmic radiation is a catch-all term for high-energy photons and particles (including nuclei of moderately high mass) coming from beyond our solar system.

Radiation Detectors

Ionizing radiation can be observed by several different types of radiation detector. Photographic film is blackened by radiation (although of course it responds to light as well), and "film badges" are worn by workers whose jobs involve possible radiation exposure. The films are developed regularly (often weekly) to monitor each worker's exposure.

The Wilson *cloud chamber* is an early device incorporating a small chamber containing a supersaturated vapor, usually ethanol. Vapor condensation is nucleated by the ions along the track left by an incoming particle, leaving a trail of droplets that can easily be seen or photographed. Cloud chamber detectors often incorporate a magnetic field, which causes curvature in the tracks of charged particles while leaving neutral particles undeflected (Figure 24.10).

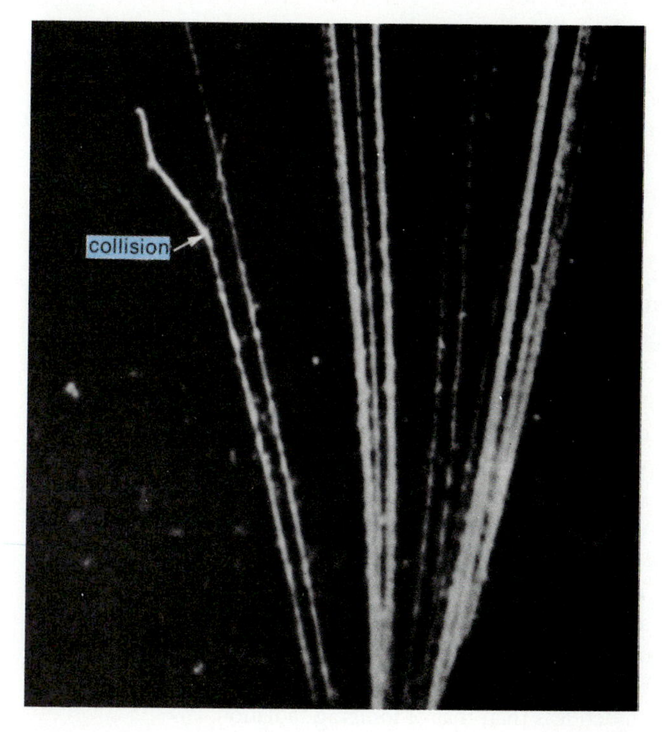

Figure 24.10 Photograph of tracks left by α-particles in a cloud chamber. The forked track (arrow) is a record of the reaction between an α-particle and a nitrogen nucleus, forming a proton and an oxygen isotope.

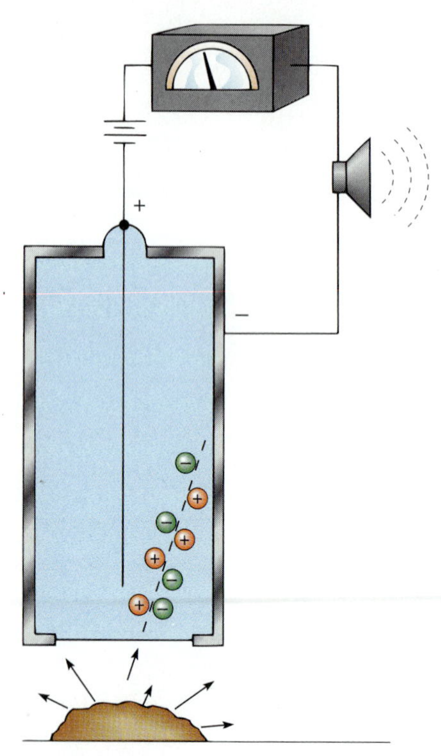

Figure 24.11 Ionizing radiation passes through the mica window and leaves a trail of ions and electrons in the gas-filled tube of a Geiger counter. Anions move to the positively charged central electrode, and anions move to the case, creating a pulse of electric current in the external circuit. The pulse is amplified and sent through a speaker, producing an audible click, or to a meter or counter.

The Geiger-Müller or **Geiger counter** (Figure 24.11) is the simplest and most common radiation detector. Its window is a strong, thin sheet of mica, through which most particles readily pass. Portable Geiger counters are widely used outside the laboratory.

The **scintillation detector** is a useful laboratory instrument, which takes advantage of the presence of electronically excited species in the track of a particle through certain materials, chiefly alkali halide crystals. As these excited species drop to the ground state, they emit light that is detected by photosensitive devices. Higher-energy particles produce more excited species, and a more intense flash of light, than lower-energy particles. Since light intensity is easily measured, one advantage of scintillation detectors is that they can be used to determine the energy of the particle causing the flash.

Penetrating Power of Radiation

Radiation penetrates materials to varying extents, depending both on the type of radiation and the type of substance. Also, the extent of chemical change they bring about in the material, the "radiation damage," varies considerably. Neutrons and γ-rays are the most penetrating: they carry no electrical charge and interact least strongly with matter. Most γ-rays can be stopped by a few centimeters of metal, but can penetrate deeply or pass right through the human body. X-rays are less energetic; they pass through soft tissue but are stopped by bone. β-particles penetrate a few millimeters into human tissue, and doubly charged α-particles are stopped by the skin or even light clothing.

Health Hazards

A great deal is known about the effects of ionizing radiation on human health, because the topic has been studied extensively for many years. There are disagreements over the interpretation of the results of these studies, however, particularly during discussions of how to minimize the risks.

Effect of Radiation on Humans

Large doses of radiation cause nausea, fatigue, and a decrease in resistance to infection. As with other illnesses, people can recover from radiation sickness. Larger doses, however, are fatal, primarily because of damage to the spleen and the bone marrow. Table 24.2 lists the specific health effects that follow from single, large doses of radiation.

TABLE 24.2 Short-Term Effects of Radiation Exposure on Humans

Dose/Rem	Effect
0–25	None observable
25–50	Decrease in white blood cell count
100–200	Nausea, fatigue
500	50% chance of death within 30 days

The units of radiation dosage, the *rem* and the *millirem*, are defined on page 1020.

There are also long-term effects, seen with much lower doses. These include several types of cancer, chiefly leukemia, and genetic damage resulting in subsequent birth defects. There is evidence that cancer is induced by free radicals formed in radiation tracks, while the genetic damage is due to changes in cellular DNA. It is estimated that, following a single, whole-body dose of 1 rem, a person's chances of developing cancer within the next 20 to 30 years increase from about 25% to 25.02%. The chances of genetic damage are less than this, but are not known with certainty.

There is growing awareness of the relationship between skin cancer and the ultraviolet component of sunlight. It is difficult to compare the hazard presented by sunlight with that of radioactivity and x-rays, because the energy carried by an ultraviolet photon is so much less than that of ionizing radiation.

α-Emitters are especially dangerous when taken into the body by respiration, eating, or drinking, since the low penetrating power of α-particles ensures that they cannot escape. They will be absorbed, and do their damage, wherever the α-emitters happen to be. Plutonium-239 and ^{90}Sr are particularly dangerous because they tend to lodge in the bones and are not excreted by the body. If the source is outside the body, neutrons and γ-radiation are more dangerous. Their greater penetrating power makes it more difficult to shield the body from them. Neutron radiation is particularly hazardous, since neutron capture can create other radioactive nuclides within the body.

Effect of Dosage: The Threshold Effect

It is generally true with radiation, as it is with most poisons and medicines, that the greater the exposure or dose, the greater the effect on the body. It is also true that human life has evolved on a planet that is awash in radiation from natural sources: the ever-present **background radiation** from radioactive nuclides in the earth's crust and cosmic radiation. Our bodies have protective mechanisms that enable us to repair damage caused by wounds, diseases caused by microorganisms, and the short-term effects of radiation. Some people believe that similar mechanisms exist that protect us from the long-term effects of background radiation. Others believe we do *not* have such mechanisms.

The controversy can be discussed in terms of the **threshold hypothesis.** Is there a threshold level of radiation exposure, below which there are no harmful effects? Or does *any* amount of radiation, however small, increase the probability of cancer and genetic damage? Experts disagree, and present impressive arguments on both sides. It is an important issue, for it helps shape our response to the many beneficial applications of nuclear chemistry in our society. One side holds that, since the existence of a threshold is unproven, the prudent course is to avoid *any* activity that increases radiation exposure above background, however slightly. The other side holds that a threshold is most likely, and in any case the harmful effects of small doses are certainly small and not of major concern. Furthermore, some activities, such as medical x-rays, have obvious and immediate benefits that outweigh the risks.

Quantitative Measurement of Radiation

Several different units are in use to describe the intensity of radiation. These units describe three different characteristics: activity, absorbed dose, and equivalent dose.

Activity

The total number of radioactive disintegrations occurring each second within a radiation source is called its **activity.** The SI unit of activity is the *becquerel (Bq),* defined as 1 disintegration per second. Another, very common, unit is the *curie* (Ci), which is defined as the number of disintegrations occurring each second in 1 g of pure radium: 1 curie $\Leftrightarrow$ 3.7×10^{10} disintegrations per second. A hospital might purchase a 10-Ci ^{192}Ir source, and use it in radiotherapy until its activity declines to 5 Ci (about six weeks).

Absorbed Dose

Those interested in radiation damage to materials are more concerned with energy than with the number of disintegrations. **Absorbed dose** is defined as the total energy received per kilogram of material. The most common unit of absorbed dose is the rad: 1 rad $\Leftrightarrow$ 0.01 J kg^{-1}. More recently the SI unit, 1 gray (Gy) $\Leftrightarrow$ 1 J kg^{-1}, is being used in addition to the rad. A slightly different unit, the röntgen, has been used since the early days of x-ray studies to measure the extent of ionization produced by radiation passing through air: 1 röntgen creates a charge (of each sign) of 2.58×10^{-4} coulombs per kilogram of air.

Equivalent Dose

The most familiar unit of radiation exposure is the **rem** (**r**öntgen **e**quivalent, **m**an) or *millirem (mrem)*. This unit is used to measure **equivalent dose,** the radiation damage to the human body. To determine an exposure or dose in *rems,* the dose in *rads* is multiplied by an experimentally determined quality factor Q that depends on the susceptibility of biological tissue to the particular type of radiation. Q also depends somewhat on dose rate and total dose. The value $Q = 1$ is assigned to γ-, β-, and x-rays, which are roughly equal in their effect on the body. Protons and slow neutrons are more harmful, and for them $Q \approx 5$. Even though it is the least penetrating, α-radiation, when absorbed, is most harmful. Q values for α-radiation are close to 20. The *sievert (Sv),* a SI unit, is also used to measure equivalent dose: 1 Sv $\Leftrightarrow$ 100 rem.

Radiation in the United States

Everyone receives a small amount of radiation each year, primarily from natural sources but some from human activities also. The amount varies: those living in the mountains receive more cosmic radiation than those at sea level; brick, concrete, and stone houses emit more radiation than wooden houses; radon from the ground is present at higher concentrations in energy-efficient buildings that exchange little air with the outside; some people have more medical and dental x-rays than others; and so on. Most published estimates of the total dose received by the average person are in the range of 100 to 200 mrem/year. As seen in Figure 24.12, more than 80% comes from natural sources. These include cosmic rays, radon and other radioactive nuclides in the soil and building materials, and ^{40}K present in the human body and foods. Most of the manmade sources are medical procedures, primarily diagnostic x-rays. Average doses from specific sources are shown in Figure 24.13. Note that the unit is millirems, not rems, as it is in Table 24.2.

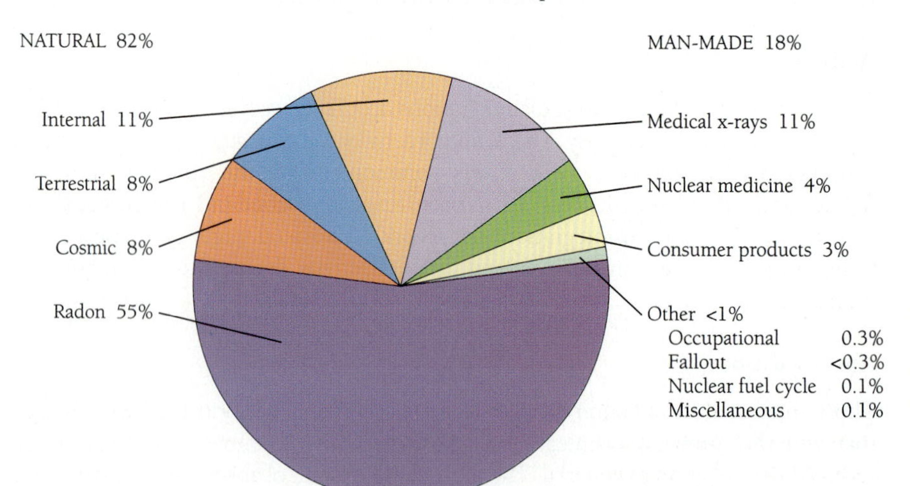

Figure 24.12 Average normal human exposure to radiation in the United States.

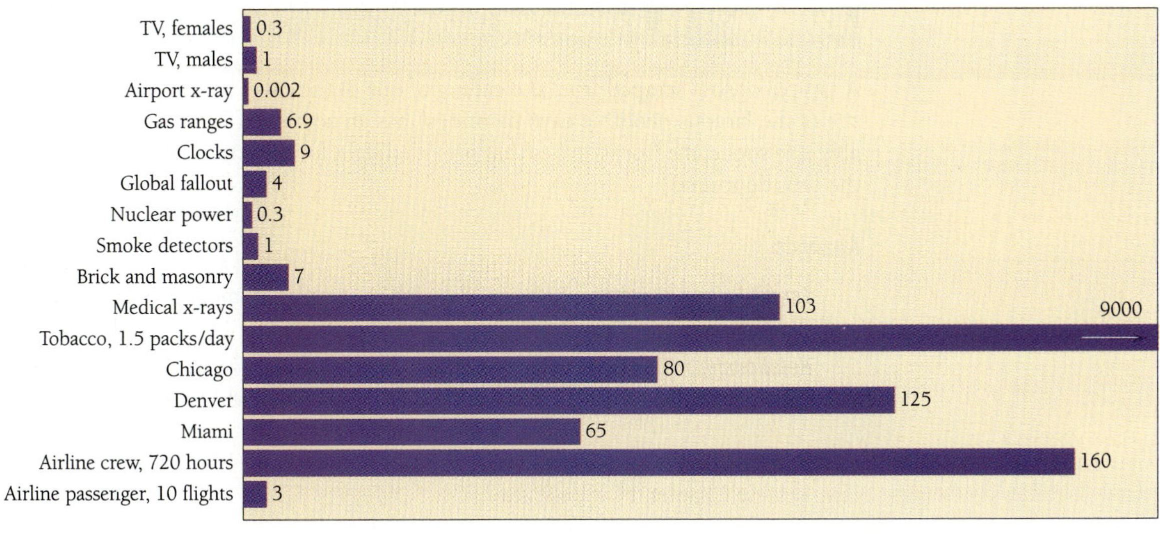

Figure 24.13 Average annual radiation dose from specific sources in the United States, natural and manmade.

24.8 OTHER APPLICATIONS OF NUCLEAR CHEMISTRY

As our knowledge of nuclear chemistry has grown over the past 60 years, so also has the number of applications of nuclear technology and radiation in our society. We make considerable use of techniques based on the half-life, energy deposition, tissue damage, and penetrating power of radiation.

Radiochemical Dating

The unchanging rate of decay of radioactive nuclides provides a natural clock by which the age of an object can be determined. **Radiocarbon dating** is widely used to determine the age of historic and prehistoric objects, artifacts, and works of art. The process begins in the atmosphere, when nitrogen atoms capture solar neutrons and are converted to radioactive carbon atoms.

The 1960 Nobel Prize in chemistry was awarded to the American chemist Willard F. Libby (1908–1980) for his work on radiocarbon dating.

$$^{1}_{0}n + ^{14}_{7}N \longrightarrow ^{14}_{6}C + ^{1}_{1}H$$

The new carbon atoms react with oxygen to form radioactive carbon dioxide. Photosynthetic uptake of CO_2 by plants, and subsequent distribution in the food chain, ensures that all living matter contains ^{14}C. Carbon-14 undergoes β-decay at a rate of 13.6 disintegrations per minute per gram of natural carbon (dpm g^{-1}). The half-life is 5730 years.

"Natural" carbon refers to the mixture of isotopes found in ordinary carbon-containing substances: ≈ 99% ^{12}C and ≈ 1% ^{13}C. Only about 1 in 10^{12} carbon atoms is ^{14}C.

$$^{14}_{6}C \longrightarrow ^{14}_{7}N + ^{0}_{-1}e$$

Uptake of ^{14}C ceases when the organism dies. After 5730 years, half of the ^{14}C has disappeared, and the activity of the remainder is down to 6.8 dpm g^{-1}. After 11,460 years, the rate is 3.4 dpm g^{-1}. The age of a sample of once-living material (wood, bone, cloth, and so on) can be determined by measuring its remaining radioactivity.

EXAMPLE 24.10 Radiocarbon Dating

A sample of soot scraped from the ceiling of one of the caves at Lascaux, France (the site of the famous neolithic cave paintings) has an activity of 2.4 dpm g^{-1}. Presumably, the soot came from torches that provided light for the artists. How long ago was the soot deposited?

Analysis

Target: t = ? years; t is the age of the carbon sample, or, more precisely, the time elapsed since the sample was part of a living organism.

Relationship: The first-order integrated rate law governs radioactive decay and connects the amount of substance remaining with elapsed time.

This problem is essentially identical to Examples 14.4 and 14.7 in Chapter 14.

Plan Determine the rate constant from the half-life, and find t from the integrated rate law. Review Chapter 14 if necessary.

Work Let N_0 (13.6 dpm g^{-1}) be the activity caused by the original amount of ^{14}C in the sample, and let N be the activity after elapsed time t. Then the integrated first-order law is

$$\ln\left(\frac{N}{N_0}\right) = -kt$$

$$t = -\frac{1}{k} \cdot \ln\left(\frac{N}{N_0}\right)$$

Using the relationship $k = 0.693/t_{1/2}$, insert the known quantities and evaluate.

$$t = -\frac{t_{1/2}}{0.693} \cdot \ln\left(\frac{N}{N_0}\right)$$

$$= -\frac{5730 \text{ y}}{0.693} \cdot \ln\left(\frac{2.4 \text{ dpm g}^{-1}}{13.6 \text{ dpm g}^{-1}}\right)$$

$$= 1.4 \times 10^4 \text{ years}$$

Check The result indicates that the Lascaux site was in use about 12,000 BCE, an age that is consistent with that of other neolithic sites in Europe.

Exercise

Wood from a chariot excavated in Bulgaria has an activity of 11.2 dpm g^{-1}. When was the chariot made?

Answer: About 400 CE.

Radiocarbon dating requires careful procedures, because the amount of ^{14}C is so very low even in modern samples. The difficulty of the measurement leads to uncertainties in the result: radiocarbon dates are normally quoted with error limits, as for instance 1630 ± 75 CE. Moreover, during periods of low solar activity, the amount of $^{14}CO_2$ in the atmosphere is less than normal. Unless this factor is taken into account, the measured age will differ from the actual age of the sample. Useful correction factors for solar activity fluctuations are obtained from tree-ring analysis of bristlecone pines, some of which have lived for over 4000 years. Radiocarbon dating is decreasingly reliable for older objects, and it cannot be used to establish dates earlier than about 40,000 BCE. After eight half-lives (46,000 years), only 1/256 of the original radioactivity remains, a count rate too small to measure.

Other radiochemical dating methods are available, based on longer-lived uranium, rubidium, and potassium isotopes, among others. These methods are used to determine the age of rocks and fossils. When several methods are used to establish the age of a sample, consistent results are usually obtained. Details are given in some of the end-of-chapter problems.

Medical Applications

The materials and techniques of nuclear chemistry have become an integral part of the modern practice of medicine.

Cancer Radiotherapy

The cells of living organisms are most susceptible to radiation during cell division. Since cancer cells divide more frequently than normal cells, a carefully administered radiation dose can destroy malignant tumors while inflicting less damage on the surrounding normal tissue. X-rays and radioisotopes are both used in radiation therapy. Radium was used very soon after its discovery, but has been supplanted by safer sources. Cobalt-60 is used for treatment of certain types of cancer, chiefly shallow lesions at or near the surface of the body. Iodine-131 is used in the treatment of thyroid cancer, because iodine is preferentially absorbed by the thyroid gland and the radiation can be delivered right where it is needed. Small pellets of ^{192}Ir are used for tumors easily reached via catheters inserted in body cavities. All of these isotopes are produced artificially in nuclear reactors, primarily at the Oak Ridge National Laboratory in Tennessee. Increasingly, hospitals are using small linear accelerators to produce high-energy electron beams for therapy.

Hair and nails also grow rapidly, and are often damaged by radiation therapy. The lining of the stomach is also fast-growing tissue, and radiation therapy patients often experience nausea after treatment sessions.

Diagnosis and Research

Thallium is preferentially absorbed by normal heart muscle, but not by damaged muscle. If radioactive ^{201}Tl is injected into the bloodstream and a photographic image of the heart is subsequently made, damaged or diseased muscle shows up as lighter (less exposed) areas on the film. There are many sites of preferential absorption in the body. A suitably chosen radioisotope can be concentrated in a specific location to yield an image, on film or produced by scintillation counters, of the organ or tissue. Other examples of preferential absorption are iodine in the thyroid, iron in the blood and marrow, and technetium in the bones. These techniques are used both for diagnosis and treatment, as well as for basic research on topics such as iron uptake in the formation of red blood cells and the role of aluminum in Alzheimer's disease.

Radiation in Industrial Society

Artificial radiation and radioactive isotopes are used in many ways in our society. Various applications depend on all the characteristics of radiation: constancy of decay rate, penetrating power, energy or heat production, detectability of small amounts (decay of a single nucleus can be detected), effect on living organisms, neutron capture, production of ion tracks, and so on. A few uses are listed in this section.

Decay Rate

This aspect of radioactivity is exploited in archaeological and geological dating, discussed previously.

Penetration

Because they are not easily stopped, γ-rays can be used to examine interior structure. Some manufacturers (of railroad tracks, for example), use γ-ray "cameras" to search for internal flaws in metal objects. β-Particles are more easily stopped, and precisely because of this they are used in the plastics, paper, and other industries in gauges that monitor and control the thickness of films, sheets, and coatings.

Energy

A thermocouple is a closed loop made from two kinds of wire. Therefore, it has two junctions between dissimilar metals. Whenever the two junctions are at different temperatures, electric current flows in the loop. The so-called "nuclear battery" uses the heat from radioactive decay to create a temperature difference in a thermocouple and to generate a small amount of electrical power. Nuclear batteries are expensive, but highly reliable and can last for thousands of years. They are ideal when periodic replacement (necessary with conventional batteries) is difficult, as in pacemaker implants and space vehicles. Nuclear batteries use α-emitters such as plutonium, because they are easily shielded.

Tracers

Perhaps the most varied uses depend on the fact that exceedingly small *traces* of radioactive material can be detected and located, as in the medical diagnostic uses described above. In other applications, radioisotopes are injected into the ground to study the flow of groundwater; into rivers, estuaries, and bays to study water movement and silting patterns; into organisms to study metabolic processes such as photosynthesis or ecological relationships such as symbiosis or food chains; into soil or agricultural chemicals to study uptake by crops; and so on.

Tracer methods are also used in chemistry. For example, use of ^{14}C-labeled reagents proved that the *terminal* carbon atom migrates to the benzene ring in the so-called *Claisen rearrangement:*

$$C_6H_5\text{—}O\text{—}CH_2\text{—}CH\text{=}{}^{14}CH_2 \longrightarrow C_6H_4(OH)\text{—}{}^{14}CH_2\text{—}CH\text{=}CH_2 \longrightarrow C_6H_4(OH)\text{—}{}^{14}CO_2H$$

Effect on Organisms

Radiation is used to kill microorganisms and to retard food spoilage. This is preferable to preservation by heating, because it does not cook the food nor change its taste or nutritional value. Either β- or γ-radiation is used, in part for their penetration characteristics and in part because their absorption by matter produces no radioactive nuclides: radiation-preserved foods are *not* radioactive! Radiation has been used to sterilize male screwworm flies. Released to the environment in large numbers, they compete successfully with fertile males, and the females (which mate only once), produce almost no offspring. This method has considerable potential for control of specific pests without introducing pesticides into the environment.

Ionization

One of several uses in this category is familiar to most of us—residential smoke detectors. A small amount of ^{241}Am continuously emits α-particles into a chamber through which room air circulates. The resulting ion tracks in the air increase its electrical conductivity so that a small current can flow between electrodes in the chamber. Smoke particles in the air impede the current flow, and the current decrease sets off the alarm.

Neutron Capture

In the technique of *neutron activation analysis,* an unknown sample is irradiated for a time with neutrons. This treatment produces a small number of radioisotopes from the different elements present. When these subsequently decay, analysis of the γ-radiation emitted allows identification of the type and quantity of the original elements in the sample. The method is sensitive to small quantities, and, with proper calibration, is quite accurate.

SUMMARY AND KEYWORDS

Radioactivity is the spontaneous, exothermic disintegration of atomic nuclei. The energy, often 10^7 or more times that from a chemical transformation, is carried away in one of three major modes: **α-particles** (helium nuclei), **β-particles** (electrons), or **γ-rays** (photons). Less common modes are **positron** emission, *electron capture* (K-capture), and spontaneous **fission.** In all modes (except γ) the **daughter** nucleus has a different number of protons and neutrons. More than 2000 **nuclides,** each with a different number of neutrons and/or protons, have been characterized. Most are unstable.

Nucleons are held together by the **strong force,** which acts between all nucleons (but only within the nucleus), and is more powerful than the mutual repulsion of the protons. Nuclear stability is enhanced by an even number of neutrons and/or protons, by **magic numbers** of nucleons, and by an *n/p* ratio that lies within the **stability belt.** No nuclides with $Z > 83$ are stable. The **binding energy per nucleon,** the average energy of cohesion of nucleons, is least for light elements. It rises to a maximum at iron-56, then decreases gradually as nuclear mass increases. All nuclei have less mass than their component nucleons, and this **mass defect** is related to binding energy by $E = mc^2$. Both **fission** of heavy nuclides and **fusion** of light nuclides lead to nuclides with greater binding energy, and are therefore exothermic.

Radioactivity is unaffected by temperature, pressure, concentration, or state of chemical combination. Each **radioisotope** (radioactive nuclide) has a characteristic and constant probability of decay in a given time period. Consequently, radioactive decay follows first-order kinetics. *Half-life* rather than rate constant ($k = 0.693/t_{1/2}$) is used to describe decay rates.

Nuclear transformations may be induced by bombardment with high-energy particles, such as protons, α-particles, and other light nuclei, or by *neutron capture.* **Transuranium elements** up to $Z = 109$ can be produced in *particle accelerators.* Very high energies, and very large accelerators, are required to overcome electrostatic repulsion between nuclei before such nuclear reactions can take place.

Elements up to iron are formed from hydrogen and helium in stars, and heavier elements are formed at much higher temperatures in *supernova* explosions. Terrestrial radioisotopes decay in one of several **decay series,** sequences of α- and β-emissions ending in one of the stable isotopes of lead.

Neutron capture by certain heavy nuclides results in **induced fission,** which releases two or more additional neutrons. If a *supercritical* mass of a fissionable isotope is assembled, as in an atomic bomb, a rapid **branched chain reaction** ensues with essentially instantaneous release of all the nuclear energy. A branched chain reaction cannot occur in a *subcritical* mass, because too many neutrons escape from, or are absorbed by, nonfissionable nuclei in the sample. A **nuclear reactor** is a device that keeps a fissionable material precisely at the **critical mass,** by means of a **moderator,** which slows down fast neutrons, making their capture more efficient, and **control rods,** which absorb neutrons. Different moderators distinguish the several types of reactor used to convert fission heat to electrical power. *Gaseous diffusion* is used to *enrich* natural uranium, bringing its ^{235}U content up from 0.7% to the $\approx$ 3% required by light-water reactors. Heavy-water reactors use natural uranium. **Breeder reactors** produce more fuel than they consume by converting nonfissionable to fissionable nuclides. **Fusion** reactors, which may use either *magnetic* or *inertial confinement* (by lasers) of a high temperature *plasma,* are under development.

Ionizing radiation, which includes x-rays and cosmic rays as well as radioactivity, leaves *tracks* of ions and free radicals when it passes through matter. Neutrons and γ-rays penetrate furthest through matter, x-rays and β-particles less so, and α-particles scarcely at all. Radiation is measured by **Geiger counters** and **scintillation detectors. Activity** (disintegration rate) of a source is measured in *curies (Ci)* or *becquerels (Bq),* while energy deposited in a substance, **absorbed dose,** is measured in *rads* or *grays (Gy).* **Equivalent dose** (often called simply *dose* or *exposure*) takes into account the varying damage done to human tissue by different forms of radiation, and it is measured in *rems* or *millirems.* Single doses greater than $\approx$ 25 rems cause symptoms of radiation sickness in humans, with death usually following within 30 days in untreated victims of dosages greater than several hundred rems. Doses less than 1000 mrem have no immediate effect, but result in a slight increase in the chance of cancer and genetic damage. Americans receive an average of 100 to 200 mrems annually, almost all of which comes from natural background and medical procedures.

Radiocarbon dating is used to establish the age of once-living objects, chiefly historical and prehistorical artifacts. The older the artifact, the less accurate the measurement, and insufficient ^{14}C remains after 40,000 years for the method to be used. Other methods, based on decay of ^{238}U, ^{40}K, or ^{78}Rb, are used to determine the age of fossils and rocks.

X-rays, electron beams, and γ-rays (chiefly from ^{60}Co, ^{131}I, and ^{192}Ir) are used to halt or hinder the growth of malignant tumors, and numerous other radionuclides are used in medical diagnostic procedures. Radiation is used in food preservation, agricultural pest control, thickness monitoring, and metal flaw detection. Small quantities of radioisotopes are used in *tracer* methods to monitor movement of materials and to elucidate the pathways of complex processes in biology, chemistry, and biochemistry. Nuclear batteries provide power for pacemakers and space-borne instruments. *Neutron activation* is a sensitive and versatile procedure for qualitative and quantitative chemical analysis.

PROBLEMS

General

1. Use your own words to describe what is meant by the term "radioactive."
2. Name and describe the major types of radioactivity.
3. What is a nuclide? What is the difference between a nuclide and a nucleus?
4. What is a nucleon?
5. What is meant by the terms "parent" and "daughter" nuclides?
6. In what type(s) of radioactive decay does the daughter nuclide have the same atomic and mass numbers as the parent?
7. For each of α-, β-, and γ-decay, and positron emission, tell whether there is an increase, decrease, or no change in (a) atomic number, (b) mass number, and (c) n/p ratio.
8. Name and describe the forces that affect nucleons within the nucleus.
9. Why is a nucleus unstable if it contains too few neutrons?
10. What evidence exists to support the theory that nucleons are arranged in energy levels, or "shells," within the nucleus?
11. Explain the significance of the stability belt and how it may be used to predict the most likely mode of disintegration of a nucleus.
12. How many elements have at least one stable isotope?
13. How many stable nuclides are there? Assuming that 2000 different nuclides have been characterized (actually the number is somewhat larger), what percentage of them are radioactive?
14. All elements with $Z > 83$ are radioactive. Are there any elements with $Z \leq 83$ that have no stable isotopes? If so, name them. (*Hint:* Most editions of the periodic table put the atomic mass of such elements in parentheses.)
15. How is it possible that an atomic nucleus can have less mass than the sum of the masses of its parts? Isn't this a violation of the law of conservation of mass?
16. What is the distinction between mass defect and binding energy?
17. Figure 24.3 (binding energy per nucleon) contains no point corresponding to hydrogen, ^1_1H. Explain this omission.
18. Explain why fission of a heavy nuclide and fusion of two light nuclides are both exothermic processes.
19. Why does nuclear fusion require high temperature?
20. Explain why fission of a heavy nucleus produces several neutrons in addition to two nuclei of intermediate size.
21. What is the difference between a chain reaction and a branched chain reaction?
22. What is meant by the term "critical mass" of a fissionable substance?
23. Why is it that nuclear transformations can be induced by capture of slow-moving neutrons, but slow-moving protons are ineffective?
24. What substances, other than fissionable isotopes, must be present in order to sustain a chain reaction at the critical level in a nuclear reactor? In a breeder reactor?
25. On what physical principle is isotopic enrichment of uranium based?
26. What is the function of the control rods in a nuclear reactor?
27. The energy released in fusion of deuterium is 6.9×10^{11} J mol^{-1}, while that released in fission of ^{235}U is 2×10^{13} J mol^{-1}. Which process produces the greater energy per gram of fuel?
28. Where and how are heavy elements formed?
29. What is a decay series? Why are there no significant accumulations of intermediate nuclides of a series?
30. What are the parent species, and the ultimate products, of the three naturally occurring decay series?
31. All of the daughter nuclides in the decay series beginning with ^{238}U have mass numbers that fit the expression $4n + 2$, where n is an integer. Explain why this is so.
32. Suppose the $^{40}K/^{40}Ar$ ratio in a rock sample is measured, and the age of the rock is determined to be 500 million years. If some argon had escaped from the rock, either by weathering or in preparation of the sample for measurement, would the true age be greater or less than that indicated by the measurement?
33. What are the immediate effects of ionizing radiation on matter?
34. Rank the following in order of increasing penetrating power: α-, β-, γ-, and x-rays, and neutrons.
35. What characteristics of radioactivity are commonly measured, and what are the common units of each? What are the SI units for these quantities?
36. Suppose you wished to reduce your personal exposure to radiation to the lowest possible level. What would be the three most effective steps you could take?
37. Without referring to the text, name as many nonmedical uses of radioactivity as you can. Reread Section 24.8 and see if you missed any. Do you know of any not mentioned in the text?
38. Uranium tetrabromide, like UF_6, is a volatile molecular compound. Would UBr_4 be as suitable as UF_6 for isotope enrichment by gaseous diffusion? Explain.

Nuclear Equations

39. Write balanced nuclear equations for the following processes.
 a. α-decay of $^{232}_{94}Pu$ b. β-decay of tritium ($^{3}_{1}H$)
 c. γ-decay of $^{60}_{27}Co$ d. positron decay of $^{170}_{72}Hf$

40. Write balanced nuclear equations for the following processes.
 a. α-decay of $^{244}_{98}Cf$ b. β-decay of $^{27}_{12}Mg$
 c. γ-decay of $^{132}_{55}Cs$ d. positron decay of $^{22}_{11}Na$

41. Write a balanced equation that describes neutron capture by each of the following stable nuclides.
 a. $^{98}_{42}Mo$ b. $^{142}_{58}Ce$

42. Write a balanced equation that describes neutron capture by each of the following stable nuclides.
 a. $^{138}_{56}Ba$ b. $^{197}_{79}Au$

43. Each of the following is produced (by neutron capture) from a stable nuclide in neutron activation analysis. Write a balanced equation that describes the process.
 a. $^{24}_{11}Na$ b. $^{121}_{50}Sn$

44. Each of the following is produced (by neutron capture) from a stable nuclide in neutron activation analysis. Write a balanced equation that describes the process.
 a. $^{90}_{38}Sr$ b. $^{77}_{32}Ge$

45. Supply the missing information to each of the following equations.
 a. $^{53}_{24}Cr + ^{4}_{2}He \longrightarrow ^{1}_{0}n + ?$
 b. $^{59}_{27}Co + ^{1}_{0}n \longrightarrow ^{56}_{25}Mn + ?$
 c. $^{187}_{75}Re + \beta \longrightarrow ?$
 d. $^{243}_{95}Am + n \longrightarrow ^{244}_{?}Cm + ? + \gamma$
 e. $^{35}_{17}Cl + ? \longrightarrow ^{32}_{16}S + ^{4}_{2}He$

46. Supply the missing information to each of the following reactions.
 a. $^{14}_{?}N + ? \longrightarrow ^{17}_{?}O + p$
 b. $^{235}_{92}U + n \longrightarrow ^{137}_{54}? + 2\,^{1}_{0}n + ?$
 c. $^{26}_{?}Mg + ? \longrightarrow ^{26}_{?}Al + n$
 d. $^{241}_{95}Am + ^{12}_{?}C \longrightarrow 4\,^{1}_{0}n + ?$
 e. $^{40}_{18}Ar + ? \longrightarrow ^{43}_{?}K + ^{1}_{1}H$

Mode of Decay

47. Predict whether electron or positron emission is the more likely mode of decay of the following radioisotopes.
 a. $^{17}_{9}F$ b. $^{56}_{28}Ni$ c. $^{8}_{3}Li$ d. $^{49}_{20}Ca$

48. Predict whether electron or positron emission is the more likely mode of decay of the following radioisotopes.
 a. $^{39}_{17}Cl$ b. $^{82}_{35}Br$ c. $^{10}_{6}C$ d. $^{59}_{26}Fe$

49. Which of the following pairs of nuclides do you expect to be more stable? Give your reason in each case. (This problem can be answered without calculating binding energy.)
 a. ^{60}Ni, ^{66}Ni b. ^{70}Se, ^{74}Se c. ^{114}Ag, ^{114}Cd

50. Which of the following pairs of nuclides do you expect to be more stable? Give your reason in each case. (This problem can be answered without calculating binding energy.)
 a. ^{25}Al, ^{27}Al b. ^{40}Ar, ^{41}Ar c. ^{47}Ti, ^{51}Ti

51. What daughter nuclide remains after uranium-234 undergoes four successive α-decays?

52. Radon-220 undergoes the following series of decays: α, α, β, α, β. What is the end product of this decay series?

Energy, Mass Defect, and Binding Energy

*53. One electronvolt is defined as the kinetic energy acquired by an electron when accelerated through a potential difference of 1 volt. Using the conversion factor 1 eV ⇔ 1.6022×10^{-19} J and the relationship $KE = mv^2/2$, calculate the velocity of (a) an electron with an energy of 1.00 eV, and (b) an electron with an energy of 1.00 keV. (*Hint:* In SI, 1 J = 1 kg m^2 s^{-2}.)

54. How fast are the electrons travelling in a beam whose energy is 6 keV?

*55. Calculate the velocity of a proton whose kinetic energy is (a) 1.00 eV and (b) 5.00 MeV. (Do not be misled by the unit "*electron* volt." The eV is an energy unit, like the joule or calorie, and may be used to describe the energy of protons and baseballs as well as electrons.)

*56. When ^{198}Bi disintegrates, the energy carried by the emitted α-particle is 5.83 MeV. Calculate its velocity. (Since the energy is known to only three significant digits, it makes no difference whether the atomic mass of helium is used in place of the mass of the α-particle, which is slightly less.)

57. Use the results of Problem 53 and the de Broglie relationship (Section 6.4) to calculate the wavelength of (a) a 1.00-eV electron and (b) a 1.00-keV electron.

58. Use the results of Problem 55 to calculate the wavelength of (a) a 1.00-eV proton and (b) a 5.00-MeV proton.

*59. Calculate the mass defect of the radioisotope ^{14}O (a positron emitter), whose atomic mass is 14.00860 amu. Express your answer in amu, then convert to binding energy and binding energy per nucleon in J nucleon^{-1}.

*60. Calculate the mass defect of the stable isotope ^{79}Br, whose atomic mass is 78.91834 amu. Express the answer in amu, then convert to binding energy in kJ mol^{-1} and binding energy per nucleon in J nucleon^{-1}.

*61. Calculate the nuclear binding energy of ^{71}Ga, whose atomic mass is 70.924700 amu.

*62. The atomic mass of the nuclide chromium-52 is 51.94059 amu. What is its nuclear binding energy?

*63. The atomic mass of ^{24}Mg is 23.985042 amu. Calculate its nuclear binding energy per nucleon, and express the result in MeV.

*64. What is the binding energy per nucleon (MeV) of ^{87}Rb, whose atomic mass is 86.909187 amu?

65. Some people prefer to express binding energy in joules per *gram*. Almost all stable nuclei have binding energies in the range 1.2 to 1.5×10^{-12} J nucleon^{-1}. Express this range in J g^{-1}.

66. The total binding energy of ^{56}Fe, the most stable nuclide, is about 4.9×10^{13} J mol^{-1}. Express this in joules per gram of nuclear matter.

67. The energy released in a nuclear transformation is the difference between the total mass of the products and the total mass of the reactants. Calculate the energy released in the α-decay of $^{235}_{92}U$. Atomic masses are $^{235}_{92}U$, 235.04394; $^{231}_{90}Th$, 231.036298; and $^{4}_{2}He$, 4.00260.

68. Calculate the energy released in the β-decay of tritium, given the atomic masses ^{3}H = 3.01605 amu, ^{3}He = 3.01603 amu.

*69. Calculate the energy released when 1.00 mole of deuterium undergoes fusion according to the overall reaction

$$3\,^{2}_{1}H \longrightarrow {}^{4}_{2}He + {}^{1}_{1}H + {}^{1}_{0}n$$

Neutron mass is 1.008665 amu; relevant atomic masses are ^{1}H, 1.007825; ^{2}H, 2.0140; and ^{4}He, 4.00260.

*70. How much energy is released when 1.00 mol ^{235}U undergoes fission according to the following equation?

$$^{1}_{0}n + {}^{235}_{92}U \longrightarrow {}^{137}_{52}Te + {}^{97}_{40}Zr + 2\,^{1}_{0}n$$

Neutron mass is 1.008665 amu. Atomic masses are ^{235}U, 235.043924; ^{137}Te, 136.925410; and ^{97}Zr, 96.910950. (Recall that this is only one of many different fission reactions that occur in ^{235}U: the average energy released is $\approx 2 \times 10^{13}$ J mol^{-1}.)

Measurement of Radioactivity

71. A γ-ray source used in food preservation has an activity of 2500 Ci. Express this in becquerels.

72. What is the activity, in curies, of a 5.0 GBq radiation source?

73. How many decay events occur in one hour in a 5.0 mCi source?

74. Calculate the number of α-particles emitted in an 8.00-hour period by a source whose activity is 7.04×10^{9} Bq.

75. A scintillation counter registers a total of 22,400 counts from a radioactive sample during a period of 10 minutes. Assuming that the counter detects all of the emitted radiation, what is the activity of the sample?

*76. A Geiger counter is positioned so that it intercepts 1/1500 of the particles emitted by a sample. Assuming that each intercepted particle registers on the counter, and that 7.6×10^{6} counts are observed in one hour, calculate the sample activity.

77. A 1.5-kg sample exposed to radiation receives a total of 33 J. What is the absorbed dose in rads?

78. If 0.015 J is deposited in a 454-g sample, what is the absorbed dose?

79. A sample receives an absorbed dose of 0.050 mrads of α-radiation. What is the equivalent dose in rems?

*80. An x-ray machine delivers an absorbed dose of 5.0 rads/h. What is the equivalent dose to a patient receiving a 20-minute treatment from this machine?

Kinetics and Radiochemical Dating

81. The rate constant for the decay of ^{78}Br, a positron emitter, is 5.0×10^{-3} s^{-1}. What is its half-life?

82. The rate constant for the γ-decay of ^{60}Co is 4.19×10^{-9} s^{-1}. What is the half-life?

83. What is the decay rate constant of ^{72}Se, which undergoes electron capture with a half-life of 9.7 d?

84. Calculate the rate constant for the α-decay of ^{247}Bk, whose half-life is 7×10^{3} y.

85. Thorium-220 emits α-particles with a rate constant of 2.7×10^{-6} s^{-1}. What is the activity, in disintegrations per second, of one mole of this isotope?

86. Krypton-87 undergoes β-decay with a rate constant of 1.5×10^{-4} s^{-1}. Calculate its activity per mole.

87. Cesium-137, a radioisotope used in cancer therapy, decays by β-emission to ^{137}Ba with a half-life of 33 y. (The product nucleus is formed in an excited state, and immediately emits a γ-ray and drops to the ground state, so that this is a "double-barreled" weapon in medicine.) Use the half-life to determine the activity (disintegrations per second) of 1.0 g ^{137}Cs.

88. Calculate the activity of 1 μg of ^{239}Pu, whose half-life is 24,500 y.

89. One of the lingering products of atmospheric testing of atomic weapons in the 1950s is ^{90}Sr. It is among the more harmful constituents of fallout because its chemical similarity to calcium causes it to be concentrated in dairy products, and, ultimately, in the bones and teeth of consumers. Given that its half-life is 25 years, what percentage of the original activity remains now, after 35 years?

90. The half-life of tritium is 12.5 years. What percentage of the original activity remains in a sample after 50 years?

91. How long a time must elapse before the activity of ^{91}Kr, a waste product of uranium fission, is reduced to 0.1% of its original level? The half-life is 10 s.

92. One of the products of fission reactors is ^{239}Pu, whose half-life is 24,500 years. How much time must elapse before its activity decreases to 0.1% of its original value?

*93. Rubidium-87 decays by β-emission (to ^{87}Sr, a stable nuclide) with a half-life of 4.9×10^{10} y. Some fossils of single-celled organisms were found in which 4% of the originally-present ^{87}Rb had converted to ^{87}Sr. Estimate the age of the fossils.

*94. Ancient rock samples from Greenland were found to contain ^{87}Rb and ^{87}Sr in the ratio 96.0 mol Rb : 4.0 mol Sr. Assuming all the Sr arose from the radioactive decay of ^{87}Rb, how long ago did the rocks form?

*95. The ultimate stable product of ^{238}U decay is ^{206}Pb. A certain sample of pitchblende ore was found to contain ^{238}U and ^{206}Pb in the ratio 68.3 atoms ^{238}U : 31.7 atoms ^{206}Pb. Assuming that all the ^{206}Pb arose from uranium decay, and that no U or Pb has been lost by weathering, how old is the rock? The half-life of ^{238}U is 4.48×10^9 years.

*96. Potassium-40 decays to ^{40}Ar with a half-life of 1.3×10^9 years. What is the age of a lunar rock sample that contains these isotopes in the ratio 34.1 atoms ^{40}K : 65.9 atoms ^{40}Ar? Assume that no argon was originally in the sample and that none has been lost by weathering.

*97. One of the problems associated with the $^{238}U/^{206}Pb$ dating method is that uranium-containing minerals frequently contain natural lead in addition to lead formed by uranium decay, so that some of the ^{206}Pb in the samples does *not* arise from uranium. However, one lead isotope, ^{204}Pb, does not arise from decay of any heavier isotope, so its presence can be used to determine the amount of lead that was present when the rock was formed. Isotopic abundances in Pb are such that there are 19.2 ^{206}Pb atoms for each ^{204}Pb atom in natural lead. Calculate the age of a rock sample containing these isotopes in the ratio 1 atom ^{204}Pb : 22.1 atoms ^{206}Pb : 3.8 atoms ^{238}U. [*Hint:* Determine x, the number of "extra" or decay-product ^{206}Pb atoms for each ^{204}Pb atom. Then the fraction of ^{238}U atoms that has decayed is $3.8/(3.8 + x)$.]

*98. A rock sample brought in for potassium-argon dating is found to contain 60 atoms of ^{40}K for every 100 atoms of ^{40}Ar. The half-life for the decay of ^{40}K is 1.3×10^9 years.
  a. What is the age of the sample?
  b. Suppose microscopic examination suggests that $\approx 20\%$ of the argon atoms may have escaped through weathering. Calculate a corrected value of its age.

APPENDIXES

Appendix A Mathematical Tools

A1 Scientific Notation

An **exponent,** also called a *power,* is the number of times a **base** number is to be multiplied by itself. It is written to the right and slightly above the base. In the notation 10^3, for example, the base is 10 and the exponent is 3; $10^3 = 10 \times 10 \times 10 = 1000$, and $10^6 = 10 \times 10 \times 10 \times 10 \times 10 \times 10 = 1{,}000{,}000$. A *negative* exponent indicates the number of times the *reciprocal* of the base is to be multiplied by itself, or equivalently, the number of times the number 1 is to be divided by the base. Thus $10^{-3} = (1/10) \times (1/10) \times (1/10) = 1/1000$, and $2^{-4} = (1/2) \times (1/2) \times (1/2) \times (1/2) = 1/16 = 0.0625$. Since any number raised to the first power is equal to itself, 10^1 is almost always written simply as 10. Also, any number raised to the zero power equals 1; remember that $3^0 = 1$ and $10^0 = 1$.

In scientific notation, the base number is 10. Numbers are written in two parts: the first is an ordinary number, often but not necessarily between 1 and 10, and the second is the base 10 raised to a positive or negative whole number exponent. The first part may or may not contain a decimal point. The number 500 is written as 5×10^2, which means $5 \times 10 \times 10$ or 5×100, and the number 0.0035 is written as 3.5×10^{-3}, that is, $3.5 \times (1/10) \times (1/10) \times (1/10)$ or $3.5/1000$.

Conversion of a large number from arithmetic to exponential notation is done by making the exponent equal to the number of places the decimal point must be moved to the left in order to achieve the desired form, which normally has a single digit to the left of the decimal point. The decimal point in the number 156.3 must be moved 2 places to the left to get 1.563, so that the exponent is 2; in scientific notation the arithmetic number $156.3 = 1.563 \times 10^2$. The decimal point is often omitted in arithmetic notation; nonetheless the number 75,298 has an implied decimal point to its right. Moving the implied decimal point 4 places to the left gives 7.5298, so that in scientific notation $75{,}298 = 7.5298 \times 10^4$.

Conversion of numbers *smaller* than 1 requires moving the decimal point to the *right,* and the corresponding exponent is *negative*. The arithmetic number 0.00000826 becomes 8.26 when the decimal point is moved 6 places to the right, so that $0.00000826 = 8.26 \times 10^{-6}$.

Conversions from scientific to arithmetic notation are done by reversing these procedures.

Exercise A1.1

Convert the following numbers to scientific notation: 180,000; 93,000,000; 747; 5,286; 10.16; 0.000725; 0.105; 0.000000000711.

Answer: 1.8×10^5; 9.3×10^7; 7.47×10^2; 5.286×10^3; 1.016×10^1; 7.25×10^{-4}; 1.05×10^{-1}; 7.11×10^{-10}.

Exercise A1.2

Convert the following numbers to arithmetic notation: 6.63×10^{-3}; 1.38×10^{-6}; 8.314×10^7; 1×10^9; 1.987×10^{-2}.

Answer: 0.00663; 0.00000138; 83,140,000; 1,000,000,000; 0.01987.

Multiplications and divisions of numbers in scientific notation are carried out by separate manipulation of the exponent and of the pre-exponential part of the number.

In multiplication, the pre-exponential parts are multiplied in the normal way, but the exponential parts are *added*. This follows from the definition: if 10^2 is 10×10 and 10^3 is $10 \times 10 \times 10$, then $10^2 \cdot 10^3$ is $10 \times 10 \times 10 \times 10 \times 10 = 10^5 = 10^{(2+3)}$. For example, $(1.5 \times 10^3) \cdot (5.0 \times 10^4) = [(1.5) \cdot (5.0)] \times 10^{(3+4)} = 7.5 \times 10^7$. Be wary of negative exponents: $(3.21 \times 10^7) \cdot (1.67 \times 10^{-11}) = [(3.21) \cdot (1.67)] \times 10^{[7+(-11)]} = 5.36 \times 10^{-4}$.

Division is carried out by dividing the pre-exponential parts and *subtracting* the exponents. For example, $(7.2 \times 10^4)/(3.5 \times 10^2) = [(7.2)/(3.5)] \times 10^{(4-2)} = 2.1 \times 10^2$. Again, be wary of negative exponents: $(9.9 \times 10^2)/(3.3 \times 10^{-12}) = [(9.9)/(3.3)] \times 10^{[2-(-12)]} = 3.0 \times 10^{+14}$.

When two numbers have the *same* exponent, they may be added to or subtracted from one another simply by adding or subtracting the pre-exponential part and leaving the exponent unchanged. Thus, $3.6 \times 10^{-6} + 7.9 \times 10^{-6} = 11.5 \times 10^{-6}$, $3.2 \times 10^{19} - 2.8 \times 10^{19} = 0.4 \times 10^{19}$, and $1.853 \times 10^3 - 5.621 \times 10^3 = -3.768 \times 10^3$. However, two numbers having *different* exponents cannot be added to, or subtracted from, one another; one of the numbers must be rewritten so that it has the same exponent as the other.* Conversion of a number to a form having a different exponential part requires moving the decimal point one place to the left for each unit increase of the exponent, or one place to the right for each unit decrease in the exponent. The number 1.52×10^3 can be written as, for example, 0.0152×10^5 or 15.2×10^2; 62.5×10^{-5} can be written as 6.25×10^{-4}, 0.00625×10^{-1}, or 625×10^{-6}. A typical addition/subtraction operation is carried out as follows.

Set up problem ⟶	Convert to same exponent ⟶	Carry out operation
-4.325×10^{-6}	-4.325×10^{-6}	-4.325×10^{-6}
$+\ 6.7 \times 10^{-8}$	$+0.067 \times 10^{-6}$	$+0.677 \times 10^{-6}$
		-4.258×10^{-6}

Exercise A1.3

a. Convert the following numbers to exponential form in which the exponent has the value +3: 7.56×10^5; 32.8×10^2; 839; 8.314; 0.08206; 1.86×10^{-1}.

b. Convert the following numbers to the form in which the exponent has the value −2: 7.56×10^{-5}; 32.8×10^{-1}; 839; 8.314; 0.08206; 1.86×10^1.

Answers: (a) 756×10^3; 3.28×10^3; 0.839×10^3; 0.008314×10^3; 0.00008206×10^3; 0.000186×10^3. **(b)** 0.00756×10^{-2}; 328×10^{-2}; 83900×10^{-2}; 831.4×10^{-2}; 8.206×10^{-2}; 1860×10^{-2}.

Exercise A1.4

Perform the indicated operations: $(75.4 \times 10^{-2}) \cdot (0.682 \times 10^{-1})$; $(3.96 \times 10^{-1}) \cdot (6.02 \times 10^{23})$; $(8.82 \times 10^{-12})/(0.0366 \times 10^9)$; $(6.63 \times 10^{-34}) \cdot (3.00 \times 10^8)/(450 \times 10^{-9})$; $650 + 1.04 \times 10^2$; $6.63 \times 10^{-34} - 8.31 \times 10^{-36}$; $(752 \times 10^2 + 8.01 \times 10^4)/(3.72 \times 10^{-6})$; $(2.87 \times 10^{-3} - 1.054 \times 10^{-2}) \cdot (0.0767 \times 10^{22})$.

Answers: 5.14×10^{-2}; 2.38×10^{23}; 2.41×10^{-19}; 4.42×10^{-19}; 754; 6.55×10^{-34}; 4.17×10^{10}; -5.88×10^{18}.

* Two such numbers *can* be added or subtracted on a calculator, but it is important to be able to do this in your head as well. Checking the results of a calculator operation by a rough mental estimate is an important aspect of problem-solving. It is all too easy to punch the wrong button on a calculator.

A2 Significant Digits

The reliability with which a quantity is known is indicated by the number of digits, called **significant digits** or **significant figures,** used to write the quantity. In this notation, the right-most digit is taken as correct to within ± 1, while the other digits are taken as exact. The greater the number of significant digits, the greater the reliability. The number 2.7, with two significant digits, indicates a quantity whose magnitude lies in the range 2.6 to 2.8; it is less precisely known that the four-digit number 2.693, which lies within the much narrower range 2.692 to 2.694. Scientific calculations must be done so as to preserve the reliability of the data, while not overstating it.

The zero poses a special problem in significant digit notation, because it serves a dual function. Not only does it mean a certain digit (not 9, or 1, or 7, but **0**), but it also serves to locate the decimal point, as in 1300 or 0.0035. As a digit it must be counted as significant, but as a decimal point locator it is not. In the number 0.0035, only the 3 and the 5 are significant. The rules for whether or not a zero is significant are:

1. Leading zeroes (those to the left of all nonzero digits) are decimal point locators, and are not significant. The following numbers all have three significant digits: 0927, 0.927, 0.000927.
2. Embedded zeroes (those lying between nonzero digits) are significant. The following numbers all have four significant digits: 1052, 3007, 9.025, 0.007502.
3. Trailing zeroes (those to the right of all nonzero digits) can be *either* significant digits *or* decimal point locators. However, if the number contains an *explicit* decimal point, the trailing zeroes are significant. The following numbers all contain four significant digits: 1.350, 7.600, 8.000, 0.009350, 4.080, 100.0, 4200., 009040., 0.6020.

The only problem arises when the decimal point is not explicitly written. In the statement, "$13 is the same amount of money as 1300 pennies," the number 1300 has four significant digits: it means not 1298, not 1304, but *exactly* 1300. In the statement, "Police estimated the crowd at 1300 people," the number 1300 probably has only two significant digits, and its meaning probably is "more than 1200 but less than 1400." Note that in scientific notation, which always contains a decimal point, no ambiguity arises: the number of pennies in $13 is 1.300×10^3, and the crowd was estimated at 1.3×10^3. Absence of ambiguity is another, very good, reason for using scientific notation.

In common or arithmetic notation there is no way to resolve the ambiguity of trailing zeroes.* The safest rule, and the one we follow in this text, is to treat trailing zeroes as significant unless there is a reason not to do so in a particular instance. Thus, in a problem or laboratory instruction, the quantity "500 mL benzene" will be taken as having three significant digits; it specifies a quantity in the range 499 to 501 mL rather than in the range 490 to 510 mL (2 significant digits) or 400 to 600 mL (1 significant digit). "About 500 mL benzene" implies less accuracy, but it is not feasible to make precise rules for inherently imprecise usages such as "about."

* Arithmetic notation can be modified by adding an overhead bar to a trailing zero if it is significant. In this system, both of the following numbers have three significant digits: $50\bar{0}$, $25\bar{0}$. The system is a good one and not difficult to learn, but it is not widely used outside of high school science courses.

There are two exceptions to the rule that the reliability of quantities is indicated by the number of digits used to express them.

1. Integers are **exact,** because they are the result of counting rather than measuring. "Dogs have 2 ears and 1 tail" does not mean that a Rottweiler could have anywhere between 1 and 3 ears, for instance 2.46 ears. Integers are to be treated as having an indefinitely large, although unwritten, number of significant digits. The context often provides clues as to whether a number is the result of a count or of a measurement. For instance, integers up to ten are frequently written out as words. However, context is not infallible. If there is any doubt, *such numbers should be presumed exact.* For example, a problem might read, "**2** liters of a liquid weighs 2.068 kg: what is the weight of **one** liter?" In this case it is likely that the number of liters involved is the integer 2 (exact) rather than perhaps 2.03. *Assume* this to be true, and express the answer as 1.034 kg, rather than with some other number of significant digits.
2. Conversion factors resulting from metric prefixes and other defined relationships are exact: 1 cm (exact) = 10 mm (exact); 1 yard (exact) = 36 inches (exact), and so on. These numbers are all integers. In addition, there are two nonintegral, but nonetheless exact, conversion factors in common use in chemistry: in the equivalence statements 1 in $\Leftrightarrow$ 2.54 cm and 1 cal $\Leftrightarrow$ 4.184 Joules, all numbers are exact.

Operations With Significant Digits

Rounding Off

Mathematical operations must be carried out so as to preserve the reliability of the quantities involved; the result of a calculation must neither exaggerate nor understate the reliability. Calculators usually express numbers to 8 or 10 digits, which is almost always more than is appropriate in a scientific calculation. Such quantities should be "rounded off" to the correct number of significant digits. The right-most significant digit should be increased by 1 (rounded up) and all digits further to the right should be dropped when the digit immediately to the right of the right-most significant digit is 5, 6, 7, 8, or 9. Otherwise the right-most significant digit should be unaltered while all the insignificant digits, further to the right, are dropped (rounding down). The 8-digit calculator result 6.5491507 is correctly rounded to 1, 2, 3, and 4 digits as 7, 6.5, 6.55, and 6.549.

Multiplication and Division

When two quantities are multiplied or divided, the product or quotient is written with the same number of significant digits as the less reliable (that is, having the lesser number of significant digits) of the two factors. An integer is treated as if it had a large number of significant digits, more than any other quantity in the calculation. The same rules apply when there are more than two factors, since longer operations can always be done on a two-by-two basis. Examples:

a. $(175.3) \cdot (62) = 1.1 \times 10^4$

b. $(0.079)/(0.00038617) = 2.0 \times 10^2$

c. $(3.14159)/[2(\text{integer})] = 1.57080$

d. $(6.25) \cdot (78.26)/[(33.31) \cdot (0.2003)] = 73.3$

Intermediate results reached in solving individual steps of a problem should not be copied from a calculator display, rounded off, and reentered before continuing; to do so introduces needless additional error into the calculation. Consider the following problem: if an object moves 4.8 meters in 2.60 seconds, how far will it move in 2.74 seconds? If we solve this stepwise by calculating first the speed and then the new time, and round off the intermediate result, the answer is:

1. $\dfrac{(4.8\text{ m})}{(2.60\text{ s})} = 1.8461538\text{ m s}^{-1}$ (calculator display) $= 1.8\text{ m s}^{-1}$ (rounded off)
2. $(1.8\text{ m s}^{-1}) \cdot (2.74\text{ s}) = 4.9320000\text{ m}$ (display) $= 4.9\text{ m}$ (rounded off)

Alternatively, we could solve the problem by a different procedure, again rounding off the intermediate result:

1. Since the ratio of the distances is the same as the ratio of the times, first calculate how much longer the second move takes than the first:

$$\frac{(2.74\text{ s})}{(2.60\text{ s})} = 1.0538462 \text{ (display)} = 1.1 \text{ (rounded off)}$$

2. Multiply the first distance by this ratio to get the second:

$$(1.1) \cdot (4.8\text{ m}) = 5.2800000\text{ m (display)} = 5.3\text{ m (rounded off)}$$

Different approaches to a problem should not give different results when expressed to the appropriate number of significant digits, so one or both of these procedures must be flawed. The error lies in rounding off the intermediate results. When the second step is carried out without rounding off the result of the first, both methods give the same final result (5.0584615 m, which is correctly expressed to two significant figures as 5.1 m).

Addition and Subtraction

When two numbers are to be added or subtracted using pencil and paper, the first step is to write them with their decimal points aligned. The operations of adding and subtracting the two numbers 21.1378×10^5 and 0.0019×10^5 are set up as:

$$\begin{array}{r} 21.1378 \times 10^5 \\ +0.0019 \times 10^5 \\ \hline 21.1397 \times 10^5 \end{array} \qquad \begin{array}{r} 21.1378 \times 10^5 \\ -0.0019 \times 10^5 \\ \hline 21.1359 \times 10^5 \end{array}$$

The result is accurate to the fourth decimal *place,* since both starting quantities are accurate to the fourth decimal place; this is true even though one number has six significant digits and the other has only two. But suppose the two quantities are not significant to the same decimal place? If we use "?" to represent an insignificant (that is, completely unknown) digit, such an operation might be:

$$\begin{array}{r} 1.13?? \\ +0.0019? \\ \hline 1.13??? \end{array}$$

The fifth decimal place of the answer is of course completely unknown: ? + ? = ?. The third and fourth decimal places of the answer are also unknown, since nothing can be known of the result of adding an unknown digit, even when it is added to a known digit: ? + 9 = ? and ? + 1 = ?. Thus, for additions and subtractions the rule is that the only decimal places that are significant in the answer are those having significant digits in *all* the addends or subtrahends.

Rounding the Result

Rounding to the appropriate number of significant digits should not be done until *after* the calculation is complete. This means that at least one insignificant digit will be retained during the operation(s). If an intermediate result must be copied on paper, it is common practice to write one or two extra, insignificant digits as subscripts. Calculators automatically retain insignificant digits in intermediate calculations.

a. Round before addition (incorrect procedure)

$$\begin{array}{r} 0.586 \\ +0.0884 \\ +0.0013 \\ \hline \end{array} \longrightarrow \begin{array}{r} 0.586 \\ +0.088 \\ +0.001 \\ \hline 0.675 \end{array}$$

b. Round after addition (correct procedure)

$$\begin{array}{r} 0.586 \\ +0.0884 \\ +0.0013 \\ \hline \end{array} \longrightarrow \begin{array}{r} 0.586 \\ +0.0884 \\ +0.0013 \\ \hline 0.6757 \end{array} \longrightarrow 0.676$$

c. Round before division (incorrect procedure) $10.4/8.8 \longrightarrow 10/8.8 \longrightarrow 1.136 \ldots \longrightarrow 1.1$

d. Round after division (correct procedure) $10.4/8.8 \longrightarrow 1.182 \ldots \longrightarrow 1.2$

Mixed Operations

Whenever a problem requires both multiplication/division and addition/subtraction, it should first be written out in unambiguous form, using parentheses. Next, each set of parentheses should be "cleared," that is, reduced to a single quantity. This is done by following the rules given above, retaining one or more extra, insignificant digits in these intermediate results. Then the remaining single quantities should be combined to give the answer, which is rounded to the correct number of significant digits in the final step. In the following examples, *in*significant digits are printed in smaller type.

$$\frac{6.25 + 8.0593}{78} \longrightarrow \frac{14.309}{78} \longrightarrow 0.183 \longrightarrow 0.18$$

$$\frac{6.25}{78} - 8.0593 \longrightarrow 0.08013 - 8.0593 \longrightarrow -7.97917 \longrightarrow -7.979$$

$$\frac{[(10.5/0.0319) + 1.3726 \times 10^{-2}]}{(2.629 + 502)} \longrightarrow \frac{329.154 + 0.013726}{504.629} \longrightarrow$$

$$\frac{329.17}{504.629} \longrightarrow 0.65230 \longrightarrow 0.652$$

A3 Logarithms

The common logarithm of a number is the power to which 10 must be raised in order to equal the number. If y is some number, its logarithm is another number z; this relationship is written as $z = \log y$. According to the definition of logarithms, the value of z is such that $10^z = y$. This can be written in more compact form as

$$10^{\log y} = y \qquad \text{(A3-1)}$$

The number y must be positive—*there is no such thing as the logarithm of a negative number.* In Equation (A3-1) the number 10 is called the "base" of the logarithm. Although logarithms can be defined with respect to some number other than 10 as base, the base 10 is common and very useful. Indeed, logarithms to the base 10 are called "common logarithms." At one time it was necessary to use a complicated table of numbers in order to find a logarithm. These days, however, all scientific calculators are equipped with a "log" key, and finding a logarithm is no more difficult than finding a square root. Try a few examples with your calculator: $\log 2.00 = 0.301$, $\log 5{,}280 = 3.7226$, $\log (6.02 \times 10^{23}) = 23.780$, $\log 10^6 = 6$, $\log (7.5 \times 10^{-8}) = -7.12$, $\log 1 = 0$.

The preceding examples illustrate some helpful rules about logarithms. If a number is greater than 1, its logarithm is a positive number; if a number is less than 1, its logarithm is a negative number; the logarithm of 1 is zero. The logarithm of a pure exponential number such as 10^{-3}, 10^8, or $10^{-4.18}$ is just the exponent itself: $\log 10^{-3} = -3$; $\log 10^8 = 8$, $\log 10^{-4.18} = -4.18$.

The operation inverse to finding a logarithm is finding a number whose logarithm is known. Suppose, for instance, we know there is some number x whose logarithm is 3.5, and we wish to know the value of x. It follows from (A3-1) that

$$x = 10^{\log x} = 10^{3.5} \tag{A3-2}$$

Finding a number whose logarithm is known is called "exponentiation," or "taking the antilogarithm" of a number.

$$\text{antilog } x = 10^x \tag{A3-3}$$

On a calculator, the antilogarithm of a number is obtained with the "10^x" key, or the key sequence "inv", "log". Try a few examples: antilog $3.5 = 10^{3.5} = 3 \times 10^3$; antilog $4 = 10^4 = 10{,}000$; antilog $(-7.428) = 10^{-7.428} = 3.73 \times 10^{-8}$; antilog $(-6) = 10^{-6}$. If $\log y = [(6.023 \times 10^{23}) \cdot (9.110 \times 10^{-28})/(0.08206)]$, then $y = 1.016$.

These examples illustrate some useful rules for finding antilogarithms: if a number is positive, its antilogarithm is a number greater than 1; if a number is negative, its antilogarithm is a number less than 1; the antilogarithm of zero is 1. The antilogarithm of an integer, either positive or negative, is 10 raised to that integer—a calculator is not needed for this.

The rule for significant figures is that the logarithm has as many digits *to the right of the decimal point* as there are significant digits in the number; $\log 2 \times 10^5 = 5.3$; $\log 2.000 \times 10^5 = 5.3010$, and so on. The digit(s) to the left of the decimal in a logarithm are not significant, as they serve only to locate the decimal point in the number.

Natural or *Napierian* logarithms use the base 2.71828 . . . , a number invariably symbolized and referred to as "e." The natural log of a number y, symbolized "$\ln y$," is some other number z such that

$$e^z = e^{\ln y} = y \tag{A3-4}$$

As with common logarithms, the following relationships hold:

if y is negative or zero, $\ln y$ is not defined

if $y < 1$, $\ln y$ is negative

if $y = 1$, $\ln y = 0$

if $y > 1$, $\ln y$ is positive

y, $\ln y$, and e^y are all dimensionless numbers

The inverse operation to taking a natural logarithm, that is, finding a number whose logarithm is known, is called "exponentiation." This operation is done on scientific calculators with the "e^x" key, or one of the key sequences "inv," "ln" or "2nd," "ln." Try these examples on your calculator: $\ln 3.14 = 1.144$, $\ln 0.082 = -2.50$, $\ln (5.2 \times 10^{-6}) = -12.17$, $e^{1.23} = 3.4$, $e^{-125} = 5.17 \times 10^{-55}$. If $\ln y = [(-296) \cdot (0.0651)/(\sqrt{2})]$, $y = 1.21 \times 10^{-6}$.

Appendix B The Gas Laws and the Kinetic Molecular Theory

The postulates of the kinetic molecular theory (Section 5.4) allow a prediction of the relationship between pressure and volume of a gas. Consider a sample of gas in a cubical container whose sides are d meters in length and whose volume is $V = d^3$ cubic meters. We choose for further consideration a molecule of mass m which happens to be traveling, with velocity v, in a direction parallel to the floor of the box and perpendicular to the right-hand wall. Soon it will collide, rebound and retrace its original path in the reverse direction. It collides with the left-hand wall, rebounds and eventually returns, after an elapsed time of t seconds, to its original position. Since such collisions are elastic, the molecule has lost no energy and is still traveling with its initial velocity. It will continue this behavior indefinitely (assuming it does not collide with another molecule), making one collision with the right-hand wall every t seconds. The round-trip path length $2d$ is related to the velocity and the time by $t = 2d/v$; here t is the elapsed number of seconds per collision with the right-hand wall.

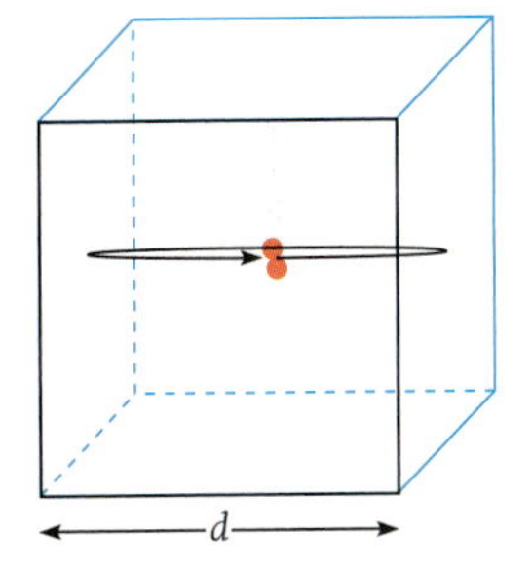

Prior to a collision the momentum of the molecule is $(+mv)$, and after the collision, since the molecule is traveling in the opposite direction, its momentum is $(-mv)$. The change in momentum is $[(-mv) - (+mv)] = -2mv$. The *rate* of momentum change is $(-2mv)/t = (-2mv)/(2d/v) = -mv^2/d$. By Newton's law (usually seen in the more familiar form $f = ma$) the rate of momentum change is equal to the force on the molecule. Also, the force on the molecule is equal and opposite to the force on the wall, which is therefore $f = +mv^2/d$. Since we do not know the value of v^2 for an individual molecule, we use the average value $\langle v^2 \rangle$ instead.

$$f = \frac{m\langle v^2 \rangle}{d}$$

Suppose that there are a total of N molecules in the container. Their motion is random, so on the average $N/3$ are traveling left-right, $N/3$ are traveling up-down, and $N/3$ are traveling front-back. The *total* force on the right-hand wall is

$$F = \left(\frac{N}{3}\right)\left(\frac{m\langle v^2 \rangle}{d}\right)$$

Since pressure is defined as force divided by the area over which the force is exerted, and the area of the wall is $A = d^2$, the pressure is

$$P = F/A = \frac{(N/3)(m\langle v^2 \rangle/d)}{(d^2)}$$

since $V = d^3$,

$$P = \frac{(N/3)(m\langle v^2 \rangle)}{V}$$

The quantity $m\langle v^2\rangle/2$ is the average energy of gas molecules, and by the postulates of the KMT the average energy is proportional to the temperature: $m\langle v^2\rangle/2 = \text{const}(T)$. Making this substitution and rearranging, we find

$$PV = (2)(\text{const})\left(\frac{N}{3}\right)T$$

Finally, the number of moles n in the container is given by N/N_A, so that $N = nN_A$ and

$$PV = n\left[\frac{2(\text{const})N_A}{3}\right]T,$$

where "const" is the constant of proportionality between temperature and the average energy of molecular motion. Except for the numerical value of this constant of proportionality, the KMT has given us not only Boyle's law and Charles's law, but the complete ideal gas law. Once the value of the constant of proportionality has been determined, the value of the gas constant can be calculated from the relationship $R = [2(\text{const})N_A/3]$. Correct prediction of the ideal gas law is a major success of the KMT.

Appendix C Thermodynamics

C1 *PV* Work in Chemical and Physical Processes

According to its fundamental definition, work is given by the product of the distance moved and the force that opposes the motion.

$$w = [\text{force}] \cdot [\text{distance}] = F \cdot D \qquad \text{(C1-1)}$$

If a chemical process in a system is accompanied by an increase in volume, expansion occurs against the opposing force of the external pressure, and the system does work on the surroundings. Suppose such a process takes place in a cylindrical container of cross-sectional area A (meters^2), and that the system is confined by a movable piston on which the surroundings exert a pressure P (pascals). When the volume increases, the piston moves out a distance D (meters) against this pressure. Since pressure is defined as force/area, the force acting to oppose the motion is given by

$$F = [\text{pressure}] \cdot [\text{area}] = P \cdot A \qquad \text{(C1-2)}$$

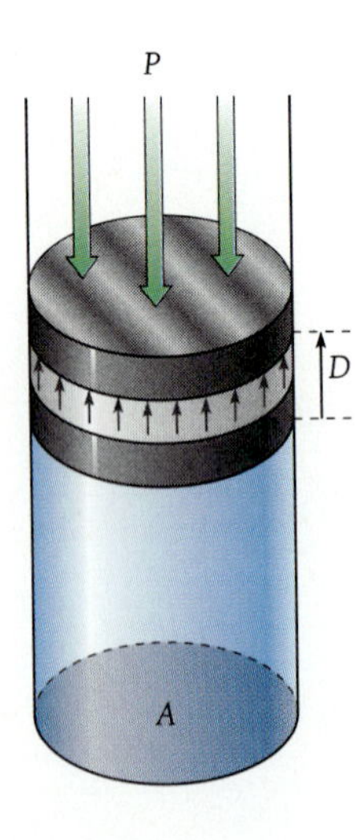

and the work done by the system in moving the piston is

$$w = [\text{pressure}] \cdot [\text{area}] \cdot [\text{distance}] = P \cdot A \cdot D \qquad \text{(C1-3)}$$

The (cylindrical) volume added to the system by the motion is equal to the piston area times the distance traveled, so that

$$\Delta V = A \cdot D$$

Substitution of ΔV for $A \cdot D$ in Equation (C1-3) gives the work done *by* the system on the surroundings as

$$w = P \cdot \Delta V$$

The negative of this is the work done *on* the system:

$$w = -P \cdot \Delta V$$

When the only type of work involved is *PV* work, the first law ($\Delta E = q + w$) may be written as

$$\Delta E = q - P \cdot \Delta V$$

When SI units are used for *P* and ΔV (Pa and m^3), the product $P \cdot \Delta V$ is in the SI unit of work and energy (J).

In SI, 1 J $\Leftrightarrow$ 1 Pa · 1 m^3.

C2 Free Energy and the Second Law of Thermodynamics

The second law of thermodynamics states that, in any spontaneous change, the total entropy of the system and the surroundings increases. The entropy change within a system, for instance a system in which a chemical reaction takes place, can be measured or calculated from known values of the entropies of elements and compounds. The entropy change of the surroundings is determined as follows.

As discussed in Section 10.5 (page 437), entropy increases when heat is added to a system, causing either a temperature increase or a phase change such as melting or vaporization (which takes place without change in temperature). The general relationship between entropy change, temperature, and heat flow (q) is

$$\Delta S = \frac{q}{T}$$

The entropy change of the surroundings is given by

$$\Delta S_{surr} = \frac{q_{surr}}{T}$$

Conservation of energy requires that

$$q_{sys} = -q_{surr}$$

The entropy change of the *surroundings* is therefore

$$\Delta S_{surr} = \frac{-q_{sys}}{T}$$

For chemical reactions (or phase changes) that take place without change in temperature, q_{sys} is given by ΔH_{sys}, equal to ΔH_{rxn} (or ΔH_{fus} or ΔH_{vap}). That is,

$$\Delta S_{surr} = \frac{-q_{sys}}{T} = \frac{-\Delta H_{sys}}{T}$$

The second law,

$$\Delta S_{sys} + \Delta S_{surr} > 0$$

can now be written entirely in terms of changes within the system:

$$\Delta S_{sys} - \frac{\Delta H_{sys}}{T} > 0$$

Multiplication by *T* gives

$$T\Delta S_{sys} - \Delta H_{sys} > 0$$

We now define the left side of this inequality as the quantity $-(\Delta G_{sys})$. That is,

$$-(\Delta G_{sys}) = T\Delta S_{sys} - \Delta H_{sys} \quad \text{or} \quad \Delta G_{sys} = \Delta H_{sys} - T\Delta S_{sys}$$

The second law becomes

$$-(\Delta G_{sys}) > 0 \qquad \text{or} \qquad \Delta G_{sys} < 0$$

The second law of thermodynamics, now defined entirely in terms of the properties of the system, is that *the free energy of a system decreases in any spontaneous change taking place at constant temperature.*

Appendix D Hybrid Orbitals

The observable and/or calculable features of the *atomic structure* of many-electron atoms are (1) the value of electron density at various points within the electron cloud, and (2) the energy associated with the various ways this electron density can be distributed. These features are understood and discussed more easily if the overall electron density is divided up into parts, and regarded as the sum of the densities associated with individual *orbitals* having the sizes, shapes, and energies discussed in Chapter 7. Such an analysis provides an excellent basis for understanding ionization energies, electron affinities, atomic radius, and the wavelengths of lines in atomic spectra. However, it does not lead to easily grasped concepts of covalent bonding, nor to good predictions of molecular shape.

In discussions of *covalent bonding,* it is more useful to divide an atom's electron density in another way, using a different set of component orbitals. These are the *hybrid orbitals* of Chapter 9. Because concrete models are easier to work with than abstract concepts, students are tempted to believe that "hydridization" is an actual physical process of electron redistribution rather than what it really is—another, more useful, way of describing the same electron density distribution in an atom's electron cloud. As an analogy, 12 objects may be described as "one dozen," "six pairs," or "four triplets." No one of these descriptions is more correct or true than the others, only more or less useful for some further purpose. An electron cloud may be considered to be built up of hydrogen-like *s*, *p*, and *d* orbitals, or of hybrid orbitals: the

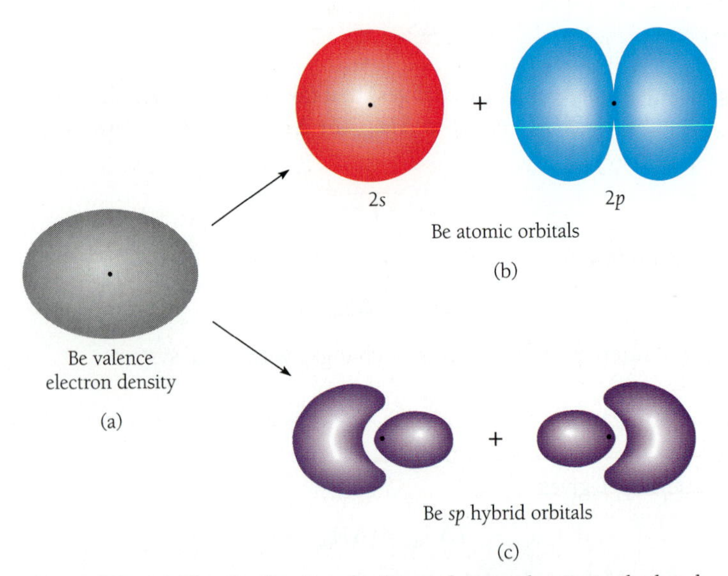

Figure D1 (a) The distribution of valence-electron density in the beryllium atom has the shape of an elongated sphere, which can be represented by either a stipple plot or a contour line. The electron cloud can be regarded as (b) superimposed *s* and *p* orbitals, or, equivalently, (c) as two superimposed *sp* hybrid orbitals.

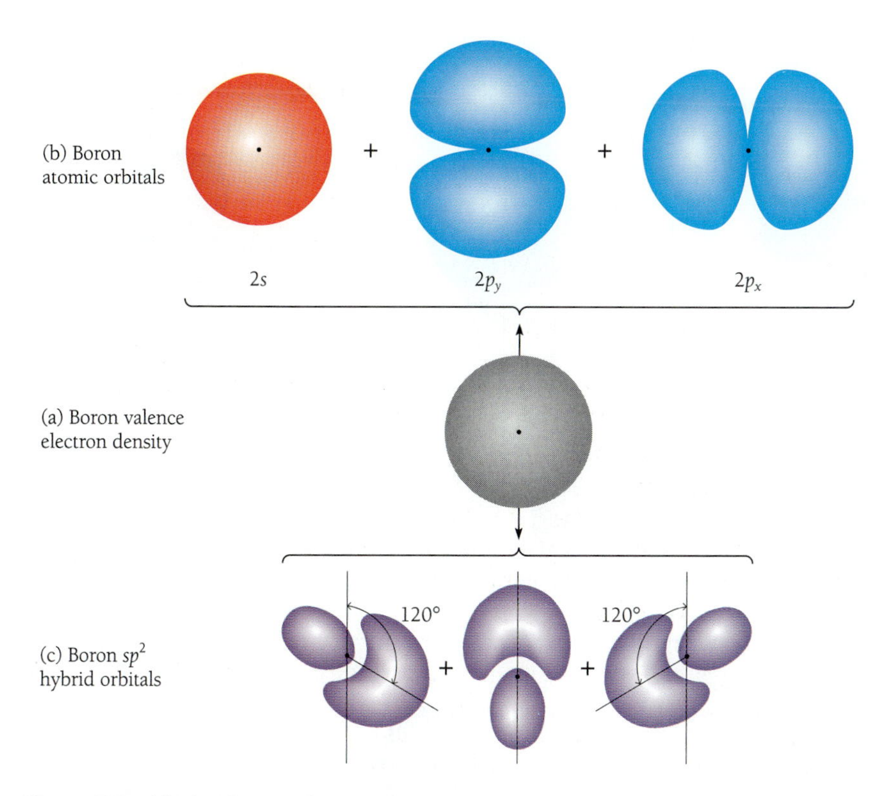

Figure D2 (a) The electron density distribution of an atom having three valence electrons has the shape of a flattened sphere, which, viewed from the top, has a circular cross-section. It may be described as consisting of (b) an *s* and two *p* orbitals, or (c) three sp^2 hybrid orbitals.

descriptions are different, but the electron cloud is the same. Figure D1 illustrates the electron density in the beryllium atom, and two equivalent ways of describing it.

The equivalence of the representations (b) and (c) in Figure D1 can be demonstrated algebraically, using the wave functions (Ψ) of the orbitals. The electron density associated with an orbital is given by the *square* of its wave function. For an atom whose valence shell consists of one *s* and one *p* electron, the total valence electron density is the sum of the densities contributed by these component orbitals.

$$D = (\Psi_s)^2 + (\Psi_p)^2$$

Hybrid orbitals are combinations of several parent orbitals. The algebraic expressions of this relationship, in the case of *sp* hybrid orbitals, are

$$\Psi_{sp(1)} = \left(\frac{1}{\sqrt{2}}\right) \cdot (\Psi_s + \Psi_p) \qquad \text{and} \qquad \Psi_{sp(2)} = \left(\frac{1}{\sqrt{2}}\right) \cdot (\Psi_s - \Psi_p)$$

The sum of the squares of these wave functions gives the overall electron density of an atom having one electron in each of two *sp* hybrid orbitals.

$$\begin{aligned} D &= [\Psi_{sp(1)}]^2 + [\Psi_{sp(2)}]^2 \\ &= \left[\left(\frac{1}{\sqrt{2}}\right) \cdot (\Psi_s + \Psi_p)\right]^2 + \left[\left(\frac{1}{\sqrt{2}}\right) \cdot (\Psi_s - \Psi_p)\right]^2 \\ &= \frac{1}{2}[\Psi_s^2 + 2\Psi_s\Psi_p + \Psi_p^2] + \frac{1}{2}[\Psi_s^2 - 2\Psi_s\Psi_p + \Psi_p^2] \\ &= (\Psi_s)^2 + (\Psi_p)^2 \end{aligned}$$

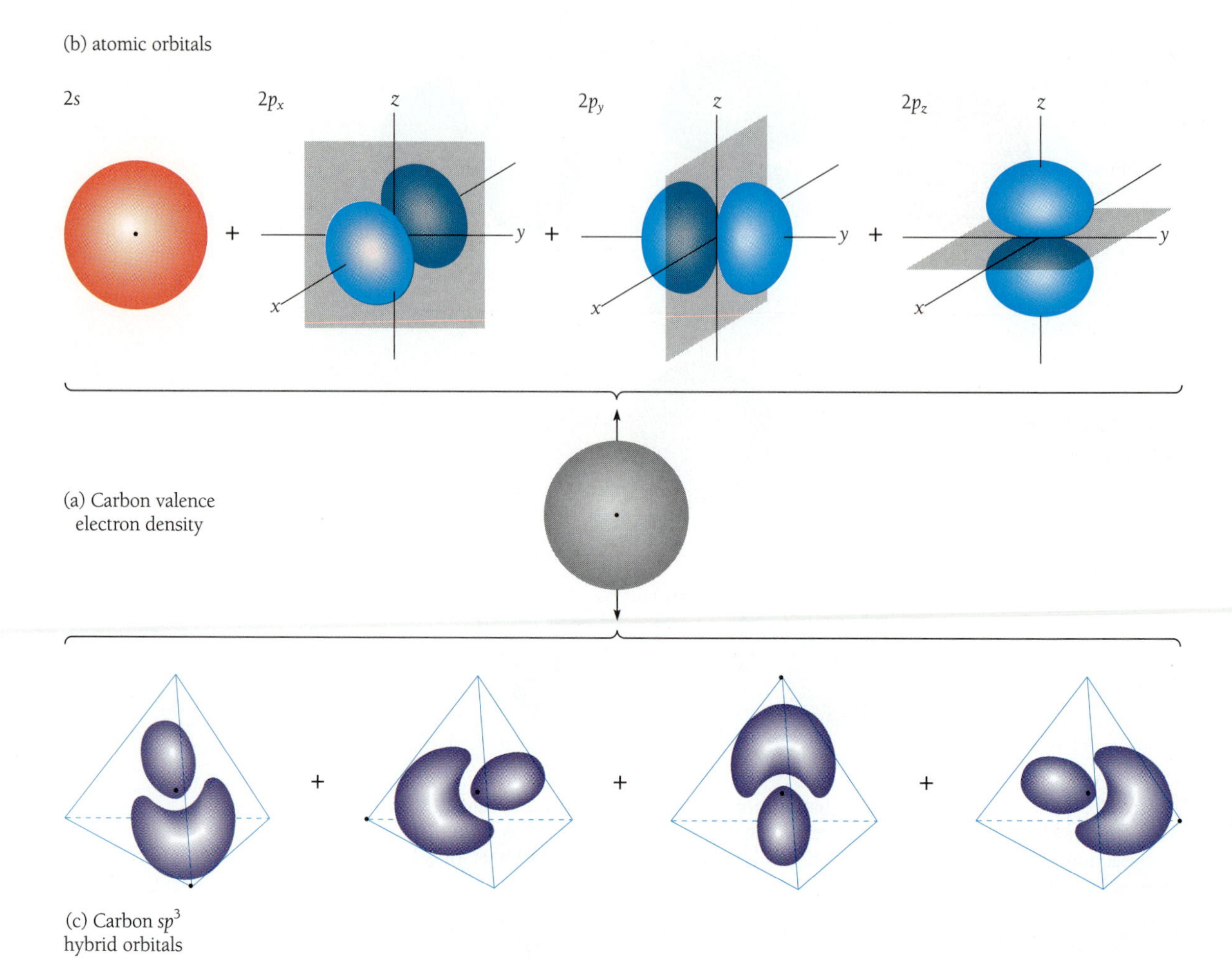

Figure D3 (a) The electron density of an atom with four electrons in its valence shell is spherical in shape. (b) In an isolated carbon atom, the electron cloud is usually described as being the sum of an *s* plus three *p* orbitals. (c) Equivalently, four sp^3 hybrid orbitals may be used to describe the electron density.

The result is exactly the same as the sum of the squares of the parent *s* and *p* orbitals: both describe the same electron density.

In the next element, boron, the electron density of the three valence electrons has the shape of a flattened sphere, as shown in Figure D2. The density may be described as the sum of the contributions of an *s* orbital plus two *p* orbitals, or, equivalently, as the sum of the contributions of three sp^2 hybrids.

Figure D3 illustrates the same principles for atoms, like carbon, that have four valence electrons. The actual electron density is spherical: it may be described as consisting of one *s* and three *p* orbitals, or equivalently as a set of four sp^3 hybrids.

Appendix E Chemical Kinetics

First-order reactions obey Equation (E-1),

$$\ln\left(\frac{C}{C_o}\right) = -kt \qquad \text{(E-1)}$$

which relates reactant concentration C at any time t after the start of the reaction to the rate constant k and the initial concentration C_o. The derivation proceeds from the rate law, which for a first-order reaction is (E-2)

$$\frac{dC}{dt} = -kC \qquad \text{(E-2)}$$

or

$$\left(\frac{1}{C}\right) dC = -kdt$$

Integration of both sides yields (E-3), which includes a constant of integration.

$$\ln C = -kt + \text{const} \qquad \text{(E-3)}$$

The constant of integration is evaluated by noting that at $t = 0$ the concentration C equals its initial value C_o

$$\ln C_o = -k \cdot 0 + \text{const}, \qquad \text{or} \qquad \text{const} = \ln C_o$$

Equation (E-3) may now be rewritten as

$$\ln C = -kt + \ln C_o,$$

or

$$\ln\left(\frac{C}{C_o}\right) = -kt \qquad \text{(E-1)}$$

Equation (E-1) is an "integrated rate law." Integrated rate laws for reactions of other orders are obtained by integrating the appropriate (differential) rate law. For example, the second order integrated rate law is obtained by integrating $dC/dt = -kC^2$.

Appendix F Normality

If the sample in a titration is a polyprotic acid, and the titrant, as is usually the case, is a monohydric base, the simple relationship "moles acid = moles base" no longer applies. For example, since one mole of a diprotic acid supplies 2 moles of H^+, two moles of base are required to neutralize it. The stoichiometry of the reaction must be taken into account in calculations. This can be done by the explicit use of the stoichiometric ratio (mol acid/mol base), as is done in the main body of this text (Section 15.8). It can also be accomplished by the implicit inclusion of the stoichiometric ratio in some newly defined quantities, "equivalent," "equivalent weight," and "normality." An "equivalent" is a quantity of matter purposely defined so as to make the statement "equivalents acid = equivalents base" hold true at the equivalence point of any titration, regardless of the stoichiometric coefficients of the balanced neutralization equation.

One **equivalent** is defined as "the amount of acid that transfers one mole of H^+ in an acid-base reaction." It follows that the number of equivalents of H^+ in a given sample is the number of moles of acid multiplied by the number of ionizable hydrogens—twice the number of moles for a diprotic acid, three times for a triprotic acid, and so on. This number is always an integer, and rarely is it larger than three. The number of ionizable hydrogens depends not only on the structure of the acid, but also on the strength of the base with which it reacts. For example, in a reaction with a strong base, carbonic acid is diprotic. The successive reactions

$$OH^- + H_2CO_3 \longrightarrow H_2O + HCO_3^-$$

$$OH^- + HCO_3^- \longrightarrow H_2O + CO_3^{2-}$$

both go to completion, so that the overall reaction is

$$2\ OH^- + H_2CO_3 \longrightarrow 2\ H_2O + CO_3^{2-}$$

In reaction with ammonia, however, only the first step goes to completion.

$$NH_3 + H_2CO_3 \longrightarrow NH_4^+ + HCO_3^-$$

Ammonia is such a weak base, and HCO_3^- is such a weak acid, that the equilibrium

$$NH_3 + HCO_3^- \rightleftharpoons NH_4^+ + CO_3^{2-}$$

lies far to the left. Thus, the second step does not take place, and the overall reaction is simply

$$NH_3 + H_2CO_3 \longrightarrow NH_4^+ + HCO_3^-$$

in which carbonic acid is monoprotic.

The concentration of a solution of a polyprotic acid is described by its **normality.** Normality is defined as "equivalents of solute per liter of solution."

$$\text{normality} = \text{equivalents/volume} \qquad \text{(F-1)}$$

The abbreviation for normality is *N*. The normality of an acid solution is given by its molarity times the number of ionizable hydrogens. The normality of a diprotic acid solution is twice its molarity, the normality of a triprotic acid solution is three times its molarity, and so on.

The **equivalent weight** is defined as "the mass in grams of one equivalent of substance."

$$\text{equivalent weight} = \text{grams equivalent}^{-1} \qquad \text{(F-2)}$$

The equivalent weight of a diprotic acid is one-half its molar mass, that of a triprotic acid is one-third its molar mass, and so on.

For a monoprotic acid, with one ionizable hydrogen, the number of equivalents in a sample is the same as the number of moles, the normality of a solution is the same as its molarity, and the equivalent weight is the same as the molar mass.

All of these concepts may be transferred directly to polyhydric bases. One equivalent of a base is defined as the amount that can neutralize one mole of H^+. The number of equivalents in a sample of a dihydric base is twice the number of moles, the normality of a solution of a dihydric base is twice its molarity, the equivalent weight of a dihydric base is one-half its molar mass, etc.

When these definitions are used to describe titrations and neutralization reactions involving polyprotic acids and/or polyhydric bases, it is *always* true that

$$\text{equivalents acid} = \text{equivalents base} \qquad \text{(F-3)}$$

at the equivalence point of the titration. It follows from Equation (F-1) that

$$\text{equivalents} = \text{normality} \cdot \text{volume} \qquad \text{(F-4)}$$

For all titrations in which both titrant and sample are solutions,

$$N_aV_a = N_bV_b \qquad \text{(F-5)}$$

N_a and V_a are the normality and volume of the acid sample, while N_b and V_b are the normality and volume of the base. In cases where the titrant is a *monohydric* base (KOH or NaOH is almost always used), the right-hand side of Equation (F-4) also equals M_bV_b. Note that, because of the way in which normality is defined, Equation (F-5) applies to all acid-base titrations, regardless of the number of ionizable hydrogens.

The stoichiometric ratio determined by the balanced neutralization reaction has not disappeared from titration calculations, but it has gone into hiding. The mole ratio is used in the calculation of the normality of the reagents, rather than in the titration calculation itself.

Remember that the number of ionizable hydrogens, and therefore the normality, of an acid is influenced by the strength of the base. It is still necessary to refer to a balanced equation for the reaction. A 0.1 M solution of carbonic acid is 0.2 *N* when reacting with KOH, but the same solution is 0.1 *N* when reacting with ammonia.

EXAMPLE F.1 Normality in Acid-Base Titrations

What volume of 0.150 *N* base is required to titrate 25.0 mL of 0.300 *N* acid?

Analysis All information necessary for the use of Equation (F-5) is given in the problem. Note that information as to whether the sample is mono-, di-, or triprotic is not needed. Insert the known values and solve.

Work

$$N_b V_b = N_a V_a \qquad \text{(F-5)}$$

$$V_b = \frac{N_a V_a}{N_b}$$

$$= \frac{(0.300 \text{ equiv L}^{-1})(0.0250 \text{ L})}{0.150 \text{ equiv L}^{-1}}$$

$$= 0.0500 \text{ L} = 50.0 \text{ mL}$$

EXAMPLE F.2 Calculation of Molar Mass

a. It is found that 23.80 mL of 0.100 *N* base are required to titrate 0.150 g of an unknown acid. What is the equivalent weight of the acid?

b. If it is known that the acid is diprotic, what is the molar mass?

Analysis Since the sample is present as a pure substance rather than as a solution, Equation (F-3) rather than (F-5) must be used. Also, after the equivalent weight is found, a second step is required to convert equivalent weight to molar mass.

Work

a.

$$\text{equivalents acid} = \text{equivalents base} \qquad \text{(F-3)}$$

$$= N_b V_b$$

$$= (0.100 \text{ equiv L}^{-1})(0.02380 \text{ L})$$

$$= 0.002380 \text{ equiv}$$

$$\text{equivalent weight} = \text{grams equivalent}^{-1} \qquad \text{(F-2)}$$

$$= \frac{0.150 \text{ g}}{0.002380 \text{ equiv}} = \frac{63.0 \text{ g}}{\text{equiv}}$$

b. The molar mass is found from the equivalent weight and the information that the acid is diprotic in this reaction.

$$\mathcal{M} = \text{(equivalent weight)(\# of ionizable hydrogens)}$$
$$= \left(\frac{63.0\text{ g}}{\text{equiv}}\right)\left(\frac{2\text{ equiv}}{\text{mol}}\right)$$
$$= 126\text{ g mol}^{-1}$$

Note: The units of "# of ionizable hydrogens" are equivalents per mole.

The concept of normality is also applicable to redox reaction stoichiometry. When a reaction involves electron rather than proton transfer, one equivalent is defined as the mass of oxidant that gains one mole of electrons in the reaction, and/or the mass of reductant that supplies one mole of electrons. All other definitions are the same: normality = (equivalents)/(volume), and normality is a small whole number times molarity. The equivalent weight is the molar mass divided by the same small whole number. In a titration or other problem in stoichiometry, the relationship "equivalents oxidant = equivalents reductant" holds, as does the titration equation $N_{ox}V_{ox} = N_{red}V_{red}$. For example, suppose a 0.15 M solution of $KMnO_4$ is used as an oxidant in a redox titration of Fe^{2+}. The balanced equation is

$$8\,H^+ + MnO_4^- + 5\,Fe^{2+} \longrightarrow 5\,Fe^{3+} + Mn^{2+} + 4\,H_2O$$

The number of electrons transferred in this reaction becomes clear when the half-reactions are written separately

$$8\,H^+ + MnO_4^- + 5\,e^- \longrightarrow Mn^{2+} + 4\,H_2O$$
$$Fe^{2+} \longrightarrow Fe^{3+} + e^-$$

Since each mole of the oxidant MnO_4^- gains five moles of electrons in the reaction, its normality is five times its molarity, or $(0.15) \cdot 5 = 0.75\,N$. The equivalent weight of $KMnO_4$ is one-fifth of its molar mass, or $158/5 = 31.6\text{ g equiv}^{-1}$. Since each mole of Fe^{2+} ions loses one mole of electrons in the reaction, its normality equals its molarity, and its equivalent weight is the same as the molar mass.

Note that the number of electrons lost by an oxidant depends on the reaction. In basic solution, permanganate ion tends to gain these electrons in the reduction half reaction

$$MnO_4^- + 2\,H_2O + 3\,e^- \longrightarrow MnO_2(s) + 4\,OH^-$$

A 0.15 M permanganate solution is 0.45 *N* in reactions that take place in basic solution.

Appendix G Selected Thermodynamic Values

Species	ΔH_f°(298 K)/kJ mol^{-1}	S°(298 K)/J K^{-1} mol^{-1}	ΔG_f°(298 K)/kJ mol^{-1}
Aluminum			
Al(s)	0	28.3	0
$AlCl_3$(s)	−704.2	110.67	−628.8
Al_2O_3(s)	−1675.7	50.92	−1582.3
Barium			
$BaCl_2$(s)	−858.6	123.68	−810.4
BaO(s)	−553.5	70.42	−525.1
$BaSO_4$(s)	−1473.3	132.2	−1362.2
Beryllium			
Be(s)	0	9.5	0
$Be(OH)_2$	−902.5	51.9	−815.0
Bromine			
Br(g)	111.884	175.022	82.396
$Br_2(\ell)$	0	152.2	0
Br_2(g)	30.907	245.463	3.110
BrCl(g)	−428.9	320.19	−350.6
BrF_3(g)	−255.60	292.53	−229.43
HBr(g)	−36.40	198.695	−53.45
Calcium			
Ca(s)	0	41.42	0
Ca(g)	178.2	158.884	144.3
Ca^{2+}(g)	1925.90	—	—
CaC_2(s)	−59.8	69.96	−64.9
$CaCO_3$(s; calcite)	−1206.92	92.9	−1128.79
$CaCl_2$(s)	−795.8	104.6	−748.1
CaF_2(s)	−1219.6	68.87	−1167.3
CaH_2(s)	−186.2	42.	−147.2
CaO(s)	−635.09	39.75	−604.03
CaS(s)	−482.4	56.5	−477.4
$Ca(OH)_2$(s)	−986.09	83.39	−898.49
$Ca(OH)_2$(aq)	−1002.82	−74.5	−868.07
$CaSO_4$(s)	−1434.11	106.7	−1321.79
Carbon			
C(s, graphite)	0	5.740	0
C(s, diamond)	1.895	2.377	2.900
C(g)	716.682	158.096	671.257
$CCl_4(\ell)$	−135.44	216.40	−65.21
CH_3Cl(g)	−80.83	234.58	−57.37
CCl_4(g)	−102.9	309.85	−60.59
CH_3Cl(g)	−80.83	234.58	−57.37
$CHCl_3$(g)	−103.14	295.71	−70.34
CH_4(g, methane)	−74.81	186.264	−50.72
C_2H_2(g, ethyne)	226.73	200.94	209.20

Species	ΔH_f°(298 K)/kJ mol^{-1}	S°(298 K)/J K^{-1} mol^{-1}	ΔG_f°(298 K)/kJ mol^{-1}
Carbon (continued)			
C_2H_4(g, ethene)	52.26	219.56	68.15
C_2H_6(g, ethane)	−84.68	229.60	−32.82
C_3H_8(g, propane)	−103.8	269.9	−23.49
C_6H_6(ℓ, benzene)	49.03	172.8	124.5
CH_3OH(ℓ, methanol)	−238.66	126.8	−166.27
CH_3OH(g, methanol)	−200.66	239.81	−161.96
C_2H_5OH(ℓ, ethanol)	−277.69	160.7	−174.78
C_2H_5OH(g, ethanol)	−235.10	282.70	−168.49
CO(g)	−110.525	197.674	−137.168
CO_2(g)	−393.509	213.74	−394.359
CS_2(g)	117.36	237.84	67.12
$COCl_2$(g)	−218.8	283.53	−204.6
Cesium			
Cs(s)	0	85.23	0
Cs^+(g)	457.964	—	—
CsCl(s)	−443.04	101.17	−414.53
Chlorine			
Cl(g)	121.679	165.198	105.680
Cl^-(g)	−233.13	—	—
Cl_2(g)	0	223.066	0
ClF(g)	−54.48	217.89	−55.94
HCl(g)	−92.307	186.908	−95.299
HCl(aq)	−167.159	56.5	−131.228
Chromium			
Cr(s)	0	23.77	0
Cr_2O_3(s)	−1139.7	81.2	−1058.1
$CrCl_3$(s)	−556.5	123.0	−486.1
Copper			
Cu(s)	0	33.150	0
CuO(s)	−157.3	42.63	−129.7
$CuCl_2$(s)	−220.1	108.07	−175.7
Fluorine			
F_2(g)	0	202.78	0
F(g)	78.99	158.754	61.91
F^-(g)	−255.39	—	—
F^-(aq)	−332.63	−13.8	−278.79
HF(g)	−271.1	173.779	−273.2
HF(aq)	−332.63	−13.8	−278.79
Hydrogen			
H_2(g)	0	130.684	0
H(g)	217.965	114.713	203.247
H^+(g)	1536.202	—	—
H_2O(ℓ)	−285.830	69.91	−237.129
H_2O(g)	−241.818	188.825	−228.572
H_2O_2(ℓ)	−187.78	109.6	−120.35

Species	ΔH_f°(298 K)/kJ mol^{-1}	S°(298 K)/J K^{-1} mol^{-1}	ΔG_f°(298 K)/kJ mol^{-1}
Iodine			
I_2(s)	0	116.135	0
I_2(g)	62.438	260.69	19.327
I(g)	106.838	180.791	70.250
I^-(g)	−197.	—	—
ICl(g)	17.78	247.551	−5.46
Iron			
Fe(s)	0	27.78	0
FeO(s)	−272.	—	—
Fe_2O_3(s, hematite)	−824.2	87.40	−742.2
Fe_3O_4(s, magnetite)	−1118.4	146.4	−1015.4
$FeCl_2$(s)	−341.79	117.95	−302.30
$FeCl_3$(s)	−399.49	142.3	−344.00
FeS_2(s, pyrite)	−178.2	52.93	−166.9
$Fe(CO)_5$(ℓ)	−774.0	338.1	−705.3
Lead			
Pb(s)	0	64.81	0
$PbCl_2$(s)	−359.41	136.0	−314.10
PbO(s, yellow)	−217.32	68.70	−187.89
PbS(s)	−100.4	91.2	−98.7
Lithium			
Li(s)	0	29.12	0
Li^+(g)	685.783	—	—
LiOH(s)	−484.93	42.80	−438.95
LiOH(aq)	−508.48	2.80	−450.58
LiCl(s)	−408.701	59.33	−384.37
Magnesium			
Mg(s)	0	32.68	0
$MgCl_2$(s)	−641.32	89.62	−591.79
MgO(s)	−601.70	26.94	−569.43
$Mg(OH)_2$(s)	−924.54	63.18	−833.51
MgS(s)	−346.0	50.33	−341.8
Mercury			
Hg(ℓ)	0	76.02	0
$HgCl_2$(s)	−224.3	146.0	−178.6
HgO(s, red)	−90.83	70.29	−58.539
HgS(s, red)	−58.2	82.4	−50.6
Nickel			
Ni(s)	0	29.87	0
NiO(s)	−239.7	37.99	−211.7
$NiCl_2$(s)	−305.332	97.65	−259.032
Nitrogen			
N_2(g)	0	191.61	0
N(g)	472.704	153.298	455.563
NH_3(g)	−46.11	192.45	−16.45
N_2H_4(ℓ)	50.63	121.21	149.34

Species	ΔH_f°(298 K)/kJ mol^{-1}	S°(298 K)/J K^{-1} mol^{-1}	ΔG_f°(298 K)/kJ mol^{-1}
Nitrogen (continued)			
$NH_4Cl(s)$	−314.43	94.6	−202.87
$NH_4Cl(aq)$	−299.66	169.9	−210.52
$NH_4NO_3(s)$	−365.56	151.08	−183.87
$NH_4NO_3(aq)$	−339.87	259.8	−190.56
$NO(g)$	90.25	210.76	86.55
$NO_2(g)$	33.18	240.06	51.31
$N_2O(g)$	82.05	219.85	104.20
$N_2O_4(g)$	9.16	304.29	97.89
$NOCl(g)$	51.71	261.69	66.08
$HNO_3(\ell)$	−174.10	155.60	−80.71
$HNO_3(g)$	−135.06	266.38	−74.72
$HNO_3(aq)$	−207.36	146.4	−111.25
Oxygen			
$O_2(g)$	0	205.138	0
$O(g)$	249.170	161.055	231.731
$O_3(g)$	142.7	238.93	163.2
Phosphorus			
P_4(s, white)	0	164.36	0
P_4(s, red)	−70.4	91.2	−48.4
$P(g)$	314.64	163.193	278.25
$PH_3(g)$	5.4	310.23	13.4
$PCl_3(g)$	−287.0	311.78	−267.8
$P_4O_{10}(s)$	−2984.0	228.86	−2697.7
$H_3PO_4(s)$	−1279.0	110.5	−1119.1
Potassium			
$K(s)$	0	64.18	0
$KCl(s)$	−436.747	82.59	−409.14
$KClO_3(s)$	−397.73	143.1	−296.25
$KI(s)$	−327.90	106.32	−324.892
$KOH(s)$	−424.764	78.9	−379.08
$KOH(aq)$	−482.37	91.6	−440.50
Silicon			
$Si(s)$	0	18.83	0
$SiBr_4(\ell)$	−457.3	227.8	−443.9
$SiC(s)$	−65.3	16.61	−62.8
$SiCl_4(g)$	−657.01	330.73	−616.98
$SiH_4(g)$	34.3	204.62	56.9
$SiF_4(g)$	−1614.94	282.49	−1572.65
SiO_2(s, quartz)	−910.94	41.84	−856.64
Silver			
$Ag(s)$	0	42.55	0
$Ag_2O(s)$	−31.05	121.3	−11.20
$AgCl(s)$	−127.068	96.2	−109.789
$AgNO_3(s)$	−124.39	140.92	−33.41

Species	ΔH_f°(298 K)/kJ mol^{-1}	S°(298 K)/J K^{-1} mol^{-1}	ΔG_f°(298 K)/kJ mol^{-1}
Sodium			
Na(s)	0	51.21	0
Na(g)	107.32	153.712	76.761
Na^+(g)	609.358	—	—
NaBr(s)	−361.062	86.82	−348.983
NaCl(s)	−411.153	72.13	−384.138
NaCl(g)	−176.65	229.81	−196.66
NaCl(aq)	−407.27	115.5	−393.133
NaOH(s)	−425.609	64.455	−379.494
NaOH(aq)	−470.114	48.1	−419.150
Na_2CO_3(s)	−1130.68	134.98	−1044.44
Sulfur			
S(s, rhombic)	0	31.80	0
S(g)	278.805	167.821	238.250
S_2Cl_2(g)	−18.4	331.5	−31.8
SF_6(g)	1209.	291.82	−1105.3
H_2S(g)	−20.63	205.79	−33.56
SO_2(g)	−296.830	248.22	−300.194
SO_3(g)	−395.72	256.76	−371.06
$SOCl_2$(g)	−212.5	309.77	−198.3
H_2SO_4(ℓ)	−813.989	156.904	−690.003
H_2SO_4(aq)	−909.27	20.1	−744.53
Tin			
Sn(s, white)	0	51.55	0
Sn(s, gray)	−2.09	44.14	0.13
$SnCl_4$(ℓ)	−511.3	258.6	−440.1
$SnCl_4$(g)	−471.5	365.8	−432.2
SnO_2(s)	−580.7	52.3	−519.6
Titanium			
Ti(s)	0	30.63	0
$TiCl_4$(ℓ)	−804.2	252.34	−737.2
$TiCl_4$(g)	−763.2	354.9	−726.7
TiO_2	−939.7	49.92	−884.5
Zinc			
Zn(s)	0	41.63	0
$ZnCl_2$(s)	−415.05	111.46	−369.398
ZnO(s)	−348.28	43.64	−318.30
ZnS(s, sphalerite)	−205.98	57.7	−201.29

* Taken from "The NBS Tables of Chemical Thermodynamic Properties," 1982.

Appendix H Solubility Products

Solubility Products of Some Slightly Soluble Salts*

Chlorides		Bromides		Iodides	
AgCl	1.8×10^{-10}	AgBr	5.0×10^{-13}	AgI	8.3×10^{-17}
$AuCl_3$	3.2×10^{-25}	$AuBr_3$	4.0×10^{-36}	CuI	1×10^{-12}
CuCl	4.2×10^{-8}	CuBr	5×10^{-9}	HgI_2	1.1×10^{-28}
Hg_2Cl_2	1.2×10^{-18}	$HgBr_2$	1×10^{-10}	Hg_2I_2	4.6×10^{-29}
$PbCl_2$	1.7×10^{-5}	Hg_2Br_2	5.6×10^{-23}	PbI_2	7.9×10^{-9}
Fluorides		**Chromates**		**Sulfates**	
BaF_2	1.7×10^{-6}	Ag_2CrO_4	1.2×10^{-12}	Ag_2SO_4	1.5×10^{-5}
CaF_2	3.9×10^{-11}	$BaCrO_4$	2.1×10^{-10}	$BaSO_4$	1.1×10^{-10}
MgF_2	6.6×10^{-9}	$CaCrO_4$	7.1×10^{-4}	$CaSO_4$	2.4×10^{-5}
SrF_2	2.9×10^{-9}	$PbCrO_4$	1.8×10^{-14}	$PbSO_4$	1.6×10^{-8}
Carbonates		**Sulfides†**		**Hydroxides**	
Ag_2CO_3	6.4×10^{-12}	Ag_2S	6×10^{-50}	$Al(OH)_3$	4.5×10^{-33}
$BaCO_3$	5×10^{-9}	CdS	8×10^{-27}	$Ba(OH)_2$	3.6×10^{-4}
$CaCO_3$	4.6×10^{-9}	CoS	4×10^{-21}	$Ca(OH)_2$	6.5×10^{-6}
$CdCO_3$	1.8×10^{-14}	Cu_2S	2×10^{-48}	$Co(OH)_3$	3×10^{-45}
$CoCO_3$	1.0×10^{-10}	CuS	6×10^{-36}	$Cr(OH)_3$	6.7×10^{-31}
$CuCO_3$	2.3×10^{-10}	FeS	6×10^{-18}	$Mg(OH)_2$	7.3×10^{-12}
$FeCO_3$	2.1×10^{-11}	HgS	4×10^{-53}	$Mn(OH)_2$	2×10^{-13}
Hg_2CO_3	8.9×10^{-17}	MnS	2×10^{-10}	$Ni(OH)_2$	6×10^{-16}
$MgCO_3$	3.5×10^{-8}	NiS	3×10^{-19}	$Zn(OH)_2$	3.0×10^{-16}
$NiCO_3$	1.3×10^{-7}	PbS	2×10^{-27}		
$PbCO_3$	7.4×10^{-14}	SnS	1×10^{-26}		
$SrCO_3$	9.3×10^{-10}	ZnS	2×10^{-24}		
$ZnCO_3$	1×10^{-10}				
Oxalates					
$Ag_2C_2O_4$	1×10^{-11}	CdC_2O_4	1.5×10^{-8}	PbC_2O_4	3×10^{-11}
BaC_2O_4	1×10^{-6}	CuC_2O_4	2.9×10^{-8}	SrC_2O_4	4×10^{-7}
CaC_2O_4	1×10^{-8}	MgC_2O_4	8.5×10^{-5}	ZnC_2O_4	1.3×10^{-9}

* Most values refer to 25 °C; a few are for somewhat lower temperature, but none is for less than 19 °C.

† Solubility products for sulfides are defined as in Equation (17-12), page 751: for the salt MS, $K'_{sp} = [M^{2+}][HS^-][OH^-]$.

Appendix I Standard Reduction Potentials

Standard Reduction Potentials in Aqueous Solution at 25 °C	
Reduction Half Reaction (acidic solution)	**$\mathscr{E}°$(V)**
$F_2(g) + 2e^- \rightarrow 2F^-(aq)$	2.87
$Co^{3+}(aq) + e^- \rightarrow Co^{2+}(aq)$	1.82
$Pb^{4+}(aq) + 2e^- \rightarrow Pb^{2+}(aq)$	1.8
$H_2O_2(aq) + 2H^+(aq) + 2e^- \rightarrow 2H_2O$	1.77
$NiO_2(s) + 4H^+(aq) + 2e^- \rightarrow Ni^{2+}(aq) + 2H_2O$	1.7
$PbO_2(s) + SO_4^{2-}(aq) + 4H^+(aq) + 2e^- \rightarrow PbSO_4(s) + 2H_2O$	1.685
$Au^+(aq) + e^- \rightarrow Au(s)$	1.68
$2HClO(aq) + 2H^+(aq) + 2e^- \rightarrow Cl_2(g) + 2H_2O$	1.63
$Ce^{4+}(aq) + e^- \rightarrow Ce^{3+}(aq)$	1.61
$NaBiO_3(s) + 6H^+(aq) + 2e^- \rightarrow Bi^{3+}(aq) + Na^+(aq) + 3H_2O$	~1.6
$MnO_4^-(aq) + 8H^+(aq) + 5e^- \rightarrow Mn^{2+}(aq) + 4H_2O$	1.51
$Au^{3+}(aq) + 3e^- \rightarrow Au(s)$	1.50
$ClO_3^-(aq) + 6H^+(aq) + 5e^- \rightarrow \frac{1}{2}Cl_2(g) + 3H_2O$	1.47
$BrO_3^-(aq) + 6H^+(aq) + 6e^- \rightarrow Br^-(aq) + 3H_2O$	1.44
$Cl_2(g) + 2e^- \rightarrow 2Cl^-(aq)$	1.358
$Cr_2O_7^{2-}(aq) + 14H^+(aq) + 6e^- \rightarrow 2Cr^{3+}(aq) + 7H_2O$	1.33
$N_2H_5^+(aq) + 3H^+(aq) + 2e^- \rightarrow 2NH_4^+(aq)$	1.24
$MnO_2(s) + 4H^+(aq) + 2e^- \rightarrow Mn^{2+}(aq) + 2H_2O$	1.23
$O_2(g) + 4H^+(aq) + 4e^- \rightarrow 2H_2O$	1.229
$Pt^{2+}(aq) + 2e^- \rightarrow Pt(s)$	1.2
$IO_3^-(aq) + 6H^+(aq) + 5e^- \rightarrow \frac{1}{2}I_2(aq) + 3H_2O$	1.195
$ClO_4^-(aq) + 2H^+(aq) + 2e^- \rightarrow ClO_3^-(aq) + H_2O$	1.19
$Br_2(\ell) + 2e^- \rightarrow 2Br^-(aq)$	1.066
$AuCl_4^-(aq) + 3e^- \rightarrow Au(s) + 4Cl^-(aq)$	1.00
$Pd^{2+}(aq) + 2e^- \rightarrow Pd(s)$	0.987
$NO_3^-(aq) + 4H^+(aq) + 3e^- \rightarrow NO(g) + 2H_2O$	0.96
$NO_3^-(aq) + 3H^+(aq) + 2e^- \rightarrow HNO_2(aq) + H_2O$	0.94
$2Hg^{2+}(aq) + 2e^- \rightarrow Hg_2^{2+}(aq)$	0.920
$Hg^{2+}(aq) + 2e^- \rightarrow Hg(\ell)$	0.855
$Ag^+(aq) + e^- \rightarrow Ag(s)$	0.7994
$Hg_2^{2+}(aq) + 2e^- \rightarrow 2Hg(\ell)$	0.789
$Fe^{3+}(aq) + e^- \rightarrow Fe^{2+}(aq)$	0.771
$SbCl_6^-(aq) + 2e^- \rightarrow SbCl_4^-(aq) + 2Cl^-(aq)$	0.75
$[PtCl_4]^{2-}(aq) + 2e^- \rightarrow Pt(s) + 4Cl^-(aq)$	0.73
$O_2(g) + 2H^+(aq) + 2e^- \rightarrow H_2O_2(aq)$	0.682
$[PtCl_6]^{2-}(aq) + 2e^- \rightarrow [PtCl_4]^{2-}(aq) + 2Cl^-(aq)$	0.68
$H_3AsO_4(aq) + 2H^+(aq) + 2e^- \rightarrow H_3AsO_3(aq) + H_2O$	0.58
$I_2(s) + 2e^- \rightarrow 2I^-(aq)$	0.535
$TeO_2(s) + 4H^+(aq) + 4e^- \rightarrow Te(s) + 2H_2O$	0.529
$Cu^+(aq) + e^- \rightarrow Cu(s)$	0.521
$[RhCl_6]^{3-}(aq) + 3e^- \rightarrow Rh(s) + 6Cl^-(aq)$	0.44
$Cu^{2+}(aq) + 2e^- \rightarrow Cu(s)$	0.337
$HgCl_2(s) + 2e^- \rightarrow 2Hg(\ell) + 2Cl^-(aq)$	0.27

Standard Reduction Potentials in Aqueous Solution at 25 °C *(continued)*	
Reduction Half Reaction (acidic solution)	**$\mathcal{E}°$(V)**
$AgCl(s) + e^- \rightarrow Ag(s) + Cl^-(aq)$	0.222
$SO_4^{2-}(aq) + 4H^+(aq) + 2e^- \rightarrow SO_2(g) + 2H_2O$	0.20
$SO_4^{2-}(aq) + 4H^+(aq) + 2e^- \rightarrow H_2SO_3(aq) + H_2O$	0.17
$Cu^{2+}(aq) + e^- \rightarrow Cu^+(aq)$	0.153
$Sn^{4+}(aq) + 2e^- \rightarrow Sn^{2+}(aq)$	0.15
$S(s) + 2H^+(aq) + 2e^- \rightarrow H_2S(aq)$	0.14
$AgBr(s) + e^- \rightarrow Ag(s) + Br^-(aq)$	0.0713
$2H^+(aq) + 2e^- \rightarrow H_2(g)$ (reference electrode)	0.0000
$Fe^{3+}(aq) + 3\,e^- \rightarrow Fe(s)$	−0.037
$Pb^{2+}(aq) + 2e^- \rightarrow Pb(s)$	−0.126
$Sn^{2+}(aq) + 2e^- \rightarrow Sn(s)$	−0.14
$AgI(s) + e^- \rightarrow Ag(s) + I^-(aq)$	−0.15
$[SnF_6]^{2-}(aq) + 4e^- \rightarrow Sn(s) + 6F^-(aq)$	−0.25
$Ni^{2+}(aq) + 2e^- \rightarrow Ni(s)$	−0.25
$Co^{2+}(aq) + 2e^- \rightarrow Co(s)$	−0.28
$Tl^+(aq) + e^- \rightarrow Tl(s)$	−0.34
$PbSO_4(s) + 2e^- \rightarrow Pb(s) + SO_4^{2-}(aq)$	−0.356
$Se(s) + 2H^+(aq) + 2e^- \rightarrow H_2Se(aq)$	−0.40
$Cd^{2+}(aq) + 2e^- \rightarrow Cd(s)$	−0.403
$Cr^{3+}(aq) + e^- \rightarrow Cr^{2+}(aq)$	−0.41
$Fe^{2+}(aq) + 2e^- \rightarrow Fe(s)$	−0.44
$2CO_2(g) + 2H^+(aq) + 2e^- \rightarrow (COOH)_2(aq)$	−0.49
$Ga^{3+}(aq) + 3e^- \rightarrow Ga(s)$	−0.53
$HgS(s) + 2H^+(aq) + 2e^- \rightarrow Hg(\ell) + H_2S(g)$	−0.72
$Cr^{3+}(aq) + 3e^- \rightarrow Cr(s)$	−0.74
$Zn^{2+}(aq) + 2e^- \rightarrow Zn(s)$	−0.763
$Cr^{2+}(aq) + 2e^- \rightarrow Cr(s)$	−0.91
$FeS(s) + 2e^- \rightarrow Fe(s) + S^{2-}(aq)$	−1.01
$Mn^{2+}(aq) + 2e^- \rightarrow Mn(s)$	−1.18
$V^{2+}(aq) + 2e^- \rightarrow V(s)$	−1.18
$CdS(s) + 2e^- \rightarrow Cd(s) + S^{2-}(aq)$	−1.21
$ZnS(s) + 2e^- \rightarrow Zn(s) + S^{2-}(aq)$	−1.44
$Zr^{4+}(aq) + 4e^- \rightarrow Zr(s)$	−1.53
$Al^{3+}(aq) + 3e^- \rightarrow Al(s)$	−1.66
$H_2(g) + 2e^- \rightarrow 2H^-(aq)$	−2.25
$Mg^{2+}(aq) + 2e^- \rightarrow Mg(s)$	−2.37
$Na^+(aq) + e^- \rightarrow Na(s)$	−2.714
$Ca^{2+}(aq) + 2e^- \rightarrow Ca(s)$	−2.87
$Sr^{2+}(aq) + 2e^- \rightarrow Sr(s)$	−2.89
$Ba^{2+}(aq) + 2e^- \rightarrow Ba(s)$	−2.90
$Rb^+(aq) + e^- \rightarrow Rb(s)$	−2.925
$K^+(aq) + e^- \rightarrow K(s)$	−2.925
$Li^+(aq) + e^- \rightarrow Li(s)$	−3.045

Standard Reduction Potentials in Aqueous Solution at 25 °C *(continued)*	
Reduction Half Reaction (base solution)	**$\mathscr{E}$°(V)**
$ClO^-(aq) + H_2O + 2e^- \rightarrow Cl^-(aq) + 2OH^-(aq)$	0.89
$OOH^-(aq) + H_2O + 2e^- \rightarrow 3OH^-(aq)$	0.88
$2NH_2OH(aq) + 2e^- \rightarrow N_2H_4(aq) + 2OH^-(aq)$	0.74
$ClO_3^-(aq) + 3H_2O + 6e^- \rightarrow Cl^-(aq) + 6OH^-(aq)$	0.62
$MnO_4^-(aq) + 2H_2O + 3e^- \rightarrow MnO_2(s) + 4OH^-(aq)$	0.588
$MnO_4^-(aq) + e^- \rightarrow MnO_4^{2-}(aq)$	0.564
$NiO_2(s) + 2H_2O + 2e^- \rightarrow Ni(OH)_2(s) + 2OH^-(aq)$	0.49
$Ag_2CrO_4(s) + 2e^- \rightarrow 2Ag(s) + CrO_4^{2-}(aq)$	0.446
$O_2(g) + 2H_2O + 4e^- \rightarrow 4OH^-(aq)$	0.40
$ClO_4^-(aq) + H_2O + 2e^- \rightarrow ClO_3^-(aq) + 2OH^-(aq)$	0.36
$Ag_2O(s) + H_2O + 2e^- \rightarrow 2Ag(s) + 2OH^-(aq)$	0.34
$2NO_2^-(aq) + 3H_2O + 4e^- \rightarrow N_2O(g) + 6OH^-(aq)$	0.15
$N_2H_4(aq) + 2H_2O + 2e^- \rightarrow 2NH_3(aq) + 2OH^-(aq)$	0.10
$[Co(NH_3)_6]^{3+}(aq) + e^- \rightarrow [Co(NH_3)_6]^{2+}(aq)$	0.10
$HgO(s) + H_2O + 2e^- \rightarrow Hg(\ell) + 2OH^-(aq)$	0.0984
$O_2(g) + H_2O + 2e^- \rightarrow OOH^-(aq) + OH^-(aq)$	0.076
$NO_3^-(aq) + H_2O + 2e^- \rightarrow NO_2^-(aq) + 2OH^-(aq)$	0.01
$MnO_2(s) + 2H_2O + 2e^- \rightarrow Mn(OH)_2(s) + 2OH^-(aq)$	−0.05
$CrO_4^{2-}(aq) + 4H_2O + 3e^- \rightarrow Cr(OH)_3(s) + 5OH^-(aq)$	−0.12
$Cu(OH)_2(s) + 2e^- \rightarrow Cu(s) + 2OH^-(aq)$	−0.36
$S(s) + 2e^- \rightarrow S^{2-}(aq)$	−0.48
$Fe(OH)_3(s) + e^- \rightarrow Fe(OH)_2(s) + OH^-(aq)$	−0.56
$2H_2O + 2e^- \rightarrow H_2(g) + 2OH^-(aq)$	−0.8277
$2NO_3^-(aq) + 2H_2O + 2e^- \rightarrow N_2O_4(g) + 4OH^-(aq)$	−0.85
$Fe(OH)_2(s) + 2e^- \rightarrow Fe(s) + 2OH^-(aq)$	−0.877
$SO_4^{2-}(aq) + H_2O + 2e^- \rightarrow SO_3^{2-}(aq) + 2OH^-(aq)$	−0.93
$N_2(g) + 4H_2O + 4e^- \rightarrow N_2H_4(aq) + 4OH^-(aq)$	−1.15
$[Zn(OH)_4]^{2-}(aq) + 2e^- \rightarrow Zn(s) + 4OH^-(aq)$	−1.22
$Zn(OH)_2(s) + 2e^- \rightarrow Zn(s) + 2OH^-(aq)$	−1.245
$[Zn(CN)_4]^{2-}(aq) + 2e^- \rightarrow Zn(s) + 4CN^-(aq)$	−1.26
$Cr(OH)_3(s) + 3e^- \rightarrow Cr(s) + 3OH^-(aq)$	−1.30
$SiO_3^{2-}(aq) + 3H_2O + 4e^- \rightarrow Si(s) + 6OH^-(aq)$	−1.70

Appendix J Ionization Constants of Weak Acids and Bases

Acids

Acid	Formula and Ionization Equation	K_a	pK_a
Acetic	$CH_3CO_2H \rightleftharpoons H^+ + CH_3CO_2^-$	1.8×10^{-5}	4.74
Benzoic	$C_6H_5CO_2H \rightleftharpoons H^+ + C_6H_5CO_2^-$	6.3×10^{-5}	4.20
Butanoic	$C_3H_7CO_2H \rightleftharpoons H^+ + C_3H_7CO_2^-$	1.5×10^{-5}	4.82
Carbonic	$H_2CO_3 \rightleftharpoons H^+ + HCO_3^-$	4.0×10^{-7} (K_1)	6.40
	$HCO_3^- \rightleftharpoons H^+ + CO_3^{2-}$	4.0×10^{-11} (K_2)	10.40
Chloroacetic	$CH_2ClCO_2H \rightleftharpoons H^+ + CH_2ClCO_2^-$	1.4×10^{-3}	2.85
Citric	$H_3C_6H_5O_7 \rightleftharpoons H^+ + H_2C_6H_5O_7^-$	7.2×10^{-4} (K_1)	3.13
	$H_2C_6H_5O_7^- \rightleftharpoons H^+ + HC_6H_5O_7^{2-}$	1.7×10^{-5} (K_2)	4.77
	$HC_6H_5O_7^{2-} \rightleftharpoons H^+ + C_6H_5O_7^{3-}$	4.0×10^{-7} (K_3)	6.40
Cyanic	$HOCN \rightleftharpoons H^+ + OCN^-$	3.5×10^{-4}	3.46
Ethanol	$CH_3CH_2OH \rightleftharpoons H^+ + CH_3CH_2O^-$	$\sim 10^{-16}$	~16
Fluoroacetic	$CH_2FCO_2H \rightleftharpoons H^+ + CH_2FCO_2^-$	2.2×10^{-3}	2.66
Formic	$HCO_2H \rightleftharpoons H^+ + HCO_2^-$	1.8×10^{-4}	3.74
Hydrazoic	$HN_3 \rightleftharpoons H^+ + N_3^-$	1.9×10^{-5}	4.72
Hydrocyanic	$HCN \rightleftharpoons H^+ + CN^-$	4.0×10^{-10}	9.40
Hydrofluoric	$HF \rightleftharpoons H^+ + F^-$	6.6×10^{-4}	3.18
Hydrogen peroxide	$H_2O_2 \rightleftharpoons H^+ + HO_2^-$	2.4×10^{-12}	11.62
Hydrosulfuric	$H_2S \rightleftharpoons H^+ + HS^-$	1.0×10^{-7} (K_1)	7
	$HS^- \rightleftharpoons H^+ + S^{2-}$	$\sim 10^{-19}$ (K_2)	~19
Hypobromous	$HBrO \rightleftharpoons H^+ + BrO^-$	2.5×10^{-9}	8.60
Hypochlorous	$HClO \rightleftharpoons H^+ + ClO^-$	3.0×10^{-8}	7.52
Nitric	$HNO_3 \rightleftharpoons H^+ + NO_3^-$	44	−1.64
Nitrous	$HNO_2 \rightleftharpoons H^+ + NO_2^-$	4.0×10^{-4}	3.40
Oxalic	$H_2C_2O_4 \rightleftharpoons H^+ + HC_2O_4^-$	5.9×10^{-2} (K_1)	1.23
	$HC_2O_4^- \rightleftharpoons H^+ + C_2O_4^{2-}$	6.4×10^{-5} (K_2)	4.19
Phenol	$C_6H_5OH \rightleftharpoons H^+ + C_6H_5O^-$	1.0×10^{-10}	10.00
Phosphoric	$H_3PO_4 \rightleftharpoons H^+ + H_2PO_4^-$	7.6×10^{-3} (K_1)	2.12
	$H_2PO_4^- \rightleftharpoons H^+ + HPO_4^{2-}$	6.2×10^{-8} (K_2)	7.21
	$HPO_4^{2-} \rightleftharpoons H^+ + PO_4^{3-}$	2.2×10^{-13} (K_3)	12.66
Sulfuric	$H_2SO_4 \rightleftharpoons H^+ + HSO_4^-$	$\sim 10^{3}$ (K_1)	~−3
	$HSO_4^- \rightleftharpoons H^+ + SO_4^{2-}$	1.2×10^{-2} (K_2)	1.92
Propanoic	$C_2H_5CO_2H \rightleftharpoons H^+ + C_2H_5CO_2^-$	1.3×10^{-5}	4.89
Thiophenol	$C_6H_5SH \rightleftharpoons H^+ + C_6H_5S^-$	3.0×10^{-7}	6.52
Trichloroacetic	$CCl_3CO_2H \rightleftharpoons H^+ + CCl_3CO_2^-$	2.2×10^{-1}	0.66
Trifluoroacetic	$CF_3CO_2H \rightleftharpoons H^+ + CF_3CO_2^-$	5.9×10^{-1}	0.23

Bases			
Base	**Formula and Ionization Equation**	K_b	pK_b
Ammonia	$NH_3 + H_2O \rightleftharpoons OH^- + NH_4^+$	1.9×10^{-5}	4.72
Aniline	$C_6H_5NH_2 + H_2O \rightleftharpoons OH^- + C_6H_5NH_3^+$	3.8×10^{-10}	9.42
Aziridine	$C_2H_4NH + H_2O \rightleftharpoons OH^- + C_2H_4NH_2^+$	1.1×10^{-6}	5.96
Benzamidine	$C_7H_8N_2 + H_2O \rightleftharpoons OH^- + C_7H_8N_2H^+$	5.4×10^{-3}	2.27
Dimethylamine	$(CH_3)_2NH + H_2O \rightleftharpoons OH^- + (CH_3)_2NH_2^+$	5.9×10^{-4}	3.23
Ethanolamine	$HO(CH_2)_2NH_2 + H_2O \rightleftharpoons OH^- + HO(CH_2)_2NH_3^+$	3.2×10^{-5}	4.50
Ethylamine	$CH_3CH_2NH_2 + H_2O \rightleftharpoons OH^- + CH_3CH_2NH_3^+$	4.3×10^{-4}	3.37
Ethylenediamine	$(CH_2)_2(NH_2)_2 + H_2O \rightleftharpoons OH^- + (CH_2)_2(NH_2)_2H^+$	8.5×10^{-5} (K_1)	6.07
	$(CH_2)_2(NH_2)_2H^+ + H_2O \rightleftharpoons OH^- + (CH_2)_2(NH_2)_2H_2^{2+}$	2.7×10^{-8} (K_2)	7.57
Hydrazine	$N_2H_4 + H_2O \rightleftharpoons OH^- + N_2H_5^+$	8.5×10^{-7} (K_1)	6.07
	$N_2H_5^+ + H_2O \rightleftharpoons OH^- + N_2H_6^{2+}$	$\sim 10^{-16}$ (K_2)	~16
Hydroxylamine	$HONH_2 + H_2O \rightleftharpoons OH^- + HONH_3^+$	9.1×10^{-9}	8.04
Methylamine	$CH_3NH_2 + H_2O \rightleftharpoons OH^- + CH_3NH_3^+$	4.2×10^{-4}	3.38
N-methylaniline	$C_6H_5NHCH_3 + H_2O \rightleftharpoons OH^- + C_6H_5NH_2CH_3^+$	6.9×10^{-10}	9.16
Nicotine	$C_{10}H_{14}N_2 + H_2O \rightleftharpoons OH^- + C_{10}H_{14}N_2H^+$	1.1×10^{-6}	5.96
Pyridine	$C_5H_5N + H_2O \rightleftharpoons OH^- + C_5H_5NH^+$	1.4×10^{-9}	8.85
Trimethylamine	$(CH_3)_3N + H_2O \rightleftharpoons OH^- + (CH_3)_3NH^+$	6×10^{-5}	4.2

Answers to Odd-Numbered Problems

You may find that your answer to a particular end-of-chapter problem differs from the one presented here by a few units in the right-most significant digit, even though you have worked a problem correctly. You should not worry, because such differences can arise from different choices of the number of digits used to represent the value of physical constants or conversion factors.

Wherever possible, we used at least two digits more than the most accurate of the quantities given in the problem. For example, if a calculation involving Avogadro's number and the molar mass of nitrogen had to be made on a quantity of 36.8 g, we used 6.0221×10^{23} and 14.0067 g mol^{-1} rather than 6.02×10^{23} and 14.0 g mol^{-1}.

Another source of small calculational differences is round-off procedure. If your habit (not a good habit, by the way) is to round off intermediate results in a calculation and later re-enter and use those rounded results, your final answers will often differ from those presented here.

Chapter 1

1. Matter is anything having mass and volume; substance is a particular type of matter.
3. Observation, hypothesis, testing, theory; useful in criminology, but not in love or religion.
5. Popular usage, having nothing to do with science.
7. **(a)** Fact, **(b)** hypothesis, **(c)** fact, **(d)** fact.
9. Solid has definite shape and volume, example rock; liquid has definite volume, indefinite shape, example water; gas has indefinite shape and volume, example air.
11. **(a)** Mixture, **(b)** mixture, **(c)** pure, **(d)** pure, **(e)** pure.
13. Hetero- and homogeneous mixtures.
15. **(a)** Homogeneous, **(b)** homogeneous, **(c)** heterogeneous, **(d)** heterogeneous, **(e)** heterogeneous.
17. Observable without formation of new substances; melting, boiling, expansion.
19. **(a)** Physical, **(b)** chemical, **(c)** physical, **(d)** chemical.
21. **(a)** Chemical, **(b)** physical, **(c)** physical, **(d)** chemical, **(e)** chemical.
23. **(a)** Element, **(b)** compound, **(c)** compound, **(d)** element, **(e)** element.
25. **(a)** Compound, **(b)** homogeneous, **(c)** heterogeneous, **(d)** heterogeneous, **(e)** compound.
27. **(a)** SI, derived; **(b)** other; **(c)** other; **(d)** SI, base; **(e)** SI, derived.
29. **(a)** 10^{-9}, **(b)** kilo, **(c)** 10^{6}, **(d)** 10^{-2}, **(e)** 10^{12}.
31. **(a)** 7.325×10^{3}, **(b)** 8.3×10^{-1}, **(c)** 1.8236×10^{4}, **(d)** 7,500,000, **(e)** 9.72×10^{-4}.
33. **(a)** 6.92×10^{2} m, **(b)** 3.151596×10^{-3} g, **(c)** 2.54 cm inch^{-1}, **(d)** 300,000,000 m s^{-1}, **(e)** 9.72×10^{-4} m^{3}.
35. **(a)** 83 g, **(b)** 150 Mm, **(c)** 15.626 kK, **(d)** 8.372 g, **(e)** 946 mL.
37. **(a)** 7.853 μm, **(b)** 285 mPa, **(c)** 1.83 m°C, **(d)** 400 kJ, **(e)** 47.2 pg.
39. **(a)** 48.2 m s^{-1}, **(b)** 3.28 mmol L^{-1}, **(c)** 50.3 μL s^{-1}, **(d)** 3.803 kJ m^{-3}.
41. **(a)** 1.5 m^{2}, **(b)** 9.03×10^{-1} g cm^{-3}.
43. **(a)** 3, **(b)** 2, **(c)** 2, **(d)** 4.
45. **(a)** 3, **(b)** 2 or 3, **(c)** 2, **(d)** 3.
47. **(a)** Exact, **(b)** 3, **(c)** ambiguous, probably 2, **(d)** 3, **(e)** exact.
49. 2-digit precision in one temperature scale cannot correspond to 4-digit precision in another; 24 °C is correct.
51. **(a)** 4.8×10^{3}, **(b)** 20. or 2.0×10^{1}; **(c)** 3.62×10^{4}.
53. **(a)** 178, **(b)** 855, **(c)** 762, **(d)** 742, **(e)** 64.9, **(f)** 11.1.
55. **(a)** 38.8, **(b)** −443, **(c)** 65.67.
57. 2
59. **(a)** 121428, **(b)** 22.
61. **(a)** 298 K, **(b)** 195 K, **(c)** 773 K, **(d)** 1507 K.
63. **(a)** 20 °C, **(b)** 4 °C, **(c)** 668 °C, **(d)** 429 °C.
65. **(a)** 50 °F, **(b)** −22 °F, **(c)** −321 °F, **(d)** 932 °F.
67. In °C: −269.0, −252.8, −195.8, −183.0; in °F: −452.1, −423.0, −320.4, −297.3.
69. ≈200 °C.
71. 0.21.
73. 8.88 g mL^{-1}.
75. No; by definition the Sp. Gr. of water = 1 (exact), at any temperature.
77. 8.62 g mL^{-1}.
79. 7.1 g mL^{-1}.
81. 112 mL.
83. **(a)** 8.93×10^{5} L, **(b)** 893 m^{3}, **(c)** 8.93×10^{-7} km^{3}.
85. 1.025 Mg.
87. 11.4 g mL^{-1}.

89. **(a)** Balsa, 17 L; fir, 4.7 L; oak, 3.2 L; ironwood, 2.10 L. **(b)** Since its density is greater than that of seawater, ironwood does not float—anchor, yes; raft, no.
91. No; the density of ethanol is 0.790 g mL^{-1}, less than that of salad oil.
93. 0.83 kg.

Chapter 2

1. Democritus; miniscule, indivisible particle of matter. Dalton: same, and one type atom (characteristic mass) for each element. Thomson: same, but composed of negatively charged particles imbedded in a positively charged sphere. Rutherford: same, but positive charge and mass concentrated in "nucleus."
3. A nuclear particle, that is, a proton or a neutron.
5. Variant types of atoms, having the same atomic number (number of protons) but different mass number, that is, different number of neutrons and therefore different mass.
7. Empirical formula gives relative numbers of atoms of different elements in a compound, molecular formula gives actual numbers of atoms in a molecule.
9. Location of bonds, that is, atom-to-atom connectivity; sometimes approximate bond angles are shown.
11. N_a = the number of atoms in a 12 g sample of pure C-12.
13. Mass spectrometry.
15. Relative amount of an isotope, expressed as a percentage of all isotopes of that element; measured in a mass spectrometer.
17. An element having only one nonradioactive isotope: F, Na, and Al are common.
19. Cation, anion.
21. **(a)** Sodium (12), **(b)** aluminum (14), **(c)** cobalt (32), **(d)** fluorine (10), **(e)** manganese (30), **(f)** arsenic (42).
23. Na-23, Al-27, Co-59, F-19, Mn-55, As-75.
25. **(a)** Oxygen, O; **(b)** magnesium, Mg; **(c)** potassium, K; **(d)** argon, Ar; **(e)** silicon, Si; **(f)** barium, Ba.
27. **(a)** $12p,12n,12e$; **(b)** $21p,24n,21e$; **(c)** $40p,51n,40e$; **(d)** $13p,14n,10e$; **(e)** $30p,35n,28e$; **(f)** $47p,61n,46e$.
29. **(a)** $7p,8n,7e$; **(b)** $76p,112n,76e$; **(c)** $74p,110n,74e$; **(d)** $17p,20n,17e$; **(e)** $16p,18n,18e$; **(f)** $37p,48n,36e$.
31. **(a)** Mn-55, **(b)** ${}^{40}_{20}Ca^{2+}$, **(c)** As-75, **(d)** ${}^{127}_{53}I^-$.
33. **(a)** 2, **(b)** 10, **(c)** 11, **(d)** 2.
35. **(a)** 6, **(b)** 5, **(c)** 6, **(d)** 18.
37. **(a)** 4, **(b)** 3, **(c)** 4, **(d)** 2, **(e)** 9, **(f)** 12.
39. **(a)** 8, **(b)** 5, **(c)** 4, **(d)** 10.
41. 24.31 g mol^{-1}.
43. 80 g mol^{-1}.
45. 50.7%, 49.3%.
47. **(a)** 0.500 mol, **(b)** 1.66×10^{-21} mol, **(c)** 26.4 mmol, **(d)** 125 mmol.
49. **(a)** 2.40×10^{21}, **(b)** 3.63×10^{24}, **(c)** 5.32×10^{4}, **(d)** 7.431×10^{23} atoms.
51. **(a)** 3.01×10^{18}, **(b)** 11 (not 11.4), **(c)** 9.94×10^{22}, **(d)** 1.5×10^{25} particles.
53. **(a)** 42.094, **(b)** 108.0104, **(c)** 175.33, **(d)** 98.999 g mol^{-1}.
55. **(a)** 44.010, **(b)** 92.0110, **(c)** 146.05 g mol^{-1}.
57. **(a)** 77.4 g, **(b)** 151 g, **(c)** 54.3 g, **(d)** 1.25 mg.
59. **(a)** 62 mmol, **(b)** 1.66 mol, **(c)** 6.5×10^{-5} mol, **(d)** 109 mmol.
61. He, 56.7 mol; Ne, 11.2 mol; Ar, 5.68 mol; Kr, 2.71 mol; Xe, 1.73 mol; Rn, 1.02 mol.
63. 18 g.
65. 0.231 mol O_2.
67. **(a)** 9.0×10^{23}, **(b)** 7.8×10^{22}, **(c)** 4.5×10^{23}, **(d)** 1.81×10^{24} H atoms.
69. **(a)** 1.8×10^{24}, **(b)** 4.4×10^{23}, **(c)** 1.2×10^{24}, **(d)** 6.3×10^{23} N atoms.
71. LiOH, 28.98%; NaOH, 57.4786%; KOH, 69.6870%.
73. CO.
75. $XeOF_6$.
77. P_2O_5.
79. NO_2, N_2O_4.
81. Formula is $ZnCl_2$; since Cl forms ions with charge (−1) $ZnCl_2$ is neutral only if the zinc cation has a charge of (+2).
83. C_6H_{12}.
85. $C_4H_{10}O$ is both the empirical and the molecular formula.
87. **(a)** RbF, **(b)** $(NH_4)_2SO_4$, **(c)** K_2S, **(d)** BaS, **(e)** $Mg(HCO_3)_2$, **(f)** $NaNO_2$.
89. **(a)** SO_3, **(b)** $HClO_2$, **(c)** CCl_4, **(d)** PCl_5, **(e)** NH_3, **(f)** HI.
91. **(a)** MnO_2 (empirical), could be either ionic or molecular compound; **(b)** SO_2 (empirical and molecular), molecular compound; **(c)** $FeCl_3$ (empirical), ionic or molecular compound; **(d)** NO_2 (empirical), N_2O_4 (molecular), molecular compound; **(e)** $Co_2(Cr_2O_7)_3$ (empirical), ionic compound; **(f)** HO (empirical), H_2O_2 (molecular), molecular compound.
93. **(a)** Ammonium hydrogen carbonate, **(b)** sodium oxalate, **(c)** magnesium nitride, **(d)** potassium dichromate, **(e)** calcium hydride, **(f)** calcium hydrogen sulfate.
95. **(a)** Chloric acid, **(b)** dichlorine heptoxide, **(c)** dinitrogen pentoxide, **(d)** carbon monoxide, **(e)** sulfur trioxide, **(f)** carbon tetraiodide.

97. (a) Iron(II) chloride, (b) hypochlorous acid, (c) germanium tetrachloride, (d) calcium hydride, (e) sodium oxide, (f) magnesium chloride.

99. Helium, He; neon, Ne; argon, Ar; krypton, Kr; xenon, Xe; radon, Rn.

101. (a) Ar, (b) Rn, (c) Xe.

Chapter 3

1. A list of formulas of reactants and products in a particular chemical reaction.

3. The same number of atoms of each element appears on both sides of a balanced equation.

5. Since the atoms are unchanged, their total mass is the same on both sides of a balanced equation—mass is conserved.

7. The ratio of the molar amounts of two reactants or products in a chemical equation.

9. Losses in purification or isolation of product; side reactions leading to other products; incomplete reaction.

11. The reactant that is consumed before the other(s) because it is present in the smallest relative amount.

13. Solvent is the solution component that, when pure, is in the same physical state as the solution; other components are solutes.

15. Concentration is any measure of the relative amount of solute in a solution, while molarity is moles solute per liter of solution.

17. (a) Balanced, (b) not balanced, (c) not balanced.

19. $HNO_3 + 3\ HCl \rightarrow NOCl + Cl_2 + 2\ H_2O$; $2\ H_2O + 2\ Cl_2 \rightarrow 4\ HCl + O_2$.

21. (a) Balanced, (b) not balanced, (c) not balanced.

23. $2\ Na + 2\ H_2O \rightarrow 2\ NaOH + H_2$; $2\ KNO_3 + S + 3\ C \rightarrow K_2S + N_2 + 3\ CO_2$.

25. (a) $Xe + 2\ F_2 \rightarrow XeF_4$, (b) $Mg + O_2 \rightarrow 2\ MgO$, (c) $N_2 + 3\ H_2 \rightarrow 2\ NH_3$, (d) $2\ Al + 6\ HCl \rightarrow 2\ AlCl_3 + 3\ H_2$, (e) $2\ Al + Cr_2O_3 \rightarrow Al_2O_3 + 2\ Cr$, (f) $Zn + 2\ HNO_3 \rightarrow Zn(NO_3)_2 + H_2$.

27. (a) $CS_2 + 3\ O_2 \rightarrow CO_2 + 2\ SO_2$, (b) $2\ BiCl_3 + 3\ H_2S \rightarrow Bi_2S_3 + 6\ HCl$, (c) $2\ RbOH + SO_2 \rightarrow Rb_2SO_3 + H_2O$, (d) $HBF_4 + 3\ H_2O \rightarrow H_3BO_3 + 4\ HF$, (e) $PCl_3 + Cl_2 \rightarrow PCl_5$, (f) $Na_2SO_4 + 2\ C \rightarrow Na_2S + 2\ CO_2$.

29. (a) $P_4O_{10} + 6\ H_2O \rightarrow 4\ H_3PO_4$, (b) $6\ P_2H_4 \rightarrow 8\ PH_3 + P_4$, (c) $3\ PbO + 2\ NH_3 \rightarrow 3\ Pb + N_2 + 3\ H_2O$, (d) $Mg_3N_2 + 6\ H_2O \rightarrow 3\ Mg(OH)_2 + 2\ NH_3$.

31. (a) $5\ AsF_3 + 3\ PCl_5 \rightarrow 3\ PF_5 + 5\ AsCl_3$, (b) $4\ NH_3 + 5\ O_2 \rightarrow 4\ NO + 6\ H_2O$, (c) $PbO_2 + Pb + 2\ H_2SO_4 \rightarrow 2\ PbSO_4 + 2\ H_2O$.

33. $2\ HMnO_4 + 3\ MnCl_2 + 2\ H_2O \rightarrow 5\ MnO_2 + 6\ HCl$.

35. (a) $Ca_3(PO_4)_2 + 3\ H_2SO_4 \rightarrow 3\ CaSO_4 + 2\ H_3PO_4$, (b) $Ca_3(PO_4)_2 + 2\ H_2O + 2\ CO_2 \rightarrow Ca(HCO_3)_2 + 2\ CaHPO_4$.

37. (a) $2\ Cu + H_2O + O_2 + CO_2 \rightarrow Cu_2(OH)_2CO_3$, (b) $KOH + NH_4Br \xrightarrow[\Delta]{} KBr + H_2O + NH_3$.

39. (a) $8\ KClO_3 + C_{12}H_{22}O_{11} \rightarrow 8\ KCl + 12\ CO_2 + 11\ H_2O$; (b) $PbO_2 + SO_2 \rightarrow PbSO_4$.

41. 0.0179 mol.

43. 0.0223 mol.

45. 0.163 g.

47. 13.4 kg.

49. 95.9 g.

51. 1.49 g.

53. 17.1 g.

55. 75 g.

57. 69.8 g.

59. (a) 116.7 g, (b) N_2O_5.

61. (a) 4.46 g, (b) 2.17 g O_2 remains unreacted.

63. (a) 106.8 g, (b) KNO_3.

65. 48%.

67. 82%.

69. 5.78%.

71. 96.4%.

73. 96 kg.

75. (a) 0.0027 mol, 0.58 g; (b) 10.0 g, 0.0340 mol; (c) 1.62 mol, 273 g.

77. (a) 5.77 L, (b) 10.2 L, (c) 518 g, (d) 1.42 kg.

79. (a) Add 12.5 g $KClO_4$ to 487.5 g H_2O; (b) to 50 mL alcohol, add water to bring the total volume to 1 L; (c) dissolve 0.0400 mol (2.24 g) KOH in $\approx$ 1 L water, then add more water until solution volume is 2 L.

81. (a) To 8.22 mL stock solution, add sufficient water to bring the solution volume to 250 mL; (b) to 31 mL stock solution, add sufficient water to bring the solution volume to 750 mL; (c) to 15 mL stock solution, add sufficient water to bring the solution volume to 300 mL.

83. 255 mL.

85. 0.0354 M.

87. 78.5 mL.

89. 0.152 M.

91. Lithium forms the oxide, $4\ Li + O_2 \rightarrow 2\ Li_2O$; sodium forms the peroxide, $2\ Na + O_2 \rightarrow Na_2O_2$; potassium, rubidium, and cesium react to form the superoxide, $M + O_2 \rightarrow MO_2$.

93. Cs is most reactive, and Li is least. The equation is 2 M + 2 $H_2O \rightarrow$ 2 MOH + H_2.
95. Melting point and boiling point decrease with increasing atomic number, while density generally increases.
97. Decomposition of trona, a mineral mixture of sodium hydrogen carbonate and sodium carbonate, and the Solvay process using NaCl and $CaCO_3$ as raw materials.
99. Ammonium ion, NH_4^+.
101. NaOH is produced by electrolysis of NaCl(aq); Cl_2 is a byproduct.
103. Sodium chloride and sodium nitrite.

Chapter 4

1. The amount of heat absorbed from or released to the environment during a chemical reaction.
3. The heat of a reaction is the same whether it occurs as a single step or as a sequence of separate steps.
5. The most stable state of a substance at 25 °C and 1 atm.
7. From the relationship $\Delta H° = \Sigma[n \cdot \Delta H_f^\circ \text{(products)}] - \Sigma[n \cdot \Delta H_f^\circ \text{(reactants)}]$.
9. Exothermic.
11. **(a)** Heat required to increase the temperature of an object by 1 °C: J °C^{-1}; **(b)** heat required to increase the temperature of one gram of a substance by 1 °C: J °C^{-1} g^{-1}; **(c)** heat required to increase the temperature of one mole of a substance by 1 °C in a process taking place with no change of pressure: J °C^{-1} mole^{-1}.
13. Fill the cup with a known mass of water at a known temperature, then add a weighed piece of metal at a known, higher temperature. From the temperature rise of the water, calculate the heat lost by the metal; divide the heat by the mass of the metal and its temperature change to get the specific heat.
15. 15.3 kJ.
17. 7.84 kJ.
19. 1.6×10^5 kJ.
21. $\Delta H = -484$ kJ.
23. +180.5 kJ.
25. $\Delta H_c = -395.4$ kJ mol^{-1}; 3.38 kJ.
27. 44.012 kJ mol^{-1}.
29. −192 kJ mol^{-1}.
31. −520 kJ mol^{-1}.
33. 2802.7 kJ mol^{-1}.
35. 4.19 kJ.
37. 0.3485 C_2H_2.
39. 25.6 kJ.
41. 12.2 °C.
43. 0.586 J °C^{-1} g^{-1}.
45. 0.947 J °C^{-1} g^{-1}.
47. 263 kJ.
49. −5 °C.
51. 37 °C.
53. 28.6 °C.
55. 4.74 kJ.
57. 1.12 g or 0.0622 mol.
59. 7.3 kcal.
61. 34 °C.
63. 6×10^6 kJ.
65. 1.1 kJ.
67. 55.8 J °C^{-1}, 0.128 J °C^{-1} g^{-1}.
69. Endothermic; 240 J g^{-1}, 18 kJ mol^{-1}.
71. 24.8 °C.
73. 4.79×10^3 J °C^{-1}.
75. 2.2 °C.
77. None. Group 2A elements are too reactive to exist in nature uncombined.
79. **(a)** Limestone, **(b)** seawater.
81. By electrolysis of the molten chlorides. $MCl_2(\ell) \rightarrow M(\ell) + Cl_2(g)$.
83. Strontium hydroxide.
85. "Quicklime" is calcium oxide, CaO; "slaked lime" is calcium hydroxide, $Ca(OH)_2$.
87. **(a)** $2\,M + O_2 \rightarrow 2\,MO$, **(b)** $3\,M + N_2 \rightarrow M_3N_2$.
89. $MO + H_2O \rightarrow M(OH)_2$, exothermic.
91. 991 g.
93. 1.16 ton.

Chapter 5

1. Solids have a definite shape and a definite volume; liquids have a definite volume, but their shape depends on the container; gases assume the shape and volume of the container.
3. Pressure is force per unit area. Pressure is how hard one substance pushes against another.
5. See Section 5.1.
7. "Constant" is a number whose value does not change as pressure and volume change.
9. The ideal gas law, $PV = nRT$, applies approximately to all gases. It is less accurate at high pressure and/or low temperature.
11. Absolute zero (0 K) is measured by extrapolating *PVT* data to low temperature. At absolute zero, the extrapolated volume of a gas is zero.

13. See Section 5.4. The postulates are reasonable, and certainly useful, but they are not all "true": gas molecules do *not* have zero volume, nor are the intermolecular forces truly zero.

15. $\langle v \rangle = (8RT/\pi \mathcal{M})^{1/2}$

17. The best evidence comes from molecular beam measurements of molecular velocity distributions.

19. See caption to Figure 5.19.

21. Real gases do not obey the ideal gas law exactly. Deviations are caused by nonzero molecular volume and nonzero intermolecular forces, both postulated as negligible by the KMT.

23. Boyle: PV = const; Charles: V/T = const; Avogadro: V/n = const; Dalton: total pressure is the sum of partial pressures in a gas mixture; Torricelli: pressure of the atmosphere; van der Waals: real gas behavior; Graham: effusion and diffusion.

25. 3800 torr, 5.07×10^5 Pa, 507 kPa, 5.07 bar.

27. 42 torr.

29. 1.013×10^5 Pa.

31. 2.21×10^3 psi.

33. 1.00 L.

35. 1.9×10^2 L.

37. By a factor of 2.

39. 99 Pa.

41. 0.277 atm.

43. Pressure increases by a factor of 3.

45. 667 mL.

47. 206 mL.

49. 465 K.

51. 590 K.

53. 0.714 L.

55. 149 K.

57. 2.35 L.

59. 3.35 atm.

61. −16 °C.

63. 27 L.

65. No change.

67. 2.5×10^3 torr.

69. 2.67 atm.

71. 1.81 atm.

73. 25.6 L.

75. 3.68 mol.

77. 224 K.

79. 841 K.

81. 2.57 g.

83. 6.52 g L^{-1}.

85. 0.556 atm.

87. 17.09 g mol^{-1}.

89. 70.9 g mol^{-1}.

91. 64.1 g mol^{-1}.

93. 5.32 g L^{-1}.

95. 454 m s^{-1}.

97. 500 m s^{-1}.

99. 21.9 g mol^{-1}.

101. 11,900 K.

103. 1.07 (N_2 is faster); 1 (energy depends on temperature only).

105. 182 g mol^{-1}.

107. 720 torr.

109. 2.22 atm.

111. P_t = 7.34 atm; P_{N_2} = 4.89 atm; P_{He} = 2.45 atm; X_{N_2} = 0.667; X_{He} = 0.333; %(w/w) He = 6.67.

113. X_{H_2} = 0.32, x_{O_2} = 0.68; P_{H_2} = 0.32 atm, P_{O_2} = 0.68 atm.

115. 253 mL.

117. 21.5 g S.

119. 300 L H_2; 200 L NH_3.

121. 6.60 mL SO_2.

123. C_7H_{14}.

125. 70 atm; 120 atm if ideal.

127. Aluminum.

129. Thallium.

131. High electrical conductivity, low density, and corrosion resistance.

133. The solvent in the electrolytic production of Al is cryolite (Na_3AlF_6); for Ga, In, and Tl, the solvent is water.

135. Al and Ga.

137. 92.56%.

139. 19.2 L.

141. **(a)** $Fe_2O_3 + 2\ Al \rightarrow 2\ Fe + Al_2O_3$, **(b)** −851.5 kJ, **(c)** 15.78 kJ g^{-1}, **(d)** 3.17 kg.

Chapter 6

1. Refraction is a change in the direction of travel of light as it passes through an interface from one transparent medium to another. Reflection also involves a change of direction at an interface between two substances, but the light does not pass through the interface, it remains in the same medium.

3. Blue light has lower wavelength, higher frequency, and higher energy per photon than green light.

5. A photon is absorbed by a metal surface, and an electron leaves the surface (with an amount of kinetic energy that depends on the wavelength of the absorbed light).

7. The spreading of light into a shadow area.
9. In *constructive* interference, two light rays of the same wavelength combine to produce one ray whose intensity is the sum of the original intensities. In *destructive* interference, the intensity of the combined beam is the *difference* between the intensities of the original beams.
11. True only in a vacuum. In other transparent media, the speed is less.
13. Spectrum: a display of frequencies (or wavelengths) at which a sample interacts with light, together with the intensity of interaction, or the set of all frequencies within a given range, as "the visible spectrum." Absorption/emission spectrum: a display of specific frequencies and intensities at which light is absorbed/emitted by a sample.
15. Most heated objects emit continuous spectra, but heated or electrically discharged gases emit discrete spectra.
17. The particle theory cannot explain diffraction or interference, and the wave theory cannot explain blackbody radiation or the photoelectric effect.
19. Light and microscopic particles as well exhibit properties of both waves and particles.
21. Surviving features: quantized energy, stationary (nonradiating) states, and interstate transitions accompanied by absorption or emission of light. Nonsurviving: planetary motion.
23. A moving particle whose momentum (mass times velocity) is such that wave characteristics are important.
25. A solution to the Schrödinger equation. The function itself has no physical significance, but its square gives negative charge density.
27. Integers that describe energy, size, shape, and orientation of orbitals.
29. Shell: set of all orbitals having the same value of the principal quantum number. Subshell: set of all orbitals having the same value of the principal and angular momentum quantum numbers.
31. Na^{10+}, K^{18+}, Fe^{25+}.
33. 4.3×10^{14} Hz to 7.5×10^{14} Hz.
35. 2.449×10^{18} Hz.
37. 0.500 m.
39. 75 cm.
41. 330 s, or $\approx 5\frac{1}{2}$ min.
43. 4.5×10^5 km.
45. **(a)** 4.41×10^{-19} J, **(b)** 6.14×10^{-26} J, **(c)** 1.64×10^{-18} J.
47. **(a)** 2.87×10^{-19} J, **(b)** 5.83×10^{-28} J, **(c)** 2×10^{-23} J.
49. 216 kJ mol^{-1}.
51. **(a)** ($\approx$0.1 nm) $\approx 10^3$ MJ mol^{-1}, **(b)** $\approx$220 kJ mol^{-1}.
53. 342 nm; ultraviolet.
55. 2.5×10^{23} photons s^{-1}, or 0.42 mol s^{-1}.
57. 12 pm, about 15% of the diameter of the H atom.
59. 3.3×10^4 m s^{-1}.
61. 7.9×10^{-33} m; hardly a problem with an object the size of a baseball.
63. 220 nm; no.
65. 141 kJ mol^{-1}, infrared.
67. 2.625 μm, infrared.
69. 30.39 nm.
71. 410 nm; borderline ultraviolet/visible.
73. 6.
75. 4, 3.
77. **(a)** n must be less than n, **(c)** $|m_\ell|$ cannot exceed ℓ, **(d)** n cannot be 0, and ℓ cannot be negative.
79. Five.
81. 3.
83. **(a)** See Figure 6.22(a), p. 251; **(b)** see Figure 6.23(a), p. 252.
85. **(a)** $n = 3$, $\ell = 2$, allowed, **(b)** $n = 3$, $\ell = 3$, forbidden, **(c)** $n = 2$, $\ell = 1$, allowed.
87. 3 in an f orbital, 1 in a p orbital.
89. 4.
91. See Section 6.5.

Chapter 7

1. Any atom or ion containing more than one electron, for example He or Li^+, is a many-electron species. Two one-electron species are H and He^+.
3. Fluorine orbitals have lower energies than the same orbitals in hydrogen.
5. No.
7. Zero. If two electrons are paired, their spins are opposite and the magnetic moments are equal and opposite.
9. The statement is false, because "configuration" refers to orbital occupancy, not just to electron pairing.
11. Ground.
13. A magnetic field **(a)** repels diamagnetic substances slightly, **(b)** attracts paramagnetic substances moderately, and **(c)** attracts ferromagnetic substances strongly.
15. Aufbau is a procedure for predicting the ground-state configuration of an atom or ion.
17. Properties and electron configurations.
19. Core electrons constitute a shield or screen that prevents the outer electrons from experiencing the full positive charge of the nucleus, causing them to be less strongly attracted.
21. Outer electrons of different ℓ penetrate the core to different extents, so that the extent to which they are shielded from the full nuclear charge differs also.

23. The energy required for the reaction $Sc^{2+} \longrightarrow Sc^{3+} + e^-$ is the third ionization energy of scandium.

25. Atoms are assumed to be hard spheres, in contact with one another in molecules. The radius of such a sphere is the covalent radius of an atom. Values of covalent radii are determined from internuclear distances, in turn measured by x-ray diffraction.

27. The anion has a larger radius because it has one or more additional electrons, while the cation is smaller because it has lost one or more electrons.

29. A set of species having the same number of electrons, for example Cl^-, Ar, K^+, Ca^{2+}.

31. A diagram shows orbital occupancy and spin within partly filled subshells, while the configuration shows only the total number of electrons in each subshell.

33. {1, 0, 0, +½}; {1, 0, 0, −½}; {2, 0, 0, +½}; {2, 0, 0, −½}; {2, 1, x, y}, where $x = -1$, 0, or +1, and $y = +½$ or −½.

35. Lowest: {1, 0, 0, (+½ or −½)}; highest: {4, 0, 0, (+½ or −½)}.

37. 10.

39. The core electrons have {1, 0, 0, +½} and {1, 0, 0, −½)}; the seven outer electrons have {2, 0, 0, (+½ and −½)}, {2, 1, (−1, 0, and +1), (two have $m_s = +½$, two have $m_s = -½$, and the fifth has $m_s = +½$ or −½)}.

41. **(b)** ℓ cannot have a negative value.

43. Since $\ell = 1$ and $n = 4$, this is a p-block element in the fourth period; it could be any of the elements Ga through Kr.

45. Five: all have $n = 3$ and $\ell = 2$; each has a different value of m_ℓ (−2, −1, 0, +1, and +2), and each has the same value of m_s, either +½ or −½.

47. Cl.

49. $7s^27p^6$: $Z = 118$.

51.

F: $1s^22s^22p^5$ F: [⇅] [⇅] [⇅][⇅][↑]
1s 2s 2p

Cl: $[Ne]3s^23p^5$ Cl: [Ne] [⇅] [⇅][⇅][↑]
3s 3p

I: $[Kr]5s^24d^{10}5p^5$ I: [Kr] [⇅] [⇅][⇅][⇅][⇅][⇅] [⇅][⇅][↑]
5s 4d 5p

53.

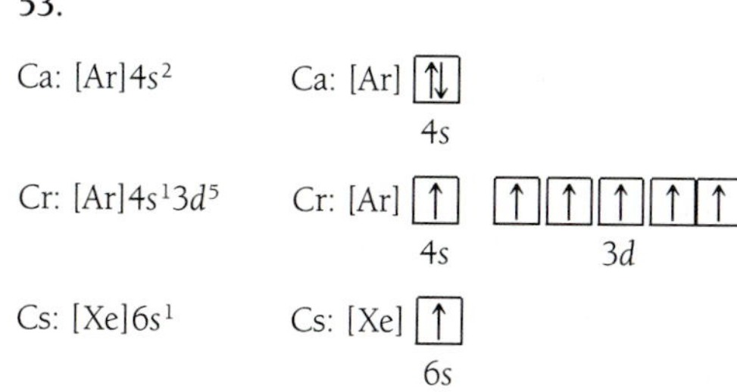

55. **(a)** Possible ground state, **(b)** possible excited state, **(c)** impossible, since three electrons cannot occupy an s orbital, **(d)** impossible, since the neon core already contains a filled $2p$ subshell.

57. **(c)** Impossible, since it shows two electrons, *unpaired*, in the same orbital, **(d)** possible excited state of carbon.

59. **(a)** Sodium, **(b)** mendelevium.

61. [He], 2; [Ar], 18; [Xe], 54.

63. a.

65. **(a)** Group 6A, fourth period, **(b)** Group 2A, third period.

67. **(a)** Transition; **(b)** main group, halogen; **(c)** main group, alkali; **(d)** main group.

69. Main-group elements, but not including Groups 1A and 2A. Any of the 25 elements in Groups 3A through 7A is an example.

71. Transition elements; third series.

73. ns^{1-2} and ns^2np^{1-6}.

75. The p-block elements are Groups 3A through 8A; the f-block elements are the inner transition series, that is, the lanthanides and the actinides.

77. Group 7A, the halogens.

79. N, P, Si.

81. **(a)** Na, **(b)** Cl.

83. **(a)** S^{2-}, **(b)** Se^{2-}.

85. **(a)** O, **(b)** Sr.

87. Se < Cl < Ne.

89. Ionization of Co^{2+} probably requires more energy than Cu^+. The IE of Cu is slightly greater than the IE of Co (due to trends in the period), but there generally is a larger difference between the IE's of successive ionizations.

91. No. Removal of an electron from a positive ion requires more energy than from a neutral atom.

93. 516 J.

95. The elements of highest atomic number.

97. HBr, 151 pm; HI, 170 pm.

99. 700 to 800 kJ mol^{-1}.

101. Al is largest, and C has the greatest ionization energy.

103. 80 mL mol^{-1}; 270 to 280 pm.

105. Mn, Fe^+, Co^{2+}, Ni^{3+}, Cu^{4+}.

107. In this case, the periodic trends work in opposite directions: Br is further to the right (larger IE) than S, but also further down (smaller IE). From Figure 7.22, Br has the larger IE.

109. Fluorine, gas; chlorine, gas; bromine, liquid; iodine, solid, astatine, presumed solid.

111. The ocean, or ancient dried seas.

113. HF < HCl < HBr < HI; all dissolve in water; all are gases at room temperature.

115. By bubbling chlorine through seawater; $Cl_2(g) + 2\ NaBr(aq) \rightarrow Br_2(g) + 2\ NaCl(aq)$.

117. HOF reacts with water.

119. A reaction in which a single species reacts with itself to give two different compounds is called a disproportionation.

Chapter 8

1. Valence: an adjective referring to chemical bonding.

3. In the solid or liquid state of an ionic compound, there are no neutral species—only ions. In the gas phase, ions form ion pairs or clusters that are electrically neutral and that might be regarded as molecules.

5. For molecular compounds, the formula unit is the molecule itself, while the empirical formula gives only the ratio of the numbers of atoms of different elements in the compound. For most ionic compounds, the two are the same, but there are some exceptions: the empirical formula of mercury(I) bromide is HgBr, while the formula unit is Hg_2Br_2.

7. Order: the number of electron pairs in the bond; length: the sum of the covalent radii, or the distance between the two nuclei of the atoms joined by the bond; energy: the energy required to separate the bonded atoms.

9. The greater the bond order, the greater the bond energy.

11. Formal charge is the electrical charge associated with an atom in a molecule, if all unshared electrons were localized on their atoms and all bonding electrons were equally shared. Yes, except in homonuclear diatomic molecules.

13. Resonance is the delocalization of a shared pair of electrons (or one electron, in odd-electron species) into more than one bond. Species for which more than one good Lewis structure can be written, for the same geometrical position of the atoms, are resonant.

15. Equal bond lengths of bonds that are of different order in the resonance contributors, and greater-than-expected stability.

17. In a general sense, delocalization is the spreading out of electron density over more than one small region of a molecule. In resonant species, an electron pair that is part of more than one bond is delocalized.

19. Bond polarity is an unequal sharing of electrons, resulting in a partial positive charge on one atom and a negative charge of the same magnitude on the other. It arises when two atoms of different electronegativity form a bond.

21. Electronegativity increases as atomic number increases within a period, and decreases as atomic number increases within a group.

23. In general, true. The measurable quantity is the molecular *dipole moment,* which is the vector sum of all bond moments in the molecule. In most molecules, individual bond moments cannot be measured. Exceptions are diatomic molecules, which have only one bond, and molecules like H_2O, which have a known geometry and only one *type* of bond.

25. Any bond between atoms of different electronegativity is polar. Since there are few elements having the same electronegativity, most bonds are polar.

27. Electronegativity measures the relative attraction of an atom for the shared pair of a covalent bond. It is not measured, but assigned on the basis of ionization energy and other physical properties.

29. 1.7 to 1.9.

31. A reaction in which bonds are broken and weaker (or fewer) bonds are formed requires energy input. That is, it is endothermic and ΔH is positive. Conversely, a negative ΔH is expected for a reaction that forms strong bonds and breaks weak (or fewer) bonds.

33. For molecules whose electron-pair geometry is trigonal bipyramidal.

35. Dipole moment, and bond moment in diatomic molecules.

37. **(a)** Li^+, **(b)** Sr^{2+}, **(c)** S^{2-}, **(d)** Al^{3+}.

39. **(a)** Mg:[Ne]$3s^2$, Mg^{2+}:[Ne], **(b)** Li:[He]$2s^1$, Li^+:[He], **(c)** I:[Kr]$5s^2 4d^{10} 5p^5$, I^-:[Xe]; **(d)** Al:[Ne]$3s^2 3p^1$, Al^{3+}:[Ne].]

41. **(a)** BaO, **(b)** Al_2O_3, **(c)** Cs_2S, **(d)** $Na_2Cr_2O_7$.

43.

```
          :F:                  :O:
           |                   //
(a) H—C—H   (b) :Cl—C
           |                    \
          :F:                  :Br:

                               H
                              /
(c) :C≡O:   (d) :Cl—N:
                              \
                               H
```

45.

```
(a) :F—Xe—F:   (b) :N=N=N: ⁻

              :O: ⁻
              //
(c) :O—Cl:        (and two others)
              \
              :O:

(d) H—N=N=N:
```

47.

```
                            :O:                 :O: ⁻
                            //                  /
(a) H—O—O: ⁻   (b) H—C        ⟷  H—C
                            \                   \\
                            :O: ⁻               :O:

       H    :O:                    :O:
       |   //                       ||
(c) H—C—C              (d) H—O—S—O—H
       |   \                        ||
       H    :O—H                   :O:
```

In answering the questions in this section, it is not necessary to include formal charges of zero in lists or diagrams.

49. **(a)** Formal charge is zero for all atoms. **(b)** C, −1; O, +1.

51. N, +1. The N atom.

53. (a) $:\overset{\oplus}{S}(=\ddot{O}:)(-:\ddot{O}:^{\ominus}) \longleftrightarrow :\overset{\oplus}{S}(-:\ddot{O}:^{\ominus})(=\ddot{O}:)$

(b) $H-\ddot{O}-C(=\ddot{O}:)(-:\ddot{O}:^{\ominus}) \longleftrightarrow H-\ddot{O}-C(-:\ddot{O}:^{\ominus})(=\ddot{O}:)$

(c) and (d) all atoms zero.

55. All single bonds except the C=O double bond in COClBr, and the C≡O triple bond in CO.

57. **(a)** All single bonds, **(b)** the NO bonds are of order 3/2, **(c)** all single bonds except for the C=O (terminal) double bond, **(d)** the H—O—S bonds are single, the S=O bonds are double.

59. The values in the table are averages, and the actual value varies from one molecule to another.

61. SO_2.

63. **(a)** −91 kJ, **(b)** −103 kJ.

65. **(a)** -408 kJ mol^{-1}, **(b)** $+181 \text{ kJ mol}^{-1}$.

67. **(a)** O, **(b)** K, **(c)** Mg, **(d)** O.

69. **(a)** H—F, **(b)** O—H, **(c)** S—Cl, **(d)** O—H.

71. **(a)** Si—O, **(b)** Cl—F, **(c)** C—B, **(d)** Li—Br.

73. **(a)** Octahedral, square pyramidal, **(b)** tetrahedral, same, **(c)** trigonal, bent, **(d)** trigonal, same.

75. **(a)** Trigonal bipyramidal, see-saw, **(b)** tetrahedral, tetrahedral, **(c)** linear, same, **(d)** trigonal bipyramidal, linear.

77. **(a)** 109°, **(b)** 120°.

79. **(a)** 90°, **(b)** 120°, **(c)** 180°.

81. **(a)** Nonpolar, **(b)** nonpolar, **(c)** polar, S end, **(d)** polar, O end.

83. b.

85. b and c.

87. The atmosphere.

88. Repeated compression, cooling, and expansion results in the condensation of air to a liquid. Pure liquid oxygen is separated from the other components of liquid air by distillation.

91. Oxide, O^{2-}; peroxide, O_2^{2-}; superoxide, O_2^-; oxygenyl, O_2^+.

93. See Table 2.3.

95. Oxides of the nonmetallic elements.

97. 10.3 g.

99. **(a)** $P_4O_{10}(s) + 6\ H_2O(\ell) \rightarrow 4\ H_3PO_4(aq)$,
(b) $3\ NO_2(g) + H_2O(\ell) \rightarrow 2\ HNO_3(aq) + NO(g)$.

Chapter 9

1. **(a)** A shared electron pair, **(b)** overlap of two singly occupied atomic orbitals, **(c)** a doubly occupied bonding orbital.

3. Farm silo, factory chimney, uncooked spaghetti, clarinet (except for mouthpiece, keys, and holes), lightbulb, rocket. . . .

5. The statement is misleading, because bonding is a continuous, rather than a stepwise, process.

7. Hydrogenlike orbitals have the general characteristics (relative size, shape, energy, orientation) of similarly named orbitals in the hydrogen atom. Parent orbitals are the set of hydrogenlike orbitals from which hybrids are formed.

9. The numbers are equal.

11. The statement is true, but misleading in that resonance, which implies delocalization, confers additional stability to molecules and ions.

13. *sp* as in O=C=O or sp^2 as in $H_2C{=}CH_2$; *sp* in triple bonds.

15. π orbitals have two lobes, with the greatest electron density off the axis, and a planar node containing the internuclear axis. π^* orbitals have four lobes and two planar nodes: one contains the internuclear axis, the other perpendicularly bisects the axis.

17. True only for bonding orbitals. σ^* orbitals have two lobes, and π^* orbitals have four.

19. In the second-period elements, the lowest available *d* orbital is the 3*d* orbital, which lies at a much higher energy than the 2*s* and 2*p* orbitals. Thus, the promotional energy cost is too high for participation of *d* orbitals in hybridization.

21. In the "sea of electrons" model, valence electrons are only weakly attracted to nuclei and so are freely mobile; in band theory, valence electrons are not mobile unless there is a vacant orbital of similar energy elsewhere in the material.

23. In the "sea of electrons" model, valence electrons are only weakly attracted to nuclei and so are freely mobile; since electrical current is the motion of charged particles, metals have high conductivity. Electrons are mobile in band theory also, but only because of the availability of low-energy vacant orbitals, in both the conduction band and the valence band, throughout the metal.

25. The orbitals in the valence band are bonding MO's, while those in the conduction band are antibonding. Conduction-band orbitals have higher energy than those in the valence band.

27. "Saturated" describes a hydrocarbon having the maximum possible number of single bonds.

29. Aromatic compounds are resonating species whose Lewis structures have alternating single and double C—C bonds. Aliphatic hydrocarbons may have one or more double bonds, but they are not resonant species.

31. Isomers are sets of different compounds that have the same molecular formula. For example, butane and 2-methylpropane.

33. **(a)** sp^3, **(b)** sp^3.

35. **(a)** sp^3d, **(b)** sp^3d^2.

37. **(a)** sp^3, **(b)** sp.

39. **(a)** sp^3d, **(b)** sp^3.

41. From left to right: sp^3, sp^2, sp^2, sp^3.

43. Both are sp^2.

45. From left to right: **(a)** sp^3, sp^2, **(b)** sp^2, sp^2, sp^3, sp^2, sp^2.

47. **(a)** sp, C—H bond is σ, C≡C bond is (σ + two π's), **(b)** sp^2, C—F bonds are σ, C=C bond is ($\sigma + \pi$).

49. **(a)** sp^2; since this is a resonant species, the closest valence bond description is that each S—O bond is a σ plus half of a π. **(b)** sp^2; σ plus one third of a π.

51. **(a)** sp, both bonds ($\sigma + \pi$), **(b)** sp^2, C—Cl bonds are σ, C=O bond is ($\sigma + \pi$).

53. The outer carbon atoms are sp^2, and the middle carbon is sp; C—H bonds are σ, and C=C bonds are ($\sigma + \pi$).

55. **(a)**, **(b)**, and **(c)**: all bonds ½ order, all paramagnetic.

57. **(a)** $F_2 < F_2^+$, **(b)** $C_2 < C_2^-$.

59. **(a)** $CO > CO^+$, **(b)** $NO < NO^+$.

61. All but O_2^{2-} are paramagnetic. Bond energy increases in the order $O_2^{2-} < O_2^- < O_2 < O_2^+$.

63. C_2.

65. C_2 is diamagnetic; C_2^{2-} has the greatest bond energy.

67. **(a)** 2-methylbutane, **(b)** 2,3-dimethylpentane, **(c)** 3,3-dimethylpentane.

69. All three compounds have the same number of bonds: 14 σ bonds (including the one that is part of the double bond) and one π bond.

71. **(a)**
```
     H   H   H   H   H   H
     |   |   |   |   |   |
  H—C—C—C—C—C—C—H
     |   |   |   |   |   |
     H   |   H   H   H   H
         |
       H—C—H
         |
         H
```

(b)
```
              H
              |
            H—C—H
              |
     H   H    |   H   H
     |   |    |   |   |
  H—C—C—C—C—C—H
     |   |    |   |   |
     H   |    H   H   H
         |
       H—C—H
         |
         H
```

(c)
```
              H
              |
            H—C—H
              |
     H   H    |   H   H   H   H   H   H
     |   |    |   |   |   |   |   |   |
  H—C—C—C—C—C—C—C—C—C—H
     |   |    |   |   |   |   |   |   |
     H   Cl   H   H   H   H   H   H   H
```

(d)
```
                 H               H
                 |               |
               H—C—H           H—C—H
                 |               |
     H   H   H   |   H   H       |   H   H   H
     |   |   |   |   |   |       |   |   |   |
  H—C—C—C—C—C—C—C—C—C—C—H
     |   |   |   |   |   |   |   |   |   |
     H   Br  H   Cl  |   H   H   H   H   H
                     |
                   H—C—H
                     |
                     H
```

73. **(a)**
```
     H   H   H   H   H
     |   |   |   |   |
  H—C—C—C=C—C—Cl
     |   |           |
     H   H           H
```

(b)
```
     H   H   H   H   O
     |   |   |   |  //
  H—C—C=C—C—C
     |           |   \
     H           H    H
```

(c)
```
     H   H   H   H   H
     |   |   |   |   |
  H—C=C—C—C—C—OH
             |   |   |
             H   H   H
```

(d)
```
     H   H   H   H   H   H
     |   |   |   |   |   |
  H—C—C—C—C—C—C—H
     |   |   |   |   |   |
     H   H   H   H   H   H
```

Chapter 10

1. Thermodynamics is concerned with heat, work, disorder, and spontaneity in chemical and physical processes, while thermochemistry is limited to consideration of the role of heat in chemical reactions.

3. *PV* work.

5. Heat is the sum of the kinetic energies of all the atoms in a substance, or alternatively, the form of energy that moves from one place to another by virtue of a difference in temperature. The common units of heat are joules and calories.

7. "State" in thermodynamics refers to any condition or arrangement of a system; unless all measurable properties of two systems are identical, the two are in different states. "States of matter" refers to the phase of a substance, that is, solid, liquid or gas.

9. Hess's law states that the heat (enthalpy) of reaction is the same whether the reaction is carried out all at once or as a series of steps. Provided that the initial and final states are the same, the change in enthalpy (or *any* state function) is independent of the path by which the process, which may be a chemical reaction, is carried out.

11. The change in internal energy of a system is equal to the sum of the work done on it and the heat added to it.

13. The negative sign means that the system has done work on the surroundings.

15. $\Delta E = q + w$.

17. The reaction releases heat.

19. Disorder or randomness.

21. The standard state of a substance is its most stable form at 25 °C and 1 atm.

23. The "formation equation" of an element in its standard state is, by definition, an equation in which the reactant is identical to the product. Since there is no change in the state of the system, the change in all properties is zero. If ΔG is zero for a formation equation of a substance, the standard free energy of formation of that substance must be zero as well.

25. $\Delta G < 0$ in spontaneous processes, and $\Delta G = 0$ in systems that are in equilibrium.

27. 16.5 J.

29. 133 m.

31. 1.8 kJ.

33. 0.14 °C.

35. +50 J.

37. 100 J.

39. 750 J.

41. -86.4 J K^{-1}.

43. 5.00 J K^{-1}.

45. +364.58 J K^{-1}.

47. -840.1 kJ.

49. -519.6 kJ mol^{-1}.

51. +50.7 kJ.

53. -413.8 kJ.

55. 2.6 kJ; 17.4 kJ.

57. **(a)** Positive—mixing of substances leads to a more disordered system; **(b)** negative—freezing a substance immobilizes its atoms, and increases the orderliness of their locations.

59. No; 1000 K.

61. Since $\Delta G = \Delta H - T\Delta S = 0$, it follows that $\Delta H = T\Delta S$ at equilibrium. That is, ΔS and ΔH must have the same sign. If the reaction becomes spontaneous, ΔG *decreases* as the temperature is changed. Since $\Delta G = \Delta H - T\Delta S$, a *positive* value of ΔS means that ΔG decreases when T *increases.* **(a)** $\Delta S > 0$ and $\Delta H > 0$, **(b)** $\Delta S < 0$ and $\Delta H < 0$.

Chapter 11

1. Fluidity is the ability to flow, viscosity is the resistance to flow.

3. Gas molecules are not in contact with one another, and can be moved closer with only a moderate increase in pressure. Liquid molecules are in contact, and their total volume can be decreased only by decreasing the size of the molecules themselves; to do so requires enormous pressure.

5. Surface tension results from an imbalance of cohesive forces in a liquid's surface layer, where molecules are attracted into the bulk liquid but not to the adjacent phase (air or vapor). The greater the cohesive forces, the greater the surface tension.

7. The best evidence is the fact that gases can be condensed by compression or cooling.

9. London force, dipole-dipole force, and the hydrogen bond are intermolecular forces. The ionic bond and the covalent bond are strong forces, but because they exist between *ions* or *atoms,* they are not inter*molecular* forces.

11. The dipole-dipole force. Dipole moments, which depend on electronegativity differences and molecular symmetry, can be very small or zero. The London force is also small, but can never be zero.

13. An induced dipole is one that is present only when a molecule is in an electric field, as when it is near another atom, molecule, or ion. An induced dipole arises because the electron cloud of a species is distorted so that it is no longer centered on the nucleus. Induced dipoles are associated with dispersion forces.

15. When a hydrogen atom, covalently bonded to an N, O, or F atom, is in the vicinity of another N, O, or F atom.

17. a.

19. b and d.

21. Ethanol is capable of hydrogen bonding, while dimethyl ether is not.

23. **(a)** CH_3Cl, **(b)** SiH_4, **(c)** CH_3OH, **(d)** $CH_3CH_2CH_2CH_3$.

25. Sublimation—transition from solid to vapor; evaporation—transition from liquid to vapor; condensation—transition from vapor to liquid.

27. No. Boiling point depends on pressure, and a liquid under a pressure less than 1 atm boils at a temperature lower than its normal boiling point.

29. Condensation is the process vapor $\rightarrow$ liquid, and deposition is the process vapor $\rightarrow$ solid.

31. 8.3 torr.

33. 64%.

35. 15 °C.

37. The heat of vaporization increases in proportion to the strength of a liquid's cohesive forces. Boiling point increases with the strength of cohesive forces.

39. No. The disorder of a system *always* increases when a liquid becomes a gas.

41. 0.0673 kJ.

43. 7.21 kJ.

45. 44.2 kJ mol^{-1}.

47. 463 J mL^{-1}.

49. 93.7 J mol^{-1} K^{-1}.

51. 53.1 J mol^{-1} K^{-1}.

53. The phase diagram of a given substance is a plot of vapor pressure vs. temperature, in which regions corresponding to solid, liquid, and vapor phases are separated by lines forming the set of (P, T) values at which two-phase equilibrium exists.

55. The triple point.

57. **(a)** liquid, **(b)** solid.

59. Initially a solid, water begins to sublime and reaches equilibrium with its vapor at a temperature slightly less than 0 °C; when the temperature rises further, only vapor remains.

61. a → b: solid melts, then the liquid heats up and boils, and finally the vapor temperature rises; b → c: vapor expands as pressure decreases; c → d: vapor cools, then deposits as solid, which cools; d → a no change, as solid is essentially incompressible.

63. The microscopic particles of an amorphous solid, for a number of possible reasons, are randomly arranged; the term "glass" is used for *covalently* bonded macromolecular solids that lack crystalline order. Glasses have a softening point and wide melting range rather than a specific melting point.

65. Crystalline substances have an orderly arrangement of microscopic particles, and form regularly shaped crystals characteristic of the substance. Amorphous substances have random arrangements of microscopic particles, and each substance forms solid particles with a large number of different shapes.

67. **(a)** ionic, **(b)** ionic, **(c)** ionic, **(d)** molecular.

69. **(a)** molecular, **(b)** ionic, **(c)** molecular, **(d)** ionic.

71. A crystal lattice is a mathematical abstraction, whereas a crystal structure is the actual arrangement of atoms, molecules, or ions in a crystal.

73. A unit cell is a brick-like solid figure that represents the repeating structural unit of a crystal. Eight lattice points define the corners of the cell, while others may lie on the faces or in the interior. No. A unit cell is an abstraction that does not correspond to any real boundaries within a crystal.

75. The Bragg equation, used to determine the interplane distances (d) in crystals from the x-ray diffraction pattern, is $2d \sin \theta = n\lambda$.

77. 7.1°.

79. 208 pm.

81. 361.5 pm.

83. 330.0 pm.

85. 143.3 pm.

87. 131.1 pm.

89. Face-centered.

91. A liquid crystal is a substance whose molecules are arranged in regular (crystal-like) order in at least one dimension, and in random (liquid-like) order in at least one dimension. The cohesive forces are similar in type to those of any liquid, but, owing to the rod-like shape of the molecules, they have different strengths in different directions.

93. According to the material in this chapter, the answer is no. However, freezing does not occur instantaneously, and with careful technique a liquid can be temporarily *supercooled* to a few degrees below its freezing point.

95. Supercritical fluids, liquid crystals, plasmas, and neutron stars.

97. Thermal and electrical conductivity both depend on the same factor, namely the mobility of electrons in a metal.

99. No. Trouton's rule applies to the vaporization of *liquids* only.

101. −78 °C.

103.

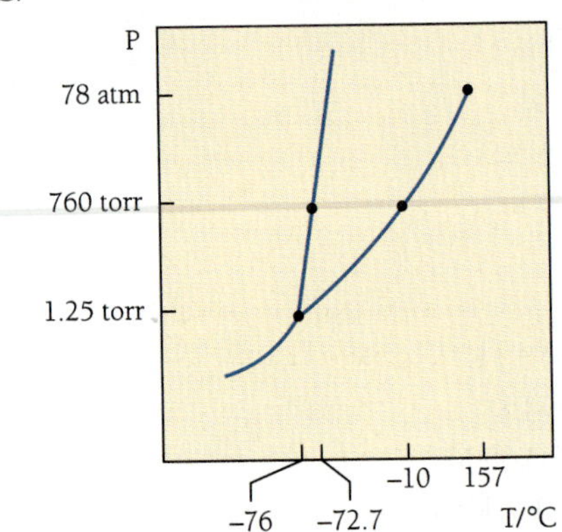

105. Conductivity depends on the motion of charged particles. Since ions are immobile in ionic solids, there can be no conductivity.

Chapter 12

1. A mixture can be heterogeneous, that is, have regions of differing composition, or can be homogeneous, with the same composition throughout; there is no difference between a solution and a homogeneous mixture. A colloid is a mixture in which one of the components exists as particles too small to be seen, yet much larger than single molecules.

3. Strength of the intermolecular forces in the pure components relative to their strength in the solution.

5. Yes. If all solid material is removed from a saturated solution, the condition of equilibrium no longer exists. However, the solution still is called saturated, because no additional solute can be dissolved.

7. Deposition means formation of a solid from a vapor; crystallization is formation of a crystalline solid from a liquid; precipitation is formation of a solid by chemical reaction of dissolved substances.

9. A supersaturated solution is one whose concentration is greater than the equilibrium value. Although it is not possi-

ble to dissolve more solute in a saturated solution, a supersaturated solution can be prepared by cooling a saturated solution.

11. One that obeys Raoult's law.
13. To gaseous solutes in liquid solvents.
15. A property of solutions that depends on the concentration, but not the identity, of the solute.
17. Less. The boiling points of solutions of volatile substances are usually intermediate between the boiling points of the pure components.
19. Since the effect of osmosis is a decrease of solute concentration, entropy is an important factor. A decrease in concentration means that the solute is dispersed in a larger volume, and is therefore in a condition of greater disorder. Enthalpy is less important, because in general very little heat accompanies the addition of more solvent to a solution.
21. An activity coefficient is the ratio of activity, or "effective concentration," to actual concentration.
23. If the particle size of one of the components is 10 to 1000 nm, a mixture is a colloid.
25. Ethanol is a polar liquid capable of hydrogen bonding, and therefore expected to be soluble in water.
27. **(a)** Soluble, **(b)** insoluble, **(c)** soluble, **(d)** soluble.
29. **(a)** Soluble, **(b)** insoluble, **(c)** insoluble, **(d)** soluble.
31. **(a)** No precipitate, **(b)** $BaSO_4$ precipitates, **(c)** no precipitate.
33. **(a)** $Pb(CN)_2$ precipitates, **(b)** MgS precipitates, **(c)** no precipitate.
35. $X(O_2) = 0.28$, 28 mole%; $X(N_2) = 0.72$, 72 mole%.
37. 0.0101.
39. 0.392 mole% NaI.
41. 0.171 mol kg^{-1}.
43. 0.271 mol kg^{-1}.
45. 2.57 g.
47. 387 torr.
49. 208 torr.
51. 98.2 torr.
53. Propyl acetate, 73.5 torr; ethyl acetate, 92.0 torr; vapor is 55.6 mole% ethyl acetate.
55. 0.20 M.
57. 0.31 M.
59. 0.00128 mol L^{-1} atm^{-1}.
61. −0.86 °C.
63. 0.28 m.
65. **(a)** −0.47 °C, **(b)** −2.8 °C.
67. **(a)** 5.2 °C, **(b)** 79.8 °C.
69. −0.85 °C.
71. 16.3% NaCl.
73. 170 g mol^{-1}.
75. 101.00 °C.
77. 180 g mol^{-1}.
79. 0.343 M; 7.68 atm.
81. 3.68 g; 13.6 mm H_2O.
83. 4.1×10^7 g mol^{-1}.
85. 31.1%; 8.46 m.
87. 9.6 m; 37% (w/w).
89. 2.
91. 2.1×10^2 g mol^{-1}.
93. 36,000 g; the density of this solution would be 36 g mL^{-1}, an unlikely value considering that no known substance has a density greater than 22 g mL^{-1}.

Chapter 13

1. Reversibility, equal rates of opposing processes, and the fact that the direction of approach does not affect the position of equilibrium.
3. "Dynamic equilibrium" means that two opposing processes occur at equal rates, resulting in no net change.
5. This is another way of saying that the condition of equilibrium is the same, no matter how equilibrium is reached.
7. When two immiscible liquids are in contact, and a substance soluble in both is added, that substance dissolves partly in one solvent and partly in the other. The ratio of concentrations in the two liquids is the "solute distribution equilibrium."
9. If the equilibrium lies far to the right, $K_{eq} \gg 1$; if far to the left, $K_{eq} \ll 1$.
11. **(a)** P_{O_2}, **(b)** $P_{H_2}P_{CO}/P_{H_2O}$, **(c)** $[HCO_2^-][H_3O^+]/[HCO_2H]$.
13. **(a)** $[Cl^-]^2[Br_2]/[Br^-]^2P_{Cl_2}$, **(b)** P_{H_2O}/P_{H_2}, **(c)** P_{H_2S}/P_{H_2}.
15. All are heterogeneous except 11c.
17. 1.63×10^8 atm^{-1}.
19. **(a)** 0.29 atm, **(b)** 1.8 $atm^{-1/2}$.
21. **(a)** To the left, **(b)** to the left, **(c)** to the right.
23. **(a)** To the left, **(b)** to the right, **(c)** to the left.
25. $K_{eq} = 5.20 \times 10^{+24}$; the huge size of this quantity means that the reaction goes to completion.
27. $\Delta G^\circ_{rxn} = -15.9$ kJ.
29. $\Delta G^\circ_{rxn} = -102$ kJ; $\Delta G^\circ_f(CH_3Cl) = -57.4$ kJ mol^{-1}.
31. 5.06×10^{17}.
33. The equilibrium will shift to the **(a)** left, **(b)** left, **(c)** right, **(d)** no effect.

35. **(a)** i & ii, shift to right; **(b)** i, shift to left; ii, shift to right.
37. **(a)** shift to left, **(b)** no effect.
39. 7.58×10^2 atm.
41. 331 K.
43. +97 kJ.
45. 0.13 atm.
47. **(a)** The extent of reaction is 0.25 atm; **(b)** $P_{CO_2} = P_{H_2} = 0.50$ atm, $P_{CO} = 0.25$ atm.
49. $P_{O_2} = 0.225$ atm, $P_{SO_3} = 0.250$ atm, $P_{total} = 0.975$ atm.
51. 0.771.
53. 8.37×10^{-3}.
55. 0.25 atm.
57. $P_{CO} = P_{Cl_2} = 1.24$ atm; $P_{COCl_2} = 0.36$ atm.
59. $P_{HI} = 0.401$ atm; $P_{H_2} = P_{I_2} = 0.497$ atm.
61. $P_{H_2} = P_{CO_2} = 0.0576$ atm, $P_{CO} = P_{H_2O} = 0.042$ atm.
63. $P_{H_2} = P_{CO_2} = 0.115$ atm, $P_{CO} = P_{H_2O} = 0.085$ atm.
65. 0.238 atm.
67. $P_{HBr} = 0.935$ atm; $P_{total} = 0.990$ atm.
69. 0.460 atm.
71. $P_{N_2} = P_{O_2} = 0.0093$ atm; $P_{NO} = 0.0015$ atm.
73. $[N_2] = [O_2] = 0.0093$ mol L^{-1}; [NO] = 0.0015 mol L^{-1}].
75. 0.0535 mol L^{-1}.
77. 1.
79. **(a)** shift to left, **(b)** no effect.
81. $P_{CO_2} = 0.314$ atm; $P_{CO} = 0.686$ atm.

Chapter 14

1. Reacting molecules must **(a)** collide, in a collision with **(b)** sufficient energy, and **(c)** proper orientation.
3. Chemical nature of reactants; concentration; pressure; temperature; presence of a catalyst; and, for heterogeneous reactions, surface type and area.
5. The order of a reaction is the sum of the exponents of the concentrations in the rate law. Yes, it is possible.
7. Rate is measured in mol L^{-1} s^{-1}, regardless of reaction order. The units of the rate constant depend on order, and are s^{-1} and L mol^{-1} s^{-1} for first- and second-order reactions, respectively.
9. $t_{1/2} = 0.693/k$.
11. The reaction coordinate describes the geometrical configuration of a collision complex as it changes from reactants to products. "Extent of reaction" describes how much material has reacted, and is appropriate to macroscopic quantities rather than to a molecular event.
13. A region on the surface of an enzyme molecule, the *active site,* has a specific and precise shape that ensures the proper alignment of reacting species.
15. A reaction mechanism is a description, by means of elementary steps, of the chemical changes undergone by the reactant(s) during a chemical reaction.
17. Intermediates are reactive substances produced in an early step, and consumed in a later step, of a reaction.
19. **(a)** Second order, first order with respect to each reactant; **(b)** first order; **(c)** first order; **(d)** second order.
21. **(b)** O_3: $\Delta[O_3]/\Delta t = -2\,k[O_3]$
    O_2: $\Delta[O_2]/\Delta t = +3\,k[O_3]$
    (c) $ClCO_2C_2H_5$: $\Delta[ClCO_2C_2H_5]/\Delta t = -k[ClCO_2C_2H_5]$
    CO_2, ClC_2H_5: $\Delta[CO_2]/\Delta t = \Delta[ClC_2H_5]/\Delta t = +k[ClCO_2C_2H_5]$
    (d) HI: $\Delta[HI]/\Delta t = -k[HI]^2$
    H_2, I_2: $\Delta[H_2]/\Delta t = \Delta[I_2]/\Delta t = -\frac{1}{2}k[HI]^2$.
23. No effect when $y = 0$; $\times 2$, $\times 4$, and $\times 8$ when $y = 1, 2,$ and 3.
25. **(a)** $\times 2$, **(b)** $\times 8$.
27. 4.1×10^{-10} mol L^{-1} s^{-1}.
29. **(a)** Second order, **(b)** first order, **(c)** third order, **(d)** $r = k[H_2][NO]^2$.
31. **(a)** First order with respect to A, second order with respect to B, and zero order with respect to C; third order overall; **(b)** 6 L^2 mol^{-2} s^{-1}.
33. **(a)** First order with respect to both A and B, second order overall; **(b)** $r = k[A][B]$, $k = 0.012$ L mol^{-1} s^{-1}.
35. 0.0078 s^{-1}.
37. 5.3 s.
39. 0.048 M.
41. 0.14 M.
43. 0.0232 M.
45. 1.31×10^{-3} s^{-1}.
47. **(a)** 1620 years, **(b)** 0.651 mmol.
49. 1.0×10^5 years.
51.

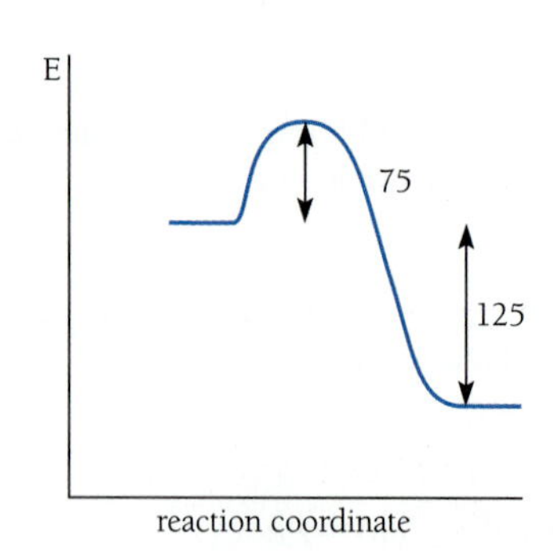

53. 350 kJ mol^{-1}.

55. Neither. $\Delta E_{rxn} = 0$: the reaction is *thermoneutral.*

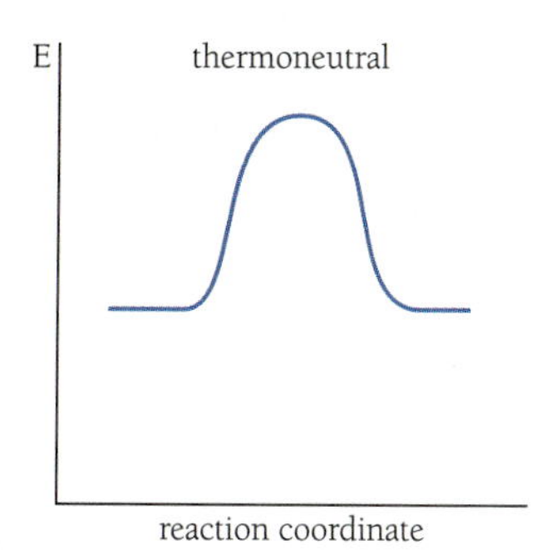

57. 123 kJ mol^{-1}.

59. **(a)** 99 kJ mol^{-1}, **(b)** $1.1 \times 10^{14}\ s^{-1}$.

61. $3.4 \times 10^{-6}\ s^{-1}$.

63. 658 K.

65. $E_a = 53{,}000\ J\ mol^{-1} = 53\ kJ\ mol^{-1}$.

67. **(a)** $r = k[H][O_2]$, **(b)** $r = k[O][O_3]$, **(c)** $r = k[Cl]^2[Cl_2]$.

69. $2\ N_2O_5 \rightarrow 4\ NO_2 + O_2$.

71. **(a)** $r = k[NO_2]^2$, **(b)** $r = k[NO_2][CO]$.

73. The single-step mechanism $2\ NO + O_2 \rightarrow 2\ NO_2$ fits the rate law.

75. 0.041 M.

77. **(a)** 11 s, **(b)** 3.2 s.

79. Exothermic.

81. Not possible. The reverse activation energy cannot be less than the exothermicity of a reaction.

83. $2.6 \times 10^{15}\ s^{-1}$.

85. 324 K.

87. By a factor of 1.35×10^5; that is, $k_{373}/k_{273} = 1.35 \times 10^5$.

89. 32 kJ mol^{-1}; no.

91.

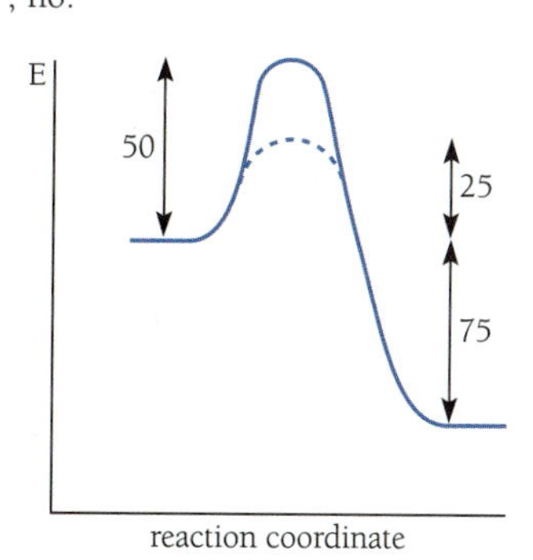

93. $r = k[CH_3CN(aq)[H^+]$.

95. **(a)** $CH_3CH_2OH + H_3O^+ \rightleftharpoons CH_3CH_2OH_2^+ + H_2O$ (fast)
$CH_3CH_2OH_2^+ \rightarrow CH_2{=}CH_2 + H_3O^+$ (slow);
(b) $k = k_2K_{eq}$, where k_2 is for the second, rate-determining step, and K_{eq} is for the pre-equilibrium step.

Chapter 15

1. An Arrhenius acid releases a proton when dissolved in water, and an Arrhenius base releases a hydroxide ion. Examples of acids are HCl, H_2SO_4, and HCN, while examples of bases are KOH, NaOH, and $Ca(OH)_2$.

3. H_3O^+. A hydronium ion is held together by covalent bonds, while a hydrated sodium ion is held together by the weaker ion-dipole force.

5. A Brønsted acid, such as HCN or CH_3CO_2H, is a proton donor; a Brønsted base, such as NH_3 or OH^-, is a proton acceptor.

7. No. NH_3 is a Brønsted-Lowry base, but not an Arrhenius base. Also, KOH is an Arrhenius base, but the corresponding Brønsted base is OH^-, not KOH.

9. $NH_3 + H_2O \rightleftharpoons NH_4^+ + OH^-$.

11. A conjugate pair, such as HSO_4^-/SO_4^{2-}, consists of a Brønsted acid together with its conjugate base, the species remaining after the acid loses a proton.

13. No acid stronger than H_3O^+ can exist in aqueous solution. All strong acids are completely ionized in aqueous solution, so, regardless of *which* strong acid is dissolved, H_3O^+ is the only acidic species in solution.

15. Autoionization is a reaction in which an ion (usually H^+) is transferred between two identical molecules. $H_2O + H_2O \rightarrow H_3O^+ + OH^-$.

17. **(a)** 2.60 M, **(b)** 1.02 M, **(c)** 1.06×10^{-3} M.

19. **(a)** 0.909 g, **(b)** 8.4 g, **(c)** 0.901 g.

21. 58.3 mL.

23. **(a)** 0.301, **(b)** 1.52, **(c)** 2.13.

25. 8.9×10^{-4}, 2×10^{-6}, 1, 10, and 3.7×10^{-3} M.

27. 9.81, 7.00, 5.75, 1.8.

29. **(a)** 14.5, **(b)** 10.41, **(c)** 13.02.

31. **(a)** 0.67%, **(b)** 1.5%, **(c)** 0.067%.

33. $[H_3O^+] = 1.5 \times 10^{-4}$ M, pH = 3.82.

35. **(a)** 5.4×10^{-4} M, 10.73; **(b)** 2.5×10^{-5} M, 9.40.

37. The strong acids on this list are HCl, HBr, H_2SO_4, and HNO_3.

39. H_2O and HCO_3^- are weak, the others are strong.

41. The acids are: **(a)** H_2O and HCN, **(b)** H_2CO_3 and H_2SO_4, **(c)** $H_2C_2O_4$ and HNO_2; all others are bases.

43. **(a)** H_2O/OH^-, HCN/CN^-, **(b)** H_2SO_4/HSO_4^-, H_2CO_3/HCO_3^-; **(c)** $H_2C_2O_4/HC_2O_4^-$, HNO_2/NO_2^-.

45. **(a)** HSO_4^-, **(b)** HF, **(c)** CH_3O^-, **(d)** H_2O.

47. (a) $H_3SO_4^+$, (b) $CH_3OH_2^+$, (c) HI.

49. (a) $HNO_2 + H_2O \rightleftharpoons H_3O^+ + NO_2^-$, (b) $HBr + H_2O \rightarrow H_3O^+ + Br^-$, (c) $CH_3CO_2H + H_2O \rightleftharpoons H_3O^+ + CH_3CO_2^-$.

51. (a) $C_{10}H_{14}N_2 + H_2O \rightleftharpoons OH^- + C_{10}H_{15}N_2^+$, (b) $N_2H_4 + H_2O \rightleftharpoons OH^- + N_2H_5^+$, (c) $(CH_3)_3N + H_2O \rightleftharpoons OH^- + (CH_3)_3NH^+$.

53. (a) ≈ 7, (b) >7, (c) <7, (d) >7, (e) >7.

55. (a) No reaction, (b) $CO_3^{2-} + H_2O \rightleftharpoons OH^- + HCO_3^-$, (c) $NH_4^+ + H_2O \rightleftharpoons H_3O^+ + NH_3$, (d) $S^{2-} + H_2O \rightarrow OH^- + HS^-$, (e) $NH_2^- + H_2O \rightarrow OH^- + NH_3$.

57. (a) $HSO_4^- + H_2O \rightleftharpoons OH^- + H_2SO_4$ and $HSO_4^- + H_2O \rightleftharpoons H_3O^+ + SO_4^{2-}$, (b) $HCO_3^- + H_2O \rightleftharpoons OH^- + H_2CO_3$ and $HCO_3^- + H_2O \rightleftharpoons H_3O^+ + CO_3^{2-}$.

59. (a) HCl, (b) HF.

61. (a) OH^-, (b) NO_2^-.

63. (a) $NO_2^- + HF \rightleftharpoons F^- + HNO_2$, to the right; (b) $HCO_3^- + HSO_4^- \rightleftharpoons SO_4^{2-} + H_2CO_3$, to the right.

65. $H_2C_2O_4 + H_2O \rightleftharpoons H_3O^+ + HC_2O_4^-$; $HC_2O_4^- + H_2O \rightleftharpoons H_3O^+ + C_2O_4^{2-}$.

67. $H_2PO_4^-$ is stronger than HPO_4^{2-}, because it is more difficult to remove a proton from the doubly charged HPO_4^{2-} than from the singly charged $H_2PO_4^-$.

69. $S^{2-} + H_2O \rightarrow HS^- + OH^-$, and $HS^- + H_2O \rightleftharpoons H_2S + OH^-$.

71. $NH_3 < PH_3 < AsH_3$; acidity of the binary hydrides increases with the size of the central atom in any group, because the greater the bond length, the more easily it is broken and the stronger the acid.

73. (a) $F_2CHCOOH$ is stronger because two F atoms have a greater inductive effect (weakening of the O—H bond) than one F atom; (b) H_2Te is stronger because, within a group, increased size of the central atom exerts a stronger influence than electronegativity on the acid strength of binary hydrides; (c) H_3PO_4 is stronger because a proton is more readily lost from a neutral species than from a negatively charged species.

75. $HOBr < HOCl < HClO_2$; HOBr is weaker than HOCl, because Br is less electronegative than Cl; HOCl is weaker than $HClO_2$, because it has fewer terminal oxygen atoms.

77. The conjugate acid H_2O is *stronger* than NH_3; because of the reciprocal relationship between the relative strengths of a conjugate pair, it must be true that the conjugate base OH^- is *weaker* than NH_2^-.

79. (a) acid: $AlCl_3$; base: Br^-; adduct: $AlCl_3Br^-$. (b) acid: $Be(OH)_2(s)$; base: $OH^-(aq)$; adduct: $Be(OH)_4^{2-}(aq)$.

81. 51.0 mL.

83. 0.1397 M.

85. 18.8 mL.

87. 93.8 mL.

89. 4.70 L.

91. 75.1 mL.

93. 0.0433 M.

95. 256 g mol^{-1}.

97. 104 g mol^{-1}.

99. (a) $NH_3 + NH_3 \rightleftharpoons NH_4^+ + NH_2^-$, (b) $NH_2OH + NH_2OH \rightleftharpoons NH_3OH^+ + NH_2O^-$, (c) $H_2SO_4 + H_2SO_4 \rightleftharpoons H_3SO_4^+ + HSO_4^-$.

101. The weak acid has the greater pH, and the strong acid reacts more vigorously.

103. $HCN < HBrO < H_2S < HNO_2 < HSO_4^-$.

105. 0.01580 M.

107. 62 mL.

109. 73.0 g $equiv^{-1}$; 146 g mol^{-1}.

Chapter 16

1. "Equilibrium expression" refers specifically to concentrations at equilibrium, while "reaction quotient" refers to both nonequilibrium and equilibrium concentrations.

3. Inverse. If the pK_a is large, the acid strength is small.

5. There is no real difference: "ionization constant" is a specialized term for the equilibrium constant of an ionization reaction.

7. $K_a = [H_3O^+][A^-]/(C - [A^-]) = x^2/(C - x)$, (since $[H_3O^+] = [A^-]$).
Rearrange to $x^2 + K_a x - K_a C = 0$, for which the positive root is

$$x = [-K_a + \sqrt{(K_a^2 + 4K_a C)}]/2$$

For acids, pH = $-\log x$; for bases, pH = $14 + \log x$.
% ionization = $100 \cdot [-K + \sqrt{(K^2 + 4KC)}]/2C$.

9. Because the acid is too weak to transfer protons to water, hence too weak to affect pH, no matter what the concentration.

11. "Hydrolysis" refers to the proton transfer reaction between an ion and water. $CN^- + H_2O \rightleftharpoons OH^- + HCN$ describes the hydrolysis of any cyanide salt, and $NH_4^+ + H_2O \rightleftharpoons NH_3 + H_3O^+$ describes the hydrolysis of any ammonium salt.

13. Amphoteric means capable of reacting with both acids and bases; amphiprotic means able to donate or accept a proton.

15. It is the amount of H_3O^+ or OH^- that can be added before a significant change occurs in the pH of a buffer.

17. Titration is an analytical procedure for determining the amount of an acid (or base) in a sample, by measuring the amount of strong base (or acid) required to neutralize the sample.

19. In the undertitrated region the sample is a buffer solution, that is, a mixture of the acid and its sodium salt; at equivalence, the sample contains only the sodium salt; when over-

titrated, the sample is a mixture of the salt and sodium hydroxide.

21. An indicator is a weak acid or base that, because it changes color over a certain pH range, is used to signal the endpoint of a titration. Good indicators are strongly colored in either protonated or deprotonated form, or both.

23. The range of pH value, typically 1.5 to 2.0 units, over which the indicator's color change is complete.

25. To the overall, first, and second ionization constants, respectively. An example is oxalic acid.

$H_2C_2O_4 + 2\ H_2O \rightleftharpoons C_2O_4^{2-} + 2\ H_3O^+$;

$$K = \frac{[H_3O^+]^2\,[C_2O_4^{2-}]}{[H_2C_2O_4]} = 3.8 \times 10^{-6}$$

$H_2C_2O_4 + H_2O \rightleftharpoons HC_2O_4^- + H_3O^+$;

$$K_1 = \frac{[H_3O^+][HC_2O_4^-]}{[H_2C_2O_4]} = 5.9 \times 10^{-2}$$

$HC_2O_4^- + H_2O \rightleftharpoons C_2O_4^{2-} + H_3O^+$;

$$K_2 = \frac{[H_3O^+][C_2O_4^{2-}]}{[HC_2O_4^-]} = 6.4 \times 10^{-5}$$

27. **(a)** 2.66, **(b)** 0.70, **(c)** 4.821, **(d)** 4.0.

29. **(a)** 0.0014, **(b)** 3.0×10^{-7}, **(c)** 0.56, **(d)** 4×10^{-4}.

31. 3.97.

33. 0.0051 M, 11.7.

35. 5.26, 0.0073%.

37. 4.2%, 2.37.

39. 0.0066 M, $[A^-] = 1.0 \times 10^{-3}$ M, $[HA] = 0.0056$ M.

41. 2.8×10^{-7}.

43. 2.65.

45. 1.6×10^{-5}.

47. 1.79×10^{-4}.

49. 6.9×10^{-10}.

51. $NO_2^- + H_2O \rightleftharpoons OH^- + HNO_2$; 2.5×10^{-11}.

53. $K_b = 3.3 \times 10^{-7}$; $OCl^- + H_2O \rightleftharpoons OH^- + HOCl$.

55. Acidic: a and d; neutral: c; basic: b.

57. 3.22.

59. 6.03, 9.3×10^{-7} M.

61. $K_a = 1.1 \times 10^{-6}$, $K_b = 9.1 \times 10^{-9}$.

63. 6.0×10^{-11} M.

65. 2×10^{-11} M.

67. 6.77.

69. 4.60.

71. [Kebz]/[Hebz] = 1.6.

73. 0.14 g acid/1 g salt.

75. Look at Table 16.1 and choose an acid whose pK_a is close to 5. Acetic acid ($pK_a = 4.74$) is a good choice. For an acetic acid/acetate buffer, pH = 5.00 when [acid]/[salt] = 0.55.

77. 0.24 mol.

79. **(a)** 8.85, **(b)** 8.72, **(c)** 8.35.

81. **(a)** 1.18, **(b)** 1.48, **(c)** 3.0, **(d)** 7.00, **(e)** 12.36.

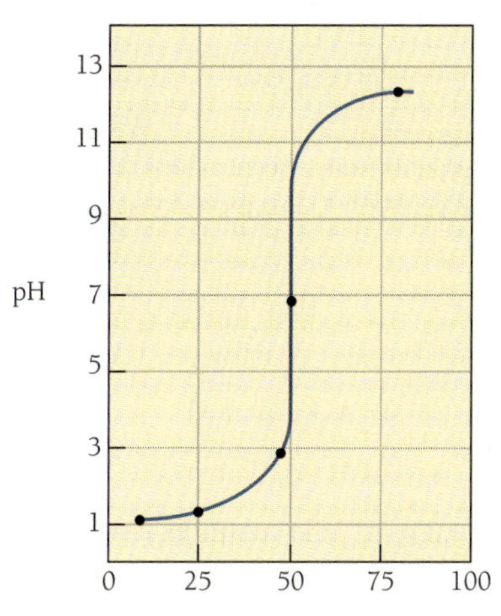

83. **(a)** 12.82, **(b)** 12.52, **(c)** 11.0, **(d)** 7.00, **(e)** 1.64.

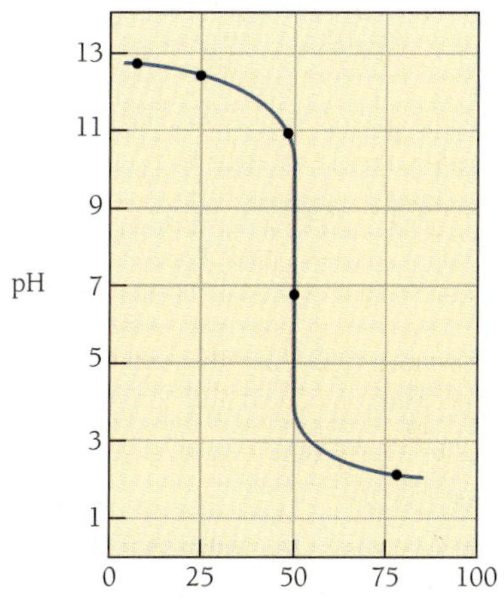

85. **(a)** $H_2SeO_4 + H_2O \rightleftharpoons HSeO_4^- + H_3O^+$,

$$K_1 = \frac{[H_3O^+][HSeO_4^-]}{[H_2SeO_4]}$$

$HSeO_4^- + H_2O \rightleftharpoons SeO_4^{2-} + H_3O^+$,

$$K_2 = \frac{[H_3O^+][SeO_4^{2-}]}{[HSeO_4^-]};$$

(b) $H_3AsO_4 + H_2O \rightleftharpoons H_2AsO_4^- + H_3O^+$,

$$K_1 = \frac{[H_3O^+][H_2AsO_4^-]}{[H_3AsO_4]}$$

$H_2AsO_4^- + H_2O \rightleftharpoons HAsO_4^{2-} + H_3O^+$,

$$K_2 = \frac{[H_3O^+][HAsO_4^{2-}]}{[H_2AsO_4^-]}$$

$HAsO_4^{2-} + H_2O \rightleftharpoons AsO_4^{3-} + H_3O^+$,

$$K_3 = \frac{[H_3O^+][AsO_4^{3-}]}{[HAsO_4^{2-}]}.$$

87. **(a)** $H_2CO_3 + 2\ H_2O \rightleftharpoons CO_3^{2-} + 2\ H_3O^+$, **(b)** 1.6×10^{-17}.

89. $[CO_3^{2-}] \approx 0$, $[HCO_3^-] \approx 0.04$ M, $[H_2CO_3] \approx 0.01$ M.

91. $[HCO_3^-] = 4.0 \times 10^{-6}$ M, $[CO_3^{2-}] = 1.6 \times 10^{-12}$ M.

93. 10.30.

95. 10.22.

97. The true equilibrium concentrations in a C molar solution of the weak acid HA (ionization constant = K_a) are found from

$$K_a = \frac{[H_3O^+][A^-]}{[HA]} = \frac{x^2}{C - x}$$

$$x^2 = K_a\,(C - x)$$

$$x = \sqrt{K_a\,(C - x)}$$

Call this value x_{true} and substitute $x = 0.1 \cdot C$ to get

$$x_{true} = \sqrt{K_a\,(C - x)} = \sqrt{K_a\,(C - 0.1C)} = \sqrt{K_a\,C \cdot 0.9} = \sqrt{K_a\,C} \cdot \sqrt{0.9}$$

Since $x_{approx} = \sqrt{K_a\,C}$, we have shown that

$$x_{true} = x_{approx} \cdot \sqrt{0.9} \quad \text{or} \quad (x_{approx}/x_{true}) = 1/\sqrt{0.9} = 1.05$$

Therefore, the approximate value is no more than 1.05 times the true value, that is, in error by no more than 5%.

99. $[ClO^-]/[HClO] = 0.00038$.

101. 10.27.

103. 7.84.

105. b, because more H_3O^+ or OH^- can be consumed by the larger concentration of acid and conjugate base.

107. **(a)** 1.30, **(b)** 12.70, **(c)** 3.02, **(d)** 4.74, **(e)** 5.29, **(f)** 9.28.

109. 122 g mol^{-1}, 4.20.

111. Phenolphthalein is a good choice for any acid, weak or strong. Its range coincides with the steeply rising part of the titration curve just past the equivalence point. The titration curves of weak and strong acids differ *prior* to the equivalence point, not after. Methyl red is a good choice for any base. Its range coincides with the steeply falling part of the titration curves, of both weak and strong bases, just past the equivalence point.

Chapter 17

1. The ion product Q can be evaluated for any concentrations of ions, while the solubility product expression K_{sp} holds only when a solid is in equilibrium with dissolved ions.

3. The number of *grams* of a compound that will dissolve in a liter of solvent is the solubility, while the molar solubility is the number of *moles* that will dissolve in one liter.

5. An undersaturated solution contains less than the equilibrium amount, a saturated solution contains the equilibrium amount, and a supersaturated solution contains more than the equilibrium amount of solute.

7. By adjusting the pH. At low pH the reaction $H_3O^+ + OH^- \rightarrow 2\ H_2O$ removes OH^- from the solution, decreasing the ion product and allowing more of the solid compound to dissolve.

9. A complex ion consists of a metal ion bonded by coordinate covalent bonds to one or more electron-donating molecules or anions. Examples are $Ag(CN)_2^-$ and $Cu(NH_3)_4^{2+}$.

11. It refers to the fact that the metal ion is joined to the surrounding molecules or anions by *coordinate covalent,* or donor/acceptor, bonds.

13. Qualitative analysis is a procedure for identification of the components of a mixture.

Appendix H should be consulted for needed values of solubility products.

15. **(a)** $K_{sp} = [Ba^{2+}][F^-]^2$, **(b)** $K_{sp} = [Bi^{3+}]^2[SO_4^{2-}]^3$, **(c)** $K_{sp} = [Cu^+][Br^-]$, **(d)** $K_{sp} = [Ba^{2+}][CO_3^{2-}]$.

17. 2.0×10^{-4} mol L^{-1}.

19. 3.3×10^{-7} mol L^{-1}; 1.0×10^{-4} g L^{-1}.

21. 2×10^{-3} g.

23. 2.7×10^{-7} g L^{-1}.

25. 5.0×10^{-13}.

27. 1.5×10^{-5}.

29. No precipitate forms.

31. No precipitate forms.

33. 0.036 mol.

35. AgBr.

37. First $PbSO_4$, then $CaSO_4$.

39. $[Pb^{2+}] = 6.7 \times 10^{-6}$ M.

41. 1.0×10^{-10} M.

43. 2×10^{-10} M.

45. 2×10^{-11} mol.

47. 2.6×10^{-3} g L^{-1}.

49. **(a)** 4.5×10^{-3} g L^{-1}, **(b)** 4.2×10^{-6} g L^{-1}.

51. **(a)** 8.4×10^{-3} M, **(b)** 1.8×10^{-8} M.

53. 8.97.

55. No. Maximum possible $[Mg^{2+}]$ in 0.01 M NH_3 is 7.7×10^{-5} M.

57. 8.1×10^{-4} M.

59. 2×10^{-9} M.

61. 6.0.

63. **(a)** $Ag^+ + 2\ CN^- \rightleftharpoons Ag(CN)_2^-$, $K_f = \dfrac{[Ag(CN)_2^-]}{[Ag^+][CN^-]^2}$; **(b)** $Cd^{2+} + 4\ NH_3 \rightleftharpoons Cd(NH_3)_4^{2+}$, $K_f = \dfrac{[Cd(NH_3)_4^{2+}]}{[Cd^{2+}][NH_3]^4}$.

65. 5.1×10^{-7}.

67. $[Ag(NH_3)_2^+] = 0.063$ M, $[Ag^+] = 1.2 \times 10^{-7}$ M.

69. Analytical Group 1 chloride salts are insoluble, while those of Analytical Group 2 are soluble.

71. Add concentrated ammonia. If the precipitate dissolves completely, it contained only AgCl. If it dissolves partially, the remaining solid is Hg_2Cl_2. Add HCl to the solution; if a precipitate reappears, it is AgCl.

73. 7.6×10^{-4} M.

Chapter 18

1. Combustion is a rapid exothermic reaction with oxygen, usually emitting light as well as heat. Oxidation is an increase in oxidation number, caused by reaction with oxygen or any other oxidant. Reactions with oxygen need not be exothermic and need not be rapid.

3. A Lewis base has an unused pair of valence electrons that it *shares* with an acceptor species. A reducing agent *transfers* one or more electrons to an acceptor species.

5. A half-reaction is either an oxidation or a reduction; it cannot occur by itself.

7. The oxidation number of an element in a compound is the number of electrons each atom of that element loses from its valence shell in the formation reaction of the compound.

9. Any ion that is unchanged in a chemical reaction is a spectator ion.

11. Disproportionation (or auto oxidation-reduction) is any reaction in which a substance reacts with itself, yielding both oxidized and reduced products. For example, $Cl_2(aq) + H_2O \rightarrow H^+ + Cl^- + HOCl$.

13. **(a)** +2, **(b)** +3, **(c)** +1.

15. From left to right: **(a)** +5, −1; **(b)** +3, −1; **(c)** +4, −2; **(d)** +4, −2; **(e)** +2, −1; **(f)** +1, +4, −2.

17. From left to right: **(a)** +6, −1; **(b)** +4, −1; **(c)** +5, −2; **(d)** +1, +3, −2; **(e)** −3, +1; **(f)** C: −2, H: +1, O: −2.

19. a, d, and e.

21. b and c.

23. **(a)** PCl_3 is oxidized, Cl_2 is reduced; **(d)** NO_2 is oxidized *and* reduced; **(e)** Mg is oxidized, HCl is reduced.

25. **(b)** F_2 is the oxidant, Xe is the reductant; **(c)** Br_2 is the oxidant *and* reductant.

27. **(a)** ionic: $3\ Na^+ + 3\ ClO^- \rightarrow 3\ Na^+ + ClO_3^- + 2\ Cl^-$, net ionic: $3\ ClO^- \rightarrow ClO_3^- + 2\ Cl^-$; **(b)** ionic: $3\ Cu(s) + 8\ H^+ + 8\ NO_3^- \rightarrow 3\ Cu^{2+} + 6\ NO_3^- + 2\ NO + 4\ H_2O$, net ionic: $3\ Cu(s) + 8\ H^+ + 2\ NO_3^- \rightarrow 3\ Cu^{2+} + 2\ NO + 4\ H_2O$.

29. **(a)** ionic: $8\ H^+ + 8\ Cl^- + H_3AsO_4(aq) + 4\ Mg(s) \rightarrow AsH_3(g) + 4\ Mg^{2+} + 8\ Cl^- + 4\ H_2O$; net ionic: $8\ H^+ + H_3AsO_4(aq) + 4\ Mg(s) \rightarrow AsH_3(g) + 4\ Mg^{2+} + 4\ H_2O$. **(b)** ionic: $K^+ + NO_2^- + 5\ H_2O(\ell) + 2\ Al(s) \rightarrow 2\ Al^{3+} + 6\ OH^- + NH_3 + K^+ + OH^-$; net ionic: $NO_2^- + 5\ H_2O(\ell) + 2\ Al(s) \rightarrow 2\ Al^{3+} + 7\ OH^- + NH_3$.

31. **(a)** ox: $Sn^{2+} \rightarrow Sn^{4+}$, red: $UO_2^{2+} \rightarrow U^{4+}$; **(b)** ox: $Hg(\ell) \rightarrow Hg_2SO_4(s)$, red: $Cr_2O_7^{2-} \rightarrow Cr^{3+}$; **(c)** ox: $Mn^{2+} \rightarrow MnO_4^-$, red: $H_2O_2 \rightarrow H_2O$; **(d)** ox: $C_2O_4^{2-} \rightarrow CO_2$, red: $IO_3^- \rightarrow I^-$.

33. **(a)** ox: $Sn^{2+} \rightarrow Sn^{4+} + 2\ e^-$, red: $UO_2^{2+} + 4\ H^+ + 2\ e^- \rightarrow U^{4+} + 2\ H_2O$; **(b)** ox: $2\ Hg(\ell) + SO_4^{2-} \rightarrow Hg_2SO_4(s) + 2\ e^-$, red: $Cr_2O_7^{2-} + 14\ H^+ + 6\ e^- \rightarrow 2\ Cr^{3+} + 7\ H_2O$; **(c)** ox: $Mn^{2+} + 4\ H_2O \rightarrow MnO_4^- + 8\ H^+ + 5\ e^-$, red: $2\ H^+ + H_2O_2 + 2\ e^- \rightarrow 2\ H_2O$; **(d)** ox: $C_2O_4^{2-} \rightarrow 2\ CO_2 + 2\ e^-$, red: $6\ H^+ + IO_3^- + 6\ e^- \rightarrow I^- + 3\ H_2O$.

35. **(a)** $Sn^{2+} + UO_2^{2+} + 4\ H^+ \rightarrow Sn^{4+} + U^{4+} + 2\ H_2O$, **(b)** $6\ Hg(\ell) + 3\ SO_4^{2-} + Cr_2O_7^{2-} + 14\ H^+ \rightarrow 3\ Hg_2SO_4(s) + 2\ Cr^{3+} + 7\ H_2O$, **(c)** $5\ H_2O_2 + 2\ Mn^{2+} \rightarrow 2\ MnO_4^- + 6\ H^+ + 2\ H_2O$, **(d)** $6\ H^+ + IO_3^- + 3\ C_2O_4^{2-} \rightarrow 6\ CO_2 + I^- + 3\ H_2O$.

37. **(a)** $4\ I_2(s) + H_2S(aq) + 4\ H_2O \rightarrow 10\ H^+ + SO_4^{2-} + 8\ I^-$, 8 electrons; **(b)** $Cl_2(g) + H_2O \rightarrow ClO^- + Cl^- + 2\ H^+$, 1 electron.

39. **(a)** $2\ I^- + H_2O_2(aq) + 2\ H^+ \rightarrow I_2(s) + 2\ H_2O$, **(b)** $IO_3^- + 2\ Cr(OH)_4^- \rightarrow I^- + 2\ CrO_4^{2-} + 2\ H^+ + 3\ H_2O$, **(c)** $Cu(s) + 2\ NO_3^- + 4\ H^+ \rightarrow Cu^{2+} + 2\ NO_2(g) + 2\ H_2O$.

41. $H_2O + 3\ CN^- + 2\ MnO_4^- \rightarrow 3\ CNO^- + 2\ MnO_2(s) + 2\ OH^-$.

Chapter 19

1. Electrochemistry is the study of interconversions between chemical energy and electrical energy in redox reactions. "Redox chemistry" need not involve electrical energy.

3. 1.6×10^{-19} C, found on electrons (negative) and protons (positive).

5. The unit of potential is the volt, and the unit of work is the joule. They are related by the equivalence statement $1\ J \Leftrightarrow 1V \cdot 1\ C$.

7. Little difference. The current is carried by Na^+ and Cl^- ions in both media, but the ions in aqueous solution are hydrated.

9. An oxidation or reduction half-reaction that occurs at the surface of an electrode.

11. An electrochemical cell is a device for converting electrical energy to chemical energy (an electrolytic cell) or chemical energy to electrical energy (a voltaic or galvanic cell).

13. Use current-time measurements to determine electrical charge; the amount of charge (in Faradays) per mole of reactant oxidized at the anode (or reduced at the cathode) is equal to the number of electrons transferred per mole of reactant.

15. A voltaic cell consisting of metallic zinc immersed in a 0.3 M solution of a zinc salt, connected via a salt bridge or other porous barrier to a 0.1 M solution of a chloride salt in which an AgCl-coated silver wire is immersed.

17. $\mathscr{E}_{red} = -\mathscr{E}_{ox}$.

19. SRP of a half-cell is the measured voltage of a voltaic cell consisting of that half-cell and a SHE. If the SHE is the cathode, the SRP is the negative of the measured voltage.

21. F_2 is the strongest, $Li^+(aq)$ is the weakest.

23. $\Delta G = -n\mathscr{F}\mathscr{E}$; if ΔG is negative, the reaction is spontaneous.

25. The lead storage battery used in automobiles consists of lead electrodes, one coated with PbO_2 and the other coated with spongy lead, immersed in aqueous sulfuric acid. The dry cell used in hardware such as flashlights has a zinc anode and a carbon cathode: the electrolyte is a moist paste of MnO_2 and NH_4Cl.

27. The faraday is the electrical charge of one mole of electrons, and is equal to Avogadro's number times 1.6×10^{-19} C, the charge on one electron.

29. Charge = 5.0 C; work = 50 J.

31. 200 s.

33. 2.2×10^5 J.

35. Current = 11 A; 1.2 faradays.

37. $Zn|Zn^{2+}(aq)||Cu^{2+}(aq)|Cu$.

39. **(a)**

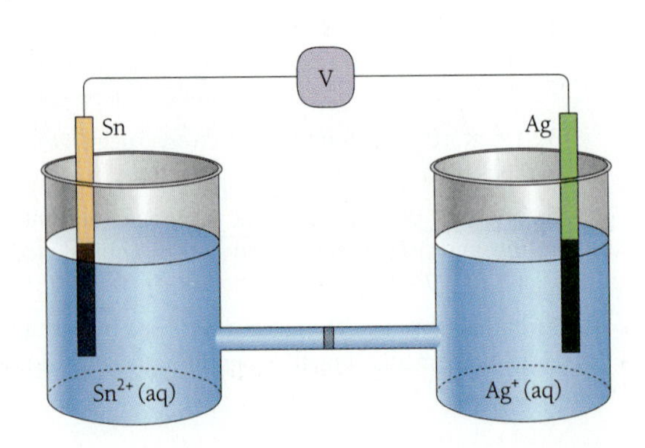

(b)

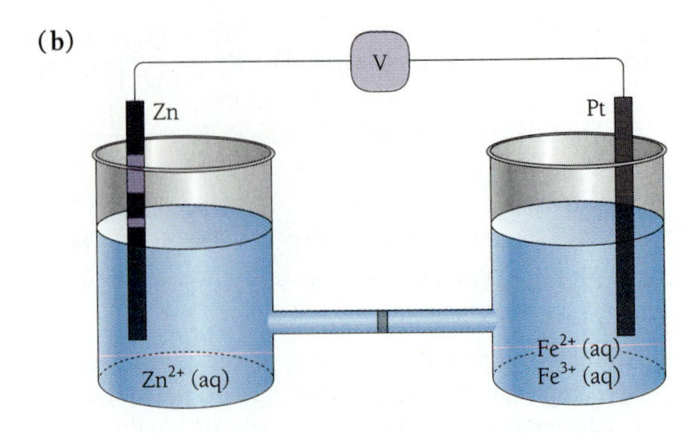

41. Anode, $Zn \rightarrow Zn^{2+}(aq) + 2\ e^-$; cathode, $Cu^{2+}(aq) + 2\ e^- \rightarrow Cu$.

43. **(a)** $Sn(s) + 2\ Ag^+(aq) \rightarrow Sn^{2+}(aq) + 2\ Ag(s)$, $n = 2$; **(b)** $Zn(s) + 2\ Fe^{3+}(aq) \rightarrow 2\ Fe^{2+}(aq) + Zn(s)$, $n = 2$.

45. a.

47. **(a)** NR, **(b)** $Cl_2(g) + 2\ KBr(aq) \rightarrow Br_2(\ell) + 2\ KCl(aq)$, **(c)** NR, **(d)** $2\ AgBr(s) + Ni(s) \rightarrow 2\ Ag(s) + Ni^{2+}(aq) + 2\ Br^-(aq)$, **(e)** NR.

49. **(a)** $Sn + Cu^{2+} \rightarrow Sn^{2+} + Cu$, **(b)** NR, **(c)** NR.

51. **(a)** Cl_2, **(b)** Pb^{2+}, **(c)** $KMnO_4$, **(d)** CuCl, **(e)** Zn^{2+}, **(f)** AgCl(s).

53. **(a)** Ni, **(b)** Fe^{2+}, **(c)** Sn, **(d)** H_2S, **(e)** Cr^{3+}, **(f)** Fe.

55. **(a)** Cu^+, **(b)** Hg^{2+}, **(c)** Fe^{3+}, **(d)** $Cr_2O_7^{2-}$.

57. +1.100 V, zinc.

59. **(a)** +0.385 V, $2\ I^-(aq) + 2\ Hg^{2+}(aq) \rightarrow Hg_2^{2+}(aq) + I_2(s)$, $Pt/I_2(s)I^-(aq)||Hg^{2+}(aq), Hg_2^{2+}(aq)|Pt$; **(b)** +0.38 V, $2\ Cu^+(aq) + H_2S(aq) \rightarrow 2\ Cu(s) + S(s) + 2\ H^+(aq)$, $Pt|S(s)|H_2S(aq)||Cu^+(aq)|Cu(s)$; **(c)** 0.834 V, $Zn(s) + 2\ AgBr(s) \rightarrow 2\ Ag(s) + Zn^{2+}(aq) + 2\ Br^-(aq)$, $Zn(s)|Zn^{2+}(aq)||Br^-(aq)|AgBr(s)|Ag(s)$; **(d)** +0.25 V, $HNO_2(aq) + ClO_4^-(aq) \rightarrow H^+(aq) + NO_3^-(aq) + ClO_3^-(aq)$, $Pt|NO_3^-(aq), HNO_2(aq)||ClO_4^-(aq), ClO_3^-(aq)|Pt$.

61. **(a)** +0.62 V, $Zn(s) + Sn^{2+}(aq) \rightarrow Zn^{2+}(aq) + Sn(s)$, diagram is correct; **(b)** +1.202 V, $Cd(s) + 2\ Ag^+(aq) \rightarrow 2\ Ag(s) + Cd^{2+}(aq)$, $Cd(s)|Cd^{2+}(aq)||Ag^+(aq)|Ag(s)$; **(c)** +0.222 V, $H_2(g) + 2\ AgCl(s) \rightarrow 2\ H^+(aq) + 2\ Cl^-(aq) + 2\ Ag(s)$, $Pt|H_2(g)|H^+(aq)||Cl^-(aq)|AgCl(s)|Ag(s)$; **(d)** +1.358 V, $H_2(g) + Cl_2(g) \rightarrow 2\ H^+(aq) + 2\ Cl^-(aq)$, diagram is correct.

63. **(a)** $5\ NO(g) + 3\ MnO_4^-(aq) + 4\ H^+(aq) \rightarrow 3\ Mn^{2+}(aq) + 5\ NO_3^-(aq) + 2\ H_2O$, $n = 15$, $Q = \{[Mn^{2+}]^3[NO_3^-]^5\}/\{P_{NO}^5[MnO_4^-]^3[H^+]^4\}$; **(b)** $2\ Ag^+(aq) + Sn(s) \rightarrow Sn^{2+}(aq) + 2\ Ag(s)$, $n = 2$, $Q = [Sn^{2+}]/[Ag^+]^2$; **(c)** $2\ Fe^{3+}(aq) + Zn(s) \rightarrow Zn^{2+}(aq) + 2\ Fe^{2+}(aq)$, $n = 2$, $Q = [Zn^{2+}][Fe^{2+}]^2/[Fe^{3+}]^2$; **(d)** $2\ AgCl(s) + H_2(g) \rightarrow 2\ H^+(aq) + 2\ Cl^-(aq) + 2\ Ag(s)$, $n = 2$, $Q = [H^+]^2[Cl^-]^2/P_{H_2}$.

65. **(a)** $Zn(s) + Sn^{2+}(aq) \rightarrow Zn^{2+}(aq) + Sn(s)$, $n = 2$, $Q = [Zn^{2+}]/[Sn^{2+}]$; **(b)** $Cd(s) + 2\ Ag^{+}(aq) \rightarrow 2\ Ag(s) + Cd^{2+}(aq)$, $n = 2$, $Q = [Cd^{2+}]/[Ag^{+}]^2$; **(c)** $H_2(g) + 2\ AgCl(s) \rightarrow 2\ H^{+}(aq) + 2\ Cl^{-}(aq) + 2\ Ag(s)$, $n = 2$, $Q = [H^{+}]^2[Cl^{-}]^2/P_{H_2}$; **(d)** $H_2(g) + Cl_2(g) \rightarrow 2\ H^{+}(aq) + 2\ Cl^{-}(aq)$, $n = 2$, $Q = \{[H^{+}]^2[Cl^{-}]^2\}/\{P_{H_2}P_{Cl_2}\}$.

67. **(a)** +0.95V, **(b)** 1.287 V, **(c)** +1.619 V, **(d)** 0.81 V.

69. **(a)** $[Ni^{2+}]/[Sn^{2+}] = 3.9 \times 10^{-6}$, **(b)** $[Sn^{2+}]/[Ni^{2+}] = 1.5 \times 10^{-13}$.

71. **(a)** 1.2×10^{-4}, **(b)** $P_{H_2} = 3.9 \times 10^{-4}$ atm.

73. **(a)** 3, **(b)** 1, **(c)** 1.

75. **(a)** 5.18×10^3 C, **(b)** 8.94×10^2 C, **(c)** 4.81×10^2 C.

77. 2.58×10^{-3} mol; 57.8 mL.

79. 0.567 C.

81. 3.25×10^4 s, or slightly more than 9 hours.

83. 0.0877 M.

85. **(a)** anode: $Zn(s) \longrightarrow Zn^{2+}(aq) + 2\ e^{-}$;
cathode: $2\ NH_4^+ + 2\ MnO_2(s) + 2\ e^{-} \longrightarrow Mn_2O_3(s) + H_2O(\ell) + 2\ NH_3(aq)$;
overall cell reaction: $2\ NH_4^+ + 2\ MnO_2(s) + Zn(s) \longrightarrow Zn^{2+}(aq) + Mn_2O_3(s) + H_2O(\ell) + 2\ NH_3(aq)$.
(b) anode: $Pb(s) + HSO_4^- \longrightarrow PbSO_4(s) + H^{+} + 2\ e^{-}$;
cathode: $PbO_2(s) + 3\ H^{+} + HSO_4^- + 2\ e^{-} \longrightarrow PbSO_4(s) + 2\ H_2O(\ell)$;
overall cell reaction: $Pb(s) + PbO_2(s) + 2\ H_2SO_4(aq) \longrightarrow 2\ PbSO_4(s) + 2\ H_2O(\ell)$.
(c) anode: $Zn(s) + 2\ OH^{-} \longrightarrow ZnO(s) + H_2O(\ell) + 2\ e^{-}$;
cathode: $AgO(s) + H_2O(\ell) + 2\ e^{-} \longrightarrow Ag(s) + 2\ OH^{-}$;
cell: $AgO(s) + Zn(s) \longrightarrow ZnO(s) + Ag(s)$.

87. The system is a voltaic cell with the reactions:
anode: $Al \longrightarrow Al^{3+} + 3\ e^{-}$;
cathode: $Ag_2S(s) + 2\,e^{-} + 2\,H^{+} \longrightarrow 2\,Ag(s) + H_2S(aq)$.

89. **(a)** $\Delta G° = -5.7 \times 10^2$ kJ, $K_{eq} \approx 10^{100}$: this is so large that it is meaningless—the reaction goes to completion; **(b)** $\Delta G° = +24$ kJ, $K_{eq} = 6.4 \times 10^{-5}$; **(c)** $\Delta G° = +32$ kJ, $K_{eq} = 2.1 \times 10^{-6}$.

91. $\mathscr{E}° = +1.36$ V.

93. Fe^{3+}.

95. $\mathscr{E} = +0.763 - 0.0592 \cdot pH$.

97. The experiment gives $\mathscr{F} = 99.5 \times 10^4$ C mol^{-1}, which is about 4% high. Either the cathode gained 4% more than 1.952 g or the current was 4% less than 0.500 A. It is not likely that the time measurement was in error.

Chapter 20

1. Organic compounds contain carbon and hydrogen, usually in combination with oxygen, nitrogen, and/or sulfur.

3. A functional group is a part of a molecule, a structural feature that confers, and is the site of, a characteristic reactivity not possessed by molecules lacking the feature. Examples are —OH, C=C, $—NH_2$, and so on.

5. The greater acid strength of the carboxylic acids, as compared to alcohols, may be ascribed to resonance stabilization of the carboxylate anion $R—CO_2^-$; resonance is not possible in the alkoxide anion, $R—O^-$; also, aromatic compounds are so stabilized by resonance that they do not undergo the addition reactions characteristic of alkenes.

7. Condensation polymers are formed in condensation reactions of bifunctional monomers. Addition polymers are formed from monomers having at least one double bond, that is, from alkenes.

9. No. It is a fact of geometry that any three points, like the carbon nuclei in cyclopropane, lie in one plane. The H-atoms of cyclopropane, of course, do not lie in the same plane. "Planarity," as commonly used, refers to the carbon skeleton only.

11. **(a)** Reduction, **(b)** oxidation, **(c)** oxidation, **(d)** neither.

13. **(a)** All C atoms are sp^3; **(b)** 2 C atoms are sp^2, and the others are sp^3 (4 sp^2 C atoms in dienes, and so on); **(c)** 2 C atoms are sp, and the others are sp^3 (4 sp C atoms in diynes, and so on); **(d)** all ring C atoms are sp^3; **(e)** all six ring C atoms are sp^2.

15. **(a)** $CH_3—C(CH_3)_2—CH_3$ (central C bearing CH_3 above and CH_3 below) **(b)** benzene ring with Cl, Cl, Cl substituents (1,2,4-trichlorobenzene)
(c) $CH_3—C(Cl)=CH—CH_2—CH_3$ **(d)** cyclopentene

17. **(a)** hexane; **(b)** chlorobenzene; **(c)** 1,1-dibromo-4-methylpentane; **(d)** 3-chlorobenzoic acid.

19. a, b, e, and f.

21. The structural formula corresponding to 3-methylpropane is $CH_3—CH_2—CH_2—CH_3$ (the terminal CH_3 drawn below the last carbon): since the longest continuous chain has *four* carbon atoms, the IUPAC name for this structure is butane.

23. Since ethane is neither cyclic nor an alkene, geometrical isomerism (implied by "*cis*-") is not possible in substituted ethanes.

25. **(a)** Cl(H)C=C(Cl)H (*cis*: Cl and Cl on the same side) **(b)** $CH_3—CH_2—C(CH_3)_2—CH_3$

(c) $CH_3—C(CH_3)_2—CH_2—CH_2—CH{=}CH—CH_3$

(d) $CH_3—C(CH_3)_2—CH_3$

27. **(a)** *cis*-1,2-dichloroethene; **(b)** 2,2-dimethylbutane; **(c)** 6,6-dimethyl-2-heptene; **(d)** 2,2-dimethylpropane.

29. The names are 1-pentene; 2-pentene (*cis*- and *trans*- forms); 2-methyl-1-butene; 3-methyl-1-butene; 2-methyl-2-butene; cyclopentane; methylcyclobutane; 1,1-dimethylcyclopropane; 1,2-dimethylcyclopropane (*cis*- and *trans*- forms); and ethylcyclopropane. The structures are given in the *Solutions Manual*.

31. c.

33. 1,1-dichlorocyclopropane; 1,1-dichloropropene; 3,3-dichloropropene; *cis*-1,2-dichloropropene; *trans*-1,2-dichloropropene; *cis*-1,3-dichloropropene; *trans*-1,3-dichloropropene; and 2,3-dichloropropene. The structures are given in the *Solutions Manual*.

35. **(a)** Amides; **(b)** 2° amines; **(c)** aldehydes, ketones, acids, esters, amides.

37. **(a)** 1° alcohols, **(b)** 2° alcohols, **(c)** aldehydes, **(d)** sulfonic acids.

39. **(a)** $(CH_3CH_2)_2NH$, **(b)** CH_3CO_2H, **(c)** CH_3CONH_2, **(d)** $(CH_3)_2CO$. The structures are given in the *Solutions Manual*.

41. 1°: $CH_3CH_2CH_2CH_2OH$; 2°: $CH_3CH_2CHOHCH_3$; 3°: $(CH_3)_3COH$. The structures are given in the *Solutions Manual*.

43. The structure of acetone is $CH_3—C(=O)—CH_3$: the carbon atom doubly bonded to O is sp^2, and the other two are sp^3.

The structure of 2-butanone is $CH_3—CH_2—C(=O)—CH_3$: all carbon atoms are sp^3 except the one sp^2 carbon doubly bonded to O. In *any* ketone of formula $CH_3(CH_2)_nCOCH_3$, all carbon atoms are sp^3 except the one sp^2 carbon doubly bonded to O.

45. One.

47. **(a)** CH_3CH_2OH, **(b)** $CH_3(CH_2)_3CO_2H$, **(c)** Na^+(soln) + $CH_3CH_2O^-$(soln), **(d)** $HCO_2CH_2CH_3$.

49. **(a)** C_6H_5—CO_2^-(aq), **(b)** cyclohexanol, **(c)** C_6H_5—O^-(aq).

51. 8.5 g acetic acid, or 14 g potassium acetate.

53. Yes. A monomer having an amino group at one end and a carboxylic acid group at the other could polymerize to a polyamide: $H_2NCH_2CH_2CO_2H \rightarrow {+}CH_2CH_2CONH{+}_n$. Other examples are $HOCH_2CH_2OH \rightarrow {+}CH_2CH_2O{+}_n$, and $HOCH_2CH_2CO_2H \rightarrow {+}CH_2CH_2CO_2{+}_n$.

55. Polyethylene is an alkane, and as such is unreactive at room temperature. The amide groups in nylon are not unreactive, and in fact are attacked by concentrated acids in the reverse of a condensation reaction.

57. See Tables 20.3 and 20.4.

59. 73%.

Chapter 21

1. In polarized light, the electric fields of all photons are parallel to one another. Optical activity is the ability of a substance to rotate the plane of polarized light passing through it.

3. A chiral molecule is optically active, that is, rotates the plane of polarized light. A chiral carbon atom is one that is bonded to four different atoms or groups. Enantiomers are stereoisomers that are mirror images of one another. An equimolar mixture of two enantiomers is a racemic mixture.

5. a and d.

7. One chiral carbon atom, two optical isomers.

9. Five chiral carbon atoms, 32 optical isomers.

11. Two enantiomers, each with a central C bearing CH_3 (top), COOH: (1) H_2N left, H bottom; (2) H left, H_2N bottom.

13. Dipole-dipole and H-bonding. In addition, a covalent bond could be formed by a condensation (esterification) reaction. Its side-chain makes serine a 1° alcohol, and the side-chain of glutamic acid contains the $—CO_2H$ group.

15. $CH_3—CH_2—CH(NH_2)—COOH$ **an α-amino acid**; $CH_3—CH(NH_2)—CH_2—COOH$ **a β-amino acid**

17.

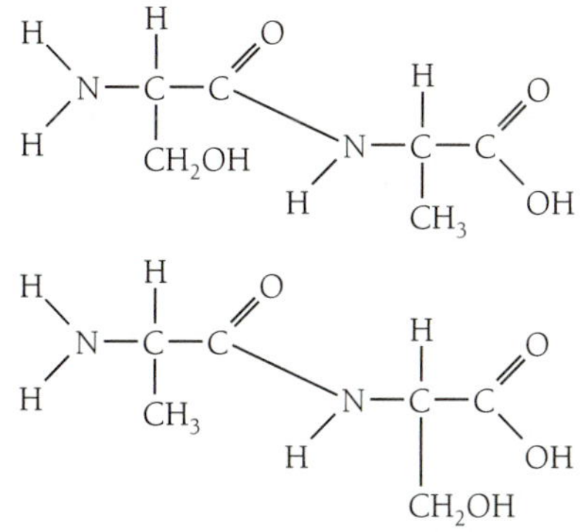

```
 H     H    O
  \    |   //
   N—C—C          H    O
  /    |    \     |   //
 H   CH2OH   N—C—C
            /    |    \
           H    CH3    OH

 H     H    O
  \    |   //
   N—C—C          H    O
  /    |    \     |   //
 H    CH3    N—C—C
            /    |    \
           H     |     OH
               CH2OH
```

19. Six: asp-ser-trp, asp-trp-ser, ser-asp-trp, ser-trp-asp, trp-ser-asp, and trp-asp-ser.

21.

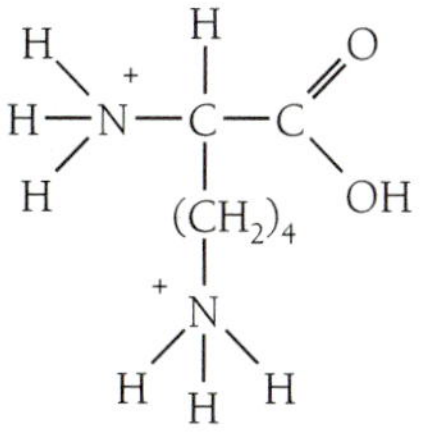

```
 H  +   H    O          H     H    O
  \     |   //           \    |   //
H—N—C—C                 N—C—C
  /     |    \          /     |    \
 H   (CH2)4   OH       H   (CH2)4   O−
        |                     |
      + N                     N
       /|\                   / \
      H H H                 H   H

     in acid              in base
```

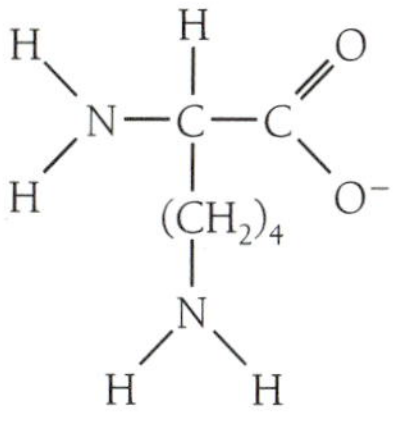

23. Amino acids not synthesized by humans are called *essential.* There are eight such amino acids.

25. They catalyze reactions, transport oxygen, produce movement in muscles, serve as connective tissue, and so on.

27. 1°: covalent bonds; 2°: H-bonding between the peptide bonds (amide linkages); 3°: dipole-dipole, electrostatic, H-bonding, and —S—S— cross-linking.

29. An aldose is a sugar whose carbonyl group is present as in

```
              O
             //
aldehydes, R—C     , while in a ketose the carbonyl group
              \
               H
```

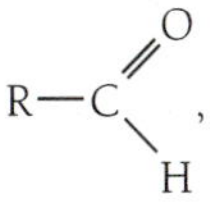

```
                             O
                            //
is present as in ketones, R—C
                             \
                              R'
```

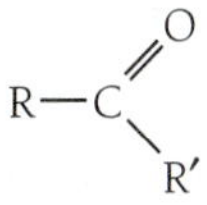

31. Following the principle of "like dissolves like," carbohydrates, which are polar substances containing the —OH group and capable of H-bonding, should dissolve in water, a polar substance containing the —OH group and capable of H-bonding.

33. When a linear monosaccharide undergoes a cyclization reaction, a carbon atom formerly bonded to three different groups becomes bonded to four different groups. That is, a new chiral carbon atom is formed, so that two stereoisomers are possible. These isomers are called the α and β forms.

35. Maltose is a disaccharide consisting of two α-glucose molecules. Therefore, a polymer of maltose is identical to a polymer of glucose in which all glycoside linkages have the α configuration.

37. In starch, the glycoside linkages are all α-1,4; in cellulose, they are all β-1,4.

39. RNA is a phosphate polyester of ribose, in which each ribose unit is bonded to one of the bases adenine, guanine, cytosine, or uracil. DNA is a phosphate polyester of deoxyribose, in which each deoxyribose unit is bonded to one of the bases adenine, guanine, cytosine, or thymine.

41. The primary structure is the sequence of bases on the sugar-phosphate backbone, while the 2° structure is the double helix, the two-strand composite molecule.

43. —CCTGT—.

45. Any compound of biological origin that is soluble in nonpolar solvents is called a lipid.

47. A saturated fat contains no C=C double bonds, while a polyunsaturated compound has at least two C=C double bonds.

49. A triglyceride is an ester produced by condensation of three moles of fatty acids with one mole of glycerol. Fats and oils are triglycerides.

51. Since they are soluble in nonpolar solvents, steroids fit the definition of lipids.

53. The literal meaning of "hydrophobic" is "water hating." Hydrophobic substances, like fats, oils, waxes, and other lipids, do not dissolve in water. All hydrocarbons are hydrophobic.

55. Fats and oils are glycerol esters of three fatty acids, while in phosphoglycerides the glycerol is esterified with two fatty acids and a phosphate group.

57. One mole of glycerol, two moles of fatty acid(s), and one mole of a phosphate ester.

59. A process for which $\Delta H < 0$ is exothermic, while if $\Delta G < 0$ the process is exergonic. Since $\Delta G = \Delta H - T\Delta S$, any process with a small, negative ΔH and a large, negative ΔS is both exothermic and endergonic.

61. 29 mol.

63. 17 mol.

Chapter 22

1. Five in B-10 and six in B-11.

3. Magnesium.

5. In many compounds, such as the boron halides, the boron atom has only six electrons in its valence shell; that is, the octet rule is not followed. In the boranes (boron hydrides) there are some bonds in which a pair of electrons is shared by three, rather than two, atoms.

7. BCl_3. All boron halides are planar trigonal molecules of formula BX_3.

9. pH = 5.22.

11. **(a)** Boron tribromide, BBr_3; **(b)** sodium borohydride, $NaBH_4$; **(c)** boron nitride, BN; **(d)** boric oxide (diboron trioxide) B_2O_3.

13. C-12 (99%) and C-13 (1%).

15. Graphite is a good lubricant because its weak interplane forces allow the layers to slide easily over one another.

17. Beryllium carbide, Be_2C. $Be_2C + 4\ H_2O \rightarrow 2\ Be(OH)_2 + CH_4$.

19. Carbon monoxide, CO; carbon dioxide, CO_2; and carbon suboxide, C_3O_2. All are molecular compounds. CO reacts with metals to form metal carbonyls such as $Ni(CO)_4(\ell)$.

21. (a) H—O—C(=O)—O^-, sp^2;

(b) H—C≡N, sp; (c) H—C(=O)—O^-, sp^2.

23. $HCO_3^- < H_2CO_3 < HCO_2H$.

25. Cave water is saturated, or nearly saturated, with $CaCO_3$. Since CO_3^{2-} is a base, the pH is greater than 7.

27. HCN.

29. 35.0 L.

31. The Si—Si bond is quite weak, so molecules containing it are not very stable.

33. A long-chain polymer whose repeating structure is $\text{+}SiR_2O\text{+}_n$.

35. The tetrahedral SiO_4 unit. In aluminosilicates, some of the silicon atoms in the SiO_4 tetrahedra are replaced by aluminum atoms. Cations of other metals, present in both silicates and aluminosilicates, are situated outside, rather than inside, the tetrahedra.

37. Long —Si—O—Si—O— chains are helically arranged in crystalline quartz. An individual crystal contains either right-handed or left-handed helices, but not both.

39. 26.8 g.

41. Germane, stannane, and plumbane.

43. Since metallic character increases as atomic number increases within a group, tin and lead are more metallic than silicon. Metals tend to form ionic compounds, while nonmetals typically form molecular compounds.

45. $PbH_4 + 2\ H_2O \rightarrow PbO_2 + 4\ H_2$.

47. The isotope is ^{15}N, which contains eight neutrons.

49. The synthesis of ammonia from the elements.

$$N_2(g) + 3\ H_2 \xrightarrow[\text{Fe}]{\text{500 atm, 450 °C}} 2\ NH_3(g).$$

51. In both ammonia, NH_3, and hydrazine, N_2H_4, the N atoms are sp^3 hybridized.

53. **(a)** $:\ddot{N}::N::\ddot{N}:^-$, linear; **(b)** $:C{\equiv}N:^-$; **(c)** $:\ddot{N}::C::\ddot{O}:^-$, linear; **(d)** $:\ddot{C}::N::\ddot{O}:^-$, linear.

55. A compound whose standard heat of formation is positive is an endothermic compound. All of the nitrogen oxides, and none of the carbon oxides, are endothermic.

57. Nitrous acid, HNO_2, is weak and unstable in aqueous solution. Aqueous solutions of the strong acid HNO_3 (nitric acid) are also unstable, but they decompose so slowly that their instability presents no problems in their use. The anhydride of nitrous acid is dinitrogen trioxide, while the anhydride of nitric acid is dinitrogen pentoxide.

59. $5\ N_2H_5^+ + 4\ IO_3^- \rightarrow 5\ N_2(g) + 2\ I_2(s) + H^+ + 12\ H_2O$.

61. The P atoms are located at the vertices of a tetrahedron; the hybridization is sp^3.

63. **(a)** $P_4(s) + 5\ O_2(g) \rightarrow P_4O_{10}(s)$, **(b)** $P_4O_{10}(s) + 6\ H_2O(\ell) \rightarrow 4\ H_3PO_4(aq)$.

65. **(a)** $H_3PO_3(aq) + H_2O(aq) + Pb^{2+}(aq) \rightarrow Pb(s) + H_3PO_4(aq) + 2\ H^+(aq)$, **(b)** no reaction.

***67.** Phosphorus, arsenic, and antimony.

69. Most is mined from deposits of almost pure elemental sulfur, and a little more than one third is recovered from stack-gas emissions from fuel combustion and smelting operations.

71. The monatomic anion, S^{2-}, is found in alkali-metal sulfides like K_2S. In addition, sulfur forms the disulfide ion, S_2^{2-}, and the polysulfide ion, S_n^{2-}, a linear species.

73. **(a)** $K_2S + H_2O \rightarrow 2\ K^+(aq) + HS^-(aq) + OH^-(aq)$, **(b)** $K_2S + 2\ HCl(aq) \rightarrow H_2S(g) + 2\ KCl(aq)$, **(c)** $SF_4 + 2\ H_2O \rightarrow SO_2 + 4\ HF$.

75. Sulfur dioxide is the anhydride of sulfurous acid: $SO_2 + H_2O \rightarrow H_2SO_3(aq)$; sulfur trioxide is the anhydride of sulfuric acid: $SO_3 + H_2O \rightarrow H_2SO_4(aq)$.

77. 9.4×10^4 g, or a little over 200 pounds.

79. **(a)** Potassium sulfate, **(b)** potassium hydrogen sulfate or potassium bisulfate, **(c)** potassium sulfite, **(d)** potassium hydrogen sulfite or potassium bisulfite.

81. **(a)** A central sulfur atom is bonded to 4 terminal O atoms; the geometry is tetrahedral, and the hybridization is sp^3; **(b)** same as **(a)**, except one of the terminal O atoms is replaced by an S atom.

83. $3\ SO_3^{2-} + 2\ MnO_4^- + H_2O \rightarrow 3\ SO_4^{2-} + 2\ MnO_2(s) + 2\ OH^-$.

85. $SO_3(g) + H_2SO_4(aq) \rightarrow H_2S_2O_7$ (pyrosulfuric acid).

87. H_2SO_4 and H_2SeO_4 are nearly tetrahedral in structure, while telluric acid, $Te(OH)_6$, is octahedral.

89. The metalloid elements, B, Si, Ge, As, Sb, Te, and At, are found in the main groups.

91. H_2Te. Among similar compounds, boiling point increases with molar mass.

93. 0.861 M.

Chapter 23

1. With the exception of Zn, Cd, and Hg, all transition elements have a partly filled *d* subshell in the neutral atom or in a common oxidation state.

3. Instead of being 10% to 15% larger then those immediately above them in the periodic table, the usual relationship, the nine elements following La are about the same size (radius) as those above them.

5. In the second series, Nb, Mo, Ru, Rh, Pd, and Ag; in the third series, Pt and Au.

7. Thorium and uranium.

9. These elements have completely filled *d* subshells in all of their oxidation states.

11. **(a)** $[Ar]4s^2 3d^2$, **(b)** $[Ar]4s^2 3d^8$, **(c)** $[Kr]5s^1 4d^7$, **(d)** $[Kr]5s^1 4d^5$, **(e)** $[Kr]5s^2 4d^1$, **(f)** $[Xe]6s^1 4f^{14} 5d^{10}$.

13. **(a)** +4, **(b)** +1, **(c)** +4, **(d)** +3.

15. Cu^{2+}.

17. +7; Mn.

19. Fe (4.7%); Ti (0.58%).

21. The greater the cohesive forces in a solid, the higher the melting point. The strength of cohesive forces increases with the number of unpaired electrons, which *(according to Hund's rule)* reaches a maximum in the middle of the group (d^5).

23. See pages 952 through 956.

25. A Lewis base is a species containing an atom having at least one lone (nonbonding) pair of valence electrons. A Lewis acid contains an atom with an incompletely filled valence shell, capable of sharing an electron pair provided by a base.

27. Both consist of a metal ion bonded to several neutral or anionic ligands. A *complex* is a charged species, while a *compound* is neutral.

29. The number of ligands bonded to a metal in a coordination complex or compound.

31. A chelate is a coordination complex or compound consisting of a metal ion and one or more polydentate ligands.

33. No.

35. a, b, and d are monodentate; the oxalate ion has four atoms with lone pairs, but is shaped so that no more than two can bond to the same metal ion: it is bidentate.

37. 4.

39. **(a)** +3, **(b)** +3, **(c)** +2, **(d)** +3.

41. **(a)** Diaquotetracyanoferrate(III), **(b)** diamminedichloroethylenediaminecobalt(III), **(c)** diamminedioxalatonickelate(II), **(d)** tetrahydroxozincate(II).

43. **(a)** +1, **(b)** +2, **(c)** +2, **(d)** +2.

45. **(a)** Pentacarbonylchlororhenium(III) chloride; **(b)** tris(1,10-phenanthroline)ruthenium(II) chloride; **(c)** sodium ethylenediaminetetraacetatoferrate(II); **(d)** diammineplatinum(II) bromide.

47. **(a)** $[Al(H_2O)_4(OH)_2]^+$, **(b)** $[FeCl_6]^{4-}$, **(c)** $[Zn(NH_3)_3Br]^+$, **(d)** $[Ni(en)_3]^{2+}$.

49. **(a)** $Fe_4[Fe(CN)_6]_3$, **(b)** $Li_2[Cr(NH_3)_2Br_4]$, **(c)** $[PtCl(NO_2)_3]Cl$, **(d)** $Ca[Ni(H_2O)_3Cl_3]_2$.

51. The atom-to-atom connectivity of the molecules differs among structural isomers, but not among stereoisomers. Stereoisomers differ only in bond orientation.

53. MWXYZ, MX_2YZ, and MX_2Y_2.

55. $M(en)_3$, M(en)FClBrI, and $M(en)_2FCl$.

57.

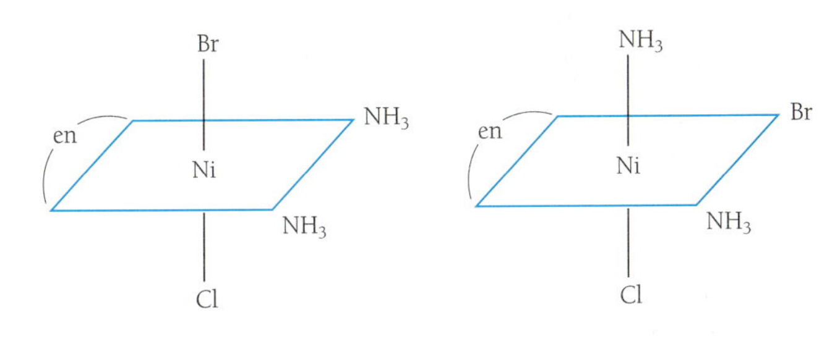

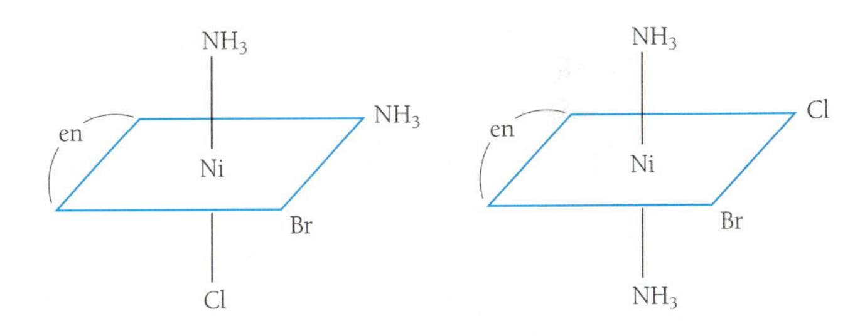

59. Valence bond; treats metal-ligand bonds as coordinate covalent, and explains the shape of complexes in terms of the metal's hybrid orbitals; usually correctly predicts magnetic properties, but not color. Molecular orbital, constructs bonding, antibonding, and nonbonding orbitals extending over metal and ligands alike. Crystal field; treats complex bonding as electrostatic attraction between metal cation and point charges (or dipoles) of ligands; gives useful explanation of color as well as magnetic properties.

61. It is the *d* orbitals, generally, that are involved in color (the absorption of visible light). Sc^{3+} has no *d* electrons, hence no color. (Ti^{3+} has one *d* electron).

63. $I^- < Br^- < Cl^- < F^-$.

65. Because of its similarity to oxygen in electronic configuration as well as size and shape, CN^- might be expected to be, like O_2, a good ligand for the iron atom in hemoglobin. If so, it could seriously (and perhaps fatally) interfere with oxygen transport by the blood. Cyanide-bound hemoglobin might well have a different color from normal blood, and this could cause a change in the color of the blood and tissue having many blood vessels, such as the lips.

67. **(a)** +5, high-spin, 0; **(b)** +5, low-spin, 0; **(c)** +4, high-spin, 1; **(d)** 4, low-spin, 1.

69. **(a)** 3, **(b)** 1, **(c)** 5.

71. Yes. The rate of ligand exchange is low in "inert" complexes, but they are not necessarily unreactive.

73. pH = 3.6, compared to 3.90 for 0.0010 M acetic acid.

75. 1×10^{36}.

77. EDTA. When the concentration ratio $[Ni^{2+}(aq)]/[Ni(complex)]$ is calculated separately for EDTA and en, it is found to be lower in the case of EDTA. Therefore, more $Ni^{2+}(aq)$ can be removed by EDTA under these conditions.

79. An aqueous solution of KCN contains $CN^-(aq)$, which is a Brønsted base by virtue of the hydrolysis reaction $CN^-(aq) + H_2O \rightleftharpoons HCN(aq) + OH^-$. The relatively high concentration of OH^- in KCN(aq) causes the precipitation of the insoluble hydroxide $Zn(OH)_2$. Higher concentration of CN^- favors the formation of the complex ion $[Zn(CN)_4]^{2-}$, and the $Zn(OH)_2$ dissolves.

Chapter 24

1. Emitting, or capable of emitting, α-, β-, or γ-radiation.

3. A nuclide is a particular type of atom, having a specific number of electrons, protons, and neutrons. A nucleus is the core of (any) atom, composed of protons and neutrons.

5. A daughter nuclide is the atom formed when a parent nuclide undergoes radioactive decay.

7. α: atomic and mass numbers decrease, n/p ratio increases. β: atomic number increases, mass number is unchanged, and n/p ratio decreases. γ: all unchanged; positron emission: atomic number decreases, mass number is unchanged, and n/p ratio increases.

9. One explanation is that, by making the nucleus larger, neutrons weaken the repulsive (destabilizing) Coulomb force between protons. If there are too few neutrons, the repulsive force is too strong and the nucleus is unstable.

11. The stability belt is significant for two reasons: it designates the n/p ratio characteristic of stable nuclei, and it is a predictive factor for the mode of decay of unstable nuclei. Unstable nuclides tend to decay so as to change their n/p ratio to a value within the stability belt.

13. There are about 280 stable nuclides; about 85% of all known nuclides are radioactive.

15. When nucleons fuse to form a nucleus, some mass is converted to energy. This is a violation of the law of conservation of mass, and a reminder that nuclear processes obey the law of conservation of mass/energy, a more general concept.

17. ${}^{1}_{1}H$ has only one nucleon, hence no binding energy.

19. High temperature, that is, high average kinetic energy, is necessary to overcome the mutual repulsion of positively charged nuclei, allowing them to make contact and fuse under the influence of the strong force.

21. In a chain reaction, an average of exactly 1 neutron per fission event goes on to induce fission in another nucleus; in a branched-chain reaction, the average is greater than 1. Chain reactions proceed at a steady rate, while a branched chain reaction accelerates as the reaction proceeds.

23. Slow-moving protons are prevented, by Coulomb repulsion, from reaching a nucleus and interacting with it. Neutrons are not repelled by nuclei, and they can approach even if slow-moving.

25. On Graham's law, which says that diffusion rate in gases is inversely proportional to the square root of molar mass.

27. Deuterium fusion (3.5×10^{11} J g^{-1}).

29. A decay series is the sequence of spontaneous decay steps by which a nucleus of uranium or thorium is transformed to a stable isotope of lead. Since all subsequent steps in a series are faster than the first step, intermediate products decay relatively quickly and do not accumulate.

31. Each step in a decay series is either β-emission, which does not change the mass number, or α-emission, which reduces the mass number by 4. The starting mass number, 238, is equal to $(4 \cdot 59 + 2)$, so, since the mass number can only change by 4, subsequent mass numbers are $(4 \cdot 58 + 2)$, $(4 \cdot 57 + 2)$, $(4 \cdot 56 + 2)$, and so on.

33. Formation of ions and free radicals along the path of the particle.

35. Energy, J (SI) or eV; activity, Bq (SI); absorbed dose, Gy (SI) or rad; equivalent dose, Sv (SI) or rem. An additional characteristic is penetrating depth, measured in meters.

37. See Section 24.8.

38. **(a)** ${}^{232}_{94}Pu \rightarrow {}^{4}_{2}He + {}^{228}_{92}U$, **(b)** ${}^{3}_{1}H \rightarrow {}^{3}_{2}He + {}^{0}_{-1}e$, **(c)** ${}^{60}_{27}Co \rightarrow {}^{60}_{27}Co + \gamma$, **(d)** ${}^{170}_{72}Hf \rightarrow {}^{170}_{71}Lu + {}^{0}_{1}e$.

41. **(a)** ${}^{98}_{42}Mo + {}^{1}_{0}n \rightarrow {}^{99}_{42}Mo$, **(b)** ${}^{142}_{58}Ce + {}^{1}_{0}n \rightarrow {}^{143}_{58}Ce$.

43. **(a)** ${}^{23}_{11}Na + {}^{1}_{0}n \rightarrow {}^{24}_{11}Na$, **(b)** ${}^{120}_{50}Sn + {}^{1}_{0}n \rightarrow {}^{121}_{50}Sn$.

45. **(a)** ${}^{56}_{26}Fe$; **(b)** α; **(c)** ${}^{187}_{74}W$; **(d)** 96, β; **(e)** p.

47. **(a)** β^+, **(b)** β^+, **(c)** β^-, **(d)** β^-.

49. **(a)** ${}^{60}Ni$, more favorable n/p ratio; **(b)** ${}^{74}Se$, more favorable n/p ratio; **(c)** ${}^{114}Cd$, even numbers of both protons and neutrons.

51. ${}^{218}Po$.

53. **(a)** 5.93×10^5 m s^{-1}, **(b)** 1.88×10^7 m s^{-1}.

55. **(a)** 1.38×10^4 m s^{-1}, **(b)** 3.09×10^7 m s^{-1}.

57. **(a)** 1.23 nm, **(b)** 38.7 pm.

59. 0.10599 amu, 9.5256×10^{12} J mol^{-1}, 1.1298×10^{-12} J nucleon^{-1}.

61. 5.9719×10^{10} kJ mol^{-1} or 1.3967×10^{-12} J nucleon^{-1}.

63. 8.2605 MeV.

65. 7.2 to 9.0×10^{11} J g^{-1}.

67. 4.70 MeV.

69. 6.86×10^{11} J mol^{-1}.

71. 9.3×10^{13} Bq.

73. 6.7×10^{11}.

75. 37 Bq.

77. 2.2 krad.

79. 1.0×10^{-3} rem.

81. 2.3 minutes.

83. 8.3×10^{-7} s^{-1}.

85. 1.6×10^{18} Bq.

87. 2.9×10^{12} Bq.

89. 38%.

91. 100 s.

93. 2.9×10^9 y.

95. 2.46×10^9 y.

97. 3.7×10^9 y.

Glossary of Terms

A

α (alpha) helix In proteins, a type of secondary structure in which a single polymer chain is coiled into a helix, 884
α particles Fast-moving helium nuclei (mass number 4, charge +2), emitted when certain nuclei undergo radioactive decay, 51, 990
absorbed dose Total energy of radiation received per kilogram of material, 1019
absorption (light) (1) Decrease in intensity as light passes through a substance, 227; (2) process in which an atom or molecule reaches a state of higher energy and a photon of energy equal to the increase disappears, 239
accuracy The degree to which the measured value of a quantity agrees with the true value, 11
acid see **Arrhenius, Brønsted,** and/or **Lewis acid**
acid salt Salt of a weak base, which, on dissolution in water, produces a solution with $pH < 7$; for example, NH_4Cl, 659
acidic oxide A nonmetal oxide that reacts with water to produce an acid, 362
acidity Concentration of $H^+(aq)$; extent to which a solution or substance exhibits acid properties, 644
actinides Members of the second inner transition series, elements 89 through 103, 276, 959
activation energy Minimum kinetic energy required for a collision to result in a chemical reaction, 596
active site A pocket or cleft on the surface of an enzyme molecule; a localized region, usually near a metal atom, at which catalysis takes place, 633, 885
activity (1) Effective concentration of a solute in a real (as opposed to an ideal) solution, 542; (2) total number of decay events per second within a source of radioactivity, 1019
activity coefficient The ratio of a solute's activity to its concentration, 542
actual yield Actual amount of product isolated from a chemical process, 102
acyclic Lacking a ring-like structure or substructure; see **cyclic**, 851
addition polymer Polymer formed through repeated addition reactions of a monomer containing a double bond, 868
addition reaction A reaction in which two parts of a reactant molecule form bonds to the two atoms of a double or triple bond, 409
adduct Compound formed by combination of two reactant compounds; for example, a Lewis acid-base adduct, 668
adsorption Adherence of atoms, ions, or molecules to a surface, caused by attractive forces other than chemical bonds, 544
alcohol Aliphatic compound containing a —OH (hydroxyl) group bonded to an sp^3 carbon atom; aromatic compounds containing this group are called phenols, not alcohols; many acids contain the —OH group, but are called acids rather than alcohols, 854
aldehyde A compound containing the —HC=O group, 857
aliphatic Not aromatic (that is, not related to benzene), 411
alkali elements (alkali metals) Elements of Group 1A, except hydrogen, 118
alkaline earth Any of the elements in Group 2A: Be, Mg, Ca, Sr, Ba, and Ra, 158
alkane Saturated hydrocarbon, containing single bonds only; all carbon atoms sp^3 hybridized, 402
alkene Unsaturated hydrocarbon having at least one double bond and two sp^2-hybridized carbon atoms, 408
alkyne Unsaturated hydrocarbon having at least one triple bond and two sp-hybridized carbon atoms, 410
alltrope One of several distinct forms of an element; for example, diamond and graphite for C, oxygen and ozone for O, 15
allowed Said of an orbital whose quantum numbers follow certain rules, and is therefore available for occupancy by one or two electrons, 247
alloy Metallic substance that is a mixture or solution of several metallic elements, sometimes containing nonmetallic elements as well, 32
aluminosilicate A mineral containing both AlO_4 and SiO_4 tetrahedra, 926
amide Compound containing the group $—CO—NR_2$, where R is an alkyl group or an H-atom, 864
amine Derivative of ammonia having formula RNH_2, R_2NH, or R_3N, 863
amino acid Molecule containing both the amine ($—NH_2$) and carboxylic acid ($—CO_2H$) groups, which, in α-amino acids, are bonded to the same carbon atom, 879
amorphous Lacking structure; glassy or noncrystalline, 481
ampere(A) Unit of electrical current, one of the SI base quantities: one ampere ⇔ 1 coulomb/1 second, 799
amphiprotic Capable both of donating and accepting a proton, 656
amphoteric Capable of reacting both with acids and with bases, 663
anabolic Describes a process, taking place in living organisms, in which large, complex molecules are built up from smaller ones, 905
angstrom(Å) Unit of length equal to 10^{-10} m, 226
anhydrous Lacking water, 57
aniline The simplest aromatic amine, $C_6H_5—NH_2$, 863
anion Any negatively charged ion, 46
anode The electrode at which oxidation takes place in an electrochemical cell, 804
antibonding orbital In MO theory, an orbital of greater energy and greater delocalization than a bonding orbital; electron occupany of an antibonding orbital creates a repulsive force between atoms, 387
aqueous Composed primarily of water; in an aqueous solution, the solvent is water, 94
aromatic Having alternating single and double bonds (Lewis description) or extensive π-electron delocalization (MO description) in a six-membered ring, 411
Arrhenius acid Compound that dissociates to give H^+ ions in aqueous solution, 644
Arrhenius base Source of OH^- ions, 644
Arrhenius equation $k = Ae^{(-E_a/RT)}$; it relates the value of a rate constant to temperature, 618
association Process in which several molecules or ions join together to form a larger molecule or ion; the opposite of dissociation, 555
atom The smallest particle of an element; atoms consist of one or more electrons surrounding a small, dense nucleus containing neutrons and protons; all atoms of the same element have the same number of protons and the same chemical properties, 2, 42
atomic mass unit(amu) A mass unit used for microscopic particles; 1 amu ⇔ $1/6.022137 \times 10^{23}$ g) = 1.660540×10^{-27} kg, 43

atomic number The number of protons in the nucleus (also, the number of electrons in the neutral atom), 44
atomic radius see **covalent radius**
atomic structure A small, dense, positively charged core (nucleus) containing neutrons and protons, surrounded by electrons, which move as a diffuse cloud of negative charge, 44
atomic theory (Dalton) Elements consist of identical atoms, compounds consist of two or more types of atom; atoms of different atoms differ in mass; atoms are indestructible, and during chemical reaction they undergo changes in arrangement only, 42
atomic weight More properly called molar mass, it is (1) the mass in grams of one mole of an element, averaged over the natural abundance of the isotopes, and (2) the average mass of an atom in amu, 64
atomization energy Energy required to break all bonds in a molecule, 339
ATP adenosine triphosphate, 906
aufbau Method of predicting the ground-state electron configuration of atoms, by placing electrons in the lowest available orbitals, 268
autoionization A reaction in which a pure compound (for example, H_2O or NH_3) undergoes a proton transfer reaction with itself, forming an anion and a cation, 566
autoionization constant The equilibrium constant of an autoionization reaction, 656
auto oxidation-reduction A reaction in which the same species is both oxidized and reduced; also called disproportionation, 788
Avogadro's number The number of particles in exactly 12 grams of C-12 (one mole); 6.022137×10^{23}, 61
Avogadro's principle Equal volumes of different gases at the same temperature and pressure contain equal numbers of molecules, 181
axial Describes the atoms or groups either (1) opposite to one another and perpendicular to the base of a trigonal bipyramid, 348; or (2) approximately perpendicular to the plane of a cyclohexane ring, 887

B

β (beta) particles Fast-moving electrons emitted when certain nuclei undergo radioactive decay; also called β rays, 51, 990
background radiation Ionizing radiation from the earth's crust and from outside the earth (cosmic radiation), 1019
balanced equation A chemical equation that gives the identity and relative number of moles of all reactants and products in a particular reaction, 95
band (energy band) In a solid, a huge number of MO's of similar energy merged into a single, continuous, broad energy level, 399
band gap Energy separation between the valence and the conduction bands, 400
barometer Instrument for measuring atmospheric pressure, 172
base see **Arrhenius, Brønsted,** or **Lewis base**
base quantity One of the seven fundamental quantities defined by SI, from which all other quantities are derived: mass, length, time, temperature, amount of substance, electric current, and luminous intensity, 9
basic oxide A metal oxide which, on dissolution in water, reacts to produce OH^- ions, 361
basic salt Salt of a weak acid which, on dissolution in water, produces a solution with pH > 7; for example, NaF, 659
basicity Concentration of OH^-; extent to which a compound or solution exhibits basic properties, 644
bidentate Literally, "having two teeth"; describes a ligand capable of forming two bonds to a metal ion in a coordination complex, 961
bimolecular see **molecularity**
binary compound A molecular or ionic compound composed of only two elements, 77
binding energy The cohesive energy of a nucleus, that is, the energy required to decompose it into its constituent nucleons, 996
binding energy per nucleon The average cohesive energy of the nucleons in a particular nuclide, 998
binding sites The set of locations that are or can be occupied by ligands in a coordination complex, 961
biochemistry Branch of chemistry dealing with compounds present in living organisms, and their reactions, 879
blackbody radiation Electromagnetic radiation emitted by a black (nonreflective) object when heated, 228
boat One of the conformations of cyclohexane and similar cyclic compounds, 852
Bohr atom The planetary model of atomic structure, including the assumption of constant-energy, quantized electron orbits; precursor to quantum mechanics, 236
boiling Vigorous vaporization that takes place throughout the body of the liquid rather than just at the surface, 470
boiling point Temperature at which boiling takes place; called the "normal" boiling point if the pressure is 1 atm, 470
boiling point constant The proportionality constant, different for each solvent, that relates the boiling point elevation to the solute molality, 531
boiling point elevation The extent to which the presence of a nonvolatile solute increases the boiling point of a liquid (one of the colligative properties), 528
bomb calorimeter Apparatus for measuring a substance's heat of combustion, 155
bond The force holding atoms or ions together, 53, 309; see also **covalent, ionic,** or **metallic bond**
bond angle The angle formed by imaginary lines joining the nucleus of an atom to the nuclei of two other atoms to which it is bonded, 345
bond character The degree to which a bond is ionic or covalent, 333
bond energy Energy required to separate two atoms joined by a covalent bond, 337
bond length Distance between two nuclei joined by a covalent bond; sum of covalent radii of two bonded atoms, 335
bond moment The electrical asymmetry of a polar bond, that is, the extent to which a bond has a positive end and a negative end, 330
bond order In Lewis theory, the number of shared pairs (1, 2, or 3), 319; in MO theory, ½ (bonding electrons−antibonding electrons), 389
bond strength see **bond energy**
bonding orbital In MO theory, an orbital of low energy, occupancy of which creates an attractive force (a covalent bond) between atoms, 388
borane A binary compound of boron and hydrogen, 915
boride A binary compound of boron and a metal, 915
Boyle's law Provided its temperature is not changed, the volume of a gas is inversely proportional to pressure, 173
branched Describes a hydrocarbon (or derivative) having one or more alkyl substituents on the chain; equivalently, having at least one carbon atom bonded to more than two other C-atoms, 404
branched chain reaction A chain reaction in which the number of chain carriers increases during the propagation step(s); occurs both with chemical and nuclear explosives, 1011

break The abrupt change in pH near the equivalence point of a titration, 707
breeder reactor Type of nuclear reactor that produces more fissionable material than it consumes, 1015
Brønsted acid A proton donor, that is, a species capable of transferring a proton to another molecule or ion, 650
Brønsted base A proton acceptor, that is, a species capable of accepting a proton from another molecule or ion, 651
Brønsted-Lowry theory An acid is a proton donor, a base is a proton acceptor, and all acid-base interactions are proton transfer reactions, 650–651
buffer capacity The amount of H^+ or OH^- a buffer can absorb without significant change in pH, 703
buffer solution A solution that resists or minimizes change in pH on addition of H^+ or OH^-; a solution containing significant and comparable concentrations of a weak acid and its conjugate base, 698

C

calorie (cal) (1) A unit of heat, related to the SI unit by the equivalence statement 1 cal ⇔ 4.184 J; (2) the nutritional calorie (C), equal to 1 kcal, 150
calorimeter Apparatus for measurement of heat involved in a chemical or physical process; a *bomb* calorimeter is used for measuring heats of combustion at constant volume, 151, 155
canal ray Early name for a beam of cations emitted and repelled by the positive electrode of a discharge tube, 49
capillary action The spontaneous creeping of a liquid into a narrow tube or channel as a consequence of strong adhesive forces, 457
carbide A binary compound of carbon with an element of lesser electronegativity, 918
carbohydrate A naturally occurring compound of approximate formula, $[C(H_2O)]_n$, 887
carbonyl group An sp^2-hybridized carbon doubly bonded to oxygen ($>C{=}O$); found in aldehydes and ketones, among others, 857
carboxylate An anion or compound containing the carboxylate group, $-COO^-$, 859
carboxylic acid A compound containing the group $-COOH$, 858
catabolic Describes a process in living organisms that results in the breakdown of large, complex molecules into smaller species, 905
catalysis The action of a catalyst, 630
catalyst Substance that increases the rate of a chemical reaction without itself being consumed, 630
catenation The formation of covalent bonds between more than two or three atoms of the same element, resulting in a chain-like structure, 921
cathode The electrode at which reduction takes place in an electrochemical cell, 803
cathode ray A beam of electrons emitted by the negative electrode of a discharge tube, 49
cathodic protection Prevention of corrosion of a metal by electrical connection to a more active metal, which corrodes first, 836
cation Positively charged ion,46
cell potential The difference in electrical potential (voltage) between the electrodes of a voltaic cell, 812
cell reaction The redox reaction that takes place in an electrochemical cell, 805
Celsius scale The temperature scale defined by assigning 0 °C and 100 °C, respectively, to the freezing and boiling points of water, 24
chain carrier See **chain reaction**
chain length The number of monomer units in a polymer, 868
chain reaction A reaction that proceeds by the stepwise mechanism of initiation (formation of reactive intermediates called chain carriers), propagation (formation of stable products), and termination (consumption of carriers), 629
chair The most stable conformation of cyclohexane and similar molecules, 852
charge A fundamental property of electrons (negative charge) and protons (positive charge), from which both attractive and repulsive forces can arise, 799
charge-dipole force The attractive force between an ion and a polar molecule, 458
Charles's law Provided the pressure is not changed, the volume of a gas is proportional to its absolute (Kelvin scale) temperature, 179
chelate A coordination complex incorporating one or more polydentate ligands, 962
chelating agent see **polydentate**
chemical bond General term for the attractive force that binds atoms together in compounds, 5, 309
chemical change Any change resulting in the formation of a different element or compound not present before the change, 6
chemical equation A list of the compounds and/or elements initially present, followed by an arrow and a list of the species present following a chemical change, 94
chemical kinetics see **kinetics**
chemical property Characteristic behavior observed during chemical change; for example, reactivity toward oxygen, 6
chemical reaction Chemical change involving a small number of different substances, 6
chiral (1) Having mirror-image or left-right asymmetry; (2) optically active, 881
cis- "Same side as"; prefix to the name of the geometrical isomer having substituents on the same side of a double bond, 408, a ring, 849, or a coordination sphere, 966
classical physics Theoretical physics prior to quantum mechanics; usually refers to theories of electricity, magnetism, and Newtonian mechanics, 231
close-packed Having the least possible free space between atoms in a crystal structure, 489
closed system A thermodynamic system that can exchange heat and work, but not matter, with the surroundings, 425
coagulation Separation (precipitation) of a solid phase from a colloid, 544
codon A three-base segment of a nucleic acid molecule, which directs the addition of a specific amino acid during protein synthesis, 899
coinage metals Cu, Ag, Au, and sometimes Pt, 32
colligative property A property of solutions that depends on concentration but does not depend on the chemical nature of the solute, 525
collision theory Theory of reaction rates stressing the importance and characteristics of bimolecular collisions, 596
colloid (or **colloidal dispersion)** A mixture in which very small particles are uniformly dispersed in a supporting medium; colloids have properties intermediate between those of true solutions and heterogeneous mixtures, 543
combustion A rapid, self-sustaining exothermic reaction of a substance with air or oxygen, often giving off a small part of its energy in the form of light, 134
common ion effect Influence, on an ionic equilibrium, of an additional substance containing an ion in common with the equilibrium system, 698, 746
complementary pairs The organic bases adenine-thymine or guanine-cytosine,

which are hydrogen bonded to one another in the double helix of DNA, 897
complex ion A charged coordination complex, either a cation or an anion, 754
composition The relative amounts of the different components of a mixture; often expressed numerically as percent composition, mole fraction, molarity, or molality, 5
compound A pure substance composed of two or more different elements, 7
compressibility factor A dimensionless number equal to PV/nRT, whose magnitude indicates the extent to which a real gas departs from ideal behavior, 207
concentrated Describes a solution containing a relatively large amount of solute, 108
concentration The amount of solute in a solution; often used as a synonym for molarity, but can refer to percent composition as well, 111
concentration cell A voltaic cell whose potential arises from a difference in concentration of the electrolyte in the two half-cells, 825
condensation (1) Process by which a vapor becomes a liquid, 453; (2) reaction in which two large molecules become coupled, usually with loss of a small molecule such as water from the site of coupling, 860
condensation polymer Polymer formed by a condensation reaction between polyfunctional monomers, 868
condensed phase Physical state in which molecules are in contact with one another, that is, the liquid or solid state, 453
conduction band A broad energy level in solids, composed of antibonding orbitals that are more delocalized and higher in energy than the valence band; partly occupied in metals and semiconductors, unoccupied in insulators, 400
conductor A material capable of carrying electrical current, 801
conformation Molecular shape having a characteristic angle of rotation about a single bond, 405, 850
conformational isomers Molecules that differ from one another only in the angle of rotation about one or more single bonds, 850
conjugate Denotes the proton-transfer relationship between Brønsted acids and bases; when a Brønsted acid loses a proton, what remains is the conjugate base, 654
connectivity Description of which atoms are bonded to each other in a molecule, 57
conservation of energy (law of) Energy is neither created nor destroyed in chemical or physical processes, 133
conservation of mass (law of) The total mass of the substances involved in a chemical change is unaffected by that change, or, more broadly, matter is neither created nor destroyed in chemical or physical changes, 47
constant composition (law of) Samples of a given compound have the same composition, regardless of source or history, 48
control rods Part of a nuclear reactor used to absorb neutrons and keep the chain reaction at exactly the critical level, or to shut down the reaction entirely for maintenance, 1014
conversion factor A quotient whose numerator is a quantity expressed in one unit (or set of units), and whose denominator is the same quantity expressed in other units; used multiplicatively to express a quantity in different units; for example, (1 mi h^{-1}/0.447 m s^{-1}), 12
coordinate bond Covalent bond in which both electrons are supplied by the same atom, 313
coordination complex An ion that is an adduct between a metal ion and one or more electron-pair donors called ligands, 669, 753, 960
coordination compound A neutral compound that is an adduct between a metal ion and one or more ligands, 960
coordination isomers Coordination compounds that differ from one another by the interchange of a ligand with the counterion, 968
coordination number (1) In an ionic crystal, the number of nearest-neighbor ions of opposite charge that a given ion has, 491; (2) in a coordination complex, the number of metal-ligand coordinate bonds, 960
coordination sphere The set of binding sites surrounding the metal ion in a coordination complex, 961
core (1) The set of filled subshells having lower energy and lying closer to the nucleus than the valence-shell orbitals, 271; (2) the central part of a nuclear reactor, in which the reaction takes place, 1014
core change The net positive charge an atom would have if all valence electrons were removed, 321
core electrons Those occupying orbitals in an atom's core, 271
corrosion Spontaneous oxidation of a metal exposed to the environment, 835
Coulomb (C) SI unit of electrical charge, 43
Coulomb force The force existing between electrically charged particles, 309
Coulomb's law Unlike electrical charges attract one another, and like charges repel; the energy of interaction is proportional to q_1q_2/r, 309
counter A device for detecting and measuring radioactivity, such as a Geiger counter or a scintillation detector, 1018
counterion In an ionic compound or in solution, the ion of opposite charge that accompanies an ion of interest, 658
covalent bond Cohesive force binding atoms together in molecules; in Lewis theory, a pair of electrons shared by, and in the valence shells of, two atoms, 313
covalent radius Effective radius of an atom when covalently bonded in a molecule (equal to half the internuclear separation in homonuclear diatomic molecules), 286
covalent solid (macromolecular solid) A solid whose particles (atoms) are held together by covalent bonds rather than van der Waals forces, 486
critical mass The amount of fissionable material required to sustain a chain reaction in a nuclear reactor or weapon, 1012
critical point The temperature above which a vapor cannot be liquified, no matter how great the pressure; or, the temperature and pressure above which there is no difference between the liquid and vapor phase of a substance, 478
critical temperature, pressure see **critical point**
crystal or **crystalline solid** Solid substance characterized by a definite, ordered arrangement of atoms, molecules, or ions in a regular geometric pattern, 481
crystal field splitting Energy difference, induced by ligands, in the *d* orbitals of the metal ion in a coordination complex, 972
crystal field theory Theory of bonding in coordination complexes that treats the coordinate bond as ionic, and it explains magnetic and optical properties of complexes in terms of the effect of the ligands on the metal ion's *d* orbitals, 971
crystal lattice (1) In solids, the regular pattern of atoms, molecules, or ions, repeated indefinitely in three dimensions, 480; (2) one of 14 ways of arranging a three-dimensional array of regularly spaced points, 492
crystallization Separation of a pure solid from a supersaturated solution, or, formation of crystals as a liquid cools, 745

crystallography X-ray Elucidation of the microscopic structure of crystalline solids by means of x-ray diffraction, 481
cubic Having a cubical unit cell, either simple (primitive), face-centered, or body-centered; describes both actual crystals and lattices, 489
current (electrical) Motion of charged particles, either electrons or ions, 799
cyclic Describes a compound whose atoms are bonded to each other so as to form a ring-like molecule, 850
cyclization Reaction in which a linear molecule (or a linear part of a molecule) rearranges into a cyclic structure, 887
cycloalkane, cycloalkene A cyclic alkane or alkene, 851
cyclotron Device for accelerating protons, electrons, or ions to high velocity, 1004

D

Δn_{gas} The change in the number of moles of gaseous species in a balanced chemical equation, 572
Dalton's law The total pressure exerted by a mixture of gases is the sum of the partial pressures of the components, 201
daughter An atom or nuclide produced in a nuclear disintegration, 991
de Broglie wavelength Wavelength of a moving particle, $\lambda = h/mv$, 240
Debye (D) A unit of dipole moment, equal to 3.3×10^{-30} C m in SI units, 356
decay series (radioactive) Stepwise process by which a naturally occurring radioactive isotope becomes a stable, non-radioactive species of lesser mass, 1011
degenerate Having the same energy, 264
delocalization Condition of covalent bonding in which a pair of valence electrons is shared by more than two atoms; see **resonance,** 324
density Mass per unit volume, an (intensive) physical property of a substance, 27
deposition A process in which a vapor becomes a solid, without the intermediate formation of a liquid; the reverse of sublimation, 472
derivative Any compound that can be prepared from a simpler, but similar, organic compound, 853
detergent A synthetic soap, usually a long-chain alkane or alkene whose terminal carbon atom is bonded to a polar, water-soluble group such as a sulfonic acid group, 902
dew point The temperature at which water vapor begins to condense when moist air is cooled, 469
diamagnetic Slightly repelled by a magnetic field; characteristic of substances having no unpaired electrons, 28
diatomic Adjective describing molecules or ions that consist of two atoms, such as O_2, HCl, or OH^-, 118
diffraction (1) Spreading of wave motion into a shadow area; (2) wavelength-dependent scattering of light by an array of closely spaced lines or points; related to interference, 224
diffusion One-by-one motion (as opposed to bulk transport) of molecules of one substance through those of another, 198
dihydric base A base that can accept two protons, or supply two OH^- ions, 663
dilute Containing relatively little solute, 108
dimensions The type of quantity, for example mass or time, 9
dipole-dipole force Attractive intermolecular force between two molecules having a permanent dipole moment, 458
dipole moment Numerical value of electrical asymmetry of a polar molecule, equal to the vector sum of individual bond moments, 355
diprotic see **polyprotic**
disaccharide Two monosaccharides bonded together by an —O— bridge; also called a *sugar,* 888
discrete Composed of separate, unconnected parts; a discrete spectrum, also called a *line spectrum,* is characteristic of the emission of light by gaseous atoms, 229
disperse To spread apart or separate, as light into its component wavelengths, 229
dispersed phase The small particles of a colloid, 543
dispersion force Attractive intermolecular force arising from instantaneous dipole moments (also called *London force*), 460
dispersion medium The substance in which the small particles of a colloid are suspended; also called the *continuous phase,* 543
displacement A reaction of the form A + BC $\rightarrow$ AB + C, in which one species (BC) forms a bond with an attacking species (A), while simultaneously losing a fragment (C), 668
disproportionation see **auto oxidation-reduction**
dissociation Separation into several parts, as a molecule into atoms or radicals, 337, or a compound into ions, 78
dissociation energy Energy required to break a bond, that is, separate a molecule into two fragments, 337
dissociative ionization Separation of a molecular or ionic compound into ions; always occurs when ionic compounds dissolve in water, 506
dissolution Process in which one substance forms a homogeneous mixture (solution) with another, 505
distillation Separation of the components of a liquid solution by vaporization and recondensation of the vapor, 522
distribution equilibrium Passage of a solute across the interface between two immiscible solvents at equal rates in both directions, so that the concentrations of the two solutions do not change, 554
divalent Having a charge of +2 or −2, 78
DNA see **nucleic acid**
doping Introduction of a controlled amount of impurity in order to modify the band structure and electrical behavior of a semiconductor, 401
dose The amount of energy deposited by ionizing radiation in a material (absorbed dose), or in human tissue (equivalent dose), 1019
dot structure see **Lewis structure**
double bond In Lewis theory, two pairs of shared electrons, 316; in valence bond theory, a σ-bond and a π-bond between the same two atoms, 374
dynamic equilibrium See **equilibrium**

E

effective nuclear charge Positive charge experienced by valence electrons; less than true nuclear charge because of shielding by core electrons, 288
effusion One-by-one passage of gas molecules through a small hole, 196
Einstein equation The mass-energy equivalence, $E = mc^2$, 996
elastomer A rubber-like polymer that regains its shape after being distorted, 874
electrochemical cell Apparatus in which either (1) electrical current is generated by a spontaneously occurring redox reaction *(voltaic cell),* or (2) an electrical current causes a chemical change *(electrolytic cell),* 804
electrochemistry The branch of chemistry dealing with aspects of redox reactions that involve electric current, 799
electrode A surface at which the mode of electric current conduction changes from

electronic to ionic, and at which either oxidation or reduction takes place, 803
electrode process The oxidation or reduction reaction that takes place at an electrode surface, 803.
electrolysis A process in which electrical current causes a redox reaction to take place, 828
electrolyte (1) Compound that, on dissolution, dissociates into anions and cations; *strong electrolytes* dissociate completely, *weak electrolytes* dissociate partially, 540; (2) in an electrolytic cell, the medium (liquid, paste, or gel) in which ionic conduction takes place, 804
electrolytic cell see **electrochemical cell**
electromagnetic radiation Periodic fluctuation in perpendicularly oriented electric and magnetic fields, travelling at 3×10^8 m s^{-1} through a vacuum and somewhat more slowly through material substances, 226
electromagnetic spectrum The collective name for electromagnetic radiation of all wavelengths, arranged in order of increasing wavelength or frequency; names are assigned to different wavelength ranges, for example, "microwaves" or "γ-rays," 226
electron Negatively charged elementary particle having a mass of 0.0005485 amu; constituent of the "electron cloud" surrounding an atomic nucleus; carrier of current in metallic conduction, 43
electron affinity ΔE_{rxn} (kJ mol^{-1}) for the reaction in which a gas phase atom acquires an electron to become a singly charged negative ion, 310
electron cloud (1) The diffuse negative charge in the region surrounding an atomic nucleus; (2) the form taken by electrons when they are part of atoms, 44
electron configuration The way in which the electrons of an atom or ion are distributed into the available orbitals, 265
electron deficient Describes an atom, in a molecule or ion, that has an incomplete valence shell; for example, B in BF_3 has only 6 valence electrons, 326
electron density (1) The value of the charge-to-volume ratio at a particular point within the electron cloud surrounding an atomic nucleus—qualitatively, the thickness of the cloud; (2) the square of the electron wave function, 246
electron pair geometry In VSEPR theory, the orientation of the bonds and lone pairs around an atom in a molecule or ion, 346
electron transfer reaction A redox reaction involving the direct transfer of an electron from one species to another, 769
electron wave As described by the Schrödinger equation, the motion or condition of an electron in an atom, 244
electronegativity The relative strength of attraction of a bonded atom for its valence electrons; Pauling electronegativities, on a scale of 0–4, are the most widely used, 331
electronvolt (eV) Energy unit used in nuclear chemistry; 1 eV $\Leftrightarrow 1.602 \times 10^{-19}$ J, 990
electroplating Coating an object with a thin layer of metal by electrolytic reduction of the metal ion, 834
element A substance composed entirely of the same type of atoms, that is, those having the same atomic number, 7
elemental symbol A one- or two-letter abbreviation of the name of an element (elements 106–109 have temporary three-letter symbols), 7
elementary charge unit (ecu) Smallest unit of electrical charge found in nature, equal to 1.60219×10^{-19} coulombs, 43
elementary particles Protons, neutrons, and electrons, 43
elementary step A chemical reaction that takes place as a single molecular event; see mechanism, 624
emission Process in which an atom, ion, or molecule drops from an excited state to a lower level, and a photon with energy equal to the energy decrease appears, 237
empirical formula Ordered list of elemental symbols, with whole-number subscripts giving the relative number of atoms or ions of each type in a compound, 56
enantiomers Two optical isomers whose structures are mirror images of one another, 881
endergonic Describes a reaction for which $\Delta G > 0$, that is, one that consumes free energy, 905
endothermic Describes a process in which a system gains heat from the surroundings; $q > 0$, 134
endpoint In a titration, the point at which sufficient titrant has been added so that the indicator changes color; see equivalence point, 715
energy Capacity for doing work, 3; energy is either kinetic or potential, 420
energy band A broad range of energy available to the valence electrons of a solid material, formed by merging a large number of atomic energy levels, 399
energy level One of the quantized states available to an atom or molecule, or a group of such states (the *p* subshell for instance) having the same energy, 236
energy of reaction see **heat of reaction**
enthalpy An extensive property of state, which measures the heat transferred to a system during a constant-pressure process, 431
enthalpy of formation The enthalpy change when one mole of a compound is formed from its standard-state elements; called the standard enthalpy of formation when all reactants and products are in their standard states at 1 atm, 432. Also called heat of formation, 141
enthalpy of reaction see **heat of reaction**
enthalpy of solution see **heat of solution**
entropy An extensive property of state measuring the extent to which a system is disordered, or randomly arranged, 437
entropy of fusion The entropy change that accompanies the melting of one mole of a solid substance, 472
entropy of reaction The change in entropy that accompanies a chemical reaction; called the *standard* entropy of reaction, ΔS°_{rxn}, when all reactants and products are in their standard states at 1 atm, 440
entropy of sublimation The entropy change that accompanies the sublimation of one mole of a solid, 473
entropy of vaporization The entropy change that accompanies the vaporization of one mole of a liquid, 472
entropy, standard Entropy of one mole of a substance at 1 atm pressure, 440
enzyme A high molecular-weight protein that catalyzes a chemical reaction within a living organism, 633, 885
equation Shorthand description of a reaction; reactant formulas on the left are separated by an arrow from product formulas on the right, 94
equatorial The three positions, or atoms bonded at those positions, that are (1) in the basal plane of the trigonal bypyramid geometry, 348; or (2) approximately parallel to the plane of a cyclohexane ring, 888
equilibrium The condition in which a reversible chemical or physical change proceeds with equal rate in both directions, resulting in no measureable change in the system, 465
equilibrium constant The numerical value of the reaction quotient evaluated at equilibrium; the equilibrium constant is characteristic of a reaction at a given temperature, 558
equilibrium expression The algebraic form of the equilibrium constant; see **reaction quotient,** 558

equilibrium sublimation pressure see **sublimation pressure**
equilibrium vapor pressure see **vapor pressure**
equivalence point In a titration, the point at which a stoichiometrically equivalent amount of titrant has been added, that is, at which all the sample has reacted, 671
equivalence statement A relationship between the magnitudes of the same quantity expressed in different units; for example, 10 km ⇔ 6.2 miles, 12
equivalent (1) In acid-base reactions, the amount of acid required to supply, or to react with, one mole of H^+ ions; (2) in redox reactions, the amount of a reductant or oxidant involved in the transfer of one mole of electrons, A.15, A.18
equivalent dose The absorbed dose of radiation, adjusted for the susceptibility of human tissue from different kinds of radiation; measured in rems or millirems, 1020
ester A compound containing the group RCO_2R', produced by reaction of an acid and an alcohol, 860
ether A compound containing the functional group C—O—C, 586
evaporation see **vaporization**
excited state Any stationary state with energy greater than the ground state, 255, 268
exclusion principle see **Pauli**
exergonic Describes a reaction for which $\Delta G < 0$, that is, one that produces free energy, 905
exothermic Describes a chemical or physical process that generates heat, 134
expanded valence shell A valence shell (of a bonded atom) containing more than eight electrons; not possible for first- or second-period elements, 328
exponential notation see **scientific notation**
extensive property A physical property, for example volume, whose numerical value depends on the amount of substance present, 28
extent of reaction The change in concentration (or partial pressure) of a reactant (or product) as a reaction proceeds from an initial condition to a later (usually equilibrium) condition, 579

F

fact A property or behavior confirmed by repeated observations by different observers; for example, "carbon monoxide is a colorless, toxic gas," or "lions hunt in packs," 3
factor-label method A problem-solving technique in which the units of quantities are treated algebraically in the same way as pure numbers, 15
Fahrenheit scale Used for nonscientific purposes in the United States, the temperature scale on which water freezes at 32 °F and boils at 212 °F, 24
faraday The charge on one mole of electrons, 1 $\mathcal{F}$ = 96,487 coulombs, 822
fat Mixture of triglycerides that is solid at room temperature; usually of animal origin, 901
fatty acid An organic compound containing a linear chain of 14 to 20 carbon atoms that terminates in a $—CO_2H$ group, 901
ferromagnetism Strong attraction to a magnetic field, due to cooperative behavior of many electron spins in a relatively large region of the structure of some metals or alloys, 282
fiber Hair-like substance of high tensile strength, usually a crystalline polymer, 870
first law of thermodynamics The change in a system's internal energy is the sum of the heat and work transferred to the system from the surroundings; $\Delta E = q + w$, 426
first order see **order**
fission Breaking apart of a large nucleus into two fragments of moderate mass, 1000; fission can be spontaneous (rare) or induced, 1011
fluid A liquid or gas; capable of flowing, 455
forbidden Said of an orbital that cannot exist because it would violate one or more of the quantum mechanical rules, 247
force Strength of a push or action, 421
force, adhesive The intermolecular force of attraction between two different materials, 456
force, cohesive The attractive force between the molecules, atoms, or ions of a pure substance, 453
formal charge Electrical charge assigned to each atom in a molecule or ion to serve as a rough guide to the actual charge distribution; evaluated for a particular Lewis structure, 321
formation constant The equilibrium constant of a ligand exchange reaction in which a coordination complex having only H_2O as ligands is converted to a complex having only a second species as ligands, 754, 978
formation reaction A reaction in which one mole of a compound is formed from its constituent elements, all in their standard states, 141
formula unit A set of atoms corresponding to the empirical formula of an ionic compound; analogous to a molecule, 56
formula weight More properly called molar mass, the mass of one mole of formula units, 67
fractional ionization The extent to which a weak acid (or base) ionizes in solution; also expressed as percent ionization, 688
fractional precipitation A separation or purification technique based on differing solubilities, 745
free energy The combination of enthalpy and entropy defined by $\Delta G = \Delta H - T\Delta S$; ΔG of a system is negative in all spontaneous processes, and zero at equilibrium, 442
free energy of formation, standard (ΔG_f°) The change in free energy that occurs when one mole of a compound is formed from its elements in their standard states; $\Delta G_f^\circ = 0$ for elements in their standard states, 443
free energy of reaction The change in free energy that accompanies a chemical reaction; called the *standard* free energy of reaction, ΔG_{rxn}°, when all reactants and products are in their standard states at 1 atm, 444
free radical A reactive fragment of a molecule, having an unpaired electron, 870
freezing Process in which a liquid becomes a solid, 463
freezing point see **melting point**
freezing point constant The proportionality constant, different for each solvent, that relates the freezing point depression to solute molality, 531
freezing point depression The extent to which the presence of a nonvolatile solute decreases the freezing point of a liquid; one of the colligative properties, 528
frequency The number of crests (or troughs) of a wave passing a fixed point per second, 226
frequency factor The pre-exponential or "A" factor in the Arrhenius equation that relates the value of a rate constant to temperature, 618
fuel cell A voltaic cell in which an externally stored fuel like H_2 or CH_4 is oxidized by O_2, 827
fuel rods Long tubes packed with uranium oxide, used in a reactor core to contain and position the fissionable "fuel," 1014
functional group A subset of atoms within a molecule that serves as the site of,

and exerts a dominant influence on, chemical reactivity, 853
fundamental quantity see **base quantity**
fusion (1) Melting of a solid, 463; (2) joining of two nuclei to form a nucleus of greater mass, 1000

G

γ (gamma) rays High-energy radiation emitted when certain nuclei undergo radioactive decay, 51, 990
gas One of the three common physical states of matter; a gas takes on the shape and volume of its container, 5
Geiger counter A device for detecting and measuring radioactivity, 1018
genetic code The set of correspondences between three-base segments (codons) of nucleic acid molecules and the amino acids the codons specify during protein synthesis, 899
geometrical isomer One of two isomers differing only in (1) placement of substituents *cis-* or *trans-* at a doubly bonded carbon atom in an organic compound, 408, or (2) in arrangement of ligands around the metal in a coordination complex, 966
glass A covalent solid whose microscopic particles are not regularly arranged; glasses do not have a distinct melting point, 487
glycoside Product of a condensation reaction in which one of the —OH groups of a monosaccharide is replaced by an —OR group from another molecule (often another monosaccharide), 888
glycoside linkage The —O— bridge joining two monosaccharides, 889
ground state In an atom, ion, or molecule, the energy level (stationary state) having the lowest possible energy, 255
group (1) The set of elements in the same vertical column of the periodic table; members of a group have the same valence electron configuration and share similar chemical properties, 8; (2) a specific and distinctive part of a molecule—for instance —CH_3, the methyl group, 404

H

half-cell Part of an electrochemical cell, in which either oxidation or reduction (but not both) takes place, 805
half-life Also called half-time, $t_{1/2}$ is the time required for one-half of the initially present reactant or radioactive nuclei to disappear, 614, 1002
half-reaction Either the oxidation or reduction part of a redox reaction, 769
halogen Member of Group 7A, having valence-shell configuration ns^2np^5, 294
heat The form of energy associated with high temperature and transferred from one place to another as a result of a temperature difference; also called *thermal energy,* 133, 424
heat capacity The amount of heat required to raise the temperature of an object by one degree Celsius (or 1 K); units are $J\ K^{-1}$, $J\ °C^{-1}$, or cal deg^{-1}, 146
heat capacity at constant pressure (c_P) Heat required to raise the temperature of one mole of a substance by one degree Celsius (or 1 K) in a process occurring without change in pressure; units are $J\ mol^{-1}\ K^{-1}$ or $J\ mol^{-1}\ °C^{-1}$, 149
heat of combustion The amount of heat released when a compound burns, that is, reacts rapidly with oxygen, 155
heat of formation, standard see **enthalpy of formation**
heat of fusion The amount of heat required to melt one mole of a solid substance, 145
heat of reaction The amount of heat released or absorbed during a chemical reaction, 134, 431
heat of solution The amount of heat released (or absorbed) when a substance dissolves, 509
heat of sublimation The amount of heat required to sublime (vaporize) one mole of a solid substance, 473
heat of vaporization The amount of heat required to vaporize one mole of a liquid substance, 145
Henry's law The solubility of a gas in a liquid is proportional to the partial pressure of the gas, 524
hertz(Hz) The SI unit of frequency: 1 hertz $\Leftrightarrow$ 1 $second^{-1}$, 227
Hess's law The heat of reaction is the same whether a chemical change occurs all at once or as a series of steps, 137
heterocyclic compound A cyclic compound having at least two different kinds of atom in the ring(s), 852
heterogeneous Having regions of different physical state or composition, 5
heterogeneous catalyst A material whose surface increases the rate of a chemical reaction, 632
heterogeneous reaction A chemical change taking place at the interface between two phases, 567
heteronuclear diatomic molecule A molecule consisting of two different atoms, 395
high-spin see **weak-field**
homogeneous Having the same composition throughout; uniform, 5
homogeneous catalyst A substance, present in the same phase as the reactants, that increases the rate of a reaction, 632
homonuclear diatomic molecule Molecule consisting of two identical atoms, 387
Hund's rule The ground-state configuration of partly filled subshells is the one that maximizes total electron spin, 269
hybrid orbital An atomic orbital, formed by mixing two or more of the *s*, *p*, or *d* hydrogen like orbitals of an atom, used to describe bonding and geometry in molecules, 376
hybridization The particular type of hybrid orbitals used to describe the bonding of an atom, as in "the hybridization of the carbon atom in H_2CO is sp^2," 376
hydrate A solid compound whose crystal structure incorporates one or more molecules of water per formula unit, 57
hydration Solvation by water, 648; see **solvation**
hydride (1) The H^- ion, or (2) an ionic compound whose anion is H^-, 79
hydrocarbon Binary compound of carbon and hydrogen, 401
hydrogen bond Inter- or intramolecular force between a hydrogen atom, covalently bonded to N, O, or F, and a nearby N, O, or F atom, 460
hydrogenlike orbitals Orbitals of many-electron atoms, very similar to the orbitals that are the solutions to the Schrödinger equation for the hydrogen atom, 264
hydrolysis Reaction with water; in *salt hydrolysis,* an ionic species donates a proton to, or accepts one from, a water molecule, producing an acidic or basic solution, 660, 694
hydronium ion H_3O^+, 649
hydrophilic "Water-loving," said of a highly polar or otherwise water-soluble molecule or group, 902
hydrophobic "Water-hating," said of a nonpolar molecule or group, 902
hydroxyl (group) The —OH group, part of a larger molecule, 854
hypothesis Tentative explanation of a relationship among a set of facts, 3

I

ideal gas law $PV = nRT$; the product of the pressure and volume of a gas is proportional to the amount and the absolute temperature, 181

ideal solution A solution whose components obey Raoult's Law, 518
indicator Dissolved substance having pH-dependent color, used to detect the endpoint of a titration, 671, 715
induced dipole Temporary dipole moment in an atom or molecule, arising from distortion of the valence electron cloud by a nearby charged or polar species, 459
inductive effect A change in bond strength, polarity, or reactivity caused by the electronegativity of an atom in a distinct part of the molecule, 665
inert Unreactive; said of a coordination complex that requires more than a few minutes to undergo a ligand exchange reaction, 980
inert gas Earlier name for noble gas
infrared Spectral region adjacent to, but with longer wavelength than, red light, 223
initial rates, method of Determination of a rate law and rate constant by measurements of reaction rates during the early part of a reaction, 600
initial step see **chain reaction**
inner electrons Electrons that are in the noble gas core of an atom; not valence electrons, 271
inner salt see **zwitterion**
inner transition element One of the lanthanides ($Z = 57$ to 71) or actinides ($Z = 89$ to 103), 276, 959
inorganic Not organic; inorganic compounds (with a few exceptions) contain no carbon, 77
insulator Material that does not conduct electricity (or heat), 40, 802
intensive property A physical property of a substance, whose numerical value is independent of the amount of substance present, 28
interference Wave interaction; in *constructive interference,* waves add to one another so that the resulting amplitude (or power, or intensity) increases; in *destructive interference* the result is a decrease in amplitude, sometimes a complete cancellation of two waves, 224
interhalogen Molecular compound consisting of two or more different halogen atoms, 300
intermediate A chemical species (usually highly reactive), produced in one step in a mechanism and consumed in a later step, 628
intermediate salt Salt whose anion is a polyprotic acid that has lost more than one (but not all) of its ionizable hydrogens, for example Na_2HPO_4; all intermediate salts are amphiprotic, 722
internal energy An extensive thermodynamic property of state, equal to the sum of kinetic and potential energies of all atoms in a system; increase in internal energy is equal to heat input at constant volume, 426
internuclear axis Imaginary line passing through the centers (nuclei) of two adjacent atoms in a molecule, 373
ion Electrically charged particle, formed by loss or gain of electron(s) by an atom or molecule, 46
ion-electron method A procedure of separately balancing the half-reactions of a redox reaction, 781
ionic bond The cohesive force, arising from Coulomb attraction, binding ions together into a compound, 309
ionic character On a scale of 0% to 100%, the extent of polarity or charge separation in a polar covalent bond, 334
ionic compound A compound consisting of ions rather than atoms or molecules; except for oxides and hydroxides, ionic compounds are also called *salts,* 54
ionic conduction Electrical current carried by moving ions, in a solution or a molten salt, 802
ionic equation A chemical equation that shows ionic compounds as separate ions; used to describe reactions in water or other polar solvents, 779
ionic radius Effective radius of an ion, determined by measuring internuclear distances in ionic compounds, 292
ionic solid A solid substance whose microscopic particles are ions, 485
ionization constant Equilibrium constant for the proton transfer reaction of an acid (K_a) or base (K_b) with water, 684
ionization energy The energy required to remove an electron from the highest occupied orbital of an atom, molecule, or ion (also called *ionization potential),* 256, 290
ionizing radiation Electromagnetic radiation or nuclear particle having sufficient energy to leave a trail of ions as it passes through matter, 1017
ion pair Two oppositely charged ions bound together by the Coulomb force into a single unit; an "ionic molecule," 311
ion product The reaction quotient for the dissolution reaction of a slightly soluble salt, 732
isoelectronic Having the same number of electrons, 293
isomer One of several different compounds having the same molecular formula but different connectivity or spatial arrangement of atoms, 404
isotonic Describes two or more solutions having the same osmotic pressure, 538
isotopes Variant forms of atoms of the same element, having the same number of protons but a different number of neutrons, hence a different mass, 45
IUPAC International Union of Pure and Applied Chemistry, an international body that oversees, among other things, systematic nomenclature in chemistry, 8

J

joule The SI unit of energy: $1 \text{ J} \Leftrightarrow 1 \text{ kg m}^2 \text{ s}^{-2}$; all forms of energy, including heat, work, enthalpy, and free energy, can be measured in joules, 133

K

kelvin scale "Absolute" temperature scale on which the theoretical limit of low temperature is 0 K, and the freezing point of water is 273.15 K, 24
ketone A compound in which two alkyl groups are bonded to a carbonyl group, $R_2C{=}O$, 857
kinetic energy Energy of motion; $KE = mv^2/2$, 420
kinetic molecular theory Matter consists of small particles in random motion, with an average energy proportional to temperature; gas-phase particles are widely separated, move independently, and undergo occasional elastic collisions, 191; liquid and solid particles are in contact, and their motion is largely vibrational, 453
kinetics Study of the rates and mechanisms of chemical reactions, 595

L

labile Easily moved; said of a coordination complex in which ligand exchange is fast, 980
lanthanide contraction The decrease in atomic radii of the sixth period elements beyond the lanthanides from the values expected in comparison to the radii of corresponding elements in the fifth period, because of the increased nuclear charge introduced by the "extra" lanthanide elements, 957
lanthanides Elements in the first inner transition series, mostly having a partly filled 4*f* subshell; also called the *rare earths,* 276, 959
lattice points Locations in a three-dimensional array characteristic of a crystalline

substance; in a crystal, all lattice points are occupied by identical particles, 480
law A concise summary statement of a set of related facts; for example, Murphy's law states, "If something can go wrong, it will," 3
Le Châtelier's principle A system at equilibrium responds to an external influence so as to reduce the effect of the disturbance, 510
leveling effect No acid stronger than H_3O^+, or no base stronger than OH^-, can exist in aqueous solutions, 661
Lewis acid An electron-pair acceptor, capable of forming a coordinate bond with an electron-pair donor, 667
Lewis base An electron-pair donor, that is, a species having a nonbonding valence electron pair, capable of forming a coordinate bond with a Lewis acid, 667
Lewis structure A representation of a molecule or ion using dots and lines to show the locations of valence electrons and covalent bonds; also called "dot structure," 313
ligand An electron-pair donor molecule or ion that has formed a coordinate covalent bond with a metal ion, 960
ligand exchange In coordination chemistry, a reaction in which one ligand is replaced by a different one, 976
light Electromagnetic radiation in the approximate wavelength ranges 10–400 nm (ultraviolet), 400–700 nm (visible), 0.7–1000 μm (infrared), 226
limiting reactant As compared to all others, a reactant present in less than stoichiometric amounts and therefore controlling the amount of product, 108
linear In a straight line, as the atoms of some polyatomic species, 346
line spectrum Emitted or absorbed light whose wavelength distribution consists of a set of discrete lines, each of which extends over a very narrow wavelength range; characteristic of atomic rather than molecular substances, 229
linkage isomers Coordination complexes that differ from one another only in that a different atom of the same ligand is bonded to the metal ion, 968
lipid A biological molecule soluble in nonpolar solvents, but insoluble in water, 900
liquid One of the three common physical states of matter, characterized by fluidity and a variable shape, but having a fixed volume, 5
liquid crystal A material in which long, rod-like molecules are well ordered in one or two dimensions but randomly arranged in the third, 496
lobe Part of an orbital, separated from other lobes by a *node,* 250
London force see **dispersion force**
lone pair A pair of valence electrons not used for bonding (unshared electrons), 315
low-spin see **strong-field**

M

macromolecular solid see **covalent solid**
macromolecule see **polymer**
macroscopic Large scale; in chemistry, any quantity referring to a large number of molecules (say an amount large enough to weigh) is called a macroscopic quantity, 59
magic number A number of neutrons, protons, or both that confers exceptional stability on a nuclide, 993
main-group element Member of a group designated "A" in the periodic table; not a transition element, 913
malleable Capable of being bent, drawn, or beaten into a new shape without breaking; most metals are malleable, 31
many-electron species An atom or monatomic ion having more than one electron, 264
mass An amount or quantity of matter, 2
mass defect The difference between the mass of a nucleus and the total mass of its component nucleons, 996
mass number The sum of the number of protons and neutrons in an atom, 45
mass percent 100 times the ratio of the mass of a particular component to the total mass (same as weight percent), 109
mass spectrometer Instrument for the precise measurement of the masses of atoms and molecules, 53
matter Anything that has mass and occupies space, 2
matter wave Term used when discussing the wave properties of particles, for instance a beam of electrons in an electron microscope, 240
Maxwell-Boltzmann distribution The probability that a randomly chosen molecule in a gaseous substance will have a certain velocity, and the dependence of that probability on molar mass and temperature, 193
mechanism Explanation of the course of a chemical reaction in terms of a sequence of molecular encounters or events called elementary steps, 624
melting The process in which a solid becomes a liquid, 463
melting point Temperature at which a solid melts, or at which solid and liquid can coexist; called the *normal melting point* if the pressure is one atmosphere; also called the *freezing point,* 463
membrane A thin layer of material separating regions of different composition; a semipermeable membrane allows the free passage of some molecules or ions, but not others, 535
metal An element, compound, or mixture exhibiting the metallic properties given in Table 1.7, 31
metallic bonding The cohesive force in metals, a type of bonding characterized by extreme delocalization of valence electrons, 398
metall character The extent to which an element has chemical and physical properties characteristic of a metal; generally increases with atomic number in a group, and decreases with atomic number in a period, 293
metallic conduction The mechanism by which electrical current is carried in metals, namely by freely moving electrons, 801
metallic solid The type of solid whose cohesive force is the metallic bond, 488
metalloid An element having properties of both metals and nonmetals, or intermediate between metals and nonmetals; one of the elements B, Si, Ge, As, Sb, or Te, 31
metallurgy The technology of extracting pure metals from their ores or preparing metal alloys, 32
metal/salt electrode A metal in contact both with a solution containing the metal cation and with an insoluble salt containing the same cation, 807
metal/solution electrode A metal in contact with a solution containing the cation of the metal, 807
metastable Describing a system or particle, such as an atomic nucleus, that is temporarily in an excited state, 991
metric system Worldwide system of decimal units, very similar to SI, 9
microscopic Small scale; no larger than a few molecules, 59
miscible Capable of forming solutions of any concentration, said usually of two liquids, 516
mixture Sample of matter containing at least two different elements or compounds, 5
mmHg same as **torr**

model A hypothesis, often in pictorial form, 3
moderator Material used in nuclear reactors to decrease the kinetic energy of neutrons, making them more susceptible to capture by fissionable nuclides, 1013
molality (*m*) Measure of solution composition, equal to moles of solute per kilogram of solvent, 528
molar heat capacity Heat capacity per mole; units are J mol^{-1} K^{-1}, J mol^{-1} $°C^{-1}$, or cal mol^{-1} $°C^{-1}$, 148
molar mass The mass of one mole of an element, compound, or ion; also called *atomic weight, molecular weight,* and *formula weight,* 64
molar volume Volume occupied by one mole of substance (for elements, a periodic property), 284
molarity (M) Measure of solution composition equal to the number of moles solute per liter of solution, 111
mole Counting unit, similar to a dozen, but equal to Avogadro's number of items; used for counting microscopic particles: one mole of an element is N_A atoms, one mole of a molecular compound is N_A molecules, one mole of an ionic compound is N_A formula units, and so on, 59
molecular compound Compound consisting of molecules rather than atoms or ions; also called *covalent compound,* 53
molecular equation A chemical equation in which all reactants and products appear as neutral compounds, even though it is their ionic components that actually participate in the reaction, 779
molecular formula Ordered list of elemental symbols, with whole-number subscripts giving the number of atoms of each type in a molecule or polyatomic ion, 56
molecular orbital A valence orbital associated with an entire molecule (rather than two atoms), 371
molecular shape Spatial arrangement of the atoms of a molecule or ion; includes approximate bond angles, either explicitly or implicitly, 344
molecular solid A solid whose microscopic particles are molecules, held together by van der Waals forces, 485
molecular weight More properly called molar mass, the mass in grams of one mole of a compound, 67
molecularity The number of reactant molecules involved in an elementary step, which is uni-, bi-, or trimolecular, 625
molecule A stable collection of atoms joined by covalent bonds; the smallest possible quantity of a molecular or covalent substance, 53
mole fraction The ratio of the number of moles of one component to the total number of moles in a mixture, 200
mole ratio An equivalence statement relating the number of moles of two species participating in a reaction, 99
monatomic Consisting of only one atom, as "monatomic gas" or "monatomic ion," 54
monodentate Describes a ligand capable of forming only one bond to a metal ion in a coordination complex, 961
monomer Compound from which a polymer is made, 867
monosaccharide A 4-, 5-, or 6-carbon polyhydroxy aldehyde or ketone; also called a sugar, 887
monovalent Having a charge of +1, or −1, 78
multiple bond In Lewis theory, a covalent bond involving more than one electron pair, 316
multiple proportions (law of) If elements combine to form more than one compound, the combining ratios (by mass) are related by small whole numbers, 48

N

natural abundance Percentage of a particular isotope in a macroscopic sample of an element, 46
Nernst equation The algebraic expression relating the potential of an electrochemical cell to electrolyte concentrations, 821
net ionic equation An ionic equation from which spectator ions have been removed, 644, 779
network solid One whose atoms are held together by covalent bonds rather than van der Waals forces; less often, the term "network solid" is used to refer to an ionic solid, 485
neutralization Reaction of an acid with a base, resulting in the disappearance of acidic and basic properties, 644
neutral salt An ionic compound formed by reaction of a strong acid with a strong base; a neutral salt has no effect on pH when dissolved, 698
neutron Uncharged particle with a mass of 1.0087 amu, a constituent of all atomic nuclei except H, 43
newton (N) The SI unit of force: 1 N ⇔ 1 kg m s^{-2}, 422
noble gases The elements of Group 8A; also called *rare gases,* and formerly called *inert gases,* 82
node (1) A point (in a one-dimensional standing wave) or a surface (in a three-dimensional wave or orbital) where the amplitude of the wave is zero; (2) in an orbital, a surface of zero electron density, 245
nonpolar Having a dipole moment of zero; electrically symmetrical, 355
normality Concentration expressed in equivalents of solute per liter of solution, A.16; see **equivalent**
nuclear reactor Device for carrying out nuclear transformations on a large scale, usually for the purpose of generating electrical power, 1013
nucleic acid A polymer whose monomeric units are a sugar, a phosphate ester, and one of five organic bases; found in and near cell nuclei; in RNA (ribonucleic acid) the sugar is ribose, while in DNA (deoxyribonucleic acid) the sugar is deoxyribose, 894
nucleon One of the two subatomic particles in an atomic nucleus; a proton or a neutron, 45
nucleoside Part of a nucleic acid, a molecule or fragment consisting of a sugar and an organic base, 895
nucleotide The monomeric unit of nucleic acids, composed of a sugar, a phosphate ester, and one of five organic bases, 894
nucleus (plural: nuclei) The small, central region of an atom, containing >99.9% of the mass and all of the positive charge, 44
nuclide An atom having a specific number of neutrons and protons, 989
nutritional calorie (C) Equal to 1 kcal or 4184 J, the Calorie is a unit of heat or energy used to describe the fuel value of foods, 150

O

octahedron (1) Eight-sided (diamond-shaped) solid; (2) electron-pair geometry of species having six electron pairs on the central atom, 346
octet rule In Lewis theory, covalently bonded atoms share electrons so as to achieve a complete valence shell containing eight electrons (s^2p^6), 313, 315; main-group atoms form monatomic ions having the noble gas, valence-shell structure of eight electrons, 311

odd-electron species A molecule having an odd number of electrons, 327
oil A mixture of triglycerides, usually of plant origin, that is liquid at room temperature, 901
oligo- Prefix meaning a small but indeterminate number, usually less than a dozen or so, 889
oligosaccharide A carbohydrate consisting of a small number (but not less than three) of monosaccharide units, 889
one-electron species H-atom, or any ion such as He^+, Li^{2+}, or Ar^{17+}, from which all electrons but one have been removed, 256
open system A system that can exchange heat, work, and matter with the surroundings, 425
optical activity The physical property that causes a compound (pure or dissolved) to rotate the plane of polarized light, 881
optical isomers Molecules that differ only in the bonding configuration of one or more chiral atoms, 887
orbital (1) A shaped region in space, near the nucleus, that defines where one or two electrons may be found; (2) the wave function of an electron in an atom, molecule, or ion, 247
orbital diagram A sketch showing the occupancy of each orbital in an atom, molecule, or ion, denoting the spins of the electrons with the symbols ↑, ↓, or ⇅, 268
order In kinetics, the sum of the exponents of concentrations in the rate law, 600
ore Rocks and soil containing usefully high amounts of a metallic element or its compound(s), 32
organic Containing carbon and hydrogen, usually in combination with oxygen, nitrogen, and/or sulfur, 77
osmosis Slow passage of solvent through a semipermeable membrane, which blocks the passage of solute, 536
osmotic pressure Pressure required to counteract and halt the flow of solvent through a membrane, 536
outer electrons Electrons in orbitals lying mainly outside of, and having higher energy than, the noble gas core, 271
overall reaction The net chemical change associated with a reaction that takes place in two or more steps, 624
overlap Constructive interference of valence orbitals on two adjacent atoms, creating an internuclear region of enhanced electron density and forming a covalent bond, 371
oxidant Reactant capable of causing oxidation of another species; also called an *oxidizing agent,* 297, 768
oxidation Loss of one or more electrons, resulting in an increase in oxidation number, 769; in organic and biochemical reactions, loss of H or gain of O by a compound, 862
oxidation number An integer that indicates the extent of electron loss (or gain) by an atom in a compound or ion, compared to its number of electrons in the pure element; also called *oxidation state,* 769
oxidation number method A procedure for balancing redox equations, beginning with the assignment of oxidation numbers, 776
oxidizing agent see **oxidant**
oxoacid A molecular acid whose acidic hydrogen is bonded to, and on dissolving dissociates from, an oxygen atom; H_2O, HNO_3, and HClO are oxoacids, while H_2S and HCl are not, 81
oxoanion A polyatomic anion whose central atom, not oxygen, is bonded to one or more oxygen atoms, 81

P

π (pi) bond In valence bond theory, a covalent bond formed by side-to-side overlap of two *p* orbitals, and characterized by two off-axis lobes and a nodal plane, 373
paired spins Two electrons in the same orbital, one with $m_s = +\frac{1}{2}$ and the other with $m_s = -\frac{1}{2}$, 267
paramagnetic Slightly attracted by a magnetic field; all substances containing at least one unpaired electron per formula unit are paramagnetic, 281
parent orbital Valence-shell, hydrogen-like orbital from which hybrid orbitals are constructed, 376
partial charge Electrical charge acquired by an atom in a molecule as a result of unequal sharing of electrons in a polar covalent bond, 330
partial pressure The pressure exerted by one component of a mixture of gases, equal to the pressure that would be exerted if all other gases were removed from the container, 201
pascal (Pa) The SI unit of pressure; 1 atm ⇔ 1.01325×10^5 Pa, 171
Pauli exclusion principle No two electrons in an atom may have the same values of the quantum numbers $\{n, \ell, m_\ell, m_s\}$; or, no more than two electrons may occupy the same orbital, 267
penetration Extent to which an outer orbital lies within the core and interacts with the nucleus, 289
percent composition (1) A list of the mass percents of each of the elements in a compound, 68; (2) the amount, expressed either as a mass or volume percent, of a solute in a solution, 108
percent ionization Extent to which a weak acid or base ionizes when dissolved, 652, 688
percent yield 100 times the ratio of actual yield to theoretical yield, 102
period Horizontal row of the periodic table, 8
periodicity, or **periodic law** In a list of elements in order of increasing atomic number, similar chemical and physical properties recur periodically, 284
periodic table Listing of elements in two-dimensional format, in order of increasing atomic number and grouped together according to similar electron configuration and chemical properties, 8
periodic trend A regular variation in a physical or chemical property as atomic number increases, 83
permanent dipole An electrical asymmetry that arises from the inherent shape of a molecule, as opposed to a temporary *induced* dipole, which is caused by environmental influences, 459
pH Measure of the acidity or $H^+(aq)$ concentration of a solution: $pH = -\log[H^+(aq)]$, 645
phase One of the components of a heterogeneous mixture, 5
phase diagram For a given substance, a graph showing which physical state (or states) is (are) stable in each region of temperature and pressure, 476
phase transformation A change of a substance's physical state from solid to liquid, solid to vapor, liquid to vapor, or *vice versa,* 5
phenol A compound having an —OH group bonded to a carbon atom in an aromatic ring, 856
phenyl (group) The —⌬ or —C_6H_5 group, 856
phosphate ester An alkyl derivative of phosphoric acid, containing the group

$$(HO)_2\text{—}\overset{\overset{\displaystyle O}{\|}}{P}\text{—OR}$$, 865

phospholipid A triglyceride formed from two fatty acids and a phosphate diester, 903

photoelectric effect Ejection of an electron from a metal surface following the absorption of a photon, 232
photoelectron An electron emitted from a metallic surface as a result of the absorption of a photon by the metal, 232
photoelectron spectroscopy Measurement of an electron's kinetic energy following absorption of a high-energy photon and ejection of an electron from any of the occupied orbitals, 279
photon A quantum or particle of electromagnetic radiation, carrying energy proportional to the frequency of the radiation, 233
photosynthesis Endothermic synthesis of carbohydrates from water and carbon dioxide by green plants, using energy from sunlight, 907
phototube A light-sensitive device, based on the photoelectric effect, which measures light intensity by converting photoelectrons into an electric current, 232
physical change Any change in form or condition that does not produce a new substance; some examples are melting, grinding, temperature change, and expansion, 6
physical property Characteristic or behavior observable without inducing chemical change, for example melting point, 6
physical quantity see **quantity**
physical state A form of matter, most commonly solid, liquid, or gas; sometimes called "phase," although strictly speaking, this is not correct, 5
planetary model An early theory of atomic structure: electrons travel around the nucleus in well-defined orbits, 236
plastic (1) Adjective describing a solid substance capable of changing shape or form; (2) a polymeric material that can be molded into a desired shape (see thermosetting and thermoplastic), 874
pleated sheet In proteins, a type of secondary structure in which many polymer chains lies parallel to one another, 883
polar bond An electrically unsymmetrical (that is, having a positive and a negative end) covalent bond in which the pair of electrons is unequally shared by the atoms, 330
polar molecule A neutral molecule that is electrically unsymmetrical, so that it has a negative and a positive end, 355
polyatomic Consisting of three or more atoms, adjective applied to molecules and ions, 54
polydentate Describes a ligand capable of forming several bonds to a metal ion in a coordination complex; a polydentate ligand is also called a *chelating agent,* 962
polyfunctional Having more than one functional group, 867
polyhydric Capable of supplying more than one OH^- ion or reacting with more than one H^+ ion, 663
polymer A substance consisting of very large molecules, called macromolecules, composed of many identical or very similar subunits, 867
polypeptide Polymer formed by condensation of amino acids; very large, naturally occurring polypeptides are called *proteins,* 882
polyprotic Capable of transferring more than one H^+ to another species; for example, H_2SO_4 is a diprotic acid, and H_3PO_4 is a triprotic acid, 662
polysaccharide A polymer of monosaccharides; some naturally occurring examples are starch, cellulose, and glycogen, 889
polyunsaturated Containing more than one multiple carbon-carbon bond per molecule, 410
porphyrin One of a class of compounds found in living organisms, whose structure incorporates four nitrogen atoms in planar geometry and which functions as a tetradentate ligand, 982
positional isomers Compounds having the same molecular formula, but differing in the location of a substituent or multiple bond, 408
positron A subatomic particle having the same mass as an electron, and bearing an equal but opposite (positive) charge, 992
potential (electrical) Driving force behind electrical current, which flows from a high to a low potential (analogous to pressure in flow of gases or liquids), 799
potential energy Energy a system has by virtue of its position, or the relative positions of the particles of the system, 420
precipitate Solid material that separates from a liquid solution during a precipitation reaction, 517, 739
precipitation reaction A reaction between two solutes that results in the formation of an insoluble product, 517, 739
precision The degree to which a set of measurements of a quantity agree with one another, 11
pre-equilibrium In a reaction mechanism, a reversible elementary step that reaches equilibrium prior to the rate-determining step, 628
pressure Force per unit surface area; some units of pressure are torr (mmHg), atm, and the SI unit, pascal (Pa), 170
primary alcohol A compound of formula $R{-}CH_2OH$, that is, with two H atoms and one $-OH$ group bonded to the same carbon atom, 855
primary amine An amine of formula RNH_2, with one alkyl group, 863
primary structure The order of amino acids in a polymer or of bases in a nucleic acid, 883, 895
primitive Describing a crystal lattice whose unit cells have only the eight lattice points at the corners, 490
product A species or substance formed in a chemical change, 94
promotion Excitation (hypothetical) of a valence electron as a preliminary to orbital hybridization, 375
propagation step see **chain reaction**
property of state see **state function**
protein A large, naturally occurring condensation polymer of amino acids, 882
proton (1) Part of atomic nuclei, a particle with a mass of 1.0073 amu and a positive charge equal to one elementary charge, 43; (2) an alternate name for the hydrogen ion, 650
proton acceptor A Brønsted base, 651
proton donor A Brønsted acid, 650
proton transfer Process in which one species donates a hydrogen ion to another, breaking one covalent bond and forming a new one, 650
purine bases Basic compounds related to purine; adenine and guanine, which occur in nucleic acids, are purine bases, 894
***P-V* work** Work of compression or expansion; work done on a system = $-P\Delta V$, 426
pyrimidine bases Basic compounds related to pyrimidine; cytosine, thymine, and uracil, which occur in nucleic acids, are pyrimidine bases, 894

Q

qualitative analysis Identification of a compound or the components of a mixture, 756
quantitative analysis Determination of the amount of a particular element or compound in a sample, 10
quantity The numerical result of a measurement, which tells how much and of what; for example, 65 miles per hour, or 4.2 m^3, 9
quantized Restricted to specific values, that is, discrete rather than continuous;

most often used to describe the energy states available to an atom or molecule, 233

quantum (plural: quanta) The smallest possible amount of a quantity; the electron is the quantum of negative charge, the photon is the quantum of light, 233

quantum mechanics The modern mathematical description of molecules, atoms, and subatomic particles that emphasizes their wave nature as well as their particle nature, 244

quantum number In quantum mechanics, a dimensionless index, usually a whole number, that characterizes and identifies a particular state or group of states, 237; rules for quantum numbers, 247

quaternary structure In proteins, the arrangement of several macromolecules into a cohesive unit, 885

R

racemic mixture Equimolar mixture of a pair of enantiomers, 882

radioactivity (1) The spontaneous, exothermic change in the structure of or disintegration of atomic nuclei; (2) the radiation or particles emitted by disintegrating nuclei, 51, 989

radiocarbon dating A technique for determining the age of once-living animal or plant tissue by means of its ^{14}C content, 1021

radioisotope Any radioactive nuclide, 989

Raoult's law The vapor-phase partial pressure of a solution component is equal to its liquid-phase mole fraction times its vapor pressure when pure; applies to ideal solutions only, 518

rare earth Synonym for **lathanide**

rare gas see **noble gas**

rate constant The constant appearing in the rate law of a reaction, numerically equal to the reaction rate when all reactants are at 1 M concentration, 600

rate-determining step In a mechanism, an elementary step that is considerably slower than preceding and subsequent steps, 625

rate law Algebraic expression relating the rate of a chemical reaction to the concentration(s) of reactant(s), for example rate = $k[NO_2]^2$, 600

reactant A compound or element present before a reaction, and converted during the reaction into a product, 94

reactant in excess The reactant present in greater than the stoichiometric amount; any reactant other than the limiting reactant, 106

reaction A specific chemical change, usually involving only a few different compounds or elements, 6

reaction coordinate The change in the geometrical relationship among the atoms of reacting molecules as they proceed from the reactant configuration to the product configuration, 616

reaction profile A graph showing how the potential energy of the species involved in a chemical reaction varies as a function of reaction coordinate, 616

reaction quotient The numerical value or algebraic form of an expression whose numerator is the product of the concentrations, each raised to the power equal to its stoichiometric coefficient, of all product species in a chemical reaction, and whose denominator is the product of the concentrations of reactant species, each raised to its stoichiometric coefficient, 558

reaction rate Speed at which a reaction takes place, expressed as a rate of decrease (increase) in concentration of a reactant (product), in mol L^{-1} s^{-1}, 598

reagent Any mixture, solution, or pure substance used to bring about chemical change, 104

reciprocal relationship The weaker an acid, the stronger its conjugate base, 655; or algebraically, $K_b = K_w/K_a$ for any conjugate pair, 695

redox electrode In an electrochemical cell, an electrode at which a redox reaction involving only solute species takes place, 809

redox reaction A reaction in which oxidation and reduction take place, 766

reducing agent A reactant capable of causing reduction of another species; also called *reductant,* 768

reduction A reaction in which a species gains one or more electrons, while one of its atoms undergoes a decrease in oxidation number, 769; in organic and biochemical reactions, a loss of O or a gain of H, 862

refraction Change in the direction of travel of light on passing from one transparent medium to another, caused by a difference in the velocity of light in the two media, 222

relative humidity Water vapor content of atmospheric air, expressed as percent of the equilibrium vapor pressure of water, 467

reliability A general term meaning either accuracy, precision, or both, 11

rem Acronym for röntgen equivalent, mankind; the unit for measuring a dose of radiation absorbed by the human body, 1020

resonance Type of covalent bonding in which π electrons are shared by more than two nuclei, that is, spread out into more than one bond, 324; see delocalization

resonance contributor One of several Lewis structures, differing only in location of valence electrons, used to describe bonding in resonant species, 324

resonance hybrid A resonant species, having an average structure intermediate among several resonance contributors, 324

resonance stabilization A measure of the greater stability, or decreased reactivity, exhibited by resonant species in comparison to otherwise similar species lacking resonance, 326

resonant species A molecule or ion having resonance, 324

reversible Describes a chemical or physical change capable of taking place equally well in either the forward or the reverse direction, 553

ring system A cyclic molecular framework, that is, a part of a molecule whose atoms form a closed chain of three or more atoms, 441

RNA see **nucleic acid**

S

σ (sigma) bond In valence bond theory, a covalent bond in which the electron density has cylindrical symmetry; a single bond, 372

σ framework The "skeletal structure," which describes the connectivity of a molecule by designation only of σ bonds and σ components of multiple bonds, 385

σ orbital Molecular orbital with cylindrical symmetry and whose electron density lies primarily between two adjacent nuclei, 387

saccharide see **mono-, di-,** or **oligosaccharide**

salt Ionic compound whose anion is not OH^- or O^{2-}, 54, 658

salt bridge A type of porous barrier between the two halves of an electrochemical cell, which provides a conducting path to complete the electrical circuit while preventing mixing of electrolytes, 806

saturated Adjective applied to either (1) an organic compound that contains no multiple bonds, hence contains no atom that can engage in any further bonding, 402; or (2) a solution in which no additional solute can be dissolved, 508
Schrödinger equation Wave equation applied to particle motion; the fundamental postulate of quantum mechanics, 244
scientific method An approach to knowledge involving observations of facts, and formulation and testing of hypotheses, 3
scientific notation System for writing numbers in two parts; a decimal number multiplying a whole-number power of 10 (indicates reliability as well as magnitude), 10
scintillation detector A device for detecting and measuring radioactivity, 1018
secondary alcohol An alcohol of formula R_2CHOH, 855
secondary amine An amine of formula R_2NH, 863
secondary structure (1) In proteins, the way in which polymer chains are coiled or disposed toward their neighboring chains, 883; (2) in DNA, the double helix formed by two complementary strands, 896, 898
second law of thermodynamics The total entropy of a system and its surroundings increases during any spontaneous change; an alternative statement of the second law is that the free energy of the system always decreases during a spontaneous change, 442
seed A small crystal of pure material, used to nucleate (initiate) crystallization of solute from a supersaturated solution, 512
semiconductor A solid with a moderate band gap, so that thermal excitation of electrons to the conduction band occurs only rarely; a material whose electrical conductivity is intermediate between a metal and an insulator, 401
SHE see **standard hydrogen electrode**
shell Set of orbitals having the same value of the principal quantum number, 249
shielding Caused by the negative charge of the core electrons, the apparent decrease in the magnitude of the nuclear charge experienced by valence electrons, 288
SI Système Internationale, a set of internationally accepted units for quantities, 9
side product A product of a side reaction, 102
side reaction A secondary reaction, usually undesired, that a reactant can undergo, 102
significant digits (figures) The reliably-known digits in the value of a quantity—the more digits used, the greater the reliability, 11
silanes Binary compounds of silicon and hydrogen, having empirical formulas of Si_nH_{2n+2}, 921
silicate minerals Naturally occurring rocks and minerals having linked SiO_4 tetrahedra as their fundamental structural unit, 922
silicone Polymeric molecule whose backbone structure is formed by alternating Si and O atoms ($—OSiR_2—$), 922
single bond A covalent bond, that is, a pair of electrons shared by, and in the valence shells of, two atoms, 316
soap A sodium or potassium salt of a fatty acid, 902
solid One of the three common physical states of matter, characterized by definite shape and volume, 5
solubility The maximum amount ($g\ L^{-1}$) of solute that can dissolve in a given amount of solvent; if expressed in units of $mol\ L^{-1}$, it is called the *molar solubility*, 508, 732
solubility product The numerical value of the ion product evaluated for a pure salt in equilibrium with its saturated solution, 732
solute Any solution component other than the solvent, 108
solution Homogeneous mixture of two or more pure substances; often, but not exclusively, applied to a pure liquid mixed with a small amount of another pure substance, 6
solvation Nonbonding interaction between solvent molecules and a solute species, 506, 648
solvent The solution component that, when pure, is in the same physical state as the solution (normally the major component), 108
specific gravity The ratio of the density of a substance to the density of water at the same temperature, 30
specific heat Heat capacity per gram; units are $J\ g^{-1}\ K^{-1}$, $J\ g^{-1}\ °C^{-1}$, or $cal\ g^{-1}\ °C^{-1}$, 147
spectator ion Any ion that is present in solution but does not participate in the chemical reaction, 644, 779
spectrochemical series A listing of ligands in increasing order of their effect on crystal-field splitting, 972
spectrograph, -scope, -meter Instruments for observation of spectra, using respectively film, the eye, or an electronic device such as a phototube for the detection of light, 229
spectroscopy Study of spectra, 228
spectrum (plural: spectra) Description of the wavelengths of light (or other electromagnetic radiation) emitted or absorbed by a substance; in a *continuous* spectrum, the emitted/absorbed intensity varies smoothly as the wavelength changes, whereas in a *discrete* or *line* spectrum, there are a number of very narrow wavelength ranges in which radiation is emitted or absorbed, 229
spin Intrinsic property of electrons, giving rise to both angular momentum and magnetic properties; spin is quantized and has one of only two possible values—equal and opposite, 265
spontaneous Having an intrinsic tendency to take place, without any external stimulation or influence, 436
square planar Molecular shape in which four atoms, groups, or ligands lie at the corners of a square, and are bonded to an atom or metal ion at the center of the square, 961
SRP see **standard reduction potential**
SSP Acronym for *standard stoichiometry problem* in which a stoichiometric ratio, together with a given mass of one species in a reaction, is used to find the mass of another, 100
stability belt The set of values of the neutron-to-proton ratio characteristic of most nonradioactive nuclides, 994
standard conditions Reference conditions for tabulation and comparison of properties: for gases, a partial pressure of 1 atm; for solutes, a concentration of 1 M; for liquids and solids, the pure substance, 812
standard hydrogen electrode (SHE) Half-cell consisting of gaseous hydrogen at 1 atm, 1 M $H^+(aq)$, and a platinum electrode, 813
standard reduction potential (SRP) Potential of a particular half-reaction, measured under standard conditions against a SHE, 814
standard state The most stable form of an element or compound at 1 atm pressure and 25 °C, 141
standing wave As opposed to a travelling wave, wave motion that oscillates in one place and does not travel through space; also called a *stationary wave*, 245
starting material Reagent from which a desired reaction product is prepared, 103
state (1) A *physical* state is solid, liquid, or gas, 5; (2) a *thermodynamic* state is a

unique condition of a system, usually described by pressure and temperature, 425; (3) in *quantum mechanics,* a (stationary) state is one of the solutions to the Schrödinger equation, 246
state function A property of a system whose value is independent of the path by which the system reaches a particular state; *P*, *T*, *E*, *H*, *S*, and *G* are state functions, *q* and *w* are not, 428
stationary state One of the quantized states or energy levels of an atom or molecule, 236
stationary wave see **standing wave**
stereoisomers Molecules having the same atom-to-atom connectivity but a different spatial arrangement, 881
steroid A lipid whose structure (see Figure 21.37a) incorporates one 5-membered ring and three 6-membered rings fused together (many hormones are steroids), 904
stock solution A concentrated solution, used as needed to prepare more dilute solutions, 114
Stock system Chemical nomenclature rules in which the oxidation state of an element is indicated by parenthetical Roman numerals, 81
stoichiometric amount(s) Amounts of reactants in exact proportion to the mole ratio in the balanced equation, that is, the amount of a reagent that can react completely, being neither limiting nor in excess, 106
stoichiometric coefficient Number preceding the formula of a species in a balanced chemical equation; indicates the relative number of moles of the species involved in the reaction, 95
stoichiometric ratio see **mole ratio**
stoichiometry The mass and mole relationships among (1) the elements in a compound's empirical formula, 79, or (2) the reactants and products of a chemical reaction, 98
STP Acronym for standard temperature (0 °C) and pressure (1 atm), 184
straight chain hydrocarbon Also called *normal,* a hydrocarbon in which no carbon atom is bonded to more than two other carbon atoms, 404
strong Describes an acid, base, or electrolyte that completely dissociates into ions when dissolved, 540, 652, 653
strong-field (case) In crystal field theory, the situation in which ligands high in the spectrochemical series cause the metal ion's *d* orbitals to be occupied with the minimum number of unpaired electrons; also called the *low-spin case,* 973
strong force The cohesive force exerted on one another by nucleons in a nucleus, 993
structural formula Pictorial arrangement of elemental symbols showing the chemical bonds and connectivity of a molecule or polyatomic ion, 57
structural isomers Molecules having the same molecular formula, but different connectivity, 404
subatomic particles see **elementary particles**
sublimation Process in which a solid, without melting, becomes a vapor; characterized by sublimation temperature and equilibrium sublimation pressure, 472
sublimation pressure The partial pressure of vapor in equilibrium with a pure solid substance, 472
sublimation temperature The temperature at which the sublimation pressure of a solid becomes equal to the confining pressure of the atmosphere; called the *normal* sublimation temperature if the confining pressure is 760 torr, 472
subshell A set of orbitals, within a designated shell, having the same value of ℓ, the angular momentum quantum number, 249
substance A particular type of matter, 5
substituent An atom or group of atoms that replaces a hydrogen atom in a molecule, 404
substitution A reaction (of aromatic compounds) in which a hydrogen atom is replaced by another atom or group, 411
substrate Molecule whose reaction is catalyzed by an enzyme, 885
sugar One of the smaller carbohydrates; a mono-, di-, or oligosaccharide, 887
sulfonic acid or **ester** A compound containing the group R—(SO_2)—O—R′, where R′ is an H-atom (sulfonic acid) or an alkyl group (sulfonic ester), 865
supercritical fluid State of matter with properties of both liquid and vapor; exists only above the critical point of a substance, 478
supersaturated Describes an unstable solution containing more than the equilibrium amount of solute; that is, a solution in which the solute concentration is greater than in a saturated solution, 512, 738
surface tension Physical property resulting from unbalanced intermolecular forces at the surface of a liquid; causes liquids to behave as if enclosed by an elastic skin, and to assume a shape of minimum surface area, 455
surroundings Everything that lies outside a system and is able to interact with it through exchange of matter, heat, and/or work, 425
symmetry An aspect of shape; an object having *spherical symmetry* looks the same when rotated about any axis passing through its center; an object having *cylindrical symmetry* looks the same when rotated about only one axis passing through its center, 373
system A limited and defined portion of the universe selected for attention in thermodynamics, 424
systematic name A compound name that uniquely describes the structure of a compound, according to a scheme developed and overseen by IUPAC, 77

T

termination see **chain reaction**
tertiary alcohol An alcohol of formula R_3COH, 855
tertiary amine An amine of formula R_3N, 863
tertiary structure In proteins, the shape assumed by the flexible α-helix, together with nonhelical portions of the molecule, 885
tetrahedron Four-sided solid whose faces are identical equilateral triangles, 344
theoretical yield The maximum amount of a product obtainable from given amounts of reactants under loss-free conditions, 102
theory The most satisfactory explanation of the known facts; a hypothesis that has survived, perhaps in modified form, considerable testing, 3
thermal energy Energy in the form of heat, due to the kinetic energy of the atoms or molecules of a substance, 133
thermochemical equation A balanced chemical equation, written together with the associated heat of reaction, ΔH_{rxn}, 136
thermochemistry Study of the heat produced or consumed in chemical reactions, 134
thermodynamics Study of heat, work, and energy in chemical and physical processes, 420
thermoplastic Adjective describing a polymer capable of repeated melting and resolidifying, 874
thermosetting Adjective describing a polymer that can be prepared in any desired shape, but cannot be melted and reformed, 874

third law of thermodynamics At absolute zero, the entropy of a perfectly ordered crystal of a pure substance is zero, 439
three-center bonding Covalent bond in which three atoms share an electron pair; found in boranes, 915
threshold hypothesis There is a dosage or level of ionizing radiation, below which there are no harmful effects, 1019
titrant In a titration, the reagent whose volume is carefully measured as it is added to the unknown being analyzed, 670
titration Determination of the amount of a substance (usually acid, base, oxidant, or reductant) in a sample by careful addition of just enough titrant for complete consumption of the substance, 670
titration curve Plot of pH vs. added titrant, beginning in the undertitrated region, passing through the region of greatest pH change ("break") near the equivalence point, and then into the overtitrated region, 707
torr A unit of pressure equal to 133.32 Pa; 1 torr = 1 mmHg = (1/760) atm, 171
trans- "Across from"; prefix to the name of the geometrical isomer having substituents on the opposite sides of a double bond, 408, a ring, 849, or a coordination sphere, 966
transition element or **transition metal** Member of a B group of the periodic table; all such elements (except Zn, Cd, and Hg) have a partly filled *d* subshell in at least one of their common oxidation states, 946, 947
transition series The three *outer* transition series have partly filled 3*d*, 4*d*, or 5*d* subshells; the two *inner* transition series, the lanthanides and actinides, have partly filled 4*f* or 5*f* subshells, 956, 959
transition state Geometrical configuration of atoms at the maximum energy of the reaction profile, between reactants and products (not a distinct chemical substance), 617
transuranium elements Elements of atomic number greater than 92, artificially produced, 1005
travelling wave Wave motion that not only oscillates but also moves through space, 244
triglyceride An ester of glycerol (1,2,3-trihydroxypropane) and three acids, usually fatty acids, 900
trigonal Having the shape or symmetry of an equilateral triangle, 346
trigonal pyramid (1) Six-sided solid whose faces are identical isoceles triangles; (2) electron-pair geometry of molecules having five electron pairs on the central atom, 346
trimolecular Older usage "termolecular"; describes an elementary step involving three atoms, molecules, or ions, 625
triple bond In Lewis theory, three pairs of electrons, shared by and completing the valence shells of two atoms, 316; in valence bond theory, a σ bond and 2 π bonds linking the same two atoms, 374
triple point Temperature and pressure at which three different phases of a substance can simultaneously exist, 478
triprotic see **polyprotic**
trivalent Having a charge of +3 or −3, 78
trivial name Nonsystematic, older name for a compound; many trivial names are still in use, 77
Tyndall effect Scattering of light by a colloid, 545

U

ultraviolet Spectral region adjacent to, but having a shorter wavelength than, violet light, 226
uncertainty principle Position and momentum of a particle cannot simultaneously be known with unlimited accuracy, 242
undersaturated see **unsaturated**
unimolecular Describes an elementary step involving only one molecule, 625
unit(s) That part of a quantity describing what the numerical part refers to; °C or grams per milliliter, for example, 9
unit cell The smallest possible repeating unit of a crystal lattice; a unit cell is a parallelepiped (a six-sided solid), and must have at least eight lattice points, 480
unit conversion factor A quotient whose numerator is a quantity expressed in one unit, and whose denominator is the same quantity expressed in another unit, 12
unpaired electron An electron not associated with another of opposite spin; any electron in a singly occupied orbital, 267
unsaturated Adjective describing either (1) a compound containing at least one multiple bond, 408 or (2) a solution containing less than the maximum possible amount of solute, that is, in which additional solute can be dissolved (same as undersaturated), 508

V

vacant orbital An orbital containing no electrons, 267
valence (1) Adjective: having to do with chemical bonding, 309; (2) noun: combining power or total number of bonds an atom can form, 309
valence band The lower of two energy bands of a solid, composed of localized bonding orbitals, 371
valence bond The theory that described a covalent bond as formed by overlap of two atomic orbitals and localized between two adjacent atoms, 371
valence electrons (1) Electrons in the valence shell; (2) electrons participating in chemical bonding, 309
valence shell The set of partly filled orbitals lying generally at higher energy and farther from the nucleus than the core electrons; also called outer shell, 309
valence shell electron-pair repulsion (VSEPR) A method for predicting molecular shape from the Lewis structure of a molecule or ion, 346
valence state Electron configuration of an atom as it forms bonds (as opposed to an isolated atom), 375
van der Waals equation $(P + an^2/V^2)(V - nb) = nRT$; an equation relating the pressure, volume, and temperature of a real (nonideal) gas, 208
van der Waals forces Nonbonding forces between molecules or atoms, including dipole, dispersion, and hydrogen bonding, 458
van't Hoff factor A measure of the extent to which colligative properties are enhanced by ionization of a solute; equal to the number of moles of dissolved particles per mole of dissolved compound, 541
vapor The gaseous state of a substance that is a liquid or a solid at room temperature, 464
vapor pressure The pressure exerted by a gaseous substance when it is in equilibrium with a liquid, 464
vapor pressure decrease The extent to which a nonvolatile solute decreases the vapor pressure of a liquid (one of the colligative properties), 465
vaporization The process in which a liquid becomes a vapor; also called *evaporation,* 464
variable-valence elements Elements that exhibit more than one oxidation number in their compounds; for example, iron forms both $FeCl_2$ and $FeCl_3$, 772
viscosity Resistance of a fluid to flow, 455
visible light Electromagnetic radiation in the wavelength range 400 to 700 nm, to which the human eye responds, 226
volatile Readily vaporized, 465

volt Unit of electrical potential; 1 volt = 1 joule/1 coulomb, 801
voltage Synonym for electrical potential difference, 801
voltaic cell see **electrochemical cell**
volume percent 100 times the ratio of pure solute volume to solution volume, 109
VSEPR see **valence shell electron-pair repulsion**

W

water of crystallization Loosely bound water molecules incorporated into the crystal structure of a hydrated compound, 57
water of hydration same as **water of crystallization**
wave function One of a set of many solutions to the Schrödinger equation describing a particular system; each wave function describes a particular stationary state, 246
wavelength The distance between two successive crests, troughs, or other particular points of a repetitive wave pattern, 224
wave-particle duality Theory that waves and particles exhibit both wave-like and particle-like behavior, 235
wax An ester of a fatty acid and a long-chain alcohol, 903
weak Describes an acid, base, or electrolyte that is not completely dissociated in solution, 540, 652, 653
weak-field (case) In crystal field theory, the situation in which ligands low in the spectrochemical series cause the metal ion's *d* orbitals to be occupied with the maximum number of unpaired electrons; also called the high-spin case, 973
weight percent same as **mass percent**
work Motion of matter against an opposing force, 3; a means of energy transfer, 421

X

x-ray crystallography Elucidation of crystal structure and the dimensions of the unit cell by study of the diffraction of x-rays by a crystal, 487

Y

yield see **actual yield, theoretical yield,** or **percent yield**

Z

zeolite Aluminosilicate mineral whose structure incorporates cavities containing metal ions; used in ion exchange processes, 926
zone refining Purification method in which impurities are swept out of a solid by creating a small, moving zone of melted material, 921
zwitterion A molecule that has two ionized groups of opposite charge; for example, $H_3N^+—CH_2—CO_2^-$; also called an *inner salt,* 879

PHOTO CREDITS

Frontmatter

p. ii, xviii, xxiii, Charles D. Winters.

p. iii, © Dr. Dennis Kunkel/Phototake NYC

p. xiii, Manfred Kage/Peter Arnold, Inc.

p. xv, Leon Lewandowski.

p. xvii, L. Brown/The Image Bank

p. xix, Charles D. Winters.

p. xxi, Nelson Max/LLNL/Peter Arnold, Inc.

Chapter 1

Chapter Opening Photo, Figs. 1.3, 1.5, 1.6, 1.11, 1.14, 1.15, unnum. figs. pp. 3, 5, 10, 15, 27, 32 *bottom,* Charles D. Winters.

Fig. 1.1(a), ROE/AAT Board 1982.

Fig. 1.1(b), Kingory/Phototake NYC.

Fig. 1.9, Leon Lewandowski.

Fig. 1.10, Marna G. Clarke.

Unnum. figs. p. 17, Phil Degginger.

Unnum. fig. p. 32 *top,* Bethlehem Steel.

Chapter 2

Chapter Opening Photo, Fig. 2.13, Fermilab.

Fig. 2.3, Michael Dalton/Fundamental Photographs, NY.

Fig. 2.7, unnum. figs. pp. 59, 71 *left,* Charles Steele.

Fig. 2.10, unnum. figs. p. 58, Charles D. Winters.

Fig. 2.11(a), Dave Pierce, Finnigan Mat, San Jose.

Figs. 2.14(a) (b), Phil Degginger.

Fig. 2.15, Argonne National Laboratory.

Unnum. fig. p. 42, Royal Institution, London.

Unnum. figs. p. 49 *bottom,* Richard Megna/Fundamental Photographs, NY.

Unnum. fig. p. 60, Richard Roese.

Unnum. fig. p. 71 *right,* UPI/Bettmann Newsphotos.

Chapter 3

Chapter Opening Photo, Figs. 3.7, 3.8, 3.10, 3.11, 3.13, 3.14, unnum. figs. pp. 95, 116, Charles D. Winters.

Fig. 3.12, Charles Steele.

Fig. 3.15, Leon Lewandowski.

Fig. 3.16(a), NASA.

Fig. 3.16(b), courtesy of Morton International Inc., Chicago, IL.

Unnum. fig. p. 123, Dresser Industries/Kellogg Company.

Chapter 4

Chapter Opening Photo, Kevin Schafer/Tom Stack and Associates.

Fig. 4.1(a), L. Brown/The Image Bank.

Fig. 4.1(b), Kent and Donna Dennen/Photo Researchers, Inc.

Fig. 4.1(c), 4.5, 4.9, unnum. figs. pp. 146, 147, 153, 158 *left,* 162 *top,* Charles D. Winters.

Fig. 4.2(a), Kristen Brochman/Fundamental Photographs, NY.

Fig. 4.2(b), Harold Sund/The Image Bank.

Fig. 4.3, NASA.

Fig. 4.6(b), Phil Degginger.

Fig. 4.7, Charles Steele.

Fig. 4.8, Michael Dalton/Fundamental Photographs, NY.

Unnum. fig. p. 158 *right,* AM General Corporation.

Unnum. fig. p. 162 *bottom,* Walter Bibikow/The Image Bank.

Unnum. fig. p. 163, Donna McLean.

Chapter 5

Chapter Opening Photo, Figs. 5.1, 5.7, 5.18, 5.19, unnum. figs. pp. 171, 183, 188, 203, 211, 212, 213, Charles D. Winters.

Fig. 5.3, NOAA Climate Analysis Center.

Fig. 5.4(a) (b), Runk/Schoenberger/Grant Heilman, Inc.

Unnum. fig. p. 214, NASA.

Chapter 6

Chapter Opening Photo, Milt and Joan Mann/CAMERAMANN INTERNATIONAL.

Fig. 6.5, J. Brousseau/The Image Bank.

Fig. 6.8, 6.11(b), 6.15(a), Charles D. Winters.

Fig. 6.10(a), Colin Molyneux/The Image Bank.

Fig. 6.10(b), courtesy of Phillips Lighting Co.

Fig. 6.14, Rob Atkins/The Image Bank.

Fig. 6.17, Deutsches Museum.

Fig. 6.18(a), courtesy of Philips Electronic Instruments Co.

Fig. 6.18(b), Kunkel/Phototake, NYC.

Unnum. fig. p. 258, courtesy of Sterling and Francine Clark Art Institute, Williamstown MA, photo by Amanda Merullo.

Chapter 7

Chapter Opening Photo, Fig. 7.11, Charles D. Winters.

Fig. 7.12, Leon Lewandowski.

Fig. 7.24, courtesy of the American Dental Association.

Fig. 7.25, Marna G. Clarke.

Fig. 7.26, Oxychem.

Fig. 7.27, copyright © Tom Till.

Fig. 7.28, Delve Communications.

Chapter 8

Chapter Opening Photo, Nelson Max/LLNL/Peter Arnold, Inc.

Fig. 8.2, Charles Steele.

Figs. 8.8, 8.9, 8.10, 8.13, 8.14, 8.15, 8.16, 8.21, 8.23, 8.24, 8.25, unnum. figs., pp. 312, 319, 341, Charles D. Winters.

Fig. 8.22, Union Carbide/Industrial Gases, Linde Division.

Unnum. fig. p. 324 *top,* NASA.

Unnum. fig. p. 324 *bottom,* Michael Melford/The Image Bank.

Unnum. fig. p. 327, David Brownell/The Image Bank.

Chapter 9

Chapter Opening Photo, Fig. 9.41, 9.42, 9.43, unnum. figs. pp. 376, 382, 383, 384, Charles D. Winters.

Unnum. figs. pp. 378, 379, 381, 386, Charles Steele.

Unnum. fig. p. 412, Archive Photos/American Stock.

Chapter 10

Chapter Opening Photo, Figs. 10.8, 10.12, 10.15, 10.17, 10.19, unnum. figs. pp. 432, 444, Charles D. Winters.

Fig. 10.2, W. Bacon/Photo Researchers, Inc.

Fig. 10.3, Kristen Brochmann/Fundamental Photographs, NY.

Fig. 10.6, Anthony A. Boccaccio/The Image Bank.

Fig. 10.10, Robin Scagell/Science Photo Library/Photo Researchers, Inc.

Chapter 11

Chapter Opening Photo, Figs. 11.5, 11.7, 11.9, 11.10, 11.20, 11.30, 11.33(b), 11.39, 11.40, 11.50, Charles D. Winters.

Fig. 11.6, Manfred Danegger/Peter Arnold, Inc.

Fig. 11.8(a) (b), Phil Degginger.

Fig. 11.18, Martin Miller.

Fig. 11.25(a), Manfred Kage/Peter Arnold, Inc.

Fig. 11.25(b), Dr. E. R. Degginger.

Unnum. fig. p. 497, Lawrence Livermore Laboratory/Mark Marten/Photo Researchers, Inc.

Chapter 12

Chapter Opening Photo, Figs. 12.10, 12.11, 12.23, 12.27, 12.28, Charles D. Winters.

Fig. 12.1, © Thomas C. Boyden.

Fig. 12.12, Konrad Wothe/The Image Bank.

Fig. 12.17, Standard Oil of Ohio.

Fig. 12.25, SIU/Photo Researchers, Inc.

Fig. 12.26(a), photo by Norman LoRusso, provided by Mount Pleasant Waterworks and Sewer Commission.

Chapter 13

Chapter Opening Photo, Leon Lewandowski.

Figs. 13.1, 13.5, 13.9, 13.10, unnum. figs. pp. 559, 569, Charles D. Winters.

Fig. 13.6, M. Digiacomo/The Image Bank.

Unnum. fig. p. 573, Marna G. Clarke.

Chapter 14

Chapter Opening Photo, Figs. 14.1, 14.4, 14.5, 14.11, 14.16, 14.18, Charles D. Winters.

Fig. 14.19, General Motors.

Unnum. fig. p. 634, North Wind Picture Archives.

Chapter 15

Chapter Opening Photo, Figs. 15.2, 15.3, 15.5(a), 15.16, 15.18, Charles D. Winters.

Fig. 15.1, Scala/Art Resource, NY.

Figure 15.5(b), Dr. E. R. Degginger.

Unnum. fig. p. 676, Oliver Strewe/Tony Stone Images.

Chapter 16

Chapter Opening Photo, Leon Lewandowski.

Figs. 16.1, 16.2, 16.8, Charles D. Winters.

Unnum. fig. p. 724, © Richard Nowitz/Phototake, NYC.

Chapter 17

Chapter Opening Photo, Fig. 17.6, Charles Steele.

Fig. 17.1, CNRI/Science Photo Library/Photo Researchers, Inc.

Fig. 17.2, Dr. E. R. Degginger.

Figs. 17.3, 17.4, 17.7, 17.8, 17.10, 17.11, 17.12, Charles D. Winters.

Fig. 17.13, James Morgenthaler.

Fig. 17.5, Robert C. Simpson/Tom Stack and Associates.

Unnum. fig. p. 760, Arthur Tress/Magnum Photos, Inc.

Chapter 18

Chapter Opening Photo, Michael Dalton/Fundamental Photographs, NY.

Figs. 18.1, 18.2, 18.8, 18.9, unnum. fig. p. 787, Charles D. Winters. Unnum. fig. p. 772, Eric L. Heyer/Grant Heilman, Inc. Unnum. fig. p. 793, courtesy of Serengeti Eyewear.

Chapter 19

Chapter Opening Photo, James Morgenthaler. Figs. 19.5, 19.6, 19.13, 19.17, 19.18, 19.23, Charles D. Winters.

Fig. 19.14, Yoav/Phototake, NYC.

Fig. 19.21, courtesy of Eveready Battery Company.

Unnum. fig. p. 821, Marna G. Clarke.

Chapter 20

Chapter Opening Photo, figs. 20.4, 20.5, 20.8(b), unnum. figs. pp. 849, 852, 853, 854, 868, 869, 871, 872, 873, Charles D. Winters.

Fig. 20.1, Robin Smith/Tony Stone, Worldwide.

Fig. 20.6, Charles Steele.

Fig. 20.8(a), John Moss/The Stock Shop.

Unnum. fig. p. 867, Union Carbide Corporation.

Unnum. fig. p. 868, Alan Pitcairn/Grant Heilman, Inc.

Unnum. fig. p. 869, Dr. E. R. Degginger.

Unnum. fig. p. 871, Irving Shapiro/The Stock Shop.

Unnum. fig. p. 871, © David R. Frazier.

Unnum. fig. p. 872, courtesy of DuPont.

Unnum. fig. p. 873, Harry J. Przekop, Jr/The Stock Shop.

Unnum. fig. p. 873, Dawson Stone/Tony Stone Images.

Chapter 21

Chapter Opening Photo, Larry Lee/Westlight.

Fig. 21.33, Leonard Lessin.

Fig. 21.35, Charles D. Winters.

Chapter 22

Chapter Opening Photo, figs. 22.5, 22.7, 22.8(b), 22.9(a), 22.9(c), 22.14, 22.17, 22.18, 22.19, 22.22, unnum. figs. pp. 913, 918, 927, 930, 934, Charles D. Winters.

Fig. 22.1, Phototake.

Fig. 22.3(a) (b), GE Superabrasives.

Fig. 22.9(b), Alabama Museum.

Fig. 22.15, Blair Seitz/Photo Researchers, Inc.

Fig. 22.16, Grant Heilman/Grant Heilman.

Fig. 22.19 *left*, Fletcher, W. K./Photo Researchers, Inc. *right*, Allen B. Smith/Tom Stack and Associates.

Fig. 22.20, C. B. Jones/Taurus Photos.

Unnum. fig. p. 917, courtesy of Leonard W. Fine.

Unnum. fig. p. 923, Warren Morgen/Westlight.

Unnum. fig. p. 925, Paul Silverman/Fundamental Photographs, NY.

Chapter 23

Chapter Opening Photo, figs. 23.6, 23.7, 23.8, 23.19(b), 23.30, 23.31, unnum. figs. pp. 946, 951, 952, 954, 955, 982, Charles D. Winters.

Fig. 23.12, © Tom Tracy/FPG International Corp.

Unnum. fig. p. 951, The Granger Collection, NY.

Unnum. fig. p. 953, Philippe Brilak/Gamma Laison.

Unnum. figs. pp. 955, 959, Runk/Schoenberger/Grant Heilman, Inc.

Unnum. fig. p. 982, Larry Lefever/Grant Heilman, Inc.

Chapter 24

Chapter Opening Photo, Will and Deni McIntyre/Photo Researchers, Inc.

Fig. 24.5, CERN/Science Photo Library/Photo Researchers, Inc.

Unnum. fig. p. 989, AIP Niels Bohr Library/W. F. Meggers Collection.

Unnum. fig. p. 990 *top*, courtesy of Patriots Point Naval and Maritime Museum.

Unnum. fig. p. 990 *bottom*, Charles D. Winters.

Unnum. fig. p. 1002, Stand Levy/Photo Researchers, Inc.

Unnum. fig. p. 1004, Argonne National Laboratory.

Unnum. fig. p. 1008, Anglo-Australian Observatory.

Unnum. fig. p. 1009, © 1979 Royal Observatory Edinburgh Anglo-Australian Telescope Board, photo by David Malin, original from U. K. Schmidt plates.

Unnum. fig. p. 1011, Mark Newman/Tom Stack and Associates.

Unnum. fig. p. 1015 *top*, Y. Arthus-Bertrand/Peter Arnold, Inc.

Unnum. fig. p. 1015 *bottom*, Tass/Sovfoto.

INDEX

Page numbers in bold type refer to definitions of terms; numbers in italics refer to illustrations; a page number followed by "t" indicates a reference to a Table, and "A" designates an Appendix page number.

Physical Constants

Electron charge	$e = 1.6021773 \times 10^{-19}$ C
Planck's constant	$h = 6.626076 \times 10^{-34}$ J s
Speed of light	$c = 2.99792458 \times 10^{8}$ m s^{-1} (exactly)
Acceleration of gravity (sea level, at equator)	$g = 9.80665$ m s^{-1} (exactly)
Avogadro's number	$N_A = 6.022137 \times 10^{23}$
Universal gas constant	$R = 8.31451$ J mol^{-1} K^{-1}
	$= 0.0820578$ L atm mol^{-1} K^{-1}
Faraday constant	$\mathcal{F} = 96{,}485.31$ C mol^{-1}
Masses of fundamental particles	
Electron	$m_e = 9.109390 \times 10^{-31}$ kg
	$= 0.00054858$ amu
Proton	$m_p = 1.672623 \times 10^{-27}$ kg
	$= 1.007276$ amu
Neutron	$m_n = 1.674929 \times 10^{-27}$ kg
	$= 1.008665$ amu

Values (except for particle mass in amu) are taken from the 1986 recommendations of the United States National Bureau of Standards, as published in the *Journal of Research of the National Bureau of Standards,* vol. 92, March–April 1987. The amu values are calculated from the relationship mass(amu) = mass(kg) $\cdot 10^3 \cdot N_A$.